IFAC 1977 Conference Proceedings

ATHERTON: Multivariable Technological Systems
BANKS & PRITCHARD: Control of Distributed Parameter Systems
CICHOCKI & STRASZAK: Systems Analysis Applications to Complex Programs
DUBUISSON: Information and Systems
GHONAIMY & YOUNIS: Systems Approach for Development
HASEGAWA & INOUE: Urban, Regional and National Planning - Environmental Aspects
LEONHARD: Control in Power Electronics and Electrical Drives
OSHIMA: Information Control Problems in Manufacturing Technology
RIJNSDORP: Case Studies in the Humanisation of Automation
SAWARAGI & AKASHI: Environmental Systems Planning, Design and Control
SINGH & TITLI: Control and Management of Integrated Industrial Complexes
SMEDEMA: Real Time Programming 1977

IFAC

International Federation of Automatic Control

CONTROL IN POWER ELECTRONICS AND ELECTRICAL DRIVES

CONTROL IN POWER ELECTRONICS AND ELECTRICAL DRIVES

Proceedings of the Second IFAC Symposium,
Düsseldorf, Federal Republic of Germany, 3 - 5 October 1977

Edited by

W. LEONHARD

Technical University of Braunschweig, Federal Republic of Germany

Published for the

INTERNATIONAL FEDERATION OF AUTOMATIC CONTROL

by

PERGAMON PRESS

OXFORD · NEW YORK · TORONTO · SYDNEY · PARIS · FRANKFURT

U.K.	Pergamon Press Ltd., Headington Hill Hall, Oxford OX3 0BW, England
U.S.A.	Pergamon Press Inc., Maxwell House, Fairview Park, Elmsford, New York 10523, U.S.A.
CANADA	Pergamon of Canada Ltd., 75 The East Mall, Toronto, Ontario, Canada
AUSTRALIA	Pergamon Press (Aust.) Pty. Ltd., 19a Boundary Street, Rushcutters Bay, N.S.W. 2011, Australia
FRANCE	Pergamon Press SARL, 24 rue des Ecoles, 75240 Paris, Cedex 05, France
FEDERAL REPUBLIC OF GERMANY	Pergamon Press GmbH, 6242 Kronberg-Taunus, Pferdstrasse 1, Federal Republic of Germany

First edition 1978

British Library Cataloguing in Publication Data

IFAC symposium on Control in Power Electronics and Electrical Drives, 2nd, Dusseldorf, 1977
Control in power electronics and electrical drives.
1. Electric motors - Control - Congress
2. Electric current converters - Congresses
I. Title II. Leonhard, W III. International Federation of Automatic Control IV. VD1/VDE - Gesellschaft Mess-und Regelungstechnik V. Energietechnische Gesellschaft im VDE
621.46'2 TK2511 78-40165

ISBN 0-08-022014-2

In order to make this volume available as economically and as rapidly as possible the authors' typescripts have been reproduced in their original forms. This method unfortunately has its typographical limitations but it is hoped that they in no way distract the reader.

Printed in Great Britain by A. Wheaton & Co. Ltd., Exeter

SECOND IFAC SYMPOSIUM ON CONTROL IN POWER ELECTRONICS AND ELECTRICAL DRIVES

Organized by
VDI/VDE-Gesellschaft Mess- und Regelungstechnik (GMR)
Energietechnische Gesellschaft im VDE (ETG)

Sponsored by
The International Federation of Automatic Control
Technical Committee on Applications
Technical Committee on Components and Instruments

International Program Committee
L. Abraham, F.R.G.
B. M. Bird, UK
A. Boehringer, F.R.G.
D. R. Bristol, USA
H. Bürskens, F.R.G.
A. De Carli, Italy
M. Depenbrock, F.R.G.
W. Harprecht, F.R.G.
R. G. Hoft, USA
R. Jötten, F.R.G.
W. Leonhard, F.R.G.
M. Marty, France
W. E. Miller, USA
A. E. Nisenfeld, USA
I. Rácz, Hungary
Ph. K. Sattler, F.R.G.
H.-Chr. Skudelny, F.R.G.
D. Ströle, F.R.G.
H. Weh, F.R.G.
A. Weinmann, Austria
R. Zwicky, Switzerland

National Organizing Committee
W. Fromme, Duesseldorf
A. Grütz, Frankfurt/Main
M. A. Kaaz, Duesseldorf
M. Rausch, Duesseldorf
A. Sedlaczek, Duesseldorf
L. Zühlke, Duesseldorf

CONTENTS

FOREWORD

Since our last IFAC-Symposium in Dusseldorf, three years ago, the field of static electronic power conversion has seen further steady progress, fed mainly from two sources:

Improvements of the high-power electronic switching components (thyristors) with regard to voltage, current and switching speed,

Application of more sophisticated control strategies, facilitated by the advances of low power integrated circuitry, eventually heading towards all digital converter control with micro-processors.

There is a continued strong tendency towards standardisation and the use of factory-built subassemblies with the aim of achieving cost reductions.

With regard to controlled electric drives, the effort of simplifying existing AC motor control schemes to make them a viable substitute for converter supplied DC drives is continuing, but it appears that no cost equivalent has been reached as yet, even though there are circuits which at least approach the simplicity of standard DC drives.

The upsurge of microcomputers will most likely have a strong impact on the control of electronic power converters, whose control signals are digital anyway. Thus most of the analogue control devices may eventually be replaced by digital signal processing elements.

This second symposium is extending the range of topics by including discussion of the effects of converter control on the design of electrical machines. This reflects the fact that the complete drive, including the converter, the machine and the control, must be dealt with as a unit. With advanced AC-drives it is no longer possible to consider the sections independently because all the connected elements are highly interacting.

It is hoped that this 2. IFAC-Symposium on

"Control in power electronics and electrical drives"

represents a timely occasion for again discussing the state of the art in this rapidly expanding field.

W. Leonhard

FOREWORD

A GENERAL DIGITAL COMPUTER SIMULATION PROGRAMME FOR THYRISTOR STATIC CONVERTERS (PROGRAMME SACSO) APPLICATION EXAMPLES

H. Foch, C. Réboulet and J. Schonek

Laboratoire d'Electrotechnique et d'Electronique Industrielle E.R.A. No 536
2 rue Camichel, 31071 Toulouse Cédex, France
Centre d'Etudes et de Recherches de Toulouse, 2 avenue Edouard Belin, 31 Toulouse, France

INTRODUCTION

Static energy converters are often studied using analytical methods. In this case their operation is broken up into a succession of elementary sequences depending on the conduction or non-conduction of the semi-conductors which are considered as perfect switches. For circuits with a low number of semi conductors these methods are completely justified and lead to good qualitative and quantitative representation of the converter under investigation.
In order to take account of a greater number of secondary phenomena (Damping, Influence of protection circuits...) or for example to investigate the effect of parameter variation, digital computer simulations based on the same principle can be used. However such methods are soon limited by the size of the circuit. Furthermore, the order of the sequences must be known perfectly and must not change during the simulation. This types of method cannot, therefore, be considered as a means of investigation in the full sense.
It thus seems indispensable to use more general methods and to resort to a global simulation for static converters (Ref. 1, 2, 3). This enables the operation of any circuit to be described without the a-priori knowledge of the order of the conducting sequences. Both the steady state and transient behaviour of complex circuits with a large number of semiconductors can be analysed. Such a method, based on a resistive model of the semiconductors, a state variable representation of the equations and the Runge-Kutta resolution method has already been presented by the authors in 1974 (Ref. 4). The difficulties encountered in the utilisation of this method originate from the excessive computing time due to the very small calculation step inherent in numerical algorithms.
Thanks to the financial help given by the D.G.R.S.T. to the L.E.E.I., to the D.E.R.A.-C.E.R.T. and to the Laboratoire d'Electrotechnique of the E.N.S.E.G.P., this work has been restarted, a critical analysis made in order to develop the positive aspects and find solutions for the disadvantages of the previous method.
It is the new method, leading to the development of the programme S.A.C.S.O., which is presented in this article (Ref. 5, 6).

DESCRIPTION OF THE METHOD

1. Semi-conductor Model

The electric model chosen to represent the semiconductors is the first fundamental aspect in the conception of a simulation method. Both the accuracy and complexity of the simulation depend upon the exactness of the model chosen.
The more elaborate models, which take into account the microscopic phenomena present in the semiconductors, have been avoided as they lead to very laborious and complex simulations because of the large number of parameters which need to be considered.
A certain number of simulation have been undertaken in which the semiconductor is considered as an open circuit when non-conducting and as an impedance when turned-on. The number of elements in the circuit is reduced by the presence of the open circuits and thus the order of the system is reduced. However, the problem is then to find, at each commutation, the topology and the equations of the resulting circuit. In order to avoid the difficulties posed by a general programme using this method, a resistive model has been chosen : the semiconductors are represented by a resistance taking two greatly different values depending on the conducting state. The value of the resistance is determined by testing the voltage appearing across it.
The introduction into the circuit of resistances of very different values causes a great dispersion in the various time constants present in the circuit. A method which overcomes this difficulty and at the same conserves a good degree of accuracy and a low computation time has consequently been chosen.

2. Resolution Method

With a view to develop a general programme, possessing a great simplicity of utilisation, the research of the system of equations is designed to be completely automatic using only a description of the position of the various components in the circuit. The method used is based on the classical results of graph theory (Ref. 7, 8).
The equations are determined in state variable form.

$$\dot{x} = Ax + Bu \qquad x\text{: state vector}$$
$$y = Cx + Du \qquad u\text{: input vector}$$
$$\qquad y\text{: output vector}$$

Since the semiconductor model is resistive, the system of equations is linear by section and the matrices A, B, C, D have constant coefficients between any two commutations. If the input vector U is constant then the solution is given analytically in the form of a recurrent equation (Ref. 7).

$$x(t+T) = e^{A.T}x(t) + A^{-1}(e^{A.T} - \mathbb{1}).B.u$$

where by definition

$$e^{AT} = \mathbb{1} + A.T + \frac{(A.T)^2}{2!} + \cdots + \frac{(A.T)^n}{n!}$$

The calculation step, T, can mathematically take any value. It can, in fact, be considered as an observation interval and, as opposed to the calculation step in interative methods, it is not necessary for it to be of the same order of magnitude as the smallest time constant in the circuit. This overcomes the disadvantages due to the great dispersion of the time constants that has been mentioned above.
Furthermore, the calculation step being relatively large, the amount of computation is reduced and the gain in computation time distinct.
In order to illustrate the performance and possibilities open to this resolution method and to analyse the influence of varying the calculation step, a single phase of a McMurray Bedford inverter operating at 400 Hz has been simulated (Fig. 1).
Figure 2 represents the results obtained with a calculation step of 12.5μs i.e. 1/200 th of the operating period. Decreasing the calculation step does not modify these results but only increases the computation time.
The observation step has to increase to almost 1/40th of the period (65 μs) before an anomaly in the programme occurs. This anomaly is seen as a fictitious commutation fault caused by the non-detection of the turn-off of a thyristor (Fig. 2).

3. Change Of Conducting State Of A Semi-conductor

The determination of the exact instant at which the semiconductors change their conducting state is one of the key points of the simulation.
Remembering that the semiconductor model used is resistive, the conducting state of a semiconductor is determined according to the voltage across it and, in the case of a thyristor, its firing control signal.
The firing of the thyristors is controlled but the other changes of conducting state (thyristor turn-off, diode turn-on and turn-off) occur when the voltage across the semiconductor changes polarity.
It is therefore necessary to :

- detect the changes in polarity of the voltage across the semiconductors

- determine if these changes actually give rise to a change of state.

This can be done with the aid of the logic flow diagram shown on Fig. 3 which represents the logical behaviour of semiconductors. In the case where a change in state should actually occur then it is necessary to determine accurately the instant at which it takes place i.e. the instant at which the voltage falls to zero. This operation is made necessary by the use of a large calculation step : two commutations may be separated by an interval of time smaller than the calculation step. The method chosen consists of using a smaller step near the voltage zero in order to obtain a greater accuracy.

APPLICATION EXAMPLE

The following example illustrates the performance of the programme SACSO. The simulation concerns a current fed inverter supplying an active load which represents approximately a synchronous machine at one of its operating points (Fig. 4). This converter is one of the most complex of the commoner circuits because of the interdependence of the 3 phases.
Furthermore it has the characteristic of possessing extremely different types of operation depending on the value of the phase shift between the load current and the voltage.
For a phase lag (the most common case which corresponds to the special case of asynchronous machines) the operationnal behaviour is extremely simple and predictable. Here the case of a phase lead is presented, an operating point perfectly accessible for a synchronous machine, but where the operational behaviour becomes very complicated, perpetually disturbed by spontaneous reconduction of the isolating diodes (Ref. 9). No less than sixty commutations by period can be observed. This makes the analysis impossible for the more usual methods of investigation.
The comparison of the simulation results with those determined experimentally (Fig. 5) shows in a rather spectacular way the effectiveness of the programme.

CONCLUSION

The problems posed by the simulation of static converters are very complex because of the basically discontinuous operation of these systems. The programme presented is specific to the analysis of the static converters but is absolutely general in this field of application. All the operational and control characteristics of these converters can be taken into account without the programme becoming too laborious and the computation time excessive. It is been shown using important examples that the programme SACSO because of its effectiveness and accuracy can be easily considered as a means of in-

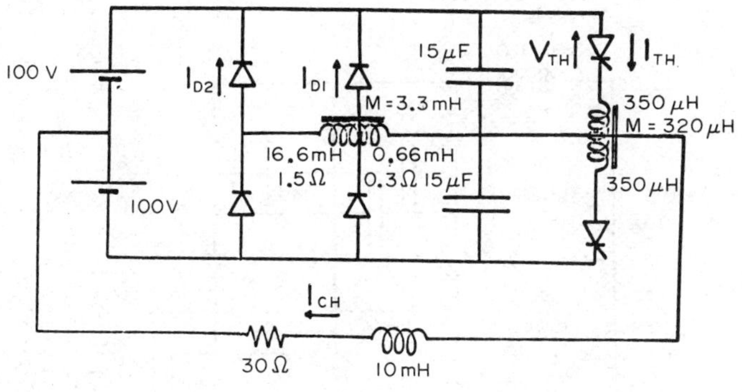

Fig.1. McMurray Bedford inverter circuit

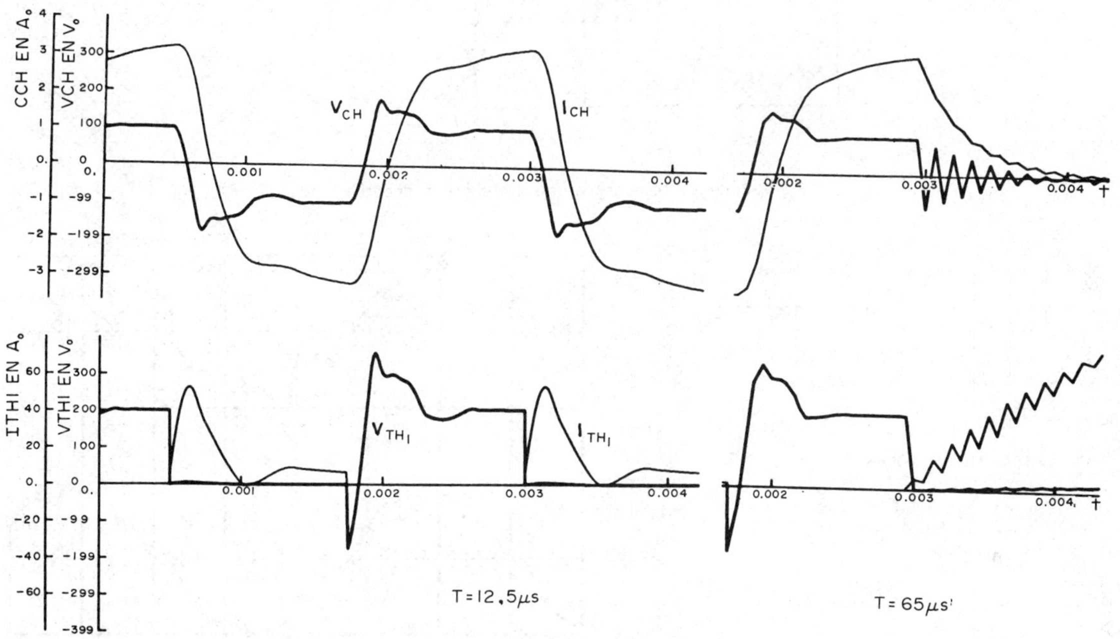

Fig.2. Simulated results obtained with two different calculation steps

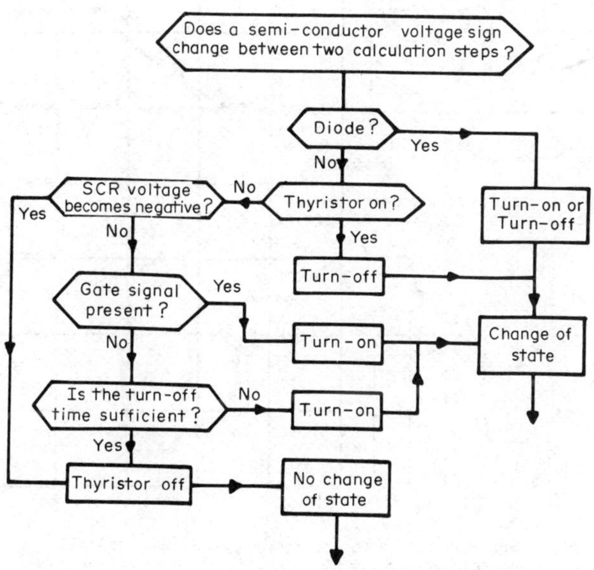

Fig.3. Flow Diagram for semi-conductors

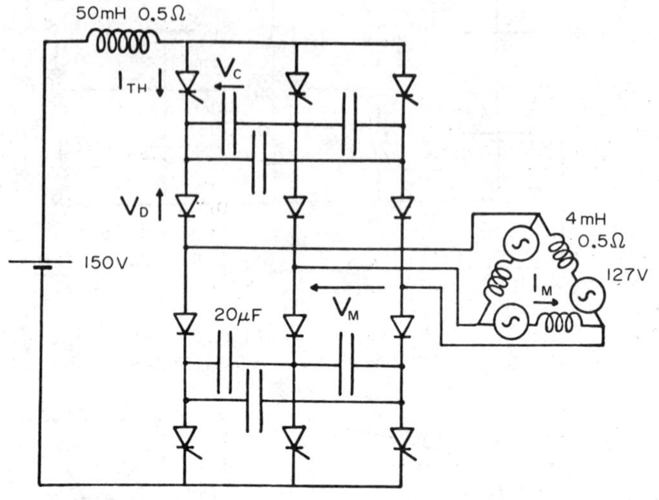

Fig.4. Current-fed commutator circuit with an active load

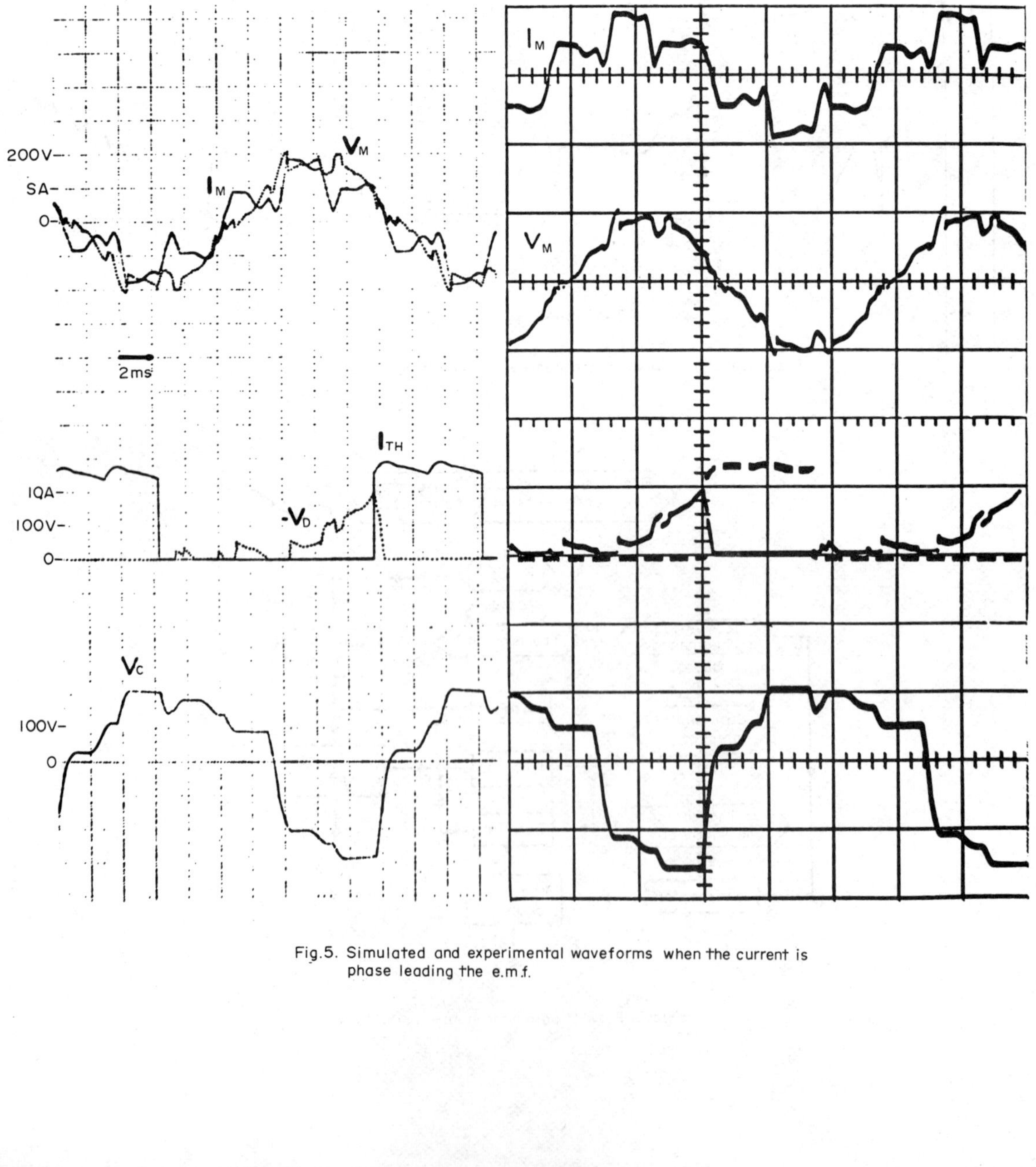

Fig.5. Simulated and experimental waveforms when the current is phase leading the e.m.f.

vestigation able to replace experimentation at all stages of the conception and development of a converter system.

REFERENCES

(1) G.N. Revankar, Computer analysis of SCR circuits, IEEE Transactions on Industrial Electronics and Control Instrumentation, February (1975)

(2) E.R.C. Da Silva, Contribution à l'étude et au calcul des circuits comportant des éléments unidirectionnels de puissance : méthode globale de simulation, identification. Application à un onduleur MacMurray Bedford, Thèse de Docteur Ingénieur, Toulouse (1972)

(3) V. Rajagopalan, K. Sankara Rao, User's manual on topological analysis of power electronic converter systems : Atosec version 1, Université du Québec à Trois Rivières, Canada

(4) H. Foch, B. Trannoy, J. Faucher, Complete simulation of a static converter by digital and analogue methods, IFAC Symposium "Control in power electronics and electrical drives", Duesseldorf, 1, 363-373, Octobre (1974)

(5) H. Foch, C. Reboulet, J. Schonek, Simulation automatique de convertisseurs statiques sur ordinateur - Programme SACSO Optimisation, Rapport DGRST, Mai (1976)

(6) J. Schonek, Simulation numérique de convertisseurs statiques. Elaboration d'un programme général. Application à la conception et à l'optimisation des convertisseurs, Thèse de Docteur Ingénieur, Toulouse (1977)

(7) N. Balabanian, T.A. Bickart, Electrical network theory, Wiley (1969)

(8) A. N'Dir, Contribution à l'étude des surtensions de manoeuvre sur les lignes à très haute tension (programme Pégase) Thèse de Docteur ès Sciences, Grenoble (1971)

(9) M. Lajoie-Mazenc, Contribution à l'étude de la commutation statique et à la conception de nouvelles structures de machines électriques, Thèse de Docteur ès Sciences Physiques, Toulouse (1969)

A DIGITAL SIMULATION OF CONVERTER CIRCUITS

A. Coffetti, G. Petrecca and A. Savini

Department of Electrical Engineering, Pavia University, Strada Nuova 106/c, 27100 Pavia, Italy

ABSTRACT

A digital simulation procedure for circuits containing semiconductor devices (diodes, thyristors) is described. A Boolean functional representation of thyristors and diodes is introduced and the state variable representation is utilized to obtain the circuit model. The state system is then resolved by computing the state transition matrix of the equivalent augmented unforced system.

In order to show the versatility of this procedure two typical examples in the field of d.c.-d.c. and a.c.-d.c. static conversion are given.

INTRODUCTION

Digital simulations of systems with converters and electromechanical components are becoming more and more important due to the increasing complexity and power range of such systems.

Converter circuit behaviour is generally investigated by cutting the duty cycle into many time intervals according to the on-off state of converter semiconductor components. This approach requires a previous knowledge of the converter modes of operation and for this reason it cannot be easily employed every time that fault or transient operations must be investigated. In addition different modes of operation of a same circuit may require a too complicated previous analysis (Ref. 1). A complete digital simulation of such systems, which utilizes semiconductor models, is then recommended.
In order to cope with the task of obtaining a good semiconductor model through a simple procedure, a Boolean functional representation of the most popular power semiconductor devices is introduced and the off state is associated with an open circuit branch (Ref. 2-3). Consequently, different topological configurations (and therefore different state circuit models) are associated with different modes of operations.

The proposed program automatically determines the state matrices connected with different topological configurations. State differential equations are organized in canonical form and their resolution is obtained by transforming the original nonomogeneous system into an unforced augmented equivalent one.
The state variables are then computed, starting from a set of initial values, by means of an iterative procedure which utilizes the state transition matrix evaluation.
In this way the computation time is reduced, if compared with the Runge-Kutta method, and instability problems are avoided (Ref.5-6).

In order to show the versatility of this procedure two typical examples in the field of d.c.-d.c. and a.c.-d.c.static conversion are given.

COMPUTER PROGRAM DESCRIPTION

A general flow diagram of the whole program is outlined in Fig. 1 where the most important blocks are pointed out. A more detailed description of these blocks is reported in the course of the paper.

Input Data (Block 1)

Data which must be introduced at the beginning of the program are reported as follows:

Network indexes. Numbers of branches (NE), nodes (NN), voltage generators (NUMG), capacitors (NUMC), resistors (NUMR), semiconductors (NS), inductors (NUML), state variables of the adjoint system (NUMGG) are introduced together with different linear combinations of the above. These indexes are intended as "parameters" because of the variable dimensions of all matrices.

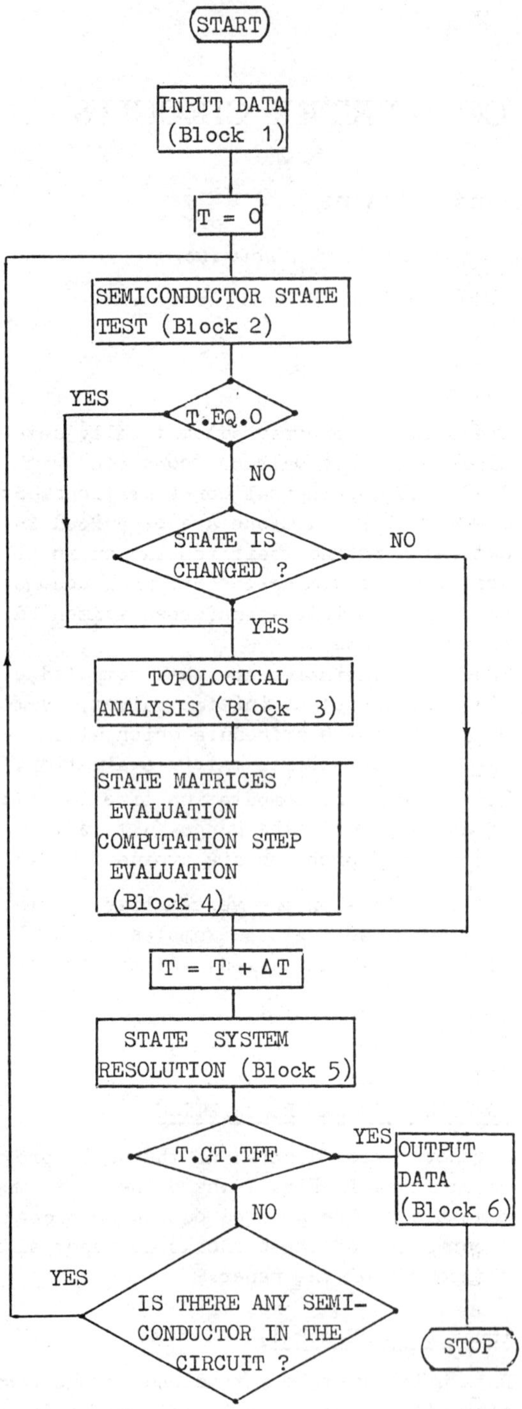

Fig. 1 - The general flow diagram of the whole program.

Network branch value vector VALNUC (1,NE).

This vector contains in the following order voltages of generators, capacitances, resistances, on-state semiconductor resistances, self-inductances . If the network contains mutual inductances two additional vectors are introduced.

If voltage generators are time varying, values at the initial instant are reported in VALNUC.

In addition, initial values of the adjoint state variables are given in DERIF (1, NUMGG) vector.

First node vector NF(1,NE) and terminal node vector NT(1,NE). These vectors contain respectively the node number from which a branch starts and the node number to which the same branch goes.

Voltage sources with a positive value must have the positive terminal in NF (and the negative in NT).

Semiconductor devices, as diodes and thyristors, must have the anode terminal in NF. In the case of diode and thyristor antiparalleled, the thyristor anode terminal must be in NF.

Branch characterizing vector TE (1,NE). Elements of this vector have the following values: 1 for voltage generators, 2 for capacitors, 3 for resistors, 4 for thyristors, 5 for diodes and thyristors antiparalled, 6 for diodes, 7 for inductors.

Initial values of currents and voltages. These values are reported in vectors EBC(1, NE) and EBV(1,NE).

Branch initial mode. The mode index has value 0 in the case of an open circuit branch and 1 in the opposite case. Its value at the beginning of computation is recorded in vector ME(1,NE).

Gate impulse sequence. In order to correctly define this sequence three vectors are utilized: TEMACC(1,NE) whose elements are the first time values at which a trigger pulse is applied to different thyristors; DELGR(1,NE) whose elements are the terminal instant values at which the first trigger pulse is applied; DSGR(1,NE) whose elements are the time interval values between two following trigger pulses.

If a thyristor is conducting at the beginning of the computation, the corresponding element of vector GR(1,NE) has value equal 1.

Additional data. On the assumption that all semiconductors have the same characteristics the holding current value (IMAN) and the forward conduction voltage value (DELV) are finally introduced into the program.

Base quantities and the last time of computation are also defined (TFF).

Semiconductor Models and Test on Their State (Block 2)

On the basis of vector TE, semiconductor branches corresponding to TE(1,i)=4,5,6 are selected in order to test their state. Models of the most popular power semiconductors (diode, thyristor, diode and thyristor antiparalled) are introduced by means of a Boolean functional representation. The on-off state of each semiconductor is dependent upon the following informations referred at the beginning of each step of computation: anode-cathode voltage, anode-cathode current, presence of trigger signal for thyristors only.

During the "on" state, anode-cathode currents and voltages are the currents and voltages of the resistor equivalent branches. If one semiconductor is on "off" state, the anode-cathode voltage is calculated by finding a loop, in which the corresponding semiconductor branch is introduced as a link, and subsequently applying the Kirchhoff voltage law. In this way anode-cathode voltages at the beginning of computation (when all semiconductors may be on "off" state) as well as at the initial stage of all following steps can easily be evaluated.

According to a digital modelling procedure on the basis of a Boolean representation (Ref. 2-3), the following models are introduced:

$$C_D=\bar{C}_D'*V_D+I_D \qquad \text{diode}$$
$$C_T=\bar{C}_T'*V_T*G+I_T \qquad \text{thyristor} \qquad (1)$$
$$C_{TD}=C_T-C_D=\bar{C}_T'*V_T*G+I_T-\bar{C}_D'*V_D-I_D \qquad \text{diode and thyristor antiparalleled}$$

where

$C_{D,T,TD}$ is the logical state indicator at the beginning of the following step of computation ;

$\bar{C}'_{D,T,TD}$ is the coniugate of the logical state indicator in the preceding step. At the beginning of the calculation $\bar{C}'_{D,T,TD}$ is the coniugate of the ME vector elements;

G has values equal 1 or 0 according to the vector GR elements;

V_{AK} = 1 if the anode-cathode voltage value is greater than the forward conduction voltage; = 0 if the anode-cathode voltage value is less than or equal to the forward conduction voltage;

I_{AK} =1 if the anode-cathode current value is greater than or equal to the holding current;

I_{AK} =0 if the anode-cathode current value is less than the holding current.

Semiconductor states thus calculated are then utilized for the following step of computation and the mode vector ME is altered according as logical state indicator $C_{D,T,TD}$ values.

Topological Analysis, State Matrices Evaluation and Resolution (Blocks 3-4-5)

The topological analysis and the state matrices evaluation is performed by the classical approach (Ref. 7-8) every time the ME mode circuit vector changes. This way the open circuit branches are removed from the circuit and a modified reduced incidence matrix is obtained.

At this stage the loop matrix which expresses the topological relation between links and tree branches can be easily found and the state system in the classical network theory formulation (Ref. 7-8-9) is obtained by introducing the circuit parameter values:

$$[\dot{x}] = [MATRA][x]+[MATRB][u]+[MATRC][\dot{u}] \qquad (2)$$

where

$[x]$ is the state variable vector containing capacitor voltages and inductor currents which are the active part of the circuit at this step of computation ;

$[u]$, $[\dot{u}]$ are the input vector and its derivative;

$[MATRA]$, $[MATRB]$, $[MATRC]$ are the state matrices which depend on the topology of the circuit and on its parameter values.

System (2) can be modified by introducing an adjoint state system, from the state variables of which input variables can be defined (see also Appendix 1) :

$$[\dot{\mu}] = [\Omega][\mu] \qquad (3)$$
$$[u] = [\Gamma]\cdot[\mu]$$

and substituting in (2)

$$[\dot{x}] = [MATRA][x] + [MATRB][\Gamma][\mu] + [MATRC][\Gamma][\Omega][\mu] \qquad (4)$$
$$[\dot{\mu}] = [\Omega][\mu]$$

so that an equivalent unforced system is obtained

$$\begin{bmatrix}\dot{x}\\ \dot{\mu}\end{bmatrix} = \begin{bmatrix} MATRA & MATRB*\Gamma+MATRC*\Gamma*\Omega \\ 0 & \Omega \end{bmatrix}\begin{bmatrix}x\\ \mu\end{bmatrix} = [S]\begin{bmatrix}x\\ \mu\end{bmatrix} \qquad (5)$$

subject to the given initial condition

$$\begin{bmatrix} x(0) \\ \mu(0) \end{bmatrix} \tag{6}$$

The solution of system (5), subject to the initial condition (6), is computed by means of an iterative method (Ref. 9)

$$\begin{bmatrix} x(n\,\Delta t) \\ \mu(n\,\Delta t) \end{bmatrix} = e^{[S]\,\Delta t} \begin{bmatrix} x((n-1)\,\Delta t) \\ \mu((n-1)\,\Delta t) \end{bmatrix} \quad n=1,2.. \tag{7}$$

The exponential matrix $e^{[S]\Delta t}$ is calculated as a matrix series by means of a subroutine which utilises a particular algorithm (Ref. 4).

The computation step time interval Δt is automatically defined every time the circuit state is altered by computing the minimum between S matrix norms evaluated by rows and by columns and then by imposing the condition $||\Delta t\ S|| < 1$. Therefore the time computation interval must satisfy the following condition $\Delta t < 1/||S||$ (Ref. 4).

The circuit resolution is then completed evaluating currents and voltages corresponding to all branches.

At the end of each computation step, semiconductor states are tested (see Block 2 of Fig. 1) and the mode vector ME is accordingly altered. If a change in vector ME elements occurs, the program goes on through Block 3 to analyse the new circuit configurations and to perform the new state system and the new computation step Δt.

If resistors or inductors are current or time dependent, the state system must be calculated again at each step by skipping the topological analysis Block 3.

Printing and Plotting Output (Block 6)

A network resolution program has a typical output represented by currents and voltages in all its branches, which can be printed or plotted according to the user's requirements (it must also be noticed that the computation is performed by introducing p.u. values).

Additional outputs, as Fourier coefficients, r.m.s. and average values, which can be computed step by step, are available if required.

EXAMPLES OF COMPUTATION

Two examples of application of the proposed program are reported in the field of d.c.-d.c. and a.c.-d.c. static conversion. Figure 2 shows a self-commutating chopper, whose output can only be controlled by varying the pulse repetition period because of the invariable pulse duration. Data input are reported in Table 1 and transient waveforms of the most interesting currents and voltages are reported in Fig. 3.

A second example is reported with reference to a three-phase rectifier bridge in order to show the flexibility of this program both in state system resolution and in the evaluation of converter operation modes. Figure 4 shows the above rectifier circuit. The corresponding input data are reported in Table 2 and current waveforms are reported in Fig. 5.

CONCLUSION

This program has proved to be a powerful method of resolution of networks containing semiconductors and it can be profitably applied to any kind of circuits with static converters and electromechanical devices. It is worth noticing that a previous knowledge of the circuit behaviour is not necessary and that the program automatically defines all modes of operation.

State system resolution is performed by evaluating the exponential matrix every time the mode circuit vector ME changes and by introducing an iterative method which allows very short computation time. In this way, instability problems, which generally arise in static converter digital simulation, are avoided.

AKNOWLEDGEMENTS

The authors wish to thank Prof. G.Cottafava, Department of Electrical Engineering Pavia University, for the insight offered with respect to the application of the network teory.

This work has been developed at the Department of Electrical Engineering, University of Pavia, with the financial support of C.N.R., Gruppo Macchine Elettriche.

APPENDIX 1

Two different kind of input waveforms are considered.

Constant Inputs

$$[u] = [K] \quad ; \quad [u]\ (NUMG,1) \tag{8}$$

The adjoint state system is assumed as follows

$$[\dot{\mu}] = [\Omega][\mu] = [0][\mu] \quad ; \quad [\mu]\ (1,1)$$
$$[u] = [K][\mu] \tag{9}$$

TABLE 1 Input Data of Fig. 2 chopper

VALNUC	100.	0.0005	1.0	0.01	0.01	0.01	0.00005	0.001
DERIF	100.							
NF	2	4	5	2	4	1	3	6
NT	1	1	6	3	5	5	4	1
TE	1	2	3	4	6	6	7	7
EBC	0.	0.	0.	0.	0.	0.	0.	0.
EBV	100.	0.	0.	0.	0.	0.	0.	0.
ME	1	1	1	0	0	0	1	1
TEMACC	0.	0.	0.	0.	0.	0.	0.	0.
DELGR	0.	0.	0.	0.0001	0.	0.	0.	0.
DSGR	0.	0.	0.	0.004	0.	0.	0.	0.
GR	0	0	0	1	0	0	0	0
IMAN = 0.001 (p.u.); DELV = 1 V; IB = 100 A; VB = 100 V; FB = 250 Hz; TB = 1/(2πFB)								

TABLE 2 Input Data of Fig. 4 rectifier bridge

VALNUC	311.	−155.5	−155.5	4.	0.01	0.01	0.01	0.01	0.01	0.01	0.001	0.001	0.001	0.01
DERIF	0.	311.												
NF	2	3	4	10	8	8	8	5	6	7	5	6	7	9
NT	1	1	1	8	5	6	7	9	9	9	2	3	4	10
TE	1	1	1	3	6	6	6	6	6	6	7	7	7	7
EBC	0.	0.	0.	0.	0.	0.	0.	0.	0.	0.	0.	0.	0.	0.
EBV	311.	−155.5	−155.5	0.	0.	0.	0.	0.	0.	0.	0.	0.	0.	0.
ME	1	1	1	1	0	0	0	0	0	0	1	1	1	1
TEMACC	0.	0.	0.	0.	0.	0.	0.	0.	0.	0.	0.	0.	0.	0.
DELGR	0.	0.	0.	0.	0.	0.	0.	0.	0.	0.	0.	0.	0.	0.
DSGR	0.	0.	0.	0.	0.	0.	0.	0.	0.	0.	0.	0.	0.	0.
GR	0	0	0	0	0	0	0	0	0	0	0	0	0	0
IMAN = 0.001 (p.u.); DELV = 1 V; IB = 100 A; VB = 311 V; FB = 50 Hz; TB = 1/(2πFB)														

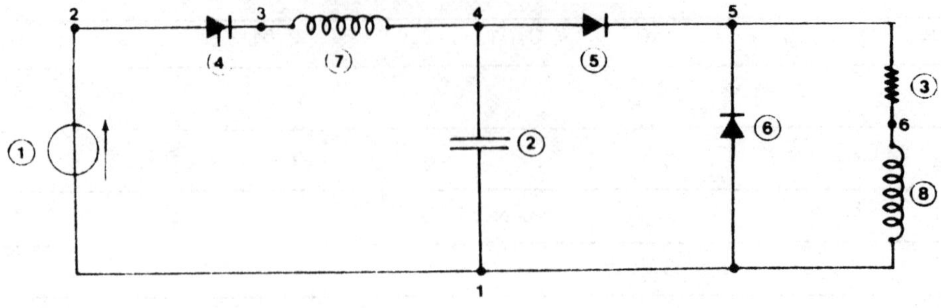

Fig. 2. Autocommutating chopper circuit (see Table 1)

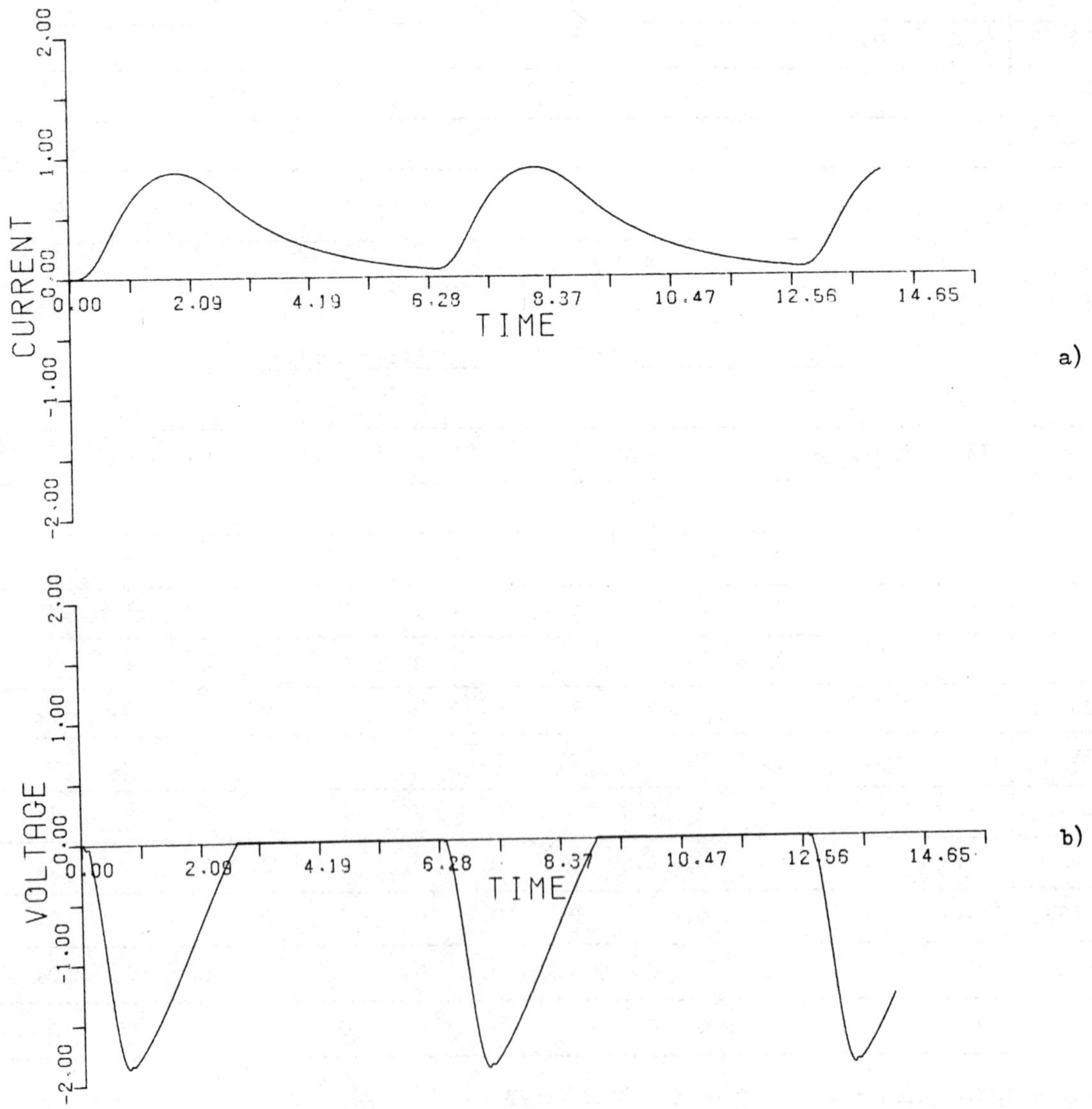

Fig. 3. a) branch 3 (or 8) current

b) branch 6 voltage

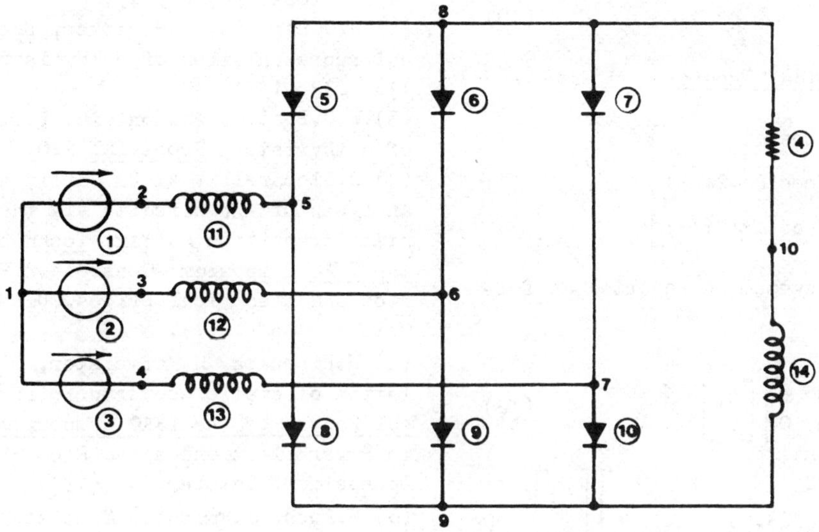

Fig. 4. Three-phase rectifier bridge (see Table 2)

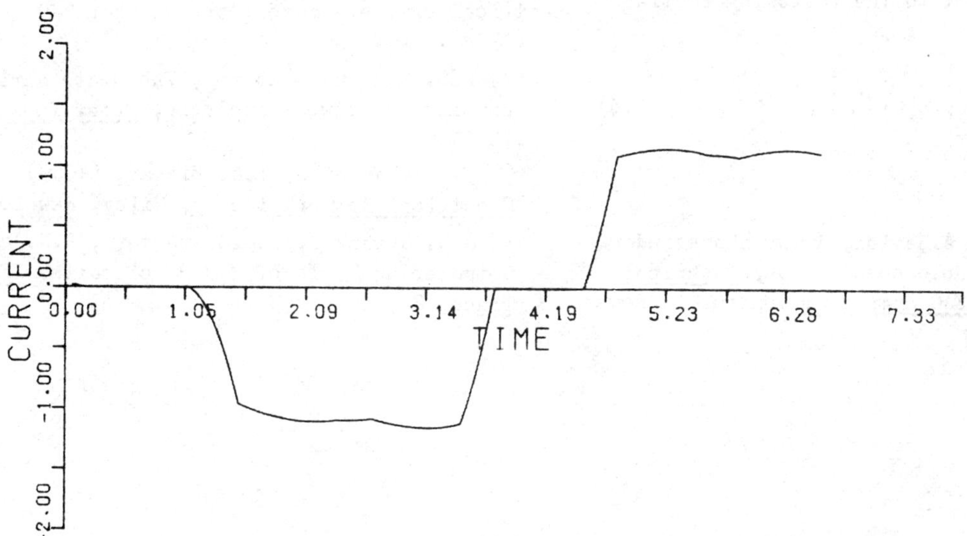

a)

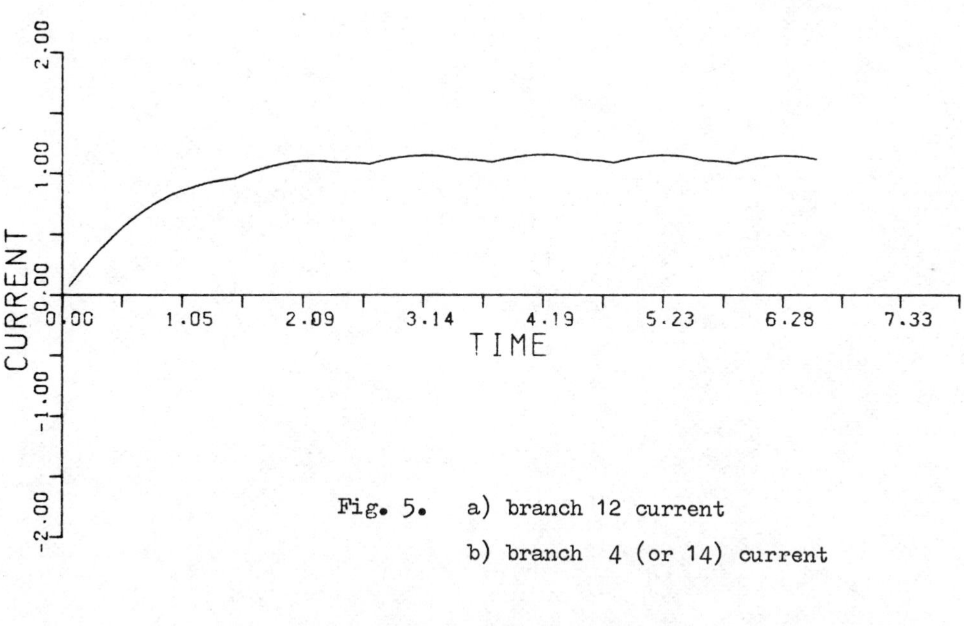

b)

Fig. 5. a) branch 12 current

b) branch 4 (or 14) current

and it is subject to the following condition

$$[\mu(0)] = [1] \tag{10}$$

Three-Phase Sinusoidal Inputs

$$\underset{(3,1)}{[u]} = \begin{bmatrix} u_1(t) \\ u_2(t) \\ u_3(t) \end{bmatrix} = \begin{bmatrix} U\cos \omega t \\ U\cos(\omega t - 2\pi/3) \\ U\cos(\omega t + 2\pi/3) \end{bmatrix} \tag{11}$$

The adjoint state system is reported as follows

$$[\dot{\mu}] = [\Omega][\mu] = \begin{bmatrix} 0 & \omega \\ -\omega & 0 \end{bmatrix} \begin{bmatrix} \mu_1 \\ \mu_2 \end{bmatrix} \tag{12}$$

$$[u] = [\Gamma][\mu] = U \begin{bmatrix} 0 & 1 \\ \sin 2\pi/3 & \cos 2\pi/3 \\ -\sin 2\pi/3 & \cos 2\pi/3 \end{bmatrix} \begin{bmatrix} \mu_1 \\ \mu_2 \end{bmatrix} \tag{13}$$

and it is subject to the following initial conditions

$$[\mu(0)] = \begin{bmatrix} 0 \\ 1 \end{bmatrix} \quad ; \quad [\mu]\ (2,1) \tag{14}$$

REFERENCES

(1) G.Petrecca, A.Savini, A two phase induction motor variable speed drive, Technical Papers of the IFAC Symposium Control in Power Electronics and Electrical Drives, Duesseldorf October 7-9 (1974).

(2) W.C.Beattie, J.H.Parker, Boolean functional representation of a thyristor, Proc.IEE 118, 232 (1971)

(3) W.C.Beattie, W.Monteih, Digital modelling of a thyristor, Proc. IEE 120, 789 (1973).

(4) G.Ciccarelli, A. De Carli, M.La Cava, Analysis of SCR circuits via the augmented state transition matrix, Technical Papers of the IFAC Symposium Control in Power Electronics and Electrical Drives, Duesseldorf, October 3-5 (1977).

(5) H.Eisenack, H.Hofmeister, Digital simulation of static converter circuits, Technical Papers of the IFAC Symposium Control in Power Electronics and Electrical Drives, Duesseldorf October 7-9 (1974).

(6) H.Foch, J.Faucher, Complete simulation of static converter by digital and analogue method, Technical Papers of the IFAC Symposium Control in Power Electronics and Electrical Drives, Duesseldorf October 7-9 (1974).

(7) E.S. Kuh, R.A. Rohrer, The state variable approach to network analysis, Proc. IEEE , 672 (1966).

(8) N. Balabanian, T.A. Bickart (1969) Electrical Network theory, Wiley, New York.

(9) G.N. Revankar, P.K.Srivastava, R.N.Jawle, Computer analysis of S.C.R. circuits, IECI Trans, 22, 48 (1975).

APPLICATION OF STIFFLY STABLE ALGORITHMS IN THE DIGITAL COMPUTATION OF THE SCR CIRCUITS

T. Kutman

Technical University of Istanbul, Faculty of Engineering and Architecture, Beşiktaş, Istanbul, Turkey

ABSTRACT

In the digital computation of the systems containing SCR circuits there arise generally two problems. One of them is the accuracy of the mathematical model used for the representation of the characteristics of the silicon switching element. The other one is the difficulty in the numerical integration of the state equations which have usually widely separated time-constants. Using a modified model for SCR elements this paper has adopted the stiffly stable algorithm, given by Gear (1), in the digital computation of the SCR circuits, and it is shown that the numerical integration problem can be solved without any fictitious components or changing the structure of the system.

INTRODUCTION

Most of the proposed models for the silicon power switches like diode, thyristor, etc., represent the element as a high ohmic or a low ohmic resistance according to the blocking and conducting states respectively (2), (3), (4). However, there arises a difficulty in the digital computation of the SCR circuits, when this resistive nonlinear model is used. Due to the different operating states the time-constants of the circuits are widely separated, because the resistances in the model differ by several orders of magnitude (e.g. 100 kΩ in blocking state and 0.01 Ω in conducting state). When the used numerical integration algorithm is not suitable for stiff differential equations, these widely separated time-constants can cause numerical instability, if the integration step size is not sufficiently small. On the other hand the number of iterations increases, if a smaller step size is chosen. In order to avoid the problem some authors have considered thyristor as an ideal switching element (5) or represented it as a high ohmic resistance connected in series with a large inductance in the blocking state (3). However, there is another case which can not be avoided. Many of the power electronic circuits are used in the control of electrical machines which have greater time-constants in comparison with the electronic part. This is another reason for the stiffness of the state equations of the systems which contain SCR circuits.

MATHEMATICAL MODEL

The mathematical model which is used for SCR in this paper, is a modified version of the model, given in (2). According to the nonlinear equivalent circuit in Fig. 1 this model

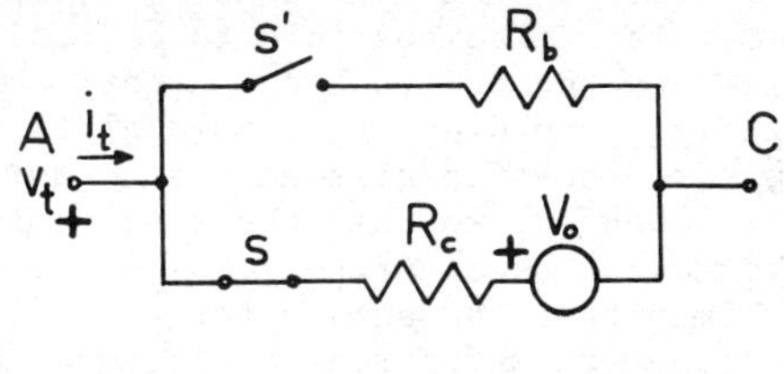

Fig. 1. Equivalent circuit for SCR

can be easily written as

$$v_t = (R_c \cdot s + R_b \cdot s') \cdot i_t + V_o \cdot s \quad ,$$

where

R_c : Resistance of the SCR in the conducting region (0.01 Ω - 0.5 Ω)

R_b : Resistance of the SCR in the blocking region (10 kΩ - 100 kΩ)

V_o : Conduction voltage drop of the SCR at the holding current or threshold voltage of the diode (0.6 V - 1.1 V)

s : Binary logic state variable representing the complementary contacts in the equivalent circuit ($s = 1$, if SCR is on ; $s = 0$, if SCR is off).

The value of the logic state variable is determined from the Boolean function

$$s = s_v \cdot (s_i + s_g) \quad ,$$

where

s_i : Binary logic variable corresponding to the current of the SCR ($s_i = 1$, if $i_t \geq I_H$; $s_i = 0$, if $i_t < I_H$; I_H: Holding current)

s_v : Binary logic variable corresponding to the anode - cathode voltage of the SCR ($s_v = 1$, if $v_t \geq V_o$; $s_v = 0$, if $v_t < V_o$)

s_g : Binary logic variable corresponding to the triggering signal of the SCR .

In order to take the turn-on and turn-off times of the thyristor into consideration the logic variable s_g is defined as follows:

$s_g = 1$, if

a) at the end of the turn-on time there is still a trigger signal at the gate of the thyristor,
b) the thyristor current is less than the holding current and the turn-off time is not yet over.

$s_g = 0$, in all other cases.

INTEGRATION ALGORITHM

The numerical integration algorithm which is proposed for the digital computation of the SCR circuits in this paper, is given by Gear (1). It is a multistep predictor-corrector algorithm and suitable for programming on digital computer. The numerical stability of the so-called stiffly stable algorithm is rather independent of the integration step size (6). Therefore it is not necessary to perform the numerical integration with smaller step sizes than the smallest time-constant of the system. This property decreases the number of iterations and shortens the computation time. It is also possible to control the step size and the order of the algorithm automatically. According to the local truncation error, the step size and the order can be easily changed, so that the transients in the system can be accurately obtained. This is an important advantage for the digital computation of the SCR circuits, because there often arise transients when a switching element turns on or off. Thus, the algorithm also offers an opportunity for the optimal selection of the circuit parameters.

APPLICATION OF THE ALGORITHM

Figure 2 shows a control system of which state

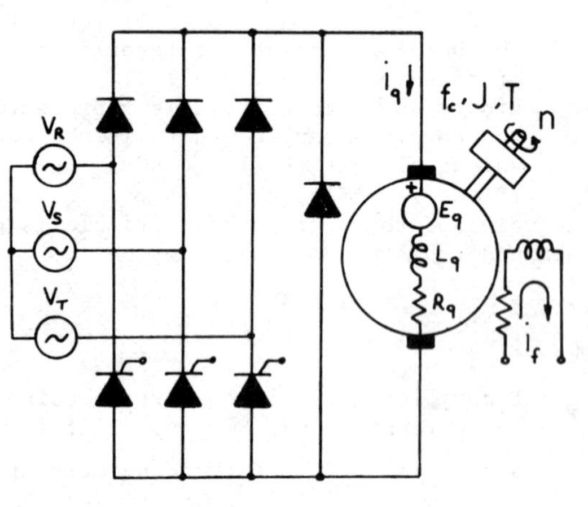

Fig. 2. A three-phase full-wave rectifier circuit supplying a DC motor

equations are computed as an example. The system consists of a three-phase full-wave rectifier circuit which supplies a separately excited DC motor. Assuming constant excitation current, the dependent voltage source E_q in the equivalent circuit of the motor is given as

$$E_q = k_v \cdot f(i_f) \cdot n = 0.126 \cdot n \quad ,$$

where

k_v : Speed-voltage constant of the motor

i_f : Excitation current

$f(i_f)$: Magnetic flux produced by the excitation current

n : Speed of the motor

Three thyristors in the rectifier circuit are triggered in sequence with a phase delay of 120^0. The trigger angle is chosen $\alpha = 0^0$. Therefore the thyristors act as diodes in the circuit. The motor is shunted with a by-pass diode in order to control the trigger angle in a wider range. The other parameters of the system are chosen as follows:

Supply voltages:

$v_R = 145 \cdot \sin(314 \cdot t + 1.57)$

$v_S = 145 \cdot \sin(314 \cdot t - 0.52)$

$v_T = 145 \cdot \sin(314 \cdot t - 2.62)$

Switching elements:
(for all diodes and thyristors)

$R_c = 17$ mΩ

$R_b = 100$ kΩ

$V_o = 0.9$ V

Motor circuit:
Electrical part:

$L_q = 10^{-2}$ H (rotor self inductance)

$R_q = 1.2\ \Omega$ (rotor resistance)

$i_{qnom} = 16.6$ A (nominal rotor current)

$i_q(+0) = 0$ A (initial value of the rotor current)

Mechanical part:

$f_c = 0.12$ J·sec (viscous friction constant)

$J = 0.208$ J·sec^2 (moment of inertia)

$T = 20.35$ N·m (load torque)

$n_{nom} = 1750$ rpm (nominal speed of the motor)

$n(+0) = 0$ rpm (initial value of the speed)

According to the values of these parameters, the greatest and the smallest time-constants of the system are equal approximately to 0.5 sec and 0.5 psec respectively. The state equations in which the rotor current i_q and the motor speed n are chosen as the state variables, are of the stiff type, because the maximum ratio among the time-constants is in the order of 10^{12}. The transient response which is obtained by the application of the model and the integration algorithm presented above, is shown in Fig. 3.

CONCLUSIONS

The results have shown that the stiffly sta-

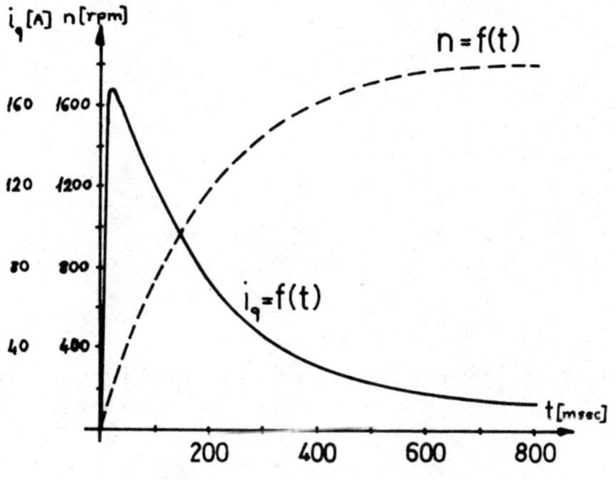

Fig. 3. Transient response of the DC motor shown in Fig. 2

ble integration algorithm, given by Gear, is very suitable in the digital computation of the systems containing SCR circuits. It is observed that the integration step size is automatically decreased when one of the switching elements changes its operating mode, and thus, it is shown that the resistive nonlinear model of the thyristor which represents nearly the same characteristics of the physical model, can be used without any fictitious components.

REFERENCES

(1) Gear, C.W. (1971) Numerical Initial Value Problems in Ordinary Differential Equations, Prentice-Hall, New Jersey.

(2) T. Kutman, A Method of Digital Computation for SCR Circuits, IEEE Trans. Ind. Electron. Contr. Instrum. 21, (1974).

(3) H. Eisenack and H. Hofmeister, Digitale Nachbildung von elektrischen Netzwerken mit Dioden und Thyristoren, Archiv für Elektrotechnik 55, (1972).

(4) G. Buja, Digital Simulation of Thyristor, Proc. European Conference on Circuit Theory and Design, (1976).

(5) G. N. Revankar, P.K. Srivastava and R.N. Jawle, Computer Analysis of SCR Circuits, IEEE Trans. Ind. Electron. Contr. Instrum. 22, (1975).

(6) Chua, L.O. and Lin, P.M. (1975) Computer Aided Analysis of Electronic Circuits, Prentice-Hall, New Jersey.

CALCULATION OF THE ELECTRICAL QUANTITIES OF CONVERTER CIRCUITS SHOWN IN THIS CASE FOR A QUENCHING, ASYMMETRICAL BRIDGE CIRCUIT (LUB)

R. Windmöller

Wissenschaftlicher Assistent am Lehrstuhl für Erzeugung und Anwendung elektrischer Energie Ruhr-Universität, 4630, Bochum, BRD

1. INTRODUCTION

When supplying d.c. traction motors from overhead lines it is possible to obtain a better power factor with single-phase, quenching asymmetrical bridge circuits than with pair of arms controlled bridge circuits /2/, /3/. Especially with respect to the cascade circuit consisting of a LUB and a pair of arms controlled bridge circuit control methods which can make much leading line currents possible are particularly suitable. Figure 1 shows the equivalent circuit of a LUB.

The capacitor C_L causes the turn-off of the main thyristors T_1 or T_2 by firing one of the turn-off thyristors T_{11} or T_{12}, while the capacitor C_S takes up the energy, which is stored in the inductances on the a.c. side during self commutation /1/, /2/. The exact calculation of the control characteristics of the circuit shown in Fig. 1 is only possible by a digital simulation, because of the inductance L_d, the value of which depends on the current, and the non-linear valve characteristics a complete, analytical calculation of the steady state is not possible. The time required for a complete simulation only permits the evaluation of a few characteristics. A good way to find the criteria for obtaining the values of the circuit parameters is to neglect all circuit parameters the influence of which on the circuit is only minor. It is then generally possible and easy to obtain a complete calculation. Only after all the important effects, which may occur in the ideal circuit have been determined, a digital simulation of the real circuit is possible. In the following three possiblities of how to calculate the control characteristics are given. All three only require a relatively short computation time. The programs for these computations were written in ALGOL 60. All the information about the computation times refer to the TR 440 (AEG-TELEFUNKEN), as it is installed at the Ruhr-Universität Bochum.

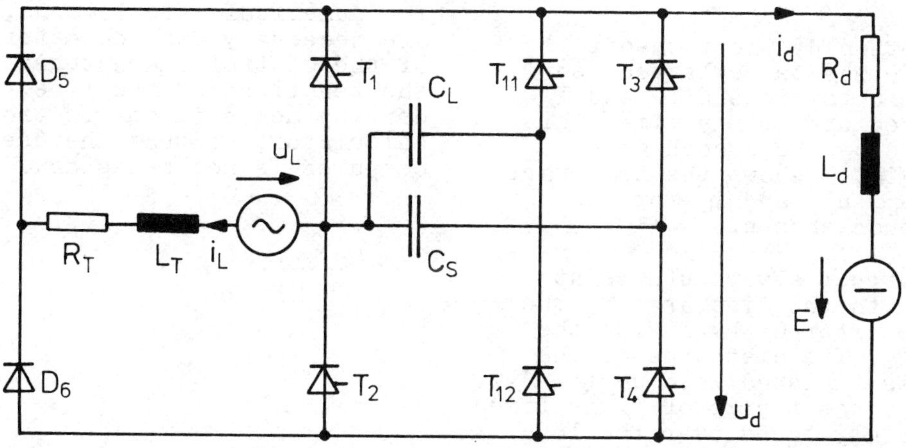

Fig. 1 Equivalent circuit of a LUB

2. CALCULATION OF THE CONTROL CHARACTERISTICS FOR THE IDEAL CIRCUIT

The calculations of chapter 2 are based on the following assumptions:

a) all losses on the a.c. and the d.c. side and the inductance L_T (Fig. 1) are neglected ($R_T = R_d = X_T = 0$)

b) the supply voltage is a true cosine wave

$u_L = \hat{u}_L \cdot \cos x \; ; \; x = 2\pi \cdot f_L \cdot t$

c) the smoothing inductance L_d is so big, that all alternating parts of the load current may be neglected ($L_d \rightarrow \infty$)

d) all valves are loss free and ideal - also with respect to their hold-off interval

e) the load voltage E contains no alternating parts.

Under these assumptions one obtains the equivalent circuit of Fig. 2.

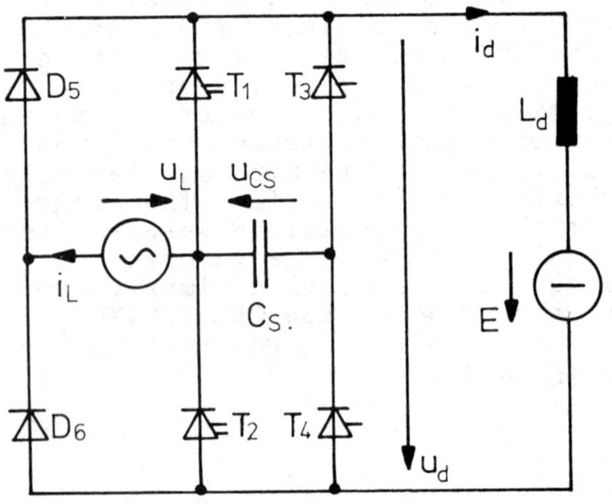

Fig. 2 Simplified equivalent circuit of the LUB

With assumption d) the turn-off capacitor C_L may be neglected, so that the main thyristors T_1 and T_2 can be turned off at any time. This is indicated by the second gate terminal. Fig. 3 shows the time variations of u_L, u_{CS} and i_L for the control method chosen.

The line current always starts at $x = \beta_2 - \pi$ with the discharge of the storage capacitor C_S by firing the thyristor T_4. The discharge of the capacitor is finished when $u_{CS}(\varepsilon - \pi) = 0$. At $x = \varepsilon - \pi$ the thyristor T_2 is fired and immediately takes over the load current. Because of the forced turn-off of T_2 and the firing of T_4 at the same time ($x = \beta_1$) the capacitor C_S is charged by the load current during the interval $\beta_1 < x < \gamma$, until the capacitor voltage u_{CS} is the same as the line voltage u_L at $x = \gamma$ and the load current immediately commutates to the pair of by-pass arms D_5, D_6. Diode D_5 conducts during the interval $\gamma - \pi < x < \beta_2$. While the capacitor is charged and discharged the time dependence of the voltage u_{CS} relative to the peak voltage $\hat{u}_L$ satisfies the equation of a straight line with the tangent of the angle of inclination ψ

$$\tan\psi = \frac{\bar{i}_d X_{CS}}{\hat{u}_L} \qquad (1)$$

The sooner the discharge of the capacitor begins the more the line current leads in phase. To avoid back commutation of the load current to the pair of by-pass arms, the capacitor voltage u_{CS} may touch the line voltage only at one point during the discharge. Only when $\tan\psi \geq 1$, all values of the turn-off angle

$$-\frac{\pi}{2} \leq \gamma \leq \frac{\pi}{2}$$

are adjustable and the control factor

$$a = \frac{\bar{u}_d}{\bar{u}_{di}} \qquad \bar{u}_{di} = \frac{2}{\pi}\hat{u}_L$$

may take values

$$0 \leq a \leq 1$$

For $\tan\psi < 1$ the control factor is limited to values /1/

$$a_{min} \leq a \leq 1$$

Curve a) in Fig. 4 shows the displacement reactive power of the line relative to the ideal d.c. power P_{di}

$$P_{di} = \bar{u}_{di}\bar{i}_d$$

in the case of $\tan\psi = \pi$.

For all values of the control factor the displacement reactive power is capacitive and shows so the desired characteristic.

The numerical calculations, which are necessary for the determination of the control characteristics and the reactions of the line, can be done by means of an electronic pocket calculator, so that the use of the computer is not necessary.

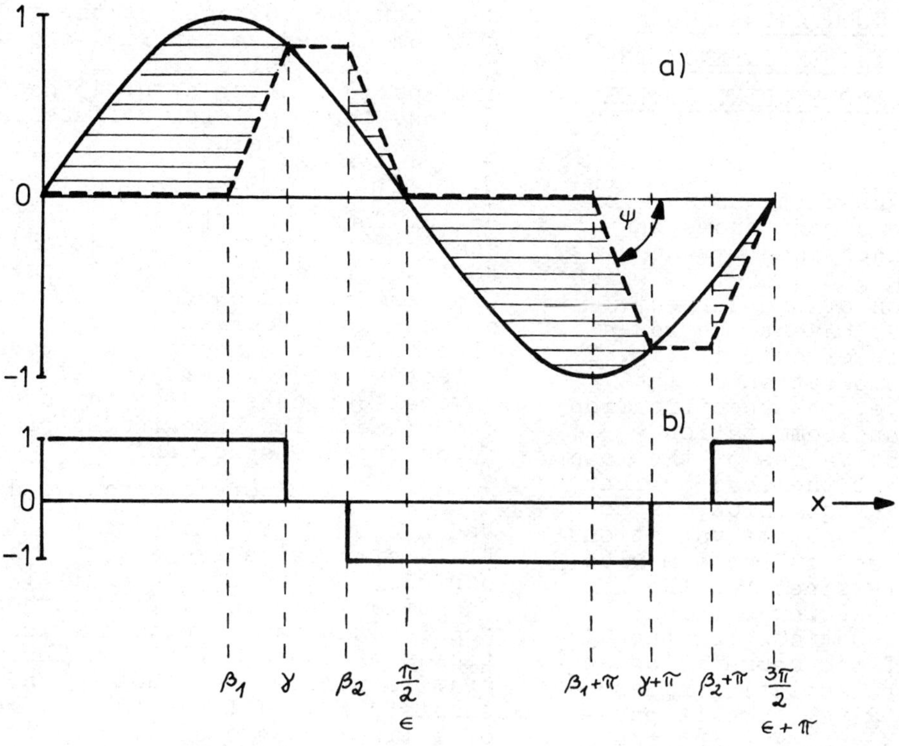

Fig. 3 Time variation of
a) $u_L/\hat{u}_L$ ——— ; $u_{CS}/\hat{u}_L$ ---------- ($u_d/\hat{u}_L$ shaded)
b) $i_L/\bar{i}_d$

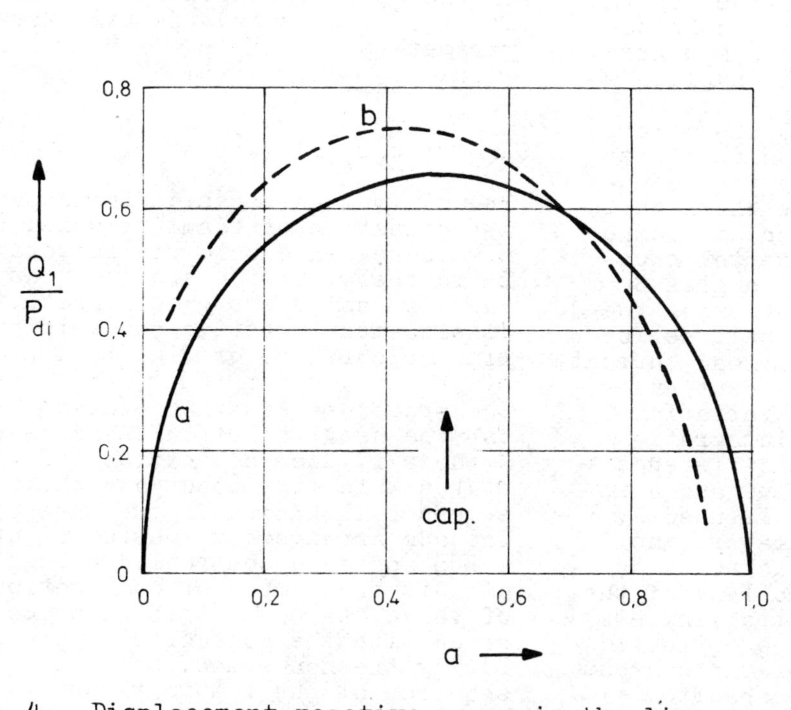

Fig. 4 Displacement reactive power in the line as a function of the control factor

3. CALCULATION OF THE CONTROL CHARACTERISTICS NOT NEGLECTING COMMUTATION INDUCTANCES AND A FINITE SMOOTHING INDUCTANCE THE LATTER BEING INDEPENDENT OF THE CURRENT

Contrary to the assumptions a) - e) in chapter 2 the circuit nows has a commutation inductance L_T and a finite smoothing inductance L_d which is not a function of the current. These inductances have to be added to the equivalent circuit of Fig. 2. An analytical computation of the steady state, i.e. the specification of all control and commutation angles and of the values of the load voltage E, so that the load current takes a given average value, is no longer possible. The time variations of all voltages and currents may be analytically determined for the separate intervals. One obtains transcendental equations for the determination of the commutation angles and the characteristic angles ε and γ. With the given circuit parameters $\hat{u}_L$, X_d, X_T and X_{CS} and the control angles β_1 and β_2 all time variations of the voltages and currents are computated for half of a period of the supply from the analytically computated equations which are valid for the different sections. Only the instantaneous values of the beginning and the end of a combination of valve states are computated. A stepped scanning which involves a long computation time is not necessary. Beginning and end of a certain combination of valve states are determined either by the commutation angles, or by the characteristic angles ε and γ, which are computated from the transcendental equations, or by the control angles β_1 and β_2. After the computation of the time variations for one half period the average value of the load current $\bar{i}_d$ is evaluated. This is done analytically from the time variation of the load current by integration over a half period. The difference between this average value and a given desired value $\bar{i}_{ds}$ is used by a time discrete PI-regulator, the effect of which is simulated by an algorithm for the controlling of the load voltage E. The control angle β_1 remains constant. Because of the varying average value of the current the final point of the capacitor discharge moves, so that also the control angle β_2 has to be regulated depending on the difference between the desired value $\varepsilon_s = +\pi/2$ and the actual value ε. After regulating β_2 and E the independent quantity x is put back to the starting point and the instantaneous values of all currents and voltages at the end of the computated half-period are used for starting values for the following half-period. After about 15 half-periods the regulator difference has diminished to values

$$|\bar{i}_{ds} - \bar{i}_d| < 1\ \%\ \bar{i}_{ds}$$

Having found the steady state the analysis of the time variations follows. This might include the plotting of all currents and voltages as well as a fourier analysis of the line current. Moreover the average, effective and extreme values of several currents and voltages are determined. The unsegmented program needs storage space for

32 K core storage words

and a computation time of about 13s for a simulation run over 100 half-periods, including the analysis of the time variations. This method enables the rapid determination of fields of characteristics. The program is written in ALGOL 60 and consists of 234 lines (max. length of a line: 80 symbols). 120 lines are necessary for the computation of the time variations in the steady state, while the remaining part is used for the analysis of the time variations. The displacement reactive power in the line as a function of the control factor is presented in Fig. 4, curve b) for the circuit parameters

$$\tan\psi = \pi$$
$$u_k = 11{,}5\ \%$$
$$X_d/X_T = 9$$

The maximum voltages on the capacitor as a function of $\tan\psi$ (1) which are now increased due to the inductance L_T on the a.c. side are shown as curve a) and the biggest capacitive displacement reactive power in the line as curve b) of Fig. 5.

The procedure described above can only be used for circuit and control methods if the combinations of valve states which may occur and their sequence is known. In the case of a cascade arrangement consisting of a LUB and an uncontrolled bridge circuit the number of combinations of valve states is tripled as compared with the case of a simple LUB. During the non-steady state the sequence of the combinations of valve states can change every half-period. This circuit was also simulated, but now the time variations which were analytically calculated for the separate intervals are numerically

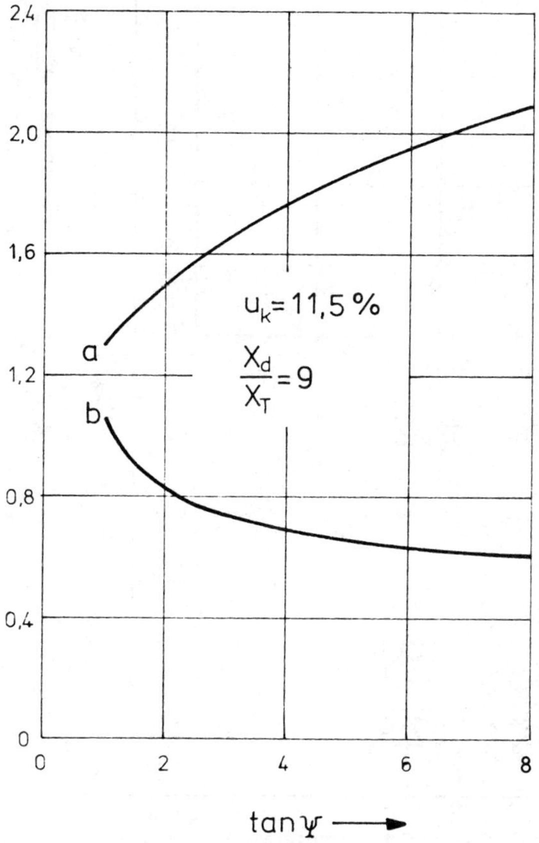

Fig. 5 Curve a): Maximum capacitor voltages
Curve b): Biggest capacitive displacement reactive power
as a function of $\tan\psi$

evaluated step by step and the appearing characteristic angles are determined directly from the time variations. Here impedances on the primary side or a current dependent smoothing inductance cannot be considered. With 4 scanning steps during each degree and 50 computated half-periods one needs a computation time of about 39s and for the unsegmented program core storage for 37 K words. The resulting displacement reactive power is shown in Fig. 7, curve c) as a function of the control factor.

4. CALCULATION OF THE CONTROL CHARACTERISTICS AND REACTIONS OF THE LINE OF THE CASCADE CIRCUIT CONSISTING OF A LUB AND AN ASYMMETRICAL, PAIR OF ARMS CONTROLLED BRIDGE CIRCUIT

The circuit regarded is shown in Fig. 6.

Besides line and transformer impedances also load losses, a finite, current-dependent smoothing inductance L_d and the turn-off capacitor C_L together with the turn-off thyristors T_{11} and T_{12} are considered. Because of the coupling of the bridge circuits on the primary side by the resistances R_P and R_L and the inductances X_P and X_N and the non-linear $(L_d - i_d)$ characteristic, an analytical solution during the separate intervals is now no longer possible. The time variations have to be evaluated step by step from the differential equations, which are defined during the separate intervals by means of numerical methods for integration. The combinations of valve states, which may appear during one half-period and their sequence have also to be known for this method. It is possible, however, to regard both bridge circuits as being decoupled, so that the total number of necessary combinations of valve states is the sum and not the product out of the possible combinations of valve states of each individual bridge circuit. With this simulation method also more complicated circuit arrangements may be computated. Just as described in chapter 3, always one half-period of the line is considered and by regulation of the control angle β_2 and the load voltage E the steady state is reached. For each scanning step a maximum of four 1st order differential equations have to be solved. For the case that the LUB contributes to the output voltage only during the higher part of the control range the variations of Q_1/P_{di} are plotted in Fig.7.

The LUB is once pair of arms controlled (Fig. 7, curve a)) and then as capacitive as possible controlled (Fig. 7, curve b)). Fig. 7 curve c) shows the case when all the losses, the turn-off capacitor and the impedances on the primary side are neglected. The value of the current-independent smoothing inductance (Fig. 7 curve c)) is equivalent to the value $X_d = f(\bar{i}_{ds})$.

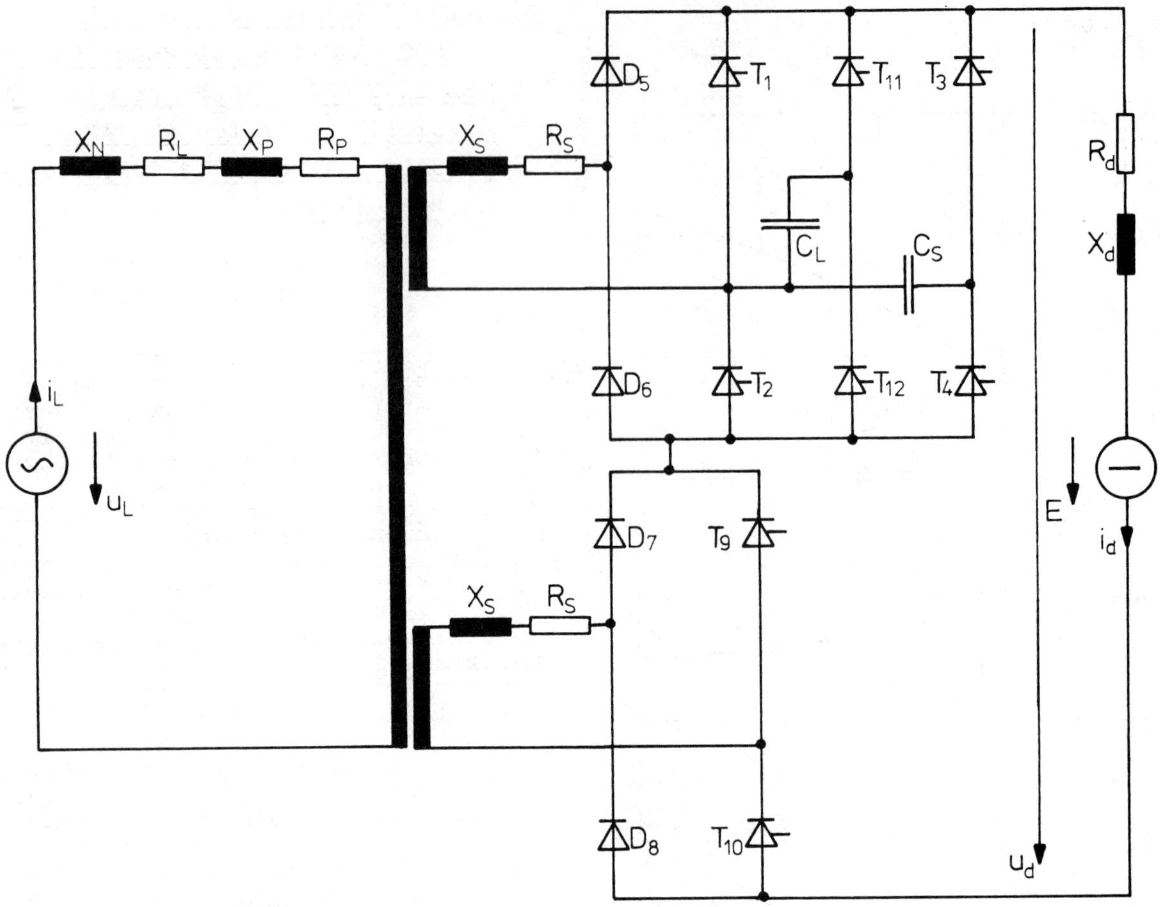

Fig. 6 Equivalent circuit of the regarded cascade circuit

The data for the simulation are based on the given technical data of the train ET 420 (DBB). With 8 scanning steps per degree and a simulation run over 50 half-periods (= 72000 steps) one needs a computation time of about 100s and the unsegmented program requires core storage space for 38 K words. This simulation run includes the evaluation of 40 average, r.m.s. and amplitude values, the fourier coefficients of the line current and the psophometrically rated disturbance current. Moreover, the time variations of all currents and voltages may be plotted. About the half of the 240 lines, which the program consists of, is only necessary for the analysis of the time variations.

5. SUMMARY

A comparison of the computation times for the methods described in this paper with times necessary for digital simulation languages like DIGSI, ASIM, ANAGOL shows, that the problem related programming by means of a higher programming language i.e. FORTRAN or ALGOL 60 is clearly better. Taking all the important circuit parameters of the cascade circuit described in chapter 4 into account, the characteristics may be determined very quickly, whereby a computation time of about 100s for the determination of one steady state is necessary. The results change only little, if practical values for common impedances on the primary side are considered. The fact that the smoothing inductance is a function of the current, however, leads to clear modifications of the power balance of the line. Also few modifications follow from the consideration of the turn-off capacitor, if its value is only small. To evaluate the exact valve voltages, the turn-off capacitor, however, has to be considered.

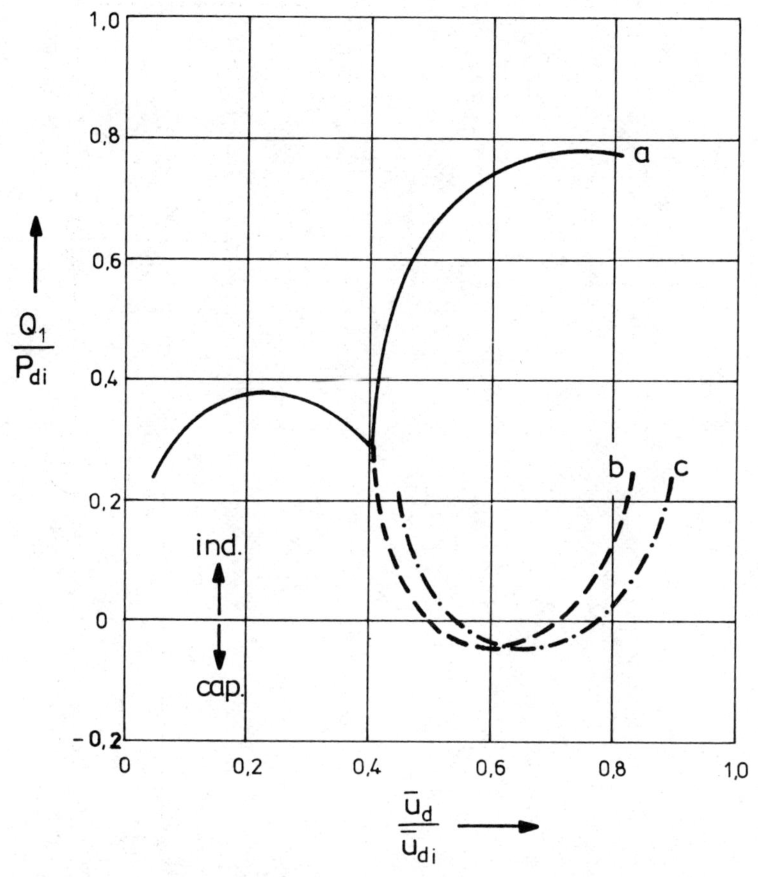

Fig. 7 Displacement reactive power relative to the ideal direct power P_{di}

References

/1/ R. Windmöller, Steuerverfahren für die LUB, ETZ-A 4, (1977)

/2/ K. H. Bezold, Löschbare, einphasige Stromrichter mit Sektorsteuerung für die Speisung von Mischstrommotoren, Elektrische Bahnen 12, (1974)

/3/ I. Förster, Grundsätzliches zur Stromrichter Netzbelastung, Wissenschaftliche Berichte AEG-Telefunken 4/5, (1976)

GENERALIZED HYBRID COMPUTER TECHNIQUE FOR SIMULATING INVERTER-FED AC MOTORS DRIVES

P. Ferraris, M. Lazzari and F. Villata

Istituto di Macchine Elettriche, Politecnico di Torino (Italy)

ABSTRACT

In the paper an investigation technique is described useful for studying steady state and dynamic performances of convertor-fed AC drives, by way of hybrid computer and with the opportunity of directly supplying the machine model with the convertor's voltage waveforms. Starting from some general considerations concerning the possibility of describing a quite general supply voltage system by way of an opportune space vector, it is pointed out that the opportune knowledge of such vector in module and position is sufficient to define the input voltages valid for any two-phase equivalent rotating machine model. It is shown that for an inverter-fed machine the module and argument space vector determination and the input model voltages are deducible from the real inverter voltages, for any reference frame, with an opportune hybrid technique. A natural hybrid simulation technique is deducible. Some examples are reported concerning the method application to the case of an inverter-fed variable speed synchronous drive when different modulation techniques are utilized.

INTRODUCTION

The actual increasing interest for invertor-fed AC motors drives makes necessary to have at disposal flexible computing techniques which should allow a system analysis under different points of view. In particular it may be opportune to easily investigate about any drive system with regards to: dynamic or control technique points of view; electrical, magnetic and mechanical motor solicitations; inverter solicitations; effects on the power supply system, etc..

The aim of the paper is then to define a hybrid computer technique for simulating inverter adjustable speed drives allowing to deduce choice criteria about the different technical solutions and project choices too. It is intended also to accept any type of inverter modulation technique and machine transformed equations; the output inverter voltage waveforms are often very complicated and sometimes it is impossible to exprime them by analytical expressions. The proposed method was then intended allowing to directly connect the machine and control system model to the output inverter terminals. Also the logic signals defining the firing time intervals of the inverter thyristors are useful at the aim. In such a way the effects of any type of modulation technique may be easily investigated. Real time operation may also allow the interconnection between computing system and other real external measuring or control components.

Another property of the simulation system should be the possibility of using a simulation approach based on a d, q, o transformation technique using any rotating reference frame. In fact an opportune reference system choice may make easier to show and to interpret some particular phoenomena. But that implies the necessity of having at disposal a transformation apparatus useful to provide the correct input voltages to the particular analog model.

In this paper such problem is examined for a quite general type of AC inverter drive in order to obtain automatically the requested model input signals in a way indipendent from the particular movement law adopted for the reference system.

GENERAL CONSIDERATIONS

Let us consider the general case of an AC rotating machine having two stator and rotor poliphase windings systems; for whatever poliphase voltages systems, not containing homopolar components, each phase voltage may be deduced by projecting two spatial vectors $\overline{V}_r$ and $\overline{V}_s$, one for the stator and the other for the rotor, on the windings axes. The stator vector module V_s and the argument δ_{so} with respect to the supplied windings are depending only on the input instantaneous voltages; V_s and $f_s = 1/2\pi \, d/dt(\delta_{so})$ represent a kind of "instantaneous voltage and frequency" of the poliphase system. V_s, V_r, δ_{so} and δ_{ro} completely describe the supply voltages.

Let us suppose that the resulting m.m.f. distribution produced at the airgap should be sinusoidal; then it is possible to consider an equivalent elementary two phases, two poles machine.

With reference to fig. 1 the following electrical equations are deducible for the stator and rotor windings pairs:

$$\overline{V}^{1,2}_{s,r} = \left[R^{1,2}_{s,r}\right]\overline{I}^{1,2}_{s,r} + p\overline{\Phi}^{1,2}_{s,r}$$
$$\Phi^{1,2}_{s,r} = \left[L^{1,2}_{s,r}\right] I^{1,2}_{s,r} \qquad (1)$$

where:

$$\overline{V}^{1,2}_{s,r} = \left[V_{s1},\ V_{s2},\ V_{r1},\ V_{r2}\right]^t$$
$$\overline{I}^{1,2}_{s,r} = \left[I_{s1},\ I_{s2},\ I_{r1},\ I_{r2}\right]^t$$
$$\overline{\Phi}^{1,2}_{s,r} = \left[\Phi_{s1},\ \Phi_{s2},\ \Phi_{r1},\ \Phi_{r2}\right]^t$$

and:

$$R^{1,2}_{s,r} = \begin{bmatrix} R_s & 0 & 0 & 0 \\ 0 & R_s & 0 & 0 \\ 0 & 0 & R_r & 0 \\ 0 & 0 & 0 & R_r \end{bmatrix}$$
$$L^{1,2}_{s,r} = \begin{bmatrix} L_s & 0 & M_1(\delta_{ro}) & M_2(\delta_{ro}) \\ 0 & L_s & M(\delta_{ro}) & M(\delta_{ro}) \\ M(\delta_{ro}) & M(\delta_{ro}) & L_r & 0 \\ M(\delta_{ro}) & M(\delta_{ro}) & 0 & L_r \end{bmatrix} \qquad (2)$$

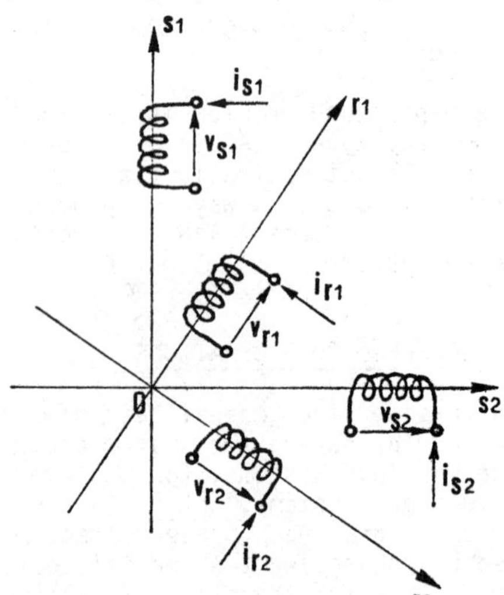

fig. 1

The equivalent voltages V_{s1}, V_{s2}, V_{r1}, V_{r2} may also be considered as the projection of the equivalent vectors $\overline{V}_s(t)$ and $\overline{V}_r(t)$ on the orthogonal axes pairs s_1, s_2 and r_1, r_2 (fig. 2).

If the equations have to be obtained valid for whatever reference frame constituted by a pair of orthogonal axes d, q rotating at a given speed ω_0 (as it is indicated in fig. 3), it is necessary to operate a reference changement defined by the transformation A :

$$A = \begin{bmatrix} \cos\delta_0 & \operatorname{sen}\delta_0 & 0 & 0 \\ -\operatorname{sen}\delta_0 & \cos\delta_0 & 0 & 0 \\ 0 & & \cos(\delta_0-\delta_{r0}) & \operatorname{sen}(\delta_0-\delta_{r0}) \\ 0 & & -\operatorname{sen}(\delta_0-\delta_{r0}) & \cos(\delta_0-\delta_{r0}) \end{bmatrix} \qquad (3)$$

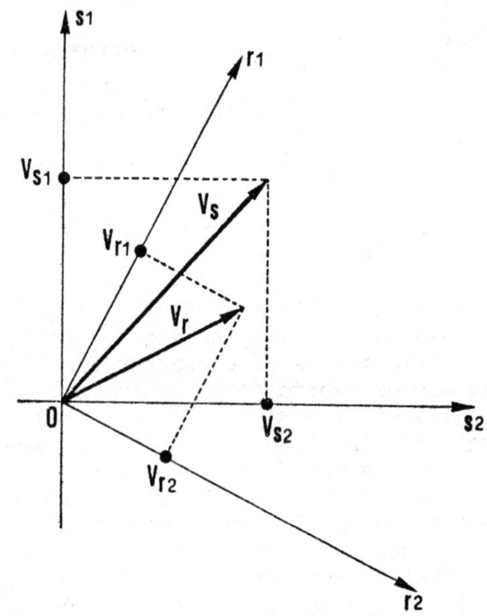

fig. 2

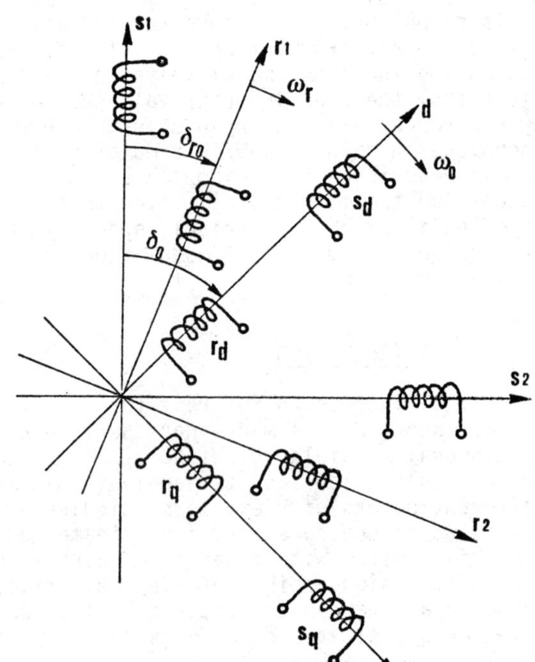

fig. 3

The deducible transformation results allow to write the machine equations in the following form:

$$\overline{V}^{d,q}_{s,r} = R^{d,q}_{s,r}\, I^{d,q}_{s,r} + p\Phi^{d,q}_{s,r} + \Omega\,\Phi^{d,q}_{s,r}$$
$$\Phi^{d,q}_{s,r} = L^{d,q}_{s,r}\, I^{d,q}_{s,r} \qquad (4)$$

where:

$$\overline{V}^{d,q}_{s,r} = \left[V_{s1}, V_{s2}, V_{r1}, V_{r2}\right]^t$$

$$\overline{I}^{d,q}_{s,r} = \left[I_{s1}, I_{s2}, I_{r1}, I_{r2}\right]^t$$

$$\overline{\Phi}^{d,q}_{s,r} = \left[\Phi_{s1}, \Phi_{s2}, \Phi_{r1}, \Phi_{r2}\right]^t$$

and:

$$R^{d,q}_{s,r} = \begin{bmatrix} R_s & 0 & 0 & 0 \\ 0 & R_s & 0 & 0 \\ 0 & 0 & R_r & 0 \\ 0 & 0 & 0 & R_r \end{bmatrix}$$

$$L^{d,q}_{s,r} = \begin{bmatrix} L_s & 0 & M & 0 \\ 0 & L_s & 0 & M \\ M & 0 & L_r & 0 \\ 0 & M & 0 & L_r \end{bmatrix} \quad (5)$$

$$\Omega = \begin{bmatrix} 0 & -\omega_0 & 0 & 0 \\ \omega_0 & 0 & 0 & 0 \\ 0 & 0 & 0 & \omega_r-\omega_0 \\ 0 & 0 & \omega_0-\omega_r & 0 \end{bmatrix}$$

The transformed voltages V_s^d, V_s^q; V_r^d, V_r^q [which may also be considered as the voltages applied to two equivalent windings pair r_d, r_q and s_d, s_q (fig. 3)], are still obtainable by projecting the spatial vector $\overline{V}_s$ and $\overline{V}_r$ on the new axes d and q.

If the machine analog model is deduced from the equations (4), the input supply voltages may be obtained from the equations (6) if the vector modules V_s, V_r and the arguments δ_s, δ_r with respect to the reference axes are known (see fig. 4):

$$\begin{aligned} V_{sd} &= -\ V_s \operatorname{sen}\delta_s \\ V_{sq} &= \ \ V_s \cos\delta_s \\ V_{rd} &= -\ V_r \operatorname{sen}\delta_r \\ V_{rq} &= \ \ V_r \cos\delta_r \end{aligned} \quad (6)$$

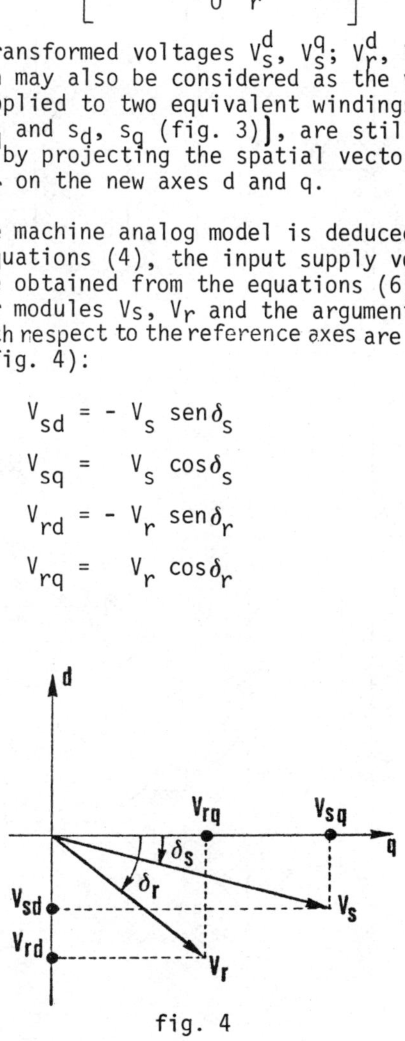

fig. 4

Such way of defining the supply voltages allows to obtain a machine model valid for whatever reference frame speed, because such information is already taken into account by the time depending values of δ_s and δ_r.

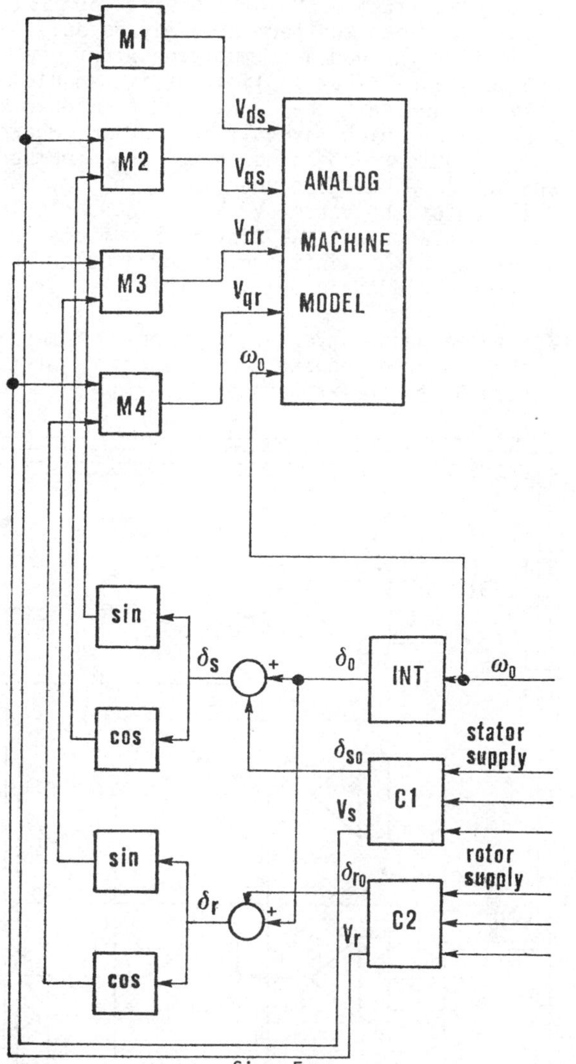

fig. 5

The realized hybrid model defining the input voltages on the basis of the equations (6) is reported in the form of a principle block scheme in fig. 5.

The necessary input signals are constituted by the realistic voltage waveforms and by the rotating axes d and q speed ω_0.

It is evident that the main problem consists in the definition of the bloks C_1 and C_2 which must operate the conversion from the actual voltage values to the module and argument of the spatial vectors $\overline{V}_s$ and $\overline{V}_r$. It has also to be observed that in the fig. 5 some quantities may indefinitely increase their value; the practical realization should provide another solution. That is possible owing to the fact that, in order to obtain the correct transformed voltages, it is sufficient to know the angles between the axes d, q and the vectors $\overline{V}_s$, $\overline{V}_r$, with regards to the value included in the interval $\pm$ 180°.

The particular characteristics of the inverters output voltages make possible a hybrid solution for the equations $V_{s,r} = f(v_i)$, $\delta_{s,r} = f(v_i)$, where v_i represent the output inverter voltages. In fact, when one inverter is supplied from a DC voltage E, the output voltage between two terminals may be only $\pm E$ or zero, with sudden commutations from one to another value. That implies, as it should be easy to show, that the corresponding module V has a value which may only be 0 or αE, where α is a number depending on the phase number and on the windings connection type too. The position of the vector $\overline{V}$ changes suddenly in correspondance of every inverter commutation, assuming only a finite number of angular positions included in the interval $\pm$ 180°.

For three phase inverters it may only assume the positions corresponding to the vertices or to centre of a regular hexagon (v. fig. 6).

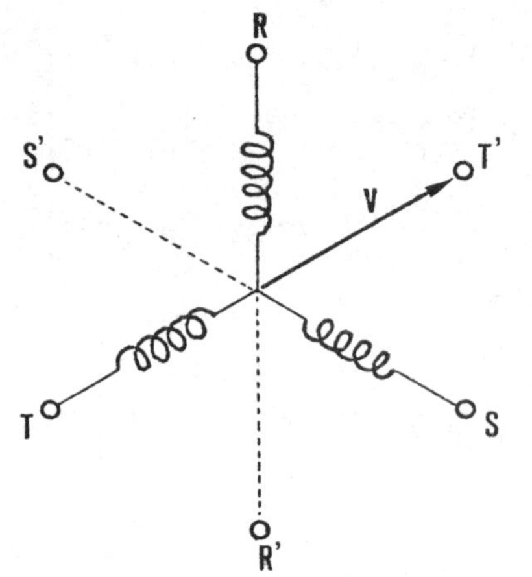

fig. 6

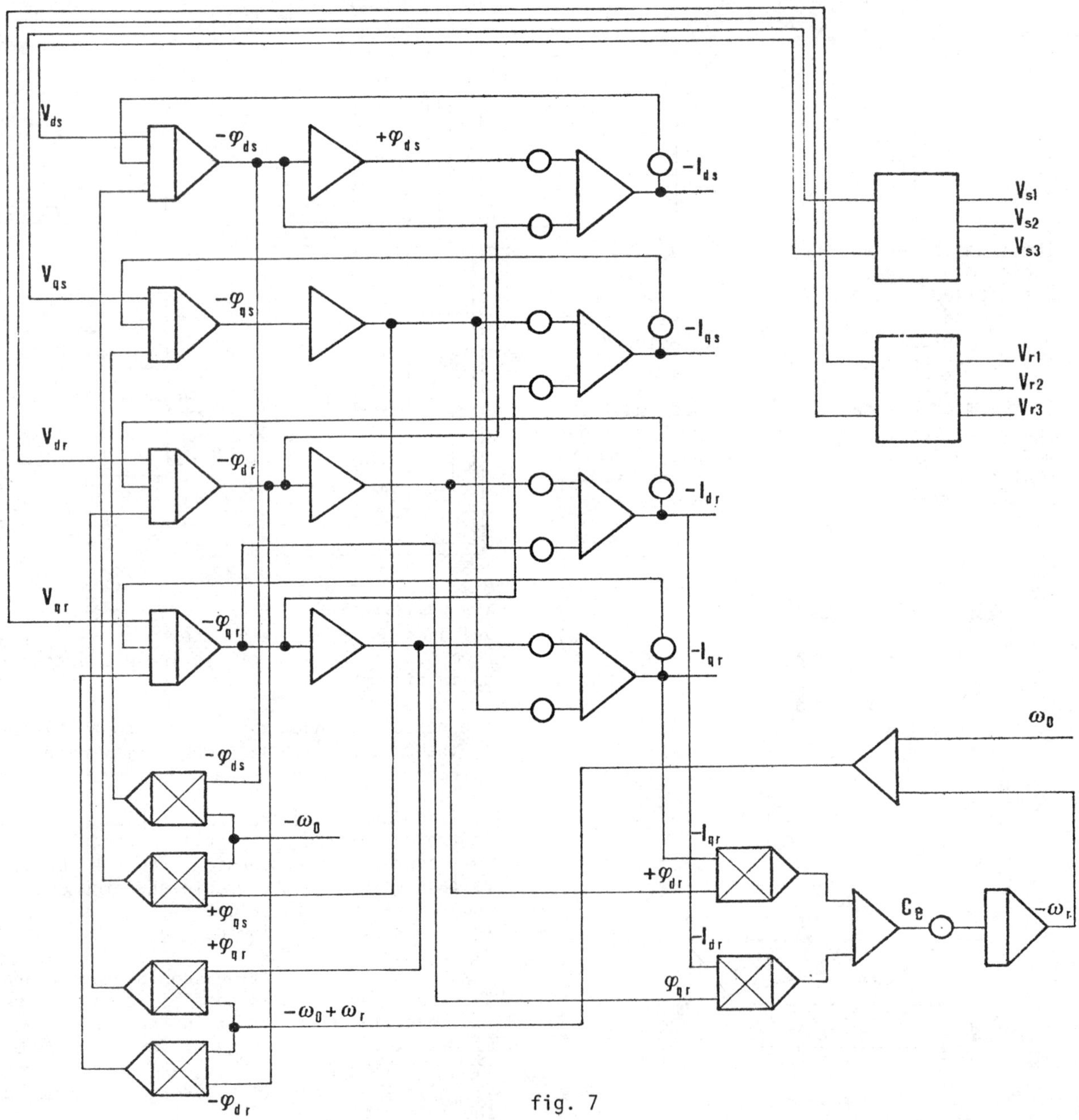

fig. 7

The vector $\overline{V}$ may then assume six positions with a constant module αE ($\alpha \neq 0$) and one position corresponding to the situation of module equal to zero.

Depending on the modulation technique type the vector's displacements may happen between adjacent angles or not. It should be possible to show that for a three-phase inverter the vector's estreme may commutate between two not adjacent points only when two static switches, belonging to different branches, are contemporany turned on; the center of the hexagon has to be considered as adjacent to all the six vertices. For an inverter-fed system, supplied with a constant voltage intermediate stage, the multipliers M_1, ... M_4 of fig. 5 may be substituded by analog gates blocks; the problem will be better discussed in the following section.

The fig. 7 shows the complete block simulation scheme valid for any AC electrical machine; stator and rotor are supposed supplied by two different inverters, whose output voltages are elaborated by two interface systems operating on the basis of the reported considerations in order to provide the necessary electrical model inputs. In the figure the general case is reported regarding any value of the reference frame speed; the particular situations $\omega_0 = 0$ and $\omega_0 = \omega_r$ are obtainable by way of evident scheme modifications. The instantaneous value of ω_0 may in general be imposed in order to make evident some particular properties by way of the obtained transformed quantities.

INTERFACE BLOCK REALIZATION

The interface system allowing to obtain a computing apparatus based on the foregoing statements was realized for the case of three phases, three terminals inverters.

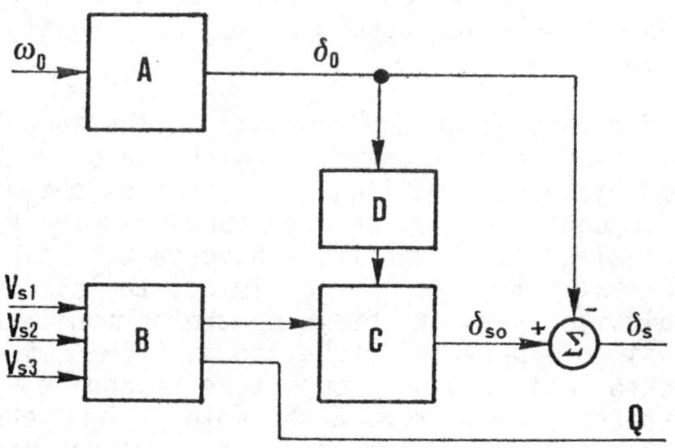

fig. 8

The fig. 8 represents the block scheme of the computing circuits which elaborates the three inverter voltages and an analog signal ω_0 proportional to the speed of the chosen reference axes; they give at the output terminals an analog signal giving in the opportune scale the angle δ_s and a logic information V_s useful to select the correct module value for the vector $\overline{V}_s$, between zero and αE. The interface device may work in real time and allow the direct connection between inverter and analog model.

The block A of fig. 8 produces an integration useful to obtain the angle δ_0 which locate the absolute reference frame position. But for $\omega_0 \neq 0$ such angle might increase indefinitely; the integration system is at the aim modified in order to give at disposal the instantaneous value of an angle δ_0' always included in the range - 180°, + 180° and given by $\delta_0' = \delta_0 \pm k\,360$, with k integer. The method used at the aim is deducible from the flow-chart reported in fig. 9.

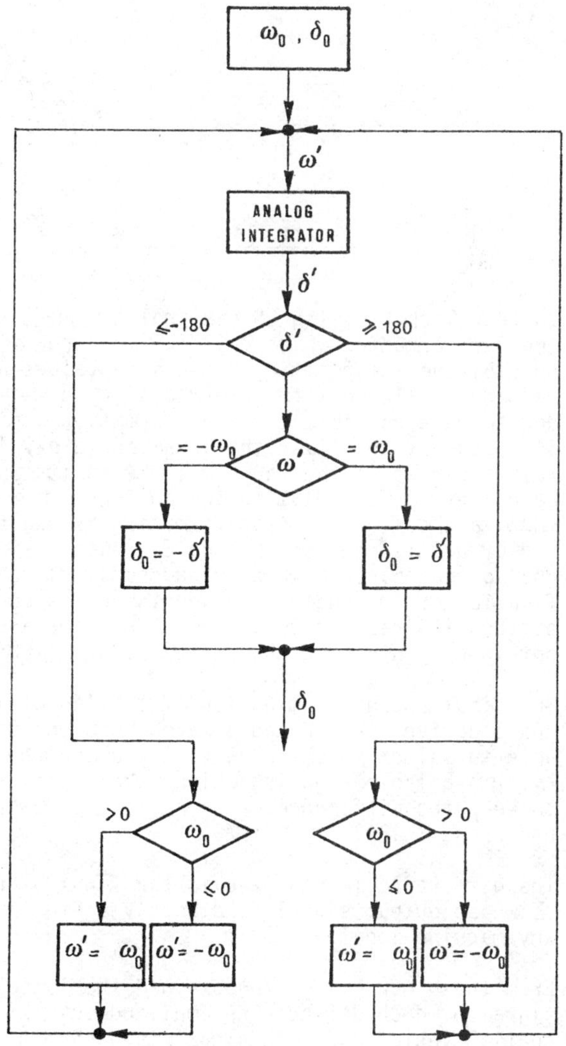

fig. 9

The block B represents a combinatorial system which gives a logical interpretation of the inverter voltages and furnishes in a logical way the instantaneous position of the vector $\overline{V}_s$ and the choice between the module values 0 or αE.

As example in fig. 10, is reported a table showing for a three-phase bridge inverter the position of the vector $\overline{V}_s$ (r,t',s,r',t, s') as a function of the output terminals potential, which is supposed assuming the values 0 and E; it is also reported the corresponding module value.

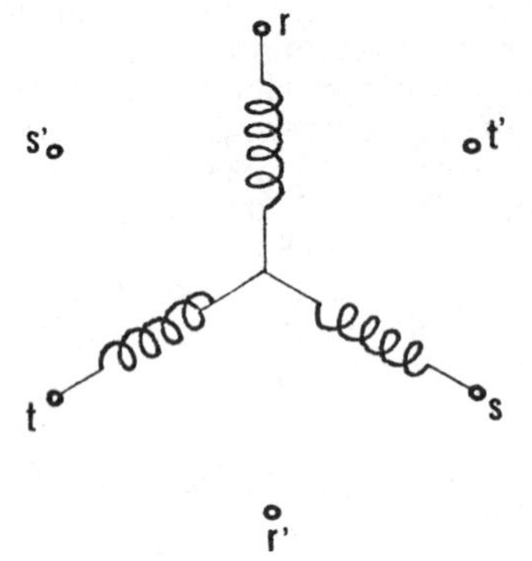

fig. 10

	r	t'	s	r'	t	s'	V_s=0	
R	E	E	0	0	0	E	0	E
S	0	E	E	E	0	0	0	E
T	0	0	0	E	E	E	0	E

In the block C of fig. 8 six analog signals values are arranged which have the meaning of possible vector positions; the same values are made available with inverted sign in order to make possible the transformation which is necessary to obtain the final angle δ_s, always included in the range ± 180°, as it is necessary for calculating the analog functions senδ_s and cosδ_s. The transformation is realized by utilizing also the logical output signal of the block D which produces the classification of the angle δ_0 by defining its position with respect to six angular intervals obtained by dividing the range - 180, + 180.

The simultaneous examination of the logic signals derived from D and B make the block C able to select that analog value, among the twelve available signals, which makes possible to keep the difference δ_{s0} - δ_0 in the desired range ± 180°.

The difference is realized in the final block Σ, whose output signal is directly useful for any machine model.

From the block B is also obtained the logic signal Q which defines the amplitude of the spatial vector $\overline{V}_s$; such signal is useful to obtain the correct value for V_{ds} and V_{qs}.

At that purpose two solutions are possible; the one (fig. 11 a) which provides analog gates for deducing V_{ds} and V_{qs} when V_s is constant; the other (fig. 11 b) which allows to calculate V_{ds} and V_{qs} by way of analog multipliers for the case of V_s variable in time. The last situation is useful for static inverters supplied by an intermediate variable voltage stage. That's the case, for example, of variable voltage square wave operation.

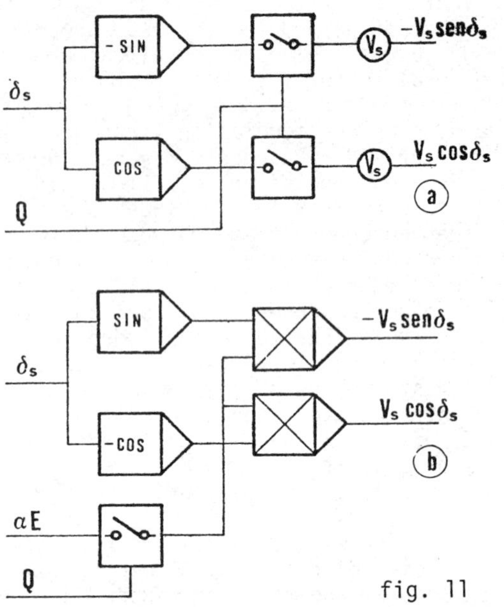

fig. 11

The realized simulation system may be adapted to more usual inverter-fed electrical machines and for investigating some other particular drive systems. It is immediate to define the simulation scheme of an inverter-fed normal induction motor: it is sufficient to suppose the input rotor signals V_{rd} = V_{rq} = 0.

A synchronous motor may be studied by supposing V_{rq} = 0 and V_{rd} = V_r.

Other situations are easily deducible with regards to drives of different type (motor with incremental motion opportunities f.e.) or to special situations regarding the supply systems (short circuit on a bridge branch, sudden variation in the intermediate DC voltage etc.).

With regards to the reference frame choice one may observe that the elasticity of the adopted technique allows to utilize the most convenient solution according to the investigation aim.

A first solution may consist in the adoption of a reference frame fixed with stator, which is obtained by making ω_0 = 0 in the scheme. Such condition may be opportune when the stator electrical quantities have to be obtained in real time (f.e. for an immediate losses evaluation, or for obtaining the harmonic analysis of currents, torque etc.). From similar reasons it should be opportune to choice a reference frame fixed on the rotor, that can be realized by introducing as input signal ω_0 the rotating speed signal ω_r.

Another possibility consists in a choice which shall produce electrical quantities having constant value in steady operating condition with sinusoidal constant frequency input voltages. That condition is obtainable with a speed ω_0 having constant value equal to the one of a spatial vector producing the same total rotating angle of the vector $\overline{V}_s$; that condition may be equivalent to the other of reference frame fixed on the rotor for a steady operating synchronous machine. Such solution may be particularly useful for investigating stability problems, evaluated by analysing the current or fluxes variations or rotor oscillations.

Another point, which will be better emphasized in the next section, consists in the absolute freedom regarding the type of modulating technique used for the inverter voltage. All the P.W.M. waveforms are accepted, in transient condition too, and waveforms which have no analytical expression may be experimentally tested

EXPERIMENTAL RESULTS

In this section some results are reported, obtained by simulating with the proposed hybrid technique a synchronous motor directly fed by a variable frequency P.W.M. inverter. No particular care was taken at the aim for the accuracy of the machine model; for example the saturation effects are not considered and no damping rotor circuit is introduced. The selected reference frame consists in two orthogonal axes d and q fixed on the rotor and with the direction d coinciding with the rotor excitation m.m.f.. The used inverter is modulated with different P.W.M. techniques depending on the frequency value; it is operating with the subharmonic method at low frequency; when the frequency increases a wave with two or four fixed position variable amplitude holes is adopted; in the highest frequency range square wave operation is used.

In the fig. 12 the block scheme is reported of the system for which the following photographs were obtained. The scheme may be obtained from the general scheme of fig. 7 with the further simplification $C = M \cdot i_q \cdot i_r$, valid for a round rotor machine with single escitation; at last it was imposed $_0 = {}_r$. The adopted parameters values are reported in p.u. in the table A.

TABLE A

$r_s = 0{,}1$

$r_r = 0{,}12$

$l_{sd} = l_{sq} = 1{,}1$

$l_r = 1{,}2$

$m = 1$

$$H = \frac{1}{2}\frac{J\omega_b^2}{V_b I_b} = 0{,}25$$

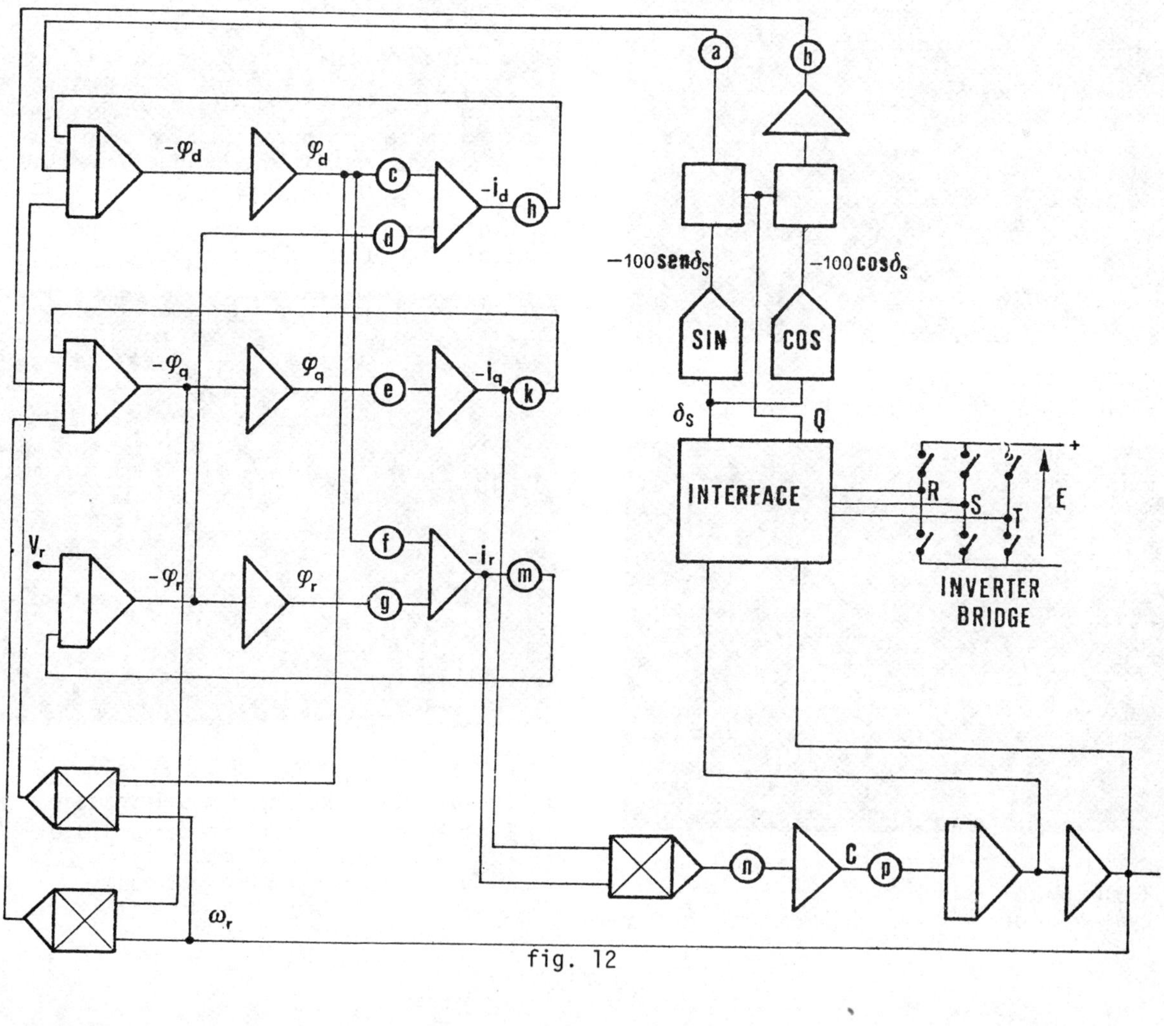

fig. 12

In the following figures the results are reported which were obtained for some different working conditions, as it is indicated for each photographic group.

Square Wave Operations

Steady-state operations at a frequency f=84 Hz

Load condition: constant rated torque

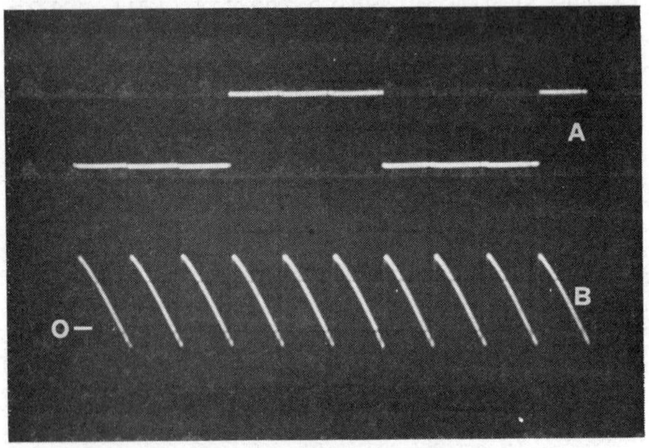

fig. 13

A) Evolution of the inverter output potential
B) Evolution of the input model voltage - V_d

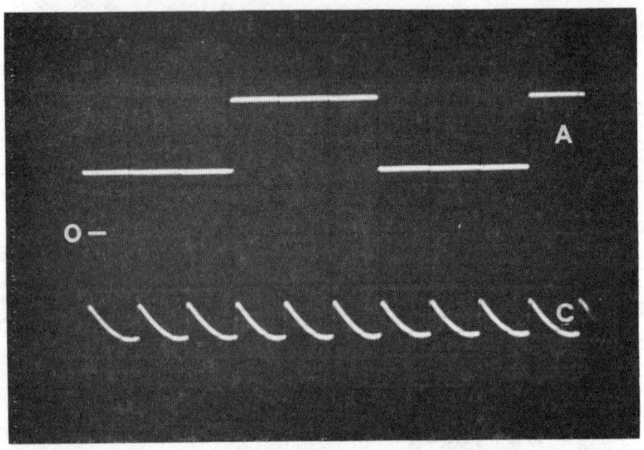

fig. 14

C) Evolution of the input model voltage - V_q

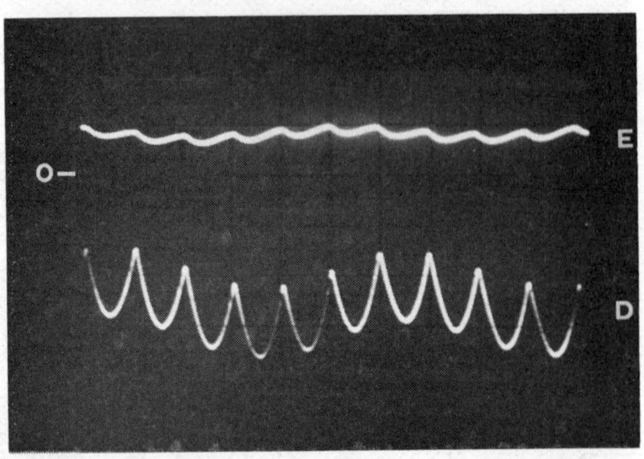

fig. 15

D) Evolution of - i_d
E) Evolution of - i_q

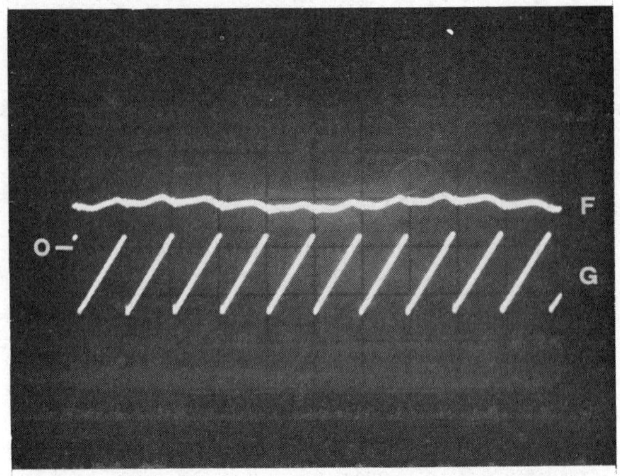

fig. 16

F) Evolution of electromagnetic torque
G) Evolution of the angle δ_s

Load condition: 100% torque overload

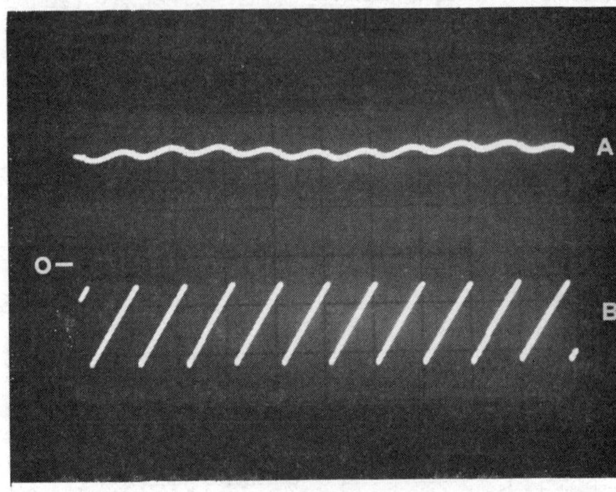

fig. 17

A) Evolution of electromagnetic torque
B) Evolution of the angle δ_s

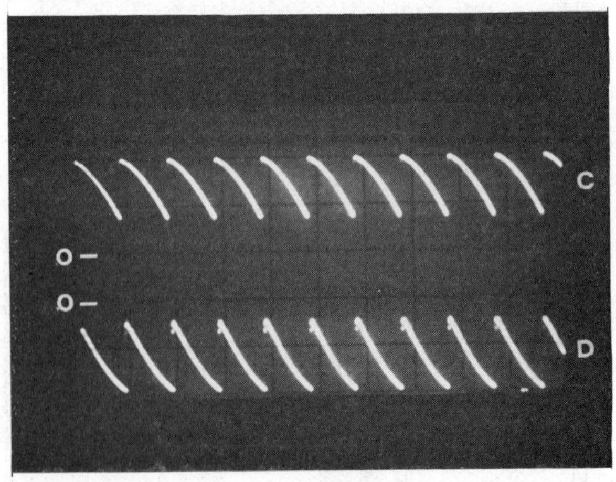

fig. 18

C) Evolution of the input model voltage - V_d
D) Evolution of the input model voltage - V_q

From the above figures it is possible to identify some peculiar characteristics of the adopted simulation technique. It is also evident the presence of an alternative current component having fundamental frequency, which should be absent; that is depending on a d.c. component in the input motor voltages.

Transient working conditions. In the figures 19, 20, 21 the evolutions are reported of some quantities in consequence of a sudden load variation.

Load condition: initial rated torque with sudden 100% overloading

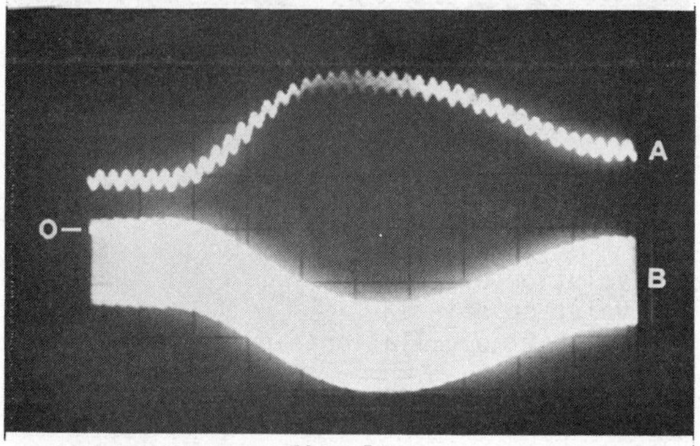

fig. 19

A) Evolution of electromagnetic torque (time scale: 50 m sec/div)
B) Evolution of the angle δ_s

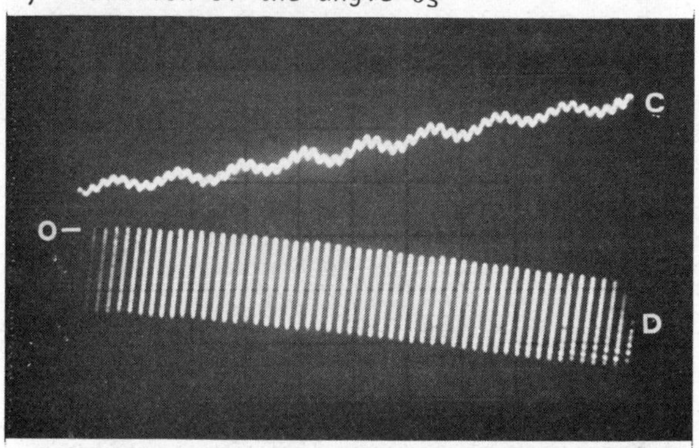

fig. 20

C) The same of A) with time scale of 10 msec/div
D) The same of B) with time scale of 10 msec/div

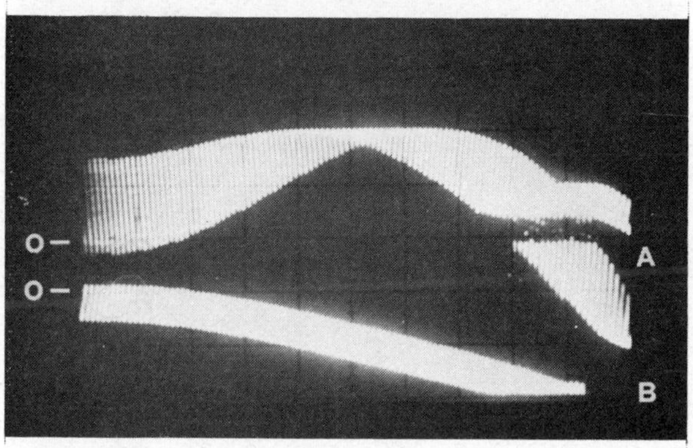

fig. 21

A) Evolution of - V_d
B) Evolution of the angle δ_s

The situation of fig. 21 represents the motor synchronism loss in consequence of the sudden overload applications.

P.W.M. Operation with One Fixed Position Variable Width Hole

Steady state operations at a frequency of f = 60 Hz

Load condition: constant rated torque

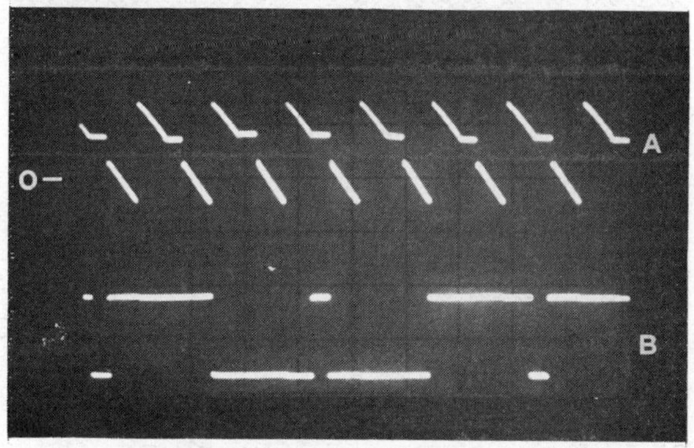

fig. 22

A) Evolution of the input model voltage - V_d
B) Evolution of the inverter output potential

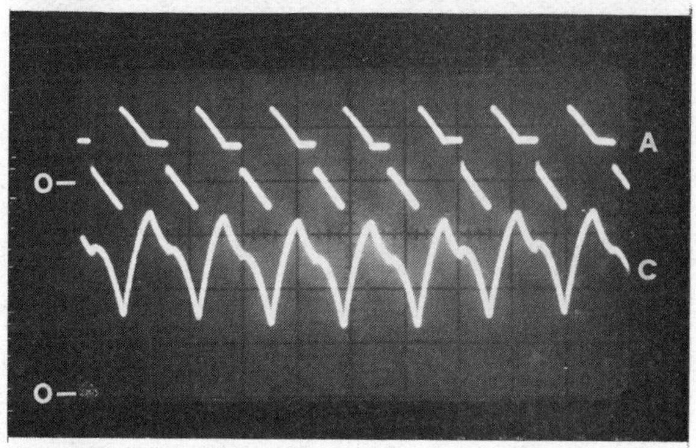

fig. 23

C) Evolution of - I_d

Transient working conditions

Load condition: initial 100% torque overload with sudden reduction to the rated torque

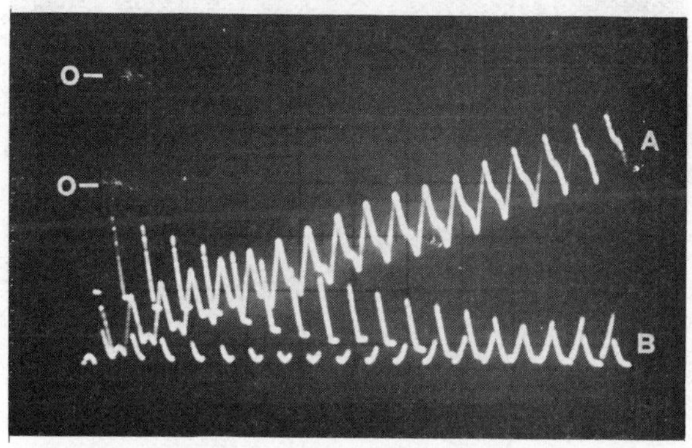

fig. 24

A) Evolution of - I_q
B) Evolution of input model voltage - V_q

P.W.M. Operation with Subharmonic Modulation Technique

Steady state operation at a frequency of 40Hz

Load condition: 100% torque overload

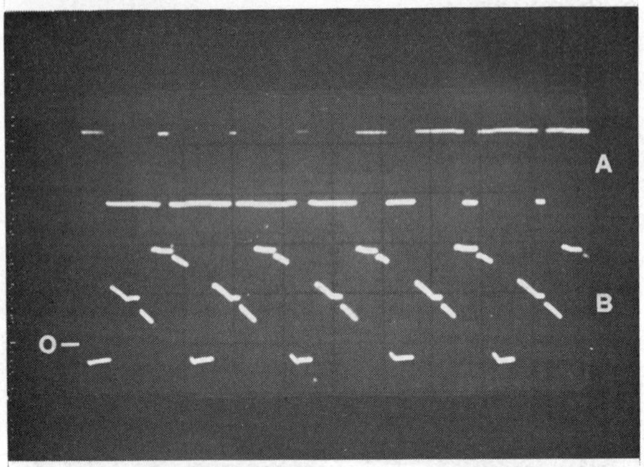

fig. 25
(time scale 2 msec/div)
A) Evolution of the inverter output potential
B) Evolution of the model input voltage - V_d

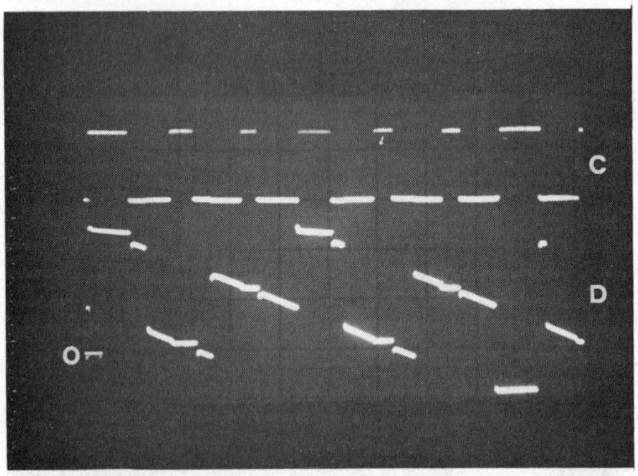

fig. 26
(time scale 1msec/div)
C) Evolution of interface output $\bar{Q}$
D) Evolution of the angle δ_s

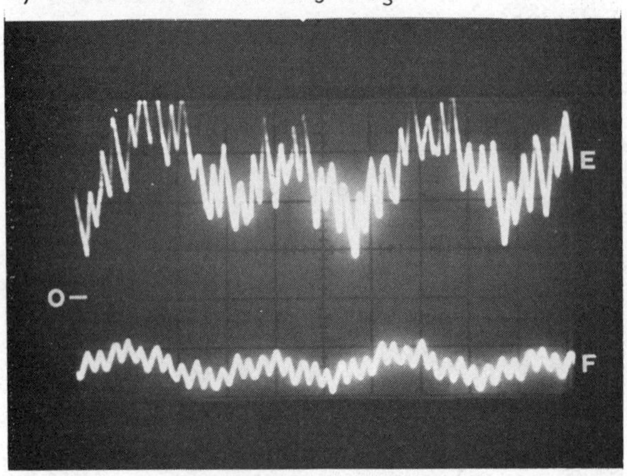

fig. 27
(time scale 5 msec/div)
E) Evolution of - I_d
F) Evolution of - I_q

Transient working condition

Load condition: initial 100% torque overload with sudden reduction to the rated torque

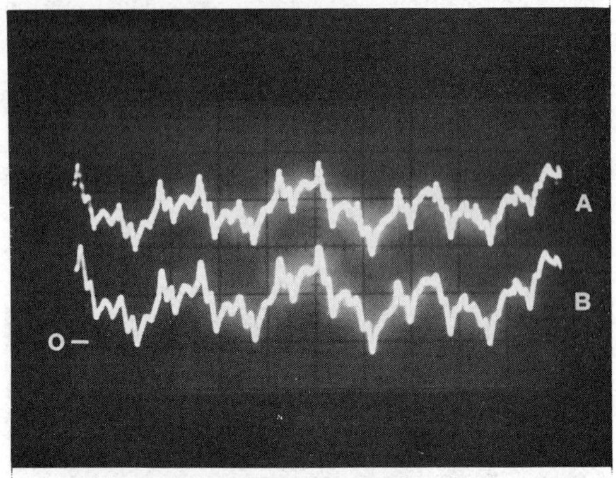

fig. 28
(time scale 5 msec/div)
A) Evolution of I_r
B) Evolution of - I_d
during shaft oscillations in consequence of load reduction

REFERENCES

P. Ferraris, M. Lazzari, A. Vagati, Contributo allo studio della dinamica dei motori sincroni alimentati da invertitori statici, con particolare riguardo ai fenomeni subtransitori, Simposio sulla Evoluzione nella Dinamica delle Macchine Elettriche Rotanti, Tirrenia, 394-402 (1975)

H. Foch, Complete simulation of a static converter by digital and analogue methods, 1st IFAC Symposium on Control in Power Electronics and Electrical Drives, Düsseldorf, 363-373 (1974)

P.C. Krause, Method of multiple reference frames applied to the analysis of symmetrical induction machinery, IEEE Transactions on Power Apparatus and Systems, 1, 218-227 PAS (1968)

ANALYSIS OF THREE-PHASE SEMICONDUCTOR CONNECTIONS BY PARK-VECTORS

István Rácz

Computer and Automation Institute, Hungarian Academy of Sciences, 1502 Budapest

ABSTRACT

The three-phase semiconductor connections are built up of elementary three-phase thyristor or diode groups. Due to the cyclical symmetry of transformers or a.c. machines connected to them the application of Park-vectors and zero sequence components is rewarding. The corresponding constraints of various conduction patterns are given in vector form for the voltages and currents. The pattern will be modified on the border of the domain permitted by the semiconductors or by the firing of a thyristor. It is rather straightforward to write the proper differential equations by this method. The Park-vector method is preferable for harmonic analysis. The vectors can be shown on a cathode-ray oscilloscope as well.

1. PARK-VECTORS

Instead of the instantaneous values of the three-phase currents $i_a(t), i_b(t)$ and $i_c(t)$ the transformed variables, the current Park-vector $\bar{i}(t)$ and the zero sequence component $i_0(t)$, are introduced:

$$\bar{i}(t) = i_x + ji_y = (1 \cdot i_a + \bar{a} \cdot i_b + \bar{a}^2 i_c) \cdot \frac{2}{3},$$

$$i_0(t) = \frac{i_a + i_b + i_c}{3}; \qquad \bar{a} = \exp\left(\frac{j2\pi}{3}\right),$$

The perpendicular components of the Park-vectors are sometimes denoted by α, β or d, q instead of x, y.

In the case of a.c. machines a physical picture can be assigned to this transformation (Fig. 1), the current Park-vector $\bar{i}(t)$ represents the resulting magnetomotive force of the three-phase winding system. Therefore we may regard the complex plane as the cross-section of a two-pole machine. Its phase axes a, b, c include angle 120°. The real axis is the phase axis a. Similar transformation is valid for the phase voltages, linked fluxes etc.

The transformation is not power invariant, because the instantaneous power is $p = u_a i_a + u_b i_b + u_c i_c = p_P + p_0$, where the Park-vector power component p_P may be expressed by the scalar product of the voltage and current vector

$$p_P = (\bar{u} \cdot \bar{i}) \frac{3}{2} = (u_x i_x + u_y i_y) \frac{3}{2}$$

and p_0 is the zero sequence component $p_0 = 3u_0 i_0$. Choosing other factors in the definition equations it would be possible to obtain power invariant transformation, but there are some reasons for this choice: (a) The tradition corresponding to the classical method of symmetrical components. (b) In normal operation of three-phase machines and networks the Park-vectors are identical with the time vectors or phasors and they may provide the initial or end conditions for transients, too. (c) Studying transient processes the instantaneous phase variables are more often used than the power formulas. It may be proved, that the projections of the Park-vectors on the phase axes a, b, c (e.g. i'_a, i'_b, i'_c, Fig. 1) provide the phase currents except their zero sequence parts: $i_a = i'_a + i_0$, $i_b = i'_b + i_0$, $i_c = i'_c + i_0$. In many instances the zero sequence components are equal zero.

Line Voltages

In wye-connection the Park-vector $\bar{u}$ of the phase voltages u_a, u_b, u_c determines also the line voltages. For instance the line voltage u_{bc} may be expressed as follows (Fig. 2)

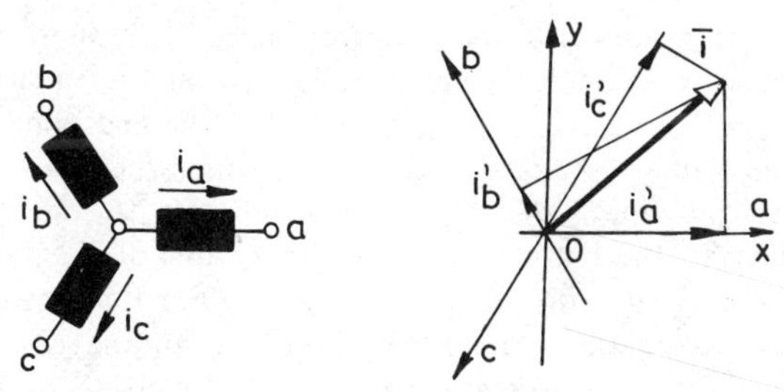

Fig. 1. Current Park-vector

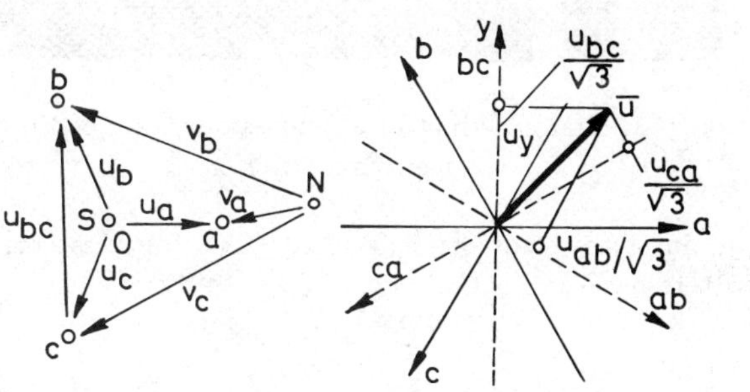

Fig. 2. Line voltages

$$u_{bc} = u_b - u_c = u_b' + u_0 - u_c' - u_0 =$$
$$= \mathrm{Re}\{\bar{a}^2\bar{u}\} - \mathrm{Re}\{\bar{a}\bar{u}\} = \mathrm{Re}\{j\sqrt{3}\bar{u}\} =$$
$$= \sqrt{3}\ \mathrm{Im}\{\bar{u}\} = \sqrt{3}u_y.$$

Therefore the projections of the phase voltage vectors on the proper axis give the instantaneous values of the line voltages with a scale factor $\sqrt{3}$. It is valid also for transients, e.g. for d.c. voltages as well.

Similar relations hold for the line and phase currents in delta-connections.

If two three-phase voltage systems have the same line voltages (u and v in Fig. 2), their voltage Park-vectors are identical, $\bar{u} = \bar{v}$, the zero sequence components differ by the voltage between the two reference points: $v_0 = u_0 + u_{NS}$.

Steady State Operation

The trajectories of the Park-vectors show a g-sided symmetry for machines constructed and controlled by thyristors symmetrically. In the case of six pulse connections g is equal six (Fig. 3) or for groups of different operation in positive and negative direction $g = 3$ (Fig. 4). In such cases it is enough to calculate the wave shapes for one g-th part of the period $\left(\tau = \frac{2\pi}{g}\right)$, the other parts of the trajectories could be determined by a rotation $\frac{2\pi}{g}$. The peridoicity conditions correspond to this fact. If there are zero sequence components the periodicity condition may be given by a factor 1 and -1 ($g = 3$ and 6), respectively.

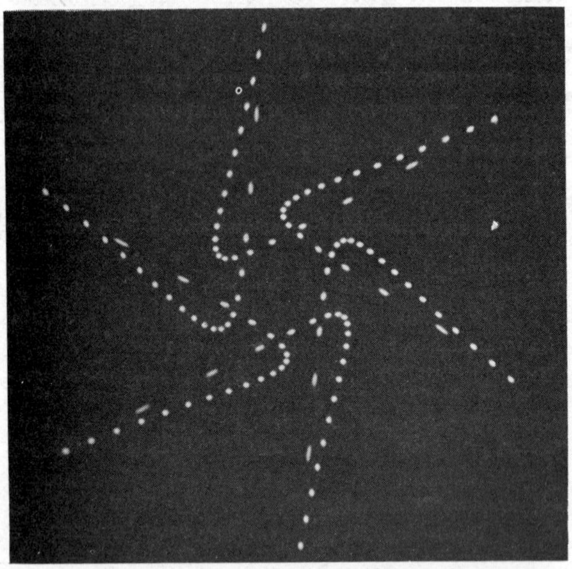

Fig. 3. Stator current of an induction motor fed by voltage source inverter

In steady-state the Park-vector time functions may be resolved in Fourier-series

$$\bar{i}(t) = \sum \bar{I}_\nu e^{j\nu\omega_1 t}$$

$\nu = 1 + gk,\ k = 0, \pm 1, \pm 2, \ldots$ the coefficients $\bar{I}_\nu$ may be calculated by an integral for the interval τ:

$$\bar{I}_\nu = \frac{1}{\tau} \int_{(\tau)} \bar{i}(t) e^{-j\nu\omega_1 t} d\omega_1 t.$$

For more details regarding of the applications of Park-vectors see (Ref. [1]-[3]).

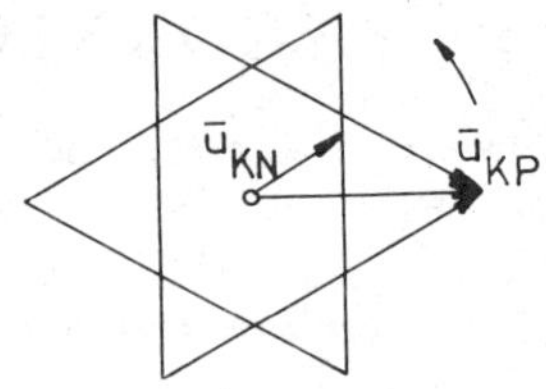

Fig. 4. Commutating condenser voltages on the positive and negative side of a current source inverter

2. THREE-PHASE VALVE GROUP

The usual circuits, bridges, current source inverters etc., can be built up of elementary three-phase thyristor or diode groups. The voltage and current vectors of a group must satisfy some restrictions depending upon the number of conducting valves. It is assumed that the semiconductor elements may be replaced by ideal switches, in conduction they are equivalent to short circuits, otherwise to open circuits.

2.1. One element is conducting. By supposing one diode in phase a to be conducting, the current $i_a = i_d$. It is usually equal to the d.c. current of the d.c. link (Fig. 5). The phase currents $i_b = i_c = 0$. In the case of

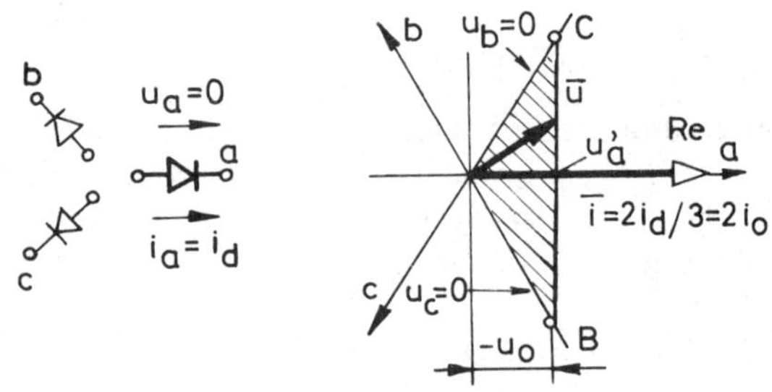

Fig. 5. Diode a is conducting

diodes the voltage restrictions are: $u_a = 0$, $u_b < 0$, $u_c < 0$. (In the case of thyristors opposite signs are necessary to be able to turn them on in the following conduction pattern). It turns out from the definition that the zero sequence component of the currents $i_0 = \frac{i_d}{3}$ can have only positive values, while that of the voltage $u_0 < 0$. The current vector $\bar{i} = \frac{i_d 2}{3} = 2i_0$ has to coincide with the positive real axis. Supposing a negative value for u_0, it follows from the equation $u_a = 0 = u_a' + u_0 = \mathrm{Re}\{\bar{u}\} + u_0$ that the end point of the voltage vector $\bar{u}$ has to be somewhere on the straight line BC being perpendicular to the real axis. In the border points B or C one of the diode voltages u_b or u_c becomes zero. Over that borders the conduction state is not possible. (In the case of thyristors the voltage vector must be outside of the section BC at the firing instant, the first possible firing instant for the thyristor b is in the position C, that is, the instant of natural commutation). If u_0 is varying, the diode voltage vector has to be in a sector including angle $\pm 60°$ with respect to the real axis.

If a diode in an other phase is conducting, the corresponding phase axis takes over the role of the real axis, or using a rotation $\pm 120°$ of the reference axes the situation remains the same.

2.2. Two elements are conducting. Let us suppose that the diodes b and c are conducting. Being now two voltages zero rather than two currents as in the previous case, and the corresponding signs of voltages and currents are reversed, this case may be regarded as a dual one. The voltage vector $\bar{u} = 2u_0$ must have its direction fixed $(u_y = 0)$ to the negative real axis, the current vector has to be in a sector $\pm 60°$ with respect to that direction (Fig. 6).

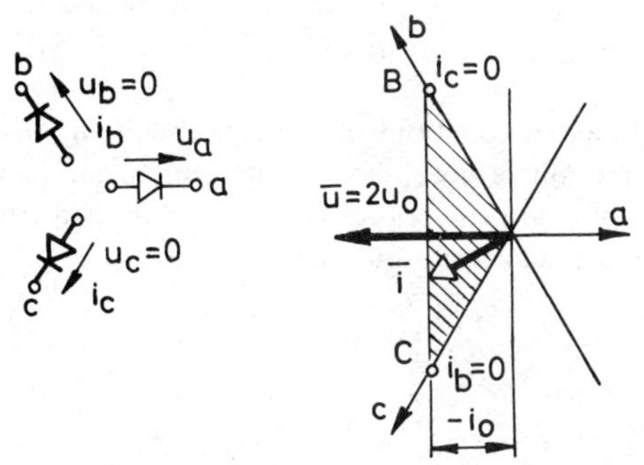

Fig. 6. Diode b and c are conducting

It is interesting to consider the power relations in both cases. The three-phase power equals 0, because either the current or the voltage is 0 in each phase, therefore $0 = p_P + p_0$. The zero sequence power is negative due to the opposite signs, and the scaler product in the Park-vector part p_P is positive due to the angle less than 60°. The zero sequence components are associated in general with the d.c. side and the Park-vectors with the a.c. side. It means, that in the case of diodes only one direction of power flow is feasible. The a.c. side delivers power to the d.c. side (rectifier). In the case of thyristors the voltage restrictions are not valid, both direction of instantaneous and average energy flow are possible.

2.3. Three or 0 elements are conducting. If all the three elements are conducting, each phase voltage is zero, therefore the voltage vector and the zero-sequence component are equal to zero, $\bar{u} = \bar{0}$, $u_0 = 0$. The currents must be positive e.g. $i_a = i_a' + i_0 = \mathrm{Re}\{\bar{i}\} + i_0 > 0$. Being $i_0 > 0$, in limit case the projection of the current vector on the real axis (phase axis a) must be equal to $-i_0$. For a given value of i_0 the permitted domain of the current vector is shown in Fig. 7.

If no elements are conducting due to the duality the permitted domain of the voltage vector for a given value of $u_0 < 0$ can be determined similarly (Fig. 8).

3. THREE-PHASE BRIDGE CIRCUITS

Two elementary three-phase groups are connected together in the well known way. The upper group (Fig. 9)

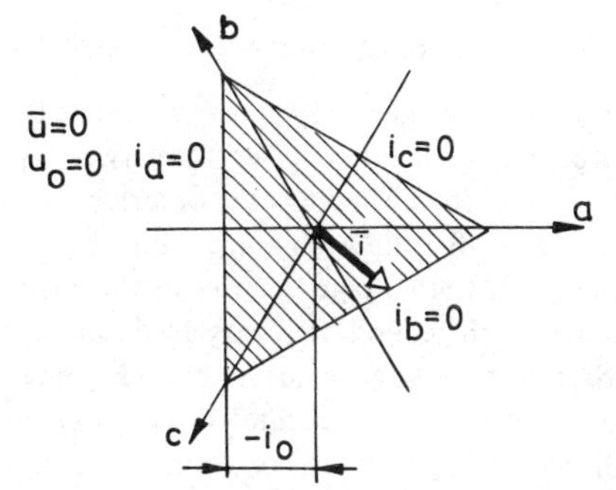

Fig. 7. Three diodes are conducting

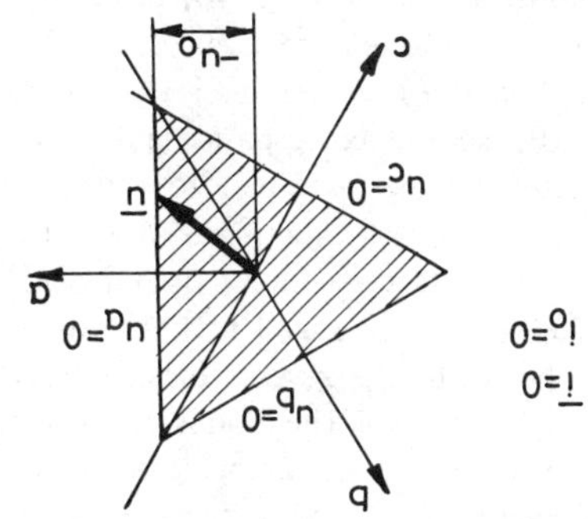

Fig. 8. No diode is conducting

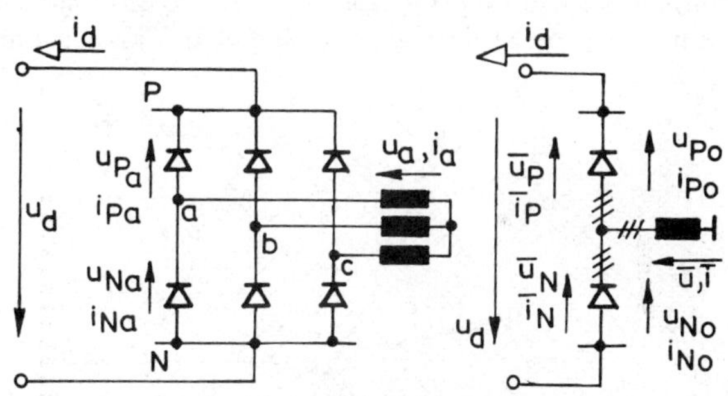

Fig. 9. Three-phase bridge

has common cathode potential on the positive d.c. bus P (in rectifier operation), the anodes of the lower group are connected to the d.c. bus N. It is supposed, as usual, that the joined three-phase winding system (transformer, stator or rotor windings of a generator or motor) has no zero sequence voltage and current component, and it is wye-connected, or we are calculating with the equivalent wye-connection.

The three-phase voltage systems of the positive and negative diode groups and of the winding system have common line to line voltages across the terminals a, b, c, therefore their phase voltage Park-vectors are identical or corresponding to the chosen reference directions: $\bar{u} = \bar{u}_N = -\bar{u}_P$. The mesh-equation with the voltages of diodes a is $u_d + u_{Pa} + u_{Na} = 0$. Similar equations hold for the phases b and c, summing them up we obtain for the zero sequence voltages: $u_d = -u_{P0} - u_{N0}$. In the case of diode bridge u_d can't have negative values. The node-equation for the terminal a is $i_{Pa} = i_{Na} + i_a$. Using the similar equations for phases b and c as well as from the definition the current vector relation is $\bar{i} = \bar{i}_P - \bar{i}_N$ and the zero sequence equation $i_0 = i_{P0} - i_{N0} = 0$ therefore $i_{P0} = i_{N0} = \frac{i_\alpha}{3}$. Due to the above relations, in general, it is favourable to consider the bridge as a whole, rather than two separate elementary groups. In the next cases of bridge conduction patterns we will follow that principle except the last case. Being all the voltage vectors in the bridge essentially identical, in general, it is easy to determine any diode (or thyristor) voltage value if one element in its group is conducting. As a result of conduction e.g. $u_{Pa} = 0$, therefore $u_{Pb} = u_{Pb} - u_{Pa}$ and $u_{Pc} = u_{Pc} - u_{Pa}$. The last two line voltages can be obtained by the proper projections and by a scale factor $\sqrt{3}$.

3.1. No diodes are conducting. All the diode voltages have to be negative. In the border case the voltage across a positive diode and a negative one is equal to zero (e.g. $u_{Pa} = u_{Nb} = 0$), so the corresponding line voltage of the three-phase windings is equal to the d.c. voltage, $u_d = u_b - u_a$. The last equation is satisfied when the end point of the voltage vector $\bar{u}$ is lying on the side P_a, N_b of the hexagon in Fig. 10. If the voltage vector is inside the hexagon, the reverse biasing effect of the d.c. voltage u_d blocks all the diodes. By increasing the absolute value $|\bar{u}|$, voltage vector $\bar{u}$ tends to step out from the hexagon. The diodes denoted on the sides begin conducting and an other conduction pattern is established. In the case of thyristors the condition for turning on two thyristors e.g. P_a and N_b is that the voltage vector be on the other side of the straight line P_aN_b.

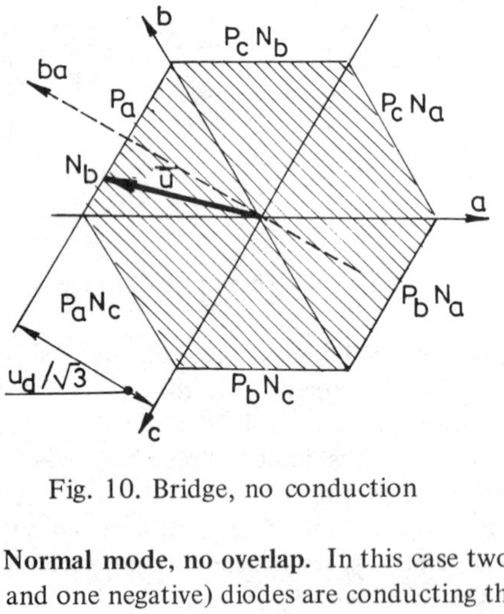

Fig. 10. Bridge, no conduction

3.2. Normal mode, no overlap. In this case two (one positive and one negative) diodes are conducting the d.c. current i_d, e.g. P_a and N_b (Fig. 11). The phase currents are: $i_a = i_d$, $i_b = -i_d$, $i_c = 0$. The current vector is drawn according to the projection law, its length is $\frac{i_d 2}{\sqrt{3}}$, and its direction determines the main, permanent direction of this conduction state. The positive and negative current vectors $\bar{i}_P$ and $\bar{i}_N$ limit the voltage vector $\bar{u}$ into sectors of ± 60°. Both conditions are satisfied in a sector of ± 30° with respect to the main direction. For the determination of the voltages of nonconducting valves the projection axes are drawn with dotted lines, a scale factor $\sqrt{3}$ is needed.

Choosing the real axis x in the direction of the

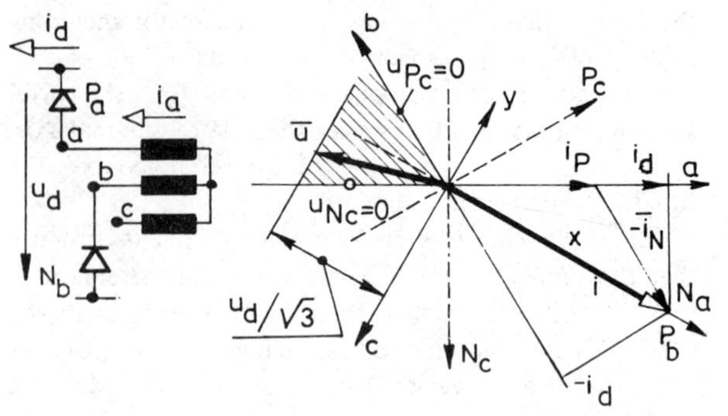

Fig. 11. Bridge, 2D, normal mode

current vector $\bar{i}$, the relations among the a.c. and d.c. side variables are the following:

$$i_x = \frac{i_d 2}{\sqrt{3}} > 0, \quad i_y = 0; \qquad u_x = -\frac{u_d}{\sqrt{3}}$$

so the differential equations of the d.c. circuit and that of the three phase machine can be combined. In a simulation program the border conditions for $\bar{u}$ are to be checked if the bridge is built up of diodes:

$$\frac{u_{Nc}}{\sqrt{3}} = \frac{u_x}{2} - \frac{u_y\sqrt{3}}{2} < 0,$$

$$\frac{u_{Pc}}{\sqrt{3}} = \frac{u_x}{2} + \frac{u_y\sqrt{3}}{2} < 0.$$

By knowing that the vectors are varying in the positive rotation sense, the second condition may be dropped out.

3.3. Overlap, three-elements are conducting. (e.g. P_a, N_b, N_c, Fig. 12). The current on the positive side

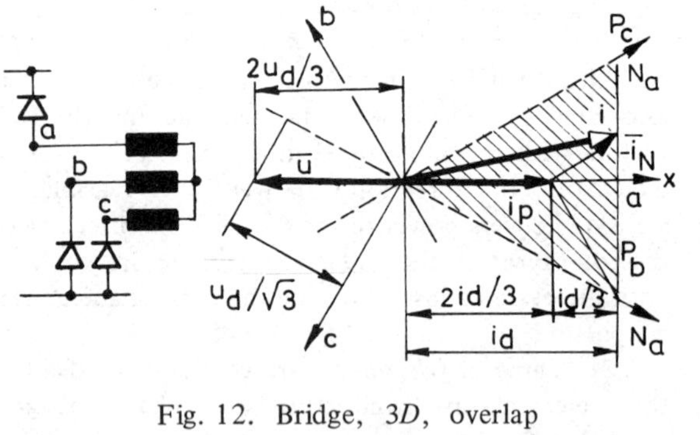

Fig. 12. Bridge, 3D, overlap

is $\bar{i}_P = \frac{2i_d}{3}$, while on the negative side $\bar{i}_N$ may vary in a sector ± 60°. The resulting three-phase current vector $\bar{i}$ has to be in a sector ± 30° around the axis a. The terminals b and c are short circuited, $u_b = u_c$. Due to the projection law the voltage vector has to have the same projection on axis b and c, that is, it coincides with axis a, its length is $\frac{u_d}{\sqrt{3}}\frac{2}{\sqrt{3}} = \frac{u_d 2}{3}$. Choosing the real axis in the main direction, the basic relations are the following: $u_x = -\frac{u_d 2}{3}$; $u_y = 0$; $i_x = i_d$ and the border conditions for the current are

$$i_b = -\frac{i_x}{2} + \frac{i_y\sqrt{3}}{2} > 0,$$

$$i_c = -\frac{i_x}{2} - \frac{i_y\sqrt{3}}{2} > 0$$

and for the voltage $u_d > 0$ in the case of a diode bridge.

3.4. Short circuit, four elements are conducting. Consider only one example (e.g. P_a, P_b, N_a, N_b). All the voltages in the bridge are equal zero, $\bar{u} = 0$. Both current vectors of the positive and negative groups may be in a sector of $\pm 60°$, so the permitted domain of the resulting current $\bar{i}$ corresponds to Fig. 13, provided,

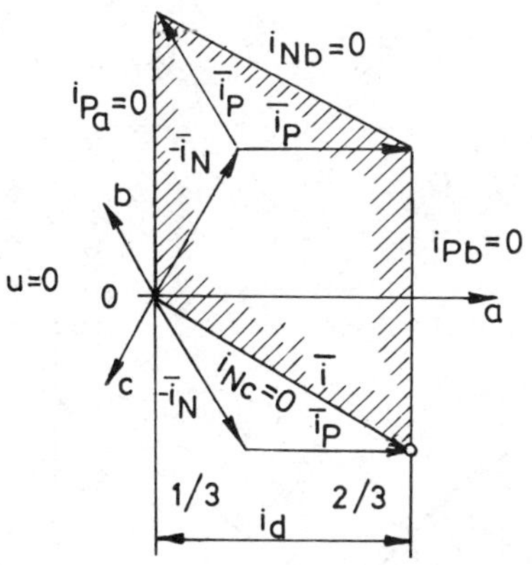

Fig. 13. Bridge, *4D*, short circuit

that the d.c. current has a given value. The border conditions denoted in the figure are valid only for thyristor bridge circuit, because in the diode bridge case the disregarded diode forward voltages may result other conduction states.

4. THE EFFECT OF SHUNT CAPACITORS ON THE COMMUTATION

It will be investigated, although only qualitatively as an application example, how the shunt capacitors connected to the three-phase terminals of a diode bridge converter (Fig. 14) modify the commutation process.

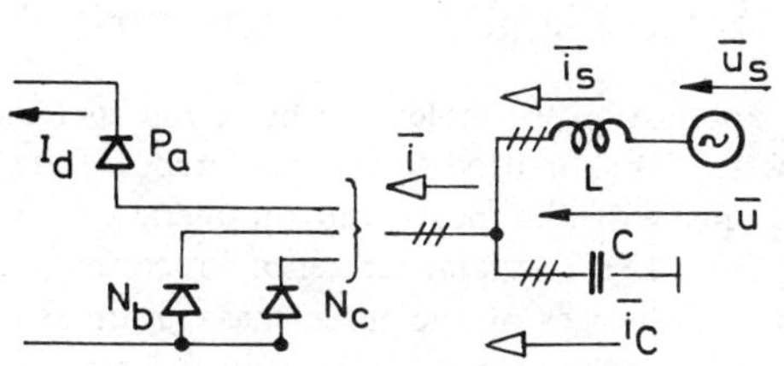

Fig. 14. Shunt capacitors

For more detailed analysis see (Ref. [4]). $\bar{u}_s$ and $\bar{i}_s$ are the Park-vectors of the three-phase supply voltage and current system. $\bar{u}_s$ is supposed to represent a three-phase positive sequence sinusoidal voltage system, therefore $\bar{u}_s$ is rotating along a circle, $\bar{u}_s = U_s e^{j\omega t}$. $\bar{u}_s$ may be the secondary no load voltage of a transformer, L is its short-circuit inductance. If a synchronous generator is supplying the bridge converter, for the calculation of waveshapes in periodical steady state $\bar{u}_s$ may be approximated by the voltages behind the subtransient reactances, and L by the subtransient inductances of the machine, $L = L_d'' \approx L_q''$. The resistances are neglected. It is assumed that the current in the *dc* link is ripple free, $i_d = I_d = \text{const.}$

One sixth of a period, that is, $\omega t_0 \leqslant \omega t < \omega t_0 + \frac{\pi}{3} = \omega t_e$ interval will be examined. The input current in phase a is constant $i_a = I_d$ and the current commutates from the diode N_b to N_c. Before commutation $(t < t_0)$ the diodes P_a and N_b were conducting. The input current vector was $\bar{i}_0 = I_d \frac{2}{\sqrt{3}} \cdot \exp\left(-\frac{j\pi}{6}\right)$. It will be $\bar{i}_e = \bar{i}_0 \exp\left(\frac{j\pi}{3}\right)$ at the end of commutation (at t_e) and because $i_a = I_d$ remains constant, $\bar{i}$ must move along a straight line perpendicular to the phase axis a (Fig. 15).

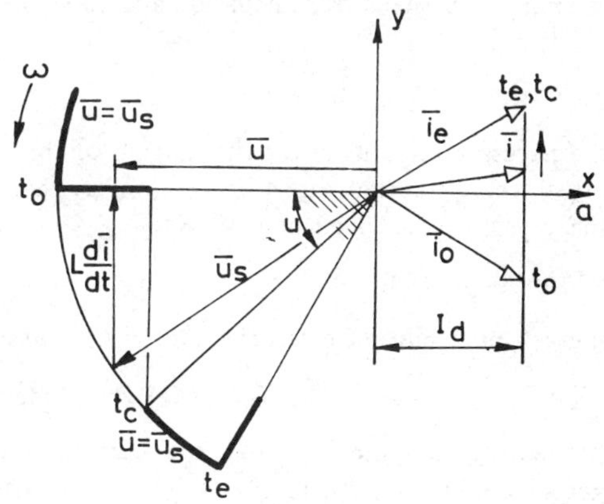

Fig. 15. Commutation, $C = 0$

The voltage equation of the three-phase side is

$$\bar{u} = \bar{u}_s + L\frac{d\bar{i}_s}{dt} \qquad (1)$$

According to the paragraphs 3.2 and 3.3 (Fig. 11 and 12) the voltage $\bar{u}$ has to be in the region above the negative real axis before the commutation (*2D*). At the instant t_0 $\bar{u}$ reaches the border line $u_{Nc} = 0$, and the commutation (*3D*) may begin. The current $\bar{i} = \bar{i}_s$ must move continuously without condenser $(C = 0)$. The transition from $\bar{i}_0$ to $\bar{i}_e$ takes a time $t_c = t_0 = \frac{u}{\omega}$, where the overlap angle u may be calculated from equation (1). After the commutation a conduction *2D* (P_a and N_c) takes place in an interval $t_e - t_c = \frac{\frac{\pi}{3} - u}{\omega}$. The voltage vector $\bar{u}$ must be in a range of $60°$ under the negative real axis (Fig. 15). Connecting a condenser bank C to the configuration, its current vector $\bar{i}_C$, as well as the bridge current $\bar{i} = \bar{i}_s + \bar{i}_C$, too, can suddenly change at the instant t_0 when the voltage vector $\bar{u}$ first coincides with the negative real axis. There are two possibilities: (a) instantaneous, and (b) mixed commutation (Fig. 16). The current vector $\bar{i}$ can change at once to the position $\bar{i}_e$, provided that the voltage vector $\bar{u}$ is moving to the region under the

kk

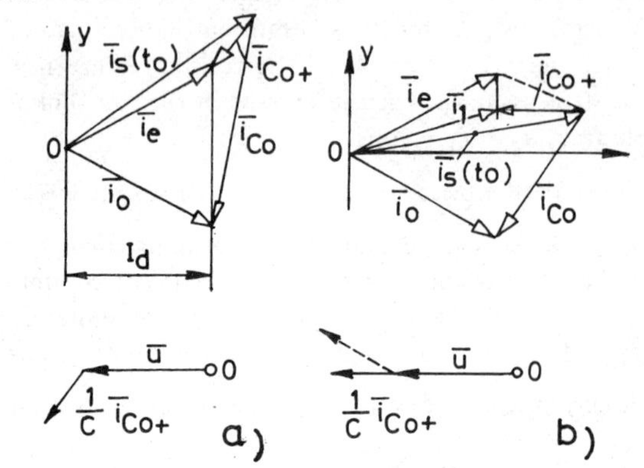

Fig. 16. Instantaneous and mixed commutation

negative real axis corresponding to the new $2D$ conduction mode (P_a, N_c). The variation speed of $\bar{u}$ will be determined by the current equation and capacity C:

$$\bar{i}_C = -\bar{i}_s + \bar{i} = C\frac{d\bar{u}}{dt} \qquad (2)$$

and the necessary condition for the motion of the voltage vector the permitted direction $(-y)$ is, that the imaginary part of the speed be negative:

$$\mathrm{Im}\{-\bar{i}_s + \bar{i}\} < 0. \qquad (3)$$

As long as this condition is satisfied at $t = t_0$ and even with $\bar{i} = \bar{i}_e = \frac{2I_d}{\sqrt{3}} \cdot \exp\left(\frac{j\pi}{6}\right)$ (Fig. 16a), the diodes are instantaneously commutated and the current vector changes suddenly from $\bar{i}(t_0)$ to $\bar{i}_e$ with $u = 0$. Therefore the current $\bar{i}$ will be constant in the whole sixth of the period, and the time functions $\bar{i}_s(t)$ and $\bar{u}(t)$ can be determined by the solution of the differential equations (1) and (2). The initial conditions can be calculated from the periodicity conditions of $\bar{i}_s(t)$ and $\bar{u}(t)$. The vectors at $t = t_0$ and at $t = t_0 + \frac{\pi}{3\omega}$ have to be equal to each other in absolute value but they have to be rotated by 60° in clockwise direction from their original position. In this case it is unnecessary to resolve the vectors into components x and y, that is, to solve four equations. The use of Park-vectors simplifies the calculation.

If the condition (3) is not satisfied at $t = t_0$ with $\bar{i} = \bar{i}_e$ the current vector $\bar{i}$ can change suddenly from $\bar{i}_0$ only to $\bar{i}_1$, when the condenser current remains real and the condenser voltage doesn't move back in the region corresponding to the previous $2D$ conduction mode (dotted lines in Fig. 16b, the lower parts show the voltage $\bar{u}$ and its speed at $t = t_0$). The conduction mode has to be $3D$ as long as the current vector takes the position $\bar{i}_e$. During this interval having overlap angle u the voltage vector $\bar{u}$ remains in the direction of the negative real axis, $u_y = 0$. Therefore this mode of operation is denoted as mixed commutation. After this a $2D$ conduction state follows without overlap. Figure 17 shows the result of the solution for a case with the overlap angle $u = 37°$ (Ref. [4]). The condenser current vectors and the $\bar{u} - \bar{u}_s = \frac{Ld\bar{i}_s}{dt}$ vectors

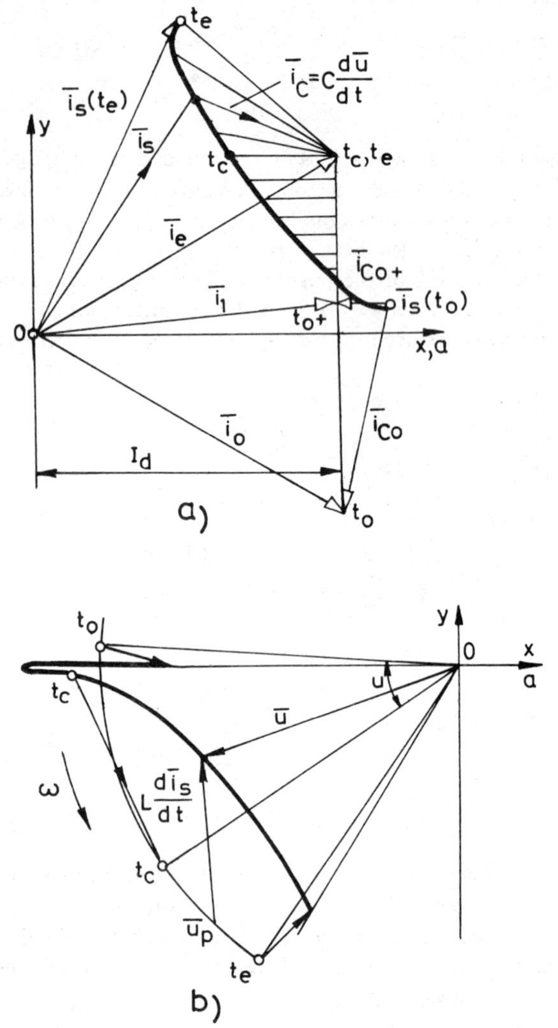

Fig. 17. Mixed commutation
a.) currents, b.) voltages

show the speed of $\bar{u}$, and that one of $\bar{i}_s$, respectively. These relations may help in the qualitative drawing of the trajectories. Using condensers the overlap angle and the voltage drop of the converter will be reduced.

5. CURRENT-SOURCE INVERTER (*CSI*)

More application examples can be found in Lázár's paper (Ref. [5]) submitted to this conference. Here we will deal only with the most common operation modes of *CSI* (Fig. 18). The same reference directions will be used for the voltages of the three-phase group as they are denoted for the currents. The shunt elements (the condensers and the stator winding of the motor) can't have zero sequence current. Therefore the zero sequence current component of each semiconductor groups are equal to $\frac{i_d}{3}$. In the more detailed figure b only the current conducting elements are indicated for one sixth of a period beginning with the firing of the thyristor *TNC* in phase c in the negative side.

5.1. Current and voltage vectors, conduction modes. The current distribution in the positive side doesn't change. The dc current is flowing in phase a, the current Park-vectors of the thyristors and diodes in the positive side are equal to $\bar{i}_{TP} = \bar{i}_{DP} = \frac{2i_d}{3}$ and point in the direction of phase a (Fig. 19a). Therefore the condenser current in the positive side $\bar{i}_P = \bar{i}_{TP} - \bar{i}_{DP} = \bar{0}$, so the condenser voltage ($\bar{u}_P$) remains constant, and it is equal with negative sign to the thyristor voltage vector.

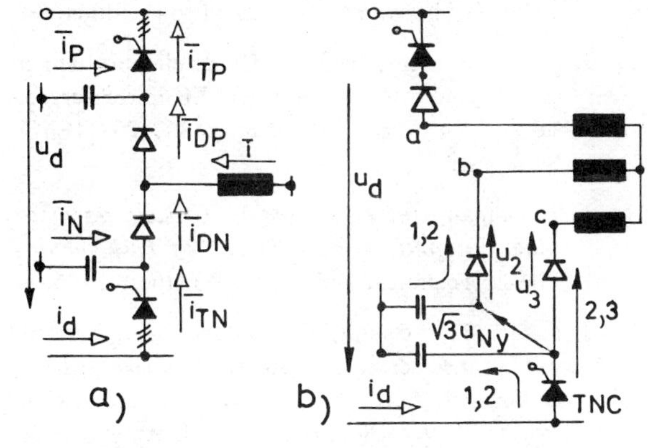

Fig. 18. One-line and detailed circuit of the *CSI*

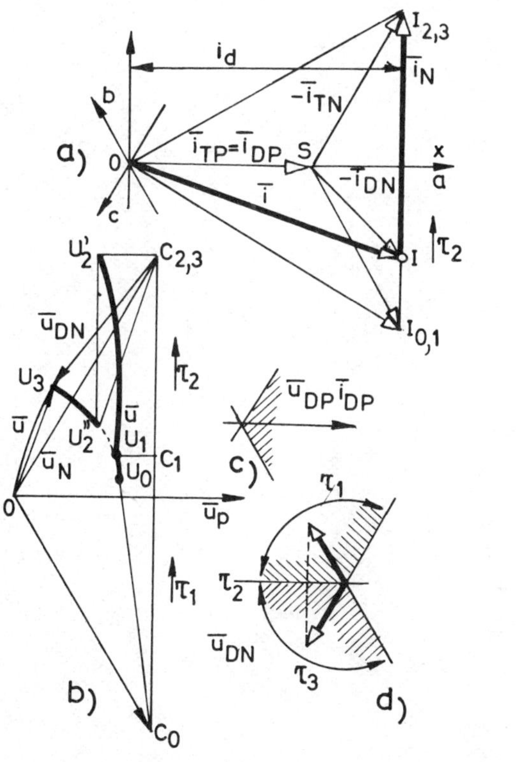

Fig. 19. Current (a) and voltage (b) vector trajectories; the permitted domains for $\bar{u}_{DP}$ (c) and $\bar{u}_{DN}$ (d)

In the negative side the condenser current vector is $\bar{i}_N = \bar{i}_{DN} - \bar{i}_{TN}$ and the motor current $\bar{i} = \bar{i}_{DP} - \bar{i}_{DN} = \overline{0I}_i$. Before firing *TNC* only the diode *DNB* and thyristor *TNB* were conducting in phase b, therefore $\bar{i}_{DN} = \bar{i}_{TN} = \overline{I_0S}$ (Fig. 19a). After firing *TNC*, the thyristor current commutates practically instantly to phase c, $\bar{i}_{TN} = \overline{I_2S}$, and the condenser current will be $\bar{i}_N = \overline{I_0S} - \overline{I_2S} = \overline{I_0I_2}$. After time t_0 denoting the firing time point of thyristor *TNC* three different conduction modes will arise with the intervals $\tau_1 = t_1 - t_0$, $\tau_2 = t_2 - t_1$ and $\tau_3 = t_3 - t_1$. In Fig. 18b the numbers 1, 2, 3 denote the current flow in these modes τ_i, and in Fig. 19a, b the indices 0-3 relate to t_i.

In τ_1 all the current vectors are varying proportional to the current i_d. i_d = const. is assumed in Fig. 19a for the sake of simplicity. Due to the equation $C\frac{d\bar{u}_N}{dt} = \bar{i}_N$ the condenser voltage $\bar{u}_N$ is moving from C_0 to C_1 with practically constant speed. The name of the mode τ_1 is "recharging". The diode voltage $\bar{u}_{dN} = \bar{u} - \bar{u}_N = \overline{C_iU_i}$ may vary in the domain shown in Fig. 19d. $\bar{u}_{dN}$ reaches the border position $\overline{C_1U_1}$ at the instant t_1. The diode *DNC* begin conducting, too, therefore in the interval τ_2 (diode commutation) $\bar{u}_{DN}$ has to point into the negative real axis direction and $\bar{i}_{DN}$ can vary in a sector ± 60° (Fig. 19a).

As $\bar{i}_{DN}$ takes the border position $\overline{I_2S}$, the conduction of the diode in phase b is terminated, and the third interval τ_3 (no overlap) starts.

Figure 19 shows that all border conditions (for $\bar{u}_{DP}$, too) are satisfied.

5.2. The differential equations of the *CSI*. The most simple equations are obtained in a simulation program, therefore the computation time is the shortest, if the reference axes are chosen in the direction of the corresponding main axes in each conduction state. Nevertheless it is sometimes desirable to use the same reference frames for each conduction state (small memory of the computer or for the sake of theoretical deductions). In this case the diodes in both forward and reverse mode can be substituted by proper voltages. In the forward mode the diode voltage is 0, while in the reverse one the voltage has to be chosen to force the diode current to zero. Such voltages are u_2 and u_3 in Fig. 18.

Let us suppose that among the elements of state-vector x x_1 and x_2 are the stator current components of $\bar{i}$, $\bar{i} = i_x + ji_y = x_1 + jx_2$; $x_3 = u_{Ny}$ is the y-component of the condenser voltage in the negative side, other elements are the rotor quantities (fluxes or currents), speed of rotation, control variables etc. The stator voltage vector $\bar{u}$ is to be expressed by this variables.

Using *TNC* (Fig. 18b) as reference point of the three-phase voltage system $\bar{u}$ we obtain $u_a = u_d$, $u_c = u_3$, $u_b = u_2 + \sqrt{3}x_3$ and with $\bar{u} = u_x + ju_y$ according to the definition of Park-vectors

$$u_x = \frac{2u_d}{3} - \frac{u_2}{3} - \frac{u_3}{3} - \frac{x_3\sqrt{3}}{3}$$

$$u_y = \frac{u_2 - u_3}{\sqrt{3}} + x_3,$$

The d.c. current can be expressed by x_1, $i_d = x_1$, the condenser current component i_{Ny} by $\frac{x_1}{\sqrt{3}} - x_2$, so the state differential equation may be written in the following form:

$$\frac{dx}{dt} = f(x, E) - k_2u_2 - k_3u_3$$

where f is in general a nonlinear function (due to the speed variation, torque equation, saturation provided that they are taken into consideration), but the terms having u_2 and u_3 are linear with constant vectors k_2 and k_3.

In the conduction mode 2 (diode commutation, τ_2) $u_2 = 0$ and $u_3 = 0$. In the mode 1 (recharging) $u_2 = 0$ and i_3 must be equal 0, the latter can be expressed as follows

$$i_3 = -i_c = \frac{x_1}{2} + \frac{x_2}{\sqrt{3}} = k_{i3}^T x = 0.$$

From the last two equations we obtain for the interval τ_1

$$u_3 = \frac{k_{i3}^T f}{k_{i3}^T k_3}; \qquad \frac{dx}{dt} = \left(I - \frac{k_3 k_{i3}^T}{k_{i3}^T k_3}\right) f.$$

The expression in the brackets is a projector matrix. The equations of the conduction mode 3 (τ_3) can be obtained similarly with the conditions $u_3 = 0$ and $i_2 = 0$.

REFERENCES

[1] K.P. Kovács – I. Rácz, *Transiente Vorgänge in Wechselstrommaschinen,* Akademie Verlag, Budapest, 1959.

[2] I. Rácz, Oszillographische Aufnahme und harmonische Analyse von Dreiphasen-Vektoren, *Periodica Polytechnica* 8, 325 (1964), Budapest.

[3] I. Rácz, Betrachtungen zu Oberwellenproblemen an Asynchronmotoren bei Stromrichterspeisung, *Periodica Polytechnica* 11, 29 (1967), Budapest.

[4] G.P. Kalman, *Investigation of Capacitor-assisted Line Commutation by Means of Park-vectors,* Thesis. Technical University, Budapest 1976.

[5] J. Lázár, The operational modes of a current-source inverter induction motor drive system, Düsseldorf, Second IFAC Symposium on Control in Power Electronics and Electric Drives, 1977.

ANALYSIS OF FORCED COMMUTATION PROCESSES OF A BRIDGE INVERTER BY MEANS OF SYMMETRICAL COMPONENTS

I. Bausz

Computer and Automation Institute, Hungarian Academy of Sciences, 1502 Budapest

ABSTRACT

In consequence of the symmetrical arrangement of the bridge-connection inverter its processes may be written in a relatively simple way by the symmetrical components of thyristor-currents. Each working point on the trajectory corresponds to processes of the inverter in the coordinate-system of symmetrical components. The working states of the inverter, depending on the initial conditions could well be seen by the aid of the above mentioned trajectory and the main parameters from the viewpoints of operation of the inverter could be got by a suitable constructional procedure.

INTRODUCTION

A single-phase bridge forced commutation thyristor inverter is shown in Fig. 1 with the d.c. and a.c. filter circuit pertaining to it. The phenomena occurring in the switching can be described, if the state of the semiconductors remains unchanged, by a matrix differential equation whose dimension is identical with the number of energy storage elements. If the conduction state of any of the 8 semiconductors of the inverter changes, the matrix equation, describing the system changes too. Even during the period of inverter operation several such changes occur: at least 4 commutations take place, every commutation consists in average of 3 processes; the circulating currents in the commutating circuit can decay between 2 commutations etc. The exact computation of the steady-state condition and transient phenomena of the circuit in Fig. 1 is complicated and difficult to survey.

In order to facilitate computations it is practical to use an approximate constraint on the one hand, and to choose, instead of plotting time functions a layout, containing in a single curve all variables, on the other. This solves also the problem of presentation of processes differing to a great extent from each other as to their length.

EQUATIONS DESCRIBING THE PROCESSES

Assuming constant d.c. input voltage reduces not only by two the number of the unknown variables according to the two energy storages abandoned but by the reduction of commutating capacitor pairs the dimension number of the matrices is reduced by four and the processes taking place in the right and left side become independent from each other.

Under the approximate constraint that the output current of the inverter remains unchanged during the commutations, the matrix differential equation describing the system is broken down to three parts independent from each other: two 3×3 matrix equations of identical form describing the inverter-part on the right and left sides and one related to the filter circuit. In the case of the interruption of the circulating currents flowing in the commutating circuit the circuits cannot be broken down to parts, they can be examined only joint-

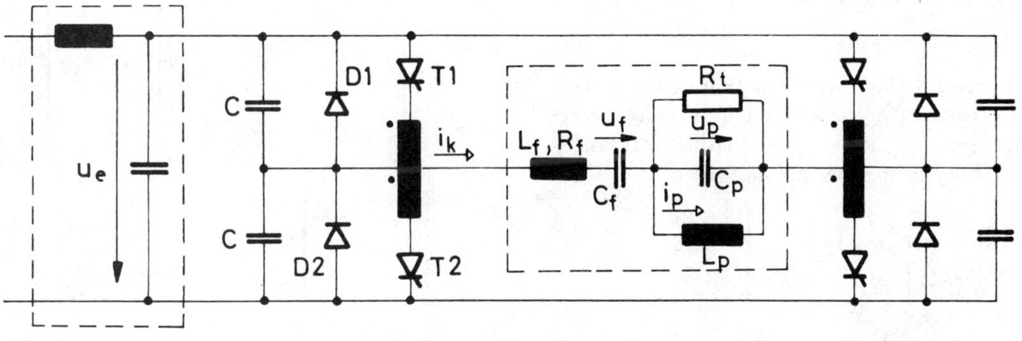

Fig. 1

ly. Due to the symmetric structure of the inverter composing the matrix equation, covering one of its parts will suffice. Fig. 2 shows the possible conduction figures of

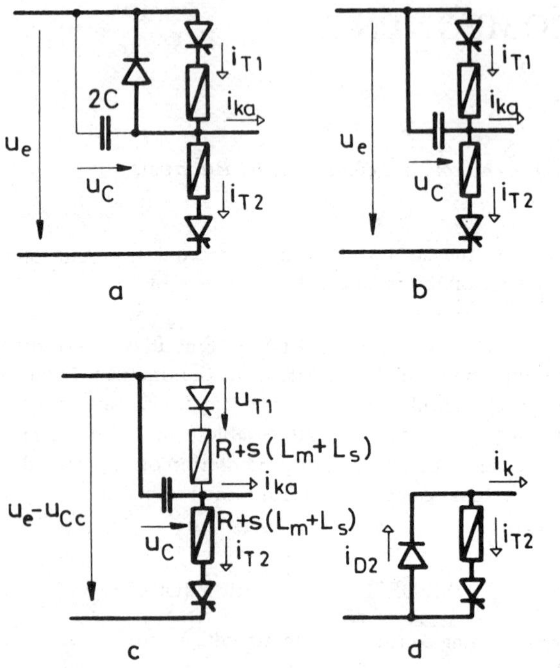

Fig. 2

the half-period of the left-side inverter part. In Fig. a, b and c the schematic equivalent figures of the thyristor-diode (*TD*) conduction, thyristor-capacitor (*TC*) conduction and capacitor (*C*) conduction sectors of thyristor 1, while in Fig. d the state with circular current between two commutations (preceding the commutation of thyristor 2) can be seen. L_m, L_s are the main and stray field inductances of the commutating transformer. R the resistance of one of its coils, $2C$ the capacity of the reduced commutating capacitors.

By introducing the vector of the electrical quantities

$$i^* = [i_{T1}, i_{T2}, u_C] \tag{1}$$

of the energy storages of the inverter, the form

$$L\frac{di}{dt} + R \cdot i = u \tag{2}$$

will be obtained for the matrix equation of the current distribution figures. The sector of the L matrices representing TD and TC sections and relating to thyristor currents is cyclic, having the canonical form

$$\begin{bmatrix} L_m + L_s & L_m \\ L_m & L_m + L_s \end{bmatrix} = \left(\tfrac{1}{2}\right)\begin{bmatrix} 1 & 1 \\ 1 & -1 \end{bmatrix} \cdot \begin{bmatrix} 2L_m + L_s & 0 \\ 0 & L_s \end{bmatrix}\begin{bmatrix} 1 & 1 \\ 1 & -1 \end{bmatrix} = \left(\tfrac{1}{2}\right) S\langle\lambda_i\rangle S. \tag{3}$$

S is the matrix consisting of eigenvectors on the right resp. left side, its square is the double of the unit matrix:

$$S^2 = 2I. \tag{4}$$

By using (3) and (4), equation (2) can be written in the form

$$S\langle\lambda_i\rangle\left(S\frac{di}{dt}\right) + R \cdot S(S \cdot i) = 2u. \tag{5}$$

By reducing matrix S with vector i we can introduce new variables instead of the thyristor currents: the symmetrical components of the thyristor currents

$$i_s = S \cdot i = \begin{bmatrix} 1 & 1 \\ 1 & -1 \end{bmatrix}\begin{bmatrix} i_{T1} \\ i_{T2} \end{bmatrix} = \begin{bmatrix} i_{T1} + i_{T2} \\ i_{T1} - i_{T2} \end{bmatrix}. \tag{6}$$

After the introduction of the symmetrical components of the thyristor currents the matrix equations for the particular sectors will be:

$$TD\colon \frac{d}{dt}\begin{bmatrix} i_{T1} + i_{T2} \\ i_{T1} - i_{T2} \end{bmatrix} = -\begin{bmatrix} \dfrac{R}{2L_m + L_s} & 0 \\ 0 & \dfrac{R}{L_s} \end{bmatrix}\begin{bmatrix} i_{T1} + i_{T2} \\ i_{T1} - i_{T2} \end{bmatrix} + \begin{bmatrix} \dfrac{u_e - 2u_d}{2L_m + L_s} \\ -\dfrac{u_e + 2u_d}{L_s} \end{bmatrix}, \tag{7}$$

$$TC\colon \frac{d}{dt}\begin{bmatrix} i_{T1} + i_{T2} \\ i_{T1} - i_{T2} \\ u_C \end{bmatrix} = -\begin{bmatrix} \dfrac{R}{2L_m + L_s} & 0 & 0 \\ 0 & \dfrac{R}{L_s} & -\dfrac{2}{L_s} \\ 0 & \dfrac{1}{2C} & 0 \end{bmatrix}\begin{bmatrix} i_{T1} + i_{T2} \\ i_{T1} - i_{T2} \\ u_C \end{bmatrix} + \begin{bmatrix} \dfrac{u_e - 2u_d}{2L_m + L_s} \\ \dfrac{u_e}{L_s} \\ \dfrac{i_k}{2C} \end{bmatrix}, \tag{8}$$

$$C:\ \frac{d}{dt}\begin{bmatrix} i_{T2} \\ u_C \end{bmatrix} = -\begin{bmatrix} \frac{R}{L_m + L_s} & \frac{1}{L_m + L_s} \\ -\frac{1}{2C} & 0 \end{bmatrix}\begin{bmatrix} i_{T2} \\ u_C \end{bmatrix} + \begin{bmatrix} \frac{u_{Cc} - u_d}{L_m + L_s} \\ \frac{i_k}{2C} \end{bmatrix}, \tag{9}$$

between two commutations:

$$\frac{d}{dt} i_{T2} = -\frac{R}{L_m + L_s} i_{T2} - \frac{2u_d}{L_m + L_s}. \tag{10}$$

u_d is the forward voltage drop of the semiconductors. The solution of the differential equation with the form

$$\frac{d\boldsymbol{i}_s}{dt} = \boldsymbol{A} \cdot \boldsymbol{i}_s + \boldsymbol{u}_s \tag{11}$$

of the particular sectors is

$$\boldsymbol{i}_s = e^{\boldsymbol{A}t} \cdot \boldsymbol{i}_{s0} + \boldsymbol{i}_p. \tag{12}$$

In equation (12) $\boldsymbol{i}_p$ is the particular solution pertaining to $\boldsymbol{u}_s$, $\boldsymbol{i}_{s0}$ is the amplitude to be determined from the initial conditions. The notations of the thyristor currents and capacitor voltages on the sector boundaries are summarized in Fig. 3.

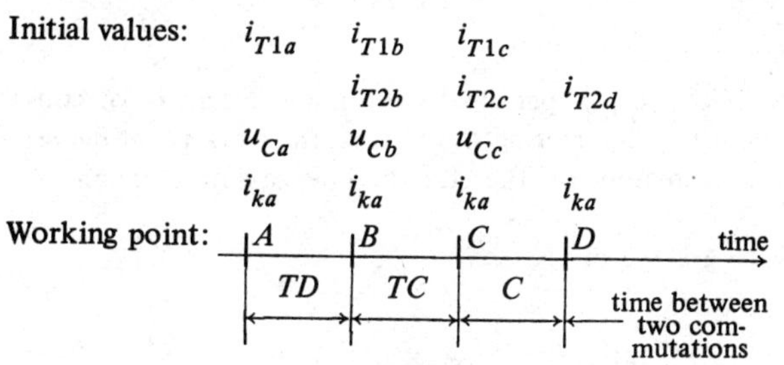

Fig. 3

In the solutions covering the particular sectors

$$T_m = \frac{L_m}{R}, \quad T_s = \frac{L_s}{R}, \quad \omega_{TC} = \frac{1}{\sqrt{L_s C}},$$

$$\zeta_{TC} = \frac{R}{2}\sqrt{\frac{C}{L_s}}, \quad \nu_{TC} = \omega_{TC}\sqrt{1 - \zeta_{TC}^2},$$

$$\varphi_{TC} = \arccos \zeta_{TC}, \quad \omega_C = \frac{1}{\sqrt{(L_m + L_s)2C}},$$

$$\zeta_C = \frac{R}{2}\sqrt{\frac{2C}{L_m + L_s}}, \quad \nu_C = \omega_C\sqrt{1 - \zeta_C^2}, \tag{13}$$

$$\varphi_C = \arccos \zeta_C$$

time constants, angular frequencies, damping factors, phase shifts serve as parameters. The solutions describing the processes are as follows:

$$TD:\ \begin{bmatrix} i_{T1} + i_{T2} \\ i_{T1} - i_{T2} \end{bmatrix} =$$

$$= \begin{bmatrix} e^{-\frac{t}{2T_m + T_s}} & 0 \\ 0 & e^{-\frac{t}{T_s}} \end{bmatrix}\begin{bmatrix} i_{T1a} - \frac{u_e - 2u_d}{R} \\ i_{T1a} + \frac{u_e + 2u_d}{R} \end{bmatrix} + \begin{bmatrix} \frac{u_e - 2u_d}{R} \\ -\frac{u_e + 2u_d}{R} \end{bmatrix} \tag{14}$$

The TD sector comes to an end when current

$$i_{D1} = (i_{T1} - i_{T2}) - i_{ka} \tag{15}$$

of diode 1 decays.

$$TC:\ \begin{bmatrix} i_{T1} + i_{T2} \\ i_{T1} - i_{T2} \\ u_C \end{bmatrix} =$$

$$= \begin{bmatrix} e^{-\frac{t}{2T_m + T_s}} & 0 & 0 \\ 0 & -\frac{e^{-\frac{t}{2T_s}}}{\sqrt{1 - \zeta_{TC}^2}} \sin \nu_{TC} t & 0 \\ 0 & 0 & -\frac{e^{-\frac{t}{2T_s}}}{\sqrt{1 - \zeta_{TC}^2}} \sin(\nu_{TC} t + \varphi_{TC}) \end{bmatrix}$$

$$\begin{bmatrix} (i_{T1b} + i_{T2b}) - \frac{u_e - 2u_d}{R} \\ u_e\sqrt{\frac{C}{L_s}} + 2\zeta_{TC} i_{ka} \\ \frac{u_e + R i_{ka}}{2} \end{bmatrix} + \begin{bmatrix} \frac{u_e - 2u_d}{R} \\ i_{ka} \\ \frac{u_e + R i_{ka}}{2} \end{bmatrix} \tag{16}$$

Decaying of the current of thyristor 1 causes the process TC to come to an end. By introducing the vector

$$x(t)^* = \frac{e^{-\frac{t}{2(T_m + T_s)}}}{\sqrt{1 - \zeta_C^2}} \cdot$$

$$\cdot [\sin \nu_C t, \sin(\nu_C t + \varphi_C), \sin(\nu_C t - \varphi_C)] \tag{17}$$

the time functions of sector C are:

$$i_{T2} = x(t)^* \begin{bmatrix} (u_e - u_{Cc})\sqrt{\frac{2C}{L_m + L_s}} \\ i_{ka} \\ -i_{T2c} \end{bmatrix} - i_{ka}, \tag{18}$$

$$u_C = x(t)^* \begin{bmatrix} \dfrac{i_{T2c} + i_{ka}}{\sqrt{\dfrac{2C}{L_m + L_s}}} - R(u_e - u_{Cc})\sqrt{\dfrac{2C}{L_m + L_s}} \\ Ri_{ka} \\ u_C - u_{Cc} \end{bmatrix} +$$

$$+ u_e + Ri_{ka}^{\cdot}. \tag{19}$$

The voltage impressed upon the thyristor, turned off by forced communication

$$u_{T1} = x(t)^* \begin{bmatrix} \dfrac{2L_m + L_s}{L_m + L_s} \dfrac{i_{T2c} + i_{ka}}{\sqrt{\dfrac{2C}{L_m + L_s}}} - \\ - R(u_e - u_{Cc})\sqrt{\dfrac{2C}{L_m + L_s}} \\ - Ri_{ka} \\ \dfrac{2L_m + L_s}{L_m + L_s}(u_e - u_{Cc}) - \dfrac{L_m}{L_m + L_s} Ri_{T2c} \end{bmatrix} +$$

$$+ u_e + Ri_{ka}. \tag{20}$$

The commutation process of thyristor 1 comes to an and when thyristor 2 starts conducting. As long as also a circulating current is flowing in the sector between two commutations, the thyristor current changes according to the relation

$$i_{T2} = \left(\frac{2u_d}{R} + i_{T2d}\right) e^{-\frac{t}{T_m + T_s}} - \frac{2u_d}{R}. \tag{21}$$

(The case of the interruption of the circulating current can be investigated only numerically regarding the great number of energy storages.)

THE PROCESSES IN THE COORDINATE SYSTEM OF SYMMETRICAL COMPONENTS OF THE THYRISTOR-CURRENTS

On basis of relation (4) S is the projector matrix. With its help we can switch over according to (6) from the thyristor currents to the symmetrical components of the thyristor currents, resp. we can recover by a relation, identical in form with

$$\begin{bmatrix} 2i_{T1} \\ 2i_{T2} \end{bmatrix} = \begin{bmatrix} 1 & 1 \\ 1 & -1 \end{bmatrix} \begin{bmatrix} i_{T1} + i_{T2} \\ i_{T1} - i_{T2} \end{bmatrix} \tag{22}$$

the doubles of the thyristor currents from the symmetrical components. Axis $2i_{T1}$ points to the direction of vectors (1, 1), while axis $2i_{T2}$ points to the direction of vectors (1, −1) of the coordinate system $(i_{T1} - i_{T2})$. In consequence of the graphic conditions created by the projector matrix the projections of a point of the plane perpendicular to the axis give at the moment pertaining to the point the symmetrical components of the thyristor currents, resp. the double of the thyristor currents.

In the coordinate system of the symmetrical components of the thyristor currents the locus in Fig. 4 describes the above processes.

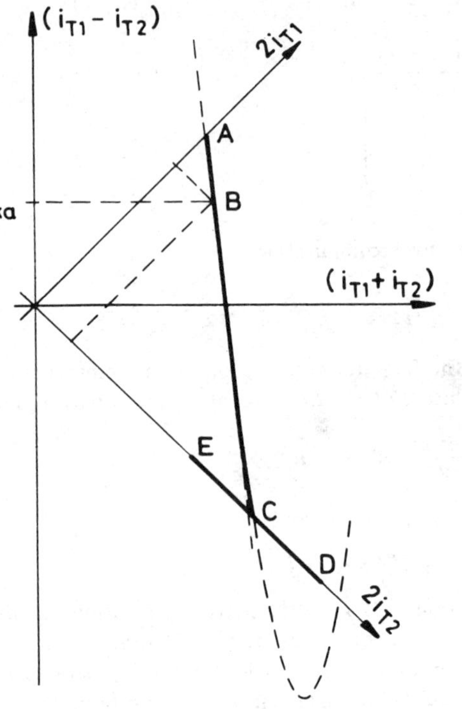

Fig. 4

$A(i_{T1a}, i_{T1a})$ pertain to the initial moment of conduction TD. During the process the work point moves approximately on the straight line passing through A with a slope of $-\dfrac{2L_m + L_s}{L_s}$,

$$i_{T1} - i_{T2} = i_{ka} \tag{23}$$

diode 1 stops conducting in point B. Conduction TC begins at point B and the work point progresses according to relation (16) along the section, drawn by solid line, of the damped sine curve drawn in the figure by a dotted line. The process lasts until thyristor current 1 becomes zero in point C. Under conduction C the working point moves in the direction of currents increasing along axis $2i_{T2}$. The process comes to an end by having attained the maximum deflection. Between two commutations the working point invariably moves along axis $2i_{T2}$ but in a diminishing direction, at the beginning of the next commutation it can be found e.g. in E. Necessary changes having been made, the results obtained hold also for the commutation of thyristor 2. This process takes place in the coordinate system of the symmetrical components as a projection on axis $(i_{T1} + i_{T2})$ of the curve in Fig. 4. The condition of periodicity: E is the mirror image of A.

Fig. 4 has the defect of representing the most important process of the commutation, conduction C as a change occurring along a coordinate-axis. This disadvantage can be removed by replacing sine functions by exponential functions.

Assuming a loss-free commutating circuit ($\zeta_{TC} = \zeta_C = 0$, $\varphi_{TC} = \varphi_C = 90°$) we can write up the approximate equations relative to the particular sectors in the form of a scalar product. For the sector TC:

$$(i_{T1} - i_{T2}) \cong \cong \mathrm{Re}\left\{[e^{j(\nu_{TC}t + \frac{\pi}{2})}, 0] \begin{bmatrix} (u_e - 2u_{Cb})\sqrt{\frac{C}{L_s}} \\ \frac{u_e}{2} \end{bmatrix}\right\} + i_{ka}, \quad (24)$$

$$u_C \cong \mathrm{Re}\left\{[0, -e^{j\nu_{TC}t}] \begin{bmatrix} (u_e - 2u_{Cb})\sqrt{\frac{C}{L_s}} \\ \frac{u_e}{2} \end{bmatrix}\right\} + \frac{u_e}{2}. \quad (25)$$

The equations of sector C:

$$i_{T2} \cong \mathrm{Re}\left\{[e^{j(\nu_C t - \frac{\pi}{2})}, e^{j\nu_C t}] \cdot \begin{bmatrix} (u_e - u_{Cc})\sqrt{\frac{2C}{L_m + L_s}} \\ i_{T2c} + i_{ka} \end{bmatrix}\right\} - i_{ka}, \quad (26)$$

$$u_C \cong \mathrm{Re}\left\{[e^{j(\nu_C t - \frac{\pi}{2})}, e^{j\nu_C t}] \cdot \begin{bmatrix} \frac{i_{T2c} + i_{ka}}{\sqrt{\frac{2C}{L_m + L_s}}} \\ -(u_e - u_{Cc}) \end{bmatrix}\right\} + u_e, \quad (27)$$

$$\frac{u_{T1}}{2} \cong \mathrm{Re}\left\{[e^{j(\nu_C t - \frac{\pi}{2})}, e^{j\nu_C t}] \cdot \begin{bmatrix} \frac{i_{T2c} + i_{ka}}{\sqrt{\frac{2C}{L_m + L_s}}} \\ -(u_e - u_{Cc}) \end{bmatrix}\right\} + \frac{u_e}{2}. \quad (28)$$

In the coordinate system of the symmetrical components considered as complex the end points of the vectors corresponding to the quantities described by the relations (24)-(28) change along arcs with shifted centres. Let us choose the scale of the currents and the capacitor voltage such as the distance OO_2 corresponds to a current of $u_e\sqrt{\frac{2C}{L_m + L_s}}$ and voltage u_e. Then distance ST resp. KN belongs to current $\sqrt{\frac{C}{L_s}}$. On basis of relations (24), (25) the process TC can be designed with the help of the circles r_1 with radius $u_e\sqrt{\frac{C}{L_s}}$ and r_2 with radius $\frac{u_e}{2}$, drawn around centre $K\left(-i_{ka}, -\frac{u_e}{2}\right)$ chosen in an appropriate way. While the vector i_s of the symmetrical components of the thyristor currents moves along the path BC, the complex vector of the difference of the thyristor currents (due to the projection on axis $2i_{T2}$) covers arc NM of radius r_1. By projecting the vector of the capacitor voltage on axis $(i_{T1} - i_{T2})$ and choosing O_2 for its neutral point, its vector changes along circle r_2 and shows at point P at the end of process TC. The distance RO_2 corresponds therefore to u_{Cc}, the distance OR to the voltage $(u_e - u_{Cc})$. The constructions relating to process C are based on equations (26)-(28). Sector FJ of axis $(i_{T1} + i_{T2})$ corresponds to current $(i_{T2c} + i_{ka})$, the distance JG perpendicular to this corresponds to current $(u_e - u_{Cc})\sqrt{\frac{2C}{L_m + L_s}}$. The distances JG, FJ give rightly also with respect to the direction the components of scalar product (26). Thus the distance FG is the complex vector of the current i_{T2} at the beginning of process C and the auxiliary work point proceeds on the arc GL of the circle r_1 drawn around the centre F, the work point is its projection on axis $2i_{T2}$. As the row vectors of the scalar product giving current i_{T2} and voltage u_C are identical and their column vectors are orthogonal to each other, the projection on axis $(i_{T1} - i_{T2})$ of the auxiliary working point proceeding on circle r_1 yields the capacitor voltage during process C. It follows from the comparison of (28) with (29) that the projection on axis $(i_{T1} - i_{T2})$ of the auxiliary working point proceeding on circle r_3 given at the same time the voltage of thyristor 1, too, the voltage scale has to be taken, however, for double and O_1 has to be chosen as neutral point. Thus distance HG is at the beginning of process C the blocking voltage of the thyristor, the angle pertaining to arc GI is the time expressed in electrical degrees of the existence of the negative blocking voltage. By assuming the output current i_{ka} as constant and by changing the initial thyristor current i_{T1a}, the end points of vectors i_{T2} take up position at the beginning of process C along the ellipse constructed with the circles r_1, r_2.

On basis of the above, all important characteristics of the processes taking place in the inverter can be read out from the curve plotted in the coordinate system of the symmetrical components of the thyristor currents and from the design which can be attached to it.

One can get the best synoptical view of which processes take place during the commutation in the inverter switching, depending on the initial i_{t1a} thyristor current, i_{ka} output current and u_{Ca} capacitor voltage, on the plane of the symmetrical components of the thyristor currents. It follows from the inverter switching in Fig. 1 that at the beginning of the commutation the output voltage cannot be higher than that of the thyristor current. Assuming an initial thyristor current of a certain extent the possible variations of the processes in the function of the output current / thyristor current ratio are:

1a) $i_{ka} = i_{T1a}$, $u_{Ca} = 0$. In this case points A and B coincide in Fig. 5, the process TD does not come about. For the given initial thyristor current the section

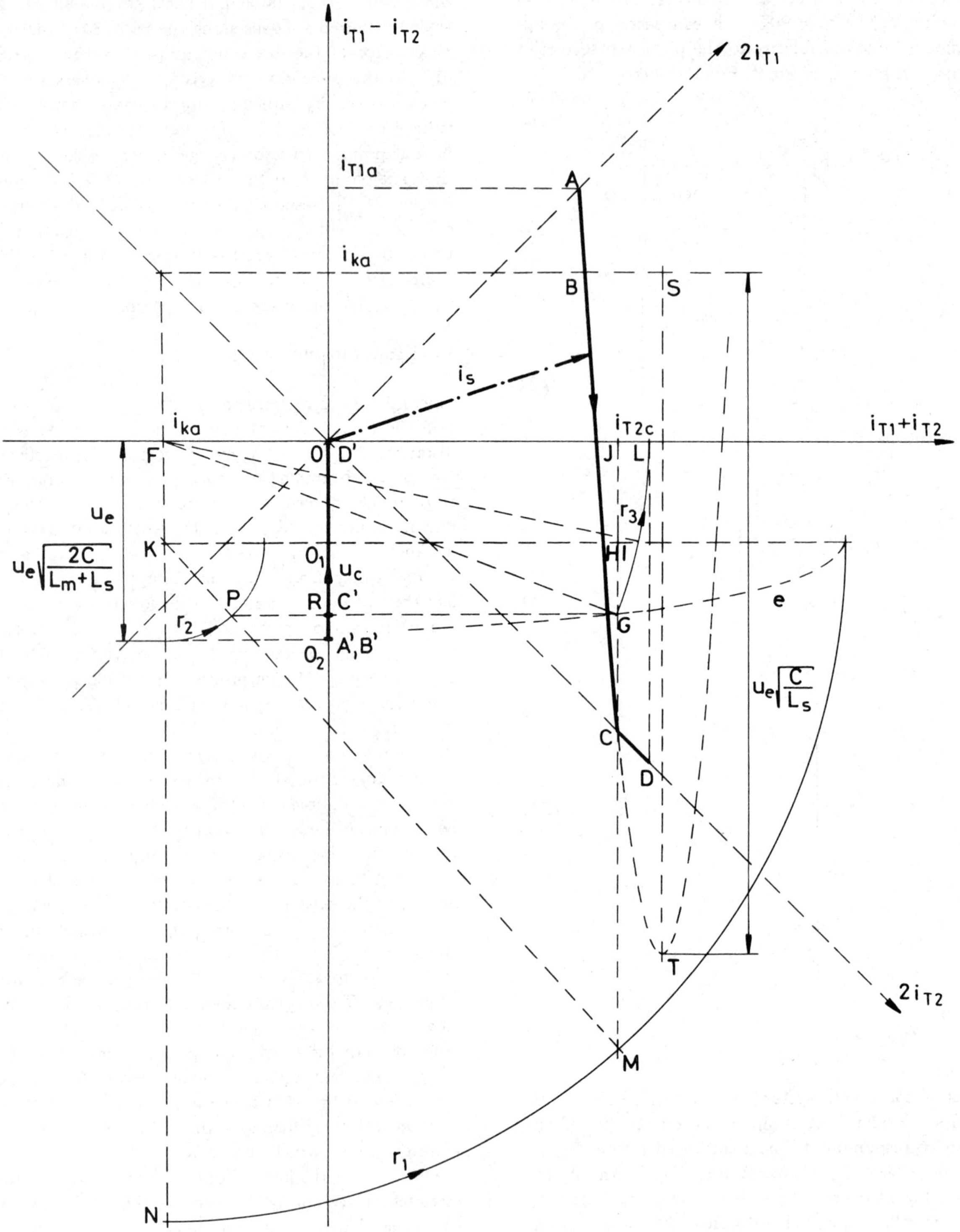

Fig. 5

TC is the longest of the possible ones, therefore the circuit turn-off time of the thyristor is the shortest in section C.

1b) $i_{ka} = i_{T1a}$, $u_{Ca} \neq 0$ ($u_{Cb} = u_{Ca}$). Neither does process TD come about at this time. The design of processes TC and C differs from that in Fig. 5 inasmuch as the radii of circles r_1 and r_2 are according to (25) and (26) smaller in a proportion of $\dfrac{u_e - u_{Cb}}{u_e}$.

In further cases when the output current is smaller than the initial thyristor current, the initial voltage of the capacitor is zero: $u_{Ca} = u_{Cb} = 0$.

2) $i_{T1a} > i_{ka} > -i_{T2c}$. The process series traceable in Fig. 5 covers this case.

3) $i_{ka} = -i_{T2c}$. Process TD is the longest of those which can come about under the given initial thyristor current. Points B and C in Fig. 5 coincide, process TD does not come about. The capacitor voltage is even at the beginning of section C zero: $u_{Cc} = 0$. Under a given initial thyristor current this case affords the maximum time available for thyristor regeneration.

4) $i_{ka} < -i_{T2b}$. Likewise to the previous case section TC has a maximum length, and section TC does not come about. After section TD the until now not investigated figure of current conduction develops in Fig. 6 the *diode* (D) *conduction*, described by the differential equation

$$\frac{d}{dt} i_{T2} = -\frac{R}{L_m + L_s} i_{T2} + \frac{u_e}{L_m + L_s} \tag{29}$$

whose solution is:

$$i_{T2} = -\left(\frac{u_e}{R} - i_{T2b}\right) e^{\frac{-t}{T_m + T_s}} + \frac{u_e}{R}. \tag{30}$$

On the plane of the symmetrical components of the thyristor currents the work point proceeds during period D along the axis $2i_{T2}$ with a nearly constant speed corresponding to (30). The end of the period is indicated by the cessation of the current of diode 1. At this moment $i_{T2c} = i_{ka}$. The duration of period D increases the time available for the regeneration of the thyristor. The process C following period D runs its course with a value $u_{Cc} = 0$ according to what has been said until now.

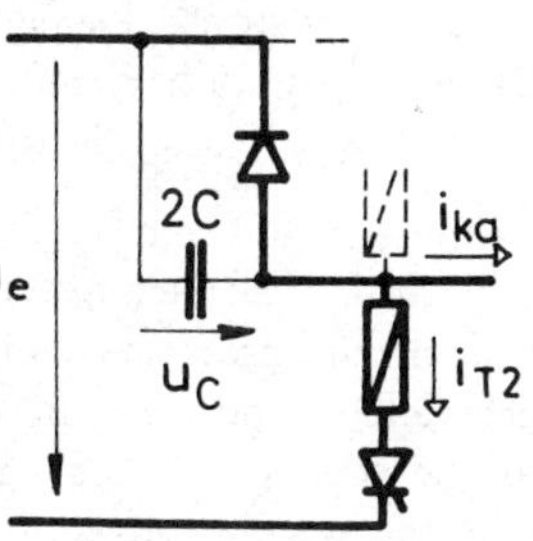

Fig. 6

THE OUTPUT CURRENT AT THE MOMENT OF THE COMMUTATIONS

Fig. 5 refers to a single commutation of the transient or stationary state of the inverter taken at random. On the other hand, the investigation of a complete transient process requires the knowledge of the series of the values of the output current during the succession of commutations, too.

In case of a given load circuit connected to the output of the inverter the values of the output current under commutation – by assuming momentary commutation and disregarding the interruption of loop currents – are relatively easy to determine. A matrix equation, identical in form with (2) can be written up for the circuit loading the inverter. The components of vector i of the matrix equation written up for the load circuit are the electrical quantities of the energy storages of the circuit, among which the output current can also be found. The inductances and capacitances of the circuit are contained in matrix L, its resistances in matrix R. The number of dimension of the matrices is equal to the number of the energy storages of the circuit. E.g. the circuit in Fig. 1 is described by the following matrix equation:

$$\begin{bmatrix} L_f & 0 & 0 & 0 \\ 0 & C_p & 0 & 0 \\ 0 & 0 & C_f & 0 \\ 0 & 0 & 0 & L_p \end{bmatrix} \frac{d}{dt} \begin{bmatrix} i_k \\ u_p \\ u_f \\ i_p \end{bmatrix} + \begin{bmatrix} R_f & 1 & 1 & 0 \\ -1 & \frac{1}{R_t} & 0 & 1 \\ -1 & 0 & 0 & 0 \\ 0 & -1 & 0 & 0 \end{bmatrix} \begin{bmatrix} i_k \\ u_p \\ u_f \\ i_p \end{bmatrix} = \begin{bmatrix} 1 \\ 0 \\ 0 \\ 0 \end{bmatrix} u_i.$$

As we do not intend to define continuous time functions, only to compute some of their discrete values, we can use Laplace-transformation later on to a z-transform domain. In the Laplace-transform domain the periodic output voltage of the inverter, composed of square waves, can also be expressed in a mathematically closed form. In the Laplace-transform matrix equation

$$(R + sL)i(s) - L \cdot i_0 = e_1 \cdot u_i(s) \tag{31}$$

i_0 is the vector of the initial values of the electrical quantities of the energy storages, $u_i(s)$ the inverter voltage, e_1 the unit vector. The Laplace-transformation of the derivative by time of the inverter voltage with the form indicated in Fig. 7 can be generated with the help of the transform formula

$$e^{s\tau} = z \tag{32}$$

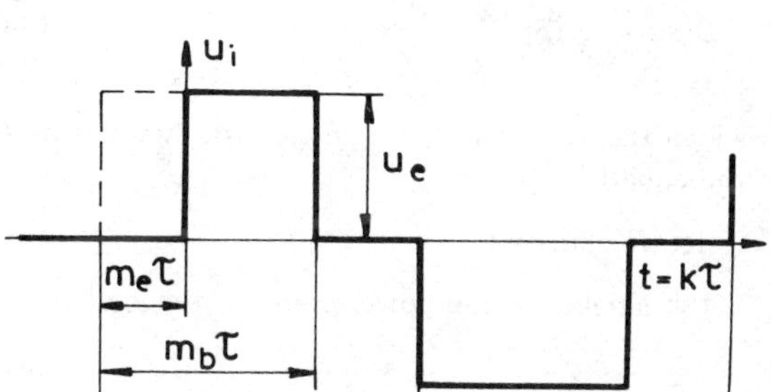

Fig. 7

as an infinite (geometric) series of Dirac pluses:

$$\mathscr{L}\left\{\frac{\partial u_i}{\partial t}\right\} = u_e(1 + [z^{-(m_b - m_e)} + z^{-(m - m_e)}] \cdot$$

$$\cdot [-1 + z^{-m} - z^{-2m} + \ldots]) =$$

$$= u_e \frac{1 + z^{-m} - z^{-(m - m_e)} - z^{-(m_b - m_e)}}{1 + z^{-m}}.$$

In the relations τ is the interval between two discrete instants, m, m_e and m_b are integers to be interpreted according to Fig. 7. The Laplace transform of the periodic voltage composed of square waves can in general be broken down to the sum of partial fractions:

$$u_i(s) = \frac{u_e}{s} \frac{P(z)}{1 + z^m} = \frac{u_e}{m \cdot s} \sum_{r=1}^{m} P(z_r) \frac{z_r}{z - z_r}, \tag{33}$$

$P(z)$ is the polinomial derived from the numerator, z_r-s are the roots

$$z_r = e^{j\frac{\pi}{m}(2r+1)} \qquad r = 1, 2, \ldots, m \tag{34}$$

of the equation

$$1 + z^m = 0.$$

By taking into account the Laplace transform of the inverter voltage and the inverse

$$(\boldsymbol{R} + s\boldsymbol{L})^{-1} = \frac{\text{adj}\,(\boldsymbol{R} + s\boldsymbol{L})}{\det\,(\boldsymbol{R} + s\boldsymbol{L})} \tag{35}$$

of the matrix $(\boldsymbol{R} + s\boldsymbol{L})$, the Laplace transform of vector $\boldsymbol{i}$ is

$$\boldsymbol{i}(s) = u_e \frac{P(z)}{1 + z^m} \frac{\text{adj}\,(\boldsymbol{R} + s\boldsymbol{L})\boldsymbol{e}_1}{s \cdot \det\,(\boldsymbol{R} + s\boldsymbol{L})} + \frac{\text{adj}\,(\boldsymbol{R} + s\boldsymbol{L})\boldsymbol{L} \cdot \boldsymbol{i}_0}{\det\,(\boldsymbol{R} + s\boldsymbol{L})}. \tag{36}$$

The passing over to z-transform is facilitated by writing up relation (36) with parcial fractions:

$$\boldsymbol{i}(s) = u_e \frac{P(z)}{1 + z^m} \left[\frac{\text{adj}\,\boldsymbol{R} \cdot \boldsymbol{e}_1}{s \cdot \det \boldsymbol{R}} + \sum_{i=1}^{n} \frac{\boldsymbol{D}(s_i)\boldsymbol{e}_1}{s_i} \frac{1}{s - s_i}\right] + \sum_{i=1}^{n} \boldsymbol{D}(s_i)\boldsymbol{L} \cdot \boldsymbol{i}_0 \frac{1}{s - s_i}. \tag{37}$$

The coefficients of the partial fractions can be calculated by the formula

$$\boldsymbol{D}(s_i) = \frac{\text{adj}\,(\boldsymbol{R} + s\boldsymbol{L})}{\frac{\partial}{\partial s} \det\,(\boldsymbol{R} + s\boldsymbol{L})\,|s_i} \tag{38}$$

in which the s_i $(i = 1, 2, \ldots, n)$ are the (simple) roots of the equation

$$\det\,(\boldsymbol{R} + s\boldsymbol{L}) = 0 \tag{39}$$

n is the number of the power stores of the load circuit of the inverter. Assuming an imaginary sampling coinciding with the moments of commutations and performed every τ time, on basis of the relation

$$\mathfrak{z}\left\{\frac{1}{s - s_i}\right\} = \frac{z}{z - z_i} \tag{40}$$

between the Laplace and z-transforms (where (32) forms a connection between s_i and z_i), the z transform of vector $\boldsymbol{i}$ is:

$$\boldsymbol{i}(z) = u_e \frac{P(z)}{1 + z^m} \frac{\text{adj}\,\boldsymbol{R} \cdot \boldsymbol{e}_1}{\det \boldsymbol{R}} \frac{z}{z - 1} + \frac{u_e}{m} \sum_{r=1}^{m} P(z_r) \frac{z_r}{z - z_r} \sum_{i=1}^{n} \frac{\boldsymbol{D}(s_i)\boldsymbol{e}_1}{s_i} \frac{z}{z - z_i} + \sum_{i=1}^{n} \boldsymbol{D}(s_i)\boldsymbol{L} \cdot \boldsymbol{i}_0 \frac{z}{z - z_i}. \tag{41}$$

The transform law

$$x(k) = \frac{1}{2\pi j} \oint x(z) z^{k-1}\, dz = \sum \text{Res}\, x(z) \cdot z^{k-1} \tag{42}$$

using the residue computation yields the value of vector $\boldsymbol{i}$ taken at the discrete moment $k\tau$:

$$\boldsymbol{i}(k) = \boldsymbol{e}_j u_i(k) + \frac{u_e}{m} \sum_{r=1}^{m} \sum_{i=1}^{n} P(z_r)\boldsymbol{D}(s_i)\boldsymbol{e}_1 \frac{z_r}{s_i} \frac{z_r^k - z_i^k}{z_r - z_i} + \sum_{i=1}^{n} \boldsymbol{D}(s_i)\boldsymbol{L}\boldsymbol{i}_0 z_i^k. \tag{43}$$

The relation yields the value of the transient process starting from the initial value $\boldsymbol{i}_0$, coming about under the influence of the inverter voltage with the shape presented in Fig. 7 and pertaining to the instant $k\tau$. The formula covers the computations of the switching on in the powerless state $(\boldsymbol{i}_0 = 0)$, the switching off in the state of the energy storage elements characterized by $\boldsymbol{i}_0$ $(u_e = 0)$ and the stationary state (because of the damping $\lim_{k \to \infty} |z_i|^k = 0$), too. Vector $\boldsymbol{i}_{(k)}$ contains not only the output current, but also the quantities of the other energy storage elements of the load circuit. The investigation of the processes taking place in the inverter, undertaken on the plane of the symmetrical components of the thyristor currents require only the determination of the output current values at the instants of the commutations.

REFERENCES

I. Rácz – J. Borka – K. Lupán – Mrs. D. Miklós: Quasi-periodic dynamic behaviour of piecewise linear multi-parameter system, Periodica Politechnica, Budapest 1969. 205-219.

J.T. Tou: Digital and Sampled-data Control Systems, Mc Graw-Hill Book Company. INC, New York, Toronto, London 1959.

A MODELLING METHOD FOR THE BEHAVIOUR OF CONVERTERS OPERATING IN CONTROL LOOPS

R. Prajoux, J. Jalade, J.C. Marpinard and J. Mazankine

Laboratoire d'Automatique et d'Analyse des Systèmes du Centre National de la Recherche Scientifique, 7, avenue du Colonel Roche, 31400 Toulouse, France

ABSTRACT

The models available to study the dynamic behaviour of control loops including power convertors can be separated in two classes : those very simplified or those very sophisticated. In this paper, we give a method which takes place between the two classes and we obtain non linear continuous models for large signal or linear continuous models for small signal. Some examples show the good agreement between the corresponding results and those obtained by using more accurate methods.

1. INTRODUCTION

In this paper, we are primarily interested in the modeling of some convertors (especially AC-DC and DC-DC convertors) with a control point of view.

A lot of work has been done in this field. This work was summarized in the survey-paper by LAGASSE-PRAJOUX at the 1st I.F.A.C. Symposium /_1_/. It appears that the various analytical approaches lead to the following classification :
- linear continuous models : they are very easy to use and permit the synthesis of correcting devices, but they are very approximate regarding the actual behaviour of the system, /_2,3,4,5,6_/ ;
- discrete linear models : they describe accurately the small signal behaviour taking into account the discrete nature of the system, but the synthesis is not straight forward /_7,8,9,10_/ ;
- discrete non-linear models : this class is not very extensive, due to the great complexity involved/_11,12,13_/.

In this paper, we give a method which is in fact a compromise between complexity and accuracy. Basically, one obtains a non-linear continuous model. This means that the non-linear properties of the actual system are preserved and that an approximation is made when passing from discrete to continuous. Furthermore, it is clear than a linear small-signal model can be derived from the previous one.

The method consists first in dividing the electrical circuit representing the power part of the convertor in two or more blocks. Each block has an invariant electrical structure but the interconnexions between the blocks are variable according to the state of the on-off components. Second, we determine which block (s) delivers the signal (s) applied to the control-loop (we call it the "fixed part"). Third, provided that some conditions are fulfilled, we replace the other blocks and their variable interconnexions by some equivalent blocks connected to the fixed part. We obtain thus a non-linear invariant circuit, i.e. a non-linear continuous model. The equivalence is based on averaging techniques and the method provides then a systematic way of achieving average-techniques and preserving the non-linear properties of the actual system.

2. GENERAL STUDY

Let us consider an electrical circuit consisting in sources, passive components and switches (fig. 2.1). The switches represent the active on-off components of the actual convertor.

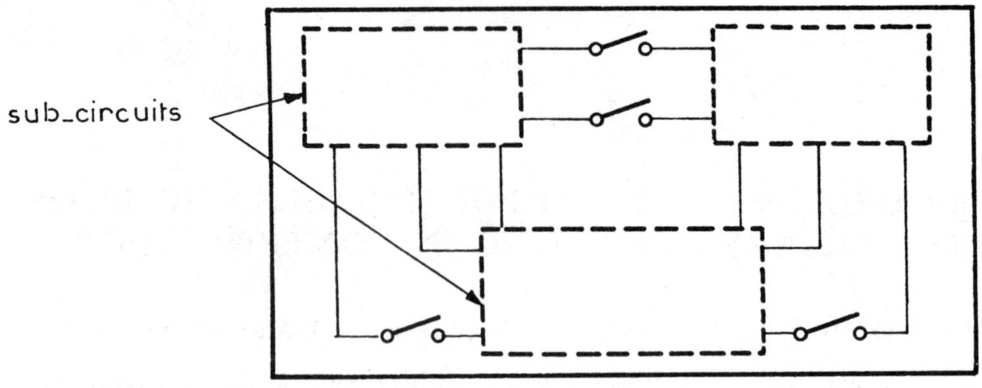

Fig. 2.1 Electrical circuit with switches

Switches provide variable interconnexions between sub-circuits. According to the state of those switches, the circuit has not an invariant topology but j possible configurations.
Let Z be a state vector of the circuit such that :

(2-1) $Z = [\vartheta c_1 \quad \vartheta c_2 \;.........\; i_{L_1} \quad i_{L_2} ....]^T$

ϑc_i being the voltages accross the capacitors and

i_{L_i} the currents through the inductors.

We suppose that Z(t) is continuous at switching instants and that there is not capacitive loops nor inductive nodes.
The circuit may then be represented by differential state equations :

$$(2\text{-}2) \quad \begin{cases} \dot{Z} = A_k Z + B_k U \\ \text{with } 1 \leqslant k \leqslant j \end{cases}$$

Equation (2-2) may be solved step by step and taking for each initial value the final value of the preceding evolution.

An alternative approach is to study the manner in which components of the state vector Z interact with each other for all the structures in the range $[1,....j]$

2.1 - Connections between components of A_k

We say that the component z_i of Z is not connected to the component z_j for the kth structure if :

$(a_{ij})_k = 0$, $(a_{ij})_k$ being an element of A_k.

One can prove that, in this case :

$(a_{ji})_k = 0$

Then, in matrix A_k, the zeros appear to be symmetrical with respect to the first diagonal.

2.2 - Variable and fixed parts

Now, we are going to separate the system in two sub-systems, represented by state vectors X and Y. Then :

$$Z = [X \quad Y]^T$$

The system X is called the fixed part and is such that :

- its structure is invariant
- the sources U are inside
- the outputs S depend on X

Y is called the variable part and is such that :

- its structure is invariant
- the sources U are not inside

- the outputs S do not depend on Y.

Then :

$$\text{(2-3)}\quad \left|\begin{array}{l} \begin{bmatrix}\dot{X}\\ \dot{Y}\end{bmatrix} = \begin{bmatrix}A & C_k\\ H_k & D\end{bmatrix}\cdot\begin{bmatrix}X\\ Y\end{bmatrix} + \begin{bmatrix}B\\ F_k\end{bmatrix}U \\ S = MX = \begin{bmatrix}M & 0\end{bmatrix}\begin{bmatrix}X\\ Y\end{bmatrix} \\ t_{n(k-1)} < t < t_{nk}\ ,\ \forall k \in [1, \ldots j],\ \forall n \end{array}\right.$$

2.3 - Equivalent system

Let us consider the system X above. We get :

$$\text{(2-4)}\quad \dot{X} = AX + Bu + C_k Y$$

Y behaves like an input applied to system X.The response related to this input is :

$$\text{(2-5)}\quad X(t) = \int_{t_o}^{t} \phi(t-\lambda)\cdot C_k \cdot Y(\lambda).d\lambda \quad \text{with} \quad \phi(t) = e^{At}$$

Since some components of X may be not connected to some components of Y, we may write :

$$\text{(2-6)}\quad C_k\cdot Y(\lambda) = \begin{bmatrix} a_k(\lambda) \\ ---- \\ \circledcirc \end{bmatrix}$$

and :

$$\text{(2-7)}\quad \int_{t_o}^{t} \phi(t-\lambda)\cdot C_k\cdot Y(\lambda).d\lambda = \int_{t_o}^{t} [\varphi_1(t-\lambda) \vdots \varphi_2(t-\lambda)]\cdot\begin{bmatrix} a_k(\lambda) \\ ---- \\ \circledcirc \end{bmatrix}.d\lambda$$

$$= \int_{t_o}^{t} \varphi_1(t-\lambda)\cdot a_k(\lambda).d\lambda$$

The above decomposition in expression (2-6) may be valid for any k, provided that the decomposition is carried out in the worst case.
Now, let us assume that :

$$\text{(2-8)}\quad \varphi_1(t-\lambda) \simeq \text{constant},\ \forall \lambda \in [t_o, t]$$

Then :

$$\text{(2-9)}\quad \int_{t_o}^{t} \varphi_1(t-\lambda)\cdot a_k(\lambda).d\lambda \simeq \varphi_1 \int_{t_o}^{t} a_k(\lambda).d\lambda$$

Let us consider a complete cycle between the switching instants t_n and t_{n+1}, i.e. a time interval involving the sequence of the j possible structures.
We get then :

$$\text{(2-10)}\quad X_{n+1} \simeq \phi(t_{n+1}-t_n)\cdot X_n + \int_{t_n}^{t_{n+1}} \phi(t-\lambda)\cdot B\cdot u(\lambda)\,d\lambda$$

$$+ \varphi_1\left[\int_{t_n}^{t_{n1}} a_1(\lambda).d\lambda + \int_{t_{n1}}^{t_{n2}} a_2(\lambda).d\lambda + \ldots + \int_{t_{n(j-1)}}^{t_{n+1}} a_j(\lambda).d\lambda\right]$$

The sequence of vectors $a_k(\lambda)$ may be considered as the evolution of a single vector $a(\lambda)$:

$$\text{(2-11)}\quad a(\lambda) = a_k(\lambda)\ ,\ t_{n(k-1)} < \lambda < t_{nk}\ ,\ \forall k \in [1, \ldots j]$$

Let be :

$$\text{(2-12)}\quad \tilde{a} \triangleq \frac{1}{t_{n+1}-t_n}\int_{t_n}^{t_{n+1}} a(\lambda).d\lambda$$

Then, (2-10) can be written as :

(2-13) $$X_{n+1} \simeq \phi(t_{n+1}-t_n).X_n + \int_{t_n}^{t_{n+1}} \phi(t-\lambda).B.U(\lambda).d\lambda + \varphi_1 .\tilde{a}.(t_{n+1}-t_n)$$

which is the solution of :

(2-14) $$\dot{X} = AX + BU + \tilde{a}$$

Then, provided that $\varphi_1(t)$ is almost constant within a time interval Δt, the behavior of the fixed part can be approximated by replacing the variable part by an equivalent input $\tilde{a}$.

Let us consider now the variable part :

(2-15) $$\dot{Y} = DY + F_k.U + H_k.X$$

We get :

(2-16) $$Y(t) = \Psi(t-t_o).Y_o + \int_{t_o}^{t} \Psi(t-\lambda).F_k.U(\lambda).d\lambda + \int_{t_o}^{t} \Psi(t-\lambda).H_k.X(\lambda).d\lambda$$

with $\Psi(t) = e^{Dt} = \mathbb{1} + Dt + \ldots\ldots$

Assuming that :

(2-17) $$X(t) \simeq X_n \text{ , for } t_n < t < t_{n+1}$$

we get :

$$Y_{n+1} \simeq \Psi(t_{n+1}-t_n).Y_n + \sum_{i=1}^{j} \left(\int_{t_{n(i-1)}}^{t_{n_i}} \Psi(t-\lambda).F_i.U(\lambda).d\lambda \right) + \sum_{i=1}^{j} \left(\left[\int_{t_{n(i-1)}}^{t_{n_i}} \Psi(t-\lambda).d\lambda \right].H_i.X_n \right)$$

We can note the above expression by :

(2-18) $$Y_{n+1} \simeq \Psi(t_{n+1}-t_n).Y_n + G(t_n, t_{n_1}, \ldots\ldots t_{n+1}, X_n)$$

Let us define a continuous variable y such that :

(2-19) $$y(t_n) = Y_n, \ \forall n$$

We have :

$$\dot{y} \simeq \frac{Y_{n+1}-Y_n}{t_{n+1}-t_n} = \frac{1}{t_{n+1}-t_n} \left[\left(\Psi(t_{n+1}-t_n) - \mathbb{1} \right).y + G(t_n, \ldots\ldots t_{n+1}, X_n) \right]$$

By denoting :

(2-20) $$\mathcal{T}_n = \left[t_n, t_{n_1}, \ldots\ldots t_{n+1} \right]^T$$

the above expression may be written :

(2-21) $$\dot{y} = D.y + G(\mathcal{T}, X_n)$$, including $\frac{1}{t_{n+1}-t_n}$ into the function G.

Combining (2-6), (2-12) and integrating (2-16) gives an expression in the form :

(2-22) $$\tilde{a}_n = F(\mathcal{T}_n, X_n, Y_n) = F(\mathcal{T}_n, X_n, y)$$

or

(2-23) $$\tilde{a} = F(\mathcal{T}, X, y)$$, using continuous variables.

Among the instants grouped into vector $\mathcal{T}$, some are given by the control system, say $\mathcal{T}_E$, others are fixed, say $\mathcal{T}_F$ and the remaining depend on the state vector, i.e. on X and Y [17].

Then :

(2-24) $$\mathcal{T} = C_E.\mathcal{T}_E + C_F.\mathcal{T}_F + C_T.T(X,y)$$

Combining (2-14),(2-21),(2-23) and (2-24) gives the set of equations :

$$(2\text{-}25)\quad \left|\begin{array}{l} \dot{X} = AX + BU + \tilde{a} \\ \dot{y} = Dy + G(\mathcal{C},X) \\ \tilde{a} = F(\mathcal{C},X,y) \\ \mathcal{C} = C_F.\mathcal{C}_F + C_T.T(X,y) + C_E.\mathcal{C}_E \end{array}\right.$$

The system (2-25) is non-linear since the functions G,F and T are non-linear. The inputs are U and $\mathcal{C}_E$, but U is known a priori and has a predetermined value, then $\mathcal{C}_E$ is the true control input.

To conclude this paragraph, we can write that :

- provided that some conditions are fulfilled, the non-linear but piecewise linear system (2-3) can be represented approximately by the non-linear continuous system (2-25) ;
- the action of the variable part on the fixed part is approximated by fictitious voltage or current sources equal to the average of the actual signals ;
- these fictitious sources are controlled by an equivalent continuous system.

3. APPLICATION TO A CLASS OF DC-DC CONVERTORS

In order to illustrate the above theoretical concepts, we are going now to deal with an application to a class of DC-DC convertors used especially in aerospace instrumentation (the so-called buck, boost and buck-boost convertors), although the method of paragraph 2 is relevant to a much wider class.

The circuit diagram of a boost convertor is given as an example on fig. 3.1.

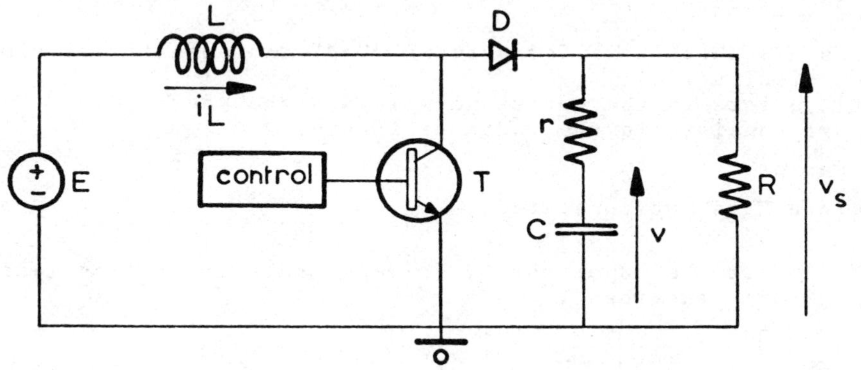

Fig. 3.1 BOOST CONVERTOR

The two other convertors differ only from the first one by the topology of the 3 devices L,D,T / 14,15,16 / and, thus, the block-diagram of fig. 3.2 can be given for the whole class.

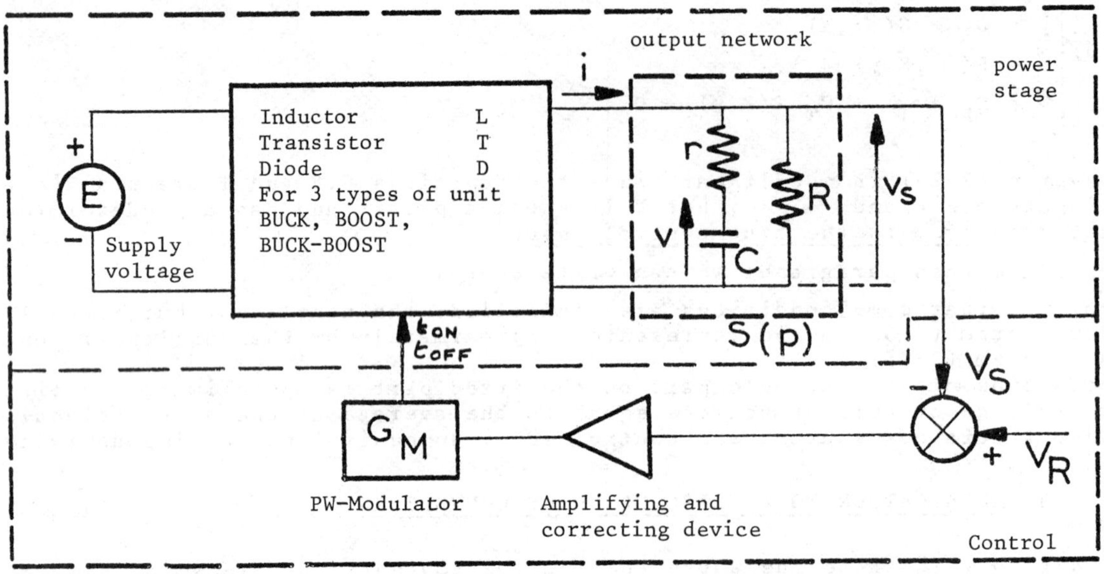

Figure 3.2 - Block-Diagram for the class of DC-DC convertors

3.1 Modeling of the class

The given class is a special case for systems (2-3) and significant simplifications can be carried out.
- The fixed part consists in the output network and the source E.
- The variable part consists in the inductor L only.

Then, for the fixed part :

$\dot{X} = AX + C_k Y$, since E is not connected.

We have
X=v, the voltage accross the capacitor C, (r represents the series resistance of an actual capacitor), and then :

$$(3\text{-}1) \quad \dot{v} = \left[- \frac{1}{(r+R)C} \right] v + C_i \left[\frac{r}{r+R} \quad \frac{1}{C} \right] i_L$$

The scalar C_i can take the values 0 or 1 and the quantity $C_i i_L$ is actually the current denoted i on fig. 3.2 :

$$(3\text{-}2) \quad i = C_i . i_L$$

It is easy to verify that the validity conditions of paragraph 2 are fulfilled.

$$(3\text{-}3) \quad \begin{cases} \phi(t) = e^{-\frac{1}{(r+R)C}t} , \text{ then} \\ \phi(t_{n+1}-t_n) = \phi(T) = e^{-\frac{1}{(r+R)C}T} \simeq 1 \\ v_{n+1} \simeq v_n \end{cases}$$

since C must have a value great enough to provide a good filtering effect.

This leads for the fixed part to the equivalent circuit diagram of fig. 3.3. The current generator i(t), whose Laplace transform is I(p), is a pulse current generator. This generator can be replaced by an equivalent generator $\tilde{i}(t)$] $\tilde{I}(p)$. The current i(t) depends on the type of convertor, the conditions of operation and on some variables.

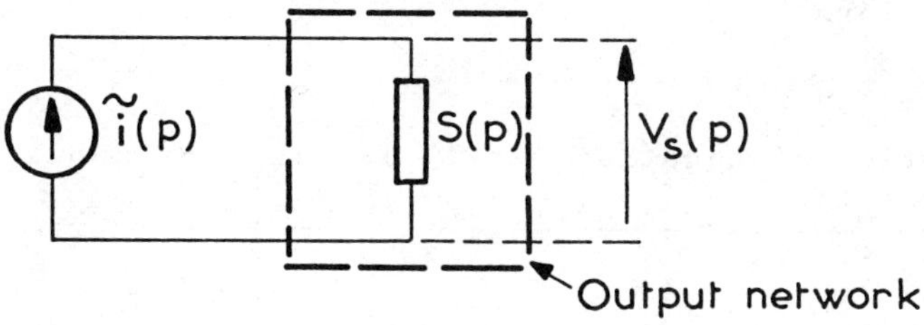

Figure 3.3 - Equivalent circuit of power stage

By definition :

$$\text{(3-4)} \quad \tilde{i}(t) = \frac{1}{T}\int_{t-T}^{t} i(t).dt \quad , \; T \triangleq t_{ON} + t_{OFF}$$

The figure 3.4 shows two typical waveforms for the current i_L, fig. 3.4a being the case of the continuous conduction and f. 3.4b the discontinuous conduction.

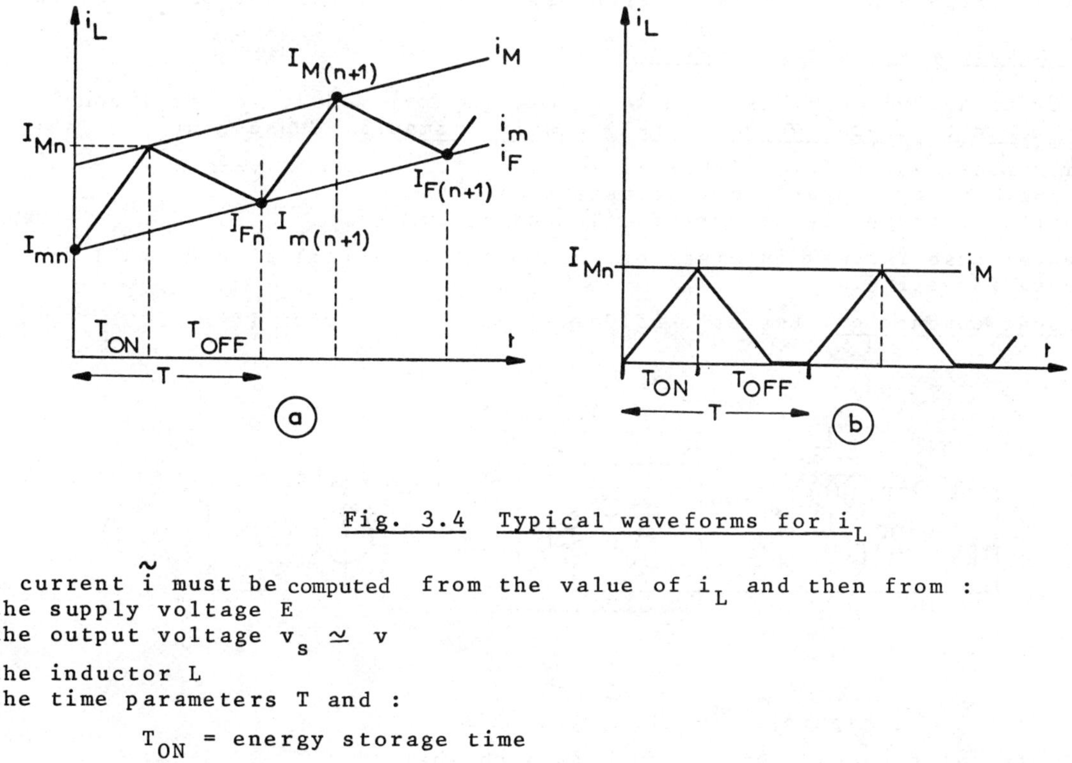

Fig. 3.4 Typical waveforms for i_L

The current $\tilde{i}$ must be computed from the value of i_L and then from :
- the supply voltage E
- the output voltage $v_s \simeq v$
- the inductor L
- the time parameters T and :

T_{ON} = energy storage time

$T_{OFF} = T - T_{ON}$

Instead of computing $\tilde{i}$ directly from the waveform of I_L, we carry out an indirect computation by using the variables I_{mn}, I_{Mn} and I_{Fn} appearing on fig. 3.4. The reason for doing this will appear later on.

Thus we define the equivalent continuous variables i_M, i_m and i_F which has the same values than I_M, I_m and I_F respectively, at the significant discrete instants.

Then, we must compute :

$$\text{(3-5)} \quad \tilde{i} = f(i_m, i_M, i_F, t_{ON}, T)$$

Direct computation of $\tilde{i}$ would have hidden the true operating conditions of the convertor (for example the transition between continuous and discontinuous conduction due to the transition of i_m through the value zero).

Knowing i_m and t_{ON}, it is possible to compute i_M and i_F. Consequently (3-5) is in fact in the form :

$$(3\text{-}6) \quad \tilde{i} = f(i_m, t_{ON})$$

The interpretation of equation (2-21) is, in continuous conduction :

$$(3\text{-}7) \quad \dot{i}_m \simeq \frac{\Delta i_m(T)}{T} = \frac{I_{m(n+1)} - I_{mn}}{T} = \frac{I_{Fn} - I_{mn}}{T}$$

It is important here to emphasize that the approximation involved in expression (3-7) does not imply that $\Delta i_m(T)$ should be small compared to i_m, but only that t_{ON} varies slowly from one period to the following one, i.e. :

$$(3\text{-}8) \qquad \frac{t_{ON(n+1)} - t_{ONn}}{T} << 1$$

In discontinuous conduction, the set (2-25) is degenerated and $\dot{y}=0$, i.e. :

$$(3\text{-}9) \qquad \begin{cases} i_{mn} = 0 & , \forall n \\ \dot{i}_m = 0 & \end{cases}$$

The explicit appearance of i_M and i_m has a further advantage. It is possible to introduce extra non-linearities, such as the current limitation in the switch.

3.2 - Example : the buck convertor

In order to derive for this example a very general model, we shall consider that t_{ON} and t_{OFF} are two completely distinct imputs. Consequently we shall obtain a model valid for, either :
- fixed-frequency convertors (T=constant)
- or variable-frequency convertors (T≠constant).

The latter case is very interesting, as most theoretical methods fail if the frequency is variable.

Continuous conduction. Let us consider a "buck" convertor (fig. 3.5).

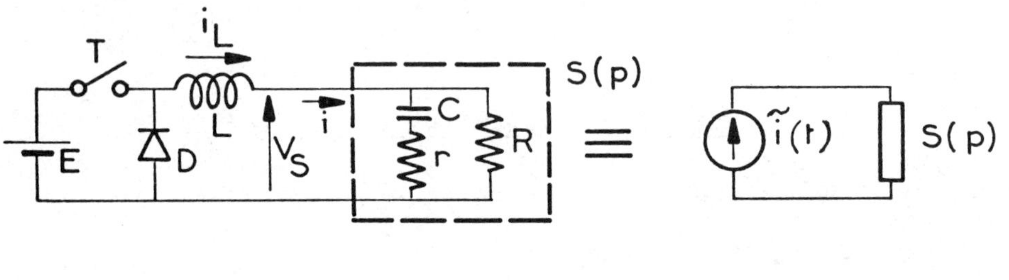

Figure 3.5 - BUCK CONVERTOR

The electrical structure of the buck is such that :

$$i(t) = i_L(t), \quad \forall t$$

In the case of a current limitation, the waveform for $i_L(t)$ is shown on fig. 3.6a.

It is easy to establish that :

$$(3\text{-}10) \quad \tilde{i}(t) = \frac{i_{ML}(t)+i_F(t)}{2} \frac{t_{ON}(t)}{T(t)} + \frac{i_M(t)+i_m(t)}{2} \frac{t_{ON}(t)}{T(t)} - \frac{(i_M(t)-i_{ML}(t))^2}{i_M(t)-i_m(t)} \frac{t_{ON}(t)}{2T(t)}$$

Discontinuous conduction. Instead of computing directly the value of $\tilde{i}$, we use the relation :

$$\tilde{i}_{DISC} = \tilde{i}_{CONT} + \mathcal{A}$$

$\mathcal{A}$ being the area appearing on fig. 3.6b.

$$\tilde{i}_{DISC}(t) = \tilde{i}_{CONT}(t) + i_F^2(t)\,\frac{L}{2v_s(t).T(t)}$$

($\tilde{i}_{CONT}$ → fictitious value)

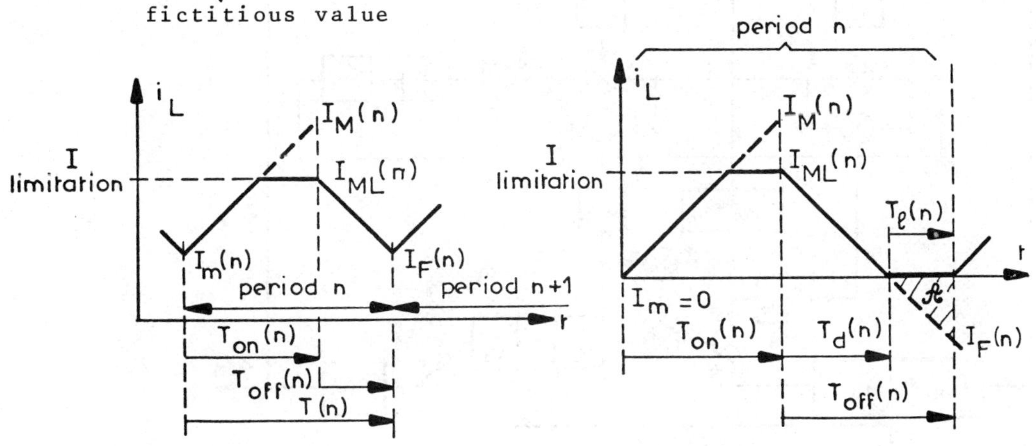

a) continuous conduction b) discontinuous conduction

Figure 3.6 - Inductor current

The above relationship is very interesting in building a simulation block diagram since it is then possible to take into account the changes from a conduction mode to the other without any initialisation problem.

Block-diagram of the buck convertor.

The block-diagram of the buck convertor obtained with the above computations is shown on fig. 3.7 . This block-diagram is valid for any amplitude of the signals, provided that the inputs t_{ON} and t_{OFF} vary slowly. The conduction mode is controlled by a flip-flop whose state changes according to the sign of the variables i_m or i_F.

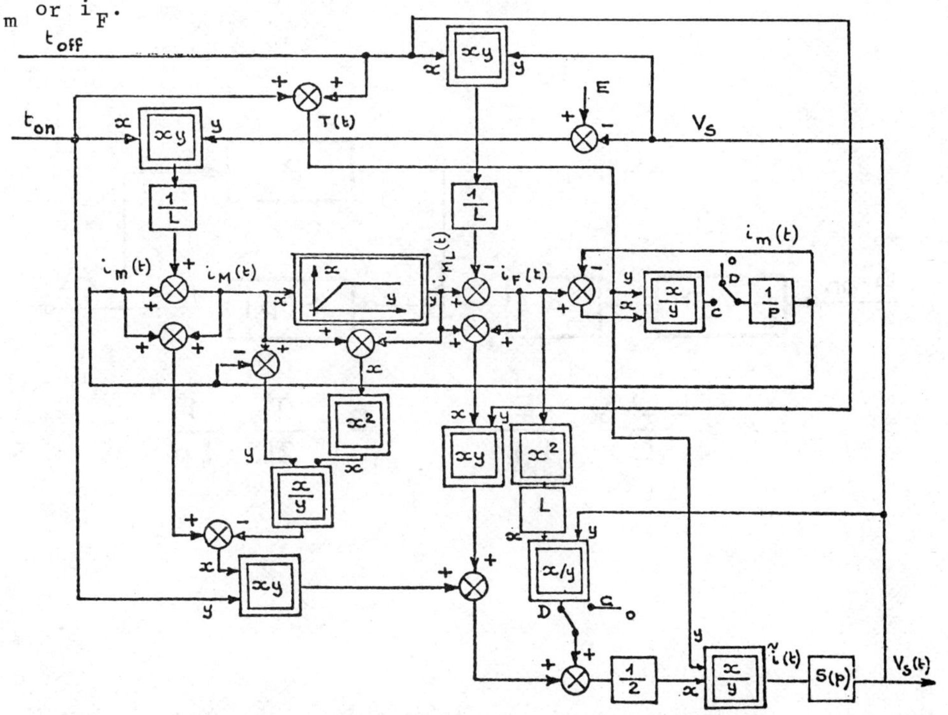

Figure 3.7 - Large-signal block diagram of the buck convertor for any mode of conduction

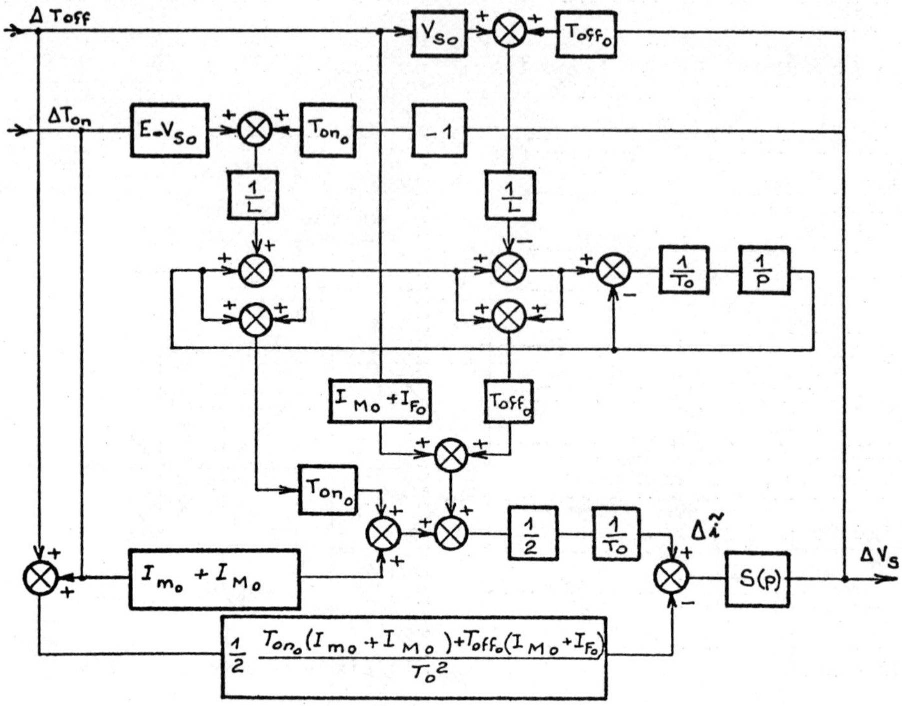

Figure 3.8 - Small-signal model for the buck in continuous conduction

-Small-signal model

Near a given operating point, it is possible to differentiate the out put value of each non-linear block and to get, then, a linear model valid, for small-signal only. This model, for the buck, is shown on fig. 3.8. After reduction, the linear block-diagram leads to the model of fig. 3.9, in the case of a fixed-frequency operation.

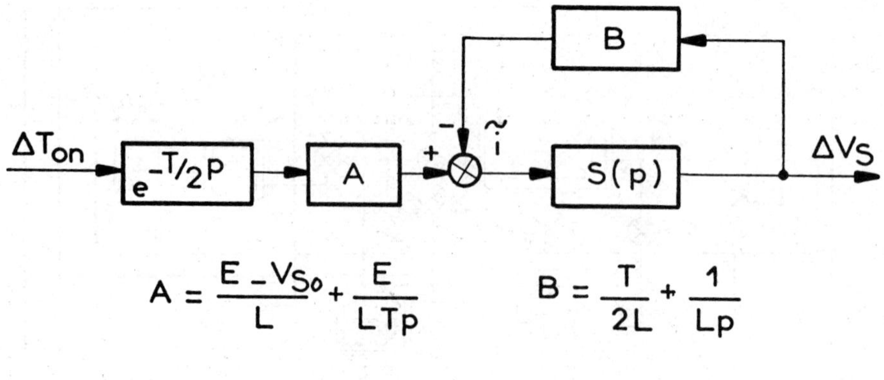

Figure 3.9 - Small-signal model for fixed-frequency operation

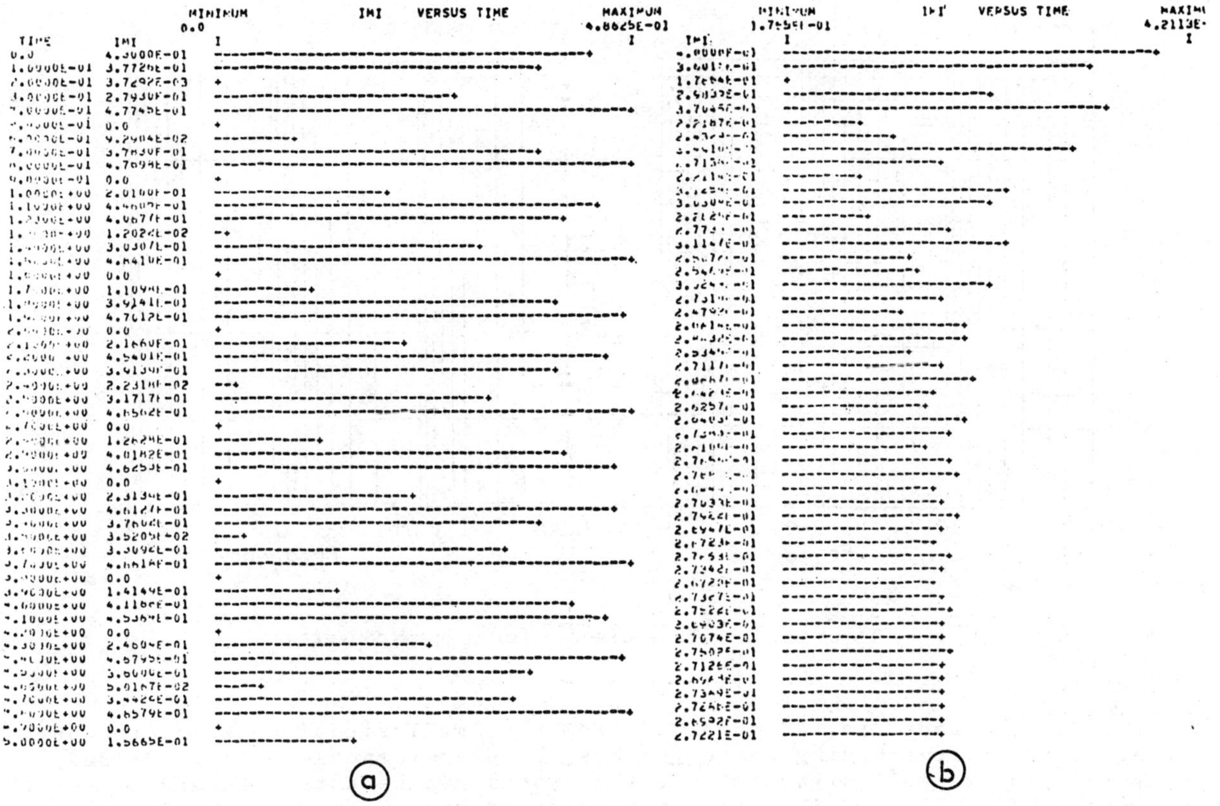

Figure 3.10 - Example of a CSMP simulation

Improvement of the models

In order to extend significantly the frequency range for which the above models are valid, it has been shown [17] that a delay T/2 must be inserted in the input connexions. This delay T/2 has the same value than BJÄRESTEN has proposed [2].

Examples using above models

The large-signal models such as the one on fig. 3.7 is sufficiently complex to lead to extremely difficult analytical study. Nevertheless, they are very useful to carry out simulations, especially using programming languages like the IBM-CSMP. There is, then, a great improvement in the computation time, according to the fact that only a few integration steps within one convertor period are needed.
An example of a CSMP simulation is shown on fig. 3.10 for a buck convertor which can be stable or unstable, depending on the initialisation. A hybrid simulation of the actual convertor is shown on fig. 3.11 and a very good agreement can be observed although the oscillations of the system occur at a frequency as high as 1/4T.

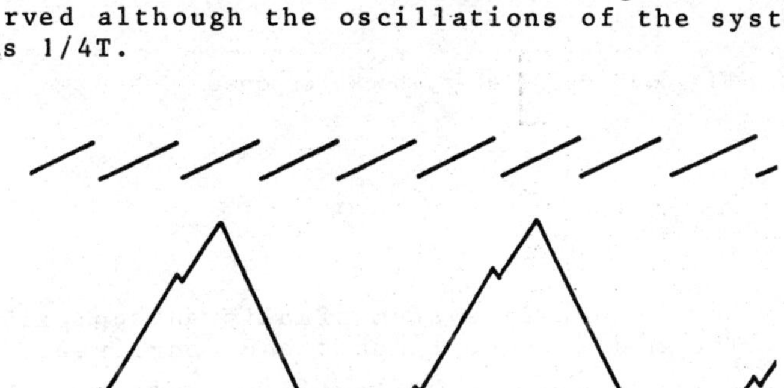

Figure 3.11 - Hybrid Simulation - Modulator Sawtooth - Inductor current

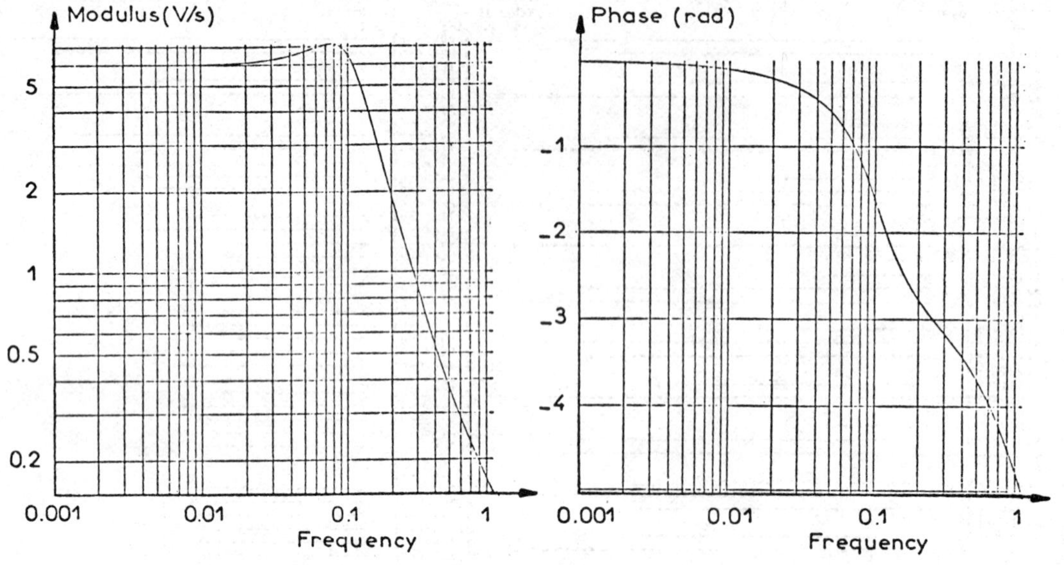

Figure 3-12 - Bode diagram for a buck convertor

On fig. 3.12, a Bode diagram obtained from the small-signal model of the buck is shown. The corresponding convertor has the step response on fig. 3.13a. Using the linear small-signal model, the system can be corrected and we get the response on fig. 3.13b. This response is as good as possible and illustrate the satisfactory accuracy of the model.

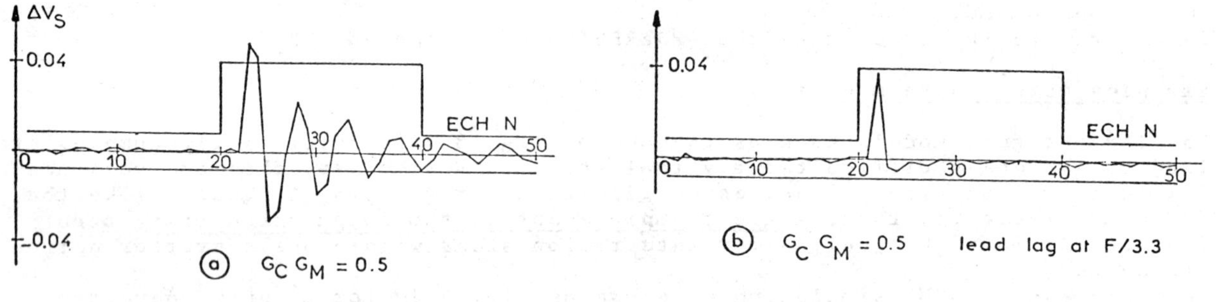

Fig 3.13 - 50% step, buck response

CONCLUSION

We have described a method to obtain systematically an approximate model for electrical convertors. The model is continuous and non-linear.

It is especially suitable to carry out digital simulations with low computation time but very good accuracy.

The linear model which can be derived from the previous one is very useful to synthesise a correcting device to obtain a satisfactory transient behaviour.

The method is then, a good compromise between complexity and accuracy.

REFERENCES

[1] - J.LAGASSE, R. PRAJOUX
Behaviour of control systems including controlled convertors, especially rectifiers : a review of existing theories
Symposium IFAC "Control in power electronics and electrical drives" (1974).

[2] - N.A.BJÄRESTEN
The static convertor as a high-speed power amplifier
Direct Current, 8, 154 (1963).

[3] - G.N.WESTER, R.D.MIDDLEBROOK
Look frequency characterization of switched DC/DC convertors
IEEE Power processing and Electronics Specialists Conference, Atlantic City (22-23 May 1972).

[4] - G.W.WESTER
Linearized stability analysis and design of a fly-back DC/DC boost regulator
IEEE Power Electronics Specialists Conference, Caltech. (June 1973).

[5] - R.D.MIDDLEBROOK
A continuous model for the tapped-inductor boost-convertor
IEEE Power Processing Electronics Specialists Conference, Los Angeles (1975).

[6] - A.K.LAHA, K.E.BOLLINGER
Power stabiliser design using pole-placement techniques on approximate power system model
Proc. IEE 122 (Sept. 75).

[7] - G.GIRALT, J.LAGASSE, Y.SEVELY
Sur la stabilité locale de la boucle de réaction de dispositifs régulateurs de tension à redresseurs contrôlés
C.R. Académie des Sciences PARIS (12 mars 1962).

[8] - R.PRAJOUX
Etablissement d'un modèle pour le comportement local d'un redresseur polyphasé utilisé en tant qu'amplificateur de puissance à réponse rapide
C.R. Académie des Sciences PARIS (22 Sept. 1969).

[9] - H.BUHLER
Investigation of a rectifier regulating circuit as a sampled data system
4th IFAC Congress, Warszawa, (June 1969).

[10] - D.SCHRODER
Analysis and synthesis of automatic control systems with controlled convertors
5th IFAC Congress, Paris, (June 1972).

[11] - R. VALETTE, R.PRAJOUX, A.GIRAUD
Systèmes à thyristors : comparaison entre un modèle avec échantillonneur et un modèle récurrent
AFCET, RAIRO Review, J3, (1974).

[12] - H.FOCH
Applied study of a fast commutation circuit with an auxiliary supply
IFAC Symposium "Control in Power Electronics and Electrical Drives" Dusseldorf (1974).

[13] - F.LEE, T.G.WILSON, S.FENG
Analysis of limit cycles in a two transistor saturable core parallel inverter
IEEE Transactions an Aerospace and Electronics Systems, AES 9 (July 1973).

[14] - T.MOORE, TG.WILSON
Basic considerations for DC to DC conversion networks
IEEE Trans. Magn. Mag. 2, n°3, (Sept. 66).

[15] - B.D.BEDFORD, R.G. HOFT
Principles of inverter circuits
John Wiley and Sons, New York (1969).

[16] - Y.ROBIN JOUAN
Modèles mathématiques de convertisseurs continu continu PWM
Power Electronics Specialists Seminar Frascati (1974).

[17] - R.PRAJOUX, J.C.MARPINARD, J.JALADE
Etablissement de modèles mathématiques pour régulateurs de puissance à modulation de largeur d'impulsion (PWM)
ESA Scientific et Technical Review 2 pages 25-43 2 pages 115-129, (1976).

STUDY OF A DC CHOPPER AS A SAMPLED SYSTEM

H. Bühler

Laboratory of Industrial Electronics, Swiss Institute of Technology, 1007 Lausanne (Switzerland)

INTRODUCTION

In this study, the theory of sampled system is used to analyze the behavior of a DC chopper in a current control circuit. The basic lay-out of the control circuit is illustrated on Fig. 1. A DC motor 1 is drived by a DC chopper 2, in a pulsed mode. The DC chopper depends on a control equipment 4 which receives a control voltage u_c from the regulator 3. The regulator is controlled by the difference between the set point i_c and the actual value i of the current to control.

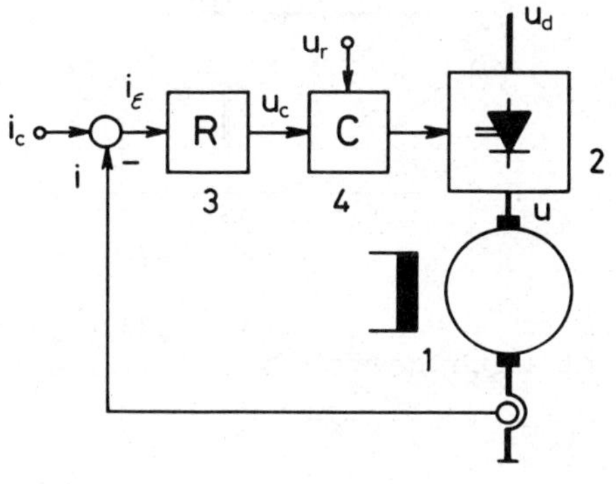

Fig. 1. Control circuit with DC chopper, driving a DC machine

The way the control equipment operates has a strong influence on the stability study of the control circuit, in particular on the setting of the equivalent circuit diagram as a sampling or continuous system. Two alternatives will thus be studied : control equipment with either a reference sawtooth voltage u_r or with a triangular one. Finally, the influence of the rippling and of the forced commutation will also be investigated.

CONTROL EQUIPMENT WITH SAWTOOTH REFERENCE VOLTAGE

Behavior of the DC Chopper for Small Variations

Figure 2 shows the reference sawtooth voltage u_r of the control equipment. When the reference voltage u_r intersects with the control voltage u_c, the DC chopper is switched on or off, with a constant pulse frequency, given by the frequency of the reference voltage u_r. At the armature of the DC motor, a pulsed voltage appears with a rectangular waveform (voltage u on Fig. 2). If the control voltage presents a variation of Δu_c, the switching off time shifts. A variation Δu appears on the voltage u with respect to the stationary working point as rectangular pulses of Δt_a duration. With the pulse period T_p, Δt_a is expressed as follows :

$$\Delta t_a = \frac{T_p}{2} \Delta u_c \tag{1}$$

For small variations, the rectangular pulses can be replaced by Dirac pulses (Fig. 2 bottom), allowing an analytical treatment as a sampled system. For this purpose, the voltage-time area must be the same for the Dirac pulses than for the rectangular ones, i.e. $(T_p/2)\ u_d\ \Delta u_c$. The Dirac pulses coincide with the sampling moments, n, n + 1, etc. The sampling period is equal to the pulse period T_p.

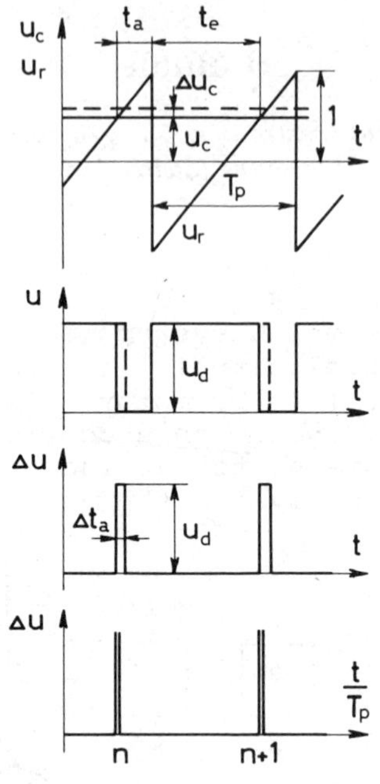

Fig. 2. Waveform of the different voltages for a control equipment with sawtooth reference voltage

Equivalent Circuit Diagram of the Control Circuit as a Sampled System

Taking into account the preceeding discussion, a block diagram can be set up as shown in Fig. 3. Figure 3 represents the equivalent diagram of the control circuit corresponding to Fig. 1 as a sampled system. $G_i(s) = \Delta i/\Delta u$ is the transfer function for the armature of the DC machine and $G_R(s)$ the transfer function for the regulator.

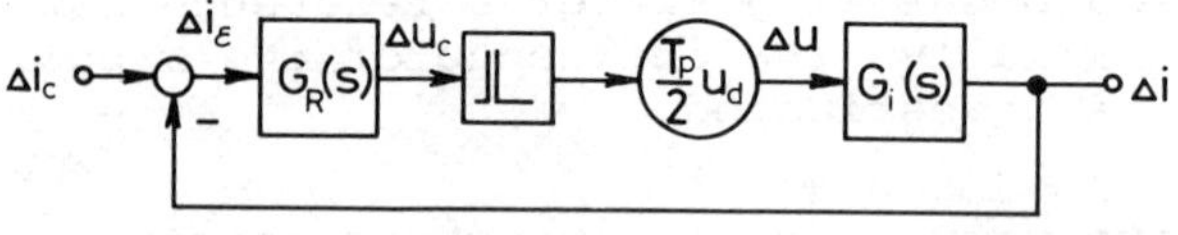

Fig. 3. Equivalent circuit diagram of the control circuit as a sampling system

In order to treat this control circuit with the z transform, it is necessary to modify the diagram until that the same representation than the one given on Fig. 4 is obtained where

$$G(s) = \frac{T_p}{2}\ u_d\ G_i(s)\ G_R(s) \tag{2}$$

considering that the pulsed sampler samples the control deviation Δx_ε just before the sampling time, the following system of equations can be set up using the z transforms

$$\Delta i(z,\varepsilon) = \frac{T_p}{2} u_d\ G_i(z,\varepsilon)\ z^{-1}\ \Delta x_\varepsilon(z,1) \tag{3a}$$

$$\Delta x_\varepsilon(z,\varepsilon) = \Delta x_c(z,\varepsilon) - \Delta x_r(z,\varepsilon) \tag{3b}$$

$$\Delta x_r(z,\varepsilon) = G(z,\varepsilon)\ z^{-1}\ \Delta x_\varepsilon(z,1) \tag{3c}$$

$\Delta x_c(z,\varepsilon)$ cannot be formulated in general but only with the help of the correspondance between the z transform and the Laplace transform :

$$\Delta x_c(z,\varepsilon) \,\blacksquare\!\!-\!\!\bullet\, G_R(s)\ \Delta i_c(s) \tag{4}$$

Equation (4) can only be solved if the set point variation $\Delta i_c(s)$ is known (e. g. unit step, harmonic variation, etc.).

From equations (3), the equation of the closed control circuit can be written as

$$\Delta i(z,\varepsilon) = \frac{T_p}{2} u_d \frac{G_i(z,\varepsilon)}{1 + z^{-1}\ G(z,1)} z^{-1} \Delta x_c(z,1) \tag{5}$$

The equivalent circuit diagram of Fig. 4 corresponds to the one of a current converter analyzed as a sampled system and the results of (ref. 1 and 2) can be used.

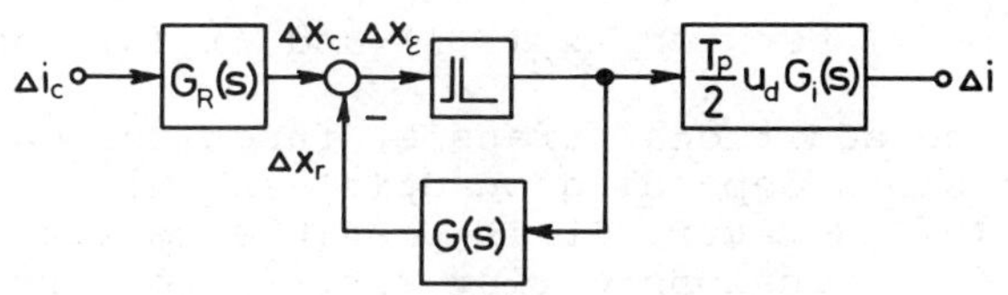

Fig. 4. Equivalent circuit diagram of the transformed Fig. 3.

Equivalent Circuit Diagram of the Control Circuit as a Continuous System

For a rapid calculation of the regulator parameters, in order to obtain a well attenuated behavior of the control circuit, the use of the z transform as a sampled system is too complicated. On the other hand, the use of the Laplace transform is better known. Using the general relationship

$$f(s) = T \int_0^1 e^{-sT\varepsilon} f(z,\varepsilon)\ d\varepsilon \tag{6}$$

where T is the sampling period (corresponding here to the pulse period T_p), one can go from the z domain to the s one. Using (6) in (5), $\Delta i(s)$ becomes :

$$\Delta i(s) = \frac{T_p}{2} u_d \frac{G_i(s)}{1 + z^{-1}\ G(z,1)} z^{-1}\ \Delta x_c(z,1) \tag{7}$$

where $z = e^{sT_p}$. $G_i(s)$ is the normal transfer function of the controlled system i.e. of the DC machine.

It is now possible to define an equivalent block diagram for the continuous control circuit, having the same transfer function as the one of equation (7). This diagram is represented on fig. 5. It shows not only the transfer function $G_i(s)$ of the controlled system but also the function $G_R(s)$ of the regulator.

The blocks with the transfer functions $G_x(s)$ and $G_c(s)$ have been introduced in order to equalize equation (8) of this continuous control circuit equivalent to the relationship (7). For this control circuit (fig. 5), one obtains

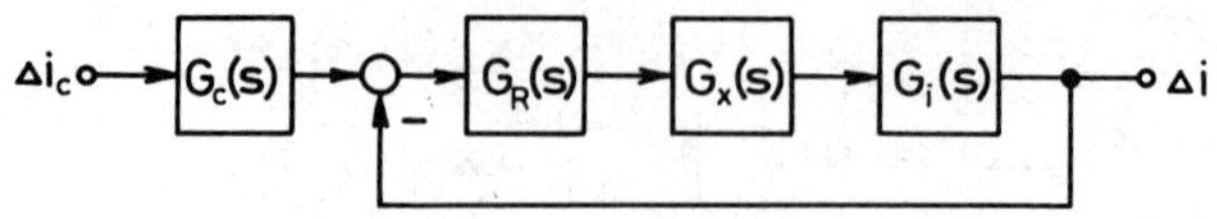

Fig. 5. Equivalent circuit diagram of the control circuit as a continuous system

$$\Delta i(s) = \frac{G_c(s)G_R(s)G_x(s)G_i(s)}{1 + G_R(s)G_x(s)G_i(s)} \Delta i_c(s) \tag{8}$$

In comparing the corresponding terms in (7) and (8), the two transfer functions $G_x(s)$ and $G_c(s)$ can be deduced :

$$G_x(s) = \frac{z^{-1}\ G(z,1)}{G_R(s)\ G_i(s)} \tag{9a}$$

$$G_c(s) = \frac{\frac{T_p}{2}\ u_d z^{-1} \Delta x_c(z,1)}{G_R(s)\ G_x(s)\ \Delta i_c(s)} \tag{9b}$$

where the relationship (4) between $\Delta x_c(z,1)$ and $\Delta i_c(s)$ has taken into account.

Rigorously speaking, the additional transfer functions $G_x(s)$ and $G_c(s)$ have very complicate expressions depending on $G_R(s)$ and $G_i(s)$, and even on the particular waveform of $\Delta i_c(s)$. However, it is possible to find with a very good approximation simplified expressions giving satisfactory results for the planned stability study (ref. 2). Therefore, $G_x(s)$ and $G_c(s)$ are expressed as follows

$$G_x(s) \approx \frac{u_d}{2}\ \frac{s\,T_p}{e^{sT_p} - 1} \approx \frac{u_d}{2}\ \frac{1}{1 + s\,\frac{T_p}{2}} \tag{10a}$$

$$G_c(s) \approx \frac{sT_p}{e^{sT_p} - 1}\ e^{sT_p} \approx \frac{1 + sT_p}{1 + s\,\frac{T_p}{2}} \tag{10b}$$

$G_x(s)$ can be seen as the transfer function of the DC chopper with its control unit having a time constant $T_p/2$. It is thus possible to study the control circuit given in Fig. 1 as a simple one using the well-known methods of the Bode diagram, the harmonic responses, etc...

CONTROL EQUIPMENT WITH TRIANGULAR REFERENCE VOLTAGE

Behavior of the DC Chopper for Small Variation

The relations obtained for the case of a symmetrical triangular reference voltage are illustrated in Fig. 6. For a change in the control voltage u_c the switching on and off times are shifted with respect to the stationary working point. The variation of the DC chopper voltage Δu presents now two rectangular pulses during a pulse period T_p. The duration of these pulses is :

$$\Delta t_a = \Delta t_e = \frac{T_p}{4}\ \Delta u_c \tag{11}$$

For small variations, Dirac pulses can be introduced with a voltage time area equal to $(T_p/4)\ u_d\ \Delta u_c$. The pulse at the switching off time of the DC chopper coincides with the sampling time, whereas the other pulse, from the switching on time of the chopper, is delayed by $\varepsilon_a = T_a/T_p$.

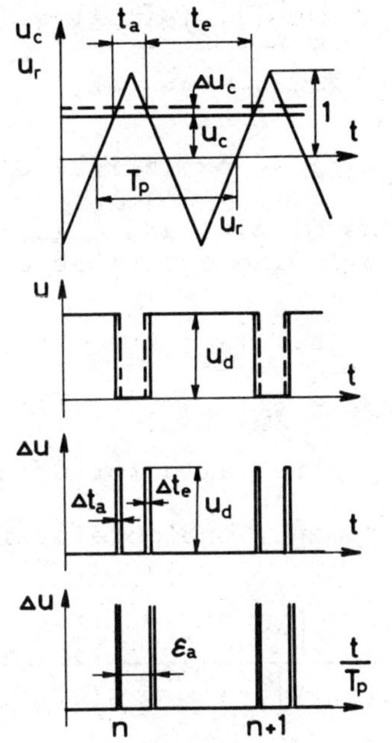

Fig. 6. Waveform of the different voltages for a control equipment with triangular reference voltages

Equivalent Circuit Diagram for the Control Circuit as a Sampled System

Because of the existence of a second pulse, the equivalent circuit diagram of Fig. 3 must be completed by a sencond pulsed sampler for which the pulse appears with a delay ε_a. This sampler must be put in parallel with the first one, as indicated in Fig. 7. The proportionality factor introduced between the DC chopper and its control is now $(T_p/4)\ u_d$.

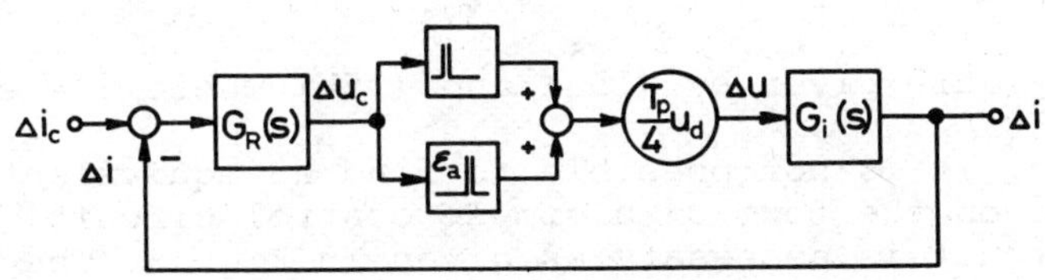

Fig. 7. Equivalent circuit diagram of the control circuit as a sampled system

The block diagram must be transformed according to Fig. 8, we have

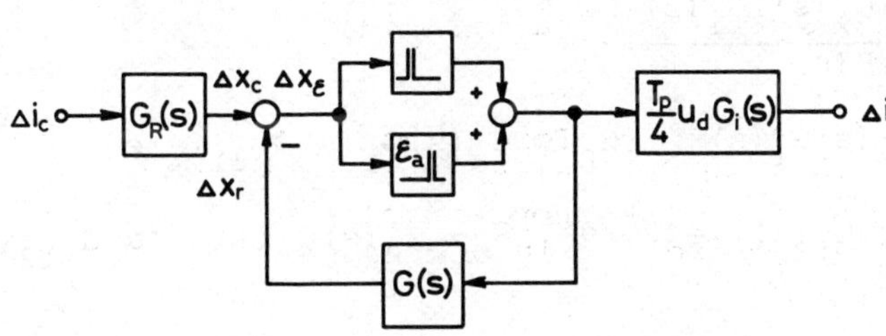

Fig. 8. Equivalent circuit diagram of the transformed Fig. 7.

$$G(s) = \frac{T_p}{4} u_d G_i(s) G_R(s) \tag{12}$$

Based on this block diagram, the following relationship can be written

$$\Delta i(z,\varepsilon) = \frac{T_p}{4} u_d \left[G_i(z,\varepsilon) z^{-1} \Delta x_\varepsilon(z,1) + G_i'(z,\varepsilon) \Delta x_\varepsilon(z,\varepsilon_a) \right] \tag{13a}$$

$$\Delta x_\varepsilon(z,\varepsilon) = \Delta x_c(z,\varepsilon) - \Delta x_r(z,\varepsilon) \tag{13b}$$

$$\Delta x_r(z,\varepsilon) = G(z,\varepsilon) z^{-1} \Delta x_\varepsilon(z,1) + G'(z,\varepsilon) \Delta x_\varepsilon(z,\varepsilon_a) \tag{13c}$$

The sampling transfer functions $G_i'(z,\varepsilon)$ and $G'(z,\varepsilon)$ are different according to the cases where $\varepsilon \gtrless \varepsilon_a$. The following equations are obtained

$$G'(z,\varepsilon) = z^{-1} G(z,1+\varepsilon-\varepsilon_a) \qquad 0 < \varepsilon \leq \varepsilon \tag{14a}$$

$$= G(z,\varepsilon-\varepsilon_a) \qquad \varepsilon_a < \varepsilon \leq 1 \tag{14b}$$

The corresponding relationships are valid for $G_i'(z,\varepsilon)$.

After a relatively long development, one obtains for the closed-loop control circuit

$$\Delta x_\varepsilon(z,1) = \frac{\left[1 + z^{-1} G(z,1)\right] \Delta x_c(z,1) - G(z,1-\varepsilon_a) \Delta x_c(z,\varepsilon_a)}{\left[1 + z^{-1} G(z,1)\right]^2 - z^{-1} G(z,\varepsilon_a) G(z,1-\varepsilon_a)} \tag{15a}$$

$$\Delta x(z,\varepsilon_a) = \frac{\left[1 + z^{-1} G(z,1)\right] \Delta x_c(z,\varepsilon_a) - z^{-1} G(z,\varepsilon_a) \Delta x_c(z,1)}{\left[1 + z^{-1} G(z,1)\right]^2 - z^{-1} G(z,\varepsilon_a) G(z,1-\varepsilon_a)} \tag{15b}$$

Then, the current variation can be deduced from equation (13a).

<u>Equivalent Circuit Diagram of the Control Circuit as a Continuous System</u>

Using the general relationship (6), equation (13a) can be transformed from the z domain into the s one. This can be written as

$$\Delta i(s) = \frac{T_p}{4} u_d G_i(s) \left[e^{-sT_p} \Delta x_\varepsilon(z,1) + e^{-\varepsilon_a sT_p} \Delta x_\varepsilon(z,\varepsilon_a) \right] \tag{16}$$

$\Delta x(z,1)$ and $\Delta x(z,\varepsilon_a)$ are given by (15a) and (15b) where $z = e^{sT_p}$.

In the present case, it is not possible to find an equivalent block diagram as a continuous system on the form of a simple control circuit. The use of a double coupled control circuit is necessary as given in Fig. 9. The two direct branches are identical and independent and the coupling appears between the two feed-back branches. For such a configuration, $\Delta i_d(s)$ are deduced as follows

$$\Delta i_d(s) = \frac{G_R(s) G_x(s) G_i(s)}{D(s)} \Big\{ \left[1 + G_R(s) G_x(s) G_i(s) - G_R(s) G_x(s) G_i(s) G_{x21}(s) e^{(1-\varepsilon_a)sT_p} G_{c1}(s) \right] + \left[\left(1 + G_R(s) G_x(s) G_i(s) \right) e^{(1-\varepsilon_a)sT_p} - G_R(s) G_x(s) G_i(s) G_{x12}(s) \right] G_{c2}(s) \Big\} \Delta i_c(s) \tag{17a}$$

with

$$D(s) = \left[1+G_R(s)G_x(s)G_i(s)\right]^2 - \left[G_R(s)G_x(s)G_i(s)\right]^2 G_{x12}(s)G_{x21}(s) \qquad (17b)$$

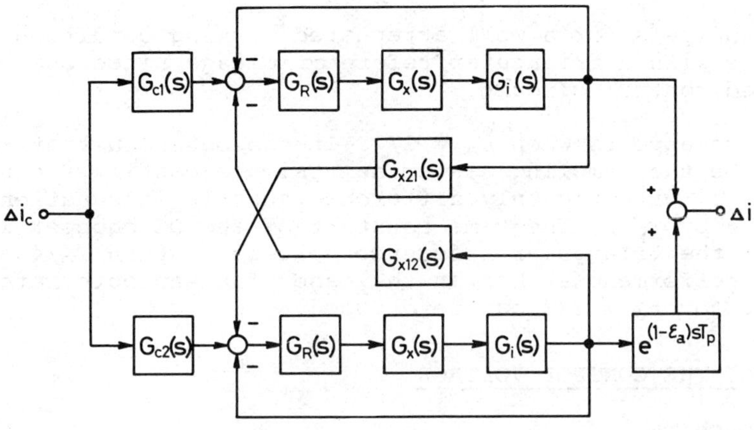

Fig. 9. Equivalent circuit diagram of the control circuit as a continuous system

In comparing the corresponding terms in (13a), (15a) and (15b), on one hand and (17a) and (17b) on the other hand, the relationship for the different additional transfer functions is obtained

$$G_x(s) = \frac{z^{-1}\ G(z,1)}{G_R(s)\ G_i(s)} \qquad (18a)$$

$$G_{x12}(s) = \frac{G(z,1-\varepsilon_a)}{z^{-1}\ G(z,1)} \qquad (18b)$$

$$G_{x21}(s) = \frac{G(z,\varepsilon_a)}{G(z,1)} \qquad (18c)$$

$$G_{c1}(s) = \frac{\frac{T_p}{4}\ u_d\ e^{-sT_p}\ \Delta x_c(z,1)}{G_R(s)G_x(s)\Delta i_c(s)} \qquad (18d)$$

$$G_{c2}(s) = \frac{\frac{T_p}{4}\ u_d\ e^{-sT_p}\ \Delta x_c(z,\varepsilon_a)}{G_R(s)G_x(s)\Delta i_c(s)} \qquad (18e)$$

With a very good approximation, simple expressions for the transfer functions can be deduced. They are

$$G_x(s) \approx \frac{u_d}{4}\ \frac{sT_p}{e^{sT_p}-1} \approx \frac{u_d}{4}\ \frac{1}{1+s\frac{T_p}{2}} \qquad (19a)$$

$$G_{x12}(s) \approx e^{sT_p} \qquad (19b)$$

$$G_{x21}(s) \approx 1 \qquad (19c)$$

$$G_{c1}(s) \approx G_{c2}(s) \approx \frac{sT_p}{e^{sT_p} - 1} e^{sT_p} \approx \frac{1 + sT_p}{1 + s\frac{T_p}{2}} \tag{19d}$$

The stability analysis for a well attenuated working condition is far too complicated here with a triangular reference voltage since one has to calculate a double coupled control circuit.

A particular case appears when $\varepsilon_a = 1/2$. It can be shown that it is then possible to describe the sampling system as a simple continuous control circuit as the one of Fig. 5, with the only difference that T_p in equations (10a) and (10b) must be replaced by $T_p/2$. The time constant of the DC chopper with its control equipment using the triangular reference voltage is then $T_p/4$ and one obtains better control performances than in the case of a sawtooth reference voltage. Similar results for $\varepsilon_a \neq 1/2$ can be deduced.

RIPPLE OF THE CONTROL VOLTAGE

General Relationships

Because of the DC chopper works in a pulsed way, the armature current shows a ripple, transmitted to the control voltage u_c according to the transfer function of the regulator. In order to obtain a well defined intersection between u_c and u_r in the control equipment, it is absolutely necessary that the control voltage u_c does not show a too high rippling. For this purpose, it is necessary to introduce a smoothing filter in the regulator. How to determine the control voltage rippling $\sim u_c$ can be solved by the sampled system theory:

Figure 10 shows the waveform of the ripple $\sim u$ of the output voltage of the DC chopper. This periodic voltage can be reproduced by the extrapolator of a sampled system as indicated in Fig. 11.

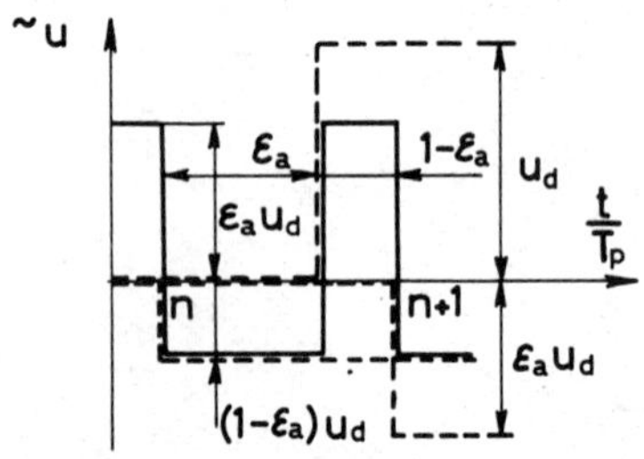

Fig. 10. Ripple of the voltage from the DC chopper

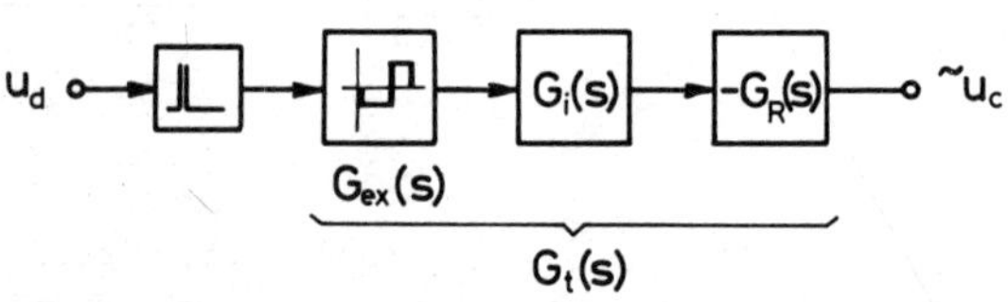

Fig. 11. Sampling system for the calculation of the ripple of the control voltage

The transfer function of this extrapolator is equal to

$$G_{ex}(s) = -\left[\frac{1 - \varepsilon_a}{s} - \frac{e^{-s\varepsilon_a T_p}}{s} + \frac{\varepsilon_a}{s} e^{-sT_p}\right] \tag{20}$$

Using the z transform, the rippling of the control voltage becomes

$$\sim u_c(z,\varepsilon) = G_t(z,\varepsilon) \frac{z}{z-1} u_d \tag{21}$$

The steady state ripple is particularly interesting, it can be obtained using the final value theorem.

$$\tilde{u}_c(\varepsilon) = \tilde{u}_c\,[\infty,\varepsilon] = \lim_{z\to 1}\,(z - 1)\ \tilde{u}_c(z,\varepsilon) = u_d \lim_{z\to 1} G_t(z,\varepsilon) \tag{22}$$

Particular Case of the PI Regulator With Smoothing Filter

In the control circuit of Fig. 1, a PI regulator with smoothing filter is usually taken. Its transfer functions is

$$G_R(s) = \frac{1 + sT_n}{sT_i}\ \frac{1}{1 + sT_f} \tag{23}$$

where T_f is the time constant of the smoothing filter. The integral action time T_n is normally chosen equal to the time constant of the DC machine armature circuit.

According to equation (22), two expressions of the rippling are obtained corresponding to the two cases $\varepsilon \lesseqgtr \varepsilon_a$. The maximum ripple occurs when $\varepsilon_a = 1/2$. Fro the peak value, the following relationship is deduced if r is the relative resistance of the DC machine armature circuit.

$$\tilde{\hat{u}}_c = u_d \frac{T_p}{r\ T_i}\ \hat{w}_c \tag{24a}$$

with

$$\hat{w}_c = \frac{1}{8} + \frac{T_f}{2\ T_p}\ \ln\ \frac{1 + e^{-T_p/2T_f}}{2} \tag{24b}$$

The factor $\hat{w}_c$ is represented in Fig. 12 as a function of T_f/T_p. For the maximum allowed ripple $\tilde{u}_c$, it is possible to determine the time constant T_f of the filter. In this determination, the integration time T_i of the regulator is also used. The time T_i depends on the stability study of the control circuit where the time constant T_f of the smoothing filter has to be taken into account (ref. 3).

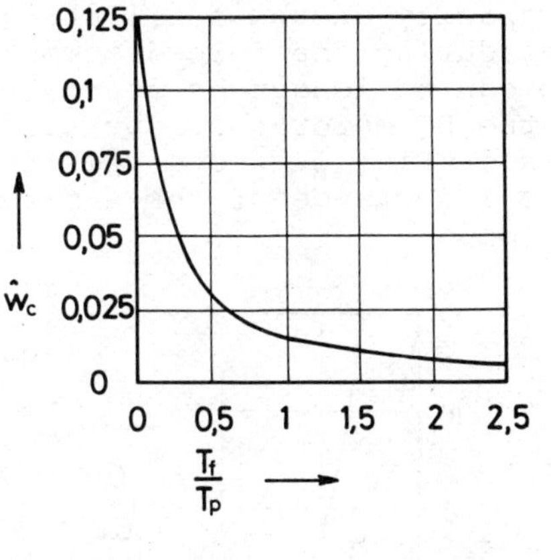

Fig. 12. $\hat{w}_c$ factor for the calculation of the maximum ripple of the control voltage

INFLUENCE OF THE FORCED COMMUTATION

Up to now, the influence of the forced commutation occurring at the switching off time of the DC chopper has been neglected. Because of this, triangular pulses are superposed on the u voltage as indicated in Fig. 13. As a result, this produces an increase in the mean value of the voltage u with regard to the ideal value $(1 - \varepsilon_a)\ u_d$.

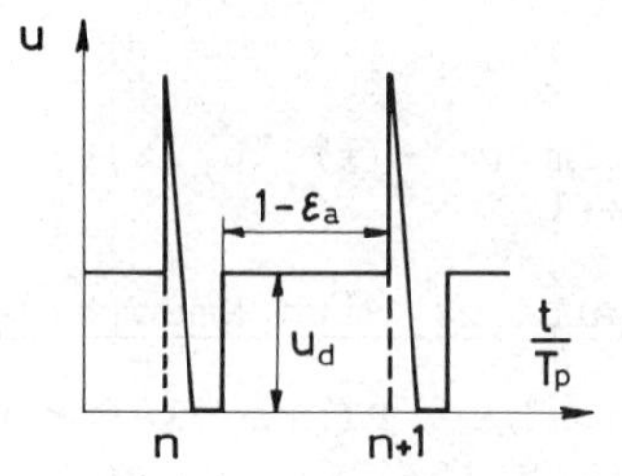

Fig. 13. Waveform of the voltage u taking into account the forced commutation of the DC chopper

It is possible to write an analytical relationship for this component, when the waveform of the current i is assumed to be smooth (ripple neglected). Per unit value, the voltage u_k is equal to

$$u_k = u_d \frac{t_{fmin}}{T_p} \, 2 \, \frac{i_{max}}{i} \, \frac{\left(1 + \frac{k_\sigma}{2} \frac{i}{i_{max}}\right)^2}{1 + k_\sigma} \tag{25a}$$

where t_{fmin} is the minimum time which the voltage at the switched off thyristor is negative. This nominal value is obtained for the maximum current i_{max}. If $i < i_{max}$, then $t_f > t_{fmin}$. The factor k_σ is given by

$$k_\sigma = \frac{1}{\omega_\sigma t_{fmin}} - 1 \tag{25b}$$

where ω_σ is the eigenvalue of the oscillator circuit formed by the leakage inductance and the extinction capacity of the DC chopper.

The component u_k due to the forced commutation and normalized with regard to $u_d \, t_{fmin}/T_p$ is represented in Fig. 14 as a function of i/i_{max}. The equation (25a) is no longer valid for small value of the current i. In this domain, it is no longer permissible to assume a smooth current waveform and one has to take into account the finite inductance of the armature circuit. The inductance forms with the extinction capacity another oscillating circuit, which determines the waveform of the voltage during the forced commutation period. In this case, an analytical relationship can no longer be found for u_k which now depends on the induced voltage u_i at the DC motor too. Curves computed numerically are represented in deshed lines in Fig. 14 where ω_0 is the eigenvalue of the oscillator circuit formed by the inductance of the armature circuit and the extinction capacity of the DC chopper.

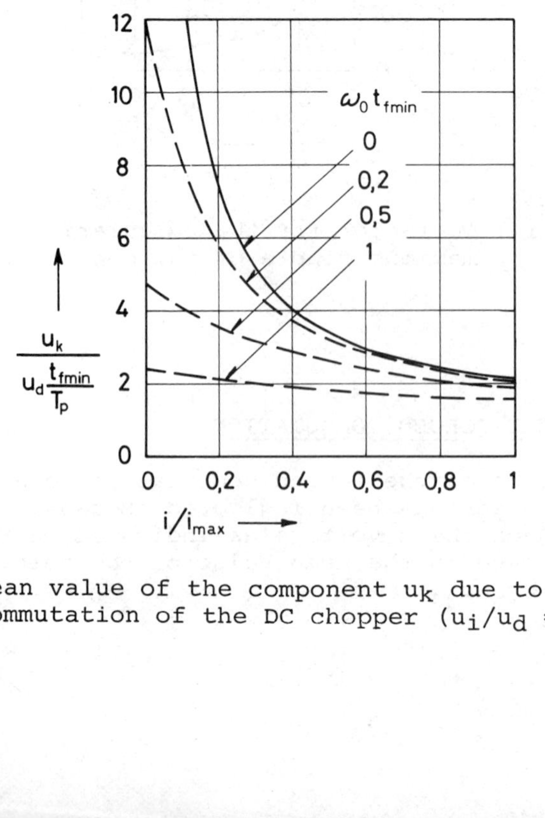

Fig. 14. Mean value of the component u_k due to the forced commutation of the DC chopper (u_i/u_d = 0,5)

In order to take into account the influence of the forced commutation on the control phenomena, the triangular pulses in Fig. 13 can be replaced by Dirac pulses with the same voltage-time surface, i.e. $u_k T_p$, and appearing at the sampling times n, n + 1, etc. As u_k is a non linear function of the current i, a linearization can be introduced for small variations using the inner resistance r_i of the DC chopper which is given by

$$r_i = - \frac{du_K}{di} = u_d \frac{t_{fmin}}{T_p} 2 \left(\frac{1}{i^2} - \frac{k_\sigma^2}{4\, i_{max}^2} \right) \qquad (26)$$

in the domain where equation (25a) is valid. For small values of the current, r_i is deduced from the tangent of the curves of Fig. 14.

The Dirac pulse appearing at the switching off time of the DC chopper can be represented by a pulsed sampling system as indicated schematically in Fig. 15.

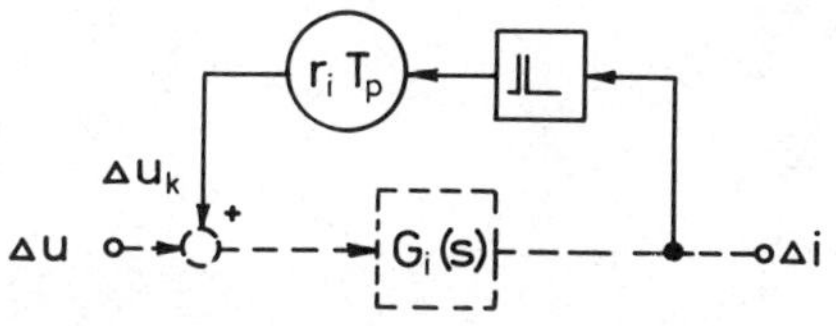

Fig. 15. Equivalent circuit diagram of the sampled system taking under consideration the influence of the forced commutation

This block diagram can be combined with the block diagram of the control circuit in Fig. 3, respectively Fig. 7, and one obtains a feedback over the block with the transfer function $G_i(s)$. For the transformation and the treatment of this configuration see (ref. 1 and 2).

In the case of an equivalent continuous diagram, the resistance r_i is to be added to the resistance r of the armature cirucit of the DC motor (ref. 3). The parameters of the transfer function $G_i(s)$, the transfer factor, and the time constant are thus varying.

REFERENCES

(1) Bühler, H., Investigation of a Rectifier Regulating Circuit as a Sampled Data System, IFAC-Congress, Warsazawa, Preprints, Tech. Sess. 64,46(1969).

(2) Yechouroun, C., Comportement dynamique des convertisseurs de courant, Thèse, Ecole Polytechnique Fédérale de Lausanne (1977).

(3) Bühler, H. (1977) Einführung in die Theorie geregelter Drehstromantriebe, Birkhäuser, Basel.

A SIMPLE METHOD FOR A QUICK-RESPONSE CHOPPER PWM SYSTEM WITH SMALL STEADY-STATE ERROR

T. Mochizuki, Y. Tanaka and M. Hyodo

Kyushu Institute of Technology, Kitakyushu 804, Japan

SUMMARY

In this paper, the dynamic behaviour of a chopper PWM system in a closed-loop is analized by using the discrete models for small signal perturbation. The extension of the model from the small signal perturbation to the large signal perturbation is discussed by graphical method. A simple method is proposed to realize a quick-response system with small steady-state error, using the dead-beat gain and the bias function.

INTRODUCTION

The dynamic behaviour of a chopper PWM system in a closed-loop has been investigated by many authors. (1),(2),(3),(4). For the investigation of dynamic behaviour of such a type of PWM regulator in a closed-loop system, it is necessary to take into consideration the actual sampling and firing mechanisms. According to the classification by Balestrino et al, (3), there are four types of PWM system; natural sampling (N.S.)-lag-type, uniform sampling (U.S.)-lag-type, N.S.-lead-type, U.S.-lead-type.

In this paper, the response of first-order system with proportional type regulator to the step change of reference input is analized by the sample-hold model and the impulse model. At these discrete models, the nature of response and the stability boundary are determined by the root of characteristic equation F(z)=0. The roots obtained by the discrete models are compared with that obtained by the linearized difference equation proposed by Leonhald (1).

When the change of reference input is large, that is, for large signal perturbation, the graphical analysis (1), (5) is useful to investigate the dynamic behaviour of system. In this paper, the graphical mapping for each type of PWM system is shown and the slope of mapping is compared with each other.

The approximate dead-beat gain is proposed depending on graphical analysis. The bias function (5), serving the purpose of eliminating the steady-state error at the approximate dead-beat gain, is proposed. A simple method to realize those performance is proposed and is confirmed by the analogue simulation.

DESCRIPTION OF THE SYSTEM

All the quantities, x, w, and V are normalized and expressed in per-unit form, where x is the output of controlled plant, w is the reference input, V is the loop gain of system and is concentrated on the gain of propotional type regulator. The transfer function of controlled plant is expressed as follows

$$G(s)= \tau_0/(s+\tau_0)$$

where $\tau_0=T/T_i$, T_i is the time constant of controlled plant, and time variable t is transformed into $\theta=t/T$.
The differential equation of x is given by

$$\dot{x}(\theta)= -\tau_0 x(\theta)+\tau_0 v(\theta) \qquad (1)$$

where $v(\theta)$ is the output of PWM,

$$v(\theta)=0 \quad (n\leqq \theta \leqq n+\alpha_n)$$
$$=1 \quad (n+\alpha_n\leqq \theta \leqq n+1)$$

for lag type

$$v(\theta)=1 \quad (n\leqq \theta \leqq n+\beta_n)$$
$$=0 \quad (n+\beta_n\leqq \theta \leqq n+1)$$

for lead type

The α_n and β_n are determined by a coincident circuit which compares the synchronous timing voltage $u_s(\theta)$ with the control voltage $u_r(\theta)$. The voltage $u_s(\theta)$ and $u_r(\theta)$ are described by

$$u_s(\theta)= 1-\theta \quad \text{for lag type}$$
$$= \theta \quad \text{for lead type}$$
$$(n\leqq \theta \leqq n+1)$$

$$u_r(\theta)= \text{Sat } y$$

where

$$\text{Sat } y= \begin{cases} y & 0 \leqq y \leqq 1 \\ 1 & y>1 \\ 0 & y<0 \end{cases}$$

$$y=V\{w-x(\theta)\} \quad \text{for N.S.}$$
$$=V\{w-x(n)\} \quad \text{for U.S.}$$

The firing pulse is produced when $u_s(\theta)$ is coincident with $u_r(\theta)$. The mathematical description of the system between any concecutive sampling periods is developed. The rigorous mathematical model is depending on the repeated uses of the above description. In what follows, unless other-wise stated, lag type system will be analized in full in the body of the paper and only a part of the results for other system will be shown.

Steady-State Characteristic

The output voltage of PWM regulator is a periodic waveform at constant firing angle α_0, so the response of system output can be easily obtained by application of the modified z-transformation. The Laplace transformation of the PWM output voltage $v(\theta)$ is

$$v(s)=\frac{1}{1-e^{-s}}\cdot\frac{e^{-\alpha_0 s}-e^{-s}}{s} \qquad (2)$$

The Laplace transform of the controlled plant output is

$$x(s)=G(s)v(s)=\frac{\tau_0}{s+\tau_0}\,v(s) \qquad (3)$$

Applying the modified z-transformation to this equation, get

$$x(z)=\frac{e^{-\tau_0 m}\{1-e^{-\tau_0(1-\alpha_0)}\}}{(z-1)(z-e^{-\tau_0})} \qquad (0\leqq m\leqq\alpha_0)$$

$$x(z)=\frac{e^{-\tau_0 m}\{1-e^{-\tau_0(1-\alpha_0)}\}}{(z-1)(z-e^{-\tau_0})} \qquad (0\leqq m\leqq\alpha_0)$$

$$=\frac{e^{-\tau_0 m}-e^{-\tau_0}+z\{1-e^{-\tau_0(m-\alpha_0)}\}}{(z-1)(z-e^{-\tau_0})} \qquad (\alpha_0\leqq m\leqq 1) \qquad (4)$$

The actual output $x(n,m)$ can be easily obtained by the inverse modified z-transformation.
The value of $x(\infty,m)$ $\{0\leqq m\leqq\alpha_0\}$ is important for determing the gain correction factor as described latter section and $x(\infty,m)$ can be obtained as follows

$$x(\infty,m)=\lim_{z\to1}(z-1)x(z,m)=\frac{1-e^{-\tau_0(1-\alpha_0)}}{1-e^{-\tau_0}}\,e^{-\tau_0 m}$$
$$=x_{10}e^{-\tau_0 m} \qquad (5)$$

where
$$x_{10}=\frac{1-e^{-\tau_0(1-\alpha_0)}}{1-e^{-\tau_0}}$$

Transient Characteristics

Under the closed-loop conditions, the differential equation governing the behaviour of the output between n-th and (n+1)th sampling periods is given by Eq.(1). Let us define $x(n)=x_n$, $x(n+1)=x_{n+1}$, then the following first order difference equation is obtained

$$x_{n+1}=e^{-\tau_0}x_n+1-e^{-\tau_0(1-\alpha_n)} \qquad (6)$$

Where α_n should be determined as follows

$$1-\alpha_n=\mathrm{Sat}\ V(w-x_n e^{-\tau_0\alpha_n}) \qquad \text{for N.S.}$$
$$=\mathrm{Sat}\ V(w-x_n) \qquad \text{for U.S.} \qquad (7)$$

Eq.(6) is essentially nonlinear, so can be solved only by numerical techniques, for example, by digital simulation or graphical analysis.
As Leonhald proposed (1), by linearising Eq. (6) in the neighbourhood of the considered steady-state operating point, it is possible to obtain clear picture of the system, at least for small signal perturbation.
According to the reference (1), if the reference input w is changed from w_0 to $w=w_0+\Delta w$, the following first order linear difference equation can be obtained by Eq.(6) and Eq. (7).

$$\Delta x_{n+1}+c_1\Delta x_n=c_0\Delta w \qquad (8)$$

where

$$c_1=\frac{\tau_0 V(1-e^{-\tau_0\alpha_0})-(1-e^{-\tau_0})}{\tau_0 V(e^{-\tau_0\alpha_0}-e^{-\tau_0})+(1-e^{-\tau_0})}\,e^{-\tau_0}$$

$$c_0=\frac{\tau_0 V(1-e^{-\tau_0})e^{-\tau_0(1-\alpha_0)}}{\tau_0 V(e^{-\tau_0\alpha_0}-e^{-\tau_0})+(1-e^{-\tau_0})}$$

for N.S

$$c_1=e^{-\tau_0}(\tau_0 V e^{\tau_0\alpha_0}-1)$$

$$c_0=\tau_0 V e^{-\tau_0(1-\alpha_0)}$$

for U.S

Applying the z-transformation to solve Eq.(8)

$$\Delta x(z)=\frac{c_0\Delta w z}{(z-1)(z+c_1)} \qquad (9)$$

$$\Delta x(n)=\frac{c_0\Delta w}{1+c_1}\{1-(-c_1)^n\} \qquad (10)$$

SMALL SIGNAL PERTURBATION

Gain Correction Factor

The dynamic loop gain of the system is constant when the control voltage u_r=const; i.e. the system operates open-loop or closed-loop for U.S. The change of firing angle $\Delta\alpha$ correspond to the change of control voltage Δu_r is given by

$$\Delta\alpha/\Delta u_r=-1 \qquad (\text{for lead type } \Delta\alpha/\Delta u_r=1)$$

In fact, under the closed-loop condition at N.S., due to the ripple component in the output waveform, a gain correction factor must be introduced (6).
It can be shown (7) that for small signal perturbation, the gain correction factor is given by

$$F=\lim_{\Delta u_r\to0}\left|\frac{\Delta\alpha}{\Delta u_r}\right|=\frac{1}{1-\left.{du_r/d\theta}\Big/{du_s/d\theta}\right|_{\theta=\alpha_0}} \qquad (11)$$

$$F=\frac{(1-e^{-\tau_0})}{\tau_0 V(e^{-\tau_0\alpha_0}-e^{-\tau_0})+(1-e^{-\tau_0})} \qquad (12)$$

for lag type

$$F=\frac{(1-e^{-\tau_0})}{\tau_0 V(e^{-\tau_0\beta_0}-e^{-\tau_0})+(1-e^{-\tau_0})} \qquad (13)$$

for lead type

For determination $du_r(\theta)/d\theta$, the modified z-transformation method mentioned at preceding section is useful.

Sample-Hold Model

For small signal perturbation, the change of PWM output is the pulse train, this is modelled by spreading the area of the train pulses over the sampling period (8),(9). As a result, the PWM is replaced by a sampler following by a zero-order circuit. The resulting model for N.S. is shown in Fig. 1.

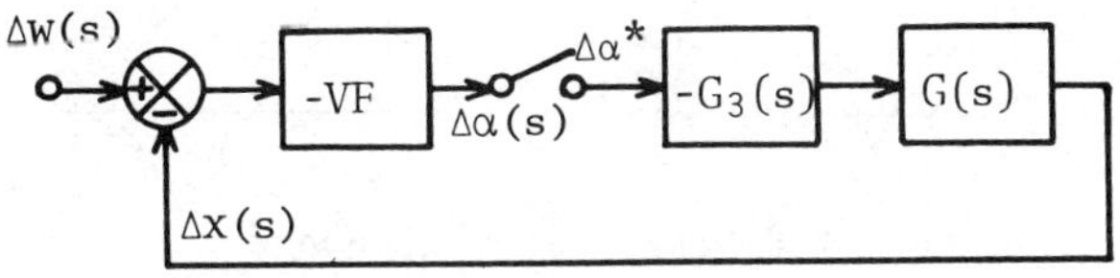

Fig. 1 Block diagram of sample-hold model

The sampler operates at the time instants $\theta=n+\alpha_0^-$ (2)

$$\Delta\alpha(s)= -VF\{w-x(s)\} \qquad (14)$$

$$\Delta x(s)= -\Delta\alpha^*(s)G_3(s)G(s) \qquad (15)$$

Where $G_3(s)$ is the transfer function of zero-order hold circuit,

$$G_3(s)= (1-e^{-s})/s$$

We apply the z-transformation to these equation,

$$\Delta x(z)= \frac{VF\Delta w(z)G_3G(z)}{1+ VFG_3G(z)} = \frac{VF\Delta wz}{(z-1)\{z-e^{-\tau_0}+VF(1-e^{-\tau_0})\}} \qquad (16)$$

The output $\Delta x(n)$ can be obtained by inverse z-transformation,

$$\Delta x(n)= \frac{VF\Delta w}{\{1-e^{-\tau_0}+VF(1-e^{-\tau_0})\}}[1-\{e^{-\tau_0}-VF(1-e^{-\tau_0})\}^n] \qquad (17)$$

In this model,the quantity of $\Delta\alpha$ would be fully characterrized from the knowledge of the state of the system at the sampling instants only (10) . For U.S, the sampler operates at the time instant $\theta=n$. Only replaing F=1, Eq.(14)—(17) can be applicable to U.S. in the same manner with N.S .

Impulse model

The "equivalent areas" concept of the sample-hold model is valid if τ_0 is small enough. But this concept breaks down when τ_0 becomes considerably large. At the impulse model, $\Delta\alpha\to 0$ the train of pulses with finite pulse width is approximated by a train of Dirac impulses (2),(10). The sampler operates $\theta=n+\alpha_0^-$ for N.S., and the resulting model is shown in Fig. 2.

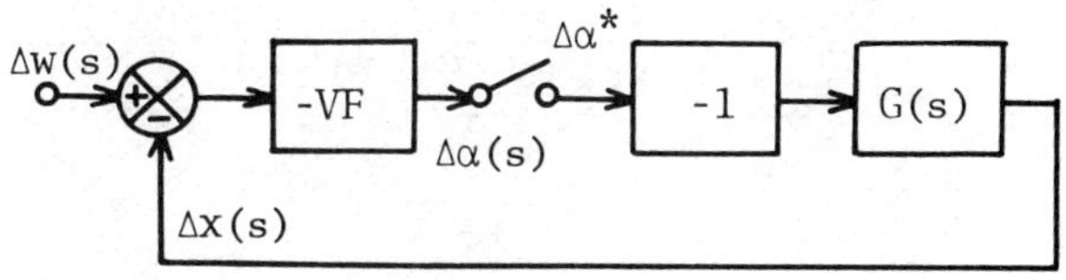

Fig. 2 Block diagram of impulse model for N.S.

$$\Delta\alpha(s)= -VF\{\Delta w(s)-\Delta x(s)\} \qquad (18)$$

$$\Delta x(s)= -\Delta\alpha^*(s)G(s) \qquad (19)$$

We apply the modified z-transformation method to above equations and get

$$\Delta x(z,m)= \frac{VF\Delta w(z,1)G(z,m)}{1+VFG(z,1)} = \frac{\tau_0 VF\Delta w e^{-\tau_0 m} z}{(z-1)\{z-e^{-\tau_0}(1-\tau_0 VF)\}} \qquad (20)$$

$$\Delta x(n,m)= \frac{\tau_0 VF\Delta w e^{-\tau_0 m}}{1-e^{-\tau_0}(1-\tau_0 VF)}[1-\{e^{-\tau_0}(1-\tau_0 VF)\}^n] \qquad (21)$$

where F is given by Eq.(12).
It should be noted that the root of characteristic equation is completely identical to the root obtained by linearized difference equation as described previous section.
For U.S., F=1 and the sampler operates at $\theta=n^-$, but the real impulse works at $\theta=n+\alpha_0^-$. So the delay function $G_2(s)=e^{-\alpha_0 s}$ should be introduced. To resulting model is shown in Fig. 3. The calculate $G_2G(z,m)$, the region $0\leqq m\leqq\alpha_0$ and $\alpha_0\leqq m\leqq 1$ should be seperated.

for $0\leqq m \leqq \alpha_0$

$$\Delta x(z,m)= \frac{V\Delta w(z,1)G_2G(z,m)}{1+VG_2G(z,1)} = \frac{\tau_0 V\Delta w e^{-\tau_0(m+1-\alpha_0)}}{(z-1)\{z-e^{-\tau_0}+\tau_0 Ve^{-\tau_0(1-\alpha_0)}\}} \qquad (22)$$

$$\Delta x(n,m)= \frac{\tau_0 V\Delta w e^{-\tau_0(m+1-\alpha_0)}}{\{1-e^{-\tau_0}(1-\tau_0 Ve^{\tau_0\alpha_0})\}} \cdot [1-\{e^{-\tau_0}(1-\tau_0 Ve^{\tau_0\alpha_0})\}^{n-1}] \qquad (23)$$

for $\alpha_0\leqq m \leqq 1$

$$\Delta x(z,m)= \frac{\tau_0 V\Delta w e^{-\tau_0(m-\alpha_0)} z}{(z-1)\{z-e^{-\tau_0}+\tau_0 Ve^{-\tau_0(1-\alpha_0)}\}} \qquad (24)$$

$$\Delta x(n,m)=\frac{\tau_o V\Delta w e^{-\tau_o(m-\alpha_o)}}{\{1-e^{-\tau_o}(1-\tau_o V e^{\tau_o\alpha_o})\}}\cdot[1-\{e^{-\tau_o}(1-\tau_o V e^{\tau_o\alpha_o})\}^n] \qquad (25)$$

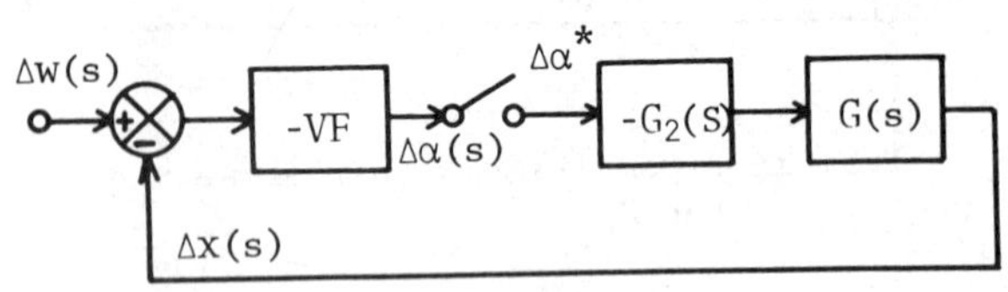

Fig. 3 Block diagram of impulse model for U.S.

In this case also the root obtained is completely identical to that obtained by the linearized difference equation. Therefore, the impulse model give the same results regarding the stability boundary and dead-beat gain as described latter section.

Numerical Example of Step Response

The performance of the models for the reference input changed from w=0.3 to w=0.4 (Δw =0.1) at τo=o.2, V=2, are illustrated in Fig. 4 (a) for N.S. and (b) for U.S.
The results of the impulse model including the ripple component are well agreed with that of the rigorous mathematical model calculated by digital computer. The degree of agreement for U.S. is better than that for N.S.
This fact shows that the root of characteristic equation of U.S. is less depend on the reference input w than that of N.S.

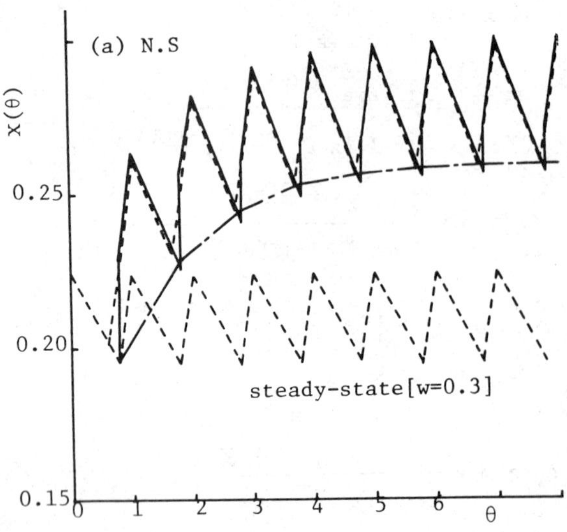

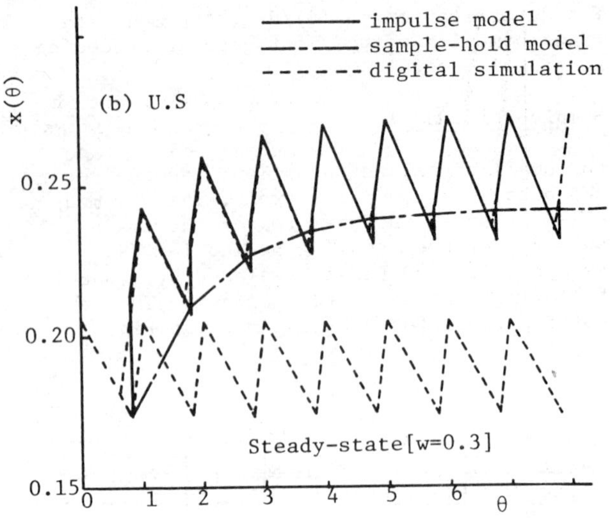

Fig. 4 (a),(b) Step response
τ_o=0.2, V=2, Δw=0.1

Stability Boundary

When increasing the loop gain V, the root of characteristic equation becomes -1, then the system becomes instable. The critical gain Vc can be obtained from each models, and the result is shown in Table 1.
Numerical examples are shown in Fig. 5 (a) and (b).

TABLE 1 Critical Gain V_c

	N.S	U.S
S-H model	$\frac{(1+e^{-\tau_o})(1-e^{-\tau_o})}{(1-e^{-\tau_o})^2-\tau_o(1+e^{-\tau_o})(e^{-\tau_o\alpha_o}-e^{-\tau_o})}$	$\frac{1+e^{-\tau_o}}{1-e^{-\tau_o}}$
I model	$\frac{(1+e^{-\tau_o})(1-e^{-\tau_o})}{\tau_o\{2e^{-\tau_o}-e^{-\tau_o\alpha_o}(1+e^{-\tau_o})\}}$	$\frac{1+e^{-\tau_o}}{\tau_o e^{\tau_o\alpha_o}}$

TABLE 2 Dead-Beat Gain V_d

	N.S	U.S
S-H model	$\frac{e^{-\tau_o}(1-e^{\tau_o})}{(1-e^{-\tau_o})^2-\tau_o e^{-\tau_o}(e^{-\tau_o\alpha_o}-e^{-\tau_o})}$	$\frac{e^{-\tau_o}}{1-e^{-\tau_o}}$
I model	$\frac{(1-e^{-\tau_o})}{\tau_o(1-e^{-\tau_o\alpha_o})}$	$\frac{1}{\tau_o e^{\tau_o\alpha_o}}$

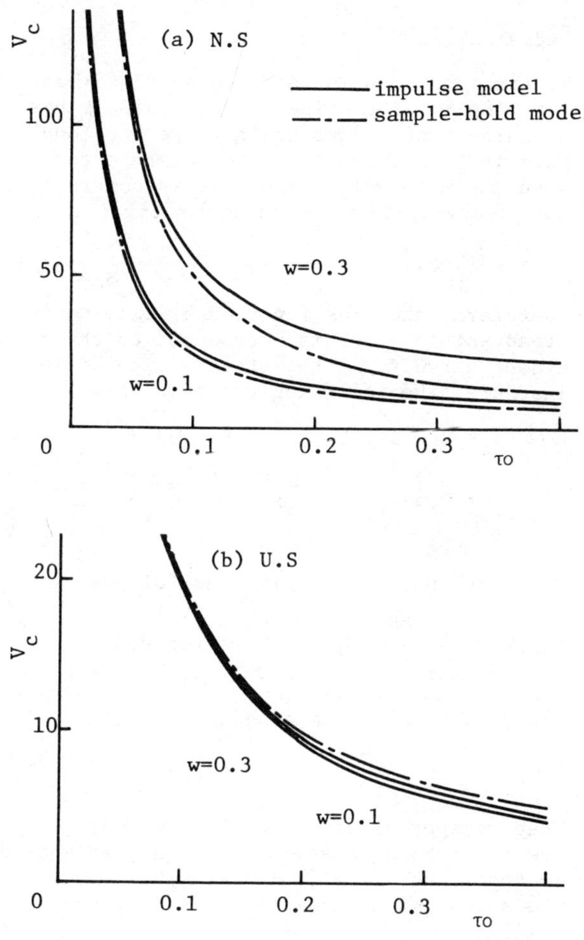

Fig. 5 (a),(b) Numerical example of critical gain V_c

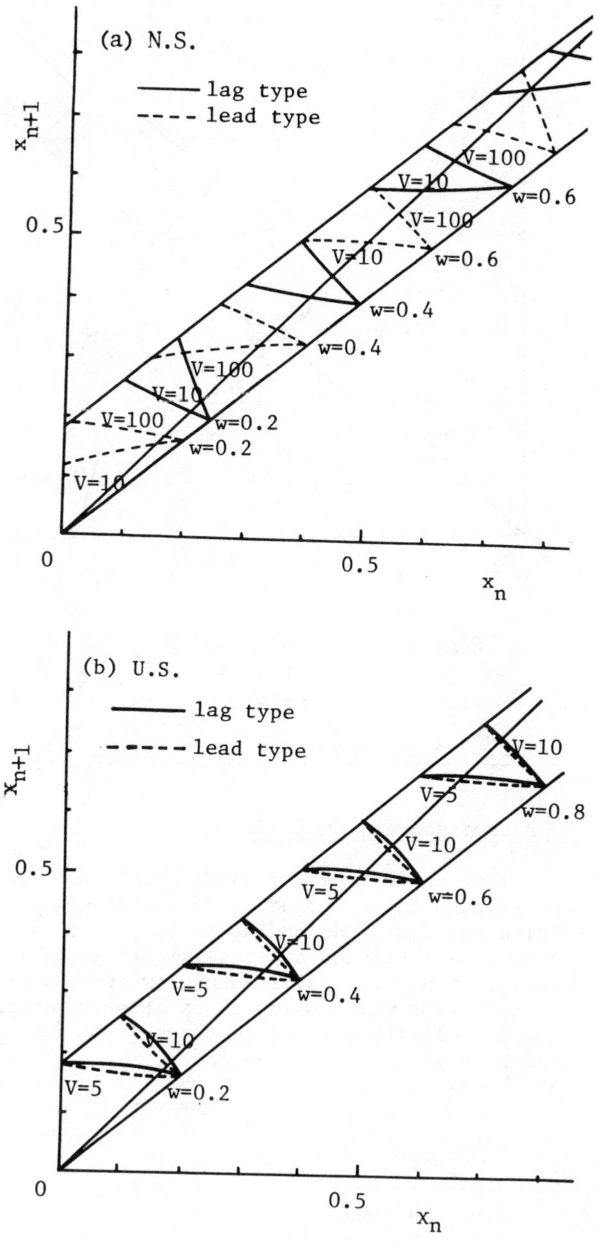

Fig. 6 (a),(b) Graphical mapping $x_n \sim x_{n+1}$ $\tau o=0.2$

Dead-Beat Gain

When the root of characteristic equation $\rho=1$, the system shows dead-beat response which is a special feature of the discrete model. The dead-beat gain V_d is shown in table 2.

LARGE SIGNAL PERTURBATION

The application of the linearized perturbational discrete models is restricted to the infinitesimal small signal. But as shown at numerical example, even for $\Delta w=0.1$ the impulse model considering the gain correction factor gives an exact result. For more large signal perturbation, the response described by the difference equation should be solved by the graphical mapping analysis (5). It is not so difficult to construct the mapping x_n-x_{n+1} . Several numerical example of mapping are shown in Fig. 6 (a) for N.S., (b) for U.S.

It should be noted that the nonlinearity of the mapping considered is not so serious. But for the large signal perturbation, the operating point moves to the saturated region where $\alpha=0$ and $\alpha=1$. At such a condition the graphical analysis is useful. The bending of two curves is opposite for lead type and lag type. This fact suggests that at the symmetrical control (lead-lag type) the degree of the linearity of the curves will be improved. Several examples are shown in Fig. 7.

The details of this system will be presented at next chance.

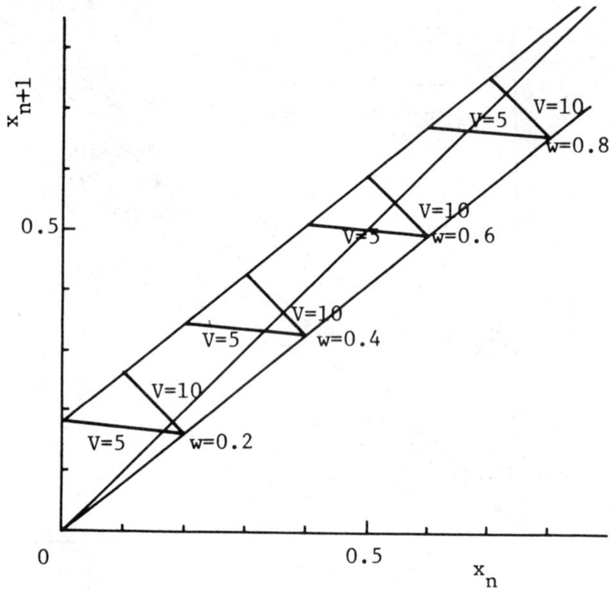

Fig. 7 Graphical mapping $x_n \sim x_{n+1}$ lead-lag type $\tau_o = 0.2$

QUICK-RESPONSE AND SMALL STEADY-STATE ERROR

Approximated Dead-Beat Gain

The rigorous dead-beat gain V_d obtained from the impulse model includes the steady-state firing angle α_o, depending on the operating points. As mentioned at large signal perturbation, if a curve (5) is approximated by a straight line which coincidents at both points $\alpha=0$ and $\alpha=1$ with a rigorous curve, then the dead-beat gain is given by

$$V_d = \frac{e^{-\tau_o}}{1-e^{-\tau_o}} \qquad \text{for U.S.}$$

$$V_d = \frac{e^{-\tau_o}}{(1-e^{-\tau_o})(1-w)} \qquad \text{for N.S.}$$

For U.S. the dead-beat gain is independent of the reference input w, whereas for N.S. the dead-beat gain is a function of w.
In analogue device, such a gain is easily realized by the method as shown in Fig. 8.
In Fig. 8, the potentiometer P_2 is coupled with the potentiometer of reference input w.

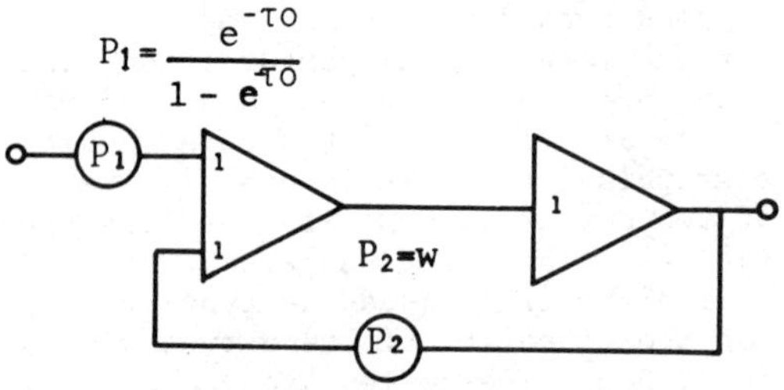

Fig. 8 Analogue method for dead-beat gain V_d

Bias Function

If the loop gain is selected as "dead-beat", for some τ_o, the value of loop gain is not so large. One object to increase the loop gain is to reduce the steady-state error. When the mean value of output is controlled, the steady-state error is given by

$$\varepsilon = w-(1-\alpha_o) \qquad (26)$$

Therefore, the bias function B_f eliminating steady-state error is introduced to the coincident circuit. At the steady-state firing instants, the following relation holds.

$$1-\alpha_o+B_f = V(w-x_{10}) \qquad \text{for U.S}$$

$$= V(w-x_{10}e^{-\tau_o\alpha_o}) \qquad \text{for N.S} \qquad (27)$$

The bias function is given as follows

$$B_f = V\left[w-\frac{1-e^{-\tau_o w}}{1-e^{-\tau_o}}\right]-w \qquad \text{for U.S}$$

$$= V\left[w-\frac{e^{-\tau_o(1-w)}-e^{-\tau_o}}{1-e^{-\tau_o}}\right]-w \qquad \text{for N.S} \qquad (28)$$

When we approximate $e^{-\tau_o} \simeq 1-\tau_o$, we can get $B_f=-w$ for both cases. But this approximation is too rough, so $e^{-\tau_o} \simeq 1-\tau_o+\tau_o^2/2$.
The approximate dead-beat gain is given as follows.

$$B_f = \frac{w(w-3)}{2} \qquad \text{for U.S.}$$

$$B_f = -(1/2)w \qquad \text{for N.S.} \qquad (29)$$

For N.S., $B_f=-(1/2)w$ is easily realized by analogue device. It is difficult to realize $B_f = -\frac{w(w-3)}{2}$ by means of analogue device.
So, $B_f = -w$ is used for N.S.
In order to confirm the effect of this approximate bias function combined with the dead-beat gain, we examined the steady-state error ε for the reference input w.
The results are shown in Fig. 9 (a) for U.S. and (b) for N.S.
The effect of this approximate bias function gives good result, i.e., the steady-state error is reduced about 10% . Especially for N.S., the steady-state error is negligible small in spite of its simple method.

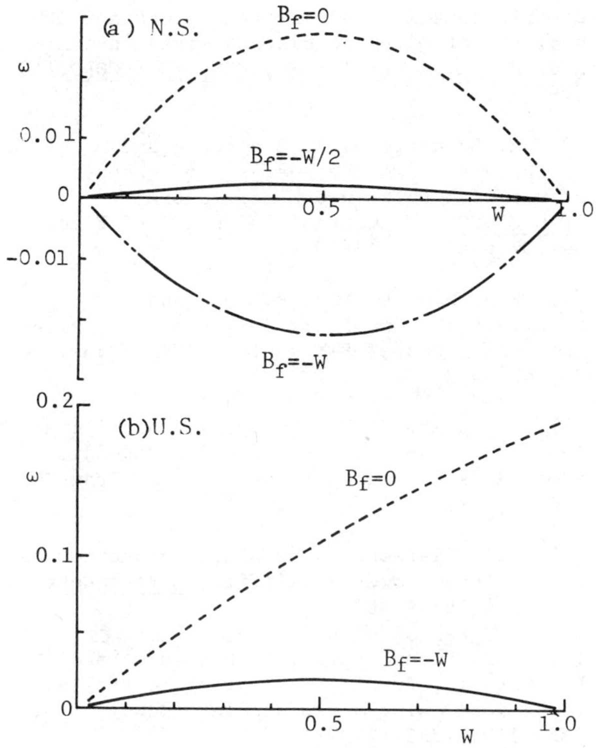

Fig. 9 (a),(b) Steady-state error ε
τo=0.2

Analogue Simulation

Using the dead-beat gain V_d along with the bias function B_f, a quick response PWM regulator with small steady-state error can be obtained. This fact is confirmed by an analogue simulation. The block diagram is shown in Fig. 10.
The results of the analogue simulation for the step change of reference input w=0.3 to w=0.4, are shown in Fig. 11 (a) for U.S. and (b) for N.S. In the case of N.S., the system nearly gives the dead-beat response, and the steady-state error is negligible small. In the case of U.S., a relatively small amount of the steady-state error remains because the bias function B_f=-w is some what small in comparison with the rigorous bias function. And also, the results obtained from the analogue simulation is well agreed with that of the digital simulation, which is not shown here.

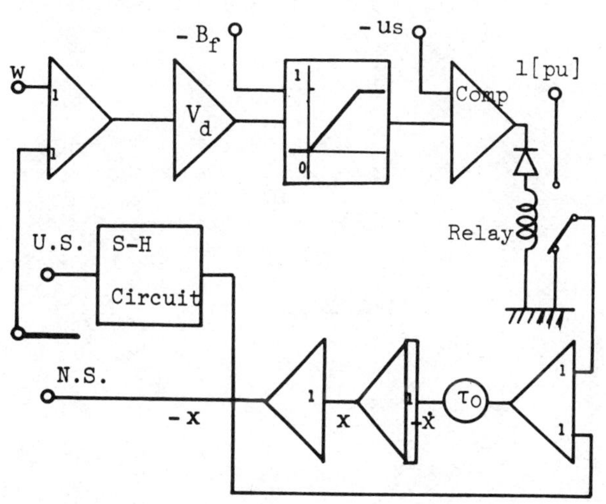

Fig. 10 Block diagram of analogue simulation

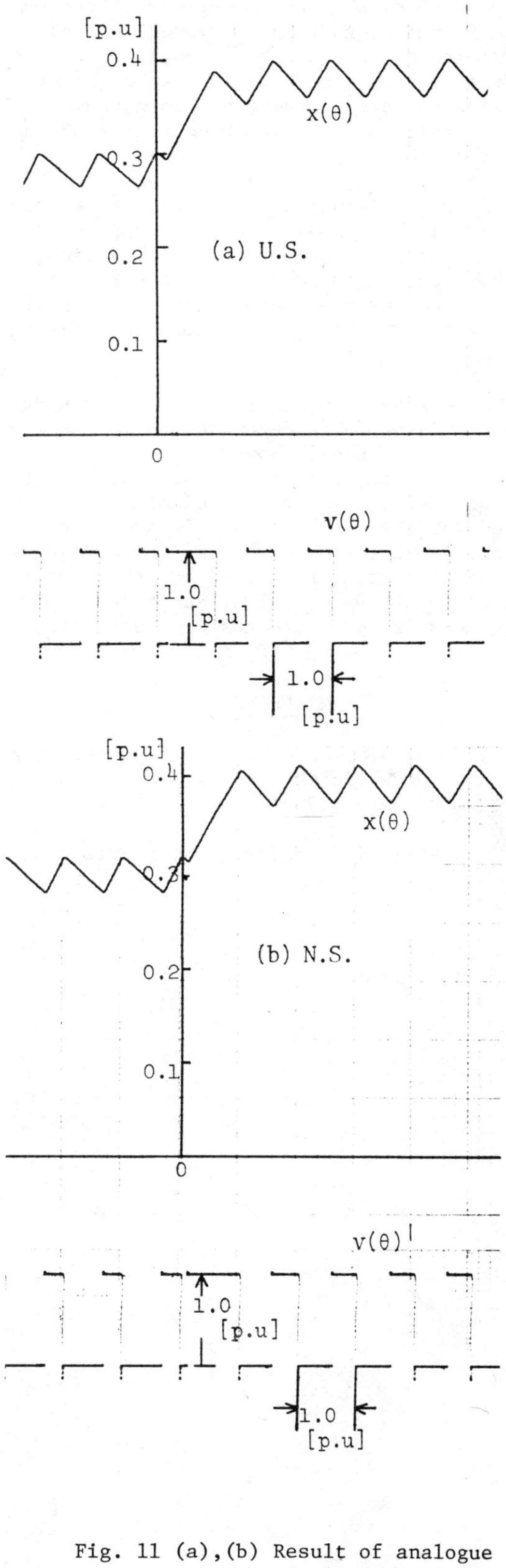

Fig. 11 (a),(b) Result of analogue simulation
τo=0.2

CONCLUSION

The root of characteristic equation obtained by the impulse model considering the gain correction factor, is completely agreed with the root obtained by the linearized difference equation. When the change of reference input is small, that is, for small signal perturbation, the result of impulse model by the modified z-transformation method is well agreed with that of the rigorous mathematical model, including the ripple component of output waveform.

As a result of graphical mapping, it is comfirmed that the impulse model is applicable to considerably large signal perturbation, if the operating points are on the linear region. And for the improvement of linearity in graphical mapping, the symmetrical firing method is suggested.

The approximate dead-beat gain to get quick response is proposed depending on the analysis of the graphical mapping.
For N.S. system, this gain should be changed with the reference input. A simple analogue method for the variable gain is proposed.
At proposed dead-beat gain, by introducing the bias function, the steady-state error is reduced about 10%. Total effect by both methods is confirmed by the analogue simulation.

REFERENCE

(1) W. Leonhald, Regelkreis mit gesteuertem Stromrichter als nichtlineares Abtastproblem, E.T.Z. A86, 513 (1965) .

(2) A. Capel, J.G. Ferrant and R. Prajoux, Dynamic behaviour and z-transform stability analysis of DC/DC regulators with a non linear P.W.M. control loop, PESC 73 RECORD.149 (1973) .

(3) A. Balestrino, A. Eisinberg and L. Sciavicco, On the analysis of chopper drives with pulse width modulators, IFAC SYMPOSIUM. Control in power and Electrical Drives. 1 , 17(1974) .

(4) B.J. Min, R. Unnikrishnan and C. Slivinsky, Stability analysis and simulation of a chopper PWM system, IEEE Trans. AC-20, 420-422, (1975) .

(5) E. Polak, Stability and graphical analysis of first order Pulse-Width-Modulated sampled-data regulator systems, IRE Trans. A.C 6 , 276-282 (1961) .

(6) N.A. Bjaresten, The static converter as a high speed power ampilifier, Direct Current 8, 154-165 (1963) .

(7) J.P. Sucena-Pavia, R. Hernandez and L.L. Freris, Stability study of controlled rectifiers using a new discrete model, Pro. IEE. 119, 1285-1293 (1972) .

(8) A.M. reider, Analysis of stability of the control systems of the Kashira-Moscow direct transmission line, Direct Current. 3, 227-240 (1957) .

(9) E.A. Parrish and E.S. Mcvey, A theoretical model for single-phase silicon-controlled rectifier systems, IEE Trans. AC-12 , 577-579 (1967) .

(10) R. Hernandez, J.P. Sucena-Pavia and L.L. Freris, Modeling of controlled rectifiers in feedback systems, IEEE. T73, 487-6 , 1-9 (1973) .

DISCRETE TIME DOMAIN MODELLING AND ANALYSIS OF DC-DC CONVERTERS WITH CONTINUOUS AND DISCONTINUOUS INDUCTOR CURRENT

R.P. Iwens*, F.C. Lee* and J.E. Triner**

**TRW Defense and Space Systems, Redondo Beach, California*
***NASA Lewis Research Center, Cleveland, Ohio*

ABSTRACT

Using discrete time state variable representation, a generalized computer-aided modelling and analysis of DC-DC converters is presented. The methodology provides exact modelling and is applicable to all types of power stages and duty-cycle control, including continuous and discontinuous inductor current operation. Converter stability, transient behavior and audio susceptibility can be analytically evaluated and predicted. The generalized theory of the proposed approach to converter modelling and analysis is presented first, followed by a demonstrative example applying the theory to a constant frequency buck converter operating in continuous and discontinuous inductor current mode. Excellent agreement with laboratory test data has been observed.

I. INTRODUCTION

Switched DC to DC power converters employing pulse modulation play an important part in industry. In order to eliminate time consuming trial and error design techniques and optimize converter performance, a thorough understanding of the converter as a system is required. One should be able to analytically predict such important behavior as stability, transient response, and susceptibility to audio frequency components in the primary DC supply. Duty-cycle control of the power switch for a converter can be achieved in a variety of ways: fixed on-time, variable off-time; fixed off-time, variable on-time; fixed frequency; or hysteresis control, to name just a few. These pulse modulation processes employed in DC-DC converters present considerable difficulties in analyzing these regulators by conventional frequency domain analysis techniques. Furthermore, in many designs, the inductor current is allowed to vanish and to dwell at zero for a portion of a cycle. In this situation the discontinuous inductor current complicates the analysis of the converter even further. Numerous analyses of DC-DC converters have been presented [1-4], however, they are all limited to only certain types of power stages using a given duty-cycle controller, and few can handle the discontinuous inductor current operation.

Recently Iwens et al.[5] presented a time domain model and analysis approach for a constant on-time buck converter. An equivalent, nonlinear discrete time system was derived that described the pulse modulation process and the converter behavior exactly. After linearization about the discrete time equilibrium solution, stability was established as a function of any arbitrary converter parameter with the information being graphically displayed in terms of the location of the system roots. Besides obtaining converter stability criteria, this analysis technique also provided information on transient behavior and susceptibility to audio frequency ripple in the primary DC supply. An important feature of this approach was that it made extensive use of a digital computer as an analysis tool, replacing many difficult and tedious analytical computations by numerical solutions.

In this paper the approach presented in Reference [5] is generalized to be applicable to all types of power stages and almost all types of duty-cycle control. The approach is extended to also include discontinuous inductor current operation. In Sections II, III, and IV the concept

---

†This work was performed under NASA Contract NAS3-18918, "Development of a Standardized Control Module, by TRW Systems, Redondo Beach, California, for NASA Lewis Research Center, Cleveland, OH.

of the approach is presented in simple but very general terms thereby demonstrating its applicability to a wide-variety of power stages and duty-cycle control methods. Even though the concepts employed are simple, the required computations are not suitable to hand analysis and a computer must be used as an analysis tool, e.g., state transition matrices and Jacobian matrices are evaluated numerically. This is considered a significant advantage of the method, as well as a contribution to converter modelling and analysis in general, for the following reasons:

(1) The use of the computer as an analysis tool makes automation in power converter design and analysis possible.

(2) Taking advantage of the tremendous and readily available computing power of present day digital computers allows the use of hitherto not seriously looked at analysis methods which produce exact results and require no dubious and restrictive approximations in order to make a problem analytically tractable.

The application of the approach to the stability analysis of a constant frequency controlled buck converter with continuous and discontinuous inductor current is demonstrated in Section V, providing a concrete example of how the presented methodology is used to solve a particular problem.

II. NONLINEAR DISCRETE TIME SYSTEM

It is important to recognize that for exact modelling of the pulse modulation process, switched DC-DC power converters should be represented as discrete time systems described by state vector difference equations. The class of switched converters primarily considered here can be illustrated by the three basic power stage configurations shown in Fig. 1. The approach to converter modelling and analysis presented in this paper is equally applicable to any of the three. An example for the buck converter presented later will illustrate a specific application of the more general theory that is developed first.

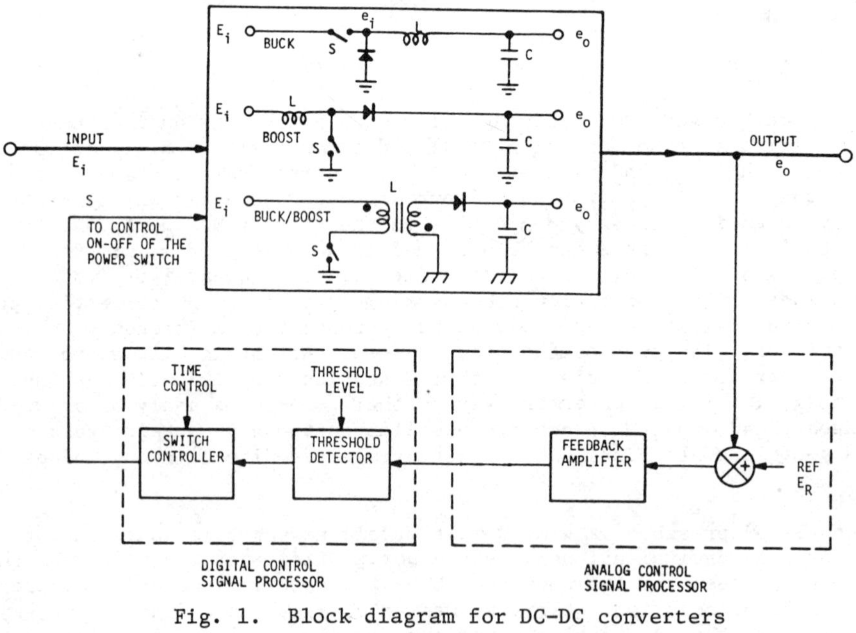

Fig. 1. Block diagram for DC-DC converters

The power converters studied all operate on the principle of controlling the duty-cycle with which a primary power source is switched to a load in a manner so that a nearly constant load voltage is maintained. Thus, for some period of time the power switch is on, and for some period it is off, repeating this duty-cycle over and over. However, unless the converter is in its steady-state and in a disturbance free environment, the on and/or off times are not necessarily constant and the analytic models presented here do not make this assumption. In fact, it is of paramount importance for successful stability and transient analysis to properly model the mechanisms producing these changes in the on and/or off times. But the circuit topological conditions, e.g.,

- power switch on, power switch off
- diode conducting, diode not conducting, etc.

repeat from cycle to cycle, and the converter need be described analytically for the conditions over one switching cycle only.

This is done by describing it by two or more linear time-invariant systems, each valid over a certain, well defined period of a cycle. A state vector $\bar{x} = (x_1, x_2, \ldots, x_n)$ is defined which typically is comprised of the converter output voltage, inductor current, and other voltages or currents describing the state of energy storing devices in the converter. Figure 2 shows the input voltage e_i (after the power switch) of an example converter and it serves here merely the purpose of establishing notation regarding time instances.

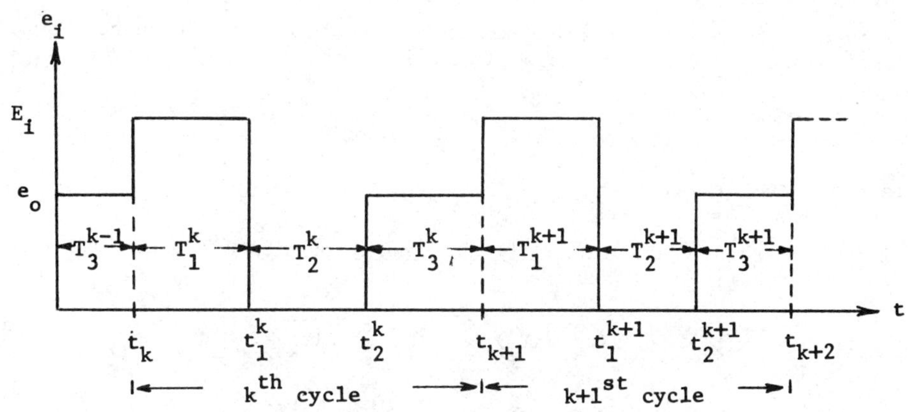

Fig. 2. Input voltage e_i as a function of time

This converter has each cycle divided into three different operational conditions, and the different time intervals could in this example describe the following circuit conditions:

$t_k \leq t < t_1^k$	power switch on, continuous current
$t_1^k \leq t < t_2^k$	power switch off, continuous current
$t_2^k \leq t < t_{k+1}$	power switch off, zero inductor current

Note that all time periods T_i in Fig. 2 are also indexed by k in the form of a superscript, denoting that they may vary from cycle to cycle, as during a transient condition, for instance. This power converter would then be described by three linear differential equations given in state vector notation as

$$\dot{\bar{x}} = F_1 \bar{x} + G_1 \bar{u}_1, \qquad t_k \leq t < t_1^k \tag{1}$$

$$\dot{\bar{x}} = F_2 \bar{x} + G_2 \bar{u}_2, \qquad t_1^k \leq t < t_2^k \tag{2}$$

$$\dot{\bar{x}} = F_3 \bar{x} + G_3 \bar{u}_3, \qquad t_2^k \leq t < t_{k+1} \tag{3}$$

F_i and G_i, i = 1, 2, 3, are constant matrices of appropriate dimensions commensurate with the order of the state vector, and $\bar{u}_i$, i = 1, 2, 3, denote input vectors comprised of piecewise constant input voltages, typically the supply voltage E_i and the reference voltage E_R to which the output is to be regulated. Note that the three systems need not be identical, since the matrices are free to differ from each other. This is particularly important for discontinuous current operation and buck-boost converter configurations. In addition, threshold conditions must be specified which define when the time instances t_1^k and t_2^k occur. For instance, suppose the state x_3 of the converter represents a voltage which causes the power switch to be turned off when it reaches a certain value E_T. Thus the threshold condition on t_1^k may be described by

$$x_3(t_1^k) = E_T = \text{threshold voltage} \tag{4}$$

Similarly, suppose the state x_2 represents the power stage inductor current. The threshold condition for t_2^k may then be described by

$$x_2(t_2^k) = 0 \tag{5}$$

since no negative current flow is allowed. The conditions on the time instances t_k and t_{k+1} depend on the type of duty-cycle control employed, i.e., constant frequency, constant off-time, etc. For constant off-time control one would have

$$t_{k+1} = t_1^k + T_{off} \tag{6}$$

where T_{off} is given and fixed.

Each of the linear systems of Equations (1) through (3) admits a closed form solution of the form

$$\bar{x}(t_1^k) = \bar{x}(t_k + T_1^k) = \Phi_1(T_1^k)\,\bar{x}(t_k) + D_1(T_1^k)\,\bar{u}_1 \tag{7}$$

$$\bar{x}(t_2^k) = \bar{x}(t_1^k + T_2^k) = \Phi_2(T_2^k)\,\bar{x}(t_1^k) + D_2(T_2^k)\,\bar{u}_2 \tag{8}$$

$$\bar{x}(t_{k+1}) = \bar{x}(t_2^k + T_3^k) = \Phi_3(T_3^k)\,\bar{x}(t_2^k) + D_3(T_3^k)\,\bar{u}_3 \tag{9}$$

where Φ_i and D_i are matrices, $\bar{x}$ is the n-dimensional state vector, and $\bar{u}_i$ are m-dimensional input vectors. In terms of their elements the matrices are defined by

$$\Phi_i = \{\phi_{jk}^i\} \qquad j,k = 1,2,\ldots,n$$

$$D_i = \{d_{jk}^i\} \qquad j = 1,2,\ldots,n \qquad k = 1,2,\ldots,m$$

and the vectors by

$$\bar{x} = (x_1,\, x_2,\, \ldots,\, x_n)^T$$

$$\bar{u}_i = (u_1^i,\, u_2^i,\, \ldots,\, u_m^i)^T$$

The Φ_i are state transition matrices and are given by

$$\Phi_i(T) = e^{F_i T}, \qquad i = 1,2,3 \tag{10}$$

and the input matrices D_i by

$$D_i(t) = e^{F_i T}\left[\int_o^T e^{-F_i s}\,ds\right] G_i\,, \qquad i = 1,2,3 \tag{11}$$

For simple, low order systems the matrices $\Phi_i(T)$ and $D_i(T)$ can often be analytically evaluated as algebraic functions of T, as was the case in the work reported in Reference [5]. But for general analysis applicable to a variety of converter configurations (some of possibly high order), the matrices $\Phi_i(T)$ and $D_i(T)$ will be evaluated numerically by the computer. The matrix exponential of Equation (10) is evaluated from its series representation, i.e.,

$$e^{FT} = I + FT + \frac{F^2T^2}{2!} + \frac{F^3T^3}{3!} + \dots \tag{12}$$

with the number of terms to be utilized being determined by an error criterion. Computation of $D_i(T)$ is then straightforward using trapezoidal or Runge-Kutta integration over the interval [0,T]. The computation of $\Phi_i(T)$ and $D_i(T)$ is programmed as a FUNCTION SUBPROGRAM with T as a formal parameter so that $\Phi_i(T)$ and $D_i(T)$ can be evaluated for any specified T.

The threshold conditions are used to define the time periods T_i^k, i = 1, 2, 3. For instance, the threshold condition of Equation (4), $x_3(t_1^k) = E_T$ becomes by Equation (7)

$$\begin{aligned}\phi_{31}^1(T_1^k)\, x_1(t_k) + \phi_{32}^1(T_1^k)\, x_2(t_k) + \dots + \phi_{3n}^1(T_1^k)\, x_n(t_k) \\ + d_{31}^1(T_1^k)\, u_1^1 + d_{32}^1(T_1^k)\, u_2^1 + \dots + d_{3m}^1(T_1^k)\, u_m^1 = E_T\end{aligned} \tag{13}$$

In effect then, all T_i^k, i = 1, 2, 3, can be represented as functions of the system state at t_k and the other known converter parameters and inputs, i.e., as algebraic functions of the form

$$\zeta_i(\bar{x}(t_k),\ T_1^k,\ T_2^k,\ T_3^k,\ \bar{u}_1,\ \bar{u}_2,\ \bar{u}_3,\ \bar{p}) = 0, \qquad i = 1,2,3 \tag{14}$$

where $\bar{p}$ is a vector of other converter parameters such as threshold voltages or certain specified constant time periods, such as constant on-time, constant off-time, or constant switching period. Note that the ith threshold condition is in general not restricted to depend only on the ith interval T_i and i^{th} input $\bar{u}_i$.

The nonlinear discrete time system that describes the converter behavior exactly, can now be obtained by combining the closed form solutions. From Equations (7)-(9) the result is the following vector difference equation

$$\bar{x}(t_{k+1}) = \Phi_3(T_3^k)\left\{\Phi_2(T_2^k)\left[\Phi_1(T_1^k)\,\bar{x}(t_k) + D_1(T_1^k)\,\bar{u}_1\right] + D_2(T_2^k)\,\bar{u}_2\right\} + D_3(T_3^k)\,\bar{u}_3 \tag{15}$$

which can be written in short form as

$$\bar{x}(t_{k+1}) = \bar{f}(\bar{x}(t_k),\ T_1^k,\ T_2^k,\ T_3^k,\ \bar{u}_1,\ \bar{u}_2,\ \bar{u}_3) \tag{16}$$

The time periods T_i^k are all functions of the current system state $\bar{x}(t_k)$ by virtue of the threshold conditions (14), and thus (14) and (16) together are a nonlinear discrete time system with the state vector x describing the exact behavior of the power converter at one instant of time for each cycle. Typically this instant represents either the peak or minimum value of the output voltage limit cycle. The nonlinear discrete time system (14)-(16) is of key importance for stability and transient analysis.

III. STABILITY AND TRANSIENT ANALYSIS

The steady-state equilibrium solution $\bar{x}^*$ of the nonlinear discrete time system (16) is defined by the conditions

$$\left.\begin{aligned} \bar{x}(t_{k+1}) &= \bar{x}(t_k) = \bar{x}^* = \text{constant} \\ T_1^{k+1} &= T_1^k = T_1^* = \text{constant} \\ T_2^{k+1} &= T_2^k = T_2^* = \text{constant} \\ T_3^{k+1} &= T_3^k = T_3^* = \text{constant} \end{aligned}\right\} \quad \text{for all } k = 0,1,2,\ldots \tag{17}$$

and it must be determined first. $\bar{x}^*$ represents the numerical value of the steady-state limit cycle solution of the converter at the time instances t_k, and usually coincides with the output voltage limit cycle minimum or maximum, depending on problem formulation. The steady-state solution is found by applying Newton's iteration method to Equations (14) and (16) until values for $\bar{x}^*$, T_1^*, T_2^* and T_3^* have been found such that Equation (17) is satisfied. The nonlinear system (16) is then linearized about its equilibrium state $\bar{x}^*$ and a linear, time-invariant, discrete time system is obtained which is characterized by the vector difference equation

$$\delta\bar{x}(t_{k+1}) = \Psi\, \delta\bar{x}(t_k) \tag{18}$$

where

$$\delta\bar{x} = \bar{x} - \bar{x}^* \tag{19}$$

and where Ψ is the Jacobian matrix of $\bar{f}$ of Equation (16) evaluated at $\bar{x}^*$, i.e.,

$$\Psi = \left.\frac{\partial \bar{f}}{\partial \bar{x}}\right|_{\bar{x}^*} \tag{20}$$

The input forcing vectors $\bar{u}_i$ are composed of nominally constant voltages (e.g., input voltage E_i, reference voltage E_r) so that the linearized system becomes autonomous (without forcing terms). Variations in the forcing vector $\bar{u}$ will be treated in the next section.

To evaluate the partial derivatives in (20) analytically is sometimes possible, but in general it is best to evaluate them numerically using difference quotients. A suitable approximation to the derivatives is

$$\left.\frac{\partial f_i}{\partial x_j}\right|_{x_j^*} = \frac{f_i(x_j^* + \Delta x_j) - f_i(x_j^* - \Delta x_j)}{2\,\Delta x_j}, \qquad i,j = 1,2,\ldots,n \tag{21}$$

which can also detect discontinuities. In order to evaluate (21) one must first determine by how much T_1, T_2, T_3 change due to a change Δx_j, j = 1, 2, ..., n, and then use the new T_1, T_2, T_3 to compute $f_i(x_j^* \pm \Delta x_j)$. The threshold conditions (14) are used to determine the change in T_1, T_2, T_3 due to a change in $\bar{x}$, and Newton's iteration method is applied to solve the threshold equations numerically. The only problem with numerical differentiation is to select the appropriate increments Δx_j. It is recommended to start with

$$\Delta x_j = 0.01\,|x_j^*|$$

but some experimentation with the increment size is advisable, since the accuracy of the partial derivatives depends on it. For instance, when the function $\bar{f}$ varies rapidly, a very small increment is clearly required. On the other hand, if the increment is chosen needlessly too small, then the accuracy degrades because of numerical problems, since in the limit a difference quotient assumes the value 0/0. Studies on the increment size and its effect on the results have also physically significant implications. If the linearized system shows high sensitivity to increment size, then this points out that the nonlinear system changes its behavior rather rapidly as it moves away from its equilibrium point, and the results obtained for the linearized system are only valid for relatively small perturbations about the equilibrium.

The linearized system (18) describes the regulator behavior about its steady-state solution and, as is well known from linear discrete time system theory, it is stable, if and only if, all the eigenvalues of the constant n x n matrix Ψ are absolutely less than unity, i.e.,

$$|\lambda_i(\Psi)| < 1, \qquad i = 1,2,\ldots,n \tag{22}$$

The eigenvalues of Ψ are evaluated by a digital computer and changes in the eigenvalues (which are the roots of the system) due to system parameter changes can be plotted in the complex plane yielding an excellent design tool, very similar to conventional root locus plots widely used in control system design. The locations of the roots do not only indicate stability or instability of the system, but also characterize its transient behavior, i.e., damping, overshoot and rapidity of response. It follows from (22) that for stability, all the roots must be inside the unit circle in the complex plane. The existing relationships between root locations inside the unit circle and corresponding system response times and damping are well known results from z-transform analysis of linear discrete time systems [6,7].

IV. SUSCEPTIBILITY TO AUDIO FREQUENCIES IN THE SUPPLY VOLTAGE

In the previous development it has been assumed that the primary DC input voltage E_i, which is a component of the input vectors $\bar{u}_i$, i = 1, 2, 3, is constant for all time. This is not necessarily so and the nonlinear discrete time system (14)-(16) is also an exact description of the converter if E_i is time-varying, but sufficiently slow so that E_i remains for all practical purposes constant over any T_{on} (i.e., T_1) period. To assume a time-varying E_i composed of the nominal DC value E_i^* plus a small superimposed AC component δE_i, i.e.,

$$\delta E_i(t_k) = E_i(t_k) - E_i^* \tag{23}$$

is a useful concept when investigating audio susceptibility of the converter. The nonlinear discrete time system (14)-(16) is then linearized about the previously defined steady-state equilibrium solution $\bar{x}^*$ and the nominal DC supply voltage E_i^*. Hence,

$$\delta\bar{x}(t_{k+1}) = \Psi\, \delta\bar{x}(t_k) + \Gamma\, \delta E_i(t_k) \tag{24}$$

where

$$\Gamma = \left.\frac{\partial \bar{f}}{\partial E_i}\right|_{E_i^*,\bar{x}^*} \tag{25}$$

with $\bar{f}$ as defined by (16). Analytical differentiation is here sometimes possible (e.g., for the constant T_{on} buck converter [5]) because E_i tends to enter the system equations linearly. In general, numerical differentiation, as described before, should be employed.

It is of interest now to examine how sinusoidal oscillations of the supply voltage E_i about its nominal DC value affect the regulated output voltage e_o in the steady-state. The z-transform method can be applied to derive a frequency domain transfer function since in the steady-state the cycle period T_p^* is constant, irrespective of the type of duty-cycle control employed. Furthermore, the amplitude of the supply voltage oscillations is constrained to be small and, therefore, the linearized system model in Equation (24) applies. Taking the z-transfrom of the vector difference Equation (24) and letting e_o be the first state in the state vector $\bar{x}$, i.e.,

$$e_o(t_k) = x_1(t_k) \tag{26}$$

one obtains

$$\delta E_o(z) = H(zI - \Psi)^{-1}\, \Gamma\, \delta E_i(z) \tag{27}$$

where H is an n-dimensional row vector given by

$$H = [1, 0, \ldots, 0] \tag{28}$$

After setting $z = e^{j\omega T_p^*}$, the frequency domain transfer function $G = \delta E_o/\delta E_i$ is given by

$$G(j\omega) = H\left(Ie^{j\omega T_p^*} - \Psi\right)^{-1}\Gamma\,, \qquad 0 \le \omega T_p^* \le \pi \tag{29}$$

which by virtue of the sampling theorem applies up to one-half the sampling frequency, i.e., up to $\omega T_p^* = \pi$. The transfer function $G(j\omega)$ of (29) can now be readily evaluated at any desired ω by a digital computer and the results plotted as a frequency response plot, yielding a measure of the audio susceptibility of the converter.

V. EXAMPLE

The series switched buck regulator shown in Fig. 3, operating in a constant frequency control mode, has been selected as the example system to demonstrate the application of the presented analysis approach to a particular problem. The critical element of the regulator is the pulse modulator that controls the power switch Q (actually a transistor) by periodically opening and closing it in such a manner that the outout voltage e_o is maintained very close to some specified reference voltage E_R. By comparing the output voltage e_o with the voltage E_R an error signal is formed which is then integrated together with a current proportional to the derivative of e_o and an AC signal obtained from a secondary winding of the power stage inductor. The power switch is turned on by a constant frequency clock pulse, and turned off whenever the output e_c of the integrator falls below a specified threshold voltage E_T. The necessary signals are generated by the Digital Control Signal Processor (DCSP).

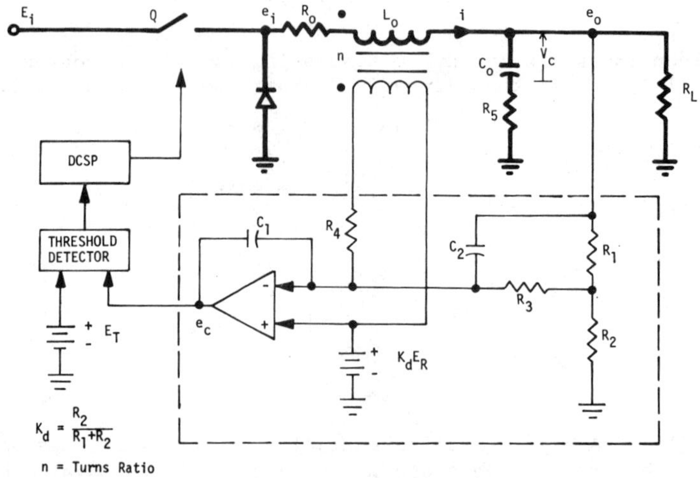

Fig. 3. Series switched buck regulator

The waveform of e_i, as shown in Fig. 4, is used to establish notation regarding the time instances when each cycle starts and each switching action occurs.

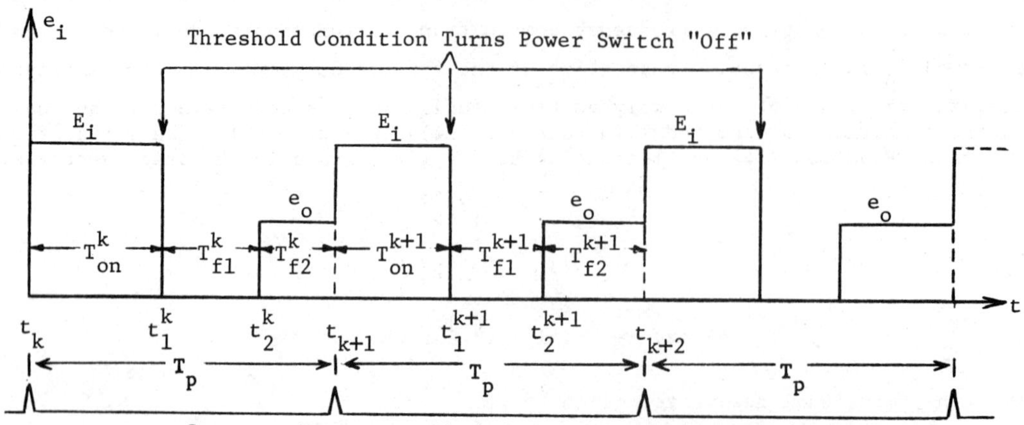

Fig. 4. Waveform of e_i for discontinuous inductor-current operation

In normal operation,

at t_k, t_{k+1}, t_{k+2} . . . . . the clock pulse turns the power switch "ON"

at t_1^k, t_1^{k+1}, t_1^{k+2} . . . . . the threshold condition turns the power switch "OFF"

at t_2^k, t_2^{k+1}, t_2^{k+2} . . . . . the zero inductor current condition turns the power diode off, if this condition exists

The time intervals $t_1^k - t_k$, $t_2^k - t_1^k$, and $t_{k+1} - t_2^k$ are defined as T_{on}^k, T_{f1}^k and T_{f2}^k, respectively, as shown in Fig. 4 and they correspond to the intervals T_1, T_2, T_3 defined earlier in Section II. These time intervals may vary from cycle to cycle. However, the time interval between t_k and t_{k+1} is, for this example, a constant equal to the period of the switching frequency, i.e.,

$$t_{k+1} - t_k = T_p \qquad \text{for all } k \tag{30}$$

Note that $T_{f2} = 0$ for continuous inductor current operation which, of course, is a permissible mode of operation for the regulator.

Referring to Fig. 3, the state vector $\bar{x}$ for the regulator is defined as

$$\bar{x} = [e_o, i, e_c]^T \triangleq [x_1, x_2, x_3]^T \tag{31}$$

and the input vector $\bar{u}$ as

$$\bar{u} = [E_i, E_R]^T \triangleq [u_1, u_2]^T \tag{32}$$

The regulator may now be described over its three different operating intervals by three differential equations in state vector form, as described previously in Equations (1)-(3). The input vector $\bar{u}$ remains here the same for all three different operating conditions, i.e., $\bar{u} = \bar{u}_1 = \bar{u}_2 = \bar{u}_3$, and the matrices F_1, F_2, F_3 and G_1, G_2, G_3 are given in terms of circuit parameters in the Appendix. The time intervals of the different operating conditions of the regulator are defined, for all k, by the modulator threshold condition

$$x_3(t_1^k) = E_T \tag{33}$$

by the zero inductor current condition (if it exists)

$$x_2(t_2^k) = 0 \tag{34}$$

and by the constant frequency condition

$$T_{on}^k + T_{f1}^k + T_{f2}^k = T_p \tag{35}$$

Equations (31)-(35) plus the matrices F_i, G_i, i = 1, 2, 3, provide all the necessary information to proceed with the analysis as outlined in Section II, and the nonlinear discrete time system for the constant frequency buck converter can be given in the form of Equations (15) or (16). The threshold conditions ζ_i, i = 1, 2, 3, of Equation (14) are derived from (33)-(35) and are here represented by the following constraint equations:

$$\phi_{31}^1(T_{on}^k)\, x_1(t_k) + \phi_{32}^1(T_{on}^k)\, x_2(t_k) + \phi_{33}^1(T_{on}^k)\, x_3(t_k) + d_{31}^1(T_{on}^k)\, E_i + d_{32}^1(T_{on}^k)\, E_R = E_T \tag{36}$$

which follows from (7) and (33);

$$\phi_{21}^2(T_{f1}^k)\, x_1(t_1^k) + \phi_{22}^2(T_{f1}^k)\, x_2(t_1^k) + \phi_{23}^2(T_{f1}^k)\, x_3(t_1^k) = 0 \tag{37}$$

which follows from (8) and (34) and the fact that $d_{21}^2 = d_{22}^2 = 0^{\dagger}$; and,

$$T_{f2}^k = T_p - T_{on}^k - T_{f1}^k \tag{38}$$

which follows directly from (35).

The steady-state equilibrium solution, as defined in (17), must be determined next. One starts with approximate expressions for T_{on}^*, T_{f1}^*, and T_{f2}^* (derived from flux conservation and duty-cycle/power considerations) which are given by

$$T_{on}^* \approx \sqrt{\frac{2L_o T_p P_o}{E_i (E_i - E_R)}} \tag{39}$$

$$T_{f1}^* \approx \sqrt{\frac{2L_o T_p P_o (E_i - E_R)}{E_i E_R^2}} \tag{40}$$

$$T_{f2}^* = T_p - T_{on}^* - T_{f1}^* \tag{41}$$

where P_o is the output power and all other symbols were previousl d fined in Fig. 3 and 4. These approximate steady-state time intervals may now be used, together with the system and threshold constraint equations, to compute an approximate equilibrium state $\bar{x}^*$.

The exact steady-state equilibrium solution must satisfy the state matching conditions, i.e., when using the nonlinear system Equations (14)-(15) to propagate $\bar{x}^*$ through one complete period T_p, then the state at t_k at the beginning of the period must match the obtained state at t_{k+1} at the end of the period. Additionally, one must satisfy the zero inductor current condition at t_2^k, if this is applicable to the mode of operation of the regulator. The state matching conditions are easy to setup. It suffices in this example to setup a state matching condition only on the state x_3 and the zero inductor current condition on the state x_2. Starting with the approximate steady-state values, iterative linearization, i.e., Newton's method [8], is then applied to the zero current and state matching conditions to determine the exact steady-state equilibrium solution $\bar{x}^*$, T_{on}^*, T_{f1}^*. The entire procedure is performed numerically by the computer and convergence from the approximate steady-state to the exact steady-state occurs usually within 1-3 iterations. The computer analysis program proceeds then with the linearization of the system, finds the system roots to determine stability and transient behavior, and evaluates the steady-state frequency response function to assess audio susceptibility of the converter.

For the present regulator the analysis program consisted of about 575 FORTRAN statements. The program has been developed in such a manner that for a given set of circuit parameters the main program first detects the mode of operation, namely the continuous current mode versus the discontinuous current mode, and then enters into the appropriate subroutine. The initial test is performed by evaluating the approximate value for T_{f2}^* using Equations (39)-(41), and if T_{f2}^* is very near or equal to zero, the continuous inductor current mode is assumed. The subsequent exact evaluation of the steady-state will then reveal whether this was the right assumption. The nature of stability of the regulator when the inductor current is at the verge from continuous current operation to discontinuous current operation is of particular interest in this phase of research and it is studied through such a program.

Stability Results

Nominal circuit parameter values for the regulator of Fig. 3 are given in Table 1. These parameters correspond to a stable, continuous current mode operation.

---

† By inspection of the matrix G_2, see Appendix, and Equation (11).

TABLE 1 **Nominal Circuit Parameter Values**

Symbol	Parameter	Units	Value
E_i	Supply Voltage	volts	50
E_R	Reference (Desired Output) Voltage	volts	20
E_T	Integrator Threshold	volts	8
R_o	Inductor Series Resistance	ohms	0.015
R_1	Part of Output Voltage Divider	ohms	28.7K
R_2	Part of Output Voltage Divider	ohms	13.5K
R_3	Op-amp DC Input Resistor	ohms	10K
R_4	Op-amp AC Input Resistor	ohms	100K
R_5	Series-Equivalent Resistance of C_o	ohms	0.077
R_L	Load	ohms	10.
C_o	Output Filter Capacitor	μF	300
C_1	Op-amp Feedback Capacitor	pF	2200
C_2	Lead Compensation Capacitor	μF	0.022
L_o	Output Filter Inductor/Transformer	μH	250
n	Transformer Turns Ratio n_2/n_1	---	0.65
T_p	Period = $T_{on} + T_{off}$	μs	30

As has often been observed in laboratory tests and other analyses to be described in [9], in the continuous current mode of operation this constant frequency converter is only stable for duty-cycles below 50 percent. This result has been reaffirmed here. The input voltage E_i was set to E_i = 40 volts corresponding to a 50 percent duty-cycle (since e_o is to be regulated to 20 volts) so that the converter is unstable. The roots of the linearized system, as obtained from the computer program, are plotted in Fig. 5 as a function of the AC loop gain resistor, R_4. Note that one of the three roots is outside the unit circle, indicating instability, and that it asymptotically approaches the unit circle at -1 as R_4 is reduced, while the two roots inside the unit circle asymptotically approach +1. Very similar pictures result when other parameters are changed, e.g., when the DC loop gain is changed by varying R_3 from 1 kΩ to 200 kΩ. The converter cannot be stabilized but merely brought to the verge of stability.

The behavior changes drastically, however, and the converter becomes stable when it is allowed to operate in the discontinuous inductor current mode. This has also been observed previously in laboratory tests and is reaffirmed here as illustrated in Fig. 6, which shows the motion of the roots as a function of the input voltage E_i, but with the power stage inductor at a new value of L_o = 50 μH. When the magnitude of the input voltage is between 25 volts and 31 volts, the converter operates in the continuous current mode with a duty-cycle greater than 50 percent and it is unstable. When the input voltage is 32 volts or greater, the converter enters the discontinuous current mode of operation and all three eigenvalues are inside the unit circle and the system is stable regardless of the duty-cycle ratio. It should be noted that during the input voltage variation, a jump phenomenon in the system eigenvalues is observed. As the input voltage varies from 31 to 32 volts, the eigenvalue λ_3 changes from -2.68 to 0 and the sudden change is definitely caused by a jump rather than by continuous motion.

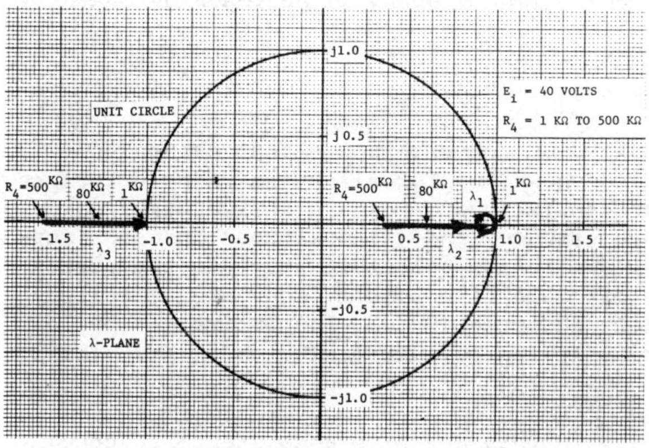

Fig. 5. Eigenvalues as a function of AC loop resistance R_4

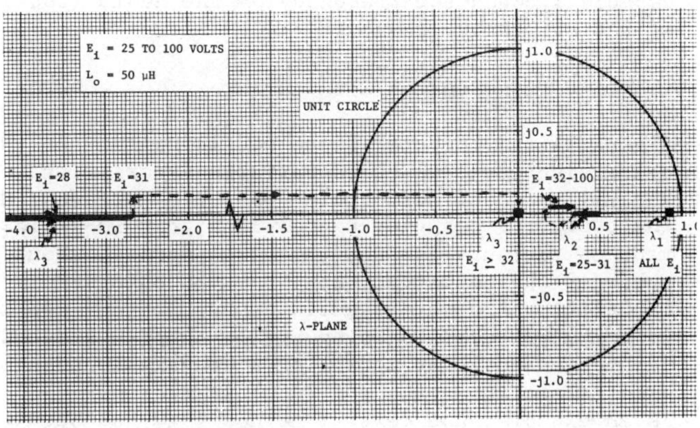

Fig. 6. Eigenvalues as a function of input voltage E_i

This observation is further supported by Fig. 7 and also by physical considerations: when changing modes, the converter equations are changed entirely from one set of equations to another, indicating a discontinuity in the physical behavior of the system.

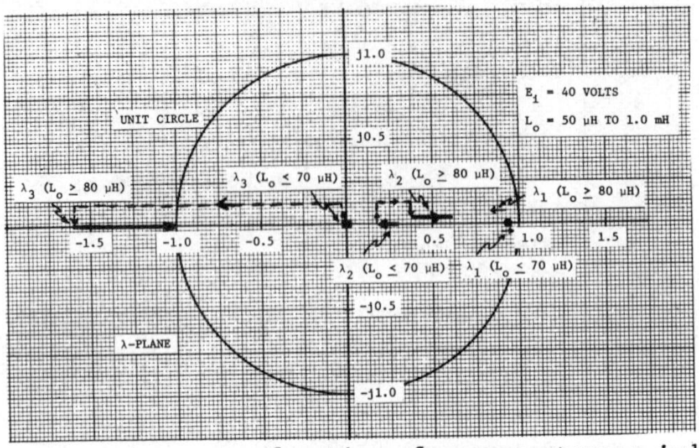

Fig. 7. Eigenvalues as a function of energy-storage inductance

Fig. 7 shows the eignevalues as a function of the energy storage inductor L_o, when the magnitude of the inductor L_o is varied from 50 μH to 1 mH with the input voltage fixed at E_i = 40 volts. The eigenvalues initially are located inside the unit circle with one eigenvalue $\lambda_3 = 0$. As the inductor L_o increases up to 80 μH, the eigenvalue λ_3 jumps outside the unit circle from the origin and the converter begins to operate in the continuous-current mode rather than the discontinuous mode. As expected, the system is unstable because the duty-cycle is 50 percent. As L_o is increased further, the converter remains in the continuous current mode of operation and, in accordance with the 50 percent duty-cycle rule, approaches the verge of stability. Thus, the behavior of the circuit as predicted by the analysis directly matches and explains laboratory observations.

Audio Susceptibility Results

A change from the continuous inductor current mode to the discontinuous inductor current mode is in practice usually brought about by a change in the load, i.e., when going from a heavy load (low resistance) to a light load (high resistance). The jump phenomenon in the roots of the linearized system could again be observed, similar to that shown in Fig. 5 through 7. The effect of a mode change on the audio susceptibility of the converter when the load resistance changes, will now be addressed. The closed loop frequency response transfer function of the converter was defined in Equation (29) as $\delta E_o/\delta E_i = G(j\omega)$. This transfer function is determined and evaluated by the computer for three different load conditions and plotted in Fig. 8. Curve (a) of Fig. 8 shows the audio susceptibility of the converter when operated in a continuous inductor current mode. While in this continuous current mode, the audio susceptibility is almost independent of the load magnitude. Curves (b) and (c), corresponding to discontinous current operations, show that the audio susceptibility does vary with load. Comparing these three curves also shows that there is a significant difference in the magnitude of audio susceptibility between continuous current mode operation and discontinuous current mode operation, especially when the frequency of the audio ripple in the primary supply is approaching one-half of the switching frequency.

VI. CONCLUSIONS

A generalized computer-aided modelling and analysis of DC-DC converters has been presented using discrete-time state variable description. It was shown how for all types of power stages and all types of duty-cycle control a nonlinear discrete-time system can be derived that describes the converter behavior exactly and that is valid for continuous and discontinuous inductor current operation. This system is linearized about its equilibrium solution, which is the steady-state of the converter, and from the obtained linear discrete time system, converter stability, transient response and audio susceptibility can readily be established. The entire methodology is computerized, replacing highly complex and tedious analyses by numerical methods and making automated converter modelling and analysis possible.

As an example, the general methodology has been applied in this paper to a constant frequency buck converter. The conducted analyses showed that this particular converter is unstable when operated at continuous inductor current at a duty-cycle ratio of 50% and larger, but that it can be abruptly stabilized as soon as the converter enters the discontinuous inductor current mode of operation. The converter is stable in either mode when operated at a duty-cycle ratio below

50%. These results are in excellent agreement with laboratory test data of the converter. The success of the generalized computer-aided methodology should make it a very valuable tool for converter modelling and analysis in the future.

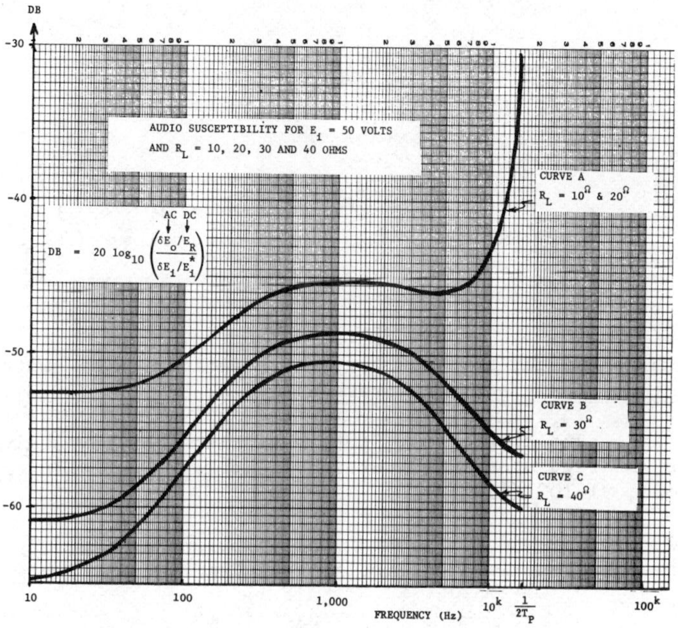

Fig. 8. Audio susceptibility for E_i = 50 volts and R_L = 10, 20, 30 and 40 ohms

APPENDIX. ENTRIES FOR MATRICES F_i AND G_i

The matrices F_i, i = 1, 2, 3, are 3 x 3 and the matrices G_i, i = 1, 2, 3, are 3 x 2. Moreover,

$$F_1 = F_2 \quad \text{and} \quad G_2 = G_3 \tag{A-1}$$

Recalling the notation $F_i = \{f^i_{jk}\}$, the entries of the respective matrices are as follows:

$$f^1_{11} = -\frac{1}{C_o(R_5+R_L)} - \frac{R_5R_L}{L_o(R_5+R_L)}, \quad f^1_{12} = \frac{R_L}{C_o(R_5+R_L)} - \frac{R_oR_5R_L}{L_o(R_5+R_L)}$$

$$f^1_{21} = -\frac{1}{L_o}, \quad f^1_{22} = -\frac{R_o}{L_o}$$

$$f^1_{31} = \frac{n}{R_4C_1} - \frac{K_d}{R_3C_1} + \frac{C_2}{C_1C_o(R_5+R_L)} + \frac{C_2R_5R_L}{C_1L_o(R_5+R_L)} \tag{A-2}$$

$$f^1_{32} = \frac{C_2R_oR_5R_L}{C_1L_o(R_5+R_L)} - \frac{R_LC_2}{C_1C_o(R_5+R_L)} + \frac{R_on}{R_4C_1}, \quad f^1_{13} = f^1_{23} = f^1_{33} = 0$$

$$f^3_{11} = -\frac{1}{C_o(R_5+R_L)}, \quad f^3_{31} = -\frac{K_d}{R_3C_1} + \frac{C_2}{C_1C_o(R_5+R_L)}$$

$$f^3_{12} = f^3_{13} = f^3_{21} = f^3_{22} = f^3_{23} = f^3_{32} = f^3_{33} = 0 \tag{A-3}$$

$$g^1_{11} = \frac{R_5 R_L}{L_o(R_5+R_L)}, \quad g^1_{21} = \frac{1}{L_o}, \quad g^1_{31} = -\frac{n}{R_4 C_1} - \frac{C_2 R_5 R_L}{C_1 L_o (R_5+R_L)}$$

$$g^1_{32} = \frac{K_d}{R_3 C_1}, \qquad g^1_{12} = g^1_{22} = 0 \tag{A-4}$$

$$g^2_{32} = \frac{K_d}{R_3 C_1}, \qquad g^3_{11} = g^3_{12} = g^3_{21} = g^3_{22} = g^3_{31} = 0 \tag{A-5}$$

REFERENCES

(1) G. W. Wester, R. D. Middlebrook, Low-frequency characterization of switched DC-DC converters, IEEE Trans. Aerospace and Electronic Systems, AES-9, 376-385 (1973).

(2) A. Capel, J. G. Ferrante, and R. Prajoux, State variable stability analysis of multi-loop PWM controlled DC/DC regulators in light and mode, IEEE Power Electronic Specialists Conference, Record, 91-103 (1975)

(3) I. M. H. Babaa, T. G. Wilson, and Y. Yu, Analytic solutions of limit cycles in a feedback regulated converter system with hysteresis, IEEE Trans. on Automatic Control, AC-13, 524-531 (1968).

(4) H. R. Weischedel, An exact method for the analysis of limit cycles in ON-OFF controlled systems, IEEE Trans. Automatic Control, AC-18, 40-44 (1975).

(5) R. P. Iwens, Y. Yu, and J. E. Triner, Time domain modelling and stability analysis of an integral pulse frequency modulated DC to DC power converter, IEEE Power Electronics Specialists Conference, Record, 80-90 (1975).

(6) Benjamin Kuo, Analysis and synthesis of sampled data control systems, Chapter 6, Prentice Hall, Englewood Cliffs, New Jersey (1963).

(7) E. I. Jury, Theory and application of the z-transform method, John Wiley and Sons, New York (1964).

(8) I. S. Sokolnikoff and R. M. Redheffer, Mathematics of physics and modern engineering, McGraw-Hill, 684-686, New York (1958).

(9) F. C. Lee, R. P. Iwens, Y. Yu, and J. E. Triner, Generalized computer-aided discrete time domain modelling and analysis of DC-DC converters, IEEE Power Electronics Specialists Conference, Record, (1977), to appear.

ANALYSIS OF SCR CIRCUITS VIA AUGMENTED STATE TRANSITION MATRIX

G. Ciccarella, A. de Carli and M. la Cava

Istituto di Elettrotecnica - Università de L'Aquila (Italy)
Dipartimento di Sistemi-Università della Calabria-Cosenza (Italy)

Summary. - A method for determining the value of the forced response, free motion, mean value and harmonic coefficients of a linear time-invariant dynamic system with canonical input is proposed. This method consists in the numerical computation of a particular augmented state system.

When the input variable is canonical and periodic, the initial state may also be calculated by elementary matrix algebra.

This method is particularly suited for analyzing electrical systems supplied by static convertors because during each conducting interval they are described by linear time-invariant differential equations with canonical inputs. An example is developed to illustrate the advantages of the method.

1. INTRODUCTION

It has been shown previously [1,2] that the behaviour of inverter fed a.c. motors can be described during each conducting state by linear time invariant differential equations. The same possibility applies to other electrical circuits supplied by static convertors, for the following reasons:

- the main power system is characterized by voltage or current waveforms of a constant or sinusoidal type;
- the static convertor applies a supply variable composed of segments of canonical waveform to the electrical circuits. Its shape depends on the main supply, on the conducting state of SCR and diodes, and in some cases on the convertor parameters as well;
- the parameters of the models describing electrical or electromechanical systems supplied by static convertors have constant values, because saturation is not reached and they are independent of the electrical variables, so that the rating of the convetor is reduced.

This paper has been developed in connection with researches programs of the Automatic Control Institute of the University of Rome with the sponsorship of the "Centro di Studio dei Sistemi di Controllo e Calcolo Automatici" of the National Researches Council.

Due to the discontinuous nature of input variables, these systems cannot be analyzed by means of the traditional methods used for electrical circuits. System theory methods should, therefore, be applied. Although the analytical solution of problems related to the analysis of these circuits can be obtained because of the canonical waveform of the input variables, its application for numerical computation is impractical due to the complexity of the analytical elaborations involved.

A new formulation of the problem should thus be worked out giving the numerical solution by simple programs which are easily implemented on a minicomputer or even on a calculator. By this procedure the output variable is determined as the free motion of a system with a suitably augmented state vector. It has not been frequently applied [3,4] for computing system behavior when the input variable has a canonical waveform. The advantage of this approach is that the augmented state system model and its state transition matrix can be easily determined. This formulation also facilitates direct evaluation of the steady state condition and its initial state. More important, by increasing the state vector and defining a suitable dynamic system the mean value and the harmonic coefficients can be calculated as the free motion of a new augmented system.

2. GENERAL FORMULATION

The problem will be formulated in relation to each time interval during which the input variable u(t) can be expressed by segments of canonical functions. The input variable may be characterized by different canonical functions or by different segments of the same canonical function. Discontinuities in u(t) usually define the width of the time interval in which the input waveform has the same shape.

For the sake of simplicity, it will be assumed that the dynamic system is subject to a scalar input variable. Let the model of the dynamic system be expressed as:

$$\dot{x} = A\,x + b\,u \qquad x(0) = x_o \tag{1}$$

in which x and b are $n \times 1$ vectors, A is an $n\times n$

matrix, and u is a scalar function.

Since u(t) is canonical, it is always possible to obtain the analytical solution of Eq.1. This is however of limited practical interest, since the analytical solution implies the computation of the eigenvalues which could require complex and occasionally inaccurate numerical procedure [5]. This drawback may be overcome by using a new formulation with which the numerical computation of the free motion and forced response can be obtained directly and simultaneously when the input is canonical. It will be proved that the free motion of a suitable dynamic system corresponds to the free motion and forced response of system given by Eq.1. The numerical solution will be obtained with a procedure in which the eigenvalues of the dynamic matrix need not be determined.

The first problem is thus to define the model of the augmented state dynamic system, obtained by aggregating the original dynamic system to an adjoined system whose free motion corresponds to the canonical input variable u(t). To obtain this it must be remember [6] that, since u(t) is canonical, it may be expressed as a linear combination of the modes of a linear dynamic system. The model of the adjoined system is, therefore, given by the following differential equations:

$$\begin{cases} \dot{\mu} = \Omega\,\mu \\ u = \Gamma'\mu \end{cases} \qquad \mu(0) = \mu_o \qquad (2)$$

in which the matrices Ω, Γ and the vector μ_o depend on the given u(t).

In order to have a minimal realization of Eq.2, the modes contained in the input variable should be worked out to connect them with canonical realization. Since the modes most frequently encountered are either exponential, polynomial or pseudoperiodic, u(t) may be decomposed into the sum of three parts respectively indicated as $u_1(t)$, $u_2(t)$ and $u_3(t)$. For exponential modes, $u_1(t)$ is usually expressed as:

$$u_1(t) = \sum_{i=1}^{r} \mu_i \exp(\lambda_i t) \qquad (2a)$$

and the following result is obtained:

$$\Omega_1 = \begin{pmatrix} \lambda_1 & 0 & \cdot & 0 \\ 0 & \lambda_2 & \cdot & 0 \\ \cdot & \cdot & \cdot & \cdot \\ 0 & 0 & \cdot & \lambda_r \end{pmatrix}; \Gamma_1 = \begin{pmatrix} 1 \\ 1 \\ \cdot \\ 1 \end{pmatrix}; \quad \mu_{10} = \begin{pmatrix} \mu_1 \\ \mu_2 \\ \cdot \\ \mu_r \end{pmatrix}$$

For any polynomial mode, $u_2(t)$ is expressed as:

$$u_2(t) = \sum_{j=1}^{s} \mu_j \; t^{j-1} \; \exp(\lambda t) \qquad (2b)$$

and the result is:

$$\Omega_2 = \begin{pmatrix} \lambda & 1 & 0 & \cdot & 0 \\ 0 & \lambda & 1 & \cdot & 0 \\ \cdot & \cdot & \cdot & \cdot & \cdot \\ 0 & 0 & 0 & \cdot & \lambda \end{pmatrix}; \; \Gamma_2 = \begin{pmatrix} 1 \\ 0 \\ \cdot \\ 0 \end{pmatrix}; \mu_{20} = \begin{pmatrix} \mu_1 \\ \mu_2 \\ \cdot \\ \mu_s \end{pmatrix}$$

For any pseudoperiodic mode,

$$u_3(t) = \exp(\alpha t)(\mu_1' \cos \omega t + \mu_2' \sin \omega t) \qquad (2c)$$

results in:

$$\Omega_3 = \begin{pmatrix} \alpha & \omega \\ -\omega & \alpha \end{pmatrix}; \; \Gamma_3 = \begin{pmatrix} 1 \\ 0 \end{pmatrix}; \quad \mu_{30} = \begin{pmatrix} \mu_1' \\ \mu_2' \end{pmatrix}$$

Therefore, the matrices Ω, Γ, and the vector μ_o of Eq.2 are:

$$\Omega = \begin{pmatrix} \Omega_1 & \cdot & 0 & \cdot & 0 \\ \cdot & \cdot & \cdot & \cdot & \cdot \\ 0 & \cdot & \Omega_2 & \cdot & 0 \\ \cdot & \cdot & \cdot & \cdot & \cdot \\ 0 & \cdot & 0 & \cdot & \Omega_3 \end{pmatrix}; \Gamma = \begin{pmatrix} \Gamma_1 \\ \cdot \\ \Gamma_2 \\ \cdot \\ \Gamma_3 \end{pmatrix}; \; \mu = \begin{pmatrix} \mu_{10} \\ \cdot \\ \mu_{20} \\ \cdot \\ \mu_{30} \end{pmatrix}$$

By combining Eq. 1 and Eq. 2, the model of the augmented state system is obtained:

$$\dot{\xi} = S\xi \qquad \xi(0) = \xi_o \qquad (3)$$

in which

$$\xi = \begin{pmatrix} x \\ \mu \end{pmatrix}; S = \begin{pmatrix} A & b\Gamma' \\ 0 & \Omega \end{pmatrix}; \; \xi_o = \begin{pmatrix} x_o \\ \mu_o \end{pmatrix}$$

In order to demonstrate that the free motion of Eq. 3 provides both the free motion and forced response of Eq. 1 with a canonical input, the solution will be obtained as a partitioned state transition matrix in terms of the complex variable s:

$$\xi(s) = (sI-S)^{-1}\xi_o = \begin{pmatrix} sI-A & b\Gamma' \\ 0 & sI-\Omega \end{pmatrix}^{-1} \begin{pmatrix} x_o \\ \mu_o \end{pmatrix} =$$

$$= \begin{pmatrix} (sI-A)^{-1} & (sI-A)^{-1}b\Gamma'(sI-\Omega)^{-1} \\ 0 & (sI-\Omega)^{-1} \end{pmatrix} \begin{pmatrix} x_o \\ \mu_o \end{pmatrix}$$

of which

$$x(s) = (sI-A)^{-1}x_o + (sI-A)^{-1}b\Gamma'(sI-\Omega)^{-1}\mu_o$$

Since from Eq. 2 it is obtained:

$$u(s) = \Gamma'(sI-\Omega)^{-1}\mu_o$$

the final result become:

$$x(s) = (sI-A)^{-1}x_o + (sI-A)^{-1}bu(s)$$

which corresponds to the solution of Eq. 1 .

To compute the numerical value of x(t) for a prefixed time instant or to work out its waveform with a prefixed step Δt , a procedure must be applied which provides the state transition matrix when only the dynamic matrix and the time interval in consideration are known.

It is known that the state transition matrix $\Phi(t)$ is given by:

$$\Phi(t) - \exp(St) = \sum_{k=0}^{\infty} \frac{(St)^k}{k!} \qquad (4)$$

but it is almost always difficult to obtain an acceptable numerical accuracy by directly applying the series expansion provided by Eq. 4, due to the truncation and round-off errors.

Expedients should, therefore, be applied to minimize the errors which are inevitable when $\Phi(t)$ is calculated by a series expansion or the analytical solution.

The first computational improvement is obtained by applying the following expression for the series expansion:

$$\Phi(t) = I + \sum_{k=1}^{\infty} \frac{St}{k} \frac{(St)^{k-1}}{(k-1)!} \qquad (4a)$$

Further improvements are had by executing the series expansion by a matrix $(S\sigma)$ whose norm is less than 1 and then calculating the actual state transition matrix by the expression:

$$\Phi(t) = \Phi(\ell\sigma) = (\Phi(\sigma))^{\ell} \qquad (4b)$$

in which σ is chosen so that $\|S\sigma\| < 1$(**). Its width σ minimizes the number of terms to be considered for calculating the series, since their norm decreases regularly from the first term on. The maximum number K^* of terms depends on the number d of digits provided by the computer and is determined by the expression:

$$K^*\,! > 10^{d} \qquad (4c)$$

To further reduce the steps entailed in calculating $\Phi(t)$, the time interval t should be subdivided according to a whole power of 2, that will be indicated by r:

$$\ell = 2^{r} \qquad (4d)$$

In this way, after $\Phi(\sigma)$ has been calculated, the following recurrent equation is applied to obtain $\Phi(t)$:

$$\Phi(\sigma) = \Phi_1 \, ; \; \Phi_p = (\Phi_{p-1})^2; \; \Phi(t) = \Phi_r$$

If applied correctly, this method gives a reliable procedure, as the result can be obtained directly, without intermediate calculations. It also provides the least significant round-off error, for a given number of digits on the computer, because it reduces to a minimum the total number of steps to be performed.

3. COMPUTATION OF THE STEADY STATE CONDITION

When the dynamic system is asymptotically stable and its input variable has a periodic waveform, the procedure described above is particularly suited for determining the initial state vector $x(T)$ in steady state conditions. It must be remembered that the state vector $\xi(t)$ of Eq.3 can be partitioned as follows:

$$\xi'(t) = \left[(x(t) \;\; \mu(t)) \right]'$$

where $x(t)$ is the state vector of the original system and $\mu(t)$ is the state vector of the adjoined system.

The periodicity conditions will then be obtained by considering the partition of $\xi(t)$. The periodicity conditions for $u(t)$, which is the output variable of the adjonied system are known *a priori* and are given by the following equation:

$$u(T) = Q\,\mu(0) \qquad (5)$$

The periodicity conditions for $x(t)$ which is the state variable of the original system are expressed as follows:

$$x(T) = Px(0)$$

Matrix P is the identity matrix for non-electromechanical systems, while it can be easily determined for electromechanical systems [7]. The periodicity conditions for the augmented state system can thus be expressed by:

$$\xi(T) = \begin{pmatrix} P & 0 \\ 0 & Q \end{pmatrix} \xi(0) \qquad (6)$$

Moreover, vector $\xi(T)$ can be expressed in terms of $\xi(0)$ by means of the following equation:

$$\xi(T) = \Phi(T)\xi(0) = \begin{pmatrix} \varphi(T) & \psi(T) \\ 0 & \chi(T) \end{pmatrix} \begin{pmatrix} x(0) \\ \mu(0) \end{pmatrix}$$

where

$$\varphi(T) = \exp(AT)$$
$$\psi(T) = \exp(AT)b{*}\Gamma' \exp(\Omega T)$$
$$\chi(T) = \Gamma' \exp(\Omega T)$$

The following equation is obtained from Eqs. 5 and 6:

$$x(0) = |P - \varphi(T)|^{-1} \psi(T)\mu(0) \qquad (8)$$

This may also be expressed as:

$$x(0) = |I - P^{-1}\varphi(T)|^{-1} P^{-1} \psi(T)\mu(0) \qquad (9)$$

where matrix P has the following properties:

$$\|P\|_2 = 1$$

$$P' = P^{-1}$$

The numerical values of the elements of matrices $\varphi(T)$ and $\psi(T)$, which appear in the partition of matrix $\Phi(T)$, can be obtained directly by computing numerically the state transition matrix of the augmented state system, when it is carried out according to the procedure described.

Considering that the norm of the eigenvalues of $\varphi(t)$ is less than one [8], since

(**) The time interval t is therefore divided in ℓ subintervals.

the original system is asymptotically stable and that $\|P\|_2 = 1$, it follows that the computation of the inverse of matrix $(I - P'\varphi(T))$, labeled as the periodicity matrix, can be carried out by the following series expansion:

$$(I - P'\varphi(T))^{-1} = \sum_{k=0}^{h} |P'\varphi(T)|^k \qquad (10)$$

The number of the terms of the series depends only on the number d of the significant digits of the computing device if $\|\varphi(T)\| < 1$, otherwise the truncation must be effected when $\|\varphi(T)^k\| < 10^{-d}$.

4. EVALUATION OF THE MEAN VALUE AND HARMONIC COEFFICIENTS

When the inputs of the original dynamic system have a canonical waveform, the mean value and the harmonic coefficient values of its state variables can be calculated by means of the free motion of a new augmented state system. Its input variable is a state variable of the original augmented state system. For this variable the mean value and the harmonic coefficients will be determined.

The new adjoined dynamic system is given by the following equation:

$$\dot{\nu} = \Omega_4 \nu + \Gamma_4 \xi \qquad \nu(0) = 0 \qquad (11)$$

where the state variable ν and the matrices Ω_4, Γ_4 may have different expressions according to whether the mean value, the harmonic coefficients, or both of these are to be determined. The new augmented state system is obtained by aggregating Eq. 11 with Eq.3 and is given by the following matrix equations:

$$\dot{\zeta} = R\zeta \qquad \zeta(0) = \zeta_0 \qquad (12)$$

where

$$\zeta = \begin{pmatrix} \xi \\ \nu \end{pmatrix}; \quad R = \begin{pmatrix} S & 0 \\ \Gamma_4 & \Omega \end{pmatrix}; \quad \zeta(0) = \begin{pmatrix} \xi_0 \\ 0 \end{pmatrix}$$

The initial condition vector ξ_0 which is given by the partition of $\zeta(0)$, may be further partitioned into two vectors, $x(0)$ and $\mu(0)$, as shown in the Eq.3. Vector $\mu(0)$ depends only on the waveform of the input variable, while vector $x(0)$ depends on the behaviour of the original system being considered. When the steady state values for a periodic input are to be determined, vector $x(0)$ is given by the equation (8).

It will now be considered the case in which the mean value and/or the harmonic coefficients of only one state variable are to be determined. When these values are to be computed simultaneously for more than one state variable it will be necessary to adjoin to the original augmented system a corresponding number of new adjoined systems given by Eq. 11.

To calculate the mean value of the k-th element of the state vector ξ in the time interval T, the following dynamic system will be adjoined to matrix Eq. 3:

$$\dot{\nu} = \Gamma_4 \xi \qquad \nu(0) = 0 \qquad (13)$$

where Γ_4 is a suitable selection vector. The latter is a $(1 \times n)$ vector and its elements are all equal to zero except the k-th which is equal to 1/T. To calculate the harmonic coefficients corresponding to the frequency $h\omega^*$ of the k-th element of the state vector ξ the new adjoined dynamic system will have the following dynamic matrix:

$$\Omega_4 = \begin{pmatrix} 0 & h\omega^* \\ -h\omega^* & 0 \end{pmatrix} \qquad (14)$$

and the following input matrix:

$$\Gamma_4 = \begin{pmatrix} 0 \cdot \frac{2}{T} \cdot 0 \\ 0 \cdot 0 \cdot 0 \end{pmatrix}; \quad \Gamma_4' = \begin{pmatrix} 0 \cdot 1 \cdot 0 \\ 0 \cdot 0 \cdot 0 \end{pmatrix} \qquad (15)$$

The elements of vector ν, evaluated at time instant T, correspond to the sinusoidal and cosinusoidal coefficients od the h-th harmonic component. This can be proved as follows. The solution in the time domain of Eq. 11 is given by:

$$\nu(t) = \exp(\Omega_4 t)\int_0^t (\exp(\Omega_4 \tau))^{-1} \Gamma_4 \xi(\tau)d\tau \qquad (16)$$

where, matrix Ω_4 being given by Eq. 14, the state transition matrix is expressed as follows:

$$\exp(\Omega_4 t) = \begin{pmatrix} \cos(h\omega^* t) & \sin(h\omega^* t) \\ -\sin(h\omega^* t) & \cos(h\omega^* t) \end{pmatrix}$$

Then, the elements of vector $\nu(t)$, as given by the Eq. 16, are the following when calculated at time instant $T = 2\pi/\omega^*$

$$\nu_1(T) = \frac{2}{T}\int_0^T \cos(h\omega^*\tau)\Gamma_4' \xi(\tau)\, d\tau$$

$$\nu_2(T) = \frac{2}{T}\int_0^T \sin(h\omega^*\tau)\Gamma_4' \xi(\tau)\, d\tau$$

These are the general expressions used for obtaining harmonic coefficients.

5. EXAMPLE

To apply the described procedure, a general purpose subroutine was first prepared to carry out the series expansion of a generic matrix. This subroutine was implemented on an HP 9820 calculator which works with 15 significant digits, only the first 12 of which are displayed. The same subroutine may be used to calculate both the state transition function and the periodicity function. For calculating the state transition function, the series expansion, considering Eqs.4c and 4d, is truncated at the twelfth term; for the periodicity function, it is automatically trucated when the norm of one term is less than 10^{-15}.

The example will be worked out with the aim of describing the application of the procedure rather than solving a particular problem. Fig. 1 shows the electrical circuit. It

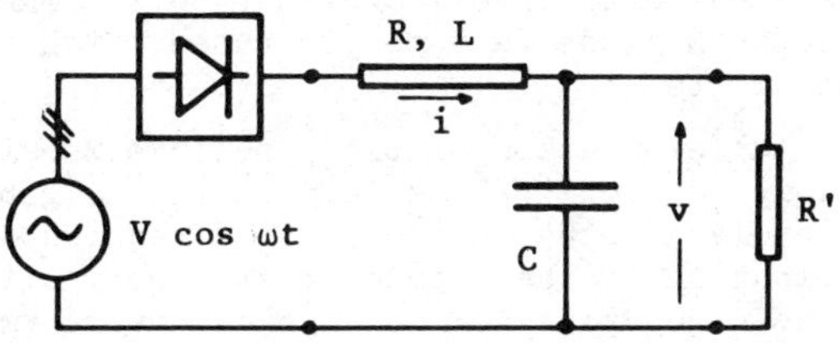

$V = 380\sqrt{2}$ V ; $R = 1\ \Omega$; $R' = 20\ \Omega$
$\omega = 100\ \pi$ s^{-1}; $L = .02$ H; $C = 500\ \mu F$

Fig. 1. Rectifier circuit layout.

is required at the steady state the computation of the current waveform, its mean value and harmonic coefficients. The numerical values of the circuit components are also indicated in the Fig. 1. The circuit behaviour is described by the following equations:

$$\begin{cases} \dfrac{di}{dt} = -\dfrac{R}{L}\, i - \dfrac{1}{L}\, v + \dfrac{V}{L}\, u(t) \\ \\ \dfrac{dv}{dt} = \dfrac{1}{C}\, i - \dfrac{1}{R'C}\, v \end{cases} \qquad (17)$$

The input variable u(t) has a periodic waveform and consists of a sinusoidal segment in the range from $-\pi/3$ and $\pi/3$. Since the frequency ω of the sinusoidal waveform is 100π, the period T of u(f) is 1/150 and therefore the corresponding frequency ω^* is 300π. Its analytical expression is the following:

$$u(t) = \cos(\omega t - \pi/3) \qquad t\varepsilon|0,T|$$

Matrices and vectors of the original dynamic systems are:

$$x = \begin{pmatrix} i \\ v \end{pmatrix};\ b = \begin{pmatrix} V/L \\ 0 \end{pmatrix};\ A = \begin{pmatrix} -R/L & -1/L \\ 1/C & -1/R'C \end{pmatrix}$$

and those of the adjoined system, considering Eq. 2c, are given by:

$$\Gamma = \begin{pmatrix} 1 \\ 0 \end{pmatrix};\ \mu_o = \begin{pmatrix} \cos(-\pi/3) \\ -\sin(-\pi/3) \end{pmatrix};\ \Omega = \begin{pmatrix} 0 & 100\pi \\ -100\pi & 0 \end{pmatrix}$$

To calculate the waveforms of variables i and v, it is necessary to calculate the free motion of the augmented state system , for which the dynamic matrix, state vector and initial conditions vector are given by:

$$\xi = \begin{pmatrix} i \\ v \\ \mu_1 \\ \mu_2 \end{pmatrix};\ \xi_o = \begin{pmatrix} i_o \\ v_o \\ \cos(-\pi/3) \\ -\sin(-\pi/3) \end{pmatrix}$$

$$S = \begin{pmatrix} -R/L & -1/L & V/L & 0 \\ 1/C & -1/R'C & 0 & 0 \\ 0 & 0 & 0 & 100\pi \\ 0 & 0 & -100\pi & \end{pmatrix}$$

The state transition matrix was computed for the time interval equal to $6.6\bar{6}$ ms by dividing it into 256 subintervals. It results

$$\Phi(T) = \Phi(.00\bar{6}) = \begin{pmatrix} -.265 & -.083 & -2.24 & 67.07 \\ 3.318 & -.348 & 308.7 & 371.3 \\ 0 & 0 & -.500 & .866 \\ 0 & 0 & -.866 & -.500 \end{pmatrix}$$

To obtain the periodicity matrix, matrix $\Phi(T)$ must be partitioned so as to obtain matrix $\varphi(T)$ expressed as follows:

$$\varphi(T) = \begin{pmatrix} -.265 & .083 \\ 3.318 & -.348 \end{pmatrix}$$

Then, seven terms of the series expansion must be computed.

The periodicity condition implies that the state vector of the original dynamic system has the same value at the beginning and at the end of each period. Therefore, applying Eq. 8 one obtains:

$$x_1(0) = x_1(T) = 18.83$$

$$x_2(0) = x_2(T) = 399.33$$

The current waveform at steady state is given in Fig. 2. It has been obtained by first

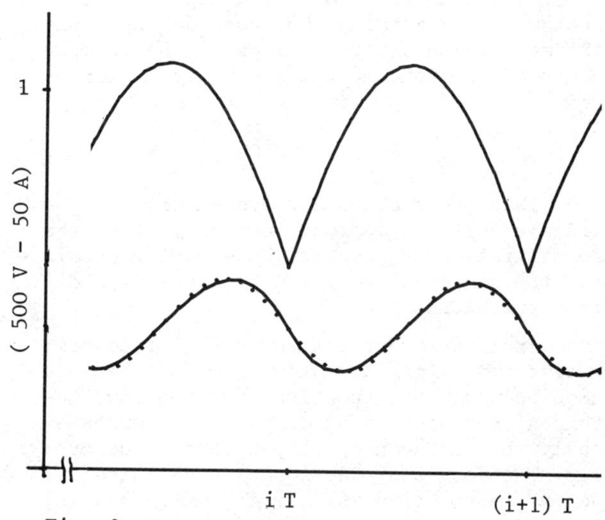

Fig. 2. Input voltage and current waveforms.

calculating the state transition matrix in correspondence with a time interval T', given by T/20 and by then using the following iterative equations:

$$\xi|(i+1)T'| = \Phi(T')\xi(iT');\ \xi_o = \begin{pmatrix} x(0) \\ \mu(0) \end{pmatrix}$$

To calculate simultaneously the mean value of the current and its harmonic coefficients of the fundamental frequency, ω* the following adjoined dynamic system is defined:

$$\begin{pmatrix} \dot{v}_1 \\ \dot{v}_2 \\ \dot{v}_3 \end{pmatrix} = \begin{pmatrix} 0 & 0 & 0 \\ 0 & 0 & 300\pi \\ 0 & -300\pi & 0 \end{pmatrix} \begin{pmatrix} v_1 \\ v_2 \\ v_3 \end{pmatrix} + $$

$$+ \begin{pmatrix} 1/T & 0 & 0 & 0 \\ 2/T & 0 & 0 & 0 \\ 0 & 0 & 0 & 0 \end{pmatrix} \begin{pmatrix} i(T) \\ v(T) \\ \cos(-\pi/3) \\ -\sin(-\pi/3) \end{pmatrix}$$

where $\omega^* = 300\pi$ and $1/T = 150$. The free motion of the last three components of the state vector of the systems having this dynamic matrix

$$R = \begin{pmatrix} -R/L & -1/L & V/L & 0 & 0 & 0 \\ 1/C & -1/R'C & 0 & 0 & 0 & 0 \\ 0 & 0 & 0 & 100\pi & 0 & 0 \\ 0 & 0 & -100\pi & 0 & 0 & 0 \\ 150 & 0 & 0 & 0 & 0 & 0 \\ 300 & 0 & 0 & 0 & 0 & 300\pi \\ 0 & 0 & 0 & 0 & -300\pi & 0 \end{pmatrix}$$

makes it possible to calculate directly the following desidered values:

$$\bar{i} = 19.29 \text{ A}$$

$$a|(i\omega^*)| = -.439\text{A}$$

$$b|(i\omega^*)| = 6.016\text{A}$$

Fig. 2 shows by a dotted line the current waveform obtained by considering only the mean value and the values of the components of the fundamental frequency harmonic.

6. CONCLUSION

This paper shows how system theory provides a suitable approach not only for the formalization of problems, but also for obtaining the numerical solution by means of a direct procedure.

Application of the augmented state transition matrix greatly simplifies the calculation because both the free motion and the forced response can be computed simultaneously when a canonical input forces the system. The fact that the input is canonical is not as restrictive as it might seem, since any type of input may generally be decomposed into segments of canonical functions. This decomposition is more flexible than the commonly used type, by which a series of pulses or steps approximates the input waveform.

This method, like the others, entails inevitable numerical errors. However, their upper bound can be evaluated more directly by applying error theory developed for the numerical solution of matrix equations [8,9].

The computation of the state transition matrix by means of a series expansion involves truncation errors which are easily determined *a priori* because they depend mainly on the number of digits used by the computing device. The value of the round-off error can also be evaluated by means of known procedures and tests have shown that this error affects only the last two or three significant digits.

Implementation of the procedure is simple, in that only elementary matrix algebra is involved. Other methods which require determination of the characteristic polynomial or the eigenvalues of the dynamic matrix may involve less memory occupation. However, this does not significantly increase the computation time since the dynamic matrix of the augmented state system has many zero elements. For these reasons, implementation of the method could be improved by using computation procedures developed for sparse matrix theory.

The procedure described in this paper is particularly appropriate for studying systems with SCR devices because of its simple formalization and implementation.

REFERENCES

[1] BELLINI A., CERRI C., DE CARLI A.: *A new Approach to the Simulation of static converter Drives*. Proc. of the International Association for Analog Computation, n.1, 1975.

[2] BELLINI A., CERRI C., DE CARLI A. : *Digital Simulation of Frequency Controlled Induction Motor Drives by a Piecewise linearization Procedure*. Proc. of the International Association for Analog Computation, n.3, 1975.

[3] MARRO G.: *Fondamenti di Teoria dei Sistemi*. Patron Editore, Bologna, pag. 306.

[4] DALLAGO E., PETRECCA G., SAVINI A.: *Impiego delle variabili di stato nello studio degli azionamenti in corrente alternata con convertitori statici*. L'Elettrotecnica, n. 4, 1976.

[5] WILKINSON J.H.: *The Algebraic Eigenvalue Problem*. Clarendon Press, Oxford 1965.

[6] ZADEH L.A., DESER C.A.: *Linear System Theory*. McGraw-Hill Book Company, 1963, pag. 293.

[7] WITHE C., WOODSON H.H.: *Electromechanical Energy Conversion*. J. Wiley & Sons, New York, 1959.

[8] WILKINSON H.J.: *Rounding errors in algebraic processes*. Her Majesty's Stationary, 1963.

[9] MANKIN J.B., HUNG G.G.: *On Round-off Errors in Computation of Transition Matrices*. Joint Automatic Control Conference, 1968, pag. 60.

D.C. CHOPPER CIRCUITS WITH A STRUCTURE DIFFERENT IN INTERVALS

M. Grötzbach

Lehrstuhl und Laboratorium für Elektrische Antriebstechnik, Technische Universität München, Arcisstraße 21, D-8000 München 2

ABSTRACT

Controlling the mean value of the energy flow by means of d.c. chopper circuits, the general case often appears that not only the input variables change periodically according to step functions, but also the structure of the system to be controlled. For such arrangements difference equations are derived to describe the behaviour of the switching points and the mean value at the moments of the largest repetition period. Presuming time invariant coefficients, we can give continuous equivalent differential equations the solution of which show the same behaviour in the discrete moments.

This exact mathematical description is rather extensive, so that its evaluation has to be done by a computer program. For the practically relevant case of a sufficiently smoothed curvature of the output signals the continuous equivalent descriptions can be found out approximately. Dominating widths of variation can be estimated or smoothing networks can be calculated out of the switching point models. Due to the rules of the d.c.-analysis, the transmission behaviour of the wanted signal can be described by the mean value model. In this way for the pulse-width modulation control a linearized model can be derived which takes into account the dynamic behaviour of the arrangement depending on the operating point. For the data of real plants the results correspond well to the exact digital calculation. As an example for the application of the approximation analysis the combined regenerative and rheostatic braking is treated in extracts.

1. INTRODUCTION

D.c. chopper circuits theoretically allow to control the mean value of the energy flow between a d.c. source and a load without a loss. In many practical arrangements the general case appears that not only the input variables change periodically according to step functions with the switching state of the semiconductor rectifiers, but also the structure of the system to be controlled.

As an example we consider the basic circuit of the combined regenerative and rheostatic braking of d.c. traction vehicles (Fig.1a), which will be referenced to in the following. Fig.1b shows the quasi-stationary behaviour of the capacitor voltage U_C, the motor current I_M and the current I_D that is delivered back; concerning the function of the circuit we refer to the literature /1,2/. The capacitor voltage serves as a control quantity for the ability of power consumption of the mains. When reaching $U_C=U_{Cmax}$, the thyristor Th is fired and the motor current flows through the braking resistor R_B.

As a result of the switch characteristic being discontinuous, the mathematically exact analysis of d.c. chopper circuits is rather extensive and not transparent. The description leads to non-linear differential equations which can be solved only iteratively by means of a digital computer /3,4/.

For the practically relevant case of a sufficiently smoothed output signal, i.e. a low pass behaviour of the system to be controlled and a switching frequency which is high, compared to the eigenfrequencies of the system, the basic circuit of the d.c. chopper is equivalent to a continuously controllable transformer of the mean value. The dynamic behaviour of a pulse-width modulation control device can f.e. be described exactly enough by means of this continuous model /5,6/.

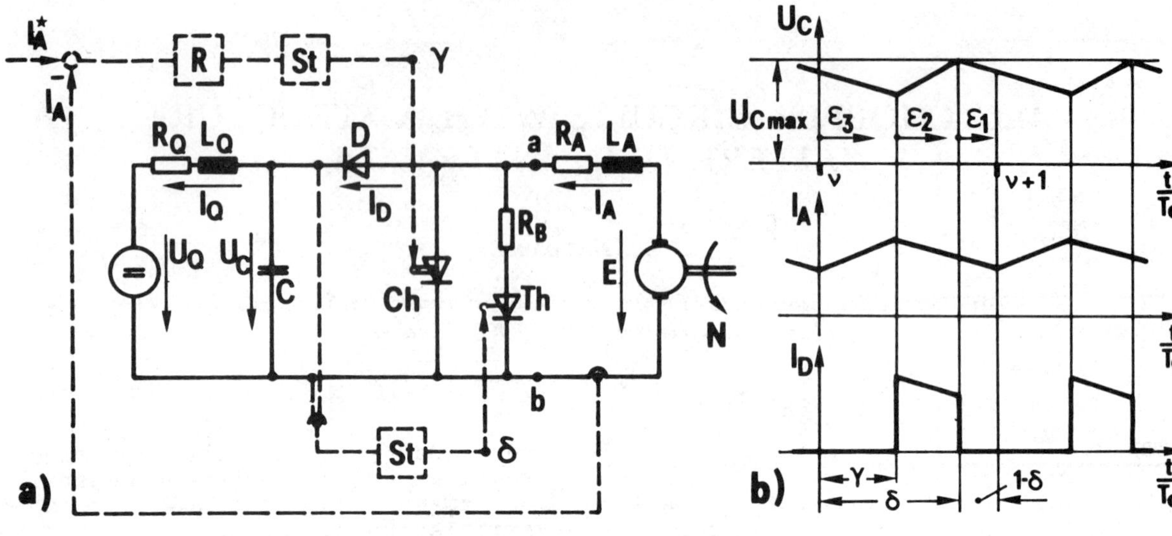

Fig. 1. Combined regenerative and rheostatic braking of d.c. traction vehicles.
a) Fundamental circuit
b) Quasi-stationary time characteristics of selected quantities

Presuming this sufficient smoothing, continuous approximation models of the behaviour of the switching points and of the mean values will be derived for <u>any</u> switched d.c. chopper circuits. In this paper the mathematical description is only considered in as much as it contributes to the understanding of the derived relations; concerning the exact realization, we refer to /7,8/.

2. OUTLINE OF THE EXACT CALCULATION

2.1 Mathematical Model

The given d.c. chopper circuit is piecewise described by linear differential equations (D.Equs.) with constant coefficients. A control device will periodically give the switching times of the switches which are assumed to be ideal. Losses of the switches can approximately be taken into account by additional linear elements, if necessary /9/.

As not only the switching state but also the structure of the circuit changes in the general case, a description with <u>physical</u> state variables seems to be advantageous. Moreover the further calculation assumes that the chosen state vector can describe <u>all</u> subintervals and is continuous in the switching points. Then the behaviour of the system between two switching points following each other, further marked with subinterval κ, is described by the following D.Equs.:

$$\frac{d\underline{v}(t)}{dt} = \underline{A}'_\kappa \underline{v}'(t) + \underline{B}'_\kappa \underline{u}'_\kappa(t) = \underline{A}'_\kappa \underline{v}'(t) + \underline{b}'_\kappa(t); \quad (1)$$

$$\kappa = 1,2,3...k.$$

Amplitude and time are suitably <u>normalized</u> in the following calculations. Using the reference magnitude $\underline{v}_{N\alpha}$, $1 \leq \alpha \leq n$ and the time axis $t=(\nu+\varepsilon)T_o$; $\nu=0,1,2...$; $0 \leq \varepsilon \leq 1$ referring to the largest period $T=T_o$, the normalized state D.Equs. are

$$\frac{d\underline{v}(\varepsilon)}{d\varepsilon} = \underline{A}_\kappa \underline{v}(\varepsilon) + \underline{b}_\kappa(\varepsilon). \quad (2)$$

If diag $(v_{N\alpha})$ signifies the diagonal matrix of the reference magnitudes, the normalized state matrix is given by

$$\underline{A}_\kappa = T_o \mathrm{diag}(1/v_{N\alpha}) \underline{A}'_\kappa \mathrm{diag}(v_{N\alpha})$$

and the input vector is defined by

$$\underline{b}_\kappa = T_o \mathrm{diag}(1/v_{N\alpha}) \underline{b}'_\kappa.$$

If any subinterval κ has time-independent input variables, i.e. $\underline{b}_\kappa \neq f(\varepsilon)$,

subinterval	circuit	state matrix $\underline{A}_\kappa$
$0 \leq \varepsilon_1 < 1-\delta$ Ch "off" Th "on" D "off"	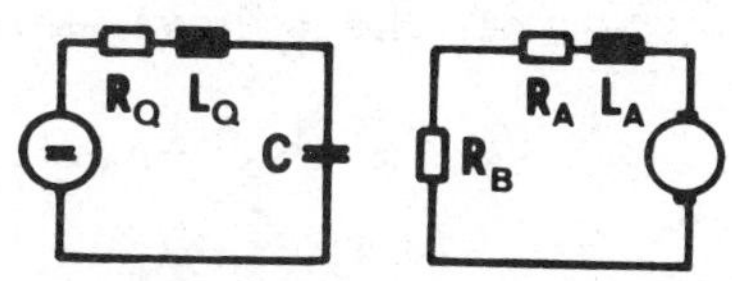	$\begin{bmatrix} -\beta_Q r_Q & \beta_Q & 0 \\ -\beta_C & 0 & 0 \\ 0 & 0 & -\beta_A(r_A+r_B-nb) \end{bmatrix}$
$0 \leq \varepsilon_2 < \delta-\gamma$ Ch "off" Th "off" D "on"	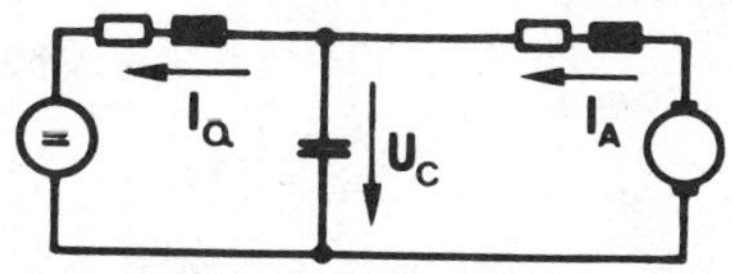	$\begin{bmatrix} -\beta_Q r_Q & \beta_Q & 0 \\ -\beta_C & 0 & \beta_C \\ 0 & -\beta_A & -\beta_A(r_A-nb) \end{bmatrix}$
$0 \leq \varepsilon_3 < \gamma$ Ch "on" Th "off" D "off"	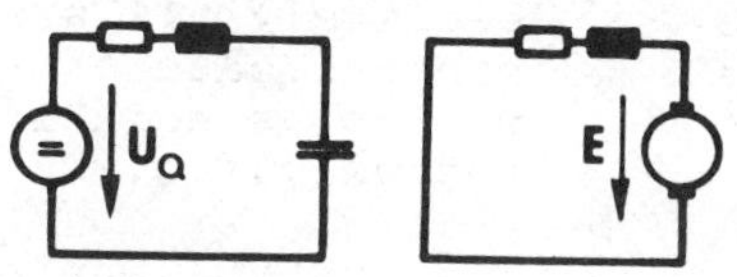	$\begin{bmatrix} -\beta_Q r_Q & \beta_Q & 0 \\ -\beta_C & 0 & 0 \\ 0 & 0 & -\beta_A(r_A-nb) \end{bmatrix}$

TABLE 1 Summary of the three state matrices $\underline{A}_\kappa$ different in intervals for the circuit of Fig. 1a.

the coordinate system of the state variables is suitably transformed into the appropriate steady state final value

$\underline{v}_\kappa(\infty)=-\underline{A}_\kappa^{-1}\underline{b}_\kappa$. With these transformed coordinates $\underline{v}_t(\varepsilon) = \underline{v}(\varepsilon)-\underline{v}_\kappa(\infty)$ the later calculation may be simplified essentially; in this connexion the following relations are applicable

$$\dot{\underline{v}}_t(\varepsilon) = \underline{A}_\kappa \underline{v}_t(\varepsilon);$$

$$\dot{\underline{v}}_t(\varepsilon) = \underline{A}_\lambda \underline{v}_t(\varepsilon) + \underline{b}_\lambda(\varepsilon) + \underline{A}_\lambda \underline{v}_\kappa(\infty); \quad (3)$$

$$1 \leq \lambda \leq k, \ \lambda \neq \kappa.$$

For the example of the combined regenerative and rheostatic braking of d.c. traction vehicles the speed of the series wound machine can be assumed constant because of the large inertia. Therefore the basic circuit of Fig. 1a is completely described by the state vector

$\underline{v}'=(I_Q,U_C,I_A)^T$; applying to the nominal values U_N and I_N of the machine,

the state vector reads: $\underline{v}=(i_Q,u_C,i_A)^T$. The piecewise valid system matrices A_κ shown in TABLE 1 can be received by the linearization of the B/H-curve corresponding to Fig. 2 and the time normalization $dt=T_o d\varepsilon$.

We will use the following abbreviations:

$$\beta_Q = \frac{T_o}{L_Q/R_N};\ \beta_A = \frac{T_o}{L_A/R_N};\ \beta_C = \frac{T_o}{CR_N};$$

$$r_\alpha = R_\alpha/R_N;\ R_N = U_N/I_N. \quad (4)$$

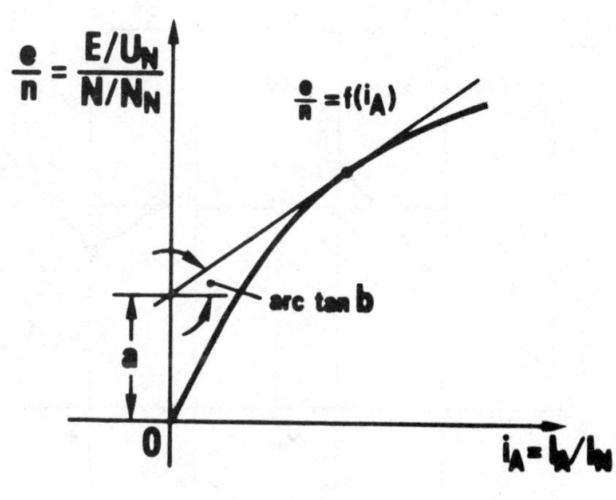

Fig. 2. Linearization of the B/H-characteristic in the stationary operating point.

The input vector $\underline{b}$ is constant in all subintervals,

$$\underline{b} = [-\beta_Q u_Q,\ 0,\ \beta_A na]^T. \qquad (5)$$

Therefore the mean value of the energy flow is only controlled by the different structure of the circuit.

2.2 METHOD OF CALCULATION

Having solved the state D.Equs. piecewise, we can set up the difference equation (ΔEqu.) of the switching points, using the condition of continuity of the state vector. The behaviour of the system is completely described by this equation in the discrete moments of the largest repetition period T_o. With the notations as in Fig. 3 and the abbreviations

$$\underline{\Phi}_\kappa(\varepsilon_\kappa) = \exp\{\underline{A}_\kappa \varepsilon_\kappa\};$$

$$\underline{\psi}_\kappa(\varepsilon_\kappa) = \int_0^{\varepsilon_\kappa} \exp\{\underline{A}_\kappa(\varepsilon_\kappa-\tau)\}\cdot \underline{b}_\kappa(\tau)d\tau$$

the ΔEqu. of the switching points applies to the case of k = 3 different subintervals

$$\underline{v}[\nu+1] = \underline{\Phi}\underline{v}[\nu] + \underline{\psi};$$

$$\underline{\Phi} = \underline{\Phi}_1(\gamma_1)\underline{\Phi}_2(\gamma_2)\underline{\Phi}_3(\gamma_3);$$

$$\underline{\psi} = \underline{\psi}_1(\gamma_1) + \underline{\Phi}_1(\gamma_1)\underline{\psi}_2(\gamma_2) + \underline{\Phi}_1(\gamma_1)\underline{\Phi}_2(\gamma_2)\underline{\psi}_3(\gamma_3). \qquad (6)$$

Integrating the time dependent state variables over the largest repetition period, we find the discrete value $\underline{v}_m[\nu_m] = \underline{M}_\nu \underline{v}[\nu] + \underline{n}_\nu$. Combined with the above-mentioned ΔEqu. this linear transformation results in the ΔEqu. of the behaviour of the mean value

$$\underline{v}_m[\nu_m+1] = \underline{K}_{\nu,\nu+1}\underline{v}_m[\nu_m] + \underline{l}_{\nu,\nu+1};$$

$$\underline{K}_{\nu,\nu+1} = \underline{M}_{\nu+1}\underline{\Phi}_\nu \underline{M}_\nu^{-1};$$

$$\underline{l}_{\nu,\nu+1} = \underline{M}_{\nu+1}\underline{\psi}_\nu + \underline{n}_{\nu+1} - \underline{K}_{\nu,\nu+1}\underline{n}_\nu. \qquad (7)$$

In this connexion the indeces ν respectively $\nu+1$ are pointing to the general case of time-variable pulse widths; a solution can be made iteratively.f.e.

Thereby the wanted exact description of the behaviour of the system in the discrete moments is found; with respect to the continuous models we search for, the transition into the continuous time axis remains to be discussed.

2.3 EQUIVALENT CONTINUOUS DESCRIPTION

With a given ΔEqu. with constant coefficients $\underline{v}[\nu+1] = \underline{\Phi}\underline{v}[\nu] + \underline{\psi}$ we associate a continuous equivalent D.Equ. $\dot{\underline{v}}_E = \underline{A}_E \underline{v}_E + \underline{b}_E$, the solution of which coincide in the discrete moments of the basic period as shown in Fig. 4. The homogeneous part of the

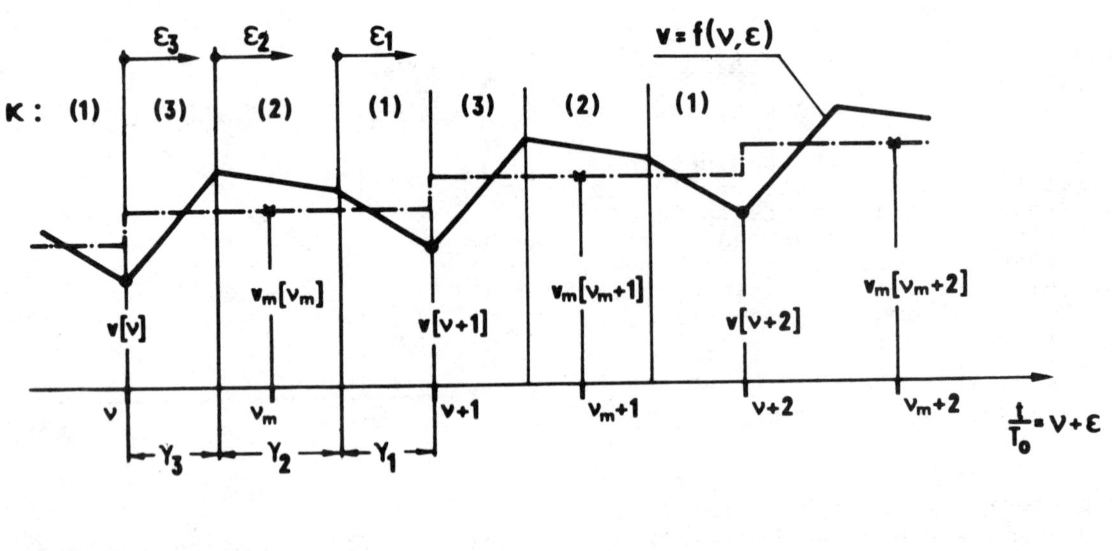

Fig. 3. Definition of the notation for k = 3 different subintervals.

continuous equivalent description $\underline{A}_E$ is given by the inverse function of the matrix exponential function; the equivalent input quantity $\underline{b}_E$ follows from the equality of $\underline{v}[\nu]$ and $\underline{v}_E(\nu T_o)$ at two arbitrary switching points. Summed up the continuous model of the behaviour of the switching points reads as follows

$$\dot{\underline{v}}_E = \underline{A}_E\underline{v}_E + \underline{b}_E;\quad \underline{A}_E = \ln\{\underline{\Phi}\};\quad \underline{b}_E = \underline{A}_E(\underline{\Phi}-\underline{I})^{-1}\underline{\psi}. \tag{8}$$

The other continuous switching point models(i.e. $t \neq \nu T_o$) are given by cyclic permutations of the subintervals; the required mean value model is found in the same way and reads

$$\dot{\underline{v}}_m = \underline{A}_m\underline{v}_m + \underline{b}_m;\quad \underline{A}_m = \ln\{\underline{K}\};\quad \underline{b}_m = \underline{A}_m(\underline{K}-\underline{I})^{-1}\underline{1}. \tag{9}$$

As the switching moments are known by presumption, a linearized model of the system to be controlled can be given for the pulse-width modulated control. For the control quantities $\underline{\gamma}_s=(\gamma,\delta)^T$, in the example of the combined regenerative and rheostatic braking in Fig. 1, we receive the D.Equ. which is linearized for small

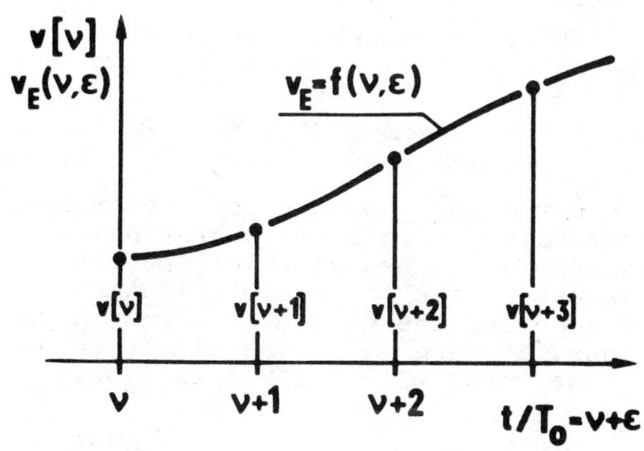

Fig. 4. Discrete time response $v[\nu]$ and the associated continuous equivalent description $v_E(\nu,\varepsilon)$.

deviations from the steady state γ_{so}, and reads in the general form

$$\Delta\dot{\underline{v}}_m = \underline{A}_m(\gamma_{so})\Delta\underline{v}_m + \underline{B}_{m\Delta}\Delta\underline{\gamma}_s. \tag{10}$$

2.4 Evaluation and Results

The matrix exponential function and its integral are evaluated by modified expansions in series, and the particular solution is determined by a corresponding polynomial set-up of the input quantities. In this connexion also piecewise singular state matrices are possibly allowed /10/. The logarithmic function $\ln(\underline{\Phi})$ is suitably determined by the appropriate equivalent polynomial, assuming for simplification the uniqueness of the mapping of the frequency plane.

With these digitally evaluated results the precision of the approximation models for given circuits which we will derive later can be tested. In contrast with the iterative calculations /3,4/, the dynamic behaviour of pulse-width modulated control is determined in the stationary steady state too.

3. ANALYTIC APPROXIMATION CALCULUS

3.1 Presumption

Controlling the energy flow by means of d.c. chopper circuits, the harmonics which are superimposed to the wanted signal are inevitable. They produce additional losses and probably influence the closed control loop. Therefore the switching frequency and the smoothing circuits are dimensioned in a way that the harmonics will remain as small as possible.

Formally this smoothing means that the dominating eigen-values of the piecewise valid state matrices are small compared to one. Thereby the expansions in series of the fundamental matrix $\underline{\Phi}=\exp\{\underline{A}\}$ and its inverse function $\underline{A}_E=\ln\{\underline{\Phi}\}$ converge quickly; an approximate description using the first terms of the series is allowed. Because of the considerably increasing expense the further calculation is restricted to a quadratic approximation. Marking the approximated value by "~", an example reads

$$\tilde{\underline{\Phi}}(\varepsilon) = \underline{I} + \underline{A}\varepsilon + \underline{A}^2\frac{\varepsilon^2}{2}. \tag{11}$$

3.2 Fundamental Relations

The computation of the continuous equivalent state matrix leads to the fundamental problem to solve the matrix equation

$\exp\{\underline{A}_E\} = \exp\{\underline{A}_1\}\cdot\exp\{\underline{A}_2\}$

for $\underline{A}_E$ with given $\underline{A}_1$ and $\underline{A}_2$. According to Hausdorff /11/, the following quadratic approximation is valid

$$\tilde{\underline{A}}_E = \underline{A}_1 + \underline{A}_2 + \tfrac{1}{2}[\underline{A}_1\underline{A}_2], \qquad (12)$$

where

$$[\underline{A}_1\underline{A}_2] = \underline{A}_1\underline{A}_2 - \underline{A}_2\underline{A}_1$$

is denoted as "alternant" and approximately takes into account the non-permutability of both state matrices.

After expanding in a series and integrating term by term we will find

$$\tilde{\underline{\psi}}(\varepsilon) = \int_0^{\varepsilon}\underline{b}(\sigma)d\sigma + \underline{A}\iint_0^{\varepsilon}\underline{b}(\sigma)d\sigma^2 + \underline{A}^2\iiint_0^{\varepsilon}\underline{b}(\sigma)d\sigma^3 \qquad (13)$$

for the particular solution

$$\underline{\psi}(\varepsilon) = \int_0^{\varepsilon}\exp\{\underline{A}(\varepsilon-\sigma)\}\cdot b(\sigma)d\sigma.$$

By means of the approximation equations stated above and the approximated inversion

$$(\underline{I}+\underline{A}_E/2)^{-1} \approx \underline{I}-\underline{A}_E/2$$

the wanted continuous approximation models for the behaviour of the switching points and the mean values can be derived, applying algebraic transformations.

3.3 Continuous Approximation Models

In the case of k = 3 different subintervals the continuous equivalent state matrices for the behaviour of the switching values and of the mean values read as follows

$$\tilde{\underline{A}}_E = \underline{A}_1\gamma_1 + \underline{A}_2\gamma_2 + \underline{A}_3\gamma_3 + \frac{\gamma_1\gamma_2}{2}[\underline{A}_1\underline{A}_2] + \frac{\gamma_1\gamma_3}{2}[\underline{A}_1\underline{A}_3] + \frac{\gamma_2\gamma_3}{2}[\underline{A}_2\underline{A}_3]; \qquad (14)$$

$$\tilde{\underline{A}}_m = \underline{A}_1\gamma_1 + \underline{A}_2\gamma_2 + \underline{A}_3\gamma_3 + \frac{\gamma_1\gamma_2\gamma_3}{2}\Big[[\underline{A}_1\underline{A}_2] + [\underline{A}_2\underline{A}_3] + [\underline{A}_3\underline{A}_1]\Big]. \qquad (15)$$

In this connexion the products put in brackets correspond to the alternants in Equ.(10), i.e.
$[\underline{A}_\kappa\underline{A}_\mu] = \underline{A}_\kappa\underline{A}_\mu - \underline{A}_\mu\underline{A}_\kappa$; $\kappa,\mu = 1,2,3$.

For the continuous equivalent input quantities we will consider only the practically important special case of piecewise time invariant input quantities. The relations are simplified by performing a coordinate transformation according to Equ.(3); using $\underline{v}_t(\varepsilon) = \underline{v}(\varepsilon)-\underline{v}_3(\infty)$ the expressions become most simple, as

$$\tilde{\underline{b}}_{Et} = \Big[\underline{I} - \frac{\gamma_2}{2}\underline{A}_2 - \frac{\gamma_3}{2}\underline{A}_3\Big]\gamma_1\underline{b}_{1t} + \Big[\underline{I} + \frac{\gamma_1}{2}\underline{A}_1 - \frac{\gamma_3}{2}\underline{A}_3\Big]\gamma_2\underline{b}_{2t} \qquad (16)$$

$$\tilde{\underline{b}}_{mt} = \Big[\underline{I} - \frac{\gamma_2\gamma_3}{2}(\underline{A}_2-\underline{A}_3)\Big]\gamma_1\underline{b}_{1t} + \Big[\underline{I} + \frac{\gamma_1\gamma_3}{2}(\underline{A}_1-\underline{A}_3)\Big]\gamma_2\underline{b}_{2t}. \qquad (17)$$

From the given relations the continuous approximation models result for k=2 different subintervals, setting γ_3=0. For the special case of constant system matrices we receive the known relations $\tilde{\underline{A}}_m=\underline{A}$ and

$$\tilde{\underline{b}}_m = \gamma_1\underline{b}_1 + \gamma_2\underline{b}_2 + \gamma_3\underline{b}_3.$$

In our example of the combined regenerative and rheostatik braking we simplify the behaviour of the mean values in the first approximation. Neglecting the quadratic terms in the Equs.(15) and (17), the simple relations reads with TABLE 1 and Equ.(5)

$$\begin{bmatrix}\tilde{i}_{Qm}\\ \tilde{u}_{Cm}\\ \tilde{i}_{Am}\end{bmatrix} = \begin{bmatrix}-\beta_Q r_Q & \beta_Q & 0\\ -\beta_C & 0 & \beta_C(\delta-\gamma)\\ 0 & -\beta_A(\delta-\gamma) & -\beta_A[r_A+(1-\delta)r_B - nb]\end{bmatrix}\begin{bmatrix}\tilde{i}_{Qm}\\ \tilde{u}_{Cm}\\ \tilde{i}_{Am}\end{bmatrix} + \begin{bmatrix}-\beta_Q u_Q\\ 0\\ \beta_A na\end{bmatrix}. \qquad (18)$$

3.4 Precision of the Approximations

To estimate the error caused by the truncation of the series by matrix norms is a method far too inaccurate for d.c. chopper circuits used in practice. Large scale matrix operations are necessary even for simple circuits to calculate the first error terms.

That is why for given d.c. chopper circuits the derived approximation models are tested by comparing them with the digitally computed results of the exact calculation. It is impossible to indicate limits of error which are valid, however, the validity of the derived second-order approximation models is fully proved by evaluating the data of practically dimensioned driving circuits. The maximum value of the absolute error of stationary control characteristics in normalized form is about 2-3%. Concerning the eigen-values of the mean value model, the relative error has an upper limit of about o,5%.

4. APPLICATION OF THE APPROXIMATION MODELS

4.1 Continuous Equivalent Circuit for the Mean Values

The continuous approximation model

$$\dot{\tilde{\underline{v}}}_m = \tilde{\underline{A}}_m \tilde{\underline{v}}_m + \tilde{\underline{b}}_m$$

can be seen as a physical equivalent circuit. Non-continuous switched elements are replaced by continuous quantities which approximately describe the behaviour of the mean value. Then we can treat this equivalent circuit by the well-known methods of d.c. analysis.

Choosing physical quantities for the state variables, the computation of the equivalent circuit belonging to the analytic approximation description is essentially simplified. Starting with a chosen output quantity v_α which corresponds to the row α of the D.Equ.-system, the conditional equations for the continuous elements of the equivalent circuit follow from the continuous equivalent description.

For the example of the combined regenerative and rheostatic braking it is suitable to choose the third row of the state D.Equ.(18) as a starting equation in regard of the armature current control which will be investigated later. Simple algebraic operations lead to the appropriate equivalent circuit of Fig. 5. By that you can calculate for example the stationary control characteristic of the armature current; the calculating rules of d.c. analysis yield

$$\tilde{i}_{Amo} = \frac{\tilde{i}_{Qmo}}{\delta_o - \gamma_o}$$

$$= \frac{n_o a - u_Q(\delta_o-\gamma_o)}{r_A - n_o b + r_B(1-\delta_o) + r_Q(\delta_o-\gamma_o)^2} \tag{19}$$

In the special case of δ=1, e.g. mere regenerative braking, Fig. 5 brings the known relations of the d.c. transformer.

4.2 Linearized Model for the Pulse-Width Control

The case of pulse-width control happens, if the energy flow is controlled by changing the relative switching times $\gamma_\kappa = T_{E\kappa}/T_o$, while the basic period T_o is constant. Since we have

$\sum_{\kappa=1}^{k} \gamma_\kappa = 1$, no more than k-1 pulse widths can change independently. Considering only small deviations from the steady state, it is possible to linearize the nonlinear D.Equs. by the rules of matrix differentiation. The received linearized models can be interpreted as a signal flow diagram and they can be treated by the known methods of control engineering.

Assuming the input vector to be

$\Delta\underline{y} = (\Delta u_Q, \Delta\gamma, \Delta\delta)^T$

for our example of the combined regenerative and rheostatik braking we get from the approximated D.Equs.(18) the linearized D.Equ. of the behaviour of the mean value

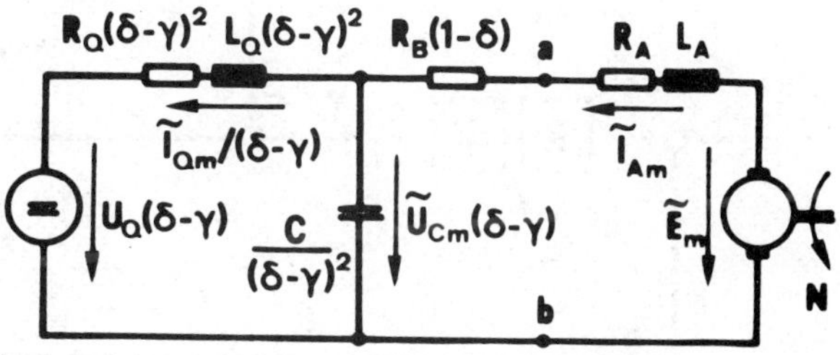

Fig. 5. Approximate mean equivalent circuit of the circuit of Fig. 1.

$$\begin{bmatrix}\Delta\tilde{i}_{Qm}\\ \Delta\tilde{u}_{Cm}\\ \Delta\tilde{i}_{Am}\end{bmatrix} = \begin{bmatrix} -\beta_Q r_Q & \beta_Q & 0\\ -\beta_C & 0 & \beta_C(\delta_o-\gamma_o)\\ 0 & -\beta_A(\delta_o-\gamma_o) & -\beta_A\left[r_A+(1-\delta_o)r_B - n_o b\right]\end{bmatrix}\begin{bmatrix}\Delta\tilde{i}_{Qm}\\ \Delta\tilde{u}_{Cm}\\ \Delta\tilde{i}_{Am}\end{bmatrix}$$

$$+\begin{bmatrix}-\beta_Q & 0 & 0\\ 0 & -\beta_C\tilde{i}_{Amo} & \beta_C\tilde{i}_{Amo}\\ 0 & \beta_A\tilde{u}_{Cmo} & \beta_A\left[r_B\tilde{i}_{Amo}-\tilde{u}_{Cmo}\right]\end{bmatrix}\begin{bmatrix}\Delta u_Q\\ \Delta\gamma\\ \Delta\delta\end{bmatrix}. \qquad (20)$$

This leads to the signal flow diagram of the armature current control dashed in Fig. 1a, compare for Fig. 6.

4.3 Estimation of Dominating Widths of Variation

The derived approximation model $\tilde{v}_E = \tilde{A}_E\tilde{v}_E+\tilde{b}_E$ describes the behaviour of the state variables in the switching points $t=\nu T_o$. After cyclic permutation of the subintervals the behaviour in all switching points can approximately be described by the appropriate D.Equs. Thereby it is possible to estimate the dominating widths of variation of the state quantities, if its extreme values appear at the switching points. For given widths of variation this estimation leads to rules for dimensioning the switching frequency and the smoothing networks. Notice, that even elements with small losses can influence the results very much /12/.

For the example of mere rheostatik braking, i.e. $\delta=1$, the stationary width of variation of the current reads after a short calculation

$$|\hat{\tilde{i}}|_o = n_o a\beta_A \frac{r_B\gamma_o(1-\gamma_o)}{r_A + r_B(1-\gamma_o) - n_o b}. \qquad (21)$$

In this case the maximum value of the stationary current variation does not appear at $\gamma_o=0{,}5$.

4.4 Control with a Variable Basic Period

The continuous approximation models of the behaviour of the switching points and the mean value are valid for time-invariant equivalent descriptions and consequently for the steady state of

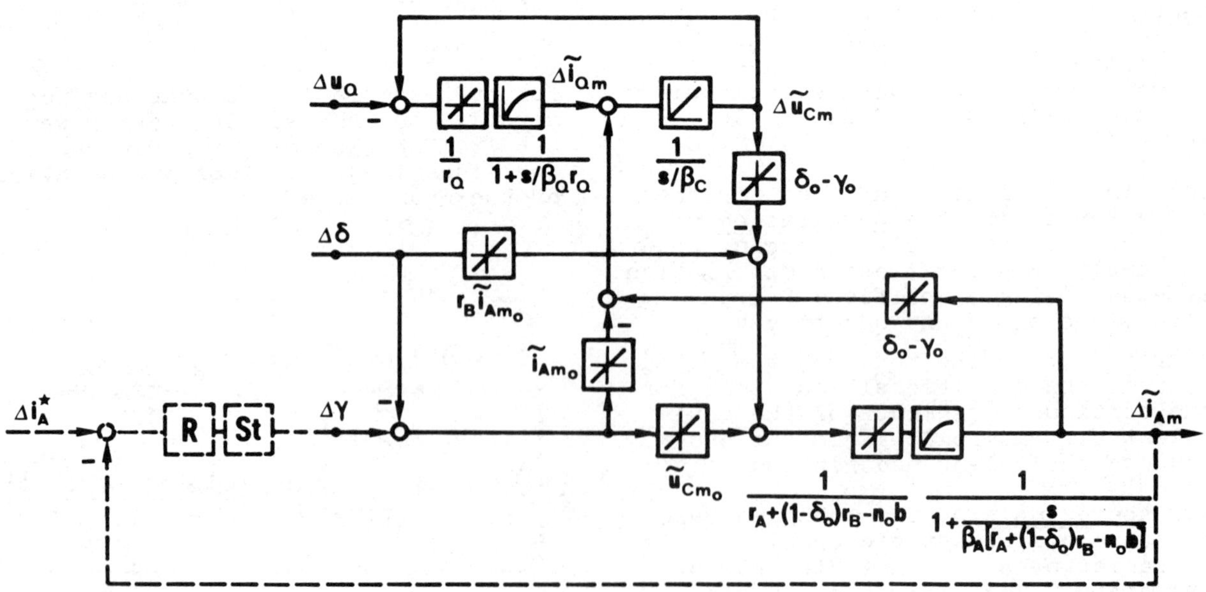

Fig. 6. Linearized signal flow diagram of the armature current control in a general form for the circuit of Fig. 1a.

two-step control and pulse-frequency control. Thus, in the case of two-step control of a state variable, the stationary switching frequency can be estimated as a function of the stationary operating point, if the width of the hysteresis curve of the controller is known. The derived results analogously apply to a pulse-frequency modulated system, if it is substituted by an equivalent pulse-width modulated system at the steady state.

5. CONCLUSIONS

The behaviour of practically dimensioned d.c. chopper circuits is exactly enough described by the derived continuous approximation models of the behaviour of the switching points and of the mean values. The essential presumptions are: piecewise linear D.Equs. with constant coefficients as shown in Equ. (1), given switching moments, sufficiently smoothed output variables and time invariant coefficients of the ΔEqus. The approximation models can further be treated by the known methods of d.c. analysis and thus allow a simple analysis of such networks.

Special cases of the approximated mean value model are already known, as f.e. the basic circuit of the d.c. chopper and the pulsing of a resistor. Therefore this model can be seen as a general linearization by periodical switching, considering the structures different in intervals. Since the dependence of the dynamic data of the operating point is given in analytical form a controlled adaptation is possible. The difference between mean value model and switching point model is significant even in the case of well-smoothed output variables and has to be taken into account for developping a model of the controlled system.

Acknowledgements - While writing this paper, the author was working at the "Lehrstuhl für Elektrische Antriebstechnik" at the Technische Universität München, Germany, as a "Wissenschaftlicher Assistent". He would like to express thanks to Prof.Dr.-Ing. G.Kessler and to the Deutsche Forschungsgemeinschaft for their generous help.

REFERENCES

/1/ H. Eisele, Gemischte Nutz- und Widerstandsbremsung bei Untergrundbahnen, ETZ-A 9o, S.7oo - 7o3 (1969)

/2/ P. Loderer, Gemischte Nutz- und Widerstandsbremsung bei Gleichstromtriebfahrzeugen, Elektrische Bahnen 44, S.17o - 174 (1973)

/3/ B. Mellitt, M. H. Rashid, Analysis of d.c. chopper circuits by computer-based piece-wise-linear technique, Proc. IEE control a.sci. 121, S.173 - 178 (1974)

/4/ D. D. Prakash, G. K. Dubey, A Digital Computer Program for Chopper-Fed DC Motors, IEEE Trans. IECI-22, S.4o8 - 412 (1975)

/5/ I. Nikoloff, Stabilitätsprobleme bei Antrieben mit Gleichstromstellern für fahrleitungsgespeiste Gleichstromtriebfahrzeuge, Elektrie 26, S.369 - 371 (1972)

/6/ H. Krug, Probleme der Regelung von Antrieben mit Gleichstrompulsstellern, VEM-Elektro-Anlagenbau 9, S.138 - 142 (1973)

/7/ M. Grötzbach, Stetige Näherungsmodelle zur Analyse von periodisch geschalteten Mehrspeichersystemen mit ausreichender Glättung, Dissertation, TU München (1976)

/8/ M. Grötzbach, Steuerung einer Tiefpaß-Regelstrecke höherer Ordnung mit periodisch geschaltetem Stellglied, Archiv für Elektrotechnik 58, Heft 6, S.321 - 331 (1976)

/9/ H. Kahlen, Vergleichende Untersuchungen an verschiedenen Gleichstromstellern für Fahrzeugantriebe, Dissertation, TH Aachen (1973)

/10/ G. Schaub, Digitale Berechnung periodisch geschalteter Systeme, Diplomarbeit am Lehrstuhl für Elektrische Antriebstechnik, TU München (1976)

/11/ R. Bellmann, Methods of Nonlinear Analysis, Vol. I Academic Press, New York, S.1o8 - 1o9 (197o)

/12/ M. Grötzbach, Geschaltete Stromversorgung, Schwankungsbreite der Ausgangsspannung, ETZ-B Heft 10 (1977)

A GENERAL MODEL FOR SWITCHED DC-DC CONVERTERS INCLUDING FILTERS

A. J. Fossard* and M. Clique**

**Ecole Nationale Supérieure de l'Aéronautique et de l'Espace (ENSAE) 10, avenue Edouard Belin - BP 4032 - 31055 Toulouse Cedex (France)*
***Centre d'Etudes et de Recherches de Toulouse (CERT), Département d'études et de recherches en Automatique (DERA) 2, avenue Edouard Belin - BP 4025 - 31055 Toulouse Cedex (France)*

ABSTRACT

A continuous, low frequency, small signal model, valid for all types of converters -both in heavy and light modes, is developed in this paper. The method, which easily allows to take into account, both input and output filters, leads to a new general canonical block diagram enhancing the respective contributions of the cell and of the filters. From it, all the desired transmittances (between the output, the duty cycle and/or the voltage source) can be derived in a straightforward general form, enlightening in particular the influence of the filters on the dynamics and the stability of the system.

The method is considered as an averaging method since it considers, at the initial point, the mean value, on one cycle, of the current injected in the filters. A judicious use of state space methods and averaging methods allows an easy implementation of these models, whatever the complexity of the input-output filters.

1 - INTRODUCTION

The modelization of switching converters, and especially of power stages, has received, in the last three years, considerable attention. This can be explained, partly by the difficulty inherent to the analysis of such systems (non linear, variable structure), partly by the very high static and dynamic performances required in spatial applications.

In fact, as it happens in many fields, a real improvement of the control loop can be viewed only through a more precise and synthetic knowledge of the behaviour of the open loop, i.e. through the obtention of models which, on the same time, should be :

- realistic enough to explain the phenomena observed on beadboards
- synthetic enough to help in the design of control laws
- simple enough to allow easy parametric studies

This explains why, although precise and accurate non linear models are valuable for final simulation runs, the bigger interest lies now on the obtention of simplified, yet sound, models, even if though this means their use is restricted to small signal, low frequency applications.

2 - MODELLING TECHNIQUES

In order to reach the above mentionned goals two main approaches have been developped :

- the first one uses the so-called "averaging techniques". In this category should be quoted the well known Middlebrook's Methods (ref. 1, 2, 3, 4) based on equivalent circuit manipulations and the more recent "injected current method" first designed by Prajoux (5).

- the other one is based on a state formulation (ref. 6,7,8,9,10) which allows though the transition function, a rigourous description while the previous methods are inherently limited to low frequency modelling on account of the hypothesis made.

- it has been objected (ref. 3) that state space methods usually "rely heavily on numerical methods" and provide little insight into the physical behaviour. This is somewhat true if one uses indeed the transition function, an exponantial matrix and if the linearization is made through gradients computation (ref. 7), but it might be considered in this case it is the price one should pay for the precision of the model and its wide range of applicability. If only a linearized low frequency model is looked for -as is the mentioned goal and the case in the first approach- a voluntary limited expansion of this matrix can be made in a litteral way by keeping a given number of terms (and not as in (ref. 7) according to a given criterion). Such a procedure considerably simplify the problem and allows the easy derivation of the useful transmittances between duty cycle and output voltage for instance. The less terms are kept in the expansion, the simpler are the expressions, at the expense of the precision, and the various methods differ in fact according to :

- the stage the resulting linearization is made (on a phase ref.10) on a cycle (ref.6, 8,9) on the open loop expression or on the closed one (ref.7)
- the number of terms which are conserved.

Table I summarizes these various approaches.

TABLE 1 : Classification of the modelization techniques

Fundamentals. Phases and associated structures recognition

State techniques
- Use of the transition matrix (ref.6)
- Differential linearization on each phase (ref.10)
- Formal linearization by limited expansion (open loop)
  First order : Robin Jouan (8)
  Second order : Owen, Capel, Ferrante (9)
- Computarized linearization (closed loop) Iwens (7)

Averaging techniques
- Equivalent "agregated" circuit method Middlebrook. Wester (ref.1 to 4)
- Injected current method Prajoux (5)- Fossard, Clique (11,12)

If, on a formal point of view, care should be taken not to separate the two methods (in fact the averaging Middlebrook"s method can be showned to be equivalent, in the heavy mode of conduction, to a first order expansion) it should be interesting to dispose of a method gathering on a practical point of view, the advantages of the two approaches, namely :

- the generality of the state approach -in particular the possibility to deal in the same way with all types of converters, and in heavy as well as in light mode of conduction
- the better physical insight provided by the averaging techniques -and in particular the possibility to singularize the respective influence of input and output filters.

It is the authors' opinion that the injected current method, here presented, is a step toward this objective.

3 - DEVELOPMENT OF THE METHOD

The method will be in the following presented in the case of an ideal boost converter shown fig.1. Needless to say however it applies to all kinds of classical converters boost, buck buck-boost and the results will be given for the three cases.

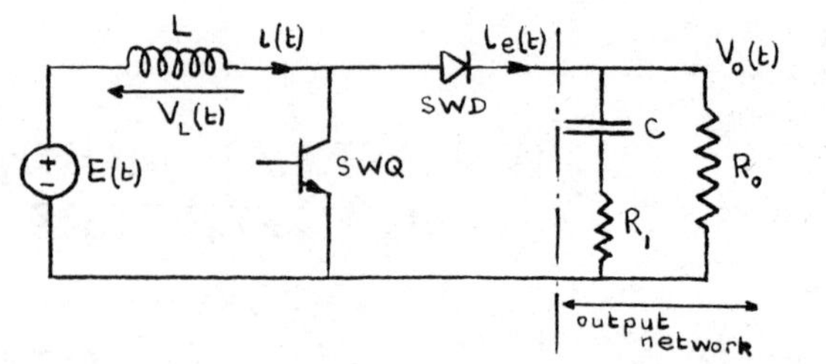

Fig. 1 Ideal boost converter configuration

It derives from the method designed by Prajoux *et al.* (5,13) but instead of considering the evaluation of a discrete derivative over a cycle (which allows obtention of large signal non linear models) it uses a direct differentiation of the "steady state" expressions (and consequently only small signal models).

3.1 - Light mode

Assume that over one cycle of length T the input and output voltages E and V_o might be considered as constant and equal to their mean values $\bar{E}$ and $\bar{V}_o$ (this hypothesis is consistant with the fact that low frequency models are looked for). Then, in the ideal case, the shapes of the currents i(t) flowing through the inductor and $i_e(t)$ injected in the output network are shown in fig.2 where t_c represents the charge time of the inductor, t_d the discharge time and t_l the free time.

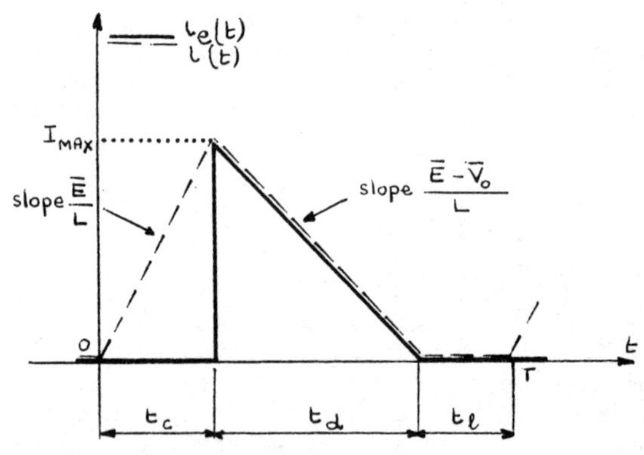

Fig.2 Characteristic shapes of the currents in light mode.

One has : $$I_{MAX} = \bar{E} t_c / L \quad (1)$$

$$t_d = t_c \bar{E} / (\bar{V}_o - E) \quad (2)$$

The output network beeing a low pass filter with a time constant much greater than the period T let us consider the mean value $\bar{i}_e$:

$$\bar{i}_e = I_{MAX} t_d / 2T = \bar{E}^2 t_c^2 / 2LT(\bar{V}_o - \bar{E}) \quad (3)$$

If the charge time t_c and/or the mean value $\bar{E}$ of the source vary $\bar{i}_e$ and $\bar{V}_o$ will vary. Noting the small variations $\tilde{t}_c$, $\tilde{e}$, $\tilde{i}_e$ and $\tilde{v}_o$ and differentiating equation (3) leads to :

$$\begin{aligned}\tilde{i}_e = {} & 2t_c\tilde{t}_c\bar{E}^2/2LT(\bar{V}_o-\bar{E}) \\ & - t_c^2\bar{E}^2\tilde{v}_o/2LT(\bar{V}_o-\bar{E})^2 \\ & + t_c^2\bar{E}(2\bar{V}_o-\bar{E})\tilde{e}/2LT(\bar{V}_o-\bar{E})^2\end{aligned} \quad (4)$$

For the nominal behavior and under energy considerations, the charge time t_c can be expressed as (ref.5,6,11) :

$$t_c = \frac{1}{\bar{E}}\sqrt{\frac{2LT(\bar{V}_o-\bar{E})\bar{V}_o}{R_o}} \quad (5)$$

Substituting (5) in (4) with the usual notations

$\bar{D}=t_c/T$ (nominal duty ratio), $\bar{D}'=t_d/T$, $\tilde{d}=\tilde{t}_c/T$ (variation of the duty ratio) gives

$$\begin{aligned}\tilde{i}_e = {} & 2E(\bar{D}+\bar{D}')\tilde{d}/R_o\bar{D}\bar{D}' \\ & - (\bar{D}+\bar{D}')\tilde{v}_o/R_o\bar{D} \\ & + (\bar{D}+\bar{D}')(2\bar{D}+\bar{D}')\tilde{e}/R_o\bar{D}\bar{D}'\end{aligned} \quad (6)$$

If S(s) is the transfer function of the output network then :

$$\tilde{V}_o(s) = S(s)\,\tilde{I}_e(s) \quad (7)$$

and if (6) is rewritten in the Laplace form one has :

$$\tilde{I}_e(s) = A(s)\,\tilde{D}(s) - B(s)\,\tilde{V}_o(s) + C(s)\,\tilde{E}(s) \quad (8)$$

where in our case A(S) B(S) and C(S) reduce to gains and

$$S(s) = R_o(1+R_1Cs)/(1+(R_o+R_1)Cs) \quad (9)$$

Equations (6) (7) and (8) can be interpreted as the block diagram shown in fig.3 :

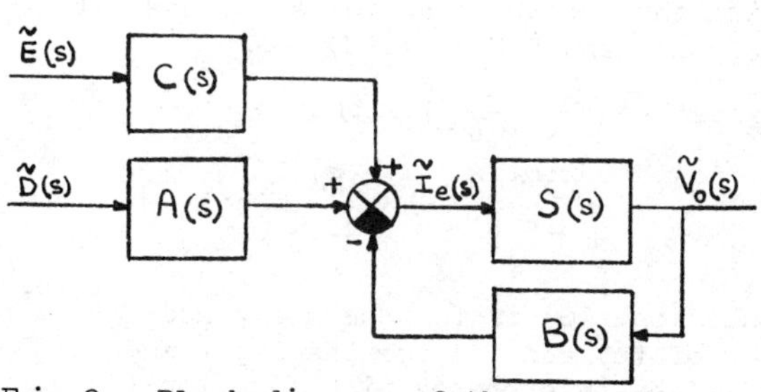

Fig.3 Block-diagram of the converter model.

For the two other types of switched dc-dc converters, the buck and the buck boost, there exist analog expressions for A(S). B(S) and C(S) which will be found in table 2 .

The associated transfer functions for the block diagram shown in fig.3 are :

$$\begin{aligned}\tilde{V}_o(s) = {} & A(s)\,S(s)\,\tilde{D}(s)/(1+S(s)B(s)) \\ & + C(s)\,S(s)\,\tilde{E}(s)/(1+S(s)B(s))\end{aligned} \quad (10)$$

3. 2 - Heavy Mode

Under the same hypothesis the shapes of the currents i(t), $i_e(t)$ and of the inductor voltage $V_L(t)$ are given in fig.4a and 4b for the norminal behavior.

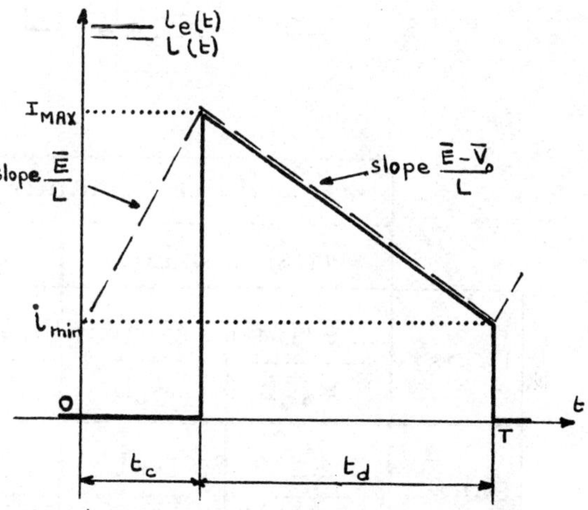

Fig. 4a Characteristic shapes of the currents in heavy mode.

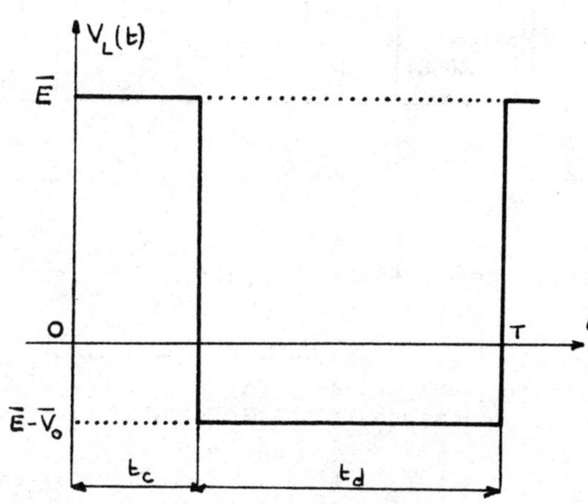

Fig. 4b Characteristic shape of the inductor voltage in heavy mode.

One has

$$t_c = T(\bar{V}_o-\bar{E})/\bar{V}_o \quad (11)$$

$$t_d = T - t_c \quad (12)$$

and under energy considerations :

$$I_{MAX} = \bar{V}_oT/R_ot_d + \bar{E}t_c/2L \quad (13)$$

TABLE 2 Ideal converter block diagram and characteristic coefficients

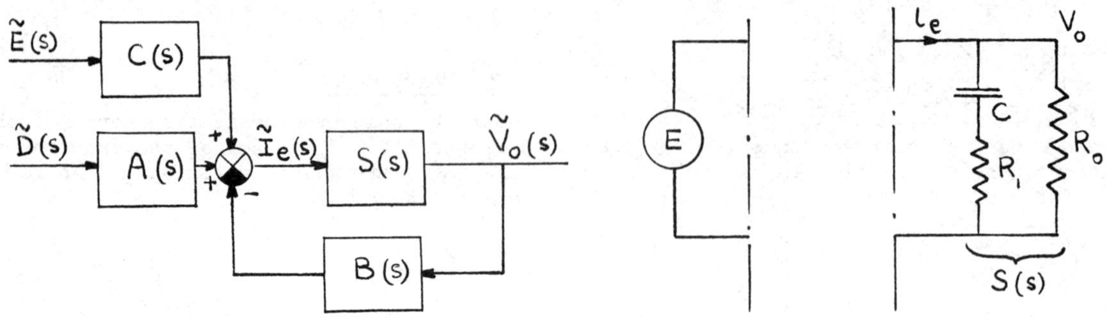

	LIGHT MODE			HEAVY MODE		
	A(s)	B(s)	C(s)	A(s)	B(s)	C(s)
BOOST	$\frac{2\bar{E}(\bar{D}+\bar{D}')}{R_o\bar{D}\bar{D}'}$	$\frac{\bar{D}+\bar{D}'}{R_o\bar{D}}$	$\frac{(\bar{D}+\bar{D}')(2\bar{D}+\bar{D}')}{R_o\bar{D}\bar{D}'}$	$\frac{\bar{E}}{Ls}-\frac{\bar{E}}{R_o\bar{D}'^2}$	$\frac{\bar{D}'^2}{Ls}$	$\frac{\bar{D}'}{Ls}$
BUCK	$\frac{2\bar{E}}{R_o(\bar{D}+\bar{D}')}$	$\frac{\bar{D}+\bar{D}'}{R_o\bar{D}'}$	$\frac{\bar{D}(\bar{D}+2\bar{D}')}{R_o\bar{D}'(\bar{D}+\bar{D}')}$	$\frac{\bar{E}}{Ls}$	$\frac{1}{Ls}$	$\frac{\bar{D}}{Ls}$
BUCK-BOOST	$\frac{2\bar{E}}{R_o\bar{D}'}$	$\frac{1}{R_o}$	$\frac{2\bar{D}}{R_o\bar{D}'}$	$\frac{\bar{E}}{Ls}-\frac{\bar{D}\bar{E}}{R_o\bar{D}'^2}$	$\frac{\bar{D}'^2}{Ls}$	$\frac{\bar{D}\bar{D}'}{Ls}$

$$i_{min} = I_{MAX} - \bar{E}t_c/L \qquad (14)$$

As in the previous case, the procedure will use the variations of the mean values :

$$\bar{i}_e = t_d\bar{i}/T \qquad (15)$$

and

$$\bar{V}_L = (t_c\bar{E} + t_d(\bar{E}-\bar{V}_o))/T = \bar{E} - t_d\bar{V}_o/T \qquad (16)$$

Differentiating equations (15) and (16) leads to :

$$\tilde{i}_e = \bar{D}'\tilde{i} + \tilde{d}'\bar{i} \qquad (17)$$

$$\tilde{v}_L = \tilde{e} - \bar{D}'\tilde{v}_o - \tilde{d}'\bar{V}_o \qquad (18)$$

For the inductor one has

$$L\,d\tilde{i}/dt = \tilde{v}_L \qquad (19)$$

and equations (17) through (19) give the relation between the variation of $\tilde{i}_e$ and those of the duty-cycle and the source voltage. Taking the Laplace transforms one has with $\bar{V}_o = \bar{E}/\bar{D}'$ and $\tilde{D}'(S) = -\tilde{D}(S)$:

$$\tilde{I}_e(s) = \left(\frac{\bar{E}}{Ls} - \frac{\bar{E}}{R_o\bar{D}'^2}\right)\tilde{D}(s) - \frac{\bar{D}'^2}{Ls}\tilde{V}_o(s) + \frac{\bar{D}'}{Ls}\tilde{E}(s) \qquad (20)$$

Equation (20) is for the heavy mode the analog of (6) and (8) for the light mode and can still be interpreted with equation (7) as the block diagram shown in fig.3.

In this case, A(S) B(S) and C(S) are no more simple gains, and even if the transfer functions are always given by equation (10), they will be different in their form.

The same procedure can be held for the buck and the buck-boost whose expressions for A(S) B(S) and C(S) will be found in table 2.

3.3 - Frequency Effect

In the previous paragraphs it has been implicitly considered that the current i_{min} was the same at the beginning and at the end of a cycle. In fact, for the perturbated behavior the current i_F at the end of a cycle is different from i_{min} at the beginning.

The use of a discrete derivative over a cycle given by $i'_{min} = (i_F - i_{min})/T$ (ref.5,13), together with a direct differentiation of the expressions then obtained, allows to compute new expressions for A(S) B(S) and C(S) ; they are slightely different from the expressions of table 2 by addition of corrective terms which are linear functions of T taking then into account the frequency effect. This modification does not exist for the light mode ($i_F = i_{min} = 0$) and even for the heavy mode, the additive terms are generally small for typical cases (it has been proved furthermore that they increase the closed loop damping for a given feedback loop).

4 - INTRODUCTION OF AN INPUT FILTER

In a pratical situation, the voltage source E(t) is the output voltage of an input filter whose input will be a voltage $\mathcal{E}(t)$ as shown in fig.5

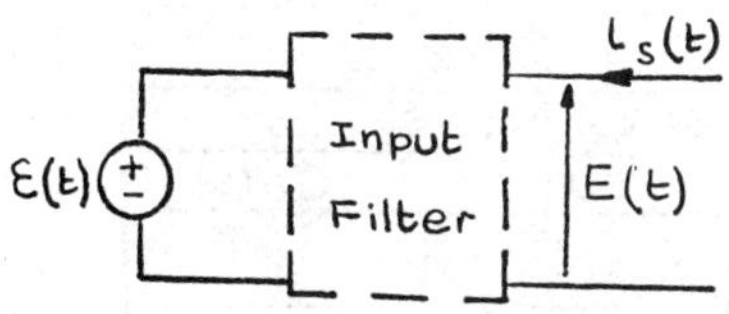

Fig. 5 Input filter

As for the ideal converter, the procedure will be given for a boost with classical output network and the same assumptions about $V_o(t)$ and E(t) will be conserved.

For both light and heavy modes, the current $i_s(t)$ shown in fig.5 is equal to -i (t) cf fig.2). Using equations (1) and (2) for light mode gives

$$\bar{i}_s = -\bar{i} = -I_{MAX}(t_c+t_d)/2T$$
$$= -\bar{E}\bar{V}_o t_c^2/2LT(\bar{V}_o-\bar{E})$$

and differentiating

$$\tilde{i}_s = -2\bar{E}(\bar{D}+\bar{D}')^2\tilde{d}/R_o\bar{D}\bar{D}'^2$$
$$+(\bar{D}+\bar{D}')\tilde{v}_o/R_o\bar{D}$$
$$-(\bar{D}+\bar{D}')^3\tilde{e}/R_o\bar{D}\bar{D}'^2$$

or more concisely in Laplace form :

$$\tilde{I}_s(s) = A_e(s)\tilde{D}(s) + B_e(s)\tilde{V}_o(s) - C_e(s)\tilde{E}(s) \qquad (21)$$

In heavy mode, equations (18) and (19) give similarly

$$\tilde{I}_s(s) = -\bar{E}\tilde{D}(s)/\bar{D}'Ls + \bar{D}'\tilde{V}_o(s)/Ls - \tilde{E}(s)/Ls$$

of the same form as (21).

If the equations of the input filter are written

$$\tilde{E}(s) = F_{e_1}(s)\tilde{I}_s(s) + F_{e_2}(s)\tilde{\mathcal{E}}(s) \qquad (22)$$

Equations (21), (22) and the previous fig.3 allow to establish the block diagram of fig. 6.

The expressions of $A_e(S)$ $B_e(S)$ $C_e(S)$ for buck boost and buck-boost in the two conduction modes are given in table 3 and the general form of the transfer functions is :

$$\tilde{V}_o(s) = \left[S(s)A(s) + S(s)F_{e_1}(s)\left(A(s)C_e(s) + A_e(s)C(s)\right)\right]\tilde{D}(s)/\Delta(s)$$
$$+ S(s)F_{e_2}(s)C(s)\tilde{\mathcal{E}}(s)/\Delta(s) \qquad (23)$$

with

$$\Delta(s) = 1 + S(s)B(s) + F_{e_1}(s)C_e(s) + S(s)F_{e_1}(s)\left(B(s)C_e(s) - B_e(s)C(s)\right)$$

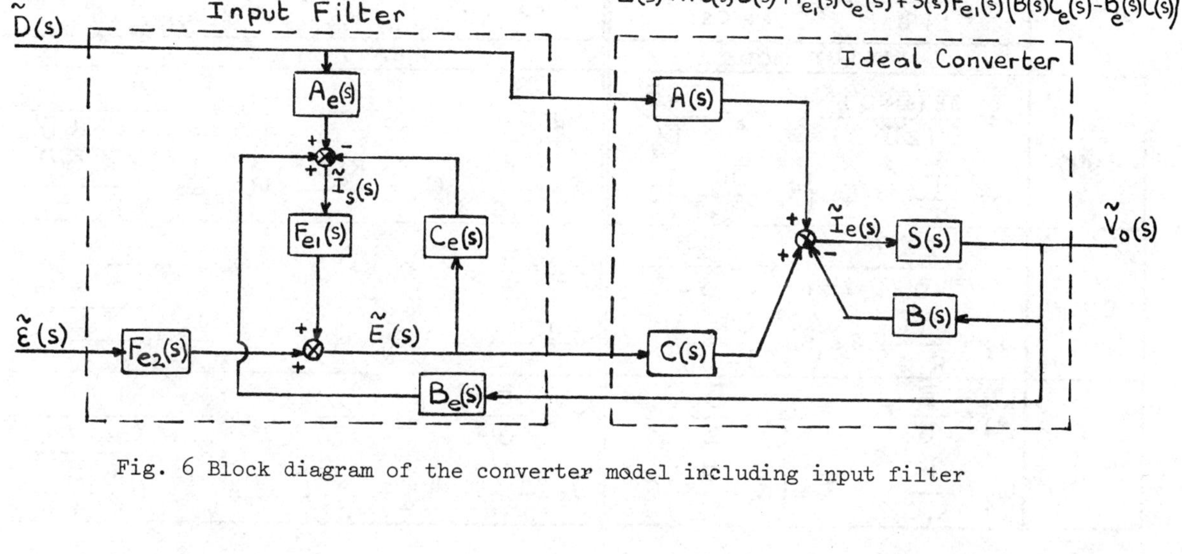

Fig. 6 Block diagram of the converter model including input filter

As well as for A, B and C, A_e, B_e and C_e reduce to gains in light mode and it is interesting to notice that in heavy mode the quantity $B(S)C_e(S) - B_e(S)C(S)$ is always zero.

Considerations about frequency effect can be made as in paragraph 3.3. by introduction of corrective terms in $A_e(S)$ $B_e(S)$ and $C_e(S)$ for heavy mode.

5 - INTRODUCTION OF AN OUTPUT FILTER

The objective of an output filter is to attenuate the ripple existing on the output voltage V_o. The configuration of the output network is given in fig. 7

Due to the capacitor C, the voltage V_e is the sum of a mean value and of a ripple. This later is effectively more important than the ripple remaining on V_o, but is relatively small compared to the mean value $\overline{V}_e$ of $V_e(t)$ and is of the same order of the ripple existing on V_o for a classical output network.

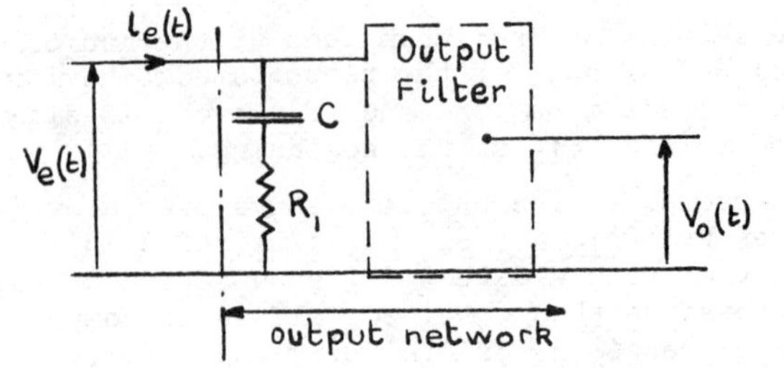

Fig 7 Output network configuration with filter

It is so possible to make the assumption that $V_e(t)$ is constant over the nominal period T and equal to its mean value $\overline{V}_e$. The used procedure is then applicable by taking into account the following considerations.

- The output mean voltage $\overline{V}_o$ which was used to establish the expressions of $\overline{D}$,

TABLE 3 Characteristic coefficients for an input filter (cf fig.6)

	LIGHT MODE			HEAVY MODE		
	$A_e(s)$	$B_e(s)$	$C_e(s)$	$A_e(s)$	$B_e(s)$	$C_e(s)$
BOOST	$-\frac{2\bar E(\bar D+\bar D')^2}{R_o\bar D\bar D'^2}$	$\frac{\bar D+\bar D'}{R_o\bar D}$	$\frac{(\bar D+\bar D')^3}{R_o\bar D\bar D'^2}$	$-\frac{\bar E}{\bar D' Ls}$	$\frac{\bar D'}{Ls}$	$\frac{1}{Ls}$
BUCK	$-\frac{2\bar E\bar D}{R_o(\bar D+\bar D')^2}$	$\frac{\bar D^2}{R_o\bar D'(\bar D+\bar D')}$	$\frac{\bar D^2}{R_o\bar D'(\bar D+\bar D')}$	$-\left(\frac{\bar D\bar E}{Ls}+\frac{\bar D\bar E}{R_o}\right)$	$\frac{\bar D}{Ls}$	$\frac{\bar D^2}{Ls}$
BUCK-BOOST	$-\frac{2\bar E\bar D}{R_o\bar D'^2}$	0	$\frac{\bar D^2}{R_o\bar D'^2}$	$-\left(\frac{\bar D\bar E}{\bar D'Ls}+\frac{\bar D\bar E}{R_o\bar D'^2}\right)$	$\frac{\bar D\bar D'}{Ls}$	$\frac{\bar D^2}{Ls}$

TABLE 4 Ideal converters transfer function with $S(s) = R_o(1+R_1Cs)/(1+(R_o+R_1)Cs)$

	$\tilde V_o = G_d\frac{1+s/\omega_z}{1+\tau s}\tilde D + G_e\frac{1+s/\omega_z}{1+\tau s}\tilde E$	$\tilde V_o = G_{cd}\frac{(1-s/\omega_a)(1+s/\omega_z)}{1+s/Q\omega_o+s^2/\omega_o^2}\tilde D + G_{ce}\frac{(1+s/\omega_z)}{1+s/Q\omega_o+s^2/\omega_o^2}\tilde E$
	LIGHT MODE	HEAVY MODE
BOOST	$G_d=\frac{2\bar E(\bar D+\bar D')}{\bar D'(2\bar D+\bar D')}$ $G_e=\frac{\bar D+\bar D'}{\bar D'}$ $\omega_z=\frac{1}{R_1C}$ $\tau=\left(R_1+\frac{\bar D R_o}{2\bar D+\bar D'}\right)C$	$G_{cd}=\frac{\bar E}{\bar D'^2}$ $G_{ce}=\frac{1}{\bar D'}$ $Q\omega_o=\frac{1}{R_1C+L/R_o\bar D'^2}$ $\omega_z=\frac{1}{R_1C}$ $\omega_a=\frac{R_o\bar D'^2}{L}$ $\omega_o=\frac{1}{\sqrt{LC}}\sqrt{\frac{R_o\bar D'^2}{R_o+R_1}}$
BUCK	$G_d=\frac{2\bar E\bar D'}{(\bar D+\bar D')(\bar D+2\bar D')}$ $G_e=\frac{\bar D}{\bar D+\bar D'}$ $\omega_z=\frac{1}{R_1C}$ $\tau=\left(R_1+\frac{\bar D'R_o}{\bar D+2\bar D'}\right)C$	$G_{cd}=\bar E$ $G_{ce}=\bar D$ $Q\omega_o=\frac{1}{R_1C+L/R_o}$ $\omega_z=\frac{1}{R_1C}$ $\omega_a=\infty$ $\omega_o=\frac{1}{\sqrt{LC}}\sqrt{\frac{R_o}{R_o+R_1}}$
BUCK-BOOST	$G_d=\frac{\bar E}{\bar D'}$ $G_e=\frac{\bar D}{2}$ $\omega_z=\frac{1}{R_1C}$ $\tau=\left(R_1+\frac{R_o}{2}\right)C$	$G_{cd}=\frac{\bar E}{\bar D'^2}$ $G_{ce}=\frac{\bar D}{\bar D'}$ $Q\omega_o=\frac{1}{R_1C+L/R_o\bar D'^2}$ $\omega_z=\frac{1}{R_1C}$ $\omega_a=\frac{R_o\bar D'^2}{\bar D L}$ $\omega_o=\frac{1}{\sqrt{LC}}\sqrt{\frac{R_o\bar D'^2}{R_o+R_1}}$

$\bar{D}'$, A(S), B(S), C(S), A_e(S), B_e(S) and C_e(S) has to be replace by $\bar{V}_e$. On the other hand, the values of t_c, t_d, t_1, I_{MAX} and i_{min} are calculated with energy considerations over the period T in nominal behavior. For the classical RC output network, the energy in the load R_o was

$$\bar{V}_o^2 T / R_o$$

It is now

$$\bar{V}_e^2 T / R_e$$

if R_e is the equivalent resistance of the output network, its inductors beeing short-circuited and capacitors in open circuits. So, in the considered expressions, it is also necessary to replace R_o by R_e. Finally, the two modifications above mentioned give new quantities, litterally unchanged which will be starred to indicate modifications of the values.

- If the equations of the output network including output filter (cf fig.7) are written :

$$\tilde{V}_e(s) = S(s)\,\tilde{I}_e(s) \qquad (24)$$

and

$$\tilde{V}_o(s) = F_s(s)\,\tilde{V}_e(s) \qquad (25)$$

(the static gain in (25) being $F_S(0)=\bar{V}_o/\bar{V}_e$) then block diagrams including output filter are immediately derived from fig.3 or 6 as shown in fig.8 (non input filter). It is interesting to notice that with the procedure used, it is particularly easy to take into account the influence of the output filter since its dynamic is only "added".

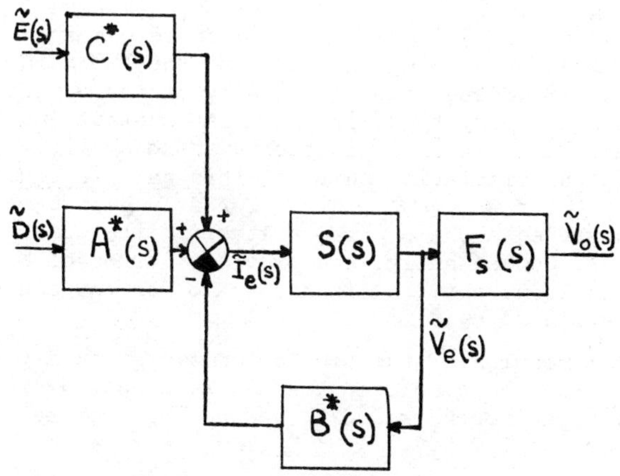

Fig. 8 Block diagram of the converter model including output filter

6 - USE OF THE OBTAINED RESULTS

6.1 - Open-Loop Behavior

For an ideal converter, the block diagram of fig. 3 and the values of table 1 allow to establish the transfer functions of the three switched dc-dc converters considered in the two conduction modes. The different expressions which characterize open-loop behavior are summarized in table 4 . Let us notice that in this case S(s) is given by equation (9) and $F_S(s) = 1$.

The results are of the same nature than those obtained by other methods (ref. 1,3,5,6,8,9, 10) and absolutely identical with those of Wester and Middlebrook (1) in the heavy mode or the results of Prajoux *et al.* (5) in the light mode.

6.2 - Design of a Feedback Compensation through State Space Methods

First of all, it is necessary to make a remark about variations of the duty cycle. It is known that its variations do not react immediately on the mean value of the injected current ; this problem frequently occurs when continuous models are considered. To take this phenomenon into account, a delay equal to half a cycle can be introduced between the variations $\tilde{d}_M$ of the PWM and $\tilde{d}$ "seen" by the converter.

This delay is approximated by a first order Pade approximation :

$$e^{-\frac{T}{2}s} \# (1-\tfrac{T}{4}s)/(1+\tfrac{T}{4}s) = -1+2/(1+\tfrac{T}{4}s)$$

For an ideal buck converter in heavy mode with $R_l = 0$ the results of table 2 lead to a state space model as

$$\begin{bmatrix} \tilde{i}_c \\ \tilde{v}_o \\ \alpha \end{bmatrix}' = \begin{bmatrix} -\frac{1}{R_oC} & -\frac{1}{L} & \frac{\bar{E}}{L} \\ \frac{1}{C} & 0 & 0 \\ 0 & 0 & -\frac{4}{T} \end{bmatrix} \times \begin{bmatrix} \tilde{i}_c \\ \tilde{v}_o \\ \alpha \end{bmatrix} + \begin{bmatrix} -\frac{\bar{E}}{L} \\ 0 \\ \frac{8}{T} \end{bmatrix} \tilde{d}_M$$

Where $\tilde{i}_c$ is the variation of the mean value of the capacitor current and α is related to the delay approximation.

Assuming $\tilde{i}_c$ and $\tilde{v}_o$ measurable and the feedback law :

$$\tilde{d}_M = k_i \tilde{i}_c + k_v \tilde{v}_o$$

the closed loop characteristic equation will write :

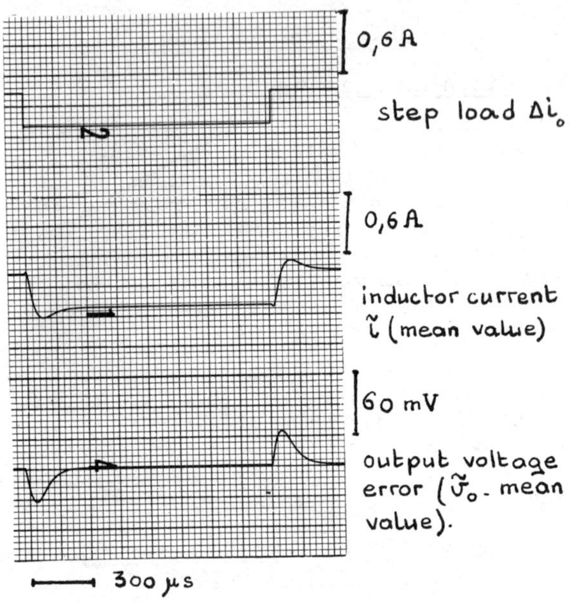

Fig 9 Simulation of the small signals continuous model : transients to a step load Δi_o

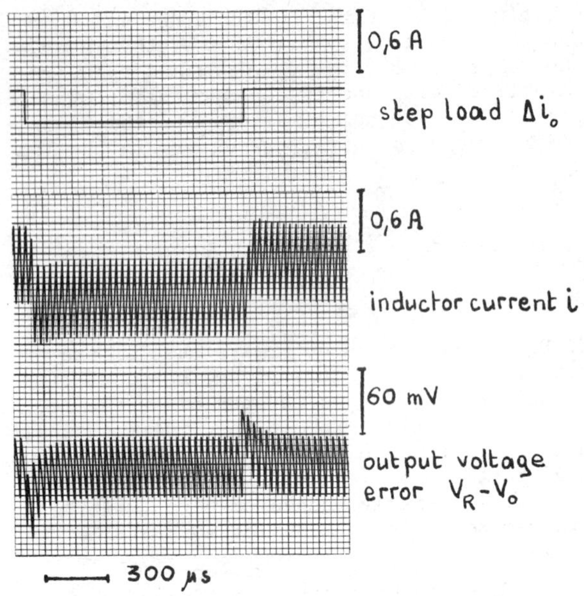

Fig 10 Simulation of the real converter : transients to a step load Δi_o

$$s^3 + \left(\frac{4}{T} + \frac{1}{R_o C} + \frac{\bar{E}}{L} k_i\right) s^2 + \left(\frac{4}{R_o C T} + \frac{1}{LC} - \frac{4\bar{E}}{LT} k_i + \frac{\bar{E}}{LC} k_v\right) s + \frac{4}{LCT}\left(1 - k_v\right) = 0$$

Having only two degrees of freedom for a third order model there will be a necessary relationship between these three modes.

Taking a typical buck regulator (ref.7) with $\bar{E} = 30$ V, $\bar{V}_o = 20$V, $L = 250\,\mu$H, $C = 300\,\mu$F, $R_o = 10\,\Omega$, $T = 30\,\mu$S, $t_c = 20\,\mu$s, $t_d = 10\,\mu$s or in the same scale as the one used in analog simulation

$\bar{E} = 1$ $\bar{V}_o = 2/3$, $L = 1/1,2$, $C = 1$,

$R_o = 10$, $T = 0,1$

the characteristic equation is

$$s^3 + (40,1 + 1,2\,k_i)\,s^2 + (5.2 - 48\,k_i + 1,2\,k_v)\,s + 48 - 48\,k_v = 0$$

with between the modes-λ_i, the relationship :

$$\lambda_1 \doteq 40\,\frac{1610,4 - [\lambda_2\lambda_3 + 40(\lambda_2 + \lambda_3)]}{(40)^2 + [\lambda_2\lambda_3 + 40(\lambda_2 + \lambda_3)]}$$

For $k_i = -7,278$ and $k_V = 22,68$ on has $\lambda_2 = \lambda_3 = 10$, $\lambda_1 = 11,3664$.

This should allow good dynamics with a time response of 3 x 0,1 sec that is a 95% response time of 9 cycles.

The analog simulation of the small signals continuous model for a step load Δi_o of 16,7% of the nominal output current verifies these results as can be seen fig.9.

More interesting are the results obtained by use of the computed feedback law on the simulation of the real regulator. The gains k_v and k_i have to be applied to the output voltage error V_R-V_o (V_R = reference voltage) and the capacitor current i_c of zéro mean value, but taking into account the PWM gain. The simulated modulator beeing of the coincidence-memory type, its gain is constant and equal to T/A if a is the sawtooth amplitude. In the simulation unities, this gain is equal to 1 for $\Delta t_c/\Delta e_c$ or 10 for $\tilde{d}/\Delta e_c$ if e_c represents the control voltage of the modulator. Consequently, the previous values of k_i and k_v have to be divided by 10 for the generation of e_c;

The results of the simulation are given fig. 10 for the same step load Δi_o, and introducing a capacitor $R_1 = 0,077\,\Omega$ and an inductor resistor $r = 0.015\,\Omega$.

It is interesting to notice the excellent prediction given by the small signals continuous model for the mean values of the inductor current i and the output voltage error $V_R - V_o$ for which the ripple can be seen on fig.10. The proposed model appears as a very practice tool to design feedback compensation through state space methods.

6.3 - Destabilizing Effect of an input Filter

When using input/output filters, closed form expressions as those given table 4 can be given only if F_s, F_{e_1}, F_{e_2} are specified.

So, consider the configuration given fig.11

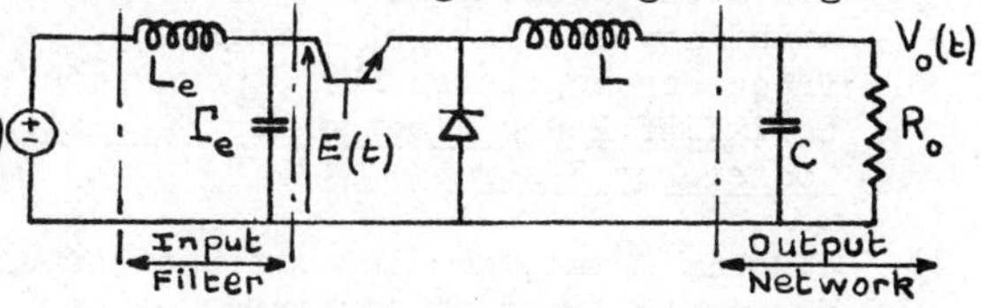

Fig. 11 Configuration of an ideal buck plus input filter

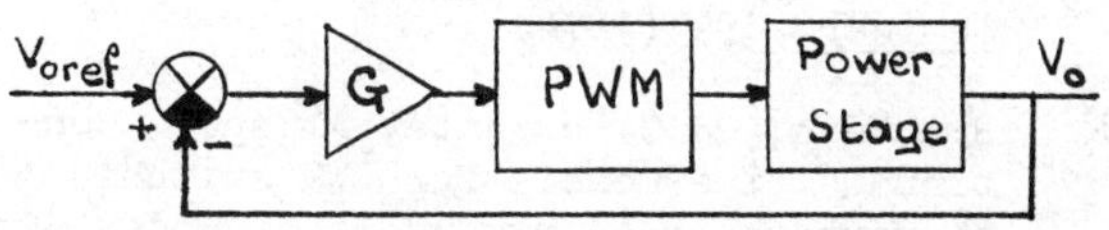

Fig. 12 Converter in feedback configuration

- In the absence of the input filter and assuming $\tilde{\mathcal{E}} = 0$, from table 3 it results that :

$$\frac{\tilde{V}_o(s)}{\tilde{D}(s)} = \bar{E}\,\frac{1}{1 + \frac{L}{R_o}s + LCs^2} \tag{26}$$

where, for simplicity, the half a cycle delay and frequency effect have been neglected.

In the closed loop configuration of fig.12 noting K the gain resulting from the amplifier G, the PWM, and the input voltage $\bar{E}$, the transfer function is :

$$\frac{V_o}{V_{oref}} = \frac{K}{K+1+\frac{L}{R_o}s + LCs^2} \tag{27}$$

and it is clear, under the previous hypothesis,that the system is always stable.

- Including now the input filter the new open-loop transfer function is (cf equ.(23) and tables 1 and 2) :

$$\tilde{V}_o(s)/\tilde{D}(s) = \left(1 - \frac{\bar{D}^2}{R_o}L_e s + L_e\Gamma_e s^2\right)/\Delta(s)$$

with (28)

$$\Delta(s) = 1 + \left(\frac{L}{R_o} + \frac{\bar{D}^2}{R_o}L_e\right)s + \left(LC + L_e\Gamma_e + \bar{D}^2L_eC\right)s^2 + \frac{LL_e\Gamma_e}{R_o}s^3 + LL_eC\Gamma_e s^4$$

which exhibits a non minimum phase zero.

The closed loop characteristic equation :

$$K+1+\left[\frac{L}{R_o} + \frac{\bar{D}^2}{R_o}L_e(1-K)\right]s + \left[LC + L_e\Gamma_e(1+K) + \bar{D}^2L_eC\right]s^2 + \frac{LL_e\Gamma_e}{R_o}s^3 + LL_eC\Gamma_e s^4 = 0 \tag{29}$$

shows that the necessary condition of stability is

$$\frac{L}{R_o} + \frac{\bar{D}^2}{R_o}L_e(1-K) > 0$$

or

$$K < 1 + \frac{L}{\bar{D}^2L_e} \tag{30}$$

The necessary and sufficient condition will be found by applying the Routh criterion to equation (29) :

$$-K^2L_e(\Gamma_e + \bar{D}^2C) + KC(L + \bar{D}^2L_e) + L_e\Gamma_e > 0 \tag{31}$$

The variation of this quantity is indicated in fig. 13 :

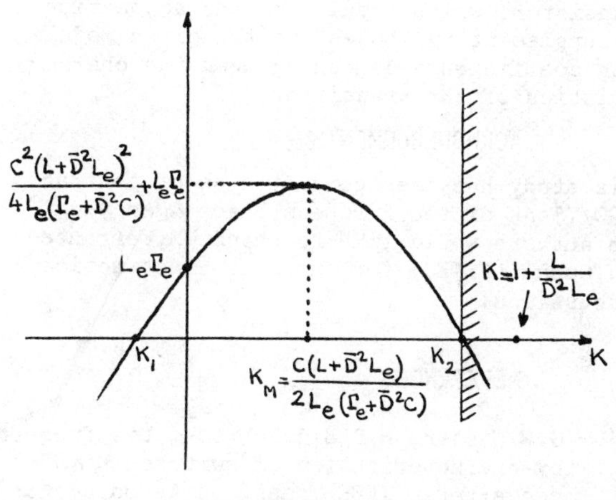

Fig. 13 Stability condition for a closed loop regulated buck with input filter.

leading to $K < K_2$ (32)

This simple example points out the destabilizing effect of an input filter, for too large value of K.

If now a resistance r_e is introduced with the inductor L_e, the same analysis shows that the necessary and sufficient condition of stability for all values of K is :

$$\frac{L_e}{\Gamma_e r_e} < \frac{R_o}{\bar{D}^2} \tag{33}$$

under the condition that the stability of the closed-loop system is non conditional (the damping decreases when increasing K).

This condition has been found by Middlebrook (4) by impedances considerations. If $r_e = 0$, this condition is never true, but not in contradiction with condition (32), depending on K.

7 - CONCLUSION

The main interesting result of the proposed method is certainly the unified presentation of switched dc-dc converters under a block-diagram form, for the two conduction modes and taking into account the use of in and output filters. The calculus required to obtain the values A, B, C, A_e, B_e, C_e characterizing each converter are very simple and this kind of approach allows to see clearly the influence of each parameter. It so becomes easy to analyse the influence of filters whose structures are known and to design feedback compensation for required performances.

Extensions of the method have been made for PWM with variable frequency and constant t_c. In this case, only the coefficients of A(S) (or $A^*(S)$) and $A_e(S)$ (or $A_e^*(S)$) are modified the new input of the open-loop converter beeing $\tilde{d}' = \tilde{t}_{OFF}/T_o$, if T_o is the nominal cycle. Adaptations have also been made to take into account the storage time of the switching transistor, which results in the adjonction of supplementary "boxes" on the given models, with coefficients depending upon the characteristics of the transistor.

ACKNOWLEDGMENTS

This study has been conducted under contract 2590/75 AK of the European Space Agency. The authors would like to thank MM.Ferrante and Capel (ESTEC) for numerous constructive discussions.

REFERENCES

(1) G.W.Wester, R.D.Middlebrook, Low frequency characterization of switched DC.DC converters, IEEE Trans. on Aerospace and Electronic Systems. AES-9 n°3, 376 (1973)

(2) R.D.Middlebrook, A continuous model for the tapped-inductor boost converter, IEEE Power Electronics Specialists Conference Record, 63 (1975)

(3) R.D.Middlebrook, S.Cuk, A general unified to modelling switching converter power stages,IEEE Power Electronics Specialists Conference Record, (1976)

(4) R.D.Middlebrook, Input filter considerations in design and application of switching regulators, IEEE Industry Applications Society Annual Meeting Chicago, (1976)

(5) R.Prajoux, J.C.Marpinard, J.Jalade, Accurate mathematical modelling of PWM regulator, II part : Continuous models. Final Report ESTEC 2026/73 AK (1975)

(6) A.Capel, J.G.Ferrante, R.Prajoux, State variable stability analysis of multiloop PWM controlled regulators in light and heavy mode, IEEE Power Electronics Specialists Conference Record, 91 (1975)

(7) R.P.Iwens, Y.Yu, Time modelling and stability analysis of integral pulse frequency modulated DC/DC power converter, IEEE Power Electronics Specialists Conference Record, 80 (1975)

(8) Y.Robin-Jouan, Modèles mathématiques des convertisseurs continu-continu à modulation de largeur d'impulsions. Applications aux études de stabilité en boucles fermées, Proceedings Frascati Seminar, 194 (1974)

(9) H.A.Owen, J.C.Ferrante, A.Capel, Continuous time models for PWM switched converters in heavy and light modes, Tech. Report ESTEC, Noordwijk, (1976)

(10) R.Prajoux, J.C.Marpinard, J.Jalade, Etablissement de modèles mathématiques pour régulateurs de puissance à modulation de largeur d'impulsions. I.Modèles discrets, ESA Sc.Tech. Review, 2 n°1, 25 (1976)

(11) A.J.Fossard, M.Clique, Modélisation des cellules élémentaires, Tech.Report 1 ESTEC 2590/75 AK, (1975)

(12) M.Clique, A.J.Fossard, Modèle linéaire d'un régulateur shunt PWM dissipatif, ESA Sc. Tech. Review, 2 n°3, 221 (1976)

(13) R.Prajoux, J.C.Marpinard, J.Jalade, Etablissement de modèles mathématiques pour régulateurs de puissance à modulation de largeur d'impulsions. II. Modèles continus. ESA Sc. Tech. Review, 2 n°2, (1976)

(14) R.D.Middlebrook, S.Cuk, Modelling and analysis methods for DC to DC switching converters, IEEE International Semiconductor Power converter conference, Orlando, Florida, (1977)

(15) M.Clique, A.J.Fossard, A general model for switching converters, IEEE Trans. on Aérospace and Electronic Systems to appear in AES 13 n°3, (1977)

(16) A.J.Fossard, M.Clique, Méthodes du courant injecté pour la modélisation des convertisseurs continu continu, Tech. Report 2 ESTEC 2590/75 AK, (1976)

(17) A.J.Fossard,M.Clique, Linéarisation et résolution numérique des équations d'état.Cas des systèmes à fréquences variables, Tech.Report 3 ESTEC 2590/75 AK

(18) A.J.Fossard,M.Clique,Représentation des cellules de puissance sous forme d'état Tech.Report 4 ESTEC 2590/75 AK, (1977)

DC AND AC ANALYSIS OF THYRISTOR CIRCUITS BY COORDINATE TRANSFORMATION AND DESCRIBING-FUNCTION METHOD

H. Haneda and T. Maruhashi

Department of Electrical Engineering, Kobe University, Kobe, Japan

ABSTRACT

A new and practical method is developed for evaluating approximately both dc and fundamental ac components of waveforms in steady-state thyristor circuits. The process is purely algebraic and is not based on simulation techniques, which utilizes a modified form of describing-function method for each thyristor. The method can be applied, at least in principle, to any configuration of linear time-invariant circuits containing several thyristors. Two simple but fundamental examples are included to show the limit and utility of the proposed method.

1. INTRODUCTION

The evaluation of dc and ac components of steady-state waveforms in thyristor circuits is a great concern for circuit engineers. Two conventional techniques are available: perturbation approach for slightly nonlinear communication circuits [1] and describing-function approach for nonlinear control circuits and systems [2],[3],[4]. Since a thyristor element can be considered as a switching device, neither of those methods can readily be applied. A direct but time-consuming method is to simulate the thyristor circuit for time solution curves and to perform Fourier analysis on the waveforms thus obtained. The difficulty arises from the fact that each time thyristors change their switching states, say from on state to off state, the differential equation for the thyristor circuits has to be reconstructed.

In this paper, we follow a describing-function approach but in a modified form, which enables us to evaluate both dc and ac components of steady-state waveforms in thyristor circuits. The proposed method is applicable at least in principle to any configuration of thyristor circuits. First, we characterize a thyristor element as a nonlinear resistor on (v,i) plane. A rotational transformation is introduced, which makes it possible to define both dc and ac describing functions. Second, those dc and ac describing functions for the transformed thyristor are computed as formulas taking the conduction angle γ into account. Those dc and ac describing functions once computed are commonly used for any other thyristors. Third, the proposed analysis method is carefully developed for thyristor circuits using those describing functions. Finally, we pick up two simple but fundamental examples to show the limit and utility of the method: a halfwave-rectifier circuit and an ac regulator controlled by back-to-back thyristor pair.

2. THYRISTOR CHARACTERISTIC AND ITS DESCRIBING FUNCTION

We assume that each thyristor is characterized on (v,i) plane as shown in Fig. 1.

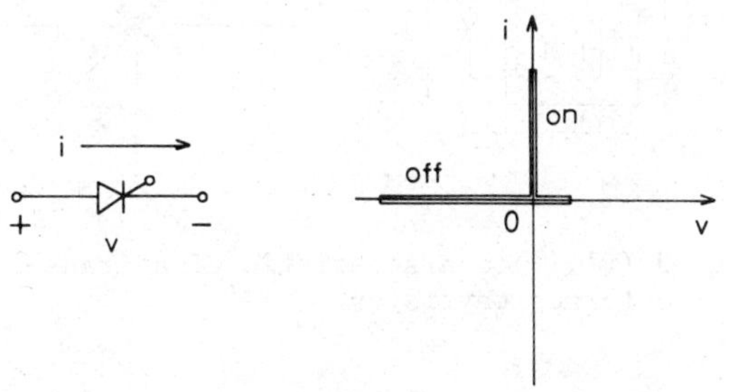

Fig. 1 (v,i) characteristic of a thyristor.

The thyristor characteristic shows the ideal switching behavior for on and off states; the transition from off to on state occurs due to the firing gate signal applied at the firing angle delay α; and the other transition ocurrs due to the nonlinear characteristic at the origin (0,0). Note that the thyristor element is a nonlinear resistor which is neither current controlled nor voltage controlled, i.e., the selection of input-output pair is not possible. One remedy for this is to introduce a regular transformation on the thyristor as shown in Fig. 2. Connect two linear two-ports, whose transmission matrices are

$$\begin{bmatrix} A & B \\ C & D \end{bmatrix} = \frac{1}{\sqrt{2}} \begin{bmatrix} 1 & -1 \\ 1 & 1 \end{bmatrix} \tag{1}$$

and

$$\begin{bmatrix} A & B \\ C & D \end{bmatrix} = \frac{1}{\sqrt{2}} \begin{bmatrix} 1 & -1 \\ 1 & 1 \end{bmatrix}^{-1} = \frac{1}{\sqrt{2}} \begin{bmatrix} 1 & 1 \\ -1 & 1 \end{bmatrix}, \tag{2}$$

to the thyristor, and consider the dotted part in Fig. 2 as a new nonlinear resistor, whose (v',i') characteristic is shown in Fig. 3.

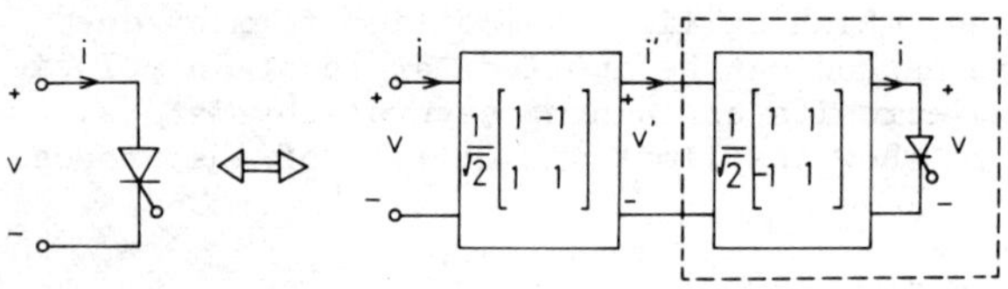

Fig. 2 Equivalent transformation of a thyristor.

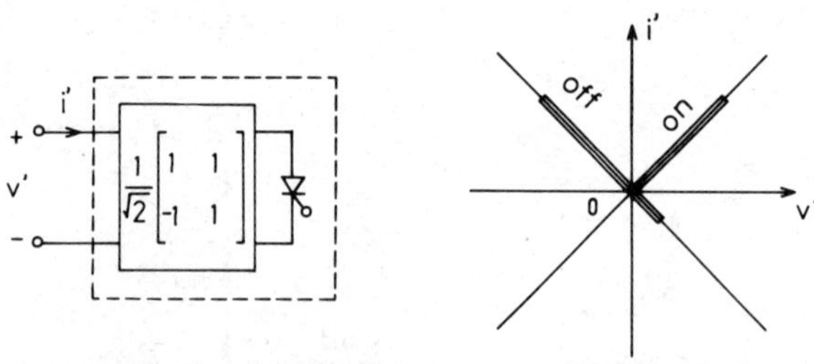

Fig. 3 (v',i') characteristic of a transformed thyristor.

The transformed characteristic is now a function of v', thus enabling us the application of the describing-function method. Note that the transformed characteristic is also graphically obtained by the clockwise rotation of the original thyristor characteristic by 45 degrees, which corresponds to a coordinate transformation from (v,i) plane onto (v',i') plane by

$$\begin{bmatrix} v' \\ i' \end{bmatrix} = \frac{1}{\sqrt{2}} \begin{bmatrix} 1 & 1 \\ -1 & 1 \end{bmatrix} \begin{bmatrix} v \\ i \end{bmatrix} \tag{3}$$

or

$$\begin{bmatrix} v \\ i \end{bmatrix} = \frac{1}{\sqrt{2}} \begin{bmatrix} 1 & -1 \\ 1 & 1 \end{bmatrix} \begin{bmatrix} v' \\ i' \end{bmatrix} \tag{4}$$

Now, apply a test input-signal

$$v'(t) = v_0' + v_1' \sin\omega t \tag{5}$$

which contains a dc component v_0' and an ac component $v_1' \sin\omega t$, to the transformed thyristor in Fig. 3, and let the output current $i'(t)$ be given by

$$i'(t) = i_0' + i_1' \sin(\omega t+\Delta) + \cdots \quad . \tag{6}$$

Then, the dc describing function $N_0(v_0',v_1')$ and the ac describing function $N_1(v_0',v_1')$ are defined by

$$N_0(v_0',v_1') \triangleq i_0'(v_0',v_1')/v_0' \tag{7}$$

$$N_1(v_0',v_1') \triangleq i_1'(v_0',v_1') \cdot e^{j\Delta}/v_1' \tag{8}$$

3. FORMULAS FOR THE DC AND AC DESCRIBING FUNCTIONS OF THE TRANSFORMED THYRISTOR

Let t denote the real time variable and suppose that the firing angle α is given a priori on the real time axis t. The actual waveforms $v'(t)$ and $i'(t)$ of a transformed thyristor would be in the form:

$$v'(t) \triangleq v_0' + v_1' \sin(\omega t+\phi') + \cdots \tag{9}$$

$$i'(t) \triangleq i_0' + i_1' \sin(\omega t+\psi') + \cdots \tag{10}$$

First, we try to compute the conduction angle γ in terms of α, v_0', v_1' and ϕ'. From now on we neglect the higher order terms in $v'(t)$. Shift the real time axis t to a new time axis $\tilde{t}$ by

$$\omega\tilde{t} \triangleq \omega t + \phi'. \tag{11}$$

Then, referring to Fig. 4, the firing angle α' on the $\tilde{t}$ time axis will be

$$\alpha' = [\alpha+\phi'+2k\pi], \tag{12}$$

where the integer k is chosen such that $0 \leq \alpha' \leq 2\pi$.

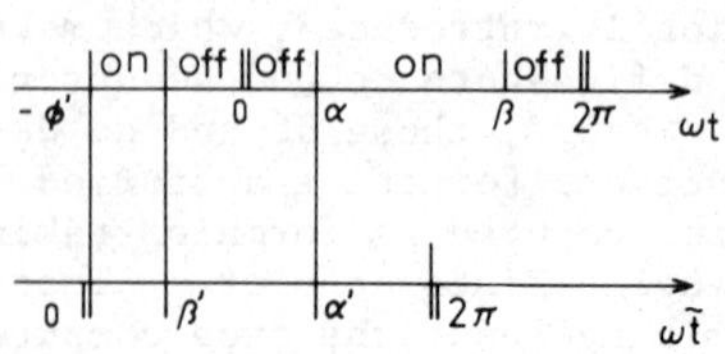

Fig. 4 Relation between the real time axis t and the transformed time axis $\tilde{t}$.

The extinction angle β' on the $\tilde{t}$ time axis should be determined on the interval $[0,2\pi)$ by

$$v'(\beta'/\omega) = v_0' + v_1'\sin\beta' = 0 \tag{13}$$

and

$$\frac{d}{d\tilde{t}}v'(\beta'/\omega) \leqq 0. \tag{14}$$

It can be shown that

$$\beta' = \pi - \mathrm{SIN}^{-1}(-v_0'/v_1'), \tag{15}$$

where $\mathrm{SIN}^{-1}(-v_0'/v_1')$ denotes the principal value of $\arcsin(-v_0'/v_1')$.
From α' and β' in eq.(12) and (15), respectively, the conduction angle γ is given by

$$\gamma = \begin{cases} \beta'-\alpha' = \pi + \mathrm{SIN}^{-1}(v_0'/v_1') - \alpha', & \text{if } \alpha'<\beta' \\ 2\pi-(\alpha'-\beta') = 3\pi + \mathrm{SIN}^{-1}(v_0'/v_1') - \alpha', & \text{if } \alpha'>\beta' \end{cases} \tag{16}$$

or

$$\gamma(v_0',v_1',\phi') = 2\pi + \mathrm{SIN}^{-1}(v_0'/v_1') - [\alpha+\phi'+2k\pi] + \pi\cdot\mathrm{sgn}\{[\alpha+\phi'+2k\pi]-\pi-\mathrm{SIN}^{-1}(v_0'/v_1')\}. \tag{17}$$

If we again shift the t time axis to another new time axis t' by $\omega t' = \omega t - \alpha'$, we may consider that for the test signal $v'(t) = v_0' + v_1'\cdot\sin\omega t'$, the transformed thyristor is in "on" state for $0 \leqq \omega t' < \gamma$ and in "off" state for $\gamma \leqq \omega t' < 2\pi$. Hence, the output $i'(t')$ will be

$$i'(t') = \begin{cases} v_0' + v_1'\sin\omega t', & \text{if } 0 \leqq \omega t' < \gamma \\ -(v_0' + v_1'\sin\omega t'), & \text{if } \gamma \leqq \omega t' < \pi. \end{cases} \tag{18}$$

By making Fourier analysis on $i'(t')$, we obtain

$$i_0' = \frac{1}{2\pi}\int_0^{2\pi} i'(t')d(\omega t') = \frac{1}{\pi}[(\gamma-\pi)v_0' + (1-\cos\gamma)v_1'] \tag{19}$$

and

$$i_1' = \sqrt{A_1^2 + B_1^2} \tag{20}$$

$$\tan\Delta = B_1/A_1, \tag{21}$$

where

$$A_1 \triangleq \frac{1}{\pi}\int_0^{2\pi} i'(t')\sin\omega t' d(\omega t')$$

$$= \frac{2}{\pi}(1-\cos\gamma)v_0' + \frac{1}{\pi}[(\gamma-\pi)-\frac{1}{2}\sin 2\gamma]v_1' \tag{22}$$

and

$$B_1 \triangleq \frac{1}{\pi}\int_0^{2\pi} i'(t')\cos\omega t' d(\omega t')$$

$$= \frac{2}{\pi}\sin\gamma\cdot v_0' + \frac{1}{\pi}\sin^2\gamma\cdot v_1'. \tag{23}$$

Hence, we obtain the formulas for N_0 and N_1:

$$N_0(v_0',v_1',\gamma) = \frac{1}{\pi}[\gamma-\pi+(1-\cos\gamma)v_1'/v_0'] \tag{24}$$

$$N_1(v_0',v_1',\gamma) = \sqrt{A_1^2+B_1^2}\, e^{j\Delta}/v_1'. \tag{25}$$

By substituting eq.(17) into eq.(24) and (25), we can observe both N_0 and N_1 are functions of v_0', v_1' and ϕ'.

4. METHOD OF WAVEFORM ANALYSIS

We consider a class of thyristor circuits which consist of independent voltage sources $E_s(t) = [E_s^{(1)}, E_s^{(2)}, \cdots, E_s^{(n)}]^T$, thyristors $Th_1, Th_2, \cdots, Th_m$, linear time-invariant resistors R, capacitors C and inductors L as shown in Fig. 5.

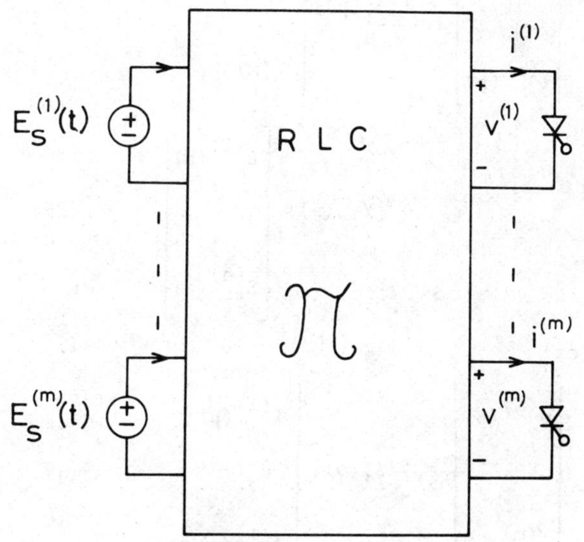

Fig. 5 A thyristor circuit.

We want to compute both dc and ac components of waveforms in the circuit, especially those of the load current. Transform each thyristor by eq.(3) or (4) and extract the transformed thyristors and independent voltage sources as shown in Fig. 6.

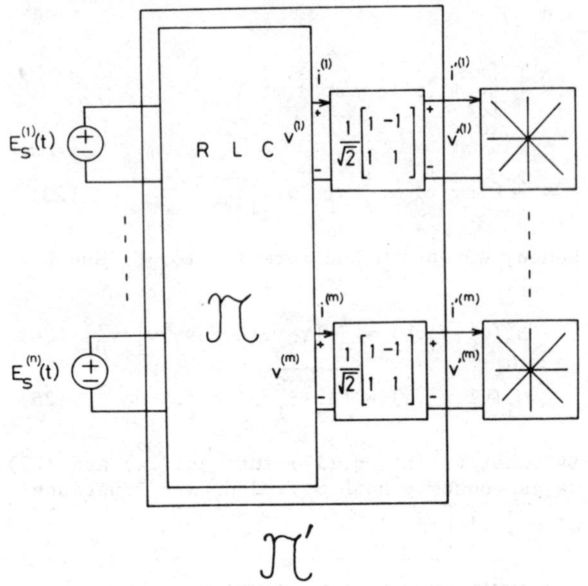

Fig. 6 A coordinate transformation of a thyristor circuit.

Suppose that from the RLC subnetwork $\mathcal{N}$ of the thyristor circuit in Fig. 5, we have a relation

$$\begin{bmatrix} i^{(1)}(p) \\ \vdots \\ i^{(m)}(p) \end{bmatrix} = [Y_{11}(p)] \begin{bmatrix} v^{(1)}(p) \\ \vdots \\ v^{(m)}(p) \end{bmatrix} + [Y_{12}(p)] \begin{bmatrix} E_s^{(1)}(p) \\ \vdots \\ E_s^{(n)}(p) \end{bmatrix} \qquad (26)$$

or

$$\begin{bmatrix} v^{(1)}(p) \\ \vdots \\ v^{(m)}(p) \end{bmatrix} = [Z_{11}'(p)] \begin{bmatrix} i^{(1)}(p) \\ \vdots \\ i^{(m)}(p) \end{bmatrix} + [Z_{12}(p)] \begin{bmatrix} E_s^{(1)}(p) \\ \vdots \\ E_s^{(n)}(p) \end{bmatrix} \qquad (27)$$

where $p \stackrel{\Delta}{=} d/dt$ is the differential operator.

Only Norton equivalent formulation, i.e., eq.(26) is explained. For the other Thevenin equivalent formulation, i.e., eq.(27), the argument is quite similar. The circuit in Fig. 6 can readily be redrawn into a block diagram as shown in Fig. 7.

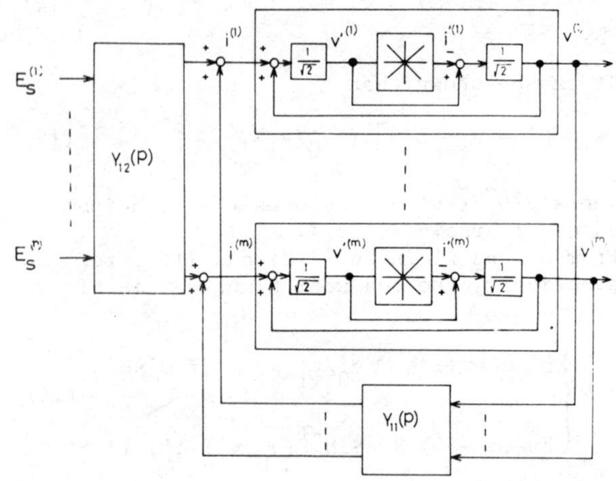

Fig. 7 Block-diagram representation of a thyristor circuit.

Replace each transformed thyristor by dc and ac describing functions eq.(24) and (25), and decompose the block diagram into dc and ac component block diagrams as shown in Fig. 8, (a) and (b), respectively.

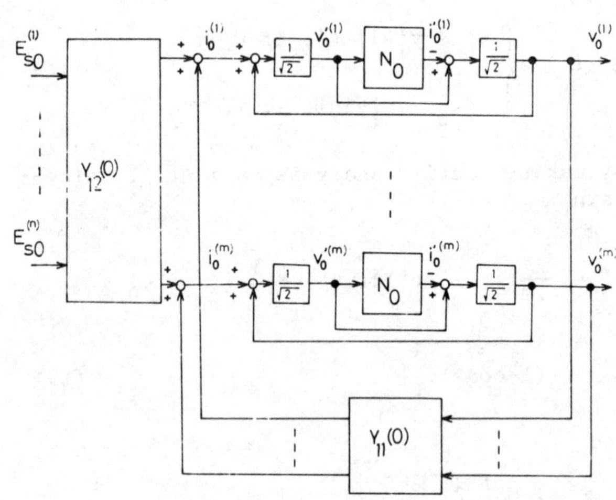

Fig. 8 (a) Dc component block diagram of a thyristor circuit.

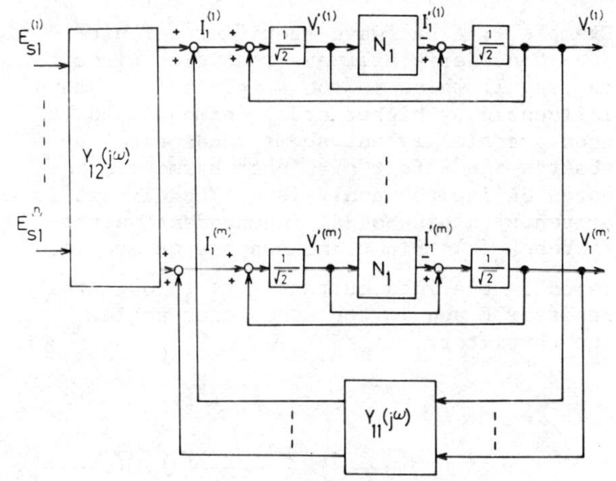

Fig. 8 (b) Ac component block diagram of a thyristor circuit.

In Fig. 8, (b), $V_1'^{(i)}$ and $I_1'^{(i)}$, $i=1,2,\cdots,m$ denote the phasors of $v_1'^{(i)}\sin(\omega t+\phi'^{(i)})$ and $i_1'^{(i)}\sin(\omega t+\psi'^{(i)})$, respectively. Let

$$[N_0] \triangleq \text{diag}\{N_0(v_0'^{(1)},v_1'^{(1)},\phi'^{(1)}),\cdots,\ N_0(v_0'^{(m)},v_1'^{(m)},\phi'^{(m)}\}, \tag{28}$$

$$v_0 \triangleq [v_0^{(1)},\cdots,v_0^{(m)}]^T, \tag{29}$$

$$i_0 \triangleq [i_0^{(1)},\cdots,i_0^{(m)}]^T, \tag{30}$$

$$[N_1] \triangleq \text{diag}\{N_1(v_0'^{(1)},v_1'^{(1)},\phi'^{(1)}),\cdots,\ N_1(v_0'^{(m)},v_1'^{(m)},\phi'^{(m)}\}, \tag{31}$$

$$V_1 \triangleq [V_1^{(1)},\cdots,V_1^{(m)}]^T, \tag{32}$$

and

$$I_1 \triangleq [I_1^{(1)},\cdots,I_1^{(m)}]^T. \tag{33}$$

Then, Fig. 8 is reduced to the one-line diagram as shown in Fig. 9. If we refer to the transformation formula eq.(3), we can express the one-line diagram in terms of the new transformed variables

$$v_0' \triangleq [v_0'^{(1)},\cdots,v_0'^{(m)}]^T, \tag{34}$$

$$i_0' \triangleq [i_0'^{(1)},\cdots,i_0'^{(m)}]^T, \tag{35}$$

$$V_1' \triangleq [V_1'^{(1)},\cdots,V_1'^{(m)}]^T, \tag{36}$$

and

$$I_1' \triangleq [I_1'^{(1)},\cdots,I_1'^{(m)}]^T \tag{37}$$

as shown in Fig. 10.

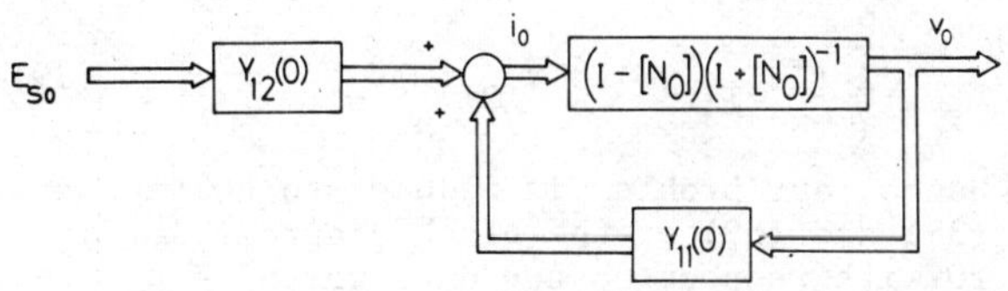

(a) Dc component one-line block diagram.

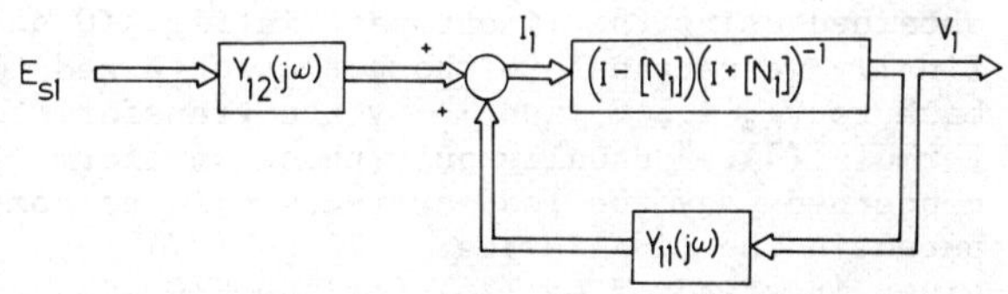

(b) Ac component one-line block diagram.

Fig. 9 One-line block diagram of a thyristor circuit.

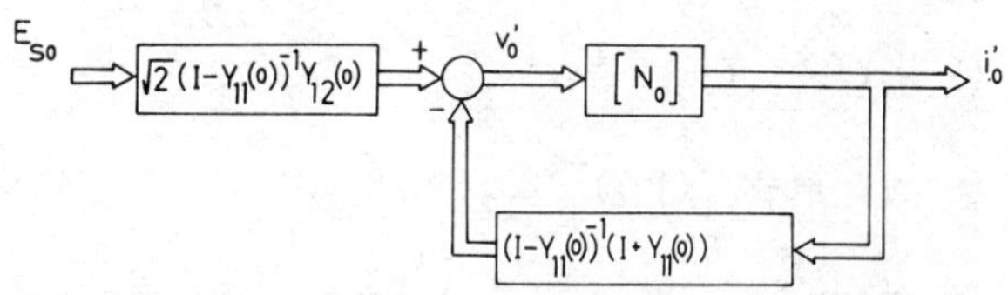

(a) Dc component one-line block diagram.

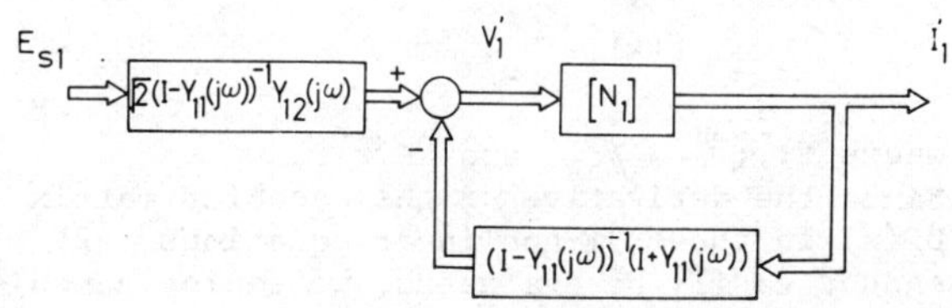

(b) Ac component one-line block diagram.

Fig. 10 A transformed one-line block diagram of a thyristor circuit.

From the dc and ac block diagrams in Fig. 10, we obtain

$$\{I + (I-Y_{11}(0))^{-1}(I+Y_{11}(0))[N_0]\}v_0' =$$

$$\sqrt{2}\ (I-Y_{11}(0))^{-1}Y_{12}(0)E_{s0} \qquad (38)$$

and

$$\{I + (I-Y_{11}(j\omega))^{-1}(I+Y_{11}(j\omega))[N_1]V_1' =$$

$$\sqrt{2}\ (I-Y_{11}(j\omega))^{-1}Y_{12}(j\omega)E_{s1}. \qquad (39)$$

Hence, our problem is reduced to solving eq. (38) and (39) or the block diagrams in Fig. 10 with respect to the unknowns $v_0' \triangleq [v_0'^{(1)},\cdots,v_0'^{(m)}]^T$ and $V_1' \triangleq [V_1'^{(1)},\cdots,V_1'^{(m)}]^T = [v_1'^{(1)}e^{j\phi'^{(1)}},\cdots,v_1'^{(m)}e^{j\phi'^{(m)}}]^T$. The transformed current variables i_0' and I_1' are obtained using the linear part in Fig. 10 and finally we retransform those thus obtained back to v_0, i_0, V_1 and I_1 by the transformation formula (4). Usually only those waveforms concerned, say the load currents will be computed in the final stage. If eq.(27) is given in stead of eq.(26) for the RLC subnetwork in Fig. 5, we can obtain the following equations to be solved similar to eq.(38) and (39):

$$\{I - (I-Z_{11}(0))^{-1}(I+Z_{11}(0))[N_0]v_0' =$$

$$\sqrt{2}\ (I-Z_{11}(0))^{-1}Z_{12}(0)E_{s0} \qquad (40)$$

and

$$\{I - (I-Z_{11}(j\omega))^{-1}(I+Z_{11}(j\omega))[N_1]V_1' =$$

$$\sqrt{2}\ (I-Z_{11}(j\omega))^{-1}Z_{12}(j\omega)E_{s1}. \qquad (41)$$

To solve eq.(38) and (39) or eq.(40) and (41), we call for a computer aid for the computation. We decompose eq.(39) or (41) in phasor form into its real and imaginary part, thus giving 3m real equations in 3m unknowns $v_0'^{(i)}$, $v_1'^{(i)}$, $\phi'^{(i)}$, $i=1,2,\cdots,m$:

$$f(x) = \theta_n, \qquad (42)$$

where $f:\mathbb{R}^{3m}\rightarrow\mathbb{R}^{3m}$ and $x\in\mathbb{R}^n$.
Since the derivative or the Jacobian matrix $Df(x)$ in those 3m nonlinear equations (42) cannot easily be evaluated, we employ a modified form of Newton-Raphson method, by approximating the Jacobian by a difference matrix:

$$Df(x) \doteq \left[\begin{array}{l}(f_i(x_1,\cdots,x_j+\Delta x_j,\cdots,x_n) - \\ f_i(x_1,\cdots,x_j,\cdots,x_n))/\Delta x_j\end{array}\right] \qquad (43)$$

5. EXAMPLES

We performed numerical experiments on the following examples.

Example 1. A halfwave-rectifier circuit.
Consider a halfwave-rectifier circuit in Fig. 11 which is too simple and too much influenced by higher order harmonics to be used practically but seems fundamental to observe the effect of higher harmonics ignored during the analysis. The circuit is driven by a sinusoidal independent voltage source $E_s(t)=E\sin\omega t$ and suppose we are interested in the load current $i(t)$ through the resistor R and in the conduction angle γ of the thyristor.

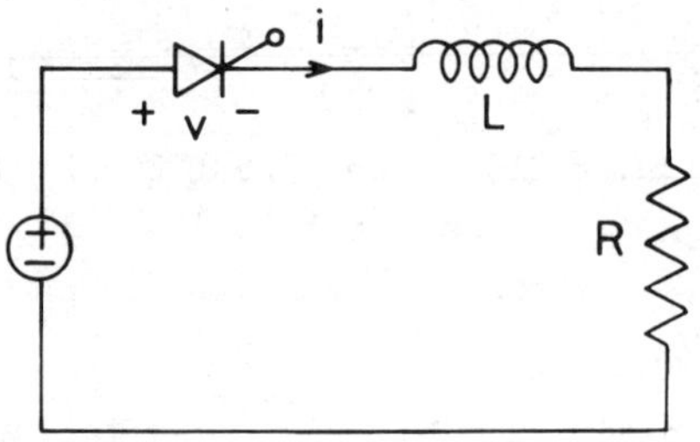

Fig. 11 A halfwave-rectifier circuit.

We chose the firing angle $\alpha=\pi/3$ (rad.), $\omega L=10.0$ (rad.henry) and $E=141.4$ (volts). As shown in Fig. 12 (b), the second order harmonics ratios to the fundamental components in voltage and current waveforms of the thyristor ranges from 104% to 58% and 31% to 54% for different load resistance R=2~14 (ohm), respectively. This shows the example circuit suffers from higher order harmonics. Figure 12 (a) shows how the computed dc and ac component of the load current $\tilde{i}_0$, $\tilde{i}_1$ and the conduction angle $\tilde{\gamma}$ differ from the exact ones with the superscript "*". Observe that the conduction angle $\tilde{\gamma}$ and the dc component $\tilde{i}_0$ show rather favorable agreement for such highly distorted signals and the errors tend to decrease as R becomes larger.

Example 2. An ac regulator [5].
Consider an ac regulator controlled by a back-to-back thyristor pair as shown in Fig. 13. Let both thyristors be triggered symmetrically. Suppose we are interested in the load current $i(t)$ through the load resistor R which contains neither dc component nor even order higher harmonics. For $\alpha=\pi/3$ (rad.), $\omega L=10.0$ (rad.henry) and R=14 (ohm), the shape of the exact load current $i^*(t)$ is drawn in dashed lines in Fig. 15. The numerical experiments showed the dc component $\tilde{i}_0$ and the conduction angle $\tilde{\gamma}$ agreed almost exactly to the exact ones and the ac component $\tilde{i}_1$ was also in good agreement as shown in Fig. 14. It is interesting to observe

that how the curve $\tilde{i}(t) \triangleq \tilde{i}_0 + \tilde{i}_1 \sin(\omega t + \tilde{\psi})$ detects the extinction angle β as shown in Fig. 15.

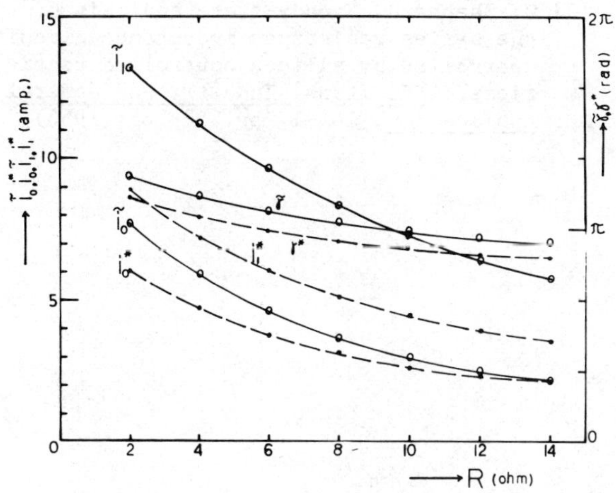

(a) Exact and computed results for the halfwave-rectifier circuit

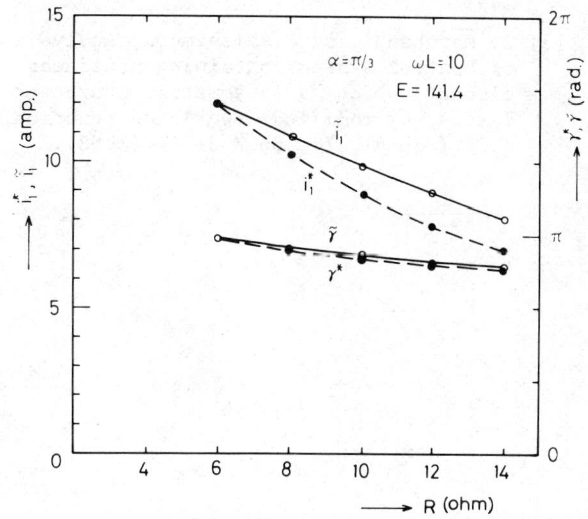

Fig. 14 Exact and computed results for the ac regulator circuit.

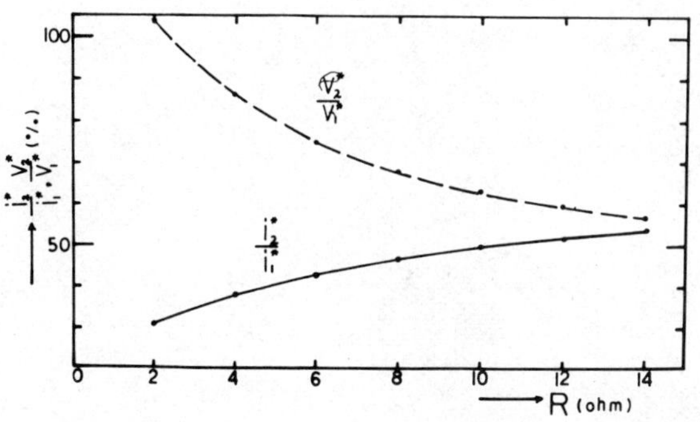

(b) Second order harmonics ratios in the voltage and current of the thyristor.

Fig. 12 Dc and ac components in the waveforms for the halfwave-rectifier circuit.

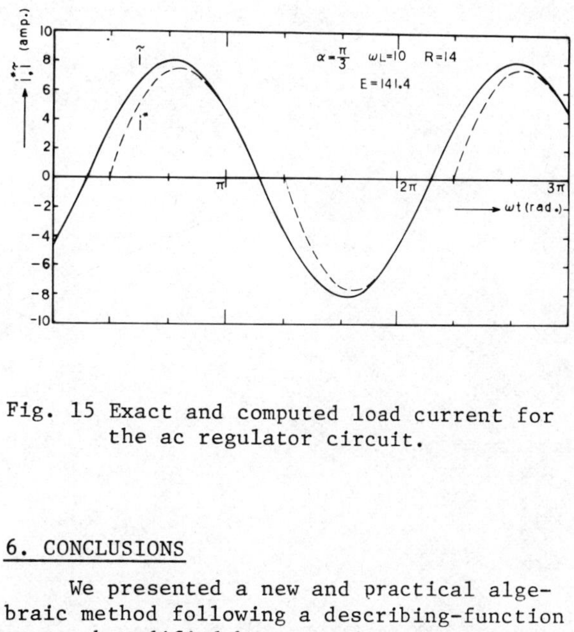

Fig. 15 Exact and computed load current for the ac regulator circuit.

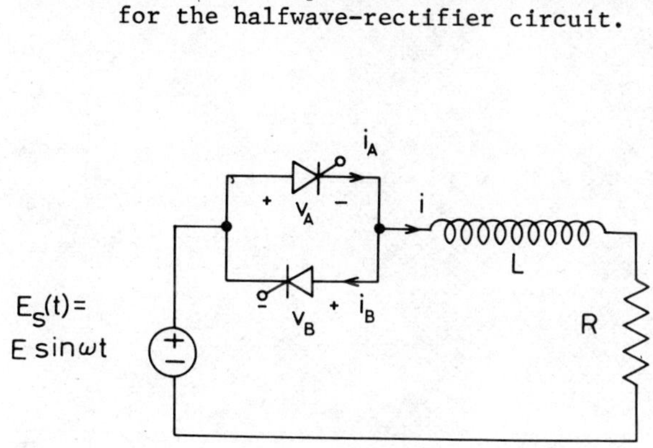

Fig. 13 An ac regulator.

6. CONCLUSIONS

We presented a new and practical algebraic method following a describing-function approach modified by a coordinate transformation. The important feature of the method is its generality with no constraints on thyristor configurations and it gives us an approximate conduction angle $\tilde{\gamma}$ as a by-product.

ACKNOWLEDGEMENTS

The authors are grateful to Mr. A. Okuno, a student for his computational assistance.

REFERENCES

[1] Y. L. Kuo, Distortion analysis of bipolar transistor circuits, IEEE Trans. Circuit

Theory, CT-20, pp.709-716 (1973).

[2] Hsu, Y. L. and Meyer, A. U. (1968) Modern Control Principles and Applications, McGraw-Hill, New York, pp.244-249.

[3] T. Maruhashi and M. Nishimura, Analysis of control system containing nonlinear element, which has asymmetric characteristics, by modified describing function, JIEE (Japan), 78, pp.733-741 (1958).

[4] Y. Murai, K. Kamimoto and Y. Tsunehiro, The application of the describing function method to the analysis of thyristor circuits, JIEE (Japan), 95, pp.177-184 (1975).

[5] W. Shepherd, Steady-state analysis of the series resistance-inductance circuit controlled by silicon controlled rectifiers, IEEE Trans. Industry and General Application, IGA-1, pp.259-265 (1965).

CONTRIBUTION TO SIMULATION AND CONTROL PROCEDURES OF BRUSHLESS DC-MACHINES

Rolf Hanitsch

Institut für Elektrische Maschinen, Technische, Universität Berlin, FRG; VDE (ETG, GMR)

ABSTRACT

The brushless d.c. motor is used in industrial equipment because of its advantages of long life, ability to eliminate brush-wear particles and arcing and high performance characteristic. In order to investigate the feasibility of some control circuits it is useful to simulate the brushless d.c. motor and its electronic commutator by means of an analog computer. In the first part of the paper the simulation aspects will be discussed. The second part of the paper deals with some control procedures because the future will require systems of high precision and reliability.

INTRODUCTION

In the past years, many different brushless d.c. motors have been proposed and developed using solid state switches to perform the function of commutation. The advantages of these motors are: low maintenance and long life, high efficiency, low acoustical and radio noise, easily obtainable speed control and a high reliability (Ref. 1,4,5,6,7). The electronic commutation circuit however has a remarkable influence on the failure rate of the whole system. It is therefore essential to consider some form of redundancy for the control circuit. This article describes a control circuit using a voltage controlled oszillator. Due to the fact that the simulation aspects of the brushless motor should be discussed first the rudimentary theory of operation and the typical construction elements are given new.

PRINCIPLE OF BRUSHLESS DC MOTOR

The rotor of the motor consists of a permanent magnet magnetized in radial direction. The armature is usually a ring winding. These airgap winding consists of four windings arranged 90° degrees apart in star-connection (Fig.1). In a different design three windings are arranged 120° degrees apart also in star-connection or delta-connection.

To derive torque from the motor, the armature field must be displaced from the rotor field. A displacement of 90 electrical degrees will give close to maximum torque. To maintain this displacement as the rotor turns, the power switches must control the position of the armature field. It is necessary to derive the winding commutation angle from the rotor position. In addition to the switching elements of the electronic commutator a special rotor position transmitter has to be incorporated. Many different methods of sensing rotor position are used (Ref. 1,3,4,7). Photoconductive cells or Hall generators are often strategically mounted to sense rotor position.

Model structure

To obtain an efficient motor drive the electronic switching must be used to connect the d.c. prime power to the motor windings with proper timing and polarity. The loss involved in the driver switches can be taken into account by an additional resistance to the winding resistance. As the resistance of the switches is included in the motor winding resistance R_j ($j = 1,2...n$),

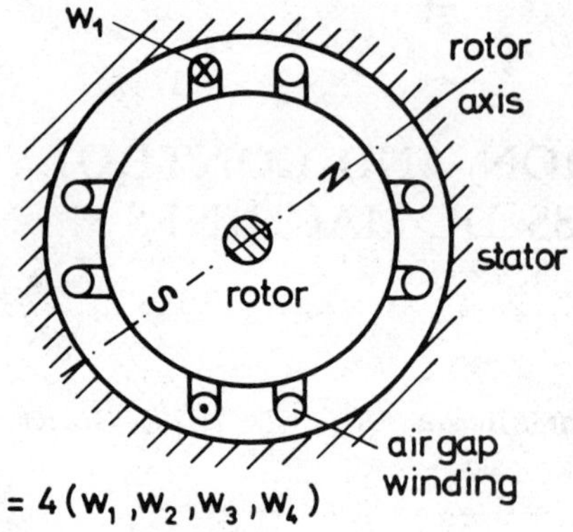

Fig.1 Cross - sectional view of brushless dc motor

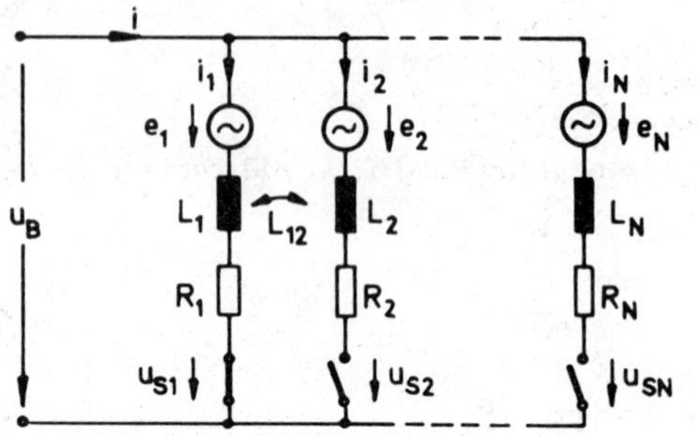

Fig. 2 Model of brushless d.c. motor

then the calculations for loss-free switches can be applied (see Fig. 2). Considering the distribution of the flux density in the air gap to be sinusoidal it is possible to obtain an expression for the instantaneous torque:

$$m_e(\alpha) = wdl \cdot f_w \hat{B} \cdot i(\alpha)\sin\alpha = Ki(\alpha)\sin(\alpha) \quad (1)$$

For a motor with one, two or four windings can be taken from Fig. 2, because the mutual inductance can be neclected for a symmetrical configuration of the windings:

$$L\frac{di_j}{dt} + Ri_j + K\omega\sin\alpha_j = u \quad \text{with} \quad \alpha_e \leq \alpha_j \leq \alpha_a \quad (2)$$

$$i_j = o \qquad \alpha_a \leq \alpha_j \leq \alpha_e$$

$$\sum_{j=1}^{N} Ki_j\sin\alpha_j = m + J\frac{d\omega}{dt} \quad (3)$$

$$\frac{d\alpha_j}{dt} = \omega$$

$$\sum_{j=1}^{N} i_j = i$$

Core losses and such losses as bearing and viscous drag losses can be incorporated into the analysis by treating them as additional torque loads on the motor: $m = m_L + m_V(\omega)$. For steady state conditions the equation (1) can be solved when $u = U =$ const. and $m = M =$ const. is assumed. The torque is given by:

$$m_e(t) = M + \sum_{\nu=1}^{\infty} M_\nu \cos(\nu N \Omega t + \varphi_\nu) \quad (4)$$

Moreover we have to consider:

$$m_e(t) = M + J\frac{d\omega}{dt} \quad (5)$$

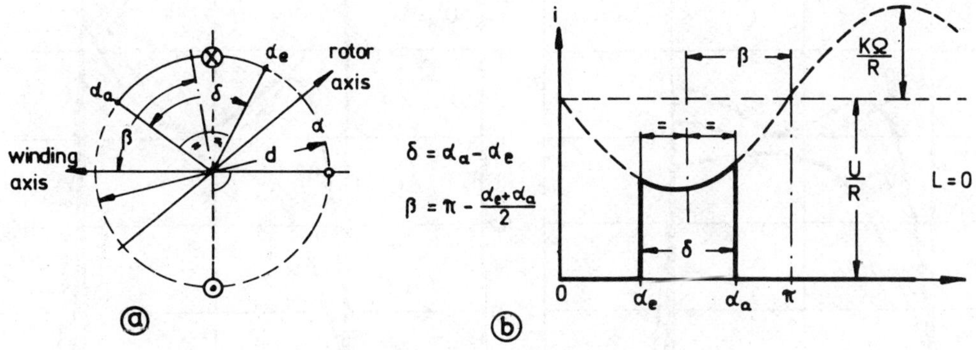

Fig.3 Representative current and voltage waveforms and characteristic quantities

A comparison between the two equations gives us the instantaneous angular velocity.

$$\omega(t) = \Omega + \frac{1}{JN\Omega} \sum_{\nu=1}^{\infty} M_\nu \cos(\nu N\Omega t + \varphi_\nu) \tag{6}$$

To reflect this fact, we have to note that $\alpha(t) \neq \omega t$ and $\alpha(t) \neq \Omega t$. Our experiments have shown that for rotational speed over 200 rpm the second term in equation (6) can be neclected, so that $\alpha = \Omega t$ can be assumed. If we consider $i(\alpha_e)=0$ (see Fig. 3) the solution of equation (1) is:

$$i(\alpha) = \frac{U}{R}\left[1-e^{-\frac{R}{X}(\alpha-\alpha_e)}\right] - \frac{K\Omega}{Z}\left[\sin(\alpha-\varphi) - e^{-\frac{R}{X}(\alpha-\alpha_e)}\sin(\alpha_e-\varphi)\right]$$

$$\alpha_e \leq \alpha \leq \alpha_a < 2\pi + \alpha_e \tag{7}$$

$$X = \Omega L$$

$$Z = \sqrt{R^2 + \Omega^2 L^2}$$

$$\varphi = \arctan(\Omega L/R)$$

When the rotor has reached the position $\alpha = \alpha_a$ the current in the corresponding winding has to be switched off by a common method in order to avoid large voltage pulses. The instantaneous current can be approximated by a straight line taking into account that a certain peak value of the voltage U_{LM} should not be exceeded.

$$i(\alpha) = i(\alpha_a) - \frac{U_{LM}}{X}(\alpha - \alpha_a)\ \mathrm{sign}\ i(\alpha_a) \tag{8}$$

$$\alpha_a \leq \alpha \leq \alpha_A < 2\pi + \alpha_e$$

$$\alpha_A = \alpha_a + \frac{X}{U_{LM}}\left|i(\alpha_a)\right|$$

$$i(\alpha) = 0 \quad \text{for} \quad \alpha_A \leq \alpha \leq 2\pi + \alpha_e$$

To generate graphs that can be used for more than one design problem, a suitable normalisation of the system parameters must be utilized. The normalized parameters are defined as follows:

$$i' = i\frac{R}{U}\ ;\ m'_e = m_e\frac{R}{KU}\ ;\ \Omega' = \Omega\frac{K}{U}\ ;\ L' = L\frac{U}{KR}\ ;\ U'_{LM} = U_{LM}\frac{1}{U} \tag{9}$$

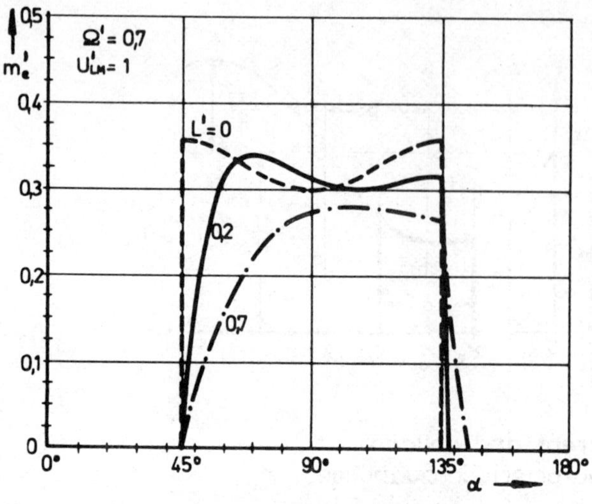

Fig. 4 Computed torque curves

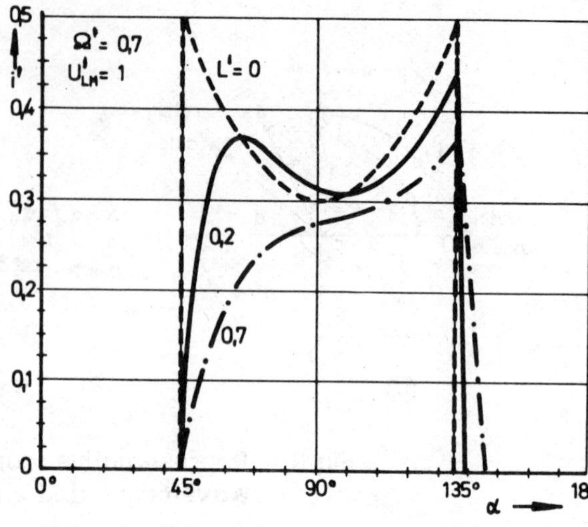

Fig.5 Computed current curves

In the curves presented in Fig. 4 and Fig. 5, the values of $\Omega' = 0{,}7$ and $U'_{LM} = 1$ are used.

For the analog computer representation it is useful to rewrite the basic equations.

$$\frac{di}{dt} = \begin{cases} \frac{u}{L} - \frac{K}{L}\omega \sin\alpha - \frac{R}{L} i & \text{for } \alpha_e \leq \alpha \leq \alpha_a \\ -\frac{U_{LM}}{L} & \alpha_a \leq \alpha \leq \alpha_A \end{cases} \tag{10}$$

$$i = 0 \qquad \alpha_A \leq \alpha \leq 2\pi + \alpha_e$$

$$\frac{d\omega}{dt} = \frac{K}{J}\, i \sin\alpha - \frac{1}{J}\,(m_L + m_V(\omega))$$

$$\frac{d\alpha}{dt} = \omega$$

The analog computer representation for a brushless dc motor with only one winding can be taken from Fig. 6 (Ref. 2). On the basis of this special case the complete circuit for a motor with four windings (symmetrical lay out) was developed (see Fig. 7). The normalization that has been chosen now relates the parameters to the nominal maximum operating conditions. The normalized parameters are defined as follows: $\tilde{x} = x/x_{max}$; $\tilde{t} = \tau t$ etc. Since the number of multipliers is restricted at our analog computer the circuit was designed to take into account a phase reference angle of $\beta = 90^o$. Note that several parameters of interest are defined in Fig. 3. Comparing the curves presented in Fig. 8 it is evident that the analog computer circuit is quite efficient.

CONTROL PROCEDURES

The design of the electronic commutator of the brushless motor has a strong influence on the motor characteristics. The conventional circuit is characterized by analog processing of all information received from the rotor position transmitter. A great progress is achieved by the incorporation of digital components. The complete description of such a system can be found in Reference (4). During startup the motor behaves like a stepper motor and running at full speed the motor has the characteristic of a dc shunt motor. In this case the stator windings are connected to the dc prime power by logical operations based on the armature induced voltage. The zero-passage of the induced voltage is used for detecting the rotor position in this special design.

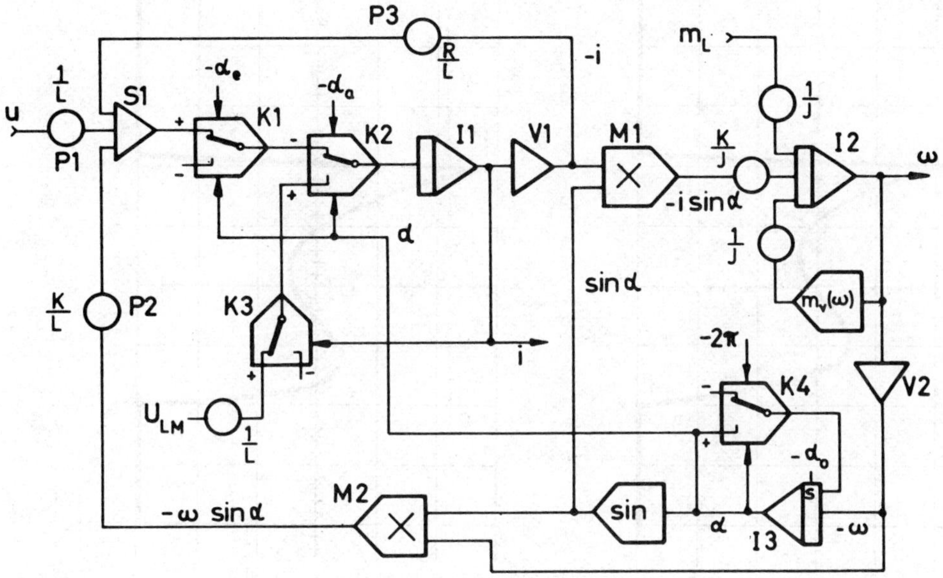

Fig. 6 Analog computer representation for one stator winding

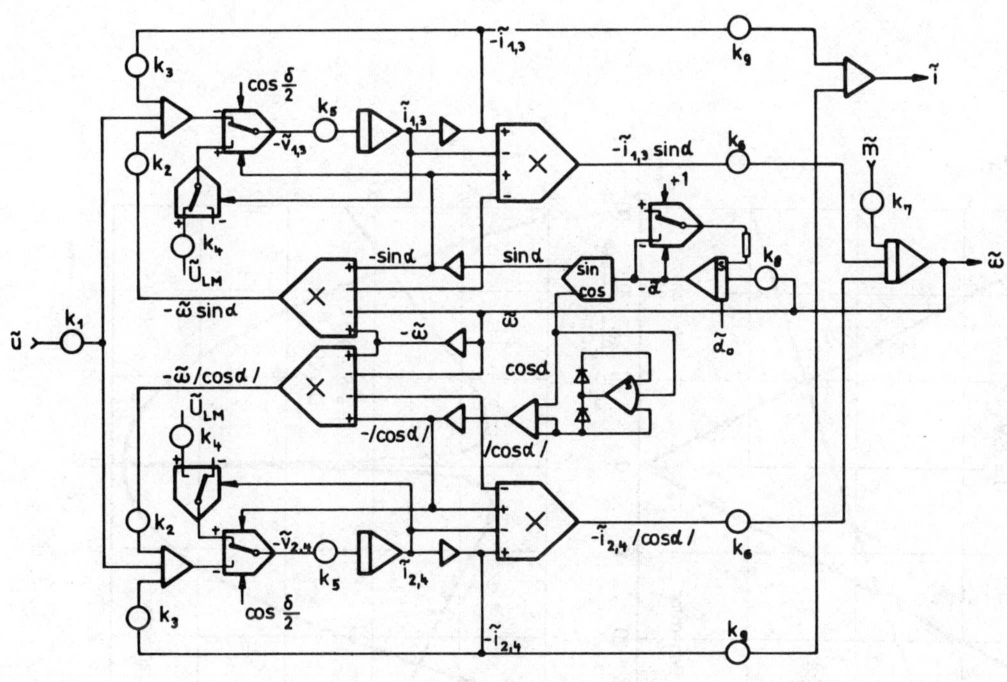

Fig. 7 Analog - computer representation for brushless d. c. motor with four stator windings

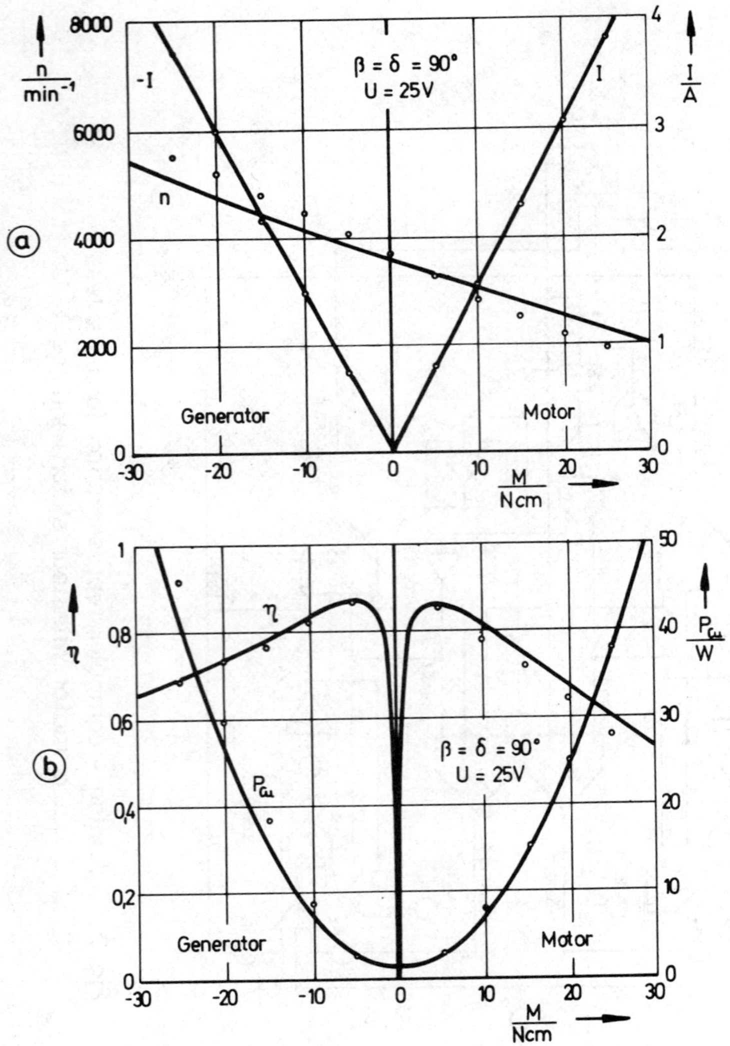

Fig. 9 Load characteristics

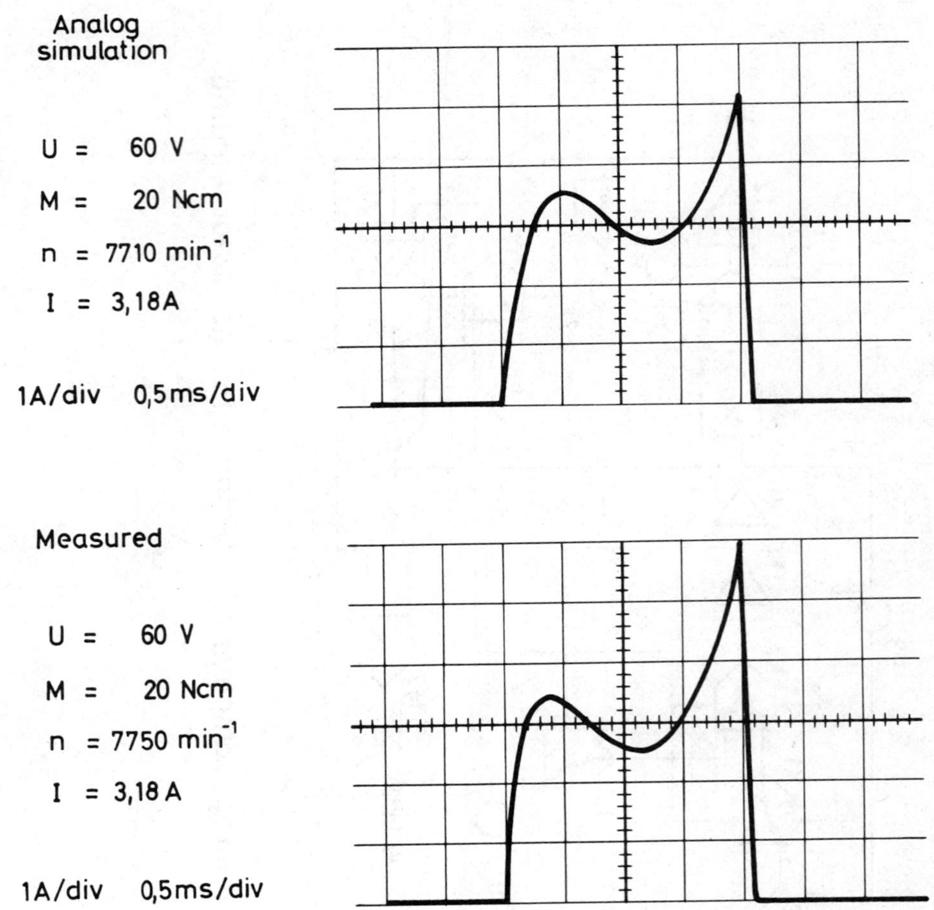

Fig. 8 Current in a winding at full load

The typical conventional load characteristics of a 120 W brushless dc motor are given in Fig. 9. Note that in this case the phase reference angle β and the current flow angle δ are equal: $\beta = \delta = 90^{\circ}$. We realized on the other hand a circuit which gave us the opportunity to change the actual phase reference angle. Measured and calculated characteristics of such a system are presented in Fig. 10. In the curves presented in this figures, a value of $\delta = 90^{\circ}$ was used. Furtheron we can see, that the phase reference angle might be used as input variable for speed control in an open-loop control system.

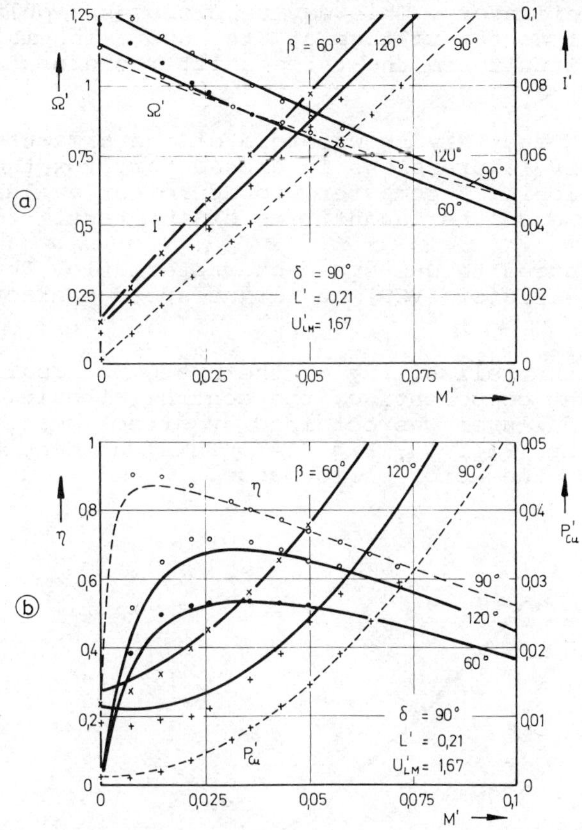

Fig. 10 Measured and calculated characteristics

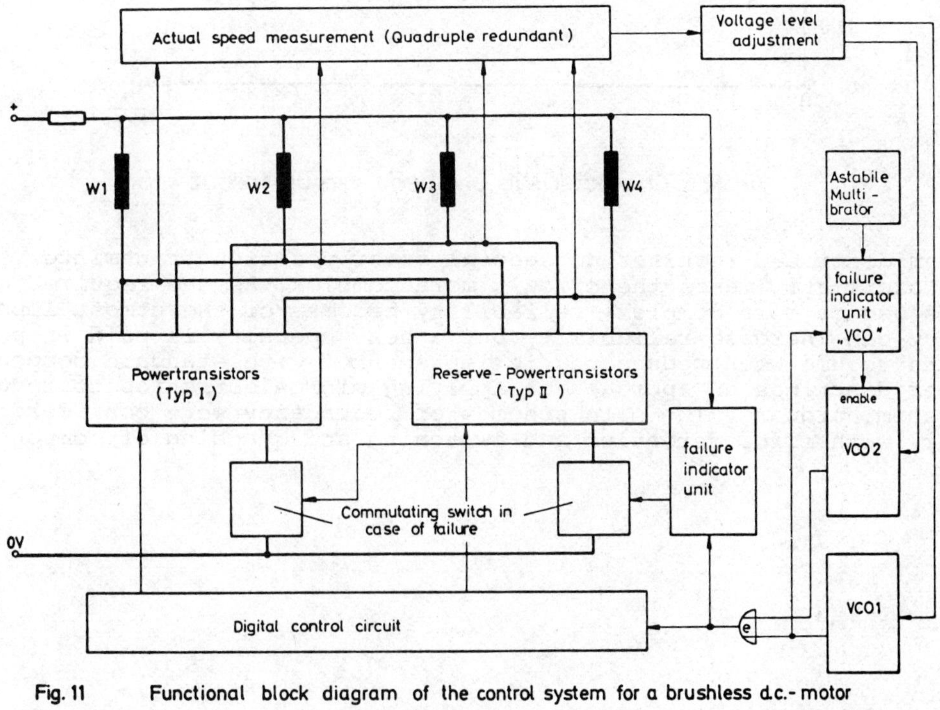

Fig. 11 Functional block diagram of the control system for a brushless d.c.- motor

A different characteristic speed versus torque can be realized by using a voltage controlled oszillator in the control circuit. The functional block diagram of the system is given in Fig. 11. In the control system, redundancy in various forms is applied to critical components and subsystems.

The machine is not equipped with an electronic position sensor to indicate the angular position of the rotor. We use the instantaneous value of the induced voltage which is coupled out by diodes and then given to the two voltage controlled oscillators. This device produces a pulse train which frequency is proportional to the voltage at its input. The main components of the digital control circuit are the three shift registers and the majority logic circuit.

Due to the complicated and varying waveform of the armature current it is very difficult to derive expressions in closed form for the torque of such machines. We are developing a computer procedure for evaluating the torque-speed characteristics and other quantities of interest.

Fig. 12 shows the measured torque-speed characteristic. The influence of the voltage controlled oszillator (VCO) on torque-speed characteristic can be seen clearly.

In order to increase the reliability of the system we realized some form of redundancy for relevant components of the control circuit. The response of the speed to typical failures was obtained by direct measurement on the prototype electronic circuit (Fig. 13; Fig. 14). This was done to study the effect of certain failures on the motor performance.

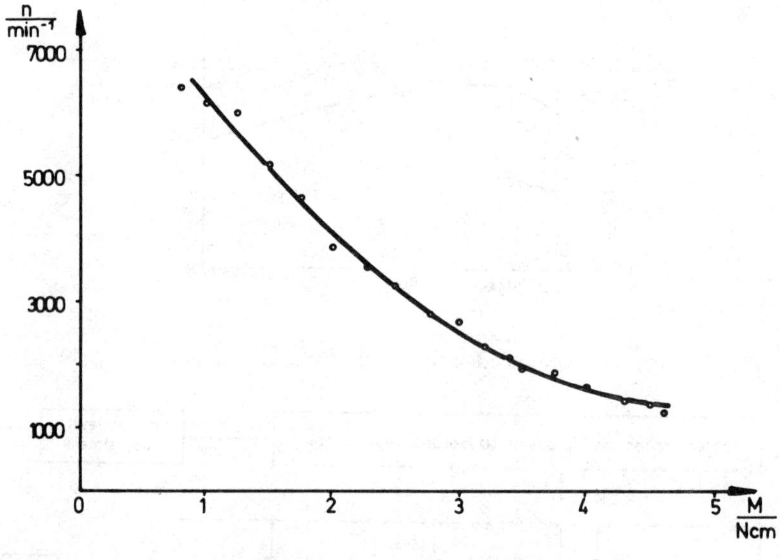

Fig. 12 Characteristic of speed versus torque

The system described requires no special rotor position transmitter, but the built-in oscillator makes the circuit more complex. As the required electronic functions become more complex, reliability becomes of the utmost importance. Redundancy can increase reliability but a heavy penalty is paid in power, weight and volume when redundancy is implemented with standard components. This paper describes an approach of applying microelectronics to redundant drive system hardware. Two main schemes of redundancy were considerd: Active redundancy with error detection and switching and quadding of components.

Conclusions

A comparison between the calculated and the measured characteristics shows that the developed mathematical model is another contribution to the theory of behaviour of brushless dc machines. To complete the discussion, therefore, a description of a prototype control system is given which uses a voltage controlled oszillator. Because this work has been undertaken in the environment of a research institute of electrical machines, development has been restricted so far to a demonstration of feasibility.

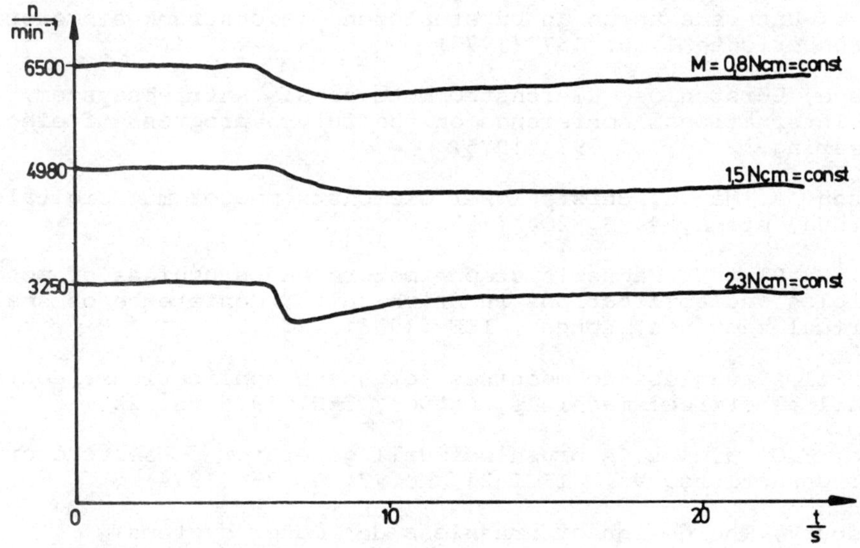

Fig. 13 The response of the speed to a failure of one powertransistor

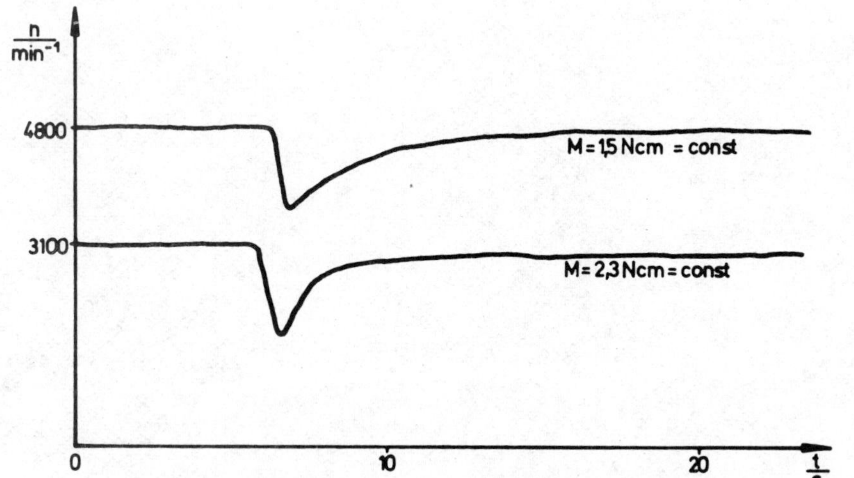

Fig. 14 The response of the speed to a failure of one voltage controlled oszillator (VCO)

Acknowledgements

The author is grateful to the Deutsche Forschungsgemeinschaft for facilities provided. The project benefited enormously from the skills of Dipl.-Ing. J. Germer, Dipl.-Ing. A. Meyna and Dipl.-Ing. K. Wegener. They are with the "Institut für Elektrische Maschinen", University of Technology, Berlin.

REFERENCES

(1) M. Liska, A new brushless variable speed dc. drive for 60 000 rev/min and torque output up to 50 Ncm, Conference on electrical variable speed drives, London, IEE, 10-12. Oct. (1972)

(2) K. Wegener, Untersuchungen an bürstenlosen Gleichstromkleinmotoren, ETG-Fachberichte 1, S. 157 (1975)

(3) R. Hanitsch, Bürstenlose Gleichstrommotoren als Antriebssystem, Proc. International conference on the future progress of electrical engineering, Prag, S. 392 (1975)

(4) R. Hanitsch, A. Meyna, Bürstenloser Gleichstrommotor mit digitaler Ansteuerung, etz-a, 4, S. 204 (1976)

(5) M. Liska, B. Ulrich, Magnetic steppermotors and brushless dc motors: principles and applications in drive units. Conference on small electrical machines, London, IEE (1976), S. 15

(6) W. Radziwill, Brushless dc machines for space applications, Conference on small electrical machines, London, IEE (1976) S. 23

(7) J.F. Wise, F.O. Simons, A brushless Hall generator dc servomotor IEEE Transactions, Vol. IECI-21, No. 2, S. 75 (1974)

(8) J.R. Woodbury, The design of brushless dc. motor systems IEEE Transactions on industrial electronics and control instrumentation, Vol. IECI-21, 2, S. 52 (1974)

DEVELOPMENTS IN SINUSOIDAL P.W.M. INVERTERS

M.G. Jayne, S.R. Bowes and B.M. Bird

University of Bristol, England

ABSTRACT

This paper reports the results of a detailed investigation into analogue controlled pulse-width-modulated inverter systems. It is shown that the natural sampling process inherent in existing control schemes, produces considerable harmonic distortion of the output modulated voltage waveforms, which limits the useable range of operating frequency. The application of regular sampling techniques to the control circuit is presented, and it is shown that this new method of control provides a considerable increase in the frequency range of operation. It is concluded that the improved harmonic spectra of the output modulated voltage waveforms, is achieved without increasing the switching frequency of the power components, the number of power components or cost.

1.0 INTRODUCTION

The main requirement of a wide range variable-speed control system which utilizes squirrel-cage induction motors, is that the magnitude and frequency of the output voltage applied to the stator windings, be independently variable over a wide range. This allows the torque-speed characteristic of the induction motor to match most required load characteristics. It is also important that the overall efficiency of the power convertor and induction motor remain above an acceptable level throughout the range of speed variation. These requirements are not easily met in practice. The cycloconvertor squirrel cage induction motor combination does not fulfil these requirements because the upper limit of output frequency is restricted to approximately half the input supply frequency. Above this limit considerable harmonic distortion is introduced in the output waveform which can cause the induction motor to crawl and cog. The overall efficiency of the convertor and induction motor also falls off rapidly when the output frequency approaches or exceeds its upper limit.

An alternative to the cycloconvertor drive system is the d.c. link inverter cage-rotor induction motor drive. This system can be made to satisfy many of the requirements of a wide range system of speed control, although many problems still remain to be solved. For example, varying the frequency of the output voltage from the convertor can cause a change in the harmonic distortion of the output waveform. Similarly, changing the magnitude of the output voltage can create commutation problems, if such a change is brought about by varying the d.c. link voltage to the inverter. Therefore, the inverter must be capable of maintaining low harmonic distortion of the output waveform when the frequency is varied, and provide a means of controlling the magnitude of the output voltage, whilst the d.c link voltage to the inverter remains constant. The overall efficiency of the inverter and squirrel-cage induction motor must also be maintained at the highest possible level, throughout the entire speed range.

The application of pulse-width-modulation (p.w.m.) to the control circuit of the inverter can provide both frequency control and voltage control without the addition of extra power components. However, increasing the frequency of the output voltage by this method of control can also lead to the introduction of unwanted harmonic voltages and switching problems of the power components, which again restricts the upper limit of frequency operation of the inverter.

2.0 THE ASYNCHRONOUS OPERATION OF ANALOGUE P.W.M. CONTROL CIRCUITS

The literature available (Refs. 1, 2 and 3) suggests that the majority of prior-art analogue controlled p.w.m. power convertors have operated on integer values of frequency ratio only. Such a mode of operation requires the carrier wave or sweep wave to be synchronised in time-phase with the modulating wave. The locking of the carrier wave to the modulating wave adds complexity to the control circuit and introduces limitations to the range of frequency variation. If, however, the carrier wave is set to a constant value which lies within the switching limitations of the power convertor, and the modulating wave is asynchronous, then varying the frequency of the modulating wave provides a range of frequency ratio variation given by:

$$\frac{2}{\pi} < \frac{fc}{fm} < \infty \qquad (1)$$

where the carrier frequency, fc, is constant and the modulating frequency fm, is variable.

Although the constant carrier-frequency asynchronous mode p.w.m. control scheme appears to be the most logical system to use for wide range variable speed induction motor drives, the literature (Ref. 3) suggests that such a means of generating p.w.m. waveforms is virturally useless for all applications, because sub-harmonic voltage components, direct voltage components and a 'beat effect' occurs in the output frequencies. However, the magnitude of these undesirable components relative to the amplitude of the wanted harmonic components does not appear to have been determined. The effects of frequency ratio variation on these unwanted components and 'beat effect' phenomenon have also avoided analysis in the literature.

The prior-art technique of generating double-edge p.w.m. control waveforms is shown in Fig. 1. It is important to note that the modulating waves and carrier wave are entirely asynchronous. This means that the triangular carrier wave and three-phase modulating waves have no known time-phase relationship. It may also be seen from Fig. 2 that the points of instantaneous intersection between the carrier wave and the modulating wave (the sampling points) coincide with the switching points of the p.w.m. output waveform which constitutes natural sampling (Ref. 4). It is also of considerable importance to note that the p.w.m. output waveform over a period of time corresponding to one cycle of the modulating wave (between t = o and t = 1/fm) is different from the p.w.m. waveform over a period of time corresponding to the succeeding cycle of the modulating wave, (from t = 1/fm to t = 2/fm) in fact, it is quite evident, in this case, that the p.w.m. waveform pattern only starts to repeat itself after a period of time, T = 2/fm, The time-period, 2/fm, is therefore the fundamental repetition time, Tr, or the frequency, 1/Tr, which in the case illustrated is equal to fm/2, is the fundamental repetition frequency, fr.

Because the carrier wave and modulating waves are not synchronised in time-phase, the value of frequency ratio, fc/fm, for any desired output frequency will never have a true integer value, although the value of frequency ratio will pass through integer values. Therefore, when the frequency spectra of the output p.w.m. waveform is determined, the frequency of the first harmonic component will be equal to the fundamental repetition frequency and not the modulating frequency. The fundamental repetition frequency of a modulated waveform is said (Ref. 5) to be equal to the highest common factor of the carrier frequency and modulating frequency. For example, if the carrier frequency is 1000 Hz and the modulating frequency is 300 Hz, then the fundamental frequency is 100 Hz, which means the wanted harmonic component is the 3rd. harmonic of frequency 300 Hz. Harmonic components of frequency less than the wanted harmonic component are therefore referred to as sub-harmonics, whereas harmonics of frequency greater than the wanted harmonic components are referred to as super-harmonics.

2.1 Effects of Asynchronism between Carrier Wave and Modulating Waves

It is said in the literature (Ref. 3) that the random time-phase displacement between the carrier wave and modulating waves can result in amplitude and phase unbalance of the wanted harmonic components when the three-phase modulating waves are compared with a single-phase carrier wave in order to produce three-phase p.w.m. waveforms. However, if the three modulating waves in the natural sampled double-edge p.w.m. process illustrated in Fig. 3 are defined by:

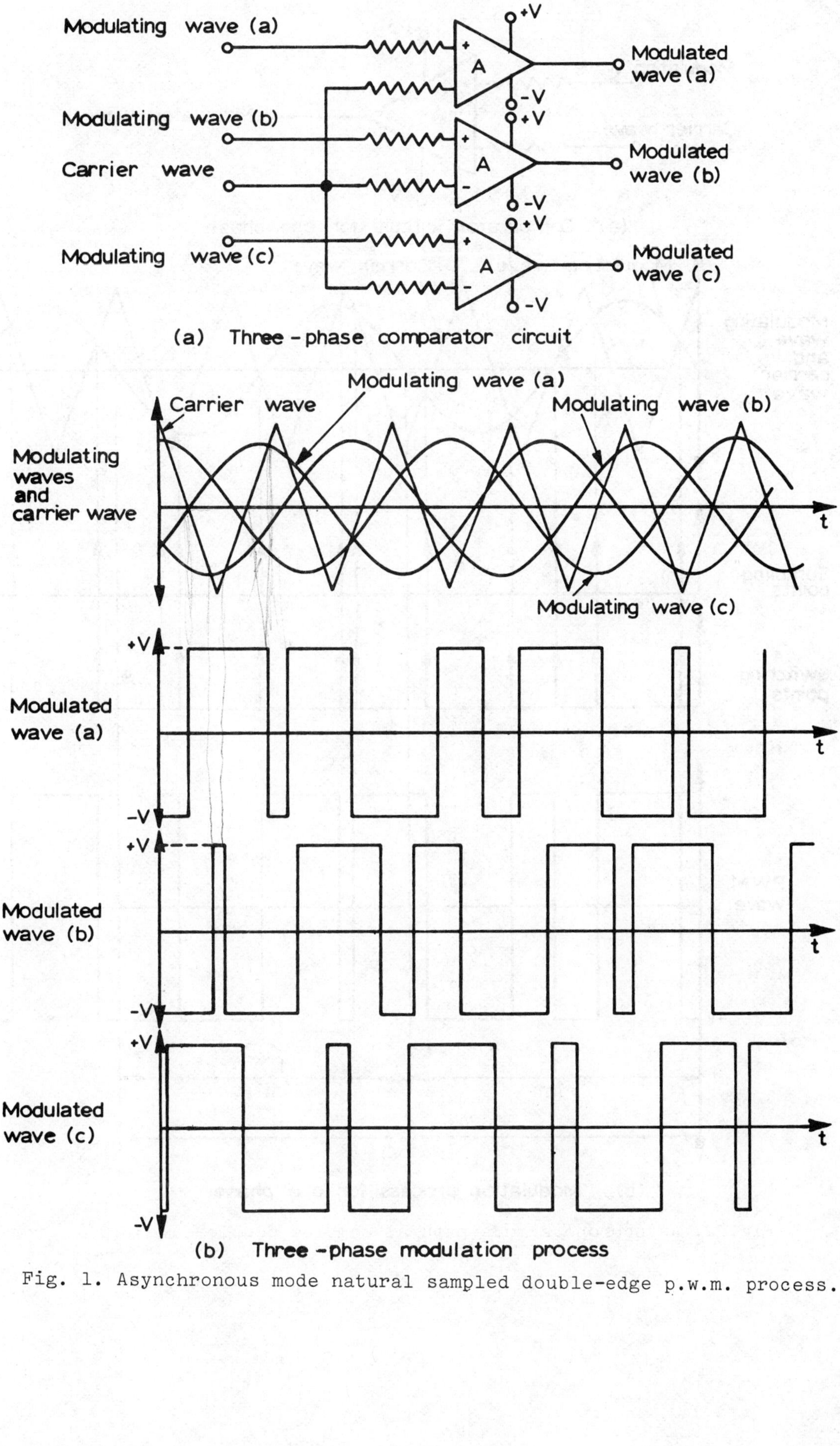

Fig. 1. Asynchronous mode natural sampled double-edge p.w.m. process.

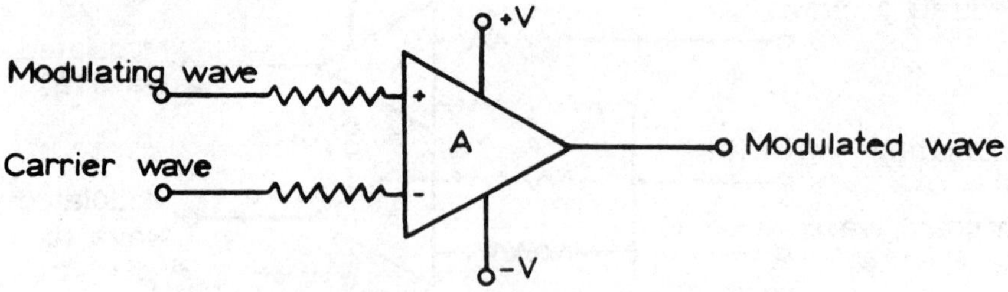

(a) Comparator circuit for one phase

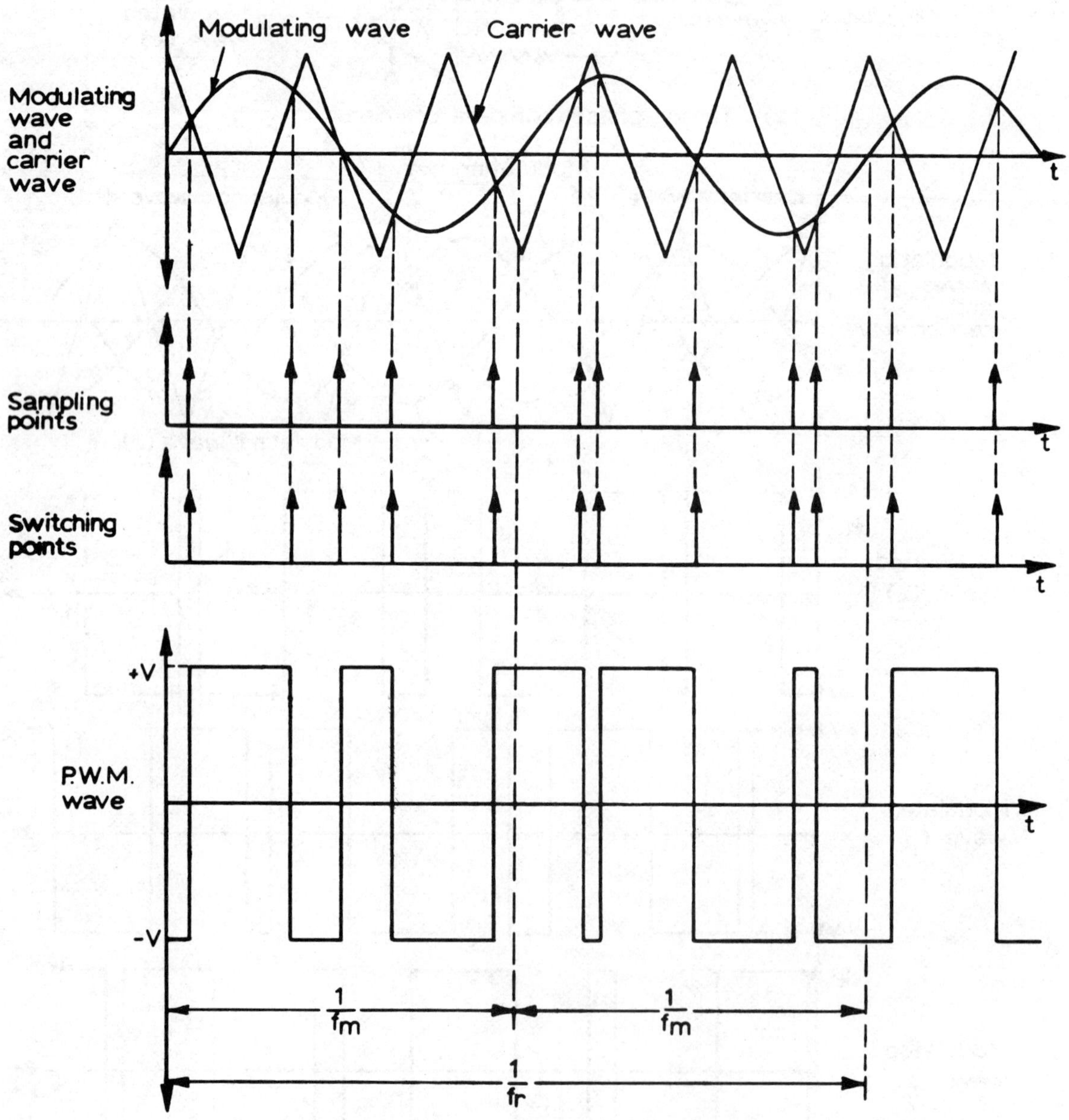

(b) Modulation process for one phase

Fig. 2. Asynchronous mode natural sampled double-edge p.w.m. process.

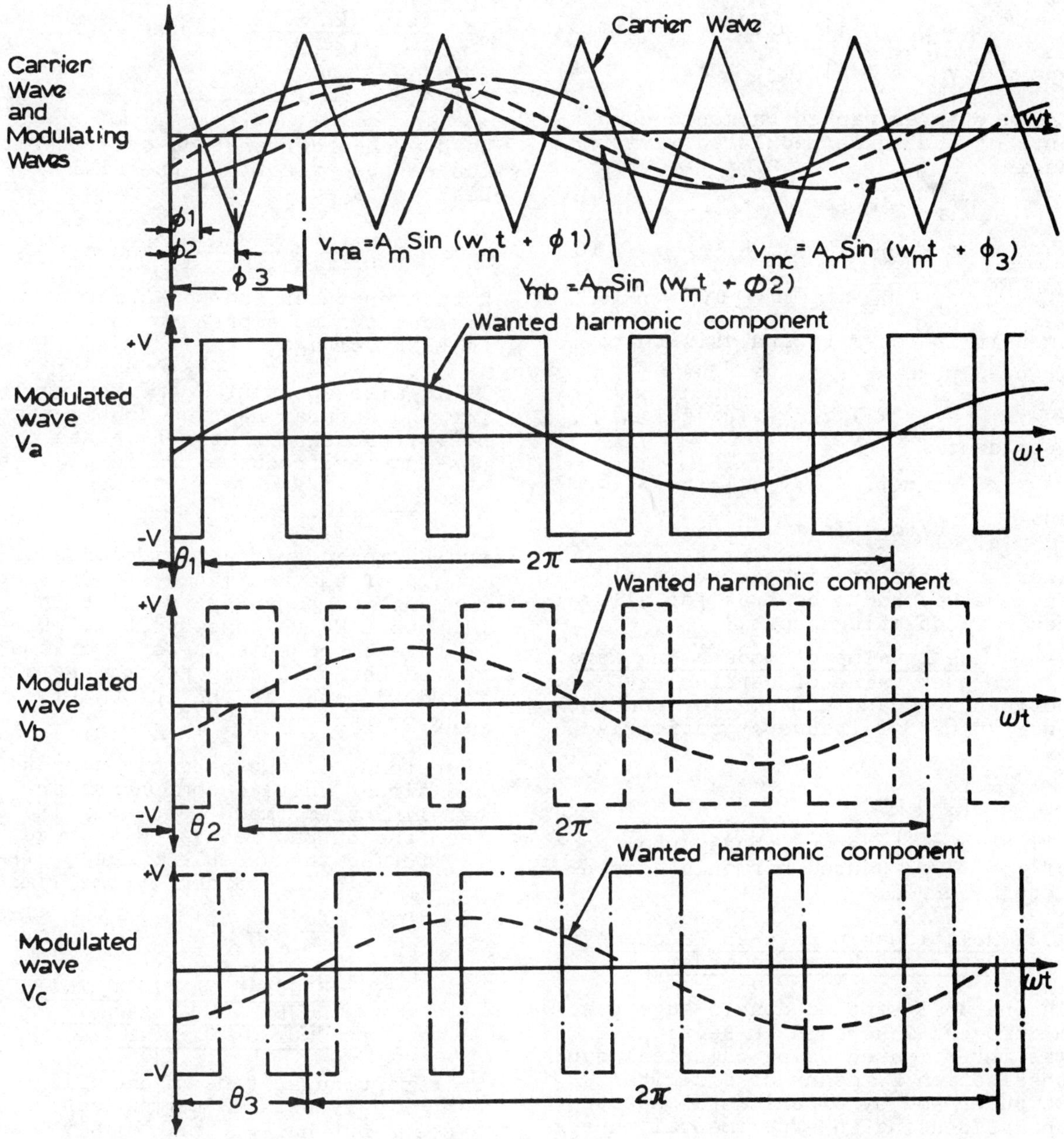

Fig. 3. The effect of the random time-phase displacement between the carrier wave and modulating wave upon the natural sampled double-edge p.w.m. process.

$$v_{ma} = A\sin(\omega_m t + \phi_1) \quad (2)$$

$$v_{mb} = A\sin(\omega_m t + \phi_2) \quad (3)$$

and

$$v_{mc} = A\sin(\omega_m t + \phi_3) \quad (4)$$

and the three wanted harmonic components of the three modulated waves defined by:

$$v_a = B_1\sin(\omega t + \theta_1) \quad (5)$$

$$v_b = B_2\sin(\omega t + \theta_2) \quad (6)$$

and

$$v_c = B_3\sin(\omega t + \theta_3) \quad (7)$$

then it is found by analysis that:

$$B_1 = B_2 = B_3$$

and

$$\theta_1 = \phi_1, \theta_2 = \phi_2, \theta_3 = \phi_3 \quad (8)$$

provided:

$$\phi_2 - \phi_1 = 120^\circ,\ \phi_3 - \phi_2 = 120^\circ$$

and

$$fc/fm = 3n/m \quad (9)$$

where n and m are integer values. This was found to be true for all values of modulation depth,

$$\frac{\text{peak value of modulating wave}}{\text{peak value of carrier wave}}$$

between zero and unity. For non-integer values of frequency ratio given by:

$$\frac{fc}{fm} \neq \frac{3n}{m} < 6$$

slight amplitude and phase unbalance of the three wanted harmonic components occurred.

2.2 Relationship between Frequency Ratio and Harmonic Order

The natural sampled double-edge p.w.m. process described in section (2.0) was investigated using numerical synthesisation by means of a digital computer and by experimental spectra analysis using an experimental three-phase inverter.

From these results the following relationships became apparent:

(1) All triplen-order harmonics (3rd, 6th, 9th, ..., 3n) always cancel in the LINE to LINE voltage, provided the numerator of the frequency ratio fraction, fc/fm, is a triple-integer multiple; that is to say:

when $\frac{fc}{fm} = \frac{3n}{m}$, and $3n > m$,

then harmonics of order, 3n, cancel in the LINE to LINE voltages.

(2) All even-order harmonics (2nd, 4th, 6th, ...2n) are eliminated from the LINE to inverter D.C. link NEUTRAL voltage and therefore from the LINE to LINE voltage when:

$\frac{fc}{fm} = \frac{(2n - 1)}{(2m - 1)}$, where $n > m$,

(3) When $\frac{fc}{fm} = \frac{(2n - 1)}{2m}$, where $n > m$,

both even-order and odd-order harmonics can be present in the LINE to LINE voltage.

(4) When $\frac{fc}{fm} = \frac{2n}{(2m - 1)}$ and $n > m$,

both even-order and odd-order harmonics can again be present in the LINE to LINE voltage.

It is also apparent from the results for the natural sampled double-edge p.w.m. process, that for values of non-integer frequency ratio given by:

$$\frac{fc}{fm} < 10$$

sub-harmonic voltage components of up to 32% of the amplitude of the wanted harmonic components can exist in the LINE to LINE voltage spectra. Direct voltage components up to 5% of the wanted harmonic components were also found to occur in the output LINE to LINE voltage.

Therefore, if sub-harmonic components and direct voltage components are to be eliminated (made less than 1%) from the output modulated voltage waveforms, the useable range of frequency ratio is bounded by the limits:

$$10 < \frac{fc}{fm} < \infty$$

3.0 REGULAR SAMPLING IN THE DOUBLE-EDGE P.W.M. CONTROL CIRCUIT

This particular type of sampling has been widely used in communication systems but has not previously been applied to the control circuits of p.w.m. power convertors.

Regular sampling basically entails the sampling of the modulating wave prior to comparison with the carrier wave as shown in Fig. 4a. However, the rate at which the modulating wave is sampled is dependent upon firstly the carrier frequency and secondly the type of p.w.m. required - symmetrical modulation or asymmetrical modulation.

3.1 Asymmetrical Double-Edge P.W.M. by Regular Sampling

This particular type of p.w.m. process is illustrated in Fig. 4b and it is immediately apparent that both edges of the width modulated pulses

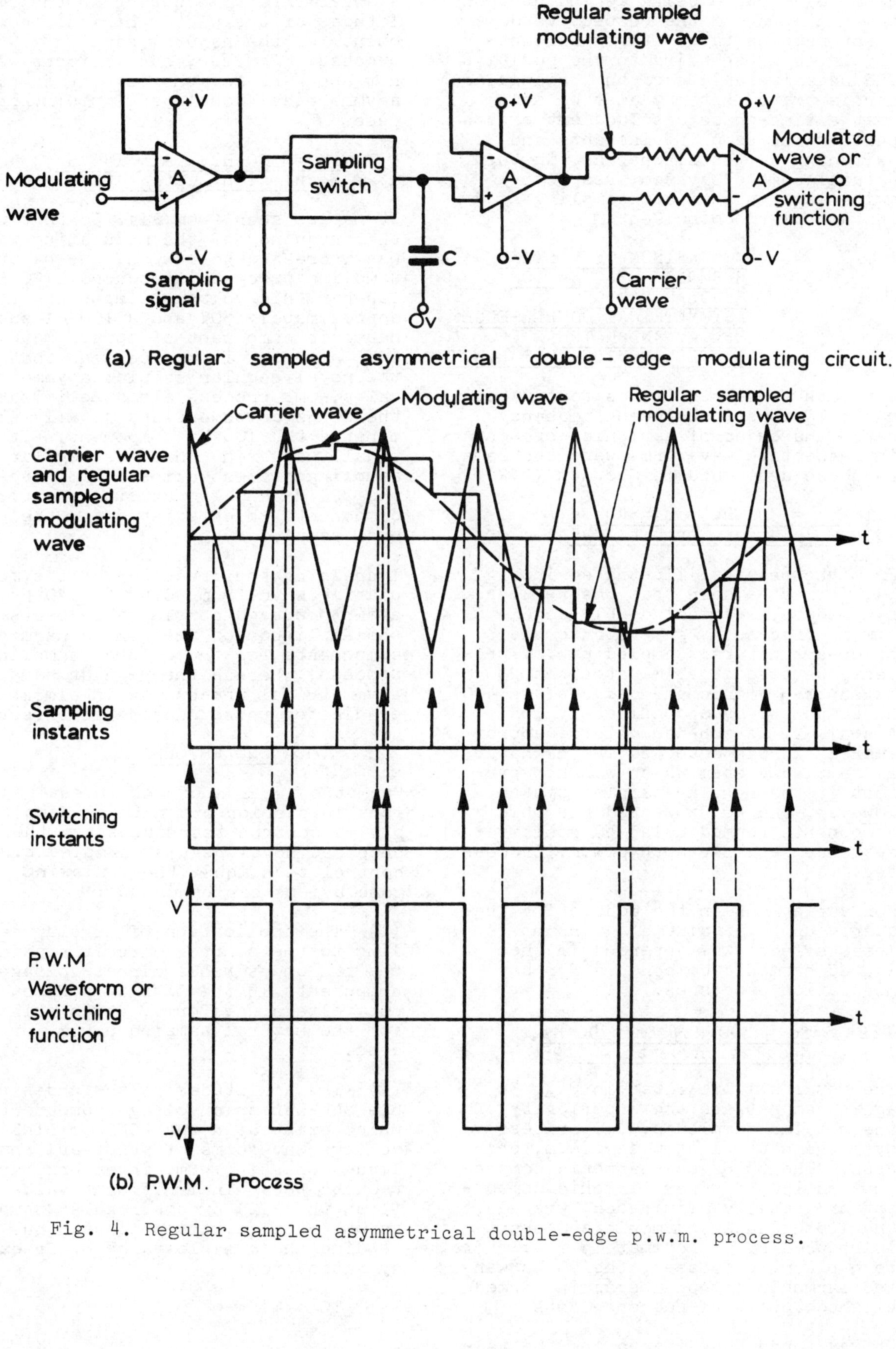

Fig. 4. Regular sampled asymmetrical double-edge p.w.m. process.

are modulated by different amounts: the amount of modulation of both edges of each pulse being dependent upon the magnitude of the respective adjacent samples taken at regular intervals, corresponding to the positive and negative apices of the triangular carrier wave. It may also be seen that a time-phase displacement exists between the sampling instants and switching instants, whereas, in the natural sampled system, respective switching instants and sampling instants were coincident.

4.0 COMPARISON OF VOLTAGE HARMONIC SPECTRA FOR REGULAR SAMPLED ASYMMETRICAL DOUBLE-EDGE P.W.M. AND NATURAL SAMPLED DOUBLE-EDGE P.W.M

For both processes it was found that the relationship between frequency ratio and order of harmonics present in respective waveforms were the same as those described in Section (2.2).

4.1 fc = 300 Hz, fm = 40 Hz, Modulation Depth = 0.95

From the results illustrated in Fig. 5a, it may be seen that the regular sampled asymmetrical double-edge p.w.m. process is superior to the prior-art natural sampled p.w.m. process, because it eliminates the 280 Hz super-harmonic and reduces the 440 Hz component by approximately 10%. A small 20 Hz sub-harmonic of approximately 1% was detected in the natural sampled system which was not present in the regular sampled system. However, the 40 Hz wanted harmonic components formed balanced positive-sequence sets for both p.w.m. processes.

For variations in the modulating frequency up to a value of 40 Hz no 'beat effect' was apparent in the wanted harmonic components for both modulation processes.

4.2 fc = 300 Hz, fm = 90 Hz, Modulation Depth = 0.95

The results illustrated in Fig. 5b again demonstrate the superiority of the regular sampled p.w.m. process over the natural sampled p.w.m. process. The 60 Hz sub-harmonic component and 150 Hz super-harmonic component are totally eliminated, whereas, the 120 Hz super-harmonic component is reduced by approximately 20%. For both p.w.m. processes, the 90 Hz wanted harmonic components again formed balanced positive-sequence sets.

It was found that increasing the value of the modulating frequency over the range 40 Hz to 90 Hz resulted in a 'beat effect' occuring in the amplitudes of the wanted harmonic components of the natural sampled p.w.m. process. However, such effects were not apparent for the regular sampled asymmetrical double-edge p.w.m. process.

4.3 fc = 300 Hz, fm = 140 Hz, Modulation Depth = 0.95

It may be seen from Fig. 5c that when the frequency of the modulating wave was increased to 140 Hz, the natural sampling process introduced a 20 Hz sub-harmonic voltage component of approximately 30% and a 100 Hz sub-harmonic component of approximately 6%. However, it may be seen that the novel regular sampled asymmetrical p.w.m. process almost eliminates the 20 Hz component and totally eliminates the 100 Hz component. It is equally apparent that the regular sampling process almost eliminates the 220 Hz and 260 Hz super-harmonic components which existed for the natural sampled system.

Increasing the frequency of the modulating wave from 90 Hz to 140 Hz again produced considerable beats in the amplitude of the wanted harmonic components for the natural sampling process, whereas the regular sampled asymmetrical process again almost eradicated this undesirable effect.

5.0 CONCLUSIONS

From the results of the investigations into prior-art natural sampled p.w.m. control techniques and the proposed novel regular sampled p.w.m. control technique, the following conclusions can be drawn :

(1) The application of regular sampling to the control circuit eradicates the undesirable direct voltage components in the LINE to LINE output voltage waveforms, which existed for the natural sampled p.w.m. process.

(2) The amplitudes of the undesirable sub-harmonic voltage components which existed in the LINE to LINE voltage waveforms of prior-art analogue control p.w.m. inverters, are reduced to an insignificant value (less than 1%) by the regular sampled asymmetrical p.w.m. process, thus extending the useable range of frequency ratio from:

$$10 < \frac{fc}{fm} < \infty$$

to $$2 < \frac{fc}{fm} < \infty$$

(3) Regular sampling to produce the asymmetrical double-edge p.w.m. waveforms reduces the amplitudes of the most significant super-harmonic components for all non-integer values of frequency ratio less than "fifteen". However, for values of frequency ratio above "fifteen" the harmonic spectra for both sampling processes became similar.

(4) The 'beat effect' phenomenon which exists in prior-art p.w.m. systems for non-integer values of frequency ratio less than "six", was almost completely eradicated by the regular sampled asymmetrical double-edge p.w.m. process for values of frequency ratio greater than "two".

It may therefore be concluded that the application of sampling techniques which are common in communication engineering but which have not previously been applied to the analogue control circuits of p.w.m. inverters, considerably increases the useable frequency range without increasing the switching frequency of the power components, the number of power components or cost.

An in-depth analysis of the 'beat effect' phenomena which occurs in prior-art analogue control asynchronous mode p.w.m. inverters has not been presented in this paper, but such an analysis will be reported in a further paper in the near future.

6.0 REFERENCES

1. S.B. Dewan and J.B. Forsythe, Harmonic analysis of a synchronised pulse-width-modulated three-phase inverter, IEEE Trans. on Ind.Apps. 1A-10 No. 1 (1974)

2. N.C. Leverance, Fourier analysis of pulse width modulated inverter waveforms, IEEE Ind. and Gen.Appls. Group Milwaukee Section (1970).

3. U. Meier and B. Mokrytzki, Voltage control thru pulse width modulation IEEE Ind. and Gen. Appls. Group Milwaukee Section (1970).

4. H.S. Black, Modulation Theory, D.Van Nostrand (1953).

5. B.M. Bird and J.S. Ford, Improvements in the phase controlled cycloconvertor using communication principles, Proc.IEE. 121, No. 10, p.1146 (1974).

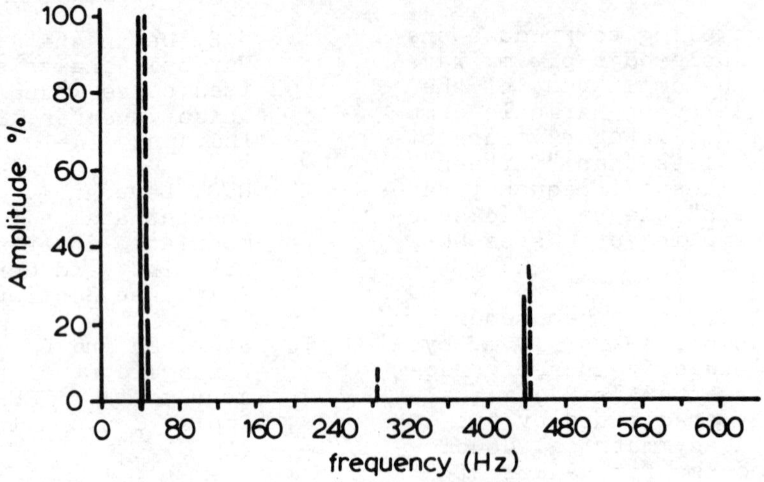

Fig. (5a) fc = 300 Hz, fm = 40 Hz, Mod. depth = 0·95

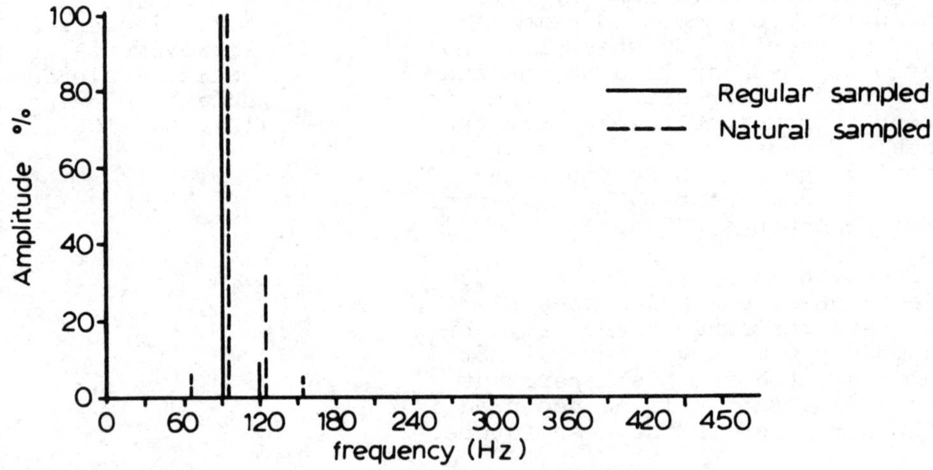

Fig. (5b) fc = 300 Hz, fm = 90 Hz, Mod. depth = 0·95

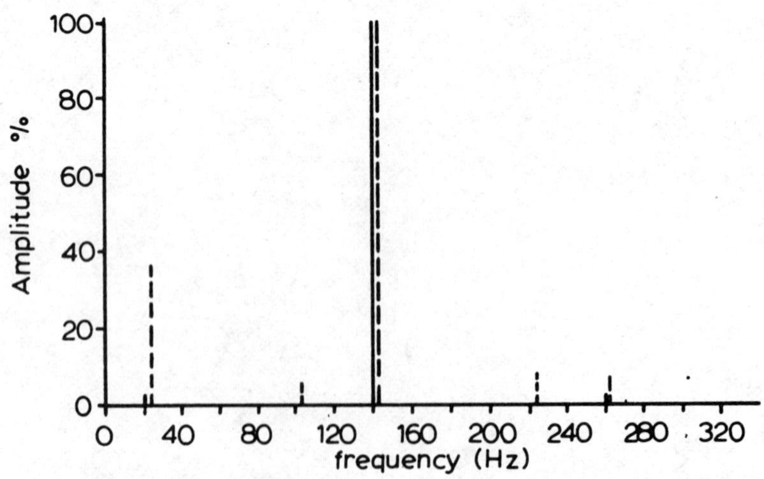

Fig. (5c) fc = 300 Hz, fm = 140 Hz, Mod. depth = 0·95

Fig. 5. Line to line voltage harmonics.

ON THE ORDINARY AND MODIFIED SUBHARMONIC CONTROL

A. Balestrino, G. De Maria and L. Sciavicco

Istituto Elettrotecnico, University of Naples, Naples, Italy

ABSTRACT

In this paper multiple pulse width modulation processes in power converters are analysed. The modulator output spectrum is given in terms of the single harmonic components. The computation heavily relies on the theory of Kapteyn series by which the implicit equations concerning the switching instants are explicitely solved.
The method of subharmonic control is thoroughly examined and subharmonics occurrence is stressed. Besides, a modified technique is pointed out, with a spectrum characterized only by odd harmonics of the modulating frequency. A comparison among different modulation techniques is worked out.

INTRODUCTION

The use of pulse modulation in static inverters has provided efficient means of controlling voltage and frequency in variable-speed a.c. motor drive system. Since harmonics generated in the modulation process affect efficiency and performance of drive system, the choice of a particular method of modulation is connected with the requirements of the application. In fact it is the balance between the inverter losses, increasing with the switching frequency, and the losses of the motor, which improves its performance in high frequency commutation, to very often bind the choice of a particular technique. Three modulation techniques trying to reduce or eliminate the harmonics are generally used: a) selected harmonic reduction (Ref. 1), b) harmonic neutralization by wave synthesis (Ref. 2), c) multiple pulse width modulation (Ref. 3). In the last class relevant interest is attached to the method of subharmonic control (Ref. 4) which belongs to the largest class of pulse-width modulation with natural or uniform sampling and two or three modulation levels.
In this paper the method of subharmonic control is thoroughly examined. The inverter output waveform is determined by the direct comparison between a sinusoidal modulating wave and a carrier wave. At first the harmonic content is determined on the basis of Kapteyn's series (Ref. 15) by which the implicit equations concerning the switching instants are explicitely solved (Ref. 16). The modulating wave subharmonics occurrence is stressed and a modification of the conventional technique is proposed to overcome such difficulties. The new technique is easily implementable by means of standard electronic circuity.
When the modified subharmonic control technique is used then the modulator output spectrum is characterized only by odd harmonics of the modulating frequency.
A comparison between conventional and modified subharmonic control is made and some numerical cases are worked out.

PULSE WIDTH MODULATION SYSTEMS

All pulse width modulation systems are characterized by a high frequency pulse train, with constant amplitude, width modulated according to the particular modulation law. Modulation process involves the sampling of modulating signal $v(t)$; the sampled information content is then converted into a series of modulated pulses $u(t)$ having pulse widths depending on amplitude of the modulating signal. The link between width and amplitude can be realized in many different ways according to the sampling type and to the modulation law. If the sampling is uniform the width of the generic pulse is determined by the value assumed by the modulating signal in coincidence with equally spaced time instants. If the sampling is natural, the width of the pulse is directly proportional to the amplitude of the modulating waveform $v(t)$ at the time when switching occurs. It is usual to implement this type of sampling through direct comparison of two waveforms: the modulating one (generally sinusoidal) and the carrier one $v_c(t)$ (triangular, sawtooth etc.) having higher frequency, switching instants and therefore the

width of generic pulse are determined by the intersection of the two waveforms.
Apart from the sampling type, pulse width modulator may be characterized as leading edge, lagging edge and double edge modulator, according to the edges which are modulated. As regards the two level double edge modulators, a modulation process is shown in Fig. 1. The modulating waveform can have a generic shape; as however, for a.c. motor drive systems, asinusoidal output waveform is required, henceforth only sinusoidal modulating waveforms will be considered: on the other hand, this class of input signals is the most suitable to analyse distortion due to the modulation process.
As far as natural sampling is concerned, at first the switching instants are determined in terms of Kapteyn series, then, through a modification of Graf formula, the spectrum is expressed in terms of the single harmonic.
As regards modulation processes with uniform sampling, it is not necessary to use special techniques to determine switching instants; the spectrum is nevertheless determined to carry out a significant comparison between the two types of sampling.

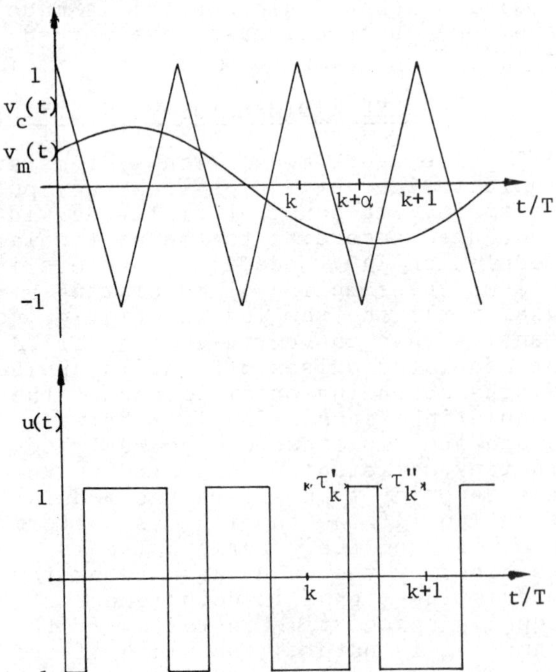

Fig. 1a Natural sampling PWM

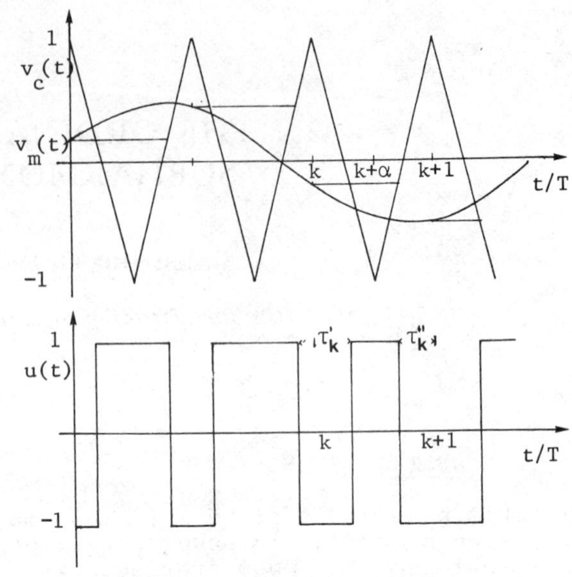

Fig. 1b Uniform sampling PWM

NATURAL SAMPLING

Switching instants determination

In the method of harmonic control with natural sampling, Fig. 1a, the output voltage switching instants are determined by the intersection of two auxiliary waveforms: triangular carrier, included between the values E and -E with frequency f_c and a sinusoidal modulating wave v(t) with amplitude: V<E, frequency f_m.
Without loss of generality assume E=1 and the modulator described by the input-output relationship:

$$u(t)=\begin{cases} 1 & V_c<V \\ -1 & V_c>V \end{cases} \tag{1}$$

The following relations are written as a function of the generic instant K+α, so that leading edge (α=0), lagging edge (α=1) and double edge modulation are treated in a unified manner. Let the modulating wave be:

$$v(t)=V\sin(2\pi f_m t+\phi) \tag{2}$$

with:

$$f_m=\frac{M}{N}f_c = xf_c \tag{3}$$

Where N and M are positive integers with N>M and N/M rational.
The output u(t) is expressed by:

$$u(t)=1 \quad K+\tau'_k \leq t/T < K+1-\tau''_k \tag{4}$$

$$u(t)=-1 \quad \begin{cases} K\le t/T < K+\tau'_k \\ K+1-\tau''_k \le t/T < K+1 \end{cases} \quad (4')$$

where τ'_k and τ''_k are defined by the equations:

$$1-2\tau'_k/\alpha=V\sin(2\pi x(K+\tau'_k)+\phi) \quad (5)$$

$$1-2\tau''_k/\beta=V\sin(2\pi x(K+1-\tau''_k)+\phi) \quad (6)$$

Note that, Fig. 1, if the modulator is a symmetrical double edge one then $\alpha=\beta=1/2$.
Let be:

$$\varepsilon=\alpha\pi xV \quad (7)$$

$$E'_k=2\pi x(K+\tau'_k)+\pi+\phi \quad (8)$$

$$M'_k=\pi x(\alpha+2K)+\pi+\phi \quad (9)$$

$$\varepsilon'=\beta\pi xV=(1-\alpha)\pi xV \quad (10)$$

$$E''_k=2\pi x(K+1-\tau''_k)+\phi \quad (11)$$

$$M''_k=\pi x(2K+2-\beta)+\phi \quad (12)$$

then eqns. (5),(6) can be rewritten:

$$M'_k=E'_k-\varepsilon\sin E'_k \quad (13)$$

$$M''_k=E''_k-\varepsilon'\sin E''_k \quad (14)$$

Those equations correspond to the analytic expression of the integral form of Kepler's second equation, where ε is the eccentricity of an elliptic trajectory, M_k is the mean anomaly, E_k is eccentric anomaly of a material particle moving along the trajectory, with centre force located in the ellipse focus.
Solving (13) in terms of Kapteyn series (Ref. 15):

$$E'_k=M'_k+\sum_{1}^{\infty}{}_n \frac{2}{n}J_n(n\varepsilon)\sin(nM'_k) \quad (15)$$

is obtained, where the series rapidly converges if $\varepsilon<1$.
Eqn. (14) and the analogue expression corresponding to the solution of (14) give:

$$\tau'_k=\frac{\alpha}{2}+\sum_{1}^{\infty}{}_n \frac{1}{\pi xn} J_n(n\varepsilon)\sin(nM'_k) \quad (16)$$

$$\tau''_k=\frac{\beta}{2}+\sum_{1}^{\infty}{}_n \frac{1}{\pi xn} J_n(n\varepsilon')\sin(nM''_k) \quad (17)$$

Harmonic analysis

The harmonic analysis for the output waveform consists in determining the expression in Fourier series:

$$u(t)=A_o+\sum_{1}^{\infty}{}_i(B_{i/M}\sin\tfrac{i}{M}\omega_m t + A_{i/M}\cos\tfrac{i}{M}\omega_m t) \quad (18)$$

If the frequency of the modulating wave is not a submultiple of the frequency of the carrier, the first harmonic in the expansion (18) corresponds to a subharmonic of frequency equal to: f_m/M.
The Fourier coefficients are represented by:

$$A_o=\frac{1}{NT}\int_0^{NT} u(t)dt \quad (19)$$

$$B_{i/M}=\frac{2}{NT}\int_0^{NT} u(t)\sin(\tfrac{i}{M}\omega_m t)dt \quad (20)$$

$$A_{i/M}=\frac{2}{NT}\int_0^{NT} u(t)\cos(\tfrac{i}{M}\omega_m t)dt \quad (21)$$

Using (4), (19) becomes:

$$A_o=1-\frac{2}{N}\sum_{0}^{N-1}{}_K(\tau'_k+\tau''_k) \quad (22)$$

By substituting for τ'_k and τ''_k eqn. (16) and (17) and summing up in K:

$$A_o=\sum_{1}^{\infty}{}_p\left\{\frac{2}{\pi Mp}\left[J_{pN}(pN\varepsilon')\sin(pN\phi-\pi\beta Mp)+ -(-1)^{pN}J_{pN}(pN\varepsilon)\sin(pN\phi+\pi\alpha Mp)\right]\right\} \quad (23)$$

is obtained.
With the position:

$$C_{i/M}=B_{i/M}+jA_{i/M} \quad (24)$$

from (12),(13),(14) one gets:

$$C_{i/M}=\frac{2}{i\pi}\sum_{0}^{N-1}{}_k\left[e^{j\frac{i}{M}(\phi+\pi)}e^{-j\frac{i}{M}E'_k}+ -e^{j\frac{i}{M}\phi}e^{-j\frac{i}{M}E''_k}\right] \quad (25)$$

By employing B4, A9, A10 and summing up in K, the generic term $C_{i/M}$ with $i\neq M$ is expressed by:

$$C_{i/M}=\frac{2}{i\pi}e^{j\frac{i}{M}\phi}\left[\sum_{-\infty}^{\infty}{}_p\frac{J_p(\omega_+)}{(\omega_+)}\frac{i\varepsilon}{M}\cdot\frac{\sin(\frac{i}{M}+p)M\pi}{\sin(\frac{i}{M}+p)x\pi}e^{-j(\frac{i}{M}\psi+p\psi-\frac{i}{M}\pi)}+ -\sum_{-\infty}^{\infty}{}_p\frac{J_p(\omega'_+)}{(\omega'_+)}\frac{i\varepsilon'}{M}\frac{\sin(\frac{i}{M}+p)M\pi}{\sin(\frac{i}{M}+p)x\pi}\cdot e^{-j(\frac{i}{M}+p)\lambda}\right] \quad (26)$$

where:

$$\omega_+=\varepsilon(p+\tfrac{i}{M}) \;;\; \omega'_+=\varepsilon'(p+\tfrac{i}{M}) \quad (27)$$

and:

$$\psi=\pi+\phi+\alpha\pi x+\pi x(n-1) \quad (28)$$

$$\lambda=\phi - \beta\pi x + \pi x(N+1) \qquad (29)$$

the generic term of Kapteyn series is significant if and only if the ratio:

$$\frac{\sin(\frac{i}{M}+p)M\pi}{\sin(\frac{i}{M}+p)x\pi} \qquad (30)$$

is different from zero,that is when the integer p satisfies the equation:

$$(\frac{i}{M}+p)x\pi=K\pi \qquad (31)$$

the term C_1 corresponding to the modulating wave frequency using the relationships B3 and B4 can be expressed as:

$$C_1/V = e^{j\phi} + \frac{2}{M} e^{j\phi} \sum_1^{\infty}{}_p \frac{1}{p}\Big[J_{pN+1}(pN\varepsilon) \cdot e^{jp\alpha_1} - J_{pN-1}(pN\varepsilon)\, e^{-jp\alpha_1} + J_{pN+1}(pN\varepsilon')\, e^{jp\beta_1} - J_{pN-1}(pN\varepsilon') \cdot e^{-jp\beta_1}\Big] \qquad (32)$$

where:

$$\alpha_1=N\phi + M\pi(\alpha-1) + N\pi(M+1) \qquad (33)$$

$$\beta_1=N\phi + M\pi(N + 1 - \beta) \qquad (34)$$

when the ratio N/M is integer. If the ratio N/M is rational,the indexes pN+1 and pN-1 must be replaced by (pN/M)+1 and (pN/M)-1 where p is an integer such that pN/M is an integer.
If the ratio N/M is irrational,the modulator output waveform is not periodic. In this case eqn. (26),(32) may be considered still valid if the irrational number N/M is considered as a limit of a sequence of rational numbers. After the passage to the limit, there is no mean value and the spectrum harmonics result only in correspondence with the angular frequencies:

$$\omega = K\omega_p - p\omega_m \qquad (35)$$

and are given by:

$$C_{i/M} = V e^{j\frac{i}{M}\phi}\Big[\alpha \frac{J_p(\omega_+)}{(\omega_+)} e^{-j(\frac{i}{M}\psi+p\psi-\frac{i}{M}\pi)} + - \beta\frac{J_{-p}(\omega_+)}{(\omega_+)} e^{-j(\frac{i}{M}+p)\lambda}\Big] \qquad (36)$$

UNIFORM SAMPLING

Harmonic analysis

Whenever a uniform sampling is used jointly with the method of subharmonic control,the values of modulating wave v(t),which are significant for determining the switching instants, are those as assumed in correspondence with the time instants kT.
The modulator output is yet expressed by eqn. (4),but the switching instants τ'_k and τ''_k are given in explicit form by the following relationships:

$$\tau'_k= \frac{1}{2}\alpha\Big[1-V\sin(2\pi kx + \phi)\Big] \qquad (37)$$

$$\tau''_k= \frac{1}{2}\beta\Big[1-V\sin(2\pi kx + \phi)\Big] \qquad (38)$$

The Fourier coefficients can be determined using eqn. (4) together with eqn. B2;by summing up in k:

$$C_{i/M} = \frac{2}{i\pi}\sum_{-\infty}^{\infty}{}_n\, e^{jn\phi}\, \frac{1-e^{j2\pi M(n-\frac{i}{M})}}{1-e^{j2\pi M(n-\frac{i}{M})}} \cdot \Big[J_n(\frac{i\pi\alpha V}{N}) e^{-ji\pi/N} - (-1)^n \cdot J_n(\frac{i\pi\beta V}{N}) e^{-ji\pi(2-\beta)/N}\Big] \qquad (39)$$

is obtained.
If the ratio N/M is not rational,the same procedure as suggested for natural sampling can be followed.
The mean value,obtained from eqn. (4), (37),(38) is given by:

$$A_0= V \frac{\sin M\pi}{\sin x\pi} \sin\Big[\pi x(M-1)+\phi\Big] \qquad (40)$$

Since M<N, A_0 is always zero.

COMPUTATIONAL REMARKS

Through eqn. (25),(26),(39) the generic Fourier coefficient $C_{i/M}$ can be evaluated both in case of natural and uniform sampling. The Kapteyn series which appear in the expression of the frequency spectrum show a fast convergence rate because of the frequency ratios generally considered (M<N).
In order to make a comparison among the different modulation techniques, a program has been set up based on the utilization of the algorithm of H. Goldstein and R.M. Thaler (Ref. 17) The fast convergence of these series allows to evaluate the generic term $C_{i/M}$ by taking into account few significant terms only;then the computation time is strongly reduced. By evaluating $C_{i/M}$ with a precision of 10^{-4} the computation time required to determine the frequency spectrum has been about 10 seconds.
In contrast with the proposed procedure it is remarkable that in the literature the spectral analysis is carried out on the basis of double Fourier series in two variables (Ref. 8, 9,10,13),when no rational relation-

ship exists between the modulating frequency and the carrier frequency, as well as of single Fourier series approach (Ref. 4,11,12,13) when the frequecy ratio is rational. These procedure give the generic harmonic by overlapping infinite partial spectra.
Then it is cumbersome to compare different modulation processes and remarkable computational work is required.

SYMMETRICAL DOUBLE EDGE MODULATION

In the case of symmetrical double edge modulation with natural sampling if the phase satisfies the relationship:

$$\phi = \pi(M \pm 1)/2N \quad (41)$$

the maximum mean value of the output is obtained;the following approximate relationship holds:

$$A_o/V = J_N(N\varepsilon)\left[1-(-1)^{N-M}\right]/N\varepsilon \quad (42)$$

The d.c. component is shown in Fig. 3 for odd N-M.
It is noted that if the difference N-M is an even number,then the mean value is zero. Moreover the mean value is zero even if:

$$\phi = x\pi/2 \pm \pi/N \quad (43)$$

Through eqn. (42) and Fig. 3 it is possible verify that d.c. component decrease when the carrier frequency is increased. In case of leading or lagging edge modulation,the mean value is zero if:

$$\phi = x\pi \pm \pi/N \quad (44)$$

whatever difference N-M is considered.
In the uniform sampling case the output mean value is always zero since N>M .
In the natural sampling pulse width modulation process,the harmonic C_1 corresponding to the modulating frequency is kept coincident with the modulating waveform if x<1/6 . If x>1/6,with N-M even,the term C_1 is attenuated when the contribution of the higher order Bessel function is opposite in phase to the term $V\cdot e^{j\phi}$. The greatest attenuation is obtainedwhen the phase satisfies:

$$\phi = x\pi/2 \pm 2k\pi/N \quad (45)$$

For instance with E=V=1, N=3 , M=1 , C_1=.66 is obtained.
If x>1/6,with N-M even,the term C_1 is magnified when the contribution of the higher order Bessel function has the same phase as the term $V\cdot e^{j\phi}$. The greatest magnification is obtained when the phase satisfies eqn. (45).

For instance,with E=V=1, N=3, M=1 C_1=1.27 is obtained.
In the uniform sampling case the harmonic C_1 is kept coincident with the modulating waveform when x<1/12 .
If x>1/12 the term C_1 is always attenuated.

SUBHARMONIC CONTENT

In the modulation process,if the ratio N/M is not integer,subharmonic terms $C_{i/M}$ (i<M) are generated. Their contribution is not negligible in correspondence with low frequency ratio. The greatest contribution is obtained in the natural sampling single edge modulation process,while the smallest one is obtained in the case of uniform sampling symmetrical double edge modulation.
In Fig. 4 and 5 frequency spectra are shown for different types of sampling and modulation; they put in evidence the above consideration in a quantitative way.
For electromechanical machine applications,where harmonics of the low operating frequencies have a significant effect on machine performance,a quantity giving the amount of the flux and therefore of the magnetising current must be considered. In Fig. 6 the terms

$$\frac{C_{1/M}\ \omega_1}{C_1\ \omega_{1/M}}$$

are illustrated in correspondence with different modulation process; spectra of this type may be an useful aid in order to select the appropriate modulation technique.
It clearly turns out that subharmonic control technique is not without troubles. Many difficulties are due to subharmonic occurrence,to d.c. components and also to harmonics of the modulating frequency. In practice, to overcome such difficulties it is necessary to choose a very high N/M ratio and,if possible, a carrier frequency multiple of the modulating frequency. With a high frequency ratio, losses due to commutating inverters considerably increase;moreover an integer frequency ratio may be realized only by locking the carrier waveform to the modulating one. Such difficulties can be overcome by suitably modifying the subharmonic control technique.

MODIFIED SUBHARMONIC CONTROL

Let us consider a symmetrical double edge pwm. At every zero crossing of the modulating waveform with positive derivative,the carrier waveform is reset to the value -1;at every zero crossing of the modulating waveform with negative derivative the carrier waveform is reset to the value 1,Fig. 2.
This modification of subharmonic control

may be implemented with a standard electronic circuity.

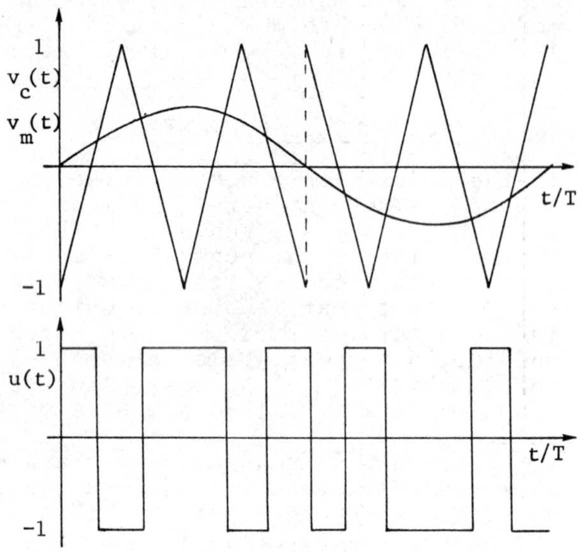

Fig. 2 Modified subharmonic control

The output waveform of this modified modulator has the same period as the modulating waveform, moreover there is a specular symmetry i.e. $u(t+T/2)=-u(t)$ In the output frequency spectrum, then there are only odd harmonics of the modulating waveform.
With modified subharmonic control tecnique there is no trouble arising from d.c. components, subharmonics or even harmonics of the modulating waveform.
The computation of harmonics of the modulating waveform is carried out as formerly. The commutation instants are determined by suitably modifying eqn. (5),(6). In short, for the symmetrical double edge case, it results:

$$C_i = \frac{2}{j\pi i}\left\{ i \sum_{1}^{\infty}{}_n \frac{1}{n}\left[J_{n-i}(n\varepsilon)e^{jn\alpha} + - J_{n+i}(n\varepsilon)\, e^{-jn\alpha}\right]\frac{1}{\sin\beta n} \cdot \left[\sin(\nu+1)n\beta + (-1)^n \sin\nu\beta n\right] + R\right\} \quad (46)$$

where: $i=2k+1$

$$\nu = \text{integer part of } (T_m/2T_c) \quad (47)$$

$$\alpha=\pi T_c(2\nu+1)/2T_m \quad (48)$$

$$\beta=\pi T_c/T_m \quad (49)$$

$$\varepsilon=\pi T_c V/2T_m \quad (50)$$

$$R' = -1-i\sum_{1}^{\infty}{}_n \frac{1}{n}\left[J_{n-i}(n\varepsilon)e^{jn\beta(4\nu+1)} + - J_{n+i}(n\varepsilon)e^{-jn\beta(4\nu+1)}\right] \quad (51)$$

$$R=\begin{cases} R' & i\neq 1 \\ R'-\nu\varepsilon & i=1 \end{cases} \quad (52)$$

if $T_m/2 \leq \nu T_c + T_c/4$

$$R=\begin{cases} 0 & i\neq 1 \\ -\varepsilon(2\nu+1)/2 & i\neq 1 \end{cases} \quad (53)$$

if $\nu T_c + T_c/4 < T/2 \leq \nu T_c + 3T_c/4$

$$R'=-1+i\sum_{1}^{\infty}{}_n \frac{(-1)^n}{n}\left[J_{n-i}(n\varepsilon)e^{jn\beta(4\nu+3)/2} + - J_{n+i}(n\varepsilon)e^{-jn\beta(4\nu+3)/2}\right] \quad (54)$$

$$R=\begin{cases} R' & i\neq 1 \\ R'-\varepsilon(2\nu+1)/2 & i=1 \end{cases} \quad (55)$$

if $T_m/2 > \nu T_c + 3T_c/4$

Fig. 7 shows some typical spectra associated with the modified subharmonic control technique so that by a comparison with Fig. 4 and 5 the improvements achieved are emphasized.

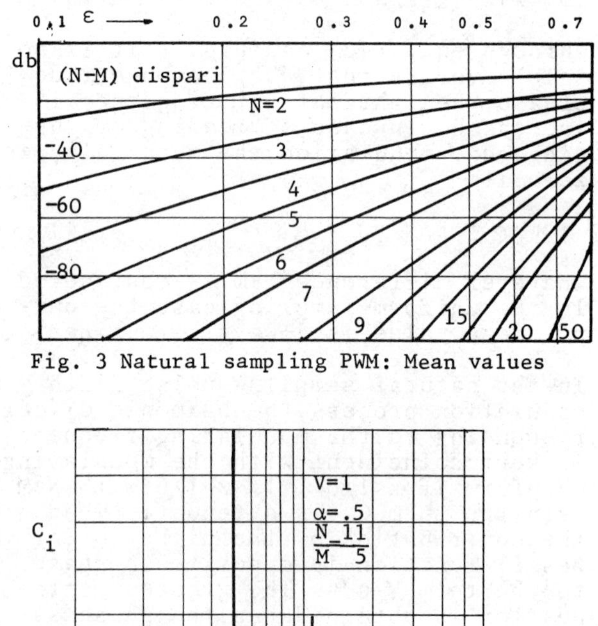

Fig. 3 Natural sampling PWM: Mean values

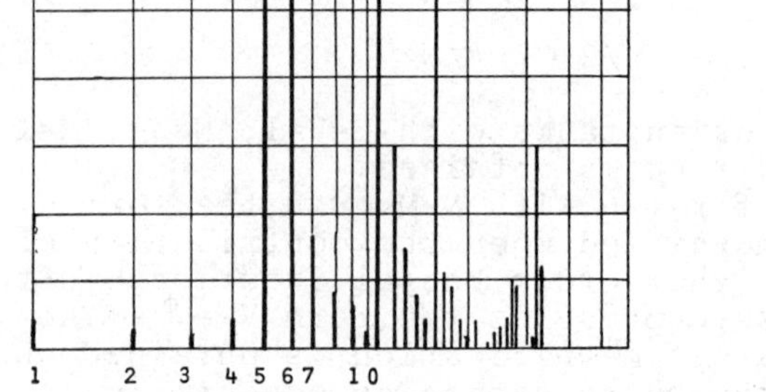

Fig. 4a Natural sampling PWM frequency spectrum

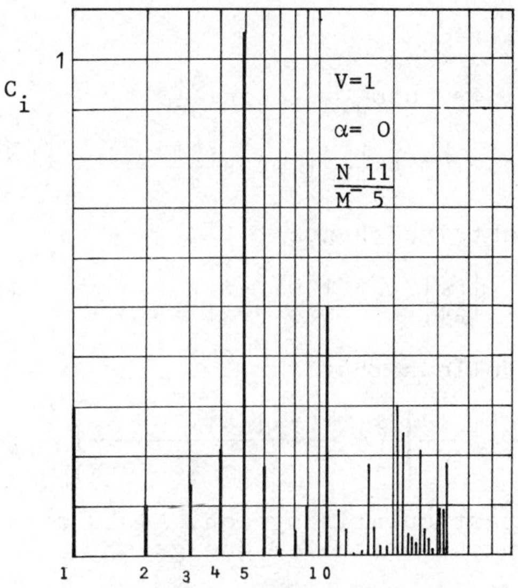

Fig. 4b Natural sampling PWM frequency spectrum

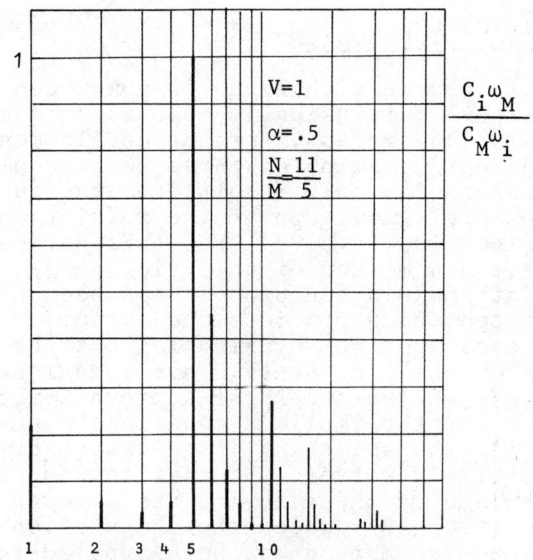

Fig. 6a Natural sampling PWM weighted frequency spectrum

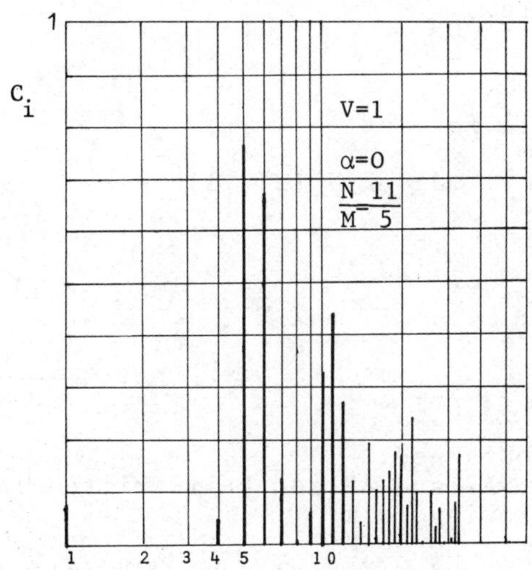

Fig. 5a Uniform sampling PWM frequency spectrum

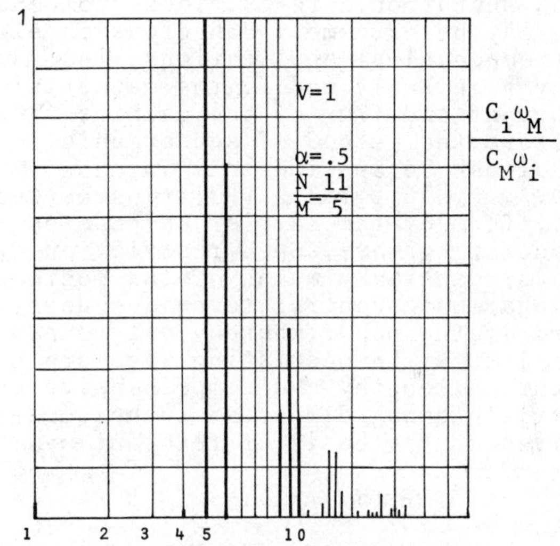

Fig. 6b Uniform sampling PWM weighted frequency spectrum

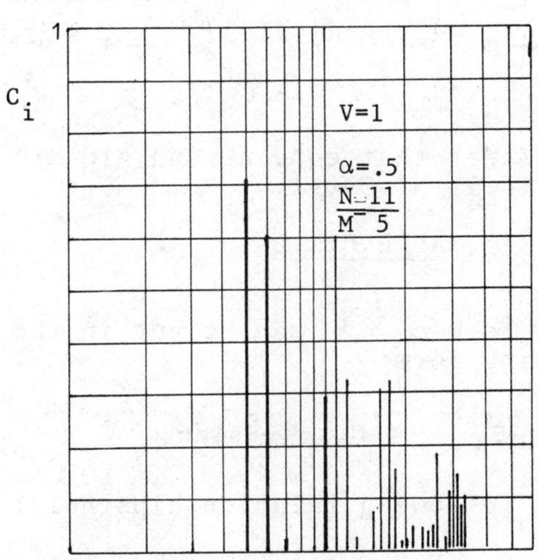

Fig. 5b Uniform sampling PWM frequency spectrum

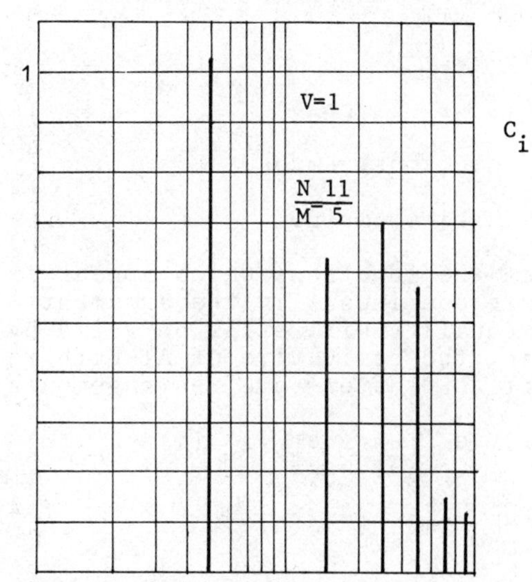

Fig. 7 Modified natural sampling PWM frequency spectrum

CONCLUSION

All frequency changing in power converters intrinsically depends on a modulation process. Further development and optimization of these power converters can only result from the fundamental knowledge of the modulation principles involved. For this purpose, a new method of analysing pulse width modulation process has been presented, which provides the output frequency spectrum by singling out the harmonics components. This technique rests on the theory of Kapteyn series, whereby the implicit equations concerning the switching instants are explicitely solved. The fast convergence rate of these series has allowed to strongly reduced the computation time. The method has been applied to significant types of converters, stressing some peculiarities, such as possible generation of d.c. components and subharmonic terms. These components may be extremely dangerous in electromechanical applications, since they may remarkably modify the magnetising current and flux of the machine. Therefore, the method of subharmonic control can be applied only in case of possibly integral high ratios between waveform of the carrier and one of the modulating wave. Such restriction can be set apart by means of the modified subharmonic control technique described in the paper, whereby only odd harmonics of the modulating wave are present. A lengthy and comprehensive comparison among different techniques is given on the basis of relevant spectra.

APPENDIX A

Let us consider Graf's formula:

$$e^{j\nu\psi} J_\nu(\omega) = \sum_{-\infty}^{\infty}{}_m J_{\nu+m}(Z)\, J_m(z)\, e^{jm\Phi} \quad \text{(A1)}$$

where:

$$\omega = (Z^2 + z^2 - 2Zz\cos\Phi)^{1/2} \quad \text{(A2)}$$

$$\psi : \begin{cases} Z - z\cos\Phi = \omega\cos\psi \\ z\sin\Phi = \omega\sin\psi \end{cases} \quad \text{(A3)}$$

Since the functions which appear in A1 are continuous in the argument Φ, the equality will still be valid by taking the derivative of A1 with respect to Φ; after some rearragements:

$$J_\nu(\omega)\nu j e^{j\nu\psi}\frac{\partial\psi}{\partial\Phi} + e^{j\nu\psi}\, J_{\nu-1}(\omega) - J_{\nu+1}(\omega) = $$

$$= \sum_{-\infty}^{\infty}{}_m\, jmJ_{\nu+m}(Z)J_m(z)e^{jm\Phi} \quad \text{(A4)}$$

is obtained.
By considering way ψ is defined, taking the derivative of A3 with respect to Φ:

$$\begin{aligned} z\sin\Phi &= \cos\psi\frac{\partial\omega}{\partial\Phi} - \omega\sin\psi\frac{\partial\psi}{\partial\Phi} \\ Z\cos\Phi &= \sin\psi\frac{\partial\omega}{\partial\Phi} + \omega\cos\psi\frac{\partial\psi}{\partial\Phi} \end{aligned} \quad \text{(A5)}$$

is obtained; hence:

$$z^2 = \left[\frac{\partial\omega}{\partial\Phi}\right]^2 + \left[\frac{\partial\psi}{\partial\Phi}\right]^2\omega^2 \quad \text{(A6)}$$

It follows that:

$$\frac{\partial\psi}{\partial\Phi} = \frac{1}{\omega}\left[z^2 - \frac{(Zz\sin\Phi)^2}{Z^2 + z^2 - 2Zz\cos\Phi}\right]^{1/2} \quad \text{(A7)}$$

By particularising eqn. A4 in correspondence with $\Phi=0$ one gets:

$$\sum_{-\infty}^{\infty} m\, J_{\nu+m}(Z)J_m(z) = \nu J_\nu(\omega_-)\,\frac{z}{\omega_-} \quad \text{(A8)}$$

where:

$$\omega_- = Z - z$$

while for $\Phi=\pi$:

$$\sum_{-\infty}^{\infty}{}_m\, mJ_{\nu+m}(Z)J_m(z)(-1)^n = $$

$$= -j\nu J_\nu(\omega_+)\,\frac{z}{\omega_+} \quad \text{(A9)}$$

where:

$$\omega_+ = Z + z$$

is obtained.
Eqn. A7 and A8 can be rewritten in the following form:

$$\sum_1^{\infty}{}_m\, mJ_{\nu+m}(Z)J_m(z) - \sum_1^{\infty}{}_m\, mJ_{\nu-m}(Z)J_m(z)\cdot$$

$$\cdot(-1)^m = \nu J_\nu(\omega_-)\,\frac{z}{\omega_-} \quad \text{(A10)}$$

$$\sum_1^{\infty}{}_m\, mJ_{\nu+m}(Z)J_m(z)(-1)^m - \sum_1^{\infty}{}_m\, mJ_{\nu-m}(Z)\cdot$$

$$\cdot J_m(z) = \nu J_\nu(\omega_+)\,\frac{z}{\omega_+} \quad \text{(A11)}$$

Consider that eqn. A9 and A10 are not valid if $\omega_-=0, \nu=1$.

APPENDIX B

The term $e^{-j\frac{i}{M}E_k}$ can be put in the following form:

$$e^{-j\frac{i}{M}E_k} = e^{-j\frac{i}{M}M_k}\, e^{-j\frac{i\varepsilon}{M}\sin E_k} \quad \text{(B1)}$$

The following relationships hold:

$$e^{\pm jz\sin\theta} = \sum_{-\infty}^{\infty}{}_n\, (\pm 1)^n J_n(z)e^{jn\theta} \quad \text{(B2)}$$

$$e^{\pm jnE_k} = n\sum_{p=1}^{\infty} \frac{1}{p}\left[J_{p-n}(p\varepsilon)e^{\pm jpM_k} + \right.$$

$$\left. \cdot \quad - J_{p+n}(p\varepsilon)e^{\mp jpM_k}\right] \qquad (B3)$$

By using B2 and B3, eqn. B1 can be re-written:

$$e^{-j\frac{i}{M}E_k} = e^{-j\frac{i}{M}M_k}\left\{J_o\left(\frac{i\varepsilon}{M}\right) + \sum_{p=1}^{\infty} \frac{1}{p}\,\cdot\right.$$

$$\cdot\left[\sum_{n=1}^{\infty} (-1)^n nJ_{p-n}(p\varepsilon)J_n\left(\frac{i\varepsilon}{M}\right) + \right.$$

$$\left. - \sum_{n=1}^{\infty} nJ_n\left(\frac{i\varepsilon}{M}\right)J_{p+n}(p\varepsilon)\right] e^{jpM_k} +$$

$$+ \sum_{p=1}^{\infty} \frac{1}{p}\left[\sum_{n=1}^{\infty} nJ_{p-n}(p\varepsilon)J_n\left(\frac{i\varepsilon}{M}\right) + \right.$$

$$\left. - \sum_{n=1}^{\infty} (-1)^n nJ_{p+n}(p\varepsilon)J\left(\frac{i\varepsilon}{M}\right)\right]\cdot$$

$$\left. \cdot\ e^{-jpM_k}\right\} \qquad (B4)$$

REFERENCES

(1) F.G. Turnbull, PWM inverters for a.c. motor drives, IEEE Trans.on Comm.and Elect. 83, 374 (1964)

(2) A.Kernick et Alii, Static inverters with neutralization of harmonics, AIEE Trans.on Appl.and Ind. 81, 59 (1962)

(3) B. Mokrytzki, Pulse width modulated inverters for a.c.motor drives, IEEE Trans.on Ind.and Gen.App. IGA3 6 (1967)

(4) A.Shonung H.Stemmle, Static frequency changers with subharmonic control in conjunction with reversible-speed a.c. drives, Brown Boveri Rev. 51, 555 (1964)

(5) H.S.Patel, R.G.Hoft, Generalized techniques of harmonic elimination and voltage control in thyristor inverters: Part.I-harmonic elimination, IEEE Trans.Ind.Appl. 1A-9, 310 (1973)

(6) H.S.Patel, R.G.Hoft, Generalized techniques of harmonic elimination and voltage control in thyristor inverters: Part.II-voltage control techniques, IEEE Trans.Ind.Appl. 1A-10 5, 666 (1974)

(7) S.P. Jackson, Multiple pulse modulation in static inverters reduces selected output harmonics and provides smooth adjustment of fundamentals, IEEE Trans.on Ind.and Gen.Appl. IGA6 4, 357 (1970)

(8) H.G.Black, Modulation theory, Van Nostrands Comp., Toronto (1953)

(9) S.R.Bowes, B.M.Bird, Novel approach to the analysis and synthesis of modulation processes in power convertors, Proc.IEE 122 5, 507 (1975)

(10) W.R. Bennet, New results in the calculation of modulation products, Bell Syst.Tech.J. 12, 228 (1933)

(11) P.F.Panter, Modulation, noise and spectral analysis, Mc Graw Hill, New York (1965)

(12) K.W.Cattermole, Principles of pulse code modulation, Iliffe Boocks, London (1969)

(13) P.Ferraris M.Migliarino F.Villata, Regolazione in frequenza di motori a induzione col metodo della sottooscillazione, analisi armonica della tensione di alimentazione, Simposio sui motori lineari, Capri (1973)

(14) B.Grzesik, On certain possibility of comparison of different methods of multiple-pulse modulation with the help of integral square error criterion, IFAC Symp. on Control in power electronics and electrical drives, Dusseldorf (1974)

(15) G.N.Watson, A treatise on the theory of Bessel function, Cambridge University Press, Cambridge (1966)

(16) A.Balestrino A.Eisinberg L.Sciavicco, Invertitori a modulazione di durata di impulsi: analisi spettrale delle forme d'onda, Simposio sulla evoluzione delle macchine elettriche rotanti, Tirrenia (1975)

(17) H.Goldstein R.M.THaler, Recurrence techniques for the calculation of Bessel functions, M.T.A.C. 13, 102

PULSE RATIO MODULATION: AN INTERESTING TECHNIQUE TO IMPLEMENT DC/DC CONVERSION

A. Balestrino, A. Eisinberg and L. Sciavicco

Istituto Elettrotecnico, University of Naples, Naples - Italy

ABSTRACT

DC/DC converters may be considered as pulse modulators; the modulation laws usually implemented are those related to pulse frequency (PFM) and pulse width modulation (PWM) techniques.
The pulse ratio modulation (PRM) technique, providing several advantages over conventional pulse modulation methods, has been developed for applications in a space vehicle attitude control system. Capable of operation with a high degree of precision and overcoming certain bandwidth problems associated with other modulation techniques, the modulator appears to hold excellent possibilities for applications in a broad range of control systems, such as DC and AC drives.
In order to investigate the dynamical properties presented by a feedback control system employing this modulation technique, in the paper a mathematical model is presented, characterizing the PRM in terms of input output relationships. By using the state variables approach, the system is described through a vector recurrence equation and linear bounds on the state characterizing the adopted modulation law. On the basis of this model both the steady state behaviour and the transient response are evaluated. As far as stability properties are concerned, the linearized model is developed in order to determine the local behaviour in correspondence with the steady state solutions, and a comparison is made with an equivalent PWM control system. Moreover a procedure, based on the use of projection operators in vector spaces, is presented to analyse the asymptotic stability in the large.

INTRODUCTION

When on-off type devices are used, for efficiency purposes, to implement DC power amplfiers, adequate modulation techniques must be used; this solution makes punctual and instantaneous correspondence impossible between input and output signals: in fact a correspondence is established between input and ouput average values in proper time intervals. The modulation techniques used fall within the pulse modulation class.
A continuous signal can be represented by the average value E of a pulse train with constant amplitude M Fig. 1a expressed by

$$E = M\zeta \qquad (1)$$

$$\zeta = \frac{t^+}{T} \qquad \text{with } \zeta \in [0, 1] \qquad (2)$$

where ζ is the duty ratio, t^+ is the pulse width and 1/T the pulse repetition frequency. A pulse modulator, therefore, can be used as an amplifier if the duty ratio depends on the input signal V through relation:

$$\zeta = k\,V \qquad \text{with } kV \in [0, 1] \qquad (3)$$

where the term kM plays the role of a level gain. Apart from the modulation technique used, i.e. the correspondence among kV, t^+ and T, these modulators are usually denoted as unipolar modulators.
With a pulse train as shown in Fig. 1b, the output average value E is:

$$E = M(2\zeta - 1) \qquad (4)$$

If a correspondence among t^+, T and V is established so that:

$$2\zeta - 1 = kV \quad \text{with } kV \in [-1, +1] \qquad (5)$$

the modulator is denoted as a bipolar modulator (Ref. 1).

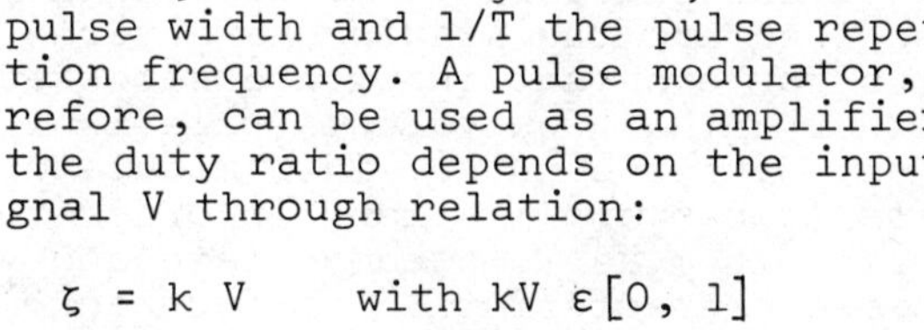

Fig. 1a. Unipolar pulse modulator.

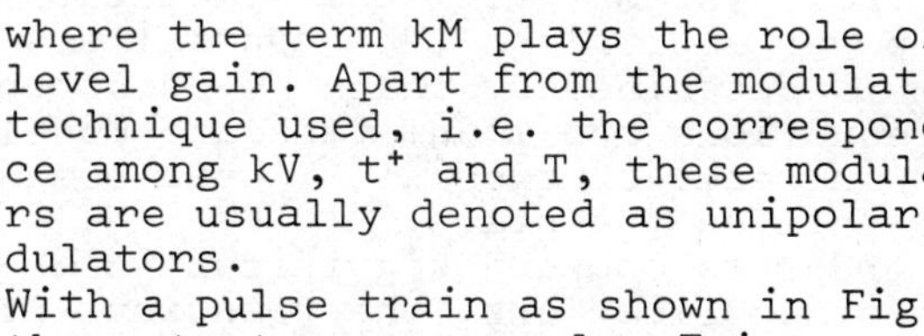

Fig. 1b. Bipolar pulse modulator.

To draw the best benefits from these amplification techniques, the key problem to face is one to work out a modulator which may control the duty ratio on the expected variation interval with sufficient accuracy. For this purpose the techniques commonly used are pulse width and pulse frequency modulation. In the former repetition period T is held constant and interval t^+ is varied; in the latter t^+ is held constant and the input signal affects repetition period T, only.
Since in applications minimum value of time intervals t^+ and $T - t^+$ are restricted by the upper cut-off frequency of switching devices, it clearly turns out that both PWM and PFM cannot cover intervals (3) and (5) completely.
In order to obtain full control in Ref. 2 and lately in Ref. 3, a modulator, denoted as pulse ratio modulator, was recommended, relying on the operating principle reported hereiafter.
A bipolar modulator is examined, considering, however, that models and remarks which follow can easily worked out to cover unipolar modulation.
First, the pulse ratio modulator is described and the static characteristics for constant input are stressed. Further on, an equivalence relation is set with the pulse width modulator, of far wider application, in order to compare them as to steady-state and dynamic characteristics of feedback control systems using these modulators as power amplifiers. The control system is described by means of recurrence equations and modulators are modelled in terms of adequate linear constraints on the state.
In order to compare significant classes of applications, the transient responses are determined as well as the domains of local stability on the plane of modulator typical parameters. The properties of asymptotic stability in the large are therefore deducted from the criteria suggested for feedback systems having generalized pulse modulators (Ref. 4).

PULSE RATIO MODULATOR

Let v(t) be the control signal, k an amplification factor, Δ the difference between two threshold levels. In correspondence with the generic interval $[t_n, t_n + T_n)$ the interval t_n^+ and the pulse repetition period T_n are defined by relations:

$$t_n^+ = \inf_{t\varepsilon[0,\infty)} t : \int_{t_n}^{t_n+t} [kv(\sigma) - 1]d\sigma = -\Delta \quad (6)$$

$$T_n = \inf_{t\varepsilon[t_n^+,\infty)} t : \int_{t_n+t_n^+}^{t_n+t} [kv(\sigma) + 1]d\sigma = \Delta \quad (7)$$

For v(t) = V = const.

$$t_n^+ = \frac{\Delta}{1 - kV}$$
$$T_n = \frac{2\Delta}{(1 + kV)(1 - kV)} \quad (8)$$

are obtained and hence (5) is satisfied. The operating laws are given on Fig. 2.

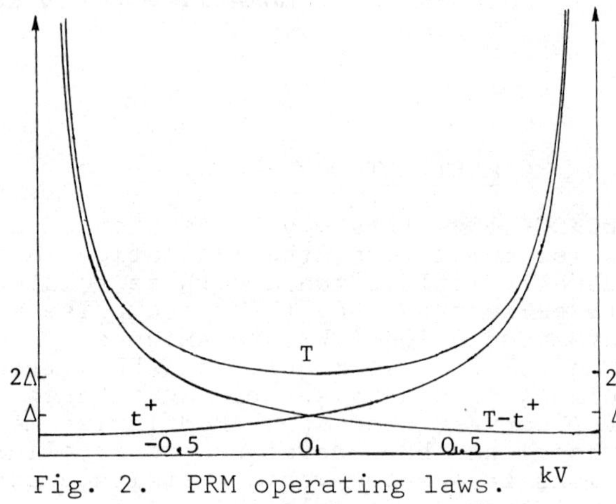

Fig. 2. PRM operating laws.

As the control signal V univocally determines the duty ratio, the denomination of this modulator is fully justified. It is also clear that the restrictions previously mentioned for PWM and PFM are overcome using PRM which, through a switching device with a finite bandwidth, can give duty ratio to fully cover intervals (3) and (5).
A possible PRM implementation scheme is given on Fig. 3, where threshold levels V^+ and V^- are such that $V^+ - V^- = \Delta$ and from the level shifter a quantity is fed back which can have values +1 and -1 only.

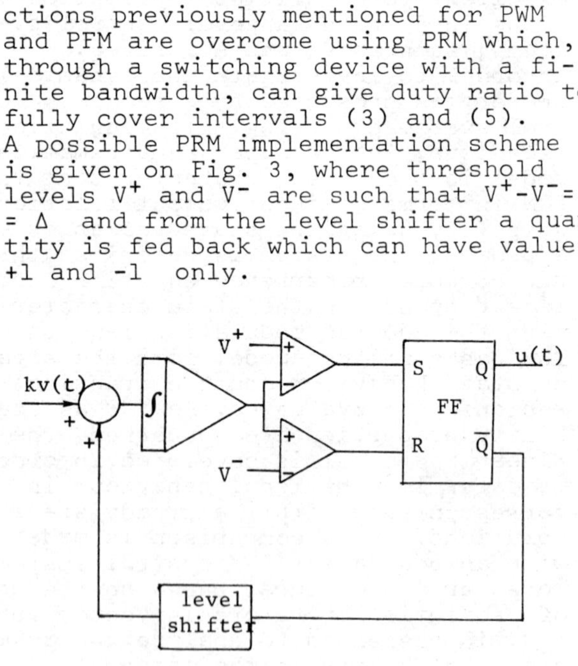

Fig. 3. PRM implementation scheme.

With a view to advantageously compare different modulation techniques, it seems useful to consider the characteristics of modulators on the plane t^+, $T - t^+$, Fig. 4.
The real characteristics of switching devices and the application of pulse modulators on servo-mechanisms and con-

trol systems imply restrictions which are usually given by:

a) minimum time values on t^+ and $T-t^+$ imposed by the finite bandwidth of the switching device;
b) minimum pulse repetition frequency set according to low pass characteristics of the controlled plant.

On the plane t^+, $T-t^+$, constant frequency loci are straight lines having negative unitary slope, whereas loci with constant output average value are straight lines

$$t^+ = \frac{\zeta}{1-\zeta}(T - t^+) \tag{9}$$

Above restrictions make all kinds of modulators to work within the triangular area on Fig. 4. In this area PFM works along straight lines having t^+ constant, PWM along constant frequency loci and PRM on a locus defined, by means of (8), through:

$$T - t^+ = \frac{\Delta t^+}{2t^+ - \Delta} \quad \text{with} \quad t^+ \geq \frac{\Delta}{2} \tag{10}$$

with maximum operating frequency equal to:

$$f_{max} = \frac{1}{2\,\Delta} \tag{11}$$

obtained in coincidence with $\zeta = 0.5$. Setting $t^+_{min} = (T - t^+)_{min} = t_m$ and $2t_m < T < T_M$ for the operating triangle, the duty ratios can be varied within the interval:

$$\zeta \in \left[\frac{t_m}{T_M}, 1 - \frac{t_m}{T_M}\right] \tag{12}$$

This interval can be fully described by PRM and PWM only if they share the

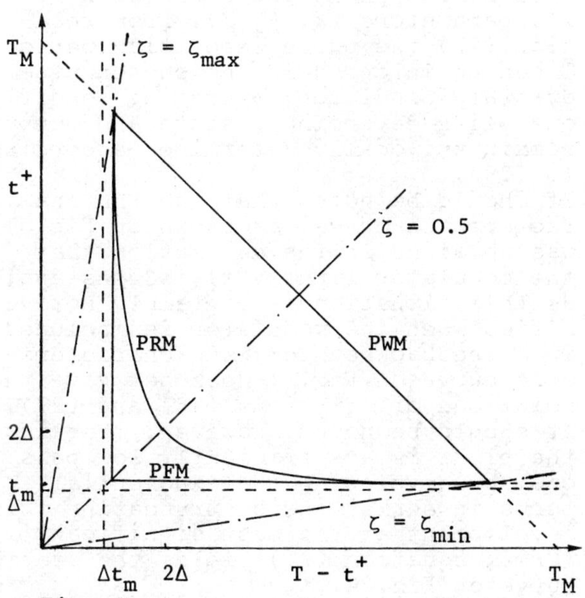

Fig. 4.
Modulators characteristic loci.

two vertices corresponding to the acute angles on Fig. 4. PFM, as can be verified, has lower performances and therefore comparison is made hereinafter between PWM and PRM only.

In order to have equivalence of PRM and PWM as to duty ratio, it is necessary that

$$\Delta = 2t_m\left(1 - \frac{t_m}{T_M}\right) \tag{13}$$

The main difference between PRM and PWM is that in coincidence with any duty ratio other than the limiting values, PRM works at higher frequencies than PWM. This characteristic intuitively improves the performances of the control system both as to increased bandwidth and decreased ripple factor. In order to make a rigorous comparison between the two types of modulation, the mathematical models of pulse modulated control systems are determined.

MATHEMATICAL MODEL OF PRM AND PWM CONTROL SYSTEMS

Fig. 5 shows the control system static and dynamic properties of which are to be investigated according to the modulation law applied. It is assumed hereinafter that modulators are not saturated ($|kv(t)| \leq 1$).

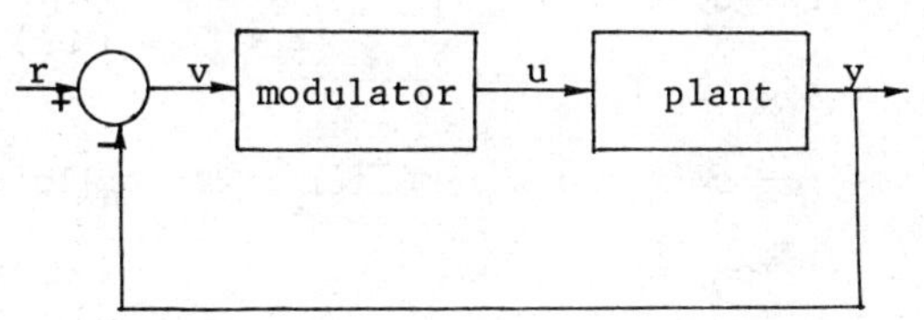

Fig. 5.
Pulse modulated control system.

With a bipolar modulator, the open loop system is described, in the generic interval $[t_n, t_n + T_n)$ by relations:

$$\begin{aligned} \dot{x}(t) &= Ax(t) + bu(t) \\ y(t) &= cx(t) \end{aligned} \tag{14}$$

$$u(t) = \begin{cases} +M & \forall\, t \in [t_n, t_n + t_n^+) \\ -M & \forall\, t \in [t_n + t_n^+, t_n + T_n) \end{cases} \tag{15}$$

where x is the m-vector state, A the mxm dynamic matrix, b and c^T input and output m-vectors, respectively; according to (14) the plant transfer function is:

$$G(s) = c(sI - A)^{-1}b \tag{16}$$

Since the input u(t) of the controlled

plant consists of a pulse train, it is convenient to consider the discrete model of the system on Fig. 5 which relates state x_{n+1} at instant $t_{n+1}=t_n+T_n$ with state x_n at instant t_n. The model made explicit by solving in the generic interval $[t_n, t_{n+1})$ equations (14) and (15) and assuming, for the sake of simplicity, that matrix A is invertible, is:

$$x_{n+1}=\Phi(T_n)x_n+MA^{-1}[I+\Phi(T_n)-2\Phi(T_n-t_n^+)]b \tag{17}$$

where

$$\Phi(\sigma) = \exp(A\sigma) \tag{18}$$

is the state transition matrix.
It is noted that if A is not invertible, it is always possible to follow the procedure indicated, bearing in mind that the solution of convolution integrals, which give the forced response of the plant, cannot be represented in a closed form, but through a series.
The model (17) is common for all types of pulse modulation. The modulation technique adopted, as referred to system on Fig. 5, is described by the relation existing between values of t_n^+ and T_n and the behaviour of error signal v(t) in the interval $|t_n, t_{n+1})$. Solving the relations (6) and (7), where v(t) is expressed by:

$$v(t)= r - c\Phi(t-t_n)(x_n+MA^{-1}b)+McA^{-1}b \tag{19}$$

$$\forall\, t\in(t_n,\ t_n+t_n^+)$$

$$v(t)= r - c\Phi(t-t_n)[x_n+MA^{-1}(I-2\Phi(-t_n^+))b]+ - McA^{-1}b \tag{20}$$

$$\forall\, t\in(t_n+t_n^+,\ t_{n+1})$$

the following constraints are obtained

$$cA^{-1}[\Phi(t_n^+)-I]x_n = \frac{\Delta}{k}+(r-\frac{1}{k})t_n^+ + McA^{-1}[It_n^+ + A^{-1} - A^{-1}\Phi(t_n^+)]b \tag{21}$$

$$cA^{-1}[\Phi(T_n)-\Phi(t_n^+)]x_n = -\frac{\Delta}{k}+(r+\frac{1}{k})(T_n-t_n^+)+ - McA^{-1}[(T_n-t_n^+)I + A^{-1}(\Phi(T_n)-\Phi(t_n^+))(I-2\Phi(-t_n^+))]b \tag{22}$$

which, along with recurrence equation (17), fully describe the static and dynamic behaviour of the control system on Fig. 5 if a PRM is used.
Since it is intended to compare PRM and PWM, it is necessary to obtain congruence equation for PWM analogous to (21) and (22). The natural sampling pulse width modulator is chosen because of better dynamic performances that the modulator shows as against PWM with uniform sampling (Ref. 5).

For the NSPWM the time interval t_n^+ is determined by the solution of implicit equation:

$$t_n^+ = [1+kv(t_n+t_n^+)]T/2 \tag{23}$$

where pedix n for T was omitted, since pulse repetition period is constant. As is remembered $v(t_n+t_n^+)$ can be derived from (19) and the following relation is obtained:

$$c\Phi(t_n^+)x_n = r + \frac{(T-2t_n^+)}{kT} + McA^{-1}[I-\Phi(t_n^+)]b \tag{24}$$

which,along with (17), describes the behaviour of control system on Fig. 5 when the modulator is a NSPWM.

STEADY-STATE BEHAVIOUR

Starting from relations (17), (21) and (22), which describe the system on Fig. 5 with a PRM, it is possible to find out steady state solutions. By setting $x_n = x_{n+1} = \hat{x}$, the following relations are obtained:

$$\hat{x} = MA^{-1}[I-\Phi(\hat{T})]^{-1}[I+\Phi(\hat{T})-2\Phi(\hat{T}-\hat{t})]b \tag{25}$$

$$r = (1-2\frac{\hat{t}}{\hat{T}})(McA^{-1}b-\frac{1}{k}) \tag{26}$$

$$2McA^{-2}[I-\Phi(\hat{t})][I-\Phi(\hat{T}-\hat{t})][I-\Phi(\hat{T})]^{-1}b + (\hat{T}-\hat{t})(McA^{-1}b-r-\frac{1}{k})+\frac{\Delta}{k}=0 \tag{27}$$

where steady-state values of T_n and t_n^+ are given as $\hat{T}$ and $\hat{t}$ respectively. After setting the desired output average value and therefore duty ratio $\hat{\zeta}$, the corresponding reference input r can be directly derived from (26) for a specified plant and a PRM with fixed parameters (k, Δ, M). From relation (27) the pulse repetition period $\hat{T}$ can be inferred in the chosen steady-state condition. Operating conditions being determined, state and output remain univocally determined according to (25).
It should be noted that the PRM characteristic curve, as shown on Fig. 4, was obtained on the assumption that the modulator imput v(t) is constant; as this situation is certainly not verified when the modulator is included in a feedback system, the characteristic curve on Fig. 4 does not give the solutions of equations (26) and (27).
It should be noted, however, that if the plant is a sufficiently low pass one, by developing the exponential terms in series and by truncating this development at the most significant terms, equation (27) yields the static curve on Fig. 4.
In order to evaluate deviations between the steady-state curves, these

curves are shown in Fig. 6 for a controlled plant represented by the transfer function $G(s) = 1/(s+1)$ and a PRM with $k = 1$, $M = 60$, $\Delta = 0.095$. As may be seen, deviations are limited, although value of M is rather high.

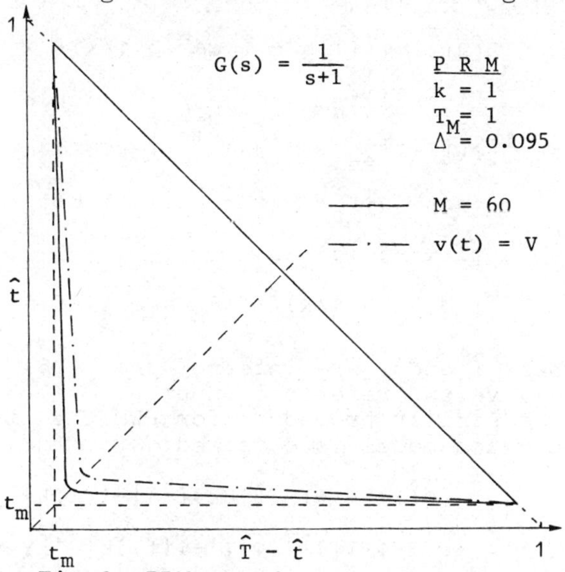

Fig.6. PRM steady-state curves.

The steady-state solution of the system on Fig. 5 with a PWM is given by

$$\hat{x} = MA^{-1}[I-\Phi(T)]^{-1}[I+\Phi(T)-2\Phi(T-\hat{t})]b \quad (28)$$

where $\hat{t}$ is given by:

$$McA^{-1}[I-\Phi(T)]^{-1}[2\Phi(\hat{t})-\Phi(T)-I]b = r - \frac{2\hat{t}-T}{kT} \quad (29)$$

When the steady-state values are known the ratio δ defined as

$$\delta = \frac{[\max y(t) - \min y(t)]_{PWM}}{[\max y(t) - \min y(t)]_{PRM}} \quad (30)$$

can be calculated, in case a PWM and an equivalent PRM are used.
For a plant with $G(s) = 1/(s+a)$, for a ratio $t_m/T_M = 0.05$, with varying ζ, curves turn out to practically coincide with the one shown on Fig. 7 for values of $a\varepsilon[0.1, 1]$.

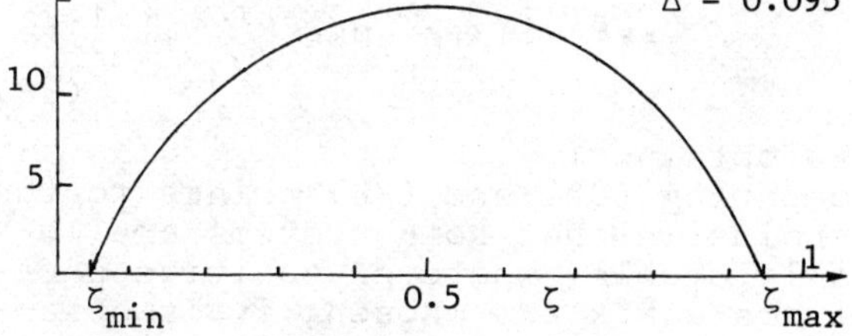

Fig. 7. Ripple comparison.

Fig. 8 shows, for a plant with $G(s) = 1/(s+1)$, the same ratio versus the characteristic parameter of modulator t_m/T_M.
This curve indicates improved performances in a quantitative way, as to ripple factor, which can be obtained by using a pulse ratio modulator.

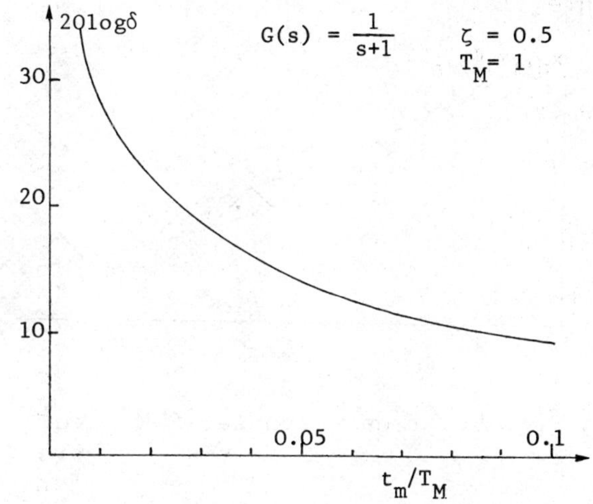

Fig. 8. Maximum ripple ratio.

DYNAMIC BEHAVIOUR

To compare dynamic properties of feedback control system with PRM and PWM modulators, dynamic responses and local stability properties for simple, yet significant, plants are evaluated.
Fig. 9 shows free responses for a system with $G(s)=1/s$.
It must be observed that, under equal equivalent gain, PRM permits lower ripple than PWM.
A comparison under equal gain conditions is not so significant as one under equal ripple conditions. Fig. 9b and 9c illustrates free responses under equal ripple conditions. In this case, with PRM, settling time is much smaller than the one with PWM and dynamic performances are far higher.

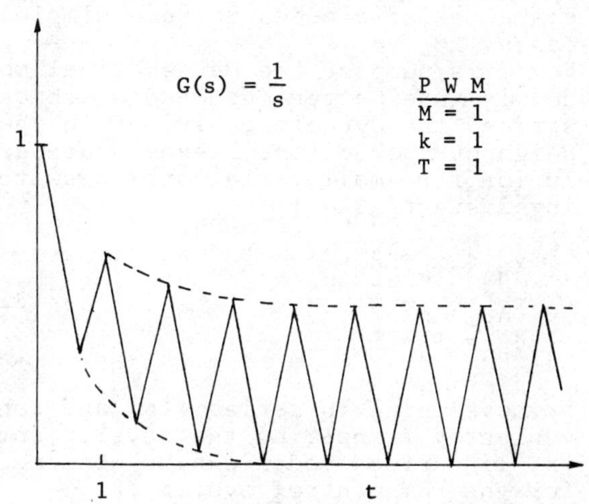

Fig. 9a. Free response: PWM case.

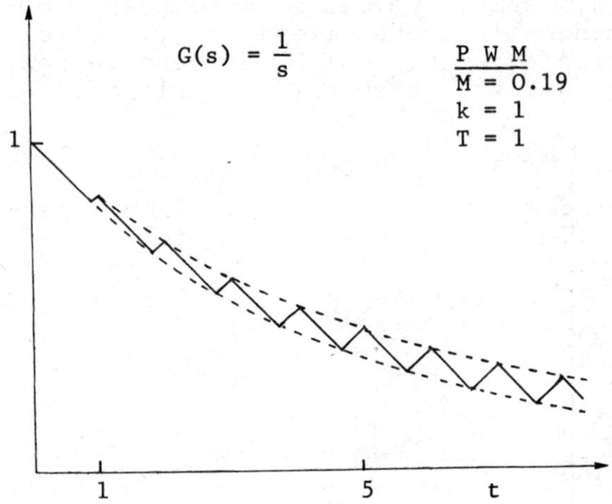

Fig. 9b. Free response: PWM case.

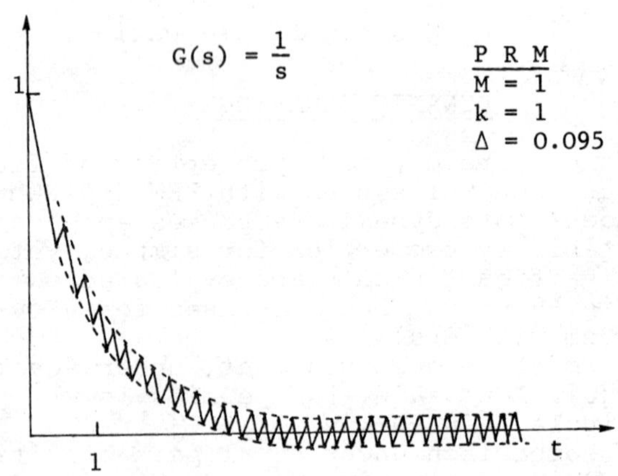

Fig. 9c. Free response: PRM case.

This can also be verified by examining the properties of local stability and aperiodicity conditions of closed-loop characteristic modes in some simple cases.
For this purpose the mathematical model is hereinafter deducted which describes the dynamic behaviour in the neighbourhood of the steady state solution for small variations of switching instants. Let:

$$\chi_n = x_n - \hat{x}$$
$$\tau_n = t_n - \hat{t} \qquad (31)$$
$$\theta_n = T_n - \hat{T}$$

by developing in series $\Phi(\sigma)$ and conveniently truncating the development, the linearized model can be obtained for the PRM control system as:

$$\chi_{n+1} = \Phi(\hat{T})\chi_n + 2M\Phi(\hat{T}-\hat{t})b\tau_n + 2M[\Phi(-\hat{T})-I]^{-1}[I-\Phi(-\hat{t})]b\theta_n \qquad (32)$$

$$cA^{-1}[\Phi(\hat{t})-I]\chi_n + 2\{McA^{-1}[I-\Phi(\hat{T})]^{-1}\cdot[\Phi(\hat{t})-\Phi(\hat{T})]b - (McA^{-1}b-1/k)\cdot(1-\hat{t}/\hat{T})\}\tau_n = 0 \qquad (33)$$

$$cA^{-1}[\Phi(\hat{T})-I]\chi_n + 2[McA^{-1}\Phi(\hat{T}-\hat{t})b+1/k+ -McA^{-1}b]\tau_n + 2\{McA^{-1}[\Phi(-\hat{T})-I]^{-1}\cdot[I-\Phi(-\hat{t})]b + +(McA^{-1}b-1/k)\hat{t}/\hat{T}\}\theta_n = 0 \qquad (34)$$

where $\hat{T}$ and $\hat{t}$ are inferred from (26) (27) versus reference input r.
By a similar procedure for PWM the linearized model is obtained:

$$\chi_{n+1} = \{\Phi(T) + \Phi(T-\hat{t})bc\Phi(\hat{t})[-1/MTk+ +c(I-\Phi(T))^{-1}(\Phi(T)-\Phi(\hat{t}))b]^{-1}\}\chi_n = W(T,\hat{t},M)\chi_n \qquad (35)$$

where $\hat{t}$ is given by (29) versus reference r.
By replacing τ_n and θ_n of (33) and (34) in (32), an equivalent recurrence equation can be obtained, in case a PRM is used, in the form:

$$\chi_{n+1} = R(\hat{T},\hat{t},M)\chi_n \qquad (36)$$

Local stability is guaranteed for PWM and PRM control system if eigenvalues of matrices W and R, respectively, are inside the unit circle; to ensure aperiodicity the eigenvalues of these matrices must belong to real interval [0,1].
Such conditions can be verified through well known methods (Ref. 7).
For example, for a controlled plant described by the transfer function $G(s) = 1/(s+a)$, with $\zeta=0.5$, according to (35) (36),

$$W(T,0.5T,M) = e^{-aT}\{1 + -[1/MTk + 1/(1+e^{aT/2})]^{-1}\} \qquad (37)$$

$$R(\hat{T},0.5\hat{T},M) = e^{-a\hat{T}}\{1 - 2[1+a(1+ +e^{-aT/2})/(Mk-Mke^{-aT/2})]^{-1}\}^2 \qquad (38)$$

are obtained.
Comparing (37) and (38) brings to the conclusion that both systems are locally stable, whatever positive parameters a,M,k are chosen; it is observed that, while aperiodicity condition is always satisfied for PRM, if:

$$MTk > 1 + e^{-aT/2} \tag{39}$$

the response is pseudoperiodical for PWM control system.
From this viewpoint, too, better performances of PRM control systems are evidentiated.

STABILITY IN THE LARGE

The problem of asymptotic stability in the large of pulse modulated systems has been widely described in literature; a criterion has been suggested (Ref. 6) to analyse a NSPWM system; an extension of this criterion (Ref. 4) synthetically presented hereinafter allows analysis of PRM feedback control systems stability.
Rewriting the recurrence equation (17) for the sake of notational simplicity as:

$$x_{n+1} = \Phi(T_n)(x_n + w_n) \tag{40}$$

and the constraints (21), (22) as:

$$K_n^T x_n + D_n = 0 \tag{41}$$

where K_n^T and D_n are 2xm and 2xl matrices, respectively, the asymptotic stability in the large of the steady state solution $\hat{x}$ for a PRM control system is investigated through the discrete analog of Lyapunov method.
Let:

$$H = \{x_n : K_n^T x_n + D_n = 0\} \tag{42}$$

and:

$$V_n = (x_n - \hat{x})^T Q (x_n - \hat{x}) \tag{43}$$

be a continuous scalar function, positive definite and radially unbounded in R^m, with Q a real, symmetric, constant matrix.
The first difference, according to (40), is:

$$V_{n+1} - V_n = - x_n^T P_n x_n - 2g_n^T x_n - h_n \tag{44}$$

where:

$$P_n = Q - \Phi^T(T_n) Q \Phi(T_n) \tag{45}$$

$$g_n = - \Phi^T(T_n) Q \Phi(T_n) w_n + [\Phi(T_n) - I] Q \hat{x} \tag{46}$$

$$h_n = 2\hat{x}^T Q \Phi(T_n) w_n - w_n^T \Phi^T(T_n) Q \Phi(T_n) w_n \tag{47}$$

It can be shown that the system is asymptotically stable in the large if:

$$\exists \min_{x_n \varepsilon H} (x_n^T P_n x_n + 2g_n^T x_n + h_n) \geq 0 \tag{48}$$

in correspondence with any t_n^+ and T_n congruent with their interval of definition.
Choosing (43) as a Lyapunov function and assuming P_n is positive definite matrix $\forall$ n, then (48) is equivalent to the following condition:

$$(D_n^T - g_n^T P_n^{-1} K_n)(K_n^T P_n^{-1} K_n)^{-1}(D_n - K_n^T P_n^{-1} g_n) + $$
$$+ h_n - g_n^T P_n^{-1} g_n \geq 0 \tag{49}$$

$$\forall\ t_n^+ \varepsilon | t_m, T_M - t_m | \ , \ \forall\ T_n \varepsilon | 2t_m, T_M |$$

The above procedure, conveniently particularized, can be followed in the case of a PWM feedback control system to determine, on the plane of characteristic parameters of modulator, regions of asymptotic stability and therefore is a useful tool to synthetize these systems (Ref. 6).
As to PRM feedback control systems it must be noted that, if Δ and k are fixed on the basis of specifications posed by the controlled plant, since steady state equations (26), (27) are to be simultaneously satisfied, $\hat{T}$ and $\hat{t}$ being fixed, the parameters M and r are univocally determined. From this point of view the procedure mentioned can be used only to check stability of the system under investigation.

CONCLUSIONS

This paper deals with an unconventional modulation technique showing the property of implementing a duty ratio proportional to average value of input signal in a convenient sampling interval; this duty ratio can vary in the interval [0, 1], although the modulator uses limited bandwidth switching devices. Such performances are obtained by suitably adjusting frequency and width of output pulses.
Examining static and dynamic characteristics of a PRM feedback control system shows remarkably improved performances as against an equivalent PWM feedback control system; such improved performances occur both for steady state ripple factor and dynamic responses, as well as for stability characteristics.
The difficulties to analyse a PRM feedback control system should not deter from using this technique in chopper fed DC drives; however, it must be noted that, at present, there is no consistent experience in this field.

REFERENCES

(1) J.Tal, Design and analysis of PWM amplifiers for DC servosystems, IEEE Trans. on Ind. El. and Cont. Instr., IECI-23, 47, (1976)

(2) R. A. Schaefer, A new pulse modulator for accurate DC amplification with linear or nonlinear devices, IRE Trans. on Instr., I-11, 34, (1962)

(3) P. Burger, Analysis of a class of pulse modulated DC/DC power converters, IEEE Trans. on Ind. El. and Cont. Instr., IECI-22, 104, (1975)

(4) A. Balestrino, A. Eisinberg, L. Sciavicco, On the stability of pulse modulated control systems, to appear on Ricerche di Automatica, (1976).

(5) A. Balestrino, A. Eisinberg, L. Sciavicco, On the analysis of chopper drives with PWM, IFAC Symp. on Control in Power Electronics and Electrical Drives, Duesseldorf, (1974).

(6) A. Balestrino, A. Eisinberg, L. Sciavicco, A generalized approach to the stability analysis of PWM feedback control systems, J. of the Franklin Inst.,298, 45, (1974).

(7) E. I. Jury, Inners and stability of dynamic systems, J. Wiley, New York, (1974).

DESIGN OF AN INVERTER WITH OPTIMUM OUTPUT VOLTAGE WAVEFORM

Mehmet Celâl Tanju

Asst. Prof. Dr. Electrical Engineering Department, Middle East Technical University Gaziantep Extension Campus, Gaziantep, Turkey

ABSTRACT

Harmonic distortion at a static inverter output has been the topic of research of many workers. Improvements in and control of harmonic distortion have been achieved by employing sampling and modulation techniques. On the other hand, there are practical difficulties in realizing the optimum inverter output voltage steps and/or control circuits. To overcome these difficulties and at the same time to obtain an inverter output having a low harmonic content, polyphase symmetrization technique is used to give the designed optimum inverter output voltage waveform.

INTRODUCTION

Harmonic distortion at a static inverter output has been the topic of research of many workers. Improvements in and control of harmonic distortion have been achieved by employing sampling and modulation techniques. These techniques, when applied to inverters for industrial drives, have their own limitations. On the other hand, optimization methods enable the researcher to get an inverter output voltage waveform that has a low harmonic content.However, there are practical difficulties in realizing the optimum inverter output voltage steps and/or control circuits. To overcome these difficulties and at the same time to obtain an inverter voltage output having a low harmonic content, polyphase symmetrization technique is used to give the designed optimum inverter output.

Optimum inverter output voltage waveform is determined by using a program written by H. G. Jacob (1) but modifying it so that the program converges to the required extremum more effectively when the constraints form a narrow boundary region.

METHOD

The method used in the design of the inverter with optimum output voltage waveform can be separated into two parts:

1. Optimization procedure which determines the variable values for optimum inverter output voltage waveform. (Optimum inverter output voltage waveform corresponds to minimum harmonic distortion of the inverter output.)

2. The polyphase symmetrization technique which enables the designed optimum inverter output voltage waveform to be properly constructed.

The Optimization Procedure

The numerical method developed by H. H. Rosenbrock (2) and thus, the static optimization algorithm written by H. G. Jacob (1) to give the extremum of a function of several variables restricted to a given region is used to obtain the optimum stepped inverter output voltage waveform. The step widths $(\theta_{k+1} - \theta_k)$ and the step heights A_{ck} of the inverter output voltage waveform are varied to give minimum harmonic distortion, k_t, for a fixed number of steps p of one cycle of inverter output voltage waveform, Fig. 1.

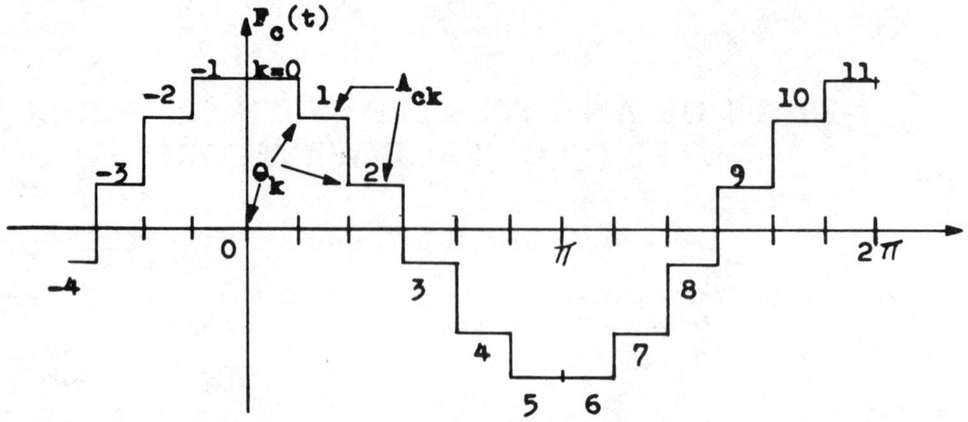

Fig. 1. The optimum inverter output voltage waveform

In order to eliminate subharmonics and even ordered harmonic components, which can occur if the inverter output voltage waveform is not symmetrical, maximum degree of symmetry is assumed. Maximum degree of symmetry means that the positive and negative half cycles of the inverter stepped voltage waveform are identical in form and inverted with respect to each other, i.e. half wave symmetry. Thus, there are only odd harmonic components in the inverter output voltage waveform. It is also known that, due to quarter wave symmetry, the Fourier series of the output voltage waveform will contain only cosine terms of odd order. In this work, the inverter output voltage waveform has both the quarter and half wave symmetry.

The aim of the optimization procedure is to minimize harmonic distortion caused by existing odd harmonic cosine series components in the inverter output voltage waveform. When the inverter output voltage waveform has both half wave and quarter wave symmetry, some of the A_{ck} and θ_k variables become dependent variables, see equations (1.7) and (1.8).

In order to preserve symmetry and at the same time avoid overlapping steps, the following constraints must be satisfied for $0 \leq \theta_k \leq \pi/2$, Fig. 1.

$$1 = A_{co} \geq A_{c1} \geq \dots \geq A_{ck} \geq \dots \geq A_{cx} \geq 0 \tag{1.1}$$

$$0 = \theta_o < \theta_1 < \dots < \theta_k < \dots < \theta_y = \pi/2 \tag{1.2}$$

$$M = p/2 \qquad \text{for p even and p/2 odd} \tag{1.3}$$

$$M = (p-2)/2 \qquad \text{for p even and p/2 even} \tag{1.4}$$

$$x = (M-1)/2 \tag{1.5}$$

$$y = (M+1)/2 \tag{1.6}$$

$$A_{c(M-k)} = - A_{ck} \tag{1.7}$$

$$\theta_{(M-k+1)} = \pi - \theta_k \tag{1.8}$$

The total number of independent variables, (x + y) = M, needed for the optimization algorithm is, thus, given by equations (1.3) and(1.4) whichever is appropriate.

The inverter output voltage waveform shown in Fig. 1. is optimized for minimum harmonic distortion (see Appendix) by using the constraints of equations (1.1), (1.2), (1.7), and (1.8). The results of optimization are as shown under column (a) of Table 1 .

In addition to this, it may be necessary to completely eliminate lower ordered harmonic components from the inverter output voltage waveform:

TABLE 1 Results of Inverter Output Voltage Optimization for p=12

Harmonic Number	(a) %	(b) %	(c) %	(d) %
1	107.25	103.74	102.39	107.22
2	2.26	0.76	0.00	2.30
5	3.26	0.96	0.00	3.25
7	2.70	0.97	0.00	2.67
9	0.68	0.99	0.00	0.69
11	2.10	6.29	9.09	2.10
13	4.69	9.16	7.69	4.69
15	7.86	1.86	0.00	7.85
17	2.81	0.98	0.00	2.82
19	0.72	0.55	0.00	0.71
21	0.60	0.31	0.00	0.60
23	0.41	1.60	4.35	0.39
25	1.36	4.81	4.00	1.37
% k_t	12.73	13.79	14.43	12.73
θ_1	41.234^o	33.213^o	30°	41.227^o
θ_2	66.195^o	61.287^o	60°	66.199^o
A_{co}	1.000	1.000	1.000	1.000
A_{c1}	0.640	0.721	0.732	0.641
A_{c2}	0.224	0.263	0.268	0.224

(a) - No constraints on harmonic magnitudes

(b) - $C_N \leq$ % 1.0 for $1 < N < p-1$

(c) - $C_N = 0$ for $1 < N < p-1$

(d) - $A_{ck} = (\mathrm{Sin}(\theta_{k+1}) - \mathrm{Sin}(\theta_k))/(\theta_{k+1} - \theta_k)$

k_t - Total harmonic distortion of inverter output voltage waveform

$$C_N = 0 \qquad \text{for} \quad 1 < N < p-1 \tag{1.9}$$

The results of optimization are shown under column (c) of Table 1 .

In some applications, it may be sufficient for the ratio of the lower ordered harmonics to the fundamental to have a certain percentage:

$$C_N/C_1 \leq \% X = \% 1.0 \tag{1.10}$$

The optimization result for this case is shown under column (b) of Table 1 .

In Table 1 , the largest step height A_{co} is normalized as $A_{co} = 1.0$. The peak value of the fundamental component as compared with A_{co} is indicated in Table 1 . It is seen from Table 1 that, the magnitude of the fundamental with %107.25 , corresponding to the optimum condition with no constraints on harmonic magnitudes, has the largest value.

Hardware realization of the optimal solutions corresponding to columns (a) and (b) of Table 1 would be quite difficult and expensive. This is due to the fact that the angles Θ_k determining the step durations must be obtained precisely. Otherwise, the harmonic distortion of the inverter output voltage waveform would depart from the optimum value. It would, thus, be pointless to devote too much expense and time in trying to implement it. It would be more easier and cheaper to realize the optimum solution corresponding to equation (1.9) .

Depending on the constraints and the initially estimated values of the independent variables A_{ck} and Θ_k , the number of stages in reaching the optimum solution varies. For p = 12 , there are 5 independent variables and the optimization program took 11 stages to reach to the optimum solution which corresponds to column (a) of Table 1 . Computer execution time for the modified optimization program of H. G. Jacob (1) took roughly 2 minutes for p=12 for each stage on the 7/32 Interdata mini computer. This means that the total optimization program takes 22 minutes of computer time. The initially estimated values for the independent variables for p=12 were:

$$A_{ck} = 1.0 \tag{1.11}$$

$$\Theta_k = 2.\pi.k/p \tag{1.12}$$

The initial step increments for all the variables, A_{ck} and Θ_k, were taken as 1/100.

With the addition of the constraints, equation (1.10), the same program after 194 trials or 7 stages entered into a search loop. The execution continued for some 3 hours after which it had to be stopped. The reason for this was found out to be that the search step increments, as the program approaches to the optimum satisfying the constraints, do not get smaller as they should. As the program by itself tries to find out the correct step increment, it can not satisfy the constraints of equation (1.10) and thus, the search goes on. In order to prevent this kind of a search loop; initially, a maximum number for search trials is allocated. If during the first stage, the current trial number exceeds the initially allocated maximum trial number, the program goes back and the inital search step increments are divided by a number n. In this work, n = 2 . The program,then, continues as for an initial search. This condition corresponds to the case of a large estimated step sizes and due to this, the constraints are not satisfied. The program, therefore, divides the estimated step sizes by 2. The division by 2 continues until the constraints are satisfied. If the first stage and any other stage is successfully completed, the search step sizes, the independent variables and the trial number are stored. The stored trial number for the first stage becomes the base for the maximum number of trials which can be made during the succeeding stages. The maximum number of trials for the following stages is taken as m (m=2) times the trial number for the first stage. If during any stage , the number of trials exceeds the maximum number of trials for the stages, the program goes back and puts in the stored values for the independent variables and divides the stored search step increments by n. The search,then, starts as for the initial stage. In this way, it was possible for the program to reach to the optimum solution effectively. Depending on the application, it is possible to assign different values for m and n.

In order to reduce the computer time, a functional relationship among the independent variables, to give the same optimal solution within acceptable limits, were sought :

$$A_{ck} = f(\theta_k) \tag{1.13}$$

The step heights being proportional to the average value of a cosine wave during the step width,

$$A_{ck} = (\mathrm{Sin}(\theta_{k+1}) - \mathrm{Sin}(\theta_k))/(\theta_{k+1} - \theta_k) \tag{1.14}$$

gives the optimal results as indicated in column (d) of Table 1 . The relationship of equation (1.14) also gives (within acceptable limits) the same results which correspond to column (c) of Table 1 . Various other functions, $f(\theta_k)$, were tried but they are not worth mentioning.

Optimizing the waveform of Fig. 1. for minimum harmonic distortion by using equation (1.14), therefore, reduces the total number of independent variables to y, equation (1.6). It is possible to express y in terms of p :

$$y = (p-2)/4 \quad \text{for p even and p/2 odd and } p \neq 2 \text{ or} \tag{1.15}$$

$$y = (p-4)/4 \quad \text{for p even and p/2 even and } p \neq 4 \tag{1.16}$$

Polyphase Symmetrization Technique

Polyphase symmetrization technique, as established by W. Fong (3), enables a balanced 3-phase vector system (V_1 , V_2 , V_3) to be obtained from 3 arbitrary vectors (V_a , V_b , V_c) by displacing each vector by $+/- a^o = +/- 120^o$. Taking a combination of the vectors give :

$$V_1 = V_a + a.V_b + a^2.V_c \tag{1.17}$$

$$V_2 = a^2.V_a + V_b + a.V_c \tag{1.18}$$

$$V_3 = a.V_a + a^2.V_b + V_c \tag{1.19}$$

Vectors V_a , V_b , V_c are the outputs of inverters. Due to economic reasons, the optimum inverter output voltage waveform corresponding to equation (1.9) has been designed. For p=12, only two voltages, V_a and V_b, are sufficient. This means that, for an optimum 3-phase inverter system, only two 3-phase bridge inverters are necessary. The outputs of the inverters are added, Fig. 2., via suitable transformers to give the 3-phase optimum inverter output voltage waveforms of Fig. 3.(a). The voltage phase relationships between V_a and V_b and V_a and V_b with respect to the optimum inverter phase voltage V_1 are shown in Fig. 3.

DISCUSSION

The output frequency and voltage of the optimum inverter can be controlled independently to give any desired V/f ratio with a constant harmonic content at all frequencies. However, the output frequency range is limited due to commutation problems at low frequencies when the input dc voltage, V_{dc}, to the inverter gets very small.

Although the optimum inverter output voltage waveform having the lowest harmonic distortion is the desired goal ; the hardware implementation of this goal must be considered from an economical point of view. If hardware implementation of an optimal solution is difficult and expensive ; then, it may be possible to change or modify the constraints so that the feasible optimal solution becomes simple and cheap. This can, ofcoarse, be achieved at the expense of degradation in performance of the inverter system.

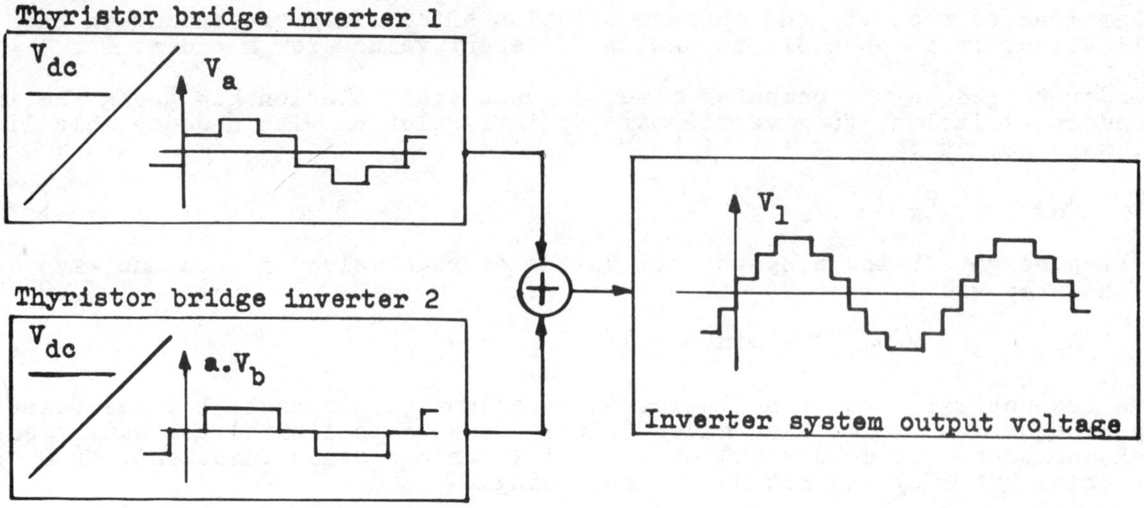

Fig. 2. Inverter system block diagram

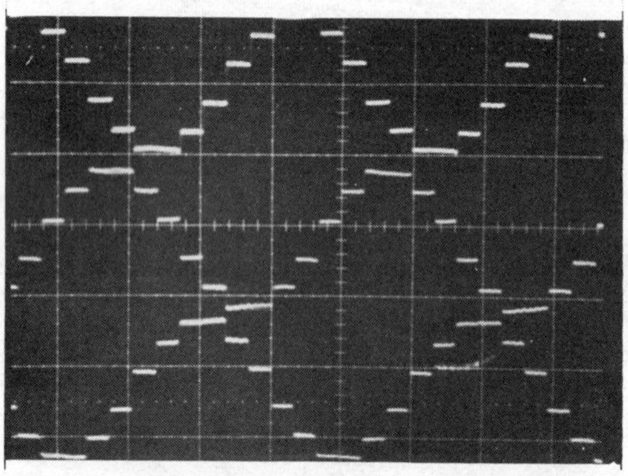

a) Inverter output phase voltages (V_1 , V_2 , V_3)

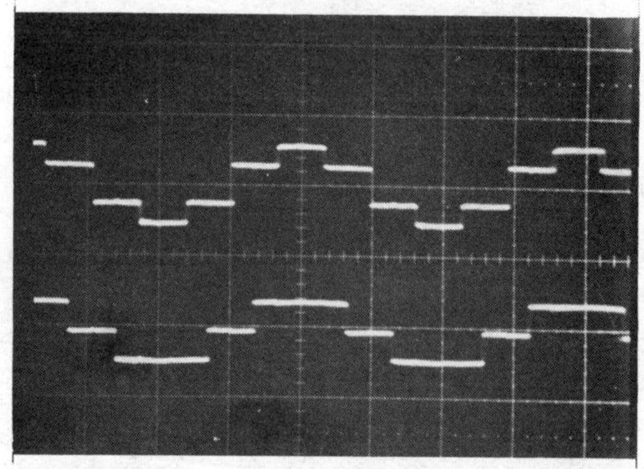

b) Top trace: V_a , bottom trace: $a.V_b$

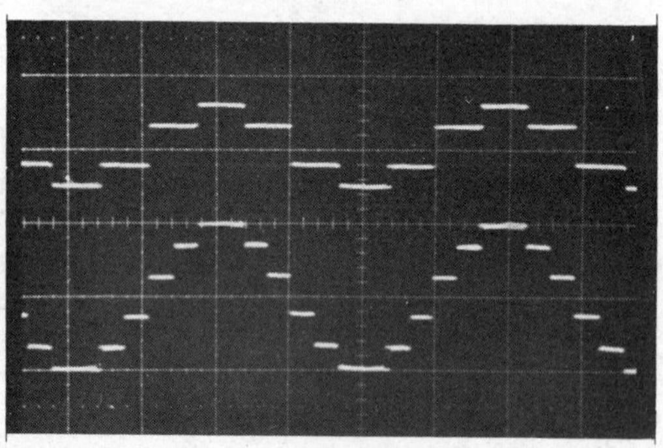

c) Top trace: V_a , bottom trace: V_1

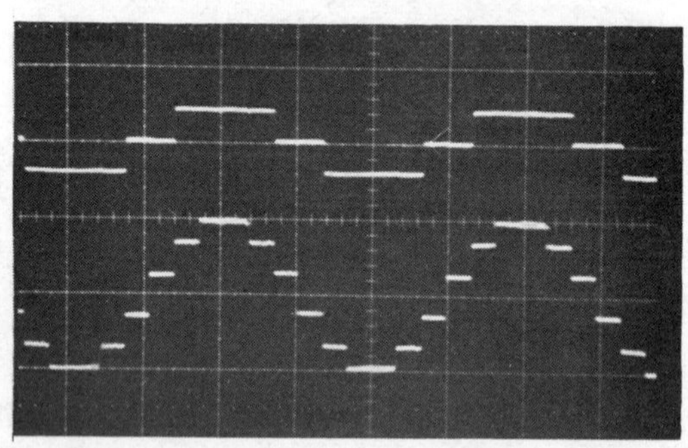

d) Top trace: $a.V_b$, bottom trace: V_1

Fig. 3. Inverter voltage waveforms

CONCLUSIONS

It is possible to get optimum inverter output voltage waveform for different step numbers p by minimizing harmonic distortion. However, economic feasibility must be considered in implementing the optimal solution.

Polyphase symmetrization technique enables the optimum inverter output voltage waveform to be properly constructed.

Although, the frequency and voltage of the optimum inverter system may be controlled independently; the problem of thyristor commutation at low voltages and high currents must be solved.

REFERENCES

1. H. G. Jacob, An engineering optimization method with application to STOL-aircraft approach and landing trajectories, N.A.S.A. Technical Note, D-6978, (1972).

2. H. H. Rosenbrock, An automatic method for finding the greatest or least value of a function, Computer Journal 6, 175 (1960).

3. W. Fong, Polyphase symmetrization, Proc.I.E.E. 115, 8, (1968).

APPENDIX

The stepped wave shown in Fig. 1., may be represented by

$$F_c(t) = \sum_{k=-\infty}^{+\infty} A_{ck}.(\, U(t-\theta_k) - U(t-\theta_{k+1})\,) \qquad (2.1)$$

where $U(t-\theta_k)$ = shifted unit step function

A_{ck} = step heights

θ_k = angles determining step durations

$F_c(t)$ can be represented in terms of complex Fourier series as

$$F_c(t) = \sum_{N=-\infty}^{+\infty} C_N.e^{jNt}.dt \qquad (2.2)$$

where C_N is the harmonic frequency coefficient and N is the harmonic number.

$$C_N=(1/\pi)\sum_{k=0}^{p-1} A_{ck}.\int_0^{2\pi} (\, U(t-\theta_k) - U(t-\theta_{k+1})\,).e^{-jNt}.dt \qquad (2.3)$$

$$(\, U(t-\theta_k) - U(t-\theta_{k+1})\,) = +1 \text{ within the limits of the integral.} \qquad (2.4)$$

$$C_N = (1/\pi).\sum_{k=0}^{p-1} A_{ck}.\int_{\theta_k}^{\theta_{k+1}} e^{-jNt}.dt \qquad (2.5)$$

The complex Fourier harmonic coefficient C_N can be separated into its real and imaginary parts :

$$C_{Nr} = (1/N.\pi).\sum_{k=0}^{p-1} A_{ck}.(\, \mathrm{Sin}(N.\theta_{k+1}) - \mathrm{Sin}(N.\theta_k)\,) = C_N \qquad (2.6)$$

$$C_{Ni} = (1/N.\pi).\sum_{k=0}^{p-1} A_{ck}.(\, \mathrm{Cos}(N.\theta_{k+1}) - \mathrm{Cos}(N.\theta_k)\,) = 0 \qquad (2.7)$$

$k_t = (\, C_3^2 + C_5^2 + \dots + C_N^2\,)^{1/2}/C_1$ becomes the harmonic distortion(2.8)

POWER FACTOR AND CURRENT HARMONICS WITH CYCLOCONVERTERS FED BY A THREE-PHASE SUPPLY

G. Möltgen, and T. Salzmann

Siemens AG Erlangen, Systemtechnische Entwicklung

The following effects are obtained when making electrical measurements with a cycloconverter of the simplest type as shown in Fig. 1. The actual power factor at the three-phase input is in practice lower than that calculated from measurements of active and reactive power. Even if the single-phase load has a leading current component, a lagging component is drawn from the three-phase supply. The spectrum of current harmonics in the three-phase supply is not the same as the usual spectrum obtained with six-pulse converters. These characteristics are difficult to understand when the simple circuit diagram is considered. The cycloconverter contains neither reactive circuit elements nor decoupling transformers, and has an identical circuit to that of a six-pulse bridge, dual-converter without circulating current. The questions arising from the power factor and current harmonics with cycloconverters will be considered in more detail in this paper.

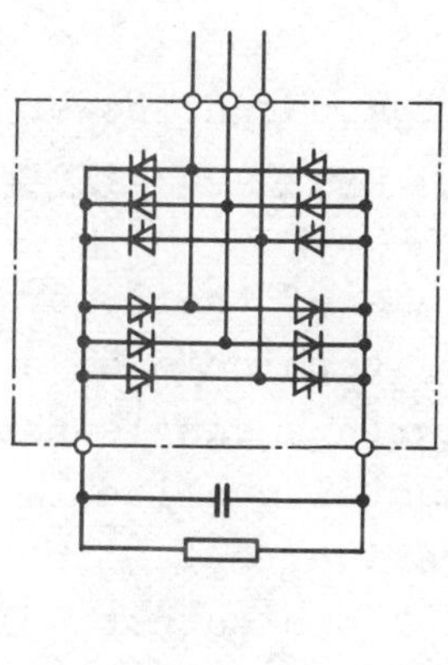

Fig. 1 Cycloconverter with a single-phase output and a three-phase supply

General consideration of power factor

The power factor λ at the terminals of a general electrical load connected to a polyphase supply is defined for steady-state operation as,

$$\lambda = \frac{P}{S_r} . \tag{1}$$

P is the sum of the powers in each phase of the polyphase system which do not necessarily have to be equal. S_r will be called the "general apparent power" (see "Rechtleistung" in Ref. 1), and is defined for a system of k phases as follows,

$$S_r = \sqrt{\sum_k U^2 \sum_k I^2} . \tag{2}$$

The r.m.s. voltages to neutral U and line currents I can contain harmonic components, and do not necessarily have to be equal in all phases. It is usually assumed that the voltages with three-phase systems (k = 3) are symmetrical and contain only the fundamental sine-wave component.

Based on these assumptions, equation (2) can be replaced by

$$S_r = \sqrt{3}\; U \sqrt{I_R^2 + I_S^2 + I_T^2}\,. \qquad (3)$$

In accordance with most practical circuits, it will also be assumed that there is no neutral conductor. The sum of the instantaneous values of current is therefore zero.

If the currents do not contain harmonic components, then equation (3) relates fundamental quantities. An unsymmetrical current system is present if the values of I_R, I_S, I_T are not equal. This can be split into two sets of symmetrical components, a current system with a positive sequence I_P and a current system with a negative sequence I_N (opposite phase sequence). For such unsymmetry, the general apparent power can be split into the fundamental apparent power S_1 and a component of apparent power D_u as follows,

$$S_r = \sqrt{S_1^2 + D_u^2}\,. \qquad (4)$$

The conventional fundamental apparent power (see DIN 40110, Quantities used in a.c. theory) is identical to the apparent power of the positive-sequence current system I_P and is made up of the components active power and reactive power. The component of apparent power D_u correspond to the negative-sequence current system I_N and does not contain any components of active power. The term "unsymmetrical power" will be used (see "Unsymmetrieleistung" in Ref. 1).

As an example, consider the case of an unsymmetrical three-phase load consisting of a linear passive load between two phases with no current flowing in the third conductor. The components I_P and I_N must have the same amplitude, and the phase angle of their instantaneous values must be such that there is no current in the third phase. It follows that S_1 and D_u have the same magnitudes, irrespective of whether or not S_1 contains an active power component. A conventional three-phase measurement of active and reactive power enables only S_1 the "geometrical apparent power" to be determined. Such a measurement with a purely resistive load between two phases results in a power factor of unity, whereas the true power factor according to equations (1) and (4) is only $1/\sqrt{2}$. With an unsymmetrical current system (can be present both in the active and reactive power components), the sum of the instantaneous powers in all phases is no longer constant, but fluctuates with a frequency which is double the supply frequency. With a non-linear unsymmetrical load, i.e. current harmonics are present, there are additional positive-sequence and negative-sequence components at multiples of the supply frequency.

The general apparent power as associated with cycloconverters

The magnitude of the power at the three-phase terminals of a cycloconverter with a single-phase output also fluctuates cyclically at steady-state operation, but with twice the frequency of the output voltage. It is also found that the magnitude of the general apparent power is higher than that of the normal apparent power S. The general apparent power S_r

contains a component "distortion power" $D_\nu = \sqrt{1 - g^2}\, S_r$ (similar to DIN 40110) due to harmonics in the supply current inherent with converter operation. This component can be eliminated for this purpose by using the notation,

$$S_{r1} = g\, S_r . \qquad (5)$$

S_{r1} can now be determined from the general apparent power obtained from electrical measurements (r.m.s. values of current and voltage) using the factor g representing the fundamental component of the supply current for the particular converter connection under consideration. This factor has a theoretical value $g=3/\pi$ for the three-phase bridge connection most commonly used in cycloconverters. It is also permissible to use this value for cycloconverters with a very low output frequency, as two successive half-waves of current can be considered with a good approximation as being symmetrical about the abscissa. If the commutation process is taken into account, or if the converter connection has a very high pulse number, then g is larger than the value $3/\pi = 0.955$, and is so close to unity that the effect of distortion power can be neglected.

The fundamental apparent power S_1 can be determined from the measured active power P and reactive power Q_1 (the average value of a half-period of output frequency). The value of S_1 so obtained is smaller than the value of general apparent power S_{r1} determined from equation (5) also using measurements. This is even though the cycloconverter acts as a symmetrical load on the supply (based on the previous approximation for low output frequencies). A component of apparent power D_m remains which, in a similar manner to equation (4), is given by

$$D_m = \sqrt{S_{r1}^2 - S_1^2} . \qquad (6)$$

This component cannot be explained by distortion power D_ν, as this is already taken account of by equation (5), nor by a fundamental component of power, as these are already fully covered by P and Q_1 based on reliable measurements. It is appropriate to interpret the quantity D_m as a type of unsymmetrical power (Ref.2), a sort of "modulation power". Instead of the previous symmetrical negative-sequence system of current phasors with supply frequency, this modulation power is a sum made up of symmetrical systems of current phasors with various frequencies.

The general apparent power for symmetrical supply voltages of a sine-wave form and without a neutral conductor is most generally given by,

$$S_r^2 = P^2 + Q_1^2 + D_\nu^2 + D_u^2 + D_m^2 . \qquad (7)$$

In this case D_u is zero, and D_ν is already taken into consideration by equation (5).

The spectrum of the sum of currents responsible for D_m depends only on the waveform of the output voltage of the cycloconverter. Some typical waveforms given in Fig. 2 show that the spectrum contains current systems with frequencies which are multiples of twice the output frequency. At steady-state operation, an integration of the current in each supply line over a half period of output frequency results in the

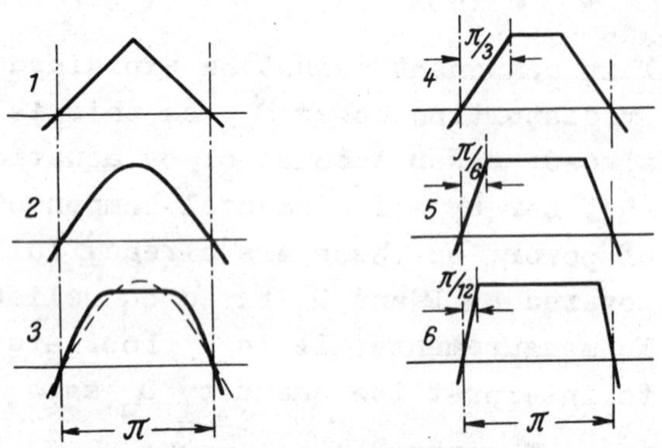

Fig. 2 Various waveforms for single-phase output voltage

same r.m.s. values for each supply line only if the supply frequency is an exact multiple of the output frequency.

From the theoretical point of view, a power fluctuation of a lower frequency than the supply frequency has an equivalent effect to that of power fluctuations of twice supply frequency caused an unsymmetrical load, or that of the higher-frequency power fluctuations resulting from harmonics due to converter operation.

It is possible to express the modulation power in a similar way to that used for distortion power, just as was the case with unsymmetrical power (Ref. 1). The circuit analogy of Fig. 3 using a three-phase rotating transformer consisting of an induction machine with a wound rotor and slip rings is useful. The rotor phases are inserted in the supply lines to a symmetrical three-phase resistive load. The connections to the machine are such that, when stalled, the phase sequence of the winding in series with the load is opposite to that of the supply. If, for example, the voltage in one phase of the supply and the associated secondary winding of the machine are of the same value but with a phase difference of half a cycle, then the load is unsymmetrical with no current in one of the resistors. If the machine is driven such that the rotor is in synchronism with the field produced by the stator winding, then the voltage across each phase of the rotor winding is zero, i.e. the load is symmetrical. If the machine is driven above synchronous speed, then the phase sequence of the rotor voltages is the same as that of the supply. The load sees a symmetrical voltage system which varies both in amplitude and phase with the "slip frequency", therefore the power in all three phases pulses in time with the slip frequency.

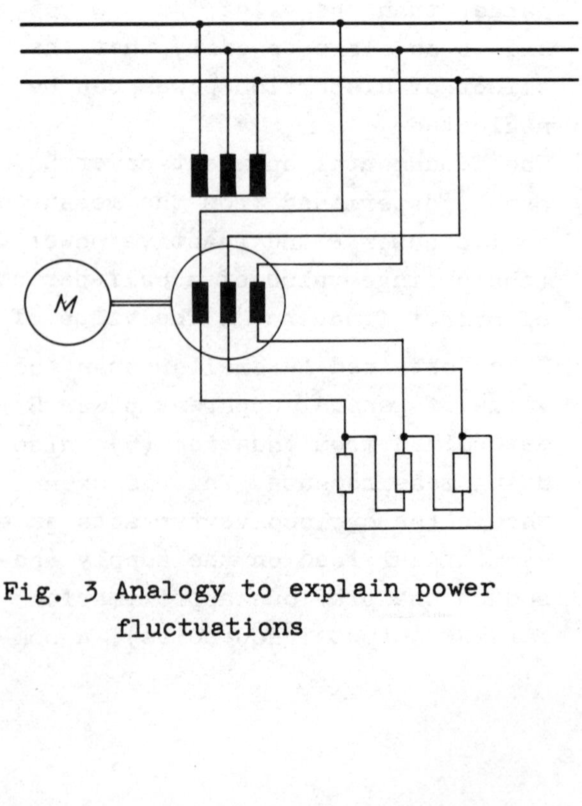

Fig. 3 Analogy to explain power fluctuations

The voltage system applied to the load contains two symmetrical components, the supply voltage with a positive sequence and an additional system with a slip frequency (compared to the first system) which is determined by the machine. Fig. 4 shows the locus of the phasors of line currents with respect to the voltages of the supply in Fig. 3. It can be seen that the magnitude and phase angle of the currents vary cyclically with the result that the general apparent power has a larger value than that of the active power. The same phenomenon occurs when the machine is driven at six times synchronuous speed, distortion power is

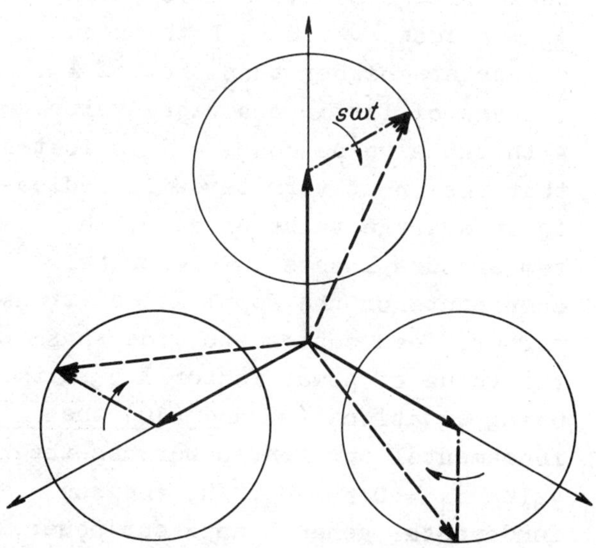

Fig. 4 Phasor diagram showing the addition of the currents in Fig. 3

- ——— Positive-sequence current system
- ——— Supply voltage
- —·—·— Secondary system of current with a slip frequency sω
- - - - - Total current

then simulated. A fifth harmonic is present in the supply current. The apparent power is again higher than the active power, but in contrast to the relationships associated with modulation power, it is permissible to determine the r.m.s. value of the line currents by integrating over a half-period of supply frequency if only distortion power is present.

Distortion power and modulation power may not be combined to a single quantity. This is because distortion power is defined in DIN 40110 as applied to harmonic components which are exact multiples of the supply frequency. This is inappropriate for the modulation power associated with cycloconverters which is determined by power fluctuations which are a function of output frequency.

Power factor with cycloconverters

If ideal cycloconverter operation is considered, i.e. at a low output frequency with negligible commutation overlap, then the active power P and reactive power Q, at the supply as well as the r.m.s. value of currents can be calculated based on a half-cycle of output frequency. The nature of the load and the output voltage waveform must be known. The simplest case to consider is that of a sinusoidal output voltage, wave shape 2 in Fig. 2. With a cycloconverter with a single-phase output feeding resistive load at maximum output voltage, the phasor of the fundamental component of current travels back and forwards along the locus shown in Fig. 5 during each half-period of the output frequency. A converter with a three-phase brid-

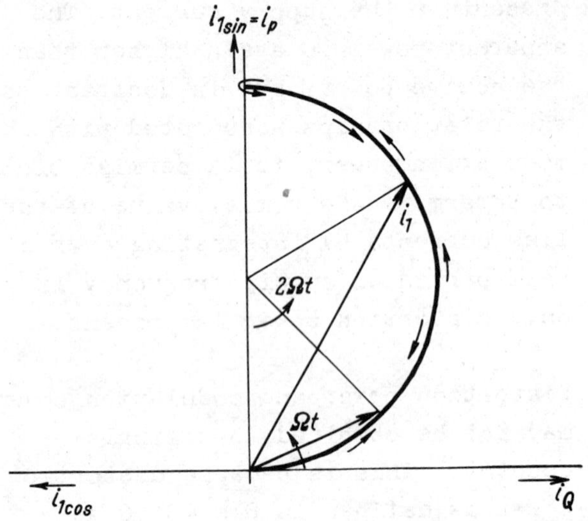

Fig. 5 Locus of the fundamental component of supply current for a cycloconverter with a single-phase output during operation with waveform 2 as in Fig. 2

ge connection and a voltage to neutral of U has the following open-circuit d.c. output voltage,

$$U_{di} = \frac{3\sqrt{6}}{\pi} U. \quad (8)$$

Using this value as the amplitude of the output voltage of the cycloconverter, the active power P and reactive power Q_1 for a resistive load R are given by,

$$P = \frac{U_{di}^2}{R}\frac{2}{\pi}\int_0^{\pi/2} \sin^2 x \, dx = \frac{1}{2}\frac{U_{di}^2}{R} \quad (9)$$

$$Q_1 = \frac{U_{di}^2}{R}\frac{2}{\pi}\int_0^{\pi/2} \cos x \sin x \, dx = \frac{1}{\pi}\frac{U_{di}^2}{R}. \quad (10)$$

The integration is conducted over a quarter-cycle of output frequency. Using the standard equations for a three-phase converter bridge, the r.m.s. value of the supply current is given by

$$I_L = \sqrt{\frac{2}{3}\frac{U_{di}}{R}\sqrt{\frac{2}{\pi}\int_0^{\pi/2} \sin^2 x \, dx}} = \frac{1}{\sqrt{3}}\frac{U_{di}}{R}. \quad (11)$$

The general apparent power is therefore given by

$$S_r = 3U_{di}\frac{\pi}{3\sqrt{6}}\frac{1}{\sqrt{3}}\frac{U_{di}}{R} = 0.740\frac{U_{di}^2}{R}. \quad (12)$$

Based on equation (1), a power factor of $\lambda = 0.675$ is obtained. The value of fundamental power factor as in DIN 40110 (Verschiebungsfaktor) calculated from equations (9) and (10) has a value $\cos\varphi_1 = 0.844$, and the power factor of the positive-sequence current system with a three-phase bridge connection is given by $\lambda_P = g \cos\varphi_1 = 0.805$. Both these values are higher than that of λ, but are of little practical value. Although a value $\cos\varphi_1 = 1$ indicates that the input varmeter will indicate an average value of zero, the temperature losses in electrical components on the supply side (transformer, feed cables etc.) depends on the value of power factor λ. Further, using equations (9) and (10) the fundamental apparent power has the value $S_1 = 0.593\ U_{di}^2/R$, and the fundamental general apparent power from equation (5) is $S_{r1} = g\ S_r = 0.707\ U_{di}^2/R$. Therefore the modulation power as given by equation (6) has the value $D_m = 0.386\ U_{di}^2/R$.

If the component Q_1 of the general apparent power is compensated by a bank of capacitors, then the modulation power remains unchanged and the power factor can be increased to a value $\lambda^x_{max} = 0.748$. If it is possible to prevent the reactive component Q_1 of general apparent power from occurring (e.g. by using force-commutated devices in the converter), then the modulation power is also reduced, as only the active component of the fluctuating apparent power remains. The power factor can then be increased to $\lambda^{xx}_{max} = 0.796$.

Table 1 shows a comparison of the performance with the various output voltage waveforms shown in Fig. 2 for resistive load with the same assumptions as before. Waveform 3 is a special case of a trapezoidal waveform containing only odd harmonics which are multiples of 3. However the output current has an approximate sine-wave shape. The first row of figures by each waveform is for a six-pulse converter, and the second row is for a pulse number which is so high that the factor g approaches unity. The more the waveform of output voltage approaches a square-wave, the higher is the power factor.

A cycloconverter with a three-phase output is obtained if three single-phase cycloconverters of Fig. 1 are connected in star and controlled with a phase displacement corresponding to a third of the period of the output voltage. With a symmetrical load, the modulated-power components of each cycloconverter are superimposed in the supply. The active component of the apparent power fluctuation is theoretically zero. In practice this fluctuation is not zero on account of the irregulari-

TABLE 1 Results of Power Factor Calculation

Wave-forms as in Fig.2	P	S_r	λ	Q_1	S_1	$\cos\varphi_1$	λ_{mit}	D_m	λ^{x}_{max}	λ^{xx}_{max}
	Single-phase output									
1	0.333	0.605 0.577	0.551 0.576	0.333	0.471	0.707	0.675 0.707	0.333	0.659 0.706	0.712 0.746
2	0.5	0.740 0.707	0.675 0.707	0.318	0.593	0.844	0.805 0.844	0.386	0.748 0.792	0.796 0.833
3	0.577	0.740 0.707	0.780 0.816	0.147	0.596	0.969	0.926 0.969	0.381	0.796 0.835	0.807 0.845
4	0.556	0.780 0.745	0.712 0.745	0.222	0.598	0.926	0.887 0.926	0.444	0.734 0.780	0.777 0.814
5	0.778	0.924 0.882	0.842 0.889	0.111	0.786	0.990	0.945 0.990	0.401	0.848 0.889	0.868 0.909
6	0.889	0.987 0.943	0.900 0.943	0.056	0.891	0.998	0.953 0.998	0.309	0.902 0.944	0.911 0.955
	Three-phase output									
2	1,5	1.815 1.779	0.805 0.843	0.955	1.778	0.844	0.827 0.844	0.042	0.972 0.995	0.98
3	1.732	1.836 1.799	0.919 0.962	0.440	1.787	0.969	0.949 0.969	0.210	0.972 0.993	0.98 1

The above powers are normalized with respect to U_{di}^{2}/R

ties of converter operation which cannot be prevented, especially by relatively high output frequencies. The modulation of the reactive power Q_1 (caused by cycloconverter operation) is also not fully eliminated (Ref. 3), so a small component of modulation power also remains with three-phase cycloconverter. If special circuits could be used to prevent the reactive power component Q_1 from occurring, and if a filter were to be used to compensate for current harmonics, then a cycloconverter with a three-phase output could be operated at a power factor $\lambda = 1$.

Table 1 also shows the performance of a cycloconverter with a three-phase output feeding a resistive load at maximum output voltage. Waveform 3 is of special interest because the output current has the same sine-wave shape as with waveform 2, but the utilization of the converter is high on account of the approximately trapezoidal, voltage waveform. The superposition of the supply currents to the three cycloconverter results in a partial compensation of the harmonics on account of the phase difference of each cycloconverter. This results in an increased fundamental component of $g \approx 0.98$ for both output waveforms with a six-pulse connection.

Analysis of the input current from the supply

The input supply current is modulated in both amplitude and phase angle with a naturally-commutated cycloconverter which has a single-phase output (Fig. 1). The envelope of the amplitude modulation corresponds to the wave-shape of the output current. It is useful to use the delay angle $\alpha = 90^{\circ}$ as a reference point for the consideration of the modulation of phase angle. The instantaneous value of output voltage is then zero provided that the current is not intermittent. Starting from this reference point, the delay angle is varied in accordance with the required waveform of output voltage. This variation is not linear, but must incorporate the inverse of the control characteristic of the converter, which in this case is an arc sin function multiplied by the sign of the output current. With negligible overlap due to the commutation process, the expression for the amplitude and phase modulation of the fundamental and harmonic components which are generated by the converter is,

$$\frac{i_\nu}{\hat{I}_a} = \frac{1}{\nu}\,\frac{2\sqrt{3}}{\pi}\left|\frac{i_a}{\hat{I}_a}\right| \times$$

$$\times \sin\nu\left(\omega t - \frac{\pi}{2} + \operatorname{sgn} i_a \arcsin\frac{u_a}{U_{di}}\right). \qquad (13)$$

This expression can be re-written as the sum of a sine and a cosine expression, both of which are amplitude modulated,

$$\frac{i_\nu}{\hat{I}_a} = \frac{1}{\nu}\,\frac{2\sqrt{3}}{\pi}\left[\frac{i_a}{\hat{I}_a}\sin\left(\nu \arcsin\frac{u_a}{U_{di}}\right)\sin\nu\omega t -\right.$$

$$\left. - \frac{i_a}{\hat{I}_a}\cos\left(\nu \arcsin\frac{u_a}{U_{di}}\right)\cos\nu\omega t\right]. \qquad (14)$$

For the fundamental component ($\nu=1$), the sine expression represents the active power and the cosine expression the reactive power. The functions with which the sine and cosine expressions of equation (14) are modulated can also be expressed with the help of Fourier analysis. A d.c.

component and a series of components with multiples of twice the output frequency are obtained. With the help of the addition theorems it can be shown that the output current is made up to the following components. Firstly components arising from the converter operation which have constant magnitudes at supply frequency ω, and frequencies $\nu\omega$. Secondly constant components with frequencies which lie at a displacements of multiples of twice the output frequency above the below the fundamental and its harmonic components,

$$\omega_t = |\nu\omega \pm 2n\Omega|. \qquad (15)$$

$\nu = 1,\ (mp\pm1);\ m,n = 1,2,\ldots;$
p pulse number

The associated amplitudes depend on the waveform of the output voltage, the phase control factor, and the reactive current of the load, each of which is a function of output frequency.

The fundamental component of current for the sine-wave output voltage as in waveform 2 of Fig. 2, with a resistive load, can be calculated using equation (14),

$$\frac{i_1}{\hat{I}_a} = \frac{2\sqrt{3}}{\pi}\left(\sin^2\Omega t \sin\omega t - \sin 2\Omega t \cos\omega t\right). \qquad (16)$$

Figure 5 shows the previously considered locus which is traced by the phasor of fundamental current during a half-period of output frequency. The average value of the active component of current (sin ωt) corresponds to the power P in equation (9), and the reactive component (cosin ωt) corresponds to the reactive power Q_1 in equation (10). The vector sum of the average values of the active and reactive components is the average r.m.s. value of the fundamental component of current, and corresponds to the fundamental apparent power S_1. As far as the general apparent power is concerned, the true r.m.s. value of the current must be used. With a cycloconverter of high pulse number, i.e. factor $g \approx 1$, the general apparent power can be calculated from the vector sum of the average value of the squares of the active and reactive components. If the converter is a components. If the converter is a six-pulse type, then the 5th harmonic of supply current given by equation (13) is

$$\frac{i_5}{\hat{I}_a} = \frac{2\sqrt{3}}{5\pi}\left[\sin\Omega t \sin(5\arcsin\Omega t)\sin 5\omega t - \sin\Omega t \cos(5\arcsin\Omega t)\cos 5\omega t\right]. \qquad (17)$$

The associated locus of the current phasor with five times supply frequency is shown in Fig. 6. Higher harmonics can be treated in a similar manner.

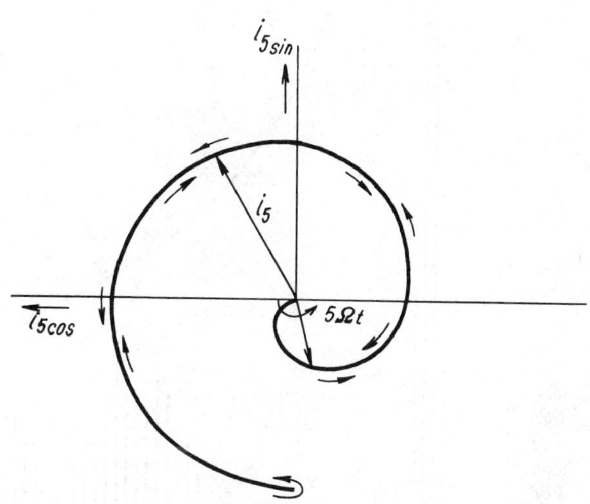

Fig. 6 Locus of the fifth harmonic of supply current for a cycloconverter with a single-phase output during operation with waveform 2 as in Fig. 2

Figure 7 shows the amplitude-frequency spectrum of supply current with a six-pulse converter operated with waveform 2 in Fig. 2. Values for other phase-control factors with loads containing capacitive and inductive components are given for waveform 2 in Ref. 4. Here the values are normalized with respect to the average r.m.s. value of the fundamental component. This must be taken into account when comparing with actual measured values.

The vector sum of the average r.m.s. value of a particular harmonic and its associated side bands is the true r.m.s. value of the associated modulating harmonic. If this value is normalized with respect to the fundamental component, then the expected value $1/\nu$ of the magnitude of each harmonic for rectifier operation is obtained. The value g of the fundamental component, which is also associated with controlled rectifiers, is obtained depending on the converter type. Relatively high output frequencies can result in the overlapping of the side bands from different harmonics. In this case the associated amplitudes must be used to determine the vector sum.

The supply currents to a symmetrically-loaded cycloconverter with a three-phase output is the sum of the supply currents to each converter. There is a considerable reduction of the depth of modulation which, as previously considered, has an influence on the spectrum of harmonics in the current drawn from the supply. Apart from the fundamental and harmonics depending on the pulse

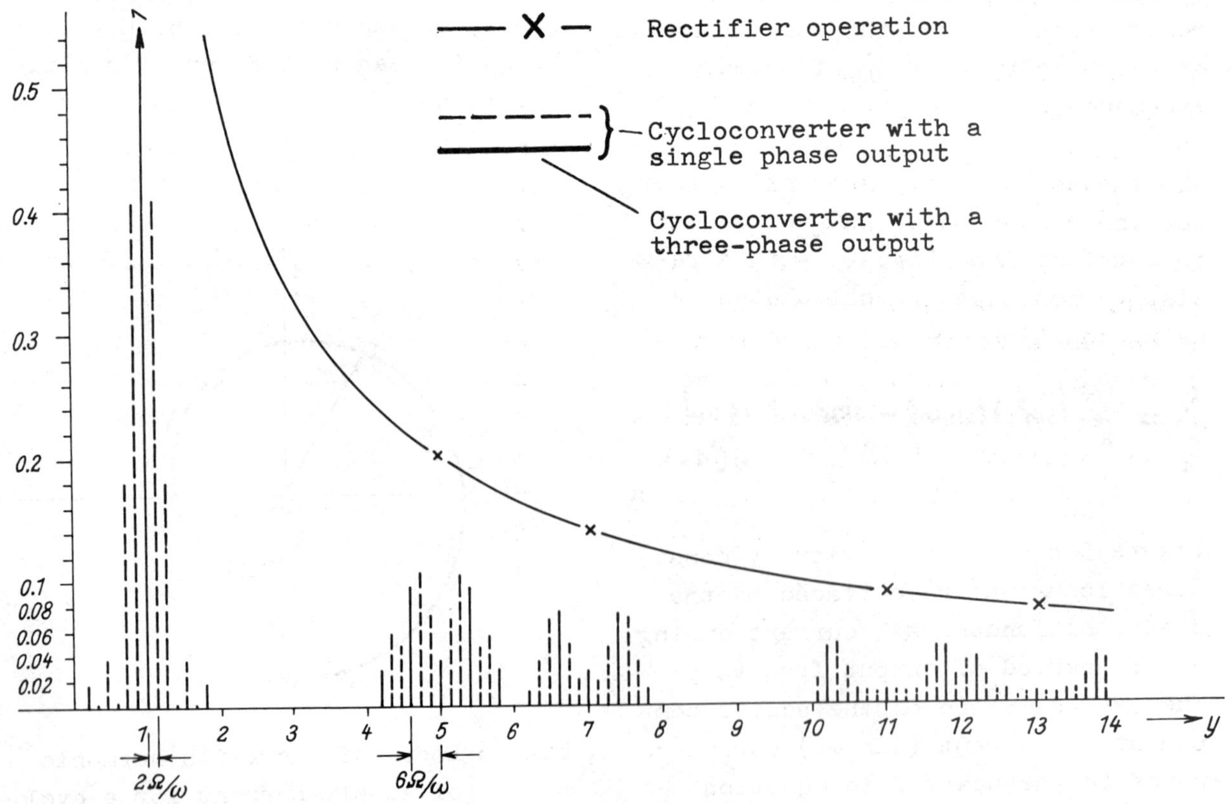

Fig. 7 Spectrum of harmonics in the supply current for a cycloconverter with an output voltage corresponding to waveform 2 in Fig. 2, ratio of frequencies $\Omega/\omega \approx 0.067$

number of the converter connection, the only other harmonics are those with side-band displacements which are multiples of six-times output frequency. Equation (15) can be replaced by

$$\omega_t = |\nu\omega \pm 6n\Omega| \, . \qquad (18)$$

The normalized amplitudes of these remaining harmonics are the same as those of a cycloconverter with a single-phase output. Fig. 7 shows the comparison between the amplitude-frequency spectrum of a cycloconverter with a three-phase output, with that of a cycloconverter with a single-phase output, both under similar operating conditions.

Conclusions

The general apparent power at the three-phase supply to a cycloconverter with a single-phase output contains a power component resulting from power fluctuations at twice the output frequency (modulation power) in addition to the usual components of active, reactive and distortion power. This modulation component of power has a great influence on power factor and can be kept within reasonable limits by arranging that the output voltage differs from a sine wave in a suitable manner, e.g. with cycloconverters for electrical slag-refining plants (Ref. 5). Cycloconverters with a three-phase output have a negligible component of fluctuating power. They are very suited to power variable-speed a.c. drives of large powers if relatively low output frequencies are required, see Refs. 6, 7 and 8.

Depending on whether a cycloconverter has a single-phase or a three-phase output, the reactive power caused by cycloconverter operation has a noticable effect on the power factor. The magnitude of this reactive power depends on the reactive component of the load, on the phase-control factor, and on the waveform of the output voltage, and can be maintained at a low value by adopting special control policies, or by using force-commutated devices in each arm of the converter.

In addition to the current harmonics dependent on the pulse number of the converter, operation as a cycloconverter results in additional a.c. currents which are a function of the frequency and the waveform of the output voltage. The effect of these currents is a maximum for a cycloconverter with a single-phase output. The use of a converter with a high pulse number in conjunction with carefully-designed filter circuits enables a fundamental component to be obtained approaching unity.

References

1. Fack, H.: Scheinleistung in Drehstromsystemen, Special Issue of ETZ (1967), No. 1.
2. Möltgen, G.: Netzgeführte Stromrichter mit Thyristoren, 3rd ed., Siemens AG, Berlin-München (1974).
3. Salzmann, Th., Schultz, W.: Cycloconverter and their reaction on supply systems, CIRED 1975, Paper 46, A.I.M. Liege.
4. Pelly, B.R.: Thyristor Phase-Controlled Converters and Cycloconverters, Wiley, New York (1971).
5. Thomas, F.W., Schmidt, W.: Einsatz von Direktumrichtern für das Elektro-Schlacke-Umschmelzverfahren, Siemens-Zeitschrift 47 (1973) pp. 676-680.
6. Riehlein, D.: Getriebeloser Antrieb für eine Zementmahlanlage, Siemens-Zeitschrift 45 (1971), pp. 181-191.
7. Spinnler, W.: Netzkupplungsumformer für die Deutsche Bundesbahn, Siemens-Zeitschrift 45, Beiheft Bahntechnik (1971), pp. 66-70.
8. Dirr, R., Neuffer, I., Schlüter, W., Waldmann, H.: Neuartige elektronische Regeleinrichtungen für doppeltgespeiste Asynchronmotoren großer Leistung, Siemens-Zeitschrift 45 (1971), pp. 362-367.

HARMONIC ANALYSIS OF PRACTICAL CYCLOCONVERTERS USING A DIGITAL SIMULATION

C. Iung and P. Therme

Laboratoire d'Electronique, d'Electrotechnique et d'Automatique de l'Ecole Nationale Supérieure d'Electricité et de Mécanique de Nancy (France)

ABSTRACT

In this paper, a digital simulation of single phase cycloconverter is presented. The programme written in FØRTRAN language is able to simulate different configurations of cycloconverter supplying various loads. The effect of several physical constraints on the harmonic frequencies of the output voltage are studied separately.

NOMENCLATURE

Fd : desired output frequency (Hz)

F_o : supply frequency (50 Hz)

$F_{i,k}$: frequency defined by
$F_{i,k}=6iF_o + (2k+1)Fd$
$F_{i,-k}=6iF_o-(2k+1)Fd$

ϕ : displacement angle of the load at frequency Fd

r : ratio between the amplitude of the control voltage and of its maximum amplitude

τ : security delay time (s)

α : rectifier and inverter angle limit

V : phase voltage

A_F : amplitude of the harmonic component at frequency Fd

X : reduced time.

INTRODUCTION

Harmonic analysis of the output voltage is commonly used for studying the performance of the cycloconverters. In addition to the desired frequency Fd there appear also other harmonic frequencies which are more or less important. These frequencies $F_{i,k}$ may be multiple of Fd or linear combination of Fd and the supply frequency F_o.

$F_{i,k}=6iF_o+(2k+1)Fd$; $F_{i,-k}=6iF_o-(2k+1)Fd$

These frequencies may be about or less than Fd, then they would be a source of nuisance when the cycloconverters supplies a motor. Previously reported studies have been considering ideal cycloconverters by supposing the following hypothesis : sinusoidal waveform of the current, unsaturated voltage reference and instantaneous reversal of the current direction /1/,/2/,/3/. In practice, these conditions can not all be realized due to certain technological difficulties. The need for a tool to study the effect of these physical constraints separately on the harmonic component is now obvious. This device would enable the determination of the behaviour of a practical cycloconverter.

A simulation programme has been written for a single phase cycloconverter which consists of a dual AC/DC converter without circulating current. The harmonic spectrum is obtained by a Fast Fourier Transform algorithm. The program is written in FØRTRAN language in a modular way to make it easy to study particular configurations.

1. THE DIGITAL SIMULATION OF CYCLOCONVERTER

1.1 General principles

The digital simulation of a cycloconverter was deduced from a simulation programme of a system composed of a load supplied by a dual AC/DC converter without circulating current /4/. This simulation took into account all the discontinuities imposed by the converter, the load and the control voltage. The closed-looped operation can be also simulated. This fact is essential to study different methods of choosing the moment when the current inverses its direction.

To avoid the use of a too small integration step, keeping in the same time a good precision, we employed the following technique : physical processes (load and controller) are simulated on an interval (X, X+DX), then the behaviour is examined to know if the simulation is correct on this step i.e. every physical law is verified (for example : there is no negative current in the thyristor) and no discontinuity occured. In this case the following step of integration is initiated by the last results, otherwise an update is made at the instant when the event occurred. When it is possible to know that a discontinuity will take place at an instant X + ΔX, the simulation is calculated up to X + ΔX. Then the treatment of this discontinuity is made at X + ΔX.

1.2 Modifications for cycloconverters simulation

To simulate correctly a cycloconverter, the determination of the control voltage is now made in a modular way. Usually the control voltage is either sinusoidal or triangular according to the form of the firing curves of the converters. Moreover, it may be interesting to search other forms of the control voltage to improve the harmonic behaviour. In case of the current should inverse its direction a discontinuity happens. If the bank selection is made when the current nullied this discontinuity cannot be predicted and it is not possible to make an interpolation with the current reference which is a logical variable. To solve this problem we choose the following solution : an indicator is set up by the bloc ANALO when X is a current cross over instant, the simulation made on (X, X + DX) is not considered, the new step (X, X + DX) is initiated, taking in account the order of the reverse of current direction.

The Fast Fourier Transform algorithm needs a constant sample. A new update is introduced at the sample instant to get the best accuracy on the physical variables.

The general flow-chart of the simulation is given by figure 1.

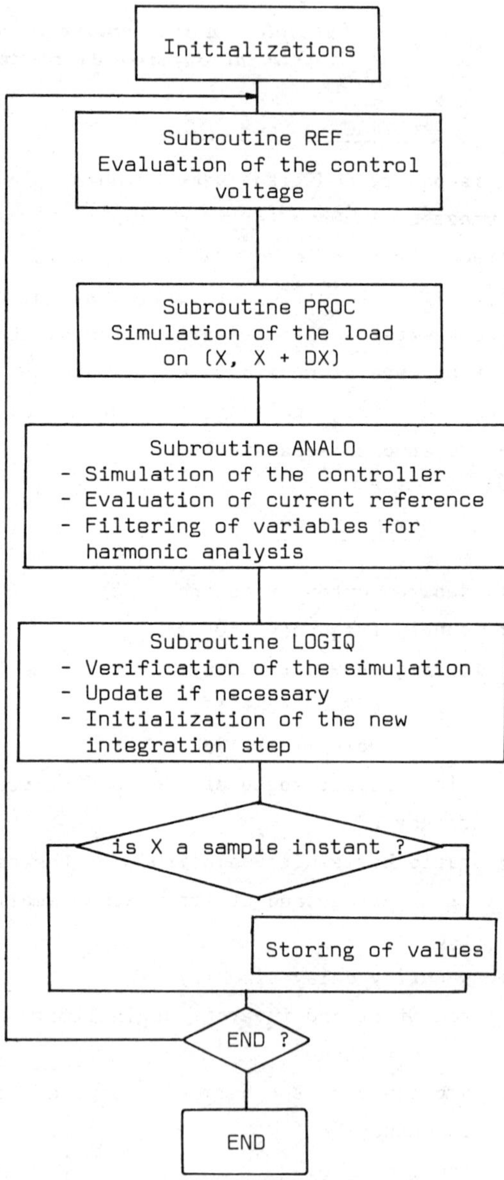

Fig. 1 Flow-chart of the simulation.

1.3 Digital harmonic analysis

A classical Fast Fourier Transform is used to compute the power spectral density function of the frequency. This spectral analysis can be employed for any electrical variable computed by the simulation : voltage, current

To obey Shannon' law, the values are filtered by a 6 order digital Butterworth's filter. It's crossover frequency is calculated such that there is no amplitude modification in the zone of study. The filtering is realized by the subroutine ANALO. The sample frequency and the observation time are given as data. To obtain the static behaviour, the system is simulated for a few periods before the harmonic analysis. The waveform of voltage and current, and the harmonic spectrum can be observed by a display unit (Fig. 2).

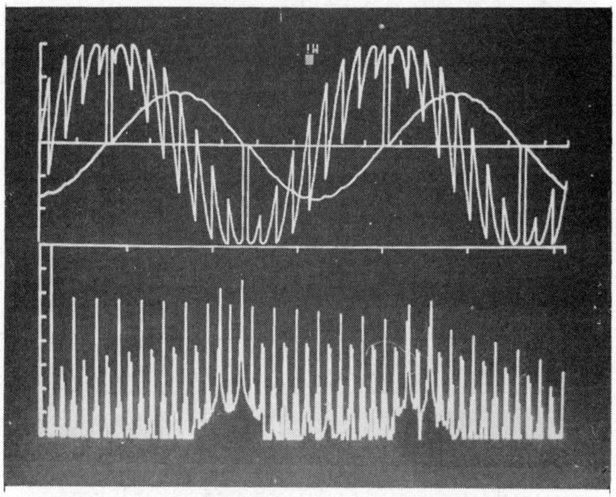

Fig. 2 Display of output voltage, current and harmonic spectrum (db).

2. RESULTS OF THE HARMONIC ANALYSIS

2.1 Ideal case

With this programme different configurations and loads were tested. To check the validity and the accuracy of results, a configuration near to the ideal case was chosen. The results were compared with the theoretical ones. To approach to this ideal configuration, the following techniques were followed.

The "Bank selection" is made at the instant when the fundamental waveform of current changes its direction. The fundamental of the current is obtained by a pass-band filter, it's resonance frequency equal to the desired output frequency Fd. Such a filter practically exists in numerical technology /5/.

. An instantaneous reversal of the current direction is realized. This is easy to obtain by simulation.

. The rectifier and inverter limits are set to their maximum values.

In the ideal case, the amplitude of the fundamental voltage is equal to $3\,V\sqrt{6}/\pi$ for a six pulses cycloconverter. To compare the simulation results with the analytical ones /2/. The harmonic amplitude is given by the percent value of the theorical fundamental. Table 1 shows that there is no great difference between the results. A Fourier series is effectively obtained. Accumulation of spectrum seems useless.

TABLE 1 Comparison between some theoretical and simulation results
A : theoretical amplitudes
B : simulation results.

Frequency (Hz)		$\phi=90°$ r=0,5		$\phi=60°$ r=1		$\phi=30°$ r=0.1	
		A	B	A	B	A	B
13	F_a	50	49.1	100	100	10	11
105	$F_{1,-15}$	1.3	<1	<1		1.4	1.2
131	$F_{1,-13}$	1.5	<1	1.1	1	1.6	1.6
157	$F_{1,-11}$	1.7	1	1.4	1.2	1.9	2.2
183	$F_{1,-9}$	2.1	1.3	1.6	1.5	2.3	2.8
209	$F_{1,-7}$	2.8	2	15.1	14	3	3.4
235	$F_{1,-5}$	<1	<1	10.1	17	4.2	4.2
261	$F_{1,-3}$	26.5	24.6	2.2	2.1	7.4	6.6
287	$F_{1,-1}$	12.3	12.2	2.3	2.2	26.2	25
313	$F_{1,1}$	8.8	8.8	2.3	2.2	16.5	15
339	$F_{1,3}$	5.4	5.5	2.2	2.2	7.1	6.3
365	$F_{1,5}$	3.6	3.8	2.1	2	4.2	3

When a pratical cycloconverter is considered, technological constraints make the ideal behaviour not possible. The study of the effect of this physical constraint was made separately in the range of the actual industrial applications. A systematical study was made when the results obtained let us think that the action of the parameter was important.

2.2 Effect of inverter on rectifier limits

For security reasons inverter and rectifier

limits are classicaly used. Table 2 gives an example of the results obtained. We conclude that the effect of these limits is small.

TABLE 2 Effect of the inverter and rectifier limits (ϕ= 90°, r=1)

Frequency (Hz)		α = 0°	α = 20°	α=30°
13	Fd	99.5	97.6	93.3
39	3Fd	<1	2.2	5.5
65	5Fd	<1	1.9	3.6
91	7Fd	<1	1.4	1.8
209	$F_{1,-7}$	13.5	12.5	11.7
235	$F_{1,-5}$	18.6	19.3	19

2.3 Effect of the overlap

The overlap angle depends on different parameters (current amplitude, short-circuit supply impedance). A systematical analysis was not made. The simulations for a pure inductance short-circuit impedance shows that the amplitude of harmonic frequency are lighly shorten (Table 3). For an actual source, there is a voltage drop on the source resistance following well-known formula /6/.

TABLE 3 Harmonic chart with a maximum overlap of 6° (ϕ = 90°, r=1)

Frequency (Hz)		Amplitude
13	Fd	99.2
209	$F_{1,-7}$	14.1
235	$F_{1,-5}$	18.8

2.4 Effect of the control voltage waveform

A triangular control voltage is commonly used when the firing curves are linear. The comparison with the sinusoidal waveform case is difficult because there is no linear relation between the maximum reference value and the amplitude of the fundamental of the output. The results of the simulation cannot lead us to a simple law. Nevertheless we have noticed that there is an important changement of the amplitudes of some harmonic components $F_{i,k}$ (Table 4). This remark makes possible to search an optimal wave of control voltage.

TABLE 4 Comparison between the sinusoidal and linear case (r=0.5, ϕ= 60°)

Frequency		sinusoidal case	linear case
13	F_d	49.5	59.9
39	$3F_d$	<1	5.2
65	$5F_d$	<1	2
131	$F_{1,-13}$	1	1.4
157	$F_{1,-11}$	1.3	2
183	$F_{1,-9}$	1.7	2.8
209	$F_{1,-7}$	2.2	3.7
235	$F_{1,-5}$	6.1	5.4
261	$F_{1,-3}$	20.1	25.2
287	$F_{1,-1}$	16.6	7.4
313	$F_{1,1}$	10.7	5.5
339	$F_{1,3}$	6	4.9
365	$F_{1,5}$	3.6	4.2

2.5 The effect of security delay

When the current reference changes it's sign, a bank selection should be made. But to allow the second bridge to conduct the current must be exactly zero. Because of the inaccuracy of zero-current transducers a delay of the charge of bridge is to be introduced. This delay "τ" is greater than the recovering time of thyristors (a few milliseconds). This security time plays an important role on the harmonic behaviour, so a systematic study is presented by figure 3 (a, b, c). It is obvious to notice that the curves obtained are approximately linear for the fundamental and the third order harmonic within a large range of frequency. Furthermore, the slope of these curves is proportional to τ. This means that the amplitudes of the fundamental and of the odds harmonics

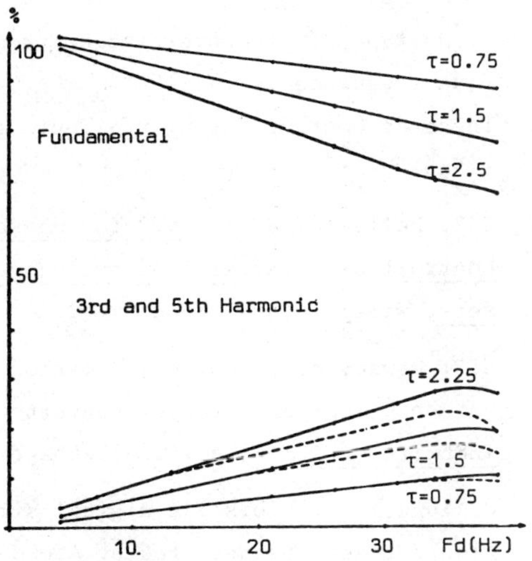

Fig. 3a (ϕ= 90°)

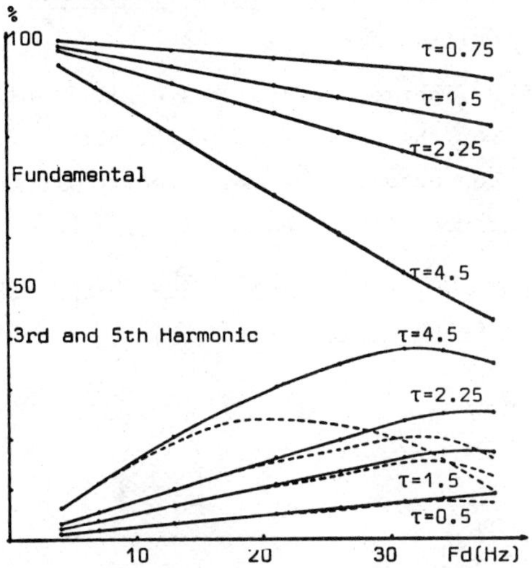

Fig. 3b (ϕ= 60°)

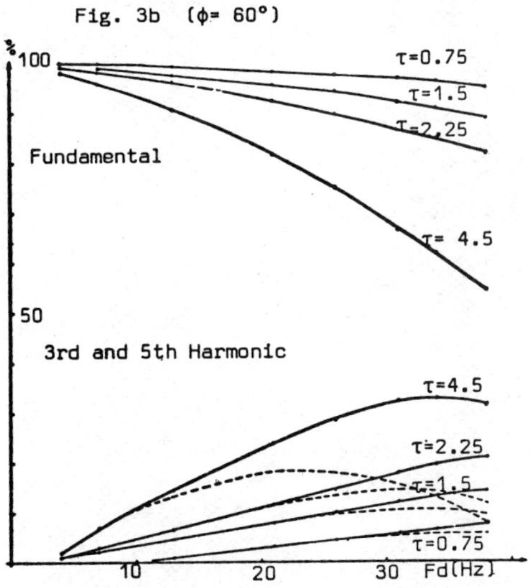

Fig. 3c (ϕ= 30°)

Fig. 3(a,b,c) represent the amplitude of fundamental and harmonic components versus the frequency for different values of τ (milliseconds). The dotted lines are the 5th harmonic component.

depend on the 'security angle' $2\pi\tau f$.
It is possible to give a rule of thumb to calculate these amplitudes.

$$A_{Fd} \simeq - K_1(\phi)\tau\ Fd + 1$$

$$A_{3Fd} \simeq K_2(\phi)\tau\ Fd$$

Table 5 gives the values of K_1 and K_2 for different load displacement angles ϕ.

TABLE 5 Values of K_1 and K_2 for different values of ϕ

ϕ	30°	45°	60°	90°
K_1	1.3	2.3	3.2	4.0
K_2	2.5	3.5	3.3	3.8

Harmonics components still exist at frequencies $F_{i,k}$ their amplitudes are relatively constant and about those obtained in the ideal case.

2.6 The effect of the bank selection method

The simulation programme is able to study different bank selection methods. An example of "First current-zero" bank selection is show, in figure 4. In the case of completely continuous mode of operation, the effect is naturally neglected, but when discontinuous mode of operation occurs, odds harmonic frequencies appear.

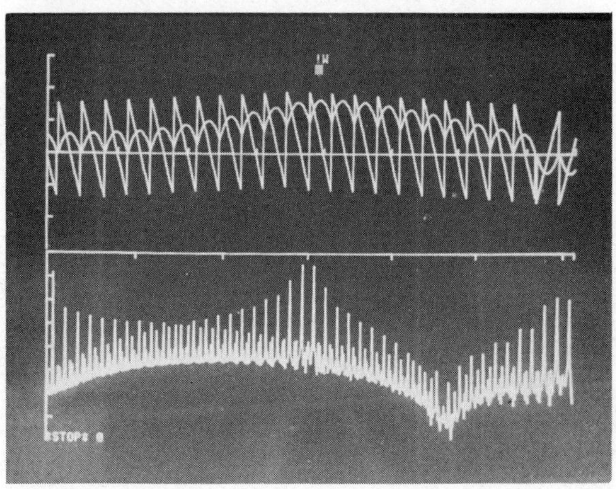

Fig. 4 Display of output voltage, current and harmonic spectrum in the case of First current-zero bank selection (r=0,1 ϕ=30°)

CONCLUSION

The studies made thanks to a digital simulation of a single phase cycloconverter have shown that the importance of the physical constraints are quite different. The main effect seems to be produced by the delay introduced when the current has to change its direction. The action of this delay angle is detailled and leads to simple empirical formula.
Furthermore, the programme used is written in a modular way such that anyone can easily simulate his proper configuration and so predict the harmonic behaviour of particular cycloconverter supplying a given load. This can be a good tool for the designer.

REFERENCES

/1/ B. de Fornel : Contribution à l'étude d'un convertisseur statique de fréquence du type "cycloconvertisseur fréquence fixe fréquence variable" Thèse de Docteur Ingénieur, Toulouse (1969).

/2/ B.R. Pelly : (1971) Thyristor phase-controlled converters and cycloconverters, Wiley Interscience.

/3/ S.B. Dewan, M.D. Kankam : A method for harmonic analysis of cycloconverters, IEEE Trans. IGA-6, n° 5, 455-462 (1970).

/4/ C. Iung, J.P. Louis : A digital simulation for some process fed by AC-DC converters. Proceeding 8th AICA Congress, Delft (1976).

/5/ P. Therme, G. Rooy : A digital solution for the bank selection problem in cycloconverter, 3rd Power Electronics Conference, (1977) Budapest.

/6/ W. Mc Murray: (1972) The theory and design of cycloconverters, MIT Press.

MODIFIED CYCLOCONVERTER WITH MINIMUM BLANKING BETWEEN BRIDGES

C. Arikan

Middle East Technical University, Ankara, Turkey

ABSTRACT

Cycloconverters employing circulating current inhibition techniques, suffer from long blanking periods.
In this research work, the circulating current inhibition logic of a cycloconverter employing "Current Zero Discrimination Method of Inhibition" (1) has been modified such that, it minimizes the blanking period. The logic operation generated can be used with any cycloconverter. A good correlation has been obtained between theoretical and practical results.

INTRODUCTION

The operation of cycloconverters at low frequency range (0-30Hz) has been discussed by various authors (1)(2)(3). The restriction to a top frequency of 30 Hz, is due to the harmonic contents of the output wave-forms at frequencies higher then 30 Hz; which is due to the nature of wave-form generation mechanism and circulating current inhibition logic employed.

INHIBITION

To study the effect of inhibition on the operation of a cycloconverter, it is necessary first to state one of the necessary conditions for a three-phase thyristorized bridge to operate. This is: In a bridge connection two thyristors, one from the upper group, and one from the lower group must operate together to bypass current from a.c. source into the d.c. load. Hence, cycloconverters are compossed of such bridges, then, for each bridge the above condition has to be satisfied.

A careful study of the firing logic of a cycloconverter, shows that the thyristors of any bridge of a three-phase cycloconverter receive pulses at different instant of time. Therefore, commutation from one bridge to the other one is a problem.

In conventional cycloconverters the commutation problem is eliminated by inserting a high value resistance between the neutral of the supply, and the upper bar or lower bar of the three-phase bridge as shown in Fig. 1.

Suppose thyristor bridge I receives a firing pulse then, there are two possibilities, this pulse is either received by one of the upper group thyristors or by one of the lower group thyristors. If the pulse is received by one of the upper group thyristors, then, one of the phases is connected to the neutral, through the starting resistance and hence, the voltage across the load is zero although the bridge has started to operate. If the pulse is received by one of the lower group thyristor, then, one of the phase is connected to the neutral through the load and starting resistance. The value of the starting resistance is choosen such that, it is very much larger compared to the load and small enough to sustain a current which is greater then the holding current of the thyristors. Therefore, the voltage across the load is very small. The second pulse will be received by one of the upper group thyristors, if the first pulse is received by one of the lower group thyristors or vice-versa. After the second pulse is received, the bridge will be at its normal operation mode.

In the cycloconverters employing "Current Zero Discrimination Method of Inhibition" the inhibition and removal of inhibition events happen 0.48 radians (50 Hz reference) after the true current zero. When the inhibition is removed from the incoming bridge then the bridge is ready to operate, but as explained above, the bridge can only start to feed current to the load, when it receives its second pulse after removal of inhibition. Therefore, the non-operative period of a cycloconverter after a true current zero can be divided into the following three intervals

- t_d = The time interval between the true current zero and removal of inhibition instant
- t_1 = The time interval between the removal of inhibition instant and first pulse received by the incoming bridge
- t_2 = The time interval between first and second pulses received by the incoming bridge

Therefore, the total non-operative time of the cycloconverter per half-cycle is

$$t_{no} = t_d + t_1 + t_2$$

The time interval t_d is constant for all frequencies and it is 0.48 radians. The time interval t_2 can be calculated for a given operating condition. However, t_1 can not be predicted easily but it can be said that it varies between 0 and t_{1max}. Therefore, the total non-operative period for half-cycle varies between following limits.

$$t_d + t_2 \leq t_{no} \leq t_d + t_{1max} + t_2$$

The percentage of non-operative time to one period of output cycle are listed in Table 1 for the whole frequency range. As it can be observed from Table 1, the non-operative period could be as high as 54% for a frequency of 40Hz, and for a frequency of 25 Hz it varies between 19% and 34%. Therefore, as the output frequency increases the percentage of non-operative period reaches to unacceptable values because,

i) The un-controlled part of the voltage is high
ii) The harmonic contents of the output voltage are beyond the acceptable limits.

Therefore, in the following section a modified cycloconverter with better characteristics is presented.

MODIFIED CYCLOCONVERTER

As explained above the main draw-back of the conventional cycloconverter at high frequency range is the length of the non-operative period, and the magnitudes of the harmonics generated. If this non-operative period can somehow be reduced then, conventional cycloconverters can be used for higher frequencies and also they will have better performance characteristics at all frequencies.

It is obvious that if the inhibition is removed from an incoming bridge then the bridge will be ready to operate if correct pulses can be generated at this instant. In order to be able to generate these pulses, information about the inhibited pulses must be available because, if the last inhibited pulses for each firing circuit of the upper and lower group thyristors can be detected. Then, these pulses can be generated after the pre-set time delay of 0.48 radians, effectively reducing the non-operative period to 0.48 radians for all frequencies.

Firing Circuit For Modified Cycloconverter

The block-diagram representation of the firing circuits for upper three thyristors of one of the bridger of the modified cycloconverter are shown in Fig.2. The chain including zero-crossing detector, differentiator, mono-stable multi-vibrator and inhibition gate with pulse transformer is the block-diagram of the firing circuit of the CZD cycloconverter (cycloconverter given in Ref 1). The additional circuits to minimize the blanking period are a bi-stable multi-vibrator and a nand gate for each firing circuit, and a mono-stable multi-vibrator and a differentiator for every three firing circuits.

The outputs of the differentiators are the pulses for triggering the appropriate thyristors of a cycloconverter. Therefore these outputs can be used to determine the correct pulses to be generated at the predicted half-cycle instant. The three pulses appearing at the outputs of the differentiators before the predicted half-cycle point are shown in Fig. 3.a. If the bi-stable multi-vibrators (I,II and III) are triggered with these pulses such that, pulses coming from the differentiator with the same group number triggers the bi-stable multi-vibrator such that its output changes for 0 to 1 and, vice-versa if the pulse is coming from a differentiator with a different group number. The outputs of the bi-stable multi-vibrators are shown in Fig.3.b. On the other hand the output of the inhibition circuit is used to obtain a pulse at the current zero (0.48 radians after true current zero) then, this pulse is used to trigger a mono-stable multi-vibrator to give a pulse starting at the half-cycle point with a width of 1 Msecs. The Bi-stable outputs and mono-stable output are then fed into nand-gates I,II and III, and corresponding outputs of the nand gates are shown in Fig.3.c. Only one of the nand gates will have an output pulse at the half-cycle point, which will trigger corresponding inhibition gate, to give a pulse train at the output of the firing circuit. The firing circuit with the pulse train at the output starting at half-cycle point, is the firing circuit whose last pulse before the half-cycle point is the nearest to the half cycle-point among the last three pulses.

A similar circuit can be used for lower group thyristors. So at the half cycle point one thyristor from upper group and, one thyristor from lower group will be triggered in the correct sequence, reducing the blanking period to 0.48 radians.

COMPUTER ANALYSIS

The harmonic content of the output voltage wave-form of the modified cycloconverter has been analysed by a digital computer for a frequency range of 0-40Hz. The typical results for 40 Hz are tabulated in Table 2 for ideal, CZD and modified cyclo converters. As it can be observed from Table 2 the harmonic contents of modified and ideal cycloconverters are very similar, and the harmonic contents of the CZD cycloconverter is much worse than the modified cycloconverter. The differences in sub-harmonic contents are obvious and expected.

EXPERIMENTAL WORK AND RESULTS

The modified firing circuit whose principle of operation has been explained in previous section has also been designed and constructed and the circuit diagram is shown in Fig.4. The modified cycloconverter is tested for a frequency range of 0-40 Hz and results obtained proved to be satisfactory. In Fig.5 the output voltage and current wave-forms of a modified cycloconverter for an inductive load is illustrated. As it can be observed from Fig.5 that the blanking period is constant, which is defined by constant pre-set time delay.

The output voltage harmonics are also obtained by a wave-analyser and a good correlation has been obtained between practical and theoretical values.

CONCLUSIONS

In this research work once again

it has been observed that one of the main draw-back of cycloconverters is the high percentage of harmonic contents, which is due to the wave-form generation mechanism. It has also been noticed that the circulating current inhibition techniques employed **further** increases the harmonic contents of the output voltage wave-form at all frequencies.

The results of this work has shown that the increase in harmonic generation due to inhibition can be minimized by modifying the firing logic of a CZD cycloconverter.

Also it has been concluded that the method for improvement can be applied to any cycloconverter employing circulating current inhibition technique.

REFERENCES

1. C.Arıkan, Inhibition of Cycloconverters Ph.D.Thesis,1971 Ankara, Turkey
2. W.McMurray, The Theory and Design of Cycloconverter, MIT Press, London
3. R.J.Bland, Factors Affecting the Operation of a Phase-Controlled Cycloconverter; Proc IEE 114,(1967)

ACKNOWLEDGEMENT

The author would like to thank to Prof. B.M. Bird from Bristol University for his vigorous encouragement through-out the research work.

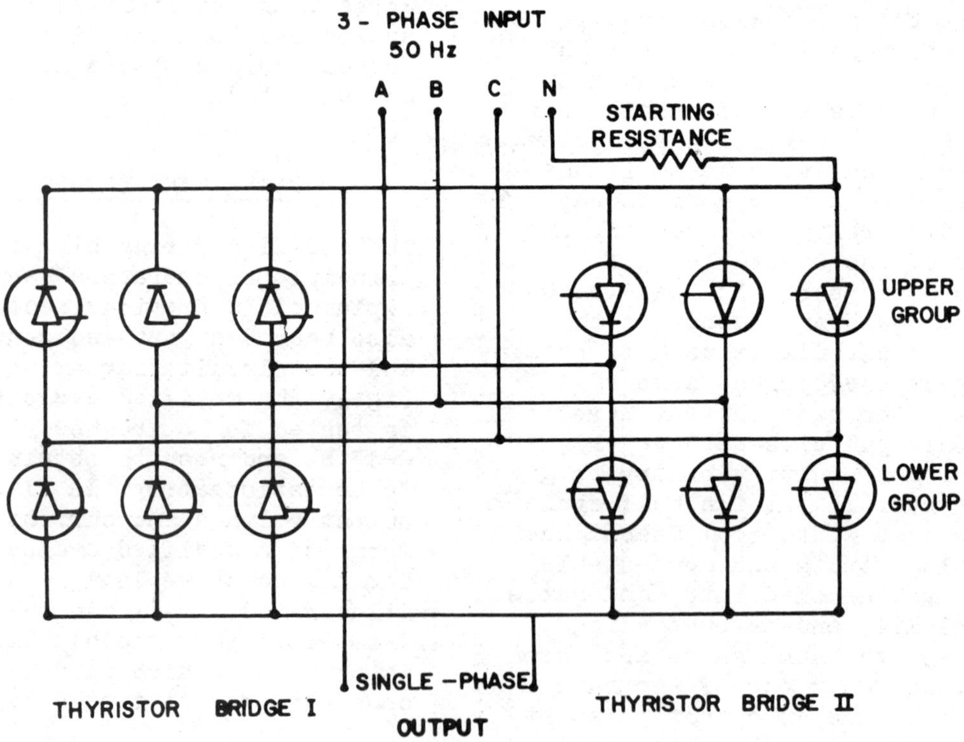

FIG.1. MAIN CIRCUIT CONNECTION FOR ONE-PHASE OF A CYCLOCONVERTOR (With starting resistance)

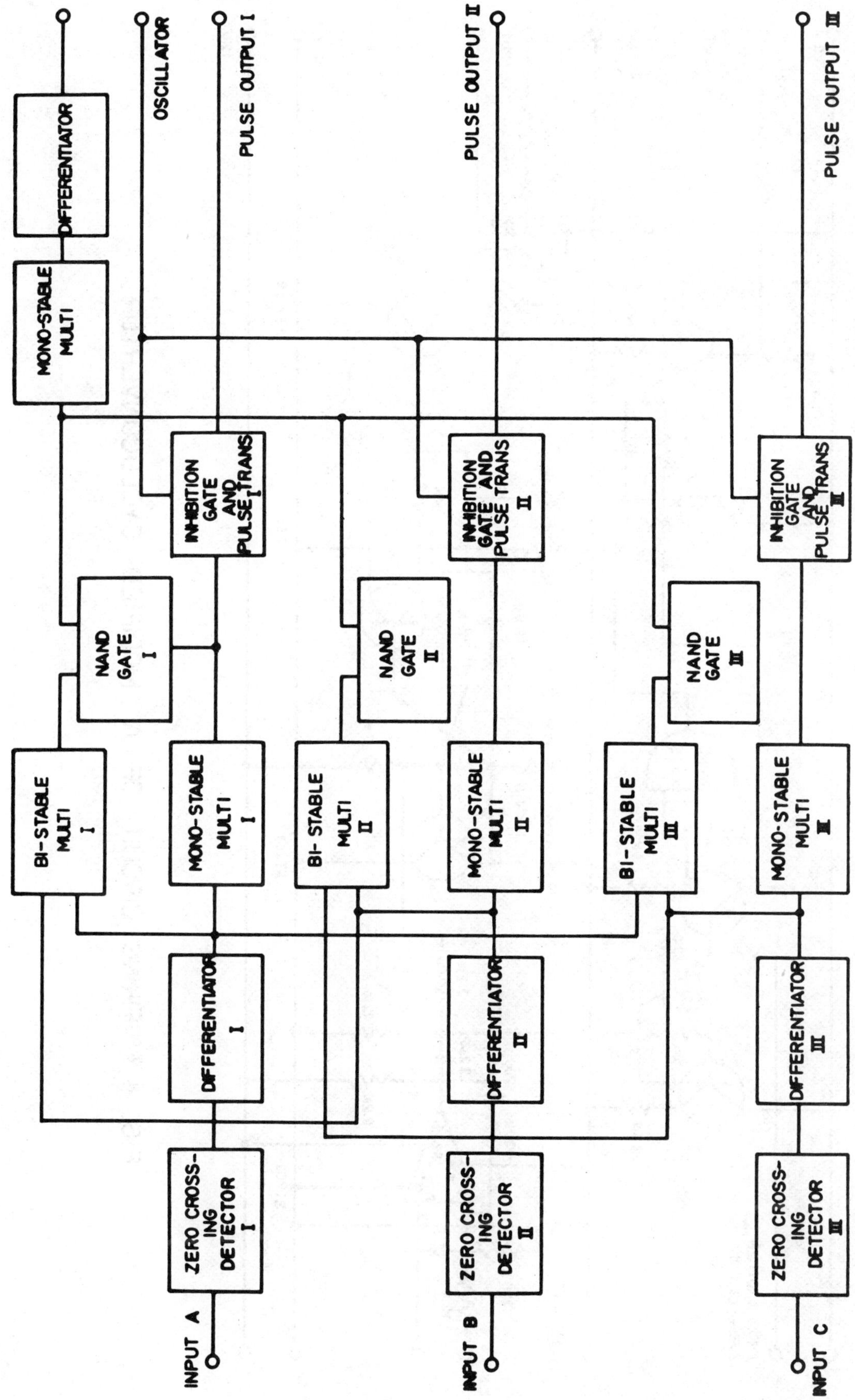

FIG. 2 BLOCK DIAGRAM OF THREE FIRING CIRCUITS OF UPPER THREE THYRISTORS OF ONE OF THE BRIDGES OF A- MODIFIED CYCLOCONVERTOR.

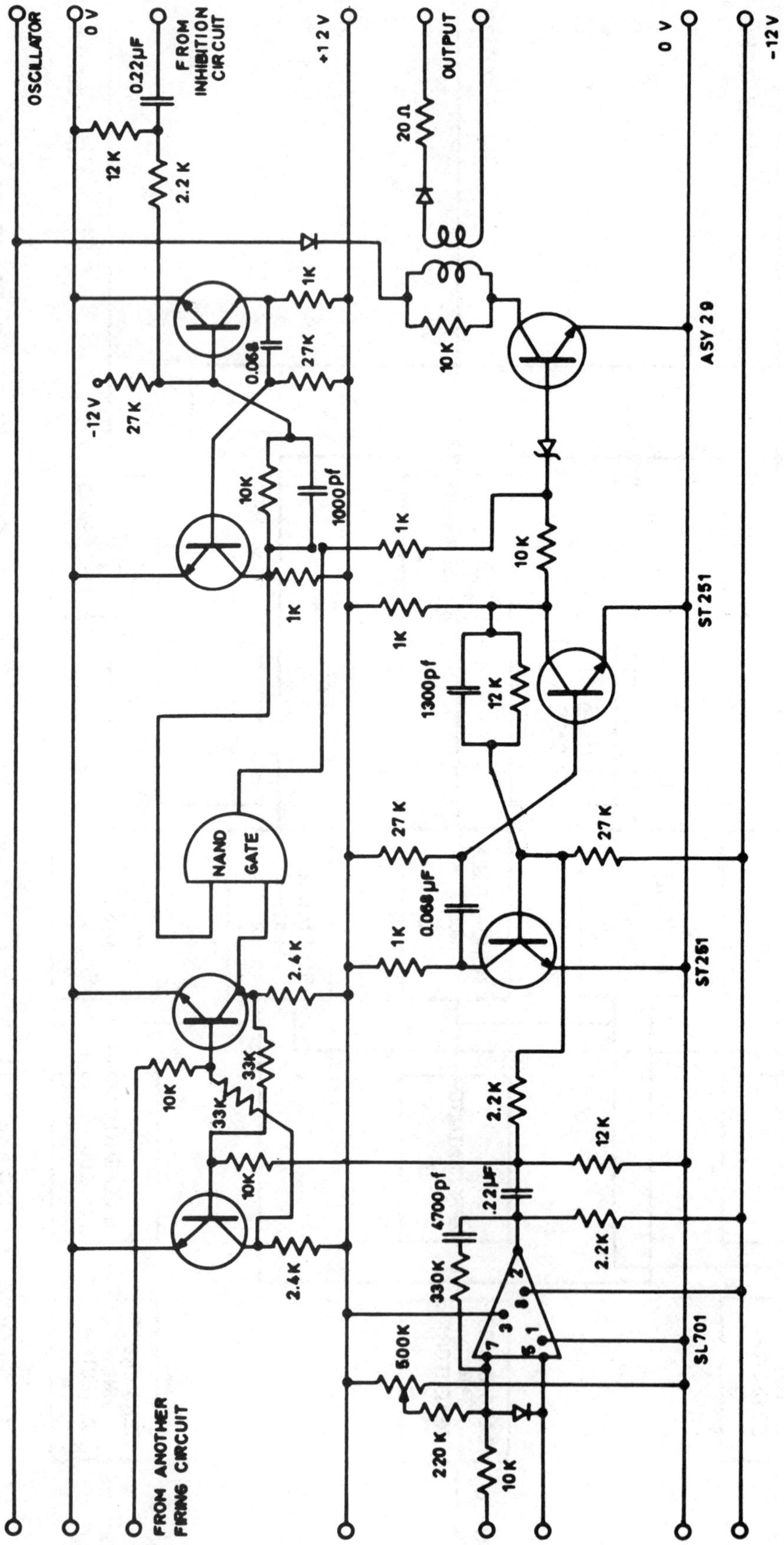

FIG. 4 FIRING CIRCUIT OF A MODIFIED CYCLOCONVERTOR.

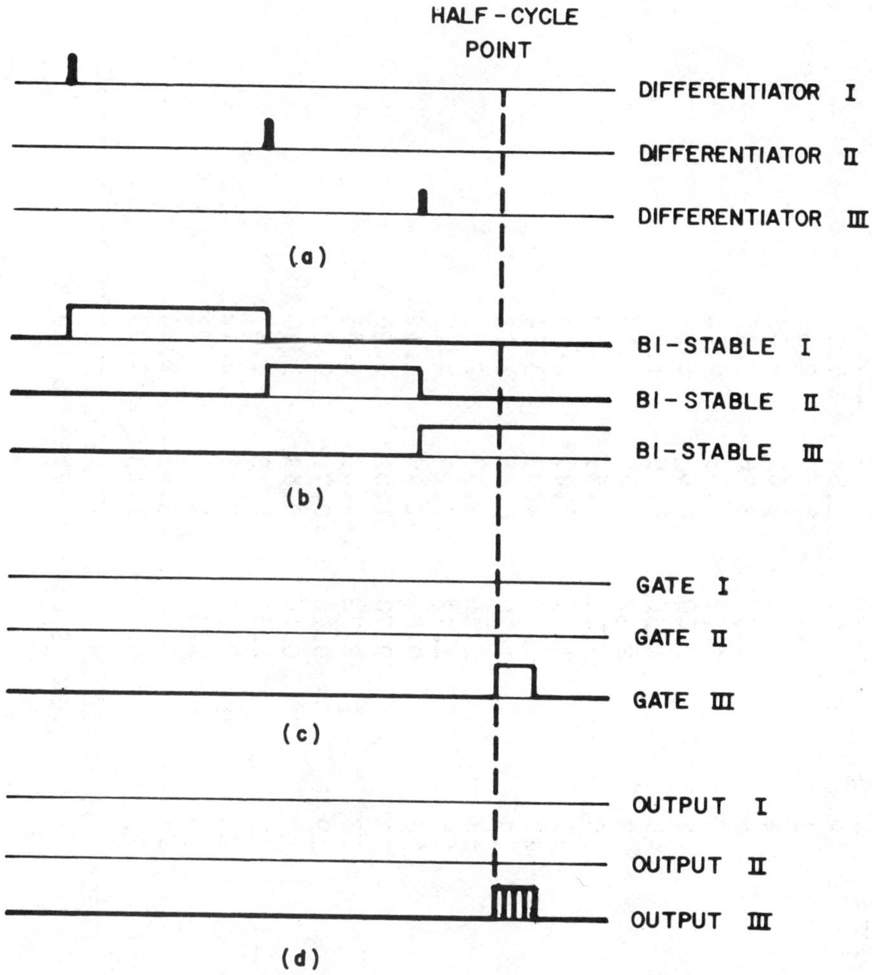

FIG.3 SOME WAVE-FORMS OF THE FIRING CIRCUIT OF A MODIFIED CYCLOCONVERTER

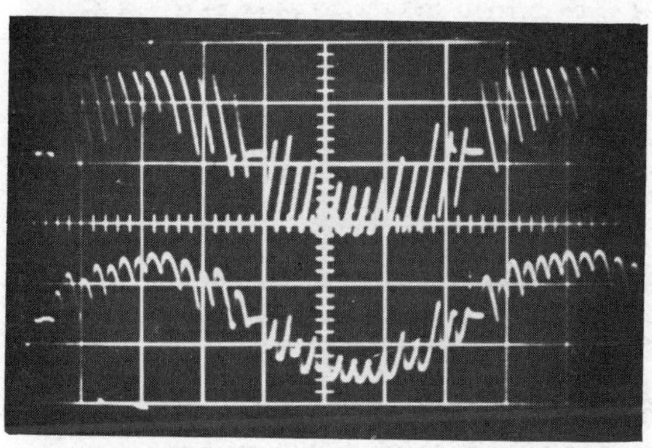

FIG.5 OUTPUT WAVE-FORMS OF THE MODIFIED CYCLOCONVERTER

Frequency	CZD Cycloconverter		Modified Cycloconverter
	min	max	
5	3.8	6.8	1.53
10	7.6	13.6	3.06
15	11.40	20.40	4.58
20	15.20	27.20	6.12
25	19.00	34.00	7.65
26	19.76	35.36	8.06
27	20.52	36.72	8.36
28	21.28	38.08	8.67
29	22.04	39.44	9.07
30	22.80	40.80	9.18
31	23.56	42.16	9.48
32	24.32	43.52	9.78
33	25.08	44.88	10.18
34	25.84	46.24	10.48
35	26.60	47.60	10.70
36	27.36	48.96	11.00
37	28.12	50.32	11.31
38	28.88	51.68	11.62
39	29.64	53.04	11.99
40	30.40	54.40	12.24
41	31.16	55.76	12.55
42	31.94	57.12	12.86
43	32.70	58.48	13.16
44	33.46	59.84	13.46
45	34.22	61.20	13.77
50	38.00	68.00	15.3

TABLE 1 PERCENTAGE NON-OPERATIVE PERIOD OF CYCLOCONVERTERS

Harmonic Number		Harmonic Contents		
		Ideal	CZD	Modified
Sub.	1	16.12	23.69	17.15
	2	8.30	12.56	7.61
Fund	1	100.00	100.00	100.00
	2	4.74	11.84	5.61
	3	0.42	18.55	2.62
	4	6.40	9.65	9.03
	5	0.59	3.76	3.45
	6	4.29	10.83	8.01
	7	2.00	6.02	1.32
	8	1.56	11.15	1.91
	9	2.32	3.85	1.10
	10	3.11	2.69	3.39
	11	1.22	5.45	1.85
	12	1.32	4.03	3.34
	13	1.30	0.56	1.73
	14	1.46	2.38	2.71
	15	1.73	2.23	1.11
	16	0.84	2.35	1.62
	17	0.70	1.81	0.96
	18	2.72	4.09	0.53
	19	0.20	0.72	0.89
	20	0.16	0.53	1.70
	21	0.83	2.12	1.37
	22	2.34	2.81	2.42
	23	1.32	1.52	1.23
	24	4.92	3.43	2.86
	25	0.71	1.13	0.85

TABLE 2 HARMONIC ANALYSIS OF THE OUTPUT WAVE-FORM OF CYCLOCONVERTERS (40 HZ)

THYRISTOR HIGH-FREQUENCY RESONANCE INVERTER WITH COMMUTATING-MARGIN ANGLE/TIME CONTROL SYSTEM BY MEANS OF PHASE-LOCKED LOOP

M. Nakaoka and T. Maruhashi

Department of Electrical Engineering, Kobe University, Rokko, Kobe, Japan

SUMMARY

This paper is concerned with a development and a system analysis on a voltage source-fed type series resonance high-frequency inverter which operates at a constant-value control system of commutating-margin angle/time. In the first place, a high-precise constant value control system of commutating-margin angle/time by means of the integrated circuit of phase(frequency)-locked loop is introduced to the voltage source-fed type inverter for induction heating and melting. This control system proposed here is usefully applicable to the current source-fed type inverter and so forth.

In the second place, a generalized system model of voltage-fed bridge type inverter designed to operate at the control policy of optimal commutating-margin angle/time is presented and discussed. And then, a system analysis of state-space approach and performance assessments for system design are described in relation to a practical analysis in a frequency domain by describing-function method.

Furthermore, an effectiveness of security control for the inverter treated here is substantiated with system theory and the experiments.

INTRODUCTION

The semiconductor static power converter equipments and systems which stably operate over an ultrasonic/audio frequency range have recently attracted special interest in the fields of power electronics. The high-frequency inverters and frequency step-up cycloconverters of voltage-fed and current fed types, respectively abbreviated to VF and CF types are suitable for the switching power controllers which drive an induction heating and melting, ultra-high speed motor and a high-intensity ultrasonic transducer system and the like.

Much additional research on such power converters and their peripheral system are practically required, and a new exploitation of security control of VF and CF types must be executed by the latest power electronics.

The authors have attempted to develop a stabilized VF type series tuned inverter operated at the security control system. In particular, a precise constant-value control of commutating-margin angle/time by phase-locked loop, abbreviated to PLL-IC, and signal processor is proposed and discussed here.

On the other hand, this sort of basic study on software of circuit analysis and performance estimations of VF type inverter system has not been minutely made so far. It is of great importance to analyse and estimate the behavior of VF type inverter circuit and system stability taking into account the normalized system parameters named as control variable and circuit constants of commutating components and load circuitry. In this respect, an analysis conducted so far are not considered from a standpoint of inverter system with feedback control loop.

Thus, the authors have made an attempt at carrying out the transient and steady-state system analyses of VF type series tuned inverter with commutating-margin angle/time(γ/tc).

This paper describes an effectual hardware technique on a precise constant-value control system of γ/tc in VF type inverter. It also illustrates a system model of the inverter mentioned above and describes the system analyses by state-space approach and a practical describing-function method on the basis of the proposed system model. Theoretical results obtained here are ascertained with a tested inverter equipment excited by the developed security control system.

Moreover, the effectiveness of the inverter system are proved and discussed by system theory and the experiments.

CONFIGURATION OF INVERTER SYSTEM

Power Circuit of Voltage-Fed Type Inverter

Figure 1 shows a VF type series tuned inverter including a commutating-margin angle/time(γ/tc) control system. This inverter is profitably designed for high-power use in the frequency range of about 3kHz to a certain extent of 10kHz. Time-variant parameters of induction heating and melting are reduced on the primary side of matching transformer.

The suppression for thyristor stresses is practically lighten by creating the overlapping-commutation mode.

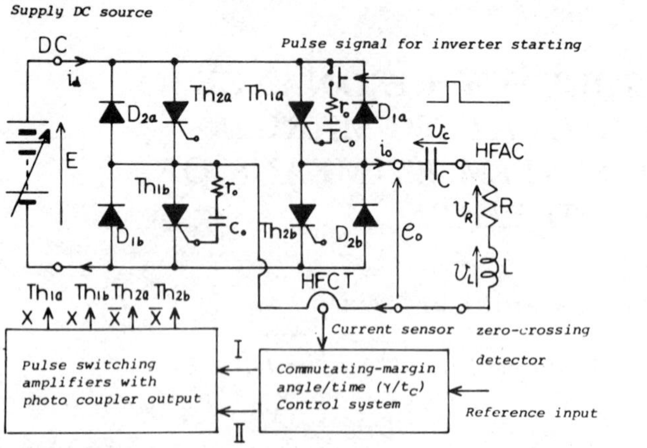

Fig. 1. Voltage controlled source-fed bridge type high-frequency inverter system including a constant-value control scheme of commutating margin angle/time.

Margin angle/time of thyristor commutation is adaptably regulated by control system which is composed of PLL-IC and a signal processor. The anti-parallel connection of reverse-blocking triode thyristor and diode is conveniently replaced by reverse-conducting triode thyristor(RTh) with excellent dynamic characteristics.

Figure 2 shows some possible modified constructions of VF type inverter. The operation of these circuits is all the same as the inverter shown in Fig.1. The voltage and current waveforms of VF type inverter by RTh, which operates at steady-state condition, are illustrated in Fig.3. The inverter operation can be optimized at the minimum leading current angle/time(γmin/tc min), i.e., minimum commutating-margin angle/time.

The VF type series resonance inverter is known as an improved type of conventional series inverter.

VF type inverter possesses the following remarkable features:
(1) The maximum voltage applied across thyristor is extremely small even in the case of high-resonance sharpness of the series tuned circuitry.
(2) The distortion factor K of output current and voltage is small. In R-C load, K is extremely small and does not vary as the parameter disturbances.
(3) The output voltage regulation is stable in order to possess the control function of reactive power in the inverter.
(4) The power regeneration into DC supply source can be attained by feedback diode path. And the power factor is smoothly adjusted by control of reference input.
(5) The stability of thyristor commutation is high even in open-loop system, and the wide range of load variations can be actualized as the stable operation region.
(6) The operation in short-circuit makes possible.
(7) For inverter system operated at the self-excited state, the adjustment of output frequency can be performed.
(8) Start-up and stop control can be performed by on-off control of selective command signal in control system.
(9) The system reliability is very high for the inverter with commutating-margin angle/time control system.

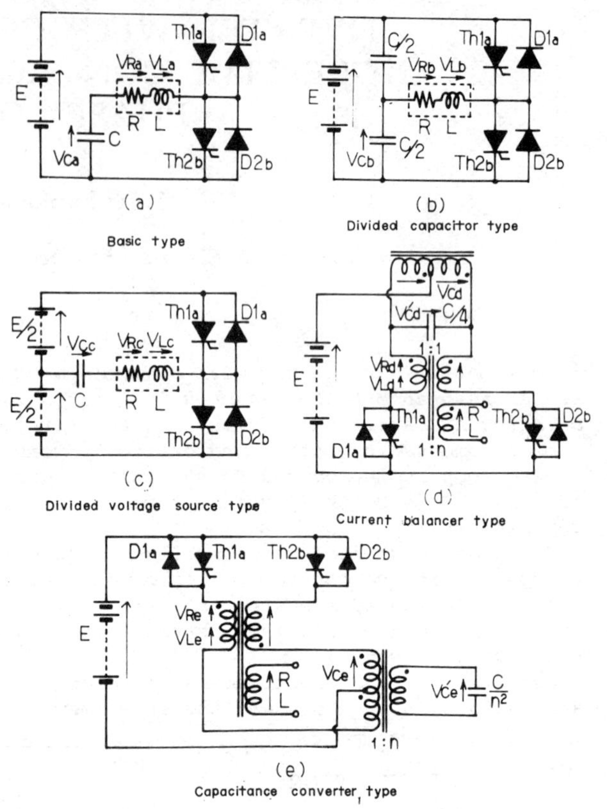

Fig. 2 Modified variations of voltage controlled source-fed bridge type series resonance inverter.

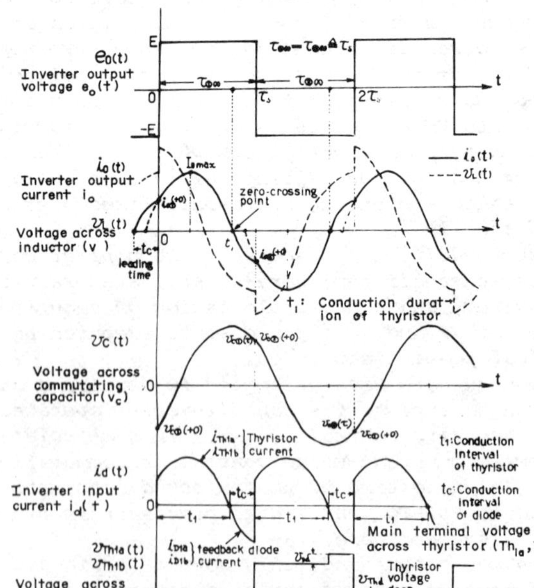

Fig. 3 Voltage and current waveforms in order to explain the inverter operation at steady-state.

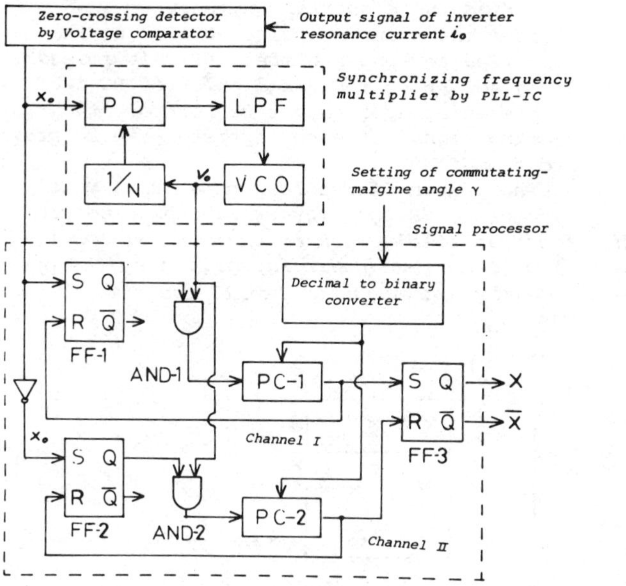

Fig. 4 A constant-value control scheme of commutating-margin angle by Phase-locked loop and counter-gate signal processor.

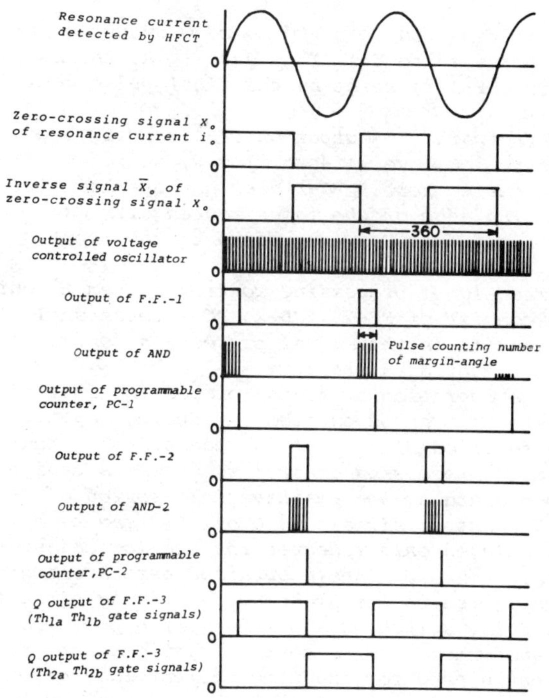

Fig. 5 Pulse sequence chart of γ/t control system shown in Fig. 4.

For convenience sake, the operating mode of the inverter is divided an output period into two main circuit modes 1 , 2 from thyristor triggering point. And each of these main modes is sub-divided into two sub-modes a , b by alternative conduction durations of thyristor and diode. The time interval of sequent sub-modes a , b respectively varies as the system parameters of series resonance circuit and output frequency. Both the triggering point and self-quenched time of thyristor are determind in operating state of the inverter with γ/tc constant control loop and also depends upon a system parameters of series tuned circuit.

Control System of Commutating-Margin Angle by Phase-Locked Loop

For a stable operation, it is required to operate at a condition of leading angle/time of resonance current i_0 with respect to output rectangular ac voltage e_0. This leading angle/time of i_0 is exactly equal to the margin-angle/time (γ/tc) of thyristor commutation, i.e., a conduction angle/time of feedback diode.

Figure 4 shows a developed constant-value control system of (γ/tc) by PLL-IC and signal processor. PLL-IC functional device is consist of digital type phase-detector PD by pulse leading edge triggering, digital-to-analog converter called charge pump, active-low pass filter LPF, DC amplifier voltage controlled oscillator VCO, programmable frequency divider (Ratio of frequency division, 1/N ; N=360) and logical signal processor with two signal channel I, II by counter-gate systems. The commutating-margin angle control system shown in Fig.4 is largely composed of synchronizing frequency multiplier by PLL-IC and pulse counting circuits including logical gate and programmable counter, and pulse switching amplifier with photo-coupler isolation outputs X, X̄.

The current transformer HFCT is used for detection and isolation of series resonance inverter current and the voltage comparator is also employed for zero-crossing detection of i_0. This part is used as a simple interface between inverter circuit and control system. The rectangular signal detected by zero-crossing circuit is supplied to the reference input of PLL-IC. The output signal V_0 of PLL-IC is synchronized with the inverter output current and the frequency of V_0 is arbitrarily multiplied for output frequency of i_0. The VCO output signal with 1/2 duty-cycle is accurately determined in a limitting resolving angle of the precision of 1 degree in accuracy. For the requirement of higher resolution of digital delayed angle γ , a smaller value of 1/N must be availably selected. The output signal of frequency multiplying equipment is directly transmitted to the next stage processor by counter-gate timing control system. Detailed operation of signal processor in Fig.4 is elucidated below.

First, reset-set flip-flop FF-1 is set at the leading edge of rectangular waveform corresponding to i_0, which is furnished from HFCT. At this instant, Q output of FF-1 becomes high-state.

Next, both Q output of FF-1 and output signal V_0 of VCO are conducted to AND-1 gate circuit. The output signal of AND-1 gate becomes a pulse number modulation type pulse group signal counted precisely from zero-crossing point of i_0. This pulse group signal is led to the input of programmable counter PC-1. The pulse counting number of PC-1 is previously set to the other input of PC-1. As soon as the expected pulse counting of PC-1 is accomplished at set-up level, FF-3 cascaded with PC-1 is instantly set *by th final pulse of PC-1 and Q output of FF-3 is provided to Th_{1a} Th_{2a} gates. At the same time, FF-1 is reset by the final pulse code of PC-1.*

Afterwards, Q output of FF-1 readily becomes to low level. Accordingly, output of AND-1 turns to zero. PC-1 does not operate till the leading edge of the input rectangular waveform detected from HFCT reappears to the input side of FF-1.

The signal processing control system II which is consist of FF-2, AND-2, PC-2 repeats all the operation as the signal processing control system I mentioned above.

The operation of signal processing control system II can be carried out during negative half-period of i_0. It is indescribable that a signal processing control system I is designed to operate at the positive half-period of i_0.

The output signals Q, $\bar{Q}$ of FF-3 can be exactly delayed only γ degree for the logical signal of i_0. These signals obtained here are individually led to the higher-gate driving circuits by pulse switching amplifiers with photo-coupler transformers. This control system can be sufficiently used for the time-variant load circuitry which fulctuates slowly such as induction heating and melting. Figure 5 shows the pulse timing chart of control system shown in Fig.4.

Control System of Commutating-Margin Time

Figure 6 shows the developed constant-value control system of commutating-margin time tc of VF type inverter. This system is not applicable to the tc-control system of CF type bridge inverter. Note that the commutating-margin angle control system treated here is effectually available for CF type bridge inverter.

This system has much simpler configuration and becomes high precise controller, which is replaced a synchronizing frequency multiplier employing PLL-IC by a rectangular pulse generator equipped separately.

As a reference oscillator in this system, a crystal oscillator which generates stably at the output frequency of 1 MHz is adopted.

The PLL type oscillator including a programmable voltage limitter shown in Fig.7 at the input side of VCO indicated in Fig.4 is also availably used in place of crystal reference oscillator shown in Fig.6.

In the control system employing PLL-IC including VCO with programmable voltage limitter, the novel control function both γ and tc is availably furnished. The γ control system can be actualized within a limitting boundary of voltage limitter. On the contrary, the tc control system can be naturally accomplished by exceeding the voltage threshhold of voltage limitter. By adjusting voltage limitting value, the γ control loop is automatically exchanged into the tc control loop.

The oscillating pulse period of crystal generator is to be exactly determined as a limitting resolving time of 1μs in accuracy. Accordingly, the commutating-margin time tc is adaptably controlled by minute resolving time of 1μs order.

From the zero-crossing point of i_0, which is detected at every half-cycle, programmable counters PC-1, PC-2 begin to count precisely the specified pulse counting number.

After the counting pulse attains to the specified set-pulse number, FF-3 is immeadiately set up by the final pulse of PC-1/PC-2. In operating at constant tc control, the triggering signal of each thyristor can be provided from FF-3.

Hence, even if the load parameters widely fluctuate, VF type inverter with tc control loop can stably work at the optimal condisition of munimum power factor. More the operating pulse sequence is all the same as Fig.5.

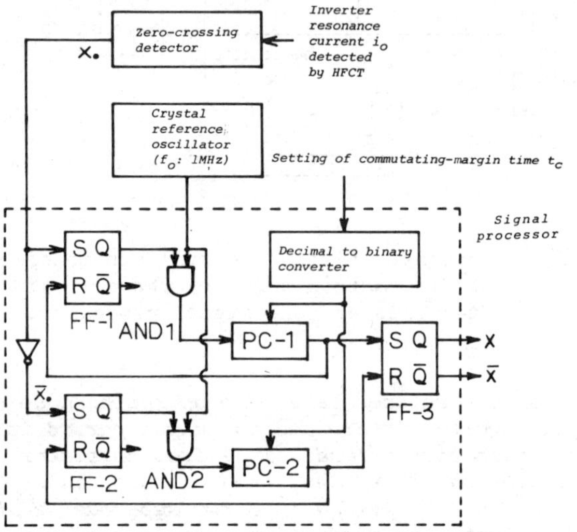

Fig. 6 Schematic system configuration of commutating - margin time control project

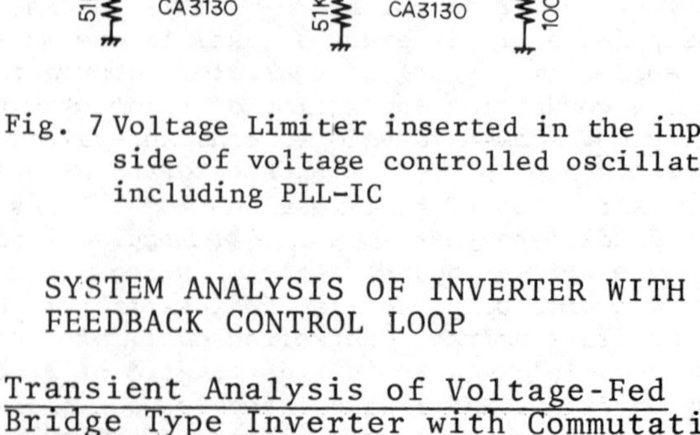

Fig. 7 Voltage Limiter inserted in the input side of voltage controlled oscillator including PLL-IC

SYSTEM ANALYSIS OF INVERTER WITH FEEDBACK CONTROL LOOP

Transient Analysis of Voltage-Fed Bridge Type Inverter with Commutating-Margin Time Control Loop

The transient analysis of VF type series resonance inverter system with constant-value control loop of tc and its theoretical considerations will be presented first.

In this inverter analysis, ac branch current i_O of L and voltage V_C of C are chosen as the state-space variables. And the rule of their positive directions and symbols are defined as indicated in Fig.1. In addition, a time origin of state variables are determined at the thyristor firing instant of each switching cycle mode. Consider the n-th cycle mode including two main switching-mode 1 , 2 after inverter starting.

Figure 8 illustrates the transient voltage and current waveforms at the n-th cycle mode after inverter starting.

This is still drawn for the case when the inverter alwalys operates at the constant-value control scheme of tc over a range of transient to stationary state.

More, figure 8 can be applied for illustrating circuit behavior in the case when the load parameters and input DC voltage are abruptly changed.

For generality, the state vector $X(t) = (i_O(t), V_C(t))^T$ of the inverter shown in Fig.1 are normalized by the rèference quantities ($I = E/2\sqrt{L/C}$, & E), and yet the real time t is normalized by $Z = t/2\tau_{\textcircled{r}n}$

n: switching interval number of main mode at the n-th cycle, r: mode number 1, 2).

Now, we define the representation of new variables of X (t) as shown below.

$$X(z) = (i_O(z)/L,\ V_C(z)/E)^T \triangleq (i_O^*(z),\ v_C^*(z))^T$$

As previously shown, each cycle mode is always divided into two main switching modes 1 , 2 including two-submodes a , b in spite of circuit operating state.

The state-vector equations with periodic input, which adaptably influences on the circuit variations, are described as follows:

For main mode 1 , we obtain

$$\frac{d}{dz}X_{\textcircled{r}n}(z) = A_{\textcircled{r}n}X_{\textcircled{r}n}(z) + B_{\textcircled{r}n}U_{\textcircled{r}n}(z) \qquad \text{-------- (1)}$$

where,

$$X_{\textcircled{r}n}(z) = (i^*_{o\textcircled{r}n}(z),\ v^*_{c\textcircled{r}n}(z))^T$$

$$A_{\textcircled{r}n} = \begin{bmatrix} -\dfrac{4\pi\lambda}{\mu_{,n}} & -\dfrac{4\pi}{\mu_{,n}} \\ \dfrac{\pi}{\mu_{,n}} & 0 \end{bmatrix} \qquad B_{\textcircled{r}n} = \begin{bmatrix} \dfrac{4\pi}{\mu_{,n}} \\ 0 \end{bmatrix}$$

$$U_{\textcircled{r}n}(z) = +1,\ \lambda = R/2\sqrt{L/C},\ f_{o,n} = 1/2\tau_{\textcircled{r}n}$$

$$\mu_{,n} = 2\pi f_{o,n}\sqrt{LC} = \pi\sqrt{LC}/\tau_{\textcircled{r}n}$$

By solving the vector differential equation under the initial condition of,

$$X^{+0}_{\textcircled{r}n} \triangleq (i^*_{on}(+0),\ v^*_{cn}(+0))^T, \quad 0 < \lambda < 1,$$

we have

$$i^*_{o\textcircled{r}n}(z) = i^*_{o\textcircled{r}n}(+0)G_1(z) + \frac{2(1 - v^*_{c\textcircled{r}n}(+0)) - i^*_{o\textcircled{r}n}(+0)\lambda}{\sqrt{1-\lambda^2}}H_1(z),$$

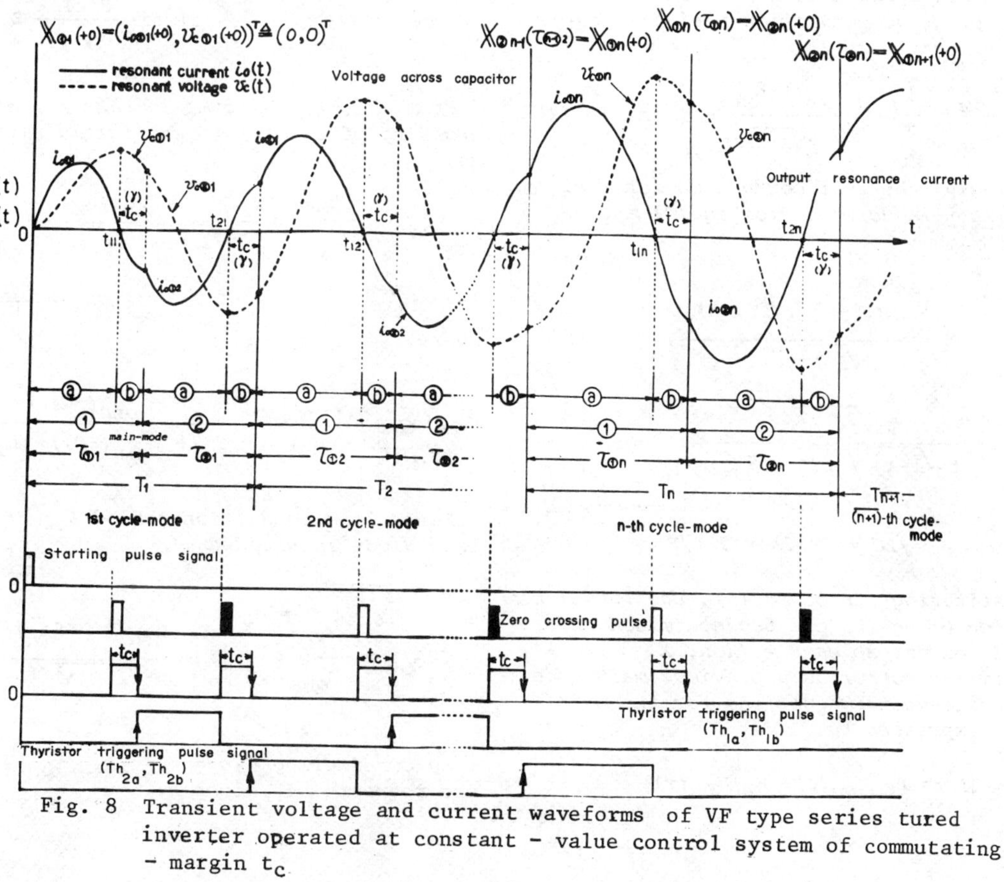

Fig. 8 Transient voltage and current waveforms of VF type series tured inverter operated at constant - value control system of commutating - margin t_c

$$V^*_{c1n}(z) = 1-(1-V^*_{c1n}(+0))G_1(z) + \frac{i^*_{c1n}(+0)-2\lambda(1-V^*_{c1n}(+0)}{2\sqrt{1-\lambda^2}} H_1(z) \quad \text{------- (2)}$$

where,

$$G_1(z) = exp(-\frac{2\pi\lambda}{\mu_1 n} z) cos \frac{2\pi\sqrt{1-\lambda^2}}{\mu_1 n} z \quad ,$$

$$H_1(z) = exp(-\frac{2\pi\lambda}{\mu_1 n} z) sin \frac{2\pi\sqrt{1-\lambda^2}}{\mu_1 n} z$$

From i n(z) = 0, a normalized quenching time Z_{1n}, at which Th_{1a} and Th_{1b} cease to conduct, is explicitly estimated by,

$$Z_{1n} = \frac{t_{1n}}{2\tau_{1n}} = \frac{\mu_1 n(\pi-\xi_{1n})}{2\pi\sqrt{1-\lambda^2}} \quad \text{---- (3)}$$

where,

$$\xi_{1n}(X^{+0}_{1n}, \lambda) = tan^{-1} \frac{i^*_{c1n}(+0)\sqrt{1-\lambda^2}}{2(1-V^*_{c1n}(+0))-i^*_{c1n}(+0)\lambda}$$

Note that Z_{1n} represets the mode-transition instant between sub-modes a , b .

By considering that the normalized switching period τ_{1n} of main-mode 1 is $Z_{1n} = \tau_{1n}/2\tau_{1n} = 1/2$ without having concerned with arbitrary cycle-number n, the normalized time Z_{c1n} of diode conduction period is expressed by,

$$Z_{c1n} = \frac{t_{c1}}{2\tau_{1n}} = \frac{1}{2}\{1- \frac{\mu_{1n}(\pi-\xi_{1n})}{\pi\sqrt{1-x^2}}\} \quad \text{------ (4)}$$

Accordingly, the real conduction interval τ_{1n} of Th_{1a}, Th_{2b} is derived from eq.(4) and $\mu_1 n = \pi\sqrt{LC}/\tau_{1n}$
Namely,

$$\tau_{1n} = t_c + \frac{p(\pi - \xi(X^{+0}_{1n}, \lambda))}{\sqrt{1-\lambda^2}}$$

$$\mu_1 n = \frac{\pi p\sqrt{1-\lambda^2}}{t_c\sqrt{1-\lambda^2} + p(\pi - \xi_{1n})} \quad \text{------ (5)}$$

where, $p=\sqrt{LC} = CR/2\lambda = 2\lambda L/R$

By substituting μ n of eq.(5), the function formulation of eq.(2) can be determined under the specified variables $P = (X\ n, \lambda, p, tc)^T$.

In a similar manner as a previous main mode 1 , the state-vector equations for the main mode 2 is expressed by,

$$\frac{d}{dz} X_{2n}(z) = A_{2n}X_{2n}(z) + B_{2n}U_{2n}(z) \quad \text{------- (6)}$$

where, $X_{2n}(z) = (\ i^*_{c2n}(+0)\ ,\ V^*_{c2n}(+0)\ )^T$,

$$A_{2n} = \begin{bmatrix} -\frac{4\pi\lambda}{\mu_2 n} & -\frac{4\pi}{\mu_2 n} \\ \frac{\pi}{\mu_2 n} & 0 \end{bmatrix} \qquad B_{2n} = \begin{bmatrix} \frac{4\pi}{\mu_2 n} \\ 0 \end{bmatrix}$$

$$U_{2n}(z) = -1\ ,\ \mu_1 n = 2\pi f_{2n}\sqrt{LC} = \pi\sqrt{LC}/\tau_{2n}$$

The solutions obtained under the initial condition of $X^{+0}_{2n} = (i^*_{02n}(+0), V^*_{c2n}(+0))^T$ are given by,

$$i^*_{c2n}(z) = i^*_{c2n}(+0)G_2(z) - \frac{2(1+V^*_{c2n}(+0))^T + i^*_{c2n}(+0)\lambda}{\sqrt{1-\lambda^2}} H_2(z)$$

$$V_{c2n}(z) = -1 + (1 + V^*_{c2n}(+0))G_2(z) \quad \frac{2\lambda(1+V^*_{c2n}(+0)) + i^*_{c2n}(+0)}{2\sqrt{1-\lambda^2}} H_2(z) \quad \text{------ (7)}$$

where,

$$G_2(z) = exp(-\frac{2\pi\lambda}{\mu_2 n} z)\ cos \frac{2\pi\sqrt{1-\lambda^2}}{\mu_2 n} z$$

$$H_2(z) = exp(-\frac{2\pi\lambda}{\mu_2 n} z)\ sin \frac{2\pi\sqrt{1-\lambda^2}}{\mu_2 n} z$$

From eq.(7), the time instant Z n of zero-crossing of i_0 n(z) is explicitly estimated as,

$$Z_{2n} = \frac{t_{2n}}{2\tau_{2n}} = \frac{\mu_2 n(\ \pi - \xi_{2n}\ )}{2\pi\sqrt{1-\lambda^2}} \quad \text{-------- (8)}$$

where,

$$\xi_{2n}(X^{+0}_{2n}, \lambda) = tan^{-1} \frac{-i^*_{c2n}(+0)\ \sqrt{1-\lambda^2}}{2(1+V^*_{c2n}(+0)) + i^*_{c2n}(+0)\lambda}$$

Hence, the conduction period Z_{c2} of feedback diode is yielded,

$$Z_{c2n} = \frac{t_c}{2\tau_{2n}} = \frac{1}{2}\{1 - \frac{\mu_{2n}(\pi - \xi_{2n})}{\pi\sqrt{1-\lambda^2}}\}$$

By considering $\mu_2 n = \pi p/\tau_{2n}$, τ_{2n} and $\mu_2 n$ are respectively given by,

$$\tau_{\textcircled{2}n} = t_c + \frac{p(\pi-\xi_{\textcircled{2}n})}{\sqrt{1-\lambda^2}}$$

$$\mu_{2n} = \frac{\pi p\sqrt{1-\lambda^2}}{t_c\sqrt{1-\lambda^2}+p(\pi-\xi_{\textcircled{2}n})}$$

-------- (9)

As a result, state-vector $X_{\textcircled{2}n}(z)$ is defined as a nonlinear function of $P' = (X^{+0}_{\textcircled{2}n},\ p,\ \lambda,\ tc)^T$.

Figure 9 shows the flow-chart in order to calculate the system dynamics; to analyse the transient and stationary behaviors of VF type inverter operated at constant-value control scheme of tc.

This flow-chart is derived from an analytical approach mentioned above.

Namely, the iteration of calculation for cycle-mode number n are successively carried out according setting values of parameters and initial values $X^{+0}_{\textcircled{1}n}$ of state vector.

Figure 10 shows the theoretical results calculated on basis of the computation algorithm shown in Fig.9. This illustrates the transient voltage and current waveforms in the case when the disturbance parameter λ_1 promptly is changed into λ_2 ($\lambda_2 > \lambda_1$).

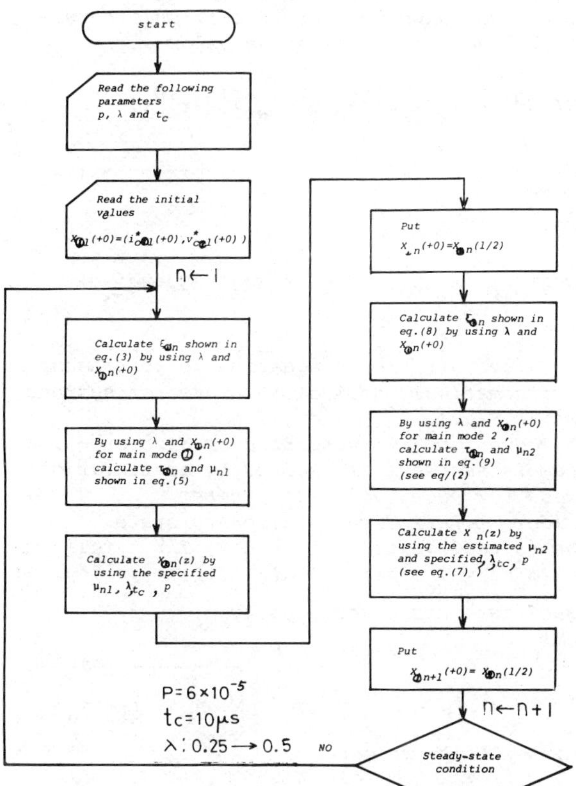

Fig.9 Flow-chart for transient analysis and steady-state analysis of voltage fed type series resonance inverter with feedback control loop shown in Fig.6.

Steady-State Analysis by State-Transition Approach

In a periodical stationary condition, a piece-wise linear theory by state-space approach is usefully applied to the periodically interrupted circuit of VF type inverter with constant-value control loop of tc.

Figure 11 gives the definition of the time division of operating-mode in the n-th cycle after starting.

The vector differential equations described by autonomous system are derived from eqs.(1) and (6) as follows;

$$\dot{X}(\delta) = A_{\textcircled{1}}X(\delta)\ :\ 0 \leqq \delta \leqq \tau_n$$

$$\dot{X}(\delta') = A_{\textcircled{2}}X(\delta')\ :\ 0 \leqq \delta' \leqq \tau_n$$

------- (10)

where,

$$X(t) = (\ E\ ,\ i_o(t)\ ,\ V_c(t)\ )^T$$

$$A_{\textcircled{1}} = \begin{bmatrix} 1 & 0 & 0 \\ 1/L & -R/L & 1/L \\ 0 & 1/C & 0 \end{bmatrix} \quad A_{\textcircled{2}} = \begin{bmatrix} 1 & 0 & 0 \\ -1/L & -R/L & -1/L \\ 0 & 1/L & 0 \end{bmatrix}$$

$$\delta = t - \sum_{k=1}^{n-1} T_k = t - t_{n-1}$$

$$\delta' = t - t_{n-1} - \tau_{\textcircled{1}n}$$

The state variables X (t) can be estimated by Laplace transformation method of eq.(10), i.e.,

$$X(t) = \Phi_1(\delta)\ X(t^{+0}_{n-1})$$

$$X(t) = \Phi_2(\delta')\ X(\overline{t_{n-1} + \tau_{\textcircled{1}n}}^{\,+0})$$

------ (11)

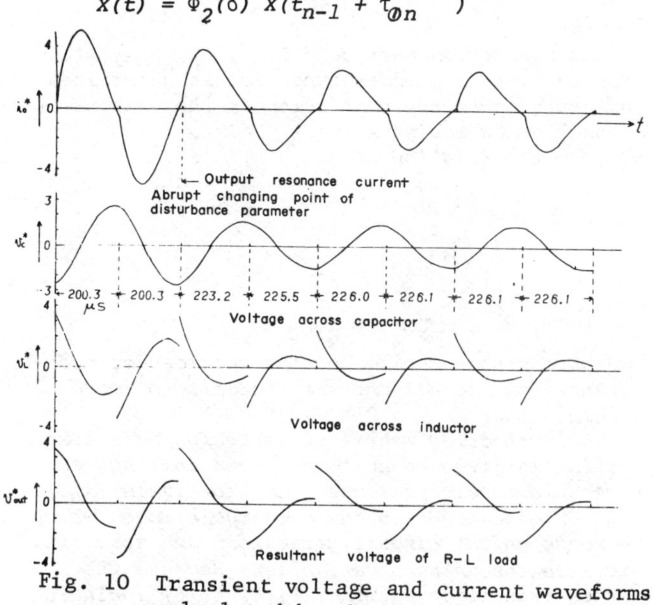

Fig. 10 Transient voltage and current waveforms calculated by the procedure of flow-chart shown in Fig. 9

where,

$$\Phi_1(\delta) = \mathcal{L}^{-1}[\, sI-A_1]^{-1} = \exp(A_1 t)$$

$$= \begin{bmatrix} 1 & 0 & 0 \\ a_{21}(\delta) & a_{22}(\delta) & -a_{21}(\delta) \\ 1-a_{31}(\delta) & a_{32}(\delta) & a_{31}(\delta) \end{bmatrix}$$

$$\Phi_2(\delta') = \mathcal{L}^{-1}[\, sI-A_2]^{-1} = \exp(A_2 t)$$

$$= \begin{bmatrix} 1 & 0 & 0 \\ -a_{21}(\delta) & a_{22}(\delta) & -a_{21}(\delta) \\ -1+a_{31}(\delta) & a_{32}(\delta) & a_{31}(\delta) \end{bmatrix}$$

$$a_{21}(\delta) = (1/L\beta)\exp(-\alpha\delta)\, \sin\beta\delta$$

$$a_{22}(\delta) = \exp(-\alpha\delta)\ (\cos\beta\delta - \frac{\alpha}{\beta}\sin\beta\delta\ )$$

$$a_{31}(\delta) = \exp(-\alpha\delta)\ (\cos\beta\delta + \frac{\alpha}{\beta}\sin\beta\delta\ )$$

$$a_{32}(\delta) = \exp(-\alpha\delta)\ \ \sin\beta\delta\ (1/\beta C)$$

$$\alpha = R/2L \qquad\qquad \beta = \sqrt{1-\lambda^2}\ /\sqrt{LC}$$

The connection matrices B_{n-1}, B_n, B_{n+1} etc. due to switching mode-transition is determined by applying to the continuity on the energy conservation law of X (t).

From the relation of

$$X(t^{+0}_{\textcircled{1}n-1}) = B_{n-1}\ X(t^{-0}_{\textcircled{1}n-1})$$

$$X(t^{+0}_{\textcircled{1}n}) = B_n\ X(t^{-0}_{\textcircled{1}n})$$

$$X(t^{+0}_{\textcircled{2}n}) = B_{n+1}\ X(t^{-0}_{\textcircled{2}n})$$

we obtain that the connection matrices of mode-transition are all the same in spite of mode cycle number.

The connection matrix of switching mode transition is given by unit matrix in this system.

As previously presented, the intervals $\tau_{\textcircled{1}n}$, $\tau_{\textcircled{2}n}$ of main-modes in the n-th cycle mode, which depends on the circuit parameters and specified tc, are the unknown variables. Because the state-transition matrices $\Phi_1(\delta)$, $\Phi_2(\delta')$ including the initial values $X\ (t^{+0}_{n-1})$, are expressed by a nonlinear function of ($X(t^{+0}_{n-1})$, R, L, C, tc), these transition matrices are not determined yet in this stage.

However, $\tau_{\textcircled{1}n}$ and $\tau_{\textcircled{2}n}$ are respectively calculated from the condition of $i_{o,n}(t)=0$ and $i_{o2n}(t)=0$ by considering the constant value control policy of tc., that is to say,

$$\tau_{\textcircled{1}n} = t_c + \frac{\pi - \xi_{\textcircled{1}n}(\ X(t^{+0}_{n-1}), R, L, C\ )}{\beta}$$

$$\tau_{\textcircled{2}n} = t_c + \frac{\pi - \xi_{\textcircled{2}n}(X\ (\overline{t_{n-1} + \tau_{\textcircled{1}n}}^{\,+0}),\ R,\ L,\ C)}{\beta}$$

where, --------- (12)

$$X\ (\overline{t_{n-1} + \tau_{\textcircled{1}n}}^{\,+0}) = \Phi_1(\tau_{\textcircled{1}n})X\ (t^{+0}_{n-1})$$

$$\xi_{\textcircled{1}n} = \tan^{-1} \frac{\beta L\ i_o(t^{+0}_{n-1})}{E-V_C(t^{+0}_{n-1}) - \frac{R}{2}\ i_o(t^{+0}_{n-1})}$$

$$\xi_{\textcircled{2}n} = \tan^{-1} \frac{-\beta L\ i_o(\overline{t_{n-1}+\tau_{\textcircled{1}n}}^{\,+0})}{E+V_C(\overline{t_{n-1}+\tau_{\textcircled{1}n}}^{\,+0}) + \frac{R}{2}\ i_o(\overline{t_{n-1}+\tau_{\textcircled{1}n}}^{\,+0})}$$

By using the analytical results obtained here, we can have a recurrence formula between $X\ (t^{+0}_{n-1})$ and $X\ (t^{+0}_n)$ as shown below.

$$X(t^{+0}_n) = \Phi_2(\tau_{\textcircled{2}n})\ B_n\ \Phi_1(\tau_{\textcircled{1}n})\ X(t^{+0}_{n-1})$$

----------- (13)

where,

$$X(t^{+0}_{n-1}) = \Phi_2(\tau_{\textcircled{2}n-1})B_{n-1}\Phi_1(\tau_{\textcircled{1}n-1})X(t^{+0}_{n-2})$$

The above difference equation is to present the mathematical model of the inverter system shown in Fig.1.

The equation (13) described as discrete-time system is successively solved by an iteration process for each cycle mode number. Next, consider the periodical stationary state.

The initial values $X^{+0}_s \triangleq (\ E,\ i^{+0}_{os},\ V^{+0}_{cs})^T = \lim_{n\to\infty} X(t^{+0}_n) = \lim_{n\to\infty} X(t^{+0}_{n-1})$ of X (t) in periodical steady-state are calculated by

$$X^{+0}_s(\ 1 - \Phi_2(\tau_s)\ B\Phi_1(\tau_s)\ ) = \Theta \qquad \text{------ (14)}$$

Where, $\tau_s \triangleq \tau_{\textcircled{1}\infty} = \tau_{\textcircled{2}\infty}$ $\quad B \triangleq B_\infty = \lim_{n\to\infty} B_n$

From eq.(14), we get,

$$X^{+0}_s = \begin{bmatrix} 1 \\ \dfrac{-2h_s}{\beta L\{(\ 1+g_s\ )^2 + h_s^2\ \}} \\ \dfrac{2h_s\alpha/\beta + (h_s^2 + g_s^2 - 1)}{(1+g_s)^2 + h_s^2} \end{bmatrix} E \qquad \text{---- (15)}$$

where,

$h_s = e^{-\alpha\tau} \sin\beta\tau_s$, $g_s = e^{-\alpha\tau} \cos\beta\tau_s$.

Substituting eq.(15) into eq.(11), we have

$$X(t) = X_{②}(t) = (E,\ i_o(t),\ V_c(t))^T$$

$$= \begin{bmatrix} E \\ \dfrac{2E\exp(-\alpha t)}{\beta L\sqrt{(1+g_s)^2 + h_s^2}} \sin(\beta t + \psi_s) \\ E - \dfrac{2E\exp(-\alpha t)}{p\beta\sqrt{(1+g_s)^2 + h_s^2}} \sin(\beta t + \psi_s') \end{bmatrix} \qquad \text{-------- (16)}$$

where,

$$\psi_s = \tan^{-1}\frac{-h_s}{1+g_s},$$

$$\psi_s' = \tan^{-1}\frac{(1+g_s) + h_s\alpha/\beta}{h_s + (1+g_s)\alpha/\beta}$$

and,

$$X(t) = X_{②}(t) = -X_{①}(t) = \Phi_{②}X_{②}(+0).$$

Oscillation Frequency and Current Amplitude

The oscillation frequency f_{os} (oscillation angular frequency ω_{os}) of self-excited type inverter in the periodical steady-state is estimated from ψ_s of eq.(16), i.e.,

$$\tan\left(\frac{\sqrt{1-\lambda^2}}{p} t_c\right) \left\{1 + \exp(\pi\lambda/\mu_s) \cos\frac{\pi\sqrt{1-\lambda^2}}{\mu_s}\right\} = \exp(\pi\lambda/\mu_s) \sin\frac{\pi\sqrt{1-\lambda^2}}{\mu_s} \qquad \text{---------- (17)}$$

From eq.(17), $\mu_s = 2\pi f_{os}\sqrt{LC}$ is determined by the specified parameters (λ, p, tc).

Namely, $f_{os} = \mu_s(\lambda, p, tc)/2\pi\sqrt{LC}$.

Figure 12 shows the parameter domain which the sustained oscillation generates stably. This domain is calculated from the constraint of $i_o(+0) \geq 0$, i.e., the condition of leading current for $e_o(t)$.

Figure 13 shows tc vs. ω_{os} curve estimated by eq.(17). Figure 14 shows μ_s vs. λ curve for the specified parameter tc(γ), p.

Figure 15 illustrates the oscillograms of voltage and current waveforms. These are throughly examined by the analytical results described here.

From eq.(12), the current amplitude at the sustained oscillation is given by,

$$I^*_{omax} = I_{omax}/I$$

$$= \frac{4\exp\left(-\dfrac{\lambda}{\sqrt{1-\lambda^2}}\right)\left\{\tan^{-1}\dfrac{\sqrt{1-\lambda^2}}{\lambda} - \tan^{-1}\dfrac{-h_s}{1+g_s}\right\}}{\sqrt{(1+g_s)^2 + h_s^2}} \qquad \text{--------- (18)}$$

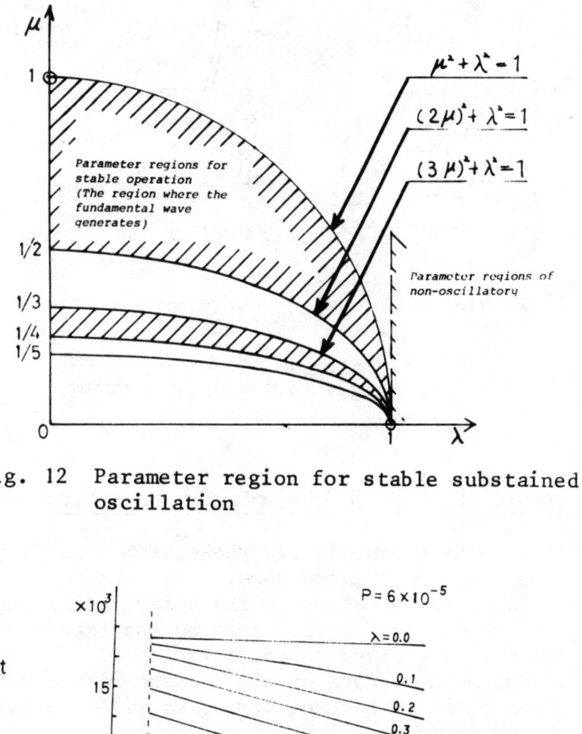

Fig. 12 Parameter region for stable substained oscillation

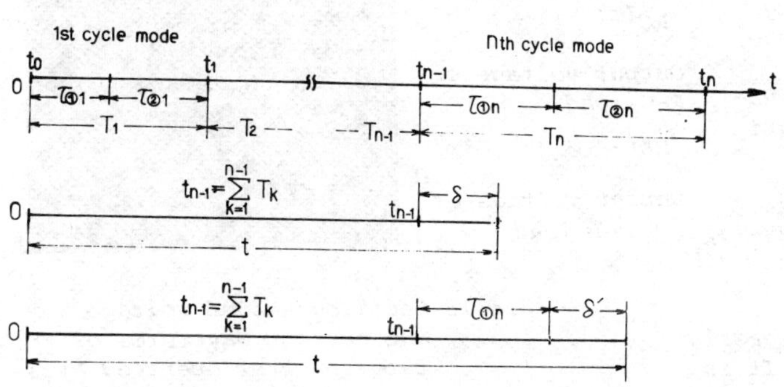

Fig. 11 Time - division of operating-mode during the n-th cycle

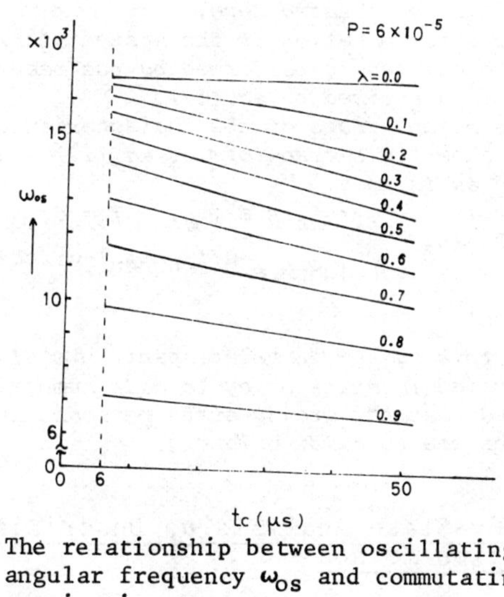

Fig. 13 The relationship between oscillating angular frequency ω_{os} and commutating-margin time t_c

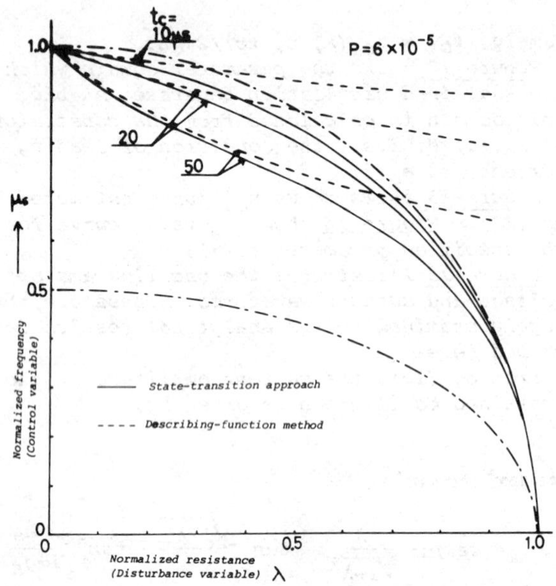

Fig. 14 (a) μ_s VS. λ curves at the constant-value control of commutating-margin time t_c.

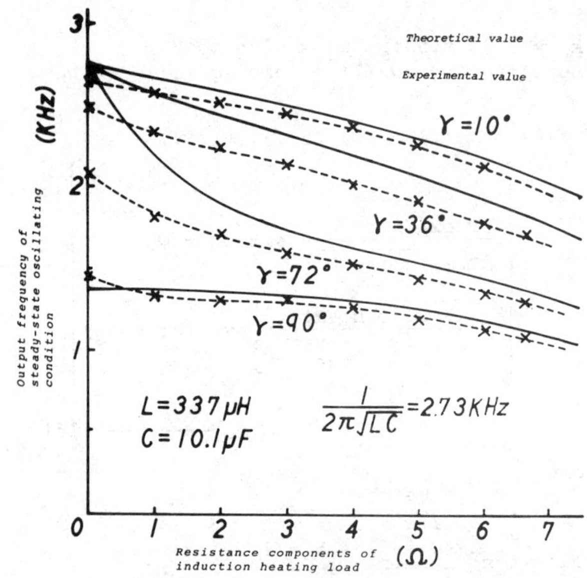

f_{os} VS. R curves at the constant-value control of commutating - margin angle γ

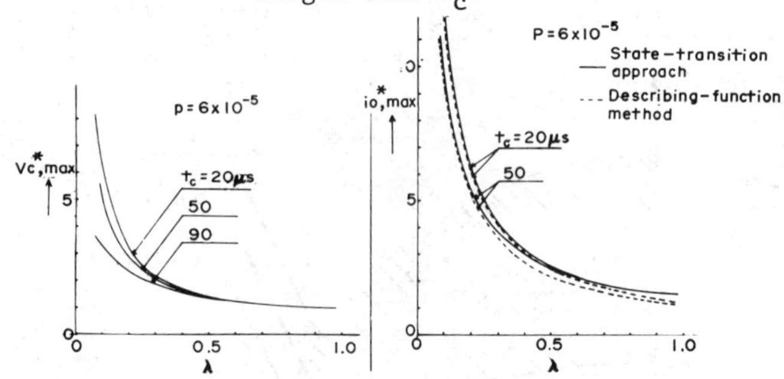

Fig. 16(a) λ vs. $V^*_{c_{max}}$ curve. $v^*_{c_{max}}$: Oscillation voltage amplitude

(b) λ vs $i_{o_{max}}$ curve. $i^*_{o\ max}$: Oscillation current amplitude

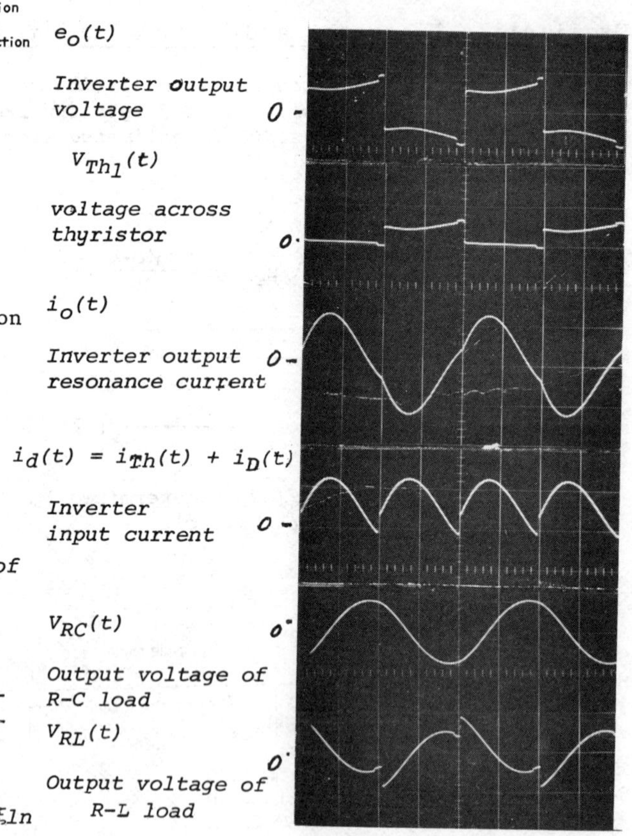

Fig.15 Oscillograms of voltage and current waveforms of VF type inverter operated at commutating-margin angle/time control system.

Vertical Scale: 50V/div., 10A/div.

Horizontal Scale: 0.8μs/div.

Stability of Self-Excited Inverter

Examine theoretically the system stability of the inverter treated here.

The discrimination of the system stability can be completely performed by the recurrence formula described by eq.(12).

The eigen values of the characteristic equation shown by z-transform of eq.(12) are estimated as follows.

$$\det [\ zI - \Phi_2(\tau_{2n})B_n\Phi_1(\tau_{1n})\ ] = 0,$$

we obtain,

$$z_1 = 1,\ |z_2| = |z_3| = e^{-\alpha(\tau_{1n}+\tau_{2n})}e^{-\alpha\{2tc+\frac{1}{\beta}(2\pi-\xi_{1n}-\xi_{2n})}$$

In this system the relationship of $|z|\leq 1$ is easily satisfied in spite of cycle mode numbers. It is proved that the steady-state periodic initial values are to exist uniquely.

Steady-State Analysis by Describing Function Method

The mathematical model of the inverter system

is described by eq.(1) and (6). Figure 17 shows the block diagram represetation of the inverter system with commutating-margin angle/time control scheme. For application of describing function method, the inverter output voltage $e_o(t)$ shown in Fig.3 is approximately assumed as

$e_{o1}(t) = (4E/\pi)\sin\omega_{os}t \quad (\ \omega_{os} = 2\pi f_{os}\ ).$

Moreover, the input current of ideal relay component is represented by $i'_{o1}(t) = I_{o1}\sin\omega_{os}t$.

The describing function of ideal relay element is accordingly given by $N(I_{o1}) = 4E/\pi I_{o1}$.

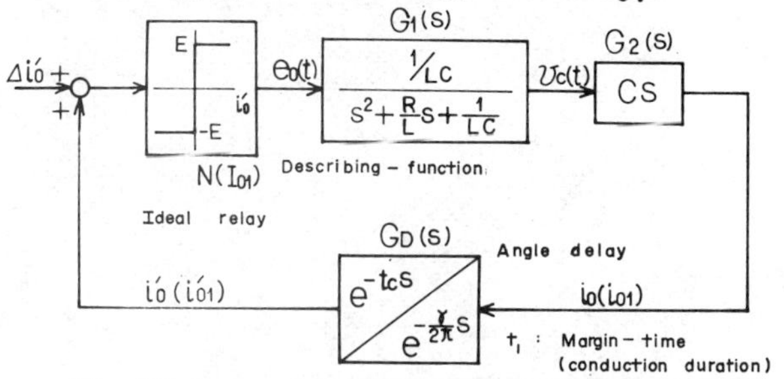

Fig.17 Generalized system mode of VF type series tuned inverter operated at commutating-margin angle/time control system

The characteristic equation of the positive feedback inverter system shown in Fig.17 is written by,

$$\frac{\cos(\omega_{os}t_c-\psi) + j\,\sin(\omega_{os}t_c-\psi)}{R\sqrt{\left(\dfrac{1}{\omega_{os}CR} - \dfrac{\omega_{os}L}{R}\right)^2 + 1}} = 1/N(I_{o1})$$

where,

$$\psi = \tan^{-1}\left(\frac{1}{\omega_{os}CR} - \frac{\omega_{os}L}{R}\right)$$

The sustained oscillation generated in this inverter system estimated by the standard criterion of Nyquist, and the oscillation frequency f_{os} is nearly calculated for the following cases;

(i) in the case of t_c constant control scheme:

$$f_{os} = \frac{\omega_{os}}{2\pi} = \frac{1}{2\pi\sqrt{LC}\sqrt{\left(\dfrac{t_c}{L/R}\right) + 1}} = \frac{1}{2\pi p\sqrt{1 + \dfrac{2\lambda t_c}{p}}}$$

(ii) in the case of γ constant control

$$f_{os} = \frac{1}{2\pi}\left\{ - \frac{R}{2L}\tan\gamma + \sqrt{\left(\frac{R}{2L}\tan\gamma\right)^2 + \frac{1}{LC}} \right\}$$

$$= \frac{1}{2\pi p}\left(- \lambda\tan\gamma + \sqrt{1 + \lambda^2\tan^2\gamma}\ \right)$$

The oscillation amplitude of i_o is approximately given by

$$\frac{I_{o1}{}^*}{I} = \frac{4E/I}{\pi\sqrt{\left(\dfrac{1}{\omega_{os}C} - \omega_{os}L\right)^2 + R}} = \frac{8\mu_s}{\pi\sqrt{(1-\mu_s^2)^2+(2\lambda\mu_s)^2}}$$

where, $\mu_s = 2\pi f_{os}p$

The theoretical results mentioned above are shown and compared with the results by state-space analysis and the various experimental results. This results by describing function method are good agreements with the exact results by state-space approach and the experimental results.

CONCLUSIONS

This paper proposed a novel system development on a commutating-margin angle/time (γ/t_c) of VF type high-frequency resonance bridge inverter by reverse-conducting thyristors.

The latest integrated circuit devices of PLL and counter-gate system were effectively applicable to a constant-value control scheme of γ/t_c for security control. The system model of the inverter operated at a security control scheme was introduced, and the transient and steady-state analysis and the performance evaluation for circuit design was elucidated by a state-space approach and a practical describing-function method was usefully applied to the system performance ascessments in steady state.

Furthermore, these theoretical results were good agreements with the experimental ones.

ACKNOWLEDGMENT

The authors would like to thank Prof. M. Nishimura of Osaka University for thoughtful suggestions, and K. Nakazawa, M. Uno and H. Okuyama of Kobe University for valuable discussions on the subject of this paper.

BIBLIOGRAPHICAL REFERENCES

/1/ J.Klapper, J.T.Frankle: Phase-Locked and Frequency Feedback Systems, Academic Press, Inc. 1972

/2/ F.M.Gardner: Phaselock Techniques, John Wiley & Sons, Inc. 1966

/3/ A.J.Viterbi: Principles of Coherent Communication, McGraw-Hill, 1955

/4/ Zivan Zabar & Alexander Yair: Guidelines for Adaptation of High-Frequency Rectangular Wave Inverter to Inductive Load, IEEE Trans. Vol.IECI-20, No.1, P.27, Feb. 1974

/5/ M. Depenbrock: Die Verknüpfungen von Frequenz, Dämpfung und Steuerwinkel beim Schwingkreiswechselrichter, 49.Band Heft-1964, P.235

/6/ Peter Pomper: Über das statische und dynamische Verhalten von Schwingkreis-Wechselrichtern, ETZ-A, Bd.92(1971),H.4, P.223

/7/ N.Revankar & D.S.Karande: Voltage-Fed High -Frequency Bridge Inverter Circuit, IEEE Trans. IECI, Vol.IECI-21, No.4, Nov. 1974 P.226

/8/ G.N.Revankar & S.A.Gadag: Analysis of High -Frequency Inverter Circuit, IEEE Trans. IECI Vol.IECI-20, Aug. 1973, P.178

/9/ T.Maruhashi, P.P.Biringer, S.B.Dewan & R.S. Segsworth: Series Inverter without Gate Control, IEEE Trans. MAG-6, No.1, March 1970, P.54

/10/ T.Maruhashi, P.P.Biringer, S.B.Dewan & R.S. Segsworth: Stability and Approximate Analysis of Series Inverter without Gate Control, IEEE Conf. Record, 1969, P.445

/11/ N.Mapham: An SCR Inverter with Good Regulation and Sine-Wave Output, IEEE Trans. IGA, Vol.IGA-3, March/April, 1967, P.176

/12/ M.Nakaoka, Y.Murakami, M.Nishimura: A Time-Sharing High-Frequency Series Capacitor Commutated Inverter Circuit and Its Performance Analysis, Electrical Engineering in Japan, S/P scripta publishing company, Vol. 94, March/April 1974, P.36

COMMUTATION MODES OF A CURRENT-SOURCE INVERTER

W. Lienau

Institut für Stromrichtertechnik und Elektrische Antriebe, RWTH, Aachen

ABSTRACT

The limits of stability of a variable-speed drive using an asynchronous machine fed by a controlled current-source inverter are investigated and confirmed by experiment. The stable range can be divided into three modes of commutation, which are single-, double- and multiple commutation. Calculated convertor voltages and currents for the three modes are shown. On the basis of single commutation mode the design for the convertor is developed.

INTRODUCTION

The convertor with interphase commutation, fed by a direct current-source, consists of a three-phase thyristor bridge and commutation equipment. The convertor circuit is shown in Fig. 1. It contains six thyristors, T_1 to T_6, each of them conducts during one third of a period. Six commutation capacitors, C_1 to C_6, provide commutation energy, and six diodes, D_1 to D_6, are included to prevent interaction of the motor load. The commutation circuits include the load. Therefore the character of the load determines an analysis of the convertor. In this paper the load is considered to be an asynchronous machine which corresponds to the circuit shown in Fig. 1 (2).

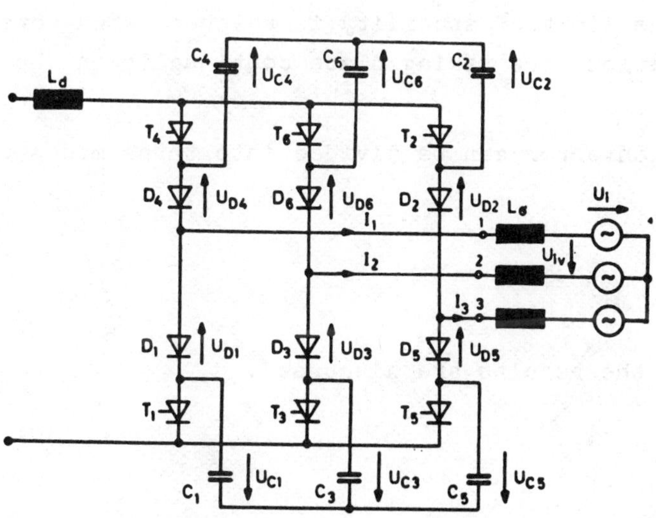

Fig. 1 Convertor with interphase commutation

An analysis of the normal operation of this type of convertor has been presented in the past (1, 3, 4). One of the properties of this convertor is the possibility of multiple pulse operations at low frequency machine currents in order to reduce the low order torque oscillation (5).

The convertor design is determined by the torque-speed diagram of the asynchronous machine (6). The design parameters can be derived from an analysis of the commutation phenomena.

In the following chapters the general design rules are specified, and the stability of this convertor in dependence on frequency are investigated.

The investigation is carried out by means of a digital simulation. The results are confirmed by laboratory measurements.

DIGITAL SIMULATION PROGRAM

The digital simulation is effected by the „Mimic-Source-Simulation-Language" (7). The convertor contains 12 semiconductor valves. Each time a valve is switched on or off, the network configuration changes. Each network configuration is described by a set of differential equations. There are $2^{12} = 4096$ possible combinations. They can be reduced if we assume, that

- one thyristor in each commutation group is conducting simultaneously,
- the indices of the two conducting thyristors differ by one,
- the proceedings in the convertor are periodical by $\frac{1}{6}$ of the period and
- the d.c. current is smooth.

The above restrictions reduce the possible configurations to 49 different combinations of conducting and nonconducting diodes.

The program is to be fed with the data of the asynchronous machine (rated voltage U_{VT}, rated current I_N, per unit short circuit voltage u_{KT}) and the voltage frequency-diagram [$0 < f < f_T$ (= rated frequency), $U_V = \frac{f}{f_T} U_{VT}$, $f_T < f < f_m$ (= maximal machine frequency), $U_V = U_{VT}$] (8, 9). The machine frequency f and the d.c. current I_d are the variables.

This program shows that the limit of stability is reached, when more than three simultaneous commutations are taking place continually in the convertor during one period.

The stable range of this convertor can be divided into three modes of commutation

- single commutation
- double commutation
- multiple commutation.

In the following chapters the results are discussed.

Single commutation

In the range of single commutation the maximum is two simultaneous commutations, one in each commutation group. In this range the process of commutation can be described by a „closed equation system". Therefore it is possible to give the design rules for this convertor.

The process of commutation in this range is well known (2, 3, 4). The convertor voltages and currents are shown in Fig. 2 for one sixth of a period. Three commutation steps are necessary to commutate the current i_d from one machine phase to the next. They are included in this section of the period. The section starts with commutation of current i_d from phase 1 to phase 2 (diode commutation $D_1 \rightarrow D_3$ see Fig. 1) and stops when the linear recharging of the capacitors C_2, C_4 is completed.

The section can be divided into three parts. For each part the most important equations are given.

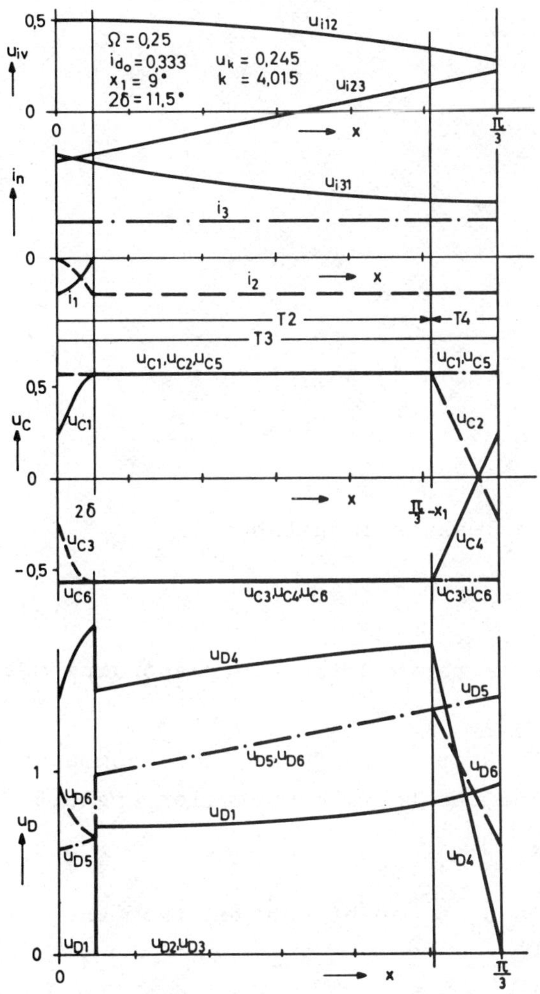

Fig. 2 Single commutation. Convertor voltages and currents at per unit frequency $\Omega = 0,25$

Diode commutation ($0 \leqq x \leqq 2\delta$)

$$i_1 = -i \cdot \cos kx - \frac{3a\Omega_T}{\pi u_K \Omega [(\frac{K}{\Omega})^2 - 1]} [\cos(x+\varphi_1') - \cos\varphi_1' \cos kx + \frac{\Omega}{K} \sin\varphi_1' \sin kx] \quad (1)$$

$$i_2 = -i - i_1$$

with $i = \frac{I}{I_N}$, $u_K = \frac{\omega_T L_\sigma I_N}{U_{1T}}$, $\Omega = \frac{f}{f_m}$, $K = \frac{f_o}{f_m}$, $f_o = \frac{1}{2\pi\sqrt{L_\sigma C}}$, $\varphi_1' = \varphi_1 - \delta$,

$\varphi_1 = \sphericalangle(U_{1n}, I_{n1})$, $a|_{0<\Omega<\Omega_T} = \frac{\Omega}{\Omega_T} = \frac{\hat{U}_{1V}}{\hat{U}_{1VT}} = 1$, $a|_{\Omega_T<\Omega<1} = 1$,

$$u_{c1} = -u_{c3} = \frac{\pi u_K}{6\Omega_T} k\cdot i + \frac{a(\frac{K}{\Omega})^2}{2[(\frac{K}{\Omega})^2 - 1]} \left\{\sin(x+\varphi_1') - \frac{\Omega}{K}\left[\cos\varphi_1' \sin kx + \frac{\Omega}{K} \sin\varphi_1' \cos kx\right]\right\} \quad (2)$$

with $u_c = \frac{U_c}{\hat{U}_{i12T}}$.

$2\delta \leq x \leq \frac{\pi}{3} - x_1$

The currents and capacitor voltages are constant.

Linear recharging ($\frac{\pi}{3} - x_1 \leq x \leq \frac{\pi}{3}$)

$$u_{c2} = -u_{c4} = u_{c1}|_{x=2\delta} - \frac{\pi u_K}{6\Omega_T} k\cdot i(x - \frac{\pi}{3} + x_1). \quad (3)$$

During the period under consideration the induced delta voltages of the machine are

$$u_{iV} = a\cdot\sin[x+\varphi_1'+(n-1)\frac{2\pi}{3}] \qquad n = 1, 2, 3. \quad (4)$$

During diode commutation the voltage on the leakage inductance is

$$u_{L\sigma}(0 \leq x \leq 2\delta) = \frac{u_{c1}-u_{c3}-u_{i12}}{2} \quad (5)$$

The diode voltages u_{Dm} (m = 1, ..., 6) can be derived from corresponding mesh equations.

The equations (1) - (3) provide an implicit statement of the diode commutation angle 2δ, the linear recharging angle x_1 and the total commutation angle β

$$\beta = 2\delta + x_1 \quad (6)$$

For frequency $\Omega \approx 0$ the minimal total angle β_{min} can be derived from the equation (3)

$$\beta_{min} = \frac{1}{K}. \quad (7)$$

The angle β_{min} is the angular hold-off interval of the thyristors, it is determined by the value of the commutation capacitor $C \sim \frac{1}{K^2}$.

Another condition for the design of the commutation capacitor C is the maximum total commutation angle $\beta_{max} = \frac{2\pi}{3}$. Assuming constant machine voltage u_{iV} during the diode commutation, the total commutation angle β can be given as follows

$$\beta = \frac{\pi}{2K} + \frac{1}{K}\left(1 + \frac{6\Omega_T\cdot a\cdot\sin\varphi_1}{\pi u_K\cdot K\cdot i}\right). \quad (8)$$

For a machine (u_K, Ω_T) the ratio $\frac{\sin\varphi_1}{i}$ and the voltage ratio $\frac{U_{iV}}{\hat{U}_{iVT}}$ are the parameters. The total commutation angle β has its top value at a given voltage

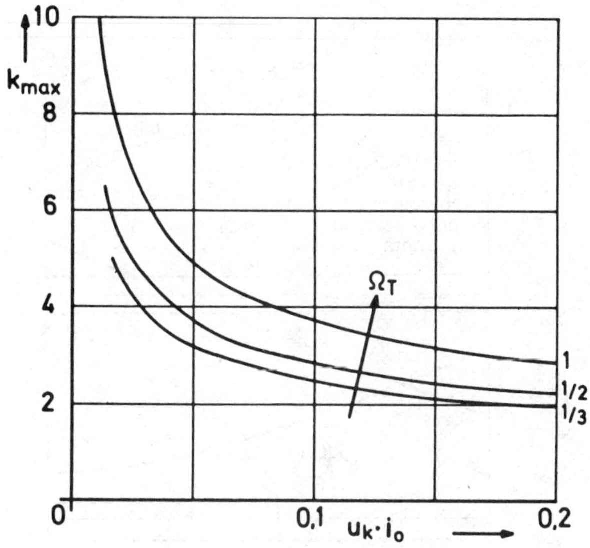

Fig. 3 Single commutation: Maximal frequency ratio K_{max} for different leakage coefficients $(u_K \cdot i_o)$ and voltage ratio $a = 1$

ratio $\frac{U_{iV}}{\hat{U}_{iVT}}$ for no-load operation $(i = i_o,\ \sin\varphi_1 = 1)$. There is only one frequency ratio K_{max} for which the top value of β reaches its maximum value $\beta_{max} = 2\pi/3$ (Fig. 3).

$$K_{max} = 0{,}614\ \left(1+\sqrt{1+2{,}42\ \frac{\Omega_T \cdot a}{u_K \cdot i_o}}\right) \quad (9)$$

It appears to be reasonable to determine the ratio K_{max} in such a manner, that for maximal machine frequency $\Omega_m = 1$ the maximum total commutation angle $\beta_{max} = \frac{2\pi}{3}$ is reached.

A small ratio K_{max} results from increasing the leakage coefficient $(u_K \cdot i_o)$. A small per unit short circuit voltage u_K is desirable for a small additional capacitor voltage $\Delta u_{cmax} = \left[\frac{\pi u_K}{6\Omega_T} \cdot K_{max}\right]$. From this follows the necessity of a great no-load current $i_o = i_o(\Omega)$. Specially developed asynchronous machines have a rated no-load current of $i_{oT} \approx \frac{1}{3}$; at machine frequency $\Omega_m = 1$ the minimum no-load current is $i_o \approx \frac{1}{6}$. They have a per unit short circuit voltage of $u_K \approx 0{,}2$.

In most cases it is necessary to design the capacitor C by taking into account the maximum total commutation angle β_{max}. This value of the capacitance is sufficient for providing the necessary hold-off time for the thyristors at frequency $\Omega \approx 0$, even if line frequency thyristors are used.

The limit of single commutation for a convertor, designed in this way, is reached at machine frequency $\Omega_m = 1$. In each commutation group of the convertor the commutations run continously. For an idling machine (idle current: $i_o(\Omega<\Omega<\Omega_T) = i_{oT}$, $i_o(\Omega>\Omega_T) = i_{oT} \cdot \frac{1}{2\Omega}$) the calculated convertor voltages and currents for one sixth of a period are shown in Fig. 2 and 4 at different frequencies $\Omega = 0{,}25;\ 0{,}7;\ 1$.

There are no mutual influences of the commutation processes of each commutation group. In the mode of single commutation three diodes are conducting simultaneously during diode commutation and after diode commutation only two diodes are conducting simultaneously.

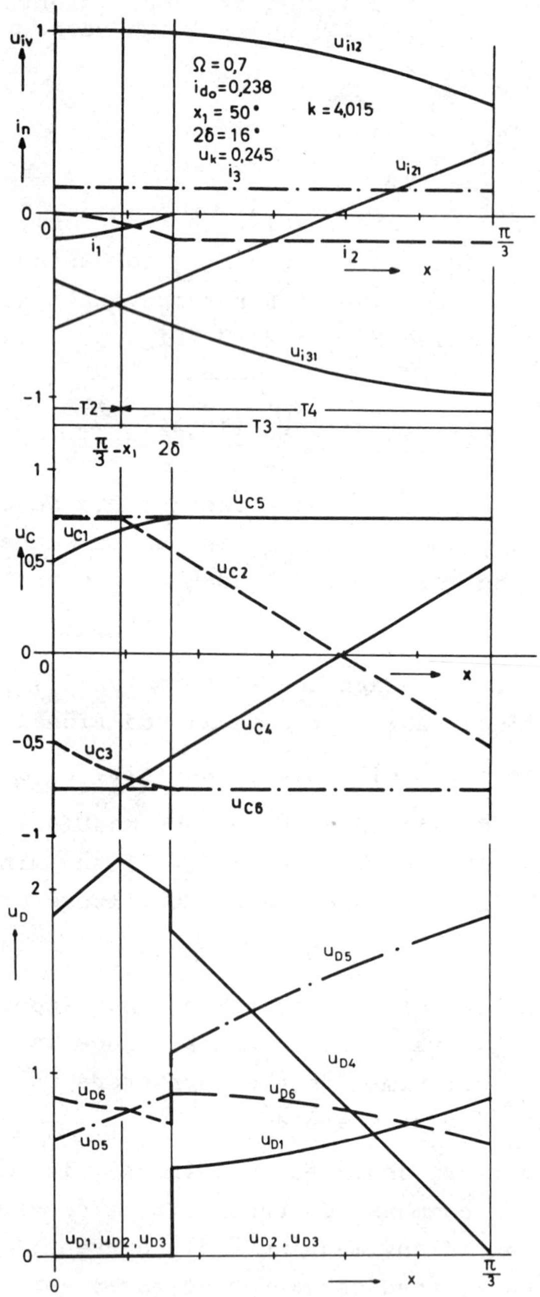

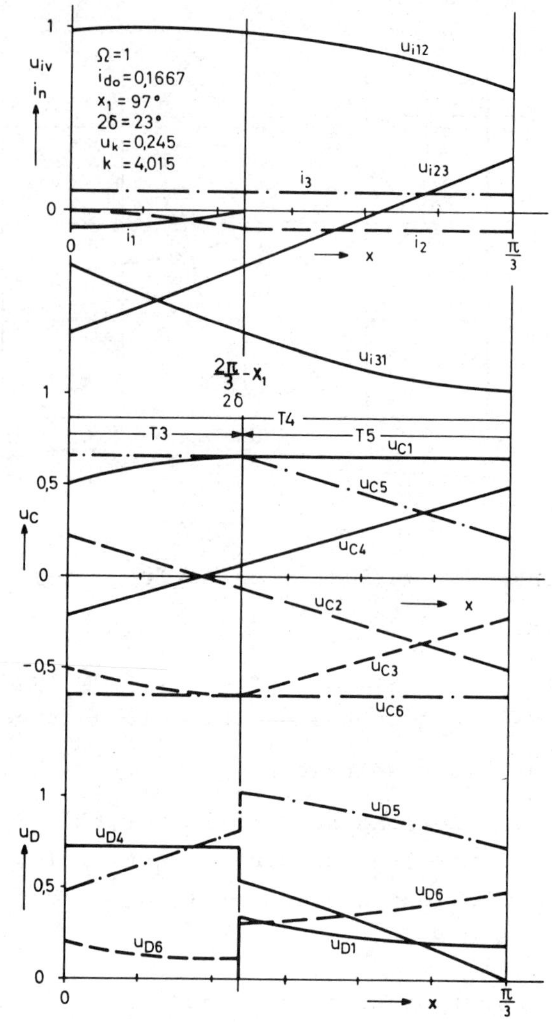

Fig. 4 Single commutation: Convertor voltages and currents at per unit frequencies Ω = 0,7 and Ω = 1 (limit frequency)

Double commutation

If the frequency Ω is raised above the limit frequency Ω_m = 1 the mode of double commutation is reached. This means that during the last interval of diode commutation in one commutation group the next thyristor of the same commutation group is fired and an overlapping of diode commutation and linear recharging takes place. The calculated convertor voltages and currents are shown in Fig. 5 at the per unit frequency Ω = 1,06.

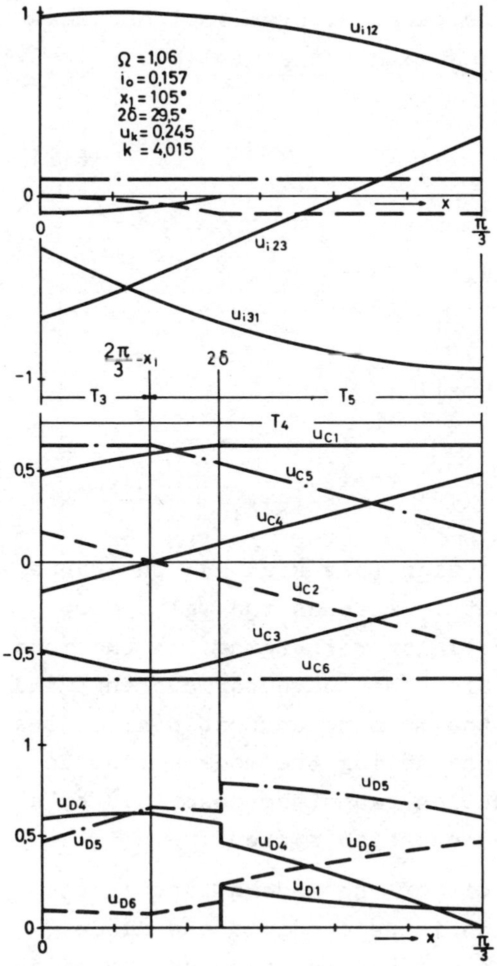

Fig. 5 Double commutation: Convertor voltages and currents at per unit frequency $\Omega = 1{,}06$

In order to explain the phenomena of double commutation the attention has to be drawn to the necessary linear recharging angle. This angle is determined by the capacitor voltage variation in dependence of the angle $\frac{du_c}{dx}$ as follows

$$\frac{du_c}{dx} = \frac{\pi \cdot u_K \cdot K^2_{max}}{6\Omega_T} \cdot i. \tag{13}$$

At no-load operation ($i = i_{oT} \cdot \frac{1}{2\Omega}$), the ratio $\frac{du_c}{dx}$ decreases for rising frequency Ω. Therefore the recharging angle x_1 increases. As the diode commutation angle 2δ is nearly independent of frequency Ω, the sum of both angles would be more than $\frac{2\pi}{3}$. But as the control is symmetrical, the following thyristor is fired during diode commutation of the previous commutation group (see Fig. 5). At this moment, a new diode commutation circuit is to be considered. Figure 6 shows the commutation circuits immediately before and after thyristor firing for the chosen section.

The differential equations for diode commutation which describe the system before the firing of the following thyristor are to be replaced by a new set of differential equations.

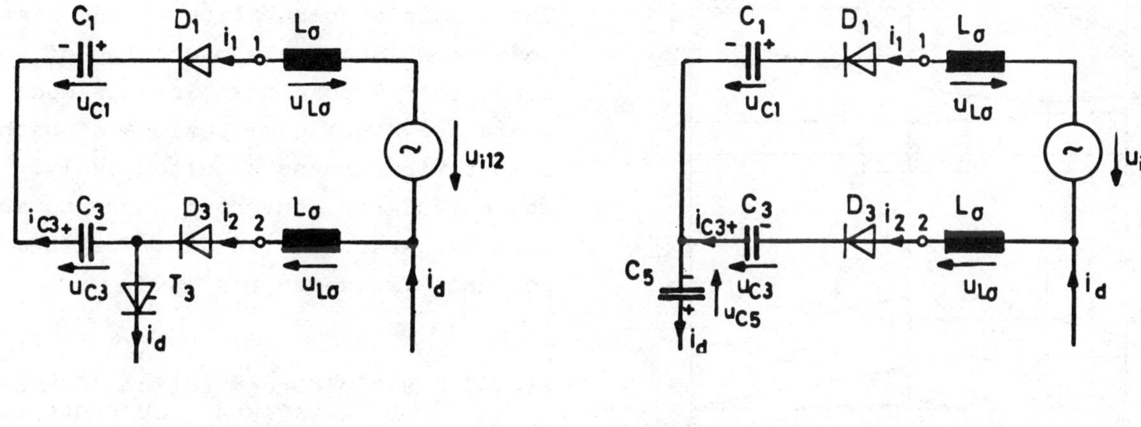

Fig. 6 Double commutation: Diode commutation circuits immediately before and after firing thyristor T_5

Assuming constant machine voltages u_{iV} during the range of overlapping commutations $x_1 \leqq x \leqq 2\delta$, the currents and voltages can be given as follows

$$i_1 = -\frac{1}{2} i_o \, [1-\cos k(x-x_1)+2\cos kx],$$
$$i_2 = -\frac{1}{2} i_o \, [1+\cos k(x-x_1)+2\cos kx], \tag{11}$$

$$u_{c1} = u_{c1}|_{x_1} + \frac{\pi u_K}{12\Omega_T} i_o K[K(x-x_1)-\sin K(x-x_1)+2\sin kx],$$
$$u_{c3} = -u_{c1}|_{x_1} + \frac{\pi u_K}{12\Omega_T} i_o K[K(x-x_1)+\sin K(x-x_1)-2\sin kx]. \tag{12}$$

The capacitor current i_{c3} changes from $i_{c3}|_{x<x_1} = i_1$ to $i_{c3}|_{x>x_1} = -i_2$.

The capacitor voltage variation $\frac{du_{c3}}{dx}$ changes its sign (see Fig. 5). The range of overlapping commutation ends, when the current i_1 reaches the value zero. The remaining angle $(\frac{2\pi}{3} - 2\delta)$ is sufficient for linear recharging. In the mode of double commutation stable operation is possible. The convertor currents and voltages are still periodical for one sixth of the machine current period. The processes during the mode of double commutation cannot be described by a „closed equation system".

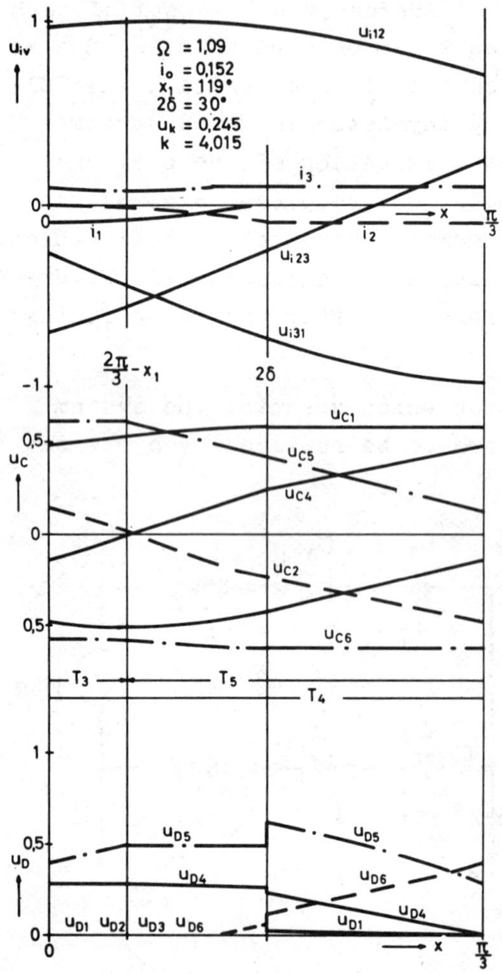

The limit of the mode of double commutation is reached, if the diode voltage u_{D6} becomes zero during diode commutation. Then the diode D_6 is conducting as an additional diode in an inverse direction to the normal cycle. Respecting the data of the asynchronous machine the limit frequency is $\Omega_2 \approx 1,08$.

Multiple commutation

The multiple commutation is the last mode in which stable operation of the convertor is possible. In this mode there are diode commutations of both commutation groups simultaneously. The calculated convertor currents and voltages are shown in Fig. 7 at the per unit frequency $\Omega = 1,09$.

Fig. 7 Multiple commutation: Convertor voltages and currents at per unit frequency $\Omega = 1,09$

Assuming that the diodes D_1, D_2, D_3 are carrying current when the diode D_6 becomes conducting, a new network configuration occurs (see Fig. 8). The new circuits include the three phases of the asynchronous machine. As this circuit contains several energy storage devices, the current i_{D6} starts oscillating. The per unit frequencies K_1 and K_2 of current i_{D6} result from the equivalent inductance of the diode commutation circuits.

$$K_1 = \sqrt{\frac{2}{3}}\,K \quad \text{and} \quad K_2 = \sqrt{2}\,K \tag{14}$$

The current i_{D6} reduces the phase currents and influences the recharging of the capacitors C_1, C_2, C_3 and C_6 in such a manner, that the total commutation process takes place within the compass of angle $\frac{2\pi}{3}$.

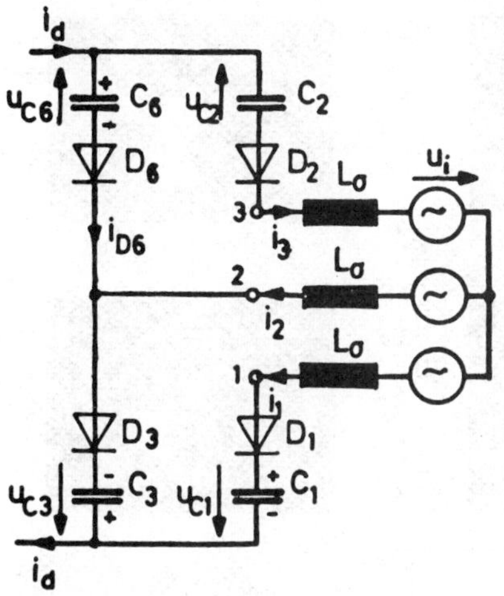

Fig. 8 Multiple commutation: Diode commutation circuits when diode D_6 starts to conduct

It amplifies the fifth harmonic of the machine currents. In order to keep the induced machine voltage u_{1T} constant in this mode of operation, the d.c. current i_d is raised by the control loops.

The limit of this mode is reached when the oscillation current i_{Dm} is no longer intermittent and changes continously from one principal arm to the next. The ultimate limit frequency Ω_G of stable operation can be given as follows

$$\Omega_G = \frac{K_2}{5} = 0{,}283\cdot K \tag{15}$$

At this ultimate limit each diode is conducting about $\frac{5}{6}$ of the period. If the frequency is raised above this limit frequency oscillations of all convertor currents result. On the basis of the data of the asynchronous machine the ultimate limit frequency is $\Omega_G = 1{,}13$.

TEST RESULTS

A convertor has been designed according to the rules described in the last chapter. The rated data of the asynchronous machine are $P_N = 10$ KW, $U_{NT} = 180$ V, $I_N = 45$ A, $f_m = f_T = 100$ Hz, $u_K = 0{,}17$; $i_o(f_T) = 15$ A. The data of the convertor are $K_{max}(\Omega_T = \Omega_m = 1) = 4{,}67$; delta connected capacitances $C_\Delta = 62$ µF, additional capacitor voltage $\Delta U_{cmax} = 106$ V, total capacitor or thyristor blocking voltage $U_{cmax} = 407$ V, hold-off interval $t_F = 340$ µs.

The measurement were made at a reduced machine frequency of $f_m = 45$ Hz. Therefore the capacitances were raised up to $C_\Delta = 430$ µF. The oscillograms in Fig. 9 represent the mode of single commutation at frequency f = 28 Hz, the beginning of the mode of multiple commutation at frequency f = 43 Hz and the ultimate limit of stable operation at frequency f = 47,5 Hz. The limits of the three modes are somewhat reduced due to the d.c. current ripple. It is no problem to vary the load between no-load (Fig. 9) and rated load. The thyristor current i_{T5} is a section of the d.c. current i_d. At the limit of multiple commutation the d.c. current is raised by 30 per cent. The harmonic analysis of a machine current i_n leads to the following amplitudes $i_{n1} = 0{,}7 \cdot i_d$ and $i_{n5} = 0{,}2 \cdot i_d$.

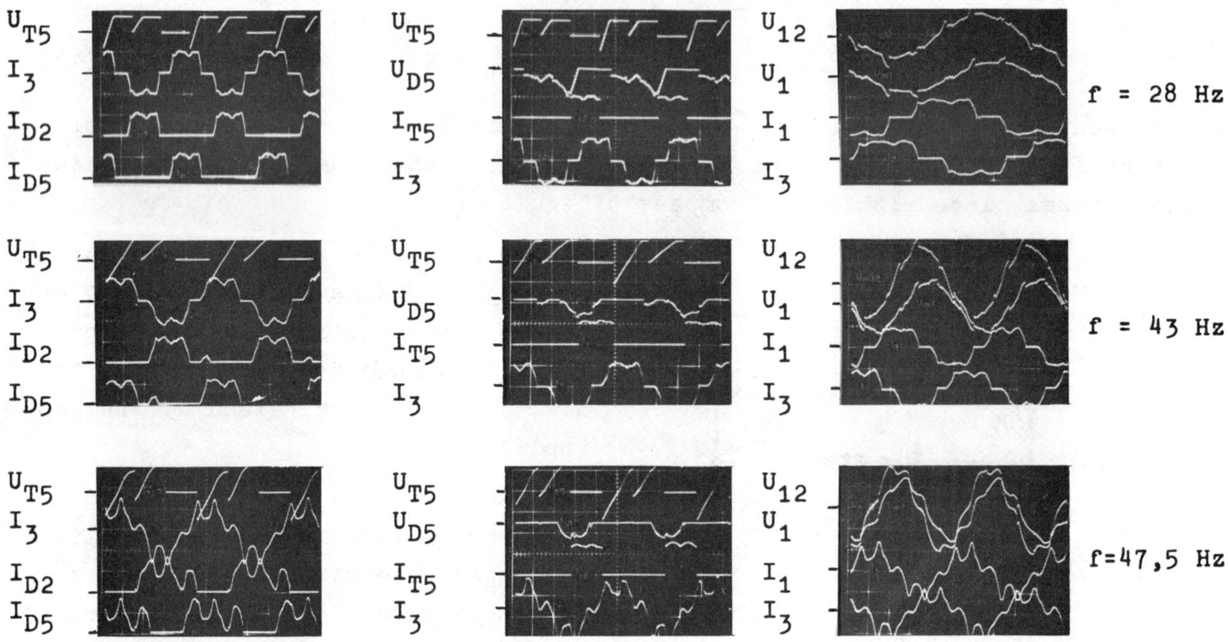

Fig. 9 Test results: Oscillograms of convertor currents and voltages at different frequencies

CONVERTOR DESIGN FOR CONSTANT ROTOR FREQUENCY OPERATION

Although the asynchronous machine has been designed especially for this application, the convertor design for no-load condition leads to a small commutation capacitance. This results to a high thyristor blocking voltage $U_T > 2U_{1VT}$ at rated load.

A different approach to convertor design is possible if the machine is operated at load-independent rotor frequency f_2 over the frequency range $0 < f < f_m$ (10). In this case the design results in a greater commutation capacitance and smaller thyristor blocking voltage $U_T < 2U_{1VT}$. The machine acquires the characteristic of a series wound d.c. machine, the torque being proportional to the square of the machine current. The total commutation angle β becomes load independent, because the ratio $\frac{a \cdot \sin\varphi_1}{i}$ is proportional to the rotor frequency f_2. If the machine is operated in this condition, the convertor is to be designed for rated load and the total commutation β_{max}, as expressed in equations (1) - (3).

REFERENCES

(1) K. P. Phillips, Current source convertor for ac motor drives, IEEE Trans. on Ind. Appl. 8, 679 (1972).

(2) H. Kleinrath, Der Kommutierungsvorgang beim Asynchronmotor mit Speisung über Umrichter mit Gleichstromzwischenkreis, Archiv f. Elektrot. 56, 12 (1974).

(3) H. Kazumo, Commutation of a three-phase thyristor bridge with commutation capacitors and series diodes, Electrical Engineering in Japan 90, 91 (1970).

(4) W. Farrer, J. D. Miskin, Quasi-sine-wave fully regenerative inverter, Proceeding IEE 120, 969 (1973).

(5) W. Lienau, A. Müller-Hellmann, Möglichkeiten zum Betrieb von stromeinprägenden Wechselrichtern ohne niederfrequente Oberschwingungen, ETZ-A 97, 663 (1976).

(6) J. Lazar, Berechnung eines Umrichters mit Gleichstromglättung, Wiss. Zeit. TH Ilmenau 22, 99 (1976).

(7) MIMIC DIGITAL SIMULATION LANGUAGE, Reference Manual Publication 446 10400 (1972).

(8) K. Heumann, K.-G. Jordan, Das Verhalten des Käfigläufermotors bei veränderlicher Speisefrequenz und Stromregelung, AEG-Mitteilungen 54, 107 (1964).

(9) J. Körber, Grundlegende Gesichtspunkte für die Auslegung elektrischer Triebfahrzeuge mit asynchronen Fahrmotoren, Elektrische Bahnen 45, 52 (1974).

(10) P. Appun, Control of rotating and linear induction motors of vehicle drives, 2. IFAC-Symposium, Düsseldorf 1977.

APPLICABLE FREQUENCY RANGE OF CURRENT SOURCE CONVERTERS

K. Moll and D. Schröder

Brown, Boveri & Cie., Mannheim, Germany

ABSTRACT

A three-phase current source converter with blocking diodes feeding an asynchronous induction machine has been analysed. The analysis illustrates the three different modes of operation of this converter, especially the operation in the full overlap mode of commutation. The results were found by a computer program.

INTRODUCTION

From many publications the existence of two current source converter types is well known: the first one needing an additional thyristor bridge for commutation, the second one operating with the commutation between the phases by blocking diodes. The commutation mechanism of both types is described by the same mathematical procedure. In the arrangement with the additional thyristor bridge there is far more requirement on the elements. As only the converter with the commutation between the phases has a region with full overlapping commutation it distinguishes from the other one. This paper wants to describe and analyse the operating range of overlapping commutation. It is demonstrated, that the general operating of the current source converter with blocking diodes is preserved.

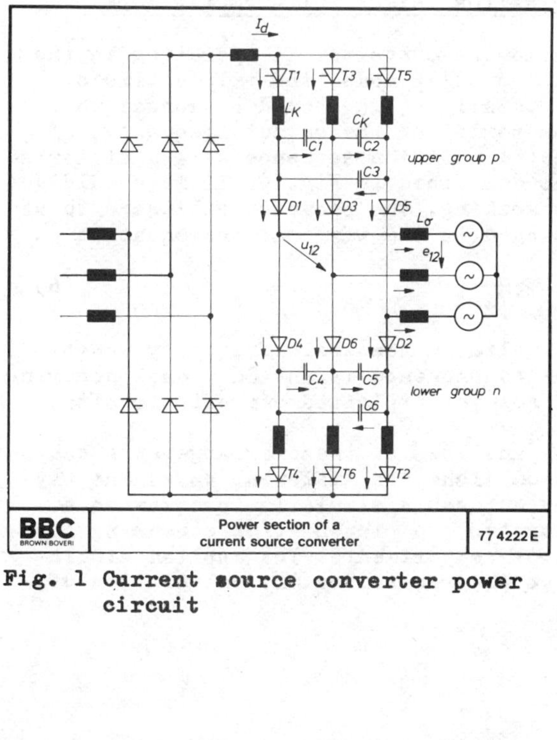

Fig. 1 Current source converter power circuit

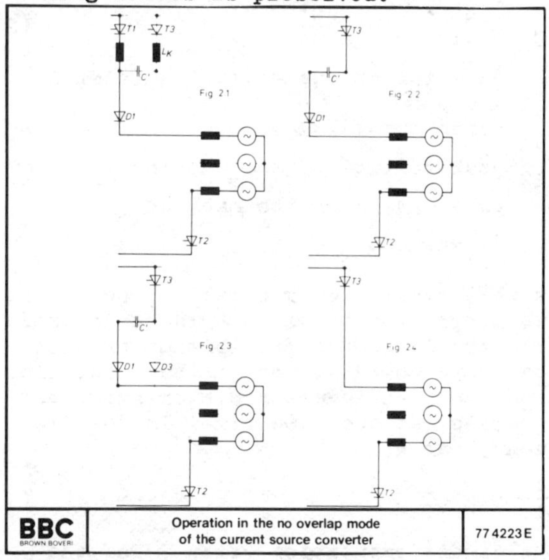

Fig.2 Commutation in the no-overlap mode

OPERATION IN THE NO-OVERLAP MODE

Fig.1 shows the entire converter power circuit, Fig.2 every single section of the usual commutation sequence. The three steps it consists of are explained by the commutation from phase 1 to phase 2 of the induction machine as follows.

Step 1: Thyristor-commutation. When the thyristor T3 turns on, T1 is biased in the reverse direction by the voltage of the upper capacitor group. By the oscillating circuit with the inductance $2 \cdot L_K$ and the capacitance $C' = 1.5 \cdot C_K$, the thyristor T1 is turned off. The whole dc current I_d flows to T3 (Fig.2.1). The duration of this step is generally of little importance.

Step 2: Linear capacitor- charging. While T3 and the diode D1 are conducting (Fig.2.2), a constant-current discharging of the commutation capacitors C1, C2 and C3 takes place. There is no effect on the induction machine. As soon as the capacitor-voltage reaches the machine-voltage e_{12}, the linear charging comes to an end.

Step 3: Diode-commutation. At the end of step 2 there is an oscillating circuit with $C'=1.5\cdot C_K$ (capacitance) and $L'=2\cdot L_\sigma$ (inductance) (Fig.2.3). The oscillation lasts until the load current has completely changed from phase 1 to phase 2 (Fig.2.4).

During the described operation, the following equations are to be considered. Certain conditions must not be neglected (Ref.3).

$$T_C = \sqrt{L'C'}\cdot(1+\frac{\pi}{2})+2C'\frac{\hat{E}_{12}\sin\varphi}{I_d} \qquad (1)$$

$$U_C = \hat{E}_{12}\sin\varphi+I_d\sqrt{\frac{L'}{C'}} \qquad (2)$$

T_C duration of the operating (step 2 and 3)

U_C capacitor-voltage

$\hat{E}_{12}$ peak-voltage of the machine

φ phase-angle of the machine

I_d dc current

As the commutation generally proceeds in two steps, the thyristor-groups uncouples from the diode-groups. This means that there are commutations in the upper diode-group and the lower thyristor-group and reversed, at the same time. In the frequency range

$$\frac{1}{6T_C} < f \leq \frac{1}{3T_C} \qquad (3)$$

the diode-commutation (step 3) of the group p overlaps the thyristor-commutation (step 1) and the linear charging (step 2) of the lower group n. Because the subsequent conducting diode during step 1 and step 2 is still biased in the reverse direction, there is no mutual influence of the upper and lower group in steady-state operation. An overlapping of commutations has no effect on the working readiness of the converter. The frequency range of the usual commutations

$$\text{region I}: \; 0 \leq f \leq \frac{1}{6T_C} \qquad (4)$$

(no-overlap mode)

can also be extended to

$$\text{region II}: \; \frac{1}{6T_C} < f \leq \frac{1}{3T_C} \qquad (3)$$

(partial-overlap mode)

The critical frequency

$$f_p = \frac{1}{3T_C} \qquad (5)$$

is reached when in the same commutation group (p or n) the end of the preceding diode-commutation coincides with the beginning of the following linear capacitor-charging. The commutation-capacitors are continually conducting, that means the times of constant capacitor-voltage ($+U_C$, $-U_C$, 0) are reduced to only a point. When exceeding the critical frequency f_p, the range of the full overlapping commutations starts, which is analysed in the following chapter. The different frequency ranges are to be seen in Fig. 3.

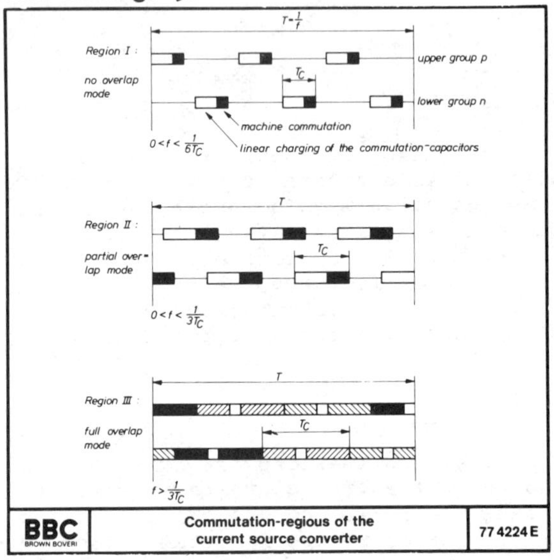

Fig.3 The three different frequency ranges of the current sources converter

Region I: Operating in the no-over-overlap mode

Region II: Operating in the partial-overlap mode

Region III: Operating in the full-overlap mode

OPERATION IN THE FULL-OVERLAP MODE

While the converter is operating in the full-overlap mode, the triple time of commutation ($3T_C$) would be longer than the period of the output frequency, if the commutation sequence were undisturbed (as described in Fig.2). If it should go on working, the sequence will have to be changed in the way that the equation

$$f = \frac{1}{3T'_C} \qquad (6)$$

is valid in the whole frequency range. The consequence is an additional occuring of not yet exploited current circuits.

The analysis of these arrangements can be accomplished be different ways. The way chosen uses a simulation program on a computer, to prevent time-consuming laboratory experiments. The applied simulation program serves the calculation of steady and dynamic states in electrical circuits with semiconductors. Using this kind of programs the special conditions of the convertertechnique must be taken into account, as for example the great differences in time constants by continually switching actions.

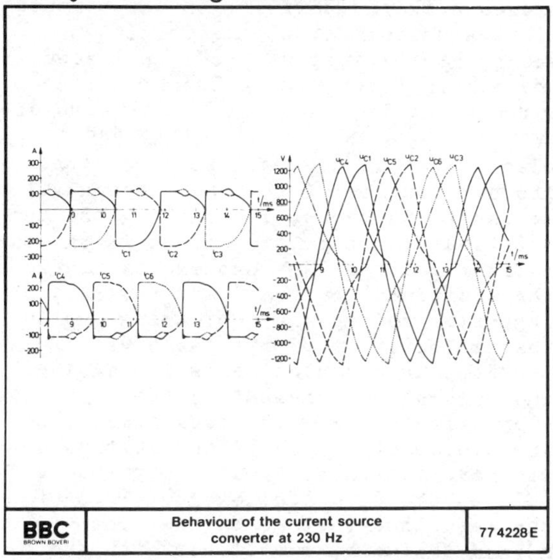

Fig.4 Operating in the full-overlap mode: capacitor-voltages and -currents of the current source converter at 230 Hz.

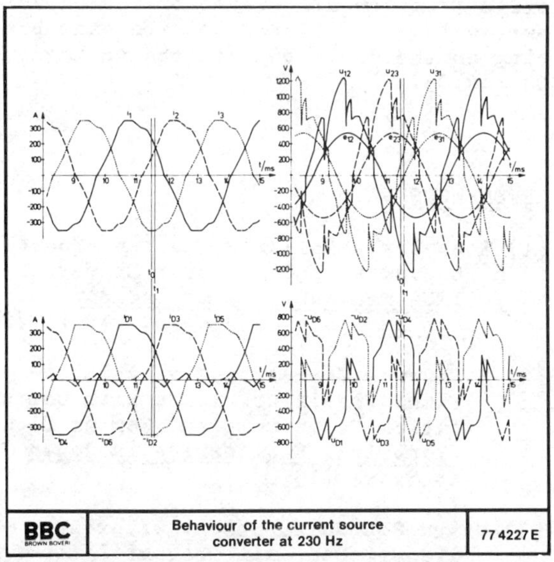

Fig.5 Operating in the full-overlap mode: diode-voltages, diode-currents, machine-voltages and machine-currents at 230 Hz.

The circuit, shown in Fig.1, is calculated at the computer. The following parameters have been employed for the analysis:

L_σ = 0,6 mH, L_K = 12 µH, C_K = 110 µF,

E_{12} = 380 V, I_d = 350 A, φ = 50°

From this operation point follows:

T_C = 1,53 msec $\qquad f_p$ = 217 Hz.

The value of the adopted frequency of the induction machine is 230 Hz (about 6 % more than f_p). The results of the simulation are to be seen in Fig. 4 and Fig.5.

The essential conclusion is the readiness for operation of the current source converter in the frequency range above f_p. A more detailed investigation shows that an additional charging takes place as suspected, as every diode conducts more than once a period. Operating in the no-overlap and partial-overlap mode a diode usually conducts only one time in a period.

At the instant t_0 (Fig.5) a diode commutation from D1 to D3 (thyristor T3 is conducting) is taking place in the upper group p, while there is a linear capacitor charging in the lower group n, diode D2 and thyristor T4 conducting. The diodes D4, D5 and D6 are biased in the reverse direction. The voltage of D6 being reduced to zero at the instant t_1

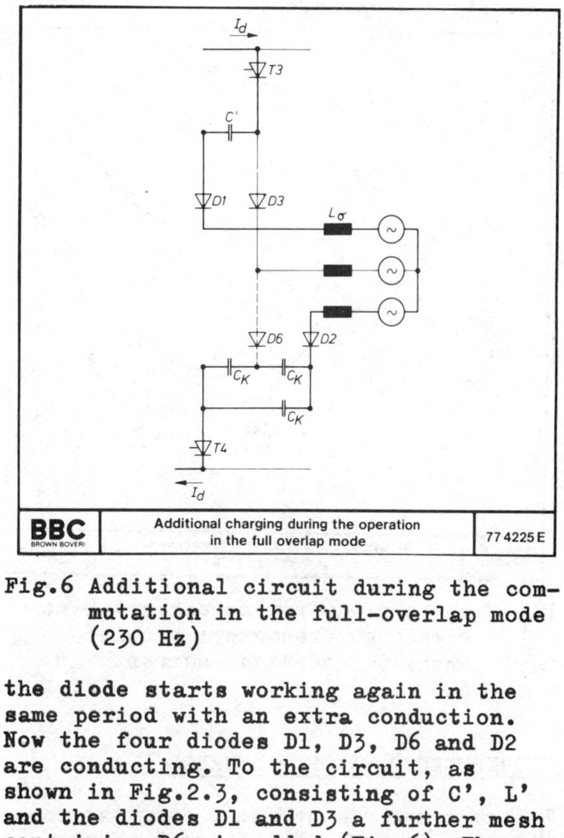

Fig.6 Additional circuit during the commutation in the full-overlap mode (230 Hz)

the diode starts working again in the same period with an extra conduction. Now the four diodes D1, D3, D6 and D2 are conducting. To the circuit, as shown in Fig.2.3, consisting of C', L' and the diodes D1 and D3 a further mesh containing D6, is added (Fig.6). The supplementary current in D6 causes an additional charging of the commutation-capacitors and a reduction of the currents in phase 1 and 3 of the induction machine. There is the advantage to the converter that an elevation of the frequency above f_p is possible and that the general working readiness is preserved. The disadvantage is the reduction of the torque-building currents, while the dc current remains constant. This fact prevents the region of full-overlap commutation from being exploited infinitely, yet there is no risk for the converter in this range as perhaps overvoltages.

The described commutation operation comes to an end at the frequency

$$f_m = \frac{1}{6T_{C2}} \qquad (7)$$

i.e.

T_{C2} duration of the machine commutation (from phase to phase).

The general conditions are maintained furtheron. An immense increasing of frequency causes a conduction of more than four diodes at the same time (Fig.7), however that is no danger for the converter either. But in this range the converter is only controllable with difficulties, as there are instabilities in the change-over of motor- and generator-operating. If the converter changes from one region to the other, the phase-angle of the induction machine will change discontinuously its value.

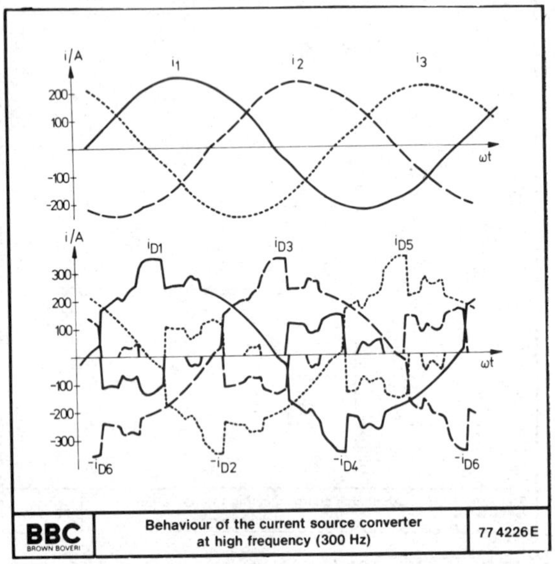

Fig.7 The current source converter at very high frequency: machine-currents and diode-currents at 300 Hz.

REQUIREMENTS ON THE ELEMENTS

In the range of the usual commutations (no-overlap and partial-overlap mode) the following requirements on the semiconductors appear:

$$\text{thyristor: } u_{Th} = U_C \qquad (8)$$
$$\text{diode: } u_{Di} = U_C + u_{12}$$

Fig.5 shows that the additional charging causes a charge-shifting of the commutation-capacitors C4, C5 and C6 (lower group) at the instant t_1:C1 is loaded in the same, C5 conducting 1/3 of the dc current I_d is loaded in the reverse direction, C6 conducts 2/3 of I_d and is also loaded in the reverse direction. By the description of the usual commutation mechanism is known, that during the commutation from thyristor T2 to T4 the capacitor-voltage $-U_C$ changes its sign. As the additional current runs reverse, the capacitor-voltage is decreasing, while the exploitation of the range with full-overlapping commutation is increasing. The general requirement on the thyristor stays with U_C. For the requirements on the diodes other conditions must be considered. In the frequency range below f_p (Fig.3) the diode voltage consists of the capacitor-voltage of one group (p or n) and the machine voltage. With the beginning of full-overlapping commutations the diode voltage is set by the voltage of both capacitor-groups (p and n). The voltage of the diode D4, for example, results from the difference of the capacitor-voltages of C4 and C1. The same bearing on the diode D5 have the voltages of C2 and C5.

REFERENCES

(1) Kenneth P. Phillips, Current-Source Converter for AC Motor Drives, IEEE Transactions on Industry Applications IA-8 No.6, 679 (1972)

(2) H.Kazuno, Commutation of a Three-Phase Thyristor Bridge with Commutation Capacitors and Series Diodes, Electrical Engineering in Japan 90 No.5, 91 (1970)

(3) Dierk Schröder, Selbstgeführter Stromrichter mit Phasenfolgelöschung und eingeprägtem Strom, Elektrotechnische Zeitschrift ETZ-A 86 H.11, 520 (1975)

ANALYSIS AND DESIGN OF A CURRENT-FED INVERTER

M.K. Parasuram* and B. Ramaswami**

**Department of Electrical Engineering, Karnataka Regional Engineering College, Srinivasanagar-574157 (India)*
***Department of Electrical Engineering, Indian Institute of Technology, Madras, Madras-600036 (India)*

SUMMARY– The current-fed inverter discussed by Philips and Maag is becoming increasingly popular for reversible speed induction motor drive applications. This paper gives the analysis and design of such an inverter for inductive loads. Experimental results have been compared with analytical results. Finally the work reported by Farrer is extended by including resistance in the motor model developed by him and the equations of performance are arrived at.

1. INTRODUCTION

Current-fed inverters for induction motor drive applications reported to in literature [1,2] prior to about 1970 required twelve thyristors, six main and six auxiliary ones. For an automatic reversible drive, another six thyristors are required for precharging the commutating capacitors. Because of the large number of thyristors needed and the complexity of the control circuitry required for reversible drive, the induction motor drive could not compete with d.c. motor drive. The current-fed inverter discussed by Phillips [3] and Maag [4] needs just six thyristors and requires only simple firing logic for reversible drive as the commutating capacitors have at each instant the right polarity for firing the inverter in either phase sequence. Further, slow turn-off thyristors could be used for the inverter. Because of these advantages and the merits of current-fed inverters over voltage-fed inverters [5], the current-fed inverter considered by Phillips and Maag is finding increasing application. Figure 1 shows this inverter feeding a delta connected induction motor. Figures 2a to 2d give different waveforms of interest.

Farrer and Miskin [5] develop a mathematical model of the induction motor for use with such an inverter. They give a detailed analysis of the commutation for a motor load. The analysis of the current-fed inverter for R-L loads is equally a significant problem as quite often thyristor converters and inverters are tested with R-L load before using them for motor load. This paper gives the analysis and design of the current-fed inverter for R-L load and also extends the work of Farrer for motor load.

Section 2 gives a detailed analysis of the commutation with R-L load and section 3, the design procedure for the inverter. In section 4, the model of Farrer is improved to include resistance in each phase of the motor model and equations of performance are arrived at. Section 5 gives experimental verification of the results developed in section 2. Section 6 gives constructional features and section 7 conclusions.

2. COMMUTATION ANALYSIS WITH R-L LOAD

Figure 3 shows the inverter feeding a R-L load. The inverter is supplied from a constant current source which invariably is nothing but a controlled rectifier with a current control loop and a large inductance in the d.c. link circuit. The inverter is shown with thyristors T_1 and T_2 conducting and the inverter thyristors are to be fired in the sequence 3,4,5,6,1,2 etc. at intervals of 60°. C_1 through C_6 are the commutating capacitors and the diodes D_1 to D_6 prevent the discharge of the capacitors during the non-commutating periods.

In Fig.3, the circuit is assumed to be in steady state with T_1 and T_2 conducting and the commutation of T_1 is considered in the analysis that follows. The positive directions of all voltages and currents of interest are shown in Fig. 3. The load currents in Fig. 3 are,

$$i_3 = \frac{2}{3} I \quad \text{and} \quad i_1 = i_2 = -\frac{1}{3} I.$$

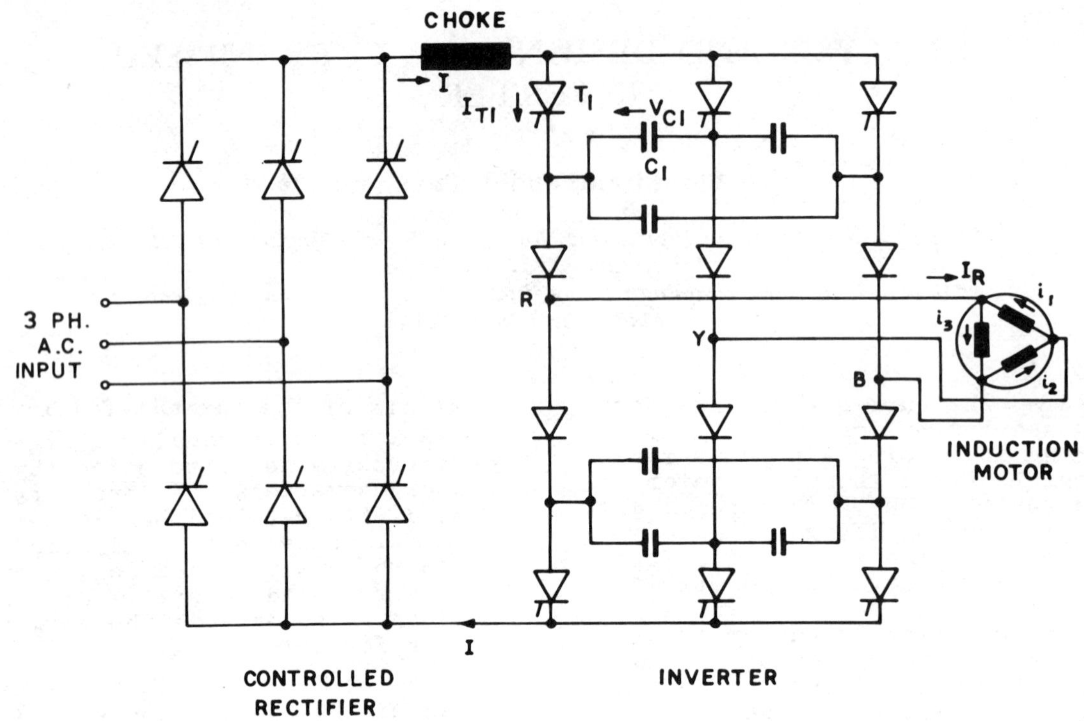

Fig. 1. Current-fed inverter supplying a motor load.

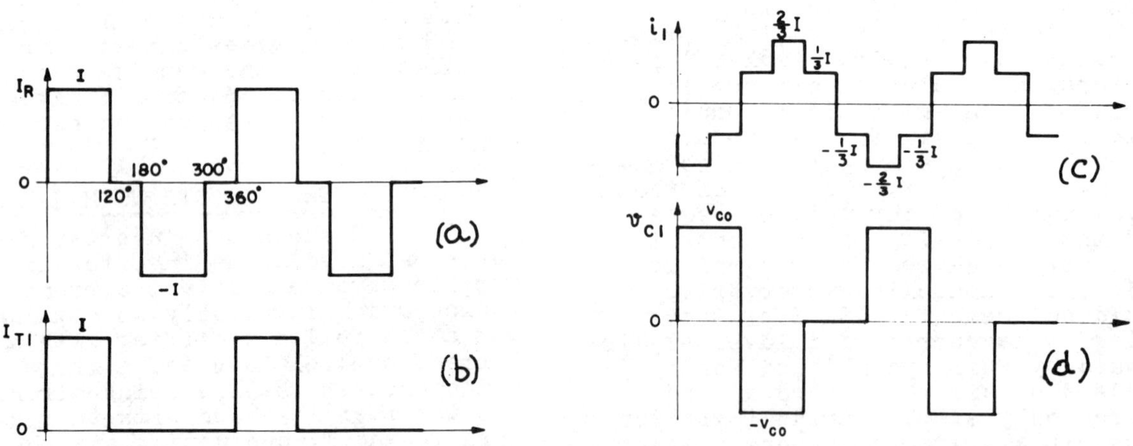

Fig. 2. Current and voltage waveforms of Fig. 1: (a) Line current I_R. (b) Current through thyristor T_1. (c) Current in phase 1 of motor. (d) Voltage of capacitor C_1.

The voltages of the capacitors in the positive leg are

$v_{c1} = V_{co}$, $v_{c5} = -V_{co}$ and $v_{c3} = o$.

T_3 is now fired to commutate T_1. Figures 4 and 5 show the intermediate stages of commutation. Figure 6 shows the situation after the commutation is over and has thyristors T_3 and T_2 conducting. The load currents in Fig. 6 are,

$i_2 = -\frac{2}{3}I$ and $i_1 = i_3 = \frac{1}{3}I$.

It is assumed that T_1 is turned off instantaneously as soon as T_3 is fired. The diode D_5 is assumed to be reverse biased throughout the commutation and the conditions to be met for this are investigated later. When T_3 is fired, D_3 will be reverse biased to start with and the d.c.

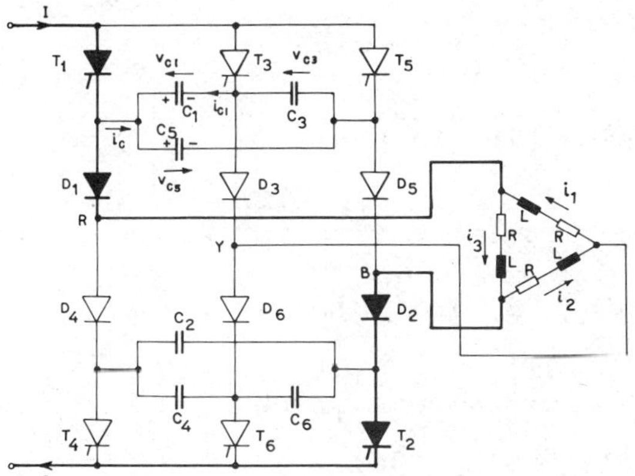

Fig. 3. Current-fed inverter with R-L load; thyristors T_1 and T_2 conducting.

link current will be conducted to the load through D_1 only. Let

t_1 = the duration for which D_1 alone conducts. (1a)

Later D_3 would conduct. The commutation is over when D_1 stops conducting. Let

t_2 = the duration for which D_1 conducts after D_3 starts conducting. (1b)

Therefore, the total commutation period, $t_c = t_1 + t_2$. (1c)

To complete the analysis, it would be necessary to compute also the reverse bias time of T_1 and the variation of the capacitor voltage v_{c1} responsible for the commutation. Once v_{c1} is determined, v_{c3} and v_{c5} could be determined easily.

2.1 Determination of t_1

The reverse bias voltage across D_3 is given by

$$v_{R,D3} = v_{c1} + L\frac{di_1}{dt} + Ri_1. \qquad (2a)$$

Till D_3 conducts, there will be no change in the load currents and hence di_1/dt will be zero. Also $i_1 = -I/3$ till D_3 conducts. Hence,

$$v_{R,D3} = v_{c1} - RI/3. \qquad (2b)$$

The diode D_3 would start conducting when v_{c1} becomes equal to $RI/3$. Also till D_3 conducts, $i_{c1} = (2/3)I$ and the voltage v_{c1} decreases linearly. Hence t_1, the period for which D_3 is reverse biased is given by

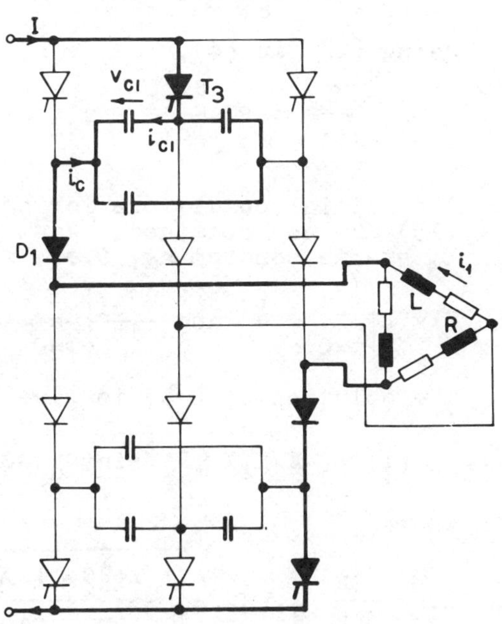

Fig. 4. First stage of commutation interval on firing T_3.

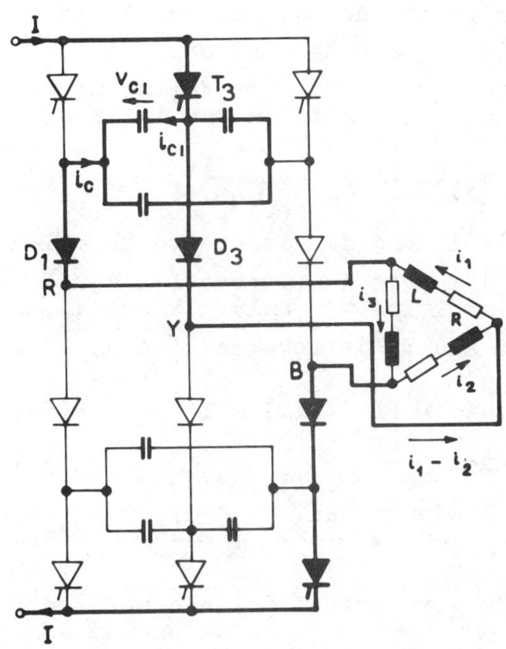

Fig. 5. Second stage of commutation interval.

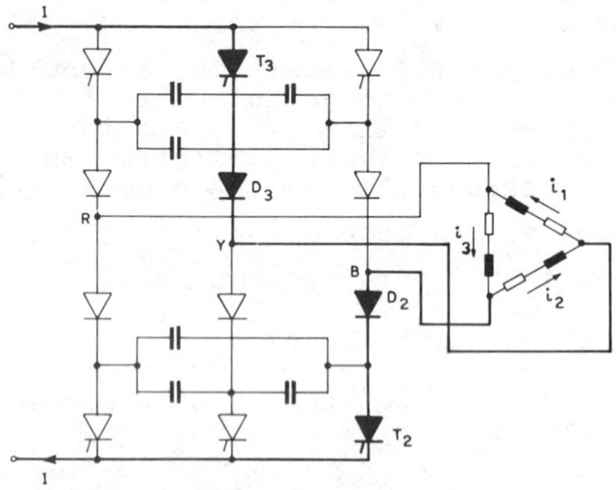

Fig. 6. Circuit at the end of commutation.

$$\frac{1}{C}\,\frac{2}{3}\,I\,t_1 = V_{co} - IR/3$$

from which,

$$t_1 = \frac{3C}{2I}\left(V_{co} - \frac{IR}{3}\right). \quad (3)$$

The determination of t_1 would be complete once V_{co} is obtained in terms of the parameters R,L and C, and the d.c. current I. This is done later.

2.2 Determination of t_2 and v_{cl}

For computing t_2, let the instant at which D_3 starts conducting be reckoned as t = 0. The following equations are valid in the interval t_2.

$$v_{cl} = -(Ri_1 + L\,\frac{di_1}{dt}). \quad (4)$$

In (4), i_1 and di_1/dt are to be eliminated to get a differential equation involving v_{cl} only. Since T_5 is off and D_5 is reverse biased,

$$-i_c + (i_1 - i_2) = I \quad (5a)$$

and hence

$$\frac{di_1}{dt} = \frac{di_2}{dt} + \frac{di_c}{dt}. \quad (5b)$$

For the symmetrical delta connected load,

$$(i_1 + i_2 + i_3) = 0 \quad (6a)$$

and

$$\frac{di_1}{dt} + \frac{di_2}{dt} + \frac{di_3}{dt} = 0. \quad (6b)$$

Also $I = i_3 - i_2$. (7a)

Hence

$$\frac{di_3}{dt} = \frac{di_2}{dt}. \quad (7b)$$

Using (7b) in (6b),

$$\frac{di_2}{dt} = -\frac{1}{2}\,\frac{di_1}{dt}. \quad (8)$$

Using (8) in (5b),

$$\frac{di_1}{dt} = \frac{2}{3}\,\frac{di_c}{dt}. \quad (9a)$$

Integrating (9a),

$$i_1 = \frac{2}{3}\,i_c + k. \quad (9b)$$

At t = 0, $i_1 = -I/3$ and $i_c = -I$. Using these initial conditions in (9b), the solution for i_1 is,

$$i_1 = \frac{2}{3}\,i_c + \frac{I}{3}. \quad (9c)$$

For the delta connected capacitor bank,

$$i_{cl} = \frac{2}{3}(-\,i_c) \quad (10)$$

and hence

$$\frac{dv_{cl}}{dt} = -\frac{1}{C}\,i_{cl} = \frac{2}{3}\,\frac{i_c}{C}. \quad (11)$$

Using (11) in (9c),

$$i_1 = C\frac{dv_{cl}}{dt} + \frac{I}{3}. \quad (12)$$

Using (12) in (4),

$$LC\,\frac{d^2v_{cl}}{dt^2} + RC\,\frac{dv_{cl}}{dt} + v_{cl} = -\frac{RI}{3}. \quad (13)$$

The initial conditions for solving (13) are now obtained. Just before D_3 starts conducting, i.e., at t = 0

$$v_{cl}\Big|_{t=0} = \frac{RI}{3} \text{ and } \frac{dv_{cl}}{dt}\Big|_{t=0} = -\frac{2I}{3C}. \quad (14)$$

The solution of (13) is given by

$$v_{cl}(t) = -K_1K_2I\,\varepsilon^{-\beta t}\sin(\omega_d t-\theta) - \frac{RI}{3}, \quad (15)$$

where

$$K_1 = \frac{2R}{3\lambda\omega_n}, \quad K_2 = \sqrt{\frac{1+2\beta\lambda + \lambda^2\omega_n^2}{1-(\beta/\omega_n)^2}},$$

$$\omega_n^2 = 1/LC, \ \beta = R/2L, \ \omega_d^2 = \omega_n^2 - \beta^2,$$

$$\theta = \tan^{-1}\left(\frac{\lambda\omega_d}{1+\beta\lambda}\right),$$

and $\frac{1}{\lambda} = \frac{1}{RC} - \frac{R}{L}$.

Using (15) in (11),

$$i_c(t) = -\frac{3}{2} C \left\{ K_1K_2 I \varepsilon^{-\beta t} [\omega_d \cos(\omega_d t-\theta) - \beta \sin(\omega_d t-\theta)] \right\}. \quad (16)$$

It can be checked that (16) satisfies the initial condition $i_c(0) = -I$. The diode D_1 stops conducting as soon as $i_c(t)$ tries to become positive. Hence, the interval t_2 is given by

$$\omega_d \cos(\omega_d t_2-\theta) - \beta \sin(\omega_d t_2-\theta) = 0. \quad (17)$$

The total commutation interval, t_c is given by $t_c = (t_1+t_2)$ where t_1 is given by (3) and t_2 by the solution of (17).

2.3 Reverse Bias Time of Thyristor, t_R

The capacitor voltage v_{c1} has an initial value of V_{co} and keeps decreasing. The thyristor is reverse biased as long as v_{c1} is positive. Hence the reverse bias period is given by

$$t_R = t_1 + t_\mu \quad (18a)$$

where t_μ is obtained by equating (15) to zero. In any practical case, t_μ would be small compared to t_1, and t_R can safely be approximated as

$$t_R \simeq t_1 = \left(V_{co} - \frac{RI}{3} \right) \frac{3C}{2I}. \quad (18b)$$

2.4 Determination of V_{co}

From (15) the capacitor voltage v_{c1} at the end of commutation is given by

$$v_{c1}(t_2) = -K_1K_2 I \varepsilon^{-\beta t_2} \sin(\omega_d t_2-\theta) - \frac{RI}{3}. \quad (19)$$

In steady state this should be equal to $-V_{co}$. Hence

$$V_{co} = K_1K_2 I \varepsilon^{-\beta t_2} \sin(\omega_d t_2-\theta) + \frac{RI}{3}. \quad (20)$$

The voltage v_{c1} varies in a linear manner upto t_1 and as a damped sinewave in the interval t_2. The voltage of C_1 decreases by $2V_{co}$ during the commutation and hence the voltage of C_3 and C_5 should each increase by V_{co} at the end of commutation since these capacitors form a closed circuit. Thus at the end of commutation

$$v_{c3} = 0+V_{co} = V_{co} \text{ and}$$

$$v_{c5} = -V_{co} + V_{co} = 0.$$

The voltage v_{c1} and the load currents during commutation are shown in Fig.7. The line voltage V_{RY} (rise from line R to line Y) is the negative of v_{c1} and is also shown in Fig. 7. The other line voltages are also sketched in the figure.

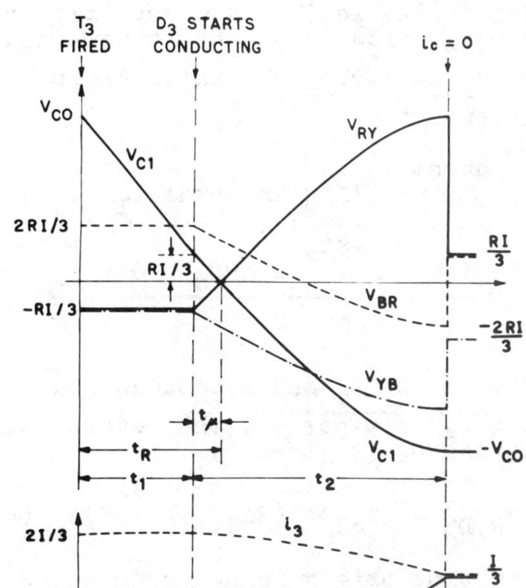

Fig. 7. Waveforms of currents and voltages during commutation.

2.5 Conditions for Proper Operation of the Inverter

The proper operation of the inverter and the entire analysis given so far is based on the assumption that D_5 is reverse biased for the whole of the commutation period. The reverse bias of D_5 for the two periods, t_1 and t_2 is considered separately.

Period t_1: D_1 alone conducting. The reverse bias voltage across D_5 is given by

$$v_{R,D5} = -v_{c5} - (Ri_3 + L\frac{di_3}{dt})$$

$$= -v_{c5} - \frac{2RI}{3} \quad (21)$$

since $i_3 = 2I/3$ during the entire period, t_1. Initially $v_{c5} = -V_{co}$. Since v_{c5} decreases in magnitude during this interval, the reverse bias is least at the end of this interval. At the end of t_1,

$$v_{c5}\Big|_{t_1} = -V_{co} + \frac{V_{co} - IR/3}{2}. \quad (22)$$

Using (22) in (21) and substituting for V_{co} from (20), the lowest value of $v_{R,D5}$ during the interval is given by

$$\text{Lowest } v_{R,D5}\Big|_{\text{Interval } t_1} =$$

$$\frac{K_1K_2I\,\varepsilon^{-\beta t_2}\sin(\omega_d t_2-\theta)}{2} - \frac{RI}{3}. \quad (23)$$

Period t_2: D_1 and D_3 conducting. During this period, the reverse bias across D_5 is given by

$$v_{R,D5} = v_{c3} + (Ri_2 + L\frac{di_2}{dt}). \quad (24)$$

The time origin is now counted as the instant at which D_3 starts conducting. Since v_{c3} at any instant should be equal to its initial value plus one half the change of v_{c1} upto that instant

$$v_{c3}(t) = v_{c3}\Big|_{t=0} + \frac{v_{c1}(0) - v_{c1}(t)}{2} \quad (25a)$$

$$= \left(0 + \frac{V_{co} - IR/3}{2}\right) + \frac{(IR/3) - v_{c1}}{2}$$

$$= \frac{V_{co} - v_{c1}}{2}. \quad (25b)$$

Using (6a), (7a) and (9c),

$$i_2 = -\frac{2I}{3} - \frac{i_c}{3}. \quad (26)$$

From (8) and (9a)

$$\frac{di_2}{dt} = -\frac{1}{3}\frac{di_c}{dt}. \quad (27)$$

Using (26), (27), (11) and (13)

$$Ri_2 + L\frac{di_2}{dt} = -\frac{RI}{2} + \frac{v_{c1}}{2}. \quad (28)$$

Using (25b) and (28) in (24)

$$v_{R,D5}\Big|_{\text{Interval } t_2} = \frac{V_{co}}{2} - \frac{RI}{2}. \quad (29a)$$

It is interesting to note that this is a constant. Substituting for V_{co} from (20)

$$v_{R,D5}\Big|_{\text{Interval } t_2} =$$

$$\frac{K_1K_2I\varepsilon^{-\beta t_2}\sin(\omega_d t_2-\theta)}{2} - \frac{RI}{3}. \quad (29b)$$

The expression on the R.H.S. of (29b) is the same as that of (23). Hence this expression must be positive for successful operation of the inverter. The reverse bias voltage being considered here, in general, refers to the diode in series with the non-conducting thyristor of the leg in which commutation is taking place.

3. DESIGN OF INVERTER FOR R-L LOAD

3.1 Current Rating of Thyristors and Diodes

Let the rms value of the line current of the load be I_L. From the waveform given in Fig. 2a, $I_L = \sqrt{(2/3)}\ I$. From Fig. 2b the rms value of the thyristor current is given by

$I_{rms,thyristor} = \frac{I}{\sqrt{3}}$. Eliminating I,

$$I_{rms,thyristor} = I_{rms,diode} = 0.707 I_L \quad (30)$$

3.2 Voltage Rating of Thyristors

It is easily seen that the peak forward blocking voltage as well as the peak reverse voltage on the thyristor is each equal to the maximum capacitor voltage. Hence the voltage rating of the thyristor is equal to V_{co}.

3.3 Voltage Rating of Diodes

The peak reverse voltage across a diode occurs just before the thyristor in the same leg is fired and is given by $(V_{co} - IR/3)$. This could be taken as V_{co} itself for a safe design.

3.4 Turn-off Time of Thyristors

The reverse bias time t_R is given by (18b). Hence the turn-off time of the thyristor chosen should be smaller than this, say by a factor of 2. The turn-off time will usually be large enough to permit the use of non-inverter grade thyristors.

3.5 Choice of C

The peak capacitor voltage is V_{co} and from Fig. 2d the rms value is given by 0.816 V_{co}. The choice of the capacitance value is closely linked with its voltage rating which is the same as the voltage rating of the thyristor. The equations here being of transcendental nature, it is difficult to see the effect of the variation of C on V_{co}. It is not difficult to imagine that a larger C would give a lower V_{co} and a higher reverse bias time. A lower value of thyristor voltage and a higher value of turn-off time are desirable. However it is not possible to increase the value of C indefinitely. With increase in C, the reverse bias voltage across D_5 given by (29b) would become less and it may become negatime with a sufficiently large value of C leading to a total failure of the inverter. Table 1 shows the interdependence of the various quantities for a numerical example and also brings out the failure of the inverter with too large a value of C.

TABLE 1. NUMERICAL EXAMPLE

R = 5 ohms, L = 25mH, I = 10A.

C μf	V_{co} volts	t_R μ sec.	$V_{R,D5}$ volts
1	1046	68.6	500
5	460	148.0	205
8	356	181.0	153
15	260	246.0	106
1000	41	1667.0	-4*

*Inverter fails for this value of C.

4. EQUATIONS FOR AN IMPROVED MODEL OF MOTOR LOAD

The model given by Farrer and Miskin [5] for the motor can be improved by including resistance in each phase of the model. The equations for such an improved model can be written by inspection from consideration of the fact that the equations developed for R-L load in this paper with R tending to zero would become the same as the ones given in [5] with the sinusoidal voltages in the model set equal to zero. This fact is established for the maximum capacitor voltage, V_{co} and could be done for other equations also.

From the equations given in [5], the maximum capacitor voltage with motor load for the model of Farrer can be written as

$$V_{co,motor} = E \sin \varphi + \frac{2I}{3\omega_n C}. \quad (31)$$

Setting $E \sin \varphi$ equal to zero,

$$V_{co} = \frac{2I}{3\omega_n C}. \quad (32)$$

From (20), V_{co} for R-L load is given by

$$V_{co} = K_1 K_2 I \,\varepsilon^{-\beta t_2} \sin(\omega_d t_2 - \theta) + \frac{RI}{3}. \quad (33)$$

It can be easily checked that as R tends to zero,

$$\frac{R}{\lambda} \rightarrow \frac{1}{C}, \quad \beta \rightarrow 0, \quad \omega_d \rightarrow \omega_n,$$

$$\theta \rightarrow 0, \; (\omega_d t_2) \rightarrow \pi/2 \text{ and } K_2 \rightarrow 1.$$

Substituting these in (33),

$$V_{co} = 2I/3\omega_n C \quad (34)$$

which is the same as (32) and hence establishes the fact.

The equations of the improved model can therefore be written by including the effect of the sinusoidal voltages of the motor model in the equations developed for R-L load. The maximum capacitor voltage would be given by

$$V_{co} = K_1 K_2 I \,\varepsilon^{-\beta t_2} \sin(\omega_d t_2 - \theta) + \frac{RI}{3} + E \sin \varphi \quad (35)$$

where t_2 is still the solution of (17). The diode reverse bias voltage that is required to be positive throughout the commutation is given by

$$v_{R,D5} = \frac{1}{2} K_1 K_2 I \epsilon^{-\beta t_2} \sin(\omega_d t_2 - \theta) - \frac{RI}{3} - 0.866E. \quad (36)$$

The peak reverse voltage across a diode is

$$V_{R,Dmax} = K_1 K_2 I \epsilon^{-\beta t_2} \sin(\omega_d t_2 - \theta) + 2E \sin \varphi. \quad (37)$$

The general expression for t_R is quite complex and this would also depend on the relative values of $RI/3$ and $E \sin \varphi$. For the case $RI/3 < E \sin \varphi$, t_R would be less and gives a safer design. Thus

$$t_R = V_{co} \frac{3C}{2I} \quad (38)$$

where V_{co} is given by (35).

Equations (35) to (38) could be used for designing the inverter.

5. EXPERIMENTAL VERIFICATION OF RESULTS FOR R-L LOAD

The delta connected R-L load for which experimental results were obtained has the values:

Resistance per phase = 6.5 ohms
Inductance per phase = 45 mH.

The commutating capacitors of the inverter are rated 22 μF. The results have been verified for two cases: one for a d.c. link current of 10 A and the other for 20 A.

Case (i); d.c. link current of 10 A.
Theoritical values of V_{max}, $v_{R,D5}$, t_R have been calculated using (20), (23) and (18b). These are, $V_{max} = 287.7$V, $v_{R,D5} = 111.3$ V, $t_R \simeq t_1 = 0.88$ msec. The total commutation period, $t_c = t_1 + t_2 = 2.52$ msec.

Case (ii); d.c. link current of 20A.
The calculated values for this case are,

$V_{max} = 574.4$ V, $v_{R,D5} = 222.6$ V,

$t_R = 0.88$ msec, and $t_c = 2.52$ msec.

From the design equations of section 2 as well as the values just calculated, the following is evident: for a given load, t_1, t_2, t_R and t_c are independent of the load current and inverter frequency; the values of V_{max} and $v_{R,D5}$ are proportional to the load current but are independent of the inverter frequency. Therefore the experimental results can be obtained for any convenient frequency of operation. The inverter was operated at different frequencies and the above was found true, but all these results are not tabulated. Table 2 shows the theoritically calculated values and the experimental results for two different link currents for an inverter frequency of 7 Hz.

TABLE 2. EXPERIMENTAL VERIFICATION

		Theoritical	Experi-mental
I = 10A	V_{max}	287.7 V	280 V
	$v_{R,D5}$	111.3 V	100 V
	t_R	0.88 ms	1.0 ms
	t_c	2.52 ms	2.7 ms
I = 20A	V_{max}	574.4 V	560 V
	$v_{R,D5}$	222.6 V	200 V
	t_R	0.88 ms	1.0 ms
	t_c	2.52 ms	2.7 ms

It can be seen that the experimental results agree well with the theoritical values. The waveforms recorded for a d.c. link current of 10A are shown in Figs. 8a to 8d. It

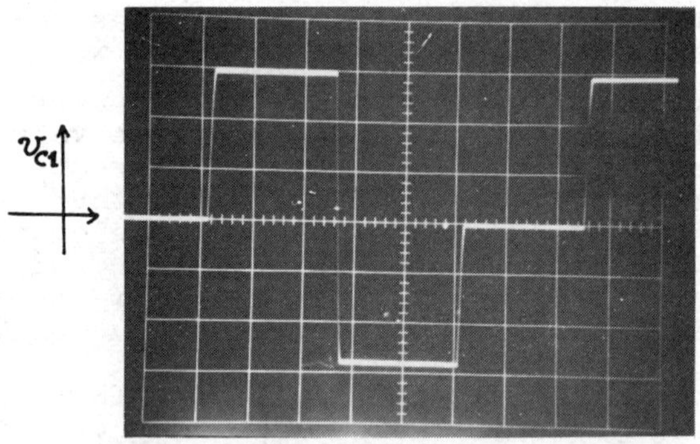

(a)

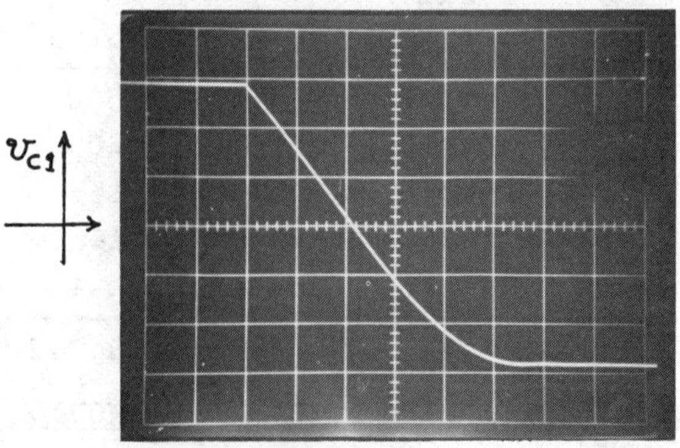

(b)

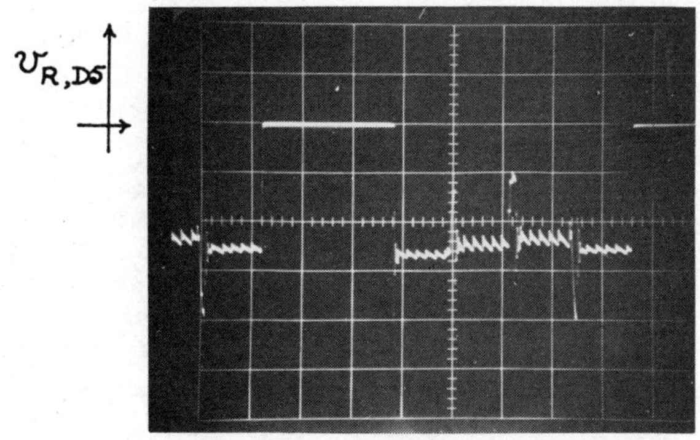

(c)

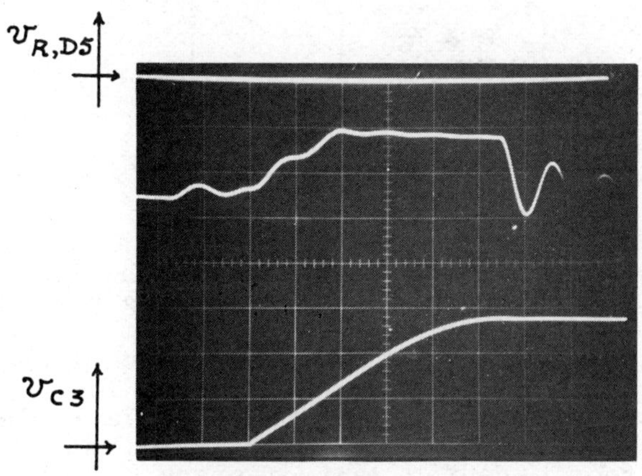

(d)

Fig. 8.

(a) Capacitor voltage v_{c1}(1 BIG DIV = 20 ms).

(b) v_{c1} with expanded time scale to show its variation during commutation (1 BIG DIV = 0.5 ms).

(c) Diode reverse bias voltage, $v_{R,D5}$ (1 BIG DIV = 20 ms).

(d) $v_{R,D5}$ and v_{c3} with expanded time scale to show their variation during commutation (1 BIG DIV = 0.5 ms).

(Y-axis scale for all - 1 BIG DIV = 100V)

may be recalled that it was shown in section 2 that the minimum value of $v_{R,D5}$ occurs at the end of the first commutation interval (t_1) and remains constant at this value during the second commutation interval (t_2). Figure 8d shows on the upper trace $v_{R,D5}$ during the commutation of T_1 and on the lower trace the variation of v_{c3} during the same period. It can be seen from the traces that $v_{R,D5}$ remains constant at its lowest value during the second commutation interval.

6. CONSTRUCTIONAL FEATURES OF THE INVERTER BUILT

The block diagram of the inverter constructed and used for experimental verification is shown in Fig. 9. Before switching on the load, the

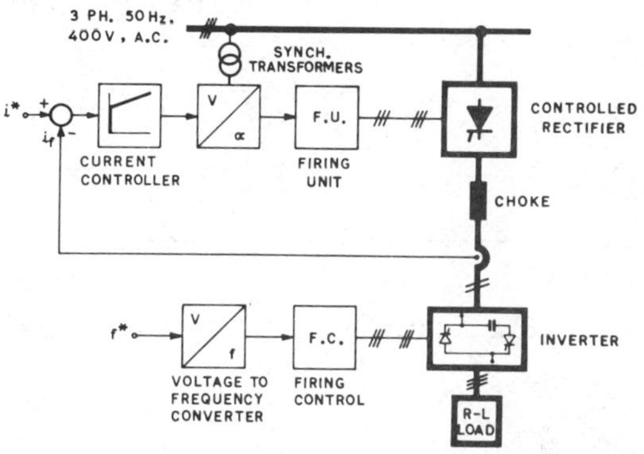

Fig. 9. Block diagram of the inverter.

commutating capacitors are to be precharged to enable the inverter take a step load. The power and control circuits are to be energised with a time delay for proper operation of the inverter. To enable the self starting of the converter and inverter, their thyristors are to be fired such that two thyristors, one on the positive leg and the other on the negative leg, receive firing pulses simultaneously. Figure 10 explains this to certain extent and also shows the sequence in which the control, precharging, power and load circuits get energised once the 'ON' button is pressed.

A photograph of the unit built is shown in Fig.11. It can be seen that the frame consists of two swinging units, one for the control circuit (upper portion) and the other for the power circuit (lower portion). The power frame itself can be opened into two separate parts, outer and inner, after it is swung out from the main frame. The outer part contains quick acting contactors, the converter and inverter thyristors and their pulse transformers. The inner part contains the saturable reactors for di/dt protection, the diodes and the commutating capacitors of the inverter. This arrangement gives good accessibility of the different terminals for measurement as well as repair work.

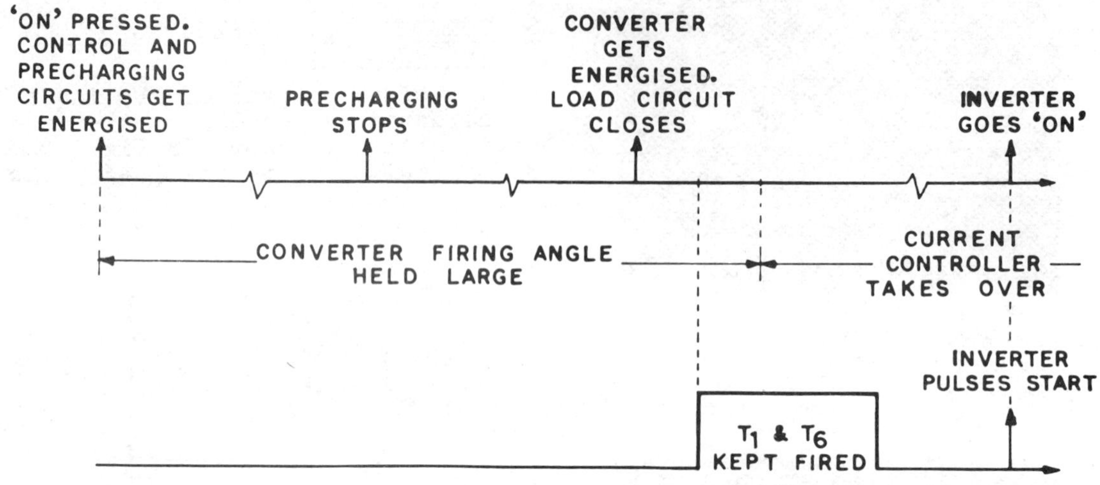

Fig. 10. Switching sequence of power and control circuits.

Fig. 11. Photograph of the inverter.

7. CONCLUSIONS

The analysis and design of a current-fed inverter for a R-L load is presented. The validity of the analysis is established by the close agreement obtained between the theoritical and experimental results. Farrer in his analysis [5] assumes that the diode in series with the non-conducting thyristor of the leg in which commutation is taking place, is reverse biased throughout the commutation period. This assumption is proved here for R-L load. The authors have proved this for Farrer's case also but have not given the proof here for want of space. Using the analysis developed here, the mathematical model given by Farrer for the motor is improved and equations for designing an inverter with this improved model are arrived at. It would be interesting to establish the validity of this improved model by comparing theoretical and experimental results. A brief description of the inverter constructed and a photograph of the equipment built are also given.

One may question the use of the analysis given in this paper as quite often one encounters only motor loads in practice. The answer lies in that the inverter and especially the commutating capacitors could be designed for motor load, but one could make sure that the converter, inverter, etc. are operating well by testing the scheme built, first with a R-L load before switching on the motor load. It is also quite possible that there are applications where the inverter is used for R-L loads and for these cases the analysis given here would be of immense value.

ACKNOWLEDGEMENT - The authors wish to thank the authorities of the Indian Institute of Technology, Madras, for providing facilities to carry out this work. Thanks are also due to the staff of the Control Engineering Laboratory, Indian Institute of Technology, Madras, and to Mr. B.S. Krishna Rao and Mr. N. Sundararajan for their help in the construction of the inverter. The second author expresses his most grateful thanks to Prof. Dr.-Ing. J. Holtz, Gesamthochschule Wuppertal, for bringing his attention to the potentialities of the current-fed inverter considered in this paper.

REFERENCES

[1] Von Gottfried Möltgen: Grundlagen einer Theorie des Stromrichters in Drehstrom-Brückenschaltung mit mehrstuffiger L-C-Kommutierung. Siemens-Zeitschrift 43, 680-685 (1969).

[2] Erwin Held, Herwig Klautschek and Lothar Schröder: Simovert single-motor drive for extruders. Siemens Review XXXVIII, 513-516 (1971).

[3] K.P. Phillips: Current source converter for a.c. motor drives. IEEE - Conference Record, 385-392 (1971).

[4] R.B. Maag: Characteristics and application of current source/slip regulated a.c. induction motor drives. ibid., 411-416 (1971).

[5] W. Farrer and J.D. Miskin: Quasi-sine-wave fully regenerative invertor. Proc.IEE 120, 969-976 (1973).

MODIFICATIONS OF COMMUTATION CIRCUITS FOR A THREE-PHASE INVERTER, CONNECTED WITH THE NEUTRAL CONDUCTOR

Degenhard Pohl

Institut für Stromrichtertechnik und Elektrische Antriebe RWTH Aachen

ABSTRACT

The self-commutated three-phase bridge connection with a common turn-off device using the neutral conductor of the three-phase system is chosen as an example to illustrate modified commutation circuits. These can extend the possibilities of operation of the three-phase bridge, although they may have other suitable applications too. A particular turn-off device is first described for the use of this bridge as inverter for three-phase drive systems, which enables this inverter to be used for multiple pulse current control. Subsequently various commutation circuits are described which are to avoid a forced charge reversal of a large highly energized commutation capacitor. Instead of the normally used resonant circuit by which the capacitor will be recharged with a high current amplitude and therefore with high losses, a small additional auxiliary capacitor will be used in these new circuits to reverse the voltage in the commutation circuit with a lower current and with lower losses. Moreover these circuits will allow a replacement of the large expensive turn-off capacitor by a cheaper d.c. capacitor.

INTRODUCTION

The interest in drive systems with three-phase machines has grown in the last years in the same measure as it has become possible to produce powerful and reliable inverters with improved components as semiconductors and commutation capacitors. Today inverter-fed three-phase drives hold a safe position in solving driving tasks. This interest is based on the possibility of producing machines with a small volume, a good ratio of power to weight and a good power efficiency. The three-phase asynchronous machine in particular is more appropriate for mass production and therefore can be cheaper than, for example, a d.c. machine. Drive systems with asynchronous machines have been built or suggested for a great number of applications, e.g. as drives in rolling or spinning mills and as traction drives in locomotives; even as drive systems of electric road vehicles they cannot be excluded from the outset.

EXAMPLES OF CURRENT-FED INVERTER CIRCUITS

Three-phase asynchronous machines for driving applications must be fed by a self-commutated convertor in the total range of operation. This convertor normally consists of a feeder device, for instance a rectifier or a d.c. chopper, and a three-phase inverter. Systems with a current source d.c. link seem to be very important in this case and they are well-known in the literature. Proposals for various inverter circuits have been published, e.g. circuits with one turn-off device for each phase (1), circuits with phase-sequence-commutation (2, 3), circuits with a common turn-off device at the inverter input (4) and circuits with a commutation device using the star-point of the three-phase system (5). Inverters with phase-sequence commutation in particular have gained a remarkable popularity in the last years because they require the smallest number of controllable semiconductors. The force-commutated inverter bridge with commutation using the neutral conductor of the three-phase machine (Fig. 1), however, combines the advantages of a small number of semiconductors and commutation elements with small power losses because, unlike phase-sequence commutated inverters, it requires no series diodes. Only two series connected semiconductors carry load current simultaneously.

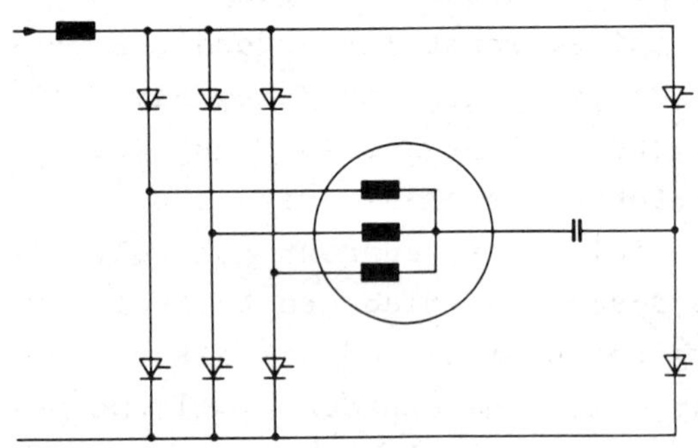

Fig. 1 Basic inverter diagram

One main problem of all these constant-current-inverters is the total commutation time t_c, which must be short in order to achieve a high limit frequency f_{max}:

$$t_c < \frac{1}{6\, f_{max}}$$

Because the commutation circuits of all these inverters include at least one winding of the machine, the single or even the double leakage inductance of the machine will always be active during commutation. Asynchronous machines can have a relatively large leakage inductance. Therefore commutation capacitors with high energy are required. This is especially true if a high current must be commutated. That means, a capacitor with a high voltage or a large capacitance is necessary. High capacitor voltage requires semiconductors with high blocking voltage (series connection if necessary) and a well insulated construction. Large capacitance implies large commutation time and low limiting frequency, especially if a low current is flowing. Therefore it seems necessary for all drives with a wide control range for speed and current to recharge the capacitor in a short time with an additional circuit, if inexpediently high voltages are to be avoided. From this point of view, inverters with a common turn-off device seem to be advantageous, because in this case only one additional circuit for a forced recharging of the single capacitor is required. In the simplest case such a circuit consists of a reactor with a small inductance and two antiparallel connected thyristors, as it is shown in Fig. 2. This circuit has already been suggested in (5) for the inver-

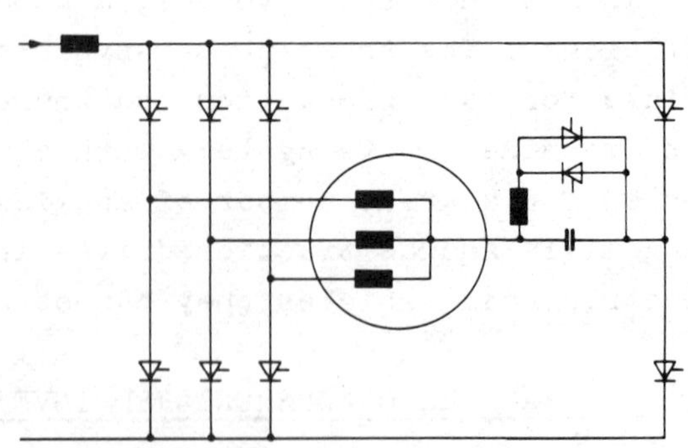

Fig. 2 Inverter with an additional device for a forced charge reversal of the capacitor

ter with commutation using the neutral conductor. In this case it is especially important to make commutation as fast as possible, because the three-phase machine biased via the star-point will be loaded with zero-phase sequence current harmonics too. The amplitudes of these harmonics are the smaller the shorter the commutation time can be made. For this inverter type the total commutation time t_c can be estimated by

$$t_c > 2\sqrt{L_\sigma C}\,\arcsin\left(\frac{I_L}{(U_{Co}+\hat{U}_1)}\sqrt{\frac{L_\sigma}{C}}\right) + 2\,t_f + 2\,\hat{U}_1\,\frac{C}{I_L}$$

t_c total commutation time

L_σ leakage inductance per phase of the machine

C commutation capacitance

U_{Co} initial capacitor voltage

$\hat{U}_1$ amplitude of induced machine voltage

t_f turn-off time of the main thyristors

I_L load current

This calculation is based on the assumption, that the capacitor voltage U_{Co} is controlled as a function of the current I_L in such a way, that it will correspond with the induced machine voltage $\hat{U}_1$ just at the end of the turn-off time of the main thyristor. An additional inductive circuit connected with the capacitor will shorten only the last part of this formula, but this part represents a significant portion of the total commutation time chiefly at small current. It can be limited to

$$t_u < \pi\sqrt{C\,L_u} \qquad (t_u\text{: recharging time, } L_u \text{ recharging inductance})$$

Moreover the total commutation time includes the double turn-off time of the main thyristors. Therefore the use of high speed thyristors in the inverter bridge may be advantageous, although, in principle, the circuit also performs when slow thyristors are used.

ABILITY FOR MULTIPLE PULSE CONTROL

Constant-current-inverters are normally switching with the basic frequency of the three-phase system. The current harmonics produce oscillating torques which are usually smoothed by the inertia factor and therefore are not noticable at high speed. Operating at low speed, however, they may cause disturbance in the drive system because there are points where the oscillating frequencies correspond to the resonant frequency of mechanical components. This can be avoided by pulsing the motor current (6). The simple inverter with commutation using the neutral conductor is not suitable for multiple pulse current control because both polarities of the capacitor voltage are used alternately for turn-off operation in either commutation group of the bridge. However, multiple turn-off operations in one half-bridge are required when pulsing the current. An addition of three thyristors to the commutation device enables this inverter to choose freely between multiple pulse or current-block control (Fig. 3). For current-block control a turn-off operation in the upper half-bridge is carried out by firing the thyristors T_1 and T_4, in the lower half-bridge with T_2 and T_6. In case of multiple pulse control several turn-off operations can be successively carried out in the upper half-bridge, for example, by firing T_1 and T_4 if the commutation capacitor is recharged by firing T_7 between every single turn-off operation. The same procedure is possible with T_5 and T_3 for turn-off operations in the lower half-bridge. A forced recharging of the capacitor during the commutation can be started by firing T_3 or T_4.

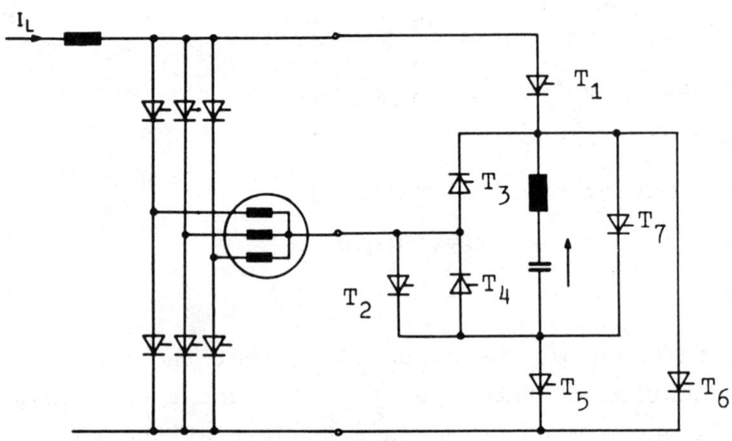

Fig. 3 Inverter with a commutation circuit suitable for current-block or multiple pulse control

This control method requires a capacitor, only one voltage polarity of which is used for turn-off operations (shown in Fig. 3). After each commutation the voltage has the reverse polarity. The capacitor must be prepared for the next commutation by recharging it in a resonant circuit with thyristor T_7 conducting. Thus the capacitor's operating frequency is six times higher than the fundamental frequency, but this fact causes no change of the capacitor design because multiple pulse control is only used at a low basic frequency.

AVOIDANCE OF THE FORCED CAPACITOR RECHARGING

It has been shown in the last but one paragraph that a forced recharging of the capacitor will be necessary wherever a high frequency limit and a wide range of current control are required. The transport of the energy stored in the large capacitor via a resonant circuit with a very short ring time produces a current pulse with a very high amplitude which results in high power losses in the semiconductors, the reactor and the capacitor. The thyristors in this resonant circuit must be able to carry a load current with high rms value, which can even be higher than that of the main thyristors. If a high power efficiency is desired, for example in cases with a limited energy reservoir, it seems to be advantageous to avoid this recharging method. The following will describe a new group of commutation circuits which enable the inverter to operate without these disadvantages. A complete charge reversal of a large highly energized capacitor only for the purpose of reversing the voltage in the commutation circuit is to be avoided. Instead, this large capacitor, which is of course necessary in this case too, has only two tasks left: to deliver reverse blocking voltage during the turn-off time of the main thyristor, and, reconnected with reverse polarity to the commutation circuit, to limit the overvoltage when commutating the current to the next winding of the machine. For this purpose it is necessary to take the constant load current out of this capacitor by commutating it to an additional branch of the commutation circuit. This branch may well be a small additional capacitor. The constant load current will recharge this auxiliary capacitor in a short time thereby reversing the voltage in the commutation circuit. Then the current is to be commutated back to the turn-off capacitor now reverse connected to the commutation circuit. This last current transfer can of course be carried out with a second turn-off capacitor too. In this case these two capacitors with the same capacitance can be used alternately for turn-off operation and voltage limitation when commutating the current to the next winding of the machine. This method of operation probably will result in smaller power losses, because the current in each semiconductor will never be greater than the load current. Another advantage of this procedure is that the large turn-off capacitor does not change its voltage polarity and therefore, instead of an expensive high quality a.c. capacitor, a cheaper d.c. capacitor can be employed. Of course, this advantage will be at the expense of an increased number of semi-

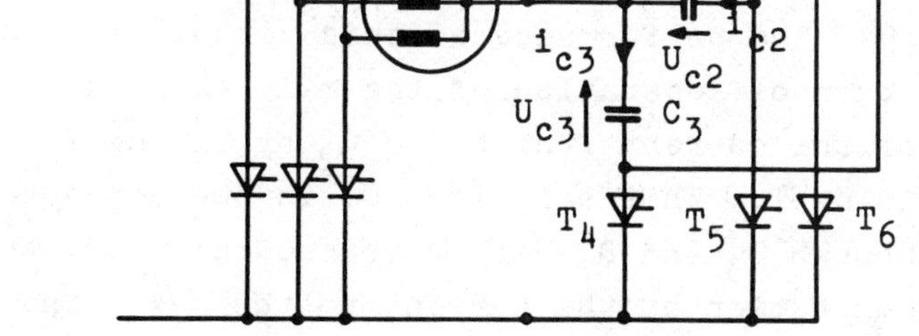

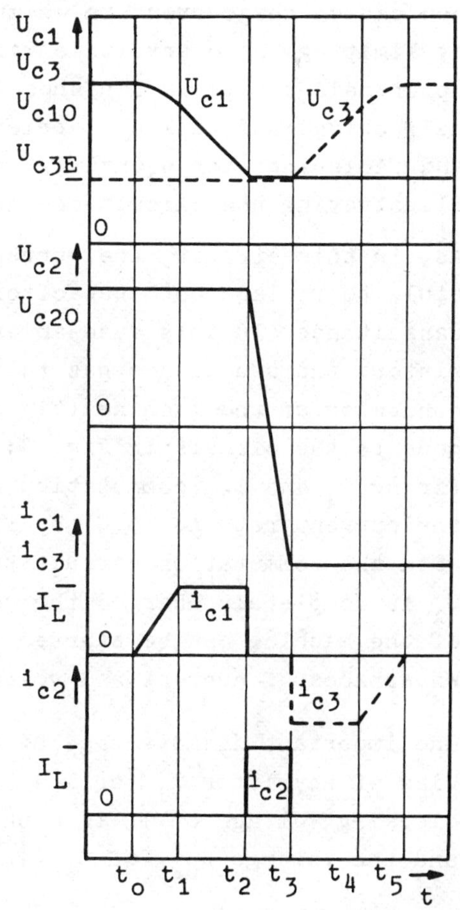

Fig. 4 Commutation circuit with two d.c. turn-off capacitors and an auxiliary capacitor and waveshapes of the voltages and the currents of the capacitors during commutation

conductors and control circuits, but thyristors for a low mean current can be chosen for these circuits. Therefore the cost of such a commutation device will probably not very much exceed that of the normal commutation circuit.

Three circuits operating with the described commutation method will be illustrated in the following paragraphs.

Figure 4 shows the simplest design of such a turn-off device using two large d.c. capacitors C_1 and C_3 with the same capacitance for turn-off. The operation of this circuit will be described in detail for the case when a thyristor of the upper half-bridge is to be turned off. Such a turn-off operation is initiated by firing thyristor T_1 (t_o). Assuming the voltage U_{C10} of the capacitor C_1 to have the polarity shown in Fig. 4 this capacitor will cause a current pulse flowing in reverse direction through the conducting main thyristor thereby turning it off (t_1). Then C_1 will be discharged with the constant load current. At the end of the turn-off time of the main thyristor thyristor T_2 is fired (t_2) causing an immediate commutation of the current from T_1 to T_2, if an appropriate voltage U_{C20} of the auxiliary capacitor C_2 is assumed. The voltage of C_1 has reached its final value U_{C1E}. The charge of the small auxiliary capacitor C_2 will be reversed within a short time by the load current. When the voltage of C_2 has reached its initial value with changed polarity, thyristor T_3 will be fired (t_3). The current will commutate immediately to T_3 because the voltage of capacitor C_3 has kept its final value U_{C3E} of the last turn-off operation in the lower half-bridge. C_3 is recharged by the load current flowing now through T_3 until the next main thyristor is fired (t_4). C_3 now causes the current to commutate to the next winding of the machine thereby limiting the overvoltage with its high capacitance. The commutation in the upper half-bridge is finished (t_5). The next commutation happens in the lower half-bridge and will be carried out in the same way, starting with T_4 and C_3 and finishing with C_1 and T_6. Waveshapes of capacitor voltage and current illustrating the circuit operation are also shown in Fig. 4.

As, in this circuit, the current has been totally taken out of C_1, it is possible to replace both capacitors C_1 and C_3 by only one capacitor with the same capacitance. In this case an artificial mid-point with two additional thyristors and two diodes has to be constructed and connected with the neutral conductor of the machine (Fig. 5). The performance of this circuit is analogous to the circuit in Fig. 4: turn-off operation of the main thyristor by firing T_1 and T_7, commutation of the current from C_1 to C_2 by firing T_2. Then the current returns to C_1 by firing T_3. This capacitor C_1 is now connected with the commutation circuit through T_3 and D_7 with reversed polarity. When C_1 is completely charged the commutation of the current to the next winding of the machine can be started by firing the corresponding main thyristor. Waveshapes of current and voltage of such a commutation are shown in Fig. 5.

One important disadvantage of these circuits (Fig. 4 and 5) is the unfavourable bias of thyristor T_1 (or T_4) concerning blocking voltage. Whereas the maximum blocking voltage of T_1 is equal to the sum of the final voltage U_{C1E} of C_1 and the voltage U_{C20} of C_2 ($|U_{C1E}| + |U_{C20}|$) (unless this value is less than

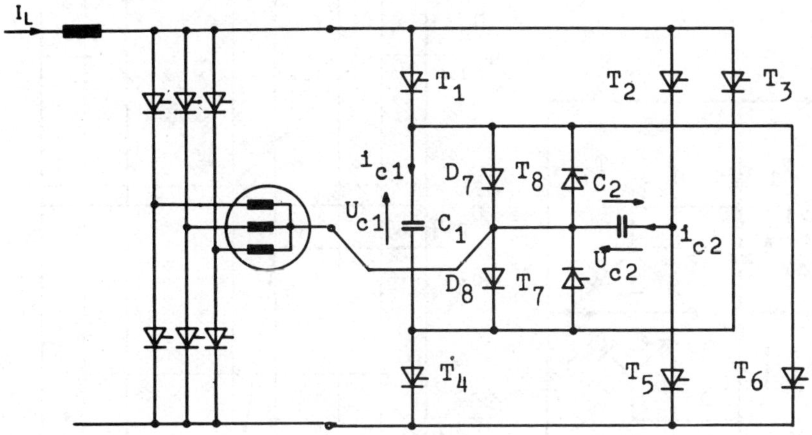

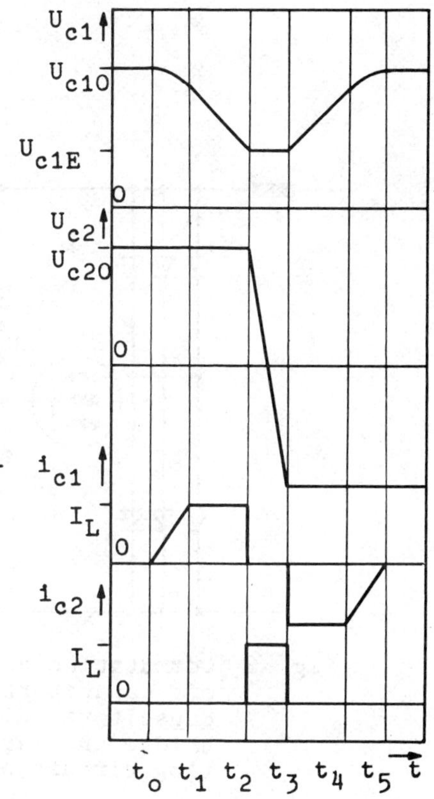

Fig. 5 Commutation circuit with only one d.c. turn-off capacitor and an auxiliary capacitor and waveshapes illustrating circuit operation

($|U_{C10}| + |\hat{U}_1|$)), the turn-off time t_{f1} of T_1 is determined by the difference of these voltages:

$$t_{f1} < \frac{C_2(|U_{C20}| - |U_{C1E}|)}{I_L}$$

If a short commutation time is desired the capacitance C_2 has to be small and T_1 must have a short turn-off time and a high blocking voltage. Therefore it will be an expensive device. This difficulty can be removed by a modified circuit shown in Fig. 6. In this case two turn-off capacitors C_1 and C_3 have to be applied again. The small auxiliary capacitor C_2 now is introduced into a thyristor bridge which is connected with the turn-off capacitors. The circuit's performance will be described for a commutation in the upper half-bridge. The conducting thyristor is turned off by firing thyristor T_1 as described earlier (t_o). At the end of the turn-off time of the main thyristor (t_2), thyristors T_7 and T_2 are fired, thereby turning off thyristor T_1 immediately, if a suitable voltage of C_2 is assumed. Now the load current is flowing through C_2, but also through C_1. When C_2 is completely recharged for the next commutation (t_3), the current is commutated to C_3 by firing T_3, thereby turning off T_2. C_3 has to be charged now, because its voltage still has the final value of the last commutation. Then, the current can be commutated to the next winding of the machine by firing the next main thyristor (t_4). The blocking voltage of T_1 is now determined by $|U_{C20}|$ only or by ($|U_{C10}| + |\hat{U}_1|$), if this value is

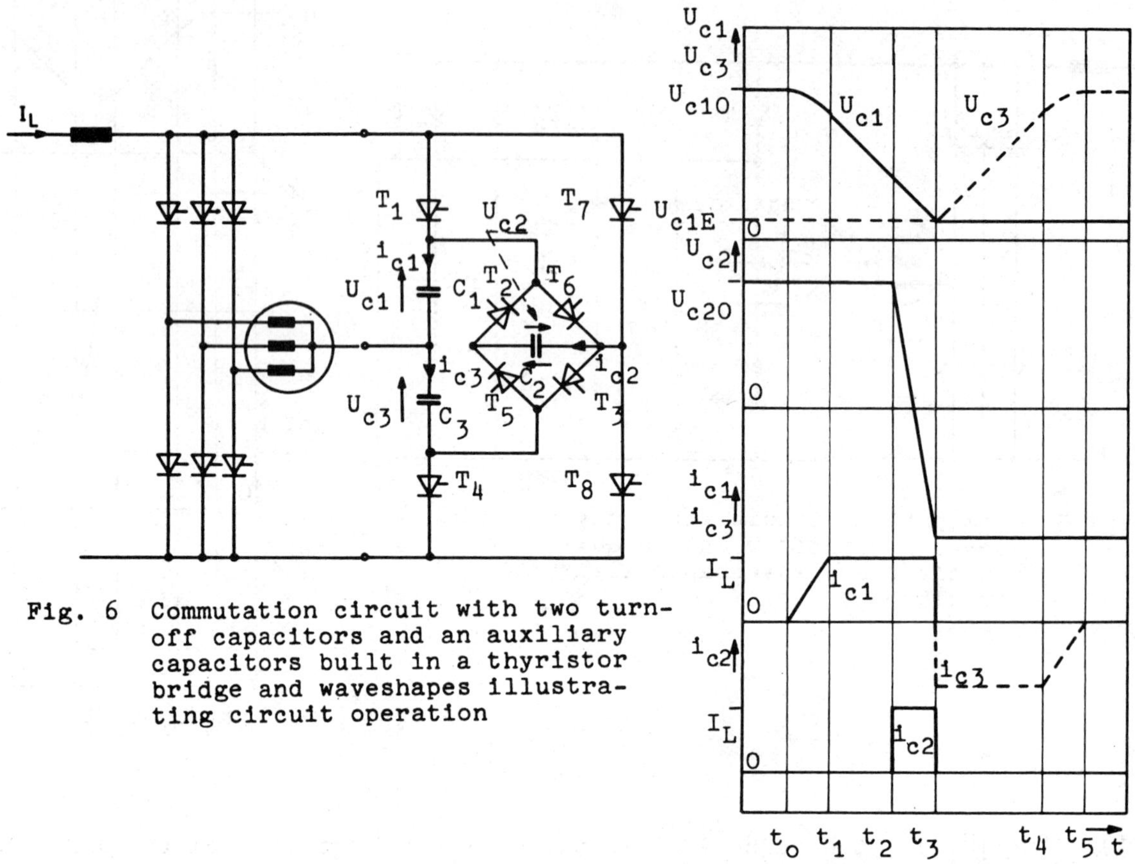

Fig. 6 Commutation circuit with two turn-off capacitors and an auxiliary capacitors built in a thyristor bridge and waveshapes illustrating circuit operation

greater than $|U_{C20}|$. However, the turn-off time can be calculated now by:

$$t_{f1} < \frac{C_2 \; |U_{C20}|}{I_L}$$

and can be a little longer than in the other circuits. But now it is impossible to replace both turn-off capacitors by a only one, because the load current is flowing through the capacitor during recharging C_2, too.

Concluding Fig. 7 shows some oscillograms of waveshapes of capacitor voltage and current during a commutation procedure of the turn-off device shown in Fig. 5. These photographs were taken during a laboratory test of this circuit. The oscillograms in Fig. 7b and 7d show no stepwise commutation of the load current from C_2 back to C_1 when firing thyristor T_3. The reason is a reactor connected in series with T_3 (and T_6). In this way it is possible to control the voltage U_{C20} of C_2 as a function of the current and no additional voltage control device for U_{C2} is necessary.

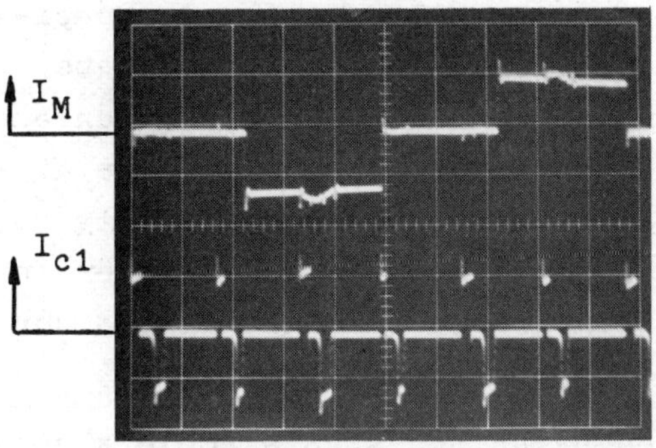

Fig. 7a
Current of the machine I_M
(50 A/div)
Current of the capacitor C_1 : I_{C1}
(50 A/div; time: 2 ms/div)

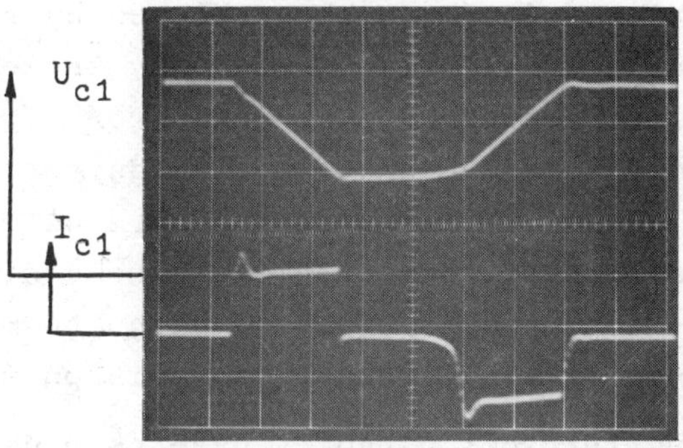

Fig. 7b
Voltage waveshape of the capacitor
C_1 : U_{C1} (100 V/div)
Current of the capacitor C_1 : I_{C1}
(50 A/div; time: 0,2 ms/div)
during one commutation

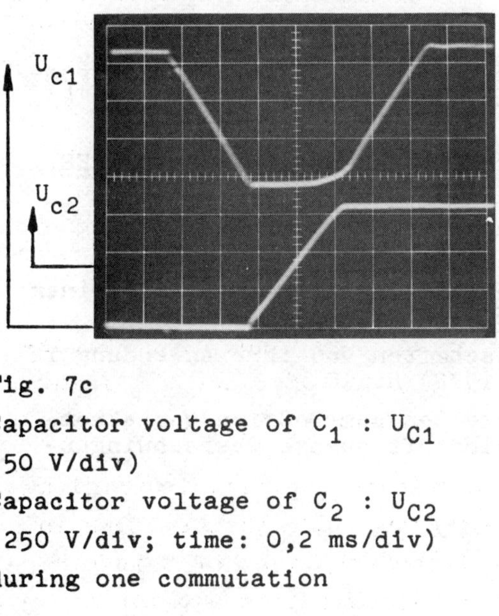

Fig. 7c
Capacitor voltage of C_1 : U_{C1}
(50 V/div)
Capacitor voltage of C_2 : U_{C2}
(250 V/div; time: 0,2 ms/div)
during one commutation

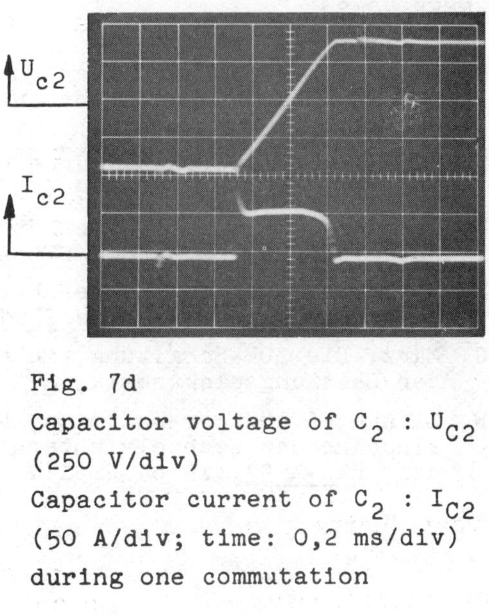

Fig. 7d
Capacitor voltage of C_2 : U_{C2}
(250 V/div)
Capacitor current of C_2 : I_{C2}
(50 A/div; time: 0,2 ms/div)
during one commutation

Fig. 7 Oscillograms of voltages and currents of the capacitors of the commutation device shown in Fig. 5

CONCLUSION

It has been shown that the described commutation circuits with an auxiliary capacitor are suitable to avoid the forced charge reversal of highly energized turn-off capacitors causing high losses. Instead of this, the reversal of the commutating voltage can be achieved by recharging a small additional capacitor with load current. In this way the expensive turn-off capacitor can be replaced by a cheaper d.c. capacitor. These improvements are at the expense of an increased number of semiconductors, which does not seem acceptable at first sight. But the semiconductors will be biased with low currents only so that small thyristors with small heat sinks can be used. If this is taken into account the costs of these circuits may be comparable to those of the normal turn-off device. But this decision depends on the specific case.

These new turn-off circuits have been described by using as an example a commutation device connected with the neutral conductor of the three-phase system, but they are not confined to this type of application. With some insignificant modifications they can be used wherever relatively large inductances are active in the commutation circuit, which require a large highly energized capacitor. If in such a case a short commutation time is to be achieved it seems to be convenient from the point of view of a good power efficiency to use these circuits instead of reversing the charge of the large commutation capacitor with high current and high losses.

REFERENCES

(1) G. Möltgen, Grundlagen einer Theorie des Stromrichters in Drehstrom-Brückenschaltung mit mehrstufiger LC-Kommutierung, Siemens Z. 43, 680 (1969).

(2) K. P. Phillips, Current-source convertor for ac motor drives, IEEE Trans. Ind. Appl. 6, 679 (1972).

(3) D. Schröder, Selbstgeführter Stromrichter mit Phasenfolgelöschung und eingeprägtem Strom, ETZ-A 96 (1975) H. 11.

(4) H. Voelker, Quasistationärer Betrieb eines Synchronmotors über einen Pulswechselrichter, Wiss.Z. TH Ilmenau 22 (1976) H. 4.

(5) G. März, Die ZDB-Schaltung, ihre Eigenschaften und ihre Anwendung in der Leistungselektronik, ETZ-A 93 (1972) H. 10.

(6) W. Lienau, A. Müller-Hellmann, Möglichkeiten zum Betrieb von strom-einprägenden Wechselrichtern ohne niederfrequente Oberschwingungen, ETZ-A 97 (1976) H. 11.

OPTIMAL DESIGN OF THE COMMUTATING CIRCUIT OF FORCED—COMMUTATED 3-PHASE-INVERTERS WITH SEQUENTIAL COMMUTATION

Helmar Fischer

Hochschule der Bundeswehr Hamburg, Fachbereich Elektrotechnik, Holstenhofweg 85, 2000 Hamburg 70, Federal Republic of Germany

Summary

When comparing other circuits with the sequential commutation at 3-phase inverters the latter excels by its minor demand of commutating means; so e.g. no further SCR's are required in addition to the main thyristors. The contribution analyses the process of commutation with regard to the necessary requirements for the commutating means and the voltage and current rating to be exspected for the thyristors and diodes. A procedure for the design of the commutating circuit is proposed.

1. Description of Commutation Process

The circuit of fig. 1 is taken as a basis; firing sequence n1, n2, ..., firing interval $\Delta t = 1/6\ f$ (f: line frequency).

For the further considerations the following simplifications are used:

- the DC-voltage and the load current during the commutating process are constant in time,
- except for the consideration of the freewheeling circuit, the commutating capacitors and chokes as well as the thyristors and diodes are considered to be lossless.

At the viewing time the thyristors n5 and n6 be conductive. At $t = t_o = 0$ the quenching of n5 is effected by firing of n1. The reactor L_1 does not participate in the commutation. The load currents $i_2(t)$ and $i_3(t)$ have only then to be considered in case they show positive sign.

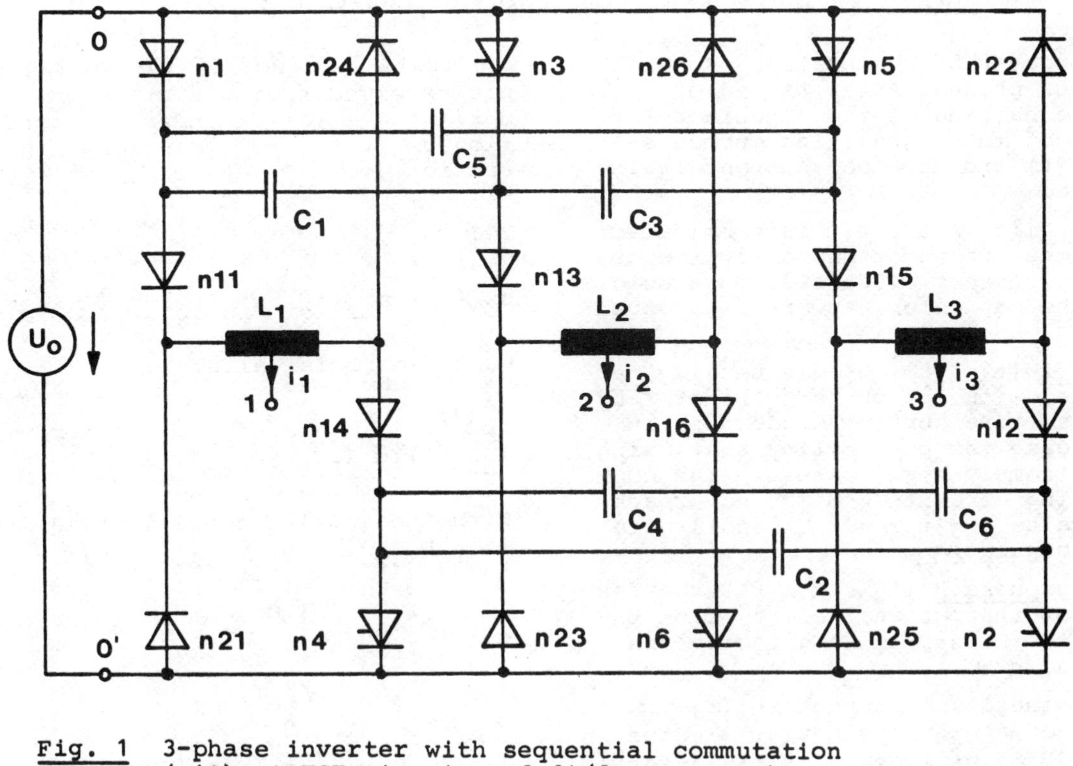

Fig. 1 3-phase inverter with sequential commutation (without TSE-circuit and di/dt-reactors)

It is set:

$$I'_2 = 0{,}5\cdot\left[|i_2(t_o)|+i_2(t_o)\right] ,$$
$$I'_3 = 0{,}5\cdot\left[|i_3(t_o)|+i_3(t_o)\right]$$

For most of the applications (no-load, passive load, reactive current operation etc.) is $I'_2 = 0$. This is assumed for the further consideration.

The commutation with the mentioned assumptions results in the equivalent circuit fig. 2.

the freewheeling circuit n23-L_2-n16--n6 and reaches its final value $i_{L2}(t'_4)$ at the latest with the firing of the thyristor n2 at the instant t_4. The reactor current $i_{L2}(t_4)$ corresponds to the initial value $i_{L3}(t_o)$ of the magnetizing current of the choke L_3.

2. Calculation of the Commutation Process

2.1 Commutation phase 1

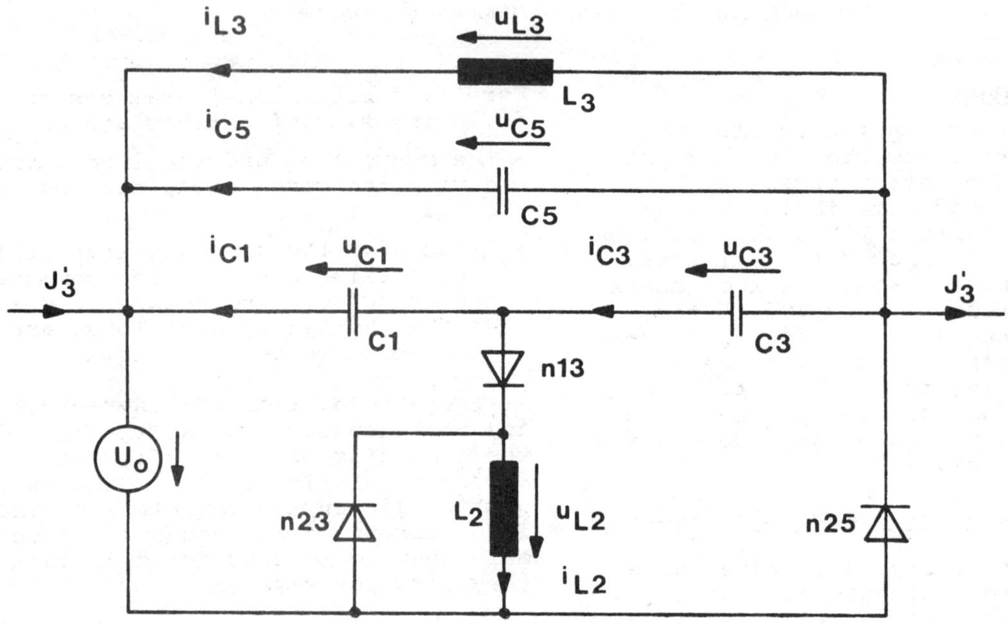

Fig. 2 Equivalent circuit of the commutating circuit

The process of commutation proceeds in four phases. After firing of n1 the recharging of the capacitors by means of the magnetizing currents $i_{L2,3}(t)$ and the load current I'_3 is started.

Phase 1 ($t_o \leq t \leq t_1$) is terminated with the current transfer by the reactive current diode n23; because of n13 the capacitor network is decoupled.

Phase 2 ($t_1 < t \leq t_2$) is usually terminated with the current transfer by the reactive current diode n25. In this case the possibility that owing to extremely high losses in the commutating circuit C_5 might not reach the value $u_{C5}(t_2) = -U_o$ shall here be excepted.

During phase 3 ($t_2 < t \leq t_3$) the feedback of the residual energy from L_3 into the DC supply is effected via n22 and n25.

The magnetizing current $i_{L2}(t_1)$, achieved at the instant t_1, decays in the course of phase 4 ($t_1 < t \leq t'_4 \leq t_4$) in

The initial values of the voltages are received from the statement $u_{C5}(t_o) = (1+a)\cdot U_o$ under the condition $u_{C5}(t_2) = -U_o$, mentioned in section 1:

$$u_{C1}(t_o) = a\cdot U_o \quad (1)$$
$$u_{C3}(t_o) = U_o \quad (2)$$
$$u_{C5}(t_o) = u_{L2}(t_o) = u_{L3}(t_o) = (1+a)\cdot U_o \quad (3)$$

Furthermore is valid:

$$i_{L2}(t_o) = 0 \quad (4)$$
$$i_{L3}(t_o) = I_{L30} = i_{L2}(t_4) \quad (5)$$

With the initial conditions (1) up to (5), $L_{1,2,3} = L$, $C_{1,2,3} = C$,

$$\omega_1 = 1/\sqrt{LC} , \quad (6)$$
$$Z_o = \sqrt{L/C} , \quad (7)$$
$$I_o = U_o/Z_o , \quad (8)$$
$$R_o = \frac{U_o}{I'_3 + I_{L30}} , \quad (9)$$

the currents $i_{L2,3}(t)$ and voltages $u_{L2,3}(t)$ of the equivalent circuit fig. 2 for the period $0 \leq \omega_1 t \leq \pi/2$ are received in close approximation to

$$\frac{u_{L2}(t)}{U_o} = (1+a)\cdot\cos\omega_1 t - \frac{Z_o}{3R_o}\cdot\sin\omega_1 t \ , \quad (10)$$

$$\frac{i_{L2}(t)}{I_o} = (1+a)\cdot\sin\omega_1 t + \frac{Z_o}{3R_o}\cdot(\cos\omega_1 t-1) \ , \quad (11)$$

$$\frac{u_{L3}(t)}{U_o} = (1+a)\cdot\cos\omega_1 t - \frac{2Z_o}{3R_o}\cdot\sin\omega_1 t, \quad (12)$$

$$\frac{i_{L3}(t)}{I_o} = (1+a)\cdot\sin\omega_1 t + \frac{Z_o}{3R_o}\cdot(2\cos\omega_1 t+1) - \frac{I_3'}{I_o} \ . \quad (13)$$

The recorvery time for the thyristor n1 is given by the zero passage of the voltage u_{L3}:

$$t_s = \sqrt{LC}\cdot\arctan\left[\frac{3(1+a)}{2(Z_o/R_o)}\right] \ . \quad (14)$$

The commutation phase 1 ends with $u_{L2}(t_1) = 0$:

$$t_1 = \sqrt{LC}\cdot\arctan\left[\frac{3(1+a)}{Z_o/R_o}\right] \ . \quad (15)$$

Eq. (15), inserted in (11) up to (13) supplies the initial conditions for the commutation phases 2 and 4.

With the help of the equations (3), (12), and (15) now the voltage factor a can be determined out of the condition

$$u_{C1}(t_1) - 0{,}5\cdot\left[U_o+u_{C5}(t_1)\right] = -u_{C5}(t_o)$$

One obtains the combitional equation

$$\left(\frac{Z_o}{R_o}\right)^2 = \frac{3\cdot(1+a)^2\cdot(1-2a)^2}{a\cdot(2-a)} \ . \quad (16)$$

Equation (16) is graphically interpreted in fig. 3.

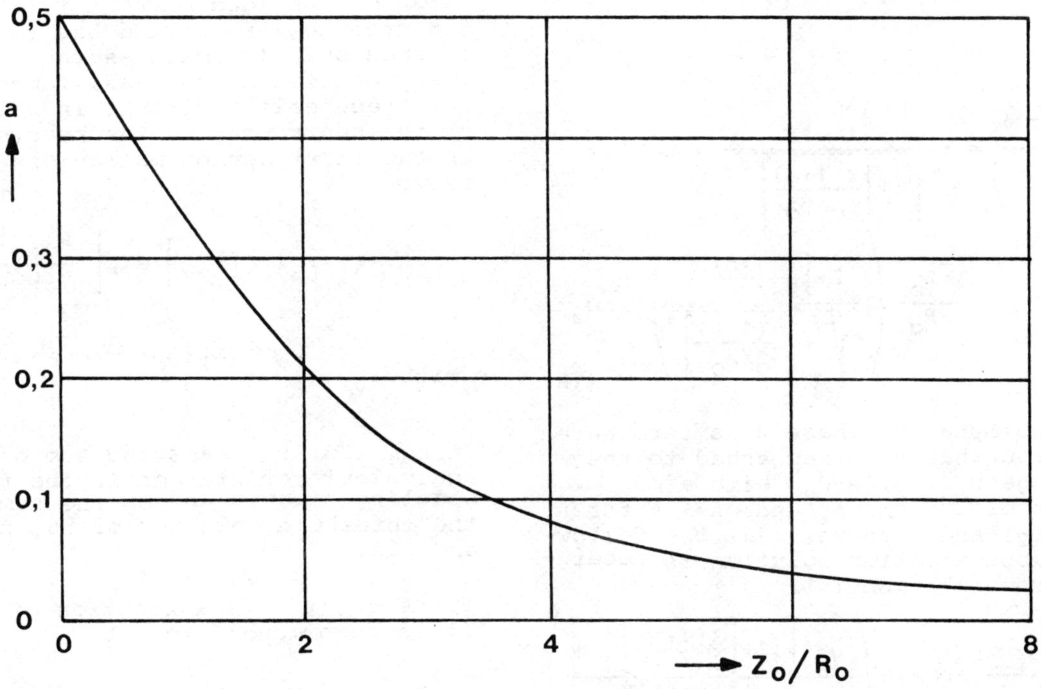

Fig. 3 Voltage factor $a = f(Z_o/R_o)$

2.2 Commutating Phase 2

The commutation phase 2 is described by the equivalent circuit fig. 4.

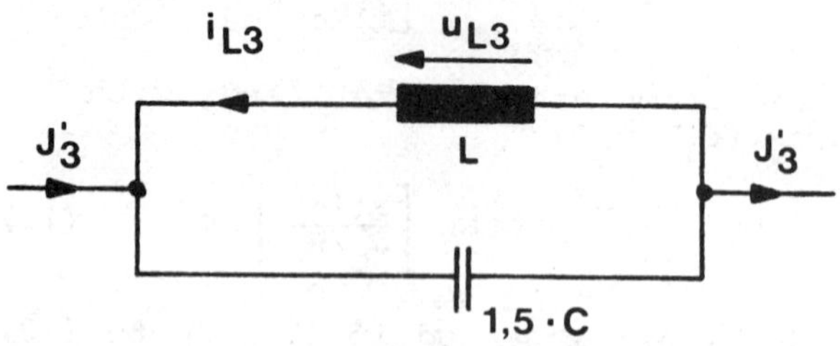

Fig. 4 Equivalent circuit for commutation phase 2

The solutions for $i_{L3}(t)$ and $u_{L3}(t)$ with $t' = t-t_1$, $\omega_2 = \omega_1 . \sqrt{2/3}$ and the initial conditions, mentioned in section 2.1, are as follows:

$$\frac{i_{L3}(t')}{I_o} = -\frac{(1+a)\cdot\sqrt{3/2}}{\sqrt{1+\left[\frac{3(1+a)}{Z_o/R_o}\right]^2}}\cdot \sin\omega_2 t' + \frac{Z_o}{3R_o}\cdot\left(1+\frac{2+\left[\frac{3(1+a)}{Z_o/R_o}\right]^2}{\sqrt{1+\left[\frac{3(1+a)}{Z_o/R_o}\right]^2}}\right)\cos\omega_2 t' - \frac{I_3'}{I_o}\ , \tag{17}$$

$$\frac{u_{L3}(t')}{U_o} = -\frac{1+a}{\sqrt{1+\left[\frac{3(1+a)}{Z_o/R_o}\right]^2}}\cdot \cos\omega_2 t' - \frac{\sqrt{6}Z_o}{9R_o}\cdot\left(1+\frac{2+\left[\frac{3(1+a)}{Z_o/R_o}\right]^2}{\sqrt{1+\left[\frac{3(1+a)}{Z_o/R_o}\right]^2}}\right)\sin\omega_2 t'. \tag{18}$$

The commutation phase 2 is terminated after C_5 has been recharged to the voltage $u_{L3}(t_2) = -U_o$. With a max. error of 3 % for t_2' and of 1 % for $i_{L3}(t_2)$ and with eq. (16) the following approximation solution is received from (17) and (18):

$$t_2' = \frac{3\sqrt{LC}}{2}\cdot \arcsin\left(\frac{4\sqrt{6}}{3}\cdot\frac{\left[\sqrt{1+\left[\frac{3(1+a)}{Z_o/R_o}\right]^2}-1\right]^2}{4\left[\frac{3(1+a)}{Z_o/R_o}\right]^2+3}\right) \tag{19}$$

$$\frac{i_{L3}(t_2)}{I_o} = 3\cdot\left(1-2a+\frac{a^2\sqrt{3}}{2-a}\right)\cdot\sqrt{1+\left[\frac{Z_o/R_o}{3(1+a)}\right]^2} - \frac{I_3'}{I_o}\ . \tag{20}$$

2.3 Commutation Phase 3

The magnetizing current, calculated in eq. (20), decays time-linearly during the commutation phase 3 until the reactor L_3 is entirely demagnetized at the instant t_3:

$$t_3 - t_2 = 3\sqrt{LC}\cdot\left(1-2a+\frac{a^2\sqrt{3}}{2-a}\right)\cdot\sqrt{1+\left[\frac{Z_o/R_o}{3(1+a)}\right]^2} - \frac{L\cdot I_3'}{U_o}\ . \tag{21}$$

2.4 Commutation Phase 4

The magnetizing current $i_{L2}(t_1)$ of the reactor L_2, mentioned in section 2.1, which with the equations (11) and (15) comes to

$$i_{L2}(t_1) = \frac{I_o \cdot Z_o}{3R_o}\cdot\left(\sqrt{1+\left[\frac{3(1+a)}{Z_o/R_o}\right]^2} - 1\right) \tag{22}$$

decays in the freewheeling circuit n23-L_2-n16-n6 down to the value $i_{L2}(t_4) = I_{L30}$ in the course of the commutation phase 4. A temporal alteration of the load current $i_2(t)$ during the freewheeling period has to be considered only inasmuch as there are dissymmetries of the resistances of the freewheeling circuit in relation to the centre-tap of the reactor L_2. In the first approximation one receives:

$$I_{L30} = \left[i_{L2}(t_1)+I_v\right]\cdot \exp\left[-\frac{R_v}{6fL}+\frac{t_i R_v}{L}\right] - I_v$$

$$\text{with } I_v = \frac{U_v}{R_v} + \frac{i_2(t_o)}{2} \tag{23}$$

In eq. (23) R_v means the sum of all equivalent resistances in the freewheeling circuit and U_v the sum of the threshold voltages of n6, n16, and n23.

Since $i_{L2}(t_1)$ is a function of I_{L30}, it is recommended to set

$$I_{L30} = k \cdot i_{L2}(t_1) \tag{24}$$

with $0 \leq k \leq 1$ and to insert an approximate value, that can be checked up later, for k.

2.5 Time Behaviour of the Currents and Voltages in the Commutation Circuit

Fig. 5 und 6 show the graphic evaluation of the current and voltage equations (10) up to (13), (17), and (18). The time is normalized to the recovery time

$$t_{so} = t_s(Z_o/R_o=0) = 0,5 \cdot \pi \cdot \sqrt{LC}.$$

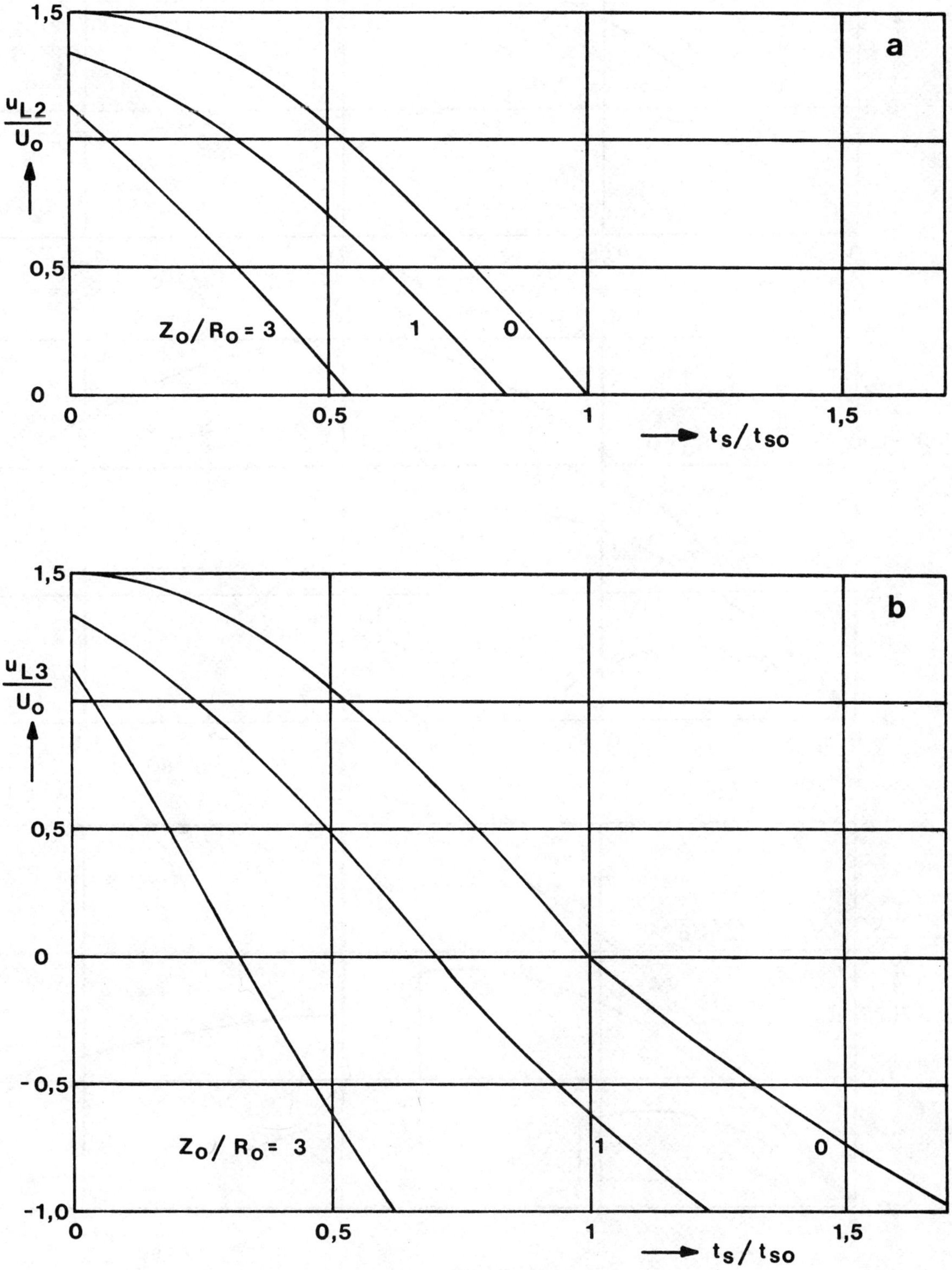

Fig. 5 Voltage Behaviour in the Commutation Circuit

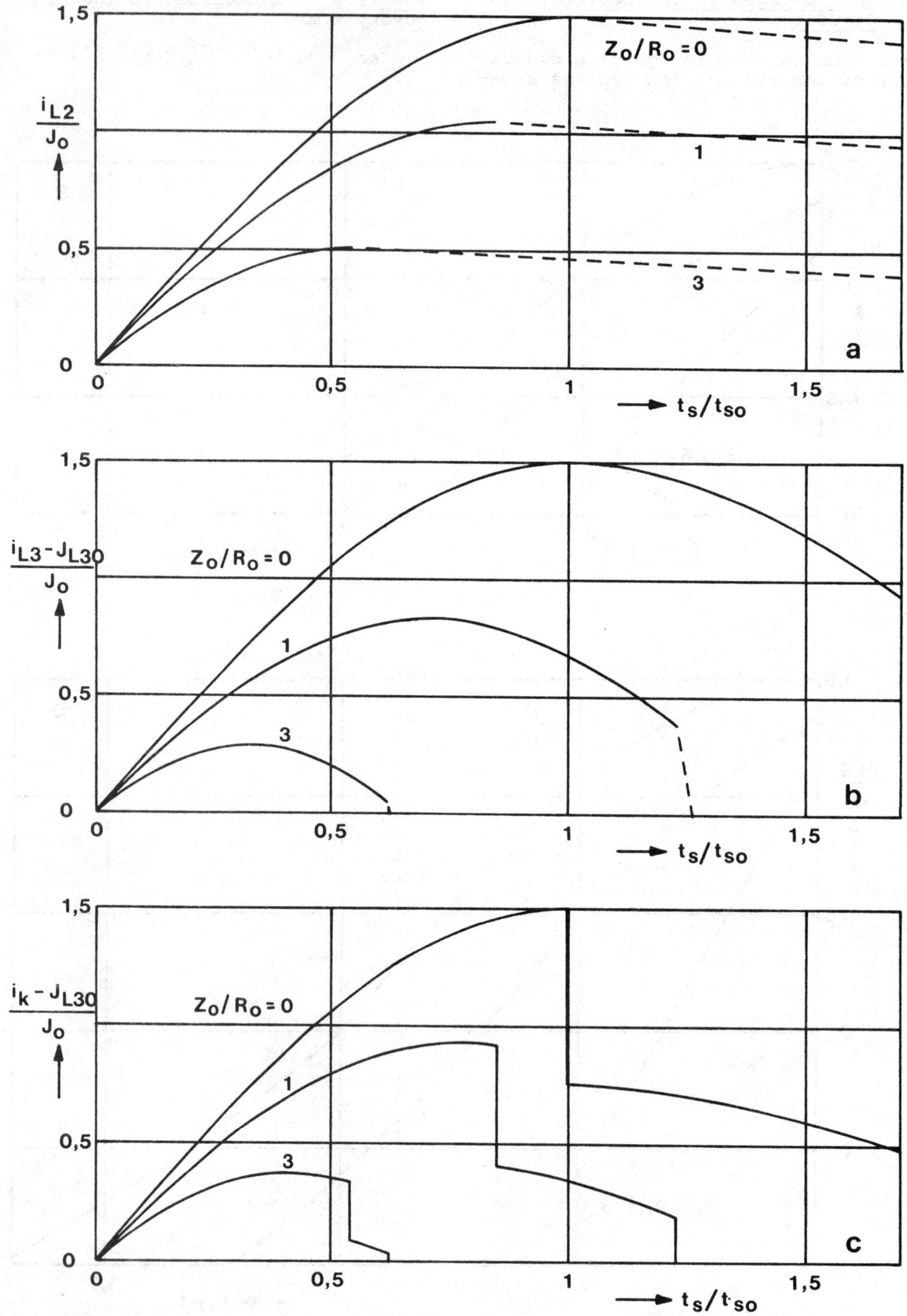

Fig. 6 Current Behaviour in the Commutation Circuit

The reactor current i_{L2} (fig. 6a) is identical with the commutation current by the thyristor n6; i_K (fig.6c) is the commutation current by the thyristor n1.

3. Dimensioning of the Commutation Circuit

3.1 Design of the commutating chokes and capacitors

On the assumption that converter characteristics and the semiconductor data are given, first the DC voltage U_o is determined under consideration of the eq. (30). The next step is to evaluate the limiting peak current $\hat{I}_{Th}$ of the thyristor current. This is either gathered as limit value from the data sheet, or it is found by means of an approximating calculation on account of the thermal conditions.

With the help of the equations (11) and (13) it follows:

$$\beta = \frac{\hat{I}_{Th}}{I_3'} = \left(1+\frac{I_{L30}}{I_3'}\right)\cdot\sqrt{1+\left[\frac{2(1+a)}{Z_o/R_o}\right]^2}\,. \quad (25)$$

(The case of $I_3' = 0$ - e.g. no-load - has to be considered separately, however, this can easely be done with the relationships, indicated in section 2.)

Furthermore, the minimum recovery time t_{smin} is determined with regard to a safe commutation.

As a fourth value the factor k is required for the freewheeling circuit (see equation (24)), which is not a mere equipment constant, but is dependent on the load and commutation current. For a first approximation it is, however, sufficient to insert an empirical or estimated value for k and to check it afterwards. With (8), (9), (22), and (24) it follows out of eq. (25)

$$\frac{U_o(1+a)}{Z_o\cdot I_3'} = K_1(\beta,k) \quad (26)$$

and out of eq. (16)

$$(1+a) = K_2(\beta,k)\,. \quad (27)$$

K_1 and K_2 may be specified in a closed form. The equations (26) and (27) supply the first determination equation for L and C:

$$Z_o = \sqrt{\frac{L}{C}} = \frac{U_o\cdot K_2(\beta,k)}{I_3'\cdot K_1(\beta,k)}\,. \quad (28)$$

The second determination equation is received out of (14) with (8), (9), and (26):

$$\sqrt{LC} = \frac{t_{smin}}{\arctan\left(\dfrac{1,5}{\sqrt{(\beta/K_1)^2-4}}\right)}\,. \quad (29)$$

With (28) and (29) the necessary values of capacitance and inductance can be calculated:

$$L = \frac{K_2\cdot U_o}{K_1\cdot I_3'}\cdot\frac{t_{smin}}{\arctan\left(\dfrac{1,5}{\sqrt{(\beta/K_1)^2-4}}\right)}\,,$$

$$C = \frac{K_1\cdot I_3'}{K_2\cdot U_o}\cdot\frac{t_{smin}}{\arctan\left(\dfrac{1,5}{\sqrt{(\beta/K_1)^2-4}}\right)}\,.$$

The normalized values

$$L' = \frac{I_3'}{U_o\cdot t_{smin}}\cdot L\,,$$

$$C' = \frac{U_o}{I_3'\cdot t_{smin}}\cdot C$$

are graphically presented in fig. 7 as function of the factor β with the parameter k.

After determination of L and C the estimated values for β and k have to be checked and corrected, if necessary. After this calculation it is advisable to check the no -load values. In relation to the calculation for $I_3' > 0$, executed in this section, for $I_3' = 0$ it is received:

$$a_L = \frac{1}{2+k}\,,$$

$$t_{SL} = \sqrt{LC}\cdot\arctan\left(\frac{\sqrt{3(3+2k)}}{2k}\right)\,.$$

Because of $0 \leq k \leq 1$ it is valid: $1/3 \leq a_L \leq 1/2$. Therewith the limiting values of the voltage in the commutation circuit are:

$$4/3 \leq \hat{U}_{Th} = \hat{U}_C = \hat{U}_L \leq 3/2$$
$$7/3 \leq \hat{U}_{n21\ldots n26} \leq 5/2. \quad (30)$$

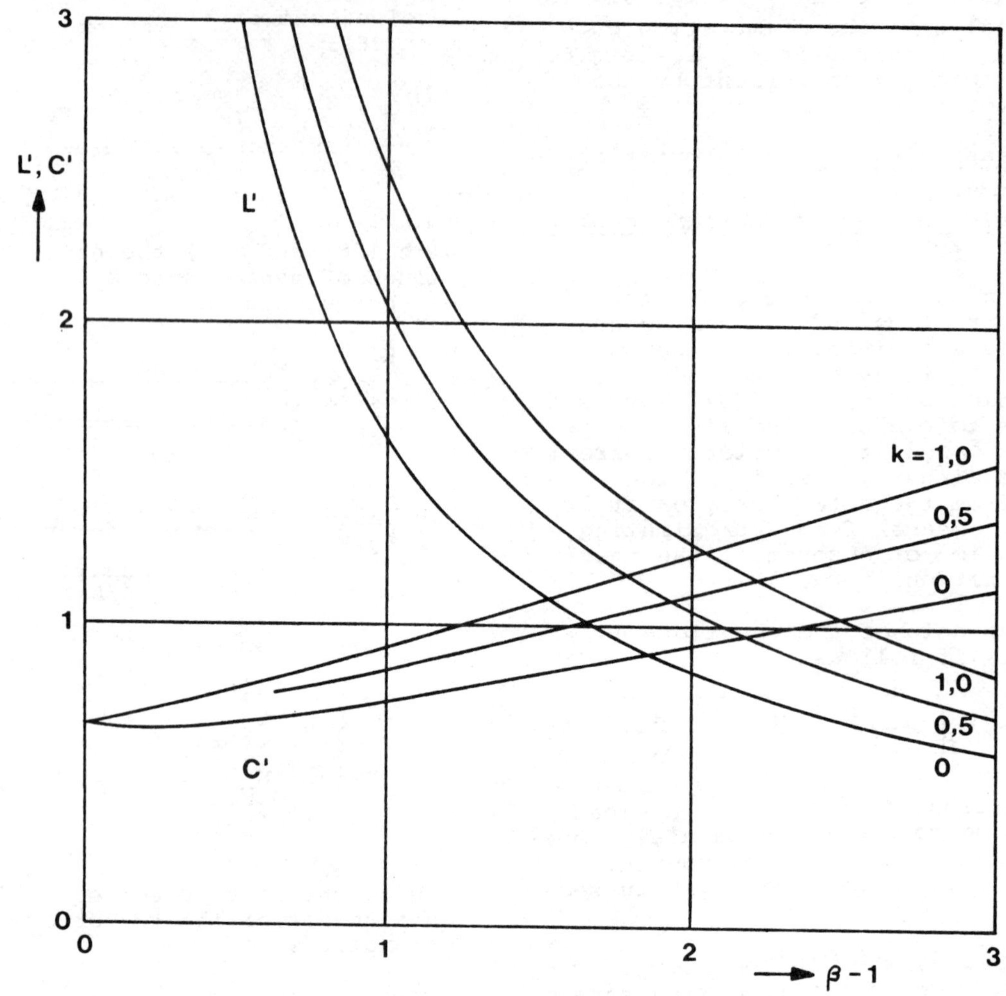

Fig. 7 Graph for the design of the commutation chokes and capacitors

3.2 Optimum Design of the Commutating Circuit

In section 3.1 a design procedure for the commutating chokes and capacitors with given use of semiconductors was proposed. In case, in doing so, β is chosen to its maximum, one receives the smallest inductance of the commutating chokes.

However, this calculation method does not necessarily lead to the most favourable design of the converter as to the cost. Especially small values of β cause high values of inductance. Indeed, also the magnetizing current of the choke is diminished; however, the load current remains the same, the physical size of the reactors is considerably increased as a whole.

For the mentioned reasons it has to be found out whether a more economical solution is possible, e.g. by connecting thyristors in parallel or by using other semiconductors. For investigations of such kind the graph of fig. 7 will serve as dimensioning help.

References:

Düll, E.H., Golde, E.: Dreiphasenumrichter mit Gleichstrom- Zwischenkreis 50/50 bis 500 Hz. AEG-Mitteilungen 54 (1964), H. 3/4

Heumann, K., Stumpe, A.C.: Thyristoren. Teubner, Stuttgart 1970

Jötten, R.: Leistungselektronik Bd. 1 Vieweg, Braunschweig 1977

A NEW APPROACH TO THE DESIGN OF ASYMMETRICALLY LOADED THREE PHASE SINUSOIDAL INVERTERS

P. Miljanić, D. Srajber, and R. Milijanović

Electrical Engineering Institute "Nikola Tesla", 11000 Belgrade, Jugoslavia

ABSTRACT

A new type of three phase inverters powered from a current source is presented. The inverter consists of only three thyristors, successively fired in equal time intervals. A three phase magnetic circuit, with windings for phase coupling and for returning surplus energy into the d-c circuit, combines with capacitors to produce: (1) a sinusoidal output voltage and (2) a symmetrical voltage system in both amplitude and phase. The three phase system, due to its nature and the way it is constructed, tends to sustain a symmetrical and sinusoid three phase voltage oscillation. The basic idea of the new design is to feed the oscillations of the system using static switches.

INTRODUCTION

In constructing sine wave three phase inverters, which are used for supplying asymmetrical three phase loads, two problems, among others, have to be sold: a separate control of the voltage amplitude of each phase and the maintaining of equal phase displacements between these voltages. The principal difficulties arise because of unequal voltage drops on the series filters. In three phase bridge inverters with feedback diodes (Ref. 1) it is not possible to regulate the phase voltages separately, making it necessary to complicate the circuit. Pulse width modulation is used to overcome this difficulty, but it does so by increasing the number of valves and using very complex electronic circuits. Futhermore, this circuit is not suited for applications requiring higher frequency a-c voltages. Another way to solve the problem of voltage control is by replacing the three phase inverter with three single phase inverters. Thus, it is possible to regulate each phase voltage separately, while the phase displacement symmetry is obtained by linking the three output voltages to a transformer whose windings are delta connected. It is obvious that if three voltages are sinusoidal and of equal amplitude and their vector sum is zero, their phase displacements should be $2\pi/3$. Voltage amplitude is regulated by the conducting angle and one way to maintain the phase displacement of $2\pi/3$ is to control the voltage by symmetrically changing the conducting angle around axes mutually displaced by $2\pi/3$. The regulation of such a system is explained in Ref. 2. The drawback of this solution is the increased price due to a large number of components - there are 12 thyristors instead of 6, with the corresponding increase in commutation components and complexity of electronic circuits. In large size inverters this drawback is not so emphasized as in small size inverters, as there exists the necessity of connecting thyristors in series and/or in parallel and because the relative price of electronics is small.

Obviously, there are sound reasons for designing three phase inverters which are much simpler than the above mentioned. In this paper one such solution will be presented: a three phase inverter with only 3 thyristors and with a very simple electronic control circuit, but which still ensures a three phase symmetrical output voltage under asymmetrical loads.

DESCRIPTION OF THE INVERTER

The main components of the inverter are shown on Fig. 1. The magnetic circuit is composed of a three-limb core type transformer with an air-gap in the middle of the windings in order to increase the magnetizing current. The primary windings N_p are star connected, while the secon-

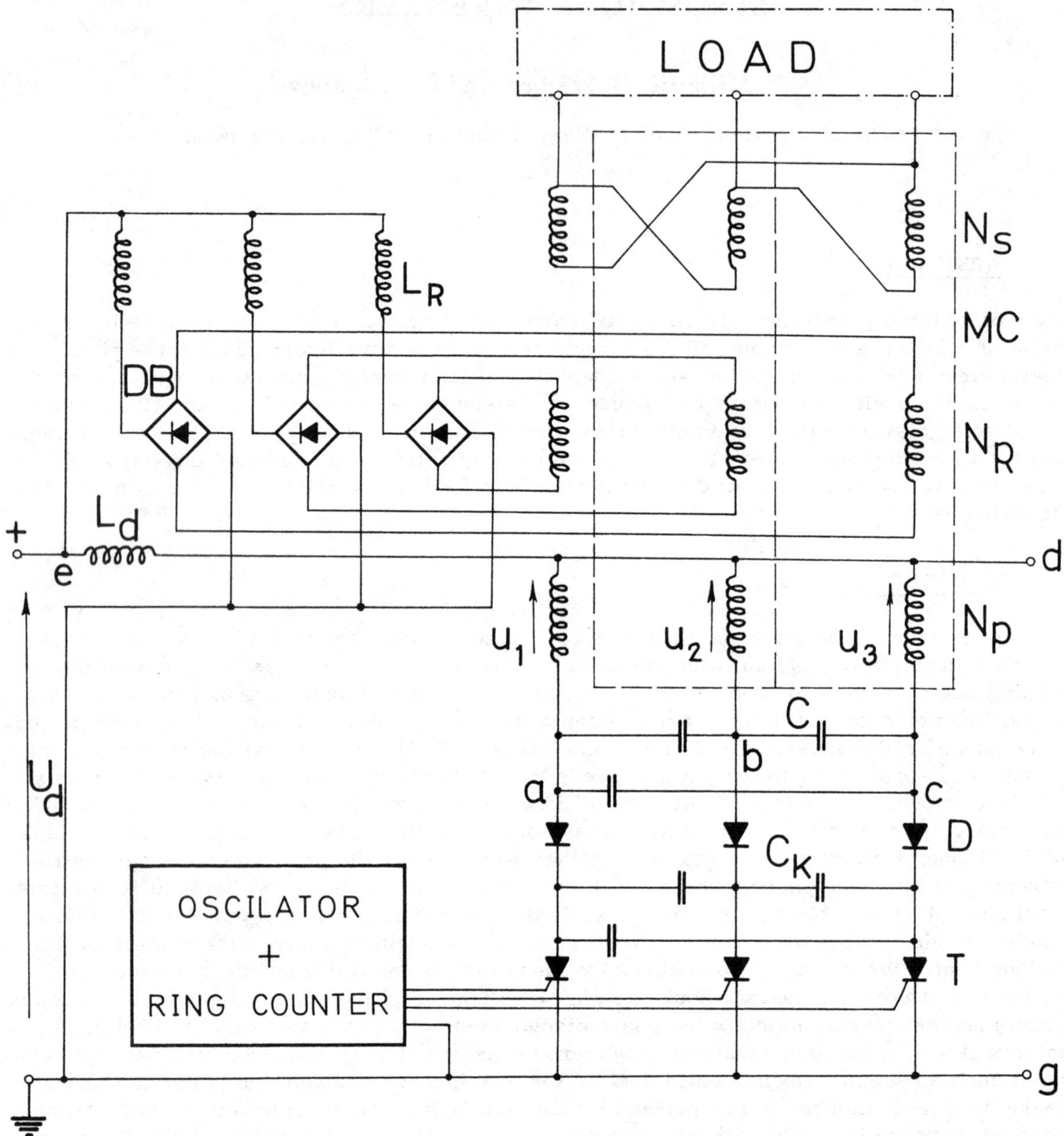

Fig. 1 The schematic diagram of the three phase sine wave inverter, showing the main power components.

dary windings N_s are delta connected. However, the secondary windings can also be star connected, if the load demands a four-lead system. On each limb of the three phase magnetic circuit there is another independent winding N_r through which a part of the energy is returned to the d-c supply.

In the circuit there are also four d-c chokes. One is found in the main circuit (L_d), while three are located in the feedback branches (L_r). Three capacitors (C) are connected between the phases on the primary side and they are in resonance with the inductances of the transformer windings. The capacitors can be connected to the secondary side as well, provided the secondary voltage is high enough. This solution has an advantage over the first one in that it results in a better form of output voltage. On Fig. 1 there is another group of three commutating capacitors C_k, which have a much lower capacitance than the first group C. Of the power semiconductor elements the inverter has three thyristors T, three blocking diodes D and three smaller single phase diode bridge rectifiers DB in the power recovery feedback circuits.

The electronic control circuit is extremely simple. An UJT oscillator drives a ring-counter, which is composed of three small thyristors. The neutral line of the power circuit to which the cathodes of all three thyristors are connected, is the same as the neutral line in the electronic circuit, which enables the firing of thyristors T, without the use of impulse transformers.

PRINCIPLE OF OPERATION

The Basic Idea

The idea of designing an oscillatory circuit and feeding the oscillations by a suitable power injection into the circuit has been known for many years. A large number of inverters have been constructed based on this principle (see, for example, Ref. 3). A single phase inverter, in which the power injection into the oscillatory circuit is realized by using a thyristor bridge, which is brought into short-circuit in regular time intervals, has been described by one of the authors of this paper (Ref. 4). Just before each inversion, by short-circuiting the bridge, the current in the d-c choke is adjusted and in that way the power to be injected into the oscillatory circuit is regulated. However, the power to be injected into the oscillatory system can be controled by recovering the surplus energy, i.e. returning this surplus to the d-c supply. The first inverter of this type the authors have described in Ref. 5. In this paper the same principle of control is presented. It allows a considerable simplification in construction and is very reliable in operation.

The Three Phase Oscillatory System

The circuit on Fig. 1 has a tendency to oscillate in such a way that the phase voltages maintain equal amplitudes and, at the same time, keep a phase displacement of $2\pi/3$. In other words, they form a stable and symmetrical three phase system. It shall be shown that the system is symmetrical, i.e. that the voltages have equal amplitudes and that their vector sum is zero. Because of the resonance effect the voltages are approximately sinusoidal. Studying the voltage situation on the choke L_r it can be seen that the d-c supply voltage U_d is found on its one side, while at the other the voltage amounts to twice the rectified voltage of the winding N_r on one of the transformer limbs. The average values of these voltages must be approximately equal. If this were not so, a very large d-c current would flow through the choke. Thus

$$U_d = \frac{1}{T}\int_0^T N_r |e_1| \, dt \qquad (1)$$

where e_1 is the value of the induced electro-motive force in one turn of the winding on the first limb. The same equations hold for the second and the third limb and so the r.m.s. value of the induced electro-motive forces in all three limbs must be equal:

$$E_1 = E_2 = E_3 = \frac{\pi}{2\sqrt{2}} \frac{1}{N_r} U_d \qquad (2)$$

On the other side, because of the fact that the magnetic core has only three limbs

$$\Phi_1 + \Phi_2 + \Phi_3 = 0 \tag{3}$$

From equation 3 it follows htat

$$\frac{d\Phi_1}{dt} + \frac{d\Phi_2}{dt} + \frac{d\Phi_3}{dt} = 0 \tag{4}$$

Thus, if the leakage is neglected, it can be stated that

$$e_1 + e_2 + e_3 = 0 \tag{5}$$

The last statement is especially true if the secondary transformer winding is delta connected, because then, apart from the magnetic, there is also an electrical connection, which dictates that the sum of induced voltages in all three windings be equal to zero. Thus the circuit is forced to maintain a symmetrical voltage system, even if the load and the power injected into its respective phases is asymmetrical.

Natural and Forced Oscillations

It can be shown that the characteristic equation, which describes the changes in the flux of a limb, is:

$$3N_pCs^2 + \frac{\theta}{N_p} = 0 \tag{6}$$

The damping due to the load and the power recovery and the reactive components of the load have been neglected in the above equation. The natural frequency of oscillation is (from equation 6)

$$f_o = \frac{1}{2\pi N_p}\sqrt{\frac{\theta}{3C}} \approx \frac{1}{2\pi N_p}\sqrt{\frac{\delta}{3C\mu_o S}} \tag{7}$$

where

N_p - is the number of turns in the primary winding
θ - is the magnetic resistance of a limb
C - is the capacitance of the capacitor
δ - is the air-gap
μ_o - is the magnetic permeability in vacuum
S - is the cross section of a limb.

However, the system will not oscillate with this frequency. It must follow the forced frequency which is defined by the UJT oscillator and the ring-counter. For the wave shapes to be sinusoidal it is necessary that the frequencies of forced and natural oscillations be approximately equal. Thus,

$$f_{osc} = 3f = 3f_o \tag{8}$$

Power Injection into the Oscillatory System

The way in which the poweris injected into the oscillatory system can best be seen by studying the wave shapes between the points g and d, which are shown on Fi. 2 for the equivalent resistive load, on Fig. 3 for the equivalent capacity load on Fig. 4 for the equivalent inductive load. At the moment one of the thyristors T is fired the point d drops suddenly to the negative potential because of the load on the commutating capacity C_k . As its capacitance is relatively small, after a short time the respective diode D begins conducting, forming a voltage difference on the inductance L_d between the supply voltage U_d and one of the three phase primary voltages u_1 , u_2, or u_3. Although the inverter could, in principle, be of the line commutating type if the equivalent load was capacitive, in order to ensure its reliability in starting and during sudden changes in the load, the diodes D and the capacitors C_k have been added to the circuit, thus securing a sure impulse commutation. The thyristors are fired successively every third of the

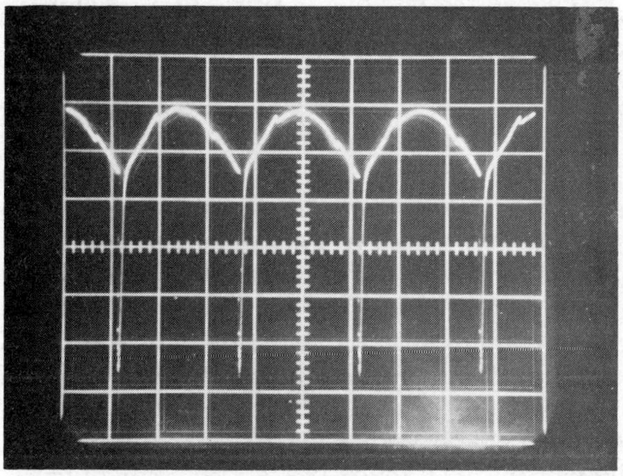

Fig. 2. Resistive load

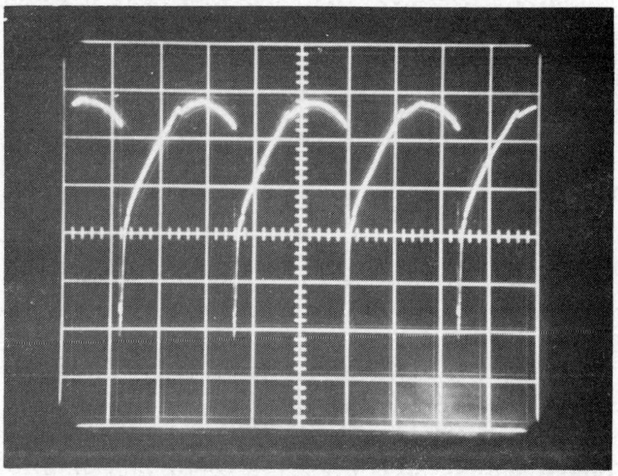

Fig. 3. Capacitive load

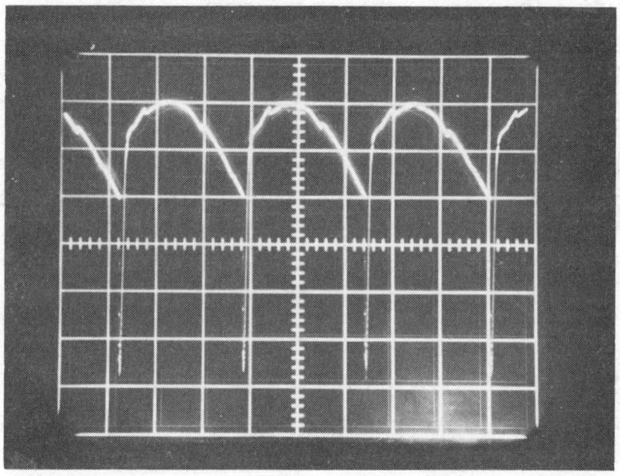

Fig. 4. Inductive load

full period (T/3) and the average value of the d-c voltage on the choke L_d has to be approximately zero, if the voltage drops are neglected. Taking this into account, the d-c supply voltage U_d can be written as:

$$U_d = \frac{3}{T} \int_{T/12-\alpha}^{5T/12-\alpha} N_p \, e dt \qquad (9)$$

α is the phase angle betweeen the first harmonic of the injected current and the voltage, and has to be equal to the phase angle of the equivalent impedance of the oscillatory circuit, feedback branches and the load. It should be noted that α is not equal to the power factor of the load φ. The angle α is smaller if the current recovery is greater or if the power rating of the oscillatory circuit is larger. If the time is counted from the moment the phase voltage passes through zero, the injection of power into that phase begins at the moment $T/12 - \alpha$ and terminates at $5T/12 - \alpha$. Taking the induced e.m.f. to be sinusoidal and of equal r.m.s. value for all three phases, it follows that

$$E_1 = E_2 = E_3 = \frac{2\pi}{3\sqrt{6}} \frac{1}{N_p \cos\alpha} U_d \qquad (10)$$

Equations (2) and (10) show that it is not possible to choose the number of turns of N_p and N_r freely. Their ratio must be approximately

$$\frac{N_r}{N_p} = \frac{3\sqrt{3}}{4} \cos \alpha \tag{11}$$

In effect, the ratio N_r / N_p defines the magnitude of the circulating current which enters the system through the choke L_d and returns through the chokes L_r. The circulating current is minimum for $\cos\alpha = 1$ and increases for both capacitive and inductive equivalent loads. Changes in the recovery current are not as large as might be supposed at first because they are limited by internal voltage drops which have been neglected up to now, of which the most important ones are the voltage drops due to leakage inductances. The recovery current helps the voltage control: it is larger in the no-load state, smaller when the inverter is loaded; it is smaller when the load is resistive than in the case of inductive or capacitive loads. It is greater in the phase where the voltage is higher and smaller in the phase where there is a decrease in voltage because of an increased load in that phase.

A more detailed analysis of the inverter presented here has not been done up to date, nor was it possible to do so easily in the limits of this paper. The workings of the process can, however, be understood notwithstanding the very complex interconnections of the reactive components: the three phase magnetic circuit, the linked capacitors, the delta connected windings, the feedback rectifiers, thyristor commutations and the currents injected into inductances L_d and L_r. However, even though a simplified analysis the designer can determine some principal parameters. For example, if the symmetry of the voltage with asymmetrical loads or the reliability of operation when sudden changes in load occur is important, the power rating of the recovery circuit should be larger. In extreme cases, this rating can be of the same order as the nominal power rating of the device. If the output voltage distortion has to be small, the reactive power of the capacitors C should be several times larger than the nominal load power. The ratio of these powers for normal demands is of the order of 5.

The Properties of the Inverter

The oscillations of energy in LC elements of the three phase system and the power recovery through feedback rectifiers enables a stable and reliable operation in the presence of sudden symmetrical and asymmetrical load changes. During a sudden load increase of the inverter, because of momentary voltage drops at the output, the recovery current decreases, while the power needed by the load is taken from the reactive components and from the slowly rising supply current. The overvoltages, due to a sudden load decrease, are damped in a similar fashion. Furthermore, the above holds true for a sudden increase or decrease in the load of only one phase. The recovery of the phase, which undergoes a load increase, decreases while the resultant inbalance, due to the existing phase couplings, is divided among all the reactive elements of the oscillatory system. When the stationary state is reached it is characterised by a larger load in one phase the other two, and by the large decrease of the recovery power of that phase.

On the basis of all that has been presented, it can be concluded that this type of inverter is immune to sudden load changes, this being one of its most positive properties.

EXPERIMENTAL RESULTS

Voltage Regulation and Control

Voltage regulation (from no-load to full load) in the inverter described in this paper is about 5 %, with a constant input voltage (see Table 1). If it is neccessary to have automatic voltage control it can be accomplished by controling the d-c input voltage. It should be noted that a very good way of doing this is by using a chopper circuit, because of the chokes at the input of the inverter. The potential of the input lines does not have to be filtered with electrolytic capacitors, as is usually done - it is enough to put a free-wheeling diode across the input. When the chopper frequency surpasses the inverter frequency, the inverter will not react to the

TABLE 1. Experimental results

LOAD CURENT PHASE			VOLTAGE MEAN	PHASE VOLTAGE DEVIATION FROM MEAN			η
1	2	3	%	1 (%)	2 (%)	3 (%)	%
0	0	0	103,5	- 0.2	+ 0.3	- 0.2	0
50	50	50	100	- 0.5	+ 0.5	0.0	0.62
100	100	100	98,6	- 0.5	+ 0.5	0.0	0.79
20	20	100	100,7	+ 1.0	+ 1.8	- 2.8	0.60
20	100	100	99,6	+ 1.8	- 0.7	- 1.1	0.71
50	50	100	99.9	+ 0.5	+ 1.0	- 1.5	0.69
50	100	100	99.4	+ 1.1	- 0.3	- 0.8	0.75

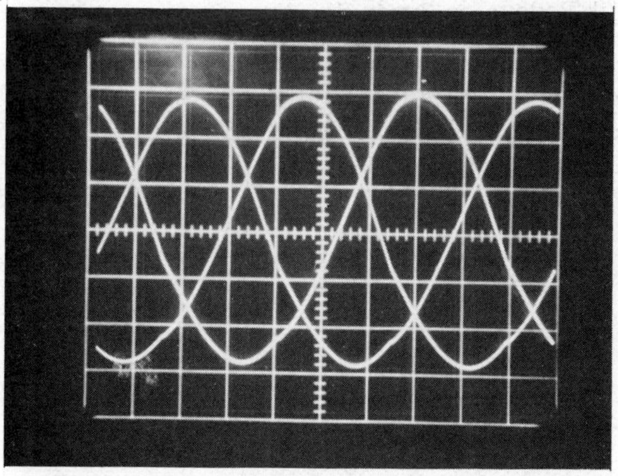

Fig. 5. Output voltages

input change of potential. The voltage balance on the chokes L_d and L_r is influenced only by the average values.

Voltage Asymmetry

The asymmetry of voltage in cases of asymmetrical loads can be seen in Table 1, where the results taken from one of the constructed inverters is shown. There it can be seen that in cases of very unbalanced loads the phase voltages deviate from the average value by not more than ± 3%. The same limits hold for the deviation of the average value of voltage from the nominal value with changing loads.

Distortion

The three phase output voltages of the inverter are shown on Fig. 5. The distortion arises because of the square form of the recovery current and because of the power injection into the corresponding phases through thyristors in the positive half-period for a time interval od T/3. Measurements show that the first reason is responsible for the distortion at no-load, while the second produces distortion at nominal loads. The distortion factor varied between 8 and 12 %

during the experiments conducted, the results of which are shown in Table 1. If the need arises, the distortion can be decreased by feeding the oscillatory circuit from two sides, in which case 6 thyristors are neccessary instead of the present 3. The distortion can be decreased, as well, by decreasing the recovery current and increasing the reactive power rating of the capacitors with respect to the nominal power.

Reliability of Operation

Several inverters working on this principle have been constructed and the experience in industry has given several years of smooth operation. Starting, stopping, sudden increase and decrease up to nominal loads are all accomplished with no sudden current surges in the semiconductor elements and the d-c supply, which is a special feature of these inverters.

CONCLUSION

The research presented in this paper and the experience acquired in exploatation have shown that it is possible to design a reliable three phase inverter using only three thyristors. The inverter can supply with sine wave voltages asymmetrical three phase loads. The output transformer has several functions. The same magnetic circuit is used for voltage transformation, for filtering higher harmonics and for the symmetrisation of voltage in both amplitude and phase displacement. The solution requires a somewhat larger amount of iron and copper, but this drawback is overcome by increased reliability in operation and simplicity in construction.

REFERENCES

(1) L. Abraham, K. Heumann, F. Koppelmann, Wechselrichter zur Drehzahlsteuerung von Käfigläufermotoren, AEG-Mitt.54, Nr. 1/2, 89 (1964).

(2) P. Miljanić, V.Vučković, Regulacija periodičnih veličina naizmeničnim poluperiodnim integraljenjem, Zbornik Radova sa Prvog Savetovanja Energetske Elektronike, Beograd, 1 (1973).

(3) SCR Manual, Fifth edition, General Electric (1972).

(4) P. Miljanić, Kratkospajajući Invertor, Zbornik Radova sa Prvog Savetovanja Energetske Elektronike, Beograd, 99 (1973).

(5) P. Miljanić, D. Šrajber, R. Milijanović, Nov Pristup Gradnji Nesimetrično Opterećenih Trofaznih Invertora, Zbornik Radova sa Drugog Savetovanja Energetske Elektronike, Beograd, 387 (1975).

A COMPARATIVE INVESTIGATION OF HIGH POWER PULSEWIDTH MODULATED INVERTER CIRCUITS

I. Topa

*University of Brasov, 2200 Brasov, Romania**

1. REQUIREMENTS OF PULSE-WIDTH MODULATED INVERTER

The maglov drive at the maximum speed of 500 km/h and a traction force of 100 kN requires a converter power of 14 to 20 MW. A linear synchronous motor with extended primary winding drive involves relatively low stator currents in order to keep reduced active and reactive losses in the train winding. Thus the converter stations along the route may be reduced in number. In order to obtain the required power, a high electromotive force of the synchronous drive is necessary and hence a high voltage in the intermediate circuit of the PWM inverter.

In this paper, the following data have been taken for the monophase load of the PWM inverter :

the resistance of the extended primary winding $R_L = 0.2125\,\Omega$

the inductance of the extended primary winding $L_L = 5.175$ mH

the electromotive force

$$\hat{u}_g = 6.3 \text{ kV}.$$

This result in a high voltage E = 16 kV in the intermediate circuit of the inverter, when the load must attain a 50 Hz maximum frequency. To investigate the different circuits, the data from a short turn-off time thyristor (Siemens B St P 44) have been taken as a basis. The turn-off time is $t_q = 25\,\mu$s and the critical rate of rise $S_{i\ krit} = 200$ A/μs. The dimensioning relations given in the paper may be generalised for the practical purpose of connecting thyristors and inverters components in series and in parallel.

For investigating the PWM inverter circuits in steady state by means of a standard program for the digital simulation of networks containing diodes and thyristors (Ref. 1), it is necessary to know the thyristor firing times. For this purpose an "ideal" PWM inverter (on-off controller) has been simulated as well as a feedback control system with a sinusoidal shape of the index value and a sinusoidal electromotive force (Fig. 1).

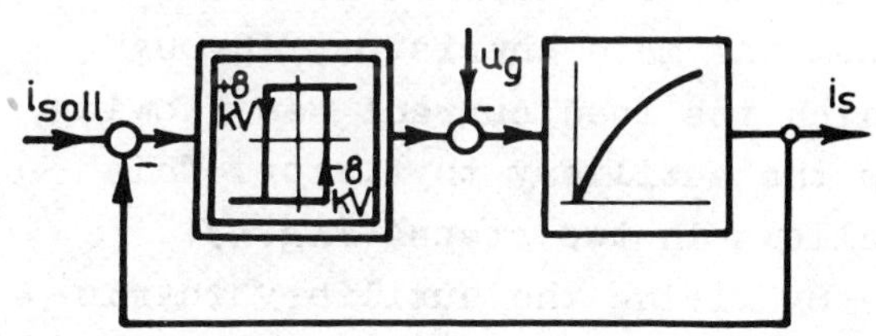

Fig. 1. The feedback control system with on-off controller.

* This paper is the result of my research work carried out in the Federal Republic of Germany, while holding a Fellowship granted by the Alexander von Humboldt Foundation.

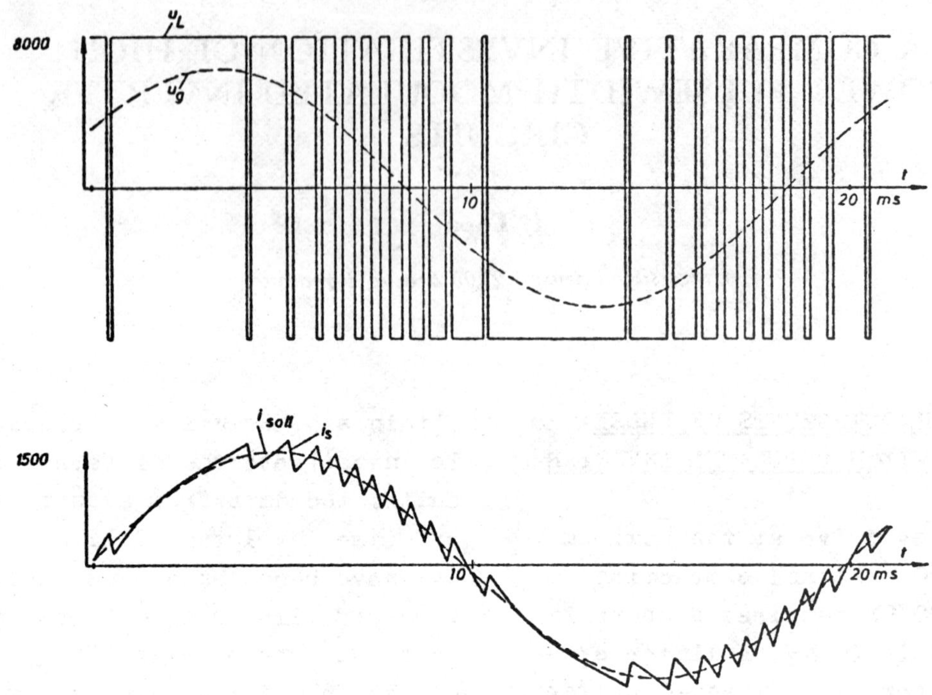

Fig. 2. Waveshapes of voltage and current of an ideal PWM inverter with active load.

$$i_{soll} = \hat{i} \sin(\omega t - \varphi_i) \qquad (1)$$

$$u_g = \hat{u}_g \sin(\omega t - \varphi) \qquad (2)$$

The results for $\varphi_i = 3^\circ$ and $\varphi = 30^\circ$ are shown in Fig. 2. The shortest time for connecting the d-c voltage to the load is thus obtained for the data provided by about $120 \mu s$.

2. GENERAL TURN-OFF PRINCIPLE

In all the circuits being examined, the load current is commutated from the main thyristor, through which the load current was flowing, to the auxiliary thyristor. This follows in two steps (Fig. 3).

By firing the auxiliary thyristor 7, capacitor voltage is applied to the main thyristor 2 in the blocking direction.

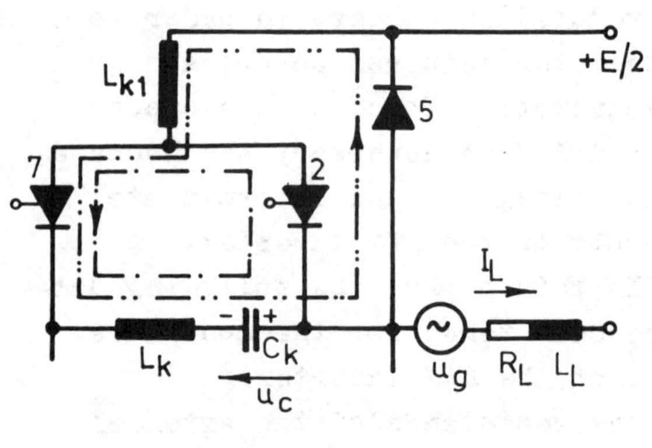

Fig. 3. Commutating circuits of a PWM inverter.

After turning off the main thyristor, the load current flows through the second oscillating circuit, through the inductance L_{k1} and diode 5, and the capacitor C_k is recharged to reversed voltage.

The rate of rise of the commutating current is limited in the first time interval by means of the commutating inductance

$$L_k \geqslant U_{co}/S_{i\ krit} \qquad (3)$$

U_{co} is the initial voltage of the capacitor.

a) In the first interval of time ($t_1 \leqslant t \leqslant t_l$) for the capacitor voltage, the oscillatory circuit resistances being neglected, the following equation is valid (Fig. 4)

$$u_c(t) = k_u E \cos \omega_{o1} t \qquad (4)$$

where $U_{co} = k_u E$ with $k_u > 1$ (5)

and

$$\omega_{o1} = 1/\sqrt{L_k C_k}\ .$$

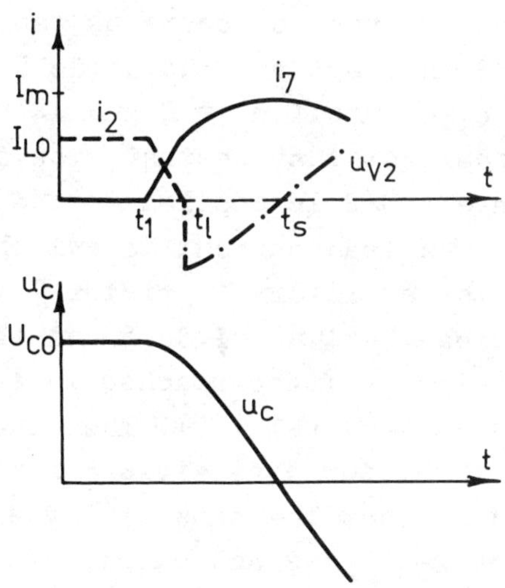

Fig. 4. Commutation process.

With the positive direction of the current indicated in Fig. 3, the capacitor current is

$$i_c(t) = C_k \frac{du_c}{dt} =$$

$$= -\, k_u E \sqrt{\frac{C_k}{L_k}} \sin \omega_{o1} t. \qquad (6)$$

The maximum time t_l, required to reduce the current of the main thyristor to zero, results from (6), provided $i_c(t = t_l) = i_7 = I_{LM}$. Here I_{LM} is the maximum load current which must be commutated. The result is

$$t_l = \sqrt{L_k C_k}\ \mathrm{arc\ sin}\left(\frac{I_{LM}}{k_u E} \sqrt{\frac{L_k}{C_k}}\right). \qquad (7)$$

At the end of this interval of time ($t = t_l$), the voltage on the capacitor has attained the value

$$U_{cl} = u_c(t_l) = k_u E \cos\left(\frac{t_l}{\sqrt{L_k C_k}}\right). \qquad (8)$$

b) For the second interval of time ($t_l \leqslant t \leqslant t_s$) the following differential equation is valid

$$L_\Sigma \frac{di_c}{dt} + \frac{1}{C_k} \int_0^t i_c(\tau)\, d\tau = 0. \qquad (9)$$

$L_\Sigma = L_k + L_{k1}$ is the inductance of the second oscillating circuit. The solution of the differential equation is

$$i_c(t) = -\, U_{cl} \sqrt{\frac{C_k}{L}} \sin \omega_{o2} t \, -$$

$$-\, I_{LM} \cos \omega_{o2} t \qquad (10)$$

where $\omega_{o2} = 1/\sqrt{L_\Sigma C_k}$ is the angular frequency of the second resonant circuit. The capacitor voltage is

$$u_c(t) = U_{cl} \cos \omega_{o2} t \, -$$

$$-\, I_{LM} \sqrt{\frac{L_\Sigma}{C_k}} \sin \omega_{o2} t. \qquad (11)$$

From (11) we obtain the safety time t_s, provided $u_c(t = t_s) = 0$.

$$t_s = \sqrt{L_\Sigma C_k}\ \mathrm{arc\ tan}\left(\frac{U_{cl}}{I_{LM}} \sqrt{\frac{C_k}{L_\Sigma}}\right). \qquad (12)$$

The safety time should be longer than the thyristor turn-off time

$t_s \geqslant k_f t_q \qquad k_f = 1.2, \ldots, 1.5$

If (12) is introduced into (10), we get the peak value of the commutating current $i_7(t = t_s) = - i_c(t = t_s) = I_m$. In order to have similar safety times, the product $L_\Sigma C_k = 1350 \times 10^{-12}\ s^2$ has been kept constant for all circuits.

3. MCMURRAY CIRCUIT

McMurray circuit represents a PWM type inverter which can be constructed with relatively few semiconductors and, therefore, it seems to be suitable for very high power loads (Ref. 2, 3). In Fig. 5 a phase of this inverter with an active load is shown. Resistances R_v and

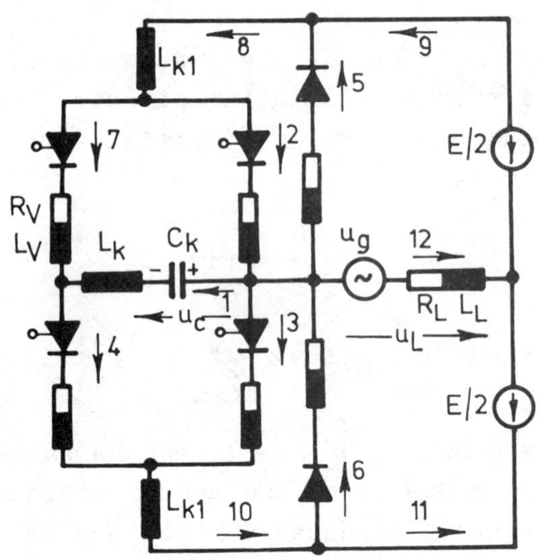

Fig. 5. McMurray single-phase PWM inverter.

inductances L_v which are connected in series to each semiconductor are required for the calculation program and represent about 3‰ of the resistance and the inductance of the load. The arrows indicate the positive directions of the currents and voltages. For an initial value of the load current $I_{Lo} = 2200$ A, the waveforms of the currents and voltages are presented in Fig. 6. The inductance L_{k1} limits the current rise slope when the load current switches from diode 6 to thyristor 2.

$$L_{k1} \geqslant E/S_{i\ krit}\ . \qquad (13)$$

The other data for Fig. 6 are $L_k = 100\,\mu H$, $L_{k1} = 125\,\mu H$, $C_k = 6\,\mu F$, $U_{co} = 1.25 \times 16 = 20$ kV and $\varkappa_m = I_m/I_{LM} = 1.58$.

Suppose that, initially, the load current flows through the main thyristor 2. For quenching it, the auxiliary thyristor 7 is fired. The quenching process occurs as was described in Chapter 2 and diode 5 is conducting until $u_c = E$ and $i_7 = I_{LM} \approx$ const. At that instant, the d-c voltage $- E/2$ is applied to the load. The load current is switched from the auxiliary thyristor 7 to the free-wheeling diode 6 and the capacitor voltage reaches the value $U_{c1} = - 30$ kV (Fig. 6). Then the main thyristor 3 receives a firing pulse so that the same firing sequence 2-7-3-4-2 and so on, can be maintained. Because of diode 6, which conducts, thyristor 3 does not fire (Fig. 6).

When the d-c voltage $+ E/2$ is to be applied to the load and the load current must be positive (the first half-cycle in Fig. 2) the main thyristor 2 must be fired again. The condition for the subsequent extinguishing of this thyristor is a positive capacitor voltage. Therefore, the next thyristor to be fired is thyristor 4. The capacitor C_k discharges and charges to the reverse polarity at the value $U_{c2} = + 30$ kV. By the firing of

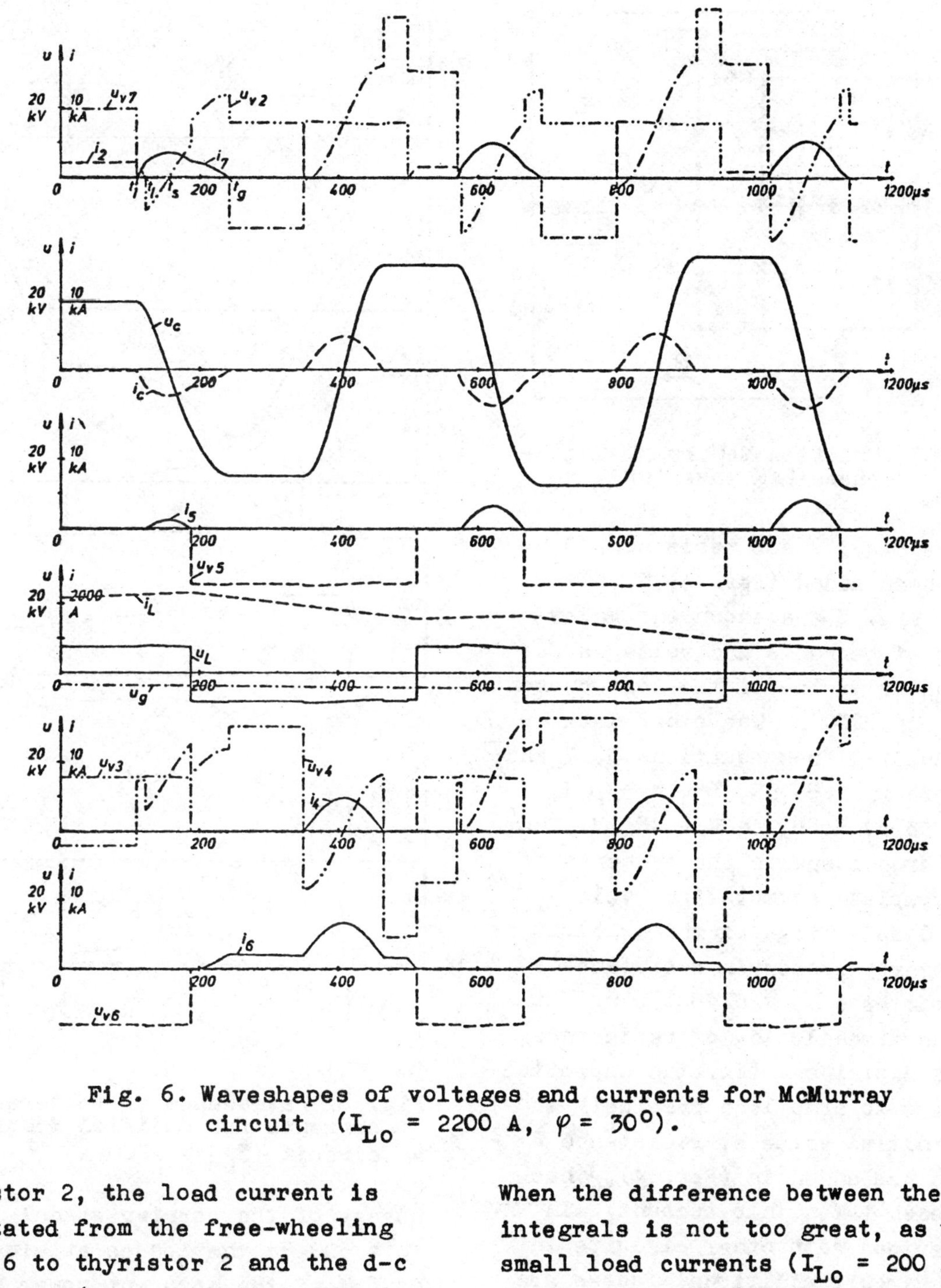

Fig. 6. Waveshapes of voltages and currents for McMurray circuit (I_{Lo} = 2200 A, φ = 30°).

thyristor 2, the load current is commutated from the free-wheeling diode 6 to thyristor 2 and the d-c voltage + E/2 is applied to the load.

A disadvantage of this circuit is that the capacitor voltage is not stabilized. The reason is that

$$\int_{t_1}^{t_s} i_c(t)\, dt \neq \int_{t_s}^{t_g} i_c(t)\, dt \quad . \qquad (14)$$

When the difference between the two integrals is not too great, as with small load currents (I_{Lo} = 200 A), the increase in capacitor voltage is not significant.

4. MODIFIED MCMURRAY CIRCUIT

The modified McMurray circuit is presented in Fig. 7. To limit the overcharging of the capacitor,

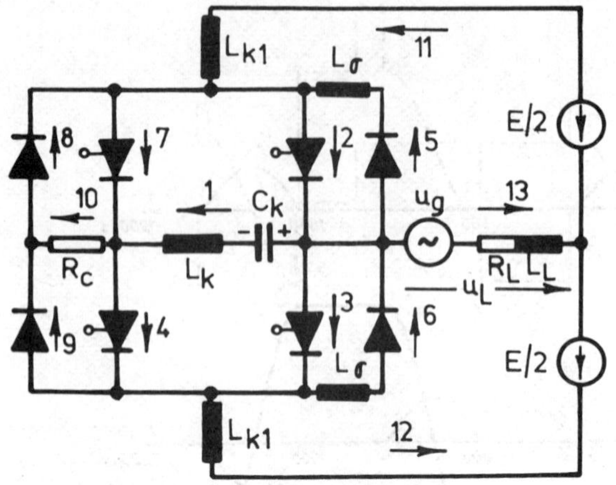

Fig. 7. Modified McMurray single-phase PWM inverter.

diodes 8 and 9 and resistance R_c have been added (Ref. 4, 5, 6).

In Fig. 8 are shown the waveshapes of currents and voltages for an initial value of the load current of I_{Lo} = 2200 A. The other data (calculated from equations 3, 7 and 13) are L_k = 85 μH, L_{k1} = 35 μH, C_k = 13.5 μF, U_{co} = E = 16 kV. There are three steps in the sequence of one complete commutation cycle.

A disadvantage of this circuit is the energy losses in resistance R_c (in our case R_c = 6.18 Ω). In (Ref. 4) the dimensioning of resistance R_c is explained. (For the capacitor voltage to stabilize precisely at the initial value E, resistance R_Q - which was added in (Ref. 4) - has not been taken into account). In comparison with other circuits in this circuit additional losses are found only in resistance R_c.

The investigations carried out so far have shown that the energy losses - calculated in acordance with (Ref. 7) - don't essentially depend on the load current. With a load current of 18% of the maximum load current, the losses attain 80% of the maximum losses. For a frequency of the carrier signal f_T = 500 Hz the losses attain 11.68% of the apparent power of the converter.

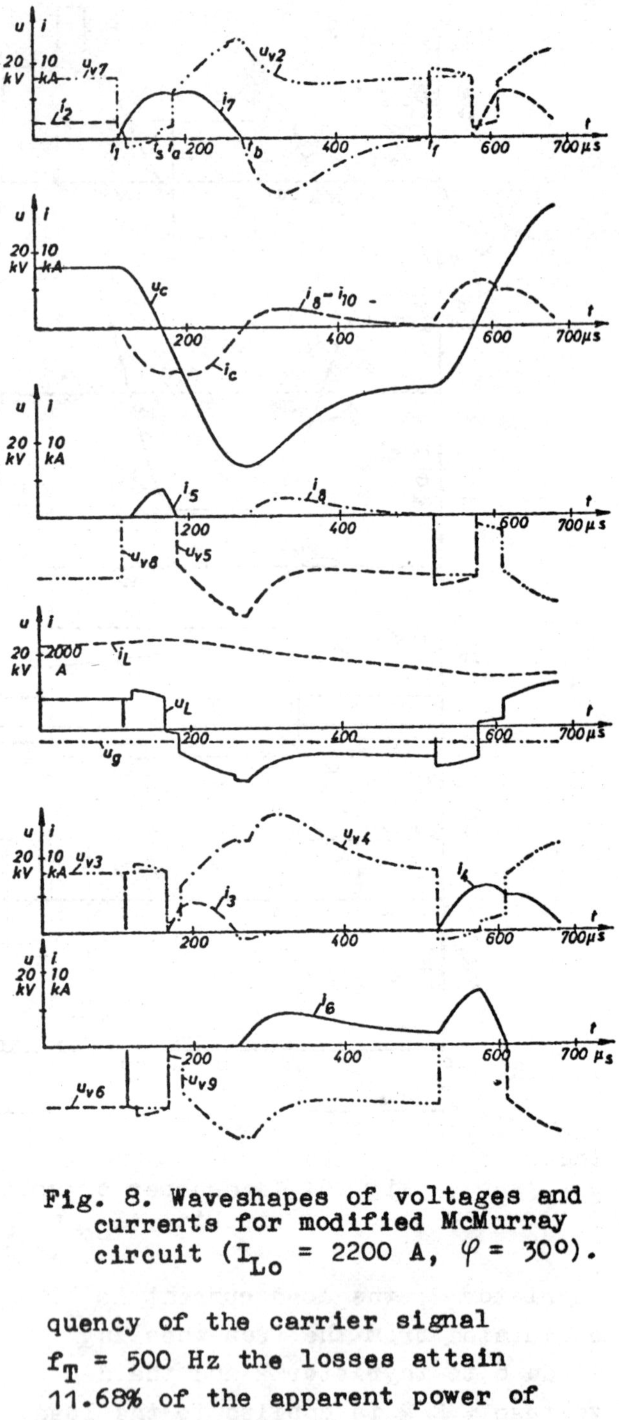

Fig. 8. Waveshapes of voltages and currents for modified McMurray circuit (I_{Lo} = 2200 A, φ = 30°).

5. SCHEME WITH OSCILLATING COMMUTATING CIRCUIT INDEPENDENT OF LOAD

Figure 9 shows a scheme with an oscillating commutating circuit independent of load (Ref. 8, 9, 10).

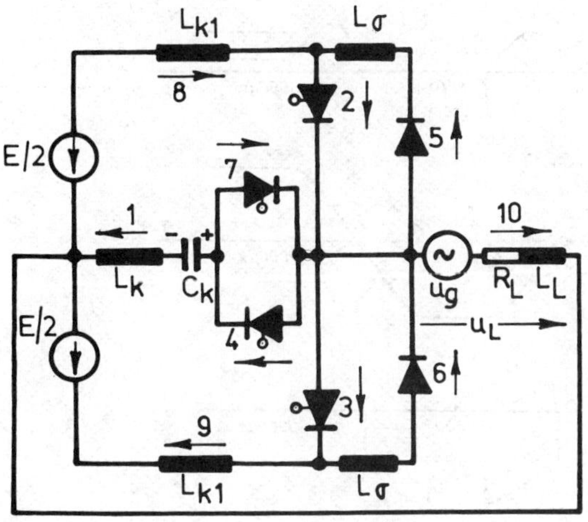

Fig. 9. Single-phase PWM inverter with oscillating commutating circuit independent of load.

The commutating inductances L_k and L_{k1} limit the rise rate of the commutating current in the first time interval.

$$L_k + L_{k1} \geqslant (U_{co} - E/2)/S_{i\ krit}. \quad (15)$$

Voltage and current waveforms for an initial value of the load current I_{Lo} = 2200 A are shown in Fig. 10. Inductances L_{k1} and L_σ limit the rate of rise when the load current is commutated from diode 6 to thyristor 2.

$$2\,L_{k1} + L_\sigma \geqslant E/S_{i\ krit}\ . \quad (16)$$

The leakage inductance L_σ increases the blocking voltage of the main thyristor (Ref. 11). The data for Fig. 10 are $L_k = 35\mu H$, $L_{k1} = 35\mu H$, $L_\sigma = 20\mu H$, $C_k = 15\mu F$, $U_{co} =$ $= 1.125 \times 16 = 18$ kV.

We assume again that the load current flows through the main thyristor 2. For turning it off, at the time $t = t_1$ (Fig. 10) the auxiliary thyristor 7 is fired.

a) In the first interval of time ($t_1 \leqslant t \leqslant t_1$), neglecting the oscillating circuit resistance, the following relation is valid

$$u_c(t) = \frac{E}{2}\left[1 + (2k_u - 1)\cos\omega_{o1}t\right] \quad (17)$$

where $L_k' = L_k + L_{k1}$ and

$$\omega_{o1} = 1/\sqrt{L_k'\,C_k}\ .$$

The capacitor current, with a positive direction according to Fig. 9, is

$$i_c(t) = -\frac{E}{2}(2k_u - 1)\sqrt{\frac{C_k}{L_k'}}\sin\omega_{o1}t. \quad (18)$$

The time t_1 results from (18) provided $i_7(t = t_1) = -i_c(t = t_1) = I_{LM}$. The result is

$$t_1 = \sqrt{L_k'C_k}\ \mathrm{arc\ sin}\left[\frac{2I_{LM}}{E(2k_u-1)}\sqrt{\frac{L_k'}{C_k}}\right] \quad (19)$$

At the end of this time interval ($t = t_1$) the capacitor voltage reaches the value

$$U_{c1}' = u_c(t_1) =$$

$$= \frac{E}{2}\left[1 + (2k_u - 1)\cos\left(\frac{t_1}{\sqrt{L_k'C_k}}\right)\right] \quad (20)$$

and the load current is commutated from the main thyristor 2 to the auxiliary thyristor 7.

b) After the main thyristor 2 is turned-off, the load current flows in the second oscillating circuit through diode 5 and the capacitor is discharged (Fig. 10 and 11a). For this interval of time ($t_1 \leqslant t \leqslant t_s$), the following differential equation is valid

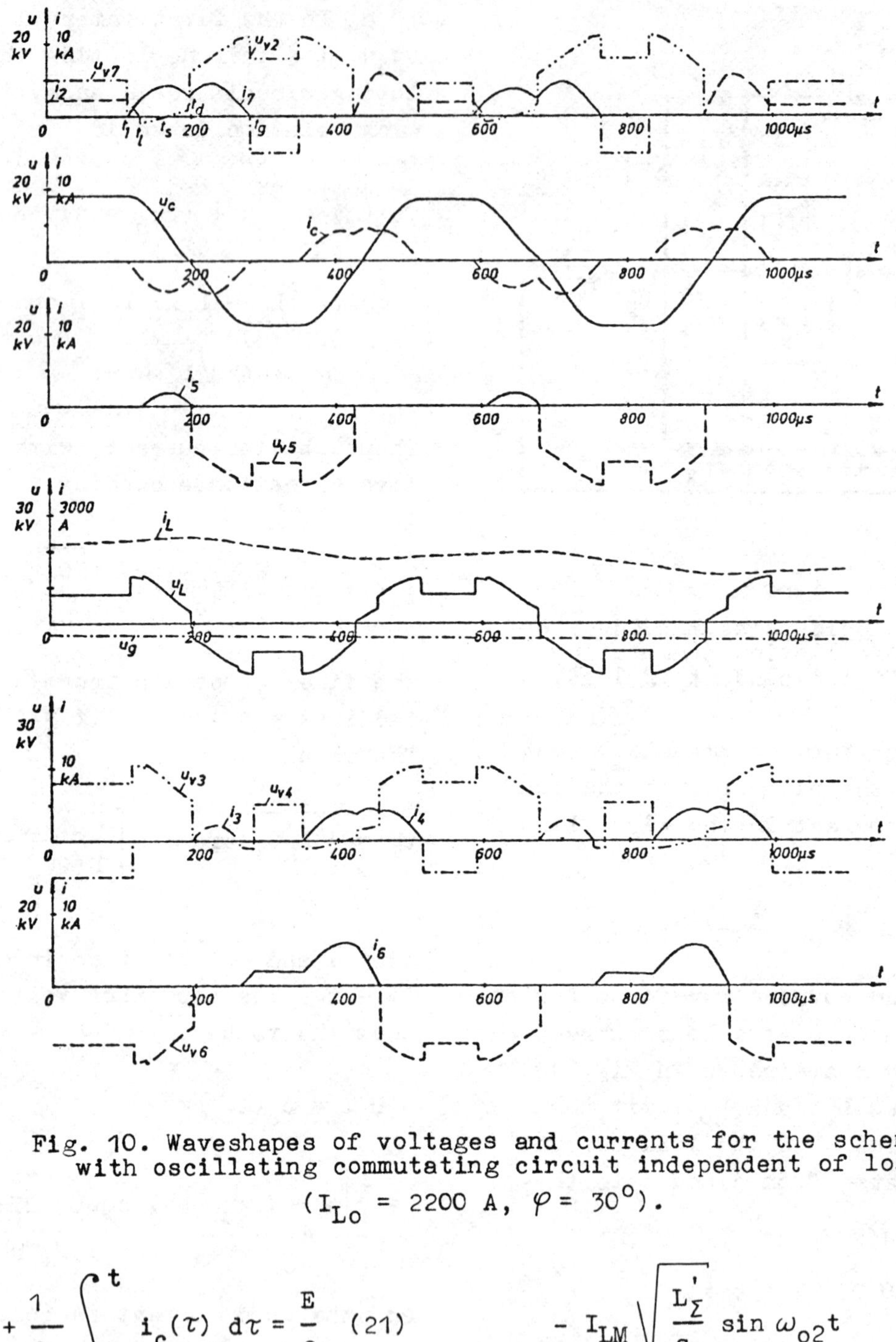

Fig. 10. Waveshapes of voltages and currents for the scheme with oscillating commutating circuit independent of load (I_{Lo} = 2200 A, φ = 30°).

$$L'_\Sigma \frac{di_c}{dt} + \frac{1}{C_k}\int_0^t i_c(\tau)\,d\tau = \frac{E}{2} \quad (21)$$

where $L'_\Sigma = L_k + L_{k1} + L_\sigma$ is the inductance of the second oscillating circuit. The solution of the differential equation is

$$u_c(t) = \frac{E}{2} + (U'_{cl} - \frac{E}{2}) \cos \omega_{o2} t - I_{LM}\sqrt{\frac{L'_\Sigma}{C_k}} \sin \omega_{o2} t \quad (22)$$

with $\omega_{o2} = 1/\sqrt{L'_\Sigma C_k}$. The capacitor current is

$$i_c(t) = -(U'_{cl} - \frac{E}{2})\sqrt{\frac{C_k}{L'_\Sigma}} \sin \omega_{o2} t - I_{LM} \cos \omega_{o2} t . \quad (23)$$

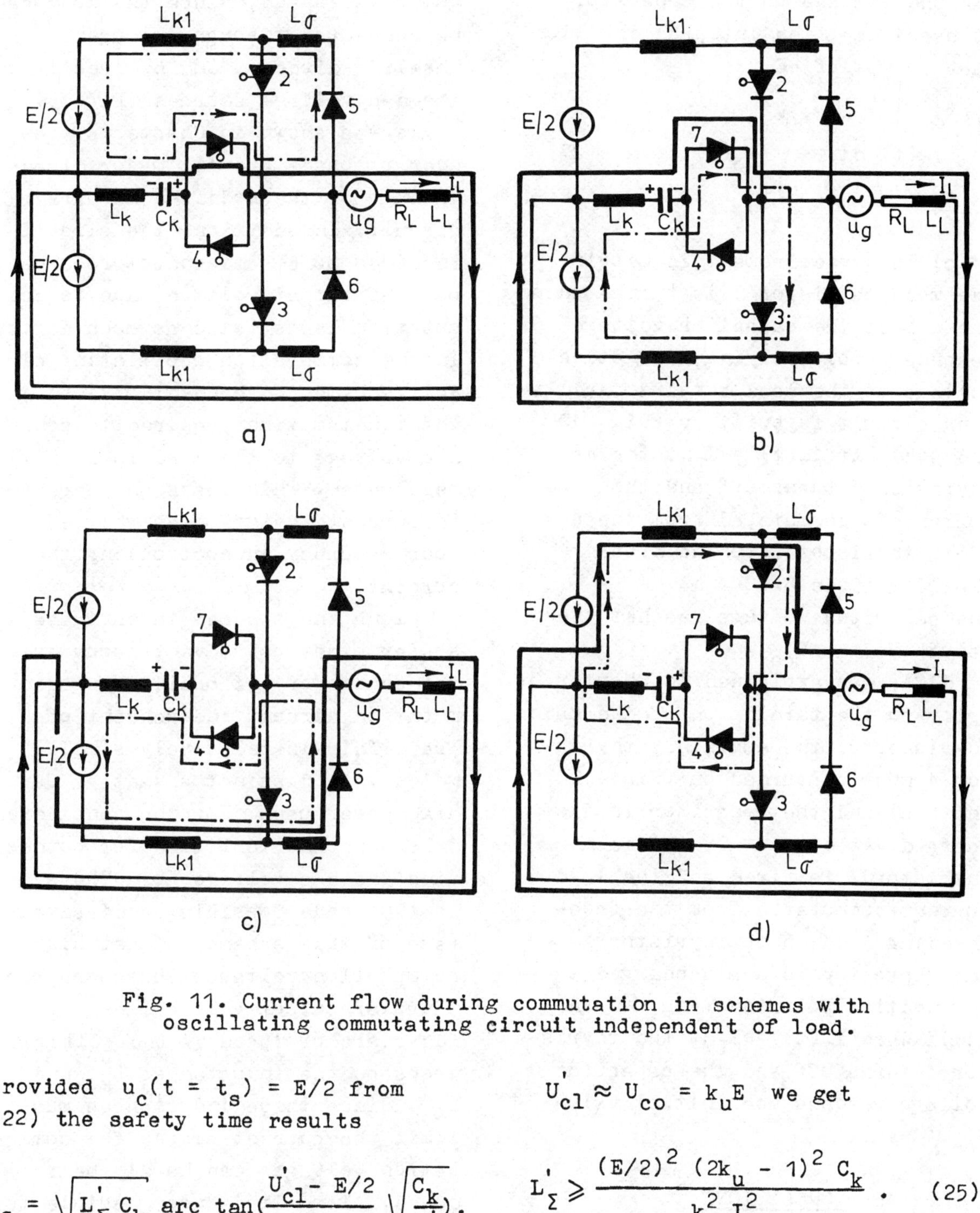

Fig. 11. Current flow during commutation in schemes with oscillating commutating circuit independent of load.

Provided $u_c(t = t_s) = E/2$ from (22) the safety time results

$$t_s = \sqrt{L'_\Sigma C_k} \text{ arc tan}\left(\frac{U'_{cl} - E/2}{I_{LM}} \sqrt{\frac{C_k}{L'_\Sigma}}\right). \quad (24)$$

If (24) is substituted in (23) we obtain the maximum value of the commutation current $i_7(t = t_s) = - i_c(t = t_s) = I'_m$. Assuming that the commutation current is sinusoidal and with $k_i = I'_m/I_{LM}$, $U'_{cl} \approx U_{co} = k_u E$ we get

$$L'_\Sigma \geqslant \frac{(E/2)^2 (2k_u - 1)^2 C_k}{k_i^2 I_{LM}^2} . \quad (25)$$

The time intervals t_1 and t_s (19, 24) depend on the load current. As the dependence of t_1 and t_s on I_{LM} is opposite, their sum remais approximately constant.

$$t_1 + t_s \approx T_{o2}/4 = \text{const.} \quad (26)$$

For the voltage on the capacitor to stabilize, the integral equation must be satisfied

$$\int_{t_1}^{t_d} i_c(t)\,dt = \int_{t_d}^{t_s} i_c(t)\,dt. \qquad (27)$$

c) This requirement is met if the main thyristor 3 is turned on at the moment the capacitor voltage reaches zero. Applying a negative voltage to the load a third oscillating circuit is built up (Fig. 10 and 11b). At $i_7(t) = I_{LM}$, the main thyristor 3 turns off and the load current is commutated from the auxiliary thyristor 7 to the free-wheeling diode 6.
The capacitor voltage reaches the value $U_c = - U_{co}$.

d) As was explained in Chapter 3, prior to the turning on of the main thyristor 2, the auxiliary thyristor 4 must be turned on (Fig. 10 and 11c) and the capacitor is discharged. At $u_c(t) = 0$, the main thyristor 2 is fired and the load current commutates from the free-wheeling diode 6 to thyristor 2. The capacitor is again charged to the initial polarity (Fig. 10 and 11d). When $i_7(t) = I_{LM}$, the thyristor 4 turns off and the capacitor voltage reaches the initial value U_{co}.

6. CONCLUSIONS

The circuit discussed in Chapter 3 cannot be utilized for feeding a linear synchronous motor with extended winding, because the voltage on the capacitor is not stabilized. Obviously, the output voltage of this scheme contains relatively few harmonics because of its rectangular waveshape, since the load can be connected through the free-wheeling diodes 5 and 6 directly to the d-c voltage sources (Fig. 5).

As was shown in Chapter 4, the energy losses are the major disadvantage of the modified McMurray circuit. In addition, the circuit requires more semiconductors. The peak of output voltage, due to commutation lasts twice as much as that in the commutation oscillating circuit scheme independent of load. The minimum time required to connect d-c voltage to the load is 2.25 times longer. This means an increase in current deviation as well as poor accuracy in controlling the current.

Among the schemes investigated so far, the most advantageous is the scheme of the oscillating commutation circuit independent of load. This is relatively simple which is a distinct advantage for high power uses. For the considered data, the shortest time for connecting the d-c voltage, of 170 μs, is thus made possible. A disadvantage of this scheme is that high commutation voltages superpose rectangular output voltage.
These are produced by the voltage drops on the inductances L_{k1} and L_σ. Since these inductances must limit the current slope, the commutation voltages can hardly be reduced. Also, it is not possible to connect the free-wheeling diodes and their inductances L_σ directly to the d-c sources (analogous McMurray scheme), since the main thyristors cannot be turned off any longer.

To build a very high power PWM inverter each of the sketched semiconductors must consist of a connection of thyristors in series and in parallel. The peak, mean and root-mean-square values of valve voltages and of commutation currents do not essentially depend on the load current.

Connecting semiconductors in parallel the critical current slope increases N_p time, where N_p is the number of semiconductors connected in parallel. The commutation inductances can be reduced (Equations 15 and 16). For a certain turn-off time t_q, the sum $t_l + t_s$ must remain constant, therefore $L_{\Sigma} C_k$ = const. (Eq. 26). In order to fulfill this condition, the capacitor must be increased. That means greater energy, since the initial voltage on the capacitor cannot be reduced. Thus the commutation currents increase. That is why the commutating inductances cannot be reduced in the same proportion.

7. REFERENCES

(1) Eisenack, H. and Hofmeister, H. Digitale Nachbildung von elektrischen Netzwerke mit Dioden und Thyristoren, Arch. Elektrotechn. 55, 32 (1972).

(2) Bedford, B. D. and Hoft, R. G. (1964) Principles of Inverter Circuits, Wiley, New York.

(3) Leonhard, W. (1974) Regelung in der elektrischen Antriebstechnik, Teubner, Stuttgart.

(4) Penkowsky, L. J. and Pruzinsky, K. E. Fundamentals of a Pulse Width Modulated Power Circuits, IEEE Conf. Record of IGA, 669 (1970).

(5) Prakash, H. N. Auxiliary Power Converter, Solid State Power Control Program Progress Report, 6. 1(1973 - 1974).

(6) Pollack, J. Advanced Pulsewith Modulated Inverter Tehniques, IEEE Trans. on IA, IA - 8, 145 (1972).

(7) Prakash, H. and Hoft, R. G. Optimizing the PWM Waverform of a Thyristor Inverter, Solid State Power Control Program Progress Report, VII, 1(1973 - 1974).

(8) Brenneisen, J. and Schönung, A. Bestimmungsgrössen des selbsstgeführten Stromrichters in sperrspannungsfreier Schaltung bei Steuerung nach dem Unterschwingungsverfahren, ETZ - A, 90, 353 (1969).

(9) Brenneisen, J. A New Converter Drive System for a Diesel Electric Locomotive with Asynchronous Traction Motors, IEEE Trans. on IA, IA - 9, 482 (1973).

(10) Stoschek, J. 5 - MVA - Umrichter für einen Linearmotorantrieb, BBC - Nachr. 4, 192 (1975).

(11) Bühler, H. Umrichtergespeiste Antriebe mit Asynchronmaschinen, N. T. 16, 121 (1974).

STEADY-STATE BEHAVIOUR OF A HIGH FREQUENCY THYRISTOR INVERTER

É. Miklós and I. Nagy

Computer and Automation Institute, Hungarian Academy of Sciences, 1502 Budapest

ABSTRACT

The paper is concerned with a special type of high frequency thyristor inverter. It can be especially applied for induction heating. The particular inverter circuit discussed in the paper uses the principle of time sharing to enable the circuit turn-off time to be extended.

The objective of this paper is to present a steady-state analysis based on an assumption concerning the load and to summarize the digital computer results of the approximate investigation. The load is a parallel oscillatory circuit. It is substituted by a sinusoidal voltage of constant amplitude and frequency. Being the inverter output current only a fraction of the condenser or coil current of the load the approximation is justified.

The numerical results are plotted in various charts in normalized form. They present the calculated relations among the inverter variables, e.g. firing angle, conduction angle, input voltage, output power etc. Various boundary curves help the orientation in the charts and encircle the possible operational areas.

The analysis revealed certain conditions in which the inverter is instable or sudden change in its steady-state operation must take place. The digital computer results help the designer to select and optimize the system parameters, to set operation point or devise control strategy of the inverter.

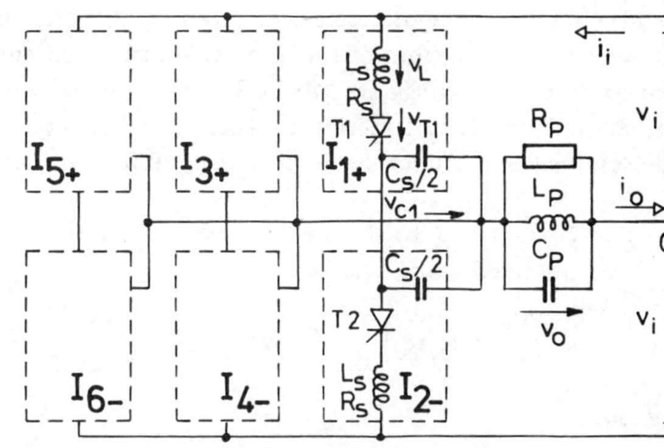

Fig. 1. Inverter configuration

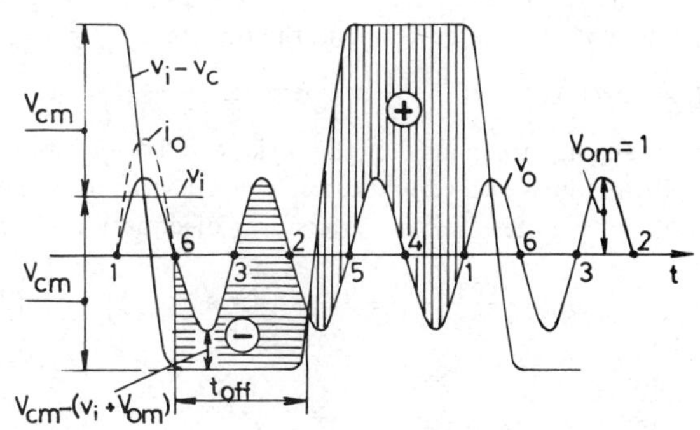

Fig. 2. Operation principle

INTRODUCTION

The paper investigates the behaviour of a high frequency thyristor inverter using the principle of time-sharing in steady state. The inverter serves for the generation of sinusoidal voltage in the frequency range between 4 kHz and some ten kHz and can be used e.g. in induction heating.

Figure 1 shows the basic inverter configuration. It is fed by a center-tapped d.c. voltage $2v_i$, its load is the parallel oscillatory circuit R_p, L_p, C_p. The inverter itself in the strict sense is built up of three similar so-called positive subinverters (I_{1+}, I_{3+} and I_{5+}) and of three similar negative subinverters (I_{2-}, I_{4-} and I_{6-}).

The principle of operation of the configuration can be traced in Fig. 2. Being the damping factor of the parallel oscillatory circuit small, the load voltage v_0 with good approximation alternates sinusoidally. At the time of every zero crossing of voltage v_0 a thyristor is fired, those of the positive subinverters at times 1, 3, 5, while those of the negative ones at times 2, 4, 6. E.g. as a result of the firing of thyristor $T1$ a current pulse of half a period long is flowing in the series oscillatory circuit $L_s - C_s$ in case of an appropriate choice of $L_s - C_s$, and it delivers power to the load on the one hand, and transpolarizes the voltage of the series condenser C_s from $-V_{cm}$ to V_{cm}. Since current is flowing again in condenser C_s from time 2 on, the condenser voltage v_c is constant from 6 to 2, that is, in a full period. The thyristor voltage $T1$ is obtained from the vertical intersection between the time functions of v_0 and $v_i - v_c$. The turn-off time t_{off} of the thyristor exceeds one period. Because of the symmetric configuration and control every subinverter works alike I_{+1}. The repetition frequency of the variables of a subinverter is one third of the frequency of the voltage v_0. The frequency division and the long t_{off} is the consequence of time-sharing operation. By building-in more subinverters instead of six, the frequency-sharing and t_{off} can be further increased.

APPROXIMATE INVESTIGATION

It is assumed in the investigation that the input d.c. voltage v_i is smooth and the output high frequency a.c. voltage v_0 is sinusoidal having constant amplitude V_{0m} and frequency f_p. The first assumption is justified by the substantial filtration, the second by the current of the reactive elements of the load being about tenfold of the output current of the inverter for inductive heating. Due to the first assumption the energy storage elements of the filter circuit, and to the second one those of the load circuit can be neglected in steady-state. On the other hand, on account of the two assumptions the subinverters can be considered to be independent of each other and the time function of their variables can be determined independently of the other subinverter variables.

The current $i_0(t)$ flowing in a positive subinverter can be obtained from equation

$$L_s \frac{di_0}{dt} + R_s i_0 + \frac{1}{C_s} \int i_0\, dt + v_0 = v_i \tag{1}$$

where

$$v_0 = V_{0m} \sin(\omega_p t + \alpha). \tag{2}$$

α is the firing angle, $\omega_p = 2\pi f_p$. Let the relative unit of the voltage, the current and the time be: $V_b = V_{0m}$; $I_b = \omega_s C_s V_{0m}$; $t_b = \frac{1}{\omega_p}$, where $\omega_s = \frac{1}{\sqrt{C_s L_s}}$ is the own angular frequency of the series oscillatory circuit.

From now on every relation will be written in per unit system. The Laplace transform of equation (1) is

$$I_0 = -\frac{1}{\Omega D(s)} \left[\frac{\cos\alpha + s\sin\alpha}{1+s^2} s - (v_i + V_{cm})\right] \tag{3}$$

where

$$D(s) = 1 + 2\xi\left(\frac{s}{\Omega}\right) + \left(\frac{s}{\Omega}\right)^2 \tag{4}$$

$$\Omega = \frac{\omega_s}{\omega_p}; \quad \xi = \frac{R_s}{2}.$$

The time function $i_0(t)$ can be found in Appendix 1. The average output power of the inverter

$$P_0 = \frac{1}{\pi} \int_0^{\alpha_e} v_0 i_0\, dt \tag{5}$$

while the average input power

$$P_i = \frac{1}{\pi} \int_0^{\alpha_e} v_i i_0\, dt = \frac{2}{\pi\Omega} v_i V_{cm} = 2 v_i I_i \tag{6}$$

where α_e is the conduction angle of the thyristor, I_i is the average value of the inverter input current. By preselecting three variables, for instance α, α_e and $V_{0m} = 1$, all other variables of the inverter, e.g. v_i, V_{cm}, P_0, P_i, furthermore all time functions of the subinverter can be computed (Appendix 1).

INVERTER STATES

Depending on the average value of output current i_0, that is, the amplitude V_{cm}, the inverter can be in three states as shown in Fig. 3. In states 1 and 2 the number of the extreme values of output current i_0 is either one (Figs. a and c) or three (Figs. b and d), while in state 3 this number is one or in limit case two (Fig. f). The turn-off time in state 1 is $1 \sim 2$ half period long, while in states 2 and 3 substantially smaller t_{off} values may occur (Figs. c and e), moreover in limit case t_{off} can in principle be zero, as well (Fig. f).

The transition between various states or between output current shapes having different number of extreme values can be continuous, but also discontinuous. The latter can be understood from Figs. b, d and f. The vertical intersection between curves v_0 and $v_i - v_c$ presents the voltage v_L developed across inductance L_s ($R_s = 0$). In Fig. f i_0 extincts with zero slope at time α_{e2} $[v_L(\alpha_{e2}) = 0]$. By increasing the firing angle α by $d\alpha$, the instantaneous value $i_0(\alpha_{e2}) > 0$ and the current i_0 will increase again at times $t > \alpha_{e2}$ following directly $t = \alpha_{e2}$, for v_L is positive. A second current wave develops. The average value of current i_0, the amplitude V_{cm} of the series condenser and most inverter quantities drasticly change.

Figures b and d also show a limit case, for here current i_0 is zero at time of the current minimum. By reducing the firing angle α by $d\alpha$ the curve $i_0(t)$ cuts the time axis in the neighbourhood of the current minimum. The thyristor turns off and the second current wave disappears. The inverter variables suddenly change in the direction opposite to the previous case.

CALCULATION RESULTS ($\Omega = 1$, $R_s = 0$)

Figures 4, 5, 6 and 7 illustrate the change of the four most important variables of the inverter: α_e, v_i, V_{cm} and t_{off} as a function of α and P_0. The parameter is in the first three figures P_0, in the last α.

In order to encircle the operation ranges and partly for orientation some of the so-called boundary curves have been plotted in the figures. The boundary curve denoting the fulfilment of the equality $V_{cm} = 1 + v_i$ and $V_{cm} = v_i$ separates the state 1 from 2 and 2 from 3, respectively. The current i_0 has two and three extreme values along the boundary curve α_{e2} and α_{e3}, respectively. Along the curve α_{e2} the current extincts with zero slope and therefore $t_{off} = 0$ (Fig. 3f), while at the points of the curve α_{e3} the current at the location of its middle extreme value just touches the time axis (Figs. 3b and d). The boundary curve α_{ei} separates the ranges pertaining to the one and to the three current extreme values. Furthermore, the curves bordering the operating ranges and pertaining to powers $P_0 = 0$ and $P_0 = \infty$ have been plotted, too. The way of determining the boundary curves is discussed in Appendix 2.

The information content of the curve sets is discussed mainly in connection with Fig. 4 plotted in the plane $\alpha_e \sim \alpha$. We have taken for the sake of explanation the path plotted in Figs. 4, 5 and 6 with heavy line and marked by points $1, 2, \ldots, 7, 8$. Starting from point 1,

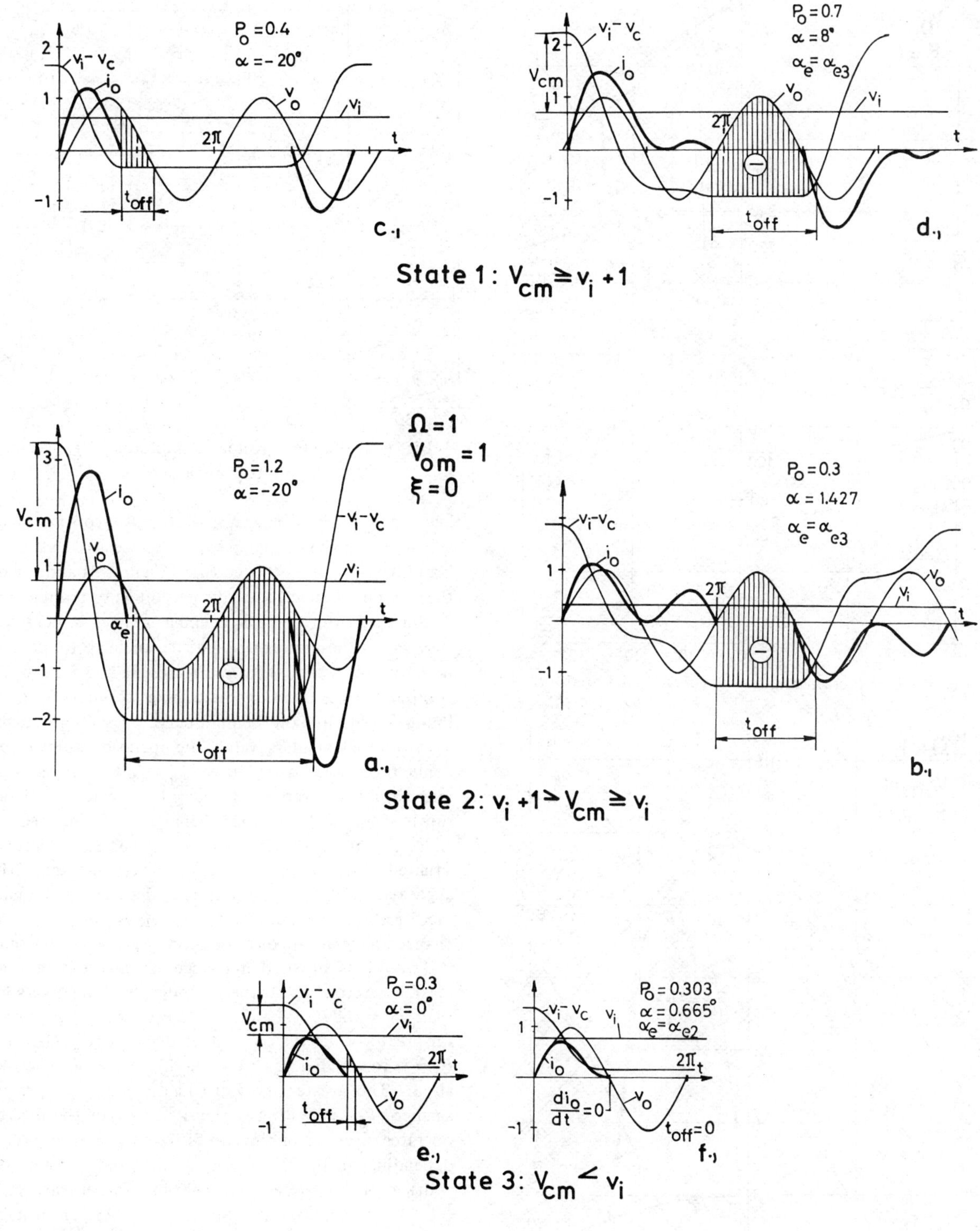

Fig. 3

Computed time functions in state 1, 2 and 3

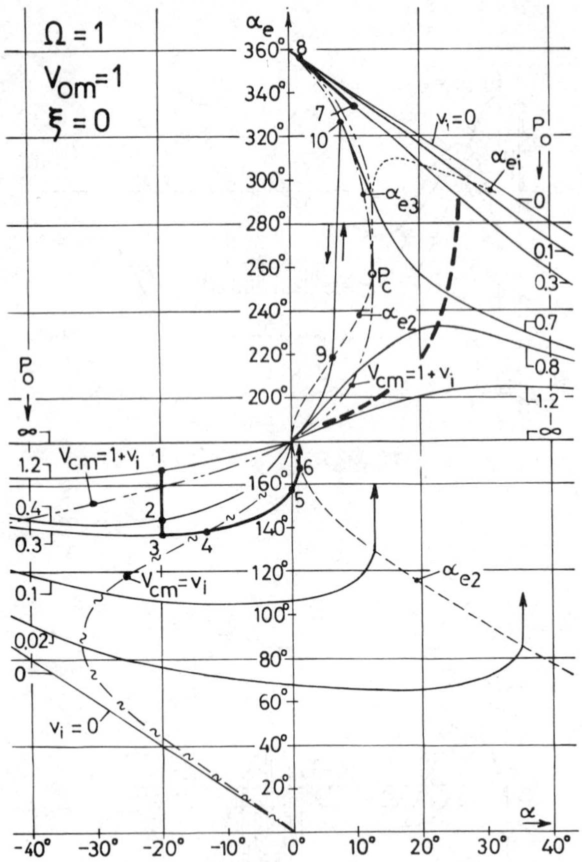

Fig. 4. Conduction angle α_e versus firing angle α

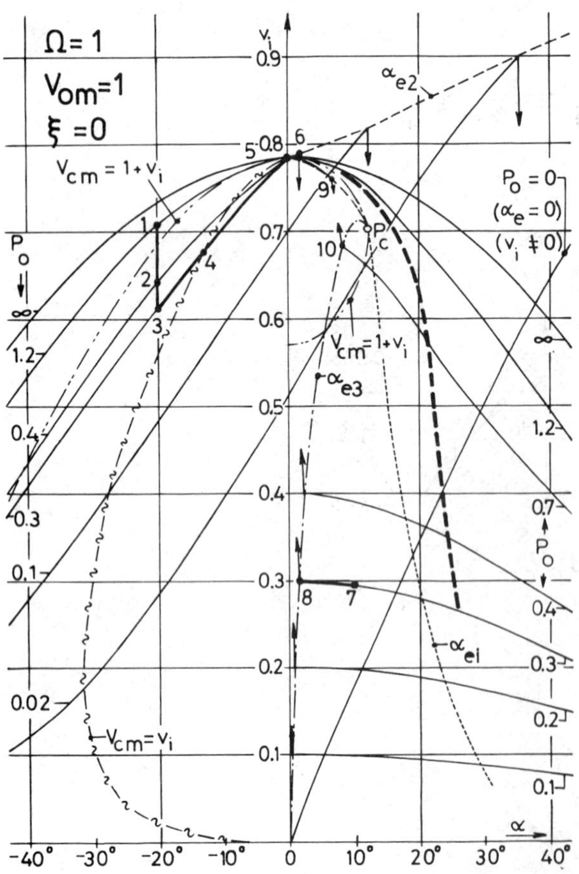

Fig. 5. Input voltage v_i versus firing angle α

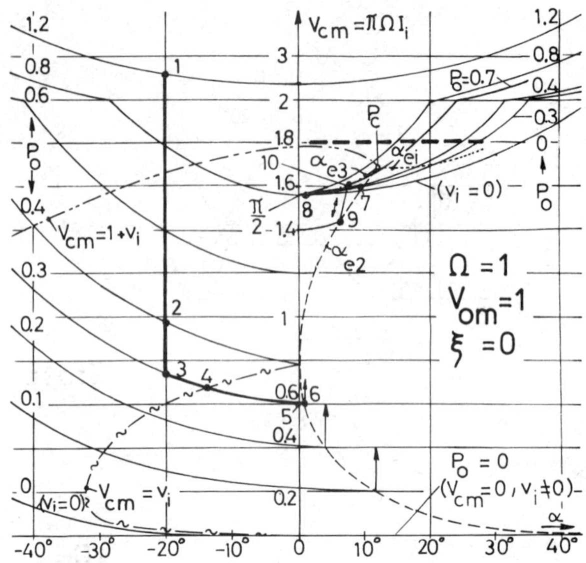

Fig. 6. Condenser voltage amplitude V_{cm} versus firing angle α

keeping the firing angle $\alpha = -20°$ at a constant value and reducing the output power P_0, we arrive at points 2 and 3. V_{cm} and P_0 decrease. Figures 3a and c show the computed time functions pertaining to points 1 and 2. Starting from point 3, keeping $P_0 = 0.3$ at a constant value and increasing the firing angle α we proceed to points 4, 5 and then 6. Figures 3e and f show the computed time functions pertaining to points 5 and 6. Point 6 is situated on the limit curve α_{e2}. By increasing α with an elementary value, the operation point is suddenly transfered to the curve $P_0 = 0.3$ running in the long conduction range, while v_i, V_{cm} and t_{off} is abruptly changed. By further changing the firing angle we arrive e.g. at either operation point 7 or 8. The latter is situated on the limit curve α_{e3}. The elementary firing angle reduction from point 8 provides again a dramatic modification in the inverter variables, but now the changes have an opposite direction to the former ones.

Two kinds of sudden change in inverter state are feasible. According to the computation for powers less than $P_0 = 0.275$ the sudden increase of α_e takes place for greater value than its sudden decrease (e.g. the curve $P_0 = 0.1$ in Fig. 5). Thus a hysteresis loop comes about. The greater its width in direction α is, the smaller P_0 is. Within the hysteresis loop the inverter operates in either of the two different operation points, depending on the initial condition of the inverter. The width of the hysteresis loop shrinks to zero at $P_0 = 0.275$ and $\alpha = 1.22$. For $P_0 > 0.275$ an unstable state comes about in a certain section of axis α instead of the hysteresis loop. Thus there are no stable inverter state e.g. for $P_0 = 0.7$ for firing angles between values α pertaining to points 9 and 10 i.e. in the range $\Delta\alpha \cong 2$.

P_c is the so-called critical operation point which is the meeting point of the limit curves $V_{cm} = 1 + v_i$, α_{e2}, α_{e3} and α_{ei}. Here $P_0 = 0.765$, $\alpha = 12.55°$ and $\alpha_e = 257.45°$ $(\alpha + \alpha_e = 270°)$. A sudden change in inverter state can take place only in case of $P_0 < 0.765$. By ensuring $\alpha > 12.55°$ and operating the inverter

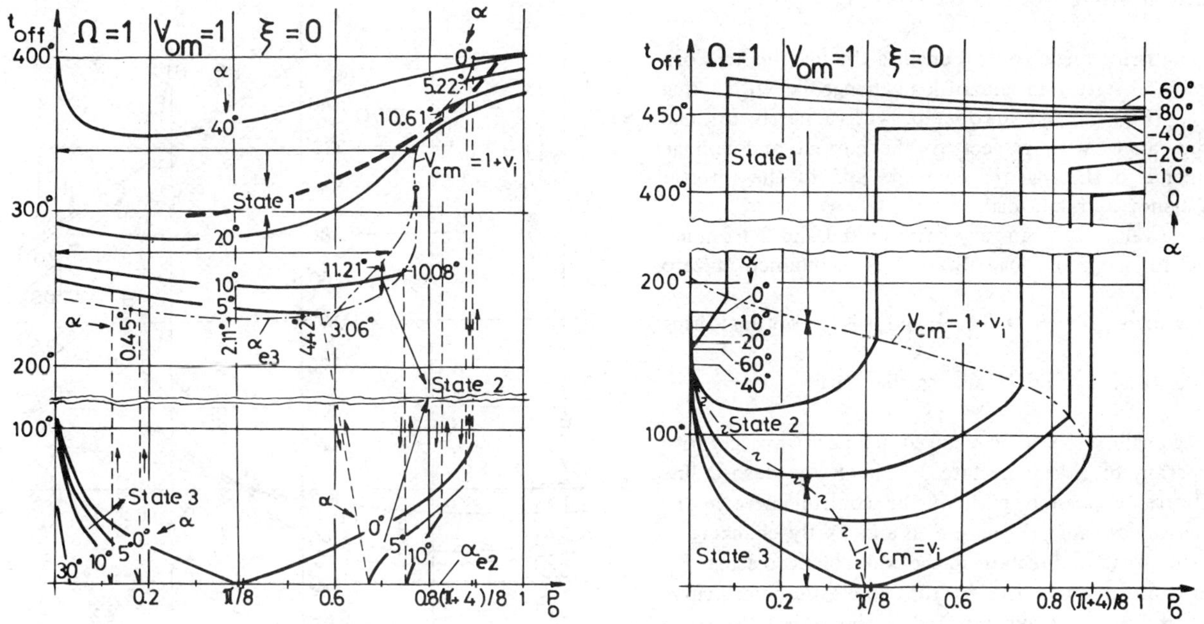

Fig. 7. Turn-off time t_{off} versus output power P_0

only in the long conduction range ($\alpha_e > 180°$) is another way for the exclusion of the sudden change in the inverter state. In the latter case the inverter always works in state 1 i.e. t_{off} is long.

Let us introduce a "figure of state" of two digits to characterize the operation of the inverter. Let the first digit characterize the state, the second the number of the extreme values of the current curve i_0. Figures 8 and 9 show the partition of the whole operation range of the inverter into areas having the same figure of state. The particular figure of state is written into each area. Two figures are required for positive α in the plane $v_i \sim \alpha$ because here the long and short conduction ranges overlap.

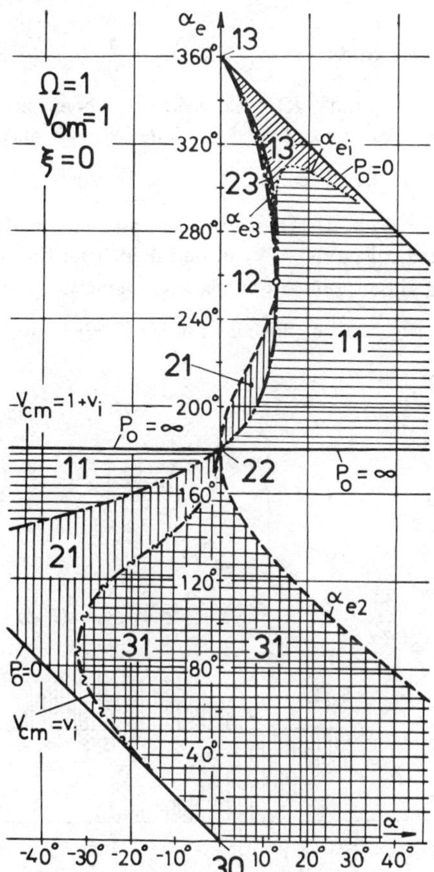

Fig. 8. Areas of various figure of state

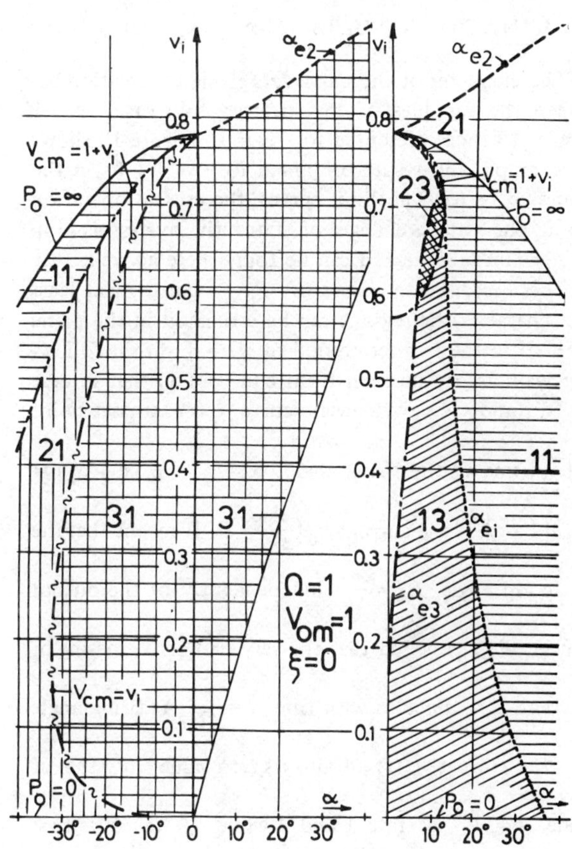

Fig. 9. Areas of various figure of state

FIRING ANGLE CONTROL ($\Omega = 1,\ R_s = 0$)

The firing angle control enables the inverter to work always in state 1 in spite of load change, i.e. with a long t_{off} (Fig. 7) and to operate with a nearly optimal firing angle with respect to the current and voltage demand in the inverter elements and to the inverter efficiency at rated load.

The value P_0 can vary between 0.3 and 2 from no-load to maximum load. State 1 is maintained at zero firing angle, for $P_0 \geqslant \frac{\pi + 4}{8}$ and it is the optimal firing angle (Fig. 6). For loads smaller than $P_0 = \frac{\pi + 4}{8}$ firing angle α has to be changed in a positive or negative direction to maintain state 1. The heavy dashed line indicates a possible path of the control curve in the positive α range (Figs. 4, 5, 6 and 7). By changing α in the negative direction in the wake of decreasing P_0, the control curve has to run above the limit curve $V_{cm} = 1 + v_i$. Figures 6 and 7 show that the latter solution is more advantageous with respect to the current and voltage demand of inverter elements and to t_{off} and neither sudden change of state, nor unstable operation region can be found in this range. State 1 can in principle be secured even at constant α for every load at firing angles $\alpha \geqslant 12.45°$ or $\alpha \leqslant -66°$. The latter, however, cannot be a practical solution because of the heavy duty of the elements and the poor efficiency.

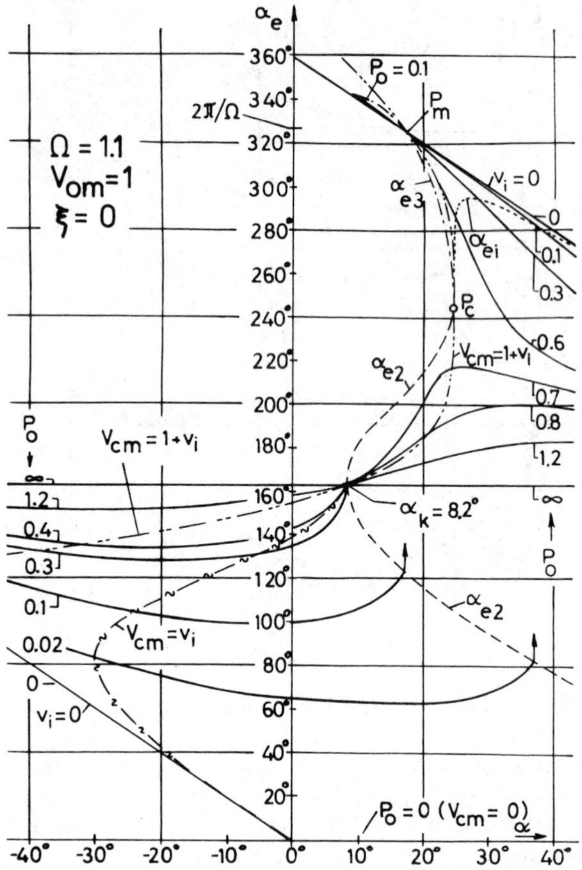

Fig. 10. Conduction angle α_e versus firing angle α

COMPUTATION RESULTS ($\Omega \neq 1,\ R_s = 0$)

The character of the curve sets giving the relation between the variables of the inverter is in case $\Omega \neq 1$ similar to those presented for $\Omega = 1$. Figure 10 shows as example the results obtained for $\Omega = 1.1$ in the plane $\alpha_e \sim \alpha$. The limit curves, the critical operation point, the sudden changes of states, the hysteresis loop and the unstable region can be found here as well.

The approximate curves of $P_0 =$ const or at least the characteristic regions can be estimated in the plane $\alpha \sim \alpha_e$ in the neighbourhood of $\Omega = 1$ in the following way: It can be seen from equ. (A1.1) that in case of α and Ω given every curve $i_0(t)$ crosses point $i_0\left(\frac{\pi}{\Omega}\right)$ and at the firing angle $\alpha = \alpha_k = \frac{1}{2}(\pi - \alpha_e) = \frac{\pi}{2}\left(1 - \frac{1}{\Omega}\right)$ the current $i_0\left(\frac{\pi}{\Omega}\right) = 0$. It means that the extiction angle $\alpha_e = \frac{\pi}{\Omega}$ independently of the output power P_0, provided that the very first zero crossing of current i_0 coincides with time $t = \frac{\pi}{\Omega}$. At firing angle α_k the current $i_0(t)$ attains its zero value truly first at time $t = \frac{\pi}{\Omega}$ in state 1 and 2 while i_0 takes its first zero value before $t = \frac{\pi}{\Omega}$ in state 3. Every curve P_0 crosses the points $\alpha = \alpha_k$ and $\alpha_e = \frac{\pi}{\Omega}$ in states 1 and 2 (Fig. 11 point K). Considering Appendix 2 the boundaries can be obtained in plane $\alpha_e \sim \alpha$ according to Fig. 11. Every curve $P_0 \geqslant \frac{2}{\pi\Omega} D_1^2 \sin^2\alpha_k$ crosses point K (App. 3). The values α_c and α_{ec} pertaining to the critical point P_c can be taken from Fig. 12. By knowing three points of curve α_{e2}, namely P_c, K and $\left(\frac{\pi}{2}, 0\right)$, it can be plotted approximately from these.

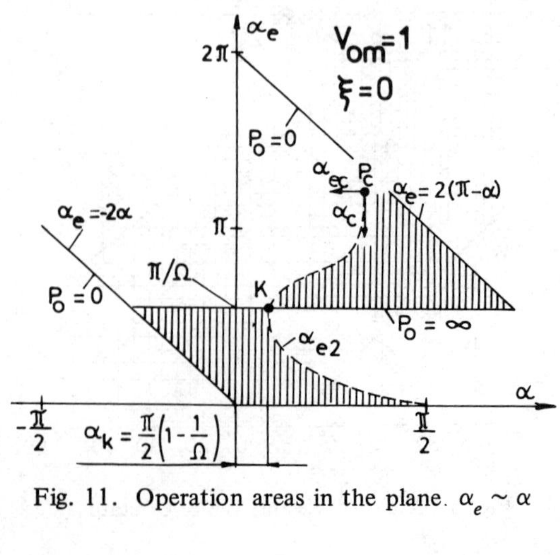

Fig. 11. Operation areas in the plane $\alpha_e \sim \alpha$

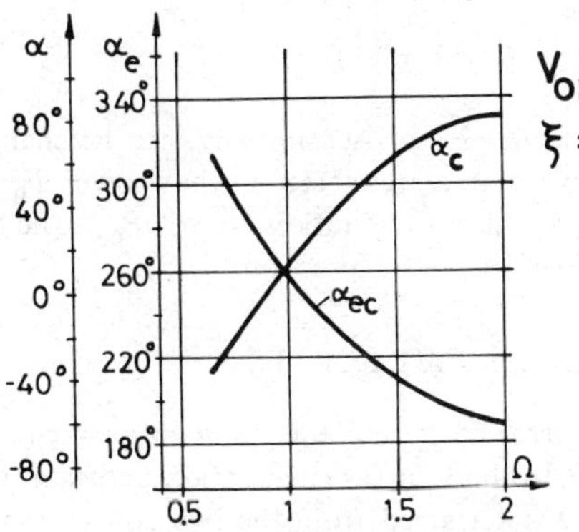

Fig. 12. Coordinate of critical point in the function of Ω

In fact, the changes of α and P_0 make the inverter frequency vary, too. [Ref. 4] Its effect can be taken into account by the numerical figures taken from the curve sets computed for different Ω-s.

COMPUTATION RESULTS ($R_s \neq 0$)

The computation was reiterated for a damping factor of the series oscillatory circuit $\xi = 0.02$. The results for $\Omega = 1$ are shown by Figures 13-16. The curves plotted in the planes $\alpha_e \sim \alpha$ and $V_{cm} \sim \alpha$ are hardly differing from those obtained for $R_s = 0$. Because of the voltage across the series resistance the voltage v_i is much higher than that of Fig. 5 and $P_i > P_0$. The inverter efficiency $\eta = \dfrac{P_0}{P_i}$ as a function of α is shown in Fig. 16. By increasing the absolute value of α, the reactive component of the fundamental harmonic of current i_0 generally increases at P_0 = const., therefore η decreases.

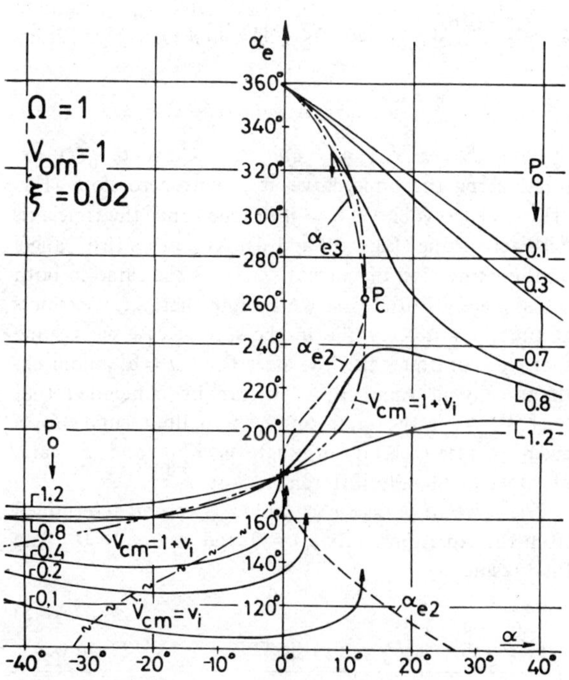

Fig. 13. Conduction angle α_e versus firing angle α

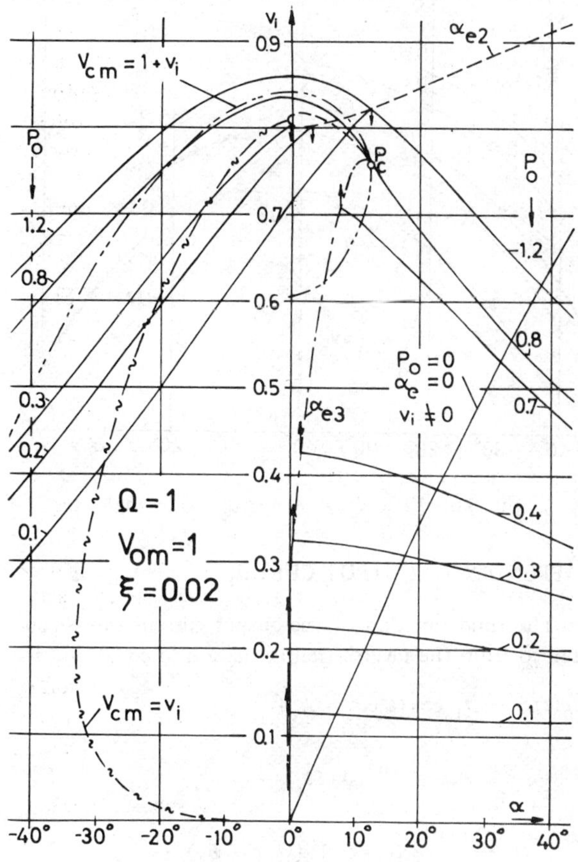

Fig. 14. Input voltage v_i versus firing angle α

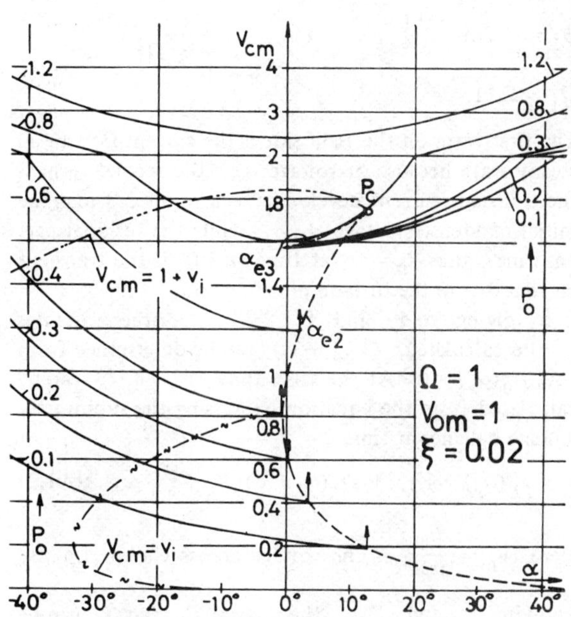

Fig. 15. Condenser voltage amplitude V_{cm} versus firing angle α

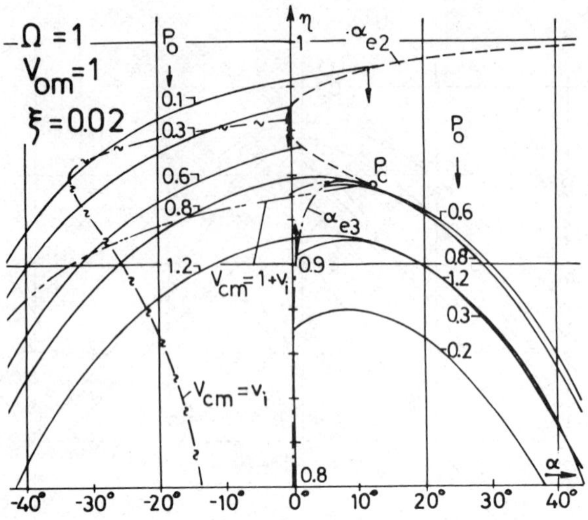

Fig. 16. Efficiency η versus firing angle α

APPENDIX 1. OUTPUT CURRENT

The time function of the output current can be obtained from the inverse transformation of equation (3):

$$i_0(t) = -D_1 \cos(t + \alpha - \psi_1) +$$

$$+ \frac{V_{cm} + v_i}{\sqrt{1 - \xi^2}} e^{-\xi\Omega t} \sin \Omega_1 t -$$

$$- D_1 \frac{e^{-\xi\Omega t}}{\sqrt{1 - \xi^2}} [\cos \alpha \sin(\Omega_1 t - \psi_2) +$$

$$+ \Omega_1 \sin \alpha \cos(\Omega_1 t - \psi_2) -$$

$$- \xi\Omega \sin \alpha \sin(\Omega_1 t - \psi_2)] \qquad \text{(A1.1)}$$

where

$$D_1^{-1} = \frac{1}{\Omega} \sqrt{(\Omega^2 - 1)^2 + 4\xi^2\Omega^2}; \quad \psi_1 = \tan^{-1} \frac{2\xi\Omega}{\Omega^2 - 1}$$

$$\psi_2 = -\tan^{-1} \frac{\sqrt{1 - \xi^2}}{-\xi} - \tan^{-1} \frac{2\xi\Omega^2(1 - \xi^2)}{1 + 2\xi^2\Omega^2 - \Omega^2};$$

$$\Omega_1 = \Omega\sqrt{1 - \xi^2}.$$

The first term on the right side is the current flowing in steady-state because of voltage v_0, the second term is the transient current developing as a result v_i and the initial condenser voltage V_{cm}, while the residual part ensuring value $i_0 = 0$ at time $t = 0$ is the transient current due to the turn-on phenomenon.

By giving α, α_e and $V_{0m} = 1$, the other variables can be calculated. $(V_{cm} + v_i)$ can be determined from equ. $i_0(\alpha_e) = 0$. At the same time $(v_i - V_{cm})$ can be calculated from the equation expressing the subinverter voltage ballance at time $t = \alpha_e$

$$v_i = v_L(\alpha_e) + V_{cm} + \sin(\alpha_e + \alpha) \qquad \text{(A1.2)}$$

Here $v_L = \frac{di_0}{d\Omega t}$ is the voltage across choke L_s. By knowing v_i and V_{cm}, the powers P_0 and P_i can be directly computed from the equations (5) and (6). With the iteration formula

$$\alpha_{en} = \alpha_{e(n-1)} - \frac{P_0 - P_{0c(n-1)}}{\left(\frac{dP_0}{d\alpha_e}\right)_{n-1}}$$

the firing angle α_e chosen initially, can be changed to match its value to a preselected output power P_0. Here n is the number of iteration steps, P_{0c} the power computed on basis of expression (5).

APPENDIX 2. LIMIT CURVES

Limit curves $V_{cm} = 1 + v_i$ *and* $V_{cm} = v_i$. In the cases of both limit curves the relation between α and α_e can be determined from the two equations of Appendix 1. Expressing v_i from $i_0(\alpha_e) = 0$ and substituting it into equation (A1.2) and taking into account $V_{cm} = v_i + 1$ the result is

$$\sin \alpha = \frac{N(\alpha_e)}{D(\alpha_e)} \cos \alpha + \frac{\sin \Omega_1 \alpha_e}{D(\alpha_e)} \qquad \text{(A2.1)}$$

while in the case $V_{cm} = v_i$

$$\tan \alpha = \frac{N(\alpha_e)}{D(\alpha_e)} \qquad \text{(A2.2)}$$

where

$$N(\alpha_e) = D_1(A_1 + A_2 + A_3) + A_4;$$

$$D(\alpha_e) = D_1(B_1 + B2 + B3) + B4$$

$$A_1 = \sqrt{1 - \xi^2} \cos \Omega_1 \alpha_e \cos(\alpha_e - \psi_1);$$

$$A_2 = \sin \Omega_1 \alpha_e \left[-\xi \operatorname{con}(\alpha_e - \psi_1) + \frac{1}{\Omega} \sin(\alpha_e - \psi_1)\right]$$

$$A_3 = -e^{-\xi\Omega\alpha_e} \cdot \sin \psi_2; \qquad A_4 = \sin \alpha_e \sin \Omega_1 \alpha_e$$

$$B_1 = \sqrt{1 - \xi^2} \cos \Omega_1 \alpha_e \sin(\alpha_e - \psi_1);$$

$$B_2 = -\sin \Omega_1 \alpha_e \left[\xi \sin(\alpha_e - \psi_1) + \frac{1}{\Omega} \cos(\alpha_e - \psi_1)\right]$$

$$B_3 = e^{-\xi\Omega\alpha_e}(-\Omega_1 \cos \psi_2 - \xi\Omega \sin \psi_2);$$

$$B_4 = -\cos \alpha_e \sin \Omega_1 \alpha_e.$$

Limit curves α_{e2}, α_{e3} *and* α_{ei}. Current $i_0(t)$ extincts along the limit curve α_{e2} with zero slope (Fig. 3f), while along curve α_{e3} it touches once the time axis with zero slope (Figs. 3b and d). At a given firing angle α the expression of current $i_0(t)$ is the same in both cases, merely in the first we assume that $i_0(t)$ extincts at time $t = \alpha_{e2}$, while in the second case we assume that i_0 continues to flow after time $t = \alpha_{e2}$ and extincts only at time α_{e3}. It can be concluded that $(v_i + V_{cm})$ is the same on the two limit curves for a given α [see (A1.1)], although v_i, V_{cm} and P_0 are, of course, generally different.

The relation between α *and* α_{e2} can be determined from the equations $i_0(\alpha_{e2}) = 0$ and $v_L(\alpha_{e2}) = 0$. The final result

$$\tan \alpha = \qquad \text{(A2.3)}$$

$$= \frac{(-F_1 \sin x + F_2 \cos x) \sin \Omega_1 x + e^{-\xi\Omega x} \Omega \sin \psi_2}{(F_2 \sin x + F_1 \cos x) \sin \Omega_1 x + e^{-\xi\Omega x} \Omega^2 F_3}$$

where

$$F_1 = \cos\psi_1 + b\sin\psi_1; \qquad F_2 = \sin\psi_1 - b\cos\psi_1$$

$$F_3 = \sqrt{1-\xi^2}\cos\psi_2 + \xi\sin\psi_2;$$

$$b = \frac{\Omega_1}{\tan\Omega_1\alpha_{e2}} - \xi\Omega$$

$$x = \alpha_{e2}.$$

The computation of α_{e3}. Firing angle α and $(v_i + V_{cm})$ can be calculated from equation (A2.3) and from $i_0(\alpha_{e2}) = 0$ at a given α_{e2}. Thereby knowing $i_0(t)$ angle α_{e3} can be determined by an iteration program around the extinction of current i_0.

The computation of α_{ei}. Along the limit curve α_{ei} the current i_0 has an inflection point at time $t = \alpha_i$ that is $\left(\frac{d^2 i_0}{dt^2}\right)_{t=\alpha_i} = 0$ and $v_L(\alpha_i) = 0$ are fulfilled.

From these two equations in case of $\xi = 0$

$$\tan\alpha = \frac{\Omega\sin\alpha_i\sin\Omega\alpha_i + \cos\alpha_i\cos\Omega\alpha_i - \Omega^2}{\sin\alpha_i\cos\Omega\alpha_i - \Omega\cos\alpha_i\sin\Omega\alpha_i}.$$

By knowing α_i and α, voltage $(V_{cm} + v_i)$ can also be calculated from the equation $v_L(\alpha_i) = 0$, thereafter α_{ei} can be determined by the iteration program around the extinction of current i_0.

Limit curves $P_0 = 0$ *and* $P_0 = \infty$. Case $\xi = 0$ is treated only.

In case of $P_0 = 0$ let us start from relation (6). P_0 can be zero if $v_i = 0$ and $V_{cm} \neq 0$ or vice versa. First let us discuss the first case. Current i_0 does not carry active power to the load but makes the capacitor voltage v_c oscillate with amplitude V_{cm}. On the one hand it can be deduced from the equation

$$I_i = \frac{1}{2\pi}\int_0^{\alpha_e} i_0\,dt = \frac{V_{cm}}{\pi\Omega} \qquad \text{(A2.4)}$$

determining the average value of the input current and, on the other hand from the expression $i_0(\alpha_e) = 0$ by using $v_i = 0$ that in case of $P_0 = 0$ the equations

$$\tan\alpha = -\tan\frac{\alpha_e}{2} \quad \text{and}$$

$$V_{cm} = -\frac{D_1}{\sin\Omega\alpha_e}[\cos\alpha\cos\Omega\alpha_e - \Omega\sin\alpha\sin\Omega\alpha_e - \cos(\alpha+\alpha_e)]$$

or in case of $\Omega = 1$

$$V_{cm} = \frac{1}{2}\sin\alpha + \frac{\alpha_e\sin(\alpha+\alpha_e)}{2\sin\alpha_e}$$

hold.

In cases of $v_i \neq 0$ and $V_{cm} = 0$ the current i_0 does not flow. This state can take place if $v_0(0) \geqslant v_i$ at the firing time $t = 0$ of the thyristor and it cannot turn on. In limit case $v_i = \sin\alpha$ and it is also valid for $\xi \neq 0$.

In limit case $P_0 \to \infty$ the current amplitude i_0 and V_{cm} are increasing beyond all limits. The current i_0 with a good approximation from (A1.1) is $i_0(t) = V_{cm}\sin\Omega t$. Therefore for an arbitrary firing angle the conduction angle $\alpha_e = \frac{\pi}{\Omega}$. Voltage v_i can be determined from the equation

$$v_i = \frac{1}{2}\frac{\Omega^2}{\Omega^2 - 1}\left[\sin\alpha\left(1 + \cos\frac{\pi}{\Omega}\right) + \cos\alpha\sin\frac{\pi}{\Omega}\right]$$

deduced from (A2.4) and $i_0(\alpha_e) = 0$. In case of $\Omega = 1$

$$v_i = \frac{\pi}{4}\cos\alpha.$$

APPENDIX 3. COMMON POINT (K POINT)

The curves $P_0 = $ const. cross common point K in the planes $\alpha_e \sim \alpha$ and $v_i \sim \alpha$ in states 1 and 2 ($R_s = 0$). Here $\alpha_e = \frac{\pi}{\Omega}$ and $\alpha = \alpha_k = \frac{1}{2}(\pi - \alpha_e)$. Let us determine the value of v_i in point K and the value of V_{cm} and P_0 on the boundary of state 2 and 3. From the expression of condenser voltage

$$v_c = \Omega\int_0^t i_0\,dt - V_{cm} = -D_1\Omega\sin(t+\alpha) +$$

$$+ D_1(\cos\alpha\sin\Omega t + \Omega\sin\alpha\cos\Omega t) -$$

$$-(V_{cm} + v_i)\cos\Omega t + v_i.$$

From equation $v_c\left(\frac{\pi}{\Omega}\right) = V_{cm}$

$$v_i = \frac{D_1\Omega}{2}[\sin(\alpha_k + \alpha_e) + \sin\alpha_k].$$

Considering the relation $\alpha_k + \alpha_e = \pi - \alpha_k$ voltage v_i

$$v_i = D_1\Omega\sin\alpha_k. \qquad \text{(A3.1)}$$

Neither α_e nor v_i change in point K.

The conduction angle $\alpha_e = \frac{\pi}{\Omega}$ is maintained as long as the slope of the current is negative or zero. In the latter limit case $V_{cm} = \frac{v_i}{\Omega^2}$ from equation $v_L(\alpha_e) = v_L\left(\frac{\pi}{\Omega}\right) = 0$ and

$$P_0 = \frac{2}{\pi\Omega^3}v_i^2 = \frac{2}{\pi\Omega}D_1^2\sin^2\alpha_k.$$

REFERENCES

[1] I. Rácz, Quasi-periodic dynamic behaviour of piecewise linear multi-parameter systems, *Periodica Polytechnika,* 13, 205 (1969).

[2] I. Nagy, Solid state high frequency power supply for induction heating, *Elektrowärme International–B,* 31, 164 (1973).

[3] I. Nagy, Investigation of a high frequency power supply for induction heating, *Elektrowärme International–B,* 31, 164 (1973).

[4] Mrs. D. Miklós – K. Lupán – I. Nagy, Start-up method in high frequency thyristor inverters, Preprints of the 1st IFAC Symposium, Duesseldorf, 2, 599 (1974).

[5] K. Lupán – Mrs. É. Miklós – I. Nagy, Digital computer investigation of time sharing inverter, 2nd Conference on Power Electronics, Budapest, (1973).

[6] Lupán Károly – Miklós Dezsőné – Dr. Nagy István, Frekvenciaosztásos inverter számítógépes vizsgálata, *Elektrotechnika,* 66, 156 (1973).

FAST RESPONSE-LOAD COMMUTATED CHOPPERS

N.T. Nicholson, S.B. Dewan and C.N. Porrelli

SPAR Aerospace Products Ltd., 825 Caledonia Road, Toronto, Ontario M6B 3X8, Canada

ABSTRACT

This paper describes two fast response load commutated choppers. Both of these current source type of choppers can be designed to operate at chopping frequencies in excess of 5 kilohertz which provides for a fast response. Also, the choppers are commutated independently of supply voltage and load variations.

The first chopper (A Chopper) operates in a single quadrant only (1, 4, 5, 7) and the second chopper (B Chopper) operates in two quadrants (2, 3, 4). The basic building blocks for the two choppers are thyristor bridges, commutating capacitors and free-wheeling diodes.

INTRODUCTION

The two choppers described in this paper are ideally suited for applications that require fast response times and current source type of operation (1, 2, 3, 4). These choppers are presently being used in dc transit applications and as current sources for current source inverters. Also, the B chopper is being used as the current source for the new class of inverters called chopper commutated current source inverters (2, 3).

These choppers offer the following advantages over most other types of choppers (1, 3, 4);

i) These choppers can be designed to operate at chopping frequencies in excess of 5 kilohertz. Thus a fast response time is easily obtained.

ii) Due to the high frequency of operation, the size of the dc supply filter components are small.

iii) The chopper thyristor commutation is independent of supply voltage variations and the magnitude of the load current.

iv) Low turn off time thyristors are not required (1, 3, 4).

v) There are no special starting or stopping procedures. The control logic is described in references (1), and (3).

For this paper, the load current is assumed to be constant and the effects of the di/dt inductors are included. Detailed mathematical simulations based on the differential equations of the choppers (3) provides "marginally" more accurate results than the analysis presented in this paper.

In this paper, the operation of the choppers are described and the choppers are analyzed by assuming a constant load current. The analysis includes the effects of the thyristor di/dt inductors. From the analysis steady state performance characteristics and the maximum ratings of the components are obtained. These results are used to design the choppers.

Due to the similarity between the two choppers, the material presented in the following sections of the paper, is combined where practical for the two choppers.

OPERATION OF THE CHOPPERS

The simplified operation of the choppers is described in the literature (1, 3, 4, 5, 7). The operation of the two choppers shown in Figs. 1 and 3 are described in Table 1. Some of the voltage and current waveforms appearing in the choppers are shown in Figs. 2 and 4.

For Table 1, the operation of the two choppers are described only for the interval of time that the thyristors T_1, T_4, T_5 and T_8 are on. For the next interval when thyristors T_2, T_3, T_6 and T_7 are on, the operation is the same (Figs. 2 and 4), (1, 3, 4).

Figures 1, 2, 3 and 4 and Table 1 show that the only differences between the two choppers are in the load voltages and the supply currents.

ANALYSIS OF THE CHOPPER

The two choppers shown in Figs. 1 and 3 are analyzed by assuming that the load current is constant and the losses in the choppers are zero. The analysis also takes into account the effects of the di/dt inductors L_1 and L_2.

Due to symmetry, it is only necessary to analyze the choppers when one set of thyristors are turned on. Also, the analysis of the choppers is very similar. Thus, for each mode of operation of the choppers, the analysis that is common to both choppers is

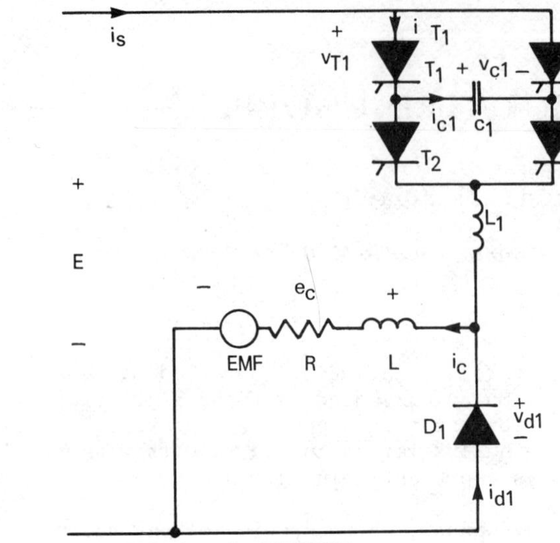

FIG. 1 A CHOPPER

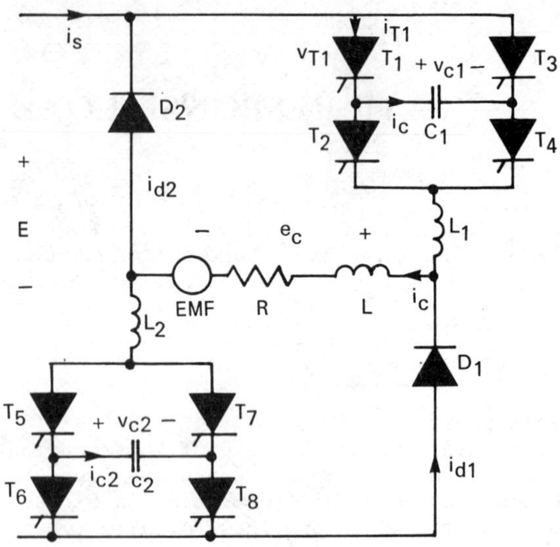

FIG. 3 B CHOPPER

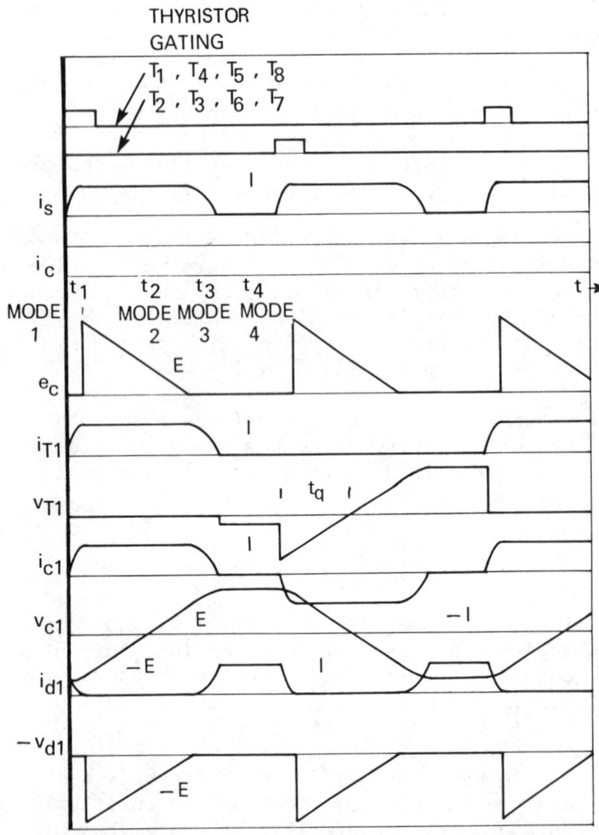

FIG. 2 A CHOPPER WAVEFORMS

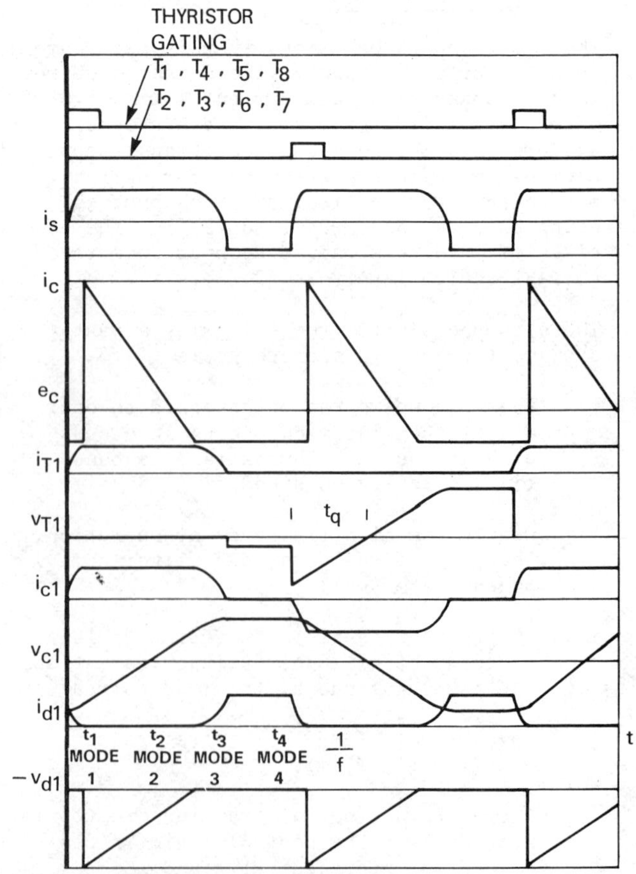

FIG. 4 B CHOPPER WAVEFORMS

TABLE 1 Operation of the Choppers (Figs. 1, 2, 3 and 4)

Mode	A Chopper Figs. 1 and 2	B Chopper Figs. 3 and 4
1	- v_{c1} is negative - thyristors T_1 and T_4 are gated - diode D_1 is carrying the load current - e_c, the load voltage is zero - i_{c1} builds up as a sine wave - mode 1 ends when i_{c1} is equal to the load current	- v_{c1} and v_{c2} are negative - thyristors T_1, T_4, T_5 and T_8 are gated - diodes D_1 and D_2 are conducting the load current back to the supply - e_c, the load voltage is -E, regeneration - i_{c1} and i_{c2} build up as sine waves - mode 1 ends when i_{c1} and i_{c2} are equal to the load current
2	- D_1 is reverse biased - e_c goes to more than 2E volts and starts to go to zero volts - i_{c1} is equal to the load current - mode 2 ends when v_{c1} has reversed polarity and becomes equal to E volts in magnitude	- D_1 and D_2 are reverse biased - e_c goes to more than 3E volts and starts to go to -E volts - i_{c1} and i_{c2} are equal to the load current - mode 2 ends when v_{c1} and v_{c2} have reversed polarity and become equal to E volts in magnitude
3	- diode D_1 is forward biased and starts to carry the load current - e_c is zero volts - i_{c1} goes to zero as a cosine wave - mode 3 ends when i_{c1} is zero	- diodes D_1 and D_2 are forward biased and start to carry the load current back to the supply - e_c is -E volts, regeneration - i_{c1} and i_{c2} go to zero as cosine waves - mode 3 ends when i_{c1} and i_{c2} are zero
4	- diode D_1 is carrying the load current - e_c is zero volts - mode 4 ends when thyristors T_2 and T_3 are gated	- diodes D_1 and D_2 are conducting the load current back to the supply - e_c is -E volts, regeneration - mode 4 ends when thyristors T_2, T_3, T_6 and T_7 are gated

presented first. The non common parts of the analysis appear at the end of each section of analysis. Finally, the time t is set equal to zero at the beginning of each mode in the analysis.

For this analysis, thyristors T_1 and T_4 of the A chopper are turned on and thyristors T_1, T_4, T_5 and T_8 for the B chopper are turned on.

Mode 1

From Figs. 2 and 4, mode 1 occurs when one set of thyristors are turned on and the capacitor currents i_{c1} and i_{c2} build up as sine waves until they are equal to the load current.

The initial capacitor voltages are;

$$V_{co} = -E - IZ \tag{1}$$

$$Z = \sqrt{\frac{L_1}{C_1}} \tag{2}$$

where V_{co} is the initial capacitor voltage.
E is the supply voltage.
I is the load current.
Z is the chopper characteristic impedance.
L_1 is the di/dt inductance in Henries.
C_1 and C_2 are the chopper commutation capacitance in Farads.

The equation describing the thyristor bridges is;

$$E = L_1 \frac{di_{c1}}{dt} + \frac{1}{C_1}\int_0^t i_{c1}\, dt + V_{co} \tag{3}$$

solving equation (3);

$$i_{c1} = i_{c2} = \left(\frac{2E + IZ}{Z}\right)\sin\omega_1 t \tag{4}$$

$$\omega_1 = \frac{1}{\sqrt{L_1 C_1}} \tag{5}$$

where i_{c1} is the C_1 capacitor current.
i_{c2} is the C_2 capacitor current.
ω_1 is the natural frequency of mode 1.

The capacitor voltages during this mode are;

$$v_{c1} = v_{c2} = \frac{1}{C_1}\int_o^t i_{c1}\,dt + v_{co} \tag{6}$$

solving equation (6);

$$v_{c1} = v_{c2} = E - (2E + IZ)\cos\omega_1 t \tag{7}$$

where v_{c1} is the C_1 capacitor voltage.
v_{c2} is the C_2 capacitor voltage.

The diode currents are;

$$i_{d1} = i_{d2} = I - i_{c1} \tag{8}$$

where i_{d1} is the D_1 diode current.
i_{d2} is the D_2 diode current.

Mode 1 ends when the capacitor currents i_{c1} and i_{c2} become equal to the load current I. Thus the time (t_1) at which mode 1 ends is obtained from equation (4).

$$t_1 = \theta/\omega_1 \tag{9a}$$

where

$$\theta \triangleq \sin^{-1}[IZ/(2E + IZ)] \tag{9b}$$

The capacitor voltages at the end of mode 1 are obtained from equations (6) and (9).

$$V_{c11} = V_{c12} = E - (2E + IZ)\cos\theta \tag{10}$$

where V_{c11} is the C_1 capacitor voltage at the end of mode 1.
V_{c12} is the C_2 capacitor voltage at the end of mode 2.

Mode 1 - A Chopper. The load voltage (e_c) during mode 1 is;

$$e_c = 0 \tag{11}$$

The supply current (i_s) is;

$$i_s = i_{c1} \tag{12}$$

Mode 1 - B Chopper. The load voltage (e_c) during mode 1 is;

$$e_c = -E \tag{13}$$

The supply current (i_s) is;

$$i_s = -I + 2i_{c1} \tag{14}$$

Mode 2

From Figs. 2 and 4, mode 2 occurs from the instant the capacitor current i_{c1} and i_{c2} are equal to the load current until the capacitor voltages v_{c1} and v_{c2} have reversed polarity and become equal in magnitude to the supply voltage E.

For mode 2, the following relationships are valid:

$$i_s = i_c = i_{c1} = i_{c2} = I \tag{15}$$

$$i_{d1} = i_{d2} = 0 \tag{16}$$

The capacitor voltages are;

$$v_{c1} = v_{c2} = \frac{1}{C_1}\int_o^t i_{c1}\,dt + V_{c11} \tag{17}$$

Substituting equations (10) and (15) into equation (17) and integrating;

$$v_{c1} = v_{c2} = \frac{It}{C_1} + E - (2E + IZ)\cos\theta \tag{18}$$

Mode 2 ends when the capacitor voltages v_{c1} and v_{c2} have reversed polarity and have charged up to a potential equal to the magnitude of the supply voltage E. Thus, the time (t_2) at which mode 2 ends is obtained from equation (18).

$$t_2 = \frac{C_1}{I}(2E + IZ)\cos\theta \tag{19}$$

The capacitor voltages at the end of mode 2 are;

$$V_{c12} = V_{c22} = E \tag{20}$$

where V_{c12} is the C_1 capacitor voltage at the end of mode 2.
V_{c22} is the C_2 capacitor voltage at the end of mode 2.

Mode 2 - A Chopper. The load voltage (e_c) during mode 2 is;

$$e_c = E - v_{c1} \tag{21}$$

substituting equation (18) into equation (21);

$$e_c = -\frac{It}{C_1} + (2E + IZ)\cos\theta \tag{22}$$

at the end of mode 2, the load voltage is equal to zero.

Mode 2 - B Chopper. The load voltage (e_c) during mode 2 is;

$$e_c = E - v_{c1} - v_{c2} \tag{23}$$

substituting equation (18) into equation (23);

$$e_c = -\frac{2It}{C_1} - E + 2(2E + IZ)\cos\theta \qquad (24)$$

at the end of mode 2, the load voltage is equal to minus the supply voltage.

Mode 3

From Figs. 2 and 4, mode 3 occurs from the instant the diodes start to conduct current until the capacitor currents go to zero. The equation describing the thyristor bridges is;

$$E = L_1 \frac{di_{c1}}{dt} + \frac{1}{C_1}\int_0^t i_{c1}\, dt + E \qquad (25)$$

Solving equation (25);

$$i_{c1} = i_{c2} = I \cos\omega_1 t \qquad (26)$$

The capacitor voltages during this mode are;

$$v_{c1} = v_{c2} = \frac{1}{C_1}\int_0^t i_{c1}\, dt + E \qquad (27)$$

Substituting equation (26) into equation (27) and integrating;

$$v_{c1} = v_{c2} = E + IZ \sin\omega_1 t \qquad (28)$$

The diode currents are;

$$i_{d1} = i_{d2} = I - i_{c1} \qquad (29)$$

The time (t_3) at which mode 3 ends is obtained from equation (26);

$$t_3 = \frac{\pi}{2\omega_1} \qquad (30)$$

The capacitor voltages at the end of mode 3 are obtained from equations (28) and (30).

$$V_{c13} = V_{c23} = E + IZ \qquad (31)$$

where V_{c13} is the C_1 capacitor voltage at the end of mode 3.
V_{c23} is the C_2 capacitor voltage at the end of mode 3.

The initial capacitor voltage (equation (1)) is obtained from equation (31).

Mode 3 - A Chopper. The load voltage (e_c) during mode 3 is;

$$e_c = 0 \qquad (32)$$

The supply current (i_s) is;

$$i_s = i_{c1} \qquad (33)$$

Mode 3 - B Chopper. The load voltage (e_c) during mode 3 is;

$$e_c = -E \qquad (34)$$

The supply current (i_s) is;

$$i_s = -I + 2i_{c1} \qquad (35)$$

Mode 4

From Figs. 2 and 4, mode 4 occurs from the instant the capacitor currents are zero until the next set of thyristors are gated.

For mode 4, the following relationships are valid;

$$i_{c1} = i_{c2} = 0 \qquad (36)$$

$$i_{d1} = i_{d2} = I \qquad (37)$$

The time (t_4) for which mode 4 exists;

$$t_4 \geq 0 \qquad (38)$$

The time t_4 is obtained from the equations describing the average load voltage which appear latter in this paper.

Mode 4 - A Chopper. The load voltage (e_c) during mode 4 is;

$$e_c = 0 \qquad (39)$$

The supply current (i_s) is;

$$i_s = 0 \qquad (40)$$

Mode 4 - B Chopper. The load voltage (e_c) during mode 4 is;

$$e_c = -E \qquad (41)$$

The supply current (i_s) is;

$$i_s = -I \qquad (42)$$

From the analysis of modes 1 and 2 of the two choppers, the thyristor turn off time (t_q) can be obtained. The previously on thyristors are reverse biased by the commutating capacitors from the instant the other set of thyristors are turned on until the capacitor voltages go to zero. Thus the turn off time (t_q) is obtained from equation (9) and from equation (18) which is set equal to zero to obtain the mode 2 contribution to the turn off time.

$$t_q = \theta/\omega_1 + C_1[-E + (2E + IZ)\cos\theta]/I \qquad (43)$$

STEADY STATE PERFORMANCE CHARACTERISTICS

In this section, the steady state performance characteristics of the A and B choppers are obtained. It is assumed that the load is an R - L - EMF and the load inductance is large so that the assumption of constant load current is valid. Also, the equations describing the load voltage and current harmonics are obtained.

From the analysis of the two choppers, normalized equations can be obtained that relate the average load voltage to the supply voltage as a function of the load current, the chopper di/dt inductor and the commutation capacitor.

For the A chopper, the load voltage is zero during modes 1, 3 and 4 and it is described by equation (22) during mode 2. Thus, the average load voltage is;

$$V_{a1} = \frac{1}{T}\int_{o}^{t_2}[-\frac{It}{C_1} + (2E + IZ)\cos\theta]dt \qquad (44)$$

$$T = t_1 + t_2 + t_3 + t_4 \qquad (45)$$

where V_{a1} is the average load voltage.
T is the time for one half cycle of operation of the chopper.

Equation (44) can be simplified by defining the following quantity X as;

$$X = \frac{E}{IZ} \qquad (46)$$

Substituting equation (46) into equation (44), dividing the resultant equation by E and integrating;

$$R_{va} = \frac{V_{a1}}{E} = \frac{t_2}{2T}\left(\frac{1 + 2X}{X}\right)\cos\theta \qquad (47a)$$

where

$$\theta = \sin^{-1}[1/(1 + 2X)] \qquad (47b)$$

Also from equations (5), (9), (19), (30), (38) and (45);

$$\frac{t_2}{T} = \frac{(1+2X)\cos\theta}{\theta + (1+2X)\cos\theta + \frac{\pi}{2} + \omega_1 t_4} \qquad (48)$$

Similarily for the B chopper:

$$R_{vb} = \frac{V_{a1}}{E} = -1 + \frac{t_2}{T}\left(\frac{1+2X}{X}\right)\cos\theta = -1 + 2R_{va} \qquad (49)$$

t_2/T in equation (49) is also defined by equation (48).

Setting t_4 equal to zero in equation (48) allows for the determination of the maximum normalized output voltage of the two choppers as a function of X. R_{va} and R_{vb} are plotted against X in Fig. 5. This figure shows that as X is increased, the maximum output voltages of the A and B choppers increase but will never be larger than the supply voltage E.

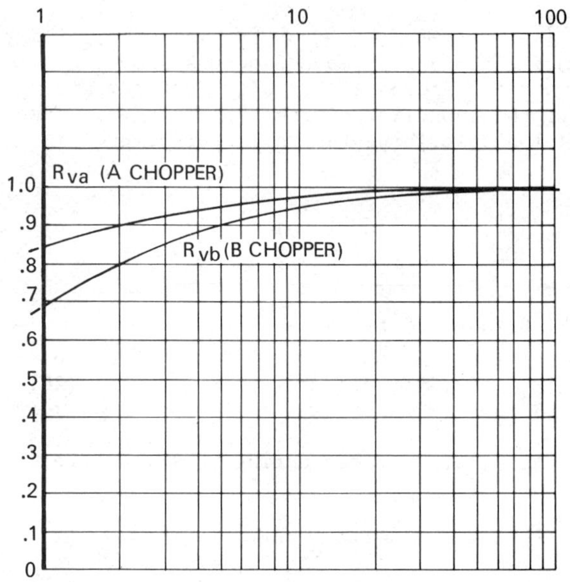

FIG. 5
NORMALIZED AVERAGE LOAD VOLTAGE VERSUS X
(Maximum Operation of the Choppers)

Also, from the equations describing R_{va} and R_{vb}, the constant load current operation of the two choppers as a function of the chopping frequency can be obtained.

The chopping frequency f of the two choppers is obtained from equations (5), (9), (19), (30), (38) and (45).

$$f = \frac{\omega_1}{\theta + (1+2X)\cos\theta + \frac{\pi}{2} + \omega_1 t_4} \qquad (50)$$

The maximum operating frequency f_{max} of the two chopper is obtained from equation (50) by setting t_4 equal to zero.

Figure 6 shows how the normalized load voltages of the two choppers vary as the chopper frequency is varied while maintaining the load current constant.

A generalized curve relating the maximum chopping frequency f_{max} of the two choppers to the load current and the rate of change of thyristor current can be obtained.

The maximum rate of change of thyristor current i_{T1} is obtained by differentiating equation (4), substituting equation (46) into

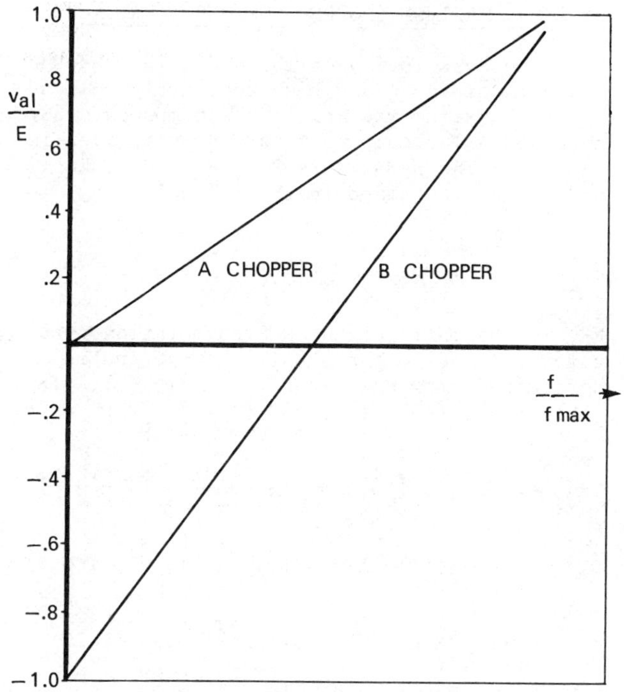

FIG. 6 NORMALIZED AVERAGE LOAD VOLTAGE VERSUS NORMALIZED CHOPPING FREQUENCY FOR A CONSTANT LOAD CURRENT

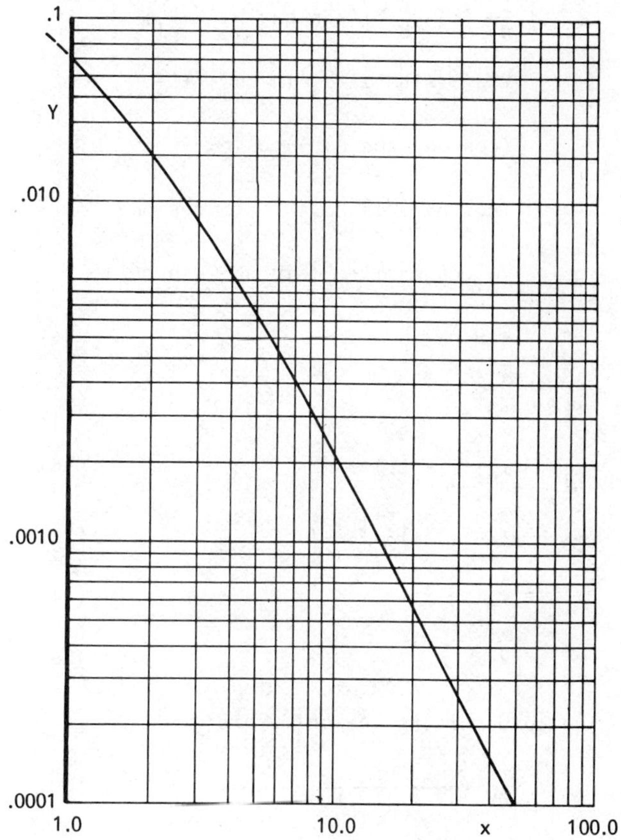

FIG. 7 Y VS. X

the resulting equation and setting t equal to zero.

$$\frac{di_{T1}}{dt} = \frac{E(1 + 2X)}{L_1 X} \tag{51}$$

Similarly, di_{T1}/dt can be obtained in terms of the commutating capacitor C_1.

$$\frac{di_{T1}}{dt} = \frac{I^2 X(1 + 2X)}{E\ C_1} \tag{52}$$

substituting equations (5), (51) and (52) into equation (50) and setting t_4 equal to zero.

$$Y = \frac{f_{max} I}{\frac{di_{T1}}{dt}} = \frac{1}{(1+2X)(\theta + (1+2X)\cos\theta + \frac{\pi}{2})} \tag{53}$$

Figure 7 shows the relationship between the maximum operating frequency f_{max} of the two choppers, the maximum load current I and the maximum rate of change of thyristor current di_{T1}/dt versus X.

Finally, the load voltage harmonics are presented for the general load voltage waveform shown in Fig. 8.

The general load voltage harmonic equation is;

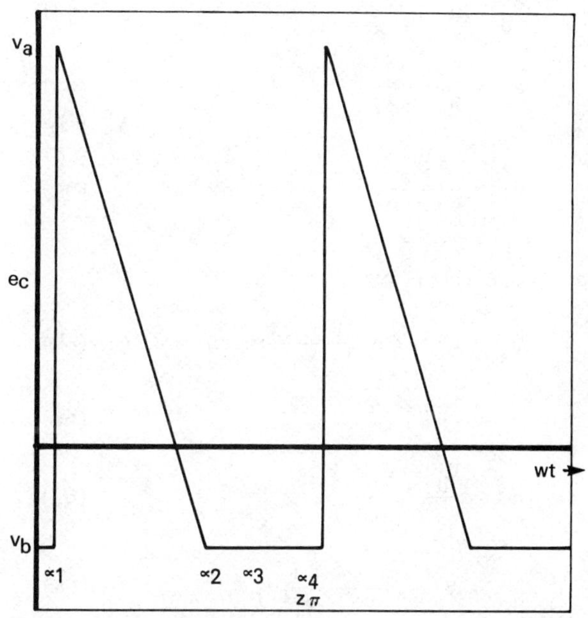

FIG. 8 GENERALIZED LOAD VOLTAGE WAVEFORM FOR THE CHOPPERS

$$e_{cn} = \frac{1}{n\pi}[\{(-V_b+b)(\sin n\alpha_2-\sin n\alpha_1) + a(\alpha_2\sin n\alpha_2-\alpha_1\sin n\alpha_1) + \frac{a}{n}(\cos n\alpha_2-\cos n\alpha_1)\}\cos n\alpha + \{(V_b+b)(\cos n\alpha_2-\cos n\alpha_1) - a(\alpha_2\cos n\alpha_2 - \alpha_1\cos n\alpha_1) + \frac{a}{n}(\sin n\alpha_2-\sin n\alpha_1)\}\sin n\alpha] \quad (54)$$

$$\alpha_1 = \omega_0 t_1 \quad (55)$$

$$\alpha_2 = \omega_0(t_1 + t_2) \quad (56)$$

$$\alpha_3 = \omega_0(t_1 + t_2 + t_3) \quad (57)$$

$$\alpha_4 = \omega_0(t_1 + t_2 + t_3 + t_4) = 2\pi \quad (58)$$

$$\omega_0 = 2\pi f \quad (59)$$

The load current harmonics are;

$$i_{cn} = \frac{e_{cn}}{R + jn\omega_0 L} \quad (60)$$

where e_{cn} is the nth load voltage harmonic.
f is the chopping frequency.
i_{cn} is the nth load current harmonic.

For the A chopper:

$$V_a = E(\frac{1+2X}{X})\cos\theta \quad (61)$$

$$V_b = 0 \quad (62)$$

$$a = \frac{-V_a}{\alpha_2 - \alpha_1} \quad (63)$$

$$b = \frac{\alpha_2 V_a}{\alpha_2 - \alpha_1} \quad (64)$$

For the B chopper:

$$V_a = -E + 2E\,(\frac{1+2X}{X})\cos\theta \quad (65)$$

$$V_b = -E \quad (66)$$

$$a = \frac{V_b - V_a}{\alpha_2 - \alpha_1} \quad (67)$$

$$b = V_a - (\frac{V_b - V_a}{\alpha_2 - \alpha_1})\alpha_1 \quad (68)$$

MAXIMUM CHOPPER COMPONENT RATINGS

From the description of the operation and the analysis of the A and B choppers (Figs. 1 and 3), the maximum ratings of the components are obtained. The component ratings are given in Table 2. The parameters R_{V1}, R_{V2}, R_{V3}, R_{I1} and R_{I2} are obtained in the following way.

R_{V1}

R_{V1} is the ratio of the RMS commutation capacitor voltage to the supply voltage and is determined from equations (5), (7), (9), (18), (19), (28), (30), (45) and (46).

$$R_{V1} = \frac{V_{crms}}{E} = [\frac{1}{\omega_1 T}\{\theta + \frac{1}{2}(\frac{1+2X}{X})^2(\theta+\frac{\sin 2\theta}{2}) + \frac{1}{3X^2}((1+2X)\cos\theta)^3 + \frac{1}{X}(1-(\frac{1+2X}{X})\cos\theta)[(1+2X)\cos\theta]^2 + (1-(\frac{1+2X}{X})\cos\theta)^2(1+2X)\cos\theta + \frac{\pi}{2} + \frac{\pi}{4X^2}\}]^{1/2} \quad (69)$$

R_{V2}

R_{V2} is the ratio of the peak thyristor voltage to the supply voltage and is determined from equations (1), (31) and (46)

$$R_{V2} = \frac{\hat{V}_T}{E} = \frac{1+X}{X} \quad (70)$$

R_{V3}

R_{V3} is the absolute value of the ratio of the peak diode voltage to the supply voltage and is obtained from equation (22) and (46).

$$R_{V3} = |\frac{\hat{V}_d}{E}| = (\frac{1+2X}{X})\cos\theta \quad (71)$$

R_{I1}

R_{I1} is the ratio of the RMS commutation capacitor current to the load current and is determined from equations (4), (5), (9), (15), (19), (26), (30), (45) and (46).

$$R_{I1} = \frac{I_{CRMS}}{I} = [\frac{1}{\omega_1 T}\{\frac{(1+2X)^2}{2}(\theta - \frac{1}{2}\sin 2\theta) + (1+2X)\cos\theta + \frac{\pi}{4}\}]^{1/2} \quad (72)$$

R_{I2}

R_{I2} is the ratio of the average di/dt inductor (L_1) current to the load current and is determined from equations (4), (5), (9), (15), (19), (26), (30), (45) and (46).

$$R_{I2} = \frac{I_{davg}}{I} = \frac{1}{\omega_1 T} \{(1+2X)(1-\cos\theta) + (1+2X) \cos\theta + 1\} = \frac{2(1+X)}{\omega_1 T} \quad (73)$$

For equations (69), (72) and (73);

$$\omega_1 T = \theta + (1+2X)\cos\theta + \frac{\pi}{2} \quad (74)$$

The component ratings are given in Table 2.

R_{v1}, R_{v2}, R_{v3}, R_{I1}, and R_{I2} are plotted in Fig. 9.

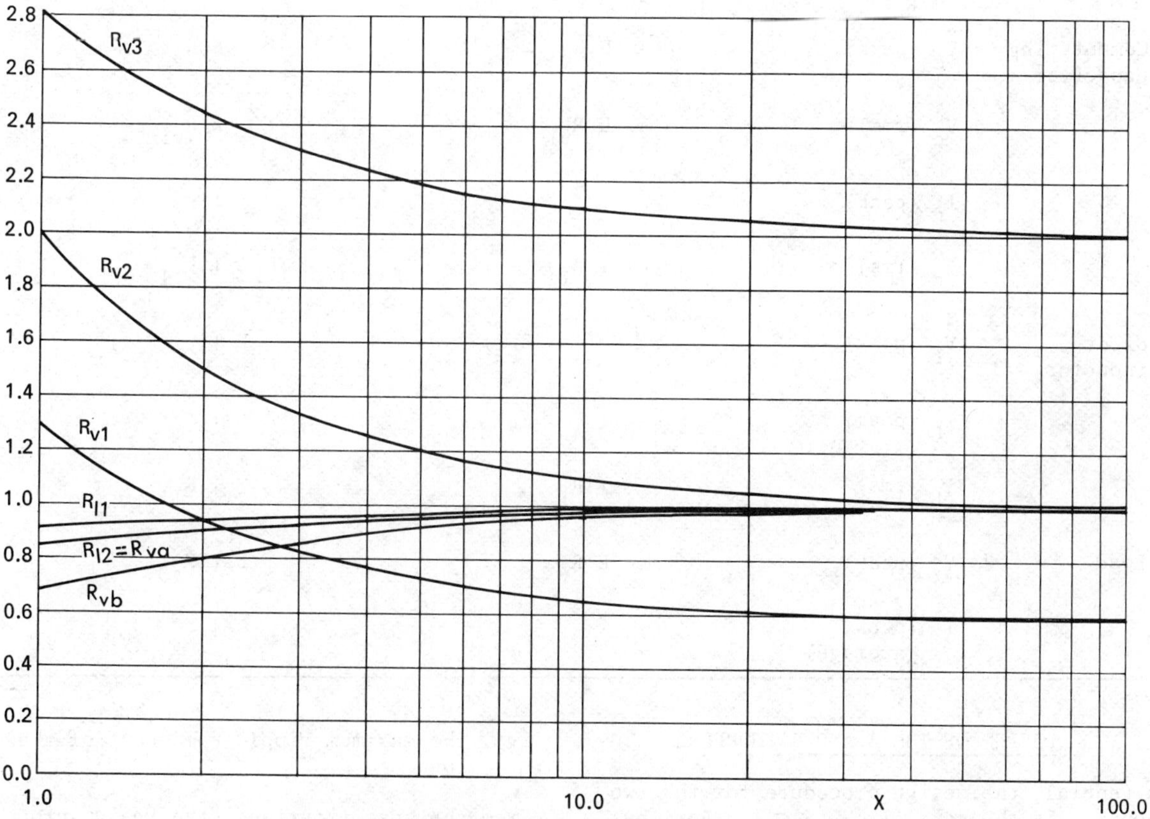

FIG. 9 QUANTITIES RELATING PEAK, RMS AND AVERAGE VALUES FOR USE IN RATING CHOPPER COMPONENTS. (TYPES A & B CHOPPERS)

TABLE 2 Component Ratings of the A and B Choppers

Component	Quantity	A Chopper	B Chopper
Thyristor	$\hat{V}_T$ (peak)	$E\ R_{v2}$	$E\ R_{v2}$
	$\hat{I}_T$ (peak)	I	I
	I_T (rms)	$IR_{I1}/\sqrt{2}$	$IR_{I1}\sqrt{2}$
	$\bar{I}_T$ (average)	$IR_{I2}/2$	$IR_{I2}/2$
	$\frac{di_T}{dt}$ (peak)	$\frac{E}{L_1}(1 + R_{v2})$	$\frac{E}{L_1}(1 + R_{v2})$
	t_q (minimum)	$\frac{\theta - X + (1+2X)\cos\theta}{\omega_1}$	$\frac{\theta - X + (1+2X)\cos\theta}{\omega_1}$
Diode	$\hat{V}_d$ (peak)	$-ER_{v3}$	$-ER_{v3}$
	$\hat{I}_d$ (peak)	I	I
	$\bar{I}_d$ (average)	$I\ (1 - R_{I2})$	$I\ (1 - R_{I2})$
Commutating Capacitor	$\hat{V}_c$ (peak)	$E\ R_{v2}$	$E\ R_{v2}$
	V_c (rms)	$E\ R_{v1}$	$E\ R_{v1}$
	$\hat{I}_c$ (peak)	I	I
	I_c (rms)	$I\ R_{I1}$	$I\ R_{I1}$
di/dt inductor	$\hat{V}_L$ (peak)	$E\ (1 + R_{v2})$	$E\ (1 + R_{v2})$
	$\hat{I}_L$ (peak)	I	I
	I_L (rms)	$I\ R_{I1}$	$I\ R_{I1}$
Load	$\hat{V}_\ell$ (peak)	$E\ R_{v3}$	$E\ (2R_{v3} - 1)$
	I (peak, average)	I	I

DESIGN OF THE A AND B CHOPPERS

In general, the design procedure for the two choppers is the same (1, 2, 3, 4). For the design the following specifications are made.

i) The maximum load current I.

ii) The maximum rate of change of thyristor current di_T/dt.

iii) The maximum operating frequency of the chopper f_{max}.

iv) The maximum ripple component of the load.

From the specifications, the design procedure is;

i) From specifications i), ii) and iii), Y, defined by equation (53), is determined.

ii) From Fig. 7 and the value of Y, X is determined.

iii) From Fig. 5 and the value of X, R_{va} for the A chopper or R_{vb} for the B chopper is determined.

iv) The minimum supply voltage E is determined from equation (47) for the A chopper or equation (49) for the B chopper.

v) The di/dt inductors L_1 and L_2 are determined from equation (51).

vi) The commutating capacitors C_1 and C_2 are determined from equation (52).

If the di/dt inductor or the commutating capacitor values obtained are not satisfactory then part of the specifications such as f_{max} should be changed and the design steps i) through vi) should be repeated. Use this iterative procedure until satisfactory values for the components are obtained.

vii) The harmonic components of the load current can be determined from equation (60) and equations (61) through (64) for the A chopper or from equations (65) through (68) for the B chopper. From these equations and specification iv), the minimum value of the load inductance is determined by using the first harmonic component of the load voltage.

viii) The ratings of the components are determined from the equations in Table 2 and Fig. 9.

ix) The snubber networks for the thyristors and diodes are for reverse recovery currents (1, 3, 4) only and the values suggested by the manufacturers of the semiconductors can be used.

CONCLUSION

In the preceding sections of this paper, the basic principles of operation of the A and B choppers were described and the two choppers were analyzed with the assumption of constant load current. From the operation and analysis of the two choppers, the performance characteristics and component ratings of the choppers were presented. Finally, a general design procedure for the two choppers was described.

Detailed mathematical simulations based on the differential equation analysis of the two choppers have been done (3, 4). The simulations do not provide any analytical design procedures and they do not provide for any more useful information than is available with the constant load current analysis of the two choppers.

Currently, designs based on these chopper configurations are being applied in dc traction and as current sources for current source inverters. Also, the B chopper is being used as the current source for a new class of current source inverters called chopper commutated inverters (2, 3).

REFERENCES

(1) S.B. Dewan, G.S. Dang, N.M. Nicholson, "A Fast Response DC Chopper", Conference Record, IEEE IAS Group Annual Meeting, pp. 922, 1975.

(2) N. Nicholson, S.B. Dewan, C. Porrelli, "Chopper Commutated Current Source Inverters", Conference Record, IEEE IAS Group Annual Meeting, pp. 1101, 1976.

(3) N. Nicholson, Chopper Commutated Inverter, Ph.D. Thesis, University of Toronto, 1976.

(4) G.S. Dang, Fast Response DC Chopper, M.A.Sc. Thesis, University of Toronto, 1973.

(5) P.R. McLellan, "Thyristor Choppers Using a Bridge-connected Capacitor for Commutation", Proc. IEE p. 514, May, 122, 1975.

(6) S.B. Dewan, A. Straughen (1975) Power Semiconductor Circuits, J. Wiley, New York.

(7) N.S. Gehlot, "Thyristor Chopper Controlled DC Motor Drive for Battery Vehicles", CCT/UFPB, Brazil, Conference Record, IEEE IAS Group Annual Meeting, pp. 239, 1976.

HIGH-CURRENT TRANSISTOR CHOPPERS

H. Knöll

Institut für Leistungselektronik und Anlagentechnik
Universität Stuttgart, FRG

SUMMARY

This article shows that transistor choppers can compete technically as well as economically with SCR-choppers.
The principle of a buck converter is briefly explained, and measures for the protection of electronic switches (single, or parallel array of transistors) against overload are stated.
An overload on an electronic switch in ON-state is prevented by detecting the voltage which is on its main current electrodes and by turning this switch to its OFF-state for a certain period of time as soon as a certain limit has been reached.
So-called relieving circuits ($\frac{di}{dt}$- and $\frac{du}{dt}$- limitations) without loss due to construction principles, which relieve the electronic switch during turn-ON and turn-OFF from undue power dissipation and which prevent the occurrence of excessive reverse voltage immediately following turn-OFF will be discussed and explained using oscillograms.
These measures result in a simple, fail-safe and cheap chopper which, compared to a SCR-chopper, can be operated at a higher pulse repetition rate, whose switching states "continuously ON" and "continuously OFF" can easily be managed, and which, finally, – due to the lack of a forced commutation circuit – can be turned OFF at any time and under any load condition.

INTRODUCTION

Choppers for power-conditioning in the propulsion systems of fork lift trucks, battery electric road vehicles and similar machinery have steadily gained importance in the field of power engineering.
These choppers, which are built for the usual application with operating voltages between 24 V and 180 V and for the switching of currents of up to 600 A, have up to now been almost exclusively realized with SCR's.
Transistor choppers can – especially with low operating voltages and high currents – compete technically and economically with SCR-choppers, provided the problems arising from the following characteristics of transistors are solved satisfactorily:

- their low transient overload capacity during ON-state,
- the existence of an "active region",
- their low capacity to dissipate power by absorbing it through the semiconductor during turn-ON and turn-OFF,
- their limited blocking capacity and
- that the parallel array of many single elements is necessary for cost reasons.

Figure 1 shows the essential components of a propulsion system consisting of a battery (d-c source) U_O and the load M, the armature of a d-c motor. The load is represented through its equivalent circuit diagram, a series connection of the armature resistance R_a, the armature inductance L_a, and the counter- e.m.f. U_q.

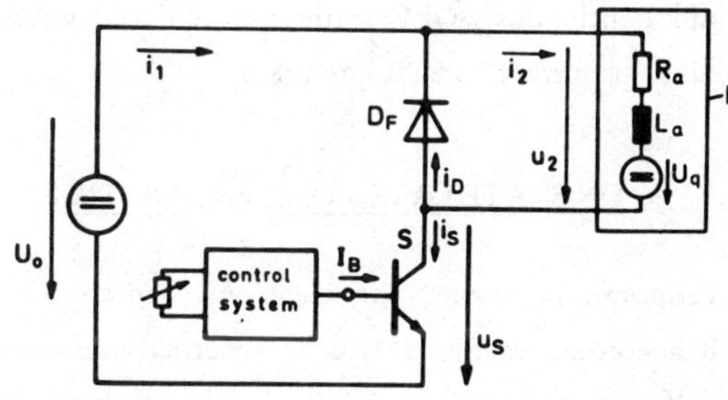

Fig. 1. Propulsion system with transistor chopper

An electronic switch S – in this case a npn-transistor – is inserted into the connecting lead between battery U_O and load M. This switch permits a quasi-continuous power control from battery U_O to load M.
If switch S is brought into ON-state by means of a

positive control-current I_B for a period of time T_g, load M is under the voltage of battery U_o. The load current i_2 is equal to current i_S through switch S.

Waveforms as shown in Fig. 2 result in steady-state condition. Both currents rise exponentially from a lower value I_{Su} to an upper value I_{So}.

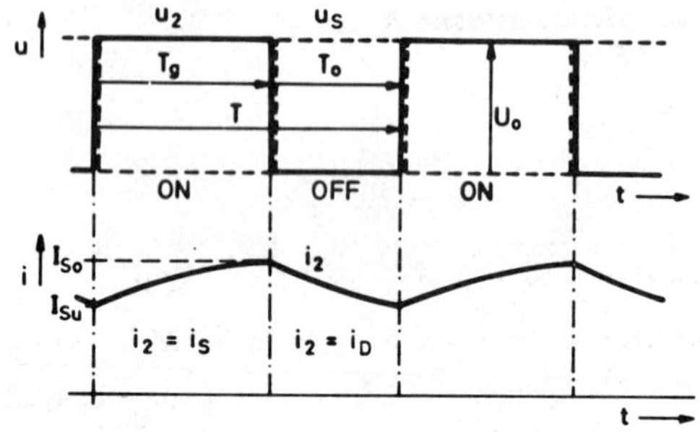

Fig. 2. Waveforms for the circuit in Fig. 1

If switch S is then brought to OFF-state by means of a negative control-current I_B for a period of time T_o, load current i_2 flows through the free-wheeling diode D_F and drops exponentially to I_{Su}. The voltage on the load is down to zero during that time, due to the conducting free-wheeling diode.

Power control to load M is effected through changing the ratio between the duration of ON-state and OFF-state appropriately. This can be done in the well-known way with the methods of pulse width modulation, pulse frequency modulation or a combination of both, or else through a bang-bang current control.

Independently from the method of control, it will prove advisable to construct the control unit in such a way that ON-state as well as OFF-state do not fall short of a minimum duration of time (T_{gmin} and T_{omin} respectively). It should be noted here that continuously-ON and continuously-OFF states are achieved without problems – contrary to SCR-circuits.

ON-STATE OVERLOAD PROTECTION

For economical reasons, the electronic switch S in a circuit according to Fig. 1 will be specified and dimensioned for the maximum required or permitted load current rather than for the maximum possible load current. As a consequence to this, the changeable counter-e.m.f. U_q and/or changeable voltage of battery U_o can lead to an overload on switch S in a circuit arrangement as shown in Fig. 1.

Contrary to already known arrangements, in the circuit which will be introduced now (according to Fig. 3), neither load current i_2 nor current i_S are detected. Instead, voltage u_S which lies between its main current electrodes, is detected (Ref. 1).

If this voltage exceeds a certain limit U_{Sg}, the switch is brought to OFF-state for a certain duration of time T_{omin}.

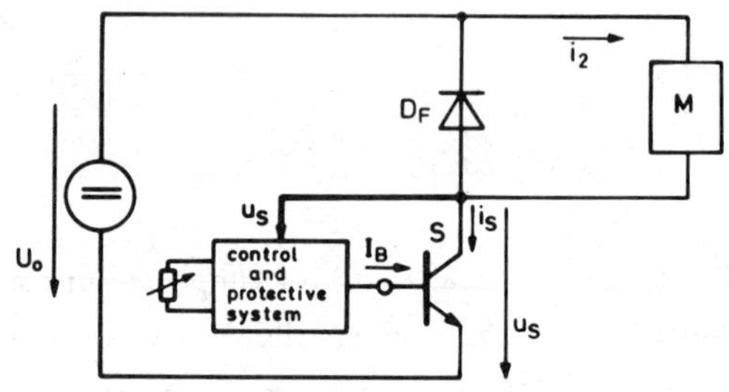

Fig. 3. Transistor .chopper with ON-state overload protective system.

This approach has several advantages.

Firstly, load condition signal u_S is free from loss and can be detected with a favourable voltage. In contrast to this, detecting of current i_S makes a shunt in the main circuit necessary. This shunt adds considerably to the dissipation of power. Reduction of these losses through lowering the resistance of the shunt results in very small voltages which have to be processed with the well-known difficulties. Excessive expenditure, signal lag and drift prohibit in most applications the use of d-c isolated, low-loss current transducers.

Voltage u_S still signifies the load on switch S far better than current i_S. This shall be explained using Fig. 4. The load limit of switch S is reached at operating point A (U_{Sg}, I_{Sg}). If many single transistors are connected in parallel, it is advisible to shift this operating point A to transition region 2 through a corresponding control-current I_{Bg}. This results in minimizing storage time to a negligible amount during turn-OFF and in avoiding a non-uniform load on single transistors due to unequal storage time.

If – due to the ever-present inaccuracy of the current transducer – switching-OFF is not made through current I_{Sg} but through the slightly larger current $I_{Sg} + \Delta I_S$, operating point A shifts to operating point A_1. This leads to a considerably higher power dissipation of switch S, which can lead to its destruction.

If, however, voltage u_S is detected, the relative inaccuracy of measuring U_{Sg} being equal to that of I_{Sg}, i. e.

$$\frac{\Delta U_S}{U_{Sg}} = \frac{\Delta I_S}{I_{Sg}} \qquad (1)$$

operating point A will shift to operating point A_2. In this case, switch S is loaded only minimally higher.

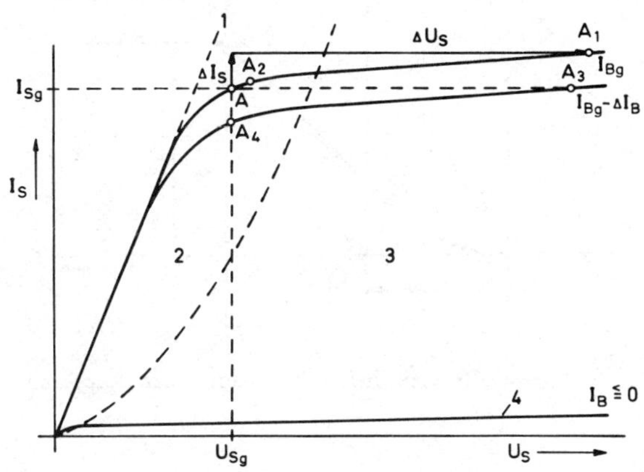

Fig. 4. Output characteristics of switch S
(for explanations see text)
1 saturation limit 3 active region
2 transition region 4 cutoff limit

If control-current I_{Bg} decreases by ΔI_B, this will lead to operating point A_3 in the case of current detection, and to A_4 in the case of voltage detection.

In the first example, the load on switch S exceeds its admissible limit by far, whereas this is not the case in the latter example.

As was mentioned before, switch S is brought into OFF-state upon reaching its load limit, and switch-back should not take place until after a minimum period of time T_{omin}, independent of the method of control.

This period of time T_{omin} of OFF-state must, on the one hand, be chosen small enough to prevent mentionable decrease of the mean value of load current i_2 upon responding to the overload protection circuit. On the other hand, a minimum limit of this period of time is set by the maximum admissible pulse repetition rate as well as the functioning ability of the relieving circuits, which will be introduced in the following paragraphs.

As was said before, a minimum period of time T_{gmin} for ON-state must not be fallen short of, either (independent from the method of control). This is essential for the functioning ability of the relieving circuits which are to be discussed later.

If the voltage u_S on switch S is used as overload detecting signal a minimum duration of ON-state is also necessary to prevent unwanted response of the overload protection circuit upon turning-ON the switch.

The ratio between the minimum duration of OFF-state (T_{omin}) and ON-state (T_{gmin}) must be of an amount which safely prevents an overload on the switch in this transitory state (ON-state for a duration of time T_{gmin}, OFF-state for T_{omin}).

All these measures guarantee fail-safe running of switch S during ON-state. Severe requirements must, however, be placed on the reliability and noise suppression of the overload protection circuit. That is why it will be useful to construct it potentially controlled, that is, signals are always processed as instantaneous values of physical quantities, and never merely as their rates of change.

SWITCHING MODE

Transition from OFF-state to ON-state, turn-ON, as well as transition from ON-state to OFF-state, turn-OFF, of switch S do not occur in a step function, but always within a finite period of time. Without further arrangements, this subjects the switch to a simultaneous load of considerable currents as well as voltages.

The resulting power dissipation should be avoided, if possible. For in an arrangement according to Fig. 1, the highest possible efficiency is intended on the one hand, and on the other hand, these switching losses very often represent the load limit of the switch – especially under high pulse repetition rates.

The following paragraphs are meant to show that power dissipation of the switch can be avoided during turn-OFF and turn-ON through so-called relieving circuits. These relieving circuits can be realized loss-free on principle which results in a chopper with a high degree of efficiency.

Turn-OFF

If turn-OFF of switch S is initiated by a negative control-current I_B in an arrangement according to Fig. 1, voltage u_S between its main current electrodes surges to the voltage on battery U_o. Load current i_2, which is almost constant, then commutates during turn-OFF within a limited period of time to free-wheeling diode D_F. Current i_S drops approximately exponentially to zero. This turn-OFF process is shown by means of an oscillogram in Fig. 5. The simultaneous voltage and current loads on switch S can be clearly seen.

The resulting considerable power dissipation during turn-OFF

$$p_S = u_S \cdot i_S \qquad (2)$$

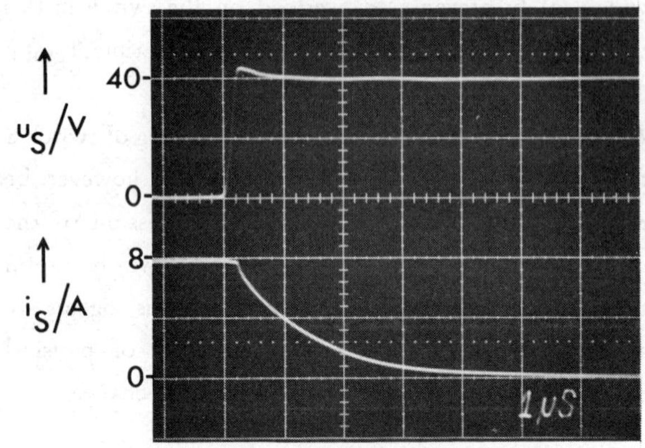

Fig. 5. Turn-OFF without du/dt-limitation (load of one transistor in a parallel array)

is avoided by an arrangement according to Fig. 6, where a series connection of a so-called relieving diode D_E and a relieving capacitor C_E is made, parallel to the main current electrodes of switch S.

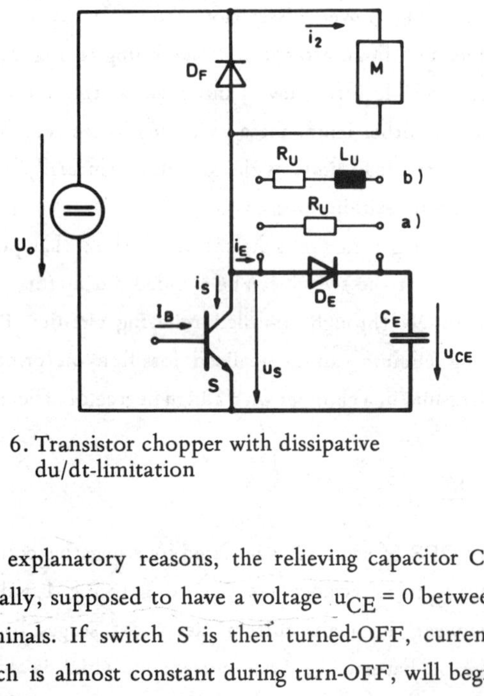

Fig. 6. Transistor chopper with dissipative du/dt-limitation

For explanatory reasons, the relieving capacitor C_E is, initially, supposed to have a voltage $u_{CE} = 0$ between its terminals. If switch S is then turned-OFF, current i_2, which is almost constant during turn-OFF, will begin to commutate to the parallel path D_E, C_E, even if voltage u_S is only slightly increased.

Whereas the current through the switch drops to zero, the current through the capacitor increases to I_{So}. This is shown in the diagrams according to Fig. 7, and in the oscillogram according to Fig. 8.

The straight line representing the fall-off of current i_S is an approximation. During fall-time t_f of current i_S according to Fig. 7, voltage u_S will increase parabolically.

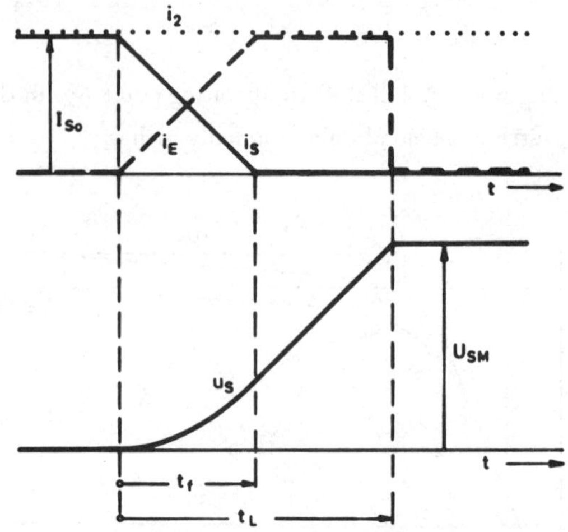

Fig. 7. Turn-OFF with du/dt-limitation (principle)

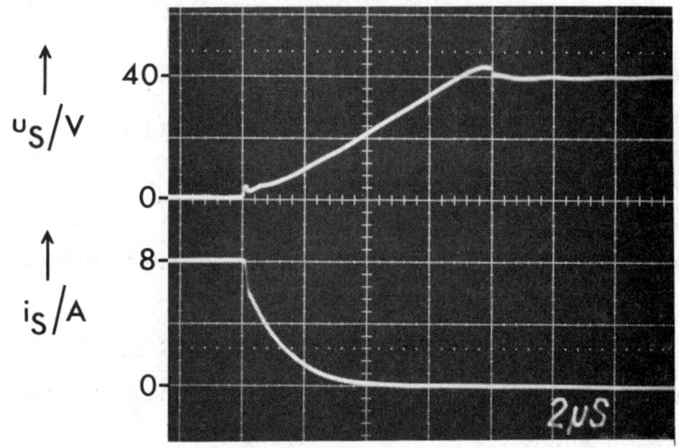

Fig. 8. Turn-OFF with du/dt-limitation (load of one transistor in a parallel array)

Consequently, current i_S has dropped to zero before there is a voltage u_S worth mentioning on the main current electrodes of switch S. Voltage u_S then rises linearily until reverse voltage U_{SM} of the switch (in the example in Fig. 6 supply voltage U_o) has been reached at a point of time t_L, and load current commutates to free-wheeling diode D_F.

Power dissipation of switch S during turn-OFF can be reduced considerably through limiting the rise of voltage u_S ($\frac{du}{dt}$- limitation). This is shown in Fig. 9 where u_S and i_S were supposed to behave according to Fig. 7.

As follows from these equations, the capacitance of relieving capacitor C_E is

$$C_E = \frac{I_{So}}{U_{SM}} \cdot \frac{t_f}{2} \cdot \left(2\frac{t_L}{t_f} - 1\right) \quad \text{and} \quad \frac{t_L}{t_f} > 1 \qquad (3a)$$

$$C_E = \frac{I_{So}}{U_{SM}} \cdot \frac{t_f}{2} \cdot \left(\frac{t_L}{t_f}\right)^2 \quad \text{and} \quad \frac{t_L}{t_f} < 1 \qquad (3b)$$

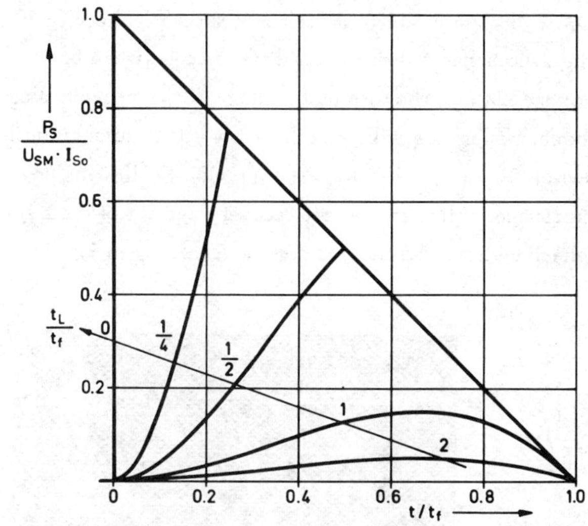

Fig. 9. Turn-Off power dissipation of switch S

$U_{SM} \cdot I_{So}$ peak power dissipation without du/dt-limitation

t_L/t_f ratio of charge time t_L of relieving capacitor C_E to fall-time t_f of current i_S

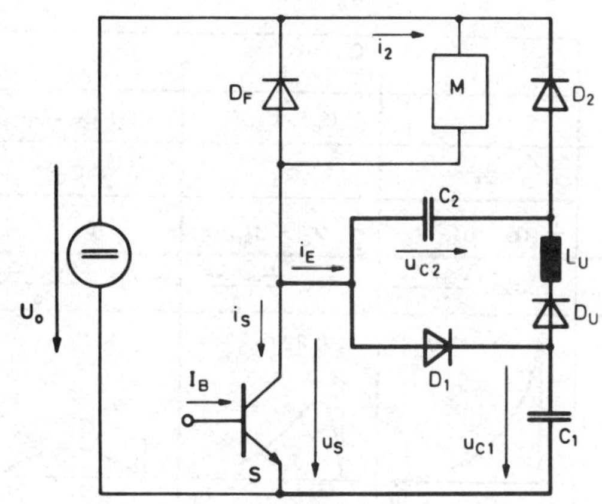

Fig. 10. Transistor chopper with non-dissipative du/dt-limitation

Relieving capacitor C_E must be under zero voltage immediately before turn-OFF in order to guarantee the functioning of the turn-OFF relieving circuit. For this reason, it will be discharged upon turn-ON of switch S through a resistor R_U connected parallel to relieving diode D_E (Ref. 2, 3).

Discharge must be finished after the minimum period T_{gmin} of ON-state, at the latest. The switch is additionally loaded during turn-ON through the discharge current of capacitor C_E. This arrangement leads to a build-up of the maximum discharge current at the beginning of turn-ON at high voltage u_S.

More favourable conditions prevail if a choke of an inductance L_U is connected in series with resistor R_U, and the resulting oscillator circuit is damped aperiodically, for example.

In both cases, however, the entire charge stored in relieving capacitor C_E after turn-OFF flows through switch S. In addition to that, the energy stored in relieving capacitor C_E is dissipated in resistor R_U.

The arrangement in Fig. 10 has substantially more favourable characteristics as far as the additional load on switch S during turn-ON is concerned. It also works without losses due to construction principles (Ref. 4).

In order to describe its relieving function, it is supposed that capacitor C_1 shall have voltage $u_{C1} = 0$ immediately before turn-OFF, while C_2 shall be charged to $u_{C2} = -U_{SM} = -U_o$. It shall also be supposed that the capacitance of the two relieving capacitors C_1 and C_2 shall be chosen identical, and each of them half the capacitance of the relieving capacitor C_E, according to Fig. 6. This choice results in waveforms shown in Fig. 7, 8 and 9, for capacitors C_1 and C_2 are effective in parallel during turn-OFF, C_1 is charged to voltage $u_{C1} = U_{SM}$, whereas C_2 is discharged to voltage $u_{C2} = 0$.

During turn-ON of switch S, capacitor C_1 will discharge itself to zero through diode D_U, choke L_U, capacitor C_2, and switch S, whereas capacitor C_2 will be charged to $u_{C2} = -U_{SM}$. This again prepares the circuit for the following turn-OFF.

From the point of view of turn-OFF relieving ($\frac{du}{dt}$ - limitation), this circuit achieves as much as the arrangement shown in Fig. 6. In contrast however, it does not show dissipation due to contruction principles. In addition to that, capacitors C_1 and C_2 are connected in series during the preparatory charging for the relieving function. Consequently, only half the charge stored in these capacitors after turn-OFF flows through switch S during the following turn-ON.

Table 1 compares the characteristics of the described circuits for reducing power dissipation of switch S during turn-OFF.

Circuits a) and b) are arrangements according to Fig. 6, where relieving capacitor C_E is discharged through a resistor R_U, as in a), or, as in b), through a series connection of a resistor R_U and choke L_U. The series resonant circuit formed with capacitor C_E is damped aperiodically ($R_U = 2\sqrt{L_U/C_E}$). Circuit c) is the arrangement shown

TABLE 1 Comparison of the circuits which reduce turn-OFF power dissipation (du/dt-limitation).
For explanations see text.

	Circuit a)	Circuit b)	Circuit c)
Q_E	$C_E \cdot U_{SM}$	$C_E \cdot U_{SM}$	$(C_1 + C_2) \cdot U_{SM} = C_E \cdot U_{SM}$
Q_U	Q_E	Q_E	$\frac{1}{2} Q_E$
P_V	$\frac{1}{2} \cdot Q_E \cdot U_{SM} \cdot f$	$\frac{1}{2} \cdot Q_E \cdot U_{SM} \cdot f$	0
$-\hat{i}_E$	$3.912 \cdot \frac{U_{SM} \cdot C_E}{t_U}$	$2.146 \cdot \frac{U_{SM} \cdot C_E}{t_U}$	$0.785 \cdot \frac{U_{SM} \cdot C_E}{t_U}$
$-i_E(t)$	$Q_U(t_U) = 0.02\, Q_E$ 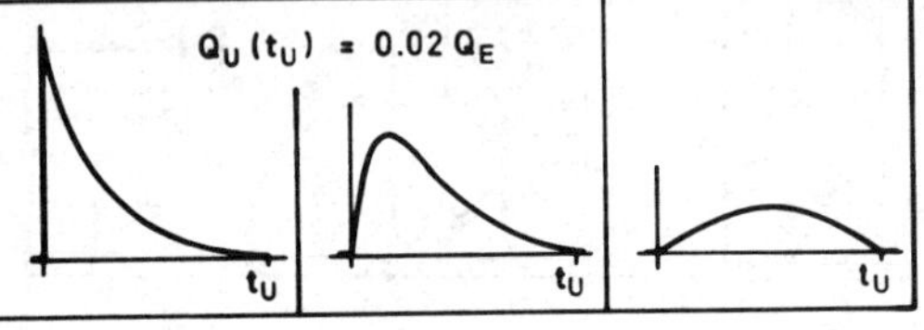		

in Fig. 10.

In all three arrangements, the value of stored charge Q_E in the capacitors is the same immediately after turn-OFF. Consequently, the same reduction of power dissipation is guaranteed by all three arrangements.

During the following turn-ON, charge Q_U flows through switch S. In circuits a) and b) this charge is the same as charge Q_E, in circuit c), however, it is only half that charge.

The discharge of the relieving capacitor C_E is dissipative in circuit a) as well as circuit b).

There is an average power of

$$P_V = \frac{1}{2} Q_E \cdot U_{SM} \cdot f = \frac{1}{2} C_E \cdot U_{SM}^2 \cdot f \qquad (4)$$

dissipated in resistor R_U, f representing pulse repetition rate. Circuit c), however, is free from dissipation.

Furthermore, switch S is loaded with an additional current $-i_E$ during turn-ON. The relieving capacitors must be discharged within a duration of time $t_U < T_{gmin}$, the definition for t_U in a) and b) being such that the charge stored in the capacitors at the end of this period is 0.02 Q_E.

With regard to the peak value $-\hat{i}_E$ of current $-i_E$ and to its waveform, circuit c) is also superior to circuit a) and b), for through suitable dimensioning, circuit c) can easily achieve that current $-i_E$ which additionally loads the switch during turn-ON, does not reach a significant value until voltage u_S between the main current electrodes of the switch has fallen to a very low value.

Turn-ON

If turn-ON of switch S according to Fig. 1 is initiated through a positive control-current I_B, load current i_2, which has previously flown through free-wheeling diode D_F, will commutate to switch S as soon as voltage u_S on the switch has fallen to the difference between battery voltage U_o and the threshold voltage of the free-wheeling diode. Voltage u_S will not fall to low values quickly until switch S has taken over the previously flowing load current as well as the reverse recovery current of the free-wheeling diode (cf. the oscillogram according to Fig. 11).

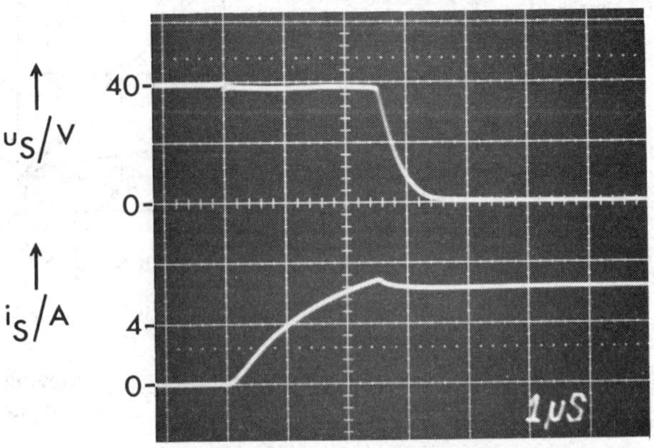

Fig. 11. Turn-ON without di/dt-limitation
(load of one transistor in a parallel array)

The resulting power dissipation of switch S can be avoided by inserting a choke with constant or current-dependent inductance into that path of the complete circuit which was without current before turn-ON, and where current flows through the switch and the supply after turn-ON. This inductance limits the rise of current i_S per time unit ($\frac{di}{dt}$- limitation). It will not rise to a considerable value until voltage u_S between the main current electrodes has become very small (Ref. 5).

There is already an inductance with the characteristics mentioned, namely the self-inductance of battery U_o and its connecting wires to the chopper.

Since switch S is exposed to a considerable load during turn-OFF through this parasitic inductance L_1 in the circuit according to Fig. 12, there is usually a smoothing capacitor added at the input terminals of the chopper. This cannot be done with high currents, since this would require a capacitor with high capacitance and extremely high current-carrying capacity. In addition to that, a relieving of power dissipation of switch S during turn-ON cannot be guaranteed without further additional circuitry.

In a circuit according to Fig. 12, however, parasitic inductance L_1 is purposely used to achieve a limitation of rise of current i_S. As a consequence, this current does not rise to considerable values until voltage u_S has become

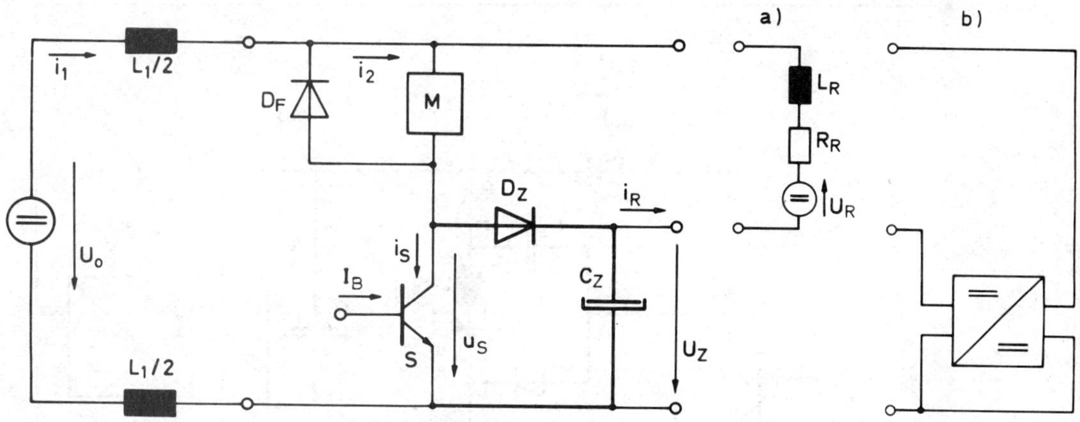

Fig. 12. Transistor chopper with di/dt-limitation
a) dissipative b) non-dissipative

very small (cf. oscillogram according to Fig. 13). Power dissipation of switch S during turn-ON has consequently fallen to negligible values.

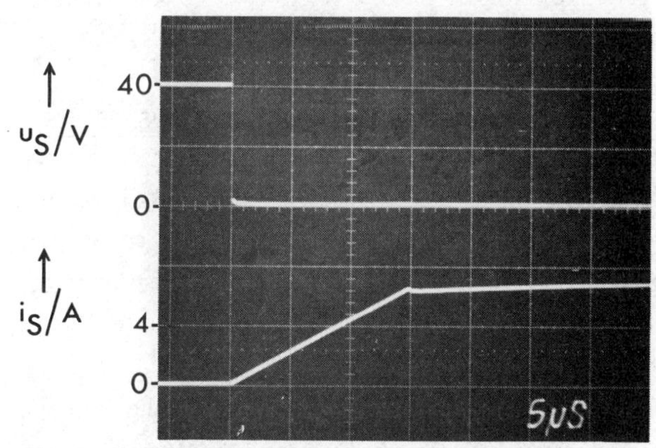

Fig. 13. Turn-ON with di/dt-limitation
(load of one transistor in a parallel array)

In order to avoid exposing switch S during the following turn-OFF to an unduly high reverse voltage, a series connection of a diode D_Z and a capacitor C_Z is connected parallel to the main current electrodes of switch S (Ref. 5).

This capacitor should have a constant voltage U_Z which is higher than that of battery U_O. It has been proved theoretically and in experiments, that U_Z should have the 1.25- to 2.0-fold value of U_O.

If switch S is turned-OFF, voltage u_S between its main current electrodes surges until it has reached the value of voltage U_Z on capacitor C_Z. Input current i_1 then commutates to the parallel path made up of diode D_Z and capacitor C_Z. If the capacitance of capacitor C_Z is chosen sufficiently high, voltage U_Z will only rise minimally, and current i_1 through the choke with inductance L_1 will fall almost linearily to zero. This guarantees that the reverse voltage on switch S remains limited, and that the energy, which is stored in the choke, immediately before turn-OFF, as well as the energy supplied by the battery during turn-OFF, are not dissipated in switch S, but stored temporarily in capacitor C_Z.

In addition to that, this capacitor is loaded with a considerably lower r.m.s. value of current than a comparable smoothing capacitor at the input terminals of the chopper. That is why a cheap electrolytic capacitor can be used.

In order to keep voltage U_Z at capacitor C_Z constant, it is necessary to take off the energy fed, through each turn-OFF of switch S.

This is suggested in two examples in Fig. 12.

Circuit a) dissipates the energy fed to capacitor C_Z through a voltage source U_R (e.g. a Zener diode or several series connected ordinary diodes), and a resistor R_R. Choke L_R serves to smooth current i_R taken off the capacitor.

Circuit b) represents a buck converter with an input voltage U_Z and an output voltage U_O. This arrangement permits re-feeding of the energy stored in capacitor C_Z after turn-OFF to the battery without losses due to principle.

Full Switching Cycle

Figure 14 shows a combination of the circuits in Fig. 10

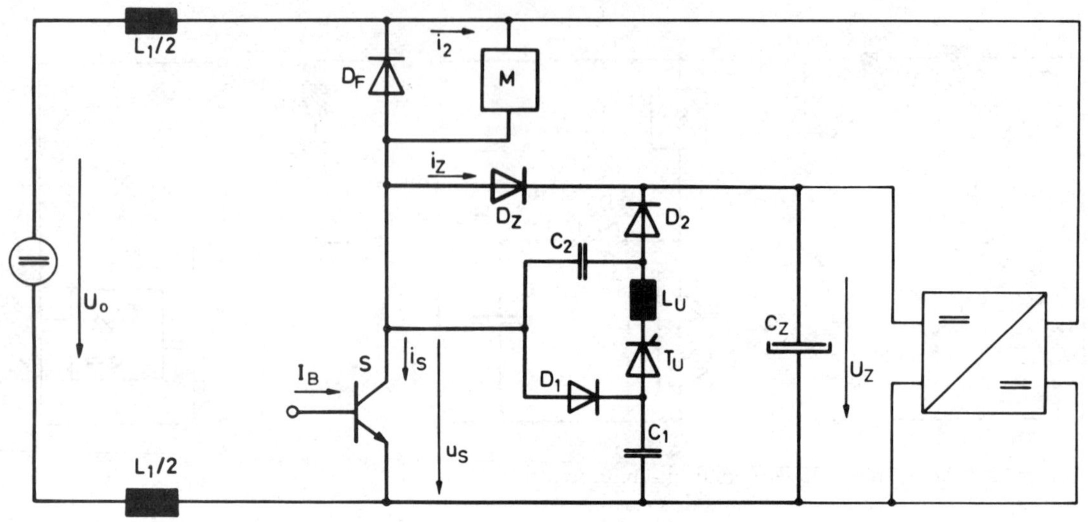

Fig. 14. Complete circuit with non-dissipative di/dt- and du/dt-limitation

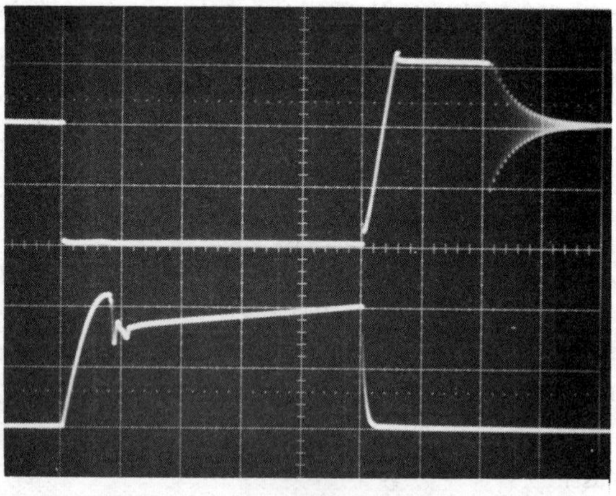

Fig. 15. Waveforms of a full switching cycle (load of one transistor in a parallel array)
upper trace: voltage u_S 20 V/DIV.
bottom trace: current i_S 4 A/DIV.
time t 20 μs/DIV.

and Fig. 12 b). Diode D_2, however, is connected to capacitor C_Z, since the reverse voltage of switch S immediately after turn-OFF is now defined by the largely constant voltage U_Z. In addition to that, diode D_U in Fig. 10 was replaced by a SCR, which is fired whenever switch S is turned-ON. This SCR prevents unwanted discharge of capacitor C_1 as well as charge of capacitor C_2 in OFF-state when load current i_2 and/or current i_Z falls to zero. This guarantees the relieving of switch S from undue power dissipation during turn-OFF even in these operating modes.

Figure 15, finally, shows the waveforms of voltage u_S and current i_S for one switching cycle. The power dissipation of switch S has declined substantially during turn-ON as well as turn-OFF. In addition to that, reverse voltage of switch S remains limited immediately after turn-OFF.

PARALLEL CONNECTED TRANSISTORS

With high currents, switch S according to Fig. 1 will, for cost reason, be constructed out of a parallel connection of many cheap single transistors with a mean current-carrying ability. These parallel arrays will operate reliably if the above mentioned measures are taken.

The single transistors can be connected parallel without emitter and/or base current sharing resistors.It may become necessary to match the single transistors according to their parameters in ON-state. It has proved sufficient to classify transistors of one type and manufacturer into three classes. If transistors from one class are connected parallel at random, the complete circuitry being such that each transistor receives 2/3 of its peak collector current under balanced current sharing, no transistor in the actual arrangement will, even at worst, exceed its peak collector current.

Power dissipation of switch S is reduced drastically during turn-ON and -OFF by using the described relieving circuits. This guarantees that single elements in parallel connections are not overloaded, even with different switching times.

OPERATING EXPERIENCE

Several buck converters making use of the principles described were installed and tested on a test stand and in various vehicles.

A fork lift truck (24 V/400 A) as well as a small electric town car (48 V/500 A) have been submitted to an endurance test for two years now. The choppers in both these vehicles have been working trouble-free up to the present day.

REFERENCES

(1) A. Boehringer, Überlastungsschutz für Leistungstransistoren, insbesondere im Parallelbetrieb, Deutsche Offenlegungsschrift Nr. 24 22 351 vom 20. 11. 75.

(2) J. M. Peter, K. Rischmüller, Anwendungshinweise für parallelarbeitende Schalttransistoren in Motorsteuerungen, Technische Information Nr. 28, Firmenschrift Thomson-CSF, München.

(3) K. Rischmüller, Motorsteuerung mit Hochvolt-Schalttransistoren, Technische Information Nr. 29, Firmenschrift Thomson-CSF, München.

(4) A. Boehringer, Einrichtung ohne prinzipbedingte Verluste zur Entlastung elektrischer oder elektronischer Einwegschalter von ihrer Verlustleistungsbeanspruchung beim Ausschalten, Deutsche Patentanmeldung P 26 50 673.0 vom 5. 11. 76.

(5) A. Boehringer, H. Knöll, Einrichtung zur Befreiung elektrischer oder elektronischer Einwegschalter von hoher Verlustleistungsbeanspruchung während des Einschaltens und überhöhter Sperrspannungsbeanspruchung beim Ausschalten, Deutsche Patentanmeldung P 26 51 492.1 vom 11. 11. 76.

A LOAD INSENSITIVE DC/DC CONVERTER FOR SUPPLY OF TELEPHONE EXCHANGES

I. Flegar

Electrotechnical Institute "Rade Končar", Zagreb - Yugoslavia

ABSTRACT

The output waveform in thyristor choppers and inverters with parallel capacitor commutation is load sensitive. This can limit the applicability of such circuits where extreme pulse width variation is needed. A solution of the problem is shown on an example of a d.c. converter for the supply of telephone exchanges.

INTRODUCTION

The commutation of the main thyristor in the majority of thyristor choppers and inverters is achieved by paralleling a capacitor to the main thyristor or to the load by aid of an auxiliary thyristor. After this auxiliary thyristor has been triggered, the capacitor takes over the load current and applies reverse voltage to the main thyristor. The main thyristor is turned off. The excess energy stored in the commutation circuit, or part of it, is delivered to the load.

The output waveform is load sensitive and differs more or less from the square wave irespectively of the configuration of the parallel capacitor commutation circuit.

The load voltage and the load current cannot be reduced below certain minimum values even in chopper circuits and inverter circuits where it is possible to turn off the main thyristor immidiately after the triggering.

These disadvantages of circuits with parallel capacitor commutation are more evident in cases where extreme pulse width variation is needed.

The extreme pulse width variation is a requirement set on d.c. converters for the supply of telephone exchanges (Ref. 1). Telephone exchanges are supplied with d.c. power by means of regulated rectifiers and stand-by batteries. With a regulated d.c. converter, connected as shown in Fig. 1, it is possible to obtain a constant load voltage irespectively of the variation of the battery voltage. Because of the wide battery voltage and load current variations, the output voltage of the d.c. converter is required to be zero at no load, or to be maximum at full load. The extreme pulse width variation is needed.

A converter circuit shown on Fig. 2 has been developed in order to satisfy these requirements.

The paper describes the operation of the circuit. The operation of the circuit is illustrated by oscillograms on the 10 V/160 A

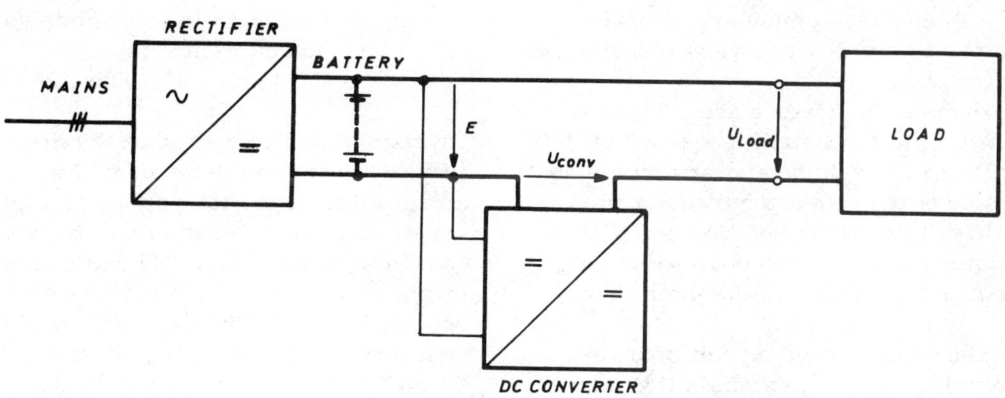

Fig. 1

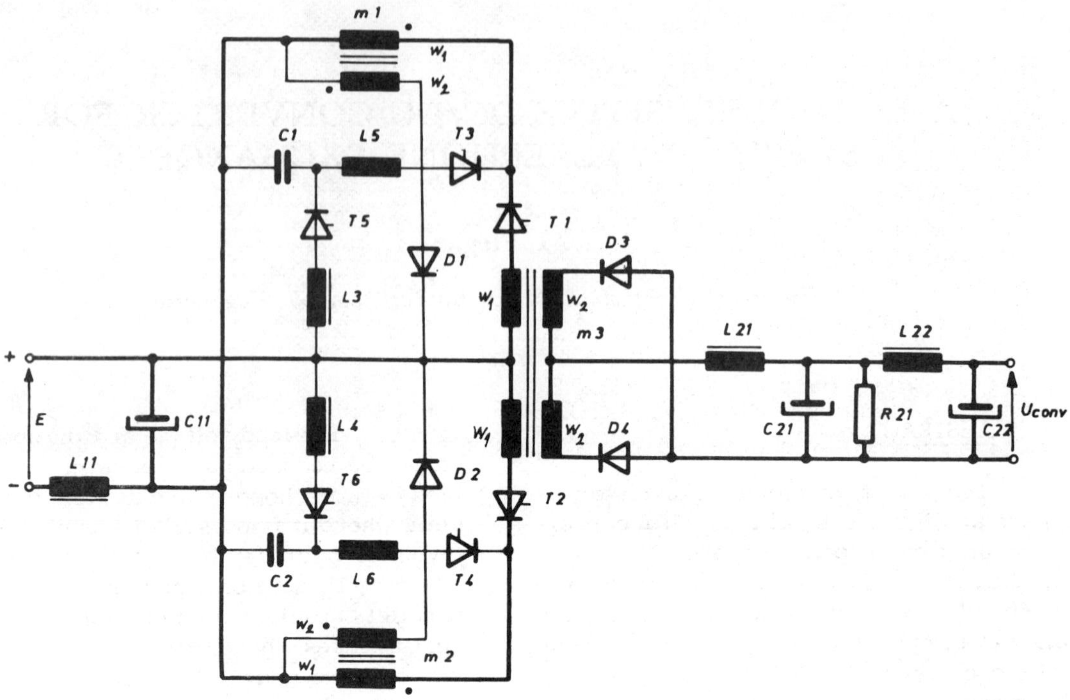

Fig. 2

d.c. converter. Some important characteristics of the circuit are also explained.

CIRCUIT OPERATION

The simplified scheme of the d.c. converter is shown in Fig. 2. The converter has two main thyristors (T1, T2), two commutation circuits with circuits for feeding back the excess commutation energy to the d.c. supply, input filter (L11, C11), output filter (L21, L22, C21, C22, R21) and converter transformer m3. First commutation circuit includes thyristors T3 and T5, commutation capacitor C1, reactors L3 and L5, transformer m1 and diode D1. Because of its function transformer m1 is called recovery transformer. Second commutation circuit includes thyristors T4 and T6, commutation capacitor C2, reactors L4 and L6, recovery transformer m2 and diode D2.

The main thyristors are triggered alternatively with a constant frequency of 400 Hz. An alternating voltage appears on the secondary side of the converter transformer. It is rectified (diodes D3 and D4) and filtered. The regulated d.c. output is obtained by varying the conduction time of the main thyristors.

The commutation circuit operates in the following way. Let the main thyristor T1 carry a current I_T (I_T is the load current reflected to the primary) and the commutation capacitor C1 be charged with the terminal connected to T3 positive.

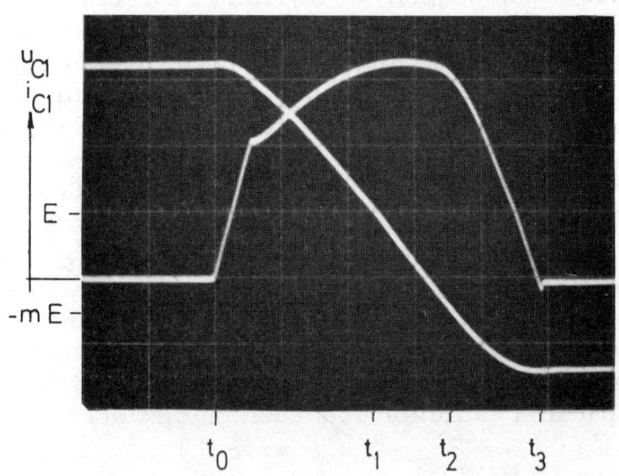

Fig. 3 Commutation capacitor voltage and current (50 V/div, 40 A/div, 20 μs/div)

Thyristor T3 is triggered at the instant t_o (Fig. 3), which is determined by a control circuit. The capacitor voltage is applied in series to the main thyristor. Thyristor T1 ceases to conduct (Fig. 4) and commutation thyristor T3 carries the load current. The reactor L5 limits the rate of rise of the current through T3. The commutation capacitor C1 and primary magnetizing inductance of the recovery transformer L1 form an oscillating circuit. The capacitor voltage must be

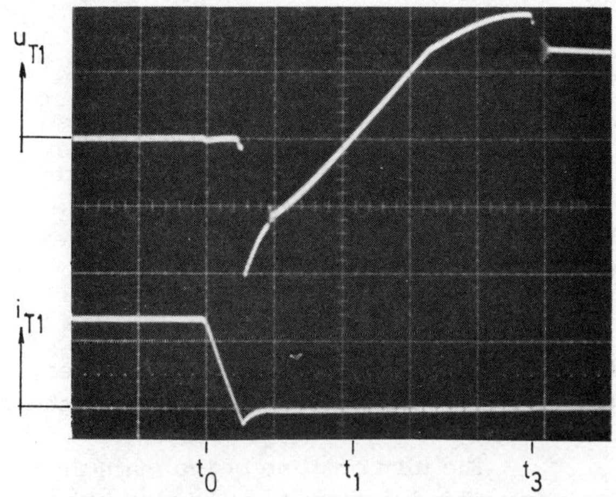

Fig. 4 Main thyristor voltage and current (50 V/div, 40 A/div, 20 µs/div)

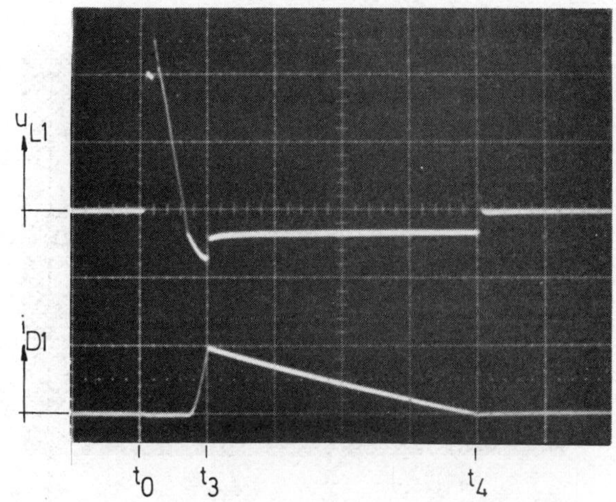

Fig. 5 Recovery transformer primary voltage and recovery diode current (50 V/div, 40 A/div, 100 µs/div)

higher than the d.c. supply voltage during a defined time interval in order to obtain a necessary turn off time for the main thyristor. The turn off time $t_1 - t_o$ can be determined from the condition:

$$u_{C1}(t_1) = E,$$

where E is the battery voltage. The voltage drop on L11 is neglected.

The commutation between rectifier diodes D3 and D4 takes place on the secondary side of the converter transformer. The commutation is over at the instant when each diode conducts half the load current. The voltage on the converter transformer is zero.

The oscillating current through the commutation thyristor reaches a maximum value and then decreases. The induced voltage on the secondary of the recovery transformer m1 reverses its polarity because of the decreasing current. The induced voltage at the instant t_2 is equal to the d.c. supply voltage and diode D1 starts conducting. The time interval $t_2 - t_o$ can be determined from the condition:

$$u_{C1}(t_2) = -mE,$$

where $m = w_1/w_2$ is a recovery transformer turn ratio.

The commutation between the commutation thyristor and the recovery diode D1 is not instantenous because of the leakage inductance of the recovery transformer. The part of the energy stored in the primary magnetizing inductance L1 is transferred during the commutation to the commutation capacitor. This means that the capacitor voltage is boosted over the value of -mE.

At the instant t_3, which can be determined from the condition:

$$i_{T3}(t_3) = 0,$$

the commutation thyristor T3 is turned off. The rest of the energy stored in the inductance of the recovery transformer at the instant t_3 is transferred during the following time interval to the d.c. supply. The time interval $t_4 - t_3$ when the recovery diode conducts only (Fig. 5), can be determined from the condition:

$$i_{D1}(t_4) = 0.$$

The commutation capacitor is held charged, but in wrong direction regarding the next turning off of the main thyristor. The commutation capacitor is recharged by aid of an auxiliary thyristor T5 (usually called recycling thyristor) after some delay time in respect to the instant t_3. The commutation capacitor is in this way ready to turn off the main thyristor T1 when required.

The same commutation process takes place in the second half period of the operation on the second half of the d.c. converter.

SOME CHARACTERISTICS OF THE CIRCUIT

Output waveform The recycling of the commutation capacitor is over before the main thyristor is triggered. The maximum possible commutation energy is always avaliable at the instant of triggering of the main thyristor. The reliable commutation is ensured in this way, even if the main thyristor must be turned off immidiately after its turning on. Neglecting the commutation time between the

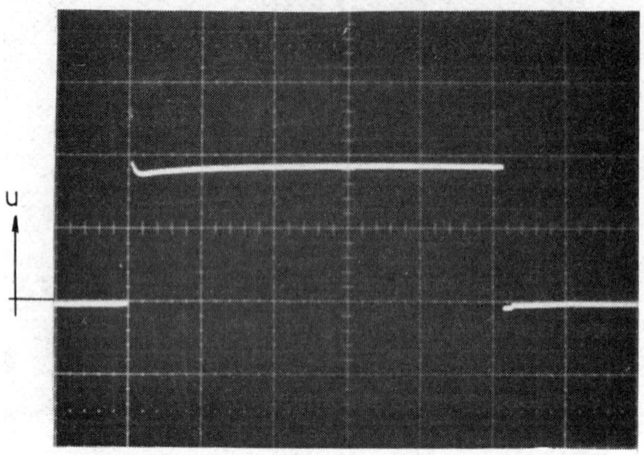

Fig. 6 The rectified secondary voltage waveform (10 V/div, 100 μs/div)

main thyristor and the commutation thyristor the load circuit and the commutation circuit are decoupled. The trailing edge of the square wave voltage on the converter transformer coincides with the trigger signal for the commutation thyristor. In this way, it is possible to reduce the load voltage to virtually zero value. The rectified secondary voltage waveform is shown in Fig. 6.

Commutation capability The voltage on the commutation capacitor is load dependent because of the leakage inductance of the recovery transformer. The capacitor voltage is increased by increasing the load current. The overshoot Δu_{C1} at the instant t_o versus load current ratio I_T/I_n is plotted on Fig. 7.

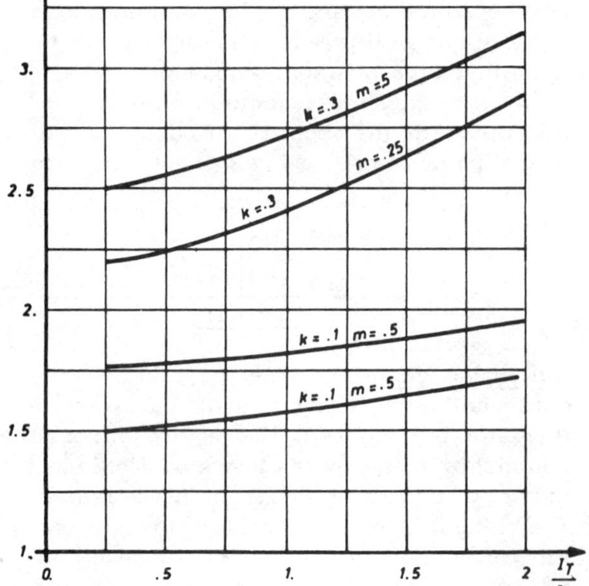

Fig. 7

The overshoot Δu_{C1} is defined according to the relation:

$$u_{C1}(t_o) = E\,(1 + \Delta u_{C1})$$

and I_n is defined as:

$$I_n = 2E/(L1/C1)^{1/2}$$

The overshoot Δu_{C1} depends on the turn ratio of the recovery transformer and on the leakage inductance coefficient k, which is defined with the following expression:

$$k^2 = m^2 L_\delta/(L1 + m^2 L_\delta).$$

L_δ is the leakage inductance of the recovery transformer obtained from a secondary short circuit test.

The turn off time of the main thyristor versus load current ratio is plotted on Fig. 8.

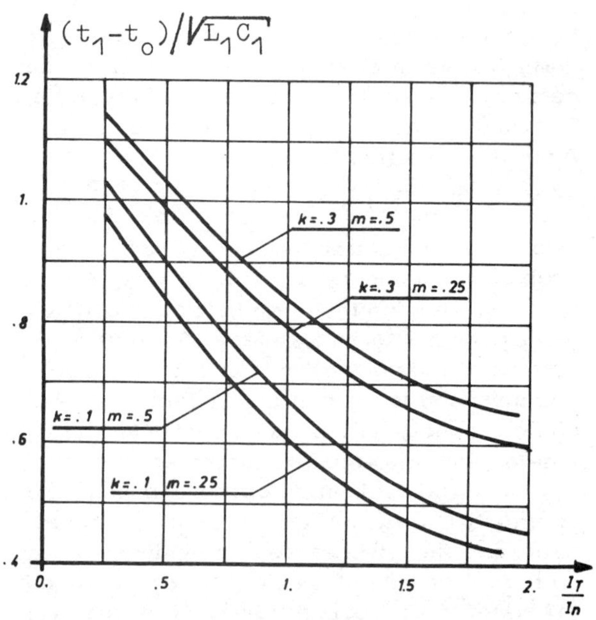

Fig. 8

The increased stored energy in the commutation capacitor, at the instant of commutation, because of the increased load current leads to a smaller capacitor for the defined load current to be commutated and safe turn off time of the main thyristor. On the other hand, the components have to be rated for the higher voltage. The extra cost for the higher voltage-rated components is normally compensated by a saving in capacitor size because of low input voltage of d.c. converters for supply of telephone exchanges (-48V).

Another fact common to the majority of commutation circuits with load dependent commutation energy must be also taken into account. Namely, the commutation energy stored in the n-th interval is used for turning off the main thyristor in n + 1-st interval.

If the variations of the load current between two successive commutation are large, the commutation can fail. This disadvantage is not important here, because of the slow variations of the load in telephone exchanges.

Starting conditions On starting, the first trigger pulse is sent to a recycling thyristor, e.g. T5. The reactor L3 and commutation capacitor C1 form an oscillating circuit. Thyristor T5 turns off when the current in the circuit reaches zero for the first time. The commutation capacitor is charged to approximately 2E with correct polarity for turning off the main thyristor T1. The trigger pulse can be applied then to the main thyristor. The same process takes place during the second half period on the second half of the d.c. converter.

It is also possible to apply trigger pulses to the thyristors T3 - T6 continuously, and to apply the trigger pulses to the main thyristors at the desired instant. The converter will switch on.

By removing the trigger pulses from the main thyristors the converter will switch off. Although the trigger pulses to thyristors T3 - T6 are still being applied, the commutation circuit and load circuit are completely decoupled. The output voltage is zero. This cannot be achieved with the d.c. converters with parallel capacitor commutation.

Maximum duty cycle The relative inductive voltage drop because of the commutation of the rectifier diodes must be taken into account in estimating the maximum duty cycle.

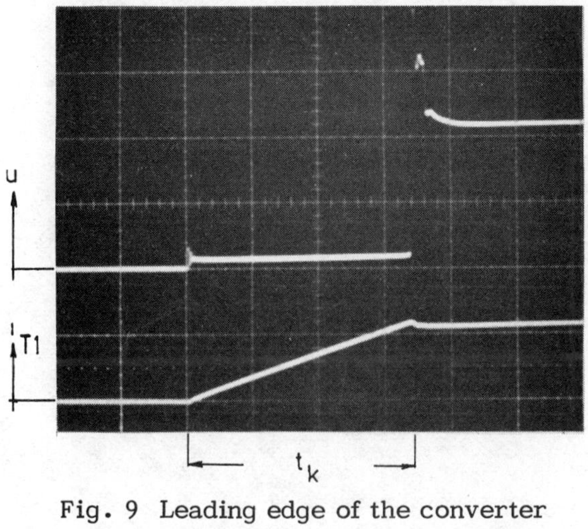

Fig. 9 Leading edge of the converter transformer primary voltage and main thyristor current (20 V/div, 40 A/div, 20 μs/div)

The leading edges of the square wave voltage on the converter transformer are delayed in respect to the trigger signals for the main thyristors. The leading edge of the converter transformer primary voltage and the main thyristor current are shown in Fig. 9.

The delay (commutation) time is equal to:

$$t_k = L1\ I_d/(nE),$$

where I_d is the load current, and $n = w_1/w_2$ is a converter transformer turn ratio. The leakage reactance of the converter transformer in respect to the primary magnetizing inductance of recovery transformer is neglected.

An inductive voltage drop arises because of the time delay. Its relative value can be obtained from the relation:

$$d_x = D_x/U_{id} = 2f\ L1\ I_d/(nE),$$

where U_{id} is the ideal mean value of the converter output voltage. It is equal to:

$$U_{id} = E/n.$$

"f" is the internal frequency of operation. The relative inductive voltage drop at maximum load current can be about 5% by appropriate choice of the value of L1. Correspondingly, the maximum duty cycle at full load yields about 95%.

REFERENCE

(1) F. Ekelund, T. Wolpert, Thyristor power converters for the supply of telephone exchanges, IEE Proc. Confer. Power Thyristors and Appl. London, May 6 - 8 (1969)

RECTIFYING SINGLE-PHASE AND THREE-PHASE A.C. WITH FORCED - COMMUTATED CONVERTERS

H. Kahlen

Brown, Boveri & Cie AG, Mannheim

Summary

Continuous controlled rectification with small reactive power by means of converters connected to single-phase and three-phase a.c. systems requires forced commutation. For the converter circuits described, a simple common turn-off facility is used. The turn-off principle is explained, citing a circuit with a diode bridge series-connected d.c. chopper as an example. By the adoption for the supporting capacitor principle, the problems associated with forced commutation on a single-phase a.c. system are solved and the influence of the mains inductance is brought under control. Forced-commutated three-phase bridge connections should be of the fully controlled type and should have free-wheeling diodes connected to the neutral of the three-phase system. A circuit for charging storage batteries provides a well-smoothed output voltage over a wide control range.

1. Introduction

When consumer are supplied from a single-phase or three-phase a.c. power system through converters using phase control, the fundamentals of current and voltage are displaced with respect to each other. In addition to the reactive power resulting from this phase displacement of the fundamentals, the current harmonics cause reactive power due to distortion. For a given load and firing angle, the reactive power caused by the displacement of the fundamentals can be reduced by static compensation (i.e.capacitors), while the reactive power due to distortion can be decreased with the aid of filters. In dynamic operation, the reactive power due to the displacement of the fundamentals can more effectively be avoided by the use of self commutated converters. The current harmonics are reduced by turning the valves on and off in several stages. In order to limit the possible connections for single-phase and three-phase a.c., a simple collective turning-off facility is used for all rectifier connections.

For the sake of clarity, the commutations are not considered for the voltage and current waveforms.

2. Series-connected converter operating from a single-phase a.c. system

An uncontrolled single-phase bridge, connected in series with a d.c.chopper, constitutes a simple form of converter that can be turned off. In Fig. 1a, the uncontrolled bridge consists of diodes D1 to D4. The operating principle of the d.c.chopper is assumed to be known (1, 2). Turning-off of the conducting thyristor T1 initiates a half-oscillation process. By firing thyristor T2, a charge on the capacitor is caused to swing over. As the charge swings back in the ensuring phase, thyristor T1 is turned off. The turn-off circuit carries the current until the capacitor is charged to the value of the mains voltage available at the instant of turn-off. The current is then commutated to the free-wheeling diode.

When being switched on for the first time, the capacitor is charged by the mains voltage via the load circuit without a thyristor being fired. Main and turn-off thyristors can be triggered simultaneously. Hence, the d.c. chopper offers a high degree of operational reliability.

However, the usefulness of this circuit configuration is limited by two criteria:

1. The capacitor, after turn-off charged to about the instantaneous value of the mains voltage.

In the case of large firing angles, the voltage approaches zero, and the capacitor can no longer be charged. Consequently, the d.c. chopper looses its turn-off capability.

2. A single-phase a.c. system - particularly one with overhead contact lines - frequently has a large inductance. On interruption of the currentflow, the energy stored in the inductance flows into the capacitor. Hence, the latter must be designed for the energy to be stored (criterion of energy) and not for the hold-off time of the thyristor to be turned off (criterion of hold-off time).

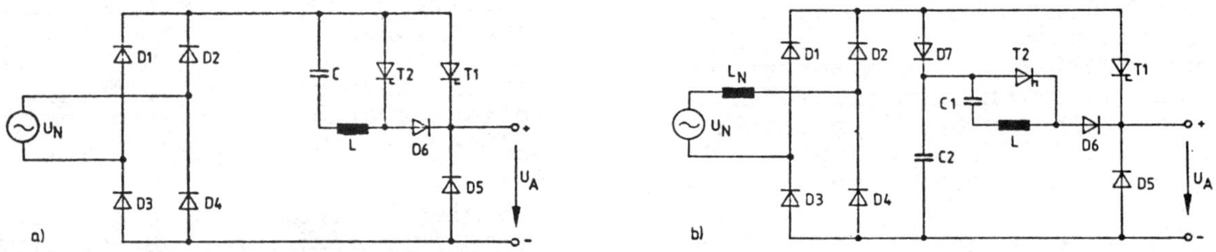

Fig.1 Series connection of a diode bridge and a d.c. chopper
a) without supporting capacitor b) with supporting capacitor

When using the criterion of energy as a basis for determining the rating of the capacitor, particularly in the case of d.c. choppers supplied from an overhead contact line, a considerably larger capacitor is obtained than using the criterion of hold-off time (3).

To keep under control these two criteria, a second capacitor C2 is used for supporting the turn-off circuit (Fig. 1b). This supporting capacitor is charged during every half-cycle approximately to the peak value of the a.c. voltage. This ensures that the turn-off capacitor always receives a definite voltage. When the thyristor is turned off the energy stored in the line inductance flows both into the supporting capacitor and into the commutating capacitor. The voltage of the supporting capacitor has only one polarity. Therefore, polarized capacitors can be used.

The voltage of the supporting capacitor increases with every turn-off when the supporting capacitor and the turn-off circuit are separated from each other or when the turn-off circuit uses the energy of the supporting capacitor only to cover its losses in the resonant circuit and its switching losses. However, by the commutating connection used, the supporting capacitor is connected in parallel with the a.c. network as turn-off thyristor T2 ist triggered. If the voltage of the supporting capacitor is higher than the instantaneous value of the mains voltage, the mains current is reduced by the capacitor discharging before the very turn-off process starts.

During the half-oscillation of the LC turn-off circuit, at maximum load current, the supporting capacitor should be discharged to such an extent that half of the current from thyristor T1 is commutated. In this interval, the supporting capacitor is discharged by the same amount by which it is recharged after commutation to the free-wheeling diode(in the steady state mode).After the thyristor has been turned off and the free-wheeling diode has taken over, the supporting capacitor and commutation capacitor are connected in parallel. (4)

It is possible to connect several parallel load circuits, each with a thyristor that can be turned off and a free-wheeling diode, to the turn-off circuit described (Fig.2).

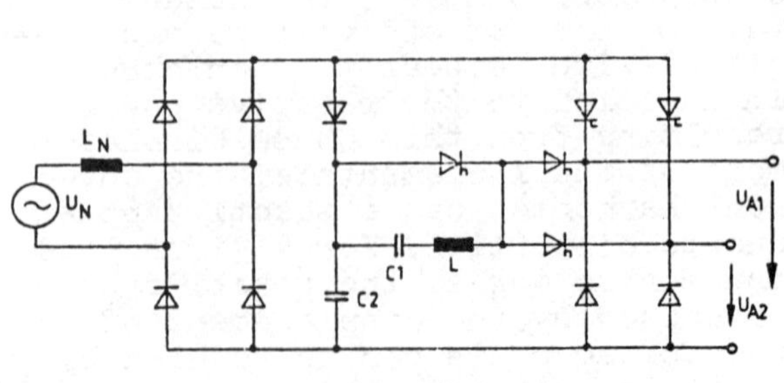

Fig.2 Converter for phase and turn-off control with two controllable outputs

When diode D6 (Fig.1a and 1b) is replaced by thyristors the individual load circuits can be switched successively. In this case, the turn-off circuit and supporting capacitor need only be rated for the maximum current of one load circuit if means are provided to ensure that only one load circuit is switched off at a time. With one load circuit, half the load current can be forced to flow in each thyristor arm by means of an interphase transformer (5).

3. Conversion of the converter - d.c. chopper series connection

3.1 On one d.c.terminal half-controllable bridge connection with provision for turning off

The series conn ection comprising the uncontrollable single-phase bridge and the d.c. chopper, together with the supporting capacitor, can be modified to give on one d.c. terminal half-controllable bridge connection that can be turned off (Fig. 3). It displays the same external behaviour as the connection according to Fig. 1b. There are only two valves in the whole circuit but the sum of mains voltage and capacitor voltage appears across the blocking thyristors whenever a conducting thyristor is being turned off. At high mains voltages, preference is therefore given to the series connection according to Fig. 1b. This connection also requires a lower number of valves when there are several parallel arms.

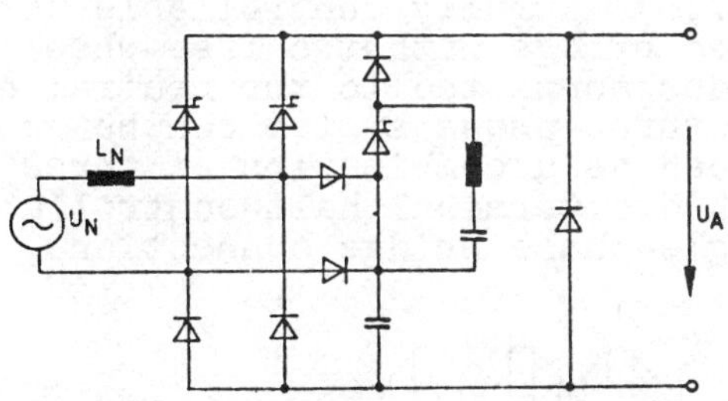

Fig.3 On one d.c. terminal half-controllable bridge for turn-off with supporting cpacitor

3.2 On both d.c. terminals half-controllable bridge connection with provision for turning off

The one d.c. terminal half-controllable bridge connection with provision for turning off according to Fig. 3 exhibits the same external behaviour as the connection shown in Fig. 1b. Its conversion into an on both d.c. terminals half-controllable bridge with turn-off feature is necessary for the three-phase bridge connection with free-wheeling diodes to be derived later. By retaining the supporting capacitor principle, a connection with two supporting capacitors and two commutating capacitors is obtained (Fig. 4). Apart from the two additional capacitors, this connection requires a somewhat higher number of valves than the on one d.c. terminal half-controllable bridge connection with turn-off feature. The voltage of the commutating capacitor appears as maximum reverse blocking voltage across the thyristors of the bridge. The diodes of the bridge are burdened with the sum of the mains voltage and the capacitor voltage.

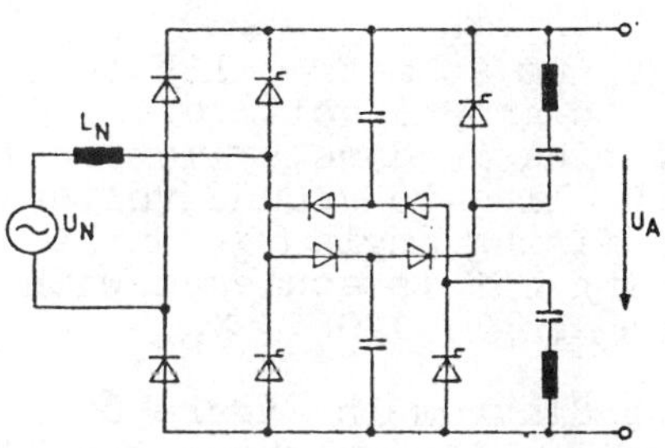

Fig.4 On both d.c. terminals half-controllable bridge for turn-off with supporting capacitors

4. Converter with turn-off feature connected to the three-phase system

4.1 Single-pole controllable converter with turn-off feature

Similarly to the connection of a diodes bridge in series with a d.c. chopper to the single-phase a.c. system, it is possible to connect a three-phase diode bridge in series with a d.c. chopper to a three-phase a.c. system (Fig. 5a). A modified connection is shown in Fig. 5b. The behaviour of this connection is to be considered first from the load side and then from the system side. The d.c. chopper can be controlled by the mains in such a way that a six-pulse voltage appears across the load, i.e. the lowest frequency of the converter output voltage is equal to six times the mains frequency. The behaviour can be clearly demonstrated with the aid of the line diagram. With this converter with turn-off, two triggering methods may be considered:

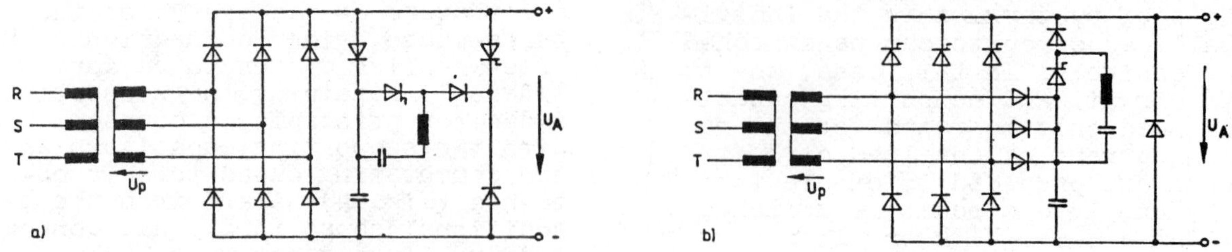

Fig.5 Single-pole controllable converter with turn-off device fed by the three phase system a) diode-bridge in series with a d.c. chopper b) half-controllable-bridge

1. Triggering symmetrically to the line-to-line voltage.
   Triggering is done symmetrically to the line-to-line voltage with firing angle α_V, while turning-off is effected with firing angle $60^o - \alpha_V$.

2. Tiggering symmetrically to the line-to-ground voltage.
   Tiggering is done symmetrically to the line-to-ground voltage with firing angle α_p, while turning-off is achieved with firing angle $120^o - \alpha_p$.

The line diagram in Figure 6 shows the line-to-line voltage, the line-to-ground voltage and the current of one phase. Compared with a d.c. voltage with phase control according to Fig. 6a, about twice the voltage time integral must be stored of in the smoothing reactor when the converter is triggered symmetrically with respect to the line-to-line voltage (Fig. 6b) or symmetrically with respect to the line-to-ground voltage (Fig. 6c). In the case of phase control (assuming a continuous flow of d.c.), the current is allowed to flow in each phase for a period of 120 electrical degrees. When the converter is triggered symmetrically to the line-to-line voltage according to Fig. 6b, two current blocks result which are symmetrical with respect to $\varphi = 60^o$ and $\varphi = 120^o$. Triggering of the converter symmetrically to the line-to-ground voltage produces three current blocks.

This method of operating a single-pole controllable converter with turn-off feature forces the fundamental of the current to be in phase with the voltage, but only at the expense of higher voltage harmonics in the load circuit and higher current harmonics in the power system. This shows that low-harmonic rectification is not possible with a circuit comprising a diode bridge with series-connected, or with a modified circuit.

4.2 Fully controllable converter with turn-off feature

The reactive power drawn by three-phase bridge conncections can be reduced by free-wheeling diodes (6,7,8). A fully controllable three-phase bridge with two free-wheeling diodes connected to the neutral of the three-phase system can be regarded as a combination of three on both d.c.terminal half-controllable single-phase bridge connections,

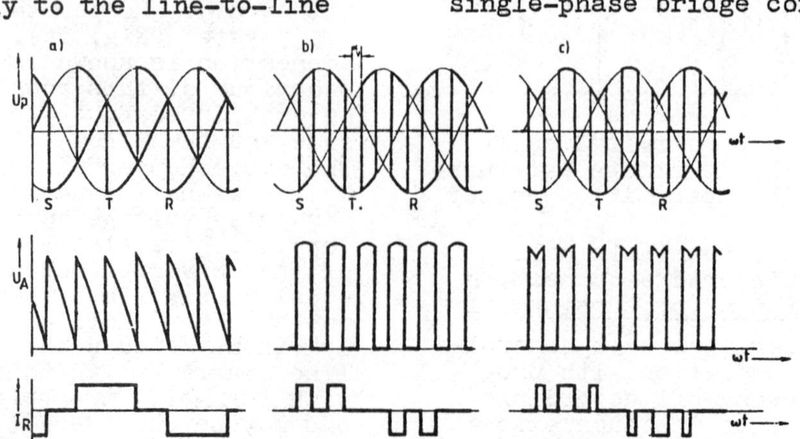

Fig.6 Voltage and current diagram for a) fully controlled converter b) converter shown in (Fig.5) symmetrically triggered to U_V c) symmetrically triggered to U_p

in which three parallel-connected diodes each are replaced by a single diode.

The three-phase bridge connection with two free-wheeling diodes is best suited for forced commutation because of the fact that the current in the phases never exceeds the phase angle 180°. Since this connection can be developed from the on both d.c.terminale half-controllable bridge connection, it is also possible to apply forced commutation. From the connection shown in Fig. 4, the connection in Fig. 7 is obtained for a three-phase a.c. system. With this three-phase bridge connection, three firing methods are possible: phase control, turn-off control, phase and turn-off control.

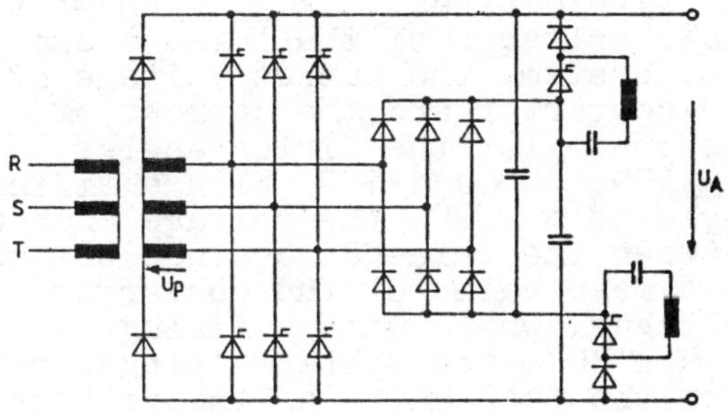

Fig.7 Fully controllable three phase converter with turn-off provision

Figure 8 shows the line diagram for a) phase control with firing angle $\alpha = 60°$ and b) torn-off control with an turn-off angle of $\beta = 60°$. The voltage and current curves as a function of time are in specular symmetry to phase angle $\varphi = 90°$. In the case of phase and turn-off control, only the line-to neutral voltage appears at the converter output in the range of $\alpha, \beta > 30°$. Neglecting the voltage peaks during turn-off, a six-pulse outpult voltage, reduced by $1/\sqrt{3}$, is obtained at the output. For firing angles $\alpha, \beta < 30°$, the line-to-line voltage appears across the output in the range where the two three-pulse converters overlap each other. Fig. 8c shows the line diagram for $\alpha = \beta = 45°$, Fig. 8d for $\alpha = \beta = 15°$. The current of a phase may be displaced between $\varphi = 0$ and 180°, depending on the trigger method.

5. Converter connections on the three-phase a.c.system with reduced harmonics

Among the converters with turn-off feature supplied form the three-phase system, only the fully controlled bridge connection with provision for turning off on both poles is capable of conducting a continuous current block in each half-cycle of a phase. However, it does not reduce the harmonics in the mains currents to any appreciable extent. Larger firing angles, i.e. narrow current pulses, give rise to a larger amount of higher-order harmonics in relation to the fundamental oscillation.

However, in the case of load circuits where a counter-voltage is always available - e.g. circuits charging accumulators (storage batteries) - a low voltage can be assumed to be available across the converter. For this reason, there is no need for the converter to commutate to the free-wheeling diode after every turn-off; instead it can commutate to an uncontrolled diode bridge connected to a transformer tap, the output voltage of which is equal to or less than the lowest load voltage. Fig. 9 shows the circuit arrangement which includes a diode bridge at transformer tappings instead of the free-wheeling diodes connected to the transformer neutral as shown for the three-phase bridge with turn-off feature in Fig. 7. Like the converter according to Fig. 7, this turn-on and turn-off type converter with fixed basic voltage can be operated with phase control, turn-off control and with phase and turn-off control.

The load current flows through the diode bridges at the transformer taps in the ranges $\varphi = 30°$ to $\varphi = 30° + \alpha$ and of $\varphi = 150° - \alpha$ to $\varphi = 150°$. The phase current is reduced by the ratio of the transformer windings between the controllable and uncontrollable bridge. In the range of $\varphi = 30° + \alpha$ to $\varphi = 150° - \alpha$, the phase current is equal to the load current. A two-step primary phase current is thus obtained which contains a smaller amount of harmonics than the phase current of a three-phase bridge with phase control. The 3rd harmonics calculated on the secondary side cannot accur in the three-phase a.c.system.

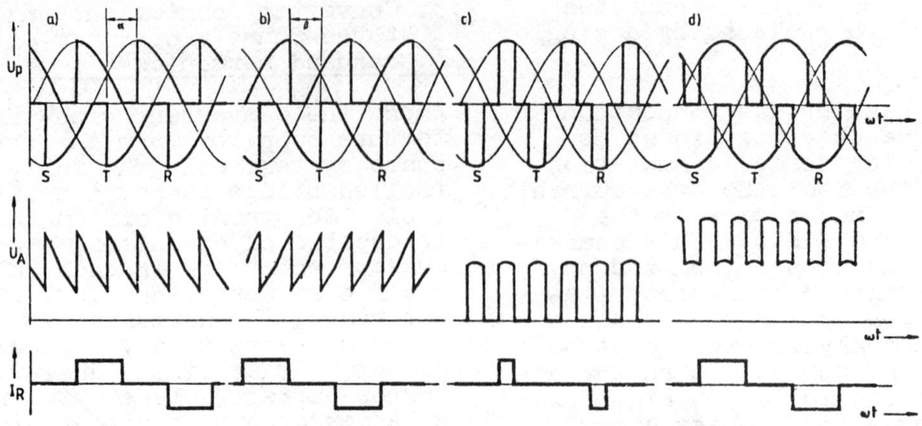

Fig.8 Voltage and current diagram for the converter shown in Fig.7
a) phase control $\alpha = 60^o$ b) turn-off control $\beta = 60^o$
c) phase and turn-off control $\alpha = \beta = 45^o$
d) phase and turn-off control $\alpha = \beta = 15^o$

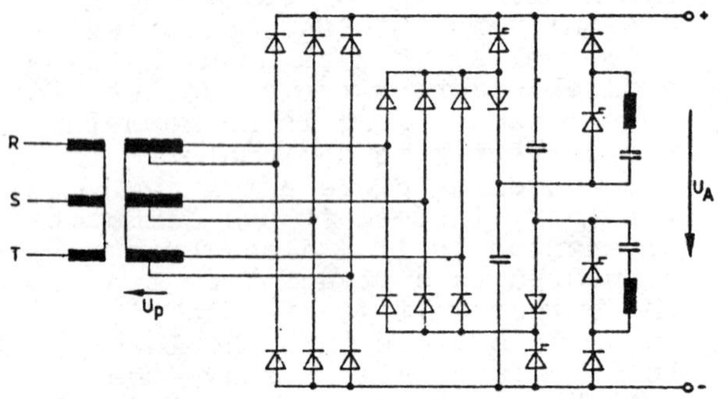

Fig.9 Controllable converter with fixed base voltage

The line diagram in Fig.10 shows the phase voltages of the three-phase a.c. system, the output voltage of the converter and the current of one phase for the firing angles $\alpha = 45^o$, 30° and 15°. With a firing angle of $\alpha > 30^o$, the voltage changes between the value of a diode bridge to a mean value of the converter voltages, while with a firing angle of $\alpha < 30^o$, the voltage changes between the mean value of the voltage and the voltage of the controlled bridge. An output voltage containing a small amount of harmonics is obtained for the firing angle $\alpha = 30^o$.

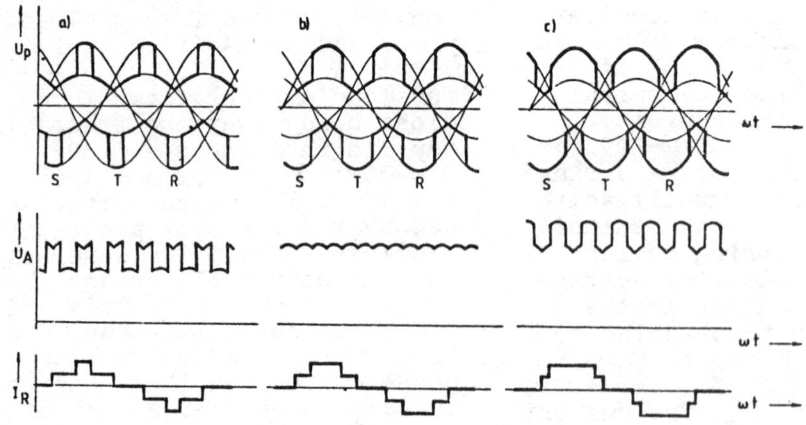

Fig.10 Voltage and current diagram for the converter shown in Fig.9 for $U_u/U_g = 0,5$
a) $\alpha = 45^o$ b) $\alpha = 30^o$ c) $\alpha = 15^o$

6. The operating characteristics of the turn-on and turn-off type three-phase converter with fixed base voltage

The converter connection as shown in Fig. 9 has three definite operating points:

1. At firing angle $\alpha = 60^o$ (corresponding to untriggered thyristors), the six-pulse basic voltage is supplied.
2. At firing angle $\alpha = 0^o$, the converter provides the full output voltage which is also of the six-pulse type.
3. At firing angle $\alpha = 30^o$, the second half of the bridge is triggered as the first half is turned off. The voltage is a twelve-pulse voltage for the ratio $U_u/U_g = 0.365$, where U_u is the voltage of the uncontrolled converter and U_g that of the controllable converter. For other voltage ratios U_u/U_g, the degree of smoothing of the output voltage is higher than in the voltage provided by a phase controlled bridge connection. Fig. 11 shows the voltage time integral for smoothing as function of the mean value of the voltage for the turn-on and turn-off converter with fixed basic voltage for $U_u/U_g = 0.5$ and, for comparison, those for a phase controllable converter.

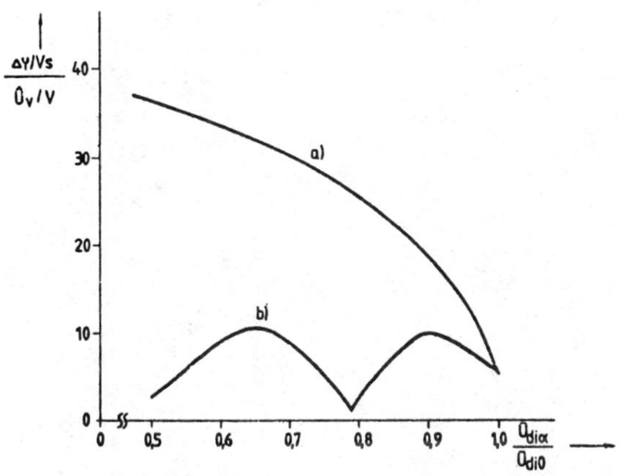

Fig.11 Voltage-time-integral vs mean value of output voltage

a) for the converter shown in Fig. 9

b) for a converter with phase-control

Assuming an completely smooth load current and neglecting commutation, in the case of phase controllable converters, the current block is conducted in each phase for a period of 120°.

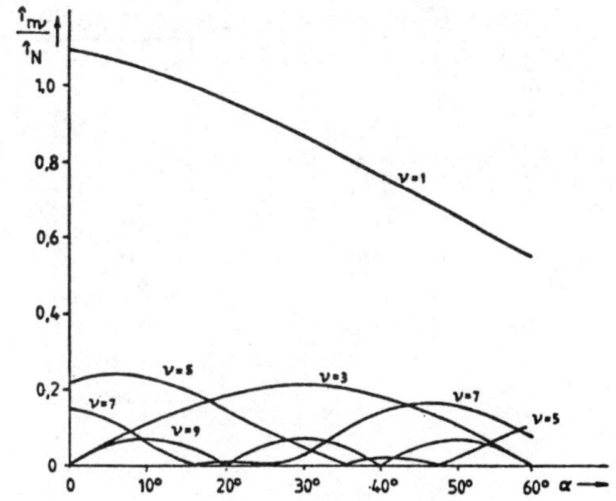

Fig. 12 Frequency spectrum of the line current vs control-angle α

For the connection according to Fig. 9 this operating point is reached with firing angle $\alpha = 0^o$. The fundamental and the harmonics of the current for a circuit designed for $U_u/U_g = 0,5$ are shown in Fig. 12 as a function of firing angle α. All harmonics vary with the firing angle with respect to their amount and phase. In this way, it is possible to reduce the amount of harmonics in a system by partly offsetting them against each other.

References

(1) U. Kunz, Ein Pulswechselrichter mit weicher Kommutierung; IFAC-Symposium Düsseldorf 1974 Vol.1, Seite 309-325

(2) H. Kahlen, Vergleichende Untersuchungen an verschiedenen Gleichstromsteller für Fahrzeugantriebe; Dissertation TH Aachen 1973

(3) J.Förster, Löschbare Fahrzeugstromrichter zur Netzentlastung und -stützung; Elektrische Bahnen Bd.43(1972) H.1, S.13-19

(4) W.Runge, Single-phase converter with forced commutation based on the supporting cpacitor principle; IFAC-Symposium Düsseldorf 1977

(5) H.Kahlen, A.Müller-Hellmann, W.Woelker, Halbgesteuerte Brückenschaltung mit unterteilter Löschung; Elektrische Bahnen Bd.46 (1975), H.12, Seite 279-285

(6) J.v.Issendorff, W.Hartel, Stromrichter mit Nullanode; Elektrotechnische Zeitschrift-A Bd.75 (1954), Seite 691-696

(7) G.Möltgen, Die Blindleistung bei Stromrichtern mit Nullanoden; Archiv für Elektrotechnik Bd.43 (1957), Seite 276-288

(8) M.Häusler, Three-phase bridge rectifiers with free-wheeling diodes; Elektrotechnische Zeitschrift-A Bd.94 (1973) Seite 230-234

REDUCTION OF THE HARMONIC CONTENT OF THE LOAD CURRENT OF THREE PHASE INVERTERS FED BY DIRECT CURRENT SOURCES

Siegfried Ditteney

Technische Akademie Esslingen, D-7302 Ostfildern 2

1. INTRODUCTION

If three phase loads are directly connected to selfcommutated inverters their input voltages and currents contain many harmonics. These harmonics are most disturbing. Therefore, they must be reduced.

In this abstract it will be shown how to half the number of harmonics in the load currents without changing the magnitudes of remaining harmonics independent of the amplitude and frequency of the fundamental. In this case it is necessary that the load can be split into two equal parts fed by two inverters with different firing angles. In addition a magnetic coupling element interconnecting both systems is necessary. This element will work as a drain for certain harmonics and has only a remarkably small equivalent rating.

2. ASSUMPTIONS AND BASIC CIRCUIT

Because of the great practical importance the study is limited to symmetrically arranged and cyclic symmetrically controlled inverter load-systems connected to a source of ideal direct current.

Nonlinearities of magnetic circuits are neglected. The thyristors are assumed to be ideal switches, which can be opened and closed at any time. Furthermore it is assumed that the direct current is not divided to two inverter output phases at the same time.

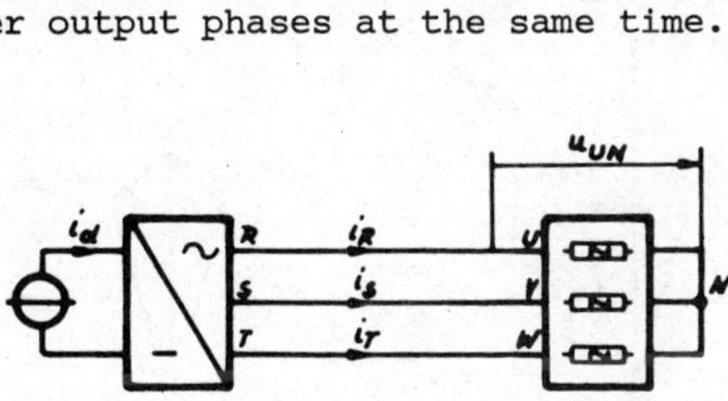

Fig. 1. Basic circuit of a 3-phase inverter load-system

The inverter output currents (Fig. 1), from now on simply called output currents, are defined with $x = \omega_1 t$ as

$$\begin{aligned} i_R(x) &= -i_R(x+\pi) \\ i_S(x) &= i_R(x-2\pi/3) \\ i_T(x) &= i_R(x+2\pi/3) \end{aligned} \quad (1)$$

The current i_R in the form of a Fourier series is

$$i_R = \sum_\nu \hat{i}_\nu \cos(\nu x + \varphi_{i\nu})$$

with $\nu = |6k \pm 1|$ and $k = 0;1;2;3;....$ (2)

The three output currents form a symmetrical 3-phase system which can be divided into two components.

1. All harmonics with the orders of

$$\nu = 6k+1 \quad \text{with} \quad k = 0;1;2;3;.... \quad (3)$$

form positive sequence systems.

2. All harmonics with the orders of

$$\nu = 6k-1 \quad \text{with} \quad k = 1;2;3;.... \quad (4)$$

form negative sequence systems.

Zero sequence systems do not exist.

A system of currents only containing harmonics of the orders of

$$\nu = |pk \pm 1| \quad \text{with} \quad k = 0;1;2;3;.... \quad (5)$$

shall be called a system with p pulses (ref. 1). Therefore, the output currents described here by harmonics according to equations (3) and (4) make up a system with 6 pulses.

3. INVERTER-FED DOUBLE LOAD WITH COMPENSATION TRANSFORMER

If it is possible to split the load into two equal parts the load currents of each part could have 12 pulses. This can be achieved if each part of the load is connected to a seperate inverter having output currents with 6 pulses. The two equal inverters work with a phase displacement of $\pi/6$ and must have a magnetic linkage with a compensation transformer as shown in Fig. 2.

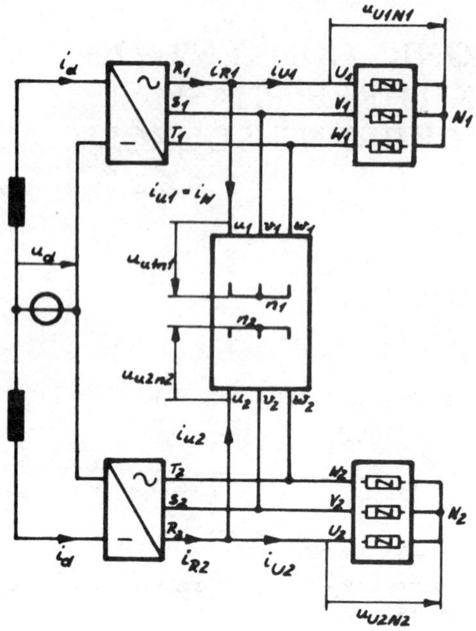

Fig. 2. Inverter load-system with compensation transformer

In the ideal case the compensation transformer has no reactance for all harmonics of the order of

$$\nu = (12k-6) \pm 1 \quad \text{with} \quad k = 1;2;3;.... \tag{6}$$

while it has an infinite inductance for all harmonics of the order of

$$\nu = |12k \pm 1| \quad \text{with} \quad k = 0;1;2;3;.... \tag{7}$$

The load has an impedance different from zero for all harmonics. Therefore, the compensation transformer is a current drain for the harmonics given by equation (6) and the current in the load will only consist of the harmonics given by equation (7).

The time variation of both output currents is as following

$$i_{R2}(x+\pi/12) = i_{R1}(x-\pi/12) \tag{8}$$

As a result of the selective separation of the output current one finds for the first system with equations (2), (6) and (7) as Fourier series

1. for the load current

$$i_{U1} = \sum_{\nu} \hat{\imath}_{\nu} \cos\left[\nu(x+\pi/12) + \varphi_{i\nu}\right] \quad \text{with } \nu = |12k \pm 1| \text{ and } k = 0;1;2;3;.... \tag{9}$$

2. for the compensation transformer current

$$i_{u1} = \sum_{\nu} \hat{\imath}_{\nu} \cos\left[\nu(x+\pi/12) + \varphi_{i\nu}\right] \quad \text{with } \nu = (12k-6) \pm 1 \text{ and } k = 1;2;3;.... \tag{10}$$

Together with formula (8) for the second system similar equations are valid.

The components of each system have to be designed for half the total apparent power and are rated equally because magnitudes and phase displacements are the same in both systems in the steady state.

The compensation transformer, as the current drain is called, can be realized with good approximation if two double windings are mounted on each limb of an iron core without an air gap. Between both three-phase systems there is only a magnetic connection so that the direct current source can easily be realized by a rectifier with smoothing inductances.

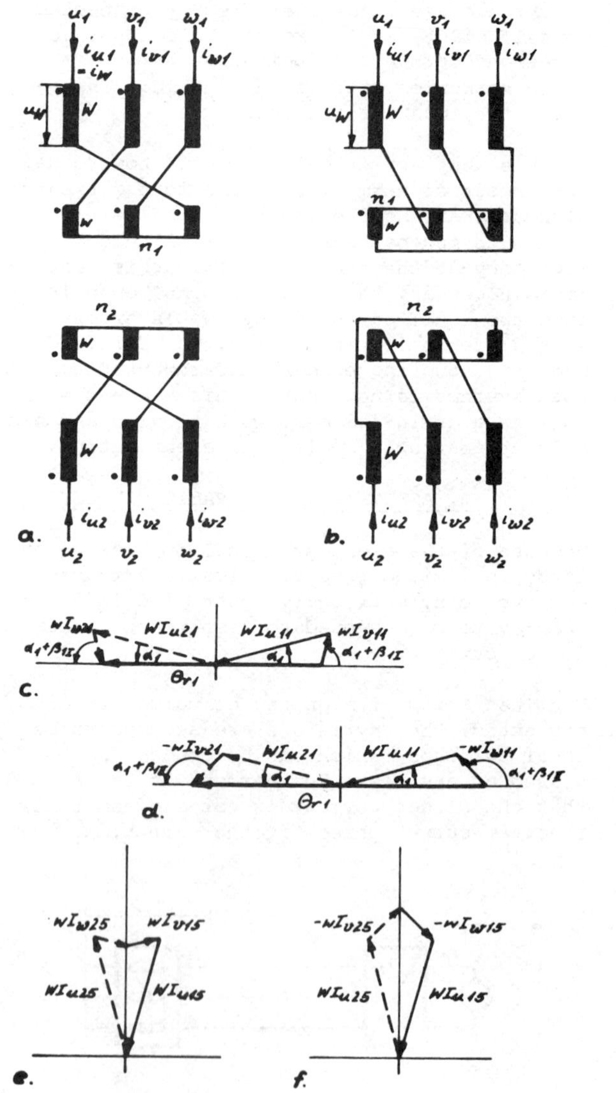

Fig. 3. Circuits of compensation transformer
a) circuit I
b) circuit II
c) phasor diagram for fundamental to a)
d) phasor diagram for fundamental to b)
e) phasor diagram for 5th harmonic to a)
f) phasor diagram for 5th harmonic to b)

4. ANALYSIS OF THE COMPENSATION TRANSFORMER

Figure 3 shows two possibilities of interconnecting the twelve windings of the compensation transformer. The phasor diagrams for the different circuits are valid for the fundamental and 5th harmonic of the flux and are shown in Fig. 3c to 3f. The winding ratios w/W must be chosen in such a way that the net sum of the 5th harmonic of the flux disappears. Using the phasor diagrams one finds the formula

$$w\cos 5(\alpha_1+\beta_1) = W\cos 5\alpha_1 \qquad (11)$$

Hence, the following is true

for case I with $2\alpha_1 = \pi/6$ and $\beta_{1I} = \pi/3$

$$(w/W)_I = 0{,}268 \qquad (12I)$$

for case II with $2\alpha_1 = \pi/6$ and $\beta_{1II} = 2\pi/3$

$$(w/W)_{II} = 0{,}366 \qquad (12II)$$

For the harmonics of the orders of

$$\nu = 12k\pm1 \quad \text{with} \quad k = 1;2;3;\ldots.$$

one obtains

$$2\alpha_\nu = (12k\pm1)\pi/6 = 2\pi k\pm2\alpha_1$$
$$\beta_{\nu I} = (12k\pm1)\pi/3 = 4\pi k\pm\beta_{1I} \qquad (13)$$
$$\beta_{\nu II} = (12k\pm1)2\pi/3 = 8\pi k\pm\beta_{1II}$$

All harmonics of the orders given by equation (7) have phasor diagrams of the same form as in Fig. 3c or 3d. The parts of the flux in a limb due to the two inverters add up to the total flux $\Theta_{r\nu}$. For these current harmonics the compensation transformer represents an infinite inductance in the ideal case. In the real case there are only small magnetizing currents if there are no saturation effects in the core.

For the harmonics of the orders of

$$\nu = (12k-6)\pm1 \quad \text{with} \quad k = 1;2;3;\ldots. \qquad (6)$$

one obtains

$$2\alpha_\nu = [(12k-6)\pm1]\pi/6 = \pi(2k-1)\pm2\alpha_1$$
$$\beta_{\nu I} = [(12k-6)\pm1]\pi/3 = 2\pi(2k-1)\pm\beta_{1I} \qquad (14)$$
$$\beta_{\nu II} = [(12k-6)\pm1]2\pi/3 = 4\pi(2k-1)\pm\beta_{1II}$$

All harmonics of the orders given by equation (6) produce phasor diagrams of the same form as in Fig. 3e or 3f with zero total flux. For these harmonics the compensation transformer has no inductance. In the real case there remains only the very small copper resistance for the symmetrical arrangement.

Therefore, the compensation transformer is a drain for the current harmonics described by formula (6). For this reason the load currents make up pure system of 12 pulses.

The three-phase voltage systems of the load and the compensation transformer do not include the 3rd harmonic or multiples of it so that they constitute only positive and negative sequence systems. Because the whole arrangement is symmetrical the points N_1 and n_1 and the points N_2 and n_2 have the same potential so that

$$u_{u1n1} = u_{U1N1}$$
$$u_{u2n2} = u_{U2N2} \qquad (15)$$

From the arrangement of the coils (Fig. 3) the voltage u_w can be derived.

For the voltage of one phase of the compensation transformer there is

$$u_{u1n1} = u_W(x) + a\frac{w}{W}u_W(x+a2\pi/3) \qquad (16)$$

and

$$u_{u2n2} = -u_W(x) - a\frac{w}{W}u_W(x-a2\pi/3) \qquad (17)$$

where

$$a = \begin{cases} +1 & \text{circuit I} \\ -1 & \text{circuit II} \end{cases}$$

Subtracting formula (17) from (16) and substituting the difference into formula (15) one gets with

$$u_W(x+a2\pi/3) + u_W(x-a2\pi/3) = -u_W(x)$$

for the coil voltage

$$u_W = \frac{u_{U1N1}-u_{U2N2}}{2-aw/W} \qquad (18)$$

The time variation of the magnetic flux of one coil Ψ_W can be calculated from

$$u_W = \omega_1 \frac{d\Psi_W(x)}{dx} \qquad (19)$$

The average of the density of losses in the windings and the maximum value of the density of the magnetic flux in the iron cores determine the dimensions of the inductive circuit components. Therefore, one finds (ref. 2) for the equivalent rating

$$P_B = \frac{\omega_1}{2\sqrt{2}} \Sigma I_W \cdot \hat{\Psi}_W \qquad (20)$$

where I_W means the effective value of the coil current.

5. INVERTER-FED 3-PHASE INDUCTION MOTOR

What has been shown above will now be used for the important special case that the load exists of 3-phase induction motors.

In good approximation the stator voltage has pure sinusoidal form when the induction motor is fed from a current source. The frequency of this voltage will be the same as that of the fundamental of the inverter output current. The three voltages form a pure positive sequence system. The following is valid

$$\begin{aligned} u_{U1N1} &= \hat{u}_1 \cos(x+\pi/12+\varphi_u) \\ u_{U2N2} &= \hat{u}_1 \cos(x-\pi/12+\varphi_u) \end{aligned} \qquad (21)$$

Then one finds for the voltage of a coil

$$u_W = -\frac{2}{2-aw/W}\,\hat{u}_1 \sin\pi/12 \cdot \sin(x+\varphi_u) \qquad (22)$$

For the maximum value of the magnetic flux one obtains from this equation

$$\hat{\psi}_W = \frac{\hat{u}_1}{\omega_1} \cdot \frac{2}{2-aw/W} \sin\pi/12 \qquad (23)$$

In the case that inverter output current consists of rectangular blocks with the hight i_d and the length $2\pi/3$ the total equivalent rating of the compensation transformer relative to the total apparent power should be calculated.

The effective value of inverter output current is

$$I_{R1} = \sqrt{2/3}\; i_d \qquad (24)$$

The effective value of the load current can be determined in the same way as that of a load voltage as explained in ref. 3.

$$I_{U1} = \sqrt{\frac{1+\sqrt{3}/2}{3}}\; i_d \qquad (25)$$

Because i_{U1} and i_W are orthogonal one obtains the effective value of the coil current of the compensation transformer as the difference between the effective values of the inverter output current and the load current.

$$I_W = \sqrt{\frac{1-\sqrt{3}/2}{3}}\; i_d \qquad (26)$$

This formula and equation (23) substituted into equation (20) provides the relative rating

$$\frac{P_B}{P_{s1}} = \frac{\sqrt{1+\sqrt{3}/2}\cdot\pi(1+w/W)\sin\pi/12}{\sqrt{18}\,(2-aw/W)} \qquad (27)$$

where P_{s1} is the total apparent power.

Substituting numbers one obtains

$$\frac{P_B}{P_{s1}} = \begin{cases} 0{,}05135 & \text{circuit I} \\ 0{,}04050 & \text{circuit II} \end{cases} \qquad (28)$$

6. REFERENCES

(1) M. Depenbrock, S. Ditteney: Spannungsverhältnisse bei gekoppelten Drehstrom-Wechselrichtern. ETZ-A 97(1976)11, S. 668-671

(2) M. Depenbrock: Fremdgeführte Zwischenkreisumrichter zur Speisung von Stromrichtermotoren mit sinusförmigen Anlaufströmen. ETZ-A 87(1966)26, S. 945-951

(3) S. Ditteney: Untersuchungen zur vollständigen mathematischen Analyse des stationären Betriebsverhaltens selbstgeführter Drehspannungswechselrichter. Dissertation, Ruhr-Universität Bochum, 1975

A NEW APPROACH TO POWER CONVERTER CIRCUITS

P. Mehta, T. Thomson and W. Un

Department of Electrical Engineering, Brunel University, Uxbridge, Middlesex, United Kingdom

ABSTRACT

The technique of 'shorting-coil' switching proposed in this paper overcomes the restrictions imposed by the voltage and current ratings of the presently available power semiconductor devices and conventional convertor circuit configurations. This is achieved by modulating flux waves in multi-limb magnetic circuits via semiconductor switches in auxiliary windings called shorting-coils.

LIST OF SYMBOLS

$Z_{(a..f,f..a)}$	mutual impedance between coils a to f
$Z_{(aa..ff)}$	self impedance of coils a to f
Z_1	mutual impedance between coils on same limb
Z_3	mutual impedance between coils on alternate limbs
Z_{LL}, Z_{HH}, Z_{JJ}, Z_{KK}, Z_0	load impedance
$i_{(a..f)}$	currents in the primitive system
$i_{(1..6)}$	currents in coils a to f
i_0	output current
i_i	input current
v_i	input voltage
v_0	output voltage

INTRODUCTION

The use of semiconductor switching techniques for the control of power flow has been restricted by the voltage and current ratings of presently available devices and by circuit limitations. As the use of power semiconductor devices grows the problems inherent in and caused by the use of conventional circuits will become more serious.

For example, phase-controlled convertors suffer from poor power factor due to the phase displacement between the input current and the line voltage at delayed triggering angle and the distortion power demanded by the harmonics in the input line current. Furthermore, the output voltage contains ripple and in some applications considerable filtering is necessary. Much work has been done in the past to alleviate some of these problems, but more recently, sequential control of series-connected bridges has been employed for power factor correction (1,2). The basic principle of this technique is to modify the waveshape of the input current and its phase relationship with the input voltage. It has been shown that to obtain maximum benefit from this approach, forced commutation of thyristors is required (1).

The output of a conventional phase-controlled cycloconvertor is limited to frequencies below that of the mains, unless forced commutation is employed. Because of this limitation d.c.-link type convertors are used but the inability of this type of convertor to transfer reactive power from the a.c. mains to the load, means that the invertor must be designed to handle reactive power; this introduces considerable complexity to the basic circuit at extra cost.

In general therefore, it can be concluded that to improve the performance of conventional convertor circuits forced commutation has to be introduced or alternatively, consideration must be given to the use of power transistors or gate-turn-off devices.

In power control applications, devices are employed as 'series' switches and therefore in the non-conducting state they must block the full system voltage, and in the conducting state they must be rated to carry the full load current. The designer, therefore, has only a limited number of basic circuits to choose from. Since the rating, both current and voltage, of the available devices is limited by either economic or technical considerations, for high voltage/current applications series/parallel connections of devices must be employed.

Some of these restriction and limitations can be overcome by employing an indirect technique called 'shorting coil' switching in which power flow in a main circuit can be regulated by switching in auxiliary circuits called shorting-coils. This introduces a new philosophy in power control which provides a unique flexibility whereby the designer, instead of selecting the voltage/current ratings of devices for a given application, designs the shorting-coil to suit the available devices. This technique should be particularly useful in utilising power transistors and gate turn-off thyristors in high-voltage applications.

The main application of the proposed technique is envisaged in the field of a.c. voltage regulation, phase-controlled convertors and direct a.c. to a.c. power conversion. Since this approach allows power transistors and gate turn-off devices to be used in high voltage applications, power factor correction and control of input current harmonics can be easily incorporated (1).

THE PRINCIPLE OF SHORTING-COIL SWITCHING

The use of short-circuited windings in the moving coil regulator (m.c.r.) (3) and as damper windings in alternators is well understood. The movement of the coil in the m.c.r. can be simulated by changing taps in a fixed winding, thereby leading to a much faster speed of operation and the elimination of magnetic circuits requiring large air-paths. In principle, because of the high switching speeds made possible by semiconductor switches, combinations of such 'static' m.c.rs. could be used to change voltage, phase and frequency.

A number of circuit configurations are possible, but the basic principle of shorting-coil switching can be illustrated best with reference to a single-phase frequency convertor which employs a three-limb magnetic circuit as shown in Fig. 1. The centre limb carries the primary winding, whilst the outer limbs carry secondary windings which are connected in series opposition, and auxiliary windings which are employed as shorting-coils. When voltage is applied to the primary winding, the resulting alternating flux divides equally between the two outer limbs. Therefore, when both the shorting coils are open circuited, the output voltage is zero because the two secondaries are connected in series opposition. The effect of closing a switch in one of the shorting coils is to reduce the rate of change of flux in that limb to nearly zero. Hence, the primary alternating flux which is maintained constant by the input voltage links only one of the other secondary windings giving the full output voltage in one direction. If the auxiliary coils are shorted alternately, the effect is to divert the flux from side to side subjecting the output coils to a modulated flux-wave. Although the shorting coils have to be rated to carry the full load VA, the turns ratio between the output and the auxiliary windings can be selected to suit voltage/current ratings of the devices to be used. Fig. 2 shows typical input and output voltage waveforms on a resistive load. Figs. 3 and 4 show respectively the corresponding shorting-coil voltage/current waveforms and the output of a search-coil wound on one of the outer limbs to observe flux. Note that during the periods when the auxiliary coil is short-circuited the rate of change of flux is approximately zero.

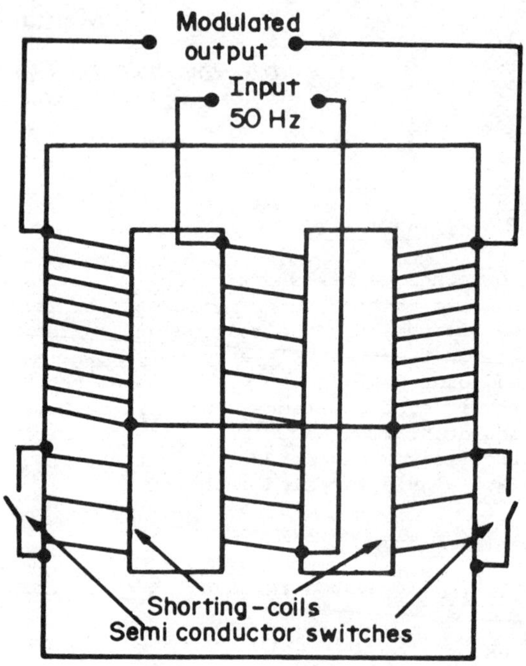

Fig l Shorting-coil modulator

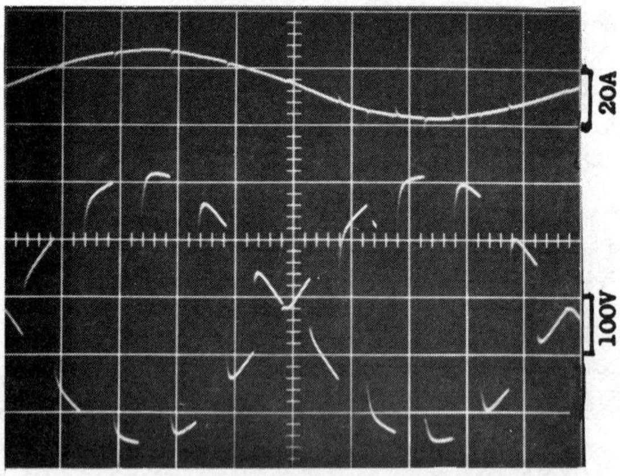

Fig. 2 Input current/output voltage waveforms - resistive load

FACTORS AFFECTING THE OPERATION OF SHORTING-COIL SYSTEMS AND CHOICE OF CIRCUIT CONFIGURATIONS

Frequency Modulator.

Ideally an auxiliary coil in conjunction with

a semiconductor switch should provide a perfect short circuit so that the rate of change of flux in that limb is maintained at zero. In practice, however, when an auxiliary coil is short-circuited, although the flux in that limb is instantaneously 'frozen', it subsequently changes at a rate dicated by the parameters of the windings and the semiconductor switch. Therefore the leakage inductance and the resistance of the shorting-coils and the 'on-state' voltage-drop of the switches produce effects which require careful consideration when evaluating the various circuit configurations and selecting semiconductor devices.

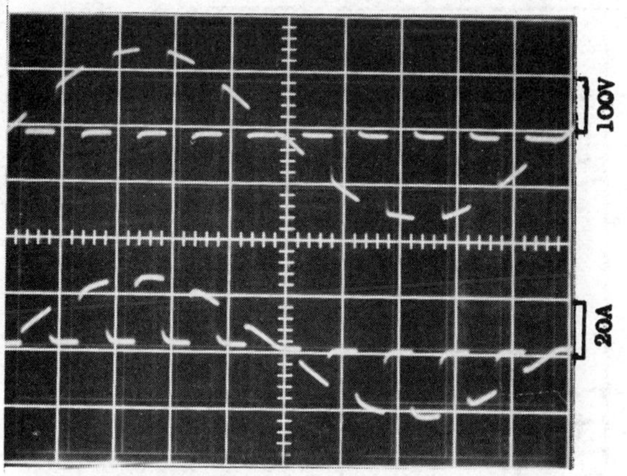

Fig. 3 Shorting-coil voltage/current waveforms - resistive load

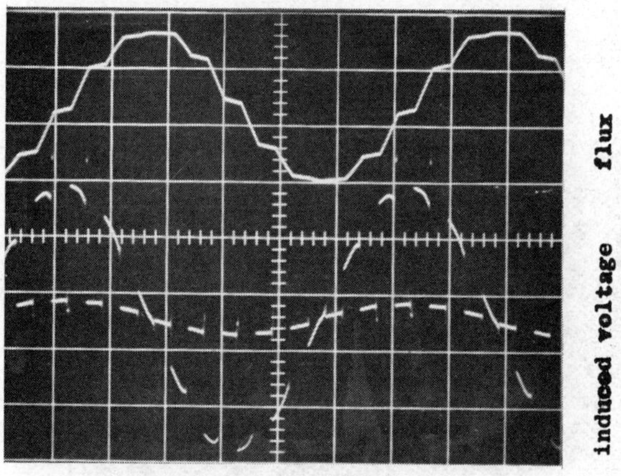

Fig. 4 Induced voltage in the search-coil and flux in an outer limb

Because of the complexity of the circuit to be analysed, Krons' (4) approach using tensors is employed. The complete analysis is presented in the Appendix. It is shown that one of the factors which affects the performance of the shorting-coil transformers is the leakage inductance of windings which must be minimised. One method of reducing the leakage inductance of the shorting-coil is to wind all the windings on the two outer limbs as shown in Fig. 5; the principle of operation remains unchanged.

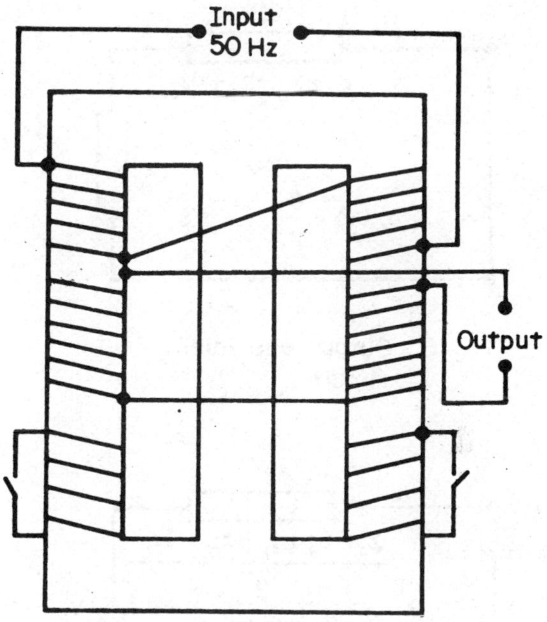

Fig. 5 Alternative single-phase modulator

Equation (7) which describes the shorting-coil transformer of Fig. 5 can be used to derive the various equivalent circuits shown in Figs. 6-8. Note that these circuits do not contain any of the mutual inductance terms. The performance of the system depends entirely on the leakage inductances and resistance values of the various windings. Of particular importance is the impedance of the shorting coils. It is evident from Fig. 6 that the impedance of the shorting-coil is a very important parameter which has a considerable affect on the output characteristics; this is discussed under Experimental Results.

Figs. 7 and 8 show the equivalent circuit as seen by the short-circuited and the open circuited auxiliary coils and they can be used to calculate the voltage and current ratings of the power semiconductor devices to be used. Note that these circuits are for overall unity turns ratio.

<u>Voltage Regulator and A.C. to D.C. Convertor.</u>

Fig. 9 shows a possible circuit arrangement for an a.c. voltage regulator which can also be used as an a.c. to d.c. convertor if a rectifier is connected at the output.

When employed as an a.c. voltage regulator,

one of the shorting-coils is employed to modulate the flux, whereas the other one provides a 'free-wheel' path for the inductive load current. Figs. 10 and 11 show typical voltage/current waveforms with and without free-wheel operation respectively.

Fig. 12 shows the equivalent circuit of the regulator. Compared with the frequency modulator, Fig. 6, the significant difference is the considerable reduction in the effective impedance of the shorting coil. This is clearly evident in the relative efficiencies of the two applications shown in Fig. 13.

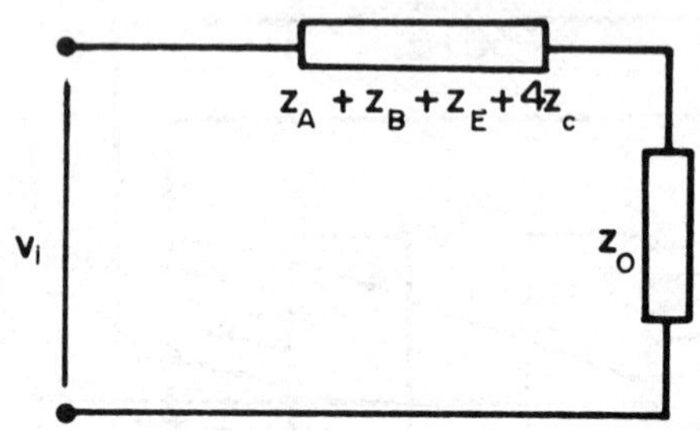

Fig. 6 Output equivalent circuit

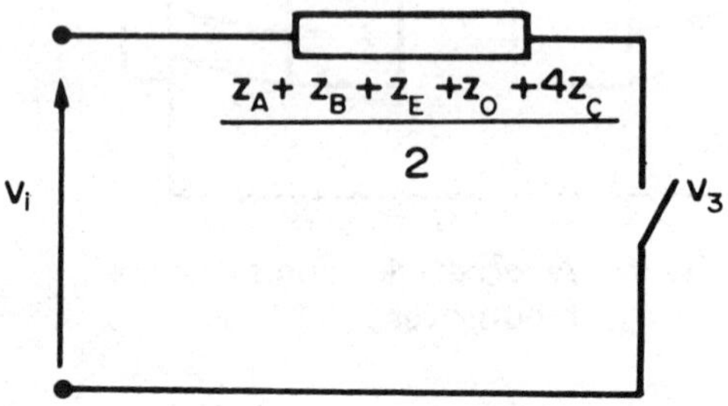

Fig. 7 Shorting - coil equivalent circuit- short circuited

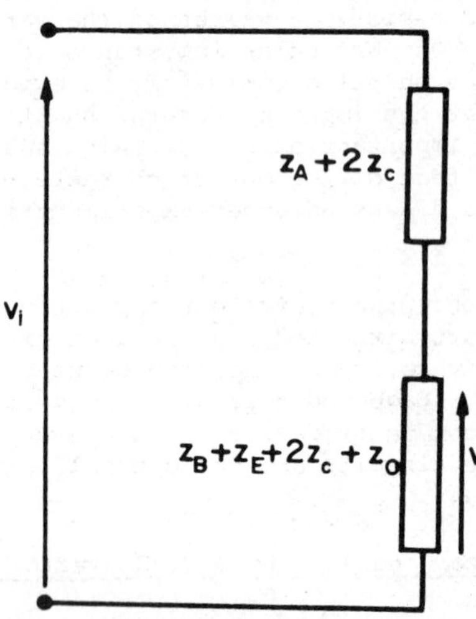

Fig. 8 Shorting-coil equivalent circuit - open circuit

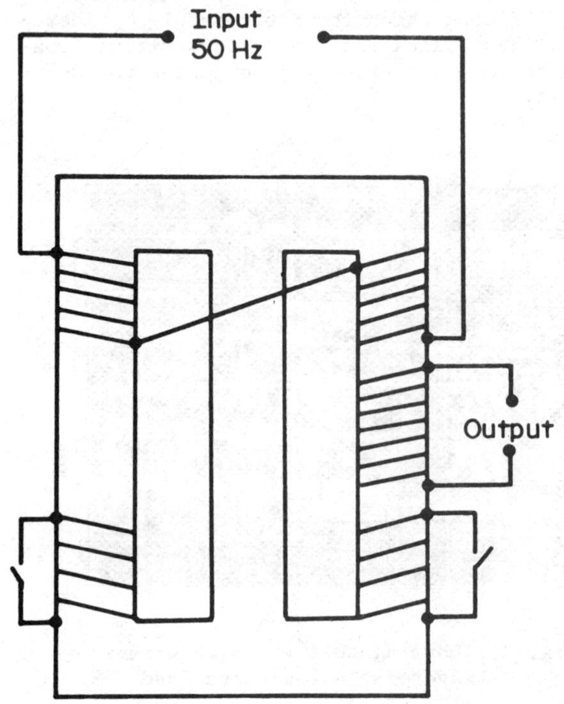

Fig. 9 A.C. voltage regulator

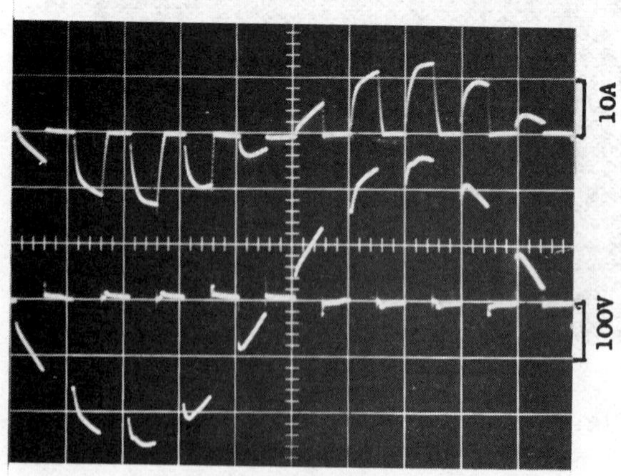

Fig. 10 A.C. voltage regulator waveforms inductive load - with free-wheel action

Experimental Results.

A standard three-limb magnetic circuit with conventional windings was employed for this work : The input/output windings were rated at 400 V and the shorting-coils at 100 V. Power transistors rated at 100 V, 80 A (160 A peak) were used as switches.
Fig. 13 shows the results of some tests carried out on the various arrangements for the two applications. For these tests, one of the shorting-coils is permanently short-circuited and the load varied. This, of course, does not take into account the switching losses which would occur under the proposed operating conditions. However, these results provide clear evidence of the efficiency of the transformer when employed in this unconventional fashion. The efficiency of the arrangement of Fig. 5 is better than that of Fig. 1 because the former employs a low leakage configuration.
The effect of the shorting-coil impedance in the two applications predicted by the equivalent circuits (Figs. 6 and 12) is also verified - the efficiency of the voltage regulator is considerably higher than that of the frequency modulator.

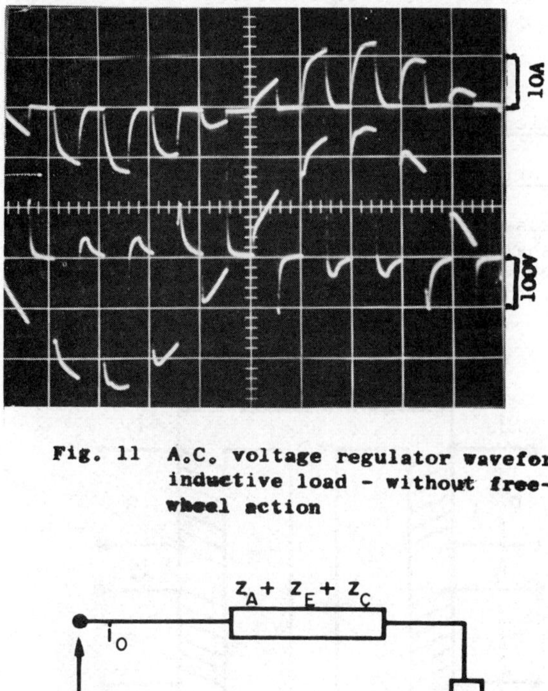

Fig. 11 A.C. voltage regulator waveform inductive load - without free-wheel action

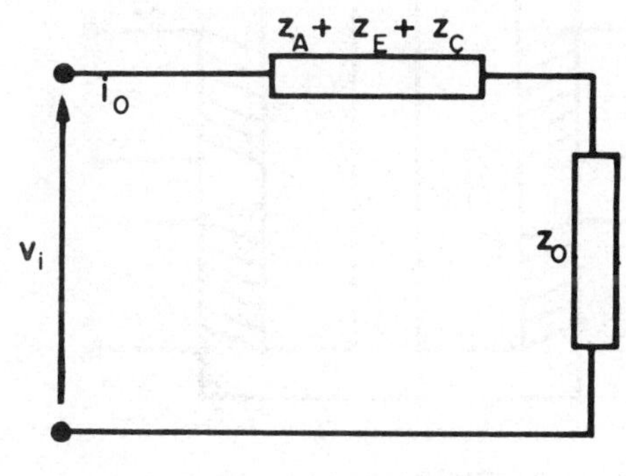

Fig. 12 Equivalent circuit of voltage regulator

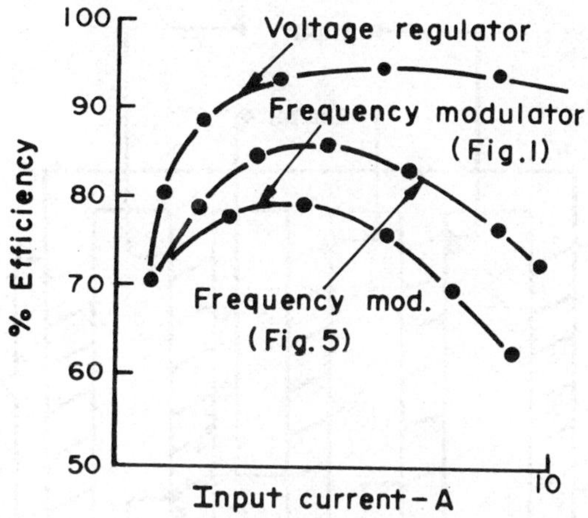

Fig. 13. Static test results

REDUCING THE RATING OF THE SHORTING-COIL AND THE ASSOCIATED SWITCHES

The forgoing has demonstrated that the designer can vary current or voltage rating of the switching devices to suit the requirements of the circuit. Full load VA, however, must be handled in all the arrangements so far considered. Some work is at present being carried out on methods of obtaining power amplification. One technique is to use the shorting-coil as a trigger in a parametric device.

The normal parametric transformer is a transfromer in which there is zero nett flux linkage between primary and secondary winding and power is transferred via cyclic variation in the reluctance of a common magnetic path (5). The cyclic variation is invariably brought about by saturation at the current peaks in the primary and the secondary windings. The output frequency is therefore limited to mains frequency or multiples thereof.

It has been shown (6) that parametric action can be induced by shorting-coil switching. The magnitude of output voltage, and the power transferred from primary to secondary is determined by the amplitude and phase of the harmonic in the cyclic reluctance wave which has been selected by tuning at the desired output frequency.

An experimental transformer has been constructed in which the shorting-coils are used to initiate parametric power transfer by causing saturation in the remainder of the circuit (Fig. 14) during the period that the coil is shorted. Thus some measure of control over the amplitdue and phase, and also the waveshape of the reluctance wave is possible. It can be shown that the shorting coils need link only a fraction of the main flux and can be reduced in rating accordingly.

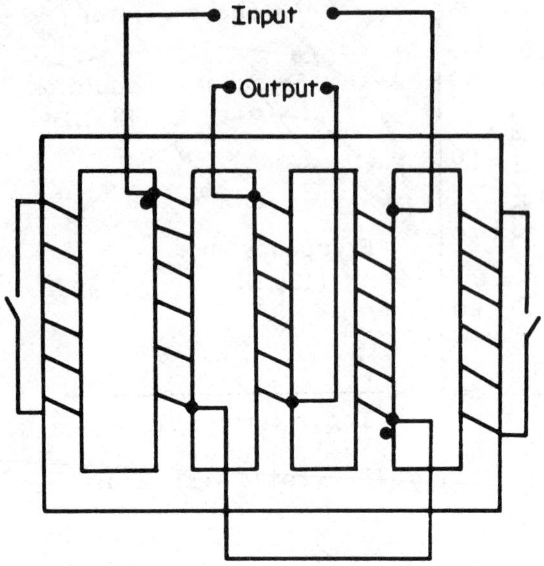

Fig. 14. Controlled parametric transformer

CONCLUSIONS

The outstanding feature of the proposed technique is the ability to perform the switching function at any desired voltage/current levels irrespective of voltage/current levels at the input/output circuits, thereby allowing optimum use to be made of presently available power semiconductor devices. In particular, this method of power control should make possible the use of power transistors and gate turn-off devices in high voltage applications.

Both theoretical and experimental work has shown that considerable improvement in the shorting-coil is necessary to reduce the leakage inductance of the shorting-coil to make the frequency modulator feasible. However, as an a.c. voltage regulator the shorting-coil transformer can operate at high efficiency.

Work is in progress to evaluate different types of windings to reduce the leakage effects. Several other configurations incorporating shorting-coils are being investigated and the results obtained so far are encouraging.

ACKNOWLEDGEMENT

The authors wish to thank the United Kingdom Science Research Council for supporting this research.

REFERENCES

1. P. Mehta, S. Mukhopadhyay, E. Orhun. Forced-commutated a.c. - d.c. convertor-controlled d.c. drives. IEE Conference Proceedings. Power Electronics - Power Semiconductors and their applications, 146-151. (1974)

2. H. Zender. Self-commutated rectifier to improve line conditions. Proc. IEE. 120, 9, 977-981. (1973)

3. G.H. Rawcliffe, I.R. Smith. The moving coil regulator: A treatment from first principles. Proc. IEE 104, 68-76. (1957)

4. G. Kron. Tensors for circuits. Dover 1959.

5. W.Z. Fam, R.P. Verma. Theory and performance of parametric transformers. IEEE Trans. 2494-2504 (1972)

6. A.J. Mullett. A new type of parametric transformer. Brunel University. M.Tech. Dissertation. (1974)

APPENDIX

Static Analysis.

The primitive system.

Because of the complexity of the circuit to be analysed tensor approach (4) is employed.

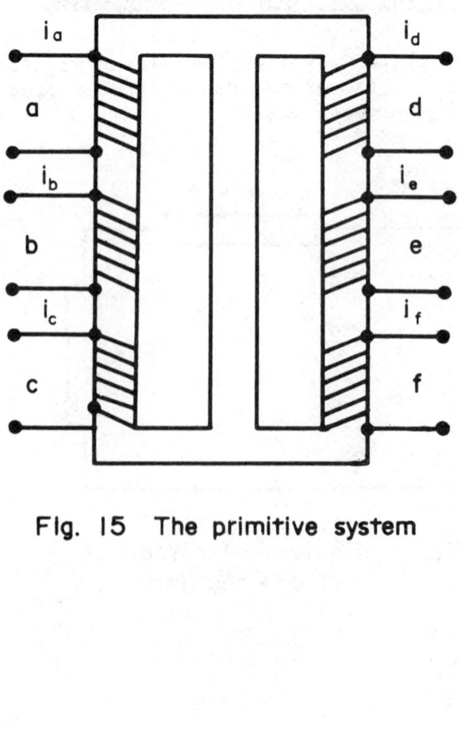

Fig. 15 The primitive system

Fig. 15 shows the primitive system with six coils a-f. Depending on the actual circuit connection employed L,H,J,K, may be load or impedances of the switches. The impedance matrix (2) of the primitive system is

	a	b	c	d	e	f	L	H	J	K
a	Z_{aa}	Z_{ba}	Z_{ca}	Z_{da}	Z_{ea}	Z_{fa}				
b	Z_{ab}	Z_{bb}	Z_{cb}	Z_{db}	Z_{eb}	Z_{fb}				
c	Z_{ac}	Z_{bc}	Z_{cc}	Z_{dc}	Z_{ec}	Z_{fc}				
d	Z_{ad}	Z_{bd}	Z_{cd}	Z_{dd}	Z_{ed}	Z_{fd}				
e	Z_{ae}	Z_{be}	Z_{ce}	Z_{de}	Z_{ee}	Z_{fe}				
f	Z_{af}	Z_{bf}	Z_{cf}	Z_{df}	Z_{ef}	Z_{ff}				
L							Z_{LL}			
H								Z_{HH}		
J									Z_{JJ}	
K										Z_{KK}

Evaluation of transformer paramters.

If the system is connected as shown in Fig. 16, the parameters of the transformer can be obtained by carrying out short-circuit and open-circuit tests.

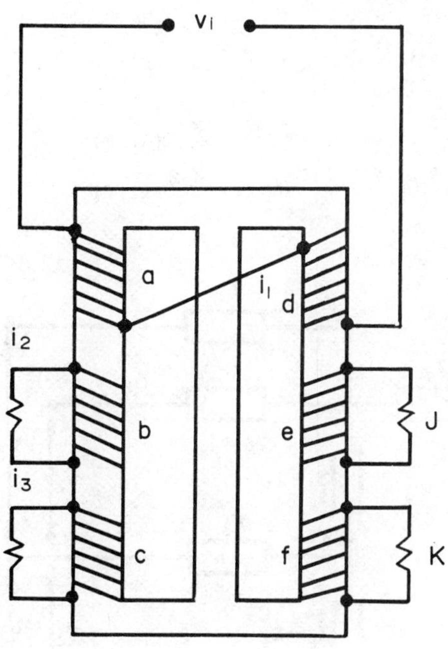

Fig. 16 Shorting-coil transformer connection for parameter evaluation

The 'new' and the 'old' currents are equated as follows:

$$i_a = i_1$$
$$i_b = i_2$$
$$i_c = i_3$$
$$i_d = i_1$$
$$i_e = i_5$$
$$i_f = i_6 \qquad (2)$$
$$i_L = i_2$$
$$i_H = i_3$$
$$i_J = i_5$$
$$i_K = i_6$$

Then the connection matrix C and current become:

	i_1	i_2	i_3	i_5	i_6
a	1				
b		-1			
c			-1		
d	1				
e				-1	
f					-1
L		1			
H			1		
J				1	
K					1

(3)

Current matrix

$$i' = \begin{bmatrix} i_1 & i_2 & i_3 & i_5 & i_6 \end{bmatrix} \qquad (4)$$

Voltage matrix is:

	a	b	c	d	e	f
e =	$\frac{1}{2}v_i$			$\frac{1}{2}v_i$		

(5)

where v_i is the input voltage.

Therefore

$$e' = C_t * e = \begin{bmatrix} v_i & 0 & 0 & 0 \end{bmatrix} \qquad (6)$$

Using the equation $e' = Z' * i'$, the equation for the transformer can be written as follows, where

$$Z' = (C_t * Z) * C$$

$$v_i = (Z_{aa} + Z_{dd} - Z_{ad} - Z_{da})\, i_1$$
$$- (Z_{ba} - Z_{bd})\, i_2 - (Z_{ca} - Z_{cd})\, i_3$$
$$- (Z_{ed} - Z_{ea})\, i_5 - (Z_{fd} - Z_{fa})\, i_6$$

$$- i_2 Z_{LL} = - (Z_{ab} - Z_{db})\, i_1 + Z_{bb} i_2 + Z_{cb} i_3 - Z_{fb} i_6 - Z_{eb} i_5$$

$$- i_3 Z_{HH} = - (Z_{ac} - Z_{dc})\, i_1 + Z_{bc} i_2 + Z_{cc} i_3 - Z_{ec} i_5 - Z_{fc} i_6$$

$$- i_5 Z_{JJ} = - (Z_{de} - Z_{ae})\, i_1 - Z_{be} i_2 - Z_{ce} i_3 - Z_{ee} i_5 - Z_{fe} i_6$$

$$- i_6 Z_{KK} = - (Z_{df} - Z_{af})\, i_1 - Z_{bf} i_2 - Z_{cf} i_3 + Z_{ef} i_5 + Z_{ff} i_6 \qquad (7)$$

putting:

mutual inductance of coil in limb 1 = Z_1

3 = Z_2

mutual inductance of coil in limbs 1 and 3 = Z_3

$$\begin{aligned} v_2 &= i_2 Z_{LL} \\ v_3 &= i_3 Z_{HH} \\ v_5 &= i_5 Z_{JJ} \\ v_6 &= i_6 Z_{KK} \end{aligned} \qquad (8)$$

$$\begin{aligned} Z_{AA} &= Z_{aa} + Z_{dd} - Z_{ad} - Z_{da} \\ Z_{bb} &= Z_{BB} \\ Z_{cc} &= Z_{CC} \\ Z_{ee} &= Z_{EE} \\ Z_{ff} &= Z_{FF} \end{aligned}$$

It can be shown that the equivalent circuit representing the transformer is as shown in Fig. 17.

$$\begin{aligned} Z_A &= Z_{AA} - (Z_1 - Z_3) - (Z_2 - Z_3) \\ Z_B &= Z_{BB} - Z_1 \\ Z_C &= Z_{CC} - Z_1 \\ Z_E &= Z_{EE} - Z_2 \\ Z_F &= Z_{FF} - Z_2 \end{aligned} \qquad (9)$$

and $Z_1 = Z_2$ if the outer limbs are identical.

If this equivalent circuit is compared with the equivalent circuit of a single-phase transformer, Z_A, Z_B, Z_C, Z_E, Z_F, are equivalent to the series impedances and $(Z_1 - Z_3)$, $(Z_2 - Z_3)$ and Z_3 are mutual impedances.

The leakage values are small compared with the mutual ones and they can be measured by carrying out short circuit tests. Similarly the mutual values are obtained by performing open circuit tests. Thus, these simple tests clearly define the steady state parameters of the transformer and can be used to evaluate the performance of any system derived from the primitive scheme of Fig. 15.

The primitive system connected as a frequency modulator.

The frequency modulator configuration is obtained by removing Z_{BB} and Z_{EE} in Fig. 17 and reconnecting the system as shown in Fig. 18 where Z_0 is the load impedance.

By writing out the new voltage equation the equivalent circuit for this application can be shown to be as in Fig. 19 where Z_{CC} and Z_{FF} represent the switch impedances.

Assuming

$$Z_{(1..3)} \times Z_{(1..3)} \times Z_{(A...F)}$$

$$\gg Z_{(1..3)} \times Z_{(A..F)} \times Z_{(A..F)}$$

$$\gg Z_{(A..F)} \times Z_{(a..f)} \times Z_{(a..f)}$$

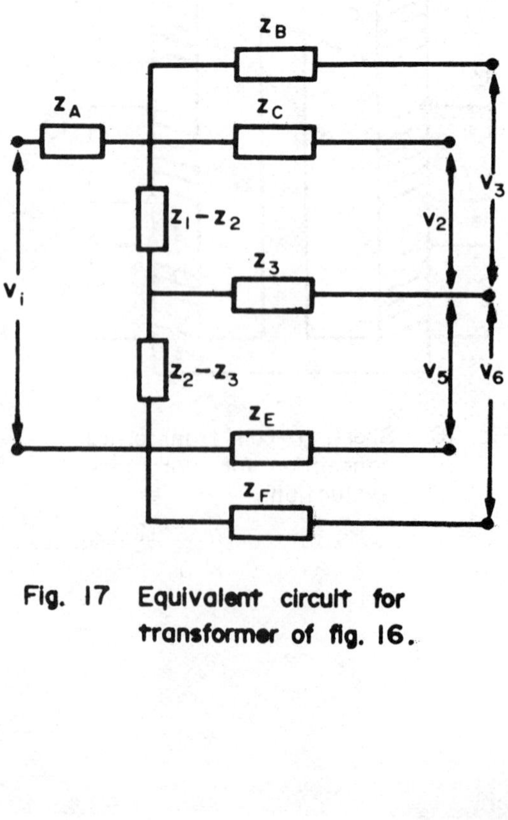

Fig. 17 Equivalent circuit for transformer of fig. 16.

and putting $Z_1 = Z_2$, since coils A, B and C are identical to coils D, E and F respectively, the above equivalent circuit can be solved to determine the equivalent impedances seen by the open and short-circuited shorting-coils and the output circuit as shown in Figs. 6,7 and 8.

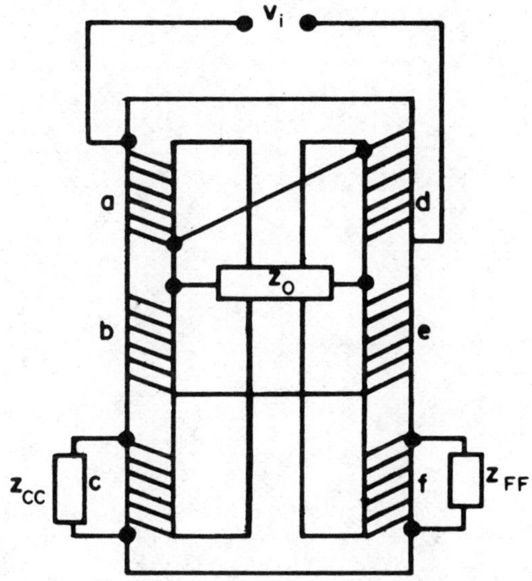

Fig.18 The primitive system connected as a frequency modulator

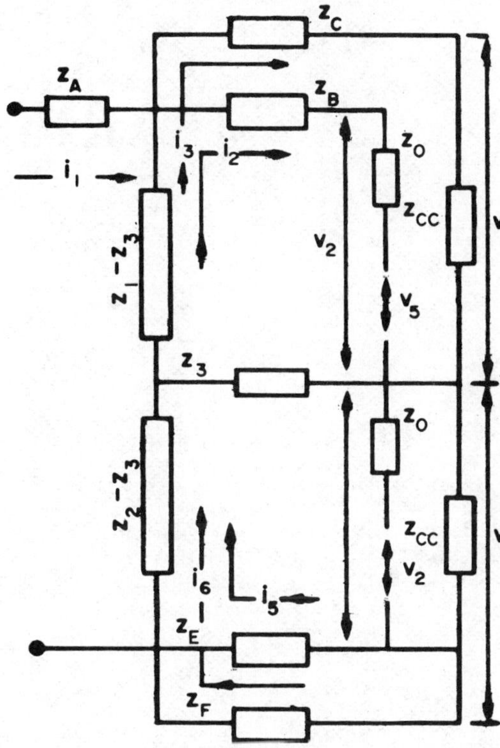

Fig.19 The equivalent circuit of modulator

SURGE SUPPRESSOR'S DESIGN BY MEANS OF NOMOGRAMS

H. Tunia and A. Wójciak

Technical University Warsaw Inst. of Control and Electronics, Koszykowa 75, Gm. E, pok. 316, PL 00-662 Warszawa, Poland

ABSTRACT

A new method for RC suppressor's design is presented, concerning commutation surges suppression in line commutated converter. This "constant-factors method" is very simple for a designer to use, and provides good results even when approximate thyristor data make base for calculations. Under additional conditions the method may be used effectively for a converter with forced commutation.

INTRODUCTION

A real commutation circuit, provided with a series connected RC suppressing network, is substituted by series connection of equivalent R_e, L_e, C_e elements. The principal assumption is, that the valve in the equivalent circuit may be considered an ideal switch, which opens after delay equal to approximate life time of minority current carriers. For a given converter this delay time is a function of firing angle and commutation angle. The set upper limit, of possible delay time values, can easily be evaluated with sufficient accuracy for design purposes.

TRANSIENT PHENOMENA IN COMMUTATION CIRCUIT

In calculation of transient phenomena the following further assumptions are also valid:

- commutation step voltage U_s is constant during the whole period of commutation
- reverse current of an SCR rises linearly up to its maximum value I_{RM}
- anode voltage of an SCR is approximately equal to zero, up to the moment when $I_R = I_{RM}$
- damping effect of the SCR resistance is neglected
- the worst-worst case makes the basis of design procedure.

These basic assumptions enable to decompose the action of the equivalent circuit Fig. 1 into two steps. In the first step the equivalent circuit can be considered as two-terminal with inductance L_e , excited by step voltage U_s. In the second step an equivalent circuit with series connection of R_e,L_e,C_e

is valid.

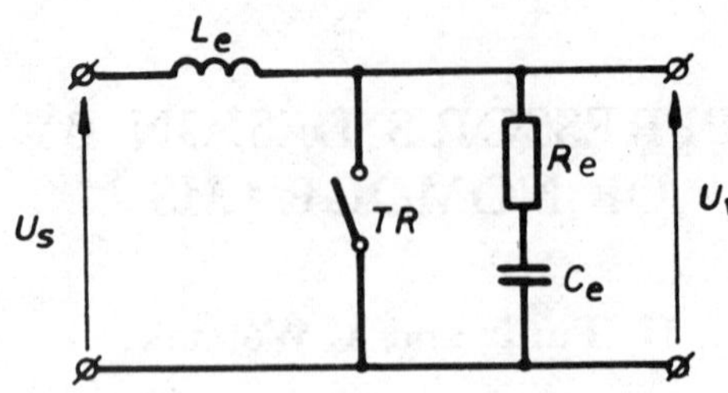

Fig. 1. Equivalent commutation circuit

The mathematical formulas for transient current $i(t)$ and voltage over the valve $U_v(t)$ for this equivalent circuit may be obtained. The only practical case which should be considered is when damping factor $\xi < 1$. Then the formulas for $i(t)$ and $U_v(t)$ are:

$$i(t) = \frac{U_s}{L_e}\left[t_r \cos(\omega_n a\ t) + \frac{1 - \xi\omega_n t_r}{\omega_n a} \sin(\omega_n a\ t)\right] e^{-\omega_n \xi t} \tag{1}$$

$$U_v(t) = U_s\Big\{1 + \Big[(2\xi\omega_n t_r - 1)\cos(\omega_n a\ t) +$$

$$+ \frac{1}{a}(\omega_n t_r - 2\xi^2\omega_n t_r + \xi)\sin(\omega_n a\ t)\Big] e^{-\omega_n \xi t}\Big\} \tag{2}$$

where: $a = \sqrt{1 - \xi^2}$; $Z = \sqrt{L_e/C_e}$; $\omega_n^2 = \sqrt{L_e C_e}^{-1}$; $\xi = R_e/2\ Z.$ (3)

Theoretical initial value of $U_v(t)$ at the moment $t = 0$ is proportional to damping factor ξ and is equal

$$U_v(0) = 2\ U_s \xi\, \omega_n t_r. \tag{4}$$

Calculations of transient phenomena in the equivalent circuit lead to three parameters Fig. 2, characteristic for the shape of surge. These are: magnitude $U_{v\,max}$, time for voltage maximum t_{max} and initial value U_{vo}. Appropriate

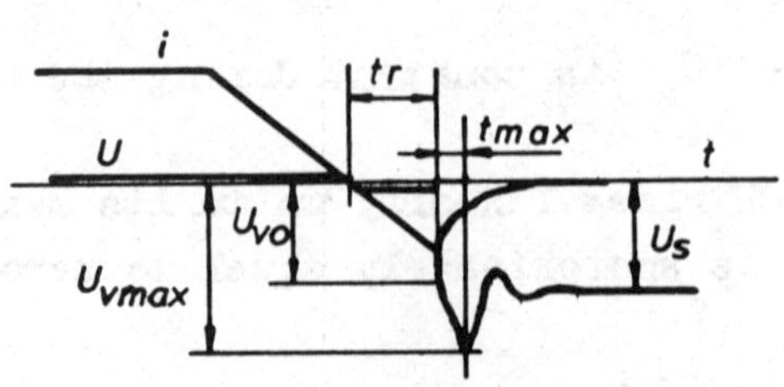

Fig. 2. Shape of surge parameters

relations are given in modificated relative form Fig. 3 to 5, with $\omega_n t_r$ and ξ as parameters.

With properly designed surge suppressor U_{vmax}, U_{vo}/U_{vmax} and t_{max} should not exceed certain assumed limites. Concerning Fig. 3, 4 and 5 we can establish three restrictions for ξ and $\omega_n t_r$ in form

$$f_i(\xi, \omega_n t_r) \leq a_i = \text{const} \qquad (i = 1,2,3) \tag{5}$$

These must be satisfied simultaneously.

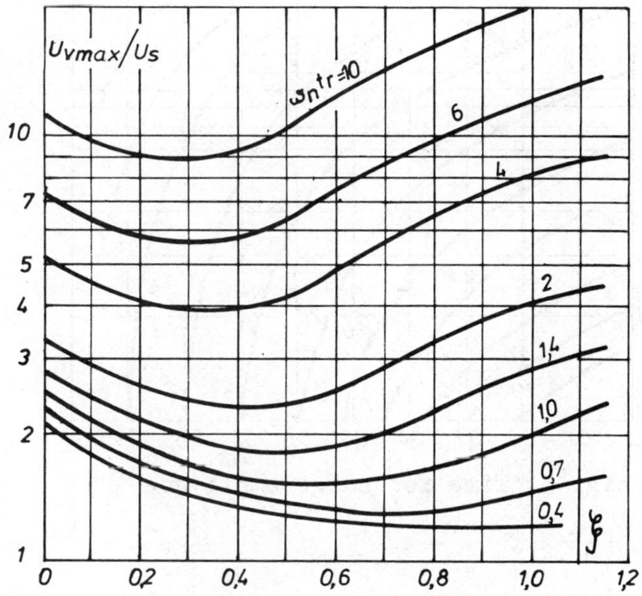

Fig. 3. Surge amplitude

In formulas (1) to (5) time t_r characterises dynamical properties of semiconductor structure during switch - off period. The value of t_r is equal to maximal possible life time of minority current carriers.

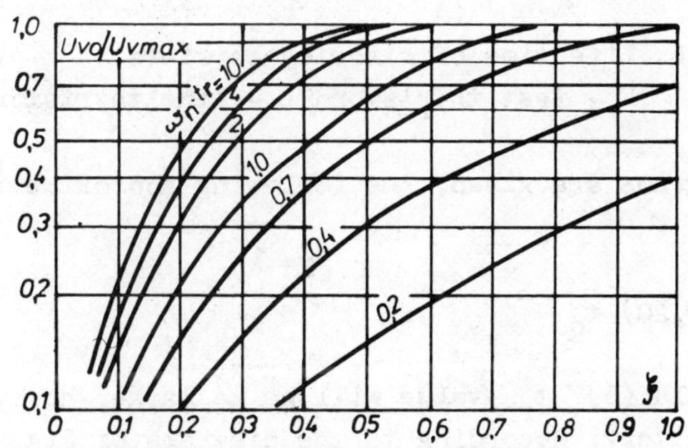

Fig. 4. Theoretical initial value

This is based on relatively small variation of t_r in different working conditions of thyristor Fig. 6, especially observed by small di/dt values. This future was not taken into account by the previously published papers, in which t_r was calculated from Q_{rr} or I_{RM} diagrams.

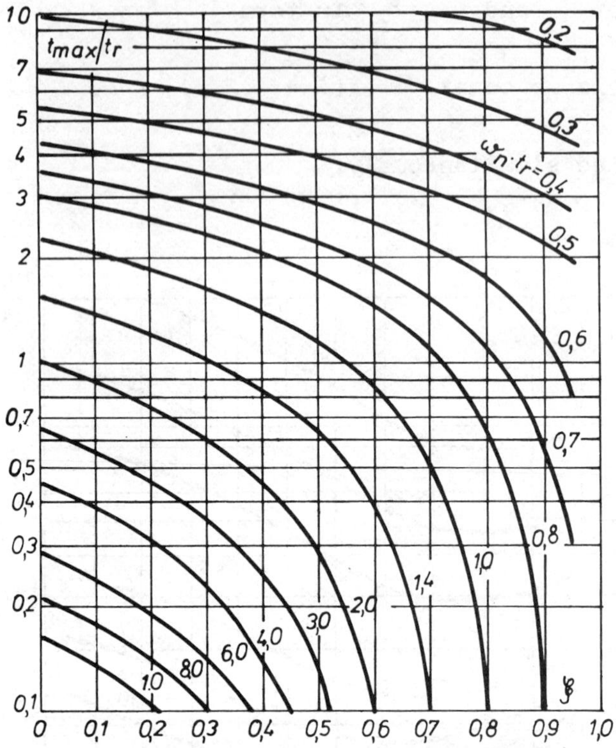

Fig. 5. Time for surge amplitude

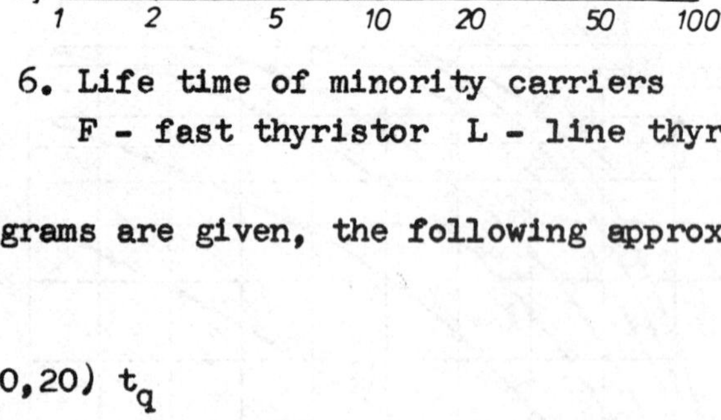

Fig. 6. Life time of minority carriers
F - fast thyristor L - line thyristor

When no adequate diagrams are given, the following approximation prooved very useful:

$$t_r = (0{,}25 \cdots 0{,}20)\ t_q \tag{6}$$

Calculated with formula (6) t_r value will be to large, however, the results of design are satisfactory. According to the diagrams of Fig.10, when the life time of minority current carriers of thyristor used is smaller then that in calculations, the surge amplitude (k_1) and its theoretical initial value (k_2) became smaller. Simultaneously relative time for U_{vmax} grows up (k_3) but its absolute value (k_4) varies very little.

Now, the new factors τ and ϑ are introduced. They are particularly convenient for design purposes and relate to parameters of equivalent commutation circuit as follows

$$\tau = \frac{t_r R_e}{2L_e} \qquad \vartheta = \frac{t_r}{\sqrt{L_e C_e}} \tag{7}$$

Restrictions (5) then become

$$\overline{t_{max}}(\tau,\vartheta) \leq a_1 = \text{const}$$
$$\overline{U_{v\,max}}(\tau,\vartheta) \leq a_2 = \text{const} \tag{8}$$
$$U_{vo}/U_{v\,max}(\tau,\vartheta) \leq a_3 = \text{const}$$

where: $\overline{t_{max}} = t_{max}/t_r$ and $\overline{U_{v\,max}} = U_{v\,max}/U_s$

A set of pairs $\tau - \vartheta$ can be specified so, that all possible combinations of τ, ϑ simultaneously satisfy inequalities (8).
On the τ, ϑ plane this set of pairs forms so called "permitted region". This prooved to be considerably small for practical cases and the autors suggest a certain pair, chosen from permitted region, be a basis for design procedure. Choice of proper vector (a_1, a_2, a_3) is based on economical and technical considerations.

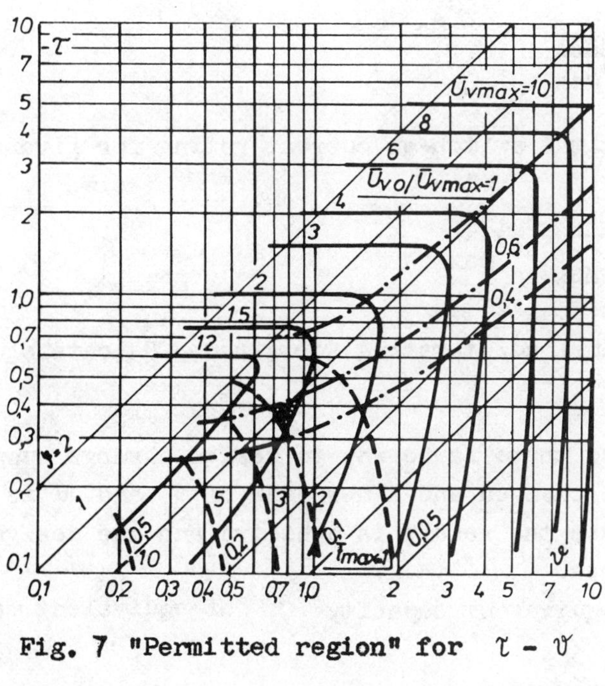

Fig. 7 "Permitted region" for $\tau - \vartheta$

However, when the higher frequency of power supply or/and forced commutation is considered, more accurate assumption of a_1 value is necessary. Although each pair of τ, ϑ chosen from permitted region provides a basis for designing, authors suggestion for line commutated converters is:

$$\tau = 0{,}4 \qquad \vartheta = 0{,}7 \tag{9}$$

and this means suggestion for damping factor $\xi = 0{,}57$.

These conclusions are basic for the "constant-factors method" algorithm Fig. 8 in protective network design.

METHOD OF DESIGN

The procedure is as follows:

- based on the technical assumptions the relations between parameters of real and equivalent circuits are to be **established**
- calculate optimal value of R_e

$$R_{e\ opt} = 2\,\tau\,L_e\,t_r^{-1} \tag{10}$$

- check conditions for equivalent resistance

$$R_{e\ min} < R_{e\ opt} < R_{e\ max} \tag{11}$$

where

$$R_{e\ min} = \frac{U_{v\ max}}{(I_{on})_{max}} \tag{12}$$

$(I_{on})_{max}$ - maximal switch-on current rating for given SCR (catalogue)

$$R_{e\ max} = \frac{L_r}{\sqrt{2}\,U_{RMS}}\left(\frac{du}{dt}\right)_{max} \tag{13}$$

$\left(\frac{du}{dt}\right)_{max}$ - maximal du/dt rating for given SCR (catalogue)

Negative result means, there is no way to design a surge suppressor for given data. These are to be changed and a new pair of τ and ϑ is to be found. For most of practical cases the result is positive and the designer will complete calculations as follows.

- calculate optimal equivalent capacity C_e of equivalent commutation circuit

$$C_{e\ opt} = t_r^2\,\vartheta^{-2}\,L_e^{-1} \tag{14}$$

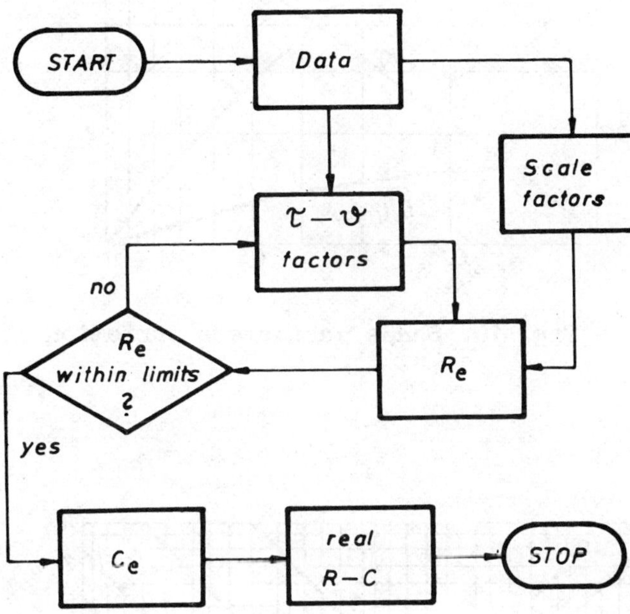

Fig. 8. Algorithm for suppressor design ("constant-factors method")

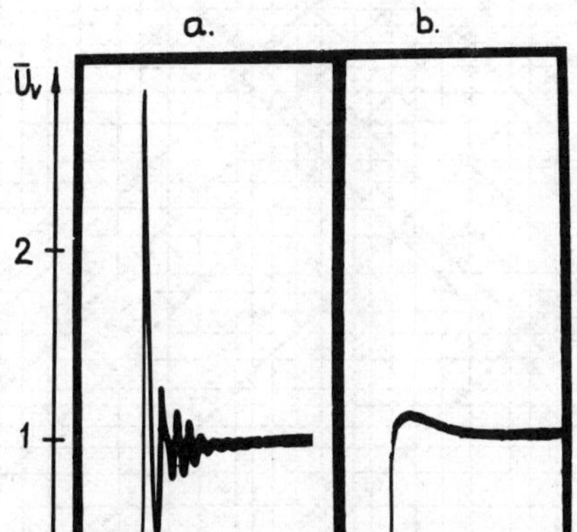

Fig. 9. An example of commutation surges in a 3-phase-bridge converter; (a) - no protection, (b) - with RC suppressors, (τ = 0,4, ϑ = 0,7)

- calculate ratings for surge suppressor components R and C. Scale factors in relations between R_e, C_e and the ratings of the real suppressor are to be found in previous papers.

In Fig. 11 a suitable nomogram based on formulas (10) and (14) is given to

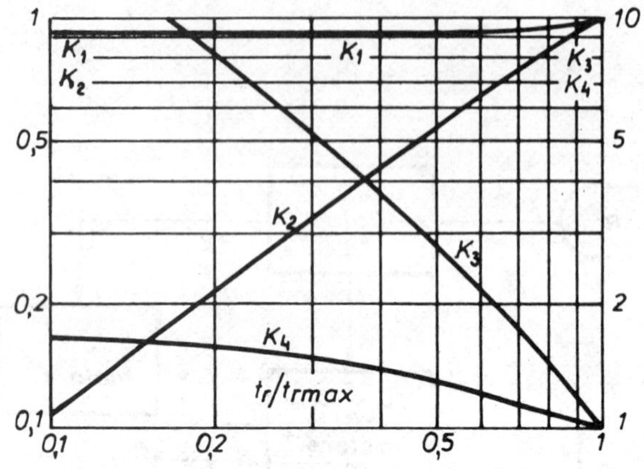

Fig. 10. Shape parameters variation

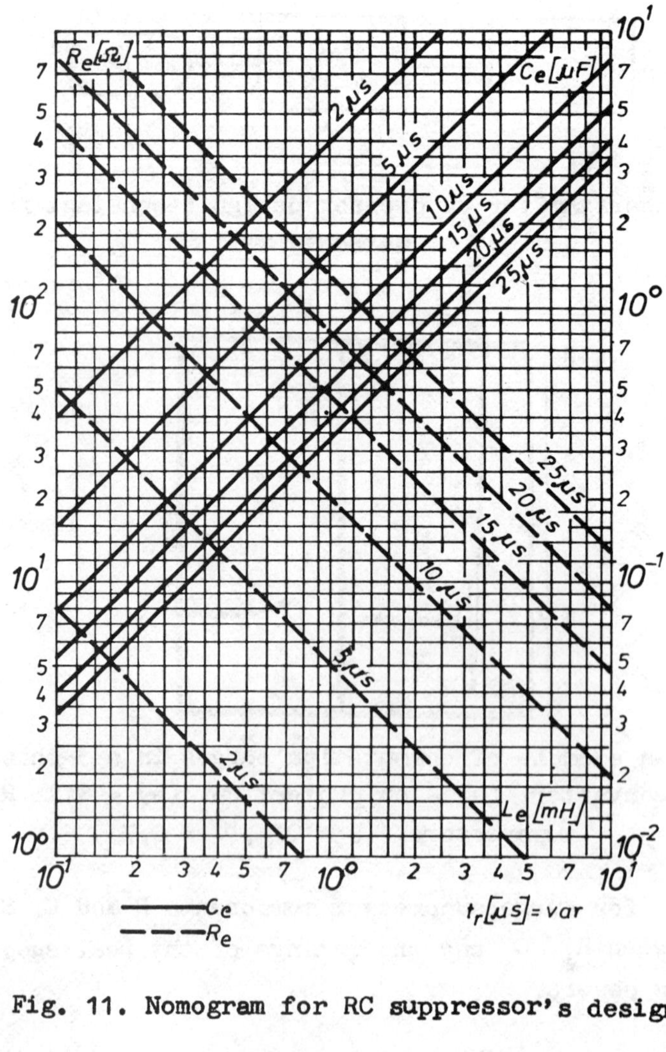

Fig. 11. Nomogram for RC suppressor's design

simplify the design procedure.

CONCLUSIONS

The presented "constant-factors method" for commutation surges suppressors design is simple in use.

The results of theoretical considerations were confirmed by effective work of real surge protection in different line commutated converters. Checks were done in the range of rated power from several hundreds VA up to several hundreds kVA.

The constant-factors method was primarily invented for line commutated converters. However, with some additional initial conditions, it may be used effectively at converters with forced commutation.

REFERENCES

(1) Hoffmann A., Stocker K. (1965) Der Thyristor als Bauelement der Leistungselektronik. Thyristor-Handbuch. Siemens-Schuckert, Berlin.
(2) Irminger G., Beschaltung von Thyristoren, BB-Mitteillungen 53, 10, 657-671 (1966).
(3) Mauersberger Ch., Möglichkeiten zur Auslegung von R-C-Gliedern zum Überspannungsschutz von Thyristoren, Elektrie 24, 1, 10-12 (1970).
(4) Samberger K., Berechnung von Trägerspeichereffekt-Beschaltungen für Silizium-Gleichrichter und Thyristoren. ETZ-A 86, 24, 781-786 (1965).
(5) Stumpe A.C., Das Schaltverhalten der Steuerbaren Silizium-Zelle, ETZ-A 83, 9, 291-298 (1962).
(6) Wójciak A., (1976) Problems of line commutated converter protection against commutation surges. Dissertation, Technical University Warsaw.

CURRENT STATUS OF POWER THYRISTORS AND RECTIFIERS

E. Spenke and P. Voss

Siemens AG, Werk Halbleiter, Frankfurter Ring 152, 8000 München 46, F.R. Germany

ABSTRACT

This review is mostly concerned with the thyristor. Recent advances in thyristor technology are described, e.g., the consequences of the introduction of silicon starting material that is homogeneously doped by means of neutron irradiation, new surface treatments that improve surface stability against electrical breakdown and electron irradiation as a means to control carrier lifetime. The turn-on of thyristors and plasma spreading are discussed with special attention given to two-stage designs, interdigitation and light firing. A section on thyristor turn-off is predominantly devoted to gate-assisted turn-off. In another section special arrangements like two terminal pnpn-structures are described. Finally the special aspects of power rectifiers are treated.

1. INTRODUCTION

Over the years power semiconductor technology has reached a high degree of maturity and now progresses at a steady pace. With the thyristor the development goes essentially in two directions: towards larger areas and towards better dynamic properties. The gain in area is exploited either to increase the current capability or to raise the blocking voltage. The development of rectifiers goes essentially in parallel except for the fast turn-off devices.

In the case of the thyristor we will go into some detail when we try to illuminate recent advances in technology and design. Emphasis in this review will be on the thyristor, but much of what is said in connection with this device applies to rectifiers as well. For a more detailed description of the topics treated in this paper we refer to the review articles of Gerlach and Köhl (1), Herlet (2), Spenke (3), Leturcq (4), Mattera (5) and Gerlach (6).

2. THYRISTORS

Even though thyristors have become very versatile, from the standpoint of applications there can be no doubt that the standard thyristor still leaves some important features to be desired. The main drawbacks are

> the lack of gate turn-off capability

and

> the slow turn-on and turn-off properties.

There is continuing effort to develop thyristors that can be turned off, but currently more work seems to be assigned to improving the dynamic properties and to bringing the other thyristor parameters closer to the principally feasible limit.

When we look at the thyristor market, we have on the one end the vast range of low-cost devices, mostly in the low and medium voltage range. Relatively recent developments in this field are the power modules and the fast turn-off (10 µs) devices with repetitive blocking voltages around 500 V at elevated junction temperatures, as desired for the use in battery operated vehicles.

At the other end we have the high-priced state-of-the-art devices which often find only very limited application. In Fig. 1 we have plotted for reverse blocking and reverse conducting devices in the current range between 100 A and 800 A the currently available maximum forward blocking voltages versus turn-off time.

Over the last two years the limit of the repetitive peak voltage of commercially available thyristors was raised beyond 3 kV. On the other hand a number of low-voltage fast thyristors was introduced that have turn-off times as low as 10 µs. Newcomers are medium-fast high-voltage reverse blocking thyristors with typical turn-off times around 50 µs and repetitive peak voltages up to 1800 V. High-voltage reverse conducting thyristors can be even faster, since thinner silicon wafers can be employed for these. Many of the devices listed in Fig. 1 are interdigitated structures.

The diagram gives an idea of the trade-offs one has to take today. The future might bring a shift of these curves to somewhat higher voltages as larger silicon wafers are introduced. Both curves could in principle shift parallel along the turn-off time axis to about half the value of the turn-off time whenever gate assisted turn-off is feasible. An extension of the curves to turn-off times beyond 800 μs appears not very likely, in view of the demands for phase-control devices.

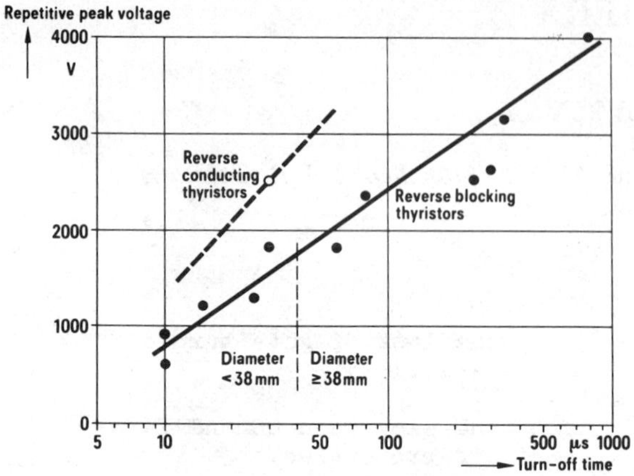

Fig. 1 Maximum blocking voltage vs. turn-off time of commercial thyristors. The curve for reverse conducting thyristors contains only one point. The drawn slope indicates that reverse conducting thyristors can have approximately twice the blocking voltage of reverse blocking thyristors for the same turn-off time (see Fig. 4 and the accompanying text in section 2.1.3).

Not included in Fig. 1 are the large and the giant phase-control high-current thyristors with diameters between 65 mm and 102 mm (de Warga and Mungenast (7)), as used in the welding and in the chemical industries. These devices with blocking voltages ranging from around 1300 V down to 600 V have current ratings somewhere between 1600 A and 3000 A, respectively.

In reviewing some of the recent advances of thyristor design and technology we will proceed at first along a line which corresponds in many ways to the path the designer has to take in order to arrive at the desired results. He would, e.g., start out by choosing the proper starting silicon that meets his needs for the intended blocking voltage for a given set of compatible current and dynamic parameters.

2.1 Blocking Voltage

2.1.1 Bulk-limited blocking behavior. The blocking voltage of a thyristor is determined either by its bulk or by its surface properties. Depending on the design of the thyristor, the bulk blocking voltage can in turn be limited either by junction breakdown or by punchthrough (Herlet (8)). Most thyristors are designed to have a blocking voltage slightly below the breakdown voltage of the junctions. This means that the space charge region supporting the voltage has to extend freely inside the low doped n-base (Fig. 2 b) and that, in addition, there has to be some separation from the opposite p-emitter to reduce the current gain α_{pnp}. Since the forward conducting characteristic and such dynamic properties as plasma spreading velocity are adversely affected by any increase in width of the bases[1], this separation has to be kept as small as possible.

The extension of the space charge region at junction breakdown and hence the breakdown voltage V_B itself are coupled to the doping of the n-base, i.e., to the resistivity ϱ_n of the starting silicon (see, e.g., SZE (9)). For diffused junction the dependence can be expressed as

$$V_B \approx c \cdot \varrho_n^{0.7}$$

where c is a proportionality constant.

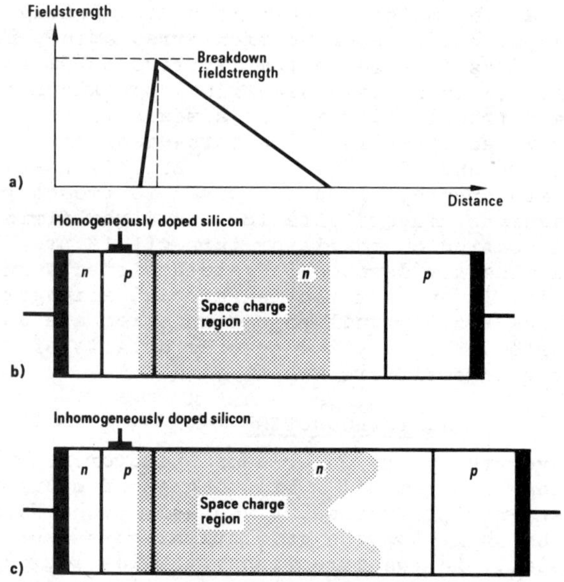

Fig. 2 Schematic representation of the influence of the doping of the n-base on the space charge region
(a) distribution of fieldstrength for case (b);
(b) homogeneously doped n-base;
(c) inhomogeneously doped n-base which has to be wider than in case (b).

Doping of the starting silicon is usually performed during the crystal pulling process. This results in nonuniformities of the resistivity across the silicon slices and in variations from slice to slice. The extension of the space charge region inside the devices varies accordingly (Fig. 2 c). In order to compensate for this, it is necessary to make the n-base overly wide.

1) See Section 2.3.2

The problem of doping variations and inaccurate doping has recently been overcome by the introduction of a new doping method which takes advantage of a nuclear reaction of the isotope Si-30 with thermal neutrons (Tanenbaum and Mills (10); Schnöller (11); Herman and Herzer (12); Janus and Malmros (13); Haas and Schnöller (14)). Silicon is transmuted into the dopant phosphorus:

$$^{30}\mathrm{Si}\ (n,\gamma)\ ^{31}\mathrm{Si} \xrightarrow{\beta} {}^{31}\mathrm{P}.$$

Using this method a very uniform and accurate doping can be achieved.

The first devices to benefit from this doping process were phase-control thyristors with repetitive peak voltages up to 3,5 kV as they are used in high-voltage dc systems and motor controls. Though similar devices had been around before (Ogawa, Kamei and Morita (15)), it was the advent of the new starting material that allowed a large scale production of such devices with high yield.

In the meantime it has been realized that the homogeneously doped silicon can also be profitably used with other devices, although it is still slightly more expensive than conventionally doped silicon. As the tendency is right now, it seems that neutron-irradiated silicon will soon be employed exclusively for most high-power devices and possibly for many low-power devices as well.

2.1.2 Surface breakdown. Besides demanding tighter controls of the diffusion processes, the increase in voltage has made the problem of surface breakdown more critical. Surface breakdown can be safely prevented for one polarity by applying a positive bevel at the respective blocking junction. For the other polarity, which is usually that in forward blocking direction, a low angle negative bevel is used (Davies and Gentry (16); Cornu (17); Bakowsky and Lundström (18)). The purpose of the bevel is to spread the space charge region at the surface and thus reduce the field strength there. The double bevel has become the standard edge contour for large-area devices (Fig. 3a).

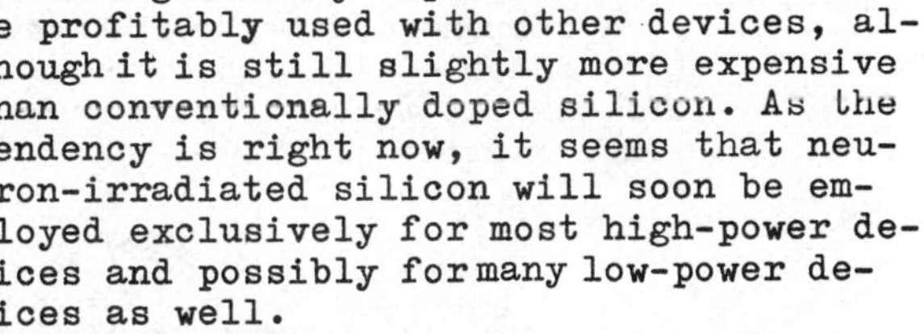

Fig. 3 Surface contours

Also very critical for the surface breakdown properties is the proper coating of the junctions. This holds true particularly for low-power devices where double beveling is economically infeasible and costs dictate a trend towards low-cost encapsulation where the device is not housed in a hermetically sealing case. Here the single or double deep moat in combination with glass passivation is more and more becoming the usual (Fig. 3b). The glass provides for the surface charge that is necessary in this case to obtain a spreading of the space charge region near the surface and it seals the junction hermetically from the environment.

2.1.3 The npinp-thyristor. If the blocking voltage of a thyristor would be the only parameter of importance, it could be significantly raised beyond the level of today's thyristors by increasing the resistivity and the width of the n-base. But there are other aspects to be considered, like forward voltage drop, current carrying capability, surge current capability and thermal stability, furthermore a string of dynamic requirements with respect to turn-off time, plasma spreading, di/dt-capability etc. Therefore, it seems unlikely that the blocking capability of today's reverse blocking thyristors will be significantly surpassed in the future.

One way to get beyond these limits is to sacrifice the reverse blocking capability. Especially in chopper circuits, the reverse blocking capability is often not needed. In this case it is expedient to replace the slightly doped n-base of the npnp-structure by a double layer consisting of an almost intrinsic i-region and a highly doped n-region (Fig. 4, Tsurushima and Kataoka (19); Gerlach and Köhl (1); Kokosa (20); Crees, Nichols and Wood (21); van Iseghem (22)). By this measure the structure of the pn-bases is changed to that of a pin-diode.[2)]

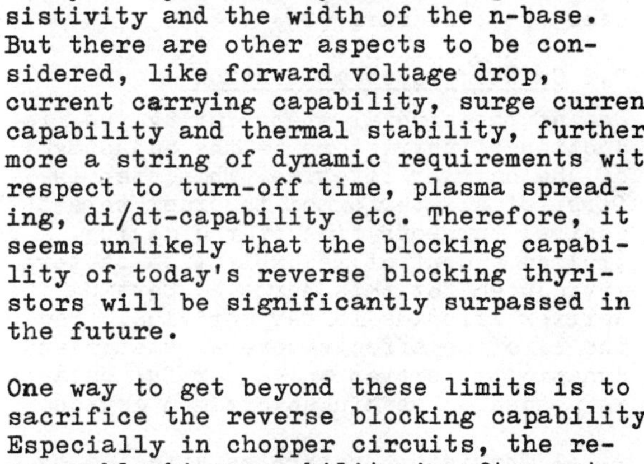

Fig. 4 Space charge regions in npnp and npinp-structures with no n-emitter shorts. In the npinp-structure the reverse blocking capability is sacrificed almost entirely in favor of an enlarged forward blocking capability.

---

2) The p-emitter is sometimes supplied with distributed shorts. This improves the blocking characteristics at high temperatures (Gerlach and Köhl (1); Kokosa (20)).

At the bottom of Fig. 4 the position of the space charge regions and the corresponding field distributions under forward and reverse blocking conditions, respectively, are indicated. The top of Fig. 4 gives the case of the npnp-structure for comparison. In the intrinsic region the high field strength remains almost constant, whereas in the slightly n-doped base of the npnp-structure it falls off about linearly to zero. Correspondingly, the forward blocking voltage is approximately doubled in the first case, while the dynamic properties stay the same. For details on the performance of npinp-structures see Gamo (23); Okamura (24).

2.2 Carrier Lifetime Control

One of the crucial steps during the production of thyristors is the adjustment of the carrier lifetime. This step is first of all performed in order to set the desired turn-off time of the device. A controlled amount of recombination centers is introduced for this purpose. Setting the carrier lifetime is very critical, because the lifetime affects several static and dynamic properties and a careful balance is needed to obtain an optimum device.

2.2.1 Thyristor properties affected by carrier lifetime. Some of the effects to be considered are:

1. A decrease of the carrier lifetime under high injection conditions accelerates the decay of stored charge at turn-off and thus reduces the turn-off time. Turn-off time amounts to approximately ten times the carrier lifetime.

2. Depending on the thickness of the base regions, carrier lifetime may strongly influence the on-state voltage drop. Fig. 5 shows the dependence of the on-state voltage drop on the carrier lifetime with the total width of the base regions as a parameter (Burtscher, Dannhäuser and Krausse (25)). At low carrier lifetime the on-state voltage drop rises very sharply.

3. The decrease of minority carrier lifetime in the neutral zone of the n-base under static blocking conditions (see Fig. 2 b) reduces the current gain α_{pnp} of the pnp-transistor section. For this reason, as the lifetime is lowered, the blocking voltage may in principle reach values closer to the junction breakdown voltage (Fig. 6, at high currents).

4. Especially at high temperatures, this beneficial effect may be more than offset by an increase of the generation current inside the space charge region (Fig. 6, at low currents).

It is essentially a matter of the tolerable current level (e.g., I_1 or I_2 in Fig. 6) which one of the cases 3. and 4. becomes more important.

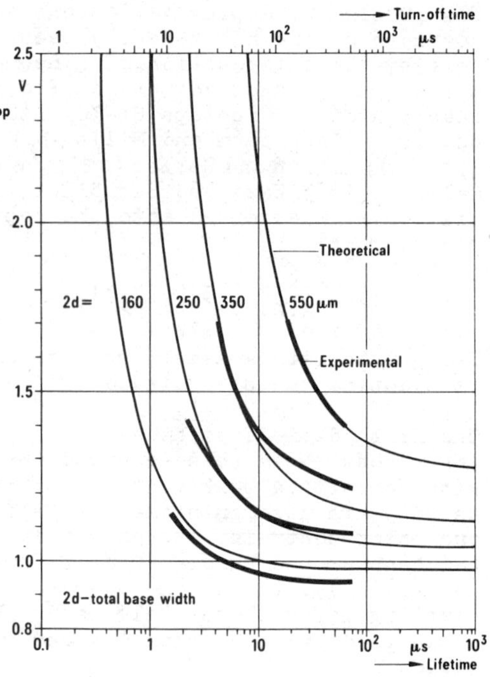

Fig. 5 Dependence of the forward voltage drop on carrier lifetime at a current density of 200 A/cm^2 (Burtscher et al. (25)). Turn-off time is taken to be seven times carrier lifetime.

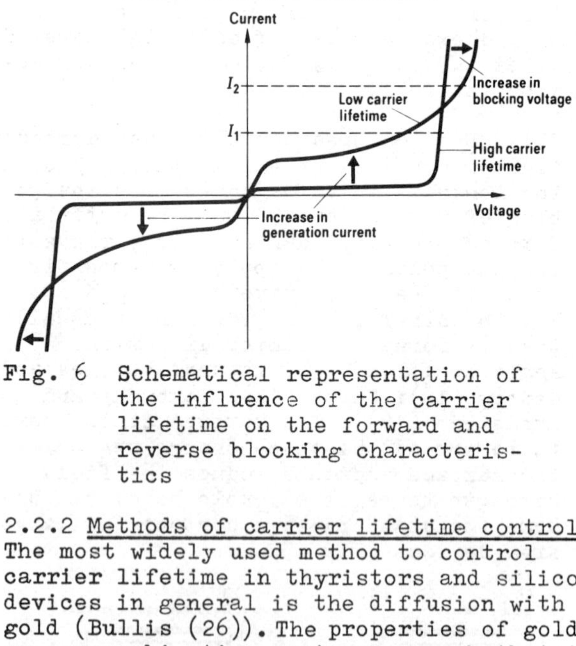

Fig. 6 Schematical representation of the influence of the carrier lifetime on the forward and reverse blocking characteristics

2.2.2 Methods of carrier lifetime control. The most widely used method to control carrier lifetime in thyristors and silicon devices in general is the diffusion with gold (Bullis (26)). The properties of gold as a recombination center are such that it allows one to build thyristors with turn-off times below 10 μs with an economically acceptable forward voltage drop. Gold does, however, have several disadvantages: it causes high generation currents, it may change the doping level and it is strongly gettered by other dopants like phosphorus and by crystal dislocations. In general this causes an uneven distribution of the gold in axial as well as in lateral dimensions and makes reproducible production rather difficult. This holds independent of whether the thyristor is a low-voltage device with a turn-off time of 10 µs or a high-voltage one with a turn-off time of 400 µs.

A strong contender to replace gold are recombination centers produced by electron irradiation damage (Tarneja and Bartko (27); Rai-Choudhury, Bartko and Johnson (28)). This method of lifetime control offers one particular advantage: lifetime adjustment can be performed on the otherwise finished device and if necessary, on an individual basis. This is a convenience that can improve the yield noticeably. Annealing of the centers starts at temperatures around 250°C. Devices manufactured in this fashion are commercially available. Turn-off times of these devices range down to 10 µs. General properties do not seem significantly different from those doped with gold, as far as one can judge from published data (Chu and Donlon (29)).

2.3 Dynamic Processes

Having treated some parameters that can be considered along one- or two-dimensional cross-cuts of thyristors, we will now turn to those thyristor properties that in all their consequences can only be considered in three dimensions. Such properties are plasma spreading, turn-on and turn-off. In this context we will also treat the shorted n-emitter.

2.3.1 The shorted n-emitter. Every modern thyristor contains some kind of local n-emitter shorts (Fig. 7). Such shorts are needed to decrease the n-emitter efficiency at low emitter currents. This is necessary in order to fix a minimum gate current below which the device does not turn on and in order to prevent an ungated turn-on that may arise from currents either thermally generated inside the space charge region or generated by rapid changes in the voltage across the device (dv/dt turn-on). The shorts provide an exit for these currents (Fig. 7) so that the lateral voltage drop underneath the n-emitter reaches at no location the value initiating turn-on. This value is of the order of 0.5 V (Frohmader (30)).

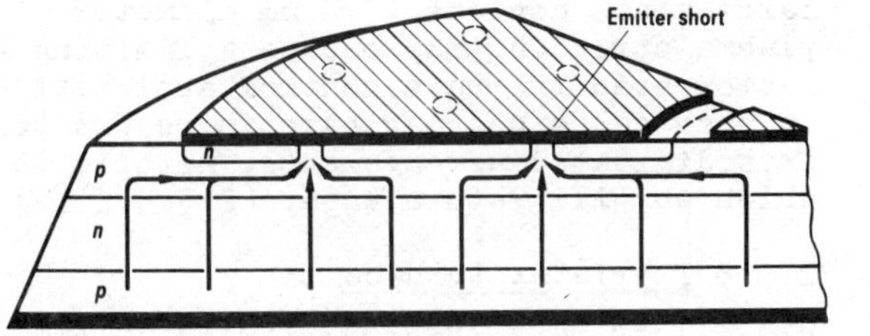

Fig. 7 n-emitter shorts

The shorted emitter has lately been given increasing attention (Chu (31); Burtscher and Spenke (32); Strack (33); Hartmann (34); Munoz-Yague and Leturcq (35)). Evidently, the more the n-emitter is shorted, the more the resistance against unwanted turn-on will improve. If we assume that a certain fraction of the total area is allotted to the shorts, a dense pattern of fine shorts is more effective than a more widely spaced pattern of large dots (Burtscher and Spenke (32)).

While the forward voltage drop at high current densities is within limits only slightly affected by the degree of n-emitter shorting, other thyristor parameters may be influenced very adversely by the shorting. This holds true primarily for the plasma spreading behavior.

2.3.2 Lateral plasma spreading. Lateral plasma spreading is a thyristor property that is implicit in several thyristor parameters, e.g., in the static and in the dynamic on-state characteristics and in the di/dt-limit. Experiments have shown that the velocity of plasma spreading increases with rising current density, increasing carrier lifetime and rising temperature. It decreases as the total width of the bases becomes larger and the amount of emitter shorting is increased (Mapham (36); Longini and Melngailis (37); Gerlach (38); Ruhl (39); Somos and Piccone (40); Terasawa (41); Matzusawa (42); Rosch (43); Yamasaki (44)).

One widely used method to detect plasma is the observation of the infrared recombination radiation emitted from the current carrying area through perforated electrodes (Gerlach (38); Somos and Piccone (40); Yamasaki (44); Voss (45)). Fig. 8 shows an example of such an observation. In this case a gated infrared converter was used (45). The six dark beams that become visible at longer times are current terminals. The device investigated was a 1300 V-thyristor with an emitter diameter of 28 mm and a turn-off time of 25 µs. The photographs show that the edge of the turned-on area is clearly defined. The edge bulges between the emitter shorts visible here as a radial pattern of dark dots. The total spreading time is above 200 µs.

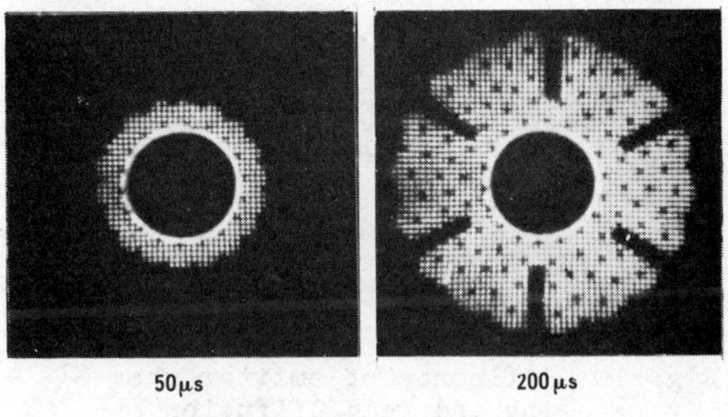

Fig. 8 Infrared observation of plasma spreading in a thyristor (V_{DRM} = 1300 V, t_q = 25 µs) with perforated electrodes for a constant current of 600 A. Distance between observation holes: 0.5 mm. The six beams visible at 200 µs are current terminals.

In thyristors with shorts, plasma spreading is an oscillating acceleration and deceleration in the areas between shorts. Fig. 9 gives an evaluation of the spreading velocity versus the radius for a thyristor with a circular arrangement of shorts (Strack (33)). Especially during the passage of the first two rows of shorts the spreading is severely delayed.

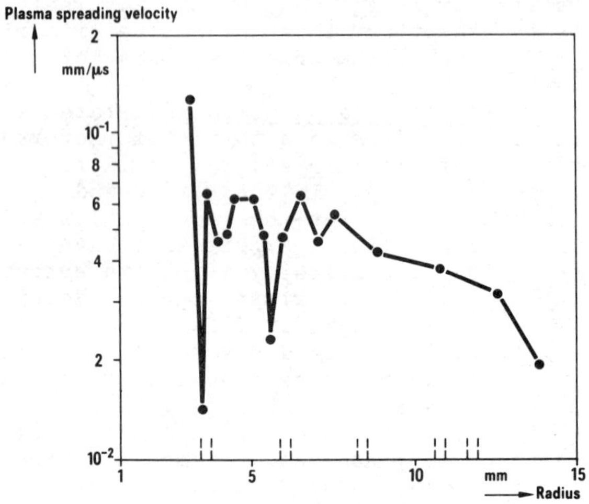

Fig. 9 Influence of n-emitter shorts on plasma spreading (Strack (41)). Double-dashed lines indicate location of shorts that are arranged in a circular fashion.

Fig. 10 gives another evaluation of a set of measurements by which the influence of the emitter shorts and that of lifetime doping was studied. Spreading velocity is plotted versus current density. These thyristors had identical thickness and went, except for the gold diffusion through identical process steps. Similar curves have been published by other investigators (39), (41), (42), (44).

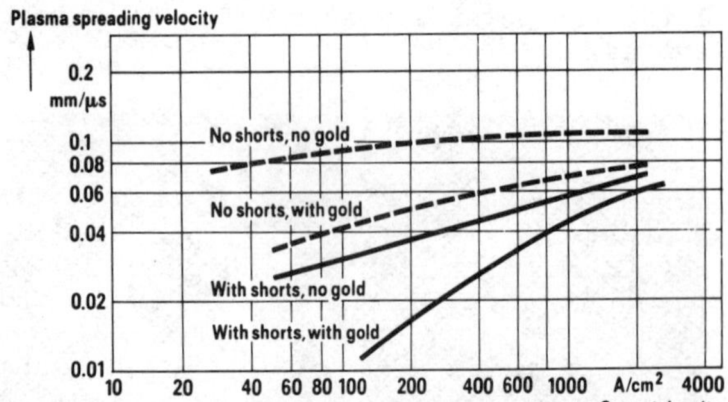

Fig. 10 Influence of emitter shorting and gold diffusion on plasma spreading. Thyristors (V_{DRM} = 1650 V) went through identical processing steps except for the gold diffusion.

The curves in Fig. 10 indicate that for these thyristors the spreading velocity almost saturates for high current densities, in this case around a value of 0.11 mm/µs. For the thyristors with shorts and with low carrier lifetime, the velocity decreases markedly. The lowest velocity is obtained for the combination of both. But it is just this combination that one has to choose if one wishes to have a short turn-off time[3].

But this correlation depends also on other parameters. In Fig. 11 on the left hand side one sees thyristors which have short turn-off times inspite of their fast spreading velocity. This is reached on account of the blocking voltage which is lower for these thyristors than for the thyristors on the right hand side.

Fig. 11 shows the result of spreading velocity measurements on a variety of thyristors with different blocking voltages, corresponding to different overall thicknesses, for a current density of 500 A/cm². Obviously, the spreading velocity decreases significantly as the blocking voltage is raised.

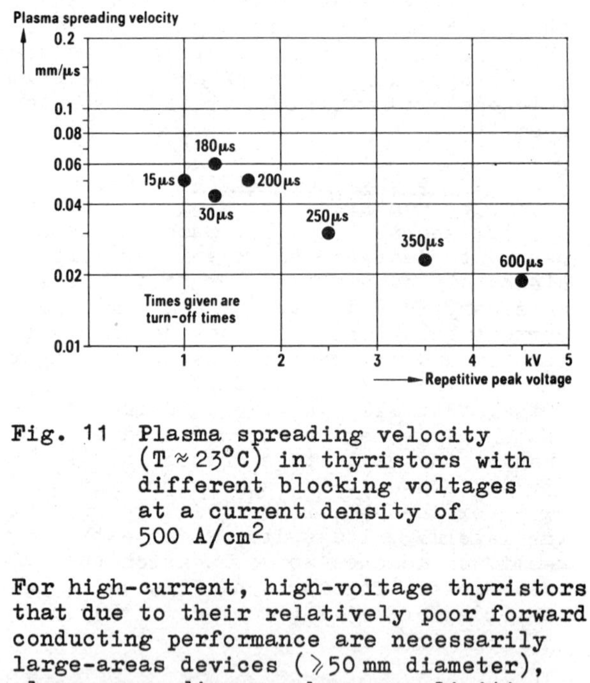

Fig. 11 Plasma spreading velocity ($T \approx 23^{o}C$) in thyristors with different blocking voltages at a current density of 500 A/cm²

For high-current, high-voltage thyristors that due to their relatively poor forward conducting performance are necessarily large-areas devices (≥50 mm diameter), plasma spreading may become a limiting factor even in phase-control applications. In such cases one may have to resort to interdigitated arrangements, a topic to which we will return later (2.3.3.2).

2.3.3 Thyristor turn-on

2.3.3.1 Primary turn-on. In its initial phases, thyristor turn-on follows its own rules and has to be considered separately from pure plasma spreading. Gated turn-on takes place at the edge of the n-emitter, adjacent to the gate. It has been established by means of different experimental techniques (Gerlach (38); Piccone and

[3] See Section 2.3.4

Somos (46); Voss (45), (48); Cordingley (47)) that the turn-on often occurs only in one or a few narrow spots. Fast rising currents, especially when reaching high levels, may overload these spots and thus cause thermal destruction.

For this reason, most high-current thyristors are nowadays equipped with some kind of two-stage gate design. All designs make directly or indirectly use of an auxiliary thyristor. The benefits from the auxiliary thyristor are two-fold: when properly designed, it turns on the main thyristor very uniformly. Furthermore, the initial turn-on area is relieved very quickly, since the auxiliary thyristor usually carries the major part of the load current only less than a microsecond.

Fig. 12 shows as an example the turn-on sequence of a high-voltage thyristor with a concentric amplifying gate, as observed with a gated image converter (Voss (45)). The corresponding voltage and current waveforms are depicted in the oscillograph. Time is counted from the end of the turn-on delay, i.e., from the time when the steep voltage fall sets in. The relative exposure times are given in parentheses.

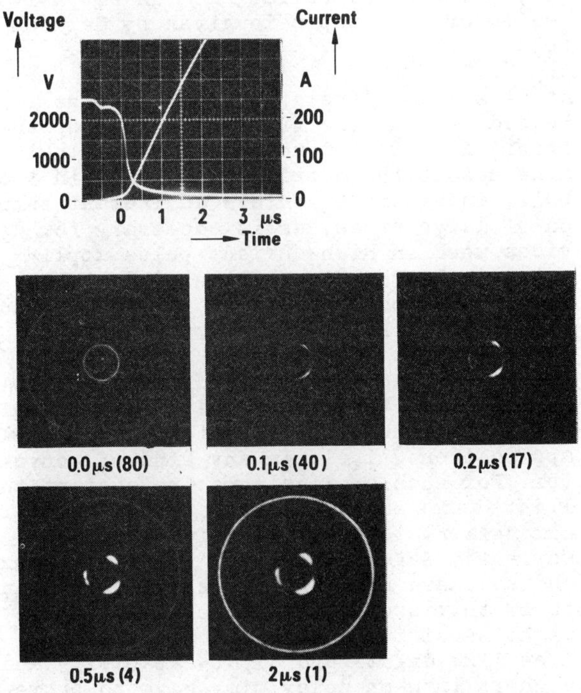

Fig. 12 Turn-on of an amplifying gate thyristor. Numbers in parentheses indicate relative exposure times.

The photographs show that during the turn-on delay phase both thyristors, the auxiliary as well as the main thyristor, draw the load current quite uniformly (0.0 μs). Current channels then form very suddenly in the auxiliary thyristor, whereas the current flow in the main thyristor remains uniform. After two microseconds the load current already flows predominantly through the main thyristor and the auxiliary thyristor begins to turn off. The design conditions that ensure such an initial turn-on at the auxiliary thyristor have been investigated by several authors (Voss (48); Kokosa and Wolley (49); Silard and Marinesch (50)).

Even though the two-stage gate arrangements have not proved to be indestructible under arbitrary gating and operating conditions, when properly laid out and operated they have made it possible to meet almost any di/dt-requirement.

2.3.3.2 Interdigitation. As we have mentioned before, the plasma spreading velocity is low in high-voltage devices and in those low-voltage devices that have a short turn-off time. Since the latter devices are to be operated at high frequencies, it became necessary to fit large-area devices with distributed gates in order to be able to make use of the full area and in order to reduce turn-on losses.

A combination of interdigitation and a gate of the amplifying gate type has proved to be of particular advantage. This is due to the fact that the auxiliary thyristor in any kind of two-stage gate arrangement is not only a source of high gate current for the main thyristor, but also effects a strong reduction of the voltage across the device before the main thyristor turns on. At reduced thyristor voltages there is no longer a tendency to form current channels. In fact, the turn-on at low voltages can be uniform for typical distributed gate lengths of 30 to 40 cm even if the gate current amounts only to a few amperes.

In Fig. 13 the turn-on of an interdigitated thyristor with a repetitive peak voltage of 900 V and a turn-off time of 10 μs is shown. Turn-on of the auxiliary thyristor is not very uniform. The turn-on of the main thyristor, however, is almost completetely uniform.

Two of the emitter finger electrodes at the bottom right of Fig. 13 are supplied with observation holes. It is apparent from the photographs in Fig. 13 that full turn-on is accomplished after about 25 μs. The same result is obtained from a measurement of the dynamic forward voltage drop under constant current conditions (Fig. 14). The static voltage drop is reached shortly after 25 μs.

Such measurements of the dynamic forward voltage drop are a very easy way of determining plasma spreading properties in general as long as one can assume homogeneity over the area. From comparing such measurements with infrared observations we have found this to be a legitimate method, except for an initial time interval of the order of 10 µs after turn-on, depending on the type of device.

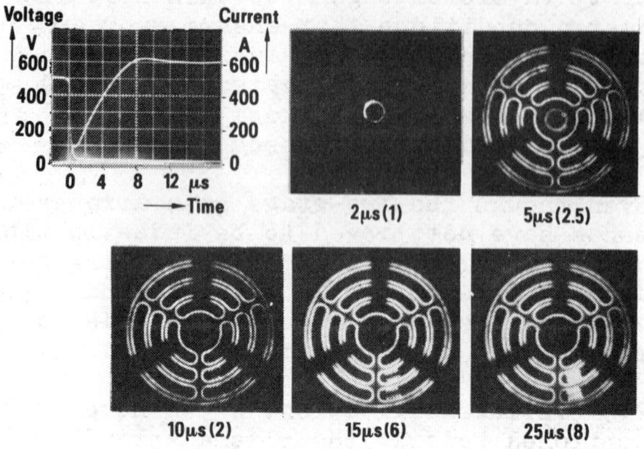

Fig. 13 Turn-on and plasma spreading in an interdigitated thyristor with amplifying gate. Two of the emitter fingers at the bottom right of this figure have observation holes in order to observe plasma spreading. Numbers in parentheses give relative exposure times.

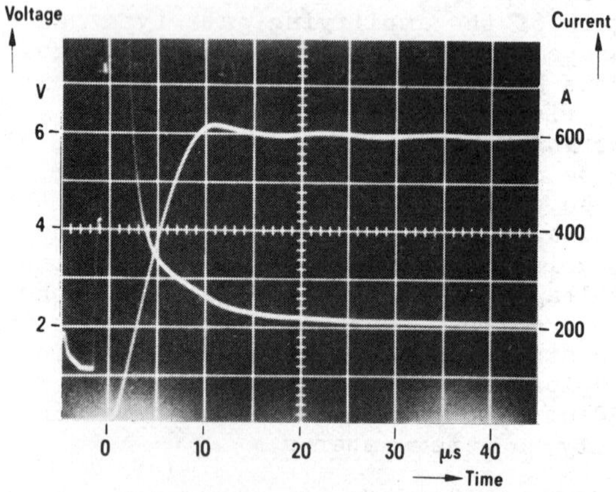

Fig. 14 Oscillographs of voltage and current vs. time for thyristor shown in Fig. 13. Turn-on with low voltage applied. Voltage approaches the static value of 2.05 V shortly after 25 µs.

Interdigitation is as yet mostly used in fast and medium-fast thyristors, i.e., in thyristors to be operated at frequencies up to around 10 kHz. At these applications where the turn-on and plasma spreading losses become very significant, the provision of sufficient data for the user poses a major problem for the manufacturer. The permissible load conditions for each application case have to be meticulously calculated on the basis of the more or less well known physical properties of the devices (Tobin and Wu (51); Somos and Piccone (52)), taking into account the scatter in each variable with good judgement of its respective influence.

2.3.3.3 Light firing. Interest has lately been revived in the direct light firing of thyristors. Direct light firing promises some major advantages: the carriers initiating turn-on are generated in the bulk of the device and the gating circuit can be completely insulated electrically from the thyristor by means of glass fibers. Such separation is particularly useful in any application where several thyristors are operated in series, as is the case, e.g., in high-voltage dc-systems and whenever there is danger of capacitive pickup in the gate circuit.

The general feasibility of direct light firing has been realized almost as long as the thyristor exists. However, the concept of the directly fired thyristor was not further pursued, mainly for lack of suitable light sources and transmitters.

This situation has now changed, as neodymium-lasers are in existence and as semiconductor light emitting diodes and lasers have become more reliable. A number of papers have recently been published on development work on light activated thyristors (Grekhov, Levinshtein and Sergeev (53); Davis (54); Davis and Roberts (55); Silber and Füllmann (56); Silber, Winter and Füllmann (57); de Bruyne and Sittig (58); Temple and Ferro (59)). An extensive review of this topic is given by Gerlach (6).

At it stands, development seems to be headed in two directions: one development effort is directed towards devices that make use of the carrier generation in the bulk, in order to achieve an instant turn-on of large areas, as is necessary for devices used in high-current pulse applications. In these cases the high light flux from neodymium lasers is utilized to swamp the bulk with carriers (53) - (55).

The main aspect of the second path of development is the electrical insulation of the gate circuit of thyristors for normal applications, i.e., in any kind of converter. For such applications the cost of the light source should be as low as possible for the arrangement to be competitive in price. Currently GaAs-LEDs with a low light flux and standard optical fibers are used. Thus, these thyristors have to be extremely light sensitive while all other properties like dv/dt- and di/dt-capability and a short turn-on delay time have to be retained. Different solutions have been published for gate structures that come close to this design goal (56) - (59). Further improvements are to be expected once semiconductor lasers have proved their long term reliability and can be reduced in price.

One possible realization of a light fired thyristor is shown in Fig. 15 (Silber (56), (57)). Not the whole arrangement is light-fired but only an auxiliary thyristor, the n-emitter of which is penetrated by the light.

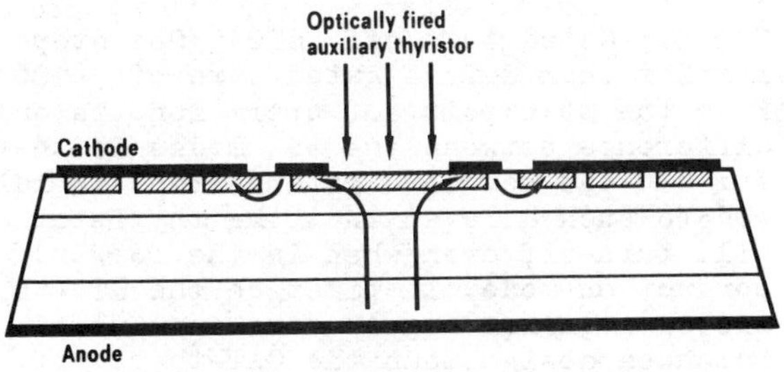

Fig. 15 Light fired thyristor (Silber et al. (56), (57))

One basic design feature of another highly light sensitive device (de Bruyne and Sittig (58)) is shown in Fig. 16. In this device the forward blocking junction is interrupted, because it is curved and thus extends to the upper surface of the silicon slice. The opening at the upper surface is kept so small that the space charge region fills it completely at voltages above approximately 100 V. Hence there is no problem with passivation. Within the opening the light can be radiated directly into the space charge region. This increases the quantum efficiency to values close to one.

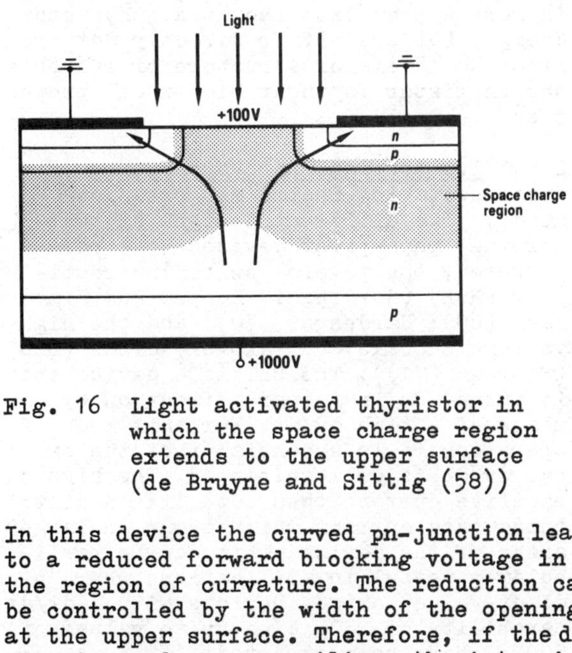

Fig. 16 Light activated thyristor in which the space charge region extends to the upper surface (de Bruyne and Sittig (58))

In this device the curved pn-junction leads to a reduced forward blocking voltage in the region of curvature. The reduction can be controlled by the width of the opening at the upper surface. Therefore, if the device is used as an auxiliary thyristor in a two-stage gate arrangement, it may be safely fired by exceeding the breakover voltage.

One of the crucial factors that will determine the long term success of highly sensitive light activated thyristors will no doubt be their turn-on delay behavior. Turn-on delay times should not be significantly longer than for gated turn-on. The published results appear promising, especially if one takes into account that one may be able to replace the LEDs by semiconductor lasers.

2.3.4 Thyristor turn-off

2.3.4.1 Non-Assisted turn-off. The normal large-area thyristor can only be turned off by reducing the applied voltage to zero or to negative values. In the latter case the current reverts until the reverse blocking junction begins to block. One of the major shortcomings of the thyristor is its relatively long turn-off time, i.e., the time that has to pass before the forward blocking capability is reestablished. This behavior is caused by the fact that not all the stored charge that is built up during the forward conducting stage can be removed by external means during the reverse blocking stage.

Once the maximum reverse blocking voltage is reached, the quasi-neutral cloud of charge carriers remaining in the base regions outside of the space charge region can only be extracted inasmuch as holes reach the space charge region by ambipolar diffusion. The rest has to disappear by means of recombination to a degree that the current generated when the forward voltage is reapplied flows predominantly into the n-emitter shorts and hence does not turn on the device (Fig. 17).

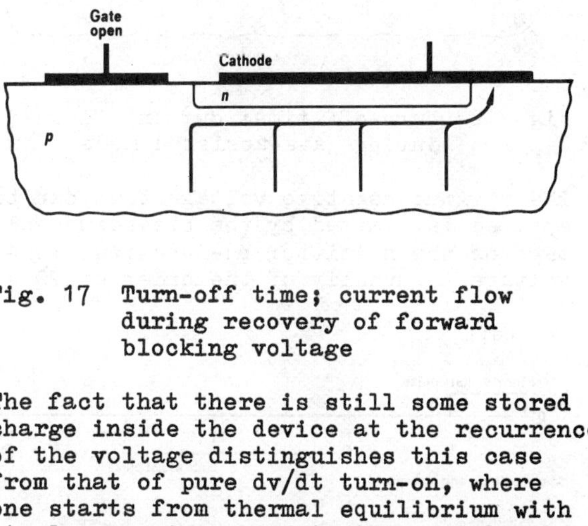

Fig. 17 Turn-off time; current flow during recovery of forward blocking voltage

The fact that there is still some stored charge inside the device at the recurrence of the voltage distinguishes this case from that of pure dv/dt turn-on, where one starts from thermal equilibrium with its low carrier concentrations. But otherwise the situation is quite similar. The more the n-emitter is shorted, the shorter the turn-off time will be. The turn-off time can thus be adjusted by means of the carrier lifetime as well as by emitter shorting. However, as pointed out before, lifetime reduction and emitter shorting have to be applied with care, if acceptable on-state characteristics[5] and good plasma spreading[6] properties are to be retained.

2.3.4.2 Gate assisted turn-off (GAT). One way that has been successfully tried to withdraw high currents out of the p-base

5) See Section 2.2 and Fig. 5

6) See Section 2.3.2 and Figs. 9 and 10

during the recurrence of the forward voltage, without impairing other thyristor properties, is the application of a negativ voltage gate pulse (New, Frobenius, Desmond and Hamilton (60); Raderecht (61); Brewster and Schlegel (62); Schlegel (63); Shimizu, Oka, Funakawa, Gamo, Iida and Kawakami (64)). This method is called gate assisted turn-off (GAT). It is commonly used for low-power thyristors that are operated in TV deflection circuits at frequencies up to around 20 kHz. Turn-off times are reduced by more than 50 %. Values as low as 2.5 µs can be obtained for 800 V devices. It should be pointed out that the negative gate pulse has little effect on withdrawing charge as long as the reverse blocking voltage is applied. Therefore, it suffices to apply the negative voltage shortly before the thyristor voltage reaches positive values (63), (64).

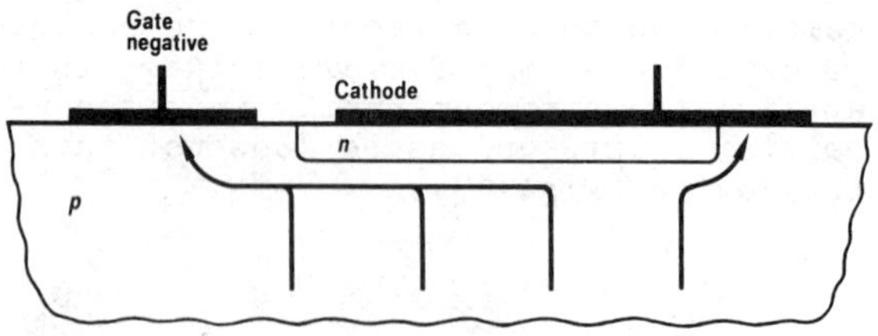

Fig. 18 Turn-off time; current flow during gate-assisted mode

The maximum negative voltage that can be applied is limited by the breakdown voltage of the n-emitter pn-junction. This voltage is usually of the order of 20 V.

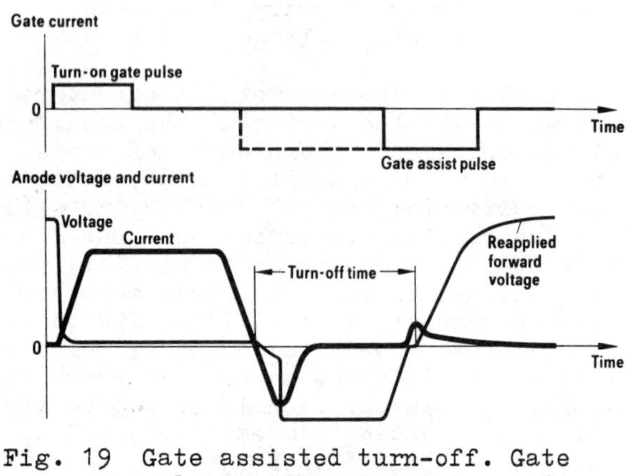

Fig. 19 Gate assisted turn-off. Gate pulses with respect to anode voltage and current transients

For large-area devices dense interdigitation is a prerequisite for this mode of operation. Gate assisted turn-off of such arrangements has been described by several authors (60) - (64). In (62) it was found that the losses in the GAT-mode at forward recovery may equal the turn-off losses. Nevertheless, as the TV-example shows, GAT has its merits where a fast turn-off is of prime importance. In radar equipment gate-assisted turn-off thyristors have been operated at frequencies up to 100 kHz (61).

2.3.4.3 Gated turn-off (GTO). One step further than GAT is gated turn-off (GTO). From the standpoint of operation the only difference between the two modes is that for GTO the negative gate current is raised to such a level that the thyristor will turn-off even when in the forward conduction mode. In practice the GTO-thyristor requires a much more carefully balanced design than the GAT-thyristor, especially when a high turn-off gain is desired (Wolley (65); Wolley, Yu, Steigerwald and Matteson (66); Kao and Brewster (67); Becke and Neilson (68)). A review of the current status of GTO-thyristors is given by Okamura (24).

Available GTO-thyristors have maximum current ratings around 10 A, whereas values as high as 200 A have been reported for laboratory samples (66). One of the main obstacles slowing the appearance of GTO-thyristors on the market appears to be their property to go into second breakdown when improperly operated.

2.4 <u>Special Arrangements</u>

We will now turn to devices that deviate in some aspect from the usual thyristor design, but are either directly derived from the thyristor structure or resemble the thyristor in their electrical properties.

2.4.1 <u>Two-terminal pnpn-structures</u> (Reverse switching rectifier, breakdown diode). Two different designs of two-terminal pnpn-power devices have been marketed, the reverse switching rectifier (RSR) (Gardenghi, Hooper and Zimmermann (69); Gardenghi (70)) and the high-voltage Shockley (breakover) diode (BOD) (Schroen (71)). The RST is a device that is turned on in forward direction by means of a high dv/dt. For this mode of operation it is essential that the critical dv/dt is well surpassed. The high capacitive current then acts like a distributed gate current which turns on large areas of the device. This can be particularly useful in applications where high di/dt-values arise. Devices with a di/dt-capability of 2500 A/µs up to values of 5000 A have been reported.

The high-voltage breakover diode is a low-power device used for over-voltage protection of high-power high-voltage thyristors. The device is turned on by exceeding the breakover voltage. In a high-voltage thyristor, breakover turn-on will usually be destructive, unless the di/dt and the current are strongly limited. Though the breakover di/dt-capability can be much improved with special thyristor designs (see Voss (72) and references there), this has so far not led to a practical device. The BOD is in principle not a high-di/dt-device, but rather profits from being connected externally to the main thyristor (Fig. 20). This allows an effective current limiting (e.g., by a resistor) in the path from the auxiliary BOD to the main thyristor gate.

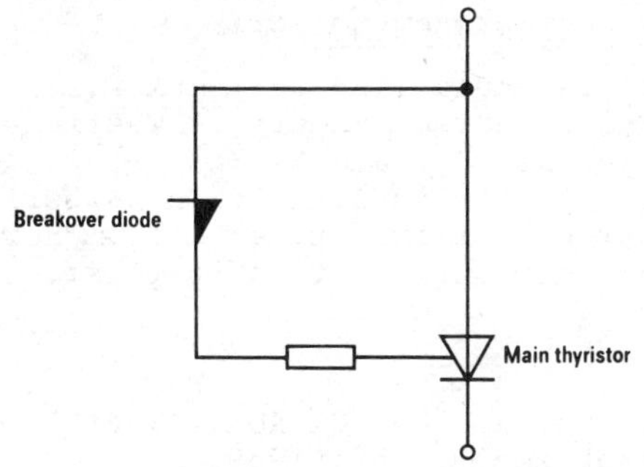

Fig. 20 Principle of externally connected Shockley breakover diode

2.4.2 Integrated devices (reverse conducting thyristor). Integration of several power devices into a single silicon chip has as yet proved to be useful in only a few cases. The most widely accepted concept is the already treated two-stage gate of various design in which an auxiliary thyristor is employed. Another concept is the integration of a thyristor and a diode in antiparallel, an arrangement which is commonly called the reverse conducting thyristor (Gerlach and Köhl (1); Kokosa (20); Yatsuo, Kamei, Terasawa, Ogawa, Wajima and Morita (73)). Since the reverse blocking capability is not exploited in this case, the thyristor section can be a npinp-structure (see Fig. 4). Reverse conducting thyristors may therefore have dynamic properties that cannot be achieved with reverse blocking thyristors. For example, a combination of 2500 V repetitive peak voltage with a turn-off time of 30 µs is possible (compare Fig. 1). A review on the state of the art of reverse conducting thyristors is given by Gamo (23).

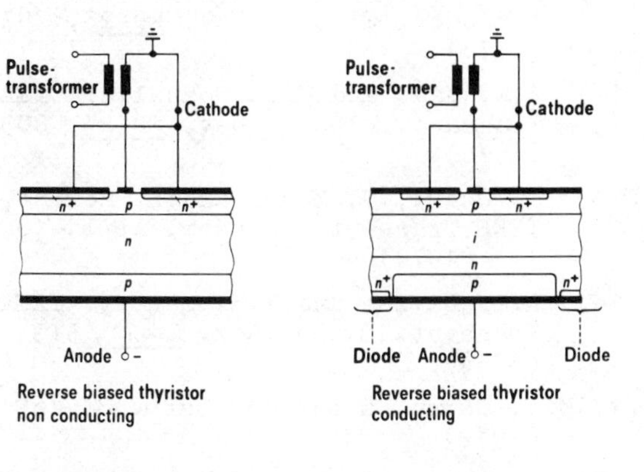

Fig. 21 Reverse conducting thyristor; npinp-thyristor with integrated diode

2.4.3 Power modules. In the current range up to 50 A, power modules with many individual devices combined in one case on an electrically insulated mounting plate have become quite common. Several manufacturers use glass passivated chips that can also be acquired as single unencapsusalted elements. Current ratings of modules are also steadily increasing and have now reached values of 120 A average current per thyristor in units containing two thyristors. It is foreseeable that this may still just be the beginning of the rise in the current ratings of modules. If the trend remains as it is, power modules are likely to displace stud-mounted thyristors in many applications.

2.4.4 Triacs. Advancement with triacs (Gentry, Scace and Flowers (74)) has as with most thyristors proceeded on the general technological level. While the triac has found widespread use in low-power phase-control applications, the use of high-power triacs is still limited, due to their inferior commutation dv/dt-capability, when compared with an arrangement of two thyristors in antiparallel. Typical maximum ratings of currently available high-power triacs are 1000 V, 200 A, in conjunction with a commutation dv/dt of 50 V/µs.

The commutation problem may be overcome by the light-activated triac (Yatsuo, Konishi, Sugawara and Wazima (75)), the schematic of which is shown in Fig. 22. The device has a central funnel-shaped groove into which the light from a LED is coupled. In this way the triac exhibits a very similar turn-on behavior for both polarities, while the two sections can be decoupled better than in the gate fired triac.

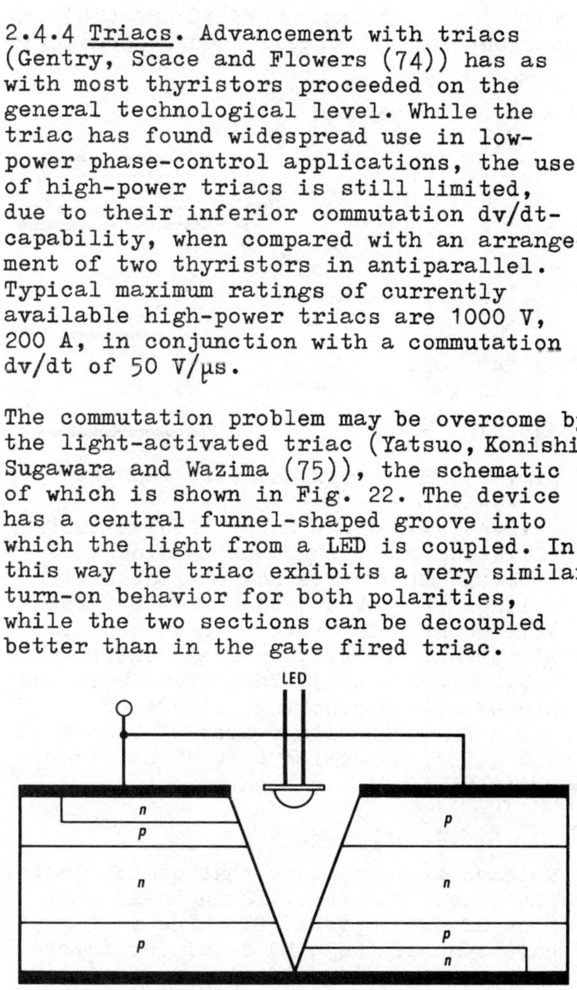

Fig. 22 Schematic representation of light activated triac (Yatsuo, Konishi, Sugawara and Wazima (75))

3. POWER RECTIFIERS

The technology of power rectifiers is very similar to that of thyristors. This is the reason why rectifier development often has become an offshoot of thyristor development. Just as with the thyristor, the trend is toward higher voltages, higher currents and higher turn-off speed. Rectifiers with voltage ratings of 5 kV are now sold by many manufacturers, in part with the extra bonus of exhibiting avalanche capability. For applications in the welding and the chemical industries rectifiers with diameters up to 76 mm are available (Rai-Choudhury, Kiggins and Pittmann (76); de Warga and Mungenast (7)).

Development of fast high-power rectifiers is impeded by an increasing demand for soft-recovery characteristics, i.e., for a lack of snap-off behavior. This requires a design that is not optimized with respect to the blocking voltage and the current carrying capability. For a p^+nn^+-rectifier exhibiting soft recovery the n-base has to be approximately twice as wide (Fig. 23) as would be necessary for a snap-off pin-rectifier (Cordingley (77); Porst (78)). The current rating decreases correspondingly.

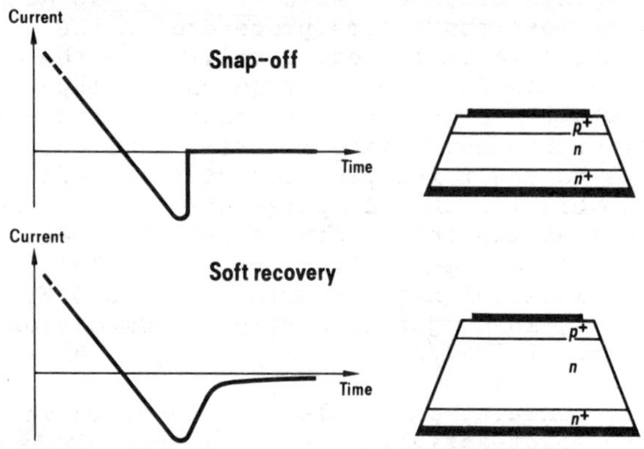

Fig. 23 Snapp-off and soft recovery

In the low-voltage range bipolar rectifiers compete with Schottky-diodes. Schottky-diodes are now available with ratings of 45 V and 60 A at 0.6 V forward voltage drop for reverse recovery times of 150 ns. However, bipolar rectifiers have become equally fast. Ion implanted diodes with a blocking voltage of 150 V and a forward voltage drop of 0.45 V at 100 A in conjunction with 75 ns reverse recovery time have been reported.

4. CONCLUSION

We have tried to show that semiconductor power devices are still in their full flow of development. Blocking voltages have reached the 4 kV level for thyristors and the 6 kV level for rectifiers. Thyristors are manufactured with diameters of 102 mm, rectifiers with diameters of 76 mm. A further increase appears possible in principle; however, the question arises where all the power to be handled by such devices is going to come from.

This question is, of course, already acute with some of devices having diameters around 75 mm. Today's market is able to absorb only relatively few such large devices, and it remains to be seen where the compromise will be between high developmental costs on the one side and the future market on the other.

Many of the developments currently under way are aimed at medium and high frequency applications. We will become accustomed to reverse conducting thyristors and probably also to gate-assisted turn-off thyristors.

ACKNOWLEDGMENT

The authors would like to thank P.L. Hower and P. Rai-Choudhury of Westinghouse Electric Co. and Y. Sato of Fuji Electric Co. for their help in gathering the material for this paper and A. Herlet for his contributions to this paper.

REFERENCES

(1) W. Gerlach and G. Köhl, Festkörperprobleme 9, 356 (1969)

(2) A. Herlet, ESSDERC 1971, Munich, Sol. State Dev. 1971, p. 123

(3) E. Spenke, Proc. 5th Conver. Sol. State Dev., Tokyo 1973, Suppl. J. Jap. Soc. Appl. Phys. 43, 375 (1974)

(4) P. Leturcq, ESSDERC 1975, Grenoble, Sol. State Dev. 1975, p. 119

(5) L. Mattera, Electronics, June 26, 1975, p. 81

(6) W. Gerlach, ESSDERC 1976, Munich, to be published

(7) J. de Warga and J.E. Mungenast, IEEE IAS Meeting 1974, p. 261 and IEEE IAS Meeting 1976, p. 69

(8) A. Herlet, Sol. State Electron. 8, 655 (1965)

(9) S.M. Sze, Physics of Semiconductor Dev., Wiley, New York (1969), p. 109 ff.

(10) M. Tanenbaum and A.D. Mills, J. Electrochem. Soc. 108, 171 (1961)

(11) M. Schnöller, IEEE Trans. Electr. Dev. ED-21, 313 (1974)

(12) H.A. Hermann and H. Herzer, J. Electrochem. Soc. 122, (1975) 1568

(13) H.M. Janus and O. Malmros, IEEE Transact. Electr. Dev. ED-23, 797 (1976)

(14) E.W. Haas and M.S. Schnöller, IEEE Transact. Electr. Dev. ED-23, 803 (1976)

(15) T. Ogawa, T. Kamei and K. Morita, IEEE Transact. Industr. Appl. 10, 112 (1974)

(16) R.L. Davies and F.E. Gentry, IEEE Transact. Electr. Dev. ED-11, 313 (1964)

(17) J. Cornu. Electron. Lett. 8, 169 (1972)

(18) M. Bakowski and I. Lundström, IEEE Transact. Electr. Dev. ED-20, 550 (1973)

(19) T. Tsurushima and S. Kataoka, Verhandl. Deutsche Phys. Ges. VI/2, 44 (1967)

(20) R.A. Kokosa, IEEE Transact. Electr. Dev. ED-17, 669 (1970)

(21) D.E. Crees, N.S. Nichols and F. Wood, IEE Intern. Conf. Power Electr.-Power Semicond. and their Appl. 1974, p. 13

(22) P.M. van Iseghem, IEEE Transact. Electr. Dev. ED-23, 823 (1976)

(23) H. Gamo, IEEE Intern. Semiconductor Power Converter Conference, March 1977, Orlando, Florida/USA

(24) M. Okamura, IEEE Intern. Semiconductor Power Converter Conference, March 1977, Orlando, Florida/USA

(25) J. Burtscher, F. Dannhäuser and J. Krausse, Sol. State Electron. 18, 35 (1975)

(26) W.M. Bullis, NBS Technical Note 788 (1973)

(27) K.S. Tarneja and J. Bartko, IEEE Power Electronics Spec. Conf. 1975, p. 269

(28) P. Rai-Choudhury, J. Bartko and J.E. Johnson, IEEE Transact. Electr. Dev. ED-23, 814 (1976)

(29) C.K. Chu and J.F. Donlon, IEEE IAS Meeting, 1976, p. 51

(30) K.P. Frohmader, PhD-Thesis, "Physik des Kurzschlußemitters", Technical University, Aachen, F.R. Germany

(31) C.K. Chu, IEEE Transact. Electr. Dev. ED-17, 687 (1970)

(32) J. Burtscher and E. Spenke, Siemens Forsch. und Entw. Ber. 3, 234 (1974)

(33) H. Strack, Siemens Forsch. und Entw. Ber. 5, 5 (1976)

(34) K. Hartmann, IEEE Transact. Electr. Dev. ED-23, 912 (1976)

(35) A. Munoz-Yague and P. Leturcq, IEEE Transact. Electr. Dev. ED-23, 917 (1976)

(36) N. Mapham, Electronics 35, 33 (1962)

(37) R.I. Longini and J. Melngailis, IEEE Transact. Electr. Dev. ED-10, 179 (1963)

(38) W. Gerlach, Telefunken Zt. 34, 301 (1966)

(39) H.R. Ruhl, IEEE Transact. Electr. Dev. ED-17, 672 (1970)

(40) I. Somos and D.E. Piccone, IEEE Transact. Electr. Dev. ED-17, 680 (1970)

(41) Y. Terasawa, Proc. 3rd Conf. Solid State Dev. Tokyo (1971) p. 198

(42) T. Matsuzawa, NEC. Res. Dev. 30, 40 (1973)

(43) R. Rosch, PhD-Thesis, "Über die laterale Zündausbreitung in Leistungsthyristoren", Technical University Aachen, FR Germany (1973)

(44) H. Yamasaki, IEEE Transact. Electr. Dev. ED-22, 65 (1975)

(45) P. Voss, VDE-Meeting, "Dynamische Probleme der Thyristortechnik", Ed. M. Depenbrock, VDE-Verlag, 1971 p. 251

(46) D.E. Piccone and I. Somos, Electr. Eng. 28, 89 (1969)

(47) B.V. Cordingley, IEEE Proceed. 58, 139 (1970)

(48) P. Voss, IEEE IAS Meeting 1974, p. 467

(49) R.A. Kokosa and E.D. Wolley, IEEE Intern. Electr. Dev. Meeting, 1974 p. 431

(50) A. Silard and V. Marinesch, IEEE Intern. Electr. Dev. Meeting, 1975, p. 359

(51) W.H. Tobin and S.J. Wu, IEEE IAS Meeting 1974, p. 477

(52) I. Somos and D.E. Piccone, IEEE IAS Meeting 1974, p. 495

(53) I.V. Grekhov, M.E. Levinshtein and V.G. Sergeev, Sov. Phys. Semicond. 8, 431 (1974)

(54) J.R. Davis, IEEE Power Electr. Specialist Conf. 1975, p. 305

(55) J.R. Davis and J.S. Roberts, IEEE Power Electr. Specialist Conf. 1976, p. 272

(56) D. Silber and M. Füllmann, IEEE Intern. Electr. Dev. Meeting 1975 p. 371

(57) D. Silber, W. Winter and M. Füllmann, IEEE Transact. Electr. Dev. ED-23, 899 (1976)

(58) De Bruyne and R. Sittig, IEEE Power Electr. Special. Conf. 1976, p. 262

(59) V.A.K. Temple and A.P. Ferro, IEEE Transact. Electr. Dev. ED-23, 893 (1976)

(60) T.C. New, W.D. Frobenius, T.J. Desmond and D.R. Hamilton, IEEE Transact. Electr. Dev. ED-17, 706 (1970)

(61) P.S. Raderecht, Intern. J. Electron. 36, 399 (1974)

(62) J.B. Brewster and E.S. Schlegel, IEEE IAS Meeting 1974, p. 663

(63) E.S. Schlegel, IEEE Transact. Electr. Dev. ED-23, 888 (1976)

(64) J. Shimizu, H. Oka, S. Funakawa, H. Gamo, T. Iida and A. Kawakami, IEEE Transact. Electr. Dev. ED-23, 883 (1976)

(65) E.D. Wolley, IEEE Transact. Electr. Dev. ED-13, 590 (1966)

(66) E.D. Wolley, R. Yu, R. Steigerwald and F.M. Matteson, IEEE IAS Meeting 1973, p. 251

(67) Y.C. Kao and J.B. Brewster, IEEE IAS Meeting 1974, p. 689

(68) H.W. Becke and J.M. Neilson, IEE Power Electr. Special. Conf. 1975, p. 292

(69) R.A. Gardenghi, E.H. Hooper and F.S. Zimmermann, IEEE Power Electr. Special. Conf. 1975, p. 313

(70) R.A. Gardenghi, IEEE Intern. Electr. Dev. Meeting 1975, p. 367

(71) W.H. Schroen, IEEE Transact. Electr. Dev. ED-17, 694 (1970)

(72) P. Voss, Sol. State Electron. 17, 655 (1974)

(73) T. Yatsuo, T. Kamei, Y. Terasawa, T. Ogawa, K. Wajima and K. Morita, IEEE IAS Meeting 1974, p. 277

(74) F.E. Gentry, R.I. Scace and J.K. Flowers, IEEE Proceed. 53, 355 (1965)

(75) T. Yatsuo, N. Konishi, Y. Sugawara and K. Wazima, IEEE Power Electr. Special. Conf. 1976, p. 267

(76) P. Rai-Choudhury, T.R. Kiggins, P.F. Pittmann, IEEE IAS Meeting 1974, p. 281

(77) B.V. Cordingley, IEE Intern. Conf. Power Electr. Power Semicond. and their Appl. 1974, p. 7

(78) A. Porst, IEEE IAS Meeting 1974, p. 645

DYNAMIC PERFORMANCE OF CONVERTER-FED SYNCHRONOUS MOTORS

F. Harashima and H. Naitoh

Institute of Industrial Science, University of Tokyo, Tokyo 106, Japan

ABSTRACT

Transfer function models of converter-fed synchronous motors have been derived, and the effects of motor parameters on the performance of the systems are discussed. The transfer function models have been derived on the basis of the average behaviors which are represented by continuous dynamic equations. The phenomena during commutation periods are taken into consideration equivalently as average voltage drops in the circuits. The effects of motor parameters such as the size of DC choke, constants of damper windings and so forth have been examined.

INTRODUCTION

Converter-fed synchronous motors, hereafter we call them Commutatorless Motors, have come into wide use in industry as a kind of variable speed drives. They can be controlled with good performance like DC drives and are easy to be maintained like AC motors.

When this system is used as a variable speed drive, it is an important matter to consider dynamic characteristics as well as steady state ones.

In this paper, transfer function models of commutatorless motors have been derived, and the effects of motor parameters on the performance characteristics of the system are discussed. The adequacy of the mathematical models has been confirmed both by simulation and experimental results.

Figure 1 shows the fundamental construction of a commutatorless motor. Two kind of dynamics should be considered for a commutatorless motor; one is electrical dynamics and the other is electro-mechanical dynamics. Since the response of electrical dynamics is much faster than that of electro-mechanical dynamics, these two dynamic characteristics can be devided into two parts as is shown in Fig. 2. The transfer function $F_2(s)$ which represent the electro-mechanical dynamics can be easily obtained from the torque-speed curves of the motor and the torque characteristics of the load. On the otherhand, the transfer function $F_1(s)$ which represent electrical dynamics can not be easily obtained, since it depends on the complicated electrical phenomena.

However, $F_1(s)$ is a very important factor when a high performance of control characteristics like a DC drive is required for a commutatorless motor. Thus, in this paper, electrical dynamic characteristics of commutatorless motor are discussed.

The transfer function models of commutatorless motors are derived on the basis of the average behaviors which are represented by continuous dynamic equations. The phenomena during commutation periods are taken into consideration equivalently as voltage drops in the circuits. The effects of motor parameters such as the size of DC choke, constants of damper windings and so forth have been examined.

DEVIATION OF TRANSFER FUNCTION MODELS

Circuits Configuration

The output voltage of the controlled rectifier in Fig. 1 is controlled by the gate signal of the thyristors. Although the output voltage includes ripple component, they do not have serious influences on the characteristics of the motor, since ripple frequencies are generally high and are smoothed by the DC choke L_0 for the most part. Therefore, the controlled rectifier

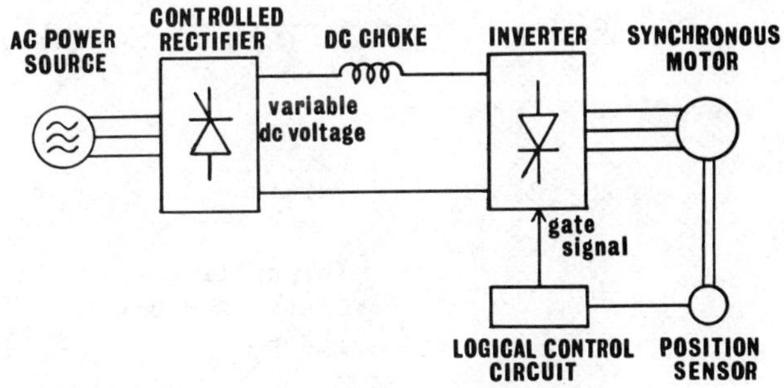

Fig. 1. Fundamental construction of commutatorless motor drive systems.

can be considered as a DC power supply with a variable output voltage. The inverter is a naturally commutated 6-pulse bridge-inverter which is commutated by the induced voltage of the synchronous motor. Then, Fig. 1 can be replaced by Fig. 3.

Equivalent Circuits and Transfer Function Models

First, we derive a simple equivalent circuit for a non-salient machine without damper windings. When only electrical dynamics are considered, the speed of the motor can be assumed to be constant. The most important problem is how to deal with the overlapping phenomena, that is, commutation modes. In the operation of a commutatorless motor, there exist two modes; single mode and commutation mode. During the single mode, the current flows through two windings of the armature windings. On the otherhand, the current flows through three windings during the commutation mode. The circuit equations for each mode are different to each other. However, we can combine these two modes into one, approximately. A line-commutated phase-controlled converter can be replaced by a variable voltage DC supply and a resistance, on the average, equivalently. The equivalent circuit is given in Fig. 4(a). In this figure, the resistance $3\omega L_c/\pi$ expresses the effects of overlapping phenomena equivalently and the amplitude of the variable DC voltage is given by $\tilde{E}\cos\gamma_0$, where $\tilde{E}$ is the induced voltage of the motor when γ_0 is set to zero, and γ_0 is the leading angle of the gate control. By replacing the overlapping phenomena by the equivalent resistance in this way, it is possible to regard that there exists only single mode throughout the operation of this system.

Thus, the equivalent circuit of the whole system is given in Fig. 4(b). The transfer function of a commutatorless motor can be derived from this equivalent circuit. In Fig. 4(b), $L=2l_1+3L_1$, where l_1 and L_1 are the leakage and effective inductances of the armature windings per phase.

From this figure, we can derive the transfer function $G_{FE}(s)$ which shows the relation between the input voltage increment ΔE_0 and the average DC current increment $\Delta\bar{I}_0$.

$$G_{FE}(s) = \frac{\Delta\bar{I}_0}{\Delta E_0} = \frac{1}{(Ls+R)} \tag{1}$$

where $L = L_0 + 2l_1 + 3L_1$

$$R = R_0 + 2R_1 + \frac{3\omega}{\pi}L_c$$

$$L_c = l_1 + \frac{3}{2}L_1 \; : \; \text{commutating inductance per phase}$$

The transfer function $G_{FE}(s)$, which denote the relation between ΔE_0 and the average torque increment $\Delta\bar{T}orq$, can be derived in the following way.

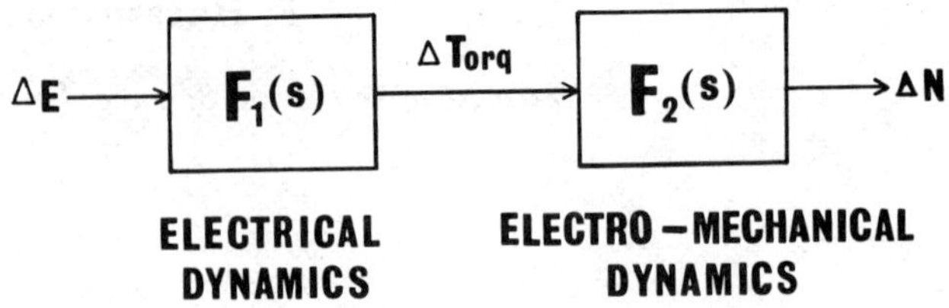

Fig. 2. Block diagram for dynamic performance of commutatorless motors.

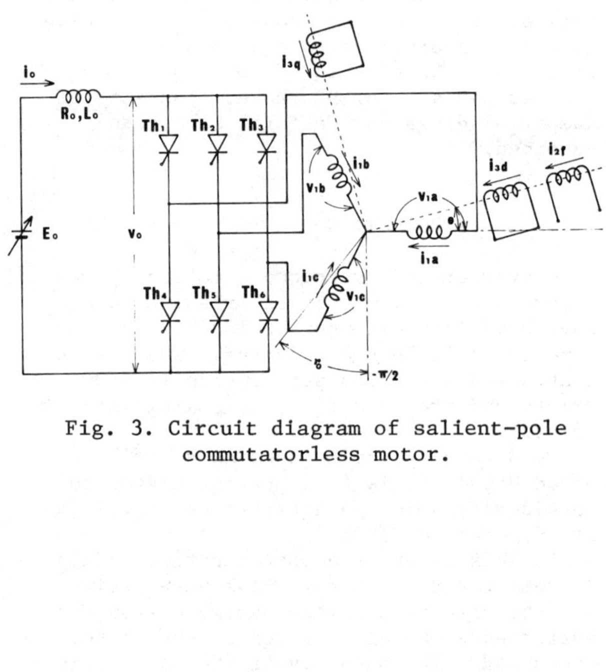

Fig. 3. Circuit diagram of salient-pole commutatorless motor.

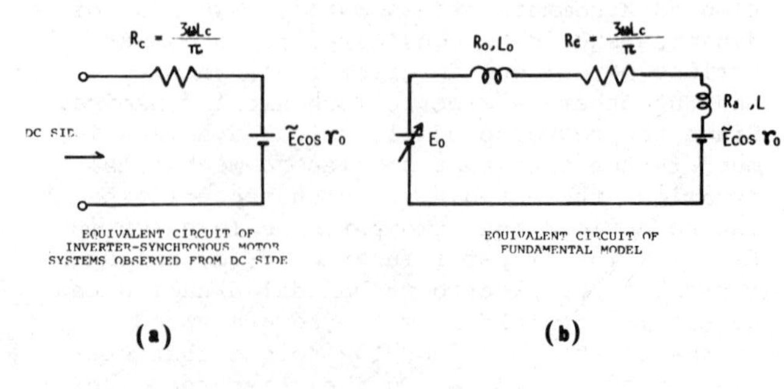

Fig. 4. Equivalent circuits.

The torque equation of this type of synchronous motor is given by Eq.(2) on the basis of two axis theory(Ref. 1) and α-β co-ordinate system (Ref. 2)(see Appendix).

$$Torq. = M_{12}'(i_{2\alpha}i_{1\beta} - i_{2\beta}i_{1\alpha}) \qquad (2)$$

For the single mode, the following relations are given.

$$i_{1a} = -i_{1c} = i_0 \quad , \quad i_{1b} = 0 \qquad (3)$$

$$i_{1\alpha} = \sqrt{3/2}\,i_0 \quad , \quad i_{1\beta} = i_0/\sqrt{2} \qquad (4)$$

where $\theta = \omega t - \frac{\pi}{2} - \gamma_0 (0 \leqq t < T,\ \omega T = \pi/3)$.

$$i_{2\alpha} = I_f\cos\theta \quad , \quad i_{2\beta} = I_f\sin\theta \qquad (5)$$

Substituting Eqs. (4) and (5) into Eq.(2), the following relation is obtained.

$$Torq. = 2M_{12}'I_f \cdot i_0\cos(\theta + \frac{\pi}{3}) \qquad (6)$$

Integrating Eq.(6) from t=0 through t=T and using the relation that the correlation between i and cos(θ + π/3) is relatively small, we obtain the following equation.

$$\overline{Torq.} = \frac{3\sqrt{2}}{\pi} M_{12}'I_f\cos\gamma_0 \cdot \overline{I}_0 \qquad (7)$$

From Eq.(1) and (7), we obtain the transfer function $G_{FT}(s)$

$$G_{FT}(s) = \frac{\Delta\overline{Torq}}{\Delta E_0} = \frac{3\sqrt{2}M_{12}'I_f\cos\gamma_0}{\pi(Ls+R)} \qquad (8)$$

The Practical Machine

Now, we consider a practical machine, that is, a non-salient pole machine with damper windings. The salient pole machine with damper windings are the most practical ones. However, we do not consider saliency for the sake of simplicity, since the effect of saliency are less important than those of damper windings.

When we derive the transfer function $G_{FI}(s)$ and $G_{FT}(s)$ in the previous section, we have derived them directly from equivalent circuit. However, when the machine has damper windings, the structure of the machine becomes more complicated. Therefore, we have to derive the transfer function for the practical machine mathematically using the basic conceptions which is obtained about the fundamental machine.

As is shown in the Appendix, the circuit equations for commutation mode are as follows.

$$A_c' \cdot pI_c = -B_c' \cdot I + D_c' \qquad (9)$$

where

$$I_c = col(i_{1\alpha}, i_{1\beta}, i_{3\alpha}, i_{3\beta})$$

$$A_c' = \begin{bmatrix} 2/3L_0+L_1' & 0 & M_{13}' & 0 \\ 0 & L_1' & 0 & M_{13}' \\ M_{13}' & 0 & L_3 & 0 \\ 0 & M_{13}' & 0 & L_3 \end{bmatrix}$$

$$B_c' = \begin{bmatrix} 2/3R_0+R_1 & 0 & 0 & 0 \\ 0 & R_1 & 0 & 0 \\ 0 & M_{13}'\dot{\theta} & R_3 & L_3\dot{\theta} \\ -M_{13}'\dot{\theta} & 0 & -L_3\dot{\theta} & R_3 \end{bmatrix}$$

$$D_c' = \begin{bmatrix} M_{12}'\,I_f\sin\theta+\sqrt{2/3}E_0 \\ -M_{12}'\dot{\theta}I_f\cos\theta \\ 0 \\ 0 \end{bmatrix}$$

From Eq.(9), we obtain eq.(10) using the relation which hold good in commutation mode.

$$A_c \cdot pI_c = -B_c \cdot I_c + D_c \qquad (10)$$

where

$$A_c = \begin{bmatrix} L_1+\frac{3}{2}L_1' & 0 & \sqrt{\frac{3}{2}}M_{13}' & 0 \\ L_1' & L_1' & 0 & \sqrt{2}M_{13}' \\ \sqrt{\frac{3}{2}}M_{13}' & 0 & L_3 & 0 \\ \sqrt{\frac{1}{2}}M_{13}' & \sqrt{\frac{3}{2}}M_{13}' & 0 & L_3 \end{bmatrix}$$

$$B_c = \begin{bmatrix} R_0+\frac{3}{2}R_1 & 0 & 0 & 0 \\ R_1 & R_1 & 0 & 0 \\ \sqrt{\frac{1}{2}}M_{13}'\dot{\theta} & \sqrt{\frac{1}{2}}M_{13}'\dot{\theta} & R_3 & L_3\dot{\theta} \\ -\sqrt{\frac{3}{2}}M_{13}'\dot{\theta} & 0 & -L_3\dot{\theta} & R_3 \end{bmatrix}$$

$$D_c = \begin{bmatrix} \sqrt{\frac{3}{2}}V_B\sin\theta+E_0 \\ -\sqrt{2}V_B\cos\theta \\ 0 \\ 0 \end{bmatrix}$$

$$V_B \equiv M_{12}'\dot{\theta}I_f$$

The circuit equation for single mode is

$$A_s \cdot pI_s = -B_s \cdot I_s + D_s \qquad (11)$$

, where

$$I_s = col(i_0, i_{3\alpha}, i_{3\beta})$$

$$A_s = \begin{bmatrix} L_0+2L_1' & \sqrt{\frac{3}{2}}M_{13}' & \frac{1}{\sqrt{2}}M_{13}' \\ \sqrt{\frac{3}{2}}M_{13}' & L_3 & 0 \\ \frac{1}{\sqrt{2}}M_{13}' & 0 & L_3 \end{bmatrix}$$

$$B_s = \begin{bmatrix} R_0+2R_1 & 0 & 0 \\ \frac{1}{\sqrt{2}}M_{13}'\dot{\theta} & R_3 & L_3\dot{\theta} \\ -\frac{3}{2}M_{13}'\dot{\theta} & -L_3\dot{\theta} & R_3 \end{bmatrix}$$

$$D_s = \begin{bmatrix} \sqrt{2}V_B\sin(\theta-\frac{\pi}{6})+E_0 \\ 0 \\ 0 \end{bmatrix}$$

Now we combine Eq.(10) and Eq.(11), and then we obtain Eq.(12).

$$A_{cs}\cdot pI_{cs} = -B_{cs} + D_{cs} \tag{12}$$

, where

$$I_{cs} = \text{col.}(i_0\ ,\ i_{1b}\Big|_0^{T_u}\ ,\ i_{3\alpha}\ ,\ i_{3\beta})$$

$$A_{cs} = \begin{bmatrix} L_{cs} & 0 & \sqrt{\frac{3}{2}}M_{13}' & \frac{1}{\sqrt{2}}M_{13}'\Big|_{T_u}^{T} \\ L_1'\Big|_0^{T_u} & 2L_1'\Big|_0^{T_u} & 0 & \sqrt{2}M_{13}'\Big|_0^{T_u} \\ \sqrt{\frac{3}{2}}M_{13}' & 0 & L_3 & 0 \\ \frac{1}{\sqrt{2}}M_{13}' & \sqrt{2}M_{13}' & 0 & L_3 \end{bmatrix}$$

$$B_{cs} = \begin{bmatrix} R_{cs} & 0 & 0 & 0 \\ R_1\Big|_0^{T_u} & 2R_1\Big|_0^{T_u} & 0 & 0 \\ \frac{1}{\sqrt{2}}M_{13}' & \sqrt{2}M_{13}'\dot{\theta} & R_3 & L_3\dot{\theta} \\ -\sqrt{\frac{3}{2}}M_{13}'\dot{\theta} & 0 & -L_3\dot{\theta} & R_3 \end{bmatrix}$$

$$D_{cs} = \begin{bmatrix} \sqrt{2}V_B\sin(\theta-\frac{\pi}{6})-(\frac{\sqrt{2}}{2}V_B\cos\theta)\Big|_0^{T_u}+E_0 \\ -\sqrt{2}V_B\cos\theta\Big|_0^{T_u} \\ 0 \\ 0 \end{bmatrix}$$

$$L_{cs} = \begin{cases} L_1+\frac{3}{2}L_1' & (0\le t\le T_u) \\ L_0+2L_1' & (T_u<t<T) \end{cases}$$

$$R_{cs} = \begin{cases} R_0+\frac{3}{2}R_1 & (0\le t\le T_u) \\ R_0+2R_1 & (T_u\le t<T) \end{cases}$$

T_u is the duration of commutation mode, and $\Big|_{T_1}^{T_2}$ indicates that the term which has this simbol is difined between T_1 and T_2 and otherwise is set to zero.

We consider the transfer function for the average behaviors of the system. The average values of state variables are defined as follows.

$$\frac{1}{T}\int_0^T i_0 dt \equiv \bar{I}_0 \tag{13}$$

$$\frac{1}{T}\int_0^T i_{3\alpha} dt \equiv \bar{I}_{3\alpha}$$
$$\frac{1}{T}\int_0^T i_{3\beta} dt = \bar{I}_{3\beta} \tag{14}$$

$$\frac{1}{T}\int_0^{T_u} i_b dt = \bar{I}_b \tag{15}$$

By integrating Eq.(12) and by the use of proper approximation, we can obtain the following equations which express the average behaviour of the system.

$$A\cdot pI = -BI + D \tag{16}$$

, where

$$I = \text{col.}(\bar{I}_0\ ,\ \bar{I}_{3\alpha}\ ,\ \bar{I}_{3\beta})$$

$$A = \begin{bmatrix} L_0+2L_1' & \sqrt{\frac{3}{2}}M_{13}' & \frac{1}{\sqrt{2}}M_{13}' \\ \sqrt{\frac{3}{2}}M_{13}' & L_3 & 0 \\ \frac{1}{\sqrt{2}}M_{13}' & 0 & L_3 \end{bmatrix}$$

$$B = \begin{bmatrix} R_0+2R_1+R_{eq} & 0 & 0 \\ 0 & R_3 & 0 \\ 0 & 0 & R_3 \end{bmatrix}$$

$$C = \begin{bmatrix} E_0-\tilde{E}\cos\gamma_0 \\ 0 \\ 0 \end{bmatrix}$$

$$R_{eq} = \frac{3\omega\sigma}{\pi}L_c$$

$$\sigma = \frac{M_{13}'^2}{L_1'L_3}$$

By the use of Laplace Transformation of Eq. (16), we can obtain the transfer function

$G_{PE}(s)$ $(\equiv \Delta\overline{I}_0/\Delta E_0)$ of this system.

$$G_{PE}(s) = \frac{L_3 s + R_3}{a_2 s^2 + a_1 s + a_0} \qquad (17)$$

, where

$$a_2 = (L_0 + 2L_1')L_3 - 2M_{13}'^2$$

$$a_1 = R_{eff}L_3 + R_3(L_0 + 2L_1')$$

$$a_0 = R_{eff} \cdot R_3$$

$$R_{eff} = R_0 + 2R_1 + \frac{3\omega\sigma}{\pi} L_c \quad : \text{effective resistance}$$

The equivalent circuit of this type of machine is given in Fig. 5.

Then, we consider the transfer function $G_{PT}(s) (= \overline{T}_{orq}/\ E\ )$. The damper currents also generate output torque, that is, damper torque. The damper torque is, however, much smaller than the torque by armature currents and its average value is considered almost to be zero (see Fig. 14). By neglecting the damper torque, the torque equation in this case is also given by Eq.(7). Then, the following equation is derived.

$$G_{PT}(s) = \frac{3\sqrt{2}}{\pi} M_{12}' I_f \cos\gamma_0 - G_{PE} \qquad (18)$$

Now, we consider the transfer function which include the electro-mechanical dynamics. In this case, the rotor speed may change.

When we examine the Eq.(17), we notice that the coefficient R_{eff} depends on rotor speed, that is, it includes $\omega(2\pi f$, f : rotating frequency). However, when we consider a small change of the input voltage, the change of R_{eff} is relatively small. Then we can regard R_{eff} to be constant. Thus the block diagram of transfer functions for the system including the electro-mechanical dynamics is given as in Fig. 6. In this

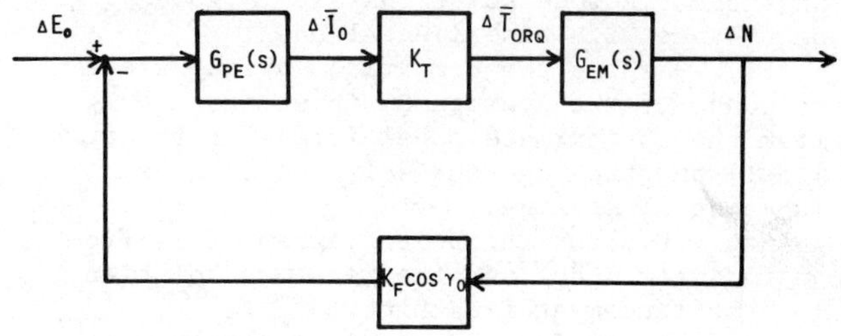

Fig. 6. Block diagram of practical system including electro-mechanical dynamics.

figure, $G_{EM}(s)$ denotes the electro-mechanical dynamics and it is determined by the torque-speed curve of the machine and the torque characteristics of the load.

NUMERICAL EXAMPLES

In this chapter, some numerical example of the dynamic characteristics of a commutatorless motor are shown. Solutions by rigorous analysis (Ref. 3) are also shown as well as solutions by the transfer function models and experimental results. The rigorous analysis is made by digital simulation based on state space method (Ref. 4).

Fig. 7 shows the comparison of DC input current wave form of the experimental result and the rigorous solution under steady state condition.

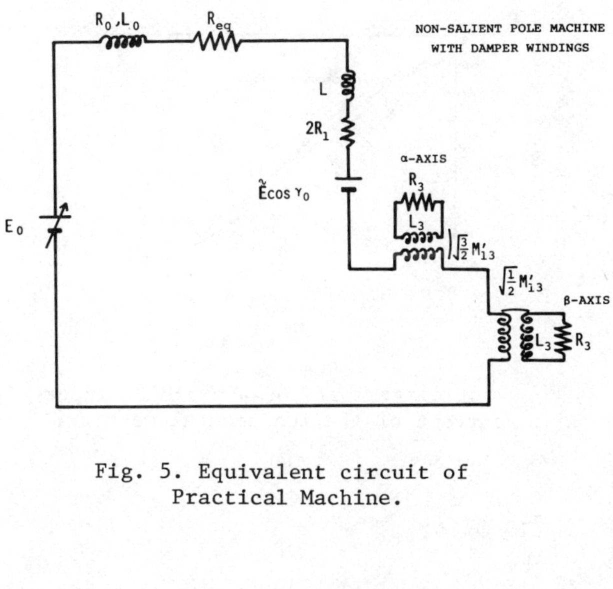

Fig. 5. Equivalent circuit of Practical Machine.

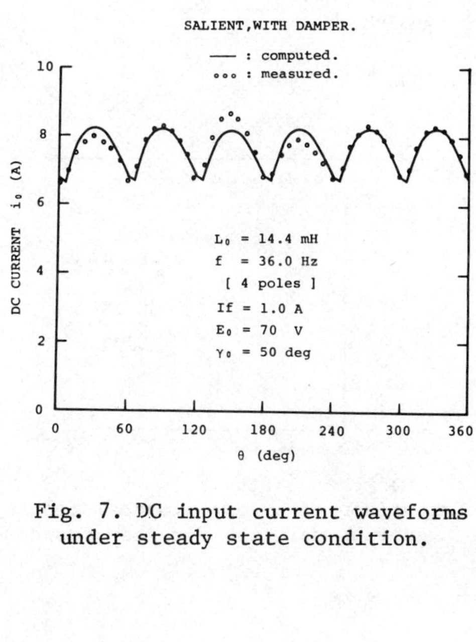

Fig. 7. DC input current waveforms under steady state condition.

Fig. 8 shows the comparison of the step response of the DC input current.

From these two figures, we can conclude that the rigorous solutions are in good accordance with experimental ones.

Fig. 9 shows the comparison of the step response of the average DC input current $\overline{I}_0$ from the approximate model (transfer function model) and the rigorous solution for the fundamental machine.

Fig. 10 shows the Bode Diagram of $\overline{I}_0$ from approximate model and the rigorous solution for the fundamental machine.

Fig. 11 and Fig. 12 show the comparison of the step responses of the average output torque $\overline{T}_{orq}$ and average q-axis damper current $\overline{I}_q$ from the approximate model and the rigorous solution for the practical machine.

Fig. 13 shows the Bode Diagram of $\overline{I}_0$ from the approximate model and the rigorous solution for the practical machine.

From the above figures, it is confirmed that the approximate models which have been proposed are in good correlation with the rigorous solution.

Fig. 14 shows the step response of the output torques from the rigorous solution.

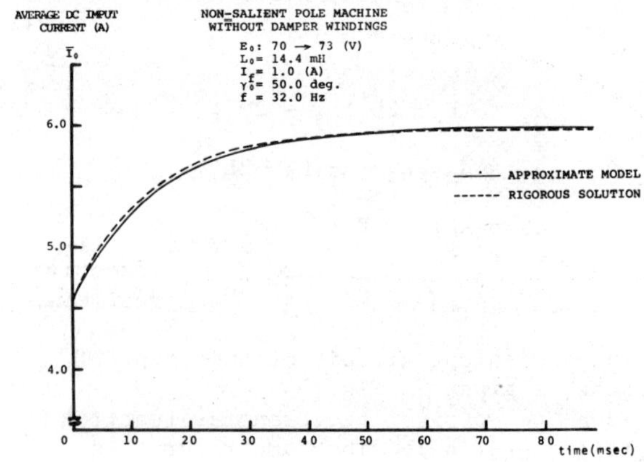

Fig. 9
Step response of average DC input current of the fundamental machine.

SOME REMARKS

In this chapter, the effects of motor parameters on the dynamic performances of the motor are discussed.

Fig. 15 shows the influence of the DC choke L_0 and damper windings on the response of the machine. From this figure, it is concluded that the existence of damper windings makes the response of the machine

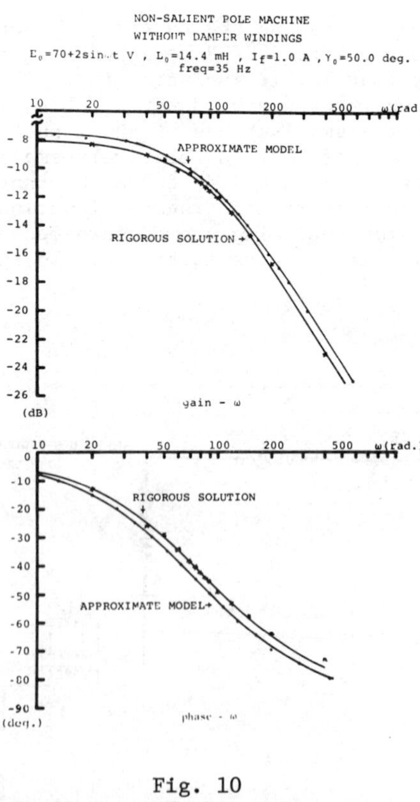

Fig. 10
Bode diagram of the average DC input current of the fundamental machine.

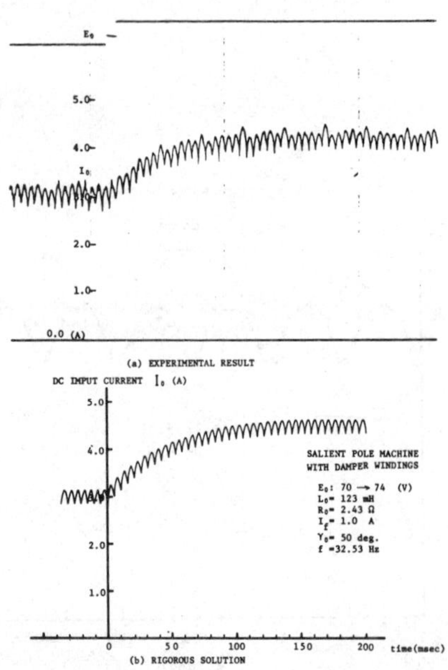

Fig. 8
Comparison of an experimental result and rigorous solution for step response.

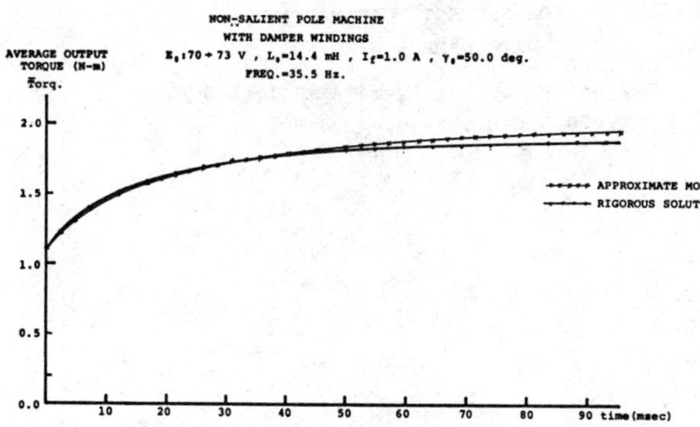

Fig. 11. Step response of average output torque of the practical machine.

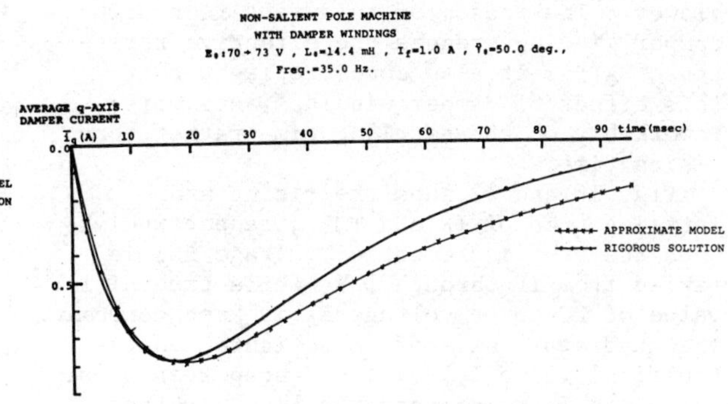

Fig. 12. Step response of average q-aixs damper current of the practical machine.

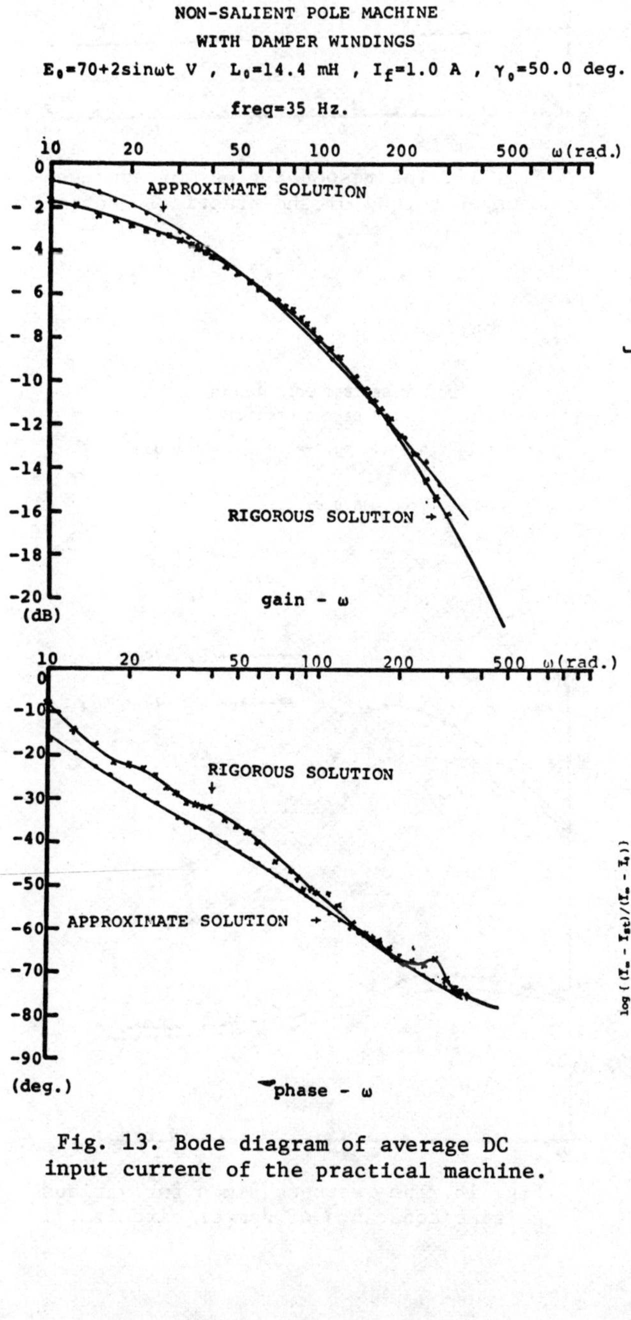

Fig. 13. Bode diagram of average DC input current of the practical machine.

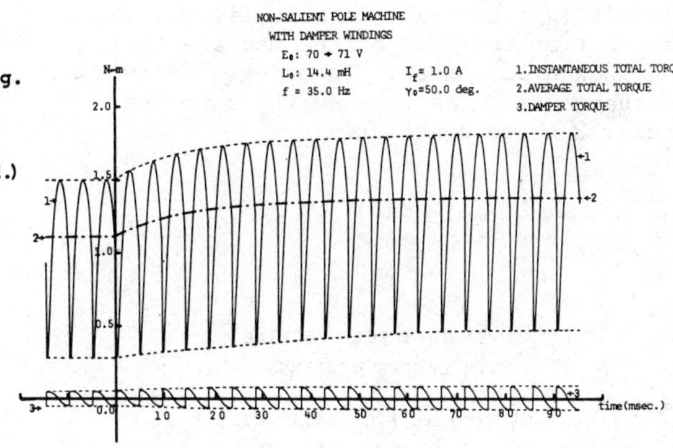

Fig. 14. Step responses of output torques of the practical machine from the rigorous solution.

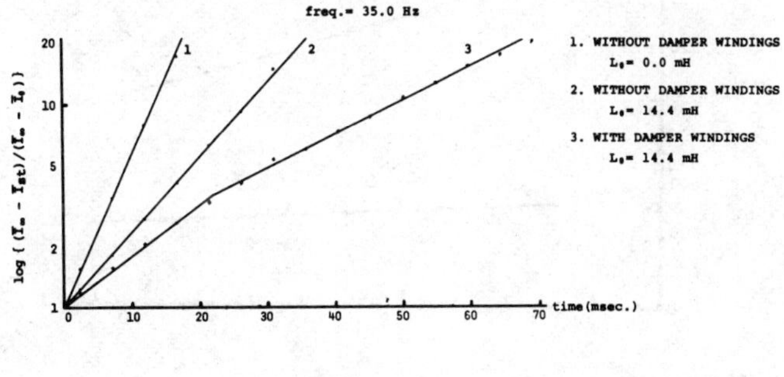

Fig. 15. Effects of DC chokes and damper windings on response speed.

slower. The reason of this fact is that damper windings reduce the effective resistance R_{eff}. It also comes to light that this effect of damper windings acts more remarkably on the settling time rather than rising time.

Fig. 16 and 17 show the rising and settling time of $\overline{I_0}$ and $\overline{T_{orq}}$, respectively, when the step increment of voltage ΔE_0 is varied from 1V through 13V, while the initial value of DC input voltage E_0 is kept constant. The quadrature axis of this figure denotes the final value($\overline{I_\infty}$) of $\overline{I_0}$ of step responses. $T_{90,I}$ and T_{90T} indicate the time for the transients of $\overline{I_0}$ and $\overline{T_{orq}}$ to reach 90% of those final value, respectively, that is the settling times. $T_{c,I}$ and $T_{c,T}$ indicate the time for the transient of $\overline{I_0}$ and $\overline{T_{orq}}$ to reach 63.2% of those final values, respectively, that is the rising times. This figure shows that this type of machine has slight nonlinearity for the input voltage.

Fig. 18 shows the response times $T_{90,I}$ and $T_{c,I}$ for various time constants of damper circuit $T_{dw}(=L_3/R_3)$. In this figure, T_{dw0} is the time constant of the machine tested and T_{dw0} = 22.5 msec. In order to vary T_{dw}, we change R_3 so as to keep the cuppling factor k to be constant.

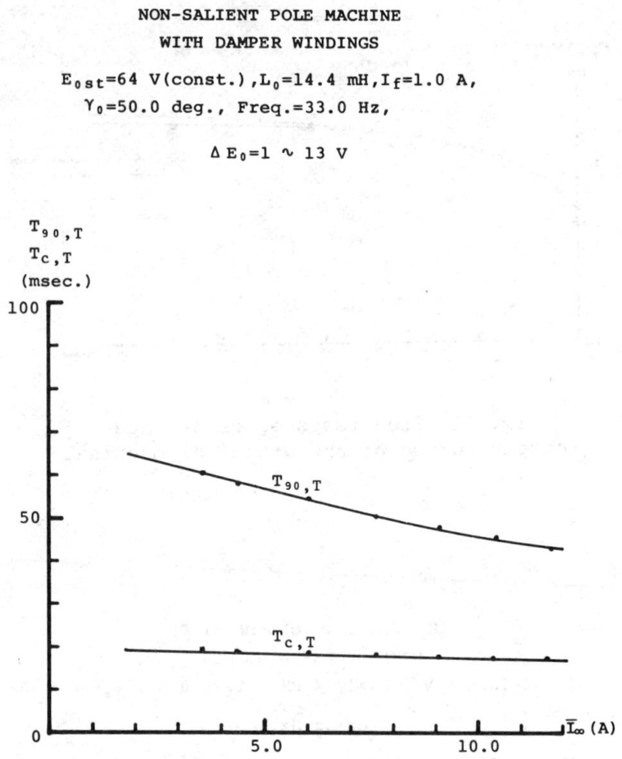

Fig. 17. The response times of average output torque of the practical machine.

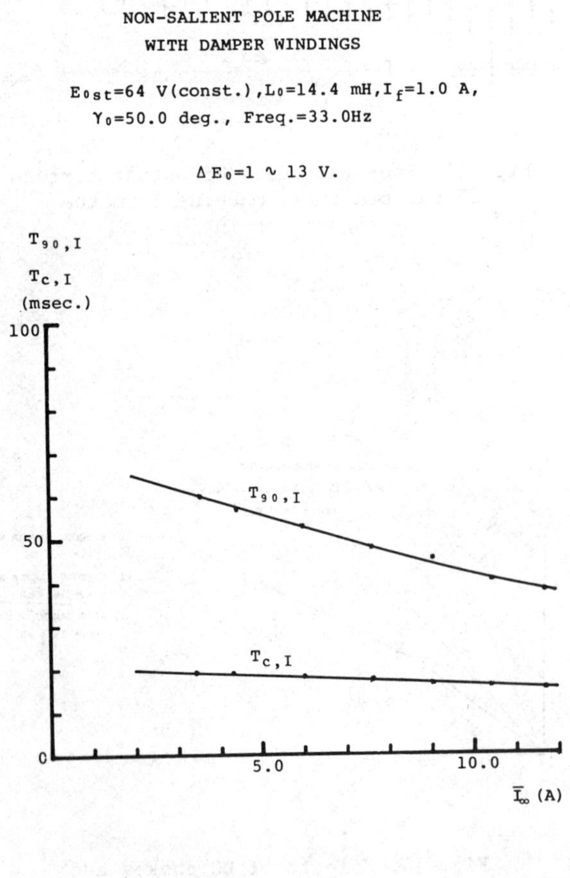

Fig. 16. The response times of average DC input current of the practical machine.

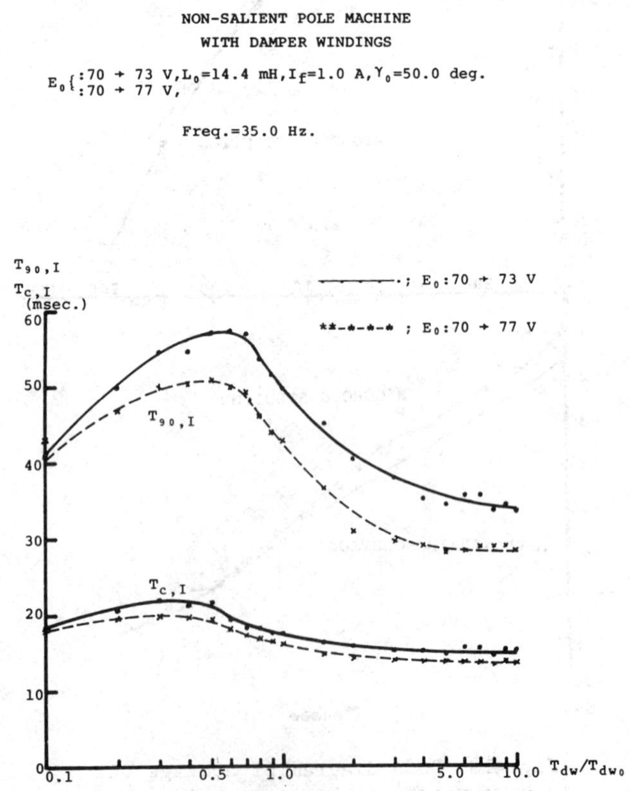

Fig. 18. The response times for various time constants of damper circuits.

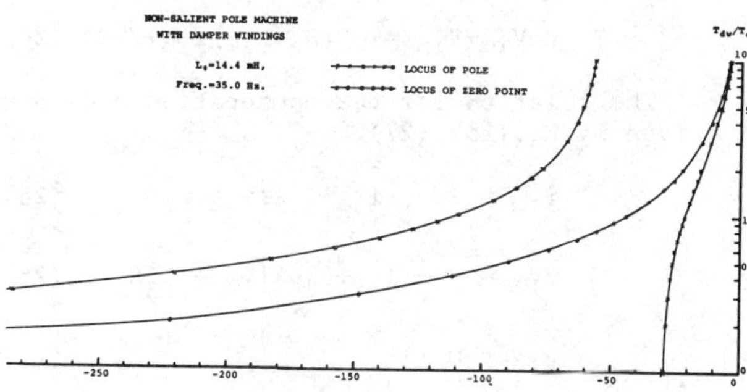

Fig. 19. The relation between the poles and zero point of the transfer function and the time constants of the damper circuit.

Fig. 19 shows the relation between the poles and zero point of the transfer function and the time constant of the damper windings. The pole nearer to origin is denoted as P_1 another pole is denoted as P_2 and the zero point is denoted as Z.

In the range where $T_{dw}/T_{dw0} = 0.1 \sim 0.4$, the response time is determined almost by P_1 since P_2 and Z are far from origin compared with P_1. When T_{dw}/T_{dw0} becomes reater than $0.6 \sim 0.7$, Z closes to P_1 rapidly and comes to cancel the effect of P_1. Then, P_2 dominates the response time of the system.

The above discussion can be also confirmed by the rigorous analysis.

CONCLUSION

Transfer function models of thyristor commutatorless motors have been proposed and the effects of motor parameters on the dynamic performances of commutatorless motors have been studied. Further works have to be done on the dynamic performances for large input voltage variations.

The optimal design from the view point of control will become possible through these studies.

REFERENCES

[1] W. J. Gibbs, Electric Machine Analysis Using Matrices, London, Sir Isaac Pitman & Sons, Ltd.(1962).

[2] F. Harashima, et al, Analysis of Inverter-induction Motor System by State Transition Method, Trans. I.E.E.of Japan, 89, 12, 27-54 (1969).

[3] F. Harashima, et al, Analysis of Thyristor Commutatorless Motors with Finite DC Reactors, Trans. I.E.E. of Japan, 94-B, 11 (1974).

[4] J. T. Tou, Modern Control Theory, New York, McGraw-Hill (1964).

APPENDIX

Co-ordinate System Used in the Analysis

In our analysis, we use the co-ordinate system based on two-axis theory. In Fig. 20, the vecotr representations of armature currents are shown. When thyristors Th_1 and Th_5 in Fig. 3 are conducting, the current flows from a-winding through b-winding. In this case, the direction of armature current vector is that of vector 1. When thyristor Th_6 turns on, the commutation starts. During the commutation period the current begins to flow through c-winding and the current which is flowing through b-winding is reduced to zero. In the meanwhile, the direction of armature current vector rotates from 1 to 2. When the current of b-winding becomes zero, the commutation is finished, and then, the direction of current vector remains in that of vector 2 until the next commutation starts. Therefore if we use this co-ordinate system. We can analyze the whole system only by considering the period from a starting time of a commutation through next starting time of it. This co-ordinate system is stationary to the armature windings except for the instant of start of the commutation. We call this co-ordinate system "α-β co-ordinates" and hereafter we analyze the commutatorless motor system using this α-β co-ordinates.

Circuit Equations of a Synchronous Motor Based on α-β Co-ordinate System

The α-β co-ordinate system is fixed to the armature windings except for the moments of commutation starting. Therefore, the circuit equations relative to this co-ordinate system are the same forms as those

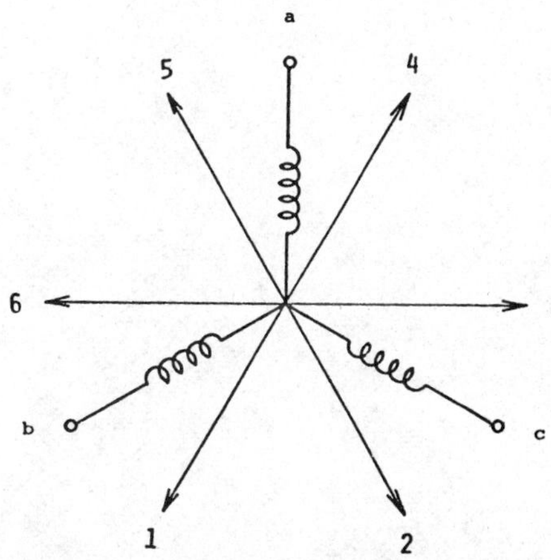

Fig. 20. Vector representation of armature currents.

relative to two-axis theory, and they are given as follows.

$$V = Z \cdot I \tag{19}$$

Eq.(19) is a fundamental equation for a commutatorless motor. Here, we show the circuit equations for each mode.

The relations common to each mode are as follows.

$$v_{3\alpha} = v_{3\beta} = 0 \tag{20}$$

$$i_{2\alpha} = I_f\cos\theta \ , \ i_{2\beta} = I_f\cos\theta \tag{21}$$

$$p\cdot i_{2\alpha} = -\dot{\theta} i_{2\beta} \ , \ p\cdot i_{2\beta} = \dot{\theta} i_{2\alpha} \tag{22}$$

$$E_0 - (R_0+L_0p)i_0 = V_{1a} - V_{1c} = v_{1a} - v_{1b} \tag{23}$$

$$(p = d/dt)$$

The relations for the single mode, are given by Eq.(3),(4) and (24).

$$\sqrt{3}V_{1\alpha}+V_{1\beta} = -2(R_0+L_0p)i_{1\beta}+\sqrt{2}E_0 \tag{24}$$

The relation for the commutation mode are given by Eq.(25) (27).

$$i_{1a} = i_0 \ , \ i_{1a}+i_{1b}+i_{1c} = 0 \tag{25}$$

$$v_{1\alpha} = -\frac{2}{3}(R_0+L_0p)i_{1\alpha} + \sqrt{\frac{2}{3}}E_0 \tag{26}$$

$$v_{1\beta} = 0 \tag{27}$$

From the above relations, the circuit equations for each mode are given by Eq.(9) and Eq.(11) for commutation mode and single mode, respectively.

OPTIMAL CONTROL OF VOLTAGE SOURCE INVERTERS SUPPLYING INDUCTION MOTORS

S. Halász

Department of Electrical Machines, Techn. Univ. 1111 Budapest, Hungary

ABSTRACT

The paper is concerned with the voltage source inverters ensuring the minimization of additional coil losses in the induction motor. Both simple and PWM inverters will be considered under non-modified control. Theoretically, the results are only valid if the skin effect in the motors is neglected. Nevertheless, the results as first approximation prove to be valid even for motors with relatively strong skin effect.

INTRODUCTION

Additional coil losses in inverter supplied asynchronous motors can be important enough to justify the application of the optimum theory. These losses can be minimized by the optimal control of three-phase inverters supplying the motors. Our analyses refer to inverters of unmodified control, hence, the motor can only be connected to the *dc* network with identical phase arrangement in the one-sixth period of the fundamental component. The asynchronous motor parameters are supposed to be constant and the inverter voltage drops are disregarded. The parameter variation as a result of skin effect will be reckoned with separately.

STATING THE PROBLEM

The inverter configuration and the locus of the stator flux Park-vector $\bar{\psi}$ as well as of its fundamental component $\bar{\psi}_1$ have been shown in Fig. 1. Losses are expected to have a minimum whenever the locus of the flux vector $\bar{\psi}$ has a six-sided symmetry over the entire period of the fundamental component. The number of order of harmonics due to the six-sided symmetry may be:

$$\nu = 1 + 6k, \qquad (1)$$

where $k = 0, \pm 1, \pm 2, \ldots$. The one-sixth period drawn by heavy line in Fig. 1 will always be discussed later. The time $t = 0$ will be at the middle of this $\frac{1}{6}$ period. The resultant vector of harmonic currents $\Delta\bar{i}$ is known (Ref. 1) to be closely approximated by:

$$\Delta\bar{i} = \bar{i} - \bar{i}_1 = \frac{\bar{\psi} - \bar{\psi}_1}{L'} = \frac{\Delta\bar{\psi}}{L'}, \qquad (2)$$

where $\bar{i}$ stator current Park-vector,

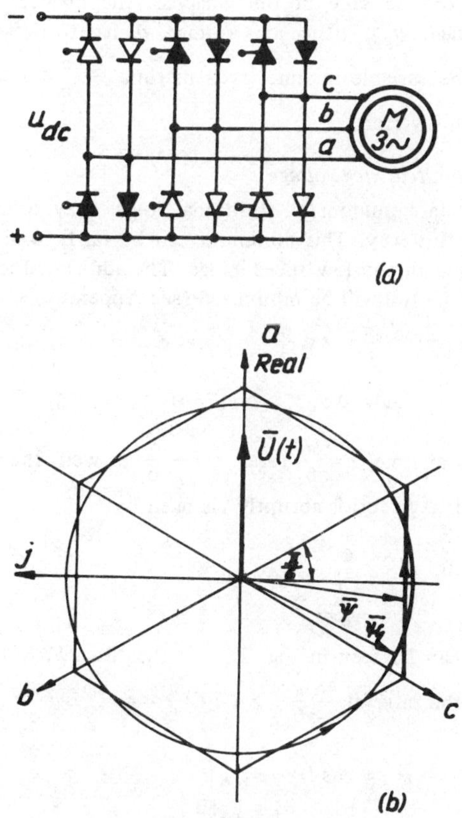

Fig. 1. Inverter configuration (a) and flux Park-vector (b)

$\bar{i}_1$ fundamental component of vector $\bar{i}$,

$\Delta\bar{\psi}$ resultant vector of harmonic fluxes,

L' stator transient inductance.

The additional losses are proportional to the square of rms value of current

$$\Delta I^2 = I^2 - I_1^2 = \frac{\Psi^2 - \Psi_1^2}{L'^2} = \frac{\Delta\Psi^2}{L'^2}. \qquad (3)$$

Be $U = \frac{2}{3} U_{dc}$,

where U_{dc} *dc* supply voltage,

t time in the angle of fundamental component.

The problem can be stated as follows:
Find $u(t)$ among the possible time functions ensuring the minimum of integral

$$\Delta\Psi^2 = \frac{3}{\pi}\int_{-\frac{\pi}{6}}^{\frac{\pi}{6}} |\Delta\bar{\psi}|^2\,dt \qquad (4)$$

with the constraint that the fundamental component of flux $\bar{\psi}$ has to remain constant (e.g. at rated value):

$$\bar{\Psi}_1 = \frac{3}{\pi}\int_{-\frac{\pi}{6}}^{\frac{\pi}{6}} \bar{\psi}e^{-jt}\,dt = \text{const.} \qquad (5)$$

SIMPLE INVERTER

For the sake of our analysis, the inverter supply voltage u_{dc}, hence also voltage u has to be assumed to be variable in time, even in time $-\frac{\pi}{6} \leqslant t \leqslant \frac{\pi}{6}$ in such inverters.

Unlimited dc voltage

The minimum of additional losses may be reached by this way. This minimum can be fairly well approached in the low speed range. The additional losses of the motor will be minimized (see Appendix) if in each interval $-\frac{\pi}{6} \leqslant t \leqslant \frac{\pi}{6}$

$$\Delta\bar{\psi} = \Delta\psi_y, \qquad \Delta\psi_x = 0 \qquad (6)$$

and at time $t = \frac{\pi}{6}$ (and $t = -\frac{\pi}{6}$ as well) the flux $\bar{\psi}$ and $\Delta\bar{\psi}$ too is abruptly changed by

$$\delta = \int_{-\frac{\pi}{6}+\epsilon}^{\frac{\pi}{6}-\epsilon} u\,dt, \qquad (7)$$

as can be seen in Fig. 2 ($\epsilon \to 0$). Be $\psi_1 = 1$, then in the interval $-\frac{\pi}{6} + \epsilon < t \leqslant \frac{\pi}{6} - \epsilon$:

$$\begin{aligned} \bar{u} &= \cos t, \\ \bar{\psi} &= \sin t - \left(\frac{\pi}{6} + \frac{\sqrt{3}}{4}\right)j, \\ \bar{\psi}_1 &= \sin t - \cos t \cdot j, \\ \Delta\bar{\psi} &= \left(\cos t - \frac{\pi}{6} - \frac{\sqrt{3}}{4}\right)j, \end{aligned} \qquad (8)$$

while $\delta = \frac{\pi}{6\sqrt{3}} - \frac{1}{4}$.

Thus:

$$\Delta\Psi^2 = \frac{3}{\pi}\int_{-\frac{\pi}{6}}^{\frac{\pi}{6}} \left(\cos t - \frac{\pi}{6} - \frac{\sqrt{3}}{4}\right)^2 dt = 0{,}001609. \qquad (9)$$

It is 74,8% of 0,00215. The later figure is proportional to the motorlosses supplied by a simple inverter of constant dc voltage (Ref. 1).

The voltage harmonics in the case of optimal control:

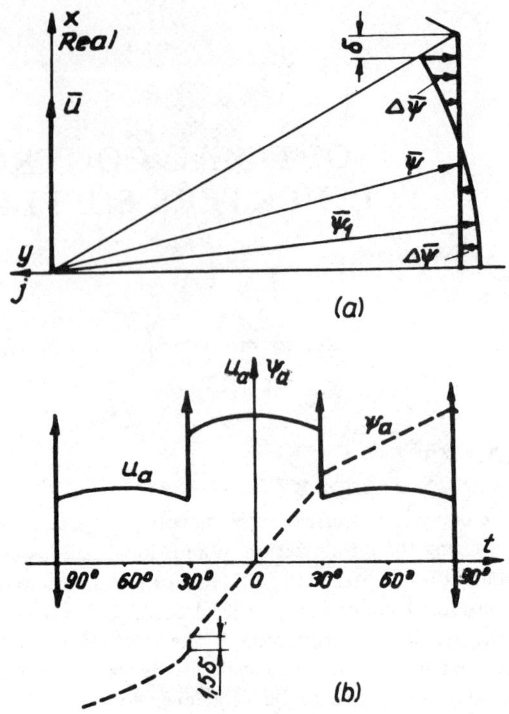

Fig. 2. Optimal control, voltage unlimited. Park-vectors (a) and phase voltage and flux (b)

$$\frac{U_\nu}{U_1} = \frac{3}{2\pi}\,\frac{\sin\left(\frac{\pi}{3} + k\pi\right)}{1 + 3k} + (-1)^k \cdot \left(\frac{1}{2} - \frac{3\sqrt{3}}{4\pi}\right), \qquad (10)$$

where $\nu = 1 + 6k$; $k = 0, \pm 1, \pm 2, \dots$. The results have been complied in Table 1, and compared with data relating to inverters of constant dc voltage and of dc voltages varying as $u = \frac{1}{\cos^2 t}$ (Ref. 2). The relative weight of harmonics for $k > 0$ is seen to have increased, namely the first and the second term in the last equation will have the same sign for these harmonics. In the cases of very high ν values the amplitude of the harmonics can be approximated by the second term in equation (10), i.e. by $(-1)^k \cdot 0{,}00865$.

The torque pulsations are obtained by:

$$\Delta m = \Delta\bar{\psi} \times \bar{i}_1 + \bar{\psi}_1 \times \Delta\bar{i} = \frac{\bar{\psi}_1 - \bar{i}_1 L'}{L'} \times \Delta\bar{\psi}. \qquad (11)$$

Assuming as an example $L' = 0{,}2$ and in no load $\bar{i}_1 = 0{,}3e^{-j\frac{\pi}{2}+jt}$

$$\Delta m = 4{,}7\sin t(\cos t - a), \qquad (12)$$

and in rated load $\bar{i}_1 = 1e^{-j\frac{\pi}{6}+jt}$

$$\Delta m = \left[5\sin t - \cos\left(t - \frac{\pi}{6}\right)\right](\cos - a), \qquad (13)$$

where $a = \frac{\pi}{6} + \frac{\sqrt{3}}{4}$.

Torque pulsations also included in Table 1.

Limited dc voltage

In the case of limited output voltage, the previously

Table 1. Control of simple inverters

		$u = \text{const}$	$u = \frac{1}{\cos^2 t}$	optimal	$U_1 = 0{,}25U_{max}$	$U_1 = 0{,}75U_{max}$
$\frac{U_\nu}{U_1}$	– 5	0,2	0,145	0,1202	0,1228	0,1405
	7	0,143	– 0,2035	– 0,1899	– 0,1929	– 0,1969
	– 11	0,091	– 0,08865	– 0,0038	– 0,0016	– 0,0045
	13	0,0769	0,1048	0,1456	0,1498	0,1453
$\Psi_1 = 1$	$\Psi(0)$	0,9497	0,9532	0,9567	0,9562	0,9540
	$\Psi(\frac{\pi}{6})$	1,097	1,101	1,1046	1,1041	1,1016
	$\Delta\Psi^2$	0,00215	0,00186	0,00161	0,00164	0,00179
Δm	no load	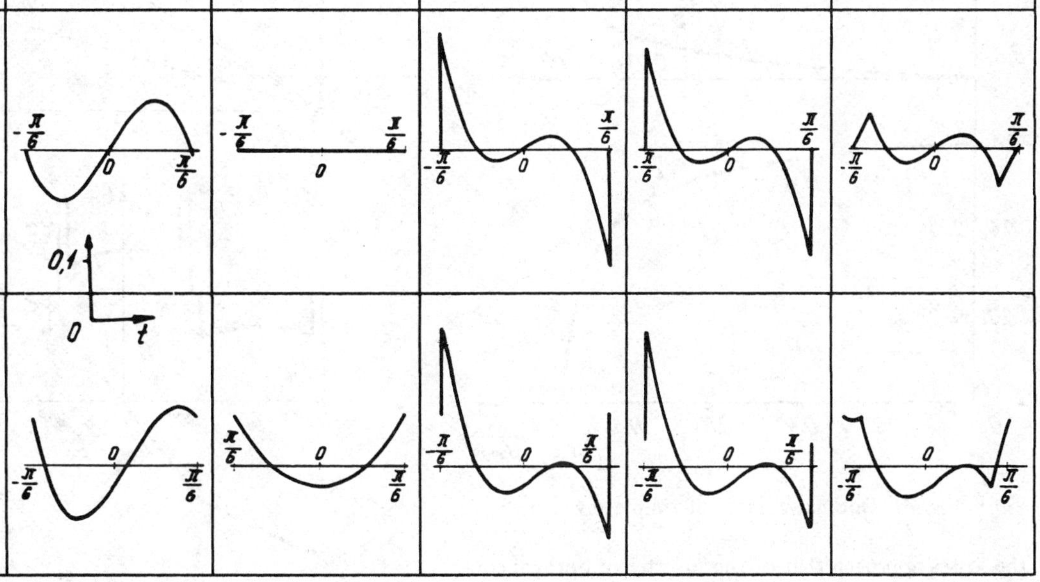				
	rated load					

deduced optimum conditions can only be approached so that in the range $-\tau \leqslant t \leqslant \tau$

$$u = U_1 \cos t, \tag{14}$$

while in the range $-\frac{\pi}{6} \leqslant t \leqslant -\tau$ and $\tau \leqslant t \leqslant \frac{\pi}{6}$

$$u = U_{max} = \frac{2}{3} U_{dc\,max}; \tag{15}$$

where $U_{dc\,max}$ is the maximum voltage of the *dc* circuit. U_1 and τ can be related to each other on the basis, that the fundamental component of the voltage given by (14) and (15) has to be U_1:

$$U_1 = \frac{1 - 2\sin\tau}{\frac{\pi}{3} - \tau - 0{,}5\sin 2\tau} U_{max}. \tag{16}$$

Relationships for $U_1 = 0{,}25U_{max}$ and $U_1 = 0{,}75U_{max}$ have been plotted in Fig. 3, while data relating to this two cases are included in Table 1.

The fluxes in the interval $-\tau \leqslant t \leqslant \tau$ are:

$$\bar{\psi} = \sin t - aj,$$
$$\bar{\psi}_1 = \sin t - \cos t \cdot j,$$
$$\Delta\bar{\psi} = (\cos t - a)j, \tag{17}$$
$$a = \left[\sin\tau + \frac{U_{max}}{U_1}\left(\frac{\pi}{6} - \tau\right)\right]\sqrt{3}$$

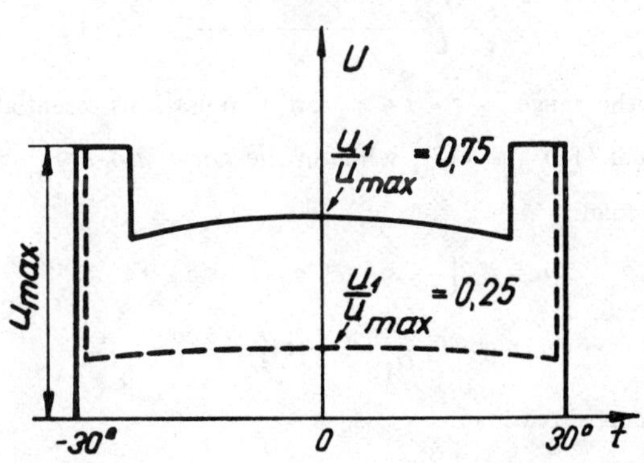

Fig. 3. Voltage time function for simple inverter

and elsewhere in the sixth of the period:

$$\bar{\psi} = \sin\tau + \frac{U_{max}}{U_1}(t-\tau) - aj,$$
$$\Delta\bar{\psi} = \sin\tau - \sin t + \frac{U_{max}}{U_1}(t-\tau) + (\cos t - a)j. \qquad (18)$$

Thus:

$$\Delta\Psi^2 = \frac{6}{\pi}\int_0^{\tau} |\Delta\bar{\psi}|^2\,dt + \frac{6}{\pi}\int_{\tau}^{\frac{\pi}{6}} |\Delta\bar{\psi}|^2\,dt. \qquad (19)$$

This result has been compared with the losses of obtained by the application of an inverter of constant *dc* voltage $\left(P = \frac{\Delta\Psi^2}{0{,}00215}\right)$ in Fig. 4. For $\frac{U_1}{U_{max}} < 0{,}4$

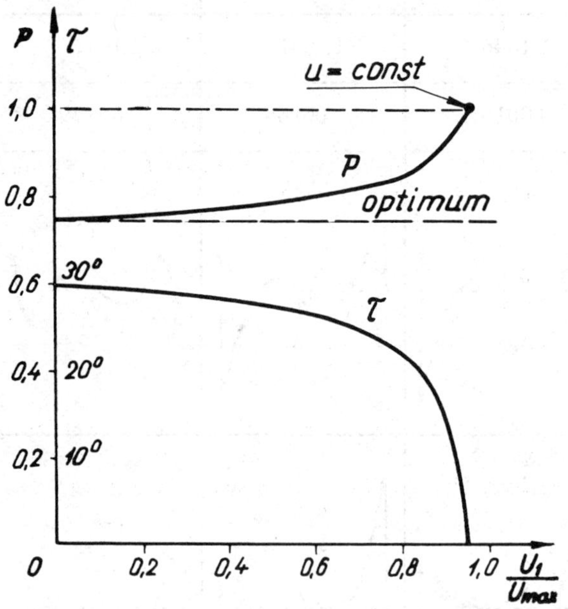

Fig. 4. Optimal control of simple inverter

the losses approach that of the inverter of optimal control. In general, results for limited voltage inverters can be stated to be in the midway of unlimited and optimized inverters and of inverters of constant *dc* voltage.

Now, voltage harmonics depend also on U_1:

$$\frac{U_\nu}{U_1} = \frac{3}{\pi}\left[\frac{\sin(\nu-1)\tau}{\nu-1} + \frac{\sin(\nu+1)\tau}{\nu+1}\right] + $$
$$+ \frac{6}{\pi}\frac{U_{max}}{U_1}\frac{\sin\frac{\pi}{6}\nu - \sin\nu t}{\nu}. \qquad (20)$$

In the range $-\tau \leq t \leq \tau$ torque pulsations essentially equal (11) and (12), while in the range $\tau \leq t \leq \frac{\pi}{6}$, at no load:

$$\Delta m = 4{,}7\Big[-a\sin t + \cos t \cdot \sin\tau + $$
$$+ \cos t\,\frac{U_{max}}{U_1}(t-\tau)\Big], \qquad (21)$$

and at rated load:

$$\Delta m = \left[5\sin t - \cos\left(t - \frac{\pi}{6}\right)\right] \cdot (\cos t - a) + $$
$$+ \left[5\cos t - \sin\left(t-\frac{\pi}{6}\right)\right] \cdot $$
$$\cdot \left[\sin\tau - \sin t + \frac{U_{max}}{U_1}(t-\tau)\right]. \qquad (22)$$

The torque pulsations at points $t = \pm\frac{\pi}{6}$ and $t = 0$ are independent of L' for a given current and can be influenced by control only in a small extent. At the same time the effect is always controversial: if the torque pulsation can somewhat be reduced at $t = 0$, then it will be accentuated by the same extent at $t = \pm\frac{\pi}{6}$. Thus, control can only reduce positive or negative areas, i.e. speed deviations due to torque pulsations. Table 1 shows

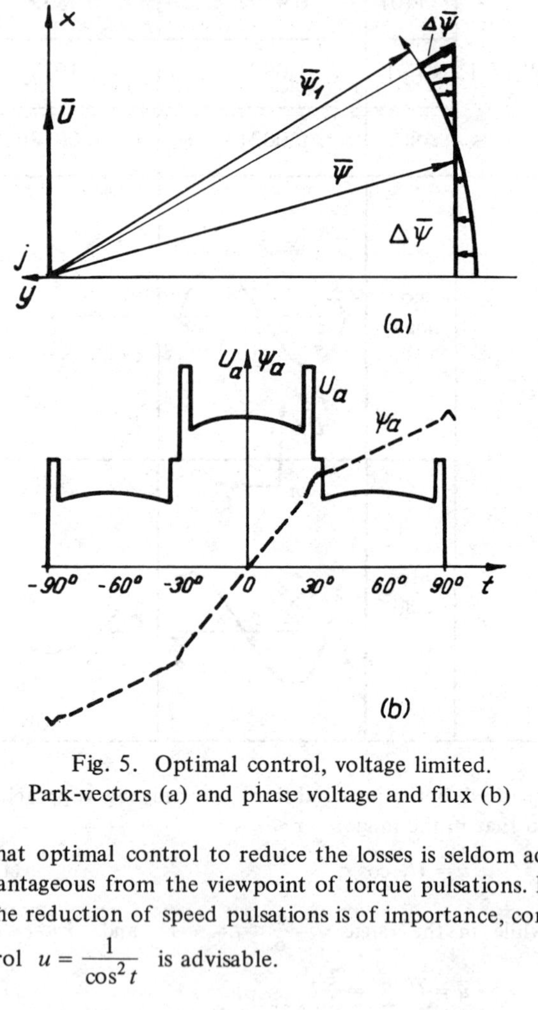

Fig. 5. Optimal control, voltage limited.
Park-vectors (a) and phase voltage and flux (b)

that optimal control to reduce the losses is seldom advantageous from the viewpoint of torque pulsations. If the reduction of speed pulsations is of importance, control $u = \frac{1}{\cos^2 t}$ is advisable.

PULSE WIDTH MODULATED (PWM) INVERTERS

The motor supply is varied by means of the number and width of pulses, while *dc* voltage is constant. Inverter control can be optimized by adequate distribution of pulses in time and by changing their width (Fig. 6).

The implementation of the results obtained for simple inverters is at present rather difficult, but they help to optimize the PWM inverters by providing important instructions. This is of practical importance, since optimal control is within the technical possibility for PWM inverters. The following conclusions can be deduced from the preceding chapters:

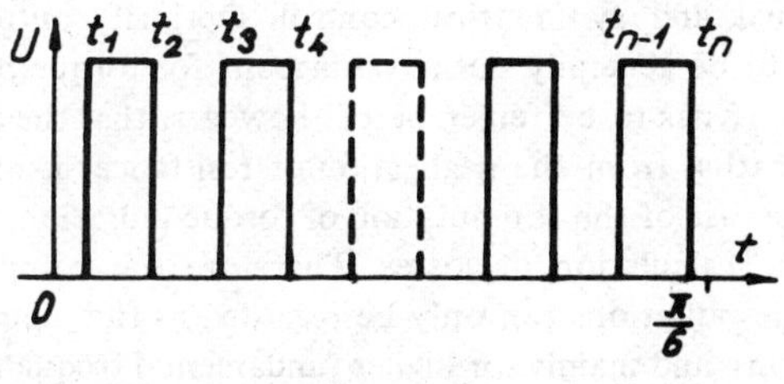

Fig. 6. Voltage time function for PWM inverter

1. The pulse density distribution in time has to correspond to Fig. 3 for a given U_1, i.e. the pulse density is higher at the ends of the one-sixth period than around its middle section.
2. The last pulse has to satisfy inequality $t_{n-1} < \tau$, where τ is determined by equation (16). For infinite number of pulses $t_{n-1} \to \tau$.
3. In conformity with Fig. 4, the last pulse width increases with U_1, hence with the speed. Therefore the locations of pulses are shifting towards the middle of $\frac{1}{6}$ periods. This is especially true for $U_1 > 0{,}8$.
4. Time t_n of the last pulse reaches $\frac{\pi}{6}$ rather soon with increasing U_1 and pulse number.

The pulse distributions minimizing additional losses can be computed by several methods.

The best from convergency aspects, proved to be in our praxis one suggested by Dr. I. Rácz. The function to be minimized in the method is consisting in minimizing the function

$$F = \Psi^2 - \Psi_1^2 + \lambda(\Psi_1^2 - 1). \qquad (23)$$

F depending on $t_1, t_2, \ldots, t_n, \lambda$. Introducing notations:

$$X = \begin{bmatrix} t_1 \\ t_2 \\ \vdots \\ t_\lambda \\ \lambda \end{bmatrix}; \quad f = \begin{bmatrix} \frac{\partial F}{\partial t_1} \\ \vdots \\ \frac{\partial F}{\partial \lambda} \end{bmatrix}; \quad D = \begin{bmatrix} \frac{\partial^2 F}{\partial t_1^2} & \cdots & \frac{\partial^2 F}{\partial t_1 \partial \lambda} \\ \vdots & & \vdots \\ \frac{\partial^2 F}{\partial \lambda \partial t_1} & \cdots & \frac{\partial^2 F}{\partial \lambda^2} \end{bmatrix};$$

$$\Psi_1 = \frac{6}{\pi} \cdot \frac{U}{U_1} \sum_1^n \sin t_i (-1)^i, \qquad (24)$$

$$\frac{\Delta\Psi^2}{\Psi_1^2} = -1 + \left(\frac{U}{U_1}\right)^2 \left(\frac{2}{\pi} T^3 + 3T^2\right) +$$

$$+ \frac{6}{\pi}\left(\frac{U}{U_1}\right)^2 \left[(t_2 - t_1)^2 (t_3 - t_2) + \ldots + T^2\left(\frac{\pi}{6} - t_n\right)\right],$$

$$T = \sum_1^n (-1)^i t_i.$$

The computation starts from a given U_1 and an assumed initial value X. First the deviation

$$\Delta x = -D^{-1} f \qquad (25)$$

is determined.

Computing $X + \Delta x$ and the computation will be repeated with new values until

$$f^T \cdot f < \epsilon_0.$$

ϵ_0 is infinitesimal (e.g. $\epsilon_0 = 10^{-23}$). The developed program takes into consideration other conditions, as well e.g. coincidence of certain pulses with the increase of U_1, or that $t_n = \frac{\pi}{6}$ is feasible. Fig. 7 presents loss

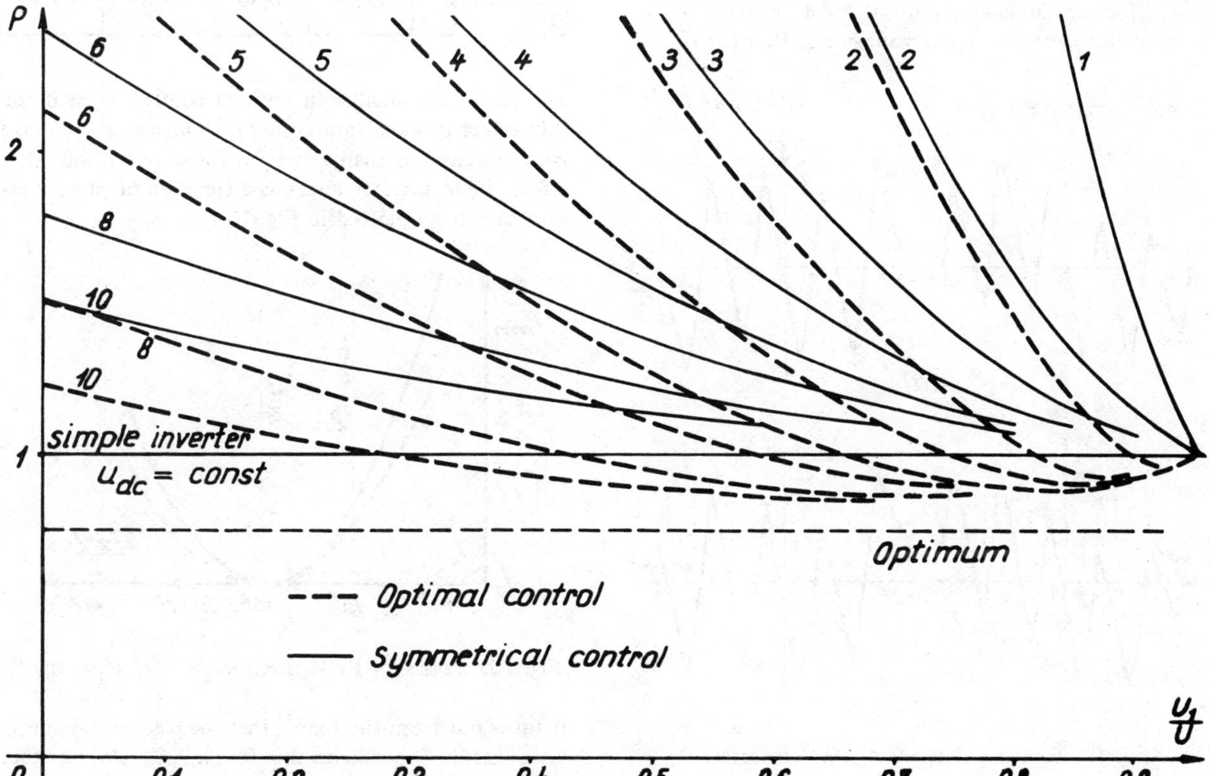

Fig. 7. Motor losses, PWM inverter

P obtained by this method for some pulse numbers n. For the sake of comparison, losses for symmetrical distribution have been plotted as well. It may be seen that the losses can be reduced by about 20% by optimal control, or keeping the losses unchanged the pulse number can be decreased by 2 or 3 by applying optimal control. It is interesting to note that the losses may be even smaller than for simple inverters of constant voltage. The time functions of certain variables are seen in Fig. 8 for pulse number $n = 5$.

As an example, Fig. 9 shows torque pulsations under

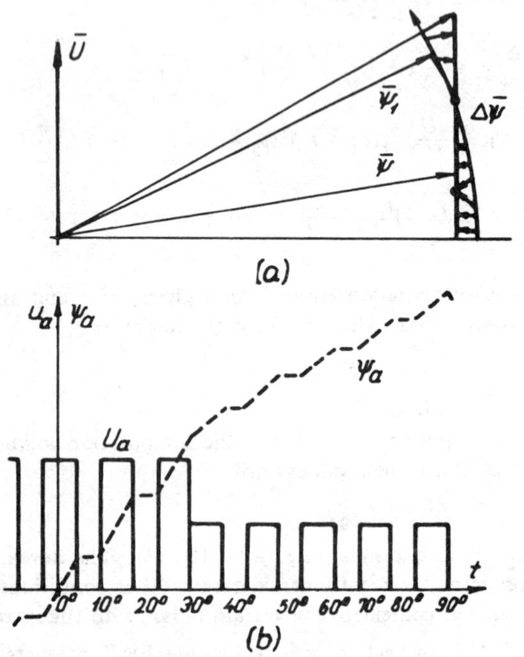

Fig. 8. Optimal control of PWM inverter. Park-vectors (a), phase voltage and flux (b)

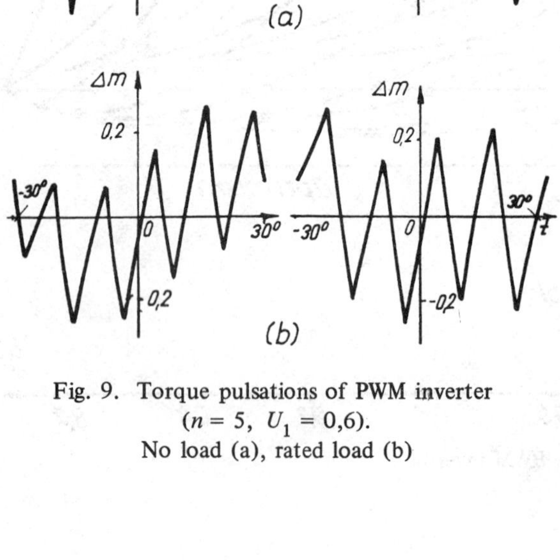

Fig. 9. Torque pulsations of PWM inverter ($n = 5$, $U_1 = 0{,}6$). No load (a), rated load (b)

optimal and symmetrical control. Optimal control is seen to be generally not advantageous for torque pulsations. It has to be remembered, however, that the error originating from the negligation of resistances is higher in the case of the computation of torque pulsation, than in the calculation of losses. Therefore the computed torque pulsations can only be regarded as first approximations, and mainly for higher fundamental frequencies.

SKIN EFFECT

The numerical analysis can not be carried out in general form in the case of the skin effect taken into account. Therefore, the computation to check the theoretical results have been made for an asynchronous motor selected as an example (Type: VZ 160 M4, 14 kW, 50 Hz, 1460 1/min, 190 V, connection delta). The method of computation and the motor data can be found in (Ref. 2., 3.). It is necessary to know the harmonics of voltage in the case of calculation, so for the simple inverters see the equations (10) and (20). In case of optimal control, computed rotor losses are seen in Table 2. The losses included in Table 2 are related to the rotor winding losses of rated operation.

Table 2. Skin effect, $\Psi_1 = 1$

	$f = 10$ Hz		$f = 50$ Hz	
		skin		skin
Optimal	0,071	0,29	0,052	0,75
U = const	0,099	0,33	0,069	0,84

The losses are smaller in optimal control, even if the skin effect is taken into account, though the advantage is somewhat less than it was in the case without skin effect. PWM inverter losses as a function of pulse location have been plotted in Fig. 10.

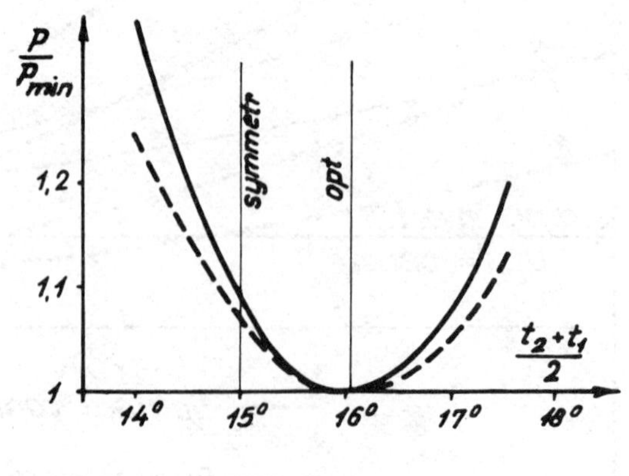

Fig. 10. Losses of PWM inverter ($n = 2$, $U_1 = 0{,}77$)

It turns out from the figure, that the place of minimum losses hardly depends on skin for $n = 2$. On the other hand, it is interesting to notice, that the value of losses is sensitive in a smaller extent to the pulse displacement originated from the control for optimal controlled inverter.

Further analyses showed that the minimum of losses in general does not differ considerable between the cases calculated with and without skin effect.

CONCLUSIONS

The relationships deduced for simple inverter for the case of optimal control can be applied to PWM inverters. The winding losses of the motor may be reduced by about 20%. Unfortunately, however, optimal control is not always favourable for torque pulsations. The numerical computations taking into account the skin effect verify that the relations deduced hold approximately for motors having considerable skin effect.

APPENDIX

Assuming $\Psi_1 = 1 = \text{const}$, (4) can be written in p.u. system:

$$\Delta\Psi^2 = \frac{6}{\pi}\int_0^{\frac{\pi}{6}} [(a - \cos t)^2 + (\psi_x - \sin t)^2\, dt$$

since $\psi_1 = \sin t - \cos t \cdot j$ and $a = -\psi_y = \cos t$

Thus:

$$\Delta\Psi^2 = a^2 - \frac{6}{\pi} a + 0{,}5 + \frac{3\sqrt{3}}{4\pi} + \Delta\Psi_x^2. \quad \text{(A1)}$$

Adequately modifying voltage u $\Delta\Psi_x^2 = 0$ may be achieved, hence:

$$\Delta\Psi^2 = a^2 - \frac{6}{\pi} a + 0{,}5 + \frac{3\sqrt{3}}{4\pi}.$$

From $\Psi_1 = 1$, $a = \frac{\pi}{6} + \frac{\sqrt{3}}{4}$, and $\Delta\Psi^2 = 0{,}001609$.

Differentiating equation (A1):

$$d\Delta\Psi^2 = \left(2a - \frac{6}{\pi}\right) da + d\Delta\Psi_x^2.$$

As a result of the infinitesimal change da the deviation of the flux at $t = \frac{\pi}{6}$:

$$d\Psi\left(\frac{\pi}{6}\right) = \frac{da}{\sqrt{3}} = d\left(\int_0^{\frac{\pi}{6}} u\, dt\right)$$

and that of the fundamental component of the inverter output voltage:

$$dU_1 = d\left(\frac{6}{\pi}\int_0^{\frac{\pi}{6}} u \cos t\, dt\right).$$

After integration

$$dU_1 = \frac{6}{\pi}\frac{da}{\sqrt{3}} \cos\frac{\pi}{6} = \frac{3}{\pi} da.$$

In order to keep the flux Ψ_1, that is, the voltage U_1 constant, the time function of $u(t)$ has to be changed, but the restriction by which the new $\Psi\left(\frac{\pi}{6}\right) = \text{const}$ has to be satisfied.

By changing voltage $u(t)$ with a Dirac impulses at times τ_1 and $\tau_2 (\tau_1 < \tau_2)$ the deviation of the fundamental component

$$-du_1 = -\frac{3}{\pi} da = \frac{6}{\pi}\delta_1(\cos\tau_1 - \cos\tau_2).$$

As a result of the first voltage impulse the flux change $d\Delta\psi_x = \delta_1$ in the interval $\tau_1 \leq t \leq \tau_2$ and the square of its rms value is:

$$d\Delta\Psi_x^2 = \frac{3}{2\pi}\left(\frac{da}{\cos\tau_1 - \cos\tau_2}\right)^2 (\tau_2 - \tau_1).$$

Its minimum is at $\tau_1 = 0{,}17$ and $\tau_2 = \frac{\pi}{6}$.

The expression of the square of the total flux deviation is:

$$d\Delta\Psi^2 = \left(2a - \frac{6}{\pi}\right) da + \frac{3}{2\pi}\frac{\frac{\pi}{6} - 0{,}17}{\left(\cos 0{,}17 - \cos\frac{\pi}{6}\right)^2}(da)^2.$$

This function has a minimum if $da = -0{,}000143$ and $d\Delta\Psi^2 = -0{,}24 \cdot 10^{-6}$. Since the deviation da and $d\Delta\Psi^2$ very small, they can be neglected in relation to the value, obtained with the assumption $\Delta\psi_x = 0$.

REFERENCES

(1) I. Rácz, Betrachtungen zu Oberwellenproblemen an Asynchronmotoren bei Stromrichterspeisung, *Periodica Polytechnika,* Electrical Engineering Vol. 11. No. 1-2. (1967), 29-57.

(2) S. Halász – I. Schmidt – S. Wahsh, Additional Losses of Asynchronous Motors Supplied by Forced-Commutation Inverter, International Conference Electrical Machines, Paper I 33-1, Vienna, (1976).

(3) S. Halász – I. Schmidt, Kényszerkommutációs inverterről táplált aszinkron motoros hajtások többletveszteségei, *Elektrotechnika,* Hungary, No. 4 (1976), 121-129.

EXPERIMENTAL AND THEORETICAL INVESTIGATIONS CONCERNING THE DESIGN OF INDUCTION MACHINES, ESPECIALLY MACHINES WITH HIGH POWER DENSITY WITH RESPECT TO THE INVERTER SUPPLY AND SPEED REGULATION*

P. K. Sattler and G. Balsliemke

Institut für Elektrische Maschinen der RWTH Aachen, BRD

SUMMARY

The special reasons concerning inverter fed induction machine design are described. The design may be quite different from the design of a network fed induction machine. Small numbers of slots with regard to high power density are possible. The ripple torque arrising from rectangular current blocks gives high mechanical stresses for certain speeds. Test models with normal laminated and solid rotor fed by inverter were measured, also equation for the steady state and dynamic behaviour were established. Test results and calculation are in good agreement.

1. INFLUENCES OF THE INVERTER SUPPLY

Variable speed drives using induction motors can be optimally realized by varying the supply frequency. The speed can be adjusted practically without loss in a wide area. The inverters chosen are usually performed as cyclo-inverters or as d.c. intermediate inverters with constant voltage or constant current in the d.c. link circuit. The frequency range goes from 2 s^{-1} to 200 s^{-1}. Higher frequencies than those conventionally used allow to develop high speed machines with a high power density. If induction machines are running at variable frequencies, different types of relations between flux and rotor frequency can be adjusted at the control system. The linked stator flux ψ_S, the air gap flux ψ_h or the linked rotor flux ψ_R are controlled on a constant value over the full range between no load and the highest load. Each design demands a separate control effort. Also the dynamic behaviour essentially depends on the chosen control system. For the experimental work described in this paper an inverter was selected, whose d.c. intermediate circuit had a great inductance, and therefore constant d.c. current was impressed. The a.c.-current of the inverter is controlled in function of the frequency of the rotor with the aim to keep the linked stator flux on its rated value for every load under steady state conditions. This results in a simple general design although for this arrangement some typical disadvantages occur (1.3). The voltage in the d.c. link is variable, the current is smoothed through the smoothing inductance. The machine impedances are placed in the commutation circuit of the artifical commutated inverter with commutation capacitors and series diodes. The stator current wave form is nearly rectangular, it is made of 120 deg. current blocks of each current

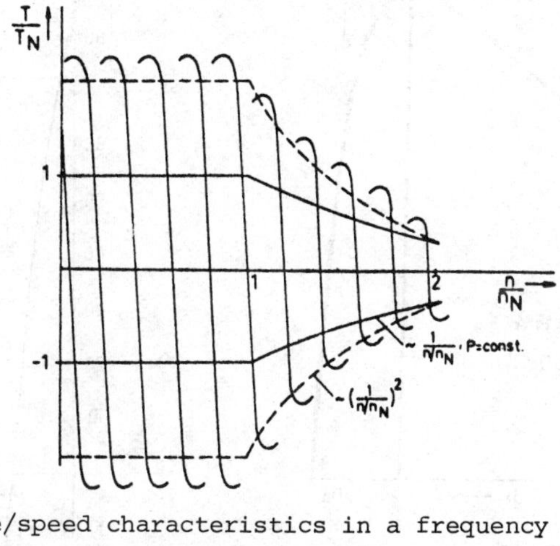

Fig. 1: Typical torque/speed characteristics in a frequency controlled induction machine

* The authors thank the Deutschen Forschungsgemeinschaft (DFG) for the financial support by this paper

direction with the amplitude of the d.c. link current, which are impressed to the machine. The voltage at the terminals gets free, it is nearly sinusoidal but contains commutation spikes. The current in the d.c. link circuit cannot change its direction because of the converter connected to the network. If the induction machine acts as a generator the voltage changes in the intermediate circuit. Therefore motor- and generator-operation is possible with only one converter connected to the network and with only one direction of current. The machine runs up to the rated point with constant flux, above the rated speed the voltage stays constant. This results in a weakening of the field analogical to the direct current motor.

The torque, which is generated by the machine, decreases square to the flux, shown in Fig. 1.

1.1 THE DESIGN OF A MACHINE FOR CONVENTIONAL OPERATION ON THE NETWORK COMPARED WITH THE DESIGN OF INVERTER FED MACHINES

Conventional machines on the network of constant voltage and frequency always have certain synchronous speeds with respect to their number of poles. The run up of those machines occurs while they are supplied by constant frequency, that means the value of the rotor frequency varies between the network frequency to zero. An inverter fed induction machine running up can be made at a constant rotor frequency and with the stator frequency following speed. Thus the torque can be kept constant during the running up. We see that for the inverter drive there exists a run up process at variable frequency in a region of the torque characteristic, where torque of the fundamental component of the air gap field is relatively high. When designing a network fed induction machine it is very important to keep small parasitic synchronous and asynchronous torques which appear in the region of greater rotor frequencies. Those torques, which are mainly caused by slot harmonics, influence the run up process on the network very intensively, while they are superposed to the relatively small torque of the fundamental component of the field. The choice of the number of slots for stator and rotor and their combination chiefly influence the parasitic torques. Therefore the number of slots per pole and phase in the stator should be as high as possible ($q \geq 3$). The relation between numbers of slots in the stator and rotor determines the behaviour of the machine in the field of high rotor frequency, but it has no significant influence in the field of low rotor frequency. This region of low rotor frequency is important at normal machine operation ($f_2 < f_{2br}$), whereas the region of high rotor frequency only has significance for the start of the motor on the network. If an induction machine is to gain as high power density as possible it must be designed for a suitably high speed. For this reason a small rotor diameter is necessary to reduce the mechanical stresses. Thus the number of slots per pole and phase must be chosen suitably low, e.g. $q=2$. The 4-pol. machine under consideration in this paper has in the stator 24 and in the rotor 18 slots, the stator winding is pitched (5/6). Thereby great field harmonics appear which are caused by the slotting. This kind of machine has on the network very bad run up characteristics - especially if they are designed with a rotor without skin effect. The machine under consideration here shows such a great synchronous torque at standstill, when starting from the network of constant frequency, that a start depending of the rotor position can be prevented. Our experiments showed, that there is no difficulty in starting the inverter fed motor, which is supplied with a suitably low stator frequency for a start at its highest torque. Then the stator frequency is nearly equal to the rotor frequency at breakdown torque. In this case it was not possible to establish any influence of the parasitic torques (Fig. 3).

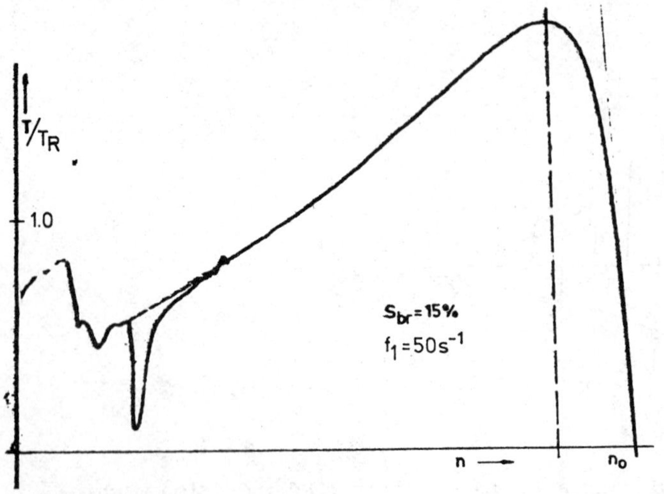

Fig. 2: Torque/speed characteristics for start on the network (without effects of transient conditions)

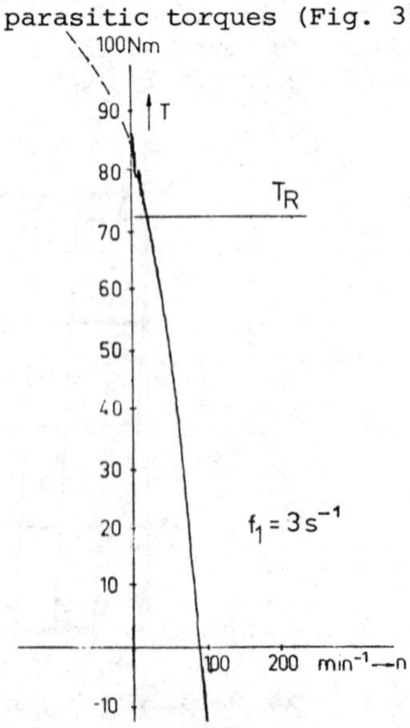

Fig. 3: Torque/speed characteristics for converter feeding in the starting area

The torque speed characteristics of the induction machine starting on the network of constant frequency shows a considerable divergence from the behaviour those without parasitic torques. This is shown in Fig. 2. Fig. 3 shows the average of instantaneous torque versus speed in the starting area for the inverter fed motor supplied by low frequencies. The effects of the parasitic torques caused by the field harmonics cannot be recognized.

It is also possible to choose the numbers of slots of stator and rotor for an inverter fed machine, that a satisfactory torque speed characteristic for a conventional network-fed machine cannot be achieved. This result of our researches is very advantageous for the design of high speed motors with small diameters.

Since the machine on the inverter is fed by nonsinusoidal voltages and currents, special phenomena are produced by current harmonics. In view of the current harmonics the machine works under short circuit conditions (Chapter 4.). The current harmonics scarcely increase the average torque. This means an increase of the losses and additional heating in the rotor bars. Therefore a low skin effect in the rotor design is to be strived for.

1.2 INFLUENCE OF THE INVERTER TYPE

A pulse modulated inverter is an ideal device for supplying the machine with a sinusoidal current. Through the switching operations in the converter considerable voltage harmonics occur, which lead to additional iron losses. In this case, to bring the current harmonics down, a higher leakage reactance of the induction motor is desired. The type of the "Impression" is most important for d.c. intermediate inverters, which deliver to the motor rectangular waves of the currents or voltages.

Rectangular voltage wave blocks are switched on the machine during voltage impression. The current depends on the impedances of the particular operation condition. A raised leakage reactance and a low no load current is desired to limit the current harmonics. Since the commutation on the d.c. side occurs it is not influenced by the leakage reactance of the motor. The flux control takes place at U/f=const. characteristics with correction in the lower frequency area. For the rotor design see Chapter 1.1.

Current blocks are impressed from the d.c. link through the inverter to the machine by current impression. The current wave of the current impressed type is little influenced by the commutation of the thyristors of the inverter. The reactance of the machine is part of the commutation circuit. Therefore the transient reactance of the induction machine is important. The transient reactance L_S is the total leakage reactance of the machine and given by $L_S = \sigma L_1$. A low leakage is desired in order to achieve favourable conditions for the commutation circuit. To obtain a stable steady state no load speed at high frequencies (particularly during the weakening of the field) a relatively high no load current is desirable. This leads to an unusual design of the machine.

1.3 THE INVERTER

For the experimental work an inverter with an impressed current was designed and built. For further specification see references (4,5,6,7,8,9).

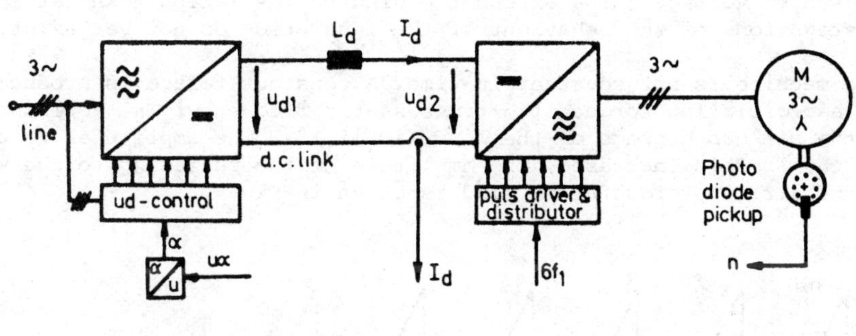

Fig. 4: Schematic block diagramm of the d.c. link with current fed inverter

During the commutation the equation

$$i = I_d\ (1-\cos \omega_e t) \qquad 0 \le t \le \frac{T_e}{4} \tag{1}$$

gives the current versus time.

ω_e marks the characteristic frequency of the commutation swinging circuit.

$$\omega_e = \frac{1}{\sqrt{C_K \; 2 \; L_S}} = \frac{2\pi}{T_e}; \qquad \begin{array}{l} C_K = \text{commutation capacity} \\ L_S = \text{the total leakage inductance of one phase} \end{array} \tag{2}$$

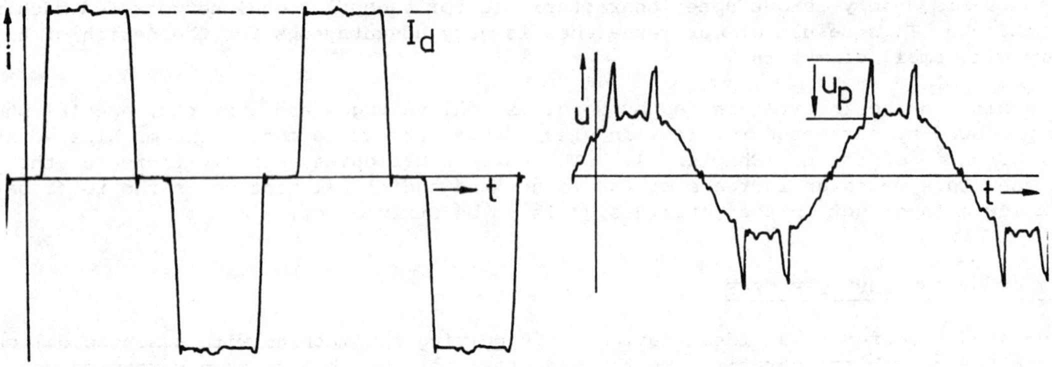

Fig. 5: Phase current and voltage

The air gap flux in the machine is nearly sinusoidal, it induces the fundamental harmonic of the stator voltage. The commutation spikes of the stator voltage are caused by current alteration of the leakage inductances. During the commutation an energy exchange takes place between the commutation capacitance and the leakage inductance of the machine.

The voltage can be evaluated by

$$u_p = L_S \frac{d_i}{dt} \tag{3}$$

$$u_p = L_S \cdot I_d \, \omega_e \cdot \sin \omega_e t \tag{4}$$

$$u_{p\,max} = I_d \cdot L_S \cdot \frac{1}{\sqrt{C_K \; 2L_S}} = I_d \sqrt{\frac{L_S}{2C_K}} \tag{5}$$

These voltage spikes are independent of the stator frequency and only dependent of the load. The maximal voltage of a phase becomes

$$U_{max} = \sqrt{2} \cdot U_R + I_d \sqrt{\frac{L_S}{2C_K}} \qquad U_R = \text{rated voltage} \tag{6}$$

The stress on the winding insulation caused by this voltage is normal with respect to the fundamental voltage wave. The spikes however have a great deal of high frequencies which depends on commutation time. The amplitude of the spikes, which is to be added to the rated voltage (Fig. 5) can reach the maximal value of the rated voltage. Measured and calculated values are in a good agreement. The sudden voltage jumps stress the winding insulation to a larger extent. At present specific investigations of the behaviour of the insulation do not yet exist.

The air gap torque of the machine is not constant in time. A constant torque is produced, which is superimposed by a oscillating torque of 6-times stator frequency. The amplitude of the pulsating torque on the air gap depends on the load (10,11,12). The amplitudes reach a maximum at no load. When the load is increased, the amplitude of the pulsating torque will descrease. The curve of the air gap torque at no load is given in Fig. 6.

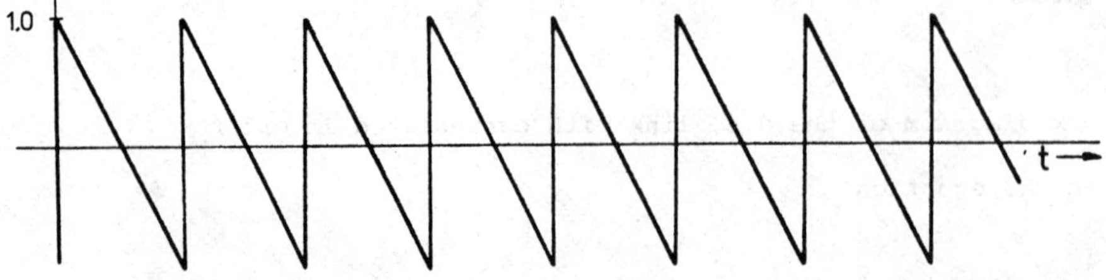

Fig. 6: The torque of the machine at no load

The analysis of this torque by FOURIER's method leads to the following series

$$T = T_A \cdot \frac{2}{\pi} \left(-\sin\omega t + \frac{1}{2}\sin 2\omega t - \frac{1}{3}\sin 3\omega t + \frac{1}{4}\sin 4\omega t \ldots\right) \qquad (7)$$

whereby is $\omega = 6\omega_1$; $\frac{\omega_1}{2\pi}$ = stator frequency.

All even and odd harmonic numbers of the 6-times supply frequencies are existent. The effect influences the whole system, which consists of inverter fed machine shaft and load. Resonances with high stress in the mechanical parts appear, they are produced by any harmonic of the motor torque. Fig. 7 shows measurements at no load different supply frequencies and speeds. Here the mechanical resonance of the system is about 176 s^{-1} (Fig. 15, Chapter 5.1).

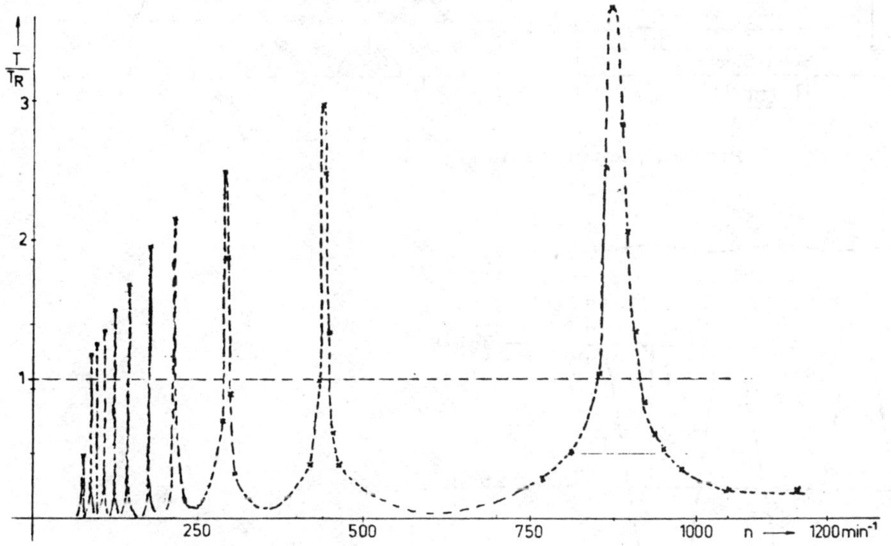

Fig. 7: The measured resonance rises, torque versus speed

These resonance rises were measured in a system as shown in Fig. 15 and described in Chapter 5.1

Under no load conditions the speed was varied by the stator frequency. The shaft torque was measured with a special measuring device, developed by our institute. The mechanical resonance frequency is constant, the fundamental wave and the harmonics of the pulsating component of the air gap torque produce at the resonance frequency high mechanical stresses. This measuring shaft was covered with resistive wire strains in the bridge circuit. This bridge voltage modulates a carrier frequency which is transmitted contactless. Fig. 7 shows on the one hand that in service of the motor it is only allowed to pass quickly through resonance areas accelerating or braking the drive while continuous operation is to be avoided. In on the other hand a continuous duty at the speeds of resonance is desired, the mechanical system must be overdimensionized.

2. TEST MODELS

The 4-pol. test models were designed for a rating of 11 kW, 50 s^{-1}, 146 V, 64 A. The machine has 24/18 slots and a two layer winding with 5/6 pitching. Thermocouples are installed in the winding overhangs and in the stator slots in the middle of the iron in the upper and under layer distributed on the circumference. For the stator described above two rotors were designed and built. The rotor A has laminated iron and a copper squirrel cage of round bars. Rotor B has a solid steel cylinder, provided with drilled slots containing copper bars of a squirrel cage. Fig. 8 shows both designs. On the surfaces of both rotors thermocouples are attached.

3. THE BEHAVIOUR OF THE MODELS ON THE NETWORK

This matter has been discussed in this paper, because on the network the machine (due to a low number of slots) shows unsatisfactory characteristics, which by no means influence the behaviour of the inverter fed machine. We mean that it would be very important to point out this fact. Both models have a high synchronous torque in the standstill.

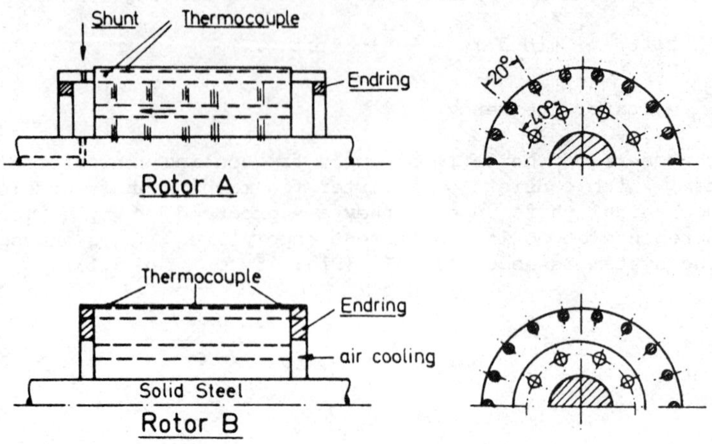

Fig. 8: Rotor design

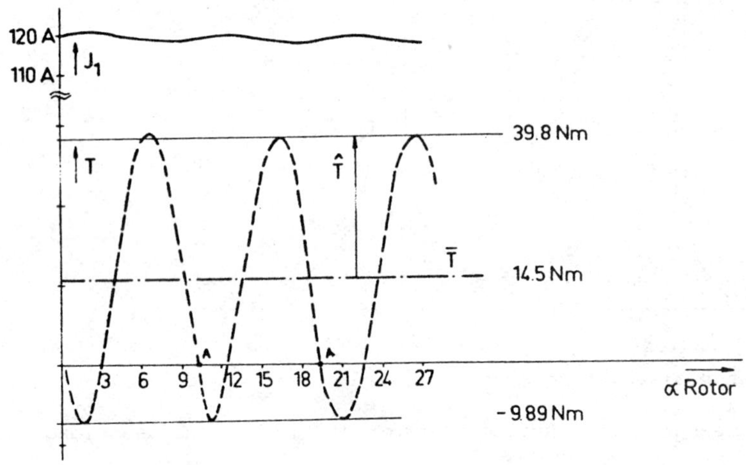

Fig. 9: Synchronous torque at standstill, Rotor A

Fig. 9 shows this torque and the effective stator current in function of the rotor angle. When starting the machine some rotor angles will lead to a stable point (A), so that the machine cannot run up.

At a speed different from zero, the synchronous torque develops into a pulsating torque, whose frequency results from the number of slots. Those pulsating torques stimulate mechanical resonance, so that the run up is influenced by the mechanical resonance frequency. This process will be described briefly. If pulsating torque frequency f_p and resonance frequency f_r are identical, it follows that

$$\frac{(1-s)f_1}{p} = \frac{f_r}{N_2 \cdot X} \qquad X = \text{wave number divided by the rotor slot pitch} \qquad (8)$$

From this investigation follows with $f_r = 176\ s^{-1}$, so that

$$(1-s)\frac{f_1}{p} = 4.89\ s^{-1} \text{ or } 293\ \text{min}^{-1}.$$

During the test we saw, that the speed sticks fast at a reduced voltage and 290 min^{-1}. Fig.10 illustrates a starting operation on the network.

Furthermore the efficiency during network operation was measured for comparison with the inverter fed motor. Fig. 12 shows the machine's efficiency of the motor for both rotors. The measurement of the power supplied by the network was carried out by electronic wattmeters, which use the time-division-method. This method also allows an exact measurement of the power during the inverter drive.

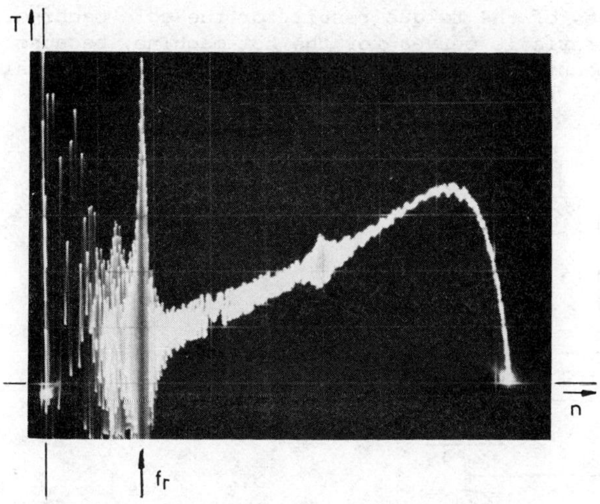

Fig. 10: Start on the network, Rotor A

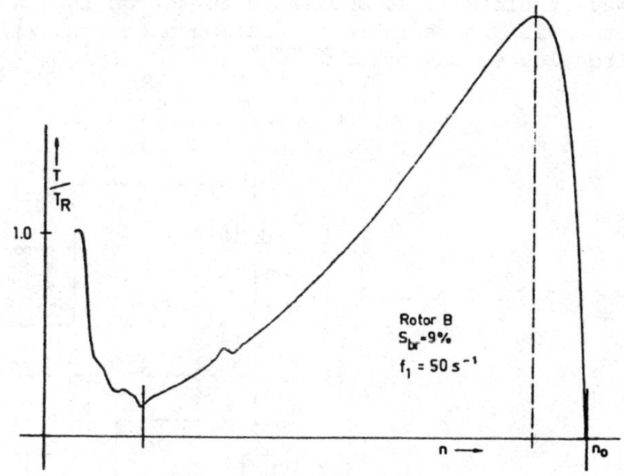

Fig. 11: Start on the network, Rotor B

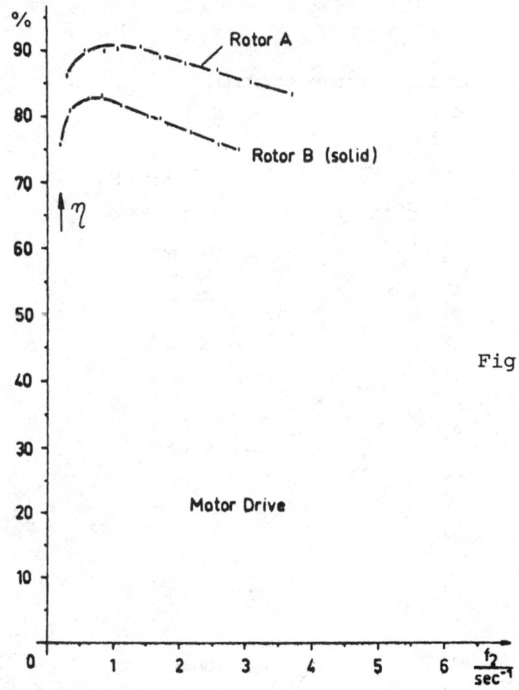

Fig. 12: The efficiency with the Rotors A, B

4. THE INVERTER SUPPLY

Indirect flux control is used. The current is given in relation to the rotor frequency,so that a constant flux linkage exists at every load and stator frequency. Therefore following equation holds

$$I_1 = I_0 \sqrt{\frac{1+(\omega_2 T_2)^2}{1+(\sigma\omega_2 T_2)^2}} \qquad (9)$$

σ = total leakage factor

T_2 = open circuit timeconstant of the rotor

$$T_2 = \frac{L_2}{R_2} = \frac{L_h + L'_{2\sigma}}{R'_2} \qquad (10)$$

I_0, σ, T_2 can be measured on the machine. σ however, must not be found by short circuit test at rated voltage, because of the saturation, which causes a smaller leakage factor. If the characteristic curve is found out by experiments the influence of R_1 has to be compensated in smaller machines. This is achieved by increasing the stator voltage in relation to the loading.

Fig. 13 shows the block circuit diagramm. A circuit with thyristors commutation capacitors and series diodes, serves as an inverter. The rotor's increased resistance by heat has considerable influence on the flux control of the machine. If the relation between current and rotor frequency for the hot machine is set up smaller values of the torque result for the cold machine. Nevertheless it is advisable to set up the characteristic curves for the hot machine, because otherwise undesirable saturation phenomena will occur during operation. Furthermore additional iron losses may occur.

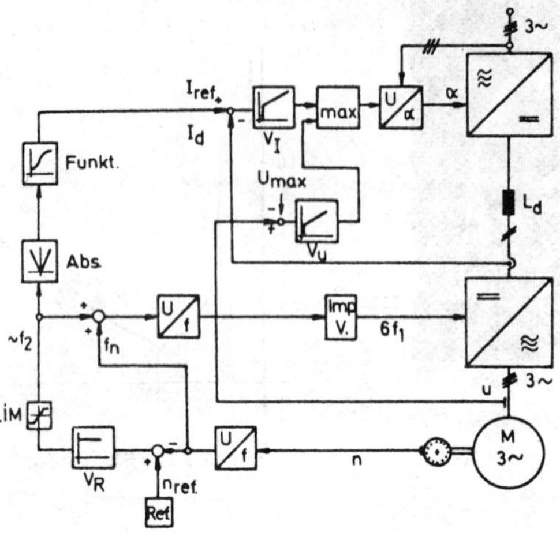

Fig. 13: System block diagramm of the inverter drive control

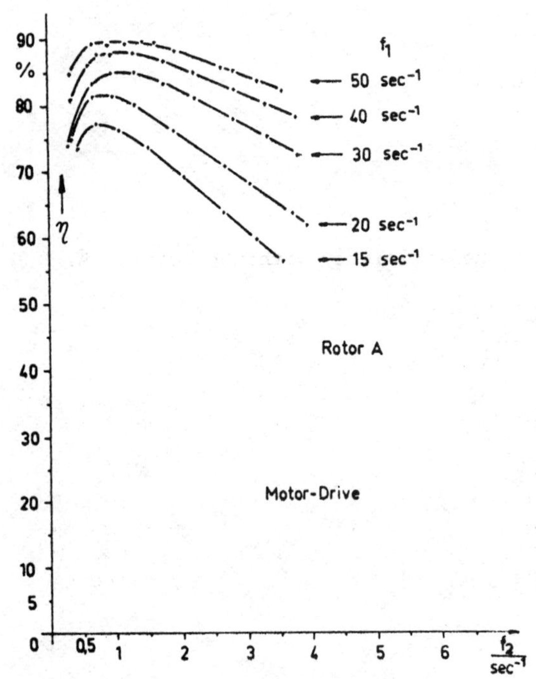

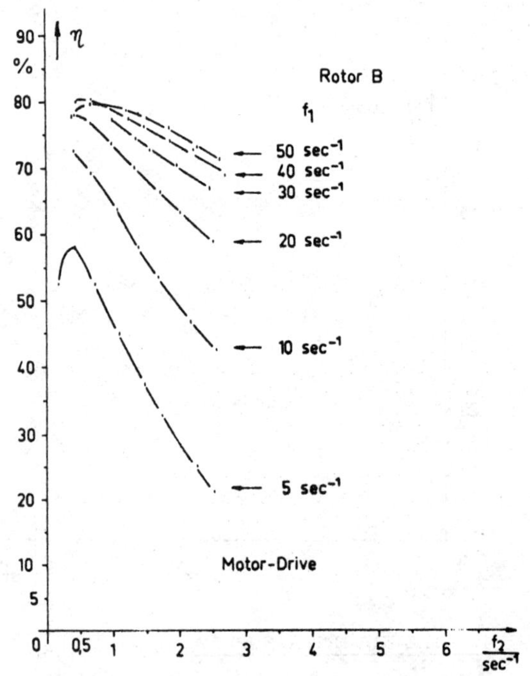

Fig. 14: Rotor A Fig. 14: Rotor B

efficiency of the inverter fed motor

Due to the harmonics the efficiency in inverter feeding is lower than in sinusoidal feeding. A first rough estimate shows this influence. Here the commutation is neglected. For a rectangular current wave of 120 degr. the relative fundamental content is

$$g_i = \frac{I_1}{I_{eff}} = \frac{\frac{4}{\pi}\,(\cos\frac{\pi}{6})\cdot I_d}{\sqrt{\frac{2}{3}}\cdot\sqrt{2}\cdot I_d} = 95.5\% \qquad (11)$$

To give the machine a certain fundamental current wave an effective current value of

$$I_{eff} = \frac{I_1}{g_i} = 1.047\, I_1 \qquad (12)$$

is necessary. This means in the stator an increase of losses of

$$1 - \left(\frac{1}{g_i}\right)^2 = 9.7\% \qquad (13)$$

A skin effect is here neglected. The current harmonics also cause losses in the rotor. The slip of the rotating field harmonic of the number $\nu = 6g+1$ which is produced by a current harmonic produced by the inverter is about 1.

$$s_\nu = 1 - \frac{1}{6g+1} + \frac{s_1}{6g+1} \approx 1; \quad g = \pm1, \pm2, \pm3, \qquad (14)$$

The total effective value of all current harmonics is defined by the relative harmonic content k_i

$$(I_{eff\,\nu})_{tot.} = k_i \cdot I_{eff}, \quad k_i = \sqrt{1-g_i^2} \qquad (15)$$

For the rotor losses follows

$$P_v = 3 \cdot R_2' \cdot I_2'^2 + 3R_2' \cdot (k_i I_{eff})^2$$

$$P_v = 3R_2' I_2'^2 \left[1 + (1-g_i^2) \left(\frac{I_{eff}}{I_2'}\right)^2 \right] \qquad (16)$$

$$I_2' = \sqrt{(I_1 \cdot \cos\varphi_R)^2 + (I_1 \cdot \sin\varphi_R - I_0)^2} \qquad (17)$$

If one takes a normal machine with $I_0/I_1 = 33\%$, $\cos\varphi_R = 0.82$ it follows $I_2'/I_1 = 85\%$.

Furthermore it follows

$$\frac{I_{eff}}{I_2'} = \frac{I_1}{g_i \; 0.85 \; I_1}$$, if one inserts this in Eq. (16) it follows

$$R_{2v} = 3 \, I_2'^2 \, R_2' \left(1 + \frac{(1-g_i^2)}{(0.85^2) g_i^2}\right) = 3 \, I_2'^2 \, R_2' \, (1 + 0.13) \qquad (18)$$

This means an increase in losses of 13% in the rotor if the skin effect is neglected. This value becomes more favourable with respect to commutation. Assuming an efficiency in the rated operation at sinusoidal supply of 87.5%, results from this calculation an efficiency reduction to 86.4% for the inverter fed machine. For the machines under consideration of the copper losses in the stator are about 45%, in the rotor about 35% of the total losses at rated conditions. These losses amount to 9.7% and 13% by the given calculation. Altogether this means increasing total loss of about 9%. The measured efficiency actually showed a loss increase lower than 10%. That explains the test result for a machine with a laminated iron rotor. For solid rotors there are higher additional losses, which are caused by surface eddy currents.

5. BEHAVIOUR DURING INVERTER SUPPLY

5.1 SPECIAL INVESTIGATIONS OF THE RIPPLE TORQUES

The system motor-shaft-load is oscillatory two mass-system (Fig. 15).

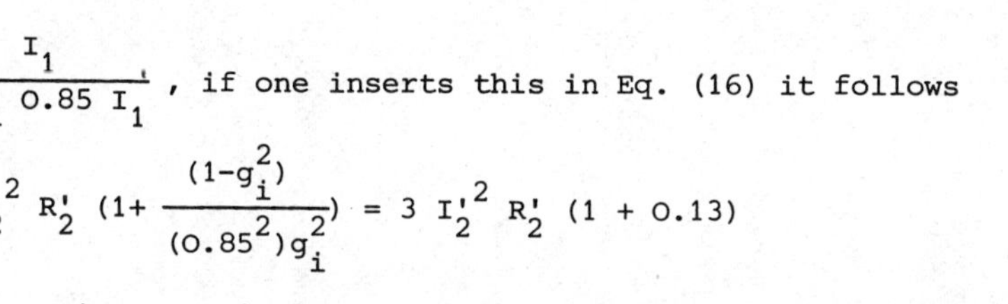

Fig. 15: Mechanical system for the complete drive

If one takes into account this oscillatory system in simulating a machine supplied by current blocks, one will find ripple torques on the shaft, which are a superimposition of the resonance frequency and the stimulation frequencies.The total system was simulated on the digital computer. A coordinate system resting relatively to the stator was used (14,15). The basic equations (19)-(31) are given, and the control design as shown in Fig. 14 was used. The commutation influence is neglected.

Current

$$\underline{i_1} = \frac{2}{3}\left[1,a,a^2\right]\begin{bmatrix} i_u \\ i_v \\ i_w \end{bmatrix} \qquad i_u, i_v, i_w = \text{instantaneous values of the stator phase currents (120 deg. blocks)} \tag{19}$$

$$a = e^{j\frac{2\pi}{3}} \tag{20}$$

$$\left|\underline{i_1}\right| = \left|\underline{i_o}\right| \cdot \sqrt{\frac{1+(\omega_2 T_2)^2}{1+(\sigma\omega_2 T_2)^2}} \tag{21}$$

Flux linkage

$$\begin{bmatrix} \underline{\Psi_1} \\ \underline{\Psi_2} \end{bmatrix} = \begin{bmatrix} L_h+L_1, & L_h \\ L_h, & L_h+L_2 \end{bmatrix} \begin{bmatrix} \underline{i_1} \\ \underline{i_2} \end{bmatrix} \tag{22}$$

$$\underline{\Psi_h} = L_h\,(\underline{i_1}+\underline{i_2}) \tag{23}$$

Voltage

$$0 = R_2\,\underline{i_2} + \frac{d\underline{\Psi_2}}{dt} - j\omega_M \underline{\Psi_2} \tag{24}$$

Speed

$$\omega_2 = \omega_1 - \omega_M \tag{25}$$

$$\omega_1 = \omega_M + V_R(\omega_{1ref} - \omega_M) \tag{26}$$

$$\omega_{e\,mech} = \sqrt{\frac{C_F}{\frac{\Theta_B \cdot \Theta_M}{\Theta_B + \Theta_M}}} \tag{27}$$

Torque

$$T_M = \frac{3}{2}\,p\,\left[\underline{i_2} \times \underline{\Psi_2}\right] \tag{28}$$

$$T_M = \frac{1}{p}\,\frac{d\omega_M}{dt} + T_F \tag{29}$$

$$T_F = \frac{C_F}{p}\int(\omega_M - \omega_B)\,dt + \frac{D}{p}\,(\omega_M - \omega_B) \tag{30}$$

$$T_F = \frac{1}{p}\,\Theta_B\,\frac{d\omega_B}{dt} + T_B \tag{31}$$

Fig. 16 and 17 show the measured and calculated torques on the shaft versus time. The resonance frequency of the mechanical system is of great importance (see Fig. 7). The extreme mechanical stress which may occur under certain circumstances must be taken into account in construction work. At the same time for a speed adjusted inverter drive it must be remembered that all kinds of stimulated frequencies will occur. Possibly other inverter types could be chosen.

5.2 DYNAMIC BEHAVIOUR

The transient behaviour of the speed is recognizable on Fig. 18 if the speed reference changes suddenly. For an acceleration process the rated torque is reached in about 50 msec growing up from zero. The dynamic behaviour is influenced by the design of the d.c. intermediate circuit.

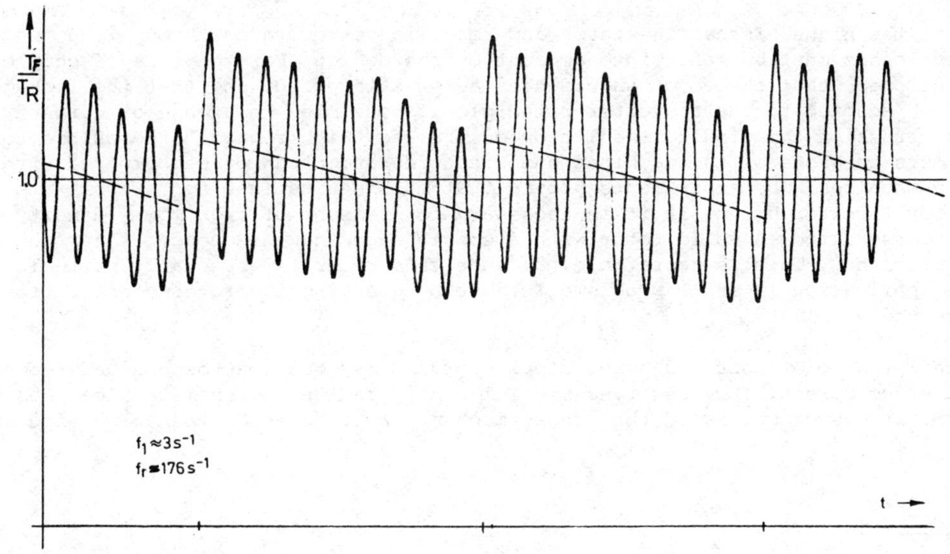

Fig. 16: Measured torque on the shaft

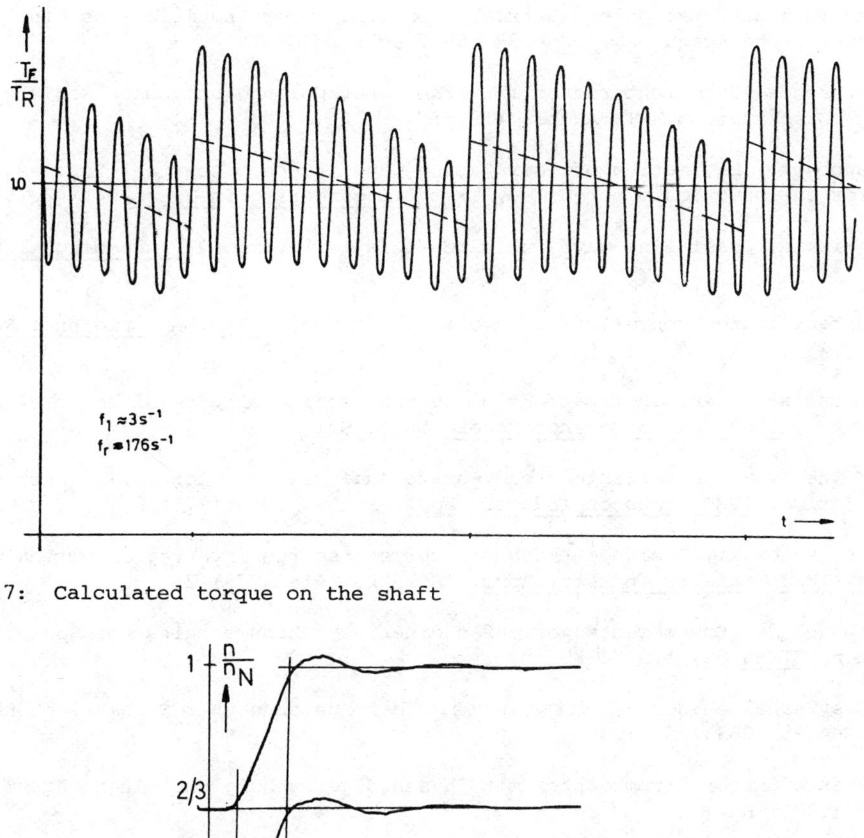

Fig. 17: Calculated torque on the shaft

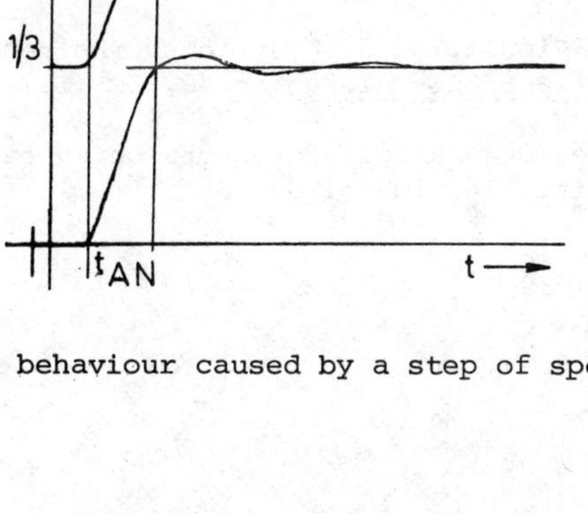

Fig. 18: Transient behaviour caused by a step of speed reference

6. CONCLUSIONS CONCERNING THE DESIGN OF SUCH MACHINES

The known fact of the higher losses in stator and rotor is corroborated through the tests. This increase in losses must be taken into account by the design. For the sake of completeness is should be mentioned, that the rotor should have a low skin effect and that the leakage inductance should be adapted to the converter principle if possible. In a case of current impression higher stress of the winding insulation must arise. With regard to parasitic torques winding and numbers of slots of the machines under consideration cannot be compared with normal network supplied machines. The experiments proved that machines useless for network are without any faults for inverter drive. If it is necessary for reasons of high power density such machines may be constructed. A solid rotor with a copper cage in consequence of surface losses has a smaller efficiency. This is regretable, because this rotor is especially suitable for high speed. The application is still problematic, but by changing the rotor construction the losses may be reduced.

For the converter principle described here, ripple torques may occur at mechanical resonance frequency as dangerous stress. This deserves special consideration in the mechanical design of the total system. It was demonstrated that these mechanical stresses can be calculated in advance.

REFERENCES

(1) K. Hasse, um dynamischen Verhalten der Asynchronmaschine bei Betrieb mit variabler Ständerfrequenz und Ständerspannung, ETZ-A 89, 77 (1968)

(2) A. Priess, Darstellung der stationären Betriebs-Kennlinien der Induktionsmaschine am Netz variabler Spannung und Frequenz, ETZ-A 88, 55 (1967) 2

(3) H. Larkamp, Asynchroner Stromrichtermotor mit Transistorspeisung und hoher Reglergeschwindigkeit, Dissertation RWTH Aachen (1973)

(4) I. Yasoka, Y. Tsunehiro, Estimation of commutating capacitance of current impressed inverter used for an induction motor-drive, Electr.Engineering in Japan 94, 48 (1974) 4

(5) W. Farrer, J.D. Miskin, Quasi-sine-wave fully regenerativ invertor, Proceedings IEEE 120, 969 (1973) 9

(6) K.P. Philips, Current source converter for a.c.motor drives, IEEE IGA Conference Record, 411 (1971)

(7) R.B. Maag, Characteristics and applications of current source slip regulated a.c. induction motor drives, IEEE IGA Conference Record 385 (1971)

(8) Th.A. Lipo, E. Cornell, State-variable steady-state analysis of a controlled current induction motor drive, IEEE Trans.on Industry Appl. IA-11, 704 (1975) 6

(9) R.H. Nelson, Th.A. Radomski, Design methods for current source inverter induction motor drive systems, IEEE Trans.on Industry Appl. IECI-22, 141 (1975) 2

(10) D. Naunin, Berechnung des Drehmomentenverlaufes einer Asynchronmaschine bei Speisung mit Rechteckströmen, ETZ-A 90, 179 (1969) 8

(11) H. Klautschek, Das Verhalten der Induktionsmaschine bei Speisung über Stromzwischenkreisumrichter, ETZ-A 95, 283 (1974) 5

(12) D. Schröder, Selbstgeführter Stromrichter mit Phasenfolgelöschung und eingeprägtem Strom, ETZ-A 96, 520 (1975) H.21

(13) U. Sobottka, Einfluß der Temperatur auf das Betriebsverhalten der drehzahlgeregelten Asynchronmaschine, Siemens Zeitschrift 43, 760 (1969) 9

(14) G. Gölz, F. Hentschel, S. Chumpitaz, Simulation des Betriebsverhaltens einer umrichtergespeisten Asynchronmaschine, Wiss.Ber.AEG-Telefunken 48, 20 (1975) 1

(15) K. Heumann, K.G. Jordan, Das Verhalten des Käfigläufermotors bei veränderlicher Speisefrequenz und Stromregelung, AEG Mitt. 54, 107 (1964) 2

A COMPARISON BETWEEN OPERATING CONDITIONS OF INVERTERFED ASYNCHRONOUS MOTORS

E. Pagano and A. Perfetto

University of Naples

1. GENERALITIES

The performances of the inverterfed asynchronous motors have been widely investigated in the last few years. The machines can be obviously operated according to different control techniques and obtain operating conditions quite different as compared with the same machines supplied with constant frequency and voltage. It seems therefore necessary to define adequate machine characteristics, which can be also considered as *reference characteristics of the machines*. Their definition contributes to state the more adequate rating of an inverterfed machine. For a traditional machine reference is generally made to the rated power and torque at rated voltage and frequency. For a variable frequency supplied a.c. machine, the maximum value must be considered of the armature current and voltage at the different frequencies and the torque-speed characteristic.

The unitary overloading capacity of the static frequency converters implies that only operations can be obtained which corresponds to armature current values not higher than the inverter rated output current. This value obviously represents the *limit value* of the armature current and is generally simply defined as the *limit current value*. The effective values of the armature impressed voltage corresponding at the different frequencies to the limit armature current gives rise to the *maximum voltage versus frequency law*. In the same way the *limit torque-speed characteristic* is defined as the set of pairs of torque and speed values the machinedevelops at the different frequencies for the maximum voltage and for the limit current.

The limit torque-speed characteristic and the maximum voltage versus frequency law have for variable frequency supplied a.c. motors the same meaning of the rated torque and voltage for a traditional machine.

We also refer to the *limit slip versu frequency law*, defined as the set of slips which at the different frequencies implies the machine carries on the limit current for the corresponding maximum value of the armature voltage.

All the operating conditions characterized by the previously defined limit values are defined as *limit operating conditions*. On the basis of these conditions we can compare the different asynchronous performances obtained with different control techniques.

2. ASYNCHRONOUS OPERATIONS

Different are the optimum operating conditions considered up to now for the inverterfed asynchronous motors. They can be briefly summerized as follows:

a. *breakdown slips for impressed current*. These operating conditions imply the machine operates at the different frequencies for the slips corresponding to the breakdown torques for a supply with impressed current, i. e. $\sigma_{bc} = R_2' f_M / [f(X_t + X_h)]$;

b. *Breakdown slips for impressed voltage*. These operating conditions imply the machine operates at the different frequencies for the slips corresponding to the breakdown torques for a supply with impressed voltage. We can approximately set $\sigma_{bv} = R_2' f_M / (f X_t)$;

c. *Optimum T/I slips*. These operating conditions imply the machineoperates at the different frequencies for the slips which gives rise to the maximum value of the ratio T/I between the electromagnetic torque and the armature current. In this case we can approximately set $\sigma_o = R_2' f_M / (f \sqrt{X_t X_h})$

The three different optimum operating conditions previously defined obtain limit torque-speed characteristics practically constant in a wide frequency range. If the minimum frequency is always adequately chosen to obtain an unitary value of the optimum slip, the limit torque-speed characteristics are increasing for the lower frequency values, where the approximated expressions previously given for the optimum slips are not more valid [1].

It is now interesting to compare the performances of the asynchronous motors which operates according to the three different optimum criteria previously recolled. The comparison can be obviously performed on the basis of different terms. For sake of simplicity we refer to approximated expressions of the electromagnetic torque and of the armature impressed voltage. We first consider three different machines adequately designed to operate at the different optimum slips with the same limit armature current. We perform the comparison on the basis of three different ratios, i. e. between the maximum effective values of the armature voltage, between the rated power of the supply converters, between the torques for unitary armature current. These ratios are simply expressed by means of the electrical machine parameters, as shown in table 1.

It is easily seen that the higher values of these three ratios correspond to the operating conditions of the point a.

The improvement of the electromagnetic torque for the case a is howewer obtained for correspondently higher values of the supply voltage and of the rated power of the converter.

Table 2 summarizes the results corresponding to a comparison for the same maximum effective value of the armature impressed voltage. The electromagnetic torques developed for unitary armature currents are now the same. The limit currents and the rated powers of the converters are different for the three different operating conditions.The operations of the type a requires the smaller value of the current and of the power.

In table 3 operations for the same electromagnetic torque are compared. It is interesting now to note that different values of the limit current and of the voltage corresponds to the three different optimum operations. The operations a therefore obtain high values of the ratio T_L/I_L.

Also if in a different way the results summarized in tables 1, 2 and 3 confirm that the operations of type a give the required output power with the lower current and with the higher voltage. To avoid the saturation phenomena,the asynchronous machines, expecially designed for the rating corresponding to type a operations,must be therefore necessarily oversized with respect to traditional machines of the same rated power.

This result is clearly understood taking into account that the impressed current machines operates with very small slip values at which unsaturated operations are obviously guaranteed by very small linear current densities.

Following the type a operations it is also very difficult to obtain satisfactory ratios T_L/I_L at the starting. To obtain a starting torque not smaller than the limit torque at full-load it is, in fact, necessary that the minimum frequency value satisfies the relation:

$$R_2' f_M / [f_m (X_t + X_h)] = 1$$

and therefore that :

$$f_m = R_2' f_M / (X_t + X_h).$$

To avoid extremely low values of f_m it is therefore necessary to realize machines with small value of X_h. This requirement agree with the requirement of a small linear current density but implies high values of the magnetizing current and gives therefore rise to values of T_L/I_L smaller than the corresponding ones to machines designed according to the conditions c.

Table 4 compares machine quantities by considering the same machine drived at variable frequency according to the three different previously considered optimum operating conditionsIf reference is made to an asynchronous machine of traditional design the rated slip is practically correspondent to the optimum slip of the type c at the maximum frequency,i.e. $\sigma_n = \sigma_0 = R_2' / \sqrt{X_t X_h}$ .The operations of the machine with the slip $\sigma_{bc} = R_2'/(X_t + X_h)$ can be therefore performed with the rated voltage if the armature current is approximately $\sqrt{X_t/X_h}$ times smaller than the rated value.The operations at the breakdown slip $\sigma_{bv} = R_2'/X_t$ can at the contrary be obtained with the the rated current if the supply voltage is $\sqrt{X_t/X_h}$ times smaller than the rated value.It follows that the maximum power developed by the machine according to the operating conditions of type a and c is therefore always smaller than the rated power.

According to the different values of the voltages and of the currents,different iron and copper losses are obtained for the three different operating conditions.Taking into account that for an asynchronous motor the ratio $\alpha_p = P_{ir}/P_{cr}$ between the iron and the copper losses is always smaller than 1,from table 4 it follows immediately that type a operations obtains p.u. losses smaller than type b operations.We can therefore finallyconclude that among the three different operating conditionspreviously considered the type c is the more useful onefor the utilization of the power of the converter and of the iron and of the copper of the machine.The type a operations obtains also satisfactory performances of the machine but they do not realize high values of the starting torque.The type b operations,compared with the other two,realizes the less satisfacoty performancesof the converter and of the machine,but obtains high starting torques and is therefore very useful for non-continuous duty operations.

On the basis of the previous considerations we can now conclude that the type a and b operations require machines adequately designed to operate at the corresponding imposed optimum slips while the type c operations can be satisfied with machines of traditional design.It must however also be taken into account that the machines properly designed according to type a and b operations always imply values of the ratio power/weight as compared with machines properly designed according to type c operations.

3.ALTERNATIVE TORQUES ANALYSIS

The evaluation of the amplitudes and of the frequencies of the electromagnetic alternative torques acting on the shaft of an inverterfed asynchronous motor was done many years ago by other authors and by ourselves [2].The studies was initially devoted to the determination of the frequencies of the alternative torquee harmonics.It is now well-known that when the supply voltage is rectangular,the frequency of these harmonics is $2m\nu f$ (we denote with m the number of the stator phases,with ν a whole number,i.e. ν=1,2,....,n, and with f the frequency of the supply voltage or current).By means of oscillographic results it was also shown that a good agreement is found between theorical and experimental results[3],both for the synchronous and the asynchronous motors.

According to the three different limit operating conditions previously considered for asynchronous motors it is now interesting to compare the amplitudes of alternative torque harmonics corresponding to the three operating conditions.For these purpose we consider an asynchronous motor operating according to the limit conditions outlined in table 4.In fig.1 the amplitudes are represented of the alternative torque harmonics as a function of the rotor speed at the left side and of the supply frequency at the right side.The amplitudes are expressed in per cent of the correspondent asynchronous torques at the different frequencies .The letters a,b and c refer to the previous considered limit operating conditions.

Examining the six different sets of curves we first note that all the alternative torque amplitudes decrease as the frequency,i.e.the rotor speed,increases.The higher values of these amplitudes are obtained at the lower frequencies,while they are constant in practice in a wide range of this last.

Type c operations obtains the higher values of the alternative torques.A meaning difference characterizes the impressed voltage or the impressed current operations.It is in fact easily seen that for type a operations the second harmonic($\nu = 2$)is many times greater than the first one ($\nu = 1$) while different relations happen for types b and c operations,when the first harmonics are many times greater than the second one.The substantial difference between impressed current and impressed voltage operations is essentially dependent on the harmonic contents of the air-gap inductions and of the rotor currents. For a voltage supplied inverterfed asynchronous motor the current and the air-gap induction harmonics are in fact practically decreasing with the square of their order. For a curren supplied asynchronous motor the current and the air-gap induction harmonics are decreasing with the their order. In the second case the first harmonics of the alternative torque are in practice compensated one with another.

This result is very important in practice,because according to the natural frequencies of the sets motor-shaft-load the higher order harmonics are more filtered than the lower ones.The alternative torques of elastic reaction are therefore many times smaller for the current supplied than for the voltage supplied asynchronous motors.No problems arise therefore in the designing the mechanical transmission of the current supplied asynchronous motors.

4.CONCLUSIONS

The asynchronous performances have been compared for the inverterfed asynchronous motors with reference to current or voltage supplied machines.It was shown that the first one requires an oversizing of the machine,because the operating slips are very small.

The operations with impressed current,compared with the oprations with impressed voltage obtain,at the contrary,lower iron losses,but require high magnetizing current.Their ratio between the developed electromagnetic torque and the stator current is therefore lower than the corresponding value for voltage supplied machines.

The alternative torque harmonics are smaller in the current supplied motors and only the second harmonic is to be considered. The amplitudes of the alternative twist angles are therefore very small in current supplied electrical drives.

5.SHORT LIST OF SYMBOLS

f_m, f_M minimum and maximum supply frequency;

R'_2 rotor resistance referred to the stator windings;

X_h main reactance at maximum frequency;

X_t stator equivalent leakage reactance at maximum frequency;

σ slip.

6. REFERENCES

1 Isastia, V.; Pagano, E.; Perfetto, A.: *"Berechnungsrichtlinien umrichtergespeister Käfigläufer -Asynchronmotoren"* ETZ-A, vol. 97, (1976), pp. 607-611.

2 Sattler, P. K.: *"Parasitäre Drehmomente von Stromrichtermotoren"* ETZ-A,vol.88, (1967), pp. 89-93.

De Carli, A.; Pagano, E.; Perez de Vera, M.: *"Coppie alternative nei motori asincroni controllati in frequenza"* L'Elettrotecnica, vol. 55, (1968), pp. 310-317.

3 Gentile, G.; Pagano, E.: *"Umrichtergespeiste Synchronmotoren"* ETZ-A,vol.93,(1972), pp. 21-26.

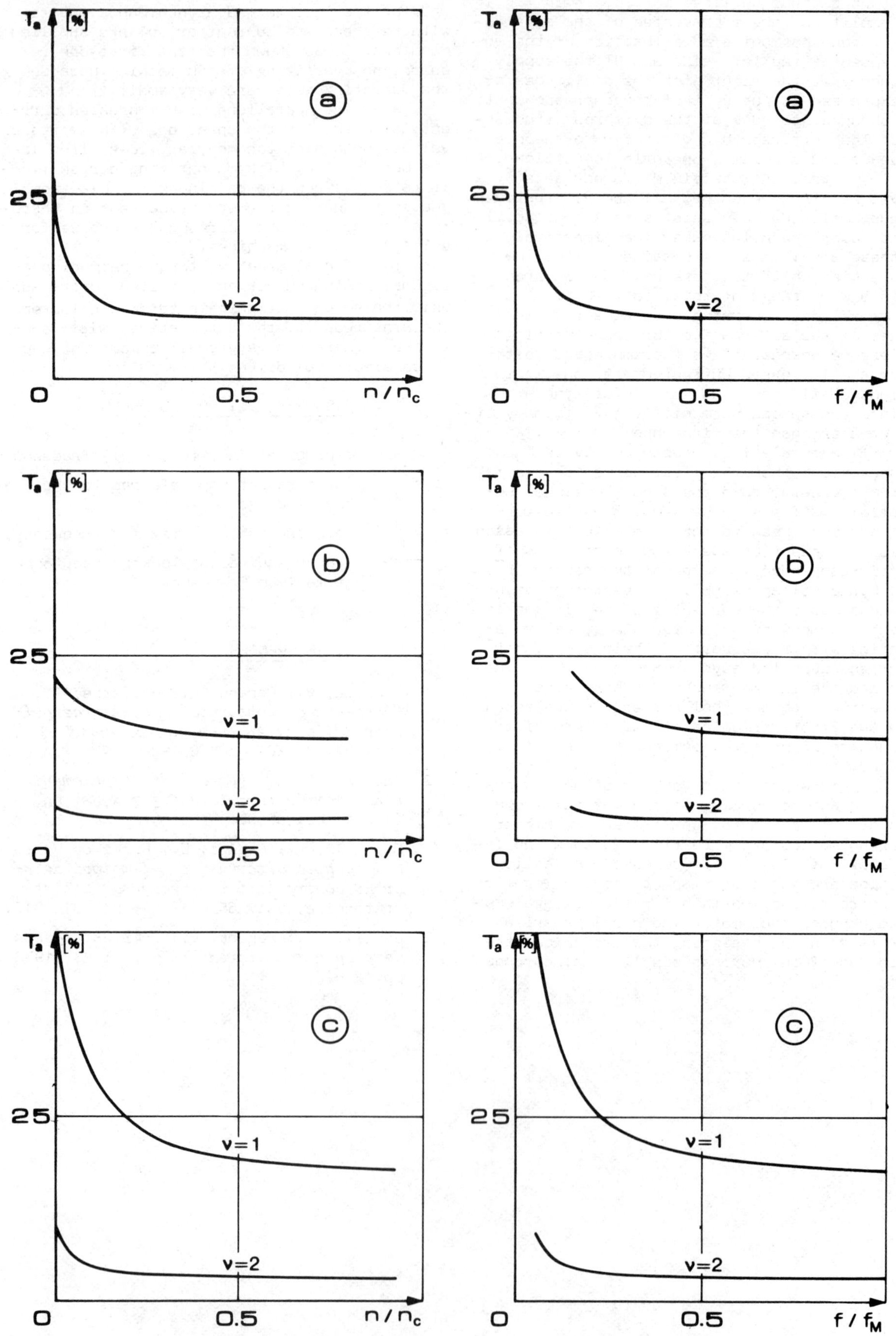

T_a [%]
a
b
c
25
$\nu=1$
$\nu=2$
0
0.5
n/n_c
f/f_M

TABLE 1 - *Comparison for the same limit current value*

Terms \ Operations	a / c	b / c	b / a
I_L	1	1	1
V_M, T_L/I_L, P_M	$\frac{\overline{X}_t + \overline{X}_h}{\sqrt{X_t^{**}X_h^{**}}}$	$\frac{X_t^o}{\sqrt{X_t^{**}X_h^{**}}}$	$\frac{X_t^o}{\overline{X}_t + \overline{X}_h}$

TABLE 2 - *Comparison for the same maximum effective voltage*

Terms \ Operations	a / c	b / c	b / a
V_M, T_L / I_L	1	1	1
I_L , P_M	$\frac{X_t^{**}X_h^{**}}{\overline{X}_t + \overline{X}_h}$	$\frac{X_t^{**}X_h^{**}}{X_t^o}$	$\frac{X_t + X_h}{X_t^o}$

TABLE 3 - *Comparison for the same limit torque*

Terms \ Operations	a / c	b / c	b / a
T_L	1	1	1
I_L	$\sqrt{\frac{\sqrt{X_t^{**}X_h^{**}}}{\overline{X}_t + \overline{X}_h}}$	$\sqrt{\frac{\sqrt{X_t^{**}X_h^{**}}}{X_t^o}}$	$\sqrt{\frac{\overline{X}_t + \overline{X}_h}{X_t^o}}$
V_M , T_L / I_L	$\sqrt{\frac{\overline{X}_t + \overline{X}_h}{\sqrt{X_t^{**}X_h^{**}}}}$	$\sqrt{\frac{X_t^o}{\sqrt{X_t^{**}X_h^{**}}}}$	$\sqrt{\frac{X_t^o}{\overline{X}_t + \overline{X}_h}}$

TABLE 4 - *Comparison for the same machine*

Operations / Terms	a / c	b / c
V	1	$\sqrt{\frac{X_t}{X_h}}$
I	$\frac{\sqrt{X_t X_h}}{X_t + X_h} \simeq \sqrt{\frac{X_t}{X_h}}$	1
P	$\frac{\sqrt{X_t X_h}}{X_t + X_h} \simeq \sqrt{\frac{X_t}{X_h}}$	$\sqrt{\frac{X_t}{X_h}}$
M	$\frac{\sqrt{X_t X_h}}{X_t + X_h} \simeq \sqrt{\frac{X_t}{X_h}}$	$\sqrt{\frac{X_t}{X_h}}$
M / I	1	$\sqrt{\frac{X_t}{X_h}}$
p	$\dfrac{\frac{X_t X_h}{(X_t + X_h)^2} + \alpha_p}{1 + \alpha_p}$	$\dfrac{1 + \frac{X_t}{X_h}\alpha_p}{1 + \alpha_p}$

A DIMENSIONING CRITERION FOR THE SHAFTING OF THE VARIABLE FREQUENCY FED INDUCTION MOTOR

I. Marongiu

Istituto di Electrotecnica, University of Cagliari, 09100 Cagliari, Italy

ABSTRACT

For the variable frequency fed induction motor the possible operating points are limited not only by the feed and motor characteristics but also by the presence of an instability region. The pattern of this instability region and the particular dependence of the extent of the frequency interval bounding it, on the torsional elasticity of the shafting, allow to reduce the extent of the instability region by appropriately dimensioning the shafting. A numerical example shows how the dominant roots locus of the characteristic equation of the linearized system, enables the numerical value of the torsional elasticity to be used in the dimensioning of the shafting, to be determined.

INTRODUCTION

Possible operating steady-state points of the variable frequency fed induction motor constitute a region of the torque-frequency plane, restricted by a boundary which represents maximum output torque at different frequencies according to feed and motor characteristics. In transient conditions not all these operating points are characterized by stable behaviour. Analysis of local stability in fact reveals that for every induction motor there exists a region of instability at low frequency operating (Refs. 1-2). At these operating conditions small sustained oscillations can be observed in the motor speed and in the stator and rotor currents. The region constituted by the unstable operating points generally assumes, in the torque-frequency plane, the configuration shown in Fig. 1, including all the interval points between f_{r1}-f_{r2}. This implies that for any variable frequency fed induction motor there exists a region which represents the possible stable operating points. The extent of the instability region and therefore the frequency interval bounding it, are determined, for a given feed voltage-frequency law, according to the characteristic parameter values of the system motor-shafting-mechanical load (motor reactances and resistances, motor and mechanical load inertia constants and torsional elasticity of the shafting. In a previous work (Ref. 3) it has been demonstrated, in particular, that the extent of the instability region and frequency interval which bounds it, diminishes

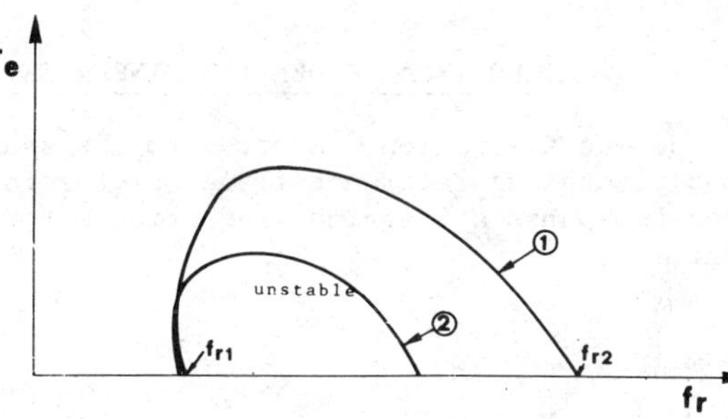

Fig. 1. Curves bounding instability region:
(1) for rigid shafting ($\bar{\alpha} = C_b/K=0$)
(2) for elastic shafting ($\bar{\alpha} > 0$)

as the torsional elasticity of the shafting increases. More specifically, the existence can be confirmed of a relationship between the extent of the frequency interval, which bounds the instability region, and the torsional elasticity of the shafting. This relationship, taking into account the fact that the extent of the frequency interval alters as the torsional elasticity varies mainly because of the shifting of its upper limit, can be substituted by a relationship between the upper limit and the torsional elasticity of the shafting. This means that it is possible, where it is necessary and required, to reduce the extent of the instability region and of the frequency interval bounding it by means of an appropriate dimensioning of the shafting without modifying the remaining parameters of the system and the feed voltage-frequency law, which are generally defined by other performance specifications. Beginning from the knowledge of the frequency interval bounding the instability region of a given system with rigid shafting, the method proposed in this paper, consists in presetting, within such an interval, the upper limit of the frequency interval outside which the system is desired to be stable. For the relationship existing between the above upper limit and the torsional elasticity, the latter turns out to be established and allows the dimensioning of the shafting.

The state variable mathematical model is deduced according to the generalized two-phase machine equation in the synchronously rotating reference frame (Ref. 4). Linearization of the non-linear mathematical model of the system is carried out using the small displacement method (Ref. 4). In the numerical example proposed in this paper, the relationship between the upper limit of the frequency interval and the torsional elasticity of the shafting is evidenced by digital computer plotting of the dominant roots locus of the characteristic equation of the linearized system. The mathematical model does not account for the damping effects of bearings, friction and ventilation. However, no further difficulty arises if these effects, which are always in favour of stability, are taken into account because the system order does not change.

MATHEMATICAL MODEL AND LINEARIZATION

The per-unit equation, referred to the synchronously rotating reference frame, of the idealized two-phase symmetrical induction motor is (Ref. 3):

$$\begin{vmatrix} v_{qs} \\ v_{ds} \\ 0 \\ 0 \end{vmatrix} = \begin{vmatrix} r_s+p\frac{X_s}{\omega_b} & f_r X_s & p\frac{X_m}{\omega_b} & f_r X_m \\ -f_r X_s & r_s+p\frac{X_s}{\omega_b} & -f_r X_m & p\frac{X_m}{\omega_b} \\ p\frac{X_m}{\omega_b} & f_r S X_m & r'_r+p\frac{X'_r}{\omega_b} & f_r S X'_r \\ -f_r S X_m & p\frac{X_m}{\omega_b} & -f_r S X'_r & r'_r+p\frac{X'_r}{\omega_b} \end{vmatrix} \begin{vmatrix} i_{qs} \\ i_{ds} \\ i'_{qr} \\ i'_{dr} \end{vmatrix} \qquad (1)$$

In (1), if the rotor speed and ω_e are constant, the voltages and the currents are constant with time.

The electromagnetic torque is given, as a function of the variables which appear in (1) by the per unit expression:

$$T_e = X_m(i_{qs}\, i'_{dr} - i_{ds}\, i'_{qr}) \qquad (2)$$

If the propogation velocity of the elastic torsional disturbances, along the shafting, is assumed to be infinite and the shafting inertia constant is not taken into account, the dynamic equation of the mechanical system induction motor rotor-shafting-load rotor shown in Fig. 2 is:

$$\begin{bmatrix} T_e \\ 0 \\ T_L \end{bmatrix} = \begin{bmatrix} 2H_1 p & \frac{1}{\bar{\alpha}} & 0 \\ 0 & -p & 1 \\ -2H_2 p & \frac{1}{\bar{\alpha}} & \frac{2H_2 P}{\omega_b} \end{bmatrix} \begin{bmatrix} \frac{\omega_1}{\omega_b} \\ \alpha \\ \dot{\alpha} \end{bmatrix} \qquad (3)$$

where:

$$\alpha = \theta_1 - \theta_2 \qquad (4)$$

$$H_i = J_i \theta_i^{\,2} / \, 2C_b \omega_b \quad (s), \quad (i = 1,2) \qquad (5)$$

$$\bar{\alpha} = C_b / K \qquad (6)$$

$$T_L = C_L / C_b \qquad (7)$$

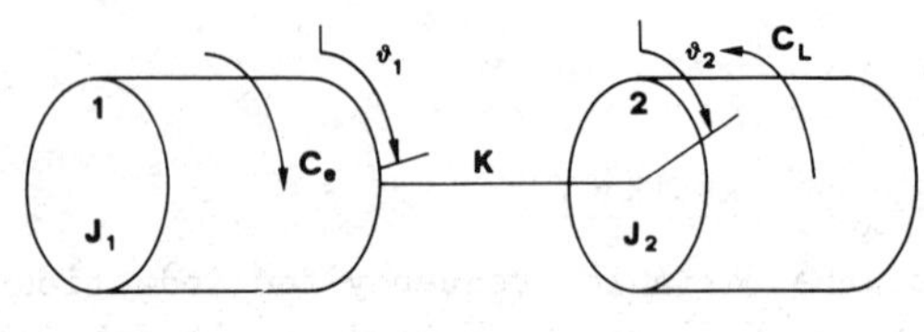

Fig. 2. Mechanical system:
(1) induction motor rotor
(2) mechanical load rotor

Equations (1), (2) and (3) constitute the overall non-linear mathematical model for the electromechanical system at hand. Linearization can be carried out applying the small displacement method (Ref. 4) which gives:

$$\frac{p}{\omega b}[\Delta I] = -[X]^{-1}.[R].[\Delta I] + [X]^{-1}.[\Delta V] \qquad (8)$$

where:

$$[\Delta V]^t = [\Delta v_{qs}, \Delta v_{ds}, 0, 0, 0, 0, \Delta T_L] \qquad (9)$$

is the input vector,

$$[\Delta I]^t = [\Delta i_{qs}, \Delta i_{ds}, \Delta i'_{qr}, \Delta i'_{dr}, \Delta \frac{\omega_1}{\omega_b}, \Delta\alpha, \Delta\dot{\alpha}] \qquad (10)$$

is the state vector

$$|R| = \begin{vmatrix} r_s & f_r X_s & 0 & f_r X_m & 0 & 0 & 0 \\ -f_r X_s & r_s & -f_r X_m & 0 & 0 & 0 & 0 \\ 0 & f_r S_0 X_m & r'_r & f_r S_0 X'_r & -(X_m i_{ds0} + X'_r i'_{dr0}) & 0 & 0 \\ -f_r S_0 X_m & 0 & -f_r S_0 X'_r & r'_r & (X_m i_{qs0} + X'_r i'_{qr0}) & 0 & 0 \\ X_m i'_{dr0} & -X_m i'_{qr0} & -X_m i_{ds0} & X_m i_{qs0} & 0 & -\frac{1}{\bar{\alpha}} & 0 \\ 0 & 0 & 0 & 0 & 0 & 0 & 1 \\ 0 & 0 & 0 & 0 & 0 & \frac{1}{\bar{\alpha}} & 0 \end{vmatrix} \qquad (11)$$

$$|X| = \begin{vmatrix} X_s & 0 & X_m & 0 & 0 & 0 & 0 \\ 0 & X_s & 0 & X_m & 0 & 0 & 0 \\ X_m & 0 & X'_r & 0 & 0 & 0 & 0 \\ 0 & X_m & 0 & X'_r & 0 & 0 & 0 \\ 0 & 0 & 0 & 0 & -2H_1\omega_b & 0 & 0 \\ 0 & 0 & 0 & 0 & 0 & -\omega_b & 0 \\ 0 & 0 & 0 & 0 & -2H_2\omega_b & 0 & 2H_2 \end{vmatrix} \qquad (12)$$

where superscript "t" denotes the transpose and subscript "o" denotes steady state values for the variables.

LOCAL STABILITY STUDY

As it is well known, the characteristic equation roots of the linearized system, that is the eigenvalues of the matrix $-[X]^{-1}.[R]$, denote the local stability or instability according to whether they all have negative real parts or not. Furthermore the system results to be at the limit of stability when the greatest real part of the eigenvalues is zero. For a given system, by resorting to a digital computer, it is possible to calculate the system eigenvalues for any f_r-S_o pair in order to determine the curve bounding the instability region. Figure 1 shows that for $\bar{\alpha} = 0$ the system, within the interval f_{r1}-f_{r2}, is unstable or can become unstable due to the accidental diminution of the torque. Outside such a frequency interval, admitting that the small instability region to the left of f_{r1} is negligible, the system is stable. As $\bar{\alpha}$ increases, the extent of the frequency interval decreases especially because of the displacement to the left of its upper limit, while the position of its lower limit remains pratically unaltered.

DIMENSIONING CRITERION FOR THE SHAFTING

According to the above, the dimensioning criterion is based on the following. Having established $\bar{\alpha}$, the feed voltage-frequency law, and the parameters of the induction motor and mechanical load, the dominant root locus of the system characteristic equation can be plotted, for S_o=0, by varying f_r. The particular f_r value which brings the dominant roots onto the immaginary axis, represents the upper limit of the frequency interval bounding the instability region. If instead of $\bar{\alpha}$, the upper limit of the frequency interval is preset, for S_o=0, the dominant roots turn out to be a function of $\bar{\alpha}$. By varying the latter parameter, the corresponding root locus can be plotted and therefore the particular $\bar{\alpha}$ value which brings the roots onto the immaginary axis, can be evaluated. This value of $\bar{\alpha}$ assures the stability of the system outside the frequency interval, the extent of which has been predetermined by means of its upper limit, while the position of its lower limit remains practically unchanged. It is obvious that, for the purpose of determining the numerical value of α the root locus is of interest only in the neighbourhood of the immaginary axis. The dimensioning of the shafting referring to the $\bar{\alpha}$ numerical value thus found is immediate.

For a shafting composed of n sections, each with uniform diameter and having length l_i, it follows that:

$$\bar{\alpha} = P_b/\omega_b \sum_1^n G_i J_{pi}/ l_i \qquad (13)$$

G_i and J_{pi} being respectively the shear modulus and the polar momentum of inertia of i-th section. Usually, for steel shafting (13) is valid only for small values of the section diameters. In such a case the transmission of the maximum torque cannot be guaranteed with sufficient certainty. Consequently it is appropriate to interpose in the shafting a flexible coupling, the dimensions of which, once the materials and sizes have been established, can be determined by the expression:

$$J_{pg} = l_g/G_g\ (\bar{\alpha} - \sum_{1\ i\neq g}^{n} G_i J_{pi}/l_i) \qquad (14)$$

where subscript "g" is referred to the flexible coupling parameters.

NUMERICAL EXAMPLE

A system composed of a tetrapolar induction motor and of a mechanical load is assumed to have the following characteristics:

rated power	P_b = 5.6 Kw
" voltage	V_b = 220 V
" frequency	ω_b = 377 s^{-1}
stator resistance	r_s = 0.025
rotor "	r'_s = 0.015
stator reactance	X_s = 3.6
rotor "	X'_r = 3.6
magnetizing "	X_m = 0.1
motor inertia constant	H_1 = 0.025 s
mechanical load inertia constant	H_2 = 0.075 s

The function:

$$v = 0.025 + f_r$$

is assumed as the feed voltage frequency law.

From Fig. 3, which shows the curve bounding the instability region for the case at hand, where $\bar{\alpha}$ = 0 (rigid shafting) we obtain f_{r1}= 0.287; f_{r2} = 0.362.

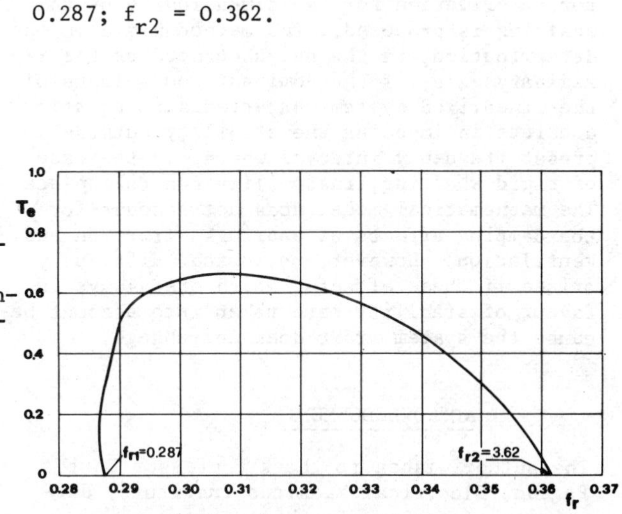

Fig. 3

By presetting for the upper limit of the frequency interval the values: 0.36, 0.35, 0.34, 0.33, 0.32, the corresponding computer plots of the dominant roots loci, in the neighbourhood of the immaginary axis, are obtained by varying $\bar{\alpha}$. Figure 4 shows these plots with their corresponding $1/\bar{\alpha}$ (rad^{-1}) values.

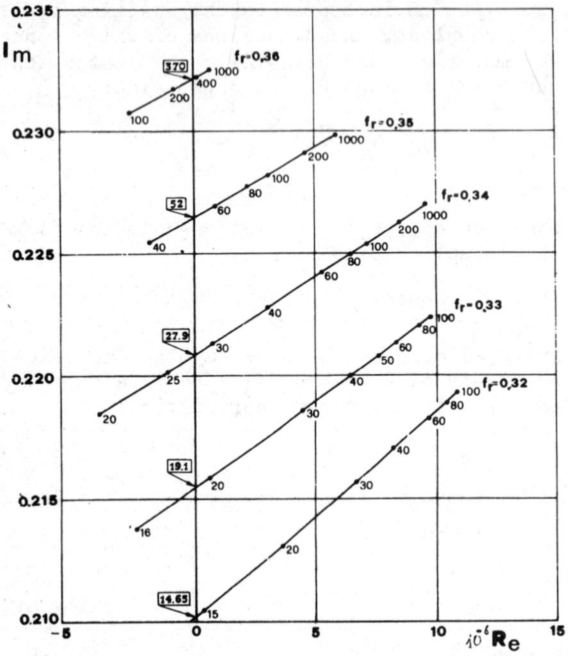

Fig. 4

The $1/\bar{\alpha}$ values 370, 52, 27.9, 19.1, 14.65 corresponding to the immaginary axis intersection, are those which ensure stability outside the frequency intervals with common lower limit f_{r1} and upper limits 0.36, 0.35, 0.34, 0.33, 0.32 respectively. Each $1/\bar{\alpha}$ value through (14) enables the dimensioning of the flexible coupling.

CONCLUSIONS

Considering the results of the stability analysis of a variable frequency fed induction motor, a criterion for the dimensioning of the shafting is proposed. The method based on the determination, in the neighbourhood of the immaginary axis, of the dominant roots locus of the linearized system characteristic equation consists in imposing the stability outside a preset frequency interval where, in the case of rigid shafting, instability can take place. The mathematical model does not account for the damping effects of bearings, friction and ventilation. However, no further difficulty arises if these effects, which are always in favour of stability, are taken into account because the system order does not change.

ACKNOWLEDGEMENT

The author wishes to thank Professor Enrico Pagano, Electrical Machines Institute, University of Naples, for his advice.

SYMBOLS

r_s	per-unit stator resistance
r'_r	" rotor " referred to the stator winding
X_s	" stator reactance
X'_r	" rotor " referred to the stator winding
X_m	" magnetizing reactance
v_{qs}	" q-axis voltage
v_{ds}	" d-axis voltage
i_{qs}	" q-axis stator current
i_{ds}	" d-axis " "
i'_{qr}	" q-axis rotor current referred to the stator winding
i'_{dr}	" d-axis rotor current referred to the stator winding
ω_b (s^{-1})	rated angular velocity of stator voltages
$f_r = \frac{\omega_e}{\omega_b}$	per unit frequency
ω_1 (s^{-1})	electric angular velocity of motor rotor
$S = \frac{\omega_e - \omega_1}{\omega_e}$	slip
$p = \frac{d}{dt}$	
$C_b(j)$	rated motor torque
$C_e(j)$	instantaneous electromagnetic torque
$C_L(j)$	instantaneous load torque
J_1 ($Kg.m^2$)	moment of inertia of the motor rotor
J_2 ($Kg.m^2$)	moment of inertia of the mechanical load rotor
θ_1	instantaneous angular position of motor rotor
θ_2	instantaneous angular position of mechanical load rotor
K (j)	torsional spring constant of the shafting

REFERENCES

(1) G.J. Rogers, Linearized analysis of induction motor transients, Proc. IEE 112, 10, 1917-1926 (1965).

(2) R.H. Nelson, T.A. Lipo, P.C. Krause, Stability analysis of a symmetrical induction machine, IEEE Trans. PAS 88, 11, 1710-1717 (1969)

(3) I. Marongiu, Influenza dell'elasticità torsionale del sistema di trasmissione del moto sulla stabilità della macchina asincrona, Int. Rep. Electrotec. Inst. Univ. Cagliari, 8 (1976)

(4) G. Kron, (1942) A short course in tensor analysis, Wiley, New York

APPROXIMATE FORMULAE FOR THE CHARACTERISTIC VALUES OF ELECTRICAL MACHINES

M. Policastro and D. Torriano

Istituto di Elettrotecnica e di Elettronica, Università di Trieste, Via A. Valerio 10, Trieste, Italia

ABSTRACT

In order to have a deep insight into the dynamics of the phenomena that occur in electrical machines the knowledge of the dependency of the modes on the parameters of the machine is required.

In this note, approximate expressions are given for the characteristic values identifying the modes with reference to the case of machines equipped also with more than one winding system either on the stator or on the rotor. The results obtained allow a critical comparison with traditional formulae to be performed.

PREMISE

For the study of electrical machines, fed through electronic solid state devices, the knowledge of the electrical dynamics in constant speed operation is of paramount importance.

An approach based upon the employment of state variables allows efficient numerical solution methods to be made available for the solution of problems associated with the determination of electrical quantities' behaviours.

For a qualitative determination of these behaviours it is however important to derive expressions of the modes of electrical dynamics in which the influence of construction parameters of the machine are displayed.

Aim of the present investigation is to obtain the above mentioned results with reference to machines supplied with more than one winding system either on the stator or on the rotor.

INTRODUCTION

In recent papers (ref.1,2) the authors have examined this problem with reference to the case of isotropic machines; in these studies the circuital parameters were assumed to be known. The determination of these parameters in presence of anisotropy will be developed here. To tackle the problem in the most general way the case of a syncronous machine is considered. Reference is made to salient pole machine equipped with both a single-phase and an n-phase windings on the stator and with a m-phase winding on the rotor. The choice of this machine, justified by lack of space, accords anyway with the requirement of obtaining a more general treatment since it allows the study of the various kinds of machines (asyncronous, direct current, repulsion) to be performed without meeting with serious obstacles for the solution of the proposed problems.

DERIVATION OF THE MODEL

The procedure for the evaluation of electrical parameters that occur in the equations of the machine under examination will be similar to that adopted in (ref. 3). In particular, the same assumption as in (ref. 3) will be adopted to represent the anisotropy of the magnetic circuit; namely the permeability will be considered to vary according to the law

$$\mu = \mu_o - \mu_2 \cos 2\phi$$

where ϕ is the angular coordinate referred to a fixed frame.

Furthermore the hypotesis of sinusoidal current sheet distribution for each belt will be assumed, with symmetrical disposition for the case of multiphase windings.

Under these assumptions the expressions of the magnetic field in the point of the air-gap corresponding to the angular position ϕ will be:

$$\begin{aligned} H = & - a_r A_o i_o Z_o \sin\phi - a_\phi B_o i_o Z_o \cos\phi - \\ & + a_r A_1 \sum_{1}^{n}{}_k i_{sk} Z_s \sin[\phi-(k-1)\delta_s] - \\ & + a_\phi B_1 \sum_{1}^{n}{}_k i_{sk} Z_s \cos[\phi-(k-1)\delta_s] - \\ & + a_r A_2 \sum_{1}^{m}{}_h i_{rh} Z_r \sin[\phi-\theta-(h-1)\delta_r] - \\ & + a_\phi B_2 \sum_{1}^{m}{}_h i_{rh} Z_r \cos[\phi-\theta-(h-1)\delta_r] \end{aligned}$$

where:

θ defines the angular position of the rotor in the same reference frame as ϕ;

Z_f ; Z_s and Z_r are the conductor densities of the three windings of the machine;

i_f , i_{sk} , i_{rh} are the instantaneous values current respectively of the field winding, of the k-th stator phase belt and of the h-th rotor phase belt;

and where $\delta_s = \frac{2\pi}{n}$; $\delta_r = \frac{2\pi}{m}$;

$$A_o = A_1 = \frac{1 + a^2/r^2}{1 - a^2/b^2} \quad ; \quad B_o = B_1 = \frac{1 - a^2/r^2}{1 - a^2/b^2}$$

$$A_2 = - \frac{1 + b^2/r^2}{1 - b^2/a^2} \quad ; \quad B_2 = - \frac{1 - b^2/r^2}{1 - b^2/a^2}$$

a and b being the inner and the outer radius of the air-gap and r the radial coordinate of the generical point of the air-gap ($a < r < b$).

By comparison of the following two expressions of the magnetic energy stored only in the air-gap

$$W_m = \frac{1}{2} \int_0^l \int_a^b \int_0^{2\pi} (\mu_0 - \mu_2 \cos 2\phi) H \times H r \, dr \, d\phi \, dz$$

$$\begin{aligned} W_m = & \frac{1}{2} \{ (L^f_\mu + L^f_{\mu_2}) i_0^2 + \sum_{1}^{n}{}_k ({}_k L^{fs}_\mu + {}_k L^{fs}_{\mu_2}) i_f i_{s_k} + \\ & + \sum_{1}^{m}{}_h ({}_h L^{fr}_\mu + {}_h L^{fr}_{\mu_2}) i_f i_{rh} + \sum_{1}^{n}{}_{k_1} \sum_{1}^{n}{}_{k_2} ({}_{k_1,k_2} L^{ss}_{\mu_2}) i_{sk_1} i_{sk_2} + \\ & + \sum_{1}^{m}{}_{h_1} \sum_{1}^{m}{}_{h_2} ({}_{h_1,h_2} L^{rr}_\mu + {}_{h_1,h_2} L^{rr}_{\mu_1}) i_{rh_1} i_{rh_2} + \end{aligned}$$

$$+\sum_{1}^{n}{}_{k}\sum_{1}^{m}{}_{h}\left({}_{h,k}L^{rs}_{\mu}+{}_{h,k}L^{rs}_{\mu_2}\right)i_{sk}\,i_{rh}\}$$

one may derive the formulae for each of the inductance parameters needed in the circuital equations of the machine.

The meaning of the symbols, that occur in the above given expression of the energy, is assumed to be known: the only remark to be made concerns the intention to distinguish the contribution of the terms depending on μ_0 (that are indicated by the symbol L affected with index μ) from the one of the terms depending on μ_2 (indicated by L_{μ_2})

With proper manipulations one finds:

$${}_{k}L^{fs}_{\mu}=L^{fs}_{\mu}\cos((k-1)\,\delta_s) \qquad {}_{k}L^{fs}_{\mu_2}=L^{fs}_{\mu_2}\cos((k-1)\,\delta_s)$$

$${}_{h}L^{fr}_{\mu}=L^{fr}_{\mu}\cos(\theta+(h-1)\delta_r) \qquad {}_{h}L^{fr}_{\mu_2}=L^{fr}_{\mu_2}\cos(\theta+(h-1)\,\delta_r)$$

$${}_{k_1,k_2}L^{ss}_{\mu}=L^{ss}_{\mu}\cos((k_2-k_1)\,\delta_s) \qquad {}_{k_1,k_2}L^{ss}_{\mu_2}=L^{ss}_{\mu_2}\cos((k_1+k_2-2)\,\delta_s)$$

$${}_{h_1,h_2}L^{rr}_{\mu}=L^{rr}_{\mu}\cos((h_2-h_1)\,\delta_r) \qquad {}_{h_1,h_2}L^{rr}_{\mu_2}=L^{rr}_{\mu_2}\cos((h_1+h_2-2)\,\delta_r+2\,\theta)$$

$${}_{h,k}L^{sr}_{\mu}=L^{sr}_{\mu}\cos(\theta+(h-1)\,\delta_r-(k-1)\,\delta_s) \qquad {}_{h,k}L^{sr}_{\mu_2}=L^{sr}_{\mu_2}\cos(\theta+(h-1)\,\delta_r+(k-1)\,\delta_s)$$

These expressions clearly shows the dependency of the self - and mutual - inductances on θ. To complete the definition of the inductances the following expressions shall be used:

$$L^{f}_{\mu}=Z_f^2\,b^2\,F\ ;\ L^{ss}_{\mu}=Z_s^2\,b^2F\ ;\ L^{rr}_{\mu}=Z_r^2\,a^2\,F\ ;$$

$$L^{fs}_{\mu}=Z_f\,Z_s\,abG\ ;\ L^{fr}_{\mu}=Z_f\,Z_r\,abG\ ;\ L^{sr}_{\mu}=Z_s\,Z_r\,abG\ ; \qquad (1)$$

$$L^{f}_{\mu_2}=2\,Z_f^2\,b^2\,E\ ;\ L^{ss}_{\mu_2}=2\,Z_s^2\,b^2E\ ;\ L^{rr}_{\mu_2}=2Z_r^2\,a^2\,E$$

$$L^{fs}_{\mu_2}=Z_f\,Z_s(a^2+b^2)E\ ;\ L^{fr}_{\mu_2}=Z_fZ_r(a^2+b^2)E\ :\ L^{sr}_{\mu_2}=Z_sZ_r(a^2+b^2)E$$

where

$$F=l\,\pi\,\mu_0\,\frac{b/a+a/b}{b/a-a/b}\ ;\ G=l\,\pi\mu_0\,\frac{2}{b/a-a/b}\ ;\ E=l\pi\mu_2\,\frac{l_n\,b/a}{(b/a-a/b)^2} \qquad (2)$$

For notational convenience the matricial form of the equations will be used hereafter.

To this aim the listed below submatrices wile be introduced:

$\mathcal{L}^{f}_{\mu}$ a 1×1 dimensioned submatrix with element $=L^{f}_{\mu}$

$\mathcal{L}^{fs}_{\mu}$ a 1×n dimensioned submatrix with element $(1,k)={}_{k}L^{fs}_{\mu}$

$\mathcal{L}^{fr}_{\mu}$ a 1×m dimensioned submatrix with element $(1,h)={}_{h}L^{fr}_{\mu}$

$\mathcal{L}^{s}_{\mu}$ a n×n dimensioned submatrix with element $(k_1,k_2)={}_{k_1,k_2}L^{ss}_{\mu}$

$\mathcal{L}^{rr}_{\mu}$ a m×m dimensioned submatric with element $(h_1,h_2)={}_{h_1,h_2}L^{rr}_{\mu}$

$\mathcal{L}^{sr}_{\mu}$ a n×m dimensioned submatrix with element$(h,k)={}_{h,k}L^{sr}_{\mu}$

and similarly for the inductance matrices with elements depending on μ_2. To simplify the notations the following rule $\mathcal{L}^{\alpha\beta}=[\mathcal{L}^{\beta\alpha}]^T$ will be assume to hold.

Setting furthermore:

$$V_f=v_f\ ;\ V_s=(v_{s1}\ .....v_{sn})^T\ ;\ V_r=(v_{r_1}\ .....v_{rm})^T$$

$$I_f=i_f\ ;\ I_s=(i_{s1}\ .....i_{sn})^T\ ;\ I_r=(i_{r_1}\ .....\ i_{r_m})^T$$

(where the v-symbols denote, obviously, the instantaneous phase voltages) the circuital differential equations of the machine may be written as follows:

$$\begin{vmatrix} V_f \\ \cdots \\ V_s \\ \cdots \\ V_r \end{vmatrix} = \begin{vmatrix} \mathcal{L}^{f}_{\mu}+L^{f}_{d} & \mathcal{L}^{fs}_{\mu} & \mathcal{L}^{fr}_{\mu} \\ \mathcal{L}^{sf}_{\mu} & \mathcal{L}^{ss}_{\mu}+L^{s}_{d}\,\mathfrak{I}_n & \mathcal{L}^{sr}_{\mu} \\ \mathcal{L}^{rf}_{\mu} & \mathcal{L}^{rs}_{\mu} & \mathcal{L}^{rr}_{\mu}+L^{r}_{d}\,\mathfrak{I}_m \end{vmatrix} \begin{vmatrix} \dot{I}_f \\ \cdots \\ \dot{I}_s \\ \cdots \\ \dot{I}_r \end{vmatrix} +$$

$$+\begin{vmatrix} \mathcal{L}^{f}_{\mu_2} & \mathcal{L}^{fs}_{\mu_2} & \mathcal{L}^{fr}_{\mu_2} \\ \mathcal{L}^{sf}_{\mu_2} & \mathcal{L}^{ss}_{\mu_2} & \mathcal{L}^{sr}_{\mu_2} \\ \mathcal{L}^{rf}_{\mu_2} & \mathcal{L}^{rs}_{\mu_2} & \mathcal{L}^{rr}_{\mu_2} \end{vmatrix} \begin{vmatrix} \dot{I}_f \\ \cdots \\ \dot{I}_s \\ \cdots \\ \dot{I}_r \end{vmatrix} +$$

$$+\begin{vmatrix} R_f & 0 & \dot{\mathcal{L}}^{fr}_{\mu} \\ 0 & R_s\,\mathfrak{I}_n & \dot{\mathcal{L}}^{sr}_{\mu} \\ \dot{\mathcal{L}}^{rf}_{\mu} & \dot{\mathcal{L}}^{rs}_{\mu} & R_r\,\mathfrak{I}_m \end{vmatrix} \begin{vmatrix} I_f \\ \cdots \\ I_s \\ \cdots \\ I_r \end{vmatrix} + \qquad (3)$$

$$+\begin{vmatrix} 0 & 0 & \dot{\mathcal{L}}^{fr}_{\mu_2} \\ 0 & 0 & \dot{\mathcal{L}}^{sr}_{\mu_2} \\ \dot{\mathcal{L}}^{rf}_{\mu_2} & \dot{\mathcal{L}}^{rs}_{\mu_2} & \dot{\mathcal{L}}^{rr}_{\mu_2} \end{vmatrix} \begin{vmatrix} I_f \\ \cdots \\ I_s \\ \cdots \\ I_r \end{vmatrix}$$

L^{f}_{d} , L^{s}_{d} , L^{r}_{d} are scalars denoting the leakage inductances of each phase-belt; R_f , R_s and R_r are scalars denoting the corresponding resistances. The symbol $\mathfrak{I}_n$ denotes the identical $(n\times n)$ matrix.

The dotted symbols denote the time derivative of the corresponding entities.

In a compact operational notation:

$$V=\begin{vmatrix} V_f \\ \cdots \\ V_s \\ \cdots \\ V_r \end{vmatrix} = (H\,p+K)\,I=(H\,p+K)\cdot\begin{vmatrix} I_f \\ \cdots \\ I_s \\ \cdots \\ I_r \end{vmatrix}$$

As well known, the characteristic values of the system are the roots of the algebraic equation in p:

$$\det|H\,p+K|=0 \qquad (4)$$

The main result of the generalized machine theory assures the coefficients of this equation to be time-invariant. This may be proved via a suitable variable transformation; namely a new current vector $\hat{I}$, and a new voltage vector $\hat{V}$ will be introduced setting

$$\hat{I}=T\,I\ ;\ \hat{V}=T\,V$$

with

$$T=\begin{vmatrix} 1 & & \\ & T_n & \\ & & P_m \end{vmatrix} \qquad (5)$$

The submatrix T_n is the unitary Clarke transformation matrix of order n (Ref. 3); P_m is the unitary Park matrix of order m whose elements are defined as follows:

$$\text{element }(j,k)\text{ of }P_m=\begin{cases} \dfrac{1}{\sqrt{m}} & \text{for } j=1 \\ \sqrt{\dfrac{2}{m}}\cos[(k-1)(j-1)\delta_r+\theta] & \text{for } 1<j<[\frac{n+1}{2}] \\ \sqrt{\dfrac{2}{m}}\sin[(k-1)(m-j+1)\delta_r+\theta] & \text{for } [\frac{m+1}{2}]<j<m \end{cases}$$

It may be useful to remind that the Clarke matrix can be obtained by setting $\theta=0$ in the Park matrix of the same order.

After proper analitical manipulation one finds the following form the equation (3):

$$\hat{V} = (\hat{H}\,p + \hat{K})\hat{I} \qquad (6)$$

$\hat{H}\,p + \hat{K}$ being the time invariant matrix shown in (7).

In the following, use will be made of the notations listed here:

$$L_0 = L^f_\mu + L^f_{\mu_2} + L^f_d \qquad L_{10} = \sqrt{\frac{n}{2}}\,(L^{fs}_\mu + L^{fs}_{\mu_2})$$

$$L_1 = \frac{n}{2}(L^{ss}_\mu + L^{ss}_{\mu_2}) + L^s_d \qquad L_{30} = \sqrt{\frac{m}{2}}\,(L^{fr}_\mu + L^{fr}_{\mu_2}) \qquad (8)$$

$$L_2 = \frac{n}{2}\,(L^{ss}_\mu - L^{ss}_{\mu_2}) + L^r_d \qquad L_{13} = \frac{\sqrt{nm}}{2}(L^{sr}_\mu + L^{sr}_{\mu_2})$$

$$L_3 = \frac{m}{2}\,(L^{rr}_\mu + L^{rr}_{\mu_2}) + L^r_d \qquad L_{24} = \frac{\sqrt{nm}}{2}\,(L^{sr}_\mu - L^{sr}_\mu)$$

$$L_4 = \frac{m}{2}\,(L^{rr}_\mu - L^{rr}_{\mu_2}) + L^r_d$$

$$R_0 = R_f \;;\; R_1 = R_2 = R_s \;;\; R_3 = R_4 = R_r$$

It is a remarkable property of the transformation to show that the dynamical characteristic of the electrical phenomena, occuring in the machine, are mainly dependent from the roots of the fifth order equation:

$$\det|\hat{H}\,p + \hat{K}|_{rid} = 0 \qquad (9)$$

where $|H\,p + K''|_{rid}$ states for the submatrix made up with the elements with row - and colum - indices equal to 1, 3, n + 1, n + 3 and n + m + 1.

That is:

$$|\hat{H}\,p + \hat{K}|_{rid} = \begin{vmatrix} pL_0 + R_0 & pL_{10} & 0 & pL_{30} & 0 \\ pL_{10} & pL_1 + R_1 & 0 & pL_{13} & 0 \\ 0 & 0 & pL_2+R_2 & 0 & pL_{24} \\ pL_{30} & pL_{13} & \omega L_{24} & pL_3+R_3 & \omega L_4 \\ -\omega L_{30} & -\omega L_{13} & pL_{24} & -\omega L_3 & pL_4+R_4 \end{vmatrix} \qquad (10)$$

where $\omega = \theta$

DETERMINATION OF THE CHARACTERISTIC ROOTS

The structure of the matrix (7) clearly shows that n-2 characteristic roots are all equal to $-R_s / L^s_d$; that one (or two if m is even) of them equals $-R_r / L^r_d$ and that other m-3 (or m-4) roots assume in pair, the complex conjugate values $(-R_r/L^r_d \pm j\,\omega)$.

Regarding the five main roots of (9) it is easily seen that the same equation may be put in the form:

$$\det|\hat{H}p + \hat{K}|_{rid} = B(p) + \omega^2 A(P) = 0 \qquad (11)$$

where

$$B(p) = \det\begin{vmatrix} pL_0 + R_0 & pL_{10} & pL_{30} \\ pL_{10} & pL_1 + R_1 & pL_{13} \\ pL_{30} & pL_{13} & pL_3+R_3 \end{vmatrix} \cdot \det\begin{vmatrix} pL_2 + R_2 & pL_{24} \\ pL_{24} & pL_{24} + R_4 \end{vmatrix} \qquad (12)$$

and

$$A(p) = \det\begin{vmatrix} pL_0 + R_0 & pL_{10} & pL_{30} \\ pL_{10} & pL_1 + R_1 & pL_{13} \\ L_{30} & L_{13} & L_3 \end{vmatrix} \cdot \det\begin{vmatrix} pL_2 + R_2 & pL_{24} \\ L_{24} & L_4 \end{vmatrix} \qquad (13)$$

The particular form (11) given to the reduced characteristic equation allows a significant meaning to be attributed to B(p) and A(p); since it is self-evident that B(p) corresponds to the characteristic polynomial pertaining the starting condition of the machine, whereas A(p) is the polynomial corresponding to the condition of the machine running at infinite speed.

In the most general situation the roots can therefore be derived by locating their intermediate position in the p-plane: this can become less troublesome if one considers the way of varying of the roots as a function of ω. This is a problem that commonly draws one to the root-locus technique.

Since in the most frequent cases ω assumes large values, it may be sufficient, for an approximate evaluation of the roots, to consider their asymptotic behaviour as functions of ω. This means that three of the roots are almost equal to the zeroes of A(p), while the other two becomes large, approaching two asymptotes whose location may be derived by the well known properties of the root-locus. With proper deductions (ref.4) one finds that the two asymptotes are vertical lines departing from the real point with abscissa:

$$x_c = -\frac{1}{2}\left\{\frac{R_3(L_0L_1 - L^2_{10})}{L_0L_1L_3 + 2L_{10}L_{13}L_{30} - L_0L^2_{13} - L_1L^2_{30} - L_3L^2_{10}} - \frac{R_4\,L_2}{L_2\,L_4 - L^2_{24}}\right\} \qquad (14)$$

Consequently the following approximate expressions may be given for three of the roots:

$$_1p_2 = x_c \pm j\,\omega \qquad (15)$$

$$p_3 = -\frac{R_2\,L_0}{L_2\,L_4 - L^2_{24}} \qquad (16)$$

The other two roots may be obtained as zeros of the first (quadratic) factor of A(p) given in (13).

These results represent a significant alternative, for the determination of the characteristic values of the machine, to their direct evaluation as solutions of the fifth-order equation (11) which, of course, must be necessarily a numerical one. In the following consideration will be given to the possibility to find out, with further semplifying assumptions, the expressions of the two last roots.

THE NORMALIZED EQUATIONS

The aim of the present analisys is to give new furtherly approximated expressions of the characteristic roots in order to establish a critical comparison with the values currently given, for the same roots, in the traditional approach (ref. 5, 6).

In this context, results are normally obtained by examining the behaviour of the machine under short,- circuit condition; comparable results will be obtained here on the basis of the preceeding analisys so overcoming the restrictive point of view of the traditional method.

By introducing new current and voltage variables, through the transformations:

$$\tilde{V} = \tau\,\hat{V} \quad \text{and} \quad \tilde{I} = \tau^{-1}\,\hat{I}$$

where:

$$
p\hat{H}+\hat{K} = \left[\begin{array}{c|cccc|cccc}
p(L^f_\mu + L^f_{\mu_2} + L^f_d) + R_f & 0 & p\sqrt{\frac{n}{2}}(L^{fs}_\mu + L^{fs}_{\mu_2}) & \cdots & 0 & 0 & p[\sqrt{\frac{m}{2}}(L^{fr}_\mu + L^{fr}_{\mu_2})] & \cdots & 0 \\
\hline
0 & pL^s_d + R_s & 0 & \cdots & 0 & 0 & 0 & \cdots & 0 \\
p[\sqrt{\frac{n}{2}}(L^{fs}_\mu + L^{fs}_{\mu_2})] & 0 & p[L^s_d + \frac{n}{2}(L^{ss}_\mu + L^{ss}_{\mu_2})] + R_s & \cdots & 0 & 0 & p[\frac{\sqrt{nm}}{2}(L^{sr}_\mu - L^{sr}_{\mu_2})] & \cdots & 0 \\
0 & 0 & 0 \;\; pL^s_d + R_s & \cdots & 0 & 0 & 0 & \cdots & 0 \\
\vdots & \vdots & \vdots & \ddots & \vdots & \vdots & \vdots & \ddots & \vdots \\
0 & 0 & 0 & \cdots \;\; pL^s_d + R_s & 0 & 0 & 0 & & 0 \\
0 & 0 & 0 & \cdots & p[L^s_d + \frac{n}{2}(L^{ss}_\mu - L^{ss}_{\mu_2})] + R_s & 0 & 0 & \cdots & p[\frac{\sqrt{nm}}{2}(L^{sr}_\mu - L^{sr}_{\mu_2})] \\
\hline
0 & 0 & \cdots & \cdots & 0 & pL'_d + R_r & 0 & \cdots & 0 \\
p[\sqrt{\frac{m}{2}}(L^{fr}_\mu + L^{fr}_{\mu_2})] & 0 & p[\frac{\sqrt{nm}}{2}(L^{sr}_\mu + L^{sr}_{\mu_2})] & \cdots & \omega[\frac{\sqrt{nm}}{2}(L^{sr}_\mu - L^{sr}_{\mu_2})] & 0 & p[L_d + \frac{m}{2}(L^{rr}_\mu + L^{rr}_{\mu_2})] + R_r & \cdots & \omega[L^r_d + \frac{m}{2}(L^{rr}_\mu - L^{rr}_{\mu_2})] \\
0 & 0 & 0 & & 0 & 0 & 0 \;\; pL^r_d + R_r & \omega L^r_d & 0 \\
\vdots & \vdots & \vdots & \ddots & \vdots & \vdots & \vdots & \ddots & \vdots \\
0 & 0 & 0 & & 0 & 0 & 0 \;\; -\omega L^r_d & pL^r_d + R_r & 0 \\
-\omega\sqrt{\frac{m}{2}}(L^{fr}_\mu + L^{fr}_{\mu_2}) & 0 & -\omega[\frac{\sqrt{nm}}{2}(L^{sr}_\mu + L^{sr}_{\mu_2})] & \cdots & p[\frac{\sqrt{nm}}{2}(L^{sr}_\mu - L^{sr}_{\mu_2})] & 0 & -\omega[L^r_d + \frac{m}{2}(L^{rr}_\mu + L^{rr}_{\mu_2})] & \cdots & p[L^{rr}_d + \frac{m}{2}(L^r_\mu - L^{rr}_{\mu_2})] + R_r
\end{array}\right] \quad (7)
$$

$$\mathcal{T} = \begin{vmatrix} Z_f & & & & \\ & \sqrt{\frac{n}{2}}Z_s & & & \\ & & \sqrt{\frac{n}{2}}Z_s & & \\ & & & \sqrt{\frac{m}{2}}Z_r & \\ & & & & \sqrt{\frac{m}{2}}Z_r \end{vmatrix}$$

the reduced impedance equations become:

$$\tilde{V} = \begin{vmatrix} pl_0 + r_0 & pl_{10} & 0 & pl_{30} & 0 \\ pl_{10} & pl_1 + r_1 & 0 & pl_{13} & 0 \\ 0 & 0 & pl_2 + r_2 & 0 & pl_{24} \\ pl_{30} & pl_{13} & \omega l_{24} & pl_3 + r_3 & \omega l_4 \\ -\omega l_{30} & -\omega l_{13} & pl_{24} & -\omega l_3 & pl_4 + r_4 \end{vmatrix} . \tilde{I} \qquad (17)$$

where:

$$\begin{aligned} l_o &= b^2(F + 2E) + l_d^f \\ l_1 &= b^2(F + 2E) + l_d^s \\ l_2 &= b^2(F - 2E) + l_d^s \\ l_3 &= a^2(F + 2E) + l_d^r \\ l_4 &= a^2(F - 2E) + l_d^r \\ l_{10} &= l_{30} = l_{13} = Gab + (a^2 + b^2)E \\ l_{24} &= Gab - (a^2 + b^2)E \end{aligned} \qquad (18)$$

and:

$$l_d^f = \frac{L_d^f}{Z_f^2} \; ; \quad l_d^s = \frac{L_d^s}{Z_s^2}\frac{2}{n} \; ; \quad l_d^r = \frac{L_d^r}{Z_r^2}\frac{2}{n} \; ; \qquad (19)$$

$$r_o = \frac{R_f}{Z_f^2} \; ; \quad r_1 = r_2 = \frac{R_s}{Z_s^2}\frac{2}{n} \; ; \quad r_e = r_4 = \frac{R_r}{Z_r^2}\frac{2}{m} \; ;$$

and E, F, G are given in (2).

A first relevant advantage of the transformation is the simplicity and the symmetry of the expressions of the inductances occuring in the equations; the direct-axis mutual-inductances (l_{10}, l_{13}, l_{30}) assume in fact the same value. A remark should also be made about the formal identity of the expression (11) and (17), which allows the correctness of the preceeding results to be confirmed, requiring only the substitution of the capital symbols with the small ones. Since the variable transformation doesn't affect the roots of the characteristic equation all of the following deductions may be drawn from the previously obtained results only by applying the forementioned substitution.

A comparison with the equations presented in (ref.5) shows some discrepancies between the two results.

In fact (Ref.5) assumes the following quantities to be identical:

$$\begin{aligned} b^2(F+2E) &\simeq a^2(F+2E) \simeq Gab + (a^2+b^2)E \\ b^2(F-2E) &\simeq a^2(F-2E) \simeq Gab - (a^2+b^2)E \end{aligned} \qquad (20)$$

These assumptions hold only if a further degree of approximation is introduced.

Namely, setting $\delta = b - a$, a Laurent expansion gives:

$$E = \frac{\pi l}{4}\mu_2 \left(\frac{a}{\delta} + \frac{1}{2} - \frac{\delta}{4a} + ...\right)$$

$$F = \pi l \mu_0 \left(\frac{a}{\delta} + \frac{1}{2} + \frac{\delta}{4a} + ...\right)$$

$$G = \pi l \mu_0 \left(\frac{a}{\delta} + \frac{1}{2} - \frac{\delta}{4a} + ...\right)$$

Neglecting all but the first term in these series expansions, one gets after substitution:

$$\begin{aligned} a^2(F \pm 2E) &\simeq \frac{a^3}{\delta}\left(\mu_o \pm \frac{\mu_2}{2}\right)\pi l \\ b^2(F \pm 2E) &\simeq \frac{a^3}{\delta}\left(\mu_0 \pm \frac{\mu_2}{2}\right)\pi l \\ Gab \pm (a^2 + b^2)E &\simeq \frac{a^3}{\delta}\left(\mu_0 \pm \frac{\mu_2}{2}\right)\pi l \end{aligned} \qquad (21)$$

which is consistent with (20).

On this basis it is possible to obtain an analytical expression for the two, till now unexpressed zeroes of the first factor of A(p) occuring (in the normalized form) in (13) which now reads:

$$[l_{md}(l_d^f l_d^s + l_d^f l_d^r + l_d^s l_d^r) + l_d^f l_d^s l_d^r]\, p^2 +$$
$$+ \{r_o[l_{md}(l_d^s + l_d^r) + l_d^s l_d^r] + r_1[l_{md}(l_d^f + l_d^r) + l_d^f l_d^r]\}p +$$
$$+ [r_o r_1(l_{md} + l_d^r)]$$

where l_{md} is the common value of the quantities (21), with the plus sign (while l_{mq} will denote the some quantities with the minus sign).

Neglecting the quantity $r_{or_1}(2l_m l_d^r)^2$, one obtains the following approximate expressions for the riciprocals of the roots p_4 and p_5, that is, for the direct axis transient and subtransient time constants of the machine:

$$\begin{aligned} \frac{1}{p_4} &= T'_d - \frac{1}{r_o}\left\{ l_d^f + \frac{l_m l_d^s l_d^r}{l_m l_d^s + l_l l_d^r + l_d^s l_d^r} \right\} \\ \frac{1}{p_5} &= T''_d - \frac{1}{r_1}\left\{ l_d^s + \frac{l_m l_d^f l_d^r}{l_m l_d^f + l_m l_d^r + l_d^f l_d^s} \right\} \end{aligned} \qquad (22)$$

THE CASE OF THE MACHINE WITH SINGLE WINDING ON EACH ZONE

It is now possible to derive from the previous analysis similar results for the machine supplied with only one winding system on each zone.

A simple reinspection of the developments presented before allows the model of this machine to be derived without any difficulty, confirming what was said in introduction.

In fact it will be sufficient to neglect the terms referring to the field winding, in the already derived expressions, to obtain the new ones fit for the present case.

The fact that some machines, of the kind here considered, present no anisotropy in their magnetic circuits, like the asyncronous ones, may be taken into account by simply neglecting the terms containing μ_2.

According to these remarks one can write down the normalized equations directly as follows:

$$\begin{vmatrix} \tilde{V}_{ds} \\ \tilde{V}_{qs} \\ \tilde{V}_{dr} \\ \tilde{V}_{qr} \end{vmatrix} = \begin{vmatrix} l_1\, p + r_1 & 0 & l_{13}\, p & 0 \\ 0 & l_2\, p + r_2 & 0 & l_{24}\, p \\ l_{13}\, p & \omega\, l_{24} & l_3\, p + r_3 & \omega\, l_4 \\ -\omega\, l_{13} & l_{24}\, p & -\omega\, l_3 & l_4\, p + r_4 \end{vmatrix} \begin{vmatrix} \tilde{I}_{ds} \\ \tilde{I}_{qs} \\ \tilde{I}_{dr} \\ \tilde{I}_{qr} \end{vmatrix} \tag{23}$$

with the same meaning for the symbols as before.

The characteristic equation is still of the form:

$$B(p) + \omega^2 A(p) = 0 \tag{24}$$

with:

$$B(p) = \det \begin{vmatrix} l_1 p + r_1 & l_{13}\, p \\ l_{13} p & l_3 p + r_3 \end{vmatrix} \det \begin{vmatrix} l_2\, p + r_2 & l_{24}\, p \\ l_{24}\, p & l_4\, p + r_4 \end{vmatrix} \tag{25}$$

$$A(p) = \det \begin{vmatrix} l_1\, p + r_1 & l_{13}\, p \\ l_{13} & l_3 \end{vmatrix} \det \begin{vmatrix} l_2\, p + r_2 & l_{24}\, p \\ l_{24} & l_4 \end{vmatrix}$$

The approximate values of the characteristic roots may be found out with the same reasoning, and are:

$$p_{1,2} = x_c \pm j\, \omega$$

with:

$$x_c = -\frac{1}{2} \left(\frac{r_3\, l_1}{l_1\, l_3 - l_{13}^2} + \frac{r_4\, l_2}{l_2\, l_4 - l_{24}^2} \right)$$

and:

$$p_3 = -\frac{r_1\, l_3}{l_1\, l_3 - l_{13}^2}$$

$$p_4 = -\frac{r_2\, l_4}{l_2\, l_4 - l_{24}^2}$$

with the same assumptions that make (20) to hold, one gets:

$$\frac{1}{p_3} = -\frac{1}{r_1} \left(l_d^s + \frac{l_{md}\, l_d^r}{l_{md} + l_d^r} \right) \; ; \; \frac{1}{p_4} = -\frac{1}{r_2} \left(l_d^s + \frac{l_{mq}\, l_d^r}{l_{mq} + l_d^r} \right)$$

A useful remark can be made about the isotropic case: from (18) one derives:

$$l_1 = l_2 \quad ; \quad l_3 = l_4 \quad ; \quad l_{13} = l_{24}$$

Since $r_1 = r_2$ and $r_3 = r_4$ it follows that the approximate values for p_3 and p_4 are equal. Concerning this conclusion attention should be given to the fact that, in this case, (24) admits only two complex conjugate pair of roots. For a more accurate expressions of p_3 and p_4 the reader should refer to (Ref.2).

CONCLUSIONS

The stationarity property of the model proposed for the case of an anisotropic machine supplied with more than one winding for each zone was shown and the consequent possibility of recognizing the modes of the electrical dynamics was investigated.

The possibility of describing the machines via a normalized equation system, the structure of which is optimal as for simplicity, was put into evidence. The coefficients can be deduced from the construction characteristic and from electrical parameters that can be easily determinated experimentally.

On this basis a procedure was suggested, wich allowed some approximate expressions of the system characteristic values to be determinated.

In this way the order of approximation and the simplifying hypothesis which must be adopted in order to recognize the validity of the expressions supplied by the traditional treatment were evinced.

REFERENCES

[1] POLICASTRO M. and TORRIANO D.; «Sulla rappresentazione di macchine elettriche con più sistemi di avvolgimenti»; Alta Frequenza N° 7 vol. XLV, pp. 439-448, July 1976

[2] POLICASTRO M. and TORRIANO D.; «Su una valutazione approssimata dei modi della macchina elettrica generalizzata»; Alta Frequenza N° 7 vol. XLV, pp. 449-456, July 1976.

[3] WHITE D.C. and WOODSON H.H.; «Electromechanical Energy Conversion,. John Wiley and Sons Inc. New York 1959

[4] POLICASTRO M. and TORRIANO D.; «Approximate expressions of the modes of generalized electrical machines». to be published on Alta Frequenza (in english)

[5] ADKINS B. «The General Theory of Electric Machines» Chapman e Hall Ldt, London 1957

[6] JONES V.C.«The Unified Theory of Electrical Machines» Butterworths-London 1967

THE INFLUENCE OF THYRISTOR CONTROL ON TRACTION MOTORS

H. Buchberger

Siemens AG - DW, D - 1000 Berlin 13

INTRODUCTION

Traction duty efforts free choice of speed and acceleration of a vehicle within certain limits. Allowing for gear and wheel-diameter the tractive effort of the vehicle responds to the motor-torque and vehicle speed to motor speed. The figurative operating range is appointed by the application-diagram of the vehicle. (Fig. 1)

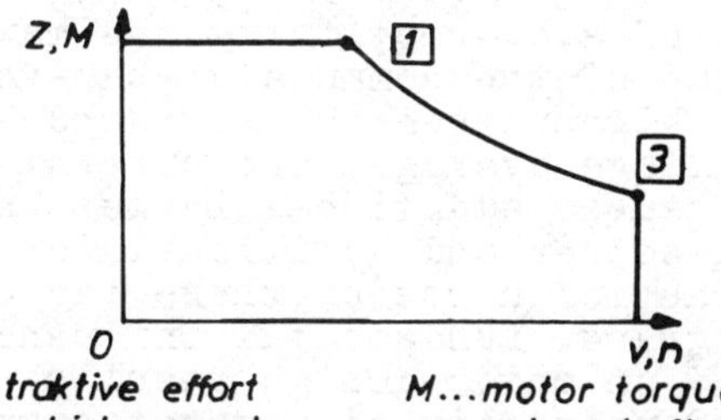

Fig. 1. Operation diagram

In the present, commutator motors and asynchronous motors with squirrel cage are in use in traction duty.

COMMUTATOR MOTOR

The main field of the commutator- is produced by the excitation winding. (Fig. 2)

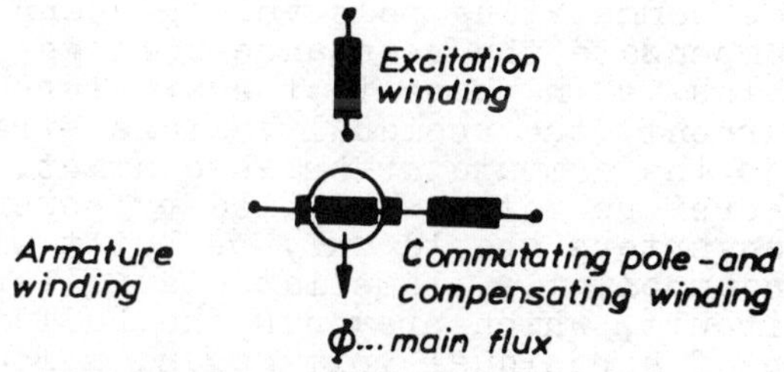

Fig. 2. Commutator motor Position of windings

The brushes connect commutator and stator winding, forming the rotor coil , whose axis is vertical to the axis of the excitation winding regardless of the rotation of the rotor.

These are the relations between electrical and mechanical quantities:

$$U = c_U \cdot n \cdot \Phi \qquad (1)$$
$$M = c_m \cdot I \cdot \Phi \qquad (2)$$
$$f = n \cdot p \qquad (3)$$

c_U ... constant of motor
c_m ... " " "
U ... voltage across brushes
I ... armature current
n ... shaft speed of rotor
Φ ... magnetic flux
p ... number of motor poles
f ... frequency of rotor conductor current

The magnetic flux is -on account of the right angle between the axes of the windings of armature and excitation- without influence by the armature circuit excited by the excitating current only. So it is -according to equ. 1-necessary to regulate the voltage for regulating the speed if the flux is constant. On the other side regulation of flux at constant voltage leads to a regulation of the speed as well. (Fig. 3)

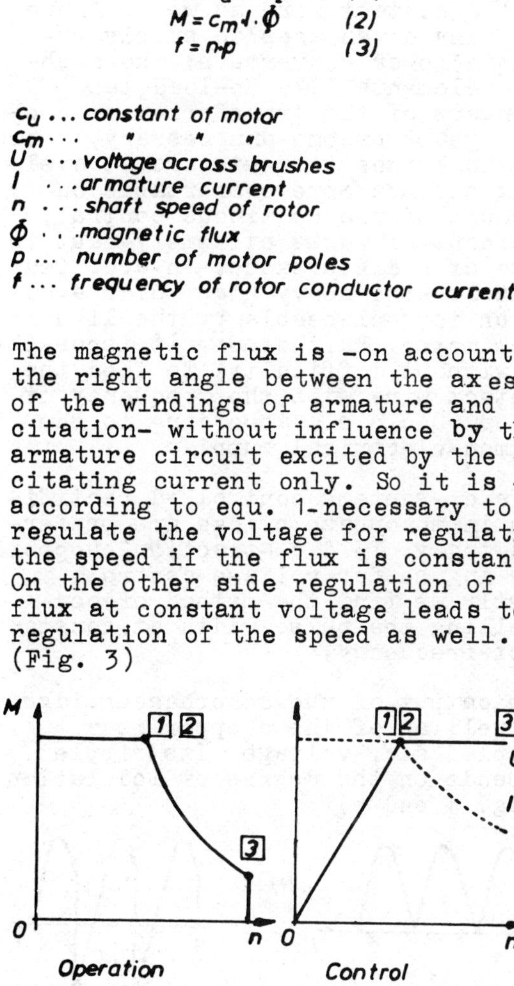

Fig. 3. Operation and control diagram

The operation diagram f.i. can be fulfilled by control of the voltage of a series-wound motor. The control diagram shows besides the controlling quantity U the corresponding functions of flux $\Phi(n)$ and current $I(n)$.

On account of this simple possibility of speed control the commutator motor has conquerred the field of electrical traction. The motor is series-wound and controlled by means of series resistances and multiple series. Field weakening by means of shunts is possible as well. On account of the possibility of energy supply with high voltage some railway companies preferred the a.c.-series-wound motor for 16 2/3 or 25 c.p.s. The voltage is controlled lossfree by means of a step-switched transformer.

The rapid development of semiconductors has opened new possibilities for control of commutator motors for both d.c. and a.c. feeding of the contact wire as well. There are many advantages in supply by static power converters: The mechanic elements like on-load tap changers of the transformer or control gears become unnecessary, there are no losses in resistors, travelling becomes more comfortable on account of the continous control, the control works extremely fast in case of a disturbance. In a.c. fed vehicles the heavy high polar a.c.-motor is replaceable by the lighter d.c. motor. Furthermore it ispossible without difficulty, to feed the contact wire with the frequency of 50 or 60 c.p.s., according to the common electrical supply.

In a.c.-systems controlled rectifiers in monophase bridge are preferred today. In d.c.-traction choppers are in use for voltage-control. Mostly we find the method of controlling the pulse width at constant tact-frequency.

The output of the monophase-bridge as well as of the chopper is a rippled d.c.-voltage. Its ripple depends on the degree of modulation (Fig. 4 and 5)

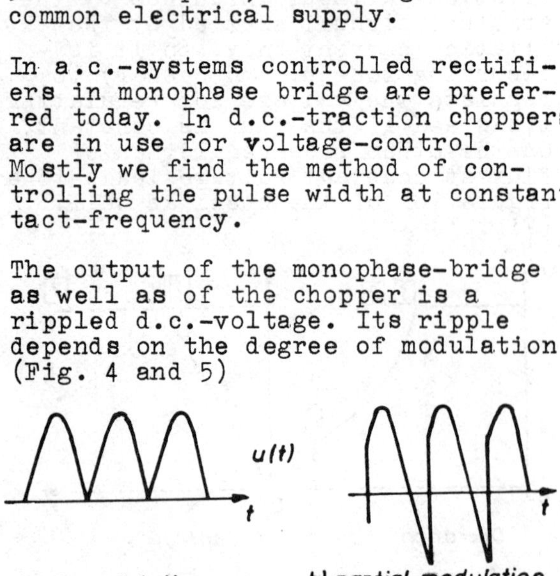

Fig. 4. Singlephase rectifier waveform of output voltage

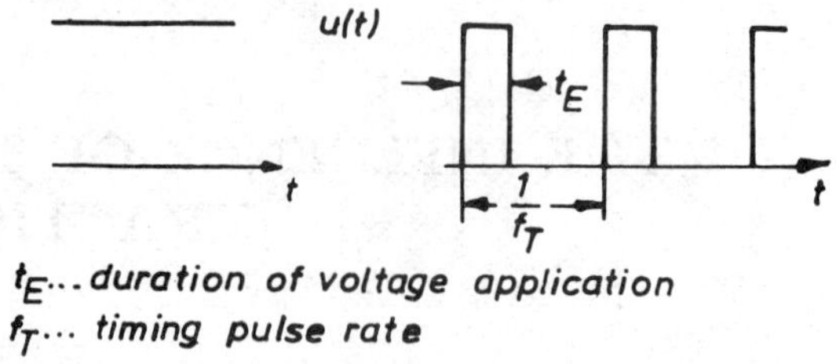

Fig. 5. Chopper, waveform of output voltage

These courses of time of the voltage are additional demands for the traction motor. Dependent on the value of the impedance between chopper, resp. rectifier and motor the motor is fed with rippled d.c.-current. (Fig. 6)

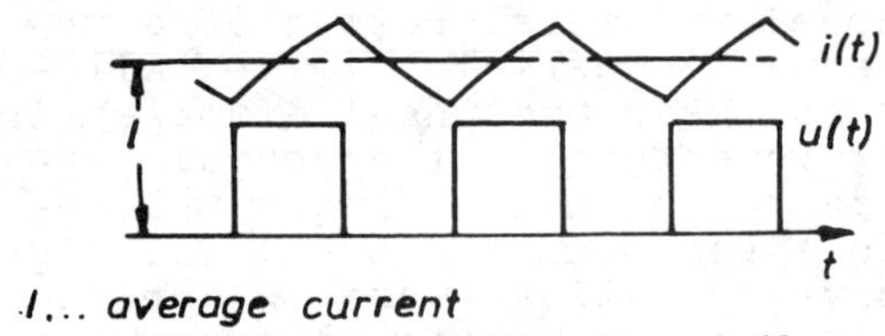

Fig. 6. Rippled d.c.-current from half modulated chopper

Influence on motor

In case of a.c.-rectifying the maximum value of the commutator-bar-voltage is higher then the corresponding voltage average. The current-ripple causes additional losses in winding-copper and additional torques. The influence of these facts is not grave. Fundamental influence, however, on motor-design results from the necessity of perfect commutation. If the commutating pole circuit is damped, f.i. by a massive yoke, the commutating field is not in phase with the exciting current. (Fig. 7)

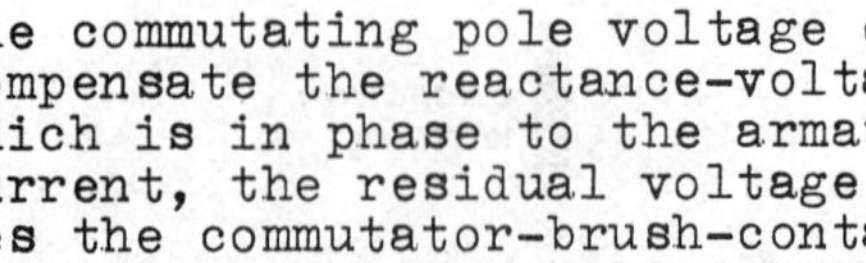

Fig. 7. Vector diagram of alternating components of reactance voltage $e_{S\sim}$ and commutating field voltage $e_{W\sim}$ if comm. field circuit is damped

The commutating pole voltage cannot compensate the reactance-voltage which is in phase to the armature current, the residual voltage straines the commutator-brush-contact. Therefore motors supplied by power converters should only be built with undamped, i.e. lamellated auxiliary circuit, which prevents the building up of a residual voltage and allows good commutation. Simultaneous the safety against fire of commutator will be improved during switching operations.

Influence on excitation circuit:

a) Series-excitation
Temporal fluctuations of the main flux induce transformer EM-Forces in the armoture coils, which are short-circuited by the brushes. These EMF's produce brush fire if they are not limited to inoffensive values.

The ripple of current is to be changed in an undangerous ripple of the flux. The alternating flux can be damped by short circuit currents in a massive yoke (internal damping). If the yoke is laminated it is possible to lead a great part of the a.c.-current through a shunt paralleled to the exitation-winding (external damping).

Figure 8 shows the complete circuit of a completely laminated series-wound chopper-fed d.c.-motor.

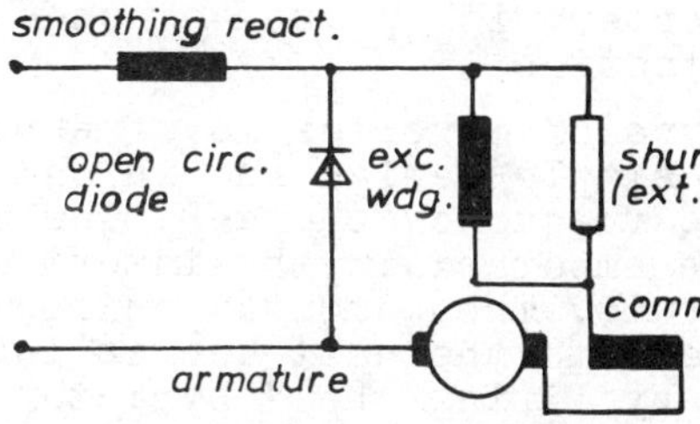

Fig.8. Series wound motor for chopper-supply, circuit scheme

b) Separate excitation
The refinement of control engineering has made it possible to equip traction motors with a separate exctition. The excitation winding is fed by an own rectifier. Thus the transformer electromotoric forоe can be kept very small.

COMMUTATOR MOTORS, SUMMARY

Static converter-fed commutator motors should be laminated in the magnetic auxiliary circuit. In the case of series-wound machines an external damping limits the transformer EMF in the commutating coils. Separate excitation is possible.

INDUCTION MOTOR

The shaft speed of a commutator motor is limited by the commutator, brushes and commutator need regular maintenance.
If it is possible to design a motor with higher speed for the same operation diagram of the vehicle, this motor will be smaller on account of the smaller torque. Squirrel cage induction motors allow more revolutions p.min. than commutator motors.
Up to the recent time is was not possible to use squirrel cage induction motors in traction, because the necessary requirement of free control of speed could not be fulfilled by simple means.

The relations for traction-duty are here:

$$U = c_U \cdot f_1 \cdot \Phi \qquad (4)$$
$$n = (1-s) \cdot \frac{f_1}{p} \qquad (5)$$
$$M = c_m \cdot I_1 \cdot \Phi \cdot \cos\vartheta \qquad (6)$$

c_U ... *constant of motor*
c_m ... *constant of motor*
U ... *terminal voltage*
n ... *shaft speed of rotor*
I_1 ... *stator current*
f_1 ... *frequency of U and I_1*
s ... *slip*
p ... *number of motor poles*
ϑ ... *load angle*

A magnetizing current with the frequency f_1 excites the flux Φ which rotates in the space at the synchronous speed of f_1/p r.p.s. The rotor follows the rotating field corresponding to equation 5. The voltage doesn't influence the speed, it determines the flux Φ at a given frequency corresponding to equation 4. A change of speed is caused by a change of frequency. The field is kept constant, if (Equ. 4) the frequency is changed proportional to the voltage. Increasing of frequency at constant voltage causes field weakening (Fig. 9).

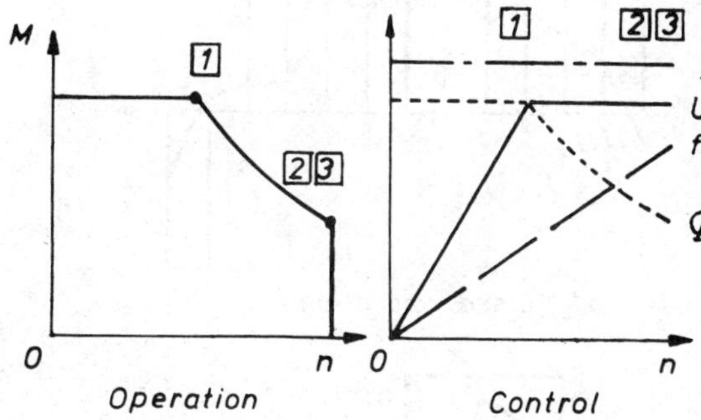

Fig.9. Operation and control diagram

An operation diagram $M(n)$ is created at its limits f.i. by a relation of the controlling variables $U(n)$ and $f_1(n)$, Flux Φ and current I_1 appear as shown in the figure.

Control of polyphase voltage in rate and frequency can be obtained by converters either of the constant voltage or the constant current type.

a) Pulse-inverter

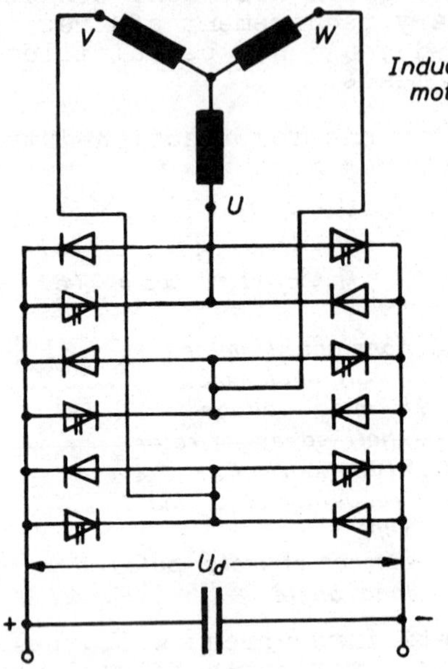

Fig. 10. Pulse inverter, circuit scheme

This inverter is one of the constant voltage type. It applies the constant d.c.-voltage cyclically in rectangular blocks of certain lengths of period to the terminals of the machine. In the sector of constant flux the rate of voltage is controlled simultaneously with the frequency by "pulsing" (Fig. 10 and 11). During field weakening the d.c.-voltage is applied to the terminals in blocks of 120 degrees el., only the frequency is changed.

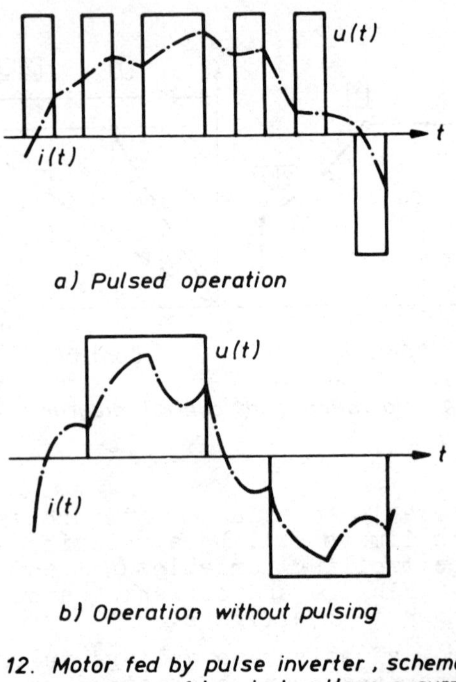

Fig. 12. Motor fed by pulse inverter, scheme of waveform of terminal voltage a. current

Fig. 11 shows the demands of the inverter which are to be fulfilled by the machine:
The maximum of the current, which must be switched by the thyristors, depends on the leakage inductance of the windings. Therfore the leakage inductance of the motor should be on a high level. The magnetizing current should be as low as possible, so that a good power factor results. The high harmonic oscillations of the currents of stator and rotor cause additional current dependent losses compared with operation at sinusoidal quantities.
The additional copper losses due to inverter feeding must be kept as low as possible. That means that the skin effect must be kept low by adequate design of the stator- and rotor-windings. The iron-losses due to the main field are not essentially increased if the motor is fed by inverter.

b) Inverter with phase-sequence-extinction

This type of inverter is of especially simple design and is only used as frequency changer. The current is smoothed in the intermediate ciruit. The commutation from one phase to the next one is initiated by firing the "thyristor in phase-sequence". The current is switched cyclically to the motor terminals in blocks of 120 degr. el. (Fig. 12)

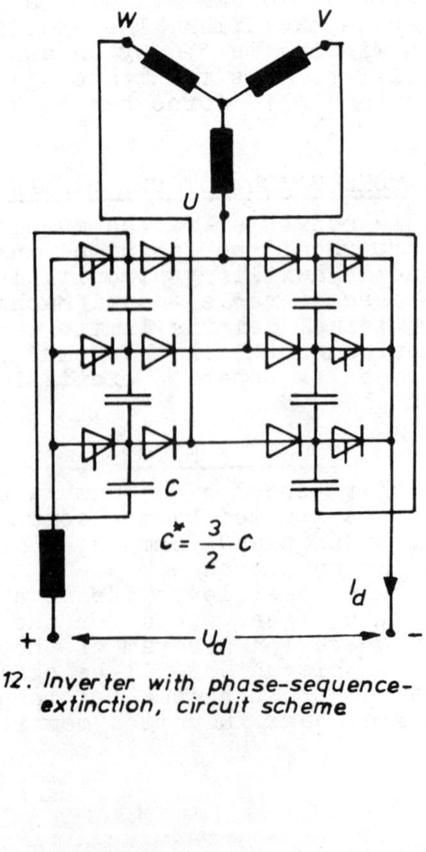

Fig. 12. Inverter with phase-sequence-extinction, circuit scheme

The current I_1 is controlled by the rate of the voltage U_d in the intermediate circuit. Regulating unit is, dependent on the available supply of the contact wire, a controlled rectifier or a chopper - just like in the case of the thyristor fed d.c. - motor. The voltage U_1 of the machine, resp. the flux ϕ appears load-dependent according to equation 7.

$$U_d = U_1 \cdot \cos\varphi \cdot 1{,}35 \quad (7)$$

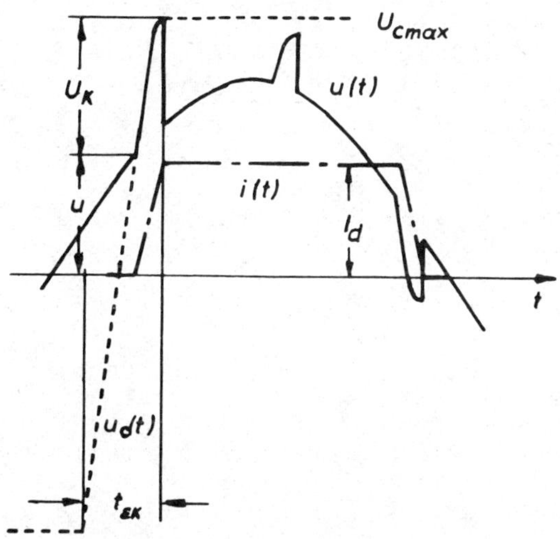

Fig. 13. Inverter with phase-sequence-extinction, time function of terminal voltage u(t) and current i(t)

The charge-reversal of the capacitors begins when the thyristor of the next phase is fired. The capacitor-voltage follows the curve $u_c(t)$ (Fig.12). When the capacitor-voltage and the momentary value of the sinusoidal Motor-E.M.F. u induced by the main field are equal, then the commutation of current from phase to phase starts. The value of u depends on the values of U_1 , I_d and $\cos\varphi$ in question. During commutation, the voltage between the motor terminals increases as the capacitor-voltage. This commutation-voltage U_K becomes the higher, the higher the leakage-inductance of the machine and the lower the capacity of the commutating capacitors are.

Another relation between inverter and motor yields form the demand, that the charge-reversal of the capacitors must have ended before it can be started anew. This reversal time $t_{εK}$ is - in a first approach - the higher, the lower the current and the higher the capacity are. It must be less than one third of the time required for one period of the frequency at maximum speed.

On account of these relations the use of a certain thyristor - for instance - requires a motor with certain values of leakage inductance as well as of magnetizing current at maximum speed. The insulation of the motor has to be provided for the maximum capacitor voltage U_{Cmax}. Similar to the operation on a pulse converter here stator and rotor currents also carry high harmonic oscillations. Therefore the windings must be designed as skin effect-free as possible in order to keep the additional copper-losses small. The shape of the field is sinusoidal and rotates at synchronous speed. Therefore the iron-losses due to the main field are not increased compared to the losses at sinusoidal operation.

INDUCTION MOTORS, SUMMARY

On account of the development in thyristor techniques it is possible now to use the commutatorless, sturdy squirrel cage induction motor for traction duty. The motor determines the work of the converter by means of its characteristic quantities like leakage inductance, rate of voltage, current and powerfactor. This must be carefully taken into consideration by the designer. The currents contain high harmonic oscillations, which increase the losses of the machine compered to the losses of sinusoidal currents.

REFERENCES

(1) K. Heintze, H. Tappeiner, M. Weibelzahl, Pulswechselrichter zur Drehzahlsteuerung von Asynchronmotoren, Siemens-Zeitschrift 45, 154 (1971)

(2) H. Klautschek, Asynchronmaschinenantriebe mit Stromzwischenkreisumrichtern, Siemens-Zeitschrift 50, 23 (1976)

(3) J. Körber, Die elektrische Lokomotive mit kommutatorlosen Fahrmotoren, ETR 24, 7/8 (1975)

(4) J. Kuhlow, Fahrmotoren für Stromrichterfahrzeuge, ETZ-A 88, 399 (1977)

(5) W. Waidmann, Drehstromantrieb für Gleichstrombahnen, Siemens-Zeitschrift 50, 493 (1976)

OPEN LOOP CONTROL OF A LINEAR VERNIER RELUCTANCE MOTOR IN A STEPPING MODE

J.W. Finch

Inter-University Institute of Engineering Control, University College of North Wales, Dean Street, Bangor, Great Britain

SUMMARY

The paper discusses the control philosophy, machine design and experimental performance of a prototype linear vernier reluctance motor used to form a conveyor system. This is intended to automatically carry samples between radio-active environment glove boxes. The important requirement was for reliability via an open loop control. The prototype supplies three glove boxes and is in modules one metre long for each, extendable up to 20 in all. A reluctance motor can satisfy these requirements since only the stator coils and rotor (attached to the underside of the trolley) are within the box. The control logic and power electronics are outside the radio-active area. A vernier motor enables a simple interchange during design of step size for excitation cycles required. The required linear speed can then be achieved with a simple resistance limited power amplifier.

INTRODUCTION

The work described forms part of an investigation into the possible replacement of an existing conveyor system by a system based on a linear reluctance actuator.

The existing system conveys components between a number of cells arranged in a line. It involves the use of a succession of motor driven belts and pulleys which carry the components in a tray from one cell to the next. The movement of the tray involves frictional drive between it and the belts. The length of the tray enables it to be passed from one belt system to the next in the adjacent cell by spanning over the gap between cells. This system was found to be generally satisfactory in the past, but the periodic maintenance required is becoming progressively more difficult because of the radio-active environment within the cells.

The proposed new system based on a linear reluctance motor was expected to offer certain major improvements over the existing arrangements. The main features which could be expected were: increased reliability and consistency of operation; greatly reduced maintenance and characteristics which make more automation easily possible.

The magnetic reluctance type of motor is particularly suitable for conveyor operation, since its method of operation is such that no relative motion is required for force to be produced. This contrasts with, for example, the linear induction motor, which depends on relative motion between the magnetic field and the conductors in which current is induced, for force to be produced. This means that a reluctance motor can be designed to hold indefinitely, if required, at any particular position, before moving on to the next fixed position.

The precise form of linear reluctance device adopted obviously depends on the required performance. The specification required low speed (50 mm/sec.) and modest positional accuracy (± 25 mm), with a load purely inertial in character of up to 5 kg. of added mass. A number of machine types were considered. The linear helical reluctance[1] motor is inherently intended for applications requiring precise positioning (fractions of a m.m.) over modest maximum excursions (say of the order of metre). The maximum excursion can be extended indefinitely simply by extending the screwed rotor (normally the moving part) or the stator (the wound portion, normally stationary). Alternatively the simple linear equivalent of a rotary reluctance motor could be used.[2] The resultant step size would be only dependent on the stator winding pitch adopted. This application does not require the very precise positioning which the helical device can offer. It is hence more economical to modify the magnetic geometry of the motor so that larger movements are achieved for each cycle of stator excitation. The screw threads of the helical motor are in effect straightened out to yield a linear reluctance device. The continuous stator can also to advantage be modified to become a series of individual coil units. By choosing differing rotor and stator pitches a vernier motor results. This enables smaller step sizes to be obtained for a given stator pitch and is the chosen motor type.[3-5]

Two series of experimental results are

presented. The first describes the behaviour of a simple test rig with a single rotor segment. The stator and rotor pitches were equal with this unit and this was used to yield the design information for the prototype conveyor motor. Static and dynamic tests are described on the prototype linear vernier motor used to propel the trolley in the installed system.

MACHINE DESIGN & EARLY TESTS

Principles of Reluctance Actuator

Reluctance motors are so called because they are made to produce force by virtue of a variation of magnetic reluctance with position. At constant excitation current the reluctance variation causes a variation in stored energy with displacement. This variation tends to force the moving part into a position of minimum reluctance and hence minimum stored energy. Any displacement from this position will produce a force which tends to restore the rotor to its equilibrium position. The precise variation in reluctance is produced in a structure which consists of ferromagnetic (usually iron or steel) parts, to guide the magnetic flux, and air gaps whose dimensions vary in some manner with displacement. These materials, iron or steel, are used because they exhibit high relative permeability. This high permeability is unfortunately a highly non-linear function of the exciting m.m.f., and reaches a peak at a certain flux-density thereafter falling rapidly. This fall is termed saturation and limits the flux-density which can be achieved in such an actuator. At flux-density levels where the relative permeability is high the contribution of the steel parts to the total reluctance may be small and the device can often initially be assumed linear. Various relations can then be shows to follow.

For a simple linear reluctance actuator with a single coil stator and constant current excitation the force produced is given by

$$F = \frac{1}{2} I^2 \frac{dL}{dx}$$

where I is the current and $\frac{dL}{dx}$ represents the variations of the stator coil inductance with the rotor position. At constant current the force magnitude and variations with displacement are governed entirely by the variation of $\frac{dL}{dx}$ with displacement, x.

A single unit of one possible form of linear reluctance actuator is sketched in Fig.1. The variation of inductance with position can be modified by variations in the size and shaping of stator and rotor iron portions and by use of selective saturation techniques. This form of geometry was deliberately chosen because the expected characteristics of the device are very suitable for use as a positioning conveyor system.

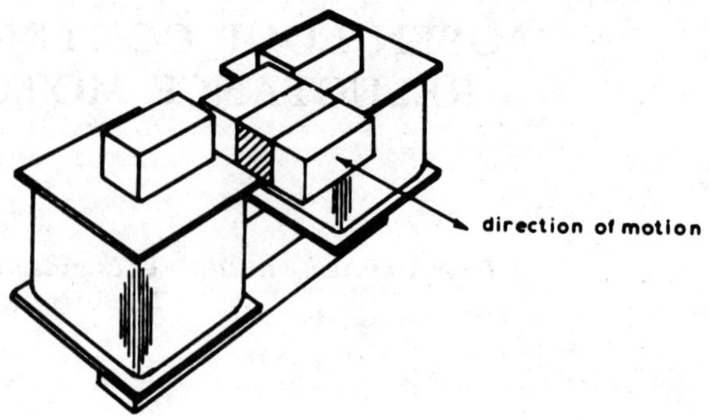

Fig. 1
Single magnet unit

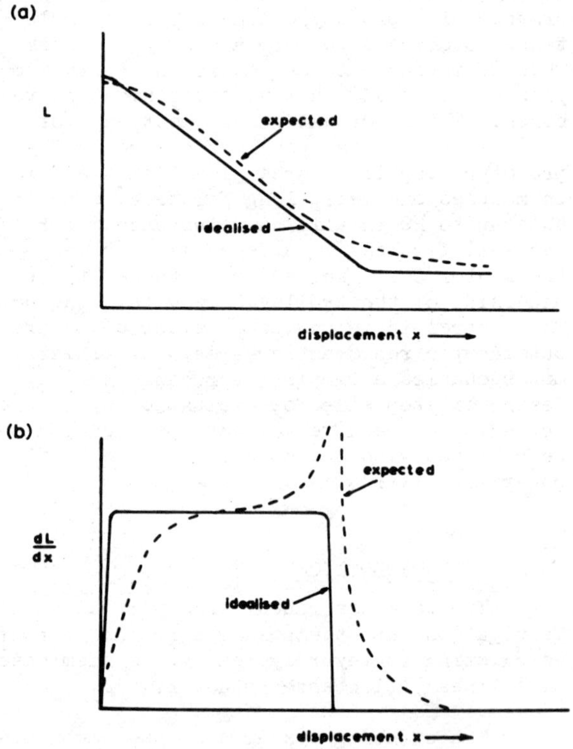

Fig. 2
Inductance and rate of change of inductance variations

If all the magnetic flux were constrained to flow normally from one stator polepiece to the other, and the iron circuit had constant reluctance, then the variation would be similar to that sketched in Fig. 2(a). The actual inductance variation will have a more 'rounded' shape because of leakage flux paths not directly crossing the main air-gap and because of successive saturation of the remaining iron circuit flux paths as the rotor is withdrawn from the equilibrium position. A likely variation is shown in Fig.2(a), also Fig. 2(b) shows the resultant idealised, and expected, variations of $\frac{dL}{dx}$ with displacement. These have the required form since the restoring force rises rapidly to a relatively high level with displacement from the equilibrium position, then stays approximately constant for much larger displacements, until finally a large peak force is realised. This 'hard' spring characteristic near equilibrium ensures accurate alignment, and the levelling off ensures controlled acceleration whilst in motion. The characteristic will be more linear with a vernier motor since the difference in pitch between rotor and stator will reduce the steepness of the curve near the origin.

Single segment testing

The form of motor shows schematically in Fig. 1 was that adopted for the initial tests. The actual design is shown in Fig. 3 which indicates the spring balance used to apply static load forces.

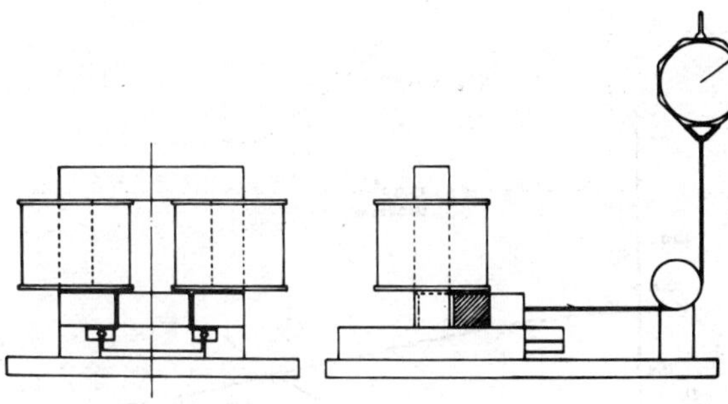

Fig. 3
Static force test rig

The form of rotor construction shown in Figs. 1 and 3 is 'Segmental' in type being composed of alternate magnetic and non-magnetic portions. This form of rotor has been conclusively shown in rotating machines to yield higher values of maximum to minimum reluctance than the alternate 'castellated' form(2) This potentially improved ratio has been effectively utilised in the design used for the conveyor to allow the air-gap, g, to be increased. This increased value of g (to 1 mm) over that normally adopted in other reluctance devices (0.1 - 0.25 mm) greatly improves the economics of the proposal, as much simpler mechanical bearing arrangements can be adopted, as described later. The interaction between separate units with a segmental rotor can be readily minimised and deduced from the particular geometry adopted. This implies that the performance of the complete system based on a number of separate units can be found by a superposition of individual performances.

The particular configuration adopted was chosen to make use of standard sizes of component parts to hasten construction, and to facilitate the extensive tests required. For this reason and to reduce costs of the final system, mild-steel has been used for the magnetic sections and aluminium for the non-magnetic. Mild-steel is a far from ideal material magnetically since it exhibits a comparatively low relative permeability and has a very wide hysteresis loop but this is of less overriding importance since switched d.c. excitation was adopted. For the same reason eddy-current losses need not be reduced and a solid, unlaminated structure can be used. The advantageous features of eddy-currents in this application are discussed later.

Standard bright mild-steel bar of 22 m.m. square section has been used in this unit. Paxolin formers and coil cheek pieces were obtained to fit this section and hand wound to form the pair of coils required. A set of precision linear bearings was used on the rig to reduce friction and mis-alignment on the rig to a minimum. In the small size required on the rig these are relatively cheap, but it is emphasised that the conveyor system was deliberately designed to use a much less sophisticated bearing arrangement.

Results were obtained on the rig at two air-gaps, 0.25 mm and 1 mm. Fig. 4 shows the variation of inductance with displacement for the smaller gap measured with 0.5A current and Fig. 5 shows the resultant force against displacement characteristic. Also shown in Fig. 4 is the desired $\frac{dL}{dx}$ curve and this can directly be compared with the shape of the force curve of Fig. 5 at the highest current. The inductance measurements were obtained by the switched d.c. technique which enables the true inductance to be determined even in the presence of coupled short-circuited paths. These paths are largely those due to the flow of induced eddy-currents in the solid structure.

Fig. 4 also shows the variation of peak to force with current and it can be seen that the force is more closely proportional to current than to current squared, beyond 0.5 A. This is due to the effect of saturation which occurs earlier with the smaller air-gap.

Fig. 6 shows the variation of inductance with positions at the design air-gap of 1 mm measured at 1A. The reduced value of maximum inductance L_d and inductance ratio (L_d/L_q) can clearly be seen by comparison with Fig. 4 for the smaller gap. Fig. 6 also shows the variation of force with displacement at both 1 and 2A, drawn with scales enabling a comparison between the shapes of the characteristics to be made. The slightly increased slope of the curve for the higher current in the "constant" force region is due to the increased successive saturation for this case as the rotor is withdrawn. These measurements show that adequate force can be produced over the wide 1 mm air-gap specified, with a conservatively rated coil unit system.

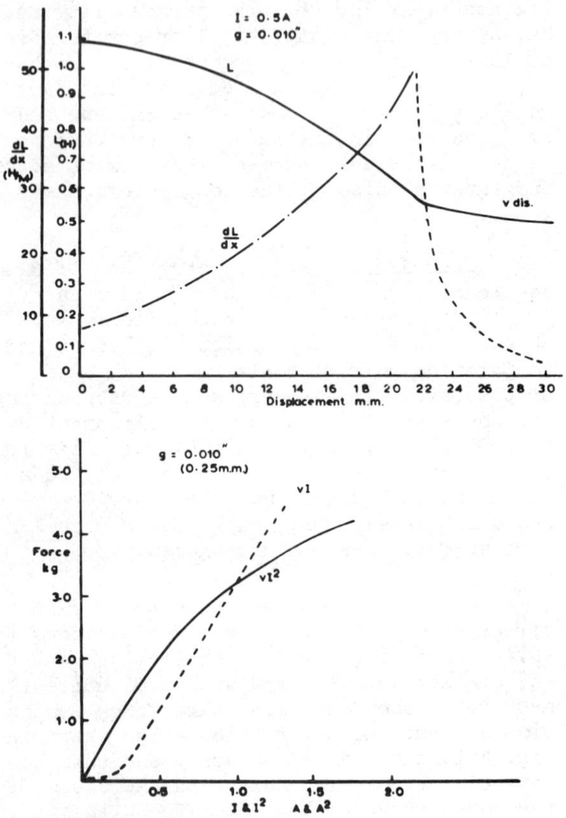

Fig. 4

Measured inductance variation with position and force against current with 0.25 mm gap

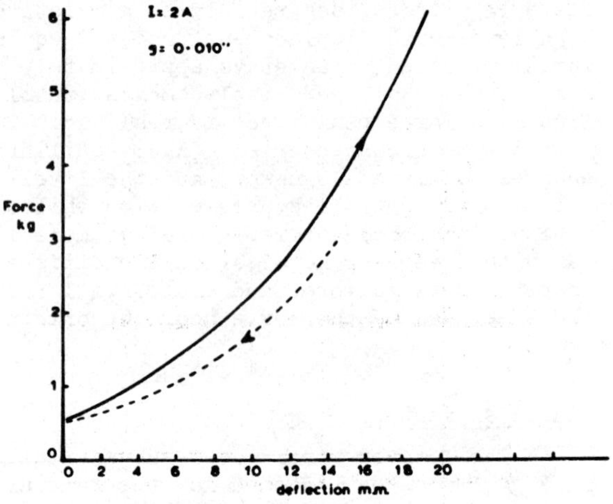

Fig. 5

Force against displacement for rig unit

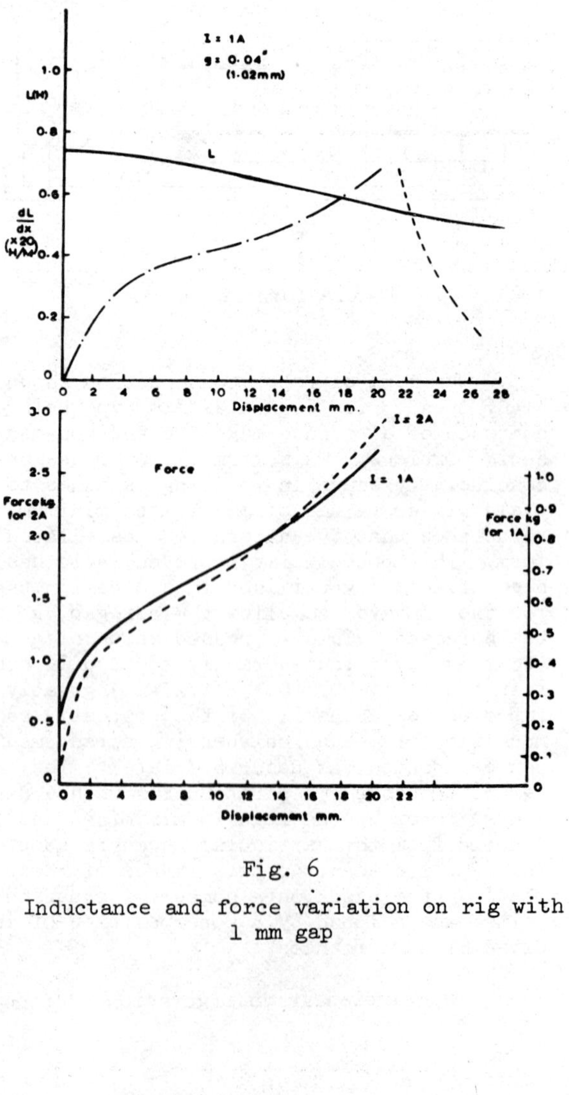

Fig. 6

Inductance and force variation on rig with 1 mm gap

VERNIER MOTOR CONVEYOR SYSTEM

Mechanical design

The conveyor system uses blocks of units similar to the test-rig within each cell. These blocks are inter-connected by a series of rigidly bolted rail sections which will serve both to align and support the conveyor trolley. The symmetry of the magnetic system used implies that with the conveyor rotor aligned between the stator units the bearings have to withstand no magnetic forces. A cross-sectional view through the rails showing the conveyor rotor is given in Fig. 7. It can be seen that this system makes full use of the magnetic structure as rail supports, ensuring accurate alignment and compactness. Other magnetic arrangements can be used to simplify the guidance rail system, but involve lateral magnetic forces which can considerably exceed the useful propulsive force.

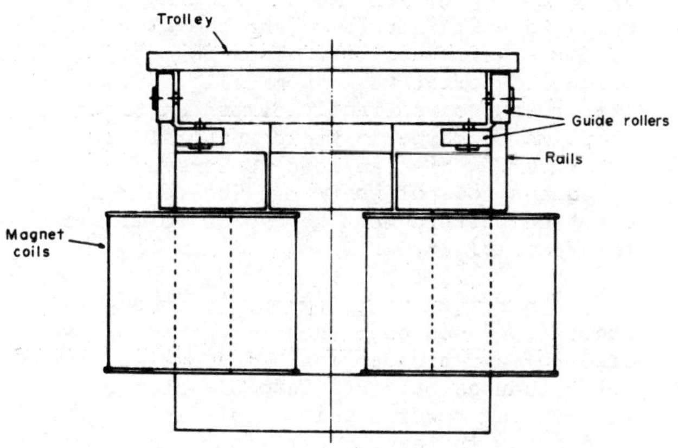

Fig. 7
End view of trolley and magnet

The trolley is moved by exciting the stator coils with current in a suitable manner, most conveniently by a switching operation. Any particular coil when excited acts to pull the nearest rotor magnetic segment into alignment with it. By arranging a rotor to consist of a number of alternate magnetic and non-magnetic sections, and making the stator and rotor pitches different, the movement produced by one switching operation can be reduced. This reduces the jumps produced by the switching operations. For example, if the relative pitch is made such that from an aligned position five stator pitches go by before another complete rotor-stator alignment, them five cycles of excitation produce one complete pitch movement, i.e. one cycle produces $\frac{1}{5}$ of a pitch movement. This is the vernier action of the motor design. This is clearly illustrated in Fig. 8 which shows a diagrammatic view, but is to scale in the direction of motion, of one complete stator. The stator consists of two blocks of four units situated within each glove-box served by the conveyor.

Variations of the distance between adjacent glove-boxes or cells are accommodated by use of connecting rail sections of approximately varied length. These rail sections ensure that the trolley is passed from one section to another without losing the accuracy of alignment required. It can be seen from Fig. 8 that the pitch ratio adopted in the conveyor design is 4 on the stator to 3 on the rotor. The stator pitch used is 50 mm which results in a full step size of 17 mm. The length of the trolley is used to span gaps in the active track within and between cells. These gaps being present because of cell geometry, and for economy. An overall view of the motor is shown in Fig. 9.

Fig. 9
Overall view of three-cell motor and its control cabinet

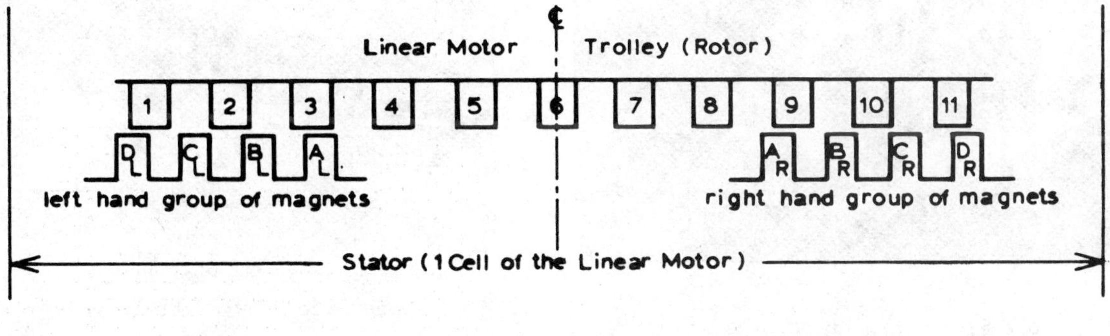

Fig. 8
Diagrammatic view of stator and rotor of vernier motor

Only one coil is used in each magnet unit in the stator (i.e. 8 per cell) to reduce the number of coils and to assist in stator packing. Figure 10 shows a detailed view of one half of the stator units for a cell. The constructional details of the trolley and rotor unit are shown in Fig. 11. The guidance and support rollers can be seen.

Control

The required operation in a radio-active environment with a minimum of maintenance placed particularly exacting requirements on the motor and its control. The performance was required to be not dependant on any transducer within the radio-active area if at all possible. This leads naturally to open-loop control for which a reluctance device is inherently suitable. A simple step-by-step switching of the motor coils in a suitable sequence can cause the motor and trolley to move from one position to another in synchronism with the switching. The single step response of the device was expected to be rather lightly damped and this has been confirmed by test. This problem is worsened in the linear motor by the large air-gap and high effective rotor mass. Control is obtained via the stator coil currents and the motor characteristics. The instantaneous coil currents are governed by the switching sequence but also by coil resistance and effective inductance. The inductance is a function of position as well as current. To move the trolley successive coil units are excited and the nearest rotor segment is pulled along in a continuous series of steps until the required position is reached. The abruptness of this procedure is greatly reduced by the staggered stator/rotor, by the inertia of the trolley and by the use of simple voltage excitation of the coils. The last point exploits the inherent slow rise of coil current after a step of voltage has been applied, owing to coil inductance, and the slow rise of force which results. This form of excitation is attractive economically. Since the step response is lightly damped multiple steps may combine to yield unsatisfactory behaviour. This can lead to de-synchronisation and loss of control. The control excitation pattern must be chosen to minimise this possibility.

In rotary stepping motors "two-phase excitation" has been shown to give improved electro-magnetic damping.[6] The motion of the rotor induces currents into the stator winding causing power dissipation. The additional damping in this linear motor was expected to be slight owing to the large air-gap used, but is still advantageous. This double excitation has a further strong advantage in the vernier motor. Since the rotor design incorporates separate magnetic segments the interaction between different sections of the motor is minimised. The action of passing current through any stator coil produces a force which will tend to pull the nearest rotor segment into line with the coil magnet structure. As the pitch of rotor and stator is different the position of alignment will be different for each stator coil in a stator unit block of four coils. A similar force pattern will be produced by each coil but with a shift in the rotor position at which it occurs. Thus the choice of the coils which are excited will vary the position at which the net force on the rotor is zero. The excitation pattern can contribute to the vernier action of the motor, already present because of the construction of rotor relative to stator. This can be used to generate an apparently better damped overall response by a suitable choice

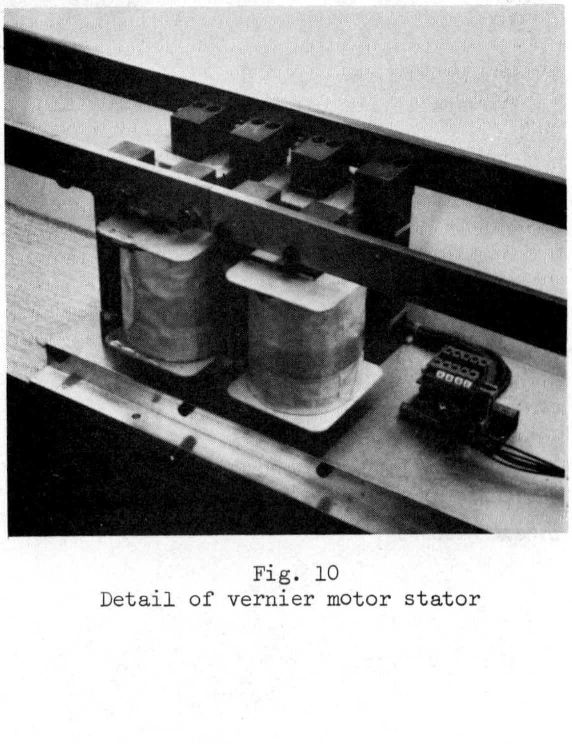

Fig. 10
Detail of vernier motor stator

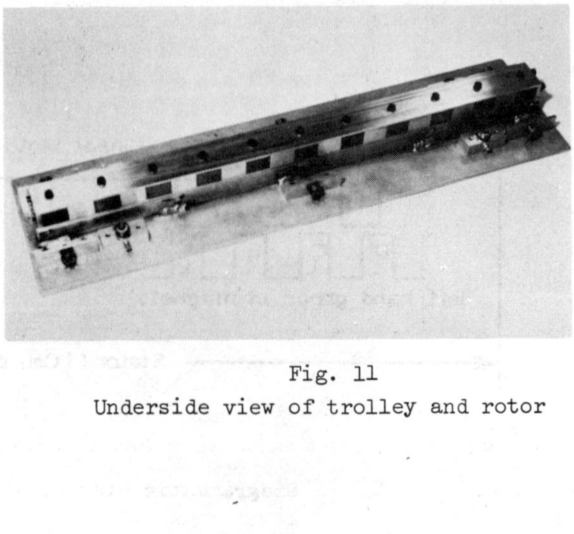

Fig. 11
Underside view of trolley and rotor

of excitation sequence. This technique is sometimes referred to as "electronic damping"[7,8]. The excitation pattern chosen results in a half-step as the first and last of the sequence of steps required to move the trolley from the centre of one cell to the centre of the next. This minimises overshoot and oscillation and provides smoother starting.[7,8] This sequence has been found to give greater immunity to parameter variations. The frictional loading is normally very light, and the inertial loading is specified to cover the range 0 - 4.5(kg) of added mass. It is this variation and the steps in the guidance rail surface at junctions that the system must encompass.

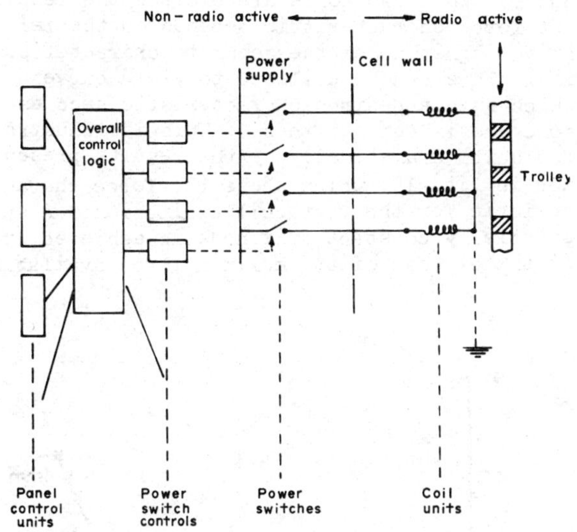

Fig. 12
Control system block diagram

Control circuitry

A simple block design of the overall circuit arrangement is given in Fig. 12. The logic arrangement adopted is based around the use of 256 x 8 bit programmable read-only-memories. Each bit in a particular memory is made to control the excitation to a stator unit. Only one memory is needed per cell since there are 8 coils within each complete stator. Thus an excitation sequence of 256 separate states can be generated by stepping through the memory contents. The logic required has been simplified by always filling the memory no matter how long the actual step sequence needed is. This is accomplished by using short sequences of identical excitation patterns within the memory. No coil switching actually occurs until a different pattern appears. The memories for adjacent cells can then effectively be connected in series. The position of the trolley is

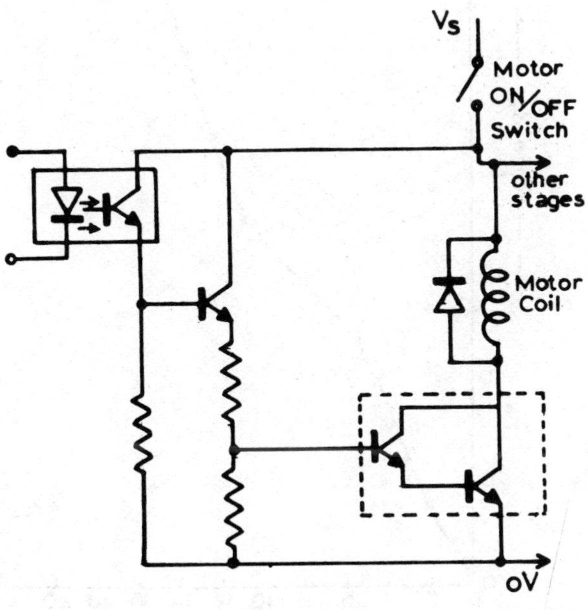

Fig. 13
Power electronic switch circuit diagram

Fig. 14
Internal view of electronics cabinet

monitored by counting from the original position. Separate control boxes with illuminated push button displays at each glove-box or cell enable the operator simply to send the trolley automatically to any desired location.

The power electronics necessary can be made both simple and robust. The circuitry adopted is illustrated in Fig. 13. A power darlington n.p.n transistor performs the switching, with a flywheel diode connected across the motor coil to clamp the voltage and take the coil current on switch-off. The circuit is optically coupled to the logic control outputs to minimise noise and earth-loop problems. Figure 14 shows the electronic cabinet which has enough logic capacity for an extension to a 20 cell conveyor.

Static Tests

The tests reported here use the two excitations which are characteristic of the control system used. Firstly, the 'half step' excitation used at the beginning and end of each cycle, and referred to as the 'rest' or holding position; for example with the coils labelled as B_L and B_R in Fig. 8 excited. Secondly, the 'regular' excitation pattern used when the trolley is in transit between cells; for example with coils A_L and D_L excited.

The results were obtained using a spring-balance loading arrangement similar to that illustrated in Fig. 3 with a linear variable differential transformer as a position transducer. When two coils are excited the resultant force characteristic depends on the relative magnitude of the separate characteristics. This can be utilised to yield curves which have a desired characteristic such as more consistent stiffness. This is illustrated quite dramatically by the results presented in Fig. 15, which shows the force characteristic for the rest coil excitation. A substantially constant stiffness is achieved, but at the expense of the maximum force available.

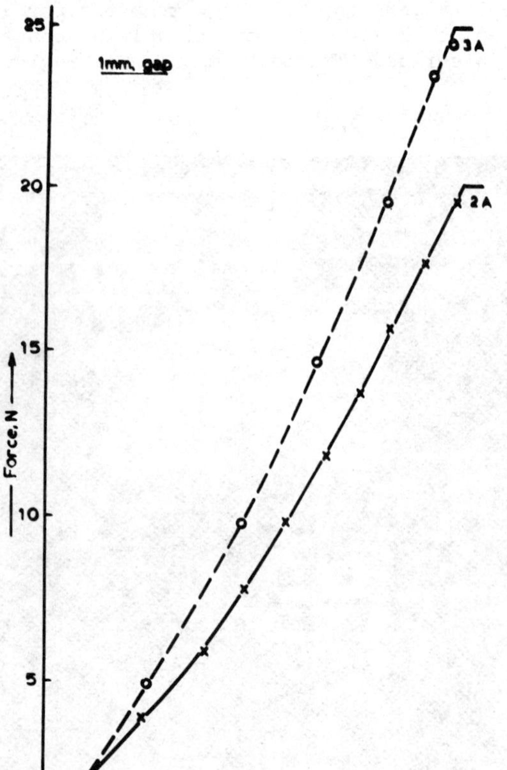

Fig. 15
Force against displacement from rest posit- of vernier motor

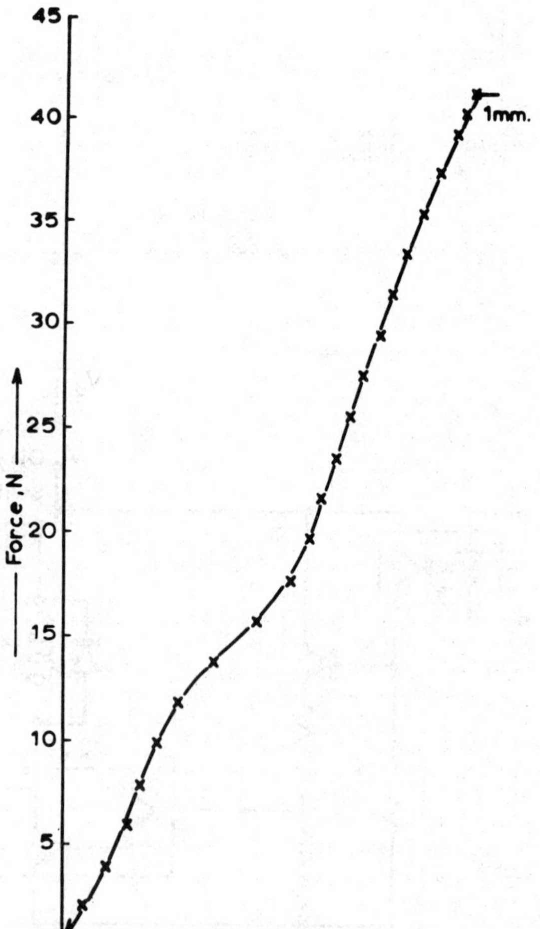

Fig. 16
Force against displacement with the regular excitation pattern

With two coils excited there will inevitably be positions where the actions of the coils are in opposition. With single coil excitation the effective spring stiffness (force/position) varies by about 3:1 over the whole curve with a maximum force of 30 Newtons. Fig. 16 presents the force characteristic measured with the regular two coil excitation. It can be seen that a rather higher stiffness than the rest coil case has been achieved but over a much wider position range.

The effect of the coil current on the motor characteristics can be considered in two areas. Firstly, the change in the shape of the force against position characteristics can be shown. Secondly, the variation in maximum force with current separately illustrated.

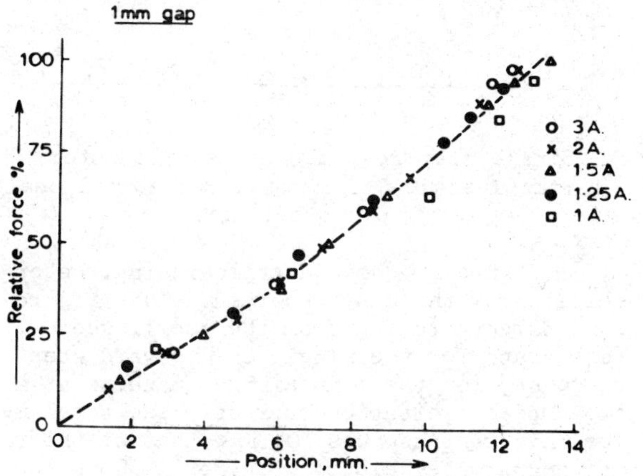

Fig. 17
Normalised force versus displacement for various currents with rest coil excitation

In Fig. 17 the first area is highlighted by plotting normalised force readings against position. The forces have been normalised by dividing the pull-out force achieved at each current. The results displayed are for rest coil excitation. Figure 17 clearly illustrates that the shape of the characteristics is unchanged by current in the range tested. The maximum deviation from the nominal curve is for currents of 1 amp. This seems to be more a variation in pullout force than in the characteristic shape. At this, the lowest current tested, any errors due to friction or to zero calibration would be most marked.

Fig. 18 shows the pull-out forces (used to normalise the results for Fig. 17) plotted against both current level and the square of current. If the motor was perfectly linear at the currents considered, the force should be directly proportional to the current squared. This is quite clearly

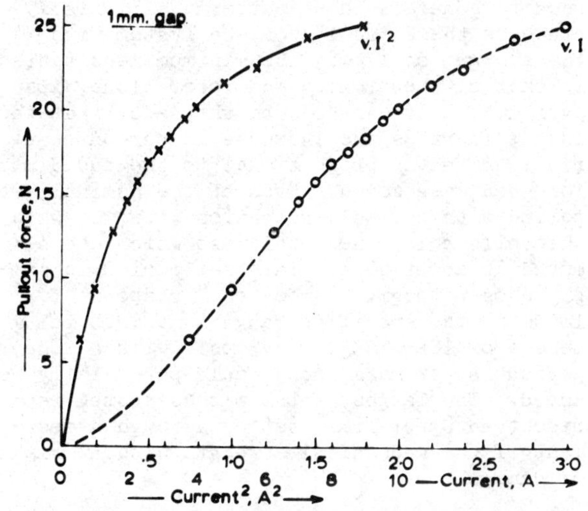

Fig. 18
Peak force against current with rest coil excitation

not the case, and a direct proportionality to current alone would be a truer representation. This implies that a higher transient current rating would be less helpful than might be supposed. Even this direct proportionality weakens as the mild steel saturates at currents of more than 2 amps, which is the continuous current rating of the coil.

A constant voltage supply such as that used in the motor will supply a gradually reducing current as the coil resistance increases with temperature. This was used in the design to build an element of self-regulation into the system. The initial coil current is set to be approximately 2 amps, but this will finally reduce to less than 1.75 amps. The coil current rating could obviously be increased by transient rating but this was not thought desirable in view of the intended motor environment.

Dynamic Tests

The dynamic performance of the motor was characterised by two classes of experiments. In the first class of test the frequency of the excitation applied to the motor was varied to determine the critical frequency at which the trolley just could not move consistently. This could then be repeated under different excitation conditions for a number of total masses of trolley plus load.

The time period recorded is that of the external variable frequency pulse generator, used in place of the internal oscillator in these experiments. The effective excitation period is four times this figure since four exactly similar excitation patterns are stored consecutively in the read-only-memory used to generate these patterns. Figure 19 presents these results on the system with the air-gap of 1 mm. The minimum mass tested is that of the trolley and rotor alone, just over 10Kg. It can be seen that except at the lowest inertias the increase in period required is directly proportional to the added mass for each case shown. Even at the nominal set period with nominal excitation (i.e. 2A with the coils cold) the total mass which can be moved is about 60Kg. This set period is simply chosen to give "smoothest" response for loads in the specified range that is total masses of 10 - 15Kg. Obviously with a longer period larger mass loads could potentially be moved. The largest total masses tested represent an upper limit set by a rough assessment of the mechanical strength without risk of damage.

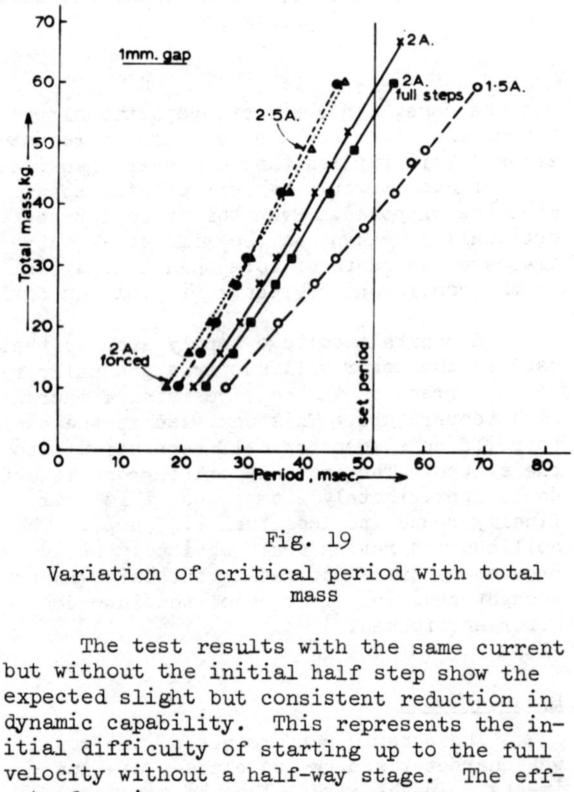

Fig. 19

Variation of critical period with total mass

The test results with the same current but without the initial half step show the expected slight but consistent reduction in dynamic capability. This represents the initial difficulty of starting up to the full velocity without a half-way stage. The effect of an increase to 2.5 amps and a decrease of current to 1.5 amps are also displayed in Fig. 19. Most interestingly the test results with forced excitation of 2 amps are also shown. These were obtained by supplying the coils from a power supply set to 40 volts on open circuit but with a current limit set to 4 amperes. This limits the current in each of the excited coils to 2 amps, but on switching the full voltage is available to rapidly increase the current to this limit. The supply acts as a current forcing amplifier, reducing the effective electrical time constant of the coils. It can be seen that the results obtained with the forced excitation of 2 amps are very similar to those with a simple voltage switching at a current of 2.5 amps.

The second series of tests were a number of recordings of instantaneous position as a function of time. These recordings were made permanent via oscilloscope photographs. The test series included single and multiple step excitations for a number of total masses with various forms of excitation.

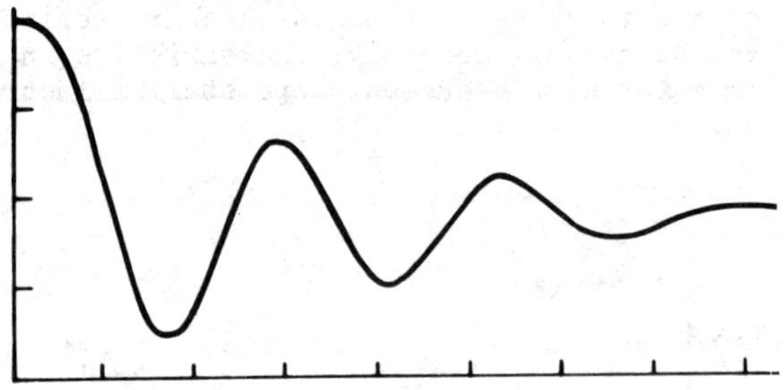

Fig. 20

Single step response of vernier motor. Vertical scale 7.2 mm/div, Horizontal scale 0.2 secs./div.

Figure 20 shows a typical single step result from the many recorded. This figure is a direct tracing from the oscilloscope photograph for the single one forward step response, for the 1 mm air-gap machine at a two ampere excitation current. The total mass for this response was 10.1Kg, that of the rotor and trolley alone. Figure 21 shows the response for the same case for the normal multiple excitation sequence chosen. It illustrates the success of the control scheme in integrating the lightly damped single-step response to give an overall response exhibiting a minimum of unnecessary oscillation. Also shown in Fig. 21 is the current in a typical coil, which causes the motion. The measured rise time constant of the current was 95 milli-seconds, with a full time constant of about 40 milli-seconds, although both these values are to some extent a function of position.

The single step responses have been compared with that for a linear second order model. The measured oscillating frequencies are in close agreement with those of the model. The model results being obtained using constant values of stiffness taken from the static force against position characteristics. Values of damping coefficient, ζ, were also estimated from the step responses. The results for the measured damping coefficients have a much greater experimental error than those for frequency. The estimated limits of error for any particular result are $\pm$ 0.02 (or approximately $\pm$ 20%), about the mean value of 0.1. Within this possible error the damping coefficient was found to be constant

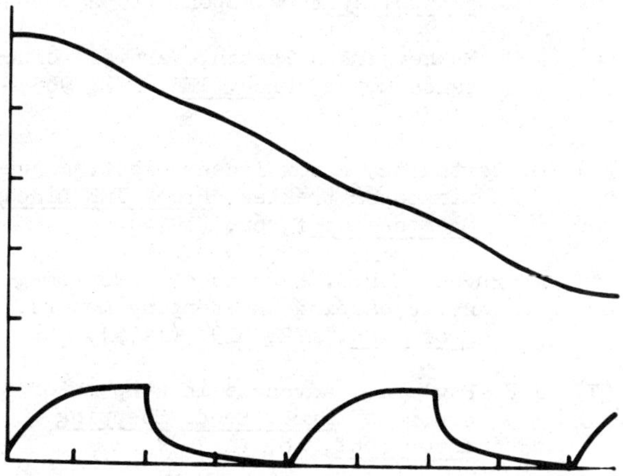

Fig. 21
Multiple step response. Vertical scale 36 mm/div. position, 2A/div. current, Horizontal scale 0.2 sec/div.

for each excitation and mass. Since ζ has been found not to vary with total mass this indicates that it is effectively non-viscous in nature. Viscous friction would yield a damping coefficient which was inversely proportional to the square root of mass. This would mean a variation of almost 1.8:1 over the range of masses used, from 10.1 to 31.2Kg. Static friction and hysteresis loss are expected causes of damping in this system which are not viscous.

Simulation

The force at any position was found experimentally to be approximately proportional to current, (see Figs. 4 and 18). The double coil excitation used also gave a characteristic which was more linear than that for single coil excitation. The rise of current with time after a coil is switched could be represented as a simple single time constant reference. A possible mathematical model would be a viscously damped mass-spring system with spring stiffness rising from zero to a constant value. In normalised form the step response of the displacements, x, would be governed by:

$$\frac{d^2x}{dt^2} + 2.\zeta\,\omega_n\frac{dx}{dt} + \omega_n^2(1-e^{-t/\tau})\,x = 1$$

Although this equation can be transformed into a form similar to Mathieu's equation results can most conveniently be obtained by simulation. Analogue and digital methods have both been used. This was found to be useful in the design as an aid to understanding the behaviour of the motor and its control. One interesting result is that with an undamped system the resultant peak to peak oscillation magnitude exceeds twice the requested step size. This results from the stiffness in the first instants of time being small, and is governed by the time-constant in relation to the natural frequency. The agreement obtained between the experimental results and this model has been discussed in the dynamic test section. Some digital simulation work was also performed using the measured static force characteristics. Work is also proceeding to attempt prediction of the force and other characteristics using numerical solutions involving a complete non-linear representation of the mild steel magnet structure.

CONCLUSION

The design and performance of a linear vernier reluctance motor conveyor system has been presented. Both static and dynamic performance experiments were described. The two coil excitation pattern adopted was seen to have a marked effect on the system behaviour. The wide variation in stiffness given with single coil excitation has been shown to superimpose to give a higher and more linear stiffness with the regular two coil excitation. The rest excitation further increases this linearity whilst reducing stiffness as expected. The maximum force was found to be fairly linearly proportional to current. The shape of the characteristic did not vary greatly with current. The results for the variation in critical period with mass emphatically show the capability of the system in moving added mass. Added masses in excess of ten times the design figure of 4.5 Kg. were moved at the nominal excitation period. An increase in period results in an increase in this capacity. The upper limits of added mass in these tests were set by structural considerations. An ultimate limit would appear to be set by bearing static friction and by the resultant component of gravity force along the track when this is not completely level. This becomes significant at very large masses. The step response results show the expected lightly damped character. The multiple step results indicate quite clearly how successful the open-loop control has been at integrating the characteristic to form a satisfactory response. The measured oscillating frequencies align closely with the values estimated from the average static stiffnesses. The damping behaviour appears to be more hysteretic than viscous in nature.

The system described has been installed and has successfully undergone evaluation trials. Work is currently proceeding in a number of areas. A larger motor is to be designed and built. Accurate prediction of the static and dynamic characteristics is being attempted from the motor dimensions incorporating the magnetic non-linearities. Considerable effort is being devoted in an attempt to use optimal control techniques to improve the behaviour of the motor and its control.

ACKNOWLEDGEMENTS

The author acknowledges the financial support and permission to publish from the UKAEA, Windscale; and the Industrial Development Unit at Bangor. He would particularly like to thank Dr. J.R. Wakefield of UKAEA, J. Moorcroft and J. Elias of ID(B), responsible for detailed mechanical and logic design respectively; and Professor P.E. Secker the Director of ID(B).

REFERENCES

(1) J. Gerard & R.J.A. Paul, Rectilinear screw-thread reluctance motor, Proc. IEE, 118, 1575, (1971).

(2) P.J. Lawrenson & L.A. Agu, Theory and performance of polyphase reluctance machines, Proc. IEE, 111, 1435, (1964).

(3) A. Tustin, Direct current machines for control systems, Spon, (1952).

(4) K.C. Mukherji & A Tustin, Vernier reluctance motor, Proc. IEE, 121, 965, (1974).

(5) M. Barthalon, A new linear electromagnetic motor traklec, Proc. IEE Linear Machines Conf, 62, (1974).

(6) A. Hughes & P.J. Lawrenson, Electromagnetic damping in stepping motors, Proc. IEE, 122, 819, (1975).

(7) J.P. Pawletko, Advances in step motor controls, Proc. Leeds Stepping motor conf, 68, (1976).

(8) P.J. Lawrenson, A. Hughes & P.P. Acarnley, Improvement and predictions of open-loop starting/stopping rates of stepping motors, Proc. IEE, 124, 169, (1977).

INVERSION OF THE ROTATING DIRECTION OF A CURRENT-FED CLOSED-LOOP CONTROLLED ASYNCHRONOUS MACHINE

B. Trannoy, B. de Fornel and M. Grandpierre

Laboratoire d'Electrotechnique et d'Electronique Industrielle, E.R.A. No 536
2 rue Camichel, 31071 Toulouse Cédex, France

Symbols

R_c	Rotor resistance
L_r	Rotor cyclic inductance
M_{sr}	Stator-rotor mutual inductance
f_s	Stator frequency
f_r	Rotor current frequency
f_N	Rotation frequency
I_s, I_r	Stator and rotor currents
T	Electromagnetic torque

INTRODUCTION

Numerous applications of a.c. machines associated with static converters are being presently developed. In particular current-fed machines possess properties very close to those of d.c. machines. These latter are very versatile and still constitute a very large proportion of variable speed drives, however at high speeds and in inflammable environments their use is limited[1,2,3,6]
It is often necessary to operate the variable speed groups in all four quadrants of the torque-speed plane (full reversibility). This necessitates an analysis of generator operation and of the reversal of the rotating direction.
The investigation deals with the reversal of the rotating direction of a current fed asynchronous machine. The behaviour of the machine and the static converter are described with the aid of an analytical study and numerical simulations. A method of accelerating the braking is deduced. It is shown how the control of the different operating sequences can be accomplished using digital control techniques. Finally, experimental results are used to confirm the theoretical calculations.

1 - SYSTEM DESCRIPTION

A three phase symmetrical asynchronous induction machine fed by a static converter with a single commutating capacitor is considered. Such a structure enables the commutating ability of the static converter to be easily controlled by prolonging, if necessary, the interval of time during which the capacitor is charged. This possibility is particularly useful during transient variations of the load current. The converter-machine group is fed from a regulated current source and the overall system operates in closed loop (autopiloting). The operating frequency of the converter is determined by the composition of the machine's rotating speed f_N and a signal corresponding to the rotor current frequency f_r. The addition of two frequencies corresponds to motor operation and their subtraction corresponds to generator operation.
Figure 1 represents the complete circuit diagram of the machine and the static converter. The role of the current inverter is to switch the current into the machine windings so as to give 120° type operation. Only two thyristors of the principal bridge should conduct simultaneously, each one being controlled to conduct during $2\pi/3$ radians, there is therefore a commutation every $\pi/3$ radians. Neglecting the commutations,the current in the windings (delta connected) takes the form of that shown on figure 2. [4,5,7]
The change to generator operation is made very simple by means of inverting the sign of the rotor current frequency. The change in the direction of rotation requires the inversion of two phases when the stator frequency falls to zero.
The system has two inputs, I_{ref} and f_r which can be either independant of each other or related in order to give, for example, constant flux operation.
It is supposed that the machine is non-saturated and that the flux density is sinusoidally distributed throughout the airgap. For the transient analysis it is supposed that the current regulated rectifier/inverter and the static converter supply the stator of the machine with perfectly sinusoidal currents of variable frequency and of an amplitude which corresponds at all moments to the input reference current. These two assumptions correspond to the current harmonics and the response time of the current regulation being neglected. In fact it has been verified that the harmonics have very little influence in the particular transients under study and as it is a constant current operation which has been studied, the characteristics of the current regulation can be ignored.

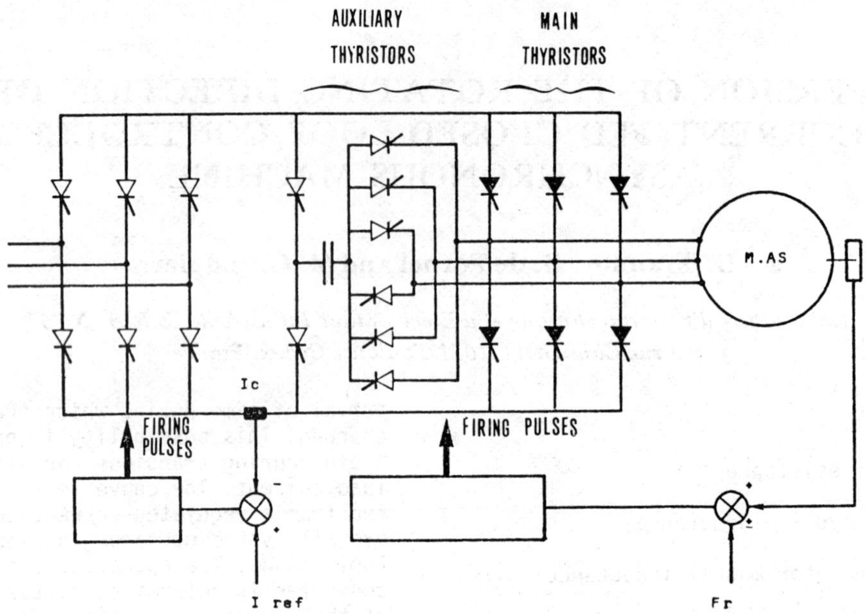

Fig. 1. System diagram

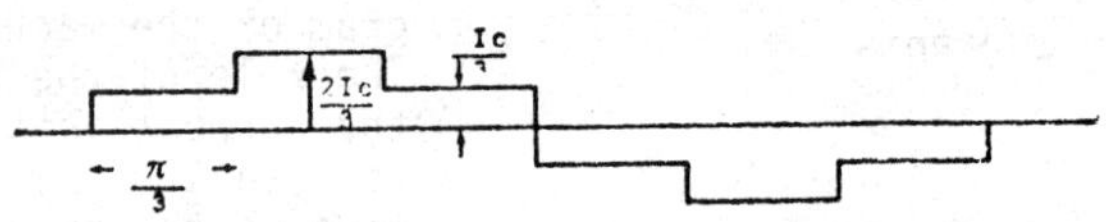

Fig. 2. Theoretical current waveform

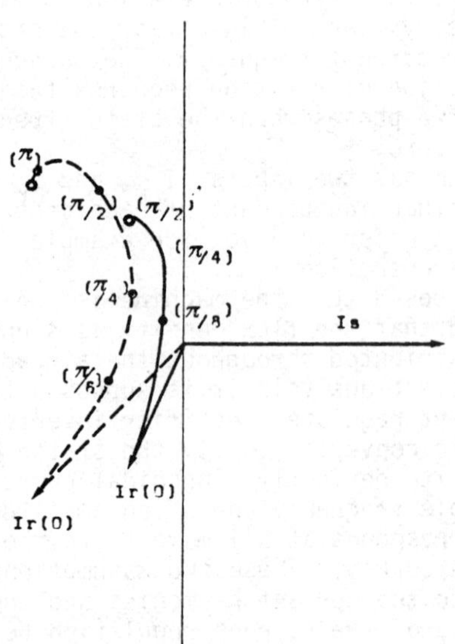

Fig. 3. Phase diagram of rotor current (for $\omega_r = \omega_{ro}$ to $\omega_r = -\omega_{ro}$)

— $\omega_{ro} = 2\pi$

--- $\omega_{ro} = R_r/L_r$

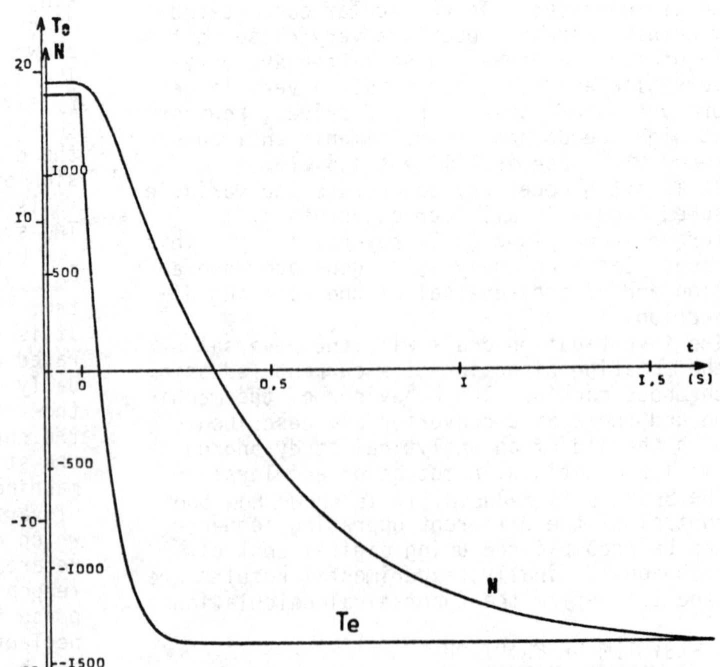

Fig. 3. Numerical simulation of reversal of the rotating direction without blocking I_s

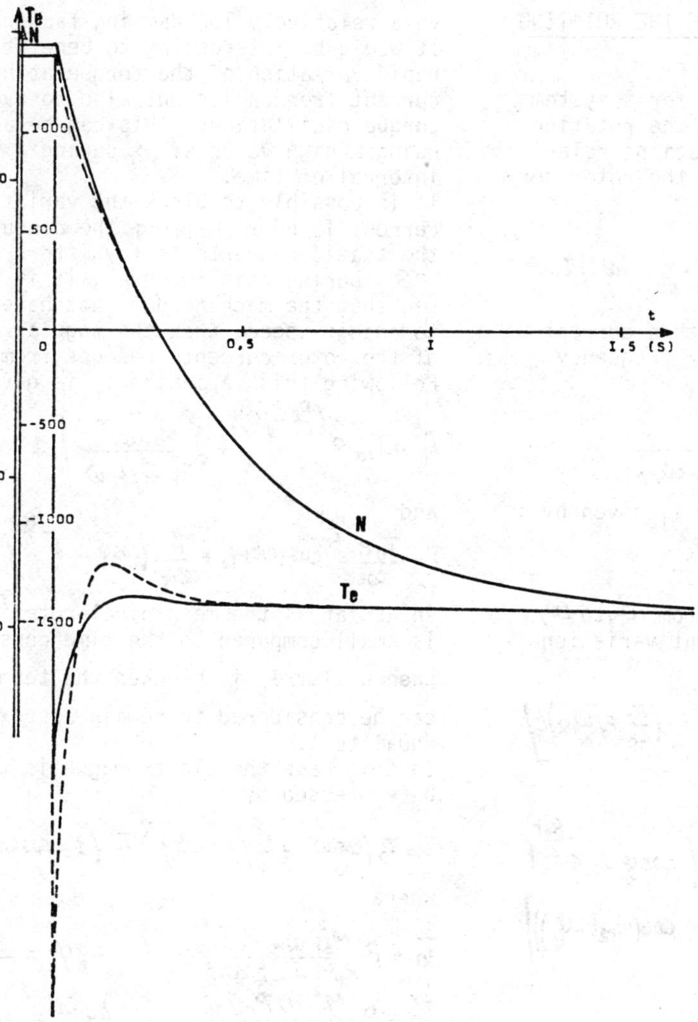

Fig. 5. Numerical simulation of the reversal rotating with blocking I_s

——— $\tau = \pi/3$

- - - - $\tau = 2\pi/3$

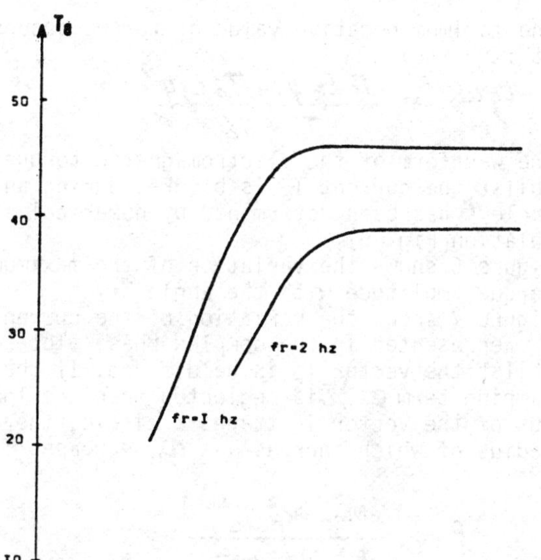

Fig. 6. Maximum torque value versus τ

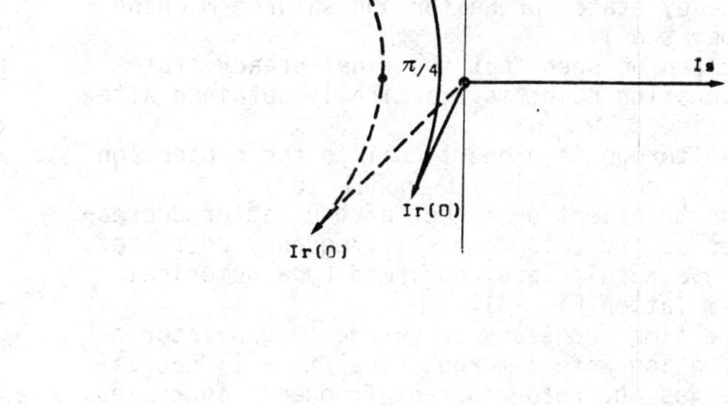

Fig. 7. Phase diagram of the rotor current for variation of ω_r
from $+\omega_{ro}$ to $-\omega$ (——— $\omega_{ro} = 2\pi$, - - - $\omega_{ro} = R_r/L_r$)
from $-\omega$ to $-\omega_{ro}$ —.—.—

2 - ANALYSIS OF THE REVERSAL OF THE ROTATING DIRECTION

With the above assumptions and for a system where the axes are referred to the rotating field, the machine's rotor equation, relating the stator current I_s and the rotor current I_r is given by [5,7]:

$$0 = M_{sr}\left(\frac{d}{dt} + j\omega_r\right)I_s + \left(R_r + L_r\frac{d}{dt} + j\omega_r\right)I_r$$

For steady state operation, with a current I_s and a rotor current angular frequency ω_{ro} the rotor current is :

$$I_{ro} = -jM_{sr}I_s\frac{\omega_{ro}}{R_r + jL_r\omega_{ro}}$$

and the electromagnetic torque is given by :

$$T_o = p\frac{M_{sr}^2\omega_{ro}R_rI_s^2}{R_r^2 + L_r^2\omega_{ro}^2}$$

Following a variation in ω_r from ω_{r1} to ω_{r2} the current and torque transient variations are expressed thus :

$$I_r = I_{r1}e^{-\left(\frac{R_r}{L_r} + j\omega_{r2}\right)t} + I_{r2}\left[1 - e^{-\left(\frac{R_r}{L_r} + j\omega_{r2}\right)t}\right]$$

$$T = \frac{T_1e^{-\frac{R_rt}{L_r}}}{\cos\varphi_1}\cos(\omega_{r2}t - \varphi_1) + \frac{T_2}{\cos\varphi_2}\left[\cos\varphi_2 - e^{-\frac{R_rt}{L_r}}(\cos(\omega_{r2}t - \varphi_2))\right]$$

$$I_{r1} = -jM_{sr}I_s\frac{\omega_{r1}}{R_r + jL_r\omega_{r1}}$$

$$I_{r2} = -jM_{sr}I_s\frac{\omega_{r2}}{R_r + jL_r\omega_{r2}}$$

$$T_1 = p\frac{M_{sr}^2R_r\omega_{r1}I_s^2}{R_r^2 + L_r^2\omega_{r1}^2} \qquad tg\varphi_1 = \frac{L_r\omega_{r1}}{R_r}$$

$$T_2 = p\frac{M_{sr}^2R_r\omega_{r2}I_s^2}{R_r^2 + L_r^2\omega_{r2}^2} \qquad tg\varphi_2 = \frac{L_r\omega_{r2}}{R_r}$$

The variation of I_r following a step change from + ω_{ro} to $-\omega_{ro}$ is represented on the complex phase diagram of Fig. 3. The variation for two values of ω_{ro} are shown : $\omega_{ro} = 2\pi$ and $\omega_{ro} = R_r/L_r$ (which corresponds to maximum steady state torque for non satured machine operation).

It can be seen that the final steady state operating point is practically obtained after a time $t \simeq \pi/\omega$

The torque is proportional to the projection of I_r on the axis orthogonal to I_s.

The transient overshoot of the latter decreases as the rotor current frequency decreases. These results are confirmed by a numerical simulation(Fig. 4). [5]

The time necessary to change to generator operation with a torque equal to - T_o decreases as the rotor current frequency increases. However the simulation of transients corresponding to high rotor current frequencies shows the presence of large amplitude torque oscillations at the frequency of the rotor currents. Furthermore these oscillations have a relatively low damping factor, R_r/L_r

It would be interesting to benefit from the rapid variation of the torque at high rotor current frequencies but also to avoid the torque oscillations. This can be achieved by using a high value of ω_r during a very short interval of time.

It is possible to block the variation of the current I_s by prolonging the conduction of the static converter's thyristors by an angle τ. During this interval, if it is considered that the machine does not have the time to change speed, then the angular frequency of the rotor currents changes from ω_{ro} to $-\omega$

Following this variation I_r is given by :

$$I_r = I_{ro}e^{-\left(\frac{R_r}{L_r} - j\omega\right)t} + j\frac{M_{sr}\omega I_s}{R_r - jL_r\omega}\left[1 - e^{-\left(\frac{R_r}{L_r} - j\omega\right)t}\right]$$

and

$$T = \frac{T_oe^{-\frac{R_rt}{L_r}}}{\cos\varphi}\cos(\omega t + \varphi) + \frac{T_o'}{\cos\varphi'}\left(\cos\varphi' - e^{-\frac{R_rt}{L_r}}\cos(\omega t + \varphi')\right)$$

In as far as the mechanical rotating period is small compared to the time constant R_r/L_r then whilst I_s is blocked the term e^{-R_rt/L_r} can be considered to remain approximately equal to 1.

In this case the electromagnetic torque can be expressed by :

$$T = T_o(\cos\omega t + tg\varphi\sin\omega t) - T_o'(1 - \cos\omega t + tg\varphi'\sin\omega t)$$

where

$$T_o = p\frac{M_{sr}^2\omega_{ro}R_rI_{so}^2}{R_r^2 + L_r^2\omega_{ro}^2} \qquad tg\varphi = \frac{L_r\omega_{ro}}{R_r}$$

$$T_o' = p\frac{M_{sr}^2\omega R_rI_{so}^2}{R_r^2 + L_r^2\omega^2} \qquad tg\varphi' = \frac{L_r\omega}{R_r}$$

The maximum negative value of torque occurs at :

$$tg\omega t = \frac{T_o tg\varphi - T_o' tg\varphi'}{T_o + T_o'}$$

The waveform of the electromagnetic torque whilst the current I_s is blocked during an angle τ has been determined by numerical simulation(Fig. 5).

Figure 6 shows the variation of the maximum torque amplitude with the angle τ.

Figure 7 shows the variation of the current I_r represented in the complex phase plot whilst the vector I_s is held fixed. If the damping term e^{-R_rt/L_r} is neglected then the locus of the vector I_r traces a circle, the radius of which increase as ω_r decreases :

$$\rho = \frac{M_{sr}^2R_r^2I_s^2}{R_r^2 + L_r^2\omega_r^2}$$

The reversal of the rotor current frequency will now be considered ; this changes from $-\omega$ to $-\omega_{ro}$. The current and torque transients are given by the same expressions as defined above but using the initial conditions imposed by the previous transient. This new variation of I_r is represented on Fig. 7

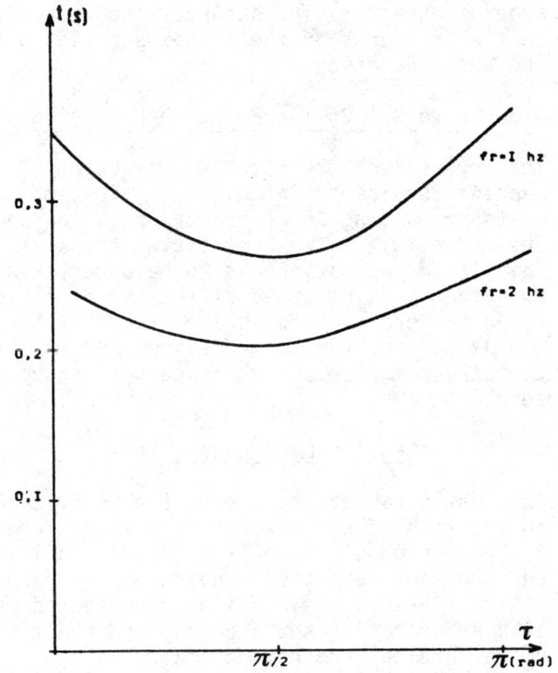

Fig. 8. Braking time versus τ

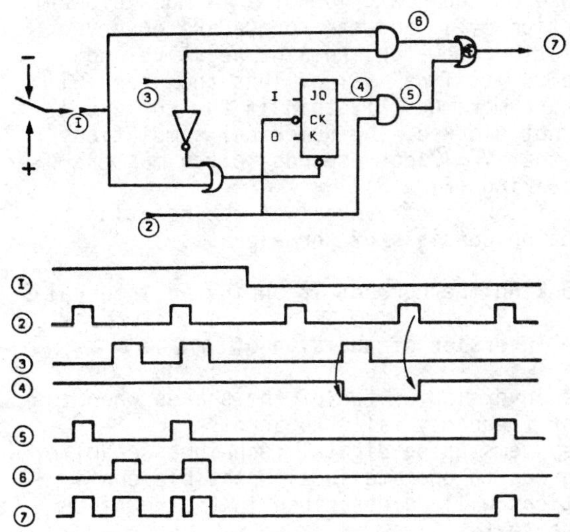

Fig. 9. Pulse adder/subtractor

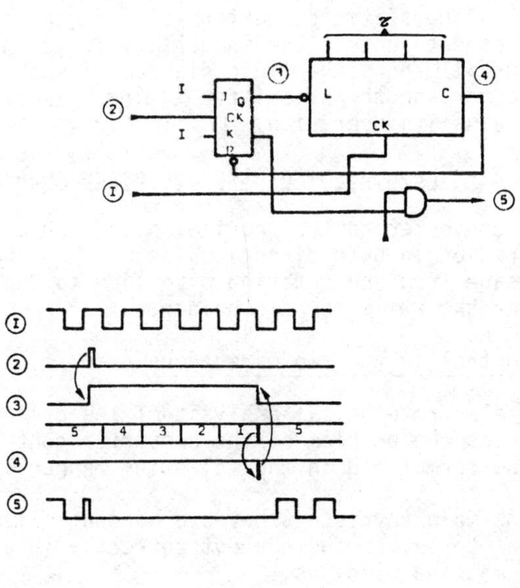

Fig. 10 Impulsions blocking

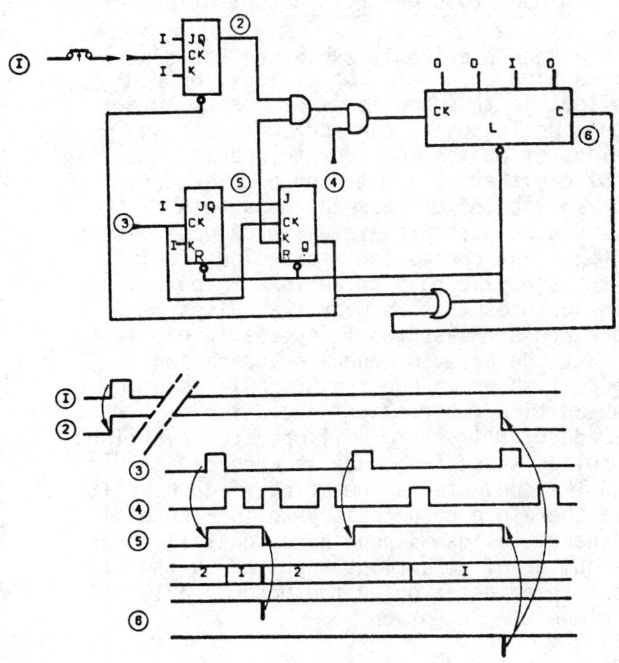

Fig. 11. Zero frequency detection

for $\omega_{ro} = R_r/4$ and for $\omega_{ro} = 2\pi$ ($\tau = 3\pi/4$)
Because of the new value of ω_r a much slower variation of I_r and the torque are observed, the latter, however, remains negative and close to its final value. Thus the speed will decrease more quickly than if the current I_s were not blocked. The numerical simulations show that $\tau = \pi/2$ corresponds to the most rapid braking (Fig. 8).
Summarizing, the reversal of the rotating direction consists of three phases :

- blocking the current I_s during an interval τ

- the inversion of the sign of ω_r

- the inversion of two of the phases when the stator frequency falls to zero.

The systems using digital techniques enabling the group to operate in all possible configurations will be described in the remainder of the text.

3 - REALIZATION

In view of using digital techniques for the conception and realization of the control of the converter it seems sensible to consider the overall problem so as to benefit from all the concepts and methods

3-1 - Closed Loop Control (Autopiloting)

As has been previously explained the converter machine group will be closed-loop controlled. To this end it is necessary to design a device which can add/subtract two strings of pulses in order to operate the machine either as a motor or a generator. The addition of two asynchronous sets of pulses uses a simple exclusive OR gate. The subtraction consists of screening a pulse of f_N each time that there is a f_r pulse. This supposes that between two pulses of f_r there are at least two f_N pulses ; this is in fact the case in generator operation. Figure 9 shows the device which is used for the addition/subtraction operation.
The adder/subtractor will operate on strings of pulses whose frequency is much greater than the converter's operating frequency. It will therefore be necessary to introduce a frequency divider depending essentially upon the number of impulsions supplied by the digital speed detector on the motor's axle during one complete turn.

3-2 - Blocking The Impulsions

The theoretical analysis has shown that it is possible to reduce the braking time, during the reversal of the rotating direction, by blocking the current in the machine windings during a certain time.
This is produced by inhibiting the control clock of the converter during a time corresponding to τ/ω_s. The circuit illustrated on Fig. 10 fulfills this operation.
To optimize the accuracy of the angle τ during which the pulses are inhibited this circuit should be placed immediately after the adder-subtractor. The accuracy thus depends on the frequency of the pulses supplied by the speed detector.

3-3 - Reversal Of The Rotating Direction

The reversal of the rotating direction of the machine considered in this investigation consists of passing to generator operation and then permutating two of the stator phases. This latter operation has to be undertaken when the rotating speed of the stator field falls to zero. The following describes the realization of the logic device used to detect the moment at which f_s passes through zero.

3-3-1 - Detection of the zero f_s

Information about the stator frequency comes in the form of a string of pulses. At zero f_s the period of the pulses becomes infinite and thus any detection requires an extrapolation. Besides the difficulty presented by such an operation the accuracy and the time of measurement are not optimal.
In the systems considered the relation between the frequencies is : $f_s = f_N - f_r$
The zero f_s corresponds therefore to $f_N = f_r$
If between two pulses of f_1 there are less than two pulses of the other frequency f_2 then $f_1 < 2f_2$
The speed detector supplies a string of pulses at a frequency much higher than that used by the static converter. It is therefore proposed to detect the inequality: $2f_N \leq 2f_r$
Figure 11 shows the logic diagram of such a detector and the associated timing diagram explains its operation.

3-3-2 - The control of the converter

The converter control device which enables operation in both directions as well as the passage from one rotating direction to the other has still to be considered.

- control in the two directions :

The electrotechnical analysis of the process enables the problem of the converter control to be formulated in the following manner :

. the main thyristors have to be controlled so as to switch the current correctly into the machine windings.

. after having conducted for 120° each main thyristor has to be turned-off using the auxiliary thyristor and the commutating capacitor.

The cyclic firing of the main bridge thyristors uses the synchronization of the gate pulses with the clock signals from the f_r and f_N frequency adder-subtractor together with a logic circuit which selects the correct thyristor. For this latter function a natural binary coding device has been used which determines the necessary sequences by simple up or down binary counting (Fig. 12).
This solution reduces considerably the hard-

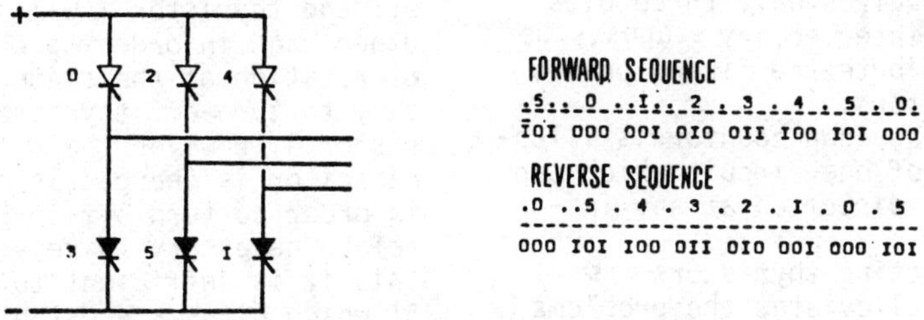

Fig. 12. Code and conduction sequences of the thyristors

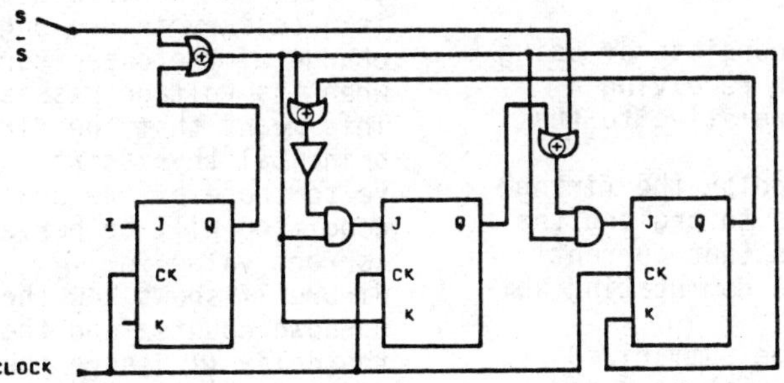

Fig. 13. An up/down modulo 6 binary counter

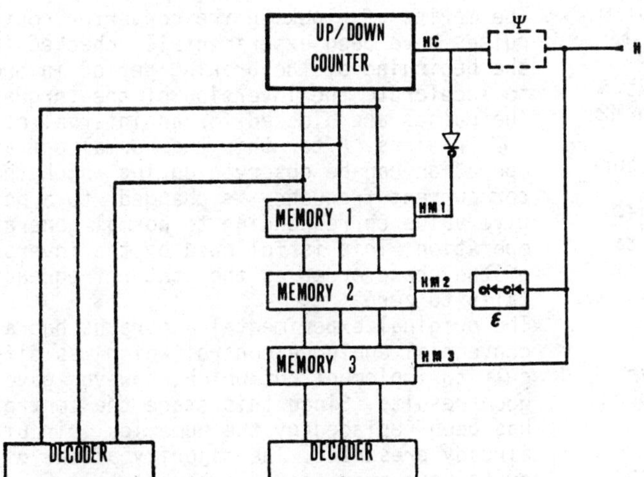

Fig. 14. Block diagram of control logic

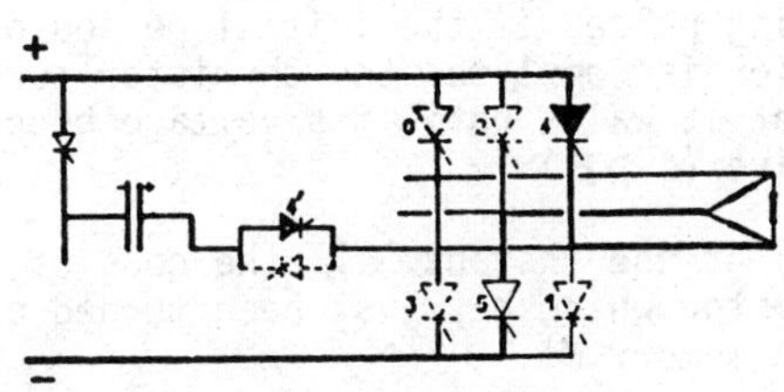

Fig. 15. Circuit during a commutation

ware used since it requires only three bits to be treated in the intermediary stages, the output decoders distribute the firing pulses to each of the thyristors.
The synthesis of the up-down counter is illustrated by an example of one circuit which can control the change in direction at any instant in the cycle.
The code of the conducting thyristors is kept in memory, this alleviates the problems associated with the coupling between principal and auxillary thyristors firing circuits, and also simplifies the inversion sequences. The control circuit are represented on Fig. 14.
Since the thyristors have to conduct during 120° it is necessary to use at least two memory stages.
In fact a very sure method consists of using 3 type D memory stages, each receiving a clock signal phase shifted relative to the others.
The delay ψ can be used to delay the firing of the main thyristors so as to prolong the charging of capacitor at constant current and therefore to improve the commutating ability of the converter.
The complete operation may be summarized by an example. Suppose that thyristors 4 and 5 are conducting and that it is desired to charge to thyristors 0 and 5 (Fig. 15).
1) Thyristor 4 which is to be turned-off was fired at the instant t_{n-2} , its code (4 = 100) appears at the output of HM2. When the closed-loop (autopiloting) signal HM3 appears the code is transfered to the input of the decoder, thyristor 4' receives a firing pulse and the commutating process is thus started.

2) At the instant $t_{n+\varepsilon}$ the code of the next thyristor to be turned off is memorized in M2.

3) At the instant $t_{n+\psi}$ the output of the counter changes to 000. By means of the decoder P the principal thyristor 0 receives its firing pulse. If the initial period of commutation is completed the thyristor turns on, if not it waits until its voltage becomes positive ($\psi \geqslant \varepsilon$).

4) At the instant t'_n the code of the thyristor which has just been turned on is found in memory M1.

- passage from one direction of rotation to another

The up-down counter defines the sequences necessary for the firing of the principal thyristors however the commutation of the thyristors in the static converter requires the adequate charging of the capacitor which depends on the control of the auxillary thyristors.
In the following example it will be shown that the control uses defines all the sequences necessary for the reversal of the direction of rotation.
If in the configuration shown on Fig. 16, the forward sequence is maintained than at the next commutation thyristor 5 will be turned off and thyristor 1 will be fired. On the other hand in order to change the direction of rotation of the stator field it is necessary to turn-off thyristor 0 and turn-on thyristor 4. However the voltage to which the capacitor is charged is of the wrong polarity in order to turn off thyristor 0 - it is therefore necessary to reverse it. To accomplish this it is sufficient to turn off 5 and fire 5' which gives a constant current change of the capacitor. When the voltage across 5 becomes positive it is possible to turn-it on again. The application of V_c turns off 5' and the system is thus brought back to the original configuration but the capacitor has now changed polarity. If it is required to maintain a certain commutating ability it is necessary to prolong the constant current charge of the capacitor beyond the instant when its voltage passes through zero.
This means that the firing pulses of the principal thyristors must be delayed relarive to those of the auxillary thyristors. This condition will be perfectly satisfied if a correct value for ψ is chosen.
Figure 17 shows how the combination of the up-down counter and the memory stages using the delay ψ assure the correct reversal of direction (Fig. 17).
It should be noted that the order for reversal can be given at a random and that it will in fact be carried out within a maximum of $1/6$ of the period.

4 - EXPERIMENTAL TESTS

The merits of blocking the converter control pulses have been experimentally checked (at the beginning of the braking period in order to accelerate the inversion of the torque). The pulses are blocked for an interval of τ radians. After this the normal operating operation can be observed during which the rotor current frequency is changed to a negative value corresponding to normal generator operation. This is followed by the inversion of the phase order as the stator frequency falls to zero.
The original experimental apparatus had a convential analogue control which was difficult to implement but which, however gave good results. Since this stage the control has been replaced by the numerical circuits already presented. The majority of the circuits have been tested experimentally and in particular those controlling the inversion of the phase order.
This type of control technology has proved much more simple and will be completed to be integrated in a programmed logic system monitoring the whole variable speed group.

5 - CONCLUSION

The operation of a.c. variable speed groups as fully reversible systems (4 quadrant operation in the torque-speed plane) requires a complex type of control. Several operations are necessary for the generator operation and the reversal of the rotating direction. It has been shown that digital techniques are

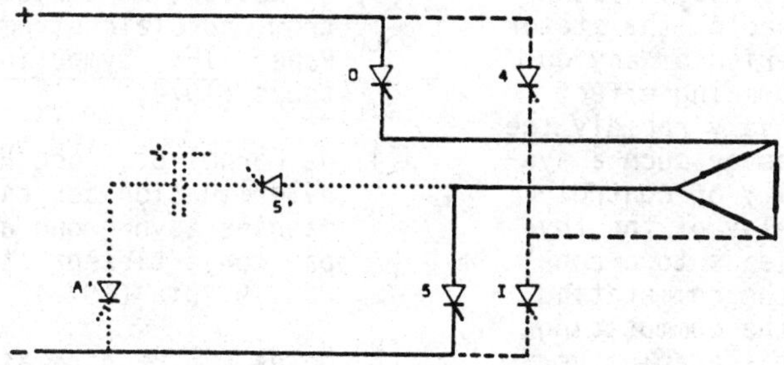

Fig. 16. Circuit configuration

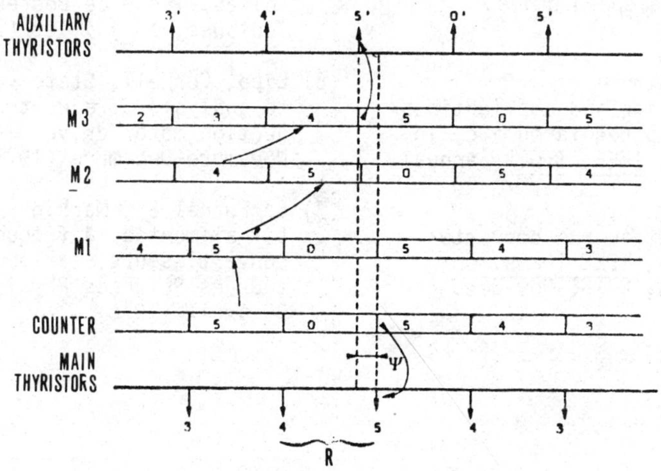

Fig. 17. Timing diagram

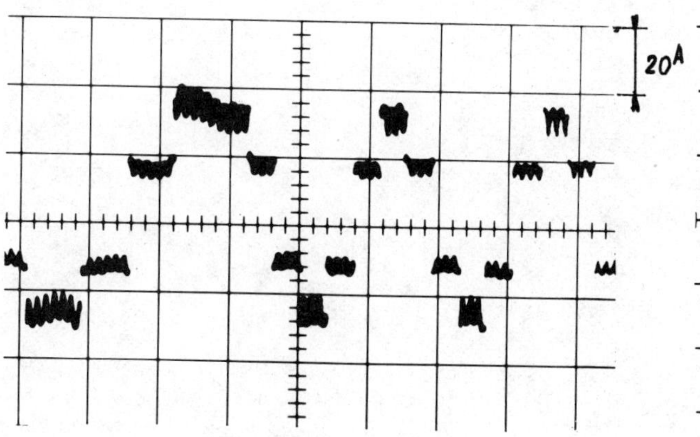

Waveform of stator current I_S due to an impulsion blocking ($\tau = \pi/3$)

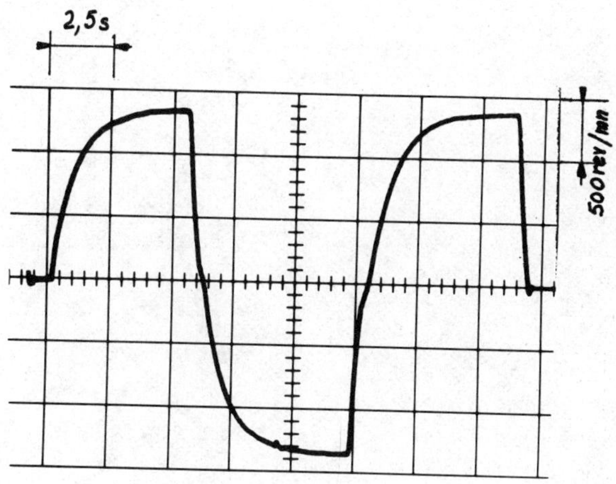

Reversal of the rotating direction with impulsion blocking

very well adapted for this purpose. Furthermore this type of control enables the stator field to be blocked for a period of any duration thus increasing the braking effect of the machine by inverting very rapidly the electromagnetic torque. Moreover such a systém also gives the possibility of controlling, at any instant, the delay of the thyristor firing pulses which leads to a control of the voltage across the commutating capacitor and therefore of the commutating ability of the converter. This enables very rapid transients with a large current associations to be considered. This realisation has been undertaken with view to using a programmed logic control for the overall monitoring of the variable speed drive.

BIBLIOGRAPHY

(1) R. Magg, Characteristics and applications of current source slip regulated a.c. induction motor drives, IEEE, I.G.A. Annual Meeting (1971)

(2) K.P. Phillips, Current source converter for A.C. motor drives, IEEE Trans. on Industry Applications, N° 5, Nov-Dec. (1972)

(3) M. Buhler, Stromrichtergespeiste drehstrom und Gleichstromantriebe, Survey Paper, IFAC Symposium, Duesseldorf, Octobre (1974)

(4) de Fornel B., Foch H., Noyes D., Sur la détermination des caractéristiques d'une machine asynchrone alimentée en courant par convertisseur statique, C.R.A.S., 280, 9 juin (1975)

(5) Noyes D., Machine asynchrone d'induction alimentée en courant par convertisseur statique à fréquence variable - Etude et mise en oeuvre des dispositifs - Caractérisation - Stabilité - Régimes transitoires, Thèse de Docteur de Spécialité, Toulouse, 10 juillet (1975)

(6) Lipo, Cornell, State variable steady state analysis of a controlled-current induction motor drive, IEEE Trans. IA, Novembre-Décembre (1975)

(7) de Fornel B., Machines à courant alternatif alimentées à fréquence variable par convertisseurs statiques, Thèse de Docteur ès Sciences Physiques, Toulouse, 30 avril (1976)

THE OPERATIONAL MODES OF A CURRENT SOURCE INVERTER INDUCTION MOTOR DRIVE SYSTEM

J. Lázár

Technical University Budapest, Department of Electrical Machines

ABSTRACT

The paper discusses the operational ranges of an induction-motor drive system supplied from a Current-Source Inverter. Using the Park-vectors, the study examines the conditions allowing the periodical steady-state to develop in the system. These conditions are determined by the conduction states of the diodes in the inverter.

In the system, an important role comes to the ratio of the output fundamental harmonic frequency to the eigenfrequency of the commutating circuit. It is demonstrated that, with a determined value of this ratio of frequencies, the system reaches a boundary state beyond which no periodical state corresponding to the known (normal) circuit modes can develop. This operational boundary is determined by the conduction states of the diodes on that side of the inverter-bridge where the commutation process is just taking place. The paper gives the condition of this boundary state.

When the boundary state is reached, the process of commutation changes and the system comes into a new periodical steady state. This is the second operational mode, which is possible up to a further boundary determined by the conduction states of the diodes of the bridge-side opposite to the previously mentioned one. The condition of this second operational mode is similarly stated in the paper. The second mode allows the semiconductors to be exploited more favourably from the view-point of voltages, as well as the upper limit of the output frequency to be increased.

INTRODUCTION

The scheme of the system in question can be seen in Fig. 1 showing an up-to-date version of the Current-Source Inverter (further on only Inverter) which is generally known and does not need any detailed explanation. The controlling and regulating circuits are not indicated in the figure, since it is not desired to discuss the problems of regulating the system. Object of the study is only to examine the commutation processes of the system.

In the examinations, the application of Park-vectors appears to be most practical, since in this way the results will be obtained in their simplest and clearest form. In accordance with [3] the motor is replaced by the circuit shown in Fig. 2. In this equivalent circuit $\bar{u}_s$ is the terminal voltage of the stator, $\bar{i}_s$ the stator current and L' the transient inductivity of the motor. Finally, $\bar{u}'$ is a voltage of fundamental harmonic with constant amplitude which has to be chosen in a manner that the current of fundamental harmonic be properly given to the working point from the equivalent circuit. The small letters with dash mean Park-vectors, those without dash are instantaneous values; the capital letters without dash stand for constant quantities.

In the equivalent circuit – which is valid for the Park-vectors and their harmonic components separately, too – the resistances of the motor with respect to the harmonics are neglected. The stator resistance will be neglected later also from the view-point of the fundamental harmonics, since the phenomena examined occur in the range of relatively higher frequencies where such a neglection is permissible. Based on Fig. 2 and using the Park-vectors, the following differential equation holds for the motor:

$$\bar{u}_s = L' \frac{d\bar{i}_s}{dt} + \bar{u}' \tag{1}$$

It is supposed that both the winding of the stator and the commutating capacitors are star-connected, whereas in the case of a delta-connected circuit we calculate with an equivalent star-connection. In this way the network shown in Fig. 3 will be obtained for the system consisting of a motor and an inverter. The two sides on the three-phase bridge-circuit constituting the inverter are marked in the figure by P and N, respectively; the three-phases are indicated by a, b and c. The positive directions of the voltages in the components and of the currents are indicated by arrows beside the respective elements. The subscripts T, D and C refer to thyristors, diodes and capacitors. The direct current I is assumed to be ideally smooth and also the speed of the motor is considered to be constant.

In periodical steady-state the processes in the system take place six times during one period of the output fundamental harmonic, therefore, owing to the periodicity, it will be sufficient to examine only a single interval lasting for $\frac{1}{6}$ of a period. For a better understanding of the following section and to introduce the symbols, the process of the well-known $L-C$ commutation will first reviewed briefly. The operational mode of the inverter with this commutation may be called the first operational mode of the inverter.

FIRST MODE

In this most common mode of operation let us choose the interval in which the commutation takes place on side N between phases b and c. This process is illustrated in the pictures of Fig. 4, where the path of the current in each circuit mode is indicated by broken lines. Picture a shows the initial state which develops at the end of the previous interval.

The interval considered begins when on side N the thyristor of phase c following according to the positive phase sequence obtains a firing pulse; then capacitors b and c having appropriate charge stop the conducting state of the thyristor of phase b. The duration of the commutation of the current from thyristor b to thyristor c is very short as compared with the other processes, and therefore it can be neglected.

The state developing when thyristor b stops conducting is shown in picture b of Fig. 4; this state is the first circuit mode of the interval examined. This lasts until the diode of phase c on side N comes to the boundary of the conducting state. At this time the first circuit mode ends and the second begins, as illustrated in picture c. This circuit mode is the overlapping during which the commutation of the current from phase b to phase c of the motor takes place. The overlapping ends when the current of phase b becomes zero and then the diode of phase b closes. After this follows the third and, at the same time, last circuit mode of the interval, which is again a two-phase conducting state. This is shown in picture d. It is from this state that the next interval begins, which is started by the firing of thyristor b on side P.

The paths of the Park-vectors for a full output period are shown in the stator coordinate system in Figs. 5, 6 and 7; the path portions belonging to the interval discussed above are marked by heavy lines in the figure.

Due to periodicity, the paths are closed and symmetrical with respect to the origo. Fig. 5 shows the Park-vector path of stator current $\bar{i}_s$, which is a regular hexagon in accordance with the assumption of $I =$ constant. In the initial moment and first circuit mode of the interval examined the current vector is in the position marked with I_0.

Fig. 6 shows the paths of the voltage vectors. Since $\bar{u}'$ is a vector of the fundamental harmonic with constant amplitude, its path is a circle around which the vector passes by at a synchronous angular speed $\omega_1 = 2\pi f_1 =$ constant determined by the stator frequency f_1. The position of vector $\bar{u}'$ is characterized by angle α at the beginning of the interval examined. The terminal voltage of the stator differs from $\bar{u}'$ by the self-induction voltage $L' \cdot \frac{d\bar{i}_s}{dt}$ only during the duration of commutation, and it is in this way that the symmetrical hexagonal path of vector $\bar{u}_s$ occurs, as can be seen in the figure.

In the figure $\bar{u}_{CN}$ indicates the vector paths of the capacitor voltages of side N while $\bar{u}_{CP}$ stands for those of side P. Considering that the periodicity recurs by 120° on both sides, but the phase is shifted by 60° between the two bridge-halves, it appears evident by the projection rules of the Park-vectors, that the loci of the capacitor voltages give equilateral triangles whose sides are perpendicular to the phase axes and their vertices lie in the axes. The two triangles are congruent, but rotated with respect to each other by 60°, in accordance with the phase shift between the bridge-sides.

The position of the vector of the capacitor voltage on side N was marked by U_C at the beginning of the interval under consideration, and this is in fact the maximum value of the phase voltages of the capacitors. In the interval examined the Park-vector $\bar{u}_{CN}$ passes along the triangle side drawn in heavy line, between points O and Q in the first circuit mode, then between Q and 3 during the overlapping. In the third circuit mode the vector remains in point 3 until the next firing of thyristor a on side N. Point Q means the time, when the line voltage of the commutating capacitors becomes equal to the line voltage $b - c$ of the stator, by which the commutation of the stator current starts. The capacitor voltages on side P remain unchanged during the whole duration of the interval, and their Park-vector is directed to the phase axis a (positive real axis): $\bar{u}_{CP} = U_C =$ constant. In the figure the duration of the first circuit mode of the interval is marked by angle γ, and the duration of the second circuit mode (overlapping) by angle δ. For later use, also the reduction $\alpha + \gamma = \beta$ is introduced.

Fig. 7 shows the vector path of the diode voltages. This path can be constructed by means of the vectors shown in Fig. 6, in compliance with the relationships

$$\bar{u}_{DN} = \bar{u}_s - \bar{u}_{CN} \tag{2}$$

$$\bar{u}_{DP} = \bar{u}_{CP} - \bar{u}_s \tag{3}$$

which can directly be written on the basis of Fig. 3. To visualize the contents of these formulae and the way of their construction, the part of Fig. 6 relative to the interval under examination is drawn separately in Fig. 8. During a full period the vectors of the diode voltages of the two bridge-sides run along the same path, but the two vectors are in different parts of the path at a given time, in conformity with the phase shift between the bridge-sides. The positions of $\bar{u}_{DN}$ and $\bar{u}_{DP}$ at the beginning of the interval under consideration as well as in the different circuit modes are indicated by the numbering used also in the previous figures.

After this qualitative analysis, the time functions of the Park-vectors can be calculated, if in addition to the above equations, also further equations of the system are used; in forms valid for the interval examined they are as follows:

$$\bar{i}_{CN} = C\frac{d\bar{u}_{CN}}{dt} \tag{4}$$

$$\bar{i}_{CN} + \bar{i}_s = I_0 \exp\left(j\frac{\pi}{6}\right) \tag{5}$$

further:

$$\bar{u}' = \bar{U}' \exp(j\omega_1 t) \tag{6}$$

and the fundamental harmonic of the stator current:

$$\bar{I}_1 = \frac{3}{\pi} \int_0^{\frac{\pi}{3}} \bar{i}_s(t) \exp(-j\omega_1 t)\, d(\omega_1 t). \tag{7}$$

This current must be equal to the current of the fundamental harmonic that can be calculated from the well-known equivalent circuit or circle diagram of the motor.

The eigenfrequency of the oscillating circuit developing during the overlapping process is

$$\Omega = \frac{1}{\sqrt{L'C}}.$$

The ratio of the output frequency of the inverter to its eigenfrequency:

$$\epsilon = \frac{\omega_1}{\Omega} = \omega_1 \sqrt{L'C}. \tag{8}$$

The calculations can be performed with the use of the per-unit system, choosing the rated quantities of the motor as basic units. Thus, e.g., the output frequency in the p.u. system will be:

$$\mu = \frac{f_1}{f_{1n}} = \frac{\omega_1}{\omega_{1n}}.$$

Instead of the stator current of fundamental harmonic it is expedient to calculate with the dimensionless quantity

$$y_1 = \frac{I_1 X'}{\left(\frac{U'}{\mu}\right)} \tag{9}$$

and to use the dimensionless quantity

$$K = \frac{\sqrt{3}}{2} \frac{U_C}{U'} \tag{10}$$

for characterizing the magnitude of the capacitor voltage. (The $\sqrt{3}$-fold value of the latter gives the peak value of the line voltage of the capacitor, see Fig. 6.) With neglection of the resistance of the stator, $\frac{U'}{\mu}$ will be a quantity proportional to the rotor flux whose magnitude in the p.u. system will be about unity. Finally: $X' = \omega_{1n} L'$.

The calculation of the first operational mode is not dealt with here, since in can be found in many places in the literature; the above relationships are meant only to state the terminology and definitions used in our study.

BOUNDARY STATE

The circuit modes developing in the first operational mode during the selected interval are shown in Fig. 4; a careful observation of these pictures gives conclusions to two important facts:

(1) During the whole duration of the interval on side P of the inverter bridge, it is only the diode of phase a that conducts, the diodes of the two other phases do not, and thus on the latter there must be a back voltage.

(2) The diode of phase a on side N does not conduct during the interval, and thus a back voltage must be present on this diode, too.

It is evident that these restrictions are necessary conditions for the circuit modes of the figures and for the first operational mode to develop. These conditions obviously raise definite requirements in respect of the diode voltages. These requirements can be examined most simply and clearly on the basis of the general theory elaborated for this conference by Prof. Dr. I. Rácz in his study [5].

Examining condition (1) one can easily demonstrate by the method of Rácz: on side P, during conduction time of diode a, the condition for diodes b and c to receive back voltage is that the Park-vector $\bar{u}_{DP}$ remains within the $\frac{2\pi}{3}$ sector closed by the negative halves of phase axes b and c during the whole interval.

Concerning condition (2) it can be demonstrated by a similar method that on side N of the inverter bridge diode a gets back voltage – while current flows only through diode b – if the Park-vector $\bar{u}_{DN}$ remains within the $\frac{2\pi}{3}$ sector closed by the negative halves of phase axes c and a.

It can be proved that this latter condition becomes first valid, and thus the first operational mode is dependent on the condition relative to the diode voltages on side N. It is the bridge side N in which the commutation takes place in the interval examined. To visualize the situation, a short portion of the path of vector $\bar{u}_{DN}$ together with the sectors mentioned above is shown in Fig. 9.

In the case corresponding to the first operational mode, as was seen in Fig. 7, the vector path intersects the negative phase axis a, and at the same time diode c gets into the conducting state and the commutation begins. This case is shown by the curve in dotted line. The curve drawn in dashed line shows a case when the vector path intersects the negative phase axis c; this means that diode a gets into conducting state and not the diode c required for the first operational mode.

Fig. 9 demonstrates clearly that in the limit case the path of $\bar{u}_{DN}$ must run into the starting point O of the coordinate system; this is the boundary of the (possible) intersection with the negative phase axis a. This case is shown by the curve drawn in heavy line. Constructing this limiting point O of the path of vector $\bar{u}_{DN}$ according to Fig. 7, one gets the rectangular triangle $O-P-Q$ from which one can write directly the following simple relationship $\frac{U_C}{2} = U' \cos\beta$ or, using the dimensionless quantities introduced above:

$$K = \sqrt{3} \cos\beta \tag{11}$$

This relationship, after some simplifications according to [4], is represented by the curve in thin lines of Fig. 14. The curves illustrate the frequency ratio ϵ defined by formula (8) as a function of the fundamental harmonic stator current interpreted according to formula (9), on the boundary of the first operational mode, for some motors with various no-load current. (The no-load current $y_{1,0}$ is given in the dimensionless form according to formula (9)). With a given load in the first operational mode of the drive system the frequency ratio must be smaller than the value read from the boundary curve corresponding to the given motor.

Since the drive must operate anywhere between no-load and a given load, therefore the maximum possible frequency ratio is determined by the values belonging to the minimum points of the boundary curves. The boundary curve of the frequency ratio determined by the minimum points is shown by the thin line curve marked 1 in Fig. 15 as a function of the no-load current of the motor. The no-load current of the asynchronous motors of conventional design is about 30 to 50 per cent, and the range corresponding to this is plotted in the figure. It can be seen that in the first operational mode the highest attainable frequency ratio for normal motors is about 0,13. It appears also that in this respect the motors of higher no-load current are more favourable.

It must be noted, that the boundary state discussed develops in the motor operation mode only, but there is no such or similar restriction in generator operation.

SECOND MODE

Overstepping the boundary of the first operational mode (which may occur also at lower frequencies if the capacitance of the commutating capacitors is chosen too high), the periodical operation of the drive system does not collapse, but the commutation process will be modified and a new periodical steady state will develop, which may be termed as the second operational mode. The possibility of the modification of the commutation has been referred to in [1], then later in [2].

The commutation process of the second operational mode, relative to the interval discussed, is illustrated in the pictures of Fig. 10. The starting position and the first circuit mode are identical with those of the first operational mode (pictures a and b of Fig. 4). However, after the first circuit mode it is not diode c but diode a that enters the conducting state on side N, as was seen in Fig. 9.

This circuit mode is shown in picture c of Fig. 10. Only after this does also diode c open, and this is how the circuit mode shown in the picture arises: all the three terminals of the commutating capacitors on side N will be connected to the terminals of the stator. After this, diode a leaves the conducting state, and in this way the state shown in picture c develops, which agrees with the overlapping period of the first mode (picture c of Fig. 4). Finally, the last circuit mode of the interval (picture f) is similarly identical with that of the first mode.

The commutation process of the first mode is thus completed by two new conducting states in the second mode (pictures c and d of Fig. 10), and so the commutation of the stator current proceeds not in one but in three circuit modes (pictures c, d and e of Fig. 10). The Park-vector paths of the state variables for an operating point of the second mode is illustrated in Fig. 11. Owing to symmetry, it is sufficient to draw the paths only for one sixth of a period. Each circuit mode of the interval examined is referred to by the numbering beside the curves, as used in Fig. 10. The duration of the three commutation periods of the stator current is characterized by the angles ϑ, δ_1 and δ_2, respectively, in Fig. 11.

The modification of the vector paths with respect to the first mode can well be observed at the triangles of the capacitor voltages; the sides of the triangle $\bar{u}_{CN}$ undergo a certain distortion and also the vertices will be somewhat displaced from the phase axes. The vector path of the diode voltage of side N will have the shape drawn in dashed line in Fig. 9. The new circuit modes will appreciably modify also the course of the stator voltage.

The Park-vector path of the stator current can be seen in Fig. 13; the deviation from the hexagon is caused by the new commutation phenomena. In the 3. circuit mode the path is approximately a portion of an elongated ellipse just tangential to the straight line of slope $-\frac{\pi}{6}$ in the starting instant.

In the point marked with $Q1$ in Fig. 11, the voltage in line $a-b$ of the capacitors and the stator becomes equal; diode a gets to the boundary of conducting state at this instant. Then the projection of vectors $\bar{u}_{CN}$ and $\bar{u}_s$ in direction $a-b$ will be equal, as drawn in Fig. 11. Based on this, one can write directly the condition determining the starting instant of the second circuit mode of the interval:

$$\mathrm{Re}\left[\bar{u}_{CN}(\gamma)\exp\left(j\frac{\pi}{6}\right)\right] = U'\cos\left(\beta+\frac{\pi}{6}\right) \qquad (12)$$

At the end of the second circuit mode also the line voltage $b-c$ of the capacitors and the stator become equal, whereby diode c starts conducting and the three-phase state develops, which means the third circuit mode of the interval. The instant of the boundary state of conduction of diode c is indicated by point $Q2$. It can be proved that for this boundary state the following equation is valid:

$$\frac{\sqrt{3}}{2}\frac{y}{\epsilon^2}\vartheta =$$

$$= \sin\left(\beta+\vartheta+\frac{\pi}{6}\right) + \sqrt{3}\cos\left(\beta+\frac{\pi}{6}\right) - 2u_{CNx}(\gamma) \qquad (13)$$

where y is equal to (9) if I_0 is substituted for I_1 therein.

Owing to the many conducting states, the calculation of the second mode is considerably more complicated than that of the first mode; even with the simplifying conditions used in this paper, it is not possible to deduce formulas so simple ant all the same giving very good approximation, as it was possible with the first mode (see. e.g. [6]).

The first circuit mode of the interval is very simple, as only u_{CNy} changes there, while all the other state variables remain constant.

In the second circuit mode, because diode a is conducting, the projection $a-b$ of $\bar{u}_{CN}$ and $\bar{u}_s$ must agree:

$$\mathrm{Re}\left[\bar{u}_{CN} \exp\left(j\frac{\pi}{6}\right)\right] = \mathrm{Re}\left[\bar{u}_s \exp\left(j\frac{\pi}{6}\right)\right]$$

and for the stator current $\bar{i}_s = i_s \exp\left(-j\frac{\pi}{6}\right)$ is valid.

In the third circuit mode $\bar{u}_{CN} = \bar{u}_s$.

In the fourth circuit mode the $b-c$-line values of $\bar{u}_{CN}$ and $\bar{u}_s$ are equal: $u_{CNy} = U_{sy}$, furthermore $\bar{i}_{CN} = ji_{CNy}$ and $i_{sx} = I =$ constant.

In the last circuit mode both $\bar{u}_{CN}$ and $\bar{i}_s$ are constant.

Considering the foregoing, the differential equations of the system can be written for each circuit mode on the basis of expressions (4) and (5). Owing to our simplifying assumptions, linear differential equation systems with constant coefficients will be obtained. It is reasonable to take the components x and y of $\bar{u}_{CN}$ and $\bar{i}_s$ as state variables. The details of the calculation can be omitted here, it has to be mentioned only, that with the use of a digital computer it is expedient to write the solution by each circuit mode, which can be performed fast and simply in the given case. The integration constants, too, can be expressed easily, for the above mentioned state variables pass across the boundary of all circuit modes continuously. After this, to determine a working point, one has to satisfy – under consideration of (12) and (13) – only the periodicity conditions by means of an iteration procedure, which can be carried out much faster than the solution of the equations step by step by the method of Runge – Kutta. The periodicity conditions are:

$$\bar{i}_s\left(\frac{\pi}{3}\right) = \bar{i}_s(0) \exp\left(j\frac{\pi}{3}\right)$$
$$\bar{u}_{CN}\left(\frac{\pi}{3}\right) = \bar{u}_{CN}(0) \exp\left(j\frac{2\pi}{3}\right) \qquad (14)$$

The Park-vector paths shown in Fig. 11 were calculated by the above method.

The second mode of the system ends when, at the end of the third circuit mode, the diode of phase c on side P of the inverter bridge reaches the boundary of the conducting state and thus capacitors a and c are connected to the terminals a and c of the motor. Such a boundary state of the second mode is visualized by the vector paths shown in Fig. 12. As can be seen, at the end of the third circuit mode vector $\bar{u}_{DP}$ is just running into the negative phase axis b; this is in accordance with what was said in connection with Fig. 9. The condition of the boundary state is:

$$\mathrm{Im}\left[\bar{u}_{DP} \exp\left(j\frac{\pi}{3}\right)\right] = 0 \qquad (15)$$

This condition can be satisfied by means of a newer iteration including also the iteration discussed above. This was the method of calculating the boundary curves drawn in heavy lines in Fig. 14, as well as the curve in heavy line marked 2 in Fig. 15, which belongs to the minimum points. These curves represent the attainable maximum frequency-ratio on the boundary of the second mode, and, as can be seen, they always run above the boundary curves of the first mode. It is easy to see from Fig. 15 that for normal motors the value of ϵ_{max} can be 10 to 12 per cent higher than the value restricted by the first mode.

A forther modification of the commutation process, and the possibility of a third or fourth operational mode are open questions at present and require further investigations.

CONCLUSIONS

In addition to the "normal" operational mode of the Current-Source Inverter Induction Motor Drive System there exists also another mode in motor operation, which arises as a consequence of the change in the conducting states of the inverter diodes at the values determined by the frequency-ratio ϵ. In this new operational mode the process of commutation has more stages than the normal one and is more complicated.

The second mode permits a higher frequency-ratio than the first, and therefore the upper frequency limit of the system can be somewhat higher. Since the quantity K defined by formula (10) is nearly inversely proportional to ϵ, (see, e.g., [4]), therefore the voltage-load of the capacitors and thyristors is somewhat more favourable in the second mode than in the first mode.

The extent of the increase of the boundary frequency can be estimated as follows: for at least a duration equal to the turn-off time following the cessation of conduction, the thyristors must receive a back voltage from the capacitors. In the interval examined, e.g., the capacitor voltage is of reverse direction with respect to thyristor b as long as the projection directed $b-c$ of vector $\bar{u}_{CN}$ is negative, i.e. until the vector arrives from its initial position in phase axis a (see Fig. 11). Based on the figure, according to equation (4) one can write: $\frac{I_0 t}{C} = \frac{\sqrt{3}U_c}{2}$ or, with the dimensionless quantities used in the paper: $\frac{y\omega_1 t}{\epsilon^2} = K$. (Since in the second mode the vertices of the voltage triangle of the capacitors are hardly displaced from the phase axes, the above relationship can be used also in the second mode with good approximation). If time t is equal to the turn-off time T_{t0}, the maximum attainable frequency can be obtained with ϵ_{max} from the previous formula:

$$f_{1\,max} = \frac{1}{2\pi}\frac{K}{y}\frac{\epsilon^2_{max}}{T_{t0}} \qquad (16)$$

Since with a given current y, the quantity K is nearly inversely proportional to ϵ, it will be obtained from (16) that with thyristors having identical turn-off-time the maximum frequency can be greater by about 10 to 12 per cent (Fig. 15) than in the system with operation restricted to the first mode.

REFERENCES

[1] H. Kazuno, Commutation of a Three-Phase Thyristor Bridge with Commutation Capacitors and Series Diodes. *Electrical Engineering in Japan,* Vol. 90, No. 5, 1970.

[2] J. Ancker, Analysis and Simulation of an Induction Motor Drive supplied from a Current Source Frequency Converter. Ph. D. Thesis, Stockholm, 1976.

[3] I. Rácz, Betrachtungen zu Oberwellenproblemen an Asynchronmotoren bei Stromrichterspeisung. *Periodica Polytechnica,* Vol. 11, No. 1-2. 1967.

[4] J. Lázár, Berechnung eines Umrichters mit Gleichstromglättung. *Wissenschaftliche Zeitschr. d. TH Ilmenau,* Heft 3, 1976.

[5] I. Rácz, Analysis of Three-Phase Semiconductor Connections by Park-Vectors. *Second IFAC Symposium on Control in Power Electronics and Electric Drives,* Düsseldorf, 1977.

[6] J. Lázár – M.M. El-Sayed, The Simplified Theory and Design of a Current Source Inverter for AC Motor Drives. *Periodica Polytechnica,* under printing.

FIGURES

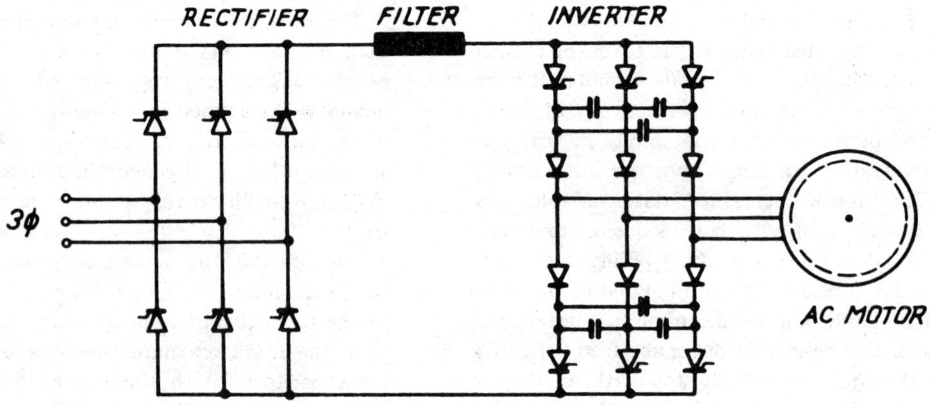

Fig. 1.

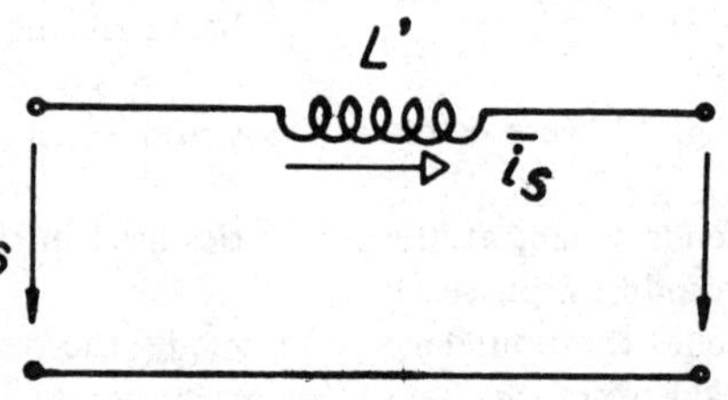

Fig. 2.

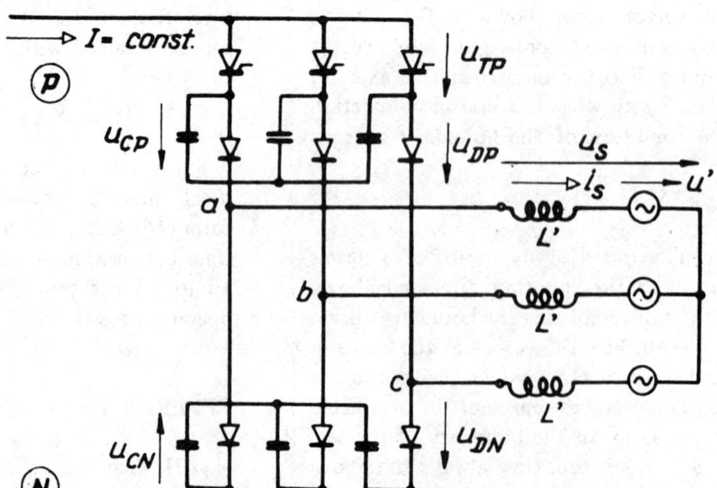

Fig. 3.

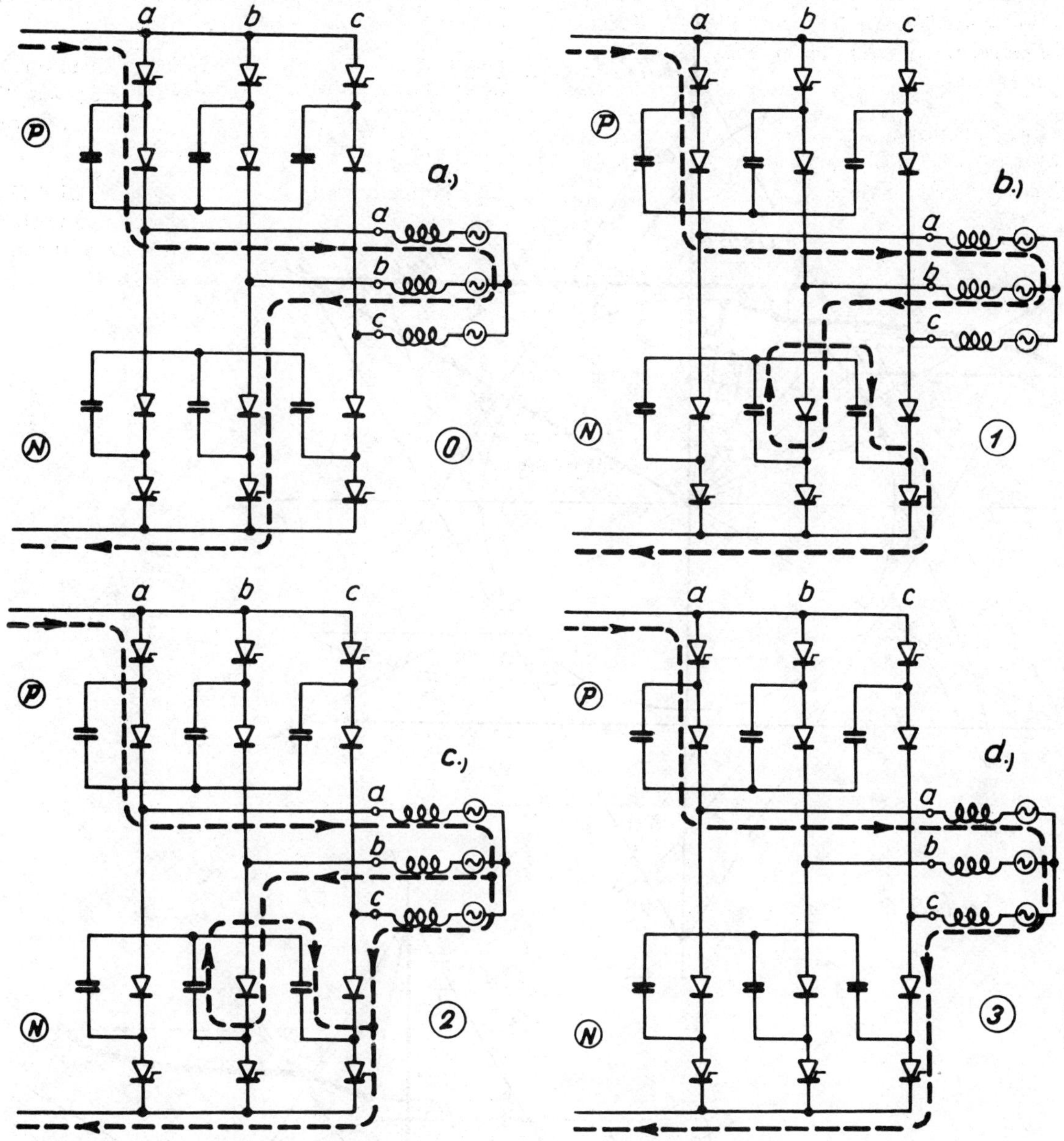

Fig. 4.

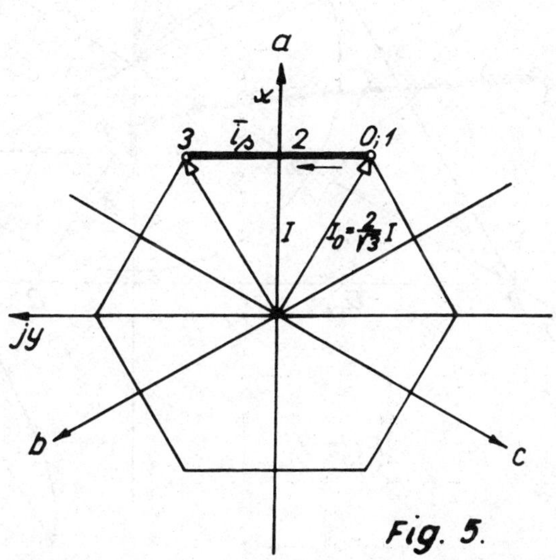

Fig. 5.

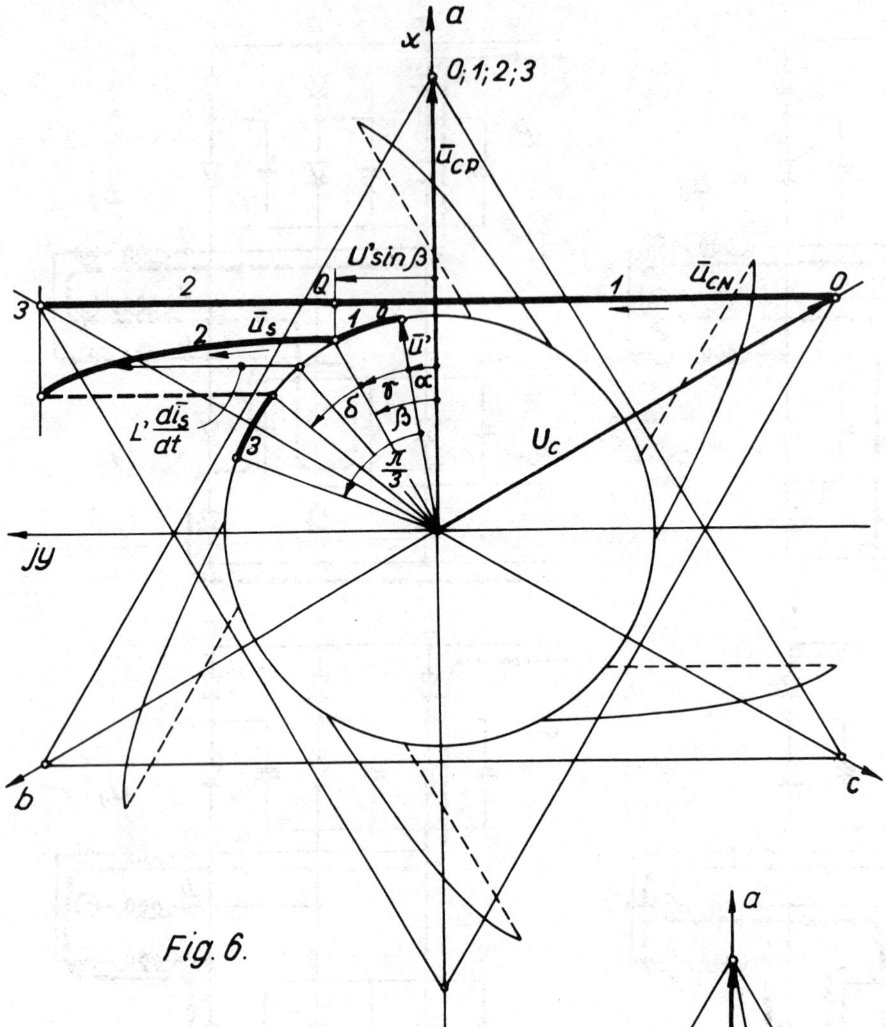
a
x
0;1;2;3
$\bar{u}_{CP}$
U'sin β
Q
2
1
$\bar{u}_{CN}$
0
3
$\bar{u}_s$
$\bar{u}'$
α
γ
δ
β
π/3
$L'\frac{d\bar{i}_s}{dt}$
U_c
jy
b
c

Fig. 6.

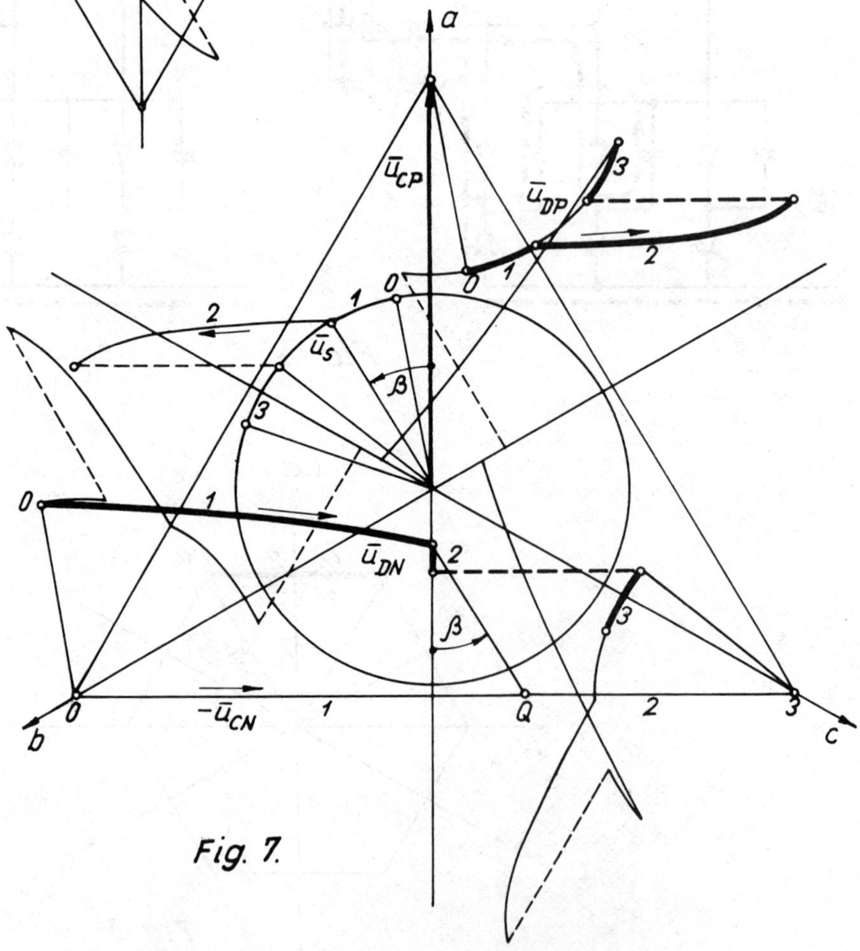
a
$\bar{u}_{CP}$
$\bar{u}_{DP}$
3
2
0
1
$\bar{u}_s$
β
$\bar{u}_{DN}$
$-\bar{u}_{CN}$
Q
b
c

Fig. 7.

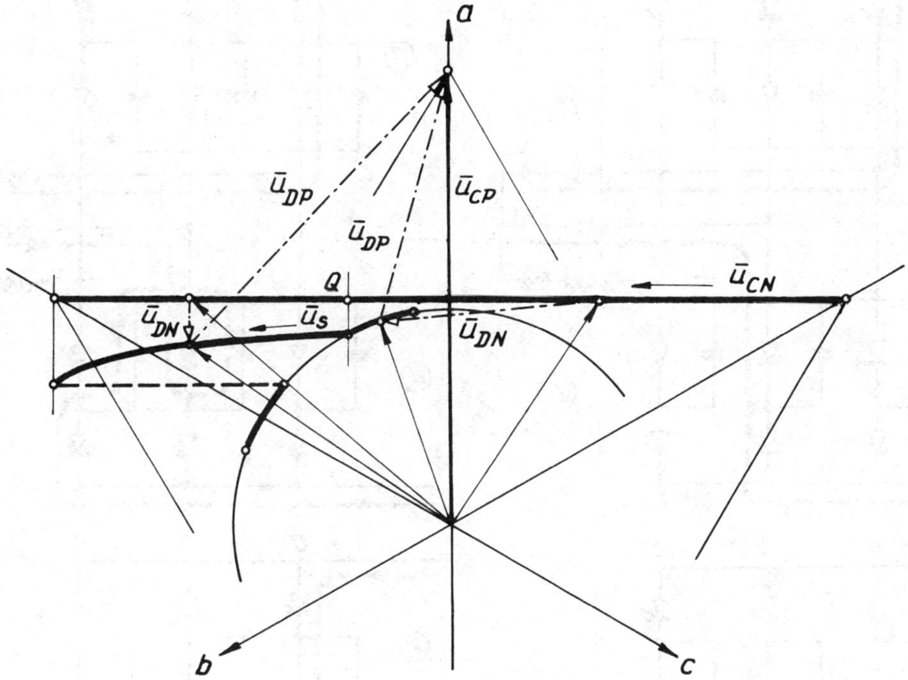
a
$\bar{u}_{DP}$
$\bar{u}_{CP}$
$\bar{u}_{DP}$
Q
$\bar{u}_{CN}$
$\bar{u}_{DN}$
$\bar{u}_S$
$\bar{u}_{DN}$
b
c

Fig.8.

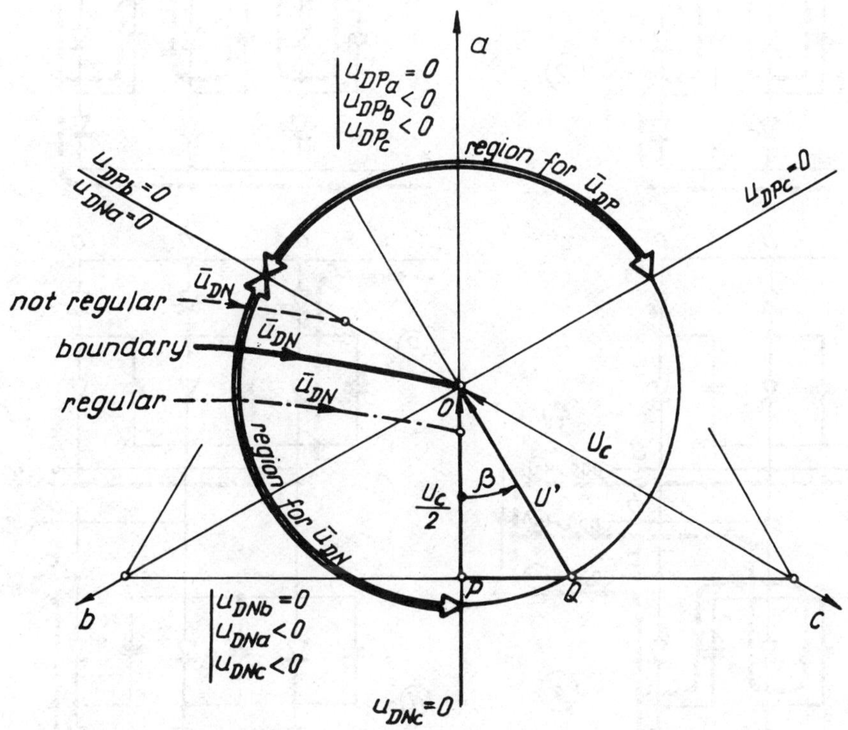
a
$u_{DPa} = 0$
$u_{DPb} < 0$
$u_{DPc} < 0$
region for $\bar{u}_{DP}$
$u_{DPb} = 0$
$u_{DNa} = 0$
$u_{DPc} = 0$
$\bar{u}_{DN}$
not regular
$\bar{u}_{DN}$
boundary
$\bar{u}_{DN}$
regular
region for $\bar{u}_{DN}$
0
U_c
β
U'
$\frac{U_c}{2}$
P
Q
b
c
$u_{DNb} = 0$
$u_{DNa} < 0$
$u_{DNc} < 0$
$u_{DNc} = 0$

Fig. 9.

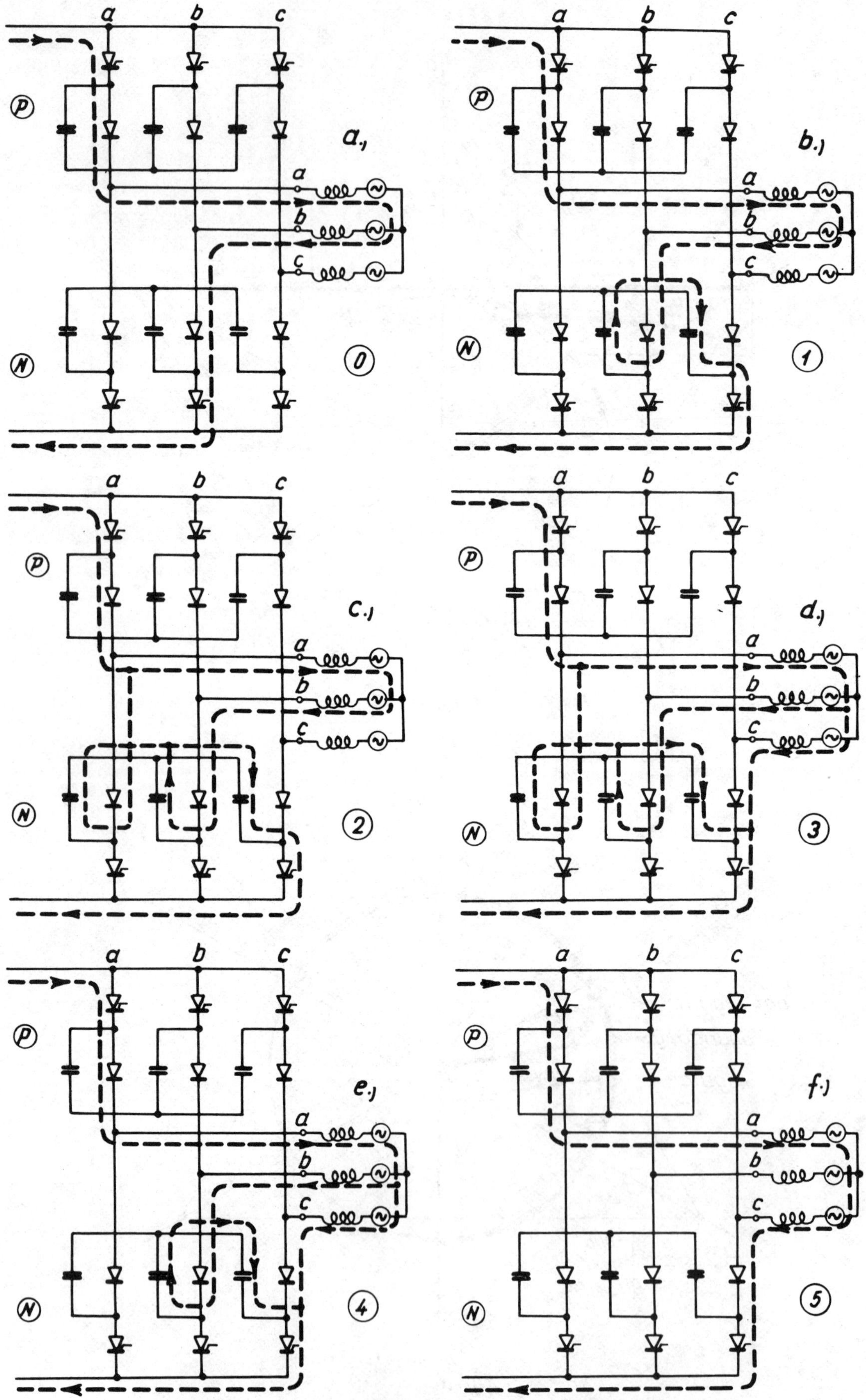

Fig. 10.

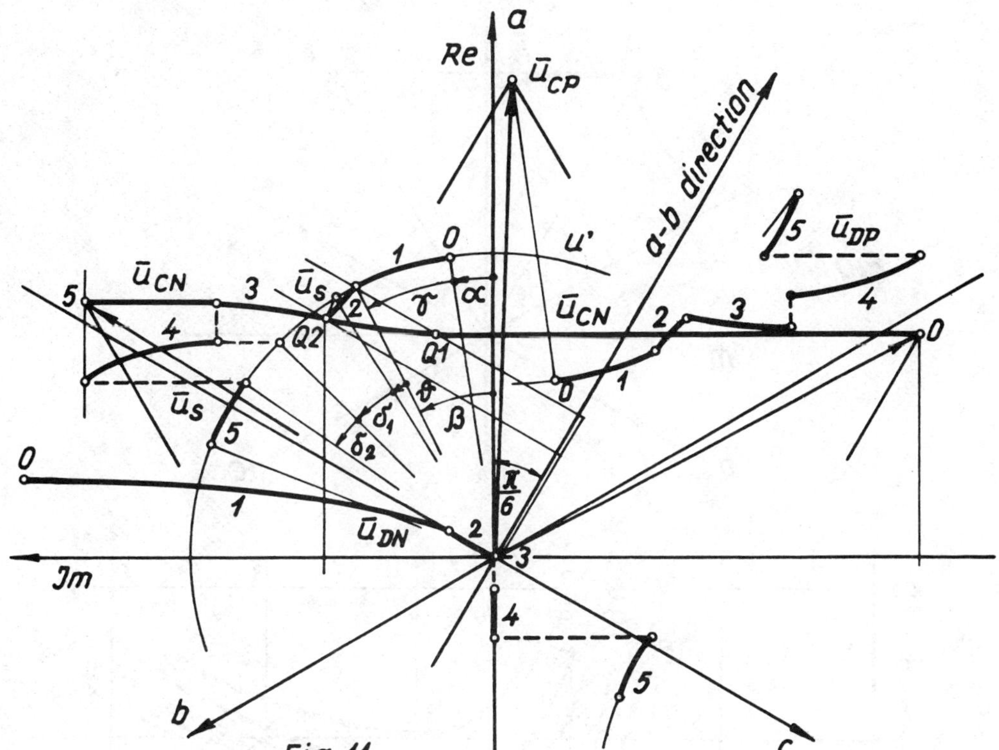
a
Re
$\bar{u}_{CP}$
a-b direction
$\bar{u}_{DP}$
$\bar{u}_{CN}$
$\bar{u}_s$
u'
Q2
Q1
α
β
γ
ϑ
δ_1
δ_2
$\pi/6$
$\bar{u}_{DN}$
Im
b
c

Fig. 11.

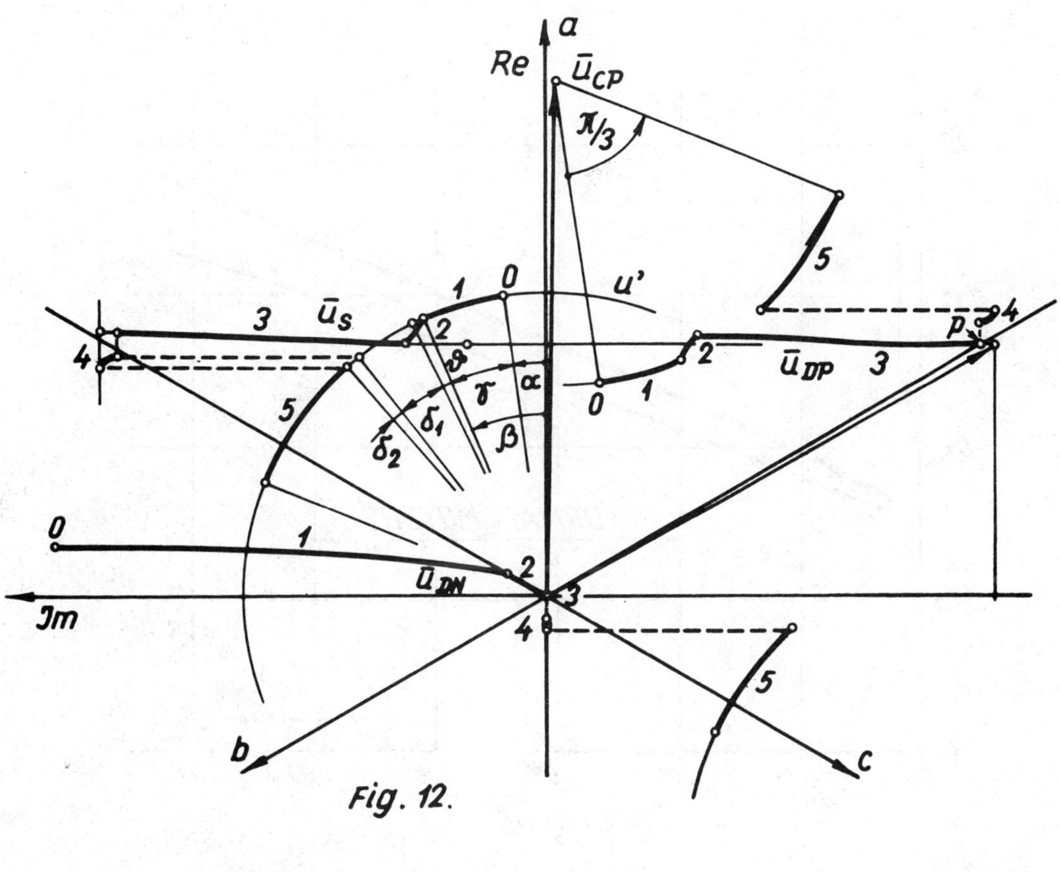
a
Re
$\bar{u}_{CP}$
$\pi/3$
u'
$\bar{u}_s$
$\bar{u}_{DP}$
p
α
β
γ
ϑ
δ_1
δ_2
$\bar{u}_{DN}$
Im
b
c

Fig. 12.

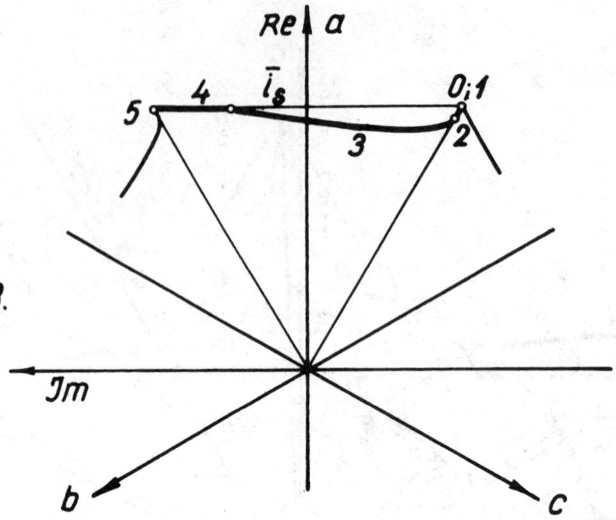

Fig. 13.

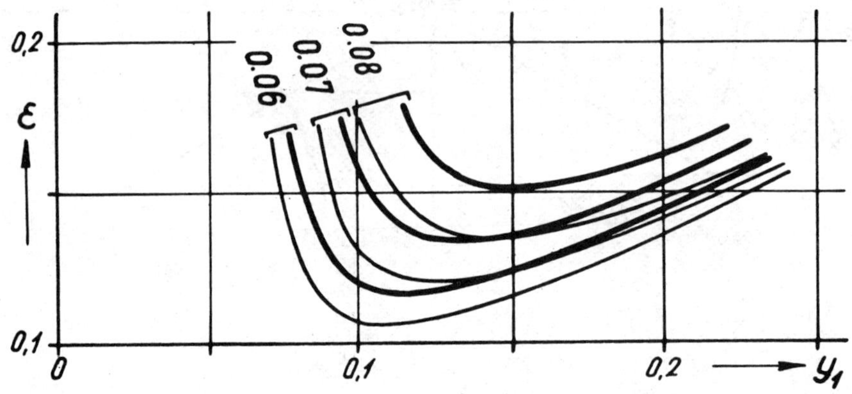

Fig. 14.

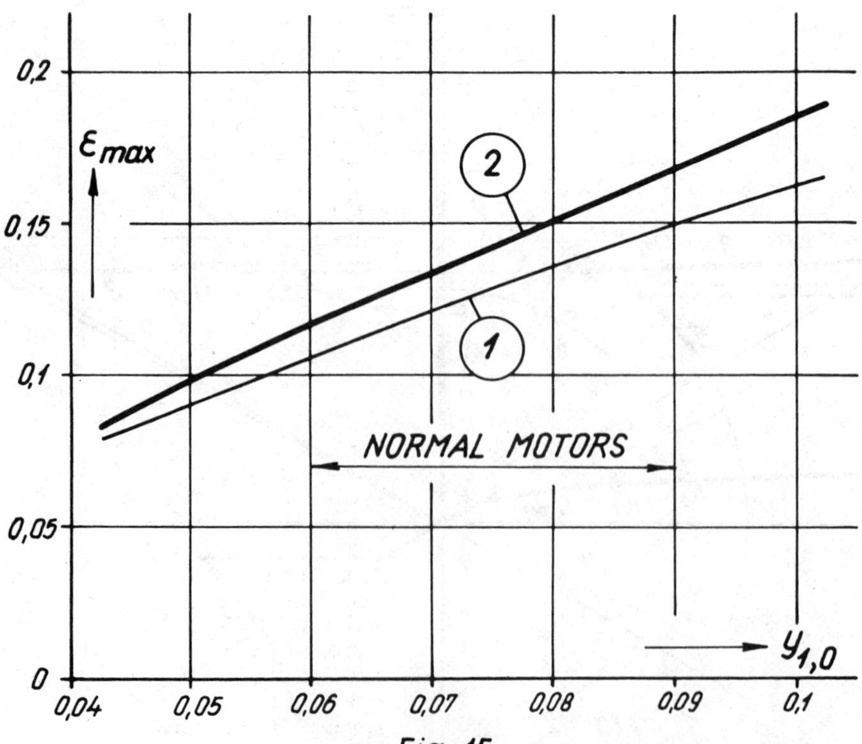

Fig. 15.

ON THE UTILITY OF SIGNAL FLOW GRAPHS IN THE ANALYSIS OF CURRENT CONTROLLED INDUCTION MOTOR

V. Subrahmanyam, D. Subbarayudu and M.V.C. Rao

Department of Electrical Engineering, Indian Institute of Technology, Madras-600036, India

Summary--A novel method is presented for the analysis of a current controlled induction motor via the state transition signal flow graph. The state transition matrix is computed in a simple, elegant and straightforward fashion by this method, with the result the solution can be obtained in a closed form. The solution automatically takes into account of the effect of different inputs thereby eliminating the necessity of evaluating the convolution integrals. Computationally too, this method is found to be superior to the other methods, as it requires lesser C.P.U. time. The solution obtained can be used to show that the locus of the rotor flux vector under steady state conditions is a circle and the rotor flux induces a sinusoidal voltage in the stator phases accompained with voltage spikes during commutation intervals, although the stator current is nonsinusoidal. It can also be shown that the motor torque pulsates with sixth harmonic oscillations, whose relative magnitude increases with the increase in rotor speed. The current-fed induction motor can also be shown to exhibit a shunt type speed-torque characteristic by suitably adjusting the stator frequency.

LIST OF SYMBOLS

$i_{\alpha s}, i_{\beta s}$	direct axis and quadrature axis components of stator current in stationery reference frame
$i_{\alpha r}, i_{\beta r}$	direct and quadrature axis components of rotor current in stationery reference frame
$I_{\alpha s}(s)$	Laplace transform of $i_{\alpha s}$
$I_{\beta s}(s)$	Laplace transform of $i_{\beta s}$
L_s, L_r	self inductance of stator and rotor
M	mutual inductance between stator and rotor
P	pairs of poles
R_s, R_r	phase resistances of stator and rotor
T_d	torque developed
$v_{\alpha s}, v_{\beta s}$	direct and quadrature axis components of stator voltage in stationery reference frame
$\varphi_{\alpha r}, \varphi_{\beta r}$	direct and quadrature axis components of rotor flux linkages in stationery reference frame
$\varphi_{\alpha r}(s)$	Laplace transform of $\varphi_{\alpha r}$
$\varphi_{\beta r}(s)$	Laplace transform of $\varphi_{\beta r}$
$\varphi_{\alpha r}(0)$	initial value of $\varphi_{\alpha r}$
$\varphi_{\beta r}(0)$	initial value of $\varphi_{\beta r}$
p	differential operator
τ	period of input current waveform

INTRODUCTION

It has been established during the last few years that the drive using a current controlled induction motor offers a variety of advantages over the drive using its voltage controlled counterpart. As is well known, the input current waveform of a current controlled induction motor is nonsinusoidal. The existing methods of approach assume the input current to vary in steps treating the waveform to be piecewise constant over a certain period.

Naunin [1,2] and Klautscheck [3] used the concept of space vector and gave a closed form solution for the machine differential equations. Lipo and Cornell [4] adopted the state variable approach and presented the solution in terms of state transition matrix obviously in the form of matrix exponential and convolution integrals. The evaluation of state transition matrix involves procedures like matrix inversion and some times determination of modal matrix. These methods become increasingly tedious and unwieldy with the increase in the order of matrix.

In the present paper an alternate method of obtaining the state transition matrix based on the state transition signal flow graph is described. A closed form solution is made available for the performance of a current-fed induction motor. Further as will be mentioned later this method has several advantages over the methods already available in the literature.

The input current of a current-fed induction motor has hitherto been assumed to have a stepped waveform, which is not strictly correct. In practice because of the machine inductance coupled with the inverter commutating capacitance, the instantaneous current transfer from one phase to the other is rather not possible. Thus the input current waveform approaches a trapezoidal one. Under these circumstances the space vector approach presented by Naunin and Klautscheck does not lead to a closed solution, since the current vector during the period of commutation cannot be easily defined. Similarly the state variable method suggested by Lipo and Cornell creates a formidable mathematical problem of requiring the solution of convolution integrals during the commutation interval. On the other hand the present method can very effectively handle this type of input and the solution can be obtained in a closed form.

MACHINE EQUATIONS AND FORMULATION OF SIGNAL FLOW GRAPH

The well known α-β-0 equations namely the two axis equations referred to stationery reference frame are used in the analysis. As the α-β components of the stator current of a current-fed induction motor are known, the order of the system is reduced to two. The rotor flux linkage components $\varphi_{\alpha r}$ and $\varphi_{\beta r}$ are chosen as the state variables. The following are the differential equations of the rotor expressed in matrix form.

$$p\begin{bmatrix}\varphi_{\alpha r}\\ \varphi_{\beta r}\end{bmatrix}=\begin{bmatrix}MR_r/L_r & 0\\ 0 & MR_r/L_r\end{bmatrix}\begin{bmatrix}i_{\alpha s}\\ i_{\beta s}\end{bmatrix}+\begin{bmatrix}-R_r/L_r & -w_r\\ w_r & -R_r/L_r\end{bmatrix}\begin{bmatrix}\varphi_{\alpha r}\\ \varphi_{\beta r}\end{bmatrix} \quad (1)$$

where

$$\begin{bmatrix}\varphi_{\alpha r}\\ \varphi_{\beta r}\end{bmatrix}=\begin{bmatrix}L_r\, i_{\alpha r}+M\, i_{\alpha s}\\ L_r\, i_{\beta r}+M\, i_{\beta s}\end{bmatrix}$$

Taking Laplace transform of Eqn.(1), we get

$$\begin{bmatrix}s\varphi_{\alpha r}(s)-\varphi_{\alpha r}(0)\\ s\varphi_{\beta r}(s)-\varphi_{\beta r}(0)\end{bmatrix}=\begin{bmatrix}K_1 & 0\\ 0 & K_2\end{bmatrix}\begin{bmatrix}I_{\alpha s}(s)\\ I_{\beta s}(s)\end{bmatrix}+\begin{bmatrix}-Q_{11} & -Q_{12}\\ -Q_{21} & -Q_{22}\end{bmatrix}\begin{bmatrix}\varphi_{\alpha r}(s)\\ \varphi_{\beta r}(s)\end{bmatrix} \quad (2)$$

The values of K's and Q's can be easily identified from Eqn.(1), $\varphi_{\alpha r}(0)$ and $\varphi_{\beta r}(0)$ are the initial values of rotor flux linkages. Solution of Eqn.(2) describes the steady state behaviour of the current-fed induction motor at a rotor angular speed of w_r. This equation can be solved via the state transition signal flow graph shown in Fig.1.

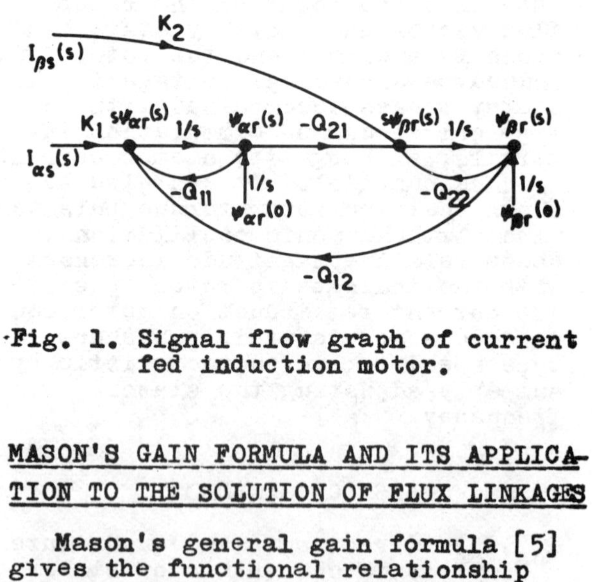

Fig. 1. Signal flow graph of current fed induction motor.

MASON'S GAIN FORMULA AND ITS APPLICATION TO THE SOLUTION OF FLUX LINKAGES

Mason's general gain formula [5] gives the functional relationship between the input and output variables of a signal flow graph. The formula is

$$AM=\frac{X_{out}(s)}{X_{in}(s)}=\frac{AM_k\,\Delta_k}{\Delta} \quad (3)$$

where

AM_k - gain of the kth forward path

Δ - system determinant
Δ_k - the value of Δ for that part of the graph not touching the kth forward path.

The Eqn.(3) is used to obtain the expressions for $\varphi_{\alpha r}(s)$ and $\varphi_{\beta r}(s)$ and they are given by the following matrix.

$$\begin{bmatrix}\varphi_{\alpha r}(s)\\ \varphi_{\beta r}(s)\end{bmatrix} = \begin{bmatrix}G_{11} & G_{12}\\ G_{21} & G_{22}\end{bmatrix}\begin{bmatrix}I_{\alpha s}(s)\\ I_{\beta s}(s)\end{bmatrix} + \begin{bmatrix}H_{11} & H_{12}\\ H_{21} & H_{22}\end{bmatrix}\begin{bmatrix}\varphi_{\alpha r}(0)\\ \varphi_{\beta r}(0)\end{bmatrix} \qquad (4)$$

where

$$G_{11} = G_{22} = \frac{MT_r(s+T_r)}{(s-\gamma_1)(s-\gamma_2)}$$

$$G_{21} = -G_{12} = \frac{MT_r\, w_r}{(s-\gamma_1)(s-\gamma_2)}$$

$$H_{11} = H_{22} = \frac{s+T_r}{(s-\gamma_1)(s-\gamma_2)}$$

$$H_{21} = -H_{12} = \frac{w_r}{(s-\gamma_1)(s-\gamma_2)}$$

and

$$T_r = R_r/L_r \qquad (5)$$

γ_1 and γ_2 are the roots of the characteristic equation of the system

$$s^2 + 2T_r s + (T_r^2 + w_r^2) = 0 \qquad (6)$$

As stated earlier, the input current waveform tends to be trapezoidal in practice (i.e.) it can be assumed that during commutation period the current falls to zero linearly. α,β components of the stator current of an induction motor fed from a current source inverter (Fig. 2.) are shown in Fig. 3.

In general α,β components of the stator current can be written as

$$\begin{bmatrix}i_{\alpha s}\\ i_{\beta s}\end{bmatrix} = \begin{bmatrix}m_\alpha t + I_{D\alpha}\\ m_\beta t + I_{D\beta}\end{bmatrix} \qquad (7)$$

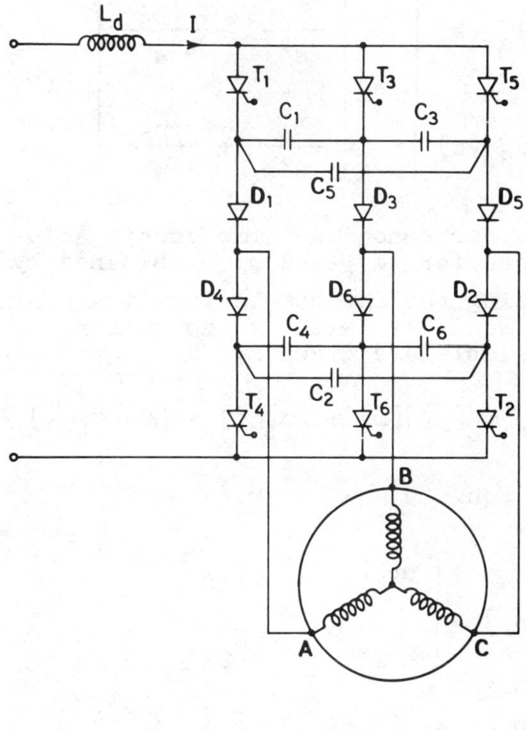

Fig. 2. Current source inverter.

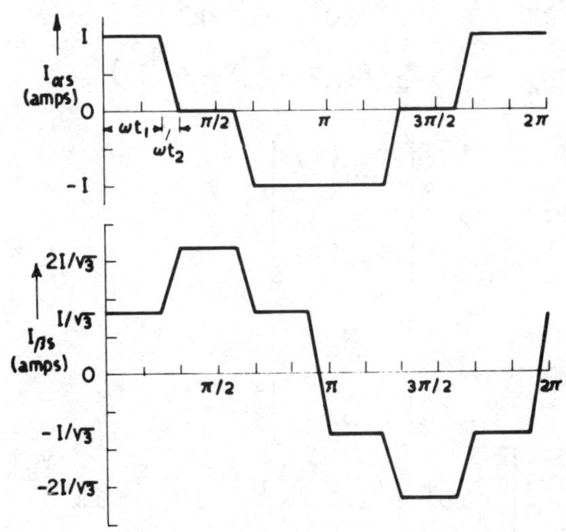

Fig. 3. α,β components of stator currents of an induction motor fed from a current source inverter.

m_α and m_β are the slopes of the current waveforms during the commutation of current from one phase to the other. They are zero during the major portion of the 1/6th cycle when the currents remain constant. The Laplace transforms of $i_{\alpha s}$ and $i_{\beta s}$ are

$$\begin{bmatrix} I_{\alpha s}(s) \\ I_{\beta s}(s) \end{bmatrix} = \begin{bmatrix} \frac{m_\alpha}{s^2} + \frac{I_{D\alpha}}{s} \\ \frac{m_\beta}{s^2} + \frac{I_{D\beta}}{s} \end{bmatrix} \quad (8)$$

The corresponding time domain solutions for $\varphi_{\alpha r}$ and $\varphi_{\beta r}$ obtained by taking the inverse Laplace transform of Eqn. (4) with k as running variable are given by

$$[\varphi] = [pp][m] + [A][I] + [\emptyset][\varphi_{(o)}] \quad (9)$$

In Eqn. (9)

$$[pp] = \begin{bmatrix} pp_{11} & pp_{12} \\ pp_{21} & pp_{22} \end{bmatrix}$$

where

$$pp_{11} = (E_1 + E_{10}t + \sum E_{1k}e^{\gamma_k t})$$

$$pp_{12} = (F_1 + F_{10}t + \sum F_{1k}e^{\gamma_k t})$$

$$pp_{21} = (E_2 + E_{20}t + \sum E_{2k}e^{\gamma_k t})$$

$$pp_{22} = (F_2 + F_{20}t + \sum F_{2k}e^{\gamma_k t})$$

$$[\varphi] = \begin{bmatrix} \varphi_{\alpha r}(t) \\ \varphi_{\beta r}(t) \end{bmatrix}$$

$$[m] = \begin{bmatrix} m_\alpha \\ m_\beta \end{bmatrix}$$

$$[A] = \begin{bmatrix} (A_{10} + \sum A_{1k}e^{\gamma_k t}) & (B_{10} + \sum B_{1k}e^{\gamma_k t}) \\ (A_{20} + \sum A_{2k}e^{\gamma_k t}) & (B_{20} + \sum B_{2k}e^{\gamma_k t}) \end{bmatrix}$$

$$[I] = \begin{bmatrix} I_{D\alpha} \\ I_{D\beta} \end{bmatrix}$$

$$[\emptyset] = \begin{bmatrix} \sum C_{1k}e^{\gamma_k t} & \sum D_{1k}e^{\gamma_k t} \\ \sum C_{2k}e^{\gamma_k t} & \sum D_{2k}e^{\gamma_k t} \end{bmatrix}$$

and

$$[\varphi(0)] = \begin{bmatrix} \varphi_{\alpha r}(0) \\ \varphi_{\beta r}(0) \end{bmatrix}$$

The constants A's, B's, C's, D's, E's and F's can be evaluated using the partial fraction expansion of Eqn. (4). The initial values of the rotor flux linkages $\varphi_{\alpha r}(0)$ and $\varphi_{\beta r}(0)$ can be calculated using the property of the rotor flux linkages vector that it varies uniformly and cannot jump from one value to the other when the current commutates.

Thus the rotor flux linkages vector at $t = 0$, $\varphi_r(0)$ and the rotor flux linkage vector at $t = \tau/6$, $\varphi_r(\tau/6)$ are related by

$$\varphi_r(\tau/6) = e^{j\pi/6}\, \varphi_r(0) \quad (10)$$

The rotor flux linkage vector φ_r can be written as

$$\varphi_r = \varphi_{\alpha r} + j\, \varphi_{\beta r} \quad (11)$$

Using this relation Eqn.(10) can be expressed as

$$\begin{bmatrix} \varphi_{\alpha r}(\tau/6) \\ \varphi_{\beta r}(\tau/6) \end{bmatrix} = \begin{bmatrix} 0.5 & -0.866 \\ 0.866 & 0.5 \end{bmatrix} \begin{bmatrix} \varphi_{\alpha r}(0) \\ \varphi_{\beta r}(0) \end{bmatrix} \quad (12)$$

Substituting Eqn. (12) in Eqn. (9), the values of $\varphi_{\alpha r}(0)$ and $\varphi_{\beta r}(0)$ can be obtained. Having determined the values of $\varphi_{\alpha r}(0)$ and $\varphi_{\beta r}(0)$, the flux linkages over the entire cycle can be calculated. If the input current waveform is assumed to be stepped one (i.e.) $m_\alpha = m_\beta = 0$ over the complete 1/6th period, the Eqn. (9) can be used even in this case by substituting zero values for m_α and m_β.

ROTOR CURRENT AND TORQUE DEVELOPED

The α,β components of the rotor currents can be expressed in terms of the rotor flux linkages as

$$i_{\alpha r} = \varphi_{\alpha r}/L_r - i_{\alpha s}\, M/L_r \quad (13)$$

$$i_{\beta r} = \varphi_{\beta r}/L_r - i_{\beta s}\, M/L_r \quad (14)$$

The electromagnetic torque developed by the machine is given by

$$T_d = 3/2\ P\ M(i_{\alpha r} i_{\beta s} - i_{\alpha s} i_{\beta r}) \quad (15)$$

or in terms of the rotor flux linkages, we can write that

$$T_d = 1.5\ PM/L_r(\varphi_{\alpha r} i_{\beta s} - \varphi_{\beta r} i_{\alpha s}) \quad (16)$$

The torque developed over a cycle can be calculated using the Eqn.(16).

STATOR INDUCED VOLTAGES

The α,β components of the phase voltages of the stator are given by

$$\begin{bmatrix} v_{\alpha s} \\ v_{\beta s} \end{bmatrix} = \begin{bmatrix} R_s & 0 \\ 0 & R_s \end{bmatrix} \begin{bmatrix} i_{\alpha s} \\ i_{\beta s} \end{bmatrix} + \begin{bmatrix} M/L_r & 0 \\ 0 & M/L_r \end{bmatrix} \begin{bmatrix} p\varphi_{\alpha r} \\ p\varphi_{\beta r} \end{bmatrix}$$

$$+ \begin{bmatrix} (L_s - M^2/L_r) & 0 \\ 0 & (L_s - M^2/L_r) \end{bmatrix} \begin{bmatrix} pi_{\alpha s} \\ pi_{\beta s} \end{bmatrix} \quad (17)$$

In Eqn. (17) the third term on the right hand side vanishes when the current does not undergo commutation. The voltage in the stator winding is only due to R_s and the rate of change of flux linkages. As it is seen from the results, the rotor flux linkages vary uniformly and have sinusoidal time variation. The third term influences the voltage during the commutation of the current from one phase to the other and is responsible for the voltage spikes in the stator voltage. Thus it is clear that the voltage induced in the stator winding is almost sinusoidal with super imposed spikes during commutation.

RESULTS

The performance equations of a current controlled induction motor derived above via the state transition signal flow graph are used to determine the steady state performance of a 3 phase slip ring induction motor at a constant rotor speed. The details of the machine are given in Table 1.

The locus of the tip of the rotor flux vector with $\varphi_{\alpha r}$ on the X-axis and $\varphi_{\beta r}$ on the Y-axis is shown in Fig. 4 for a rotor speed of 1460 rpm, and a supply frequency of 50 Hz.

TABLE 1 Details of Induction Motor

	Stator	Rotor
No. of phases	3	3
No. of poles	4	4
Connection	Star	Star
Voltage	380V	105V
Resistance/ phase	1.35 ohms	2.07 ohms
Self inductance	0.1688 H	0.1655 H
Mutual inductance between stator and rotor	0.156 H	0.156 H

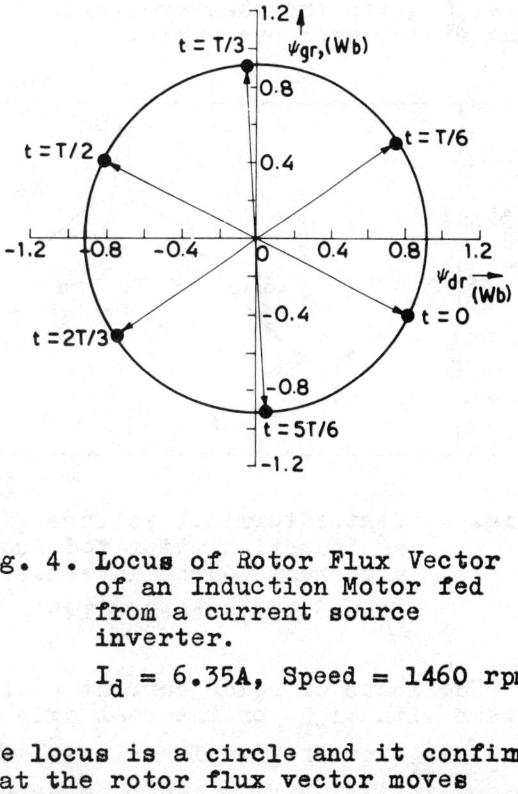

Fig. 4. Locus of Rotor Flux Vector of an Induction Motor fed from a current source inverter.

I_d = 6.35A, Speed = 1460 rpm

The locus is a circle and it confirms that the rotor flux vector moves uniformly with synchronous angular velocity w_s. The variation of $\varphi_{\alpha r}$ with time is shown in Fig. 5, from which it is clear that it is sinusoidal eventhough the input stator current is nonsinusoidal.

The sinusoidal rotor flux induces sinusoidal voltage in the stator winding. This voltage is superimposed by voltage spikes, which are brought into existence by the machine

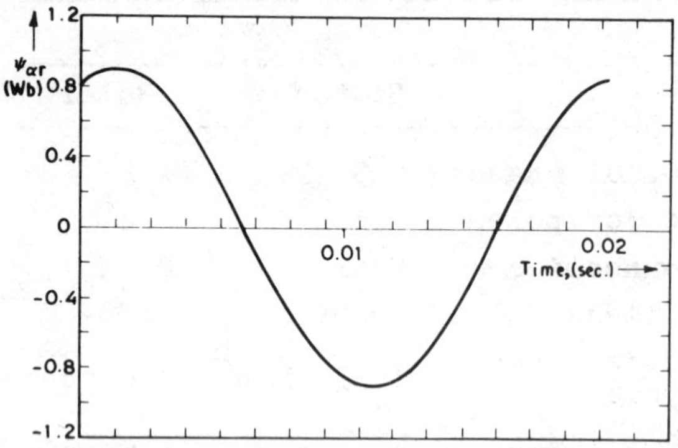

Fig. 5. Rotor flux linkages of an induction motor fed from a current source inverter.

I_d = 6.35A, Speed = 1460 rpm

inductance and rate of change of current during commutation of current from one phase to the other. Fig. 6 gives the phase voltage of the stator over one cycle.

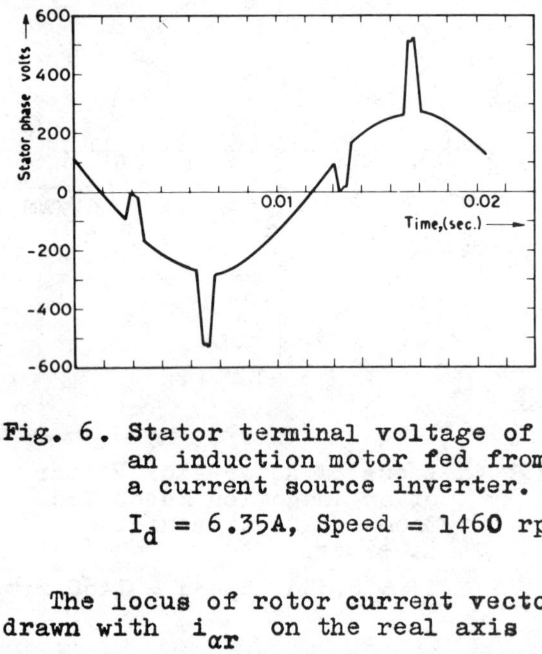

Fig. 6. Stator terminal voltage of an induction motor fed from a current source inverter.

I_d = 6.35A, Speed = 1460 rpm

The locus of rotor current vector drawn with $i_{\alpha r}$ on the real axis and $i_{\beta r}$ on the imaginary axis is shown in Fig. 7. It can be observed that the rotor current varies uniformly during the major portion of the 1/6th period and the end of commutation interval jumps to a different value from which it varies uniformly again. It can be noticed from the figure that the magnitude of the rotor current vector at the begining of each 1/6th period is the same in magnitude with a phase difference of 60° between the consecutive 1/6th periods. The time

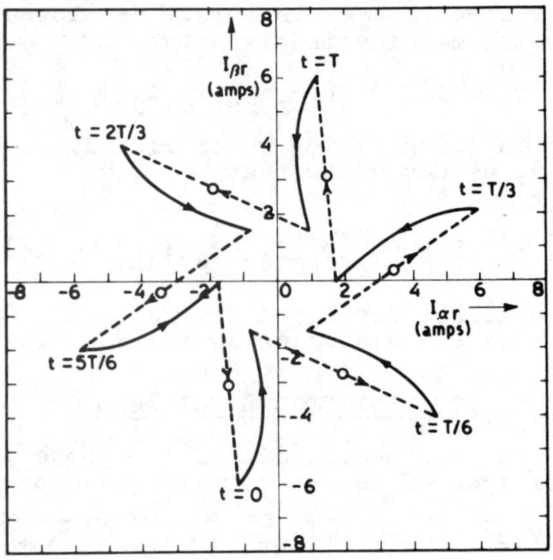

Fig. 7. Locus of the rotor current vector of an induction motor fed from a current source inverter.

I_d = 6.35A, Speed = 1460 rpm

variation of $i_{\alpha r}$ is shown in Fig. 8, which is found to have rich harmonic content.

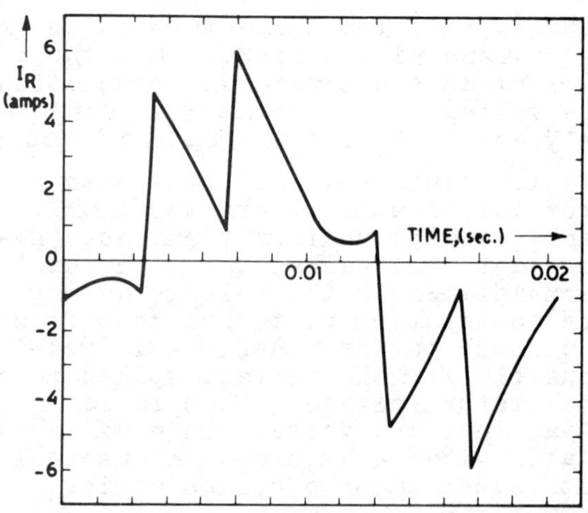

Fig. 8. Rotor current waveform of an induction motor supplied from a current source inverter.

I_d = 6.35A, Speed = 1460 rpm

The torque developed during one cycle is calculated and given in Fig. 9. The torque oscillations present have predominently sixth harmonic content which is brought into existence by the fundamental interacting with fifth and seventh harmonics. The relative amplitude

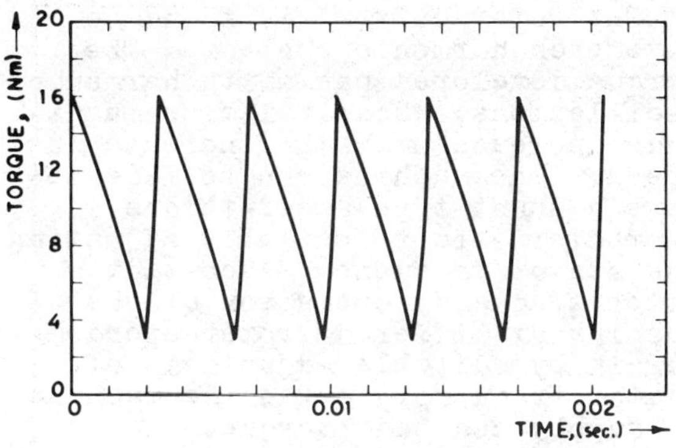

Fig. 9. Torque developed in an induction motor supplied from a current source inverter.

I_d = 6.35A, Speed = 1460 rpm

of the harmonic torque is found to increase as the rotor speed increases for a given supply frequency. The torque developed by the current-fed induction motor over 1/6th period at different speeds is shown in Fig. 10 for a supply frequency of 50 Hz. From the above considerations it can be concluded that the performance of the machine repeats every 1/6th cycle.

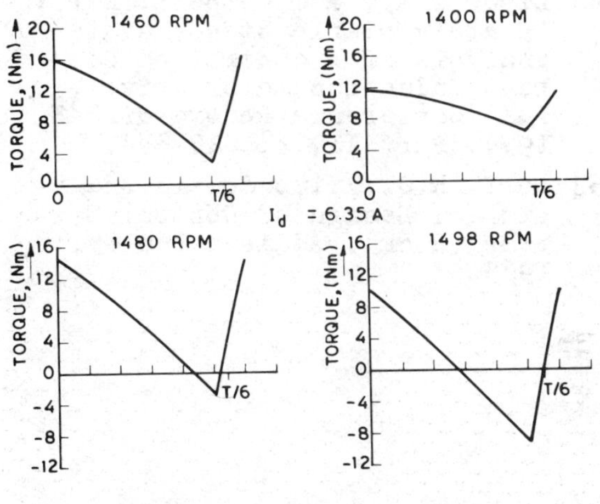

Fig. 10. Torque developed by an induction motor excited by a current source inverter.

I_d=6.35A.

The average torque developed by the machine can be controlled by controlling the rotor frequency. In Fig. 11 is shown the torque developed by the machine over 1/6th period for a constant rotor frequency at different rotor speeds. From the figure it is clear that the torque developed is the same at any rotor speed for a given rotor frequency. This is obtained by changing

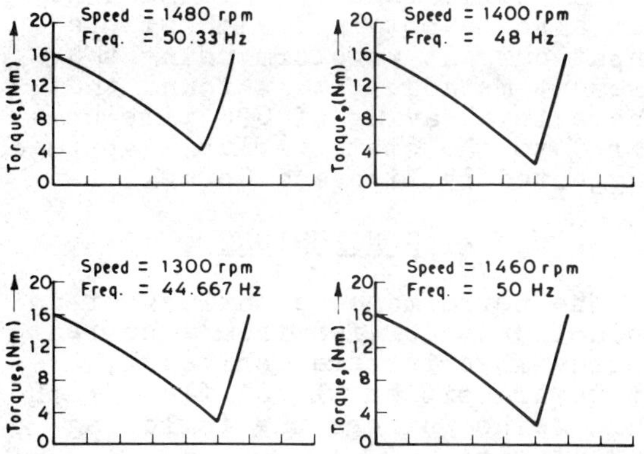

Fig. 11. Torque developed by an induction motor fed from a current source inverter for constant rotor frequency.

I_d = 6.35 A.

correspondingly the stator frequency. The average torque produced at each of these speeds is the same. Fig. 12 shows the average torque Vs speed characteristic of the machine under constant rotor frequency operation condition. Thus it is also possible

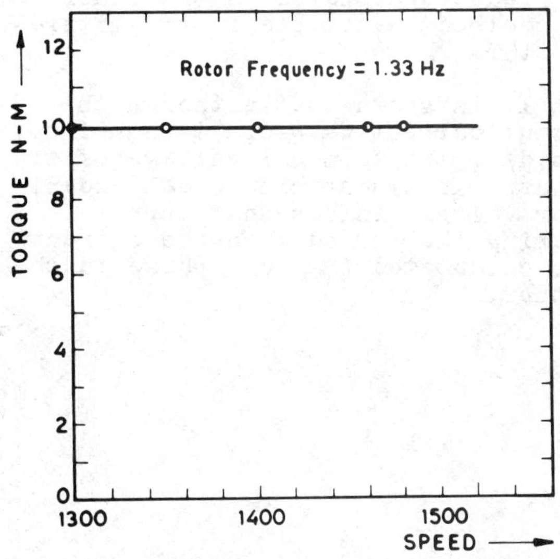

Fig. 12. Average torque Vs speed of an induction motor fed from current source inverter with rotor frequency as parameter.

I_d = 6.35 A.

to obtain a speed-torque characteristic similar to that of a d.c. shunt motor by controlling the stator frequency suitably so that the machine operates at constant rotor frequency.

The performance of the machine can also be obtained with stepped input current waveform using the present method. It is found that there is a saving of CPU time compared to the state variable approach suggested by Lipo and Cornell.

CONCLUSIONS

The performance equations of an induction motor fed from a current source inverter are derived by a state transition signal flow graph. This method offers the following advantages:

a) It illustrates intuitively the physical foundation of the state transition equation.

b) It provides a simple and straightforward method for determining the state transition matrix.

c) It eliminates the necessity of evaluating the convolution integrals as the effect of input signal is automatically accounted for. The input can have any wave shape.

d) The solution is provided in a closed form.

e) The CPU time taken is very much reduced compared to the other methods available in the literature.

It is shown that although the input current waveform is nonsinusoidal, the terminal voltage of the stator is almost sinusoidal except for voltage spikes that appear during the period when the current is commutated from one phase to the other.

The rotor current is found to have rich harmonic content. The torque developed has sixth harmonic oscillations, whose relative magnitude increase with the increase in speed. The machine can be made to have a shunt type speed-torque characteristic by suitably adjusting the stator frequency. Constant rotor frequency operation of the machine at different rotor speeds (again by suitable adjustment of stator frequency) makes the machine to supply constant torque.

REFERENCES

[1] Naunin, D.: 'Berechnung des Drehmoment Verlaufs einer Asynchronmaschine bei Speisung mit Rechtechströmen', ETZ-A, 1969, Bd.90, pp.179-182.

[2] Naunin, D.: Der Einfluss der Läuferfrequenz auf das Drehmomentverhalten von frequenzgestenerten Asynchron maschine'. Bull. SEV. 1970, Vol.61, pp.118-123.

[3] Klautschek, H.: 'Das Verhalten der Induktionsmaschine bei Speisung über Storm-Zwischenkreis Umrichter, ETZ-A, 1974, Bd.95, pp.283-287.

[4] Lipo, T.A., and Cornell, E.P.: 'State variable steady state analysis of a controlled current induction motor drive', IEEE Conference Record of IAS, 1974, Part II, pp.227-234.

[5] Dorf, R.C.: Time domain Analysis and Design of Control Systems (Book), Addison-Wesley, 1965.

STATIONARY AND DYNAMIC BEHAVIOUR OF A SPEED CONTROLLED SYNCHRONOUS MOTOR WITH COSφ - OR COMMUTATION LIMIT LINE CONTROL

Josef Leimgruber

Institute for Automatic Control and Industrial Electronics, The Swiss Federal Institute of Technology, Zurich, Physikstrasse 3, ETH-Zentrum, 8092 Zurich, Switzerland

1. ABSTRACT

In actual realizations of thyristor-controlled synchronous machines for high speed ranges, converters with D.C. link are usually applied. Normally, the converter for the synchronous motor is realized in a three phase thyristor bridge, and the gating of the thyristors is controlled by a position transducer on the rotor shaft. The reactive power for the commutation of the rectifier is supplied by the synchronous machine driving the synchronous motor in an overexcited state. The control angle of the synchronous motor, which is defined as the angle between the armature field and excitation field, is applied as constant value in actual implemented equipments, so that the synchronous motor is working by nominal power on the commutation limit line of the converter, according to the best possible cosφ of the synchronous motor. In all other states it is working with a rather worse cosφ than by nominal power, which represents a disadvantage of these methods.

In this paper a control method is presented allowing to control the synchronous motor with constant resulting flux in the machine and cosφ = 1 in the whole speed and torque range, by changing the control angle in function of the state of the machine. The condition for such an operation is a self-commutated converter on the machine side of the D.C. link.

For applications, in which for reason of cost the converter with forced commutation cannot be used, a further control method is proposed, wherein the cosφ of the synchronous motor is optimized by maintaining the resulting flux of the machine on a constant value, so that the converter is operating on the commutation limit line within the whole speed and torque range.

The control characteristics for the variation of the control angle and the excitation in function of the state of the machine are derived from the computation of the stationary characteristics of the synchronous motor. The dynamic behaviour of the synchronous motor controlled by means of these methods is depending, as shown in this paper, on a large scale on the working point of the machine.

Based on the computation of the linearized transfer functions of the thyristor-controlled synchronous machine it is shown how the control parameters can be optimized for different working points by means of the state-variable representation of the process.

2. INTRODUCTION

By the alimentation of a synchronous machine through a thyristor converter with D.C. link and variable frequency, a traction system is realized by which the speed is changeable in a large range. If the gating of the thyristor of the converter is controlled by a position transducer on the rotor shaft of the synchronous machine, the converter produces an armature field of which the electrical frequency is in synchronism with the speed of the excitation wheel of the synchronous machine. Thus, the synchronous machine loses the overturning nature which it has, when the machine is alimented by an A.C. power supply system. Now its characteristics correspond with a commutatorless D.C. machine. Figure 1 shows a self-commutated thyristor-controlled synchronous motor.

The tension of the D.C. link $U_{di\alpha 1}$ is adjustable by means of the control tension U_{st1}. The rectifier (1) is connected to the A.C. power supply. With the D.C. tension $U_{di\alpha 1}$

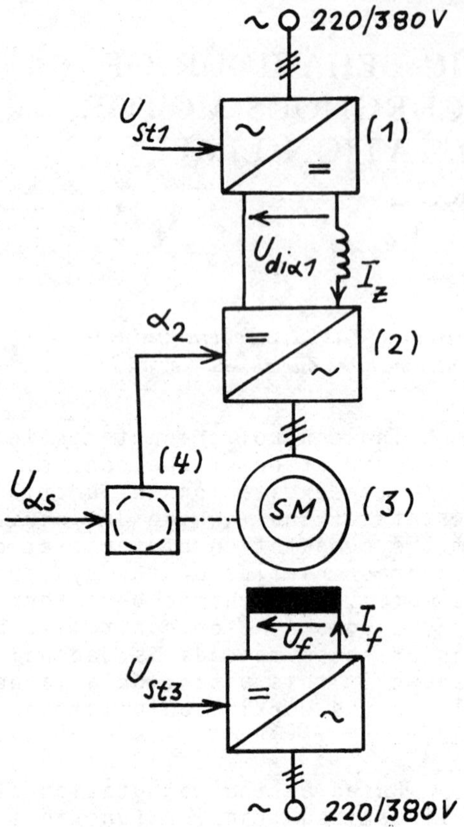

Figure 1: Thyristor controlled synchronous motor

the torque and the real power of the machine can be regulated. The exchange of the reactive power between the converter (2) and the synchronous machine (3) is depending on the firing angle $_2$ of the converter. The position transducer on the rotor shaft cascaded with a control electronic for the rectifier (4) produces the necessary gate impulses for the thyristors. The firing angle α_2 can be controlled by the control tension $U_{\alpha s}$ in such a way, that the machine is able to operate in all conditions with a power factor $\cos\varphi = 1$. This kind of operating needs the changing of the excitation current and the control tension U_{st3} in a function of the state of the synchronous machine. The excitation current can be adjusted by means of an excitation rectifier (5).

3. STATIONARY BEHAVIOUR

The stationary behaviour of the thyristor-controlled synchronous machine can be represented by means of the vector diagrams of the tensions, the currents and the magnetic flux vectors in the machine. Figure 2 shows the vector diagram of an overexcited synchronous machine by neglecting the ohmic resistance in the machine. According to figure 2 the following three variables are composing the vector diagram for the fluxes: armature flux ψ_s, excitation flux ψ_f, control angle α_S. From this vector diagram for the value α_s the following equation can be derived, where α_2 represents the firing angle of the rectifier:

$$\alpha_S = 90° - (\vartheta + \varphi) \qquad (1)$$

$$\varphi = 180° - \alpha_2 \qquad (2)$$

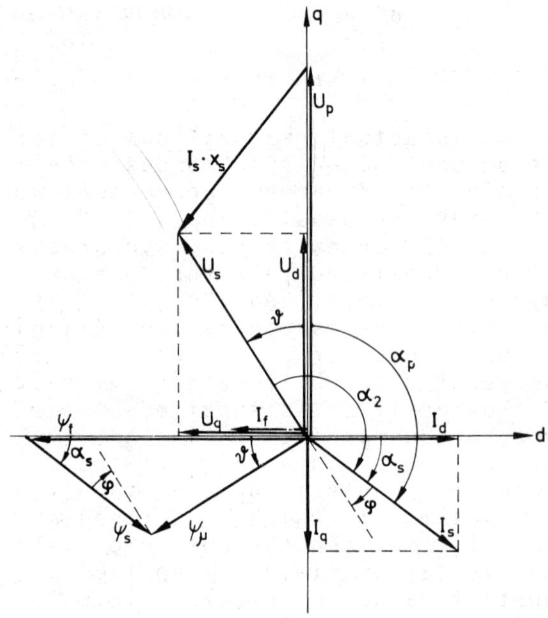

Figure 2: Vector- diagrams of the synchronous motor in a over excited state

Equation (1) and (2) describe the main equation for the $\cos\varphi$ - control of the thyristor-controlled synchronous motor. Thereby, ϑ and α_s represent the power angle and the control angle. With the control angle α_s and the excitation current i_f the exchanging of the reactive power between the synchronous motor and the rectifier is controllable. An optimal solution regarding the utilization coefficient of the process is a control method by means of which the air-gap flux ψ_μ, the speed n and the power factor $\cos\varphi$ can be given as nominal values. For the implementation of this control method a measuring electronic to measure the power angle ϑ is needed.

For the stationary state of the synchronous machine the following equations in per unit according to figure 2 can be noted:

$$i_f = i_\mu \cdot \cos\vartheta + X_d\, i_s \cos\alpha_s \qquad (3)$$

$$\vartheta = \mathrm{arctg}\, \frac{\cos\varphi}{\dfrac{i_\mu}{i_s X_d} + \sin\varphi} \qquad (4)$$

$$m_{el} = i_f \cdot i_s \sin\alpha_s \qquad (5)$$

The numerical evaluations of the equations (3) to (5) are represented for a 10 MVA synchronous motor in figures 3a) to 3c). These figures describe the adjusting function for the excitation current i_f and the control angle $\alpha_p = \alpha_s + 90^0$ as a function of the armature current i_s, in the case where the air-gap flux ψ_μ and the power factor $\cos\varphi$ are the nominal values of the control system. At this place it must be sa d, that the change of speed by a constant mechanical load does not modify the flux vector diagram. For changing the speed of the synchronous motor, we must only vary the tension $U_{di\alpha 1}$ in the D.C. link. So, the results in figures 3a) to 3c) are valid for all speed ranges.

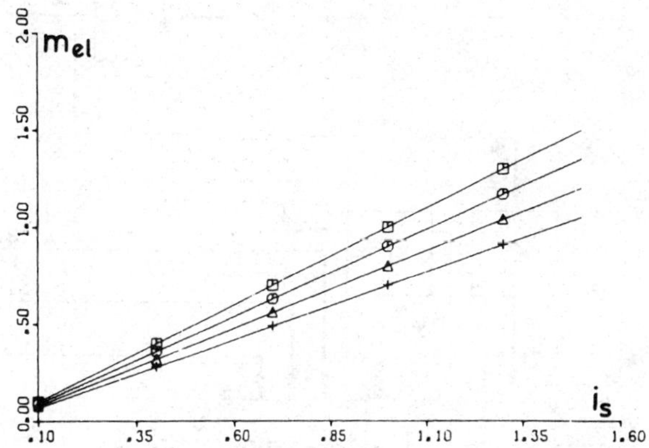

Figure 3c: Electrical torque as a function of the stator current i_s

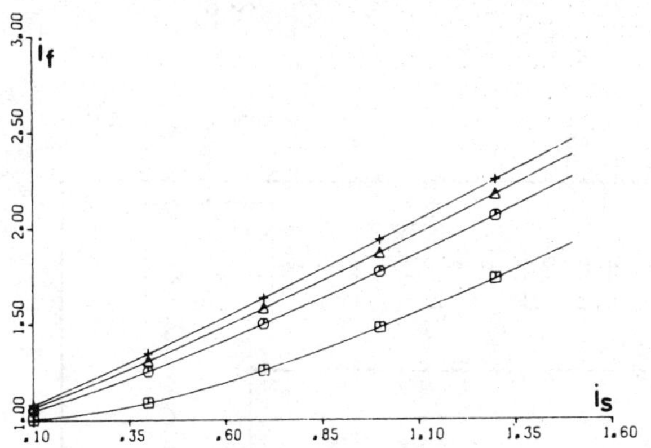

Figure3a: Excitation current i_f as a function of the statorcurrent i_s. Parameters: $\cos\varphi = 1{,}0$ □/ $\cos\varphi = 0{,}9$ ○/ $\cos\varphi = 0{,}8$ Δ / $\cos\varphi = 0{,}7$ +/

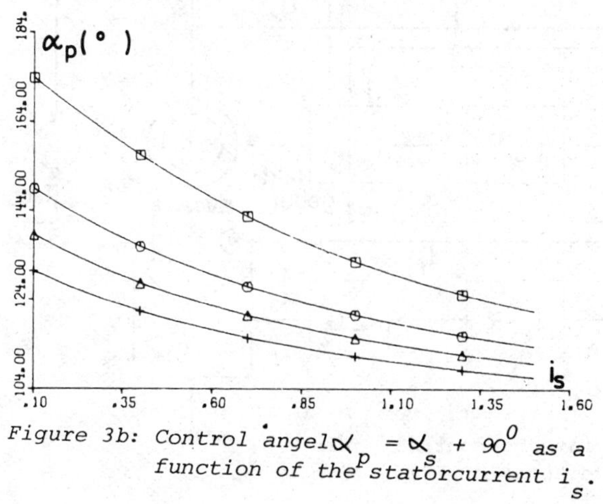

Figure 3b: Control angel $\alpha_p = \alpha_s + 90^0$ as a function of the statorcurrent i_s.

4. CONTROL EQUIPMENTS

In technical application the following two control methods are of practical interest:

A) - speed control
- $\cos\varphi = 1$ control
- flux control, constant air-gap flux $\psi_\mu = 1$

As to this control method a converter with forced commutation is needed for the alimentation of the synchronous motor.

B) - speed control
- flux control, constant air-gap flux $\psi_\mu = 1$
- commutation limit line control

If the control angle α_s (equation 1) is given as a constant value, the power factor $\cos\varphi$ is depending on the load of the synchronous motor. Provided, that the converter is consisting of a thyristor equipment for natural commutation, it is possible to drive the converter on its commutation limit line by changing the control angle α_s in function of the power angle ϑ according to the following equation:

$$\alpha_s^* = 90^\circ - (\vartheta + \varphi_{min}) \qquad (6)$$

$$\varphi_{min} = 180^\circ - \alpha_{2max} = 180^\circ - (\mu + \gamma) \qquad (7)$$

γ: turn off time of the thyristor

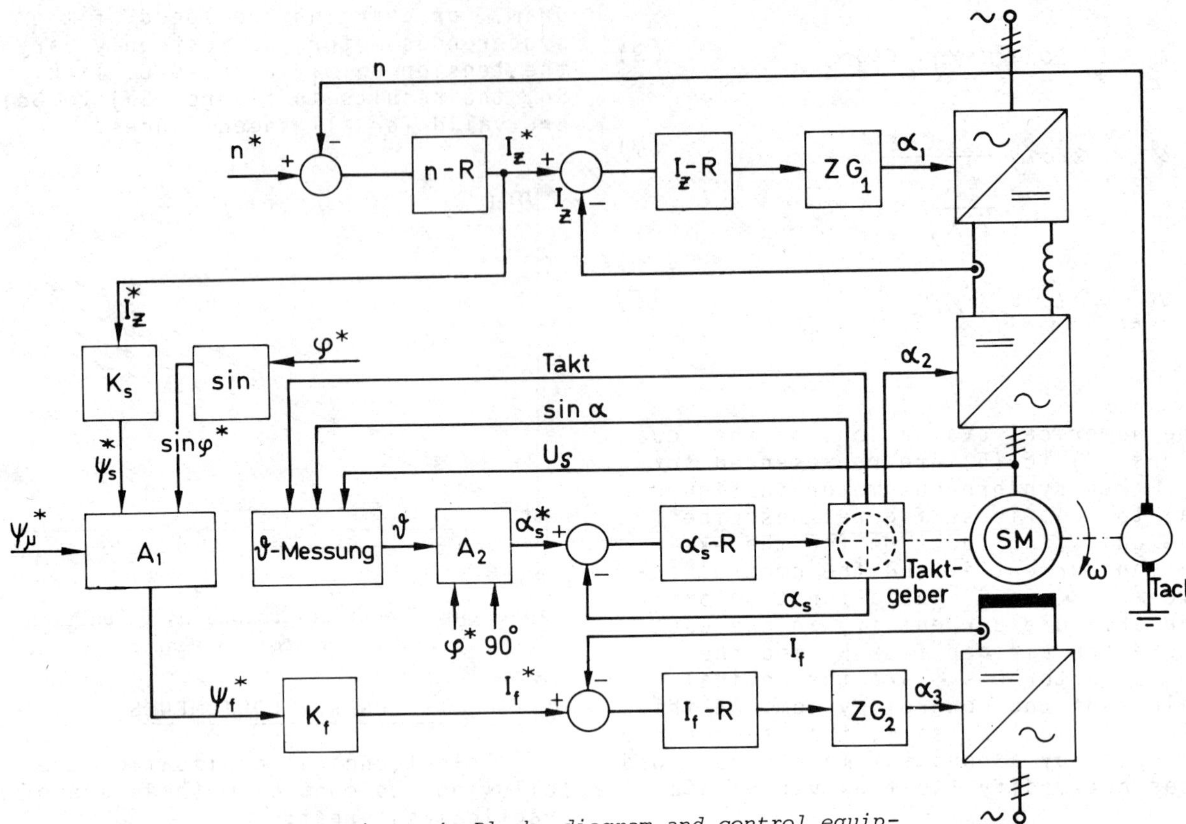

Figure 4: Block- diagram and control equipment for the cos φ -control

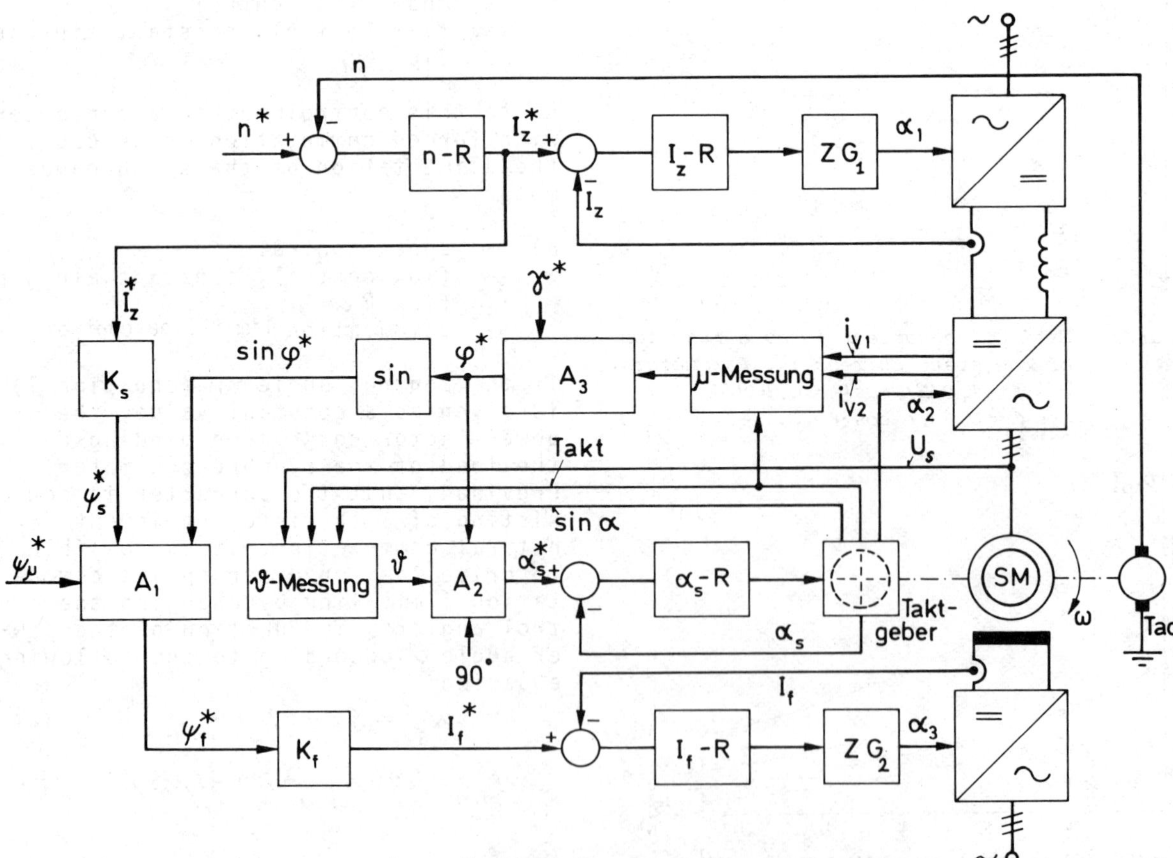

Figure 5: Block diagram and control equipment for commutation limit line control

μ : angle of overlap

By satisfying the conditions of the equations (6) and (7), the synchronous motor is working in each working point with its optimal power factor $\cos\varphi$.

The block diagrams for the control equipments of these two control methods are shown in figures 4 and 5. The power electronic in both control methods are the same and are consisting of the converter with D.C. link and the rectifier for the excitation. The control equipments for the $\cos\varphi$ - control shown in figure 4 are the following:

1) speed controller: (n - R)
2) armature current controller: (i_z - R); $i_z = i_s$
3) excitation current controller: (i_f - R)
4) calculators for the nominal values ψ_f^* and α_s^*: (A_1, A_2)
5) ϑ- calculator
6) firing equipments: ZG1, ZG2
7) position transducer equipment and rectifier firing electronic

The calculator A_1 has the task to calculate the nominal value for the excitation flux ψ_f^* according to the nominal values ψ_μ^*, ψ_s^* and $\sin\varphi^*$ as input values of the calculator. The equation to be realized is the following:

$$\psi_f^* = \sqrt{\psi_\mu^{*2} + \psi_s^{*2} + \psi_s^{*2} \cdot \psi_\mu^{*2} \cdot 2\sin\varphi^*} \qquad (8)$$

This nominal value is transformed into the nominal value i_f^* for the excitation current controller. The calculator A_2 computes the nominal values for the control angle α_s^* by means of the following equation:

$$\alpha_s^* = 90^\circ - (\vartheta + \varphi^*) \qquad (9)$$

The measurement of the power angle for computing the value of α_s^* is realized by measuring the zero passage of the armature tension u_s and the absolute position $\sin\alpha$ of the magnetic wheel measured by means of an 8-bit cray-code shaft position encoder. Power angle measurement is available each half electrical period by measuring the negative and positive zero passage of the armature tension u_s and the absolute position of the magnetic wheel.

Figure 5 shows the block diagram of the control equipments for the speed controlled synchronous motor with commutation limit control. In addition to the control equipment of the $\cos\varphi$- control a measurement equipment is needed to measure the angle of overlap μ and a calculator A_3 to compute the nominal value of φ^* as a function of overlap angle μ and the turn off time γ^* of the thyristors. For the start up of the synchronous motor with this control method an additional electronic equipment is needed to extinguish the current I_Z in the D.C. link by commutating the stator current i_s from one phase to another.

5. DYNAMIC BEHAVIOUR

In figure 6 the thyristor-controlled synchronous motor is shown as a control system.

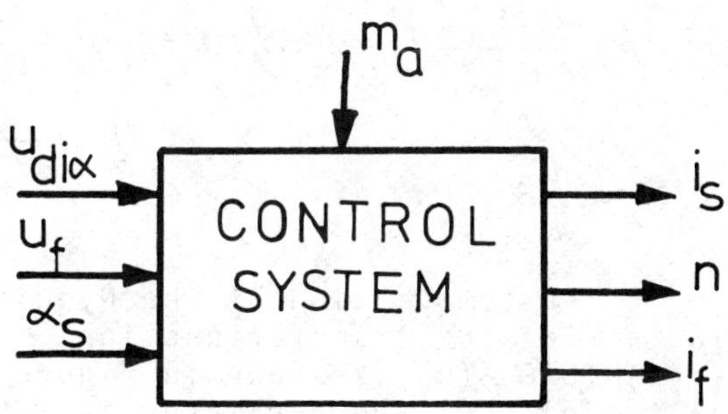

Figure 6: The thyristor controlled synchronous motor discribed as a control system

The input values of the system are given by the D.C. link tension $U_{di\alpha 1}$, the excitation tension u_f and the control angle α_s. The mechanical torque m_a represents the disturbance value of the system. On the output of the control system we have the state variables of the D.C. current i_Z, the excitation current i_f and the actual value of the speed. Provided that the excitation field time constant T_f is much greater than the time constant T_D of the damping windings in the d - axes, the following transferfunctions are derivable, by linearizing the differential equation of the synchronous machine in a determined working point.

$$G_{11}(p) = \frac{1}{r_z(1+pT_z) + \frac{p}{\omega_n}\left[\sin\alpha_p^2 \cdot X_d(p) + \cos^2\alpha_p \cdot X_q(p)\right] + \ldots}$$

$$\ldots \frac{1}{+\sin\alpha_p \cdot \cos\alpha_p\, n\left[X_d(p) - X_q(p)\right] + r_s} \quad (10)$$

$$G_{12}(p) = G_{11}(p)\, \frac{n \cdot \cos\alpha_p + \frac{p}{\omega_n} \cdot \sin\alpha_p}{1 + pT_f} \quad (11)$$

$$G_{21}(p) = G_{11}(p)\, \sin\alpha_p\, (X_d - X_d') \cdot \frac{pT_f}{(1 + pT_f)} \quad (12)$$

$$G_{22}(p) = \frac{1 + G_{21}(p)\left[n \cdot \cos\alpha_p + \frac{p}{\omega_n} \cdot \sin\alpha_p\right]}{1 + pT_f} \quad (13)$$

$$G_{nf}(p) = \frac{-\, i_s \cdot \cos\alpha_p}{pT_m} = \frac{k_{nf}}{pT_m} \quad (14)$$

$$G_{nz}(p) = \frac{-\cos\alpha_p}{pT_m}\left[i_f - i_s \sin\alpha_p \left(2X_d + X_d'' - X_d' - X_q - X_q''\right)\right] = \frac{k_{nf}}{pT_m} \quad (15)$$

$$T_z = \frac{L_z}{R_z} \quad (16)$$

The computation of the Nyquist diagrams of the transferfunction $G_{11}(p)$ to $G_{22}(p)$ is shown in figure 7a) to 7d). For the parameter $i_s = 1{,}5$, $\cos\varphi = 1$, $\psi_M = 1$ and speed $n = 1{,}5/1/0{,}5$ (parameters in per unit) the transferfunctions have an unstable behaviour, with $n = 0{,}1$, however, all

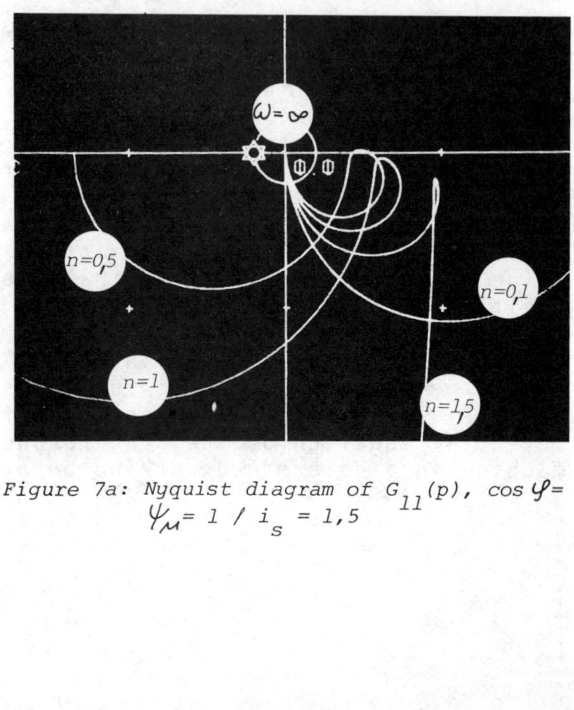

Figure 7a: Nyquist diagram of $G_{11}(p)$, $\cos\varphi = 1$ $\psi_M = 1$ / $i_s = 1{,}5$

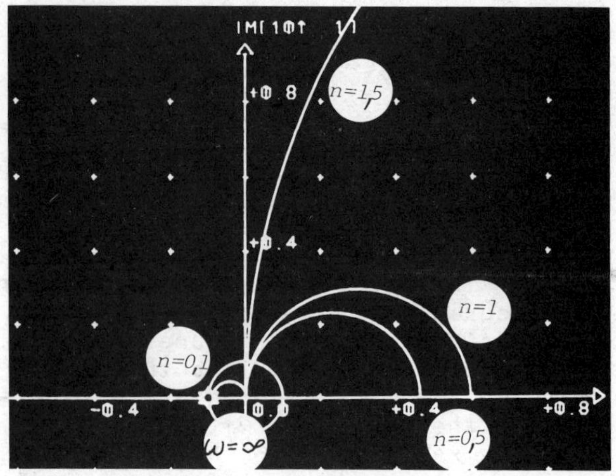

Figure 7b: Nyquist diagram of $G_{21}(p)$, $\cos\varphi = 1$ $\psi_M = 1$/ $i_s = 1{,}5$

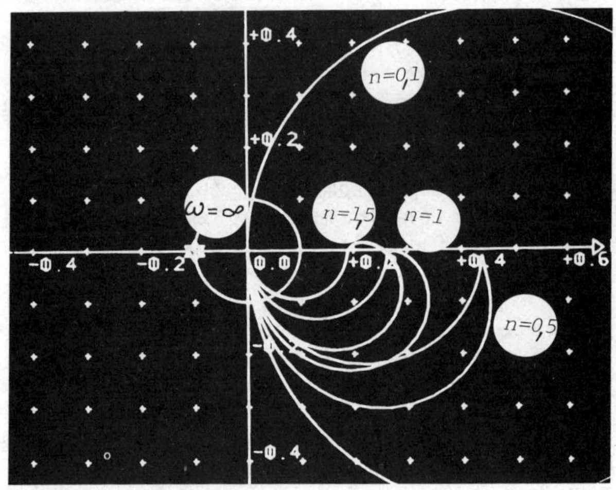

Figure 7c: Nyquist diagram of $G_{12}(p)$, $\cos\varphi = 1$ $\psi_M = 1$ / $i_s = 1{,}5$

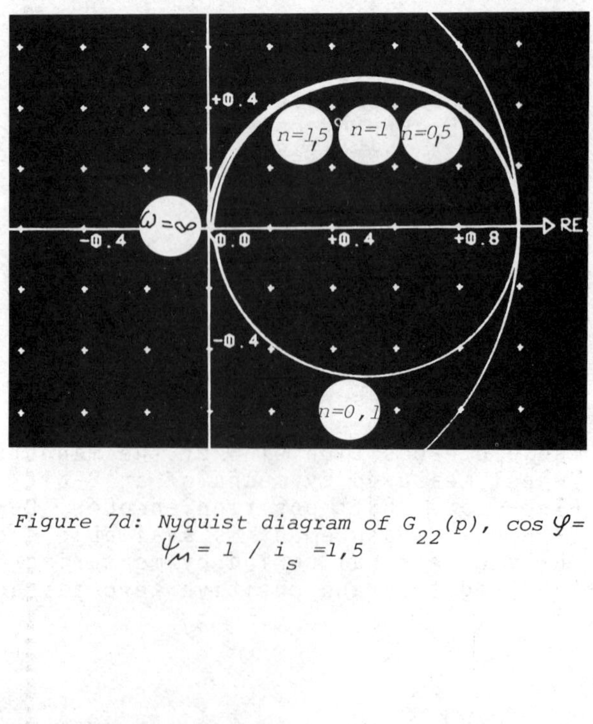

Figure 7d: Nyquist diagram of $G_{22}(p)$, $\cos\varphi = 1$ $\psi_M = 1$ / $i_s = 1{,}5$

transferfunctions are stable. By changing the parameter of the stator current from the value i_s = 1,5 to i_s = 0,1 the transferfunctions $G_{11}(p)$ to $G_{22}(p)$ show a stable behaviour independent of the speed value n. The analysis of the root locus of the transferfunctions demonstrates that $G_{11}(p)$ and $G_{22}(p)$ can be stabilized by means of a feedback system, which is shown in figure 8. On the other hand, $G_{21}(p)$ and $G_{22}(p)$ cannot be

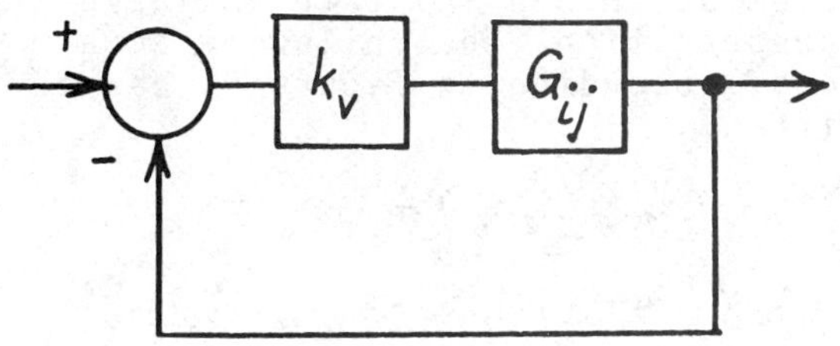

$$G_{ij}^{*} = \frac{k_V \cdot G_{ij}}{1 + k_V \cdot G_{ij}}$$

Figure 8: Stabilization of $G_{ij}(p)$ by means of a feedback system k_v^{ij}

stabilized by increasing the amplifying factor k_v in the feedback system, because there is a pole of $G_{21}^{*}(p)$ and $G_{22}^{*}(p)$ in the right half plane for all values k_v = 0 to k_v =∞.

For this reason it is necessary to test the behaviour of the whole control system on stability and to optimize the control parameters by means of a digital computer.

6. SIMULATION AND PARAMETER OPTIMIZATION

Proceeding from the transferfunction $G_{11}(p)$ to $G_{22}(p)$, $G_{nZ}(p)$ and $G_{nf}(p)$ the speed controlled synchronous motor can be described by the structure diagram of figure 9.

To simulate this control system on a digital computer, the state variable representation is used for the system description. For this reason, the transferfunctions in figure 9 are written in the polynomial form.

$$G_{ij} = \frac{A_0 + A_1 \cdot p + A_2 \cdot p^2 + \dots + A_n \cdot p^n}{B_0 + B_1 \cdot p + B_2 \cdot p^2 + \dots + B_m \cdot p^m}$$

$$m > n$$

The numerical values of the polynomial A(p) and B(p) are presented at

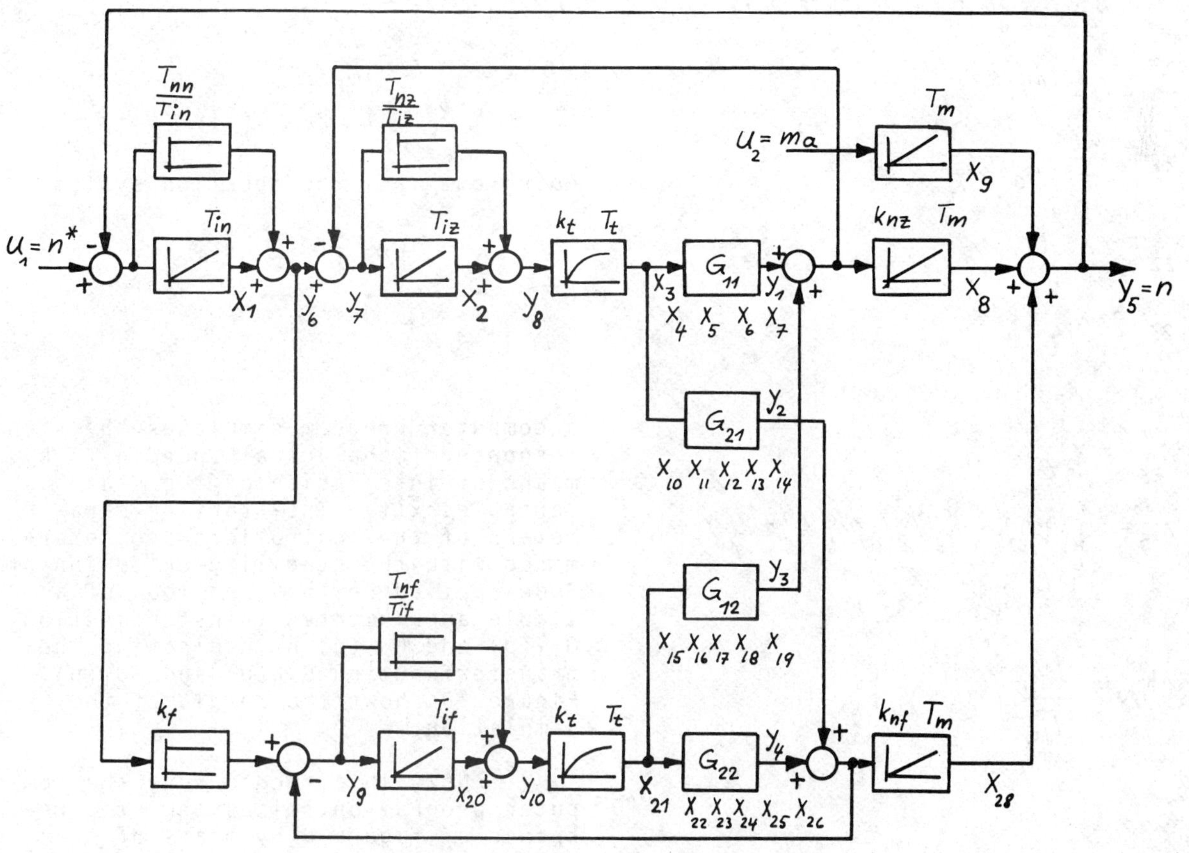

Figure 9: Structure diagram of the speed controlled synchronous motor

the end of this paper for a 10 MVA thyristor-controlled synchronous motor. Proceeding from these values, the following equations can be written for the state variables X_1 to X_{28}, which are introduced according to the structure diagram in figure 9.

$$\dot{X}_1 = \frac{1}{T_{in}}(U_1 - Y_5)\,;\quad \dot{X}_2 = \frac{1}{T_{iz}} \cdot Y_7\,,$$

$$\dot{X}_3 = \frac{kt}{T_t} \cdot Y_8 - \frac{1}{T_t} \cdot X_3$$

$$\dot{X}_4 = X_5\,;\quad \dot{X}_5 = X_6\,;\quad \dot{X}_6 = X_7$$

$$\dot{X}_7 = -\frac{1}{B_4}\left(B_0 \cdot X_4 + B_1 \cdot X_5 + B_2 X_6 + B_3 \cdot X_7 - X_3\right)$$

$$\dot{X}_8 = \frac{k_{nz}}{T_m}\left(Y_1 + Y_2\right)$$

$$\dot{X}_9 = \frac{1}{T_m} \cdot U_2$$

$$\dot{X}_{10} = X_{11}\,;\quad \dot{X}_{11} = X_{12}\,;\quad \dot{X}_{12} = X_{13}\,;\quad \dot{X}_{13} = X_{14}$$

$$\dot{X}_{14} = -\frac{1}{B_5}\left(B_0 \cdot X_{10} + B_1 \cdot X_{11} + B_2 \cdot X_{12} + B_3 \cdot X_{13} + B_4 X_{14} - X_3\right)$$

$$\dot{X}_{15} = X_{16}\,;\quad \dot{X}_{16} = X_{17}\,;\quad \dot{X}_{17} = X_{18}\,;\quad \dot{X}_{18} = X_{19}$$

$$\dot{X}_{19} = -\frac{1}{B_5}\left(B_0 X_{15} + B_1 X_{16} + B_2 X_{17} + B_3 \cdot X_{18} + B_4 X_{19} - X_{21}\right)$$

$$\dot{X}_{20} = \frac{1}{T_{if}} \cdot Y_9$$

$$\dot{X}_{21} = \frac{kt}{T_t} \cdot Y_{10} - \frac{1}{T_t} \cdot X_{21}$$

$$\dot{X}_{22} = X_{23}$$

$$\dot{X}_{23} = X_{24}$$

$$\dot{X}_{24} = X_{25}$$

$$\dot{X}_{26} = X_{27}$$

$$\dot{X}_{27} = -\frac{1}{B_6}\Big(B_0 \cdot X_{22} + B_1 \cdot X_{23} + B_2 \cdot X_{24} + B_3 \cdot X_{25} + B_4 \cdot X_{26} + \dots B_5 X_{27} - X_{21}\Big)$$

$$\dot{X}_{28} = \frac{k_{nf}}{T_m}\left(Y_2 + Y_4\right)$$

The controllers of the current i_Z in the D.C. link, the speed n and the excitation current i_f are chosen in PI-structure. The control angle α_s is kept as a constant value in the following considerations. This supposition is allowable, because the α_s -controller in the practical implementation is working much quicker than the other control circuits of the process.

For the output values y of the system described in state variable representation, the following equations, according to figure 9, can be written.

$$Y_1 = A_0 \cdot X_4 + A_1 \cdot X_5 + A_2 \cdot X_6 + A_3 X_7$$

$$Y_2 = A_0 \cdot X_{10} + A_1 \cdot X_{11} + A_2 \cdot X_{12} + A_{13} \cdot X_{13} + A_{14} \cdot X_{14}$$

$$Y_3 = A_0 \cdot X_{15} \cdot A_1 \cdot X_{16} + A_2 \cdot X_{17} + A_3 \cdot X_{18} + A_4 \cdot X_{19}$$

$$Y_4 = A_0 \cdot X_{22} + A_1 \cdot X_{23} + A_2 \cdot X_{24} + A_3 \cdot X_{25} + A_4 X_{26} + \dots \dots + A_5 \cdot X_{27}$$

$$Y_5 = X_8 - X_9 + X_{28}$$

$$Y_6 = X_1 + (U_1 - Y_5) \cdot \frac{T_{nn}}{T_{in}}$$

$$Y_7 = Y_6 - Y_3 - Y_1$$

$$Y_8 = X_2 + Y_7 \cdot \frac{T_{nz}}{T_{iz}}$$

$$Y_9 = Y_6 \cdot k_f - Y_4 - Y_2\,;\quad Y_{10} = X_{20} + Y_9 \cdot \frac{T_{nf}}{T_{if}}$$

Going out from the equation system

$$\dot{\vec{X}} = \vec{A} \cdot \vec{X} + \vec{B} \cdot \vec{U}$$

$$\vec{Y} = \vec{C} \cdot \vec{X} + \vec{D} \cdot \vec{U}$$

a computer program simulates the step response of the actual speed n(t) by means of integration of the state variables $\dot{x}(t)$. The starting parameters of the controllers are determined with the stability criterion of Bode applied on the open loop of a stable approximated transferfunction $G_{11}(p)$ and $G_{22}(p)$ by neglecting the transferfunction $G_{12}(p)$ and $G_{21}(p)$. Figure 10 shows the result of the simulation.

Now, in a second run, the computer program optimizes the step response of figure 9 by means of the ISE-criterion. Figure 11 shows the step response of the control system

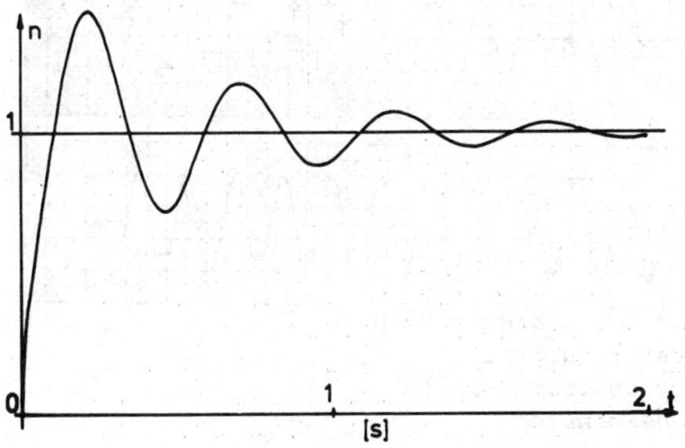

Figure 10: Simulation of the step response n(t) with the controllparameters determined by the stabilitiy criterion of Bode

with optimized parameters.

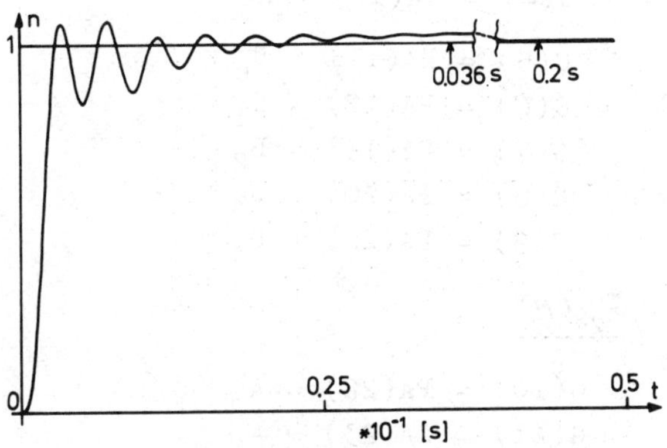

Figure 11: Step response of the speed n(t) with the optimized control parameters by means of the ISE-criterion; $\cos\varphi = 1$; $\psi_m = 1$

For the working point indicated in figure 11 the speed controlled system shows a stable behaviour. With the optimized parameter set of the working point (n = 1,5 / i_s = 1,5) the step response for n = 0,5 / i_s = 1,5 and n = 0,1 / i_s = 1,5 has been simulated. The result was an absolutely stable behaviour, but with a little quantitative difference in the characteristic values of the step response. By the variation of the stator current from the value i_s = 1,5 to i_s = 0,5, the step response with the parameter set of the working point n = 1,5 and i_s = 1,5 shows an unstable behaviour. After a computer run determinating the new optimal control parameter the step response can also be stabilized in this new working point. The influence of changing the value of speed n in the new working point (i_s = 0,5) shows no quantitative change in the optimal control parameters. Finally, one comes to the conclusion, that for an optimal dynamic behaviour an adaptive control of the control parameters in function of the stator current i_s may be a better realization than the implementation of PI-controllers with constant parameters.

7. RESULTS

The speed controlled synchronous motor described above has been built-up with a 3 kW synchronous machine. Figure 12 shows an oscillogram of the stator tension u_s, the stator current i_s and the speed n during the start up procedure of the synchronous motor. The dynamic behaviour of the step response for a positive speed step is shown in the oscilloscopes of figure 13 and 14 for the cases, where the nominal value of speed is changed from 1400 up to 1600 revolutions per minute, and where the process by a nominal speed of 1500 revolutions per minute is charged in one step from zero to its nominal mechanical load. Both experiments show a satisfactory dynamical behaviour.

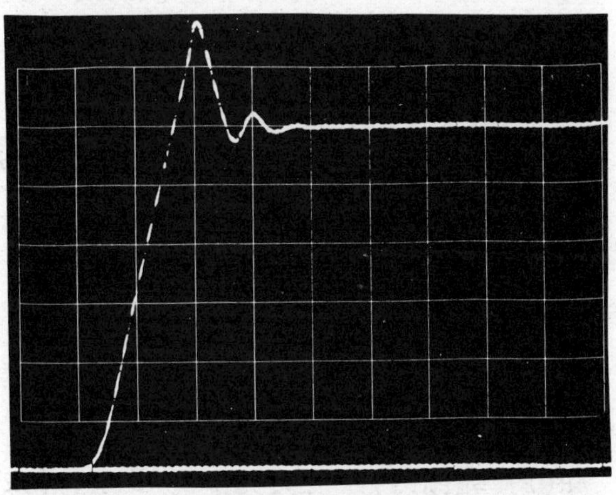

Figure 13: Step response of the speed n by changing the nominal value of speed from 1400 up to 1610 r/min X = 200 ms/Div/ y = 0,2 V/Div/ 1 mV in y- axes = 0,18 r/min/ Load: 3kW

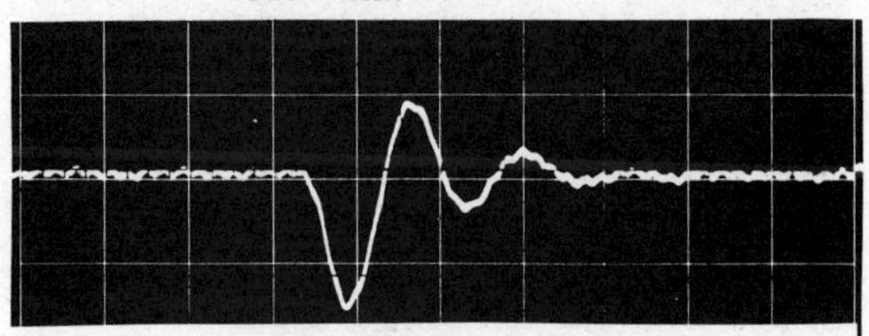

Figure 14: Dynamic behaviour of the speed by charging the machine from no- load to nominal load (3kW) in one step. Nominal speed = 1500 r/min/ y = 50 mV/Div/ x = 100ms/Div/ 1 mV in y- axes = 0,18 r/min

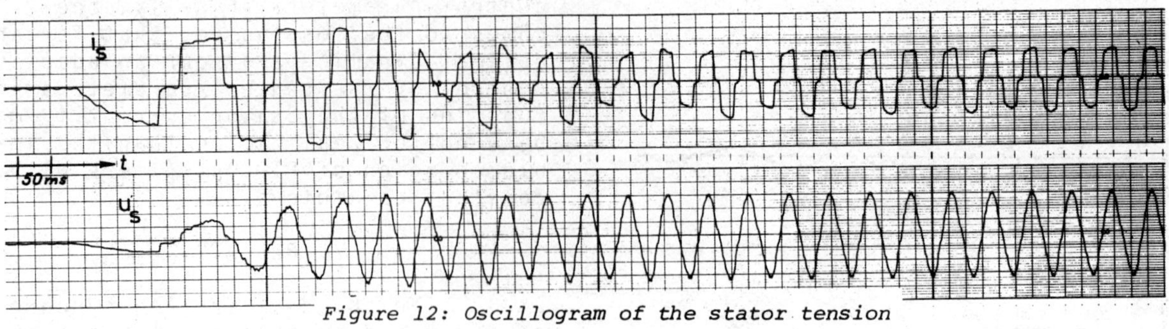

Figure 12: Oscillogram of the stator tension u_s and the stator current i_s during a start up from zero to 500 revolutions/minutes

8. ACKNOWLEGEMENT

I wish to express my gratitude towards Prof. Dr. R. Zwicky from the Institute of Automatic Control and Industrial Electronics at the Swiss Federal Institute of Technology Zürich, who gave me the opportunity to carry out the research work described in this paper. Furthermore I thank Dr. D. Rufer for the disposition of the identification program and his Information concerning the application of the program.

9. Datas of the Control System

Synchronous machine:

$S = 10$ MVA / $x_d = 1{,}29$
$x'_d = 0{,}25$ / $x''_d = 0{,}145$/ $x_q = 0{,}75$/ $x'_q = 0{,}75$
$x''_q = 0{,}156$ / $T_f = 2{,}7$sec/ $T'_d = 0{,}52$/ $T''_d = 0{,}01$
$T''_q = 0{,}01$ $\sin\alpha_p = 0{,}79$/ $\cos\alpha_p = -0{,}61$
$r_s = 0{,}01$ / $I_{n\lambda} = 577$ A/ $U_{n\lambda} = 5770$ V

Start parameters for the simulation of n(t)

R(1) = PA(1) = 13.36E-03 = T_{nn} [sec]
R(2) = PA(2) = 13.36E-05 = T_{in} "
R(3) = PA(3) = 45.53E-03 = T_{nz} "
R(4) = PA(4) = 51.10E-03 = T_{iz} "
R(5) = PA(5) = 2.71 = T_{nf} "
R(6) = PA(6) = 9E-03 = T_{if} "

S(1) = PA(7) = k_t = 2,7
S(2) = PA(8) = T_t = 1,6 ms
S(3) = PA(9) = T_m = 2,5 s
S(4) = PA(10) = k_{nz} = 0.04
S(5) = PA(11) = k_{nf} = 0.92
S(6) = PA(12) = k_f = 0.92

$G_{11}(p)$ / $\psi_m = 1$ / $n = 1{,}5$ / $i_s = 1{,}5$ / $\cos\varphi = 1$ /

G(1) = PA(13) = A_o
G(2) = PA(14) = A_1
G(3) = PA(15) = A_2
G(4) = PA(16) = A_3

G(5) = PA(17) = B_o
G(6) = PA(18) = B_1
G(7) = PA(19) = B_2
G(8) = PA(20) = B_3
G(9) = PA(21) = B_4

$G_{21}(p)$

G(10) = PA(22) = A_o
G(11) = PA(23) = A_1
G(12) = PA(24) = A_2
G(13) = PA(25) = A_3
G(14) = PA(26) = A_4
G(15) = PA(27) = B_o
G(16) = PA(28) = B_1
G(17) = PA(29) = B_2
G(18) = PA(30) = B_3
G(19) = PA(31) = B_4
G(20) = PA(32) = B_5

$G_{12}(p)$

G(21) = PA(33) = A_o
G(22) = PA(34) = A_1
G(23) = PA(35) = A_2
G(24) = PA(36) = A_3
G(25) = PA(37) = A_4
G(26) = PA(38) = B_o
G(27) = PA(39) = B_1
G(28) = PA(40) = B_2

G(29) = PA(41) = B_3
G(30) = PA(42) = B_4
G(31) = PA(43) = B_5

$G_{22}(p)$

G(32) = PA(44) = A_o
G(33) = PA(45) = A_1
G(34) = PA(46) = A_2
G(35) = PA(47) = A_3
G(36) = PA(48) = A_4
G(37) = PA(49) = A_5

G(38) = PA(50) = B_o
G(39) = PA(51) = B_1
G(40) = PA(52) = B_2
G(41) = PA(53) = B_3
G(42) = PA(54) = B_4
G(43) = PA(55) = B_5
G(44) = PA(56) = B_6

Parameter of the controllers optimized with the ISE criterion:

T_{nn} = 3,07 ms/ T_{in} = 0,03 ms/ V_{Rn} = 100

T_{nz} = 65,75 ms/ T_{iz} = 49,57ms/ V_{Rz} = 1,33

T_{nf} = 2,2 sec/ T_{if} = 0,69 ms/ V_{RF} = 3200

10. References

Prof. Bühler: Stromrichtergespeiste Drehstrom- und Gleichstromantriebe, IFAC Düsseldorf 1977

Laible Th.: Die Theorie der Synchronmaschine im nicht stationären Zustand, Springer, Berlin (1952)

Leimgruber, J.: Regelverfahren bei selbstgeführten Stromrichtermotoren, IFAC Düsseldorf 1974.

Leimgruber, J: Direct dicital control of a th thyristor controlled synchronous motor, IFAC Fredericton 1977, Canada.

TRANSIENT CHARACTERISTICS AND LIMITS OF A SQUIRREL-CAGE MOTOR FED FROM A FREQUENCY-CONVERTER WITH D.C. CURRENT LINK

L. Abraham

Brown, Boveri & Cie., Mannheim, Germany

ABSTRACT

The dynamic behaviour of a drive system consisting of converter with d.c. current link and a squirrel-cage motor is discussed. After modelling a suitable simplified structure of the system by elimination of six-pulse-harmonics it is shown that the parameters of the d.c. link essentially influence the transfer characteristic of the system. In normal operating conditions the usual control circuits are sufficient for adequate running. If the input d.c. voltage has reached maximum limit however, it may occur transient characteristics in the drive, which require complicated control circuitry and restrict dynamic quality.

INTRODUCTION

The transient behaviour of inverter fed squirrel-cage motor has already been studied in detail, e.g. in (1)...(8). In such studies, the motor voltage or the motor current are mostly assumed to be freely controllable. This is justified in the case of a frequency converter with d.c. voltage link, since the storage capacitor permits quick current changes, and quick voltage variations can also be achieved by pulse-width-modulation methods in the frequency converter; in many cases, the dynamic behaviour of the supplying system is sufficient to eliminate the required power gradients within a tolerable change of the d.c. link voltage.

In frequency converters with d.c. current link, the rate of change of the current is limited by the d.c. choke so that this has to be considered when investigating the dynamic response. In the following, the transient characteristics of such a drive are outlined and discussed.

Fig.1 shows the basic arrangement of a drive with d.c. current link. The supply voltage U_V is rectified by the controlled rectifier G. The d.c. link includes the smoothing reactor with inductance L_d and resistance R_d. The selfcommutated inverter W switches the d.c. link current i_d to 2 phases of the connected squirrel-cage machine each. By suitable control of rectifier G , the magnitude of the direct voltage u_e can be adjusted, while the gate pulses of the inverter W determine the phase position of the motor current. At the motor shaft, which rotates with an angular frequency of ω/p_z, a load with the reaction torque M_g is effective.

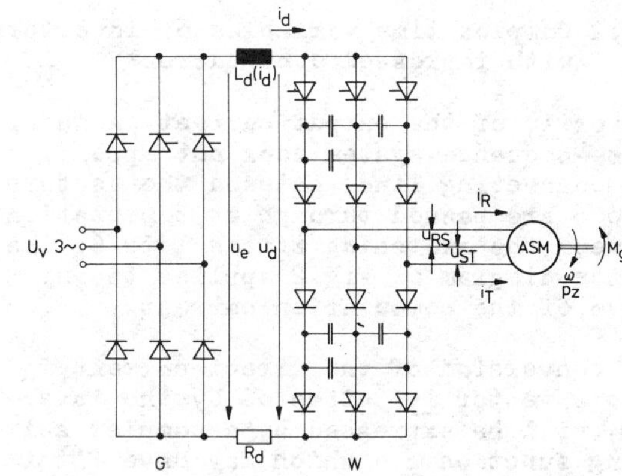

Fig.1 Frequency-converter drive with d.c.-current-link

BEHAVIOUR OF THE INVERTER AND MODELLING

The inverter W in Fig.1 with impressed d.c. current generates roughly square-wave output currents at constant direct current i_d (see Fig.2). In each of the switching states 1 to 6 the d.c. current is flowing in two phases of the a.c.side. Between these states, there are commutations during which all three phases are simultaneously conducting. When the system of the phase currents i_R, i_S, i_T is transformed into a vector system with complex time variables (see (9)), the curve shown above in Fig.2 results for complex

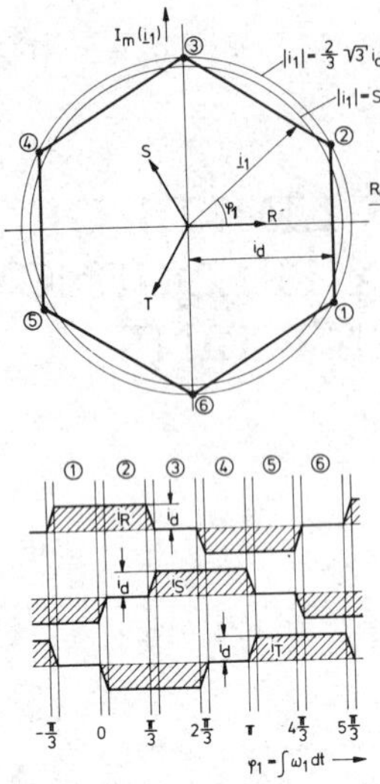

Fig.2 Complex time variables of inverter with impressed d.c. current

vector $\underline{i}_1$ of the output current. A zero phase-sequence system does not appear. The connecting lines between the vectors 1 to 6 are passed through at commutation between the switching states 1 to 6. The vector diagram of Fig.2 applies to any curve of the commutation current.

The conversion of the direct current i_d into a vector $\underline{i}_1$, effected by the inverter, will be expressed by a complex switching function $\underline{s}$ - which may have first any figure:

$$\underline{i}_1(\varphi_1) = \underline{s}(\varphi_1)\cdot i_d \qquad (1)$$

In this equation, φ_1 is the actual angle of the inverter control which initiates the commutation as soon as the switching points $\varphi_1 = 0, \pi/3, 2\pi/3 \ldots$ are reached. In the case of six-pulse operation according to Fig.2, the Fourier analysis of the switching function gives

$$\underline{s} = \underline{S}_1 e^{j\varphi_1} + \underline{S}_5 e^{-j5\varphi_1} + \underline{S}_7 e^{j7\varphi_1} + \ldots \qquad (2)$$

With very short commutation times (square-wave current curve) results

$$\underline{S} = \frac{\sqrt{12}}{\pi}\cdot\frac{1}{\nu} = 1{,}10\,\frac{1}{\nu} \qquad (2a)$$

With linear commutation over $\pi/3$ (trapezoidal current curve), the following results:

$$\underline{S} = \frac{3}{2}\,\frac{\sqrt{12}}{\pi}\,\frac{1}{\nu^2}e^{-j\nu\pi/6} = 1{,}05\,\frac{1}{\nu^2}e^{-j\nu\pi/6} \qquad (2b)$$

Multiple commutations(see (10)) bring about deviations from the curve shown in Fig.2 since in this case a part of the d.c. current is short circuited within the inverter and so does not pass over the output terminals. This leads to a reduction of the amplitude of the switching function. However, these phenomena are not to be discussed here.

The harmonics of the switching function, which occur according to equation (2), cause hunting torques in the connected induction machine. The problems associated with this have been repeatedly investigated. For the following basic considerations of the transient response of the drive system, only the fundamental oscillation (ν=1) is considered. In addition to that, only the fundamental of the square-wave curve according to equation (2a) will be considered, since equation (2b) shows that the amplitude of the fundamental of the switching function is only sligtly reduced by extended commutation, and the resultant phase displacement can be compensated by suitable control measures. Consequently,

$$s(\varphi_1) = \underline{S}_1 e^{j\varphi_1} \text{ with } \underline{S}_1 = S_1 = \sqrt{12}/\pi \qquad (3)$$

The effect of the inverter on the d.c. link is taken into account by the assumption that no active or reactive power is consumed in the inverter and that, Therefore, the d.c. power is equal to the three-phase a.c. power.

$$u_d\cdot i_d = \frac{3}{2}(\underline{u}_1\cdot\underline{i}_1) = \frac{3}{2}(\underline{u}_1\cdot\underline{S}\ i_d)$$

$$u_d = \frac{3}{2}(\underline{u}_1\cdot\underline{s}) = \frac{3}{2}\mathrm{Re}(\underline{u}_1\cdot\underline{s}^*) \qquad (4)$$

This assumption is certainly not permissible when investigating the commutation occurance since then also the commutating capacitors have to be taken into account with the power balance. These capacitors, however, only cause energy pulsations with a six-fold and higher frequency so that the simplification according to equation (4) exactly corresponds to the neglected harmonics and losses in the inverter.

Equations (1), (3), (4) allow a fully sufficient model formation of the inverter for the operating range without multiple commutations, control being effected by adjustment of phase angle φ_1.

MODELLING OF THE DRIVE SYSTEM

For developing the major properties of the complete system, the following simplifying assumptions are made:

a) The supply system and the supplying rectifier in Fig.1 are replaced by a fast-response controlled voltage source; the voltage drops on the d.c. side are substituted by an appropriate increase of R_d.

b) The inductance of the d.c. link choke is constant.

c) The squirrel-cage machine is described with the aid of complex time variables (vectors) according to (9). Field harmonics, current displacement, core losses and core saturation are neglected.

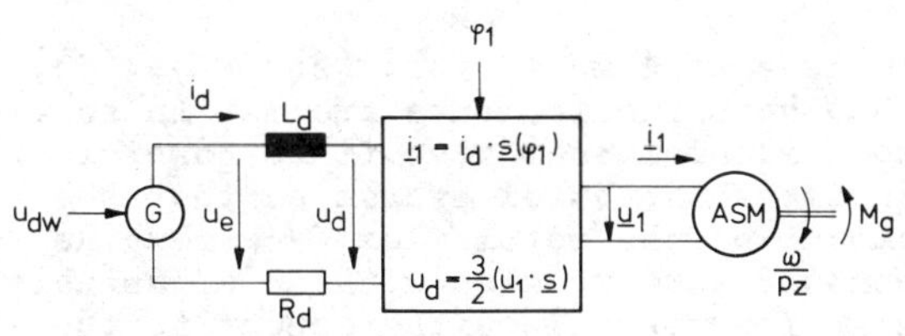

Fig.3 Simplified diagram of the drive sytem

On the basis of these assumptions, the diagram shown in Fig.3 is obtained.

For a description of the behaviour of the machine it is profitable to introduce suitable polar coordinates since amplitude and phase position of the stator current are preset separately to the inverter.

The rotor equation

$$0 = \underline{i}_2 R_2 + \dot{\underline{\psi}}_2 - j\omega\underline{\psi}_2 \qquad (5)$$

and the coupling between stator-flux and rotor-flux immediately give the two vector diagrams in Fig.4, and the following equations apply:

$$\omega_s = \frac{R_2}{1+\sigma_2}\, i_1 \sin\alpha/\psi_2 \qquad (5a)$$

$$\dot{\psi}_2 = \frac{R_2}{1+\sigma_2}\, i_1 \cos\alpha - \frac{R_2}{L_2}\,\psi_2 \qquad (5b)$$

The angular frequency ω_s of the rotor flux with respect to the rotor is referred to as "slip angular frequency". In the case of drives with high dynamic properties, the rotational frequency ω of the drive is mostly fed back to the inverter control, and a preset frequency ω_v and, if necessary, a angle correction $\Delta\varphi$ is added to it so that the following is obtained for the control angle φ_1 of the inverter

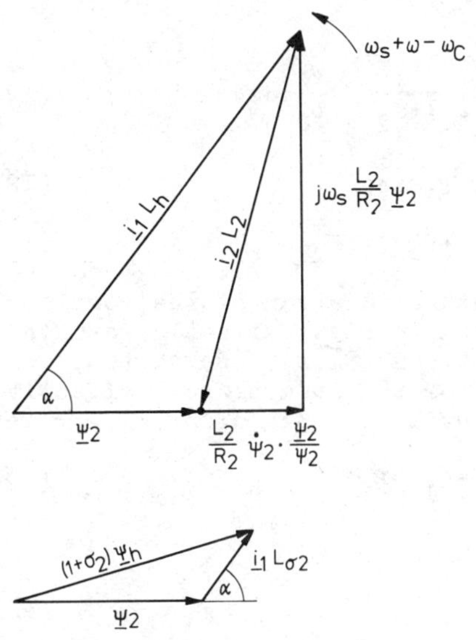

Fig.4 Diagram of rotor-flux-vectors

$$\varphi_1 = \int (\omega+\omega_v)dt + \Delta\varphi$$

The angle α between the stator current vector and the rotor flux vector can then be described by the equation

$$\alpha = \varphi_1 - \int(\omega+\omega_s)dt = \int(\omega_v-\omega_s)dt + \Delta\varphi \qquad (6)$$

For the torque and speed of the machine, the following applies:

$$M_d = \frac{3}{2} p_z\, \psi_2 \sin\alpha\, i_1/(1+\sigma_2) \qquad (7)$$

$$\dot{\omega}_m = (M-Mg)/\Theta \qquad \omega = \omega_m \cdot p_z \qquad (8)$$

The stator equation of the machine

$$\underline{u}_1 = \underline{i}_1 R_1 + \dot{\underline{\psi}}_1 \qquad (9)$$

can be transformed into

$$\underline{u}_1 = \underline{i}_1 R_1 + \frac{1}{1+\sigma_2}\left(\frac{R_2}{1+\sigma_2}\underline{i}_1 - \frac{R_2}{L_2}\underline{\psi}_2 + j\omega\underline{\psi}_2\right) + \sigma L_1 \dot{\underline{i}}_1 \qquad (10)$$

Using equation (4) and considering that

$$(\underline{s}\cdot\underline{s}) = 1 \quad \text{und} \quad (\underline{s}\cdot\underline{s}^*) = 0 \qquad (11)$$

applies for the simplified switching function according to equation (3), the reaction of the machine on the d.c. circuit is as follows:

$$u_d = \frac{3}{2} s_1^2 R_1 i_d + \frac{3}{2} s_1^2 L_1 \dot{i}_d + u_{d2} + u_M \qquad (12)$$

In this equation the reaction components (u_{d2} and u_M) of the rotor losses and the torque are as follows:

$$u_{d2} = \frac{3}{2}S_2R_2\left[\frac{i_1}{1+\sigma_2} - \frac{\psi_2}{L_2}\cos\alpha\right] \qquad (12a)$$

$$u_M = \frac{3}{2}S_2\omega\,\psi_2\sin\alpha \qquad (12b)$$

$$S_2 = S_1/(1+\sigma_2) \qquad (12c)$$

The structural diagram of the complete drive shown in Fig.5 directly results from equations (5a,b), (6) to (8), (12a,b). For the inductance L effective in the d.c. link, the following applies

$$L = L_d + \frac{3}{2}S_1{}^2\sigma L_1 \qquad (12d)$$

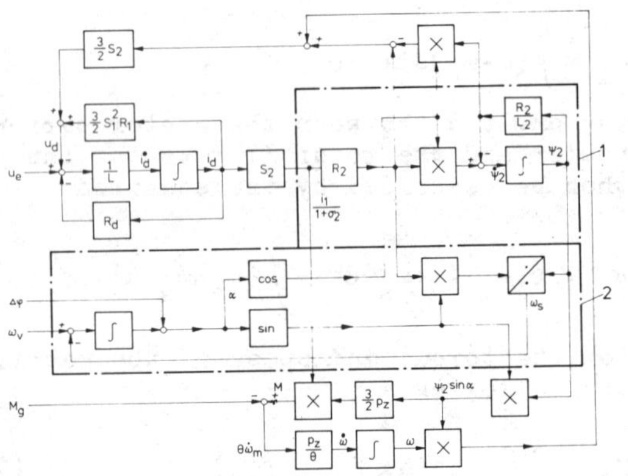

Fig.5 Structural diagramm of the drive with d.c.-current-link

If the saturation of the main field is to be considered, the main flux can be determined in accordance with Fig.4 by

$$(1+\sigma_2)^2\psi_h{}^2 = \psi_2{}^2 + i_1{}^2L_{\sigma2}^2 - 2i_1L_{\sigma2}\sin\alpha\cdot\psi_2 \qquad (13)$$

and the resultant L_h can be included in the structural diagram as variable factor.

CONTROL LOOPS FOR SQUIRREL-CAGE MOTORS

For the control of squirrel-cage machines - as for the control of d.c. drives - it is common to set up a hierarchy of superordinate control loops. The inner loop mainly serves for adjusting the torque. The speed control loop and, if necessary, the position control loop are superposed on that loop. The dynamic response of the complete system is largely determined by the transient response of the inner control loop. Therfore, this loop must meet the followingrequirements:

- The speed as disturbance variable should have a minimum effect on the torque.
- The torque should follow the reference input with a minimum time constant or good damping factor.

It is a well-known fact (see (3), (5)) that both requirements are met in an almost ideal manner when it is possible to devise the control system so that the amount of the rotor flux Ψ_2 remains constant and rotates with an adjustable slip frequency ω_s with respect to the rotor. This leads to the well-known control condition

$$\underline{i}_1 = \frac{\Psi_2}{L_h}(1+jb_v)e^{j(\varphi+\varphi_v)} \qquad (14)$$

where the "slip control"

$$b_v = 2\pi f_v L_2/R_2 \qquad (14a)$$

can be preset as reference input and where the "slip control frequency" f_v is fed in the inverter and the angle

$$\Delta\varphi = \mathrm{atn}\ b_v \qquad (14b)$$

must be put in as phase correction.

As long as condition (14) is satisfied, no transient phenomena will occur in the machine, and the torque is proportional to b_v.

LIMITS OF THE CONTROL BEHAVIOR OF THE DRIVE

When the control system for the drive is so designed that equation (14) is fulfilled, then continually appears

$$f_s = f_v \qquad \text{and} \qquad \alpha = \Delta\varphi$$

In this case, the effect of the machine on the d.c. link corresponds to the effect of a machine operated with ω, ω_s, i_1 under steady-state conditions. Fig.6 shows the direct input voltage required in this operation mode for an example.*)

*) In the examples here and below, the following drive parameters are used: R_d=35mΩ, L_d=4,5mH, R_2=13mΩ, L_2=4,6mH, σ_1=0,067, σ_2=0,02, i_{1N}=450A, f_N=50Hz, $\tan\alpha_N$=2

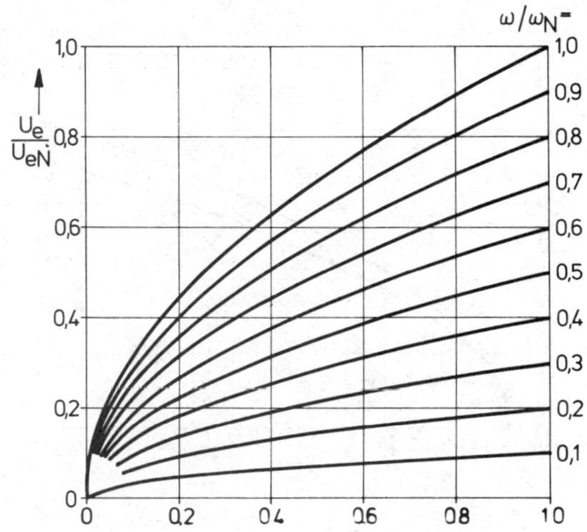

Fig.6 Required d.c. voltage at steady state operation and constant rotor flux

For changing the torque, an additional dynamic component

$$u_L = L\,\dot{i}_d = L\,\frac{L_2}{R_2}\,i_{d\mu}\,\dot{\omega}_v\,\sin\alpha \qquad (15)$$

must be added to the direct input voltage, $i_{d\mu}$ being the direct current at no load in accordance with

$$i_{d\mu} = \Psi_2/L_2\,S_2 \qquad (16)$$

For estimating the necessary relative voltage reserve, u_L is related to the direct voltage required for the torque, according to equation (12b):

$$\frac{u_L}{u_M} = \frac{2}{3}\,\frac{1}{S_2^2}\,\frac{L}{L_2}\,b_v\,\frac{1}{\omega}\,\frac{\dot{\omega}_v}{\omega_v} \qquad (17)$$

By substituting the rating of various drives it was found that with a voltage reserve of 10 % at rated frequency, rates of change in the preset frequency (i.e. in the torque) of

$$\frac{\dot{\omega}_v}{\omega_{vN}} = \frac{\dot{M}}{M_N} = 0{,}5\ \ldots\ 2\ \%/\mathrm{ms}$$

can be obtained. The required steady-state d.c. voltage decreases rapidly with diminishing of speed or torque as shwon in Fig.6; therefore the feasible change rate of torque rises accordingly. This characteristic is sufficient for most drive applications. The control behaviour of the pulse-controlled frequency converter with impressed voltage, however, is considerably better as the voltage reserve is only required for current variations in the leakage inductance (see (5)).

BEHAVIOUR AT CONSTANT SLIP CONTROL FREQUENCY

With regard to the improved rating conditions with d.c. current link convertors there are also control modes used keeping constant the slip control frequency (11). The steady-state value of d.c. current i_d and rotor flux Ψ_2 then both rise proportional to the square root of the torque. The d.c. voltage required in this case is shown in Fig.7. The curves do not differ much from those in Fig.6. The decrease of d.c. voltage, i.e. the increase of voltage reserve, at reducing speed is poorer in Fig.7. But the characteristic allows also quick current control.

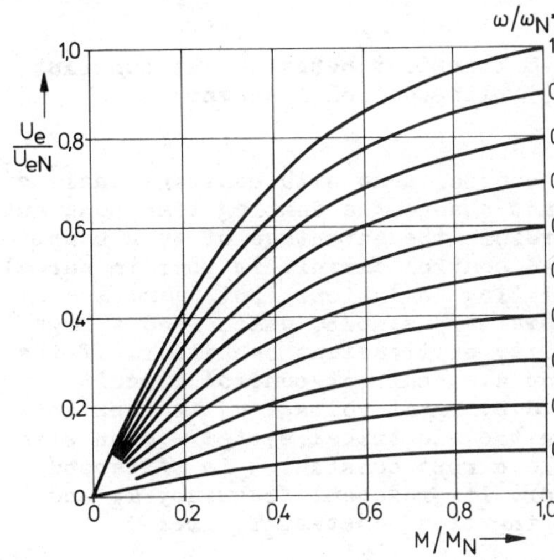

Fig.7 Required d.c. voltage at steady state operation and constant slip-control-frequency

If the slip control frequency in transient operation is kept constant without any phase correction, the wellknown transient behaviour, shown in Fig.8, appears in reaction to quick change of motor current. The torque turns to the new state by oscillating with the slip frequency and a damping time constant equal L_2/R_2. During this time the phase angle α oscillates too, which results from Fig.8.

It is possible to eliminate the torque oscillating by introducing control loops, which keep constant the angle α even in transient conditions.A suitable control instruction is given by the equations (5a) and (5b). It may be e.g. realized by modelling the machine using Fig.5 (only parts 1 and 2 are necessary) and delivering the real current to the model at constant phase angle α. The frequency ω_s produced by the model can be used as slip control frequency.

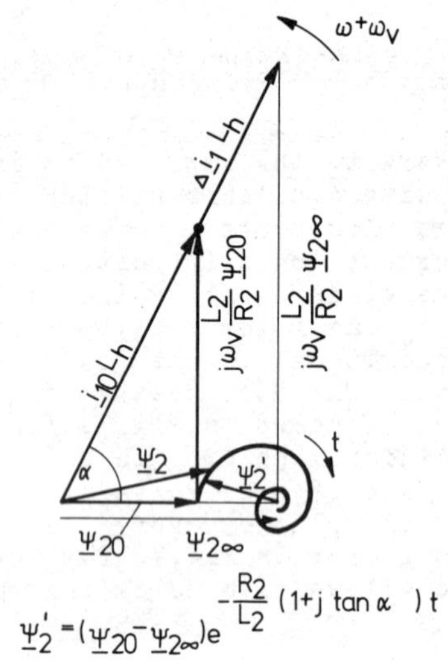

Fig.8 Transient behaviour at constant slip-control-frequency

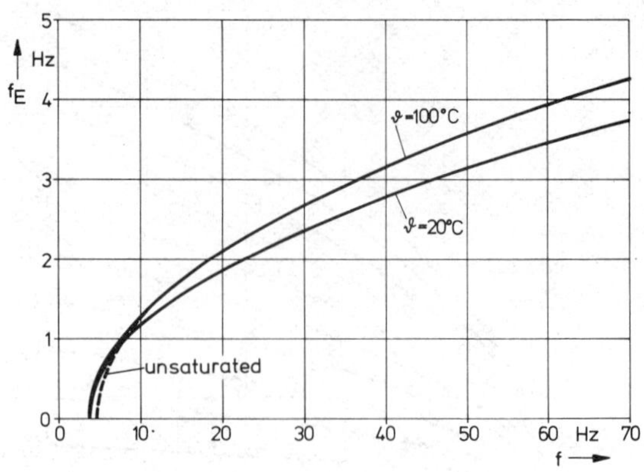

Fig.9 Resonant frequency for u_e = const, α = const, ω = const

The control mode with constant angle α cannot change the damping time constant, therefore the advantage of such complicated control circuit is poor in normal operating conditions. But there are extraordinary states, which need a high quality of transient behaviour. If there is no d.c. current control circuit and the d.c. input voltage u_e is constant, then the electrical system - even with angle α kept constant - is of second order. Its resonant frequency ω_E and damping time constant T_E are:

$$\omega_E^2 = \frac{R_2 R}{L_2 L} - \frac{3}{2}s_2^2\frac{R_2^2}{L_2 L}\cos^2\alpha + \frac{3}{4}s_2^2\omega\frac{R_2}{L}\sin 2\alpha - \frac{1}{T_E^2} \tag{18a}$$

$$\frac{1}{T_E} = \frac{1}{2}\left(\frac{R_2}{L_2} + \frac{R}{L}\right) \tag{18b}$$

$$R = R_d + \frac{3}{2}s_1^2 R_1 + \frac{3}{2}s_2^2 R_2 \tag{18c}$$

Fig.9 shows the resonant frequency of the exemplary drive as a function of speed. The value of f_E is in the range of the value of slip frequency. So the slip control frequency may excite undamped oscillations, which was proved by simulation. The system then will be stabilized by the control of phase angle α.

SPEED CHANGE AT LIMITED D.C.VOLTAGE

The disturbance of the electric system caused by speed change effects against the input d.c. voltage (see Fig.5). It may be eliminated by feeding back the speed to the reference input voltage. So the system is able to reach extremely high change rates of speed.

Transgressing the nominal speed however, the d.c. voltage is limited and the flux is weakened. In drives with high acceleration rate it may happen, that the field weakening cannot follow the rapid increase of rotation. Than the voltage reserve will be consumed and the current control loop therefore will get out of operation. In such cases only the slip control frequency f_s including phase angle correction have a bearing on this system. Then the control of angle α may be necessary for stable operation.

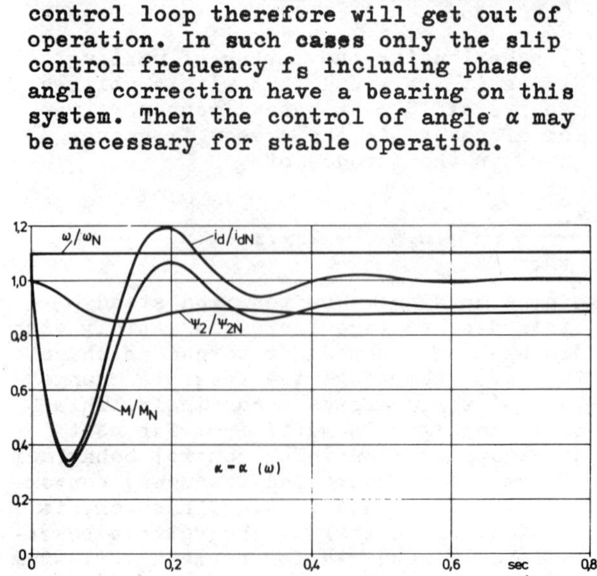

Fig.10 Transient behaviour at speed change using ideal α control

In Fig. 10 the transient characteristic is demonstrated by assuming an instantaneous speed change of 10 % and ideally operating of α-control. The transient characteristics are the same as described in (18a) to (18c).

The characteristic shown in Fig.10 may additionally cause electro-mechanical oscillations in the system. By suitable modulation of angle α it is possible, to improve the transient behaviour, but the breakdown of torque M remains. Because of the influence of speed, temperature and harmonics it will be difficult to get satisfactory operation at any condition. In practice the acceleration rate will be limited to such values that the field weakening can follow the speed.

SYMBOLS

b_v	slip control,(14a)
f_s	slip
f_v	slip control frequency
i_1	amplitude of stator current
$\underline{i}_1$	vector of stator current
i_d	direct current
$i_{d\mu}$	fictitious no load d.c.current, (16)
L	inductance effective in d.c.link, (12d)
L_d	inductance of d.c. link
L_1	inductance of stator
L_2	inductance of rotor
M,Mg	torque, load torque
p_z	number of poles
R_d	d.c. link resistance
R_1	stator resistance
R_2	rotor resistance
S_1	fundamental of switching function
S_2	see (12c)
u_e	d.c. input voltage
u_d	d.c. voltage of inverter
u_{d2},u_M	see (12a), (12b)
u_L	transient required d.c.voltage, (15)
α	see Fig.4
φ_1	state angle of inverter
Δφ	phase angle correction, Fig.5
σ	leakage parameter
σ_2	rotor leakage parameter
ω	angular frequency of speed (electrical angle)
ω_m	angular frequency of speed (mechanical angle)
$\omega_E = 2\pi f_E$	see (18a)
ω_s	angular frequency of rotor flux vector with reference to rotor
ω_v	slip control angular frequency
suf.N	marks nominal value

REFERENCES

(1) F. Blaschke, H. Ripperger, H.Steinkönig, Regelung umrichtergespeister Asynchronmaschinen mit eingeprägtem Ständerstrom, Siemens-Zeitschrift, 42, 773 (1968)

(2) D. Naunin, Die Darstellung des dynamischen Verhaltens der stromrichtergespeisten Asynchronmaschine durch ein komplexes VZ1-Glied, Wiss.Ber. AEG-TELEFUNKEN 42, 48 (1969)

(3) K. Hasse, Drehzahlregelverfahren für schnelle Umkehrantriebe mit stromrichtergespeisten Asynchron-Kurzschlußläufermotoren, Regelungstechnik und Prozeß-Datenverarbeitung 20, 60 (1972)

(4) F. Blaschke, Das Verfahren der Feldorientierung zur Regelung der Asynchronmaschine, Siemens Forsch.- und Entwickl.-Ber. 1, 184 (1972)

(5) L. Abraham, W. Kuhn, Möglichkeiten und Grenzen bei der Regelung von umrichtergespeisten Kurzschlußläufermaschinen, IFAC-Symposium Düsseldorf 1974 Vol.1 pp 587 VDI/VDE-Gesellschaft Meß- und Regeltechnik, Düsseldorf

(6) G. Gölz, F. Hentschel, S. Chumpitaz, Simulation des Betriebsverhaltens einer umrichtergespeisten Asynchronmaschine, Wiss.Ber.AEG-TELEFUNKEN 48, 20 (1975)

(7) D.W. Novotny, J.H.Wouters, Induction Machine Transfer Functions and Dynamic Response by means of Complex Time Variables, IEEE Trans. PAS-95, 1325 (1976)

(8) A.B.Plunkett, T.A.Lipo, New Methods of Induction Motor Torque Regulation,IEEE Trans. IA-12 (1976)

(9) K.P. Kovacs, I.Racz, Transiente Vorgänge in Wechselstrommaschinen, Verlag der Ungarischen Akademie der Wissenschaften, Budapest, (1959)

(10) K.Moll, D.Schröder, Applicable Frequency Range of Current Source Converter, IFAC-Symposium Düsseldorf 1977,Pergamon Press,Oxford

(11) P. Appun, Control of rotating and linear induction motors for vehicle drives, IFAC-Symposium Düsseldorf 1977, Pergamon Press, Oxford

STABILITY PROBLEMS WITH THE CONTROL OF INDUCTION MACHINES USING THE METHOD OF FIELD ORIENTATION

K.-H. Bayer and F. Blaschke

Erlangen, Siemens AG, Systemtechnische Entwicklung

Summary

The method of field orientation for the control of induction machines has become well-known in the last few years. With this method, the vector of magnetic flux is used to provide self-control of the machine. As a result the angle of the flux vector operates on the actuating input to the machine as positive feedback. Stability problems can occur due to this positive feedback. This paper will show how these stability problems can be easily overcome.

Introduction

The method of field orientation provides the induction machine with similar advantageous control characteristics to those obtained with a d.c. machine. This is achieved by providing a fixed coupling between the operating axes of electrical current and of magnetic flux in the machine. The coupling is such that the angle of the vector of stator current is formed principally by the angle of the vector of air-gap flux. Considering the induction machine as a control system, there is positive feedback of a state variable to the input of the machine.

The effect of this positive feedback on the stability of the resulting system will be investigated in this paper. A signal-flow analysis shows that the induction machine can be represented as a system of four integrators with coupling including a negative feedback path. Oscillations are therefore possible. A simplified consideration of the system shows that the positive feedback inherent with field orientation compensates the internal negative feedback in the machine. The coupling between the integrators is therefore in effect neutralized, i.e. the system is decoupled and no longer susceptible to oscillations.

However, a more exact investigation of the system shows that full compensation of the coupling only occurs with a machine which has no leakage. Residual coupling remains if there is a leakage in the machine. A stability analysis of the resulting system shows that this coupling is present as positive feedback, the magnitude of which depends on the loading of the induction machine. Instability can occur above a certain critical load. This undesirable effect can be overcome if the so-called rotor flux is used to accomplish field orientation instead of the air-gap flux. The two fluxes differ by an amount related to the leakage of the machine. Full compensation of the internal negative feedback is possible using the rotor flux for orientation, regardless of the amount of leakage in the machine. The complete decoupling of the system mentioned above is therefore achieved for any machine.

Properties of the induction machine

The signal flow diagram of the induction machine will be firstly considered as this forms the basis for the investigation of the stability problems previously described. For the sake of clarity, it will be assumed that the stator current is directly controllable. There are several methods of achieving this, e.g. by using a current-source inverter (Ref. 1), or by using a pulse-width-modulated inverter in conjunction with suitable control of the instantaneous values of the phase currents (Ref. 2). A machine without rotor leakage $L^{\sigma L} = 0$ will be firstly considered. The signal flow diagram of such a machine is shown in Fig. 1, see Ref. 3. Fig. 1 describes the relationship between the vectors of stator current i^S, of magnetic flux Ψ (will also be referred to as the field vector), and the electrical torque M_{el}, using mainly polar coordinates, see also Fig. 2. This method of representation was chosen because it lends itself well both to the consideration of the method of field orientation, and to the problems associated with this method.

The inverter enables the vector of stator current to be applied to the system with variable magnitude i^S and angle ε^S_S, see Fig. 1. The difference angle ε^S_φ between the current vector and the field vector is obtained by subtracting the angle

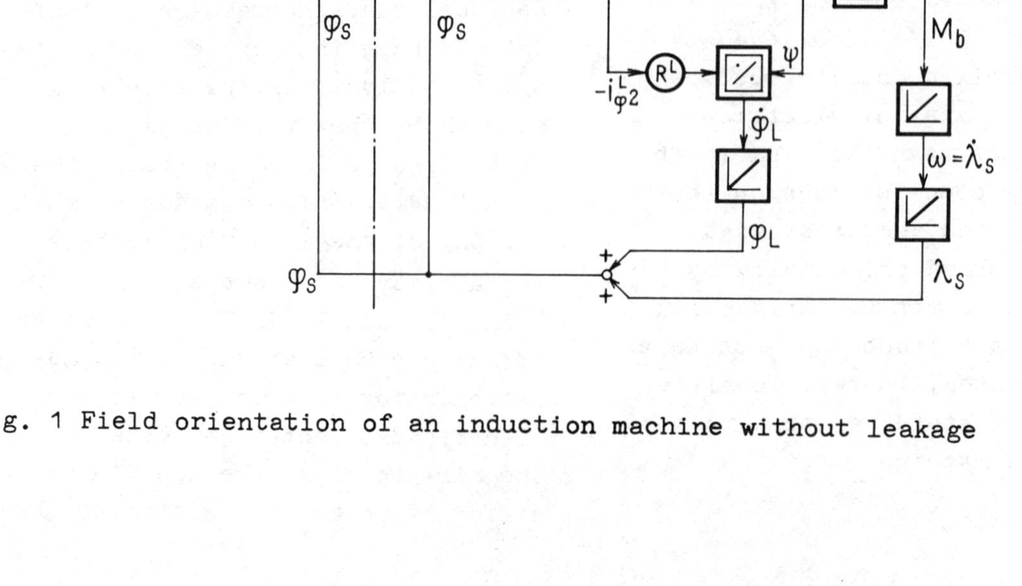

Fig. 1 Field orientation of an induction machine without leakage

φ_S of the field vector from ε^S_S . A polar-cartesian coordinate converter P/K converts i^S and ε^S_φ into cartesian coordinates which are the so-called field-orientated coordinates $i^S_{\varphi 1}$ and $i^S_{\varphi 2}$ of the stator current, the projections of the vector of stator current parallel and perpendicular to the field vector, see Fig. 2. The magnitude of the field Ψ is formed from $i^S_{\varphi 1}$ via a first-order lag network. The product of $i^S_{\varphi 2}$ and Ψ is the electrical torque M_{el}. Signals

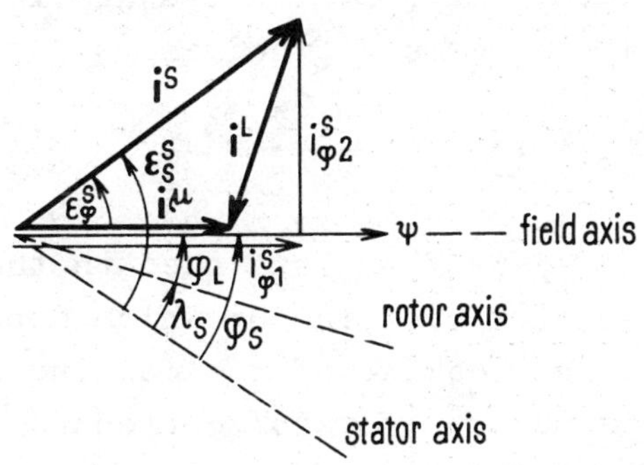

Fig. 2 Vector diagram of current and flux

$i^S_{\varphi 2}$ and Ψ also determine the angle φ_L between the field and the rotor via an integrator. The rotor angle λ_S is determined by M_{el} via two integrators. These two angles added together give the field angle φ_S relative to the stator which is fed as negative feedback to the adding point at the input of the machine. The induction machine is a potentially oscillatory system because of the four integrators and the negative feedback path. In practice, the electrical torque and field adjust themselves to each operating point with a slow transient. At full load, this transient is usually underdamped.

Method of field orientation

The aim of the method of field orientation (Ref. 3, 4) is to enable electrical torque and field to be quickly-adjusted with good damping. The first step is to eliminate the underdamped transient by compensating the effect of the negative feedback of φ_S by positive feedback of this angle at the input of the system, see left-hand side of Fig. 1. The angle ε^S_S of stator current is "precontrolled" by the field angle φ_S , i.e. angle ε^S_S is formed principally by the field angle. This compensation orientates the vector of stator current to the vector of field, hence the name for this method of control. The second step is to compensate the P/K converter by an inverse converter, a K/P converter. Fig. 3 shows the signal flow diagram of the system obtained.

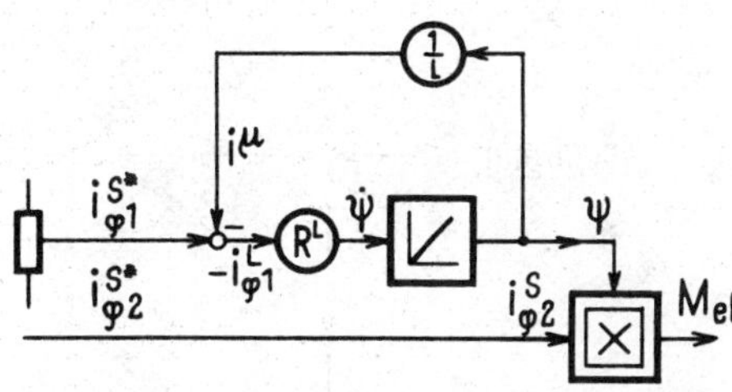

Fig. 3 Resulting signal-flow diagram with field orientation $L^{\sigma L} = 0$

The field-orientated components $i^S_{\varphi 1}$, $i^S_{\varphi 2}$ track the inputs to this converter $i^{S*}_{\varphi 1}$, $i^{S*}_{\varphi 2}$ without delay or oscillation. Direct adjustment of electrical torque and field is therefore possible. The field angle φ_S necessary for this method of control-by field the air-gap flux is referred to - can be determined by several methods, e.g. by a direct measurement of the flux in the air-gap of the machine (Ref. 5), or by using an electronic model (Ref. 6).

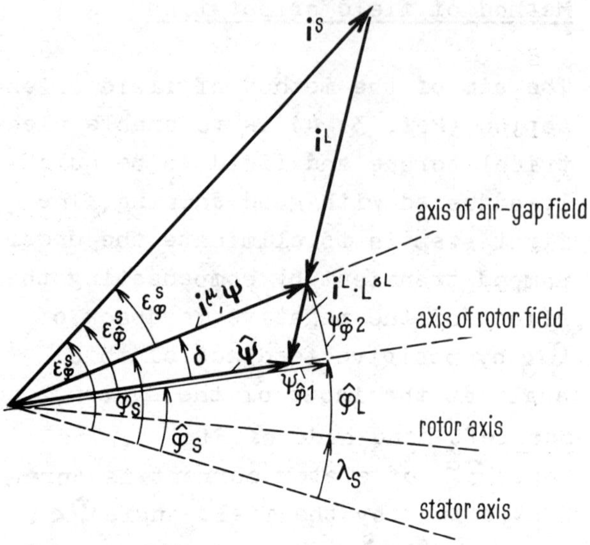

Fig. 4 Vector diagram for $L^{\sigma L} \neq 0$

Effect of rotor leakage

The previous considerations held for a machine without rotor leakage. This limitation will now be overcome. It is shown in Ref. 4 that the signal flow diagram of an induction machine with leakage $(L^{\sigma L} \neq 0)$ is very similar to that of an induction machine without leakage if the vector of rotor flux $\hat{\psi}$ is chosen as a state variable instead of the vector of air-gap flux ψ. Fig. 5 shows the resulting signal flow diagram. The unshaded part of the diagram, with the exception of unimportant factors, is the same as that of Fig. 1 if all the symbols referring to the air-gap flux are replaced by symbols for the rotor flux (^). The leakage factor used here is defined as

$$\sigma^L = \frac{L^{\sigma L}}{L} \tag{1}$$

The rotor flux is linked to the air-gap flux by a factor which depends on the leakage of the machine and on the vector of rotor current i^L in the following manner,

$$\hat{\psi} = \psi + i^L L^{\sigma L} \tag{2}$$

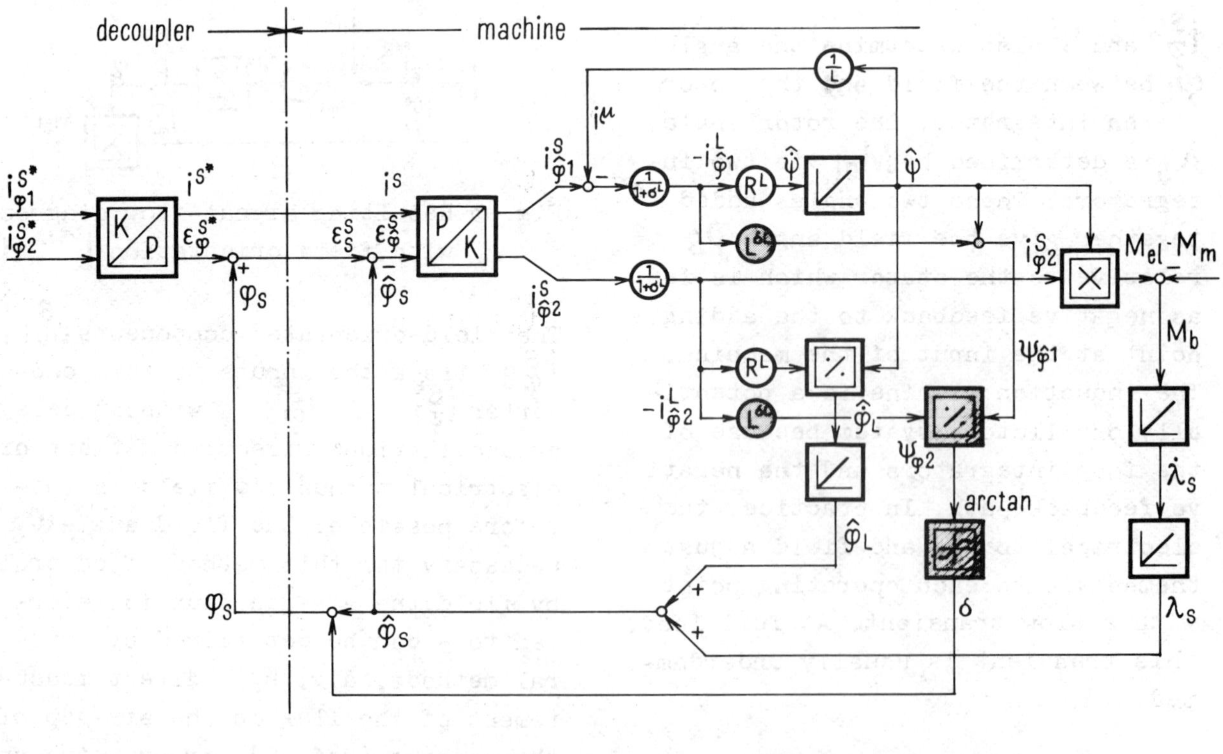

Fig. 5 Field orientation of an induction motor with rotor leakage

The resulting triangle of flux vectors is shown in Fig. 4. It can be seen that the angle $\hat{\varphi}_S$ of the vector of rotor flux differs from the angle of the vector of air-gap flux φ_S by an angle δ. This angle can be calculated from the rotor-field orientated coordinates of the vector of air-gap flux, see Fig. 4 and equation (2). The latter are the projections of the vector of air-gap flux parallel and perpendicular to the vector of rotor flux,

$$\Psi_{\hat{\varphi}1} = \Psi - i^L_{\hat{\varphi}1} L^{\sigma L} \qquad (3)$$

$$\Psi_{\hat{\varphi}2} = \quad - i^L_{\hat{\varphi}2} L^{\sigma L} \qquad (4)$$

The angle δ is given by,

$$\delta = \arctan \frac{\Psi_{\hat{\varphi}2}}{\Psi_{\hat{\varphi}1}} \qquad (5)$$

These relationships are represented as the shaded functions shown in Fig. 5.

The angle δ has no effect on the signal flow path and on the properties of the uncontrolled machine (without field orientation), but has a direct effect on angle φ_S of the vector of air-gap flux. Angle φ_S is the only variable which can be directly measured, and therefore must be considered as an output variable.

Nature of the problem

If field orientation is accomplished by pre-controlling the current angle ε^S_S by the angle φ_S of air-gap flux as before, then it is apparent that an undesired coupling still exists due to the angle δ associated with the rotor leakage. Fig. 6 shows the resulting signal flow diagram.

The coupling acts as positive feedback. This leads to an error in the desired angle between current and

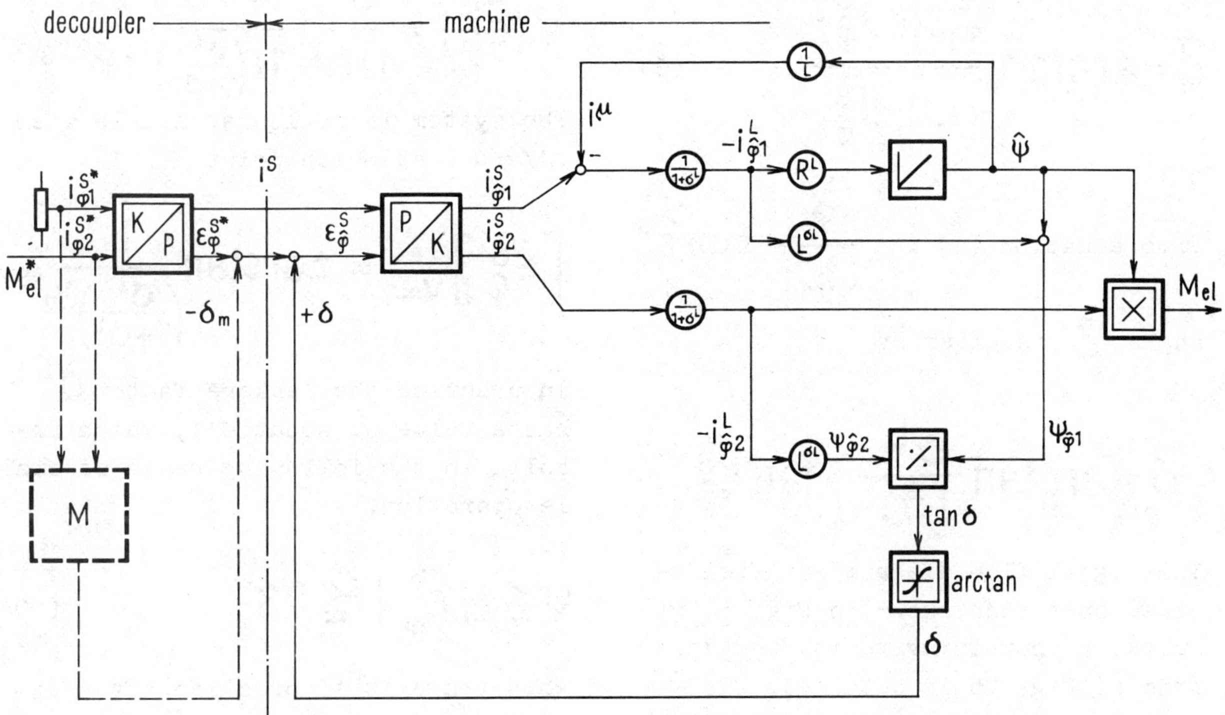

Fig. 6 Resulting signal-flow diagram with field orientation for $L^{\sigma L} \neq 0$

flux, as well as affecting the stability of the complete system.

The amplification V of the positive feedback governs the degree of stability. As soon as $V > 1$ the positive feedback becomes over-critical and the system is unstable (exponential function). The amplification factor will be determined for steady-state operation ($\bar{}$). The function

$$\bar{\delta} = f(\bar{\varepsilon}^S_{\hat{\varphi}})$$

will be firstly considered. Considering Fig. 5 for steady-state operation, i.e. with

$$\bar{\dot{\hat{\Psi}}} = 0$$

which gives

$$\bar{\hat{\Psi}} = L\, \bar{i}^S_{\hat{\varphi}1}$$

the following relationship is obtained,

$$\bar{\delta} = \arctan \frac{L \frac{\sigma^L}{1+\sigma^L}\, \bar{i}^S_{\hat{\varphi}2}}{L\, \bar{i}^S_{\hat{\varphi}1}} \qquad (6)$$

From equation (1) and $\frac{i^S_{\hat{\varphi}2}}{i^S_{\hat{\varphi}1}} = \tan \hat{\varepsilon}^S_{\varphi}$ angle $\bar{\delta}$ is given by

$$\bar{\delta} = \arctan \frac{\sigma^L}{1+\sigma^L} \tan \bar{\varepsilon}^S_{\hat{\varphi}} \qquad (7)$$

The signal flow diagram for steady-state operation shown in Fig. 7a results. By combining terms, the diagram of Fig. 7b is obtained. The amplification of the functions within the positive feedback loop must now be determined to assess stability. From equation (7) with $V = \frac{d\bar{\delta}}{d\bar{\varepsilon}^S_{\hat{\varphi}}}$ amplification $V(\bar{\varepsilon}^S_{\hat{\varphi}})$ is given by

$$V(\bar{\varepsilon}^S_{\hat{\varphi}}) = \frac{\sigma^L}{1+\sigma^L} \cdot \frac{1+\tan^2 \bar{\varepsilon}^S_{\hat{\varphi}}}{1+\left(\frac{\sigma^L}{1+\sigma^L}\right)^2 \tan^2 \bar{\varepsilon}^S_{\hat{\varphi}}} \qquad (8)$$

The system is no longer stable when $V = 1$, at which point

$$\left[\bar{\varepsilon}^S_{\hat{\varphi}}\right]_{V=1} = \pm \arctan \frac{1}{\left(\frac{\sigma^L}{1+\sigma^L}\right)^{1/2}} \qquad (9)$$

In practice the leakage factor σ^L has a value of about 0.1, which results in the following range of stable operation,

$$0 \leqq |\bar{\varepsilon}^S_{\hat{\varphi}}| \leqq 73° \qquad (10)$$

This range is in practice often exceeded, especially if operation with field weakening is required. It follows that the field orientation of

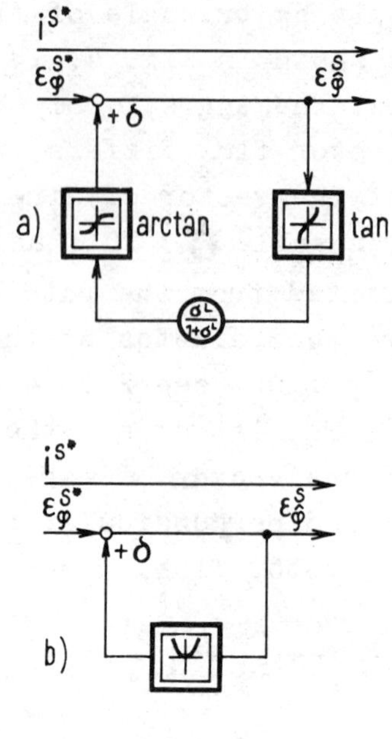

Fig. 7 Simplification of Fig. 6 for steady-state operation

induction machines using the angle of air-gap flux can not be universally used.

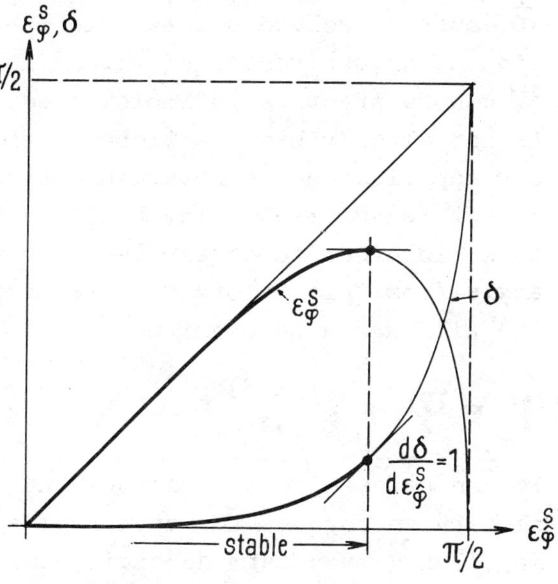

Fig. 8 Angles δ and ε^S_φ plotted against $\varepsilon^S_{\hat{\varphi}}$

The area for stable operation will also be considered with the help of the Heyland circle diagram, see Fig. 9. From Fig. 4 it follows that

$$\bar{\varepsilon}^S_\varphi = \bar{\varepsilon}^S_{\hat{\varphi}} - \bar{\delta} \tag{11}$$

Angles $\bar{\varepsilon}^S_\varphi$ and $\bar{\delta}$ are plotted in Fig. 8 as a function of $\bar{\varepsilon}^S_{\hat{\varphi}}$. By differentiating $\bar{\varepsilon}^S_\varphi$ and $\bar{\delta}$ with respect to $\bar{\varepsilon}^S_{\hat{\varphi}}$, the following result for the gradient of the function $\bar{\varepsilon}^S_\varphi = f(\bar{\varepsilon}^S_{\hat{\varphi}})$ is obtained,

$$\frac{d\bar{\varepsilon}^S_\varphi}{d\bar{\varepsilon}^S_{\hat{\varphi}}} = 1 - \frac{d\bar{\delta}}{d\bar{\varepsilon}^S_{\hat{\varphi}}} = 1 - V \tag{12}$$

The gradient of $\bar{\varepsilon}^S_\varphi = f(\bar{\varepsilon}^S_{\hat{\varphi}})$ is positive in the area of stable operation $V < 1$, and is zero at the limit of stability $V = 1$. The function $\bar{\varepsilon}^S_\varphi = f(\bar{\varepsilon}^S_{\hat{\varphi}})$ therefore always has a positive gradient in the area of stable operation.

As the vector of stator current advances along the Heyland circle in the direction of increasing current, see Fig. 9, the angle of this function $\bar{\varepsilon}^S_{\hat{\varphi}}$ increases continuously*). Therefore the range of the Heyland circle corresponding to stable operation is the range with in which the angle ε^S_φ continuously increases while advancing round the circle. This holds for the range of the circle between the synchronous point S and the point of contact T of the tangent, which therefore forms the limit of stability.

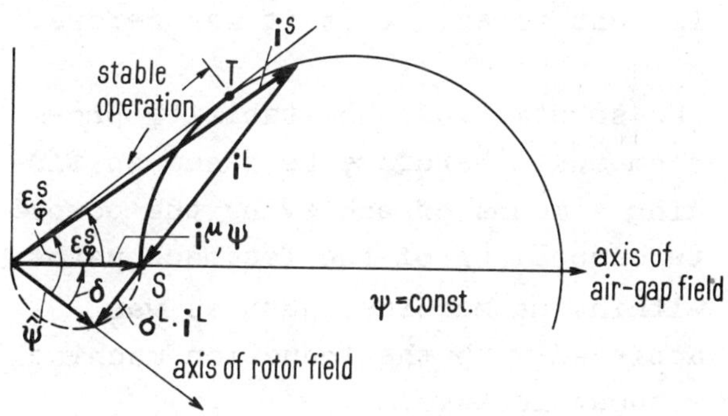

Fig. 9 Heyland circle diagram

Solution of the problem

It has been shown that compensation by pre-controlling with the air-gap flux results in a residual coupling of the system which leads to an error in the orientation and can lead to instability.

*) At steady-state operation the vector of rotor flux is perpendicular to the vector of rotor current. The adjacent ($\hat{\psi}$) of the right-angled triangle formed by these two vectors always decreases, and the opposite $i^L + \sigma^L i^L$ always increases. The tangent of the angle $\varepsilon^S_{\hat{\varphi}}$, and thus the angle $\varepsilon^S_{\hat{\varphi}}$ itself, therefore also always increases.

Whereas an error in the orientation can in principle be compensated for by injecting the angle δ_m (a model can be used to determine the angle δm based on the desired-value signals, see Ref. 2), an improvement in the stability of the system is not possible in this manner. The reason for this can be seen from the signal flow diagram of Fig. 5 in which such a model M is shown with dotted lines. The orientation error would be eliminated, but the loop within the machine with positive feedback which determines stability is just as active as it was before.

The solution of the stability problem must therefore be based on finding a means of achieving the complete decoupling of the feedback paths within the machine, just as was achieved with the induction machine without leakage.

It can be seen in Fig. 5 that complete decoupling, and therefore perfect stability over the complete range of operation, can be achieved by simply using the angle of rotor flux $\hat{\varphi}_S$, instead of the angle of air-gap flux φ_S as the orientation angle. This orientation of current to the rotor flux means the impression of angle $\varepsilon_{\hat{\varphi}}^S$ instead of the angle ε_{φ}^S .

Practical implementation

The problem of determining the angle $\hat{\varphi}_S$ must be solved before orientation using the vector of rotor flux $\hat{\Psi}$ can be practically implemented. As has already been mentioned, the air-gap flux can be determined by a direct measurement using Hall detectors. In order to obtain the required angle from this information, equation (2) will again be examined,

$$\hat{\Psi} = \Psi + i^L L^{\sigma L} \qquad (2)$$

It can be seen that it is only necessary to correct the vector of air-gap flux Ψ by a term depending on the rotor current in order to obtain the vector of rotor flux $\hat{\Psi}$. The vector of rotor current i^L cannot usually be measured, but can be eliminated by the following relationship,

$$i^S + i^L = \frac{\Psi}{L} \qquad (13)$$

The vector of rotor flux $\hat{\Psi}$ is then given by,

$$\hat{\Psi} = \Psi(1+\sigma^L) - i^S L^{\sigma L} \qquad (14)$$

This result shows that the required vector of rotor flux $\hat{\Psi}$ can be determined by subtracting the stator current dependent term $i^S L^{\sigma L}$ from the product of the measured vector of air-gap flux Ψ with the factor $1+\sigma^L$. The stator current must also be measured in addition to the air-gap flux. Both sets of signals should be measured in the natural stator-orientated coordinates Ψ_{S1}, Ψ_{S2} and i_{S1}^S, i_{S2}^S respectively. Fig. 10 shows the circuit of the necessary measurement arrangement with the associated calculation circuit in

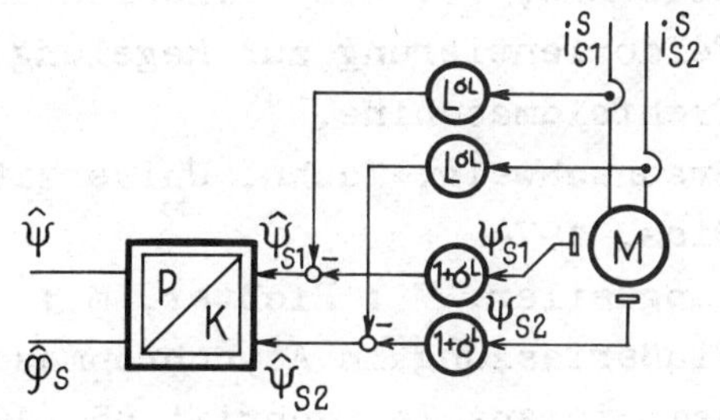

Fig. 10 Determination of the vector of rotor flux $\hat{\Psi}$

cartesian components, as well as the following coordinate converter which determines the signals $\hat{\Psi}$ and $\hat{\varphi}_S$ in polar coordinates.

The adjustment of the factor $L^{\sigma L}$ can be very simply accomplished in practice as follows. With a constant set-value $i_{\hat{\Phi}1}^{S*}$ of the coordinate of stator current, the value of $L^{\sigma L}$ should be adjusted until the magnitude of the field given by equation (13) (signal $\hat{\Psi}$ in Fig. 10) remains constant, independent of load.

Fig. 11 shows a few typical oscillograms using field orientation with the air-gap flux b) and with the rotor flux a) for an impulse-shaped disturbance of the impressed current for no-load and full-load operation.

Both methods of orientation have a similar good performance at no-load, i.e. the fall in the magnitude of flux is only temporary. However at full load using orientation with the air-gap flux (Fig. 11b), the machine undergoes a complete fall-off of electrical torque and flux. This undesirable effect is eliminated by using orientation with rotor flux, see Fig. 11a.

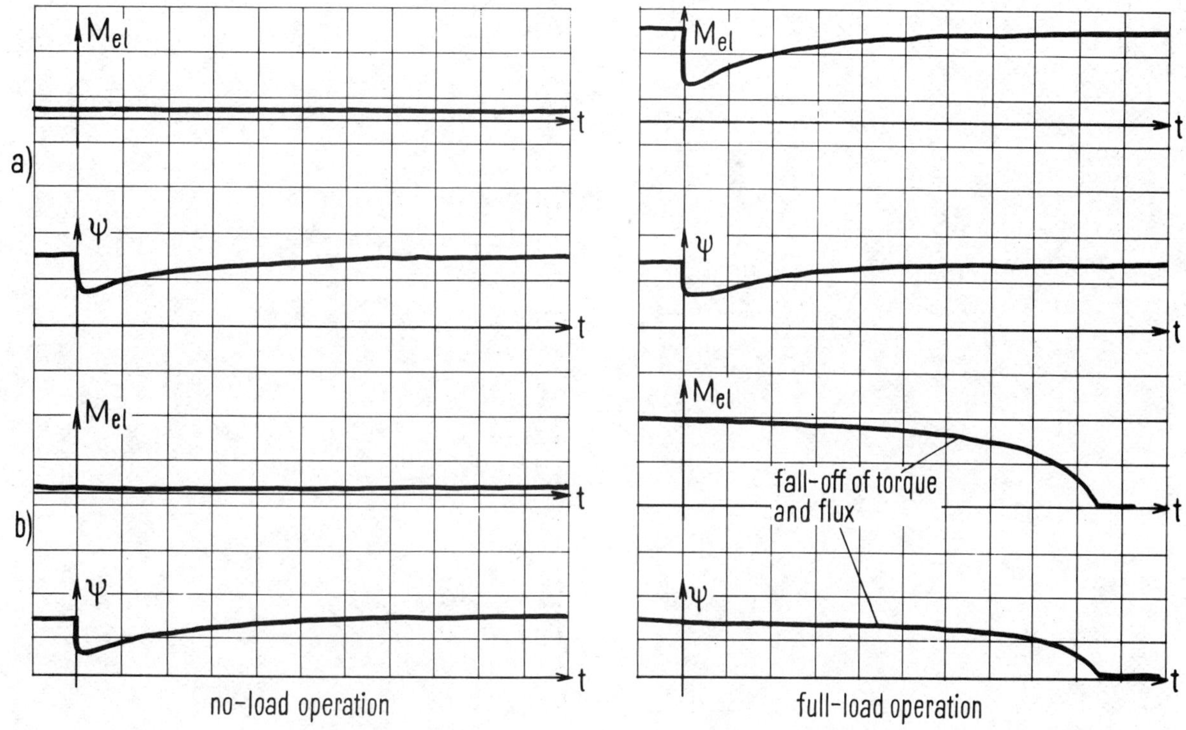

Fig. 11 Field orientation using a) rotor flux $\hat{\Psi}$ b) air-gap flux Ψ

Reference

1. Becker, H.: Dynamisch hochwertige Drehzahlregelung einer umrichtergespeisten Asynchronmaschine. Regelungstechnische Praxis und Prozeßrechentechnik 1973, No. 9, pp. 217 - 221
2. Blumenthal, M.: Die feldorientiert betriebene umrichtergespeiste Asynchronmaschine mit Statorstromeinprägung.

   Karlsruhe, Techn. Universität, Diss. 1975
3. Blaschke, F.: Das Verfahren der Feldorientierung zur Regelung der Asynchronmaschine, Siemens Forsch. und Entwicklungsberichte No. 1 (1972), pp. 184 - 193
4. Blaschke, F.: Das Verfahren der Feldorientierung zur Regelung der Drehfeldmaschine. Braunschweig, Techn. Universität, Diss. 1974
5. Langweiler, F.; Richter, M.: Flußerfassung in Asynchronmaschinen, Siemens-Zeitschrift 45 (1971), pp. 768 - 771
6. Blaschke, F.; Böhm, K.: Verfahren der Felderfassung bei der Regelung stromrichtergespeister Asynchronmaschinen. Messen, Steuern , Regeln, 18, automatisierungspraxis (1975), No. 12, pp. 278 - 280

TIME-RESPONSE ANALYSIS OF INVERTER-FED INDUCTION MOTORS WITH CONTROLLED STATOR CURRENT

G. Pfaff

Inst. f. Elektrische Antriebe u. Steuerungen, Universität Erlangen, FRG

Summary

An important operating mode of inverter-fed induction motors is attended with impressed stator currents. The stator currents are imposed on the motor either by feedback control or by using a current source inverter. If assuming the stator currents as input variables of the motor, the dynamic behaviour is described by a third order system of nonlinear differential equations. It is shown how suited transformation techniques have been used to minimize the number of system parameters. By this way it is possible to investigate the dynamic behaviour of the motor in the time domain and to calculate transients, valid not only for a special case but generally valid. As an example the method is applied to analyze the control performance of the current-fed induction motor and to compare two different control strategies. Subsequently a sensitivity analysis of the system is carried out with regard to variations of the remainded parameters. The sensitivity equations of the system are formulated in the time domain. An interesting problem in this connexion is the influence of variations in rotor-resistance, which is dependent on the operating temperature of the machine.

INTRODUCTION

Probably the most important type of converter-fed induction motor control is an operating condition with impressed stator currents. In this case the currents are imposed to the stator of the motor by means of feedback control with a pulsewidth modulated inverter. Another possibility to realize impressed stator currents is to supply the motor from a current source inverter with a large series reactor in his dc-link. In a first approach the stator currents in that case may be interpreted as independent variables of the considered system and the equations of the induction motor are reduced in order. For the investigation remains a third order system of differential equations which is very much nonlinear due to several products of system variables. Therefore it is not possible to obtain a closed solution of these equations.

One method available to study the transient performance of electrical drives in such cases is small signal linearization. Each variable is considered to be composed of a steady-state and a small time-varying component. In this form the variables are substituted into the original nonlinear equations. All purely steady-state terms drop out and all second order perturbations are ignored. The remaining equations are linear in the perturbation variables and so it is possible to determine transfer functions of the system and to study the dynamic response in the frequency domain. There is the feasibility to investigate the pole-zero configurations of the transfer function in the s-plane /Ref. 3/, or an alternative is the projection of Bode plots for several operating points /Ref. 4/.

This paper describes a method to analyze the transient performance of current-fed induction motors in the time domain with regard to the existing nonlinearities. Therefore solutions of the system equations are only possible by means of analog or digital computers. The difficulty, that these solutions are only valid for the special combination of the applied parameters, is to get over by suited transformations, which reduce drastically the number of parameters. In this manner the computer results obtain an universally valid character.

SYSTEM ANALYZED

It is well known that machine dynamic behaviour regarding torque and speed control is determined primarily by the fundamental components of excitation variables. The idealized model used for this analysis contains the conventional assumptions as sinusoidal distribution of MMF in the airgap, negligible saturation and negligible iron losses. Furthermore sinusoidal stator currents are assumed but this assumption is not indispensable. The model uses a reference frame that rotates around the airgap in synchronism with the stator MMF at a speed corresponding to stator excitation frequency. If all harmonics are neglected, then the machine variables become constants in the synchronously rotating reference during steady state operation. We suppose that the reference frame α-axis and motor phase a-axis are aligned at $t=0$, and also the current maximum in phase a occurs at $t=0$.

The stator currents in the three phases of the motor are

$$i_a = \hat{i}_1(t)\cos(\lambda(t) + \delta(t))$$
$$i_b = \hat{i}_1(t)\cos(\lambda(t) + \delta(t) - \frac{2\pi}{3}) \quad (1)$$
$$i_c = \hat{i}_1(t)\cos(\lambda(t) + \delta(t) + \frac{2\pi}{3})$$

Here describes $\lambda(t)$ the angular frequency of the stator current system

$$\omega_1(t) = \frac{d\lambda}{dt} \quad (2)$$

while $\delta(t)$ serves for the representation of steps in phase shift. The components of the resulting space vector /Ref. 1/ of stator current in the α- and β-axis of the rotating reference frame are

$$i_{1\alpha} = \hat{i}_1\cos\delta$$
$$i_{1\beta} = \hat{i}_1\sin\delta \quad (3)$$

The equations of the current-fed induction motor, using per unit values and state space notation, are:

$$\frac{d\psi_{2\alpha}}{dt} = -\frac{R_2}{X_2}\psi_{2\alpha} + \psi_{2\beta}(\omega_1 - n) + R_2\frac{X_h}{X_2}i_{1\alpha}$$

$$\frac{d\psi_{2\beta}}{dt} = -\frac{R_2}{X_2}\psi_{2\beta} - \psi_{2\alpha}(\omega_1 - n) + R_2\frac{X_h}{X_2}i_{1\beta} \quad (4)$$

$$\frac{dn}{dt} = \frac{1}{T_A}\left[\frac{X_h}{X_2}(\psi_{2\alpha}i_{1\beta} - \psi_{2\beta}i_{1\alpha}) - d_L\right]$$

The electromagnetic torque d_M of the motor is given by

$$d_M = \frac{X_h}{X_2}(\psi_{2\alpha}i_{1\beta} - \psi_{2\beta}i_{1\alpha}) \quad (5)$$

In these equations the parameters are the rotor resistance R_2, the mutual reactance X_h and the self-reactance X_2 of the rotor to base frequency. T_A is the inertia constant of the system in per unit. $\psi_{2\alpha}$ and $\psi_{2\beta}$ are the flux linkages of the rotor in the two axes of the reference frame, d_L is the load torque and n means the rotor speed. In per-unit values the relation between angular frequency ω_1 of the stator currents and angular frequency ω_2 in the rotor is given by

$$\omega_1 - n = \omega_2 \quad (6)$$

REDUCTION OF PARAMETERS

In order to reduce the number of system parameters the following new variables are introduced:

$$\psi^*_{2\alpha} = \frac{R_2}{X_2}\psi_{2\alpha} \qquad \omega^* = \frac{X_2}{R_2}\omega$$

$$\psi^*_{2\beta} = \frac{R_2}{X_2}\psi_{2\beta} \qquad n^* = \frac{X_2}{R_2}n$$

$$i^*_{1\alpha} = \frac{X_h}{X_2}R_2\, i_{1\alpha} \qquad d^*_M = \frac{R_2^2}{X_2}d_M$$

$$i^*_{1\beta} = \frac{X_h}{X_2}R_2\, i_{1\beta} \qquad d^*_L = \frac{R_2^2}{X_2}d_L$$

$$t^* = \frac{R_2}{X_2}t$$

These new variables are substituted into the equations of the former section:

$$\frac{d\psi^*_{2\alpha}}{dt^*} = -\psi^*_{2\alpha} + \psi^*_{2\beta}\omega^*_2 + i^*_{1\alpha}$$

$$\frac{d\psi^*_{2\beta}}{dt^*} = -\psi^*_{2\beta} - \psi^*_{2\alpha}\omega^*_2 + i^*_{1\beta}$$

$$\frac{dn^*}{dt^*} = P(\psi^*_{2\alpha}i^*_{1\beta} - \psi^*_{2\beta}i^*_{1\alpha} - d^*_L) \quad (7)$$

$$d^*_M = \psi^*_{2\alpha}i^*_{1\beta} - \psi^*_{2\beta}i^*_{1\alpha}$$

The only one remaining parameter in the system is

$$P = \frac{X_2^3}{R_2^4 T_A} \quad (8)$$

Fig. 1 shows the dynamic structure of the system in a block diagram.

Now it is possible to solve the equations with various values of the parameter P by an analog or digital computer and to obtain a general view of the dynamic performance of current-fed induction motors. So two different control strategies will be analyzed and compared in the next section of the paper. Another problem that could be treated in this way is the computation of torque and speed at rectangular current excitation and other wave forms of stator currents.

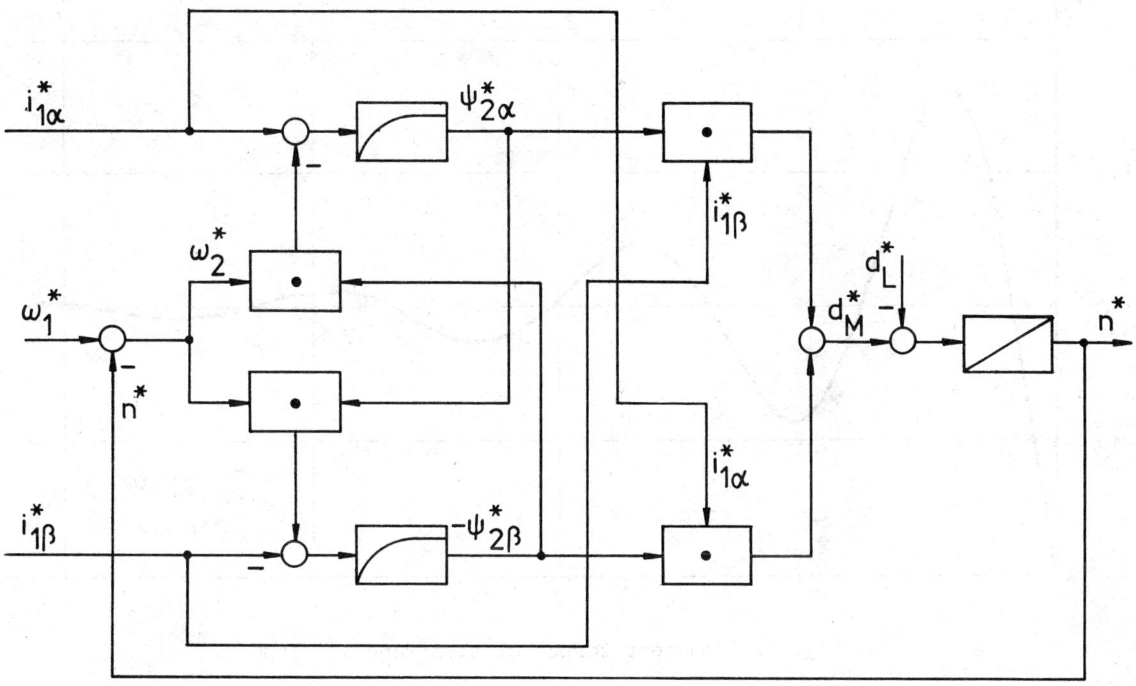

Fig. 1. Block diagram of current-fed induction motor

METHODS OF SPEED CONTROL

Considering the equations (4) or (7) it can be noted that the magnetic flux and the torque of the current-fed induction motor are dependent only on the slip frequency ω_2 and the components of the stator current vector. They are independent of the stator frequency ω_1. Tprque regulation is required to achieve the desired motor speed, beyond it the method must involve regulation of the flux level in the machine. The flux in the motor can be indirectly determined by controlling the stator current and the rotor slip frequency ω_2. When constant-flux operation is required, prescribed values of stator current and rotor frequency must always be suitably related. In a closed-loop control system the actual motor speed is compared with the desired speed. The magnitude and polarity of the speed error are used to determine the rotor frequency. This demanded rotor-frequency ω_2 is added to, or subtracted from, the measured rotational frequency n in order to determine the stator frequency ω_1, which must be supplied to the motor. A slip limit is incorporated in the control circuit so that the demanded rotor frequency can never exceed the breakdown value in the motoring or generating regions. In the so-called controlled-slip drive the stator current must be suitably related to the rotor frequency in order to maintain constant flux in the machine. Two different functions of stator current are investigated in the following.

The first method demands constant flux for only stationary operation of the motor. The stator current is influenced only in magnitude not in phase. /Ref. 5 and 4/ The magnitude of the rotor flux is

$$\psi_2 = \sqrt{\psi_{2\alpha}^2 + \psi_{2\beta}^2} \stackrel{!}{=} \text{const.} \qquad (9)$$

and control of stator current only in magnitude is described by

$$i_{1\beta} = 0. \qquad (10)$$

With these assumptions the equations (4) for stationary operating conditions yield the control strategy

$$i_{1\alpha} = I_{1\alpha o}\sqrt{1 + \frac{X_2^2}{R_2^2}\,\omega_2^2} \qquad (11)$$

The constant value $I_{1\alpha o}$ is the magnetizing current of the motor in no-load operation. In the parameter-free system equation (11) results in

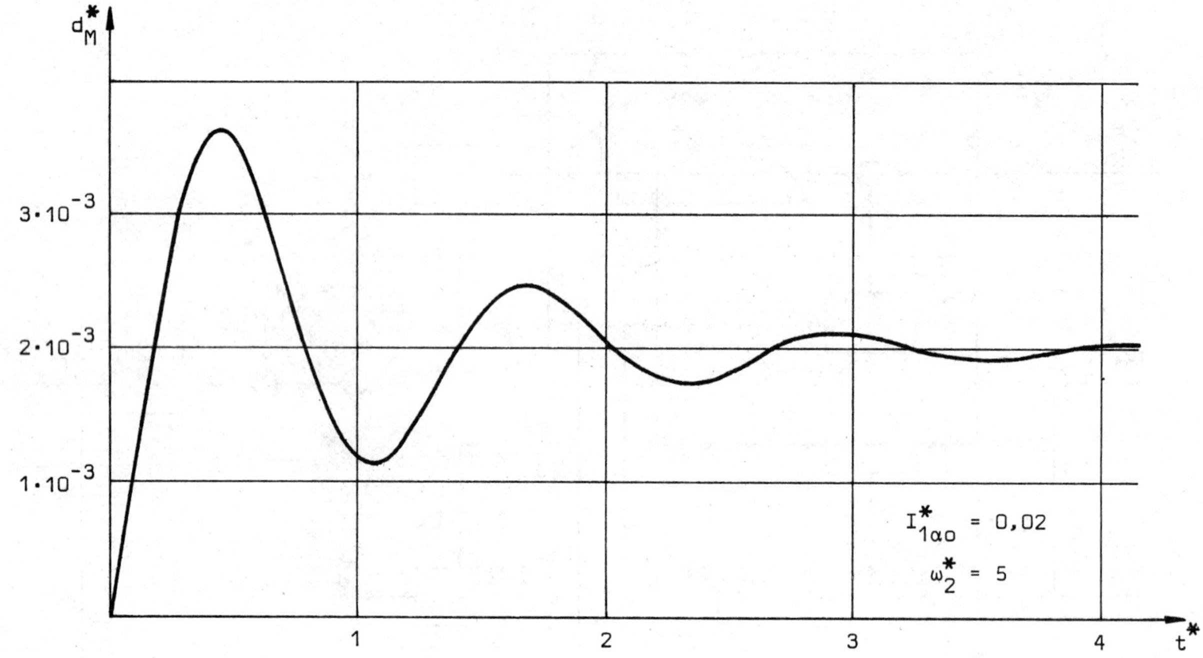

Fig. 2. Transient torque at step-function input ω_2^*

$$i_{1\alpha}^* = I_{1\alpha o}^* \sqrt{1 + \omega_2^{*2}} \qquad (12)$$

Now the only input variable of the system is the rotor angular frequency ω_2^* and the interesting output variable is primarily the motor torque d_M^*. Fig. 2. shows the torque response of the system at step-function input. The transient response of flux magnitude given by equation (9) at the same input is demonstrated in Fig. 3. For constant values of ω_2^* the model contains a linear second order system as can be seen in the block diagram of Fig. 1. The damping coefficient and the time constant of this system depend only on ω_2^* and are given by

$$d = \sqrt{1 + \omega^{*2}} \qquad T = \sqrt{1 + \omega^{*2}} \qquad (13)$$

On the basis of these results the design of the closed loop control system for speed regulation can be completed and analyzed by computer simulation in the usual range of ω_2^*.

Another concept of torque and speed control which was first explained in /Ref. 6/ demands constant rotor flux for all operating conditions. To meet this requirement the stator current is directed not only in magnitude but also in phase. With

$$\frac{d\psi_{2\alpha}}{dt} = \frac{d\psi_{2\beta}}{dt} = 0 \qquad (14)$$

and

$$\Psi_{2\alpha o} = X_h\, I_{1\alpha o} = \text{const}, \quad \Psi_{2\beta o} = 0 \qquad (15)$$

equations (4) result in the control law for this operating mode:

$$i_{1\beta} = \frac{X_2}{R_2}\, I_{1\alpha o}\, \omega_2 \qquad (16)$$

Here is also $I_{1\alpha o}$ the magnetizing current of the motor in no-load operation. In the parameter-free system the above control strategy is described by

$$i_{1\beta}^* = I_{1\alpha o}^*\, \omega_2^* \qquad (17)$$

It can easily be shown that in this case the torque follows directly the input signal ω_2^* without time lag. The rotor flux remains constant here in contrast to Fig. 3. In this case the design of closed-loop control system for speed control can easily be done. An integral plus proportional controller is used to give a good speed response to reference input and to disturbance input d_L^*. The parameters of the controller are always adapted to the system parameter P according to equation (8) and so we obtain a uniform transient response of the speed control, which is demonstrated in Fig. 4. and 5. The reference input signal is provided with a time lag in order to avoid overshoot in motor speed.

It is seen that in the second case for control strategy II the dynamic behaviour of the system is better than in the first case. On the other hand the type of current processing and signal processing is more complicated. /Ref. 7/ But on principle for both

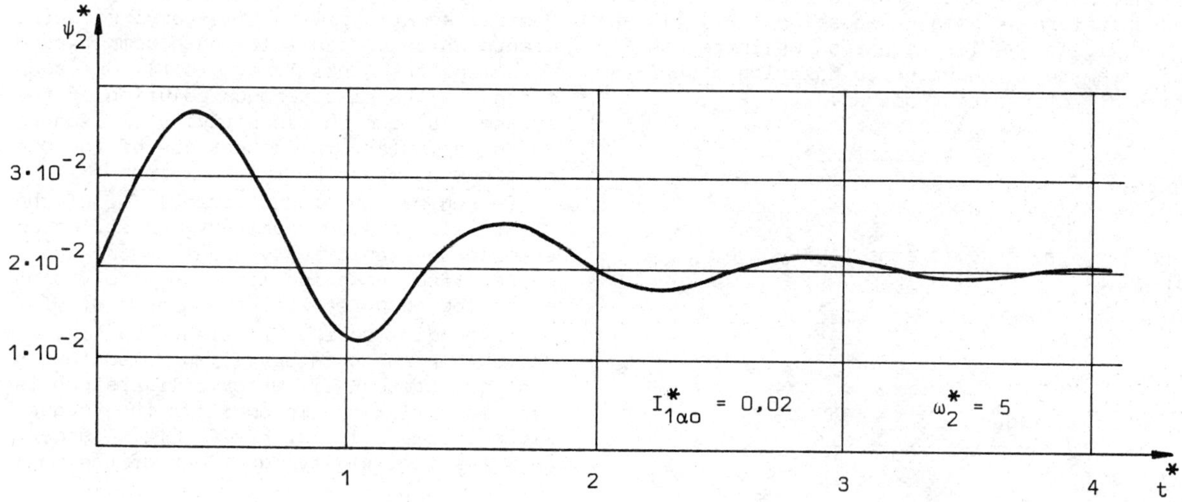

Fig. 3. Transient flux at step-function input ω_2^*

concepts the control strategy is adapted to the special value of rotor resistance R_2. The influence of mismatching caused by variations of R_2 with temperature will be investigated in the next section.

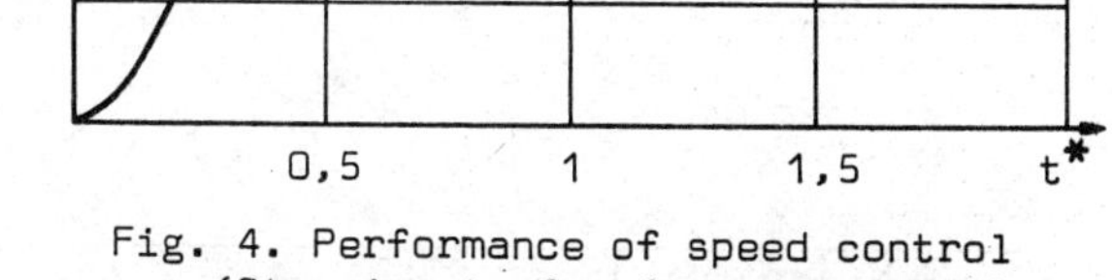

Fig. 4. Performance of speed control (Step-input of reference signal)

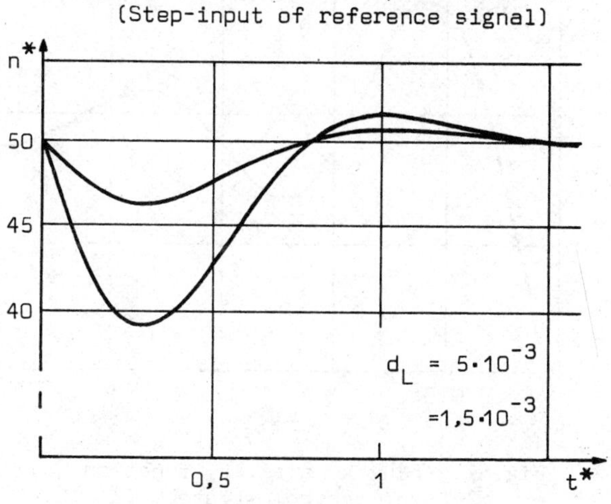

Fig. 5. Performance of speed control (Step-input of load torque)

SENSITIVITY ANALYSIS

Much of the early work about sensitivity in control theory was concerned with the sensitivity of systems to changes in plant parameters. We consider the system equation

$$\frac{d\underline{x}}{dt} = \dot{\underline{x}} = \underline{f}\left[\underline{x}(t),\underline{u}(t),\underline{p},t\right] \tag{18}$$

where $\underline{x}(t)$ is an n-dimensional state vector, $\underline{u}(t)$ is the control vector and $\underline{p}$ is an r-dimensional parameter vector. We define the dynamic sensitivity coefficient as the change in the state variable $\underline{x}$ due to variations in one of the parameters p. For small changes in $\underline{p}$, a first-order approximation for the corresponding change in the state vector $\underline{x}$ is

$$\underline{\Delta x} = \sum_{j=1}^{r} \underline{v}_j \, \Delta p_j \tag{19}$$

where $\underline{v}_j$ is the sensitivity vector

$$\underline{v}_j = \left[\frac{\partial x_1}{\partial p_j} \; \cdots \; \frac{\partial x_n}{\partial p_j}\right]^T \tag{20}$$

By partial differentiation of equation (18) with regard of one parameter p_j we obtain the sensitivity equation of the system

$$\dot{\underline{v}}_j = \frac{\partial \underline{f}}{\partial \underline{x}} \cdot \underline{v}_j + \frac{\partial \underline{f}}{\partial p_j} \tag{21}$$

Now we will employ these considerations for the investigation of rotor resistance variations. First we study the control strategy I and therefore we combine equations (7), (10) and (12). In the last equation we introduce the parameter F in order to describe a variation of rotor resistance R_2.

$$i^*_{1\alpha} = I^*_{1\alpha o}\sqrt{1 + F^2\omega_2^{*2}} \qquad (22)$$

We define the dynamic sensitivity coefficients

$$v_1 = \frac{\partial\psi^*_{2\alpha}}{\partial F} \qquad v_2 = \frac{\partial\psi^*_{2\beta}}{\partial F}$$
$$v_3 = \frac{\partial d^*_M}{\partial F} \qquad v_4 = \frac{\partial\psi^*_2}{\partial F} \qquad (23)$$

and obtain the following sensitivity equations:

$$\dot v_1 = -v_1 + \omega^*_2 v_2 + \frac{I^*_{1\alpha o}\, F\, \omega_2^{*2}}{\sqrt{1 + F^2\omega_2^{*2}}}$$
$$\dot v_2 = -v_2 - \omega^*_2 v_1 \qquad (24)$$

Analogous we obtain from equations (7.4) and (9)

$$v_3 = -I^*_{1\alpha o}\sqrt{1+F^2\omega_2^{*2}}\cdot v_2 - I^*_{1\alpha o}\frac{F\,\omega_2^{*2}}{\sqrt{1+F^2\omega_2^{*2}}}\cdot\psi^*_{2\beta}$$
$$v_4 = \frac{1}{\psi^*_2}(\psi^*_{2\alpha}v_1 + \psi^*_{2\beta}v_2) \qquad (25)$$

In the second case for control strategy II we employ in place of equation (22) in accordance with equation (17)

$$i^*_{1\beta} = FI^*_{1\alpha o}\,\omega^*_2 \qquad (26)$$

With the same definitions for the sensitivity coefficients as above we obtain for this case the following sensitivity equations

$$\dot v_1 = -v_1 + \omega^*_2 v_2$$
$$\dot v_2 = -v_2 - \omega^*_2 v_1 + I^*_{1\alpha o}\omega^*_2$$
$$v_3 = I^*_{1\alpha o}\left[\omega^*_2(Fv_1 + \psi^*_{2\alpha}) - v_2\right] \qquad (27)$$
$$v_4 = \frac{1}{\psi^*_2}(\psi^*_{2\alpha}v_1 + \psi^*_{2\beta}v_2)$$

The sensitivity equations are linear differential equations with time-varying coefficients which we can solve on a computer to determine the sensitivity v(p,t). One such method is the simultaneous solution of the system (18) and the sensitivity (21) on an analog computer. This makes use of the connections between the two equations.

In general the input variable ω^*_2 of the plant will have an arbitrary course. As an example the sensitivity equations of the current-fed induction motor were solved on an analog computer with step input of ω^*_2 corresponding to the transients in Fig. 2 and 3 (control strategy I) and also for the control strategy II, which delivers constant dynamical flux and in addition the torque without time delay at F = 1. Fig. 6 and 7 show the sensitivity functions of the mag-

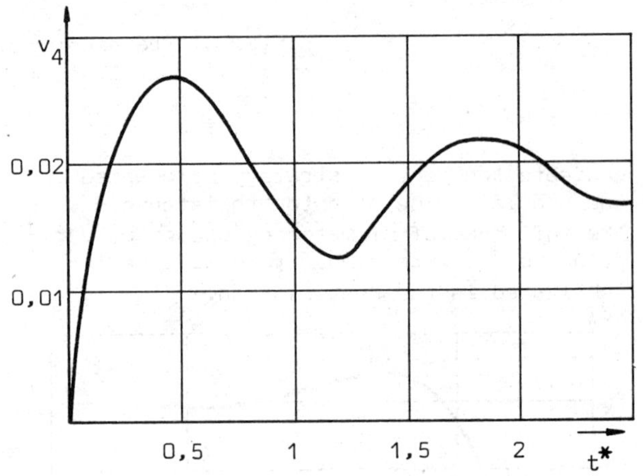

Fig. 6. Flux sensitivity function for control strategy I

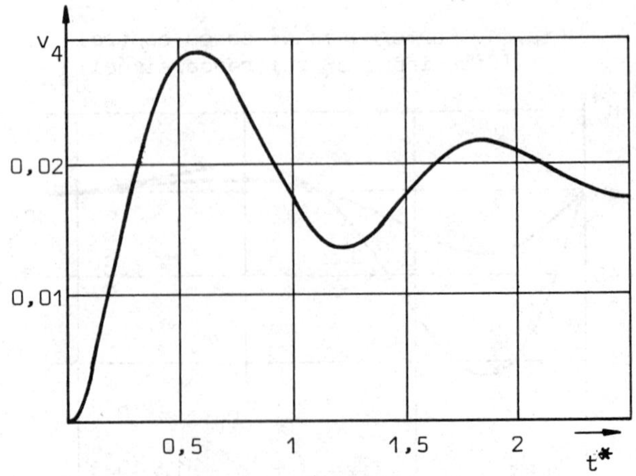

Fig. 7. Flux sensitivity function for control strategy II

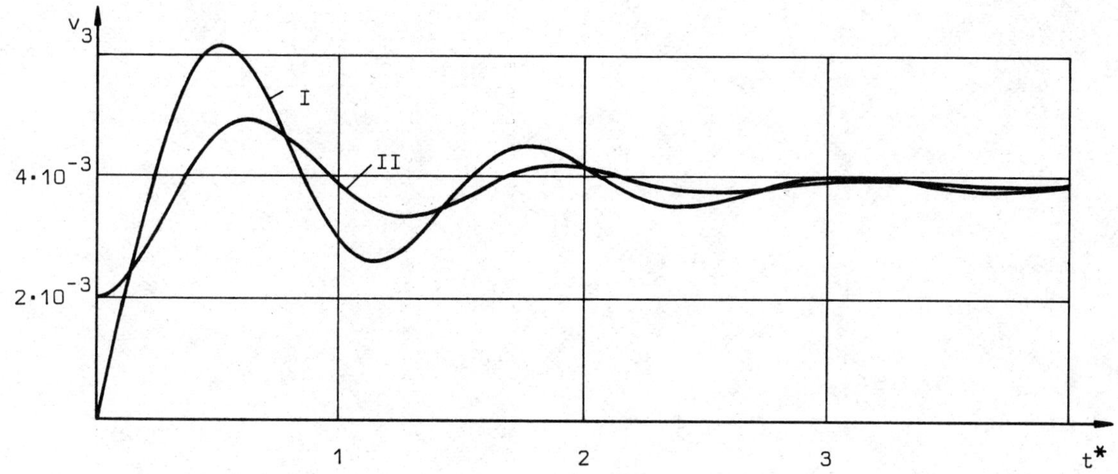

Fig. 8. Torque sensitivity functions for control strategy I and II

netic flux for both control strategies due to changes of the parameter F during the step response of the motor system. A greater difference between the curves is observed for the sensitivity functions of torque which are demonstrated in Fig. 8. The stationary final value must be the same in both cases. In the same way it is possible to calculate sensitivity functions of the current-fed induction motor for arbitrary courses of ω_2^* as they occur in the closed-loop control system.

REFERENCES

/1/ Kovacs, K.P. (1959) Transiente Vorgänge in Wechselstrommaschinen
Racz, J.
Verlag d. Ungar. Akademie d. Wissenschaften, Budapest

/2/ Pfaff, G. (1971) Regelung elektrischer Antriebe I
R. Oldenbourg, München

/3/ E.P. Cornell Design of controlled current ac drive systems using transfer function techniques
T.A. Lipo
Symposium "Control in power electronics and electrical drives" 1, 133 (1974)

/4/ H. Bühler Stromrichtergespeiste Drehstrom- und Gleichstromantriebe
ibid., Survey Papers 63 (1974)

/5/ A. Schönung Geregelter Drehstrom-Umkehrantrieb mit gesteuertem Umrichter nach dem Unterschwingungsverfahren
H. Stemmler
BBC-Mitt. 51, 555 (1964)

/6/ K. Hasse Zur Dynamik drehzahlgeregelter Antriebe mit stromrichtergespeisten Asynchron-Kurzschlußläufermaschinen
Dissertation TH Darmstadt (1969)

/7/ R. Jötten Signalverarbeitung für die Regelung umrichtergespeister Drehfeldmaschinen
Symposium "Control in power electronics and electrical drives" 1, 681 (1974)

/8/ A.B. Plunkett New methods of induction motor torque regulation
T.A. Lipo
IEEE Trans. Industry Appl. 12, 47 (1976)

COMPUTATION OF OPTIMUM CONTROL FUNCTIONS FOR TRANSIENTS OF INDUCTION MOTORS

C. Meier

Inst. f. Elektrische Antriebe u. Steuerungen, Universität Erlangen, FRG

Summary- In this paper, there is presented the strategy for computation of optimum control functions to obtain an optimum transient behaviour of the induction motor. At first, there is described the principle of the iteration process basing on the Maximum principle of Pontrjagin and the gradient technique. After the preparation of the equations there is presented the computation of the optimum control functions for time optimum acceleration processes of the speed. This optimization is made with consideration to necessary limitations of state variables in the machine. In the examples the good convergence of the iteration process is demonstrated.

INTRODUCTION

The motion of the induction motor at any speed and good dynamics requires a supply with a three phase voltage system with variable frequency and amplitude. The right temporal guide of the control variables, the frequency $\omega(t)$ and the amplitude $\hat{u}_1(t)$ of the voltage system is of great significance to the quality of the transient process.

The quality of a transient process is measured by a cost function, where the requirements on the transient process are defined and rated. The transient process, which performs these requirements in the best manner, is the optimum process. The cost function will become extremely. The control functions $\omega^*(t)$ and $\hat{u}_1^*(t)$ belonging to this optimum process are the optimum control functions we searched.

In this paper, there is described a method, which allows the computation of the guide of the control variables to give the transient process of the induction motor an optimum behaviour, measured by a cost function.

THE MOTOR - EQUATIONS

For the description of the induction motor we take the Space - vector - description in the per - unit system /Ref. 1/. The necessary assumptions associated with this idealized model are: sinusoided voltage and current exitation, sinusoidal distribution of MMF in the airgap, negligible saturation and iron losses.

The orientation of the space vectors is given in a reference frame with the α- and β-axis. This system rotates in synchronism with the space vector of the stator voltage with the speed or frequency $\omega(t)$ /Ref. 1/. The space vector of the stator voltage is fixed in the direction of the α-axis. The amplitude $\hat{u}_1(t)$ and the frequency $\omega(t)$ of the symmetrical three phase supply may be movable continuous in the operating range and there may be no load torque. Now we can describe the induction motor with the system of equations (1):

$$\frac{d}{dt}\Psi_{1\alpha} = -\frac{R_1}{\sigma X_1}\Psi_{1\alpha} + \frac{R_1 X_h}{\sigma X_1 X_2}\Psi_{2\alpha} + \omega\Psi_{1\beta} + \hat{u}_1$$

$$\frac{d}{dt}\Psi_{1\beta} = -\frac{R_1}{\sigma X_1}\Psi_{1\beta} + \frac{R_1 X_h}{\sigma X_1 X_2}\Psi_{2\beta} - \omega\Psi_{1\alpha}$$

$$\frac{d}{dt}\Psi_{2\alpha} = -\frac{R_2}{\sigma X_2}\Psi_{2\alpha} + \frac{R_2 X_h}{\sigma X_2 X_1}\Psi_{1\alpha} - (n-\omega)\Psi_{2\beta}$$

$$\frac{d}{dt}\Psi_{2\beta} = -\frac{R_2}{\sigma X_2}\Psi_{2\beta} + \frac{R_2 X_h}{\sigma X_2 X_1}\Psi_{1\beta} + (n-\omega)\Psi_{2\alpha}$$

$$\frac{d}{dt} n = \frac{X_h}{T_A \sigma X_{1.2}}(\Psi_{1\beta}\Psi_{2\alpha} - \Psi_{1\alpha}\Psi_{2\beta}) = \frac{d_M}{T_A} \qquad (1)$$

THE STRATEGY OF OPTIMIZATION

For realization of a compact notation, which is used in publications above optimization - techniques /Ref. 2,3,4,5/ too, the following expressions are introduced:

$$\text{state vector} \quad \underline{x}(t) = \begin{pmatrix} \Psi_{1\alpha}(t) \\ \Psi_{1\beta}(t) \\ \Psi_{2\alpha}(t) \\ \Psi_{2\beta}(t) \\ n(t) \end{pmatrix} \tag{2}$$

$$\text{control vector} \quad \underline{u}(t) = \begin{pmatrix} \hat{u}_1(t) \\ \omega(t) \end{pmatrix} \tag{3}$$

So we can write the equation (1):

$$\text{system equation} \quad \dot{\underline{x}}(t) = \underline{f}(\underline{x}(t),\underline{u}(t)) \tag{4}$$

$$\text{initial condition} \quad \underline{x}(0) = \underline{x}_0 \tag{4a}$$

We can do without the formulation of end-conditions in the time t_1. (There are indications for the treatment of such requirements by Sage /2/.

Now we write down the cost function

$$J = \int_0^{t_1} \Phi(\underline{x}(t),\underline{u}(t))dt \tag{5}$$

This functional describes the requirements, which are postulated at the transient behaviour in the whole interval of the time t: (0, t_1). (t_1 is the fixed end - time). (A quality criterion $\Theta(\underline{x}(t_1),t_1)$ which estimates only the terminal state at t_1 isn't interesting here).

Now we wish to find an optimum control function $\underline{u}^*(t)$ so that the cost function J becomes a minimum. With the Maximumprinciple of Pontrjagin, there is shown a way, which makes possible the computation of optimum control functions. /Ref. 2-6/. Let us define the scalar Hamiltonian, using equation (4) and (5) introducing the multiplier $\underline{\lambda}(t)$:

$$H = \Phi(\underline{x}(t),\underline{u}(t)) + \underline{\lambda}^T(t)\underline{f}(\underline{x}(t),\underline{u}(t)) \tag{6}$$

Now the Maximumprinciple of Pontrjagin declares, that the Hamiltonian will become a minimum, if we use the optimum control function $\underline{u}^*(t)$:

$$H(\underline{x}(t),\underline{u}^*(t),\underline{\lambda}(t)) < H(\underline{x}(t),\underline{u}(t),\underline{\lambda}(t)) \tag{7}$$

But using this statement a direct computation (as shown for linear 2nd-order systems in /Ref. 4/) of the optimum control function $\underline{u}^*(t)$ is not possible in the case of the large non-linear system of the induction machine.

With the evaluation of the requirement, that the cost function J (5) falls to a minimum in the optimum, we will get the canonic system of equations:

$$\text{system} \quad \dot{\underline{x}}(t) = \frac{\partial H(\underline{x}(t),\underline{u}(t),\underline{\lambda}(t))}{\partial \underline{\lambda}(t)} = \underline{f}(\underline{x}(t),\underline{u}(t)) \tag{4}$$

$$\text{adjoint system} \quad -\dot{\underline{\lambda}}(t) = \frac{\partial H(\underline{x}(t),\underline{u}(t),\underline{\lambda}(t))}{\partial \underline{x}(t)} \tag{8}$$

$$\underline{0} = \left.\frac{\partial H(\underline{x}(t),\underline{u}(t),\underline{\lambda}(t))}{\partial \underline{u}(t)}\right|_{\underline{u}(t)=\underline{u}^*(t)} \tag{9}$$

Because we can't solve this equations (4), (8) and (9) in a close form or in one step, we must try to find an iteration method, which searches the optimum control function $\underline{u}^*(t)$ and fulfils these equations.

To solve this problem, we use now the gradient-technique /Ref. 2,3/. In this technique, we solve the equations (4) and (8) exactly in each step of the iteration. So we can observe the state variables $\underline{x}(t)$ and the convergence of the computation very good. Equation (9) will be satisfied after convergence of the computation.

At first we estimate a control function $\underline{u}^o(t)$. Then we compute with this function the state variables $\underline{x}^o(t)$ and the adjoint system $\underline{\lambda}^o(t)$. With this functions we get the first value of the cost function J^o. With the gradient technique we can improve the control function $\underline{u}(t)$ by $\Delta\underline{u}(t)$ so that we will get a smaller value of J after a new computation of $\underline{x}(t)$. J will become smaller with a maximum decrease in each step, if we choose

$$\Delta\underline{u}(t) = - K(t) \frac{\partial H(\underline{x}(t),\underline{u}(t),\underline{\lambda}(t))}{\partial \underline{u}(t)} \tag{10}$$

with $K(t) > 0$.
The control function for the step (N + 1) is found then:

$$\underline{u}^{N+1}(t) = \underline{u}^N(t) + \Delta\underline{u}^N(t) \tag{11}$$

With K(t), for example $K(t) = \text{const} > 0$ we can control the quantity of the iteration step.

For the computation of the iteration process a hybrid computer is very suitable. With this computer it is possible to compute the system and adjoint equations very fast at the analogue computer. We need the digital computer to store the functions and to compute the control functions in each step of iteration. It is not possible to use a particular analogue computer for the whole computation.

We must observe a characteristic at the computation of the adjoint system $\underline{\lambda}(t)$. We compute $\underline{\lambda}(t)$ not in the normal direction $0 \to t_1$, because the adjoint system (8) is then an unstable system at the analogue computer. If we compute with $\tau = t_1 - t$, in backward direction of t: $t_1 \to 0$, we will get with

$$-\dot{\underline{\lambda}}(t) = +\frac{d}{d\tau}\underline{\lambda}(\tau)$$

a stable system at the computer. The initial values for this computation are the values $\underline{\lambda}(t_1)$. The time direction of the computation is determined by the direction of the storage of the functions at the digital computer.

TIME OPTIMUM CONTROL OF THE SPEED

Now it shall be presented the computation of the optimum control functions $\hat{u}_1^*(t)$ and $\omega^*(t)$ to get a time optimum transient behaviour. There are the following requirements at the transient processes:

We will obtain a time optimum transient process of the speed n(t) from the initial condition n_0 to the endspeed n_1. (R1)

When the endspeed n_1 is reached, we demand: $\frac{d}{dt}\, n(t) = 0$ that means steady state with this speed n_1. (R2)

The magnitude of the rotor flux shall be bounded by the value of the nominal rotor flux all the time: $\hat{\Psi}_2(t) \approx \hat{\Psi}_{2N}$. This requirement is very important and it is necessary for the limitation of the state variables. (In all optimizations we must take care of any boundary conditions.) (R3)

At first it must be formulated the cost function. We will compute with a fixed time interval $(0,t_1)$. If we find a suitable formulation of the cost function which satisfies the requirements (R1), (R2) and (R3) we need no end conditions in t_1 (i.e. $\underline{N}(\underline{x}(t_1),t_1)=\underline{0}$) which are not easy to verify in this form and we need no further boundary conditions.

We can write a cost function which satisfies these requirements:

$$J = \int_0^{t_1} \Big[K_1(n(t) - n_1)^2 + K_2(d_M(t) - d_{Ms}(t))^2 + K_3(\hat{\Psi}_2^2(t) - \hat{\Psi}_{2N}^2)^2 \Big] dt \qquad (12)$$

with $\hat{\Psi}_2^2(t) = \Psi_{2\alpha}^2(t) + \Psi_{2\beta}^2(t)$.

The part $K_1(n(t) - n_1)^2$ is the essential part of the cost function. With this function the iteration runs fast into the proximity of the optimum (with the condition, that the starting functions $\hat{u}_1^0(t)$ and $\omega^0(t)$ are not very "bad").

The part $K_2(d_M(t) - d_{Ms}(t))^2$ improves the convergence. This part is of great importance for the movement into the steady state with $n(t) = n_1$. $d_{Ms}(t)$ is a function which depends on the time, shown in Fig. 1. The maximum of this function is given by the maximum of the admissible torque of the motor.

The part $K_3(\hat{\Psi}_2^2 - \hat{\Psi}_{2N}^2)^2$ performs the limitation of the magnitude of the rotor flux.

With the factors K_1, K_2 and K_3 the parts are weighted and the influence of each part is defined. These factors may be varied during the iteration process and they may depend on the time. To find the right relative size of these factors, there we need some test computations.

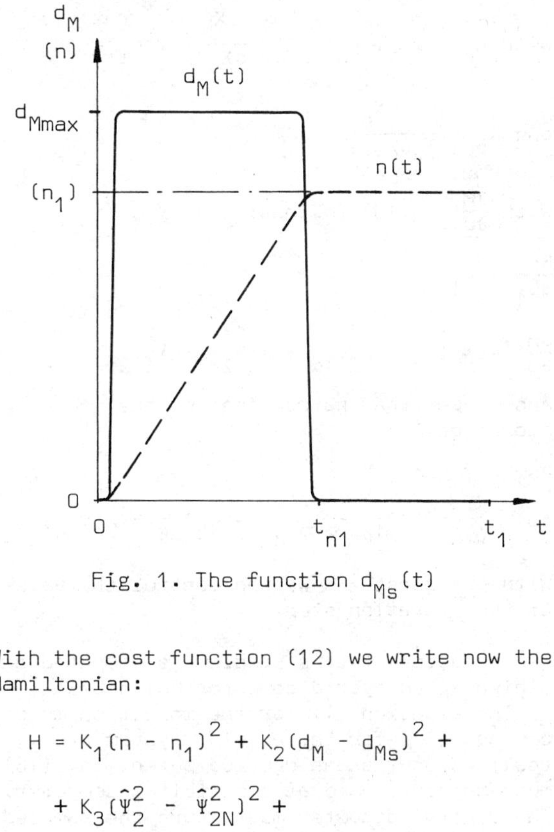

Fig. 1. The function $d_{Ms}(t)$

With the cost function (12) we write now the Hamiltonian:

$$\begin{aligned} H = {} & K_1(n - n_1)^2 + K_2(d_M - d_{Ms})^2 + \\ & + K_3(\hat{\Psi}_2^2 - \hat{\Psi}_{2N}^2)^2 + \\ & + \lambda_1\Big(-\frac{R_1}{\sigma X_1}\Psi_{1\alpha} + \frac{R_1 X_h}{\sigma X_1 X_2}\Psi_{2\alpha} + \omega\Psi_{1\beta} + \hat{u}_1\Big) \\ & + \lambda_2\Big(-\frac{R_1}{\sigma X_1}\Psi_{1\beta} + \frac{R_1 X_h}{\sigma X_1 X_2}\Psi_{2\beta} - \omega\Psi_{1\alpha}\Big) \\ & + \lambda_3\Big(-\frac{R_2}{\sigma X_2}\Psi_{2\alpha} + \frac{R_2 X_h}{\sigma X_2 X_1}\Psi_{1\alpha} - (n - \omega)\Psi_{2\beta}\Big) \\ & + \lambda_4\Big(-\frac{R_2}{\sigma X_2}\Psi_{2\beta} + \frac{R_2 X_h}{\sigma X_2 X_1}\Psi_{1\beta} + (n - \omega)\Psi_{2\alpha}\Big) \\ & + \lambda_5 \frac{X_h}{T_A \sigma X_1 X_2}(\Psi_{1\beta}\Psi_{2\alpha} - \Psi_{1\alpha}\Psi_{2\beta}) \end{aligned} \qquad (13)$$

and the adjoint system:

$$\frac{d}{d\tau}\lambda_1 = \frac{\partial H}{\partial \Psi_{1\alpha}} = -K_d(d_M-d_{Ms})\Psi_{2\beta} - \frac{R_1}{\sigma X_1}\lambda_1 - \omega\lambda_2 + \frac{R_2 X_h}{\sigma X_2 X_1}\lambda_3 - K_5\Psi_{2\beta}\lambda_5$$

$$\frac{d}{d\tau}\lambda_2 = \frac{\partial H}{\partial \Psi_{1\beta}} = K_d(d_M-d_{Ms})\Psi_{2\alpha} + \omega\lambda_1 - \frac{R_1}{\sigma X_1}\lambda_2 + \frac{R_2 X_h}{\sigma X_2 X_1}\lambda_4 + K_5\Psi_{2\alpha}\lambda_5$$

$$\frac{d}{d\tau}\lambda_3 = \frac{\partial H}{\partial \Psi_{2\alpha}} = K_d(d_M-d_{Ms})\Psi_{1\beta} + K_\Psi(\hat{\Psi}_2^2-\hat{\Psi}_{2N}^2)\Psi_{2\alpha} + \frac{R_1 X_h}{\sigma X_1 X_2}\lambda_1 - \frac{R_2}{\sigma X_2}\lambda_3 + (n-\omega)\lambda_4 + K_5\Psi_{1\beta}\lambda_5 \quad (14)$$

$$\frac{d}{d\tau}\lambda_4 = \frac{\partial H}{\partial \Psi_{2\beta}} = -K_d(d_M-d_{Ms})\Psi_{1\alpha} + K_\Psi(\hat{\Psi}_2^2-\hat{\Psi}_{2N}^2)\Psi_{2\beta} + \frac{R_1 X_h}{\sigma X_1 X_2}\lambda_2 - (n-\omega)\lambda_3 - \frac{R_2}{\sigma X_2}\lambda_4 - K_5\Psi_{1\alpha}\lambda_5$$

$$\frac{d}{d\tau}\lambda_5 = \frac{\partial H}{\partial n} = K_n(n - n_1) - \Psi_{2\beta}\lambda_3 + \Psi_{2\alpha}\lambda_4$$

---

with $K_n = 2K_1$, $K_d = 2K_2\dfrac{X_h}{\sigma X_1 X_2}$, $K_\Psi = 4K_3$

and $K_5 = \dfrac{X_h}{T_A\sigma X_1 X_2}$.

With $\dfrac{\partial H}{\partial \underline{u}}$ (eq.10) follows:

$$\frac{\partial H}{\partial \hat{u}_1} = \lambda_1$$

$$\frac{\partial H}{\partial \omega} = \lambda_1\Psi_{1\beta} - \lambda_2\Psi_{1\alpha} + \lambda_3\Psi_{2\beta} - \lambda_4\Psi_{2\alpha} \quad (15)$$

and we get the instruction for the iteration:

$$\Delta\hat{u}_1 = -K\lambda_1$$

$$\Delta\omega = -K(\lambda_1\Psi_{1\beta} - \lambda_2\Psi_{1\alpha} + \lambda_3\Psi_{2\beta} - \lambda_4\Psi_{2\alpha}) \quad (16)$$

With K = const > 0 we can control the value of the iteration step.

The solution of this problem is computed at a high speed hybrid computer EAI 680/640.

The equation (1) for the induction motor and eq. (14) for the adjoint system are realized at the analogue computer. Eq. (16) must be programmed at the digital computer. The digital computer must store the sampled control and system functions and control the transfer of data between the analogue computer and the storage.

For the starting functions $\hat{u}_1^o(t)$ and $\omega^o(t)$ we take for example constant functions, we get out of the computation of the steady state with $n = n_1$: $\hat{u}_1^o(t) = \omega^o(t) = n_1$. The initial conditions for the backward computation of the adjoint system are then set to zero: $\underline{\lambda}(t_1) = \underline{0}$. (Other functions may be generated.)

In a short form we can write the scheme of the computation for the iteration shown in Fig. 2. The computation stops if $\Delta\hat{u}_1$ and $\Delta\omega$ become so small, that there is no developement from step to step of the iteration. Then the optimum is reached.

RESULTS OF THE COMPUTATION

In the following figures there are demonstrated characteristic results of the optimization process in some examples.

Example 1 shows the optimization of the switching-on process of a 10kW machine. All the initial conditions (machine flux and speed) are zero at the time t = 0. In Fig. 3 the control functions $\hat{u}_1(t)$, $\omega(t)$ and the speed n(t) are shown. The iteration process starts with the functions $\hat{u}_1^o(t) = \omega^o(t) = 1$ and the normal switching on speed $n^o(t)$. Then there are plotted the functions marked with 1, 2 and 3 which are computed in the 20th, 40th and 60th step of iteration. With this curves the good convergence of the process is demonstrated.

The factors K_n and K_Ψ were constant and K_d had been increased during the optimization process.

The factor K which determines the size of each iteration step was very small in this example. If we take K greater, we don't need so much steps. But if K is too great, the iteration process will become unstable and we get no convergence.

The time t is not normalized in all figures. There are the following maximum values of some variables in the optimum process: $d_{Mmax} \approx 4d_{MN}$, $\hat{\Psi}_{2max} \approx \hat{\Psi}_{2N}$ and $\hat{\Psi}_{1max} \approx 1{,}5\hat{\Psi}_{1N}$.

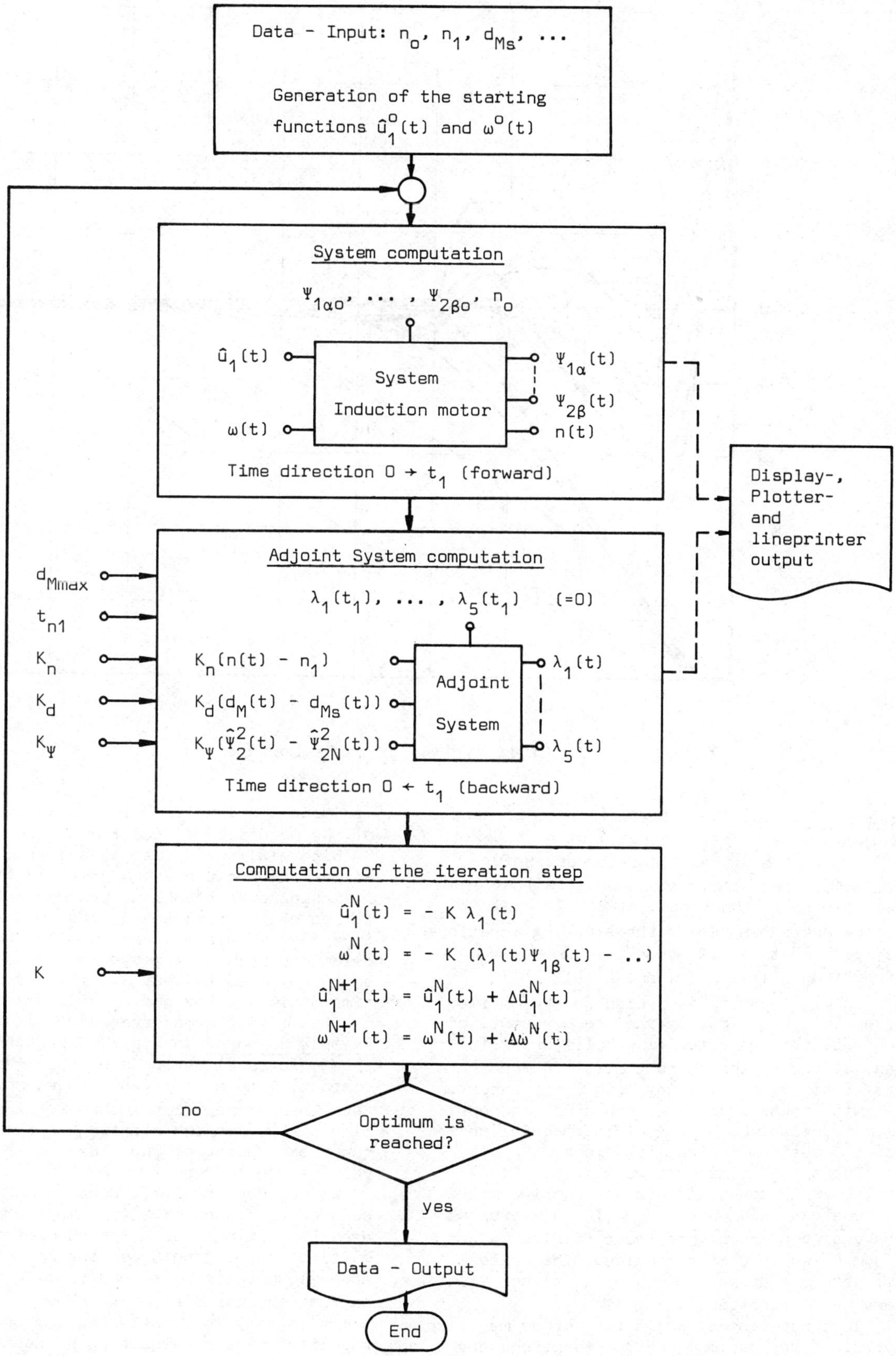

Fig. 2. Scheme of the iteration process

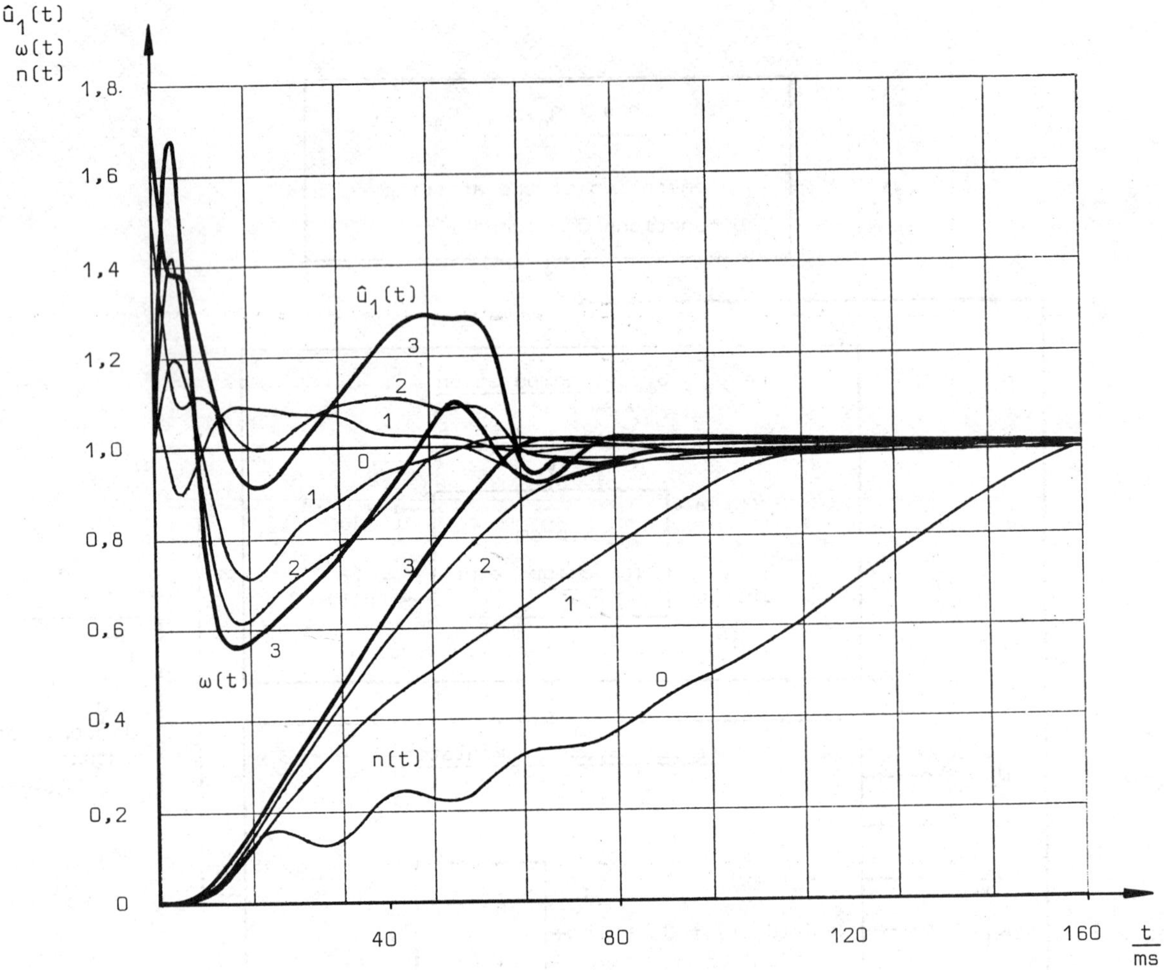

Fig. 3. Optimization of the switching - on process

Example 2 shows the optimization of an acceleration process of the speed from $n_o = 0{,}2$ to $n_1 = 1{,}2$ also without load torque. Before and after the acceleration process the machine is in the steady state. In Fig. 4 there are shown beside the starting functions $\hat{u}_1^o(t) = \omega^o(t) = 1{,}2$ and $n^o(t)$ the optimum functions $\hat{u}_1^*(t)$, $\omega^*(t)$ and $n^*(t)$.

In Fig. 5 there is presented the optimum transient behaviour of the space vectors of the stator- and rotor flux $\Psi_1(t)$ and $\Psi_2(t)$ and of the stator current $i_1(t)$ in the reference frame rotating with $\hat{u}_1(t)$ synchron.

With these results we can study the optimum transient behaviour of the machine and of the optimum control functions.

During the time when we have an acceleration with constant torque, we drive with a constant slip $\omega_2 = \omega - n$. In this area we have a good analogy to those results we get from known regulating devices. The angles between the space vectors of $\hat{u}_1(t)$ and $\Psi_{1,2}(t)$ are nearly constant.

In those ranges, where the torque moves up or down, the optimum control functions are more complicated. There are two problems which are solved by $\hat{u}_1(t)$ and $\omega(t)$.

The main problem for $\hat{u}_1(t)$ is the regulation of the magnitude of the flux in the machine.

With $\omega(t)$ there must be adjusted the angles between the space vectors. In the beginning of a transient process this regulation of the flux and of the angles is very important. There must be adjusted the right relations between the system variables in the machine as fast as possible to get a time optimum behaviour.

That may be demonstrated with example 3 and 4. In Fig. 6 and 7 there are shown two acceleration processes from the standstill to $n_1 = 1$. In both examples we have the nominal flux in the machine at the time $t = 0$. But the starting angles between the space vector of the flux and the stator voltage are different.

We can see the difference in the optimum control functions of $\omega^*(t)$. In example 4 we need the maximum value (bounded at 1,8) of $\omega^*(t)$ at the beginning of the process to get an angle between $\hat{u}_1(t)$ and $\Psi_1(t)$ of about 80° as fast as possible. (Like shown in Fig. 5)

It is very interesting to compare the results of the acceleration processes out of the standstill with different initial conditions shown in Fig.3, 6 and 7.

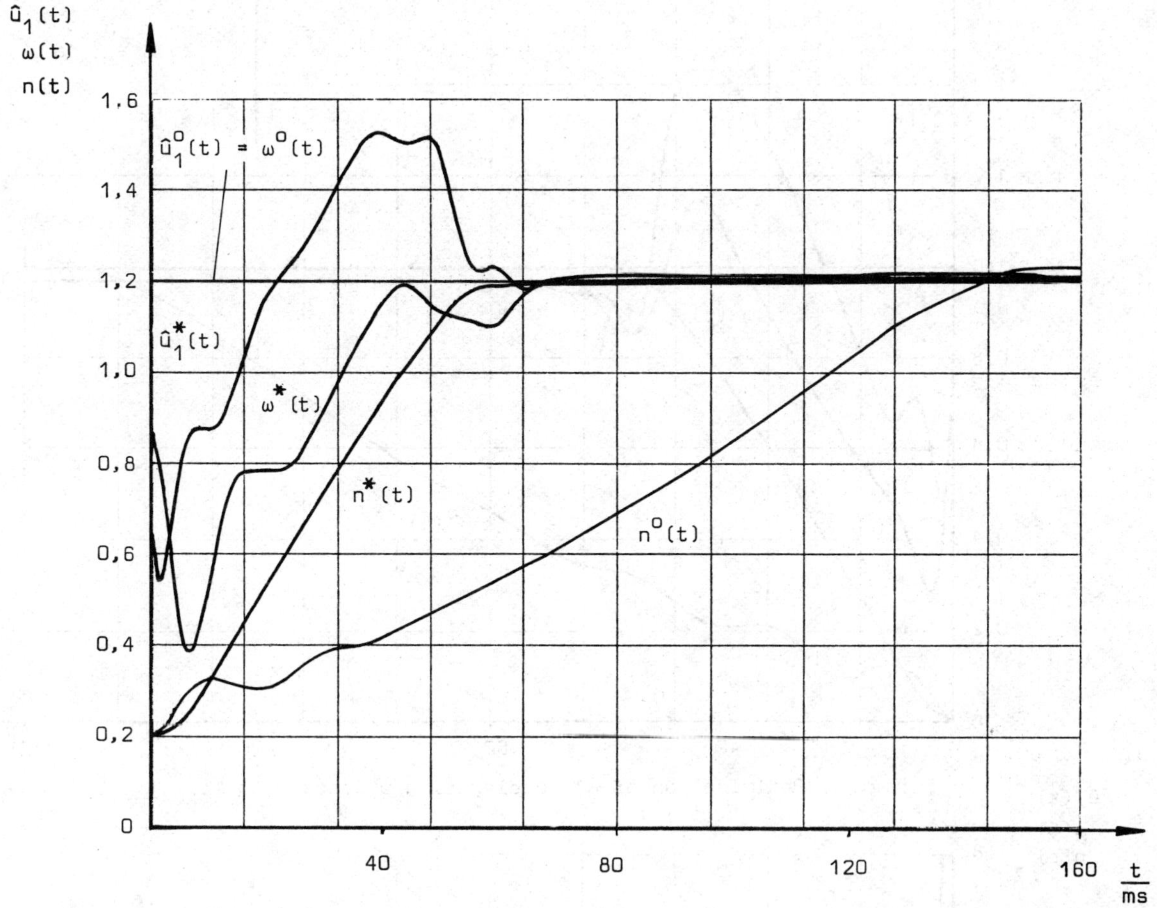

Fig. 4. Optimization of an acceleration process

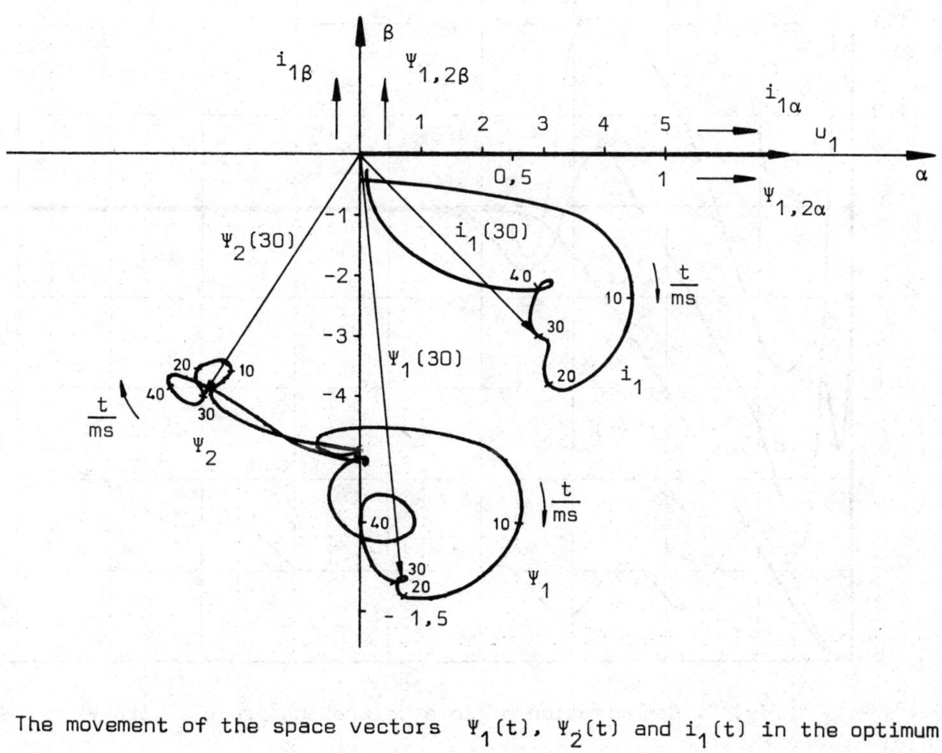

Fig. 5. The movement of the space vectors $\Psi_1(t)$, $\Psi_2(t)$ and $i_1(t)$ in the optimum process

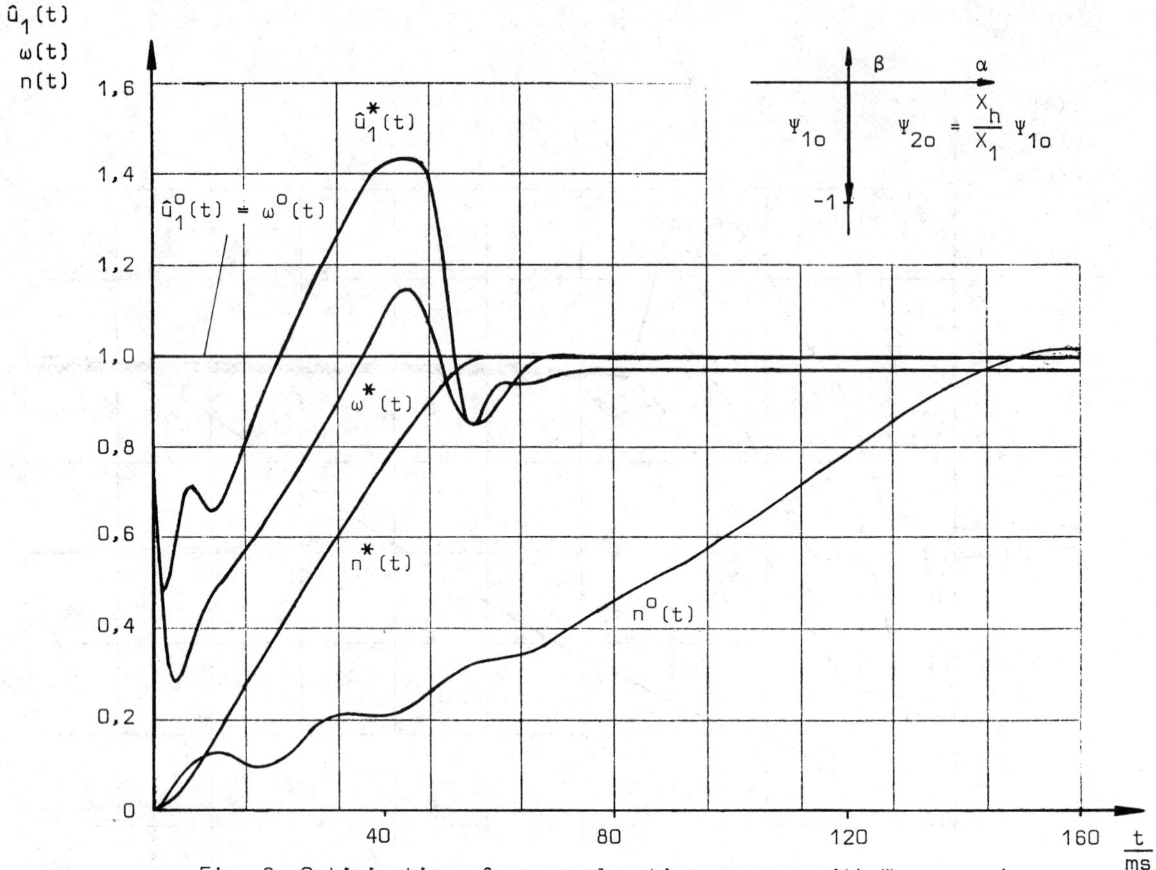

Fig. 6. Optimization of an acceleration process with $\Psi_{1\beta o} = -1$

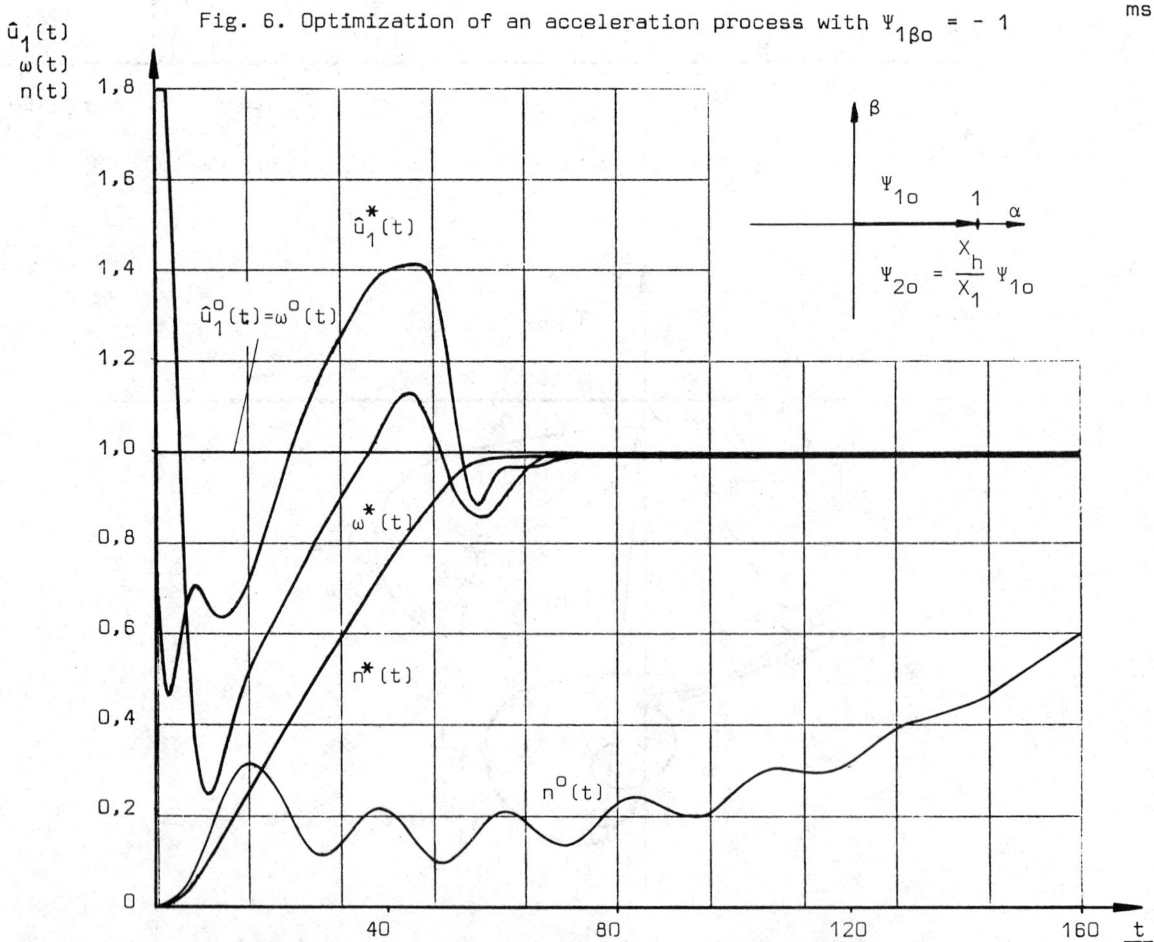

Fig. 7. Optimization of an acceleration process with $\Psi_{1\alpha o} = +1$

With these examples there should be demonstrated the principle way of optimization of the transient behaviour of the induction motor using the Maximum principle and the gradient technique.

With other quality functions it is possible to compute solutions for different optimization problems. This optimization can be done in consideration of given boundary conditions of state and control variables.

It is possible to test given control systems with these results to find how good the control systems use the possibilities of the drive system to realize an optimum automatic control.

SYMBOLS AND MOTOR DATA

u_1 stator voltage

i_1 stator current

Ψ_1 stator flux

Ψ_2 rotor flux

α,β axis of the reference frame

α,β index: components of the space vectors

n speed

d_M motor torque

ω frequency

t time

$\underline{x}$ system variable

$\underline{u}$ control variable

$\underline{\lambda}$ adjoint system variable

H Hamiltonian

J cost function

* upper index: optimum function

o upper index: starting function

o lower index: initial condition

^ value or magnitude of a space vector

N index: nominal value

P_N = 10kW

U_N = 220V

I_N = 21A

Z_p = 2

f_N = 50cps

R_1 = 0,04

R_2 = 0,0505

T_A = 55

$X_1 = X_2$ = 2,61

X_h = 2,5

σ = 0,0825

REFERENCES

/1/ G. Pfaff: Regelung elektrischer Antriebe I.
R. Oldenbourg Verlag München Wien (1971)

/2/ A. P. Sage: Optimum systems control.
Prentice-Hall, Inc.
Englewood Cliffs (1968)

/3/ E. Hofer, R. Lunderstäd:
Numerische Methoden der Optimierung.
R. Oldenbourg Verlag München Wien (1975)

/4/ G. Weihrich: Optimale Regelung linearer deterministischer Prozesse.
R. Oldenbourg Verlag München Wien (1973)

/5/ D. E. Kirk: Optimal Control Theory.
Prentice-Hall Inc. Englewood Cliffs (1970)

/6/ M. Athans, P. L. Falb:
Optimal Control
McGraw-Hill Book Company New York 1966

ANALYSIS AND COMPARISON OF DIFFERENT P.W.M. TECHNIQUES FOR INDUCTION MOTOR DRIVES

A. Bellini and G. Figalli

Istituto di Automatica Via Eudossiana 18 00184 Roma

Summary — In the paper the modulation techniques usually employed in variable frequency induction motor drives are examined. A general procedure is shown for the computation of the most important phenomena connected with the non-sinusoidal waveform of the voltages applied to the motor and a comparison between the performances obtained by each modulation technique is effected. This makes it easier to select the most suitable modulation technique.

NOMENCLATURE

R_s R_r	stator and rotor resistances
L_s L_r L_{sr}	stator, rotor and mutual inductances
Φ_s Φ_r	stator and rotor fluxes
p	pole pairs

1. INTRODUCTION

In drives using as actuator an induction motor supplied by a variable frequency inverter, the amplitude of the voltage applied to the motor must be changed with the frequency.

When the inverter is supplied by a d.c. constant voltage, the a.c. voltage variation can be achieved by two different modulation techniques. In both cases the nonsinusoidal waveform of the inverter supply voltage produces several unwanted phenomena. The most important ones are the pulsating torques and the a.c. currents superimposed on the d.c. supply. To select the more suitable technique for an actual drive, also these phenomena must be taken into account.

In the paper, the peculiarities of both modulation techniques are examined and a general procedure for the computation of the torque and current harmonics is shown. Finally a comparison between the performances obtained by the two modulation techniques is effected.

2. MODULATION TECHNIQUES

In drives, in which the three-phase bridge inverter is supplied by a d.c. constant voltage, two different modulation techniques are usually employed: the first consists in the classical pulse-width-modulation with a triangular carrier voltage; the second is typical of the three-phase systems [1,14].

2.1 First modulation technique

The classical modulation technique consists in comparing reference voltage v_{ri} with carrier voltage v_{ci} and imposing that output voltage v_i is equal to $+E_a$ when $v_{ri} > v_{ci}$ and equal to $-E_a$ when $v_{ri} < v_{ci}$.

Carrier voltage v_{ci} is an isosceles triangular wave of fixed amplitude E_c and frequency f_c, while reference voltage v_{ri} is a sinusoidal wave of frequency f_r equal to, and amplitude E_r proportional to, those desired to supply the motor.

The peculiarities of output voltage in terms of ratio K between carrier frequency f_c and reference frequency f_r are described in [15]. In particular, when K is irrational, voltage v_i is aperiodic and its frequency spectrum is continuous. When K is rational but not integer, voltage v_i is periodic, with a period higher than $1/f_r$, and the frequency of its first harmonic is lower than f_r. When K is integer, voltage v_i is periodic with a period T equal to $1/f_r$. In addition, if K is odd the following recurrent condition, which reveals the absence of the even harmonics is true:

$$v_i\left(t + n\frac{T}{2}\right) = (-1)^n\, v_i(t) \qquad n \text{ integer} \tag{1}$$

In variable speed drives frequency f_r changes continuously; so, if f_c is assumed constant, ratio K is continuously variable and all the previously examined situations can take place. This modulation technique produces, therefore, a bad operation of the motor, specially when the speed is slow, owing to the presence of subharmonics in connection with not integer values of K.

To avoid it, it is therefore necessary to change frequency f_c in such a way that ratio K is always integer (*).

The waveform of the a.c. voltage depends on ratio $V = E_r/E_c$ and phase ϕ of the reference voltage measured when the carrier voltage is zero and its slope is positive.

The amplitude of the voltage harmonics can be computed in various ways. In particular, in [16] is presented an interesting computation method, based on a rapidly convergent Kapteyn series. If K is odd

(*) If it is desired that at low frequencies K should assume a high value and at high frequencies the maximum number of commutations per second be sufficiently small, it is necessary to modify discontinuously K according to frequency f_r.

3. HARMONIC COMPUTATION PROCEDURE

The computation of the harmonic content of the motor torque and the d.c. supply current can be worked out on the basis of the system theory results assuming that the influence of the pulsating torques on the motor speed is negligible. It is also assumed that the magnetic circuits are linear, the motor has an isotropic structure and the reference axes are fixed with the stator windings.

Selecting as state variables the stator and rotor fluxes expressed as complex variables, the electromagnetic model of the motor is given by the following matrix equation:

$$\dot{\underline{\Phi}}(t) = \underline{A}(\omega)\,\underline{\Phi}(t) + \underline{v}(t) \qquad (6)$$

in which

$$\underline{\Phi} = \begin{bmatrix} \Phi_s \\ \Phi_r \end{bmatrix} \qquad \underline{v} = \begin{bmatrix} v_s \\ 0 \end{bmatrix} \qquad \underline{A} = \begin{bmatrix} -\alpha & \alpha K_r \\ \beta K_s & -\beta + j\omega \end{bmatrix}$$

$$v_s = \sqrt{2/3}\,[v_1 + v_2\exp(j2\pi/3) + v_3\exp(j4\pi/3)]\exp(j\,\psi)$$

$$K_s = \frac{L_{sr}}{L_s} \qquad K_r = \frac{L_{sr}}{L_r} \qquad \alpha = \frac{R_s}{\sigma L_s} \qquad \beta = \frac{R_r}{\sigma L_r}$$

$$\sigma = 1 - K_s K_r$$

and ψ is the electrical angle between the real axis and that of winding 1.

Since a.c. voltages v_1, v_2 and v_3 have the same waveform and a time displacement equal to T/3, recurrent condition (1) imposes that the steady-state behaviour of the motor can be completely defined, if the shapes of the state variables are known during an observation interval T^* equal to T/6.

Furthermore the following steady-state condition can be deduced:

$$\underline{\Phi}(t_0 + T^*)_- = \underline{P}\;\underline{\Phi}(t_0)_+ \qquad (7)$$

in which matrix $\underline{P}$ represents a rotation of the reference axis equal to $\pi/3$ and is:

$$\underline{P} = \begin{bmatrix} \exp(j\pi/3) & 0 \\ 0 & \exp(j\pi/3) \end{bmatrix}$$

Observation interval $(0\ ,\ T^*)$ can be divided into M subintervals, in which voltage v_s remains constant. Therefore matrix equation (6) makes it possible to determine the state variables during each subinterval $(t_{k-1}\ ,\ t_k)$ in terms of their value $\underline{\Phi}_{k-1}$ at instant t_{k-1} and of value v_{sk} of v_s:

$$\underline{\Phi}(t_{k-1} + \lambda) = \exp(\underline{A}\lambda)\,\underline{\Phi}_{k-1} + \underline{A}^{-1}[\exp(\underline{A}\lambda) - \underline{I}]\underline{v}_k \qquad (8)$$

in which

$$\underline{v}_k = \begin{bmatrix} v_{sk} \\ 0 \end{bmatrix}$$

Since the order of matrix $\underline{A}$ is only 2 it is possible to obtain $\exp(\underline{A}\lambda)$ in a closed form [18]:

$$\exp(\underline{A}\lambda) = \underline{R}_1\exp(z_1\lambda) + \underline{R}_2\exp(z_2\lambda) \qquad (9)$$

The flux at final instant $t_k = t_{k-1} + \tau_k$ of the subinterval is therefore:

$$\underline{\Phi}_k = \exp(\underline{A}\tau_k)\,\underline{\Phi}_{k-1} + \underline{A}^{-1}\left[\exp(\underline{A}\tau_k) - \underline{I}\right]\underline{v}_k \qquad (10)$$

Applying Eq. (10) to each subinterval of the observation interval, flux $\underline{\Phi}_M$ at final instant T^* can be obtained in terms of flux $\underline{\Phi}_0$ at initial instant, i.e.:

$$\underline{\Phi}_M = \underline{\Phi}(t_0 + T^*) = \exp(\underline{A}T^*)\,\underline{\Phi}_0 + \underline{s} \qquad (11)$$

in which

$$\underline{s} = \underline{A}^{-1}\sum_{k=1}^{M}\exp[\underline{A}(T^* - \sum_{i=1}^{k}\tau_i)]\cdot[\exp(\underline{A}\tau_k - \underline{I})]\underline{v}_k$$

Eq. (11) and steady-state condition (7) make it possible to compute the value of flux $\underline{\Phi}_0$:

$$\underline{\Phi}_0 = \left[\underline{P} - \exp(\underline{A}T^*)\right]^{-1}\cdot\underline{s} \qquad (12)$$

Therefore Eq. (8) makes it possible to compute the value of the flux at the final instant of each subinterval and then its instantaneous value.

Motor torque t_q and d.c. supply current i_L are periodic with period T^*. Their instantaneous values can be determined, in terms of the state variables and voltage v_s, and are given by the following equations:

$$t_q(t) = \underline{\Phi}(t)^{*T}\;\underline{G}\;\underline{\Phi}(t) \qquad (13)$$

$$i_L(t) = Re\left\{\underline{a}^T\,\underline{\Phi}(t)\underline{v}_s^*(t)\right\} \qquad (14)$$

in which

$$\underline{G} = j\,\frac{pL_{sr}}{2\sigma L_s L_r}\begin{bmatrix} 0 & 1 \\ -1 & 0 \end{bmatrix}$$

$$\underline{a}^T = \frac{1}{2\sigma L_s E_a}\begin{bmatrix} 1 & -K_r \end{bmatrix}$$

It is then possible to obtain the explicit equations which give the values of amplitudes A_{tn} and A_{in} of the n-th harmonic of the motor torque and d.c. supply current [17].

4. COMPARISON BETWEEN THE TWO MODULATION TECHNIQUES

The computation procedure has been employed to compare the two modulation techniques, taking into consideration a drive in which the three-phase bridge inverter is supplied by a d.c. voltage of 3000 V (E_a = 1500 V), and the induction motor is characterized by the following parameters:

$p = 3$ $\quad R_s = .3121$ $\quad R_r = .2006$

$L_s = .1792$ $\quad L_r = .1772$ $\quad L_{sr} = .1724$

For the second modulation technique all three modes have been examined. The comparison has been effected selecting for each techniques the same number of commutations per period corresponding to K = 15. Amplitude A_1 of the first harmonic of the a.c. voltage has been selected so that, for any value of frequency f_r, the mean value of the motor torque is equal to rated value T_{qr} = 2580 Nm, when the absoluted slip assumes its rated value. The maximum value of the frequency has been selected equal to 39Hz, which corresponds to a value of ratio V prac-

and $0 \leqslant V \leqslant 1$, amplitude A_n of the n-th harmonic is (**):

$$A_n = X_n + \frac{4E_a}{\pi} \sum_{p=1}^{\infty} \frac{1}{p} F_p \qquad (2)$$

in which

$$F_p = J_{pK+n}\left(p\frac{\pi}{2}V\right) \exp(jpK\phi) - J_{pK-n}\left(p\frac{\pi}{2}V\right)\exp(-jpK\phi)$$

$$X_n = \begin{cases} VE_a & \text{for } n=1 \\ 0 & \text{for } n\neq 1 \end{cases}$$

$J_p(\cdot)$ is the Bessel function of first kind and n is odd.

As shown in [17], when K is higher than, or equal to seven, the amplitude of the first harmonic is roughly proportional to V:

$$A_1 \cong E_a\, V \qquad (3)$$

and the amplitude of the harmonics of order $n \leqslant K+2$ is practically independent of ϕ and can be computed by the following simplified equation:

$$A_n \cong \frac{4E_a}{\pi} J_{K-n}\left(\frac{\pi}{2}V\right) \qquad (4)$$

Eq. (4) shows that the amplitude of the harmonics of order $n < K-4$ is practically negligible and moreover those of the harmonics of order $n = K-4$, $n = K-2$, $n = K$ and $n = K+2$ are independent of the value of K and have the waveforms shown in fig. 1.

In a three-phase bridge inverter, in order to obtain that a.c. voltages v_1, v_2 and v_3 have the same shape and a time displacement equal to T/3, it is necessary that also the three reference voltages and the three carrier voltages should have a displacement equal to T/3. In this situation the harmonics with a frequency multiple of $3f_r$ are in phase and do not affect the behaviour of the motor. It is then convenient to assume a value of K multiple of three. In this way the effect of the K-th harmonic is eliminated and it is possible to use the same carrier voltage.

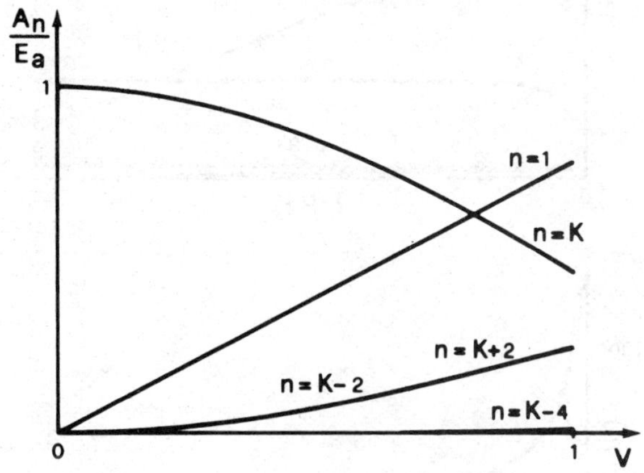

Fig. 1. Voltage harmonics.

2.2. Second modulation technique

Fig. 2 shows a typical waveform of voltage v_i obtained by the second modulation technique. Time interval (T/6 , T/3) is constituted by N subintervals in which voltage v_i is constant and alternately equal to $+E_a$ and $-E_a$, while during time intervals (0 , T/6) and (T/3 , T/2) v_i is equal to $+E_a$. Also with this modulation technique, to avoid the even harmonics it is necessary that recurrent condition (1) is true.

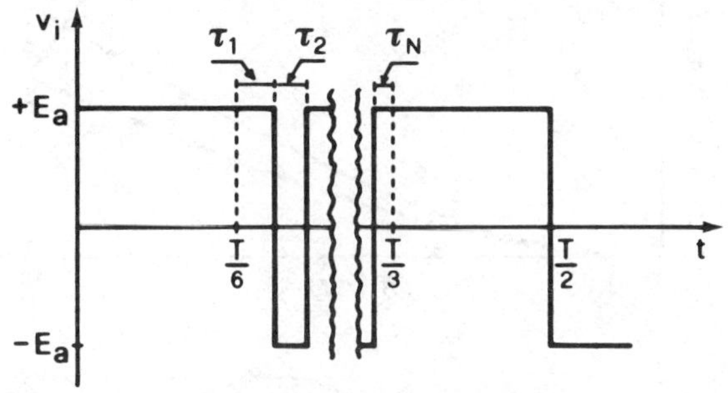

Fig. 2. Phase voltage waveform.

Therefore the modulated voltage is fully characterized by number N and duration $_i$ of the subintervals constituting time interval (T/6 , T/3). The amplitude of the odd harmonics of voltage v_i is then:

$$A_n = \frac{4E_a}{n\pi} \left| 1 + \sum_{i=1}^{2\left[\frac{N}{2}\right]} (-1)^i \cos\left[n\pi\left(\frac{1}{3} + \frac{2}{T}\sum_{k=1}^{i} \tau_k\right)\right] \right| \qquad (5)$$

The harmonic content of the a.c. voltage depends on the modulation mode which gives the selection of the durations of the N subintervals. This choice can be effected either to simplify the inverter control circuit or to reduced the unwanted phenomena connected with the nonsinusoidal waveform of the a.c. voltage.

In this paper three different modulation modes are considered:

Mode 1. In this mode, which requires a very simple inverter control circuit, the number of the subintervals is even. The duration of all the subintervals, in which $v_i = +E_a$, is assumed equal to τ_1, whereas that of the others to $\tau_2 = T/(3N) - \tau_1$.

Mode 2. In this mode, which requires a little more complex control circuit, the number of the subintervals is odd. The duration of the first and the last subinterval is assumed equal to half of that of the other subintervals, in which $v_i = +E_a$. On the contrary all the subintervals, in which $v_i = -E_a$, have the same duration.

Mode 3. In this mode, which requires a much more complex control circuit, all the durations of the subintervals are assumed different and are selected on the basis of a suitable criterion. For example in this paper it is assumed that:

$$\tau_i = \begin{cases} \frac{B}{2}\left[C - \sin(N\gamma)\right] \cdot X & \text{for } i=1 \text{ and } i=N \\ B\left\{1 - \left[C - \sin(\delta_i)\right]\right\} \cdot X & \text{for } i \text{ odd} \\ B\left[C - \sin(\delta_i)\cos(\gamma)\right] \cdot X & \text{for } i \text{ even} \end{cases}$$

in which

$$B = \frac{T}{3(N-1)} \quad C = 1 + \frac{\sqrt{3}}{2} \quad \gamma = \frac{\pi}{3}\frac{1}{N-1} \quad \delta_i = (N+i-2)\gamma$$

and X depends on the value selected for the first harmonic and is never greater than one.

(**) In general, when V is higher than one, the described modulation technique is not used. In fact, in this situation the amplitude of the first harmonic is not proportional to V.

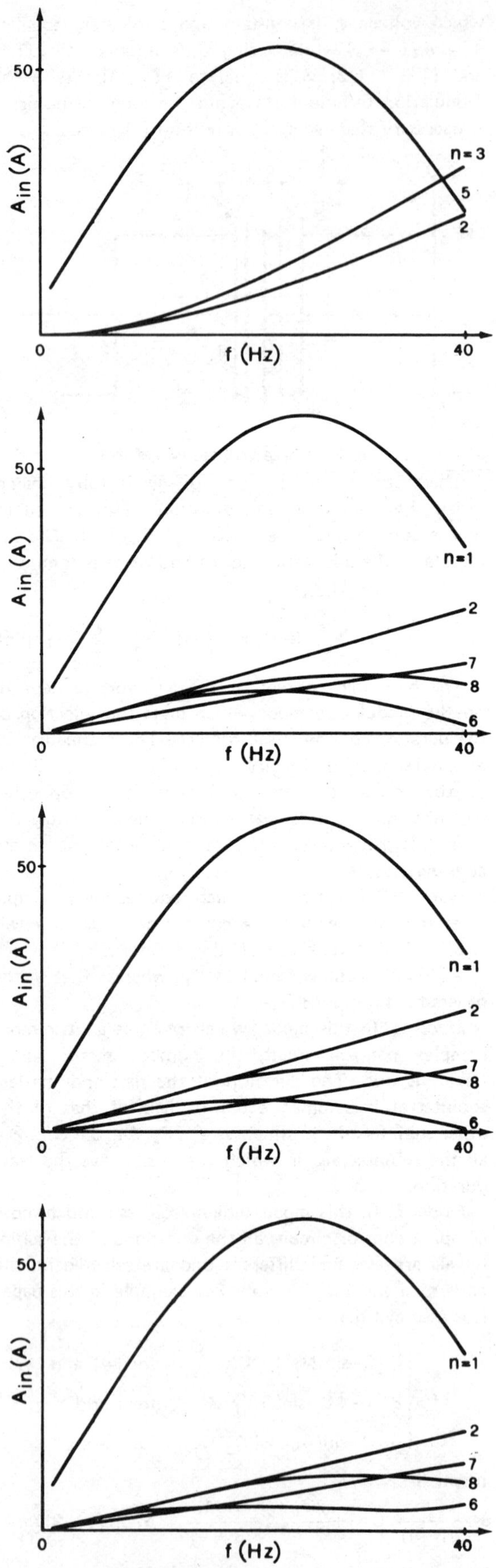

Fig. 3. Current harmonics.

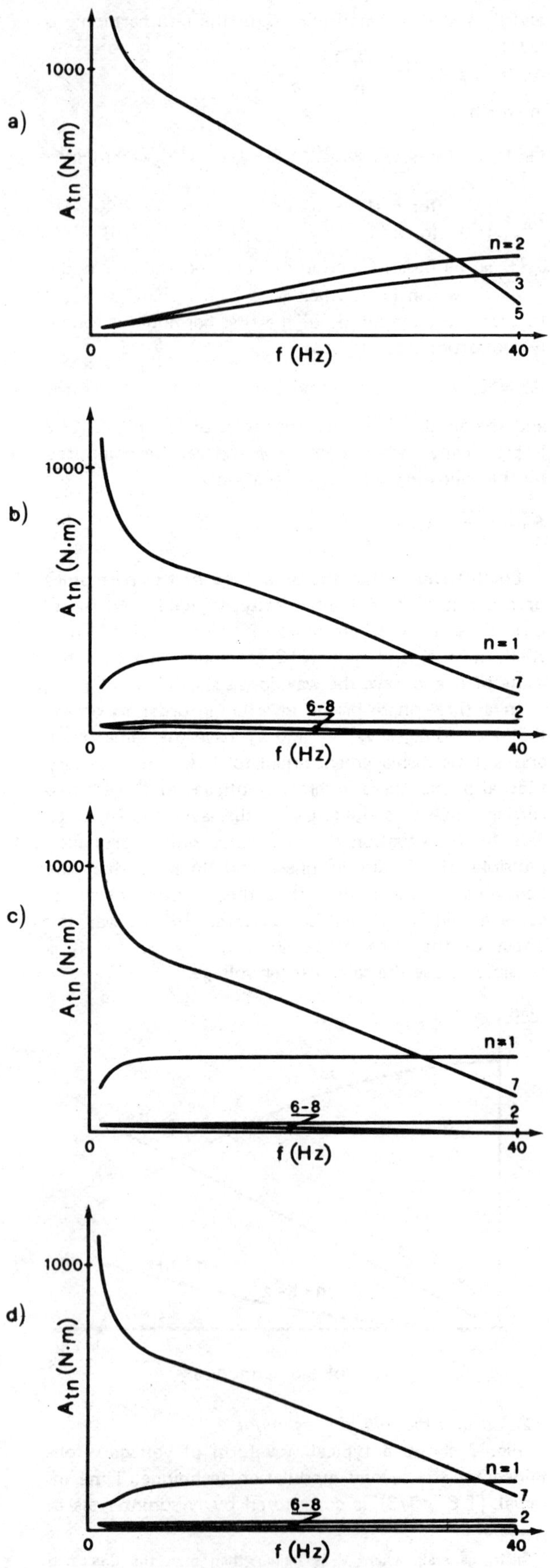

Fig. 4. Torque harmonics.

tically equal to one.

Figs. 3 and 4 show the value assumed by the first harmonics of the motor torque and the d.c. supply current, when the mean value of the motor torque is equal to its rated value. Letter *a* refers to the first modulation technique, letters *b, c* and *d* to the modes 1,2 and 3 of the second one. These figures show that in the first technique the first three harmonics, the amplitude of which is not negligible, have an order equal to (K – 3)/6, (K + 3)/6 and K/3. On the contrary in the second technique the more weighty harmonics have an order equal to 1,2 and (K – 1)/2.

As regards the second technique, it is possible to deduce that modes 1 and 2 are practically equivalent, whereas in mode 3 the first harmonic both of the motor torque and of the d.c. supply current is considerably reduced.

The selection of the modulation technique depends on the drive requirements. In particular it is convenient, when it is possible, to use modes 1 and 2 of the second technique owing to the more simple control circuit. On the other hand the selection between the first technique and mode 3 of the second is quite difficult. By analysing the harmonic content with a suitable weighting function, it has been possible to deduce that the first technique produces a better operation except for the lower frequencies.

With regard to the other performances of the drive, i.e. the efficiency and the r.m.s. value of the motor phase current, the two modulation techniques are practically equivalent.

5. CONCLUSIONS

The results obtained in the paper have shown the peculiarities of the modulation techniques usually employed in drives using as actuator a frequency controlled induction motor.

Their comparison makes it easier to select the most suitable modulation mode on the basis both of the requirements of the actual drive and of the complexity of the inverter control circuit.

Further investigations carried out with several motors of different rated power have confirmed the generality of these results.

REFERENCES

[1] A. BELLINI and G. CIOFFI: Induction machine frequency control: three-phase bridge inverter behaviour and performance. *IEEE Trans. Ind. Gen. Appl.* IGA-7, N. 4 (1971).

[2] A. DE CARLI, E. PAGANO and M. PEREZ: Coppie alternative nei motori asincroni controllati in frequenza. *Istituto Elettrotecnico, University of Rome* Int. Rep. 42 (1967).

[3] D.T. ROBERTSON and K.M. HEBBAR: Torque pulsation in induction motors with inverter drives. *IEEE Trans. Ind. Gen. Appl.* IGA-7, N. 2 (1971)

[4] A. BELLINI and A. DE CARLI: On the selection of the supply voltage waveform in induction motor drives. *IFAC Symp. on Control in Power Electronics and Electrical Drives*, Düsseldorf (1974).

[5] J.M.D. MURPHY: *Thyristor control of a.c. motors.* Pergamon Press (1973).

[6] A. WIART: Association d'onduleurs et de machines asynchrones. *Revue Générale de l'Electricité*, Tome 84, N. 11 (1975).

[7] J.J. POLLACK: Advanced pulse-width modulated inverter techniques. *IEEE Trans. Ind. Appl.* IA-8, N. 2 (1972).

[8] T.A. LIPO, P.C. KRAUSE and H.E. JORDAN: Harmonic torque and speed fluctuactions in a rectifier-inverter induction motor drive. *IEEE Trans. Power App. Syst.* PAS-88, N. 5 (1969).

[9] S.B. DEWAN and G.B. FORSYTHE: Harmonic analysis of a synchronized pulse-width modulated three-phase inverter. *IEEE Trans. Ind. Appl.* IA-10, N. 1 (1974).

[10] T.A. LIPO and F.G. TURNBULL: Analysis and comparison of two types of square-wave inverter drives. *IEEE Trans. Ind. Appl.* IA-11, N. 2 (1975).

[11] P.H. NAYAK and R.G. HOFT: Optimizing the PWM waveform of a thyristor inverter. *IEEE Trans. Ind. Appl.* , IA-11, N. 5 (1975).

[12] H.S. PATEL and R.G.HOFT: Generalized techniques of harmonic elimination and voltage control in thyristor inverters. *IEEE Trans. Ind. Appl.* IA-9, N. 3 (1973).

[13] H. STEMMLER: Ein-und mehrpulsige Unterschwingungswechselrichter, Steuerverfahren, Strom-und Spannungsverhältnisse. *IFAC Symposium on Control in Power Electronics and Electrical Drives*, Düsseldorf (1974).

[14] H. KERMANN, W. LIENAU and R. NILL: Der Vierquadrantenstellerein Pulsstromrichter zur Einspeisung von Triebfahrzeugen mit Drehstromantrieb aus dem Einphasen-Bahnnetz. *IFAC Symposium on Control in Power Electronics and Electrical Drives*, Düsseldorf (1974).

[15] F. CANE, P. FERRARIS and F. VILLATA: La regolazione in frequenza di motori trifasi mediante invertitori statici comandati col metodo della sotto-oscillazione. Criteri di scelta dei parametri di modulazione. *Simposio sui motori lineari* Capri (1973).

[16] A. BALESTRINO, A. EISENBERG and L. SCIAVICCO: Invertitori a modulazione di durata di impulsi: analisi spettrale delle forme d'onda. *Simposio sulla evoluzione nella dinamica delle macchine eletriche rotanti*, Tirrenia (Pisa) (1975).

[17] A. BELLINI and G. FIGALLI: Harmonic analysis in induction motor drives with a pulse-width-modulated inverter. *Istituto di Automatica, University of Rome* Rep. 76-15 (1976).

[18] A. BELLINI, C. CERRI and A. DE CARLI: Digital simulation of frequency controlled induction motor drives by a piecewise linearization procedure. *Proceedings of AICA*, N. 3 (1975).

INTERACTION BETWEEN CONVERTER, SIGNAL-PROCESSING AND CONTROL FOR A CONVERTER-FED SQUIRREL CAGE MOTOR

K.-F. Warnecke

Institut für Stromrichtertechnik und Antriebsregelung, Technische Hochschule, Darmstadt, Bundesrepublik Deutschland

SUMMARY

A 4-quadrant induction motor drive, its converter and its control are described. The features of the investigated system are the following: A converter with intermediate constant d.c. voltage, forced commutation with the possibility of controlling the instantaneous values of the phase currents at the machine side, superimposed speed control, measurement of the rotor position angle as a basis for deriving the appropiate current reference values ensuring constant rotor flux linkage. The system is designed for high quality dynamic response. The behaviour of the converter and its interaction with the drive is also considered. The used signal processing avoids a measurement of the air-gap field, which has been used by other authors. This simplification has as a consequence a dependency on the rotor resistance which is discussed in this paper.

The wish to utilize the stator-currents as input to the machine is important for the choice of the converter. It has a constant voltage bus, and its output-currents are "bang-bang" controlled (Fig. 1). The most significant characteristics of this converter type are:

a) Quick changing of the output current,

b) low reaction on the feeding a.c. system,

c) variable frequency.

The converter's power-part is divided into a d.c. voltage source and a forced commutated inverter.

Fig. 2 shows the circuit of this forced commutated inverter. The inverter consists of three phases, each of which splits into two groups. Each group works as a d.c.-chopper. In Fig. 2 also the smoothing capacitor in the intermediate circuit and the short-circuiter are shown. The short-circuiter protects the inverter from voltage surges. Fig. 3 shows the circuit diagramm of the used d.c.-chopper /3/.

A forced commutation cycle occurs in the following sequence: The main thyristor T_H conducts the load-current i. The voltage of the charged quenching capacitor C is $U_C = U_B$. If the load current reaches its limit, the current control fires the quenching-thyristor T_L. The capacitor-voltage U_C changes its polarity in a cosine oscillation. This causes the current in the thyristor T_L to have a sinus-wave shape. T_L blocks during the zero-passage of the current. The quenching capacitor's polarity is now negative. The negative current half wave passes by D_L and T_H and brings i_H to zero with a definite di/dt. Then the load-current i continues the rechanging of the capacitor's polarity. Simultaneously, a great part of the rechanging current flows through the branch D_Z-L_Z independent of the load-current. Diode D_F limits the voltage u_c to the value u_B. With the connection of two d.c.-choppers to a phase of the inverter, see Fig. 2, the circuit D_Z-L_Z is not necessary. Its task is taken care of by Diodes and reactors of the two choppers. If the capacitor voltage reaches the magnitude of the voltage U_B, the diode D_F will take over the load-current i. The time for the described quenching-cycle amounts to 240 μs. In the following, the d.c.-choppers are regarded as ideal switches. The load is star-connected with isolated neutral point. Therefore the current-control is over-appointed as three currents are controlled, but because of the condition $\Sigma i = 0$ only two currents can be freely chosen. The investigation reported in /4/ discusses this problem. Summarizing the result it should be mentioned here that the shape of the current differs from the form to be expected of a single of bang-bang controlled quantity but no difficulties for the overall control occur.

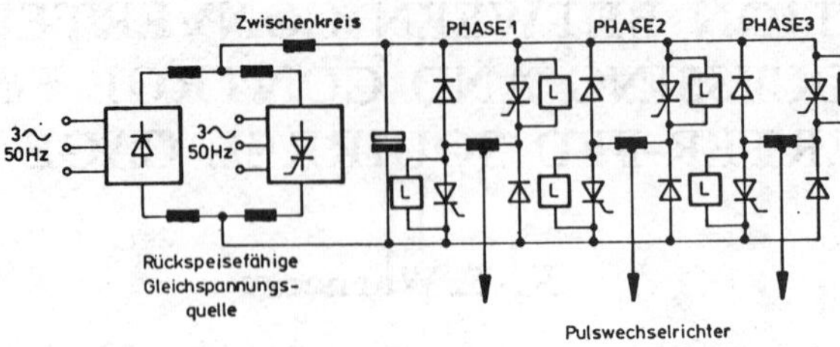

Fig.1. Pulse converter

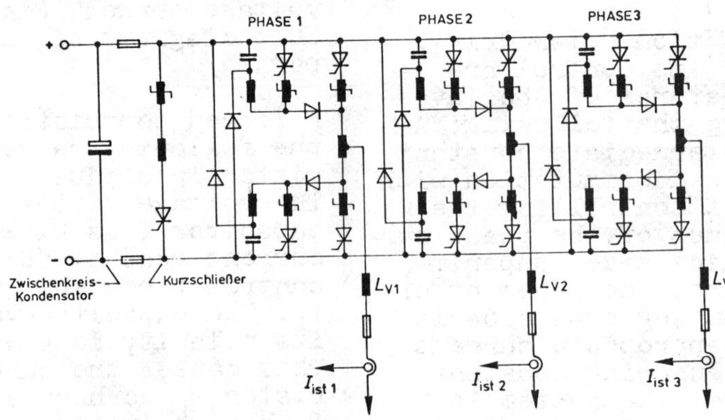

Fig.2. Pulse inverter

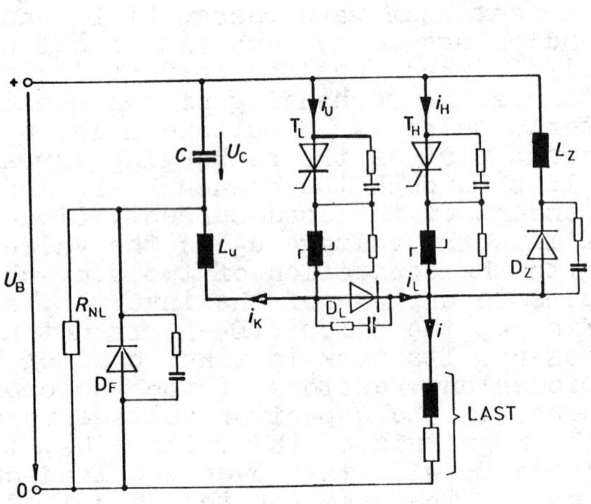

Fig.3. D.C. chopper

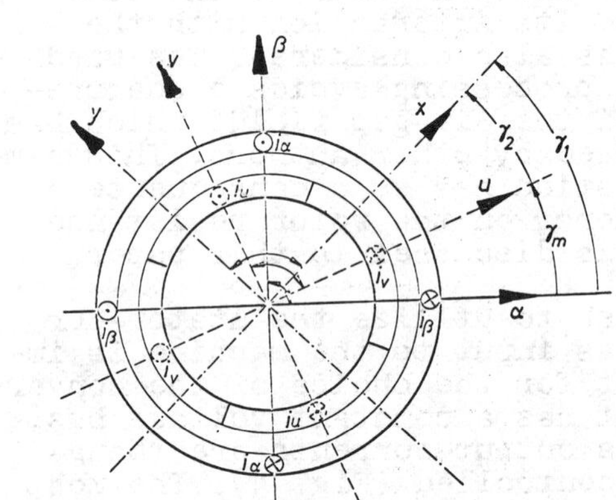

Fig.4. Used Coordinate systems

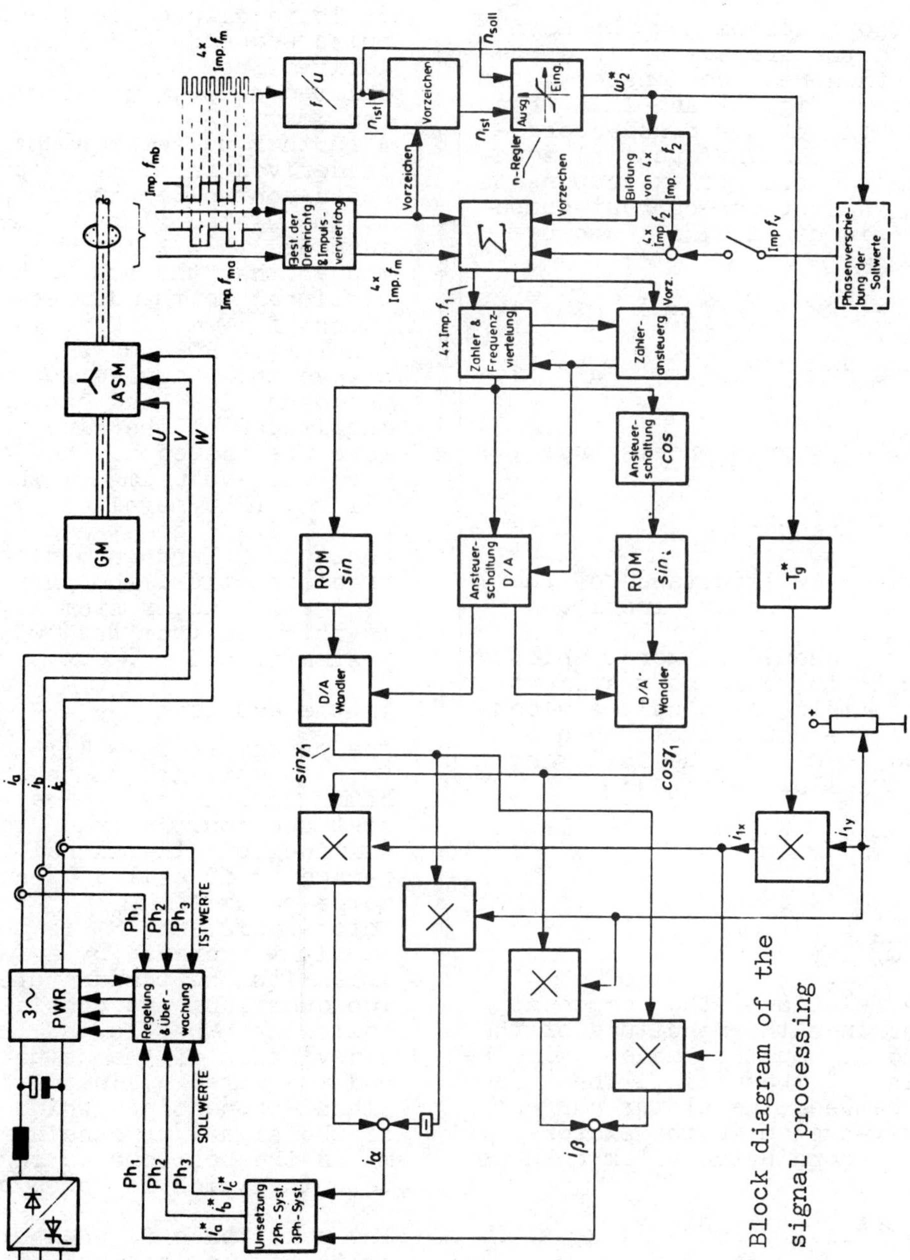

Fig.5. Block diagram of the signal processing

The generation of the current references will now be considered. The usual simplifying assumptions are made.

a) The regarded 2-pole induction motor is built up symmetrically.

b) Saturation and core losses are neglected and only the fundamental wave is considered.

The equations describing the asynchronous motor are set up in a rotating orthogonal coordinate system (x-y-system, fig. 4) and then transferred to the stator-fixed-coordinates α - β. The differential equations for the rotor flux components ψ_{2x}, ψ_{2y} and stator-current-components as inputs i_{1x}, i_{1y} can be written:

$$\dot{\psi}_{2x} = -\omega_g \psi_{2x} + \omega_2 \psi_{2y} + \omega_g L_h u_{1x}$$

$$\dot{\psi}_{2y} = -\omega_2 \psi_{2x} - \omega_g \psi_{2y} + \omega_g L_h i_{1y} \qquad (1)$$

$\omega_g = \frac{R_2}{L_2}$, R_2 = rotor resistance

ω_2 = rotor frequency,

$L_2 = L_h + L_2$ = Inductance of the rotor circuit.

$\dot{\psi}_{2x}$ and $\dot{\psi}_{2y}$ should be zero, which means that there is no transient in the rotor field. Also only a y-component of the rotor field should exist ($\psi_{2x} = 0$). So the following equations follow from (1):

$$\psi_2 = \psi_{2y} = i_{1y} L_H$$

and $\qquad (2)$

$$i_{1x} = -\frac{\omega_2}{\omega_g} i_{1y}$$

In these relations, the component i_{1y} determines the magnitude of the field and i_{1x} must be made $\sim \omega_2$. The torque is then $\sim i_{1x}$. The relations between the stator currents in the x-y-system in the stator-fixed α-β-coordinate-system read as follows:

$$i_\alpha = i_{1x}\cos\gamma_1 - i_{1y}\sin\gamma_1$$

$$i_\beta = i_{1x}\sin\gamma_1 + i_{1y}\cos\gamma_1 \qquad (3)$$

$$\gamma_1 = \gamma_m + \gamma_2$$

Finally the three current-references are:

$$i_a^* = i_\alpha$$

$$i_b^* = -\frac{1}{2} i_\alpha + \frac{3}{2} i_\beta \qquad (4)$$

$$i_c^* = -\frac{1}{2} i_\alpha - \frac{3}{2} i_\beta$$

The calculation of the current references following equations (2) through (4) is realized in fig. 5 /1.4/. The signal-processing is based on a measurement of the mechanical rotor-position. The pulse converter with its current controls and the asynchronous motor with its incremental pulser for the measurement of the rotor position are shown in fig. 5. The analogue speed value and the pulse sequence 4 X Imp. $\{f_m\}$ are derived from the two 90° degrees shifted square-wave signals Imp. $\{f_{ma}\}$ and Imp. $\{f_{mb}\}$.
A further pulse sequence 4 X Imp. $\{f_2\}$ is derived from the output of the speed-controller and is proportional to the wanted slip-frequency ω_2^*. It is added to 4 X Imp. $\{f_m\}$ in such a manner that the sum of both is proportional to the demanded stator-frequency $f_1 \sim \omega_1 = \omega_m + \omega_2$. A binary-counter, a ROM and a D/A-transducter change this impulse sequence in an analogue sine- and cosine-wave. The magnitudes of the sine- and cosine-wave are fitted to the desired slip-frequency-value ω_2^* with four multipliers. (The magnitude is determined by ω_2^*, i_{1y} = const and $T_g = 1/\omega_g$ = const.) Through a coordinate-transformation, the three-phase current-reference-system i_a^*, i_b^* and i_c^* is obtained from the two phases system i_α and i_β.

In the equation $i_{1x} = -\frac{\omega_2}{\omega_g} i_{1y}$ (2) the parameter $\omega_g = R_2/L_2$ appears. Some investigations on the influence of its variations are reported in the next section. In order to make a distinction, the signal-processing parameter is called ω_g^* and the actual motor-parameter ω_g. Especially the rotor-resistance R_2 is not constant; it might increase by a factor of 1.5, with the temperature.
Two questions are considered in the following. What is the influence of a deviation of the actual value ω_g and the fixed value ω_g^* in the signal-processing? Which adjustement of the signal processing parameter ω_g^* is the best choice regarding the detuning which is to be expected?

The next three figures give the results of the theoretical and experimental investigations on these questions: Fig. 6 shows the transient response of the torque when the rotor-frequency suddenly jumps from zero to a magnitude corresponding to 1/4, 1/2 and 3/4 nominal torque and the associated current reference changes are properly made. (Analogue computer studies). The ideal response would be a step function. The following is characteristic in the transient:

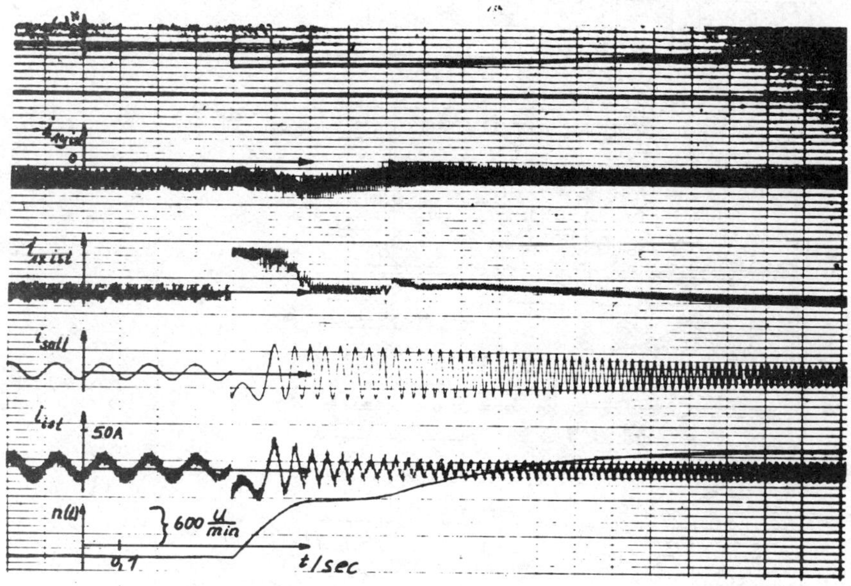

Fig.9. Behaviour of $i_{1x\ ist}$ and $i_{1y\ ist}$ without increasing ω_2 in a case of speed change.

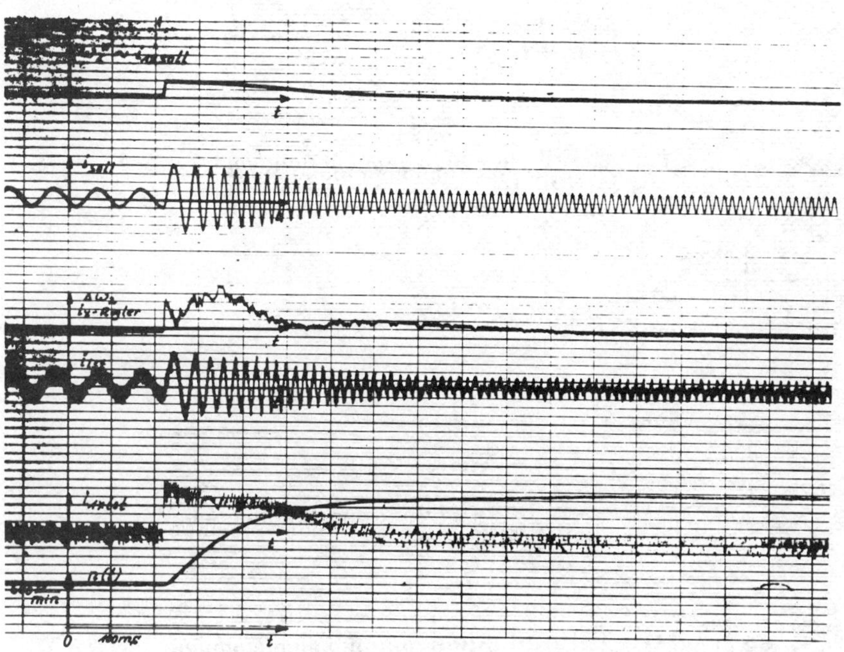

Fig.11. Speed change with a control of $i_{1x\ ist}$.

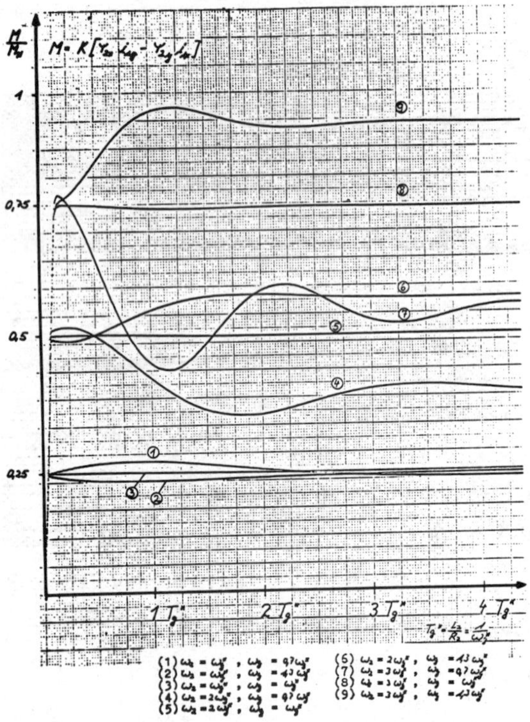

Fig.6. Transient behaviour of the torque

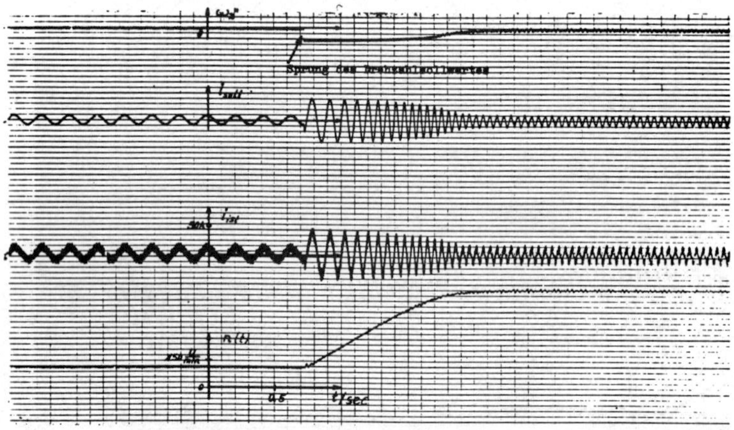

Fig.7. Speed change

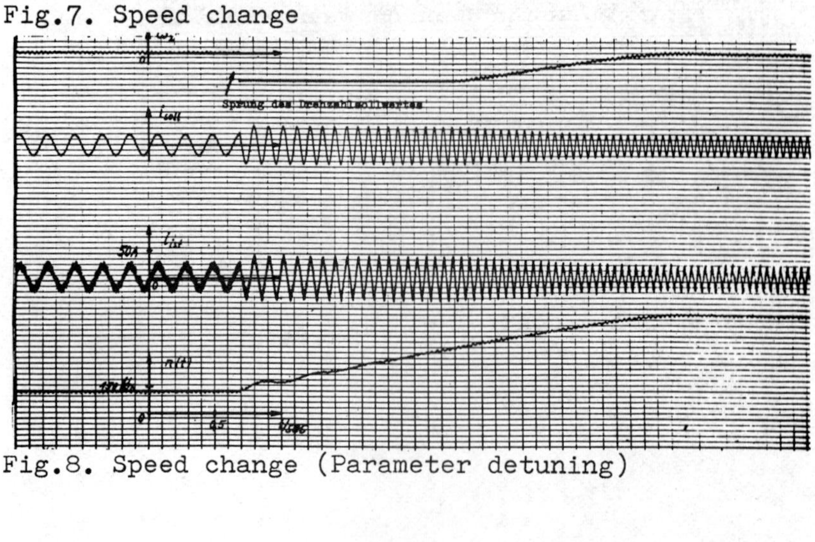

Fig.8. Speed change (Parameter detuning)

a) The amplitude of the ocurring oscillation increases with the magnitude of the torque step.

b) The final value of the torque decreases if $\omega_g/\omega_g^* \ll 1$.
- In this case, the transient torque oscillates strongly with the rotor-frequency as oscillating frequency.

The condition $\omega_g/\omega_g^* \ll 1$ is fulfilled if in the signal-processing, the set value of the rotor-resistance is greater than the actual value. On the other hand, if an inverse detuning occurs, the transient torque's will not deviate so much from the desired behaviour; This is the more the case if the influence of magnetic saturation is considered.

Fig. 7 and 8 show two oscillograms of a speed change. In both of them, the motor accelerates a flywheel. During the first test (7), the signal processing parameter ω_g^* is equal to the actual value. This speed change is therefore time-proportional as long as the speed-controller gives a constant torque reference. In the oscillogram 8 ($\omega_g/\omega_g^* \ll 1$) the inclination of the speed-curvature oscillates and the final value decreases.

The investigations on the dynamic behaviour of the speed-control need not be regarded here, as they are explained in /2/.

The final section of this paper discusses some investigations on the influence of the limited range in the current control. If the stator-frequency increases, the actual current values will begin to deviate from its references.
Therefore the demanded relation between the stator-current components i_{1x} and i_{1y} is no longer fulfilled ($i_{1x} \neq - \omega_2/\omega_g \cdot i_{1y}$). A torque transient results. For experimental investigations, an inverse coordinate-transformation splits up the actual current values in its x- and y-components.

The effect of a deviation of the controlled currents during a speed change is demonstrated in Fig. 9. Upon regarding the behaviour of the actual components $i_{1x\ ist}$ and $i_{1y\ ist}$, one can recognize that the regulation reserve is missing. To avoid this disadvantage, a way to increase the regulation margin had to be found.

Fig. 10 shows the equivalent circuit of an induction motor. One can see that the distribution of the stator-current I_1 on the main inductance L_h and on the parallel rotor-circuit determines the main-field and in this manner the e.m.f. as well. A large e.m.f. diminishes the regulation margin of the current control. In consequence the parallel rotor-resistance must be decreased. This can be handled by increasing ω_2 for a given ω_1 and I_1.

In Fig. 11, a control of the stator-current-component i_{1x} makes sure that ω_2 is raised if $i_{1x\ ist}$ deviates from the set value i_{1x}.

A comparison between Fig. 9 and Fig. 11 demonstrates the advantage of the enlarged rotor-frequency.

LITERATURE

/1/ R. Jötten: Signalverarbeitung für die Regelung umrichtergespeister Drehfeldmaschinen, IFAC-Symposium, Düsseldorf (1974).

/2/ K. Hasse : Zur Dynamik drehzahlgeregelter Antriebe mit stromrichtergespeisten Asynchron-Kurzschlußläufermaschinen, Diss. T.H. Darmstadt (1969).

/3/ U. Kunz : Ein Pulswechselrichter mit weicher Kommutierung, IFAC-Symposium, Düsseldorf (1974).

/4/ K.-F. Warnecke : Wechselwirkung zwischen Umrichter, Signalverarbeitung und Regelung bei einem Stromrichtermotor mit Käfigläufer, Diss. T.H. Darmstadt (1976).

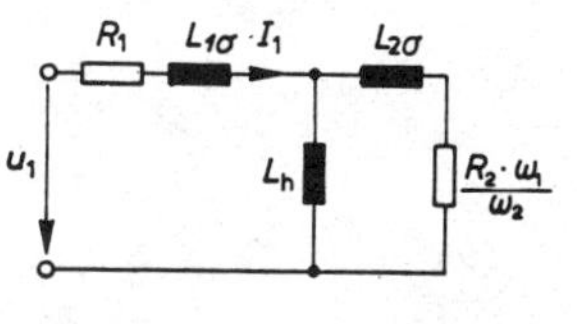

Fig. 10. Equivalent circuit of an induction motor

ANALYSIS AND DESIGN OF PULSE COMMUTATED INVERTER CIRCUITS FOR MULTIMOTOR DRIVES

M. Nowak

Technical University Warsaw, Inst. of Control and Electronics, Koszykowa 75, Gm. E PL 00-662 Warszawa, Poland

ABSTRACT

Analysis of a commutating process as well as methods of computation of inductance and capacitance in terms when a supply voltage changes its value in a large range and value of commutating load current increases rapidly during the start of one of the motors, is rather complicated.
In the paper, a certain calculation method of factors which describes commutating process and are very useful for precise analysis, is presented. Computation algorithm is based on numerical solutions of four differential equations which describe equivalent inverter circuits during commutating interval. In the paper, a way of determining commutating circuit parameters, (inductance and capacitance), in terms when commutating load current and dc supply voltage change their value, is given.

INTRODUCTION

Puls commutated inverters which, are supplied from a regulated voltage source are used in a large scale in a variable multimotor drives. The important features of these invertors are: supply voltage changes within the range forced by the output frequency range and a rapid increase of commutated load current reaching 30 % of steady state value. This current increase occurs during transits which happen at the switch-on of one of the motors sections. Such conditions of the pulse commutated circuits need precise design and appropriate choice of inductance L and capacitance C .

The goal of this paper is to present the method of analysis of pulse commutated circuit which enables the description of its properties in the above mentioned conditions.
The subject of this analysis is a voltage inverter circuit shown in Fig. 1. Inductors L2 which are in the main thyristor circuit, ensure the right performance of the circuit during rapid increases of load current and wide range of supply voltage changes. Inductor L1 limits di/dt of thyristor current. The influence of commutated circuit quality factor Q and inductance division factor $m = L2/L1$, is taken into account in this method. The computation method

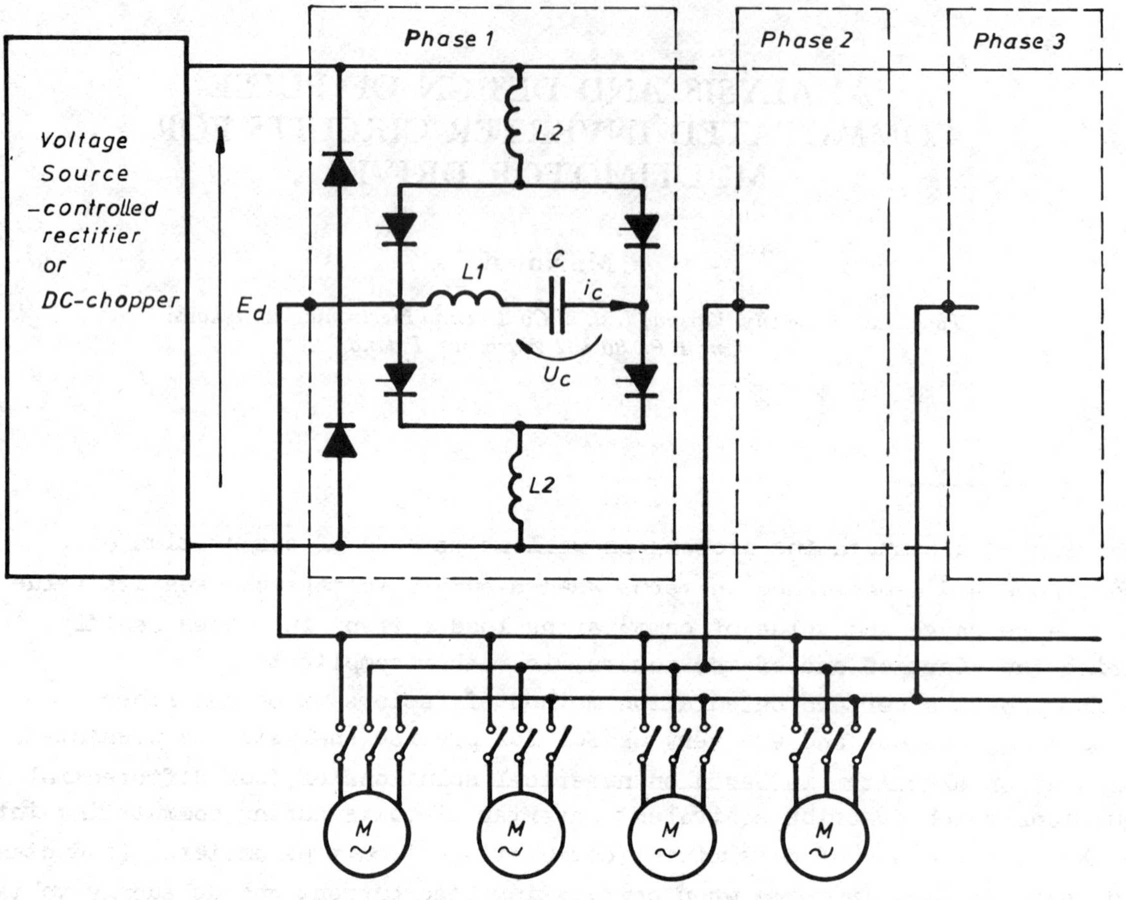

Fig. 1. The scheme of the voltage inverter utilized in multi-mitor drives

of LC parameters which permits for limitation of maximal capacity voltage, when the supply voltage and load current are changeable, is also considered. The paper includes diagrams showing a method of capacitor energy minimalization through the appropriate choice of LC parameters.

METHOD OF ANALYSIS OF COMMUTATION CIRCUIT

The principle and basic properties of commutation circuit are presented in literature (Ref. 1) . The capacitor voltage u_c and circuit turn-off time of main thyristors depend on the supply voltage value, commutated load current and circuit parameters (quality factor Q and inductance division factor L2/L1). During commutation, four characteristic time intervals, indicated in Fig. 2 by symbols A, B, D and E, should be taken into consideration. Each of these time intervals refers to appropriate differential equation describing the commutating process. The following, relative quantities are being introduced into these equations:

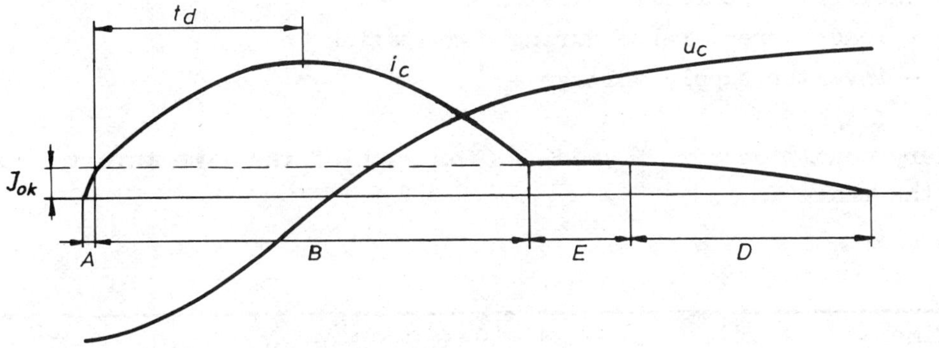

Fig. 2. Capacitor voltage - u_c and capacitor current - i_c during the commutation interval

- relative variable of capacitor voltage:

$$X = \frac{u_c}{E_d}$$

- relative current variable:

$$Y = \frac{dX}{d\tau}$$

- relative time variable:

$$\tau_d = \frac{t}{(L_1 + L_2)\, C}$$

Differention equations take form:
- in the interval A:

$$X + \frac{1}{1+m}\frac{dY}{dX} Y + \frac{1}{Q} Y = 0 \qquad (1)$$

- in the interval B:

$$X + \frac{dY}{dX} Y + \frac{1}{Q} = 0 \qquad (2)$$

- in the interval D:

$$X + \frac{dY}{dX} Y + \frac{1}{Q} Y = 1 \qquad (3)$$

- in the interval E

$$Y = Y_o \quad \text{for} \quad X < 1$$

For these equations following relations are valid:

$$m = \frac{L_2}{L_1}, \quad Q = \frac{R}{\frac{L_1+L_2}{C}}, \quad Y_o = \frac{I_{ok}}{E_d} \frac{L_1+L_2}{C}$$

where:

R - oscillation circuit resistance

I_{ok} - load current value during commutation

E_d - inverter supply voltage

The boundary condition for X and Y for each of the time intervals are given in the table I.

Table I

Time interval	A	B	E	D
Initial condition	$Y = 0$ $X_{pa} = X_o$	$Y = Y_o$ $X_{pb} = X_{ka}$	$Y = Y_o$ $X_{pe} = X_{kb}$	$Y = 0$ $X_{pd} = X_{kb}$ $X_{pd} = 1$
Terminal condition	$Y = Y_o$ X_{ka}	$Y = Y_o$ X_{kb}	$Y = Y_o$ $X_{ke} = 1$	$Y = 0$ X_{kd}
Interval validity condition	$Y < Y_o$ $X < 0$	$Y > Y_o$	$Y = Y_o$ $X_{kb} \leqslant X < 1$	$Y < Y_o$ $X \geqslant 1$

Remark: The subscript p stands for the beginning of the interval and subscript k for the end of the interval.

This form of differential equations was used in a computer program which enabled determining the relative capacitor voltage X_o and relative circuits turn-off time τ_d versus relative commutated current Y_o. The flow-chart of this program is presented in Fig. 3. The obtained characteristics of $X_o = f(Y_o)$ and $\tau_d = f(Y_o)$ are showed in Fig. 4.

Presented method of X_o and τ_d computation is specially useful for LC parameter evaluation when voltage supply and load current are being changed in a wide range.

THE METHOD OF CALCULATION OF LC PARAMETERS

The inductances L1 and L2 and capacity C in commutation circuit designed for a given range of supply voltage and, when step changes of commutated load current are predicted, should satisfy the following requirements:

a) circuit turn-off time must be greater than given in catalogues the critical value $t_{d\,min}$

b) the capacitor voltage should not exced the maximal value $U_{c\,max}$.

The principle of computation of LC parameters is explained in Fig. 5. For calculations a computer programme, which enables to determine the values X_o

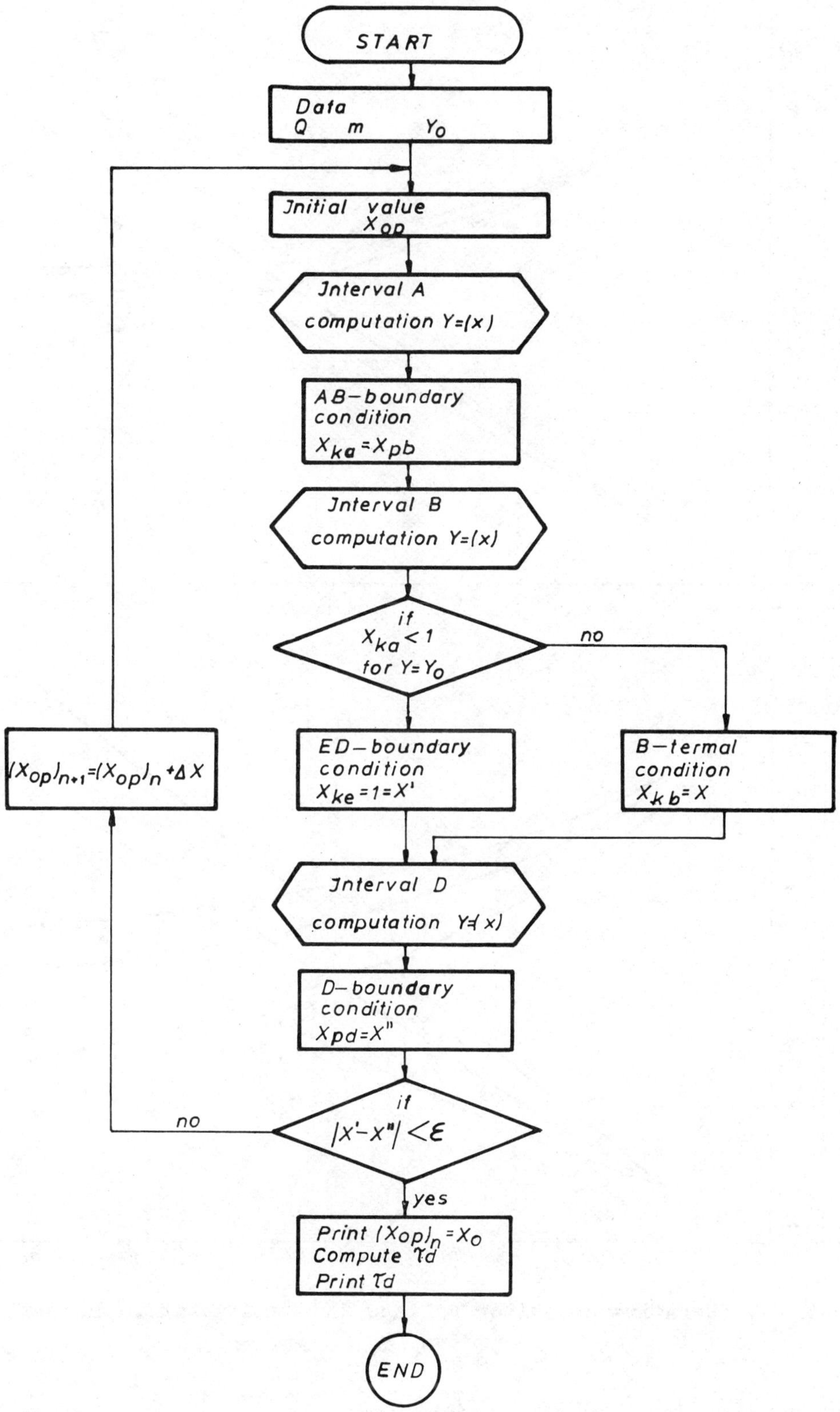

Fig. 3. Flow chart for the programme of $X_o(Y_o)$ and $\tau_d(Y_o)$ computation

and τ_d for given Y_o, was used. The procedure contains the following steps for computing LC parameters, as listed below:

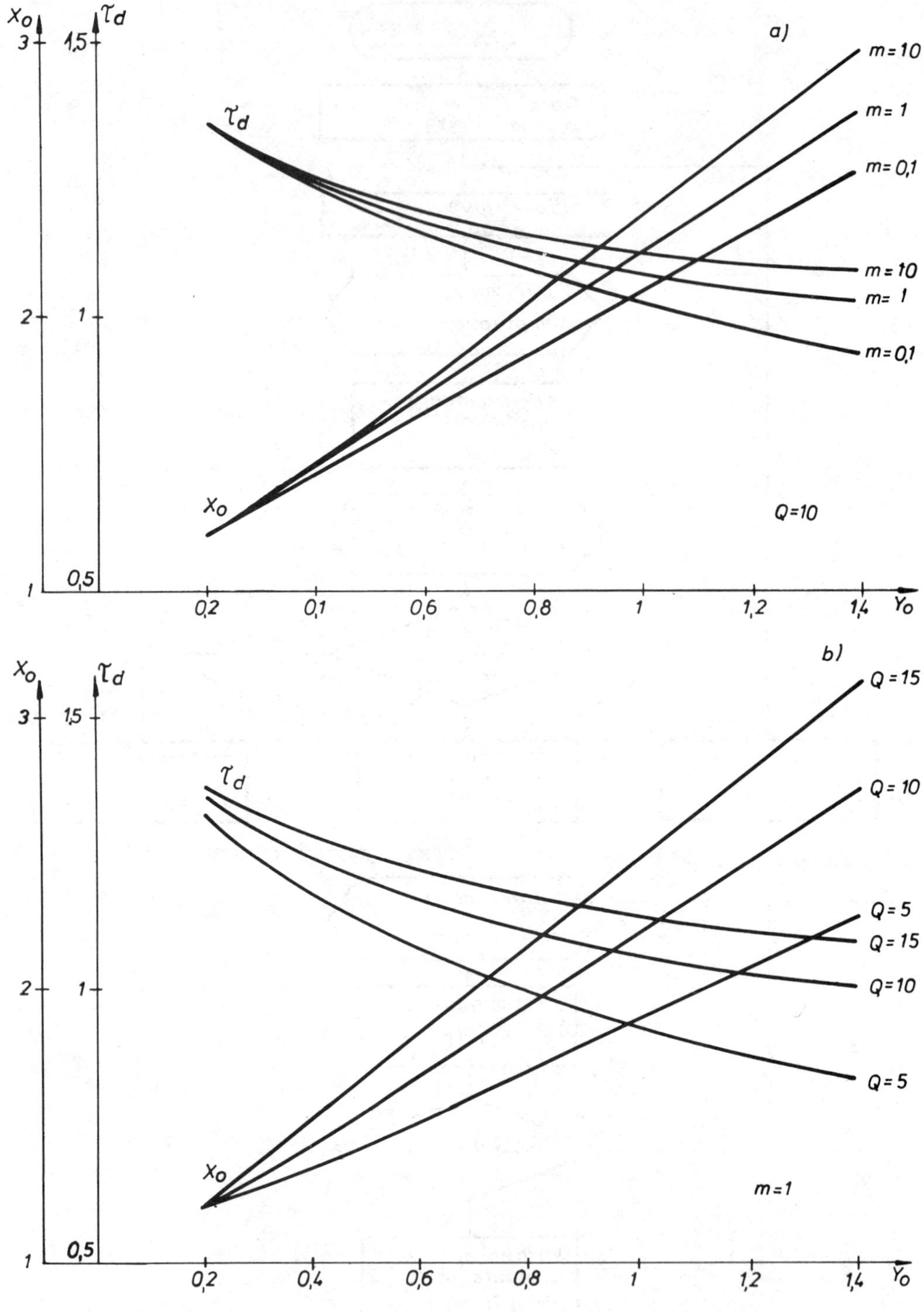

Fig. 4. Relative capacitor voltage X_o and relative turn-off time τ_d vs. Y_o parameter

1) computation of

$$X_{o\ min} = \frac{U_{cm}}{E_{d\ max}}$$

where U_{cm} - given maximal value of capacitor voltage

2) Computation of $Y_{o\ min}$ by an appropriate subroutine for given X_{omin} and Q_{max}.

3) Computation of

$$Y_{o\ max} = \frac{I_2\ E_{d\ max}}{I_1\ E_{d\ min}}$$

where: I_1 - commutated current value during the start of one of the motors sections and when maximum supply voltage is applied

I_2 - steady state value of commutated current for minimum supply voltage.

4) Computation by an appropriate subroutine of: $X_{o\ max} = f(Y_{o\ max})$ for assumed minimal value of quality factor Q_{min}.

5) Computation of

$$Y_o^* = \frac{I_3}{I_2} Y_{o\ max}$$

where: I_3 - commutated current value in transient when the minimal supply voltage is applied.

6) Computation of τ_d^* for given Y_o^* and $X_{o\ max}$.

7) Computation of $(L_1 + L_2)$ and C on the basis of relation:

$$C = \frac{t_{dn}}{\tau_d^*} \frac{I_3}{Y_o^*\ E_{d\ min}}$$

$$(L_1 + L_2) = \frac{t_{dn}}{\tau_d^*} \frac{Y_o^*\ E_{d\ min}}{I_3}$$

where: t_{dn} - the nominal turn-off time obtained through the cataloque thyristor data.

The values of L_1, L_2 and C obtained through this computer program satisfy the requirement of minimal circuit turn-off time and maximal capacitor/value, provided that the current quality factor Q is within the given range ($Q_{min} \div Q_{max}$).

The critical value of thyristors di/dt can cause additional correction of division factor L_2/L_1 and recomputation. The minimalization of the energy losses in commutated circuit can be obtained through optimal choice of maximal capacitor voltage value U_{cm}. In this case, it is necessary to evaluate the condenser capacitance for which the minimum of stored energy takes place for given U_{cm}.

RESULTS OF COMPUTATION AND EXPERIMENTS

On the basis of characteristics of commutated load current within the range $E_{d\ min} \div E_{d\ max}$ (Fig. 6) (obtained via experiments), some features diagrams were determined which illustrate the influence of maximal capacitor voltage value U_{cm}, the quality circuit factor Q ($Q_{min} \div Q_{max}$) and division factor m

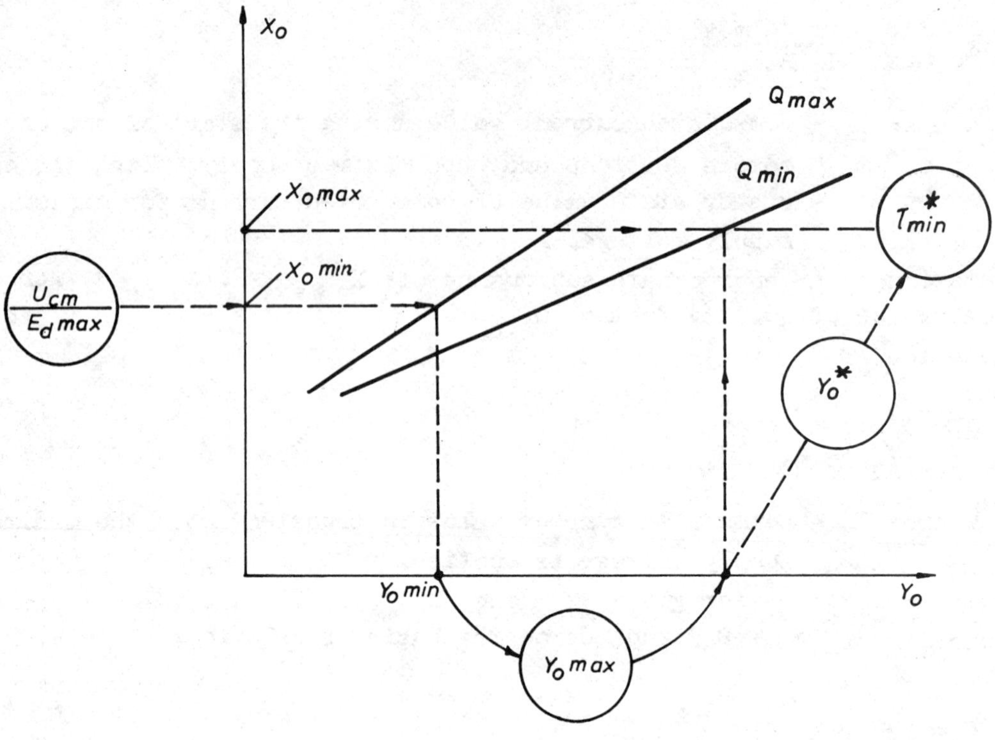

Fig. 5. Illustration of the method of LC-parameter computation

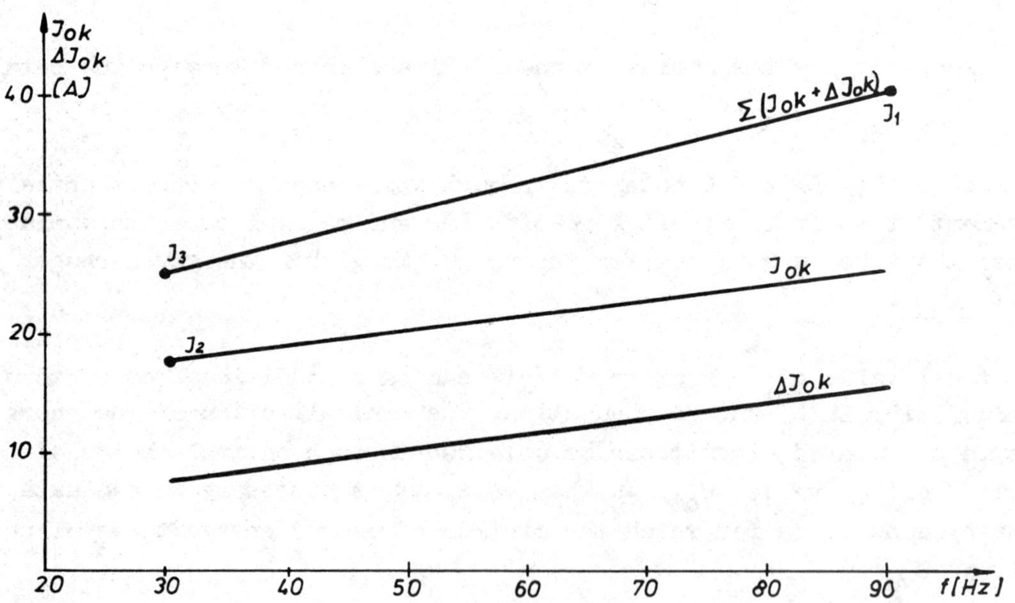

Fig. 6. Steady state I_{ok} and transient ΔI_{ok} values of commutated load current

on obtained, by computation, the values of parameters L_1, L_2 and C. These characteristics are presented in Fig. 7, Fig. 8 and Fig. 9. Relation between energy stored in capacitor C and voltage U_{cm} value, is shown in Fig. 10.
It indicates the possibility of loss minimalization. The characteristics taken

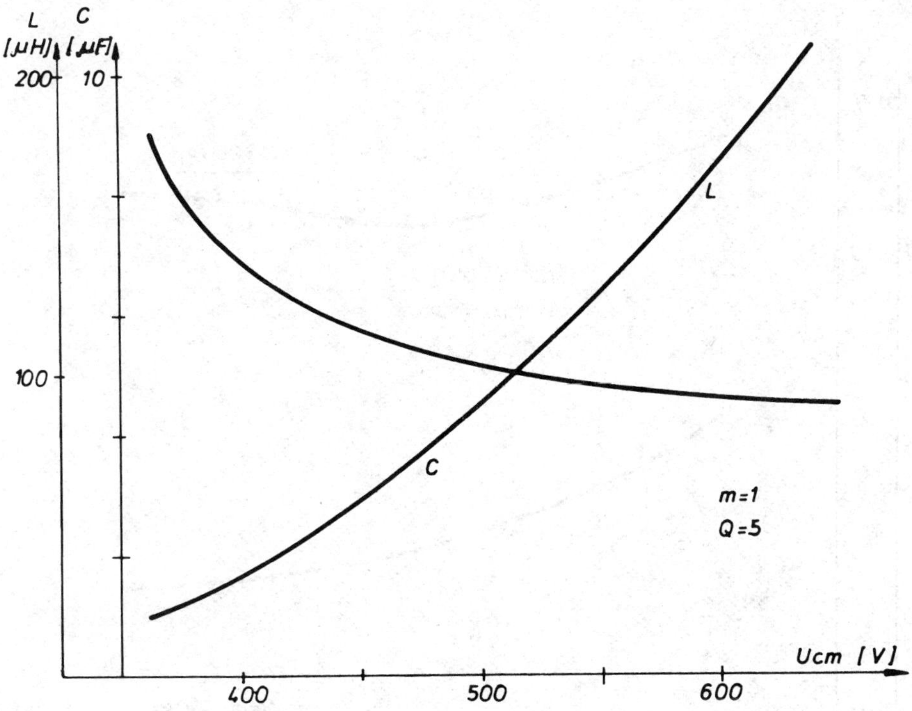

Fig. 7. Capacity C and inductance L versus U_{cm}

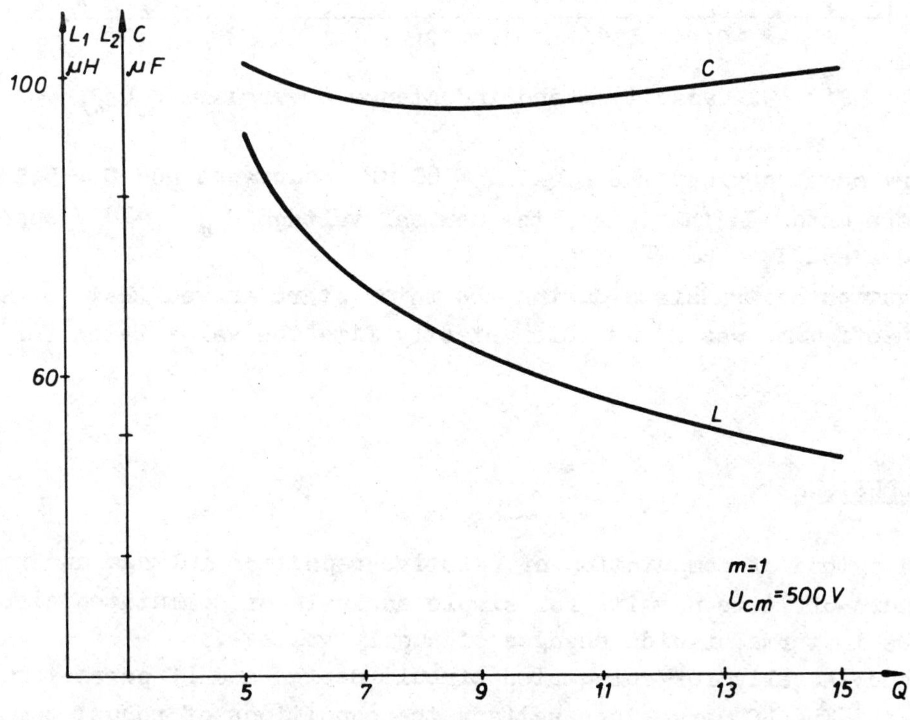

Fig. 8. Capacity C and inductance L versus Q

during the laboratory investigation prove the method validity and its sufficient accuracy. For given in Fig. 6 load current characteristics and for Q = 5, U_{cm} = 450 V, t_{dn} = 20 μs, the following values were obtained:
$L_1 = L_2$ = 60 μH, C = 5,65 μF

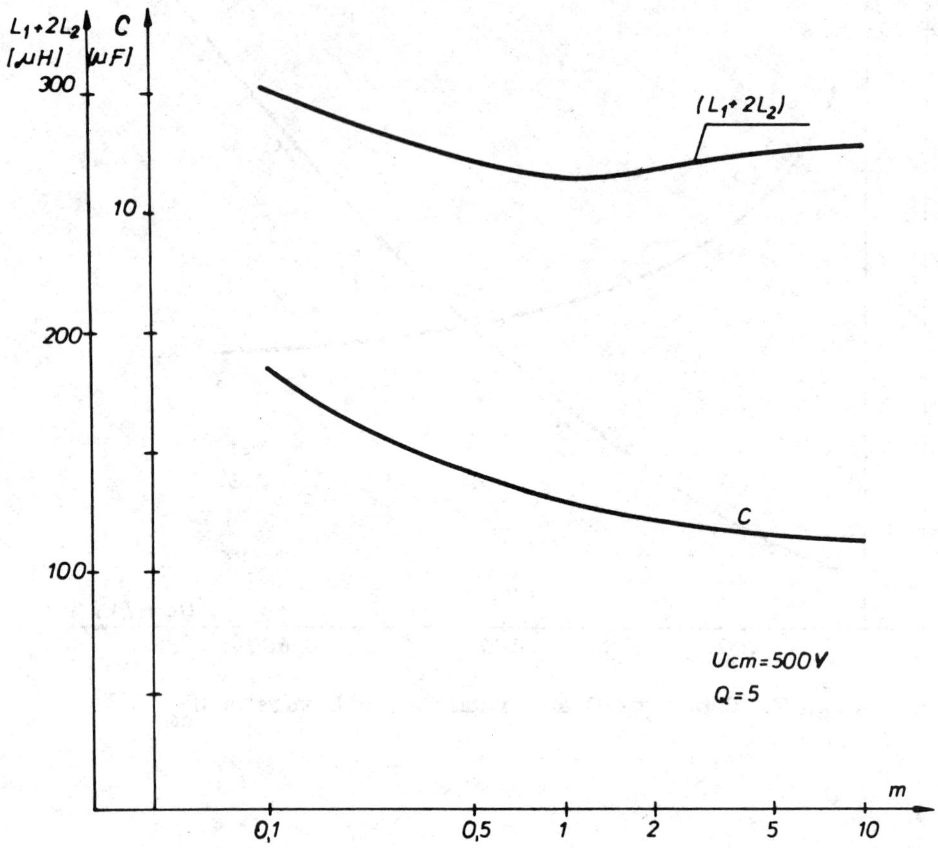

Fig. 9. Capacity C and inductance L versus $m = L_2/L_1$

In laboratory model device, the $L_1 = L_2 = 60\ \mu H$ inductance and $C = 5{,}8\ \mu F$ capacitor were used. In this case, the maximal voltage $U_{cm} = 420$ V appears for $I_{ok} = 40$ A and $E_d = 289$ V.
The investigation of transists during the motor start showed that the minimal circuit turn-off time was 20 μs which exactly fits the value taken for computation.

CONCLUSIONS

1) Presented method of computation of relative capacitor voltage and relative circuit turn-off time permits for simple analysis of commutated circuit properties in terms of wide changes of supply voltage.
2) This method is valid for commutated circuit design and LC parameter calculation. The LC parameters satisfy the conditions of robust commutation in which the maximal capacitor voltage value and minimal circuit turn-off time in transits are limited.
3) The computation results of LC - parameters are influenced by the values of Q and $m = L_2/L_1$.
4) The minimalization of the commutation losses can be performed via an appropriate choice of maximal voltage value.

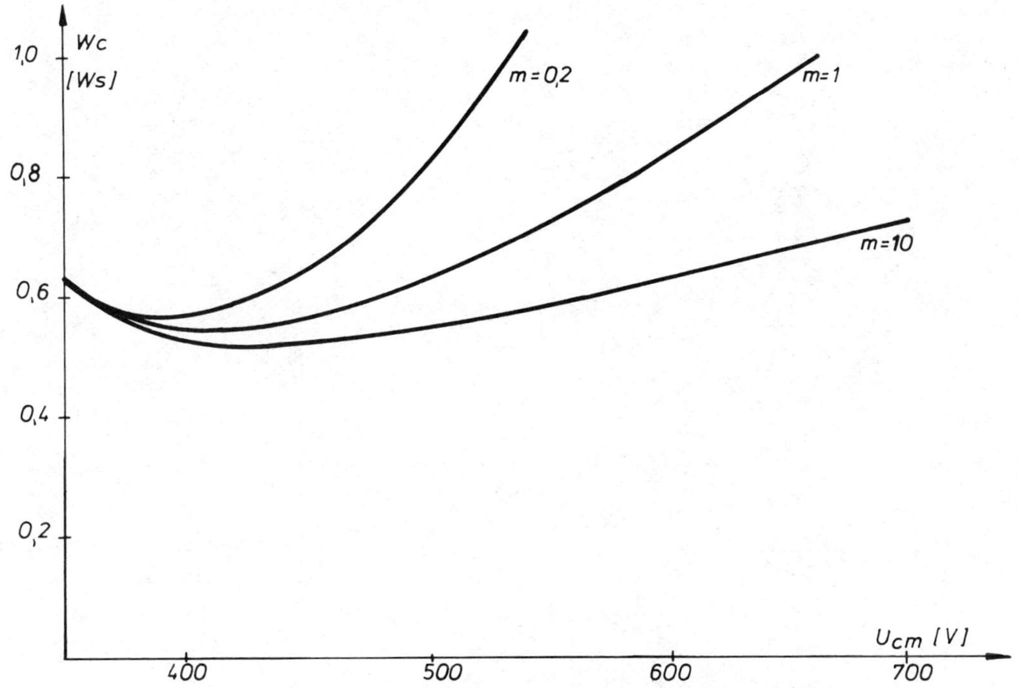

Fig. 10. Maximal energy stored in the capacitor circuit versus U_{cm} and m

REFERENCES

(1) Bedford B. D., Hoft R. G. (1964) Principles of inverter circuits. John Wiley & Sons New York

(2) Klaassens J.B., Analysis of Forced Commutation Circuit for Design of a Closs of Thyristor Inverters, IEEE Trans. on Ind. El. and Contr. Instr. 3, (1973)

(3) Glazienko T. A., Goncażenko R. B. (1969) Poluprowodnikowyje preobronowatieli castoty w elektropriwodach, Energia Leningrad

CONTROL OF CYCLOCONVERTERS FOR FEEDING OF ASYNCHRONOUS MACHINES

K. Hasse

Institut für Stromrichtertechnik und Antriebsregelung, Technische Hochschule, Darmstadt/Bundesrepublik Deutschland

SUMMARY

For the cycloconverter operation of thyristor bridges in anti-parallel connection, control techniques by which the output frequency can be considerably increased have been published. In the present investigation such an improved cycloconverter is used for the stator current control of a speed-controlled induction motor. A method of indirect flux control is applied. It is based on former investigations on the dynamic behaviour of the induction motor fed by PWM-controlled inverters. The performance of the cycloconverter brings steady-state errors in the current control. They can be eliminated by a co-ordinate transformation of the actual values of stator currents into the d,q-system and by an overlaid d,q component feedback control. Experimental results of the model plant are shown.

INTRODUCTION

For the speed control of squirrel cage induction motor drives by regulation of voltage and frequency, cycloconverters with natural commutation, instead of converters with intermediate d.c. circuit and forced commutation, can be used as frequency changers.

The expenditure for the commutation equipment and the smoothing circuits of intermediate circuit converters has to be compared with the great number of thyristors needed for the cycloconverters, although for the latter the requirements concerning turn-off time, current rise time, and du/dt sensitivity are low. As to operation characteristics, especially the high reactive power taken from the mains is inconvenient, whereas the possibility to work at lowest frequencies and even at zero frequency has to be mentioned as an advantage.

The principal disadvantage is seen in the fact that, with the 50 Hz-three-phase system as source, the maximum frequency usually has to be limited to about 25 Hertz. However, studies published in 1971 /1/, /2/ show that by use of improved control techniques for line commutated converters with anti-parallel connection of thyristor bridges substantially higher output frequencies are attained. The application of these methods includes a fast current control loop which is worked by sine wave reference signals and thus supresses the subharmonics in the output voltage.

One of these papers /1/ refers to antiparallel converters with circulating current, whereas the other /2/ describes a new equipment for work without circulating current. The essential improvement of this equipment is a technique of measurement which immediately indicates the forward blocking of the thyristors at zero current by means of a high frequency signal, so that the non-current delay during the crossover can be reduced to about 100 microseconds.

In both papers oscillograms demonstrate that output frequencies even higher than the mains frequency can be reached. However, the published results concern inductive load only. The present paper is to give a report about investigations of the speed-controlled induction motor fed by a cycloconverter. For these investigations a model plant using the method of /2/ was developped.

STATOR CURRENT CONTROL SYSTEM

For the feeding of a three-phase induction motor three pairs of antiparallel six-pulse bridges are applied, as shown in Fig. 1. The galvanic separation of the stator windings makes it possible to work without a converter transformer. The stator currents (i_a, i_b, i_c) are controlled by three separate feedback control systems, the desired time functions are effected by a three phase system of reference values (i^*_a, i^*_b, i^*_c), which have sine wave form in steady-state operation and whose amplitude and frequency are variable. These reference values are generated by a two phase system (i^*_α, i^*_β ; see Fig. 2)

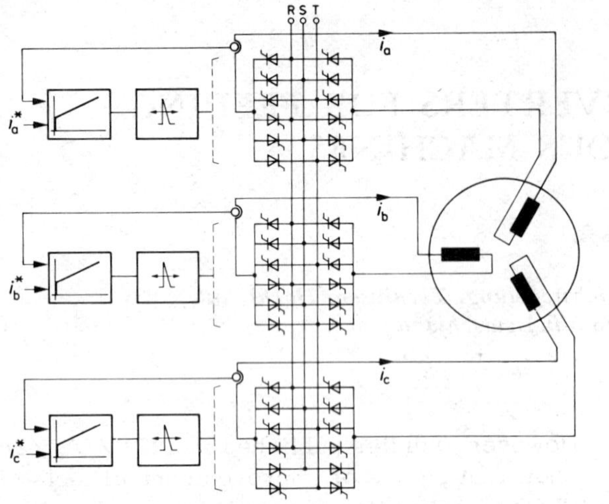

Fig. 1. Schema of the main power circuits

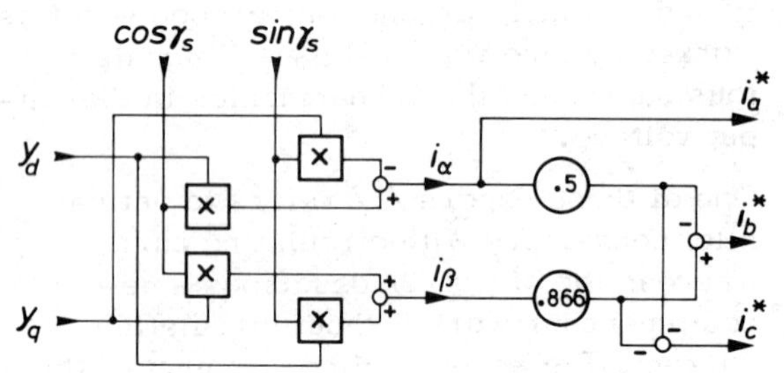

Fig. 2. Formation of stator current references

according to

$$
\begin{aligned}
i_a^* &= i_\alpha^* \\
i_b^* &= -\frac{1}{2} i_\alpha^* + \frac{\sqrt{3}}{2} i_\beta^* \\
i_c^* &= -\frac{1}{2} i_\alpha^* - \frac{\sqrt{3}}{2} i_\beta^*
\end{aligned} \tag{1}
$$

This two-phase system is produced by a stator angle reference system ($\cos\gamma_S$, $\sin\gamma_S$) and by the component amplitudes (y_d, y_q) in a so-called co-ordinate changer, which carries out the transformation

$$
\begin{aligned}
i_\alpha^* &= y_d \cos\gamma_S - y_q \sin\gamma_S \\
i_\beta^* &= y_d \sin\gamma_S + y_q \cos\gamma_S
\end{aligned} \tag{2}
$$

y_d and y_q represent the components of the resultant stator current vector along the perpendicular axes (d, q) of a co-ordinate system, which rotates at the speed of the revolving magnetic field. The angle γ_S marks the actual position of this d, q co-ordinate system in relation to the stator-field α, ß system. In steady-state operation y_d and y_q are constant and the time derivative of the angle γ_S is the stator frequency:

$$\omega_S = \frac{d\gamma_S}{dt} \tag{3}$$

The stator current values have to be set in a way that guarantees an economically resonable operation of the motor and, at the same time, an optimal dynamic behaviour. For this purpose, it has been proved as a most favourable method to control and to limit the slip frequency and to keep the rotor flux constant. If the latter requirement is met, the torque becomes linear with the slip frequency and follows its changes respectively changes of the stator current without delay nor transients. So, as long as the needed stator voltage is available, the dynamic behaviour is independent of the speed of the motor.

Systems of immediate feedback control of flux have been put into practice /3/,/4/. However they need installation of Hall probes in the air gap. In the present investigation an indirect method, which was published by the author /5/ and which avoids this expenditure, is used. It comprises the above-mentioned control of slip frequency and the setting of the stator current components as a function of slip frequency in such a way, that the rotor flux comes to the desired value and remains constant. These considerations are based on the differential equations of the rotor flux d, q-components at given stator current components i_d, i_q. These equations are

$$T_R \frac{d\Psi_{Rd}}{dt} = -\Psi_{Rd} + \omega_R T_R \Psi_{Rq} + L_h i_{Sd} \tag{4}$$

$$T_R \frac{d\Psi_{Rq}}{dt} = -\Psi_{Rq} - \omega_R T_R \Psi_{Rd} + L_h i_{Sq} \tag{5}$$

with $T_R = L_R/R_R$. The slip frequency ω_R means the angular velocity of the d, q co-ordinate system in relation to a rotor-fixed axis. It is convenient to give the d,q system such a position that the d-axis coincides with the resulting rotor flux vector, i.e.

$$\Psi_{Rd} = \Psi_R \tag{6}$$

$$\Psi_{Rq} = 0 \tag{7}$$

The additional requirement that the time derivatives of the rotor flux components have to be zero results in the following rule for the setting of the stator current components as functions of the desired value of rotor flux and of the slip frequency:

$$i_{Sd} = \frac{\Psi_R}{L_h} \tag{8}$$

and

$$i_{Sq} = \omega_R T_R \frac{\Psi_R}{L_h} = \omega_R T_R i_{Sd} \quad (9)$$

This means, that the current component along the rotor flux (i_{Sd}) has to be kept constant, whereas the component i_{Sq}, by which the torque is set, has to be linear with the slip frequency. A disadvantage of this very simple method is the sensitivity of the performance to the change of the parameters, especially of the rotor resistance.

For the slip frequency control a voltage-controlled oscillator with a two-phase sine wave output, as described for instance in /5/, is applied. Its input voltage is the analog signal for the desired slip frequency $\omega_R = d\gamma_R/dt$, the output signals are $\cos\gamma_R$ and $\sin\gamma_R$, the angle γ_R marking the wanted actual position of the rotor flux vector in relation to a rotor-fixed axis. As shown in Fig. 3, a co-ordinate transformation of this two phase system with the measured rotor position angle γ_m generates the stator angle (γ_S) reference system, which according to equ. (2) is needed for the formation of the sine wave stator current reference values, (see Fig. 2). The co-ordinate transformation results in the summation

$$\gamma_S = \gamma_R + \gamma_m \quad (10)$$

resp. $$\omega_S = \omega_R + \omega_m \quad (11)$$

The stator angle γ_S therefore indicates the rotor flux position in relation to a stator-fixed axis (α axis or a axis). The frequency addition (11) acts as a decoupling measure, by which the influence of the speed on the electromagnetic transients is eliminated.

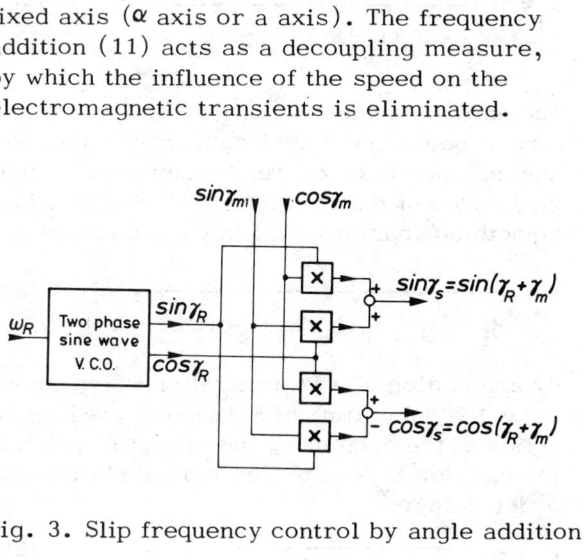

Fig. 3. Slip frequency control by angle addition

OPERATING CHARACTERISTICS OF CURRENT CONTROL USING THE CYCLOCONVERTER

In drives using inverters with forced commutation and with voltage control by pulse-width modulation, the accuracy of current control can be made good enough to adjust the stator current components by mere setting of the reference values (y_d, y_q in Fig. 2) according to the equ. (8) and (9). The prerequisite is a closed loop frequency response which, in the area of the stator frequency values, remains near to 1. However, if a cycloconverter is applied as power amplifying device, this prerequisite cannot be fulfilled.

For the study of the dynamic structure of the induction machine as plant of the instantaneous current control system it is proper to consider the differential equations for the electrical quantities in the stator-fixed co-ordinates (α, β).

So the stator currents of the two-phase equivalent machine ($i_{S\alpha}$ and $i_{S\beta}$) can be regarded as controlled variables, and the transformer coupling of the three-phase windings need not be taken into account. By application of the Laplace transformation without notice of the initial conditions the differential equations for the stator flux components will be

$$p\,\Psi_{S\alpha} = U_{S\alpha} - I_{S\alpha} R_S \quad (12)$$

$$p\,\Psi_{S\beta} = U_{S\beta} - I_{S\beta} R_S \quad (13)$$

and for the rotor flux components

$$p\Psi_{R\alpha} = -\,\omega_m \Psi_{R\beta} - I_{R\alpha} R_R \quad (14)$$

$$p\Psi_{R\beta} = \omega_m \Psi_{R\alpha} - I_{R\beta} R_R \quad (15)$$

The relation between flux and current components is given by

$$\Psi_{S\alpha(\beta)} = L_S\, I_{S\alpha(\beta)} + L_h\, I_{R\alpha(\beta)} \quad (16)$$

$$\Psi_{R\alpha(\beta)} = L_h\, I_{S\alpha(\beta)} + L_R\, I_{R\alpha(\beta)} \quad (17)$$

From these equations the block diagram of the controlled system (Fig. 4) can be evaluated, where the controlled conditions can be expressed by

$$I_{S\alpha} = F_1\,(U_{S\alpha} + F_2\,\omega_m\,\Psi_{R\beta}) \quad (18)$$

and

$$I_{S\beta} = F_1\,(U_{S\beta} - F_2\,\omega_m\,\Psi_{R\alpha}) \quad (19)$$

The transfer functions are

$$F_1 = \frac{1 + pT_R}{R_S + p(R_S T_R + L_S) + p^2 \sigma L_S T_R} \qquad (20)$$

$$F_2 = \frac{L_h}{L_R} \cdot \frac{pT_R}{1 + pT_R} \qquad (21)$$

$$\text{with } T_R = \frac{L_R}{R_R} \text{ and } \sigma = 1 - \frac{L_h^2}{L_S L_R}$$

The diagram shows the crosswise coupling of the stator circuits as a consequence of the induced voltages ($\omega_m \Psi_{R\alpha(\beta)}$) which are caused by the rotor motion. Therefore the rotor angular velocity (or the speed) affects the controlled system as a disturbance. As shown in Fig. 4, the rotor flux components are formed by

$$\Psi_{R\alpha} = \frac{1}{1 + pT_R}(L_h I_{S\alpha} - \omega_m \Psi_{R\beta}) \qquad (22)$$

$$\Psi_{R\beta} = \frac{1}{1 + pT_R}(L_h I_{S\beta} + \omega_m \Psi_{R\alpha}) \qquad (23)$$

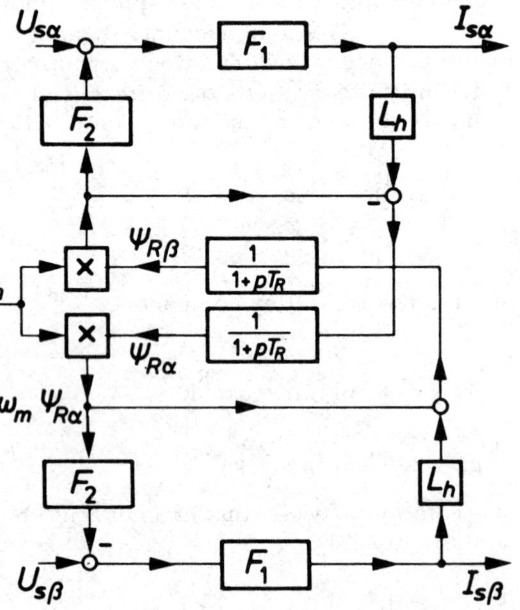

Fig. 4. Block diagram of the stator circuits

The mentioned coupling is eliminated, if the rotor flux Ψ_R (i.e. the amplitude of $\Psi_{R\alpha}$ and $\Psi_{R\beta}$) is kept constant by the slip frequency dependent current control described above.

In this case the rotor flux components as functions of time are

$$\Psi_{R\alpha} = \Psi_R \cdot \cos \gamma_S = \Psi_R \cdot \cos \omega_S t \qquad (24)$$

$$\Psi_{R\beta} = \Psi_R \cdot \sin \gamma_S = \Psi_R \cdot \sin \omega_S t \qquad (25)$$

and the amplitude of the voltage induced by rotor motion $\omega_m \Psi_R$ is determined only by the speed. Based on this consideration it is correct to describe the dynamic structure of the stator circuits for any function of time by two separate equivalent circuits as done in Fig. 5.

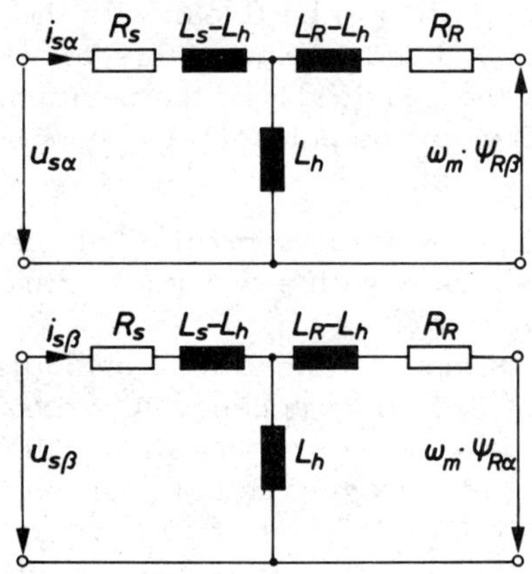

Fig. 5. Equivalent circuits for the equations (12) to (17)

For the design of the current controllers the generated speed voltage is not taken into account, that means a standstill of the rotor ($\omega_m = 0$) is assumed. For stator frequencies above $\omega_S = 1/T_R$ the transfer function F_1 (20) may be approximated by

$$F_1 \approx \frac{1}{R_S + R_R(L_h/L_R)^2 + p\sigma L_S} \qquad (26)$$

The entire transfer function F_s of the stator circuit between the cycloconverter output voltage and the stator current comprises the inner resistance of the converter (R_i) and additional smoothing reactors (L_z) in the motor links:

$$F_s = \frac{1}{R_i + R_S + R_R(L_h/L_R)^2 + p(\sigma L_S + L_z)} \qquad (27)$$

By application of a PI controller which compensates the delay time of F_s and effects integral action of the open loop, the closed loop transfer function F_w can be approximated by a second order delay:

$$F_w = \frac{1}{1 + pT_o + p^2 \frac{1}{2} T_o^2} \qquad (28)$$

In respect of the dead time of the six-pulse rectifier and of a small time lag caused by a smoothing element for the signal of the measured current, the time constant T_o cannot be set much lower than about 3 milliseconds.

Since in steady-state operation the controlled variables are sine wave functions, there is a current deviation of amplitude and phase which, for passive load, can be calculated by the expression (28), where $p = j\omega_S$. Under induction motor load condition the necessary error increases with rising speed by the influence of the generated speed voltage. This error depends not only on the speed but also on the slip frequency, because the phase angle between the stator current and the generated speed voltage is a function of the latter.

The closed loop frequency response of the current control system at steady-state operation of the induction motor with constant rotor flux was measured. Fig. 6 shows the results obtained from the model plant under the following operating conditions: positive rated slip frequency (full load motor action), negative rated slip frequency (full load generator action), and zero slip (idle motion). The measurements were carried out in such a way that the equations (8) and (9) for the d, q components of the stator current (i_d, i_q) were fullfilled by suitable setting of the input signals y_d and y_q. The response curves (Fig. 6) represent, in complex plane, the function

$$F(j\omega) = \frac{i_d + j\, i_q}{y_d + j\, y_q} \tag{29}$$

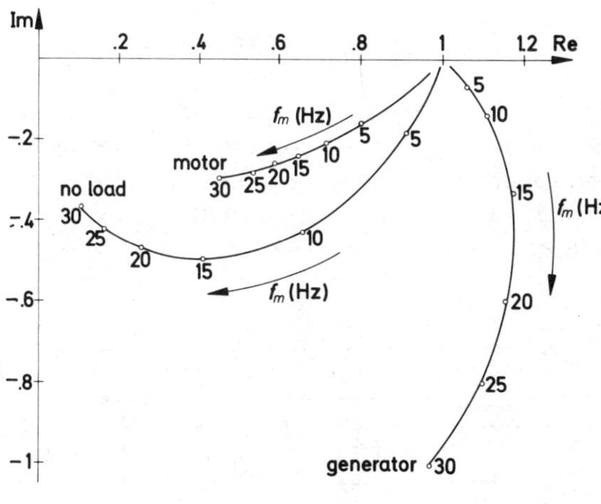

Fig. 6. Frequency response of $F(j\omega)$ on complex plane

OVERLAID COMPONENT CONTROL AND DECOUPLING

From the closed loop performance as expressed by the transfer function $F(j\omega)$ it is evident that, in order to attain the desired current values, offset signals of y_d and y_q which increase with speed are required in steady-state operation. Therefore the arrangement is supplemented by overlaid feedback control loops, whose controllers produce the required signals y_d and y_q. It is suitable to choose the d,q components of the stator currents (i_d, i_q) as controlled variables of these overlaid control loops. Since in steady-state condition i_d and i_q are constant, the final deviation can be eliminated by application of controllers with integral action. So the requirements for constant rotor flux can be met by setting the references (i_d^*, i_q^*) of the components according to equ. (8) and (9).

The feedback signals i_d and i_q are evaluated by the computation of the components (i_α, i_β) from the measured values of the phase currents (i_a, i_b, i_c)

$$i_\alpha = \frac{2}{3}\left(i_a - \frac{1}{2} i_b - \frac{1}{2} i_c\right) \tag{30}$$

$$i_\beta = \frac{1}{\sqrt{3}}\left(i_b - i_c\right) \tag{31}$$

and by a subsequent co-ordinate transformation by means of the stator angle reference system ($\sin\gamma_S$, $\cos\gamma_S$):

$$i_d = i_\alpha \cos\gamma_S + i_\beta \sin\gamma_S \tag{32}$$

$$i_q = -\, i_\alpha \sin\gamma_S + i_\beta \cos\gamma_S \tag{33}$$

Fig. 7 shows the block diagram of the resultant component control system.
As a consequence of the performance of the inner current control there exists a coupling between the two overlaid control loops. Their interaction complicates the layout of the controllers and impairs the dynamic behaviour. The decoupling network included in the arrangement (Fig. 7) is meant to reduce this mutual influence.

The design of this decoupling network is based on the mentioned measurements of the closed loop frequency response of the inner current control. It only comprises static characteristics which were calculated for steady-state condition. Test results obtained from the plant showed that this simplified way of decoupling is also sufficient for dynamic performance. Each of the references

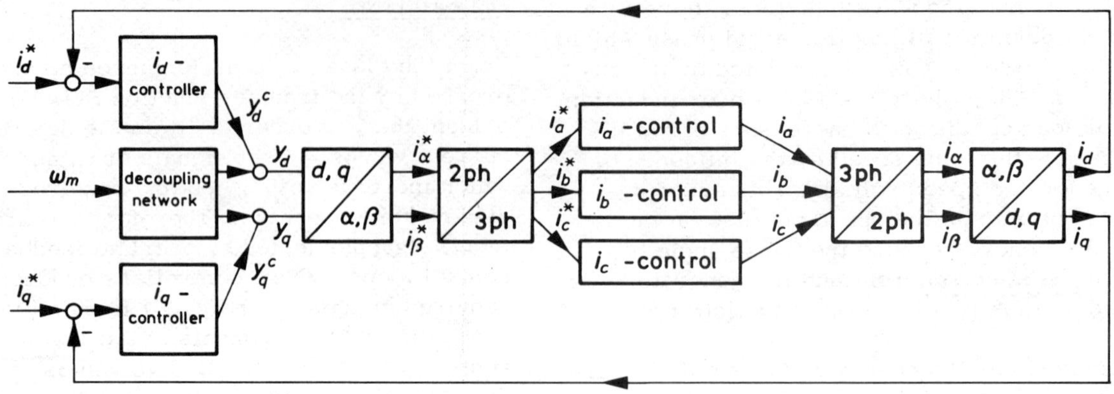

Fig. 7. Block diagram of the component control system

(y_d and y_q) consists of two items, one being required for the component i_d, the other for the component i_q:

$$y_d = y_{d1}(i_d) + y_{d2}(i_q) \tag{34}$$

$$y_q = y_{q1}(i_d) + y_{q2}(i_q) \tag{35}$$

As the results of the measurements show, these items are nearly linear with the components (i_d, i_q), so that equ. (34) and (35) can be approximated with satisfactory accuracy by

$$y_d = K_{11} i_d + K_{12} i_q \tag{36}$$

$$y_q = K_{21} i_d + K_{22} i_q \tag{37}$$

The coefficients (K) are functions of speed only and are calculated from the mean values of i_d and i_q. The functions K_{11} and K_{21} are obtained from $y_d = y_{do}$ and $y_q = y_{qo}$ at no load condition of the motor ($i_q = 0$):

$$K_{11}(\omega_m) = \frac{y_{do}}{i_d} \tag{38}$$

$$K_{21}(\omega_m) = \frac{y_{qo}}{i_d} \tag{39}$$

From measurements under load condition the functions

$$K_{12}(\omega_m) = \frac{y_d - y_{do}}{i_q} \tag{40}$$

and

$$K_{22}(\omega_m) = \frac{y_q - y_{qo}}{i_q} \tag{41}$$

are computed.

The equations (36) and (37) can be converted into

$$i_d = F_{11} y_d - F_{12} y_q \tag{42}$$

$$i_q = - F_{21} y_d + F_{22} y_q \tag{43}$$

where the static gain factors F of the inner closed loops as functions of speed are

$$F_{11}(\omega_m) = \frac{K_{22}}{D} \qquad F_{21}(\omega_m) = \frac{K_{21}}{D}$$
$$F_{12}(\omega_m) = \frac{K_{12}}{D} \qquad F_{22}(\omega_m) = \frac{K_{11}}{D} \tag{44}$$

with $D = K_{11} K_{22} - K_{12} K_{21}$

Fig. 8 shows the equivalent block diagram of the controlled system for steady-state operation together with the decoupling equipment, in which the component controller output signals are multiplied by suitable functions of

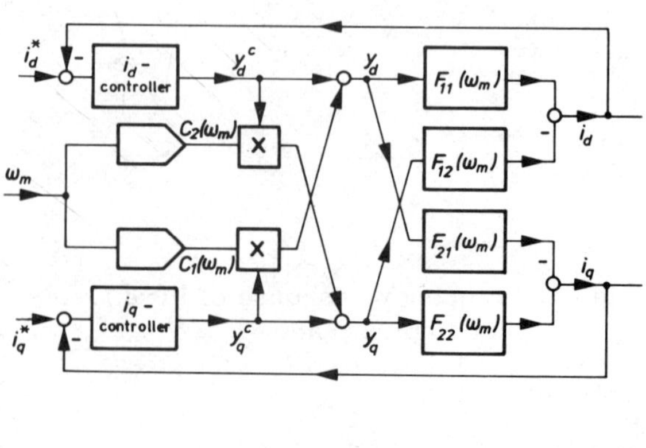

Fig. 8. Equivalent block diagram of the component control

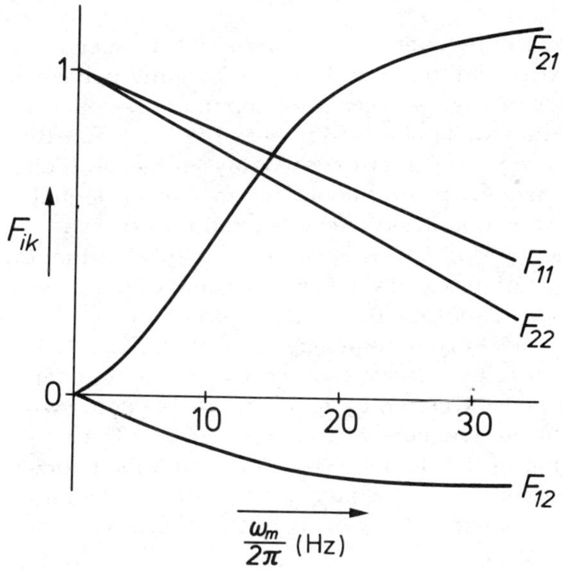

Fig. 9. Coefficients F_{ik} of equ. (42) and (43) versus speed

speed (ω_m). Fig. 9 contains the functions $F(\omega_m)$. The decrease of the static gain factors F_{11} and F_{22} with speed is caused by the influence of the induced speed voltage on the inner control loops as discussed above. Without decoupling measures, the layout of the component controllers would have to be set, for stability reasons, on the basis of the maximum gain, which appears at low speed. Therefore, the delay of the closed loop would increase considerably with speed according to the decrease of the gain factor. To attain a satisfactory dynamic behaviour measures must be taken to keep the dependence of the resultant gain on the speed as small as possible. The decoupling network used here is also apt to take over this function.

From Fig. 8 the required functions C_1 and C_2 which fulfill the decoupling condition can be read immediately:

$$C_1(\omega_m) = \frac{F_{12}}{F_{11}} \tag{45}$$

$$C_2(\omega_m) = \frac{F_{21}}{F_{22}} \tag{46}$$

C_1 and C_2 can be expressed by the measured functions (see equ. (44)) as

$$C_1(\omega_m) = \frac{K_{12}}{K_{22}} \tag{47}$$

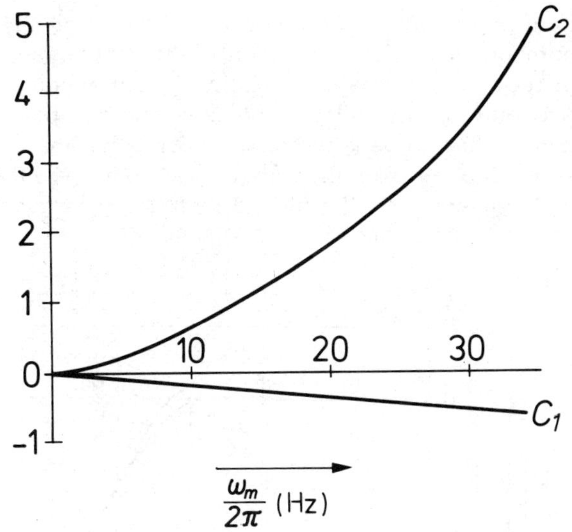

Fig. 10. Functions $C(\omega_m)$ of the decoupling network

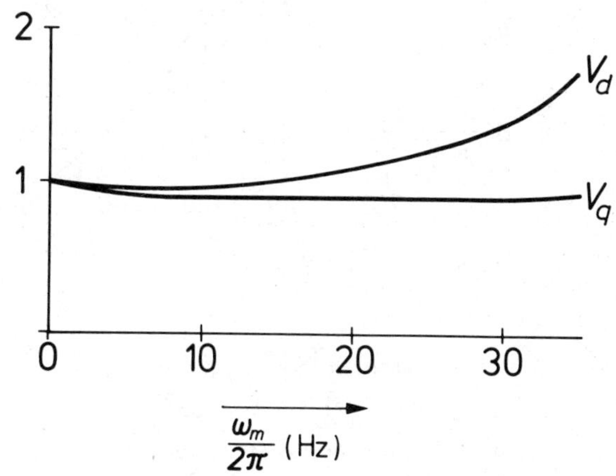

Fig. 11. Resultant open-loop gain factors versus speed

$$C_2(\omega_m) = \frac{K_{21}}{K_{11}} \tag{48}$$

These decoupling functions, shown in Fig. 10, are executed in the plant by inserted non-linear elements with the speed as the input signal (see Fig. 8). Then the resultant gain factors from the controller outputs (y_d^c, y_q^c) to the controlled variables will be

$$V_d(\omega_m) = \frac{i_d}{y_d^c} = F_{11}(1 - C_1C_2) \tag{49}$$

$$V_q(\omega_m) = \frac{i_q}{y_q^c} = F_{22}(1 - C_1C_2) \tag{50}$$

As to be seen in Fig. 11, the application of the decoupling network brings about the desired constancy of the resultant open loop gain factor (V_q) in the control of the current component i_q. In the i_d control the increase of the resultant gain factor (V_d) with speed presents no difficulties, because the component i_d has to be kept constant for constant rotor flux as explained above (see equ. (8)) and consequently the closed loop delay is not of such importance.

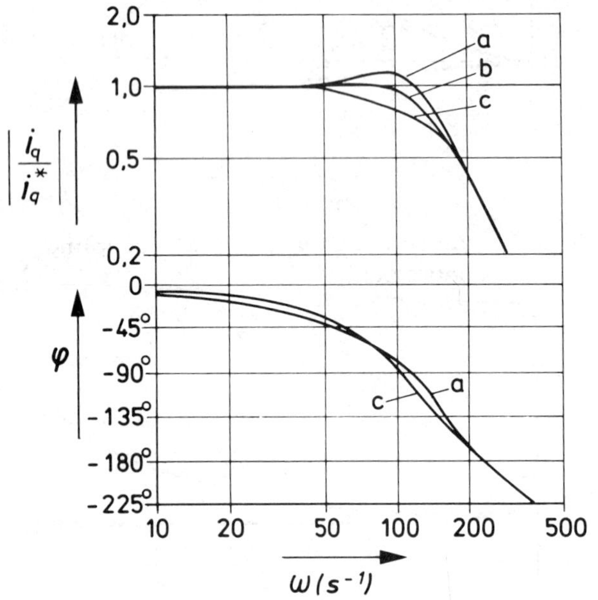

Fig. 12. Closed-loop frequency response i_q/i_q^* at the speed f_m = 25 Hz
a) full load generator b) no load
c) full load motor

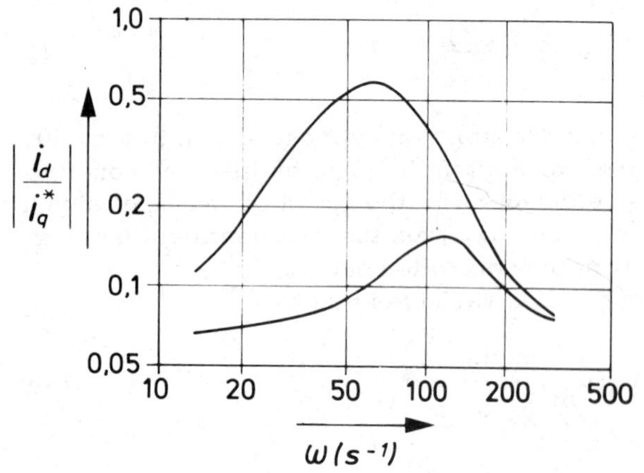

Fig. 13. Magnitude of the closed-loop frequency response i_d/i_q^* at full-load motor operation and f_m = 25 Hz

EXPERIMENTAL RESULTS

For the performance of the plant as speed-controlled drive mainly the dynamic behaviour of the i_q-control is important becouse i_q is the component which sets the torque. With respect to a first order delay element which is necessary to smooth the measured signal of the controlled variable, the transit frequency was set in a way that at full load motor operation a satisfactory damping of the closed loop was attained. As the Bode diagram of the measured frequency response (Fig. 12) shows, the closed loop transfer function can be approximated by a second order delay with a break frequency of $\omega_o = 125 \ s^{-1}$. The alteration of the damping with changing load conditions is a consequence of a certain deviation of the open loop gain from the average curve (Fig. 11).

The efficacy of the decoupling device was tested under closed loop condition by measuring the influence of a sinusoidal i_q reference signal on the i_d control. The Bode diagram Fig. 13 contains the steady-state frequency responses for the magnitude of i_d/i_q^* at full motor load and at a speed of 25 Hz. It shows the improvement attained by the application of the decoupling network.

Fig. 14 and 15 show the transient responses of the controlled variables i_d and i_q and of one of the phase currents to a step function of the reference i_q^*. The response of i_d in Fig. 14 indicates the disturbing effect of the coupling in dynamic operation. Fig. 15 demonstrates to what extent the performance can be improved by the decoupling.

The model plant tests proved that the overlaid component control loops together with the decoupling equipment yield a reliable operation and a good dynamic behaviour of the current control system as a part of a speed-controlled induction motor drive. Figure 16 illustrates the performance at a fast speed reversal of the unloaded motor with an overlaid speed control.

The characteristics of transient operation are comparable to those of speed-controlled dc-motor drives. Faultless working is possible up to a stator frequency of 35 Hertz. At higher frequencies there arise major subharmonics of the stator currents and, therefore, discontinuities of the torque. In cases in which these inconveniences may be tolerated, stator frequencies close to the power frequency can be reached.

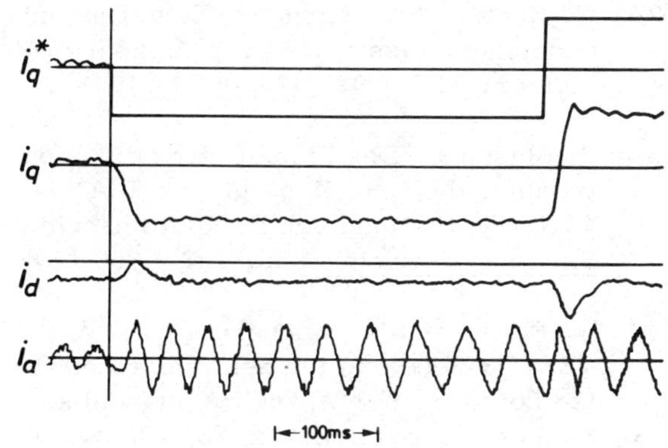

Fig. 14. Time characteristics of the component control without decoupling network

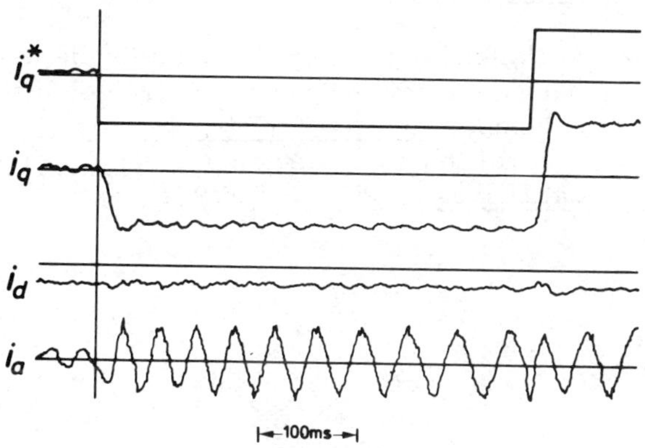

Fig. 15. Time characteristics of the component with decoupling network

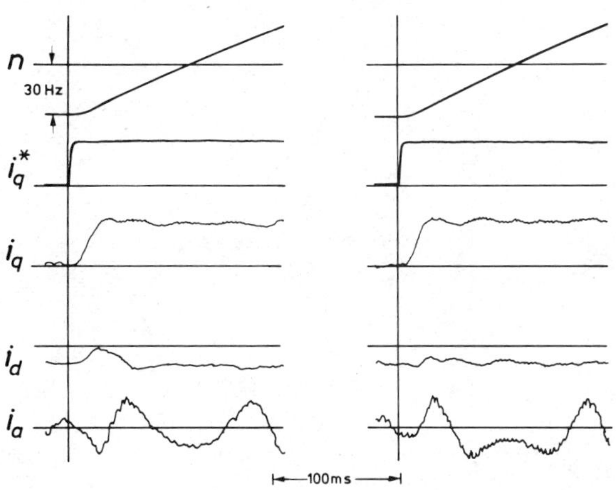

Fig. 16. Time characteristics at a speed reversal

left: without decoupling network
right: with decoupling network

SYMBOLS

f_m	frequency corresponding to the speed, $f_m = z \cdot n$
F, F_1, F_2	transfer functions
F_s	transfer function of the controlled system
F_w	closed loop transfer function
i_a, i_b, i_c	stator phase currents
$i_\alpha, i_\beta, i_{S\alpha}, i_{S\beta}$	components of the stator current vector in stator-fixed axes
$I_{S\alpha}, I_{S\beta}$	Laplace transforms of $i_{S\alpha}, i_{S\beta}$
$I_{R\alpha}, I_{R\beta}$	Laplace transforms of the components of the rotor current vector in stator-fixed axes
i_d, i_{Sd}	direct-axis component of the stator current vector
i_q, i_{Sq}	quadrature-axis component of the stator current vector
L_h	mutual inductance
L_S	total stator inductance
L_R	total rotor inductance
L_z	inductance of additional reactors
n	speed
p	Laplace operator
R_i	inner resistance of source
R_S	stator resistance
R_R	rotor resistance
T_R	rotor time constant
$u_{S\alpha}, u_{S\beta}$	components of the stator voltage vector in stator-fixed axes
$U_{S\alpha}, U_{S\beta}$	Laplace transforms of $u_{S\alpha}, u_{S\beta}$
y_d, y_q	references of the inner current control loops
y_d^c, y_q^c	component controller output signals
z	number of pole pairs
γ_m	angle of the rotor position related to a stator-fixed axis
γ_S	angle of the direct axis related to a stator-fixed axis
γ_R	angle of the direct axis related to a rotor-fixed axis
σ	leakage coefficient
$\psi_{S\alpha}, \psi_{S\beta}$	components of the stator flux linkage in stator-fixed axes

$\psi_{R\alpha}, \psi_{R\beta}$	components of the rotor flux linkage vector in stator-fixed axes
ψ_R	rotor flux linkage
ψ_{Rd}	direct-axis component of the rotor flux linkage vector
ψ_{Rq}	quadrature-axis component of the rotor flux linkage vector
ω	angular frequency
ω_m	mechanical angular velocity of the rotor, $\omega_m = 2\pi \cdot z \cdot n$
ω_S	stator frequency
ω_R	slip frequency

REFERENCES

/1/ D. Schröder, Dynamische Eigenschaften von Stromrichter-Stellgliedern mit natürlicher Kommutierung, Regelungstechnik und Prozeß-Datenverarbeitung 19, 155 - 162 (1971).

/2/ U. Kunz and H. Zürneck, Steuerung und Regelung kreisstromfreier Umkehrstromrichter, ETZ-A 92, 418 - 421 (1971).

/3/ F. Blaschke, Das Prinzip der Feldorientierung, die Grundlage für die TRANSVEKTOR-Regelung von Drehfeldmaschinen, Siemens-Zeitschrift 45, 757 - 760 (1971).

/4/ W. Flöter and H. Rippberger, Die TRANSVEKTOR-Regelung für den feldorientierten Betrieb einer Asynchronmaschine, Siemens-Zeitschrift 45, 761 - 764 (1971).

/5/ K. Hasse, Drehzahlregelverfahren für schnelle Umkehrantriebe mit stromrichtergespeisten Asynchron-Kurzschlußläufermotoren, Regelungstechnik und Prozeß-Datenverarbeitung 20, 60 - 66 (1972).

/6/ R. Jötten, Signalverarbeitung für die Regelung umrichtergespeister Drehfeldmaschinen, Proc. 1. IFAC-Symposium Control in Power Electronics and Electrical Drives 1, 681 - 695 (1974).

CONTINUOUS OPERATION METHODS OF STATIC SCHERBIUS CONTROL SYSTEM AT INSTANTANEOUS POWER FAILURE

M. Honbu*, T. Hori*, Y. Hiro, H. Ichikawa***, K. Iwata** and Y. Fukuda****

**Hitachi Research Laboratory, **Hitachi Works, ***Omika Works of Hitachi Ltd., Hitachi-shi, Ibaraki-ken, Japan*

ABSTRACT

The static Scherbius control system, widely used for speed control of large-capacity induction motors, is incapable of smooth continuous operation at instantaneous power failure lasting about 0.1 to 0.6 second, because it is then subject to (a) commutation failure of inverter due to a lack of commutation voltage or to (b) overcurrent based on an overvoltage inducing in the secondary winding of the motor.

This paper proposes a method to overcome this defect.

(1) Transient voltage and current phenomena in the induction motor at instantaneous power failure are first made clear.

(2) On this basis a method that permits continuous operation even at instantaneous power failure was developed whereby, upon instantaneous power failure, shift and block control of inverter gate firing signals is effected and a resistor is inserted in the secondary winding of the induction motor.

(3) The effect of this method was verified experimentally.

INTRODUCTION

The static Kraemer control system and static Scherbius control system, in which a static frequency converter is connected to the secondary winding of a wound-rotor induction motor to effect speed control, have been widely used to drive large-capacity pumps because they offer high efficiency, good control performance, and a wide range of speed control (1)-(5). In recent years, speed control systems for large-capacity pump drives have been required to provide higher reliability to prevent any water shut-off and turbidity, to secure steady water supply and better service to the user, and maintenability as well as ease of servicing and saving of labor.

To meet this requirement, the authors developed a commutatorless Kraemer control system in which the direct current motor is replaced by a commutatorless motor (6),(7). In this system, inverter commutation is carried out by the counter electromotive force of the commutatorless motor, so that, even at instantaneous power failure, commutation failure of inverter is prevented and normal operation is continued without interruption.

The static Scherbius control system, which is employed more frequently than the static Kraemer control system, has presented reliability problems because it cannot operate through instantaneous power failure. To make the continuous operation possible a new static Scherbius control system was developed.

This paper first points out problems involving motor and inverter at instantaneous power failure with regard to a conventional static Scherbius control system. To grasp these problems quantitatively, transient phenomena at instantaneous power failure are analyzed for the induction motor operating with and without the static Scherbius control system. Methods to detect instantaneous power failure, prevent inverter commutation failure, and suppress transient secondary voltage of induction motor are described as means of enabling the system to operate through instantaneous power failure. Results of our consideration are discussed with oscillograms from tests.

PROBLEMS ON CONVENTIONAL STATIC SCHERBIUS CONTROL SYSTEM AT INSTANTANEOUS POWER FAILURE

Figure 1 shows a conventional static Scherbius control system. Phenomena that occur at instantaneous power failure are as follows: An abnormal condition develops by, for instance, lightning, wind or snow, and causes a voltage drop. The abnormal condition is detected and power transmission is stopped. After a short period of power shut-off, the circuit is closed again to resume a conducting state, and normal conditions are recovered. Instantaneous power failure can occur in different modes, such as single-phase open circuit, three-phase open circuit, line-to-ground short circuit, line-to-line short circuit, three-phase short circuit, and voltage unbalance. Simulation tests in these modes revealed the following problems for the conventional static Scherbius control system.

Transient Secondary Voltage of Induction Motor

During instantaneous power failure and at recovery, an extremely high over-voltage is induced in the secondary winding of the induction motor. When the power source voltage is unbalanced, a negative-sequence voltage is induced. The transient secondary voltage that is induced at instantaneous power failure is the largest at the recovery from a three-phase open circuit and is equal to the secondary voltage at slip S = 1. Because of this transient secondary voltage, a current larger than the commutatable maximum current of the inverter flows, resulting in commutation failure of the inverter. Further this transient voltage causes an overvoltage breakdown of diodes in the rectifier and thyristors in the inverter because their allowable maximum voltages are based on steady state operation.

Primary Residual Voltage of Induction Motor

To develop the new static Scherbius control system capable of operating through instantaneous power failure, a means of quick and sure detection of instantaneous power failure must be found. The most common method of detecting instantaneous power failure is by a low-voltage relay, which responds to a voltage drop. The relay generates a signal when the power source voltage falls to 80 or 85 % of the rated value. In the static Scherbius control system, power-factor-correcting capacitors are connected to the power source. Owing to these capacitors, the primary voltage of the induction motor falls very slowly at instantaneous power failure, or the induction motor undergoes self-excitation to raise the primary voltage. As a result, depending on the mode of instantaneous power failure, a low-voltage relay will be delayed in the response or will even fail to respond at all.

Commutation Failure of Inverter

The static Scherbius control system uses a separate excited inverter in which commutation is carried out by the power source voltage. As power source voltage drops at instantaneous power failure, commutation voltage of the inverter also drops and causes commutation failure. Another possible cause of inverter commutation failure at instantaneous power failure is the unstable operation of control circuits and gate pulse generator.
In a mode where power source voltage is unbalanced, a negative-sequence voltage causes the induction motor secondary voltage to rise and the rectifier DC voltage to increase. In a phase where power source voltage drops, inverter commutation fails and inverter DC voltage drops. Therefore, the DC voltage between the rectifier and the inverter is not in the state of equilibrium, overcurrent flows, and the high-speed circuit breaker for overcurrent protection is tripped.

Vanish of Power Source for Control and Sequential Circuit

Operation of the automatic current regulator, automatic speed regulator, gate pulse generator, etc. becomes unstable and this leads to inverter commutation failure and overcurrent.

ANALYSIS OF TRANSIENT PHENOMENA AT INSTANTANEOUS POWER FAILURE

Transient secondary voltage, inverter commutation failure and other phenomena have been pointed out above as problems on the conventional static Scherbius control system at instantaneous power failure. Now these transient phenomena will be analyzed in an attempt to quantify them.

First, to determine the magnitude of transient secondary voltage, the transient phenomena of the induction motor with secondary resistor will be analyzed. Figure 2 shows a model of the induction motor expressed with α-β axis. The relationships of the primary and secondary currents expressed by three-phase components, viz. i_a, i_b, i_c, and i_u, i_v, i_w, to the primary and secondary currents expressed by α-β axis, viz. I_α, I_β, and i_α, i_β, are given in the following equations:

$$\begin{bmatrix} I_\alpha \\ I_\beta \end{bmatrix} = \sqrt{\frac{2}{3}} \begin{bmatrix} 1 & -\frac{1}{2} & -\frac{1}{2} \\ 0 & \frac{\sqrt{3}}{2} & -\frac{\sqrt{3}}{2} \end{bmatrix} \begin{bmatrix} i_a \\ i_b \\ i_c \end{bmatrix}, \quad \begin{bmatrix} i_\alpha \\ i_\beta \end{bmatrix} = \sqrt{\frac{2}{3}} \begin{bmatrix} 1 & -\frac{1}{2} & -\frac{1}{2} \\ 0 & \frac{\sqrt{3}}{2} & -\frac{\sqrt{3}}{2} \end{bmatrix} \begin{bmatrix} i_u \\ i_v \\ i_w \end{bmatrix} \qquad (1)$$

Similar relations are given for voltages. The basic equation of the induction motor by the α-β axis components is given as follows (8),(9):

$$\begin{bmatrix} L_s & 0 & M\cos\theta & M\sin\theta \\ 0 & L_s & -M\sin\theta & M\cos\theta \\ M\cos\theta & -M\sin\theta & L_r & 0 \\ M\sin\theta & M\cos\theta & 0 & L_r \end{bmatrix} \begin{bmatrix} PI_\beta \\ PI_\alpha \\ Pi_\beta \\ Pi_\alpha \end{bmatrix} = \begin{bmatrix} E_\beta - M\omega_r(-i_\beta\sin\theta + i_\alpha\cos\theta) - R_sI_\beta \\ E_\alpha - M\omega_r(-i_\beta\cos\theta - i_\alpha\sin\theta) - R_sI_\alpha \\ e_\beta - M\omega_r(-I_\beta\sin\theta - I_\alpha\cos\theta) - R_ri_\beta \\ e_\alpha - M\omega_r(I_\beta\cos\theta - I_\alpha\sin\theta) - R_ri_\alpha \end{bmatrix} \qquad (2)$$

where $\omega_r = (1 - S)\omega_s$, $\theta = \omega_r t + \theta_0$

ω_r = Electrical angular velocity of the rotor
ω_s = Electrical angular velocity of the power source voltage
S = Slip
P = Operator $(\frac{d}{dt})$

If it is assumed that the power source voltage is in three-phase balance then:

$$E_\alpha = E_s\cos(\omega_s t + \phi), \quad E_\beta = E_s\sin(\omega_s t + \phi) \qquad (3)$$

where E_s = Line-to-line voltage of power source (RMS value)
External resistor R_{2F} is connected to the secondary winding. Therefore:

$$e_\alpha = -R_{2F}i_\alpha, \qquad e_\beta = -R_{2F}i_\beta \qquad (4)$$

Substituting Equations (3) and (4) into Equation (2), and performing numerical analysis of Equation (2), we obtain the transient phenomena of the induction motor with secondary resistor.
Table 1 shows the maximum voltages and currents measured and calculated in a transient state of recovery from a three-phase open circuit. The measured and calculated results are in good agreement. Constants of the motor used for the calculations are presented in Table 2.
It can be predicted that the transient secondary voltage of the induction motor that is induced at instantaneous power failure is related to the resistance connected to the secondary winding. By varying R_{2F} in Equation (4) to seek its relationship to the maximum transient secondary voltage obtained by numerical analysis of Equation (2), we get Fig. 3. It is a matter of course that transient secondary voltage decreases as the resistance decreases. When the resistance is less than 10 times the secondary winding resistance, the transient secondary voltage decreases at a rapid rate. It is saturated over the ten times.
Next, to obtain the current that flows in the rectifier and inverter at instantaneous power failure, the transient phenomena of a static Scherbius control system will be analyzed. In the static Scherbius control system shown in Fig.1, a rectifier is connected to the secondary winding of the induction motor. Commutation phenomenon of this rectifier is represented by the single mode and commutation mode shown in Fig. 4. In the single mode, in Fig. 4(a), current flows from the w phase to the v phase; therefore:

$$i_u = 0, \quad i_v = -i_w = I_d \qquad (5)$$

By substituting Equation (5) into Equation (1), we get:

$$i_\alpha = 0, \qquad i_\beta = \sqrt{2}\, I_d \tag{6}$$

Therefore, in the single mode, it is sufficient to obtain i_β only for the secondary current. e_β here is, from Equations (1) and (4),:

$$e_\beta = \frac{1}{\sqrt{2}}(e_v - e_w) = -\frac{1}{\sqrt{2}} E_{di} = -\frac{1}{\sqrt{2}}(E_{ds} + L_d P I_d + R_d I_d)$$

$$= -\frac{1}{\sqrt{2}} E_{ds} - \frac{1}{2}(L_d P i_\beta + R_d i_\beta) = e'_\beta - \frac{1}{2}(L_d P i_\beta + R_d i_\beta) \tag{7}$$

where E_{di} = Rectifier DC voltage, E_{ds} = Inverter DC voltage,
L_d = Inductance of smoothing reactor, R_d = Resistance of DC circuit.

E_{ds} can be expressed as follows by denoting the inverter control angle of advance by γ and neglecting the overlapping angle:

$$E_{ds} = \sqrt{2}\, E_{t2} \cos(\omega_s t - \gamma - \tfrac{\pi}{6}), \quad 0 \le \omega_s t < \tfrac{\pi}{3} \tag{8}$$

where E_{t2} = Secondary line-to-line voltage of inverter transformer (RMS value)

By eliminating the i_α and e_α terms from Equation (2) and substituting Equation (7), and rearranging the result, we get a relationship equation that holds in the single mode.

$$\begin{bmatrix} L_s & 0 & M\cos\theta \\ 0 & L_s & -M\sin\theta \\ M\cos\theta & -M\sin\theta & L_r + \frac{1}{2}L_d \end{bmatrix} \begin{bmatrix} PI_\beta \\ PI_\alpha \\ Pi_\beta \end{bmatrix} = \begin{bmatrix} E_\beta + M\omega_r i_\beta \sin\theta - R_s I_\beta \\ E_\alpha + M\omega_r i_\beta \cos\theta - R_s I_\alpha \\ e'_\beta + M\omega_r(I_\beta \sin\theta + I_\alpha \cos\theta) - (R_r + \frac{1}{2}R_d) i_\beta \end{bmatrix} \tag{9}$$

The single mode is under the following conditions:

$$|\sqrt{3}\, e_\alpha| < -e_\beta, \; e_\beta < 0 \quad (\because e_v < e_u < e_w) \tag{10}$$

where $e_\alpha = M\{\omega_r(-I_\alpha \sin\theta + I_\beta \cos\theta) + \cos\theta \cdot PI_\alpha + \sin\theta \cdot PI_\beta\}$

When the conditions given in Equation (10) are no longer satisfied, the single mode transfer to commutation mode. For instance, when $\sqrt{3}\, e_\alpha$ becomes equal to $-e_\beta$, viz. $e_u = e_w$, then commutation begins from the w phase to u phase and the commutation mode shown in Fig. 4(b) is realized. In Fig. 4(b), the u, v and w phases are replaced by virtual (w), (u) and (v) phases respectively. Here:

$$i_{(u)} = I_d \,(\because i_v = I_d), \qquad e_{(w)} = e_{(v)} \,(\because e_u = e_w)$$

By substituting these into Equation (1):

$$i_\alpha = \frac{\sqrt{3}}{2} I_d, \qquad e_\beta = 0$$

$$e_\alpha = -\sqrt{\tfrac{2}{3}}\, E_{ds} - \frac{2}{3}(L_d P i_\alpha + R_d i_\alpha) = e'_\alpha - \frac{2}{3}(L_d P i_\alpha + R_d i_\alpha) \tag{11}$$

By substituting Equation (11) into Equation (2) and rearranging the result, we get a relationship equation that holds in the commutation mode.

$$\begin{bmatrix} L_s & 0 & M\cos\theta & M\sin\theta \\ 0 & L_s & -M\sin\theta & M\cos\theta \\ M\cos\theta & -M\sin\theta & L_r & 0 \\ M\sin\theta & M\cos\theta & 0 & L_r + \frac{2}{3}L_d \end{bmatrix} \begin{bmatrix} PI_\beta \\ PI_\alpha \\ Pi_\beta \\ Pi_\alpha \end{bmatrix} = \begin{bmatrix} E_\beta - M\omega_r(-i_\beta \sin\theta + i_\alpha \cos\theta) - R_s I_\beta \\ E_\alpha - M\omega_r(-i_\beta \cos\theta - i_\alpha \sin\theta) - R_s I_\alpha \\ - M\omega_r(-I_\beta \sin\theta - I_\alpha \cos\theta) - R_r i_\beta \\ e'_\alpha - M\omega_r(I_\beta \cos\theta - I_\alpha \sin\theta) - (R_r + \frac{2}{3}R_d) i_\alpha \end{bmatrix} \tag{12}$$

Replacing the v phase by virtual (u) phase is, as is obvious from Fig. 4(b), equal to advancing by $2\pi/3$ from the actual value. This means substituting $\theta + 2\pi/3$ for θ in Equation (12). The initial conditions of Fig. 4(b) are:

$$i_{\alpha s} = \sqrt{\frac{3}{2}}\, I_{ds}, \quad i_{\beta s} = -\frac{1}{\sqrt{2}}\, I_{ds} = -\frac{1}{\sqrt{3}}\, i_{\alpha s}$$

(I_{ds} = DC current at the start of commutation mode)

The terminal conditions are:

$$i_{\alpha e} = \sqrt{\frac{3}{2}}\, I_{de}, \quad i_{\beta e} = \frac{1}{\sqrt{2}}\, I_{de} = \frac{1}{\sqrt{3}}\, i_{\alpha e}$$

(I_{de} = DC current at the end of commutation mode)

Other single and commutation modes can be considered in the same way. Thus, by distinguishing between single and commutation modes, and solving either Equations (9) or (12) accordingly, the transient phenomena of the static Scherbius control system can be obtained.

Figure 5 shows the measured and calculated waveforms at recovery from a three-phase open circuit. They show good agreement. The equipment used is a static Scherbius control system incorporating induction motor II shown in Table 2. A small capacity system was used because, with a large one, the high-speed circuit breaker (HSCB) for overcurrent protection would be tripped and this would make it impossible to measure waveforms suitable for comparison with calculated ones.

In the equipment used, the maximum value of DC current of measured waveforms was 94A, as against the rated value of 100A. Therefore there was no commutation failure of the inverter. This is obvious from the fact that the inverter DC voltage waveform E_{ds} has not been reduced to zero.

To predict the maximum current in the case of large-capacity equipment, calculation was performed for a static Scherbius control system incorporating induction motor I in Table 2. In this case, the maximum value of DC current is 3,310 A, viz. 9 times the rated value of 367 A. Thus, the inverter will certainly fail in commutation. The trip level of the HSCB of this equipment is 625 A, viz. about 1.7 times the rated value. In Fig. 6, this level is reached at 4 ms after recovery from a three-phase open circuit, so that the HSCB is expected to trip. This was verified in the experiment, in which the maximum value of DC current was 1,250 A at 10 ms after recovery. In calculations, the value at 10 ms after recovery is 1,145 A.

CONSIDERATION OF CONTINUOUS OPERATION OF STATIC SCHERBIUS CONTROL SYSTEM AT INSTANTANEOUS POWER FAILURE

In enabling the control system to operate through an instantaneous power failure, it must be studied how to: (1) detect instantaneous power failure, (2) suppress transient secondary voltage of induction motor, (3) prevent inverter commutation failure, and (4) secure power sources for the control and sequential circuit. These will be discussed below in detail.

Detection of Instantaneous Power Failure

To take proper steps, it is necessary and important that instantaneous power failure is detected quickly and surely. Because instantaneous power failure is usually accompanied by a voltage drop, a low-voltage relay has been used to detect the voltage drop. However, as stated before, there are cases where a voltage drop does not occur on account of residual voltage or self-excitation of the induction motor. Instantaneous power failure in such cases must be detected somehow.

In a static Scherbius control system, primary current I_s of the induction motor and direct current I_d flowing in the inverter are in a fixed functional relationship as shown in Fig. 7. In other words, irrespective of slip S, the primary current and DC current are generally proportional. As the power source is in an abnormal state at instantaneous power failure, primary current decreases. Also, as speed drops, the automatic speed regulator operates to compensate for the speed drop. As a result, DC current increases, and the proportional relationship between primary current and DC current is broken. This phenomenon can be utilized to detect instantaneous power failure. By combining a detector using this proportional relationship and a low-voltage relay, instantaneous power failure can be detected quickly and surely.

Suppression of Transient Secondary Voltage of Induction Motor

The transient secondary voltage induced in the secondary winding of the induction motor during instantaneous power failure or in the course of recovery can cause overvoltage breakdown to the diodes and thyristors in the rectifier and inverter. To prevent this, this voltage must be suppressed below a specified level.
In the induction motor with the secondary resistor the transient secondary voltage and resistance stand in the relationship shown in Fig. 3. As shown, the transient secondary voltage is heavily dependent on resistance and is hardly affected by slip, or speed. Therefore, by connecting a resistor to the secondary winding of the induction motor at instantaneous power failure, the transient secondary voltage can be suppressed below a given value. In the case of Fig. 3, in order to hold the voltage to 257.5 V, which is the maximum voltage induced normally in the speed control range of a typical static Scherbius control system, viz., the voltage induced normally at $S = 0.5$, a resistor with a resistance equal to four times that of the secondary winding of the induction motor must be connected.
To prevent the transient secondary voltage of the induction motor from being applied to the rectifier or inverter, the contactor between the secondary winding and the rectifier in Fig. 1 must be opened at instantaneous power failure. To open this contactor it will take 0.1 to 0.2 second and the transient secondary voltage induced during this time will affect the rectifier and inverter. Therefore, a resistor must be connected to the secondary winding simultaneously with instantaneous power failure. This is accomplished by means of a static switch.

Prevention of Inverter Commutation Failure

In the initial stage of instantaneous power failure, commutation voltage for the inverter is usually secured owing to the residual voltage of the induction motor. It is expected that this voltage can be used to bring the current in the inverter to zero. A possibility here is to shift the inverter control angle of advance to minimum upon detection of instantaneous power failure, and then, after verifying a zero current, to block the gate firing signals. By this method, which will be called gate shift and block control, inverter commutation failure after extinction of inverter commutation voltage can be prevented.
Gate shift and block control is effective against overcurrents caused by transient secondary voltages of the induction motor, particularly by those at recovery from a three-phase open circuit. Figure 8 shows calculated results for a static Scherbius control system incorporating induction motor I in Table 2, under the condition that gate shift and block control is released 0.3 second after recovery from a three-phase open circuit. The maximum value of DC current is 332 A. In Fig. 6, which shows calculated results for a case where no gate shift and block control is effected, the maximum value is as high as 3,310 A. This demonstrates the effect of this control method.
At times the power source voltage drops so abruptly that the inverter current can not be brought to zero although the inverter control angle of advance is shifted to a minimum, and this leads to inverter commutation failure, with the result that overcurrent flows and the HSCB for overcurrent protection is tripped. Conventionally, this HSCB tripping has been taken as a sign of serious trouble, which requires stopping of the equipment. Our close study of phenomena associated with HSCB tripping at instantaneous power failure revealed that no breakdown of equipment, particularly rectifier or inverter, occurs at HSCB tripping which, consequently, should not be regarded as serious trouble but rather as slight trouble. On this belief, we developed a method to continue operation by reclosing after recovery of the power source, so long as there is no abnormality in the rectifier or inverter. If there is an abnormality, the secondary winding of the induction motor is short-circuited to carry on single operation of the induction motor.

Securing Power Source for Control and Sequential Circuit

By receiving supplies from an uninterrupted power source, the control and sequential circuit will be able to secure a power source even at instantaneous power failure.

The above control methods can be realized by means of a circuit such as shown in Fig.9. Figure 10 is a flow chart showing the operation of the circuit in Fig.9.

EXPERIMENTAL RESULTS AND DISCUSSION

The circuit diagrammed in Fig. 9 was applied to a static Scherbius control system incorporating induction motor I in Table 2. The results will be described. Oscillograms in every conceivable modes on instantaneous power failure are shown in Fig. 11. The modes are: (a) line-to-line short circuit, (b) three-phase short circuit, (c) single-phase open circuit, and (d) three-phase open circuit. In (a), the inverter fails in commutation, an overcurrent flows, and the HSCB is tripped. Although tripping occurs upon instantaneous power failure, the HSCB is reclosed after power source recovery, and operation is continued without interruption. In (b), (c) and (d), there is no inverter commutation failure, so that the HSCB is not tripped. It is seen that during instantaneous power failure and recovery, no overcurrent flows. Neither does an overvoltage that causes insulation breakdown occur, and smooth operation is continued.
In the single-phase open circuit shown in (c), inverter commutation failure can occur in the range of low revolution velocity, causing an overcurrent to flow and the HSCB to trip. In the single-phase open circuit and line-to-line short circuit, the primary voltage of the induction motor is unbalanced, so that a negative sequence voltage is induced in the secondary winding and DC voltage of the rectifier increases. Meanwhile simultaneously with the occurrence of instantaneous power failure, the inverter fails in commutation at the arm connected to the phase where power source voltage drops, and inverter DC voltage drops. As a result, it is presumed, the DC voltage between the rectifier and the inverter is not in the state of equilibrium, overcurrent flows, and the HSCB is tripped.
In the case of a three-phase open circuit or three-phase short circuit, inverter commutation voltage also decreases but no negative-sequence voltage is induced in the secondary winding of the induction motor. Therefore, secondary voltage also drops in proportion to the decrease in power source voltage. Since the secondary voltage and inverter commutation voltage decrease while being in the state of equilibrium, even if inverter commutation failure occurs, overcurrent is not large enough to trip the HSCB.
The transient secondary voltage induced in the secondary winding of the induction motor at instantaneous power failure is the largest at recovery from a three-phase open circuit, as in (d). The voltage is 370 V. Because a resistor is connected to the secondary winding of the induction motor by means of a static switch upon detection of instantaneous power failure, the transient secondary voltage is held as low as 370 V. Without such a resistor, the voltage would reach a maximum of 515 V.
Thus it was shown that operation can be continued without interruption in any conceivable mode of instantaneous power failure.

CONCLUSIONS

A new static Scherbius control system capable of operating through instantaneous power failure was developed. Control methods to make this continuous operation possible and experimental results are summarized below.

1) The proportional relationship between primary current of the induction motor and inverter DC current does not exist at instantaneous power failure. By combination of a detector utilizing this phenomenon and a conventional low-voltage relay, instantaneous power failure can be detected quickly and surely.
2) The transient secondary voltage induced in the secondary winding of the induction motor at instantaneous power failure is suppressed by connecting a resistor to the secondary winding by means of a static switch.
3) To prevent inverter commutation failure, gate shift and block control is performed. This method is effective also against overcurrents caused by transient secondary voltages, particularly by those at recovery from a three-phase open circuit. This is obvious from the results of analysis of transient phenomena of the static Scherbius control system.
4) Despite control as in 2) and 3) above, inverter commutation may fail, overcurrent flow, and the HSCB for overcurrent protection trip, in which case operation is continued by reclosing after power source recovery.
5) Experiments revealed the following: In instantaneous power failure modes of three-phase open circuit and three-phase short circuit, there is no inverter commutation failure and smooth operation can be continued. On the other hand, in modes where power source voltage is extremely unbalanced, as in a single-phase open circuit or line-to-line short circuit, inverter commutation failure occurs with the result that overcurrent flows and the HSCB is tripped. Nevertheless, operation can be continued by reclosing after power source recovery.

Now that it is capable of operating continuously through instantaneous power failure, the new static Scherbius control system has been enhanced in reliability. As a speed control system for large-capacity induction motors, the new static Scherbius system is expected to find increasing applications in driving water supply pumps and other fields where labor-saving, energy-saving and high reliability are required.

The authors wish to express their gratitude to Mr. Watanabe, Department Manager, First Department, Hitachi Research Laboratory, Hitachi, Ltd., Mr. Kawakami, Department Manager, Rectifier Equipment Design Department, Hitachi Works, Hitachi, Ltd., and Mr. Iwaki, Department Manager, Industrial Control Design Department, Omika Works, Hitachi, Ltd. for their expert guidance and generous cooperation in carrying out this research project.

REFERENCES

(1) K. Honda, et al., 6200 kW Raw-Water Transfer Pumping Equipment of the Asaka Purification Plant, Tokyo Metropolitan Water Works Bureau, Hitachi Review, 15, 60 (1966).

(2) Y. Hiro, et al., Speed Control of Induction Motors by the Secondary Voltage Excitation Method, THE HITACHI HYORON, 47, 1841 (1965).

(3) K. Craf, AC Pump Drives with Static Converter Slip-Power Recovery System for the Lake Constance Water Supply Scheme, Siemence Review, XL, 539 (1973).

(4) Y. Hiro, et al., Brushless Scherbius Control of Induction Motor, IEEE. Conf. of the IAS, 9th Annual, 111 (1974).

(5) T. Hori, The Characteristics of an Induction Motor Controlled by a Scherbius System (Application to Pump Drive), IEEE. Conf. of the IAS, 7th Annual, 775 (1972).

(6) T. Hori, et al., Commutatorless Kraemer Control System for Large-Capacity Induction Motors for Driving Water Service Pumps, IEEE. Conf. of the IAS, 11th Annual, 822 (1976).

(7) H. Shimasaki, et al., Pumping Plant for Water Supply in Misono Filtration Plant of Water Works Bureau, Tokyo Metropolitan Government, Hitachi Review, 25, 229 (1976).

(8) A. K. DE. Sarkar, et al., Digital Simulation of Three-Phase Induction Motors, IEEE. Trans. on PAS, PAS-89, 1031 (1970).

(9) P. C. Krause, Simulation of Symmetrical Induction Machinery, IEEE. Trans. on PAS, PAS-84, 1038 (1965).

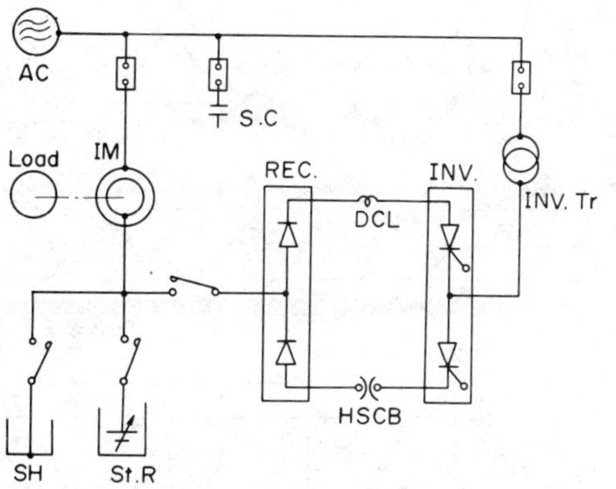

AC : AC Power Source IM : Induction Motor
HSCB : High Speed Circuit Breaker
S.C : Power-Factor-Correcting Capacitor
REC. : Rectifier INV. : Inverter
INV. Tr : Inverter Transformer
DCL : Smoothing Reactor St. R : Starting Resistor
SH : Short-Circuit Equipment

Fig.1. Basic schematic diagram of conventional static Sherbius control system.

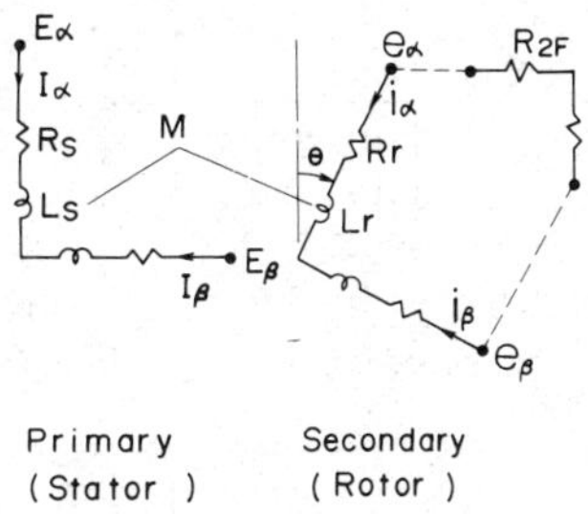

Rs : Primary Winding Resistance
Rr : Secondary Winding Resistance
Ls : Primary Winding Inductance
Lr : Secondary Winding Inductance
M : Mutual Inductance
R2F : External Resistance

Fig.2. Representation of α-β axis for induction motor.

Table 1 Comparison of Measured Values and Calculated Values

Item \ Motor		Motor I measured	Motor I calculated	Motor II measured	Motor II calculated
Maximum Value of Primary Current	ia	162 A	106.5 A	42 A	44.4 A
	ib	113	90	67.5	64.5
	ic	99	82.5	50	48.8
Maximum Value of Secondary Current	iu	781 A	807 A	90 A	91.3 A
	iv	683	689	105	102.3
	iw	671	680	80	92.6
Maximum Value of Secondary Voltage	euv	400 V	400 V	104.4 V	105 V
	evw	340	357	107.9	108
	ewu	300	310	106.7	105.8
		(S=0.427, R2F=0.273 Ω)		(S=0.1, R2F=0.64 Ω)	

Table 2 Constants of Motor

Item \ Motor	Motor I	Motor II
Capacity	150 KW	11 KW
Number of Poles	6	6
Rated Voltage	3 kV	220 V
Primary Winding Resistance Rs	1.42 Ω	0.136 Ω
Secondary Winding Resistance Rr	211 mΩ	499 mΩ
Primary Winding Inductance Ls	386 mH	44.7 mH
Secondary Winding Inductance Lr	4.93 mH	11.6 mH
Mutual Inductance M	42 mH	22 mH

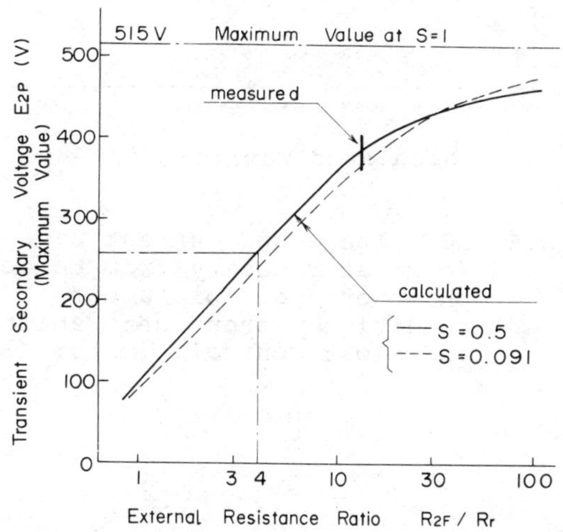

Fig.3. Illustration of suppression effect of transient secondary voltage by external resistor.

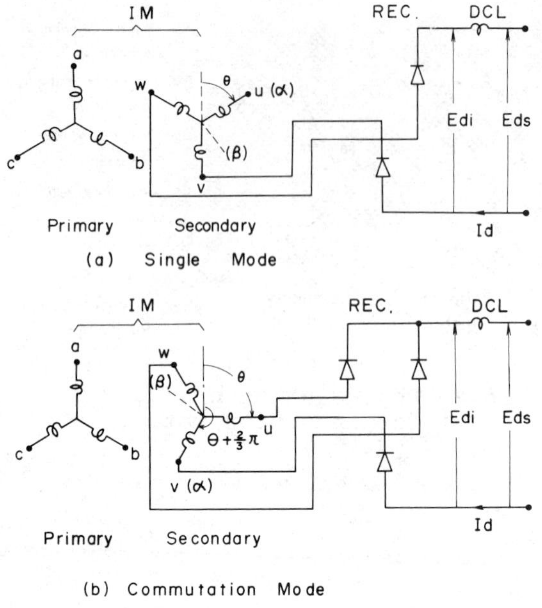

Fig.4. Secondary circuit modes of induction motor in static Scherbius control system.

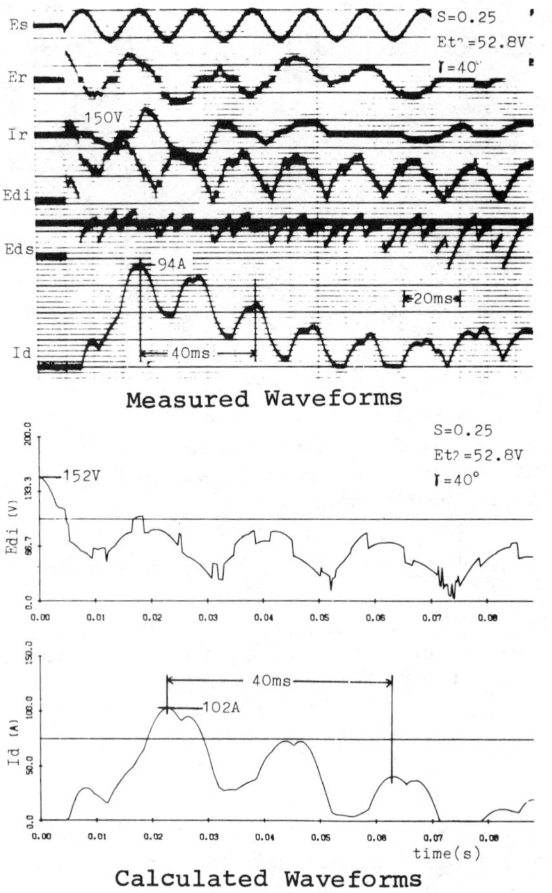

Fig.5. Measured and calculated waveforms at recovery from three-phase open-circuit. (motor II)

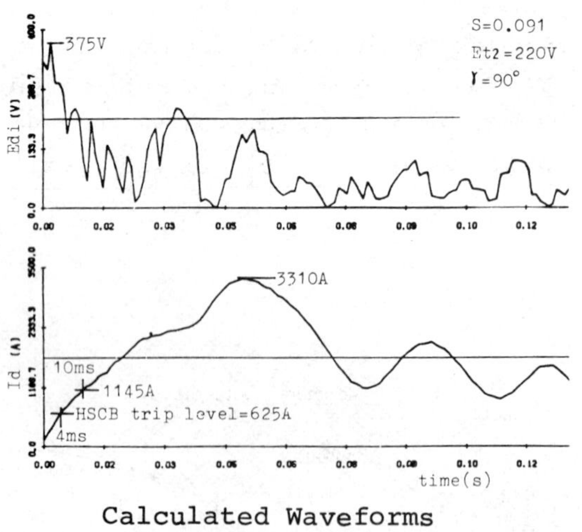

Fig.6. DC voltage and current waveforms at recovery from three-phase open-circuit under condition without gate shift and block control. (motor I)

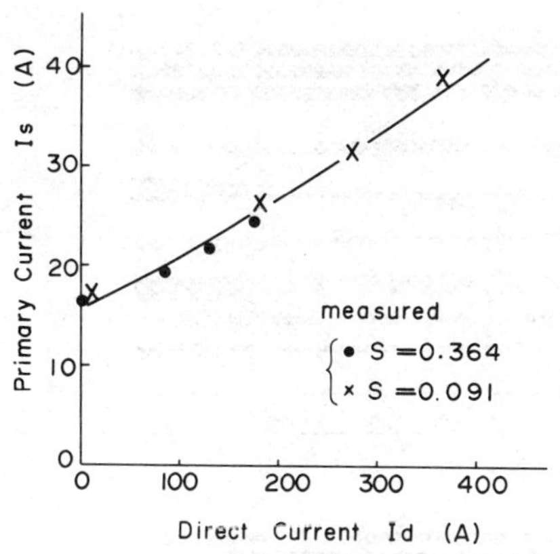

Fig.7. Direct current-primary current characteristics. (motor I)

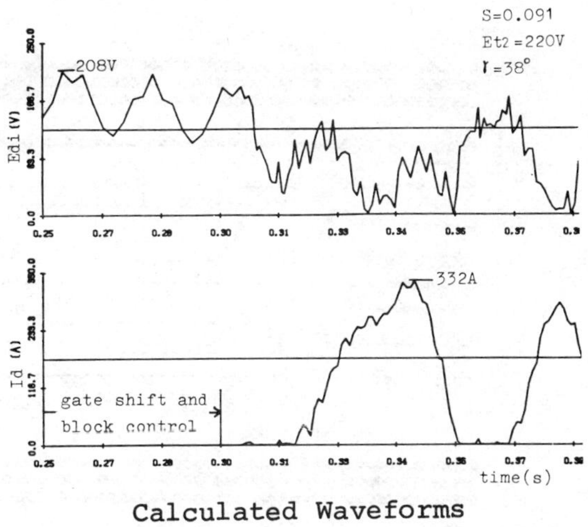

Calculated Waveforms

Fig.8. DC voltage and current waveforms at recovery from three-phase open-circuit under condition with gate shift and block control. (motor I)

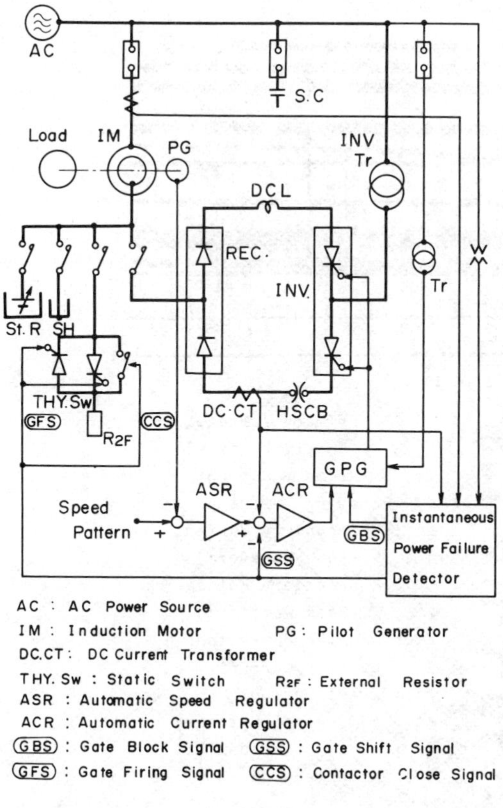

Fig.9. Basic schematic diagram of new static Scherbius control system.

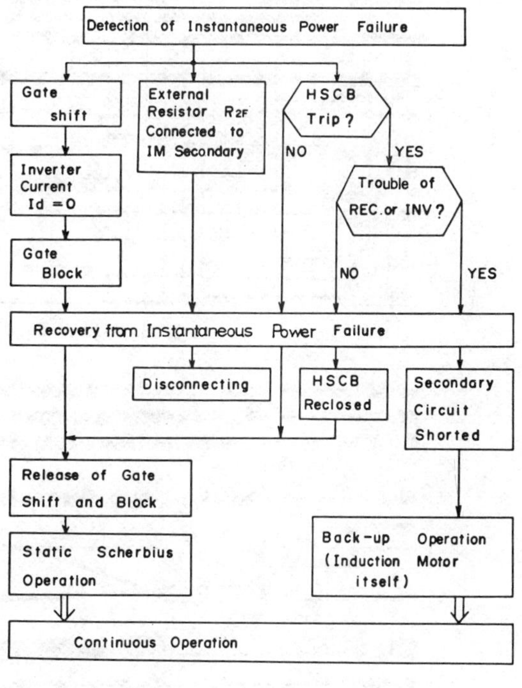

Fig.10. Flow chart of new static Scherbius control system during instantaneous power failure.

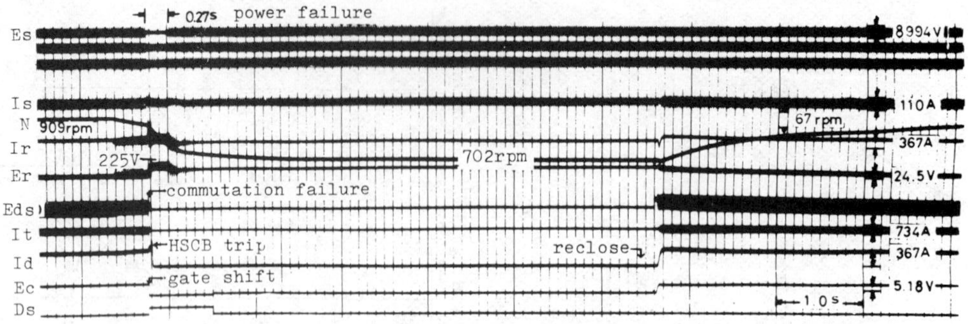

(a) Line-to-Line Short-Cirucit

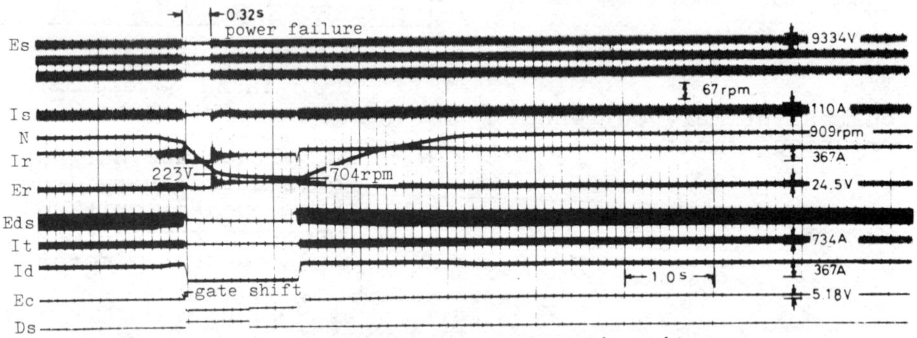

(b) Three-Phase Short-Cirucit

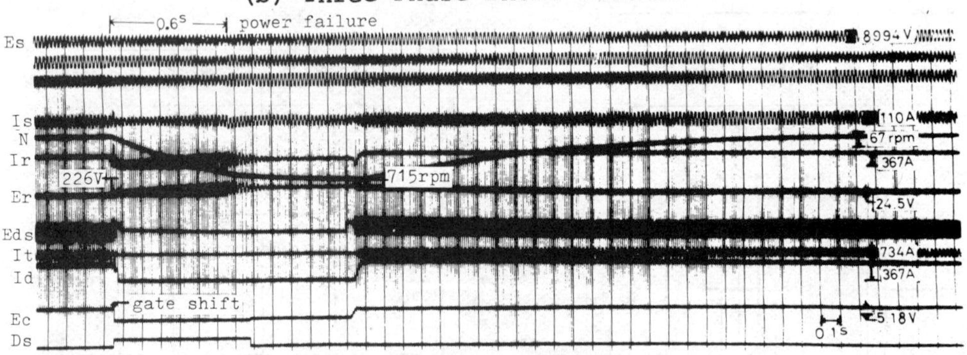

(c) Single-Phase Open-Circuit

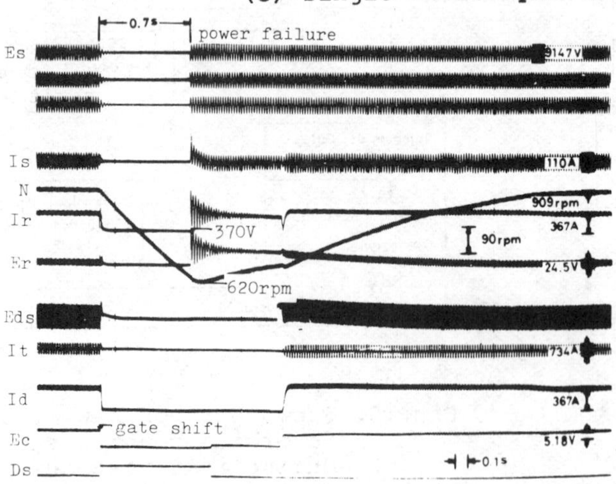

(d) Three-Phase Open-Circuit

Es :IM Primary Voltage (Supply Voltage)
Is :IM Primary Current
N :Rotating Velocity
Ir :IM Secondary Current (Peak Waveform)
Er :IM Secondary Voltage (Peak Waveform)
Eds:Inverter DC Voltage
It :Inverter Current
Id :Direct Current
Ec :GPG Control Voltage
Ds :Detection Signal of Instantaneous Power Failure

Fig.11. Operation oscillograms at instantaneous power failure.

SUPER-SYNCHRONOUS STATIC CONVERTER CASCADE

P. Zimmermann

Institut für Stromrichtertechnik und Antriebsregelung, Technische Hochschule, Darmstadt, FRG

SUMMARY

Up to this time only the sub-synchronous cascade has been used for the speed control of a slip ring motor. It is also possible to obtain super-synchronous speed by reversing the direction of the power flow in the cascade circuit. In this operation mode the converter at the slip ring side must be able to function as an inverter.

The proposal of this system made in (1) claims the following advantages: If twice the synchronous speed is chosen as maximum (rated) speed, the power of the machine is doubled,only half the rated power must be transmitted by the converter, and the maximum of the reactive power occurs at synchronous speed and accounts to about half the rated power It can be reduced by further measures. These measures at the line side converter and at the slip-ring side converter are described. The control and signal processing are also covered in this paper.

INTRODUCTION

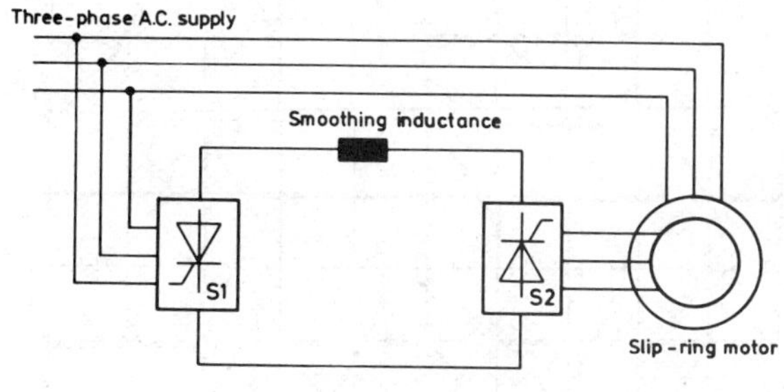

Fig.1. Block diagram of a sub- and super-synchronous static converter cascade.

The block diagram of a super-synchronous static converter casade is shown in fig. 1. The diode rectifier used at the slip-ring side of a subsynchronous static converter cascade is replaced by a controlled converter. Super-synchronous speed may now be obtained by reversing the direction of the power flow in the cascade circuit, thereby feeding additional power into the rotor.

In a sub-synchronous cascade, the static converter must have the same rating as the induction motor. If the slip is zero, the reactive power is maximal and approximately as great as the synchronous power of the machine, that is half the newly defined rated power.

Due to the fact that at twice the synchronous speed only one half of the maximal shaft power is fed into the rotor, the reactive power is reduced to about 50 per cent. In order to further reduce the reactive power, a three-phase controlled bridge rectifier with two controlled free-wheeling valves can be used. If a forced commutated converter is inserted at the slip-ring side, it can be controlled with negative firing angles and thus avoid the reactive power caused by commutation e.g. near zero speed or twice synchronous speed.

With natural commutation the necessary commutation margin has as a consequence an unavoidable amount of reactive power. With the described measures the total reactive power is about a quarter of the maximal shaft power. The reactive power associated with the magnetizing current of the machine has been neglected in the consideration up to this point. It could be compensated by inserting capacitors in parallel.

With twice the synchronous speed as rated speed the machine can be constructed with twice the number of poles. This is advantageous in regard to the weight and the size of the machine.

THREE-PHASE CONTROLLED BRIDGE RECTIFIER WITH TWO CONTROLLED FREE-WHEELING VALVES

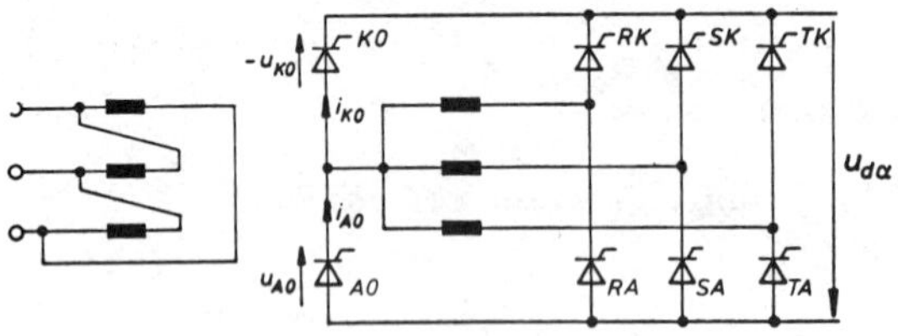

Fig.2. Three-phase controlled bridge rectifier with two controlled free-wheeling valves

The reduction of the reactive power and the advantage of having a 6-pulse reaction on the feeding A.C. system instead of a 3-pulse reaction (which is achieved with a half-controlled bridge) can be obtained with a three-phase controlled bridge converter and two additional controlled free-wheeling valves (2). Thus the fundamental power factor is improved. The changing of polarity of the D.C. link would not be possible with free-wheeling diodes.

The main problem in controlling this converter is to fire the main valves and the free-wheeling valves in such a way that, in the whole range of control, the commutation of the free-wheeling valves is ensured.

Taking this into consideration and by inserting two properly shaped characteristics between the current controller and the gating circuits of the main valves, respectively the free-wheeling valves, (see fig.10) a nearly linear function between the controlling input voltage and the inverter output voltage is reached(see fig. 3).

Figure 4 shows some typical voltage waveforms of the converter. The upper graphs shows $-u_{ko}$ (see fig. 4) the voltage of terminal K against the transformer neutral, which also represents the anode-cathode voltage of the free-wheeling valve, The lower graph shows the corresponding bridge output voltage.

The experience has been that the described connection is as safe in operation as the ordinary bridge.

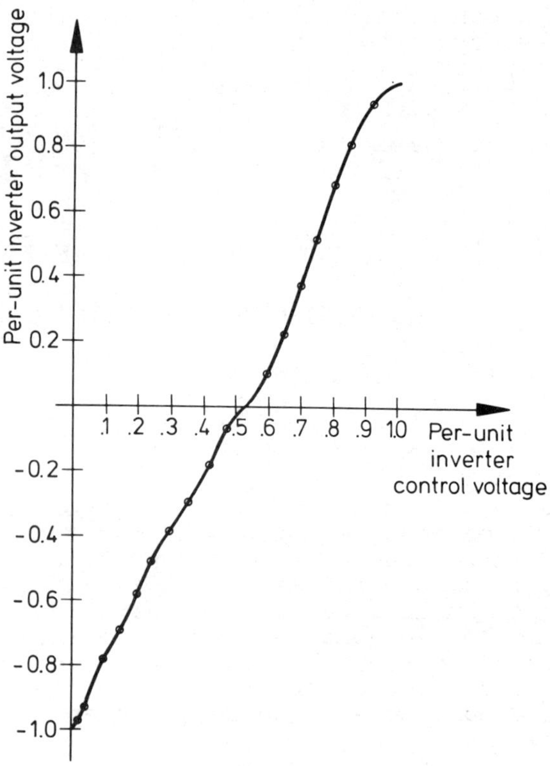

Fig.3. Typical output characteristic of a bridge converter with controlled free-wheeling valves.

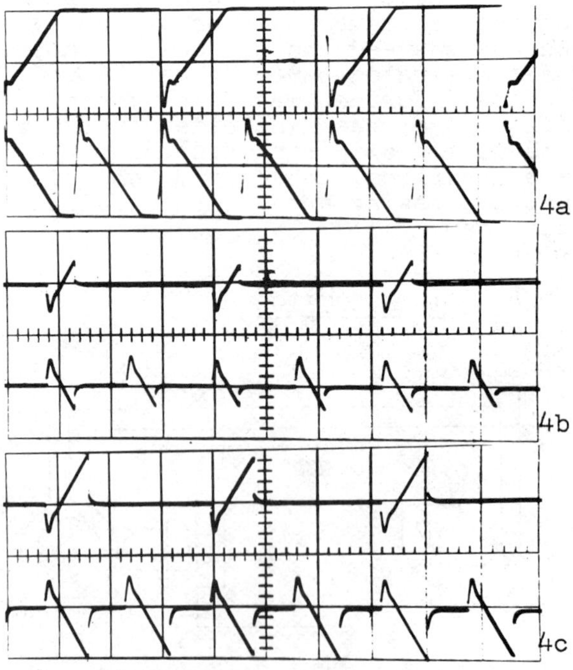

Fig.4. Influence of the free-wheeling valves:
4a. in rectifying operation
4b. at an average output voltage of zero
4c. in inverter operation
Scale: 2 millisec./div.

This converter can be used also for other applications, wherever the additive cost for the two controlled valves is bearable regarding the advantages of reducing reactive power and maintaining a 6-pulse reaction on the feeding A.C. system.

A FORCED COMMUTATED INVERTER TO BE USED IN A VARIABLE VOLTAGE D.C. LINK

The forced commutated inverter can be used favourably at the slip-ring side of an induction motor in a super-synchronous cascade as well as at the machine side of the intermediate D.C. circuit feeding a synchronous motor.

The inverter must be able to work under the following conditions:

- at a variation of the frequency between zero and some output frequency which might be considerably higher than the input line frequency.
- at variable e.m.f. of the machine between zero and U_{max} .
- at a variable voltage D.C. link of both voltage polarities with great and small currents.
- with a negative firing angle to avoid reactive power caused by the commutation.
- at both phase sequences.

Only the asynchronous cascade is considered in the following. Fig.5 shows the basic circuit diagram of the converter. Figure 6 shows, as a part of fig. 5, its functioning.

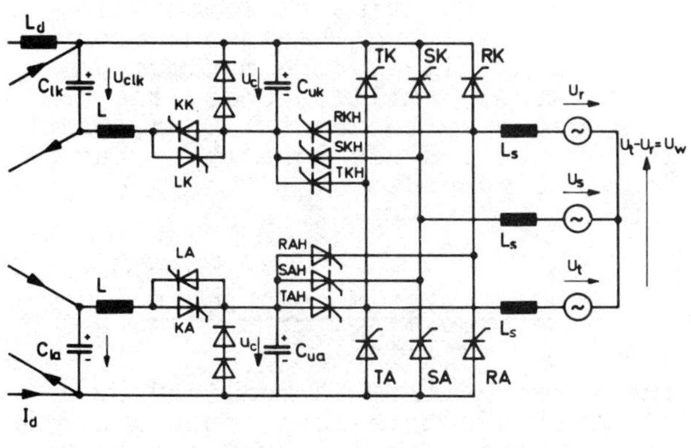

Fig.5. Basic circuit diagram of the inverter

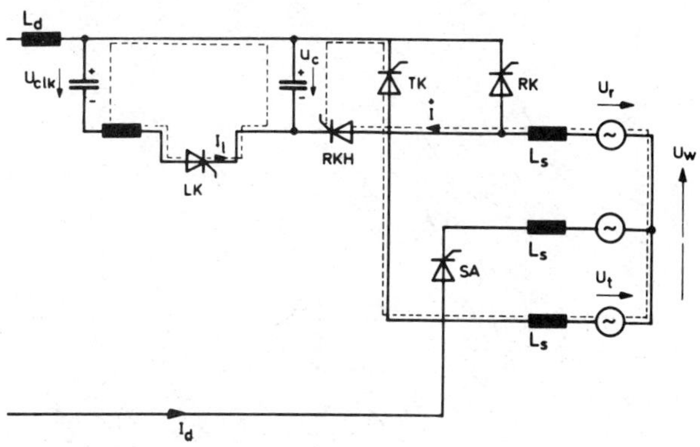

Fig.6. Functioning of the forced commutation of the connection in fig. 5

A conducting valve can be forced commutated by switching a charged capacitor parallel to the valve in such a way that the capacitor voltage appears as a negative anode-cathode voltage. The current, assumed constant at first, discharges and reloads the capacitor to the reverse polarity. Only after the capacitor voltage is high enough and more than counterbalances a phase to phase voltage of the wrong polarity for commutation, the forced commutation can begin.

In the newly proposed method a loaded capacitor is switched on in such a way that the current i in the following phase increases immediately*. In the same time the current of TK which is to be commutated, decreases to zero and the thyristor passes over into the blocking state. The commutating capacitor is fed by a seperate source in order to be independent of the tension of the D.C. link. At small currents, small commutation times can be obtained by adding a bypass to reload the commutating capacitor (L, KK, LK) .

By investigations using computer simulation it was found, that a favourable dimensioning is reached if L is twice L_S . U_C is about 1.5 times the average D.C. voltage that is 1.43 times the maximal amplitude of the phase to phase voltage of the machine.

*This special feature is due to a proposal made to the author by U. Kunz .

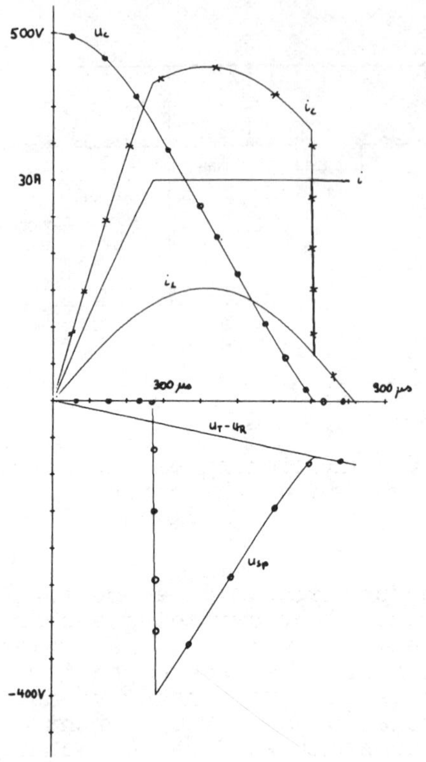

$C = 50\ \mu F \quad I_d = 30\ A = 0{,}75\ I_{dN}$

Slip $s = 1 \quad \alpha = 0^{\circ}$ U_{sp} represents the anode-cathode voltage of the commutating valve.

Fig.7. Voltage and current waveforms at the forced commutated inverter with a high current.

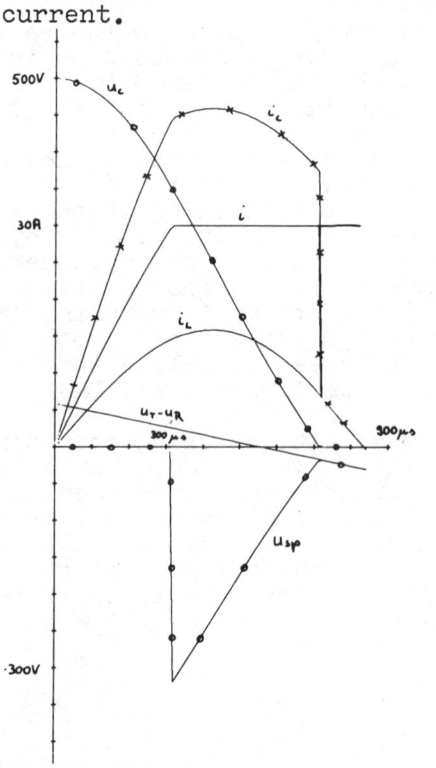

$C = 50\ \mu F \quad I_d = 2\ A = 0{,}05\ I_{dN}$

$s = 1 \quad \alpha = 0^{\circ}$

Fig.8. Voltage and current waveforms of the forced commutated inverter with a low current.

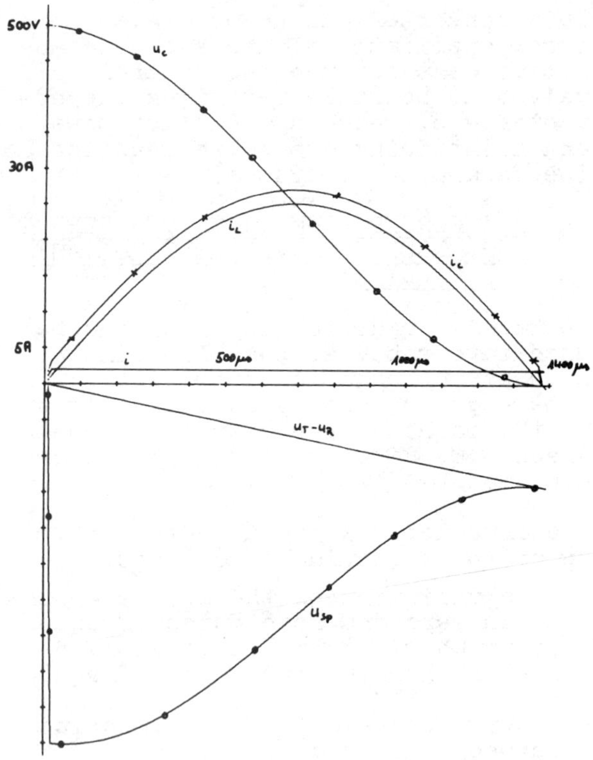

$C = 50\ \mu F \quad I_d = 30\ A = 0{,}75\ I_{dN}$

$s = 1 \quad \alpha = -10^{\circ}$

Fig.9. Voltage and current waveforms of the forced commutated inverter with a negative firing angle.

The time interval which the inverter needs for commutation is always less than 1.5 milliseconds. Adding the time of about 1 millisecond for the reloading of the capacitor means that the total time of 2.5 milliseconds, occuring when the current is minimal, limits the minimum period to 7,5 milliseconds, the maximum frequency to about 130 Hz, which is for the cascade more than what is reasonably needed.

CONTROL SYSTEM FOR CLOSED LOOP OPERATION

The speed control is realized by means of a double-loop control system. The speed controller, which measures the difference between set speed and actual speed, sets the current value (torque proportional). The current control loop adjusts the current by variation of the firing angles of the line side converter with its two controlled free-wheeling valves. If the set speed greatly exceeds the actual speed, the machine will accelerate rapidly with its maximum torque and current (e.g.110 per cent).

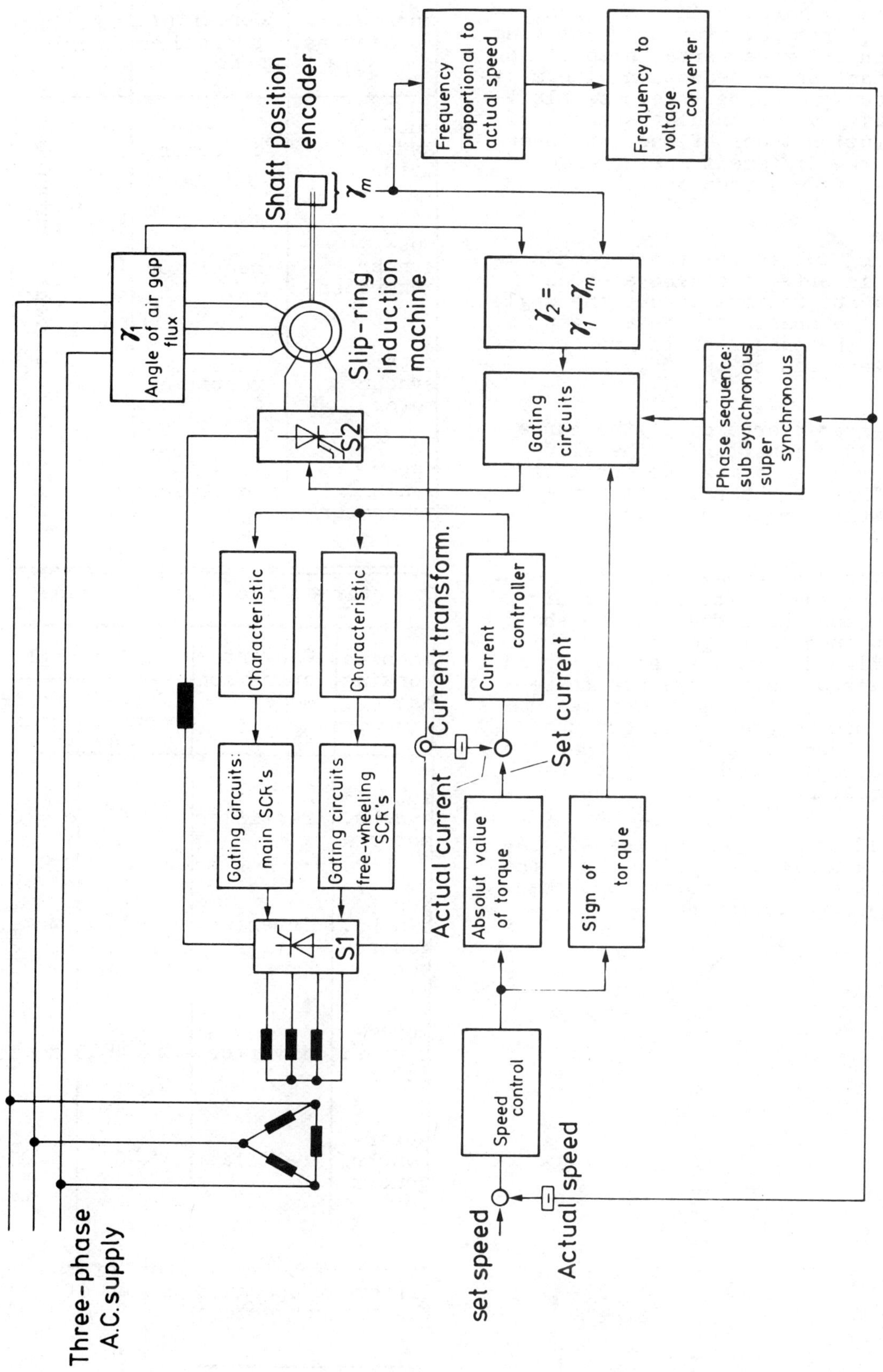

Fig.10. Block diagram of the control system for closed loop operation

When a sudden reduction in the demanded speed occurs, the sign of the set current changes. Both inverters change automatically into the contrary state, from inverter to rectifying operation and vice versa. Also at no load a fast speed reduction with maximum torque and current is possible. Figure 11 gives a tabular summary of the operation modes of the two inverters in the different speed ranges and power flow directions.

The firing pulses for the forced commutated inverter are generated as follows: For maximum torque the angle between the phasor of the air gap flux and the phasor of the rotor current always has to be 90°.

The six-step operation of the three phase bridge inverter at the slipring side gives a maximum error of $\pm 30^{\circ}$. This is the reason for the unavoidable torque pulsations.

The difference between the signal representing the position of the air gap flux and the signal of the shaft position encoder is generated. This information determines, at any speed, which pair of thyristors has to be in the conducting state. The measured average D.C. current corrects the firing pulses in such a way that the reactive power which is caused by the overlap is avoided.

Figure 12 as a result of the simulation on a digital computer shows the acceleration from zero to more than twice synchronous speed.

Three-phase controlled bridge rectifier (line side)

machine working as:	converter operation mode	firing angle
sub-synchr. motor	inverter	adjusted by current controller
sub-synchr. generator	rectifier	
super-synchr. motor	rectifier	
super synchr. generator	inverter	

Converter at the slip-ring side

machine working as:	Converter operation mode	firing angle	
		forced comm.	phase* comm.
sub-synchr. motor	rectifier	$\alpha \leq 0^{\circ}$	$\alpha = 0^{\circ}$**
sub-synchr. generator	inverter	$\alpha \leq 180^{\circ}$	$\alpha = \alpha_{max}$***
super-synchr. motor	inverter	$\alpha \leq 180^{\circ}$	$\alpha = \alpha_{max}$
super-synchr. generator	rectifier	$\alpha \leq 0^{\circ}$	$\alpha = 0^{\circ}$

Fig.11. Tabular summary of the different operation modes of the two converters

* see next chapter
** zero firing angle
*** extinction angle

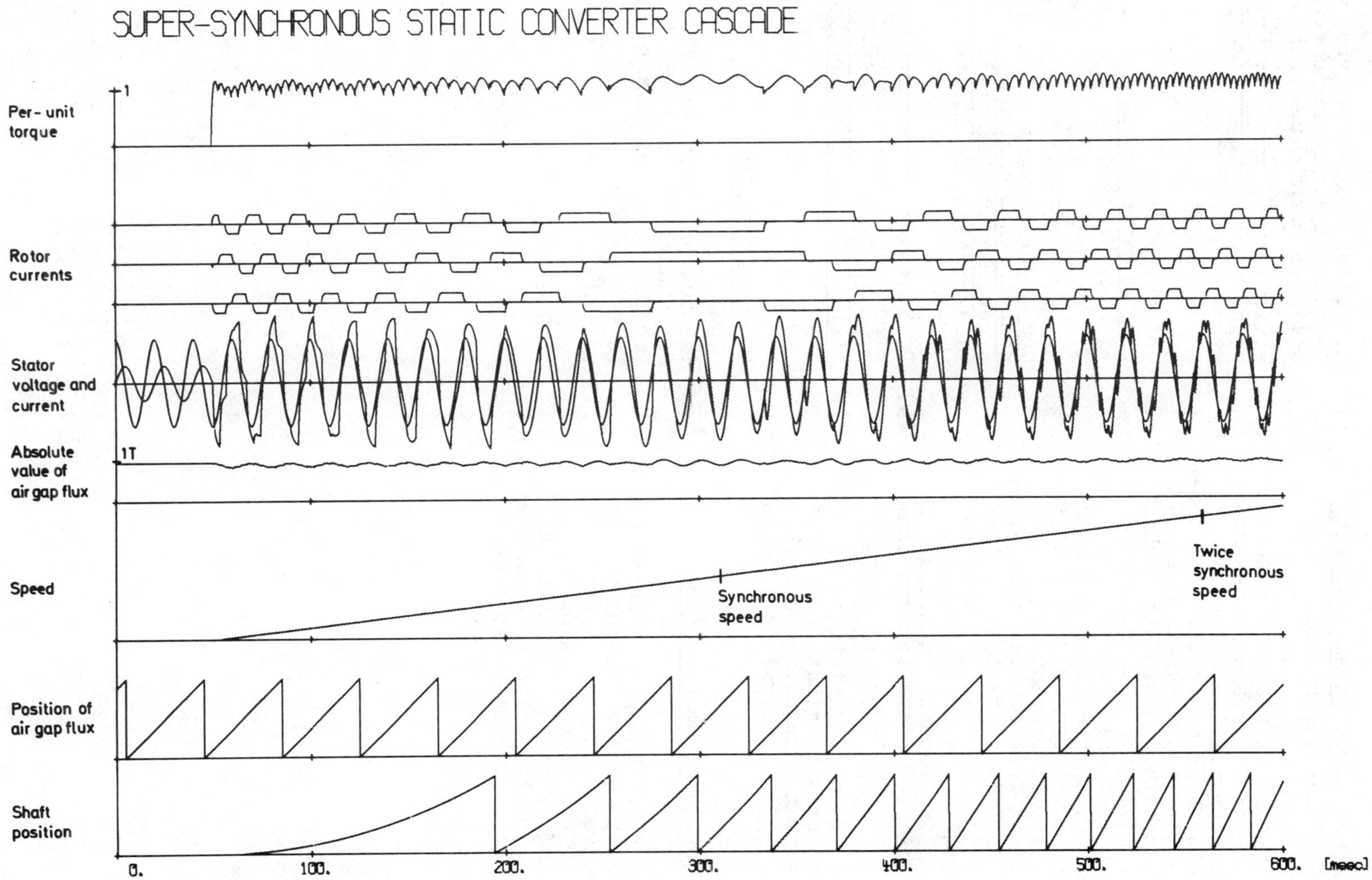

Fig.12. Acceleration from zero to more than twice synchronous speed

VARIATIONS OF THE DRIVE CONCEPT

A Phase Commutated Converter at the Slip-Ring Side

The consumption of reactive power in this case becomes greater. However, it can be economically attractive, because the phase converters are considerably less expensive than the forced commutated ones. When working in the rectifying operation, the valves are triggered at the zero firing angle. In the inverting operation the valves can only be triggered at a specific instant which must be calculated from several parameters such as the D.C. current, slip-frequency, slip-ring voltage, commutating reactance. Near synchronism, when the slip-frequency e.m.f.s. are insufficient for phase commutation, D.C. current is brought to zero by the current controlled inverter. After the conducting state of the valves at the slip-rings is changed, the current and torque are brought automatically to the values required by the load.

Four-Quadrant Operation

An operation with negative sense of the speed can easily be achieved by changing two of the three phases of the stator A.C. supply using A.C. solid state switches. In the signal processing in this case care must be taken of the reversed phase sequence.

REFERENCES

(1) R. Jötten, Signalverarbeitung für die Regelung umrichtergespeister Drehfeldmaschinen, Proc. 1. IFAC - Symposium Control in Power Electronics and Electrical Drives 1, 687-695 (1974).

(2) Jötten, R. (1976) Leistungselektronik I, Vieweg, Wiesbaden.

(3) Maier, R. (1974) Blindstrom- und Drehzahlregelung der doppeltgespeisten Drehstrommaschine mit eingeprägten Läuferströmen, Dissertation, Karlsruhe.

(4) D. Schröder, Untersynchrone Stromrichterkaskade, VDI/VDE Gesellschaft Meß- und Regelungstechnik 2. Jahrestagung, 90 (1976).

ACKNOWLEDGEMENT

The author thanks the "Deutsche Forschungsgemeinschaft" (DFG) for financial support in this investigation.

A MODIFIED ROTOR CHOPPER FOR SPEED CONTROL OF SLIP RING INDUCTION MOTORS

S.S. Kelkar and S.K. Pillai

Department of Electrical Engineering, Indian Institute of Technology, Bombay, Powai, Bombay 400076, India

ABSTRACT

This paper presents the analysis and design of an improvised rotor chopper circuit, which can be used to control the speed of slip ring induction motors. Not only the disadvantages of the circuits, which have been reported earlier, have been removed, but also the design procedure has been simplified to a great extent. A scheme for feedback speed control is also discussed.

NOMENCLATURE

E	Totor rectified voltage
E_c	Voltage across the capacitor as defined in the text
I	Rotor rectified current
i_c	Capacitor charging current
i'_c	$-i_c$
I_L	Full load rotor rectified current
$i_{L_1}, i_{L_2}, i_{R_1}, i_{S_1}, i_{S_2}$	Instantaneous currents in L_1, L_2, R_1, S_1 and S_2
L_d	Inductance used for smoothing rectified current
R_1	External resistance
T	Total time period of chopper
t_{o1}	Turn-off time provided for S_1
V	Auxiliary supply voltage
v_c	Instantaneous voltage across the capacitor
v'_c	$-v_c$

INTRODUCTION

With the increase in availability of high current thyristors, smooth and quick variation of the external resistances introduced in the rotor circuit of slip ring induction motor to control its speed, can be accomplished electronically. Schemes employing chopper controlled resistance can be used to obtain a constant torque, constant speed or any desired characteristic by using a proper feedback circuit along with it. Such circuits are widely used in industrial application where the drive operation is intermittent, for example, hoists, conveyors, cranes, lifts, excavators and where good power factor and high starting torque are more important and the efficiency.

The most commonly used chopper circuit for controlling the speed of slip - ring induction motors is given in [1], but the absence of a systematic and complete analysis of the above makes it difficult to design the circuit for a particular application. Hence in [2], a detailed analysis and procedure to determine the ratings of the various elements of the chopper circuit had been presented. A simple modification of the circuit, which slightly reduces the power consumed by the auxiliary supply, had also been suggested and analysed in [3]. But both the circuits suffer from the following disadvantages :

1. The design procedure is complicated
2. The power drain from the auxiliary supply is high
3. The commutating circuit elements have high current ratings

In [4] the author had suggested six modified versions as shown in Fig.1 of the basic circuit. The analysis and design of the above circuits presented in [5] had established that the above mentioned disadvantages can

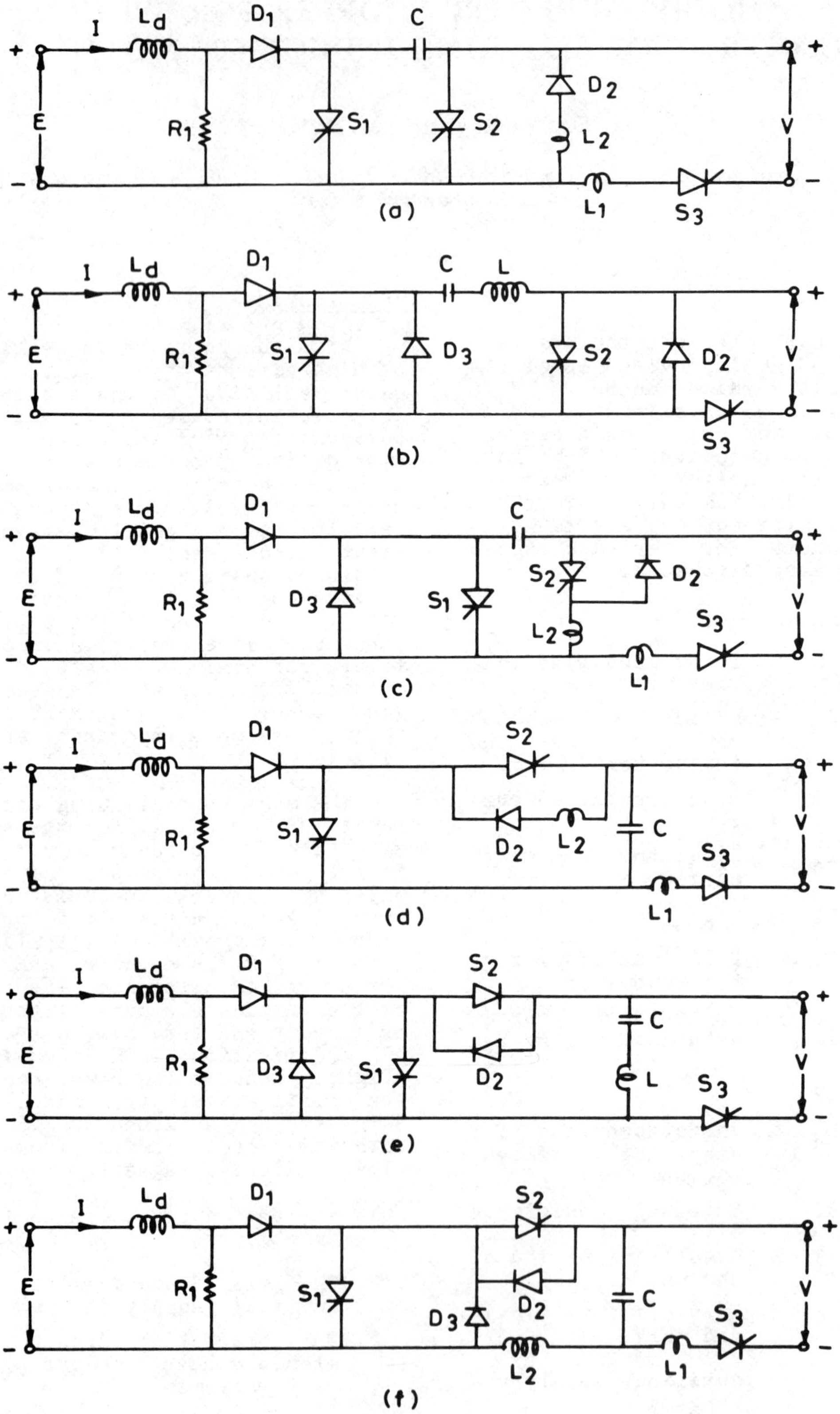

Fig.1 Modified Chopper Circuits.

be removed. This paper presents the analysis and design of the most simple and economical circuits amongst them. A speed control scheme is also given and the results obtained are discussed.

CIRCUIT UNDER STUDY

The chopper circuit under investigation is given in Fig. 2. The magnitude of E varies with the slip. However, the current is to be controlled all throughout. Hence a separate source represented as V is necessary to ensure that commutation takes place at all values of slip. The effective value of R_1 is given as

$$R_1^* = \frac{R_1 T_2}{T_1+T_2} \qquad 0 \leq R_1^* \leq R_1$$

with T_2 the OFF time of S_1 and $T_1 + T_2 = T$. The actual rotor resistance per phase R_2 is given as

$$R_2 = \frac{\pi^2}{18} R_1^* \doteq 0.5 R_1^*$$

I varies with the load but I is assumed constant during the cycle. Actually when S_1 is put ON the current I will increase from a value I_{min} to a value I_{max}, but this variation I could be kept small with a high value of L_d and the variation is neglected except in finding the value of L_d for a chosen value of I.

At t = o, S_1 is put ON. The initial conditions are

$$i_{S_1} = i_{L_2} = i_c = i_{S_2} = i_{L_1} = o$$

$$i_{R_1} = I \qquad v_c = -E_c.$$

The combined resistance of the SCR, diode and the inductance is represented as r,

$$i_c = \frac{E_c}{w_2' L_2} e^{-\alpha t} \sin w_2' t = i_{L_2} \qquad (1)$$

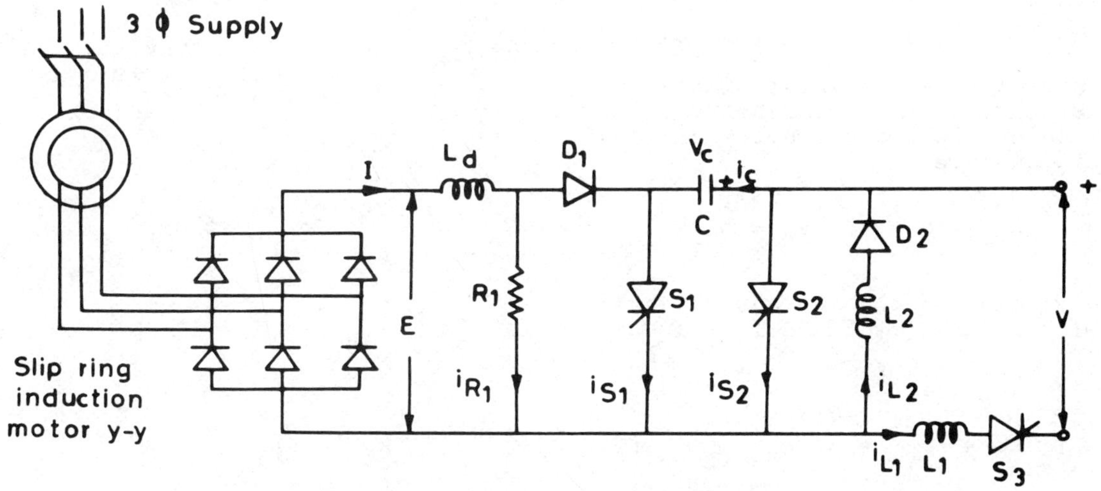

Fig.2 Circuit under Investigation.

ANALYSIS

The following assumptions have been made in the analysis :

1) The SCRs are perfect switches except for the limitation of turn off time and the diodes are ideal diodes

2) The rotor rectified voltage E together with the large inductance L_d form a constant current (I) source to the chopper. The value of

$$i_{S_1} = i_c + I \qquad (1a)$$

$$v_c = -E_c \, e^{-\alpha t}(\cos w_2' t + \frac{\alpha}{w_2'} \sin w_2' t) \qquad (2)$$

where $\alpha = \frac{r}{2L_2}$, $w_2'^2 = w_2^2 - \alpha^2 \doteq w_2^2$

and $w_2^2 = \frac{1}{L_2 c}$, $Q_o = \frac{w_2 L_2}{r}$.

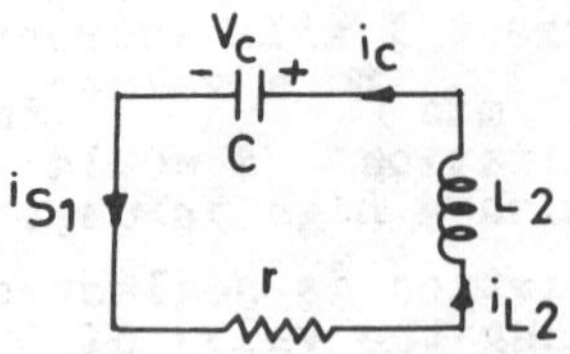

Fig.3 Equivalent Circuit S_1ON.

At time t when $w_2t = \pi$, $i_c = o$. D_2 turns OFF and $v_c = E_{c_1} = E_c\, e^{-\pi/2Q_o}$.

It is assumed that S_3 is fired at t when $i_c = o$, the reasons for this are discussed later. Two possibilities arise -

Case 1 : $E_{c_1} < V$.

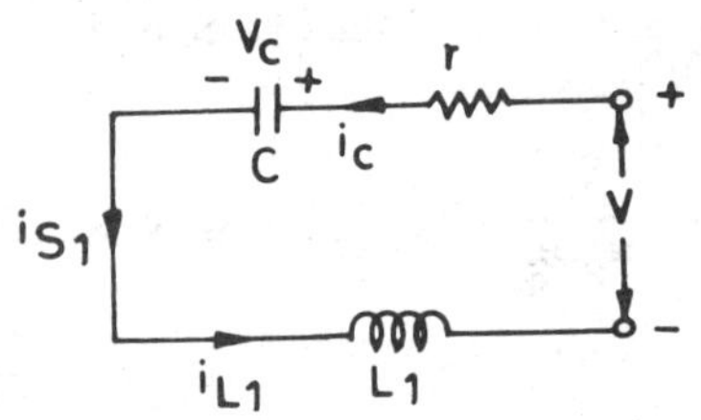

Fig.4 Equivalent Circuit S_3ON.

The symbol r has the same significance as earlier - it represents all the resistance in the circuit. Initial conditions are

$$i_{L_1} = o = i_c \quad v_c = E_{c_1}$$

with origin at the instant of firing S_3 with $i_{s_1} = i_c + I$ and $i_c = i_{L_1}$, we have

$$i_c = \frac{V - E_{c_1}}{w_1' L_1} e^{-\alpha_1 t} \sin w_1' t \qquad (3)$$

$$v_c = V - (V - E_{c_1})\, e^{-\alpha_1 t} \left[\cos w_1' t + \frac{\alpha_1}{w_1'} \sin w_1' t\right] \qquad (4)$$

where

$$\alpha_1 = \frac{r}{2L_1}, \quad w_1'^2 = w_1^2 - \alpha_1^2 \doteq w_1^2$$

$$w_1^2 = \frac{1}{L_1 C} \quad \text{and} \quad Q_1 = w_1 L_1 / r$$

when

$w_1 t = \pi$ $i_c = o$ and S_3 turns OFF with

$$v_c = V + (V - E_{c_1})\, e^{-\pi/2Q_1} = E'_{c_2}$$

Case 2 : $E_{c_1} \geq V$

No further charging from the auxiliary supply takes place.

After the duty cycle of the chopper is over S_2 is turned ON to commtate S_1 OFF. By specifying a minimum ON time for S_1 it is ensured that S_2 is fired only after the charging of the capacitor through the auxiliary supply, if required, is over.

In the above equivalent circuit the resistance can be ignored since no inductance is present in the circuit. Initial conditions are :

$$i_c' = o = i_{R_1}, \quad v_c' = -E_{c_2}$$

with origin at the instant of firing of S_2

E_{c_2} = the final capacitor voltage and it is either equal to E'_{c_2} or to E_{c_1} ($\geq V$).

$$i_c' = \left(I + \frac{E_{c_2}}{R_1}\right) e^{-t/CR_1} = i_{S_2} \qquad (5)$$

$$i_{R_1} = I - i_c' \qquad (5a)$$

$$v_c' = I_{R_1} - (I_{R_1} + E_{c_2})\, e^{-t/CR_1} \qquad (6)$$

At a value of $t = 5CR_1$, i_c' is assumed to be zero and then the capacitor voltage will be $+I_{R_1}$. Diode D_1 prevents capacitor discharge. The circuit is now ready for the duty cycle when S_1 is turned ON and it is noted that the capacitor voltage E_c at the start of the analysis equal I_{R_1}. Approximate waveforms are as in Fig.6

Design

A simple design for the motor whose specifications are given below, is presented. The design procedure remaining the same for any other motor.

Output power	2.8 kW
Power factor	0.8
Speed	1370 rpm
Stator/Rotor connection	Y/Y
Stator	400 V, 6.8 A
Rotor	84 V, 22.5 A

The value of R_1^* is varied by varying the OFF time of S_1 i.e., T_2. The frequency i.e., $1/T$ is selected on the same basis as in [2]. Let the chopper frequency be 300 Hz. A value for L_d is chosen the same way as in [2]. The value of I_L is found to be 27.1A, and the value of the no load rotor rectified current is not required in this design which is an advantage over the circuits in [2] and [3]. A value for R_1 is selected in the same way as in [2]. The maximum forward voltage across S_1 when S_1 is OFF is $I_L R_1$ which would increase if R_1 is increased and this consideration also affects the choice of $R_1 = 5$ ohms, thus giving L_d = 350 mH, 30 A.

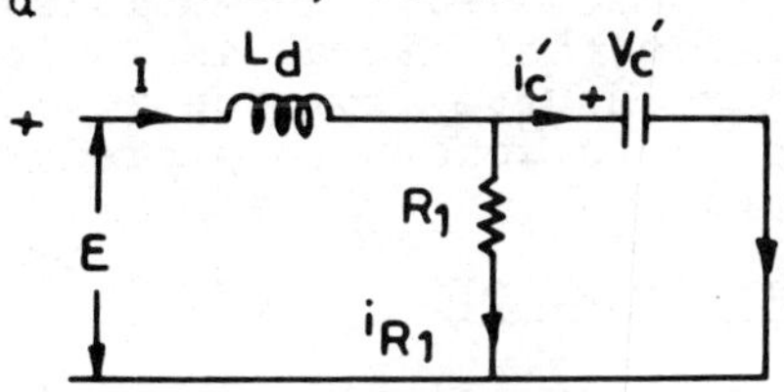

Fig.5 Equivalent Circuit S_2ON.

It is assumed that S_3 is fired after the resetting of the capacitor voltage through L_2 D_2 is complete . The advantages are :-

(1) The power drain is considerably reduced. Charging takes place only if $E_{c_1} < V$. The maximum value of E_{c_1} is $I_L R_1$ i.e., 135.5 V. If V = 100 V then at or near full load there is no power drain.

(2) The maximum capacitor voltage is reduced, to a maximum of 200 V if V = 100 V. Firing S_1 and S_3 at the same instant would give a maximum capacitor voltage of 335.5 V. Reverse voltage ratings are also thus reduced.

(3) The reduction in the capacitor voltage results in a reduction of the currents carried by all the elements of the chopper.

(4) The di/dt for the SCR's is reduced.

(5) An important advantage is that the analysis and design become simpler.

The two disadvantages of the assumption are :

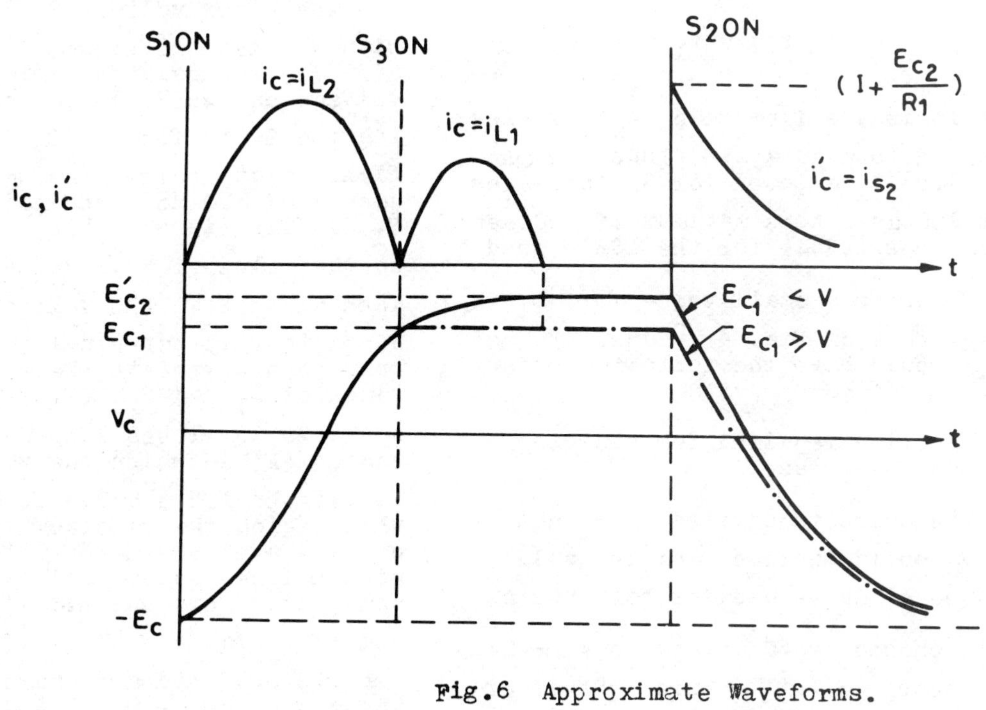

Fig.6 Approximate Waveforms.

(1) The value of the minimum ON time of S_1 increases, since the charging of the capacitor through the auxiliary supply has to be included in the duty cycle. Thus the maximum value of T_2 is reduced.

(2) The firing circuit becomes slightly complicated.

It is assumed in this design that S_3 is fired 200 μsec. after S_1 thus giving $w_2 = 1.57 \times 10^4$ rad/sec. Assuming that $v_c = o$ at $t = t_{o1}$ in (6), we get

$$e^{t_{o1}/CR_1} = \frac{IR_1 + Ec_2}{IR_1} = 1 + \frac{Ec_2}{IR_1}.$$

Under the worst conditions $E_{c_2} = V$ and $I = I_L$. Since t_{o1} for the SCR's used should be at least 30 μsec, using $V = 100$ V and $c = 20$ μF gives $t_{o1} = 55$ μsec which is safe even under the worst conditions.

The value of L_2 is calculated as

$$w_2^2 = \frac{1}{L_2 c} = 2.46 \times 10^8 \quad \text{and} \quad c = 20\mu F$$

Hence $L_2 = 200$ μH.

From (2), $v_c = o$ at time $t = t_1$ when,

$$\cos w_2 t_1 = -\frac{\alpha}{w_2} \sin w_2 t_1 \quad \text{or}$$

$$\tan w_2 t_1 = -\frac{w_2}{\alpha}$$

As α increases from zero, $w_2 t_1$ varies from $\pi/2$ to π at $\alpha = \infty$. Thus the turn-off time t_1 allowed for S_2 increases from 100 μsec to a maximum of 200μsec which is suitable for the SCR's used.

Choosing a small value for L_1 in the circuit of Fig. 4 [Eqns.(3) and (4)] would have the following effects :

(1) The time required for charging would decrease.

(2) The currents carried by S_1 and S_3 would increase and so would the di/dt values for both SCR's.

L_1 is chosen as 50 μH and so $w_1 = 2w_2$ and hence S_3 will turn OFF after 100 μsec.

The minimum ON time for S_1 is thus 350 μsec, which includes the 100 μsec required for charging through the auxiliary supply. Hence maximum value of $T_2 = 2.98$ msec at the chopping frequency of 300 Hz and so the maximum value of $R_1^* = 0.9\ R_1$.

The values of di/dt and dv/dt for the SCR's can easily be found from the equations. For the complete cycle all the pulses of current flowing through any component are drawn and the average and rms values of this current will give the current rating of that component. The equations for capacitor voltage give the voltage ratings.

The other circuits have been analysed in [5] in an exactly similar manner except for taking into account of the fact that in the last three circuits of Fig.1, S_3 is fired after S_2 and the delay in firing is fixed, whereas in the first three circuits of the same figure S_3 is fired after S_1.

It is seen that the disadvantages associated with the earlier circuits have been removed. An additional advantage is the natural isolation of the auxiliary source from the motor when the motor is loaded to a predetermined level.

FEEDBACK CONTROL

The block diagram for feedback speed control is given in Fig. 7.

The error voltage V_1 is compared with a fixed reference amp voltage of 9 V. The point at which the ramp voltage equals V_1 is the point at which the S_1 is fired. S_2 is fired at fixed points corresponding to a frequency of 300 Hz. The point at which S_1 is fired is variable; it depends on the value of V_1. S_3 is fired 200 μsec after S_1. The voltage V_1 is allowed to vary between 1.5 to 7.8 V only to accommodate the minimum ON times of S_1 and S_2. Thus the system operates in closed loop only in the region AA' in which the value of V_o is between 1.5 and 7.8 V. Outside this region the feedback loop has a control over the motor which operates with a fixed rotor resistance of 4.35 ohms, if V_o exceeds 7.8 V, or with 0.75 ohm if V_o is less than 1.5 volts. The results obtained are shown in Fig. 8 using a value of K = 8 and $K_1 = 200$.

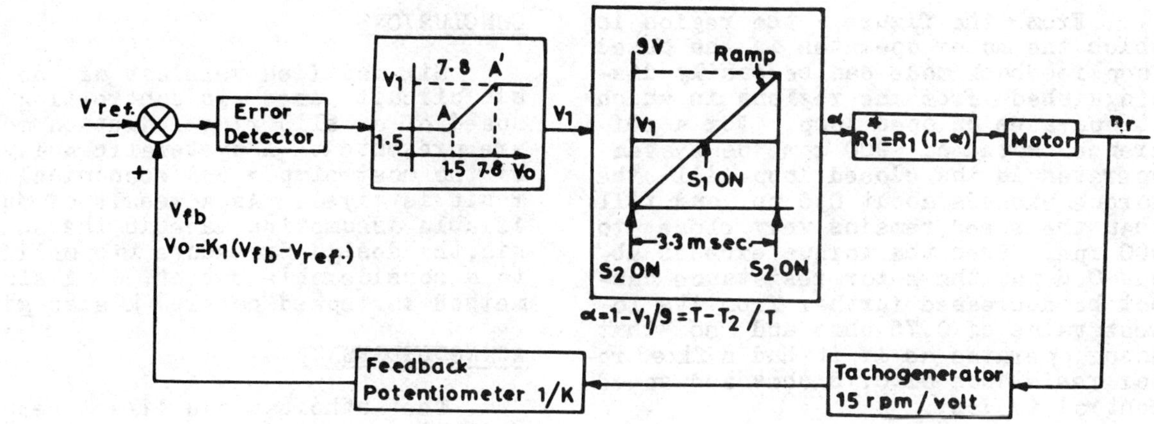

Fig.7 Block Diagram for Feedback Speed Control.

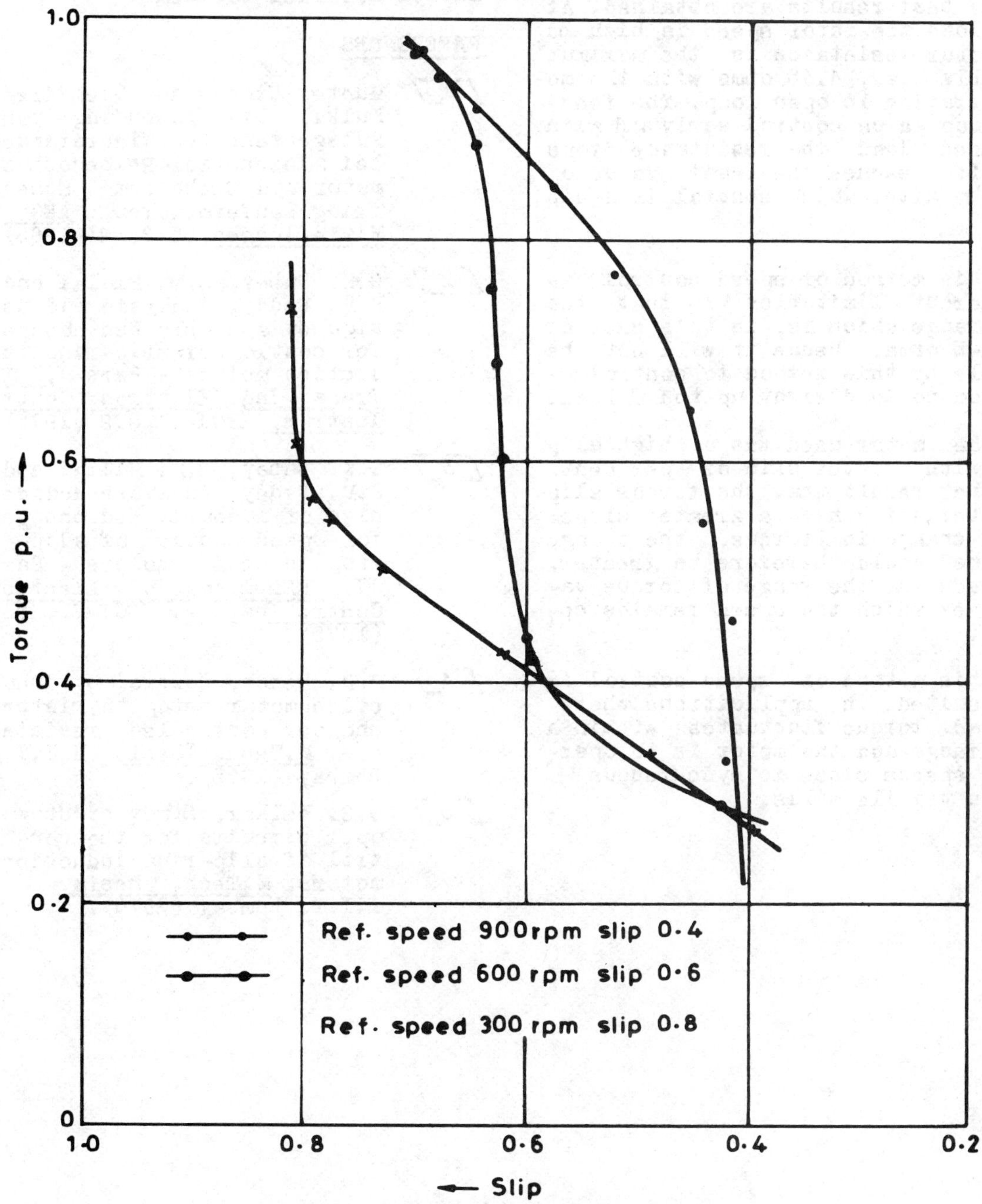

Fig.8 Torque-speed Characteristic.

From the figure, the region in which the motor operates in the closed loop feedback mode can be easily distinguished from the regions in which it operates on open loop. For a reference speed of 900 rpm the system operates in the closed loop till the torque exceeds about 0.6 pu and till then the speed remains very close to 900 rpm. When the torque exceeds about 0.6 pu, the rotor resistance cannot be decreased further from its lowest value of 0.75 ohms and so that motor operates as if it had a fixed rotor resistance of 0.75 ohms, and speed control is lost.

For the reference speed of 300 rpm the best results are obtained. At light load the motor speed is high and the rotor resistance is the maximum available i.e., 4.35 ohms with the motor operating in open loop. The feedback loop gains control early and with increased load the resistance drops till it reaches the least value of 0.75 ohm after which control is again lost.

This method of speed control has an inherent limitation in that the rotor range which is, in this case at best 0-5 ohms. Hence it will not be possible by this method to control speed from no load right up to full load.

The motor used was a high slip motor with a rates slip of 9per cent, with the result that the torque slip characteristics have a greater slope. With a change in torque, the change in speed would therefore be greater, thus reducing the range of torque values over which the speed remains constant.

This method of speed control is best suited in applications where the load torque fluctuates within a small range and the motor is to operate at speeds close to synchronous i.e. in textile mills.

CONCLUSIONS

Six modified versions of the basic circuit used in controlling the speed of a slip ring induction motor are presented. A systematic analysis of the most simple and economical circuit is given. As a result of justifiable assumption made in the analysis, the design procedure is simplified to a considerable extent. A simple method for speed control is also given.

ACKNOWLEDGEMENT

The authors would like to express their gratitude to the Director, I.I.T. Bombay, for the help and facilities afforded to them.

REFERENCES

[1] Gunter Ciesow and Siegfired Kulka, Die Anwendung von Pulsgesteuerten Widerstanden bei Gleichstrom-Reihenschusmotor und Drehstrom - Scheifring Laufermotoren, AEG - Mitteilungen, 55, 2, 123 (1965).

[2] G.K. Dubey, S.K. Pillai and P.P. Reddy, Analysis and design of a doubly fed chopper for control of slip-ring induction motors - Part I, IEEE Trans. Ind. Electron. Contr. Instrum. IECI-21, 522 (1975).

[3] G.K. Dubey, S.K. Pillai and P.P. Reddy, Analysis and design of a doubly fed chopper for speed control of slip-ring induction motors - Part II, IEEE Trans. Ind. Electron Contr. Instrum. IECI-21, 531 (1975).

[4] P.P. Reddy, Control of Induction motor using thyristor chopper controlled resistance, M.Tech. Thesis, I.I.T. Bombay (1975).

[5] S.S. Kelkar, Study of Chopper circuits for the control of slip-ring induction motors, M.Tech. Thesis, I.I.T. Bombay (1976).

ANALYSIS OF SELF-EXCITATION OF VOLTAGE CONTROLLED INDUCTION MOTOR

M. Hunyár

Department of Electrical Machines, Tech. Univ. 1111. Budapest, Hungary

ABSTRACT

The three phase induction motor, fed by inverse-parallel thyristor-diode pairs, shows self-excitation at $\frac{50}{2}$Hz =25Hz subharmonic pulsating frequency in the region of plugging operation. The self-excitation can be verified also by exact calculation, however by this way it is hard to understand the physical phenomena, and long computation time is required. Present paper investigates the above phenomena by applying an equivalent circuit closely analogue to that of the induction machine, consisting of a three-phase resistance,inductance and voltage source. As it is proved in the paper,self-excitation is realized by the modification of extinction time. In the calculation the method of small deviations around an operating point is used. The boundary curve of the unstable region calculated by using the approximate equivalent circuit is in close agreement to these results obtained from exact calculations and test.

INTRODUCTION

As it was already proved in the previous papers /Ref. 1,2/ the steady state of the three-phase induction motor fed by inverse-parallel thyristor-diode pairs can be calculated relatively exactly on the basis of the equivalent circuit shown in

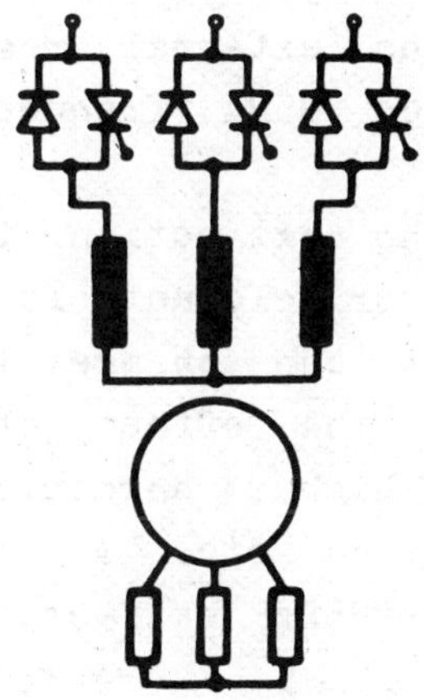

Fig. 1.

Fig. 2. The equivalent circuit is to be considered for three-phase or Park vectors,L is the transient inductance of the stator, R is the sum of the stator and reduced rotor resistances, $\bar{U}'_{50}$ counter EMF having fundamental harmonic frequency is determined by that the fundamental harmonic /50Hz/ current is to be calculated exactly:

$$\bar{U}'_{50}=\bar{I}_{50}(\bar{Z}_{m50}-\bar{Z}_{50}), \qquad /1/$$

where $\bar{Z}_{m50}$ is the impedance of the

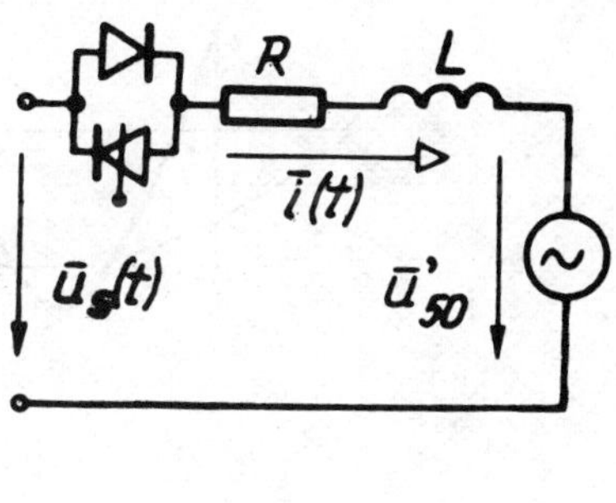

Fig. 2.

motor, $\overline{Z}_{50}$ the impedance of R-L circuit, related to 50Hz currents.

FUNDAMENTALS OF CALCULATING SELF-EXCITATION AT 25Hz

In ideal case the power-supply voltage is three-phase symmetrical and sinusoidal, the firings are repeated regularly and periodically. Therefore there is no external reason which would force 25Hz frequency on the system.

However the extinction time of the semiconductor elements is not fixed rigidly, it can change to a small extent on the effect of the slot harmonics, higher harmonics of the mains voltage etc. As it will be proved, the 25Hz self-excitation of the examined circuit, is realized by the modification of the extinction time of the doides.

The three phase-currents /i_a, i_b, i_c/ are shown in Fig. 3, in such a mode of operation, when three-phase /3Ph/ and two-phase /2Ph/ conducting modes occur alternately. Further it is assumed that due to some kind of previous effect, current of indefinite small amplitude /Δi_a, Δi_b, Δi_c/ is superposed on the phase-currents, which has a fundamental harmonic of 25Hz and of negative sequence /-25Hz/. For the sake of better illustration the scale of the additional current naturally differs from the scale of the phase-current. If the phase-current is substituted with its tangent in the region of zero crossing and if in the moment of

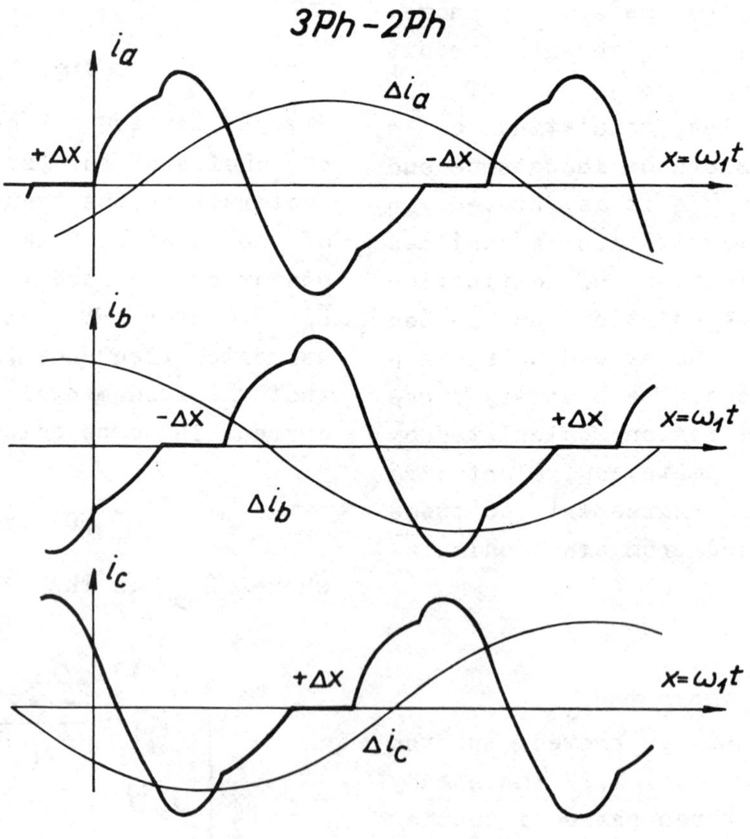

Fig. 3.

extinction the instantenous value of the superposed current-part is $\Delta i/\delta/$, then the modification of the extinction angle is:

$$\Delta x = \frac{\Delta i/\delta/}{\left(\frac{di}{dx}\right)_{\delta}} \qquad /2/$$

Where Δx is positive, if the conducting time increased $/x=\omega_1 t$, t is the time, and ω_1 is the angular frequency of the supply voltage/.

All the further calculations will be carried out on the margin of the stability, therefore it can be assumed, that Δi remains constant, its amplitude is neither increased nor decreased, As a consequence of periodicity and symmetrical relations of Fig. 3, modification of extinction time with identical absolute value but with varying sign $/\pm\Delta x/$ can be obtained. The form of the additional current really arising differs from the form assumed in Fig. 3, but its fundamental harmonic fulfils the above two conditions /is of -25Hz/, however the higher-harmonics do not alter our statements concerning $\pm\Delta x$.

Periodicity enables us also in the forthcoming analysis to apply approximate equivalent circuit consisting of R-L-$\overline{U}'$ elements. According to our calculations the results obtained correspond to the physical reality, if the induction motor is considered correctly also for -25Hz currents, i.e. a voltage source of -25Hz frequency is inserted in the circuit. By analogy to /1/:

$$\overline{U}'_{25} = U'_{25} e^{j\varphi_u} = \overline{I}_{25}\left(\overline{Z}_{m25} - \overline{Z}_{25}\right) \qquad /3/$$

/In the followings firing moment of phase "a" is chosen for the time origin, so φ_u is the phase-angle of $\overline{U}'_{25}$ in t=0 moment./

The complete equivalent circuit can be seen in Fig. 4a. Indication $\delta\pm\Delta x$ written beside the semiconductor-pair, symbolizes the modification of the extinction time. The circuit can be divided into two superposed parts, counter EMF with 50Hz can be found in Fig. 4b and with -25Hz can be found in Fig. 4c. In Fig. 5a voltage vector locus of R-L circuit in equivalent circuit Fig. 4b is shown, variation

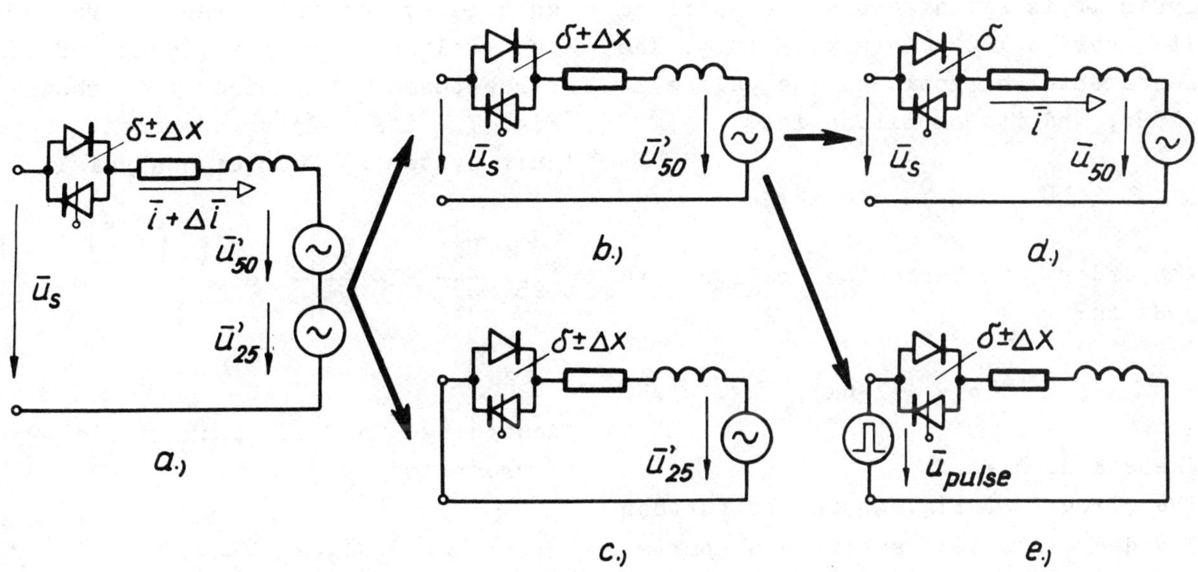

Fig. 4.

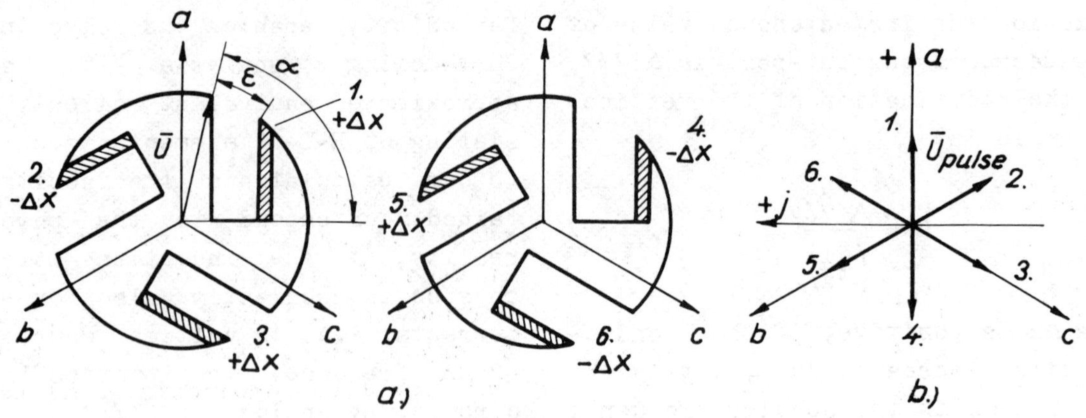

Fig. 5.

related to the operating point is shaded. As $\Delta x \to 0$, the absolute value of the amplitude of pulse like variations:

$$U_{pulse} = U \sin(\alpha - \varepsilon), \qquad /4/$$

where $\bar{U} = \bar{U}_s - \bar{U}'_{50}$, $\bar{U}_s$ is the supply voltage, α is the firing angle measured from the zero crossing of this voltage, ε is the conduction time of 2Ph. In Fig. 5 serial numbers have been given for the pulse like variations, and the sign of Δx is also shown. The process repeats itself only after two periods $/\omega_1 t = 720^\circ/$. Nevertheless in case of Fourier analysis it is satisfactory to perform it over a 120° interval only. The fundamental harmonic of the pulses is -25Hz, and its amplitude is:

$$\bar{U}_{25} = U \frac{3.\Delta x}{2\pi} \sin(\alpha - \varepsilon) e^{-j\frac{\varepsilon}{2}} \qquad /5/$$

the order of harmonics related to 50Hz is:

$$\nu = -\frac{1}{2}(1+6k), \qquad /6/$$

where $k = 0, \pm 1, \pm 2 \ldots$

The circuit of Fig. 4b can be further divided into two superposed parts, Fig. 4d is the equivalent circuit of the normal operating point /see Fig. 2/, the voltage source to be found in circuit Fig. 4e provides the voltage pulses according to Fig. 5b. Superposed parts Fig. 4c and Fig. 4e are reducible according to Fig. 6, the whole additional current $/\Delta \bar{I}/$ can be

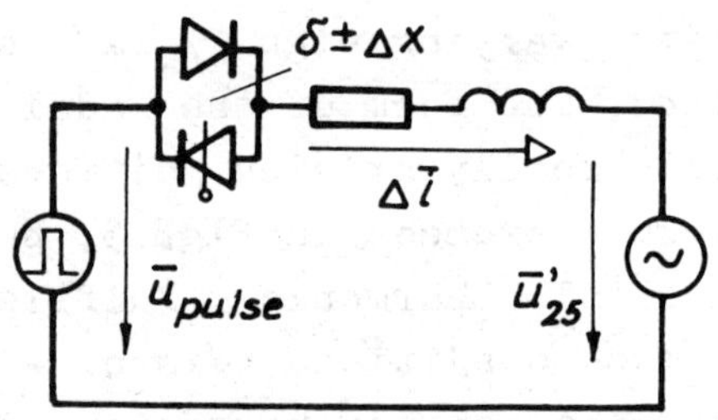

Fig. 6.

calculated on this basis. From $\bar{U}'_{25}$ the voltage vector locus of the three-phase R-L circuit is shown in Fig. 7, its -25Hz component in per unit system by Fourier analysis:

$$\bar{F}_{25} = \frac{\bar{U}^{*\prime}_{25}}{\bar{U}'_{25}} = 1 - \frac{3}{2\pi}\left[\frac{\varepsilon}{2} + e^{-j\left(2\varphi_u + \frac{\varepsilon}{2}\right)} \sin\frac{\varepsilon}{2}\right] \qquad /7/$$

According to Fig. 6, the whole -25Hz current is:

$$\bar{I}_{25} = \frac{\bar{U}_{25} - \bar{U}^{*\prime}_{25}}{\bar{Z}_{25}} \qquad /8/$$

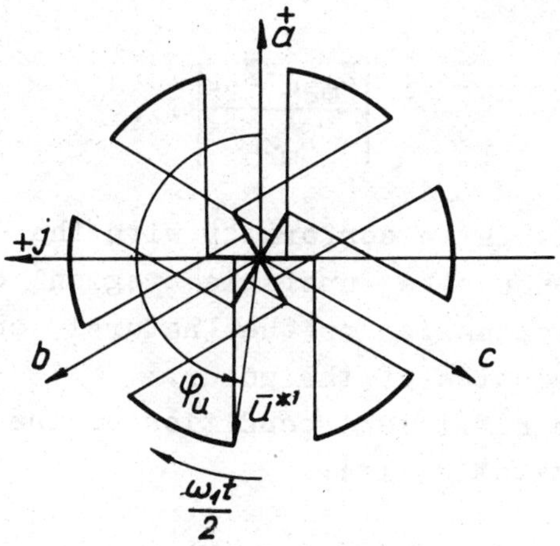

Fig. 7.

In case of firing of the thyristors normal three-phase transient phenomena of the R-L circuit occurs concerning $\Delta\bar{I}$ too /3Ph-2Ph mode of operation/:

$$\Delta\bar{I}=\bar{I}_{st25}e^{-j\frac{x}{2}}+\bar{C}_{25}e^{-\frac{x}{tg\varphi}} \qquad /9/$$

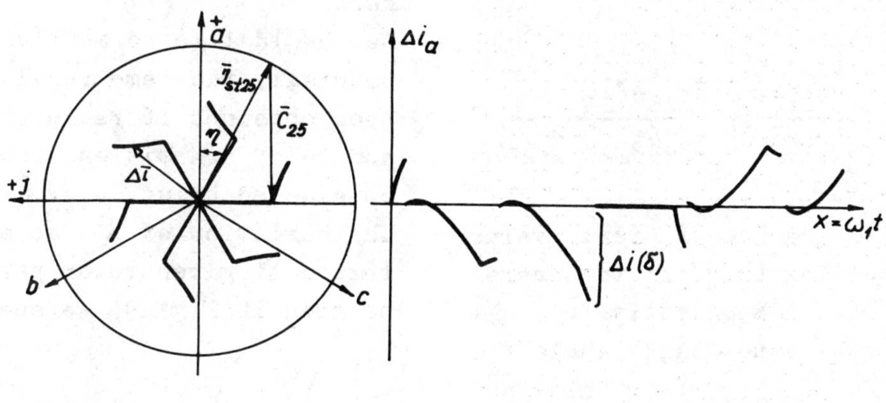

Fig. 8.

/Where $tg\varphi=\omega_1 L/R$./ The stationary current with the assumed positive directions:

$$I_{st25}e^{j\eta}=-\frac{U'_{25}e^{j\varphi_u}}{Z_{25}e^{j\varphi_{25}}} \qquad /10/$$

The transient component can be calculated from the conditions of symmetry shown in Fig. 8:

$$\bar{C}_{25}=-I_{st25}\cos\eta\left[1+j\frac{\sin\pi/3}{e^{\frac{2\pi}{3tg\varphi}}-\cos\pi/3}\right] \qquad /11/$$

The explanation for the unusual sudden change in the current is that in case of extinction of the diode currents as a result of U_{pulse} during a very short Δx time the "a", "b" and "c" direction components of $\Delta\bar{I}$ become zero. Namely in $\delta\pm\Delta x$ moments the total current has to become zero.

By means of /9/, /10/, /11/ the instantaneous value of the additional current in the phase turning off /e.g. in phase "a"/:

$$\Delta i_a/\delta/=-I_{st25}\cdot y \qquad /12/$$

where:

$$y=\sin\left(\eta-\frac{\delta}{2}-\frac{\pi}{6}\right)-e^{-\frac{\delta}{tg\varphi}}\cos\eta\left[\frac{1}{2}+\frac{3/4}{e^{\frac{2\pi}{3tg\varphi}}-1/2}\right] \qquad /13/$$

/δ is the conduction time of 3Ph/.

Reverting back to /2/, the slop of the current in the moment of zero crossing can be defined from that $i_a=0$ and after extinction $U\sin/\alpha-\varepsilon/$ voltage will appear across the thyristor:

$$\left(\frac{di}{dx}\right)_\delta = \frac{U.\sin/\alpha-\varepsilon/}{X_{50}} \qquad /14/$$

Let us designate with $\overline{U}_{25new}$ that -25Hz voltage which occurs as a result of variation of existing Δx conduction time, and with $\overline{U}_{25old}$ that voltage which previously was necessary for establishing Δx. The ratio λ^* of these two determines the stability, if is $\lambda^*>1$, then self-excitation of -25Hz will be produced, otherwise the additional current will be zero.

For presentation of demonstrative final result let us define from point of view of -25Hz currents the impedance equivalent with the semiconductor pairs:

$$\overline{Z}_{t25}=\overline{Z}_{25}\frac{1-\overline{f}_{25}}{\overline{f}_{25}}, \qquad /15/$$

thus:

$$\lambda^* = \frac{\overline{U}_{25new}}{\overline{U}_{25old}} = -\frac{3X_{25}}{\pi Z_{25}}\, y\, \frac{\overline{Z}_{m25}-\overline{Z}_{25}}{f_{25}(\overline{Z}_{m25}+\overline{Z}_{t25})} \qquad /16/$$

Formula /16/ can be of real value only, as according to /5/ the operation point $/\varepsilon/$ independently from Δx determines the same phase angle for $\overline{U}_{25new}$ and for $\overline{U}_{25old}$ too. This fact is utilized when calculating η or φ_u angle.

CALCULATED RESULTS, CONCLUSIONS

From /13/ it can be calculated that:

$$|y|\left|-\frac{3X_{25}}{\pi Z_{25}}\right|<1, \qquad /17/$$

thus only due to the second fraction can be $\lambda^*>1$. In case of full voltage $\overline{f}_{25}=1$ and $Z_{t25}=0$, therefore the second fraction simplifies, and:

$$\left|\frac{\overline{Z}_{m25}-\overline{Z}_{25}}{\overline{Z}_{m25}}\right|<1 , \qquad /18/$$

that is in conformity with the experience, the unstable region cannot reach as far as the natural characteristics of the motor.

The sufficient condition of the self-excitation is:

$$|\overline{Z}_{m25}+\overline{Z}_{t25}|\to 0 \qquad /19/$$

Condition /19/ means the approach of series resonance, as $\overline{Z}_{m25}$ is ohmic-inductive, $\overline{Z}_{t25}$ is of ohmic-capacitive characteristics in a part of the region of plugging. The real part of $\overline{Z}_{m25}$ becomes negative only in the slip range of s>1,5, for the -25Hz component the generator region starts here.

We should like to mention, that for λ^* naturally the same result would have been obtained if ratio of other "new" and "old" quantities would have been considered by us, e.g. $\lambda^*=\Delta x_{new}/\Delta x_{old}$. The variation of λ^* as a function of torque at given motor parameters, can be seen in Fig. 9. We should like to

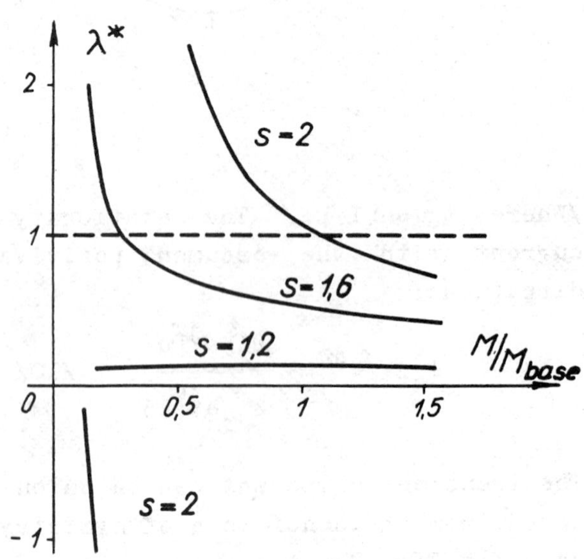

Fig. 9.

remind to that our calculations are only valid for $\lambda^{*}=1$. The boundary curve lying in the unstable region of 3Ph-2Ph mode of operation, has been illustrated in Fig. 10 on speed-

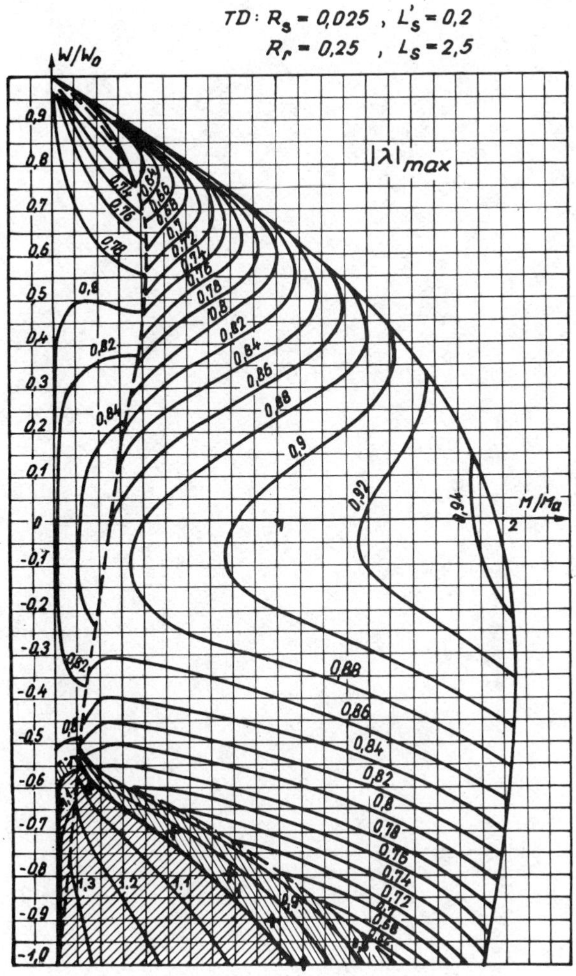

Fig. 10.

torque plane, with small crosses. The basic figure has been adopted from /Ref. 3/ in which the outhor determined with exact calculation method the maximal absolute value of eigenvalues of the transfer-matrix. The analogy is evident betwen λ^{*} and λ_{max}, as in the latter case too $|\lambda_{max}|=1$ determines the boundary curve of the unstable region. The negligible daviation of the two boundary curves can be explained first of all by calculating on basis of approximate equivalent circuit the harmonic of $\nu=-2$ order /with 100Hz/ and its torque do not give accurate result. The calculation of the boundary curye also in 2Ph-0Ph or 3Ph-2Ph-0Ph mode of operation, can be carried out similarly as mentioned above.

We should like to remark that the unstable region does not extend until region of infinite slips. Especially in case of lower rotor resistances relatively narrow region is obtained above a slip of s=1,5. As a thumb rule it can be stated, that the lower and similarly the upper limit of the unstable region will occur at such slips where the phase angle of $\overline{Z}_{m25}$ remains unchanged independently of the rotor resistance.

REFERENCES

/1/ I.Rácz,Theorie des Betriebs von Asynchronmotoren mit antiparallelen Thyristoreen, XIII. Intern. Wiss. Koll. T.H. Ilmenau /1968/.

/2/ M.Hunyár, Mit antiparallelen Thyristorpaaren gesteuerter Asynchronmotor, I. Konferenz für Leistungselektronik,Budapest /1970/.

/3/ J.Borka,A.Varga,Computer Analysis of Voltage Controlled Induction Machines, IFAC Symposium, Control in Power Electronics and Electrical Drives,Duesseldorf /1974/.

COMPARISON OF SPEED CONTROLLED DC DRIVES WITH AND WITHOUT SUBORDINATE CURRENT LOOP

H. Grotstollen

Inst. f. Elektrische Antriebe u. Steuerungen, Universität Erlangen, FRG

Summary

Two different configurations for controlling the speed of armature-fed dc drives are compared in this paper. At the first concept a subordinate current control is used to compensate the first-order factor of the armature circuit and to limit the armature current. At the other concept the speed controller directly leads the power amplifier and the armature and the machine. These drives generally have a current controller, too, which normally is not in action but relieves the speed controller when the current is exceeding its limit. By comparisons in the frequency domain it is shown that the relation between the equivalent time constant of the subordinate current control and the time constant of the armature circuit is a criteria that demonstrates which of the compared control concepts yields a faster transient response with fixed amplifier and motor. Some possibilities of current limitation for direct speed-control are presented that join with simple and inexpensive current measuring devices.

INTRODUCTION

Speed controlled electric drives are usually designed with dc motors fed by thyristor converters with natural commutations. Two reasons are decisive of all electric machines the seperately excited dc motor is best suited for control in general and for speed control especially. Besides this nearly all requirements related to steady-state accuracy and transient response can be met by feeding this machine with thyristor rectifiers commutated by the main voltages i.e. without special commutating devices. Thus the control concepts as well as the equipment for control and power supply of dc drives described are simple compared with the control concepts and the equipment required for ac drives. In consequence the special merits of dc drives are very good reliability and low cost.

Thyristor converters feeding dc motors in a speed controlled mode are mostly operated in a closed current control loop subordinated to the speed control loop: the reference signal for the current control is generated by the speed controller. In accordance to the German expression "Stromleitverfahren" this control concept shall be called current leading speed control in this paper. By means of the subordinated current control one first-order factor of the electro-mechanical system, the time constant of the armature circuit, can be compensated by use of an integral plus proportional controller (PI-controller). Furthermore the armature current can be limited simply by limiting the current reference signal to an admissible value. The current leading speed control is well known and often described in literature on electric drives /Ref. 1 to 4/.

In the last time another concept for speed control has been applied to thyristorized dc drives. Typical for this configuration is the absence of a subordinate current control loop: the speed controller directly leads the power stage and by this the electromechanical system. With respect to this fact this configuration is called direct speed control in this paper. For the purpose of current limitation direct speed controlled drives must have a current control, too. This current control device does not affect the speed control under normal operation conditions. Only when the speed control causes the motor current to exceed its limit, set at the current controller, this unit determines the input of the power amplifier. Speed control and current control relief one another when the current is reaching or leaving its limitation. At first direct speed control was applied to low-power servo drives with transistorized amplifiers generating a continuous voltage. The field of application where thyristorized drives can be found today are servomechanisms, too.

ANALYTIC COMPARISON OF CURRENT LEADING AND DIRECT SPEED CONTROL

Methods and Assumptions

In this chapter the speed control concepts described above are compared in the frequency domain. The decisition for use of Bode diagrams was taken for the frequency response characteristics offer the most convenient alternative for determination of the controllers and for comment of the steady-state accuracy as well as the transient-response.

To make the frequency response method applicable to the speed control systems their components are supposed to be linear and time invariant.

The controllers are required to provide zero steady-state error and a fast response. Nevertheless they must not amplify the ripple voltages contained by the feedback signals. In case of current feedback the current itself contains the ripple caused by unsufficient smoothing inductance in the armature circuit. In case of speed feedback the ripple is produced by the measuring device, the tachometer. With regard to these requirements proportional plus integral controllers (PI-controller) have been chosen as commonly used in drive controls.

The parameters of the controllers are designed using the symmetric optimum in its original form /Ref. 5/. Modifications are possible /Ref. 6/ but are of no importance for the qualitative results presented in this paper.

The symmetric optimum as well as the magnitude optimum /Ref. 7/ are applicable to a system with an overdamped second-order plant. Such a plant is characterized by a damping ratio d 1 and can be seperated to two first order factors. This assumption is met with the plant to be controlled in the current leading speed control version. In contrast the speed controller of a direct speed controlled dc drive normally has to deal with an underdamped plant represented by the electromechanical system. In this case the controller can be designed by a rule which is justified as follows.

If the transfer function of the plant is

$$F_P = \frac{1}{1 + 2dT\cdot s + T^2\cdot s^2} \qquad (1)$$

and if the transfer function of the controller is

$$F_C = K_C\,\frac{1 + T_C\cdot s}{T_C\cdot s} \qquad (2)$$

then the transfer function of the closed loop made up by the plant and the controller is given by

$$F = K_C\,\frac{1 + T_C\cdot s}{K_C + (1+K_C)T_C\cdot s + 2dTT_C\cdot s^2 + T^2T_C\cdot s^3} \qquad (3)$$

The magnitude of the frequency response characteristic is obtained as

$$|F| = K_C\sqrt{\frac{1 + T_C^2\cdot\omega^2}{N}} \qquad (4)$$

where $N = K_C^2 + \left[(1+K_C)^2T_C^2 - 4K_CdTT_C\right]\omega^2 + \left[(2dT)^2T_C^2 - 2(1+K_C)T^2T_C^2\right]\omega^4 + T^2T_C^2\omega^6$

Now the criteria is applied on which symmetric optimum and magnitude optimum base, too: The parameters of the controllers are to be chosen in such a manner that the magnitude $|F|$ approaches a constant as far as possible with growing frequency. From equ. (4) can be seen, that two coefficients of the denominator might be equalled to zero. By evaluation of

$$(1 + K_C)^2T_C^2 = 4K_CdTT_C$$
$$(2dT)^2T_C^2 = 2(1+K_C)T^2T_C^2 \qquad (5)$$

the following design rule is obtained

$$K_C = 2d^2 - 1$$
$$T_C = \frac{2d^2 - 1}{d^3}\,T \qquad (6)$$

As equs. (6) demonstrate the optimization is possible only when the damping ratio of the plant is d > 0,707, that is to say when the amplitude curve of the plant has no resonant peak.

When the optimal parameters are substituted in equ. (3) this equation becomes

$$F = \frac{1 + T_C\cdot s}{1 + \frac{2}{d}\frac{T}{d}\cdot s + \frac{2}{d}\left(\frac{T}{d}\right)^2\cdot s^2 + \frac{1}{d}\left(\frac{T}{d}\right)^3\cdot s^3} \qquad (7)$$

where $d > \frac{1}{\sqrt{2}}$.

By optimization only the denominator of the transfer function is affected. Hence the numerator causes the amplitude curve to deviate from the desired optimal curve with a resonant peak. The numerator undamps the system and is therefore compensated by filtering the reference signal with a first-order factor.

Although the design of controller parameters is based on extensive simplifications the Bode diagrams presented in these paper describe the transfer behaviour of the systems shown by the block diagrams exactly.

The Plant to be controlled

The arrangement and the block diagram of the plant to be controlled is shown in Fig. 1. The power amplifier which is assumed to be a thyristor converter feeds a constant excited dc motor driving a mechanic load. In the block diagrams and equations all voltages, currents and torques are referred to their rated values. The angular speed n and the back emf e are referred to the values that

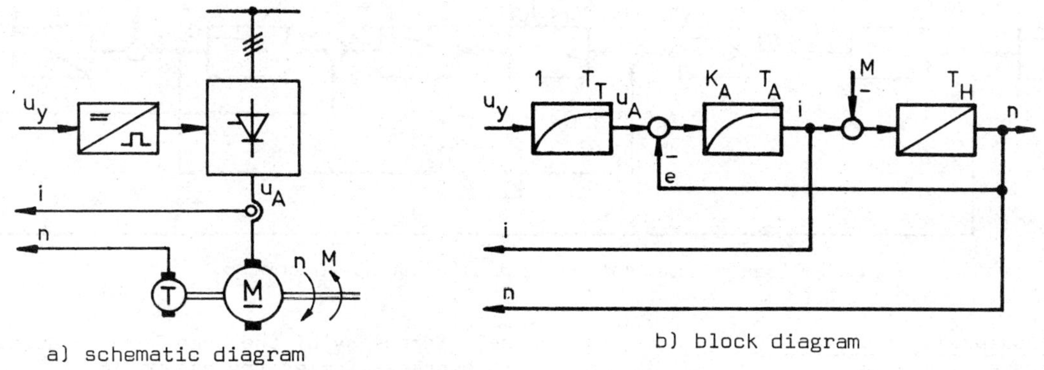

Fig. 1. Plant to be controlled

appear when the unloaded machine is fed with the rated voltage.

Fig. 1b shows that the transfer behaviour of the power amplifier is approximated by a first-order factor. In the case of a thyristor converter the time constant equals to half the period between two consecutive firing pulses. The electro-mechanic system is characterized by the gain and the time constant of the armature circuit and by the integration time of mechanic inertia. As the presence of the electric time constant shows the current is assumed to be continuous. The integration time equals the time the machine needs to accelerate from stillstand to the reference speed defined above at rated torque. This constant can be substituted by another constant

$$T_M = \frac{T_H}{K_A} \tag{8}$$

which is called mechanical time constant because of its physical background: this time constant characterizes all transients that appear when the electrical time constant is neglected and the motor thus becomes the behaviour of a first-order factor. By use of this substitution the transfer functions of the plant for leading the speed and the current become

$$\frac{n}{u_y} = \frac{1}{1 + T_T \cdot s} \cdot \frac{1}{1 + T_M \cdot s + T_M T_A \cdot s^2} \tag{9}$$

$$\frac{i}{u_y} = \frac{1}{1 + T_T \cdot s} \cdot K_A \frac{T_M \cdot s}{1 + T_M \cdot s + T_M T_A \cdot s^2} \tag{10}$$

The denominator of the motor transfer functions can be written like equ. (1) and equ. (9) becomes

$$\frac{n}{u_y} = \frac{1}{1 + T_T \cdot s} \cdot \frac{1}{1 + 2dT \cdot s + T^2 \cdot s^2} \tag{11}$$

where

$$d = \frac{1}{2}\sqrt{\frac{T_M}{T_A}} \qquad T = \sqrt{T_M T_A} \tag{12}$$

The Bode diagrams shown in this paper refer to an existing electro mechanical system with the rated values and the referred constants listet in table 1. The armature current was experimentally found to be continuous when exceeding 20 % of the rated value with use of a six pulse converter.

TABLE 1 Data of electro-mechanical system

Data of equipment:		
Rated power		9,9 kW
Rated torque		64 Nm
Rated speed		1500 rpm
Reference angular speed		167 rad/s^2
Rated armature voltage		400 V
Resistance (total)		1,4 Ω
Inductance (total)		25,2 mH
Torque constant		2,4 Nm/A
Inertia (total)		0,33 kgm^2
Data of block diagram:		
Gain of armature circuit	K_A	7,27
Electrical time constant	T_A	18 ms
Acceleration time	T_H	325 ms
Mechanical time constant	T_M	44,7 ms
Damping ratio	d	0,79
Corner time constant	T	28,4 ms

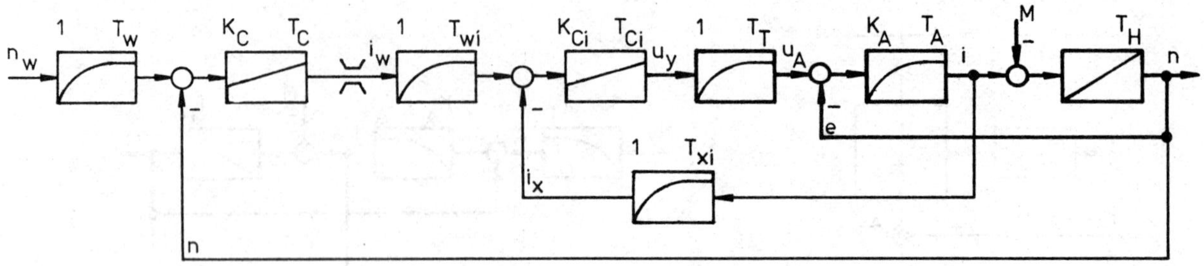

Fig. 2. Block diagram of current leading speed control

Current leading speed control

The configuration of the current leading speed control is commented by Fig. 2. The block diagram shows the plant as known from Fig. 1b now completed by the two control loops mentioned in the introduction. Besides the two PI-controllers there are two first-order factors for delaying the reference signals with respect to the lead caused by the numerator of the closed loop transfer function. An additive first-order factor is provided that can be used to filter the current feedback signal. This action is necessary when the control action is disturbed by high ripple of the armature current caused for instance by a small pulse number of the converter.

TABLE 2
PARAMETERS FOR CURRENT LEADING SPEED CONTROL

	Case A1	Case A2
T_T	1,65 ms	3,3 ms
T_{xi}	0	3,3 ms
K_{Ci}	0,76	0,19
T_{Ci}	6,6 ms	26,4 ms
T_{wi}	6,6 ms	29,6 ms
K_C	16	3,74
T_C	40 ms	172 ms
T_w	46 ms	200 ms

Table 2 shows the controller parameters for two cases that differ with respect of the pulsenumber i.e. delay of the converter as well as in the absence respectively the presence of a filter for the current feedback signal.

The open loop gain of the current loop is determined by the plant -equ. (10)- and by the first-order factor of the current feedback. As verified a little later the crossover frequency of the open loop frequency response satisfies the assumption

$$T_M \cdot \omega_c \gg 1 \tag{13}$$

Hence equ. (10) can be reduced to first-order. Thus the loop to be controlled consists of three first-order factors. By combining the two smallest of them the open loop gain becomes

$$F_{0i} = K_{Ci}\,\frac{1+T_{Ci}\cdot s}{T_{Ci}\cdot s}\cdot K_A\,\frac{1}{1+T_A\cdot s}\cdot\frac{1}{1+(T_T+T_{xi})\cdot s} \tag{14}$$

with $T_A > T_T + T_{xi}$

The open loop frequency responses for case A1 and A2 of table 2 are shown in Fig. 3 where equ. (12) is verified.

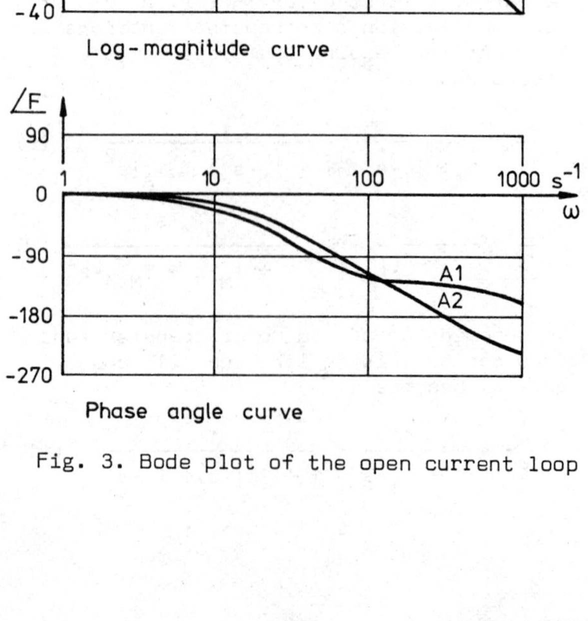

Fig. 3. Bode plot of the open current loop

By simplifying equ. (10) the influence of the back emf was neglected. Therefore equ. (14) does not describe the low frequency behaviour of the real system shown by Fig. 3.

For calculating the parameters of the speed controller by means of the symmetric optimum the closed current control with the reference filter included is assumed to be a first-order factor. Thus the transfer function of the open speed loop is made up by the controller with this first-order factor having the equivalent time constant T_{ei} and with the mechanical integration.

$$F_0 = K_C \frac{1 + T_C \cdot s}{T_C \cdot s} \cdot \frac{1}{1 + T_{ei} \cdot s} \cdot \frac{1}{T_H \cdot s} \qquad (15)$$

With the controller parameters calculated by means of the symmetric optimum the transfer function of the closed speed loop is

$$F = \frac{1 + T_C \cdot s}{1 + 4T_{ei} \cdot s + 8T_{ei}^2 \cdot s^2 + 8T_{ei}^3 \cdot s^3} \qquad (16)$$

where $T_C = 4T_{ei}$

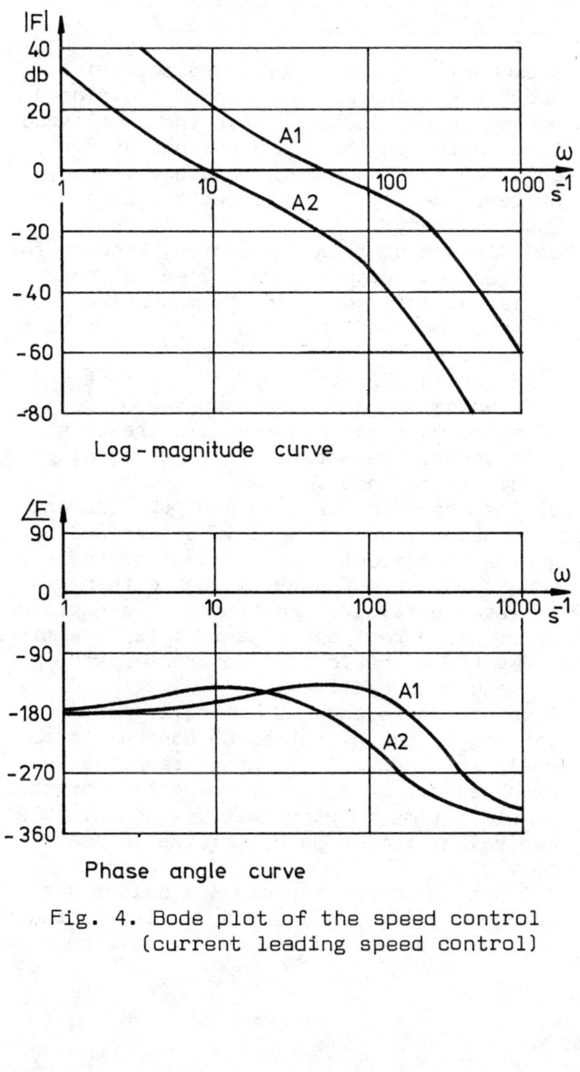

Fig. 4. Bode plot of the speed control (current leading speed control)

The frequency response of the whole speed control including the reference filter can be seen below at Fig. 7. Here the dynamics of the closed loop can already be critizised by the crossover frequency of the open loop gain which is a good measure for the corner frequency of the closed loop. See Fig. 4.

The essential result of the control design described above is the fact, that the electro-mechanical system -electrical and mechanical time constant- does hardly influence the transient response of the current control and of the overlied speed control. Of deciding influence however are the small time constants inside the current loop that can not be compensated.

Comparing case A1 and case A2 in table 2 shows:

a) If there is no reference filter: the integral times of the controllers, the reference filtering time constants and the equivalent time constants are proportional to the time constant of the power amplifier i.e. converter.
b) If a reference filter is added the time constant influences the transient response twice: At first it increases the summation time constant in equ. (14) what is equivalent to an increase of the converter time constant. Besides this the time constant situated in the feedback causes an increase of the numerator of equ. (3) and by this an additional increase of the required reference filter time constant.

Direct speed control

The block diagram of the speed control loop is shown by Fig. 5. The frequency response function of the plant to be controlled is given by equ. (11). For the first-order factor of the converter is normally negligible this function equals to equ. (1). In contrary to the second-order systems that are to control in current leading systems the plant of direct speed control is normally of low damping ratio and often underdamped. When the damping ratio satisfies $d > 0{,}707$ the system can be optimized by equs. (6) which combined with equs. (12) give

$$K_C = \frac{T_M}{2T_A} - 1$$

$$T_C = 4T_A\left(1 - \frac{2T_A}{T_M}\right) \qquad (17)$$

With these controller parameters the transfer function of the optimized closed control loop given by equ. (7) is obtained

$$F = \frac{1 + T_C \cdot s}{1 + 4T_A \cdot s + 8T_A^2 \cdot s^2 + 8T_A^3 \cdot s^3} \qquad (18)$$

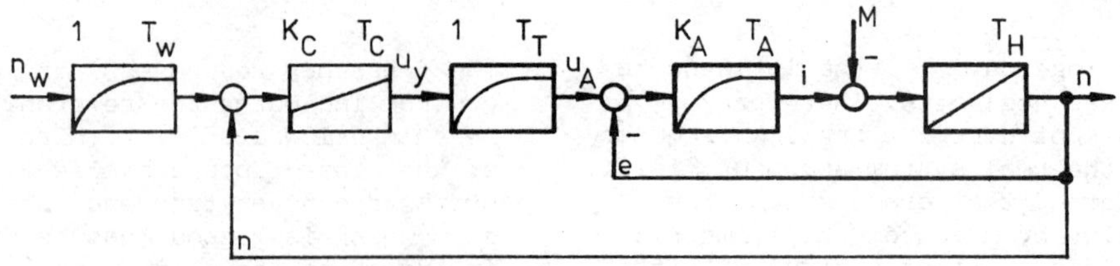

Fig. 5. Block diagram of the direct speed control

In the case of the drive specified in table 1 the optimization criteria is applicable. The result is shown in table 3 and Fig. 6. For

TABLE 3 PARAMETERS FOR DIRECT SPEED CONTROL

	Case B1	Case B2
T_T	1,65 ms	3,3 ms
T_{xi}	0	3,3 ms
K_C	0,25	0,25
T_C	14,4 ms	14,4 ms
T_w	14,4 ms	14,4 ms

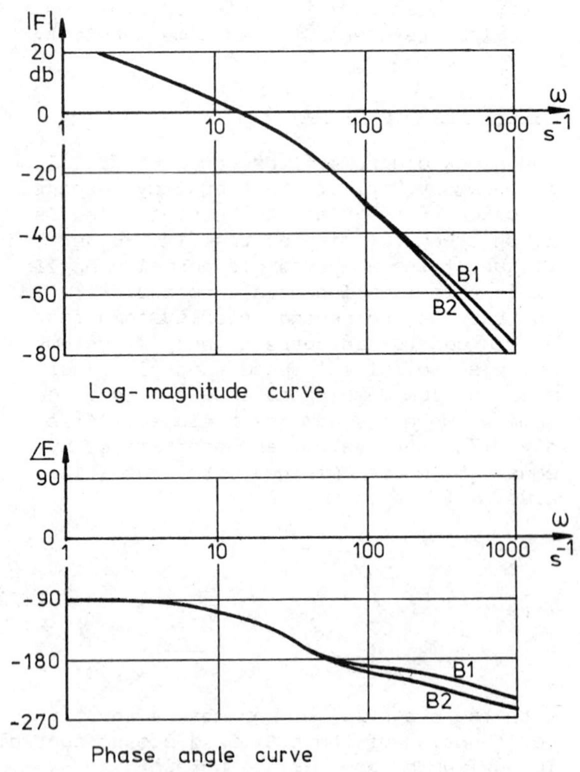

Fig. 6. Bode plot of the speed control (direct speed control concept)

the optimization rule does not depend on the time constant of the converter and for the current feedback filter is not part of the speed control loop the controller parameters do not depend on these time constants. All frequency characteristics referring to direct speed control i.e. those from Fig. 6 demonstrate how small the influence of the converter time constant on the behaviour of the real system is.

It is worth to be mentioned that at direct speed control, too, the mechanic inertia does not influence the transient response of the speed control.

Comparison in regard to the transient response to the reference signal

A comparison of equ. (16) and equ. (18) shows the transfer function of the closed speed control loops to have the same structure. Although these expressions do not describe the transient behaviour of the systems exactly it can be established with good accuracy:

a) Independent from the control concept the transient response of the speed control is not influenced by the mechanical time constant that means by the mechanical inertia.
b) Regarded qualitatively the current leading speed control and the direct speed control show the same transient response characteristic. The deciding difference however is:
c) The transient behaviour of the current leading speed control is determined by the equivalent time constant of the closed current control loop with reference filter, the transient behaviour of the direct speed control is determined by the electric time constant of the armature circuit.

Thus the decision whether the current leading or the direct speed control is to apply when optimal transient response is required can be taken by comparison of the armature time constant with the obtainable equivalent time constant of the current control.

When the converter has six pulses per period and if no filter is applied to the current feedback an equivalent time constant of about 10 ms can be obtained /Ref. 1/. In this case current leading speed control will give the better result when the time constant of the armature circuit is greater than this value of 10 ms.

Direct speed control however becomes lucrative when the relation between the deciding time constants changes. This can be caused by reducing the time constant i.e. the inductance of the armature circuit as well as by an increase of the equivalent time constant of the current control. In practice the first modification causes the second modification: When a motor with a small armature inductance is used it can be advisable not to spoil the good transient response of the armature circuit by smoothing the current with an inductor. So the armature current and the current feedback signal have high ripple especially when a converter with a low number of firing pulses is used. With respect to this ripple the current control will not operate satisfactory without a feedback filter smoothing now the current feedback signal.

This consideration makes necessary to combine the increase of the converter's delay with a filter delay when passing over from case1 to case 2 at the special system specificated by tables 1 to 3. The transient behaviour of the two different plants 1 and 2 controlled in the two different modes A and B are shown by the frequency characteristics of the speed controls including the reference filter. The curves shown in Fig. 7 demonstrate the small influence of the converter's delay on direct speed control that only appears at high frequencies that in practice are without importance for the transient response.

Comparison with regard to the transient response to load disturbance

For the systems specified by the tables 1 to 3 comment on the transient response to the load disturbance can be given by means of the frequency characteristics shown at Fig.8. Here, too, the current leading control proves to be better only when the current control has a good transient response. This fact can be explained as follows:

The dc-motor has the natural feature to

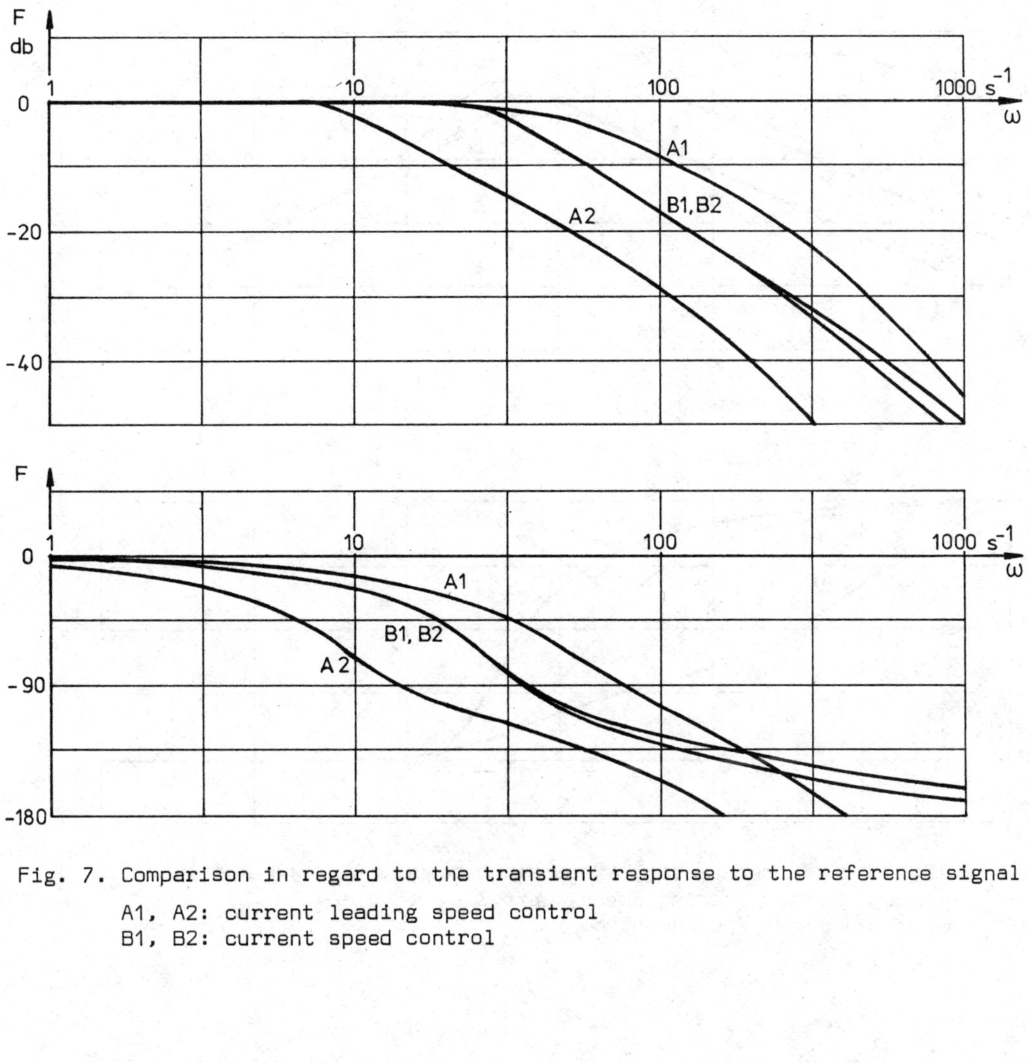

Fig. 7. Comparison in regard to the transient response to the reference signal

A1, A2: current leading speed control
B1, B2: current speed control

work against the load disturbance: When for example the load disturbance causes the machine to slow down the reduce of back emf causes the armature current and the generated torque to increase.

With direct speed control this reaction occurs independent of the dynamics with which the speed controller recognizes the change of speed and corrects the armature voltage. With current leading speed control however the natural reaction of the machine against the load disturbance is hindered by the current control: Immediately after a change of load torque this controller tries to keep the armature current constant and by this the generated torque, too. In this way the transient response of the current leading speed control can be slower than the transient response of the electro-mechanic system to be controlled. Hence the compensation of armature time constant by current control is disadvantageous if the speed controller, setting the current reference, has no adequate dynamics.

In Fig. 8 no difference can be recognized between the curves of the direct speed controls compared. The reason is that altering the converter time constant does not influence the low frequency behaviour and the high frequency behaviour is exclusively decided by the mechanical inertia: The curves of all controls approach for high frequencies to the curves given by the integration factor T_H.

CURRENT LIMITATION AT DIRECT SPEED CONTROL

When dc drives are operated in a speed controlled mode precautions have to be taken to prevent the equipment from undue strain. Generally the most critical apparatus is the converter for its semiconductors have a small overload capacity and low thermal time con-

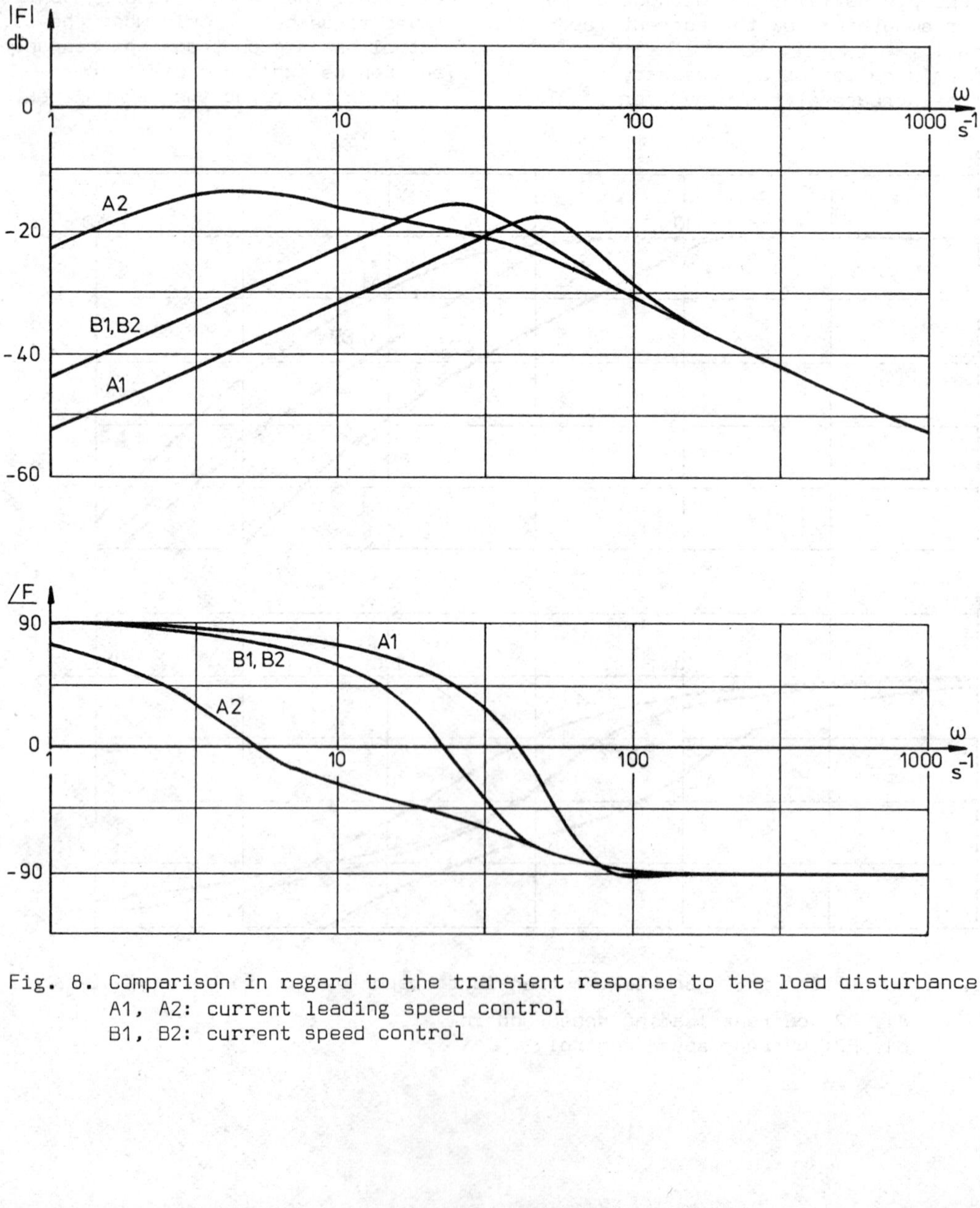

Fig. 8. Comparison in regard to the transient response to the load disturbance
A1, A2: current leading speed control
B1, B2: current speed control

stant. Undue current will occur the more the faster the control reacts to changes of the reference and of the disturbance. That is why speed controlled drives must not be operated without current limitation.

At current leading speed control the current reference signal can be simply limited at the output of the speed controller, and the armature current will not exceed this limitation when the current control is overdamped.

At direct speed control there is no current reference signal, therefore another solution must be found to limit the current.

Besides the direct speed control loop a current control loop can be made up. The controller of this loop must not influence the speed control under normal operation conditions. This controller however has to reduce the input signal of the converter, when the speed controller would cause the armature current to exceed its rated value. In referred values the input of the converter is

$$u_y = u_A = e + R_A \cdot i \tag{19}$$

and with respect to the permissible current u_y must not exceed

$$u_{ymax} = e + R_A \cdot i_{max} \tag{20}$$

Fig. 9 illustrates the configuration need to limit the current in one direction. To limit the other polarity of current, too, a second loop of the same kind can be made up with the only difference that the reference of the controller and the nonlinear characteristic are of opposite polarity. The current transducer can be a simple unipolar version,

a) when the armature current can have only one polarity,
b) when there is only one firing unit for two converter bridges (circular current-free mode of operation) or
c) when two converter bridges are fed by two identical firing units.

Limiting the current behind the speed controller causes this device to overdrive when being relieved by the current control. This effect is avoided when the current li-

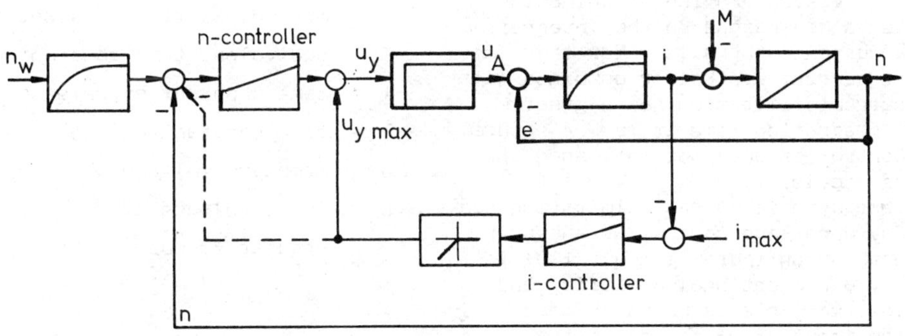

Fig. 9. Direct speed control with current limitation (only one polarity represented)

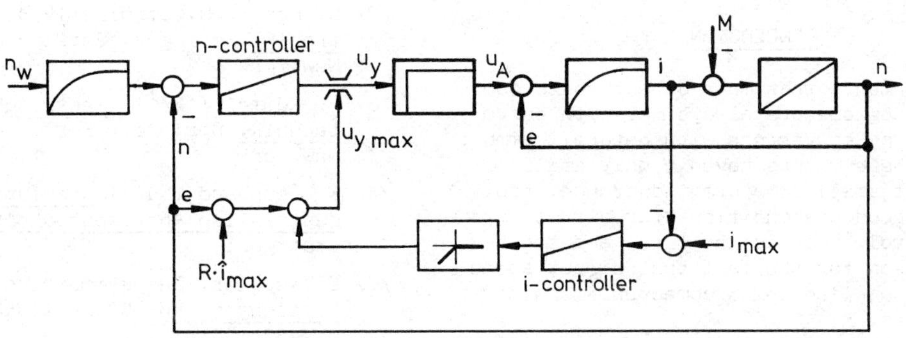

Fig. 10. Direct speed control with static and dynamic current limitation (only one polarity represented)

miting signal is entered in the speed control loop at the input of the speed controller as indicated by the dotted line in Fig. 9.

Especially with servomotors, having a high overload capability it is advisable to have a dynamic current limitation besides the steady-state current limitation. Thus up to four current controllers can be found in servo amplifiers. When dynamic and steady-state current limitation are combined they must have different time delays. Delay time is designed by the integration time of the controller and by the voltage to which the amplifier is clamped when out of action. When the current exceeds its limit the controller has to integrate the output voltage from this clamping voltage to zero before it relieves the speed control in leading the drive. It must be considered that the integral behaviour of the current controller causes the current not to reach its reference value during acceleration. The reason is the need of changing the output of the current controller demands an error signal at the input. This effect is inevitable at current leading speed control. In the cases A1 and A2 examined above the current will reach only 97 % respectively 70 % of the current reference during acceleration of unloaded motor. These values depend on the integration time of the current controller. For the steady-state current controller combined with a dynamic current limitation has a great delay i.e. integration time it must be ensured that this controller does not influence the drive during acceleration.

A very fast dynamic current limitation can be made up by open loop control as shown in Fig. 10. In this configuration equ. (20) is determined by a proportional amplifier and the determined variable is used as limit u_{ymax} for the input u_y of the converter. This limit works without any delay when the current goes to exceed the dynamic current limit $\hat{i}_{max}$ and is reduced by the current controller later.

CONCLUSION

The requirements that make direct speed control advantageous are always met with servomotors fed by converters with natural commutations. These motors haven't only small inertia but small armature inductance, too, so giving good preconditions for direct speed control. That is the first and the deciding reason for the fact that direct speed control is applied in servomechanisms with good results.

Besides this, direct speed control offers the possibility to perform specific tasks with reduced cost for the converter and the current limiting device. Direct speed control with current limiting control as described can be found in very different configuration. The control concept was found at transistorized and at thyristorized converters, at midpoint circuits and at bridge configurations, at amplifiers and converters with and without circular current.

SYMBOLS

d	damping ratio of plant
e	back emf
F	transfer function
i	armature current
i_{max}	current limit value
K_A	gain of armature circuit
K_C	gain of controller
n	angular speed
s	complex frequency
T	time constant of 2nd-order plant
T_A	electric time constant
T_C	integration time of controller
T_H	acceleration time
T_M	mechanical time constant
T_T	equivalent time constant of converter
T_w	time constant of reference filter
T_x	time constant of feedback filter
u_A	armature voltage
u_y	input voltage of plant (converter)
ω	angular frequency

REFERENCES

/1/ A. Buxbaum, K. Schierau (1974) Berechnung von Regelkreisen der Antriebstechnik, Elitera, Berlin

/2/ D. Ernst, D. Ströle (1973) Industrieelektronik, Springer, Berlin-Heidelberg-New York

/3/ F. Kümmel (1971) Elektrische Antriebstechnik, Springer, Berlin-Heidelberg-New York

/4/ W. Leonhard (1974) Regelung in der industriellen Antriebstechnik, Teubner, Stuttgart

/5/ C. Kessler, Das symmetrische Optimum, Regelungstechn. 6, 395 (1958) and 6, 432 (1958)

/6/ W. Leonhard, Regelkreise mit symmetrischer Übertragungsfunktion, Regelungstechn. 13, 4 (1965)

DESIGN OF AN OPTIMAL, AUTOADAPTATIVE CURRENT LOOP FOR D.C. MOTOR. REALIZATION WITH AN HYBRID DEVICE INCLUDING A MICROPROCESSOR

A. Oumamar, J.P. Louis and A. El-Hefnawy

Laboratoire d'Electronique, d'Electrotechnique et d'Automatique de l'Ecole Nationale Supérieure d'Electricité et de Mécanique de Nancy, France

ABSTRACT

The paper deals with the current loop for D. C. motor regulation using a PI controller. Analytic formulas are presented, which give the optimal response as well in continuous mode as in discontinuous one. The gain of the controller must be autoadaptative, because it depends of the operating point (current reference, speed). A realization is presented using a microprocessor.

INTRODUCTION

It is well-known that the current loop of a D.C. motor supplied by a dual converter (two thyristor bridges in antiparallel) must have an optimal response (that is as fast as possible) in continuous and discontinuous operations. This current loop is discrete and nonlinear and its dynamic behaviour depends on the operating point. The dynamic behaviour in discontinuous operations is completely different from that in continuous operation. Thus it is necessary to build an autoadaptative gain which must follow a particular law. Some methods have been used (/1/ to /4/), which are interesting because of their simplicity.

For optimal behaviour, we choose the dead-beat response and we establish the analytical formulas giving the optimal values of a proportional integral controller type. The optimal time constant of the controller is effectively constant. But the value of the optimal gain depends on the operating point, which is imposed by the current reference and by the speed of the motor. This gain is realized by an hybrid device including a microprocessor which determines the mode of operation, imposed by the reference and the back E.M.F. of the motor and also the corresponding optimal gain (one law for the continuous mode and another for the discontinuous one).

1. DESCRIPTION OF THE SYSTEM-NOMENCLATURE

A current control system of a DC motor will be considered (Fig. 1). The load consists of a resistance "r" and an inductance "l" and the back electromotive force "E" due to the rotation of the motor (supposed constant) supplied from an AC/DC polyphase converter (three phase thyristor bridge or two bridges connected in antiparallel). The thyristors are fired from pulses generated when the control voltage equals the timing voltage. The control voltage is the output of a P.I. controller.

i) Nomenclature (see Fig. 2)

ω, m : pulsation of supply, number of bridge controlling phase.

l, r : inductance and resistance of the load

$Z = \sqrt{r^2+(l\omega)^2}$, ϕ = Arctan $(l\omega/r)$: modulus and phase of impedance

E : back f.e.m.

$v_s = V_m \cos(\omega t-\pi/m)$ converter output voltage

$I_m = V_m/Z$, $I_o = E/r$

i, $\bar{I}$: load current and its average value

v_a, $a(t) = A(-\omega t+\delta)$ control voltage, timing voltage

(the linear case is considered in this paper)

v_2 : reference voltage

k_1, τ : controller coefficients (gain, time constant)

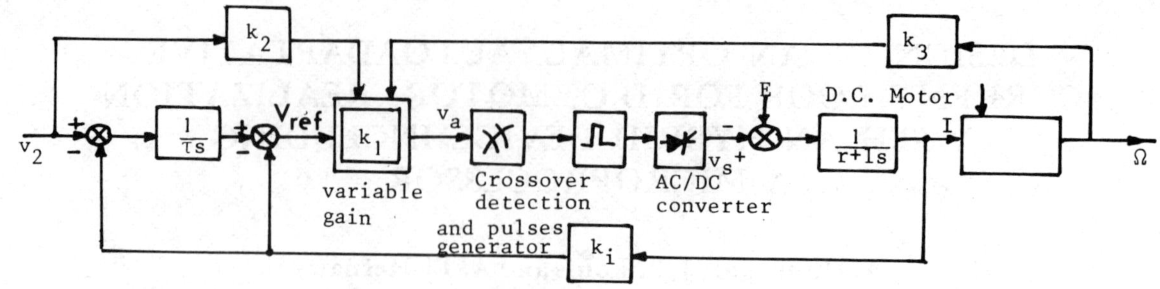

Fig. 1 Block-diagram of the process.

k_i : current transducer gain

n, ∞: index associated to the nth firing, the steady state operation.

ii) Reduced variables and coefficients

$T_e=1/r$, $\Delta=\omega T_e = \tan\phi$, $e_1=\exp(-2\pi/m\Delta)$

$x = \omega t$: reduced time (angular)

$\psi_n = \omega t_n$: n^{th} firing angle

$\psi'_n = \omega t'_n$: n^{th} extinction angle (discontinuous operation)

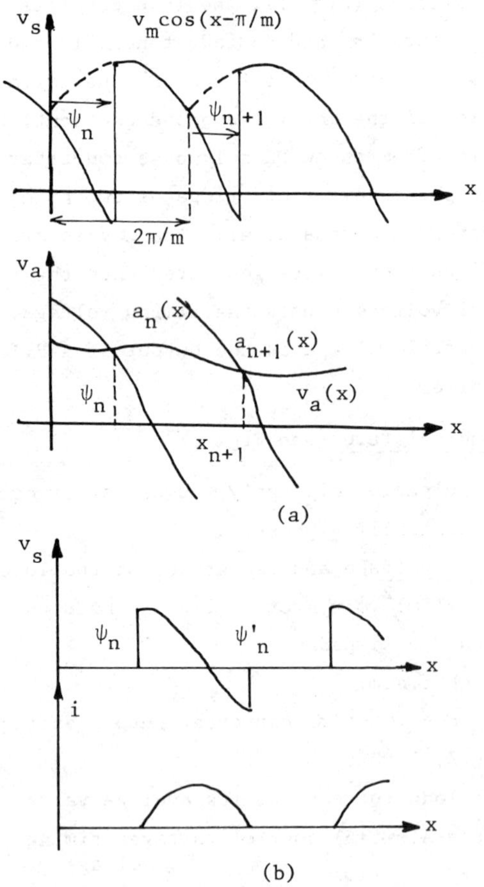

Fig. 2 Definition of some variables.
(a) continuous mode
(b) discontinuous mode

$\xi = (i+I_o)\cos\phi/I_m$: reduced current

$\nu = (v_2/k_i+I_o)\cos\phi/I_m$: reduced set-point

$\nu_o = (v_2/k_i)\cos\phi/I_m$: reduced reference current

$\xi_o = I_o \cos\phi/I_m$: reduced back f.c.e.m.

$k_a = k_1 k_i I_m/A \sin\phi$: reduced gain

$\gamma = 1/(\nu_o+\tau/k_a T_e)$: reduced coefficient.

iii) Numerical values

The system considered as an example in this investigation has the following coefficients, corresponding to a 5 KW DC Motor :

$\omega = 100\pi$, $l = 18$ mH, $r = 1\Omega$, $V_m=312$ V, $m = 6$, $A = 3.14$, $\delta = 172°$, $k_i = 1/30$.

2. EQUATIONS AND STEADY STATE CHARACTERISTICS OF THE SYSTEM

. Load equation :

$$ri + l\frac{di}{dt} = V_m \cos(\omega t - \frac{\pi}{m}) - E$$

during conduction, and : $i = 0$

during extinction (discontinuous operation).

. Controller equation :

$$\frac{dv_a}{dt} = k_1\left(\frac{v_2}{\tau} - \frac{k_i}{\tau} i - k_i \frac{di}{dt}\right)$$

Initial and final conditions :

$$v_a(\psi_n) = a(\psi_n)$$

$$v_a(\psi_n+1) = a(\psi_n+1)$$

By integrating these equations between n^{th} and $(n+1)^{th}$ firing instants, we obtain the recurrence equation which discribes the dynamic behaviour of the system. We use reduced variables to give generality to the results.

i) The case of continuous operation /6/

We get a 2nd order recurrence equation

defined by the following system :

$$\left.\begin{aligned} &f(\xi_{n+1},\psi_{n+1}+\pi/m)=f(\xi_n,\ \psi_n-\pi/m)\\ &g(\xi_{n+1},\psi_{n+1}+\pi/m)=g(\xi_n,\psi_n-\pi/m)+2\pi/m \end{aligned}\right\} \quad (1)$$

with : $f(\xi,\theta)=[\xi-\cos\phi\ \cos(\theta-\phi)]\exp(\theta/\Delta)$

$g(\xi,\theta)=\theta(1+k_a T_e \nu/\tau)-k_a \sin\theta\ T_e/\tau+k_a\Delta(T_e/\tau-1)\xi$

From these equations we can find the steady-state characteristics of the system. Practically the operating point is defined by ψ_∞ steady-state firing angle which is imposed by the reference as shown by the following equations :

$$\nu=(v_2/k_i+I_o)\cos\phi/I_m=(\bar{I}_\infty+I_o)\cos\phi/I_m = \frac{\sin\pi/m}{\pi/m}\cos\psi_\infty \quad (2)$$

and the steady-state current is defined by :

$$\xi_\infty=\frac{\cos\phi}{1-e_1}[\cos(\psi_\infty-\phi+\frac{\pi}{m})-e_1\ \cos(\psi_\infty-\phi-\frac{\pi}{m})] \quad (3)$$

ii) <u>The case of discontinuous operation</u>

In this case the current is always zero at the firing instants ($i_n=0$ or $\xi_n=\xi_o$). The recurrence which describes the system is reduced to 1st order :

$$\psi_{n+1}=\psi_n-2\nu_o\gamma\ \pi/m+\gamma[g(\psi_n-\pi/m)-g(\psi'_n-\pi/m)] \quad (4a)$$

with : $g(\theta)\ =\ (\theta+\ \pi/m)\ \xi_o-\sin\ \theta$

This equation depends on the angle of extinction which is defined by :

$$f(\psi_n-\pi/m)=f(\psi'_n-\pi/m) \quad (4b)$$

with : $f(\theta)=\ [\xi_o-\cos\ \phi\cos(\theta-\phi)]\exp\ (\theta/\Delta)$

Only solutions which have a physical meaning will be considered. In this case the steady-state operation is defined by :

$$\left.\begin{aligned} &g(\psi_\infty-\pi/m)-g(\psi'_\infty-\pi/m)=2\pi\nu_o/m\\ &f(\psi_\infty-\pi/m)=f(\psi'_\infty-\ \pi/m) \end{aligned}\right\} \quad (5)$$

iii) <u>Limit between continuous and discontinuous operations</u>

The critical mode of operation can be found from discontinuous operation by supposing : $\psi'_n=\psi_n+2\pi/m$ or from continuous operation by supposing : $\xi_\infty=\ \xi_{oc}$. Then the critical operation is defined by the following equation :

$$\xi_{oc}=\frac{\cos\ \phi}{1-e_1}[\cos(\psi_\infty-\phi+\pi/m)-e_1\cos(\psi_\infty-\phi\ -\pi/m)]$$

Figure 3 gives the characteristic $\xi_{oc}(\psi_\infty)$ for the considered system.

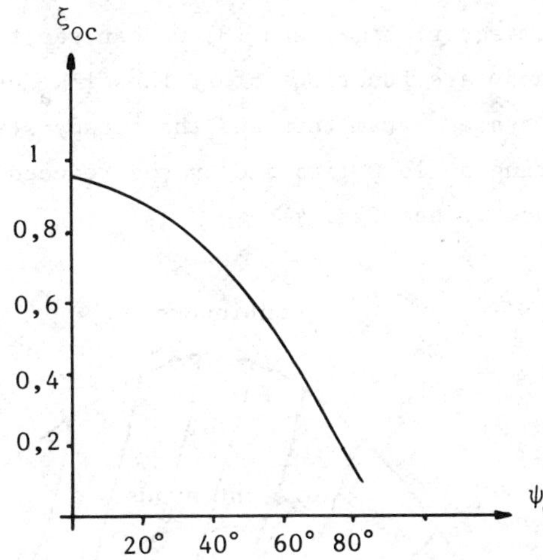

Fig. 3 Critical mode.

3. DETERMINATION OF THE OPTIMAL GAIN

For two principal response (current security, dynamic acceleration) the current loop is considered optimal if the dynamic response is as fast as possible without overshoot ; for a sampled data system like this, the best response will be the deadbeat operation; so the transient period is equal to pT, where T is the discrete period and p the system order.

The deadbeat response is obtained if the roots of the characteristic equation are zero. The system characteristic equation can be determined by linearising the system equation around the operating point (ψ_∞,ξ_∞).

i) <u>The case of continuous mode</u>

By linearising the equations (1) we can determine the following system characteristic equation

$$Z^2+[K_T[\frac{T_e}{\tau}+e_1(1-\frac{T_e}{\tau})]-(1+e_1)]\,Z+e_1(1-K_T)=0 \quad (6)$$

with : $K_T=\dfrac{2\sin\psi_\infty\sin\pi/m}{1/k_a+\xi_\infty(1-T_e/\tau)+T_e\nu/\tau-\cos(\psi_\infty\pm\pi/m)}$

The equation (6) can be reduced to : $Z^2=0$

if : $\tau\ =\ T_e(1-e_1)$ (7)

$K_T\ =\ 1$ (8)

Condition (7) shows that τ is constant ; we can determine the reduced optimal gain k_a from condition (8) :

$$1/k_a=\cos(\psi_\infty-\pi/m)\ -\xi_\infty\ (1-T_e/\tau)\ -\nu T_e/\tau \quad (9)$$

From equation (2) and (3) we can see that ξ_∞ and ν are functions of ψ_∞. Thus $1/k_a$ depends on system parameters and the steady-state firing angle ψ_∞ (imposed by the reduced reference ν) (see Fig. 4).

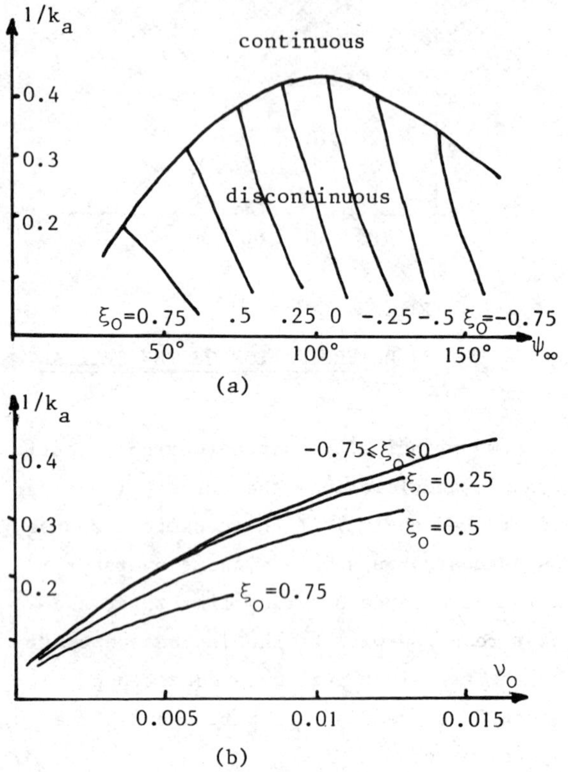

Fig. 4 Optimal Gain
(a) Continuous and discontinuous mode vs (ψ_∞).
(b) Discontinuous mode vs (ν_o).

ii) <u>The case of discontinuous mode</u>

By linearising the equation (4) we can determine the following characteristic equation :

$$Z - [1+\gamma[\xi_o - \cos(\psi_\infty - \frac{\pi}{m})][1-\exp(-\frac{\psi'_\infty - \psi_\infty}{\Delta})]] = 0 \quad (10)$$

Equation (10) becomes : Z = 0 if :

$$\frac{1}{k_a} = \frac{T_e}{\tau}[[\cos(\psi_\infty - \frac{\pi}{m}) - \xi_o][1-\exp(-\frac{\psi'_\infty - \psi_\infty}{\Delta})] - \nu_o] \quad (11)$$

This formula depends on ψ_∞, ψ'_∞, ξ_o and ν_o. The steady-state operation (imposed by ν_o and ξ_o) can only be described by two parameters, for instance ψ_∞ and ξ_o, if we solve the equations (5). These equations can be solved graphically or numerically by the Newton method. Then we can draw the curve $1/k_a$ vs ψ_∞ or vs ν_o for different values of ξ_o (figure 4). We can conserve the same value τ for the two modes of operation because there is only one condition for k_a and τ in discontinuous operation.

iii) <u>Critical case</u>

The optimal gain is a continuous function, because in the critical case, the gain deduced from equ. (9) or from equ. (10) is :

$$\frac{1}{k_a} = \cos(\psi_\infty - \frac{\pi}{m}) - \xi_o - T_e\nu_o/\tau \quad (12)$$

4. CHARACTERISTICS OF THE AUTO-ADAPTATIVE GAIN USED IN PRACTICE

Formulas (9) and (11) show that we need a variable gain k_1, which is realized by a device including a microprocessor. We notice that the microprocessor has two inputs ξ_o and ν_o ; it must determine :

. The mode of operation (continuous or discontinuous)

. The optimal gain.

It is necessary to compute numerically the characteristics as functions of real measured parameters ν_o and ν ($\nu = \nu_o + \xi_o$).

i) Figure 5 gives the relation between ξ_{oc} and ν. If $\xi_o > \xi_{oc}$ we have discontinuous operation, otherwise we have continuous operation.

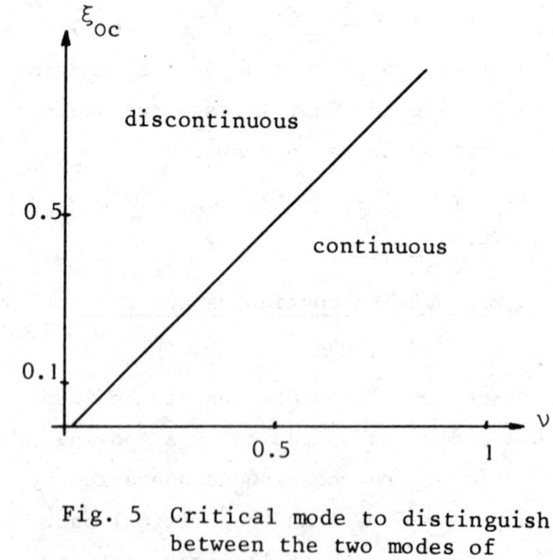

Fig. 5 Critical mode to distinguish between the two modes of operation.

ii) Figure 6 gives the relation between the optimal gain (k_{1c}) and ν for continuous operation.

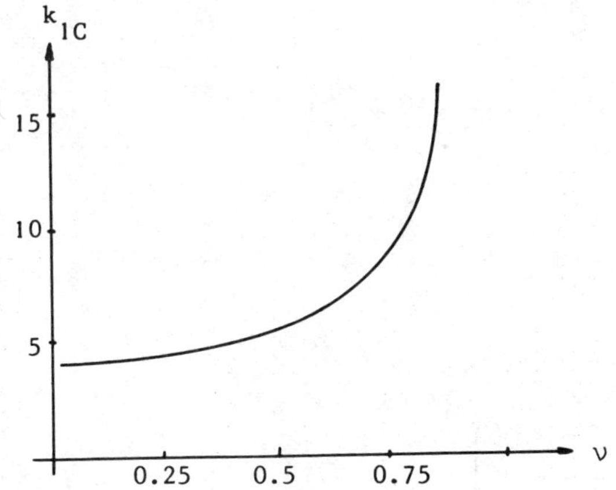

Fig. 6 Optimal Gain (Continuous mode)

iii) For the discontinuous operation, the optimal gain depends on ν_o and ξ_o. To simplify the realization we consider only the curve corresponding to $\xi_o=0$. Figure 4 (b) shows that is a good approximation for : $-0.75 \leqslant \xi_o \leqslant 0.25$. For ξ_o larger that 0.25 the gain is only sub-optimal. Figure 7 gives the relation between the optimal gain (k_{1D}) and ν_o in discontinuous operation.

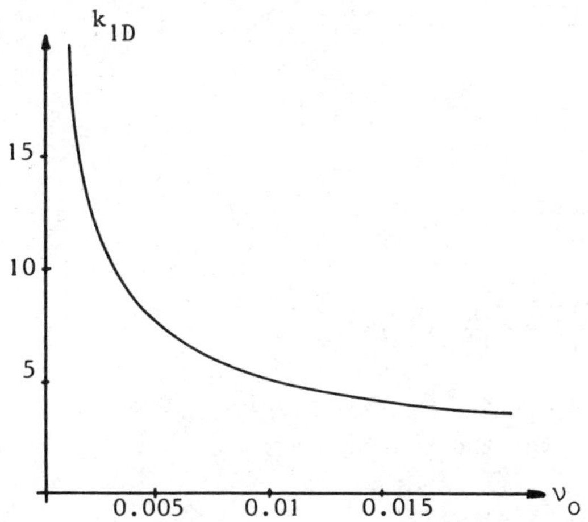

Fig. 7 Optimal Gain (Discontinuous mode).

iv) Figure 8 gives the flow-chart of the operations required from the microprocessor.

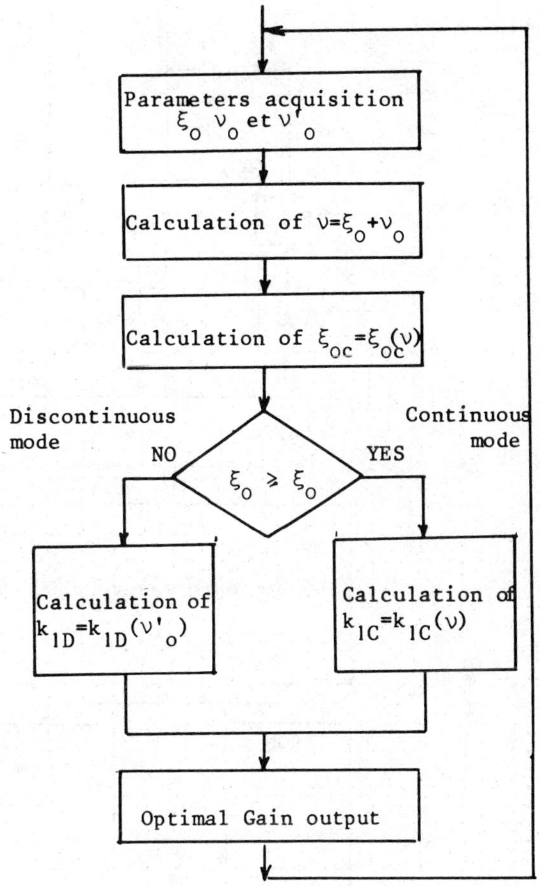

Fig. 8 Microprocessor Flow-Chart.

5. CHOICE AND FUNCTION OF THE MICROPROCESSOR

i) Choice

The choice of a microprocessor for this application was self-evident. On the one hand it allowed us to treat a relatively complex problem with a simple device, and on the other the very flexibility of the mounting allowed us to consider other cases of motors with parameters different from those of our own motor.

ii) Function of the microprocessor

The system which determined the optimal gain was based on a MOTOROLA M6800 microprocessor. Its function was simplified by external analogical devices (see Fig. 9). We shall give in detail the different stages and calculation methods used by the microprocessor.

6. MOUNTING AND WORKING

i) Mounting schema (Fig. 9)

For this application, we have used an M6800 microprocessor equipped with two parallel interfaces PIA M 6820 /7/. The output D.A.C.

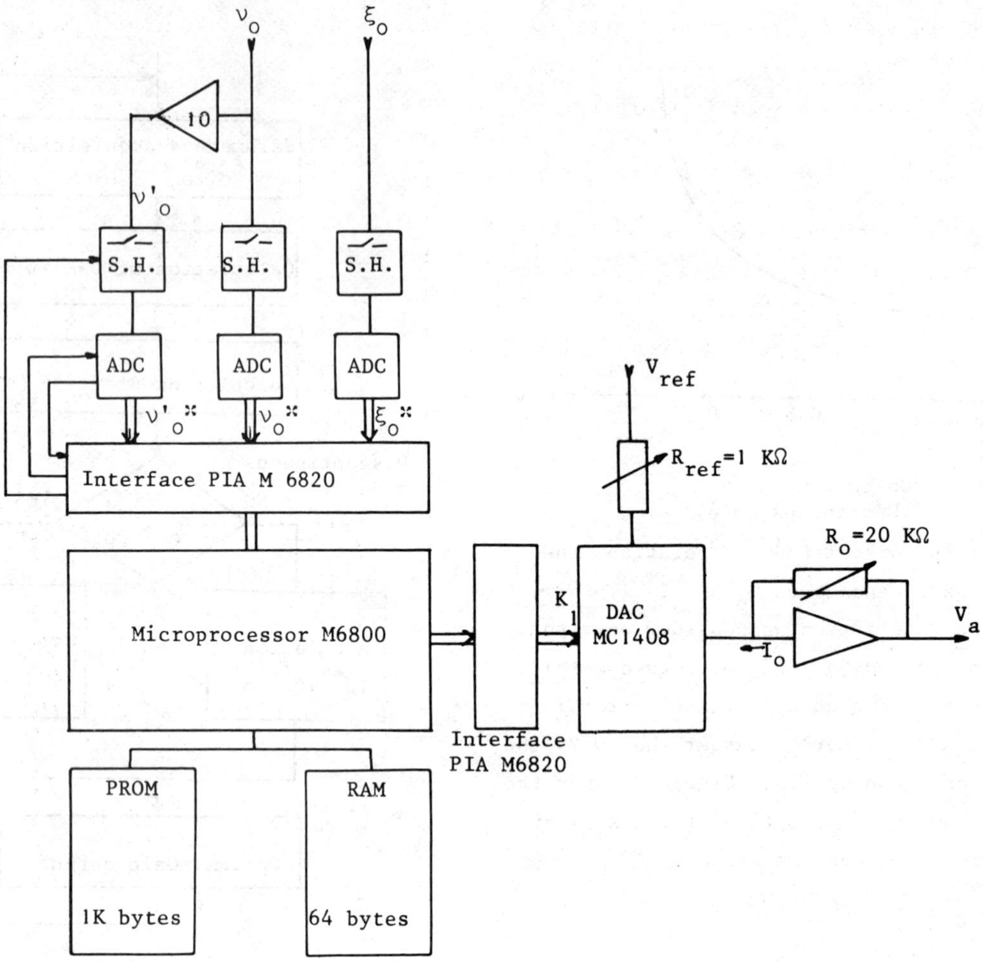

Fig. 9 Autoadaptative gain Block-Diagram.

is mounted as a multiplier. We shall see the reason for this later.

ii) Calculation method

We consider the functions : $\xi_{oc} = \xi_{oc}(\nu)$ and $k_{1C}=k_{1C}(\nu)$ (Fig. 5 and 6) where ν is a multiple of 2^{-8}, $\nu\in[0,1]$, and $k_{1D}=k_{1D}(\nu'_o)$ where ν'_o is a multiple of $2^{-8}/10$. ν_o was chosen so that we had a greater number of values when $\nu_o\in[0,0.02]$ (see Fig. 7 and 9).

We then tabulated these functions on 3 look-up tables, each having 256 bytes. To do this the 3 functions will now be considered as digitalised in the following way :

$$\xi_{oc} = \xi_{oc}(\nu) = \sum_{i=1}^{8} A_i 2^{-i}$$

$$\frac{k_{1C}}{10} = \frac{k_{1C}}{10}(\nu) = \sum_{i=0}^{7} B_i 2^{-i}$$

$$\frac{k_{1D}}{10} = \frac{k_{1D}}{10}(\nu'_o) = \sum_{i=0}^{7} C_i 2^{-i}$$

with A_i, B_i, C_i binary values (0 or 1).
Idem for the two inputs :

$$\xi_o = \sum_{i=1}^{8} D_i 2^{-i}$$

$$\nu_o = \sum_{i=1}^{8} E_i 2^{-i}$$

with D_i, E_i : binary values.
When ξ_{oc} and ξ_o are thus tabulated, they are directly comparable in binary.

. Calculation of $v_a=k_1 V_{ref}$ (see Fig. 9)
It is well known that the output of a D.A.C. (in this case the current output) is such that :

$$I_o = \frac{V_{ref}}{R_{ref}} \sum_{i=1}^{8} \frac{d_i}{2^i}$$

where V_{ref} is the reference tension of the

converter and R_{ref} the reference resistance which determine the voltage variation range. If we consider the D.A.C. in figure 9, we obtain :

$$v_a = \frac{R_o}{R_{ref}} V_{ref} \sum_{i=1}^{8} \frac{d_i}{2^i}$$

Since the numerical value at the input of the D.A.C. is equal to optimal $k_1/10$, we may write :

$$v_a = \frac{R_o}{R_{ref}} V_{ref} \frac{k_1}{10} \frac{1}{2}$$

Then if we take $\frac{R_o}{R_{ref}} \cdot \frac{1}{20} = 1$,

we obtain : $v_a = k_1 V_{ref}$

(In practice, we take R_o=20KΩ, R_{ref}= 1KΩ). Evidently some precautions are necessary when this D.A.C. is used as a multiplier. Such precautions are given in /8/.

. Working : The microprocessor will acquire the 3 parameters ξ_o, ν_o, ν'_o, will compute $\nu = \xi_o + \nu_o$ and will look for the corresponding value $\xi_{oc} = \xi_{oc}(\nu)$. Then it will compare ξ_{oc} and ξ_o, and depending on the result of the test will look for either k_{1C} or k_{1D}, and will put out the value on its interface. This value is instantaneously multiplied by input V_{ref}. The 3 tables as well as the program are placed in PROM. A memory buffer RAM is used to store the intermediary results.

7. PRACTICAL RESULTS AND CONCLUSION

In industrial environnement, the microprocessor works correctly. The case of the D.A.C. makes the task of the microprocessor much easier, for instead of wasting time doing the multiplication $k_1 V_{ref}$ by software, we prefered to allow a device to do it as quickly as possible. This takes 400 ns (compared with 200 µs needed with a microprocessor).

The error margin made on the gain is of the order of 2^{-7}. This error is mainly due to the components, for example the converters and the resistances.

When the microprocessor works at 1MHz, we obtain a sampling period of 67 µs, which gives us reason to be optimistic, because future microprocessors will be much quicker and thus the sampling period will be very much reduced.

The autoadaptative gain makes it possible to obtain very good performances with variable operating points.

Figure 10 shows a current loop with a constant gain. This gain is optimal for a particular continuous current reference (10a). But in discontinuous mode the system is unstable and we observe an oscillation(Fig. 10b).

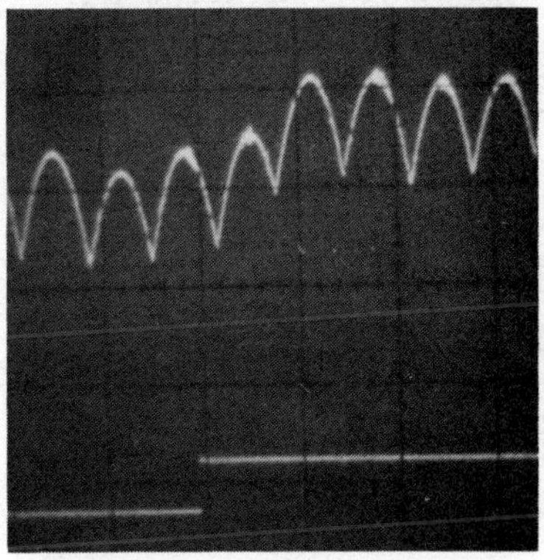

(a)

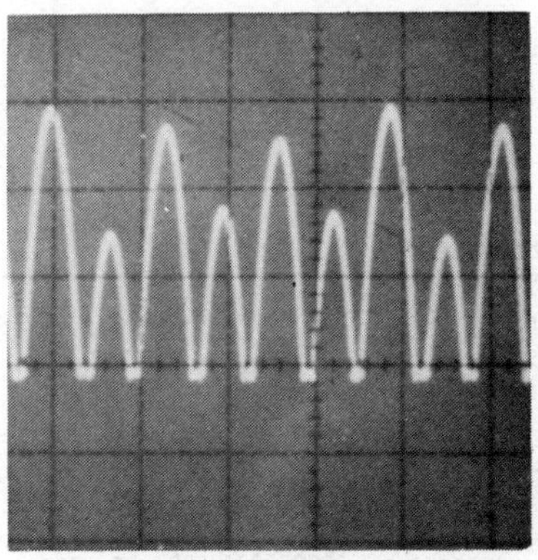

(b)

Fig. 10 Examples of Dynamic behaviour of the current loop with constant gain for two different references current.

(a) = 15 A
(b) = 3 A

With the autoadaptative gain the current response is always optimal (or sub-optimal) for different operating points and different conducting modes. Figures 11 a, b, c, d give the dynamic behaviour for different current reference : 15A (continuous mode), 10A (idem), 5A (from the discontinuous mode, near the critical case to the continuous mode), 3.5 A (discontinuous mode).

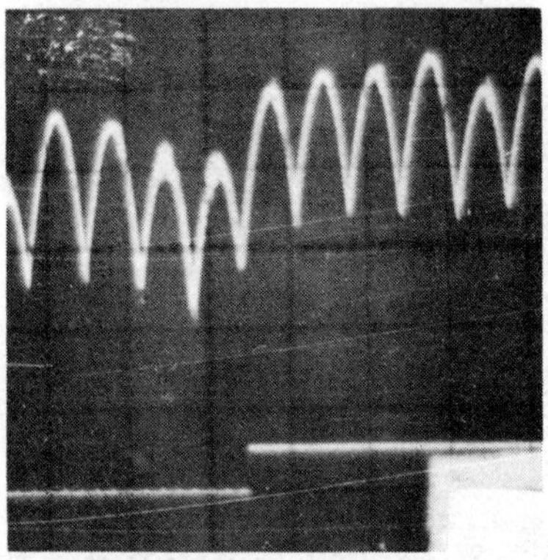

(a)

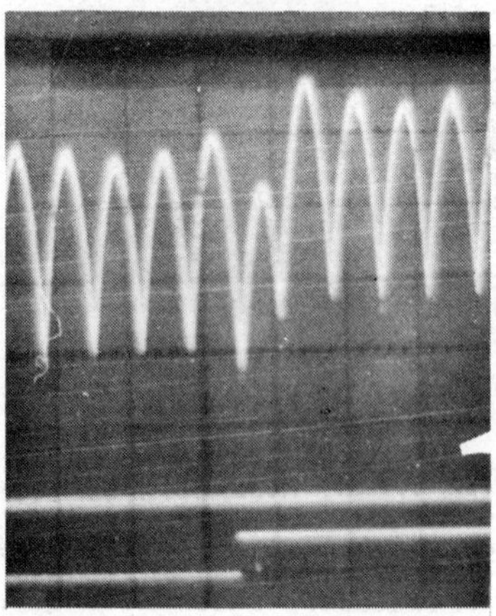

(b)

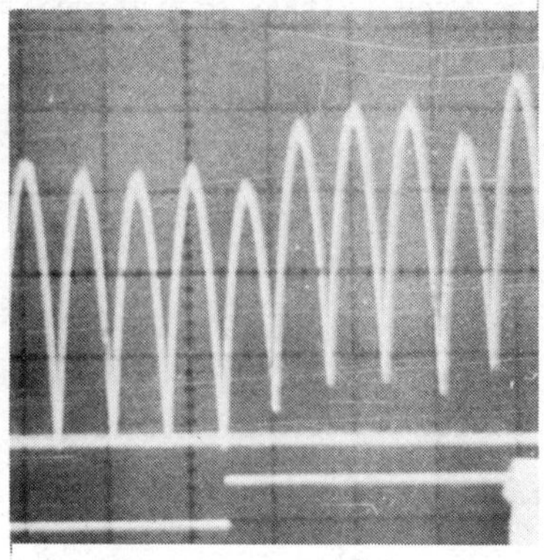

(c)

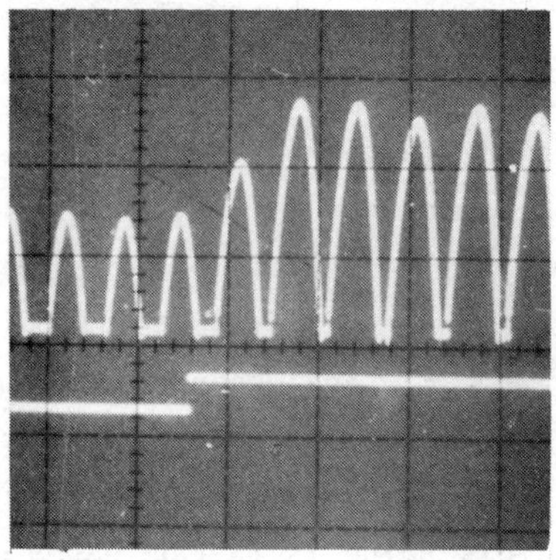

(d)

Fig. 11 Examples of dynamic behaviour of the current loop with auto-adaptative gain for four different references current.
(a) = 15 A
(b) = 10 A
(c) = 5 A
(d) = 3.5 A

We can see that theoretical results can be practically applied by means of a microprocessor. This device is destined to be included in a complete current and speed regulation system /9/.

BIBLIOGRAPHY

/1/ Magyar P. "Adaptative Control in the applications of electrical Drive Technics", Electrical Engineering Techn. University Budapest, Vol 19, n°3, (1975), pp 203-227.

/2/ Buxbaum A. "Regelung von Stromrichterantrieben bei Tückendem und nichtlückendem Ankerstrom", Techn. Mitt. AEG-TELEFUNKEN, 59 (1969), 6, pp 348-352.

/3/ Buxbaum A. "Aufban und Funktionsweise des adaptiven Ankerstromreglers", Techn. Mitt. AEG-TELEFUNKEN, 61, (1971)7, pp 371-374.

/4/ Buxbaum A. "Stromregler für Gleichstromantriebe mit Stromrichter Stellgliedern", IFAC-Symposium "Control in Power Electronics and Electrical Drives", Düsseldorf, October 7-9, (1974), Vol. 2, pp 71-87.

/5/ Courtiol B., Landau I.D. "High Speed Adaptation System for controlled Electrical Drives", Automatica, Vol. 11, (1975), pp 119-127.

/6/ Gudefin E.J., Louis J.P., El-Hefnawy A. "Non-linear Analysis of a 2nd Order Thyristor Device Control System", Int. Conf. Elec. Machines, Vienna, (1976), part 1, n° I-26.

/7/ MOTOROLA : Microprocessor Applications Manual.

/8/ MOTOROLA : Application Note on MC 1408 Motorola. Specifications and Application's information.

/9/ Oumamar A., Rooy G. "Direct Digital Control of a Thyristors Converter". Application for D.C. Motor regulation, (1977, to be published).

CONTROL OF DC-DRIVES BY MICROPROCESSORS

E. Schnieder

Institut für Regelungstechnik, Technische Universität Braunschweig, Hans Sommer-Str. 65/66, 3300 Braunschweig, Germany

1. Summary

The possibility of employing microprocessors for direct digital control of thyristors depends on the required dynamics and the attainable resolution. This paper describes a firing-control-algorithm for a three phase thyristor AC/DC convertor. The scheme has been implemented on a commonly available microprocessor; it operates sufficiently fast, so that there is enough spare time to carry out the computations for superimposed current and speed control loops as well. Without further pulse forming the output of the processor initiates directly the gate firing amplifier. Current and speed control algorithms are of proportional-integral type with output limiting. The current signal is the only analogue variable requiring A/D conversion, since speed measurement can be performed by digital techniques thus disposing of analogue transmitters such as DC tachometers.

2. Introduction

Control of thyristor fed DC-drives is usually performed by analogue techniques. Fig. 1 shows the block-diagram of a conventional separately excitated DC-drive controlled by cascaded current-speed-control loops (Ref. 1).

The present rapid development of digital technology opens a wide perspective for new concepts of controlling DC-drives replacing the conventional analogue control and measurement by digital components, i. e. microprocessors, memories, D/A- and A/D-converters. Applying digital techniques to thyristor fed drives entails a number of advantages. As a pulse-controlled thyristor circuit represents in itself an A/D-converter a separate conversion can be omitted. Additional advantages can be obtained by digital processing of speed and angular position, because digital methods are more exact and today inherently less expensive then analogue techniques such as tachogenerators or position transmitters. Digital input and storage of the set point is also possible.

With these components the analogue control unit can be replaced by a compact digital control and measurement unit. Comparison of the current signal remains the only additional task.

Fig. 2 shows the structure of this

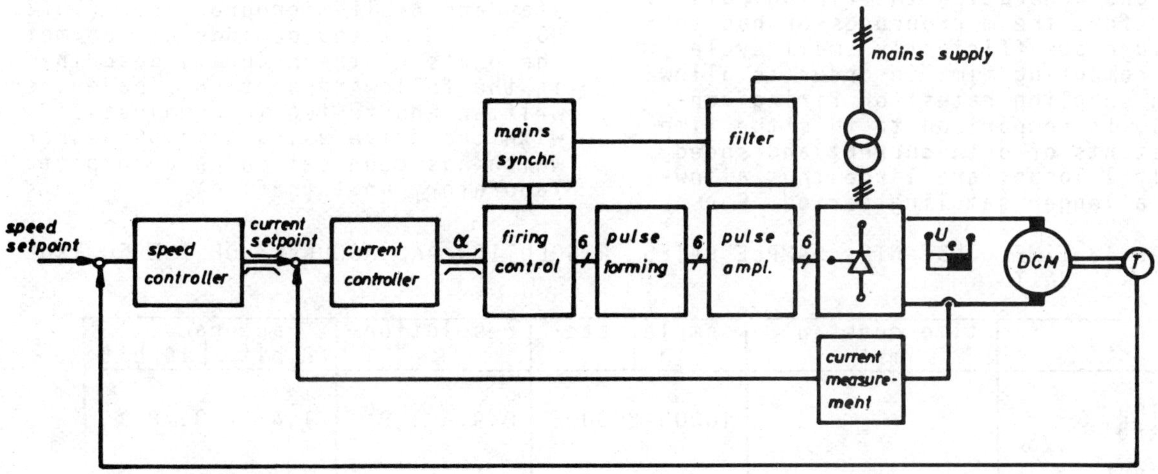

Fig. 1. Block diagram of a separately excited DC-drive with an analogue control scheme

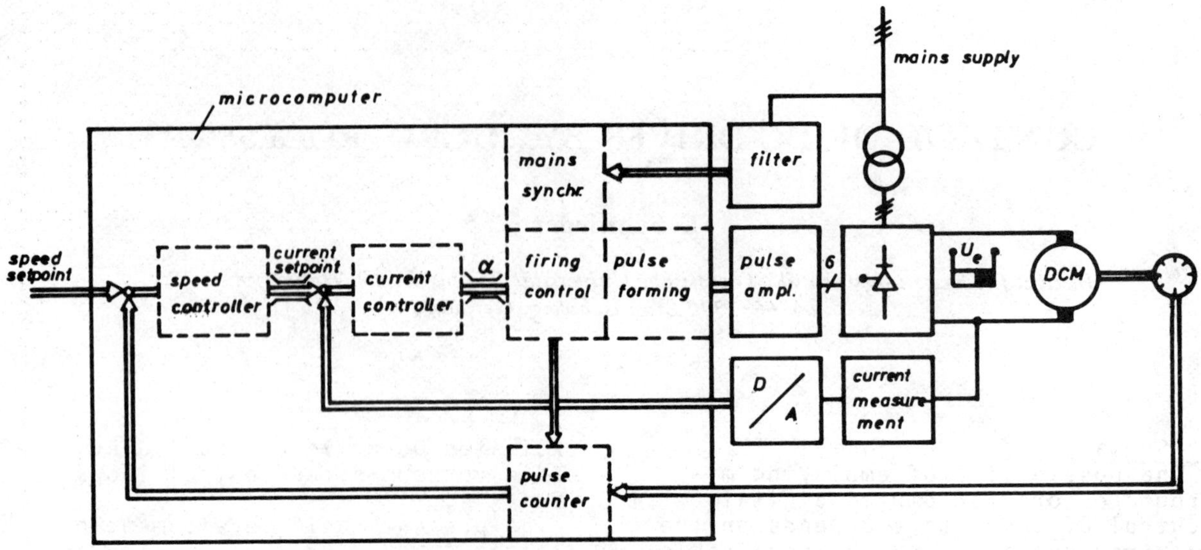

Fig. 2. Block diagram of a separately excited DC-drive with a digital compact controller

concept. This design causes a shift of emphasis from hardware to the development of the necessary software and control algorithms. To adapt this concept to different plants one only has to modify parts of the program or to enter new parameters, which is much easier done than modifying hardware.

3. Criteria for selecting a suitable microprocessor

The design of a compact control unit for a DC-drive includes the firing control for the thyristor bridge operating with a three phase 50 Hz power supply. This imposes the highest dynamic demand on the processor. The limit of allowable asymmetry requires a resolution for the firing angle to about 1° - 2°, i. e. 50 - 100 µs. This necessitates a sampling frequency of 10 - 20 kHz when computing and generating the firing pulses. Therefore the microprocessor has to feature a sufficiently small cycle and computing time in order to allow high sampling rates for firing control. In comparison to this the time constants of both current and speed control loops, are large thus allowing a longer sampling period. For these loops sampling intervals of about 1/5 ... 1/10 of the respective time constants can be estimated. For current control an accuracy of about 1 ... 3 % of the maximal current and for speed control an accuracy of about 0,5 ... 3 % of the rated speed has been aimed for. When selecting the accuracy of the speed loop, another aspect of microcomputer control must be kept in mind. Upgrading the accuracy requires longer measuring periods and larger numbers to be processed; thus double precision computations may have to be employed. The resulting data of a small experimental DC-drive with a rated speed of 1500 RPM are seen in Table 1. These sampling rates can only be realized by a combination of fast microprocessor and carefully optimized algorithms. In this context, the numbers of registers, flags and interrupts are important. Applying a standard 8-bit-microprocessor (INTEL 8080) the stated demands can be met. The parts of the program, described in the following sections, have been written and tested in combination with the drive using this processor which has been set up as a complete laboratory unit (Ref. 4).

Table 1. TIME CONSTANTS, SAMPLE RATES, RESOLUTION AND ACCURACY OF THE CONTROL UNITS

	time constant |ms|	sample rate |Hz|	resolution	accuracy 8 bit	accuracy 16 bit
firing control		10000-20000	$0{,}9 - 1{,}8^\circ$	$1{,}4^\circ \hat{=} 0{,}78$ %	
current control loop	10 - 20	500- 1000		0,39 %	
speed control loop	100 - 200	50- 100		0,78 %	0,03 %

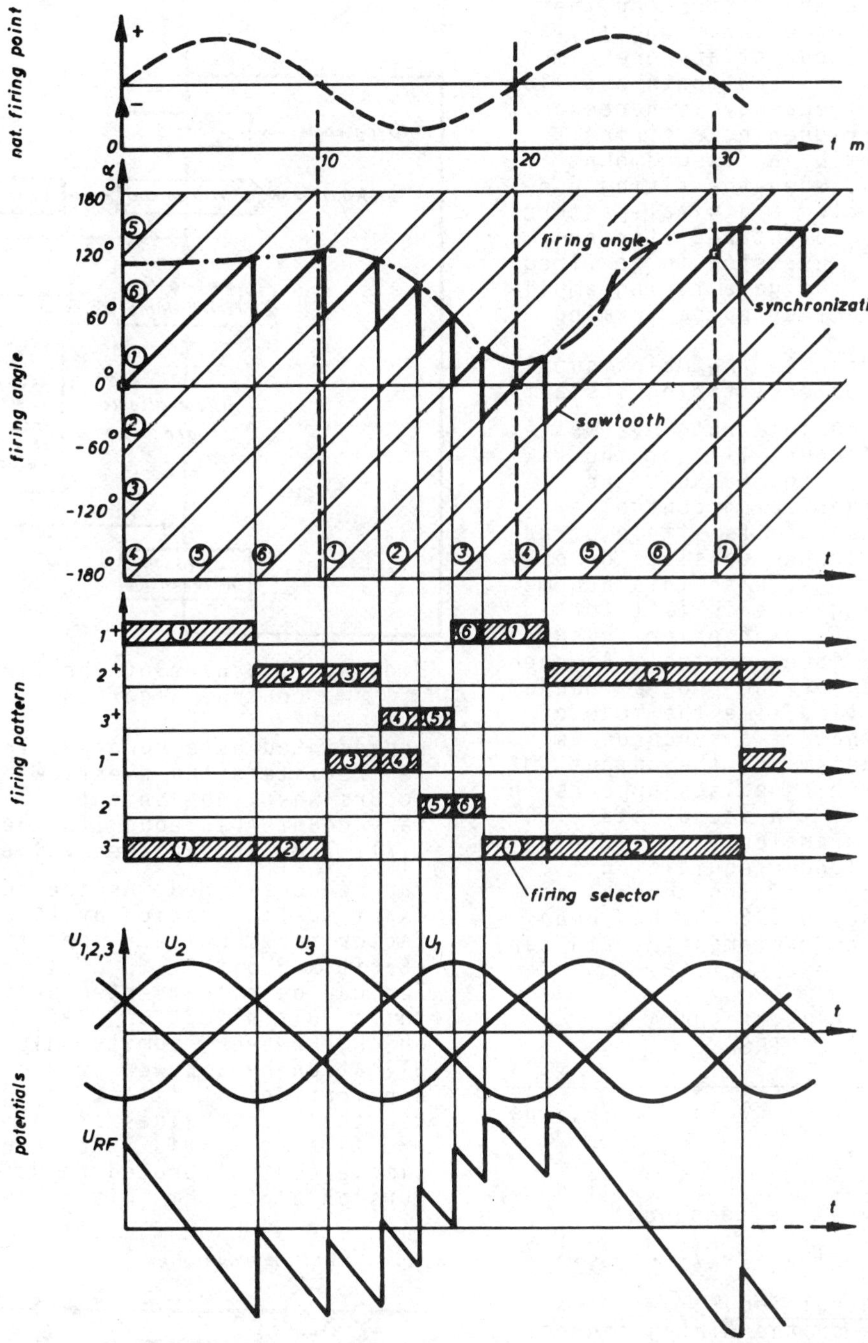

Fig. 3. Time arrangement of mains voltage, synchronization, sawtooth, firing pulses and output voltage

4. Firing control

Firing control of line commutated converters is based on the principle that firing occurs with a certain delay after the natural firing instant. Implementation of the conventional firing control strategy is obtained by comparing a linearly increasing periodic voltage with a voltage corresponding to the desired delay time (Ref. 5). Therefore three saw-tooth-signals are needed to control a three phase bridge consisting of 6 thyristors. The actual firing pulses are generated by a subsequent amplifier performing suitable pulse shaping.

A firing control algorithm has been designed along these lines and a microprocessor program was developed approximating the saw-tooth by a staircase signal. By doing so a digital firing control with a sample rate of 10 kHz could be realized with the INTEL 8080. However, since all available registers have been used and no spare computing time had been left, a completely new concept employing a more efficient algorithm had to be designed.

It is based on the fixed firing

sequence of the thyristors and the fact, that firing of only one thyristor needs to occur at any one time. Therefore only one saw-tooth oscillator of triple frequency is necessary, which, after triggering a firing pulse, is reset by a fixed amount (Ref. 5, 6, 7). When the firing condition is detected a digital pattern appears at the output register according to the thyristor to be fired. This initiates the gate firing amplifier without further pulse forming (Fig. 3).

Synchronization to the mains supply occurs at the natural firing instant every 180° by an interrupt overwriting the actual count list of the sawtooth register. Fig. 4 shows the flowchart of the firing control algorithm. Because of its simple structure the main branch consists of only a few instructions. With this scheme enough computing time is left for program parts of lower priority such as current and speed control. As each possible branch of the program needs the same computing time the mode of operation is that of a synchronous stepping mechanism and the output word for the firing pulses appears in a fixed interval in steady state. Of course during transients, the firing pulses are no longer equidistant. With this type of firing control a sample frequency of 12 kHz has been achieved. The characteristic data can be found in Table 2 (Ref. 8).

Table 2. CHARACTERISTIC DATA OF FIRING CONTROL

max. computing time/cycle ...	79,5 µs
accuracy	$1,4^{o}$
resolution	$1,4^{o}$
shift of d-c voltage component due to sampling ($\alpha = 90^{o}$).....	2,4 %

5. Current control loop

Superimposed on the firing control is the slower current control loop. In this stage the microprocessor has to perform subtraction between current set point and actual current value, dynamic processing of the input error, calculating and limiting of the firing angle. Measurement of the current is both possible on the DC or AC side of the power converter. The signal, which is proportional to the current, is fed to the microprocessor through an A/D converter.

In order to achieve a satisfactory performance a proportional-integral controller is provided. Implementing the controller by a program two different ways can be chosen. Fig. 5a shows the block diagram of the controller, the integrator of which is

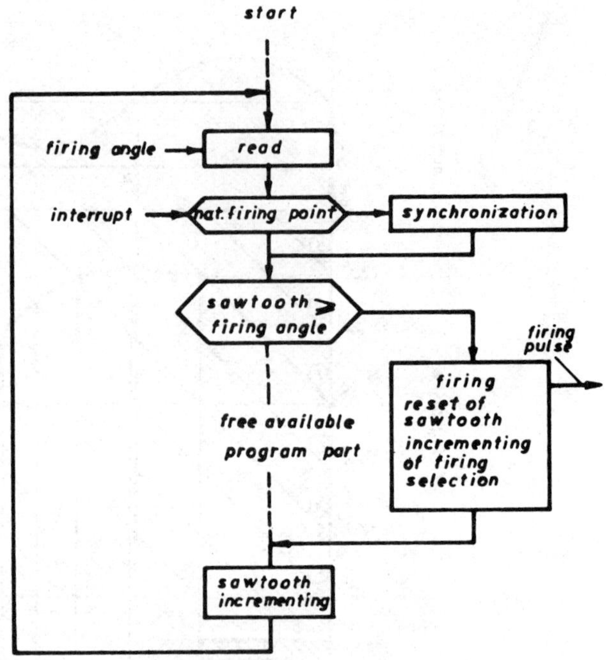

Fig. 4. Flowchart of the firing control algorithm

represented by a counter. This counter registers the overflow, which occurs when the content of an auxiliary counter is equal to the ratio T_i/T. Thus any integer value of T_i/T can be considered. As the additional exact multiplication by the gain factor takes quite a long time (8 bit x 8 bit ≙ 125 µs), this problem can be circumvented by the fast shift multiplication (Ref. 10), which, however, admits only a multiplication by a power of 2. In many applications it is not necessary to use the exact values of gain factor and time constant; this results in a simple control procedure using a recursion algorithm (Fig. 5b).

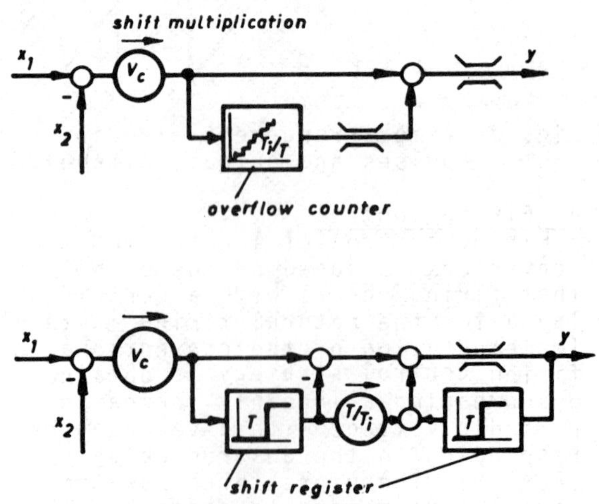

Fig. 5. Proportional-integral-controller
a) realization by counter
b) realization by recursive algorithm

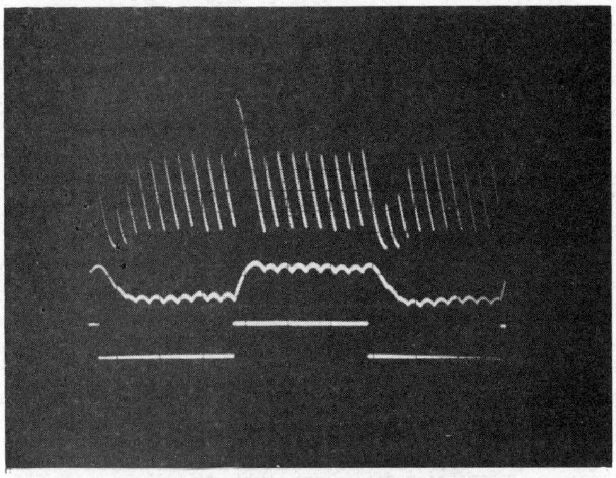

R = 6 Ω L = 30 mH U_{DC} =18V ΔI=0,25 A

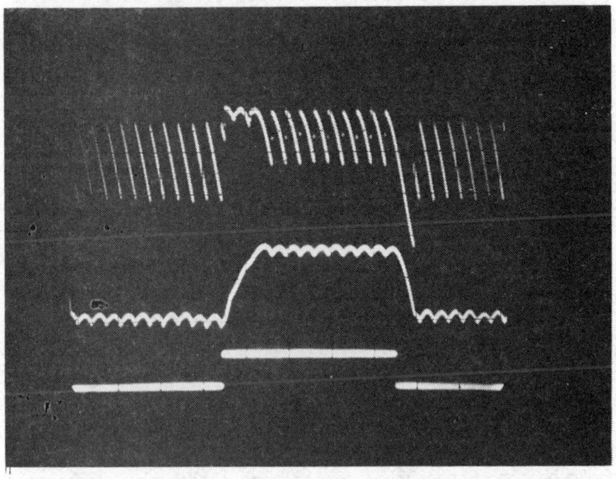

R = 3 Ω L = 3 mH U_{DC} = 5V ΔI = 2,5 A

Fig. 6. Step response of the current control loop

To achieve a quasi-analogue mode of operation a sampling period of 5 kHz has been chosen for the experimental drive. This means that after 6 cycles of the firing control loop a new firing angle must have been computed by the current control branch. This is also necessary in order to be able to compute the firing angle during the fast change from inverter to rectifier operation. Effective programming permits a computation of the firing angle during the spare time of two cycles through the firing control loop. In addition there is enough time to implement additional control loops. Fig. 6 shows the step response of the current control loop employing a current controller as shown in Fig. 5a.

6. Speed control system

Like the current controller, the speed controller has been designed in proportional-integral mode. Since this control loop imposes the least dynamic demands, a sample rate of 50 Hz is provided. Remaining computing and storage capacity is ample, but as all immediately available registers are blocked, the slower memory has to be referred to.

Digital processing of speed control brings with it two additional advantages. If the speed set-point is given as a numerical value, which is usually the case, no A/D conversion is required. The digital speed measurement by pulse tachometer disposes of any analogue transmitters.

Depending on the application one of several speed measurement methods may be employed. If the drive is to be operated with constant speed, time measurement appears as the simplest method. The drive shaft is equipped with an optical or magnetic pick-up device; between two gate pulses fed to the interrupt, the processor counts with an internal clock frequency. As the relation of speed and period is a reciprocal one, the control loop operates linearly only in a small range around the rated speed.

If the drive is to operate over a wide speed range, it is preferable to measure speed by counting the pulses fed to the microprocessor from a pulse tachometer. The speed then corresponds to the number of the pulses received during a certain intervall.

As the speed of a universal drive operating in all four quadrants of the speed-torque-plane can assume both positive and negative values, the resulting speed of processing sign and magnitude by counting and logically comparing of pulses is too slow to be handled by the microprocessor. Therefore an external logic and digital circuit has to be supplied. For the mentioned experimental four-quadrant-drive therefore the latter method has been chosen. The characteristic data of the speed control system can be found in Table 3 (Ref. 9).

Table 3. CHARACTERISTIC DATA OF SPEED MEASUREMENT

rated speed	± 1500 min^{-1}
pulse-frequency	10-20 kHz
period of measurement	
8 bit	~ 10 ms
16 bit	~500 ms

7. Outlook

Applying faster microprocessors with shorter cycle time (for example INTEL 8080 A) the direct digital control of four-quadrant-drives will be possible. The conventional parallel-push-pull circuit can be replaced by a single ended three-phase-converter with subsequent reversing of current by two pairs of thyristors-similar to the reversing of the armature current by contactors (Fig. 7). By doing so the 8-bit-structure can be retained, because

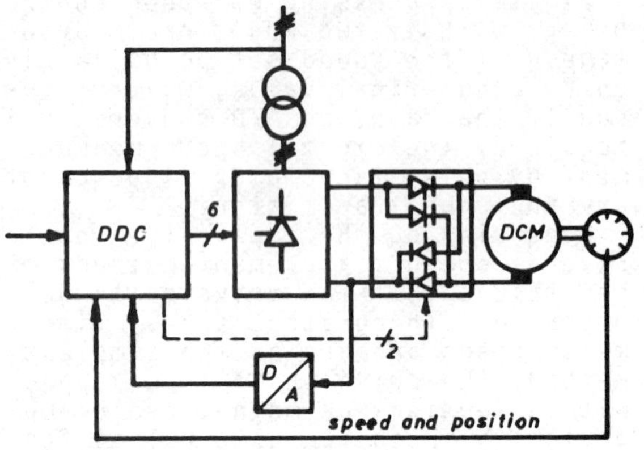

Fig. 7. Structure of a four-quadrant DC-drive controlled by a microprocessor

the number of different ignition states is no more than 8.

Another aspect is the replacement of present AC filters for the synchronizing voltages by a suitable algorithm in the microprocessor. Thus the disturbed mains voltages can be reconstructed to be utilized for synchronization.

At last the possibility of extinction-angle control for high power converter applications such as HVDC should be mentioned, which could be realized by applying faster microprocessors, already on the market.

Another extension is the superposition of a digital position control loop on the speed controlled four quadrant drive.

References

|1| W. Leonhard: Regelung in der elektrischen Antriebstechnik. Teubner, Stuttgart (1974)

|2| S. G. McLaren a. o.: Direct digital control of thyristors. IFAC-Symposium, Duesseldorf (1974)

|3| H. Brick: Digitale Regelung durch Mikroprozessoren. Elektronik, H. 3, (1976)

|4| G. Fromme: Digitale Filter mit Mikroprozessor. PDV-Fachtagung, Karlsruhe (1976)

|5| F. Maurer: Vergleich verschiedener Zündsteuerverfahren für netzgeführte Stromrichter. ETZ-A, Bd. 95, H. 1

|6| J. Holtz: Ein neues Zündsteuerverfahren für Stromrichter am schwachen Netz. ETZ-A, Bd. 91, H. 6

|7| P. Schultze: Zündsteuergerät für geregelte Stromrichter. Diplomarbeit, Institut für Regelungstechnik, TU Braunschweig (1968)

|8| N. Fotiadis: Realisierung eines digitalen stromgeregelten Zündsteuergerätes mit einem Mikroprozessor. Studienarbeit, Institut für Regelungstechnik, TU Braunschweig (1976)

|9| N. Schmitt: Thyristorgespeister Gleichstromantrieb mit Steuerung und Regelung durch einen Mikroprozessor. Diplomarbeit, Institut für Regelungstechnik, TU Braunschweig (1976)

|10| B. Itschner: Vorstellung einer neuen Klasse von DDC-Algorithmen Regelungstechnik, H. 9 (76)

SIMULATION OF THE CHOPPER CONTROLLED DC SERIES MOTOR

T. Fujimaki, K. Ohniwa and O. Miyashita

Tokyo Electrical Engineering College, Tokyo, Japan

Summary

The authors describe in this paper the equivalent circuit for a dc motor in consideration of both magnetic saturation and eddy current and the evaluation of its parameters in detail. On the basis of these results the dynamic characteristics of a chopper controlled dc series motor are investigated.

1. Introduction

In recent years thyristor choppers and phase-controlled rectifier circuits have been widely used to control dc motors in many practical applications. When a dc series motor is controlled by those solid-state circuits, the current in the field and armature circuits ripples and then eddy current arises in the iron core. Moreover, in practice the motor is often operated in the saturated region of magnetic flux. Under these conditions, the inductance of the electrical circuits is no longer constant. Such nonlinearities are significant factors affecting transient performance of the motor [4],[5],[6],[7].

However, in many conventional analyses [1], [3] the field and armature circuits have been treated as a simple RL circuit, with consideration of magnetic saturation if it is not negligible. Also, the effects of eddy current have been ignored in many cases.

In 1951, the late Professor H. Takai of Tokyo Institute of Technology suggested, in his paper [2], an excellent idea about the equivalent circuit which represented accurately the effects of eddy current in the field of a dc generator. He emphasized that the equivalent circuit was more convenient as concerns the system dynamics.

On the other hand, with rapid developments of an electronic computer, we can easily simulate even more complicated systems. Therefore, it seems to be more important to construct an appropriate mathematical model for the motor system and to determine its parameters.

2. Equivalent circuit for dc motor

In order to construct the equivalent circuit for the electrical system of a dc series motor, it is necessary to take the attention of the factors which have considerable influence on the dynamic performance. That is, the magnetic saturation and eddy current play an important role in the dynamic response of the motor. The effects of hysteresis in the iron core usually used for electric machines, however, may be assumed small enough to be negligible in comparison with the others.

2.1 Field winding

Consider first a separately excited dc generator with operation restricted to the linear range, as shown in Fig.1, in order to investigate the effects of eddy current.

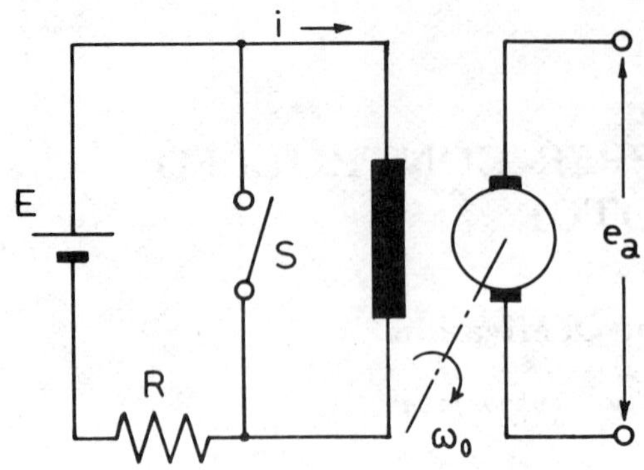

Fig.1. Schematic diagram of a separately excited dc generator

Now, let the switch S be closed at t = 0. The field current i and the armature terminal voltage e_a as a function of time are shown in Fig.2. There is no similarity between them evidently and this difference may be considered to be caused principally by eddy currents.

Consequently, the field current can not be represented as a single RL time constant. However, it is difficult to concern strictly the eddy current because they are distributed in a complex manner [2],[4].Then, the authors suggest the equivalent circuit of Fig.3 with effects of eddy current for practical use. In Fig.3, Z_e is the equivalent circuit for eddy currents, which consists of several RL circuits in parallel. This circuit has obviously no effects for direct current. Note that the magnetic flux directly proportional to the generated armature voltage is produced not by the current i but by i_m ; i_m is the so-called magnetizing current. One must distinguish between the field current i and the magnetizing current i_m.

The generator of Fig.1 may be modified by using the equivalent circuit of Fig.3, as shown in Fig.4 . If the switch S is suddenly closed at t = 0, the voltage equation is given, under conditions of negligible saturation

$$L_f \frac{d\, i_m}{d\, t} + R_f\, i = 0 \qquad (1)$$

with initial conditions $i(0)=i_m(0)=I_0$ and

$$w\,(\Phi_f - \Phi_r) = L_f i_m \qquad (2)$$

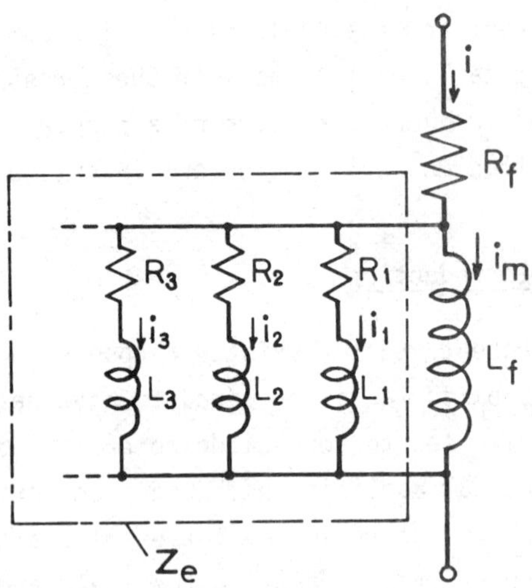

Fig.3. Equivalent circuit for field winding

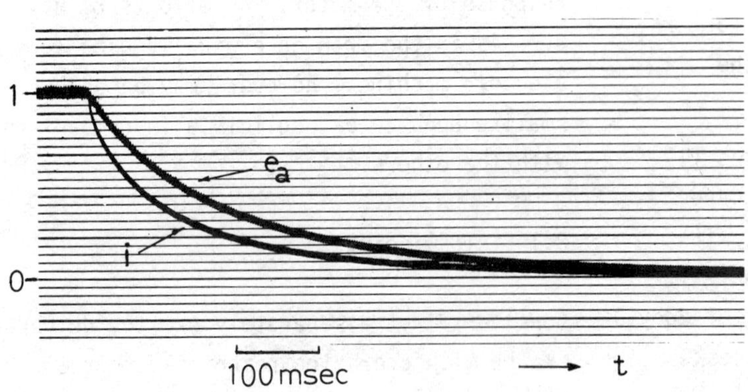

Fig.2. Armature voltage and field current following field short-circuit(in normalized form)

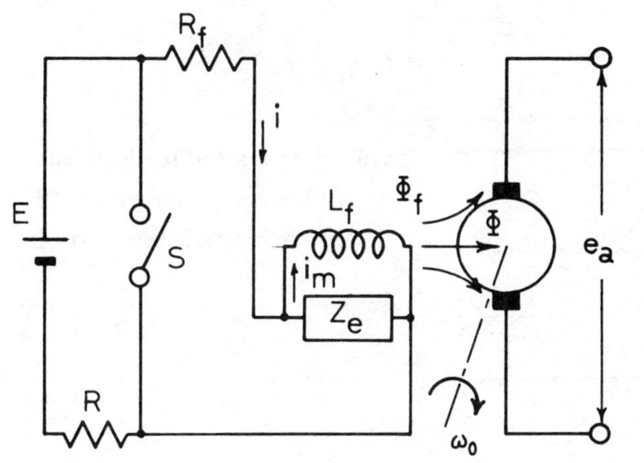

Fig.4. Modified diagram for a separately excited dc generator

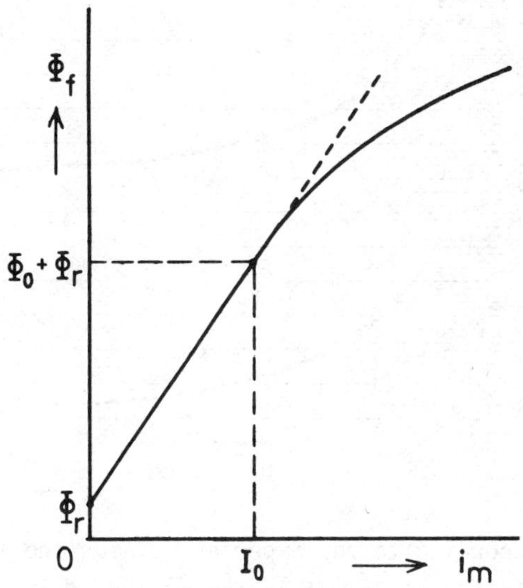

Fig.5. Magnetization curve

where L_f and R_f are the inductance and resistance of the field circuit, w the number of turns in the field winding, and a small residual flux Φ_r is assumed which has no effects on the transient response, corresponding to a small generated voltage at zero excitation.

From Eqs.(1) and (2), the magnetic flux yields

$$\Phi_f(t) = \Phi_0 - \frac{R_f}{w}\int_0^t i(\lambda)\, d\lambda + \Phi_r \qquad (3)$$

where $\Phi_0 = L_f I_0/w$ is the produced flux corresponding to the initial current I_0 in the steady state. The air-gap flux Φ is somewhat smaller than the field flux Φ_f, because of field leakage– flux; thus

$$\Phi(t) = \frac{1}{\sigma}\Phi_f \qquad (4)$$

where $\sigma$is the leakage coefficient, which we shall assume to be constant.

On the other hand, at a constant speed ω_0 the induced armature voltage for a dc generator can be expressed as

$$e_a = K_c \Phi\, \omega_0 \qquad (5)$$

where K_c is a constant.

It is necessary to compare Φ_f with e_a in order to find whether the flux given by Eq.(4) is proportional to the air-gap flux or not. That is, the equivalent circuit of Fig.3 is available, if the similarity between Φ_f and e_a exists. The experimental results are shown in Fig.6, and then the above condition is evidently satisfied.

Finally, it is possible to consider the effects of eddy current, if the parameters of Z_e are determined so that the field current i may be represented as closely as possible.

2.2 Approximation of the field current

Consider the short-circuit field current of a dc generator, as shown in Fig.2. It may be represented by a sum of several exponential terms as follows.

$$i(t) = I_A e^{-t/T_A} + I_B e^{-t/T_B} + I_C e^{-t/T_C} + I_D e^{-t/T_D} + \cdots \qquad (6)$$

where $i(0) = I_0 = I_A + I_B + I_C + I_D + \cdots$.

Usually the assumption $T_A \gg T_B \gg T_C \gg T_D \gg \cdots$ is satisfied. The effects of eddy current have influence, of course, on the number of the right-hand terms in Eq.(6). It is evident that, let n denote the number of the exponential terms, n-1 denotes that of the parallel branches in Z_e.

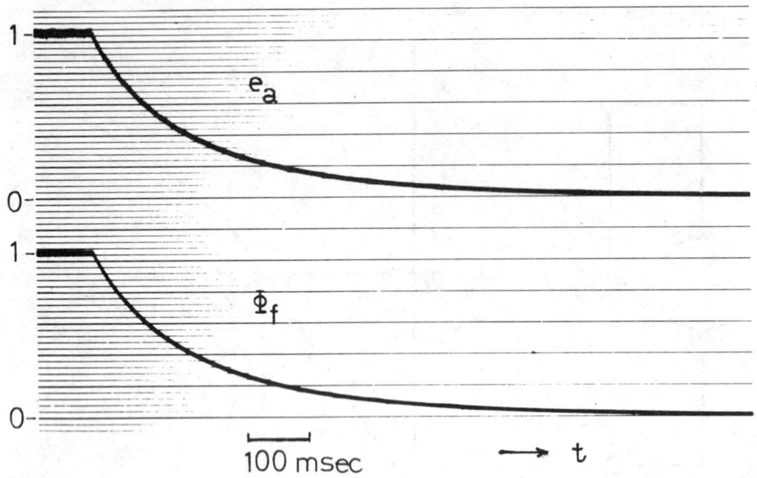

Fig.6. Voltage build-down and flux Φ_f given by Eq.(3) (in normalized form)

According to our experiences, for the field winding of dc motors the number of the branches would be less than four practically and these parameters may be assumed to be constant, though they depend upon the current, frequency and so on, in the strict sense.

Suppose that the field current i is approximated by the summation of three exponential terms; that is

$$i(t)=I_A e^{-t/T_A}+I_B e^{-t/T_B}+I_C e^{-t/T_C} \tag{7}$$

where $I_0=I_A+I_B+I_C$ and $T_A \gg T_B \gg T_C$.

The Laplace transform of Eq.(7) is

$$I(s) = \frac{I_A}{s+1/T_A}+\frac{I_B}{s+1/T_B}+\frac{I_C}{s+1/T_C} \tag{8}$$

or

$$I(s) = \frac{T_AT_BT_C(I_A+I_B+I_C)s^2+[\,I_AT_A(T_B+T_C)+}{T_AT_BT_C\,s^3+(T_AT_B+T_AT_C+T_BT_C)s^2+}$$

$$*\frac{I_BT_B(T_A+T_C)+I_CT_C(T_A+T_B)\,]s+I_AT_A+I_BT_B+I_CT_C}{(T_A+T_B+T_C)s+1} \tag{9}$$

Then the constants of the equivalent circuit as shown in Fig.3 can be evaluated in the following manner.

If $t \gg T_B$, Eq.(7) is given approximately

$$i(t) \simeq I_A e^{-t/T_A} \tag{10}$$

Dividing both sides by the initial current I_0 and taking the natural logarithm yields

$$\ln\frac{i(t)}{I_0} \simeq \ln\frac{I_A}{I_0}-\frac{t}{T_A} \tag{11}$$

Equation(11) is plotted to a logarithmic scale as a function of time as shown in Fig.7. Accordingly, T_A can be determined from the slope of the straight line which is an asymptote as $t\to\infty$ and I_A from the intersection of the line with the vertical axis.

In the similar process, I_B and T_B are obtained. If $T_A > t \gg T_C$, Eq.(7) is approximately

$$i(t) - I_A e^{-t/T_A} \simeq I_B e^{-t/T_B} \tag{12}$$

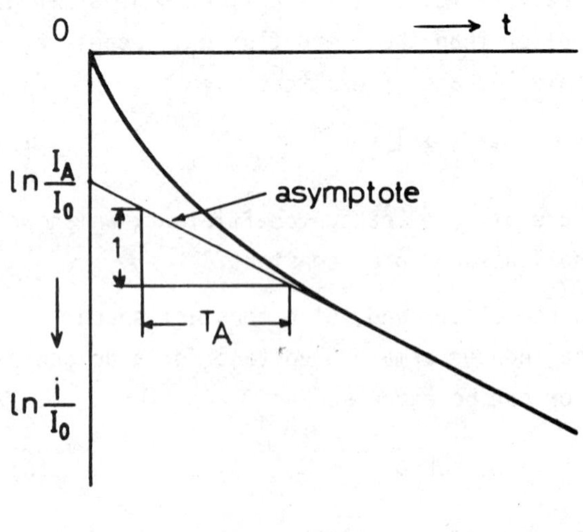

Fig.7. Current plotted to semilog coordinates

Hence

$$\ln \frac{i(t)-I_A e^{-t/T_A}}{I_0-I_A} \simeq \ln \frac{I_B}{I_0-I_A} - \frac{t}{T_B} \quad (13)$$

Furthermore, during the period of $t \ll T_B$

$$i(t)-I_A e^{-t/T_A}-I_B e^{-t/T_B} = I_C e^{-t/T_C} \quad (14)$$

and then

$$\ln \frac{i(t)-I_A e^{-t/T_A}-I_B e^{-t/T_B}}{I_C} = -\frac{t}{T_C} \quad (15)$$

where $I_C=I_0-I_A-I_B$.

From a plot of Eq.(15) I_C and T_C are determined easily.

After the work is done carefully, if I_C, i.e. generally the final term of the current in Eq.(6), is not less than about 5 percent of the initial current I_0, then more exponential terms for representing the current should be considered; in other words, the number of parallel branches in Z_e correspondingly increases. The reason is due to our experience.

2.3 Determination of parameters

As the current waveform of Eq.(7) may be expressed on the equivalent circuit with two parallel RL circuits in Z_e, we shall determine the values of R_1, L_1, R_2 and L_2 as follows. In this case the current in s-domain is

$$I(s)= \frac{I_0 [T_f T_1 T_2 s^2 + T_f(T_1+T_2)s + + T_f]}{T_f T_1 T_2 s^3 + (T_f T_1+T_f T_2+T_1 T_2+T_1 T_{f2}+ T_2 T_{f1}) s^2 + (T_f+T_1+T_2+T_{f1}+T_{f2}) s + 1} \quad (16)$$

where $T_f=L_f/R_f$, $T_1=L_1/R_1$, $T_2=L_2/R_2$, $T_{f1}=L_f/R_1$ and $T_{f2}=L_f/R_2$.

Identifying Eq.(16) with Eq.(9) gives

$$I_0 T_f = I_A T_A+I_B T_B+I_C T_C \quad (17)$$

$$I_0 T_f(T_1+T_2) = I_A T_A(T_B+T_C)+I_B T_B(T_A+T_C) +I_C T_C(T_A+T_B) \quad (18)$$

$$T_f T_1 T_2 = T_A T_B T_C \quad (19)$$

$$T_f+T_1+T_2+T_{f1}+T_{f2} = T_A+T_B+T_C \quad (20)$$

$$T_f T_1+T_f T_2+T_1 T_2+T_1 T_{f2}+T_2 T_{f1} = T_A T_B+ + T_A T_C+T_B T_C \quad (21)$$

Accordingly, the process of evaluating the circuit constants is the following:

(a) From Eq.(17)

$$T_f = \frac{I_A T_A+I_B T_B+I_C T_C}{I_0} \quad (22)$$

(b) From Eqs.(18) and (19)

$$T_1+T_2 = \frac{I_A T_A(T_B+T_C)+I_B T_B(T_A+T_C)+I_C T_C(T_A+T_B)}{I_0 T_f} \quad (23)$$

$$T_1 T_2 = \frac{T_A T_B T_C}{T_f} \quad (24)$$

T_1 and T_2 are determined from Eqs.(23) and (24).

(c) From Eqs.(20) and (21)

$$T_{f1}+T_{f2} = (T_A+T_B+T_C)-(T_f+T_1+T_2) \quad (25)$$

$$T_1 T_{f2}+T_2 T_{f1} = (T_A T_B+T_A T_C+T_B T_C) -(T_f T_1+T_f T_2+T_1 T_2) \quad (26)$$

T_{f1} and T_{f2} are determined from Eqs.(23) and (24).

(d) The constants of the equivalent circuit are easily determined

$$\begin{aligned} L_f &= T_f R_f \\ R_1 &= L_f/T_{f1} \\ L_1 &= T_1 R_1 \\ R_2 &= L_f/T_{f2} \\ L_2 &= T_2 R_2 \end{aligned} \quad (27)$$

2.4 Effects of magnetic saturation

In previous section, the parameters of the equivalent circuit with the effects of eddy current were determined on the linear basis with magnetic saturation neglected. However, the field inductance L_f inherently depends on saturation and then is function of current in general.

On the other hand, the estimation of the parameters of the eddy-current circuit Z_e in the saturated region is extremely difficult in practice. Hence, we assume that these are regarded as constant in both saturated and unsaturated regions for the sake of simplicity.

In this case, from the flux linkages $\psi_f = L_f(i_m)i_m$ in the field winding the voltage equation corresponding to Eq.(1) is given

$$\frac{d(L_f i_m)}{d\ t} + R_f\ i = 0 \tag{28}$$

or

$$L_f^* \frac{d\ i_m}{d\ t} + R_f\ i = 0 \tag{29}$$

where $L_f^* = L_f + i_m \frac{dL_f}{di_m}$, which is called "operational inductance" for the field circuit. The operational inductance L_f^* obviously coincides with unsaturated one L_f for linear operation and also is similar to the first derivative of the magnetization curve of machines with respect to the current. Thus the inductance L_f^* is easily obtained.

2.5 Armature circuit

To an equivalent circuit of the armature circuit the procedure mentioned above can be applied similarly. The measuring circuit is illustrated in Fig.8. The decaying armature current i_a would be approximated by a single time constant in many cases. However, saturation is not negligible when a larger current flows. Unlike the field winding, the saturation curve in armature circuit can not be obtained by the same test as the magnetization curve of a separately excited dc generator.

Then the authors suggest the following method. In Fig.8 the voltage equation in terms of the flux linkages can be expressed

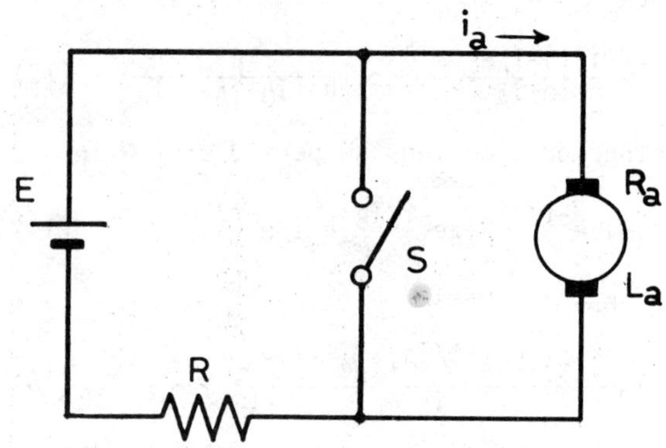

Fig.8. Measuring circuit of armature inductance

$$\frac{d\ \psi}{d\ t} + R_a\ i_a = 0 \tag{30}$$

with $\psi = \psi_0$ and $i_a = I_0$ at t=0.

Integrating of Eq.(30) gives

$$\psi_0 = R_a \int_0^{\infty} i_a(t)\ dt \tag{31}$$

From Eq.(31) the magnetic saturation curve is obtained. Practically, this integral can be carried out easily by an analog integrator. In general, the operational armature inductance L_a^* is calculated in the same manner as the field one.

3. Computer simulation

Digital simulation of the dynamic behavior of a chopper controlled dc series motor as illustrated in Fig.9, is studied in this chapter. It is also applicable to the analog simulation.

Suppose that the chopper acts as an idealized switch. In addition, we make the assumptions that the armature reaction and hysteresis will be neglected, that brush voltage-drop e_b is constant, and that the main thyristor in the chopper circuit and the free-wheeling diode give the same voltage drop denoted as a constant e_d. By using the equivalent circuits for the field and armature circuits, the schematic diagram of Fig.9 can be represented as in Fig.10.

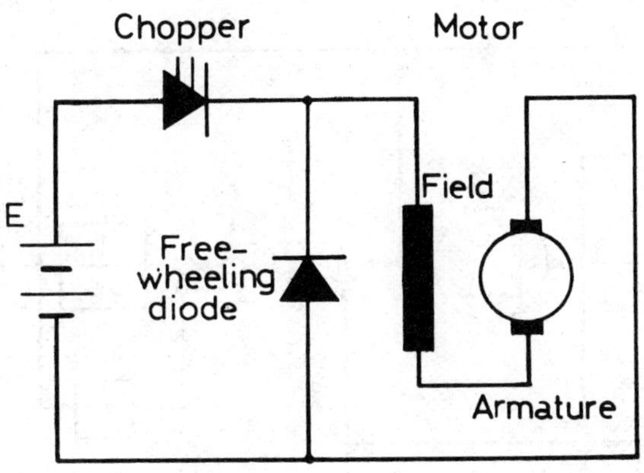

Fig.9. Schematic representation of chopper-controlled dc series motor

Accordingly, the dynamic equations are the following:

(a) Conducting period

For the electrical system the voltage equations are

$$L_f^* \frac{d\,i_m}{d\,t} + L_a^* \frac{d\,i}{d\,t} + (R_f + R_a) i + K_c \Phi(i_m)\,\omega + e_b + e_d = E \qquad (32)$$

$$L_f^* \frac{d\,i_m}{d\,t} = L_1 \frac{di_1}{d\,t} + R_1\, i_1 \qquad (33)$$

$$L_f^* \frac{d\,i_m}{d\,t} = L_2 \frac{di_2}{d\,t} + R_2\, i_2 \qquad (34)$$

$$i = i_m + i_1 + i_2 \qquad (35)$$

where K_c is a constant and ω is the angular velocity.

The torque is given

$$\tau \doteq K_t\, \Phi(i_m)\, i \qquad (36)$$

where K_t is a constant.

For the rotational system the torque equation is

$$\tau = J \frac{d\,\omega}{d\,t} + f\,\omega + \alpha + \tau_L \qquad (37)$$

where J is the moment of inertia, f the coefficient of viscous friction, α Coulomb friction, and τ_L the load torque.

(b) Free-wheeling period

For this operation, in Eq.(32) E=0.

Under some condition, the motor current would be discontinued. In Eq.(37), then the torque is equal to zero, due to i=0.

Finally, the block diagram for a motor system is shown in Fig.11. For example, a 1.3 kw, 100 volt, 17 amp, 1500 rpm dc series motor has the constants as listed in Table 1. The saturation curve and the operational field and armature inductances are shown in Fig.12 and Fig.13, respectively.

Figure 14 shows the results of the simulation for the motor at no load, during starting and in the steady state, compared with the experiments.

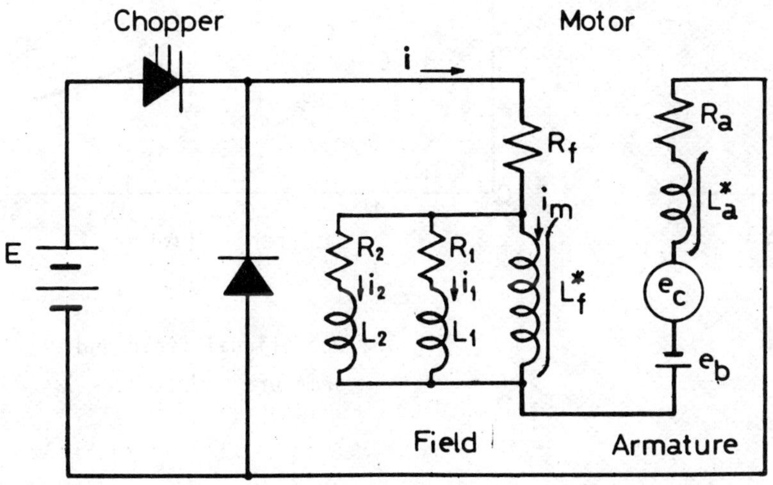

Fig.10. Equivalent circuit of Fig.9

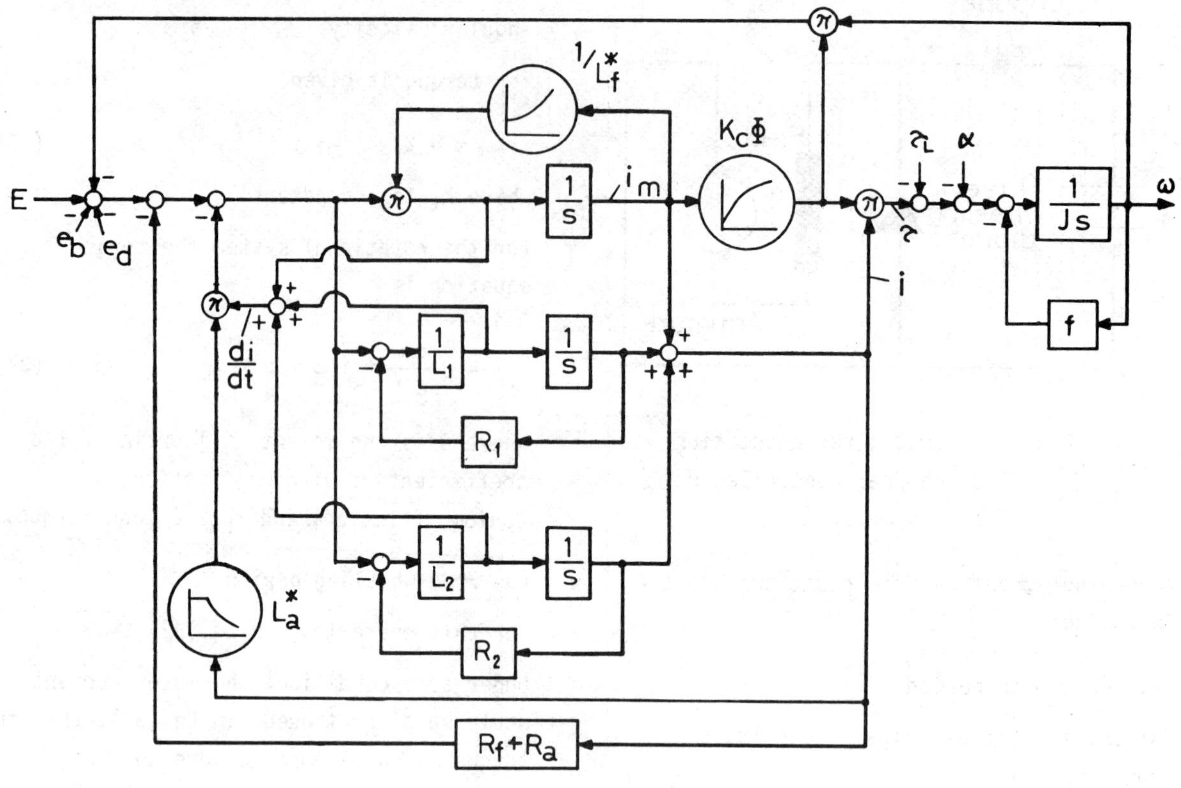

Fig.11. Block diagram for a d-cseries motor

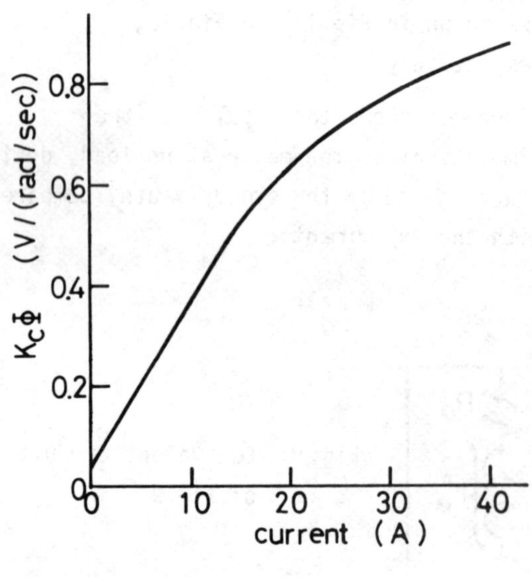

Fig.12. Saturation curve

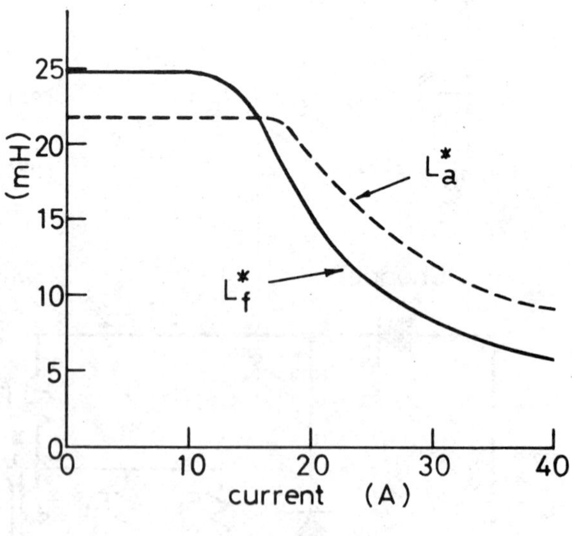

Fig.13. Operational field and armature inductances

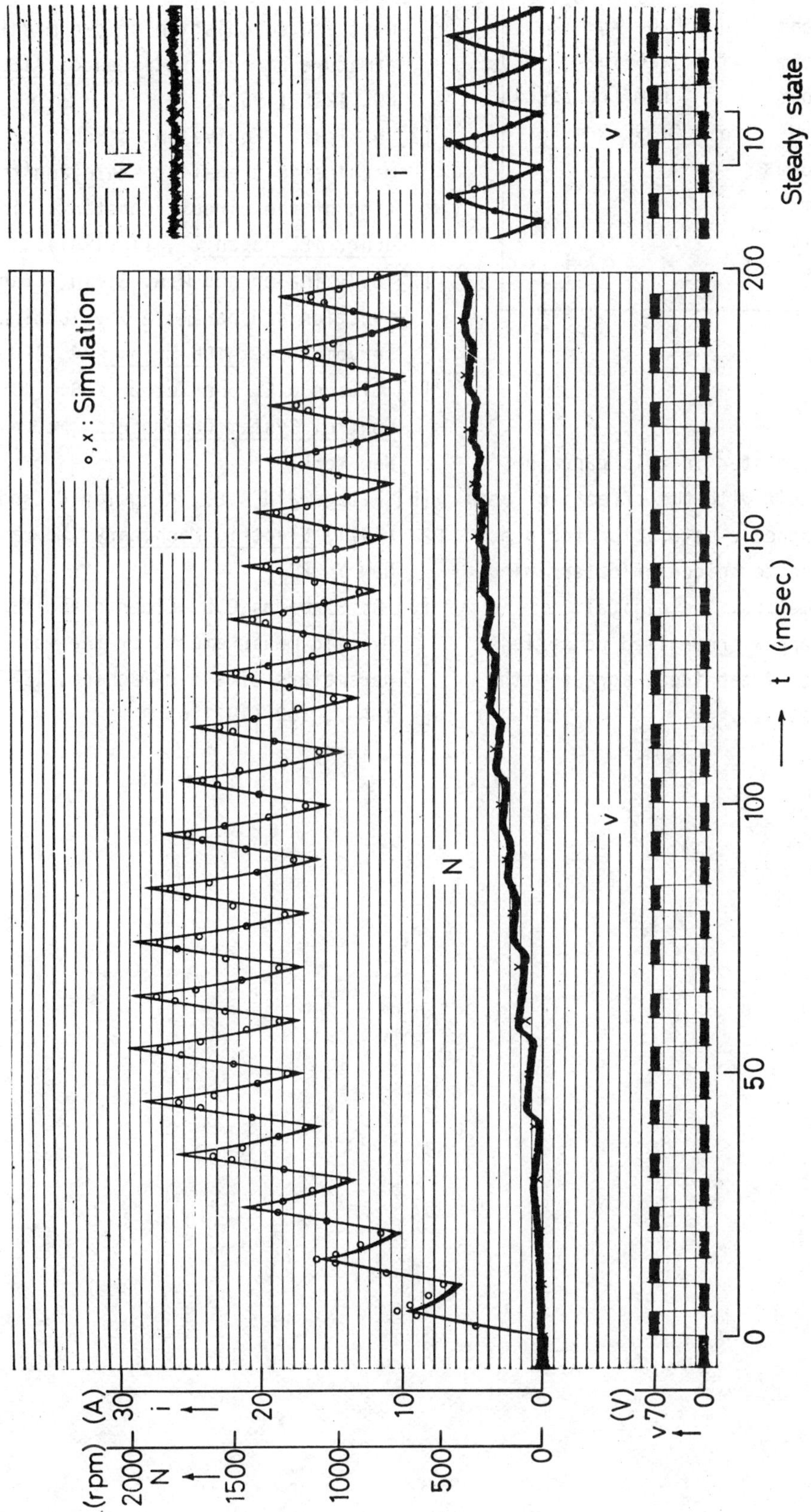

Fig.14. Starting characteristics of a chopper controlled dc series motor(E=71 volts, chopper frequency= 100 Hz and duty factor=0.474)

Table 1

R_f = 0.10 ohm	J = 0.04 kg-m^2
R_a = 0.77 ohm	f = 1.26×10^{-3} N-m/(rad/sec)
R_1 = 0.77 ohm	
L_1 = 42.7 mH	α = 0.392 N-m
R_2 = 1.98 ohms	
L_2 = 5.5 mH	
e_b = 2 volts	
e_d = 1 volt	

4. Conclusion

The authors have suggested in this paper a new equivalent circuit with the effects of both eddy current and magnetic saturation for a dc series motor and the technique of determining these circuit parameters. The digital simulation of a chopper controlled dc series motor, on the basis of the idea, agrees with the experimental results.

References

[1] H.D. Snively and P.B. Robinson, Measurement and calculation of d-c machine armature circuit inductance, AIEE Trans. 69, 1228 (1950).

[2] H. Takai, Equivalent circuit of a dc machine in transient state and the estimation of the circuit constants(in Japanese), Jour. IEE Japan 71, 11 (1951).

[3] G.J. Thaler and W.A. Stein, Transfer function and parameter evaluation for d-c servomotors, AIEE Trans. 74, 410 (1956).

[4] Gibson,J.E. and Tuteur,F.B. (1958) Control System Components ,McGraw-Hill, New York.

[5] Fitzgerald,A.E., Kingsley,C. and Kusko,A. (1971) Electric Machinery ,McGraw-Hill, New York.

[6] B. Szabados,C.D. Dicenzo and N. Sinha, Dynamic measurements of the main electrical parameters of a d-c machine, IEEE Trans. IGA-7, 109 (1971).

THE COMMUTATORLESS D-C MOTOR WITH THREE-PHASE CURRENT EXCITATION

A. Läuger

Institut für Leistungselektronik und Anlagentechnik, Universität Stuttgart, FRG

SUMMARY

The commutatorless d. c. motor with three-phase current excitation is a new, simple and variable speed drive which has a speed-torque characteristic similar to that of a d. c. shunt motor.

The circuitry of the static converter consists of a line-commutated rectifier controlled by line voltage and a load-commutated inverter controlled by motor stator voltage. As motor, a rotor-excited wound-rotor induction motor is used, which is particularly designed for usage with above mentioned converter. It guarantees a safe commutation of the load-commutated inverter under any condition of operation. There are no starting problems. The favourable speed-dynamic characteristics which are typical for this drive are also given at very low speeds.

This paper presents the characteristics of the new drive and shows possible application fields. Its range of rated power can extend from about 5 kW to more than 1 MW.

In many applications, the new drive is an obvious alternative to the converter-fed d. c. shunt motor. A comparison of these two drive systems shows, from an economical point of view, that the commutatorless d. c. motor with three-phase current excitation is clearly superior.

INTRODUCTION

The definitely most frequent problem in drive technology asks for a one directional drive. The motor has to accelerate the load from standstill to the operative speed required, drive it for a certain period of time, and then brake it down to zero speed.

Here are some examples for loads with only one or a privileged direction of rotation: centrifuges, electrically driven vehicles, and test beds for a wide variety of industrial fields.

This means that in most cases a motor is needed whose operating range only includes quadrants I and II of the speed-torque plane. Drive technology offers a variety of speed variable drives for these purposes, which are, however, suited for the driving problem mentioned above and occuring very frequently to a very varying degree.

The motor which is to be introduced here is an extension of the range of drives offered so far. It claims to be especially suited to solve the most frequently occuring driving problem.

In order to show this quite clearly, three well-known speed variable drives will be used to solve the above mentioned driving problem, so that the most important differences between these and the new drive shall become apparent.

In doing this, those characteristics of a drive are to be called "positive" which are useful for the solution of this frequently occuring driving problem. Characteristics which do or might have disadvantageous effects are called "negative".

First, a maintenance-free, dynamically excellent but also very expensive drive is to be considered.

CONVERTER-FED CAGE-ROTOR INDUCTION MOTOR

Figure 1 shows the basic arrangement of this drive and its typical operating quadrants.

positive:

- operating range in all quadrants
- favourable speed dynamic over the entire operating range
- high operation speed possible
- maintenance-free, robust motor
- small over-all length

negative:

- high expenditure on converter equipment, since a forced-commutated inverter is needed (e. g. a sub-harmonic inverter) and a double converter must be provided for power flow back to the a. c. supply.

This converter-fed drive can even fulfil higher requirements than those that were made for the driving problem described initially. The wish for a speed drive whose construction expenditure for the converter is clearly

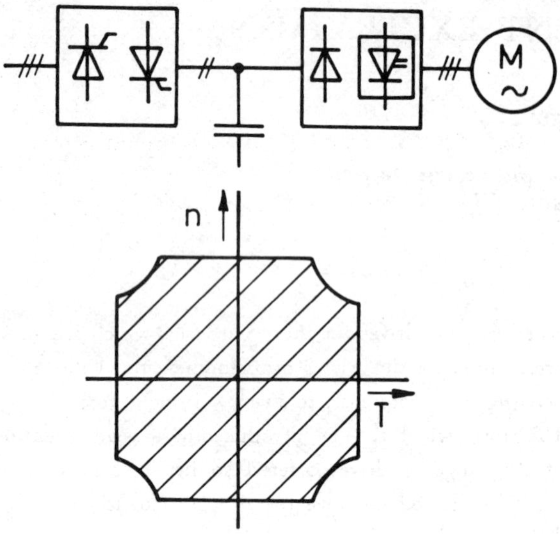

Fig. 1. Cage-rotor induction motor fed by a forced-commutated inverter

smaller, must, however, be maintained.

From this point of view, those motors are suited which are fed through line-commutated and/or load-commutated converters. The most important example for this type of drives is the double-converter-fed d. c. motor which shall be discussed briefly.

DOUBLE-CONVERTER-FED D. C. MOTOR

The basic construction principle of this drive and its operating quadrants are shown in Fig. 2. In order to guarantee a dynamically high-quality transition from motor to regenerative braking mode, neither a mechanical switching of the armature connections, nor a reversal of the field, respectively, were planned.

positive:

- only two single converters
- favourable speed dynamic over the entire operating range
- operating range in all four quadrants

negative:

- motor with commutator
- commutator maintenance
- higher over-all length through the commutator
- output limit, and rotational speed limit subject to commutator and rotor winding
- under high demands on speed dynamic, circulating current operation of the double-converter can become necessary

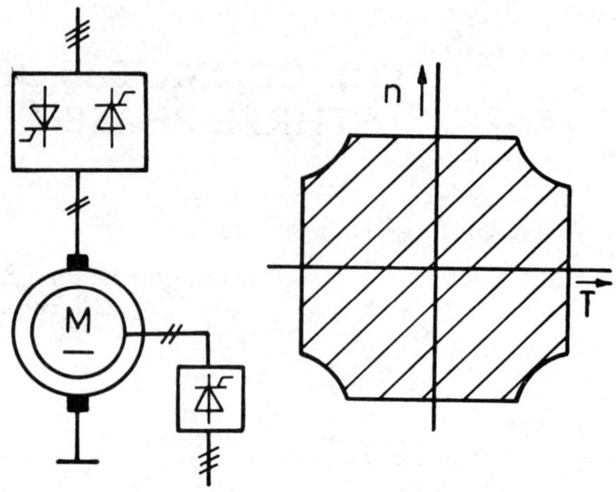

Fig. 2. Double-converter-fed d. c. motor

- additional exciting rectifier necessary

Through this drive, the initially mentioned driving problem can be solved rather satisfactorily.

Quite frequently, however, there are technical and/or economical marginal conditions which stand against the use of a commutator motor. That is why a solid-state commutator is to be considered in the following example.

COMMUTATORLESS D. C. MOTOR WITH D. C. EXCITATION

Figure 3 shows the d. c. excited commutatorless d. c. motor and its typical operating quadrants.

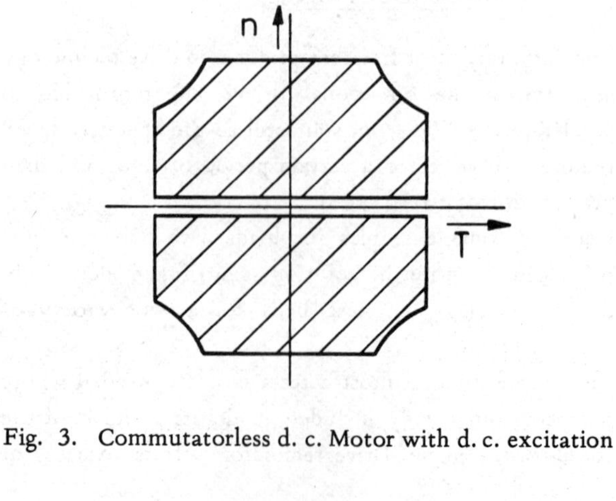

Fig. 3. Commutatorless d. c. Motor with d. c. excitation

Here the drive is a synchronous motor fed by two single converters. At least during the starting process, the load-commutated inverter is controlled by a position transducer on the shaft.

positive:

- two single converters which are not in inverse-parallel connection
- operating range over all quadrants
- high operating speed possible
- favourable speed dynamic
- built without slip rings, almost maintenance-free

negative:

- critical operating region at low speeds resulting in the following problems:
  1. commutation of the load-commutated inverter is no longer possible since stator voltage and stator frequency approach zero
  2. low frequency harmonics in the air-gap torque
- additional expenditure for converter equipment is necessary to effect inverter commutation in the critical speed region
- higher expenditure for circuitry of the control electronics; in the upper starting range a switching from controlling by shaft transducer to controlling by voltage takes place
- a shaft position transducer is necessary; this results in only one free shaft end; in a double motor arrangement (one shaft), this leads to dynamic problems due to controlling with one shaft position transducer only

As a consequence, this motor is only used for solving the initially mentioned driving problem, where the problems arising from the shaft-position transducer must be accepted due to technical marginal conditions (e. g. free of maintenance), and wherever there are no high demands on speed dynamic in the range of low speeds.

The characteristics rated as "negative" with this drive do not come up if a poly-phase motor is used whose rotor is excited by a poly-phase a. c. supply. This motor and its feeding converter shall be called "commutatorless d. c. motor with three-phase current excitation".

THE COMMUTATORLESS D. C. MOTOR WITH THREE-PHASE CURRENT EXCITATION

A complete circuit diagram of this type of drive is given in Fig. 4. The operating quadrants of this drive are shown in Fig. 5. The circuit arrangement of the converter according to Fig. 3 was retained. As a consequence, firstly, the expenditure of converter equipment is also rather low and, secondly, there is also the possibility of a dynamically high-value reversing of the power flow direction.

Instead of a synchronous motor, a special rotor-excited wound-rotor induction motor is used here. The rotor carries a poly-phase winding which is connected through slip rings to an a. c. supply with fixed voltage and fixed frequency f_2. As a consequence, at zero speed of the rotor on the stator side, a three-phase voltage system with a sufficient voltage amplitude and the frequency f_2 are induced. This guarantees the commutation of the load-commutated inverter at zero speed and allows starting of the motor without additional components.

positive:

- two single converters which are in series connection
- operating range in quadrants I and II; no critical operating region
- if use is restricted to low negative speeds only, highly effective four quadrant operation is possible; zero speed is no singularity
- small expenditure on converter equipment; no auxiliary commutation circuits
- no commutator
- nearly maintenance-free

negative:

- rotor with windings, due to these, maximum speed is limited by the stability of the rotor windings
- slightly higher converter rated power and physical size of the motor

A comparison with the previously listed drives shows that the new drive is virtually predestinated to solve the frequent driving problem mentioned initially. It has all the necessary characteristics which permit the solving of drive problems at minimal expenditure.

A circuit arrangement of the converter with only two single converters, in which the thyristors are subject to a definitely lower voltage stress compared to forced-commutated converter circuits, and the commutatorless motor form the essential prerequisites for the creation of an economical drive.

In addition to that, the complete operating range of the new drive includes the range of low negative speeds. They are effected without reversing the direction of the air-gap field. Even in this range, a favourable speed dynamic is guaranteed, so that – restricted to low negative speeds – there is also a dynamically optimal four quadrant operation possible. This is a prerequisite for the exact load positioning ability of the motor. This opens a new field of application for this drive.

We will go on showing how this rather unusual operating region is made possible, and which technical arrangements lead to the favourable characteristics of this drive.

Controlling the Load-commutated Inverter

The rotor-excited wound-rotor induction motor is generally fed by a three-phase SCR bridge on the stator side. The thyristors of this inverter switch the d. c. link current from one stator phase winding on to the next.

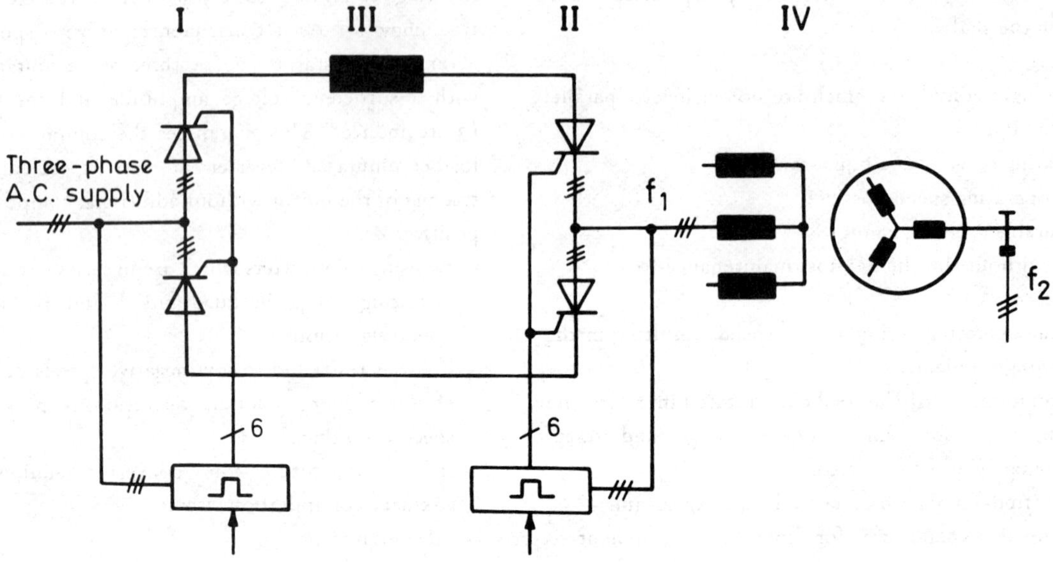

I PHASE-CONTROLLED RECTIFIER
II INVERTER CONTROLLED BY STATOR VOLTAGE
III D. C. LINK
IV WOUND-ROTOR INDUCTION MOTOR
f_1 STATOR FREQUENCY
f_2 ROTOR FREQUENCY

Fig. 4. Commutatorless d. c. motor with three-phase current excitation

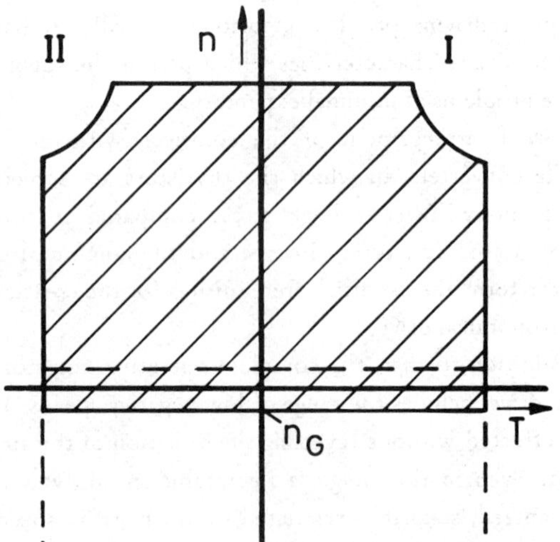

Fig. 5. Operating quadrants of the commutatorless d. c. motor with three-phase current excitation

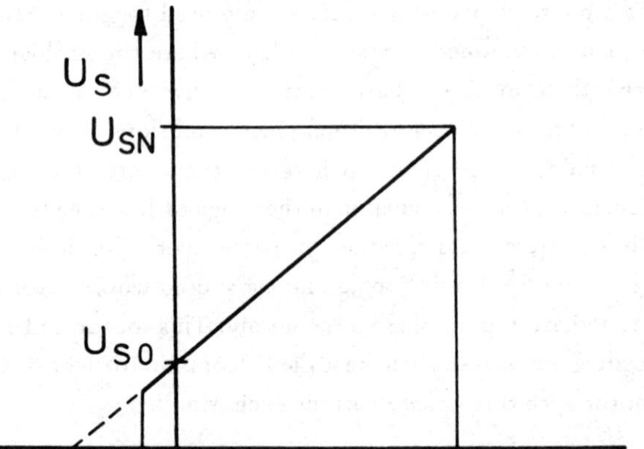

Fig. 6. Stator no-load voltage U_S vs. shaft speed n

This generates the required rotating ampere conductors per cm in the motor.

For the commutation of the thyristor current the motor has to generate a sufficiently high commutation voltage. The r. m. s. value of this voltage is shown in Fig. 6 as a function of the shaft speed.

As can be seen from this, the motor can generate the necessary voltage over the entire operating range. In comparison to the commutatorless d. c. motor with d. c. excitation, here, even at zero speed (n = 0), a three-phase voltage system with the voltage amplitude U_{SO} is induced transformator-like. The frequency of this voltage is 50 Hz, provided that the rotor winding is connected to a three-phase 50 Hz a. c. supply. The magnitude of the voltage U_{SO} is defined by the magnitude of a. c. supply voltage, by the ratio of the number of series turns per phase and winding factors of the motor.

In order to achieve fast commutation processes the demand is for a small commutation reactance. In this case, it equals the leakage reactance of the motor. If it is kept small through adequate lay-out of the motor, an accordingly high current can be commutated, and consequently, a sufficiently high shaft torque can be reached (e. g. twice the rated shaft torque). Therefore the rotor can start from standstill even if there is a great load and/ or a great inertia torque.

That this can be achieved to a sufficient degree has been proved by means of a 25 kW output prototype motor.

If the converter is controlled in such a way (line-commutated rectifier and load-commutated inverter) that the direction of energy flow goes from the converter to the motor, the motor speeds up in its main direction of rotational speed. In this case, the rotor and the air-gap field rotate in the same direction. The stator frequency follows this equation:

$$f_1 = f_2 + p \cdot f_{mech} \qquad (1),$$

wherein f_{mech} signifies mechanical frequency, p the number of pole-pairs and f_2 the electrical rotor frequency. The stator frequency f_1 and, at the same time, the amplitude of the stator voltage show a linear rise with the speed of the motor (cf. Fig. 6). At the rated speed n_N the rated stator frequency f_{1N} as well as the rated stator voltage U_{SN} are reached. For speeding up the rotor in reverse direction (negative speed), the converter operation mode changes (line-commutated inverter, load-commutated rectifier). The direction of power flow reverses and is now directed from the motor through the converter to the a. c. supply. Now the rotor speeds up in the rotational direction opposite to that of the air-gap field. The stator frequency f_1 and the amplitude of the stator voltage decrease with increasing (now negative) speed. (cf. Fig. 6, $n < o$). The operating limit for this drive lies at the negative speed n_G, as regenerative braking mode of the motor (load-commutated inverter mode) leads to same commutation problems as they occur with the commutatorless d. c. motor with d. c. excitation at low speeds. This region will be avoided completely, and does, therefore, not belong to the operating range of this drive. This fact has also been considered in Fig. 5, in which the operating range ends with the negative speed n_G. This speed limit is established at fixed rotor frequency and rotor voltage through the lay-out of the motor. It lies within a range of about 5 – 10 % of the rated speed.

By changing the phase sequence of the rotor a. c. supply, additonal operating points in quadrants III and IV can be realized (cf. Fig. 5, operating region marked by dotted lines).

Figure 7 shows an oscillogram which was registered at a prototype motor, the upper part showing the phase winding voltage of the stator, the lower part showing the corresponding current. The motor is developing 1.6-times rated torque and is at zero speed.

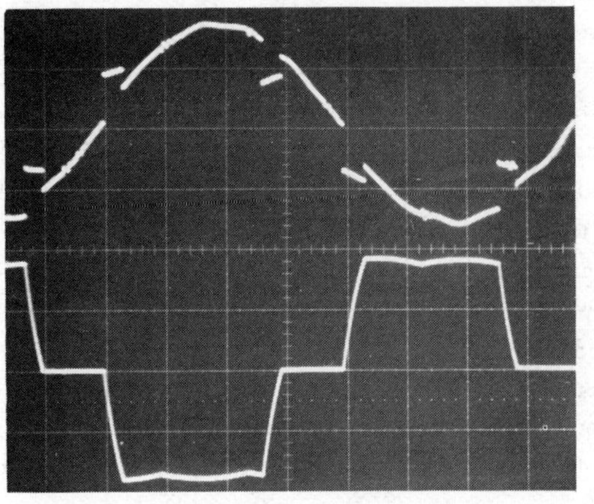

Fig. 7. Top: Phase winding voltage vs. time
Vert. 30 V/DIV; Horiz. 2,5 ms/DIV
Bottom: Phase winding current vs. time
Vert. 40 A/DIV; Horiz. 2,5 ms/DiV
Both oscillograms are made at zero speed and 1.6-fold rated torque

For the purpose of a safe commutation of the inverter at a two quadrant drive operation, this is the least favourable condition. The oscillogram shows convincingly that a perfect commutation process is guaranteed even at zero speed of the rotor.

Controlling Inverter Frequency by Stator Voltage

Controlling of the load-commutated inverter by stator voltage can be effected over the entire range of operation of the motor, since there is always a sufficiently high stator voltage available. For this purpose, the voltage zero-axis crossings are first picked off by means of a simple electronical circuit and then the firing of the corresponding thyristors is carried out.

As a result of this principle, a position transducer on the rotor shaft is not required. Thus, all problems are avoided completely that generally arise from the use of transducers on the shaft.

Exciting of the Motor

The rotor winding is connected to a poly-phase a. c. supply through slip rings. A 50 Hz three-phase current supply is most advantageous. Consequently line voltage, line frequency and therefore also magnetic flux in the motor are fixed. There are no other auxiliary devices necessary for motor excitation. At this operation mode the complexity of the power electronics is least. This is one of the basic requirements for an economical drive. For special applications, where there is no three-phase a. c. supply available for motor excitation (e.g. in applications for traction), a poly-phase supply has to be generated. This can be done by means of an additional converter, which can have a considerable smaller rated power than that of the converter on the stator side. Now there is a way of controlling the magnetic flux in the motor through changing the rotor voltage so, that a field weakening can be carried out.

In addition to that, exciting frequencies less than 50 Hz can be realized now, which creates favourable conditions for the lay-out of the motor. This shall be illustrated by an example. A poly-phase a. c. supply of 30 Hz is to be supplied through a converter on the rotor side. At a given rated speed of 1500 min^{-1} and a chosen number of pole-pairs p = 4 the stator rated frequency is 130 Hz. This frequency is completely acceptable for the motor as well as the inverter equipped with thyristors for line application.

The Apparent Power to be Conducted over the Slip Rings

The apparent power which has to be conducted over the slip rings is composed of the magnetizing reactive power for motor excitation, and of an apparent power P_L dependent on the operating condition of the load-commutated inverter.

At zero speed (n = o, the motor operates like a transformer) the apparent power P_L equals the apparent power of the load-commutated inverter. At his operating state, the a. c. supply exciting the motor supplies the required reactive power and harmonic power of the inverter. When the rotor moves, the reactive power required for the inverter can be additionally supplied by the motor.

The active power P_W resulting in the rotor is determined by the basic power division process in the rotor circuit of an induction motor. According to that, rotor active power P_W dependent on internal mechanical rotor power $P_{mech\ i}$ can be written as follows:

$$P_W = \frac{f_2}{p \cdot f_{mech}} P_{mech\ i} \qquad (2)$$

wherein f_2 represents the electrical rotor frequency, f_{mech} the mechanical shaft frequency, and p the number of pole-pairs.

Though the rotor active power P_W flows back to the a. c. supply (rotor losses shall be neglected), it must, however, be lead through the converter and the motor (and the slip rings). As a consequence, converter rated power and physical size of the motor are increased.

That is why rotor active power should be kept rather small at the rated point of the motor (at $f_{mech\ N}$). This can be achieved through a small ratio $f_2 : (p \cdot f_{mech\ N})$. If f_2 is fixed, (e. g. by excitation through a 50 Hz supply) the product $p \cdot f_{mech\ N}$ must be chosen accordingly, which means that at a fixed shaft frequency $f_{mech\ N}$ of the motor, these conditions have to be created by choosing the appropriate number of pole-pairs.

The product $p \cdot f_{mech\ N}$ does, however, also define the magnitude of the rated stator frequency (cf. equation (1)). The latter must not be chosen too high with regard to the iron losses in the motor stator and to the recovery time of the thyristors. This leads to an upper limit for $p \cdot f_{mech\ N}$. Both requirements on the ratio $f_2 : (p \cdot f_{mech\ N})$ have to be weighed and brought to a reasonable compromise.

To add an example, the frequency and power ratios of the 25 kW output prototype motor shall be given as follow:

The motor is excited from the rotor side by a three-phase a. c. supply. Consequently the electrical rotor frequency is 50 Hz. At a rated rotational speed of 3000 min^{-1} and a number of pole-pairs p = 6, the rated stator frequency is 350 Hz. Rotor active power P_W amounts to about 17 % of rated shaft power. This results in an apparent current – to be lead over the slip rings – of 11 A (total r. m. s. current).

A relatively small rotor active current (proportional to rotor active power P_W), is the main prerequisite for a small physical size of the required slip rings. Thereby the over-all length of the motor is also reduced.

Figur 8 shows the rotor of the above mentioned 25 kW prototype motor. The photograph shows the very small slip rings.

Fig. 8. Wound-rotor with very small slip rings. Prototype motor with 25 kW output

This drive has quite favourable conditions regarding brushes stress during no-load operation. In practice, there will be a ratio in the range of 0.4 to 0.7 between magnetizing current and rated rotor current. This leads to the fact that there is a sufficiently large current through the brushes during no-load motor operation (where P_L is small). This means that they are not subject to underloading at any motor operation point. Consequently brushes wear and their claims for maintenance are small.

The Harmonics of the Air-Gap Torque

Due to the non-continuously rotating ampere conductors per cm in the motor, harmonics of the air-gap torque arise as in the drive according to Fig. 3.

The first harmonic in the air-gap torque amounts to the six fold of stator frequency f_1, provided there is a three-phase inverter bridge. The lowest stator frequency results from a negative shaft frequency $f_{mech} = f_G$ (cf. equation (1)). It is supposed that at this point stator frequency is not substantially lower than the exciting frequency. Supposing a drop to 80 % of this exciting frequency, the frequency of the first harmonic of the air-gap torque under 50 Hz excitation amounts to 240 Hz. It does consequently, in most cases, differ sufficiently from the range of rotation speed frequencies critical because of torsional vibration of the mechanical system. That is why this drive has small amplitudes of harmonics in shaft torque even at low speeds. Thus, starting is very smooth. The user is provided with a drive which can also solve very demanding positioning problems.

Detection of Rotational Speed

In this motor, there is no need for a tachometer for measuring shaft rotational speed. Thus, the dynamic or possible spatial problems caused by a detection of rotational speed at the shaft, do not exist.

According to equation (1), a signal proportional to rotor frequency f_{mech} can be obtained by subtracting rotor frequency from stator frequency. These values can be detected through simple and reliable electronic circuits – disconnected locally from the motor. The subtraction of the frequencies is realized digitally with consequent accuracy.

The necessary information for positioning the rotor can be gained from a comparison between stator- and rotor frequency. There is no need for a special position transducer on the rotor shaft. As a consequence, this drive always has two free shaft-ends which are at full disposal for drive uses.

The Natural Performance Characteristics

The load-commutated inverter and its gating circuits controlled by stator voltage have the effect of an electrical, non contacting commutator.

Under excitation at a fixed frequency the motor shows operating characteristics which are similar to those of a d. c. shunt motor (including the field weakening possible in both cases).

Speed control is effected by varying the rectifier delay angle, since this alters the direct voltage delivered to the solid state commutator. Loading of the motor leads to a speed drop, according to its shunt characteristic. Due to inverter controlling by stator voltage, there are no stability problems. (Stator frequency f_1 of the motor is a function of the shaft frequency, according to equation (1). There is no external reference oscillator. The commutatorless d. c. motor with three-phase current excitation is definitly no doubly-fed wound-rotor induction motor which presents the well-known stability problems).

The 25 kW Prototype Set

The characteristics of the commutatorless d. c. motor with three-phase current excitation quoted in this article have been proved by measurings and tests on a 25 kW prototype motor.

The set construction can be seen in Fig. 9. The new drive (right-hand side of the photograph) is connected to a converter-fed loading machine (d. c. motor, left).

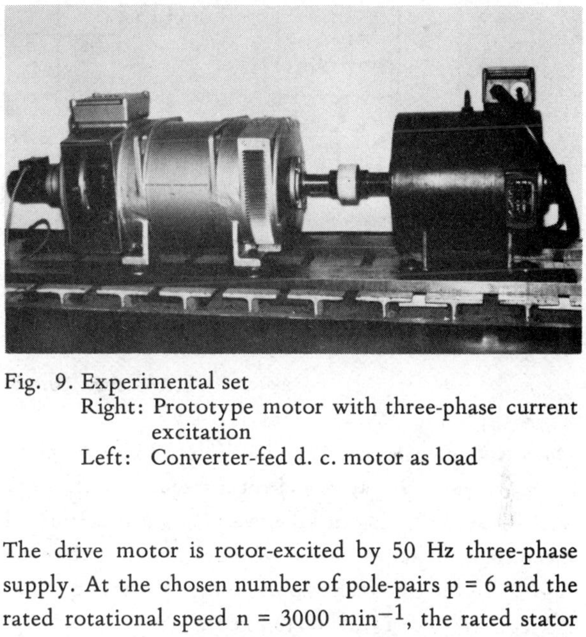

Fig. 9. Experimental set
Right: Prototype motor with three-phase current excitation
Left: Converter-fed d. c. motor as load

The drive motor is rotor-excited by 50 Hz three-phase supply. At the chosen number of pole-pairs p = 6 and the rated rotational speed n = 3000 min^{-1}, the rated stator frequency is 350 Hz. The rated shaft power is 25 kW at an efficiency of $\eta = 0.89$. At rated torque the active power flow back to the a. c. supply through the slip rings is about 17 % of the rated shaft power. To achieve the smallest possible leakage reactance of the motor, its rotor was not unskewed (cf. Fig. 8). Sufficiently fast commutation processes are achieved with a leakage reactance per unit of 0,1.

The maximum negative speed (without changing the phase sequence of the rotor a. c. supply) is 6 % of the rated rotational speed. The main direction of speed of the drive can be chosen freely by changing the phase sequence of the rotor by means of SCR-circuits.

Speed Control System

Two special characteristics of the new drive can be used to full advantage when planning the lay-out of a control system.

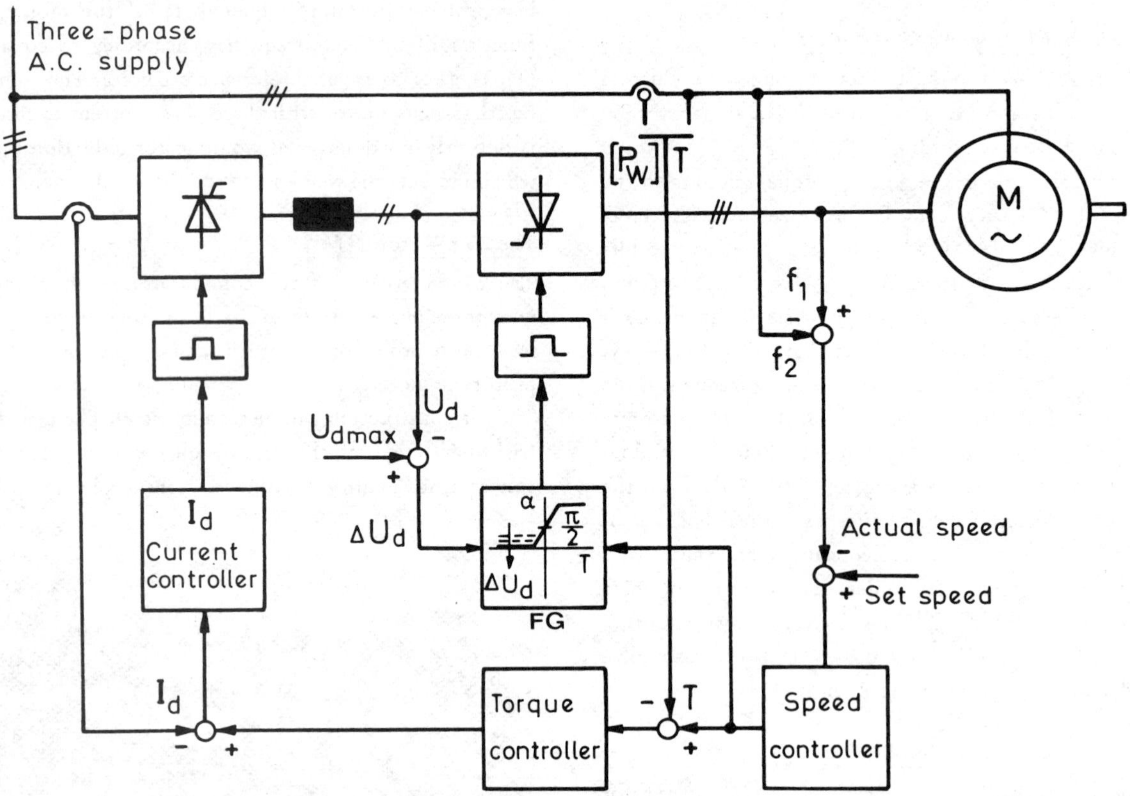

Fig. 10. Speed control system

One of them is above mentioned possibility of speed detection through stator- and rotor-frequency. The other is the easy measuring of effective motor torque through measuring the active power on the rotor side (cf. equation (2), the internal motor torque is the quotient $P_{mech\,i} : \omega_{mech}$; f_2 and p are fixed).
This even registers the influence of a possibly effected field weakening on the motor torque automatically.
Given the possibility of torque measuring, an underlying torque control loop can now be introduced. As an example for one of the many possible variations, Fig. 10 shows the block diagram of a speed control system which makes full use of all the listed characteristics of the motor.
In addition to the outer speed loop, there are two more underlying control loops, one for d. c. link current I_d and the other for motor torque T. Besides, the output signal of the speed controller effects the necessary changing of the operation mode of the load-commutated inverter during transfer from motor- to regenerative braking mode and vice versa. This is done by a function transmitter FG, which is also used to prevent d. c. link voltage from exceeding a fixed magnitude $U_{d\,max}$. This also guarantees that the line-commutated rectifier can always control the d. c. link current.

APPLICATION FIELDS

There are some examples for possible applications of the new drive:

- drives for veneer peeling installations (power about 100 kW to 300 kW) this kind of drive must permit a dynamically high-quality two quadrant operation)
- drives for test beds in a wide variety of industial fields (power 50 kW to 400 kW)
- drives for centrifuges (e. g. as they are used in the sugar-refining industry)
- pull-controlled drives for take-up and feed-spools in paper-making and sheet metal industries (this requires the control of a wide range of manipulated pull)
- a. c. traction drives (e. g. urban traction vehicles)
- drives for industrial cranes (which must be able to reach a position in optimal time and without shock)
- drives for transfer cars and big converters in steel mills (with high demands on positioning)
- main spindle drives for machine tools with possibility for positioning of the tools (e. g. automatic tool changing)

- wide total range of speed adjustment (e. g. 3000 : 1)
- high operation speeds (they are only limited by the stability of the rotor winding)
- small amplitudes of shaft torque harmonics even at low and zero drive speeds
- a large zero speed torque must be maintained for a considerable period of time
- commutatorless motor
- lowest possible expenditure of converter equipment (favourable power/weight – and power/space-ratio)
- motor shall be nearly maintenance-free
- no transducers on rotor shaft needed for rotational speed and rotor position detection
- simple motor torque measuring possibility
- considerably smaller over-all length than motor with commutator

REFERENCE

A. Boehringer, Stromrichter-Synchronmaschine mit Drehfelderregung, Deutsche Offenlegungsschrift Nr. 2413266 vom 2. 10. 75.

NOTE

The research-works are being sponsored by the „Deutsche Forschungsgemeinschaft".

- drives for the sheet metal industry (where metal slips must be manufactured continuously at a defined prestress. A shaft torque with a small amplitude of harmonics is required with regard to a mechanically induced vibration of the metal strip).

SUMMARY OF DRIVE CHARACTERISTICS

The commutatorless d. c. motor with three-phase current excitation represents a drive for versatile use. Its range of power from about 5 kW to over 1 MW makes it fit for a wide range of applications. Its use is especially favourable whenever one or more of the following typical characteristics of this drive are required:

- there is only one direction of rotation; it requires motor and regenerative braking modes; favourable speed dynamic
- there is one main direction of rotation; only small rotational speeds with favourable speed-dynamic in reverse direction (four quadrant operation, which is necessary for positioning)
- there are both positive and negative directions of rotational speed with a wide total range of speed adjustment (this can be realized through admitting a short pause with zero torque for changing the phase sequence of the rotor a. c. supply during the required changing of rotational speed direction)

HYBRID LINEAR ELECTRIC DRIVE FOR INDUSTRIAL APPLICATIONS

M. Hájek, O. Roubíček and Z. Zežulka

VÚSE-Research Institute of Electrical Engineering, Prague 9-Běchovice, Czechoslovakia

ABSTRACT

The article elucidates the concept and structure of a new linear drive for mixed, continuous and stepped motion. It characterizes the properties of a linear hybrid motor of both the induction-reluctance and the pure induction type. It includes connections diagrams, describes the principles of operation of power converters for the drive, indicates the measuring methods used in its control circuits and the ways and problems of controlling its operation. Research prototypes of the new drive are described and suggestions for its application presented.

INTRODUCTION

In the technique of industrial manipulation and transport we frequently require a rapid linear dislocation of loads and their rigid positioning at predetermined places. New prospects are offered by the use of the hybrid linear electric drive. It has the advantages common to all types of linear electric drives (Ref. 1, 2 and 3).

STRUCTURE OF THE DRIVE

The drive is based on the use of an electronically controlled hybrid linear motor which fenerally is constructionally integrated with the driven working mechanism. The motor combines the features of a continuous motion linear induction motor and a reluctance-type or induction-type linear stepping motor for positioning. The operation in the continuous mode with high tractive forces is used for the acceleration and braking of inertial masses and for a rapid transport path travel. The stepping mode follows up the continuous motion braking phase and provides for taking the desired rigid position. The position rigidity is obtained by a mere fixing action of the motor without any external positional feedbacks.

The operation of the drive is automatically controlled to obtain the desired speed or the continuous-motion acceleration and the desired position. Fig. 1 will help to interpret the operation as exemplified for the case of material handling. The material is distributed on ramp 4 and is

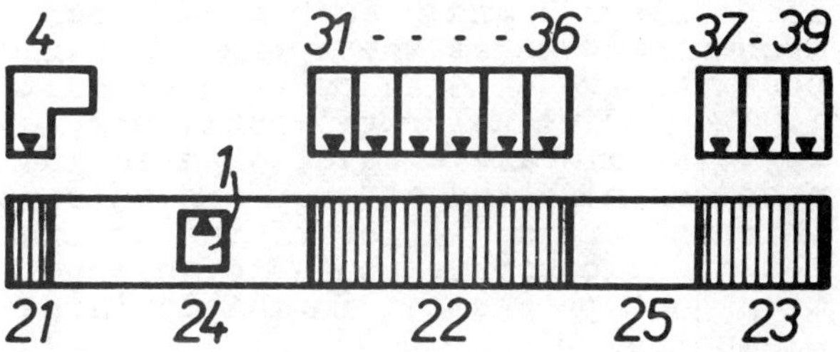

Fig. 1. Example of application.

to be transported by conveyer 1 to racks 31-39 according to its kinds. Conveyer 1 connected with the moving part of the motor travels along path 2 described by the fixed part of the motor. This part consists of sections 21-23 where conveyer 1 must provide rigid positions with respect to sections 4 and 31-39, and of sections 24 and 25 where a rapid and only continuous travel of conveyer 1 is required. The continuous motion is possible along the entire length of path 2, the positioning only in sections 21-23.

The basic block diagram of the drive is shovn in Fig. 2. The continuous-mode operation of motor 1 is controlled by the continuous-mode controller 2 according to the desired values v_w and a_w of the continuous speed $\underline{v}$, respectively, and the acceleration $\underline{a}$ of the driven mechanism 4. The speed $\underline{v}$ is measured directly by sensor 3 and the acceleration $\underline{a}$ is evaluated from it in controller 2. The control of the stepping-mode and positioning is provided by the stepping mode controller 5. Information on the steps performed and on the position orientation are evaluated from

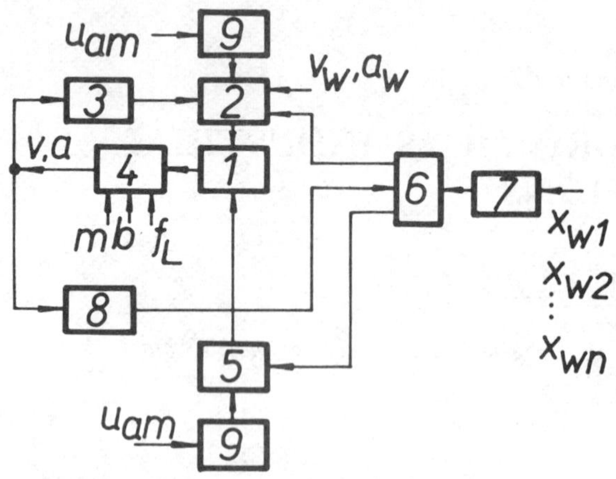

Fig. 2. Basic block diagram.

the signal of code-mark sensor 8 at the position $\underline{x}$ of mechanism 4 in the decision element 6. This element switches the control of motor 1 to controllers 2 and 5 such a way that mechanism 4 takes the optimum desired positions x_w entered by program unit 7. The individual rigid positions x_{w1}, x_{w2}, ... x_{wn} are always taken in the stepping mode and at such instants at which a zero deviation of the controlled position corresponds to the position just programmed in the decision element 6.

The drive is energized from mains 9 and the main control disturbances affecting its operation are the following: variations of the mains voltage amplitude u_m and changes in the inertial mass $\underline{m}$ and passive resistors $\underline{b}$ of the working mechanism. To these cases sometimes add changes in the active forces f_L of the load.

PROPERTIES OF THE MOTOR

General aspects

The electromagnetic circuit of the hybrid linear motor is common for both oterating modes and is usually fully utilized in both of them. The circuit, therefore, constitutes a physically integrated hybrid motor unit. Its design is based on the continuous-motion linear motor.

In the continuous mode the principle of the linear induction motor is directly used. The generation of the stepped motion and the action of the fixing tractive force in the positions taken is based on the formation of a nonhomogeneous, magnetically orientable reactive part of this motor. There are two possibilities that may be taken into consideration. The first is the application of the reluctance principle leading to a mixed-type of motor (designated in the following text as the IR type). The second is the use of induction activation of the reactive part of the motor which in this case will be a pure induction motor (designated as type II).

The IR type of motor

The active (energized) part of this motor wholly corresponds to the known active (flat of tubular) parts of the usual linear induction motors. The reactive (not energized) part is positionally orientable by a slotted ferromagnetic, the number of slots per pole being by one larger or smaller than for the active part. In this way a nonius-like effect producing the stepping capability of the motor is obtained. The slot pitches τ_{d1} and τ_{d2} of the active and reactive part of the motor, respectively, are related by the expression

$$\tau_{d2} = \tau_{d1} m_1/(m_1 \pm 1), \qquad (1)$$

where m_1 is the number of phases of the active part. For the length of step we then obtain

$$\Delta = \tau_{d1} - \tau_{d2} = \tau_{d1}/(m_1 \pm 1). \qquad (2)$$

For short-step motors the alternative with the positive sign is chosen and vice versa.

The stepping based on the reluctance principle is obtained by exciting every m_1-th tooth and switching the excitation cyclically in m_1 steps. This is achieved by energizing two phases of the winding of the active part (Fig. 3) at a time by direct current. The method of feeding the most

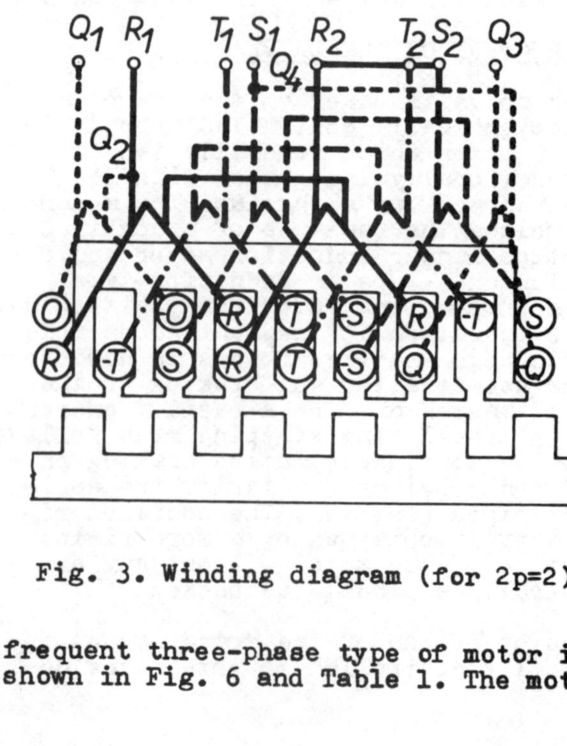

Fig. 3. Winding diagram (for 2p=2).

frequent three-phase type of motor is shown in Fig. 6 and Table 1. The motor has three phase windings with the terminals R, S, T (Figs. 3 and 6) and the ends of its active part are provided with compensations coils (with the terminals Q) compensating for the excitation unsymmetry at the edge regions.

The stepping-mode static tractive force F as a function of the mutual position x of the active and reactive parts of the motor is shown in Fig. 4.

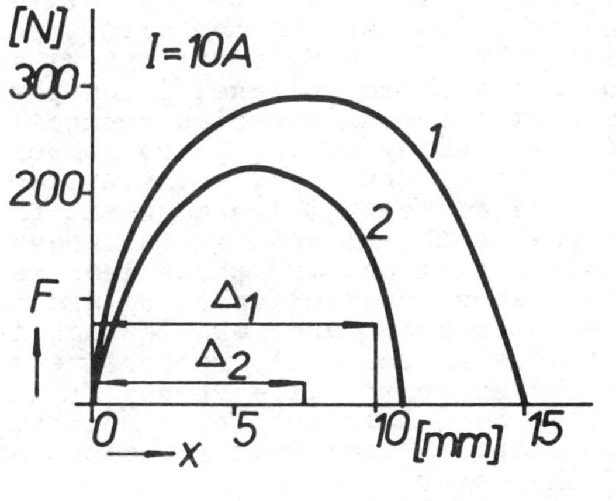

Fig. 4. Static stepping characteristic of motors shown in Figs. 14 (1) and 15 (2).

The reactive part of the flat-one-sided motor is designed in the form of a rack of ferromagnetic with a bar cage used for the continuous mode. As regards the two-sided variant the reactive part consists of a band made of compact ferromagnetic with teeth in the form of bars separating the slot openings (Fig. 14). The continuous-mode operation may be improved by inserting blocks of highly conductive material into the slots and connecting them longitudinally along both ends to form a bar cage. During the stepped motion the cage increases the damping effect.

The tubular motor has its reactive part in the form of a ferromagnetic bar with peripheral slots. The bar is plated with a highly conductive material (Fig. 15), which acts similary to the cage in the continuous mode. In the stepping mode it behaves a damper.

<u>The II type of motor</u>

This alternative type of motor is reasonably practicable only as a flat model. The positional orientability of the reactive part is provided by single-turn coils with a pitch equal to the pole pitch. The stepping effect occurs with single-phase a.c. feeding of the active part of the motor by the action of currents induced in the coils of the reactive part. The cyclic connection of the winding phases of the active part to a.c. voltage produces a stepped motion having the step length

$$\Delta = \tau_p / m_1, \quad (3)$$

where τ_p is the pole pitch.

In this case the step is longer (about 2 to 4 times longer than for the IR type) so that it is practical to refine it by intermediate positions. If an intermediate cycle is introduced into the switching schedule between the switching from the phase just excited and the following intermediate position, consisting in a simultaneous connection of both these phases (one of them with opposite polarity to that for the continuous operation in the given direction) the length of step will be reduced, according to (3), to one half. The step may be additionally reduced by extending the intermediate switching cycle to several graded ratios of currents flowing in the two phases.

As shown in Fig. 5 the mean stepping-mode static tractive force F in dependence on the mutual position x of the active and reactive part of the motor, is less favourable than

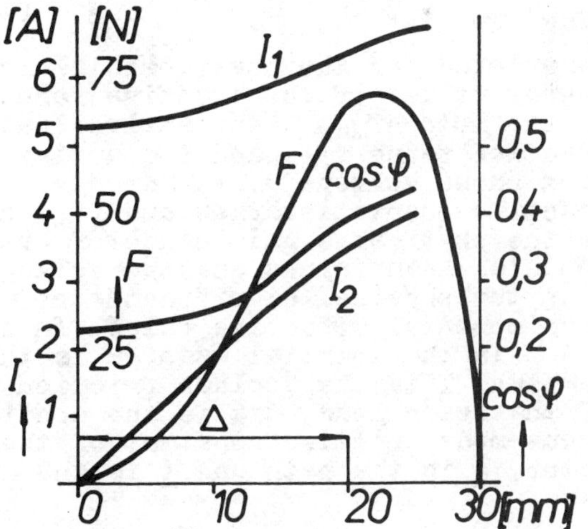

Fig. 5. Static stepping characteristic of the motor shown in Fig. 16.

for the IR type. The diagram corresponding to harmonic voltage feeding also includes curves for the effective active part phase current I_{f1}, the reactive part phase current I_{f2} reduced to the active part and the power factor cos φ.

The constructional design of the reactive part of the two-sided type of motor is shown in Fig. 16. A band of highly conductive material is provided with rectangular openings separated by bars having the pitch τ_p/m_1. The bars at the pitches τ_p connected longitudinally by the material of the band edges constitute the above mentioned single-turn coils corresponding to the winding of each phase of the active part. In the continuous operation the reactive part behaves as a cage.

The one-sided motor requires to underlay the reactive part with a ferromagnetic plate.

<u>Dynamic equations of the motors</u>

Considering the character of operation of hybrid linear drives it may be assumed that the duration of the mechanical transient states is substantially longer than of the electromagnetic transient effect in the motors. Taking this into account it will be sufficient to consider in the continuous mode the electromechanical relation

$$M \frac{d^2x}{dt^2} + B \frac{dx}{dt} + T \operatorname{sign} \frac{dx}{dt} + f_{Lav} = f_{Mav}, \quad (4a)$$

where

$$f_{Mav} = K_{FM}\, u_{1ef}^2 , \quad (4b)$$

formulated for the quasimomental integral values of the tractive force of the motor f_{Mav}, the active load tractive force f_{Lav} and the active part phase voltage u_{1ef}. Here the tractive quantities f_{Mav} and f_{Lav} have the physical significance of quasimomental mean values and the voltage u_{1ef} the physical significance of a quasimomental effective value. In eqn. (4) M is the inertial mass, B is the damping, T is the Coulomb friction of the motor load, K_{FM} is the continuous-mode tractive constant of the motor, <u>x</u> is the path and <u>t</u> is the time.

In the stepping mode, more detailed descriptions are unavoidable because we are concerned with a quasistationary effect comprising a set of transient processes. The state at the end of each step defines the initial conditions for the next step.

The basic system of equations for the instantaneous values of the quantities of the IR type of motor is given by the relations

$$u_1 = R_1 i_1 + L_{1\sigma} \frac{di_1}{dt} + w_1 n \frac{d\phi}{dt}, \quad (5a)$$

$$0 = R_2 i_2 + L_{2\sigma} \frac{di_2}{dt} + w_2 \frac{d\phi}{dt}, \quad (5b)$$

$$M \frac{d^2x}{dt^2} + B \frac{dx}{dt} + T \operatorname{sign} \frac{dx}{dt} + f_L = f_M , \quad (5c)$$

$$f_M = k_{FIR}\, i_1^2 \sin(2\pi x/\tau_{d2}). \quad (5d)$$

The subscripts 1 and 2 relate the variables in eqns. (5) to the active and reactive part of the motor, respectively. In these equations u_1 denotes the phase voltage, <u>i</u> the phase current assuming a series connection of the winding coils, <u>w</u> the number of turns of one coil, <u>n</u> the number of excited teeth, ϕ the magnetic flux in one tooth, R_1 and $L_{1\sigma}$ the phase ohmic resistance and phase leakage inductance, respectively, R_2 and $L_{2\sigma}$ the ohmic resistance and leakage inductance of one coil, respectively, f_M the motor tractive force, f_L the active load tractive force and k_{FIR} the stepping-mode tractive constant of the motor.

With regard to the arrangement of windings the usual value of <u>n</u> in eqn. (5a) is n=2p where <u>p</u> is the number of pole pairs whereas in eqn. (5b) describing the effect of the equivalent damping winding of the reactive part, we generally consider that w_2=1.

In solving the system (5) for i_1, i_2, f_M and <u>x</u> we substitute into eqns. (5a) and (5b) the expression

$$\frac{d\phi}{dt} = w_1 \frac{di_1}{dt} + w_2 \frac{di_2}{dt} [\varepsilon + \psi \cos(2\pi x/\tau_{d2})] + (w_1 i_1 + w_2 i_2)(2\pi\psi/\tau_{d2}) \cdot \frac{dx}{dt} \sin(2\pi x/\tau_{d2}) \quad (5e)$$

where ε and ψ are partial magnetic conductances of the motor air gap defined as $\varepsilon + \psi = \Lambda_{max}$, $\varepsilon - \psi = \Lambda_{min}$, where Λ_{max} and Λ_{min} are the maximum and minimum magnetic air-gap conductances, respectively, within one step.

For the stepping-mode of the type II motor it is possible to establish the system of equations

$$u_1 = R_1 i_1 + (1 + \sigma_1) L_1 \frac{di_1}{dt} + \alpha L_1 \frac{di_2'}{dt}, \quad (6a)$$

$$0 = R_2' i_2' + (\alpha + \sigma_2) L_1 \frac{di_2'}{dt} + \alpha L_1 \frac{di_1}{dt}, \quad (6b)$$

$$f_M = k_{FII}\, i_1 i_2' \qquad (6c)$$

to which adds the electromechanical relation (5c). In eqns. (6) the symbols σ_1 and σ_2 denote the coefficients of inductive leakage of the active and reactive part, respectively, L_1 the main inductivity of a phase of the active part for the lowest inductive coupling with the reactive part, k_{FII} the stepping-mode tractive constant of the motor and α the relative path expressed as

$$\alpha = 2x/\tau_p \ . \qquad (6d)$$

The primes of i_2' and R_2' denote that the corresponding quantities of the reactive part refer to the active part.

Even in this case the system (6) and (5c) allows to solve the time dependence of the variables i_1, i_2, f_M and $\underline{x}$.

The calculation using eqns. (5) or (6) are most conveniently programmed for a hybrid computer.

POWER CONVERTER OF THE DRIVE

General aspects

The power converter is an active member providing for the feeding and control of the motor in both operating modes.

Considering its physical properties the motor shows to have a resistive shape of the continuous-mode mechanical characteristic. It is, therefore, very well suited for a simple voltage control of the continuous speed. This may be used to advantage in feeding the motor from a symmetric m-phase (usually three-phase) system of harmonic voltages through a reversing connection of triacs, or antiparallel pairs of thyristors in each phase. The star-connected motor with a brought-out neutral point (for simplicity in the stepping mode) has in this case a relatively little distorted terminal voltage containing only odd harmonics. A wide control range at good stability may be obtained.

The motor is rather sensitive to the content of the higher harmonics in the supply voltage spectrum. That is to say, a hybrid type tends to develop parasitic tractive components which are strongly supported by the harmonically distorted supply voltage and deteriorate the continuous-mode control stability.

For the control of the stepping mode of the motor the converter is equipped with a digital part in the form of a d.c. or a.c. sequence thyristor switch (for the type IR or type II motor). The switch ensures pulse feeding of the motor windings in the stepping mode and steady current feeding in fixed positions.

Converter for the IR type of motor

The basic connection diagram is shown in Fig. 6. The motor windings Q_1 through T are denoted in accordance with Fig. 3.

The continuous operation control of the motor is effected by triacs T_1 to T_5 phase-controlled by the control pulses u_{g1} to u_{g5} of pulse generator 1 according to the signal u_v from the continuous-speed controller. The logic signal Y of the motion direction selector determines the instants of reversal of the triac unit. Before transition to the stepped motion the continuous control is locked by the logic signal U_{b1} from the motion direction selector and the triacs T_1 to T_5 separate the a.c. mains supply from the motor windings.

The stepping mode is controlled by a sequence switch consisting of power switching units SQ_1 to ST and controller 2. The power switching units controlled by the output logic channels q_1 to t of controller 2 switch the d.c. current source 4 to the individual motor windings. Before transition to the continuous mode the power units SQ_1 to ST are locked by the logic signal U_{b2} of the motion direction selector.

The structure of the power switching units is apparent from the detailed diagram of the block ST. It represents a usual switching connection of the thyristors T_h and T_p ; thyristor T_b is a blocking element for the separation of d.c. in continuous operation.

Controller 2 controls the switching of the current from source 4 to two winding phases at a time (according to Table 1). This process takes place at a frequency f_{d2} (Fig. 6) applied from the output from the output of the stepping frequency generator. The current distribution must be ensured according to the positional states of the active part of the motor with respect to the reactive part. In the case of the usual three-phase motor there are three states characterized by the switching data according to Table 1. The corresponding switching schedule may be constructed by assigning the combination of the values of two logic variables A and B to each state, so that the

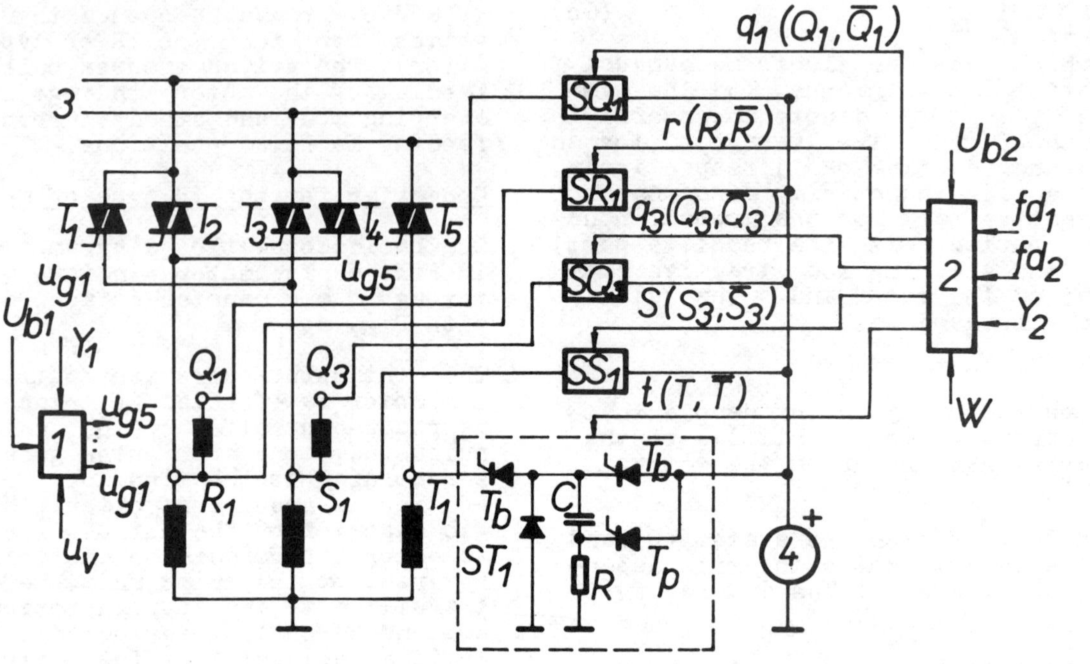

Fig. 6. Converter used for feeding the IR type of motor.

Table 1 Switching and state schedule for the control of type IR motors

			Winding terminal											
			Switching					Output logic variable						
			Q_1	Q_3	R_1	S_1	T_1	Q_1	Q_3	R_1	S_1	T_1	A	B
Stepping mode	State	1	+	+				1	1	0	0	0	0	0
		2				+	+	0	0	0	1	1	0	1
		3			+		+	0	0	1	0	1	1	0
Continuous mode					~	~	~							

individual windings (Fig. 6) will be identified by the terminal codes

$$Q_1 = \bar{A}\,\bar{B}\,,$$
$$Q_3 = \bar{A}\,\bar{B}\,,$$
$$R_1 = A\,\bar{B}\,, \qquad (7a)$$
$$S_1 = \bar{A}\,B\,,$$
$$T_1 = \bar{A}\,B + A\,\bar{B} = S_1 + R_1\,,$$

and with regard to the circuit realization of the last relation by the NAND type of elements we used the form

$$\bar{T}_1 = \overline{S_1 + R_1} = \bar{S}_1\,\bar{R}_1\,. \qquad (7b)$$

The combinations (7) may be realized by a modulo "3" bi-directional synchronous counter whose configuration of the intrinsic variables A and B is changed by clock pulses according to Table 1. Thus the conditions for a synchronous stepped motion in one direction are established. A change in direction implies a reversal of the state sequence and is executed under the logic stimuls Y_2 (Fig. 6) by a reverse switching program.

The state orientation of controller 2 in continuous operation is maintained by the input f_{d1} with a logic reference signal from the sensor of the positional orientation of the drive. The recurrence frequency f_{d1} of this signal is usually much higher than the control frequency of the stepped motion (depending on the speed difference of both motional modes).

The basic diagram of the controller

generating the relation (7) is shown in Fig. 7. The outputs A to $\bar{B}$ of the

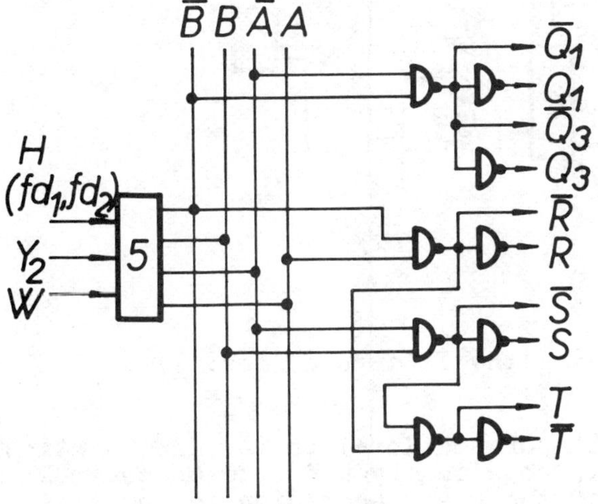

Fig. 7. Connection diagram of the controller.

modulo "3" bi-directional synchronous counter are connected to a logic network producing the signals $\bar{Q}_1$, Q_1 to $\bar{T}$, T for the control channel q_1 to t in Fig. 6. The inputs H or Y_2 of counter 5 (Fig. 7) are used to introduce two types of clock pulses, or the logic information on the direction of the stepped motion; input W serves for resetting the counter to the initial state corresponding to the reference positional state of the motor.

The starting circuits of power switching units SQ_1 to ST of Fig. 6 are connected as shown in Fig. 8. This

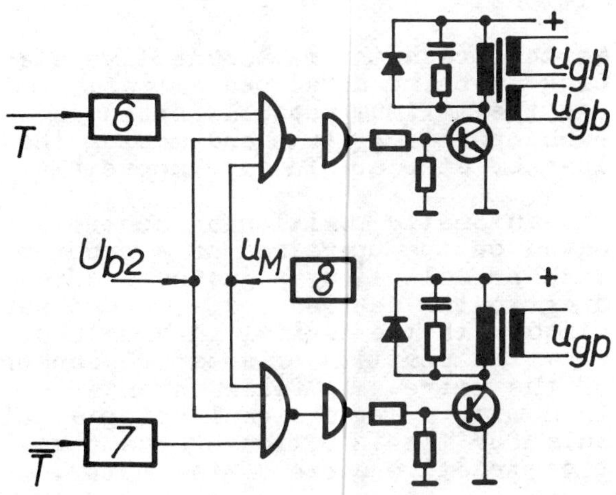

Fig. 8. Connection diagram of the starting circuits.

figure gives the connection for starting the unit ST drawn in detail in Fig. 6, controlled by channel t with logic signals T and $\bar{T}$ (Fig. 8). The front edges of these signals release the monostable flip-flop circuits 6 and 7 generating the pulses u_{gh}, u_{gp} and u_{gb} for the thyristors T_h, T_p and T_b in Fig. 6. The frequency modulation of these pulses is provided by the signal u_M of multivibrator 8 (Fig. 8). The input U_{b2} is uded to introduce the locking input of the motion direction selector on transition of the motor to continuous motion.

Converter for the II type of motor

The continuous operation of the motor is voltage controlled in the same manner as teh IR type. The stepping mode requires a cyclic connection of one phase at a time to a.c. voltage. The basic block diagram of the converter providing for the mentioned functions is shown in Fig. 9, the corresponding switching schedule is given in Table 2. The symbols in Fig. 9 correspond to those of Fig. 6; the control methods and the control circuits connections for the range and channels used are the same as in the case of the converter for the IR type of motor. The difference is mainly in the arrangement of generator 1 producing the pulses u_{g1} to u_{g5} for the triacs T_1 to T_5, which is in operation for both operating modes. Its inputs 11, 12 and 13 are switched by switch 9 either to the control of the signal u_v from the continuous-speed controller or to the channels q_1, s and r of the stepping mode controller 2 (reduced analogy of the diagram in Fig. 7).

MEASUREMENTS IN THE CONTROL CIRCUITS

The continuous speed is measured by means of an analogue or a digital rotary tachogenerator with a linear-to-rotary motion converter. It is usually constituted by a friction gearing with an intermediate gear whose driven pulley rolls on the reference surface of the path. If slip cannot be avoided a finer Gall s chain, a rack or a measuring gear is placed on the reference surface and the motion is transmitted to the tachogenerator by means of a toothed wheel.

The measurement of the path length required to ensure the positional orientation of the drive is digital.

In less exacting cases the path is sampled by means of relay sensors with addres codes. The sensors are equipped with a starting element carried by the moving part of the motor. However, they permit positioning only at the places where they are mounted.

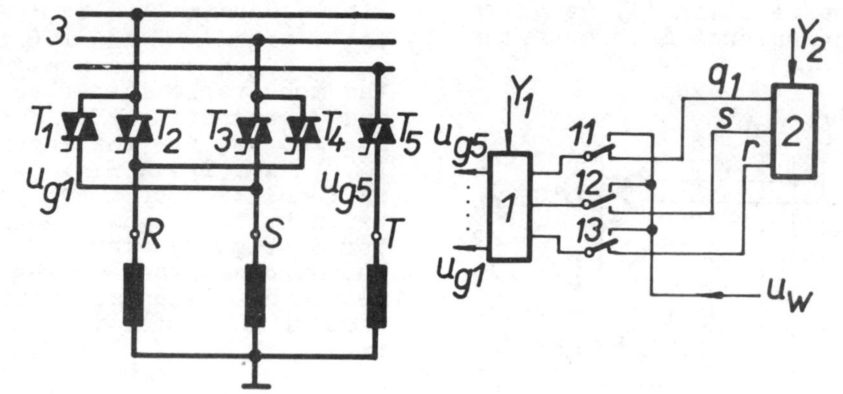

Fig. 9. Converter used for feeding the II type of motor.

Table 2 Switching schedule for the control of type II motors

			Winding terminal		
			R	S	T
Stepping mode	State	1	~		
		2		~	
		3			~
Continuous mode			~	~	~

A universal and centrally programmable positioning can be realized only by means of a continuous sampling by code marks indicated by the sensor carried by the moving part of the motor. Only the contactless (inductive or photoelectric) indication has proved to be successful. The path is evaluated by a counter from the number of pulses corresponding to the number of code marks indicated. From the instantaneous pulse frequency it is also possible to evaluate the instantaneous speed of motion by means of a frequency-to-voltage converter.

When the motion is transmitted by means of a toothed wheel it is under certain conditions possible to use the same method for the evaluation of the path and speed from the signal of the digital rotary tachogenerator.

CONTROL OF THE DRIVE OPERATION

Continuous motion control

The control is the same as for linear continuous-motion induction drives. The automatic speed control to the desired value is provided by an integral action analogue controller (usually of the PI type). The desired value of acceleration (or deceleration) is obtained by a follow-up control of the speed according to the linearly programmed time dependence of the change in its desired value.

The direction of motion is controlled by the logic signals from the motion direction selector which is part of the plant programming system.

Stepped motion control

In the stepping mode is used the control of the stepping frequency, i.e. the rate of the stepped motion, and the control of the positioning.

As already stated the stepping is synchronous, controlled by an independent stepping frequency generator. This frequency is determined from the analysis of a mathematical model of the drive on a hybrid computer to the end that the step may not be lost even under the most unfavourable conditions.

At the present time an adaptive stepping is being developed allowing to use the maximum possible speed in each operating state and making the loss of step absolutely impossible.

The automatic positioning control is based on the operation of a counter with preselection, adjusted according to the desired position and supplied with the digital information from the position orientation sensor of the drive. The filled counter transmits a logic signal being a stimuls for the initation of the fixation tractive force of the motor.

The transition from the continuous to the stepping operation is effected on the basis of a logic signal from the position control block of the drive. The signal is transmitted at the moment of approach within a given number of steps to the desired position.

The synthesis of the stepping controller of the drive is a complex problem since the controlled system is discontinuous and nonlinear in character and has variable parameters. Practically the only rational procedure is to use a hybrid computer. The calculations start from the solution of the electromechanical equation (5c) with the variable initial conditions $x(i_0)$ and $\dot{x}(i_0)$ at the beginning of every i-th step. The tractive force of the motor, f_M, is determined as the function

$$f_M = \varphi(x, t, i) \quad (8)$$

according to the pathwise distribution of the tractive force along the active part and to the switching sequence of the individual phases of the motor winding in each i-th step. The tractive force path dependence has the form of a periodic function being odd with respect to the pole axis of one phase of the winding and having the period $(3/2)\tau_p$ for a three-phase motor. Its form depends on the design of the motor and on the required degree of approximation to reality. For the other two phases these functions are displaced by multiple of the step length Δ (i.e. by Δ or 2Δ). To avoid modelling the entire three-phase system the transition to the mentioned two further functions at times corresponding to the stepping frequency is solved by a transformation of the path $\underline{x}$ into the definition interval of the basic function in dependence on the sequence $\underline{i}$ of the step.

The simplified diagram of the hybrid model for B=0 showing the division of the calculation algorithm into the analogue and the digital part is illustrated in Fig. 10. The analogue integration is controlled by the signals TA, TB, TC of a system of independent time sources and the continuous integration after the last step is ensured by the signal D. Information on the stepping process is transfered to the digital part through the sense line. The digital program allows a change of the parameters and initial conditions in eqn. (5c), a choice of the number of steps and a desired change of the active tractive force f_L of the load before an arbitrary step. The valid dependence of the tractive force of the motor, f_M, on the path $\underline{x}$ in the definition length region of its active part is written as a subroutine called in the hybrid loop of the algorithm. The model shown in Fig. 10 is used for finding the stability region of the stepping mode and for the determination of the optimum stepping frequency for various values of the parameters and the tractive force f_L in eqn. (5c), for the determination of optimum initial conditions of transition from the continuous to the stepping mode and for the ascertainment of the reproducibility of the desired position.

Synchronization of transition

The synchronization will be exemplified by a drive using the IR type of motor.

To prevent a position error (due to loss or rise in step) the controller must be at any moment correctly state oriented according to Table 1. In the continuous mode this is done by the

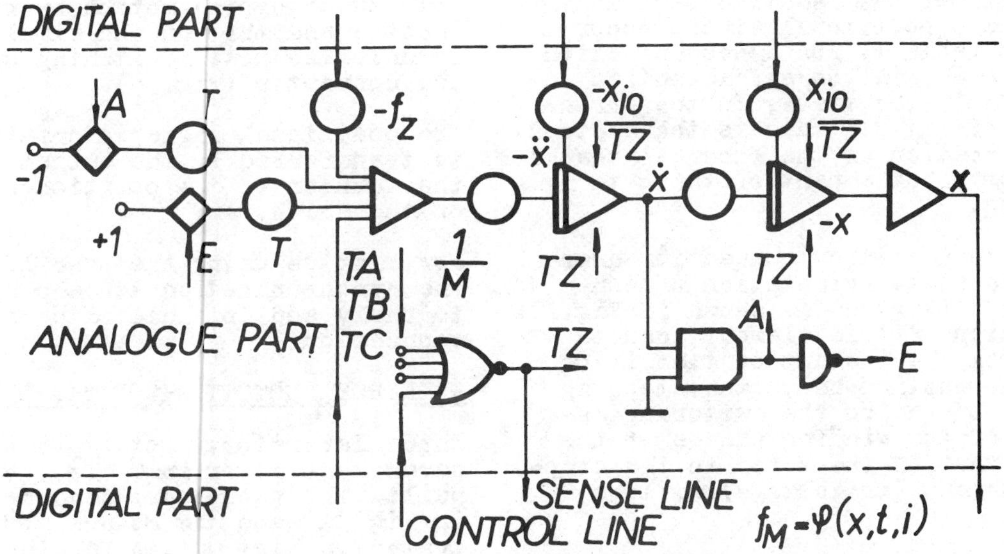

Fig. 10. Basic hybrid model of the drive.

reference logic signal f_{d1} controlling the combinations of the intrinsic variables of counter 5 (Fig. 7) through the clock pulse input H. The logic signal f_{d1}, transmitted by the sensor of the positional orientation of the drive, contains the logic information C and D, satisfying the state conditions according to Table 3.

Table 3 State schedule of path sampling

State	C	D
1	1	0
2	1	0
3	0	1

The way of obtaining them by sensing equal code marks of the path is shown schematically in Fig. 11. This applies to a three-phase motor with the

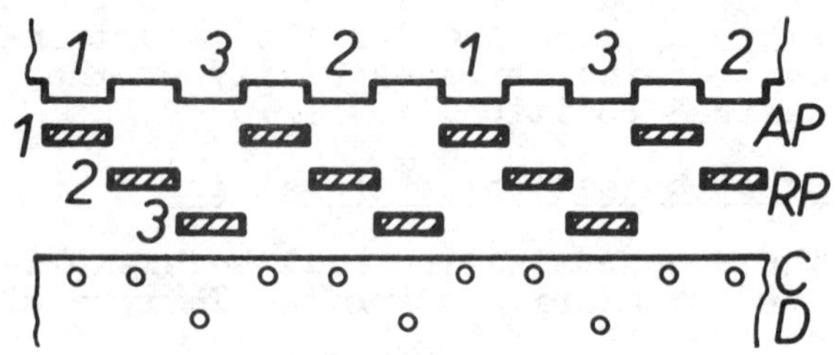

Fig. 11. Code marks and states.

active part AP and reactive part RP. The numbers 1 to 3 relate the excited teeth of the active part to the engaged teeth of the reactive part in the individual positional states according to Table 1. For synchronization purposes it will be sufficient to indicate only one state. In the arrangement of Fig. 11 this is the state 3. The indication of the other states is henceforth but a part of the path sampling.

The synchronization connection ensuring the state orientation of controller 2 (Fig. 6) is shown in Fig.12. The action of this element ensures that from the instant of transition from the continuous to the stepping mode of operation the switching sequence of the winding phases of the active part of the motor to the stepping current (or voltage) will be correct.

When the drive starts its stepping motion controller 2 is adjusted by the logic information W in such a way

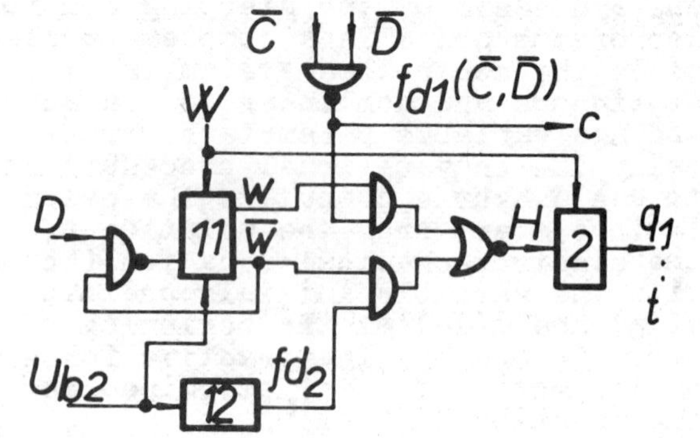

Fig. 12. Synchronization element.

that its output variables correspond to the state 3 of the motor (see Table 1). At the same time the bistable flip-flop circuit 21 is reset so that $\overline{w}$=1. The output of generator 22 generating the stepping frequency f_{d2} is locked by the logic signal U_{b2} from the motion direction selector (together with the power switching units of the switch of Fig. 6). On arrival of the first logic information D (Figs. 11 and 12) about state 3 circuit 21 (Fig. 12) is triggered. The logic signal w=1 opens the gate for the logic signal f_{d1} with the position sampling components $\overline{C}$ and $\overline{D}$ applied by input H in the form of clock pulses into controller 2, which begins to operate synchronously with the motion of the motor. On releasing the stimuls for the transition to the stepping motion the logic signal U_{b2} changes, the bistable circuit 21 locks the entrance of the signal f_{d1} to the input H of the clock pulses of controller 2 and unlocks the output of the stepping frequency generator 22. This output then brings the motor into the stepping motion by way of input H and through the simultaneously unlocked power switching units of the converter (Fig. 6).

The positional logic information f_{d1} is transferred by the output c into the counter of the position control of the drive.

For a drive using the type II motor the synchronization takes place analogously and is achieved by the same connection.

RESEARCH PROTOTYPES OF THE DRIVE

Three laboratory prototypes with the parameters given in Table 4 vere built. The flow-chart used is shown in Fig. 13 and the motors are illustrated in Figs. 14 to 16. In Fig. 14 the cage of the IR motor is removed from the reactive part. The small circular openings on the reactive

Table 4 Basic characteristics of the research prototypes of motors

		Flat type IR of Fig. 14	Tub. type IR of Fig. 15	Flat type II of Fig. 16
Continuous mode	Path length [m]	0,7	0,7	0,7
	Synchronous speed [m/s]	6	12	6
	Minimum speed [m/s]	0,05	0,05	0,05
	Thrust at rest [N]	450	240	500
	Power input at rest [W]	4700	4770	3800
	Power factor at rest [1]	0,52	0,69	0,5
	Supply voltage [V,Hz]	3x380, 50	3x380, 50	3x380, 50
Stepping mode	Path length [m]	0,7	0,7	0,7
	Length of step [mm/imp.]	10	7,5	20
	Reproducibility [mm]	±1,5	±1	±2,8
	Stepping frequency [Hz]	2-15	2-15	2-15
	Adjustable thrust [N]	150	200/80	45/20
	Holding thrust [N]	200	300/120	160/70
	Holding power input [W]	640	660/240	1750/760
	Supply voltage [V,Hz]	32, d.c.	32/20, d.c.	3x380/250, 50

part are the code marks of the positional orientation of the drive for photoelectric indication.

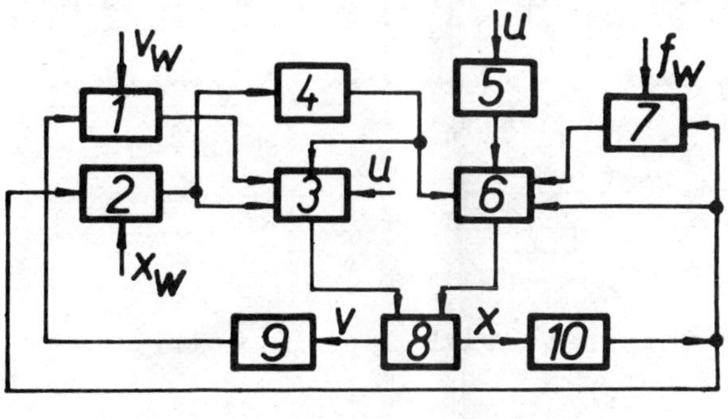

Fig. 13. Block diagram variant. 1 controller of the continuous speed $\underline{v}$ to the desired value v_W, 2 controller of the position $\underline{x}$ to the desired value x_W, 3 reversing converter for continuous-mode voltage control, 4 motion direction selector, 5 current source, 6 power switch for stepping operating mode, 7 stepping frequency generator with adjustable value of f_W, 8 motor and working mechanism, 9 speed sensor, 10 position orientation sensor, $\underline{u}$ feeding

The rest characteristics for the continuous mode follow from Table 4.

The stationary characteristics for the stepping mode were shown in Figs. 4 and 5.

The dynamic behaviour of the drives is illustrated by the oscillograms of Figs. 17 and 18 showing the time dependences of the phase voltage u_R, phase currents i_R to i_T and the speed

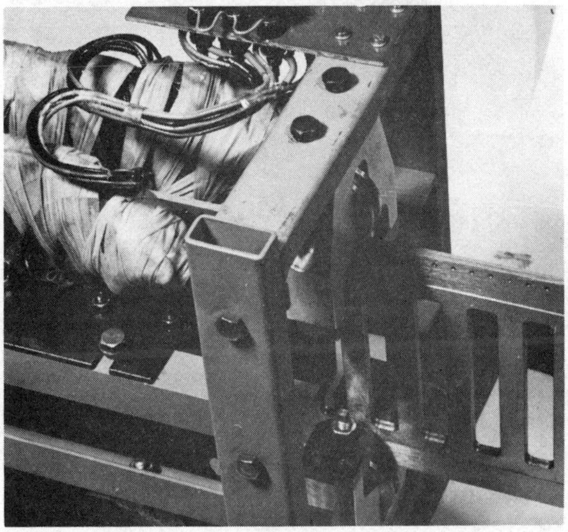

Fig. 14. Flat two sided IR type of motor with the cage of the reactive part removed.

Fig. 15. Tubular IR type of motor.

v in the continuous mode (left portions of the records) and in the stepping mode (right portions of the records). From the speed curve of the drive using the IR motor (Fig. 17) the high damping of the moving part is apparent and, conversely, in the case of the drive with the type II motor the damping is low. The records evidence good control stability of the hybrid linear drive under the two mechanically quite diverse operating conditions.

CONCLUSION

In this work a new type of linear electric drive, i.e. the hybrid linear drive, has been described. For the continuous mode of operation the drive may be designed to have tractive forces up to several thousand of newtons and speeds ranging from a few dm/s to several m/s; the length of path is almost unlimited. As regards power, this may reach the order of

Fig. 16. Flat two-sided II type of motor.

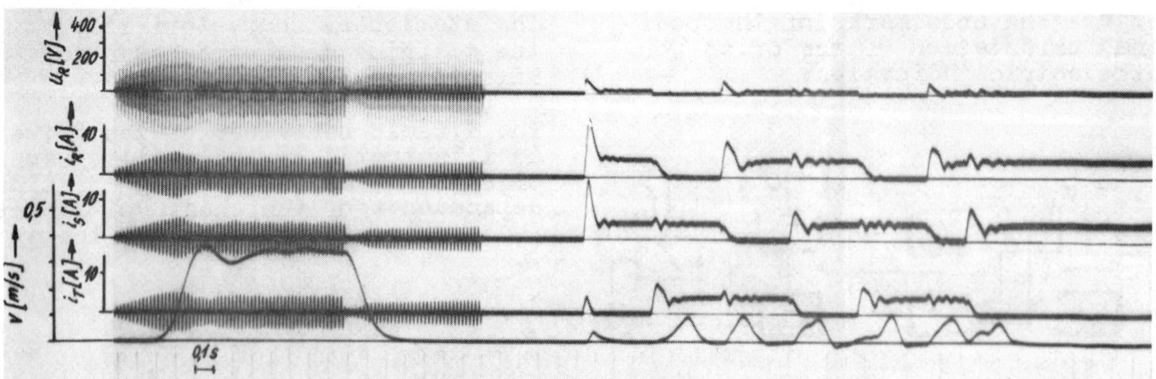

Fig. 17. Operation oscillogram for the drive with type IR motor.

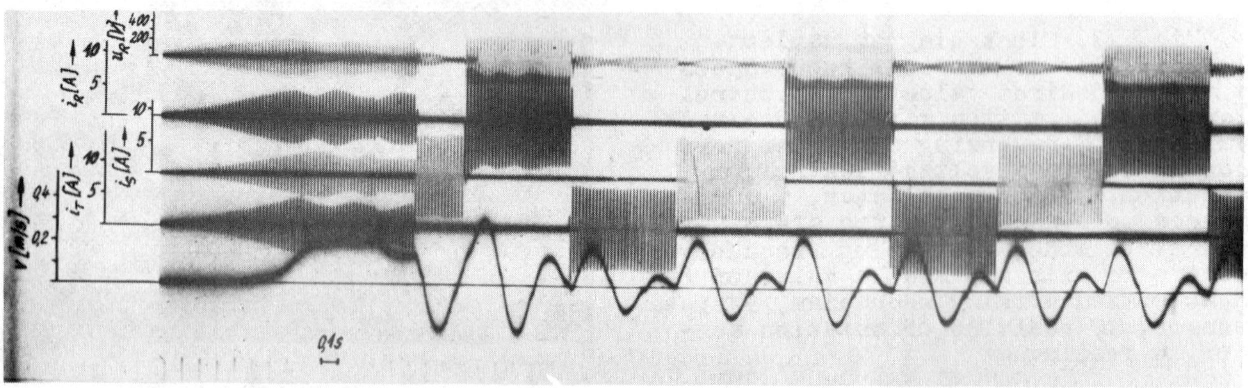

Fig. 18. Operation oscillogram for the drive with type II motor.

several tens of kilowatts. In the stepping mode the tractive forces may reach several hundred newtons at speeds up to several m/min and a stepping frequency of about 2 to 15 Hz depending on the load; the step length ranges from a few milimeters to several centimeters per pulse with a reproducibility better than about ±15% of the length of step. The length of one stepping section of the path may reach up to several meters.

The type IR motor yields in the stepping mode under equal conditions higher values of traction force and a more favourable positional tractive force dependence than type II. In the stepping mode it also operates more economically since with d.c. feeding only ohmic losses occur in the winding of the active part. The stepping-mode stability of the IR type is higher than that of type II, the stepping is finer and the positional reproducibility is better. Conversely, the tractive force, particulary at higher continuous-motion speeds is lower than that of type II. The reactive part imposes higher demands on the production and is more expensive, mainly with regard to closer production tolerances.

It should be noted that in the stepping mode the type II motor is relatively uneconomical. This is particulary true dor the states of rest because the imitation of traction and consequently also of the fixation traction is always connected with the development of losses in the winding and in the ferromagnetic of both parts of the motor. The stability of stepping operation is lower because with faster stepping there is a danger of starting into continuous operation in a single-phase induction mode. The manufacture of the reactive part is simple and inexpensive and grants the motor better properties in continuous operation than in the case of the IR type.

The drive with the IR motor is suitable for exacting applications with higher demands on the dynamic properties and on the accuracy of positioning and in cases where the positioning operation dominates over the continuous mode.

The drive with the II motor may be recommended if less severe demands are imposed on dynamics and positioning accuracy when driving such equipment where continuous operation along longer paths with infrequent positioning places predominates.

As an example of the application possibilities of the hybrid linear drive we mention especially the various equipment in spatially extensive arrangements and with a larger number of positioning places. Under these applications belong charging, palletization and sorting mechanisms, rack-stock stowing machines, assembly belts, traversing mechanisms for industrial haulage and interplant material handling, some types of material handling equipment for elastic production systems, various transport and positioning mechanisms for hazard workplaces where the presence of actuators with external positional feedbacks is not desirable.

The authors wish to extend their sincere thanks to Mr. Miroslav Exner for his co-operation as regards modelling of the drive on the hybrid computer and to the management of the Research Institute of Electrical Engineering for their permission to publish the work.

REFERENCES

(1) O. Roubíček, Contribution of the Czechoslovak research to the development of linear electrical controlled drives, IEE Conference Publication No. 120, p. 31 (1974).

(2) O. Roubíček, Linearantriebe für Industriezwecke, ETZ-B (in print).

(3) O. Roubíček, J. Holan, J. John, Progress in the field of less common electrical controlled linear drives, Proceedings of the IFAC-Symposium on Control in Power Electronics and Electrical Drives Vol. 2, p. 175 (1974).

THE INFLUENCE OF DRIVE CIRCUIT PARAMETERS ON THE PERFORMANCES OF STEPPING MOTORS

A. Kelemen and V. Trifa

Polytechnic Institute of Cluj - Napoca, Romania

ABSTRACT

The parameters of current waveforms in the phases of stepping motor are discussed. They are introduced into the equation of motion of rotor, which is used for an energy approach of stepping motor operation. A method for computation of maximum stepping rate for given values of load torque is presented.

INTRODUCTION

The stepping motor has the basic property to change a command pulse into a determined angular displacement, named stepping angle, or step. For this purpose it must be supplied with current pulses from a power source in a sequence with standard command pulses. Thus the conventional concept of electric machine attributed to the stepping motor may be associated or even replaced with that of electro-mechanic pulse-displacement converter. This conception is based on the fact that stepping motor performances do not depend only, and especially not firstly, upon its mechanic construction, but rather upon its control and supplying electronic circuit, which can ensure high performances in terms of start, stop and running rate limits, the smooth of its movement, the stability etc.

The electronic circuits for pulse supplying of stepping motors contain power transistors in series with motor phases. In the most cases the use of a simple transistor as switch coupled with a power return diode is not sufficient to obtain high performances. Due to low constant time of motor phase, the rising time of current is to large and thereby the motor is able to follow only a low command pulse rate without losing considerably its output torque.

In order to reduce time constant or rising time of the current, that is to obtain the required waverform, the following techniques are used (Ref. 4):

1. Series resistors are added to motor phases.
2. Series resistors coupled with parallel capacitors are mounted across the motor phase or transistor.
3. Dual-voltage drive involves a high voltage source used during pulse initiation, with a low voltage source during the flat top of the pulse.
4. A "Chopped" waveform is generated by applying a high voltage at the beginning of the step until the current reaches a predeterminated level, then the high voltage is time modulated into pulses to sustain current at constant level.
5. A programmed voltage source is used to deliver the voltage required to maintain the current constant at any pulse rate of stepping motor.

The present work hasn't as purpose to examine in detail these techniques, but to define some parameters of drive circuit, which serve for determination of a computation method of running limit characteristics of the stepping motors.

DRIVE CIRCUIT PARAMETERS

Figure 1 shows some typical waveforms of supplying voltage and current known in drive circuit technique.

The shape drawn in Fig. 1 a is for driver with series resistors added to motor phases. T_{m1} and T_{m2} are time constants refered to charging and discharging circuits and I is the final value of current, which depends upon applied voltage U and total resistance of series circuit. The rissing time of current is the duration for the current to reach 95 % of final level I, that is (Ref. 2)

$$t_1 = 3T_{m1} = 3L_m / R_t \qquad (1)$$

where L_m is average inductance of motor phase and R_t is total resistance of circuit. This time normally is constant, but the time interval between command pulses t_q (inverse of

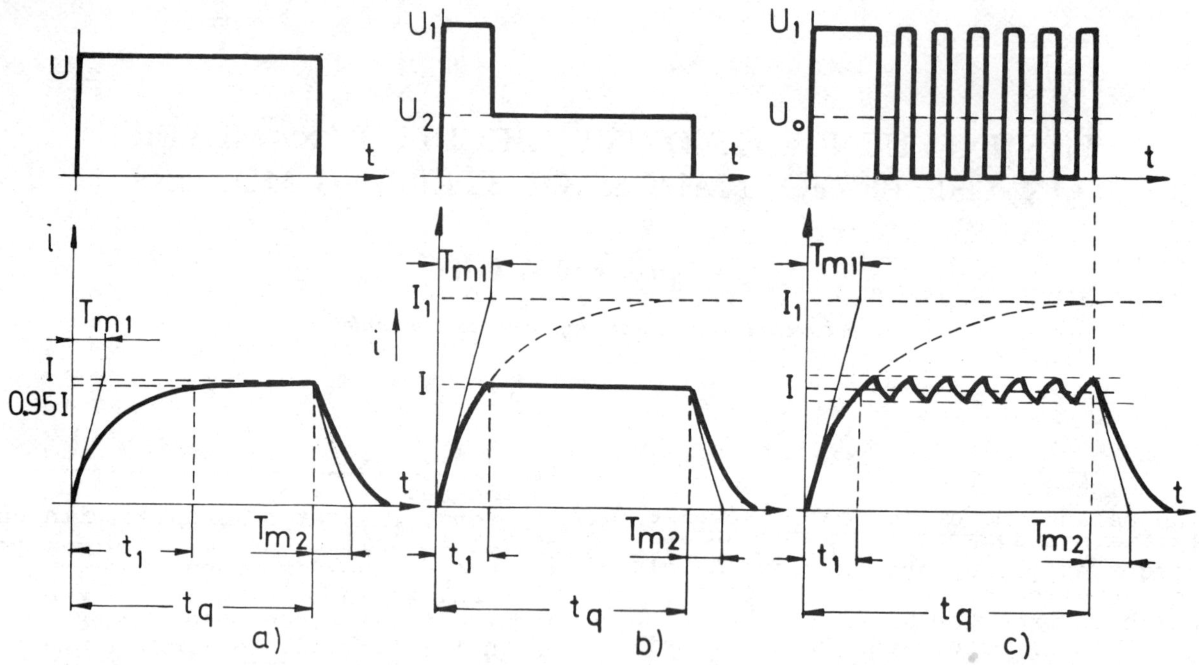

Fig. 1. Waveforms of the voltage and current

stepping rate) decreases at high stepping rate. Thus the rising time t_1 limits the maximum stepping rate, for a given load of the motor.

In dual-voltage technique illustrated in Fig. 1 b the shape of current resembles with a trapezium due to rapid rise of the current. The flat level I depends upon voltage U_2, thus

$$I = U_2 / R_t \qquad (2)$$

One can define a forcing factor as ratio of the two voltages U_1 and U_2, thus

$$k_f = U_1 / U_2 \qquad (3)$$

which can reach value of 30. With this factor the rising time of current becomes

$$t_1 = T_{m1} \ln \frac{k_f}{k_f - 1} \qquad (4)$$

and may be reduced considerably with the grows of k_f.

The "chopped" waveform drawn in Fig. 1 c is similar to that of dual-voltage drive. The rising time is

$$t_1 = T_{m1} \ln \frac{U_1}{U_1 - U_o} \qquad (5)$$

where U_o is steady state motor voltage. For this case the forcing factor $k_f = U_1 / U_o$.

In both previous cases $T_{m1} = L_m / R_t$, where R_t is total series resistance of circuit(phase, transistors, sensor). It is evident that an ideal waveform of the current consisting in a squarewave shape may be not realised. Therefore the current rate of rise in motor phases has a great influence on the maximum stepping rate of the motor. A method for computation of these limits, known as running limit characteristics will be presented further by introducing an energy approach of stepping motor operation.

ENERGY APPROACH OF STEPPING MOTOR OPERATION

As is known, for simulation of stepping motor operation, mathematical models are used. These models consist of voltage equations associated with the equation of motion of the rotor. The complexity of these models, based on many assumptions, does not justify always their use to determine the performances of stepping motor in given conditions (Ref. 3). The use of motion equation of the rotor only, offers a very clear and simple analysis of energy transformation processes of stepping motor in connection with its driver and load characteristics.

First, the normalized equation of motion around the equilibrium position of the rotor is considered, thus

$$\frac{d^2\theta}{d\tau^2} + 2\xi \frac{d\theta}{d\tau} + \mu_r = \mu_e \qquad (6)$$

where

θ electrical stepping angle
τ normalized time
ξ normalized total damping ratio
μ_r load torque, normalized
μ_e developed torque, normalized.

By introducing normalized angular velocity $\omega = d\theta / d\tau$, equation (6) becomes

$$\omega \frac{d\omega}{d\theta} = -2\xi\omega - \mu_r + \mu_e \quad (7)$$

The damping ratio $\xi = 0$ is considered and (7) can be written as follows

$$\omega \frac{d\omega}{d\theta} + \frac{dE}{d\theta} = 0 \quad (8)$$

where E is potential energy of rotor. By integrating (8), the equilibrium of energies is found

$$\frac{\omega^2}{2} + E(\theta) = h \quad (9)$$

where h is the constant of energy and $\omega^2 / 2 = E_c$ is kinetic energy. The potential energy of the rotor is

$$E(\theta) = \int_0^{\theta} (-\mu_e + \mu_r)\, d\theta \quad (10)$$

which has two component, as follows

$$E_d(\theta) = \int_0^{\theta} (-\mu_e)\, d\theta$$ - developed energy

$E_r(\theta) = \mu_r \theta$ - energy for compensating load torque

With (9) and (10) the equilibrium of energies is

$$E_c + E_d + E_r = h \quad (11)$$

In order to implement this equation on stepping motor operation, a four-phase variable-reluctance (VR) stepping motor, one phase on, is considered and a sinusoidal waveform of developed torque with respect to rotor position θ is assumed. Figure 2 shows the energy portrait of the motor, along a full step ($\pi / 2$).

The developed torque, around the equilibrium position 0 , is

$$\mu_e = -\sin\theta \quad (12)$$

and with (10), equation (11) becomes

$$\frac{\omega^2}{2} + 1 - \cos\theta + \mu_r\theta = h \quad (13)$$

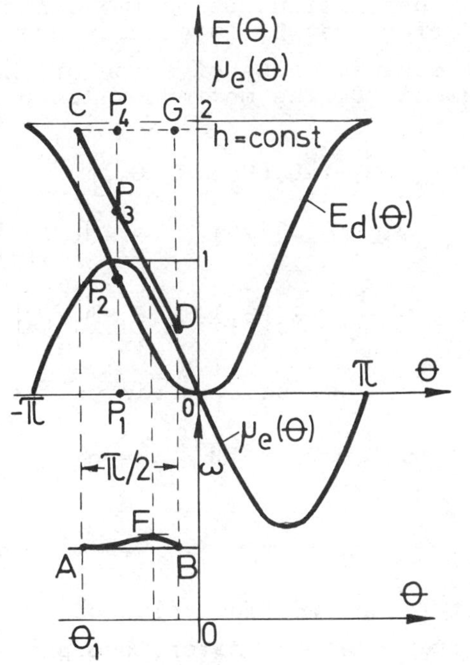

Fig. 2. Equilibrium of energies

If θ_1 is rotor position at the initiation of step, for an arbitrary position $\theta_1 \leq \theta \leq \theta_1 + \pi / 2$, the seqments pointed by P_1, P_2, P_3, P_4 indicate:

$P_1P_2 = 1 - \cos\theta$ developed energy

$P_2P_3 = \frac{\omega^2}{2}$ kinetic energy

$P_3P_4 = \mu_r\theta$ energy for load (energy loss)

It is evident that if μ_r = const the energy loss will be delimited by the vertical distance between segments CD and CG, and the value of normalized load torque μ_r will be the slope of segment CD. For running operation the average angular velocity is constant, therefore kinetic energy is constant at the beginning and at the end of the step, that is $E_c(\theta_1) = E_c(\theta_1 + \pi / 2)$.

A small variation of ω between θ_1 and $\theta_1 + \pi / 2$ will exist yet and its maximum F takes place for an angular position determined by maximum vertical distance between segment CD and curve $E_d(\theta)$.

A method of computation the maximum value of μ_r for a given ω will result if one can find the maximum slope of segment CD so that its projection on horizontal axis θ be $\pi / 2$, that is a full step. With this definition, using (13), and considering that θ_1 is angular position for maximum value of slope of the segment CD, the maximum value of μ_r is

$$\mu_{rm} = \frac{2}{\pi}\left\{\left[E_d(\theta_1)+E_c(\theta_1)\right] - \left[E_d(\theta_1+\pi/2)+E_c(\theta_1+\pi/2)\right]\right\}$$

$$= \frac{2}{\pi}\left[E_d(\theta_1)-E_d(\theta_1+\pi/2)\right]$$

$$= \frac{2}{\pi}(\cos\theta_1 - \sin\theta_1) \qquad (14)$$

This is for $\theta_1 = -\pi/4$ and

$$\mu_{rm} = \frac{2\sqrt{2}}{\pi} = 0.9 \qquad (15)$$

that is the maximum value of μ_r for a given stepping motor, one phase on.

COMPUTATION OF RUNNING LIMIT CHARACTERISTICS

The approach presented was for a square-wave current in the motor phases and the developed torque was a function of angular position θ only. In fact the torque depends upon the current and this dependence is marked by time constant or rising time of the current. For an exponential shape of current, the expression of developed torque, for a four-phase VR stepping motor, one phase on, is (Ref. 1)

$$\mu_e(\theta) = -(1-e^{-\tau/T})^2 \sin\theta \qquad (16)$$

where T is normalized time constant. In this expression the exponential shape of current at the end of supplying pulse is neglected. For dual-voltage or "chopped" waveforms the shape of current is just like a trapezium. A similar expresion can be written for this case if the real shape of current is replaced with an equivalent exponential waveform how is shown in Fig. 3.

The real current is

$$i = i_f = I_1(1-e^{-t/T_{ml}}), \quad \text{for } t<t_1, \text{ and}$$

$$i = I, \quad \text{for } t \geq t_1 \qquad (17)$$

where $I_1 = k_f I$. The equivalent current i' is the current which, for $t = T_f < t_1$, has the same value as the real current, that is 0.637 I. Therefore

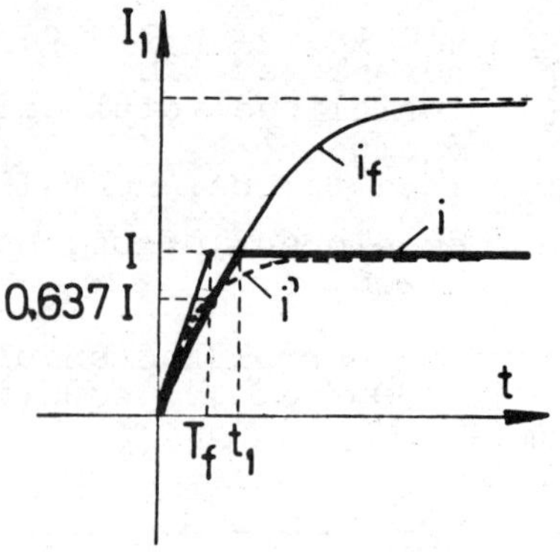

Fig. 3. Equivalent waveform of current

$$i = I(1-e^{-t/T_f}) \qquad (18)$$

where the new time constant Tf is obtained from $i'(T_f) = i(T_f)$ and is

$$T_f = T_{ml} \ln \frac{k_f e}{(k_f-1)e-1} \qquad (19)$$

Thus the expression (16) may be considered valid for any pulse supplying technique and will be used for determination of maximum values of μ_r for given velocities ω, that is a running limit characteristic.

In order to simplify the computation the following assumptions must be taken into account:

1. Stepping motor is a four-phase VR type, one phase on.
2. The developed torque depends sinusoidally upon the angular position.
3. The shape of current in the motor phase is exponential.
4. The dependence of inductances upon angular position and magnetic saturation are neglected.

In running operation of stepping motor the variation of angular velocity along a full step can be neglected, therefore the normalized time from (16) can be written as

$$\tau = \frac{\theta - \theta_1}{\omega} \qquad (20)$$

where θ_1 is the initial position of the rotor within a full step. With (16) and (20) the developed energy is

$$E_d(\theta) = \int_0^{\theta} (-\mu_e)\, d\theta$$

$$= \int_0^{\theta} \left[1-e^{-(\theta-\theta_1)\sigma}\right]^2 \sin\theta\, d\theta$$

$$= 1-\cos\theta - \frac{2e^{\sigma\theta_1}}{1+\sigma^2}$$

$$+ \frac{2e^{-(\theta-\theta_1)\sigma}}{1+\sigma^2}(\cos\theta + \sigma\sin\theta)$$

$$+ \frac{2e^{\sigma\theta_1}}{1+4\sigma^2} - \frac{e^{-2(\theta-\theta_1)\sigma}}{1+4\sigma^2}$$

$$\times (\cos\theta + 2\sigma\sin\theta) \qquad (21)$$

where

$$\sigma = 1/\omega T \qquad (22)$$

In concordance with definition (14) the maximum value of μ_r for given ω and T is

$$\mu_{rm}(\theta_1) = \frac{2}{\pi}\left[E_d(\theta_1) - E_d(\theta_1 + \pi/2)\right]$$

$$= \frac{2}{\pi}\Big[- \cos\theta_1 - \sin\theta_1$$

$$+ \frac{2}{1+\sigma^2}(\cos\theta_1 + \sigma\sin\theta_1)$$

$$- \frac{2e^{-\pi\sigma/2}}{1+\sigma^2}(-\sin\theta_1 + \sigma\cos\theta_1)$$

$$- \frac{1}{1+4\sigma^2}(\cos\theta_1 + 2\sigma\sin\theta_1)$$

$$+ \frac{e^{-\pi\sigma}}{1+4\sigma^2}(-\sin\theta_1 + 2\sigma\cos\theta_1)\Big] \qquad (23)$$

and angle θ_1, which corresponds to μ_{rm}, is obtained form $d\mu_{rm}/d\theta_1 = 0$, thus

$$\mathrm{tg}\,\theta_1 = \frac{1 - \dfrac{2\sigma + 2e^{-\pi\sigma/2}}{1+\sigma^2} + \dfrac{2\sigma + e^{-\pi\sigma}}{1+4\sigma^2}}{1 - \dfrac{2 - 2\sigma e^{-\pi\sigma/2}}{1+\sigma^2} + \dfrac{1 - 2\sigma e^{-\pi\sigma}}{1+4\sigma^2}} \qquad (24)$$

Giving various values for σ, at T = const, with (23) and (24) the dependence $\omega(\mu_r)$ can be obtained and hence a running limit characte - ristic.

STEPPING MOTORS SPECIFICATIONS AND PERFORMANCE DATA

The stepping motor used for this ap- plication study is of AE - 265 type, made in I. P. Cluj - Napoca /Romania (Ref. 1). The specifications of the motor are

type
  variable reluctance (VR)
stepping angle
  $\theta_p = 2.65^o$
number of phases
  m = 4
number of rotor teeth
  $z_r = 34$
rated current
  I = 5 A per phase
resistance
  $R_m = 2.5\ \Omega$ per phase
d - axis inductance
  $L_d = 0.072$ H
q - axis inductance
  $L_q = 0.038$ H
moment of inertia of the rotor
  $J_m = 1.5 \times 10^{-4}$ Nms^2
peak holding torque at 5 A, one phase on
  $M_{max} = 14.5$ Nm

Other specifications are

average inductance
  $L_m = (L_d + L_q)/2 = 0.055$ H
time constant of motor phase (rezistance of transistors and sensors are neglected)
  $T_{ml} = L_m / R_m = 0.022$ s
load inertia
  $J_L = 0.5 \times 10^{-4}$ Nms^2
reference time for normalization
  $T_b = \sqrt{\dfrac{J_m + J_L}{M_{max}\, z_r}} = 6.36 \times 10^{-4}$ s
normalized time constant
  $T_n = T_{ml} / T_b = 34.59$
steady state voltage
  $U_o = R_m I = 12.5$ V

Normalized time constant T from (22) is obtained making use of definition (19), where T_{ml} becomes T_n, thus

$$T = T_n \ln \frac{k_f e}{(k_f - 1)e - 1} \qquad (25)$$

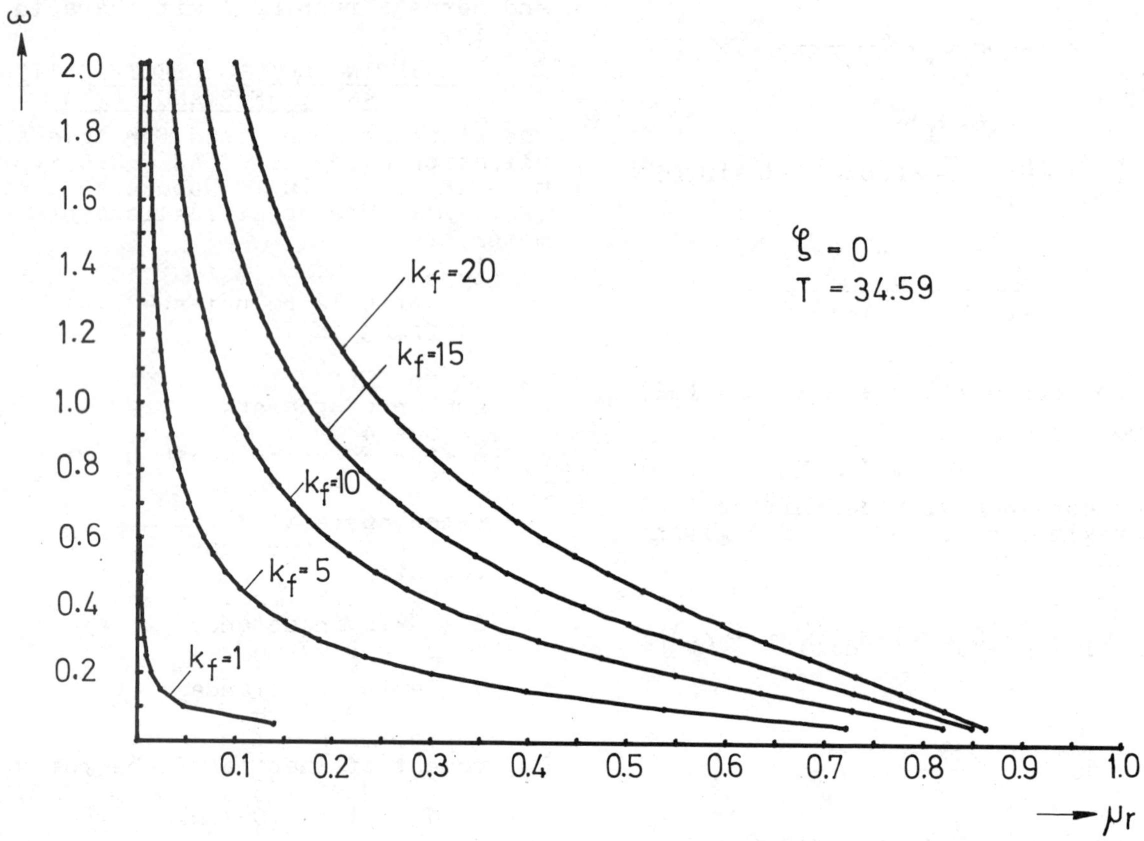

Fig. 4. Running limit characteristic plotts

The input data for computation of $\omega(\mu_r)$ dependence are k_f, T_n and various values for ω. With (23), (24) and (25) a running limit characteristic will result. Figure 4 shows running limit characteristic plotts for damping ratio $\zeta = 0$ and forcing factor $k_f = 1 \ldots 20$. For a given μ_r critical stepping rate for running operation increases with forcing factor k_f.

In order to compare these theoretical results with experimental stepping rate / load torque curve of the motor, the following test conditions are taken

frictional factor

$\beta = 0.094$ Nms

applied voltage (chopped waveform)

$U_1 = 125$ V

Theoretical curves of Fig. 4 are for $\zeta = 0$. Examining equation (7), it is evident that term $2\zeta\omega$ (that was neglected), is directly proportional to ζ for ω =const, and may be added to the value of μ_r. Hence a simple method to include the influence of ζ will result, by translation of each μ_r left with $2\zeta\omega$ at a given ω. That helps to extend theoretical curves previously obtained for real cases of $\zeta \neq 0$.

With test conditions shown above and the motor specifications, are obtained:

damping ratio

$$\zeta = \frac{\beta T_b}{2J} = 0.15$$

forcing factor

$$k_f = U_1 / U_o = 10$$

stepping rates (steps/s)

$$f_p = \frac{180^o \omega}{\pi \theta_p^o z_r T_b} = 1000 \times \omega$$

load torques (Nm)

$$M_r = 14.5 \times \mu_r$$

Figure 5 shows the two curves $\omega(\mu_r)$, theoretical and experimental, for $k_f = 10$ and $\zeta = 0.15$. They drawn near especially at high velocity, that is a confirmation of present method of computation, based on the consideration of ω = const.

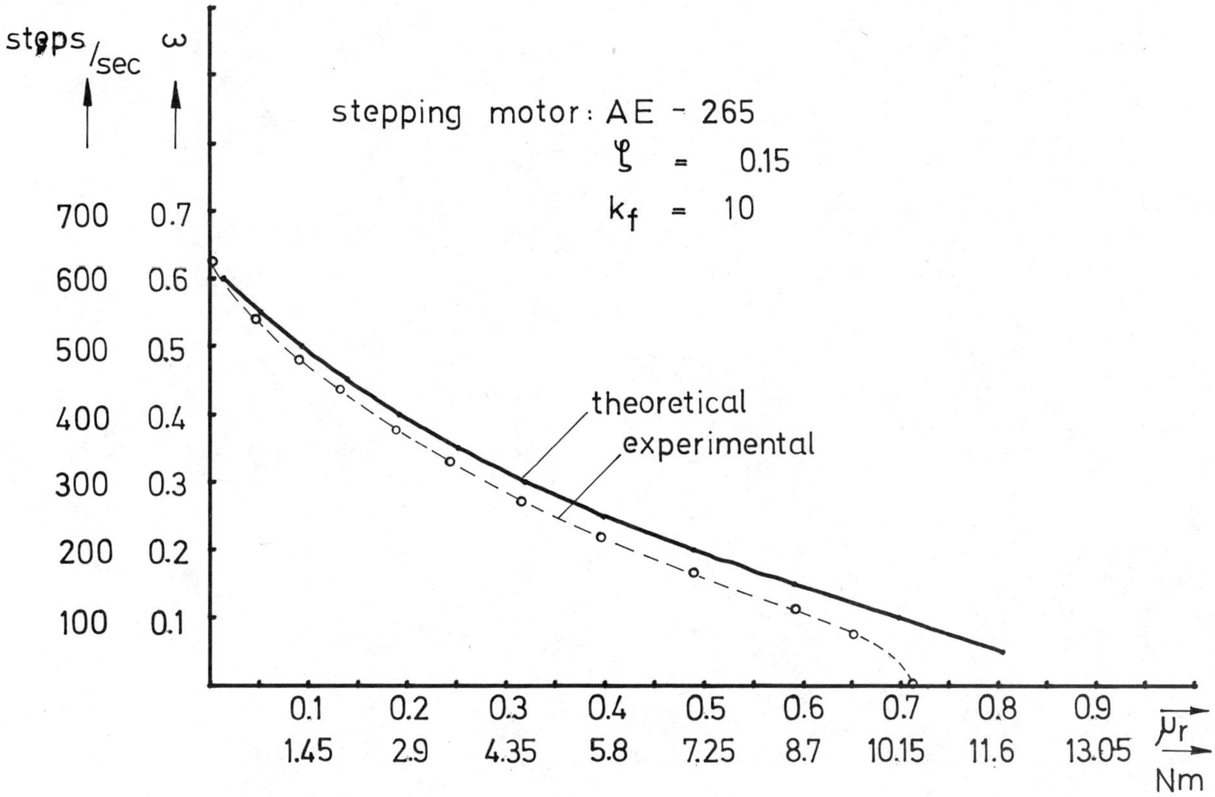

Fig. 5. Variation of maximum stepping rate with load torque

CONCLUSIONS

A method of computation of running limit characteristics for stepping motor, using an energy approach of its operation has been presented. The influence of drive circuits parameters is marked by using of an equivalent exponential waveform of motor phase current.

An interesting remark can be made about the significance of angle θ_1 as rotor position at the initiation of the step, for maximum load torque. This angle can be interpreted as an optimal switch angle, dependent on time constant and velocity. Thus the method discussed here can be used for study of various closed-loop control of stepping motors.

The theoretical curves are verified with experimental results for typical test conditions. The implementation of theoretical results to any VR stepping motor is very easy due to normalized magnitudes used in this study.

REFERENCES

(1) Kelemen, A.,Crivii, M. (1975) Motoare electrice pas cu pas, Ed. Tehnica, Bucarest.

(2) Leonenko, L., I. (1974) Polupro-vodnicovie forsiruscie shemi, Iz. Energia, Moskva.

(3) Julovian, V., V., Lidengoltz,I., F., C rascetu ciastotnih haracteristic reactivnih sagovih electrodvigatelei, Electrotehnika, 11, 41 (1973)

(4) * * * (1974) Sigma Stepping Motor Handbook.

THE APPLICATION OF CONTROLLED STATIC CONVERTERS IN TRACTIVE UNITS

Hans Heinrich Weber

Swiss Federal Railways, Motive Power and Workshops Division, 3030 Berne

1. INTRODUCTION

The present article does not claim to offer a complete survey. Its intention is to point out the many advantages offered by the application of controlled static converters, mainly in a.c. traction, but also the problems encountered with various circuits.

The conventional direct-motor drive, i.e. the traction motor fed directly from the power supply system through appropriate controls and regulators for speed and tractive force, meets the traction requirements to a fairly high degree. The following is a summary of the most important traction requirements:

1. High tractive force corresponding to the physical factors between wheel and rail, from standstill to a speed as high as possible.

2. At maximum speed (e.g. V = 200 km/h or more) there shall be sufficient tractive force to overcome the high air resistance.

3. Between the two extreme operating conditions, that of maximum tractive force at a given speed and the tractive force at maximum speed, the absorbed power should be rather constant, i.e. the curve of the tractive force as a function of the speed should, in this range, ressemble a hyperbola.

4. The alteration of the torque of the traction motor should be continuous (no jerks).

5. Optimum use should be made of the adhesion between wheel and rail at any speed.

6. The drive should be designed so as to minimize the stress on the track, despite of high axle loads, i.e. to keep to a minimum the forces between wheel and rail.

7. The electric drive should be designed for low maintenance requirements and accessibility to those parts requiring periodic maintenance.

8. The interference of the electrical system with the signalling, telecommunication and power station installations should require but minor alterations in the fixed installations.

A further point we have to bear in mind is that the most important of the present traction current systems, at the time of the electrification, were selected in view of the available traction motors, in other words, the traction motor and the traction current system are closely interconnected in the conventional direct-motor drive (power supply fed directly to the traction motor). This resulted in the following traction current systems:

d.c. 1,5 kV
d.c. 3 kV
a.c. 15 kV, 16 2/3 Hz
a.c. 25 kV, 50 Hz

It was the latter system only which required a conversion on the locomotive, initially by means of mercury-vapour rectifiers, and more recently with semi-conductor rectifiers. The separation of the traction current system from the traction motor was inevitable for the 50 Hz system. Technically the design of a 50 Hz single-phase series-wound commutator motor is feasible, but its weight and volume would be far higher than any d.c. motor, pulsating current motor or single-phase current motor with 16 2/3 Hz. It was, therefore, impossible to develop an economic 50 Hz direct-motor traction, not even with the considerable progress made in that area.

Figure 1 is a comparison of various equally rated motors: one each for 16 2/3 Hz and 50 Hz direct-motor traction, a pulsating current motor and an asynchronous three-phase motor. The advantage of the present conversion on the locomotive is evident.

2. STATIC CONVERTER SYSTEMS WITH PULSATING CURRENT TRACTION MOTORS

The time-honoured steam engine alredy allowed a continuous control of the tractive effort, except for a certain unbalance in the piston forces, and so do the modern diesel locomotives with electric or hydraulic transmission of torque. For the conventional direct-motor fed with a.c. or d.c. from the catenary, the control of the tractive force required a considerable effort which resulted in a quasi-continuous control only. The steps are quite noticeable, even in modern direct-motor designs. Comfort as well as utilization of adhesion are clearly affected. Figur 2 shows the conventional control of speed and tractive force by tap changer used in a.c. traction. Figure 3 shows an example of the vehicle characteristics (speed versus tractive force) for 11, 29 and 39 steps and for continuous control, whereby the damping effect of the moving masses has been neglected. The importance of the continuous control with respect to acceleration and jerking is evident. It is the controlled converter only which offers continuous control of speed and tractive force for catenary fed locomotives. In other words, they allow us to regain a feature which the steam engine already offered. Continuous control is a characteristic common to all traction circuits using controlled converters.

Before considering the traction requirements listed in paragraph 1, which are materially better achieved in comparison with direct-motor techniques, we are going to discuss the basic circuits In particular, we shall deal with the optimization of the various circuits with respect to interference with the signalling and telecommunications equipment, and interference in substations and power station.

2.1 Basic circuits with firing angle control (phase angle control)

The discussion will be restricted to the bridge circuits in common use today, neglecting the push-pull circuits.

2.11 The fully-controlled symmetric bridge circuit

In the following considerations it is assumed that a very large smoothing chocke has been installed in the traction motor circuit. The leakage inductance of the transformer has also been neglected. It is thus easier to follow the process, since the individual rectified currents and the primary a.c. current will appear as rectangular current. Figure 4 shows the circuit of the fully-controlled converter. The currents i_1 and i_2 can be identified, together with the negative instantaneous d.c. voltages occurring in this type of circuits. The primary current i_L, the fundamental harmonic, is out-of-phase by the angle φ compared to the voltage U_d. The phase shift is equal to the firing angle α. The voltage regulation with the above firing angle procedure results in an increased reactive power. When the firing angle is increased up to $\alpha = 90^o$, the average d.c. voltage is $U_d = 0$; the rectangular current continues, and therefore, we practically have reactive power demand only on the a.c. supply. If we increase α beyond 90°, we obtain a negative average d.c. voltage at positive current. This happens when the traction motor changes to generator; regenerative braking is therewith possible [11].

2.12 The semi-controlled asymmetric bridge circuit

The circuit shown in Figure 5 offers the possibility of reducing the reactive power consumption. Instead of thyristors, two arms are equipped with diodes. This results in a freewheeling through diodes 1 and 2 for the current caused by inductance in the traction motor circuit. Thyristors 1 and 2 are turned off immediately on voltage crossover and negative d.c. voltage portions are no longer produced. Thus, the currents i_1 and i_2 are flowing only during the positive portions of u_d, and during the gaps of u_d the freewheeling current i_D is flowing, which however is not transmitted to the primary side of the transformer. As a result the rectangular currents on the primary side are shorter and the fundamental wave shows a phase shift which is reduced to $\varphi = \frac{\alpha}{2}$. Thus, the reactive power can be materially reduced compared to the controlled symmetric converter circuit [11].

With respect to harmonics, both types of circuits are equal.

2.13 Measures to reduce the influence of the firing angle control on the power supply, signalling and telecommunication installations

The influence on the grid system can be defined as follows:

- occurence of even-numbered and odd-numbered harmonics in the grid and feeder systems, in substations and power stations
- additional load for the power supply through increased reactive power consumption.

In addition, special attention is required with respect to:

- influence on the signalling circuits
- influence on the telecommunication circuits
- resonances in the feeder lines

In particular, there must be no interference with the signalling and telecommunication circuits which might affect the safety. The solution must be sought in decreasing the "interference power" of the tractive unit, and in reducing the reception sensitivity of the "disturbed" fixed installations. Accordingly, the following objectives have emerged concerning the circuits on the tractive units:

- reduction of the amplitude of the upper harmonics in the power supply in view of the interference with the telecommunication installations
- reduction of the primary current harmonics of very low order in view of the interference with the a.c. track circuits and the additional load on the substations and power stations. 125 Hz track circuits are common in Switzerland; they are influenced in particular by the 7th and 9th harmonics
- improvement of the power factor to relieve the substations and power stations.

2.131 Multi-sequence circuits

2.131.1 Influence of the commutation on the power supply

In Switzerland, thorough investigations have proved the controlling of the commutation phase, particularly in the phase cut range, to play an important part for the behaviour towards the grid, and mainly for the upper harmonics [1]. To obtain a realistic picture of the commutation behaviour, we can no longer neglect the leakage reactance of the feeding transformer, on the other hand, we still may assume an infinite smoothing. The corresponding voltage and current diagrams are presented in Figure 6, both for retarded firing operation and diode like operation (firing angle at minimum), where u_o is the overlap angle (crossover to free-wheeling), u_{oo} overlap angle at diode like operation, and u_α overlap angle in the retarded firing range. For diode like operation the following approximation formula can be derived for the amplitude of the odd-numbered grid current harmonics of the nth order [1]:

$$\hat{I}_{Ln} \approx \frac{1}{n^2} \cdot \frac{2}{\pi} \cdot \bar{I}_d \cdot \frac{\sin u_{oo}}{1 - \cos u_{oo}}$$

$\hat{I}_{Ln}$: grid current harmonics of the nth order

$\bar{I}_d$: linear average d.c. current value

u_{oo} : overlap angle at diode like operation

For retarded firing operation the corresponding approximation formula is [1]:

$$\hat{I}_{Ln} \approx \frac{1}{n^2} \cdot \frac{2}{\pi} \cdot \bar{I}_d \left[\frac{\sin u_o}{1 - \cos u_o} + \frac{\left| \sin (n \frac{u_\alpha}{2}) \right|}{\sin \frac{u_\alpha}{2}} \right]$$

u_o : overlap angle after crossover of the grid voltage

u_α : overlap angle in the retarded firing range

From the two approximation formulas we recognize the important part played by the overlap angles u_{oo}, u_o and u_α for the amplitude of the harmonics. These various overlap angles can be influenced to a high degree by the corresponding reactances X_{KO} and $X_{K\alpha}$. The commutation reactance at the current commutation after the grid voltage crossover X_{KO} is determined by the grid reactance and the leakage reactance of the feeding transformer; for the given semi-controlled converter circuit, the commutation reactance on commutation in the firing angle is $X_{K\alpha} = X_{KO}$. The above two approximation formulas suggest that the amplitudes of the various harmonics are higher in the retarded firing range than in diode like operation. This is confirmed in Figures 7 and 8, where the amplitudes for switching through the converter at practically the same parameters of u_{oo} take higher values than in the diode like operation. This fact forces us to keep the portion of the current, resp. voltage, which is obtained by retarded firing operation as low as possible, i.e. the power controlled by retarded firing has to be reduced. This may be done with multi-sequence circuits, which gives the possibility - as will be shown later - of favourably influencing the harmonics with the help of the commutation reactance $X_{K\alpha}$.

The basic effect of sequence operation of the converter on the momentary form of the grid current is presented in Figure 9, together with the appropriate approximation formula [1]. By selecting the number of sequences N, we can determine the size of the retarded firing portion $\Delta I L_\alpha$. It is

$$\Delta I L_\alpha = \frac{\bar{I}_d}{N}$$

By providing as many sequences N as we possibly can, we are in a position to decrease the height of the current portion in the retarded firing range and thus the term in the approximation formula which determines the contents of harmonics generated by the firing angle control. Thanks to the lower weighting of this second term for the creation of harmonics, the spectrum of the harmonics for retarded firing operation approaches that of diode like operation.

2.131.2 Circuit design

Multi-sequence circuits can be installed either on the d.c. side or on the a.c. side. Figure 10 shows the two possibilities. It is evident that the desired increase in the number of sequences results in a proportional increase in the number of thyristors. In sequence circuits on the d.c. side, the rectifiers however are loaded with N times less voltage than in a corresponding single-sequence bridge circuit. On the other hand, the voltage at the rectifiers is higher for a sequence circuit on the a.c. side; but, you need less rectifier arms and less connections between converter and transformer.

The expenditure for multi-sequence circuits being quite high, the next step was the development of more economic circuits. The basic principle is shown in Figure 11, where a four-sequence and an eight-sequence economic converter circuit are illustrated. In a four-sequence circuit, thyristors 1 and 1', resp. 2 and 2', each always assume control by retarded firing for a quarter of the voltage, until half of the total voltage u_{dI} is controlled. At that moment the diode like controlled bridge of voltage u_{dII} is switched in, and simultaneously the bridge with voltage u_{dI} is switched off. The procedure is similar for the economic eight-sequence converter circuit. Figure 11 shows that an additional choke L_K in the multi-sequence circuit enables the reactance in the phase cut range to be increased; this results in a reduced steepness of the current increase. Figures 12 and 13 represent the power frequency spectra on switching through the highest voltage stage, for one-sequence to eight-sequence circuits, and varying ratios of the reactances $X_{K\alpha}/X_{KO}$. On switching through an eight-sequence circuit with a ratio $X_{K\alpha}/X_{KO} = 16$, the harmonics spectrum reaches nearly that of a diode like controlled converter.

To sum up, an optimum behaviour towards the power supply grid, i.e. that of a diode locomotive, can be obtained with the following measures:

- reduction of the portion of current controlled by retarded firing angle, which can be effected with:
- the highest possible number of sequences
- increasing the overlap angle u_α in the phase cut range by inserting a special commutation inductance which slows down the commutation, mostly in the phase cut range only.

The new four-car suburban units RABDe8/16 of SBB and the new Bo'Bo' locomotives of the Rhaetian Railway have been equipped with the four-sequence economic circuit, whereby the field weakening can be taken advantage of in an economical way through the regulation of the separate excitation of the traction motor by means of controlled converters (Fig. 14 and 15).

2.131.3 Effect of d.c. smoothing on the power supply

As mentioned earlier, ideal d.c. smoothing has been assumed in the previous paragraphs. For 16 2/3 Hz tractive units in particular, there is an interest in installing smoothing chokes of limited performance. Without going into details for the calculation of the converter-power supply behaviour in case of not ideal d.c. smoothing, the result of such a calculation can be summarized as follows [1]:

- the psophometrically weighted disturbing currents are practically not influenced by an increasing d.c. ripple
- the fundamental oscillation is increased, independently of the number of sequences; for a d.c. ripple of 60% by 14 to 18%
- of the lower harmonics, the 3rd und 5th increase with growing d.c. ripple at small angles in two-sequence circuits; in four-sequence and eight-sequence circuits, they decrease if there are large overlaps. The harmonics of the 7th to the 11th order, however, are practically unaffected in all circuits
- the upper frequency range is practically unaffected by d.c. ripple.

To sum up, we can establish that an increasing d.c. ripple leads to somewhat smaller power supply harmonics of the lowest order, and simultaneously to a noticeable deterioration of the power and displacement factors. The upper harmonics of the power supply, however, are much less dependent on the d.c. smoothing reactance than on the commutation reactance and the number of sequences in the converter described in the previous paragraphs.

2.131.4 Some results of measurements made in Switzerland with various thyristor controlled tractive units

From the many comprehensive measurements made by various railway administrations, we notice that, generally, the interference voltages measured in track circuits and telecommunication cables reflect the power supply harmonics in the section concerned. As mentioned before, the "reception sensitivity" of the disturbed installations can be reduced by special measures, otherwise it is possible to develop safety installations which are not affected by a.c. currents. Depending on the type of cable and the safety system in use, the obstacle for the immediate introduction of converter techniques with thyristor control may be the safety system rather than the communication equipement, or vice-versa. In any case it is essential that the thyristor controlled tractive units to be built have an optimized behaviour towards the power supply system, including of course, a good power factor. Figure 16 shows the circuit diagram of the BLS Re4/4 No. 161 thyristor-controlled test locomotive with eight-sequence circuit. This engine was used some years ago to measure and confirm the calculations of the influence on power supply of the various multi-sequence circuits with and without commutation choke. The circuit shown is symmetric for the two bogies, i.e. two parallel connected traction motors each could be switched in the tests for two, four or eight-sequence operation. The two commutation chokes are magnetically coupled. In addition to the saving of weight, thanks to the elimination of a second iron frame, this measure results in a slight improvement of the power factor for part-load control.

Figure 17 illustrates the maximum values of the amplitude of the harmonics in the power supply current measured with this locomotive for two, four and eight-sequence converter circuits with and without commutation choke. The figure shows that the eight-sequence circuit with commutation choke attains practically the same quality as the converter in diode like operation. Figure 18 shows how the increase of the number of sequences results in an improvement of the power factor at low speeds already. The relative fall-offs as a function of the switching time are also much smaller. From about 11 km/h upwards, the power factor of the two and four-sequence circuits is better than that of the diode locomotive. The psophometrically weighted disturbing current I_p - reflecting also the interference in the telecommunication cables - decreases considerably with the number of sequences. With an additional commutation choke it can again be materially reduced.

Figure 19 shows the influence of the multi-sequence converters on the track circuits and telecommunication cables during the tests with the BLS locomotive No. 161. The effect of the multi-sequence control is evident:

- the effective track interference voltage is reduced; the increase is in smaller jumps.
- the interference voltage U_{LP} in the telecommunication cables can be noticeably reduced.

Resonance phenomena for direct-motor, as well as for static converter locomotives, find their expression in Figure 20. They were noticed for the first time during measurements on the SBB section Winterthur-Gossau in island feeding.

The following tractive units were being tested:

Type	Use	1-hour rating	Converter circuit
SBB Re4/4^{II} No. 11101-11304	mainline locomotive	4'780 kW (at 100 km/h)	no converter, a.c. motors
BLS-Re4/4 No. 162-189	mainline locomotive	4'980 kW (at 75 km/h)	uncontrolled rectifier (with tap changer)
BLS-Re4/4 No. 161	mainline locomotive	4'980 kW (at 75 km/h)	4-sequence semi-controlled circuit
ET - 420 DB	suburban multiple unit (12 axles)	2'300 kW (at 72 km/h)	2-sequence semi-controlled circuit
Eem6/6· No. 17001-17006	heavy shunter	788 kW (at 24 km/h)	fully-controlled circuit (push-pull circuit) without sequence operation

This particular island feeding represented an extremely weak power supply arrangement which in reality would be temporary only in case of a breakdown. Figure 20 shows clearly that the contents of harmonics in the power supply depends on the reactive power consumption of the converters; the locomotive Eem6/6 with its fully controlled converter generates the relatively highest harmonics. The lowest values are generated by the BLS diode locomotive. The oscillograms of the catenary voltages (top, Fig. 20) infer that compensating oscillations are excited due to the converter commutation. In the lower illustration in Figure 20 they are clearly expressed by the resonant rises. According to calculations, the feeding system of this line section has a natural frequency of approx. 500 Hz. This natural frequency is depending on the series inductance between the power station and the traction motors of the tractive units. Decisive for the capacitance is the total transverse capacitance with the catenary capacitance in the foreground. If these factors vary - the number of substations of course also plays its part - the natural frequency can alter considerably.

2.132 The quenchable asymmetric bridge circuit (LUB) for an improved power factor, phase angle control, turn-off control, sector control

In the previous paragraphs it was explained how means and ways could be found for the semi-controlled converter circuits with phase angle control to improve the power factor even at low speeds. It must be understood, however, that further efforts were made to increase the power factor to the level of direct-motor tractive units. Figure 21 illustrates the power performance for various control modes of single-phase converters. The conventional firing angle control for converters with semi-controlled bridge circuits (Fig. 21a) results in a lag of the supply current over the voltage. In the so-called turn-off control the output voltage is controlled via angle β, causing the quenching angle γ. This requires quenchable rectifier arms. The displacement factor is then made capacitive. With a combination of two converters, one working in the firing angle mode, and one in the turn-off mode, it is possible to obtain a displacement factor nearing 1. This is also the case for the so-called sector control, which combines firing angle and turn-off control, see Figure 21c.

The appropriate circuits were shown in several publications [2], [3]. A common feature is that the grid current is quenched before the grid voltage passes zero. The magnetic energy stored in the grid and transformer inductances is transferred to a capacitive energy store in the circuit. This store is later discharged into the load to prepare it for a new charge. The required output voltage is set by the angles β and γ.

Two types of circuits have emerged, depending on the number of capacitors used, the 2C and the 3C circuit. Figure 22 illustrates the general arrangement of the quenchable asymmetric bridge circuit (LUB). The highest power factor for a two-bridge sequence circuit with LUB can be obtained, when the "gravity-centre" of the current is continuously towards the front. The objective, therefore, is to obtain firing at the start of the half-wave and turn-off control over the whole half-wave range. This is possible with both the 2C LUB and the 3C LUB. We have to take into consideration, however, that the quenching of the primary current by turn-off control results in overshoot voltages due to catenary and transformer inductances. These overshoots are damped in the 2C circuit through the commutation capacitors C1 and C2. If turn-off is required at the peak of the pulsating current, for instance in starting, we need relatively large commutation capacitors. To avoid this, the turn-off angle for starting current must be limited at approx. 30° in order to turn off after the current has slightly decreased. The voltage adjustment is then made through firing angle control. For smaller currents it is not necessary to limit the turn-off angle. Depending on the current, the turn-off angle can be increased to approx. 80° in the part-load range. The 3C LUB has an additional capacitor C3 for the purpose of more effectively damping the overshoot voltages. It also allows even at maximum starting current a liberal turn-off of the voltage over the whole half-wave.

Essentially, the selection of the type of circuit depends on the intended use of the vehicle. For suburban trainsets running mostly with starting current, the 3C LUB is more suitable. For mainline locomotives, where starting plays a minor part, compared to the normal running in the part-load range, the simpler und cheaper 2C LUB seems to present the more economic solution.

Test runs of DB during the past two years with the ET 420 suburban trainset, as well as with the two-frequency locomotive 181.2, showed encouraging results after some alterations, amongst others due to high transient overshoot voltages at the main thyristors at current turn-off [3]. In particular, they confirmed the improved power factor λ as shown in Figure 23. During a normal passenger run from Frankfurt to Saarbrücken, an average power factor λ = 0,91 was established. It appears that for a two-bridge sequence circuit the reduced reactive power consumption of the quenching circuit can be fully obtained, even if one bridge only is equipped with LUB. In which of the two bridges it is installed depends on the use of the unit and the desired curve of the power factor in the firing or turn-off angle range. In tractive units which have to accelerate very often (e.g. suburban trains), the energy absorbed in the lower speed range on switching through the first bridge is negligible compared to the total starting energy. Therefore, the LUB is inserted in the second bridge, whilst main-line units in the part-load range are controlled with the first bridge and consequently the LUB is installed in the first bridge.

But still, the problem of the harmonics is not completely solved with the LUB arrangement. It appears that the 3C circuit with turn-off angle control only, shows better results than the 2C circuit with firing angle and turn-off control. The ORE experts committee A 122 conducted tests with a suburban trainset ET 420 of DB. Measurements were made in a normal, two-sequence semi-controlled asymmetric bridge circuit, on 1C LUB, 2C LUB and 3C LUB. The 1C LUB and the 2C LUB were in the first bridge, the 3C LUB in the second bridge. The best values of LUB by far were obtained with the 3C circuit. The results of the frequency analysis are shown in Figure 24, comparing the normal circuit and the 3C LUB with diode like operation of the second bridge. The amplitudes of the 5th upto the 21st harmonics produced by the 3C LUB are clearly higher, some even considerably so, than those of the two-sequence firing angle control. From the 22nd harmonic upward they are lower, but even those produced by the two-sequence firing angle control are harmless. With respect to the first harmonic current, the psophometrically weighted disturbing current of the 3C circuit reaches about the same peak value as that of the normal ET 420 circuit.
(ORE = Office for Research and Experiments of the UIC)

2.2 Summary of converter systems with pulsating current traction motors

It may be said that for the aforementioned reasons, application is now generally made only of the semi-controlled asymmetric bridge circuit with phase angle control and two-sequence or multi-sequence switching. The optimization of the commutating reactance in the phase cut range of multi-sequence circuits, as well as the ripple in the d.c. circuits, has led to results which are equal to those of an uncontrolled diode bridge. The power factor in particular can be improved by turn-off angle control, or a combination of firing angle control and turn-off control. The railway administrations having safety and telecommunication equipement of various designs (d.c. or a.c. safety circuits of various frequencies, old or new cables), the emphasis of the development - particularly with respect to the harmonics - differs from administration to administration. On the other hand, the efforts for obtaining a high power factor are of equal importance everywhere. However, it would be unwise to improve the power factor at the expense of dead weight in the tractive unit (e.g. capacitors, filters, etc) or reactive power compensators in the power supply stations. It is therefore essential to support all the efforts leading to a small reactive power consumption through appropriate control of the thyristors.

Attention must also continue to be paid to the problem of superimposing the influence of the simultaneous operation of tractive units taking into account their different position at different times. In all these matters, and with respect to the problems discussed earlier in this report, ORE Committee A 122 is performing very valuable research work.

In the investigations made so far the even-numbered harmonics have been rather neglected. They can take their origin through faulty, aperiodic firing from half-wave to half-wave, or through very fast displacement of the firing angle, as can happen in suburban trainsets with high acceleration. The former can be controlled by supervision of the firing operation, the latter must be tolerated to some extent, bearing in mind that even-numbered harmonics of higher amplitudes may be generated when switching on a transformer from the no-load condition.

Regenerative braking is possible with phase angle control. When driving, the corresponding bridge circuit must be semi-controlled, when braking, it must, however, be fully controlled.

3. CONVERTER TRACTION SYSTEMS WITH ASYNCHRONOUS MOTORS FREQUENCY CONVERSION TECHNIQUES

3.1 Background

The traction with the three-phase asynchronous motor is an old dream of the vehicle designers. Efforts in this respect date back to the electrification era of several railway companies. A few examples are the three-phase traction of the Burgdorf-Thun-Railway and the Simplon Railway in Switzerland, and the Italian State Railways in Northern Italy, where the unfavourable speed-traction force regulation of the asynchronous motor fed with constant frequency directly from the grid was revealed, and together with other difficulties, e.g. a two-phase catenary, that technique had no chance of survival. The idea however of using the asynchronous motor as traction motor lingered on. Rotating machines were used in tractive units for conversion into three-phase current (e.g. synchronous phase converters of Kandó, mono-three phase converter locomotive of SNCF). All these developments resulted in a few units being built only.

The breakthrough came with the development of the static converter which gave the opportunity of using three-phase techniques with less expenditure than with the rotating frequency converter. Compared to the simple thyristor control, the separation of power supply and traction motor ist now even more pronounced.

During the severe winter of 1962/63, SBB suffered a rather high number of traction motor defects. (Then, insulating material of class B was used, whilst now material of class H is used for new motors and overhauls with considerably improved characteristics against humidity, etc). The defects which occurred in that winter induced SBB, in collaboration with the AG Brown, Boveri & Co in Baden, Switzerland, to develop the new frequency converter technique with static converters. The result of years of calculations and laboratory tests was in 1971 the testing of the first experimental unit Be4/4 12001 (figure 25). It proved the feasability of and promised success for this new technique. The experimental locomotive Be4/4 12001 of SBB has an hourly rating of 1'000 kW at 28 km/h and a top speed of 75 km/h [6]. It is at present being equipped with the latest converter elements, and after upgrading, it will remain in regular service. Figure 26 shows the basic circuit diagram of the primary and the traction motor circuits. The transformer, a simple diode bridge and a semi-controlled bridge circuit connected in series, feed an intermediate d.c. circuit, which in turn feeds the three-phase asynchronous traction motors through the inverter. The intermediate d.c. circuit should have a constant voltage over the whole driving range to save the inverters. The fluctuations of the supply voltage are, therefore, balanced in the semi-controlled converter. It will be noticed right away that no improvement could then be found for the behaviour towards the grid, compared to the simple thyristor control. The objective of SBB, however, was to combine the advantages of the three-phase asynchronous motor with a good power performance, comparable to that of the direct-motor traction, with respect to harmonics and power factor.

The four-quadrant setter developed by BBC Mannheim now offers a device which should meet all these requirements to a great extent. Measurements, however, will have to be made by SBB to confirm this in detail.

Encouraged by the results of the experimental locomotive Be4/4 12001, SBB ordered the diesel-electric shunter Am6/6 from BBC Baden and Thyssen-Henschel AG Kassel, in 1973. The technical data are:

- Continuous output at the rim (driving and braking)	1'470 kW at 13,5 km/h
- Max. tractive effort at the rim	400 kN
- Max. brake effort (electrical) at the rim	140 kN
- Max. speed	85 km/h

Figure 27 shows the general circuit diagram, and Figure 28 is a picture of this locomotive.

When driving, the three-phase generator feeds, through contactors (3) the intermediate circuit rectifiers (4), which in turn feed the inverters (10) through the intermediate d.c. circuit. The intermediate d.c. circuit contains the choke (5) and the stand-off capacitor (7). The voltage of the intermediate circuit remains at 1'500 V over the whole operating range of the diesel engine. The inverters (10) convert the d.c. voltage into a three-phase a.c. voltage which is connected to the three-phase asynchronous traction motors (12). With the inverters, or rather with their controls, it is possible to vary the amplitude and frequency of the voltage within a wide range. When braking, the inverters feed the energy generated by the traction motors into the intermediate d.c. circuit, where it is converted into heat in the brake resistance (9). The four sub-resistors are switched in through staggered firing of the appropriate thyristors. The thyristors are quenched by opening of the contactors which brings the intermediate circuit momentarily to a voltageless state [7].

Pursuing the further development of this technique, SBB placed an order in late 1976 for an electric shunter Ee6/6 with BBC Baden and SLM[1], having similar performance data as those of the Am6/6. This is a frequency converter locomotive collecting the energy from the catenary. Figure 29 shows a type drawing of the locomotive Ee6/6, and Figure 30 is the basic circuit diagram. It will be noticed that both its exterior and its circuit represent a logical development of the frequency converter technique, the emphasis of the development being placed on an advantageous behaviour of the engine with respect to the power supply grid.

The first of three diesel-electric main-line locomotives DE 2500 applying similar equipment has been put into operation in 1971 by BBC Mannheim and Rheinstahl AG Transporttechnik (now Thyssen-Henschel AG). Subsequently, this locomotive was equipped with a four-quadrant setter for the feeding of the intermeditate d.c. circuit from the catenary. The four-quadrant setter, feeding the intermediate d.c. circuit was installed in a driving trailer. Figure 31 shows this

1) SLM = Schweizerische Lokomotiv- und Maschinenfabrik
Winterthur, Switzerland

locomotive set. For a short time it was also tested on SBB lines to check the influence on the power supply system. On that occasion it was discovered for the first time that the grid resonances generated by the pulse repetition rate of the four-quadrant setter have to be closely watched. And finally, it should be noted that frequency converter locomotives were taken into service by the connecting railways of the Ruhrkohle AG, approx. nine months ago.

3.2 The general design of a frequency converter motive power unit of good behaviour towards the grid and having three-phase asynchronous traction motors

A number of articles have been published lately on this subject [4], [5], [8]; the present article can, therefore, be confined to a short description of the functioning of the inverter and the four-quadrant setter.

3.21 Inverter control for the feeding of the three-phase asynchronous traction motors

In the starting range, the inverters are operated under the sub-harmonics system, and at higher speed, i.e. higher frequencies, with staggered fundamental frequency pulsing. Figure 32 shows the flow of the various voltages in the circuit of a motor phase. The basic single-phase converter units consist of two sub-inverters each; the two sub-inverters have a two-pulse circuit for each phase. In the sub-harmonics range the two sub-inverters are switched several times during each half-wave. The switching sequence is adjustable so that the average value of the winding voltage is sinusoidal. The difference between the voltages U_2 - U_3 is the output voltage U_4 of inverter 1. The motor current follows the voltage surges with a delay due to the leakage inductance of the windings; however its average is also sinusoidal. This control process sets the amplitude and frequency of the fundamental frequence of the motor voltage, independent of each other. The three windings of the in-phase choke GD are magnetically closely coupled and they form an equal voltage in the three windings which is proportional to the linear average of the three converter output voltages. The choke replaces the missing star-point of the motor windings. It suppresses all in-phase harmonics of the three-phase net (3rd, 9th, 15th harmonics, etc). The phase voltage U_6 then appears as the sum of the output voltage U_4 and the choke voltage U_5.

As soon as a certain fundamental frequency is reached, which has to be fixed from case to case, switchover occurs to the rectangular mode with staggered pulse. The two voltage-time blocks are staggered in relation to each other; this also allows to obtain a sinusoidal phase voltage U_6.

The rectangular mode with synchronous pulses is not generally used due to the harmonics content of the motor current; there remains then a certain minimum staggered angle.

Figure 33 shows the curves of the most important electrical values for the Am6/6 SBB locomotive operating at full torque. We have to distinguish between the following frequency, resp. speed ranges:

- in the range 0 - 15,6 Hz (0 - 13,5 km/h) the voltage is increased roughly in a linear way at constant current, with the result that the traction motor operates with practically constant flux. The performance also increases in a linear way.
- from 15,6 Hz (13,5 km/h) upwards the traction motors absorb the maximum available output of the diesel engine. Up to the maximum voltage at 24 Hz (21,5 km/h) current and slip are decreased roughly in a linear way.
- in the range above 24 Hz (21,5 km/h) the maximum voltage is constant and it is the frequency only which is increased; the flux decreases. The motor now operates in the "field weakening range". For the current to remain constant in the first approximation, the rotor frequency is increased in a roughly linear way.

3.22 The four-quadrant setter

For the size of the basic inverter unit to be kept low, it is important to keep the voltage in the intermediate circuit rather constant. The data for the Am6/6, for instance, are the following:

- intermediate circuit voltage, nominal value Ug = 1'500 V
- permitted static voltage deviation ± 5%
- permitted dynamic voltage deviation - 10%, + 17%

The existing voltage fluctuations in the catenary exceeding these tolerances of the intermediate circuit, it would in fact be necessary to use phase angle control for the feeding rectifier, as installed in the locomotive Be4/4 12001. But that would again result in harmonics.

The four-quadrant setter is a pulse-controlled inverter consisting of similar single-phase basic inverter units as described under 3.21. It allows the keeping in phase of voltage and current of the power supply, and the lower harmonics in the supply remain in the order of those generated by direct-motor tractive units.

Figure 34 shows the basic circuit diagram of a two-pulse four-quadrant setter. The converter proper consists of a bridge circuit with an uncontrolled rectifier as well as an ignitable and quenchable rectifier per arm. On the d.c. side we find the booster circuit (adjusted to double the supply frequency) and the smoothing capacitor. The converter connects the terminals on the supply side with either the positive or the negative pole of the intermediate d.c. circuit, depending on the switched state of the rectifiers. The circuit can be illustrated by means of two time lag free switches S_1 and S_2; their positions can indicate the functions rectifying, pulsing and feedback of the four-quadrant setter. Figure 34 shows in a table the positions of the "switches" S_1 and S_2 at positive and negative halfwave of u_T, and in illustration b we see the momentary shape of voltages and currents at the input and output of the converter for ideally smoothed d.c. voltage. The input voltage u_S at the supply side assumes either the value $+u_-$, resp. $-u_-$, or zero, depending on the position of the "switches" S_1 and S_2. Their working frequency with the modern "fast" thyristors is (also for economic reasons) approx. 183 Hz. Through pulse width modulation, the mean value u_S of the input voltage is given a sinusoidal form. The a.c. current i_T is given its form by the power supply choke L_N. In accordance with the pulsing of the pulse width modulation, i_T is composed of a quick succession of blocks with rising and falling shape. As shown in the table of Figure 34, the current i at the output of the converter is either i_T, $-i_T$ or zero, depending on the position of the switches. It can be clearly recognized that for instance a positive a.c. voltage is fed to the negative pole of the intermediate d.c. circuit, which generates a negative d.c. current -i (example a), or similarly, a negative a.c. voltage u_T is fed to the positive pole of the intermediate d.c. circuit, which again generates a negative current -i (example c'). Thus, reactive energy flows from the d.c. to the a.c. side and creates a small phase shift between a.c. current and voltage.

The grid inductance L_N can be generated by a large leakage inductance in the transformer. This may result in a short circuit voltage of up to approx.30%.

The harmonics in the power supply can be considerably reduced through multi-pulsation. Figure 34 shows the two-pulse circuit. By dividing the secondary winding of the transformer into several sub-windings which feed the units as per Figure 34, connected in parallel, it is possible to design four, six or eight-pulse circuits which can be operated staggered. For the Ee6/6 locomotive ordered by SBB, the calculated psophometrically weighted current is for instance 6 A in two-pulse operation, 3,08 A in four-pulse operation, 1,558 A in six-pulse operation and 0,71 A in eight-pulse operation. The order was placed for a six-pulse circuit.

As implied by its description, the four-quadrant setter can work in all four quadrants: driving inductive and capacitive, as well as braking inductive and capacitive. The regenerative brake - a brake installed in practically all direct-motor locomotives of SBB, the usefulness of which one has grown accustomed to - is therefore possible in frequency converter operation.

3.23 Measurements made on SBB lines with the locomotive set DE 2500

Figure 35 shows the power supply frequency analysis in relation to the first harmonic of the DE 2500, compared to that of the new SBB suburban trainset RABDe8/16 (four-sequence phase angle control with economic circuit), and the SBB direct-motor locomotive $Re4/4^{II}$. The amplitudes up to the third harmonic are much lower for the DE 2500. For the 7th, 23rd and 41st harmonics there are, however, marked amplitude increases. The pulse repetition frequency of 720 Hz resulting on this vehicle with staggered four-pulse circuit corresponds to approx. the 43rd harmonic, and is rather identical with the amplitude rise of the 41st harmonic. The three amplitude increases mentioned have a frequency spacing of approx. 280 Hz. It must be emphasised that in various respects the locomotive DE 2500 with four-pulse circuit was not optimised with respect to the creation of harmonics in the power supply current.

The higher the frequencies of such grid resonances, the easier it is to check them on the vehicle itself. However, these problems cannot be surveyed exactly until measurements have been made with an optimised vehicle, for instance the Ee6/6.

The optimum phase position with λ towards 1 between catenary current and voltage could be proved in the above tests.

4. CONCLUSIONS AND PROSPECTS

Initially, we listed the major traction requirements for electric traction. We shall compare them now with the known and the extrapolated results of thyristor locomotives with particular emphasis on the question of how the application of converters can effectively improve the traction.

Points 1 and 2 in paragraph 1 should be discussed in the light of Figure 36. With the possible and purposeful utilization of the field weakening in vehicles with controlled or uncontrolled converters, the tractive force at maximum speed can be considerably increased (illustration 1 in Figure 36).

The elimination of the commutator in the inverter traction makes space available which can be filled with torque-creating iron, which together with higher rpm (the maximum rpm for pulsating current motors or 16 2/3 Hz direct-motors is determined by the commutator) results in a considerable increase in the available motor output. Illustration 2 in Figure 36 tries to show the development of the present inverter/field weakening circuit to a nearly rectangular speed/tractive force graph which today, however, is but a theoretical possibility. The real limit will be determined amongst others by:

- the losses of the three-phase asynchronous motor with increasing speed (stator frequency)
- the inverter output which possibly can be built into the vehicle
- the optimum tuning of inverter and traction motor output
- the ratio motor speed/driving wheel speed
- the design of the motor bearings
- the admissible axle load
- the capacity of the feeding grid

In this context we can mention the project of DB of which Figure 37 shows a sketch and a graph. As a logical development of their frequency converter locomotives Be4/4, Am6/6 and Ee6/6, SBB are likely to order a mainline locomotive Re4/4^{IV} sometime in 1978.

The requirement contained in point 4 for continuous torque control can be obtained in an electric tractive unit only if it is equipped with controlled converters. The improvement of passenger comfort and utilization of adhesion (point 5) are marked. The tests performed by SBB in 1965 [9] and subsequent tests of other administrations showed already that a steep characteristic Z = f (V) and a continuous control of the tractive effort are major factors for a full utilization of the adhesion. The pulsating current motor, as well as the asynchronous motor (the latter at constant frequency and voltage) have a steep characteristic and continuous control is possible, as described earlier. Figure 38, as an example of the investigation, shows the slip probability on starting on wet rails, measured on an SBB Re4/4^{II}, the traction motors of which could be fed alternately with undulatory current (uncontrolled bridge circuit) or single-phase a.c. for these adhesion tests. Parallel connected traction motor fields (PH) (in use on all modern mainline locomotives of SBB) provide for a field strengthening of the slipping traction motor, i.e. for an increased steepness of its Z = f (V) characteristic. It will be realized that this process results in practically the same utilization of the adhesion as can be obtained with a diode circuit. Appropriate tests (continuous increase of the brake power of the braking locomotives) allowed the simulation of a continuous traction force control, and it was established that this increased the adhesion coefficient by another 10%. The indication was then already that a continuously controlled tractive unit should result in markedly increased utilization of the adhesion, which has in the meantime been fully confirmed. But, thorough investigations are necessary in the form of tests and measurements with frequency converter locomotives, in order to study the dynamic phenomena occurring during sudden loss of adhesion of one axle or another (e.g. momentary opposed driving-braking operation of individual traction motor groups).

It may be pointed out here that the aforementioned investigations in the year 1965, as well as later measurements made by SBB, proved that the adhesion coefficient as a function of the speed- at least up to approx. 100 km/h - need not decrease, provided the mechanical part of the locomotive has been designed for an optimum utilization of the adhesion. The adhesion curve by Curtius-Kniffler, which shows a steep decrease, and which is quoted again and again in various publications, is no longer applicable for a modern bogie locomotive. In this light there are no limits with respect to the maximum useful range of tractive force and speed, as discussed under point 1 and 2. This is illustrated in Figure 39 with the adhesion coefficient as a function of speed measured under very bad conditions (rails wetted with soapy water) on the SBB Re6/6 (Bo'Bo'Bo') locomotive. Under better conditions μ_{Lok} remains constant at 0,33 over the whole speed range. The test section being low powered, measurements could be made only at speeds of up to V = 85 km/h.

D.c. tractive units with direct-motor traction have a very flat characteristic Z = f (V) on the resistance notches, which leads to the known adhesion difficulties. The chopper is a valuable instrument for an increased utilization of the adhesion of d.c. vehicles, and in addition it provides higher passenger comfort and more economic traction, since no energy needs to be wasted in starting resistances. (In city traffic, a vehicle equipped with chopper control needs approx. 15 - 16% less energy under operating conditions in a place like St.Gallen for instance than a conventional vehicle with d.c. direct-motor).

Point 6: With pulsating current motors, and in particular with the three-phase asynchronous motor, the weight of the bogies can be much reduced. The dynamic guiding forces Y and the vertical wheel loads Q can be noticeably decreased. The rails and their fastenings are subjected to less stress and the danger of distortion and lateral displacement of the track is considerably less. Since the traction motors can be of small size, particularly in three-phase operation, the wheel-base in the bogie can be decreased, which again contributes to a reduction of the quasi-static Y forces (guiding forces) [10].

Major economies can be expected in the maintenance (point 7) due to the elimination of the tap changers and - on frequency converter locomotives - the elimination of commutator and brushes. DB expect a cost saving of 10 - 15% for the frequency converter locomotive.

Point 8 has already been extensively discussed. The main problem with converter-type locomotives is the interference with the signalling and telecomunication installations, the sub-stations und power stations. A solution must be found, at reasonable expenditure, in view of the advantages of this type of traction and the progress which can be expected in future.

Finally it should be mentioned that the future may bring an interesting solution for multi-system tractive units in the sense that the intermediate d.c. circuit of a frequency converter locomotive can be fed from a chopper in d.c. operation and from the four-quadrant setter in a.c. operation.

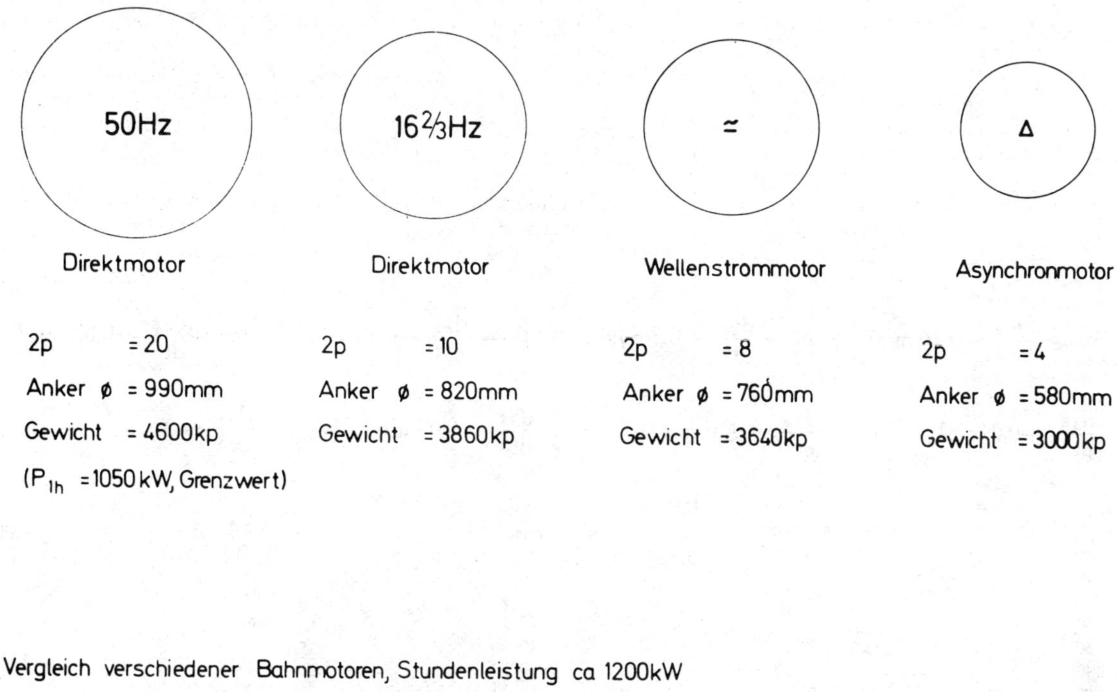

Fig. 1

Number of poles, diameter of the armature and weights of various traction motors (one hour rating approx. 1'200 kW)

Direktmotor	=	single phase a.c. motor
Wellenstrommotor	=	pulsating current motor
Asynchronmotor	=	asynchronous motor
Anker	=	armature
Gewicht	=	weight
Grenzwert	=	upper limit

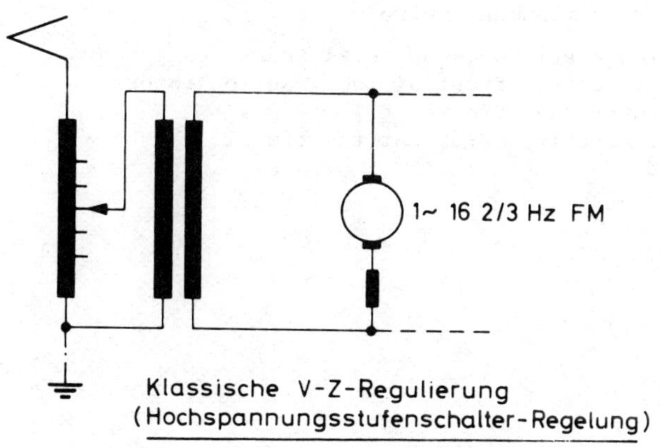

Fig 2

Conventional control of speed and tractive force by means of high-voltage tap changer for a.c. direct-motor traction

$p_m = \frac{d^2s}{dt^2} = 0{,}46\ m/s^2$ $p_m = \frac{d^2s}{dt^2} = 0{,}5\ m/s^2$ $p_m = \frac{d^2s}{dt^2} = 0{,}53\ m/s^2$ $p_m = \frac{d^2s}{dt^2} = 0{,}63\ m/s^2$

$\frac{d^3s}{dt^3} = 0{,}2\ m/s^3$ $\frac{d^3s}{dt^3} = 0{,}16\ m/s^3$ $\frac{d^3s}{dt^3} = 0{,}07\ m/s^3$ $\frac{d^3s}{dt^3} = 0$

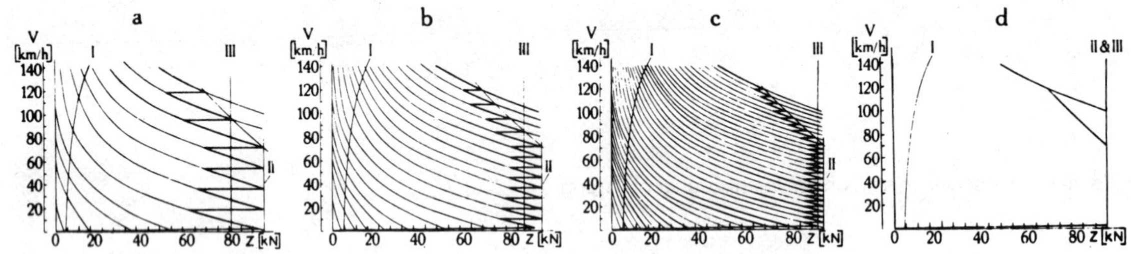

a Diagramm mit 11 Stufen
b " " 29 "
c " " 39 "
d " für stufenlose Regelung

I Fahrwiderstand auf ebener Bahn
II Höchstzulässige Anfahrzugkraft der Fahrmotoren
III Mittlere Anfahrzugkraft
Z Gesamte Zugkraft in kN am Radumfang
V Fahrgeschwindigkeit in km/h

<u>Fig. 3</u>

Vehicle graphs (speed/tractive force) for various numbers of steps

diagram with 11 steps
" " 29 "
" " 39 "
" for continuous control

I running restistance on level track
II max. tractive effort of the traction motors
III mean starting effort
Z total tractive effort at the rim
V speed

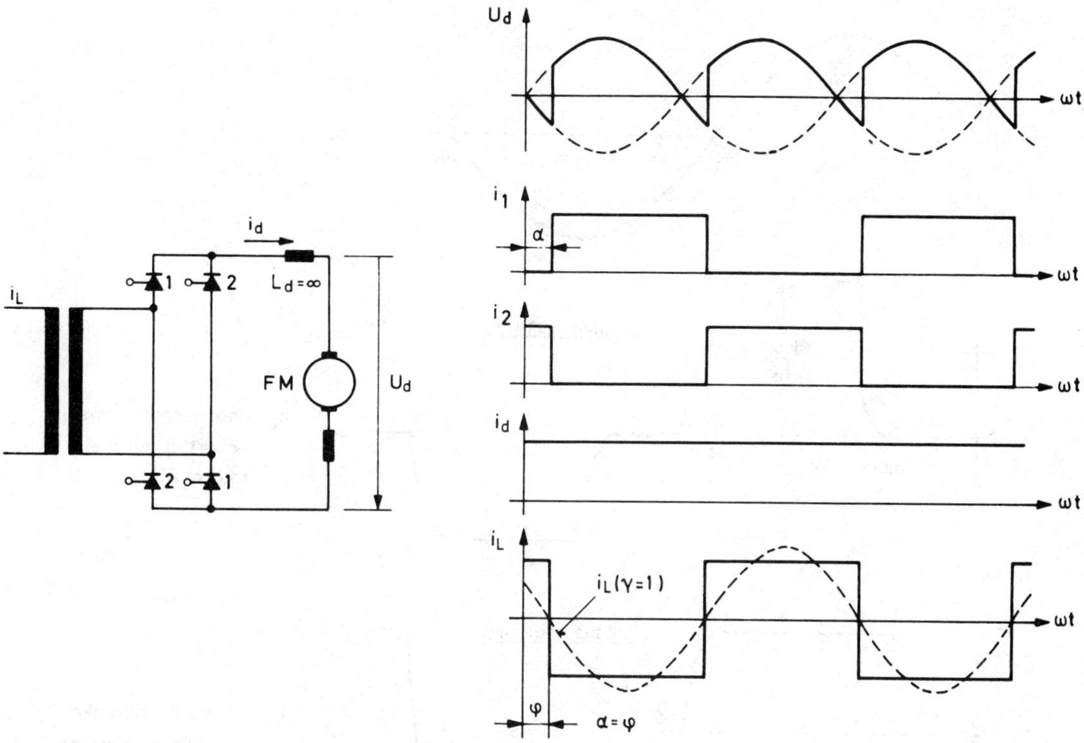

Fig. 4

Fully-controlled symmetric bridge circuit

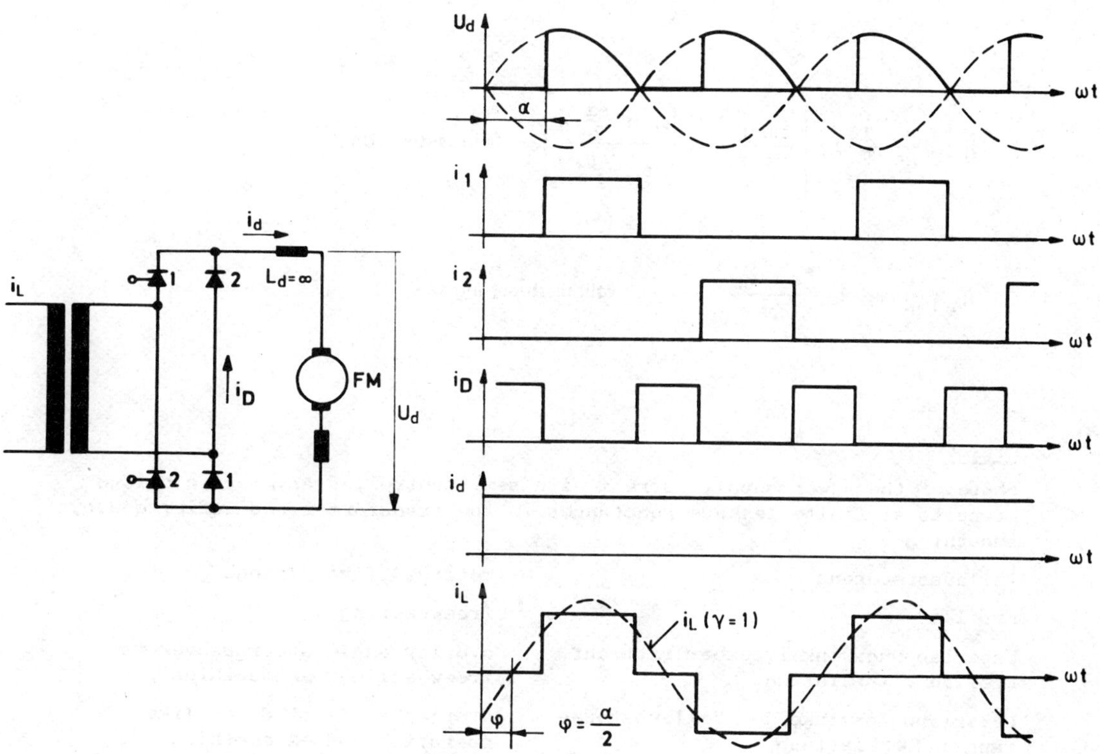

Fig. 5

Semi-controlled asymmetric bridge circuit

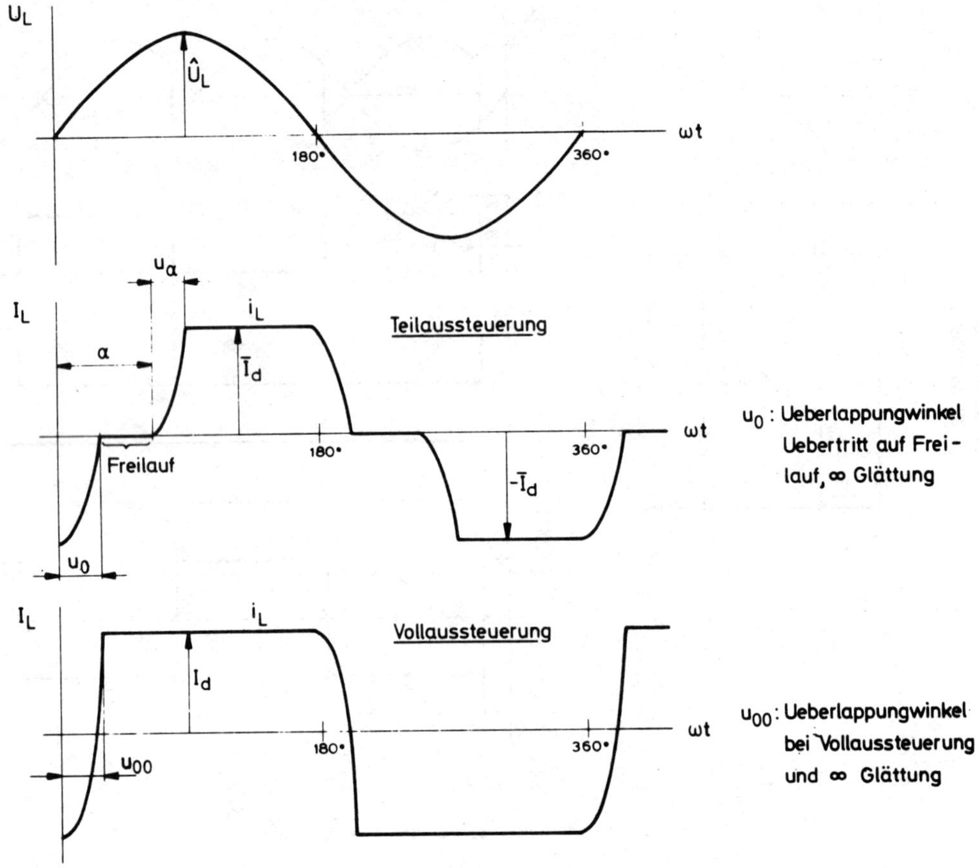

$$\hat{I}_{L_n} \cong \frac{1}{n^2} \cdot \frac{2}{\pi} \cdot \bar{I}_d \left[\frac{\sin u_0}{1-\cos u_0} + \frac{\left|\sin \left(n \frac{u_\alpha}{2}\right)\right|}{\sin \frac{u_\alpha}{2}} \right] \quad \text{Teilaussteuerung}$$

$$\hat{I}_{L_n} \cong \frac{1}{n^2} \cdot \frac{2}{\pi} \cdot \bar{I}_d \cdot \frac{\sin u_{00}}{1-\cos u_{00}} \quad \text{Vollaussteuerung}$$

Fig. 6

Shape of the power supply current with semi-controlled asymmetric bridge circuits at finite leakage reactances in the transformer and infinite d.c. smoothing

Teilaussteuerung	=	retarded firing range
Freilauf	=	freewheeling
Ueberlappungswinkel, Uebertritt auf Freilauf, ∞ Glättung	=	overlap angle at crossover to freewheeling, ∞ smoothing
Ueberlappungswinkel bei Vollaussteuerung und ∞ Glättung	=	overlap angle at diode like operation and ∞ smoothing
Vollaussteuerung	=	diode like operation

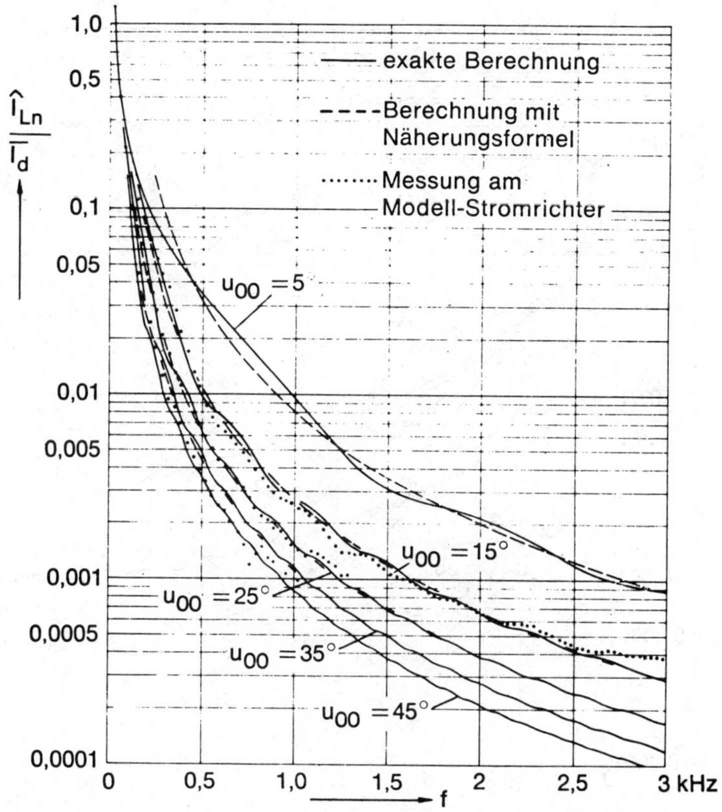

Netzstrom-Frequenzspektrum des vollausgesteuerten Stromrichters

Fig. 7

Computed power supply current frequency spectrum of the diode like operated converter at various overlap angles

——— exact calculation
----- calculation with approximation formula
.... measurement at model converter

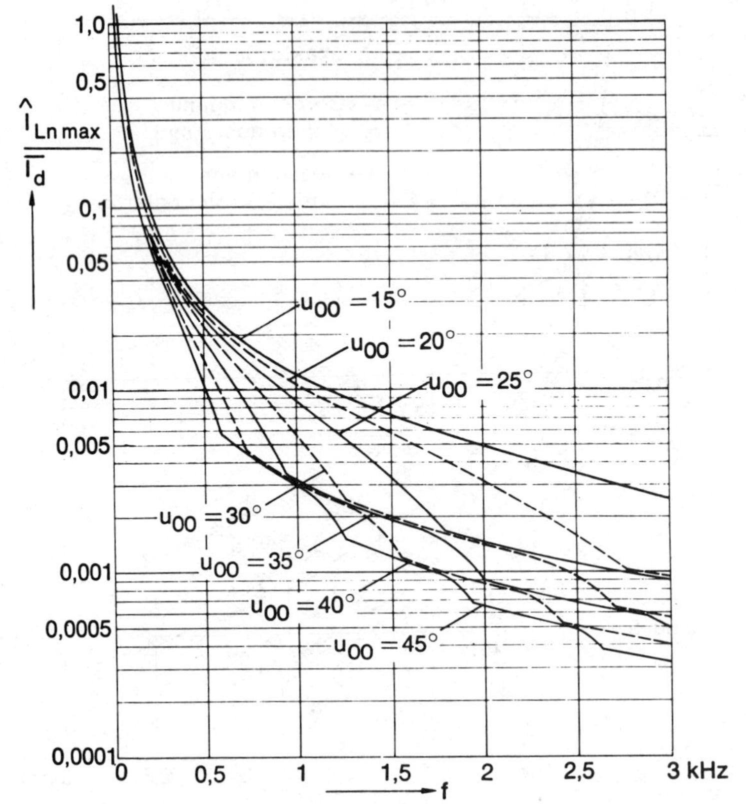

Netzstrom-Frequenzspektren der halbsteuerbaren Brückenschaltung
Maximale, beim Durchsteuern des Stromrichters auftretende Amplituden

Fig. 8

Computed power supply current frequency spectrum of the semi-controlled converter bridge circuit; maximum amplitudes on switching through the converter

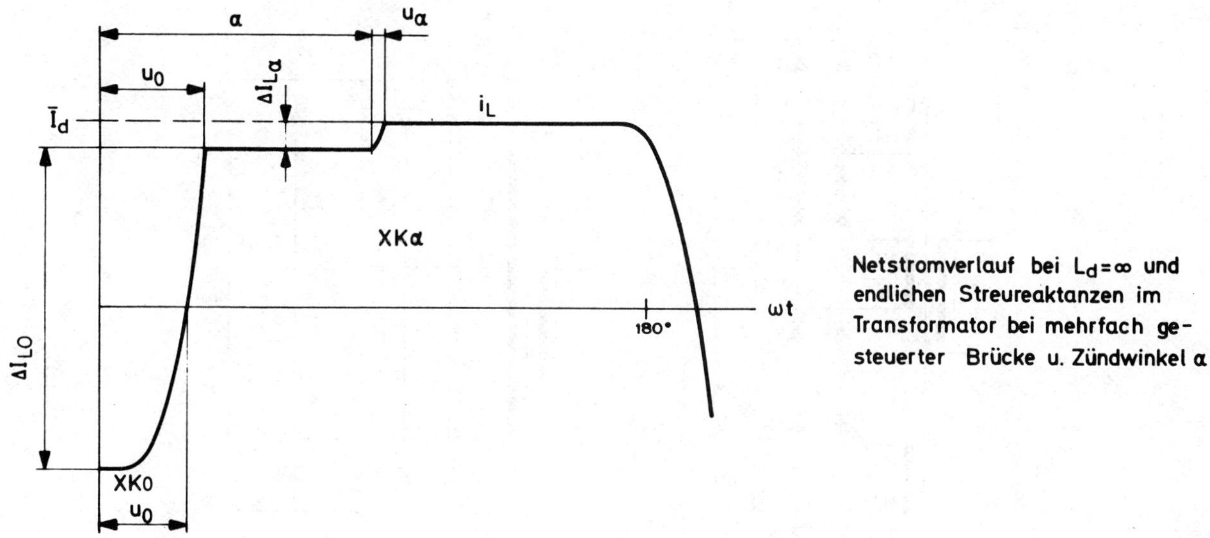

$$\hat{I}_{Ln} \cong \frac{1}{n^2} \cdot \frac{2}{\pi} \left[\Delta I_{LO} \frac{\sin u_0}{1 - \cos u_0} + \Delta I_{L\alpha} \frac{\left| \sin\left(n \frac{u_\alpha}{2}\right) \right|}{\sin \frac{u_\alpha}{2}} \right] \qquad \Delta I_{L\alpha} = \frac{\bar{I}_d}{N}$$

(N = Anzahl Stufen)

Fig. 9

Shape of the power supply current and approximation formula for the amplitudes of the odd-numbered harmonics in multi-sequence bridges with finite leakage reactances in the transformer and ideal d.c. smoothing. firing angle α

N = number of sequences

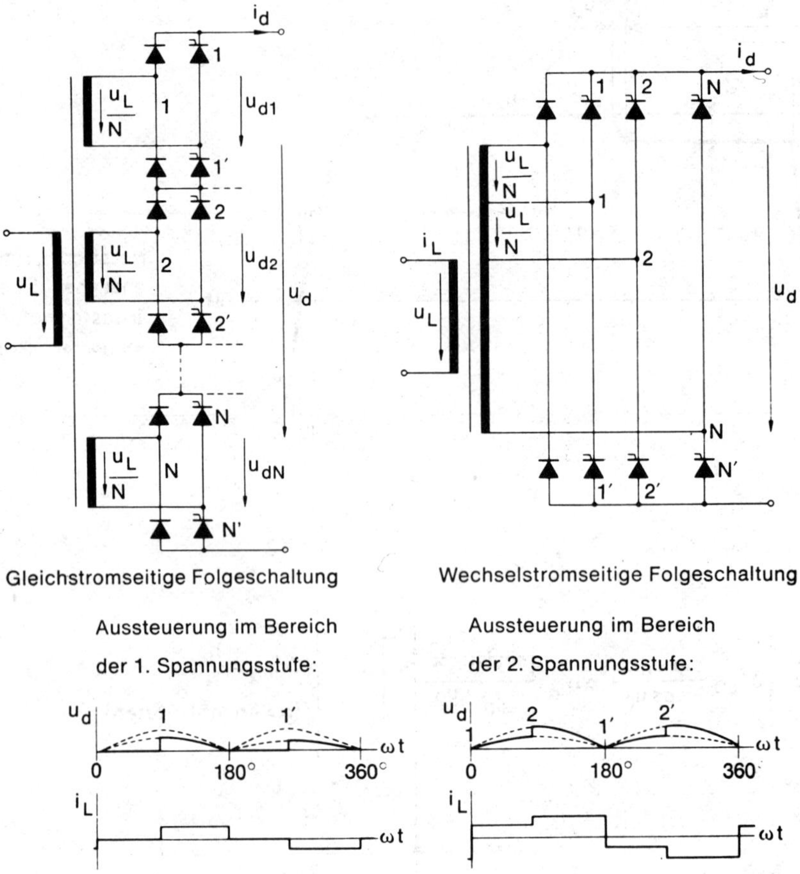

Konventionelle Mehr-fach-Folgeschaltungen

Fig. 10

Conventional multi-sequence circuit (d.c. side and a.c. side) firing angle range in the 1st voltage sequence 2nd

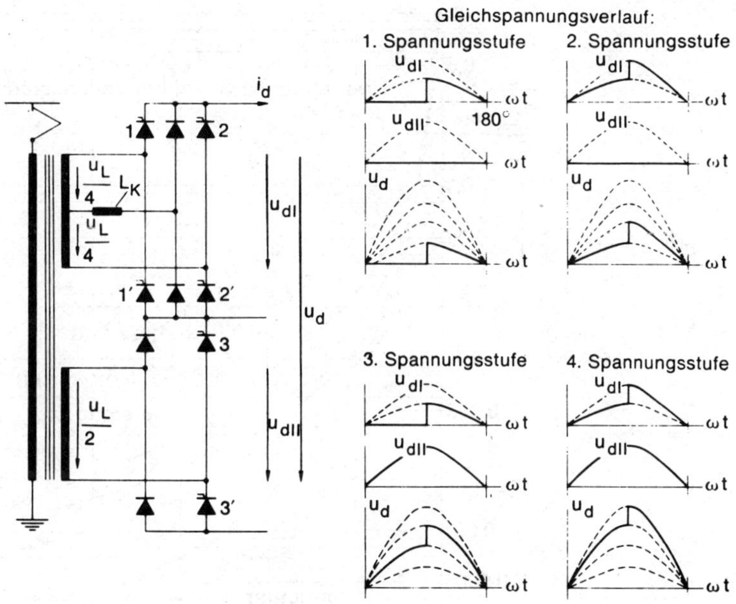

Vierstufige Stromrichter-Sparschaltung mit Zusatz-Kommutierungsdrossel

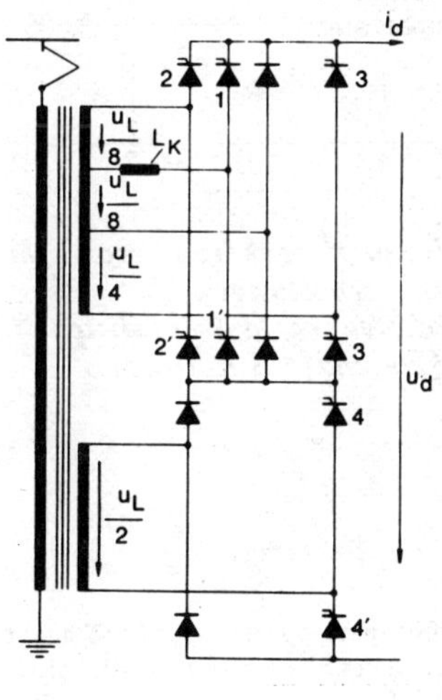

Achtstufige Stromrichter-Sparschaltung mit Zusatz-Kommutierungsdrossel

Steuerschema:

Spannungsbereich $u_d / u_{d\,max}$	Kontinuierlich aufsteuernde Ventile	vollausgesteuerte Ventile
0 ... 1/8	1,1'	–
1/8 ... 2/8	2,2'	1,1'
2/8 ... 3/8	1,1'	3,3'
3/8 ... 4/8	2,2'	1,1'; 3,3'
4/8 ... 5/8	1,1'	4,4'
5/8 ... 6/8	2,2'	1,1'; 4,4'
6/8 ... 7/8	1,1'	3,3'; 4,4'
7/8 ... 8/8	2,2'	1,1'; 3,3'; 4,4'

Fig. 11

Multi-sequence economic converter circuit with additional commutation choke

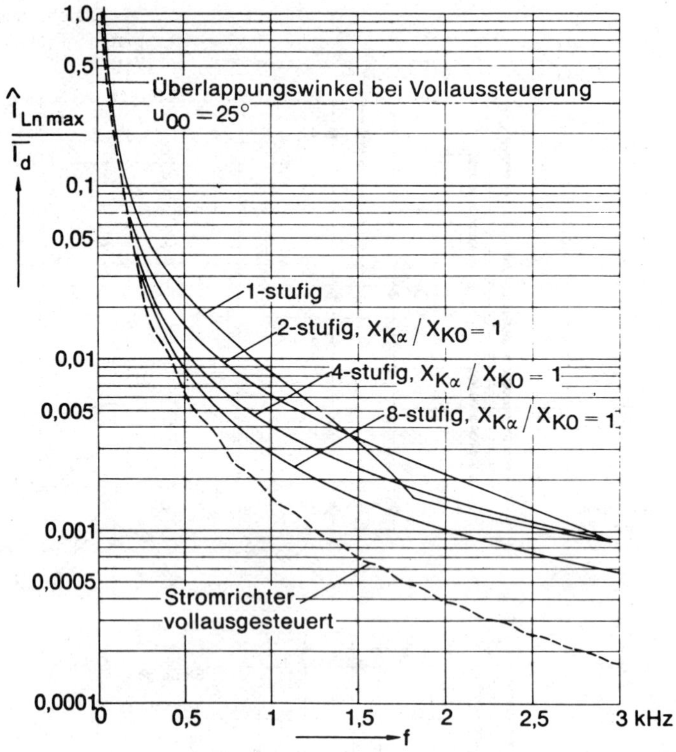

Netzstrom-Frequenzspektren von Mehrfach-Folge-schaltungen
Maximale, beim Durchsteuern der höchsten Spannungsstufe auftretende Amplituden

<u>Fig. 12</u>

Computed power supply current frequency spectra of multi-sequence circuits; maximum amplitudes occurring on switching through the highest voltage stage.

$X_{K\alpha} = X_{KO}$

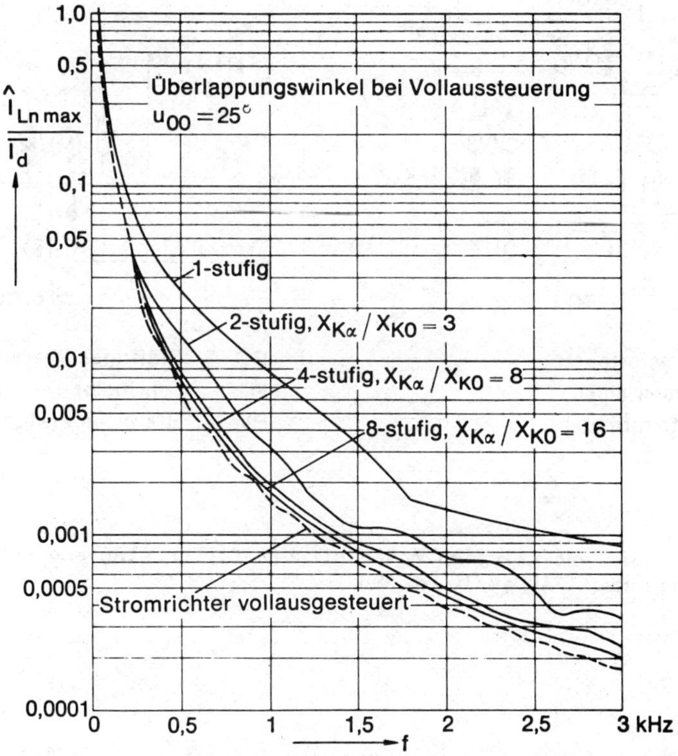

Netzstrom-Frequenzspektren von Mehrfach-Folgeschaltungen
Maximale, beim Durchsteuern der höchsten Spannungsstufe auftretende Amplituden

Fig. 13

Computed power supply current frequency spectra of multi-sequence circuits; maximum amplitudes occurring on switching through the highest voltage stage

$X_{K\alpha} \neq X_{KO}$

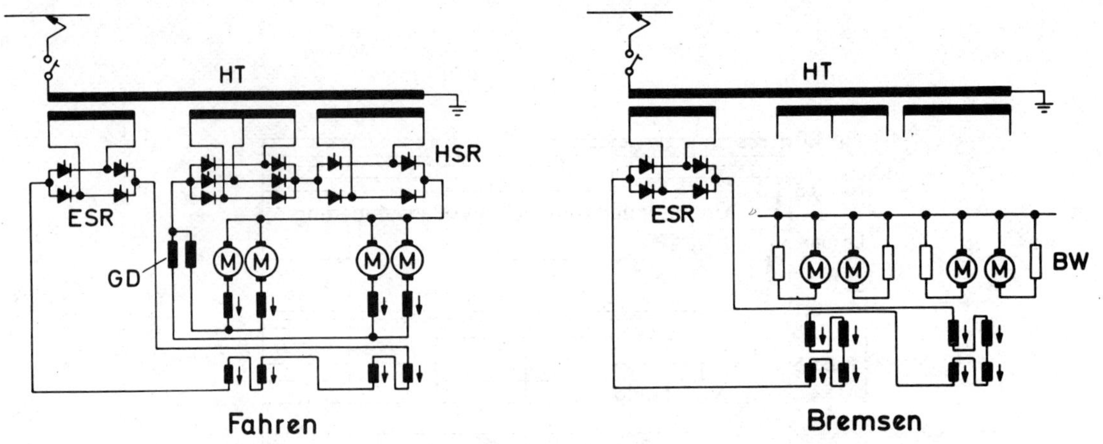

HT	Haupttransformator	GD	Glättungsdrosselspule
HSR	Hauptstromrichter	M	Fahrmotor
ESR	Erregerstromrichter	BW	Bremswiderstand

Fig. 14

Primary circuit and traction motor circuit for driving and braking of the suburban trainset RABDe8/16 of SBB

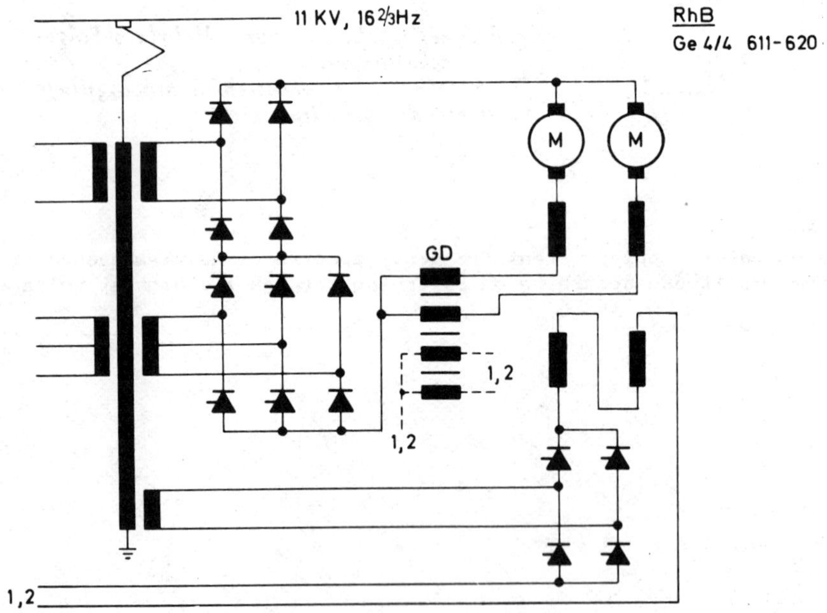

Fig. 15

Primary circuit and traction motor circuit of the Bo'Bo' Ge4/4 locomotive of the Rhaetian Railway

BLS - Stromrichterlokomotive Re 4/4 Nr 161

Prinzipschaltbild der 8-stufigen Schaltung

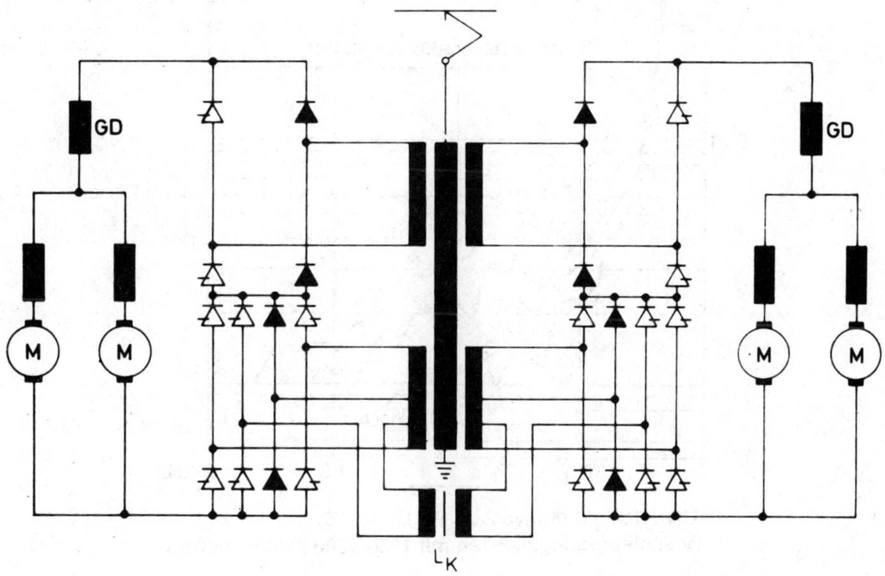

Drehgestell I

Drehgestell II

GD Glättungsdrosselspule
M Fahrmotor
L_K Kommutierungs-Drosselspule

Fig. 16

Primary circuit and traction motor circuit of the BLS Re4/4 locomotive No. 161 (experimental locomotive) with economic eight-sequence circuit

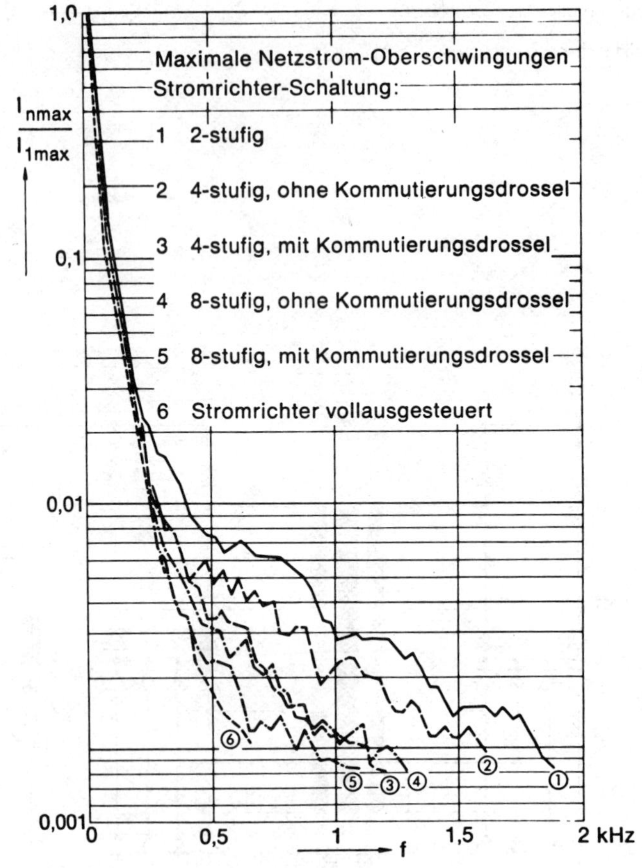

Thyristorlokomotive BLS-Re 4/4 Nr. 161
Beschleunigungsfahrten mit 1100/1300 A-Motorstrom

Gemessene Maximalwerte der Netzstrom-Oberschwingungsamplituden

Fig. 17

Maximum values of the power supply current harmonics for two, four and eight-sequence circuits with and without commutation choke, measured on BLS locomotive No. 161

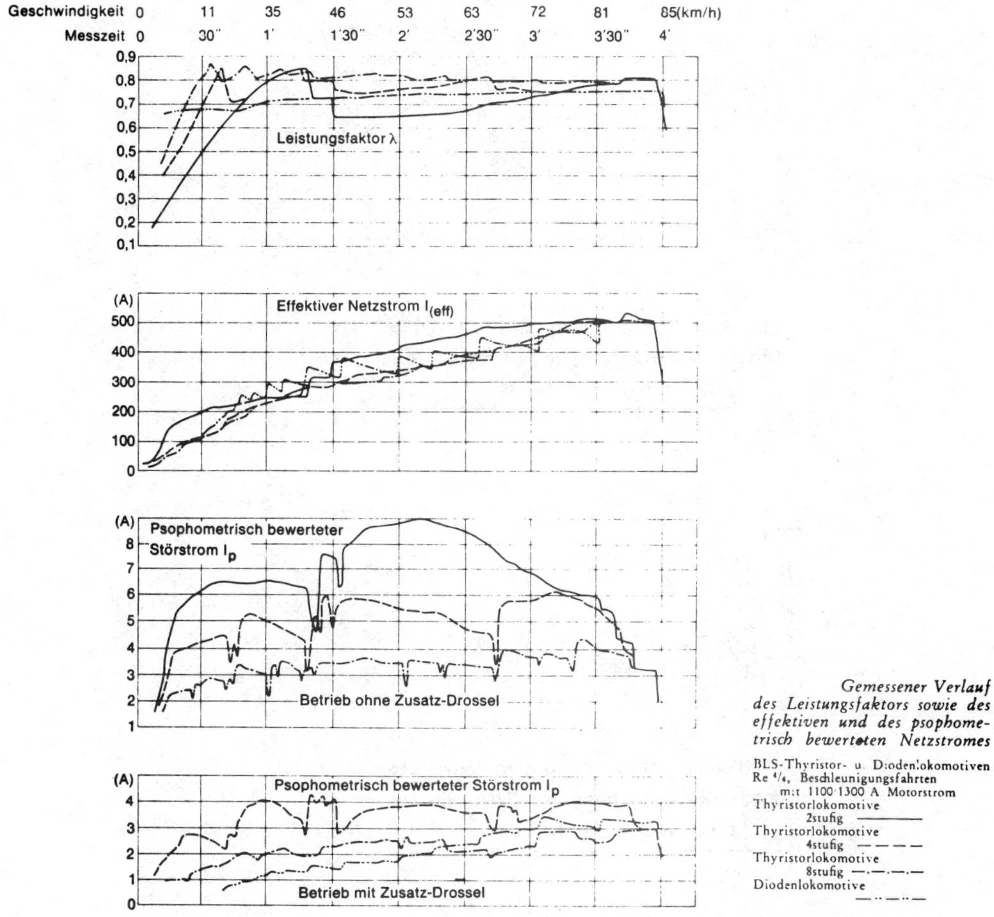

Fig. 18

Measured power factor, effective and psophometrically weighted supply current for two, four and eight-sequence circuit and for a diode locomotive

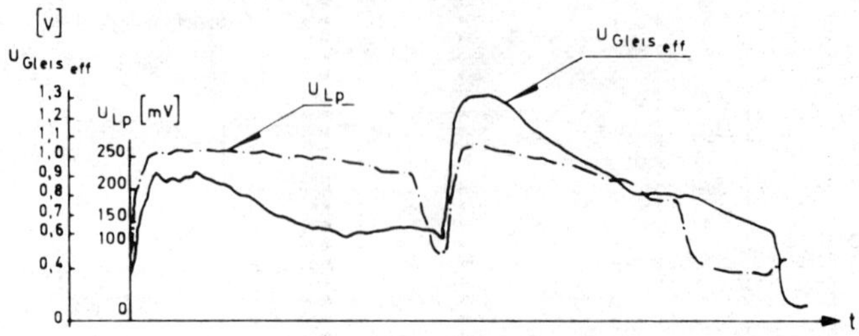

Eff. Gleis - Störspannung und bewertete
Längs - Störspannung im Fernmeldekabel
2 - Stufig B L S Lok. Nr. 161

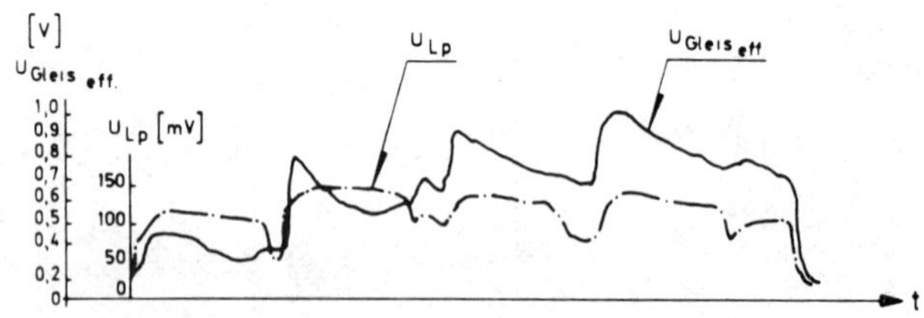

Eff. Gleis - Störspannung und bewertete
Längs - Störspannung im Fernmeldekabel
4 - Stufig B L S Lok. Nr. 161

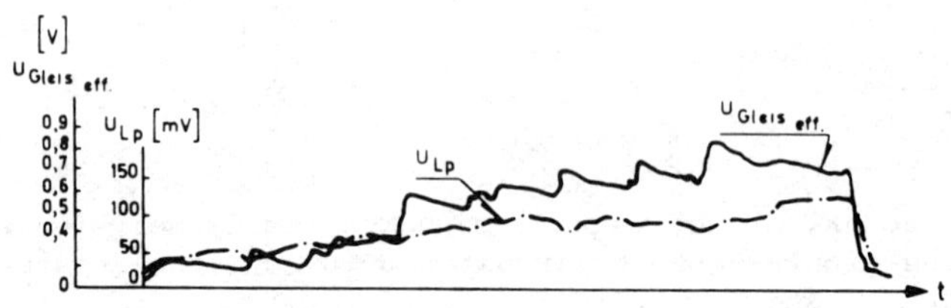

Eff. Gleis - Störspannung und bewertete
Längs - Störspannung im Fernmeldekabel
8 - Stufig B L S Lok. Nr. 161

Fig. 19

Effective track interference voltage and weighted interference voltage along the telecommunication cable for two, four and eight-sequence circuit

Masstäbe: Spannung (oben): 17,5 kV/Einh. Strom (unten): 800 A/Einh. (Bilder 1,2,3) bzw.
Zeit : 10 ms/Einh. 400 A/Einh. (Bild 4) bzw.
145 A/Einh. (Bild 5)

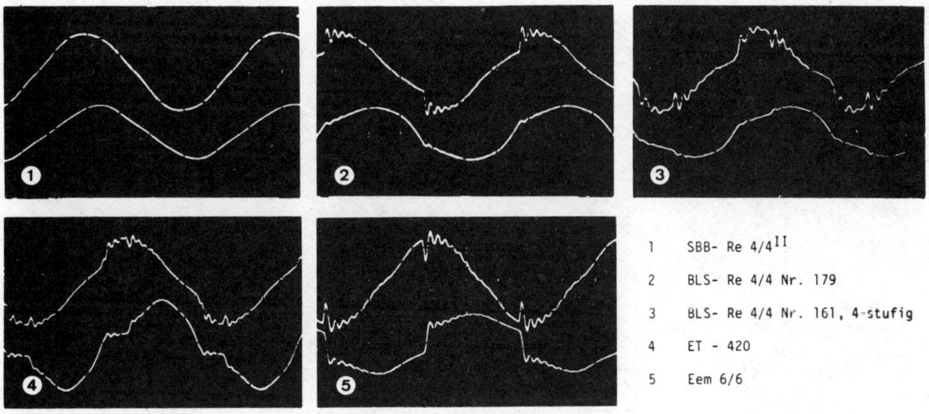

Abb. ① Oszillogramme von Fahrleitungsspannung und -strom

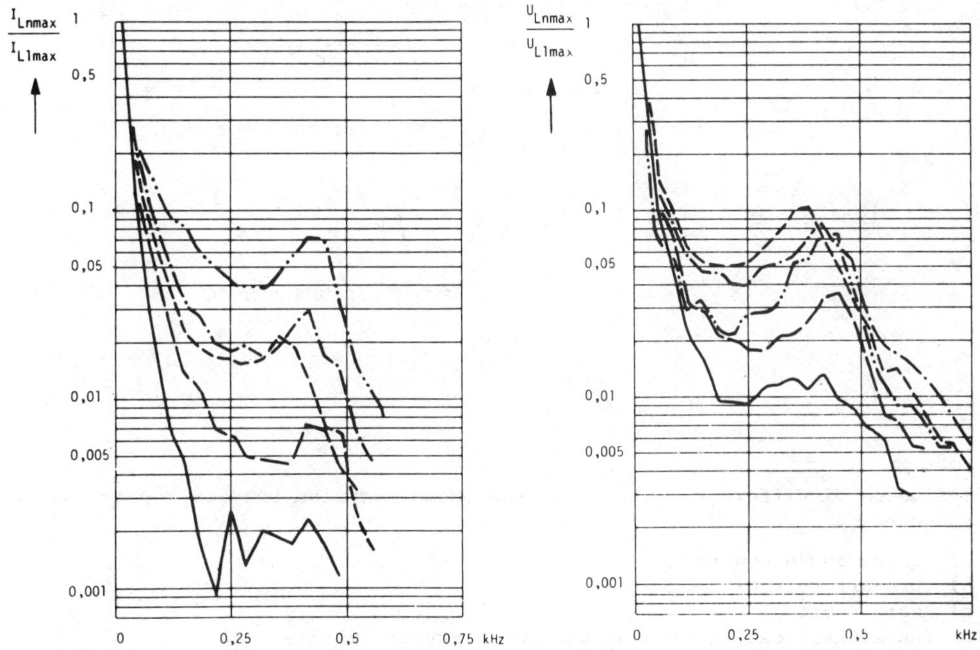

Abb. ②

—— SBB- Re 4/4II (I_{L1max} = 450 A)

— — BLS- Re 4/4 Nr. 179 (I_{L1max} = 500 A)

- - - BLS- Re 4/4 Nr. 161, 4-stufig (I_{L1max} = 570 A)

— · — ET 420 (I_{L1max} = 358 A)

— ·· — Eem 6/6 (I_{L1max} = 97 A)

Frequenzspektren von Fahrleitungsspannung und -strom (Höchstwerte)

Vergleichsmessungen mit verschiedenen Triebfahrzeugen auf der Strecke Winterthur - Gossau (Insel-Schaltung mit 1 Speisetransformator à 3 MVA und 45 km Fahrleitung; alle Grössen im Unterwerk Winterthur-Grüze gemessen).

__Fig. 20__

Comparative measurements with various tractive units of DB, BLS and SBB on the SBB line Winterthur - Gossau in island feeding (1 transformer of 3 MVA, 45 km of catenary, all values measured in the Winterthur-Grüze substation)

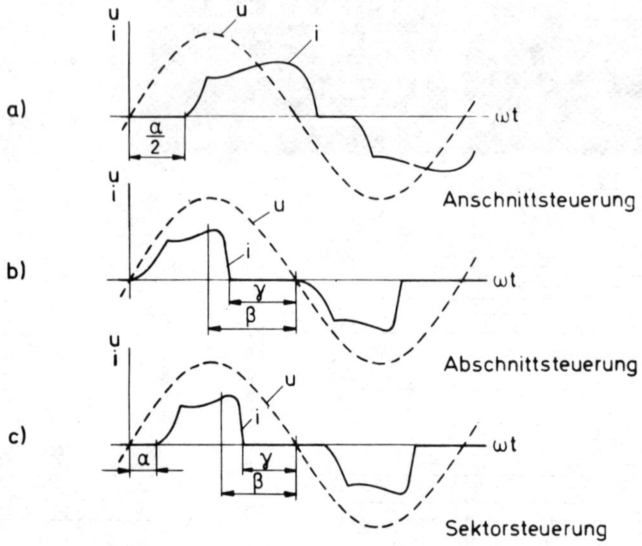

Fig. 21

Influence of different controls on the phase position between supply voltage and current

a) firing angle control
b) turn-off control
c) sector control
u fundamental oscillation of the power supply voltage
i supply current

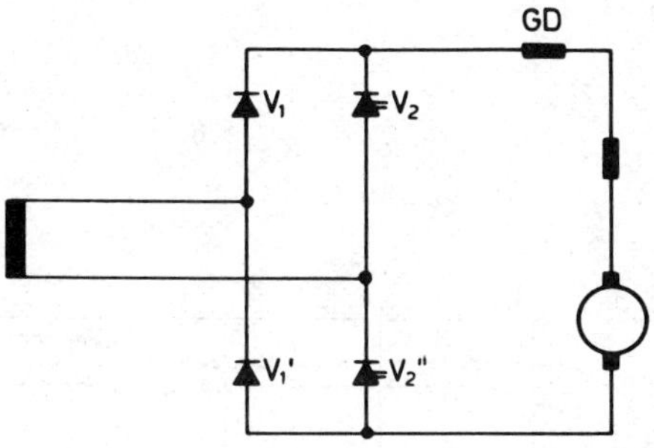

Genereller Aufbau der LUB

GD Glättungsdrosselspule
V1-V2" Stromrichterventile

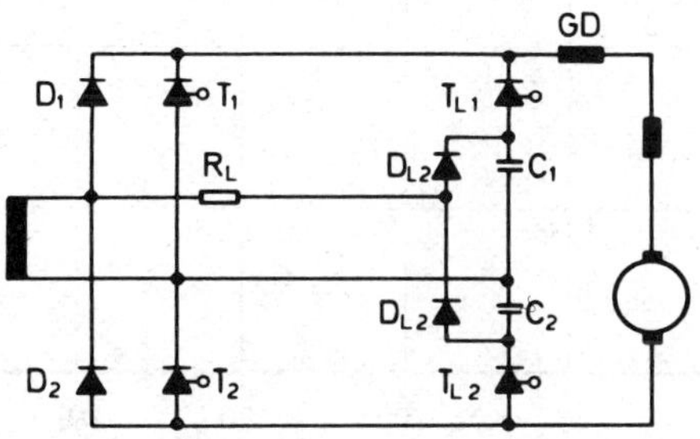

2C - Schaltung

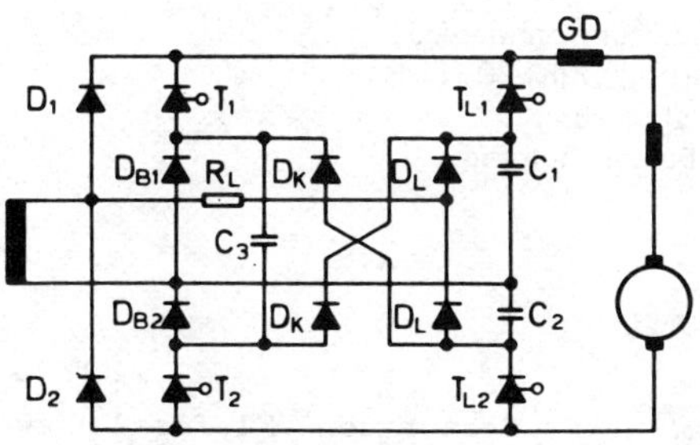

3C-Schaltung

GD	Glättungsdrosselspule
T	Hauptthyristor
T_L	Löschthyristor
D	Freilaufdiode
D_B	Blockdiode
D_L	Ladediode
D_K	Koppeldiode
R_L	Ladewiderstand
C_1, C_2	Löschkondensator
C_3	Speicherkondensator

Fig. 22

Quenchable asymmetric bridge circuit, basic diagram 2C circuit, 3C circuit

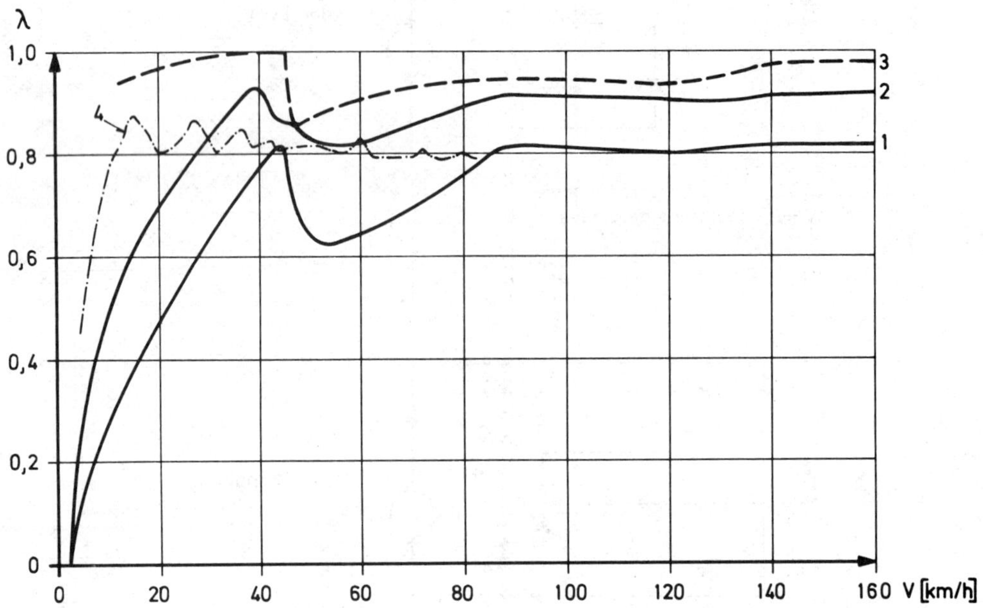

1 λ bei Anschnittsteuerung ohne LUB
2 λ bei Sektorsteuerung mit 2C-LUB
3 cos φ_1 bei Sektorsteuerung
4 λ bei 8-stufiger Folgeschaltung

Fig. 23

Power factor λ and fundamental power factor cos φ1 for locomotive 181.208 of DB on starting with 138% supply current

1) with firing angle control (without LUB)
2) with sector control (with 2C LUB)
3) cos φ1 with sector control
4) λ with firing angle control in eight-sequence circuit

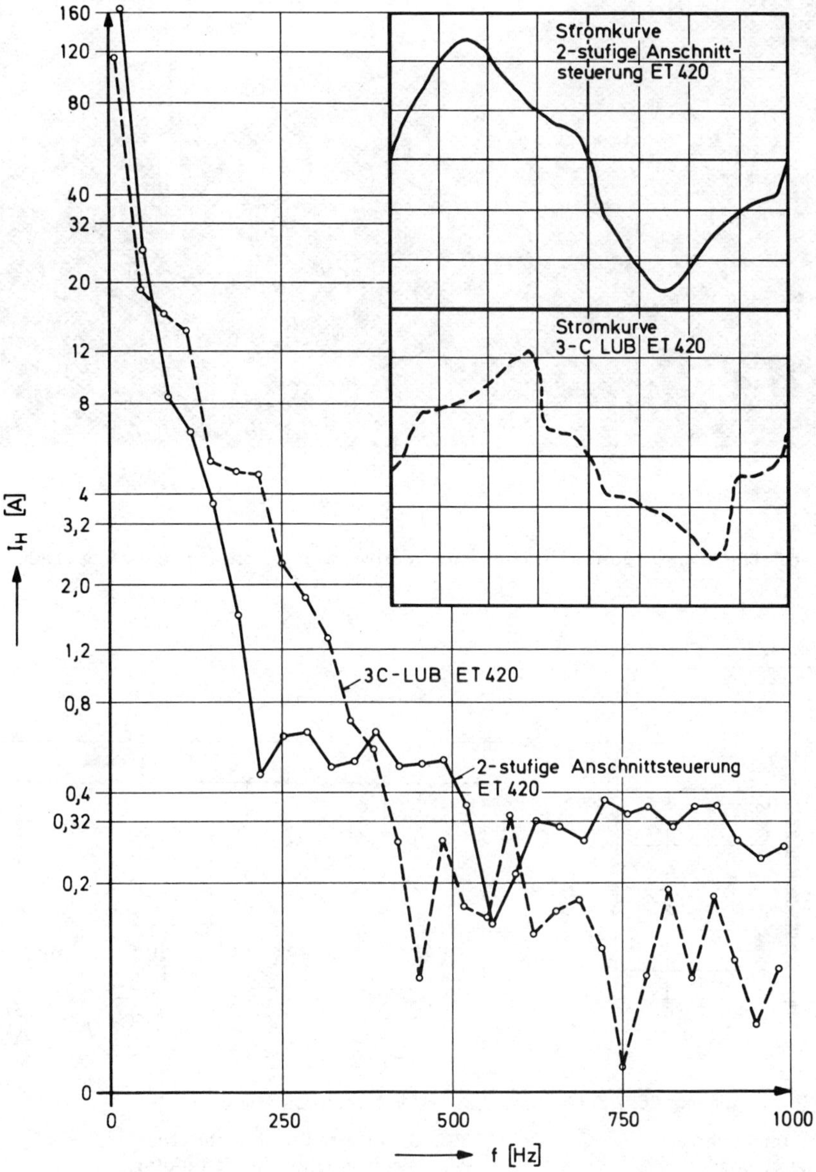

<u>Fig. 24</u>

Comparison of two frequency analyses, measured with normal two-sequence semi-controlled asymmetric bridge circuit and with 3C LUB, on DB trainset ET 420.075

Fig. 25

Picture of the experimental frequency converter locomotive Be4/4 12001 of SBB

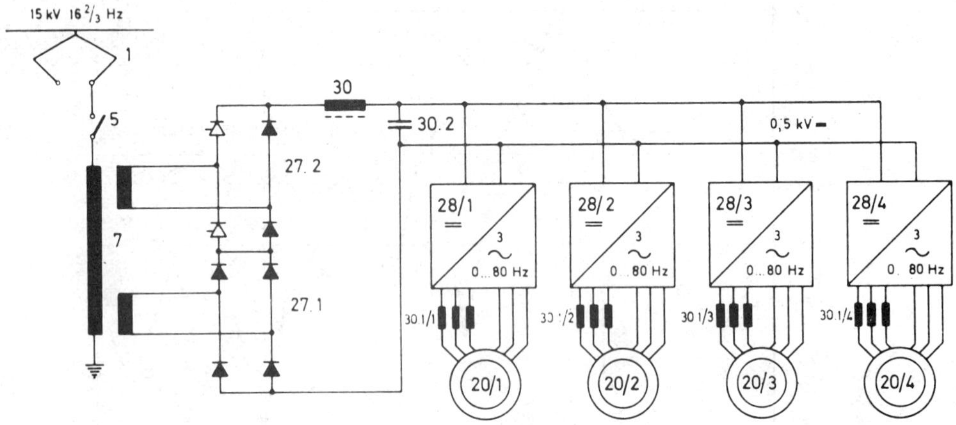

Fig. 26

Primary circuit and traction motor circuit of the experimental locomotive Be4/4 12001 of SBB

1 pantograph
5 main circuit breaker
7 transformer
20 three-phase asynchronous traction motor
27.1 uncontrolled converter bridge
27.2 controlled converter bridge
28 inverters for traction motors
30 smoothing choke for intermediate circuit
30.1 in-phase choke
30.2 capacitor for intermediate circuit

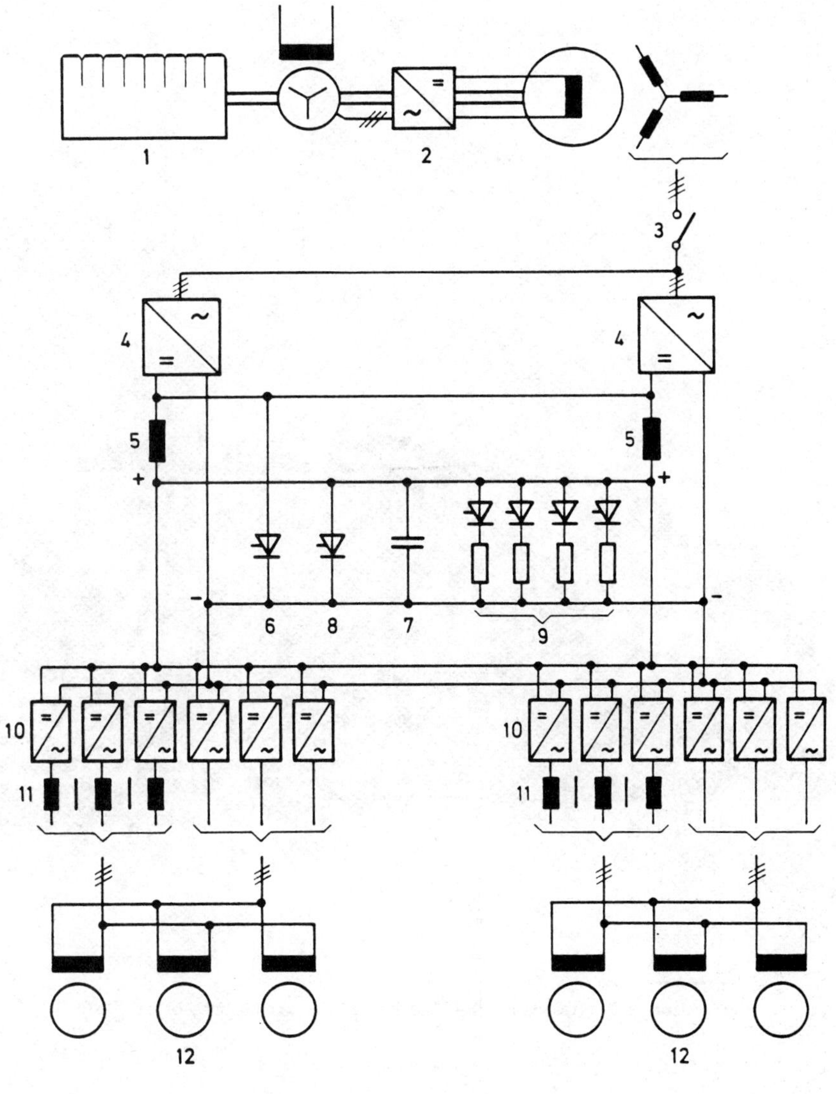

Fig. 27

Basic circuit diagram of the diesel-electric shunter Am6/6 of SBB

1 diesel engine
2 three-phase generator with three-phase exciter and rectifier for excitation
3 contactor
4 rectifier for the intermediate circuit
5 choke
6 short-circuit thyristor
7 stand-off capacitor
8 short-circuit thyristor
9 brake sub-resistances with thyristors
10 inverter
11 in-phase chokes
12 three-phase asynchronous traction motors

Fig. 28

Picture of the diesel-electric shunter Am6/6 of SBB

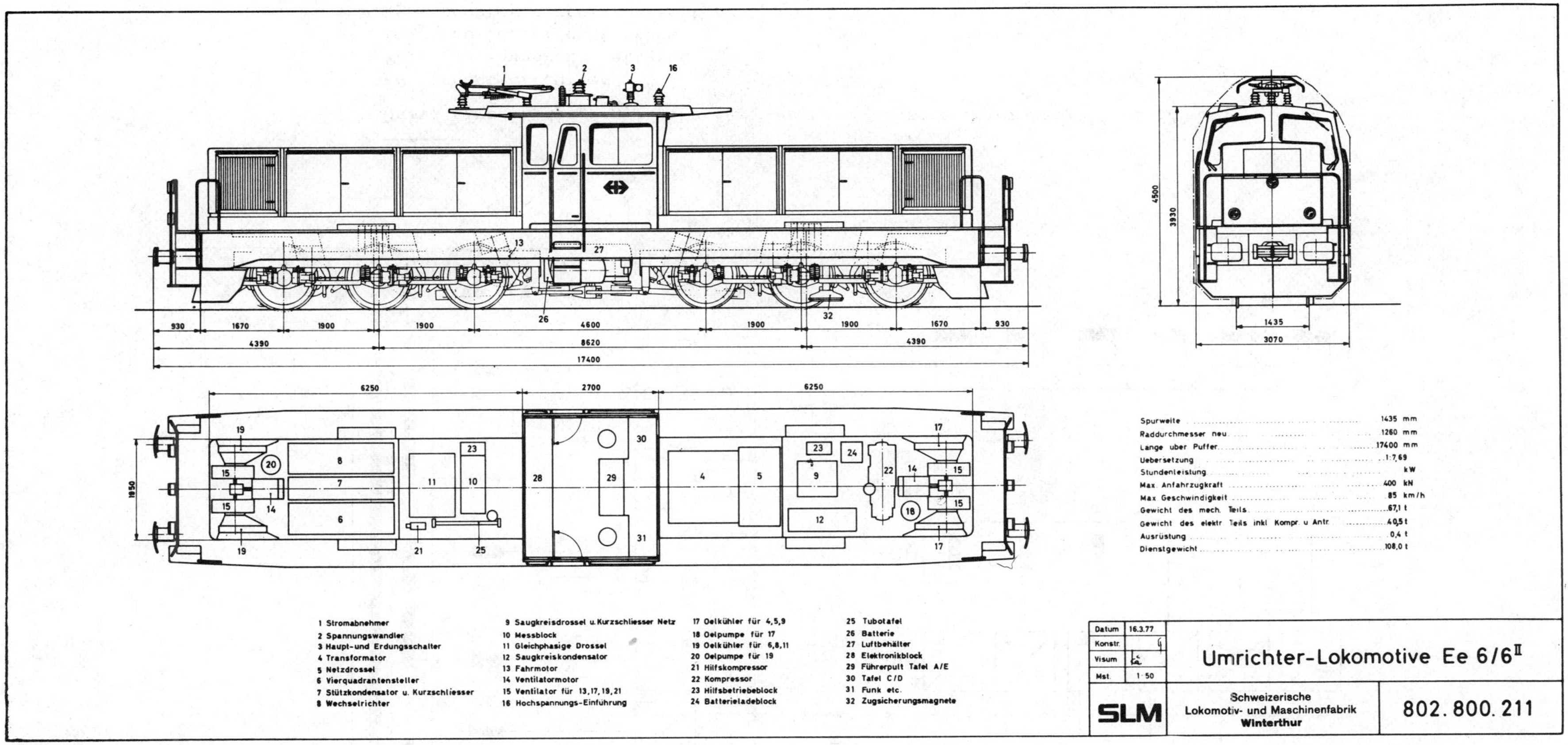

Fig. 29

Type drawing of the frequency converter shunter Ee6/6^{II} of SBB

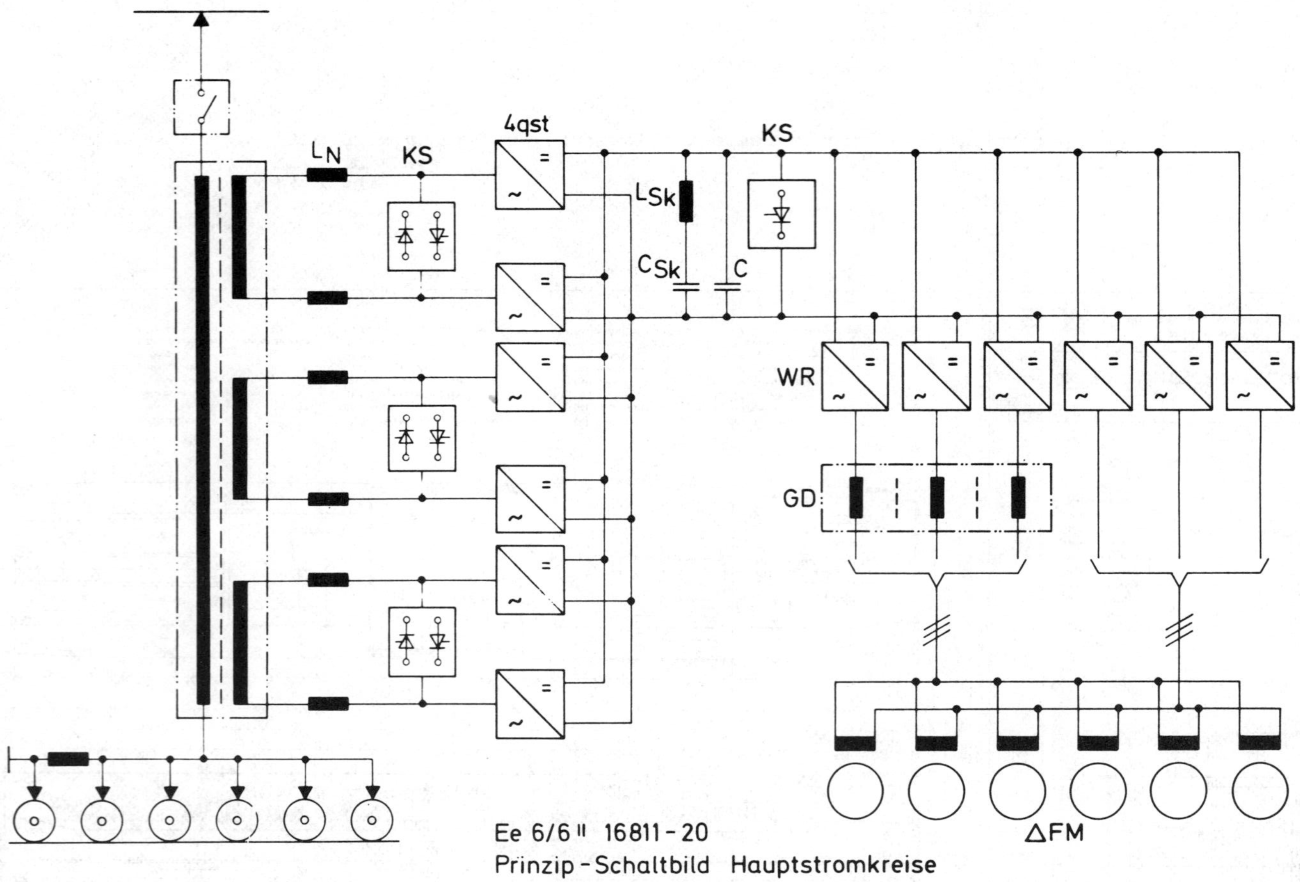

Fig. 30

Basic circuit diagram of the frequency converter shunter Ee6/6^{II} of SBB

KS	short-circuiter
4qst	four-quadrant setter
L_{Sk}	booster circuit choke
C_{Sk}	booster circuit capacitor
C	stand-off capacitor
WR	inverter
GD	in-phase choke
FM	three-phase asynchronous traction motors

Fig. 31

DE 2500 with driving trailer, where the feeding of the intermediate d.c. circuit through four-quadrant setters was investigated for the first time on DB and SBB lines

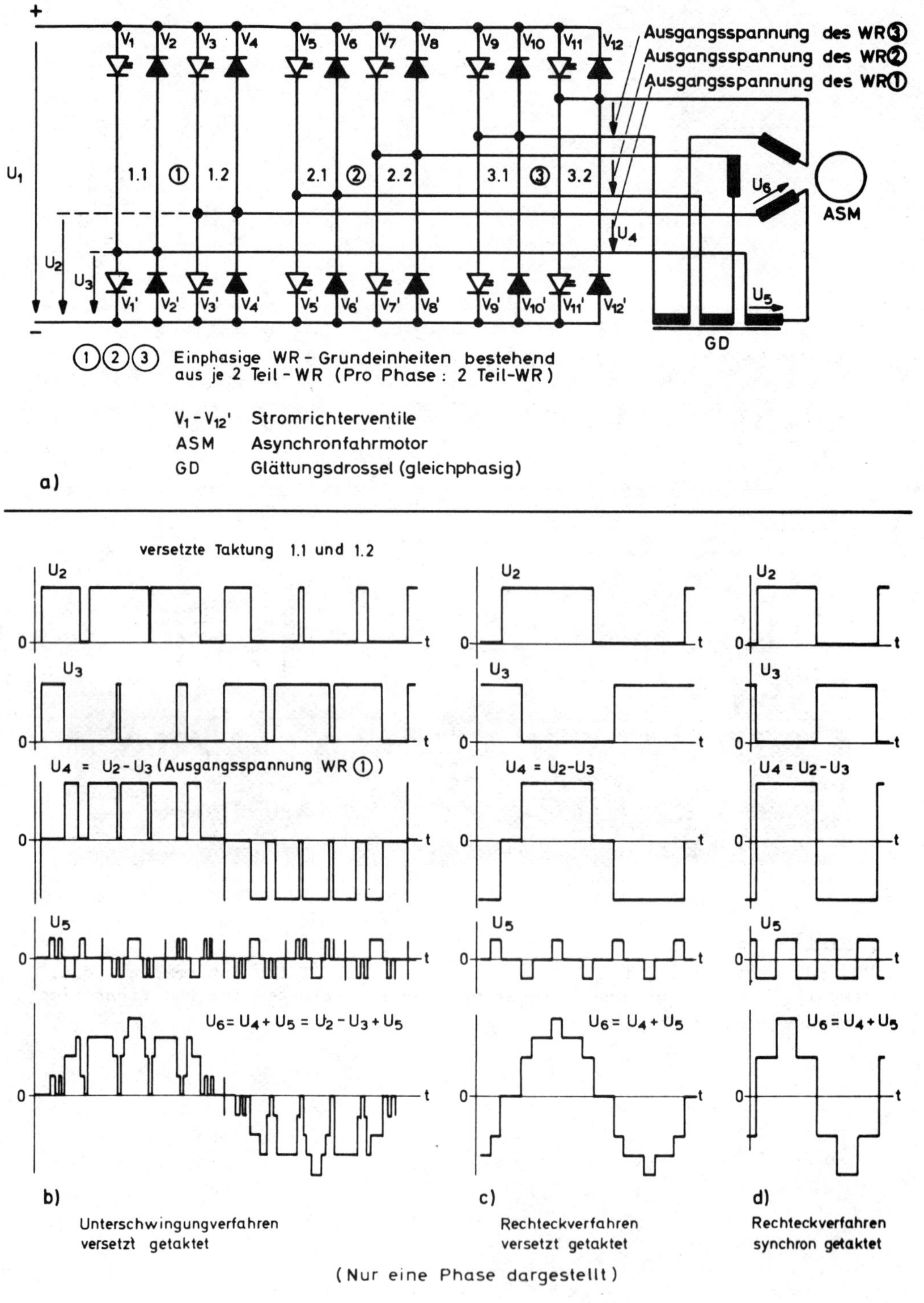

Fig. 32

Basic design of two-pulse inverters in three-phase circuits with the corresponding voltages for the sub-harmonics mode, the staggered rectangular mode and the synchronous pulsed rectangular mode

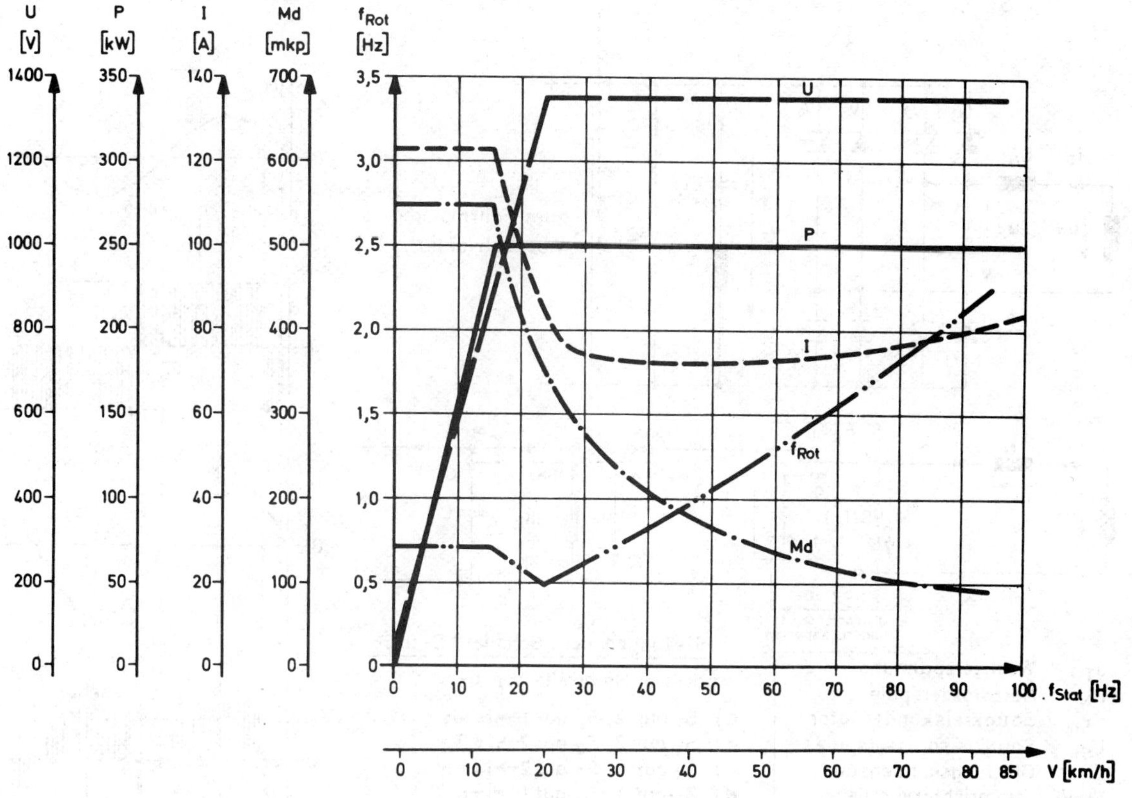

P Leistung an der Motorwelle (Maximalwert)
U Netzspannung / Phase
I Strom (maximaler Effektivwert)
Md Drehmoment an der Motorwelle (Maximalwert)
f_{Rot} Schlupffrequenz am Rotor (Maximalwert)
f_{Stat} Statorfrequenz
V Geschwindigkeit

Fig. 33

The most important electrical and mechanical values at operation with maximum torque. Diesel-electric shunter Am6/6 of SBB

P output at the traction motor shaft
U supply voltage (phase)
I motor current
Md torque at the traction motor shaft
f_{Rot} slip frequency at rotor
f_{Stat} stator frequency
V speed

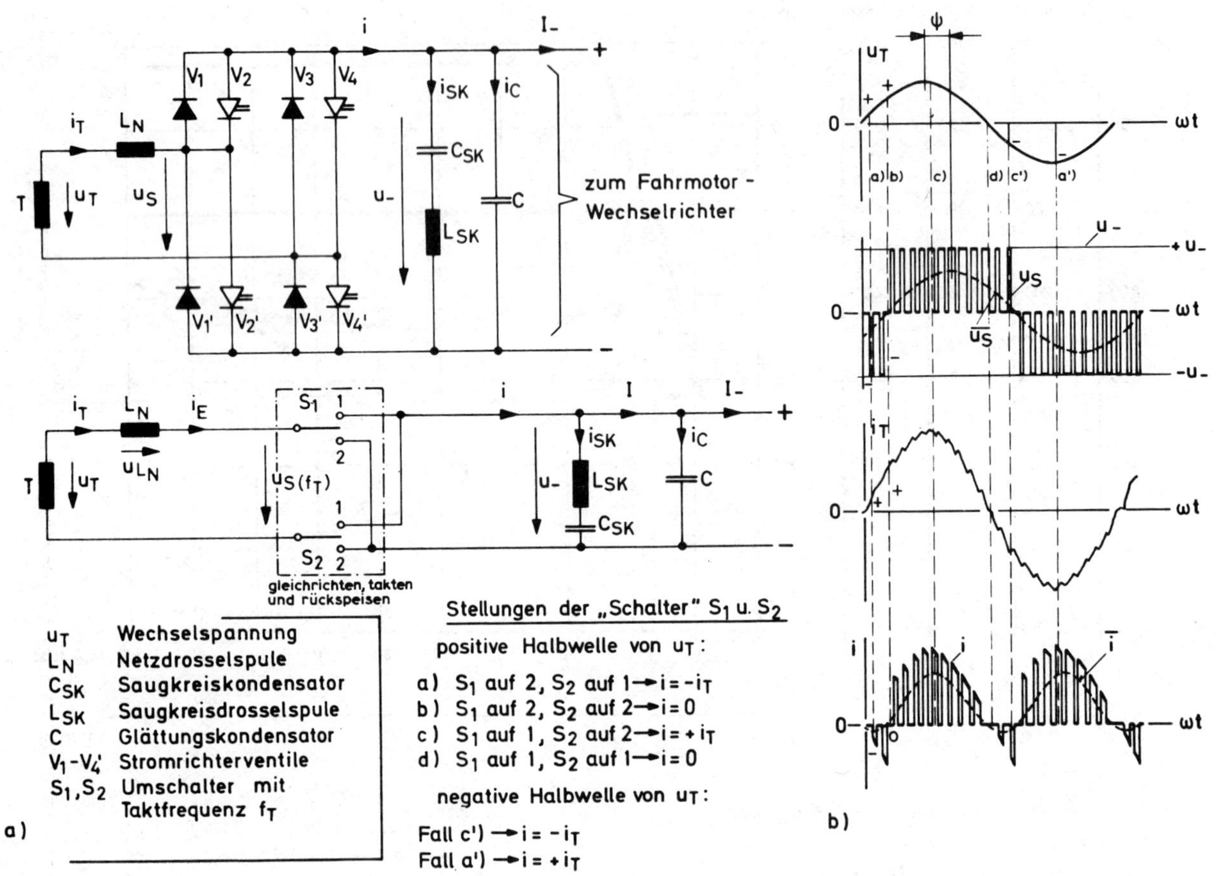

Fig. 34

Basic circuit of the four-quadrant setter

u_T	a.c. voltage
L_N	power supply choke
C_{SK}	booster capacitor
L_{SK}	booster choke
C	smoothing capacitor
V_1-V'_4	semi-conductor valves
S_1, S_2	switches with pulse frequency f_T
u_S	input voltage at four-quadrant setter
i_T	supply current

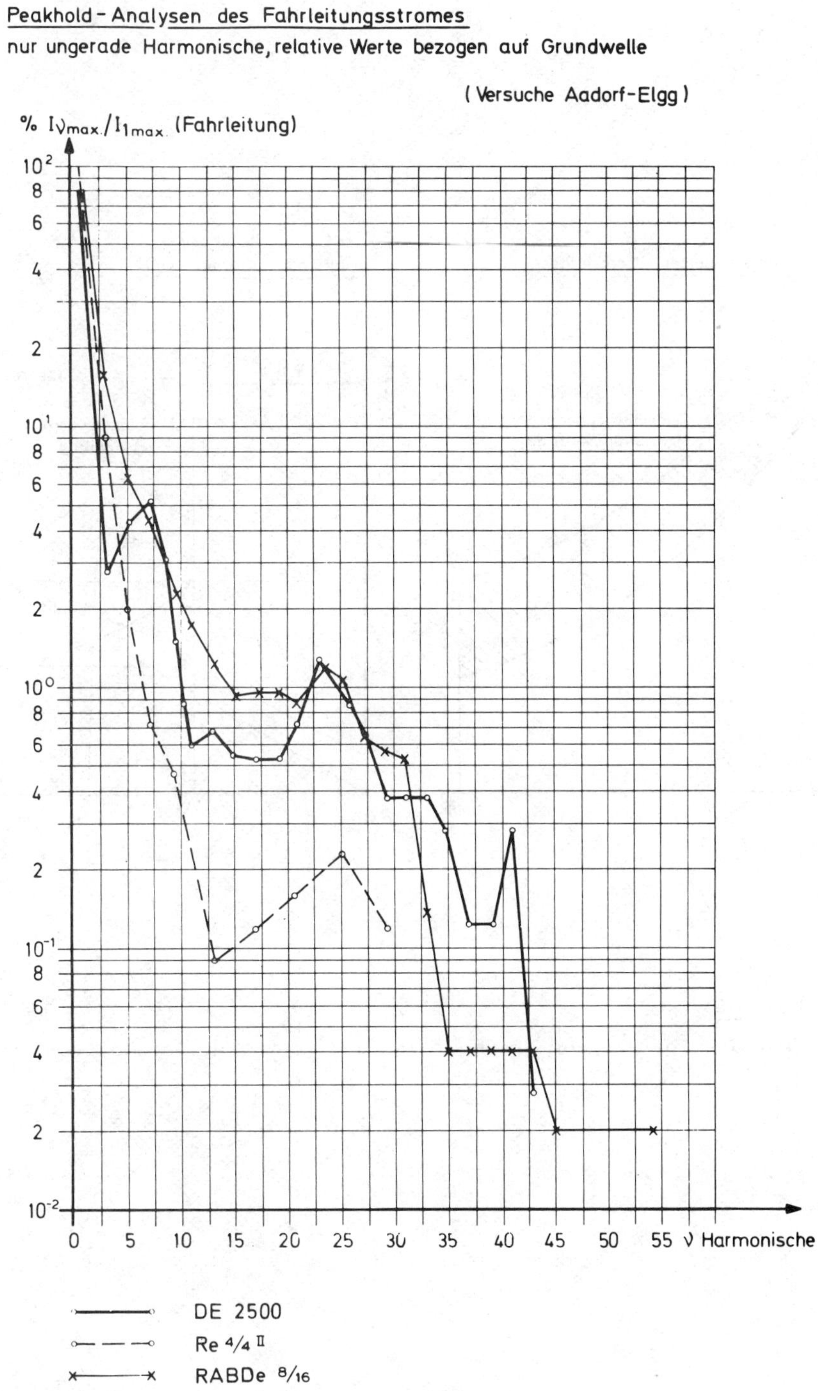

Fig. 35

Power supply current frequency analysis related to the first harmonic. DE 2500 locomotive set in comparison with suburban trainset RABDe8/16 of SBB and direct-motor locomotive $Re4/4^{II}$ of SBB

Anschnittsteuerung oder ungesteuerter Stromrichter

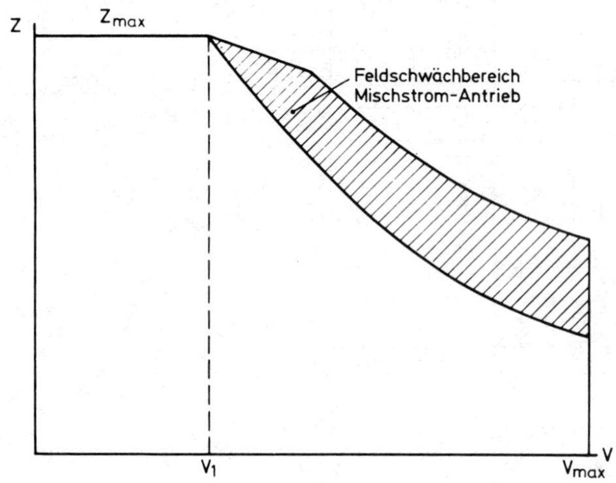

Abb. 1

Umrichter - Antrieb

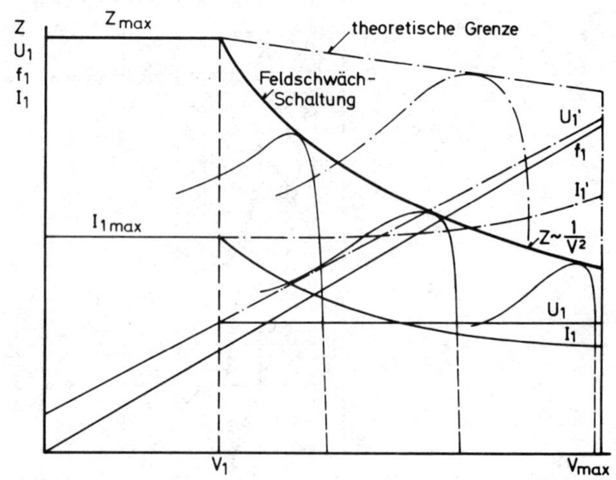

Abb. 2

Fig. 36

Tractive force/speed graph of various converter locomotives

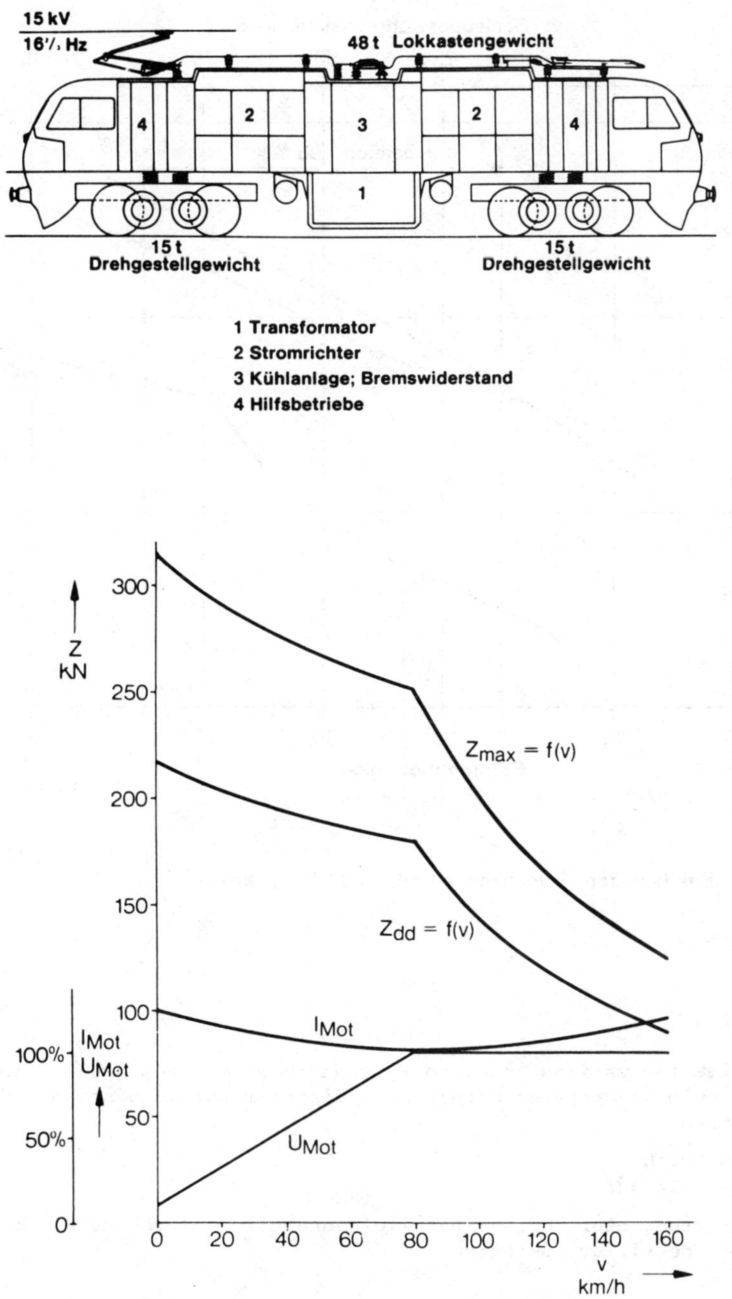

Fig. 37

Type drawing and characteristics of the DB project for an electric locomotive with asynchronous traction motors. Total weight 78 t, output 5,6 MW, **maximum speed 160 km/h**

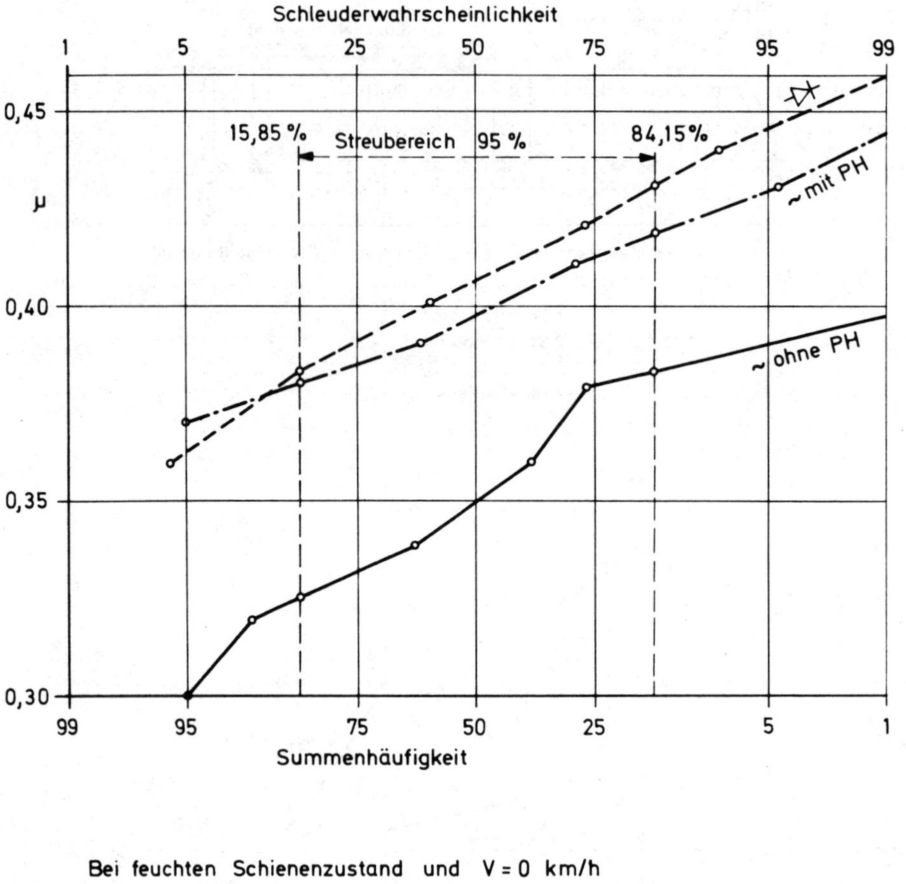

Fig. 38

Slip risk for various traction systems measured on a $Re4/4^{II}$ locomotive which could be operated either as a direct-motor locomotive or rectifier locomotive

mit = with
ohne = without

PH = contactor for the parallel connection of the motor fields
-▷|- = rectifier operation

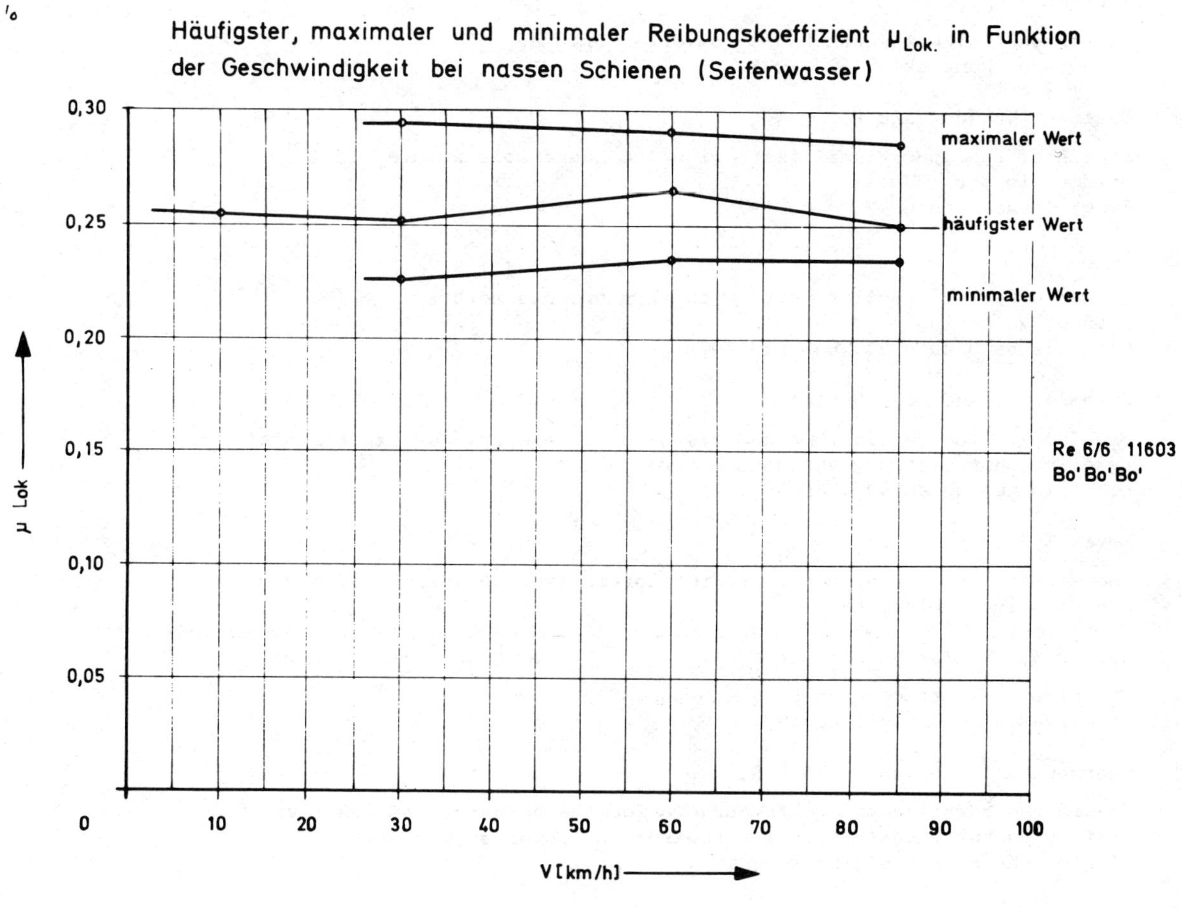

Fig. 39

Most frequent, maximum and minimum adhesion coefficient μ_{Lok} as a function of speed on wet rails (soapy water) of the SBB locomotive Re6/6 (Bo'Bo'Bo')

Bibliography

[1] Winter P.

The influence of smoothing and commutation reactances on the behaviour towards the power supply grid of multi-sequence converters in single-phase semi-controlled circuits
Diss. Nr. 4945 der ETH Zürich, 1973

[2] Förster J.

Sector control with quenchable converter bridges
Technische Rundschau 560, 1973

[3] Dreimann K. and Falk P.

Status of the operational tests with the quenchable asymmetrical bridge circuit (LUB)
Elektrische Bahnen 47, H 6, 1976

[4] Teich W.

Diesel-electric tractive units with asynchronous motors without slip-ring
Elektrische Bahnen 43, Heft 4, 1972

[5] Stemmler H. and Brechbühler M.

Problems in the development and design of a catenary-fed experimental tractive unit with asynchronous motors
Elektrische Bahnen 43, Heft 5, 1972

[6] Meyer K.

The experimental frequency converter locomotive Be4/4 12001 of SBB
SBB-Nachrichtenblatt 4/73

[7] Gerber M., Müller M., Winter P.

The diesel-electric locomotive Am6/6 of SBB
Schweizerische Bauzeitung 1977

[8] Kehrmann H., Lienau W., Nill R.

The four-quadrant setter - favourable for the behaviour of tractive units with three-phase traction towards the power supply grid
Elektrische Bahnen 45, Heft 6, 1974

[9] Weber H.H.

Investigations and findings on the adhesion behaviour of electric locomotives
Schweizerische Bauzeitung, Heft 48, 2.12.1965

[10] Weber H.H.

The wheel arrangements Co'Co' and Bo'Bo'Bo' and their stresses on the track
Glasers Annalen 7/8, Jahrgang 98, Juli/August 1974

[11] Zwicky R.

Theoretical principles of the influence of thyristor-controlled tractive units on low current circuits
Bulletin SEV 62 (1971) 1 9.1.1971

THE ANALYSIS OF DYNAMICS OF INDUCTION DRIVING MOTORS OF MINE WINDERS WITH FREQUENCY CONTROL

L.M. Szklarski and A. Dziadecki

Department of Mining Automation
Mining and Metallurgical Academy, Cracow, Poland

1. Introduction

The induction a.c. machine and especially one with a squirrel cage rotor shows serval advantages when applied in electric drives. The main advantages are as follows: robustness, low costs, minimum maintenance costs, etc. These advantages are especially valuable in big drives with heavy duty cycles. The most successful means for achieving good speed regulation with induction motors is the use of variable frequency supply at low freqencies. This enables the direct coupling of motors to the mine winders with no need of the reduction gear.

Recent progress in thyristor technique enables the construction of rather complex supply systems of such motors, but designing of above systems and precise analytical evaluation presents very considerable difficulties.

In the considered case the motor is being supplied which controlled low frequency, from zero to 10 or 20 percent of nominal frequency. These conditions immediately imply the use of cycloconverter which fulfills the above requirements. Since the output voltage of the cycloconverter contains several harmonics, the precise analysis of the a.c. machine behaviour and its charakteristics is very difficult. The most suitable way of analysis is that by means of digital computer, with programmed cycloconverter output voltage. The analysis in this paper is carried out by this metod as well as the simulation of the a.c. squirrel cage motor.

The computer used by the authors was CDC CYBER 72 with the simulation by CSSL III language.

2. Induction a.c. machine

The model used to the simulation is shown in Fig.1.

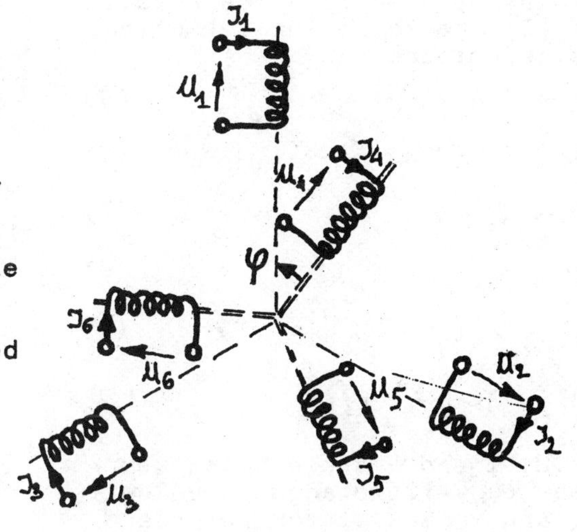

Fig.1.

From the functional of energy conversion for this machine the following equations may be derived:

$$[u] = [R][J] + \frac{d}{dt}\left\{[M][J]\right\} \qquad (1)$$

$$\mathcal{M}_{el} = [J]^T \left\{\frac{d}{d\varphi}[M]\right\}[J] \qquad (2)$$

$$J\frac{d\omega}{dt} = \mathcal{M}_{el} - M_m \qquad (3)$$

where

$$[u] = \begin{bmatrix} u_1 \\ u_2 \\ \vdots \\ u_6 \end{bmatrix}, \quad [J] = \begin{bmatrix} J_1 \\ J_2 \\ \vdots \\ J_6 \end{bmatrix},$$

$$M = \begin{bmatrix} M_{11} & M_{12} & \cdots & M_{16} \\ M_{21} & & & \vdots \\ \vdots & & & \\ M_{61} & \cdots & \cdots & M_{66} \end{bmatrix} \qquad (4)$$

The mutual inductances M_{ij} in equations (1) to (4) are the functions of currents and angle φ, they are also time dependent. One may see that due to strongly nonlinear relations the system of equations (1) to (3) is practically useless to analytical of solution of those equations. Even when assumed that both rotor and stator have balanced treephase windings, the magnetic field in the air gap is sinuesoidal and neglecting the saturation effect, the situation will not improve. Several cases may be discussed with an assumption being made that the stator supply 3-phase voltage is of purely sine waveform. After transformation to the rotating reference axes of the system associated with the rotating field [3] the following set of equations is obtained :

$$u_s = r_s i_s + \frac{1}{\omega_n}\frac{d\Psi_s}{dt} + jf_s\Psi_s \qquad (5)$$

$$u_r = r_r i_r + \frac{1}{\omega_n}\frac{d\Psi_r}{dt} + jf\,\Psi \qquad (6)$$

$$f_r = f_s - \nu \qquad (7)$$

$$\Psi_r = x_r i_r + x_h i_s \qquad (8)$$

$$\Psi_s = x_s i_s + x_h i_r \qquad (9)$$

$$M_{el} = \mathrm{Im}\left[\Psi_s^{*} \cdot i_s\right] \qquad (10)$$

where U, i and Ψ are complex quantities, ω_n -field angular velocity, x_s, x_r and x_h are stator, rotor and main magtetization reactances, respectively.

The above set of equations enables the analysis of several cases of machine performance, such as starting, braking, etc, as well as deduction of transfer functions. But one must keep in mind that the relations (5) to (10) are valid only when the supply voltage is sinusoidal, without higher harmonics, i.e. at the magnetic field rotating with constant speed.

When the machine is being supplied from the cycloconverter the output voltage contains higher harmonics which make the application of this method impossible. On the other hand the detailed analysis of harmonics is unrealistic due to changing of harmonics with the variation of output voltage and frequency. Thus the only suitable method of analysis is that of digital computer simulation. The simulation model applied was that given by the equations (1) to (3) with changes of saturation being neglected, both rotor and stator were assumed to have balanced trhee-phase windings and magnetic field being perfectly sinusoidally distributed in the air gap. Since the rotor has squirrel cage windings $U_4=U_5=U_6=0$. The stator windings are supplied from a thyristor cycloconverter. One may keep in mind that at low frequencies the motor impedance is low and can not supress and filter out the harmonics.

3. The cycloconverter

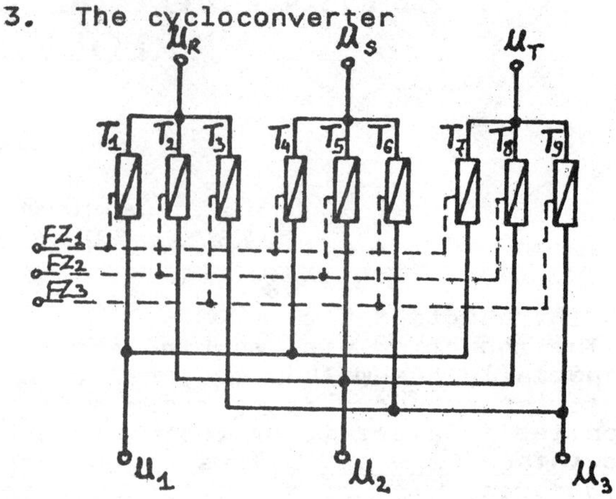

Fig. 2.

The equivalent circuit of the cycloconverter is shown in Fig. 2. Each of the elements T_i of the cycloconverter may have one of two states : on and off /which may be represented by logic quantities either 1 or 0 /.

I/. Triggering takes place at the moment when waveform of supply voltage /normalized to the interval [0,2π] / exceeds the value FZ_i / which exactly corresponds to the firing of thyristors /.

II/ Triggering of any given element causes the turning off of all remaining elements of that group.

The quantities FZ_1 to FZ_3 may be varied arbitrarily in the course of time ; at the problem discussed they were periodic time functions symmetrical in respect to zero and shifted by one thrird of a complete cycle each.

In the example discussed the machine parameters are as follows : rated power P_N=2000 kW, rated voltage U_N=3000 V, rated synchronous speed at nominal frequency f_N=50Hz n_o=500 rpm, rated speed n_N=490 rpm. The following cases were discussed :

1. Starting at constant supply frequency 10 Hz, modulation functions FZ_i being sinusoidal.

The supply voltages, currents, rotational speed and motor torque within the selected time intervals are shown in Figs.3-18

$U_{RST}=1/5U_N$

2. Starting at constant frequency 10 Hz when modulation functions FZ_i are rectangular. The same output functions as previously are shown in Fig. 19-34

$U_{RST}=1/5U_Y$

3. Starting at constant frequency 10 Hz, modulation functions FZ_i being sinusoidal, $U_{RST}=U_N$.
The supply voltages, currents, rotational speed and motor torque within the selected time intervals are shown in Figs.35-50
4. Starting at constant frequency 10 Hz, modulation functions FZ_i being rectangular, $U_{RST}=U_N$.
The same output functions as previously are shown in Figs.51-66

The stabilization of main magnetic flux was not used, but U/f was kept constant.

IA1 is the geometric sum of all current vectors projections on the axis bound with the stator.

Let us consider the obtained results. In all discused cases the motor supply voltage obtained from the cycloconverter is of a very crude waveform and contains several harmonics, it differs very much from the sine wave. It is even discontinuous as well, which is due to thyristors commutaton. But on the other hand the motor phase currents I_1, I_2, I_3 are continous though they shape differ from sine wave. It is important to notice that despite such deformations due to the rectangular shape of modulation voltage, the magnetic field is nearly sinusoidal, after the transients have decayed. The motor speed-time curve at the motor acceleration is smooth /figs.15,31,47,63/,without oscillations; except at the very short initial starting period; this lasts lessthan 0.5 sec. One may notice that speed-time curve at given no-load start conditions is nearly linear.

4. Conclusions

The analysis of a.c. induction machine performance supplied from the frequency changer carried out by means of digital computer simulationis relatively simple, though it requires theapplication of speed computer with big storage and rather rich software /simulation languages/. The method described may be very handy for the designer of driving systems since it enables to calculate the output values, e.g. currents, and provides the facility of checking the influence of several parameters on the system performance; e.g. application of several kinds of modulation functions FZ_i - sinusoidal, rectangular,staircase type, etc.

Some additional advantages of the method is that once prepared software may be used for several versions of supply voltage and the drive load.

References

[1] Meisel,J. : Principles of Elektromechanical Energy Conversion. McGrawċHill,1966.

[2] Puchała,A. : Theory of Electromechanical Converters./in Polish/ textbook,Cracow,1968.

[3] Bühler,H. : Strorichtergespeiste Drehstrom- und Gleichstromantriebe. IFAC Symposium, duesseldorf,1974

[4] Hancock,N.N. : Matrix Analysis of Electric Machinery Pergamon Press, 2-nd Ed.,1974

[5] Figaro,B.,Gotovsky,B.,Liss,Z. : Thyristor Cycloconverters./in Russian/ Moscow,1973

[6] Pelly,B.R. : Thyristor Phase-Controlled Converters and Cycloconverters. John Wiley and Son,1971

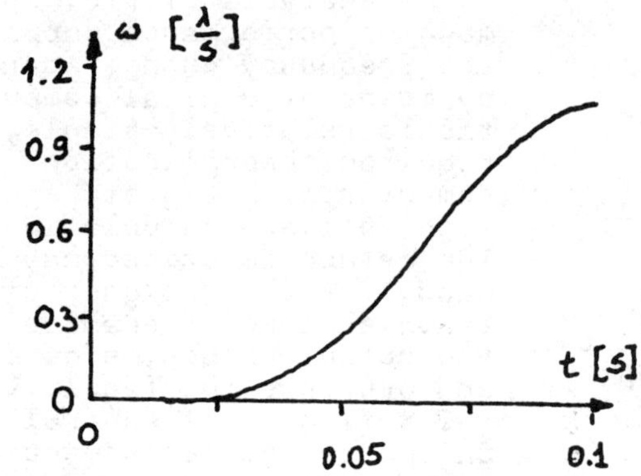

Fig.3.

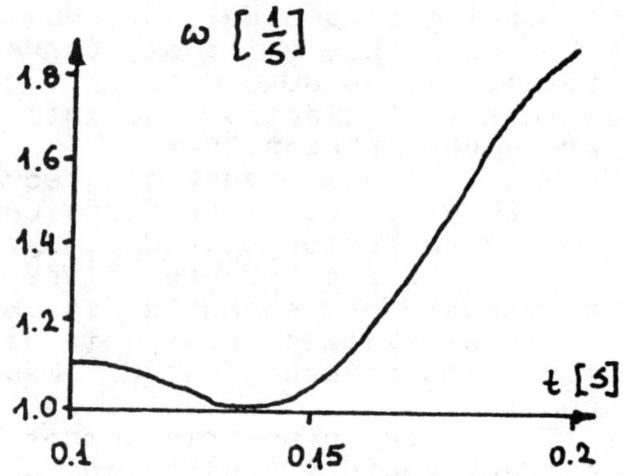

Fig.4.

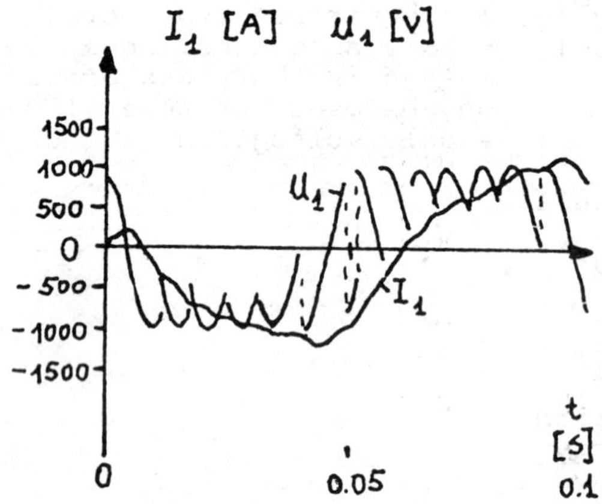

Fig.5.

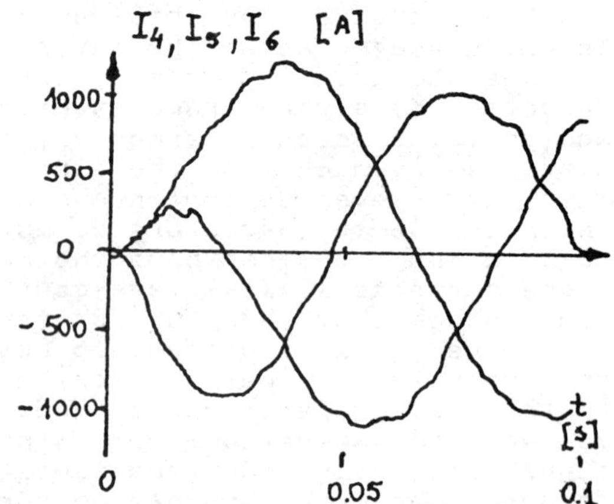

Fig.6.

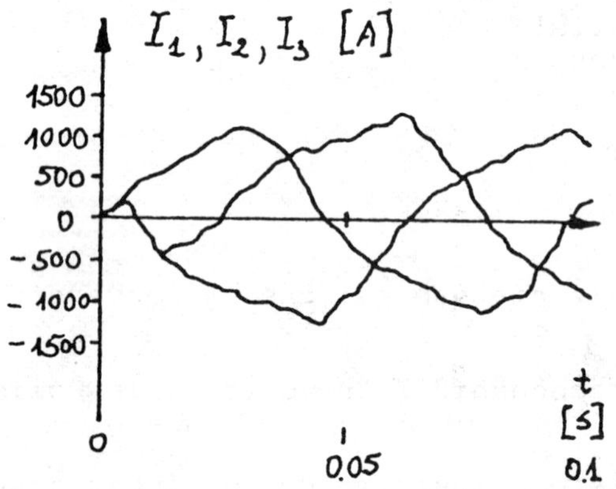

Fig.7.

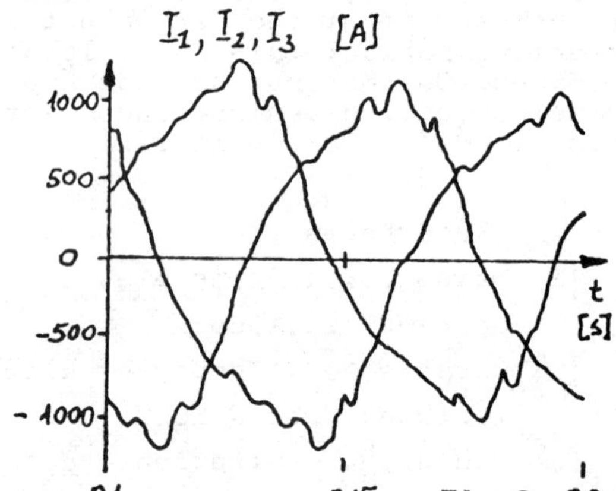

Fig.8.

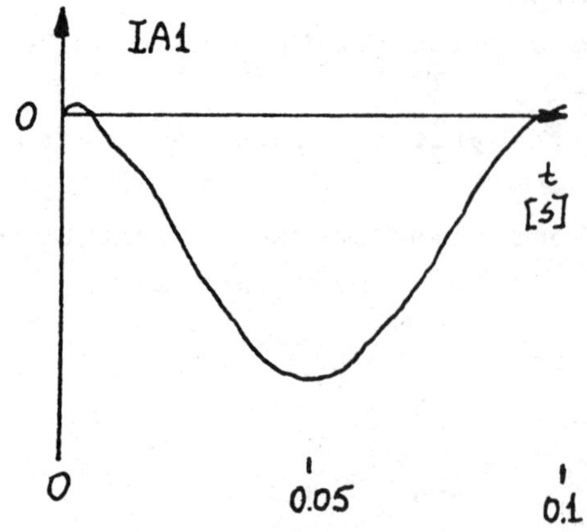

Fig.9

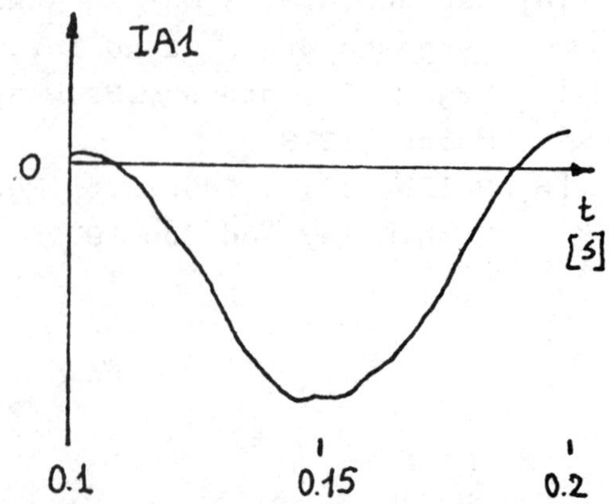

Fig.10.

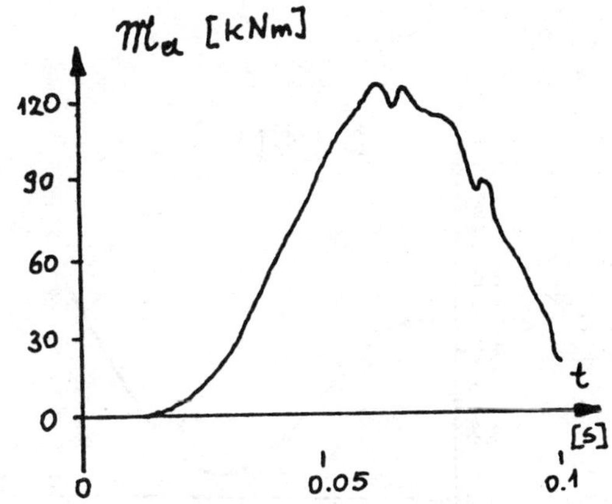

Fig.11.

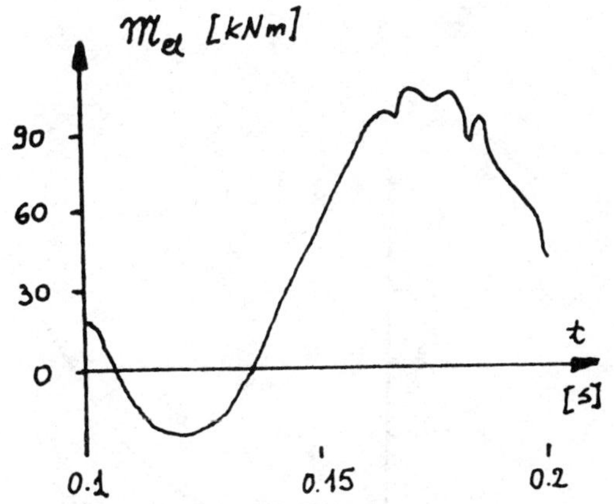

Fig.12.

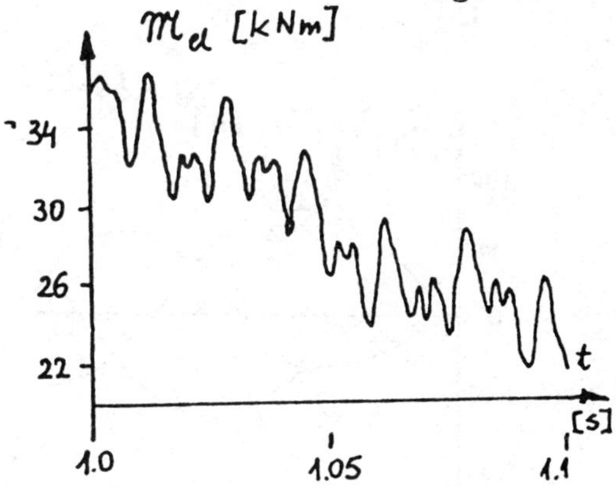

Fig.13.

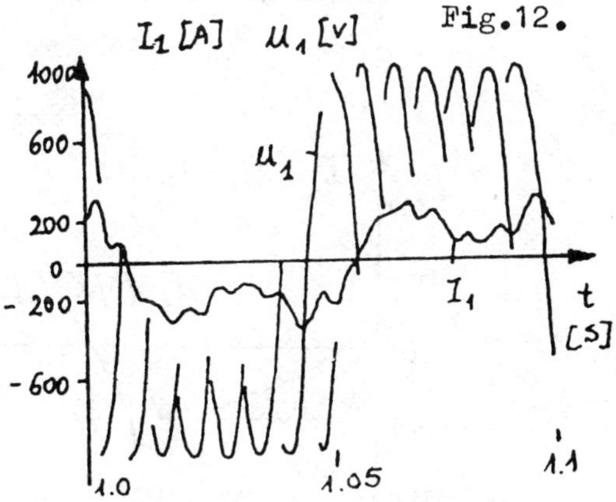

Fig.14.

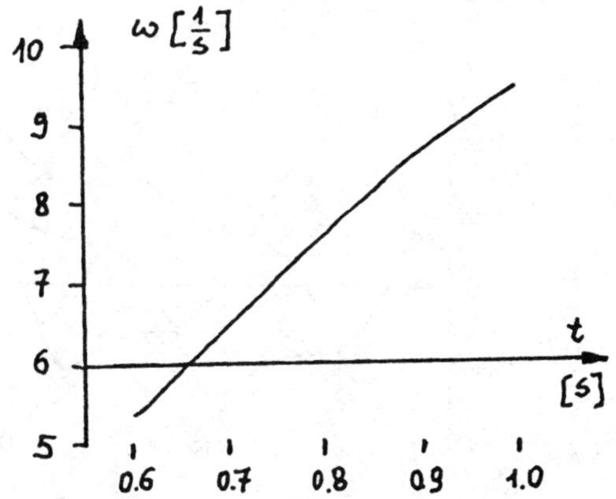

Fig.15.

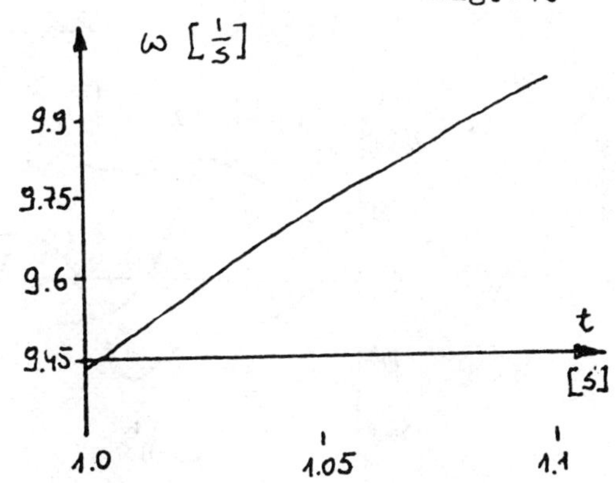

Fig.16.

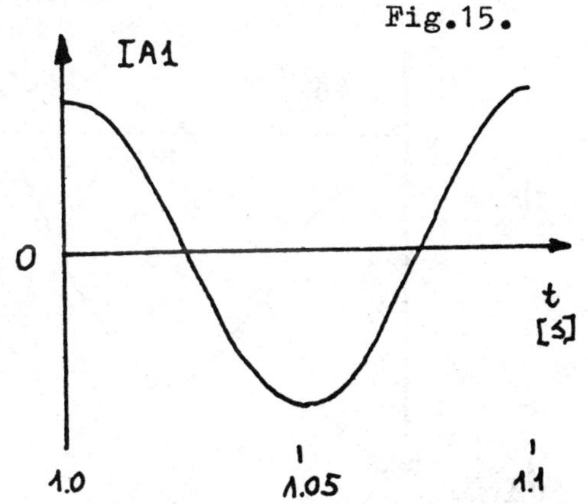

Fig.17.

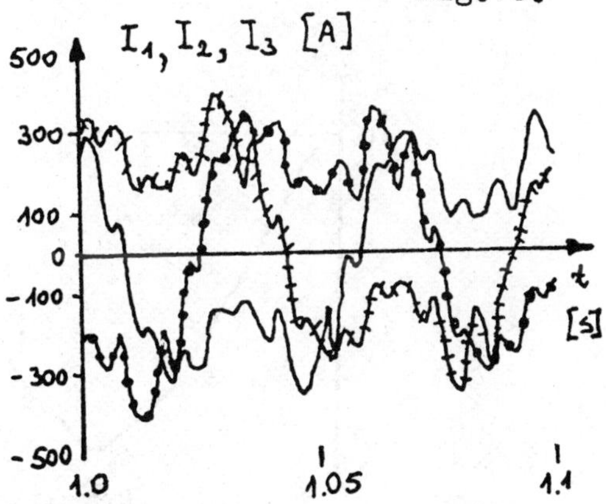

Fig.18.

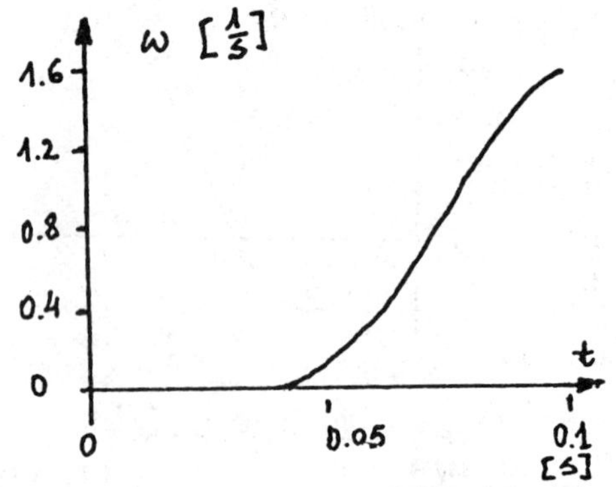

Fig.19.

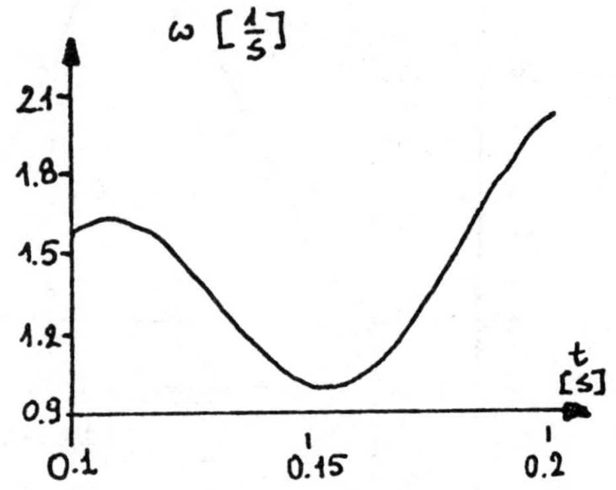

Fig.20.

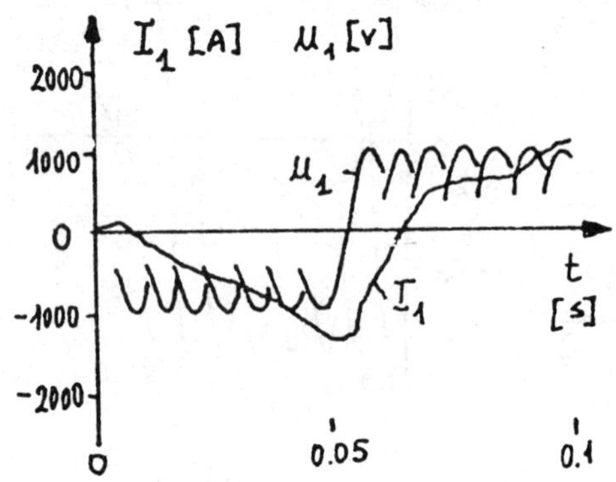

Fig.21.

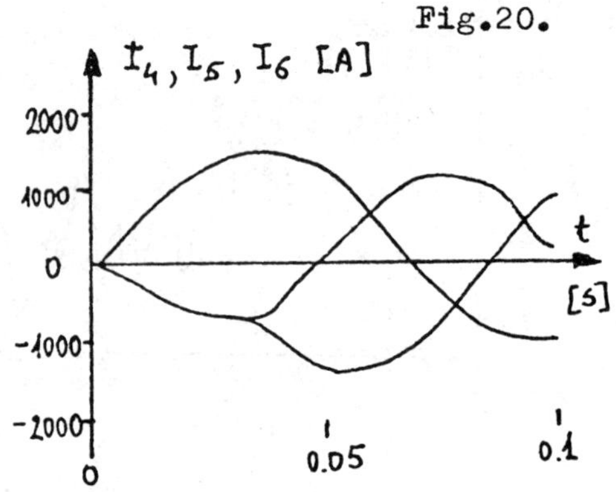

Fig.22.

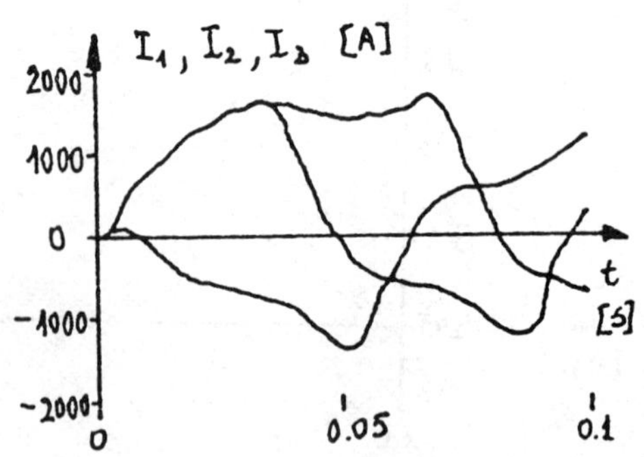

Fig.23.

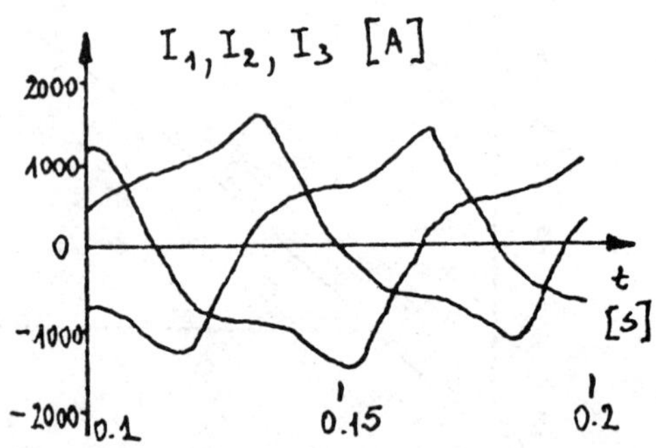

Fig.24.

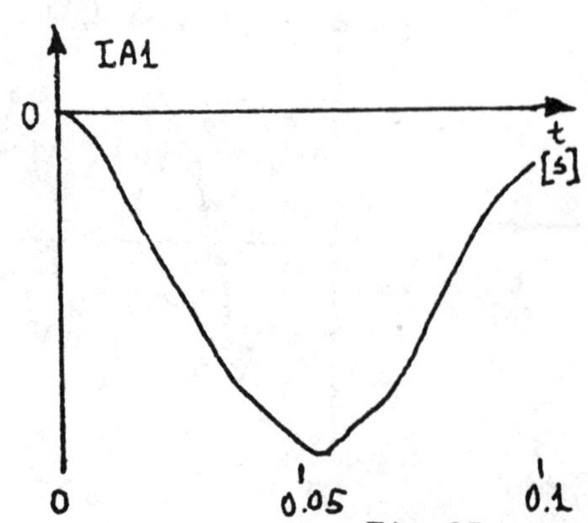

Fig.25.

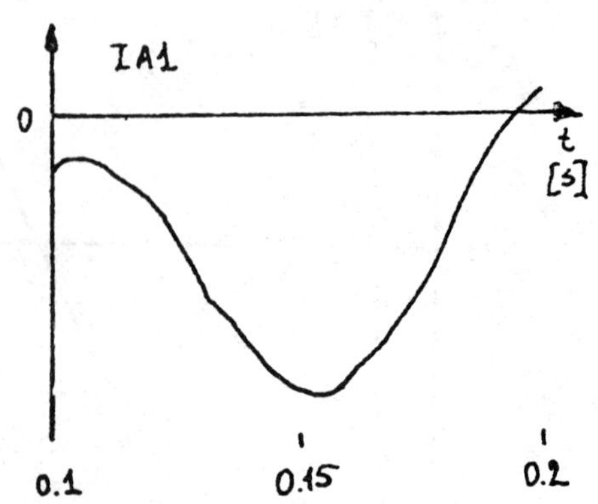

Fig.26.

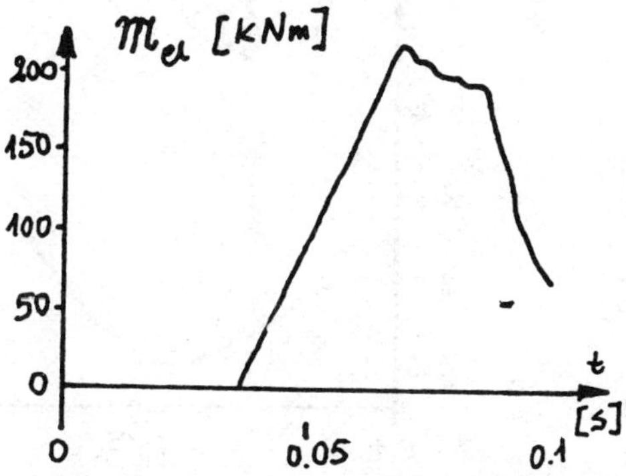

Fig.27.

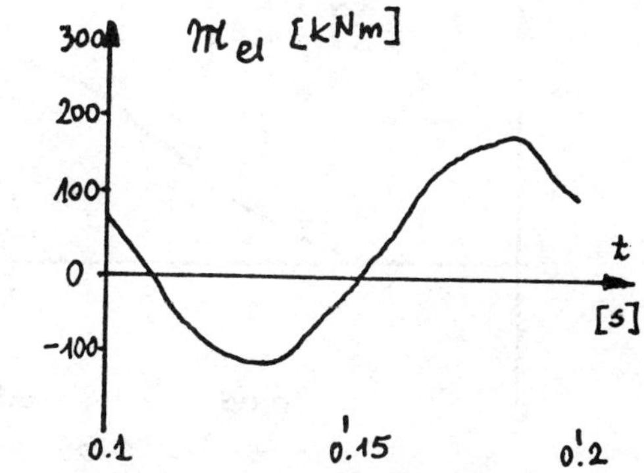

Fig.28.

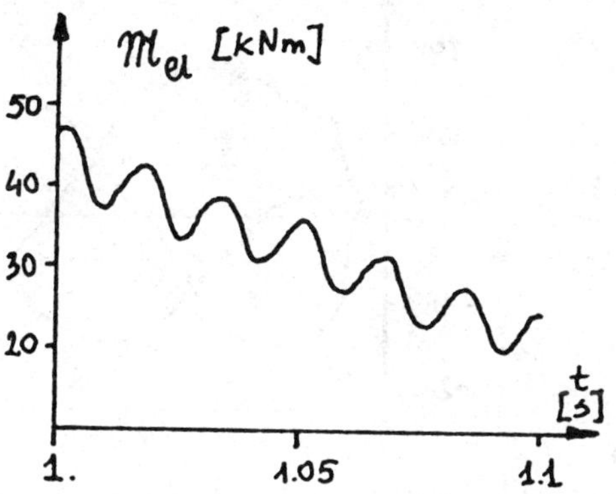

Fig.29.

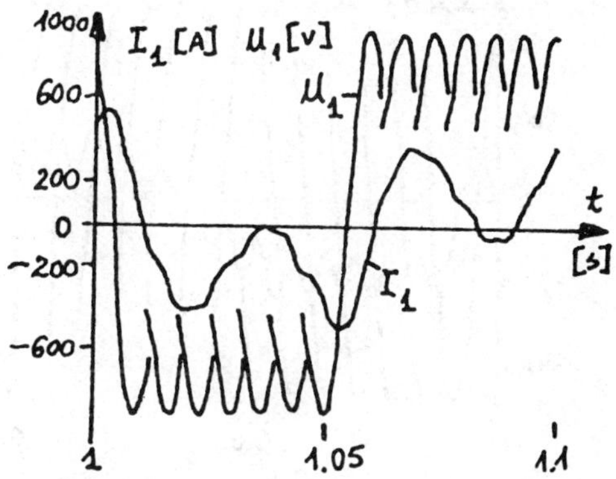

Fig.30.

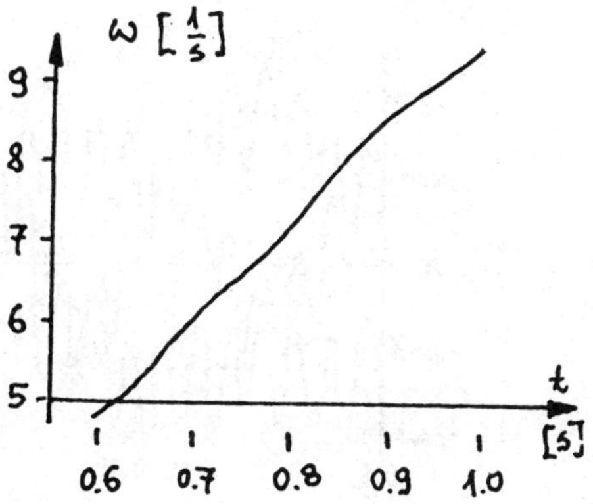

Fig.31.

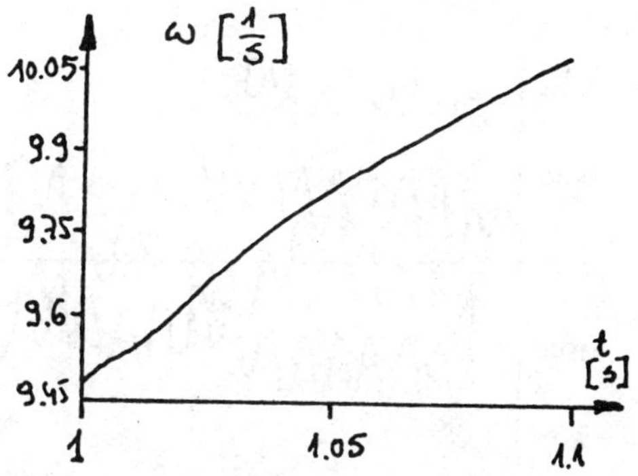

Fig.32.

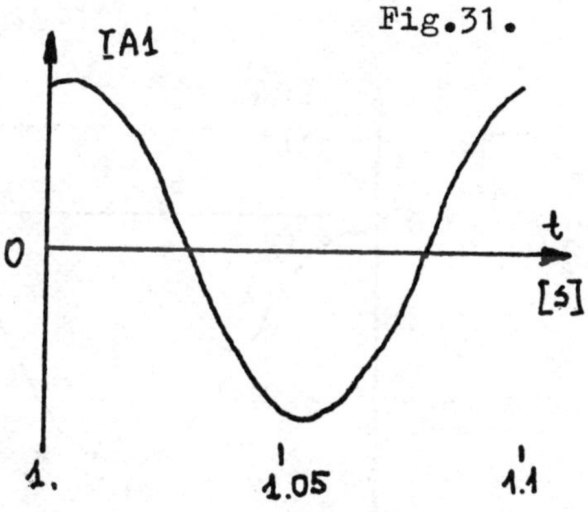

Fig.33.

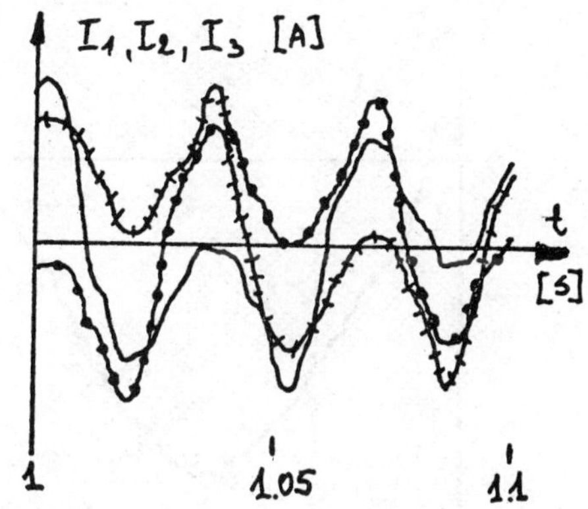

Fig.34.

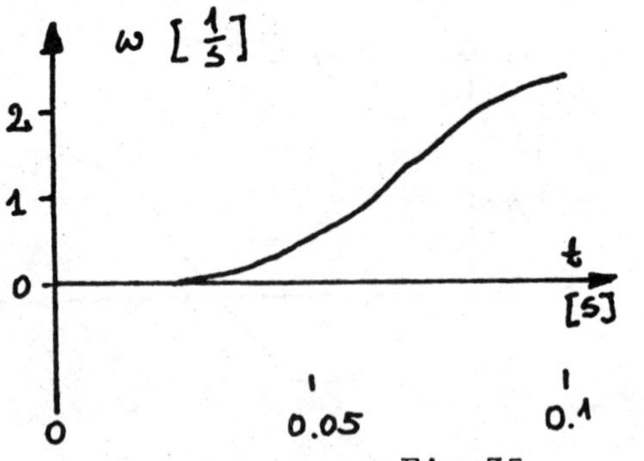

Fig.35.

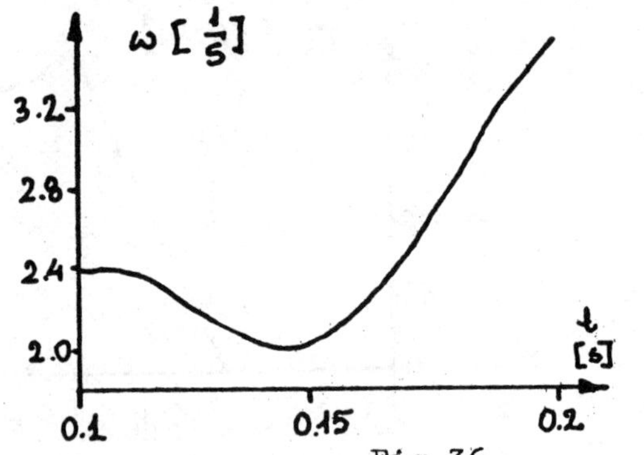

Fig.36.

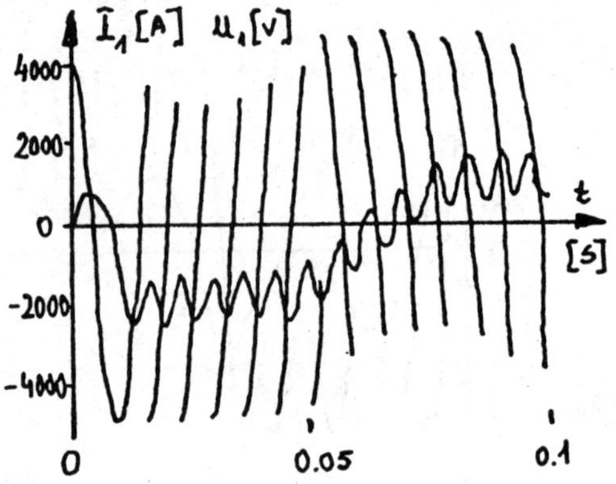

Fig.37.

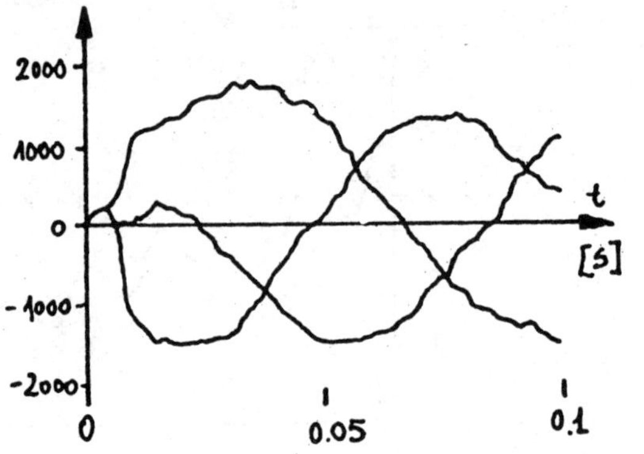

Fig.38.

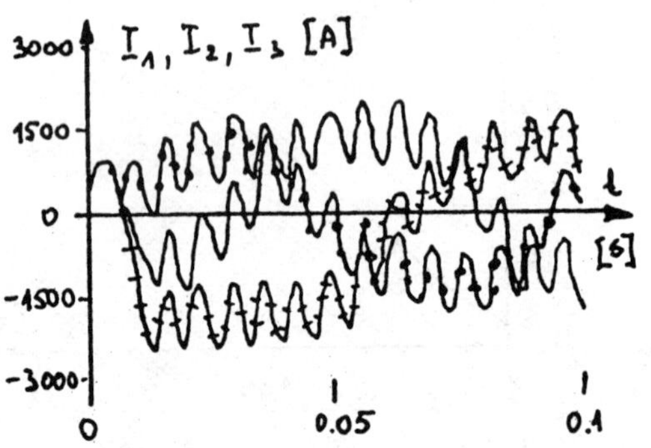

Fig.39.

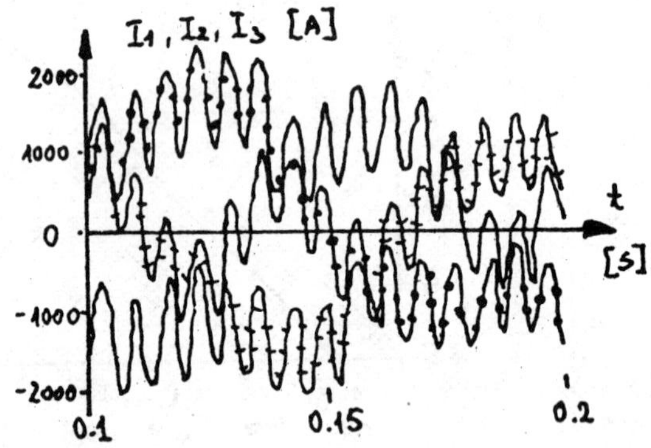

Fig.40.

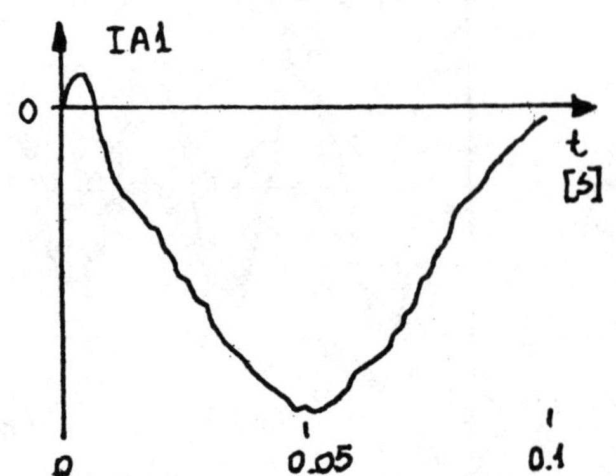

Fig.41.

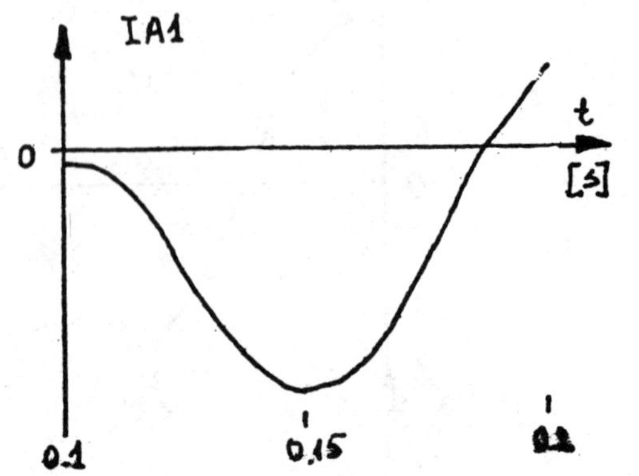

Fig.42.

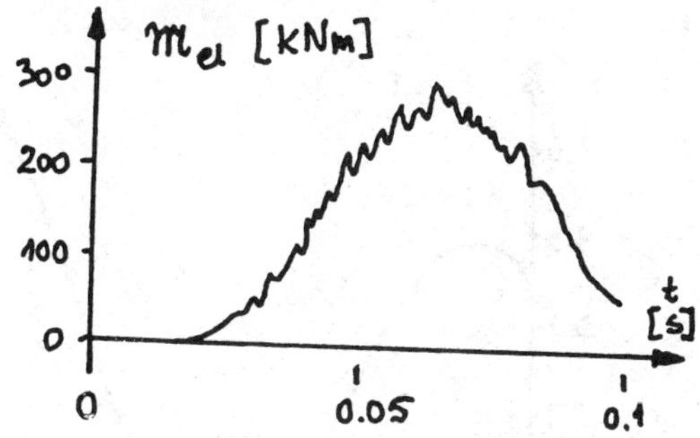

Fig.43.

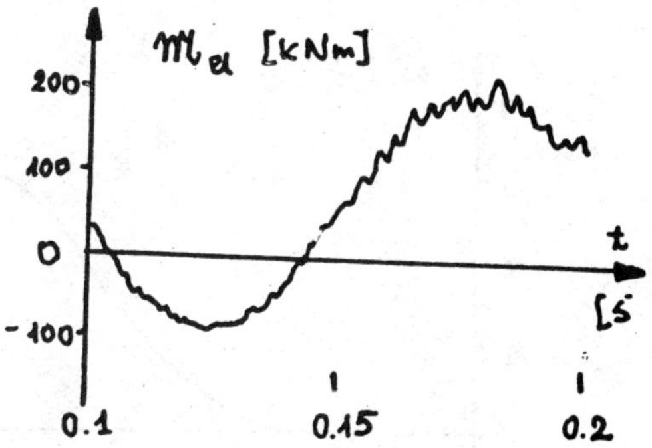

Fig.44.

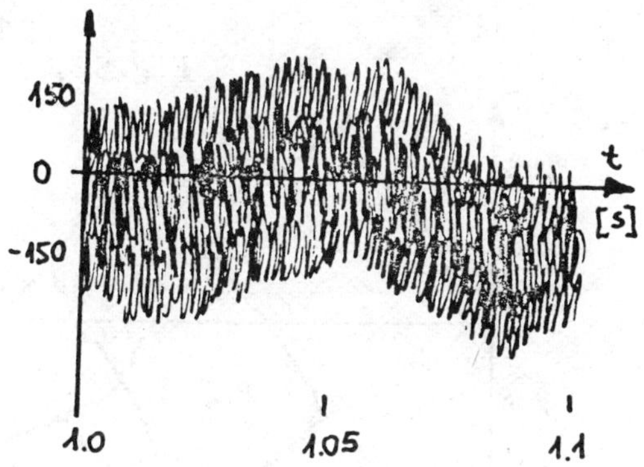

Fig.45.

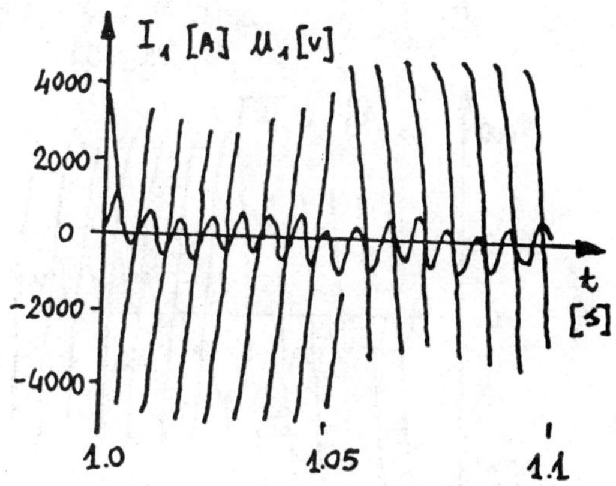

Fig.46.

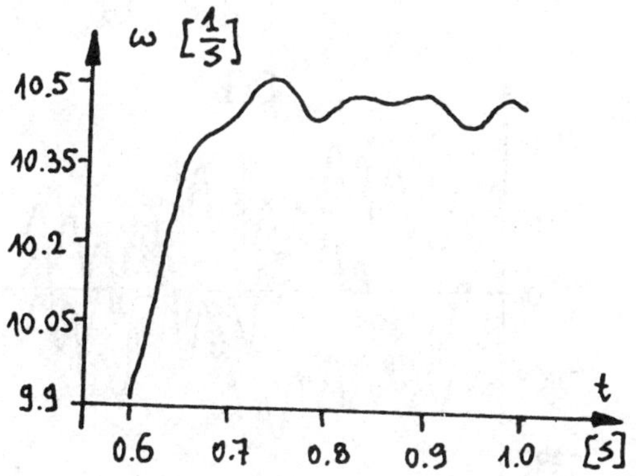

Fig.47.

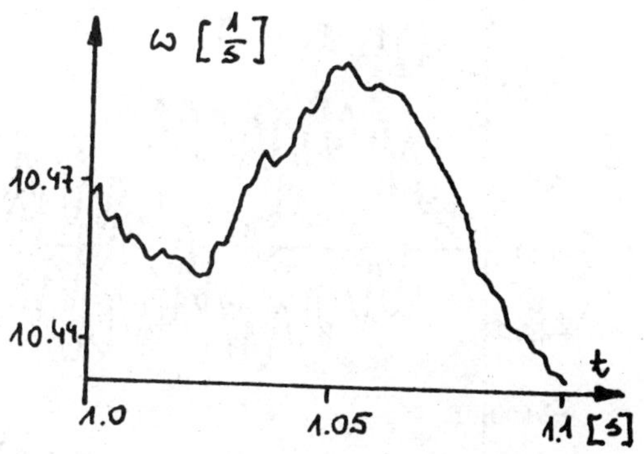

Fig.48.

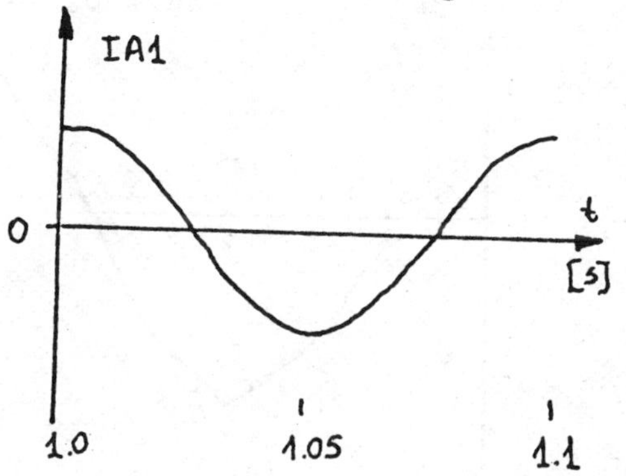

Fig.49.

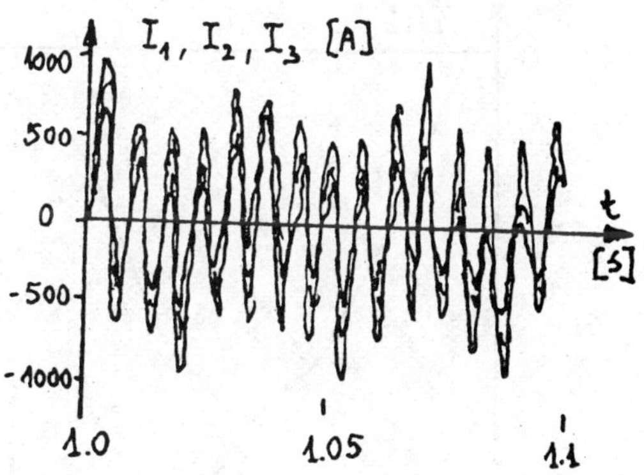

Fig.50.

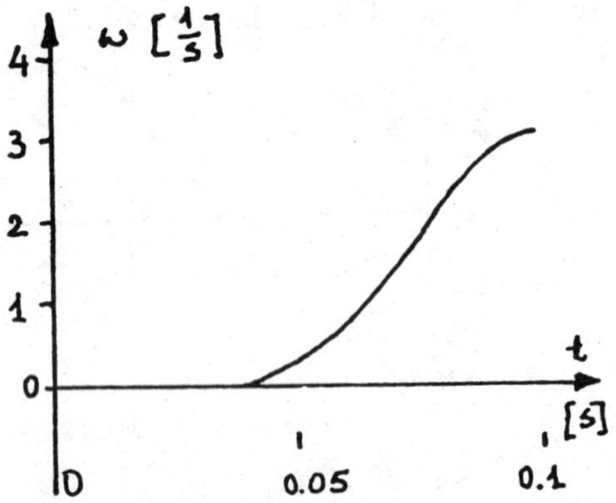

Fig.51.

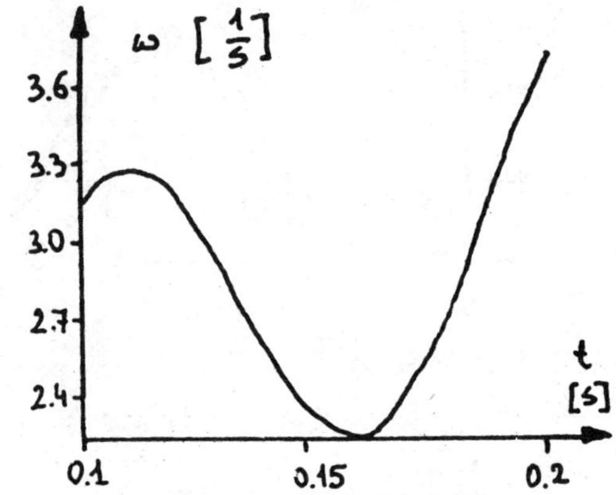

Fig.52.

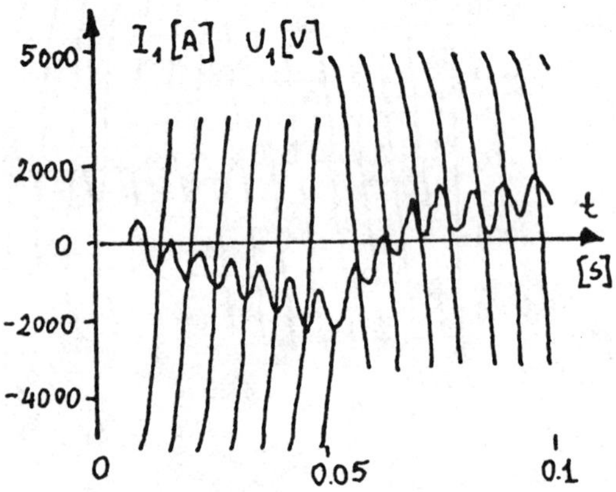

Fig.53.

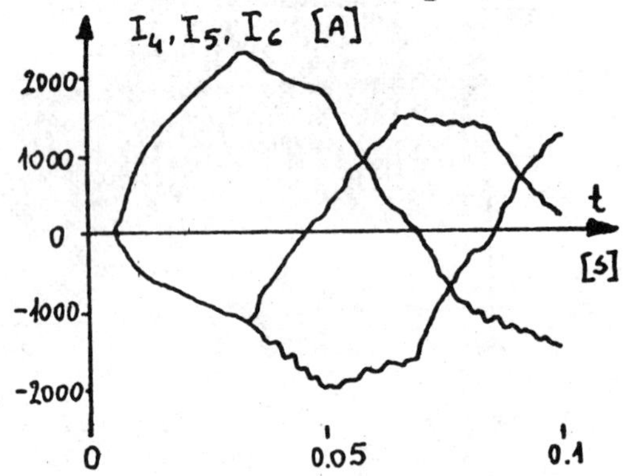

Fig.54.

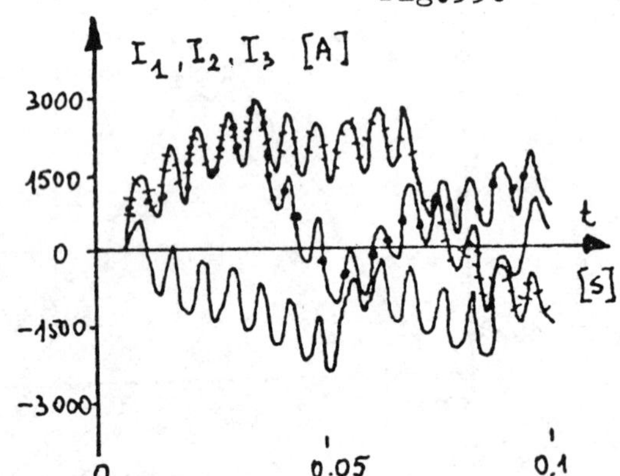

Fig.55.

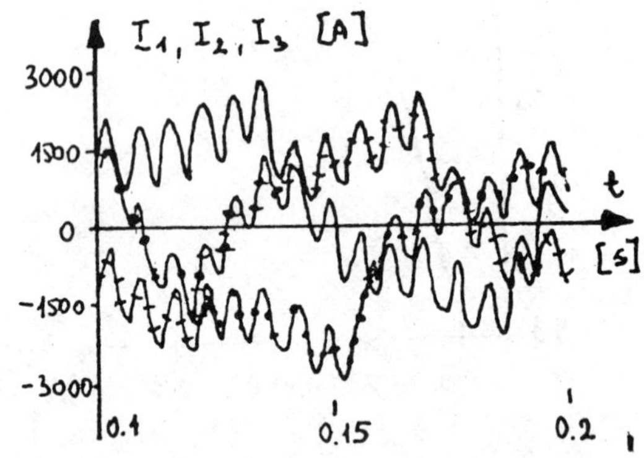

Fig.56.

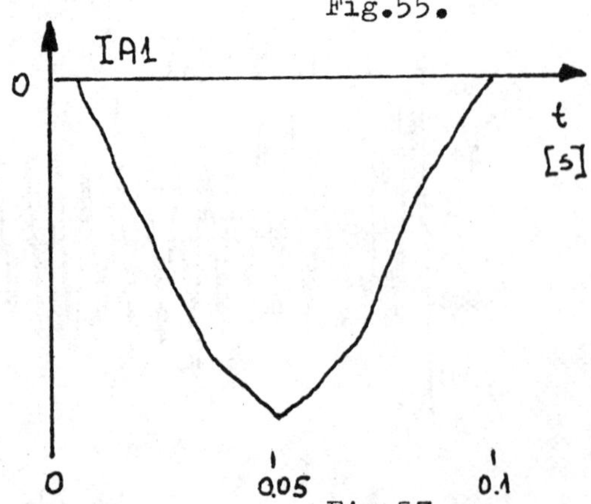

Fig.57.

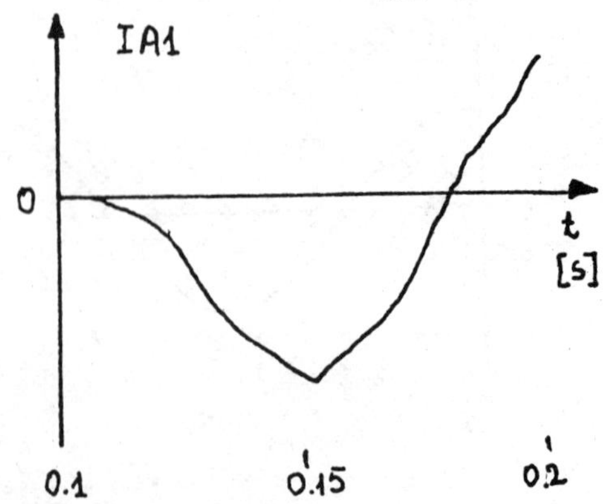

Fig.58.

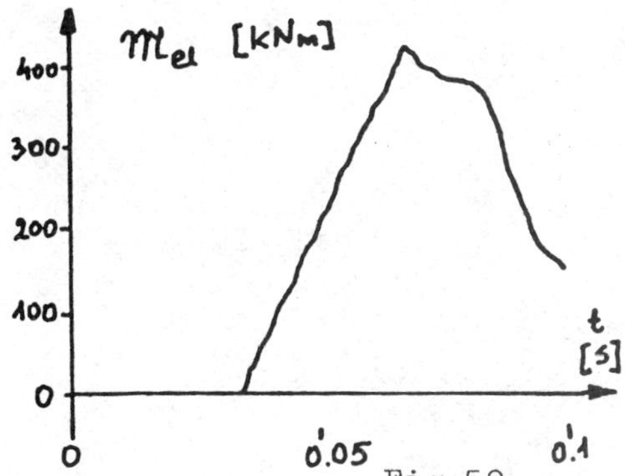

Fig.59.

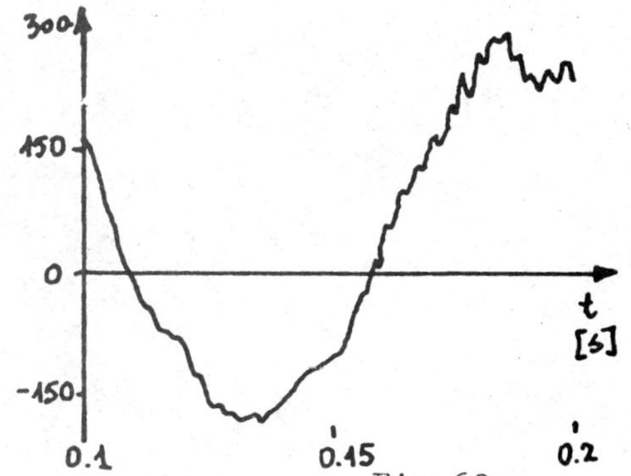

Fig.60.

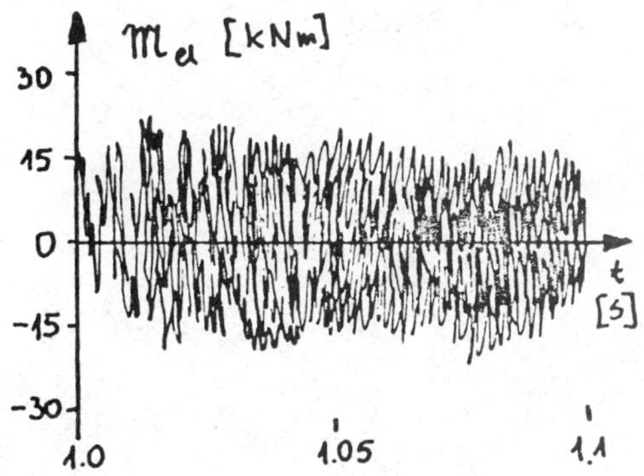

Fig.61.

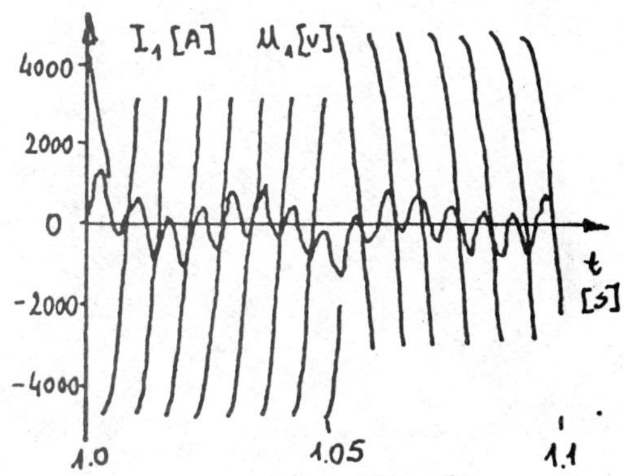

Fig.62.

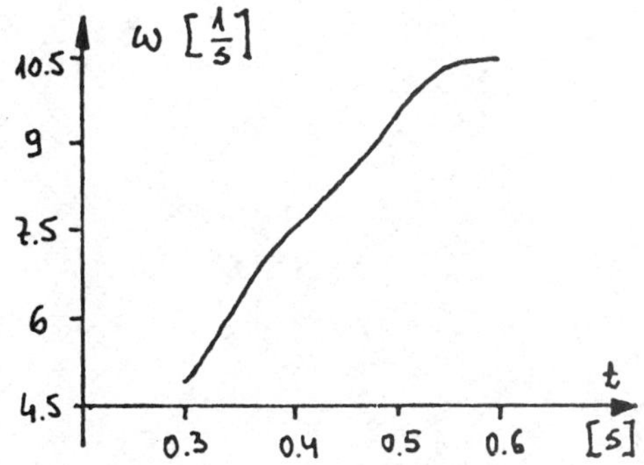

Fig.63.

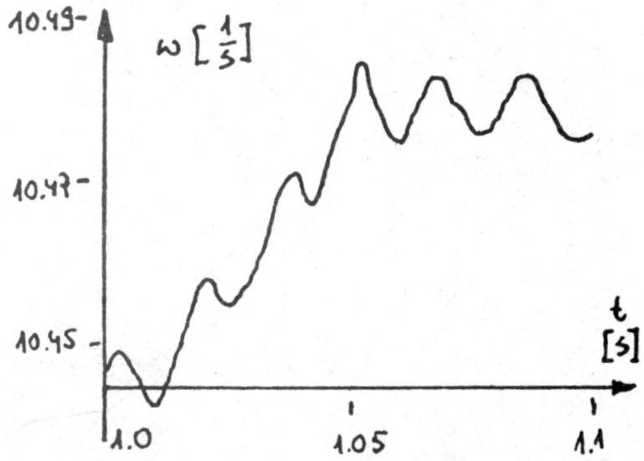

Fig.64.

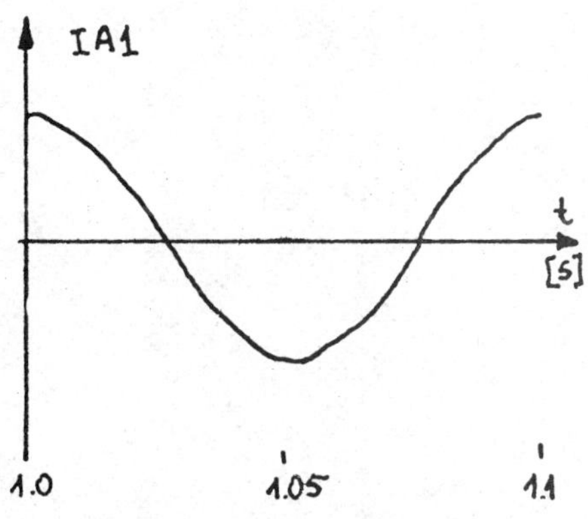

Fig.65.

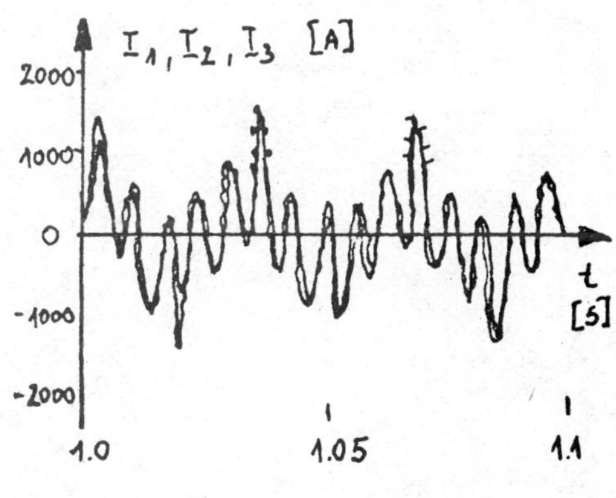

Fig.66.

ON – LINE RECORDING WITH A STORAGE DISPLAY AND A PROCESS COMPUTER FOR CURRENT LOCUS AND TORQUE-SPEED CHARACTERISTIC OF AN INDUCTION MACHINE AT RATED VOLTAGE

Fr. Depping and J. Weigel

Arbeitsgruppe Prozeßrechentechnik auf dem Gebiet der Erzeugung und Anwendung elektrischer Energie, Ruhr-Universität, Bochum, Germany

ABSTRACT

Hardware and software implementation within the PRISMA-System [1)] are discussed for recording current locus and torque-speed characteristic of an induction machine at rated voltage. A dc shunt machine at constant field current is coupled with an induction machine via a torque measuring device. The dc drive system is built up with a 4-quadrant thyristor converter for speed control and armature current control within the internal feedback loop. The speed reference signal is delivered by a process computer (set point control). The analoguous values for the mains-voltage and -current, torque, and speed are fed into the process computer via integrating A/D conversion. Phase measurement between the mains-voltage and -current is achieved by fast A/D conversion and direct memory access. The programs for measurement and processing of data allow the display of the result at a certain speed setting immediately after the measurement on the screen of a storage cathode ray tube. The necessary time for measurement is small enough to avoid excessive heating even at high slip values opening hereby new vistas for machine testing and offering valuable help for the development of new electrical machines including the necessary control equipment.

INTRODUCTION

Current locus of the stator current and torque-speed characteristic of an induction machine are useful for the evaluation of the machine's performance. Since a long time (Ref. 1) efforts have been made to get rid of the lenghthy procedures necessary for the determination of the current loci of induction machines when using traditional pointer instruments for the measurement of power, voltage, and current. The big investment of time when applying conventional methods is accompanied by the necessity of a certain expenditure for personnel required for the simultaneous reading of the measuring instruments and keeping the minutes. Moreover there is a drawback which results from the requirement that the machine should not be overheated at high slip values. The determination of the current locus at a reduced stator voltage limits the validity of the measurement result, since nonlinearities and saturation effects are difficult to be allowed for.

The availability of small process computers makes it feasible to use digital techniques in those fields, where mostly for reasons of cost analoguous methods have been exclusively applied before. Moreover those capabilities of a digital computer system have to be considered which open new possibilities that cannot or unsatisfactorily be implemented by conventional means.

The aspect of cost is extremely important for industrial testing of electrical machines. The initially high investment of a process computer system is coupled with the expectation, that it is amortized in a relatively short time by reduced testing times and savings of personnel.

SYSTEM DESIGN CONSIDERATIONS

When starting the first design studies it had to be kept in mind that the application of a digital computer must be considered as one part of a whole system. The question of data interchange between the process computer and the process, i.e. the induction machine with its associated auxiliary equipment, had to be deliberately planned to achieve hardware and software harmony. The necessary instrumentation has to fit the requirements of converting the interesting physical analoguous quantities into digital form to be processed in the computer and to transfer data and commands to the machine test set-up. All this has to be done before the background of an electrically noisy environment and the problem of fighting system disturbances of various nature. Last not least there is the question of man-machine interfacing comprising those topics

---

1) Prozess-Rechner Instrumentierungssystem zur Steuerung, Messung und Automatisierung

like ease of operation and highly informative computer output. From the experience gained so far an alphanumeric graphical CRT storage display in combination with a keyboard for interactive operation fits the demands (Ref. 2).

In the following the various components are described so far as it is necessary to explain the operation of the whole system. The abbreviations used in the general circuit diagram of Fig. 1 refer to the components described in the following text. The overall view of the experimental set-up is shown in Fig. 2.

SYSTEM CONFIGURATION AND MEASURING TECHNIQUES

The induction machine (IM) to be investigated has a 10 kW nominal rating with a synchronuous speed of 1500 r.p.m.. The machine is equipped with a wound rotor. The ends of the star-connected rotor windings are led to slip rings, allowing to close the rotor circuit with a contactor (C). To avoid contact wear by arcing antiparallel thyristors (APT) are connected to each contact. A firing circuit assures the release of gate pulses to these thyristors before closing and their blocking after the mechanical contacts have opened.

For protecting the system from abnormal or critical operating conditions the design philosophy was followed that computer control independent equipment had to be installed, which is housed in the auxiliary units AU1 and AU2. This concerns such topics like switching off commands caused by blowing fuses, failure of cooling fans or an illicitly high number of r.p.m. of the machines ($>$ n). This doesn't exclude, however, that computer commands are inserted into interlock chains.

A four quadrant thyristor converter (TC) with two fullwave, fully controlled 3-phase bridges in antiparallel connection (Ref. 3) was selected to meet the demands of accurate closed loop control according to the requirements of the experiment. The dc drive system was built up following existing theories (Ref. 4) and design rules (Ref. 5, 6, 7). The closed-loop control system uses combined proportional and integral action for the speed regulation and for the internal current feedback loop.

A dc shunt machine (M) of 28.5 kW rating with a separate regulated field excitation (FE) is coupled with the induction machine.

A current reversal logic circuitry detects zero current by means of two current transformers in the mains connection and the polarity of the output voltage of the speed regulator for selecting one of the thyristor bridges. The gate control set is switched to either one of the bridges with a no current interval of approximately 5 ms between reversal of current direction in the dc machine's armature circuit, allowing to operate the converter without circulating current. During the currentless interval the shift of the firing pulses is tracked by the EMF control, thereby improving the dynamic behavior of the control system.

Because of inherent asymmetries of the mains and the non-ideal performance of gate control sets a subharmonic 100 Hz-oscillation in the thyristor control circuit (Ref. 8) was detected during the first test runs. For the prevention of this hunting effect a slight modification was successfully executed within the commercial converter unit. The change provided the pick-up for the current signal from a shunt (S) in the armature circuit by bringing the signal to regulator reference potential through a chopper type isolation amplifier (IAC) and a subsequent absolute value former instead of taking the rectified ac signal from the mains current transformers.

A choke with L = 50 mH nominal inductance permits smooth running conditions with a low limit for the onset of a discontinuous current.
Besides of the dc current signal the armature voltage of the dc machine is acquired by another isolation amplifier (IAV) providing means for making efficiency calculations with the process computer.

For the galvanic separation between the drive system and the process computer inputs and outputs optically coupled isolators are used to tackle the problem of common mode signals and disturbances produced by switching transients of the power circuits including the thyristors.

Remembering that the root-mean-square value of an alternating voltage or current is basically an average value the application of an integrating analogue-to-digital conversion method is quite natural. In addition, however, there arises the question of squaring the a.c. quantities before integrating them. For this reason an analogue multiplier module is used. The multiplication function is realized after a pulse modulation principle. A pulse train is modulated in that fashion that one input controls the pulse duty cycle, the other the amplitude of the pulse in a linear manner. The resultant output pulse train is fed to a lowpass filter to extract the product of the input signals to be multiplied. With both signals connected together as a squarer the output voltage of the module is $\frac{u^2}{10V}$, with u as the instantaneous value of the input voltage. To measure the phase voltage of the mains a voltage transformer (VT) with a relative voltage error of $\pm$ 0.5 % and a phase error of $\pm$ 0.33 °el at 50 Hz is used to adapt the voltage to the input of the squaring unit. The requirement of adequate accuracy can easily be met since the phase voltage changes only in a rather limited range. The adaptation of the current signal, however, is a bit more difficult for several reasons. 1st: The physical quantity of the phase current of the induction machine has to be converted into a proportional voltage. 2nd: The phase current covers a considerable range with the parameter slip varied through the motor, generator, and braking region. 3rd: The squaring result is affected by an untolerable error if the input voltage is less than 2 V_{RMS}. It follows from Fig. 3 that the interesting range of the phase current of the induction machine is acquired by three current transformers having turn ratios of the secondary windings to the primary windings of 100 : 1, 30 : 1, and 20 : 1. The current transformers are of the toroidal core type (Ref. 9) with a nominal burden of 30 VA at 50 Hz with an over current ratio of 5. Accuracies are given by a relative current error of $\pm$ 0.2 % and a phase error of $\pm$ 0.17 °el at nominal current. The burden for the current transformers are wirewound resistors with a noninductive winding (DALE Type NH), mounted in an aluminium housing (inductance app. 0.2 μH measured at 1 MHz). The resistors are mounted on profile heat sinks to provide thermal stability with a temperature coefficient of $5 \cdot 10^{-5}/^{\circ}C$.

The secondary current of the current transformers produces a voltage drop across the burden which is fed to the analogue to digital converter of the computer interface within the PRISMA-System (Ref. 10). Since all voltages to be converted have to be confined to the standard level of $\pm$ 10 V, means are provided to shorten the burden resistances of the 20 : 1 and 30 : 1 current transformers with a relay contact as soon as the RMS value of the voltage across them reaches 7 V. For transient states a fast acting combined Zener diode-transistor voltage limiter protects the input of the converter circuitry until the shorting relay contacts commutate the secondary current from the electronic components after some milliseconds.

The voltages for determining the switching level of the shorting relays are measured at the 100 : 1 current transformer burden. Under consideration of a certain hysteresis region the choice of the primary to secondary current ratios of the current transformers and the associated burden resistances permit the measurement of phase currents with adequate accuracy. To avoid an appreciable error (>1 %) it has been provided that at least one of the input voltages that are fed to the squarers for subsequent coding in a voltage to frequency converter is not less than 2 V_{RMS} within the interesting phase current region under symmetrical conditions. The time base of the integrating analogue to digital converter has a 20 ms duration, eliminating hereby the ac component of the squaring result for the ac quantities phase voltage and current. The superimposed harmonic ripple components of the armature voltage of the dc machine and the shunt signal of the armature current are likewise suppressed in the coding process.

The average value, which is coded in the analogue-to-digital converter after squaring a sinusoidal voltage with the RMS value U, is determined by

$$\bar{u} = \frac{(\sqrt{2}\,U)^2}{V} \cdot \frac{1}{10} \cdot \frac{1}{2} \,.$$

Coding is performed with a 10 bit (pure binary) resolution plus sign with a 10 V dc signal corresponding to a decimal value of 1000.

By determining the RMS values according to the above method slight harmonic distortions of the mains phase voltage and the phase current are not critical for the measuring result. The error caused by a higher harmonic, which for example has an amplitude of 0.1 of the fundamental harmonic, would only cause an increase of the measured RMS value by app. 0.5 % compared with the undistorted case.

The signals for determining the phase angle between the mains phase voltage and the associated phase current are derived from the voltage transformer and the 100 : 1 ratio current transformer. Each of the two instantaneous signals are fed to separate track and hold amplifiers. A multiplexer samples each output of these amplifiers with a 50 kHz sampling frequency. The sampled signals are coded with a 10 bit resolution (binary offset code) by an analogue-to-digital converter. The sampling frequency of 50 kHz permits a phase angle resolution of 0.36 °el for a 50 Hz sinusoid. The coded values of the voltage and current signals are brought into the central memory of the process computer by direct memory access in the cycle stealing mode (Ref. 11). The evaluation of the stored data in a 500 words buffer is performed by a program that starts counting the number of current signal samples from the zero crossing of the phase voltage signal until it detects the change of sign of the current samples (Ref. 10). The number of counted current samples up to the zero crossing point of the phase current multiplied by the sampling period of 20 μs $\hat{=}$ 0.36 °el delivers the phase angle between the phase voltage and the phase current of the induction machine.

The torque measuring device (TMD) (Ref. 12) connects the induction machine with the dc shunt machine as a hub with a flexible flange mount. A torque applied to the measuring hub produces a torsion, which is detected in a strain gauge bridge circuit mounted on the measurement shaft. The strain gauge bridge is energized by a L.F.-F.M. oscillator with a central frequency of 6.75 kHz and a maximum sweep of $\pm$ 30 %. The bridge unbalance controls the oscillator frequency deviation, which is directly proportional to the acting torque. The arrangement provides contactless transmission of the measured torque signal from the shaft to the stationary part of the measuring system. Moreover power is also transmitted to the rotating device without sliding contacts by transformer action. A combined signal and power induction coil serves for supplying power from the mains to the mobile part of the system and transferring the measuring signals from the rotating to the fixed section. A discriminator recovers the measuring signal from the modulated transmission signal. A voltage of $\pm$ 10 V corresponds to a torque of $\pm$ 50 kpm (50 · 9.81 Nm). The linearity error is less than 0.5 %. The dynamic response is characterized by a -3dB corner frequency of about 1 kHz.

COURSE OF MEASUREMENT, EVALUATION AND SOFTWARE CONSIDERATIONS

The course of the measurement for the interesting quantities when recording the current locus and the torque-speed characteristic of the induction machine is consistent with the asynchronuous (query-response basis) hand shaking mode of digital computers. The speed reference signal is delivered by the process computer via a digital-to-analogue converter (set point control). A supervision circuitry detects the state when the actual speed value taken from the tachometer (T) has reached the reference value within a 1 %-tolerance band. When running to a new speed setting the induction machine is already connected with the stator windings to the mains. The rotor circuit, however, is still open. Upon reaching the new speed value within the tolerance band the rotor circuit is closed by a process computer command. After the transient state, following the closing of the rotor circuit (Ref. 13), an interrupt signal, derived from the tolerance band supervision in conjunction with a time delay in the interface is sent back to the process computer with the information that the induction machine has reached a quasistationary state at a certain speed or slip. This causes the measurement program to be started. Upon completion of the measurement its evaluation is immediately followed with a corresponding measuring point eventually displayed on the screen of the storage CRT with the associated slip values listed on the right side of Fig. 4. The photograph of Fig. 4 was taken with a diapositive film. In order to have a hard copy the acquired data of the measurements are used to be processed in a FORTRAN plot program. The torque-speed characteristic of Fig. 5 is the result of such a plot. Table 1 is the print-out of an electrical typewriter with a golf-ball type writing head. The machine had been adapted to the process-computer interface providing a cheap means for writing a measurement protocol as a useful completion to the graphical displays.

Assuming a speed setting which belongs to the motor region of the induction machine, the operation of the dc drive system is as follows. With the rotor circuit of the induction machine open the thyristor converter works as a rectifier delivering energy to the dc machine running as a motor. Upon closing the rotor circuit the induction machine is developing a torque causing the dc machine to become a generator and the thyristor converter with the other bridge to go into inverter operation. After execution of the measurements of the interesting physical quantities the rotor circuit is reopened to avoid excessive heating of the induction machine. This causes again a change of the thyristor converter's state into the initial rectifier operation.

The computer programs for the acquisition and the evaluation of data are built up by separate modules, each of them following the rules of structured programming. Before starting the various parameters are transferred into storage by writing them into so called instruction data files (Ref. 10) via the keyboard. Parameters are speed ranges and stepwidths for going from one speed or slip setting to the next. The application of instruction data files makes it very convenient, even for unskilled system users to activate programs from the keyboard with the CRT screen for an interactive operation.

The program sequence diagram of Fig. 6 is self-explanatory according to the preceding sections. The implementation of the individual program modules for the on-line operation was performed by using symbolic assembly language for the process computer used (Ref. 11). It should be mentioned, that the programming effort must not be underestimated. There is hope, however, that the availability of problem-oriented computer languages will have a growing importance in the future by readily allowing the description of a system and the activation of the process interface. This would help the user of a process computer to concentrate his efforts more on the process itself rather than to become heavily involved in programming as a major job.

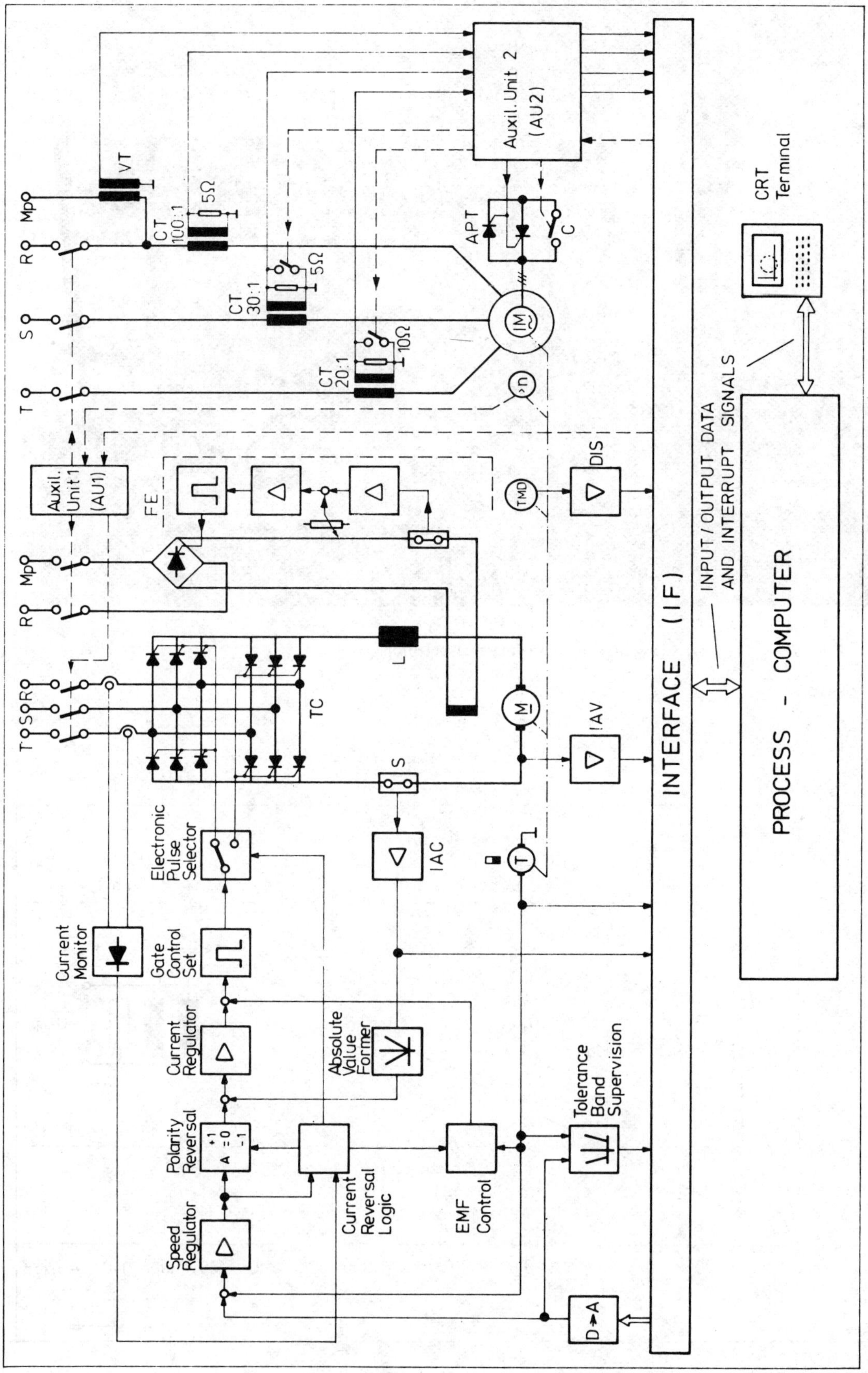

Fig. 1. General circuit diagram of the experimental set-up

Fig. 2. Overall view of the experimental set-up

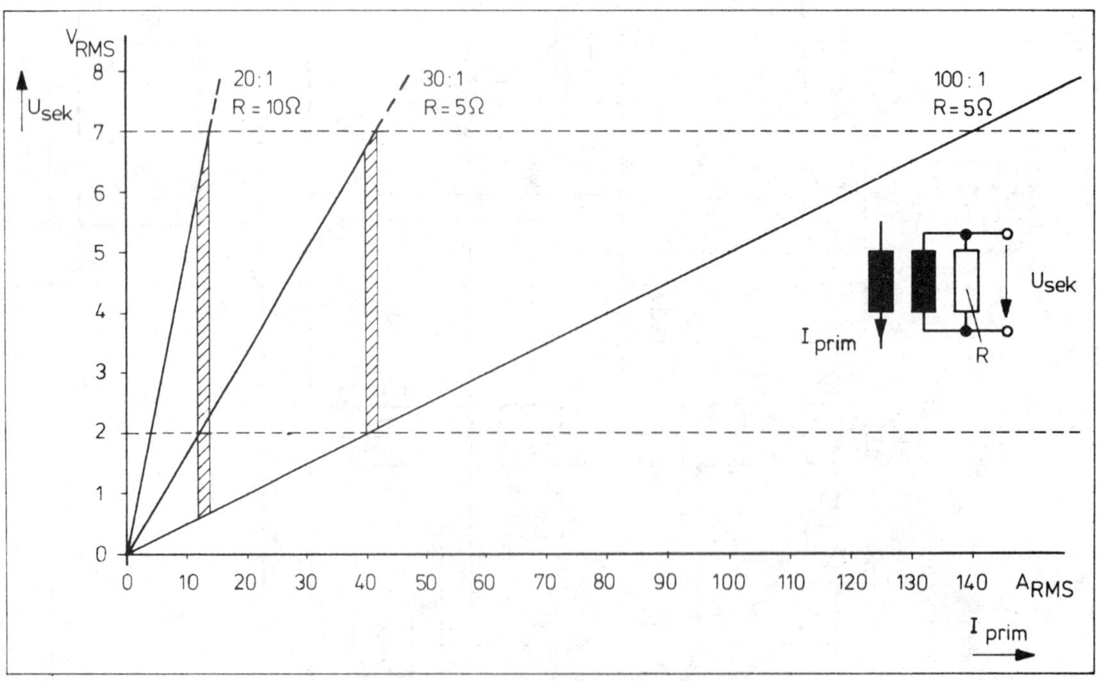

Fig. 3. Output voltage of the current transformers for the measurement of the phase currents of the induction machine

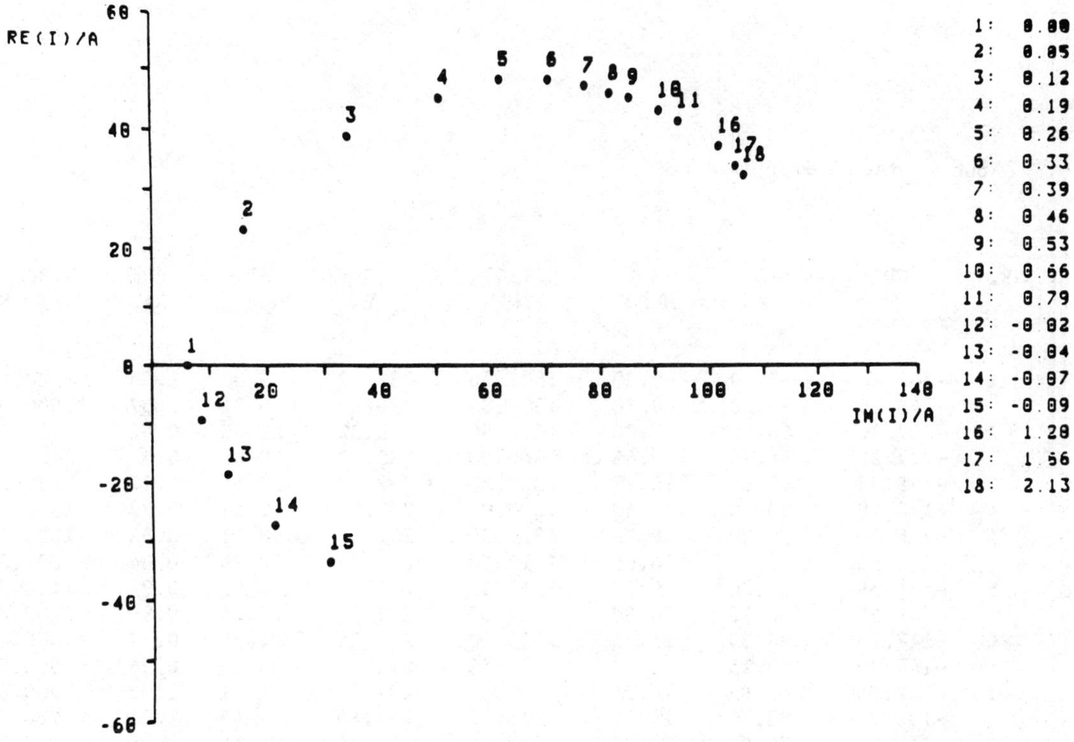

Fig. 4. Photograph of the CRT storage display after recording the measuring points for the current locus of the induction machine

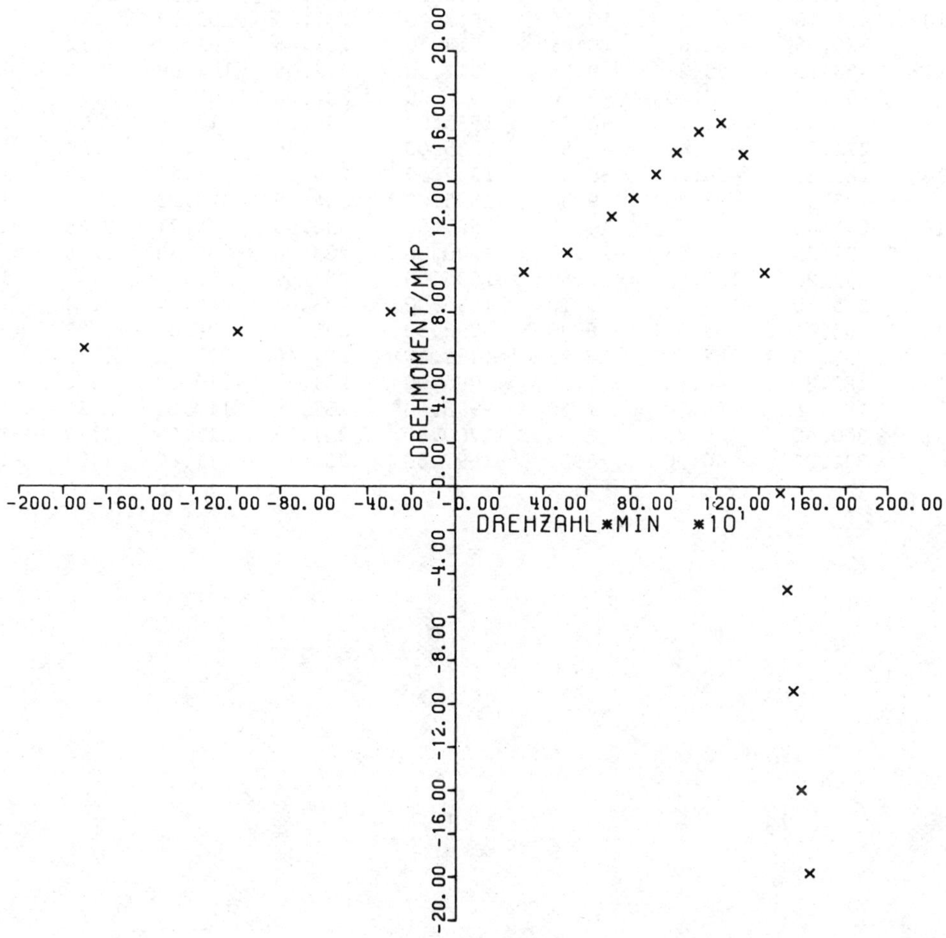

Fig. 5. Torque-speed characteristic of the induction machine (computer plot)

TABLE 1 Measurement protocol

NR.	UD V	ID A	MD MKP	N(IST) 1/MIN	U V	I A	PHI RAD	N(SOLL) 1/MIN
1	-262.80	-3.00	-0.30	1502.50	235.17	6.32	1.58	1500.00
	-263.25	-3.00	-0.30	1500.00	235.17	6.32	1.57	1500.00
2	-227.70	56.10	9.80	1422.50	233.32	28.40	0.61	1400.00
	-227.25	56.70	9.85	1420.00	233.63	28.65	0.60	1400.00
3	-195.75	90.60	15.25	1322.50	232.39	52.54	0.72	1300.00
	-196.20	90.90	15.40	1320.00	232.08	52.54	0.72	1300.00
4	-176.40	99.60	16.70	1220.00	231.46	68.41	0.83	1200.00
	-175.50	99.75	16.65	1217.50	231.77	68.70	0.84	1200.00
5	-161.55	97.20	16.30	1117.50	231.14	78.99	0.92	1100.00
	-161.55	96.75	16.30	1117.50	231.14	79.25	0.90	1100.00
6	-147.15	90.30	15.35	1015.00	231.14	86.26	0.97	1000.00
	-146.70	90.15	15.30	1012.50	231.46	86.26	0.96	1000.00
7	-132.75	83.85	14.35	917.50	231.14	91.21	1.02	900.00
	-132.75	83.70	14.35	915.00	231.46	91.43	1.01	900.00
8	-118.80	76.65	13.25	812.50	231.14	94.66	1.06	800.00
	-118.35	76.35	13.20	812.50	231.14	94.66	1.05	800.00
9	-103.50	71.10	12.40	715.00	231.46	97.37	1.09	700.00
	-103.50	71.10	12.35	712.50	231.46	97.37	1.07	700.00
10	-71.10	61.95	10.75	512.50	231.46	101.39	1.14	500.00
	-70.65	61.65	10.85	512.50	231.46	101.39	1.12	500.00
11	-38.70	55.20	9.85	312.50	232.08	104.69	1.15	300.00
	-38.70	55.20	9.80	312.50	232.08	103.92	1.15	300.00
12	-274.50	-29.10	-4.75	1535.00	234.86	12.59	2.38	1540.00
	-275.85	-29.40	-4.80	1535.00	234.86	12.70	2.39	1540.00
13	-285.30	-56.40	-9.40	1562.50	234.56	22.61	2.53	1580.00
	-285.30	-57.30	-9.50	1565.00	234.25	22.93	2.53	1580.00
14	-291.60	-85.50	-13.95	1597.50	233.63	34.31	2.48	1620.00
	-292.05	-86.85	-14.15	1600.00	233.94	34.73	2.48	1620.00
15	-295.20	-113.55	-17.80	1635.00	233.32	45.17	2.41	1660.00
	-295.20	-114.45	-18.10	1635.00	233.01	46.04	2.40	1660.00
16	63.45	52.35	8.00	-292.50	231.46	108.63	1.23	-300.00
	63.45	52.50	8.25	-292.50	231.46	109.00	1.21	-300.00
17	181.35	44.85	7.10	-995.00	231.77	110.82	1.26	-1000.00
	181.35	44.70	7.30	-995.00	231.46	110.82	1.25	-1000.00
18	300.60	40.20	6.35	-1700.00	231.77	111.89	1.29	-1700.00
	301.05	40.50	6.35	-1697.50	231.77	111.89	1.27	-1700.00

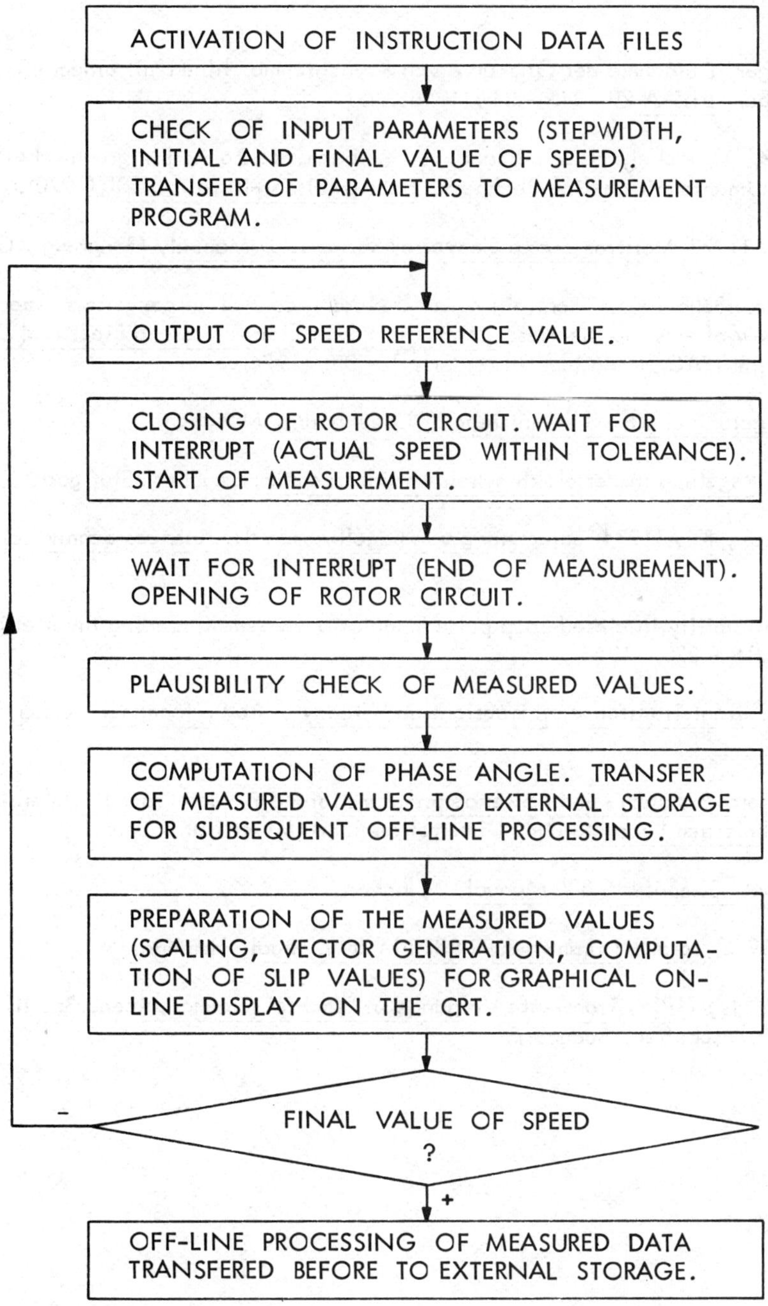

Fig. 6. Program sequence diagram

References

(1) D. Schäfer, A. Rüdiger, Aufnahme der Ortskurve von Asynchronmaschinen mit einem elektronischen Ortskurvenschreiber, ETZ-A 79, 345 - 349, (1958).

(2) P. Abend, R. Bublitz, D. Jackel, H. Klessmann, A. Kolbach, J. Zahn, Ein graphisches Sichtgerät für nukleare Experimente mit on-line Rechnern, Hahn-Meitner-Institut B 100, (1970).

(3) Brown, Boveri & Cie, (1973) Veritron Static Converter Series CAD Manual, Mannheim, Germany.

(4) I. Lagasse, B. Prajoux, Behaviour of control systems including controlled convertors especially rectifiers. A review of existing theories, Control in Power Electronics and Electrical Drives, Survey papers of the IFAC Symposium Düsseldorf, 1 - 38, (1974).

(5) Pfaff, G., (1971) Regelung elektrischer Antriebe I, Oldenbourg, München.

(6) Leonard, W., (1973) Regelung in der elektrischen Antriebstechnik, Teubner, Stuttgart.

(7) Buxbaum, A., Schierau, K., (1974) Berechnung von Regelkreisen der Antriebstechnik, Elitera, Berlin.

(8) F. Fallside, Ripple instability in closed-loop pulse modulation systems including inverter drives, Proc. IEE, 115, 218 - 228, (1968).

(9) Wright, A., (1968) Current Transformers, Transient and Steady State Performance, Chapman & Hall, London.

(10) Weigel, J., (1976) Betriebssystem- und Anwenderprogrammierung für den Prozeßrechnereinsatz in einem elektrotechnischen Laboratorium, Dissertation Ruhr-Universität Bochum.

(11) Siemens, (1971) Process Computer S 305 Manual, München.

(12) Philips Elektronik, (1972) Torque Transmitter PR 9372 M/50 Manual, Hamburg

(13) Kovaćs, K. P., Rácz, I., (1959) Transiente Vorgänge in Wechselstrommaschinen, Bd. II, Ungarische Akademie der Wissenschaften, Budapest.

HIGH ACCURACY AND FAST RESPONSE DIGITAL SPEED MEASUREMENT FOR CONTROL OF INDUSTRIAL MOTOR DRIVES

V. Török and J. Valis

The Royal Institute of Technology, Department of Electric Machinery, Stockholm

ABSTRACT

This paper concerns digital speed control for high performance industrial drives (paper machines, rolling mills etc.) based on a novel principle for frequency measurement.

In sections 1-2 the principle and theory for digital measurement of angular velocity by means of a low resolution pulse transmitter and the new frequency-to-digital convertor is presented. This system gives high measurement accuracy and resolution combined with short measuring time within a wide speed range at a low ripple content. Further, in section 3 it is shown how this F/D convertor, consisting of a binary rate multiplier, an up/down counter and a register can be built with a few digital integrated circuits.

The next section describes the design of the speed control system for a thyristor controlled DC motor drive based on the new F/D convertor. Significantly better steady state accuracy and linearity (better than 0.01 %) compared with present analogue systems has been achieved. The dynamic control performance is also better than that of most analogue systems available (response time better than 30 ms for a 100 kW drive). These characteristics are documented by experimental performance curves.

In the last section, two industrial applications of such drives are described - one for a paper machine and one for a wire rod mill. In both cases microprocessor-based program-controlled electronics is used for digital overriding (higher level) control.

1. INTRODUCTION

A considerable, and ever increasing proportion of electrical drives are designed for variable speed operation. A great many of these are speed controlled either in order to ensure accurate control over the speed or to enable the speed controlled motor to be used as "actuator" in an overriding control system or a hierarchy of overriding control systems.

The demand for accuracy and fast response in speed regulation has been increasing constantly, particularly with respect to sectional drives for processes with continous tracks such as paper machines, where small differences between the speeds of the motors have considerable influence on the process.

Increased production rates increase the demand for fast control since, in principle, the response time should be reduced to keep up with the reduced period spent by the product in the controlled sections.

That which at present limits the performance of the variable speed drive system most prevalent today - the power convertor fed DC motor - lies purely within the field of measuring technique.

It can be shown [1,2] that the theoretically shortest response time for speed control with respect to the periodic mode of operation of the convertor is the time for 1.5 convertor pulses, i.e. 5 ms in Europe for the most usual 6-pulse convertor. In practice the control time is 5 - 200 times longer.

Most of the blame for this great gap between theory and practice can be laid on the measurement noise inherent in the tachometers now available (see Table 1 on page 11).

Other defects - non-linearity, temperature drift, long-term drift (see Table 1) may also play an important part, especially in sectional drives without much process feedback to the drive system, such as paper machine drives. Combined analog and digital measuring systems have been used for a long time for such applications, the purpose of the digital part being to compensate the drift in the analog measurement. The digital part thus acted either as a slow but accurate speed transmitter (Fig. 1) or as a fast-acting transmitter of the angular deviation (Fig. 2) in relation to an imaginary reference axis.

The previously known digital speed-measuring methods, based on pulse transmitters on a shaft in combination with pulse counting or period time measurement, were unable to combine in a practical manner the speed and accuracy of measurement required for speed control with a large measurement range. The analog tachometers could not therefore be replaced by purely digital systems.

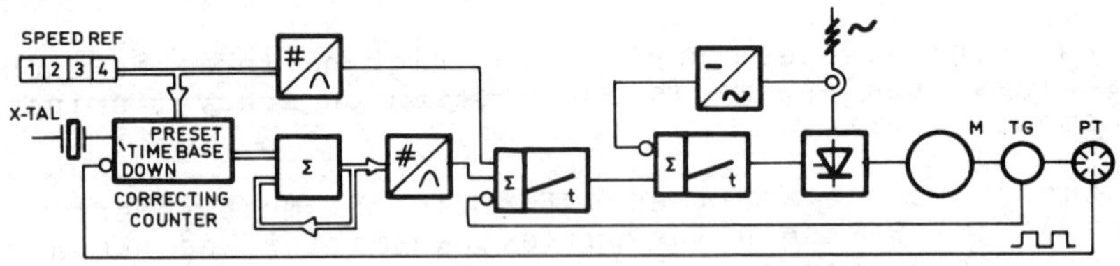

Fig. 1. Analog speed control with slow-acting digital correction

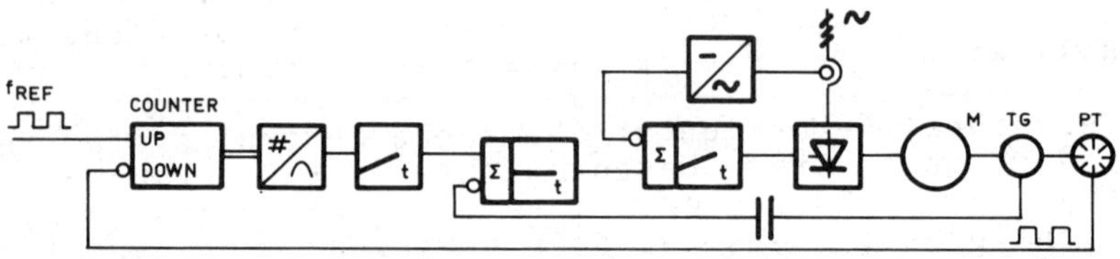

Fig. 2 Fast-acting digital position control with analog stabilizing feedback

2. NEW METHOD OF DIGITAL SPEED MEASUREMENT

It is well known that speed cannot be altered suddenly due to Newtonian forces. There is consequently no reason to start from scratch for each measuring interval - as is done both in the case of pulse counting and period time measurement. The same applies to speed as to the weather: the weather tomorrow is most likely to be the same as today! The new measuring method operates exactly on this principle: each time there is an expected moment at which the new transmitter pulse should occur, the expected length of the new interval being inversely proportional to a stored "measurement number" corresponding to the last (estimated) value of the speed. When the new transmitter pulse arrives, we adjust that number by an amount depending on the difference between the observed and the predicted length of the interval, thus improving the measurement iteratively. Figure 3 shows that the abovementioned difference is measured by subdividing the expected pulse distance $T_{p,E}$ into a large number (N) of parts by generating an internal pulse train f_R having a frequency N times higher than the expected value of the transmitter frequency $f_{p,E}$.

The internal pulse train f_R is generated by a digital rate multiplier (Fig.4), the pulse input of which is driven by a known precise clock frequency and the rate inputs of which are connected to a register containing a number Z. The internal frequency f_R is thus proportional to this "measurement number" Z in accordance with the equation

$$f_R = \frac{Z}{Z_0} fc \qquad (1)$$

where f_c = the clock frequency Z_0 = the divisor of the digital rate multiplier

The measurement number Z is adjusted by the difference between the expected number of internal pulses/interval, N, and the actual number, P_k, so that the new measurement number will be

$$Z_{k+1} = Z_k + (N - P_k) \tag{2}$$

where Z_k - the measurement number for the kth measured interval, i.e. the old measurement number

and Z_{k+1} - the measurement number for the $k+1^{th}$ measured interval, i.e. the new measurement number.

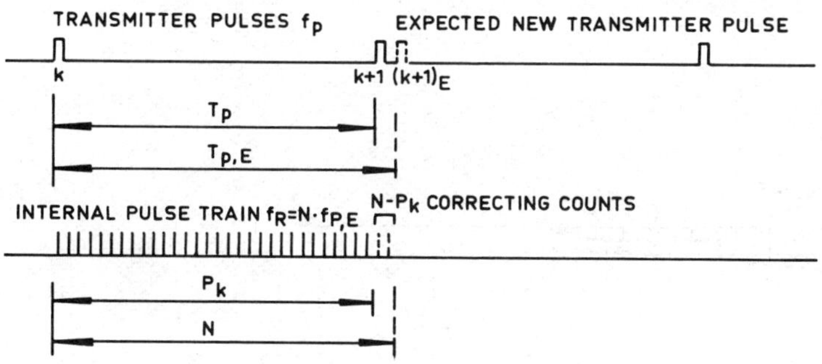

Fig. 3

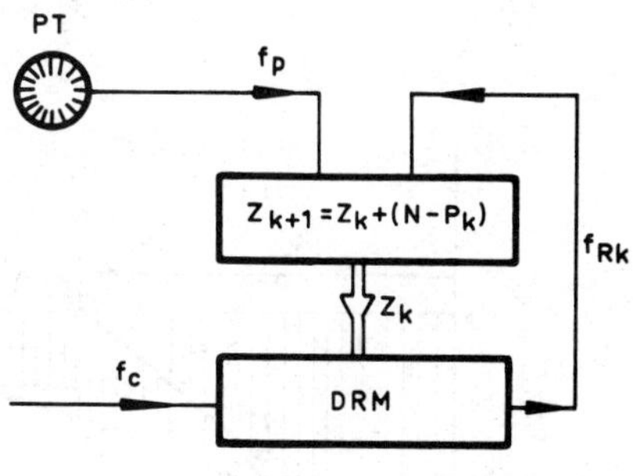

Fig. 4

The exact value of the measurement number Z_p shall fulfil the condition N-P = 0 (see equation (2)), equivalent to the condition

$$f_R = N \cdot f_p \tag{3}$$

which inserted in equation (1), gives after simplification

$$\frac{Z_p}{Z_0} = N \frac{f_p}{f_c} \tag{4}$$

The actual number of internal pulses in a measuring interval can be expressed with respect to equation (1) (see also Fig. 3)

$$P_k = \frac{Z_k}{Z_0} \cdot f_c \cdot T_{p,k} \tag{5}$$

where $T_{p,k} = \frac{1}{f_{p,k}}$ - the time interval between two transmitter pulses.

If (4) and (5) are inserted in equation (2) the following is obtained after simplification

$$Z_{k+1} = Z_k + (Z_{p,k} - Z_k) \frac{f_c \cdot T_{p,k}}{Z_0} \tag{6}$$

We assume that the transmitter frequency, and thus Z_p, remain unchanged during the $k+1^{th}$ interval. The deviation in the measurement number expressed with respect to equation(6) will be

$$Z_p - Z_{k+1} = (Z_p - Z_k)(1 - \frac{f_c T_p}{Z_o}) \tag{7}$$

The coefficient

$$a = 1 - \frac{f_c T_p}{Z_o}$$

is positive and less than 1 within the measuring range. The equation (7) can be used recoursively and for the k+xth interval this gives

$$Z_p - Z_{k+x} = (Z_p - Z_k)a^x \tag{8}$$

i.e. the measurement error vanishes exponentially (see Fig. 5).

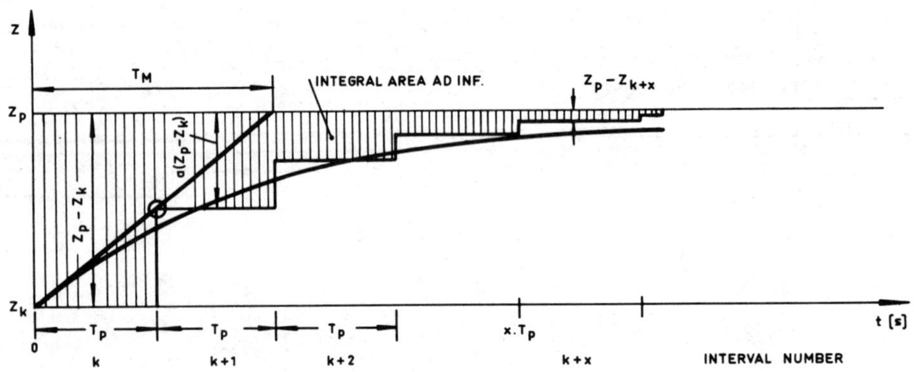

Fig. 5

The deviation of the measurement number integrated ad infinitum can be expressed with the sum of a geometric series

$$\int_0^\infty (Z_p - Z)dt = (Z_p - Z_k)T_p \cdot \sum_{x=0}^{\infty} a^x \qquad [s] \tag{9}$$

where

$$\sum_{x=0}^{\infty} a^x = \frac{1}{1-a} = \frac{Z_o}{f_c T_p} \tag{10}$$

An equivalent time constant of the measurement method can be defined as

$$T_M = \frac{1}{Z_p - Z_k} \int_0^\infty (Z_p - Z)dt \qquad [s] \tag{11}$$

The equations (9),(10)and (11) combined give

$$T_M = \frac{Z_o}{f_c} \qquad [s] \tag{12}$$

i.e. *the time constant is unchanged within the entire range of measurement* and is equal to the cycle time of the digital rate multiplier.

A "constant" time constant is extremely desirable in the field of control engineering as well as meeting the requirements of constant measurement accuracy throughout the entire measurement range.

The latter can be understood as follows: the peripheral length of the pulse wheel entailed in determining a measured value, i.e. the length passing a pickup during the measuring time T_M, is proportional to the speed. The manufacturing tolerances of the position of the marks on the pulse wheel are constant irrespective of the length of the periphery embraced (for ground toothed wheels, for instance, which are best for the purpose, the position error of the teeth is less than $\pm$ 2 μ). Let us designate this ΔL. If the peripheral speed of the pulse wheel is V_p, then the speed measurement is impaired by a relative error of

$$\Delta L/(V_p T_M) \tag{13}$$

This measurement error, related to the maximum speed within the range of measurement, corresponding to a peripheral speed V_0 or the measurement number $Z_p = Z_0$, can be expressed with respect to equation (12)

$$\frac{\Delta V}{V_0} = \left(\frac{V_p}{V_0}\right) \frac{\Delta L}{V_p T_M} = \frac{\Delta L}{V_0 T_M} \tag{14}$$

The measurement accuracy is thus constant through the entire range of measurement. A somewhat perplexing feature from the point of view of measuring techniques is that the *resolution Z_0 is independent of the number of teeth on the pulse wheel.* Expressed from equation (12) the resolution is

$$Z_0 = f_c T_M \tag{15}$$

and at a certain desired time constant T_M it is only dependent on the clock frequency. On the basis of equation (15) it can be said that the entire measuring procedure is substantially equivalent to a pulse counting procedure applied to the internal frequency f_R, i.e. *N times more advantageous than pulse counting applied to the transmitter pulse train.* In practice N is in the order of magnitude 10 - 1000.

An invariable time constant (equation (12)) will therefore include a varying number of intervals T_p within one measuring time constant. The step response of the measurement procedure is convergent for

$$|a|<1 \tag{16}$$

(compare equation (8)), monotonic for

$$0 < a < 1 \tag{17}$$

and oscillatory for

$$-1 < a < 0 \tag{18}$$

Equations (8) and (12) combined result in

$$a = 1 - \frac{T_p}{T_M} \tag{19}$$

which means that the borderline case between monotonic and oscillotory step response occurs when

$$T_p = T_M \tag{20}$$

i.e. the period time of the transmitter pulse train is equal to the measurement time. Under these circumstances the step response is of "dead beat" character, the measurement method behaves in principle as a pure period time measurement. *Equation (20) defines a certain limit frequency f_{pL} and a corresponding limit speed* for specified pulse transmitter and measurement time.

The stability limit for oscillatory step response (equation (18) occurs at *half the limit speed* and *it is here that the abolute limit of the measuring method lies*. The practical limit may be set at about 70 - 100 % of the limit speed. In equipment available so far the lower limit of the measurement range has been set at the limit speed and was in the order of magnitude 0.4 - 6.25 % of the upper limit of the measurement range.

So far we have not taken into account the influence of quantization, i.e. that P and Z are integers. Furthermore, when f_R is generated by means of a binary rate multiplier (BRM), the pulse train becomes somewhat irregular. In practice these factors result in a superimposed digital noise which usually has an amplitude of $\pm$ 1 unit in the number Z. This is of no significance from the control point of view since the lowest frequency it may have is determined by BRM's cycle time = T_M. However, this frequency is outside the cutoff frequency of the speed control.

Figures 6A - 6B show the step responses of the measurement method.

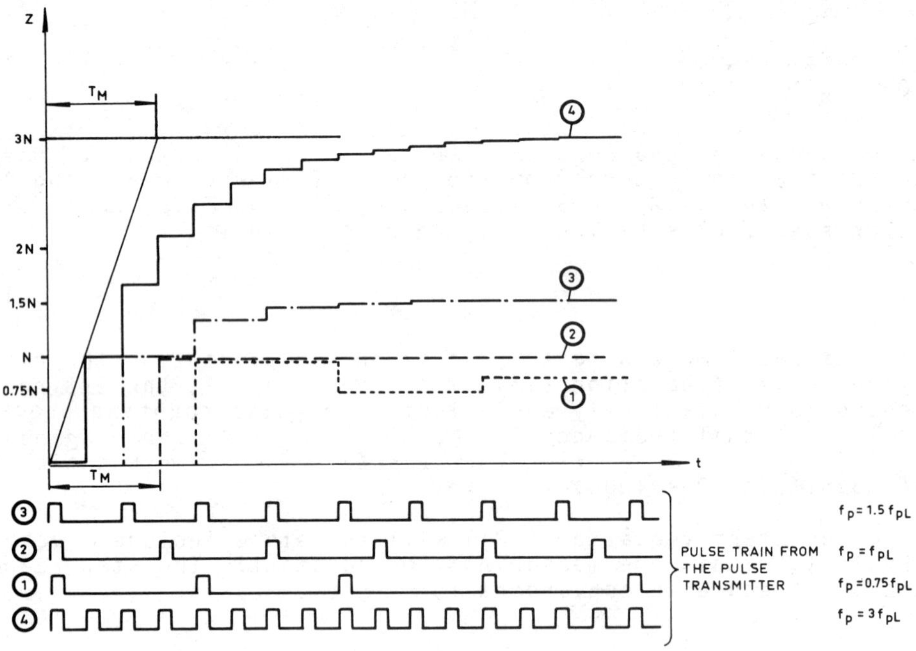

Fig. 6A

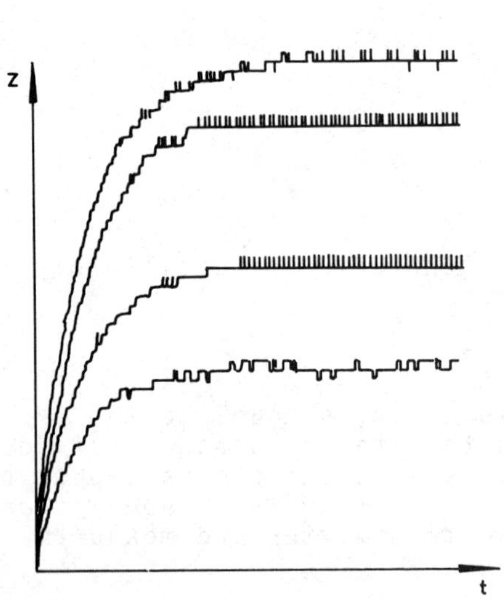

Fig. 6B Step response containing digital noise

Possible extensions

The internal pulse train f_R in effect constitutes fictitious marks on the pulse wheel and is thus equivalent to a fictitious pulse wheel with N times as many "teeth" as the actual wheel. This enables an angular measurement with extremely high resolution, which can be used in position measurement and control for various purposes.

An extremely large measuring range is necessary for certain applications, i.e. it is desirable to be able to measure at almost zero speed. In such cases the frequency range of the transmitter can be shifted in accordance with the resolver principle, for instance, so that the transmitter frequency is greater than zero within the entire speed range.

3. PRACTICAL DESIGN OF A FREQUENCY TO DIGITAL CONVERTOR BASED ON THE METHOD

As described in the preceding section, the F/D convertor comprises three main parts:

- a digital rate multiplier, where the internal frequency $f_{R,k}$ during each transmitter pulse interval is generated as the product of a known clock frequency f_c and the measurement number Z_k
- a memory for the measurement number Z_k
- an artithemetic unit where the new measurement number Z_{k+1} is calculated in accordance with equations (2) and (5)

$$Z_{k+1} = Z_k + (N - \frac{Z_k}{Z_0} f_c \cdot T_{p,k}) \qquad (21)$$

These functions can be implemented as an algorithm in a microprocessor or as hard wired logic. The Fig.7 shows a complete diagram for a 12 bit binary F/D convertor using a few ordinary TTL ICs.

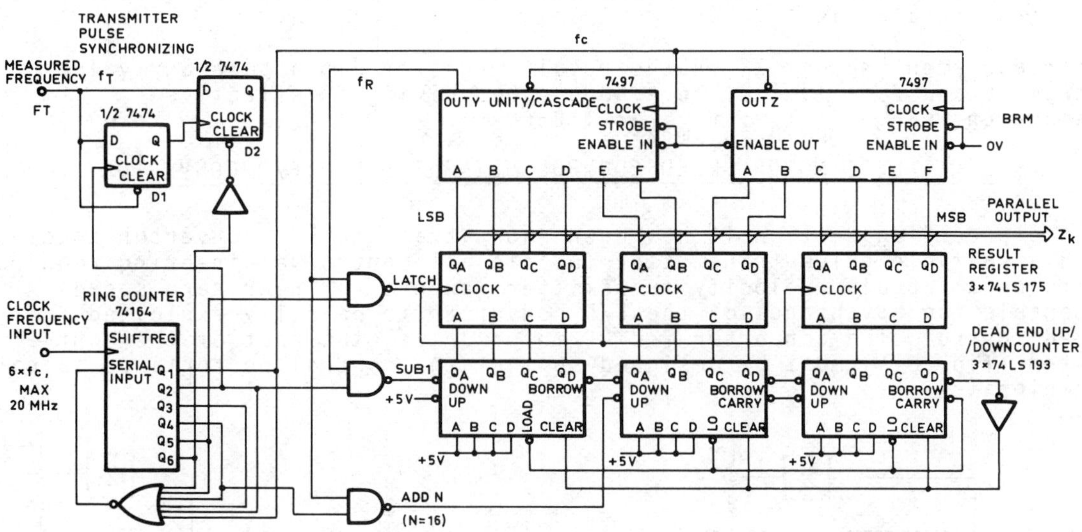

Fig. 7. 12 bit binary F/D convertor

This arrangement has been used for laboratory experiments with digital speed control and - with certain modifications enabling the parameters (resolution, measuring range and measuring time constant) to be easily varied as required for a certain application - in industrial applications [3].

Two cascade-connected 6 bit BRM IC's SN7497 are used as a 12 bit binary rate multiplier. The old measurement number Z_k is stored in a 12 bit parallel register consisting of three SN74LS175. A 12 bit up/down counter consisting of three SN74LS193 is used as the arithmetic unit to calculate the measurement number Z_{k+1}. This counter is a "dead end counter" in order to prevent overflow of the measurement number and to set lower limit of the frequency range to the limit frequency f_{pL} (see equation (20)).

The operation of the F/D convertor is controlled by a six-phase control signal Q_1 - Q_6 which is generated in a ring counter - a series in parallel out shift register and a NOR gate, where a "1" is shifted around with a shift frequency 6 times higher than the desired clock frequency f_c.

The BRM is stepped forward one step during the Q1 phase. Depending on its previous state and the current contents of the result register Z_k, the output signal of the BRM will be logical "1" or "0". If the result is "1", the current contents of the counter will be decreased by 1 during the Q2 phase. At the same time the flip-flops D1, D2 are set if the input signal FP has changed from "0" to "1", i.e. if a new period has started for the transmitter pulse train whose frequency is being measured. If flip-flop D2 has been set, the number N (N = 16 in the circuit shown in the fig. 7) is added to the contents of the counter during Q4 phase and thereafter the contents of the counter is copied into the result register during phase Q5. During Q6 flip-flop D2 is reset. If flip-flop D2 has not been set during phase Q2, nothing happens during phase Q3 - Q6.

These 6 phases are repeated cyclically with the frequency equal to the clock frequency f_c of the F/D converter. This means that during each period $T_{p,k}$ of the measured frequency the contents Z_k of the counter will gradually decrease by a total of

$$T_{p,k} \cdot \frac{Z_k}{Z_o} f_c$$

and at the end of the period will be increased by the number N = 16 (signal ADD N) and thereafter the contents of the counter will be copied into the register as Z_{k+1}.

The measurement number Z_k in the result register can assume any value between N and Z_o - 1, i.e. 16 - 4095, thus giving an effective measurement range $f_{pmin} : f_{pmax}$ = 1:256.

4. DIGITAL DC MOTOR SPEED CONTROL USING THE NEW F/D CONVERTOR

In the preceding section it has been shown that the F/D convertor together with a suitable pulse transmitter fulfils all control engineering requirements on an angular velocity transmitter (except for near zero speed).
To enable the advantages of the F/D converter to be fully exploited for speed control of a converter fed DC motor drive, the rest of the control system (fig. 8) must be designed so that the control performance is not deteriorated.

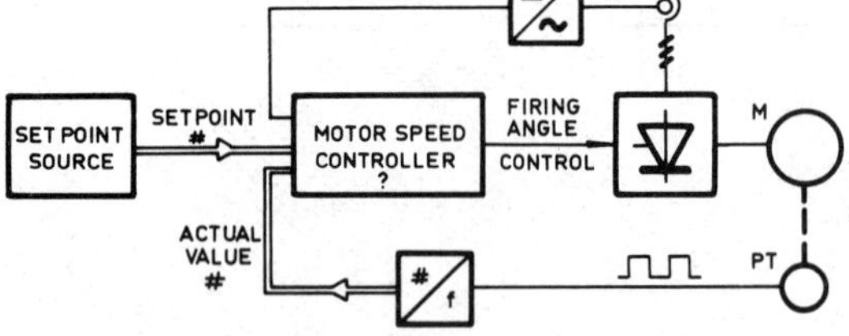

Fig. 8. Speed control using a F/D convertor

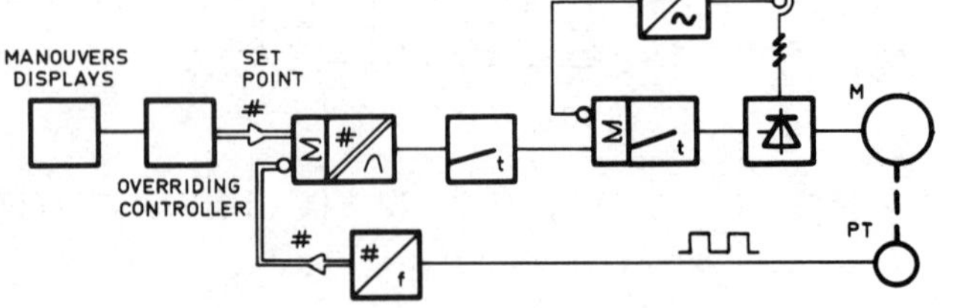

Fig. 9. DC-motor drive with digital speed control

There would not be much point in D/A converting the measurement number Z when connecting the F/D convertor to a conventional analog control system since a considerable portion of its accuracy would then be lost. On the other hand, a completely digital control system with its tempting possibilities to implement some "optimal control law" would - despite the present state of the art of digital IC technology, with inexpensive microprocessors and solid state memories - still be too exclusive for industrial applications.

Most convertors nowadays already include complete speed and current regulators and firing puls devices in analog technique and it would be practically advantageous to be able to retain these as far as possible in the control system. On the other hand, in most industrial applications the set point comes from an overriding control system in which extensive and complex data processing of various control and process signals and calculation of displayed values takes place and in which digital electronics such as mini and microcomputers has already been used. It would therefore be practical if the speed control system could work using a set point in digital form.

The above considerations have resulted in the control structure being shown on Fig. 9. Each individual speed controlled motor drive consists of a DC motor with built-on pulse transmitter, a standard thyristor convertor with its current and firing angle control circuitry and two printed circuits boards, one with electronics for the frequency/digital convertor and one with digital difference unit and digital/analog convertor. (Both PCBs are normally located in frames containing the overriding digital control system.)

The output of the pulse transmitter is connected to the frequency/digital convertor, whose output signal - the measurement number Z - is a binary number in parallel form, proportional to the actual speed. The control deviation, i.e. the difference between the set point, given by the overriding controller, and the actual value, is formed in a digital difference unit (a parallel subtractor) and then converted into an analog signal. Since the speed controller in any case is driven into saturation even at relatively small values of control deviation, it is sufficient for the D/A convertor to be connected to the difference unit so that it will reach its maximum output for a deviation of 6.25 per cent which allows a very high signal gain (16 times) in this stage. This enables the output signal of the D/A convertor to be easily amplified in an analog amplifier without deteriorating its accuracy by the amplifier offset voltage and temperature drift. This means that the D/A convertor can be directly connected to the normal PI speed control amplifier (controller) of the thyristor convertor. The same procedure and the same setting elements are therefore used for tuning the digital speed control as for analog control.

The overriding digital control system may include speed control loops for a large number of motors. The speed control loop for each motor drive is closed via an external cable connection between the two PCBs containing the F/D convertor and the difference unit, and no software of the overriding controller is involved in the speed control loop. Hence the software, program cycle time and controls and displays used in the overriding control may be entirely devoted to process control.

Dynamic control performance of a typical DC motor drive with the digital speed control is shown on fig. 10. The steady state control performance is given in Table 2.

TABLE 2 Steady state control accuracy

Source of disturbance	Steady-state deviation
Supply voltage change ± 5 % Supply frequency change ± 3 % Temperature change ± 5°C (rate of change <1°C/min) Load variation 0-100 %	Better than 0.01 %

5. INDUSTRIAL APPLICATIONS

Paper machines

The first industrial application of digital speed control was for a combined sack paper and cardboard machine PM5 at Korsnäs Marma AB, near Gävle, Sweden, having a machine wire width of 7700 mm and a wide speed range, 100 - 750 m/min. The drives for wire part, press part, drier part, FLAKT drier, on-machine coater and reeler - in all 40 drives - are equipped with digital speed control. An overriding digital micro-processor based control system contains the complete multiplicative speed reference system for draw-setting and measurement and automatic web tension control between certain sections, a measuring system for absolute and difference speed measurement, measurement of produced reeled length, speed monitoring, a storage for draw settings and communications with a supervising computer.

The digital speed control enables the draw variations between two sections to be kept within less than ± 0.005 % throughout the entire speed range.

This paper machine has been producing paper since December, 1976 and the digital control and speed measuring system is appreciated by the customer because of the simple and reproducible settings.

Hot wire rod mills

Drive systems with digital speed control have also been used in a new hot wire rod mill at Boxholm AB in Boxholm, Sweden. The mill comprises two parts, a roughing section with three stands and a finishing section with 16 stands. The final speed of the wire rod is 15 m/s. 19 out of a total of 26 motors are provided with digital speed control. The overriding digital control system comprises a multiplicative cascade system for the speed set points for all stands and auxiliary drives (e.g. shears) and automatic loop control using an optimal control law on eight stands (see Fig. 12) as well as manual loop control on the remaining stands. The operation of the mill is to a considerable extent automatic. The loops between stands are controlled with a sampling time of 50 ms and the rest of the system with 250 ms. The entire overriding digital control system is built around one microprocessor (8080).

The rolling mill was commissioned during February, 1977 and since then rolling experience has been good.

6. ACKNOWLEDGEMENT

The method of F/D conversion described in this paper was invented by the authors and the industrial applications were developed together with numerous highly qualified collaborators while the authors were employed by ASEA AB in Västerås, Sweden.

The authors wish to express their gratitude to the ASEA AB for permission to publish the results.

7. REFERENCES

[1] V. Török, J. Valis, Samplat reglersystem, Swedish pat.appl.No 74 08069-8

[2] V. Török, J. Valis, Prediktiv varvtalsregulator, internal report, ASEA 72-12-18

[3] K. Jönsson, R. Lockley, J. Valis, Digital speed control, using ASEA's DS-8 system, ASEA Journal, 49, nr. 5, 123 (1976).

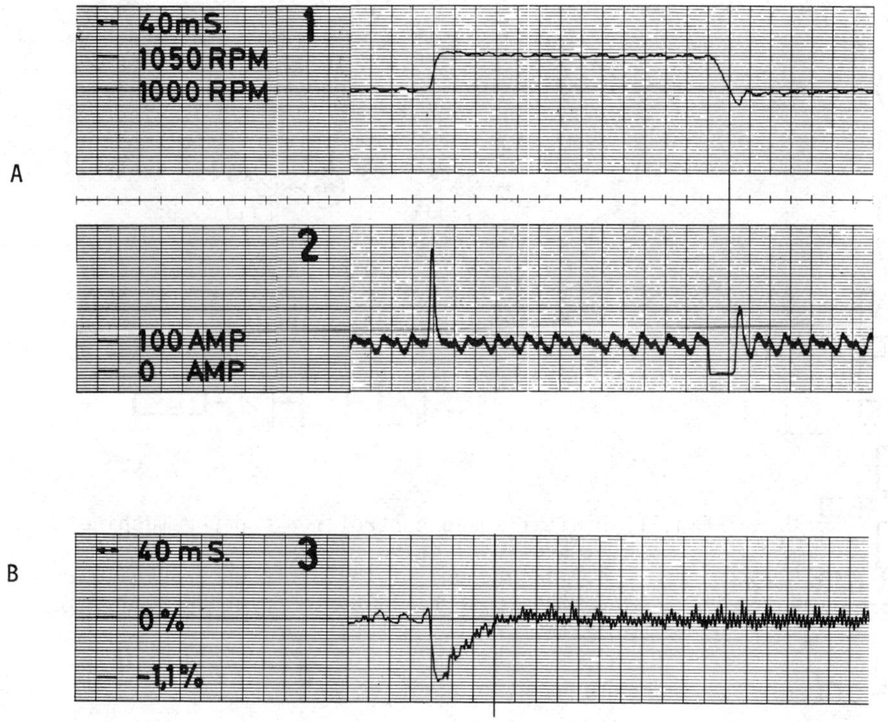

Fig. 10. Step response of a 100 kW DC motor drive with digital speed control for
A. set-point step change (motor speed - 1, rotor current - 2)
B. load-torque step change 0-100 % (motor speed - 3)

Table 1 Measured performance of high quality DC tachogenerators
All values are stated in $^{o}/oo$.

Type	Deviation of F.S. voltage from spec. value	Ripple at 1500 RPM Amplitude	Ripple at 1500 RPM Dominating frequency Hz	Linearity error One direction of rotation	Linearity error Both directions of rot.	Temperature error 1500 RPM per 10°C	Long-term stability 1500 RPM 3 months
Make 1	9.0	4	450	± 1.4	± 2.0	- 0.5	- 7.2
Make 2	-3.0	8	500	± 1.0	± 1.2	- 1.4	-
Make 3	1.0	5	600	± 0.5	± 1.0	- 1.1	- 2.2
Make 4	73.0	18	150	± 0.8	± 2.0	- 1.5	-

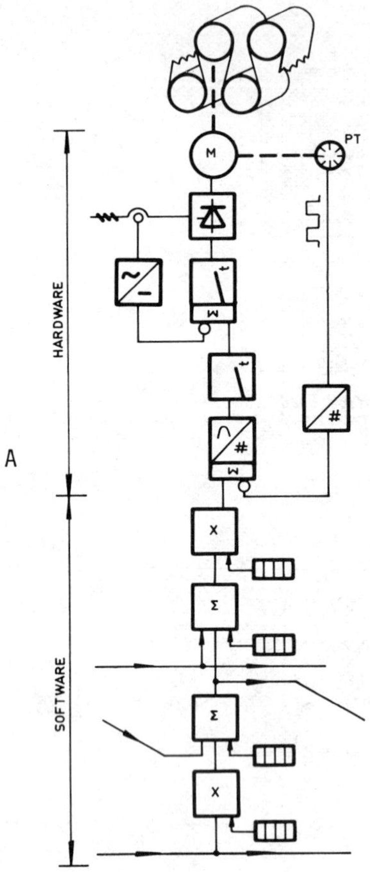

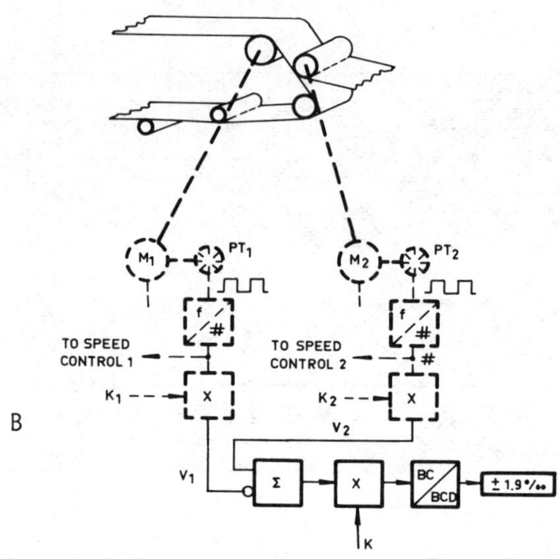

Fig.11. Digital speed control for a paper machine
A. sectional drive
B. measurement of the difference speed between the couch roll and the pick-up roll using two F/D convertors

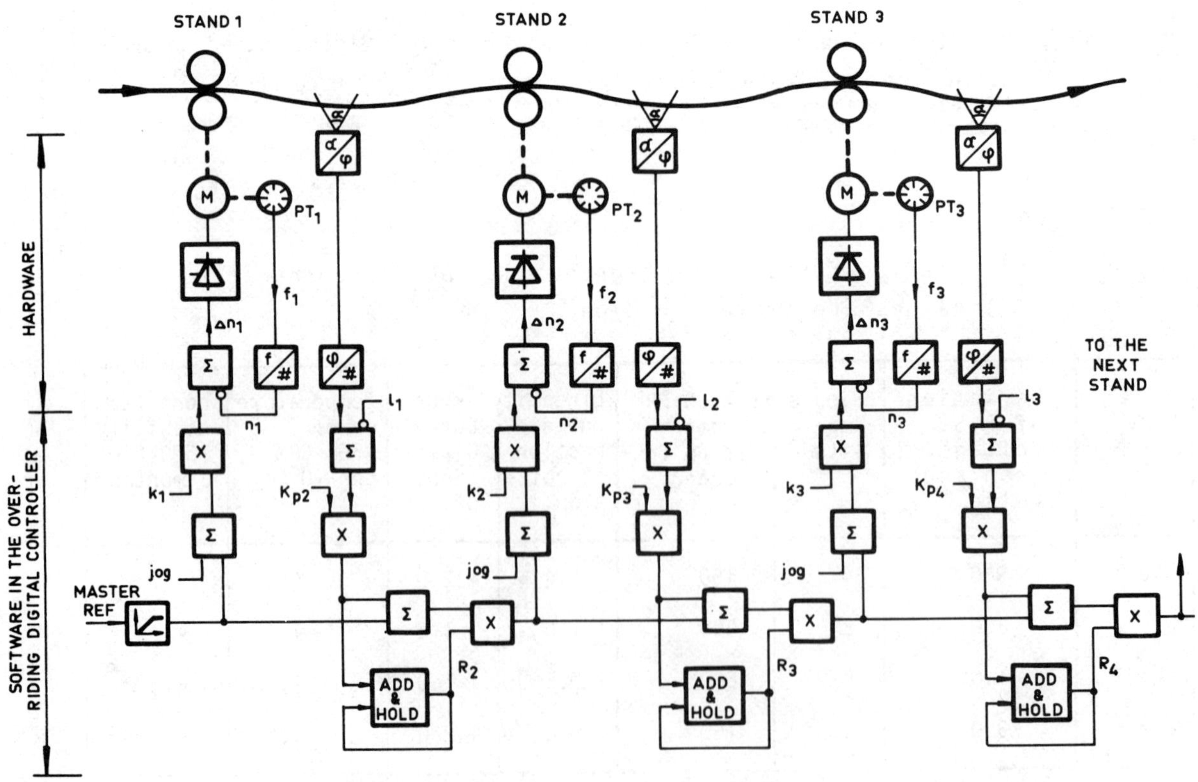

Fig. 12. Digital speed control of a hot wire rod mill, comprising overriding loop control and a multiplicative cascade system for set point calculations.

MULTI-STEP LINE- AND SELF-COMMUTATED SINGLE-PHASE BRIDGE CONVERTER

A. Müller-Hellmann

Institut für Stromrichtertechnik und Elektrische Antriebe, RWTH Aachen, FRG

ABSTRACT

By means of fully controllable multi-step two-pulse bridge convertors d.c. loads as inverters or d.c. machines can be fed with variable voltage and current at increased line side power factor. These connections can be used for both rectifier- and inverter operation. Resistance braking is possible.

INTRODUCTION

It is well known that the optimum energy conversion from a.c. to d.c. can be achieved by means of convertors, the transformation of which can be varied sinusoidally (1). They lead to a sinusoidal wave shape of the line side current. This conversion relationship can be approximated very well by using the PWM technique. A further improvement is a 4 step convertor with PWM control (2). The use of 2 PWM convertors parallel connected with displaced control is also well known (3, 4).

A different approach to this optimum conversion relationship is to have several conversion stages in conjunction with connections, which allow a single on and off switching action of each step per half cycle of the supply voltage (1, 5).

In this paper basic circuits are discussed, which permit up and down commutation of the supply current in several steps. An examination of turn-off methods, which are suitable for single-phase convertors follows. When applying these turn-off methods to the discussed basic circuits a multitude of circuit variations are obtained. A description follows of the operation of a single-phase bridge convertor, which can be fired and switched off in two steps. Some resistance-braking circuits are discussed.

BASIC CIRCUITS

In order to obtain a multi-step approach to a sinusoidal line side current a combination of two-pulse bridge convertors can be chosen. The convertors are parallel connected on the a.c. side, the d.c. terminals may be connected in series or in parallel. For parallel connection on the d.c. side another two variants can be distinguished.

The partial convertors can be operated on different transformer taps with alternating firing or they are both connected to the secondary winding of the transformer and carry current simultaneously with a different phase control factor. They are parallel connected by means of an interphase transformer (6). In Figure 1 the possibilities of combination using two two-pulse bridge convertors are represented.

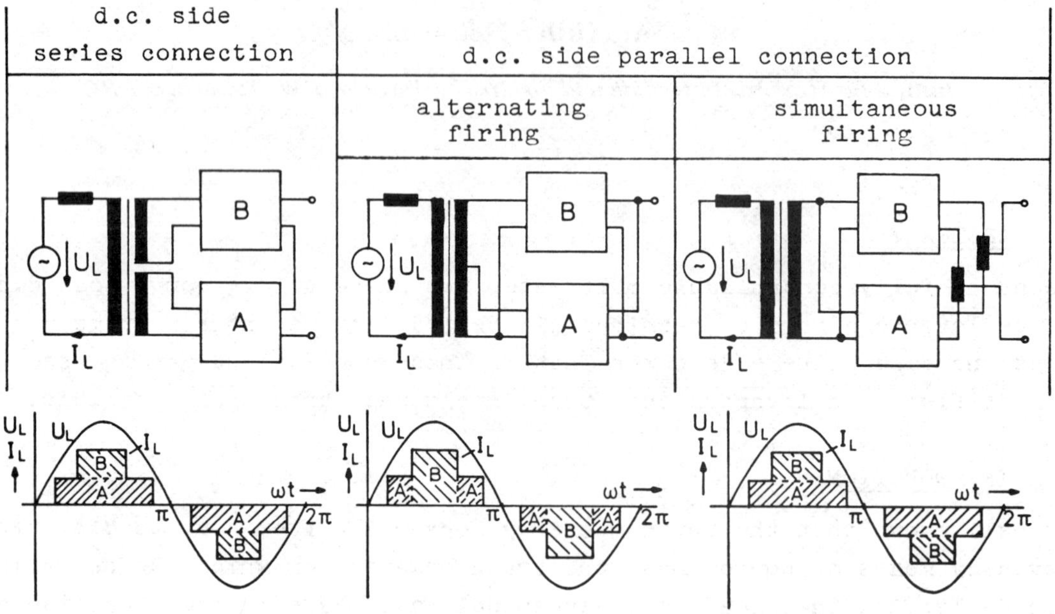

Fig. 1 Possibilities of combination of two two-pulse bridge convertors with a.c. side parallel connection

Figure 2 shows examples for the construction of half controllable convertors which permit up and down commutation of the supply current in one step as well as in two steps. The three circuits in the lower line of Figure 2 need more sophisticated control techniques than those in the upper line, as the thyristors which can be turned off have to fulfill both the free-wheeling operation and the phase control. Another disadvantage of these connections is that they are susceptible to conduction through.

If the convertors also have to regenerate energy to the supply system fully controllable connections become necessary. That means, the diodes have to be replaced by thyristors.

In the rectifier operation the multi-step line side current is switched on by means of line commutations and switched off by means of self commutations. In inverter operation the order of the commutations is reversed, the switch on commutations are effected by self commutation and the switch off commutations are effected by line commutation.

Figure 3 shows examples of the construction of connections able to generate a one- or two step line side current in the rectifier operation as well as in the inverter operation.

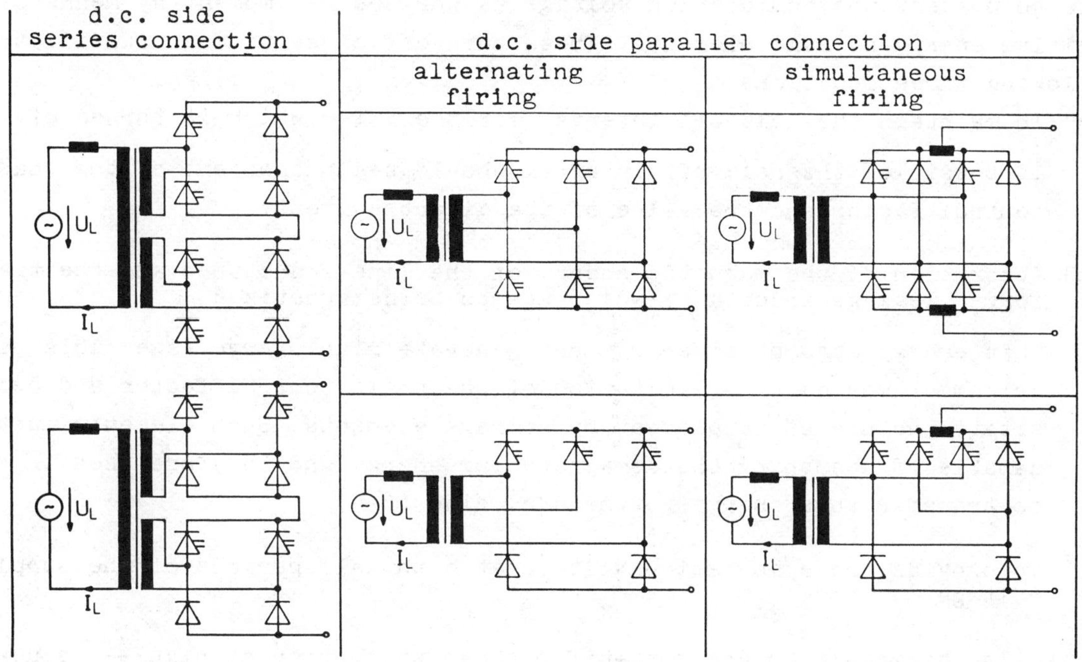

Fig. 2 Principle construction of half controllable connections for one- or two-step line side current

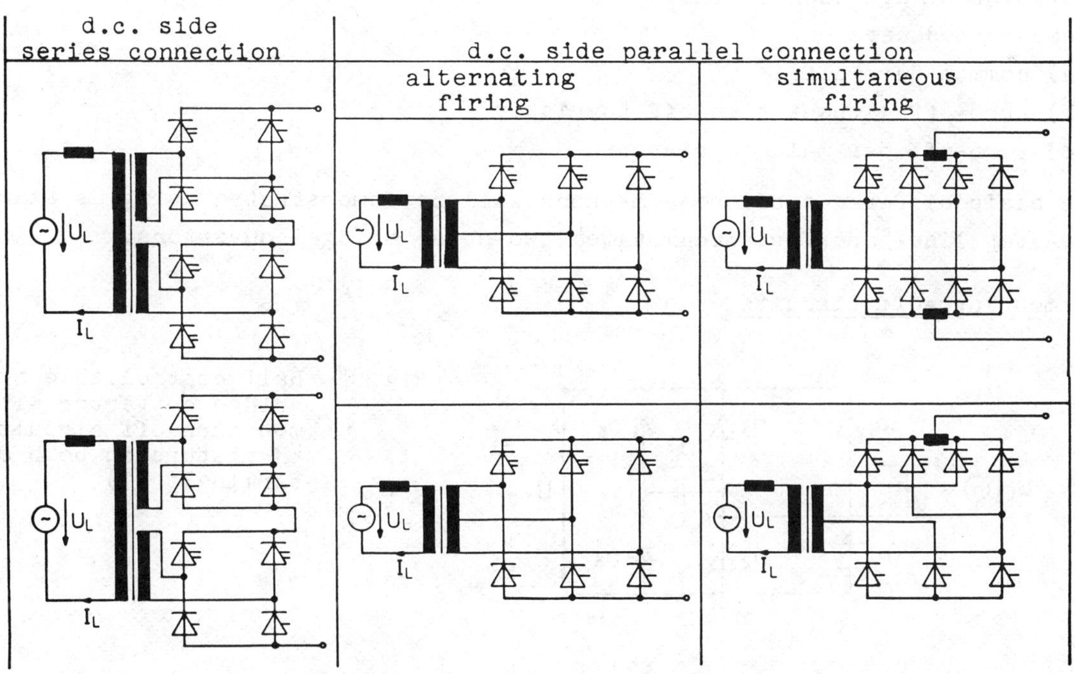

Fig. 3 Principle construction of fully controllable connections for one- or two-step line side current

FUNCTIONS OF THE TURN-OFF CIRCUITS

As self commutations are necessary, turn-off circuits are needed, which are able to present the commutating voltage at the desired moment by means of capacitive energy storage elements. These turn-off circuits must fulfill the following three functions:

a) To maintain the hold-off interval for the valve which is turned off.

   If possible, the hold-off interval should be independent of the phase control factor and the value of the direct current.

b) Absorption of the magnetic energy of the line-inductance and the transformer leakage inductance which are to be demagnetized.

   This energy absorption should not generate high overvoltage. This requirement may cause a limitation of the phase control factor and necessitate the use of supplementary storage elements. Such elements must be capable of absorbing the demagnetizing energy and they are used in order to transfer this energy to the d.c. side (6-9).

c) To provide the commutation voltage at both half periods of the supply voltage

It is also desirable to use turn-off methods which prevent high $\frac{di}{dt}$ values of the line current. Otherwise the optimisation of the power factor can only be achieved at the cost of an increase of the noise current (9).

TURN-OFF METHODS

The turn-off methods, well known from inverter technique, can also be used for two-pulse bridge connections.

These methods are

a) common turn-off
b) turn-off without turn-off thyristor
c) turn-off per main thyristor

The basic difference of these methods will be demonstrated using as examples one-step line- and force commutated two-pulse bridge convertors.

Common Turn-off Circuits

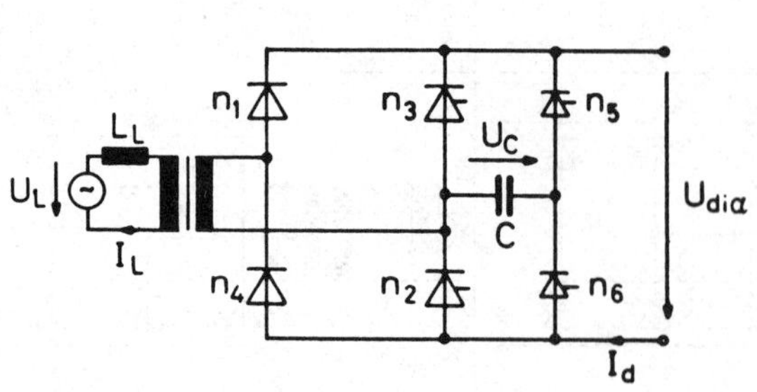

Fig. 4 Half controllable two-pulse bridge convertor with common turn-off circuit for thyristors on both d.c. terminals (7)

To follow one cycle of operation, assume the following initial conditions: a constant d.c. current I_d is flowing through the main thyristor n_1 and n_2. By firing thyristor n_6 the capacitor voltage is connected in reverse direction parallel to the value to be turned off. After recharging the capacitor to the instantaneous value of the supply voltage, the diode n_4 goes into conduction. This results in free wheeling operation. After the supply voltage has reversed, the conducting valve n_3 is turned off by firing the commutating thyristor n_5.

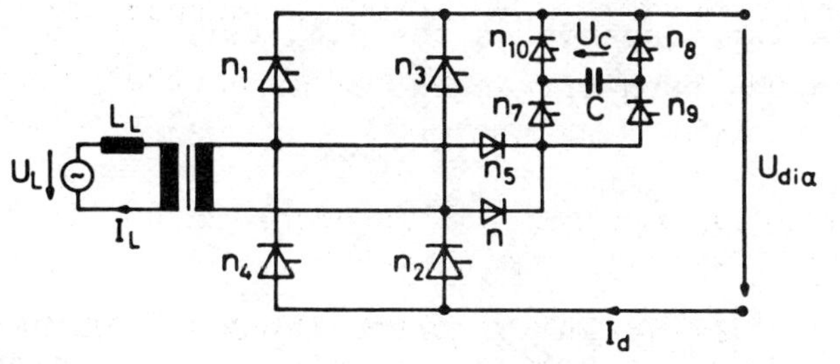

Fig. 5 Fully controllable two-pulse bridge convertor with common turn-off circuit for thyristors on one d.c. terminal

To illustrate the operation assume that a constant d.c. current I_d is flowing through the main thyristors n_1 and n_2 the line side voltage U_L being positive. By firing the turn-off thyristors n_7 and n_8 the capacitor voltage U_C is connected parallel to the thyristor n_1 which is turned off. After recharging the capacitor voltage to the instantaneous value of the line voltage the thyristor n_3 can be fired resulting in the free wheeling operation of the thyristors n_2 and n_3. The value of the capacitor voltage can be adjusted in dependence on the load by delayed triggering of thyristor n_3. As the polarity of the capacitor voltage has changed, the main thyristor n_3 can be turned off by triggering the commutating thyristors n_9 and n_{10}.

Characteristics. It is possible to reduce the expense of the turn-off operation by means of common turn-off circuits, especially if these circuits can be used for multi-step connections. In case of failure or small current the convertors can work without triggering the turn-off circuit. For the design it must be remembered that the recharging time of the turn-off capacitor depends on the d.c. current.

Other kinds of common turn-off circuits are described in (10, 11).

Turn-off Circuit without additional Turn-off Thyristors

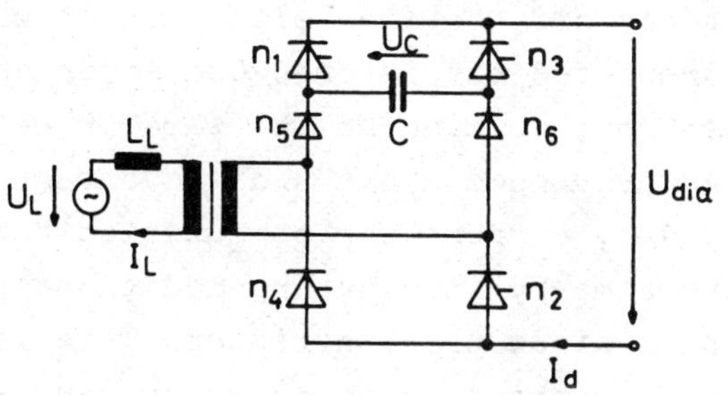

Fig. 6 Fully controllable two-pulse bridge convertor with turn-off circuit without turn-off thyristors

Assume that smooth d.c. current I_d is flowing through the main thyristors n_1 and n_2. The firing of the main thyristor n_3 initiates the turn-off operation. When the capacitor has been recharged the current I_d commutates from n_5 to n_6. In the other half period of the supply voltage the switch on commutation of the line current is started by firing thyristor n_4 and the turn-off operation by firing thyristor n_1.

Characteristics. As this circuit has no commutating thyristors it does not need additional firing equipment. Once a self commutation has been started it will be finished automatically. For the design it must be remembered that the recharging time of the turn-off capacitor depends on the d.c. current and the voltage of the turn-off capacitor on the turn-off angle.

Turn-off Circuit per Main Thyristor

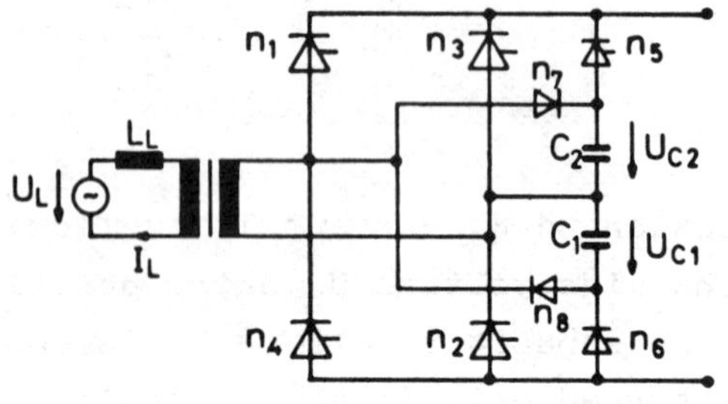

Fig. 7 Fully controllable two-pulse bridge convertor with turn-off circuit per main thyristor (12)

The d.c. current I_d, flowing through the main thyristors n_1 and n_2, is commutated by the turn-off operation from n_2 to n_6 and C_1. After the recharging of C_1 to the instantaneous value of the line voltage, n_4 can be fired in order to achieve the free wheeling operation. When thyristor n_5 is fired, the valve n_3 will be turned off. After the recharging of the capacitor, the direct current commutates into the free wheeling operation of the diodes n_4 and n_1.

Characteristics. The connection is able to operate without triggering the turn-off circuit. The turn-off voltage does not depend on the turn-off angle because of the load diodes n_7 and n_8. The recharging time of the turn-off capacitors depends on the d.c. current.

APPLICABILITY OF THE TURN-OFF METHODS

As shown in Figure 6, the turn-off circuit without additional turn-off thyristors is applicable only for those connections in which the force commutated thyristors are connected at one d.c. terminal. In rectifier operation every commutation of these thyristors switches off the line side current which had been switched on by one of the line commutated thyristors. Convertor configurations, for both one- and two-step control, working on the same principle are shown in the lower line of Figure 3. As mentioned above a disadvantage of this commutation principle is that for low d.c. current the commutation takes too much time. This disadvantage can be overcome by introducing additional recharging arms which contain a reactance and a diode or a thyristor. This additional expense can be saved by using turn-off methods a) or c) which can be operated without triggering the turn-off circuit at low d.c. current.

For connections with force commutated thyristors at both d.c. terminals these thyristors have to fulfill both switching on and switching off the line side current. Three examples of this type of convertor are given in the upper line of Figure 3. For these configurations the turn-off methods a) or c) are applicable.

ARRANGEMENTS OF SELF-COMMUTATED SINGLE-PHASE CONVERTORS

When using these turn-off methods according to their applicability to the discussed basic circuits a variety of convertor configurations is obtained. A few of them, which are suitable for transfering energy in either direction are presented in Figure 8.

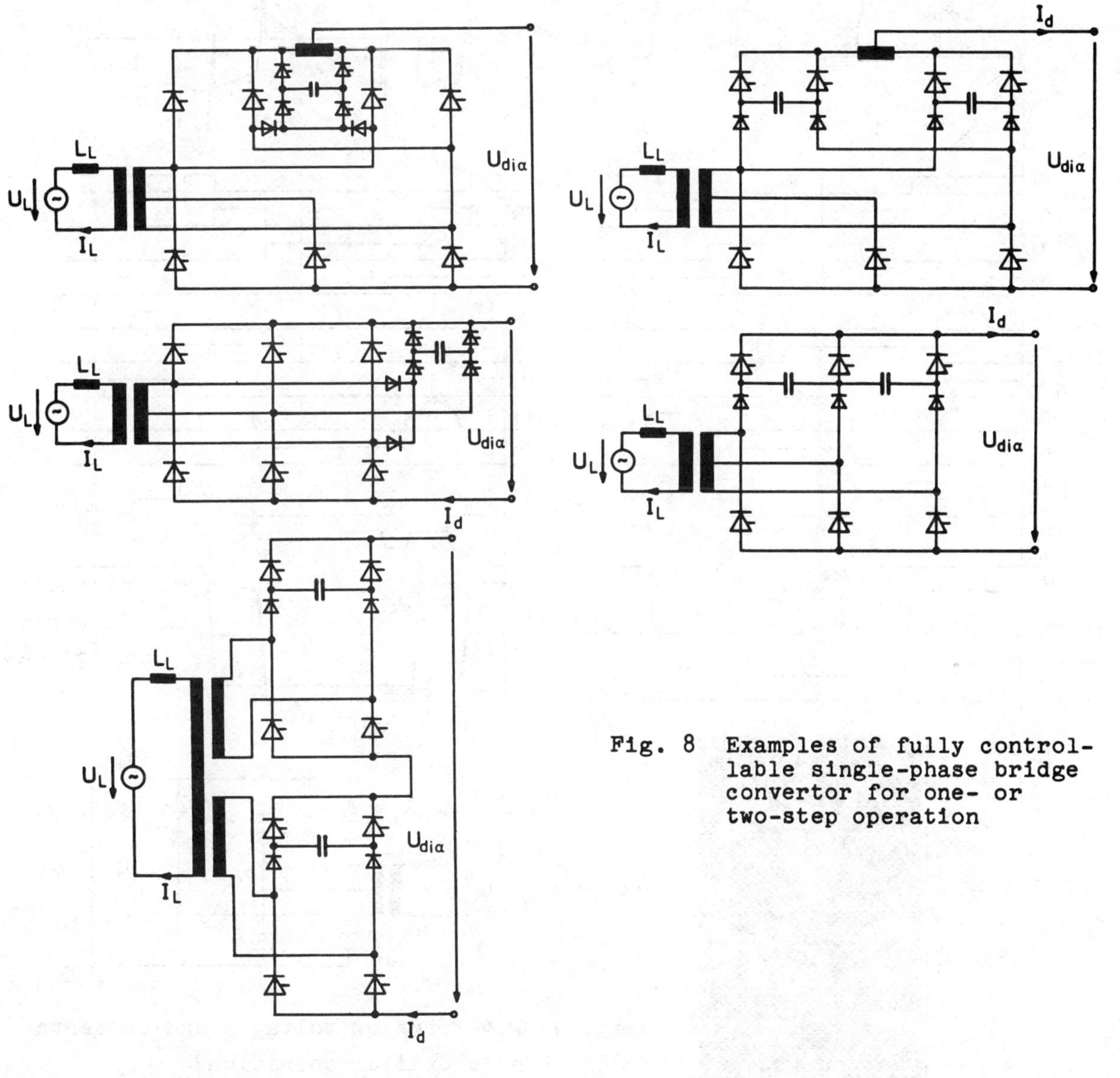

Fig. 8 Examples of fully controllable single-phase bridge convertor for one- or two-step operation

The configuration as given in Figure 8 can operate in a one-step or a two-step mode. The number of steps can be increased by either adding additional parallel branches which are connected to further transformer taps (13) or by series connecting of several configurations (1).

WAVEFORMS OF VOLTAGES AND CURRENTS

Figure 9a), b) gives the calculated waveforms of the voltages and currents in rectifier- and inverter operation for the convertor given in Figure 9d). Oscilloscope records of the input and output quantities for three different phase control factors are shown in Figure 9c).

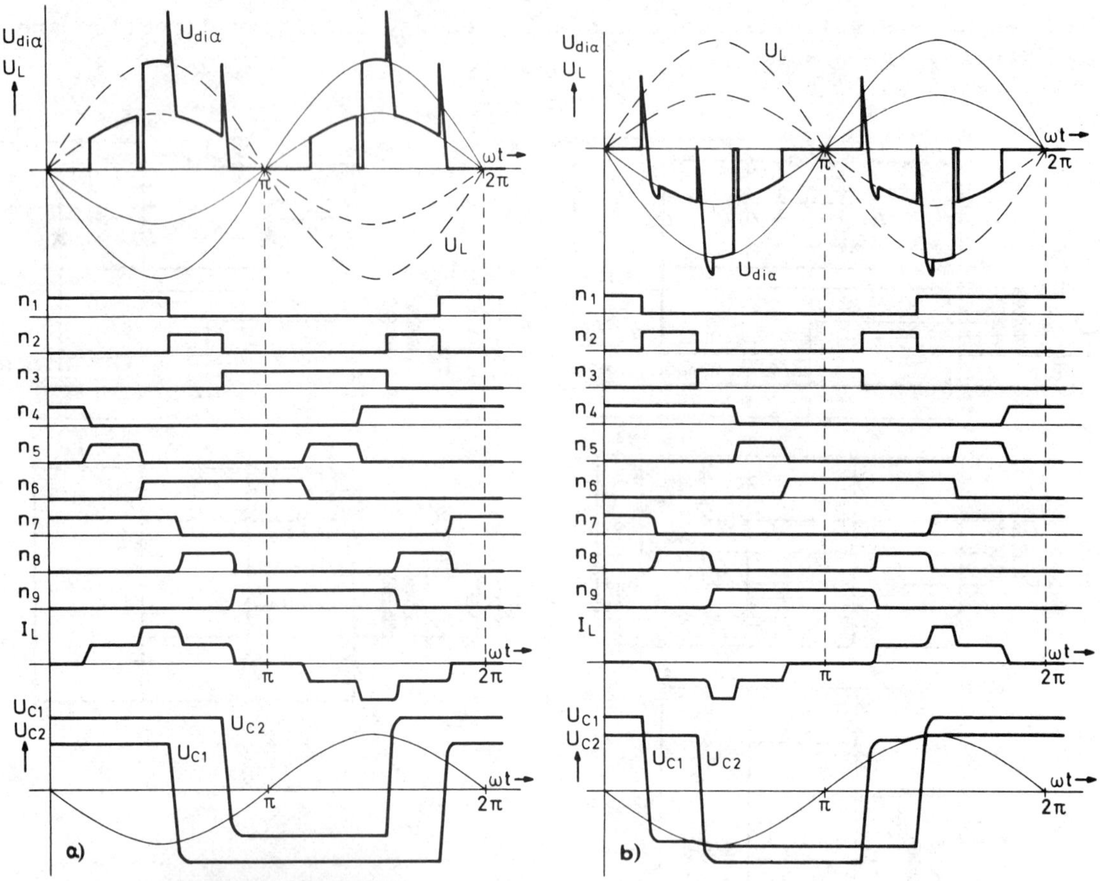

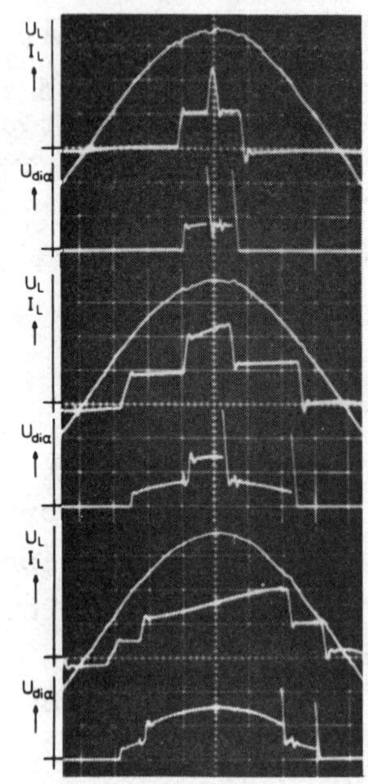

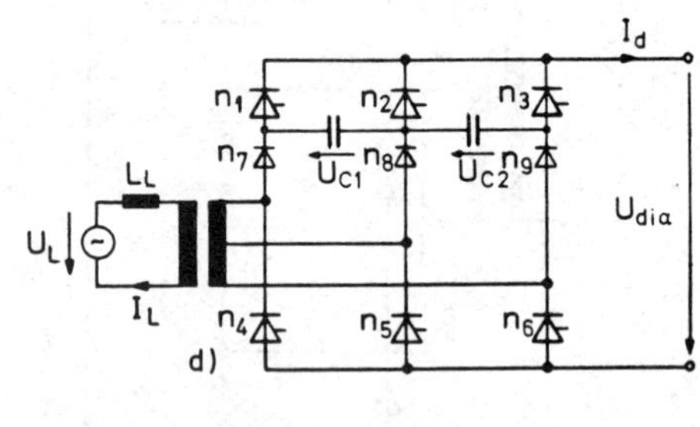

Fig. 9 Waveforms of voltages and currents

a) rectifier operation
b) inverter operation
c) Oscilloscope records on an experimental set up

$$\frac{U_{di\alpha}}{U_{dio}} = 0{,}15;\ 0{,}45;\ 0{,}75;$$

U_L = 160 V; I_d = 30 A; f = 50 Hz

d) Convertor connection

Dependence of the commutation voltage on the turn-off angle can be avoided by using a supporting capacitor (9). Such capacitors can be inserted without additional semiconductors into connections which have common turn-off circuits as shown in Figure 5 (6).

Figure 10c) shows how a supporting capacitor C_s is inserted in a configuration consisting of two parallel connected simultaneously operating two-pulse bridge convertors. The oscillograms shown in Figure 10a), b) demonstrate the influence of the supporting capacitor on the voltage of the turn-off capacitor for different phase control factors. An additional advantage of the supporting capacitor is its absorption of the demagnetizing energy of the line- and transformer leakage inductances.

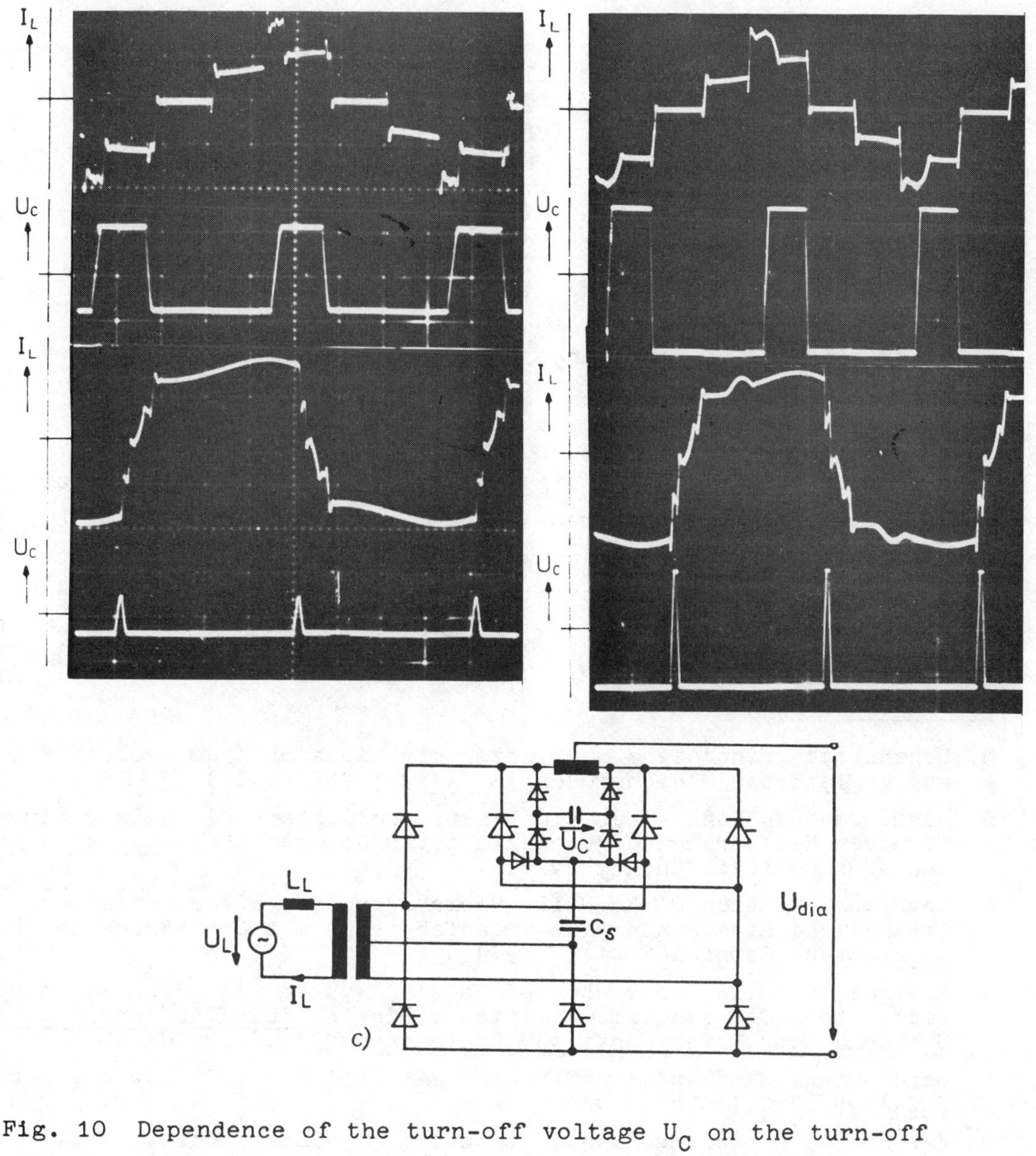

Fig. 10 Dependence of the turn-off voltage U_C on the turn-off angle for two different phase control factors

a) without supporting capacitor
b) with supporting capacitor
c) circuit configuration

EXAMPLES OF RESISTANCE BRAKING EQUIPMENTS

In normal operation the fully controllable convertors are able to regenerate energy to the supply by dynamic braking operation. If the line cannot accept the fed back energy the convertor may be switched over to resistance braking. As shown in Figure 11 the braking resistance can be connected at one of the following points:

a) the resistance is connected parallel to the a.c. terminals of the convertor

b) the resistance is series connected in one of the convertor branches, the convertor operating as a d.c. chopper

c) the resistance is connected parallel to the d.c. terminals of the convertor, the convertor acting as a d.c. chopper.

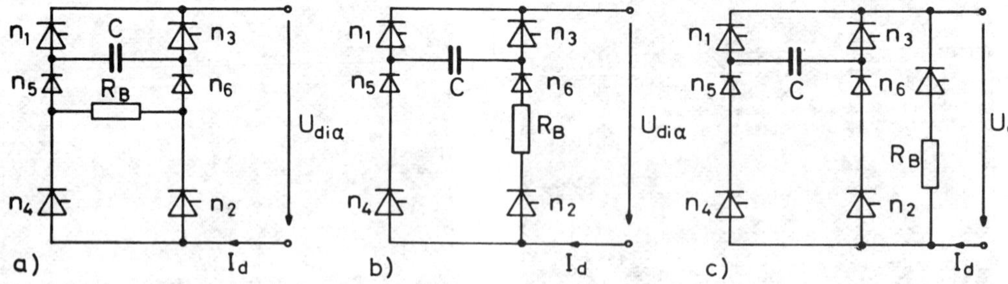

Fig. 11 Possibilities of resistance braking connections

CONCLUSION

From a detailed analysis of the configurations as presented in this paper it follows that a significant improvement of the line side power factor can be achieved by adding turn-off equipment with reasonable expense. Therefore additional tests and further investigations will be carried out in the future.

REFERENCES

(1) M. Depenbrock, Einphasenstromrichter mit sinusförmigem Netzstrom und gut geglätteten Gleichgrößen, ETZ-A 94, 466 (1973).

(2) D. Daum, Untersuchung eines Einphasenstromrichters mit nahezu sinusförmigem Netzstrom und gut geglätteten Gleichgrößen, Dissertation Ruhr-Universität, Bochum 1974.

(3) H. Kehrmann, W. Lienau, R. Nill, Vierquadrantensteller -eine netzfreundliche Einspeisung für Triebfahrzeuge mit Drehstromantrieb, Elektrische Bahnen 45, 135 (1974).

(4) H. Kielgas, R. Nill, Convertor propulsion system with 3-phase induction motors for electric traction vehicles, IEEE/IAS International Semiconductor Power Convertor Conference 1977, Florida.

(5) M. Depenbrock, Einphasenstromrichter mit optimiertem Leistungsfaktor, ETZ-A 95, 360 (1974).

(6) H. Kahlen, A. Müller-Hellmann, W. Woelker, Halbgesteuerte Brückenschaltung mit unterteilter Löschung, Elektrische Bahnen 46, 279 (1975).

(7) J. Förster, Netzrückwirkungen von Stromrichtern, Techn. Rundschau 66, 45 (1974).

(8) H. Fick, V. Fleckenstein, P. Loderer, Bahn-Stromrichter mit gutem Leistungsfaktor, ETZ-A 96, 239 (1975).

(9) W. Runge, Single-phase convertor with forced commutations based on the supporting capacitor principle, 2. IFAC Symposium Düsseldorf 1977.

(10) M. Schulze, Der Pulsgleichrichter, ein einfaches Stellglied für Gleichstromantriebe im Einrichtungsbetrieb, Elektrie 22, 432 (1968).

(11) H. Kahlen, Rectifying single phase and threephase a.c. with forced commutated convertors, 2. IFAC Symposium Düsseldorf 1977.

(12) J. Förster, Löschbare Fahrzeugstromrichter zur Netzentlastung und Netzstützung, Elektrische Bahnen 43, 13 (1972).

(13) W. Lienau, A. Müller-Hellmann, Folgegesteuerte löschbare Brückenschaltung, Offenlegungsschrift 2 530 961 vom 29. 1. 1977.

(14) W. Lienau, A. Müller-Hellmann, H.-Ch. Skudelny, Power convertors for feeding asynchronous traction motors of single phase a.c. vehicles, IEEE/IAS International Semiconductor Power Convertor Conference 1977, Florida.

SINGLE-PHASE CONVERTERS WITH FORCED COMMUTATION BASED ON THE SUPPORTING CAPACITOR PRINCIPLE

W. Runge

Brown, Boveri and Cie AG, Mannheim

SUMMARY

The reactive power due to phase displacement caused by controlled line commutated convertors can be avoided by the use of forced commutation whereby the line current is cut off before the voltage reverses. The basic problems of forced-commutation devices are presented by examples of well-known circuits. Using these circuits as a basis convertors are described and analysed, whose forced commutation is assisted by a supporting capacitor. It is paralleled to the AC side during the whole commutation period. This effects a soft "cut-off" of the line current with comparatively low over-voltage or markedly lower capacitor expenditure. The reactive power loading of the supporting capacitor permits the use of smoothing capacitors. The principle is equally suitable for application in convertors during inverter operation.

0. INTRODUCTION

With the development of increasingly efficient semiconductor valves, single-phase power convertors are becoming more and more important in the field of AC railway traction. By the transition from simple diode rectifiers to controlled rectifiers, the way was opened for the rapid and stepless control of the DC voltage. This advantage was achieved at the expense of unfavorable effects of the convertor on the mains:

a) The control causes the fundamental of the current to lag behind the mains voltage, i.e. the displacement factor cos φ_1 deteriorates; the mains is loaded with a larger amount of reactive power.

b) Whereas the commutation of the diode rectifier proceeds with relatively slow change of line current this can be significantly accelerated in controlled convertors due to the larger instanteneous value of the mains voltage during the commutation. The resulting steep slopes of the line-current curve signify an increased harmonic content and worsen the problems of electromagnetic compatibility (EMC) with communications and signalling installations.

The disadvantage of a) can be reduced by using free-wheeling diodes or half-control. In this case only the initial angle of the DC voltage pulses and the AC current blocks varies with the delay angle, whereas the end angle remains constant at voltage zero (phase control).

The disadvantage of a) can be completely eliminated by equipping the convertors with a device for forced commutation (Ref. 1). This permits the arbitrary cut-off of the current blocks before the line voltage reverses. A displacement factor of 1 can be achieved if phase and turn-off control are combined (sector control).

Partial these measures have an additional unfavorable effect on the harmonic content. Certain improvements can be achieved with the use of sequential control of a number of convertors and by employing multi-stage arrangements.

Another method is the use of pulsed convertors. By means of rapid pulsing they cause the mains current fo follow a sinusoidal reference value. Using this method, a displacement factor of 1 can be achieved with only a very small deviation from the sinusoidal shape, the deviation being due to the finite elementary frequency. Especially favorable is the design of a pulse-controlled convertor, when it operates at a DC link with fixed voltage, which fixes also the voltages of the valves and the commutating capacitors (Ref. 2). The traction motors are supplied from the DC link via pulse inverters.

The circuits which are the subject of this article belong to the group of convertor with turn-off control or sector control. They can supply variable voltage to commutator motors through a smoothing choke, or to current type inverters which are used to feed 3-phase traction motors. A large number of circuits are known, these differ mainly in their forced commutation devices. These devices have two main tasks to fulfil :

1. They must provide the hold-off interval for those thyristors which have to be turned off;
2. After a valve has been turned off they must take over the line current which continues to be driven by the inductances of the line and the transformer, and must bring it to zero with controlled over-voltage.

The supporting-capacitor principle dealt with in this article is especially suited to fulfilling both tasks and has, up to now, been little heard of.

1. THE COMMUTATION PRINCIPLES OF SOME WELL-KNOWN CIRCUITS

In order to explain the particularly interesting points about the supporting-capacitor principle, the most important details of some well-known circuits must be presented first. The coded abbreviations for the circuits give information on the number of capacitors, and the letters LVB or LUB stand for turn-off type fully-controllable (symmetrical) bridge, and for turn-off type half-controllable (asymmetrical) bridge, respectively.

1.1 Circuits with one capacitor

Figure 1 shows a symmetrical, turn-off type, fully-controllable bridge (1C-LVB). Due to the facts that the circuits using supporting capacitors are developed from the

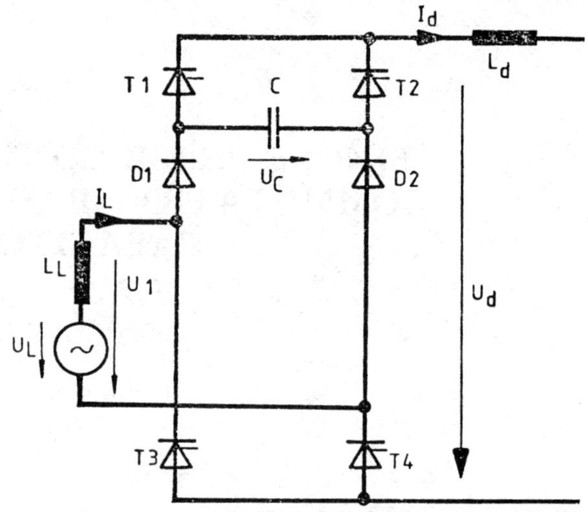

Fig. 1. Fully controllable turn-off type bridge, fundamental circuit (1C-LVB). L_L=inductance of line and transformer.

1C-LVB circuit, and that the 1C-LVB demonstrates a number of fundamental aspects concerning forced-commutated single-phase convertors, we will examine this circuit in more detail.

The upper commutating group of the 1C-LVB, with the thyristors T1 and T2, can be forced-commutated with the turn-off device composed of capacitor C and the blocking diodes D1 and D2. Its delay angle is β. The lower group with T3 and T4 commutates naturally by the line voltage U_L at delay angle α. (U_L is already referred to the secondary side of the vehicle transformer). The curves of the most important quantities during one period in steady-state operation are shown in Fig. 2. The direct current I_d which, due to a very large smoothing choke, can simplifying be regarded as being constant, flows up till $\omega t = \alpha$ via T3 - D1 - T1. By triggering of T4 it commutates between α and α_1 from the free-wheeling state onto the mains. At $\omega t = \beta$ T2 is triggered. Because $U_c = -\hat{U}_c$, T1 turns off immediately and I_d linearly reverses the charge on C. During the hold-off interval

$$t_s = \hat{U}_c C / I_d \qquad (1)$$

reverse blocking voltage is applied to the thyristor to be turned off, thus the turn-off device fulfilling its first task.

---

^: peak, ~: RMS value

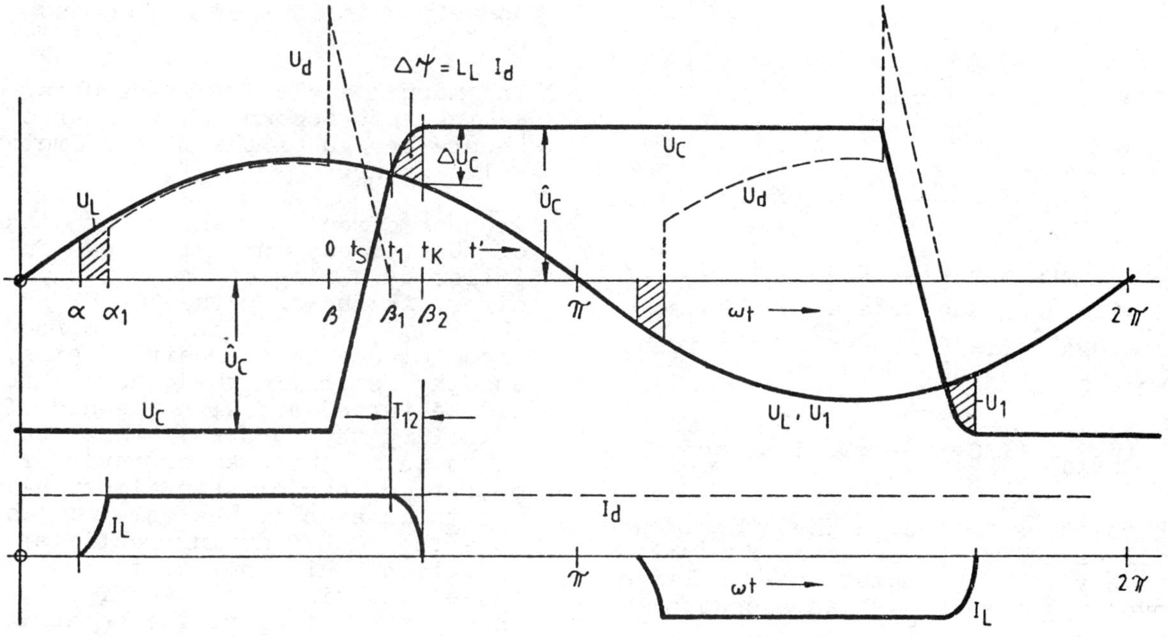

Fig. 2. Voltage and current curves of the 1C-LVB shown in Fig. 1 during one cycle.

If, at the instant

$$t_1=(\hat{U}_c+\hat{U}_L \sin \beta_1)\, C/I_d \quad (2)$$

U_c has become equal to U_L, diode D2 starts to conduct. The line and transformer inductance L_L together with C form a resonant circuit having the image impedance $Z_K=\sqrt{L_L/C}$. The initial current I_d in this resonant circuit decays with a 1/4 cosine oscillation having a period of

$$T_{12} = \frac{\pi}{2} \sqrt{L_L\, C} \quad (3)$$

This means that the turn-off device has fulfilled its second task. The capacitor voltage U_c, which from β_1 to β_2 is also the input voltage U_1, rises with a 1/4 sine oscillation by the amount:

$$\Delta U_c=I_d\, Z_K=I_d \sqrt{L_L/C} \quad (4)$$

The change in the mains voltage U_L is ignored during this period. The capacitor voltage $+\hat{U}_c$ at the end of the commutation is the initial state for the next forced commutation in the negative half-cycle. The value $\Delta\Psi$ of the hatched voltage-time integral in the interval from β_1 to β_2 between U_1 ($=U_c$) and U_L is determined by the mains-current change and the mains inductance:

$$\Delta\Psi = \Delta I_L\, L_L=I_d L_L \quad (5)$$

This applies to all circuits.

In periodic operation the following result from equations (1,2,4) and with

$$\hat{U}_c=\hat{U}_L \sin \beta_2+ \Delta U_c \quad : \quad (6)$$

$$t_s=\frac{C\hat{U}_L}{I_d} \sin \beta_2+ \sqrt{L_L C} \quad (7)$$

$$t_1=\frac{2\, C\hat{U}_L}{I_d} \sin \beta_2 + \sqrt{L_L C} \quad (8)$$

In order to evaluate the conditions we will substitute a relative short-circuit voltage for the line and transformer inductance L_L.
Since the turn-off circuit must be rated for the maximum current (including ripple, if I_d is not totally smoothed), this current I_{dmax} is used as reference:

$$u_{xm} = \frac{L_L\, I_{dmax}}{\tilde{U}_L} \quad (9)$$

If, furthermore, one specifies the relationship

$$\Delta U_{cmax} / \tilde{U}_L = k_u \quad (10)$$

which may occur at I_{dmax}, then the equations (7,8) can be converted with equations (4,9) to give:

$$t_s = \frac{u_{xm}}{\omega k_u}\left(\frac{\sqrt{2}\, I_{dmax}}{k_u I_d} \sin \beta_2 + 1\right) \quad (11)$$

$$t_1 = \frac{u_{xm}}{\omega k_u}\left(\frac{2\sqrt{2}\, I_{dmax}}{k_u I_d} \sin \beta_2 + 1\right) \quad (12)$$

If we presume e.g. that $k_u = \Delta U_{cmax}/U_L = 1$ or 0.5 and take $u_{xm} = 10$ % and $\omega = 105 s^{-1}$ (f=16 2/3 Hz), this results (at $\beta_2 = \pi$) in:

$$t_{smin} = \sqrt{L_L C} = \frac{u_{xm}}{k_u} = 0.95 \text{ ms or } 1.9\text{ms}$$

This means that even when the turn-off capacitor is selected for relatively rapid commutation with large over-voltage ΔU_{cmax}, an adequate minimum hold-off interval even for slow thyristors still results. At $\beta_2 = \pi/2$ and $I_d = I_{dmax}$, we obtain the maximum hold-off interval t_{smax} =2.29ms or 7.27ms, the period of the linear charge reversal t_1=3.64 or 12.65ms, and the total commutation period $t_K = t_1 + T_{12}$=5.12 or 15.63ms. If $I_d < I_{dmax}$, then these values increase accordingly.

A disadvantage of the 1C-LVB circuit is the long forced-commutation period and its dependence on the turn-off angle . It is only worthwhile using this circuit with a high over-voltage ΔU_{cmax}. This though is included in the maximum voltage $\hat{U}_{cmax} = \hat{U}_L + \Delta U_{cmax}$ that occurs at the capacitor and at the thyristors. The blocking diodes are loaded with max. $U_{cmax} + \hat{U}_L = \Delta U_{cmax} + 2\hat{U}_L$. The product $C\hat{U}_{cmax}^2$ is a measure for the dimensioning of the commutating capacitor. It is:

$$C\hat{U}_{c\,max}^2 = \frac{u_{xm}\, \tilde{U}_L\, I_{dmax}}{\omega}\left(1 + \sqrt{2}/k_u\right)^2 \quad (13)$$

Of considerable interest is the fact that the 1C-LVB can also be sector-controlled during inverter operation. In this case, in every half-cycle, at turn-off angle β (with $\beta<\alpha$) the current is forced from the free-wheeling state to the line by means of forced commutation. At delay angle α the current goes back again into the free-wheeling state due to natural commutation. The maximum control factor during inverter operation is limited by β=0 and $\alpha = \pi - \gamma$.

The margin angle γ must be observed, taking into account the danger of shoot-through of the line-commutated valve group.

Another circuit in which both tasks of the forced commutation are fulfilled by a single turn-off capacitor, is shown by the 1C-LUB (Ref. 1,3) in Fig. 3. It incorporates diodes in one pair of arms, and thyristors in the other, which can be turned-off with the aid of auxiliary thyristors. This circuit, which is not usable for inverter operation, performs in almost all points in exactly the same way as does the 1C-LVB during rectifier operation. This applies to the curves vs. time shown in Fig. 2, and to the rating of the capacitor as well as to the maximum voltages across the thyristors and diodes. Differences only appear in the conducting period of the main valves. Neglecting the commutations, all the valves of the 1C-LVB have an on-time of 50 %. On the other hand, in the 1C-LUB the on-time of the diodes is 100 ... 50 % and the on-time of the thyristors is 0 ... 50 %, depending on the control factor.

In the 1C-LUB, the long duration t_1 of linear recharge particularly at low currents has a disturbing effect as the flow interval of the line current is extended beyond the turn-off angle β by the time t_1. This makes it impossible to apply turn-off and phase-control angles approaching $\pi/2$ to control the output voltage and/or power below a certain value. This limitation is not present with the 1C-LVB because it can reverse its voltage.

1.2 Circuits with two and three capacitors

The disadvantages of those circuits using only one capacitor (designed for high over-voltage, very long commutation periods with forced commutation at the peak of the mains voltage) are avoided in the following circuits.

The 2C-LUB circuit (Ref. 1,3) shown in Fig. 3b contains two identical capacitors which are charged up to a value of at least $\hat{U}_L$ via a diode and a damping resistor in the half-cycle before the turn-off of the respective valve. In the first phase

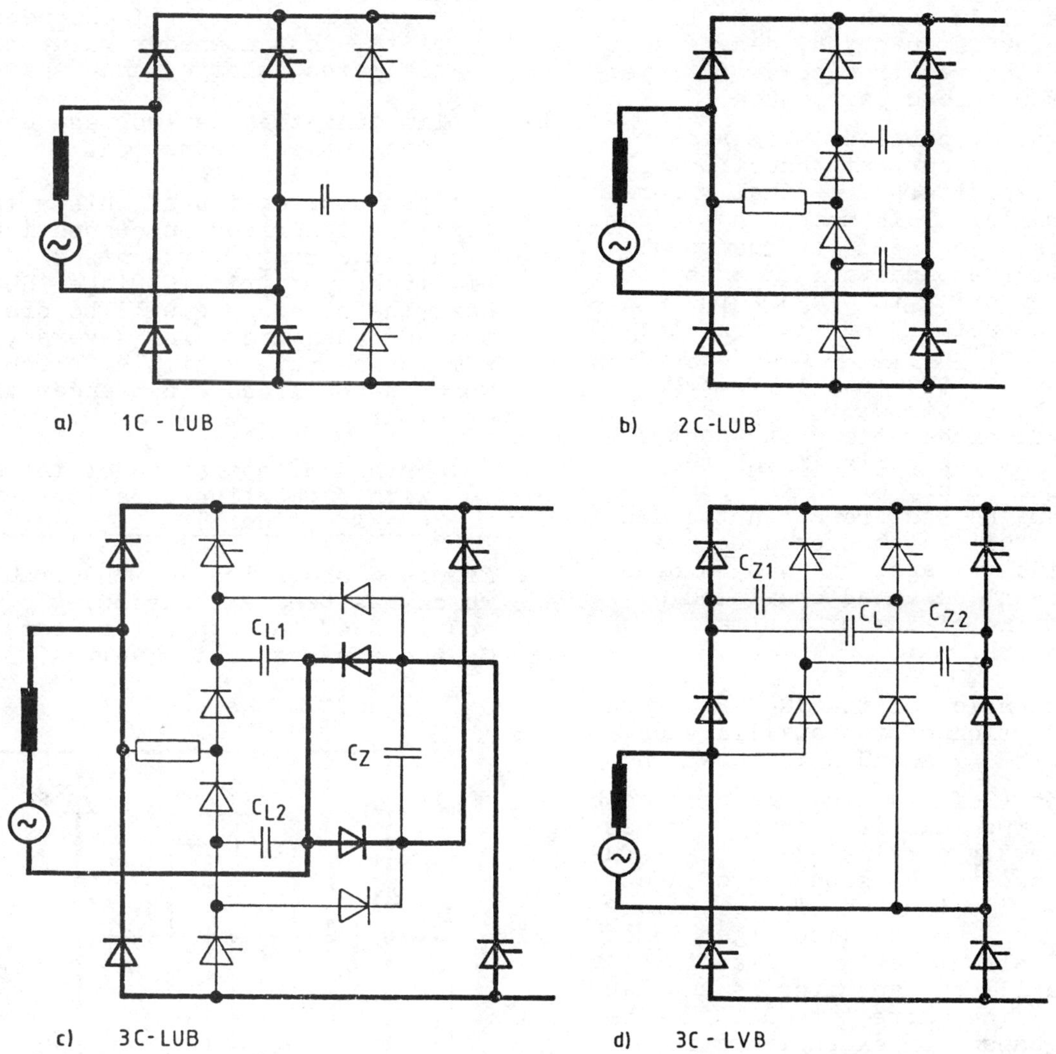

Fig. 3. Several well-known single phase converters with phase- and cut-off - control

of the forced commutation the voltage on one of the capacitors linearly reverses to the line voltage. In the second phase the mains current begins to fall. As soon as the voltage of the capacitor concerned up to now reaches that of the other capacitor, which is already charged to approx. $\hat{U}_L$, it also takes part in current reduction via the two charging diodes. Both capacitors enter into the rating formula using the over-voltage criterium. The measure for the total capacitor expenditure ΣCU^2_{cmax} is just as large as in the 1C-circuits as per equation (13). The charge-reversal period t_1 is, under otherwise identical conditions, only half that of the 1C-circuits at $\beta = \pi/2$, and due to the pre-charging of the capacitors only slightly dependent on turn-off angle β.

In order to reduce the recharging period still further while the overvoltage is kept small, both forced-commutation tasks have to be taken over by capacitors having rather different values. Fig. 3c shows the 2C-LUB extended to the 3C-LUB by the addition of an auxiliary capacitor C_Z (Ref. 1,3). Here applies $C_{L1} = C_{L2} < C_Z$. The auxiliary capacitor is arranged, with the aid of four further diodes, so that it is connected in parallel with the particular turn-off capacitor which is in operation as soon as both voltages have become the same. Due to the fact that the auxiliary capacitor is discharged in its initial state, this parallel operation starts when the voltage of the turn-off capacitor passes through zero.

The first part of the linear charge-reversal, in which only one turn-off capacitor is involved, can be adapted to the required hold-off interval. The second part, when C_Z is paralleled, proceeds much slower. During line current reduction all three capacitors are finally working in parallel. This means that equation (13) is also valid for the total capacitor expenditure of this circuit. In order to prepare the auxiliary capacitor for the next turn-off operation, it is discharged through the two main thyristors into the DC circuit. This can take place at either the phase-control angle at the beginning of the voltage sector, or in the pause between the sectors. Then the discharge must be completed before the line voltage reverses in order to prevent unwanted continued conducting of one of the main thyristors.

The extension of the 1C-LVB through the addition of two auxiliary capacitors C_{Z1} and C_{Z2} is shown in Fig. 3d (Ref. 4). In this case each of the two auxiliary capacitors is alternately paralleled to the turn-off capacitor as soon as voltage equality is reached. The auxiliary capacitors can be discharged into the DC circuit at any time. As only one auxiliary capacitor is involved in the line current reduction, the total capacitor expenditure is larger, by the second auxiliary capacitor than in the previous circuits.

The complete charge and discharge of the auxiliary capacitors in the 3C-LUB and the 3C-LVB necessitate a certain power throughput which effects the capacitor loading and the minimum control factor. The discharge into the DC circuit during the pause between the sectors reduces the voltage ripple during rectification; in the inverter operation which is possible with the 3C-LVB it increases the ripple.

2. COMMUTATION WITH SUPPORTING CAPACITOR

The idea behind the supporting-capacitor principle was that of supporting the input voltage of the circuit during commutation by the use of a large capacitor charged up to $\hat{U}_L$ or higher, for the purpose of:

1) making the voltage shift of the turn-off capacitor independent of the instantaneous value of the mains voltage, and

2) limiting the over-voltage during line-current reduction.

A first application of this principle is described in connection with a two-stage turn-on and turn-off type symmetrical bridge (Ref.5). Here the priniciple will be presented and analyzed using several, in some cases rather simple, circuits which have already been under discussion.

2.1 Method of operation of the LVB with supporting capacitor (Basic circuit)

Figure 4 shows the 1C-LVB circuit which has been extended by the addition of the supporting capacitor C_s and the respective valves D3, D4, T3 and T4 (Ref. 6).

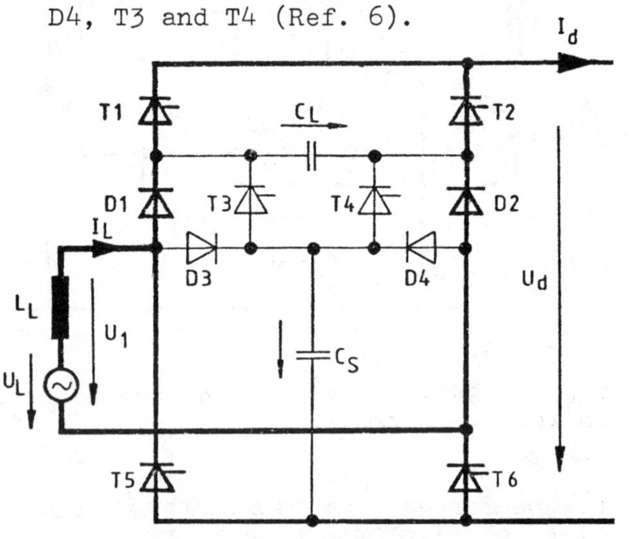

Fig. 4. Circuit from Fig.1 fitted with a supporting capacitor C_s. Operation of C_s only in the rectifier-mode

The delay angle of the line-commutated group T5, T6 is selected smaller than $\pi/2$ in order that C_s can charge up to at least $\hat{U}_L$. Before the forced commutation of the upper group T1, T2 the load current has to flow via T6, mains, D1 and T1. The forced commutation at delay angle $\omega t = \beta$ is started due to firing of the discharge thyristor T3 at the instant t'=0, and the firing of the main thyristor T2 at $t'=t_o$. At the same time as the familiar linear charge reversal of the turn-off capacitor C_L takes place, the supporting capacitor (charged to $\hat{U}_{cs} > U_L$) forces a discharge current into the mains via T3, D1 and T6, against the line current (Fig. 5a). If the line voltage U_L is regarded as constant during commutation, the discharge current of C_s increases according to the function:

$$I_{cs} = -(\hat{U}_{cs} - U_L)\sqrt{C_s/L_L}\,\sin\omega_s t' \quad (14)$$

where

$$\omega_s = 1/\sqrt{L_L C_s} \quad (15)$$

As shown in Fig. 6, this current reduces the mains current from the very beginning. In the meantime the difference between U_{cs} and U_L decreases according to a particular cosine function. This state is maintained until $t'=t_1$, at which point the voltage reversal of the turn-off capacitor C_L has proceeded to $U_{CL}=U_{cs}$. Then the blocking diode D2 becomes conducting. This means that C_L is also parallel to the input voltage. The remaining line current splits between C_L and C_s according to their capacitance values (Fig.5b). That causes I_{CL} to drop suddenly from I_d to a rather low value, while I_{cs} increases by the same amount (Fig. 6). This commutation between the two capacitors is only spontaneous if no inductance is present in the C_L-C_s commutation circuit (shown dotted in Fig. 5b).

From time $t'=t_1$ both capacitor voltages increase to the peak at t_K, following a further sine function, and the line current decays. The over-voltage ΔU_c of both capacitors with reference to the instanteneous line voltage U_L at the end of the process is composed geometrically of the initial values of current and over-voltage at t_1:

$$\Delta U_c^2 = (U_{cs}(t_1) - U_L)^2 + I_L^2(t_1)\frac{L_L}{C_L+C_s} \quad (16)$$

During periodic operation:

$$U_{cs}(t_1) - U_L = \Delta U_c \cos(\omega_s t_1) \quad (17)$$
$$= \Delta U_c \cos\delta$$

$$I_L(t_1) = I_d - \Delta U_c\sqrt{C_s/L_L}\,\sin\delta \quad (18)$$

Equation (17,18) substituted in (16) gives:

$$\Delta U_c^2 = (\Delta U_c\cos\delta)^2 + (I_d - \Delta U_c\sqrt{C_s/L_L}\sin\delta)^2\frac{L_L}{C_L+C_s}$$

$$\Delta U_c \sin\delta = (I_d - \Delta U_c\sqrt{C_s/L_L}\sin\delta)\sqrt{\frac{L_L}{C_L+C_s}}$$

$$\Delta U_c = \frac{I_d\sqrt{L_L/C_s}}{(\sqrt{1+C_L/C_s}+1)\sin\delta} \quad (19)$$

Equation (19) shows that in order to achieve minimum over-voltage, the angle $\delta = \omega_s t_1$ should be about $\pi/2$. This can be achieved by either selecting a suitable interval t_o between the firing of the discharge thyristor and of the main thyristor, or, where $t_o=0$ for the most unfavorable case in which $I_d=I_{dmax}$ and $U_L=\hat{U}_L$, by appropriate choice of C_L and C_s. This means though that the scope in the choice of hold-off interval and over-voltage is reduced, whereas both values, in the case of sequential firing, can be chosen independently of each other.

Compared with the circuits discussed in section 1, whose over-voltage at turn-off angle $\beta \approx \pi/2$ with the sum of the capacitors concerned equals

$$\Delta U_{cmax} = I_{dmax}\sqrt{\frac{L_L}{2C}}$$

the supporting-capacitor principle reduces the over-voltage considerably with the same capacitor expenditure. If for the most unfavorable case $\delta \approx \pi/2$ is realized, and if $C_s \gg C_L$, then the equation (19) is simplified as follwos:

$$\Delta U_{cmax} \approx \frac{I_{dmax}}{2}\sqrt{\frac{L_L}{C_s}} \quad (20)$$

In this way half the mains current is reduced due to the discharge of the supporting capacitor, and the other half by the re-charge. Permitting the same over-voltage during periodic operation, as in the circuits mentioned above, the supporting-capacitor principle can function, in extreme cases, with only a 1/4 of the total capacitance. Furthermore it is to be noted that the supporting capacitor, in contrast to the auxiliary capacitors of the 3C-circuits, is only partly discharged and charged again.

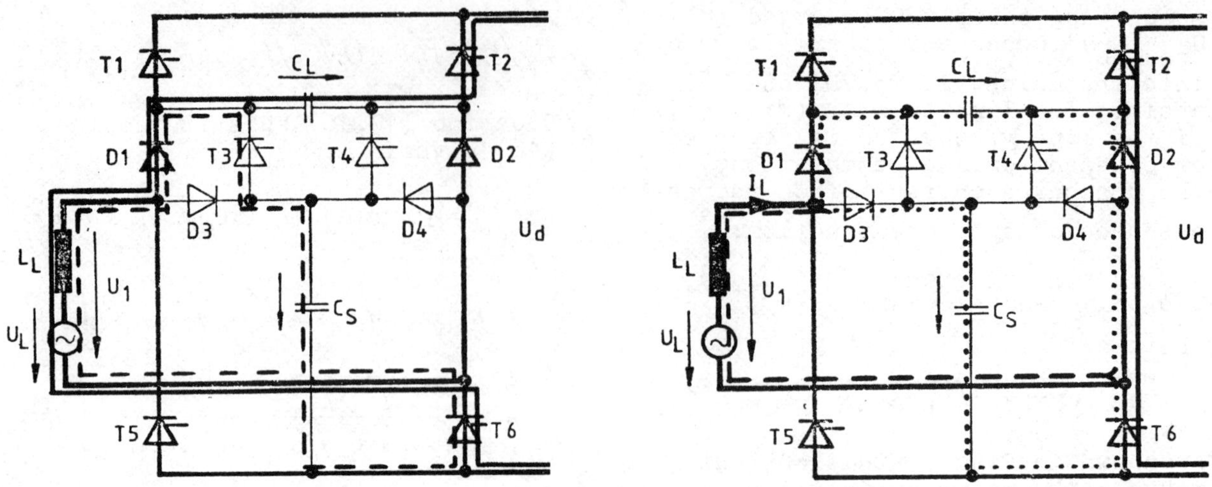

Fig. 5. Current paths during forced commutation.
a) First interval $t_o < t' < t_1$, b) second interval $t_1 < t' < t_K$
Dotted: Commutation circuit between capacitors

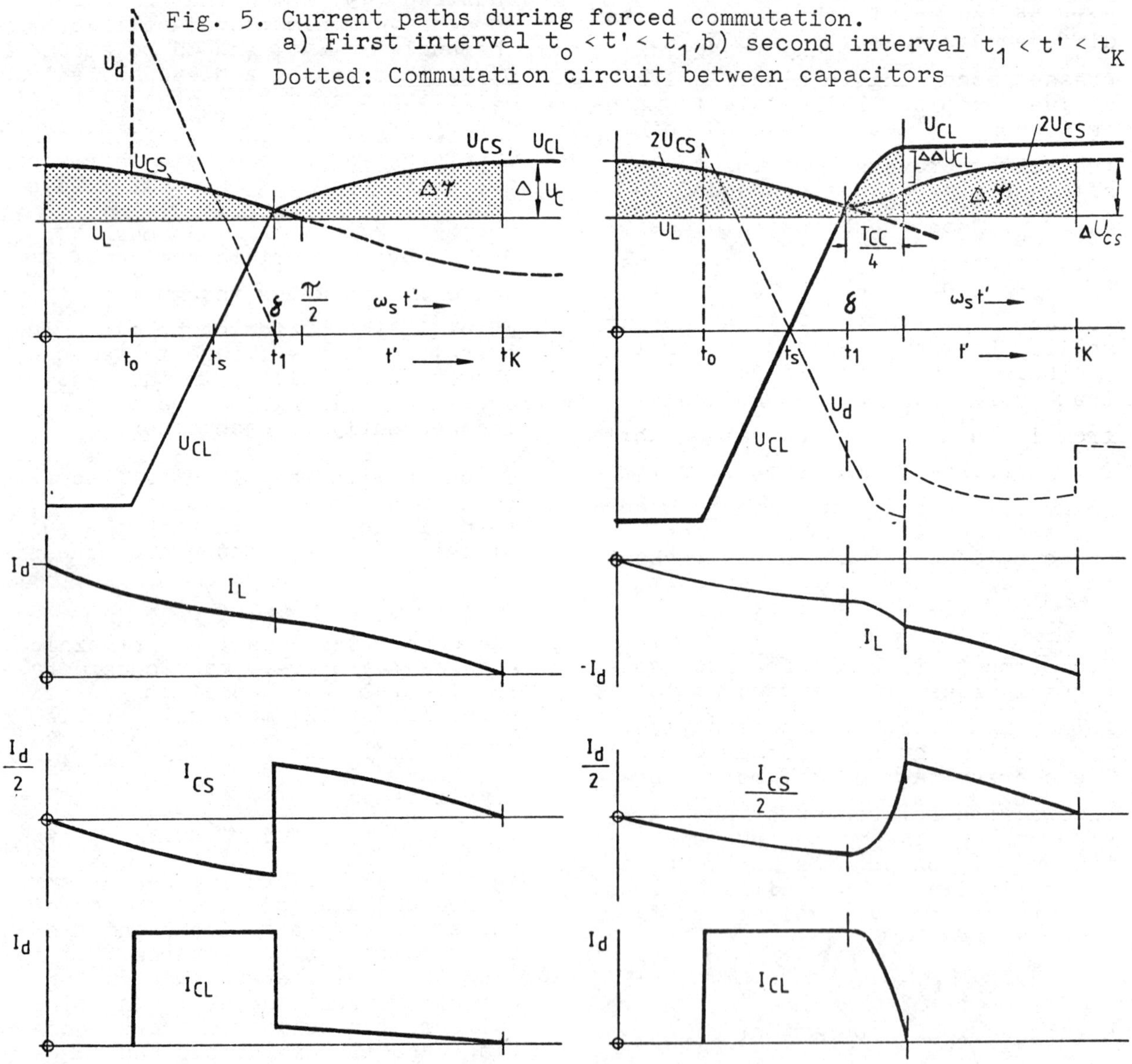

Voltage and current courses of line and of both capacitors during forced commutation.

Fig. 6. Circuit according to Fig. 4, no inductance in the C_L-C_s-circuit, rectifier mode.

Fig. 8. Circuit according to Fig.7, inductance L_1 of the center tap in the C_L-C_s-circuit, inverter mode

The low stresses due to reactive power and RMS current permit the use of smoothing capacitors whose weight referred to $C\hat{U}_c^2$ is only about 1/4 of that of commutating or reactive-power capacitors.

A further advantage is the fact the line current reduction is stretched out across the whole of the commutation period and can be arranged to proceed very gently. This has a positive effect on the interference current of the installation.

In the circuit shown in Fig. 4 the supporting capacitor fulfills an important protective function against line over-voltage transients in the period between the delay angle of the naturally commutated group and the following reversal of U_L.

This protection is present permanently in the circuits to be dealt with above where C_s is connected completely via uncontrolled valves.

The following disadvantages of the supporting-capacitor principle must be mentioned:

- If necessary, measures must be taken against excessive overshooting of the supporting-capacitor voltage when the installation is switched on.
- The blocking diodes D1 and D2 are subjected, at maximum, to a voltage of $\hat{U}_{csmax}+\hat{U}_{CLmax}=2\,\Delta U_{cmax}+2\,\hat{U}_L$, comparable diodes in the other circuits are stressed with only max. $\Delta U_{cmax}+2\,\hat{U}_L$.

All other valves are stressed, at maximum, with the voltage $\Delta U_{cmax}+\hat{U}_L$. The relative conduction period of the main thyristors and the blocking diodes is 50 %, ignoring the commutation.

Although the circuit shown in Fig. 4 is fully controllable, it is not suitable for inverter operation because in this case the supporting capacitor in ineffective. It cannot be paralleled to the mains in the forced commutation from the freewheeling state necessary during inverter operation.

However, like the basic circuit 1C-LVB, it is suitable for inverter operation if a braking resistor is used. The discharge thyristors are then blocked so that the supporting capacitor is not involved in the operation. The braking resistor R_B can be connected instead of the mains at the input terminals, or at the output terminals on the DC side. Current pulses of adjustable frequency and width are applied to the resistor at alternating or constant polarity. Due to the fact that the turn-off capacitor charges up to $\hat{U}_{cL}=I_dR_B$, the hold-off interval is independent of the current:

$$t_s = C_LR_B.$$

If the possibility of using a braking resistor is of no interest, the thyristors T5, T6 of the mains-commutated group can be replaced by diodes. This means though that the adjustable phase-control angle α is no longer available. It can be provided again by the use of a freewheeling diode on the DC side. This diode is stressed, as are the blocking diodes, with $2\,\Delta U_{cmax}+2U_L$.

2.2 Modification of the circuit for regenerative braking

In order to be able to connect the supporting capacitor parallel to the mains, independent of the state of the line-commutated group, its negative terminal can no longer be connected to the negative terminal of the circuit. Instead, one can connect it to both input terminals by using a center-point connection consisting of diodes and anti-parallel-coupled thyristors (Ref. 6). Another possibility is shown in Fig. 7a. The negative terminal of C_s is connected with the center tap of the supply transformer. In the case of a transformer with ideal coupling (L_1=0), the supporting capacitor C_s, which is connected in parallel with half the winding when it comes into operation, functions exactly as $C_s/4$ across the whole winding. The product $C_s\hat{U}_{cs}^2$ does not change at all.

The non-ideal coupling between primary and secondary winding is, as previously agreed, taken into account in the line inductance L_L.

An important precondition for the practical realization of the circuit is as close a coupling as

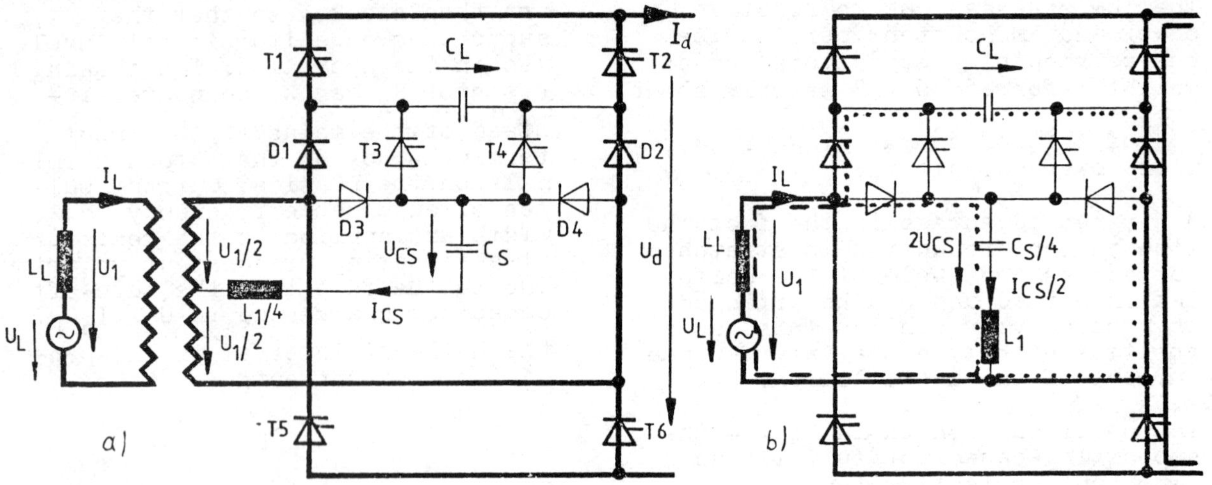

Fig. 7a. Modified circuit suited to rectifier and inverter operation. Supporting capacitor can be paralleled to each half of the transformer secondary winding. L_1 = short circuit inductance between both halfs.

Fig. 7b. Equivalent circuit to Fig. 7a if $U_L > 0$ and $I_{cs} \leq I_d$

possible between the halves of the secondary winding. That means that the inductance L_1 measurable in the short-circuit test between both halves (primary winding open) should be as small as possible. It is present as $L_1/4$ at the center tap.

The method of operation will be discussed using the forced commutation of T1 to T2 with $U_L > 0$. Fig. 7b represents an equivalent circuit for this case, in which the transformed elements $C_s/4$ and L_1 are applied to the full voltage. T5 is presumed to conduct in the line commutated group; that means inverter operation. It is of no importance for the functioning of the forced commutation whether T5 or T6 conducts, merely the course of U_d and I_L is affected.

With the firing of the thyristor T3 at the instant t'=0 the supporting capacitor, which has been charged up to $(U_L + \Delta U_{cs})/2$, starts to discharge as shown in Fig. 8 (aside from Fig. 6). The line current, which was idling, is increased in negative direction as a result. Analogous to equation (14), the current $I_{cs}/2$ of the transformed supporting capacitor $C_s/4$ follows the equation:

$$\frac{I_{cs}}{2} = -\left(2\hat{U}_{cs} - U_L\right)\sqrt{\frac{C_s/4}{L_L + L_1}}\,\sin(\omega_s t') \quad (21)$$

where

$$\omega_s = 1/\sqrt{(L_L + L_1)\,C_s/4} \quad (22)$$

The linear voltage reversal of the turn-off capacitor takes place between the firing of the thyristor T2 at $t'=t_o$ and $t'=t_1$ when the voltage U_{cL} is equal to the transformed supporting capacitor voltage $2U_{cs}$. Up to this point, the conditions are in principle exactly as in the previously described arrangement. At $t'=t_1$, D_2 begins to conduct and closes the commutation circuit between C_L and $C_s/4$ (shown dotted in Fig. 7b). According to the inductance L_1 included in this circuit the shifting of the current from C_L to $C_s/4$ no longer takes place spontaneously. If we assume that the line current remains nearly constant during this commutation, then I_{cL} decays following a cosine oscillation around the reference line

$$(I_d + I_L)\frac{C_L}{C_L + C_s/4} \quad ; \quad I_L < 0$$

The oscillation ends after slightly more than 1/4 cycle at $I_{cL}=0$. Its period is:

$$T_{cc} = \frac{1}{2\pi}\sqrt{\frac{\frac{1}{C_L} + \frac{4}{C_s}}{L_1}} \approx \frac{1}{2\pi\sqrt{L_1 C_L}} \quad (23)$$

Meanwhile, U_{cL} shoots over by the amount $\Delta\Delta U_{cL}$ and remains steady at

$$\hat{U}_{cL} = U_L + \Delta U_c \cos\delta + \Delta\Delta U_{cL} \quad (24)$$

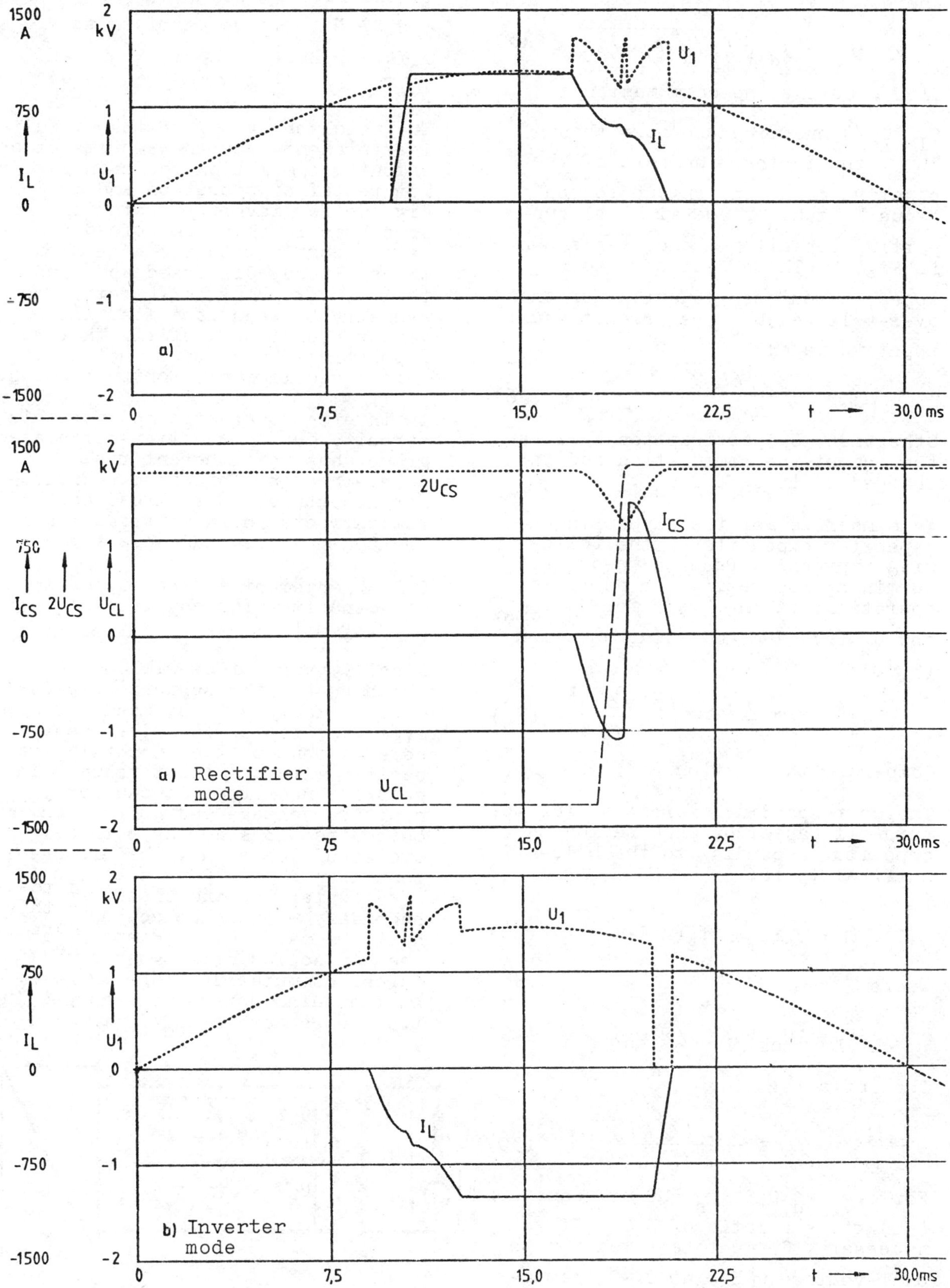

Fig. 9. Input voltage U_1, line current I_L, capacitor voltages U_{cL}, U_{cs} and current I_{cs} during one half cycle of the circuit shown in Fig. 7a as results of a computer simulation. Data: $\tilde{U}_L$=1000 V, f=16 2/3 Hz, I_d=1000 A, u_{km}=10 %, L_1=0.1 L_L, C_s/4=1mF, C_L=0.3mF

where

$$\Delta\Delta U_{cL} \approx I_d \sqrt{L_1 (1/C_L + 4/C_s)} \quad (25)$$

If U_{cL} during the commutation $C_L \rightarrow C_s/4$ has risen to more than $2\hat{U}_{cs}$, the further increase of the negative mains current up to the value I_d takes place with the supporting capacitor alone, as shown in Fig. 8. The effect of C_L disappears from the equation for the over-voltage ΔU_{cs} (referred to full input voltage):

$$\Delta U_{cs} = \frac{I_d \sqrt{4 L_L / C_s}}{2 \sin \delta} \quad (26)$$

Otherwise, C_L participates again at the end of the commutation and is charged up to $\hat{U}_{CL} \geq 2\, \hat{U}_{cs}$.

As a measure for the supporting-capacitor expenditure, the following expression results when determining the rating for periodic operation, taking $\delta \approx \pi/2$ for $I_d = I_{dmax}$ and $U_L(\beta) = \hat{U}_L$ and using equation (9,10):

$$C_s \hat{U}_{cs}^2 = \frac{u_{xm} \tilde{U}_L I_{dmax}}{4\, \omega} \left(1 + \frac{\sqrt{2}}{k_u}\right)^2 \quad (27)$$

Compare with equation (13)!

The corresponding expression for the turn-off capacitor when rating the capacitor according to the hold-off interval t_s is:

$$C_L \hat{U}_{cL}^2 = \tilde{U}_L I_{dmax}\, t_s (\sqrt{2} + k_u) \quad (28)$$

where

$$\hat{U}_{cL} = \hat{U}_L + \Delta U_{cs} = \hat{U}_L\, (\sqrt{2} + k_u) \quad (29)$$

The ratio is:

$$\frac{C_s \hat{U}_{cs}^2}{C_L \hat{U}_{cL}^2} = \frac{C_s/4}{C_L} = \frac{u_{xm} (\sqrt{2} + k_u)}{4\, \omega\, t_s\, k_u^2} \quad (30)$$

For e.g. $k_u = 0.5$, $t_s = 500\ \mu s$, equation (30) gives a ratio of 3.64. If necessary, C_L must be rated for the over-voltage $\Delta\Delta U_{cL}$ at C_L-C_s commutation. Using equation (25,26) the following results:

$$\frac{C_s/4}{C_L} = \frac{L_L}{4 L_1} \left(\frac{\Delta\Delta U_{cL}}{\Delta U_{cs}}\right)^2 - 1 \quad (31)$$

If one calculates with $L_1 = 0.1\ L_L$ and $\Delta\Delta U_{cL}/\ U_{cs} = 1.3$ one arrives at $\frac{C_s/4}{C_L} = 3.22$.

The total behaviour of the circuit is represented by the results of a computer simulation shown in Fig.9a for rectifier operation and in Fig. 9b for inverter operation. The processes involved in forced commutation comply with the considerations already discussed and find their interpretation therein. The symmetry between rectifier and inverter operation confirms that the supporting-capacitor principle is suited equally well for both cases.

In inverter operation however this circuit, due to the line-commutated group, has the inherent risk of shoot-through. This is of particular importance in railroad operation where short interruptions of the contact can take place.

2.3 Arrangement of the supporting capacitor for any type of circuit

Finally, variations will be shown which enable the supporting-capacitor principle to be used for any type of circuit. It is necessary here to connect the supporting capacitor with the input voltage in such a manner that it can carry positive voltage and current in both directions during both half cycles of the mains. The arrangement as per Fig. 10 can be applied if a center tap and closely-coupled secondary winding halves are present. Of particular advantage is the connection of a two-stage forced-commutated bridge (Ref.7). As the mains current is changed in two stages each of $I_d/2$, the

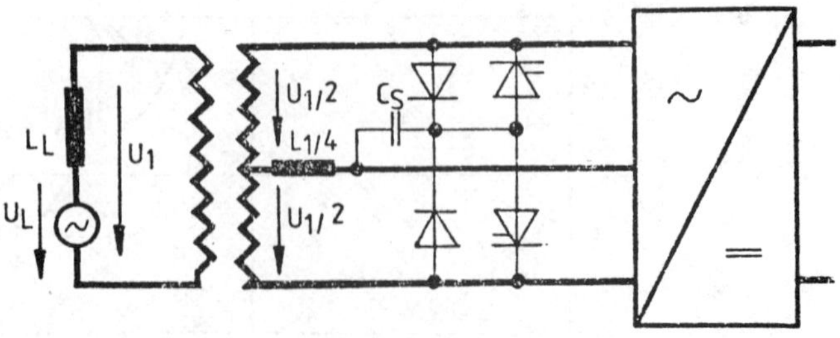

Fig. 10. Generalized arrangement of the supporting capacitor, specially suited for two-step bridges. Possible, if $L_1 \ll L_L$.

over-voltage ΔU_c is reduced by 50 % compared to a single-stage circuit. Or, if equal ΔU_c is permitted, the supporting-capacitor expenditure is reduced to 25 %.
A disadvantage is the danger of a direct short-circuit when one of the discharge thyristors fires in the false half-cycle.

If a center tap is not available, the supporting capacitor can be arranged in a bridge circuit comprising anti-parallel connected diodes and thyristors.

REFERENCES

(1) Förster,J.: Zur Stromrichter-Netzbelastung. ETZ-A 96 (1975) H. 1, pp. 52-57.

(2) Kehrmann,H.; Lienau,W.; Nill,R.: Vierquadrantensteller - eine netzfreundliche Einspeisung für Triebfahrzeuge mit Drehstromantrieb. Elektrische Bahnen 45 (1974) H. 6, pp. 135-141.

(3) Bezold,K.-H.: Löschbare, einphasige Stromrichter mit Sektorsteuerung für die Speisung von Mischstrommotoren. Elektrische Bahnen 45 (1974) H. 12, pp. 281-287.

(4) Fick,H.; Fleckenstein,V.; Loderer,P.: Bahn-Stromrichter mit gutem Leistungsfaktor. ETZ-A 96 (1975) H. 5, pp. 239-242.

(5) Kahlen,H.; Müller-Hellmann,A.; Woelker,W.: Halbgesteuerte Brückenschaltung mit unterteilter Löschung. Elektrische Bahnen 46 (1975) H. 12, pp. 279-286.

(6) Lienau,W.; Müller-Hellmann,A.: Folgegesteuerte, löschbare Brückenschaltung. Deutsche Offenlegungsschrift 2 530 961.

(7) Müller-Hellmann,A.: Mehrstufige zünd-und löschbare Einphasen-Brückenschaltungen (Multi-step line-and forced-commutated single-phase bridge convertors) IFAC-Symposium 1977.

DIGITAL SIMULATIONS OF A FORCED-COMMUTATED CONVERTER FOR SINGLE-PHASE FOR AC LOCOMOTIVES

H. Schlunegger

Swiss Federal Institute of Technology, Zurich, Switzerland
Since 1.1.1977 with Brown, Boveri & Company Limited, Baden, Switzerland

Summary

A static convertor for feeding an intermediate d.c. circuit in a traction vehicle forms the basis of this paper, in which a digital program used to simulate this convertor will be described. The curves of the electrical quantities with respect to time for two different types of circuit will also be dealt with. Results important for the behaviour of the power system will be shown in addition to the spectra of the power system voltage and current. It was found necessary to include the catenary in the simulation.

Zusammenfassung

Diesem Beitrag liegt ein Triebfahrzeug-Stromrichter zur Einspeisung eines Gleichspannungszwischenkreises zugrunde. Es wird ein digitales Programm zur Simulation dieses Stromrichters beschrieben. Für zwei verschiedene Schaltvarianten mit einem und mit 3 Stellern, werden die zeitlichen Verläufe der elektrischen Grössen dargestellt und die Spektren von Netzstrom und Netzspannung sowie die Leistungsfaktoren berechnet. Der Einsatz eines Störstromfilters bringt eine erhebliche Reduktion der Harmonischen höheren Ordnungszahl. In einem Falle wurde auch die Fahrleitung in die Simulation einbezogen.

1. Introduction

When employing asynchronous machines for traction applications, forced-commutation inverters are required for varying the frequency and voltage of the power supply. For most of these inverters a constant direct voltage is required at the input. Thus, for single-phase systems a convertor which has no adverse effects on the power system and which can be used to regulate the direct voltage in the intermediate circuit so that it stays constant, must be installed between the power system and the intermediate d.c. circuit. To examine this convertor it is necessary to simulate the behaviour of the electrical quantities with respect to time. Either a hybrid or a digital computer can be used for the simulation. Due to the complexity of the problem, however, it is not possible, generally speaking, to obtain analytical expressions for the chronological functions, and these of numerical integration. Such integrations are often built into digital computer programs. Digital programs for network analysis exist in various forms [1...5] . When used for convertors, and particularly in the case of forced commutation, the computing times for these programs are relatively long. Thus, a program written in FORTRAN, specially designed for single-phase convertors installed in a.c. locomotives, will be used for the computations used in this paper.

First order differential equations, written in the form shown below, are used to describe the system:

$$\dot{\underline{x}} = A\ (x,\ t) \cdot \underline{x} + b \cdot \underline{u}$$

Where $\underline{x}$ = vector of state variables

A = coefficient matrix (a function of $\underline{x}$ and t)

b = input vector

$\underline{u}$ = input values

A and b describe the state of the convertor circuit and must be computed again after each change (e.g. when a thyristor is triggered). Vector $\underline{x}$ moves continuously during changes of state. The differential equations are integrated according to the Runge-Kutta-Fehlberg method with step size control. Termination conditions (e.g. the crossing of a given level by a state variable) are detected by a special subroutine and computed using the secant method. Each termination is followed by the new circuit state being determined from the previous one. The initial step size is also the same as that used in the previous integration. On termination of the desired integration interval (1 or 2 cycles of the line voltage), the root mean square values, power factors and spectra of the voltages and currents are calculated and the curves plotted. Fig. 1 shows the flow chart for the program used. The computing time for one cycle of the line voltage is approx. 20 seconds when a CDC 6400-6500 system is employed. [6, 7]

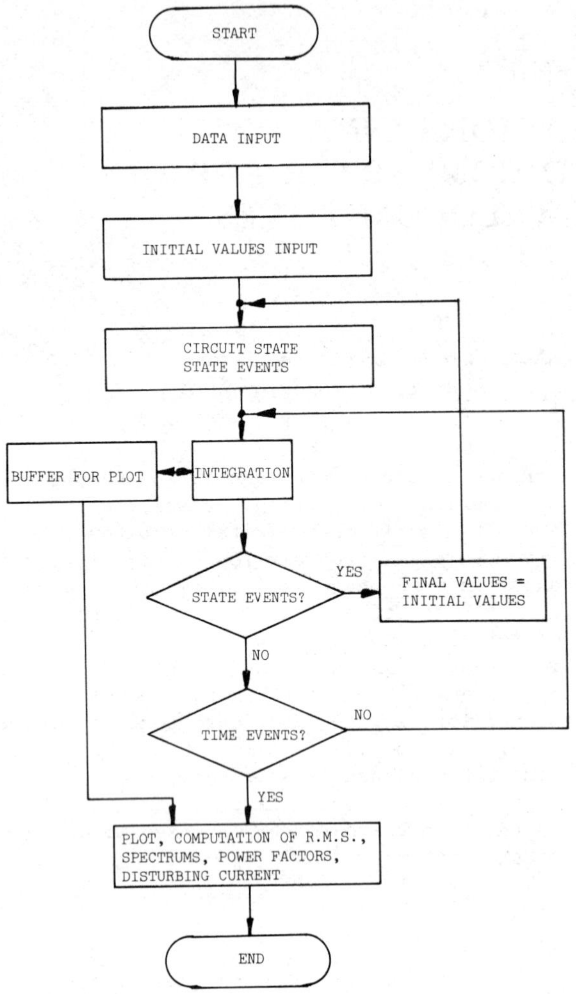

Fig. 1 Flow chart of the simulation program

2. Convertors for supplying an intermediate d.c. circuit

Requirements which must be met by a convertor supplying an intermediate d.c. circuit in a single-phase a.c. locomotive are as follows:

- Favourable line compatibility,
- minimum filter requirements in the intermediate circuit, and
- minimum control system requirements.

In the interests of the operation of the input and output convertor a constant voltage should be applied to the intermediate circuit. Thus, a filter must be provided to suppress the power fluctuations arising from the single-phase supply line. This may comprise a series resonance circuit tuned to the second harmonic of the line voltage (Fig. 2). The operation of the convertor and filter is described in [8, 9, 10, 11] . At the line end, the convertor can operate in either two or four quadrants of the power plane (u_L - i_L plane), irrespective of its structure (Fig.3) In this paper a convertor is described which operates in two quadrants, and whose line current is controlled by a two-step control

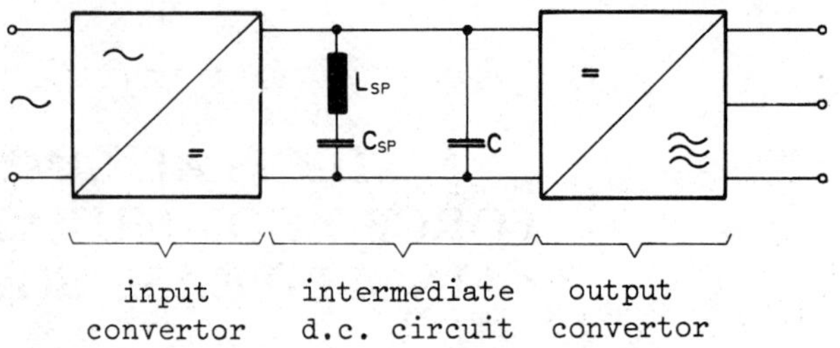

Fig. 2 Input and output convertors with filters and intermediate d.c.circuit

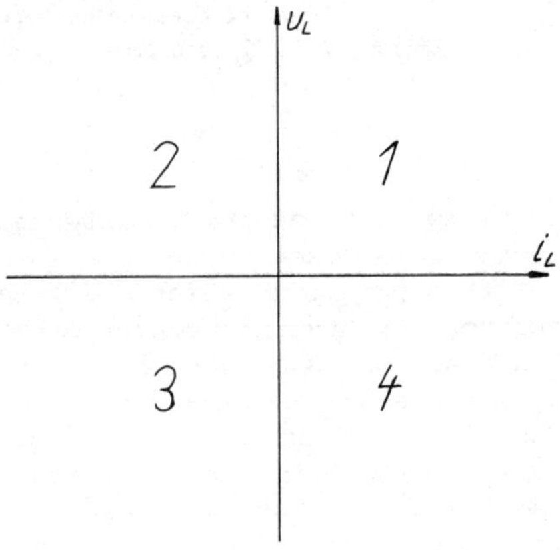

Fig. 3 u_L-i_L plane
The "Zweiquadrantensteller" convertor operates in quadrants 1 and 3, and the "Vierquadrantensteller" in all quadrants

system with a sinusoidal reference band.

3. Single-chopper convertor

Fig. 4 shows the convertor circuit, comprising a diode rectifier and a chopper. As shown in Fig. 5 the line current is controlled by means of a sinusoidal reference band. Besides the relative straightforwardness of this circuit, it also has the advantage that when a quenching failure occurs in the chopper the intermediate circuit does not discharge through the latter. The result of this is a simplified protection concept for the input convertor. However, the inverter mode of operation (regenerative braking) is not possible. The various states of the convertor circuit must be determined before the differential equations can be deduced. Equivalent circuits for this purpose are shown in Fig. 6.
All simulation results were obtained using the following data:

U_L = 15 kV,	f_L = 16 2/3 Hz
P_N = 7.5 MW	U_{ZN} = 1200 V
I_N = 6250 A	

Winding ratio of transformer $\ddot{u}$ = 20.

These results are shown in Fig. 7. All values are referred to the transformer secon-

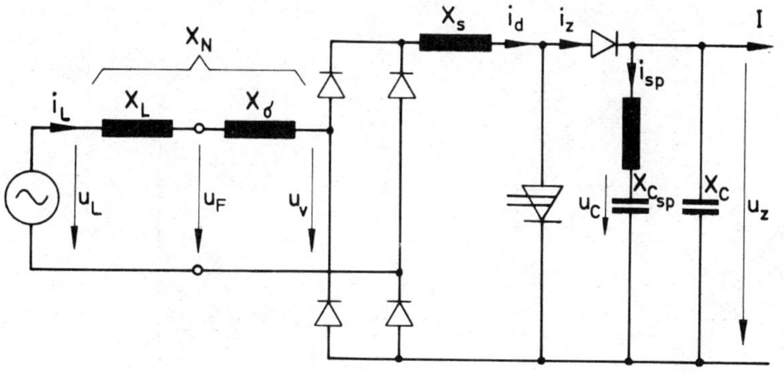

Fig. 4 Convertor comprising a diode rectifier and a chopper

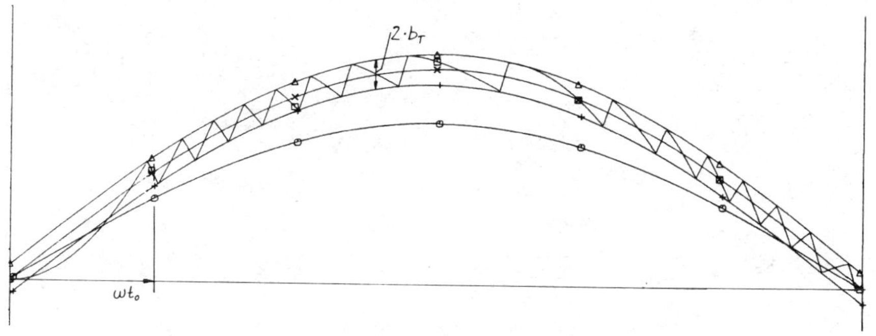

Fig. 5 Two-step control for the line current during a half-cycle

○ line voltage
△ + upper and lower limits of the reference band
X reference current i_{Lref}
□ line current i_L

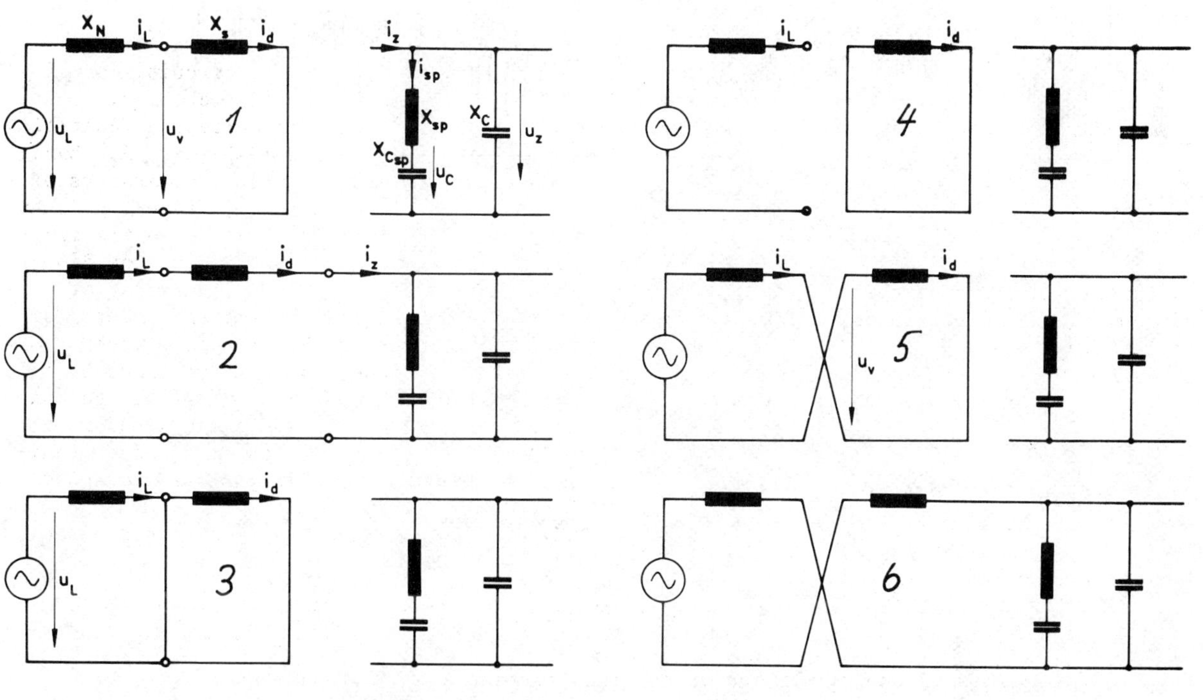

Fig. 6 Equivalent networks for the simulation of the convertor

1) chopper conducting
2) chopper non-conducting
3) commutation of the diode rectifier
4) line current is zero
5) state 1) with inverted sign
6) state 2) with inverted sign

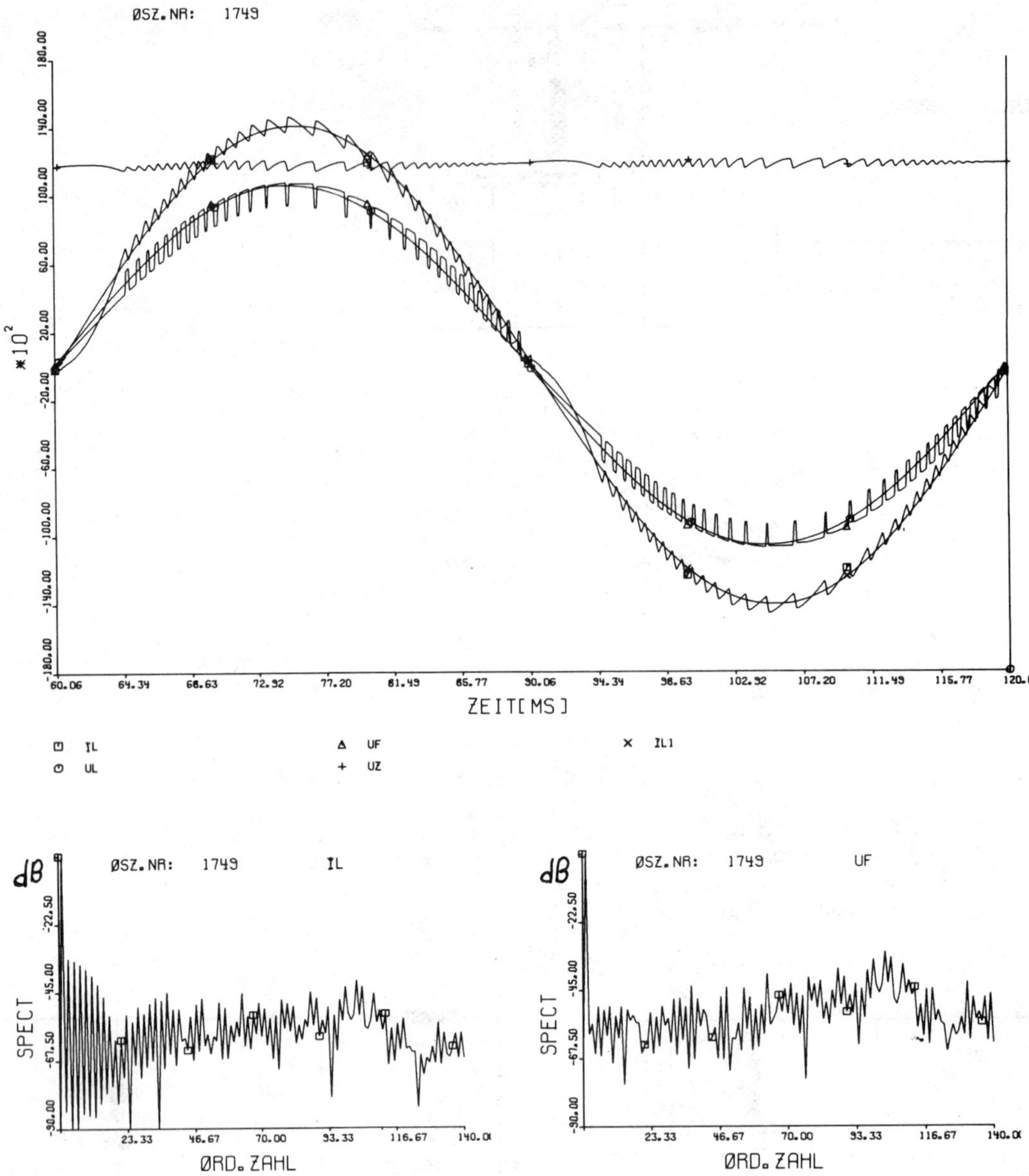

X_L = 2 mΩ	X_σ = 6 mΩ	I = 6360 A	P = 7480 kW
X_S = 9 mΩ	X_{Sp} = 96 mΩ	I_L = 499 A	U_F = 15042 V
X_{Csp} = 384 mΩ	X_C = 318 mΩ	λ_L = .9987	λ_F = .9964
β = 4.41	b_T = 5oo A	$\cos\varphi_L$ = .9999	W_U = 3.11 %
		I_{St} = 16.29 A	

Fig. 7. Results of the simulation of the line current and its fundamental harmonic, the voltage at the current collector and the intermediate circuit voltage. For symbols see fig.4 and chapter 6.

dary , with the exception of the line current, the voltage at the current collector and the equivalent disturbance current which are referred to the primary. A noticeable feature is the relatively high disturbance current, for which reason a filter had to be included in the simulation. This RC filter and the convertor are shown in Fig. 8. When using such filters it is important that the low-order harmonics are not accentuated. A correction of the displacement factor cos $_L$ as described in [12] is not necessary in this case as the cos φ of the simulated convertor is practically unity. Data for the filter are:

R_F = 40 Ω

C_F = 7.5 μF referred to 15 kV.

The spectra of the line current and voltage at the current collector with and without the filter are shown in Fig. 9. A reduction of the high-order harmonics is clearly discernible.

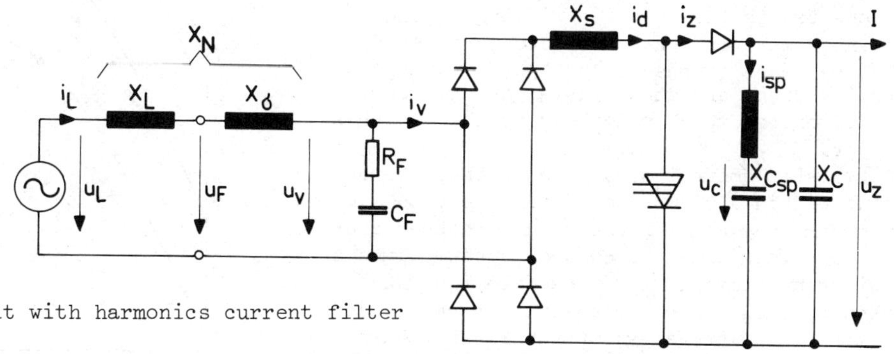

Fig. 8 Convertor circuit with harmonics current filter

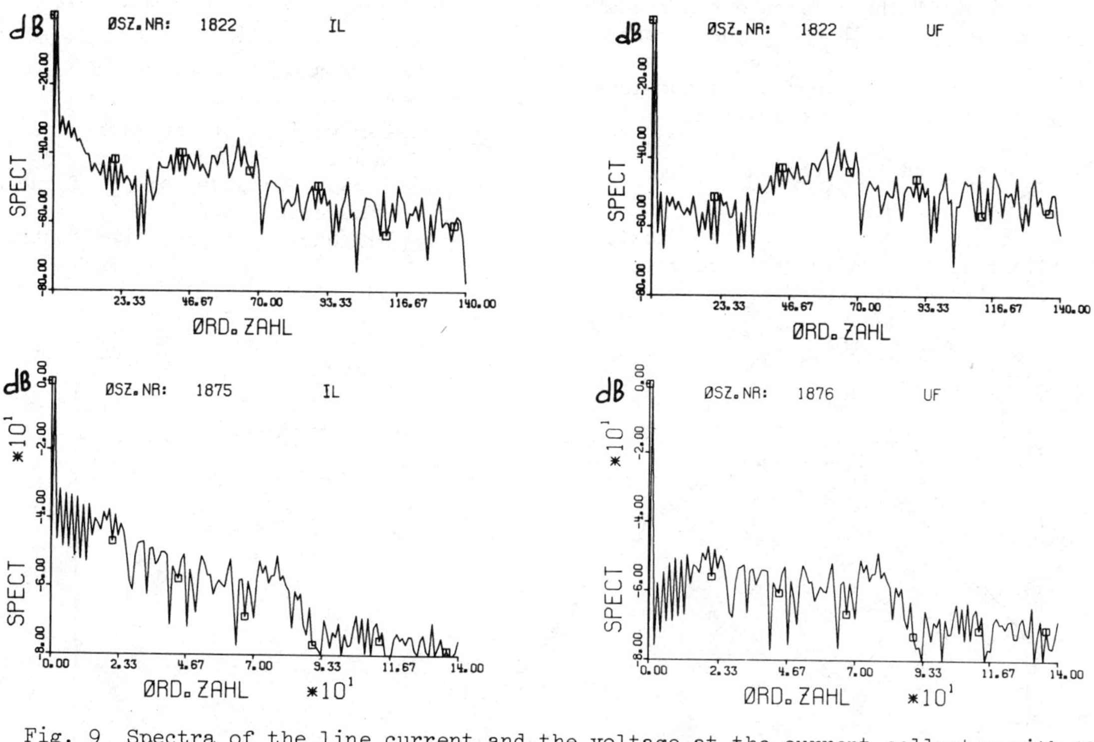

Fig. 9 Spectra of the line current and the voltage at the current collector with the harmonics filter connected (below) and disconnected (above).

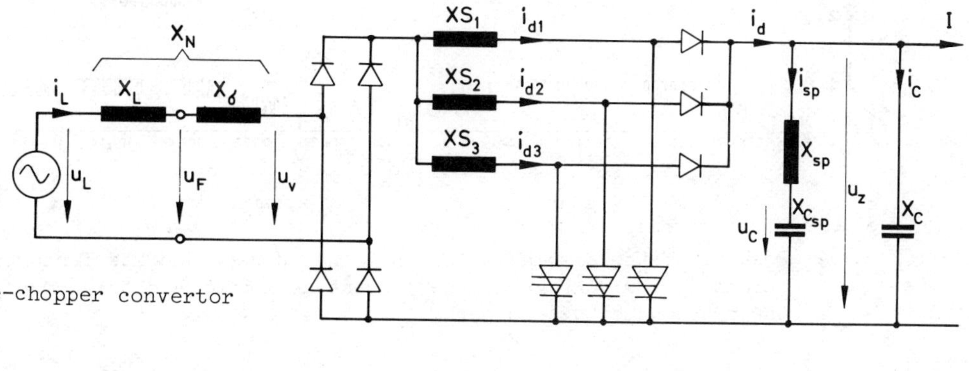

Fig. 10 Three-chopper convertor

4. Three-chopper convertor

Very short switching intervals are required to obtain favourable line compatibility when a convertor having a single chopper is employed. The minimum switching interval being governed by the type of thyristors used in the convertor. For this reason a convertor having three choppers is propounded and examined. When high-rated powers are involved, it is anyway necessary to divide the convertor into several units (see Fig.10). The proposed circuit comprises a diode rectifier connected to the traction winding of the transformer with the three choppers decoupled by means of reactances X_{sn}. Various control systems are conceivable for this circuit, one of which will be described in this paper.

A two-step control system holds the line current within a sinusoidal reference band, in the same way as with the single-chopper convertor. Each time the control system gives a switching command, i.e. when the line current (actual value) crosses a reference band limit (setpoint value) or turned off. When the upper limit of the reference band is attained, the chopper with the highest current is turned off. Similarly, the chopper with the lowest current is triggered on the lower limit being reached. The derivatives

$$\frac{di_{L\,act}}{dt} \quad \text{and} \quad \frac{di_{L\,ref}}{dt}$$

are now compared in this new state. Should the gradient of the line current be such that the latter might leave the reference

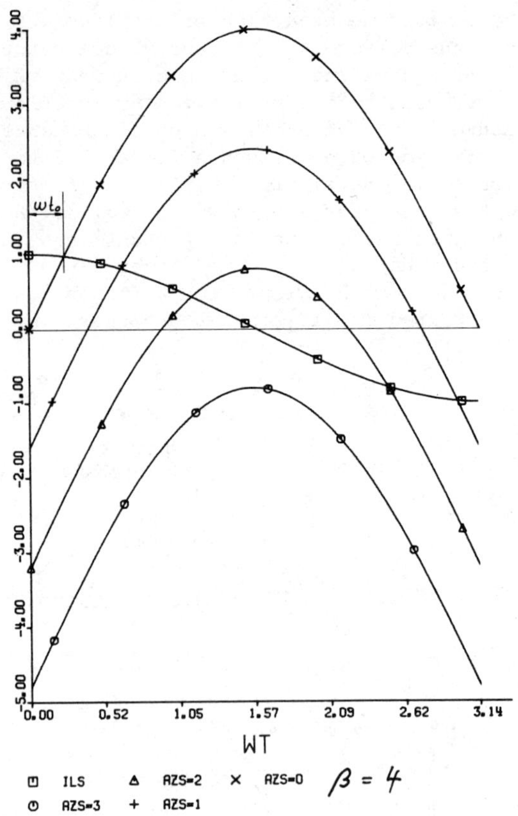

Fig. 11 Derivatives of the reference current and line current.
Parameter: azs= number of the non-conducting choppers.
All curves are reffered to the peak value of $\frac{di_{L\,ref.}}{d\omega t}$

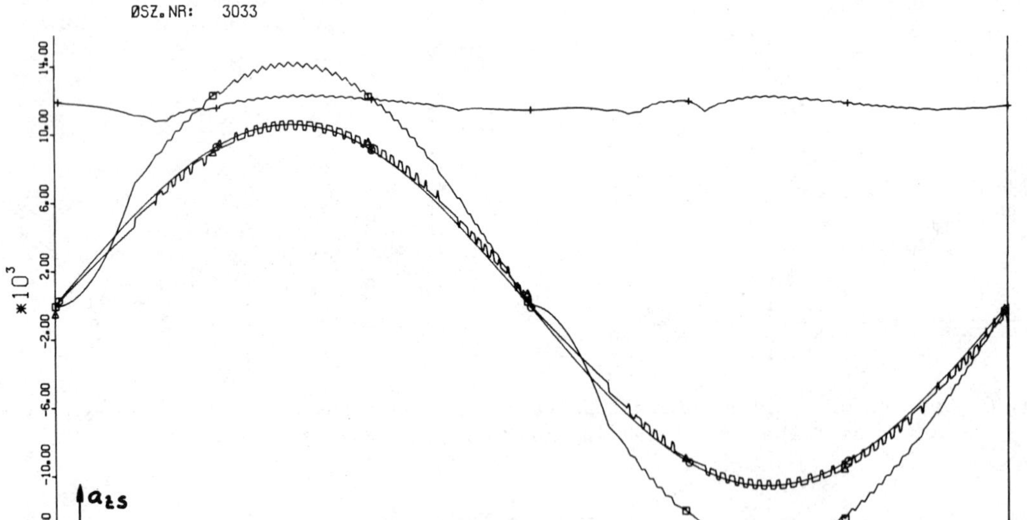

Fig. 12 Results of the simulation of the line current and its fundamental harmonic, the voltage at the current collector and the number of the non-conducting choppers.

band, the state of the chopper with the next highest current, or the next lowest as the case may be, is changed to prevent this happening. If this second change of state is still inadequate, a third change takes place based on the same principles. As it is shown in Fig. 11, however, the state of only one chopper need be changed at each switching point. The derivative of the line current is given by the differential equation:

$$\frac{di_L}{d\omega t} = \frac{q \cdot \hat{U}_L \cdot \sin\omega t - U_Z \cdot azs}{q \cdot X_N + X_S}$$

Where q = total number of choppers (here 3)
azs = number of non-conducting choppers

Fig. 11 shows the gradient of the reference value and actual values of the current for all possible states. It is possible to maintain the required current form with the exception of the interval between 0 and $\omega$to, by switching only one chopper each time the control system gives a command. The results of the simulation are given in Fig. 12. Curves of the line current and the three currents through reactances X_S are shown in Fig. 13. It can be seen that providing the reference band is sufficiently narrow, even distribution of the current between all choppers will be guaranteed.

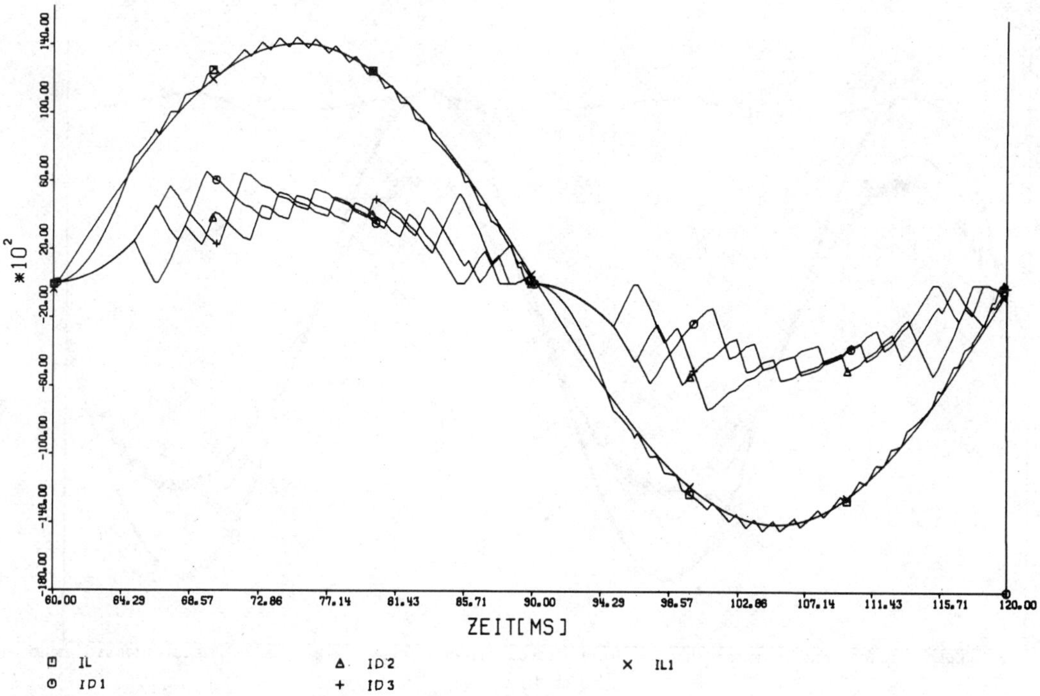

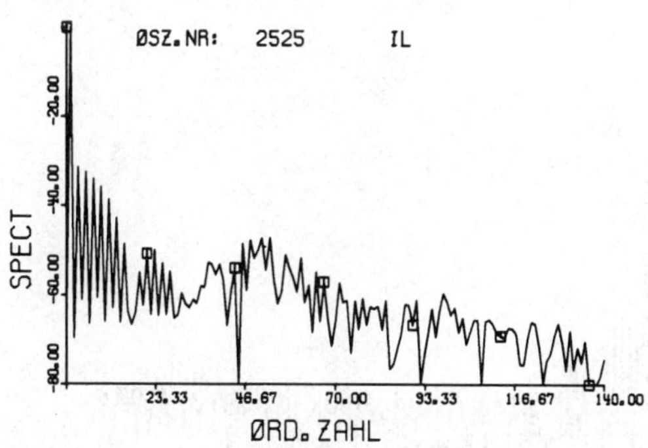

Fig. 13 Results of the simulation of the line current and its fundamental harmonic, and the three currents through the reactances X_{Sn}.

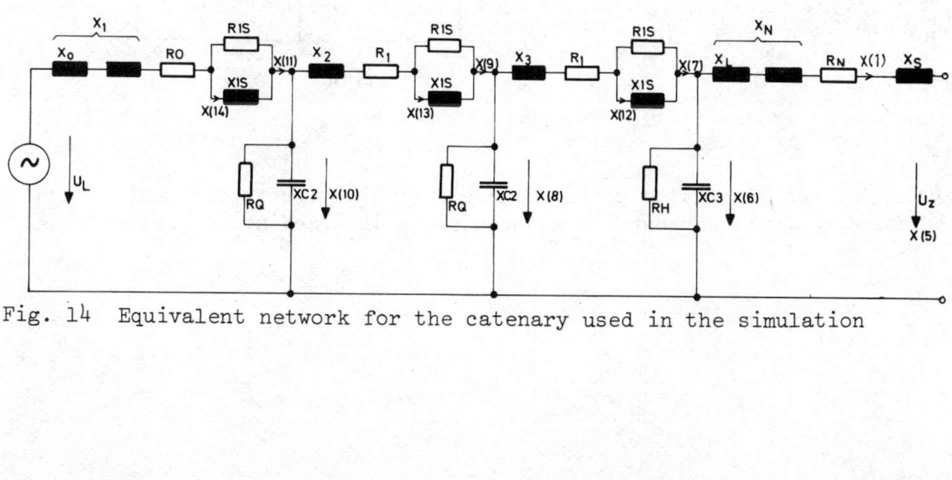

Fig. 14 Equivalent network for the catenary used in the simulation

5. Convertor with line simulation

Railway power supply systems are capable of resonance. The switching operations taking place in the static convertor can lead to considerable oscillation on the line. It is therefore necessary to include the catenary in the simulation. The equivalent network chosen is illustrated in Fig. 14. Up to 2kHz the series impedance follows the frequency as desired. Fourteen state variables are used to describe the system. This high order and the exceptional diversity of the eigenvalues result in a relatively long computing time, approximately 60 seconds being required for one line cycle. The results of the simulation are shown in Fig. 15. To prevent oscillations on the catenary a narrow reference band must be chosen so that the frequency range of the oscillations generated by the convertor are as high as possible. The data for the catenary were taken from [14, 15] . It could be established that the resistances connected in parallel with the catenary capacitances have a considerable influence on the magnitude of the oscillation amplitudes, and must therefore be included in the simulation.

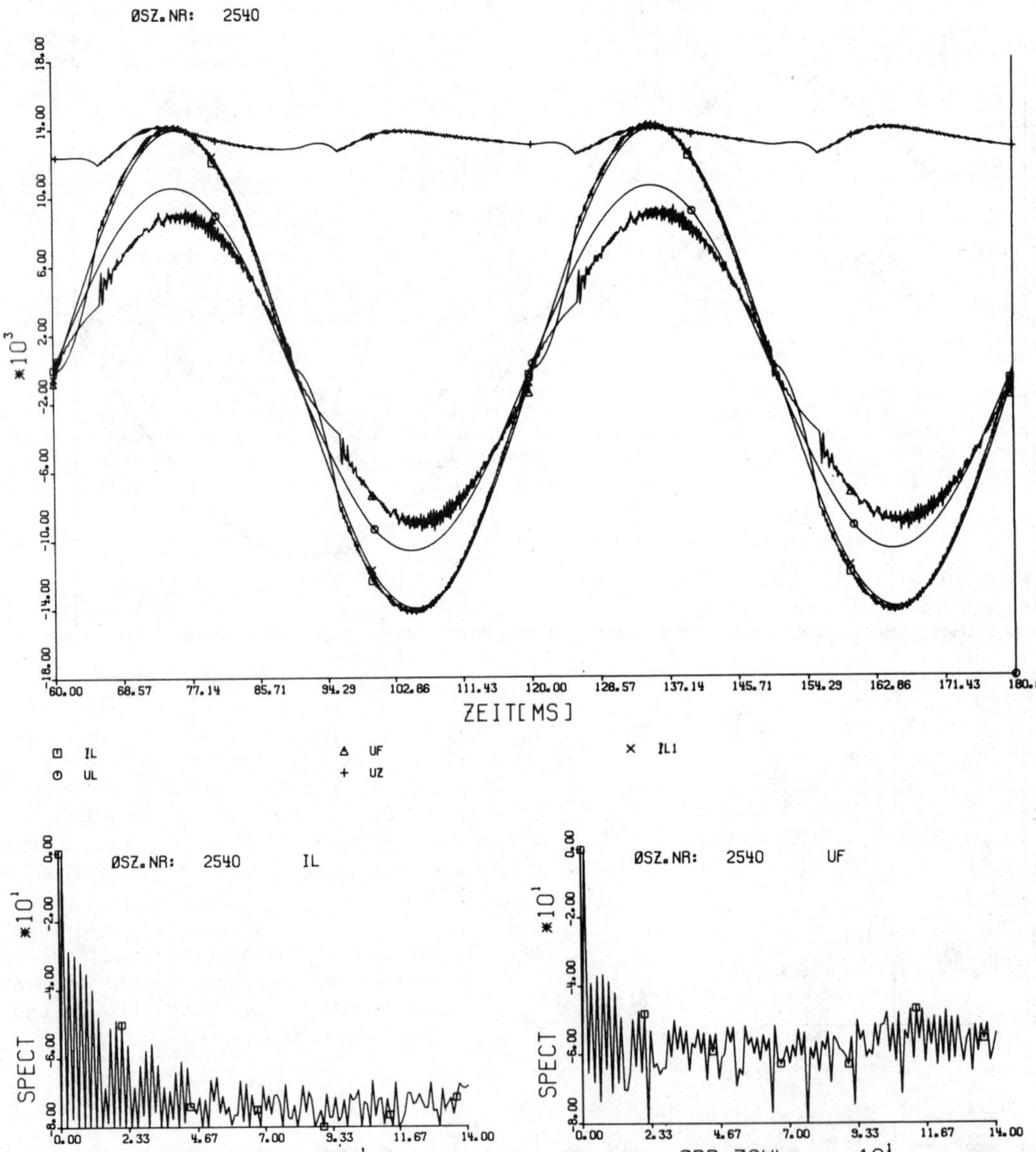

Fig. 15 Results of the simulation of the line current and its fundamental harmonic, the voltage at the current collector and the intermediate circuit voltage.

6. Key to symbols, not explained elsewhere

i_{Lact}: Actual value of the line current

i_{Lref}: Reference value of the line current

I_{St}: Equivalent disturbance current

W_u: Ripple of the voltage on the intermediate d.c. circuit

$$W_u = \frac{U_{Zmax} - U_{Zmin}}{U_{Zmax} + U_{Zmin}}$$

β: Ratio of the peak values of the short circuit current and the peak value of the reference current

$$\beta = \frac{\hat{I}_L}{\hat{I}_{Lref}}$$

Index L: substation
Index F: current collector

References

1 H. Eisenack, D. Cordes:
Digitale Nachbildung der Vorgänge in Stromrichterschaltungen.
ETZ-A 92, 120 (1971)

2 A. Schöne (Herausgeber):
Simulation technischer Systeme Bd. 2.
Carl Hauser Verlag München Wien (1976)

3 P. Mutschler:
Verfahren zur digitalen Simulation beliebiger Stromrichterschaltungen.
ETZ-A 95, 610 (1974)

4 F. Vogt:
Die Simulation von Stromrichtern.
ETZ-A 94, 479 (1973)

5 H. Eisenack, H. Hofmeister:
Use of Digital Simulation in Designing Power Electronic Circuits.
Siemens Forschung- und Entw. Ber. 3 (1974)

6 E. Fehlberg:
Classical Fifth-, Sixth-, Seventh- and Eight Order Runge-Kutta Formulas with Stepsize Control.
NASA-TR-287, (1968)

7 F. Cellier, D. Rufer:
Algorithm Suited for the Solution of Initial Value Problems in Engineering Applications.
Proceeding of "Simulation 75",160-165

8 M. Depenbrock:
Einphasen-Stromrichter mit sinusförmigem Netzstrom und gut geglätteten Gleichgrössen.
ETZ-A 94, 966 (1973)

9 H. Kehrmann, W. Lienau, R. Nill:
Vierquadrantensteller - eine netzfreundliche Einspeisung für Triebfahrzeuge mit Drehstromantrieb.
El. Bahnen 45 (1974)

10 H. Kehrmann, W. Lienau, R. Nill:
Der Vierquadrantensteller - ein Pulsstromrichter zur Einspeisung von Triebfahrzeugen mit Drehstromantrieb aus dem Einphasennetz.
IFAC-Symposium, Düsseldorf, 7.-9.10.1974 2, 357

11 D. Daum:
Untersuchung eines Einphasenstromrichters mit nahezu sinusförmigem Netzstrom und gut geglätteten Gleichgrössen.
Diss. Ruhr-Universität, Bochum 1974

12 T. Nordin:
Der Weg zu serienmässigen Thyristorschienenfahrzeugen in Schweden.
ZEV-Glasers Annalen 92, 198 (1968)

13 J. Förster:
Zur Stromrichter-Netzbelastung.
ETZ-A 96, (1975)

14 "Versuche Aadorf-Elgg":
Interner Bericht über Beeinflussungsmessungen des Elektrotechnischen Studienbüros der Bauabteilung der Generaldirektion der Schweizerischen Bundesbahnen.
September 1975

15 F. Krittian:
Messung und Berechnung der Fahrleitungsimpedanz elektrifizierter Eisenbahnstrecken.
El. Bahnen 42, 151 (1971)

16 H. Schlunegger:
Untersuchung eines netzrückwirkungsarmen, zwangskommutierten Triebfahrzeug-Stromrichters zur Einspeisung eines Gleichspannungszwischenkreises aus dem Einphasennetz.
Diss. ETH Zürich, 1977

TOLERANCE BAND CONTROLLED SINGLE PHASE CONVERTER CIRCUIT WITH MINIMUM INTERACTIONS BETWEEN CONVERTER AND SUPPLY BY OPTIMUM CONTROL PARALLEL OPERATION

G. Klinger

Lehrstuhl für Erzeugung und Anwendung elektrischer Energie Ruhr-Universität, 4630 Bochum, BRD

ABSTRACT

In this paper a single phase convertor circuit is described in which the a.c. current is kept within a given tolerance band of a sinusoidal current input. With a smoothed d.c. voltage on the secondary side d.c. power can be supplied to a constant voltage convertor for three-phase traction motors. It will be shown that the interactions between convertors and supply can be minimized by the optimum parallel operation of two independent convertors, i.e. by the simultaneous extinguishing and firing of different valve arms. Besides describing the control mechanisms, the operation of the convertor circuits and the results are illustrated by oscillograms. These oscillograms were obtained by simulation carried out on an analog computer combined with on actual control network.

1. INTRODUCTION

In the field of tractions systems three phase motors become more and more common. The three phase motors can be supplied from d.c. intermediate circuits via convertors. In order to achieve this conversion of a. c. power from the single phase traction line to d.c. power is necessary. A conversion with minimum line interaction is possible with the circuit shown in Fig. 1. See also (Ref. 1,2, 3,4)

The transformer has a time variable transmission ratio. Ideally the transmission ratio oscillates at the same frequency $f_\sim$ as the supply and at a constant phase.

$$\ddot{u} = \frac{i_2}{i_1} = \frac{u_1}{u_d} = k \cdot \cos\left(2\pi \cdot f_\sim \cdot t - \delta\right) \quad (1)$$

With the d.c. voltage u_d being constant, the voltage $u_1 = u_d \cdot k \cdot \cos(2\pi f_\sim t - \delta)$ can be determined by choosing certain values for k and δ. With $u_\sim = \hat{u}_\sim \cdot \cos(2\pi f_\sim \cdot t)$

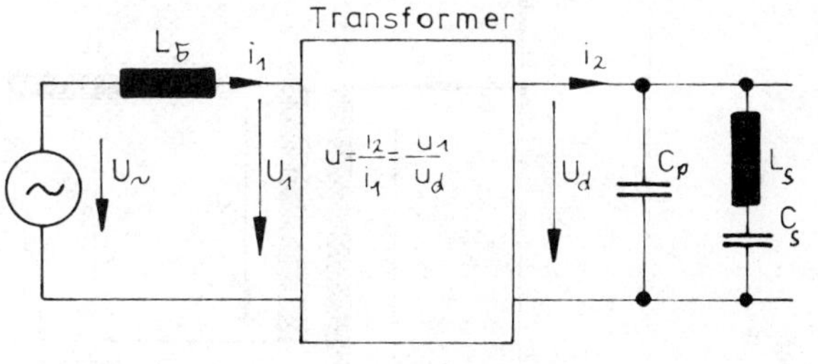

Fig. 1. Convertor arrangement

the line current becomes $i_1 = \hat{i}_1 \cdot \cos(2\pi f \cdot t - \varphi)$. For the secondary current i_2 one obtains $i_2 = \ddot{u} \cdot i_1 = i_d + \hat{i}_s \cdot \sin(4\pi f_\sim \cdot t - \varepsilon)$. The part of i_2 which oscillates at twice the line frequency flows into the tuned series resonant circuit (L_s, C_s).

With u_d being constant, the d. c. component of i_2 becomes the d.c. load current.

The transmission ratio which is assumed to be ideal can only be approximated with convertor circuits using pulse width modulation (p.w.m.) (Ref. 3) describes a method of the tolerance band control of the p.w.m. with optimum control action.

The time variation of the sinusoidal reference quantity of the line current i_1 is derived from the d.c. load current i_d, a sinusoidal quantity at line frequency and the output of the d.c. voltage regulator.

A tolerance band of $\pm\Delta i_{Gmax}$ is formed around this reference quantity. The line current is kept within this tolerance band by using discrete transmission ratios.

2. BASIC TRANSFORMER CIRCUIT

A basic transformer circuit is the four quadrant convertor (4 Q - St) shown in Fig. 2.

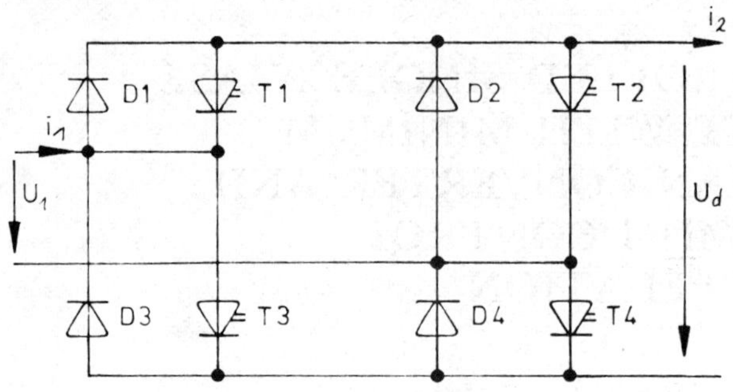

Fig. 2 Four quadrant convertor (4 Q - St)

The valve arms T1 to T4 which can be turned off and the diodes D1 to D4 are anti-parallel. The following current and voltage ratios may be set by firing or extinguishing the valve arms T1 to T4.

$$
\begin{array}{lll}
i_1 = i_2 & u_1 = u_d & ü = +1 \\
i_1 = -i_2 & u_1 = -u_d & ü = -1 \\
i_2 = 0 & u_1 = 0 & ü = 0
\end{array}
\qquad (2)
$$

A harmonic transmission ratio may be achieved only with a small error by using the three possible values ü = +1; 0; -1 of the transmission ratio of the p.w.m.

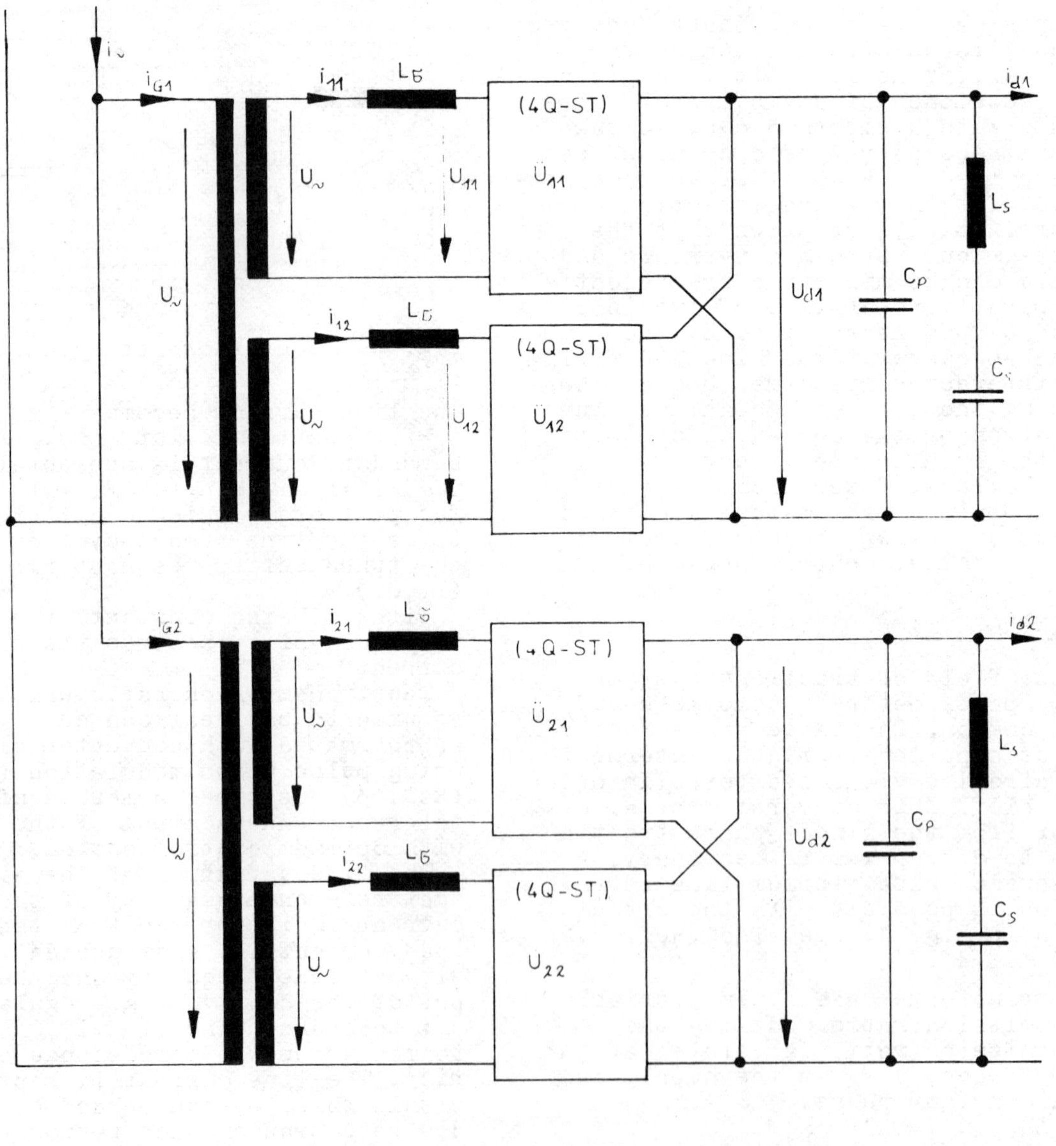

Fig. 3 Main circuit with two inedependent d.c. voltage intermediate circuit

3. MAIN CIRCUIT

Fig. 3 shows the main circuit for two independent d.c. voltage intermediate circuits (2groups).

Each group consists of two four quadrant convertors (4 Q - St) operating parallel on the primary and secondary side.

The currents of the different 4Q-St add and become the group current i_G in the main transformer. The sum of the two group currents form the line current $i_\sim$.

4. REGULATOR AND CONTROL UNIT

4.1 Regulation of a group

Each group consists of two 4Q-St and supplies a d.c. voltage intermediate circuit followed by a d.c. convertor.

A reference quantity i_{GSoll} for the group current is formed in the regulator system of Fig. 4.

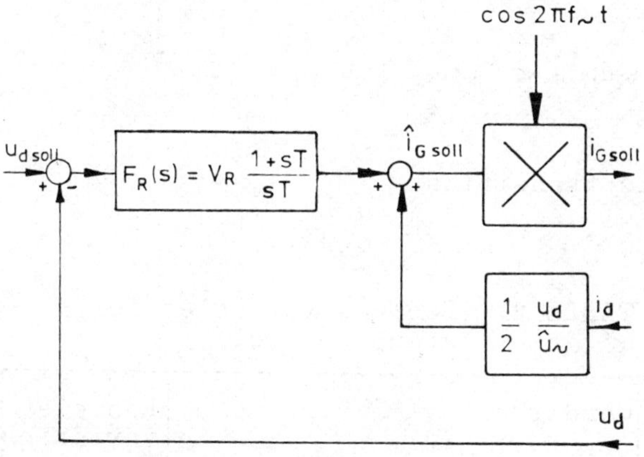

Fig. 4 Regulator system

A tolerance band with the upper limit $l1 = i_{GSoll} + \Delta i_{Gmax}$ and the lower limit $l2 = i_{GSoll} - \Delta i_{Gmax}$ with $\Delta i_{Gmax} \approx (0{,}01 \div 0{,}2)\, i_{Gmax}$ is formed around this reference quantity. The two currents $i_{\nu 1}$ and $i_{\nu 2}$ $(\nu = 1,2)$ of the two 4 Q - St are measured. If the sum $i_{\nu 1} + i_{\nu 2} > l1$ or $i_{\nu 1} + i_{\nu 2} < l2$ the switch position of one of the 4Q-St is changed. The group current i_G then flows back into the tolerance band.

4.2 Control system of a group

Because the 4Q-St may take three different transmission ratios $\ddot{u} = +1;\ 0;\ -1$, $3^2 = 9$ different switch positions for one group are possible. From Fig. 3 one obtains for group I

$$\begin{aligned}
i_{G1} &= i_{11} + i_{12} \\
\frac{d}{dt} i_{11} &= \frac{1}{L_\sigma} \cdot (u_\sim - u_{11}) \\
\frac{d}{dt} i_{12} &= \frac{1}{L_\sigma} (u_\sim - u_{12}) \\
u_{11} &= \ddot{u}_{11} \cdot u_{d1};\quad u_{12} = \ddot{u}_{12} \cdot u_{d1} \qquad (3) \\
\frac{d}{dt} i_{G1} &= \frac{1}{L_\sigma} (2 \cdot u_\sim - (\ddot{u}_{11} + \ddot{u}_{12}) \cdot u_{d1}) \\
\ddot{u}_{ges.} &= \frac{1}{2} (\ddot{u}_{11} + \ddot{u}_{12}) \\
\frac{d}{dt} i_{G1} &= \frac{2}{L_\sigma} (u_\sim - \ddot{u}_{ges} \cdot u_{d1})
\end{aligned}$$

From equation (3), it can be seen that only the total transmission ratio $\ddot{u}_{ges}$ is of significance for the differential $d/dt(i_G)$. For $\ddot{u}_{ges}$, five different values are possible by which the gradient of the group current can be influenced.

In order to obtain a uniform distribution of current from the two 4Q-St the combinations ($\ddot{u}_{11} = +1$; $\ddot{u}_{12} = -1$) and ($\ddot{u}_{11} = -1$; $\ddot{u}_{12} = +1$) of the nine possible switch positions can be eliminated. These two combinations have the same effect on the group current i_G as the combination $\ddot{u}_{11} = \ddot{u}_{12} = 0$. In Fig. 5 the notation for the remaining seven switch positions is in the form Z ($\ddot{u}_{11}$; $\ddot{u}_{12}$).

The transitions from one position to another are called W1 to W18 in the same figure.

The task of the control unit is to switch to the different positions by using the paths W1 to W18 in such a way that the sum of the two 4Q-St currents stay within the tolerance band of i_{GSoll} . The following conditions should be fulfilled as far as possible.

1. The two 4Q-St should both provide the same current.
2. The tolerance band should be narrow in order to keep the harmonic content of the group current small.
3. The amount of switching which has an effect on the supply should be kept small in order to keep the psophometric disturbance current small.

In order to fulfill these conditions the times and paths of transitions from one position to another has to be optimized. In order to be able to choose the best transmission ratios an auxilary voltage

$$u_{1Soll} = u_\sim + L_\sigma \cdot 0{,}5 \cdot \frac{d}{dt} i_{GSoll} \qquad (4)$$

is formed. Since $\Delta i_G = i_{G1Soll} - i_{G1}$ one obtains from equ. (3) and (4).

$$\frac{d}{dt} \Delta i_G = \frac{2}{L} (u_{1Sol} - \ddot{u}_{ges} \cdot u_{d1}) \qquad (5)$$

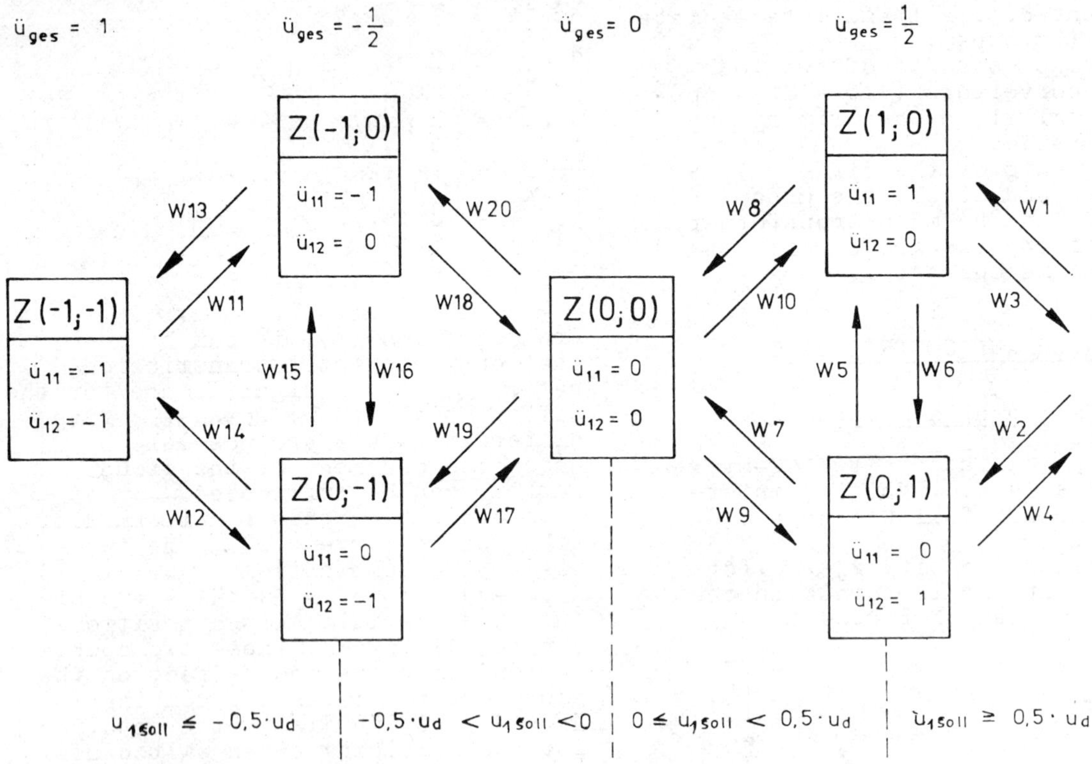

Fig. 5 Switchpositions and their possible transitions

To fulfill the conditions 2) and 3) $|d/dt(\Delta i_G)|$ has to be a minimum. According to the varies transmission ratios the voltage u_{1Soll} can be in four different ranges. $u_{1Soll} \geq 0{,}5 \cdot u_d$; $0{,}5 \cdot u_d > u_{1Soll} \geq 0$; $0 > u_{1Soll} \geq -0{,}5 \cdot u$; $-0{,}5\ u_d > u_{1Soll}$. If the group current reaches the tolerance limit 11 the minimum total transmission ratio is set greater than or equal to u_{1Soll}/u_d $\ddot{u}_{ges} \geq u_{1Soll}/u_d$. When $\ddot{u}_{ges} = \pm 1/2$ the 4Q-St with the more positive current is switched to obtain current symmetry. When the negative tolerance limit is reached similar things happened but with opposite signs.

Beside switching at the tolerance limits four other transitions (W5, W6, W15, W16) take place in order to obtain current symmetry. The latter do not influence the group current.

The Boole's variables which are necessary for optimum control are defined in Table 1. In Table 2 the transition paths W1 to W18 are given as functions of switch position variables defined in Table 1.

TABLE 1. Boole's variables for group control

Condition	Boole's variables
$u_{1Soll} \geq 0$	a
$\lvert u_{1Soll} \rvert \geq 0{,}5\ u_d$	b
$\Delta i_G = i_{GSoll} - i_{11} - i_{12} > \Delta i_{Gmax}$	α
$\Delta i_G < -\Delta i_{Gmax}$	β
$\lvert i_{11} \rvert > \lvert i_{12} \rvert$	f
$\lvert i_{11} \rvert > \lvert 0{,}5\ i_{GSoll} \rvert + \Delta i_{11\,max}$	e
$\lvert i_{12} \rvert > \lvert 0{,}5\ i_{GSoll} \rvert + \Delta i_{12\,max}$	g
failure of one 4Q-St	i
$h = ((A \vee B) \wedge f) \vee (\overline{(A \vee B)} \wedge \overline{f})$	
$\Delta i_{11max} = \Delta i_{12max} \approx (1 \div 2)\,\Delta i_{Gmax}$	

For the control of one group, $c = \alpha_1$ and $d = \beta_1$ (6)

TABLE 2. Possible transitions

Path	Function
W 1	$Z(1,1)\wedge h\wedge((a\wedge d\wedge(b\vee i))\vee\bar{a}\vee(\bar{b}\wedge\bar{i})$
W 2	$Z(1,1)\wedge\bar{h}\wedge((a\wedge d\wedge(b\vee i))\vee\bar{a}\vee(\bar{b}\wedge\bar{i})$
W 3	$Z(1,0)\wedge((a\wedge c\wedge(b\vee i)$
W 4	$Z(0,1)\wedge((a\wedge c\wedge(b\vee i)$
W 5	$Z(0,1)\wedge h\wedge(e\vee g)$
W 6	$Z(1,0)\wedge\bar{h}\wedge(e\vee g)$
W 7	$Z(0,1)\wedge((a\wedge d\wedge(\bar{b}\vee i))\vee\bar{a})$
W 8	$Z(1,0)\wedge((a\wedge d\wedge(\bar{b}\vee i))\vee\bar{a})$
W 9	$Z(0,0)\wedge\bar{h}\wedge((a\wedge c\wedge(\bar{b}\vee i)\vee(a\wedge b\wedge\bar{i}))$
W10	$Z(0,0)\wedge h\wedge((a\wedge c\wedge(\bar{b}\vee i)\vee(a\wedge b\wedge\bar{i}))$
W11	$Z(-1,-1)\wedge h\wedge((\bar{a}\wedge d\wedge(b\vee i))\vee a\vee(\bar{b}\wedge\bar{i}))$
W12	$Z(-1,-1)\wedge\bar{h}\wedge((\bar{a}\wedge d\wedge(b\vee i))\vee a\vee(\bar{b}\wedge\bar{i}))$
W13	$Z(-1,0)\wedge((\bar{a}\wedge d\wedge(b\vee i))$
W14	$Z(0,-1)\wedge(\bar{a}\wedge d\wedge(b\vee i))$
W15	$Z(0,-1)\wedge h\wedge(e\vee g)$
W16	$Z(-1,0)\wedge\bar{h}\wedge(e\vee g)$
W17	$Z(0,-1)\wedge((\bar{a}\wedge c\wedge(\bar{b}\vee i))\vee a)$
W18	$Z(-1,0)\wedge((\bar{a}\wedge c\wedge(\bar{b}\vee i))\vee a)$
W19	$Z(0,0)\wedge\bar{h}\wedge((\bar{a}\wedge d\wedge(\bar{b}\vee i)\vee(\bar{a}\wedge b\wedge\bar{i}))$
W20	$Z(0,0)\wedge h\wedge((\bar{a}\wedge d\wedge(\bar{b}\vee i)\vee(\bar{a}\wedge b\wedge\bar{i}))$

The transmission ratios which are relevant for the line current can be doubled by the controlled parallel operation of two 4Q-St as opposed to the operation of a single 4Q-St.

By choosing the optimum transmission ratios the amount of switching which has an effect on the supply is diminished, the tolerance band being the same. The improvements which are obtained by the parallel operation of two 4Q-St can also be achieved for the parallel operation of two independent groups.

4.3 Master control

The master control influences the group controls in such a way that the variable c_1, d_1 and c_2, d_2 not only contain the variable α_1, β_1 and α_2, β_2 as in equ. (6) but also contain variables of a master tolerance band control. To do this, the master control uses the variable α, β, the switch positions Z ($\ddot{u}_{\nu1}$; $\ddot{u}_{\nu2}$) and the difference between the actual and desired value of the group current Δi_G of each group.

In Table 3 the additional Boole's variables are defined and Table 4 gives the functions for c_1, d_1 and c_2 and d_2.

TABLE 3. Boole's variables for master control

Condition	Boole's variables
$\Delta i_{\sim}=\Delta i_{G1}+\Delta i_{G2} > \Delta i_{\sim max}$	γ
$\Delta i_{\sim} < -\Delta i_{\sim max}$	ψ
$\Delta i_{G1} > \Delta i_{G2}$	ϑ
$(\frac{d}{dt}\Delta i_{G1}>0)\wedge(\frac{d}{dt}\Delta i_{G2}<0)$	Z1
$(\frac{d}{dt}\Delta i_{G1}<0)\wedge(\frac{d}{dt}\Delta i_{G2}>0)$	Z2

TABLE 4. Master control functions

Function
$c_1=\alpha_1\vee(\beta_2\wedge Z1)\vee(\gamma\wedge(Z1\vee(\overline{Z2}\wedge\vartheta)))$
$d_1=\beta_1\vee(\alpha_2\wedge Z2)\vee(\psi\wedge(Z2\vee(\overline{Z1}\wedge\bar{\vartheta})))$
$c_2=\alpha_2\vee(\beta_1\wedge Z2)\vee(\gamma\wedge(Z2\vee(\overline{Z1}\wedge\bar{\vartheta})))$
$d_2=\beta_2\vee(\alpha_1\wedge Z1)\vee(\psi\wedge(Z1\vee(\overline{Z2}\wedge\vartheta)))$

The master tolerance band $\Delta i_{\sim max}$ has to be smaller than Δi_{Gmax}.

Advantages of the group control are also valid for the master control. Because of the master control the power of the two groups is independent of each other while the reactive power in the two groups is not independent. This keeps total reactive power of the supply to a minimum.

5. RESULTS

The oscillograms of an analog computer simulation are shown in Fig. 6 to Fig. 7. The control circuit for the simulation was built with industrial components. The other components were simulated on the analog computer. The dates of the simulated components for one group (Fig. 3) are given by

P	= 3,6 MW	L_σ	= 6,3 mH
$U_{\sim N}$	= 1876 V	u_k	= 33 %
$f_\sim$	= 16 2/3 Hz	L_s	= 3,1 mH
U_{dN}	= 2800 V	C_s	= 7,36 mF
I_{dN}	= 1286 A	C_p	= 11 mF

The distortion $k = \sqrt{\frac{\sum_{\nu=3}^{300} I_\nu^2}{\sum_{\nu=1}^{300} I_\nu^2}}$ and the psophometric disturbance current

$$I_{St} = \frac{1}{p_{(f=800\,Hz)}} \cdot \sqrt{\sum_{\nu=1}^{300} (k_\nu \cdot p_\nu \cdot I_\nu)^2}$$

are computed up to the frequency f = 5 kHz. With $k_\nu = 1$, the factors p_ν are given in VDE 0227/20K/Table 6.

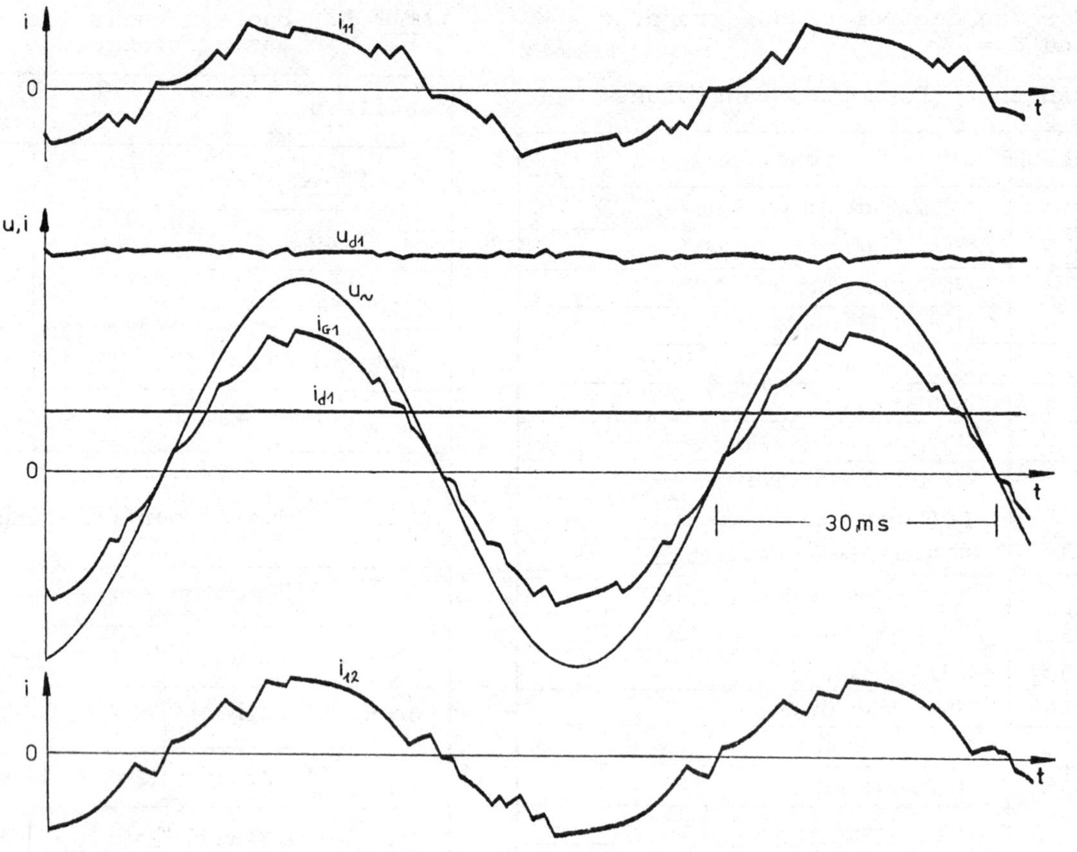

Fig. 6 Operation of one group

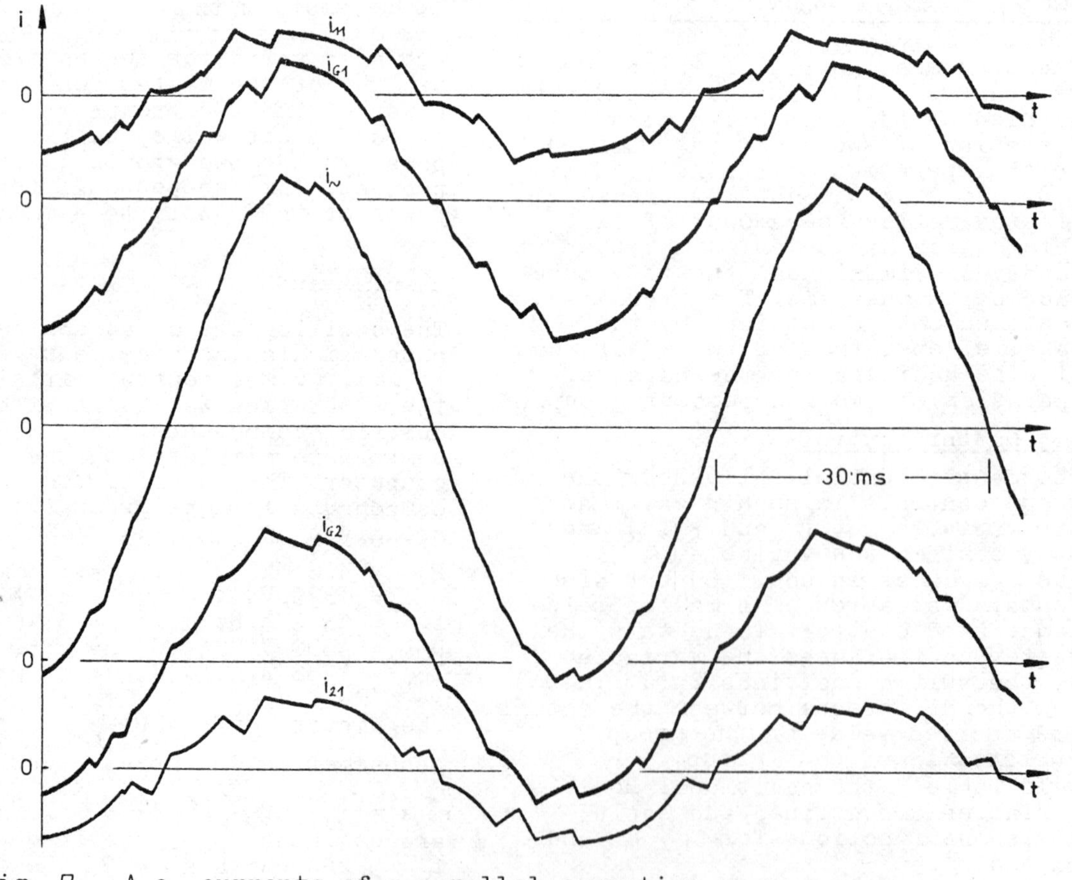

Fig. 7 A.c. currents of a parallel operation

Fig. 6 shows the operation of one group with the smoothed d.c. voltage u_{d1} , d.c. current i_{d1} , the line voltage $u_{\sim}$ and the a.c. currents of the two 4 Q-St i_{11} and i_{12} which add and become the group current i_{G1}. The amount of switching which has an effect on the supply is 50 % of the amount of switching the two 4 Q-St. The distortion k and the relative psophometric disturbance current $i_{st} = I_{st}/I_N$ for the group current i_G

are $k \simeq 4\%$ $\qquad i_{st} \simeq 6‰$

Fig. 7 shows the currents for parallel operation of the two groups under normal operating conditions as in Fig. 3. For each group the current i_{r1} of one 4 Q-St and the group current i_{Gr} is also shown as the sum of the two group currents, i.e. the line current $i_{\sim}$. The master control reduces the amount of switching which has an effect on the supply by 50 % as compared with the sum of switching the two group currents. The relative improvement can be seen from $k \simeq 2\%$ $\qquad i_{st} \simeq 4‰$

Fig. 6 and Fig. 7 show that it is possible to control the four 4 Q-St in such a way that the line current is nearly sinusoidal with a minimum of interactions between convertors and feeding line. The currents of the 4 Q-St may not be sinusoidal but the sum of the four 4 Q-St currents, the line current $i_{\sim}$ is nearly sinusoidal with the amount of switching beeing a minimum.

REFERENCES

/1/ M. Depenbrock, Einphasen-Stromrichter mit sinusförmigem Netzstrom und gut geglätteten Gleichgrößen. ETZ-A Bd. 94 (1973) S. 466 - 471

/2/ D. Daum, Untersuchung eines Einphasen-Stromrichters mit nahezu sinusförmigem Netzstrom und gut geglätteten Gleichgrößen. Diss. Ruhr-Universität Bochum 1974

/3/ G. Klinger, Toleranzbandgeregelte Einphasen-Stromrichterschaltung mit optimaler Stellgrößenauswahl. ETZ-A Bd. 97 (1976) S. 87 - 90

/4/ H. Kehrmann; W. Lienau; R. Nill, Vierquadrantensteller - eine netzfreundliche Einspeisung für Triebfahrzeuge mit Drehstromantrieb. Eletr. Bahnen Bd. 45 (1974) S. 135 - 142

/5/ G. Klinger, Regelungstechnische Untersuchung einer Einphasen -Stromrichterschaltung mit harmonisch zeitvariantem Übertrager. IFAC Symposium in Düsseldorf vom 7. - 9.10.74 Bd. 1 S. 443 - 456

/6/ D. Daum, Steuerung von Einphasenstromrichter mit Pulsbreitenmodulation. IFAC Symposium in Düsseldorf vom 7. - 9. 10. 74 Bd. 1 S. 403 - 415

THREE-PHASE A.C. MOTOR VEHICLES FROM D.C POWER SUPPLY FEED-IN CIRCUIT AND INTERFERENCE CURRENT REDUCTION

G. R. Nill and W. Daumann

Brown, Boveri & Cie AG, Mannheim

Summary

Asynchronous motors of traction vehicles are preferably fed by pulse-mode operated inverters with a constant d.c. input voltage. Also in the case of a contactwire with a d.c. supply voltage an input-converter is favourable.

An advantageous circuit of the input converter is described, which transform the line voltage to a higher value of the d.c. link voltage (step-up-chopper). It consists of two choppers with phase displaced chopping. They have a common commutation-circuit. In the case of dynamic braking the power-conversion in the resistance is controlled by the chopper. The chopper takes from the line a rippeled d.c. current. Its ripple-components could disturb the signalizing systems and communication-circuits. This a.c. components are caused mainly by the chopper, and in a smaler part by the inverter and the interference of both. The mode of generation of these harmonic components in the case of constant chopping-ratio will be described. In addition the a.c. components are influenced by the ripple of the line voltage and the converter control system.

The harmonics can be reduced by special filtering circuits. In case of very low frequencies this would result in large filter elements. A reduction of the currents by a special control -circuit will be described.

1. Introduction

The propulsion of motor vehicles with three-phase induction motors offers major advantages, particularly with respect to

- low maintenance requirements due to the absence of the commutator in the traction motor
- light-weight bogies due to the low weight of the traction motors, reducing the rail wear caused by high speed vehicles
- favourable traction properties owing to the good controllability of the converter.

These advantages have already been discussed in detail in several publications (Ref. 1, 2, 3, 4).

The three-phase induction motor is supplied from an inverter which generates a three-phase a.c. voltage with variable amplitude and frequency. A favourable type of inverter for this application is the pulse inverter connected to a d.c. voltage link (Ref. 5, 6, 7).

The technical expenditure of the inverter is the lower the closer are the limits within which the static and dynamic voltage variations in the d.c. link can be maintained since both the maximum voltage supplied to the motors and the maximum inverter output current that can be commutated are directly dependent on the voltage of the d.c. link. Although it is in principle possible to supply the inverter directly or via a filter from the d.c. contact wire, it is more convenient to connect a feeding converter with filter elements ahead of the inverter. The overall expenditure of this arrangement is in most cases more favourable than an appropriately rated inverter. It offers the additional advantage that a d.c.link voltage most suitable for the converter and motor ratings can be chosen.

The feeding converter must meet the following requirements:

- It must convert the widely varying line voltage with a high harmonic content into a constant direct voltage with a low harmonic content

- The current drawn from the line must have a low harmonic content to avoid disturbing of telecommunication and signalling equipment
- Since the line voltage already contains harmonics, the input reactance of the vehicle must not fall below predetermined limits for definit frequencies
- The efficiency of the plant should be high.

It goes without saying that the feed-in circuit and the inverter should be as light, small and inexpensive as possible, in addition to being reliable in operation and requiring little maintenance.

2. Design and operating principle of the feeding converter

Figur 1 shows the schematic diagram of feeding converter particularly suited for this application. The converter is a d.c. chopper for stepping up the voltage. Choke L is connected between the d.c. contact wire and the converter. The converter consists of an electronic switch, shown as thyristor St with provision for turn-on and turn-off, in the cross leg and a decoupling diode D, decoupling the d.c. link form the input terminals. Capacitor C_d in the d.c. link simultaneously serves as smoothing capacitor for the inverter. Fig. 2 shows the

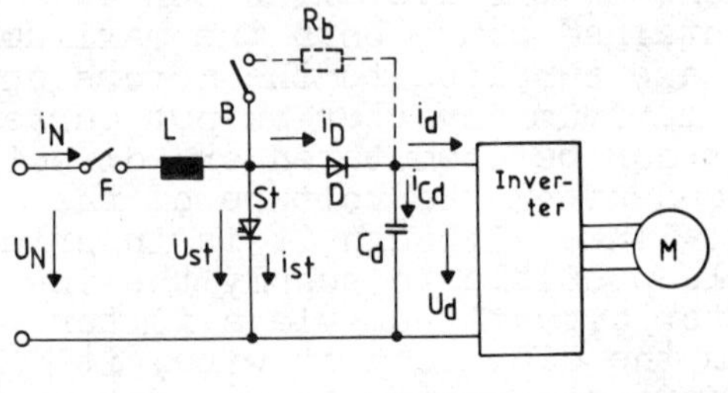

Fig.1: schematic diagram of the 1q-s

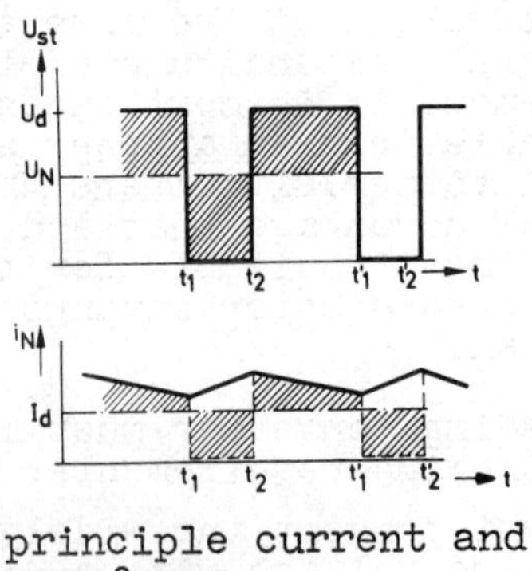

Fig.2: principle current and voltage waveforms

current and voltage waveforms at the d.c. chopper input. Thyristor St is conducting in the time interval from t_1 to t_2. The line voltage is available across choke L, and the current is increasing.

At instant t_2, the thyristor is turned off by a turn-off circuit not shown. The line current continues to flow into the d.c. link via diode D driven by choke L. Available across the choke is the difference between the line voltage U_N and the voltage of the d.c. link U_d. Since, in the steady state, the d.c. link voltage is higher than the line voltage, the line current is reduced during this period. When the diode is conducting, the d.c. link voltage is available at the d.c. chopper input. At instant t_1 the thyristor is fired again and the cycle of operations is repeated. The input voltage to the d.c. chopper consists of square-wave pulses of the d.c. link voltage, the current in the d.c. link consists of pulses of the line current. Thus, for low frequency the feeding converter works like a transformer with the transformation ratio

$$K = \frac{t_1' - t_2}{t_1' - t_1}$$

In case of a realizable choke L the input current consists of a d.c. component with superposed triangular harmonics. Since only a pure alternating voltage can appear across the choke, it is essential to preset the turn-on and turn-off instants of the d.c. chopper in such a way that, in static operation, the mean value of the d.c. chopper input voltage becomes equal to the line voltage. Furthermore, the d.c. component of the line current must be controlled in such a manner that the mean value of the current flowing over the diode into the d.c. link during the period between t_2 and t_1' is equal to the load current I_d taken up by the inverter. Consequently, the time integrals of the choke voltage and of the capacitor current must be identical in both switching states, as shown in Fig. 2 by the hatching.

The current and voltage at the d.c. chopper input always have the direction indicated in Fig. 1. For this reason, the converter was called one-quadrant chopper (1q-S).

In braking operation, the 1q-S can be used to adjust the power converted in the braking resistor. In Fig. 1, the connection for braking operation is indicated by broken lines. The supply system is disconnected by way of switch F and the braking resistor is connected in parallel with the decoupling diode by switch B. When the chopper is switched on, a current equal to the d.c. link voltage, divided by the value of the braking resistor, flows away from the d.c. link. When the chopper is blocked, the current becomes zero. By the controller, the on-period of the chopper is predetermined in such a way that the mean value of the current flowing through the resistor is equal to the current supplied by the inverter.

As mentioned in the introduction the harmonics in the line current must not be allowed to assume excessive values. Particularly critical are harmonic currents with the low frequencies of the signalling circuits (e.g. 75 Hz). The limit values are considerably below 1 percent of the direct current. Harmonics in the frequency range of 500 to 2000 Hz are essential to the interference on telephone transmissions. The harmonic current produced by the 1q-S are dependent on the control ratio, i.e. the ratio of d.c. link voltage to line voltage, inversely proportional to the pulse frequency and the inductance L. However, it would be too ineffective to reduce the harmonic currents to the required values only by increasing the choke and the pulse frequency.

A more convenient method is adding a choke in serie and a small capacitor in parallel to the input tabs of the 1q-S. A further reduction of the harmonics is achieved by splitting the 1q-S into several smaller choppers which a connected in parallel via separate chokes. The choppers are pulsed displaced in phase, so that the harmonic currents of the individual choppers partly cancel each other. In section 3, the problem of harmonic currents on the supply system is discussed in greater detail.

Fig. 3 shows a circuit in which two individual choppers have a common commuating circuit. The extended input filter comprising the serie choke L_N and the filter capacitor C_1 is already indicated. Chopper 1

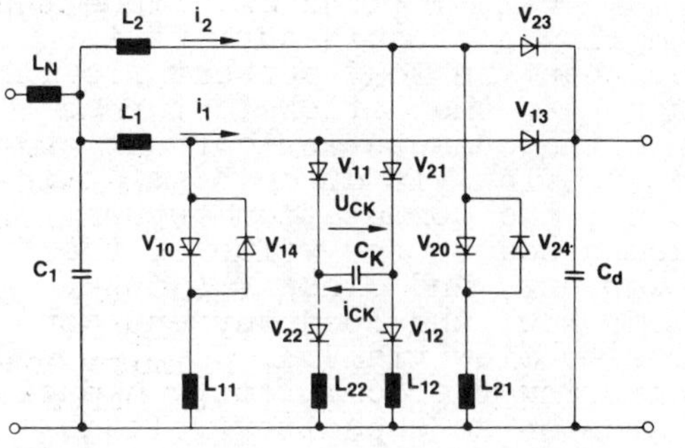

Fig. 3: circuit of the 2 phase 1q-S

consists of input choke L_1, the main thyristor V_{10} with the antiparallel-connected swing-back diode V_{14}, the decoupling diode V_{13} to the d.c. link, the turn-off thyristors V_{11} and V_{12}, and the split-up commutating chokes L_{11} and L_{12}. The elements of chopper 2 are marked accordingly (V_{20}, V_{24}, V_{23}, V_{21}, V_{22}, L_{21}, L_{22}). The commutating capacitor C_k is common for both choppers.

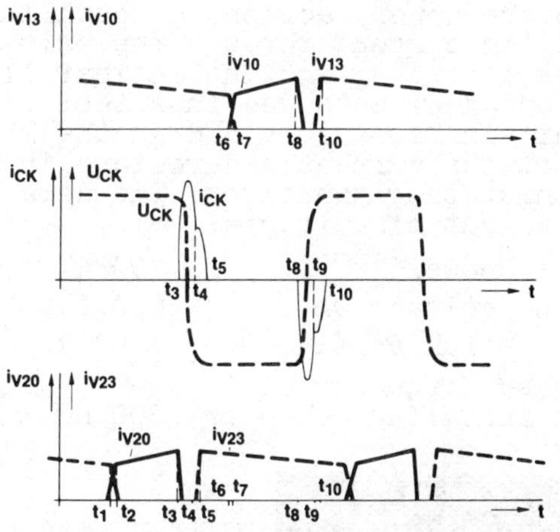

Fig. 4: current and voltage waveforms of the 2 phase 1q-S

The operating principle of the circuit is illustrated by Fig. 4. It is assumed that at instant t_o both main thyristors V_{10} and V_{20} are blocked. Each choke L_1 and L_2 forces about half of the line current into the d.c. link via diodes V_{13} and V_{23}. Voltage U_{ck} of the commutating capacitor C_k is assumed to have the polority shown. At instant t_1, thyristor V_{20} is fired. The load current commutates from diode V_{23} to thyristor V_{20}. At instant t_3, turn-off thyristor V_{21} and V_{22} are turned on causing a sinusoidal current pulse to flow through V_{22},

L_{22}, L_{21}, V_{20}, V_{21} and C_k. Since the the amplitude of this current is larger than the load current flowing through V_{20}, the latter is turned off and the commutation current continues to flow via diode V_{24}. At instant t_4, the commutation current has decreased to the value of the load current, and diode V_{24} is blocked. Now, the load current continures to flow via V_{21}, C_k, V_{22} and L_{22} charging the commutating capacitor. As soon as capacitor voltage U_{ck} has reached the value of the d.c. link voltage, diode V_{23} starts conducting, taking over the load current from the commutating circuit. In consequence of energy stored in choke L_{22} the commutating capacitor is charged up to a voltage which is higher than that of the d.c. link. The capacitor now has the voltage of the magnitude and polarity necessary to turn-off the other main thyristor V_{10}. This can be turned on at any desired instant t6 providing that it must not be turned off before instant t_5. Commutation procedure of thyristor V_{10} is effected analogously in the interval from t_8 to t_{10}. Fig.5 shows, in the upper section, an oscillogram of the current through the main thyristor V_{10} and the antiparallel diode V_{14}, measured in a test circuit in laboratory and in the lower part the current and voltage in the commutating capacitor. The data of the test circuit are:

max. power	150 kW
line voltage U_N	600...900 V
d.c. link voltage U_d	1100 V
pulse frequency of an individual chopper	800 Hz

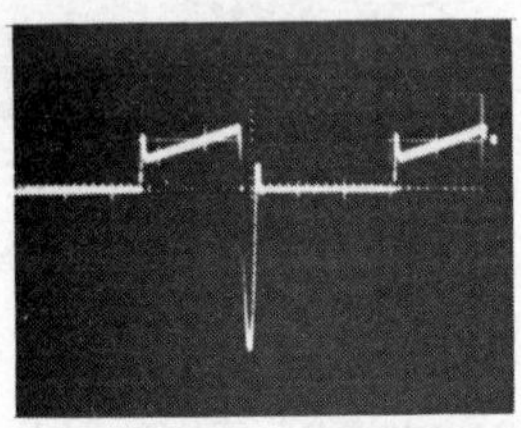

Current in V_{10} (positiv) and V_{14} (negativ)
Scale : 100A/cm
200 µs/cm

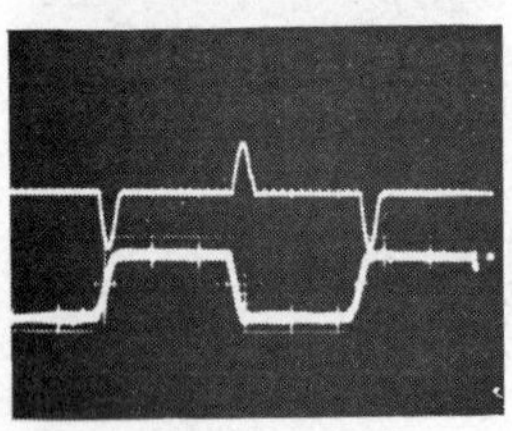

Current i_{ck}
Scale : 500A/cm
Voltage : U_{ck}
Scale : 2000 V/cm
200 µs/cm

Fig.5: current and voltage waveforms measured in the 2 phase 1q-S

3. Effects of the input converter on the supply system

The harmonic current flowing in the supply system consists of two components:

a) the line voltage contains harmonics which are mainly caused by the three-phase rectifier bridges in the substations. The voltage with maximum amplitude has a frequency of six times that of the supplying three-phase a.c. system. In addition to this, harmonics can also be provoked by unsymmetries and resonance phenomena in the supply system and harmonic currents provoked by other vehicles. These harmonic voltage cause a harmonic current through the reactance of the line and the filter circuit.

b) As mentioned in Section 2, a square-wave voltage with amplitude of the d.c. link voltage appears at the input of the 1q-S. At constant d.c. link voltage U_d and constant transforming ratio K, the following equations apply for their Fourrier components:

$$u_{ST} = K \cdot U_d + \sum_{\nu=1}^{\infty} C_\nu \sin(\nu \omega_T + \varphi_\nu) \quad (1)$$

$$\text{with } C_\nu = \frac{U_d}{\nu \cdot \pi} \sqrt{2 - 2 \cdot \cos \nu k 2 \pi} \quad (2)$$

$$\text{and } \tan\varphi_\nu = \frac{\sin \nu k \cdot 2 \cdot \pi}{1 - \cos \nu \cdot k \cdot 2 \cdot \pi} \quad (3)$$

Choppers with the same voltage ratio, which are pulsed displaced in phase, have harmonics with the same amplitude, but an additional phase displacement of

$$\psi = \frac{2 \cdot \pi \cdot \nu}{n} \quad (4)$$

where n is the number of the successeively pulsed choppers, the ordinal number of the harmonic and ω_T the cyclic pulse frequency of a d.c. chopper.

The harmonic current drawn from the supply system is calculated from the harmonic components of the voltage and from the reactances fo the supply system and the filter determined at the same frequency. For a first assessment of the harmonic behaviour, a simplified equivalent circuit diagram as shown in Fig. 6 will do. For calculation of the harmonics caused by the choppers, the line is consireded at the substation. The harmonics of the chopper input voltage ar fed into the output taps of the equivalent circuit. For calculating the input reactance in the frequency range below the chopper pulse frequency with the chopper working, the transformed capacity of the smoothing capacitor C_d is connected to the output taps of the equivalent circuit. An extension of the equivalent circuit diagram to cover more complicated supply system conditions, discontinuous operation of the choppers and sophisticated filter circuits is possible, but we will not be considered here.

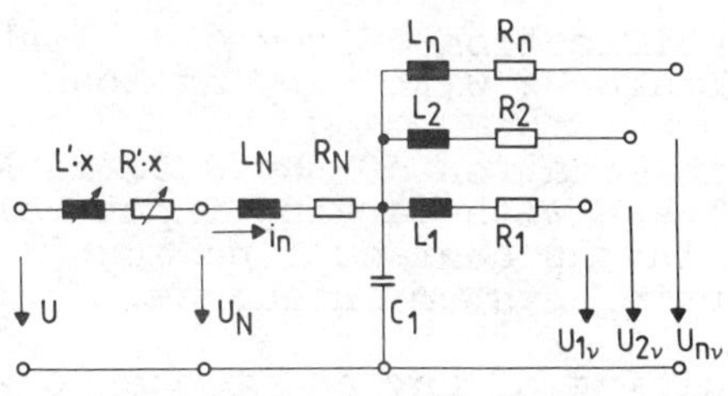

Fig.6: equivalent circuit diagram for calculation of harmonic currents

Fig.7 shows in the upper section, for arbiterarily chosen filter parameters, the absolute value of the input reactance the chopper beeing in operation and blocked and in the lower section the line current, referred to the converter input voltage, as a function of the harmonic frequency.

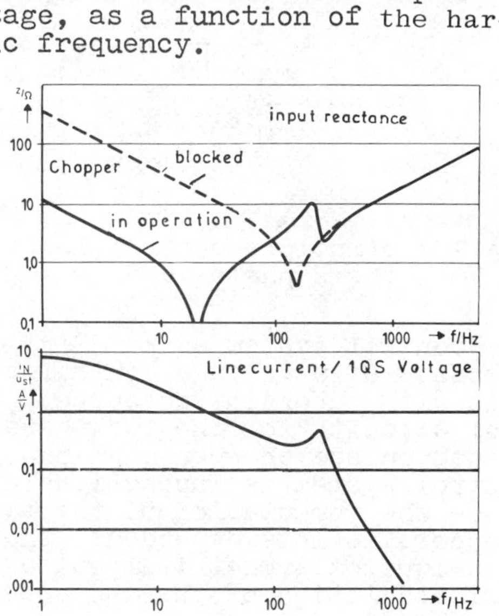

Fig.7: Input reactance and line current as a function of the frequency

For determining the higher-frequency harmonic current, which are mainly responsible for the magnitude of the posphometrically evaluated interference current, the preceding method of calculation is fully sufficient.

For the calculation of the lowfrequency currents having signalling frequency, also the interaction with the motor inverter must be taken into consideration.

In calculating the harmonic voltage of the chopper, a constant d.c.link voltage was so far taken as a basis. However, since for signal frequencies even very small harmonic currents of several hundred milliamps may be critical, it is convenient to consider also the case of the rippled d.c. link voltage.

As described in (8), the pulse inverter operates like an assembly of three change-over switches, operated with motor or pulse frequency, which apply the motor terminals to the positive or negative pole of the d.c. link voltage. Hence, the motor voltage consists of blocks of the d.c. link voltage. The resultant 3 motor currents will flow through the positive or negative pole of the d.c. link, depending on the switch positions. The sum of the currents drawn by the inverter form the d.c. link consists of a d.c. component corresponding to the active power of the motor, and of harmonics flowing over the smoothing capacitor. The frequencies and amplitudes of these harmonics are dependent on the motor current, voltage and frequency, as well as on the pulsing method of the inverter. These current harmonics now give rise to harmonics of the d.c. link voltage.

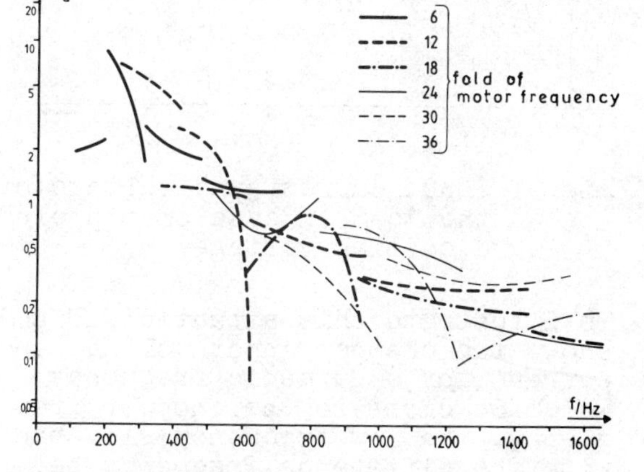

Fig.: harmonics of the d.c. link voltage caused by a pulse inverter

Fig.8 shows, for a special inverter installation, the effective value, referred to the d.c. link voltage, of the harmonic voltage for a tractive effortspeed characteristic of the drive. The harmonics with the same multiple of the motor frequency are taken together and represented as characteristics. When selecting the d.c. link capacitor C_d as usual, their amplitudes are mostly far below 1% of the d.c. link voltage. For one of the harmonics of the d.c. link voltage the following equation applies:

$$u_{d\mu} = \hat{u}_{d\mu} \cdot \sin \omega_{d\mu} \cdot t \qquad (5)$$

The effect of this harmonic on the chopper input voltage can be obtained by substituting equation (5) in (1):

$$U_{St} \approx k(U_d + \hat{u}_{d\mu} \cdot \sin(\omega_{d\mu} \cdot t) + \sum_{\nu=1}^{\infty} C_\nu \sin(\nu\omega_T t + \varphi_\nu) \qquad (6)$$
$$+ \sum_{\nu=1}^{\infty} C_{\nu\mu} \cdot \{\cos[(\nu\omega_T - \omega_{d\mu}) \cdot t + \gamma_{\nu\mu}] - \cos[(\nu\omega_T + \omega_{d\mu}) \cdot t + \partial_{\nu\mu}]\}$$

In addition to the d.c. component and the harmonics having the pulse-frequency of the chopper, also components with sum and difference frequency and with the harmonic frequency f_μ in the d.c.link appear in the input voltage to the 1q-S.

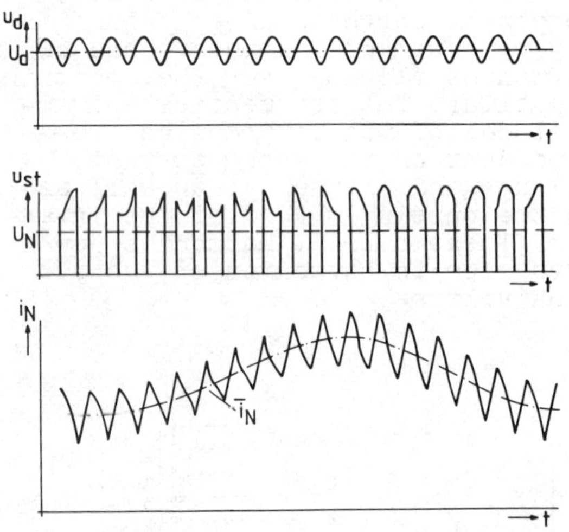

Fig.9: input current and voltage of the 1q-S in case of rippeled d.c. link voltage

To illustrate this situation, Fig.9 shows the chopper input voltage and current for a harmonic frequency which is close to the chopper frequency. The low-frequency side-band with the difference frequency between the chopper pulse frequency and the d.c.link frequency can be clearly seen. For clarity, the amplitude of the harmonic of the d.c. link voltage has been exaggerated. In fact, all low-frequency harmonics appearing in the line current are in the range below 1‰, referred to the d.c. component, for constant chopper firing angles already. A further reduction can be achieved by appropriate control measures.

4. Control

The control of the 1q-S converter must perform several tasks which include:

a) stabilization of the d.c. link voltage U_d with varying load.

c) Stabilization of the d.c. link voltage with varying input voltage in the contact line and varying system inpedance.

d) Avoiding of any additional excitations in the supply system input current.

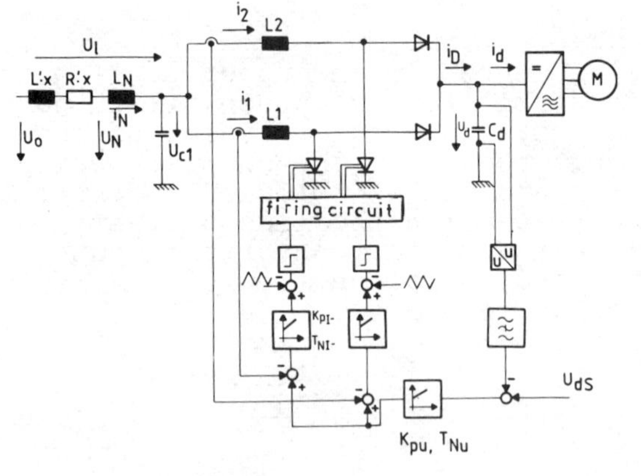

Fig.10: principle control loop of the 2 phase 1q-S

The control system as per Fig. 10 consists of a voltage control circuit with subordinated current control circuits for the two phase displaced pulsed choppers. An on-off control system is impracticable since the commutation of the two choppers is interdependent. The technique of symmetrical pulse width modulation of the choppers is here used.

The 1q-S is designed to increase the voltage. In control actions using frequencies well below the pulse frequency, the converter can be regarded as a transformer with the transformation ratio $K = U_{C1}/U_d$.

The output of the voltage controller provides a quantity which is proportional to the reference value of the supply current. The two current control circuits consist of the two PI current controllers with the controller characteristics $K_{PI1,2}$, $T_{NI1,2}$. The output voltages of the current controllers correspond to the normalized transformation ratio K. In the comparators, these output voltages are compared with the phase-displaced triangular voltage of the associated generators. The resultant voltage ist used in the gate control unit for forming the firing pulses for the main and turn-off thyristors. At the same time, the correct sequence and the timing of the turn-off pulses is supervised to ensure proper turning-off of both choppers even in the case of fast dynamic changes in the firing-angle.

The controlled system consists of the line-side circuit, i.e. the converter inductors L_1 and L_2, the input filters L_N nd C_1, and the d.c. system with the attenuation line constants per unit L' and R', as well as of the 1q-S and the d.c. link filter capacitor C_d.

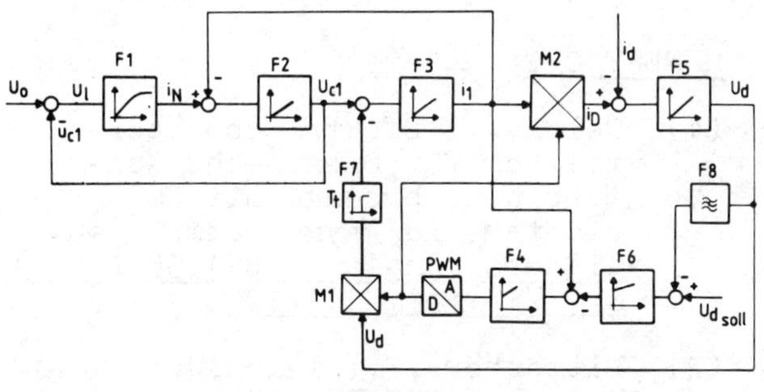

Fig.11: equivalent circuit of the control loop

Fig.11 shows the equivalent circuit diagram of a 1q-S on the d.c.system with special reference to the control aspect.

Here, F_1 is the frequency response of the input filter choke with connected d.c. system of line length X. The following applies:

$$u_L = L_N \frac{di_N}{dt} + X\left(L' \cdot \frac{di_N}{dt} + R' i_N\right) \qquad (1)$$

$$F(j\omega) = \frac{i_N(j\omega)}{u_c(j\omega)} = \frac{U_{Max}/I_{Max}}{X \cdot R'\left(1 + j\omega \frac{L'}{R'}\left(1 + \frac{L_N}{L'} \cdot \frac{1}{X}\right)\right)} \qquad (2)$$

The quantities are normalized to the maximum values of current and voltage occurring. (U_{max} = 4 kV, I_{max} = 1 kA, L' = 2 mH/km, R' = 60 mΩ/km)

$F_1(j\omega)$ has the frequency response shown in Fig.12 (paramenter X = line length per unit)

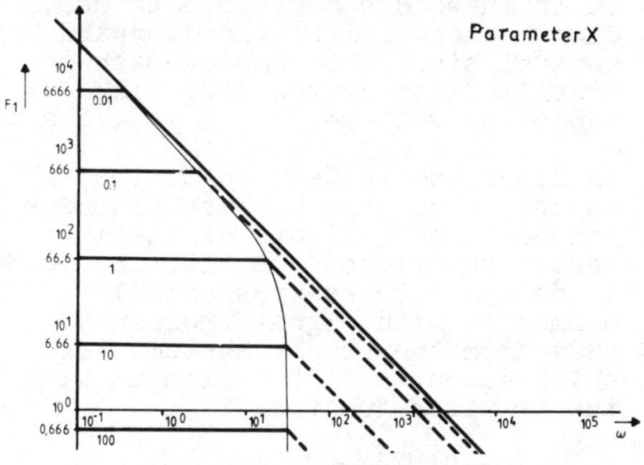

Fig.12: frequency response of line reactance L'· x and R'· x and filter choke L_N

Filter capacitor C_1 essentially froms an integral component with frequency response F_2.

$$\frac{U_{c1}(j\omega)}{i_{c1}(j\omega)} = \frac{1}{j\omega C_1 U_{Max}/I_{Max}} = \frac{1}{j\omega T_1} \qquad (3)$$

Choke L_1 or L_2 relevant for the chopper has the frequency response $F = 1/(R_3/4\,\Omega + j\omega \cdot T_3)$ in the equivalent circuit diagram.This is essentially an integral component when neglecting the resistive losses. The change in the inductane isstill very small with the currents considered and can therefore be neglected. The commutation of the converter causes a dead time. Related to the time constant of the converter choke, the dead time roughly has the magnitude $T_d \approx 0.1 \cdot T_3$.

The current controller consists of a PI controller. In the case on well-damped current control circuits ($D \geqslant 0.5$), the influence of the pulse width modulator can be neglected (Ref.9). From the equivalent circuit diagram of the current control loop,adjusting instructions are derived for the control characteristics K_{PI}, T_{NI} which are obtained with linear control circuit elements but can here be adopted without any appreciable modifications (Ref.10).

This is particularly true of the frequency range $0 \leqslant f \leqslant 300$ Hz considered here.

The voltage controller is also a PI controller with the frequency response $F_6 = K_{pu} (1 + 1/T_{NU})$. Its action is slower than that of the current controller and it cannot follow fast variations in the power drawn from the d.c. link.

It is therefore advisable to use disturbance-variable feedforward control since this improves the stabilisation of the d.c. link capacitor voltage.

As described in Section 3, the capacitor voltage U_d has harmonics caused by the effects of the inverter superimposed on it. The 1q-S is designed to keep especially harmonics with signal frequency away from the supply system. For this reason, a filter element with the transfer function $F_{(p)}$

$$F(p) = \frac{p^2 + p\omega_z/Q_z + \omega_z^2}{p^2 + p\omega_p/Q_p + \omega_p^2} \quad (6)$$

is connected behind the device measuring the actual value of the capacitor voltage. Components with signal frequency f_o are removed from the capacitor voltage U_d. The attenuation for components with signal frequency $f_o = f_z = f_p$ is about 40 to 43 db. These superimposed frequency components are practically eliminated at the output of the voltage controller. As shown in Section 3, at a constant firing angle, components with the frequency of the d.c. link harmonics appear at the converter input. Since the current reference value does not contain any components with signal frequency and since the time constant of the current control is faster than the period of the signal frequency by at least a factor of ten, the disturbing components of the converter input voltage are eliminated by displacing the turn-off instants.

5. Future Trends

After the 1q-S having proven the expected results in te circuit, the next step will be its testing in practical railway service. Therefore a test vehicle has been constructed in cooperation with the NS (Netherlans State Railways). In order to get an equivalent model to a 4400 KW locomotive projected for the future, the asyncronous motor and inverter have aproximatly the data of that of one axle of the future locomotive. The input filter and 1q-S are dimensioned in such a manner that the harmonics of the line current are equal to that of the whole future locomotive. The test vehicle consists of the frame of a DE 2500 type diesel-electric locomotive with only one traction motor.

The principal ratings of the test vehicle are:

max. power	1400 kW
DC line voltage	1500 V (1050-1800V)
DC link voltage	2800 V
max. speed	140 km/h
low speed tractive force	91 KN

The test unit will start working in summer 1977. Besides testing its traction properties there will be measured mainly its behaviour at the dc-line and the harmonic current drawn from the source.

REFERENCES

(1) Teich,W.: Elektrische Übertragung für Diesel-und Gasturbinenfahrzeuge mit Umrichter-und Asynchronmotoren. Glasers Annalen (ZEV) 95 (1971) Nr. 718, pp. 181-190

(2) Bitterberg,F.; Teich,W.: Herschel-BBC-DE 2500 - Ein Wendepunkt in der Lokomotivtechnik. ETR 20 (1971) H. 11 pp. 441-459

(3) Wolters,H.: Die Drehstromtechnik, neue Möglichkeiten für die elektrische Zugförderung. Die Bundesbahn 11/1975

(4) Brenneisen,J.; Futterlieb,E.; Müller,Ed.; Schulz,M.: A new Converter Drive System for a Diesel Electric Locomotive with Asynchronous Traction Motors. IEEE-Transactions on Industry Applications Aug.1973

(5) Körber,J.: Grundlegende Gesichtspunkte für die Auslegung elektrischer Triebfahrzeuge mit asynchronen Fahrmotoren. Elektrische Bahnen 45 (1974) H. 3, pp. 52-59

(6) Körber,J.: Die elektrische Lokomotive mit kommutatorlosen Fahrmotoren. ETR 24 (1975) H. 7/8, pp. 251-259

(7) Bohli,U.; Brechbühler,M.: Erfahrungen mit der Versuchslokomotive Be 4/4 12001 der Schweizerischen Bundesbahnen mit statischen Umrichtern. BBC-Mitteilungen 1973 H. 12, pp. 581-588

(8) Müller,E.; Ricke,F.: Die Auswirkungen verschiedener Steuertechniken des Unterschwingungsverfahrens auf die Wechselrichter-Ausgangsspannung. BBC-Mitteilungen 1973 H. 1, pp. 35-44

(9) Leonard,W.: Diskrete Regelsysteme. Bibliographisches Institut Mannheim, pp.205-220

(10) Pressler,G.: Regelungstechnik Bibliographisches Institut Mannheim, pp. 146-151

STEADY-STATE BEHAVIOUR OF INDUCTION MACHINES FED BY A FREQUENCY CONVERTER WHICH IS SUPPLIED BY A DC-LINK WITH AN AC-COMPONENT

W. Kuhn

Brown, Boveri & Cie., Mannheim, Germany

1. INTRODUCTION

Three-phase drives consisting of a self-commutated inverter and a frequency controlled squirrel-cage motor are gaining more and more significance in the electric traction. As Fig. 1 shows, it is possible, as a general principle, to supply the drive machine from a frequency convertor using either a constant voltage or a constant current d.c. link. The d.c. link supplied from an alternating voltage supply by means of an input rectifier, the level of control ß of which is generally set by a voltage regulator in the case of a convertor with a constant voltage d.c. link, and by a current regulator in the case of a frequency convertor with a constant current d.c. link. These regulators attenuate the voltage fluctuations and the voltage drop. The filter has two functions. Firstly, it has be the function together with the regulator, of reducing the internal resistance during dynamic operation for the direct voltage d.c. link case, or of holding the current constant for the direct current d.c.link case. Secondly, the filter reduces the voltage harmonics, which are caused by the input rectification, to such an extent that the voltage and current at the input to the inverter are constant. The 33,3 Hz power pulses of double mains frequency, which occur when the input supply is taken from a 16 2/3 Hz single-phase mains, is kept away from the three-phase drive which is optimally operated with a constant power input. For filter structures corresponding to Fig.1 these demand necessitate a very large supporting capacitance C_d for a constant voltage d.c. link, and a very large series inductance L_d for a constant current d.c. circuit. However, due to restrictions on volume and on weight, a limit has to be set to the increase in size of the filter elements.

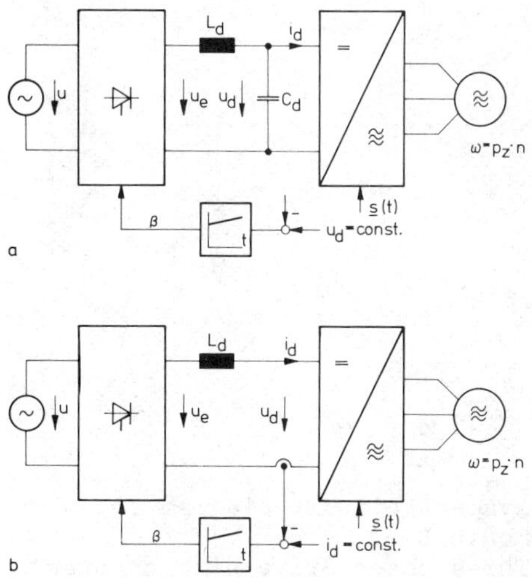

Fig.1 Three-phase drive with input supply from single phase mains
a) with constant voltage d.c.link
b) with constant current d.c.link

Consequently, a.c. quantities appear at the input to the inverter in addition to the d.c. quantities.

$$u_d(t) = U_0 + \sum_{L=1}^{\infty} \hat{u}_{dL} \cdot \cos\{L \cdot \Omega t + \varphi_L\} \tag{1.1}$$

$$i_d(t) = I_0 + \sum_{L=1}^{\infty} \hat{i}_{dL} \cdot \cos\{L \cdot \Omega t + \varphi_L + \alpha_L\} \tag{1.2}$$

Because of these a.c. quantities, additional polyphase current or voltage systems are developed on the output side of the inverter. These rotate with the interference frequencies of the harmonics in the d.c. link, and with the pulse timing of the inverter, and at certain drive speeds reach amplitudes which are overproportionally large in comparison with the originating a.c. component at the inverter input. The current or voltage systems react, for their part, upon the d.c. link. This interaction is to be described for a particular harmonic L, with the aid of a speed dependent equivalent conductance $\underline{G}_{dL}$ or equivalent resistance $\underline{Z}_{dL}$ as depicted in Fig.2. The values of $\underline{G}_{dL}$ and $\underline{Z}_{dL}$ are such that, for a single-phase input supply of 16 2/3 or 50 Hz, they are of the order of the filter impedances and filter admittances to the 2nd harmonic. They must therefore be taken into consideration in the calculation and design of the drive.

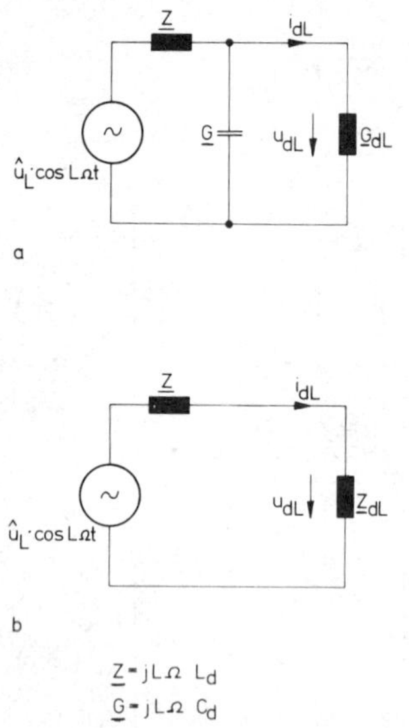

Fig.2 Equivalent circuit diagram for harmonic L
a) Three phase drive with constant voltage d.c. link
b) Three phase drive with constant current d.c. link

2. CALCULATION OF THE EQUIVALENT CONDUCTANCE $\underline{G}_{dL}$

For the calculation of the equivalent conductance $\underline{G}_{dL}$ the following assumptions are made:

1. For frequencies other than the fundamental, the saturation of the machine is neglected.
2. The inverter will be assumed to be free of losses and without any storage of energy.
3. The drive is in a steady state.

Assumption 1 makes it possible to determine the equivalent conductance for each harmonic L independently, and to superimpose current and voltages in accordance with the linear network theory. Assumption 2 means that the momentary value of the inverter input power is equal to the momentary value of the inverter output power; i.e. equal to the motor input power. By equating the corresponding powers on the three-phase side with the power conversion on the equivalent conductance, the equivalent conductance can be determined.

2.1 Transformation of the Inverter Input Voltage into the Machine Plane

The values on the three-phase side, i.e. at the machine input, are determined in the machine plane using vectors, and are represented by complex variables(Ref.1).

Fig.3 shows the transformation of the inverter input voltage $u_d(t)$ into the machine plane. The action of the inverter is described using the transformation vector $\underline{s}(t)$, which can be derived from the switching functions of the inverter phases. As these functions are inconstant, $\underline{s}(t)$ is a vector which changes its position in the machine plane in steps. If the switching functions are periodic, then the changes in the complex plane are also periodic. Analogous to the Fourier analysis of real numbers, the transformation vector can be resolved as the sum of rotating vectors $\underline{s}_K(t)$ rotating with constant speed $K\cdot\omega$

$$\underline{s}(t) = \sum_{K=-\infty}^{\infty} \underline{s}_K(t) = \sum_{K=-\infty}^{\infty} \underline{S}_K \cdot e^{jK\cdot\omega t} \qquad (2.1)$$

The coefficients $\underline{S}_K = S_K e^{j\varphi_K}$ are determined through the Fourier coefficient of the switching function $s_R(\omega t)$, $s_S(\omega t)$ and $s_T(\omega t)$. For a 120° symmetrically displaced system of switching functions and for a symmetrical three-phase winding, it is sufficient to determine the Fourier coefficient $\underline{B}_K$ of the switching function for the inverter phase co-ordinated to the phase winding, whose axis coincides with the real axis of the machine plane.

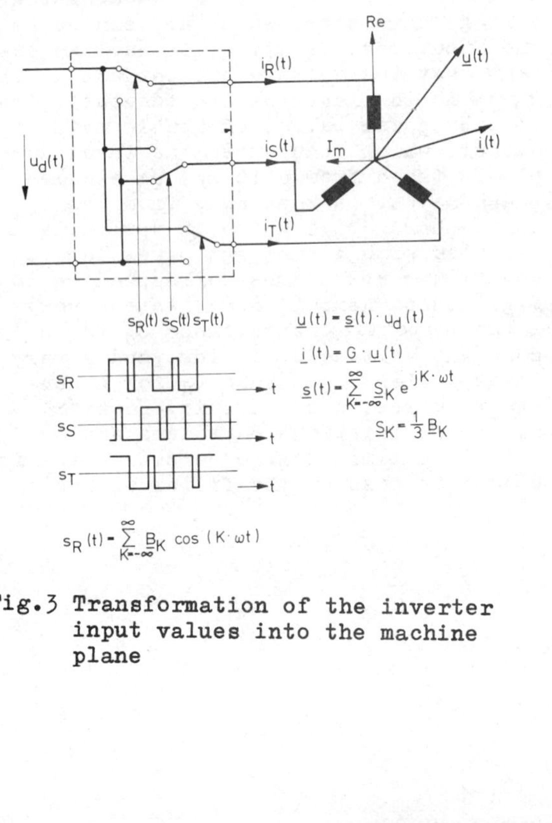

Fig.3 Transformation of the inverter input values into the machine plane

Due to the invariable transformation from the constant voltage d.c. link to the machine plane, the coefficient $\underline{S}_K$ emerges as

$$\underline{S}_K = \frac{1}{3} B_K \cdot e^{j\varphi_K} \quad ; \quad K \neq n \cdot 3 \qquad (2.2)$$

whereby the harmonics $\underline{S}_K$ with the ordinal number divisible by three are, because of the switching arrangement, not contained in the $\underline{S}_K$.

If the inverter input value is d.c. quantity U_0, then the vector corresponding to equation 1.3 is proportional to the transformation vector $\underline{s}$ (t).

$$\underline{u}_0 = U_0 \sum_{K=-\infty}^{\infty} |\underline{S}_K| \cdot e^{j\{K \cdot \omega t + \varphi_K\}} \qquad (2.3)$$

The transformation of the a.c. component, from equation

$$u_{dL}(t) = \hat{u}_{dL} \cdot \cos\{L \cdot \Omega t + \varphi_L\} \qquad (2.4)$$

gives, on the other hand, a sum of vectors, which rotate with the frequency $K \cdot \Omega$ (2.5)

$$\underline{u}_L = u_{dL}(t) \cdot \underline{s}(t) = \hat{u}_{dL} \cdot \sum_{K=-\infty}^{\infty} \underline{S}_K \cdot \cos\{L \cdot \Omega t + \varphi_L\} \cdot e^{jK \cdot \omega t}$$

The amplitude modulated vector $\underline{u}_L$ may be transformed with the aid of the relationship

$$\cos\{y\} \cdot e^{j\{x\}} = \frac{1}{2}\left[e^{j\{x+y\}} + e^{j\{x-y\}}\right] \qquad (2.6)$$

into two vectors $\underline{u}_{L1}$ and $\underline{u}_{L2}$, which rotate with different frequencies, but whose amplitude is constant.

$$\underline{u}_L = \underline{u}_{L_1} + \underline{u}_{L_2} \qquad (2.7)$$

$$\underline{u}_{L_1} = \frac{1}{2} \hat{u}_{dL} \sum_{K=-\infty}^{\infty} |\underline{S}_K| \cdot e^{j\{(K \cdot \omega + L \cdot \Omega)t + \varphi_K + \varphi_L\}} \qquad (2.8)$$

$$\underline{u}_{L_2} = \frac{1}{2} \hat{u}_{dL} \sum_{K=-\infty}^{\infty} |\underline{S}_K| \cdot e^{j\{(K \cdot \omega - L \cdot \Omega)t + \varphi_K - \varphi_L\}} \qquad (2.9)$$

The individual parts $\underline{u}_{L1K}$ and $\underline{u}_{L2K}$ produce currents in the machine windings, which are represented by the vectors $\underline{i}_{L1K}$ and $\underline{i}_{L2K}$. Magnitude and phase position of these stator current vectors may be depicted through the conductances $\underline{G}_{L1K}$ and $\underline{G}_{L2K}$

$$\underline{i}_{L1K} = \underline{G}_{L1K} \cdot \underline{u}_{L1K} \qquad (2.10)$$

$$\underline{i}_{L2K} = \underline{G}_{L2K} \cdot \underline{i}_{L2K} \qquad (2.11)$$

The values $\underline{G}_{L1K}$ and $\underline{G}_{L2K}$ are dependent not only on the machine data but also on its speed n. The conductance for the fundamental oscillation of the motor voltage is calculated from the fundamental oscillation current and cosφ, being proportional to the machine load.

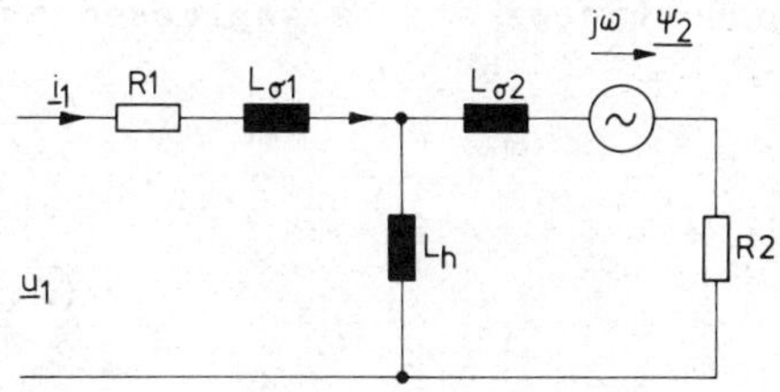

$$\underline{u}_1 = R_1 \cdot \underline{i}_1 + L_1 \cdot d\underline{i}_1/dt + L_h \cdot d\underline{i}_2/dt$$

$$\underline{0} = R_2\, \underline{i}_2 + L_2 \cdot d\underline{i}_2\ dt + L_h \cdot d\underline{i}_1\ dt - \omega \cdot \underline{\Psi}_2$$

$$\underline{\Psi}_1 = L_1 \cdot \underline{i}_1 + L_h \cdot \underline{i}_2 \qquad L1 = L_{\sigma 1} + L_h$$

$$\underline{\Psi}_2 = L_h \cdot \underline{i}_1 + L_2 \cdot \underline{i}_2 \qquad L2 = L_{\sigma 2} + L_h$$

$$\sigma_1 = L_{\sigma 1} / L_h \qquad \omega_{K1} = \frac{1}{\sigma} \cdot \frac{R1}{L1}$$

$$\sigma_2 = L_{\sigma 2} / L_h \qquad \omega_{K2} = \frac{1}{\sigma} \cdot \frac{R2}{L2}$$

$$\sigma = 1 - \frac{L_h^2}{L_{\sigma 1} \cdot L_{\sigma 2}}$$

Fig.4 Equivalent circuit diagram of the three phase machine

For the other frequencies the equivalent circuit-diagram of the asynchronous machine in Fig.4. is used. From the applied definitions and equations in this figure, the conductance is given by

$$\underline{G} = \underline{u}_1 \cdot \underline{i}_1 \qquad (2.12)$$

$$= \frac{1}{R_1} \cdot \frac{\frac{\omega_1}{\omega_{K2}} - \frac{\omega}{\omega_{K2}} - j\sigma}{\frac{\omega_1}{\omega_{K1}} + \frac{\omega_1}{\omega_{K2}} - \frac{\omega}{\omega_{K2}} + j\left\{\frac{\omega_1^2 - \omega_1 \cdot \omega}{\omega_{K1} \cdot \omega_{K2}} - \sigma\right\}}$$

ω_1 is the speed of rotation of the vector of the stator voltage $\underline{u}_1$. If it is assumed that the machine is in the stable working range i.e.

$$\omega_1 - \omega < \omega_{K2} \qquad (2.13)$$

and that for all ordinal numbers where L = 0 the condition

$$L \cdot \Omega \gg \omega_{K2} \qquad (2.14)$$

is valid, then the speed of the vectors $\underline{u}_{L1K}$ and $\underline{u}_{L2K}$ with the ordinal numbers K and L, due to the a.c. component in the d.c. link as well as of the mechanical speed ω, and of the frequency Ω of the a.c. voltage of the power supply, may be determind from the following equations.

$$\omega_1 = K \cdot \omega + L\Omega \quad \text{für } \underline{U}_{L1K} \qquad (2.15)$$

$$\omega_1 = K\,\omega - L\Omega \quad \text{für } \underline{L1}_{L2K} \qquad (2.16)$$

With the introduction of relative magnitudes

$$X = \frac{\omega_0}{\Omega} \qquad (2.17)$$

both conductances may be expressed as a function of X, K and L

$$\underline{G}_{L1K} = \frac{1}{R_1} \frac{\frac{K-1}{\omega_{k2}} \cdot X + \frac{L}{\omega_{k2}} - j\frac{\sigma}{\Omega}}{\left(\frac{K}{\omega_{k1}} + \frac{K}{\omega_{k2}} - \frac{1}{\omega_{k2}}\right) \cdot X + \left(\frac{L}{\omega_{k1}} + \frac{L}{\omega_{k2}}\right) - j\left\{\frac{\Omega}{\omega_{k1} \cdot \omega_{k2}} \cdot \left[(K^2-K)X^2 + (2K \cdot L - L)X + L^2\right] + \frac{\sigma}{\Omega}\right\}} \tag{2.18}$$

$$\underline{G}_{L2K} = \frac{1}{R_1} \frac{\frac{K-1}{\omega_{k2}} X - \frac{L}{\omega_{k2}} - j\frac{\sigma}{\Omega}}{\left(\frac{K}{\omega_{k1}} + \frac{K}{\omega_{k2}} - \frac{1}{\omega_{k2}}\right) \cdot X + \left(\frac{L}{\omega_{k1}} + \frac{L}{\omega_{k2}}\right) - j\left\{\frac{\Omega}{\omega_{k1} \cdot \omega_{k2}} \cdot \left[(K^2-K)X^2 - (2K \cdot L - L)X + L^2\right] + \frac{\sigma}{\Omega}\right\}} \tag{2.19}$$

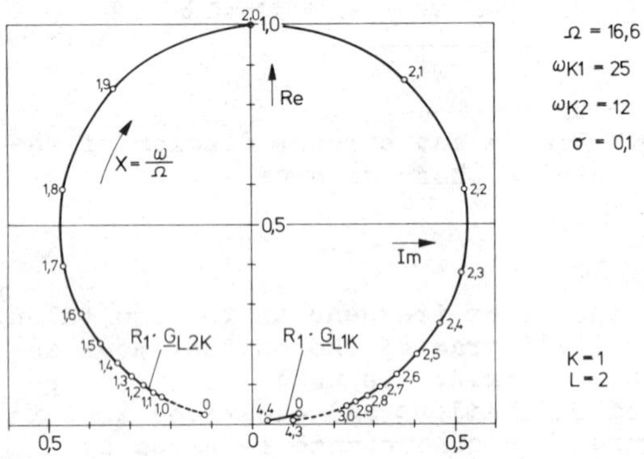

Fig.5 Relative conductance of the three-phase machine

The locus diagram in Fig.5 of the conductances for the fundamental oscillation of the switching function (K = 1) and for the 2nd harmonic in the d.c. link (L = 2), both relative to the reciprocal of the stator resistance R1, shows that the conductance $\underline{G}_{L2K}$ as derived from equation 2.19, for the value of a speed n equal to the harmonic in the d.c. link

i.e. for $\omega = \frac{L}{K}\,\Omega$

is equal to the reciprocal of the stator resistance.
For smaller values of X, the imaginary part of the conductance is positive, while for speeds greater than L/K·Ω it is negative. The conductance $\underline{G}_{L1K}$ is small compared to $\underline{G}_{L2K}$ and always has a negative imaginary part.

2.2 Power Calculation

To determine the momentary value of the power $p_L(t)$ which is assigned to the a.c. part of the input voltage u_{dL}, the scalar procuct of all the parts of the vectors $\underline{u}_{L2K}$ as well as $\underline{i}_{L1K}$ and $\underline{i}_{L2K}$ must be formed together.

From the voltage vector

$$\underline{u}_{LK_X} = \underline{u}_{L1K_X} + \underline{u}_{L2K_X} \tag{3.1}$$

and the current vector

$$\underline{i}_{LK_Y} = \underline{i}_{L1K_Y} + \underline{i}_{L2K_Y} \tag{3.2}$$

there ensues a power component

$$p_{LK_x, K_y} = Re\,\frac{2}{3}\left\{(\underline{u}_{L1K_x} + \underline{u}_{L2K_y}) \cdot (\underline{i}_{L1K_y} + \underline{i}_{L2K_y})\right\} \tag{3.3}$$

Using equations 2.10 , 2.11 and also 3.1 and 3.2 the power part $P_{LK_XK_Y}$, which is composed of four terms, is calculated

$$\begin{aligned} p_{L.K_x.K_y} = {} & \hat{U}_{dL}^2 \cdot |\underline{S}_{K_x}| \cdot |\underline{S}_{K_y}| \\ & \cdot \Big[|\underline{G}_{L2K_y}| \cdot \cos\{((K_x-K_y) \cdot \omega - 2L \cdot \Omega) t + \varphi_{K_x} - \varphi_{K_y} - 2\varphi_L - \alpha_{L2K_y}\} + \\ & + |\underline{G}_{L1K_y}| \cos\{(K_x-K_y) \cdot \omega t \pm \varphi_{K_x} - \varphi_{K_y} - \alpha_{L1K_y}\} + \\ & + |\underline{G}_{L1K_y}| \cdot \cos\{((K_x-K_y) \cdot \omega + 2L \cdot \Omega) t + \varphi_{K_x} - \varphi_{K_y} + 2\varphi_L - \alpha_{L1K_y}\} + \\ & + |\underline{G}_{L2K_y}| \cos\{(K_x-K_y) \omega t + \varphi_{K_x} - \varphi_{K_y} - \alpha_{L2K_y}\}\Big] \end{aligned} \tag{3.4}$$

The power P_L, which is created at the input of the squirrel-cage motor by the a.c. component u_{dL} of the input voltage at the inverter, is given by the sum of all components as

$$p_L = \sum_{\substack{K_x = -\infty \\ K_y = -\infty}}^{\infty} p_{L \cdot K_x \cdot K_y} \tag{3.5}$$

The power of $P_{dL}(t)$ of the harmonic with ordinal L at the inverter input is depicted using the sought after conductance $\underline{G}_{dL}$.

$$i_{dL} = |\underline{G}_{dL}| \cdot \hat{u}_{dL} \cdot \cos\{L \cdot \Omega t + \varphi_L + \alpha_L\} \tag{3.6}$$

$$p_{dL} = u_{dL}(t) \cdot i_{dL}(t) \tag{3.7}$$

resulting in a constant power component and power pulses of frequency 2LΩt

$$p_{dL} = \frac{1}{2}\,\hat{u}_{dL} \cdot |\underline{G}_{dL}| \cdot \left[\cos\alpha_L + \cos\{2 \cdot L\,\Omega t + 2\varphi_L + \alpha_L\}\right] \tag{3.8}$$

To determine the momentary value of the equivalent conductance $\underline{G}_{Ld}$ the power part with the frequency 2LΩ is compared at the input and output of the inverter. This means that, for the formation of the power part on the output side, only those parts of the power will be considered which originate from the scalar product of a current and voltage vector with the same ordinal number, i.e.

$$K_X = K_Y = K$$

The required power component with the frequency 2LΩ is given from the equations 3.5, 3.6, 3.9

$$p_L\Big/_{2L\Omega} = \left(\tfrac{1}{2}\right)^2 \cdot \hat{u}_{dL}^2 \sum_{K=-\infty}^{\infty} |S_K|^2 \cdot \Big[|\underline{G}_{L1K}| \cdot \cos\{2L\Omega t + 2\varphi_L + \alpha_{L1K}\} + |\underline{G}_{L2K}| \cdot \cos\{2L \cdot \Omega t + 2\varphi_L - \alpha_{L2K}\} \Big] \qquad (3.10)$$

The comparison of the coefficients between the equation 3.10 and the power component with the frequency 2LΩt in equation 3.8 results in the required value $\underline{G}_{dL}$.

$$\underline{G}_{dL} = \frac{1}{2} \cdot 2 \sum_{K=-\infty}^{\infty} |\underline{S}_K|^2 (\underline{G}_{L1K} + \underline{G}_{L2K}^*) \qquad (3.11)$$

As the derivation shows, the equivalent conductance $\underline{G}_{dL}$, whose reciprocal is equal to the equivalent resistance $\underline{Z}_{dL}$, is determined by

- **the pulse technique and the pulse voltage ratio control of the inverter**
- **the asynchronous system data of three-phase machine**
- **the speed of the three-phase motor.**

Above all, the temperature of the stator winding is to be taken into consideration as the change in ohmic resistance is entered linearly equation 2.18 and 2.19.

3. CALCULATIONS AND MEASUREMENT

Using the conductance $\underline{G}_{dL}$, which must be determined individually for each harmonic, the a.c. component of the inverter input quantity can now be calculated for every operating point with the aid of the equivalent circuit diagram in Fig.2a and the quantities in the d.c. link.

The further studies are carried out on the basis of a particular three-phase drive. In Fig.6 the functional diagram of this drive with its constant voltage d.c. link is reproduced. The energy is drawn from a single phase supply with a frequency of 16 2/3 Hz. The d.c. link is fed via a transformer with two secondary windings feeding a diode bridge which is connected in series with a half-controlled bridge. The Fourier coefficient of the filter input voltage in Fig.7 shows that the second harmonic can amount to up to 70 % of the d.c. component. The filter, consisting of a series inductance L_d and a supporting capacitor C_d reduces this harmonic, neglecting the machine, to approx. 5 %, and the 4th harmonic to 0,2 % of the d.c. value.

The drive electrical data for the fundamental oscillation of the motor and for the d.c. quantities in the d.c. link can be seen from Fig.8. The fundamental oscillation of the motor voltage is raised by the subharmonic method (pulse width modulation) up to the rated frequency f_N.

Fig.9 shows the amount of the Fourier coefficient $\underline{B}_K$ of a switching function of the subharmonic technique . For the coefficient of the fundamental, according to increase of the machine voltage in proportion with the speed the following applies:

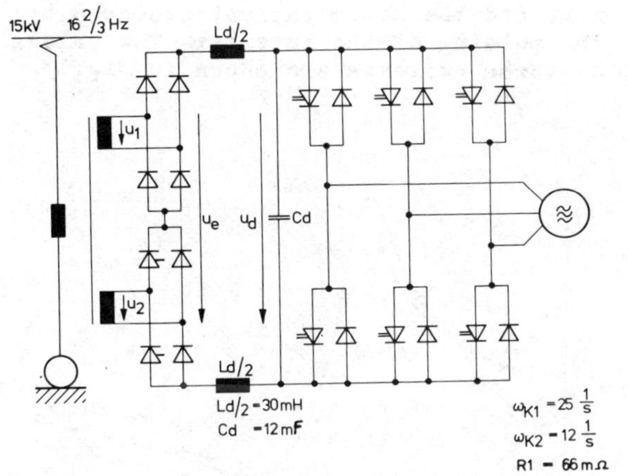

Fig.6 Three phase drive: Function diagramm

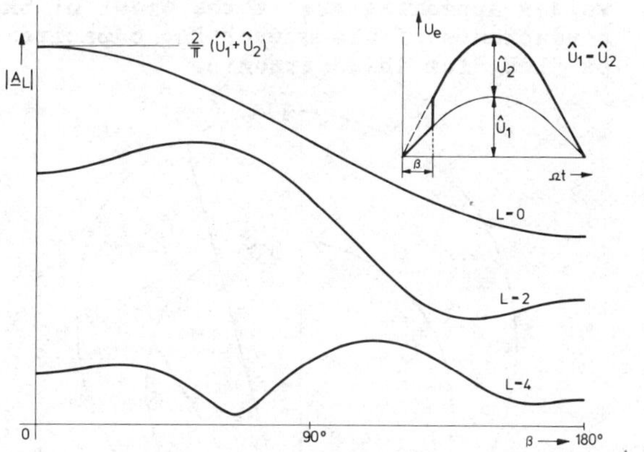

Fig.7 Three phase drive: Amount of Fourier coefficient at the output side of the supply rectifier

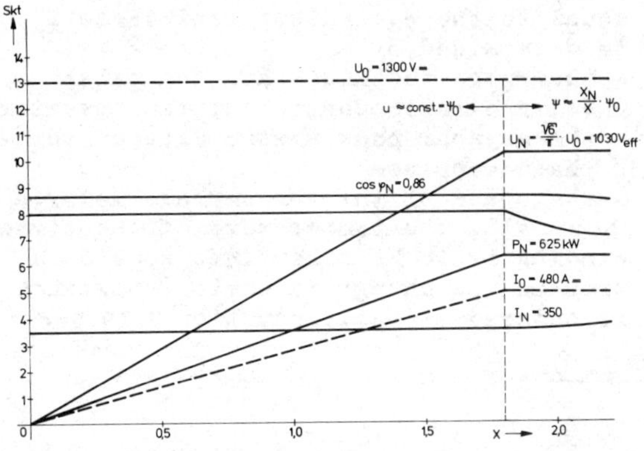

Fig.8 Electrical and mechanical quantities at maximum load torque

$$0 < X < X_N = \frac{2\pi \cdot f_N}{\Omega} \tag{4.1}$$

$$B_{K=1} = B(X) = \frac{X}{X_N} \cdot \frac{4}{\pi}$$

For the field weaking range the voltage is constant, i.e.

$$X_N < X < X_{max}$$

$$B_{K=1} = B(X) = \frac{4}{\pi} \tag{4.2}$$

The equivalent conductance $\underline{G}_{Ld}$ of the three-phase component described above, is plotted for harmonic L = 2 and for the fundamental of the switching function K = 1 in Fig.10. Nevertheless, as for all the following calculations, only the fundamental of the switching function has been taken into consideration. The copper temperature ϑ_{R1} of the stator winding has been given as parameter. The values appearing are in the order of the conductance of the supporting capacitance $jL\Omega C_d$ for this harmonic.

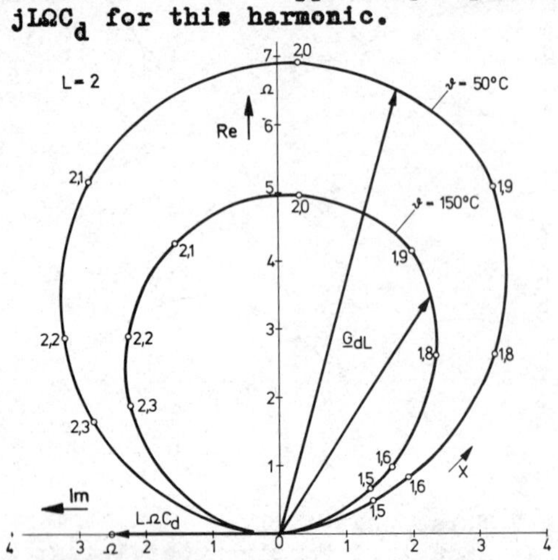

Fig.10 Equivalent conductance for inverter and three phase machine

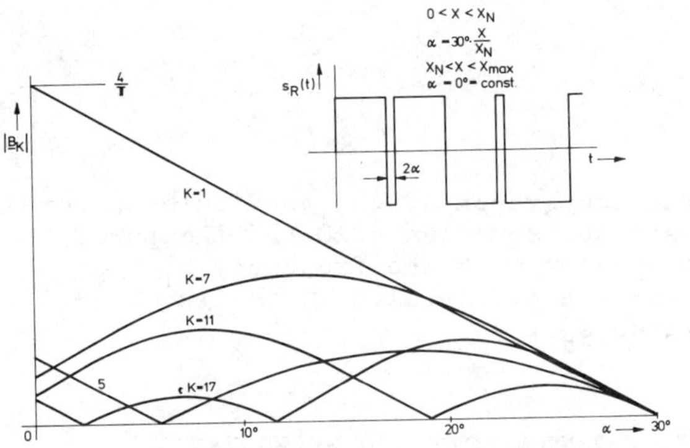

Fig.9 Amount of Fourier coefficient B_K of a switching function for synchronous pulse width modulation

Fig. 11 shows, for the equipment considered, the influence of the LC-Filter damping, as in Fig. 2, through the three-phase drive, dependent upon the refered speed and upon the stator copper temperature ϑ_{R1}. As to be expected from the course of the curve of the equivalent conductance $\underline{G}_{dL}$, the effect of the filter is decreased for speed less than 1,6 x due to the negative imaginary part, while for higher speeds the three-phase drive has a positive bearing on the damping.

Fig. 12 shows the curve of the motor current vectors $\underline{i}_{L1}$ and $\underline{i}_{L2}$, which, because of the a.c. component of the inverter input voltage $u_{dL}(t)$, is superposed by the fundamental oscillation current and the motor current produced by the pulsing of the inverter. The values of these currents are shown in Fig.13

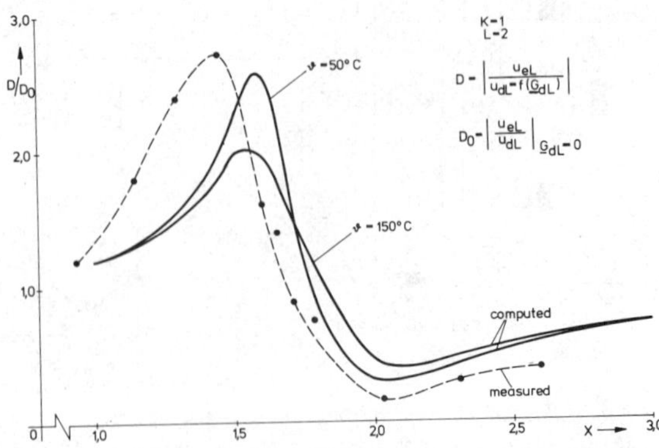

Fig.11 Influence of the d.c. link filter caused by the frequency controlled three phase drive

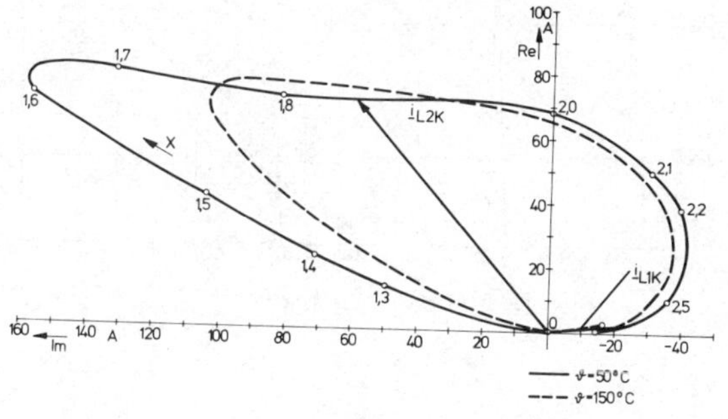

Fig.12 Curve of motor currents qualified by the a.c. component of the d.c. link

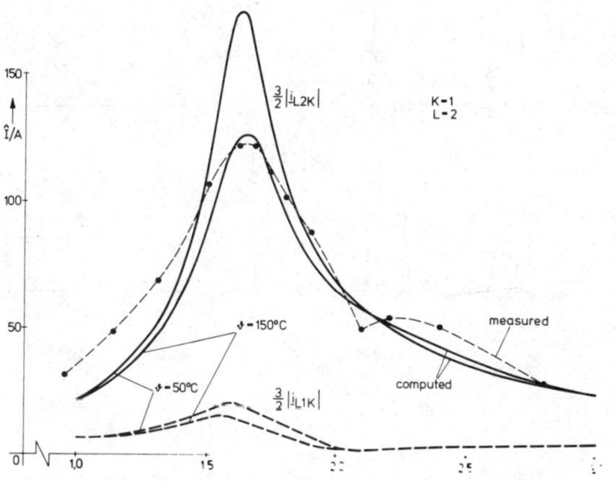

Fig.13 Amount of additional motor currents

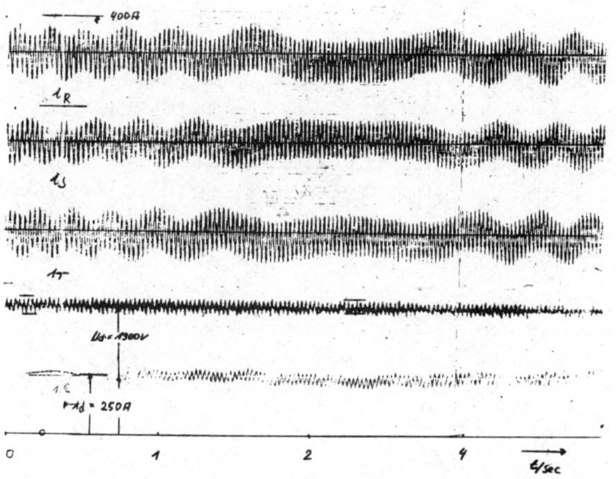

Fig.14 Run-up against 70 % load torque (Field Weakening Range)

for 2 temperatures. Calculations and measurements show in the critical range, these additional currents, which can amount to 30 % of the motor current at low temperatures, must be taken into consideration, particularly in the design of self-commutated inverters.

Fig.14 shows the inverter input quantities as well as the inverter output currents, while passing through a speed range corresponding to 1.6 $< X <$ 2.1 with a load torque of 70 % rated torque. It is seen that for a speed corresponding to 2 Ω, the frequency of the additional polyphase system ist equal to zero, and its direction of rotation changes. This change of direction of rotation corresponds to the change of sign of the imaginary part of the equivalent conductance $\underline{G}_{dL}$ in Fig.10.

Fig. 15 and 16 show a spectral analysis of the quantities at the inverter input and at the inverter output, for the speed at which the damping of the 2nd harmonic is a minimum. The amplitudes of the additional power parts, which are caused by the a.c. component in the d.c. link, are indicated in Fig.17 for a maximum load torque over the reference speed on the above mentioned machine. The interaction of the fundamental oscillation of the voltage and current with the difference frequency $\omega - 2\Omega$ is the main reason that, at the input to the three-phase machine, power pulses of double mains frequency appear and which amount to up to 50 % of the rated power at lower temperatures ϑ_{R1}.

From the curve of the filter current, the capacitor current and the inverter input current in Fig.18, it is seen that the series inductances for this power pulsation effectively decouples the drive from the feed in, and that the power pulsation shown in Fig.17 is an interaction between the supporting capacitance and the three-phase drive.

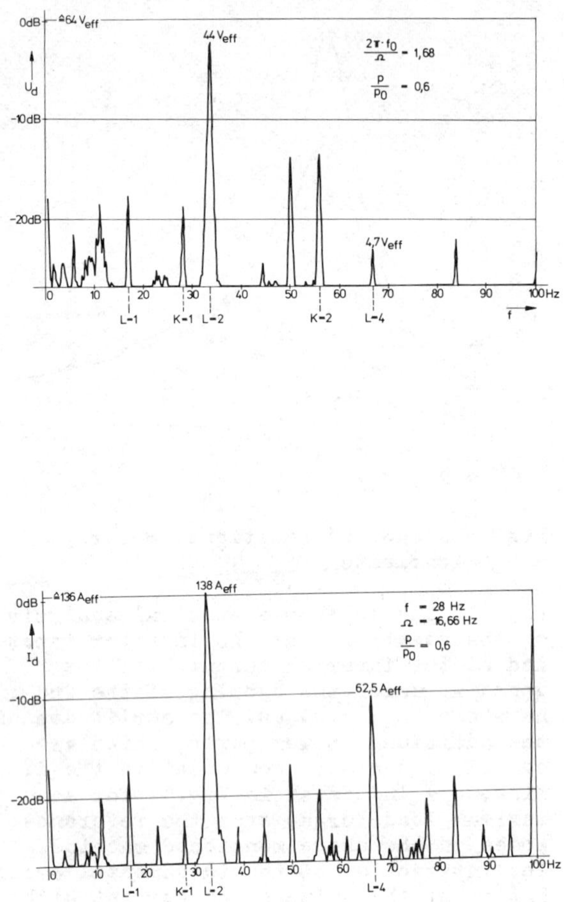

Fig.15 Spectral analysis of the inverter input quantities

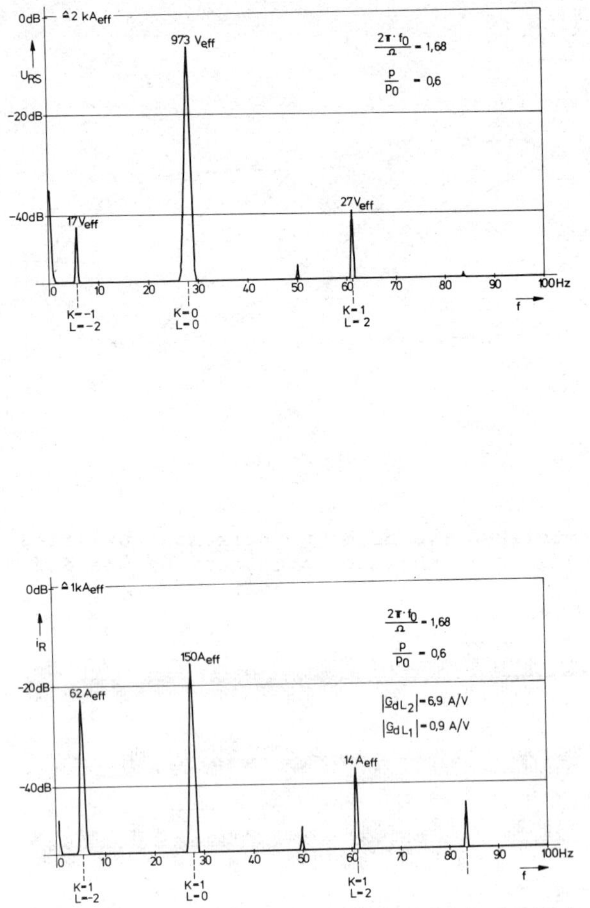

Fig.16 Spectral analysis of the inverter output quantities

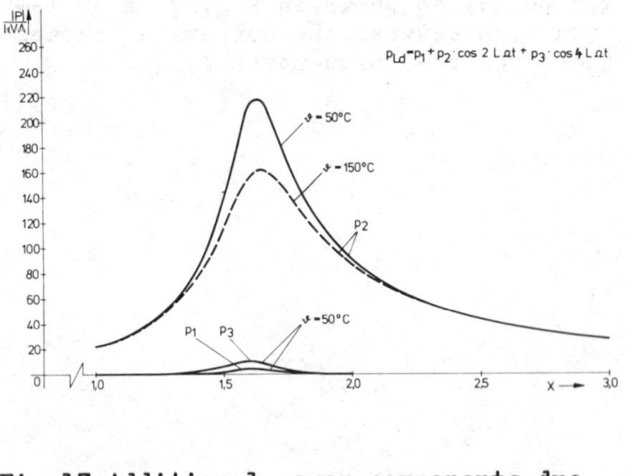

Fig.17 Additional power components due to the a.c. component in the d.c. link of the inverter

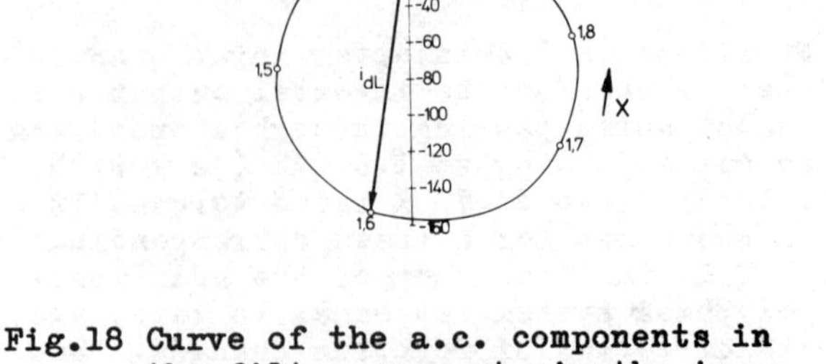

Fig.18 Curve of the a.c. components in the filter current, in the inverter and in the supporting capacitance

4. CALCULATION OF THE STEADY-STATE CONTION FOR A FREQUENCY CONVERTER WITH A D.C. LINK

The steady-state behaviour of a three-phase drive with a d.c. link for the harmonic L will be determined for the equivalent circuit-diagramm b in Fig.2. In the calculation the equivalent resistance $\underline{Z}_{Ld}$ is similarly constant, as the voltage adjustment, on which the speed depends, is effected at the input rectifier. Fig.19 shows the curve of the equivalent resistance for the drive described, for a 120° pulsing of the inverter phases for the harmonic L = 2.

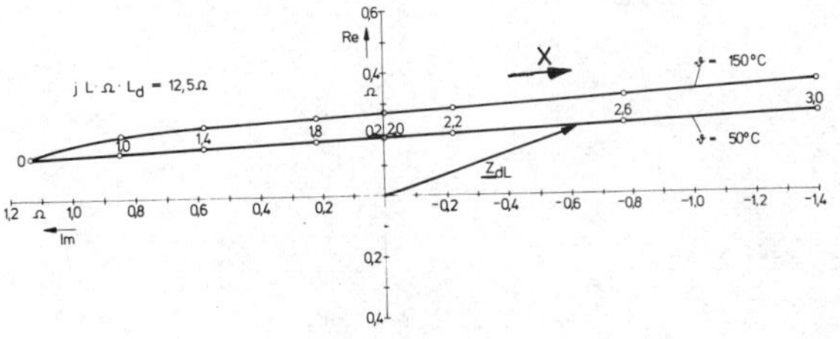

Fig.19 Curve of the equivalent resistance $\underline{Z}_{Ld}$ on operation using constant current d.c. link

5. METHODS OF DECREASING THE FILTER EXPENDITURE

A comparison of the volume and weight of the drive described with the maximum additional motor currents appearing, and with the power pulsation at the machine input, indicates that, for high powers, other solutions must be found for supplying a three-phase drive from a single phase supply. While one possibility would be to keep the power pulsation from the drive using a tuned suction circuit, the machine can, on the other hand, be so controlled that the equivalent conductance $\underline{G}_{dL}$ remains constant, and even magnifies the effect of the supporting capacitor. This means a controlled power pulsation with the squirrel-cage motor.

With an additional transformation vector $\underline{s}_X(t)$

$$\underline{s}_{KX}(t) = \frac{\hat{u}_{dL}}{u_0} \cdot B_K \cdot e^{j\{(K\cdot\omega + L\cdot\Omega)t - \varphi_K + \varphi_L\}} \tag{5.1}$$

this gives, instead of the three exciting vectors given from the equations 2.3, 2.8 and 2.9.

$$\underline{u}_{0K}{}' = B_K \cdot u_0 \left[1 - \frac{1}{2}\left(\frac{\hat{u}_{dL}}{u_0}\right)^2\right] \cdot e^{j\{K\cdot\omega + \varphi_K\}} \tag{5.2}$$

$$\underline{u}_{1LK} = B_K \cdot \hat{u}_{dL} \cdot e^{j\{(K\cdot\omega + L\cdot\Omega)t + \varphi_K + \varphi_L\}} \tag{5.3}$$

$$\underline{u}_{2LK} = B_K\ \hat{u}_{dL} \cdot \frac{\hat{u}_{dL}}{u_0} \cdot e^{j\{(K\cdot\omega - 2L\cdot\Omega)t + \varphi_K - 2\varphi_L\}} \tag{5.4}$$

While the additonal transformation vector hardly influences the fundamental oscillation of the output voltage, and the rotating voltage system with the frequency K·ω+LΩ remains unchanged, the rotating field with the difference frequency Kω − LΩ no longer exists. Instead, there is a rotating field whose amplitude is smaller by the factor $\hat{u}_{dL}/U_0$, and whose frequency is equal to Kω − 2LΩ. The motor currents resulting from this rotating field by an order of magnitude smaller than the motor currents which are caused by the system $\underline{u}_{L2}$.

The technique was tried out on a test plant, and the additional transformation vector was realized with the aid of a phase modulation of the inverter switching functions. For steady-state operation, this showed the expected results. However on dynamic operation, this resulted in instability. This instability is caused by the reaction of the phase position of the a.c. component at the inverter input on the action of the transformation vector.

REFERENCES

(1) K. P. Kovacs, I. Racz, Vorgänge in Wechselstrommaschinen, Verlag der Ungarischen Akademie der Wissenschaften, Budapest, 1959.

INVERTER-FED ASYNCHRONOUS MOTORS FOR TRACTION SYSTEMS

C. Rossi*, G. Stopiglia and E. Tortello*****

**Divisione Transporti, Ansaldo S.p.A.*
***Direzione Ricerca e Sviluppo, Ansaldo S.p.A.*
****Divisione Grandi Macchine, Ansaldo, S.p.A.*

INTRODUCTION

Variable-frequency inverter-fed asynchronous motor constitutes, in the field of electric traction, an effective option to traditional d.c. motor.
The success of this equipment is connected to the advantages the squirrel cage asynchronous motor may provide (lower weight and small dimensions, higher power at high speed, less maintenance requirements, higher mechanical rotor strength, lower cost, greater travelling comfort) and is a direct consequence of the enormous progress obtained by power electronics during recent years, both as concerns costs and reliability.
During the latest few years our Company too has been devoting itself to develop this for employ in electrical Traction.
An equipment with continuative power of 120kW and maximum power of 180kW was built and tested for an experimental trolley-bus; a motor with continuative power of 280kW is now being built for an E 323 electric shunting locomotive of Italian Railways.
At the same time many higher-rating motor designs have been completed (also exceding 1000 kW) that have confirmed even more the advantages of a motor of this kind, specially as concerns weight and dimensions.
The aim of this paper is to report the results of a complete tests series performed on the prototype trolley-drive.

TEST EQUIPMENT

The system and instrumentation employed for performing tests are indicated in diagrams shown in fig. 1 and 2. They allow great testing possibilities and analysis and also any kind of behaviour simulation:

a) Inverter feeding, as well as recovery of power delivered by the load d.c. generator, are obtained by two controlled rectifiers with an adeguate power and voltage (275 kW to 1000 V).

b) By acting on the two converters regulation it has been possible to reproduce transients both in the inverter feeding voltage and in the motor load.

c) The instruments chosen for measurements of electric feeding of the motor have optimum frequency response, and consent to consider correctly all the harmonics actually important for the motor.

d) Also the torque measure is obtained by an instrument with a very good dynamic performance. Its rotating part consists of an oscillator and a strain-gages bridge bonded on the load coupling/shaft.
The torsional strain determines the bridge unbalance and consequently a frequency modulation of the oscillator. Such frequency is transmitted, without any contact, to a frequency-voltage transducer, and its exit may be measured, recorded or acquired on a minicomputer.

e) Rotor temperature measurement is possible in different positions. Thermoresistors are employed and the signal is transferred from the rotating part by a collector.

f) The voltages, currents and couples waveforms may be acquired both under steady and transient state conditions, on a minicomputer which may be connected contemporaneously with a maximum of 8 analogic and 4(16 bit) digital channels. It is thus possible to obtain experimentally both instant values and harmonic analysis first dise storing and then printing or plotting.

ANALYSIS OF TEST RESULTS

Before analysing asynchronous motor tests, we report in Tab. I and fig. 3 its main characheristics compared to the d.c. motor it replaces [1, 2, 3]. The mechanical characteristic of the asyncronous motor (fig. 3) was obtained with a linear increase of voltage and frequency in the range 0÷75 Hz, and a constant voltage in the range 75÷190 Hz.
Considering efficiencies and power factors constant, current absorbed is also constant in the whole working range.
The motor design is related to the limit between the two feeding laws as flux, torque and voltage attain in it simultaneously the respective maximum values.
Fig. 4 shows asynchronous motor efficiency versus speed with inverter and sinusoidal feeding. On the same diagram the d.c. motor efficiency is reported.
In the higher frequency range the two motors efficiencies are similar, being the asynchronous one higher.
The analysis of the asynchronous motor behaviour in both cases, of sinusoidal and inverter supply is very effective.
In the constant power range, in which the inverter feeds the motor with a square wave the efficiency decrease is less than 1.5%
In the low frequency range we must keep in mind that this relates to starting and so it is not too significant from the efficiency point of view. In this range too, however, the asynchronous motor behaviour is considerably better than the d.c. motor one.
The difference with the sinusoidal supply efficiency is on the contrary greater, as the harmonic content of the supply is higher and the motor filtering is less effective.
The phenomenon is emphasized by the reduced value of the rotor leakage inductance, and its effect evident in the stator winding heating diagrams (fig. 5) and in the data reported in table II, relative to an 80 Hz test with both inverter and sinusoidal supply.
Separation of iron losses from additional ones was not carried out as a sufficiently satisfactory method does not exist in every situation.
The motor thermal behaviour is described in full taking ventilation diagram into account (fig.6). The measured overtemperatures are under the insulation class limits in the whole working range.
Harmonics, which we saw cause a loss increment, have an effect as well on the magnetizing characteristic.
No load tests (fig. 7), in fact, indicate that the fundamental component of current is higher with inverter supply, that is the magnetizing inductance decreases.
When the motor runs near the saturation, the presence of time harmonics determines an increase in the maximum induction value, and consequently an increase in the current.
An even greater effect is due to the saturation of those parts of the magnetic circuit in which harmonic fluxes close.
A complete calculation of this additional saturation effect is very complex and it is preferable to adopt the semi-empirical method suggested by Klingshirn, Jordan [4], based on the assumption that such additional saturation is proportional to the instantaneous sum of the current harmonics and particularly to its maximum. The adoption of this method in our case has given good results.

RELATION BETWEEN SUPPLY WAVEFORM AND JOULE LOSSES [4, 5, 6, 7]

Traction asynchronous motors, owing to required performances, must be dimensioned in terms of torque and for this reason are particularly affected by the flux weakening working range amplitude. The low inductance values of such motors favour harmonic currents circulation and this makes the importance to improve the supply waveform at low frequencies evident.
The inverter chops the d.c. voltage according to the following diagram:

Frequency range	Chopping type
0÷42 Hz	sub-harmonic PWM
42÷60 Hz	5 pulses PWM
60÷75 Hz	3 pulses PWM
75÷190 Hz	square wave

Joule losses measured for the stator and semi-experimentally deduced for the rotor, are diagrammed in fig. 8 and 9, fig. 10 and 11 report the r.m.s. current values for the entire frequency range and, for some frequency values, the harmonic analysis of both tension and current.
In the analysis of such values rotoric parameters variation related to frequency was taken into account (fig. 12). The correlation between Joule losses and supply waveforms was clarified by the analysis carried out on every chopping condition used in the 0÷75 Hz frequency range.
The results of such analysis are given in diagrams shown in fig. 13 and 14.
The sub-harmonic PWM chopping is preferable up to frequencies of 20÷30 Hz, beyond which the square wave is better.
With PWM chopping the higher the impulses number and the frequency, the better are the results obtained.
The curves may be of help in the choice of the frequency range for every type of supply modulation.

TABLE I	D.C.MOTOR	INDUCTION MOTOR
TYPE	CV 1227	M-6-120
Rated voltage	550 V	430 V (line to line)
Rated current	190 A	215 A/phase
Continuous rating	105 kW	120 kW
One-hour rating	120 kW	150 kW
Rated speed	1950 rpm	1500 rpm
Maximum speed	3600 rpm	3800 rpm
Number of poles	6	6
Supply frequency	-	0÷190 Hz
Weight	660 kg	600 kg
One-hour rating /weight	0.182 kW/kg	0.25 kW/kg

TAB.1 -

TABLE 2	f = 80 Hz	V = 430 V
LOSSES	SINUSOIDAL WAVE	SQUARE WAVE
STATOR COPPER	2.2 kW	2.3 kW
IRON+STRAY	1.3 kW	1.9 kW
MECHANICAL	0.8 kW	0.8 kW
I^2 R ROTOR	1.2 kW	2.1 kW
TOTAL	6.1 kW	7.6 kW

TAB. 2 - Motor losses (full load) 80 Hz/430 V.

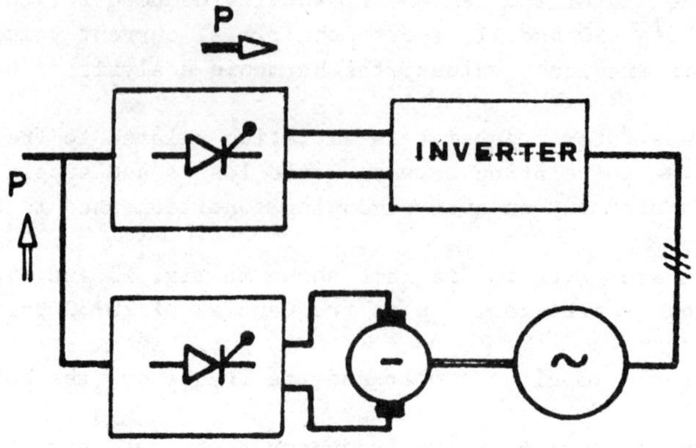

FIG. 1 - Test equipment.

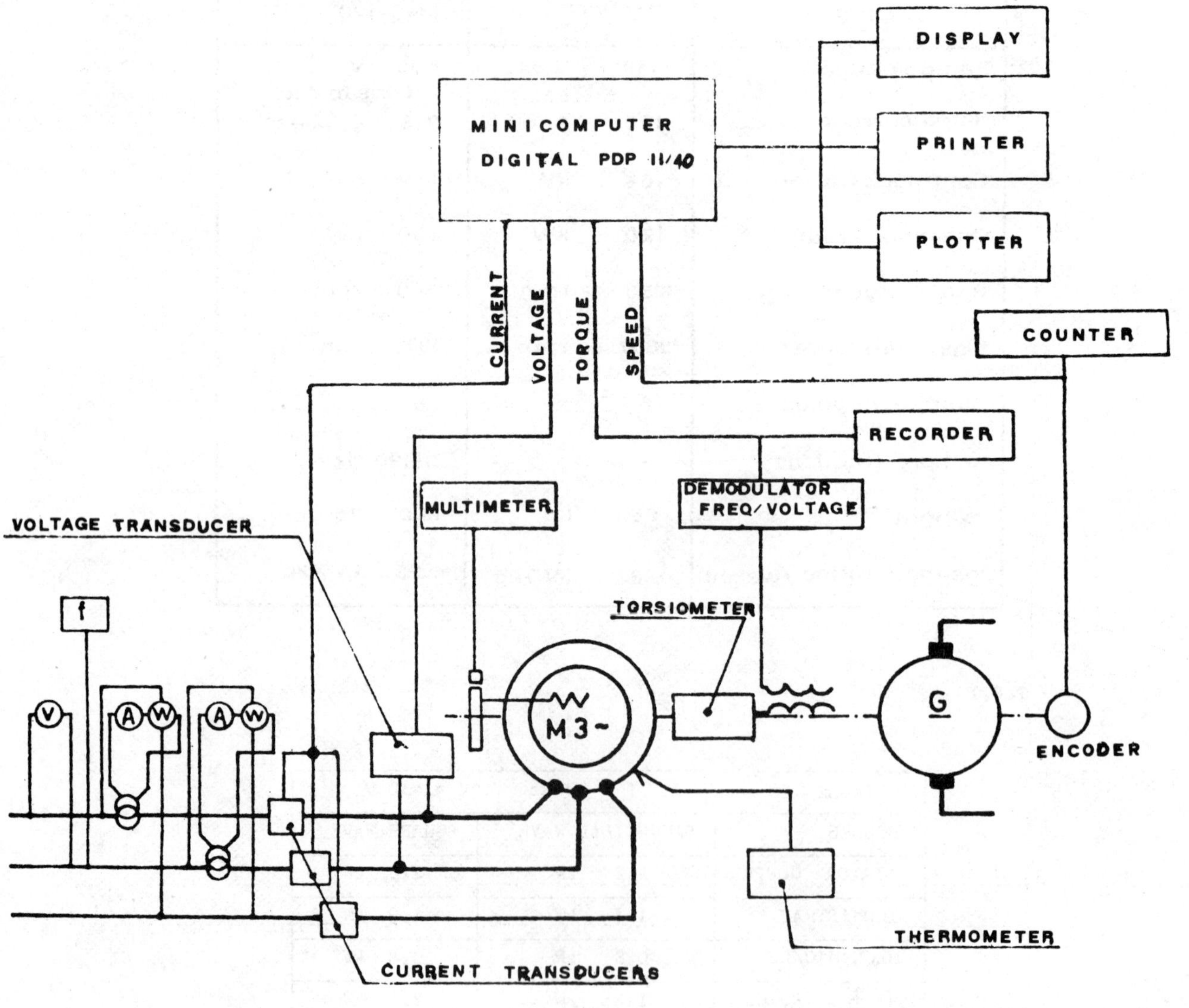

FIG. 2-Instrumentation .

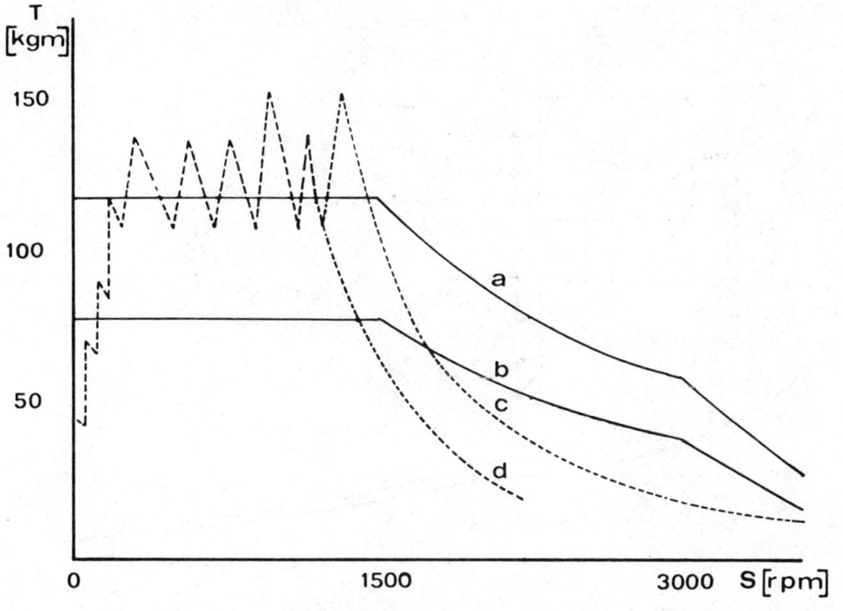

FIG. 3 - Torque vs. speed characteristics of M/6/120 and CV 1227 motors
a) M/6/120 maximum rating
b) M/6/120 continuous rating
c) CV 1227 weak field
d) CV 1227 full field

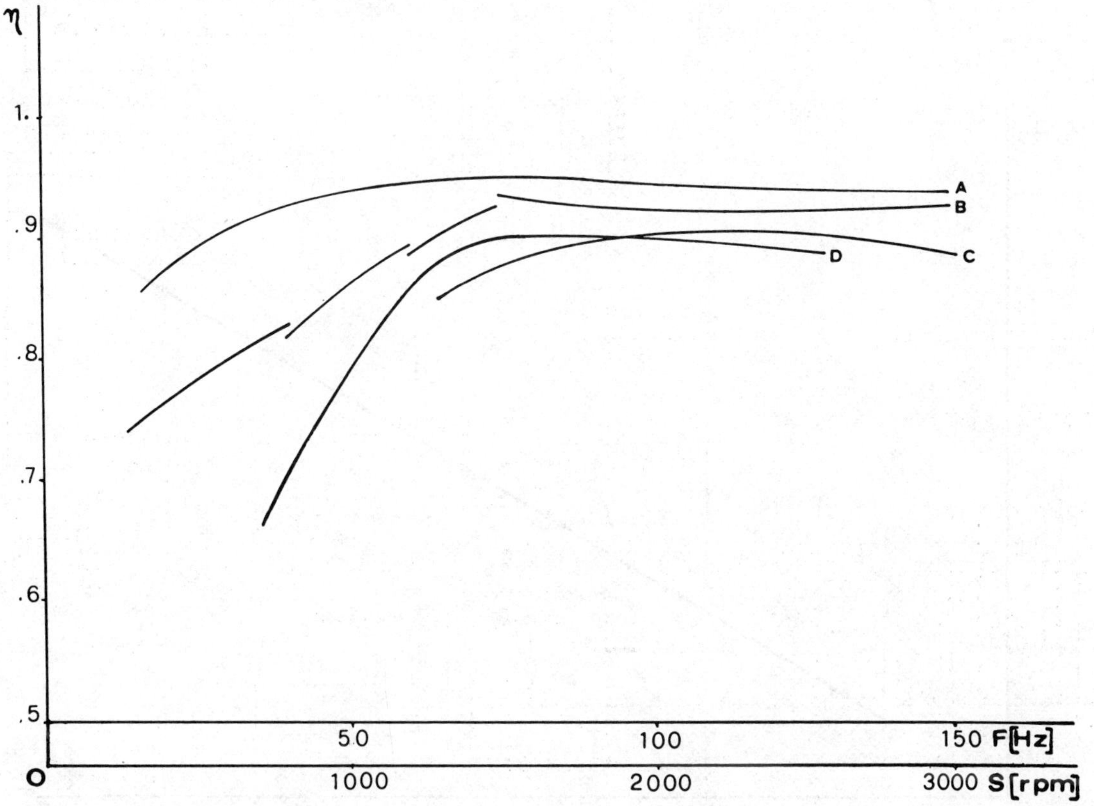

FIG. 4 - Efficiency vs. speed and frequency.
A) Sinusoidal - M-6-120
B) Inverter - M-6-120
C) Weak field - CV 1227
D) Full field - CV 1227

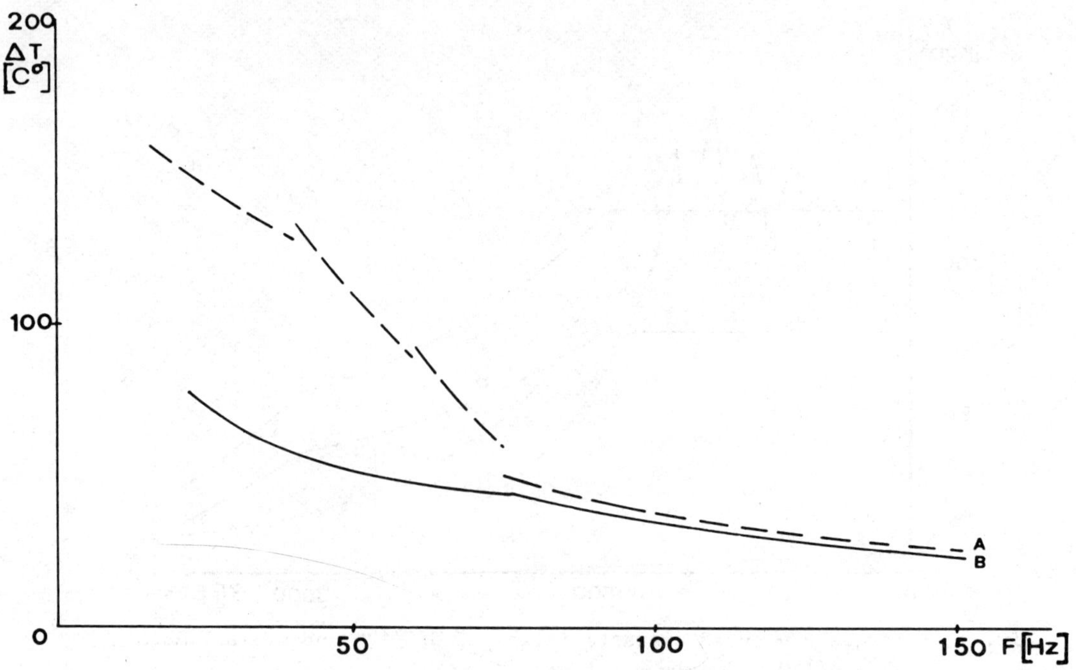

FIG. 5 - Stator copper overheating vs. frequency
a) inverter
b) sinusoidal

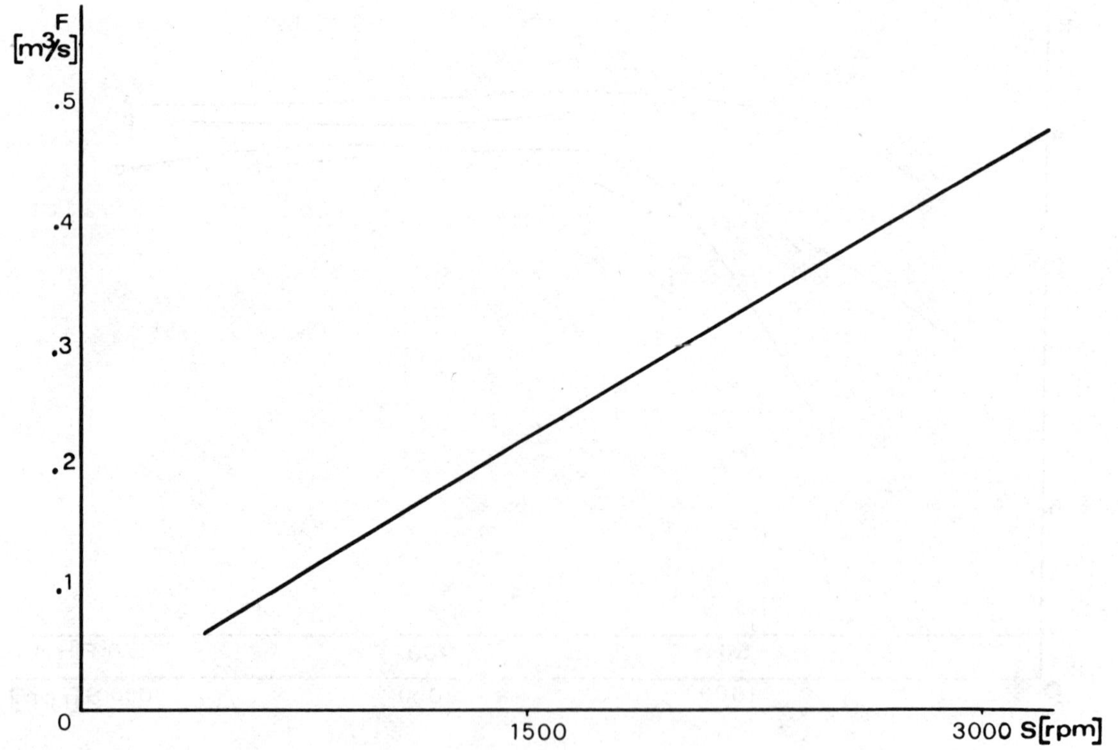

FIG. 6 - Ventilation curve. Open circuit, self-ventilated.

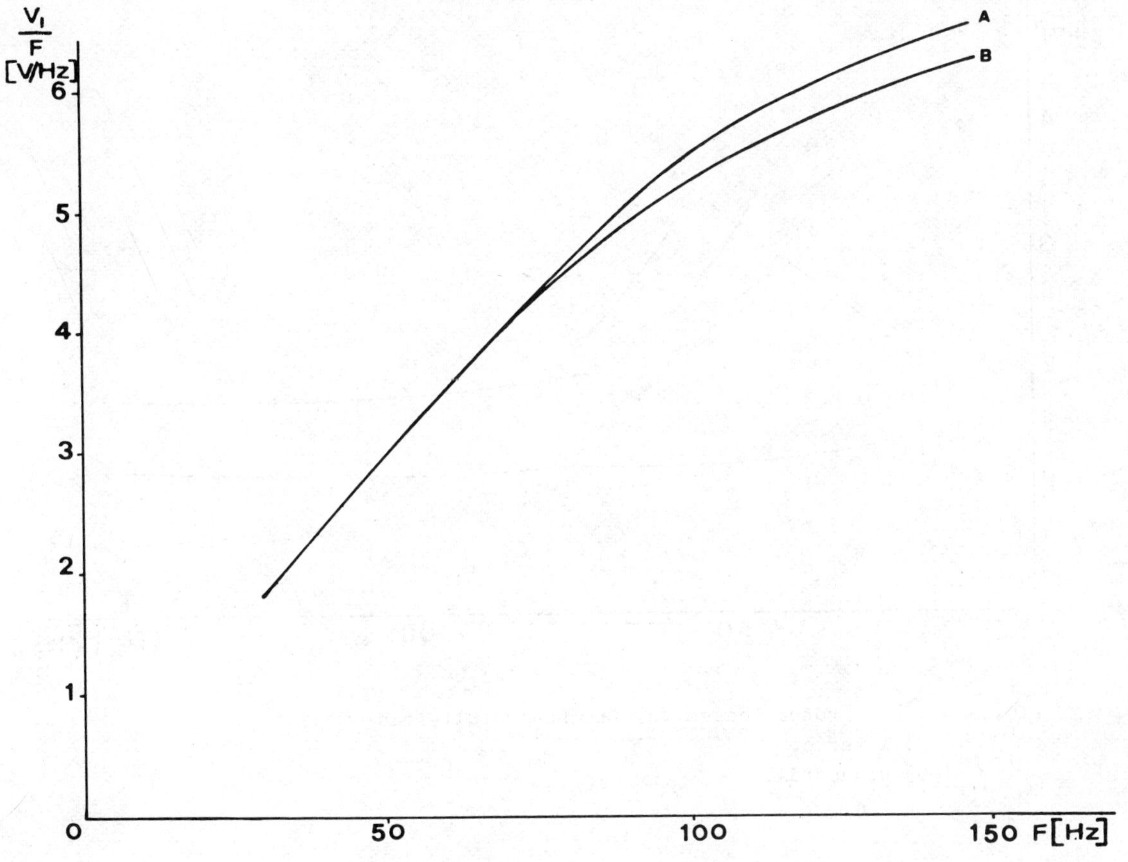

FIG. 7 - Magnetization curve, a)sinusoidal - b) square wave

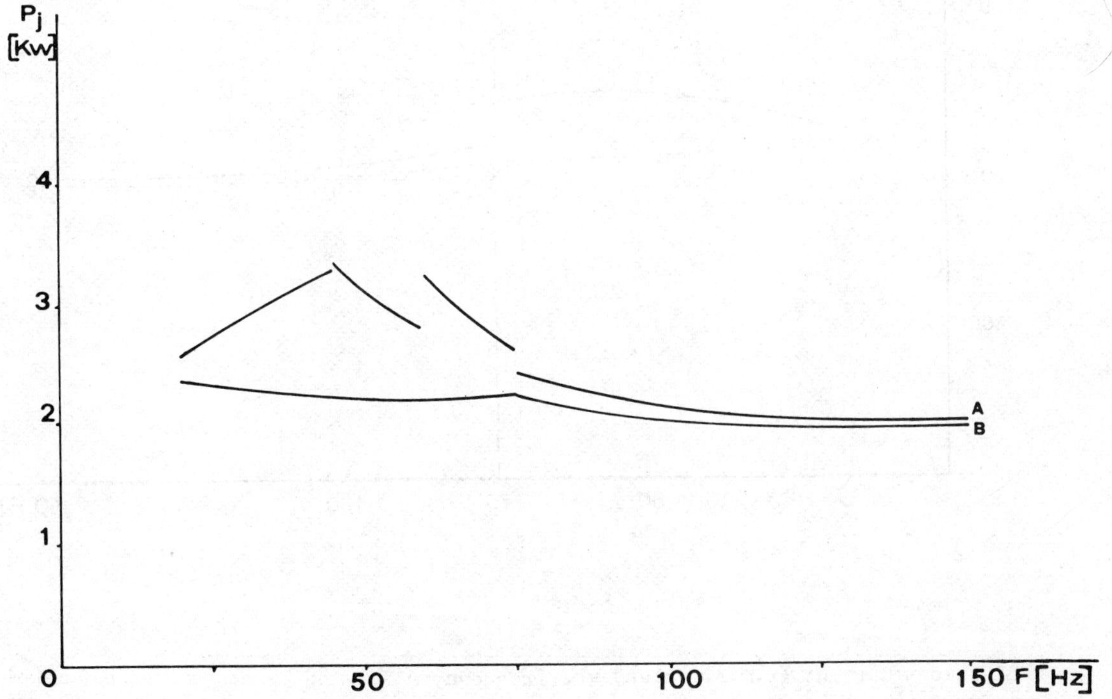

FIG. 8 - Copper stator losses vs. frequency (full-load)
a) inverter
b) sinusoidal

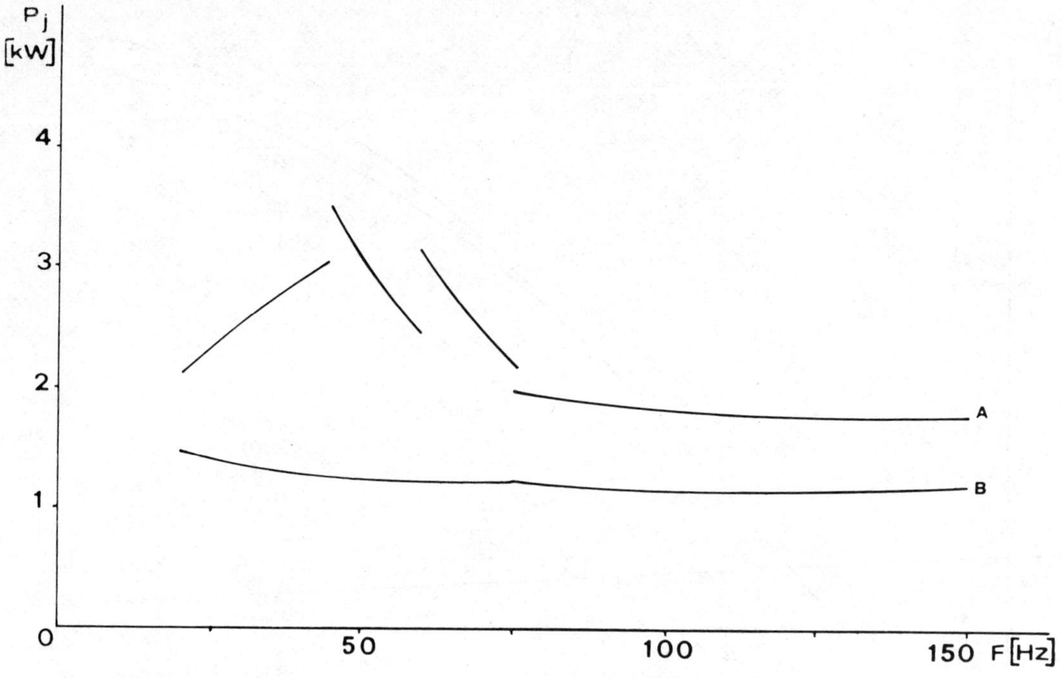

FIG. 9 - $I^2 R$ rotor losses vs. frequency (full-load)
a) inverter
b) sinusoidal

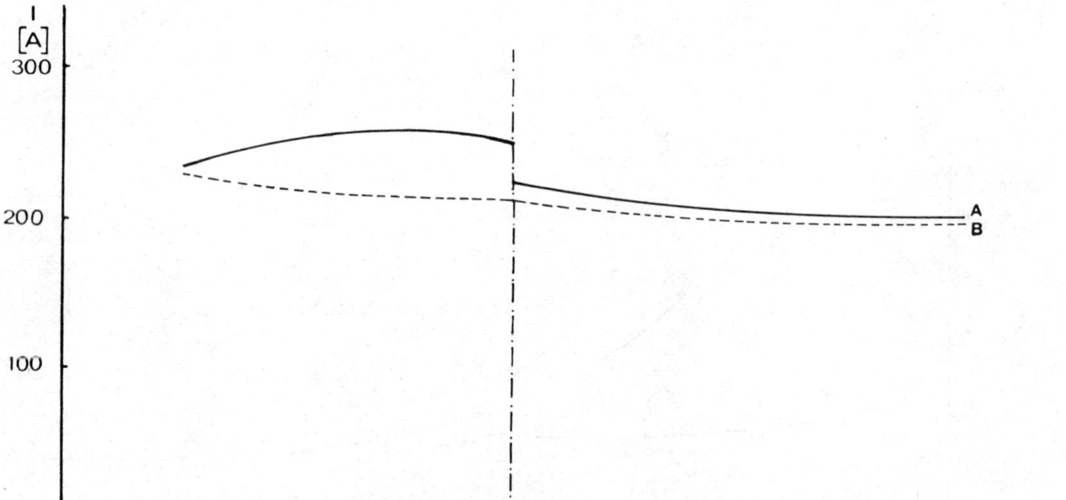

FIG. 10 - Current (r.m.s. value) vs. frequency
a) inverter
b) sinusoidal

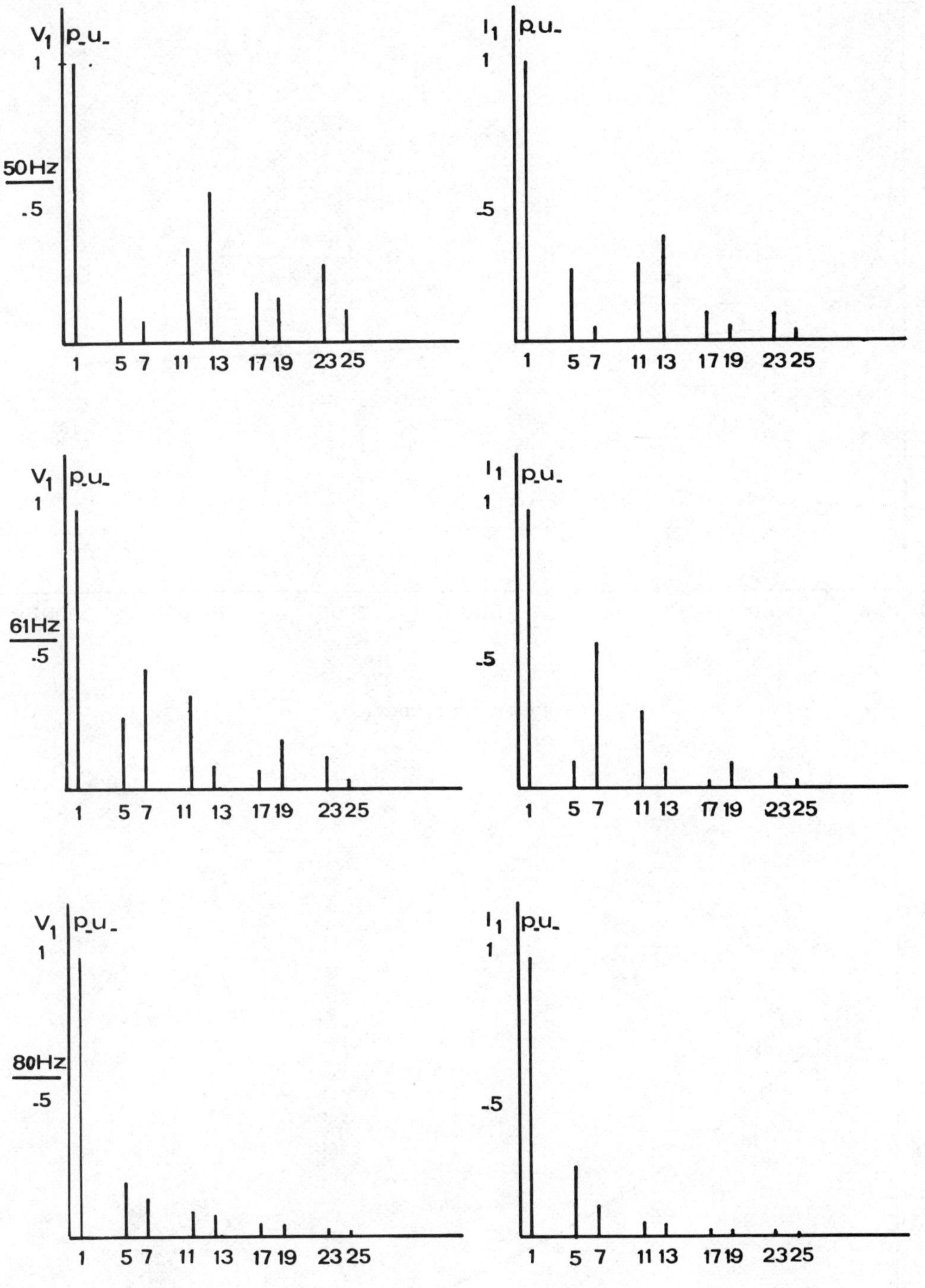

FIG. 11 - Voltage and current harmonic spectra.

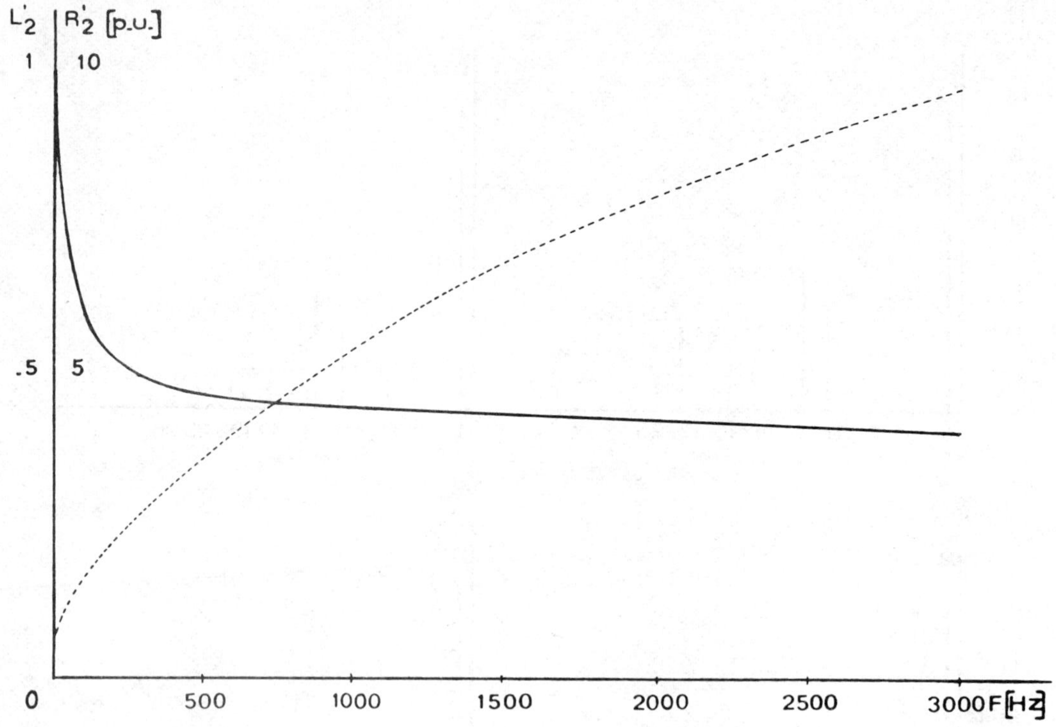

FIG. 12 - Rotor parameter vs. rotor frequency.

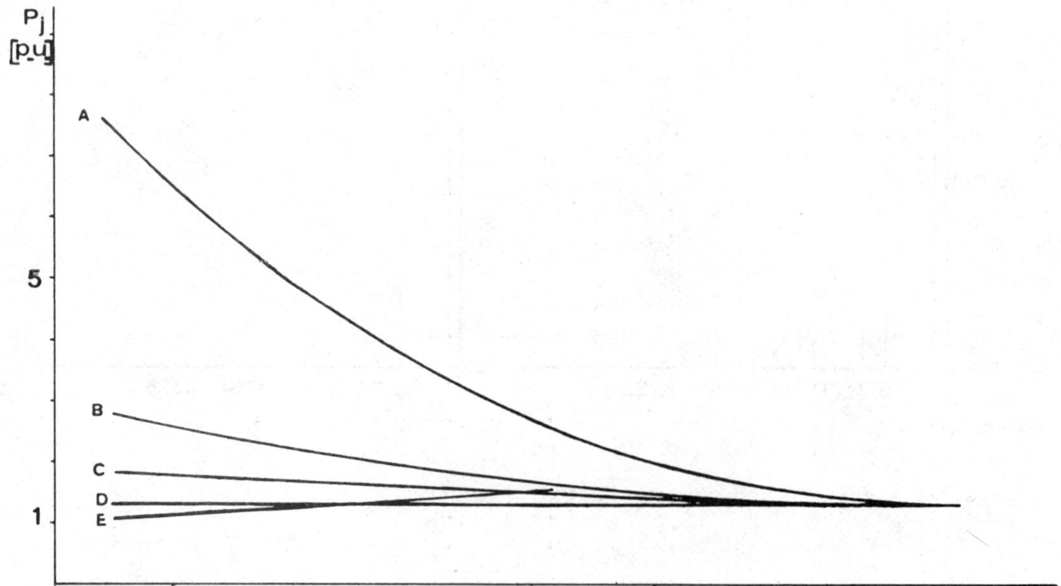

FIG. 13 - Copper stator losses frequency
a) PWM - 3 pulses
b) PWM - 5 pulses
c) PWM - 7 pulses
d) Square wave
e) Subharmonic PWM

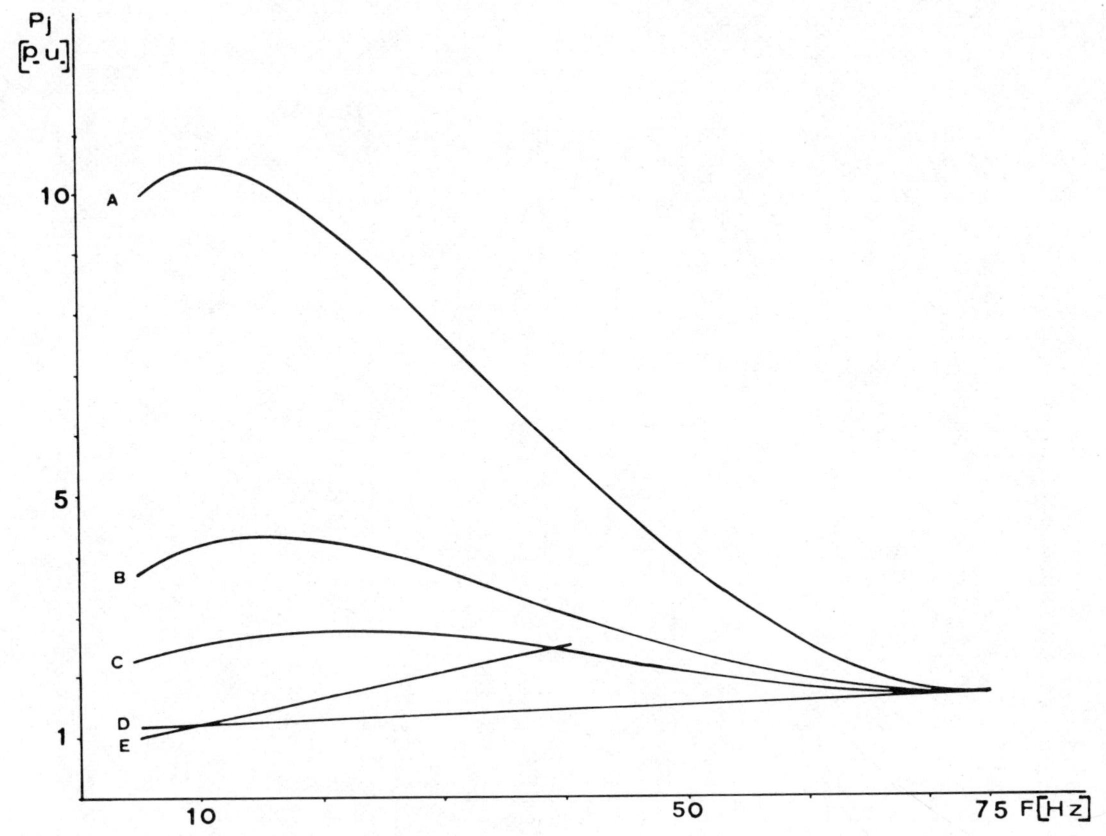

FIG. 14 - I^2 R rotor losses versus frequency
a) PWM - 3 pulses
b) PWM - 5 pulses
c) PWM - 7 pulses
d) Square wave
e) subharmonic PWM

BIBLIOGRAPHY

1 G. Giovanardi - "L'impiego del motore asincrono alimentato a tensione e a frequenza variabili quale motore di trazione per locomotive ed automotrici ferroviarie" LXXVII Riunione annuale A.E.I. di Sorrento 1976.

2 G. Stopiglia - E. Tortello - "La macchina asincrona a gabbia come motore di trazione" LXXVII Riunione annuale A.E.I. di Sorrento 1976.

3 F. Ghislanzoni- S. Gho - "Filobus sperimentale a frequenza variabile" LXXVII Riunione annuale A.E.I. di Sorrento 1976.

4 E.A.Klingshirn- H.E. Jordan - "Poliphase Induction Motor Performance and losses on Sinusoidal Voltage sources". IEEE P.A.S. vol. 87.

5 H. Largiadér - "Quelques aspects du dimensionnement des moteurs asynchrones alimentés par convertisseurs statiques de fréquence pour la traction électrique". Rev. Brown Boveri - 4-70.

6 Svante Von Zweygbergk E. Sokolov - "Verlustermittlung im stromrichtergespeisten asynchron motor" ETZ -A - Bd. 90 H 23

7 A. Bellini A. De Carli M. La Cava "Parameter Identification for Induction Motor Simulation" IFAC Symposium on Control in Power Electronic and Electrical Drives. Duesserldorf, October - 7-9, 1974.

CONTROL OF ROTATING AND LINEAR INDUCTION MOTORS FOR VEHICLE DRIVES

P. Appun

Brown, Boveri & Cie AG, Mannheim

Summary

For frequency-controlled induction motors control characteristics for the dependence of important quantities on the speed as well as control methods for the torque are derived. The main object is not a good dynamic behavior, but - as is most essential to vehicle drives - the minimum weight and volume of the drive consisting of converter and motor. Furthermore, the influence of particular features of linear induction motors on the control characteristics and methods is shown. Series resistors used to reduce the high voltages during braking operation necessitate special control characteristics, too.

1. Introduction

The requirements put by a vehicle on the performance of the drive are generally given in the form of a tractive effort vs. velocity diagram. A first velocity range of constant tractive effort which is limited by the coefficient of adhesion or by reasons of comfort, is normally followed by a second range of constant power in order to keep the expenditure for the power supply and conversion equipment within certain limits. When induction motors are used, the required high tractive effort on starting, the frequent operation in the partial load range and the demand for a continuous controllability of effort and velocity in most cases necessitate a frequency control by means of static converters which is the only subject of the following statements.

The conventional control methods for rotating induction motors aim at a constant magnetic flux in the machine to ensure a good dynamic behaviour of the drive. As the mass of the vehicle is relatively large, the speed changes slowly at normal operation. For reasons of comfort a jerk limitation is necessary which means that normally no fast torque changes are required. The highest requirements on the dynamic behaviour are imposed by extraordinary operating conditions such as wheel slipping or sliding. In this case the energy flow must be reduced quickly.

Particularly with vehicle drives a low weight and small volume is of major importance in addition to the requirements imposed on the dynamic behaviour. The weight and volume required can be influenced by the selection of the control characteristics. A difference has to be made between control characteristics for voltage, current and secondary frequency as a function of the speed along the boundary line in the torque - speed diagram and the control methods for adjusting the torque between zero and the maximum value at a definite constant speed. By using different control characteristics vs. speed the expenditure can be shifted between motor and converter. With regard to the total expenditure, optimum control characteristics can be found. Control methods for the torque have a direct influence on the dimensioning of the converter, especially in the case of a current source inverter.

Due to their particular features linear induction motors necessitate special control characteristics. Contrary to rotating machines linear induction motors have speed-dependent characteristics on account of the end effects increasing with the speed which cause additional losses and a weakening of the magnetic field. When deriving control characteristics vs. velocity, these effects have to be considered. A particular feature of single-sided linear motors are normal forces which have an attractive effect within the range of small secondary frequencies which is of interest here. Often, a limitation of these normal forces is required necessitating special control methods for the tractive force.

With drives for suburban vehicles, however also for very high velocities, the required braking power is often considerably higher than the tractive power. In order to avoid a dimensioning of the converter with respect to the braking power, braking resistors can be inserted between inverter and motor which are modifying the control characteristics versus velocity as well as the control methods for the braking force.

2. Control characteristics vs. speed

2.1 Rotating induction motor

In the first section of the torque - speed diagram where an approximately constant torque is required, current and secondary frequency are constant, too. The voltage increases approximately proportional to the primary frequency. In this section there is no possibility of reducing the expenditure for motor or converter by selecting different control characteristics. If at all, the expenditure may be varied within certain limits by the degree of the magnetic utilization of the motor. In the field weakening range where a constant power is required it is, however, possible to shift the expenditure from the motor to the converter and vice versa (Ref. 1,2) by means of different control characteristics. If one succeeds in defining the motor and converter expenditure an optimum control characteristic can be found resulting in a minimum total expenditure for the drive. Depending on the application expenditure means, first of all, weight, volume or cost.

To explain these means of influencing Fig. 1 shows examples of two different control characteristics. All parameters are per unit quantities. A given machine and a defined size of the field weakening range are assumed. In Fig. 1a a control characteristic with constant voltage has been assumed. This characteristic is of advantage when a pulse-width-modulated inverter is used, because one can operate with a pulse frequency equal to the fundamental output frequency and the required maximum pulse frequency is small. The mechanical output to be obtained is limited by the pull-out torque at maximum speed. At the beginning of the field weakening range the available torque and thus the capacity of the motor is not fully utilized, since the pull-out torque is inversely proportional to the square of the frequency and thus also approximately to the speed. The converter is used rather efficiently, even if at the beginning of the field weakening range a relatively unfavourable operating

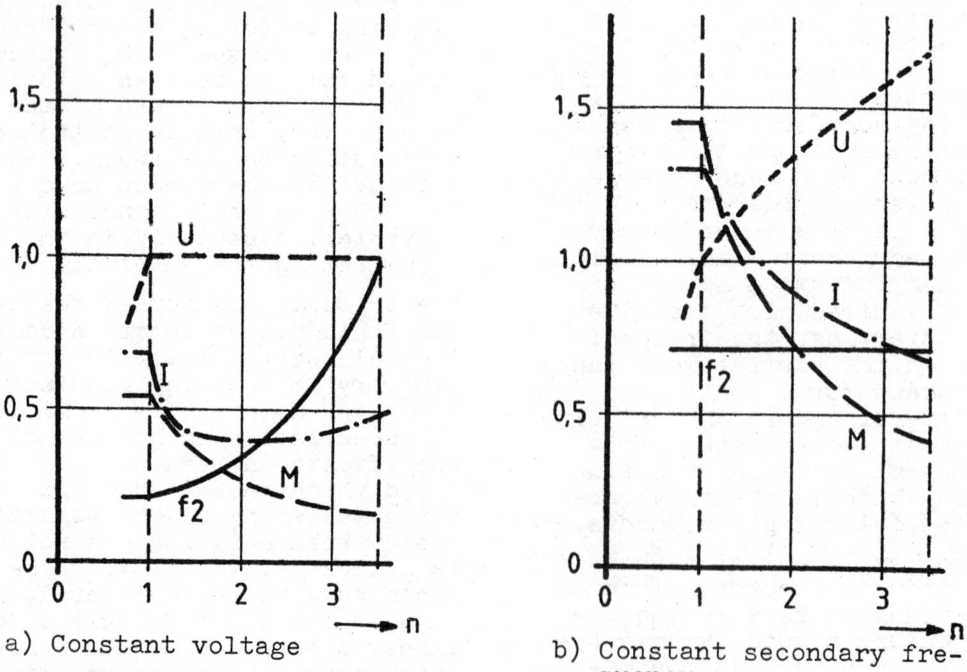

a) Constant voltage

b) Constant secondary frequency

Fig.1. Torque M, voltage U, current I and secondary frequency f_2 vs. speed in the field weakening range

point with a poor power factor and high current results on account of the parameters selected.

As other extreme Fig. 1b shows a control characteristic with constant high secondary frequency by means of which a considerably higher mechanical output can be obtained, since the machine is always operating close to its pull-out point. As compared with Fig. 1a the thermal load on the motor will also be higher which fact has normally to be considered with regard to the motor expenditure. Contrary to the motor the converter is not used efficiently, since it must be rated for the maximum values of current and voltage occurring not commonly.

When looking for an optimum control characteristic resulting in a minimum expenditure, the main problem is the definition of the expenditure. Starting from a given motor, the expenditure for the motor is constant:

$$A_{Mot} = K_1 \qquad (1)$$

The converter is defined to the first approximation by the maximum occurring voltage and current:

$$A_{Con} = K_2 \cdot U_{max} \cdot I_{max} \qquad (2)$$

If this expenditure is referred to the mechanical output supplied by the drive at a certain control characteristic, an objective function is obtained whose minimum

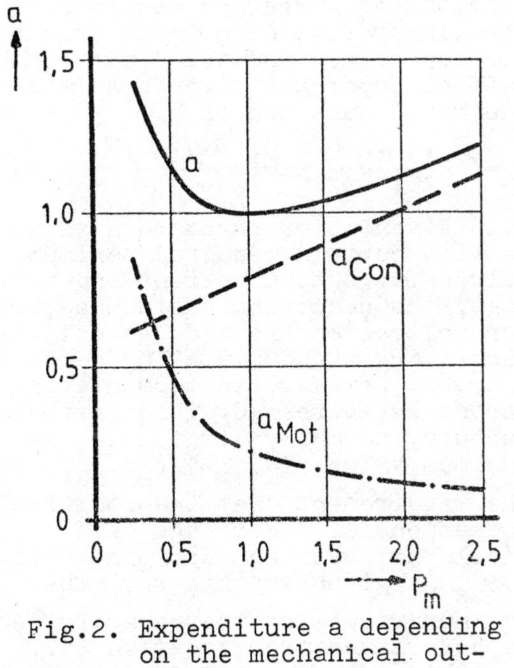

Fig.2. Expenditure a depending on the mechanical output P_m

value has to be found:

$$a = a_{Mot} + a_{Con} = \frac{A_{Mot}}{P_m} + \frac{A_{Con}}{P_m} \qquad (3)$$

The factors K_1 and K_2 permit a varying mutual evaluation of converter and motor.

Fig. 2 shows as an example the result of an optimization with such an objective function and a certain evaluation. The same motor paramerters and size of the field weakening range have been used as in Fig. 1. To begin with, one tried to find the control characteristic resulting in the minimum expenditure for different given mechanical outputs. By changing the mechanical output, one obtains the curves for converter, motor and total expenditure as shown in Fig. 2, by means of which the optimum mechanical output and thus also an optimum control characteristic can be given. The position of the optimum is dependent on the evaluation factors K_1 and K_2 in the objective function, the configuration of the objective function, the machine parameters and the size of the field weakening range. At a more detailed investigation, the thermal load on the motor as well as drive components, if any, besides converter and motor have to be considered.

2.2 Linear induction motor

The relations described above apply, in principle, also to linear induction motors. When deriving control characteristics, one has to consider a few deviations in the performance as compared with rotating induction motors. On account of the finite length of the linear motor velocity-dependent end effects (Ref. 3,4) are obtained. In addition to a deterioration of the efficiency, these end effects cause a weakening of the magnetic field resulting in velocity-dependent motor characteristics.

Fig. 3 shows characteristics which were measured on a linear motor test assembly with rotating disc. The linear motor was supplied by a self-commutated converter with variable frequency. A difference is made between range (I) limited by the current and range (II) with the additional limitation by the voltage. In range (I) the respective operating point was selected which provided the maximum tractive force at a given current. The influence of the end effects is to be seen in a decrease of the attainable maximum tractive force with growing speed.

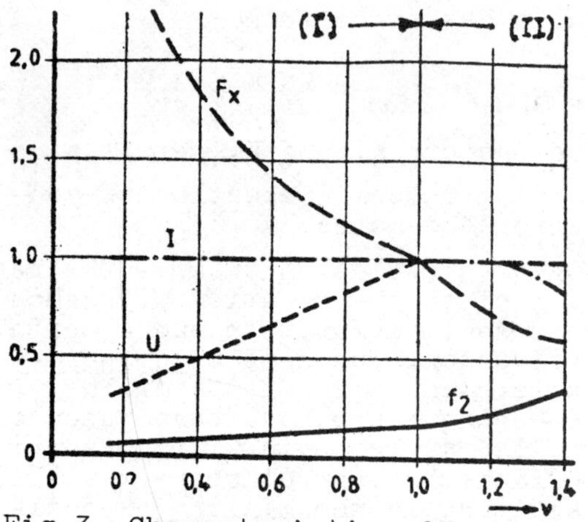

Fig.3. Characteristics for tractive force F_x, voltage U, current I and secondary frequency f_2 of a frequency-controlled linear motor vs. velocity

The appertaining secondary frequency increases simultaneously (Ref. 5). In the range (II) the secondary frequency has still to be increased as in the case of a rotating induction motor in the field weakening range.

3. Control methods for the tractive effort

3.1 Control methods for the torque of rotating induction motors

Besides the control characteristics for voltage, current and secondary frequency along the boundary line in the torque-speed diagram, the control of the size of the torque is an essential feature for vehicle drives. Investigations carried out to find a good dynamic behaviour resulted in control methods with constant flux, e.g. (Ref. 6,7,8). Such a control method may, however, lead to unfavourable dimensioning conditions for the converter. A pulse-width-modulated inverter for instance, which is operating in the field weakening range with a pulse frequency equal to the fundamental output frequency, supplies a constant output voltage necessitating respective control methods (Ref. 9). A further example which will be discussed in detail is the current source inverter (Fig. 4).

The function of the inverter is described in (Ref. 10,11). The equivalent circuit diagram used for the induction motor has been derived in (Ref. 12,13). The stator resistance has been neglected, the leakage inductance is constant and the induced voltage is assumed to

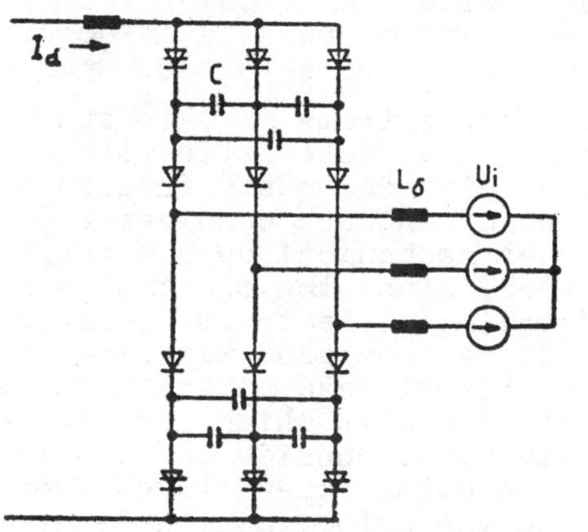

Fig.4. Current source inverter

be sinusoidal. The quantity which is decisive for the voltage stress on the semiconductors, is the voltage occurring on the commutating capacitors. The capacitor voltage is applied, at the same time, to the thyristors and - increased by the line-to-line motor voltage - also to the diodes. According to (Ref. 10,11) the capacitor voltage is calculated as follows

$$U_C = \hat{U}_{iRS} \cdot \sin\varphi_i + 2 \cdot \dot{I}_d \cdot \sqrt{\frac{L_\sigma}{3 \cdot C}} \qquad (4)$$

where φ_i is the phase angle between U_i and the fundamental of the stator current. This relation presupposes, above all, that no changes of induced voltage and dc current occur during the commutation and that the commutating operations do not influence each other. The commutating operation consists of the linear reversing the charge of the capacitors and the diode commutation. Under the given conditions the total duration is

$$T_K = (1 + \frac{\pi}{2}) \cdot \sqrt{3 \cdot L_i \cdot C} + 3 \cdot C \cdot \frac{\hat{U}_{iRS} \cdot \sin\varphi_i}{I_d} \qquad (5)$$

With given motor parameters as well as with certain required maximum values for output current and voltage, the occurring maximum capacitor voltage can be reduced only by increasing the commutating capacity whereby, however, the commutating period is increased. This period is, in turn, normally limited to a maximum value $T_{Kmax} = 1/(3f_{1max})$ by the requirement that the commutating operations of one bridge half should not overlap each other. If T_{Kmax} is given in this way, the maximum value for the capacity can be determined from equation (5). For this purpose, the data of the

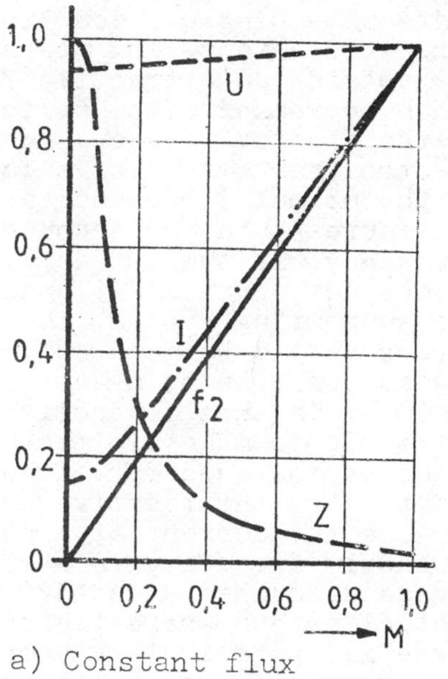

a) Constant flux

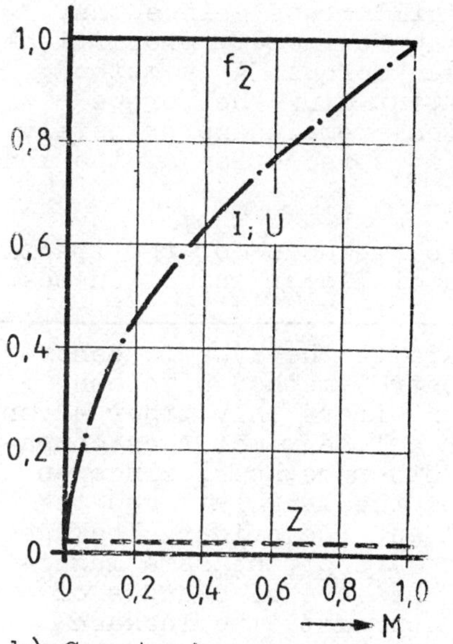

b) Constant secondary frequency

Fig.5. Curve of voltage U, current I, secondary frequency f_2 and abbreviation $Z = (\hat{U}_{iRS} \cdot \sin\varphi_i)/I_d$ vs. torque at constant speed

most unfavourable operating point have to be substituted which depend on the selected control method.

Fig. 5a shows the curves for some essential quantities vs. torque at constant maximum speed for a control method with constant flux. The data of a 4-pole 130 kW-motor with a field weakening range of f_1 = 50 Hz to 100 Hz have been used. According to equation (5) the quantity $\hat{U}_{iRS}\sin\varphi_i/I_d$, abbreviated by Z and shown in Fig. 5a, is decisive for the dimensioning of the capacitor. The dimensioning is obviously dependent on the no-load point with a value for Z which is higher by approximately the factor 50 than during rated operation. As a contrast, Fig. 5b shows a control method with constant secondary frequency at which Z remains unchanged as against the rated operation.

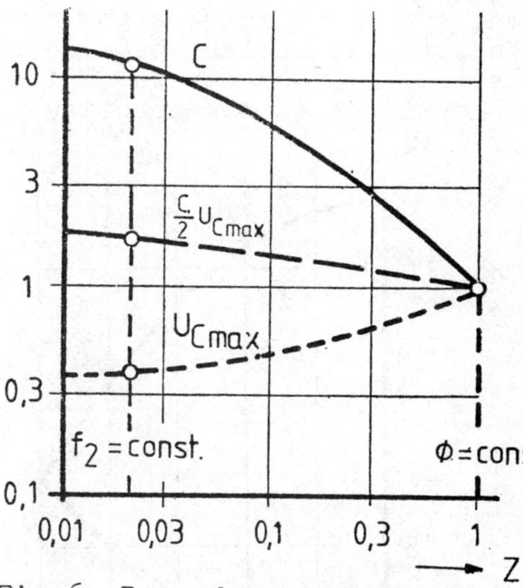

Fig.6. Dependence of capacity, capacitor voltage and size on quantity $Z = (\hat{U}_{iRS} \cdot \sin\varphi_i)/I_d$

From Fig. 6 it can be seen which maximum value can be obtained for the commutating capacity depending on the quantity Z. At the same time, the maximum occurring capacitor voltage is given as well as the value $\frac{C}{2}U^2_{Cmax}$ forming a measure for the capacitor size. With the two control methods described the values to be obtained are accentuated. At a rated voltage of the motor of U_{RS} = 380 V, the maximum capacitor voltage is U_{Cmax} =1370 V at a control method with constant flux. It is only U_{Cmax}=525 V at a method with constant secondary frequency causing a considerable reduction of the semiconductor expenditure.

With regard to the application in a vehicle a control method with constant secondary frequency is favourable also with regard to the torque oscillations, since they decrease approximately proportional to the mean torque. At a method with constant flux the torque oscillations remain approximately constant at a reduction of the mean torque.

3.2 Control methods for the tractive force of linear induction motors

With regard to the dynamic behaviour a control method with constant flux offers only minor advantages for a linear motor as compared with other methods, since on account of the large air gap the principal and the primary leakage inductance are of the same magnitude. In the case of changes of force and current, the leakage field has to be changed, too, even if the principal field is constant. Special requirements may be imposed on the control method by the normal forces which occur between the primary and secondary part of single-sided motors with secondary parts which are wholly or partly of ferromagnetic material. Within the normal operating range with low secondary frequencies these forces are attractive and may be of the same magnitude as the tractive force.

Fig. 7a shows results of measurements on a single-sided linear motor. At constant flux operation one obtains an attractive force which decreases with the increase in the tractive force, since the electrodynamic repulsive component of the normal force increases with the increase in the secondary frequency. If the attractive normal forces act as an additional load on the supporting system, the normal forces should be generally kept as low as possible by means of the control. In this connection, it is advisable to maintain the maximum value of the secondary frequency and to vary the tractive force by varying the current at a constant secondary frequency in a similar way as described under 3.1. This control method where the normal force is approximately proportional to the tractive force is shown in Fig. 7b.

If the single-sided linear motor is installed in the vehicle in such a way that its normal force contributes to the suspension of the vehicle, a control method could be required where the normal force remains constant. In such a case a method is obtained with which the primary current as a function of the secondary frequency has to be

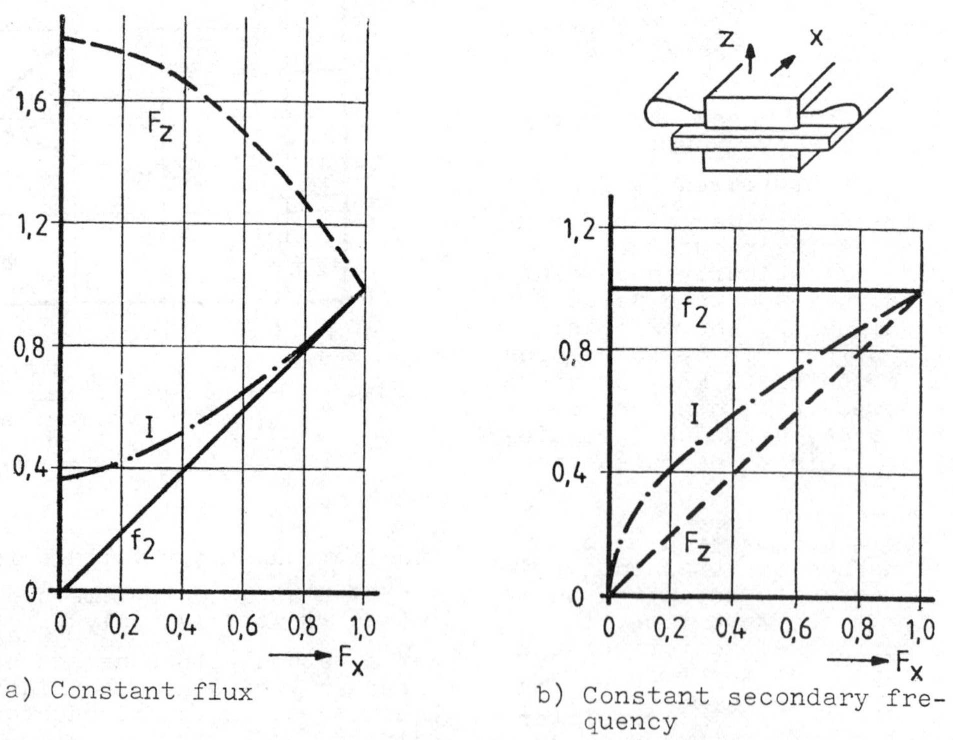

a) Constant flux b) Constant secondary frequency

Fig.7. Curve for current I, secondary frequency f_2 and normal force F_z vs. tractive force F_x for a single-sided linear motor.

set in such a way that the magnetic flux increases at a certain degree with the increase in the secondary frequency.

4. Braking with series resistors

With drives for suburban vehicles, however also for very high velocities, the required braking power is often considerably higher than the tractive power. In this connection, the higher power is generally not provided by a higher maximum torque during braking, but by a greater range with constant torque. This means approximately the same motor current as during traction, however higher voltages. In order to avoid a dimensioning of the converter for these higher voltages, one can try to reduce the voltages by series resistors to the motor windings. To keep the expenditure as low as possible, one intends to use fixed resistors which must, however, be capable of being bridged.

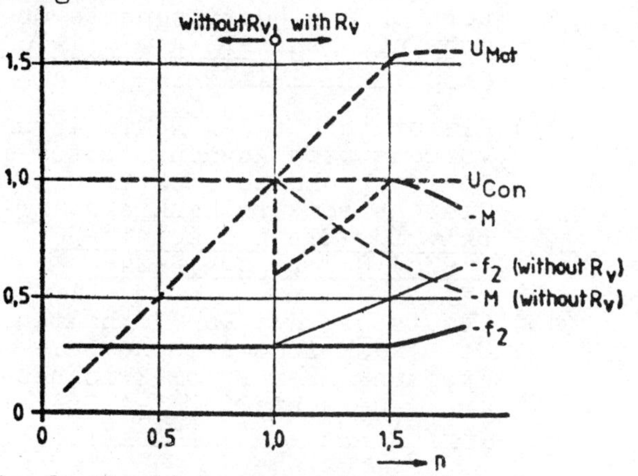

Fig.8. Control characteristics vs. speed for braking with series resistors R_V.

Fig. 8 shows the curves for some quantities vs. speed. With a decreasing speed the braking resistors have to be bridged when the power transmitted by the converter becomes zero. With series resistors below this point positive active power had to be transmitted by the converter which would result in an energetically unfavourable feature forbidding, at the same time, a mains-independent braking. In Fig.8 a limitation of the converter output voltage has been assumed. With the same converter voltage a braking power which is approximately 50 % higher can be obtained in this example by means of the series resistors. The power factor of the motor has a strong influence on this percentage.

Fig. 9 shows a comparison of different control methods for the braking torque at constant speed and operation with series resistors. In Fig. 9a it is to be seen that with a method with constant flux the converter voltage increases with decreasing torque. Such a control method would not offer any advantages with regard to the dimensioning on account of the series resistors. Fig. 9b shows curves resulting from a control method with constant converter voltage which represents a borderline case. Control methods may also be possible where the converter voltage decreases with the decreasing torque as, for instance, a method with constant secondary frequency (Fig. 9c).

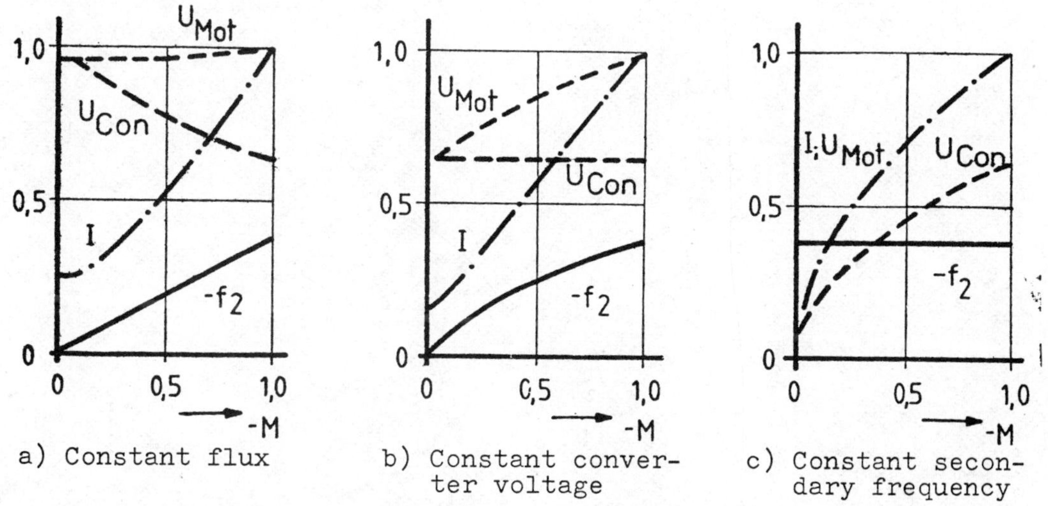

a) Constant flux b) Constant converter voltage c) Constant secondary frequency

Fig.9. Control methods for the braking torque at constant speed and operation with series resistors

5. References

(1) Leitgeb,W.: Zur Bemessung drehzahlveränderbarer Antriebe konstanter Leistung mit stromrichtergespeisten Drehfeldmaschinen. ETZ-A 94 (1973) H. 10, pp. 584-588.

(2) Teich,W.: Grundsätzliche Abhängigkeiten der Fahrmotor- und Wechselrichter-Dimensionierung beim Asynchronmotorantrieb. Eisenbahntechnische Rundschau 24 (1975) H. 1/2, pp. 44-50.

(3) Mosebach,H.: Effekte der endlichen Länge und Breite bei asynchronen Linearmotoren in Kurzständer-und Kurzläuferbauform. Dissertation TU Braunschweig 1972.

(4) Deleroi,W.: Der Einlaufvorgang beim asynchronen Linearmotor. ETZ-A 94 (1973) H. 8, pp. 488-493.

(5) Feldmann,U.; Sattler,Ph.K.: Steuerung von asynchronen Linearmotoren mit großen magnetischen Reynoldszahlen. ETZ-A 96 (1975) H. 7, pp. 311-316.

(6) Naunin,D.: Ein Beitrag zum dynamischen Verhalten der frequenzgesteuerten Asynchronmaschine. Dissertation TU Berlin 1968.

(7) Hasse K.: Zur Dynamik drehzahlgeregelter Antriebe mit stromrichtergespeisten Asynchron-Kurzschlußläufermaschinen. Dissertation TH Darmstadt 1969.

(8) Blaschke,F.: Das Verfahren der Feldorientierung zur Regelung der Asynchronmaschine. Siemens Forschungs-und Entwicklungsberichte 1 (1972) H. 1, pp. 184-193.

(9) Abraham,L.; Kuhn,W.: Möglichkeiten und Grenzen bei der Regelung von umrichtergespeisten Kurzschlußläufermaschinen. IFAC-Symposium Düsseldorf (1974), Vol.1, pp. 587-602.

(10) Schröder,D.: Selbstgeführter Stromrichter mit Phasenfolgelöschung und eingeprägtem Strom. ETZ-A 96 (1975) H. 11, pp. 520-523.

(11) Lienau,W.; Müller-Hellmann,A.: Möglichkeiten zum Betrieb von stromeinprägenden Wechselrichtern ohne niederfrequente Oberschwingungen. ETZ-A 97 (1976) H. 11, pp. 663-667.

(12) Kleinrath,H.: Der Kommutierungsvorgang beim Asynchronmotor mit Speisung über Umrichter mit Gleichstromzwischenkreis. Archiv für Elektrotechnik 56 (1974) H. 1, pp. 12-24.

(13) Klautschek,H.: Das Verhalten der Induktionsmaschine bei Speisung über Strom-Zwischenkreis-Umrichter. ETZ-A 95 (1974) H. 5, pp. 283-287.

CONTROL SYSTEM WITH CONVERTER FOR IMPROVING THE DYNAMIC RESPONSE OF MAGNETICALLY SUSPENDED VEHICLES*

E. Koopmann and U. Wiedemann

Brown, Boveri & Cie AG, Mannheim

SUMMARY

Rapid transit vehicles with magnetic suspension using the electrodynamic principle require additional damping forces that become effective on the primary side, i.e. between the supporting and guidance magnets and the track, in order to improve the travelling comfort in the passenger compartment.

These damping forces are produced through a damping control circuit; in this circuit, a damping coil arranged below the supporting and guidance magnets and supplied from a converter serves as actuator. The damping control system acts on the subordinated current control circuit of the converter. An adaptive controller in the subordinated current control circuit ensures a uniform and constant control gain and dynamic control response also in case of low actual current values when the double converter changes to intermittent current flow despite the presence of inductances in the system.
The requirement for a quick reversal of current flow can also be met by the provision of a high-speed current-zero detector. On the basis of simulations and experimental work, the results of which are described, it was possible to prove the effectiveness of active primary damping, i.e. of the system comprising damping coil, converter and regulator.

1. INTRODUCTION

For physical reasons, the natural damping of the electrodynamic suspension system is only low. This fact, combined with the maximum speed of the suspended vehicle of v = 140 m/s, leads to very high accelerations in the passenger compartment even at small irregularities in the track, which may be due to the elastic deflections of the track supports or the settling of support-mounted tracks, or to installation tolerances in the case of tracks run on the ground. On the other hand, the passenger can tolerate only certain acceleration limits which are defined in the so-called travel comfort curves.
To meet these two requirements, i.e. technically feasible minimum track tolerances and reasonable passenger comfort, it is necessary to have damping facilities on the primary side, in addition to a secondary-side suspension and guidance system as used already in other tracked transport systems such as the rail-wheel system. In principle, the vehicle is here subdivided into a magnet carrier and a passenger compartment. Arranged between the track and the magnet carrier, the primary side, are magnets which ensure, by an appropriate lifting force - suspension height characteristic in the direction of lift, and a lateral force - lateral deflection characteristic in the direction of guidance, that the vehicle fairly smoothly follows the track. An additional system that becomes effective on the primary side is an active damping system which produces damping forces exclusively on the primary side through normally-conducting coils located below the supporting and guidance magnets.

2. PRINCIPLE OF ELECTRODYNAMIC SUSPENSION

The principle of electrodynamic suspension is based on a simple induction process: A current-carrying coil which moves with the vehicle over a conducting plate embedded in the track, induces eddy currents in that plate; these eddy currents are coupled with a magnetic field which opposes that of the coil. This gives rise to a repulsive force which consists of two components. The one component acts in a vertical direction with respect to the plate, the other one is a force impeding the movement of the vehicle. The lifting and braking forces always occur jointly.

* This work was sponsored by the Ministry of Research and Technology within the framework of the research projects TV 7428 and TV 7519.

The decisive advantage offered by electrodynamic suspension is its inherent stability due to the repulsive forces. In this respect, it differs from the electromagnetic suspension principle with which, owing to the attractive forces, the levitation process can only be achieved with the aid of a stabilizing control. A disadvantage associated with the electrodynamic principle is the fact that a certain initial speed must first be reached before levitation starts. For this reason, starting and stopping facilities are required.

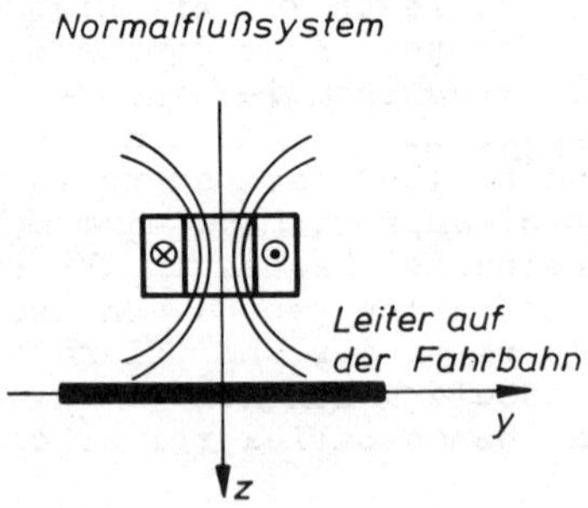

Fig. 1 Principle of electrodynamic lifting, normal flux system

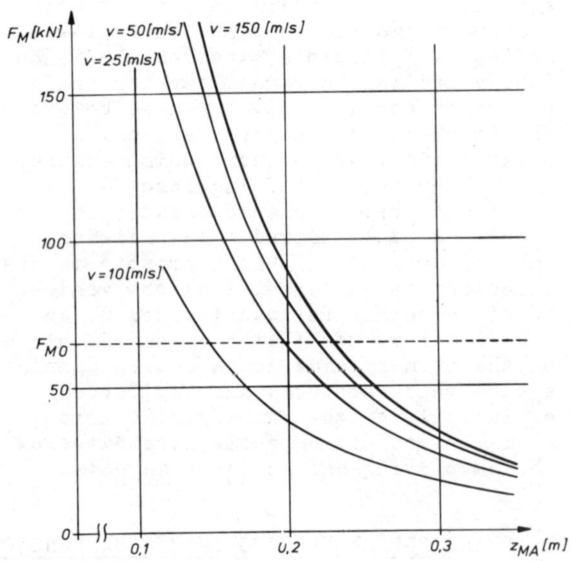

Fig. 2 Lifting force - suspension height characteristic of a normal flux magnet

The very high magnetic fields which are required to obtain high suspension heights are produced by superconducting magnets. The coils cooled to the necessary low temperatures permit the flow of heavy currents in a small conductor cross-section; the current flows in the magnetic coil practically without meeting any resistance.

With the normal flux suspension method (1) which is here used in the direction of lift, the magnets are located above the conductor surface of the track (Fig. 1). The curve of the lifting force plotted against the suspension height at different travelling speeds of a normal flux supporting magnet, calculated for a prototype test vehicle, is shown in Fig. 2.

The forces required in the direction of guidance are produced through a difference flux guiding facility (Fig. 3). The excitation coil arranged in the vehicle

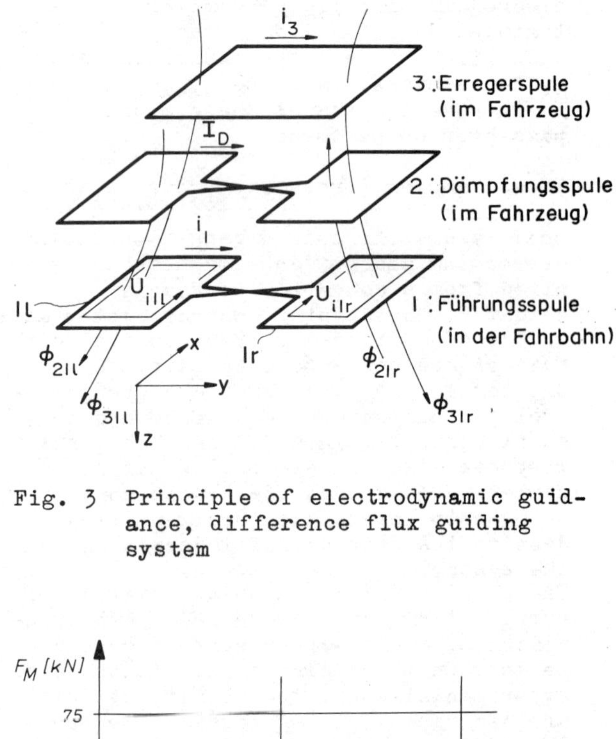

Fig. 3 Principle of electrodynamic guidance, difference flux guiding system

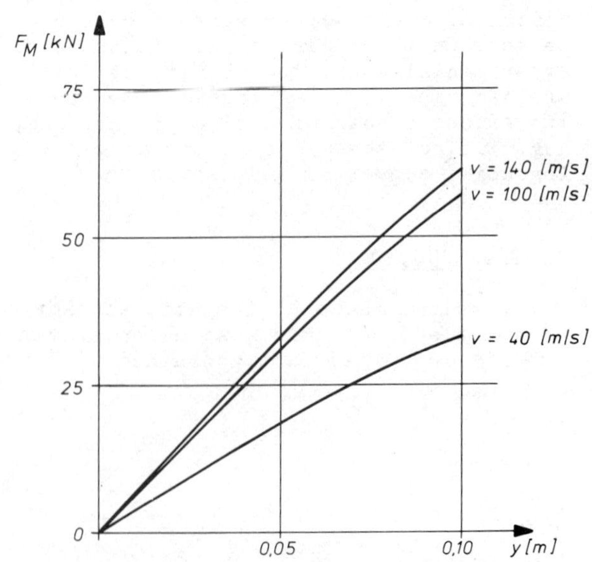

Fig. 4 Lateral force - lateral deflection characteristic of a difference flux guiding system

produces in the track-mounted, eight-shaped guiding coil the partial magnetic fluxes ϕ_{311} and ϕ_{31r}. These magnetic fluxes have the same value in the absence of a lateral deflection y_o and cancel each other because of the opposite directions of flow in the coil halves 1l and 1r. If the excitation coil is displaced in the y-direction by y_o, a flux difference of $\Delta\phi = \phi_{311} - \phi_{31r}$ and, consequently, a guiding force between coils 1 and 3 will result. The curve of the lateral force as a function of the lateral deflection at different travelling speeds is shown in Fig. 4. A controllable additional force for the active primary damping in the direction of guidance can be obtained by changing the flux difference $\Delta\phi$. If a normally-conducting, eight-shaped damping coil is arranged in the vehicle below the excitation coil and supplied with control current I_D, the values of the partial fluxes ϕ_{11} and ϕ_{1r} will change. This gives rise to an additional force in the direction of guidance which is dependent on current I_D and can be used for damping the system.

3. PRIMARY-SIDE LIFTING OR GUIDING AND DAMPING COMPONENTS IN PROTOTYPE TEST VEHICLES

Suspended vehicles of track-guided rapid transit systems mainly consist of the passenger compartment and a magnet carrier (Fig. 5). On the secondary side, i.e. between the passenger compartment and the magnet carrier, an air-type-spring system separated into the direction of lift and the direction of guidance takes over the function of maintaining a prescribed travelling comfort in the passenger compartment and to position the latter in relation to the track. On the primary side, the supporting magnets and the excitation coils for propulsion and guidance are mounted on the magnet carrier. The track incorporates the conducting plates for the supporting system and the windings for guidance and propulsion. The propulsion system operates on the principle of the coreless synchronous oblong stator motor (2); the excitation coil takes over the additional function of a guidance magnet. The normally-conducting damping coils are mounted below the supporting and guidance magnets. The damping system is active, only primary-side damping forces being produced through damping coils. These additional coils do not have the function of taking up additional stationary lifting and guiding forces, but are rather intended to serve as actuator for compensating any irregularities in the track. Owing to their function, these additional coils can be rated for a low value in relation to the supporting and guidance magnets. The concept underlying the control system is that the damping coils are controlled in groups by individual control circuits both in the direction of lift and in the direction of guidance.

The separation of the primary system into a passive supporting and guiding system on the one hand and the active damping system on the other offers an essential advantage. Even if the damping control fails, the vehicle still fully performs the functions of support and guidance, though at the expense of the travelling comfort.

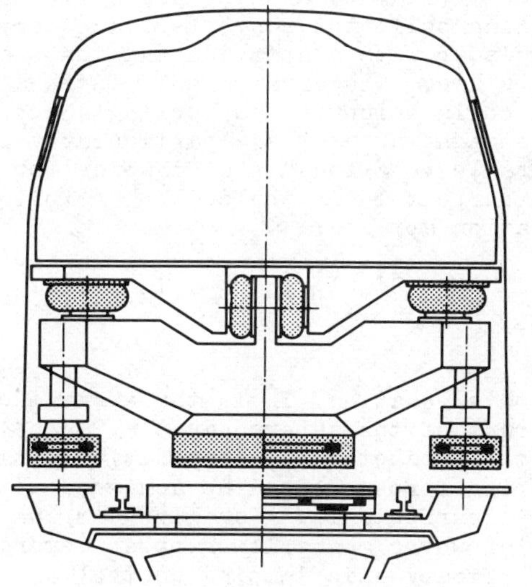

Fig. 5 Prototype test vehicle

By the passive behaviour of the supporting and guidance magnets the air gap between the magnet and the track changes according to the non-linear lifting force - suspension height characteristic of the supporting magnet and the nearly linear lateral force - lateral deflection characteristic of the guidance magnet. Consequently, steady-state position deviations will occur which are within a permissible range determined by the rating of the magnets. Owing to changes in the air gap and the speed, the operating points move on the magnet characteristics, thus influencing the dynamic response of the primary system. The adaptation of the control parameters of the damping control circuit to the respective system parameters by a superordinated adaptation ensures optimum control response.

The good dynamic behaviour of electrodynamic suspension systems is due to the relatively soft suspension and guidance systems on the primary side. In relation to the total mass of the vehicle, the natural frequencies in the direction of lift are 1.5 Hz, and in the direction of guidance 2 Hz (at a travelling speed of v = 140 m/s). This degree of softness is only possible because of the fact that the changes in the vehicle's position which are due to the influence of external forces such as aerodynamic loads, varying inertia forces caused by the vehicle following the track and by changes in the pay load, can be taken up by the large air gap.

4. DAMPING CONTROL CIRCUIT

Major components of the damping control circuit (Fig. 6) are the normally-conducting coil, the converter, the damping regulator with adaptation and the measuring system. The arrangement of the damping coils below the supporting and guidance magnets ensures a particularly favourable relationship of damping force F_D and current I_D in the coil. The following equation applies:

$$F_D \sim 2 I_M I_D + I_D^2 \ , \quad I_M \gg I_D \ .$$

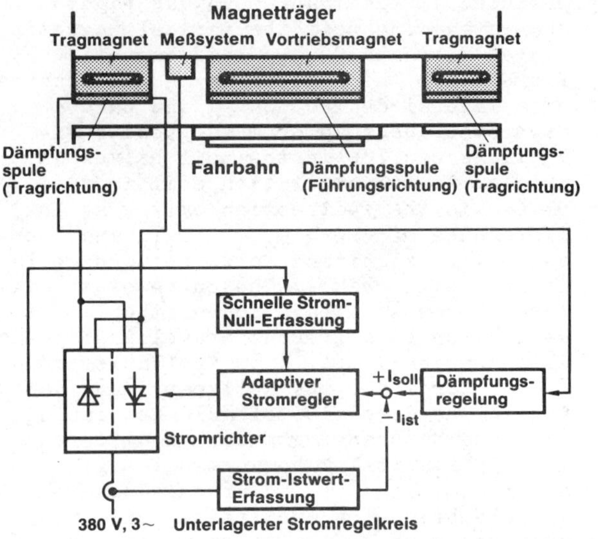

Fig. 6 Primary-side damping control circuit

In this equation, I_M is the excitation current of the superconducting magnets. By the product $I_M I_D$, comparatively high damping forces F_D can be achieved with small currents I_D. A smaller damping current however means lower power demand for the primary-side damping control and a smaller winding cross-section for the additional coil. By the arrangement of the damping coils below the supporting and guidance magnets the magnetic field of the latter can be increased or reduced by the field required for the damping force F_D.

The primary-side damping control circuit consists of a subordinated current control circuit and a superordinated damping control circuit. The controller in the superordinated damping control circuit acts on the subordinated current control circuit in accordance with the signals of the measuring system attached to the magnet carrier. The acceleration of the magnet carrier in the directions of lift and guidance is used as the only measuring quantity.

The control concept has to meet requirements which result directly from the dynamics of the movement of track-bound vehicles: On the one hand, the controller must compensate any track irregularities acting on the magnets, thus damping the magnet carrier, and on the other cause the vehicle to follow the track in the lower frequency range, i.e. when running through curves and at up-grades. From this it follows that the primary-side control circuit must exhibit both a good response with respect to irregularities in the track, and a good following characteristic. Following characteristic in this context means that, on the primary side, the vehicle can position itself relative to the track according to the characteristic curves of the magnets. These requirements are fulfilled by a controller with PDT_1 action and a high-pass filter which suppresses the low frequencies of the acceleration signal. To ensure that the controller operates optimally at different operating points, the controller parameters are adapted to the changed system parameters as, for instance, different rigidity of the supporting and guidance system as a result of a change in the air gap and travelling speed. The adaptation can be achieved in an especially simple manner by a digital system, e.g. by a microcomputer. This computer calculates the control parameters in the individual operating points on the basis of the system characteristics stored, as a function of the air gap between the track and the magnet, and different travelling speeds.

5. CONVERTER

In the case of irregularities in the track, a magnetic field is set up in addition to the stationary field produced by the superconducting supporting magnets. The force of this field counteracts the irregularities, thus reducing the acceleration of the passenger compartment considerably.

This magnetic field is produced by the subordinated current control circuit consisting of a converter, the current controller and the damping coil.

It is necessary to use a double converter

as actuator for the damping coil
since both attractive and repulsive forces have to be produced in the case of track irregularities. These forces are produced by increasing or reducing the magnetic field generated by the supporting magnets, i.e. the control element must permit reversal of the current direction.

For reasons of weight - dead weight of the current limiting reactors - and in view of the energy consumption, the use of a double converter with circulating current is not possible.

The circulating current, which is of the order of 10 to 20 % of the rated current, requires too much energy which contributes only a constant proportion to the lifting force.

The advantages of a double converter carrying circulating current, i.e. the prevention of intermittent current flow by the base load and quick reversal of the direction of current flow due to the fact that both converters are always in operation, are obtainable with circulating-current-free double converters only by the provision of additional electronic trigger and control equipment.

The demand for a short reversing interval - the period during which both converters are blocked - necessitates the use of a fast-response current-zero detector. An electronic detector with a response time of $t < 400\ \mu s$ makes it possible to reduce the dead time to a minimum. The dead time of $t \approx 2$ ms adjusted on this converter is necessary to ensure observance of the reverse recovery time and the circuit-commutated recovery time of the thyristors.

The fast-response current-zero detector measures the voltages across 6 antiparallel thyristors at a time. In the case of current values differing from zero, a low voltage drop is noticeable across the thyristors. In the blocked condition ($I = 0$) the voltage rises abruptly. Potential isolation between the converter section and the electronic trigger and control equipment is effected by means of optocouplers. A signal emitted when the current passes through zero, is logically combined with other boundary conditions in the current direction reversing control. The schematic arrangement of the current control circuit is shown in Fig. 7. The double converter used here consists of two antiparallel six-pulse three-phase bridge connections. Each bridge takes over current conduction only for one direction, this bridge being blocked upon reversal of current flow. A comparator provides a logic signal, as a function of the direction of the current reference, which is combined in two AND gates with the signal of the fast-response current-zero detector. Consequently, for changing over of the thyristor firing pulses to the other bridge, the following conditions must be fulfilled:

1) The current reference must change its sign, and

2) the current actual value in the coil, which lags the reference value because of the inductance, must become zero.

A subsequent OR circuit drives a dynamic flip-flop. Through signal switches, the firing pulses for the thyristors are then transmitted to the conducting converter bridge.

Unlike a double converter carrying circulating current, a circulating-current-free design needs only one trigger unit whose pulses are supplied to the appropriate bridge by the logic of the current direction reversing feature. Therefore, there is no need for synchronizing the two trigger units, an operation that presents difficulties in the case of a double converter carrying circulating current. Another disadvantage associated with the circulating-current-carrying double converter is the necessity of having an additional load-dependent control of the circulating current.

With this circulating-current-free double converter, observance of the circuit-commutated recovery time of about 2 ms during reversal of the direction of current flow requires that a pulse and controller blocking signal is given to the pulse generating stage and the controller unit with the aid of time elements. No firing pulses are released during this period.

The pulse generator, or trigger unit, has the function of shifting the firing pulses in dependence on the d.c. control

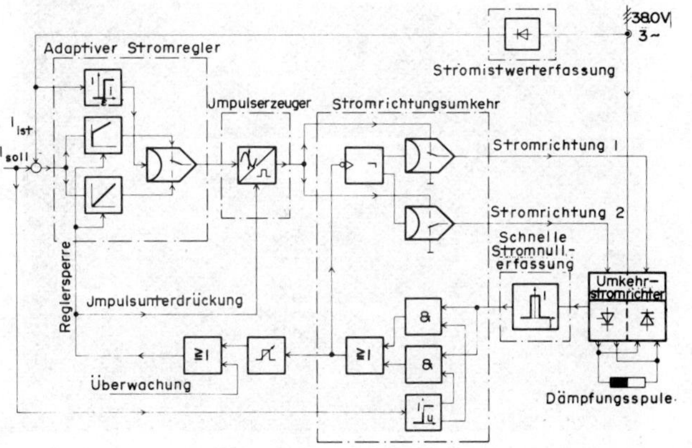

Fig. 7 Subordinated current control circuit

voltage in such a manner that as uniform as possible a voltage amplification results over the whole control range. Furthermore, the trigger unit is designed to confine the control range between the rectifier limit and the inverter limit to prevent conduction through of the converter.

For pulse displacement, the phase control principle has been adopted. The sinusoidal a.c. control voltage is here compared with the d.c. control voltage derived from the control deviation. As soon as the a.c. control voltage falls below the d.c. voltage, the firing pulse is generated.

The amplification of the converter is V = 51,3 for a rated value of the mains voltage by U_N = 380 V. This amplification factor applies only for continuous current flow. However, since the application requires a constant reversal of the current direction, an adaptive current controller must ensure during intermittent current flow that the amplification remains the same over the whole control range.

The intermittent current flow detector which operates an analog switch constantly monitors the mean value of the actual current value to find out whether it is above or below the limit of intermittent current flow. This limit is dependent on the firing angle of the converter and the time constant $\tau = L/R$ in the load circuit. In rectifier and inverter operation, the current controller operates as PI controller, in intermittent operation as I controller. The limit of intermittent current flow must be adjusted in dependence on the inductance in the load circuit. The limit between intermittent and continuous current flow is determined by an adjustable current threshold. The amplifications and the time constants of

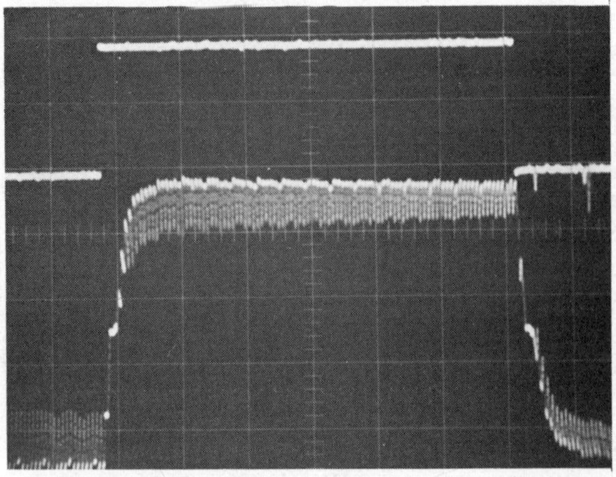

Fig. 8 Curve of the current actual value with a rapid change in the reference value

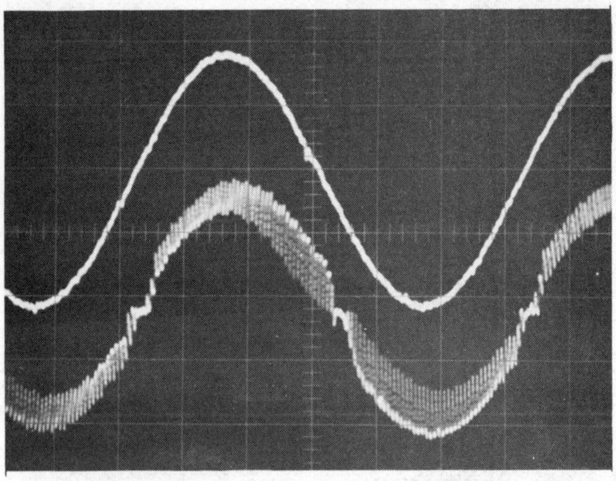

Fig. 9 Curve of the current actual value when a sinusoidal reference value (frequency f = 1 Hz) is preset

the PI and I controllers are adjusted to optimal values when sinusoidal reference values are preset. Sinusoidal reference values were chosen since the vehicle, in continuous operation, is caused to perform sinusoidal oscillations as a result of periodic track irregularities. The system behaviour of the subordinated current control circuit in response to an abrupt change in the reference value is shown in Fig. 8. Figures 9 and 10 show the curve of the current actual value when a sinusoidal current value of the frequency of 1 Hz and 5 Hz is preset.

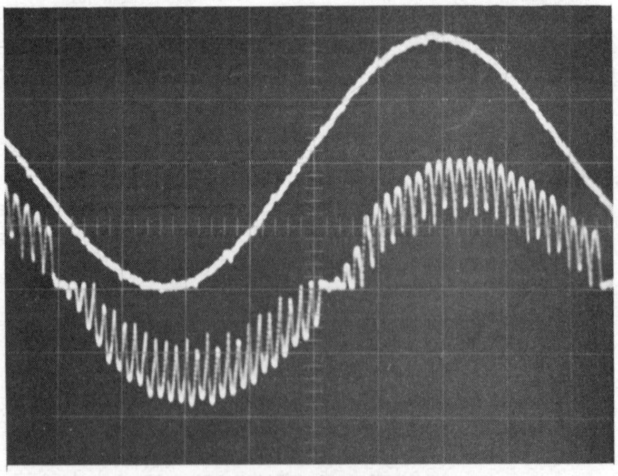

Fig. 10 Curve of the current actual value when a sinusoidal reference value (frequency f = 5 Hz) is preset

In normal operation, the continuous current of the damping coil is not more than I_D = 10 A. The converter is capable of carrying a current of I_{rated} = 33 A. It is rated for a higher value to provide an adequate voltage reserve in the event that also non-sinusoidal track disturbances occur. For the current reference and actual value curves for different frequencies, shown here, small amplitudes have been assumed so that the ripple appears comparatively large. However, these harmonics are irrelevant for the system behaviour since the frequency of the 1st harmonic of a six-pulse three-phase bridge is at f = 300 Hz. For comparison, the natural frequency of the supporting system is 1.5 to 2 Hz.

Figs. 11 and 12 show the behaviour of the converter in the range of current direction reversal, again for the frequencies of 1 Hz and 5 Hz.

The lower curve represents the respective current actual value while the upper signal represents the controller control signal.

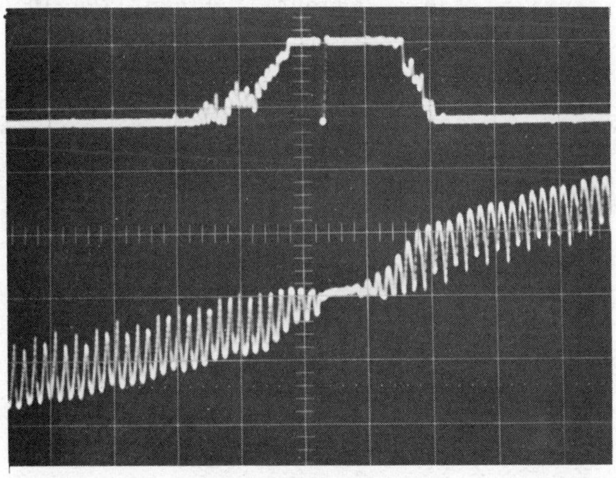

Fig. 11 Curve of the controller control signal and of the current actual value in the range of current direction reversal (frequency f = 1 Hz)

First, the conducting bridge is operating as a rectifier; when the actual value falls below the limit for intermittent current flow, the current controller is switched from PI action to I action. Owing to the ripple content of the actual value, the latter repeatedly exceeds or falls below the threshold value; accordingly, the control signal switches the controller from PI action to I action and vice versa. When the actual value passes through zero, the controller is blocked for 2 ms and switched to PI operation. The firing pulses for the next converter bridge are released by a change-over to I operation since at first current flow is still intermittent. After the limit for intermittent current flow has been

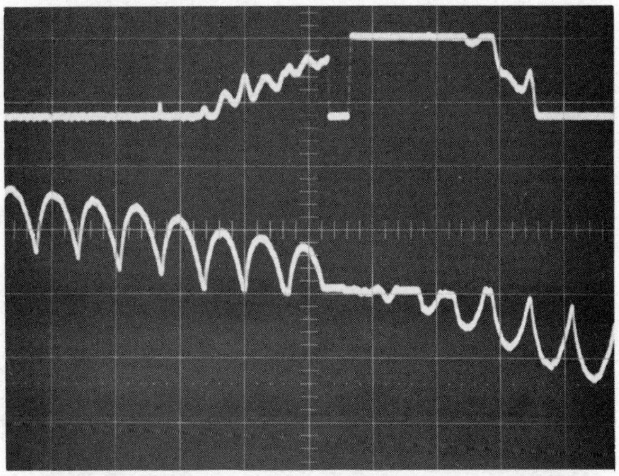

Fig. 12 Curve of the controller control signal and of the current actual value in the range of current direction reversal (frequency f = 5 Hz)

exceeded, the current controller acts again as PI controller and the conducting bridge now changes to rectifier operation. These figures clearly show the reversing interval of about 2 ms which is independent of the excitation frequency. At the end of the reversing interval, the current is immediately caused to rise to its full value as a result of the current reference curve.

6. TEST STAND FOR EXPERIMENTAL TESTING OF THE DAMPING CONTROL CIRCUIT

Before the damping control system was tried in test vehicles, its basic behaviour and effectiveness were tested in a stationary preliminary test. Since, with this test, the parameter "travelling speed" essential for electrodynamic suspension could not be realized, the suspension process could only be initiated through additional measures. The conductor plates arranged in the track as reaction parts, were replaced by normally-conducting, water-cooled mirror coils, which produced a magnetic field acting in the direction of lift. Fig. 13 shows the test arrangement. The mirror coil is mounted on a base frame and has a so-called disturbance coil arranged on it. This last mentioned coil serves for producing definite track irregularities of different amplitudes and frequencies. A magnet carrier supported on one side carries the superconducting liquid-cooled magnet which has the normally-conducting damping coil attached to its underside. By varying the ampere-turns of the mirror coils it is possible to adjust different suspension heights within a wide range. A current of 800 A, with 41 turns, corresponds to a flux of 352.8 k ampere-turns

Fig. 13 Test stand for experimental testing of the active primary damping

in the superconducting magnet. The flux of 32 k ampere-turns (32 turns) in the mirror coil gives a free suspension height of 100 mm. Figure 14 shows the dependence of the free suspension height, i.e. the distance between the bottom edge of the damping coil and the top edge of the disturbance coil, on the energization of the mirror coil. The smallest suspension height is limited on the test stand by the value z_{LO}. The wide operating range permits the investigation of the primary damping control at different suspension heights. The larger the air gap, the lower is the rigidity and, consequently, the natural frequency of the oscillation system.

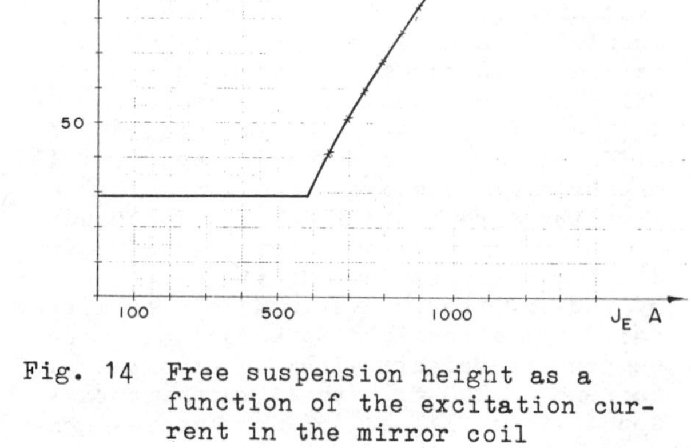

Fig. 14 Free suspension height as a function of the excitation current in the mirror coil

7. SIMULATION AND MEASURING RESULTS OF THE DAMPING CONTROL CIRCUIT

The physical data, such as rigidity and natural damping of the magnetic suspension, are obtained from the so-called die-down test. With this test, the magnet carrier is excited with its natural frequency without active primary damping. After removal of the disturbance, the system comes to rest as a result of its natural damping properties. The damping factor of this arrangement is determined from the amplitude ratios of the die-down test. It is between D = 0.1 and 0.2 for the various suspension heights but is not sufficient for maintaining the travelling comfort in the range of the natural frequency.

A simple simulation model of the damping control circuit permits a comparison of the test results with the simulation results. The signal flow diagram of the model is shown in Fig. 15. For examining the dynamic response of the magnetic suspension, the behaviour is linearized in a definite operating point.

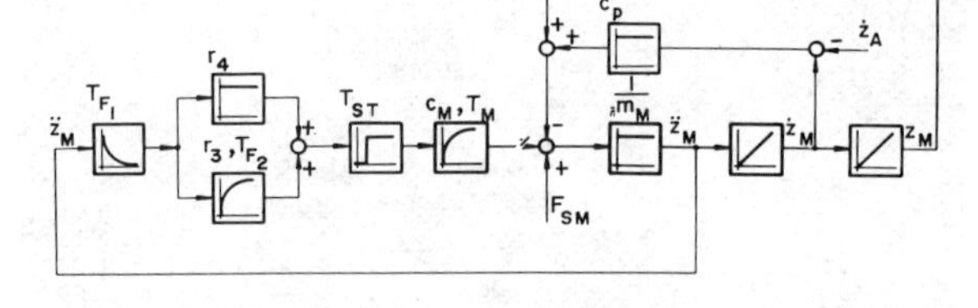

Fig. 15 Signal flow diagram of the damping control circuit

Fig. 16 shows the time behaviour of the important quantities air gap, damping current and magnet carrier acceleration with and without active primary damping. The excitation of the system is near the natural frequency of the magnetic suspension. The acceleration of the magnet carrier $\ddot{z}_M$ is reduced by the active primary damping, by a factor of 5. The middle curve shows the time behaviour of the damping current that falls to zero after elimination of the damping effect, i.e. disconnection of the damping control circuit.

Fig. 16 also shows a comparison of various system quantities between the test stand and the analog computer model for a track disturbance lying in the natural frequency range of the system. It can be seen that the model adequately describes the behaviour of the test stand.

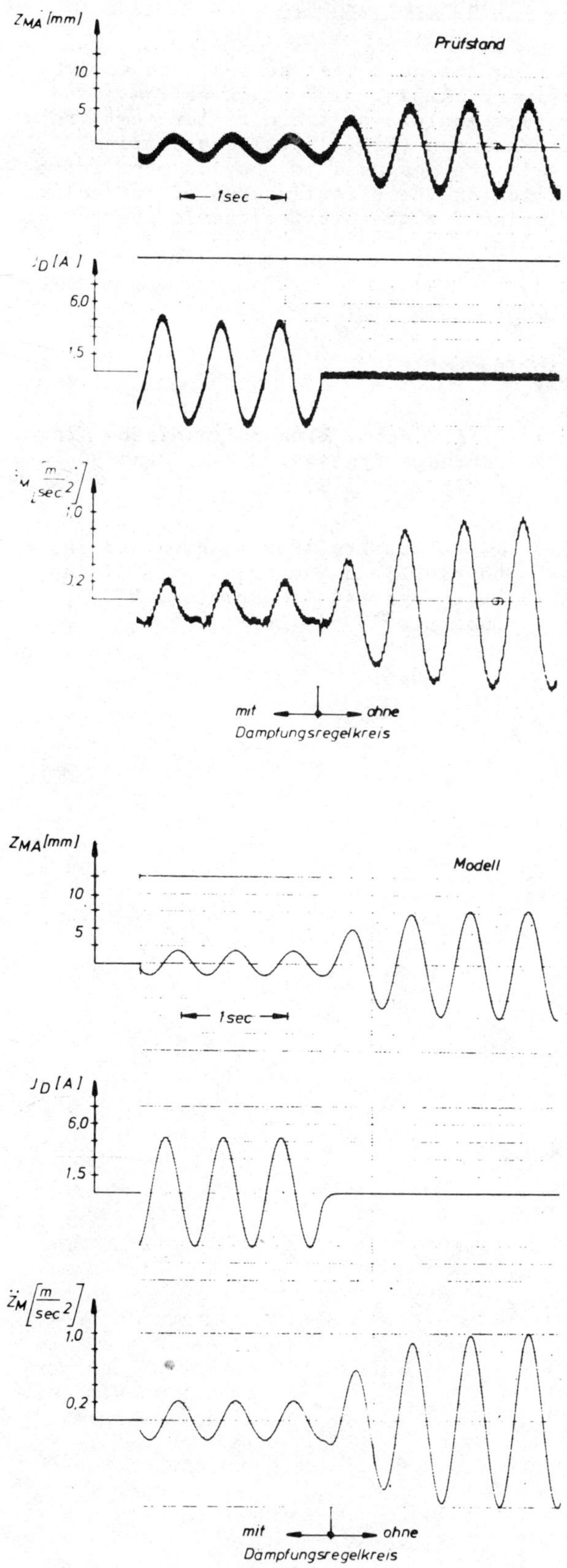

Fig. 16 Comparison of the system quantities of test stand and simulation: air gap z_{MA}, damping current I_D and magnet carrier acceleration $\ddot{z}_M$

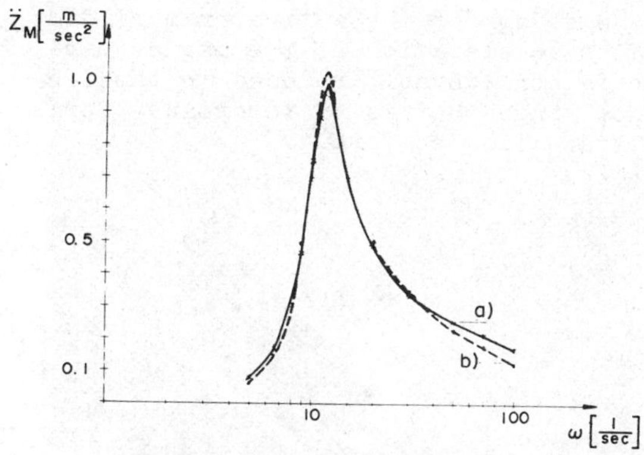

Fig. 17 Frequency response of the magnet carrier acceleration of the undampened system

a) test stand
b) simulation

Fig. 17 shows the frequency response of the magnet carrier acceleration of the undampened system of the test stand and that of the analog computer model. Here, the acceleration of the magnet carrier is measured for force disturbances of constant amplitude but different frequencies. Agreement is ensured also over a wide frequency range. The deviations in the lower and upper frequency range are due to the fact that the resolution of the acceleration pick-ups is inadequate in this range.

Very good agreement between the model and the test stand can also be observed over a wide frequency range when the active primary damping is effective (Fig. 18). By inaccuracies in the overall amplification between the model and the test stand, the frequency response is somewhat displaced but allows description of the system behaviour by the simple model used.

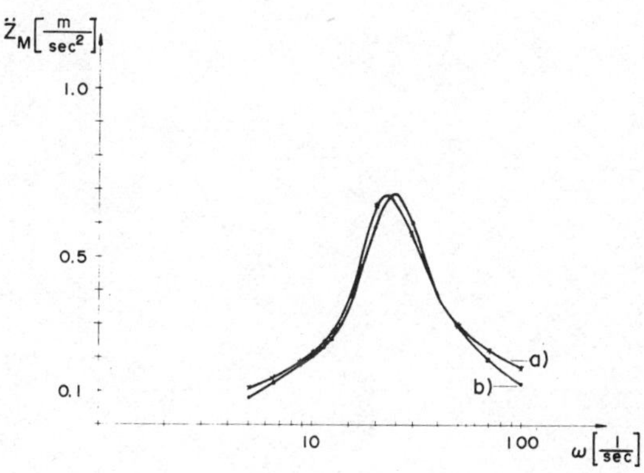

Fig. 18 Frequency response of the magnet carrier acceleration with active primary damping

a) test stand
b) simulation

By the effect of the active primary damping the acceleration of the magnet carrier is considerably reduced in the frequency range that is of importance for the travelling comfort.

Fig. 19 Test vehicle at Erlangen

8. FUTURE DEVELOPMENTS

During the next test series, the active primary damping will be investigated on a test vehicle on the circular test track at Erlangen (Fig. 19). These damping tests are designed to provide more information on the effectiveness of the active damping for the electrodynamic suspension system.

Bibliography

(1) C. Albrecht, Elektrodynamische Trag-Führungs-Systeme, ETZ-A, Band 96 (1975), Heft 9

(2) J. Holtz, Kraftkomponenten und deren betriebliche Steuerung beim eisenlosen Synchronlinearmotor, ETZ-A, Band 96 (1975), Heft 9

MICROCOMPUTER BASED DATA ACQUISITION AND PROPULSION CONTROL FOR A TRACK-POWERED LINEAR SYNCHRONOUS MOTOR FOR HIGH-SPEED GROUND TRANSPORTATION

B. Böning

Institut für Regelungstechnik, Technische Universität Braunschweig, Hans Sommer-Str. 65/66, 3300 Braunschweig, Germany

SUMMARY

Track powered synchronous motors are being developed for driving magnetically levitated high speed tracked vehicles (maglev). The sections of the track winding are supplied by variable frequency thyristor invertors (14 MW, 0 - 50 Hz), located in trackside invertor stations about 15 km apart. In order to obviate the need for a continuous data link between the vehicle moving at high speed and the wayside invertors, the information on the state of the vehicle with respect to the track winding is to be derived from the variables observed at the invertor station. This is illustrated by the block diagram in Fig. 1. At the invertor stations the instantaneous values of active and reactive power are calculated on the basis of the terminal voltages and the pulsating currents. After the ohmic and reactive losses of the track-winding have been eliminated, the pole position angle, the thrust and the velocity are computed and used for closed-loop control. The computations required are performed by microcomputers.

INTRODUCTION

Among the various possible propulsion schemes for maglevs the iron-free linear synchronous motor (LSM) with superconducting excitation coils is of special interest, as it allows a large gap and a flat-bottom vehicle, when combined with a null-flux guidance system and an electrodynamic suspension system. Compared with a short-stator motor the payload of the vehicle is increased and the problems of high-speed track side power collection are avoided. Compared with other synchronous track-powered motors the capital expenditure for the active guideway is relatively low.

In addition to the thrust the LSM produces also considerable lift and guidance forces, because all three spatial components of the vehicle-borne magnetic field take effect at

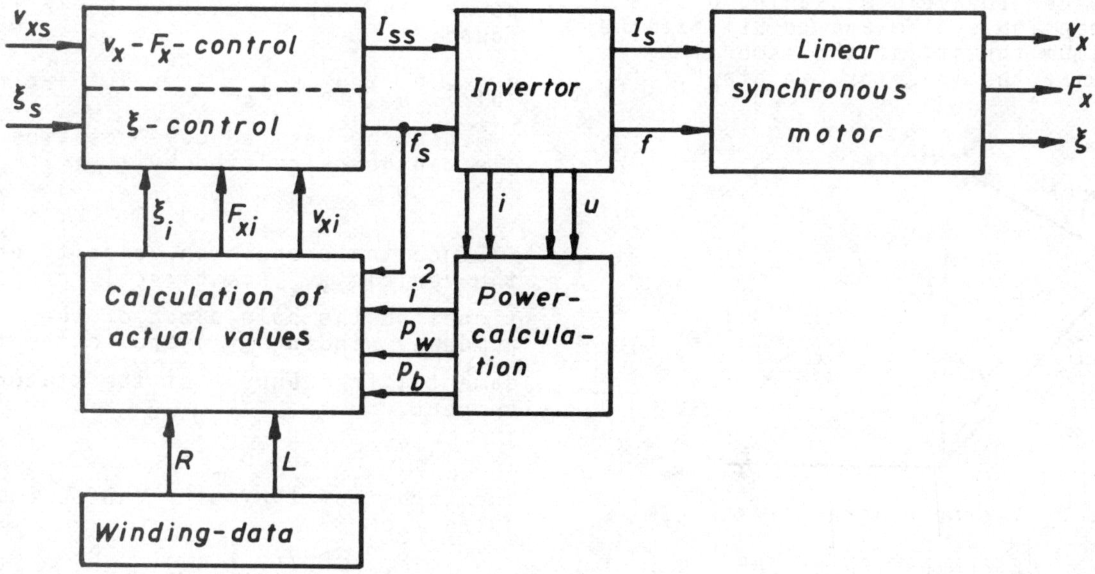

Fig. 1. Block diagram of the control system with calculation of the actual values

the two-dimensional stator-winding. The direction of the force vector is determined by the phase displacement between the vehicle poles and the stator travelling field (pole position angle ξ, Fig. 2). Only if the stator current is fed at $\xi = 0$, does the force vector point in the direction of the velocity. At $\xi = \pi/2$ the motor develops only guidance- and lift-forces. Since the weight-dependent drag forces of the electrodynamic suspension system are relatively high, the motor should be operated at a small positive pole position angle, in order to utilize the lift force of the motor.

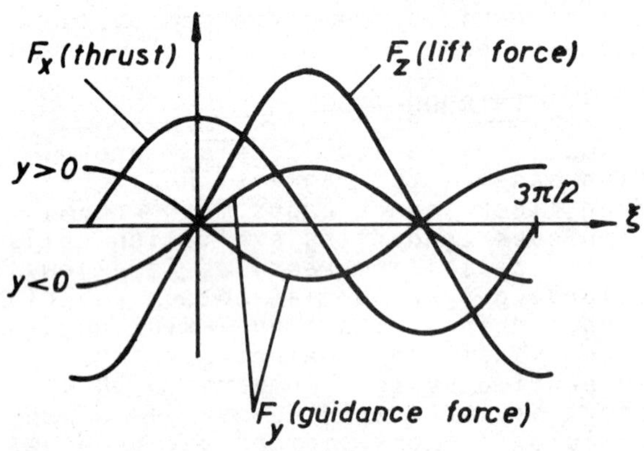

Fig. 2. Motor forces as function of the pole position angle

Another reason for operating the motor in the unstable range can easily be derived from the vector diagram of the LSM (Fig. 3): A positive pole position angle reduces the required reactive power and allows a lower invertor voltage. Control of the pole position angle is necessary, in order to avoid weakening of the suspension system and to utilize the maximum thrust, at the same time keeping the invertor "in step".

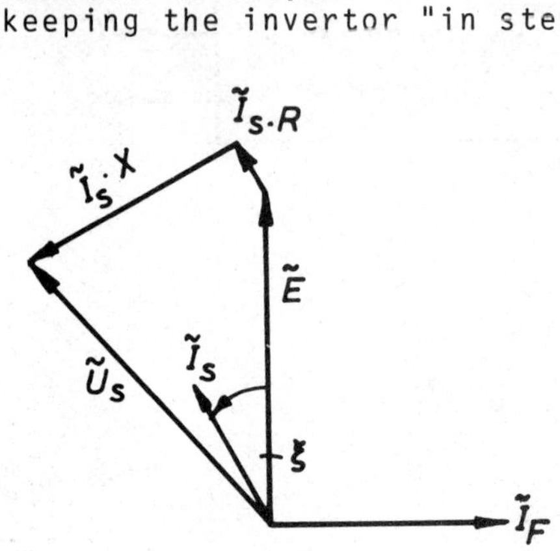

Fig. 3. Vector diagram of the LSM

DETERMINATION OF THE POLE POSITION ANGLE

The acquisition of the feedback quantities for motor control (velocity, thrust, pole position angle) is difficult with the track powered linear motor. A direct measurement of the pole position angle for example is not feasible, because the magnetic flux of the stator winding in the vicinity of the magnets amounts to less than 10^{-3} of the flux of the superconducting vehicle magnets. As an additional factor, a secure high frequency vehicle to ground communication link would be necessary.

The sources listed in the appendix describe details of control systems requiring no such communication link. Here another method is proposed, where the operating state of the motor is calculated at the invertor from data easily available there. The idea of this method can be deduced from the vector diagram of the drive (Fig. 3). E represents the fundamental harmonic of the voltage induced in the guideway winding, the vector of the stator current $\tilde{I}_s$ leads the voltage $\tilde{E}$ by the pole position angle ξ. With the knowledge of the resistance (R) and inductance (L) of the guideway winding, the power fed into the motor can be written as

$$P = P_w + j\,P_b$$
$$= 3\,(RI_s^2 + EI_s\cos\xi + j(EI_s\sin\xi - \omega LI_s^2)). \qquad (1)$$

Subtracting the resistive losses from the active power results in the mechanical power,

$$P_m = P_w - 3R\,I_s^2 = 3E\,I_s\cos\xi. \qquad (2)$$

By adding the inductive losses of the guideway winding to the reactive power, the inner reactive power is found,

$$P_{bi} = P_b + 3\omega\,L\,I_s^2 = 3E\,I_s\sin\xi. \qquad (3)$$

Hence the desired pole position angle can be calculated by

$$\xi = \arctan\,(P_{bi}/P_m). \qquad (4)$$

Since the propulsion velocity of the vehicle (v_x) is directly proportional to the pole pitch of the guideway winding (τ_p) and to the fundamental frequency f of the stator current,

$$v_x = 2\,f\,\tau_p, \qquad (5)$$

the thrust of the motor can be computed from

$$F_x = P_m/(2\,f\,\tau_p). \qquad (6).$$

Fig. 4 shows the principle of the

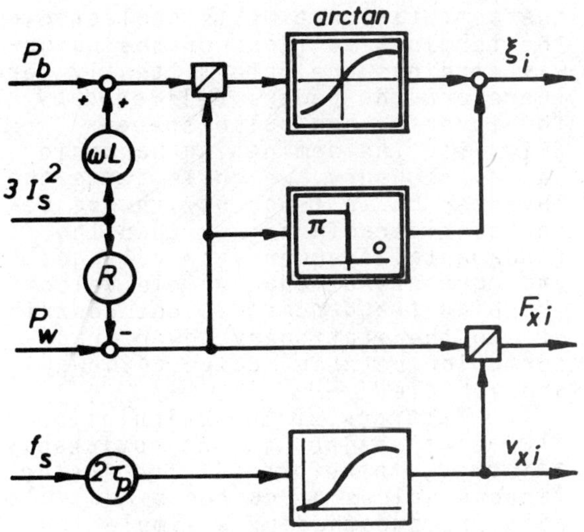

Fig. 4. Structure of the calculation of actual values

proposed procedure for calculating the important state variables of the motor; the ambiguity of the tangent function is included.

ACQUISITION OF THE VARIABLES

For reasons of clarity the principle of the calculation has been illustrated by means of vectors and r.m.s.-values. In order to exactly realize such a measurement, fundamental-frequency filters would be required, the dynamics of which would be detrimental to closed-loop control. The cut-off frequency of these filters, moreover, would have to be adjusted depending on vehicle velocity.

Instead of this it is more appropriate to compute the power values by means of the instantaneous values of voltages and currents. Since the neutral point of the guideway winding is isolated, only two terminal voltages (u_{RS}, u_{ST}) and two phase currents (i_R, i_T) have to be taken into account. From the quantities

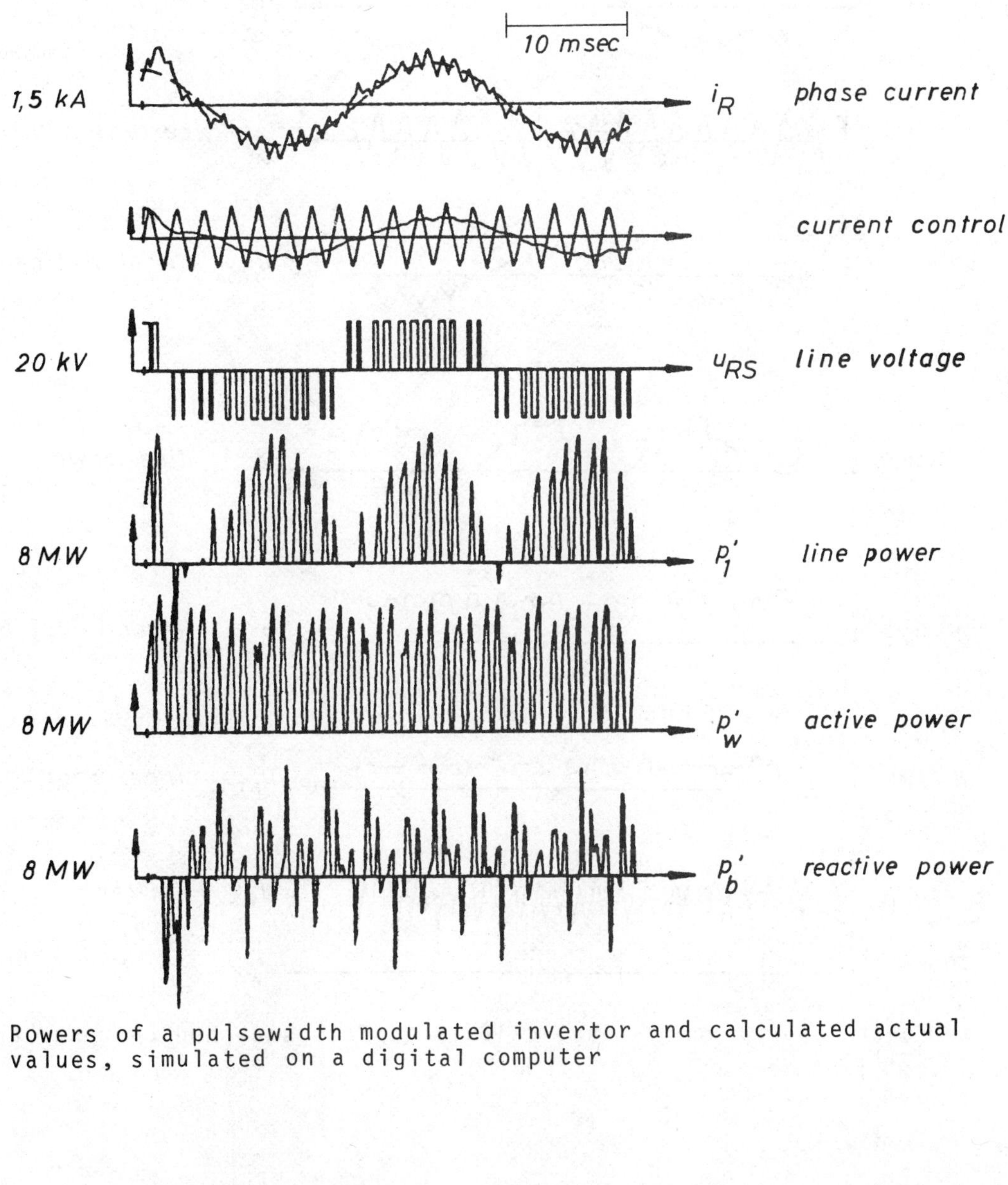

Fig. 5. Powers of a pulsewidth modulated invertor and calculated actual values, simulated on a digital computer

$$i_p(t) = i_R + i_T$$
$$p_1'(t) = u_{RS}\ i_R$$
$$p_2'(t) = u_{ST}\ i_T \qquad (7)$$
$$p_3'(t) = 2(u_{RS} + u_{ST})\ i_p$$

the inputs of the scheme in Fig. 4 can be calculated:

$$p_w'(t) = p_1' - p_2'$$
$$p_b'(t) = (p_1'+p_2'-p_3')/\sqrt{3} \quad (8)$$
$$3\ i_s^{2'}(t) = 2\ (i_T^2+i_p \cdot i_R).$$

The primed quantities indicate instantaneous values which include a high harmonic content, caused by the operation of the pulsewidth modulated invertor. Since the output voltages of this invertor change only the sign and not the magnitude, if the commutation time is neglected, the terminal voltages of the guideway windings are intermittently zero. Therefore the powers delivered by the invertor are pulse-shaped (Fig. 5). The dominating harmonic has a frequency, which is twice the invertor pulse frequency, hence being considerably higher than the fundamental frequency of voltages and currents at any vehicle velocity. The high frequencies do not contribute to the stationary power transformation between active guideway and vehicle.

The accuracy of the calculation of the actual values is not noticeably affected, therefore, if the instantaneous values of currents and voltages are smoothed by a simple low-pass filter with a time-constant of about 1 msec. (Such a delay may already be involved in the transducer.) Fig. 6 shows the inputs filtered in

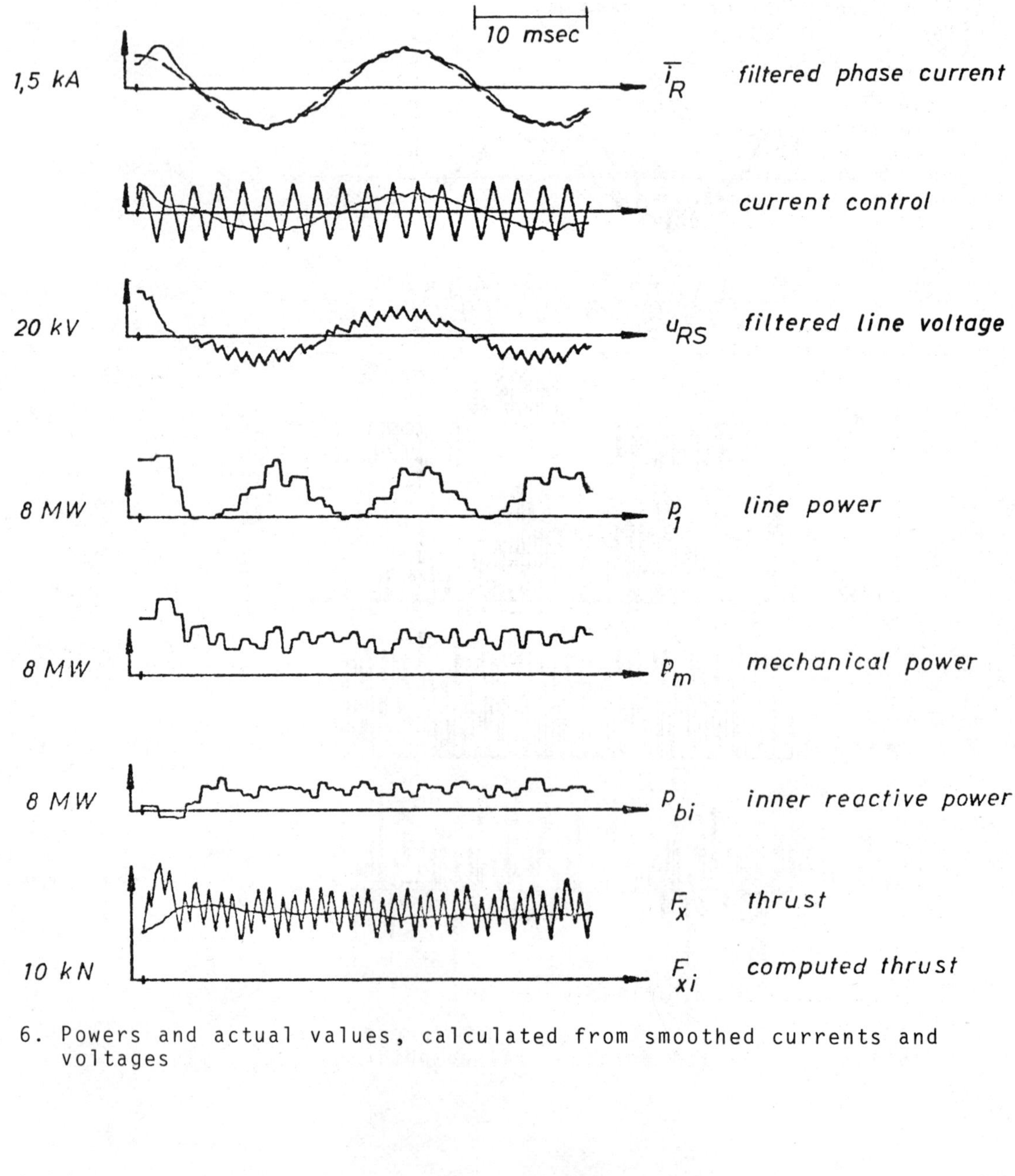

Fig. 6. Powers and actual values, calculated from smoothed currents and voltages

this way and the calculated power quantities, while the other parameters are the same as in Fig. 5.

Since the power data are proportional to the widely varying vehicle velocity, digital computation of the powers and actual values is advisable. In the simulation for Fig. 6 digital computing has been considered by a time delay and by sampling of the computed values. The thrust determined and filtered in this way corresponds to the average value of the true thrust.

If the calculations are made by a microprocessor in NMOS-technology (e. g. type INTEL 8080) and the multiplications and divisions carried out externally, sample rates of 1000 Hz may be achieved. The microprocessor system for acquiring the actual values is at present being built and is to be tested on a rotating test equipment of the "Projektgruppe Magnetschwebebahn" at Erlangen, using a 6 m diameter wheel and an actual size superconducting magnet.

PROPULSION CONTROL

Only the translatorial degrees of freedom in propulsion- and lift-direction are considered in the single-phase equivalent circuit diagram of the iron-free track-powered synchronous motor in Fig. 7. The forces accelerating the vehicle produce, after double integration, the position of the vehicle along the track. The spacial lag with regard to the stator travelling field determines the forces according to a sin- or cos-function.

As the excitation flux (I_E) of the superconducting magnets cannot be modified without heavy eddy-current losses the motor can only be controlled by magnitude (I_s) and frequency (f_s) of the stator current. Two parallel closed-loop controls have been assigned to the two corresponding regulating inputs of the invertor. The thrust is controlled by the magnitude of the stator currents. The reference value of the thrust is determined by a superimposed velocity-controller. The phase of the travelling field and the pole position angle can be changed quickly by means of the frequency control of the invertor.

In this way a coupled multivariable control system is formed. An increase of the velocity by increased current and the ensuing thrust results in interaction with the angle controller which holds the pole position angle constant by raising the frequency according to equation (6). On the other hand changing of the

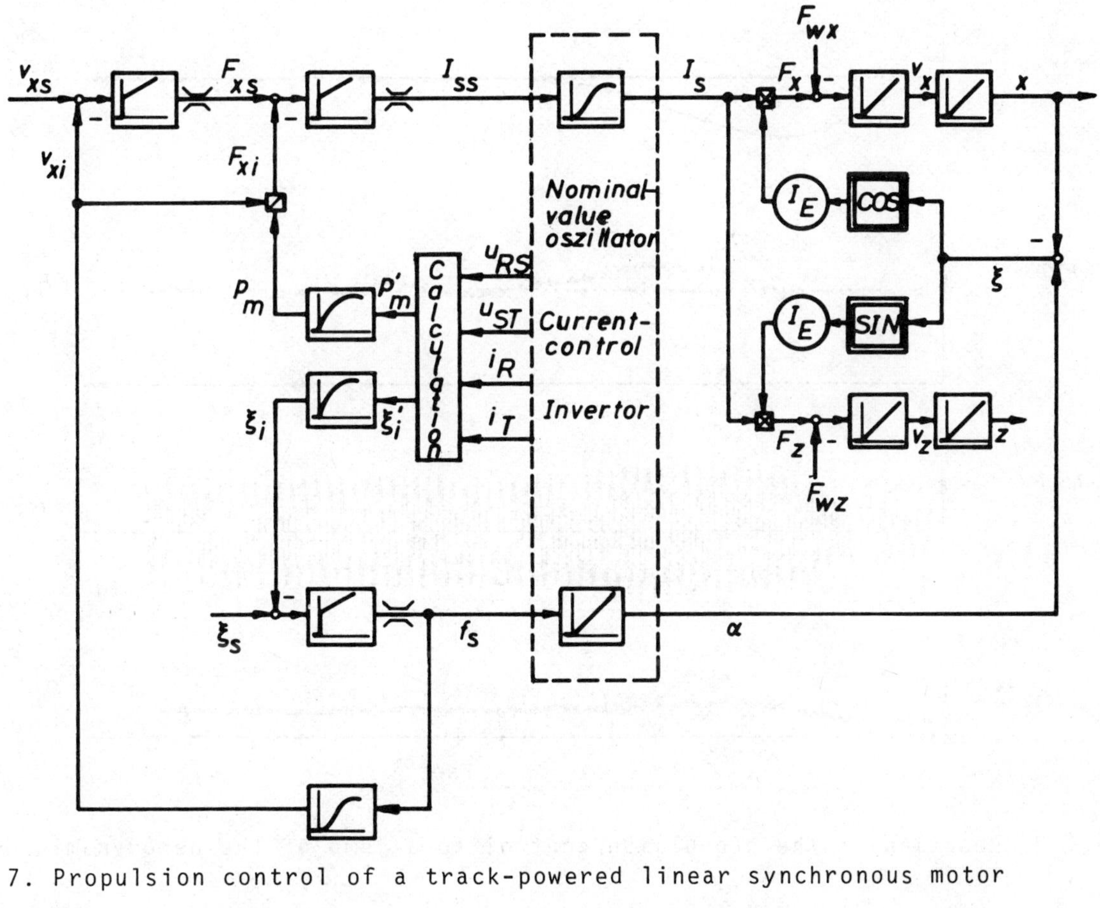

Fig. 7. Propulsion control of a track-powered linear synchronous motor

angle reference value results in a new operating point affecting the thrust, which means that the magnitude of the stator current has to be readjusted.

The calculations described in the preceeding chapters are carried out at the invertor stations, which are installed along the track at distances of about 10 - 15 km. The computed values of thrust and pole position angle produce, after simple filtering, the feedback signals for the control system. Without regard to brief alterations of the pole position angle (for example transition from driving to braking) the controller output signal f_s is directly proportional to the vehicle velocity. Hence, this actuating variable can, after some filtering, be fed to the velocity controller in place of an actual measurement.

SIMULATION OF THE CONTROL SYSTEM

The proposed control system and the calculation of the feedback quantities have been simulated on a digital computer, whereby functioning of the pulsewidth modulated invertor was approximated by an on-off controller neglecting the commutation intervals.

In Fig. 8 the initial thrust developed by the vehicle corresponds to the aerodynamic drag force of 40 kN at 400 km/h.Then the drag

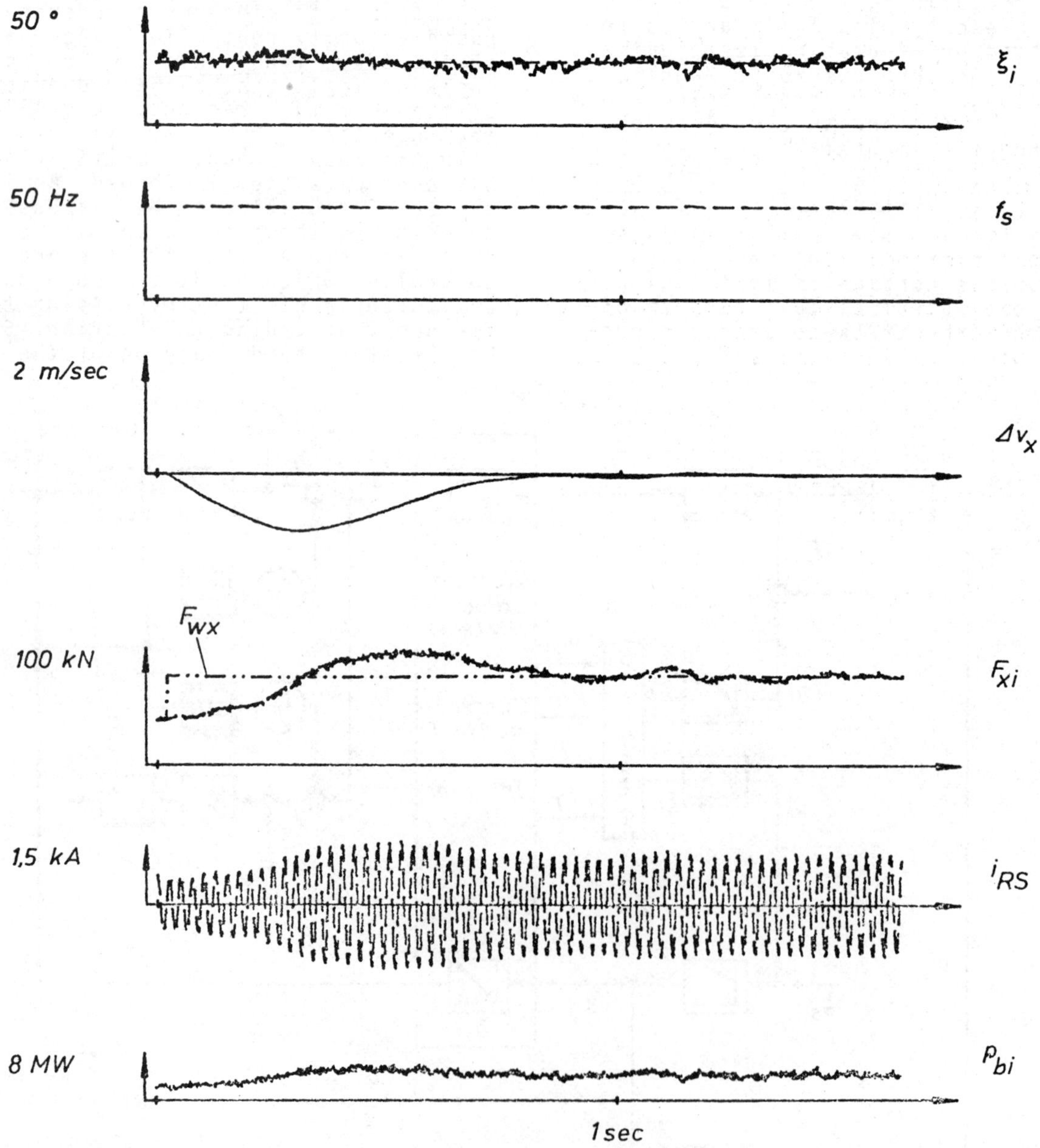

Fig. 8. Reaction of the propulsion control to a jump of the aerodynamic drag force

force suddenly increases to twice this amount, as would be caused by entering a tunnel.

First the controller of the pole position angle reacts, by reducing the frequency according to the decreasing velocity. The diminution of f_s causes the velocity controller to raise the reference thrust. After the velocity loss has been compensated for, the thrust is adapted to the new drag force. The maximum deviation of the velocity from its nominal value amounts to less than 1 %. The pole position angle is held within close tolerances in the unstable range.

OPERATION AT LOW SPEED

The control systems, as described in (1), (2) and (3), cannot be operated in the very low velocity range. Therefore the vehicle has to be started according to a fixed open loop procedure which means that it must be accelerated at less than the maximum thrust.

The lower velocity limit of the closed loop control system is determined by the accuracy achievable in current and voltage measurements, power calculations and numerical compensation of the losses. If the calculations of powers and actual values are performed with a word length of 16 bit, the cut-off velocity would to less than 10 % of the top speed.

The effect of the temperature of the track winding is less important than with most indirect measurement methods applied to rotating machines. Before the vehicle enters a section of the active guideway, the calculation of actual values can be adjusted to the present temperature of the winding by measuring its resistance.

The effect of temperature rise of the winding can most likely be neglected, since the sections are in use only for a short-time, for example 100 sec at 500 km/h.

CONCLUSION

The operating state of a synchronous track-powered linear motor can be deduced from the instantaneous values of stator currents and voltages and from power calculations. The harmonics caused by the basic frequency are eliminated by the symmetric calculation of the power quantities. As a result of the operation of the invertor considerable harmonics remain, which have to be reduced by a low-pass filter. The resultant delay in the calculation is, however, not a fundamental disadvantage for a closed-loop control system, because the velocity-coordinates of the maglev are continuous. A power measurement is, by virtue of its reliability, preferable to a phase angle measurement of the highly distorted current. The practical test of the microcomputer system for acquiring the actual values is due to be completed in 1977.

LITERATURE

|1| J. Holtz: Der Zweikomponentenmotor-Schnellbahnantrieb und Stellglied zur aktiven Dämpfung des Schwebesystems. 3. Statusseminar des Bundesministeriums für Forschung und Technologie, Referat Nr. 27, Berlin, (1974)

|2| C. P. Parsch: Regelkonzept und Erprobungseinrichtung für den eisenlosen Synchronlinearmotor. 5. Statusseminar des Bundesministeriums für Forschung und Technologie, Referat Nr. 13.2, Bad Kissingen (1976)

|3| A. R. Eastham: Superconducting Magnetic Levitation and Linear Synchronous Motor Propulsion for High Speed Guided Ground Transportation. Canadian Institute of Guided Ground Transportation, Queen's University of Kingston, Ontario (1975)

ANALYSIS OF DRIVES FOR ELECTRIC ROAD VEHICLES—REQUIREMENTS, BASIC DESIGN, PROPERTIES, VALUE AND OPTIMIZATION CRITERIA

A. Cupsa

GES Gesellschaft für elektrischen Straßenverkehr mbH, 4 Düsseldorf, Germany

ABSTRACT

In electric road transportation and its attendant special features, particular importance must be attached to the drives. Following an outline of typical requirements, the basic design and general properties are analysed using a block diagram. The drives are classified into straight electric types and hybrid types. In conclusion, technical and economic optimization criteria and results as well as present trends are mentioned.

INTRODUCTION

Electric road vehicles, just as other modern transport alternatives, can only be developed as part of an overall system. Figure 1 shows

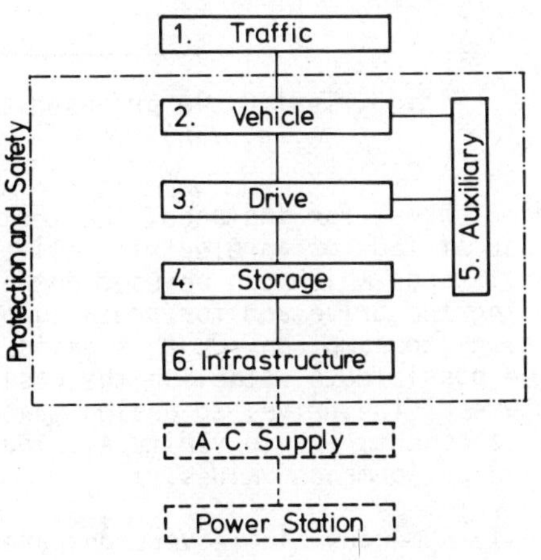

Fig. 1. General Structure of the System "Electric Road Traffic"

shows the general system structure of transportation by electric road vehicles. This comprises essentially six different groups which nevertheless influence each other. This paper deals with group III, the drive, but feedback to or logical relationships with other groups have to be taken into account. In other words: one cannot simply replace the internal combustion engine with an electric motor or the fuel tank with a storage battery and then refer to the propulsion machinery as an electric drive.

Moreover, it should not be forgotten that even in the future, electric road vehicles can only be put to good use for short-distance traffic. For one reason it is extremely necessary to reduce atmospheric pollution and secondly because the limited capacity of electrical storage batteries and the permissible loads on the low-voltage supply systems restrict the range per charge to about 1oo - 12o km which has, however, proved adequate. The requirements and limitations are important factors in the design of the drive.

Finally, mention should be made of the fact that the class of vehicle also determines important features of the drive. The wide variety of vehicles will therefore have to be classified into three major groups:

Class A: light-weight vehicles, gross weight 1 - 2 ton (e. g. passenger cars)

Class B: medium-weight vehicles, gross weight 2 - 5 ton (e. g. delivery van)

Class C: heavy vehicles, gross weight 5 - 2o ton (e. g. buses, lorries)

GENERAL REQUIREMENTS PLACED ON THE DRIVE

As electric vehicles are mainly used in short-distance service, their road performance should be designed to meet requirements in such duty so that in conurbations they do not interfere with the traffic of other motor vehicles [5]. Apart from a minimum range this includes a specified maximum speed, gradeability and acceleration. Numerous observations and tests ultimately led to the approximate values given in Table 1.

TABLE 1 Drive Performances for Electric Road Vehicles

Drive Performances \ Vehicle Class		A: 1-2 t (Car)	B: 2-5 t (Van)	C: 5-20 t (Bus)
Max. Vehicle Speed [km/h]		70 (100)	70	70
Vehicle Gradeability [%] at Full Payload	About 30 min at Max. Speed	≥1.5	1	1
	About 10 min at Half Max. Speed	≥ 7	7	7
	About 1 min after Starting	18 - 20	16 - 17	10 - 12
Average Acceleration Time [s] at Half Payload on Level Road	0 to 18 km/h	2.8	3.3	4.5
	0 to 30 km/h	5	7	10
	0 to 50 km/h	12	15	22

These have also been classified into three categories and at the same time represent the basic requirements to be met by the drive. This is explained in detail in Fig. 2 which shows the values for a delivery van.

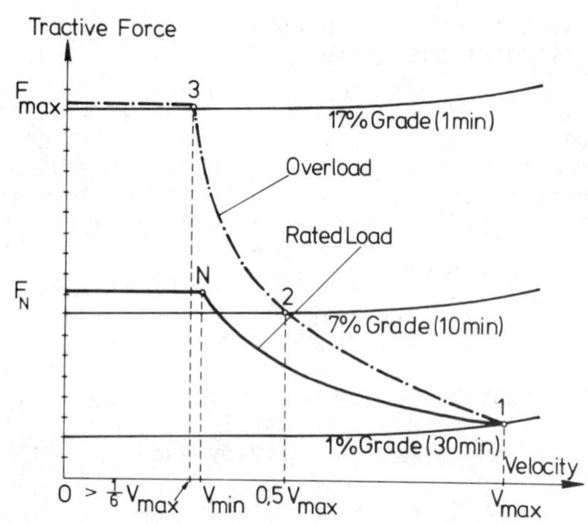

Fig. 2. Requirements for Drive Units (e. g. Van)

There are three curves representing, in the case of the delivery van, the resistance to motion for three typical gradients. Using the speeds mentioned in Table 1, we arrive at three reference points 1, 2 and 3. These three points as well as stability or overload limits determine the characteristics of the drive for rated load (continuous line) and overload (dotted line). In the range from minimum to maximum speed the rated load characteristic is, as a rule, a hyperbola; however, the overload characteristic is not generelly a hyperbola and its shape depends mainly on the thermal and electromagnetic limitations of the motor and its control. In the range from standstill to minimum speed the aim is usually to obtain constant tractive force. With a view to a comfortable ride, it is desirable to make the transition from one speed range to the other and from part-load to rated-load and overload operation as smooth as possible.

Further requirements result from the conditions of loading of the vehicle in short-distance traffic. They include a plurality of stochastic accelerating, braking and cruising operations, as well as periods of standstill, which depend on service and topographic conditions.

Experience has shown, however, that at least in urban transport actual vehicular movements can be well simulated by a suitable combination of cyclic and constant trips.

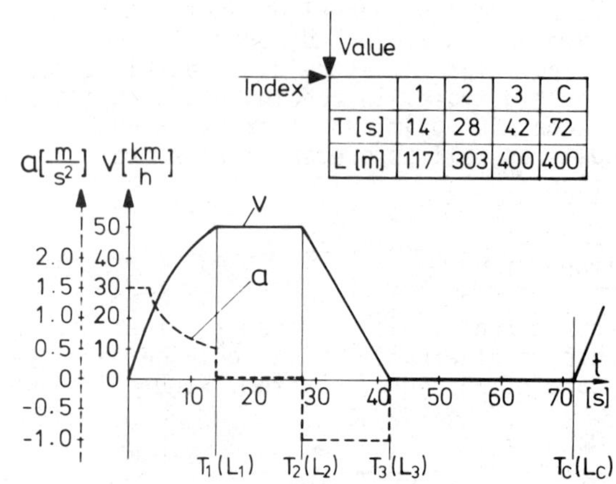

Fig. 3. Test Cycle proposed at ISO (e. g. Van)

Figure 3 shows, for instance, the test cycle proposed at ISO for an electric delivery van on level roads which can be used both for designing the drive and for precalculating the energy consumption. On this basis it will be possible to establish the basic structure of the drive, to design major drive components and to verify all loadings or drive performance values.

Naturally numerous single versions are conceivable. It is therefore recommendable to classify electric drives into two large groups:

- straight electric drives
- hybrid electric drives.

STRAIGHT ELECTRIC DRIVES

These are characterized by using electricity as the sole propulsion source. A general block diagram and the most important components of such drives are shown in Fig. 4.

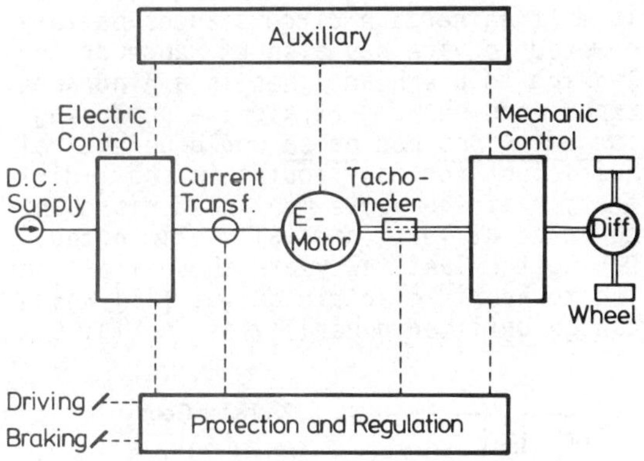

Fig. 4. Block Diagram for Electric Drive Systems

The question of choice is as follows:

- power supply: overhead lines or storage battery?

- traction motor: induction or d. c. motor?

- control: electric (e. g. electronic), mechanical or a combination of the two?

The question of power supply is fairly easy to answer. For reasons of cost the use of overhead lines is today restricted to public transport services in cities where trolley systems already exist or where such overhead lines need only be installed on main arteries. In all other cases storage batteries will have to be used.

Before answering the second question let me state that drives for battery-powered vehicles differ considerably from those used for stationary installations or even for rail-mounted vehicles supplied from the overhead lines; this is partly due to the fact that the capacity of the storage battery and the space available for the drive (traction motor plus controls) are very limited. As a whole, drives for electric vehicles have to meet the following requirements: small power-to-weight ratio, small size, large speed and overload range, reliability, good resistance to wear, low maintenance requirements, good dynamic characteristics, loss-free control of simple design and as inexpensive as possible (for auxiliary units too).

However, if one considers that a battery-powered induction motor can only operate with a static inverter [8] then it should come as no surprise that it is just the sturdy and reliable induction motor which compares unfavourably with others in battery-powered electric road vehicles. In spite of the undeniable advantages of the induction motor itself, the expenditure of power electronics for a controlled inverter is so high that a drive unit using an induction motor is up to approx. 5o % heavier and more expensive than one with a d. c. motor. Of course this applies to today's prices for power electronics and to the vehicle power range considered here.

Moreover, the d. c. motor can be controlled by other means (e. g. battery switching, gearbox, torque converter etc.), which are considerably cheaper than a d. c. chopper. Therefore - in spite of the necessary maintenance - the d. c. drive has a good chance of becoming even cheaper and more suitable for electric vehicles.

And now to the third question, motor control. Apart from high-loss control with additional resistors and contactors the following possibilities are available:

a) Mechanical control

In the automobile industry there are numerous control elements for internal combustion engines which can be used with only a few alterations: hydraulic torque converters, manual or automatic gear-boxes, continuously variable transmissions etc. These elements are mass produced and therefore relatively cheap. This type of control offers interesting possibilities although, at present, it is not entirely loss-free or suitable for electric motors.

b) Electric control

This includes the parallel series switching of batteries or motors, field weakening of the d. c. motor etc. These means are relatively inexpensive but insufficient for covering the whole speed range. Moreover, the switching cannot be effected without momentaneous loss of traction. They are therefore mostly used in conjunction with other control elements.

c) Electronic control with d. c. chopper

This modern control was made possible by advancements in thyristor research and is characterised by a high degree of driving convenience, high efficiency, electric regenerative breaking up to standstill and low maintenance requirements. The disadvantages are excessive cost and certain negative effects on other drive components. Recently there has been a new trend whre the chopper is combined with power transistors [9]. For relatively low power ratings a reduction in co st can now be achieved.

d) Electro-mechanical control

This type represents a combination of a) and b) or a) and c) and should offer good prospects for electric vehicles. This possibility was recognised in all the countries whre work is being carried out on electric vehicles and various concepts were developed. The main point about this type of control is that the electric components should be adapted, to a large extent, to the existing mechanical components thus permitting the use of numerous conventional components [1], [3], [4], [6].

Most of the electric road vehicles which have been designed and are being tested have, at least in Germany, a drive with a shunt-wound d. c. motor and an electronic d. c. chopper. It would therefore be interesting to mention a few characteristics of these drives. Taking the proposed ISO test cycle as well as some representative values of the gross vehicle weight as a basis the data summarised in Table 2 can be calculated [2].

TABLE 2 Typical Data for DC Drives with Shunt Motor and Chopper (ISO-Test-Cycle)

Gross Vehicle Weight	Rated Drive Data (in respect to Wheel)		Effic. of the Regener Braking	Peak Electric Power Required		Elec. Energy Required for 60km Range	
G_V [t]	P_{D_N} [kW]	η_{D_N} [%]	η_{RB} [%]	$\hat{P}_{el}$ [kW]	$\hat{P}_{el}^*$ [W/kg]	W_{el} [kWh]	W_{el}^* [Wh/kg]
1	5.5	71	16	17	68	8.8	35.2
2	10.8	71	20	31.5	63	16.3	32.6
5	26	76	23	70	56	37	29.6
10	50	79	25	128	51.2	68	27.2
20	93	81	26	235	47	125	25

*) Battery Weight 0.25 G_V

They agree well with operational experience and permit the following to be said:

- the necessary rated and maximal drive power related to the gross vehicle weight is higher with light-weight vehicles but the efficiency and regererative braking effect is lower than with heavy vehicles;

- the drive must be capable of being overloaded up to twice its rated power for short periods;

- bearing in mind the power density permissible in modern batteries, and leaving aside the range, the ISO test cycle can only be fulfilled by light-weight vehicles if the battery weight amounts to at least 25 % of the permissible gross vehicle weight.

HYBRID ELECTRIC DRIVES

Although the straight electric drive is entirely nonpolluting, it is at present still too expensive. A further disadvantage is that in certain circumstances battery-powered service has a short range or the linking to overhead lines is a hindrance. Experience shows that a drive producing low emissions and noise would be entirely sufficient for many routes in short-distance traffic if the above-mentioned disadvantage could be at least partially eliminated. During the last few years these ideas have led to hybrid electric drives [11] which can be outlined generally as in Fig. 5.

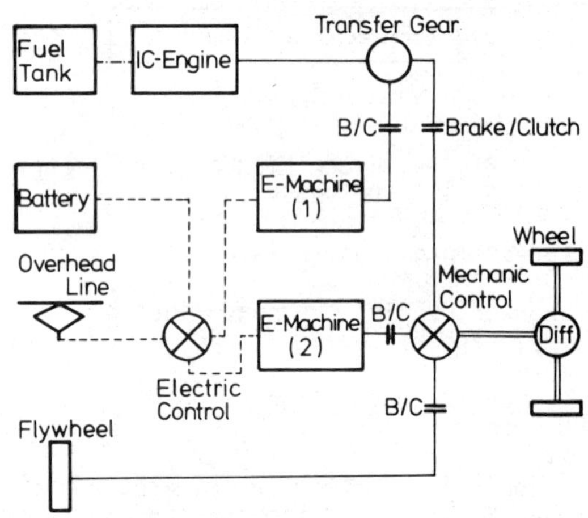

Fig. 5. Block Diagram for Hybrid Drive Systems

This shows all conceivable combinations with the following limitations [10]:

a) There are two or more power sources. At least one of them is electric.

b) Therefore, two or more different ways of storing energy are necessary, one of which could be replaced by overhead lines.

c) Under certain conditions there must be several motors present (at least one being electric).

d) Several control elements adapted to the differing power sources and motors are necessary.

e) Bearing in mind the emissions, energy consumption and protection of the units the power peaks should only be covered by particular components designed for this purpose (e. g. flywheel, electric motor etc.).

It is neither possible nor desirable to discuss every variant which could be derived from Fig. 5. I will merely give a brief outline of one promising solution. In principle, it consists of an internal combustion engine and an electric motor, a battery and the associated controls. Theoretical and experimental studies have shown that the following principles must be observed:

- In view of emissions and fuel consumption the internal combustion engine should only operate within its most favourable load and speed range;
- The power of both motors are to be determined by the driving requirements, the range in the battery mode as well as the permissible duty cycle of the internal combustion engine;
- It is particulary advantageous to link both motors by a clutch and use the existing mechanical control elements;
- The type and design values of the storage means should be chosen with particular regard to the necessary power density. Moreover, it must be specially designed for intermittent service and regenerative braking;
- With regard to efficiency and oil substitution it is advisable to charge the battery from the mains supply.

According to V. Wouk [12] this design, when used in passenger cars in urban traffic, provides a fuel saving of up to 4o % and sharply reduces the emissions to below the levels valid in America from 1977.

I would now like to say a few words about a hybrid electric drive using the socalled "Super Flywheel", which has caused a lot of discussion recently. This is a flywheel made of new types of light alloys or fibres which permits energy densities of around 1oo Wh/kg at peripheral speeds of roughly 85o m/s and therefore is supposed to make the battery, at least to some extent, superfluous. However, this is overlooking the fact that its practical application in road vehicles is linked with certain difficulties:

- The kinetic energy cannot be stored for a long period and the necessary equipment is not extremely simple.
- Owing to the relatively small internal losses huge "short circuit" peak power results, which can scarcely be controlled.
- Considerable expenditure for control equipment is necessary for charging the flywheel over a relatively long period (8 - 12 hours) without unduly overloading the electric supply system.

From this one can assume that, in the foreseeable future, a flywheel could be used mainly as a shorttime storage means to meet peak power requirements in urban traffic [7].

TECHNICAL AND ECONOMIC DECISION-MAKING AND OPTIMIZING CRITERIA

The foregoing reflexions outline some of the special features and conceivable solutions regarding electric drives for road vehicles. However, one important question remained unanswered:

"Which type of drive should be used at a given stage of development for a particular vehicle?"

This question simply could not be answered because this is only possible through a decision-making and optimization process for the system as a whole. It would then be interesting to know how this decision would in fact be made and an optimization programme carried out. This includes the following stages:

- Determining in detail the service conditions and parameters for the vehicle class in question.
- Establishing the minimum drive performance or other requirements (e. g. payload, quality of ride etc.).
- All auxiliary units including heating and secondary power supply must be considered in selecting the optimum version.
- Careful study of the safety and reliability.
- The overall specific costs (e. g. DM/km) for all versions under consideration are calculated (according to the Schöpke method) and compared with one another. A decision is then made in favour of the cheapest version also because, in the final analysis, the electric drive can only be conceived as a part of a total system.

One should now mention some of the optimization results which should apply in the short and medium term:

a) The power transistor controlled d. c. motor seems to be a satisfactory solution for vehicles in class A.

b) With class B vehicles an electro-mechanical control with a shunt-wound d. d. motor gives the best results.

c) However, the electronically controlled shunt-wound d. c. motor currently proves to be the optimal solution for heavy vehicles of the class C.

It must be stressed, however, that optimization represents a largely dynamic process in which each case must be considered on its own merits.

- In this respect the type of vehicle and its use as well as the quantity are very important.
- The optimization results depend on the stage of development of the individual components.
- Ecological, economic and political factors must always be taken into account.

CONCLUSIONS

From the user's point of view, this paper deals with a topical and concrete example of controlled drives for electric road vehicles. The task of the engineer is to recognise the aims and trends of a particular stage of development and, in doing so, to bear in mind both technical and economic considerations. It should therefore come as no surprise that, at a symposium on power electronics, other versions of and additions to electronic control of these special drives were mentioned.

On the contrary, this future-oriented field of application not only offers electronics a large market potential but also the opportunity to have a positive effect on the further development of electric road vehicles. Power electronics can not only be applied to motor control but can also assist in designing a charger suitable for vehicles. This is where the irreplaceable role of power electronics begins.

In conclusion I would like to summarise some development objectives for power electronics that are desirable with a view to electric vehicles:

- Further reduction in costs by standardisation and adaptation to the special feature of electric vehicles.
- Designing less expensive motor controls using power transistors.
- Adapted electronics for electro-mechanical controls.
- Research into the possible use of certain motor control components as battery charger.

REFERENCES

(1) Bader, C., Stephan, W.: Vergleich elektrischer Antriebe für Straßenfahrzeuge, etz-a, Bd. 98, H. 1 (1977).

(2) Cupsa, A., Görlach, U.: Grundlagen zur Vorausberechnung von Elektro-Straßenfahrzeugen, etz-a, Bd. 98, H. 6 (1977).

(3) Gaulon, J.: Description et essais de deux véhicules électriques à variateurs à courroies, Fourth International Electric Vehicle Symposium, Paper No. 212.1, Düsseldorf (1976).

(4) Gutberlet, H., Härer, H., Schneider, H.: Einfachelektroantriebe mit nachgeschaltetem Getriebe, etz-a, Bd. 98, H. 1 (1977).

(5) Harding, G.: High Performance Electric Commercial Vehicles for City Use, SAE-Paper No. 76o073, USA (1976).

(6) Kahlen, H.: Transporterantrieb mit elektronischem Steller und mechanischem Schaltgetriebe, etz-a, Bd. 98, H. 1 (1977).

(7) Locker, D., Miller, M. L.: Flywheel-Electric Hybrid Vehicle, Fourth International Electric Vehicle Symposium, Paper No. 212.4, Düsseldorf (1976).

(8) Naydowski, C. O., Pohl, D., Skudelny, H. C.: Untersuchungen an elektrischen Straßenfahrzeugantrieben, etz-a, Bd. 98, H. 1 (1977).

(9) Sornicle, C.: Commandes électroniques à transistors, Fourth International Electric Vehicle Symposium, Paper No. 211.4, Düsseldorf (1976).

(1o) Stephenson, R. R.: Should we have a new engine? JET Propulsion Laboratories, Chapter 9, California, USA (1975).

(11) Strifler, P.: Hybride Antriebe für Elektro-Straßenfahrzeuge, etz-a, Bd. 98, H. 1 (1977).

(12) Wouk, V.: An experimental ICE/battery-electric Hybrid with low Emissions and low fuel Consumption Capability; SAE-Paper No. 76o123,USA (1976).

THE APPLICATION OF TRANSISTORIZED SWITCHES TO D-C AND A-C MACHINES FOR THE CONTROL OF BATTERY VEHICLES UP TO 30kW

J.D. van Wyk and J.J. Schoeman

Department of Electrical and Electronic Engineering, Rand Afrikaans University, Johannesburg

ABSTRACT

This paper examines important parameters of transistors when used as high level power switches. The influence of the geometrical arrangement of switching elements in composite power switches is examined, determining the influence of the structural inductance on switch-off behaviour. Linear, rectangular and concentric arrangements are evaluated experimentally, indicating methods for the optimization of switch-off behaviour. Practical applications to choppers and inverters are discussed and illustrated by means of 1kW, and 10kW battery-fed chopper control of d.c. machines and a 5kW battery-fed inverter application for an induction traction machine. It is shown that present transistor switch technology can achieve construction of controllers in excess of 30kW with transistor chips not exceeding 50A maximum rating as basis.

1. OUTLINE OF INVESTIGATED PROBLEM

The application of power electronic switches for low loss control of power flow has in the past achieved widespread application to d.c. machines, lately extending also to a.c. machines. For battery-fed vehicles the availability of a low loss controller is important, as this directly influences range of operation through the limited supply of energy carried on board the vehicle. It is well known that thyristor technology has been successfully applied to the control of series and separately excited d.c. machines. This subject has been studied exhaustively in the past (1 to 4).

However, the use of thyristor technology for controllers has its own range of problems to be solved, these problems becoming increasingly severe with the increase of the number of force commutated power switches included in the controller. Solutions to these problems conducive to mass production, aimed at achieving low component cost, small physical volume, high efficiency and high reliability is much more difficult for thyristor inverter controllers then for thyristor chopper controllers - due to the increased circuit complication of inverters. Requirements of forced commutation tends to make circuits more bulky and complicated at the lower voltages, used in present experimental battery vehicles, while battery maintenance experience in large fleets (mine locomotives and underground traction for instance) seems to indicate that even the voltages of some 120V to 200V used in the said present experimental vehicles could prove to be unrealistically high in the long run.

The past few years have seen an increase of interest in the development of larger power switching transistors. (5, 7) Margrain has for instance indicated what growth curves can be expected for power transistors in the near future. (6). Based on the considerations of higher possible switching speed, absence of large energy storage circuits and possibly lower loss, the application of transistorized switches for lower voltages found in battery vehicles (48V, 72V, 96V) has been studied in our laboratory for the past two years.

This investigation has indicated experimentally that properly designed switches using transistors have low switching losses, conduction losses, small volume and can achieve switching speeds much in excess of thyristorized switches. The mechanical design can be such that automated mass production is much easier than with thyristorized equipment. Achieving higher frequencies of operation of the switches is important, since some of the past investigations have indicated that especially the lower switching speeds achievable with thyristorized inverters has the consequence of on the one hand either increasing the weight of the drive by necessitating inclusion of extra inductance or of a machine with higher leakage, or on the other hand decreasing the efficiency of the drive due to larger current ripple or higher switching losses. (3)

The problem areas have been shown to be the high speed protection of the transistor switching elements, the difference between thyristor switch technology and transistor switch technology, small thermal time constants of the transistors, and the lack of regenerative characteristics in the conducting state.

2. CONSIDERATIONS REGARDING TRANSISTORS AS POWER SWITCHES

The characteristics of transistors as power switches in general have been covered in the existing literature (5, 7). A few observations on those aspects important for the design of the power switches investigated shall be touched upon in this section, since this differs from what has been published elsewhere. (5,6,7). The devices used were all made in triple diffused planar technology (Silicon), had breakdown voltages below 150V and chip sizes of 5mm by 5mm. All were constructed in state of the art technology with emitter multifinger structure, soldered to appropriate headers with aluminium bonded wire of 0,5mm diameter to the external contacts. This meant that in all designs a minimum current amplification of 10, maximum switching time of 1μs between on and off states and base-emitter saturation voltage of 1,2V could be used.

Some remarks regarding saturation voltage and switching speed must be made. It is customary for device manufacturers to quote the peak collector current of devices with the above specified chip size as anything between 30A and 50A. Our work has confirmed that for the applications to follow, especially when keeping good current sharing for parallel devices in mind, these devices should be typically run at 15A, and at the most 20A. This will depend on the saturation voltage, and cannot be related to chip size only. Chip mounting on the header, emitter connections and chip metallization should also be taken into account.

In specifying switching speed, it is furthermore customary to relate this to current waveforms through the transistor. This has only meaning in circuits where magnetic field energy may be neglected, such as in circuits with either sufficiently small dimensions or low enough current levels, or both. The subsequent treatment in this paper will indicate that this is not the case for power switches, so that for these transistors switching speed specifications should be related to collector-emitter voltage waveforms, specifying the time it takes for the base-collector junction to revert from its forward injection state to its normal reverse-biased state. In this paper switching speed shall only have this meaning.

During design of the power switches the devices have been utilized at currents where the saturation voltage does not exceed 0,8V. For the chip sizes used, this amounts to 15A per chip, thus two chips have to be connected in parallel for obtaining 30A, the manufacturer's chip rating. The conduction loss in the parallel case will amount to 24W, whereas it would have been 54W in the case utilizing the chip at 30A. The driving power (current gain 10) will be two times 1,95W and 6W respectively. The difference between the total power dissipation of 28W and 60W respectively is not only related to the efficiency of the switch, but linearly to the switching speed achievable in the eventual structure, as will be explained in the following section.

3. COMPOSITE POWER SWITCHES BASED ON TRANSISTORS

There are a number of considerations dictating the use of paralleled devices in high current switches, one of the most important being the limited current capability of the chips used in the devices freely available. Existing high current transistors (5) are at present still expensive. Furthermore the use of a number of devices in parallel offers the attractive features of possibly a redundant design, thermal sharing in case of short circuits (especially when each chip is not run at maximum current capacity), sharing of energy to second breakdown amongst the chips, and the possibility of using chips with a higher inherent switching speed (in theory at least).

The currents to be handled by these controllers will be 20A per kilowatt at 50V level and 40A per kilowatt at the 100V level. This yields 200A even for a 10kW controller. At these current levels one cannot expect the magnetic field energy associated with the circuit to be negligible, and consequently this aspect has to be investigated.

3.1 Theoretical Considerations

For any conducting structure the impedance may be noted as

$$Z_{struct} = Z_{int} + Z_{ext} = R_{int} + jX_{int} + R_{ext} + jX_{ext} \qquad (1)$$

To model the behaviour of the switch accuratly during transients is a difficult electromagnetic problem, and will not be investigated here. However in order to achieve at least an estimate of the structure's behaviour in the microsecond switching region, the impedance of a long cylindrical structure will be considered. Although this would not take into account all the end effects, conductor shapes, etc, it should at least yield an order of magnitude answer, and indicate which parts of this impedance could be neglected and what should be taken into account.

Solution of the wave equations for the said structure (8) under the conditions of a non-conducting, perfect dielectric medium outside the structure and a good conductor inside, yield the following approximate expressions for the components of the total structural impedance:

$$Z_{int,hf} \simeq \left[\frac{1}{2\pi a}\sqrt{\frac{\omega\mu}{2\sigma}} + j\,\frac{1}{2\pi a}\sqrt{\frac{\omega\mu}{2\sigma}}\right] \Omega/\text{m} \qquad (2)$$

under the condition that $|\gamma a| >> 1$. When $|\gamma a| << 1$, the approximation becomes,

$$Z_{int,\ell f} \simeq \left[\frac{1}{\pi a^2 \sigma} + j\omega\,\frac{\mu}{8\pi}\right] \Omega/\text{m} \qquad (3)$$

For both cases $\sigma >> \omega\varepsilon$. As long as the condition prevails that $\beta a << 1$, the external impedance may be written as:

$$Z_{ext} \simeq \left[\frac{60\pi^2}{\lambda} + j\frac{\eta}{\lambda}\ ln\left(\frac{\lambda}{2\pi a}\right)\right]\ \Omega/m \qquad (4)$$

From the above the real and imaginary parts lead to the respective internal resistances and inductances,

$$R_{int,hf} \simeq \left[\frac{1}{2\pi a}\sqrt{\frac{\omega\mu'}{2\sigma}}\right]\Omega/m\ ;\ R_{int,\ell f} \simeq \left[\frac{1}{\pi a^2\sigma}\right]\Omega/m \qquad (5)$$

$$L_{int,hf} \simeq \left[\frac{1}{2\pi a}\sqrt{\frac{\mu}{2\sigma\omega}}\right]\ H/m\ ;\ L_{int,\ell f} \simeq \left[\frac{\mu}{8\pi}\right]\ H/m \qquad (6)$$

while the external resistance and inductance is approximated by:

$$R_{ext} \simeq \left[\pi\omega\ 10^{-7}\right]\ \ \Omega/m \qquad (7)$$

$$L_{ext} \simeq \left[\frac{\mu}{2\pi}\ ln\left(\frac{\lambda}{2\pi a}\right)\right]\ \ H/m \qquad (8)$$

Taking the resistivity between that of copper and aluminium as used in these switches, the low frequency approximations in relations 5 and 6 will be valid for frequencies lower than 10^{-1}Hz in structures with a radius of some 10^{-2}m, while the high frequency approximation will be valid for frequencies higher than 10^5Hz. Unfortunately the frequency band of interest for transistor power switches is somewhere between these two values, and to obtain more exact answers the expressions will become much more involved. However, one is in any event not modelling the real structure of the power switch, but merely attempting to obtain an order of magnitude answer by using the expressions for a conducting cylinder, so that the above should suffice.

Calculation shows that for the frequency band and dimensions of interest (10Hz - 10^5Hz, 10^{-3}m - 10^{-1}m)

$$L_{int,\ell f} << L_{ext},\ L_{int,hf} << L_{ext} \qquad (9)$$

so that only the external inductance is to be considered further. The difference is typically two orders of magnitude. Comparison of the numerical values of relations 5 and 7 for normal conductive material ($\sigma = 5.10^7$mho/m) also shows that at a radius of 10^{-2}m:

$$\left.\begin{array}{ll} \text{at}\ \ 10kHz : R_{ext}/R_{int} & \simeq 44 \\ 100kHz : R_{ext}/R_{int} & \simeq 140 \\ 1MHz : R_{ext}/R_{int} & \simeq 441 \end{array}\right] \qquad (10)$$

indicating that only the external resistance due to radiation should be taken into account. Seeing furthermore that the ratio of radiation resistance to reactance for the same structure would be:

$$\left.\begin{array}{ll} 10kHz : X_{ext}/R_{ext} & \simeq 8,3 \\ 1MHz : X_{ext}/R_{ext} & \simeq 5,4 \end{array}\right] \qquad (11)$$

only the reactive effects are to be considered further. However especially at higher currents the radiative losses have to be kept in mind, since at 500A calculation again indicates that some 50W per cm length may be radiated at 10kHz.

From relation 8, valid below 100MHz for $a = 10^{-2}$m, the external inductance may be estimated as

$$L_{ext} \simeq 26nH/cm \ \text{ for } \ f = 10kHz, \qquad (12)$$

giving a magnetic field energy at 100A of

$$\tfrac{1}{2}L_{ext} \cdot I^2 = W_m = 0,13mJ/cm \qquad (13)$$

To evaluate the effect of this structurally dependent magnetic field energy on the behaviour of the composite power switch, let it be assumed that the same order of magnitude of magnetic energy will be involved with a switch having the same linear dimensions.

With reference to figures 1 and 2, the magnetic energy inherent in the structure of the switch will be represented in the model by

L_{struct}, R_{struct} being the resistance associated with the structure of the switch. After turning the switch of at $t = t_o$, the magnetic energy associated with the switch structure forces a circulating current around the switch-supply-freewheel diode loop. The current decays until it becomes zero at $t = t_1$. The time $(t_1 - t_o)$ represents dissipation of a quantity of energy in the switch quite independent of any switching characteristics of the device itself.

At $t_o < t < t_1$

$$v_s = v_L + V_B \qquad (14)$$

with $R_{struct} \simeq 0$

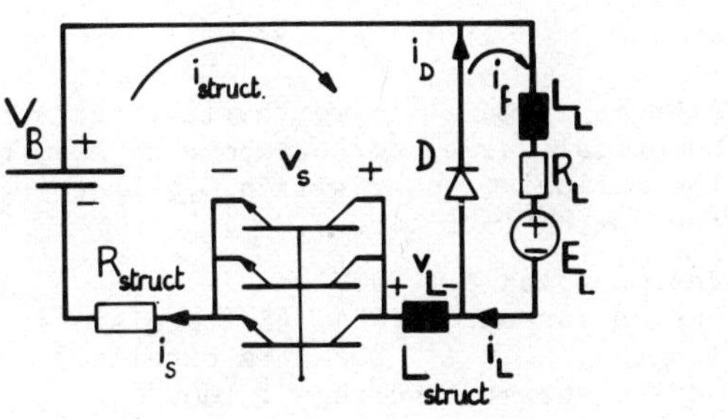

Fig. 1. Composite switch model during current switch-off

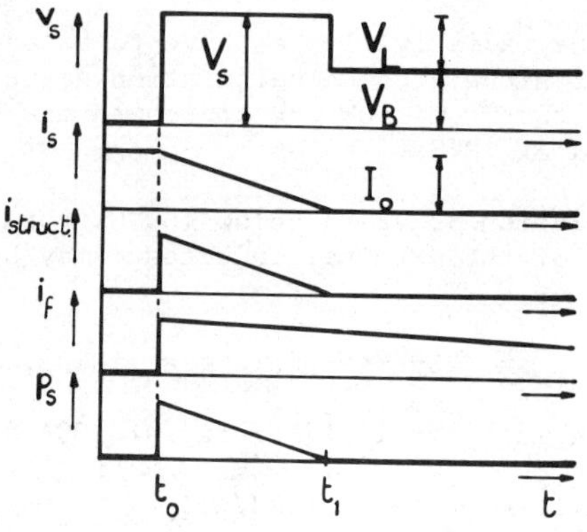

Fig. 2. Waveform definitions during switch-off

$$\int_{t_o}^{t_1} v_s i_s \, dt = \tfrac{1}{2} L_{struct} I_o^2 + V_B \int_{t_o}^{t_1} i_s \, dt \tag{15}$$

$$W_s = W_{struct} + W_B \tag{16}$$

Since the time $(t_1 - t_o)$ does not only represent the dissipation of the magnetic energy from the structure of the switch, but also dissipation from the source into the switch, it is evident that this time has to be kept to a minimum. From relation 16 the time to zero current may be calculated as:

$$\left[t_1 - t_o\right] = \frac{L_{struct} I_o}{V_L} \tag{17}$$

and the energy in the power switch during switch-off as:

$$W_s = \tfrac{1}{2}\left[V_L + V_B\right] \cdot I_o \cdot \left[t_1 - t_o\right] \tag{18}$$

giving the power dissipated in the switch as

$$P_s = W_s f \tag{19}$$

At this stage it is worthwhile to attach some numerical values to the above by examining the following power switch: (something like the switch in figure 3)

Design voltage : 60V
Design current : 100A (5 transistors, TO-3,
length : 20cm in parallel)
Max. transistor voltage : 100V
Repetition frequency : 5kHz
Assume V_L : 20V
Consequently

$$(t_1 - t_o) = 2\mu s$$
$$W_s = 8mJ$$
$$\text{and} \quad P_s = 40W$$

Using the same type of design for a 200A switch the length would be doubled, and the switch-off time increases to 8μs, W_s to 64mJ and P_s to 320W, indicating that the frequency of operation have to be decreased to 625Hz to have the same dissipation.

The above theoretical considerations all indicate that the mechanical layout of the transistor switches may be of great importance. The structure as such, chiefly determined by the layout of the heat sinks used, should be done so as to minimize the magnetic field energy.

3.2 Experimental Investigation of Layout

In order to investigate the predictions, a power switch using 5mm chips in TO-3, as shown in figure 3, was placed in an experimental set-up as indicated in figure 4. This type I switch was rated for 80A, 64V, 5kW and a typical switching waveform is given in figure 5. The power dissipated in the switch may be then be determined as shown in figure 6. The area under this power curve will yield W_s, the total power dissipated in the switch, consisting of the two components contributed respectively by the magnetic energy of the switch and by the source W_{struct} and W_B. By employing relations 14 and 15, the first of these two components, as related to the lay-out of the switch, may be determined for this type of switch. Since the initial current is also known, the equivalent inductance

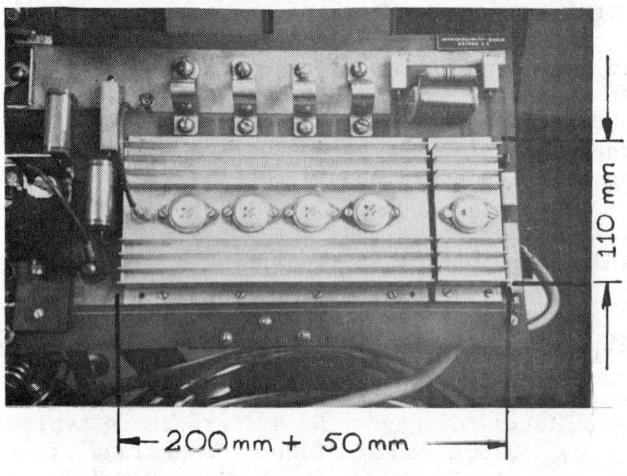

Fig. 3. Layout of type 1 transistor power switch (5kW) giving dimensions

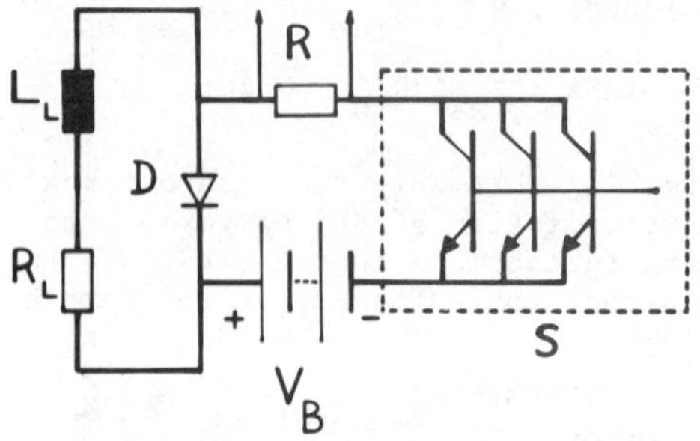

Fig. 4. Experimental layout for testing of power switches

of the type I switch may be calculated as

$$L_{struct,I} \simeq 0,33\mu H \qquad (20)$$

Influence of the structural layout has been further investigated by constructing switches type II and III as shown in figures 7 and 8. The type II switch was rated for 160A, 64V (10kW) and type III switch for 120A, 72V (8kW). By employing the same measurement techniques as

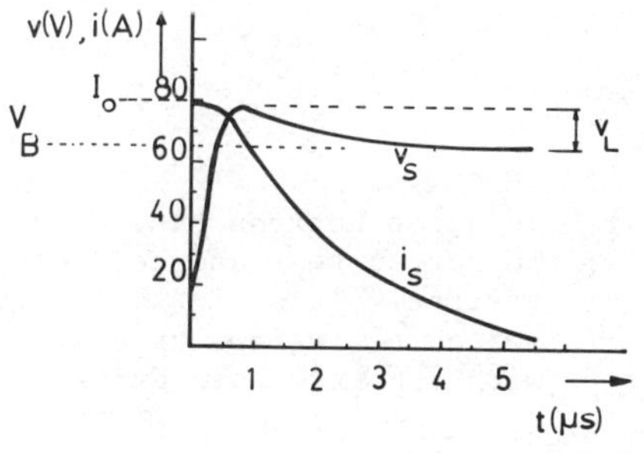

Fig. 5. Switch-off characteristics for type I switch at rated current

for type I, the equivalent structural inductances for these two switches were found respectively as

$$L_{struct,II} \simeq 0,28\mu H \qquad (21)$$

and $$L_{struct,III} \simeq 0,34\mu H \qquad (22)$$

It can therefore be seen that, in spite of the doubling of rating between switches of type I and type II, there has been an effective decrease in structural inductance. According to design, type III, being of concentric geometry, should have the lowest structural inductance of the three types of lay-out. It can be seen clearly from figure 8 that this type has the smallest overall dimensions. As may be seen from 22 above, the experimental results did not live up to these expectations. Unfortunately the different type of layout of this switch did not facilitate incorporation in the same experimental set-up as the first two types, and it is suspected that the experimental set-up contributed considerably to the result given above. A better comparison between the three types of switch may be made by expressing the structural inductance in per rated ampere values.

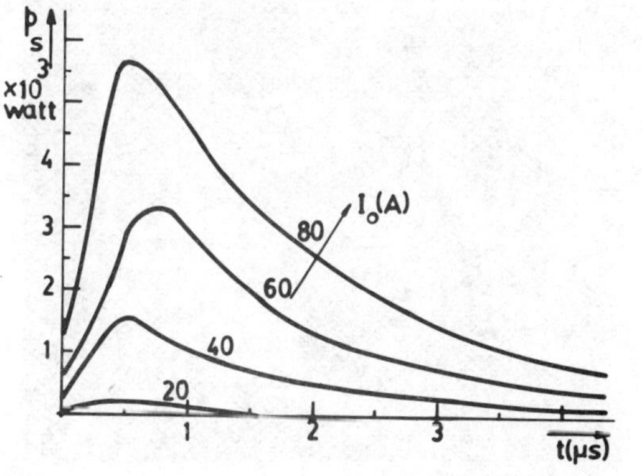

Fig. 6. Power dissipated in type I power switch as a function of time during switch off, with initial current as parameter

Fig. 7. Layout of type II transistor power switch. (10kW), indicating dimensions

Therefore these values are:

$$\left.\begin{array}{lcl} L'_{struct\ I} & \simeq & 4,125nH/A \\ L'_{struct\ II} & \simeq & 1,750nH/A \\ L'_{struct\ III} & \simeq & 2,833nH/A \end{array}\right] \qquad (23)$$

It is also interesting to note that when the linear length of type I switch (16 cm) is taken into account, the value of

$$L''_{struct,I} \simeq 20nH/cm$$

is obtained in good correspondence to relations 12 and 13.

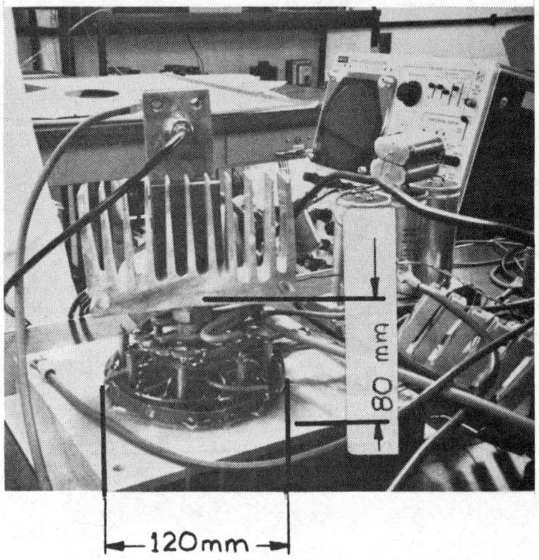

Fig. 8. Layout of type III transistor power switch. (8kW), indicating dimensions

3.3 Experimental Investigation of Other Switch Characteristics

Further experimental investigation concerning these switches treated the usual switch characteristics such as saturation voltage, switching speed and mean switching dissipation as function of current. The suppression of the voltage spikes appearing across the power switches was also studied.

The behaviour of saturation voltage as function of current presented no new or unexpected behaviour. The switching speed in this context shall only be related to the speed with which the voltage across the collector-base junction rises, as the foregoing investigation has indicated clearly that switching behaviour related to the current waveforms in the circuit is as much a function of the switch layout as of the device itself. To this, something should be added however. In the previous analyses, no consideration has been given to the form of the base drive, and its influence on the switching characteristics. Obviously, when the transistor is in saturation, the delay time between base current reduction and voltage rise shall be determined by the charge state of the device-influenced by the magnitude of both the base and collector currents.

Fig. 9(a) gives an example of the relation between collector and base current and collector-emitter voltage for type I switch, the base drive being chosen to accentuate the effects being discussed. This can be related to the energy dissipation in the switch by looking at the different components of the dissipation.

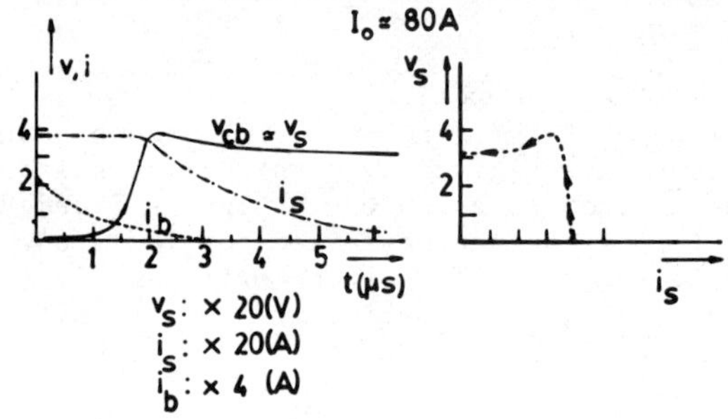

Fig. 9(a). Relation between base drive, collector current and collector-emitter voltage in the active switching region, type I switch with separate base drive

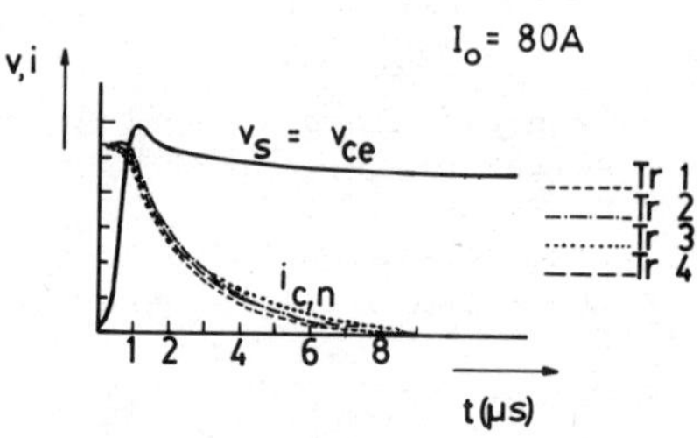

Fig. 9(b). Dynamic current sharing among four chips during switch-off in a type I power switch

From a study of relation 15,16,18, and figures 5 and 6, it can be calculated that at 60A, for example:

$$W_{s,I} = 6{,}100\text{mJ}$$
$$W_{struct,I} = 0{,}594\text{mJ}$$
$$W_{B,I} = 5{,}506\text{mJ}$$

During transistor power switch design, this aspect has to be optimised by tailoring of the base drive to minimise dissipation. During this process it has to be kept in mind that the switch voltage during this phase v_s, is directly controlled by the form of the base drive. Total shut-off at $t = t_0$ of the base drive will merely lead to breakdown of the transistor, and its possible destruction. When structural energy has to be dissipated in the chip, it is much better to do this in the active mode, as dissipation of structural energy in the chip in the breakdown mode leads to hot spots, probably secondary breakdown and consequent chip destruction.

On the other hand, the energy dissipated by the source into the chip

$$W_B = \tfrac{1}{2}V_B I_o(t_1 - t_o) \qquad (24)$$

must be minimised by keeping the time $(t_1 - t_o)$ as short as possible. Since from relation 17 it may be seen that this time is fixed by the geometry of the switch and the rated current of the switch, the only way to reduce this turns out to be to increase the di/dt after switch-off through tailoring the base drive. The minimum time is then represented by

$$(t_1 - t_o)_{min} = \left[\frac{L_{struct} \cdot I_o}{V_{CEO} - V_B}\right] \qquad (25)$$

When applying transistor choppers to low voltage drives, the difference $(V_{CEO} - V_B)$ may be large enough to reduce W_B to an acceptable level, without undue increase of cost for higher voltage transistors. With an increase in supply voltage, the situation becomes increasingly unfavourable, and eventually external R-C circuit elements across the switch has to be used to (i) keep the voltage down
(ii) divert some of the dissipation from the chip to the external circuit.

Our work has shown that this leads to a new optimisation problem, but this will not be discussed further in the present context.

During these experimental investigations the current sharing between the different chips in the switch was also examined. As far as the steady-state current sharing is considered, we have found current sharing to be within 5% and can therefore support the view of Lataire et. al. (7) that chips may be connected in parallel without any further measures, as long as the construction is such that no undue resistances are included in base and emitter leads.

Dynamic current sharing, determining the sharing of the switch-off dissipation previously discussed, have to be approached with more care, however. Following the lines of the previous paragraphs in minimising structural inductance and keeping the structures symmetrical, also results in dymanic current sharing within 5%, as shown clearly in figure 9(b) at rated current of 80A for a type I switch. The same may be concluded for the switches of type II and type III.

4. CONSIDERATIONS FOR APPLICATION OF CHOPPERS TO D.C. MACHINES AND INVERTERS FOR A.C. MACHINES

The considerations for application of transistor choppers to d.c. machines in the armature circuit may be divided into considerations, related to devices and layout for this specific application, considerations related to the control of d.c. machines in battery vehicles and considerations related to the application of choppers for inverters to drive a.c. machines in battery vehicles. They will be discussed in this order. However, since reports have already been published on transistor choppers, only those aspects we believe are different will be reported on.

4.1 Considerations related to devices and layout

Source inductance may be troublesome in any application of a chopper circuit to the control of d.c. machines by causing voltage spikes at switch-off. In the case of transistor power switches, this effect is of special importance, since the voltage rating of transistors is not yet comparable to thyristors, while the current levels in transistor power switches are higher for the same power level, due to the lower voltages generally used. As mentioned previously, one has to allow for 200A even for a 10kW controller at 50V battery voltage. Referring to figure 10, this would mean keeping the cabling stretches AB and HG as short as possible as this influences size and cost of C_B. On the other hand the magnetic energy stored in loop BGFC adds to the structural energy of the composite power switch, and also has to be kept to a minimum to increase efficiency, operating frequency and decrease cost and weight. As a result of our experimental work it has been found advantageous in all these respects to construct the chopper on top of the battery power supply when using transistor technology.

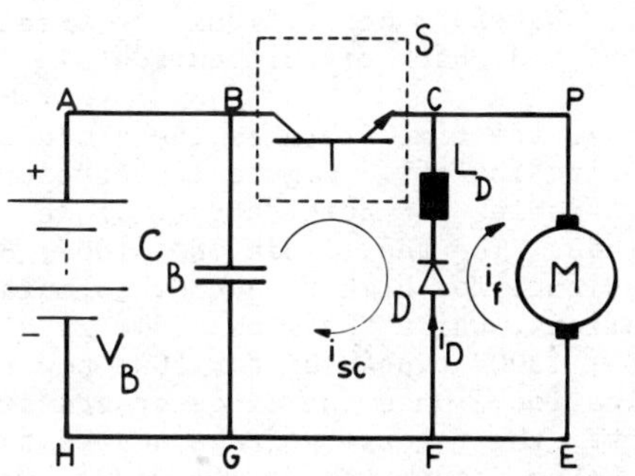

Fig. 10. Transistor chopper for d.c. machine control

When the power switch S takes the freewheeling current i_f over from the circuit CPEF (figure 10), a short circuit current i_{sc} is being built up in the circuit BCFG. This has also been studied by other investigators for this type of application (6,7). To investigate this further, let the circuit be characterized by the inductance of the branch CF, the branch BC (approximately the structural inductance of the switch), and the inductance of BGF.

$$L_{TOT} = L_{struct} + L_D + L_{BGF}$$

The diode current and short circuit currents are related by:

$$i_D = \left[I_D - i_{SC}\right] = \left[I_D - \frac{V_B}{L_{TOT}} \cdot t\right], t_2<t<t_4 \qquad (26)$$

where the maximum short circuit current will be determined by the duration of $(t_4 - t_2)$, and is given by

$$I_{SC} = I_D + I_{DRM} \qquad (27)$$

The circuit di/dt is given by the fixed supply voltage and circuit inductance, so that it may be expected that I_{DRM} is a function of

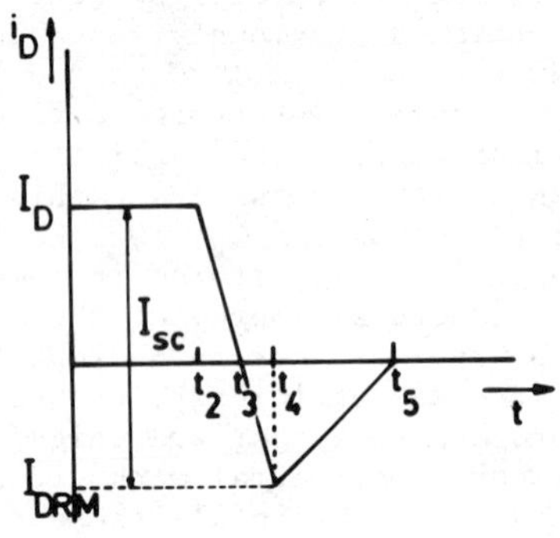

Fig. 11. Waveform definitions for freewheel and short circuit currents

$(t_4 - t_3)$, the time it takes the diode to start switching off. Figure 12 represents these variables for both a fast diode and a slow diode. The fast diode is a 100A, 600V type designed for high frequency commutation applications, while the slow diode is a normal 150A, 1200V type used for 50Hz power rectification. Once the diode starts switching off, the reverse voltage across the diode, being determined by the di/dt, may

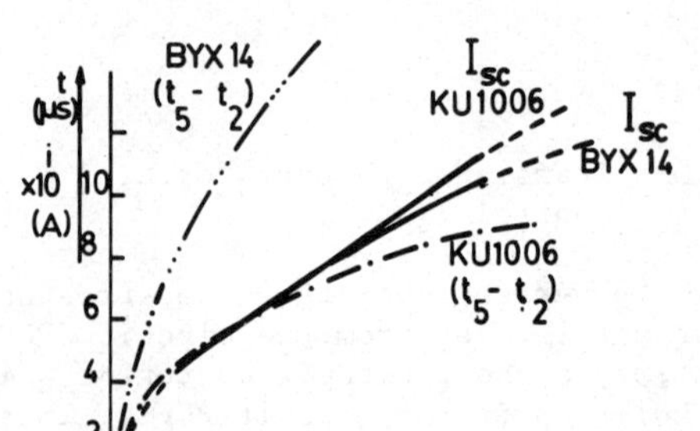

Fig. 12. Freewheel circuit variables for slow and fast diodes in exactly the same set-up (type II switch)

rise to very high levels, necessitating well designed protection for the freewheel diode. In our applications it was found that these voltages could reach 1000V at 48V supply voltage when using the fast diode. With an adequate R-C protection circuit these reverse voltages are under good control. Figure 12 indicates that the short circuit current that has to be carried by the power switch of the chopper is very nearly the same for both diodes up to a switch-off current of 60A (this would typically correspond to a mean current of 80A, half the rated current). Even after this there is not so much difference percentage wise, the peak of the short circuit current being somewhat lower for the slower diode. However, the power switch has to carry the short circuit current for nearly double the time, i.e. 17µs for the slow diode as against 9µs for the fast diode. From the graph it also follows from extrapolation that even the current in the fast diode might endanger the switch at rated current, the peak amounting to some 340A. It is therefore advisable with these bigger choppers to include a carefully optimized, small saturable core around the freewheel diode lead.

When applying transistor power switches as high-low choppers to d.c. machines, the active nature of the load can lead to base overdrive and underdrive. While the former leads to slower switching and a lower efficiency, the latter leads directly to an increase in on state voltage and destruction of the switch. Development of protective active circuitry was therefore started as part of our programme from the start. Subsequently it has been found that other investigators have shared this philosophy. (7). These active protective circuits have a twofold function: In the first instance the current through the switch is measured, and the base drive of the switch adjusted accordingly (see figure 13). This means that in effect circuitry has been developed that emulates the regenerative action inherent in thyristor switches. The base drive circuitry naturally has only a limited capacity, and taking the current carrying capability of the power transistor chip into account, this base drive levels off at a preset point. The voltage across the switch is measured, and at a certain level the switch is shut off through the base drive circuitry BD in figure 13. This is a capability not inherent in a thyristor switch. Seeing that the sensing and comparison has a limited bandwidth, and the circuit a finite reaction time, it is still necessary to include passive protection for the switch in the form of R, C, elements, connected by a diode to the main circuit, and in parallel with some form of VDR. The problem also arises whether each power transistor has to be protected individually or all collectively, but we are not yet clear on the merits of these two ways of approaching the problem.

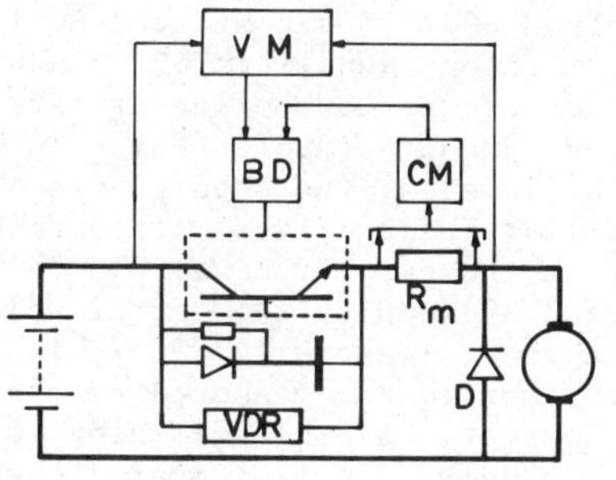

Fig. 13. Active protective circuits for application of choppers to d.c. machines. VM: voltage measuring circuitry; CM: current measuring circuitry; B.D. base drive circuitry, RC: passive protection with VDR, R and C.

4.2 Considerations related to battery vehicle applications for the choppers

Due to its current source effect the well known two level current control system has been found best for application to choppers for battery vehicles. This system is considered known, and will not be discussed further.

The electronic command system in the battery vehicle choppers has been isolated from the transistor power switches by optical isolation. This is a very important safety aspect in these systems, when applied to battery vehicles, and needs emphasis. In addition this opto-isolation allows a design of the electronic command system much less sensitive to noise and voltage level shifting. The details of this has been discussed elsewhere. (9,12).

In the foregoing discussion no decision as yet has been made as to what type of circuit configuration is to be used for the output stages. Some previous workers have preferred the Darlington configuration on different grounds (6,7). However, some arguments may be put forward to use a separate supply for the base drive, as has been done in our designs.

The base-emitter voltage drop in saturation of the power transistors used for power switches is slightly more than the collector-emitter saturation voltage, so that it may be taken that, for equal currents, the Darlington configuration will provide at least double the saturation voltage, and consequently double the conduction loss (A typical 5mm chip rated at 30A will have a 1,3V base-emitter drop at 15A and a driver chip 0,7V collector-emitter drop at 1,5A giving 2,0V total drop at current amplification of 10. Of the newer generation of transistors the BUX20, rated at 50A, will have 1,3V base-emitter voltage drop at 40A, with collector-emitter drop of 0,3V at 16A, all under assumption of current amplification of 10. Therefore, even when the BUX20 is used as a driver for four output transistors of the same type, the voltage drop will be 1,6V, that is double that of the same switch with a separate base supply).

Our work has indicated that a separate supply for the driver should have at least four times the saturation value of the main transistors (that is a minimum of 3,2V). Taking this into account, the comparison between the dissipation for the Darlington configuration and separate base drive configuration can be made.

$$P_{M,DAR} = 2V_{CE,SAT}(I_{CM} + 0,1I_{CM}) = 2,2V_{CE,SAT} \cdot I_{CM}$$

$$P_{M,SEP} = V_{CE,SAT} \cdot I_{CM} + 4V_{CE,SAT} \cdot 0,1I_{CM}$$

$$= 1,4V_{CE,SAT} \cdot I_{CM}$$

$$\therefore \Delta P_M = 0,8V_{CE,SAT} \cdot I_{CM} \qquad (28)$$

Seeing that this difference in dissipation - under the assumption of not using chips at "rated" current as stated earlier in the paper- would amount to some 64W per 100A in current technology, it has been decided to adopt designs with a separate base drive supply. This not only increases efficiency (peak conduction loss for systems with separate base drive is 80W per 100A in current technology) but enables more compact switch design (refer to enclosed figures of applications), reducing structural switch inductance, transient dissipation and consequently enabling higher frequencies of operation. Furthermore, in a battery vehicle already containing an enormous number of secondary cells, it seems to be a small price to pay to carry one more auxiliary battery to supply this base drive, in stead of using a down converter.

Previously, arguments has also been presented for a Darlington configuration in view of the fact that the base current tends to adapt to the load current, whereas the separate base drive supplies approximately constant base current. Understandably, this would lead to excessive base drive at small load currents, reducing the advantage given in relation 28. (7). This argument has to be applied with care, since the application of choppers as control elements in the armature for d.c. battery drives at present tends to be confined to the accelarating region, where the switch continuously works at maximum rated current, being shorted out afterwards during normal running with field weakening to reduce losses (13). This has been incorporated in many prototypes of electric vehicles running at present. Furthermore, as has been pointed out in connection with figure 13, the base drive may also be adapted to the load current when using a separate base drive supply.

4.3 Considerations related to the application of choppers to battery fed inverters

Application of battery fed inverters to induction machine drives in transistor technology, has been approached as an extension of the chopper technology. All the considerations discussed in the previous parts of this paper on composite switch design and transistor chopper circuits apply to inverter switches as well, and therefore a detailed discussion of the matter is to be found elsewhere. (10).

Figure 14 gives a schematic representation of the inverter system. The following remarks concern this type of application:

In order to have easy access to a separate battery for the base drive of the power switches, both PNP and NPN switches were developed, being optically isolated from the command electronics as in the previous case. Being a voltage-fed bridge, pulse-width modulation has to be used to adapt the output voltage to the frequency. The maximum switching frequency is determined by the switching losses as previously discussed, and is considerably higher (3kHz) than with corresponding thyristor bridge inverters. Voltage peaks and switching losses due to structural inductance of the bridge arrangement and the switches have to be kept to a minimum in the same fashion as discussed in section 3.

In the bridge arrangement, however, the protective measures per switch, as discussed in section 4.1, are extremely important to protect each cell. These protective measures are augmented with an extremely fast current measurement and overriding switch-off in the d.c. link to detect any short circuit currents through switches across the d.c. busbar.

5. EXAMPLES OF APPLICATION OF TRANSISTOR POWER SWITCHES TO BATTERY TRACTION

In this section the application of the ideas discussed in the preceding paragraphs to practical traction in terms of a drive for an electric scooter, for an electric motorcar with d.c. traction motor and for a low voltage a.c. traction machine will be treated.

A chopper using a type I switch for 80A, 24V was applied as control to a Solo Electra Electric Scooter, (see figure 15) using two level current control for the d.c. machine rated at 1,0kW short term, with supply from two 12V, 30A-h traction batteries and one 4V auxiliary supply. Switching losses at 1000Hz, 80A amount to 10,5 watt, and conduction and base drive losses to 90W. This gives the control system an efficiency of only 90%, illustrating clearly the disadvantages of such a low voltage supply and an overdimensioned controller.

As a next example figure 16 shows a chopper using a type II switch for 160A, 64V applied to the control of a 10kW separately excited d.c. machine for an electric car. The supply is from 60V, 190A-h Willard EV 190 batteries, with one auxiliary 6V, 45A-h supply. Efficiency of the controller at rated current and 1000Hz amount to 96%. This can be improved by lowering the supply voltage of the auxiliary base drives, since this amounts to 180W in the present case. In the case of using a 3V supply, the efficiency should increase to at least 97%. This has not yet been implemented and measured at the present stage, however. As the switching losses amount to only 50 watt at rated current and 1kHz, it can be seen that in this application the chopper frequency is limited to 3kHz, as it rises linearly with frequency.

As a final example of application to battery traction, figure 17 shows a 5kW, 80A, 64V transistor inverter developed to drive a low voltage induction machine, the stator of the machine being wound for 100V, and the short term rating of the machine being 10kW. The

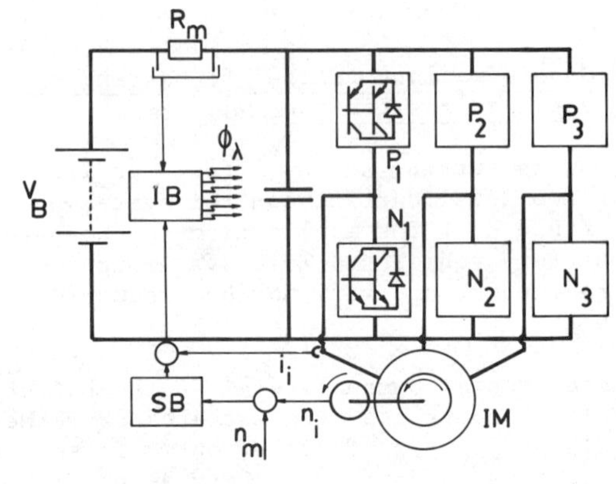

Fig. 14. Application of transistor inverter for control of 5kW battery-fed induction motor drive system

Fig. 15. Transistor (80A,24V) chopper control applied to an electric scooter with permanent magnet motor, 1,0kW. a: power switch; b: main battery; c: drive motor

Fig. 16. Transistor chopper for 160A, 64V applied to 10kW separately excited d.c. machine for electric car (R4L). a: power switch; b: freewheel diode; c: auxiliary battery; d:fan for forced cooling

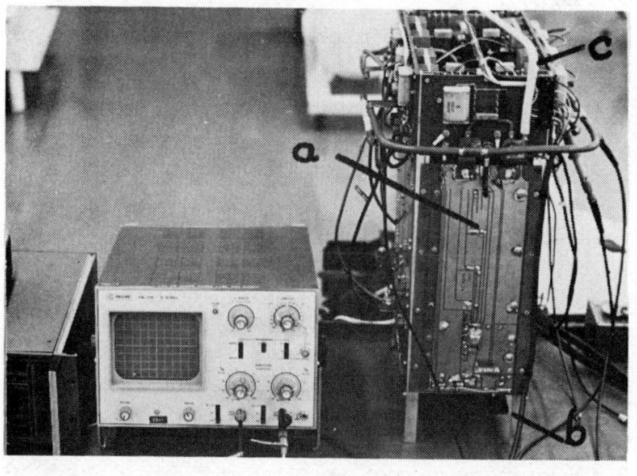

Fig. 17. Transistor inverter for 80A, 64V applied to 5kW induction motor for electric car drive. a: type I power switches; b: fans for cooling; c: protective circuits and feedback diodes

inverter has been built up by using 3 type I switches (80A, 64V) with npn-devices (N_1 to N_3) and 3 with pnp-devices (P_1 to P_3) as shown in figure 14. In contrast to the previous applications, this application incorporates both current control and speed control by maintaining the slip of the induction machine constant. Naturally the system is also fully regenerative, while the previously cited chopper application is only regenerative by field strengthening, through its field chopper circuit, which has not been discussed. As in all previous cases, optical isolation is used between the command and power electronics, while two auxiliary 6V supply batteries supplies the separate base drive circuits for the top and bottom switches.

At present unfortunately no thorough efficiency measurements on the bridge has been completed. The maximum switching frequency per switch for pulse width modulation purposes is 5kHz, switching losses appear to be 20W per switch (5kHz), conduction and base drive losses 50W per switch, although the latter may still be decreased to 35W per switch.

6. EVALUATION OF TRANSISTOR POWER SWITCH APPLICATIONS FOR BATTERY FED TRACTION

It has been indicated in this paper that it is possible to construct choppers for d.c. motor control and inverters for a.c. motor control in battery vehicles with repetion frequencies of switching of at least 5kHz. Although much remains to be done for the refinement of measurement techniques in this field the efficiencies appear to be notably higher than for thyristor converting equipment. The higher switching frequencies feasible appear to make higher system efficiencies possible through reduced losses in the electrical machines due to current ripple. The chopper systems developed and tested range up to 10kW, this also being the power level of inverters assembled from these switches.

The smaller thermal time constants, susceptibility to second breakdown and proportionality between base and load (collector-) currents, higher switching speeds and smaller voltage margins necessitate a technology quite different from conventional thyristor technology, as far as switch layout, design, fast active protection, and regenerative base drive are concerned. From the investigations conducted, part of which has been described in this paper, it would appear that pursuing the development of this technology might be extremely rewarding in eventually yielding d.c. and a.c. controllers with efficiencies in excess of 95%, with dimensions smaller than 20% of that of the drive motor.

It has to be realized that the power levels up to 10kW for the applications discussed has been dictated by laboratory circumstances such as funding and equipment, but does not represent the limit with the presented technology. With reference to figures 7 and 16, this can be explained. In this type II power switch, without any mechanical alterations whatsoever, the power transistors can be replaced by 50A chips in modified TO-3 with V_{CEX} of 160V, at present commercially available. In correspondence with previous discussions in this paper, the chips are worked at 40A each to obtain a saturation voltage of 0,8V at maximum rated current. This then yields a power switch for 400A, 100V (40kW) in a package of 250mm by 20mm x 100mm, excluding the fan for forced air cooling. With the methods presented in this paper the maximum frequency of operation of this switch is estimated at 2,5kHz, and would approach the limit of present technology. Understandably the limits of this technology will be pushed further upward by new devices becoming available.

7. ACKNOWLEDGEMENT

It is a pleasure to acknowledge the able assistance of G.W. Loots, H.G. Smith and D.A. Viljoen for parts of the experimental work. Funds have been made available through the National Committee for Battery Driven Vehicles of the Council for Scientific and Industrial Research for research and development on drive systems for battery vehicles.

LITERATURE REFERENCES

1. Weigel, W.D.
Untersuchungen an Gleichstromantrieben für Strassenfahrzeuge.
Wiss. Mitt. ISEA II, RWTH Aachen, 1973, 210 pp.

2. Kahlen, H
Vergleichende Untersuchungen an verschiedenen Gleichstromstellern für Fahrzeugantriebe.
Diss. Fak. Elektrotechnik, RWTH Aachen, 1973

3. Naydowski, C.-O, Pohl, D; Skudelny, H.-Ch.
Untersuchungen an elektrischen Strassenfahrzeugantrieben, dargestellt am Beispiel eines Kombiwagens für den Stadtverkehr.
Viertes Internationales Elektro-Fahrzeug-Symposium, Düsseldorf, 31 Aug. - 2 Sept. 1976, 24 pp.

4. Wagner, R.
Antriebstechnik für Elektroautos.
ETZ -A, 94(11), 1973, 691 - 694.

5. Polner, A
High current transistor output systems for motor controls.
Paper 2.1.2, Electric Vehicle Conference, Düsseldorf 31 Aug. - 2 Sept. 1976.

6. Margrain, P.
Evolution technique et economique du variateur a transistors: son application a un vehicle specifiquement electrique.
Paper 2.1.3, Electric Vehicle Conference, Düsseldorf, 31 Aug. - 2 Sept. 1976.

7. Lataire, Ph.; Maggetto, G.; Varkas, M.; Van Eck, J.L.
Developpement d'une serie hacheurs a hautes performances utilisant le transistor de puissance en regime de commutation.
Paper 2.1.9, Electric Vehicle Conference, Düsseldorf, 31 Aug. - 2 Sept. 1976.

8. Jordan, E.C.
Electromagnetic waves and radiating systems.
Constable, London, 1953.

9. Van Wyk, J.D.; Schoeman, J.J.
Electronic control system for an electric scooter.
Research Report, Department of Electrical and Electronic Engineering, Rand Afrikaans University, Johannesburg, 1976.

10. Van Wyk, J.D.; Schoeman J.J.; Loots, G.W.
Study of a transistorized electronic inverter for induction machine control.
Research Report, Department of Electrical and Electronic Engineering, Rand Afrikaans University, Johannesburg, 1977.

11. Van Wyk, J.D.
Composite transistor power switches
Research Report, Department of Electrical and Electronic Engineering, Rand Afrikaans University, Johannesburg, 1977.

12. Van Wyk, J.D.; Schoeman, J.J.
High current transistor chopper for d.c. machine control.
Research Report, Department of Electrical and Electronic Engineering, Rand Afrikaans University, Johannesburg, 1976.

13. Kahlen, H.
Transporter drive with electronic controls and mechanical transmission.
Paper 2.1.16, Electric Vehicle Conference, Düsseldorf, 31 Aug. - 2 Sept. 1976.

SYMBOLS USED

a : radius of conducting structure.
f : frequency
i : instantaneous current
I : constant or effective value of current
p : instantaneous power
L : inductance
P : average power
R : resistance
t : time
v : instantaneous voltage
V : constant or effective value of voltage
W : energy
X : reactance
Z : impedance

β : wave constant
γ : propagation constant
ε : electric permittivity
η : wave impendance
λ : wavelength
μ : magnetic permeability
μs: microseconds
σ : conductivity
ω : angular frequency

STEADY STATE PERFORMANCE OF CHOPPER CONTROL SYSTEM OF A SERIES MOTOR ON THE TRACTION MODE

C.P. Bottura and H.M.F. Tavares

Department of Electrical Engineering, UNICAMP Campinas, São Paulo, Brazil

ABSTRACT

In this work, the steady state performance of pulsewidth torque control system of a series motor for electric traction is analysed for the following cases:

a) open loop
b) closed loop and proportional controller
c) closed loop and proportional plus integral controller

Through a double point of a recurrence, on a Lamerey diagram, the periodic steady state is characterized; some variables are calculated with the objective of evaluation and comparison of the steady state performances.

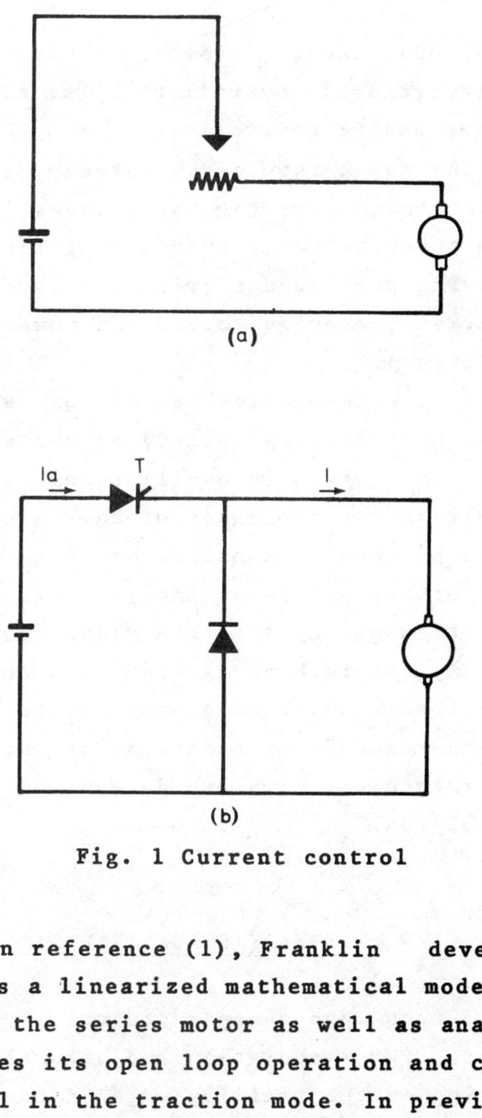

Fig. 1 Current control

1. INTRODUCTION

Current control of series motor for electric traction purposes is classicaly done through a variable resistance in series with the motor, with low power efficiency, Fig. 1.a.

By the use of pulsewidth modulation it is possible to control the motor current with high efficiency. Through a thyristor chopper, pulsed constant source voltage V is applied to the motor, whose current depends on the width of the applied pulses, Fig. 1.b.

In reference (1), Franklin develops a linearized mathematical model for the series motor as well as analyses its open loop operation and control in the traction mode. In previous works (2, 3, 4), through the method of point transformation of Poincaré-Andronov, nonlinear implicit recurrent mathematical models are obtained, and the stability of periodic oscillations for different types of controllers, as well as the consesequences of instability, and how to suppress it are studied.

The objective of the present work is to analyse the steady state performance of a series motor for electric traction, feed by pulses of continuous voltage, with pulsewidth modulation on the trailing edge of pulses, using natural sampling, in the traction mode. The control circuit includes a clock and the frequency of the pulses is constant. The following cases are analysed:
(a) open loop; (b) closed loop and proportional controller; (c) closed loop and PI controller.

The comparison of the steady state performance for the three cases is done through the calculation of the following performance criteria: average current, average torque and power efficiency.

From a theoretical point of view, the problem of stability of the steady state periodic oscillations as well as the avoidance of the chattering phenomenum constitutes the most important points of analysis; in this work, supposing that the above considered problems have already been solved so that a periodic steady state has been reached, expressions for the performance criteria are determined and analysed.

2. STEADY STATE ANALYSIS

The open loop chopper control system with PWM modulation and natural sampling on the trailing edge of the pulses can be represented by Fig. 2, where Z in the voltage signal that is applied to the series motor, P and θ are constants, the first beeing the period of the pulses and the second one beeing the conduction angle, with θP beeing the width of a pulse during the steady state.

According to Franklin (1), linearizing the magnetization curve of the machine, the following basic equations can be written for the pulsed series motor:

$$Z = L\frac{dI}{dt} + RI + F_1 \quad , I \geqslant 0$$

$$\phi = \phi_0 + C_1 I$$

$$F_1 = K_1 \phi \Omega = F + K_1 C_1 \Omega I \quad ,F=K_1\phi_0\Omega$$

where ϕ is the magnetic flux, ϕ_0, C_1 and K_1 are constants, F, is the counterelectromotive force and Ω is the speed of rotation of the motor shaft, that is supposed practically constant relatively to the frequency of the pulses, for the purposes of this work.

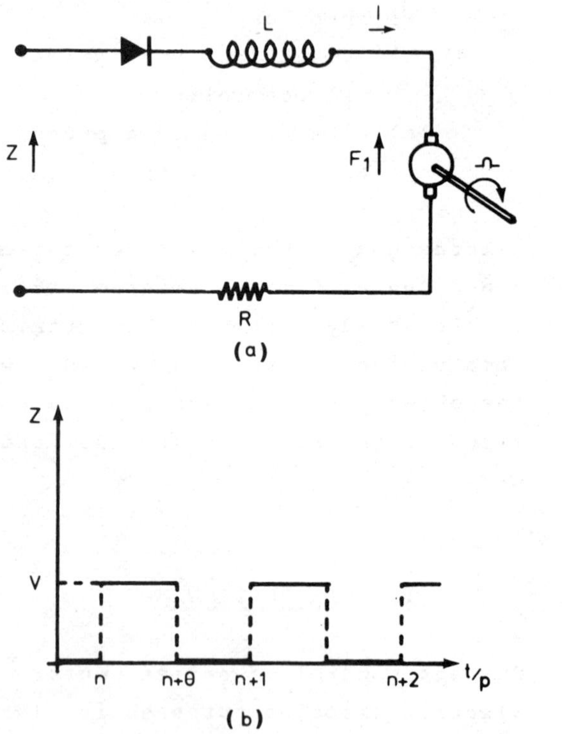

Fig. 2 Basic scheme for the motor

Thus it can be written for the series motor

$$\tau\frac{dI}{dt} + I = B\,(Z - F)$$

with

$$B^{-1} = R + K_1 C_1 \Omega$$
$$\tau = LB$$

that has the following solution

$$I(t) = I_1 + (I_n - I_1)\exp\left(\frac{-t+nP}{\tau}\right) \quad (1)$$

for $nP \leqslant t \leqslant (n+\theta)P$

$$I(t) = -I_2 + (I_{n+\theta} + I_2)\exp\left(\frac{-t+nP+\theta P}{\tau}\right) \quad (2)$$

for $(n+\theta)P \leqslant t \leqslant (n+1)P$

with $I_1 = B(V - F)$

$I_2 = BF$

where I_1 and I_2 are positive, by hypothesis, I_n, $I_{n+\theta}$ and I_{n+1} are respectively the currents at the instants nP, $(n+\theta)P$ and $(n+1)P$.

From (1) and (2), the Lamerey's diagram (5) of Fig. 3 is obtained; by imposing the condition for a double point, that characterizes the periodic steady state for the chopper controlled series motor on the Lamerey diagram, we have

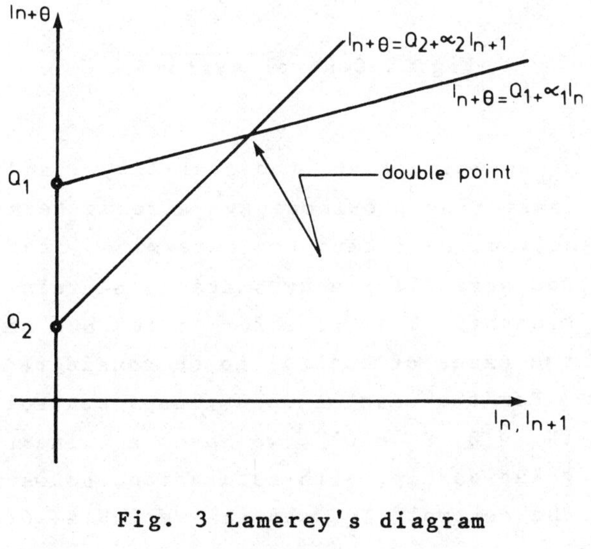

Fig. 3 Lamerey's diagram

$$\alpha_2 = \exp\left(\frac{(1-\theta)P}{\tau}\right) > \alpha_1 = \exp\left(-\frac{\theta P}{\tau}\right)$$

$$Q_1 = I_1\left\{1 - \exp\left(-\frac{\theta P}{\tau}\right)\right\}$$

$$Q_2 = I_2\left\{-1 + \exp\frac{(1-\theta)P}{\tau}\right\}$$

If $Q_1 > Q_2$ we will have a double point type I, that is characterized by

$$I_n = \frac{Q_1 - Q_2}{\alpha_2 - \alpha_1}, \quad I_{n+\theta} = \frac{\alpha_2 Q_1 - \alpha_1 Q_2}{\alpha_2 - \alpha_1}$$

On the opposite case we will have a double point type II, that is characterized by

$$I_n = 0, \quad I_{n+\theta} = Q_1$$

For the double point type I, the motor current will be positive during all of the cycle.

In this way we determine $I(t)$ as a function of θ, for the steady state, after which we can calculate some performance measures (1, 4):

- motor average current:

$$\bar{I} = I_1\theta - I_2\beta$$

where

$\beta = 1 - \theta$ for the type I

$$\beta = \frac{\tau}{P}\ln\frac{I_{n+\theta} + I_2}{I_2} \quad \text{for the type II}$$

- feeding line average current

$$\bar{I}_a = I_1\theta - \frac{\tau}{P}(I_{n+\theta} - I_n)$$

- average torque

$$\bar{S} = K_t\phi_0\bar{I} + K_t C_1\bar{I}^2$$

The instantaneous torque is given by $S = K_t\phi I$, K_t constant, and the mean square value $\bar{I}^2$ of the motor current is very easily calculated.

- efficiency

$$\eta = \frac{\bar{S}\,\Omega}{V\bar{I}_a}$$

In Figs. 4a, b, c, the above criteria are plotted for the case of $\frac{L}{R} = 40P$, $K_1 C_1 \Omega = 19R$, $K_1\phi_0\Omega = 0{,}3V$.

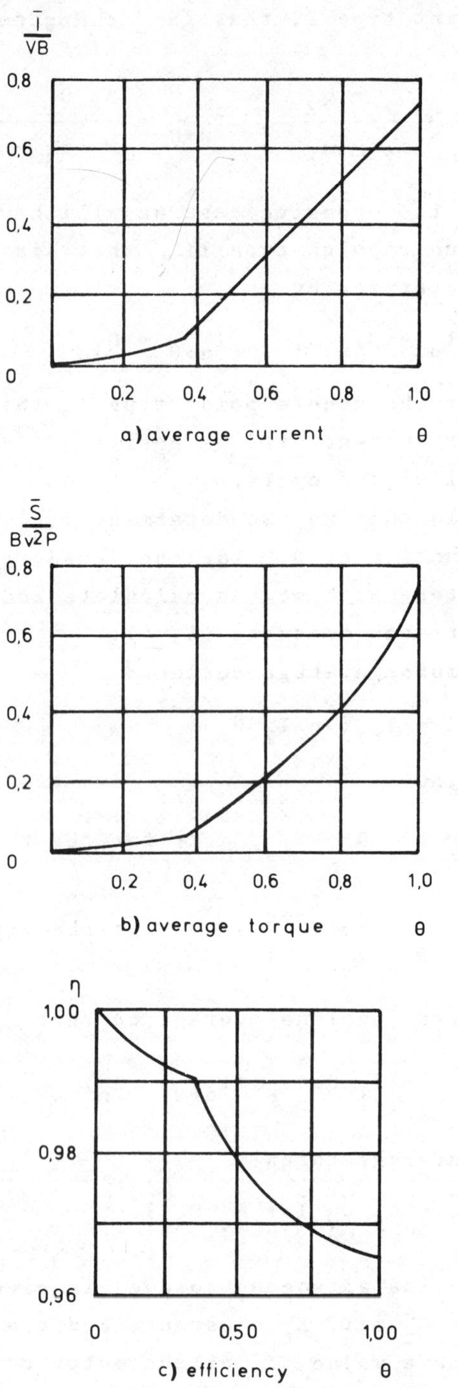

Fig. 4 Performance criteria as functions of θ

3. STEADY STATE CONTROL STUDIES

The chopper control system with proportional plus integral controller and PWM modulation with natural sampling on the trailing edge of the pulses can be represented by Fig. 5, where $I_r > 0$ is the reference current, K_2 and K_3 are constants, D is the differential operador,

$$V_c = -A(t-nP) \quad nP \leq t \leq (n+1)P;\ n=0,1,2,\ldots$$

where t is the time and A a positive constant .

Depending on the sign of the control signal $m = C + V_c$, there is chopper conduction and correspondly the tension Z applied to the motor is

$$Z=1 \quad \text{for } m>0 \ ,\ nP \leq t \leq (n+\theta)\ P$$

$$Z=0 \quad \text{for } m<0 \ ,\ (n+\theta)\ P \leq 1 \leq (n+1)\ P$$

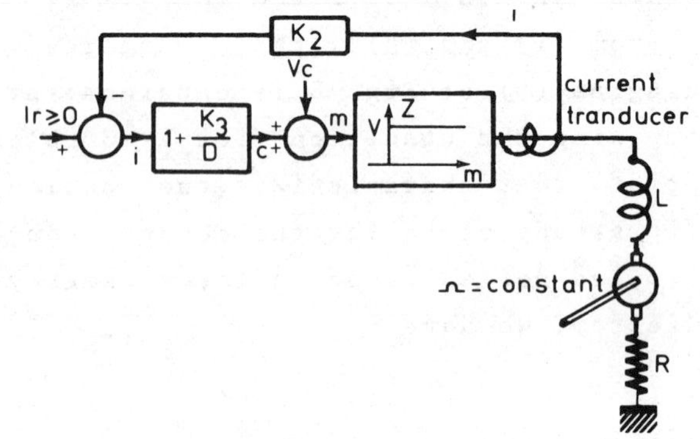

Fig. 5 Control system

Supposing that the stability and chattering problems have already been solved, we intend to determine, for the periodic steady state, the relationship $\theta = \theta(I_r)$ for each one of the cases of control to be considered.

For the case of open loop control $(K_2 = 0, K_3 = 0)$, we have a linear relationship, with saturation, between the control I_r and the variable θ:

$$\theta = \frac{I_r}{AP} \qquad 0 \leq I_r \leq AP$$

Thus, curves of the type presented previously in Fig. 4 are adequate for steady state performance determination as a function of the command I_r.

For the case of proportional control $(K_2 = 1, K_3 = 0)$, we obtain θ as implicit function of I_r:

$$I_r - AP\theta - I_1 + (I_1 - I_n)\, e^{-\frac{\theta P}{\tau}} = 0$$

For the case of proportional plus integral control ($K_2 = 1$, $K_3 > 0$), the mathematical formulation of the problem becomes more complicated, due to the introduction of an additional state variable (4), but the results are simplified due to the integral action on the error; a linear relationship for double point type I.

$$\theta = \frac{I_r + I_2}{I_1 + I_2}$$

and an implicit relationship for double point type II:

$$I_r = I_1\,\theta - I_2\,\frac{\tau}{P}\,\ell n\,\frac{I_2 + I_1\left(1 - e^{-\frac{\theta P}{\tau}}\right)}{I_2}$$

are obtained

It should be noted that the results do not depend on K_3.

In Fig. 6 we present the corresponding curves for VB = AP and remaining values as used for Fig. 4

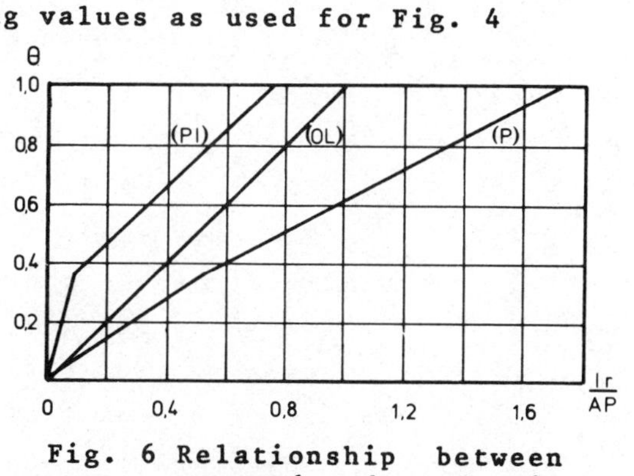

Fig. 6 Relationship between command and conduction angle

4. CONCLUSIONS

The combination of the curves in sections 3 and 2 allows the discussion of the steady state performance of the series motor with PWM control on the trailing edge of the pulses with natural sampling.

From the analysis that was done, the high efficiency of the system receives confirmation, and we can conclude in addition that in terms of average torque, the solution with PI controller is the best and that the solution with P controller is the worst of the three considered.

We emphasize that this work was limited to the calculation of some measures for the steady state performance, without studying the stability problems that may represent serious limitations to the operation of the system.

From the mathematical model adapted for the motor (1), we concluded about the existence of double point type II, and probably it lead to considerable inaccuracies for the small values of θ. We suggest that this study should be continued by the improvement of the linear approximation of the magnetization curve of the machine.

REFERENCES

1. P.W. Franklin, Theory of the D. C. motor controlled by power pulses , part I: motor operation, IEEE Transactions on Power Apparatus and Systems, PAS-91, 249-255, (1972).
2. C.P. Bottura and Y. Burian Jr., Local stability of proportional plus integral chopper control systems, Proceedings of the 1977 IEEE International Symposium on Circuits and Systems, 25-27 April, Phoenix, Arizona, USA (to appear).
3. C.P. Bottura and Y. Burian Jr.,Stability of periodic oscillations in proportional chopper control of a series motor, Proceedings of the 1977 Joint Automatic Control Conference, 22-24 June, San Francisco, California, USA (to appear).
4. C.P. Bottura, Contribuição ao Estudo do Controle de Torque em Máquina Série por Modulação em Largura de Pulso, Ph.D. Thesis, Department of Electrical Engineering, Universidade Estadual de Campinas, Campinas, Brasil (1973).
5. A.A. Andronov, A.A. Vitt and S.E. Khaikin (1966) Theory of Oscillators, Pergamon Press.

A FAIL-OPERATIONAL STEERING SERVO FOR VEHICLES

J.L. Feaver

Department of Cybernetics, University of Reading, England

INTRODUCTION

Automatic guided vehicles running on existing roads and motorways have been proposed as part of future land transportation systems. A buried guidance wire should be cheaper than railway track and in theory any existing steerable vehicle can be adapted to run under automatic control.

Fundamental to the practical implementation of a guided vehicle scheme is the provision of a high reliability steering servomechanism, preferably of such design that can be fitted readily to a production car or bus so that the vehicle can be driven either automatically or manually at the touch of a switch.

Over the past few years several vehicles have already been adapted in this way, notably by the Transport and Road Research Laboratory at Crowthorne, England.

Originally the research was directed towards improving motorway safety but it was soon realised that a reliable steering servo would have an immediate application as an aid for disabled drivers, particularly those handicapped by weak or missing limbs.

Therefore, attention was turned to the development of a fully drive-by-wire steering system with an electrical input provided by a tiny joystick control requiring only a small movement of hand or foot to replace the relatively heavy demands imposed on a driver's arms by the conventional steering wheel. A BLMC Mini with automatic transmission has been fitted with a prototype duplex steering servo powered by small lead-acid batteries.

RELIABILITY

The term "drive-by-wire" is derived from the aerospace industry's "fly-by-wire" which is used to describe fully electrical control which may have mechanical reversion available in the event of failure.

The technical problems of drive-by-wire are similar in some ways to those of fly-by-wire; both must have high integrity, but the vehicle system must be produced at automotive rather than aerospace costs. At the same time we must remember that a severely disabled driver will not be able to use mechanical reversion if electrical or electronic faults occur. Therefore, the vehicle steering actuator must be fail-operational which implies at least one redundant control channel.

Two basic methods of achieving this were considered. Standby redundancy, that is duplicated control channels, and active redundancy, that is duplex or triplex working where multiple channels share the load.

Standby redundancy was rejected for a number of reasons, including the difficulty of ensuring a smooth switchover to the standby channel, and the all important fact that there is no way of ensuring that an idle standby channel will be in working order when called upon.

Active redundancy on the other hand has a number of advantages. Both channels of a duplex system are working all the time so that they can be monitored for correct performance at any time. Both are sharing the load and so by running well below their maximum rating have an enhanced MTBF.

If a faulty channel can be switched out of service rapidly, discontinuity in control at switchover is more easily avoided.

Statistics show that in Great Britain an average driver is involved in an injury accident approximately once in every 10,000 hours of driving.(Ref.1). Assuming that loss of steering will give rise to an injury accident we must ensure that any increase in the probability of accidents arising from this cause will be relatively insignificant, say less than 1%. This implies that the MTBF of the steering system must be better than 1 x 10^6 hours.

With a duplex system, for example if each channel has an MTBF of 1,000 hours, then a single failure can be expected every 500 hours of driving. Thus, after the first failure, the probability of a second fault arising within half an hour is one in two thousand. The mean time between two failures arising within the half hour is then 500 x 2000 = 1 x 10^6 hours.

In practice it should always be possible to bring the vehicle safely to a halt within six minutes (1/10th of an hour) so that even with a single fault occurring every 500 hours, the mean time between two faults arising within this short period is 500 x 10,000 = 5 x 10^6 hours. MTBF calculations for the present system indicate that these figures will be exceeded by at least a factor of 2.

The Prototype System

System reliability decreases with increasing complexity and the MTBF of a parallel redundant system is a function of the MTBF of each constituent path and the number of parallel paths. (Ref. 2). If the reliability of a single channel is 80% and each channel is identical, then the reliability of duplex channels is

$$1 - (1-R)^2 = 96\% \qquad (1)$$

the reliability of triplex channels is

$$1 - (1-R)^3 = 99.2\% \qquad (2)$$

Thus duplex working realises the greatest gain in reliability. If each channel has a reliability of 80% or over there is little to be gained from increasing the number of channels to three, particularly with regard to cost effectiveness. However, it is important to ensure that the reliability of each channel is adequately high.

By comparing the outputs of two channels it is easy to detect a failure but not which of the channels has failed. To do this it is necessary to add a third reference channel against which to compare the other two. This is cheaper than using a full triplex system because there need be no power controller and drive motor in the reference channel. Failure of the reference channel is not potentially dangerous so the reliability predictions remain those of a duplex system.

However, satisfactory results depend upon achieving complete independence between channels so that there is no possibility of a common mode fault affecting two or more channels, and devising an automatic fault monitoring system which will reliably switch out the faulty channel quickly and warn the driver to stop at the next opportunity. The system must not give false alarms when no real fault exists. At the same time, it must be fail-safe and tested from time to time for correct functioning. These last two are conflicting requirements but whereas the former is mandatory, the latter is relatively less important.

A fully effective self testing facility can only be allowed to function when the car is stationary, and even then would cause nuisance alarms, so testing of the complete monitoring system is done manually before setting off on each journey.

A block diagram of the duplex monitored system fitted to the Mini is shown in figure 1.

The driver's input joystick is fitted with a triple ganged potentiometer which provides command signals to the three channel controllers. Two of these supply the electric motors which are geared to the steering column by two disengageable worms. Three servo feedback potentiometers are coupled to the column by anti-backlash gears. The electronic monitor compares the output from the three channels and feeds the logic switching circuits, which control the declutching solenoids and warning horns.

A small control panel with on/off and reset switches and power indicator lamps completes the system.

THE STEERING SERVO-MECHANISM

Steering effort is supplied by two ironless rotor D.C.machines coupled to the steering column mounted on a gearbox assembly fitted just below the steering wheel. This arrangement minimises the alterations required to the car during installation.

The motors drive the column through

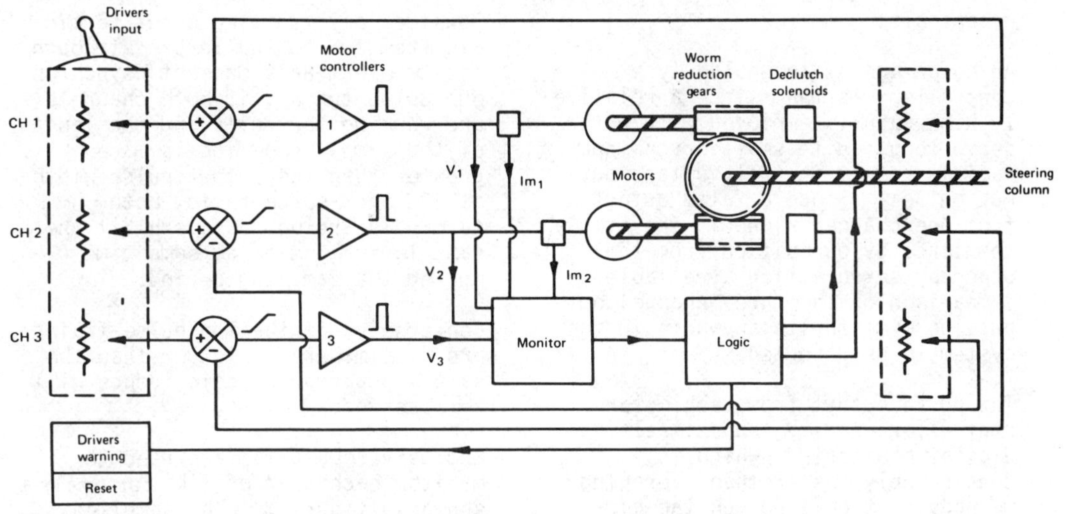

Fig.1. Block Diagram of the Steering System.

worm gears, each having a separate worm and carrier which engages with a common worm wheel on the column. The worm carriers are held in mesh by simple linkages operated by car overdrive solenoids, so that either channel can be disconnected by switching off the current to the appropriate solenoid. Worm gears allow the necessary 16/1 reduction between motor and column to be obtained in one stage and have the advantage of providing a mechanical lock to stop immediate steering run-away if one channel should fail to a hard on state. A corresponding disadvantage, however, is that if one channel fails dead the steering is locked until the failed channels worm has been disengaged. This apparent disadvantage has been turned to an advantage by making use of the fact that whatever failure mode is encountered, the failed channel sets up a locked situation so that the remaining working channel will try to oppose the lock by servo action. The electronic monitoring circuit switches out the failed channel by releasing the solenoid.

In normal running both motors are in use together and this provides useful reserve torque for parking loads. Three potentiometers are driven via antibacklash gears to provide steering position feedback to the electronic control channels. The whole assembly is contained in a die-cast aluminium housing and comprises a "class 1" mechanical component which is assumed to have an infinite MTBF in the context of the life of the vehicle justified by robustness, simplicity of design and high manufacturing quality control.

THE ELECTRONICS

Since a high integrity servo motor controller had already proved successful in various simplex controlled vehicles it was decided to use two of these, one for each working channel of the duplex system. (Ref. 3). A third reference channel utilises a low powered circuit board only without the high power output stages so that the output signals under normal working conditions correspond with those of the working channels.

In order to avoid common mode faults, each channel is supplied from a separate battery contained in the boot of the vehicle. It is envisaged that in production, one channel could be supplied directly from the vehicle battery.

For the same reason it was decided to avoid, if possible, synchronising the output pulse trains, since if synchronism was adopted it could not be relied upon in the event of failure.

Failures fall naturally into two categories; those which are potentially catastrophic, for example disconnections, short circuits, or power failures; and those which are non-catastrophic in character, for example transducer offset or component deterioration which may cause mismatch between the working channels.

Some experimental work had shown previously that there were severe practical difficulties in attempting to compare accurately the analogue input signals to each control channel. Small differences between channels during normal working are inevitable and to some degree acceptable.

An advantage of the pulse type motor controllers is that since a relatively high gain is employed within the servo loop, quite small error inputs rapidly result in full voltage output pulses. Since a false output from one channel alone is inevitably countered by opposition from the other by servo action a reliable comparison of the three channel outputs permits faults anywhere in the system to be detected.

The pulse output from each motor controller readily lends itself to digital monitoring, which is considerably faster than averaging methods. At only 50 Kph the car will have travelled 14 metres in one second so that speed of fault detection is paramount if hazardous situations are to be avoided. To be useful averaging would have to take place over relatively large portions of a second. However, a simple digital comparison of the outputs is not possible as they are unsynchronised.

The Comparator

When a motor is not being driven both controller output terminals are positive, corresponding to a logical 1. When a motor is driven, one of the terminals is grounded corresponding to a logical 0. The truth table for one motor is shown in Fig. 2.

As stated before it is not possible under practical working conditions to equate precisely the three channels due to finite tolerances associated with potentiometers, summing amplifiers, resistors, etc., but since a potentially catastrophic fault condition will necessarily entail a difference between two like controller output terminals which persists for more than a certain predetermined time it is possible to make a meaningful comparison during a short sampling period.

Consider any two like ouput terminals from two controllers. Although the two terminals present asynchronous pulse trains, if both channels are working correctly and the length of the sampling period is made greater than the output pulse width at 50% mark-space ratio, both terminals are bound to exhibit the same logical state at some time during the sampling period.

Conversely, if the two pulse trains are dissimilar for longer than the sampling period no coincidence will be observed.

Exclusive NOR gates are used to monitor each pair of like terminals as shown in figure 3. The sampling periods are defined by a series of clock pulses which are gated to a retriggerable monostable. If the motor pulses differ for longer than the sampling period, clock pulses are prevented from retriggering the monostable, which commences to time out. If more than two consecutive retrigger pulses are missing the monostable times out completely indicating a fault. The three channel outputs are compared in pairs in this way, each pair having a separate monostable. A single fault then results in two of the monostables timing out. By combining the output from pairs of monostables using AND gates, a register in the form of a D-type flip-flop, indicates precisely which channel has failed, and initiates the switching out of the faulty channel at the same time alerting the driver by means of the warning horns. The output from the registers associated with the two working channels are interlocked on a first-come-first-served basis so that subsequent fault indications on other channels are ignored.

Typical sampling pulse period is 13ms repeated every 17ms allied to a typical motor pulse period at 50% mark-space ratio of 12ms.

MOTOR DRIVE	TERMINAL A	TERMINAL B
NIL	1	1
CLOCKWISE	0	1
ANTICLOCKWISE	1	0

Fig. 2. Motor Terminal Truth Table.

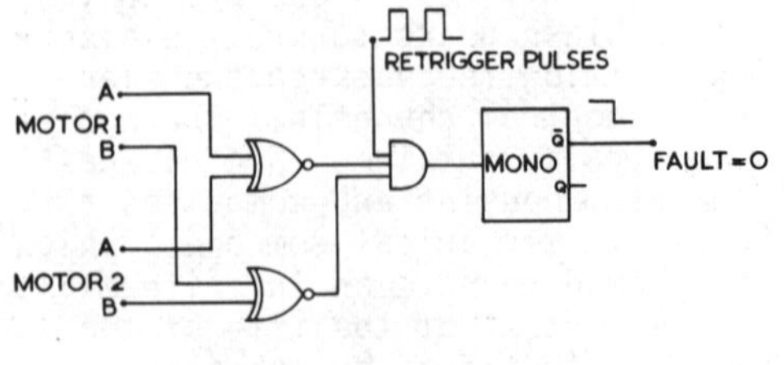

ig. 3. The Basic Comparator.

On the prototype system the monostable time-out was set at 60ms. This results in an inherent and controllable safeguard against the registration of spurious nuisance faults, whilst dealing adequately with real faults which are effectively registered within 75ms.

It is interesting to note that the particular circuit used to generate the motor drive pulses has the peculiarity that the pulse width at 50% mark-space ratio is double the on-pulse width at low mark space ratios. This imposes a limitation that errors due to unbalance between channels, for example, must exceed 50% in order to be registered.

In practice,however, a moderate degree of mismatch at controller inputs increases the stiffness seen by the motors which tend to fight each other so that on a statistical basis there is an increased probability that smaller mismatch errors will be detected, although the actual 50% mismatch situation appears only fleetingly at the controller output terminals. If a constant pulse width, variable repetition rate motor control circuit was used it would theoretically be possible to detect much smaller degrees of mismatch. The complete electronics are housed in a small rack under the glove compartment on the passengers side of the car. Four separate plug-in modules contain the two servo motor controllers, a reference and monitor module, and the solenoid and warning horn drive circuits.

POWER SUPPLIES

In order to avoid common mode faults between channels, three separate batteries are used, and these are charged via a transistor D.C. converter. The original intention was to isolate the three batteries by using silicon diodes, but in computer simulations of reliability the diodes showed up as contributing a major source of potential unreliability. Three separate output windings on the converter transformer give much enhanced protection against coupling two of the batteries together. At present the prototype converter charges each of the working channel batteries at 1.5 amps maximum, regulation being provided by the vehicle alternator control.

Apparatus was constructed to measure the ampere-hours required by the steering motors while the vehicle was driven over a typical road circuit including negotiating roundabouts, sharp turns at low speed, etc. The maximum power taken was only 1.2 ampere-hours per motor, nearly one of which was attributable to the worm-gear holding solenoids. A block diagram of the power supplies is shown in figure 4.

The regulators which supply the low level electronics and the servo potentiometers at 8.8 volts maintain regulation down to 10.5 volts input which gives an adequate margin of control in the event of a complete charging system failure. If a regulator loses control this is detected by the monitor system due to the false potentiometer signal which appears at the input of the faulty channel.

SOLENOID DRIVE

The car overdrive solenoids used to

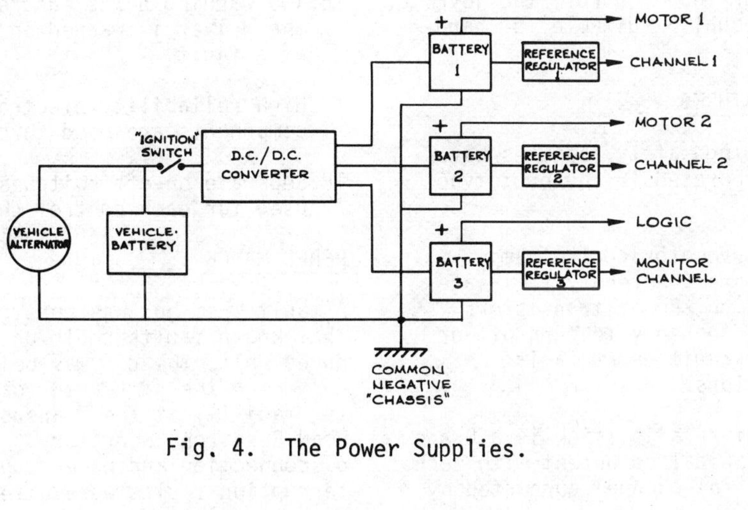

Fig. 4. The Power Supplies.

engage the worm gears were chosen because they contain two windings, one high current winding to provide a good pull-in force, and the other a low current holding coil. In order to protect the solenoids from burn-out if they remain switched to the high current winding should the gears not engage properly, a pulsed drive is provided via a transistor switch. When the solenoid is properly engaged, pulsing ceases automatically.

AUDIBLE WARNINGS

Two warning horns are fitted to ensure that the driver is adequately informed of a fault occurring. These are energised by more than one chain of command and in particular do not refrain from working until the entire system is switched on and functioning normally. Upon hearing one of these warnings the only action required by a driver is to continue driving to the nearest convenient stopping place, whereupon he can investigate the fault by trying to reset the system. If this is unsuccessful he must summon aid.

THE JOYSTICK

A high quality triple ganged potentiometer is used to convey command signals to the control channels. This is actuated by a control stick carried on substantial bearings, the whole being mounted in a robust diecast housing. The mechanical components are "class 1" as in the case of the gearbox Separate three cored cables connect the joystick to the electronics via three separate locking type plugs and sockets to avoid the possibility of common mode faults.

Although the prototype car has been adapted for foot control, the joystick is equally suitable for hand control.

SAFETY FEATURES

The following safety features have been incorporated in the prototype system.

1. The servomotor control amplifiers are inherently fail-safe by design. Power transistors have a tendency to turn off or open-circuit under fault conditions.

2. The driver's joystick is a Grade 1 mechanical component with each electrical channel connected by a separate wiring loom and separate plugs and sockets.

3. The servomotor and gearbox assembly is also a Grade 1 mechanical component.

4. Worm gears prevent a motor runaway situation developing.

5. Electronic monitor system failure leaves both servomotors active.

6. An electronic interlock prevents simultaneous disengagement of both worm gears.

7. Worm engagement solenoids fail-to-off (Disengagement = off).

8. Electronic monitor uses fail-safe pulse and latch C-mos logic.

9. Servomotor drive voltages are monitored for channel mismatch.

10. Servomotor drive current is monitored for disconnection faults.

11. Channel isolation is ensured by the use of three separate batteries, each having a separate connecting wire loom.

12. An isolating D.C.converter maintains batteries by float charging and further ensures channel isolation.

13. Battery capacity allows for several hours use in the event of a complete charging system failure.

14. Servomotors are high efficiency disc type. (Load-sharing further enhances reliability.)

15. Each motor is capable of steering the car alone.

16. Two warning horns ensure that the driver is warned in the event of a fault.

17. High reliability electronic components are used throughout.

18. Separate on-off switches are used for each control channel.

PERFORMANCE

A fault test box was constructed so that known faults could be introduced while the car was being test driven on the simulated road testing facility at the Transport and Road Research Laboratory. Various disconnection and power supply interruption faults were often barely perceptible to the driver and produced a minimal effect on the steering performance of the car.

The car has proved to be very manoeuvrable and to have generally acceptable handling characteristics. It is probable that if the system were fitted to a larger and more powerful car some change of servo gain with speed may prove necessary to reduce the steering sensitivity at high speed. This has already been tried with some success on single channel systems and a method of adapting the duplex system has been devised

Some controversy has arisen between various people who have driven the car; some like to see the steering wheel gyrating in front of them, and others prefer driving with the steering wheel removed. Since nearly half the servo power is needed to overcome steering wheel inertia there is some improvement in response to be gained from removing it.

It has been suggested that able bodied drivers used to controlling a normal car obtain some directional information from the attitude of the wheel. Disabled subjects more familiar with joystick control of wheelchairs manage very well without a wheel, which merely constitutes a hazard in the event of a collision.

SUMMARY

The viability of a duplex-monitored electronically controlled steering system for vehicles has been demonstrated. Reliability predictions have shown that an adequate margin of safety can be achieved using a duplex system, provided that the mechanical interfaces are of a sufficiently high standard.

A fail-safe method has been devised for monitoring the performance of the working channels. It has a fast response to real faults and inherently guards against spurious faults causing false alarms. At the time of writing the car has completed nearly 100 hours of testing both on the track and the public roads.

In addition to the development staff, a number of severely disabled people have driven the car and all have commented upon the ease of control as compared to using the conventional steering wheel.

Some further work is required on the ergonomic aspects of the system. Throttle and brake control could be combined in a two axis joystick.

Production costs are estimated to place the system well within the reach of the majority of disabled people who could derive great benefit from a new found mobility.

REFERENCES

(1) J.L.Feaver, S.Penoyre, B.G. Stoneman.
A Drive-by-Wire Vehicle Control System for Severely Disabled Drivers.
International Conference on Automobile Electronics.
I.E.E. July 1976.

(2) Electronics, Reliability - Calculation and Design.
Pergamon Press.
G.W.A.Dummer, N.Griffin.

(3) J.L.Feaver.
A High Integrity Single Source Reversible D.C.Motor Controller
IFAC Symposium on Control in Power Electronics and Industrial Drives.
Dusseldorf, October 1974.

INVESTIGATION OF CONTROL PROBLEMS DURING HVDC INVERTER OPERATION

K. Fay

Institut für Stromrichtertechnik und Antriebsregelung, Technische Hochschule, 61 Darmstadt, Germany

ABSTRACT

A hybrid computer simulation was arranged to investigate the main reasons for control problems during HVDC inverter operation. Three different control methods were investigated, all of which use special theories of operating the inverter in const.γ- mode, as well as different gate control devices with a proportional or integral characteristic.

INTRODUCTION

This paper supplements a contribution on HVDC control at the 1974 IFAC meeting (Ref. 1), regarding some special problems in the inverter control. The behaviour of HVDC inverter operation depends on the special system conditions as well as on the methods applied for inverter control.

DESCRIPTION OF THE CONTROL METHODS

Three different control methods for inverter operation were investigated. They are listed in Table 1. Method (1) and method (2) are presently adopted for HVDC schemes. Method (3) was included, because it seemed to be an interesting alternative at a small expense.

Method (1) is characterized by a closed loop γ- control: The actual margin angle γ is measured and controlled by a firing angle adjustment, using a feedback system. In order to avoid harmonic instability during in verter operation, the deviation $(\gamma_{act} - \gamma_{ref})$ is valued by a nonlinear characteristic (Fig. 1) which results in a high gain of the control loop for small values of γ and a slow reaction for great values.

control method	(1)	(2)	(3)
operation mode γ=const.	closed loop γ-control with nonlinear characteristic	instantaneous γ-prediction with "symmetrizer"	steady state γ-prediction
operating mode I_d=const.	PI- current controller	PDT_1- current controller	PI- current controller
gate control system	pulse-phase control (proportional)	pulse-frequency control (integral)	pulse-phase control (proportional)

TABLE 1 Components of the different control methods

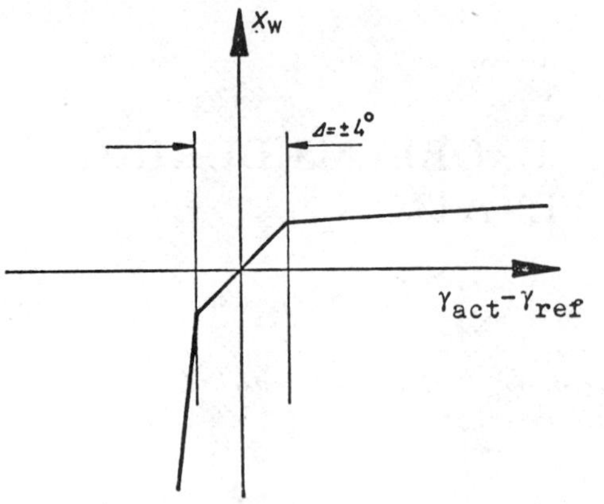

Fig. 1. Nonlinear characteristic in the γ-control loop

The method of γ-prediction, which is used with control method (2) is described in Ref. 2: The extinction angle of each valve is continuously predicted in advance from the measured commutation voltage, the direct current and the measured elapsed time from the previous zero crossing of the commutation voltage. If the predicted value of the extinction angle is equal to or less than a minimum reference value, a firing pulse will be generated immediately.

In order to avoid harmonic instability, this predictive type is combined with a so called "symmetrizer": If a predictive firing is indicated, the symmetrizer causes the gate control system to generate equidistant firing pulses corresponding to the pulse number of the converter. The symmetrized interval has the duration of one period and is affected by a new predictive firing.

The above described type of inverter control will be called "instantaneous γ-prediction".

Another possibility of operating an inverter in const. γ- mode is applied in method (3): The firing angle α_{max} which results in a given extinction angle γ_{ref} can be predicted by using the well known steady state equation:

$$\cos \alpha_{max} = -\cos \gamma_{ref} + 2.d_x . \frac{U_{\sim N} . I_d}{U_{\sim} . I_{dN}}$$

The effective value U~ of the AC voltages can be measured by applying the equation:

$$3.U_{\sim}^2 = u_R^2(t) + u_S^2(t) + u_T^2(t)$$

where $u_{R,S,T}(t)$ are the momentary values of the three phase voltages.

In the following, this method of inverter control will be called "steady state γ-prediction". If gate control systems with a proportional characteristic are used, this type of control can easily be implemented by limiting the output of the current controller to a value, which corresponds to the predicted α_{max}.

The methods (1) and (3) are combined with a proportional characteristic "pulse phase" gate control unit (PPC, Ref. 3). In order to obtain a mere integral characteristic of the open current control loop (Ref. 1), a PI-current control amplifier is used. Method (2) is combined with a "pulse frequency" gate control unit (PFC, Ref. 3), which is similar to the well known "phase locked oscillator" described in Ref. (4). This integral characteristic device can be used only in the closed current control loop. It is combined with a PDT_1 -current control amplifier.

In both, the PPC and PFC methods, the firing angle is not directly influenced by the shape of the commutation voltages. Thus, the well known feedback loop over the gate control unit, which can cause harmonic instability is eliminated.

DESCRIPTION OF THE POWER SYSTEM

The power circuit of the system, which was represented on the hybrid

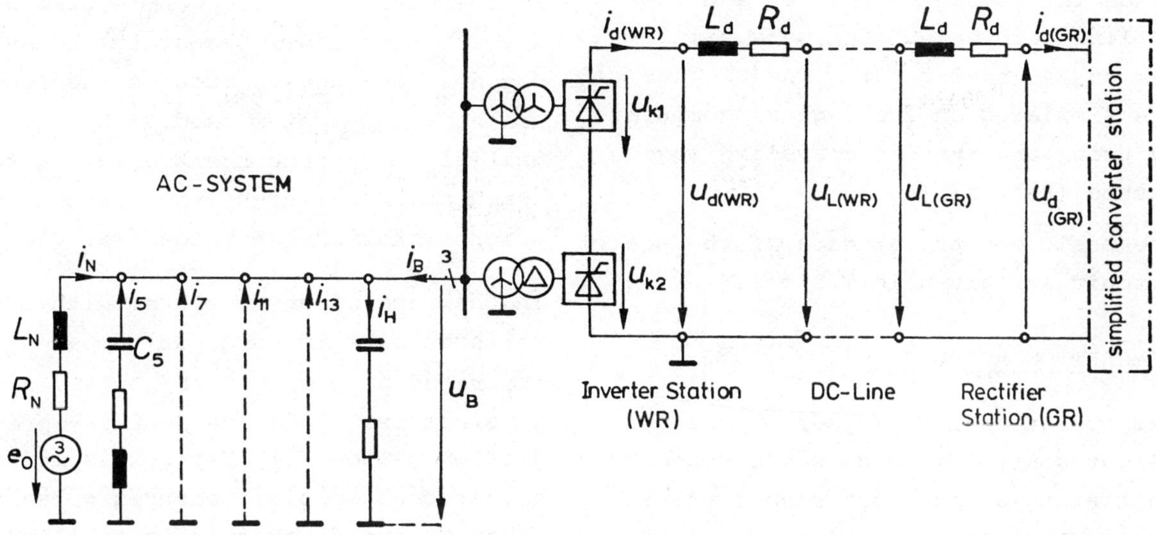

Fig. 2. Power circuit of the system investigated

computer, is shown in Fig. 2. A twelvepulse inverter unit was simulated in detail on the analog computer. On the inverter side, the impedance of the AC system as well as the filter circuits for the 5th, 7th, 11th and 13th harmonics and a high pass branch are represented.

The rectifier station was simulated in a simplified manner, using a controlled voltage source instead of the converter. This modell was refined by representing the equivalent resistance due to commutation, the time delay of the converter, and the nonlinear dependence of the DC voltage on the firing angle α, especially when α increases within a wide range. The rectifier station was equipped with control method (3) from Table 1, the AC-system impedance on the rectifier end was assumed to be zero. This model was simulated on the digital computer. The DC branches of both stations in-

DC system:		
direct voltage	U_{dN}	533 kV
direct current	I_{dN}	1800 A
converter transformer	$X_{ct} = 2.d_x$	0,2
smoothing reactor	L_d	800 mH
DC line length	l_L	700 km
Short circuit ratio:		
rectifier end	S_k''/P_{dN}	∞
inverter end	....	6 (2)
Inverter AC system:		
AC voltage	U_N	230 kV
impedance angle	φ	-85°/50 Hz
AC filter quality factor	$Q_{5,7,11,13}$	100

TABLE 2 Basic electrical data of the power circuit

clude the smoothing reactor and the DC line terminals. They were simulated on the analog part. The DC line was simulated on the digital computer by using an extended traveling wave method (Ref. 5).

The basic electrical data of the power circuit are given in Table 2.

SIMULATION RESULTS

The operational behaviour of the different control methods was checked by initiating a short but severe voltage dip on the generator side of the AC-system impedance. During all simulation runs, the time and duration of the fault was kept constant while the short circuit power of the AC system was varied.

In the case of a strong AC system on the inverter side, the system behaviour does not depend much on the special control system used. After an initial DC voltage break down due to commutation failures, the system recovers within 2-3 periods (Ref. 6).

In the case of an AC system with medium short circuit power as compared to the rated DC power ($S_k''/P_{dN} = 6$), some difficulties did arise in the γ-prediction method (3). Using this method, hitherto no special measures were taken to avoid harmonic instability as it was done in method (1) and (2). These methods did show satisfactory behaviour under the same given system conditions. Fig. 3a shows the behaviour of the unmodified method (3).

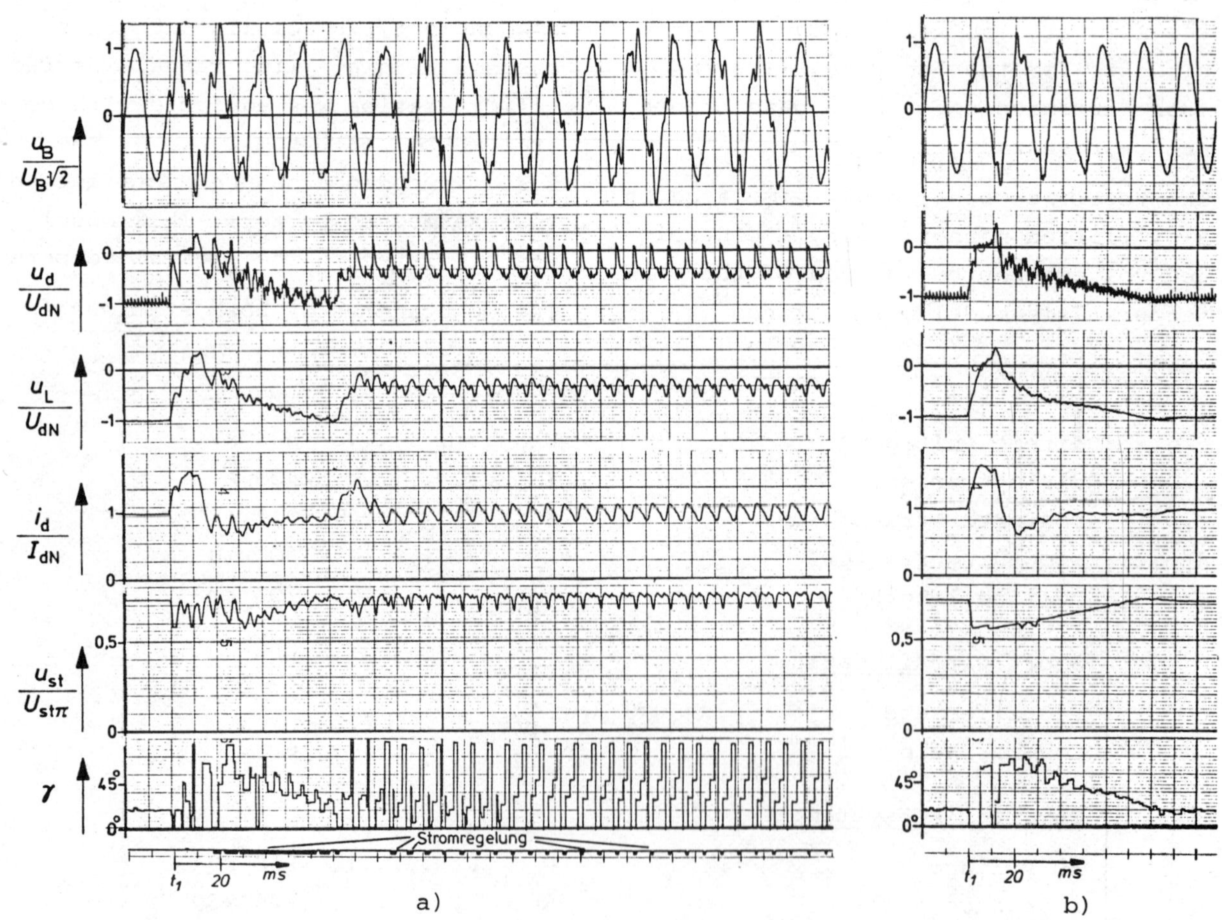

Fig. 3. Inverter operation with steady state γ-prediction (control method (3)). $S_k''/P_{dN} = 6$.

a) behaviour with unmodified control method (3)
b) behaviour with modified control method (3)

After the initial disturbance was initiated, an instability occured which led to repetitive commutation failures. With the given data of the system there was a parallel resonance of about 200 Hz on the AC side, caused by the resulting impedance of the AC-system and the AC filter circuits. Thus a strong 4th harmonic appeared on the AC side which led to a corresponding 3rd and 5th harmonic on the DC side due to the frequency transformation of the converter. Furthermore, a strong 2nd harmonic appeared on the DC side which is caused by the characteristic development of the repetitive commutation failures (Ref. 6).

The cause of this instability was traced out to be a closed loop over the mathematical model which was used for operating the inverter in const.γ mode: The predicted firing angle α_{max} depends directly on the momentary values of the AC voltages and the DC current which in return is influenced by the firing angle. In order to avoid these difficulties, the steady state γ-prediction type of control was supplied with a special nonlinear characteristic (Ref. 6), which allows the firing angle to decrease rapidly corresponding the predicted value and subsequently to increase with a very low gain. With this modification, the system behaviour Fig. 3b was achieved. Instability was avoided during all operation modes.

The following results show the operational behaviour of an inverter station, connected to a very weak AC system.

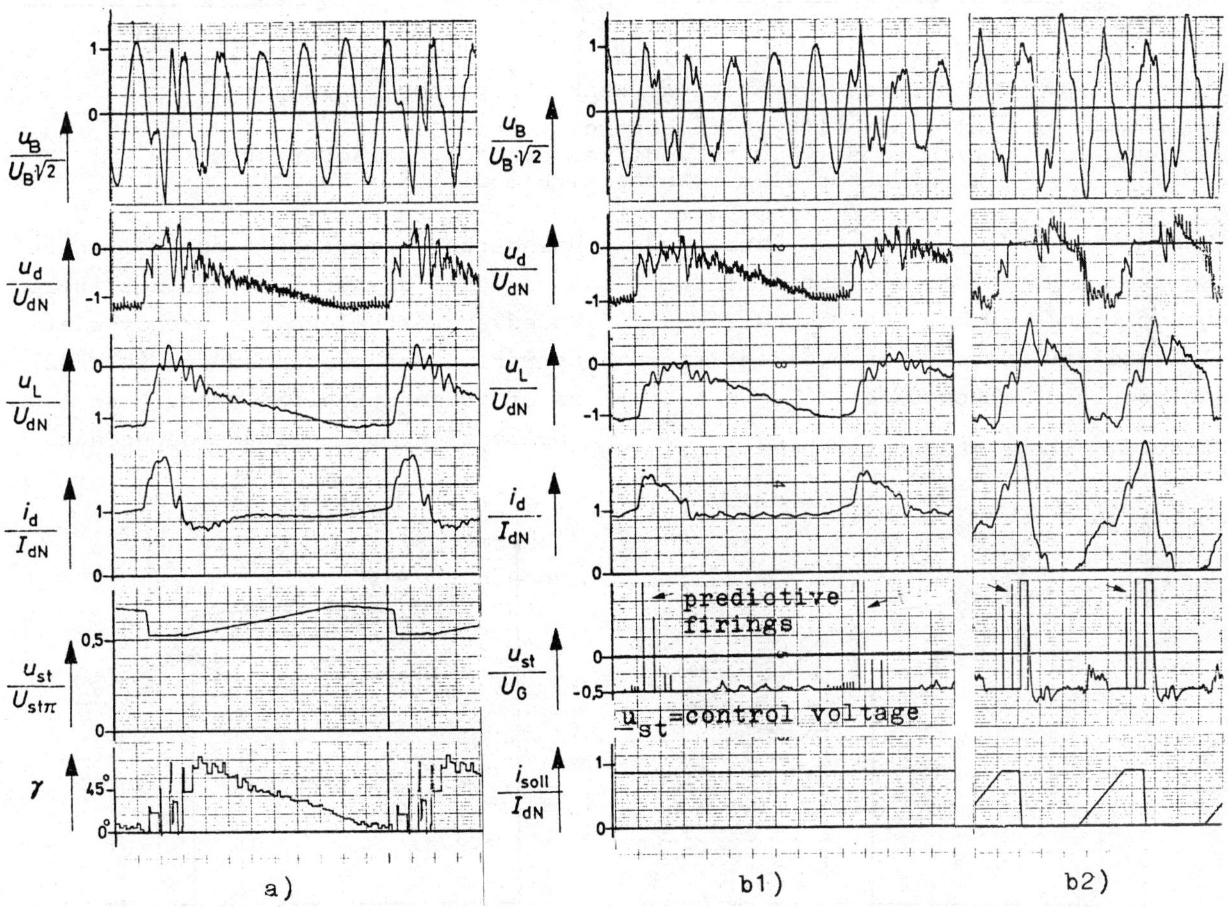

Fig. 4. Inverter operation at weak AC-systems ($S_k''/P_{dN} = 2$)

a) instability with control method (3)

b) instability with control method (2) b1) without LVDCL

b2) with LVDCL

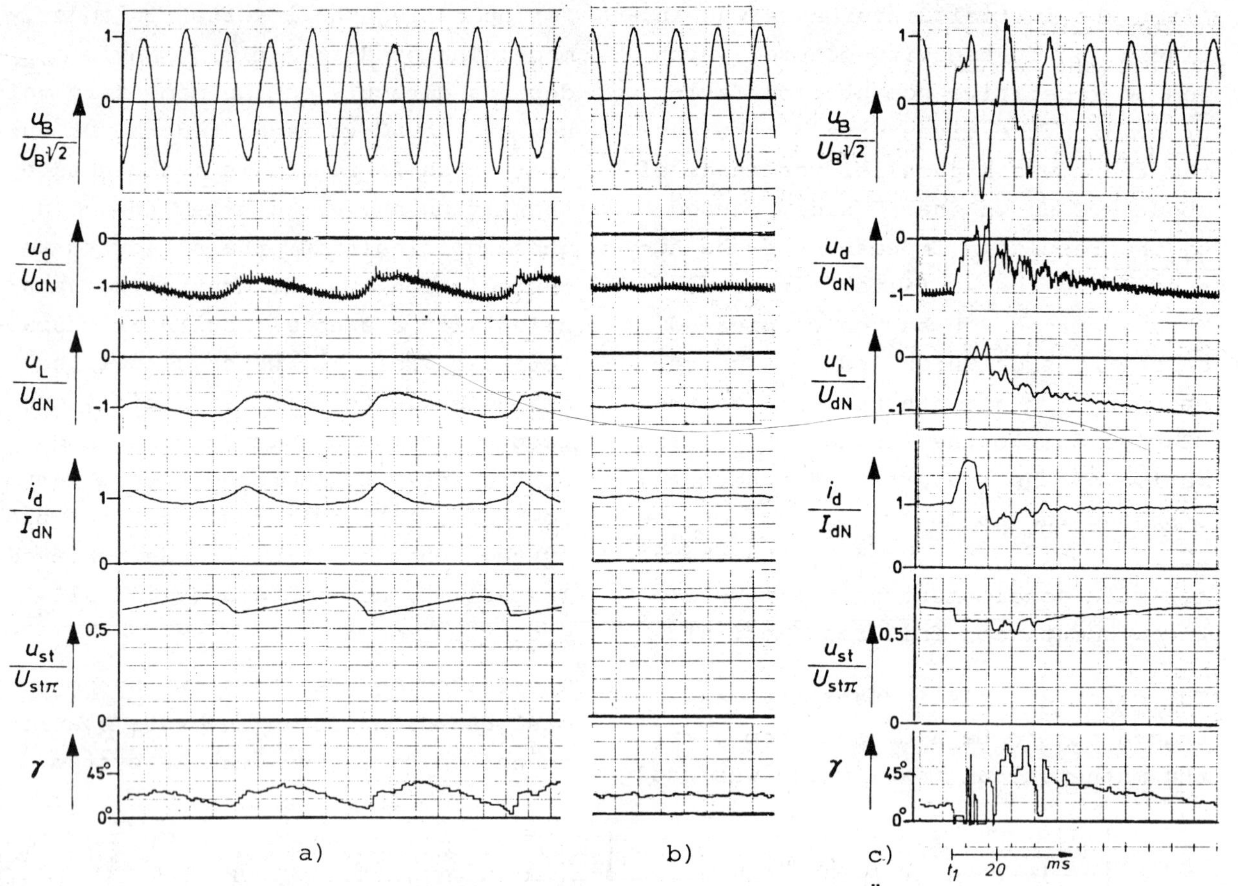

Fig. 5. Inverter operation at weak AC-systems ($S_k''/P_{dN} = 2$)

a) instability with control method (1)
b) method (1) with modified control characteristic
c) inverter disturbances with modified method (1)

The AC system impedance was increased so that a short circuit ratio of two was obtained. Under these conditions, instability appeared independent from the control method used.

Figure 4 shows the system behaviour, using the γ-prediction control methods. An initial disturbance was not introduced. Nevertheless the steady-state operation was prevented due to a series of commutation failures. Also the application of a voltage dependant

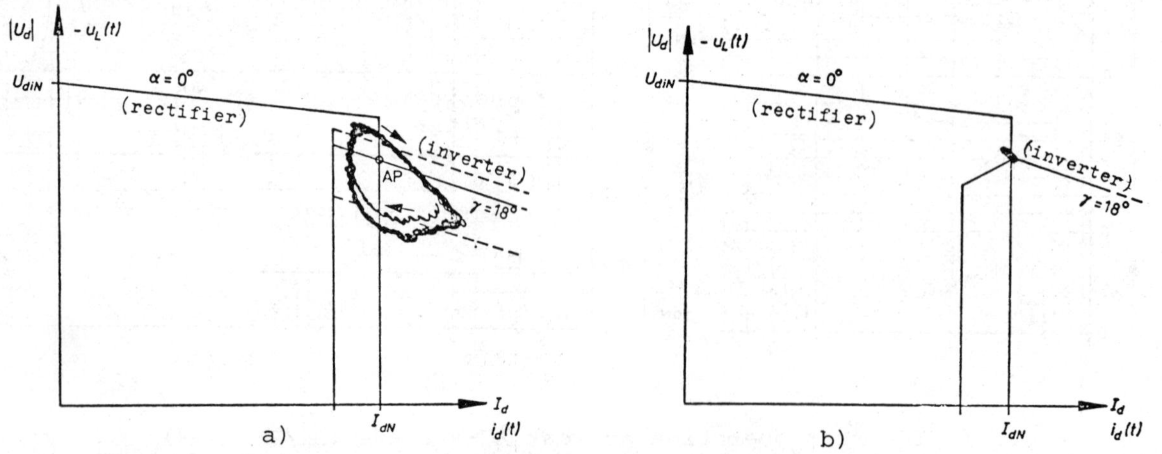

Fig. 6. Limit cycles and control characteristics

a) corresponding to Fig. 5a b) corresponding to Fig. 5b

current limit (LVDCL, Ref. 7) did not improve the system behaviour.

Using the closed loop γ-control, an oscillation of the DC- and AC- voltages and currents with a frequency of about 14 Hz appeared (Fig. 5a). Corresponding to Fig. 5a, a limit cycle of the oscillation is represented together with the steady-state control characteristics of the system in Fig. 6a. The γ=const. characteristic of the inverter is drawn in dotted lines for the respective maximum and minimum values of the AC voltages during the disturbance.

It was supposed, that the instability was caused by the negative resistance characteristic of the inverter during γ=const. operation and the dynamic interaction between the inverter control and the current control of the rectifier: Due to the weak AC-system on the inverter side, the slope of

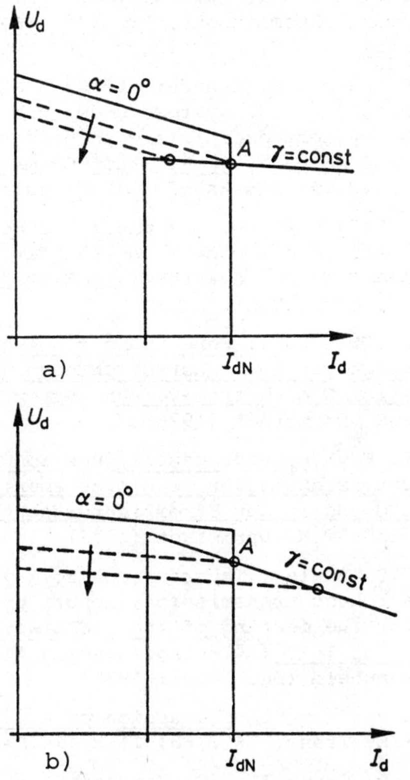

Fig. 7. Control characteristics
a) positive total resistance
b) negative total resistance

the γ=const. characteristic is steeper than the α=const. characteristic of the rectifier as it is shown in Fig. 7b. In this case, the operation point AP is unstable because an accidental increase in direct current will cause the current controller of the rectifier to increase the firing angle which will further increase the direct current. This results in monotonous instability (Ref. 8). In the other case, if the slope of the γ=const. characteristic is less steep than the slope of the α=const. characteristic (Fig.7a), an increase of the rectifier control angle will reduce the current and the operation point will be stable.

An effective damping of the oscillation could be obtained by introducing a simple modification of the inverter control characteristic as it is shown in Fig. 6b. The limit cycle in Fig. 6b corresponds to the steady-state operation in Fig. 5b.

In the following investigations, all control methods listed in Table 1 were provided with this modification. The figures Fig. 5c and Fig. 8a,b show the behaviour of the three different control methods after initiating the AC-fault on the inverter side. Good results could be obtained with the methods (1) and (3). Also method (2) performed quite well but the recovery time was increased. This disavantage is primary caused by the integral characteristic gate control unit used in this method.

SUMMARY

In this paper, two different types of instability are described:

Due to the interaction between the rectifier- and inverter control, instability can appeare which is characterized by frequencies lower than the

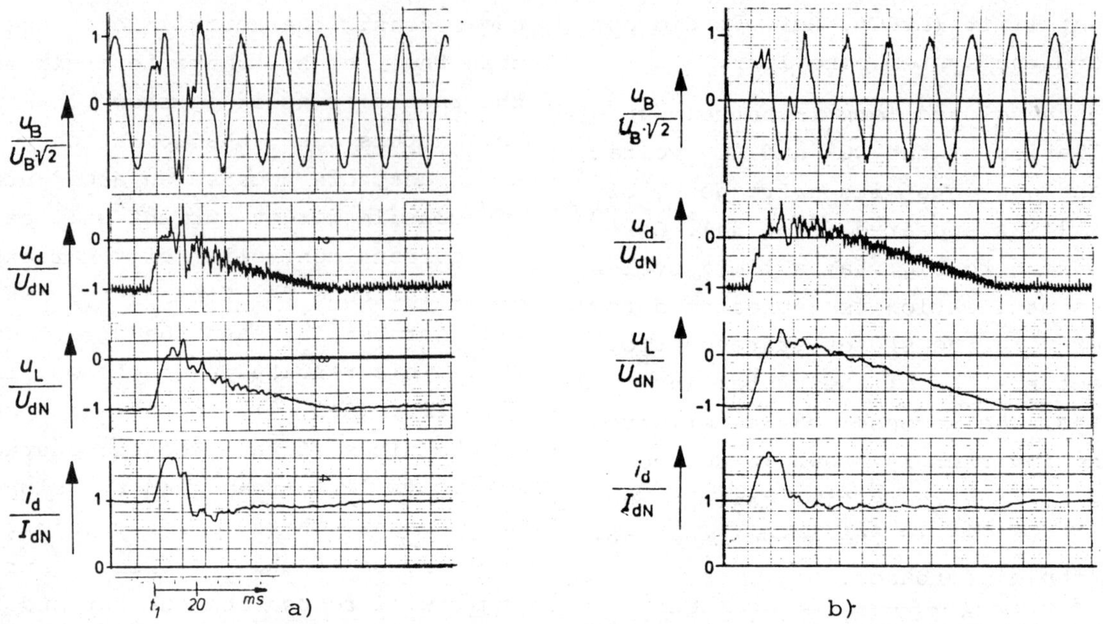

Fig. 8. Inverter disturbaces at a weak AC-system ($S_k''/P_{dN} = 2$)
a) modified control method (3) b) modified control method (2)

fundamental. The bacic cause for these difficulties is the negative resistance characteristic of the inverter during γ=const. operation. Although this type of instability in principle depends only on the shape of the control characteristics, there is nevertheless some influence of the special control method used. This is caused by he more or less intensive γ=const. behaviour of the different control methods during inverter disturbances. This type of instability can be avoided by introducing a modification of the control characteristics.

Another type of instability is characterized by frequencies higher than the fundamental and is caused by a closed loop over the extinction angle measurement or over the mathematical model used for γ-prediction. This depends primary on the type of control method used and must be avoided by applying such special modifications as nonlinear characteristics or symmetrizing devices. In both cases, the system behaviour is influenced by the given system conditions.

REFERENCES

(1) R. Jötten, R. Müller, E. Rumpf: Einfluß der Netzparameter auf die Regelkreise einer HGÜ, Proc. 1. IFAC-Symposium 2, 527, Düsseldorf (BRD) Okt. 1974.

(2) A. Eckström, P.O. Jackson, G. Liss: Firing control of thyristor valves in hvdc transmissions, Proc. 1. IFAC-Symposium 2, 509, Düsseldorf (BRD) Okt. 1974.

(3) E. Rumpf, S. Ranade: Steady-state performance of hvdc systems under unbalanced ac system conditions with different control methods, CIGRE Study Committee 14-71(SC)14, Los Angeles (USA) Juli 1971.

(4) J.D. Ainsworth: The phase-locked oscillator. A new control system for controlled static converters, IEEE Trans. PAS. 87, 859 (1970).

(5) E. Groschupf, Berechnung von Ausgleichsvorgängen in Drehstromsystemen und Drehstrom-Gleichstrom-Verbundsystemen, Diss. T.H. Darmstadt (1976).

(6) K. Fay: Hybridrechnersimulation zur Untersuchung von Regelproblemen in der HGÜ und in der Stromrichtertechnik, Diss. T.H. Darmstadt (1977)

(7) J.P. Bowles: Inverter control systems for hvdc transmission, Report prepared for the meeting of the CIGRE working group 14-02: "Control systems for hvdc", Mannheim (BRD) April 1975.

(8) H.P. Lips, in discussion on Ref. 9.: IEEE Trans. PAS. 89, 1118 (1970).

(9) D.P. Carroll, P.C. Krause: Stability a analysis of a DC power system. IEEE Trans. PAS. 89, 1112 (1970).

IMPROVING THE PERFORMANCE OF A PARALLEL AC/DC TRANSMISSION SYSTEM USING AN ADAPTIVE OPTIMAL CONTROL POLICY

B. Cuno

Institut für Stromrichtertechnik und Antriebsregelung, Technische Hochschule Darmstadt
Darmstadt, Germany

ABSTRACT

In this paper a new adaptive control policy improving the performance of a.c./d.c. power systems is described. The proposed control scheme is applied to an a.c./d.c. parallel transmission as an illustrative example. A hybrid simulation of the system is developed and used to study the operation of the system.

1. INTRODUCTION

In the two last decades, a number of high-voltage-direct-current-(h.v.d.c.) transmission links have gone into commercial operation in various parts of the world. Several other h.v.d.c. links are now under construction or consideration (1).
As demonstrated through model plant tests (2,3), analogue (4,6) and digital computer studies (7) as well as through theoretical investigations (4,8), the h.v.d.c. can improve the stability and the transient behaviour of connected a.c. networks. For this purpose supplementary signals are added to the reference values of the h.v.d.c. power- respectively current-control.
The weighting factors of these signals - derived from network values - must be acquired through model plant tests or digital simulations. In order to achieve an optimal system performance, these factors have to be adapted when a change occurs in structure, parameters, or in the operating point of the transmission system. In order to avoid the above mentioned disadvantages, a systematic procedure had to be found. In this case, of all the possible design methods, the optimal regulator with a quadratic performance index has been chosen.

2. SYSTEM UNDER CONSIDERATION

The system investigated is illustrated by a single line diagram in Fig. 1. It consists of a generation plant connected to a large receiving system through a parallel a.c./d.c. transmission link. The generating source is represented by a voltage controlled single synchronous machine equivalent and a hydroturbine with governor, whereas the receiving a.c. system was assumed to be an infinite bus. The bipolar h.v.d.c. system is represented by a monopolar network and one station-half by a single 12-pulse converter unit connected to the a.c. buses by star-star and star-delta tap changing transformers. For the purpose of this paper, the h.v.d.c. system is operated with marginal current control (9).

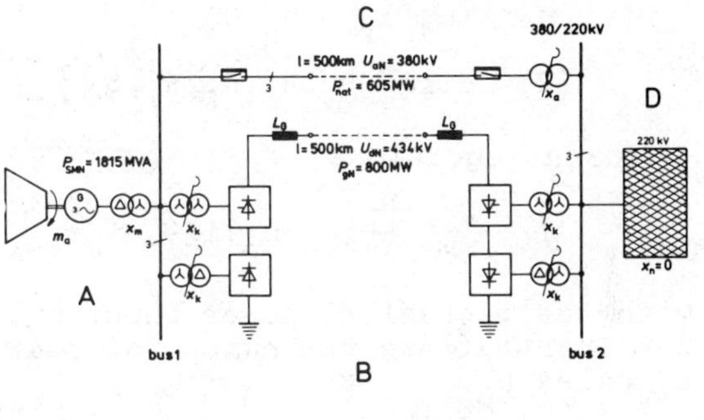

Fig. 1 Single-line diagram of the power-system investigated
(A) generating plant
(B) h.v.d.c. system
(C) a.c. transmission line
(D) network of infinite capacity

3. SYSTEM MODELING

In studying the given power system, a detailed representation would result in an extremely complex mathematical model. So it became necessary to develop a model which would describe the interesting transient performance of the system. The basis of the required modeling techniques is briefly summarized below.

3.1 Synchronous Generator

Assuming an idealized synchronous machine with the frame of reference fixed to its rotor, its performance may be described in the following equations in state space form (10).

- flux linkage equations

$$\underline{\psi} = (L)\underline{i} \qquad (1)$$

with

$$\underline{\psi} = (\psi_d\, \psi_f\, \psi_D\, \psi_q\, \psi_Q)^T$$

$$\underline{i} = (i_d\, i_f\, i_D\, i_q\, i_Q)^T$$

$$(L) = \text{diag}\left\{(L_d),\ (L_q)\right\}^T$$

$$(L_d) = \begin{bmatrix} L_{dd} & L_{df} & L_{dD} \\ L_{fd} & L_{ff} & L_{fD} \\ L_{Dd} & L_{Df} & L_{DD} \end{bmatrix}$$

$$(L_q) = \begin{bmatrix} L_{qq} & L_{qQ} \\ L_{Qq} & L_{QQ} \end{bmatrix}$$

- voltage equations

$$\underline{u} = \dot{\underline{\psi}} + (R)\underline{i} + \omega_m(-\psi_q\ 0\ 0\ \psi_d\ 0)^T$$

where

$$\underline{u} = (u_d\ u_f\ 0\ u_q\ 0) \qquad (2)$$

$$(R) = \text{diag}\left\{R_1, R_f, R_D, R_1, R_Q\right\}$$

- torque equations (3)

$$\ddot{\gamma} = \dot{\omega}_m = \frac{n_p}{\Theta}(m_a + n_p\left[i_q\psi_d - i_d\psi_q\right])\ )$$

with the mechanical power input m_a, the inertia Θ and the number of pairs of poles n_p.

- excitation system (voltage controlled)
  the voltage regulator for the field winding and the excitation system are represented by one time constant (11). Its transfer function is

$$\frac{u_f}{U_{Ref} - u} = \frac{K_E}{1 + sT_E} \qquad (4)$$

with the generator terminal voltage given by

$$u = \sqrt{u_d^2 + u_q^2} \qquad (5)$$

The time constant T_E and the gain K_E are selected depending on performance constraints and stability considerations.

3.2 Hydroturbine and governor

Figure 2 shows a simplified scheme of a hydrogovernor-turbine representation developed in (12). Assuming an idealized turbine operating at full load, the turbine-penstock transfer-function results

$$G_T(s) = \frac{1 - T_w s}{1 + \frac{T_w}{2}s} \qquad (6)$$

with the water starting time T_w.

The transfer function of the governor is approximated by

$$G_G(s) = \frac{\frac{1}{\sigma}(1 + T_r s)}{\left(1 + \frac{T_g + T_r(\sigma+\delta)}{\sigma}s\right)\left(1 + \frac{T_r T_g}{T_g + T_r(\sigma+\delta)}s\right)} \qquad (7)$$

with the governor response time T_g, the dashpot time constant T_r, the permanent speed droop σ and the temporary speed droop δ.

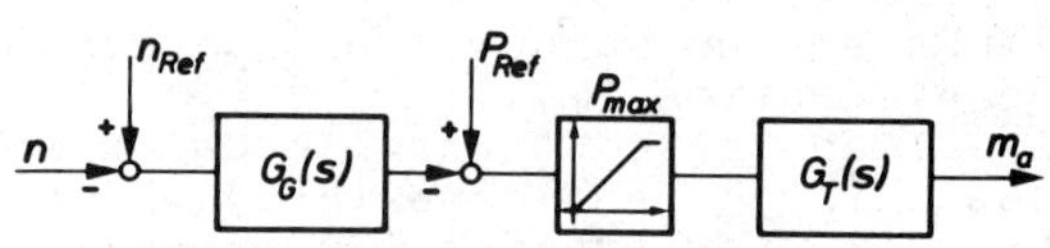

Fig. 2 Block diagram of the hydraulic turbine-governor system

3.3 A.c. transmission line

The a.c. transmission line is represented by a lumped parameter double T-section-model described by Fig. 3. The parameters of the line are obtained by parameter programs developed in (13). By star-delta transformation the line-to-line capacities are transformed into line-to-earth capacities and then are added to the respective values. The transformer is represented by its series inductance L_{Tr} and the transformer ratio $\ddot{u}_a$.

In order to facilitate the analysis, the phase variables are transformed to a rotating reference frame attached to the rotor of the synchronous machine.

Neglecting 0-quantities we get (8)

$$\begin{pmatrix} \dot i_{da1} \\ \dot i_{qa1} \\ \dot u_{dC1} \\ \dot u_{qC1} \\ \dot i_{da12} \\ \dot i_{qa12} \\ \dot u_{dC2} \\ \dot u_{qC2} \\ \dot i_{da2} \\ \dot i_{qa2} \end{pmatrix} = \begin{pmatrix} -\frac{R_a}{L_a} & \omega_N & \frac{-1}{L_a} & 0 & & & & & & \\ -\omega_N & \frac{R_a}{L_a} & 0 & \frac{-1}{L_a} & & & & & & \\ \frac{1}{C_a} & 0 & 0 & \omega_N & \frac{-1}{C_a} & 0 & & & & \\ 0 & \frac{1}{C_a} & -\omega_N & 0 & 0 & \frac{-1}{C_a} & & & & \\ & & \frac{1}{L_a} & 0 & -\frac{R_a}{L_a} & \omega_N & \frac{-1}{L_a} & 0 & & \\ & & 0 & \frac{1}{L_a} & -\omega_N & \frac{R_a}{L_a} & 0 & \frac{-1}{L_a} & & \\ & & & & \frac{1}{C_a} & 0 & 0 & \omega_N & \frac{-1}{u_a C_a} & 0 \\ & & & & 0 & \frac{1}{C_a} & -\omega_N & & 0 & \frac{-1}{u_a C_a} \\ & & & & & & \frac{u_a}{L_a+L_{Tr}} & 0 & \frac{-R_a}{L_a+L_{Tr}} & \omega_N \\ & & & & & & 0 & \frac{u_a}{L_a+L_{Tr}} & -\omega_N & \frac{-R_a}{L_a+L_{Tr}} \end{pmatrix} \begin{pmatrix} i_{da1} \\ i_{qa1} \\ u_{dC1} \\ u_{qC1} \\ i_{da12} \\ i_{qa12} \\ u_{dC2} \\ u_{qC2} \\ i_{da2} \\ i_{qa2} \end{pmatrix} + \begin{pmatrix} \frac{1}{L_a} & & & \\ & \frac{1}{L_a} & & \\ & & & \\ & & & \\ & & & \\ & & & \\ & & & \\ & & & \\ & & \frac{-u_a^2}{L_a+L_{Tr}} & \\ & & & \frac{-u_a^2}{L_a+L_{Tr}} \end{pmatrix} \begin{pmatrix} u_{da1} \\ u_{qa1} \\ u_{da2} \\ u_{qa2} \end{pmatrix}$$

3.4 Converters and d.c. transmission line

When considering transient problems, the exact simulation of the phenomena associated with the discrete nature of the converter operation is not necessary. Therefore, only the fundamental frequency behaviour of the converters will be described by a model given in (14). This consists of a voltage source in series with a resistance - including the ohmic resistance of the bridge R_{ir} and an equivalent resistance due to commutation R_{ix} -, smoothing and commutation-inductance L_G and L_K, and the voltage drop D_v in the valves. Neglecting the voltage drop (see Fig. 3), the converter current i_g is given by

$$\dot i_g = -\frac{R_{ix}+R_{ir}}{L_G+L_K} i_g - \frac{1}{L_G+L_K}(u_g - u_{gi\alpha}) \tag{9}$$

Under normal operating conditions, the rectifier station controls the current while the inverter station is operating in the extinction angle control mode. In order to avoid a great change of current and power caused by a moderate dip in alternating voltage, the inverter is also equipped with a current regulator, which is, however, set to a lower current than the rectifier regulator. The difference between the current command of the rectifier and that of the inverter is called the current margin and is generally 10 % of the rated current. Assuming that the rectifier is current controlled and the converter characteristic, the arccos nonlinearity, the firing system and the control amplifier with proportional-integral characteristic are combined, the cosine of the firing angle α is given by the following transfer function

$$\frac{\cos\alpha}{I_{gRef} - i_g} = \frac{K_{Ri}}{1 + sT_{Ri}} \tag{10}$$

Using the extinction angle control, the inverter average voltage is

$$u_{gi\alpha} = u_{gi}\cos\alpha = 2R_{ix} i_g - u_{gi}\cos\alpha \tag{11}$$

The dynamics of the extinction angle control system will be neglected as the time constants involved in this loop are much smaller than those of the current control loop. This means that the cosine of the extinction angle is assumed to be equal to its reference value (7).
The d.c. transmission line is represented by a T-section model. The lightning-protection condensators of the converter stations are added to the line-to-line capacities. The converter currents and the direct voltage in the middle of the line are shown in the following state space equation

$$\begin{pmatrix} \dot i_{g1} \\ \dot u_{g12} \\ \dot i_{g2} \end{pmatrix} = \begin{pmatrix} -\frac{R_{ix1}+R_{ir1}}{L_{G1}+L_{K1}} & \frac{1}{L_{G1}+L_{K1}} & 0 \\ \frac{1}{C_L+2\,C_B} & 0 & \frac{-1}{C_L+2\,C_B} \\ 0 & \frac{1}{L_{G2}+L_{K2}} & -\frac{R_{ix2}+R_{ir2}}{L_{G2}+L_{K2}} \end{pmatrix} \begin{pmatrix} i_{g1} \\ u_{g12} \\ i_{g2} \end{pmatrix} + \begin{pmatrix} u_{gi1}\cos\alpha_1 \\ 0 \\ u_{gi2}\cos\alpha_2 \end{pmatrix}$$

3.5 Transformations

The necessary transformation which relates the quantities in the generating plant 1 to those in the receiving network 2 is given by the transformation matrix

$$(D_{12}) = \begin{pmatrix} \cos(\gamma_{m1}-\gamma_{m2}) & \sin(\gamma_{m1}-\gamma_{m2}) \\ -\sin(\gamma_{m1}-\gamma_{m2}) & \cos(\gamma_{m1}-\gamma_{m2}) \end{pmatrix} \tag{13}$$

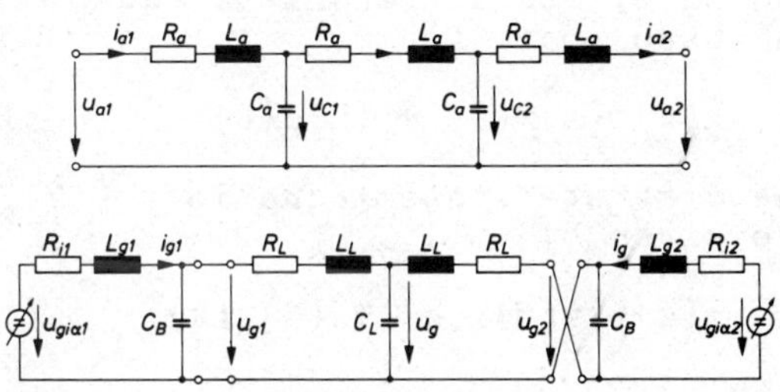

Fig. 3 Representation of a.c. line (one phase) and h.v.d.c.

4. STABILISATION METHODS

4.1 Conventional Control Methods

It has been shown in several papers (see chap. 1) that the dynamic stability of a parallel a.c./d.c. power system can be improved by modulating the nominal value of the d.c. current reference. The following three types of signals were mainly used:

- a method to increase the steady state stability was proposed by Uhlmann (15). As an additional signal the slip (frequency difference between rotor and network) is used.

$$\tilde{I}_{gRef}=I_{gRef}(1+V_a \frac{1}{\omega_N} \frac{d\vartheta}{dt}) \quad (14)$$

- to increase the transient stability - ability of the system to revert to a synchronous state following a fault -, Machida (16) had the idea of applying a signal which is derived from the difference between the driving power P_{SMN} and the active electrical output P_{SM} during a fault case at the a.c. system.

$$\tilde{I}_{gRef}=I_{gRef}\left(1+\begin{cases} V_b(P_{SMN}-P_{SM}) \\ 0, \text{ if } P_{SM}\geq P_{SMN}\end{cases}\right) \quad (15)$$

- a third method employs another signal which increases the h.v.d.c. current if the rotor angle becomes greater than 90 degrees (3).

$$\tilde{I}_{gRef}=I_{gRef}\left(1+\begin{cases} V_c(\vartheta-90^\circ), & \vartheta>90^\circ \\ 0, & \vartheta\leq 90^\circ\end{cases}\right) \quad (16)$$

4.2 Optimal Control

4.2.1 Basic Principle For the design of a closed-loop optimal system with a process computer as regulator, we need the discrete system equations, which describe the system at the sampling points t = k T. If we consider a general linear system represented by its differential equations in state space form

$$\dot{\underline{x}}(t)=(A)\underline{x}(t) + (B)\underline{u}(t) \quad (17)$$

the discrete system equations (17) are given by

$$\underline{x}(k+1)=(M)\underline{x}(k) + (N)\underline{u}(k) \quad (18)$$

with

$$(M)=\exp\{(A)T\},$$

and

$$(N)=\exp\{(A)T - (I)\}(A)^{-1}(B)$$

A useful criterion for measuring the performance of a control system is the quadratic performance index. For the given final value k_f it has the form

$$J = S(k_f) + \sum_{k=0}^{k_f-1} L(k) \quad (19)$$

with

$$S(k_f)= \frac{1}{2}\underline{x}^T(k_f)(S)\underline{x}(k_f)$$

$$L(k)=\frac{1}{2}[\underline{x}^T(k)(Q)\underline{x}(k)+\underline{u}^T(k)(R)\underline{u}(k)]$$

The weighting matrices (Q) and (S) must be non-negative definite and (R) must be positive definite.
The design problem is now to determine a control sequence $\underline{u}(k)$ that minimizes the performance index for the system specified by its linear differential equations in state space form. By use of a set of Lagrange multipliers $\lambda(k)$ the problem is recast in minimizing the augmented performance index

$$\tilde{J}=J+\lambda^T(k+1)[(M)\underline{x}(k)+(N)\underline{u}(k)-\underline{x}(k+1)] \quad (20)$$

As the solution of the above mentioned problem is well documented in the literature - see e.g. (17,18)- it will only be summarized in the following.
Utilizing the so-called Riccati-transformation

$$\lambda(k) = (P(k))\underline{x}(k) \quad (21)$$

the matrix (P(k)) must be solved by the recursive relationship

$$(P(k))=(Q)+(M)^T(P(k+1))[(I)+ +(N)(R)^{-1}(N)^T(P(k+1))]^{-1}(M) \quad (22)$$

starting at the final stage obtained by

$$(P(k_f)) = (S) \quad (23)$$

The desired optimal control law is then

$$\underline{u}(k) = (K(k))\underline{x}(k) \quad (24)$$

with

$$(K(k))=-(R)^{-1}(N)^T((M)^T)^{-1}[(P(k))-(Q)]$$

The matrix of "Kalman gains" is solved backwards in time from $k = k_f$ to k = 0 and is stored and applied to the physical system as it runs forward in time.

4.2.2 System Reduction and Linearization (19) As shown above, the arrangement of the linearized power system model equations in normal state space form is a first step in the investigation of linear optimal control system design techniques. To reduce the order of the system given in chap. 3, some further neglections in system modeling are introduced:

- line capacities are not represented
- the terms $\dot{\psi}_d$ and $\dot{\psi}_q$ in the armature circuit voltage equations and the subtransients of the damper windings of the generator are neglected
- the turbine and governor are only represented by a first order differential equation.

If the nonlinear system equations are expressed in the form

$$\underline{x}(t) = \underline{F}(\underline{x}(t), \underline{u}(t)) \tag{25}$$

the system is linearized by taking the total differential of $\underline{F}$ at the desired operating point $\underline{x}^o$. For small deviations

$$\Delta\underline{x} = \underline{x} - \underline{x}^o \tag{26}$$

we then get the linearized small displacement model in state space form

$$\Delta\underline{x}(t)=\underline{f}(\Delta\underline{x},\Delta\underline{u})=\left[\mathrm{grad}\ \underline{F}(x,u)\right]^T\Big|_{x^o}\Delta\underline{x} \tag{27}$$

The synchronous generator together with the exitation system and the turbine can be represented in linear state space form by the equations

$$\begin{aligned}\Delta\underline{\dot{x}}_m&=(A_m)\Delta\underline{x}_m+(B_{m2})\Delta\underline{i}_m+\underline{b}_{m1}\Delta U_{1\,Ref}\\ \Delta\underline{u}_1&=(C_{m1})\Delta\underline{x}_m+(C_{m2})\Delta\underline{i}_m\end{aligned} \tag{28}$$

where

$$\Delta\underline{x}_m=(\Delta\psi_f\ \Delta u_f\ \Delta m_a\ \Delta\omega_m\ \Delta\gamma_m)^T$$

(all matrices of the state space model are listed in the Appendix)

i_m is the vector of d- and q-axis generator currents and $\underline{u}_1$ is the vector of d- and q-axis generator voltages (voltages at bus bar 1) referred to the maschine axis

$$\begin{aligned}\Delta\underline{i}_m&=-(\Delta i_{dm1}\ \Delta i_{qm1})^T\\ \Delta\underline{u}_1&=(\Delta u_{d1}\Delta u_{q1})^T\end{aligned} \tag{29}$$

The a.c. line equations may be written as

$$\Delta\underline{\dot{x}}_a=(A_a)\Delta\underline{x}_a+(B_a)\Delta u\ +\underline{b}_{a2}\Delta\vartheta_1$$

with the small displacements of line currents at bus bar 1.

$$\Delta\underline{x}_a = (\Delta i_{da}\ \Delta i_{qa})^T$$

and the angle $\Delta\vartheta_1$ given by

$$\Delta\vartheta_1=\left[\mathrm{grad}(\arctan(u_{d1}/u_{q1})\right]^T\Delta\underline{u}_1$$

$$= \underline{c}^T(C_{m1})\Delta\underline{x}_m+\underline{c}^T(C_{m2})\Delta\underline{i}_{dm}$$

The h.v.d.c. transmission system may be realized by

$$\begin{aligned}\Delta\underline{\dot{x}}_g&=(A_g)\Delta\underline{x}_g+\underline{b}_{g1}\Delta I_{gRef}+\underline{b}_{g2}\Delta u_{gi1}\\ \Delta i_g&=\underline{c}_g^T\Delta\underline{x}_g\end{aligned} \tag{31}$$

where

$$\Delta u_{gi1}=\left[\mathrm{grad}(k_q\sqrt{u_d^2+u_q^2}\right]^T\Delta\underline{u}_1=\underline{q}^T\Delta\underline{u}_1$$

Assuming the primary transformer current has no harmonics except the fundamental one, it is given in d-q-quantities by

$$\begin{aligned}\Delta\underline{i}_g&=(\Delta i_{dg1}\ \Delta i_{qg1})^T\\ &=(R)(\Delta\vartheta_1\ \Delta\varphi_1)^T + \underline{s}\Delta i_{g1}\end{aligned} \tag{32}$$

with the displacement angle (angle by which the fundamental current lags the source voltage)

$$\Delta\varphi_1 \simeq c_{g1}\Delta I_{gRef}+\underline{c}_{g2}^T\Delta\underline{x}_g$$

Coupling the different sub-systems we get

$$\Delta\underline{\dot{x}} = (A)\Delta\underline{x} + (B)\Delta\underline{u} \tag{33}$$

where

$$\Delta\underline{x} = (\Delta\underline{x}_m\ \Delta\underline{x}_a\ \Delta\underline{x}_g)^T$$

$$\Delta\underline{u} = (\Delta U_{1\,Ref}\ \Delta I_{gRef})^T$$

4.2.3 <u>Application on the given system</u> The calculations required to determine the feedback-gain matrix (R(k)) were carried out by a subroutine for DDC-synthesis (20).

An interesting property of the resultant linear discrete regulator was observed by changing k_f. The results showed that if k_f is made sufficiently large, the feedback gains become constants for low values of k and decrease rapidly to zero as k approaches k_f-1, whereas the responses of the state variables remain identical. This permits a closed loop control system design with constant feedback gains. Experimentations with different combinations of weighting matrices (R), (S) and (Q) were performed in order to study their influence on stability. Eigenvalues of the system matrix (A) represent the roots of the characteristic equation of the system and indicate its stability. Therefore, the eigenvalues of the open-loop systemgiven in equation (33) and those of the closed-loop system

$$\Delta\dot{\underline{x}} = [(A) + (B)(K)]\,\Delta\underline{x} \qquad (34)$$

were calculated and compared with each other. So, beginning with chosen values for the weighting matrices, a suitable choice of (R), (S) and (Q) can be made by an trial-and-error method.

4.2.4 Implementation of the adaptive control-system As the feedback gains for the chosen performance index are dependent on the operating point of the system and on the fault case, an adaptive control must be realized.
As the necessary mathematical operations demand much computing time, the "on-line" calculation of the feedback matrix after a fault case is impossible to execute. Therefore, after each change in the operating state or structure of the network occurs, the process computer calculates - through background processing - the state space model and solves the Matrix-Riccati-Equation for various given probable failure cases likely to emerge from this state.
For cleared faults (e.g. short circuit on the a.c. line with autoreclosure) the system will - provided that system stability is maintained - return to its previous state $\underline{x}^o$, whereas the state $\underline{x}^o$ changes after permanent faults, switching operations and changes in load flow.
In Fig. 4 a modified phase-plane technique is applied to demonstrate this fact. As it is difficult to visualize and construct the complete plot of the "phase-state" diagram, only the coordinate of frequency- and load angle-deviation ($\Delta\omega_m(t_1)$, $\Delta\gamma_m(t_1)$) for the time from the given event to infinity.
To ensure that the system runs on a nearly optimal trajectory, the system equations are linearized according to the new probable point of operation. All the matrices calculated in this way are stored in the core-memory of the process computer.

Once a special fault in the transmission system has been identified by a fault-detecting-program, the respective data-set is recalled and used to improve the stability of the system. In the case of pre-programmed changes of a load flow or the switching commands, there is in most cases enough time to calculate the new feedback coefficients "on-line". Fig. 4 presents the general lay out of the adaptive optimal control-policy on the process computer. The precise description of the software used is not the purpose of this paper. It will be explained in a following paper.

5. SIMULATION TEST

5.1 Methods of investigation

As a tool for testing the effectiveness of the control schemes, a simulation had to be developed. The investigation of power systems can be undertaken with the aid of physical or mathematical models of the network components.
The physical simulation in a model simulates the power circuits by using components which exercise the same physical functions but at a lower power level. The disadvantages of such a model are high costs of construction and the fact that the variation of parameters over a wide range and the changes in network configuration are complicated to introduce.
Mathematical models permit the use of analogue or digital computers. The hybrid computer as a combination of both, permits good computing efficiency if the simulation is arranged in such a way that the best features of both computers are utilized.

5.2 Hybrid computer facilities

The digital computer used in the simulation as a process computer is a CII 100 20 with a 16-bit wordlength and core-memory of 64 K. The analogue part consists of two Telefunken RA 770's, each equipped with about 90 amplifiers inclusive 30 integrators. An additional digital expansion system (DEX) is available too. The analogue and digital parts are linked together by a Dornier HCS 712 Interface. Control-lines, Sense-lines, DA-converters and AD-channels are available, 32 of each. The operating mode of each integrator can be controlled separately. Two real-time clocks can be programmed and controlled by the digital computer. Furthermore there are 3o external interrupt entries for the real-time control of the digital computer.

Graphics display, plotting, and printing facilities are available as research aids. As the CII 100 20 is considerably slower (no floating point arithmetic by hardware) than the presently available process computers, the increased calculation time will represent the telecommunication delay and the time for the state estimation of non-measurable variables.

5.3 Implementation of the system at the hybrid computer

The mathematical basis for the analogue computer simulation is given by the equations in chap. 3. The synchronous machine is hereby set up using the fluxes as state variables (10,21). With the equations (1) and (2) we get

$$\dot{\underline{\Psi}} = \underline{u} - (R)(K)\underline{\Psi} + (\omega_m \Psi_q \;\; 0 \;\; 0 \;\; -\omega_m \Psi_d)^T \tag{35}$$

with

$$(K) = (L)^{-1} = \mathrm{diag}\left\{(L_d)^{-1},\ (L_q)^{-1}\right\}$$

The elements of the matrix (K) are calculated from any given set of machine data by using a digital program. The machine transformer values are hereby included in the parameters of the generator.
At the sending bus bar, the models for the synchronous generator, the a.c. line, and the h.v.d.c. transmission system are coupled together utilizing an inverting programming (22) technique. A fault in the system may be introduced by manual command. In the case of faulty operation a logical circuit is controlling the operation mode of integrators and the action of binary switches. The duration of fault and circuit breaker action are determined by monoflops. An extern sine-cosine-generator and two intern electronic resolvers are uased for the necessary transformations (see equ. 13). Investigations were carried out to demonstrate the effectiveness of the model and the results were compared with already published ones, gained by analogue (3) and digital (7,23) simulations and model plant tests (2,3). It was found that the developed mathematical model is sufficiently accurate to permit its use for stability investigations.
As the digital simulation of the investigated power system (7) allowed a time scale factor of $\beta = 2400$ for the detailed model (ICL 1907) and $\beta = 350$ for a simplified model (Siemens 305), the hybrid simulation was done in real-time ($\beta = 1$).

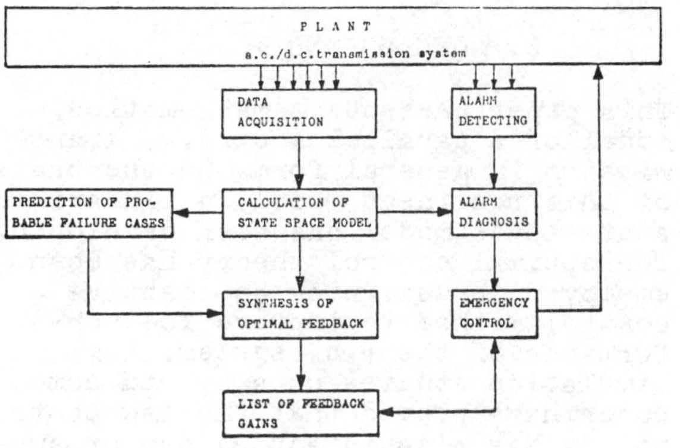

Fig. 4 Program flow chart for adaptive optimal control policy

5.4 Results of simulation studies

A brief description of simulation results is now given.
To be able to compare the simulation results with other already published investigations, the system parameters were chosen to be equal to those given in (3) and (7).
In order to make a comparison of system responses under various conditions the examples of a step change in the power command setting and of a three-phase-short-circuit (with autoreclosure) as the severest fault on the a.c. line were regarded. The step change in power was 30% of the nominal value. The time to interruption of the fault and the succeeding dead-time were 150 and 300 ms. As an important additional requirement for the effectiveness of the stabilizing control methods of the h.v.d.c. it was found that the generator voltage must regain its reference value very quickly after a fault (see Fig. 6,(a),(b) and (2)). For this reason only the d.c. reference current was modulated with supplementary signals.
The weighting matrices were selected to be diagonal.
Experimentations with different combinations of weighting factors were performed. In the case illustrated in Fig. 6 (c) the frequency was weighted very strongly. So, as a result of the regulator synthesis, mainly this state was used as an additional signal. Fig. 6 (d) demonstrates that the optimal feedback of all the state variables permits the h.v.d.c. to intervene even before any marked change in the frequency occurs.
As the effect of the conventional stabilization methods is well documented in the literature (2-4 and 6-8), the concerning traces are not

shown in this paper.

6. CONCLUSIONS

This paper presents a mathematical model of a parallel a.c./d.c. transmission in general form. On the basis of this nonlinear model, a linearized state space model has been developed. The optimal control theory has been employed in designing an adaptive control method to improve the performance of the a.c. system. Simulation studies on a hybrid computer have proved that the new control policy has a satisfactory performance capability.

ACKNOWLEDGEMENT

Financial support for this research was provided by Deutsche Forschungsgemeinschaft (DFG)

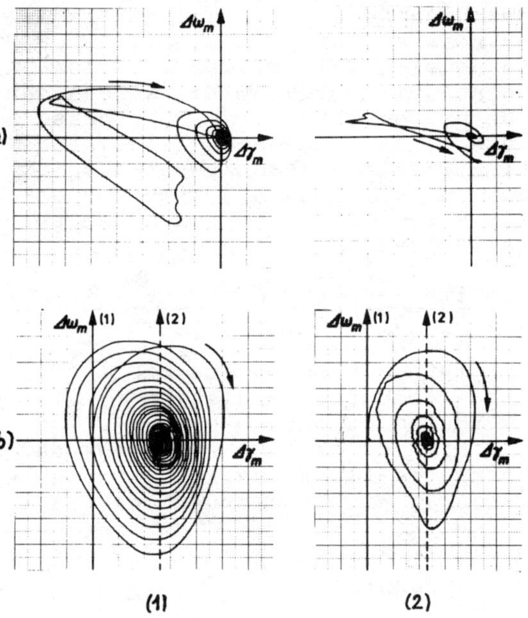

Fig. 5 Trajectory $\Delta\gamma_m(\Delta\omega_m)$
(a) three-phase-to-ground fault on a.c. line
(b) Step demand in power
(1) no feedback signals
(2) supplementary feedback

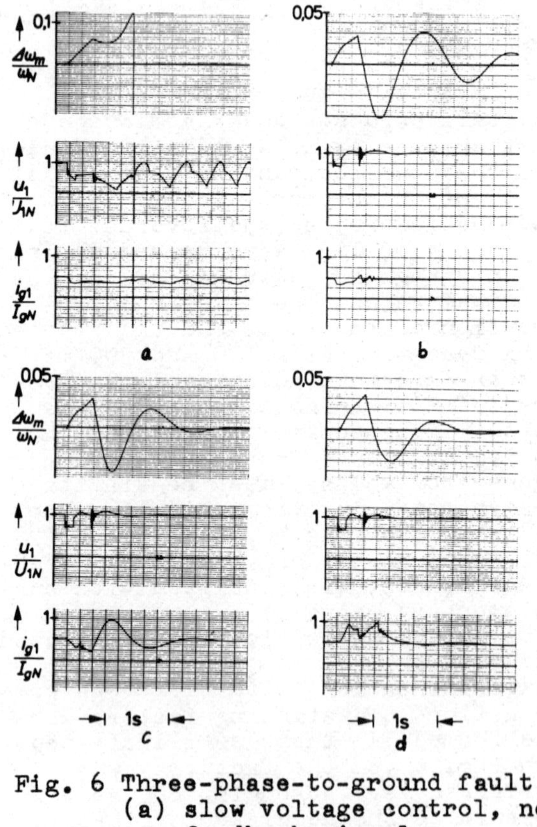

Fig. 6 Three-phase-to-ground fault
(a) slow voltage control, no feedback signals
(b) fast voltage control, no feedback signals
(c) fast voltage control, adaptive optimal control, case 1
(d) fast voltage control, adaptive optimal control, case 2

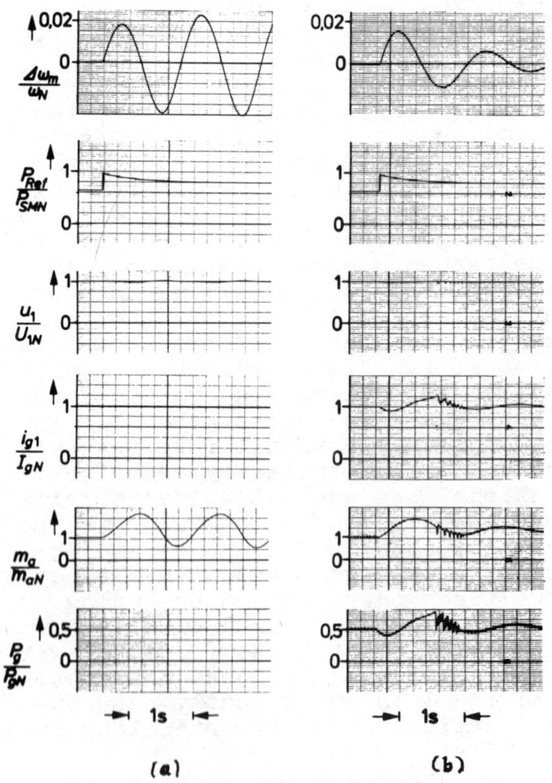

Fig. 7 30% step demand in power at the generation plant
(a) no feedback signals
(b) adaptive optimal feedback

REFERENCES

(1) R. Jötten, Hochspannungs-Gleichstrom-Übertragung - Stand zur Jahreswende 1973/1974, ETZ-A 95,1 (1974)

(2) R. Jötten and R. Foerst, Model Plant Tests on the Behaviour of a hvdc Transmission during ac and dc-System Disturbances, Cigré Report 43-08, (1968)

(3) J. Käuferle and E. Rumpf, The Influence of hvdc Transmission Systems on the Stability of associated ac Networks, Cigré Report 32-05, (1970)

(4) J. Käuferle and E. Pröll, Stabilität von Drehstromsystemen unter Einwirkung von HGÜ-Anlagen, Brown Boveri Mitteilungen 56, 112 (1969)

(5) H. A. Peterson, P. C. Krause, J. F. Luini and C. H. Thomas, An Analog Computer Study of a Parallel AC and DC Power System, IEEE Trans. on Power Apparatus and Systems 85, 191 (1966)

(6) W. A. Patterson, G. Désilets and J. N. Dalzell, More effective Damping on A.C. Systems through the Addition of external Control Signals to a HVDC Link - the Ed River Case, Paper presented at the Spring Meeting of the Canadian Electrical Association, Montreal (1972)

(7) Schmidt, H. H. (1973) Digitale Nachbildung umfangreicher dynamischer Systeme am Beispiel eines Netzmodelles mit Drehstrom- und Gleichstrom-Übertragung, Dissertation, TU Braunschweig

(8) A. M. El-Serafi and H. S. Khalil, Dynamic stability analysis of parallel ac/dc power systems, Proc. IEEE PES Summer Meeting, Portland (1976)

(9) H. Forsell, The Gotland d.c. link: The grid control and regulation equipment, Direct Current 2, 166 (1955)

(10) R. Jötten, Beschreibung dynamischer Vorgänge in Drehfeldmaschinen, lecture course, TH Darmstadt (1975)

(11) H. Glavitsch, Über die Wahl der Regelparameter von Erregersystemen für große Turbogeneratoren, Brown Boveri Mitteilungen 61, 207 (1974)

(12) D. G. Ramey and J. W. Skooglund, Detailed Hydrogovernor Representation for System Stability Studies, IEEE Trans. Power Apparatus and Systems 89, 106 (1970)

(13) Groschupf, E. (1976) Simulation transienter Vorgänge auf Leitungssystemen der Hochspannungs-Gleichstrom- und -Drehstrom-Übertragung, Dissertation, TH Darmstadt

(14) Jötten, R. (1976) Leistungselektronik, Band 1 Stromrichter-Schaltungstechnik, Vieweg, Braunschweig

(15) E. Uhlmann, Stabilisation of an A.C. link by a parallel D.C. link, Direct Current 9, 89 (1964)

(16) T. Machida, Improving transient stability of ac system by joint usage of dc system, IEEE Trans. on Power Apparatus and Systems 85, 226 (1966)

(17) Cadzov, J. A. and Martens H. R. (1970) Discrete-Time and Computer Control Systems, Prentice-Hall, Englewood Cliffs

(18) Kuo, B. C. (1970) Discrete Data Control Systems, Prentice-Hall, Englewood Cliffs

(19) Cuno, B. (1976) Entwurf des Zustandsgrößenmodells einer HGÜ/HDÜ-Parallelübertragung, Report No. 318, Inst. für Stromrichtertechnik und Antriebsregelung, TH Darmstadt

(20) Cuno, B. (1976) Ein Programmpaket zur Synthese digitaler Regelalgorithmen, Report No. 331, Inst. für Stromrichtertechnik und Antriebsregelung, TH Darmstadt

(21) Urbanke C. (1977) Die Berechnung eines analogen Synchronmaschinenmodells in d-q-Koordinaten, Report No. 344, Inst. für Stromrichtertechnik und Antriebsregelung, TH Darmstadt

(22) Fay, K. (1977) Hybridrechnersimulation zur Untersuchung von Regelproblemen in der HGÜ und in der Stromrichtertechnik, Dissertation, TH Darmstadt

(23) Mutschler, P. (1975) Berechnung von Ausgleichsvorgängen in Drehstromsystemen und Drehstrom-Gleichstrom-Verbundsystemen, Dissertation, TH Darmstadt

APPENDIX

Matrices of state space model

$$(A_a) = \begin{pmatrix} \frac{-lR'_a}{lL'_a+kL_{Tr}} & \omega_N \\ \omega_N & \frac{-lR'_a}{lL'_a+kL_{Tr}} \end{pmatrix} \qquad (B_a) = \begin{pmatrix} \frac{1}{lL'_a+kL_{Tr}} & 0 \\ 0 & \frac{1}{lL'_a+kL_{Tr}} \end{pmatrix} \qquad \underline{b}_a = \frac{k\ddot{u}_a u_2^o}{lL'_a+kL_{Tr}} \begin{pmatrix} -\cos\delta^o \\ \sin\delta^o \end{pmatrix}$$

for k = 0 a three-phase short circuit in 1 km distance from bus bar 1 is given

$$\underline{c}_m = \begin{pmatrix} \frac{u_q^o}{\sqrt{(u_d^o)^2+(u_q^o)^2}} \\ \frac{-u_d^o}{\sqrt{(u_d^o)^2+(u_q^o)^2}} \end{pmatrix} \qquad (C_{m1}) = \begin{pmatrix} 0 & 0 & 0 & -\psi_q^o \\ \omega_m^o \frac{L_{hd}}{L_f} & 0 & 0 & \psi_d^o \end{pmatrix} \qquad (C_{m2}) = \begin{pmatrix} -R_d & \omega_m^o L_q \\ -\omega_m^o (L_d - \frac{L_{hd}^2}{L_f}) & -R_q \end{pmatrix}$$

$$\underline{q} = \frac{k_q}{\sqrt{(u_d^o)^2+(u_q^o)^2}} \begin{pmatrix} u_d^o \\ u_q^o \end{pmatrix} \qquad (R) = \sqrt{\tfrac{2}{3}}\, u_{g1} i_g^o \begin{pmatrix} \cos(\vartheta^o+\varphi^o) & \cos(\vartheta^o+\varphi^o) \\ -\sin(\vartheta^o+\varphi^o) & -\sin(\vartheta^o+\varphi^o) \end{pmatrix} \qquad \underline{s} = \sqrt{\tfrac{2}{3}}\, u_{g1} \begin{pmatrix} \sin(\vartheta^o+\varphi^o) \\ \cos(\vartheta^o+\varphi^o) \end{pmatrix}$$

$$(A_g) = \begin{pmatrix} 0 & 1 \\ \frac{-K_{Ri} u_{gi1}^o}{T_{Ri}(L_{G1}+L_{G2}+L_L)} & \frac{-(R_{gi}+K_{Ri}u_{gi1}^o)}{L_{G1}+L_{G2}+L_L} \end{pmatrix} \qquad \underline{b}_{g1} = \frac{K_{Ri} u_{gi1}^o}{L_{G1}+L_{G2}+L_L} \begin{pmatrix} 1 \\ \frac{1}{T_{Ri}} - \frac{R_{gi}+K_{Ri}u_{gi1}^o}{L_{G1}+L_{G2}+L_L} \end{pmatrix} \qquad \underline{b}_{g2} = \frac{(\cos\alpha_1^o)}{L_{G1}+L_{G2}+L_L} \begin{pmatrix} 1 \\ \frac{-R_{gi}-K_{Ri}u_{gi1}^o}{L_{G1}+L_{G2}+L_L} \end{pmatrix}$$

$$\underline{c}_g = (1 \quad 0)^T \qquad R_{gi} = R_{ix1} - R_{ix2} + R_{ir1} + R_{ir2} + R_L$$

$$(A_m) = \begin{pmatrix} -\frac{R_f}{L_f} & 1 & 0 & 0 \\ -\frac{K_E}{T_E}\frac{L_{hd}}{L_f}\frac{\omega_m^o u_q^o}{\sqrt{(u_d^o)^2+(u_q^o)^2}} & \frac{-1}{T_E} & 0 & \frac{K_E}{T_E}\left(\frac{\psi_q^o u_d^o - \psi_d^o u_q^o}{\sqrt{(u_d^o)^2+(u_q^o)^2}}\right) \\ 0 & 0 & \frac{-1}{T_T} & -\frac{K_T}{T_T} \\ \frac{n_p^2}{J}\frac{L_{hd}}{L_f} i_g^o & 0 & \frac{n_p}{J} & 0 \end{pmatrix} \qquad (B_m) = \begin{pmatrix} -R_f \frac{L_{hd}}{L_f} & 0 \\ -\frac{K_E}{T_E}\frac{R_1+\omega_m^o(L_d-L_{hd}^2/L_f)}{\sqrt{(u_d^o)^2+(u_q^o)^2}} & -\frac{K_E}{T_E}\frac{R_1-\omega_m^o L_q}{\sqrt{(u_d^o)^2+(u_q^o)^2}} \\ 0 & 0 \\ \frac{n_p^2}{J}(i_q^o(L_q-L_{hd}^2/L_f)-\psi_q^o) & \frac{n_p^2}{J}(\psi_d^o - L_q i_d^o) \end{pmatrix}$$

$$\underline{b}_m = (0 \quad K_E/T_E \quad 0 \quad 0)^T$$

$$(A) = \begin{pmatrix} (A_m)+(B_{m2})(F_i)\underline{r}_1\underline{c}^T(C_{m1}) & (B_{m2})[(I)+(F_i)\underline{r}_1\underline{c}^T(C_{m2})] & (B_{m2})(F_i)[\underline{r}_1\underline{c}_{g2}^T+\underline{s}\,\underline{c}_g^T] \\ \{(B_{a1})+(C_{m2})[(B_{a1})+\underline{b}_{a2}\underline{c}^T](F_i)r_1\underline{c}^T+\underline{b}_{a2}\underline{c}^T\}(C_{m2}) & (A_a)+[(B_{a1})+\underline{b}_{a2}\underline{c}^T](C_{m2})(F_i)\underline{r}_1\underline{c}^T(C_{m2})+(C_{m2})[(B_{a1})+\underline{b}_{a2}\underline{c}^T] & [(B_{a1})(C_{m2})+\underline{b}_{a2}\underline{c}^T(C_{m2})(F_i)](\underline{r}_1\underline{c}_{g2}+\underline{s}\,\underline{c}_g^T) \\ \underline{b}_{g2}\underline{q}^T[(C_{m1})+(C_{m2})(F_i)\underline{r}_1\underline{c}^T(C_{m1})] & (A_g)+\underline{b}_{g2}\underline{q}^T(C_{m2})(F_i)[\underline{r}_1\underline{c}_{g2}^T+\underline{s}\underline{c}_g^T] & \underline{b}_{g2}\underline{q}^T(C_{m2})[(I)+(F_i)\underline{r}_1\underline{c}^T(C_{m2})] \end{pmatrix}$$

$$(B) = \begin{pmatrix} \underline{b}_{m1} & (B_{m2})(F_i)\underline{r}_1 c_{g1} \\ 0 & [(B_{a1})(C_{m2})+\underline{b}_{a2}\underline{c}^T(C_{m2})(F_i)]\underline{r}_1 c_{g1} \\ 0 & \underline{b}_{g2}\underline{q}^T(C_{m2})(F_i)\underline{r}_1 c_{g1}+\underline{b}_{g1} \end{pmatrix}$$

$$(F_i) = [(I)-\underline{c}^T\underline{r}_1(C_{m2})]^{-1}$$

$$c_{g1} = -K_{Ri}$$

$$\underline{c}_{g2} = K_{Ri}\underline{c}_g^T$$

UNIT CONNECTION OF GENERATOR AND HVDC CONVERTER, LINK BETWEEN POWER GENERATION AND POWER TRANSMISSION

M. Häusler and P. Mutschler

*Brown Boveri & Cie AG., *Käfertaler Str. 250-254, 6800 Mannheim 1, FRG*

ABSTRACT

The structure and operating conditions of the unit connection of generator and HVDC converter are described. The specific requirements of the plant being isolated from other AC networks are considered both for normal operation and for operation at transient or persistent faults.

A control concept for unit connections is proposed which is mainly based upon the control system for an isolated hydro plant feeding an HVDC transmission. The mutual interactions between turbine, generator and HVDC are analyzed. The behaviour of the unit connection in the occurrence of severe faults has been simulated by a digital program. The results of these simulator studies are represented in diagrams for the most interesting quantities frequency, generator voltage and fault current.

INTRODUCTION

The sending end of long distance HVDC transmissions sometimes is connected to plants which are electrically isolated from AC networks. The specific mode of operation resulting from the interdependence of plant and rectifier station justifies to take both as a whole: The HVDC power station (1).

HVDC transmissions offer technical advantages even for rather short transmission distances e.g. for infeed of new power into a network close to the limits of short-circuit capacity or for the transport of power from an offshore generating station to the continental grid. Up to now in HVDC power stations all generators are allocated to common AC busses to which the converters and loads are connected. If, however, the generated power shall be supplied directly and exclusively to the HVDC transmission, another arrangement becomes economically attractive: The unit connection of generator and converter (2, 3).

In view of these alternatives attention is drawn to the operational behaviour of the unit connection. The intent of this paper is to give more information to the utility engineer on the overall control system of the unit connection and to show to which extent simplification of the existing control concept is possible for unit connections.

THE UNIT CONNECTION

Figure 1 is a one-line diagram showing one unit connection with its principal controllers. The unit connection is composed of the turbine, generator, generator breaker, auxiliary supply, converter transformer, converter and DC reactor. The receiving end of the transmission is represented in dotted lines.

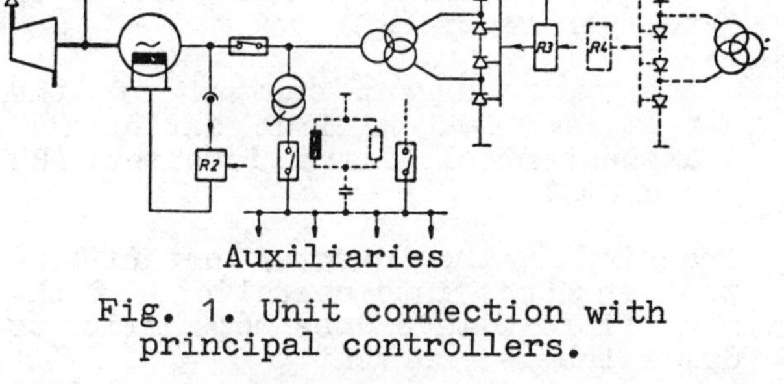

Fig. 1. Unit connection with principal controllers.

AC filters, a separate unit transformer and thus one transformation stage and the AC switchyard are eliminated in the unit connection. As compared to conventional HVDC stations, costs and space are substantially reduced.

* Member of the HVDCT Working Group, formed by AEG, BBC and Siemens

Since the generator voltage control can be used to limit the ideal no-load DC voltage and to maintain the firing angle of the rectifier in the desired range, no tap changer is required for the converter transformer, thus allowing the use of a simple and hence reliable transformer design. The converter consists of two three-phase bridges operating in twelve-pulse mode only, in order to fully utilize the generator capacity (4).

Two alternatives have been proposed for the converter:

Controlled rectifiers (as represented in Fig. 1) or simple diode rectifiers combined with an HVDC circuit breaker (5). For the control concept of the latter a complete new approach is necessary whereas for the "conventional" controlled rectifier the existing concept needs only to be simplified.

OPERATING CONDITIONS AND SPECIFIC REQUIREMENTS ON THE CONTROL SYSTEM OF THE UNIT CONNECTIONS

Under normal conditions the unit connection has to transmit bulk power which varies only slowly. After transient line faults the system shall restore normal operation as soon as possible.

The requirements for supplying the auxiliaries reliably and economically have to be considered for the design of the control concept.

As mentioned earlier the principal control loops of the unit connection are represented in Fig. 1. Four control loops are more or less interacting with each other: (the related controllers are set in brackets)

- the slow turbine speed control (R1)
- the rather fast voltage control of the generator (R2)
- the fast current control of the rectifier and minimum extinction angle control of the inverter (R3 and R4).

Transiently the inverter current controller comes into operation, if the rectifier voltage does not suffice to drive the set current.

The reference for the rectifier current is normally given by the superimposed power controller, which operates in open loop in order to avoid detrimental effects from telecommunication failures.

The operating conditions of the unit connection are to a certain extent similar to those of conventional HVDC power stations: the rectifiers constitute in both arrangements the preponderant if not only load of the generators, a load which is practically inertialess and frequency insensitive in the constant power control mode. Under normal conditions generated and transmitted power are balanced with the speed governor controlling the frequency and the HVDC controlling the transmitted power. However, disturbances in the transmission system can result in severe and rapid load changes. In order to help stabilize the system under these circumstances the HVDC must temporarily control the generator frequency. This is achieved by means of frequency limit controllers (6).

Simplification and easiness of the control of unit connections can be expected due to the following reasons:

- Much larger temporary frequency deviations can be tolerated than in stations with AC filters. Successful restoration of power transmission after transient faults is correspondingly facilitated.
- There is no risk of selfexcitation of the generator after loss of load. The requirements on the performance of the generator voltage control are therefore reduced.
- Power flow oscillations between generators are excluded on the AC side. No special measures are necessary for balancing of real and reactive power between units.

If the power of different units has to be transmitted by the same transmission line, the units may be connected in series (3) or in parallel (7) on their DC side. The definite choice depends upon a lot of factors of the power equipment such as the size of generators, DC transmission voltage etc. From the control point of view the first alternative is characterized by a common current controller for all series connected units, whereas in the second alternative all units operate at the same DC voltage. It seems as if the mutual interactions are less for parallel connected units. The load distribution during taking in operation or shutting down a unit can be changed without the need for rectifier operation at large firing angles thus minimizing the reactive power variations in the generators.

REQUIREMENTS OF AUXILIARY POWER SYSTEMS

In order to permit functioning of structures, systems, and components important to safety, two independent supplies are generally required for the auxiliaries. In Fig. 1 the first source of auxiliary power has been indicated. All auxiliaries for one complete unit, right through from the turbine to the converter are supplied from the unit auxiliary transformer. Because of the distorted generator voltage a damped highpass filter may be connected to the auxiliary network. It is less frequency sensitive than tuned AC harmonic filters and therefore allows for larger transient frequency excursions. Its influence on the generator can be neglected.

A transfer from the unit auxiliary transformer to a second source will be provided in the unlikely event of a failure of the unit auxiliary transformer, in case of a fault causing to trip the generator breaker, or after a complete loss of load due to persistent DC line faults where the generator frequency is expected to deviate largely from the nominal value.

The second source of auxiliary power could be the unit auxiliary transformer of another unit feeding a second transmission line. If, however, only one unit or transmission line is available at all, a separate standby supply is required for the essential auxiliary loads, which can be provided by an emergency diesel set.

STARTING A UNIT CONNECTION

The start-up power of a unit considerably exceeds the power needed for the essential loads. A diesel generator of adequate size is usually provided for the start-up of hydro power plants. For the same purpose isolated thermal power plants need even a house turbine. Both solutions are rather expensive. Therefore, the question arises whether means could be provided to start the unit connection from the DC line less costly than with the afore mentioned equipment.

The given arrangement of generator, generator breaker and unit auxiliary transformer principally allows to start the unit from the DC line with reversed power flow direction. For this purpose provision for inverter operation must be made. The reactive power for the auxiliaries and the "inverter" has to be delivered by a source independent from the unit generator, e.g. a synchronous condensor. The size of the synchronous condensor could be moderate compared with the unit generator, but the controls for inverter infeed into such a weak AC have still to be developed. There are promising results of investigations which indicate that inverter operation is feasible even for these low short-circuit to DC power ratios (12).

The power needed at the very beginning to drive the synchronous condensor and to control the "inverter" is small compared with the total consumption of all auxiliaries and could be delivered by the station diesel.

PROPOSAL FOR THE OVERALL CONTROL SYSTEM OF THE UNIT CONNECTION

The master power controller (MPC) of the HVDC power station receives the main power setting via telecommunication from the central load dispatcher of the AC network. This signal is stored and divided between the available units. The unit power order is limited according to the actual capability of the turbine depending on its operating point (block 1 in Fig. 2).

The rate of change of power demand is limited in order to respect the behaviour of the turbine control (block 2). At the output of the MPC the power order signal is given to both the turbine and the rectifier control.

The turbine power order passes a function generator with inverse turbine characteristic (block 3) and then is added to the output of the speed governor (block 4). The sum of both signals acts upon the turbine gate or steam valve.

For the rectifier power order a time lag (block 5) may be desirable in order to adapt the HVDC to the turbine behaviour. The actual power order of the rectifier is reduced or increased by the f_{min} and f_{max} controllers, if the measured frequency deviation exceeds a selected value (blocks 6, 7).

The division of the power order by the measured DC line voltage yields the current reference for the rectifier current controller (blocks 8, 9). The output of block 9 normally determines the firing angle of the rectifier. For the rare event of an open line fault a line voltage controller (block 14) is provided which may eventually determine either the firing angle (by 10) or the generator voltage (by 13).

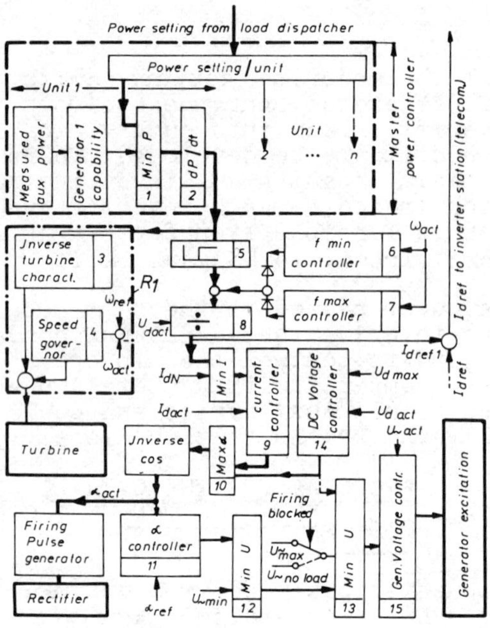

Fig. 2. Control circuits.

The actual firing angle α is compared with its desired value. The difference is acting on the α - controller (block 11), the output of which is the reference signal for the generator voltage controller (block 15). MIN and MAX selection devices (blocks 12, 13) limit the range of the generator voltage reference.

If the firing pulses of the rectifier are blocked, the α -controller (block 11) can no longer determine the generator voltage reference. For this case a constant voltage reference is provided at the input of the selecting device (block 13).

If only the ideal no-load voltage of the rectifier has to be limited, the α - controller can be replaced by a constant voltage reference. This concept probably allows to eliminate the tap changer of the unit auxiliary transformer.

THE MUTUAL INTERACTIONS

Although the unit connection is less complex than a system with several HVDC converters connected to an AC busbar, there are many interactions both between turbine, generator, and converters and between their associated control loops. Prior to try to simulate the interacting processes a sufficient understanding of the system is necessary. The most important interactions are shown in Fig. 3. In the right half of Fig. 3 the simplified diagram of the processes in turbine, generator, rectifier and inverter is given. Fig. 3 is to present an overall view on the mutual interactions between the system components, not details of the components. The output of the voltage controller determines the exciter flux. The stator current can be substituted by the direct current I_d and the firing angle α in Fig. 3.

The stator current determines the stator flux in d- and q-axis. The resulting flux yields after multiplication by the frequency ω the terminal voltage $U_\sim$ of the generator. The mutual flux linkages within the generator are not shown in Fig. 3, but they are respected in the simulation studies. The output voltage $U_\sim$ of the generator represents one of the inputs for the rectifier. The other input is the firing pulse train which is obtained from the firing angle α. Output of the rectifier and inverter are I_d and U_d. The interactions between rectifier and inverter are not shown in Fig. 3. At first there are three interacting loops in the power equipment. Two of them are very similar. The voltage $U_\sim$ and the stator current form the basic loops between generator and rectifier: loop 1 (a-b-c-d-a) and loop 1'(a-b-c-e-f-d-a). For the HVDC connected to an AC bus, the corresponding loops have been investigated in (10) and (11).

It has been shown that this loop causes even harmonics on the AC side if there are odd harmonics on the DC side and vice versa. If such harmonics in $U_\sim$ coincide with a resonant frequency of the combination of AC system and filters, then instability may arise under certain circumstances. From the control point of view, it is advantageous for the unit connection that the loop 1 (1') does not contain AC filters and that previously mentioned difficulties are thus avoided.

There exists another loop which contains rectifier, turbine and generator. The output power of the rectifier $P_d = I_d \cdot U_d$ is fed back to the turbine. The turbine produces the accelerating torque t_a. For detailed turbine representation see e.g. (8) (hydro) or (9) (thermal). The decelerating torque t_d is obtained from P_{load} and ω. The angular frequency is obtained by integrating the torque difference $t_a - t_d$ (block 2 in Fig.3). Thus loop 2 (c,g-h-k-m-a-b-c,g) contains P_d, ω and $U_\sim$.

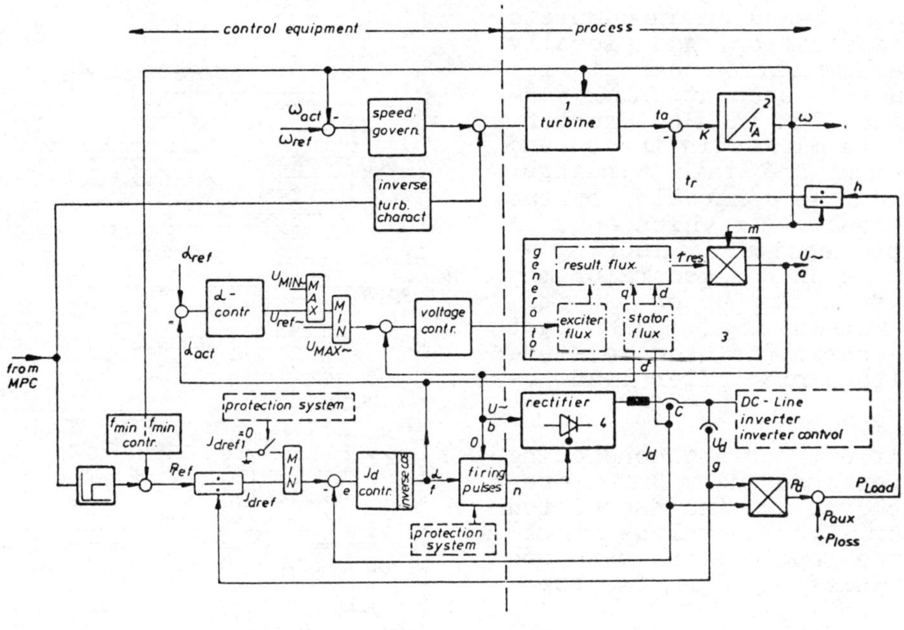

Fig. 3. Mutual interactions.

There are other interacting loops which are directly influenced by the controllers. Fast transient interactions take mainly place between rectifier, its control and generator. A fast acting loop is given by loop 3 (n-c-d-a-o-n), where o denotes the synchronizing input for the firing pulse device. The distorted generator voltage $U_{\sim}$ may cause harmonic instability if the firing pulses are directly synchronized by $U_{\sim}$. The negative effect of loop 3 can be avoided by using an equidistant firing pulse device.

REQUIREMENTS ON THE PROTECTION SYSTEM OF THE UNIT CONNECTION

Faults in the AC network of the inverter station can cause persistent commutation failures. In this case a voltage dependent device reduces the current reference signal of the rectifiers of the corresponding pole.

Faults on the DC line are mostly transient ones and can be cleared by a short de-energization of the faulted pole, as it is practice in conventional HVDC schemes (5). If several unit connections are feeding into the same transmission line, a synchronization of restarts is necessary which is performed by the MPC.

Faults in the rectifier station can be dealt with by blocking the converter valves. Under certain circumstances periodic emergency firing of the valves may be initiated. In this case the fault will be cleared by the generator breaker. Field suppression would only be necessary in the unlikely event of a short-circuit between generator and generator breaker.

SIMULATOR STUDIES

In this chapter, the performance of the unit connection during faults on the DC line and in the rectifier will be investigated. Due to the complexity of the system (Fig.3), the investigations are carried out using a digital simulation program, specially developed for HVDC studies. The intention of the following study is to obtain reliable information about

- voltage stress of the valves during faults and breaker operation
- voltage rise after fault clearing
- current stress of the valves during faults.

The used program is based on the state variable approach. The set of nonlinear, discontinuous differential equations is solved by numerical integration. While Fig. 3 gives only a survey over the plant, in the simulation study the generator, converter transformer, and the rectifier are treated in detail. The generator representation is based on the complete set of Park's equation. Additionally the nonlinear saturation characteristic is taken into account. State variables are the flux in the exciter winding and the fluxes in the d- and q-axis of damper and stator windings. The voltage $U_\sim$ is controlled by the P-I voltage controller which acts on a fast static exciter. Thus, the exciter voltage of the generator may be changed from its upper to its lower limit within 10 ms. This fast acting $U_\sim$ -control is used to minimize the voltage rise after load rejection.

In the rectifier, current conducting valves are represented by their foreward voltage drop and the associated valve reactor. Blocked valves are removed from the topology in order to speed up computation. Thus, the topology of the rectifier changes with every firing and blocking of a valve. Since the inverter and its control is not the subject in this study, it is represented in a simplified form as a controlled voltage behind the equivalent resistance of the inverter. Also the representation of the DC line is simplified. The protection system of the rectifier will reduce I_{dref} to zero when detecting a line fault. Therefore, with fault studies it is not necessary to include the MPC and the f_{max}, f_{min} controllers since during the first time after a failure their action is overridden by the protection system. The current controller is of P-I type and its parameters are obtained according to (10).

When studying faults and their consequences for a short time (some hundred ms), the turbine can be simplified, because of its relatively large time constant. The speed governor is of P-I type.

RESULTS

Load Rejection

Fig. 4 shows the load rejection without previous fault. After setting $I_{dref} = 0$, the direct current will become zero within 5 ms. The fast acting voltage controller reaches its lower limitation within 10 ms. According to loops 1 (1'), the sudden change in I_d and α influence $U_\sim$ and thus the valve side transformer voltage shown in Fig. 4. The single voltage peak occuring 1.1 ms after setting $I_{dref}=0$ on the valve side reaches 1.06 pu *.

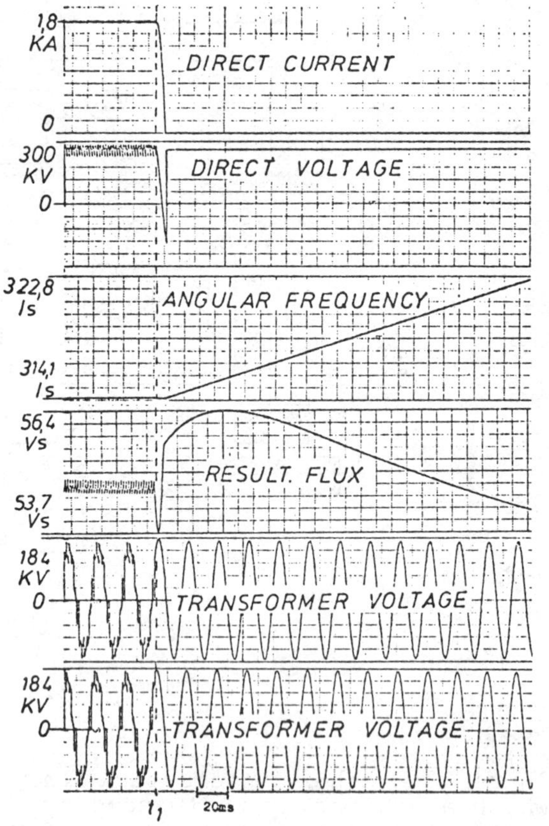

Fig. 4. Load rejection

The frequency increase with 11,5% /s acts via loop 2 on $U_\sim$. The fast voltage controller almost eliminates the voltage rise due to the frequency rise ($\Delta \hat{U} = 0,03$ pu). The above data were obtained using a salient pole generator. When simulating a typical turbogenerator, the results have been very similar.

DC Line Fault

Detecting a line fault, the protection system reduces I_{dref} to zero. Fig. 5 shows the fastest possible control action after a short-circuit on the line of the DC-reactor. Starting from worst case steady state conditions with $I_d = I_{dN}$, $\alpha = 0^\circ$ and coincidence of short-circuit with the beginning of a commutation, the peak value of the direct

*The peak value of voltage in steady state operation with $I_d=I_{dN}$ and $\alpha=12^\circ$ has been taken as the base for the pu voltage quantities.

current reaches only 2 pu*. The peak values of the valve side transformer voltages are in the same order as in case of load rejection.

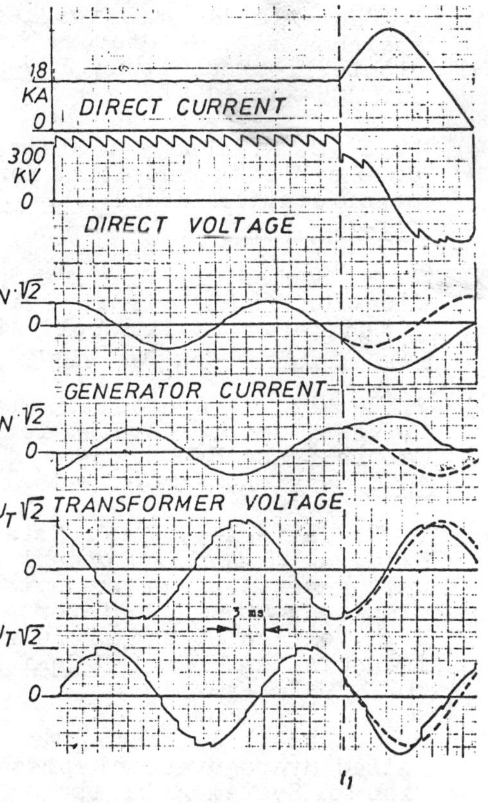

Fig. 5. DC line fault.

In practice, the protection system keeps $I_{dref} = 0$ for the de-ionization time of e.g. 150 ms and then restores power transmission via a ramp fuction for I_{dref} (duration e.g. 200 ms) (6). If the first attempt to deionize the fault location is successful, the frequency at the end of this procedure will be increased by approx. 3 %, if the unit connection was previously operated at full load.

Internal Rectifier Fault

In case of internal rectifier faults the protection system blocks the firing pulses and, if necessary, initiates generator breaker tripping. The highest current stress for the valves occurs during a short circuit across a valve.

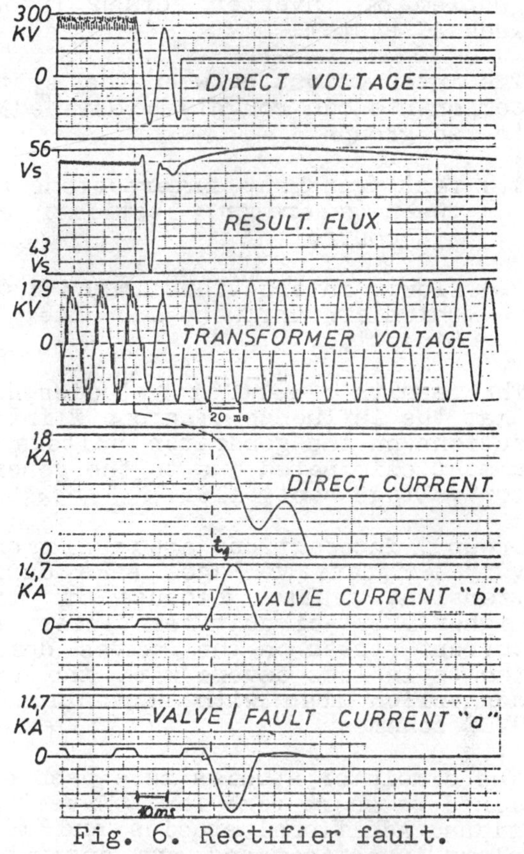

Fig. 6. Rectifier fault.

In Fig. 6, the short-circuit across valve "a" is initiated at t_1. Valve "b" then carries a peak current of 8.2 pu, where the conduction time reaches 325°. The direct current decreases rapidly. All currents in the rectifier become zero approximately 33 ms after blocking the firing pulses. During this time, no overvoltages occur at the valve side of the transformer. The voltage rise after fault clearing is of the same order as in case of load rejection. In case of periodic emergency firing of valve "b", the fault will be cleared by the generator breaker. Currently available generator breakers need 60 ms to open. During this time, valve "b" has to withstand three current peaks (max. 8.3 pu) with a conduction time increasing to nearly 360°. No overvoltages occur during breaker operation.

CONCLUSIONS

The direct connection of generator and HVDC converter simplifies the control concept in many respects. E.g. the generator voltage control can be integrated in the rectifier control replacing the tap changer control for the converter transformer.

The incentive to provide the power needed for start-up of the unit connection from the DC line may stimu-

*The peak current is less than with conventional HVDC schemes, since the commutation reactance is increased by the generator's subtransient reactance.

late investigations how to solve the problem of inverter infeed into a weak AC system.

The interactions between the system components can roughly be devided into two groups

1.) fast transient interactions between generator, rectifier and its control,

2.) slower interactions in the system which are influenced by the turbine, too.

The results of simulator studies show, that the influence of group 2 interactions on the generator voltage is almost eliminated due to the generator voltage controller.

Group 1 interactions are causing only very moderate voltage peaks during transient faults. Because of the higher short-circuit reactances, the current stress on the valves during faults is less severe with the unit connection than with conventional HVDC schemes.

The simulator studies have been confined to short time fault investigations. Additional studies and model plant investigations are desirable with respect to the overall control system including e.g. parallel operation of converters, the MPC and detailed speed governing system representation.

The presented results are mainly based upon the operational conditions and experiences with hydro plants. Because of the partly differing and more stringent requirements of thermal plants, especially of nuclear plants further work on the control and protection of HVDC unit connection is necessary.

REFERENCES

(1) D. G. Chapman, F. A. Jost, Operation of an isolated hydro plant supplying an HVDC transmission load, Trans IEEE, PAS-95, 1099-1104 (1976)

(2) K. W. Kanngießer, F. Hölters, W. Ziegler, Technik und Einsatzmöglichkeiten der Hochspannungs-Gleichstrom-Übertragung , ETZ-A 89, 170-176 (1968)

(3) J. Käuferle, HGÜ-Kraftwerkseinspeisung mit Generator-Stromrichterblockschaltung, BBC-Mitt. 60, 205-211 (1973)

(4) P. S. C. Krishnayya, Block and double-block connections for HVDC power station infeed, IEEE Conference Paper C 73227-6

(5) J. P. Bowles, L. Vaughan, Specification of HVDC circuit-breakers for different system applications CIGRE 1976 Session, Paper 13-09

(6) W. Becker, S. Ranade, E. Rumpf, T. Wess, Control system for the Cabora Bassa HVDC Scheme, CIGRE 1976 Session, Paper 14-06

(7) T. E. Calverley, C. H. Ottaway, D. H. A. Tufnell, Concept of a unit generator converter transmission system, Conference on High Voltage DC and/or AC Power Transmission 1973, IEE Conf.Publ. 107, 19-23 (1973)

(8) D. G. Ramey, J. W.Skooglund, Detailed Hydrogovernor Representation for System Stability Studies, Trans.IEEE PAS-89, 106-112(1970)

(9) G. Mathias, Das dynamische Verhalten der Kraftwerks-Dampfturbine bei Frequenz-Leistungs-Regelung, Thesis T.U. München 1973

(10) R. Jötten, R. Müller, E. Rumpf, Einfluß der Netzparameter auf die Regelkreise einer HGÜ, IFAC-Symposium 1974, 527-541

(11) R. Müller, Gegenseitige Rückwirkungen zwischen Oberschwingungsresonanzen im Drehstromnetz und der Stromregelung einer HGÜ, ETZ-A 95, 30-34 (1974)

(12) H. P. Lips, H. Ring, The performance of AC systems with predominant power supply by H.V.D.C. inverters, Trans IEEE PAS 94, 408-415 (1976)

COMPARISON OF INSTALLATIONS FOR COMPENSATION OF REACTIVE POWER

G. Loocke

Hochschule der Bundeswehr Hamburg, Fachbereich Elektrotechnik, Holstenhofweg 85, 2000 Hamburg 70, FRG

ABSTRACT

Among installations for the compensation of continuously changing reactive power, the conventional rotating synchronous condensers are more and more replaced by static installations of power electronics.

This paper comprises a technical and economic comparison of the different alternatives of power electronic components as well as rotating synchronous condensers. The most important technical criteria for comparison are:
The dynamics of energy flow and the symmetrizing characteristics at unsymmetrical phase load.

The comparison of costs is carried out for installation costs and for installation costs plus consideration of losses which vary among the different alternatives.

The losses are calculated for a standard duty cycle, and the cost of losses thus calculated are added to the costs of installation.

1. INTRODUCTION AND PROBLEM AREA

In general the compensation of reactive power in industrial networks is effected by stepwise switchable *CAPACITOR BATTERIES*, which might be executed for high-, medium-, and low voltage. (In the following the medium voltage type with a connection voltage of 30 kV will be termed as comparison *version A)*. For the compensation of continuously variable reactive power the *ROTATING SYNCHRONOUS CONDENSER* has preferably been used up to now (comparison *version B)*. During the passed years the rotating condensers are more and more replaced by static equipments of the power electronics. This applies to compensation equipments of arc furnaces as well as to those of heavy-duty rolling mills.
At present the following variations of static solutions are under discussion:

C. FIXED CAPACITOR PLANTS IN PARALLEL OPERATION WITH CONTROLLED REACTIVE CURRENT LOAD

By means of this principle the problem of the control of capacitive reactive power is led back to the technically simplier task of the control of inductive reactive power.
The fixed capacitor battery can be designed for the economically favourable medium voltage range. If necessary with regard to power harmonics, the capacitor batteries are equipped with series reactors (effect of filtering circuit). The control of the inductive reactive power via semiconductor controlled elements requires voltage transforming down to a semiconductor-compatible level.
Reactive current load can be as shown in fig. 1:

a) Inductive reactive converters:
Line-commutated converters, which are short-circuited on direct current via smoothing reactors.

b) d.c. premagnetized choke reactors:
Principle of the ampereturn control at the transductor.

c) Choke reactors in connection with semiconductor three-phase-regulators:
The reactive current is controlled via the firing angle of the 3-phase thyristor regulator. Here the idea is obvious to connect reactances and transformer in one component by installing a leakage-reactance transformer (approx. 100 %).

d) Thyristor switched reactors:
Based on their physical function the variants C.a), C.b), and C.c) cause a harmonic content in the controlled inductive reactive current, which depends on the modulation conditions. This can be avoided by mere switch operation of the reactances. By means of suitable gradations of the single thyristor switched reactances the switching process can be transformed into an almost continuous adjusting process. Transients have to be avoided by synchronous actuation of the thyristor switch.

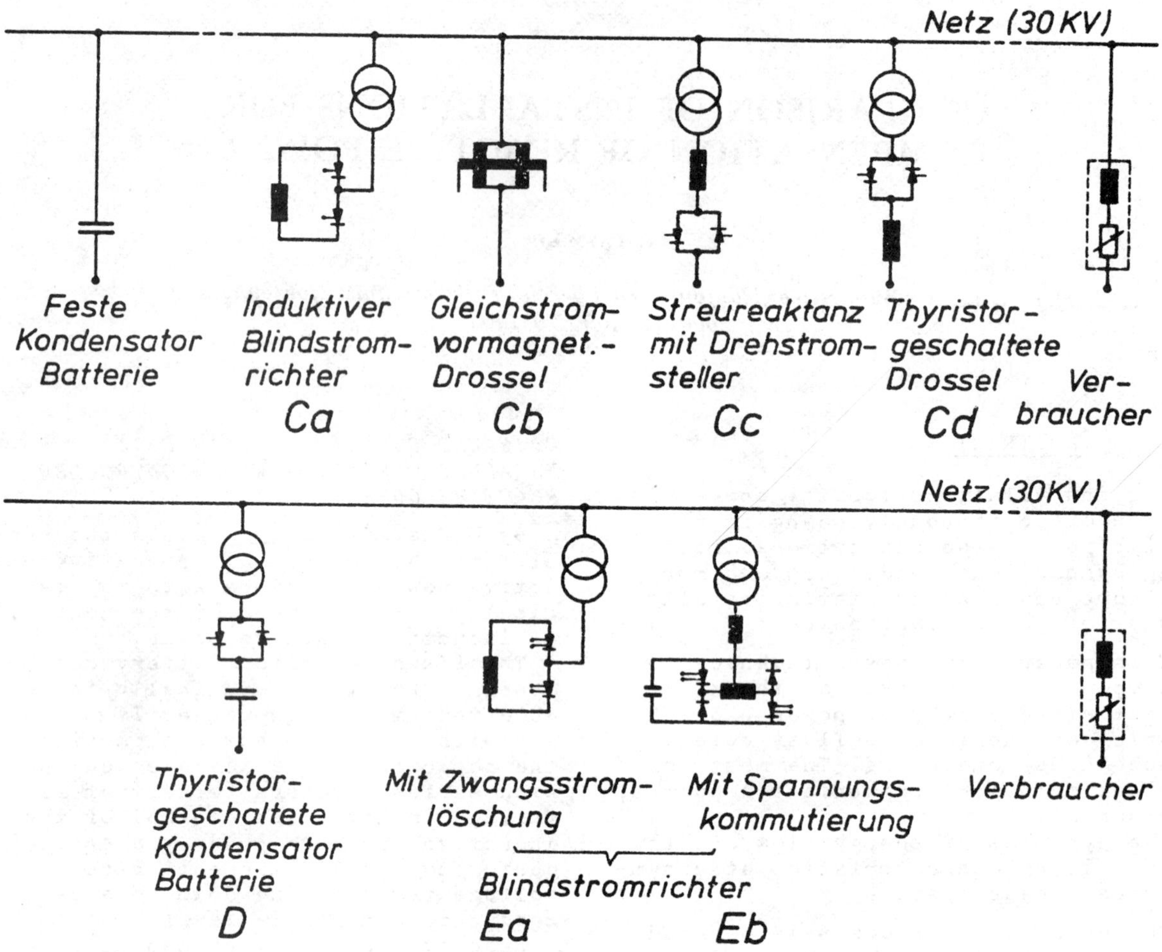

Fig. 1 Static equipment for the compensation of continuously variable reactive power

D. DIRECT WITH THYRISTORS CONNECTED CAPACITORS

On principle it is possible to adjust directly the capacitive reactive current almost continuously by a respective gradation of thyristorswitched capacitor batteries of the variant C.d. However, this technical solution is subject to the well known problems of a control of capacitor batteries on duty. This includes: precise line-synchronous switching, in order to avoid the transient currents, which are critical especially when switching capacitors; preventive measures to protect the semiconductor switches against the discharge current of the capacitors in case of line short-circuit, e.g. protection reactors as indicated in the literature. In addition, there are the special problems to have a.c. capacitors stand under d.c. voltage.

E. CAPACITIVE CONVERTERS WITH FORCED COMMUTATION

As in the variants C. and D. semiconductor control elements in the order of nominal power have to be installed, it suggests itself to equip the power electronic components with forced extinguishing and thus to employ the converters for the direct supply of capacitive reactive power. Two circuits are possible (1):

a) The converter with forced current commutating on alternating current.

Its operation principle is the same as that of the inductive reactive converter (C.a), however, additional capacitors allow a forced extinguishing of the current.

b) The converter with forced voltage commutating on direct current.

The principles of construction correspond to those of an inverter with d.c. intermediate circuit, as it is used for feeding of asynchronous motors with variable frequency (2). The different wave forms of voltage-sinusoidal voltage in line cycled/d.c. voltage at the inverter-cause the connection of an inverter to the line via a coupling reactor. Also here the idea is obvious to connect reactances and transformer in one component by installing a leakage-reactance transformer (15-30 %). In the operation mode as reactive converter a d.c. capacitor battery is used in lieu of the d.c. intermediate circuit. It is possible to continuously pass the zero range and to supply capacitive as well as inductive reactive power. By this means the parallely working with a capacitor battery is basically possible. In such case the reactive converter need only be rated for half of the required capacitive power.

New technologies inevitably demand from the manufacturer and operator of a plant a critical comparison of the new variants with the hitherto applied procedure with respect to their techniques and economy. The present contribution represents an attempt to compare firstly the new variants with one another and secondly with the conventional rotating condensers.

2. GENERAL COMPARISON ASPECTS

The technique of this new compensation procedure as well as the technology of the compensation process - especially with regard to arc furnaces - has been treated in the literature in detail during the passed years. Therefore, reference can be made to such publications when comparing the technical criteria.

As far as the literature comprises already comparisons of the individual systems, mainly the different physical operation modes have been compared (3). However, for the following study in the first place the comparison according to the view point of the different technical operation features of the individual variants are needed. For this purpose the criteria, indicated in section 3, have been chosen, whereby especially the dynamic behaviour of the plant, i.e. how fast the reactive power can be made available, yield to be the most important criterion. Only by such comparison it becomes possible to judge whether at all the individual plant variants can be estimated one by another and how their components have to be dimensioned in order to undertake a comparison of cost on the basis of equal operation characteristics.

No doubt, the statement of cost for such comparisons will always be strongly influenced by the field of view of the respective author. Other manufacturers and operators might well start from other data. In order to take this into consideration the author tried first to indicate the criteria for comparison of cost and the theoretical principles, according to which costs develop. After that comparative figures for the individual cost statements will be given for discussion.

Based on the matter to be investigated, such comparison will never be generally valid in its intirety. There will always be a certain cost part, which will allow comparison criteria in an individual case only, and that with regard to installation costs (e.g. erection, foundation) as well as to operating expenses. Other technical features can only conditionally be used as comparative criteria or cannot at all be referred to, owing to the fact that either no data are available yet from the manufacturers, or their effects are not yet obtainable quantitively (e.g. availability).

The numerical data have been elaborated for plants with the power of 45 Mvar, 24 Mvar, and 7,5 Mvar, based on concrete projects.

3. CRITERIA FOR THE TECHNICAL COMPARISON

3.1 Limiting output

All mentioned technical solutions can also be executed for smaller and larger powers as those on which the study is based.

3.2 Dynamics ot the energy flux

According to their dynamic behaviour on the energy side the following four subdivisions of this group can be made:

α *Stepwise switchable capacitor batteries*
The times for action are within the range of seconds to minutes.

β *Slow* -
Synchronous condensers and d.c. premagnetized reactors without impulse excitation with simple arrangement of excitation: They provide nominal reactive power in the range of 0,2-1 secs.

γ *Moderately rapid* -
Synchronous condensers and d.c. premagnetized reactors with impulse exitation: They provide according to the presumption mentioned in the literature (4, 5) nominal reactive power within 50-100 msecs.

δ *Very rapid* -
Variants with thyristor-control are able to supply nominal reactive power within a line period (20 msecs.) at the most; with the majority of the variants even within half a line period (10 msecs). As far as multipulse converter-connections are used, the time for action goes down to nearly two pulse lengths (6,6 msecs with 6-pulse and 3,3 msecs with 12-pulse rectifier bridges).

The problem of a rapid reactive power transfer for given reactive currents can be evaded with a synchronous condenser as active voltage source. By interlocking of the power supply with a buffer reactor a current-dividing behaviour between power supply and condenser in the ratio of the participating reactances can be forced. With regard to the stability conditions the size of this buffer reactor is limited upwards (6). Within this limit also the necessity of a rapid data logging and handling according to the following section 3.3 is eliminated.

3.3 Data logging and handling, output of the operating commands

For a rapid control on the energy side logically a respectively rapid data logging and handling for the issue of the operating commands is necessary. For symmetrical systems electronic reactive power regulating devices are available. Thereby power drifts are measured, handled and issued as analogue operating commands practically without delay. For asymmetrical operation it is possible to refer to equipment developed for respective tasks in single phase lines e.g. railway-lines (7).

In case the power control element is not actuated continuously but switched delicately stepped, as e.g. variant *Cd-thyristor-connected reactor* -a computing unit for the determination of the string reactive power and a recording logic for the string operating commands is required additionally. In addition, the control equipment has to include the logic for line-synchronous switching.

Furthermore, for thyristor switched capacitors a circuit subunit has to be provided, effecting a cyclic recharging of the capacitors which are in standby position under d.c. voltage (at least 2 recharges per second).

3.4 Balancing

With the help of its damper cage the synchronous condenser can produce a certain voltage balancing with asymmetrical phase load only.

The variants Cc, Cd and D, however, can symmetrize reactive currents and - when applying the Steinmetz principle - also active currents. The same applies to d.c. premagnetized reactorsCb when designed on single-phase operation.

3.5 Harmonic behaviour

Synchronous condensers, thyristor-connected reactors and capacitors (Cd, D) do not cause harmonic contents. With the d.c. premagnetized reactor Cb the harmonics can be kept down with symmetrical operation by the measures well kwown from high-power transformers (approx. 3 %). With unsymmetric mode of operation, for the purpose of symmetrization - at the variants Ca, Cc and E also with symmetric operation - appear harmonics on account of the physical function.

On principle, this could be taken into consideration when comparing the economy, by providing additional reactors for the parallely working capacitor batteries, in order to form filtering circuits. However, this would not do justice to the real economic situation of the plants: The consumers, for whose account the aformentioned reactive compensating plants are installed, bring a high harmonic distortion reactive power into the line by their own. That means that filtering circuits are necessary anyhow. This applies to arc furnaces in the same way as it does for rectifier drives of high power.

3.6 Overload capacity

Peaks heterodyne the basic reactive power to be compensated; at the arc furnace for instance caused by the recharge of scrap during the melting process.

Electrical machines and transformers are able to absorb such peaks within their thermal time constant. Thyristor control elements and capacitors do not dispose of such a thermal overload capability for the time duration, in which such reactive power peaks might occur.

For all installations, operating according to the variant C, the possibility of overloading is eliminated anyhow: on the one side the reactive power peaks have to be fully covered by the compensation equipment, and on the other side the plant must be ready for operation, even during longer periods, with the reactive power 0.

4. CRITERIA FOR THE COMPARISON OF INSTALLATION COSTS

4.1 General conception

With regard to the installation costs it can be considered to a close approximation that the power dependent costs are dominant. For this reason the comparison is executed on the basic of power-specific costs, which are indicated as ratio values (ratio cost figure/kvar).

The costs of a 45 Mvar synchronous condenser in relation to 1/kvar are chosen as comparison figures. Costs, which are conditionally dependent on the power, are considered as surcharges.

In the following calculation there will be investigated:

B three variants of the rotating synchronous condenser

Ba synchronous condenser with simple arrangement of excitation

Bb synchronous condenser capable of impulsive overload, with rectifier excitation
  1 related to the nominal reactive power
  2 related to impulsive reactive power

Cb two variants of the d.c. premagnetized reactor
  1 with simple excitation eqipment
  2 with rectifier impulse excitation

D thyristor-switched capacitors in the following version:
Fully controlled semiconductor switch, i.e. thyristors are provided for both current directions. Therefore, the action time for the energy flow comes to half a line period at the latest. This corresponds to the dynamic behaviour of the other very rapid variants.
The power of the installed capacitors corresponds to that of the reactive power to be compensated. In order to utilize the possibility of action time within half a line period, the then to be switched capacitor power with the two possible polarities has to be in standby position. With halved division thus only half of the installed reactive power could be in standby position during half of a line period. The solution to install double the capacitor power and to have standby each one half with the two polarities, cannot be accomplished for reason of economy.

E merely the inverter type b with parallely working capacitor battery.
The inverter type Ea does not offer the possibility of a continous zero passage and therefore it cannot be compared with the type Eb regarding a comparable economy.

The comparative calculation is executed for phase symmetrical compensation. As far as data are available, the additional costs, which might arise from asymmetrical operation, are estimated.

4.2 Specific costs of installation, dependent on the power - C_A

Comperative figures for the specific basic costs for the elements, which are used in the individual circuits in the power ranges of 45, 24, and 7,5 Mvar are given in table 1.

The specific costs of installation, dependent on the power C_A result from the basic costs by means of the design methods of the plant. For the different variants these costs are:

A Medium voltage capacitor plant
$C_A = c_{CAM}$

TABLE 1 Specific basis costs for elements

power range		45		24	7,5	Mvar
synchronous condenser	c_{PH}	1		1,26	1,55	per kvar
transformer plant	c_{TR}		0,68		0,9	
thyristor cabinet	c_{TH}		0,59		0,68	
diode cabinet	c_{DI}	0,18			0,2	
reactor	c_{DR}		0,72		0,72	
medium voltage capacitor plant	c_{CAM}	0,93		1,14	1,5	
low voltage capacitor plant	c_{CAN}	1,71		1,93	2,27	
d.c. premagnetizing choke reactor	c_{SDR}	1,09		1,31	2,61	
low voltage- capacitors	c_{CON}			1,58		
plastic-foil capacitors	c_{COK}			1,4		
medium voltage- capacitors	c_{COM}			0,86		
power electronic section of the inverter	c_{BR}			1,35		

Ba *synchronous condenser with transformer; without regarding any possibility of overloading*

$C_A = c_{PH} + c_{TR}$

Bb *impulsive overloading synchronous condenser with transformer, designed for rectifier excitation*

1 *related to the nominal reactive power*

$C_A = 1{,}25\ c_{PH} + C_{TR}$

(The usual impulsive overload of 1,6./.1,8 requires an by 15-25 % oversized model)

2 *related to impulsive reactive power*

$C_A = \frac{1}{1{,}6 - 1{,}8}\,(1{,}25\ c_{PH} + c_{TR})$

Ca *inductive reactive converter*

$C_A = c_{CAM} + 1{,}05\ c_{TR} + c_{TH} + 0{,}15\ c_{DR}$

(The factor $\frac{\pi}{3} = 1{,}05$ for the transformer follows out of the designing as converter transformer)

Cb1 *d.c. premagnetized reactor*

$C_A = c_{CAM} + c_{SDR}$

In case, for a better dynamics behaviour, the dead time, which occurs only when up-magnetizing the d.c. circuit from the precise zero-value, shall reliably be avoided, a limitating of the adjusting range and a basic excitation according to a basic reactive power of 5 % is necessary. From here results an oversizing of the adjusting range of 1.05 to 0.05 = 1; and therefore

Cb2 *d.c. premagnetized reactor, designed for rectifier excitation*

$C_A = 1.05\ (c_{CAM} + c_{SDR})$

(When designed on single-phase operation, costs for the reactor arise to 1.75 c_{SDR})

Cc *leakage-reactance transformer in connection with three-phase semiconductor regulator*

$C_A = c_{CAM} + 1{,}5\ c_{TR} + c_{TH}$

(The costs of a transformer with a leakage of 100 % are between 140 and 160 % of those of a line transformer. For the comparison an average value 150 % is chosen)

Cd *thyristor-switched reactors*

$C_A = c_{CAM} + c_{TR} + c_{TH} + c_{DR}$

D *thyristor-switched capacitor batteries*

$C_A = c_{TR} + 1{,}7\ c_{TH} + 1{,}15\ c_{CO} + 0{,}15\ c_{DR}$

(The factor 1,7 at the thyristors is caused by the fact, that the voltage stress is larger than at the other variants. The thyristor switch has to insulate the capacitors under potential safely from the line)

Eb *Capacitive reactive converter with parallely working fixed capacitor batterie*

$C_A = 0{,}5\left[1{,}15(1{,}05 c_{TR} + c_{BR} + c_{ZC}) + c_{CAM}\right]$

(c_{CZ}, the costs for the accessories of commutation and the dc-capacitors depend on characteristic values of the used thyristors. More details are given in the following)

4.3 Costs, accessories, engineering, conditionally dependent on the power

The following indicated costs are valid in the sequence of the plant sizes 45; 24; 7,5 Mvar. They are given as a multiple of the basic comparison figure in according to pos. 4.1

For a simple excitation unit at the *synchronous condenser Ba* without impulsive excitation there will be: 1577; 1351; 1126.

For a *d.c. premagnetized reactor Cb1* operating under the same conditions, the expenses are within the same dimension.

For a rectifier-impulsive excitation with conventional reactive power control at the *synchronous condenser Bb* the following costs will arise: 8108; 6306; 3604.

The *d.c. premagnetized reactor Cb2* requires for rectifier excitation and conventional reactive power control 15 766; 11 261; 7207. When designed on single-phase operation, 3 rectifier units of each **1/3** power output are needed.

The fixed costs for the control and circuit accessories are independence of the power range:

Inductive reactive converters Ca.
Leakage-reactance transformer plantCc.
Capacitive reactive converter Eb 901; when variant Cc designed on single-phase operation 1802.
Thyristor-switched reactors Cd 3604.
Thyristor-switched capacitor batterries D 7208.

For auxiliary equipment, protection devices and controls at the rotating synchronous condenser B 6757; 5856;4505 will arise.

Furthermore, engineering costs of 3604 have to be added at installation with power electronic control. At the simple plants Ba and Cb1 only 901 have been considered for engineering costs, and 450 have been calculated for standard capacitor plants A.

4.4 Other expenditure that can be evaluated quantitatively

The synchronous condenser requires

a machine foundation, the costs of which for the 45 Mvar model are indicated with at least 1937, and for a 7,5 Mvar model with at least 676.

The remaining set-up conditions can be quantitatively evaluated for an individual case only, and for this reason only a factural comparison can be made.

At the d.c. premagnetized reactor the reactor and the capacitor plants can be erected in the open air, with the exception of the excitation accessories and the control. At installations with semiconductor control elements also the power element of the thyristor has to be excepted. For thyristor connections and accessories of the low-voltage type an erection in the open air will not be economic. Also for the plant with rotating synchronous condenser the calculation of cost has been made under consideration of an erection for indoor location. However, when installating the synchronous condenser in the open air it can be equipped with a cover, the cost of which will come up to 8-12 % of the net machine price.

Owing to the lack of data, differences in the maintenance costs of the individual variants (average value 1 % of the investment amount per year) cold not be considered quantitatively.

5. EVALUATION OF THE LOSSES

5.1 General conception - specific plant losses V_A

With plants of this high power and utilization time the varying height of the power losses of the individual variants are of importance for the total evaluation. Nowadays it es common usage to evaluate the power losses of the individual plant variants with a certain cost factor per Watt-power losses, to add these loss costs to the prime costs and then to judge the individual variants according to the total of prime costs and loss costs. Also in the following comparison of reactive power it is handled in this way.

When calculating the balance of losses it is useful to start from the specific losses of the used elements, as given in table 2.

With the help of the design methods of the plant (pos. 4.2) the specific plant losses V_A can be calculated herefrom. In case the specific losses of the component parts are available divided into no-load and load losses, V_A for every state of modulation control of the plant can be found. However, the different specific plant losses can only then be used for the direct evaluation of the cost, in case the two plant variants to be

TABLE 2 Specific power losses of the elements in W/kvar

power range			45	24	7,5 Mvar	
synchronous condenser	v_{PH}	no load	10	13	16,5	
		load	5	7	8,5	($\sim I^2$)
transformer plant	v_{TR}	no load		1,2		
		load		3,8		($\sim I^2$)
thyristor cabinet	v_{TH}	no load		1,3		
		load		3,7		($\sim I/\sim I^2=1$)
diode cabinet	v_{DI}			4		
reactor	v_{DR}			6,5		
medium voltage capacitor plant	v_{CAM}			1		
low voltage capacitor	c_{CAN}			4		
d.c. premagnetizing choke reactor	v_{SDR}	no load	1,1	1,3	1,4	
		load	7,9	9,2	10,6	($\sim I^2$)
leakage reactance transformer equipment	v_{STR}	no load		0,7		
		load		12		($\sim I^2$)
low voltage- capacitor	v_{CON}			4		
plastic foil capacitor	v_{COK}			2		
medium voltage capacitor	v_{COM}			1		
inverter	v_{BR}	no load		5		
		load		10		

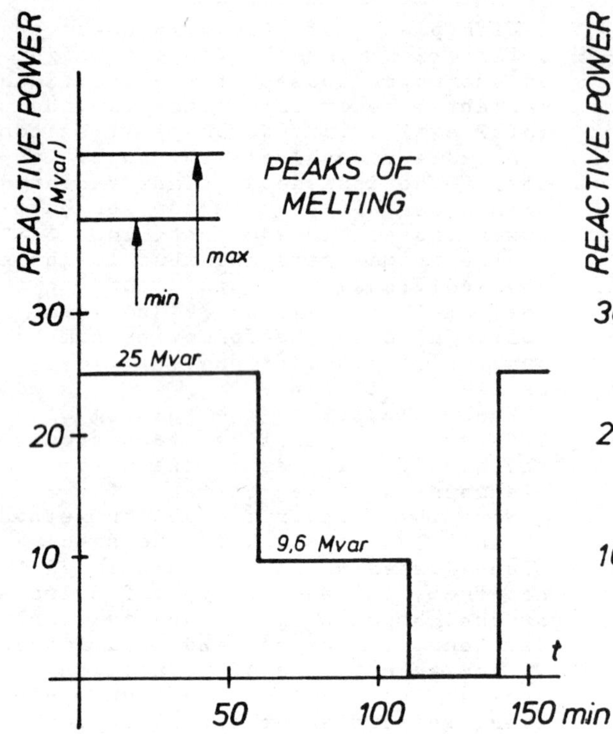

2. at one 60 tons arc furnace

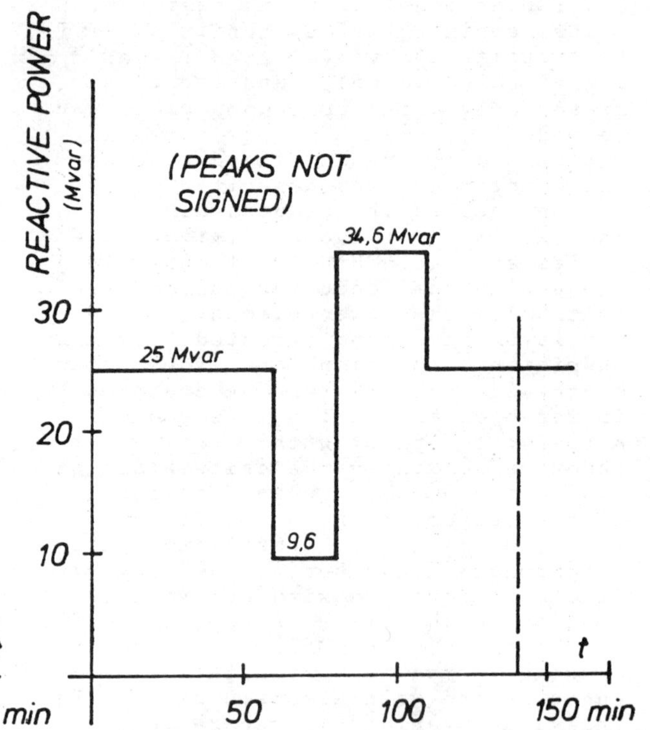

3. overlapping operation of two 60 tons arc furnaces

Fig. 2/3 Diagram of reactive power cycle, corrected for calculation

t_1 = 60 min ; Q = 25 Mvar } basic reactive power
t_2 = 50 min ; Q = 9,6 Mvar } basic reactive power
t_3 = 30 min interval time

Reactive power pulsation when melting
max $\pm$ 1 · 18 Mvar
min $\pm$ 0,7 · 18 Mvar

compared are operating according to the same principle. This does not take place in this case:

So, for instance, synchronous condensers and capacitive reactive converters only then suffer load losses, in case they have to compensate reactive power; no-load losses only occur during the intervals. On the other side, with the variants C the maximum loss take place during the intervals, while the minimum losses occur with utmost reactive power compensation. Furthermore, the variants differ by varying proportions of no-load losses, current-proportional and square current-proportional-dependent losses.

5.2 Programm losses

Comparison of losses for the individual plant variants can only be made on the basis of concrete cycle programmes. For this purpose the following cycles have been checked:

5.2.1 Operation of one 60-tons UHP-arc furnace

Fig. 2 shows the diagram of reactive power, corrected for the calculation:

Reactive power compensating periods alternate with no-load periods or those of minor power output of nearly the same length. Longer intervals have been considered by variation of the time t_3.

The calculation has been done alternatively:

5.2.1.1 for the following of a programme of the quasi stationary reactive power flow.

5.2.1.2 for the compensation of additional reactive power peak when melting.

5.2.2 Optimum overlapping operation of two 60-tons-UHP-arc furnaces

TABLE 3 COST EVALUATION FIGURES OF INSTALLATIONS FOR COMPENSATION OF REACTIVE POWER

VARIANT	SYMBOLS	POWER RANGES (Mvar) 45	24	7,5	45	24	7,5	45	24	7,5	P_s W/kvar
A STEPWISE SWITCHABLE CAPACITOR PLANT		0,94	1,16	1,56	WITHOUT INTEREST			WITHOUT INTEREST			1
SLOW											
Ba SYNCHR. COND. - SIMPLE EXCITATION	(S)	1,99	2,45 (1)	3,78 (1)	3,32 (2)	4,22 (2)	6,0 (2)	4,05 (2)	5,18 (2)	7,21 (2)	13
Cb1 D.C. PREM. REACTOR - SIMPLE EXCIT.		2,07	2,54 (2)	4,38 (2)	2,02 (1)	2,47 (1)	4,32 (1)	1,99 (1)	2,44 (1)	4,28 (1)	11,5
MODERATELY RAPID											
Bb1 {SYNCHR. COND. - } RELATED TO NOMINAL P.	(S)	2,38 (2)	2,97 (2)	4,5 (2)	4,24 (4)	5,27 (4)	7,5 (3)	5,24 (3)	6,79 (4)	9,27 (3)	22,4
Bb2 {RECT. EXCITAT. - } " " IMPULSIVE P.	(S)	1,43 (1)	1,78 (1)	2,7 (1)	2,12 (1)	2,82 (1)	4,09 (1)	2,51 (1)	3,37 (2)	4,83 (1)	13
Cb2 D.C. PREM. REACTOR - RECTIFIER EXCIT.		2,49 (2)	3,08 (2)	5,41 (3)	2,49 (2)	3,08 (2)	5,41 (2)	2,49 (1)	3,08 (1)	5,41 (2)	21,6
DESIGN. ON SINGLE PHASE LOAD	S H	(3,66) (3)	(4,59) (3)	X	(3,66) (3)	(4,59) (3)	X	(3,66) (2)	(4,59) (3)	X	
VERY RAPID											
Ca INDUCTIVE REACTIVE CONVERTER	H	2,43 (1)	2,73 (1)	3,83	2,87 (2)	3,04 (2)	4,0	3,11 (2)	3,19 (2)	4,08	21
Cc LEAKAGE - REACTANCE TRANSF - PLANT	S H	2,63 (2)	2,93 (2)	4,13 (3)	3,41 (3)	3,56 (3)	4,63 (3)	3,82 (3)	3,91 (3)	4,9 (3)	30,4
Cd THYRISTOR - SWITCHED REACTORS	S	3,07 (3)	3,42 (3)	4,76 (4)	4,08 (4)	4,28 (4)	5,48 (4)	4,61 (4)	4,74 (5)	5,87 (5)	30,1
D THYRISTOR - SW. CAPACITOR BATT	S	3,83 (4)	4,04 (4)	5,41 (5)	4,19 (4)	4,26 (4)	5,49 (4)	4,39 (4)	4,42 (4)	5,53 (4)	5
Eb CAPACITIVE REACTIVE CONVERTER	H	2,35 (1)	2,8 (1)	X	2,48 (1)	2,79 (1)	X	2,55 (1)	2,79 (1)	X	12,1
		INSTALLATION COSTS (PRIME COSTS)			INSTALLATION COSTS + LOSS EVALU. 2,08 DM/W			INSTALLATION COSTS + LOSS EVALU. 3,20 DM/W			

(S) (SYMMETRIZES ASYMMETRICAL LOAD)

H (HARMONICS OF REACT. POWER ARE CAUSED)

(SPEZIFIC POWER LOSSES (24 Mvar - RANGE) IN INTERVALS P_s)

(X DATA NON AVAILABLE)

Fig. 3 shows the diagram, corrected for the calculation. The calculation was done alternatively, too.

5.2.2.1 for the following of the programme of the quasi reactive power flow,
5.2.2.2 for the compensation of additional reactive power peak when melting.

5.3 Selection of a representative programme

After detailed discussions with manufacturers and operators the case 5.2.1.2 with shortest possible interval time has been selected for the loss evaluation of rapid compensation equipment.
Motivation:

Plants for continuous compensation of reactive power in electrical networks will only then be procured when impulsive reactive power can be compensated as well. Otherwise the mechanically switched capacitor battery will be preferred with the aim of compensation of a medium reactive power, which is much more favourable in costs.

In electrical steelworks the operation with one furnace and a shortest-possible interval time will make up a large part of the utilization time, even in case at least two furnaces are installed.

To a certain extent the programme is representative for the unfavourable cycle of a hot strip mill, too.

The reactive power programme of a tandem cold strip mill also approaches the representative programme, but with prolongated time t_3.

From the basically different operation method of the variants B, D, and E on the one side, and C on the other it follows for the balance of losses when programmes are altered:

Longer duty cycle e.g. overlapping operation of several UHP arc furnaces results in less favourable values for B, D, and E; more favourable values for C.

Shorter duty cycles e.g. tandem cold strip mill result in more favourable values for B, D, and E; less favourable values for C.

The quantitative values of the reactive power cycle coresponds to the power range 24 Mvar. The programme losses in the 45- and 7,5 Mvar range have been computated by means of the alternative specific plant losses. For slow plants the programme 5211 was taken as a basis for the evaluation of losses.

6. EVALUATION OF ECONOMY

6.1 Finding of the cost evaluation figures

For the total evaluation of the individual variants numerical data are to be found. Such an evaluation figure is composed of:

I) the specific plant costs C_A
II) the share of conditionally power dependent costs for accessories, engineering or special installation conditions C_Z
III) the valuable costs of losses

(To III: The variant Cb "d.c. premagnetized reactor" produces the lowest losses in the 45 Mvar range. For this reason the losses, occuring at the other variants, are evaluated in relation to Cb2 and added to the costs of installation).

When evaluating the loss, e.g. for machines and transformers, it is common practice in Germany to distinguish between no-load and load losses. In this case, however, it is not advisable to act accordingly, owing to the fact that by the varying modes of operation of the individual plant variants the no-load and the load losses in the single operational sections are different. Therefore, for the entire duty cycle the calculation is executed with a constant evaluation factor. In addition, indications are made on the losses for the time of no-load intervals, during which the plant has to be ready for operation, e.g. unforeseen prolongation of the time t_3 in fig. 2.

For the application of the examination results in practice it turned out to be useful to present the cost evaluation figures in dependence on the loss evaluation factor. This is of special importance for export projects where the cost evaluation of the individual plant variants are based on quite different criteria, viz. ranging from the mere comparison of prime cost up to the evaluation of losses in the order of magnitude of 4 DM/W. The result is shown in table 3.

6.2 Summarized presentation of the evaluation results

A classification in lines according to very rapid, medium-rapid and slow equipment acc. to section 3.2 has been made. Further technical features are marked by symbols. In the colums the evaluation numbers of the individual variants are given for the investigated power ranges of 45, 24, and 7,5 Mvar. Cost evaluation figures are given for mere prime costs, as well as an evaluation of the losses in relation to the variant Cb2 of 2.08 and 3.20 DM/W power loss. Numbers in circles ◯ are to easen the sequence of evaluation.

The following is remarkable:
At the *VERY RAPID PLANTS* the favourable evaluation of the reactive converters is clearly shown. The costs of the reactive converter with paral-

lelly working capacitor battery Eb are nearly the same as that of the moderately rapid variant "d.c. premagnetized reactor Cb2". That only applies, when frequency-thyristors are used. If line thyristors are used the costs for the accessories of commutation C_{ZC} rise considerably.

Under the plants with the feature to symmetrize phase-asymmetrical loads the leakage-transformer plant Cc is a most favourable. Only when large power loss evaluation factors are taken the moderately rapid variant "d.c. premagnetized reactor Cb2" is on the same level.

Of all *MODERATELY RAPID PLANTS* the rectifier-excited synchronous condenser principally represents the most favourable solution as to the costs, in all power ranges and independent of the loss evaluation, in case it can be rated according to its special characteristic of producing an impulsive reactive power Bb2.

If for (plant-)technical reasons, the synchronous condenser has to be dimensioned for the same nominal power as the static installations, it remains of interest only with regard to the prime cost Bb1.

In *SLOW PLANTS*, where only the mere prime costs are taken into consideration, synchronous condenser plants are the most favourable solution as to the cost. Even with a loss evaluation of 2.08 DM/W the simple d.c. premagnetized reactor Cb1 proofs to be the most favourable solution as to the costs.

7. REFERENCES

(1) L. ABRAHAM und M. HÄUSLER: Blindstromkompensation über Halbleiterschalter oder Umrichter. VDE-Fachtagung Elektronik 1969 II/9, S. 100-114.

(2) E.H. DÜLL und E. GOLDE: Dreiphasenumrichter mit Gleichstrom-Zwischenkreis 50/50-500 Hz. AEG-Mitteilungen 54, S. 165 (1964).

(3) A. CHIT und W. HORN: Neuzeitliche ruhende Blindleistungskompensation für Industrienetze. Techn. Mitt. AEG-TELEFUNKEN 66, S. 266 (1976).

(4) R. GÄRTNER und H. ZANDER: Die Bemessung der Erregerdeckenspannung einer Synchronmaschine für Stossblindleistung. Elektrotechn. Z.-A 87, S. 551 (1966).

(5) H. BECKER: Beherrschung von Blindlastgrössen in Verteilernetzen durch statische Kompensationseinrichtungen. VDE-Fachber. 27, S. 16 (1972).

(6) A. CHIT und W. HORN: Stabilitätsbetrachtung an Synchron-Blindleistungsmaschinen mit vorgeschalteter Netzreaktanz und parallel liegendem Verbraucher veränderlicher Last. Wiss. Ber. AEG-TELEFUNKEN 48, S. 118 (1975).

(7) G. BREITENBERGER: Verzögerungsarmer Wechselspannungs-Gleichspannungsumsetzer zur Darstellung von Effektivwert, Wirk- und Blindleistung. Techn. Mitt. AEG-TELEFUNKEN 61, S. 327 (1971).

HIGH-SPEED VAr CONTROL USING STATIC CONVERTERS IN SHORT-CIRCUIT

E. Schmid

Siemens AG, Systemtechnische Entwicklung, Erlangen, BRD

Abstract

A new circuit is described which enables the compensation of rapid changes in the reactive loading of a supply system or of disturbances on the system. The circuit has a very fast response time and involves the use of two controlled rectifier bridges, one of which is force-commutated. The circuit can accept leading currents from the supply, allowing a reduction in size of the capacitor banks normally required. Some design considerations are given and results obtained from a 30 kVA experimental circuit are discussed.

Introduction

Rapid changes in the reactive power and unbalance of loads such as large convertor drives or arc furnaces, cause voltage variations when operated with a power supply system of relatively low short-circuit power. To reduce these voltage variations, load changes must be compensated for as fast as possible. In the spectrum of random load-parameter variations, frequencies around 10 Hz are most troublesome. This is because the human eye is most sensitive towards light-intensity variations (flicker) at this frequency (Ref. 1). Many solutions to this problem exist, some involving rotating machinery, but static compensation is normally preferred. Static VAr control is possible with reactor compensators (Ref. 2) or with circuits using thyristor-switched capacitors (Ref. 3). An alternative approach is to use power electronics. Figure 1 shows some of the arrangements which have been suggested or successfully used (Refs. 4, 5). Circuit a) employs a line-commutated rectifier having a reactor connected across its d.c. terminals. A lagging 3-phase current is drawn from the supply system, the magnitude of which is dependent on the delay angle of the rectifier bridge. As virtually no active power is taken from the supply, the delay angle is always very close to + 90° so that the rectifier always draws a purely inductive current. The maximum dead-time (3.3 ms for a 6-pulse circuit) determines the response time of the system. The rectifier is

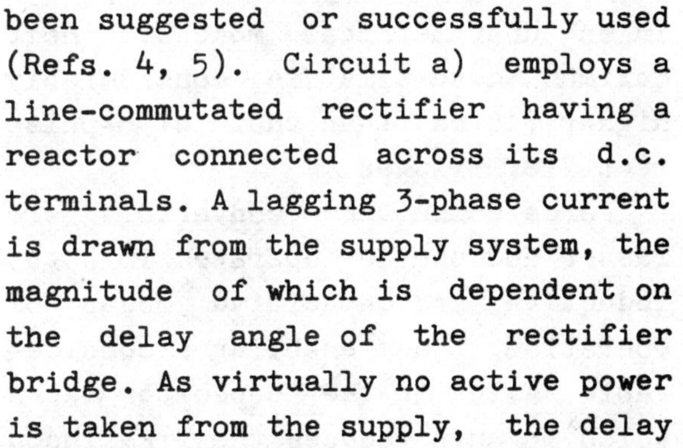

Fig. 1. Some methods of static VAr control.

operated with a fast current-control circuit which ensures that the sum of the reactive powers, i.e. that of the load and that of the rectifier is constant. The resultant lagging current is then compensated for by a fixed capacitor bank which is arranged so as to serve additionally as a filter for the current harmonics produced by the load and the rectifier, see Fig. 1b.

Since a rectifier bridge always links two phases of the supply, it is only able to provide symmetrical reactive power. More versatile in this respect and also more simple are reactors, the current of which is adjusted by thyristor controllers, see Fig. 1c. Leakage transformers having a short-circuit ratio near unity are often used instead of the reactors. Such circuits can operate in an unsymmetrical mode but their maximum dead-time is considerably higher (10 ms) than that of 6-pulse rectifier bridges.

Force-commutated convertors are faster and can be operated in both, inductive and capacitive modes of operation, thus enabling a considerable saving in the capacitor banks which in turn reduces the likelihood of supply oscillations at some critical operating conditions. Invertors of the voltage-source type (Fig. 1d) for instance, can change over smoothly from inductive to capacitive operation, and vice versa , by simple control of the output voltage of the invertor. High frequency modulation of the invertor results in a low harmonic content of the supply current. A rectifier bridge with force-commutated arms feeding a reactor load, see Fig. 1e,is capable of operation in both, the inductive mode (see Fig. 1a) and the capacitive mode. In the latter, the rectifier draws a leading current from the supply. Before changing over from one mode to the other, the d.c. current must be made zero, i.e. the energy stored in the smoothing reactor must be fed back into the supply system. The transition to and from inductive and capacitive operation is thus barely possible in the desired short time of less than a period of the supply voltage. In addition the smoothing reactor must be large. As the circuit operates at zero phase-control factor, the reactor must insure continuous current at, say 1/10 of the rated value, if a 1:10 range of available reactive power is required.

These problems can be avoided by the application of the circuit in Fig. 2 which combines the advantages of the voltage-source invertor (smooth change-over, capacitive currents) with those of the "short-circuited" rectifier bridge of Fig. 1a. This new type of circuit will be considered in more detail.

Only the reactive power of balanced loads can be compensated for by this method. For unsymmetrical loads, additional problems occur with the supply system, which can be overcome by simultaneous balancing. This is possible by impressing a three-phase current system with a negative phase-sequence into the supply lines, depending on the degree of unbalance of the load. Different solutions to this problem are under consideration.

Principle of operation

The circuit in Fig. 2 consists of two controlled rectifiers (1) and (2). Rectifier (1) is line-commutated and operates at positive delay angles, while rectifier (2) uses so-

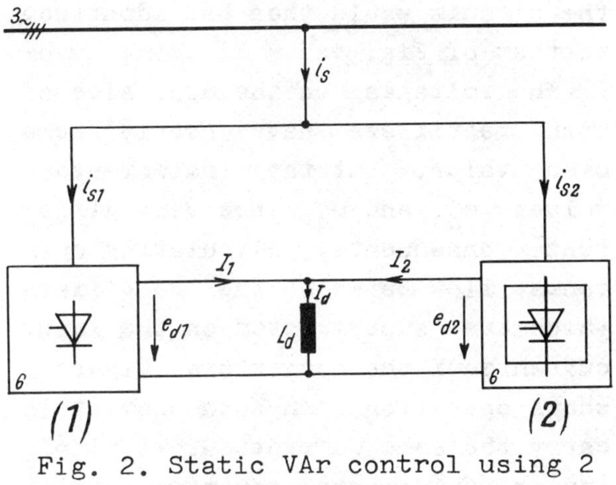

Fig. 2. Static VAr control using 2 rectifiers one of which is force-commutated.

me kind of force-commutation and is therefore able to operate at negative delay angles so that it can commutate before the instant of natural commutation. It should be pointed out in this context that rectifier (2) is not truly force-commutated, as for leading delay angles, commutation takes place not only under the influence of the capacitor voltages (characteristic of force-commutation), but also under the influence of the supply voltages (characteristic of line-commutation). This type of commutation has been discussed in detail (Ref. 6).

The two convertors (1) and (2) are connected in parallel and loaded by a smoothing reactor L_d which is common to both rectifiers and carries the current I_d. Current I_d is made up of two components I_1 and I_2, where I_1 is the d.c. current of the line-commutated rectifier (1) and I_2 is the current due to rectifier (2):

$$I_1 + I_2 = I_d = \text{const.}, \qquad (1)$$

The reactor current I_d is maintained at a constant value, the respective contributions of each rectifier bridge towards I_d being varied by adjusting the delay angles of the bridges. In this way the magnitude of the combined input current $i_s = i_{s1} + i_{s2}$ of both rectifiers is determined, and is proportional to the difference in the magnitudes of both d.c. current contributions. If both contributions are identical, only a small current i_s flows corresponding to the total loss as the currents i_{s1} and i_{s2} cancel. Neglecting this active current, rectifier (1) always draws a purely inductive current from the supply while rectifier (2) takes a purely capacitive current. Therefore, the phase displacement between the supply voltage and the combined input current i_s can be set to a leading or lagging angle (-90° or $+90^o$) depending on which rectifier carries the larger current. The change-over from one mode of operation to the other is smooth as long as I_d is kept constant. A further advantage of the circuit is that the reactor need not be "charged" or "discharged" whenever changes in the delay angles occur. The transient behaviour of the system is therefore no longer influenced by L_d.

The maximum attainable inductive or capacitive power is determined by the magnitude of I_d, and is obtained if the appropriate rectifier operates at minimum or zero current. It should be noted that the reactor in no way influences the maximum attainable reactive power. Its sole purpose is to determine the mean value of I_d for which the current waveform becomes discontinuous. In theory the reactor must be just large enough to prevent the currents I_1 and I_2 from becoming zero, but in practice it is chosen to be somewhat larger in order to reduce the current harmonics in the supply current (Ref. 7).A suitable size is that for which the d.c. current becomes discontinuous at approx. $0.5 \cdot I_{d\ max}$.

The ratio of the rated power of convertor (1) to that of convertor (2) need not necessarily be 1:1. It is determined by the power factor of the load to be compensated, and by economic considerations relating the size of the capacitor bank to the rating of rectifier (2). The larger the capacitor bank, the smaller is the required rated power of rectifier (2). Large fixed capacitors may cause problems with oscillations in and stability of the supply system. However, the rated power of rectifier (2) should not be made zero, as

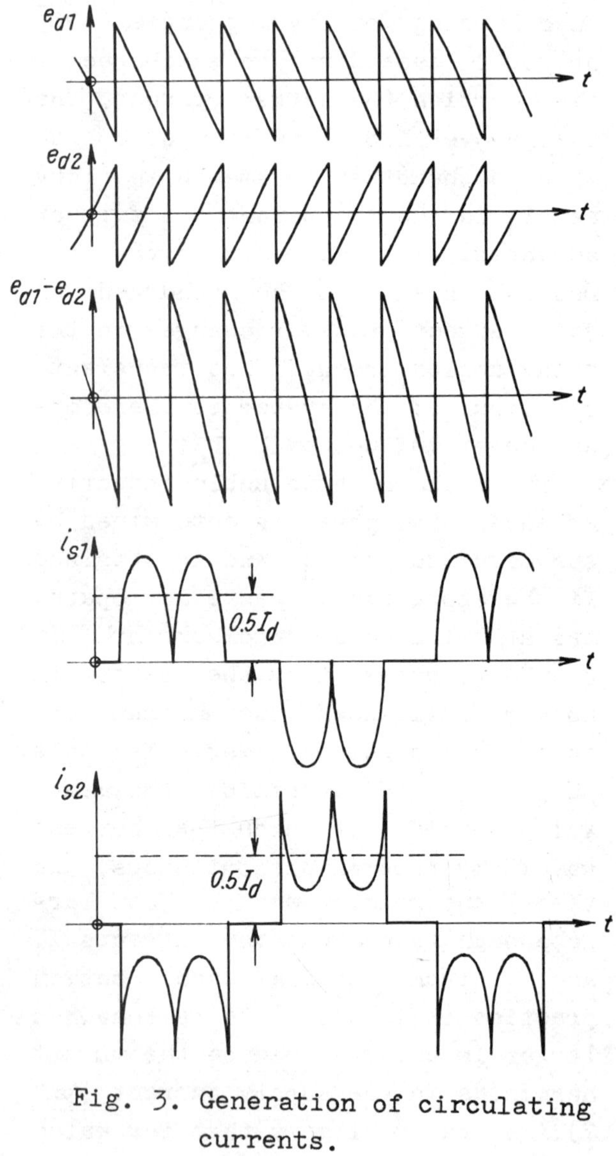

Fig. 3. Generation of circulating currents.

the circuit would then be identical to that of Fig. 1a.

The voltages on the d.c. side of both rectifiers have nearly zero mean value, but the instantaneous values e_{d1} and e_{d2} are very different. Consequently, circulating currents flow between the rectifiers which are superimposed on the input currents of the convertors. Figure 3 shows operation when both convertors carry the same current $0.5 \cdot I_d = I_1 = I_2$ and no fundamental reactive current flows in the supply lines. The circulating currents can cause the direct currents I_1 and I_2 to be discontinuous, with a result that full reactive power output can no longer be attained. As the difference of d.c. voltages $(e_{d1} - e_{d2})$ is responsible for these currents, they can only be limited to reasonable values by inductances inserted either in the d.c. path or in the a.c. paths. The latter method is more favourable since an increased stray inductance of the line transformer can be used for this purpose. It is not important where the inductances are situated in the circuit, and whether and to what extent they differ for each rectifier. However, it is useful to have the series inductances at the input of convertor (2) as small as possible in order to keep the energy of the commutation capacitors, and also the maximum voltage in rectifier (2) within reasonable limits. The reactances on the a.c. side of rectifier (1) can have any size. As rectifier (1) operates with nearly zero phase-control factor, the voltage reduction due to the a.c. side inductances is of no significance. It can be shown that reactors of the size $L_k = 0.5 \cdot L_d$ in the a.c. lines of rectifier (1) are an economical compromise. The problems of discontinuous and circulating currents are obviously less significant if 12-pulse circuits are used.

Circuit details

Figure 4 shows the circuit of Fig. 2 in more detail. For the force-commutated rectifier, a three-phase thyristor bridge of the current-source type, with commutation capacitors and series diodes, is most suitable (Refs. 8 and 9). This results in an economical circuit of high operational safety.

The experimental circuit was designed for about 30 kVA. No filters or capacitor banks were used at this stage. The most important data are: I_d=100 A, U_{s1}=U_{s2}=123 V, L_d=10 mH, L_k=2 mH. Inductance L_d does not correspond to the value mentioned previously, and therefore relatively high circulating currents are obtained.

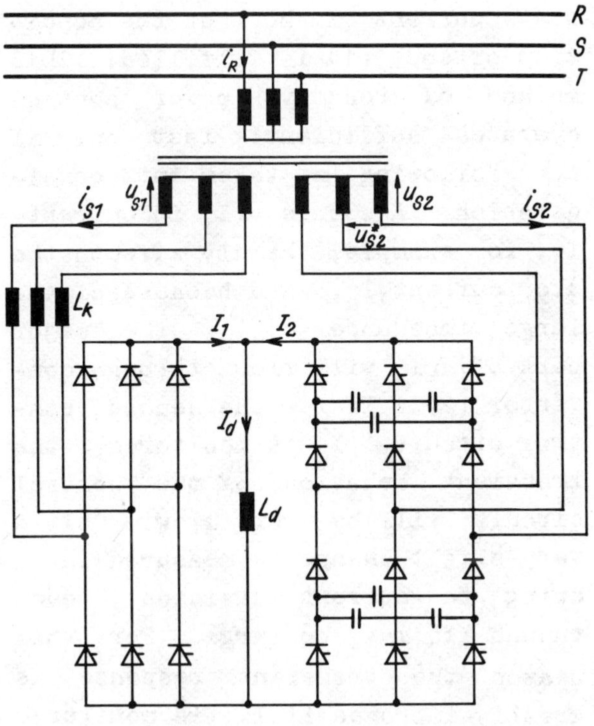

Fig. 4. Practical design of the circuit in Fig. 2.

Results from the experimental circuit

Simulation. The performance of the circuit was investigated with the help of digital simulation (Ref.10). The current and voltage waveforms and transient behaviour of rectifier (1) and (2) were investigated. The results coincided well with those abtained from the experimental circuit, some of which are shown in Figs. 5, 6 and 7.

Limited range of control. To insure a cyclic commutation in convertor (2), the current I_2 can not be decreased below a minimum value $I_{2\,min}$ determined by the size of the commutation capacitors, the turn-off time of the thyristors and the supply voltage U_{s2}. This limits the range of available capacitive VAr. A ratio of

$$I_{2\,min}/I_{2\,max} = I_{2\,min}/I_d = 0.1 \quad (2)$$

can be attained without difficulties. If for a given current I_d the range of control must be increased, then the size of the commutation capacitors can be reduced, as safe commutation is guaranteed for smaller currents i_{s2}. However, this measure increases the peak voltages within the circuit since the capacitors can

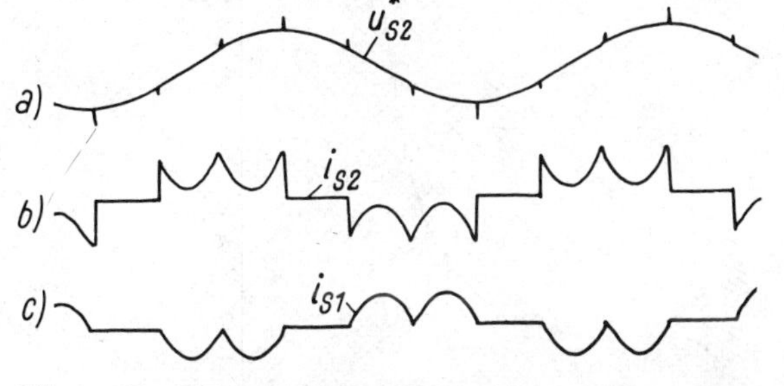

Fig. 5. Simulation results for near max. capacitive operation. a) voltage u^*_{S2} b) current i_{S2} and c) i_{S1}.

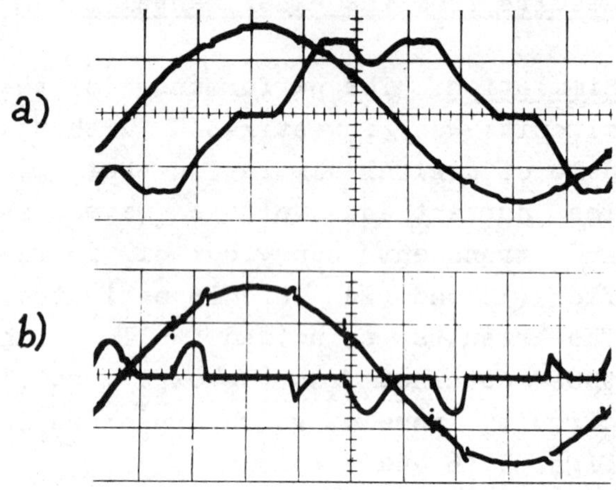

Fig. 6. Currents and voltages at max. inductive operation. a) i_{S1} and u_{S1} b) i_{S2} and u_{S2}. 100 V/div, 80 A/div, 2ms/div.

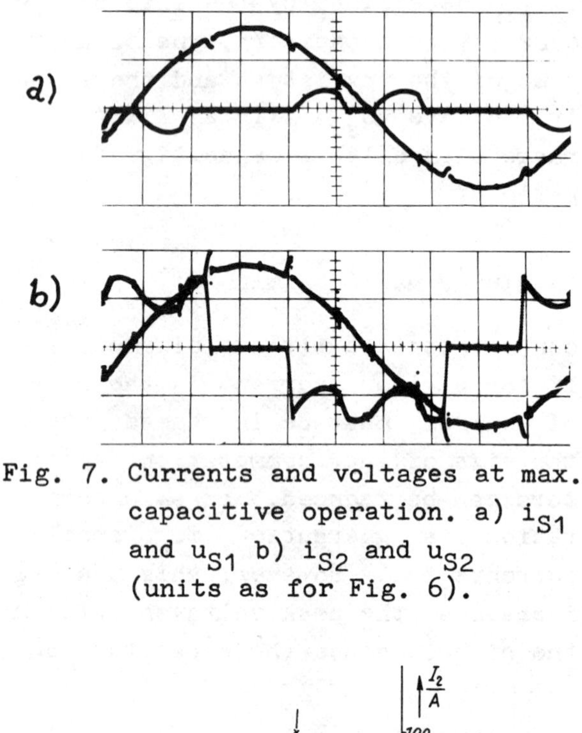

Fig. 7. Currents and voltages at max. capacitive operation. a) i_{S1} and u_{S1} b) i_{S2} and u_{S2} (units as for Fig. 6).

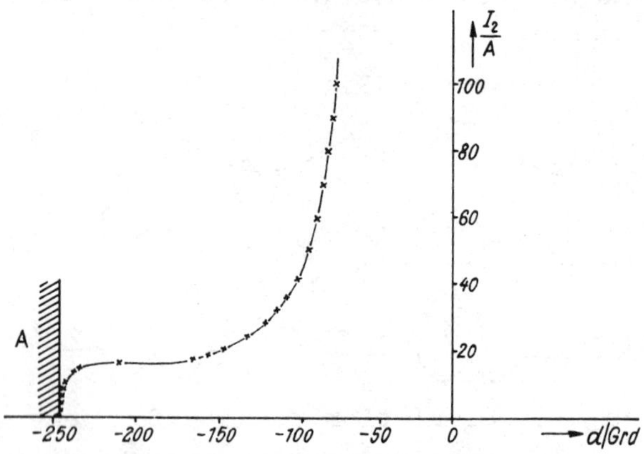

Fig. 8. Control characteristic of rectifier (2), A: range of unstable operation.

be charged up to higher voltages. Another method of extending the range of available reactive power would be to turn to "multiple commutation". This involves a reduction of i_{s2} by a further increase of the delay angle, so that the interval of the commutation extends beyond 120°. More than two bridge arms are then involved in the commutation. This very complex process results in a quasi non-linear control characteristic (Fig. 8). Zero a.c. current i_{s2} can be attained, but due to the non-linear behaviour, the control circuit of convertor (2) is very problematic.

For maximum capacitive supply current i_s, the current I_1 can be made zero as no capacitors are involved in the commutation.

<u>Control</u>. Control of the reactive power "output" of the circuit is achieved by controlling the current I_1 in accordance with the VAr demand. A second control circuit then adjusts current I_2 so that the condition of equ. (1) is fulfilled. This method of reactive power control operates sufficiently fast only if the following is taken into consideration. A change ΔI_1 in current I_1, for example, hardly affects the d.c. current I_d, and because of the large impedance of L_d, the major part of ΔI_1 will flow through convertor (2). If for the second control circuit, I_d is monitored, the transient behaviour of the control circuit will be very slow. Only a very slight change is measured in L_d owing to current division , even though ΔI_1 may be large. For this reason the transient response is greatly improved if I_2 is monitored and the desired value I_2^*, for I_2, is computed by a simple circuit using

the equation $I_2^* = I_d - I_1^*$, where I_1^* is the desired value for I_1 and is given by the VAr control circuit.

It is advantageous to use some kind of adaptive control to allow for the operation at discontinuous rectifier currents. As the gain of the control loop varies considerably when changing from continuous to discontinuous currents, the controller parameters must be varied. These can be changed in accordance with the magnitude of the d.c. currents in each bridge.

Dynamic performance. The dynamic performance of the complete circuit is determined by two factors:

a) the dead-time of up to 3.3 ms for 6-pulse circuits which enables a change in the delay angle only every 60°, and
b) the influence of the reactors at the a.c. side of rectifier (1).

The dead-time cannot be avoided. The size L_k of these reactors does not greatly affect the dynamic performances. They give rise to an inductive d.c. voltage drop, the reduction in voltage being proportional to the current I_1. Since rectifier (1) normally operates at very low phase-control factors, the value of the inductive direct voltage drop is not significant. More important is the time interval needed to supply the voltage-time integral necessary to enable a fast rise (ΔI_1) in the current I_1 of the reactances L_k. As the phase-control factor is very low under stationary conditions, this time is dependent on the magnitude of the transformer secondary voltage U_{s1} and on the required change in the delay angle. Neglecting the dead time at this point, the delay angle can be varied almost instantaneously. By a superimposed anticipatory control, the required voltage-time integral can be supplied even more quickly, practically from one firing pulse to the next.

Figure 9 shows the performance of the experimental circuit (Fig. 4) without anticipatory control. Both d.c. currents, I_1 and I_2, and the supply current I_R are presented for changes from 100 % inductive current to 100 % capacitive current at a modulation frequency of approx. 7 Hz.

Current harmonics. Figure 3 shows that the circulating currents are determined by the difference ($e_{d1} - e_{d2}$). The waveform of the supply currents, i_{s1} and i_{s2}, consists of

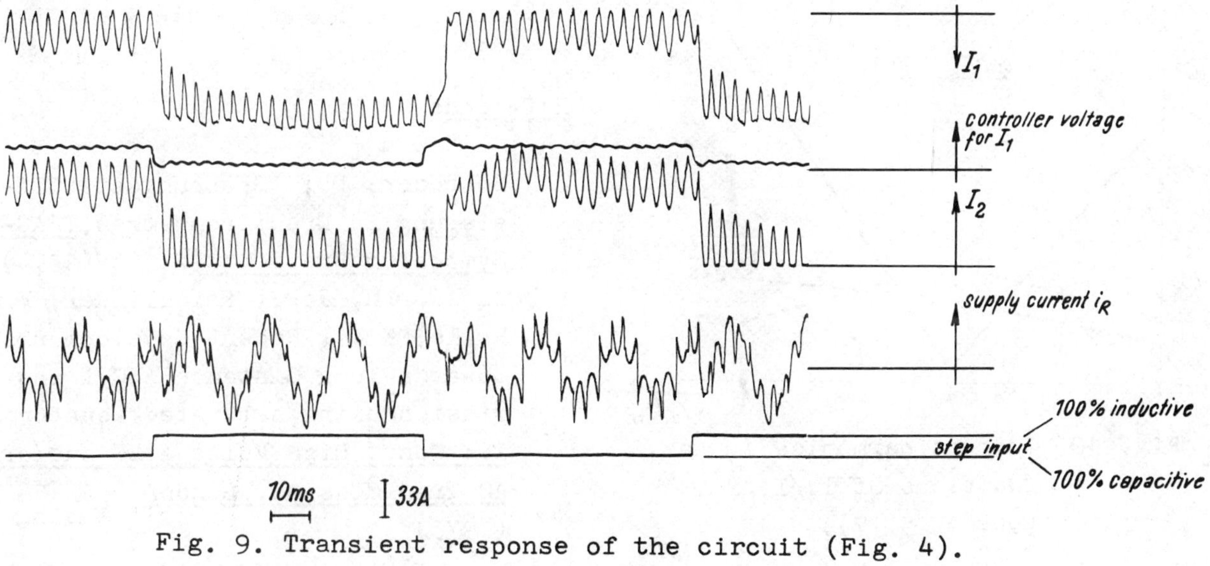

Fig. 9. Transient response of the circuit (Fig. 4).

120° blocks responsible for the current fundamental, and a superposed circulating current. Dependent on L_d and the delay angle of the individual rectifiers, the harmonics components of the circulating currents and the 120° blocks can be added linearly for each rectifier. In this way, the harmonic content of the total supply current i_R can be determined. Figure 10 is a sketch illustrating the most important components (ν = 1, 5, 7) of i_R as functions of the factor I_1/I_{1D}, where I_{1D} is the value for which I_1 becomes discontinuous. In the experimental circuit, $I_{1D} = 0.5 \cdot I_d$. Only three points were determined for each curve. Figure 10a is based on calculated results, and Fig. 10b on measured results for $I_{1D} \approx 0.5\ I_d$. The reactors L_k not only limit the circulating currents, but also account for "softer" commutation in rectifier (1), reducing the current harmonics. This effect was neglected for the calculation of values shown in Fig. 10a.

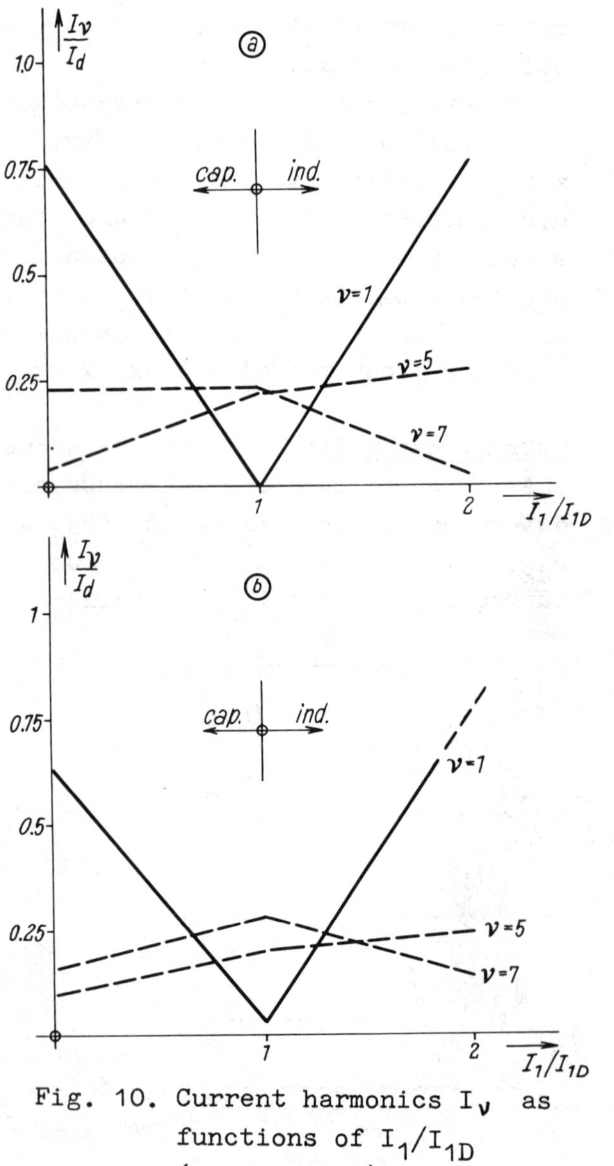

Fig. 10. Current harmonics I_ν as functions of I_1/I_{1D} (ν = 1, 5, 7).

Conclusions

The circuit of Fig. 4 has an excelent dynamic performance and seems well suited for compensation of balanced inductive loads. Although the commutation in rectifier (2) involves additional complexity, the problems associated with it need not necessarily be larger than those associated with line commutation. Line commutated rectifiers present problems particularly at high supply voltages and large units. The advantage of high-speed control together with a reduction in the capacitor banks may well be worth the additional expense of a "force"-commutated rectifier.

References

(1) Schröder, D.; Spannungsstabilisierung in Drehstromnetzen, IFAC-Symposium Preprints 2, 633(1974)

(2) Ainsworth, J.D.; Friedlander, E.; Ralls, K.J.; Recent developments towards long distance A.C.transmission using saturated reactors, IEE Conf. High Voltage DC and/or AC Transmission, London, 242 (1973)

(3) Frank, H.; Landström, B.;Faskompensering med tyristorkopplade kondensatorer, ASEA TIDNING 63, 56 (1971)

(4) Heumann, K.; Knuth, D.; Stromrichter als Energieumformer, Techn.Rundschau Nr. 37 (10.9.74)

(5) Jäger, S.; Beispiel von Anlagen zur Blindstromkompensation, Symmetrierung und Spannungsunterstützung industrieller Elektrowärmeanlagen, elektrowärme internat. 32, 326 (1974)

(6) Möltgen, G.; Grundlagen einer Theorie des Stromrichters in Drehstrom - Brückenschaltung mit mehrstufiger LC-Kommutierung, Siemens-Z. 43, 680 (1969)

(7) Möltgen, G.; (1974) Netzgeführte Stromrichter mit Thyristoren, Siemens AG, Berlin/München

(8) Kazumo, H.; Commutation of a three-phase thyristor bridge with commutation capacitors and series diodes, Electr. Eng. in Japan 90, 91 (1970)

(9) Schröder, D.; Selbstgeführter Stromrichter mit Phasenfolgelöschung und eingeprägtem Strom, ETZ-A 96, 520 (1975)

(10) Arremann, H.; (1976) Ein Verfahren zur digitalen Simulation von Stromrichterschaltungen mittels blockweiser Programmierung, Diss. TH Darmstadt

CONTROLLABLE STATIC REACTIVE-POWER COMPENSATORS IN ELECTRICAL SUPPLY SYSTEM

H. Achenbach, W. Hanke and W. Hochstetter

Siemens Aktiengesellschaft, 8520 Erlangen, Germany

ABSTRACT

Long-distance transmission of electrical power through extra-high voltage overhead lines and power distribution through cable system call for controllable reactive-power compensating circuits on an increasing scale. The transmission losses are thereby reduced, voltage deviations kept within permissible limits and the system stability is improved. The achievements in the field of thyristor techniques made it possible to build steplessly controllable compensating reactors which are combined with a three-phase thyristor power controller. Selection of optimum voltages, current and reactor reactance of the static compensator is dealt with and the current and voltage ratings of the thyristors are explained. The problem of harmonics, the influence of unbalanced conditions, and the protective and monitoring arrangements are also referred to.

INTRODUCTION

Many problems associated with environmental protection preclude the possibility of installing the generating plant required to cover the increasing energy demand in conurbated areas, and the electrical energy must therefore be transmitted over large distances in many cases. One possibility is to utilize hydropower, which is abundantly available in some countries, but requires transmission of the power generated through long extra-high-voltage overhead lines. In densely populated areas, a cable system have to be installed on an increasing scale for transporting and distributing the power. In both cases, transmission losses and a strongly varying reactive-power demand result.

As in power-station construction, and improvement of the efficiency is an important requirement for the operation of transmission systems.

Since the losses P_v of a high-voltage line are proportional to the square of the transmitted power P and inversely proportional to the square of the transmission voltage U, the losses can be expressed as

$$P_v = k \frac{P^2 \cdot l}{U^2 \cos^2 \varphi} \qquad [1]$$

where

k Proportionality factor
l Line length
cos φ Power factor of the transmitted power P

With increasing transmission power P, a correspondingly higher value of transmission voltage U is selected to keep the losses P_v in an economically justifiable relationship to the power P. The proportional increase of the losses with increasing line length l is also allowed for in the selection of a suitable transmission voltage. Furthermore, equation [1] indicates that the line losses P_v are inversely proportional to the square of the power factor of the transmitted power. Reduction of the reactive power is therefore the main prerequisite for obtaining efficient operation of the high-voltage transmission system with minimum losses and voltage fluctuation.

Owing to the operating capacitance and the inductance of the line, the reactive-power balance of the transmission system depends on the loading conditions, i.e. the amount and power factor of the transmitted power. When the high-voltage transmission system is operated at the surge-impedance loading, the capacitive charging power is just compensated by the magnetizing reactive power. When the line is at no load, the full charging power occurs and would result in considerable voltage increases if no compensating means were employed. Therefore, the charging power during no-load operation of the transmission line has to be fully compensated and the compensating equipment should be controlled automatically. It would then be possible to operate the line at any partial load between no load and rated load with the voltage not exceeding the permissible limits.

Compensating reactors installed at the line ends and in intermediate substations can compensate the charging power to up to about 70 %. The remaining 30 % compensating power required at no load would have to be provided by underexcited operation of the generators supplying the line. When the load increases, the generator excitation is increased to overexcited operation while the reactor units are gradually cut out. Another possibility is to switch in capacitor banks step by step, as a function of the voltage increase, the capacitors could be installed in the load centres and in the intermediate substations. Continuous control of the compensating power to improve the voltage and system stability has so far been effected by means of synchronous compensators installed in the intermediate substations and/or at the line ends. An advantage of the synchronous compensators is that they can be controlled steplessly within a reactive-power range corresponding to the range between overexcited operation (synchronous machine acts as a capacitor) and underexcited operation (synchronous machine acts as a reactor).

Like the generators supplying the system, the compensators are thus capable of carrying out dynamic reactive-power compensation, i.e. to make available the required reactive power within a short time in the case of sudden load variations or in the event of faults, thus keeping the voltage within permissilbe limits.

Since the charging power of a high-voltage line can be compensated by reactors, it appears to be a logical step to assign the automatic control function to the reactor circuits. Today, a three-phase thyristor power controller is available which permits this idea to be put into effect. One of the circuit arrangements presently known (1 to 7) is described in the following by reference to Fig. 1.

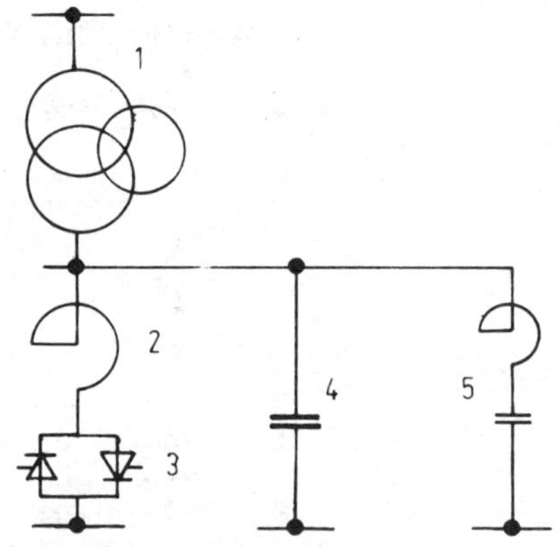

1 Transformer
2 Linear reactor
3 Thyristor controller
4 Capacitor banks
5 Filters

Fig. 1 Controllable static reactive-power compensator, basic circuit

SELECTION OF OPTIMUM VOLTAGES AND CURRENTS

The capacity can be increased by series and/or parallel connection of several units. With parallel connection, each individual thyristor should have a series-connected fuse. In the case of the three-phase power controller shown in Fig. 1, failure of one thyristor causes a current which corresponds to that at zero-delay firing angle and is not fully transferred to the faulted circuit before the control setting of the thyristors in the parallel circuit of the same and opposite polarity is automatically reduced. Selective operation of the fuses is not possible before the number of parallel circuits or the ratio between the overload current and the continuous operating current is adequately high. Uniform distribution of the current during normal operating conditions over all parallel-connected thyristors is obtained through careful selection of the components and a suitable circuit arrangement.

Fuses are not employed if the thyristors are connected in series. Here, the required redundancy, i.e. the means required to avoid service interruptions on the failure of one thyristor, is obtained by connecting additional thyristors in series. If one thyristor fails, the blocking function is assumed by the series-connected thyristors. Interruptions have not been experienced, but it may occur that the firing circuits of individual thyristors in the series circuit will be blocked. In this case, an electronic auxiliary circuit comes into operation through which firing of the thyristors is initiated without endangering these thyristors. RC circuits ensure uniform distribution of the voltage over the series-connected thyristors.

For a given thyristor type, a capacity increase results in a higher voltage in the case of the series connection and in a current increase with parallel connection. This has, of course, effects on the equipment connected in the external circuits of the thyristor controller, such as the reactor, switchgear and transformer secondary winding. A three-phase arrangement without parallel circuit using the highest-power thyristors presently available would involve currents of the order of 1000 A, and very high currents would result if the power were increased by parallel connection. This is undesirable, also in view of the connecting leads. With the series arrangement, isolation of the firing circuits from the power circuit presents problems. Today, this problem is solved by the use of signal transmission through light-optical conductors. The energy for the firing pulses has to be drawn from the voltage across the thyristor through a capacitor store.

Depending on the capacity, unusually high voltage values, e.g. between 30 kV and 110 kV, result with the series connection. This is of no importance in the case of the system transformers, since selection of a suitable secondary voltage has only a slight effect on the costs. In the case of the capacitor banks, which are made up of small units arranged in series and parallel circuits, only the insulators between the capacitors involve a slight increase of the costs. Switchgear is available for certain insulation ratings and the slight increase in costs for the equipment for a higher insulation rating is compensated by the advantage of lower currents and smaller conductor cross sections, particularly if isolating switches are employed.

A reduction of the overall expense may, under circumstances, be obtained through a combination of series and parallel circuits. If, for instance, two thyristors are connected direct in parallel, a saving may be obtained from the less complicated auxiliary and firing circuits.

SELECTION OF REACTOR REACTANCE

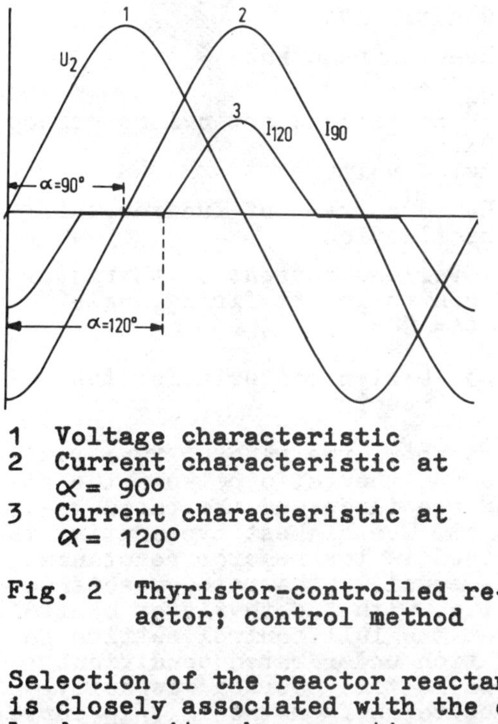

1 Voltage characteristic
2 Current characteristic at $\alpha = 90^{\circ}$
3 Current characteristic at $\alpha = 120^{\circ}$

Fig. 2 Thyristor-controlled reactor; control method

Selection of the reactor reactance is closely associated with the overload capacity when reactive power is required, and with the firing angle selected for the rated current. The reactor itself with its copper winding can be designed for a high overload capacity without appreciable additional expense. The type rating is in this case reduced, since the reactor is designed for $I^2 . X$ in continuous duty. For economical reasons, it is therefore a logical step to operate the equipment at firing angles $\alpha > 90^{\circ}$. A disadvantage of this method is that higher harmonic currents result. The control method for the thyristor-controlled reactor is illustrated in Fig. 2.

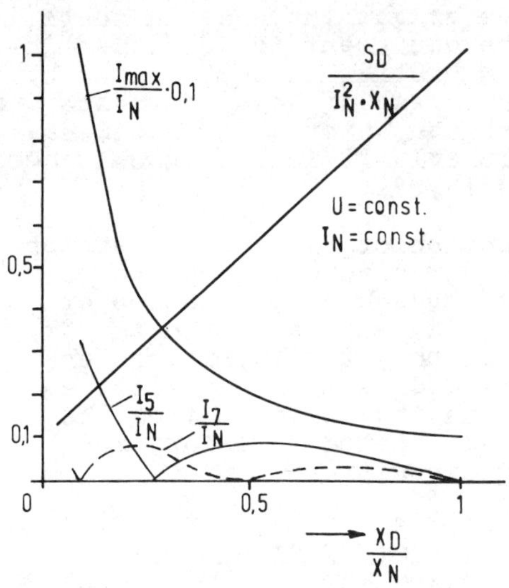

S_D Reactor kVA

X_D Reactor reactance

X_N $\frac{U_N}{I_N}$ fictitious nominal reactance

U_N Rated voltage

I_N Rated current of fundamental oscillation

I_{max} Overload current of thyristor controller at firing angle $\alpha = 90°$

Fig. 3 Design criteria for the reactor

Fig. 3 shows the reactor type rating S_D versus the ratio between the reactor reactance and the rated reactance. The highest type rating is obtained if the reactor reactance X_D is equal to the rated reactance X_N, i.e. with the thyristor controller at the full control setting in operation under rated conditions. The short-time overload capacity can be taken from the I_{max}/I_N curve; also shown are the currents due to the 5th and 7th harmonic. The harmonic currents transferred to the system can be reduced by filter circuits or other means.

CURRENT RATING OF THE THYRISTORS

The main factor determining the current rating of the thyristors is the temperature at the thyristor junction. Owing to the limited thermal capacity of the thyristors, short-time current overloads have to be allowed for in addition. There are various approximation methods for calculating the junction temperature. If high powers are involved, an accurate method, e.g. that described in (8) is employed for the calculation. In addition to the exact thyristor data, transient conditions, e.g. those occurring when the controllers are connected to the system, must be known. The basis is the continuous operating current, which determines the initial temperature for the following cases of overload, but certain overload conditions have to be allowed for as well. It may be assumed that the X_D/X_N ratio in operation under rated conditions will be decisive factor determining the losses.

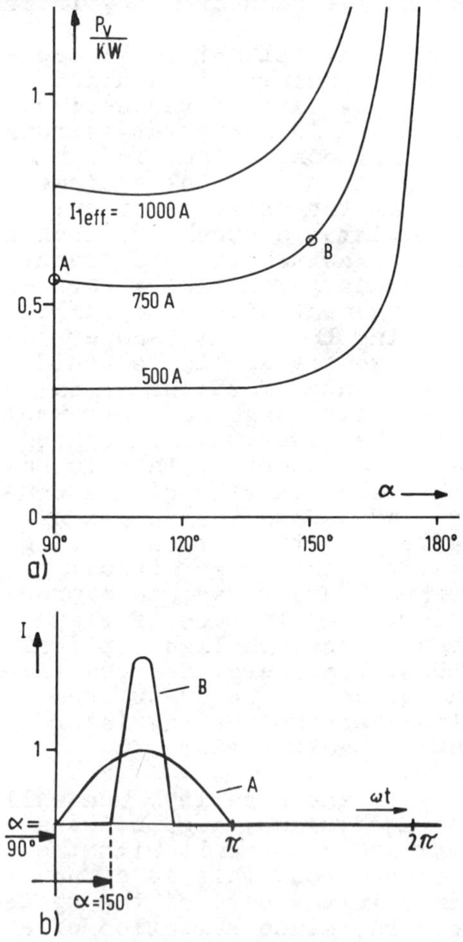

P_V Average losses during one cycle

I_{1eff} R.M.S. value of fundamental component current

I Instantaneous value of thyristor current

Fig. 4 Losses in thyristor
a) as a function of firing angle α
b) Comparative current curves for points A and B at the same fundamental-component current

Fig. 4a shows these losses P_V in a thyristor at an operating current with a constant fundamental component and for different firing angles α. In Fig. 4b, the strongly differing current curves at $\alpha = 90^o$ and $\alpha = 150^o$ are shown for points A and B, the fundamental current component being the same in both cases. A surprising fact is that the losses hardly vary within a wide range of firing angles (up to approx. 140°). Selection for a high overload without appreciable increase of the

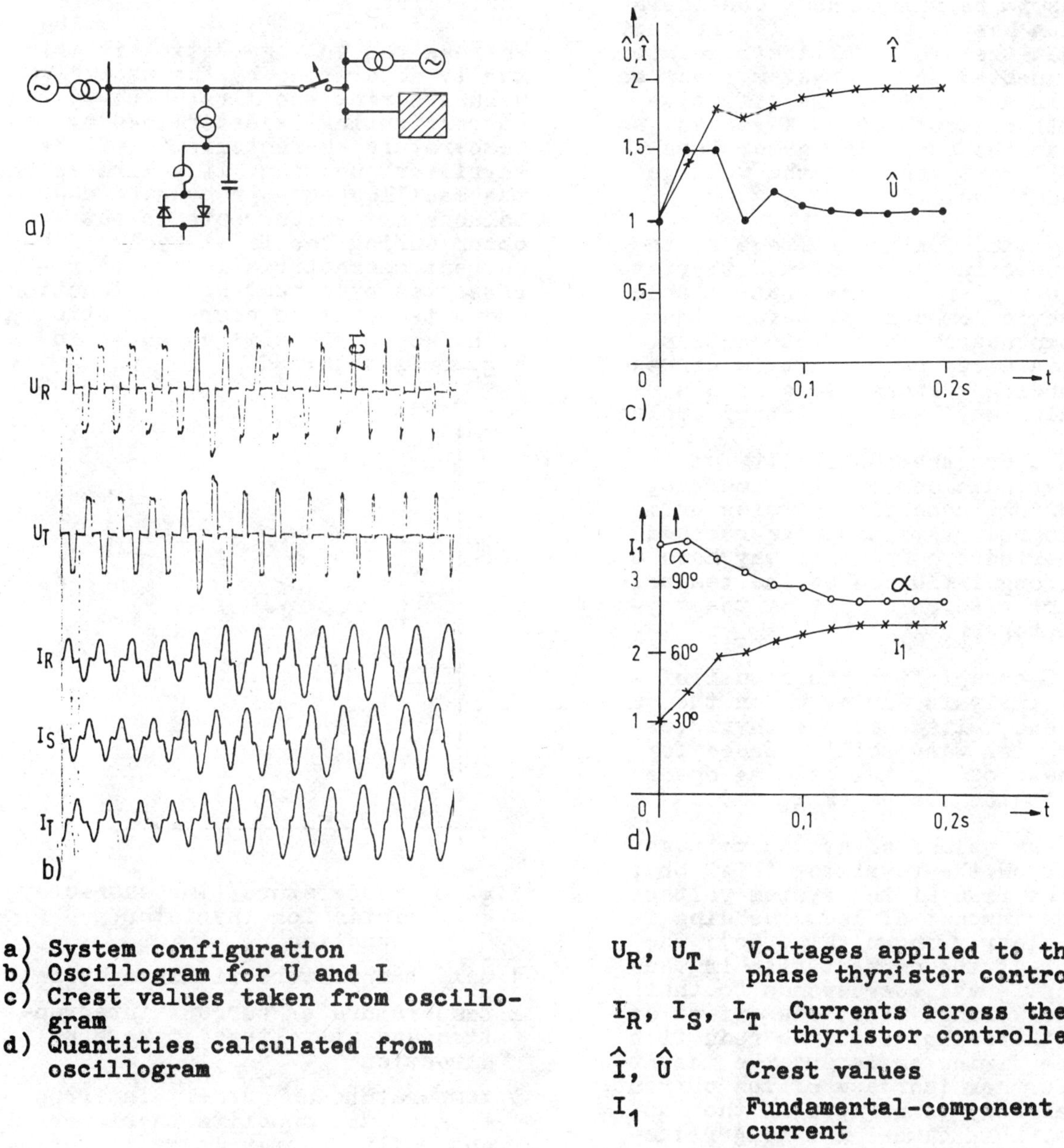

a) System configuration
b) Oscillogram for U and I
c) Crest values taken from oscillogram
d) Quantities calculated from oscillogram

U_R, U_T Voltages applied to three-phase thyristor controller
I_R, I_S, I_T Currents across the thyristor controller
$\hat{I}$, $\hat{U}$ Crest values
I_1 Fundamental-component current
α Firing angle

Fig. 5 Model investigation; load shedding in system; characteristics of voltage and current at thyristor controller

losses in continuous operation is thus possible. For determining the overload conditions, the following design criteria have to be found:

a) Overcurrent required to keep the system voltage within permissible limits on system disturbances and period for which this overcurrent has to be maintained. Consideration has to be given to the fact that the other voltage regulators connected to the system react as well and will become effective within about 0.5 to 1 second, so that the supplying generators will contribute to the voltage reduction.

b) Magnitude of the voltage at the connecting point of the thyristor controller and the controllable static compensator before the compensator comes into action. This delay may result in correspondingly increased currents which may last for 1 to 3 cycles.

c) Is there the possibility of aperiodic components occurring during these first cycles until the new state has been reached. Aperiodic components may have a strong influence on the temperature rise on a part of the thyristors.

Fig. 5 exemplifies the result of a model analysis during which the current and voltage at the thyristor controller was oscillographed for the case of a load-shedding operation in the system (Fig. 5a).

The peak values of system voltage appear on the thyristor (Fig. 5b), and the rise of the system voltage at the instant of load shedding is thus clearly shown (Fig. 5c). The amount of the current rise in the 1st half-wave corresponds to that of the voltage rise. The effect of the controller, i.e. the reduction of the firing angle and the resulting further increase of the current, can already be noticed in the 2nd cycle. This causes a strong reduction of the voltage rise rate. As a result of the system optimization of the controller, a further reduction of the firing angle now follows i.e. to the smallest possible value (Fig. 5d) in this case, and the maximum current across the thyristor controller is obtained after about 5 cycles, corresponding to 100 ms. If the system voltage is thereby reduced to a value near the setpoint, the current must be maintained (Figs. 5c and d) until all the remaining regulating gear in the system can become effective. This will be the case after 1 second at the latest. If the thyristor controller carried the maximum continuous current before the load-shedding operation, this oscillogram can demonstrate the maximum loading of the controller.

Whereas the voltage rating is primarily determined by the peak values during the first cycles, the current rating is determined by the temperature characteristic of the thyristor junction. If - like in the oscillogram - pronounced unbalance or overcurrents do not occur during the first cycles, the current characteristic may be represented by a rectangular function, and a temperature characteristic with respect to time as shown in Fig. 6 is obtained.

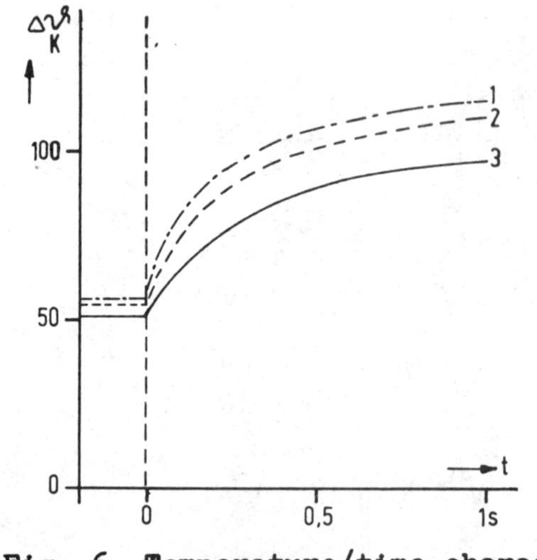

Fig. 6 Temperature/time characteristic for thyristor junction

1 max. temperature within a cycle

2 temperature on current interruption and at voltage in reverse direction

3 temperature on current interruption of the opposite thyristor and built-up of voltage in the forward direction

Since an alternating current is applied to the thyristor, unequal temperatures result at discrete instants within a cycle (Fig. 6). The maximum temperature and the temperature resulting when the thyristor has to carry the inverse voltage are of particular importance. There is a certain difference between the blocking function in the positive and negative directions.

If critical overloads are to be expected during the first cycle, the maximum permissible temperature may occur even during the first cycles.

In addition to the design of the heat sink, these conditions also depend on the overload conditions (9). It will be expedient to choose a heat sink with a low thermal resistance through which the initial temperature is kept low even before the overload occurs. If, however, the limit temperature is reached at the end of the overload period, a heat sink with a higher thermal capacity is employed. The initial coolant temperature has, of course, a decisive influence. Depending on the climatic conditions, it will be expedient to provide for artificial cooling, e.g. by evaporation coolers or refrigeration machinery, instead of using additional thyristors. Finally, unbalance conditions and aperiodic components should be allowed for in the current rating of the thyristors.

VOLTAGE RATING OF THYRISTORS

For high powers, thyristors having the highest inverse-voltage ratings are employed. Since- with a series circuit - a protective firing of the thyristors is provided, an effective protection will be available against high overvoltages. However, the protective firing arrangement must not come into action too early, otherwise the control capability of the thyristor controller would be affected. This condition is eased by the fact that simultaneous firing will also be required at high voltages and that the delay angle of the controller is anyhow changed into that particular direction. The rating for a high overload capacity takes into account that reduction of the current, e.g. by a current limiting control, must still be possible.

Limitation of the rate of voltage rise is normally effected by the series-connected reactor. An undesirably high rate of voltage rise may only occur under certain fault conditions, e.g. on incoming lightning surges or on earth faults. Such stresses can be eliminated by a suitable circuit arrangement of the equipment and its protection by buildings or other screening means. For operation at the maximum continuous service voltage, the voltage design factor for the thyristors is chosen so as to allow for non periodic overvoltages to be expected. The number of redundant series-connected thyristors can be strongly reduced if suitable means are provided for indicating the failure of individual series thyristors.

ELIMINATION OF HARMONIC CURRENTS FROM THE SYSTEM

In the case of balanced operation, which is the normal case if the controllers are operated in conjunction with a supply system, the current due to the triplen harmonics can be kept away from the system by using a delta connection of thyristor controllers. The harmonics of the other modal numbers are effectively suppressed by a permanently connected capacitor and coupling through a reactance provided by the system transformer. A filter circuit will, however, be necessary in most cases for eliminating the 5th harmonic (and the 7th in some cases).

In any case, possible points of resonance must be investigated to prevent amplification of certain harmonic oscillations. Here, the properties and configuration of the system to which the compensator is connected plays an important role. The reactance of the transformer, i.e. its reactance drop, may be adapted to the particular requirements, but this is not normally done on account of the economical advantage obtained through the use of standard-type transformers.

If very high powers are involved, suppression of the 5th and 7th harmonics may be effected by two separate compensators whose supply phases are displayed by 30°. In this case, the system transformer is provided with two secondary windings, one connected in delta and the other one in star. Balancing to the two thyristor controllers connected is by means of the closed-loop and open-loop control.

To prevent radio interference by the thyristor firing circuits, a screen is installed in the building housing the thyristor controllers. An important requirement is that the series-connected reactor should be installed close to the building; the connecting cable should be properly screened.

INFLUENCE OF UNBALANCE DURING STEADY-STATE OPERATION AND TRANSIENT CONDITIONS

The individual phases of the static reactive-power compensator can be controlled separately, which is not possible with a synchronous machine. This possibility is maily used on reactors in industrial plants where unbalance resulting from the operation of arc furnaces and other loads have to be compensated (10). In electrical supply systems, the compensator is normally operated in symmetry, the applied system voltage being normally symmetrical as well. Care should be taken to see that the compensator symmetry is inherently maintained; this is ensured by suitable balancing controls.

In addition to the possibility of unbalanced control of the three phases, manufacturing tolerances of the components of the thyristor firing circuits may cause different control settings for the positive and negative half-waves, which may result in the generation of an aperiodic component. It can therefore not be excluded that - with high powers and a comparatively low magnetizing current - the transformer is driven into saturation (11). To avoid this, suitable balancing controls are employed.

The characteristics of the magnetizing current of a model transformer with and without balancing control are shown in Fig. 7.

PROTECTION AND MONITORING

Where the rating does not provide for overload capacity the series-connected reactor protects the thyristor controller against current overloads. If the overload capacity is included, the thyristor controller may be subjected to high currents, e.g. if the limiting control or the firing circuits fail with the thyristors being continually conductive. In such a case, an overcurrent protector must isolate the compensator from the system. Standard-type protective arrangements are used for the system transformer, the reactor, the capacitor bank and the filter circuits. The entire equipment arrangement may also be protected by a differential protection.

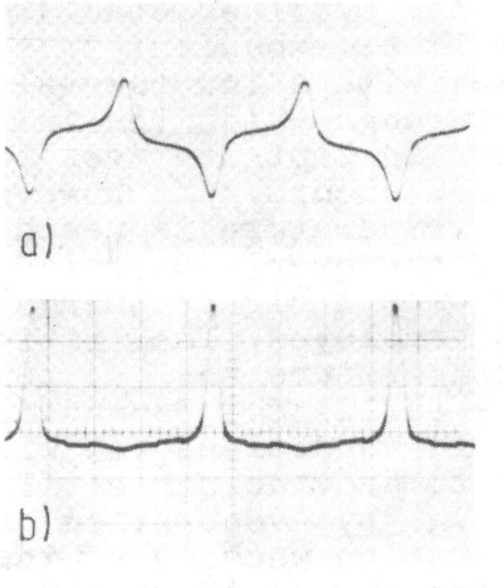

Fig. 7 Transformer magnetizing current; characteristic with respect to time

a) with
b) without balancing control

Monitoring of all the components of the compensator is of particular importance. The thyristors of the series-connected groups, for instance, are monitoried for conductivity and readiness for gating. If unbalance results in the capacitor banks, appropriate indications are given as a first step. Partial disconnection under load of plant sections, e.g. capacitor banks or filter circuits, is not provided for, since such an operation may entail undue stressing of other parts.

CONCLUSIONS

The conventional reactive-power compensating arrangements for high-voltage systems, such as fixed-value reactors, capacitors and synchronous compensators, have now been supplemented by the controllable static reactive-power compensator. This consists essentially of a thyristor-controlled three-phase reactor which is distinguished by the possibility of continuous controllability within a kVA range covering both the supply of reactive power to the system and the drawing of such power from the system. The control range depends on the system conditions to which the static compensator can easily be adapted by suitable selection and rating of its components. It is superior to any other type of compensating equipment as it is capable of making available the reactive-power demand of the high-voltage system practically without delay.

It is expected that the controllable static reactive-power compensator will, in future, be used on an increasing scale in addition to the conventional compensating equipment, and will partly replace the latter.

REFERENCES

(1) L.O. Barthold et al., Static Shunt Devices for Reactive Power Control, CIGRE Paper 31-08 (1974)

(2) W. Herbst, J. Käuferle, F. Peneder, K. Reichert, Statische regelbare Blindleistungskompensation für Hochspannungsnetze, Brown Boveri Mitt. 9/10, 433-439 (1974)

(3) H. Frank. S. Ivner, Blindleistungskompensation von Lichtbogenöfen mit Hilfe von thyristorgeschalteten Kondensatoren System TYCAP, ASEA-Z. 19, 11-16 (1974)

(4) I.D. Ainsworth, E. Friedlander, K.J. Ralls, Recent developments towards long distance A.C. transmission using saturated reactors, IEE Conf.,London, 242 (1973)

(5) A. Chit, W. Horn, Neuzeitliche ruhende Blindleistungskompensation für Industrienetze, Techn. Mitt. AEG-TELEFUNKEN 66, 7, 286-290 (1976)

(6) M. Erche, H.P. Hettler, D. Povh, Performance of Static Compensators, Canadian Communications & Power Conference, Montreal, 350-353 (1976)

(7) W. Hochstetter, Eigenschaften der regelbaren statischen Blindleistungskompensatoren in der Elektrizitätsversorgung, Siemens-Zeitschrift 51, 3, 141-145 (1977)

(8) R. Müller, Untersuchungen zur thermischen Beanspruchung der Thyristoren in einer Hochspannungs-Gleichstrom-Übertragung, Elektrotechnische Zeitschrift A 97, 7, 422-426 (1976)

(9) T. Salanki, Berechnungsmethode für Luftkühlkörper von Thyristoren hoher Leistung, Siemens-Zeitschrift 50, 3, 172-177 (1976)

(10) W. Meusel, H. Waldmann, Coordinate Transformations of Multi-Term Regulation Systems for the Compensation and Symmetrization of Three-Phase Supplies, Siemens Forsch.- und Entwickl.-Ber. 6, 1 (1977)

(11) H. Bold, Stromrichtertransformatoren für die Hochspannungs-Gleichstromübertragung Cabora Bassa-Apollo, Brown Boveri Mitt. 7, 432-441 (1976)

CONVERTER-CONTROLLED REACTIVE POWER COMPENSATORS FOR STABILIZING EHV THREE PHASE TRANSMISSION SYSTEMS

H. Stemmler

Brown, Boveri & Company Limited, Dept. EIE, CH-5401 Baden/Switzerland

1. INTRODUCTION

As the lengths of three-phase e.h.v. transmission lines increase, so too does the difficulty experienced in operating them. There are two main reasons for this:

- The voltage profile along the line is dependent to a large degree on the load conditions. Suddem load rejection and line energization have the effect of raising the voltage at the end of the line to a dangerously high level.
- An increase in length reduces the maximum power transfer capability of the line, thereby reducing the margin between the planned power transfer and the maximum limit at which the line is susceptible to transient and dynamic instability.

These problems are overcome by sectioning the line and providing reactive power infeed to the individual sections.

Conventional methods of reactive power compensation are:

- control of the reactive power of the generator,
- installation of components, such as shunt reactors or capacitors, capacitors for series compensation, and
- employment of synchronous compensators.

However, with these methods the control feature is either not available or is too slow. The shunt reactor tends to even reduce the maximum power transfer capability still further. Besides being inherently slow, synchronous compensators store kinetic energy in their rotors and thereby exhibit a tendency to produce undesirable oscillations.

The usual method of raising the transmission power is to increase either the number of lines or the transmission voltage - measures inevitably leading to higher costs and increased space requirement.

However, due to recent advances in power electronics and the falling cost of component production, reactive power compensators can be constructed which are:

- fully static
- continuously controllable, and
- capable of very fast response.

By limiting overvoltages and raising the power transfer level, these compensators ensure stable operation and make fullest use of the transmission capacity of the line.

It is the object of this contribution to show how converter-controlled reactors can be used in three-phase e.h.v. transmission systems for

- controlling voltage,
- improving the static stability, and
- the dynamic stability of the power transmission

and consequently, for making optimum use of the line.

Diagrams and equations will be used to explain their operation.

The simplified three-phase e.h.v. transmission system taken as an example in the explanation, is illustrated at the top of Table 1 (and also in Fig. 3 and 4). It comprises the source of power (a generator), two 500 km sections of line with controlled, reactive power compensators at the section junctions, and the link to the load network.

2. BASIC DESIGN AND OPERATION OF THE CONTROLLABLE REACTIVE POWER COMPENSATOR

The descriptions are kept brief and are confined to the main principles only (for details see (1)).

Fig. 2.1 shows the circuit diagram of a controllable reactive power compensator. This comprises a bank of capacitors C and a reactor R, the latter being in the form of a transformer with high leakage inductance, whose secondary can be operated continuously between no-load and short-circuit by means of a convertor control unit SR. Thus, by using a controlled convertor the reactor R assumes the characteristics of a continuously controllable inductance, and the capacitive current of the capacitor bank C can be partially, fully or overcompensated. Hence, as a whole, the controllable, reactive power compensator can draw both capacitive and inductive current from the transmission line.

The inductances L extend the function of the capacitor bank C to that of a filter, so that the harmonics produced by the convertor do not reach the power system.

The operation of the compensator is illustrated in Fig. 2.2. This shows the effect of a sudden control adjustment from zero to 75% on the current $i_{R,S,T}$, and on the active power absorption P(t) and the lagging reactive power absorption Q(t) of the reactor. The reactive power Q(t) responds to the control command with a delay of only approx. 6 msec, while the active power P(t) remains zero as desired.

3. STABILIZATION OF THE VOLTAGE ALONG THE TRANSMISSION LINE

In section 3.1, for the purpose of elucidation, the current and voltage ratios at the junction of the individual line sections are represented by loci. Following this, in section 3.2, equations are used to show to what degree the currents of the controllable, reactive power compensator affect the interface voltages U_{1A}, U_{2A}, U_{2B}.

3.1 Locus representation

To construct the loci, the equations for the individual sections of the transmission line are required (it will be assumed for the sake of simplicity that these are free of losses). The equations are given in Table 1.

The current and voltage loci shown in Fig. 3.1, 3.2, 3.3 and 4.1, 4.2, 4.3 can be drawn with the aid of the quadripole equations 8/11 from Table 1.

Fig. 3.1, 3.2, 3.3 show the ratios for the arrangement with only one controllable, reactive power compensator AB at the centre of the line. It is assumed in Fig. 4.1, 4.2, 4.3, that two additional compensators A, B are installed, one at each end of the line.

In Fig. .1, .2 and .3, voltage vectors $\underline{U}_{1A}$, $\underline{U}_{2A}$ ($\equiv \underline{U}_{1B}$) and $\underline{U}_{2B}$, respectively, are placed on the real axes (Re) of the co-ordinate system.

The heavy, full-line loci in Fig. 3.1, 3.2, 3.3 show, assuming the voltages at the ends and centre of the line to be equal and constant ($U_{1A} = U_{2A} \equiv U_{1B} = U_{2B}$),

- how, in this particular case, the source voltage $\underline{U}_0$ (curve a) and the power system voltage $\underline{U}_N$ (curve k) should behave,
- which currents must be supplied by the source ($\underline{I}_0$, curve b) and which delivered to the power system ($\underline{I}_N$, curve h),
- how the current $\underline{I}_{AB}$ (which equals the difference between the reactive powers at the output of line section A, curve f, and the input of section B, curve e) must be set at the controlled, reactive power source.

The locus parameter ist the transmission power P as a percentage of the surge impedance load P_n. Parameter curves for 0%, 50% and 100% are shown by dashed lines.

Also shown directly by the loci are the active and reactive power ratios at the beginning, centre and end of the line.

It will be seen in Fig. 3.1, that input current $\underline{I}_{1A}$ (curve b) of section A is purely capacitive at no load (leading $\underline{U}_{1A}$ by 90°). With increasing transmission power the current becomes higher, the capacitive component being reduced. The result on attaining the surge impedance load is a purely active current, which becomes inductive when the power increases still further.

By analogy, the same applies to the input current $\underline{I}_{1B}$ (curve e) of section B with reference to voltage $\underline{U}_{2A} \equiv \underline{U}_{1B}$ (Fig. 3.2).

The output current $\underline{I}_{2B}$ (Fig. 3.3, curve h) at the end of line section B (and, by analogy, output current $\underline{I}_{2A}$ of section A) (Fig. 3.2, curve f) is purely inductive at no-load (lagging $\underline{U}_{2B}$ by 90°), increases in value with increasing transmission power (the inductive component becoming correspondingly smaller), is purely active on the surge impedance load being attained, and becomes capacitive with an even further increase in power.

The current $\underline{I}_{AB}$ which the compensator AB must draw (given by the horizontal line connecting curves e and f in Fig. 3.2) is inductive (54%) for no-load operation. This current must decrease as the transmission power increases, be zero at the surge impedance load and have a leading power factor when the surge impedance load is exceeded.

Moreover, Fig. 3.1 and 3.3 show quite clearly how the angles between the voltage vectors $\underline{U}_0$ (curve a), $\underline{U}_{1A}$, $\underline{U}_{2A} \equiv U_{1B}$ (curves d and g, respectively), $\underline{U}_{2B}$, $\underline{U}_n$ (curve k) increase along the line as the transmission power increases.

Finally, in Fig. 3.2 it is shown how, to raise line voltage U_{2A} at the centre of the line the current of the controlled, reactive power source must be reduced or - the higher the power, the more so - even made capacitive, and vice versa, how voltage U_{2A} can be reduced by increasing the lagging reactive power.

These control measures also have an effect - as shown clearly by the thin, full-line curves in Fig. 3.1 and 3.3 - on the reactive power component of the line input and output current, this being also true when the voltages are kept constant (assumed in Fig. 3.1 and 3.3).

The above explanation sets out clearly the procedure for regulating voltage by means of controllable reactors.

Voltage regulation by controlled compensators is also of significance for suppressing load oscillations: by using controllable compensators to raise and lower the voltage at the centre of the line oscillations can be counteracted (see sect. 5).

The ratios shown in Fig. 4 differ from those in Fig. 3, in that controlled, reactive power sources are also provided at both ends of the line.

Installation of a controllable reactive power source between the generator and the line could also be of value should - in the distant future - superconducting generators be employed whose excitation current can be adjusted only very slowly. The controlled reactive power source could, in such a case, also perform as a generator-voltage regulator. The current $\underline{I}_A$, which is to be delivered by the reactive power source to keep the initial line voltage U_{1A} constant (for a constant source voltage U_0) despite variations in the transmission power P, is shown in Fig. 4.1 as being the difference between the initial line current $\underline{I}_{1A}$ (curve c) and the source output current $\underline{I}_0$ (curve b) (horizontal connecting line). It must also be capable of becoming capacitive at high transmission powers.

A further reason for installing additional reactive power sources at the beginning and end of a line is the ease with which the end voltages can be kept constant when, for example, loads are rejected, compared with when there is only one reactive power source at the line centre.

Controlled reactive power sources can likewise be employed for reactive load compensation.

3.2 Mathematical representation

The above explanations were based on Fig. 3 and 4. It shall now be shown by equations upon which quantities line voltages $\underline{U}_{2A}$, $\underline{U}_{1A}$ and $\underline{U}_{2B}$ at the reactor back-up points are dependent. It will be seen that these are influenced not only by the reactive current of the respective reactive power source, but also by a whole number of "disturbance variables".

The equations required are referred to below:

- The difference between the reactive power at the output of one line section and the input of another is compensated by the reactive power source at that point: see equations 12, 13 and 31, equations 14, 15 and 32, equations 16, 17 and 33 in Table 1.
- At the point where two line sections are connected, the active currents at the output of the one section and at the input of the other are equal: See equations 18, 19 and 28, equations 20, 21 and 29, and equations 22, 23 and 30 in Table 1.
- In addition, the sum of the angles between the input and output voltage vectors of the individual sections is equal to the angle φ_{N0} between the source and system voltage vector. See equation 7, Table 1.

If equations 12 to 17 are substituted in equations 31 to 33, and equations 18 to 23 in equations 28 to 30 (Table 1), a set of equations is obtained which, together with equation 7 (Table 1), enable the desired relationships, i.e. those between the voltages at the back-up points U_{1A}; U_{2A}; U_{2B} and their dependence upon the compensator currents I_A; I_{AB}; I_B, the source and system voltage U_0; U_N and the angle φ_{N0}, to be calculated:

$$U_{1A(2A,2B)} = U_{1A(2A,2B)}\,(I_A; I_{AB}; I_B; U_0; U_N; \varphi_{N0})$$

Because of the trigonometric functions, only deviations Δ... from the steady-state condition are written.

As simple, linear equations are involved, the calculations are fundamentally straightforward. The expressions obtained, however, are very long and are not easy to interpret simply. Thus, it is not considered expedient to give the results here in full, but to express these in the following general form:

$$\Delta U_{1A\,(2A)\,(2B)} = \Delta I_{Z\,A0}\,(\partial U_{1A\,(2A)\,(2B)}:\partial I_{Z\,A0}) + \Delta I_{Z\,AB0}\,(\partial U_{1A\,(2A)\,(2B)}:\partial I_{Z\,AB0}) + \Delta I_{Z\,B0}\,(\partial U_{1A\,(2A)\,(2B)}:\partial I_{Z\,B0}) + \Delta U_0\,(\partial U_{1A\,(2A)\,(2B)}:\partial U_0) + \Delta U_N\,(\partial U_{1A\,(2A)\,(2B)}:\partial U_N) + \Delta\varphi_{N0}\,(\partial U_{1A\,(2A)\,(2B)}:\partial\varphi_{N0}) \qquad \text{(a)}$$

Taking data from Table 1 as an example:

$$\Delta U_{1A\,(2A)\,(2B)} = \Delta I_{Z\,A0}\cdot 0{,}339\,(0{,}188)\,(0{,}018) + \Delta I_{Z\,AB0}\cdot 0{,}188\,(0{,}462)\,(0{,}069) + \Delta I_{Z\,B0}\cdot 0{,}018\,(0{,}069)\,(0{,}099) + \Delta U_0\cdot 0{,}777\,(0{,}314)\,(0{,}008) + \Delta U_N\cdot 0{,}152\,(0{,}643)\,(0{,}983) - \Delta\varphi_{N0}\cdot 0{,}006\,(0{,}008)\,(0{,}002) \qquad \text{(a1)}$$

$$P = P_n;\; U_0 = U_{1A} = U_{2A} = U_{2B} = U_N$$

$$\Delta U_{1A\,(2A)\,(2B)} = +\Delta I_{Z\,AB0}\cdot 0{,}188\,(0{,}459)\,(0{,}068) + \Delta U_0\cdot 0{,}787\,(0{,}335)\,(0{,}013) + \Delta U_N\cdot 1{,}152\,(0{,}636)\,(0{,}981) - \Delta\varphi_{N0}\cdot 0{,}006\,(0{,}008)\,(0{,}002) \qquad \text{(a2)}$$

$$P = P_n;\; U_{1A} = U_{2A} = U_{2B};\; I_A = I_B = \Delta I_A = \Delta I_B = 0$$

The numeric equation (a1) applies when the steady-state transmission power equals the surge impedance load ($P = P_n$), when controllable, reactive power sources are installed in the middle and at both ends of the line and when, for the steady-state condition, the voltages along the line all have the same value ($U_0 = U_{1A} = U_{2A} = U_{1B} = U_{2B} = U_N$). The numeric equation (a2) is also valid when the steady-state transmission power equals the surge impedance load ($P = P_n$), however for the case when a controllable compensator is installed at the line centre only. In the steady-state condition the voltages along the line are equal ($U_{1A} = U_{2A} = U_{1B} = U_{2B}$).

The numeric example illustrates to what extent the individual correcting variables I_A, ΔI_{AB} and ΔI_B and the "disturbance variables" ΔU_0, ΔU_N and $\Delta\varphi_{NO}$ affect the voltage U_{1A}, U_{2A} or U_{2B} being considered.
Voltage U_{2A} in the middle of the line, for example, can be influenced most by variation ΔI_{AB} of the respective reactive power source. Reactive power sources at the beginning and end of the line, however, also affect voltage U_{2A} at the centre with their changes in current ΔI_A and ΔI_B. Voltage U_{2A} is clearly also influenced by voltage changes at the source (ΔU_0) and in the power system being fed (ΔU_N), and by variation in the difference between the angles of these two voltage vectors ($\Delta\varphi_{NO}$).

4. IMPROVEMENT OF THE STATIC STABILITY OF THE POWER TRANSMISSION

A line, like a synchronous machine, can only transmit power power up to a maximum limit, at which the line falls out of step.
The critical power $\hat{P}$ of a line is given by the following equation b (see equations 25 and 26 in Table 1).

$$P = \frac{3U_1U_2}{2Z_0 \sin\beta l} \sin\varphi_{12} \rightarrow \hat{P} = \frac{3}{2}\frac{U_1U_2}{Z_0 \sin\beta l} \qquad (b)$$

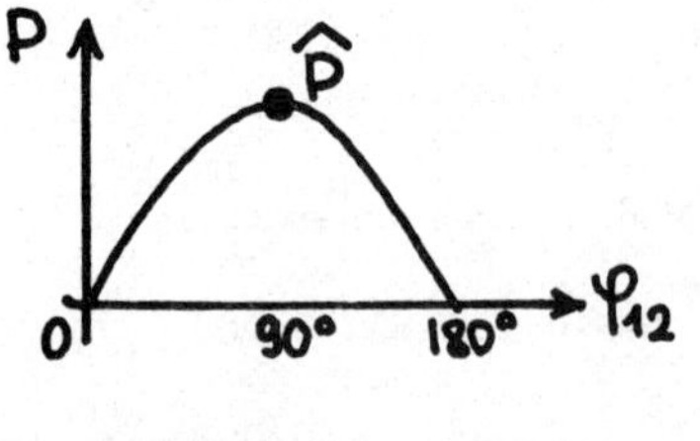

In equation (b), U_1 and U_2 are the amplitudes of the input and output voltages, respectively, β is the phase-shift coefficient and Z_0 the characteristic impedance of the line. The length of the line is given by l and the angle between the input and output voltage vectors by φ_{12}.
From equation (b) it can be seen that as the length of the line increases (i.e. with an increase in $\sin\beta l$), the critical power and consequently the stability of the power transmission decreases.
To counteract this, long lines are divided into two or more suitable sections of approximately the same length, and controllable, reactive power compensators installed at their junctions. If now - as mentioned in section 3.1 - the nominal voltages at the compensator back-up points are kept constant irrespective of transmission power P, the critical power of the entire transmission increases to the same value as that of a single section (the smaller $\sin\beta l$ and also the incoming and outgoing voltages of the section are to be inserted in equation b). In other words, the line appears to be shortened electrically, which gives the power transmission increased static stability.

5. IMPROVING THE DYNAMIC STABILITY OF POWER TRANSMISSION

Besides keeping the voltage along the line constant, the controllable reactor-compensator can also be utilized for suppressing load oscillations which may cause the line to fall out of step. This is achieved by controlling the compensator back-up voltage such that the transmission power counteracts the load swing.
It is assumed in the following that, for the simplified arrangement shown at the top of Table 1, though with only one controlled compensator (AB) at the line centre, rotating field oscillations of the synchronous machine (source) lead to oscillation of the transmission power. Equations are used to explain the oscillation phenomena, how it is instigated and how it can be suppressed.
The following equation (c) is obtained direct from equations 34 - 37 in Table 1. It is represented in non-dimensional form (characterized by ') and uses the referred variables and relationship given in Table 1.

$$\Delta m'_{an} = p\Delta\omega_N T'_n + p'^2 \Delta\varphi_{NO} \cdot T'_n + \Delta P' \qquad (c)$$

where: m_{an} = driving torque of the turbine; p = differential operator; $p\omega_N$ = variation of angular frequency of the power system with time; T_n = starting time constant for generator (including turbine); $p^2\varphi_{NO}$ = second derivative of angle between power system and source voltage vectors (rotor position) P = transmission power

Only deviations Δ... from the steady state have been taken into account in the equation (c). For the sake of simplicity, it was assumed in the equations that the entire power transmission is free of losses and that the electromagnetic dynamics of the machine and line are negligible compared with the mechanical phenomena. (However, the transients of the synchronous machine are taken into account for the calculation of the source impedance ωL_0).

The transmission power P' is found in equation (c). It shall now be determined upon what this depends, and especially if and how it can be influenced by the compensator voltage.
Beginning with power equations 24, 25, 26, 27 (Table 1), equation 7 (Table 1) involving the angles of the individual line sections and taking-because of the angle functions - only deviations Δ... from the steady state into consideration, we obtain by simple calculation the following equation (d). This is likewise written in non-dimensional form, using the referred variables given in Table 1 (shown by '):

$$\Delta P' = \Delta U'_{2A}(\partial P':\partial U'_{2A}) + \Delta U'_{0}(\partial P':\partial U'_{0}) + \Delta U'_{N}(\partial P':\partial U'_{N}) + \Delta\varphi_{NO}(\partial P':\partial\varphi_{NO}) \quad (d)$$

$$I_A = I_B = \Delta I_A = \Delta I_B = 0$$

This equation is given here in a general form only, and shows that the transmission power P' is a function of line voltage U'_{2A} at the back-up point, of source voltage U'_0 and power system voltage U'_N, and of angle φ_{NO} between the source and power system voltage vectors.
The following example with data from Table 1 shows to what extent the individual variables U_0, U_{2A}, U_N, φ_{NO} influence the transmission power P:

$$\Delta P = \Delta U_{2A}\cdot 1{,}000 + \Delta U_0\cdot 0{,}539 + \Delta U_N\cdot 0{,}342 + \Delta\varphi_{NO}\cdot 0{,}600$$

$$P = P_n;\ U_{1A} = U_{2A} = U_{2B};\ I_A = I_B = \Delta I_A = \Delta I_B = 0 \quad (d1)$$

It is seen that the transmission power can be acted upon effectively by means of line voltage U_{2A}, which is controlled by the reactor-compensator. It is assumed in numeric equation (d1) that a controllable reactive power source is installed at the line centre only, that the surge impedance load ($P = P_n$) is transmitted in the steady state, and that in the steady state the voltages along the line are equal ($U_{1A} = U_{2A} \equiv U_{1B} = U_{2B}$).

By substituting equation (d) in equation (c) we obtain the differential equation given below in operator notation:

$$\Delta P\left\{1 + p'^2\frac{T'_n}{\left(\frac{\partial P'}{\partial\varphi_{NO}}\right)}\right\} = \Delta m'_{an} - p'\Delta\omega_N T'_n + p'^2\frac{T'_n}{\left(\frac{\partial P'}{\partial\varphi_{NO}}\right)}\left\{\Delta U'_{2A}\left(\frac{\partial P'}{\partial U'_{2A}}\right) + \Delta U'_0\left(\frac{\partial P'}{\partial U'_0}\right) + \Delta U'_N\left(\frac{\partial P'}{\partial U'_N}\right)\right\} \quad (e)$$

The equation shows that load swings can be described by means of a 2nd order linear differential equation without damping coefficient ($p'\Delta P'$ missing). This means that once instigated, oscillations do not decay (in practice: only very slowly).
Oscillations are instigated by variation of the driving torque $\Delta m'_{an}$ of the turbine, of the source voltage $\Delta U'_0$ and power system voltage $\Delta U'_N$, of the line voltage $\Delta U'_{2A}$ at the back-up point and of the power system angular frequency $\Delta\omega'_N$ at the load end.

Of the variables mentioned, however, it is possible to control voltage U'_{2A} at the back-up point. This is done in such a way that a damping effect is obtained. To achieve this objective, the respective reactor is used to control $\Delta U'_{2A}$ as a function of the measured transmission power $\Delta P'$ in accordance with the following equation:

$$\Delta U'_{2A}\cdot\left(\frac{\partial P'}{\partial U'_{2A}}\right) = \frac{p'T'_D}{1+p'T'_D}\Delta P' \quad (f)$$

By substituting this control equation (f), in which damping time constant T'_D can be set as desired in the reactor controller, in the differential equation (e), we obtain:

$$\Delta P'\frac{\left\{1 + p'T'_D + p'^2\frac{T'_n}{(\partial P':\partial\varphi_{NO})}\right\}}{\left\{1 + p'T'_D\right\}} = \Delta m'_{an} - p'\Delta\omega'_N T'_n + p'^2\frac{T'_n}{\left(\frac{\partial P'}{\partial\varphi_{NO}}\right)}\left\{\left(\frac{\partial P'}{\partial U'_0}\right)\Delta U'_0 + \left(\frac{\partial P'}{\partial U'_N}\right)\Delta U'_N\right\} \quad (g)$$

Damping coefficient D and the self-resonant angular frequency $'_0$ of equation (g) are given by:

$$D = \frac{T'_D}{2}\sqrt{\frac{(\partial P':\partial\varphi_{NO})}{T'_n}} \quad (h)$$

$$\omega'_0 = \sqrt{\frac{(\partial P':\partial\varphi_{NO})}{T'_n}} \quad (i)$$

It is seen from equation (h) that damping coefficient D can be set by varying time constant T'_D of control equation (f), e.g. D = 1 for aperiodic damping.
Fig. 5 illustrates the frequency responses to load swing occurring when a controlled compensator is not used for suppression (as is the case in equation (e)), the frequency response of control equation (f), in accordance with which the back-up voltage U'_{2A} at the line centre is to be controlled as a function of transmission power P' to

TABLE 1

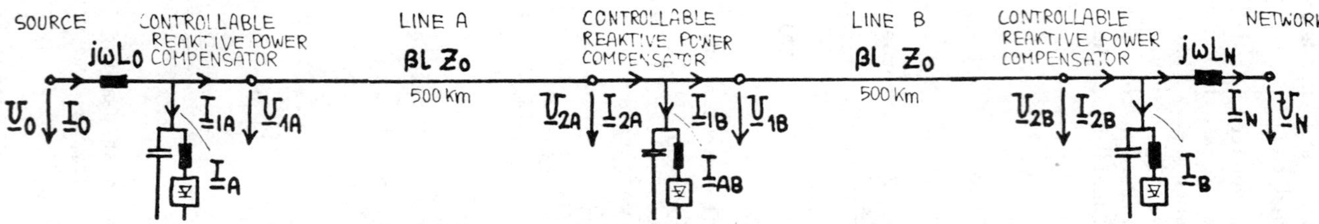

EQUATIONS	Source	Line A	Line B	Network
voltages	$\underline{U}_0=U_0e^{+j\varphi_{1A0}}$, $\underline{U}_{1A}=U_{1A}$ (1)	$\underline{U}_{1A}=U_{1A}e^{+j\varphi_{2A1A}}$, $\underline{U}_{2A}=U_{2A}$ (2); $\underline{U}_{2A}=U_{2A}e^{-j\varphi_{2A1A}}$, $\underline{U}_{1A}=U_{1A}$ (3)	$\underline{U}_{1B}=U_{1B}e^{+j\varphi_{2B1B}}$, $\underline{U}_{2B}=U_{2B}$ (4); $\underline{U}_{2B}=U_{2B}e^{-j\varphi_{2B1B}}$, $\underline{U}_{1B}=U_{1B}$ (5)	$\underline{U}_N=U_Ne^{-j\varphi_{N2B}}$, $\underline{U}_{2B}=U_{2B}$ (6)
angles	$\varphi_{N0}=\varphi_{1A0}+\varphi_{2A1A}+\varphi_{2B1B}+\varphi_{N2B}$ (7)			
quadripole equations of the sections	$\underline{U}_0=\underline{U}_{1A}+j\underline{I}_0Z_0\frac{\omega L_0}{Z_0}$ (8)	$\underline{U}_{1A}=\underline{U}_{2A}\cos\beta l+j\underline{I}_{2A}Z_0\sin\beta l$; $j\underline{I}_{1A}Z_0=-\underline{U}_{2A}\sin\beta l+j\underline{I}_{2A}Z_0\cos\beta l$ (9)	$\underline{U}_{1B}=\underline{U}_{2B}\cos\beta l+j\underline{I}_{2B}Z_0\sin\beta l$; $j\underline{I}_{1B}Z_0=-\underline{U}_{2B}\sin\beta l+j\underline{I}_{2B}Z_0\cos\beta l$ (10)	$\underline{U}_{2B}=\underline{U}_N+j\underline{I}_NZ_0\frac{\omega L_N}{Z_0}$ (11)
reactive currents at the junctions	(12) $\frac{U_{1A}-U_0\cos\varphi_{1A0}}{\omega L_0:Z_0}=\mathrm{Im}(\underline{I}_0Z_0)$ (1;8)	(13) $\mathrm{Im}(\underline{I}_{1A}Z_0)=\frac{-U_{1A}\cos\beta l+U_{2A}\cos\varphi_{2A1A}}{\sin\beta l}$ (3;9); (14) $\frac{U_{2A}\cos\beta l-U_{1A}\cos\varphi_{2A1A}}{\sin\beta l}=\mathrm{Im}(\underline{I}_{2A}Z_0)$ (2;9)	(15) $\mathrm{Im}(\underline{I}_{1B}Z_0)=\frac{-U_{1B}\cos\beta l+U_{2B}\cos\varphi_{2B1B}}{\sin\beta l}$ (5;10); (16) $\frac{U_{2B}\cos\beta l-U_{1B}\cos\varphi_{2B1B}}{\sin\beta l}=\mathrm{Im}(\underline{I}_{2B}Z_0)$ (4;10)	(17) $\mathrm{Im}(\underline{I}_NZ_0)=\frac{-U_{2B}+U_N\cos\varphi_{N2B}}{\omega L_N:Z_0}$ (6;11)
active currents at the junctions	(18) $\frac{U_0}{\omega L_0:Z_0}\sin\varphi_{1A0}=\mathrm{Re}(\underline{I}_0Z_0)$ (1;8)	(19) $\mathrm{Re}(\underline{I}_{1A}Z_0)=\frac{U_{2A}}{\sin\beta l}\sin\varphi_{2A1A}$ (3;9); (20) $\frac{U_{1A}}{\sin\beta l}\sin\varphi_{2A1A}=\mathrm{Re}(\underline{I}_{2A}Z_0)$ (2;9)	(21) $\mathrm{Re}(\underline{I}_{1B}Z_0)=\frac{U_{2B}}{\sin\beta l}\sin\varphi_{2B1B}$ (5;10); (22) $\frac{U_{2B}}{\sin\beta l}\sin\varphi_{2B1B}=\mathrm{Re}(\underline{I}_{2B}Z_0)$ (4;10)	(23) $\mathrm{Re}(\underline{I}_NZ_0)=\frac{U_N}{\omega L_N:Z_0}\sin\varphi_{N2B}$ (6;11)
active power	$P=\frac{3}{2Z_0}\frac{U_{1A}U_0}{\omega L_0:Z_0}\sin\varphi_{1A0}$ (24)	$P=\frac{3}{2Z_0}\frac{U_{1A}U_{2A}}{\sin\beta l}\sin\varphi_{2A1A}$ (25)	$P=\frac{3}{2Z_0}\frac{U_{1B}U_{2B}}{\sin\beta l}\sin\varphi_{2B1B}$ (26)	$P=\frac{3}{2Z_0}\frac{U_{2B}U_N}{\omega L_N:Z_0}\sin\varphi_{N2B}$ (27)
comparison of active and reactive currents at the junctions	$\mathrm{Re}(\underline{I}_0Z_0)-\mathrm{Re}(\underline{I}_{1A}Z_0)=0$ (28); $\mathrm{Im}(\underline{I}_0Z_0)-\mathrm{Im}(\underline{I}_{1A}Z_0)=I_AZ_0$ (31) (1;3)	$\mathrm{Re}(\underline{I}_{2A}Z_0)-\mathrm{Re}(\underline{I}_{1B}Z_0)=0$ (29); $\mathrm{Im}(\underline{I}_{2A}Z_0)-\mathrm{Im}(\underline{I}_{1B}Z_0)=I_{AB}Z_0$ (32) (2;5)	$\mathrm{Re}(\underline{I}_{2B}Z_0)-\mathrm{Re}(\underline{I}_NZ_0)=0$ (30); $\mathrm{Im}(\underline{I}_{2B}Z_0)-\mathrm{Im}(\underline{I}_NZ_0)=I_BZ_0$ (33) (4;6)	
mechanical dynamic equations of the generator	$m_{an}-m_{el}=p\varphi_0\Theta$ (34); $P=m_{el}p\varphi_0$ (35); $\varphi_0Z_P=\varphi_N+\varphi_{N0}$ (36); $p\varphi_N=\omega_N\gg p\varphi_{N0}$ (37)			

Z_0 : characteristic impedance
βl : phase-shift coefficient
L : inductance
T_n : starting time constant
T_D : damping time constant
P : transmission power
$\underline{I}$ $\underline{U}$: current-, voltage-vector
I U: current-, voltage amplitude

per unit variables: $m'_{an}\equiv m_{an}:M_n$; $P'\equiv P:P_n$; $U'\equiv U:U_n$; $p'\equiv p:\omega_n$; $\omega'_N\equiv\omega:\omega_N$; $T'_n\equiv T_n\omega_n$; (0) (0) (D)(D)

reference values: P_n: surge impedance load; U_n: rated voltage; M_n: rated torque at U_n and ω_n; ω_n: rated angular frequency;

relationship between the reference values: $M_n=P_nZ_P:\omega_n$; $\omega_nT_n=\omega_n^2\Theta:Z_PM_n$; $P_n=U_n^2:Z_0$

m_{an}: torque of the turbine
m_{el}: torque of the generator
Θ : moment of inertia
Z_P : number of pole pairs

φ_0; φ_N; φ_{N0}: angles, see above
$\varphi_{1A0}\varphi_{2A1A}\varphi_{2B1B}\varphi_{N2B}$: angles, see above
Δ: deviations from the steady-state condition
p: differential operator

Fig. 2.1

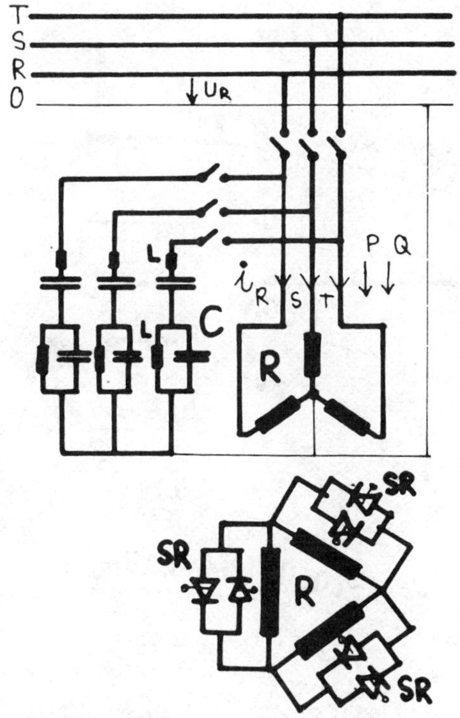

Schematic diagram of the convertor-controlled reactive power compensator

Fig. 2.2

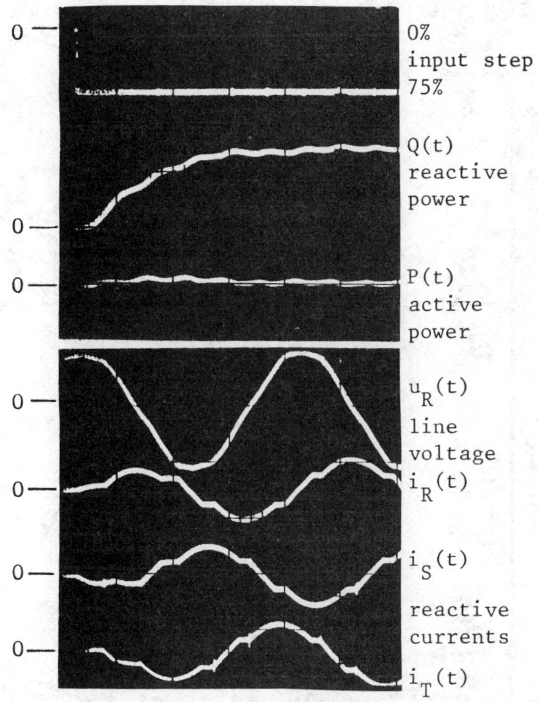

Operating principle (step response) of the convertor-controlled reactor

Fig. 5. Frequency responses of the power transmission

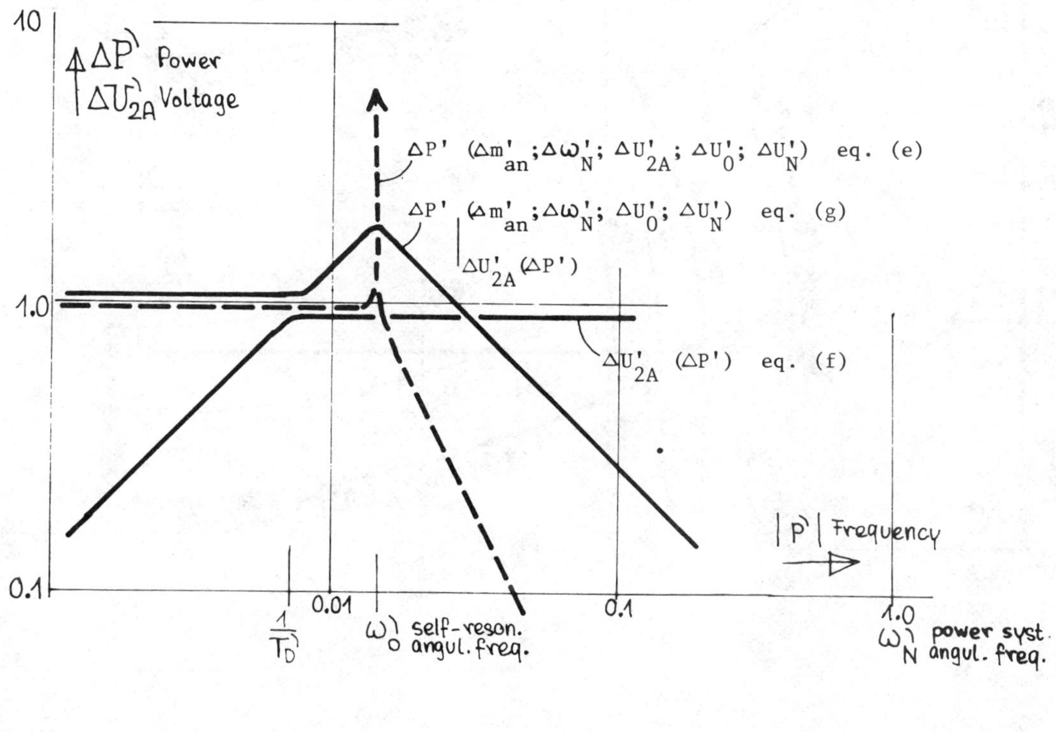

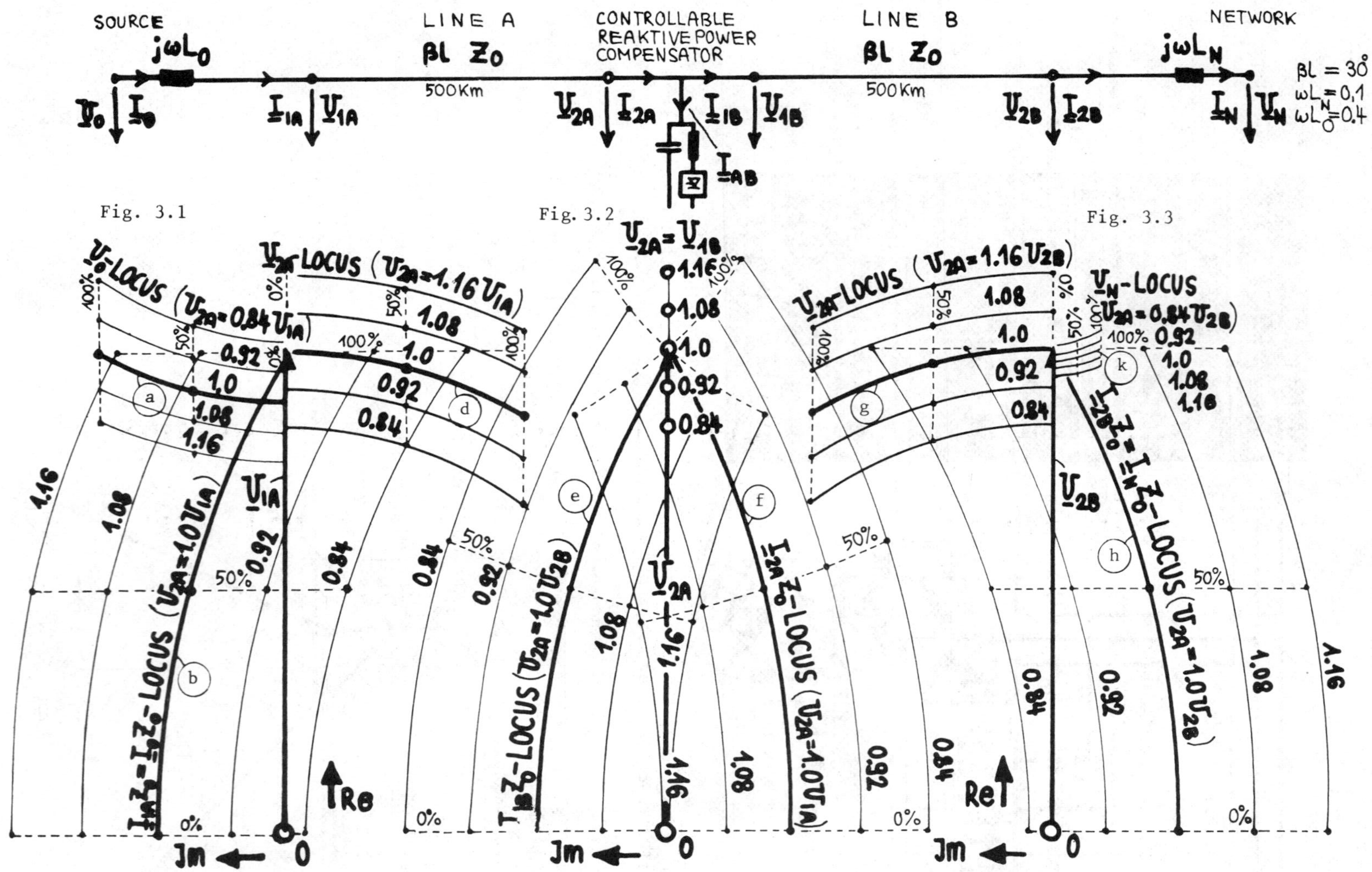

Fig. 3
Current and voltage ratios at the junctions between sections of a three-phase e.h.v. transmission system, with a controllable reactive power compensator at the line centre

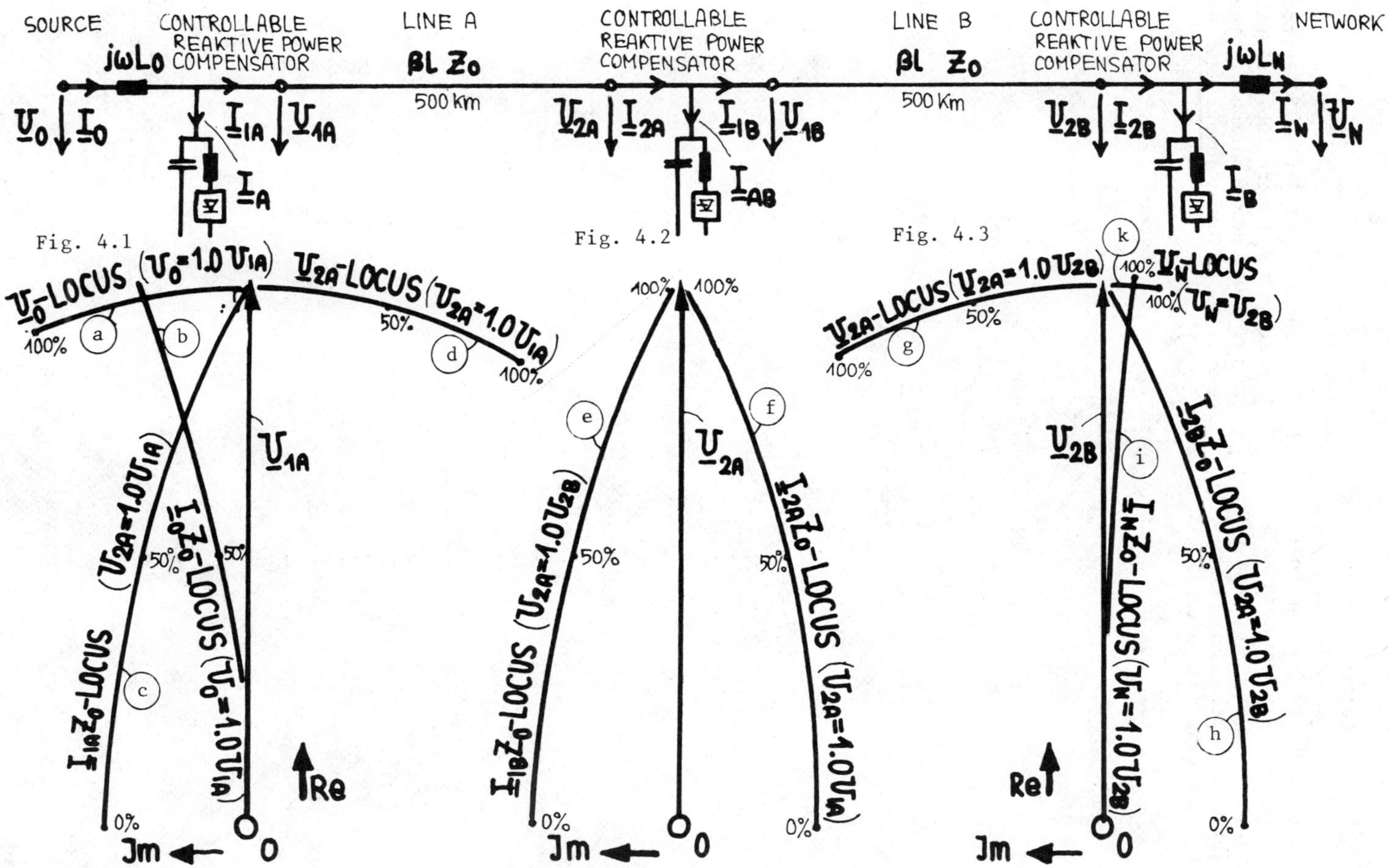

Fig. 4
Current and voltage ratios at the junctions between sections of a three-phase e.h.v. transmission system, with controllable reactive power compensators at the beginning, centre and end of the line

suppress load swing, and the frequency response of the load swing damped in this way (as per equation (g)). Fig. 5 is true for the numeric data given in Table 1 and for an aperiodic suppression of D = 1. It shows also that the self-resonant angular frequency ω_0' lies about 2 Orders of magnitude below the angular frequency ω_N' of the power system ($\omega_0' \ll \omega_N'$).

It is evident that for the suppression of oscillations allowance must be made for appropriate fluctuation of the back-up voltage (at the reactor), and this the more so the greater the suppression is desired to be. The greater the allowed fluctuation, the greater the oscillation amplitude which can be fully suppressed. Conversely, the attainable fluctuation of the back-up voltage U_{2A} is given by the ratio of the installed compensator reactive power to the transmission power (equation (a) and Fig. 3). The higher the installed reactive power is the wider the range becomes within which the back-up voltage can be set.

Reference:

(1) W. Herbst, J. Käuferle, F. Peneder and K. Reichert, Controllable static reactive power compensation of high-voltage systems, Brown Boveri Rev. 9/10, (1974).

OPERATING RESULTS OF A COMPENSATING EQUIPMENT

D. Schröder

Department of Technique Industry, Brown Boveri & Cie, Mannheim

SUMMARY

Electricity supply networks are increasingly supplying loads which have objectionable effects on the supplying network. Such effects may be caused by load currents containing harmonic components, by quick changes in the active and reactive power of the load or by unbalanced power consumption. One of the most objectionable loads in low-power supply systems is the arc furnace used for scrap melting. The arc furnace gives rise to all three phenomena leading to undesirable effects on the mains.

The reduce the effects of such loads, the compensating equipment must have excellent static and dynamic characteristics. The requirements on the compensating equipment had been shown in (1) with respect to the static and dynamic characteristics.

The results of this investigations are:

1. Necessity of the suppression of harmonics.
2. The reactive components of the compensator must be controlled independently in the three phases.

   Argument: Compensation of unsymmetrical reactive power and symmetrization of unsymmetrical active power.
3. The reactive power of the compensator should be controlled without steps.

   Argument: The static precision must be high because of the sensitivity of the human eye to flicker. For example a disturbance of the main-voltage with 3 $^{o}/oo$ in amplitude, a frequency of f = 7 Hz and a rectangular waveform can be detected easily by the human eye.
4. The dynamic characteristics of the compensating equipment must be exellent. The over-all response time should be half a period of the mains or shorter.

   Argument: A reduction of the length of time of a voltage-variation from for example 500 ms to 50 or 35 ms will increase the flicker-objection. A reduction of the flicker-objection begins when the reduction is to less than 20 ms.

This paper shows the static and dynamic characteristics of an installed equipment under various aspects.

SAME THEORETICAL ASPECTS

The installed compensating equipment is of the typ "indirect compensation". Indirect compensating equipment consists of a filter batterie for generating a reactive power with leading power factor and a inductive load for the variable reactive power with lagging power factor.

The control of the variable inductive power source requires reference value calculation and a control circuit to balance set values w and the actual values x (Fig. 1).

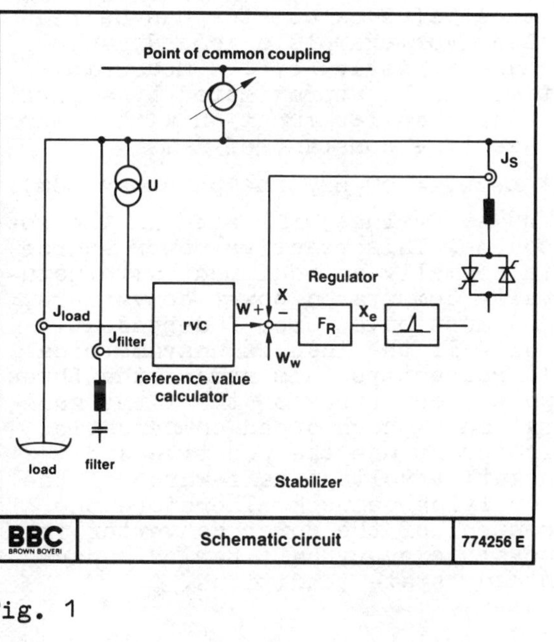

Fig. 1

The reference value calculator is necessary for the determination of the operating point of the load and for the evaluation of the required set values for the variable inductive power source. The input signals for the reference value calculator are the voltages, the currents of the load and the currents of the filter.

For example the equation for set value $\frac{1}{X_{12}}$ or Q_{12} is:

$$-\frac{1}{X_{12}} = -\frac{1}{\sqrt{3}\,R_{23}} + \frac{1}{\sqrt{3}\,R_{31}} + \frac{1}{X_{12}} + \frac{\varepsilon}{3}\left(\frac{1}{R_{12}} + \frac{1}{R_{23}} + \frac{1}{R_{31}}\right)$$

or

$$Q_{12} = \left(\frac{\varepsilon}{3} - \frac{1}{3}\right)\cdot\left(\frac{1}{T}\int_0^T u_{23}\, i_2 dt\right) - \left(\frac{\varepsilon}{3} + \frac{1}{3}\right)\cdot\left(\frac{1}{T}\int_0^T u_{31}\, i_1 dt\right)$$

The equations for the other set values can be derived by continuous changing the indices of the signals. The first equation use the admittances of the load, which can be determined by sampling the currents at distinct aquidistant moments. This method is advantageous if the signals have no or little harmonic content.

Under the aspect of the harmonic content the second equation with the power components fits more. But for the sake of quick calculation of the set values the calculation methods will be favorably a method with phase multiplication (1). Under this circumstances good static and dynamic characteristics of the reference value calculator can be reached. For example a good dynamic characteristic of the measuring time can be approximated by a first order transfer function with a worst case time constant of 2 ms ($f_{netw.}$ = 50 Hz, one phase system).

The set values will used in the power source. This reactive power source is normally an equipment with naturally commutated power converters and must have three independent units if the load is unsymmetrical. In respect to this demand the three power converters of the first generation of high speed compensation equipment use the positive and negative voltage-half-waves of the supplying network. Therefore the control of the power converter is possible every half period under steady state conditions.

If the power controller is a sampled system as for example thyristor switched condensers the above stated dynamic characteristic remain under static and dynamic conditions.
In Fig. 2 the installed variable reactive power source is shown.

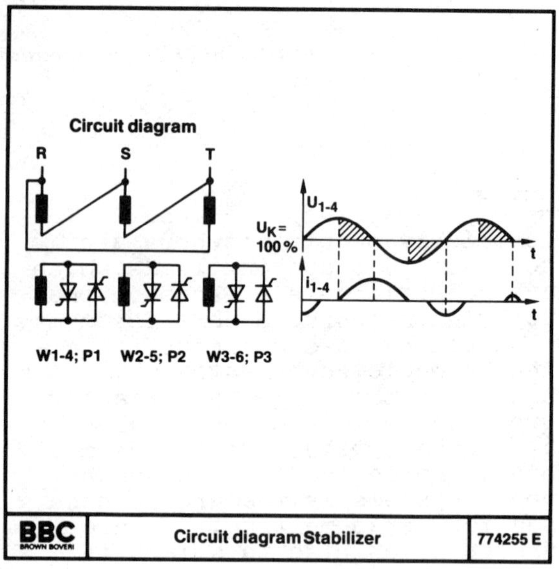

Fig. 2

It is composed of a transformer with high leakage ($u_K \approx 100$ %) and three independent pairs of antiparallel thyristor switches. The reactive power varies in the range maximum $\geqslant Q_L \geqslant 0$ when the firing angle varies in the range $90^0 \leq \alpha \leq 180^0$, thus a continuous variation of the reactive power is achieved.

Because of the variation of the firing angle the dynamic characteristics of this system differ from a system which is aquidistant sampled. An mathematical approximation shows clearly that this system with puls-phase modulation and puls-amplitude demodulation has a better dynamic behaviour as an aquidistant sampled system. For example: The dynamic characteristic is approximated by the statistical value of the time delay between variation of the input signal and corresponding variation of the output signal. The result is a statistical value in the hyperspace with the coordinates statistical value T_W, operating point of the converter α_0, amplitude of the variation $\Delta\alpha$ and moment of the variation t_i.

$$T_W = f(\alpha_0, \Delta\alpha, t_i)$$

The mean value of this statistical value gives a good approximation of the dynamic characteristic (1). Mean value T_w, aquidistant sampled system in every half-period of the mains $T_W = T/4$ (one phase); Mean value T_W of the installed system $T_{W1} = T/8$ (one phase, $\Delta\alpha \rightarrow 90°$) and $T_{W3} = 3T/16$ (one phase, $\Delta\alpha \rightarrow 180°$).

An interpretation of these results T_{W1} and T_{W3} are:

a) T_{W1}: The greater the amplitude of the variation $\Delta\alpha \rightarrow 90°$ the shorter the delay time (puls-phase modulation)

b) T_{W3}: The greater the amplitude of the variation $\Delta\alpha \rightarrow 180°$ the longer the delay time (puls-phase modulation, but additional the longer the period of the interval without any current (non continuous current). This interval without any current represents an overshoot and thus shortens the effective delay time.

Consequently the continuous controlled reactive power source has a better dynamic quality as an aquidistant sampled reactive power source.

STATIC AND DYNAMIC CHARACTERISTICS OF THE COMPENSATING EQUIPMENT

The equipment shown in Fig. 3 is connected to the point of common coupling with a rated short-circuit capacity of 4.8 GVA by way of a line and a transformer.

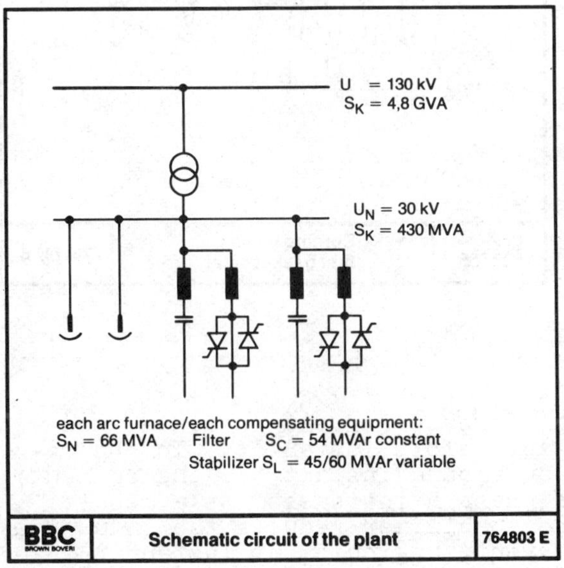

Fig. 3

The "load" consists of two arc furnaces each with nominal rating of S_N = 66 MVA and a power factor of cos N = 0,8 to 0.6 during melting. In the case of a three-phase short-circuit between the electrodes the short circuit capacity is S_K = 132 MVA and the power factor is cos φ = 0.1.

The compensating equipment comprises two identical plants. Each plant consists of filters which are tuned to the harmonics of the complete system, and the stabilizer which constitutes the source of variable lagging réactive power. To ensure complete balancing and compensation by the compensating equipment, the inductive currents of the stabilizer are independently and continously variable in all three phases. The compensating equipment thus meets all requirements placed on the static attributes of high-quality compensators.

However, the effectiveness of compensating equipment is determined both by its static and by its dynamic qualities. So far, only the static attributes have been considered, for instance, whether the equipment is capable of "correcting the power factor to any desired value" or "balancing unsymmetrical loading".

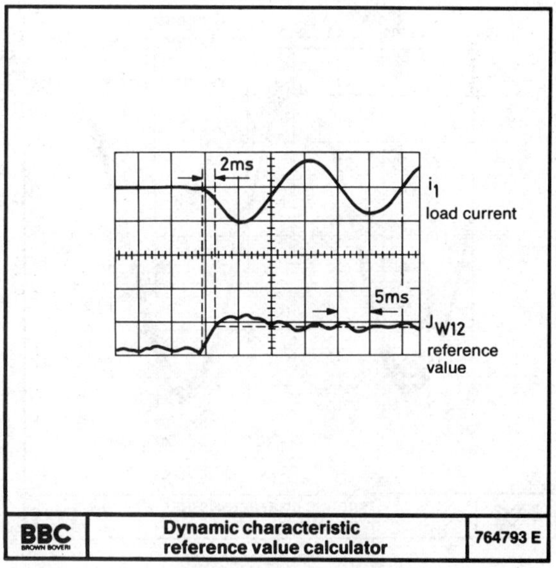

Fig. 4

The dynamic characteristic of the equipment consists of two components, i.e. the measuring time of the reference value calculator and the adjusting time of the power element. The dynamic response of the reference value calculator is shown in Fig. 4. In this experiment, a resistive-inductive load was connected between phases RS which causes a current to flow in addition to the load current. This additional, unsymmetrical current changes the reference values for the stabilizer currents. The measuring time of the reference value calculator is very short and is of the order of about 2 ms. The d.c. component in the load current, which is caused by the switching operation is included in the reference value.

One of the most important requirements placed on compensating equipment for flicker reduction - fast response of the complete system with a short measuring time for reference value calculation - is fully satisfied by the system, compared with the measuring time of 20 ms for the integration over one period.

The dynamic quality of the complete system in a test, with all three phases evenly loaded, is shown in Fig. 5. In this test, sinusoidal current reference values with a frequency of f = 10 Hz and an amplitude of Q_L = 20 MVAr are preset for the power element. The reference value calculator detects the current changes at the 30 kV busbar, the output signal constitutes the change in the reactive power.

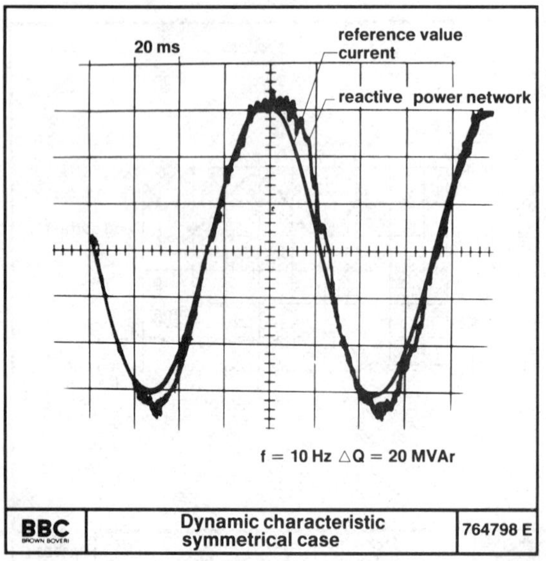

Fig. 5

In this test, the sequence "measuring time - adjusting time" for normal operation was thus reversed for experimental operation. The test result therefore shows the overall dynamic attribute of the equipment.

The good dynamic attribute can be seen from Fig. 5. The response time is less than 10 ms. Owing to the three single phase power regulators operating independently of each other and to the statistic changes of the dead times of the power converters, differences in amplitude and phase are liable to occur between the reference value and the actual value; these differences must change statistically.

In the most unfavourable case, the instances at which the reference and actual values pass through zero, are displaced with respect to each other by about 4 ms. The amplitudes of the reference and actual values differ momentarily on reduction of the power.

An oscillographic record of a furnace duty cycle during scrap melting is shown in Fig. 6. The upper curve represents the furnace current in phase R, the lower curve the stabilizer current in phase R which serves both for compensation between the phases RS and for balancing between phases ST and TR.

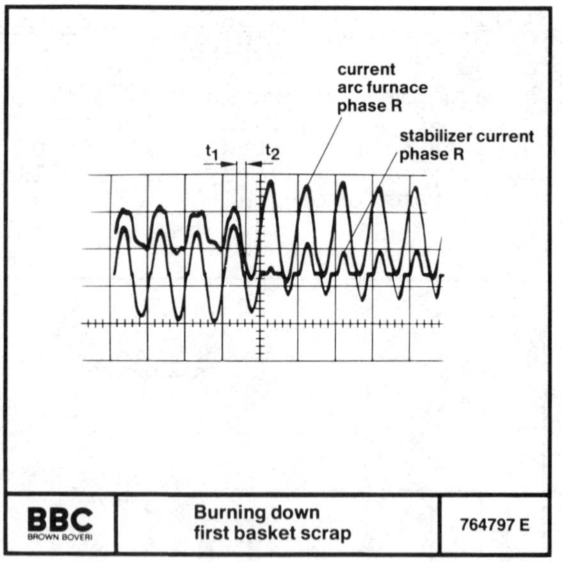

Fig. 6

The furnace current is almost constant before instant t_1. In the region around instant t_1, the current in phase R increases. As a result. the stabilizer reduces the inductive current, beginning at instant t_2, so that a capacitive current is available for compensation. The time difference $t_2 - t_1$ is less than 5 ms.

Since the second large furnace current half-cycle (positive) is larger than the first large furnace current half-cycle (negative), the stabilizer must continue to reduce the current. Since the third large furnace current half-cycle (negative) is smaller than the second large half-cycle, but larger than the first large half-cycle, the stabilizer must increase the current again in the third half-cycle. In addition to the quick power variation of the furnace, a slow power variation is noticeable in the last three cycles recorded. The stabilizer current follows this variation.

The dynamic characteristics of the compensating equipment can also be seen from Fig. 7. In this figure, the following parameters are recorded over a period of about two seconds: the reactive power of phase R of the furnace, the reactive power of the compensating equipment in phase R, the reactive power in phase R of the mains, the furnace current in phase R and the current of the compensating equipment in phase R. The reactive powers were measured with a measuring system corresponding to the measuring system in the reference value calculator. The period of measurement is at the beginning of the scrap melting cycle of the first basket.

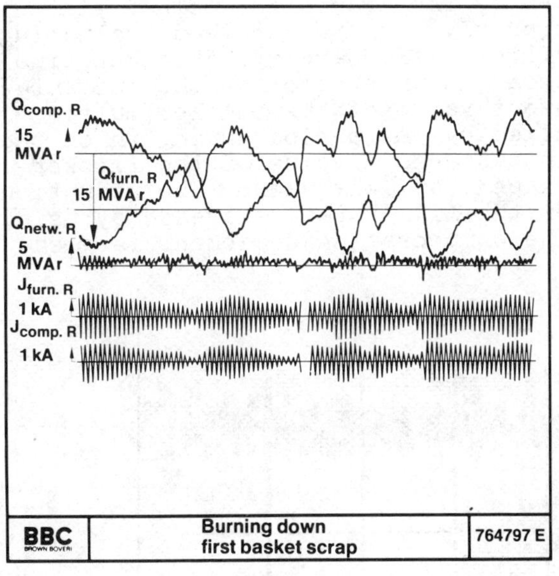

Fig. 7

It can be seen from Fig. 7 that the furnace produces considerable reactive power swings with very steep reactive power changes. Despite the very steep reactive power changes in both directions and reactive power swings up to 15 MVA in one phase, the mean reactive power consumption from the feeding power system remains almost constant and capacitiv due to $\varepsilon = R/X$ of the supplying network. The signal for the system reactive power has a high-frequency noise, due to the measuring method, superposed on the mean value. The low-frequency changes in the system reactive power are less than 2 MVAr in phase R. Fig. 7 also shows the good dynamic quality of the compensating equipment for the reactive power changes.

INVESTIGATION ON THE VOLTAGE STABILIZATION

Information on the voltage stabilization and flicker reduction can be obtained in two different ways. The first one takes account of the stochastic nature of the processes taking place in the furnace.

The information on flicker reduction is obtained by comparing tests made with different data of the compensating equipment. During the test the data of the compensating equipment are kept constant, and the duration of the test may be a week or so.

With the second method, the various tests are performed successively, with the complete system in the same operating state. This method offers the advantage that the effectiveness of the different test conditions can be compared quickly. This comparison is easier when there are major differences in the results.

In order to facilitate comparisons between different test conditions the second method - the direct chronological succession of the tests at the beginning of a charge - was adopted. The tests show the situation at the beginning of "Melting of scrap - first basket".

The first test begins with the melting phase and takes about five minutes. In this first test, the compensating equipment, independently of the active power, stabilizes the reactive power in the supply system to a desired constant value.

In the second test, the compensating equipment is completely switched off during a period of about four minutes. The furnace draws active and reactive power from the supply system.
In the subsequent third test, the compensating equipment is reconnec-

ted. The equipment stabilizes the reactive power consumption to an adjustable, constant power factor. During the third test, the second furnace momentarily starts with melting the first basket with scrap. Fig. 8 shows the results of the various tests.

A comparison of the deflections of the FGH flickermeter for comparable variations in the total reactive and total active power swings gives the following ratio between the tests when making an unfavourable estimate

$V\,1 : V\,2 : V\,3 \approx 0.45 : 1 : 0.5$

In reality, the flicker reduction is higher since, in the second test, the FGH flickermeter was temporarily overmodulated. For instance, the ratio of the flicker between test 1 and test 2 shortly before and after change-over is

$V\,1 : V\,2 \approx 0.35 : 1$

This proves that a considerable flicker reduction can be achieved thanks to the static and dynamic attributes of the compensating equipment used. In particular, it should be borne in mind that values below 0.3 are no longer perceptible. This can mean that any flicker that is objectionable when no compensating equipment is used, is below the flicker perception borderline when the compensating equipment is connected.

A comparison of the results of the Era or EdF method with those obtained with the FGH methods is not possible. Both the Era and the EdF method are methods permitting a yes-no decision to be taken as to whether or not the voltage fluctuations at the point of common coup-

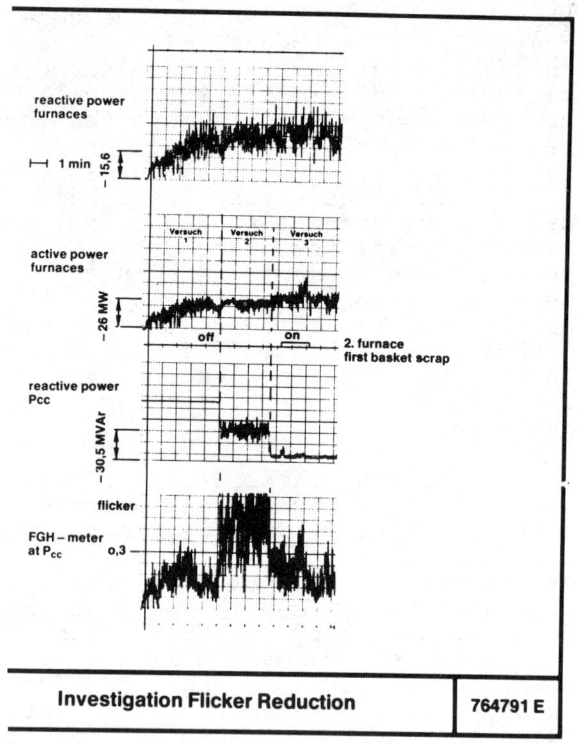

Fig. 8

ling give rise to complaints through flicker disturbances. The FGH method, however, indicates the instantaneous value of the flicker perception. An extension of the FGH method with a view to obtaining information provided by the Era or EdF method will probably lead to a method evaluating "flicker-time areas". "Flicker-time area" is understood as the instantaneous value of the flicker multiplied by the period during which the instantaneous value of the flicker exceeds a limit value. The reduction of the "flicker-time areas" by using the compensating equipment is grea-

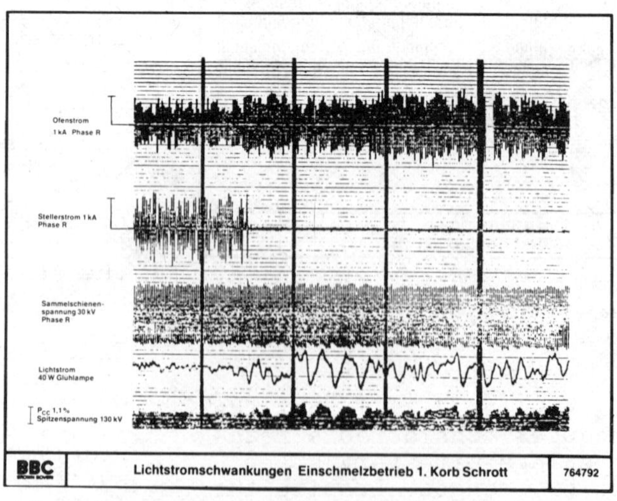

Fig. 9 Fluctuations of the Light Intensity

ter than the reduction of the flicker instantaneous values because of the fact that the compensating equipment reduces both the instantaneous value and the duration of the flicker.

To complete the description of the effectiveness of the compensating equipment, the furnace current of phase R, the stabilizer current of phase R and the light intensity of a 40-watt incandescent lamp have been recorded in Fig. 9 at a high recording speed. The period of observation covers a total of three seconds during change-over from test 1 to test 2.

From Fig. 9, the following can be seen: During test 1 with connected compensating equipment small high-frequency fluctuations predominate in the light intensity. When the compensating equipment is disconnected, large low-frequency fluctuations can be clearly noticed in the light intensity. The ratio of the amplitude variations of the light intensity is

V 1 : V 2 = 0.25 : 1.

(1) D. Schröder
Spannungsstabilisierung in Drehstromnetzen
IFAC congress 1974 S. 633 - 654

REACTIVE POWER STATIC COMPENSATION AND HARMONIC FILTERING IN A METAL INDUSTRY PLANT

C. Boisdon and M. Boidin

Jeumont Schneider, 31, 32 Quai National, 92806 Puteaux, France

SUMMARY

In a metal industry plant, the installation of a new blooming and its thyristor power supply sat the probleme of disturbances caused on the mains by this equipement. These disturbances are minimized by the installation of a static compensator.
The same industrial network includes other variable loads, amongst which a 18 T furnace.
The present paper recalls the disturbing characteristics of such loads and the permissible limits of disturbances.
Then the design and size of the compensating device will be defined.
The satic compensator, its control and performances will be analysed in detail.
The technology and performances of that installation are then presented as an exemple.

A. INTRODUCTION

The growing use of thyristor electronic converters allows to carry out variable speed drives with high dynamic performances.
Butthe high gradients of the current as well as the converter operation itself entail increased reactions on the mains. The level of those disturbances can be too high to be permissible ; it leads to the need of a compensating system which we shall describe here : the "stato-compensator".

B. DISTURBING LOADS IN A METAL INDUSTRY PLANT

B.1 Rectifier reactive power demand

In up to date rolling-mills, DC drive are supplied by thyristor converter units.
Seen from the mains, these converters operate with an apparent power proportional to the load direct current. Therefore their power factor (cos ϕ) is directly related to the control angle (α), according to the following relation :

$$\cos\phi = \cos\alpha - \frac{E_x}{E_{do}}$$

Where : $\frac{E_x}{E_{do}}$ is the reactive voltage drop the divided by the rectified voltage (no load ideal voltage).
The absorbed reactive power fluctuates with the current together with the thyristor converter delay angle. One method used to reduce this reactive power demand is the asymmetrical gating applied to two series-connected converters, for instance.

Figure 1 represents the diagram : active power vs. reactive power in a normally connected converter (Fig. 1a) and in a delayed gating converter (Fig. 1b) for given value of the DC load.
The full line drawing does not take into account the commutating effect. This effect is represented by the dotted line.
In inverter operation the reactive power increases with thefiring advance angle, but it decreases slightly when the commutation is considered.

B.2 Harmonics resulting from rectifier operation

Due to its operation, the supply current of a rectifier it not purely sinusoïdal.
Practically it operates as a high impedance harmonic current source. Those harmonics are odd ones, not multiple of 3, i.e. 5th, 7th, 11th, 13th order... : $6\,k \pm 1$ for a 6 phase connection.
For a 6 phase rectifier unit, if the commutating effect is not taken into account, the alternating current can be expressed as follows :

Let : I_d : direct current in the load
E_s : secondary line to neutral AC voltage
E_{LN} : primary line to neutral AC voltage.

Then :

$$i = \frac{2\sqrt{3}}{\pi} I_d \cdot \frac{E_x}{E_{LN}} \{\sin(\omega t-\alpha)-\phi\} - \frac{1}{5}\sin\{5(\omega t-\alpha) - 6\phi\} - \frac{1}{7}\sin\{7(\omega t-\alpha) - 6\phi\} + \frac{1}{11}\sin\{11(\omega t-\alpha) - 12\phi\} + \frac{1}{13}\sin\{13(\omega t-\alpha) - 12\phi\}$$

The harmonic current phase delay is proportional to the rectifier delay angle (α) and to the harmonic order. Conversely, a geometrical phase delay ϕ resulting from the transformer primary to secondary connection will involve a $6\,k\,\phi$ phase delay for the $6\,k \pm 1$th harmonic.

B.3 Arc furnaces

In an arc furnace, the arc resistance can rapidly fuctuate from 0 (short-circuit) to infinity (non-conduction or unidirectional conduction of an electrode). The furnace requires a large quantity of electrical energy with poor lagging power factor, to optimize production {2}.
Thus, in the melting critical phase, the arc furnace will demand a rapidly varying reactive power and draw unbalanced currents.

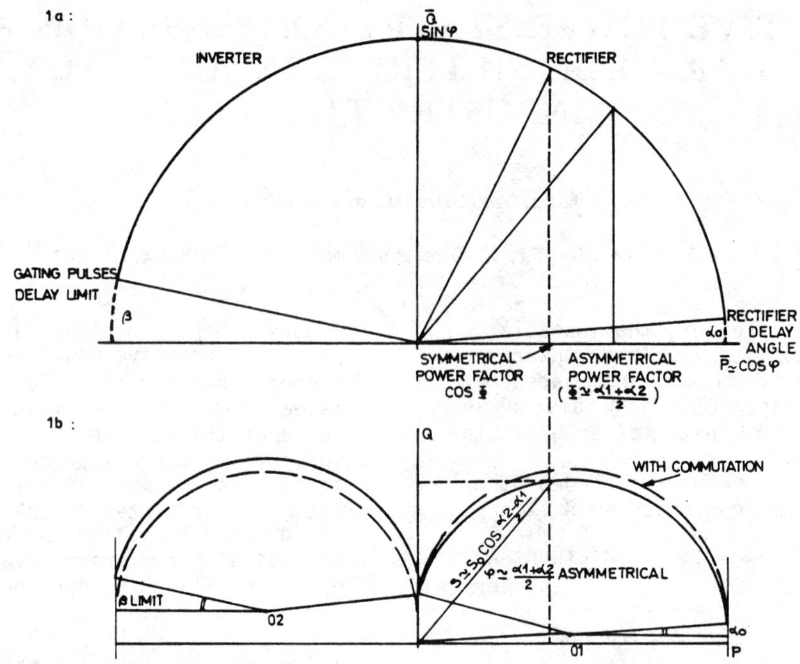

Fig. 1a : Symmetrical converter

Fig. 1b : Asymetrically gated converter reactive power VS active power diagram

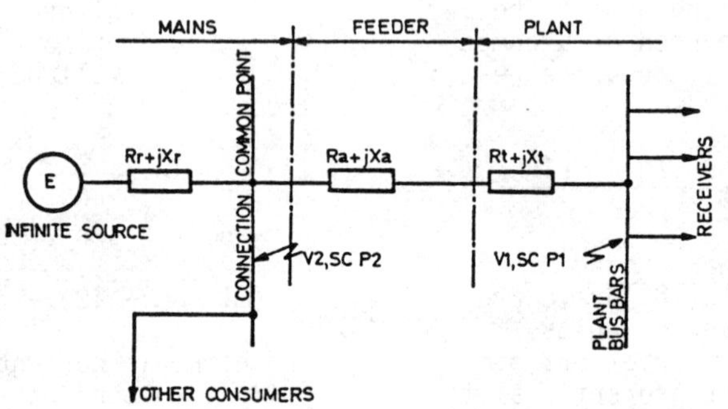

Fig. 2 : **Plant power supply diagram critical busses**

The arc furnace non-linear load and the presence of erratic no-conduction half-cycles explain the appearance of even and odd harmonic currents with low order and high amplitude.

C. DISTURBANCES AND THEIR PERMISSIBLE LIMITS

On the diagram represented on figure 2, two interesting points will be defined.

- firstly, the connection point to the mains, where other users require a good quality of voltage (voltage fluctuation, harmonic distorsion).
- plant busbars where the user's concern is to avoid disturbances on his own loads only.

C.1 Voltage drop

The flow of active power P and reactive power Q in a network with a given short-circuit power : P_{cc}, entails a relative voltage drop as follows :

$$\frac{\Delta U}{U} = \frac{1}{P_{cc}} \left(Q + \frac{P}{tg\psi}\right) + \frac{1}{2P_{cc}^2} \left(\frac{Q}{tg\psi} + P\right)^2$$

Where $tg\psi = \frac{X}{R}$

In fact, the second term is negligible, therefore the voltage fluctuation in a given point of the network is, with a good approximation :

$$\overline{\Delta U} \simeq \frac{1}{P_{cc}} \left(Q_{max} - Q_{mini} + \frac{P_{max} - P_{min}}{tg\psi}\right)$$

Voltage drop is increased by the possible unbalance between line currents.

C.2 Voltage drop limit values

On an industrial set of busbars, repetitive voltage fluctuations shall normally be less than or equal to 5 %. The power supplier is anxious to avoid any flicker (variations in the light output of electrical bulbs) at the connection point to the mains {4}

The flicker is due to periodic or erratic rapid fluctuations of the RMS voltage in the 0,5 to 25 Hz frequency band.

The disturbance is proportional to the integral of the squared voltage fluctuation amplitude. It depends upon frequency, the perception level being maximum at 10 cycles/second. The flicker perception threshold is 0,3 % squared minute. This value is the limit slope of permissible "flicker dose".

As regards the current unbalance, for sophisticated installations, the reasonable limit would be :

$$\frac{|I_i|}{|I_d|} < 0{,}05$$

Where $|I_i|$ = module of the negative sequence current component

$|I_d|$ = module of the positive sequence current component.

C.3 Harmonic effects

The flow of harmonic currents into the network impedances results into harmonic voltages superimposed to the fundamental voltage {4}

The RMS value of residual harmonic voltage is defined as :

$$r = \frac{\sqrt{\Sigma E_h^2}}{E_1}$$

If this residual voltage does not exceed 5 %, the disturbance for the users is considered as permissible. The power supplier shall have to take into account the cumulative effect of several user's harmonics sources. Thus in France allowance is such as the harmonic currents, injected into the mains by a user, may generate an harmonic voltage lower than or equal to 1 % per odd frequence, and lower than or equal to 0,5 % per even frequence, the total residual harmonic voltage being lower than 1,5 %

D. DESIGN AND RATINGS OF THE COMPENSATING SYSTEM : THE STATO-COMPENSATOR

D.1 Design

The stato-compensator is composed of two units connected to the mains (Fig. 3) :

- a bank of static capacitors, connected as a harmonic filter, generating a fixed reactive power Q_C. It increases the plant power factor to the required value and presents a low impedance to the harmonic currents generated by the rectifiers.
- a unit of three delta-connected reactors, each one being connected to the mains by a thyristor controller. They absorb three independently variable reactive powers on each of the three lines, their sum being Q_L.

The reactive power demand to the mains will be maintained constant while the load reactive power demand Q_R is fluctuating, if :

$$Q_L + Q_R + Q_c = \text{constant}$$

D.2 Stato-compensator characteristics

The use of the stato-compensator allows an accurate, continuous, high speed, phase independent compensation, of reactive power disturbances.

It can meet with any type of load, whether balanced or unbalanced.

It allows an automatic control of the power factor, or of the reactive power at a preset constant value. Thus, it reduces,

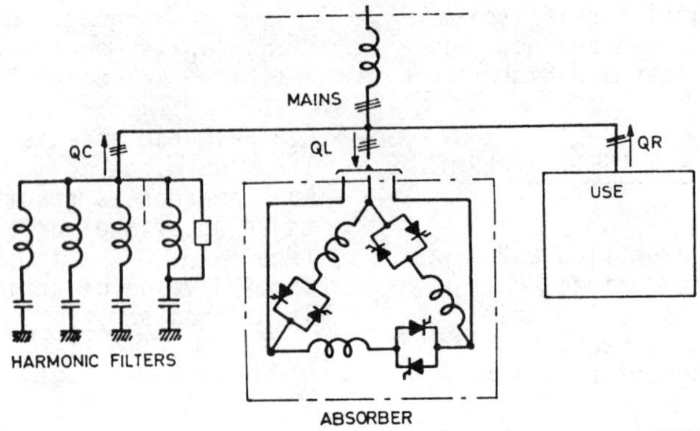

Fig. 3 : Stato-compensator - single line diagram

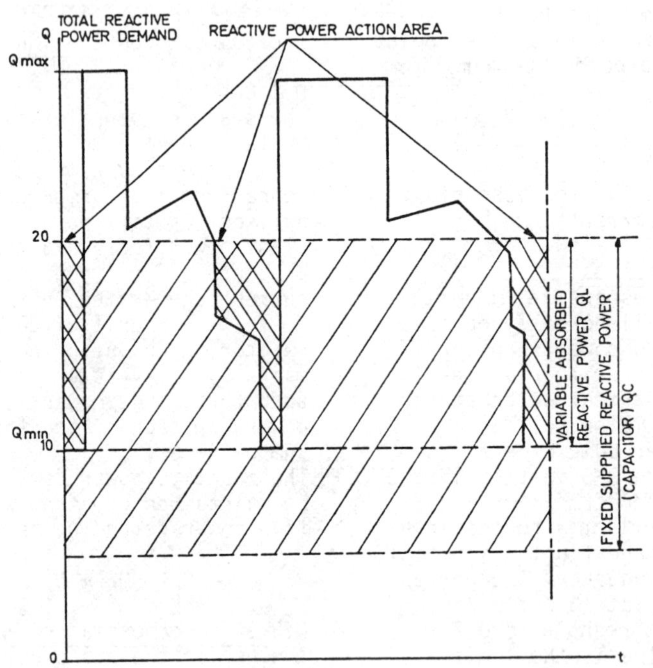

Fig. 4 : Fluctuations of the total reactive power demand
Operating range of the compensator during 2 typical passes of the blooming

or cancels, any voltage fluctuation resulting from load surges. Its operation is independent from the supply voltage variations : so it needs no kind of tap changing under load.
It ensures the filtering of harmonics generated by the load and prevents any parallel resonance with the network impedance.
According to its rating, it gives to the busbar, which it is connected to, the characteristics of a network with a nearly infinite power characteristics, wihout taking a part in the short-circuit power endured by the switchgear in case of fault.
Being designed according to a high voltage technology, for high powers, the stato-compensator allows to avoid the use of any transformer.

D.3 Rating
The absorber - the adjustable element of the compensator - is rated in relation to the reactive power fluctuation to be compensated :

$$\Delta Q_L = - \Delta Q_R$$

As a general rule it is not necessary to compensate entirely reactive power fluctuations, but the voltage fluctuations should be maintained within reasonable limits, on the compensated busbar set.

If $\frac{\Delta U}{U}$ is the p.u. voltage drop without compensation, and $\frac{\Delta U_1}{U}$ the rated permissible p.u. voltage drop, the maximum absorber power is :

$$Q_L = \left(\frac{\Delta U}{U} - \frac{\Delta U_1}{U}\right) \cdot P_{cc}$$

P_{cc} being the short-circuit capacity of the compensated busbar set.
Figure 4 illustrates the operating range of the stato-compensator and the total reactive power demand of the hereabove mentioned loads.

D.4 Plant power factor improvement ; capacitor bank rating
In any point of the operating cycle of the plant, the calculation gives the active power and the total reactive power demand to the mains, taking into account the reactive absorber operation.
The average value of P and Q for a cycle will be deduced. To obtain a power factor at least equal to the required value, the condition is :

$$Q_c = Q - P\,tg\phi$$

This result will be reduced by the mill duty factor.

D.5 Harmonic neutralization
The capacitor bank, whose power has been determined, is split into an appropriate number of parallel branches, each one being tuned to one of the major harmonics encountered in the network.
The design of each harmonic filter is done, once the following elements being known :
- the harmonic current existing in the filter
- the required degree of harmonic attenuation
- the best condition on the economical point of view, taking into account the cost of the investment and of the capitalized losses. Moreover the total RMS current of the bank shall be maintained below 1,3 times its rated fundammental current. The rated voltage of the bank will be :

$$U_c = U \cdot \frac{h^2}{h^2-1} + \Sigma\, I_h \frac{1}{hC\omega}$$

A low capacitance will bear a high voltage rise, owing to harmonic current flow.
The harmonic voltage attenuation is another rating factor. For $h \geqslant 5$, it can be expressed as :

$$\frac{U_h}{U} \# 3U \cdot \frac{I_h}{h} \cdot \frac{1}{q\, Q_c}$$

U, U_h = line to neutral fundamental and harmonic voltages.
Q_c = reactive power generated by the filter under consideration
$q = \frac{Lh\omega}{\Sigma R}$ = quality factor.

Bus frequency fluctuations, together with the filter elements constructive deviations (or tolerances), lead to foresee a detuning factor of the filter :

$$\delta = \frac{\omega_h - \omega_{o_h}}{\omega_{o_h}}$$

ω_{o_h} = filter tuning frequency
ω_h = frequency of the harmonic to be filtered.
For small frequency tuning variations

$$\Delta\delta = \frac{\Delta\omega}{\omega} + \frac{1}{2}\left(\frac{\Delta C}{C} + \frac{\Delta L}{L}\right)$$

The filter impedance is no longer minimum, and can be expressed as :

$$Z_h = R_h\,(1 + 2jQ\delta)$$

The quality factor varies between 30 and 50, which corresponds to a band value homogeneous with the foreseen variations.
As seen from the terminals of harmonic current sources, filter impedance is in parallel with the network impedance : therefore a parallel resonance, with ω_p = angular frequency lower than filter series resonance frequency, may develop :

$$\omega_p = \frac{1}{\sqrt{(L + L_R)\, C}}$$

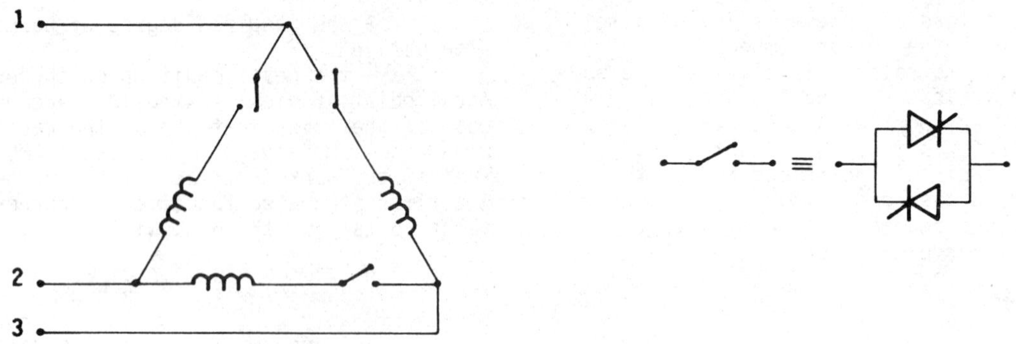

Fig. 5

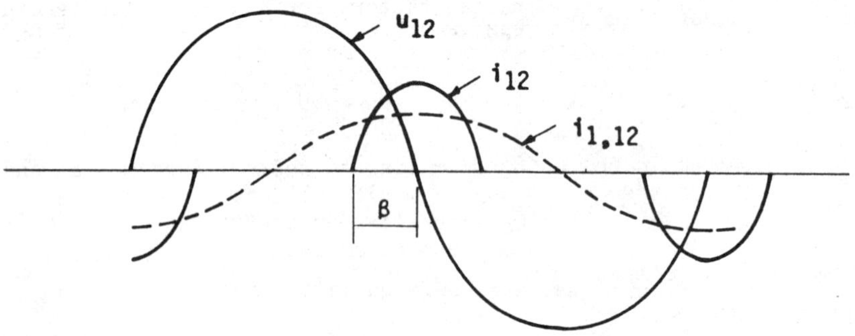

Fig. 6

Fig. 9

Fig. 7

Fig. 8

A complete analysis of the curve impedance versus frequency, is necessary.
The selection of filter orders and ratings must be carried out in such a way that there should be no inserted harmonic currents foreseen at parallel resonance frequencies.

E. STATIC COMPENSATOR ADJUSTING ELEMENT : REACTIVE ENERGY ABSORBER

E.1 Principle of operation

The absorber is a unit including 3 delta-connected reactors.
A thyristor static switch, series-connected in each branch of the delta, allows to control the amplitude of the fundamental frequency current for each branch. So appears one of the advantages of this connection : an independent control of reactive energy circulating in a line.
Taking into account the selected axes the Fourier expansion (crest value) gives :

$$n = 1 \qquad b_1 = \frac{1}{\pi}\,(2\beta - \sin 2\beta)$$

$n \neq 1$
(odd figure)

$$b_n = \frac{2}{\pi}\{(-1)^{\frac{n-1}{2}} \cdot \frac{\sin(n-1)\beta}{n(n-1)} + (-1)^{\frac{n+1}{2}} \cdot \frac{\sin(n+1)\beta}{n(n+1)}\}$$

To get a proper study of harmonics that are injected into the line, it must be distinguished between balanced and unbalanced loads.
With balanced loads "3n" order harmonics disappear on line wire, the other harmonics remain with an amplitude which is $\sqrt{3}$ times the branch amplitude and with the usual sequence.
In those conditions, this connection reinserts into the line a small content harmonics, about 4 times less than a Graëtz bridge of equal rating.
With unbalanced loads, all terms of the Fourier expansion lead to unbalanced harmonic systems, generating, in the line, balanced systems with positive and negative sequence components. This justifies the presence of a third harmonic filter.

E.2 Description of absorber

Two technologies are used, according to a low or to a medium voltage use.

E.2.1 Low voltage. As a general rule, for ratings lower than or equal to 10 MVAR approximately, low voltage technology, - a classical one, - is cheaper even though it requires a step-down transformer. Fig. 7.
Reactors have no magnetic core. There is a reactor for each branch of the absorber. Fig. 8.

E.2.2 Medium voltage. With this technology, the absorber is directly connected to the medium voltage network (10 to 33 kV). The thyristor switch is of modular construction : it consists of two medium voltage valves connected back-to-back. Each one consists of a series parallel connection of modules. Each module has the following essential characteristics :

- rated RMS voltage 6 kV
- maximum peak voltage 25 kV
- rated RMS current 800 A
- maximum peak current 1500 A

Each self-contained module consists of 10 thyristors, pressure-tightened between heat-sinks.
It is cooled by oil circulation. Fig. 9.
As it is the case for the low voltage technology, the reactor coils have no magnetic core. However, to get a better distribution of potentials in relation to earth, and to minimize fault currents, each reactor is divided into two elements located on either side of the thyristor switch.

E.3 Process control and regulation

E.3.1 Principle. Let us consider a network, "ideally perfect" in point A, defined by its positive sequence voltage $\overline{E}_d$. Fig. 10.
The line impedance, as seen from B, is defined by $\overline{Z}_d$, $\overline{Z}_i$. Should any load be connected to point B, positive and negative component currents $\overline{I}_d$ and $\overline{I}_i$ will result. Assuming that the system is linear, the following equations may be derived : {6}

$$\overline{V}_d = \overline{A}\,\overline{I}_d + \overline{B}\,\overline{I}_i$$

$$\overline{V}_i = \overline{C}\,\overline{I}_d + \overline{D}\,\overline{I}_i$$

Network relations are :

$$\overline{E}_d = \overline{V}_d + \overline{Z}_d\,\overline{I}_d$$

$$0 = \overline{V}_i + \overline{Z}_i\,\overline{I}_i$$

In order to get a voltage and current balanced network, i.e. to cancel $\overline{I}_i$ and $\overline{V}_i$, it is necessary and sufficient to get $\overline{C} = 0$.
This condition can be achieved by the use of two control parameters ; therefore, a third parameter is sufficient to control the reactive power absorbed in the balanced network.
Those three parameters are the currents of the absorber branches. The hereabove mentioned control rules do not consider the load nature whatever it may be : balanced or unbalanced.
However the load nature has an effect upon the size on steady state conditions. This fact has already been noted by several authors.
Let us assume a variable balanced purely

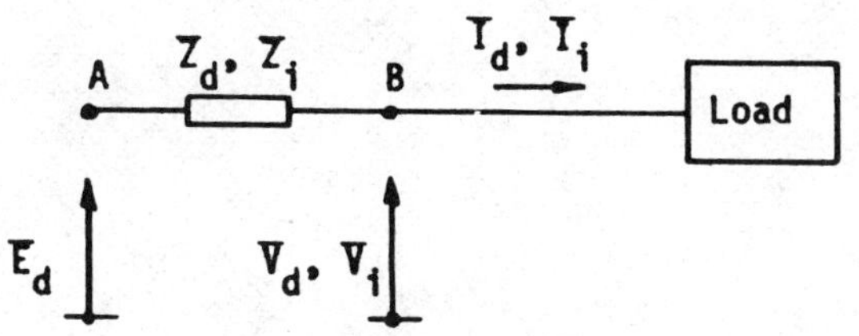

Fig. 10

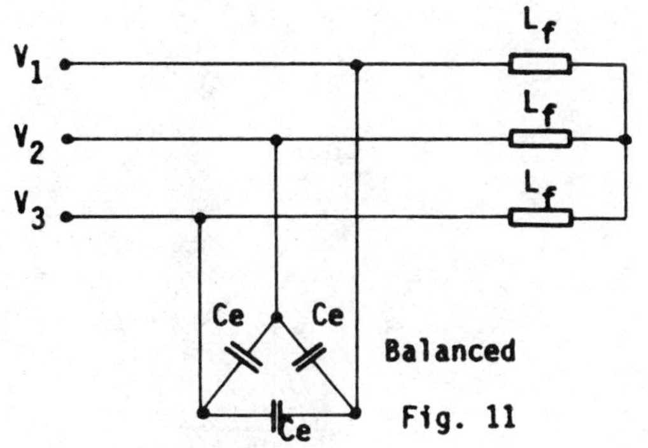

Balanced

Fig. 11

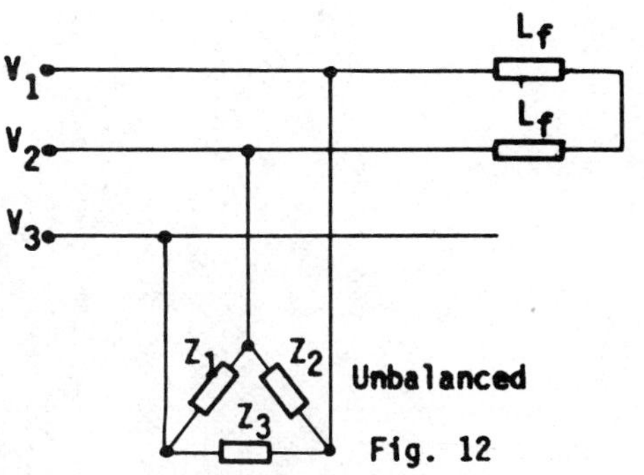

Unbalanced

Fig. 12

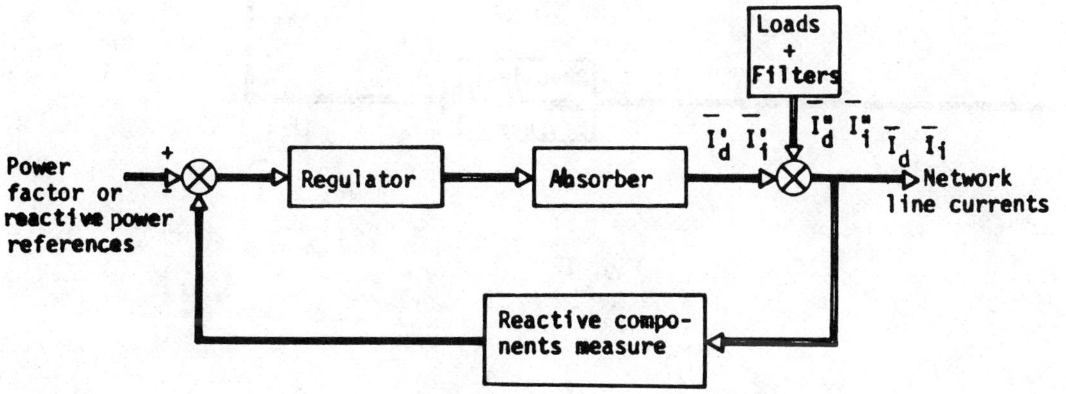

Fig. 13

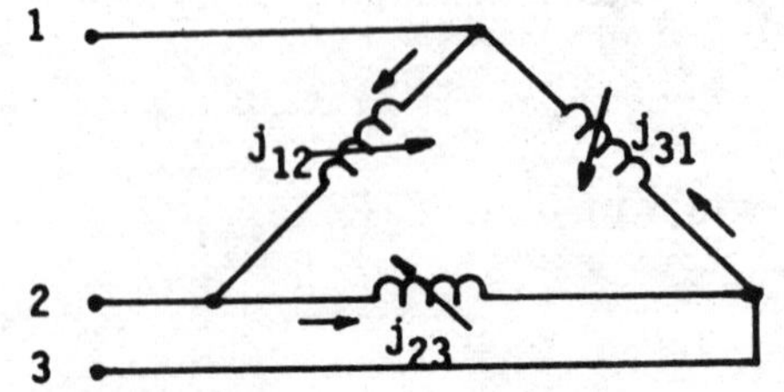
1
j_{12}
j_{31}
2
j_{23}
3

Fig. 14

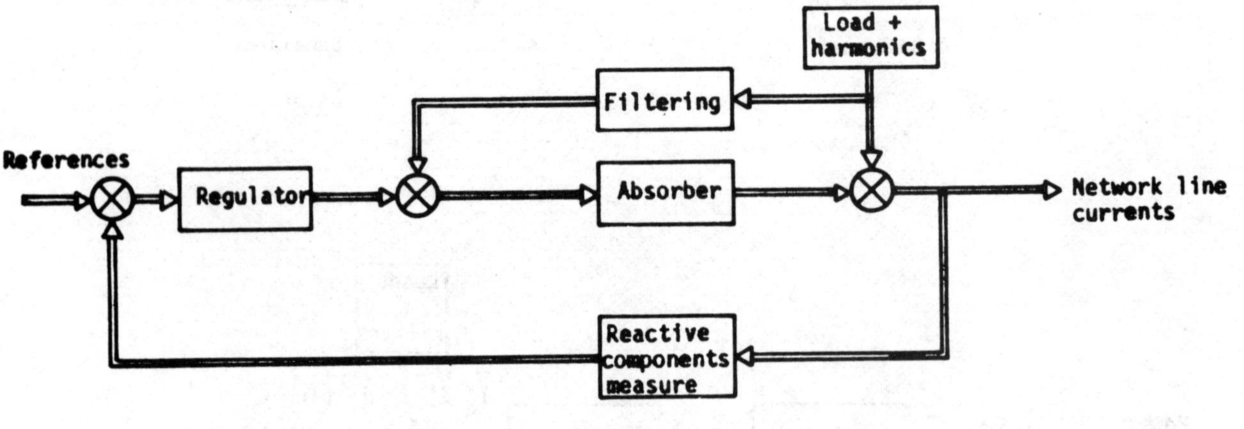
Load +
harmonics
Filtering
References
Regulator
Absorber
Network line
currents
Reactive
components
measure

Fig. 15

reactive load "Lf" : it is necessary to install 3 adjustable capacitors of "Ce" value, to compensate for the power factor.
Now, let us assume that this load is unbalanced : a line to line two phase short-circuit for instance.
Fig. 11 and Fig. 12.
In those conditions, to get a balanced network current system and to control the power factor, those devices will have to be as follows :

- Z_1 will be a capacitive reactance
- Z_2 will be an inductive reactance
- Z_3 will be a capacitive reactance.

It follows that :

- the absorber will have to absorb more reactive power than is delivered by the bank of capacitors,
- the bank of capacitors should be 1 1/2 more important than it would be necessary to compensate for balanced fluctuation.

Practically the size of capacitors and reactors is reduced by taking in account the tolerance of voltage fluctuation.

E.3.2 Regulation. The control block-diagram will present two different outlines according to the presence of rapidly fluctuating loads (blooming, rolling-mill) or very rapidly fluctuating loads (arc-furnaces).
Fig. 13 shows the bloc-diagram of the process. The controlling device is the reactive power absorber.
If we assume that the process is correctly operating, the network voltage does not vary significantly ; the vectorial sums of the load current and the capacitor currents, used as filters, can be considered as a disturbance.
In that case, the analysis of the line currents is sufficient to control the process.
As it is known, the regulation mode is based upon the cancelling of the negative sequence component of the active and reactive line currents components, and upon the control of the positive sequence reactive component.
j_{12}, j_{23} and j_{31}, the absorber branch currents, are the selected adjusting parameters. Fig. 14.
If $\overline{u}$ is a unit vector, colinear to $\overline{V}_1$, - the supply voltage - the symmetrical components of the absorber line currents are :

$$\overline{I'}_d = - j \frac{\overline{u}}{\sqrt{3}} (j_{12} + j_{23} + j_{31})$$

$$\overline{I'}_i = \frac{\overline{u}}{\sqrt{3}} \{\frac{\sqrt{3}}{2}(j_{12}-j_{31}) + j(j_{23}- \frac{1}{2} j_{12}- \frac{1}{2} j_{31})\}$$

Where $j = \exp j \frac{\pi}{2}$

Should the system neutral be ungrounded, the symmetrical components of the line currents can be expressed in relation to their reactive components. {3}

In those conditions :

$$j_{12} = - \frac{1}{\sqrt{3}} (I'_{dw_1} + I'_{dw_2} - I'_{dw_3})$$

$$j_{23} = - \frac{1}{\sqrt{3}} (- I'_{dw_1} + I'_{dw_2} + I'_{dw_3})$$

$$j_{31} = - \frac{1}{\sqrt{3}} (I'_{dw_1} - I'_{dw_2} + I'_{dw_3})$$

It can be seen that one branch current reacts upon the three line reactive components. Therefore if the line current reactive components are measured :
j_{12} will be applied to the linear combination :

$$I_{dw_1} + I_{dw_2} - I_{dw_3}$$

j_{23} will be applied to the linear combination :

$$- I_{dw_1} + I_{dw_2} + I_{dw_3}$$

j_{31} will be applied to the linear combination :

$$I_{dw_1} - I_{dw_2} + I_{dw_3}$$

E.3.3 System performances. After this demonstration, the system stability is to be studied.
It should be noted that this system presents a certain quantity of delays.
The absorber thyristor gating delay has a statistical value equal to 1/4 of the fundamental period. The line current reactive component is sampled when the corresponding voltage passes through zero. Time delay due to this sampling is approximatly 1/4 of the fundamental period, if the samplings are stored in memory and held by means of a zero order holder device.
These time delays lead to the control of a not-minimum phase shift system. A cut off frequency in open loop of about 70 rd/s is obtained by the addition of an integral corrective network with phase lead.
In these conditions, the response time to a disturbance is about 1 and 1/2, or two, fundamental periods.
This response time is sufficient to control unbalanced or balanced loads, rapidly variable (essentially drives with adjustable speeds). In case of fast varying loads (such as arc-furnaces during melt-down), to reduce flicker it is necessary to use a prevision loop from load current fluctuations. In those conditions, the response time that will be obtained is about one cycle, and this seems to be sufficient.
Thus the regulator has to ensure accuracy in steady state rating.
This type of control can be used for balanced

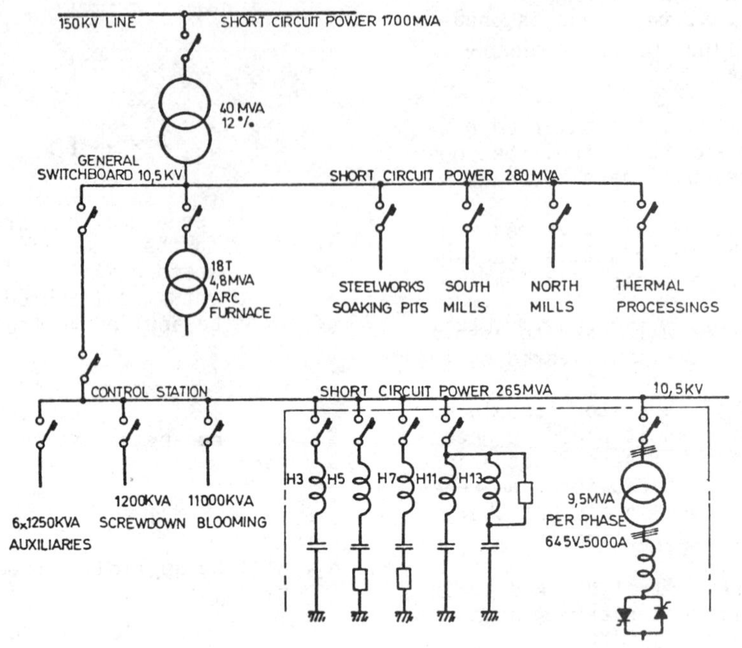

Fig. 16 : Blooming power supply general diagram

Fig. 17 : General view of harmonics filters

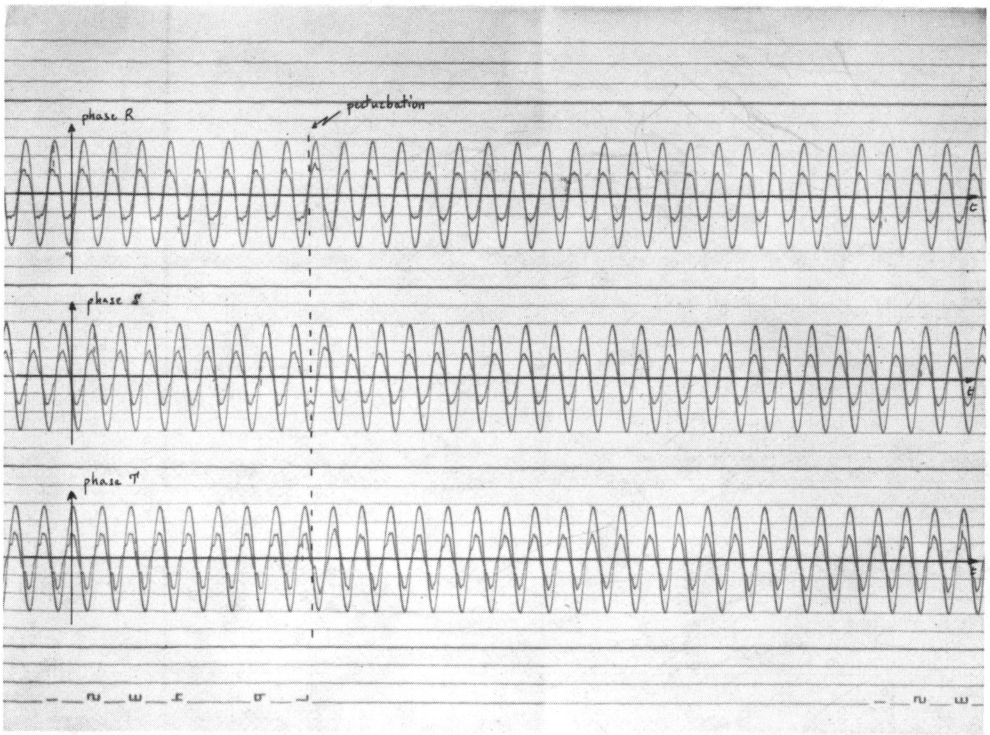

Fig. 18 : Response of system to a **disturbance**

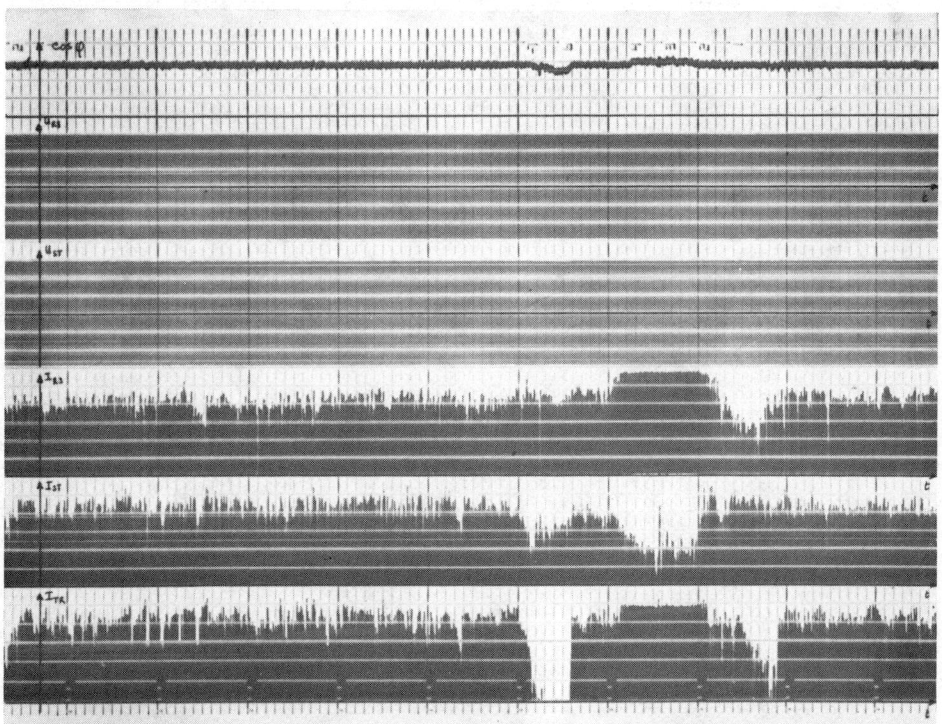

Fig. 19 : Operation of stato-compensator effect on power factor and line to line voltage

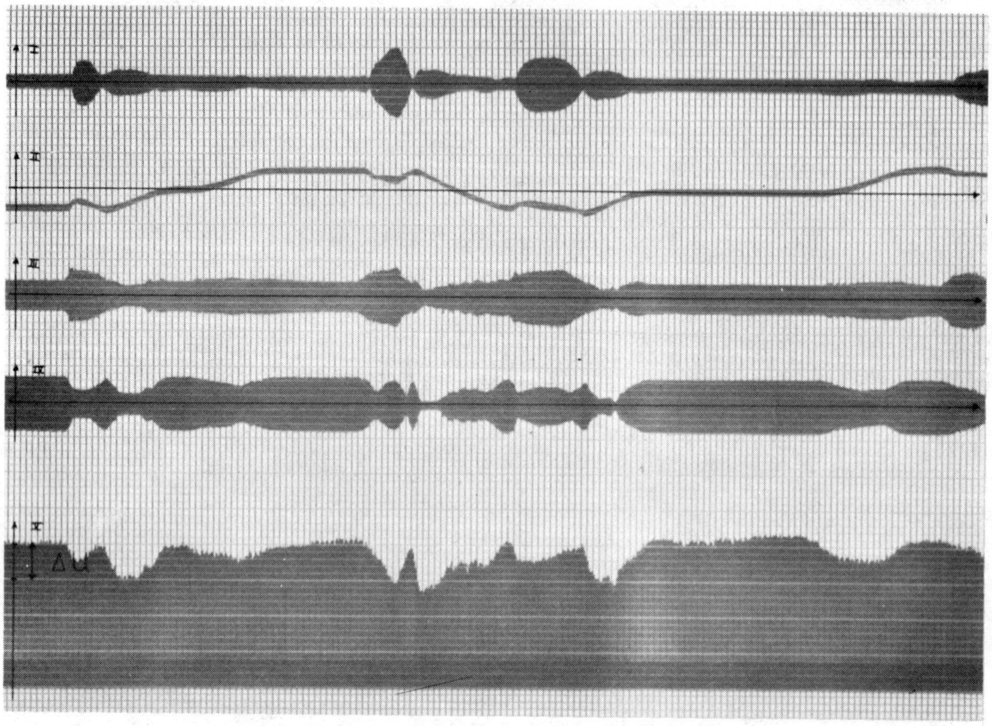

Fig. 20 : Reduction of blooming voltage fluctuation

or unbalanced loads, with a response time close to one cycle of the supply voltage and a significant static accuracy.
The control block-diagram is as follows. Fig. 15.

F. POMPEY STEEL-WORK BLOOMING MILL COMPENSATION

The Pompey blooming mill is supplied by a 150 kV network through a 40 MVA step-down transformer 150/10,5 kV (Fig. 16).
Before the installation of this mill, the reactive power consumed by the plant already exceeded the permissible limit and an average reactive power of 5 MVAR was invoiced by the mains supplier.
Besides, the 150 kV plant feeder load was close to the limit capacity and a reactive power overload was to be avoided.
The selected rating of the absorber was 9,5 MVAR. Taking into account the initial deficit and the maximum blooming load, the capacitor bank rating was stated at 14,5 MVAR.
The filters are tuned to the 3rd, 5th, 7th and 11th harmonics, and there is a high pass filter for the 13th harmonic and above. (Fig. 17).
The capacitors are of the plastic film type, with pyralen dielectric ; their losses are low : less than 1 watt/kilovar. An adjusting element is provided for each bank. The design of the reactors, made of laminated aluminum, takes into account the transient stresses appearing at time of capacitor energization.
The absorber is of the low voltage type, with two parallel 5 MVA dry transformers.
The stato-compensator control devices of the absorber, together with the capacitor bank control, are located in a control cabinet.
The control system, installed in Pompey plant, takes into account the blooming mill disturbances only, the arc furnace being considered during refining as a balanced and slowly varying load ; therefore the control does not include an advance loop.
Figure 18 allows to estimate the response of the system to a disturbance (filter 11 and 13 switched off) and its harmonic filtering efficiency. It shows three line to neutral voltages of the network, and line currents at 40 MVA transformer secondary.
Figure 19 illustrates the plant power factor, two line to line voltages of the network, and the three rectified currents of the absorber branches. The load consists of the blooming in operation and the 18 T arc furnace at the end of melt-down. It should be noted that the control system carries out a voltage balance owing to an independent conduction of the three branches.
The blooming voltage fluctuation, without stato-compensator is seven per cent.
With the stato-compensator, this voltage fluctuation is three and half per cent, as shown by the figure 20.
On this figure
Line I shows the motor current.
Line II : the motor speed.
Line III : the current on the mains.
Line IV : the absorber current.
Line V : the voltage fluctuation (scale : $\Delta U = 2{,}1\ \%$).

CONCLUSIONS

This compensation system the "stato-compensator" has been found in operation to be a very fast efficient and accurate system for the network compensation.

REFERENCES

{1} WILLIAM MAC MURRAY A study of asymetrical gating for phase controlled converters. IEEE Vol. IA-8 n° 3 May/June 1972

{2} MM. CAUSSIN - X. HENRY Four a arc de très grande puissance. Revue Jeumont-Schneider juillet 1972.

{3} C. CAUSSIN A propos des perturbations apportées aux réseaux par certains appareils consommateurs d'énergie électrique.

R. CHAUPRADE Le stato-compensateur. Une solution moderne de compensation de l'énergie reactive. Journée SEE de Lille 22/23 Avril 1976. Revue Jeumont-Schneider n° 23.

{4} P. MEYNAUD Flicker et conditions de raccordement au réseau d'appareils produisant des variations. RGE Tome 80 n° 11 Novembre 1971.

{5} M. LEMOINE Quelques aspects de la pollution des réseaux par les distorsions harmoniques de la clientèle . RGE Mars/Avril 1976 Tome 85 n° 3 et 4.

{6} B. TRANNOYS Contribution à l'étude des dispositifs statiques d'équilibrage des réseaux polyphasés. Doctorat es Sciences Physique 1964.

THYRISTOR CONTROLLER FOR INDUCTION SMELTING FURNACE

A. Dmowski, K. Stańkowski, R. Zajac and P. Fabijański

Technical University of Warsaw, Inst. of Control and Electronics, Koszykowa 75, Gm. E, pok. 316, Pl 00-662 Warszawa, Poland

ABSTRACT

In the paper the principle and the way design of thyristor power controller (75 kVA - 50 Hz) for 50 Hz electric heating are presented. Described controller is a model of industral controller for 650 kW smelting furnace. This controller enables compatible work with minicomputer which controls whole technological process.

INTRODUCTION

The introduction of computers for controlling industrial processes made necessary for conventional control equipment to obtain a greater compatibility with the new computer technique requirements. An example here can be the automatic control system for the induction smelting furnace when a specialized minicomputer is used. Such a solution makes possible to attain the optimal utilization, both energetical and technological, of the furnace.

In the present paper the thyristor controller model of 650 kVA, 50 Hz, device designed as 75 kVA, 50 Hz, is described. The model connected with the electronic interface system Camac enables a computer control of metal smelting processes. An industrial controller (650 kVA) is being built at present.

Before experimental work the following assumptions have been made:

1. The controller should make possible to supply the furnace with the controlled sinusoidal voltage 50 Hz, with a high watt-hour efficiency. It enables to control the supplied power in every moment of the industrial process.
2. The power coefficient of supply system for the whole technological cycle should be close to one.
3. The system should enable the supply of single-phase surface from three-phase mains by symmetrical load of all phases.
4. The control networks of the controller enables interfacing with the computer.

In Fig. 1 the block diagram of the controller is presented.

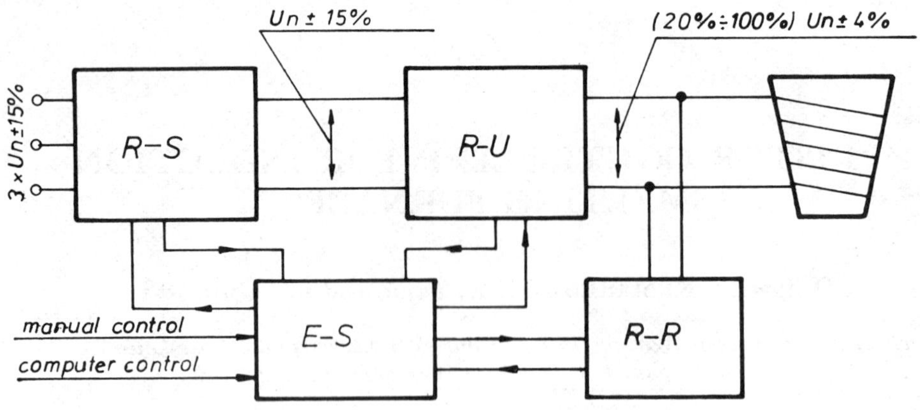

Fig.1. Schematic diagram of the Power Regulator

The controller consists of the following blocks.

1. Thyristor controllers of symmetrical load of three-phase mains R-S
2. Thyristor controller of alternating voltage R-U
3. Thyristor controllers of reactive power R-R
4. Central network for interfacing with the Camac module E-S.

The principle of operation of the power controller is as follows. At first the furnace is supplied with the voltage of 20 % U_n which permits for the action of the reactive power controller R-R and the three-phase mains symmetrical load controller R-S. When the two controllers reach the assumed compensation coefficients ($\cos\varphi = 1 \pm 0.5$ % and phase load unbalance is less than 10 %), the voltage controller R-U is automatically switched on. The R-U controller increases the voltage to the required value. Any rise in the load voltage U_o is possible only in the case when both the R-R and the R-S reached the assumed coefficients of compensation and symmetrisation.

The negative feedback loop makes the whole controller insensitive to the changes of mains supply voltage. The additional current feedback loop prevents from overloading the controller and supply mains.

CURRENT SYMMETRIZATION NETWORK R-S

The supply of one-phase induction furnace requires the load balance of three-phase transformers. This symmetrization can be reached by means of a static network which consists of the inductance coil and the condenser battery.

One of the most popular networks is shown in Fig. 2.
The condenser battery sections are switched with the aid of thyristor inverse parallel switches.
The whole network is composed of the following blocks:

I. The main circuit

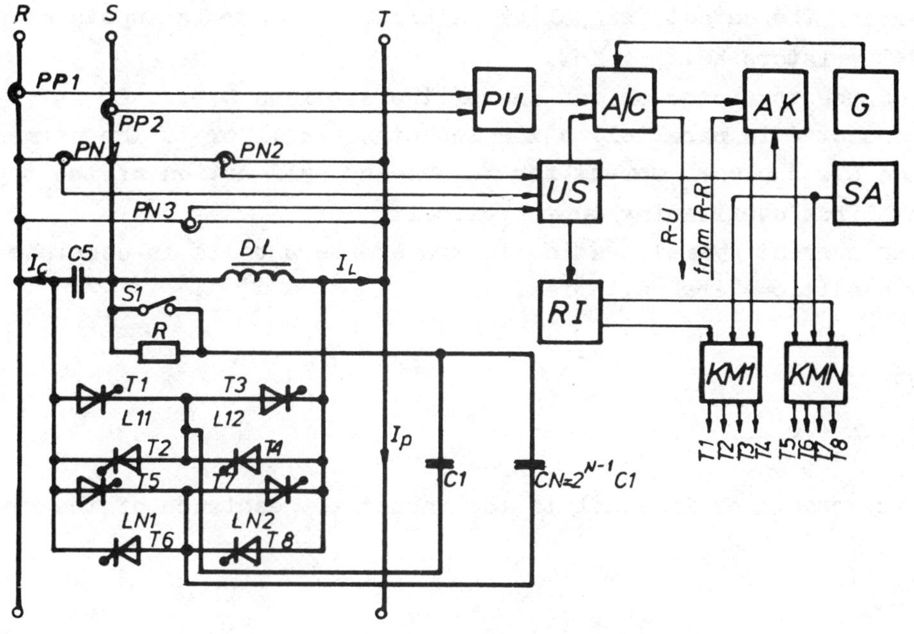

Fig.2 Schematic diagram of R-S regulator

1. thyristor switch L11 (T1; T2), L12 (T3; T4), LN1 (T5; T6), LN2 (T7; T8)
2. permanent battery section C5
3. switched condensers sections C1, CN
4. coil for symmetrization DL
5. resistor for preliminary battery load R

II. The control circuit
1. error measurement circuit PU
2. A/C converter
3. the circuit which memorize digits of measurement error - AK
4. the block of timeliness measurement US
5. clock pulse generator G
6. the distribution circuit of thyristor gate pulses RI
7. the circuit for setting in which prepares switched sections for steady state conditions SA
8. gate control circuits of thyristor switches KM1,...,KMN.

The operation principle of the device is as follows. The permanent battery section symmetrises the supply mains. The other sections are loaded via resistor R and thyristors T1, T5 by positive half waves of the line-voltage U_{RS}. The transit is about 0.2 sec. and after that time the resistor R is short circuited by the thyristor S1 and actual action is started then.
The output voltage error appears as a signal which is a difference between the two phase currents I_R and I_S. The signal is changed by the A/C converter into a digital signal in the 8421 code. The error signal sign determines whether a certain number equal to that of clock pulses synchronized at the instant by

the US circuit is added or subtracted in the reverse counter of the A/C converter. After the measurement action the AK circuit memorizes digits of measurement error. The output signal of AK circuits controls inputs of pulse amplifiers of thyristors KM1 ... KMN.

On switching on the thyristor switch one of the sections C1 ... CN is immediately connected in parallel to the inductance coil or to the permanent battery section C_5. A proper moment for the on-and -off action of the thyristor switch prevents from overloading in the circuit.

The appropriate current symmetrization in transformers coils is possible when the following conditions are satisfied:

$$I_P = \sqrt{3}\, I_L \qquad (1)$$

$$I_P = \sqrt{3}\, I_C \qquad (2)$$

where $U_{RS} = U_{ST} = U_{TR}$

and the capacity reactance is equal to the inductive reactance of the symmetrization coil.

$$\sum_{n=1}^{n=N} X_{Cn} = X_L \qquad (3)$$

If (3) is fulfilled, the conditions (1) and (2) are fulfilled when $I_R = I_S$ or $I_T = I_S$. (This relation is present in the circuit). From (3) appears that the total reactive power of the condenser battery must satisfy:

$$\sum_{n=1}^{n=N} Q_{Cn} = Q_L \qquad (4)$$

where Q_L is the reactive power of inductance coil.

The reactive power of the smallest section of condenser battery which ensures more than 5 per cent accuracy is given by:

$$Q_{C1} = \frac{0{,}1}{\sqrt{3}} P_o \qquad (5)$$

where P_o is a load active power.

The smallest power of permanent section is given by:

$$Q_{C5} = \frac{P_{omin}}{\sqrt{3}} \qquad (6)$$

where $P_{o\,min}$ is the minimum value of load active power.

By diminishing the value of Q_{C5} (5) one can obtain a smaller asymmetrization of phase currents but a number of condenser sections will then increase.

The tests of the model 75 kVA, 50 Hz have confirmed the validity of the theoretical assumptions which were the basis for designing the current symmetrization circuit.

In Fig. 3 an oscillograph record of the following signals in the A/C converter and synchronization circuit are presented:

1. the voltage signal for comparison with the control error P2;
2. the signal of measurement time of the control error P;
3. the output signal of the comparator C;
4. the control signal of reverse counter C3;

In Fig. 4, respectively

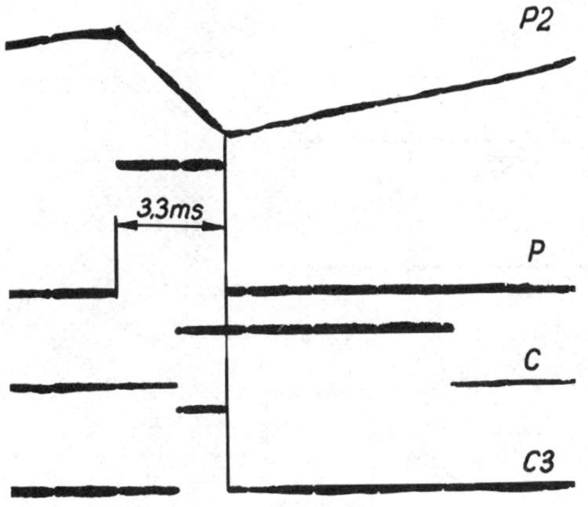

Fig.3.

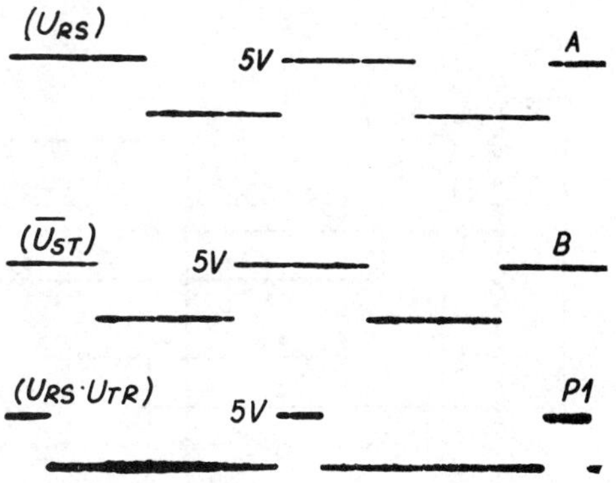

Fig.4.

1. the signal synchronizing the circuit with the U_{RS} - A voltage;
2. the signal synchronizing the circuit with the U_{ST} - B voltage;
3. the signal proportional to the measurement time of the control error P1;

In the presented model an asymmetry of the phase circuit is less than 5 per cent.

THE R-U VOLTAGE CONTROLLER

The main task of R-U controller is to control voltage supply U_o of the load. This voltage should be controlled within 20 % - 100 % of U_n. The block diagram of the controller is presented in Fig. 5.

This controller consists of

1. Autotransformer AT with five taps A, B, C, D, E (in which $U_A = U_B = U_C = U_D = U_E = 20\ \%\ U_n$)
2. Booster transformer T_d
3. Eleven double thyristor switches T1, T2,... T21, T22
4. Electronical circuit of regulator control.

The control block consists of:

a) A circuit which controls the difference between the set point voltage (U_{st}) and the controller output voltage (U_o)

b) two electronical reverse counters which control the switch of appriopriate thyristor pairs (L_1 - thyristor pairs T1, T2, ... T9, T10 and L_2 - thyristor pairs T11, T12, ... T21, T22)

c) L_1 and L_2 counters switching generator

d) a circuit for synchronization and distribution of thyristor control pulses.

The principle of controller: The supply voltage U_n switches on the pairs of thyristors T1, T2 and T21, T22. This stage refers to the zero state of both counters L_1, L_2. In the case when the voltage control value U_{st} increases and coefficients of symmetrization and compensation have appropriate values

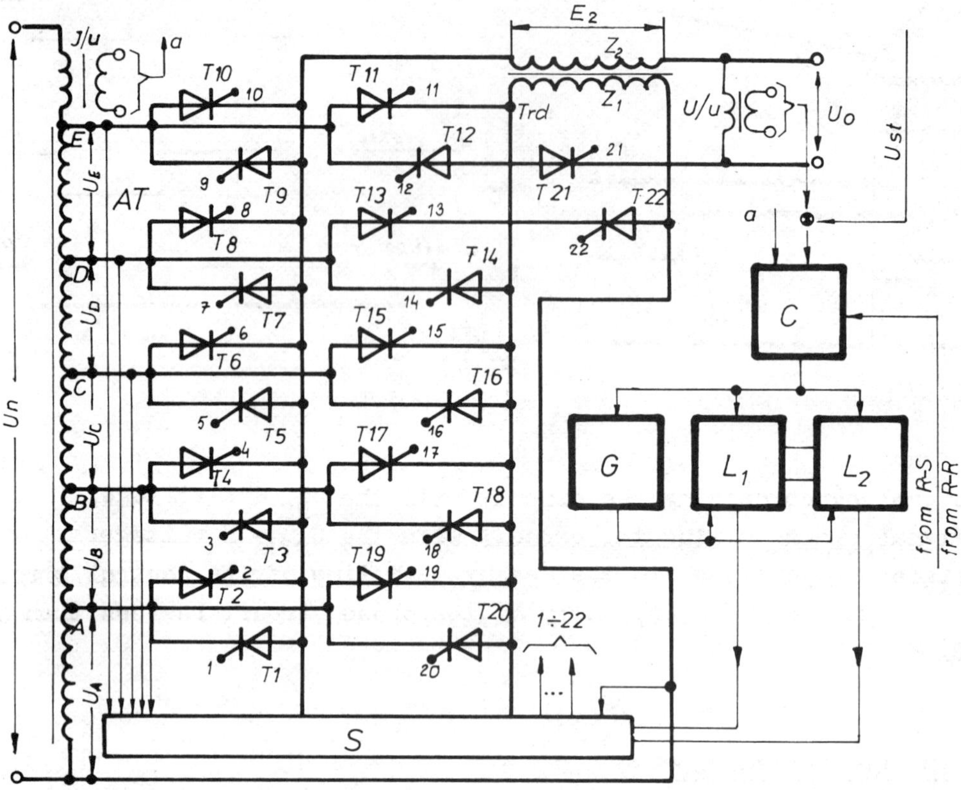

Fig.5 Voltage Regulator R-U

the generator G is unblocked.

At each end of positive half - wave of supply voltage U_n on the generator G output a pulse for counter switch appears. Each state of these two counters causes the conduction of an appriopriate thyristor pair. These thyristors conduct unless the output voltage U_o reaches the value of the control voltage. In the case when $U_o = U_{st}$ the pulse train is subtracted up to the moment when $U_o = U_{st}$. Thus each change of counter state causes the change of U_o voltage of $\frac{U_n}{5}\frac{z_2}{z_1}$ value. The principle of the regulator is illustrated in Fig. 6 in which the oscillograph of U_o & E_2 voltage versus time when step unit signal of magnitude (20 % U_{st} - 120 % U_{st}) was applied. (The load current = 50 A and $\cos\varphi$ = 0.9 ind.).

Each moment of thyristor switch causes a high harmonic dystribution in U_o voltage. The magnitude of these harmonics can be evaluated on the base of the equivalent circuit given in Fig. 7.

In this diagram 1T and 2T are on the state of conduction and 3T, 4T will be switched in the next instant.

On the basis of Fig. 7 the following set of equations can be written

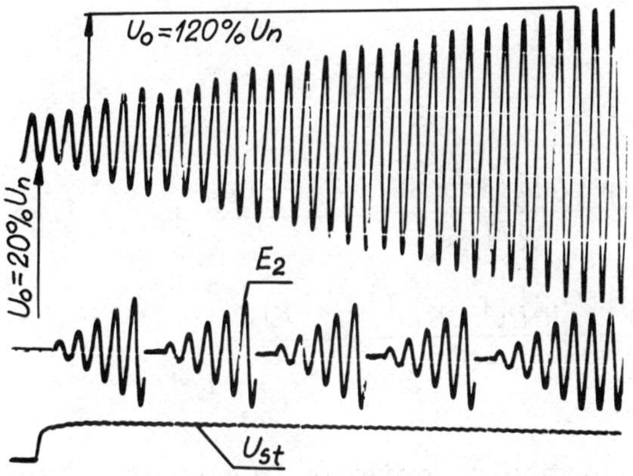

Fig.6.

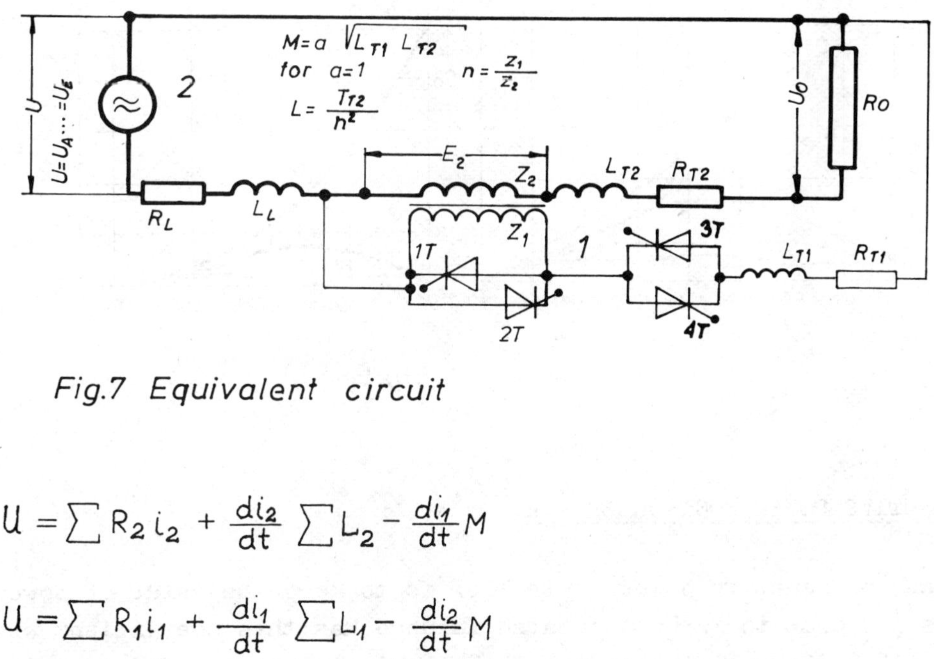

Fig.7 Equivalent circuit

$$U = \sum R_2 i_2 + \frac{di_2}{dt} \sum L_2 - \frac{di_1}{dt} M \tag{7}$$

$$U = \sum R_1 i_1 + \frac{di_1}{dt} \sum L_1 - \frac{di_2}{dt} M \tag{8}$$

this set of equations can be rewritten in a matrix form

$$\dot{\underline{x}} = \underline{A}\, \underline{x} + \underline{B} f(t) \tag{9}$$

The solution of (9) for non-sin inputs takes form (when $\varphi \neq 0^\circ$el.)

$$X_n = \sum \left[\underline{A} - j\omega k' \underline{1} \right]^{-1} \underline{B}\, \underline{F}\, mk' e^{k'\omega t} \tag{10}$$

when:

$$
F_m = \begin{cases} \frac{1}{2}(a_1 - jb_1), \; a_1 = 0, \; b_1 = \frac{U}{\pi}\left[-\frac{1}{2} + \pi - \varphi + \frac{1}{2}\cos\varphi\right] \quad \text{– first harmonic} \\ \text{other wise} \\ \frac{1}{2}(a_k - jb_k) \\ a_k = 0 \\ b_k = \frac{U}{\pi}\left\{ \frac{\sin(1+k)\varphi + \sin[(1+k)\pi] + (1-k)\varphi}{2(1-k)} + \right. \\ \qquad \left. + \frac{\sin(1-k)\varphi + \sin[(1-k)\pi] + (1-k)\varphi}{2(1-k)} \right\} \end{cases} \tag{11}
$$

Fig. 8 and Fig. 9 give the values of harmonic contains in output voltage caused by the switches of thyristors.
These results permit for evaluation of the number of autotransformer AT taps (A,B,C,D,E) and thyristor pairs for given harmonic contain in controller output voltage (for assumed commutation angle).

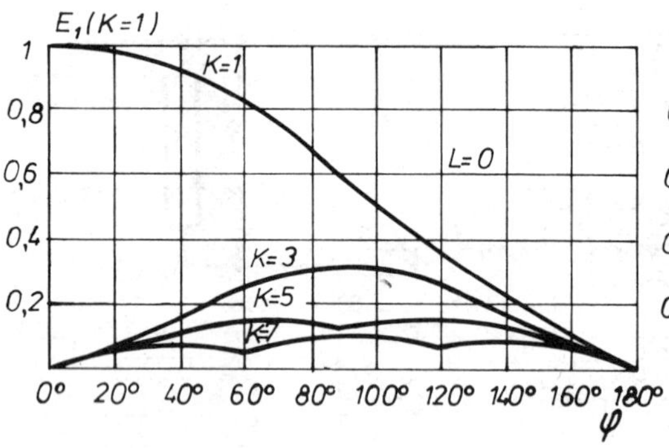

Fig. 8

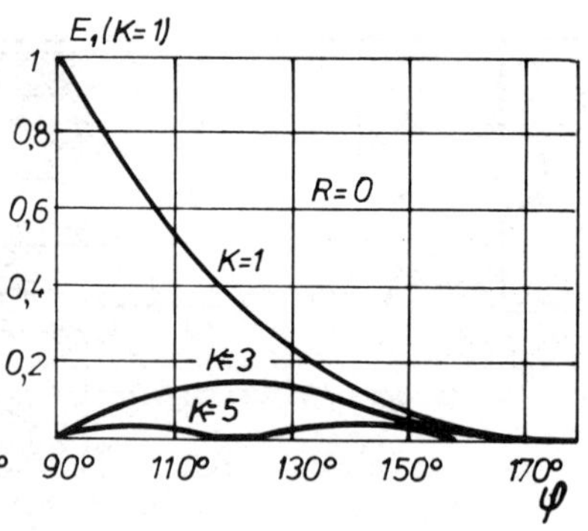

Fig. 9

REACTIVE POWER CONTROLLER R-R

The main task of reactive power controller is to keep the value of power factor (cos φ) dose to 1. Uncompenated furnace has this coefficient within the range 0.15-0.22 depending on the technological process. The reactive power is compensated by an additional condenser battery the capacity of which should be adjustable in the case when the value of cos φ close to 1 is desired.
A diagram of reactive power controller is presented in Fig. 10.

This controller consists of

a) the condensor battery divided into n sections with capacities $C_1, \ldots C_n$ weighted binary i.e. C_1, $C_2 = 2C_1, \ldots, C_n = 2^{n-1}C_1$
b) semiconductor swithes (dioda - thyristor) for bettery switching
c) voltage meassurent block with adjustable on 50 Hz filter and phose-shift

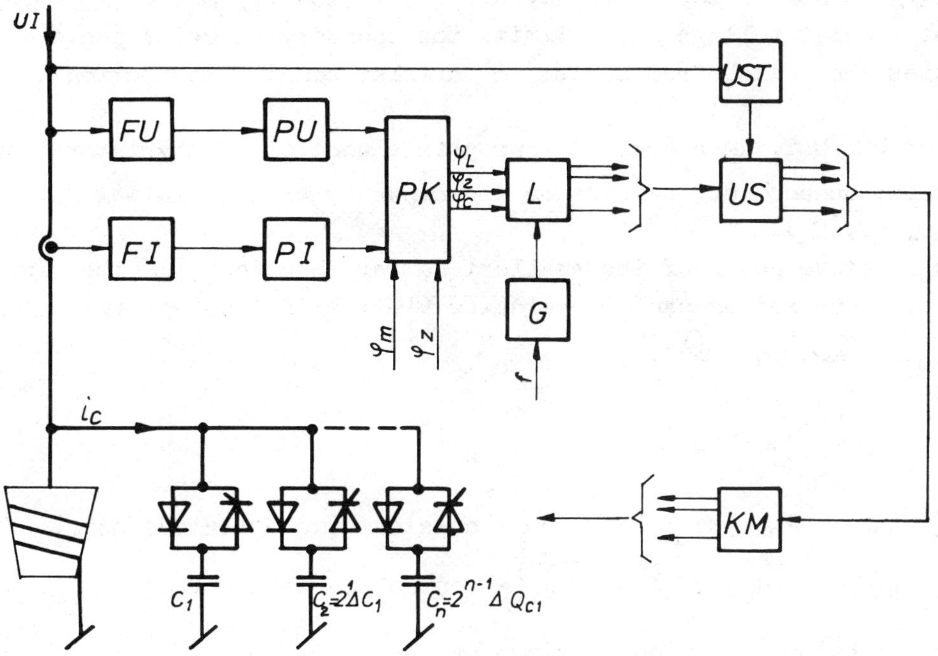

Fig.10 Schematic diagram of R-R regulator

circuit FU and PU

d) current measurement block with 50 Hz filter and phase - shift circuit FI and PI

e) the phase - shift angle PK measurement block (the angle is measured between voltage and current supplying the furnace)

f) the reverse counter block L

g) the clock pulse generator G

h) the block of synchronization ignition pulses with mains voltage UST

i) the block of synchronization block L output pulses with UST-US

j) the amplifiers of thyristors KM ignition pulses

The principle of the controller is as follows:
Signals proportional to the first harmonics of the load voltage and current are lead from the output of PU and PI blocks to the input of PK block.
The block PK output rectangular pulses have a width proportional to the phase shift between supply voltage and load current.
In the case when the load is inductive the output signal appears on the φ_L output, otherwise in the φ_C output.
The block PK output signals φ_L and φ_C cause addition or sub-traction of counter digits, respectively.

The block L consists of two counters first of which controls the switch the battery with capacity from C_1 to C_k, the second from C_{k+1} to C_n.

These counters for large values cos φ are connected in series otherwise the paraller connection is used (in the case of cos φ small values). This counter control enlarges the dynamics of whole controller R-R. The set-point voltage φ_m controls the tolerancy of $\cos\varphi$ ($\cos\varphi_{ca} = 1 \pm \varphi_m$ %), the second set - point voltage φ_z limits the boundary value of phase - shift which causes the switch from series to paraller counter connection.

The task of US, UST, KM blocks is appropriate control of thyristors ignition. The compensation accuracy depends on the capacity of the smalest battery section ($C_1 \ldots C_n$).
For given reactive power of the smallest battery section Q_{C1} the minimal accuracy of compensation can be evaluated via the following expressions

$$\varphi_{ca\,min} = \mathrm{arctg}\,\frac{Q_{C1}}{P_{min}} \qquad (12)$$

$$Q_{c\,max} = P_{min}\,\mathrm{tg}\,\varphi_{L\,max} \qquad (13)$$

$$Q_{c\,min} = P_{min}\,\mathrm{tg}\,\varphi_{L\,min} \quad \text{- constant (unregulated) term} \qquad (14)$$

$$Q_{c\,reg} = Q_{c\,max} - Q_{c\,min} \quad \text{- controlled term} \qquad (15)$$

the number of battery section is given by

$$n = \frac{\lg\left[\frac{Q_{creg}}{Q_{C1}} + 1\right]}{\lg 2} \qquad (16)$$

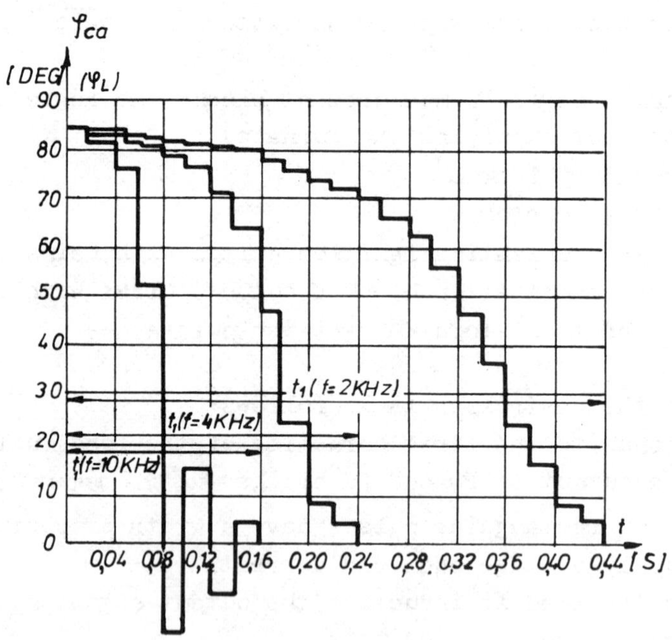

Fig. 11

In Fig. 11 the computed transients of the regulation excited by unit step are presented (time t_1) for the following set of parameters:

$$P = 4KW, \quad \varphi_L = 84\,^{\circ}\mathrm{el}, \quad Q_{C1} = 225\,\mathrm{Var}, \quad f = 2, 4, 10\,\mathrm{KHz}, \quad Q_{c\,min} = 0$$

In Fig. 12 to 15 the oscillographs of the mains current for the parameters listed above are presented. (The controller action start at the moment t = 0).

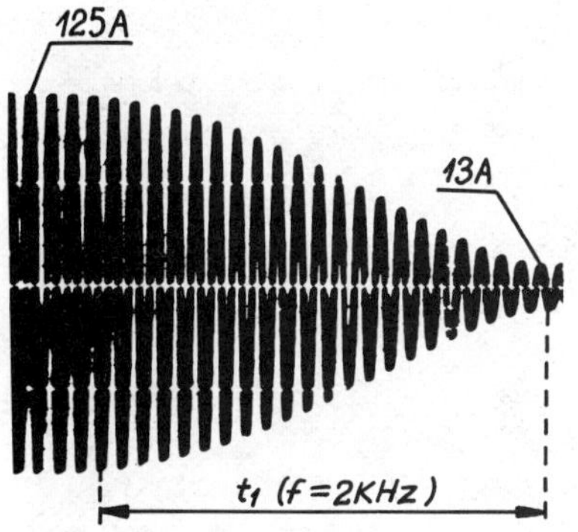

Fig. 12

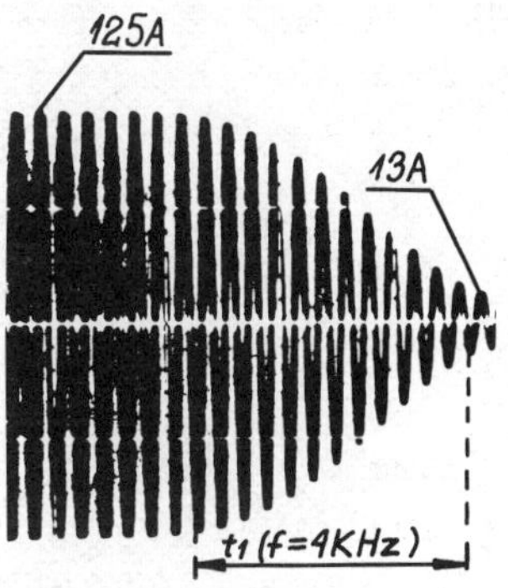

Fig. 13

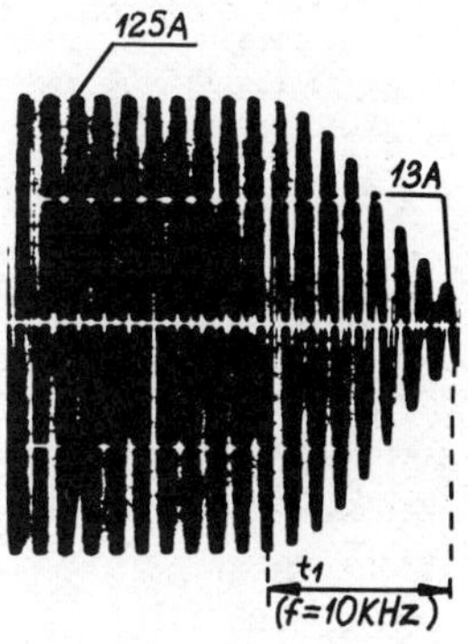

Fig. 14.

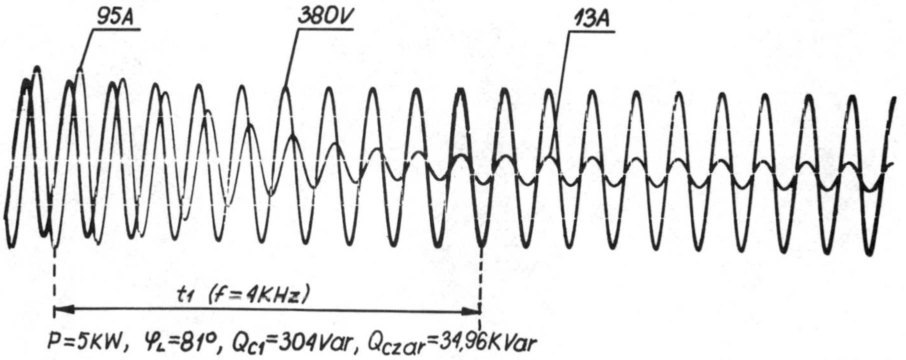

Fig. 15.

CONCLUSIONS

The results of modelling and experimental works prove the possibility of thyristor circuit application for 50 Hz electric heating devices.
The digital inputs of thyristor controller are compatible with the minicomputer outputs which make possible to control the whole group of thyristor controller by one minicomputer, simultaneously in technological process. Different parts of power controller (R-U, R-S) can be utilize in other application, for example, mains compensation, their voltage stabilization etc.

REFERENCES

(1) H. Frank and B. Landstrom, Electronics Division ASEA, Power-factor correction with thyristor - controlled capacitors ASEA Journal 6 (1971)

(2) A. K. Szibłowski, Simmetrirowanie riezima mnogofaznej sistemy pri pitaniu adnofazowych nagruzok, Energia (1974

(3) S. Lloyd, P. Marshall, Dinamical performance of a single-phase thyristor a.c. regulator, Proc. Instn. Electr. Engrs. 122, 12 (1975)

DIGITAL CONTROL SYSTEM FOR UNINTERRUPTIBLE POWER SUPPLY

F. Benvenuto, C. Di Tomaso, L. Fiore Donati, D. Sbragia and A. Valcada

ANSALDO Società Generale Elettromeccanica, Genova Cornigliano, Italy

ABSTRACT

The article is concerned with the control procedures of a IV generation Uninterruptible Power Supply (U.P.S.) system.
The system to be controlled is here described in detail; the reasons why some of the most important choices have been made are shown, and finally some hints are here mentioned about the simulation method used to verify the main features of the system.

SYSTEM DESCRIPTION

The U.P.S. to be controlled is composed by the following main parts:

1) Static rectifier and battery charger
2) Buffer battery
3) Inverter
4) Output filter
5) Transfer switch

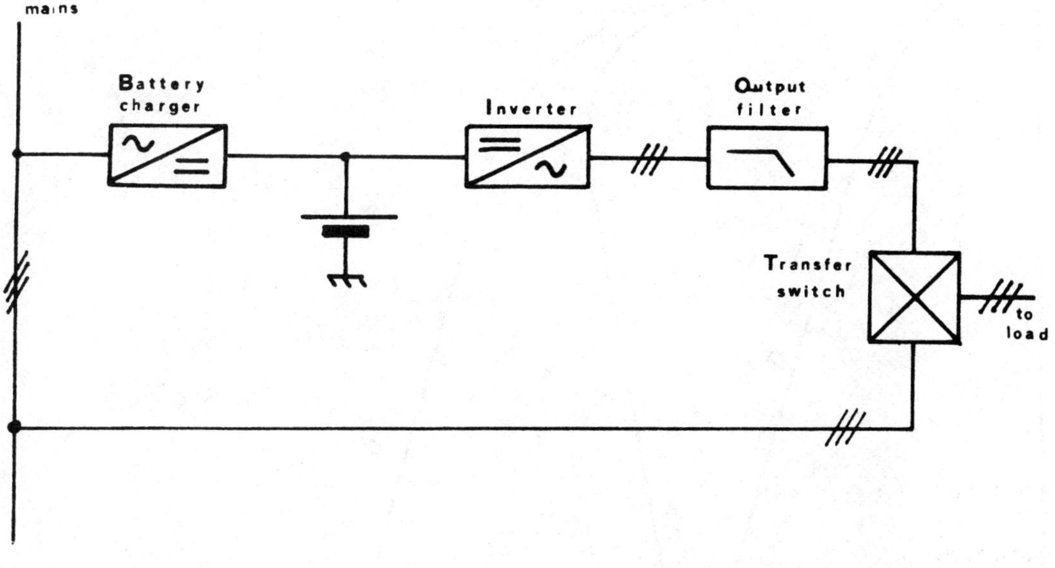

Fig. 1. Power Block Diagram

The main features this system should be able to provide are:

1) uniform output voltage, independent of load and battery variations
2) Total Harmonic Distortion (T.H.D.) below 5%
3) settling time after load step variation must be held within 3 or 4 periods of the waveform.

To obtain such features we chose a kind of modulation that would allow us to use a small - sized low-pass output filter.
This filter has a very little response time and moreover it does not overload excessively the inverter output. We have chosen the Pulse Width Modulation (P.W.M.) with harmonic cancellation and three independent angles (Fig. 2).

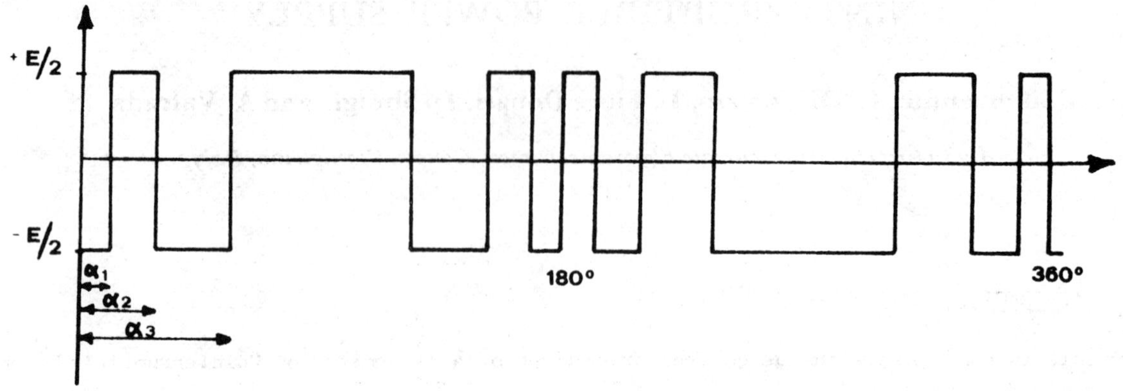

Fig. 2. Single-phase inverter voltage

The values of angles α_1, α_2, α_3 have been chosen to cancel the V and VII harmonics and to vary the fundamental at will. In this kind of P.W.M., as we take the line voltage of a three-phase system, the first significant harmonic that is encountered is the XI one.
The graphic of Fig. 3 shows the value of the ratio between the fundamental value (A_1) and the battery voltage (E) versus the value of the angles $\alpha_1, \alpha_2, \alpha_3$ (partialization ratio $\rho = A_1/E/2$) (Ref. 1).

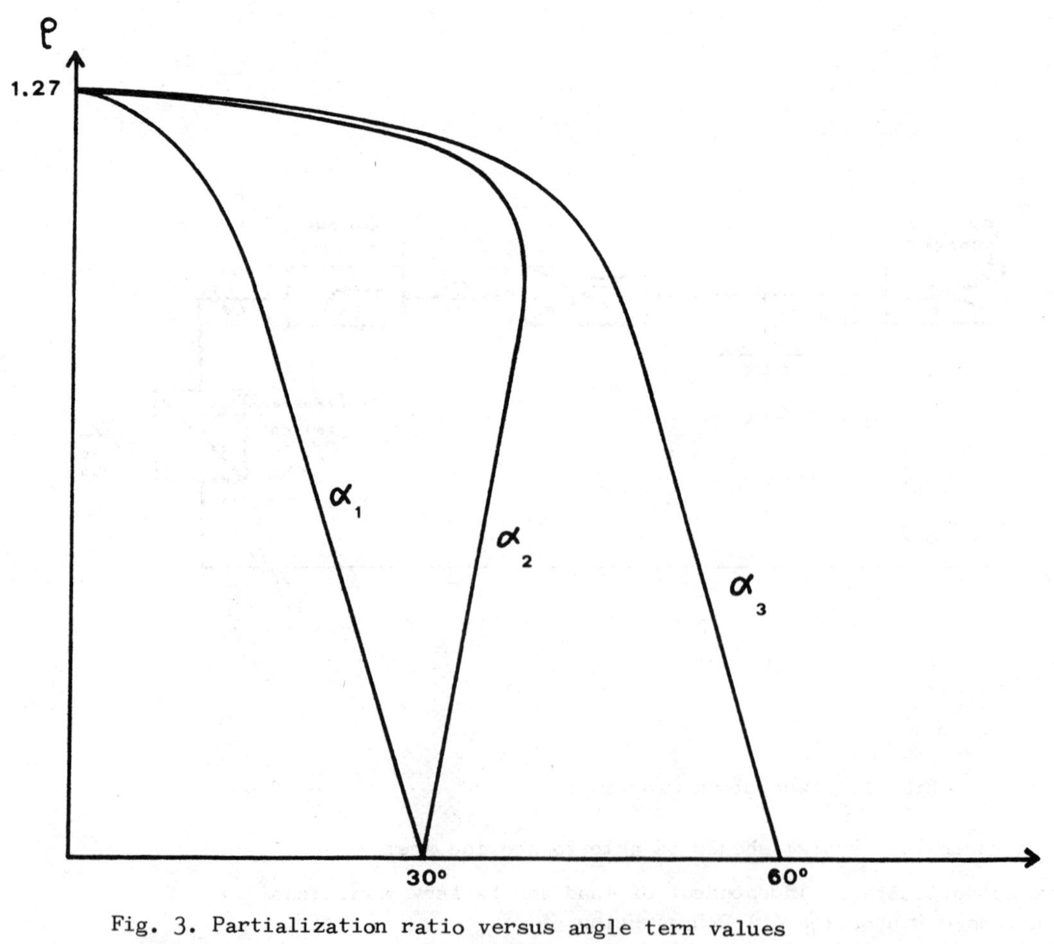

Fig. 3. Partialization ratio versus angle tern values

The range of ρ is determined by the maximum voltage battery variations together with the changes due to load - no-load transitions.
The values of ρ range from 0.6 to 1.1. The top limit of ρ value is imposed by the turn-off time of the SCR's used in the inverter.
The control must guarantee a load voltage static error below $\pm$ 1%.

CONTROL DESCRIPTION

The following figure shows the regulation system block diagram.

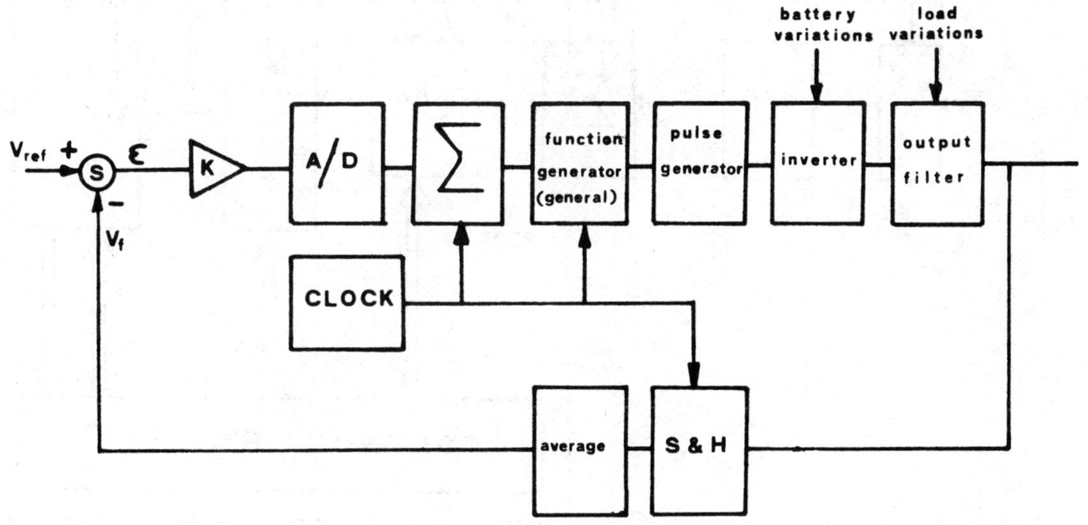

Fig. 4. System Block Diagram

We have used an integral regulator. We have chosen this type of regulator keeping in mind the absolute necessity to contain the response time and the static error within the specifications. As we have verified, by simulating the whole system, that this type of regulation can satisfy the needed features, we have abandoned the idea of inspecting more elaborate but super - abundant solutions, such as the PI or PID regulators.
The regulation acts by modulating properly the partialization angles to which the amplitude of the fundamental is evidently connected.
While designing the control we decided to generate a function like that shown in Fig. 5, to obtain the desired waveform, by properly inverting, with appropriate synchronism signals, and by repeating it.

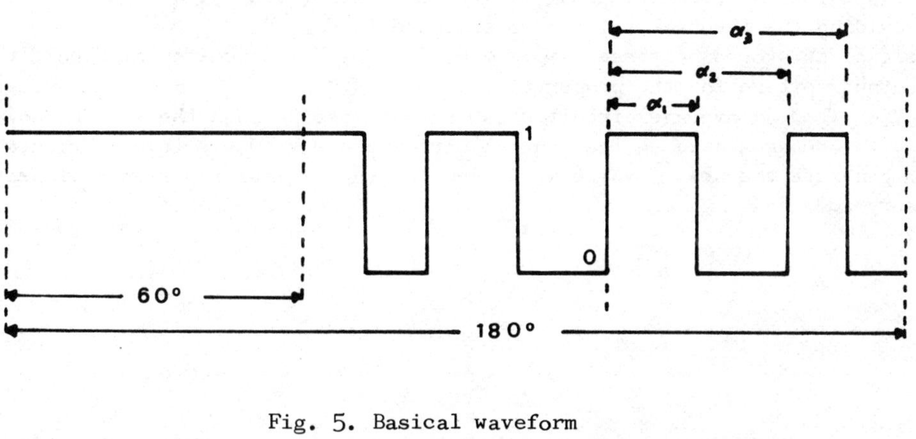

Fig. 5. Basical waveform

The desired function has been obtained starting from a clock frequency sufficiently high, in order to have a truncation error not significantly modifying the theoretical waveform features. Figure 6 shows the block diagram representing the logic circuitry.

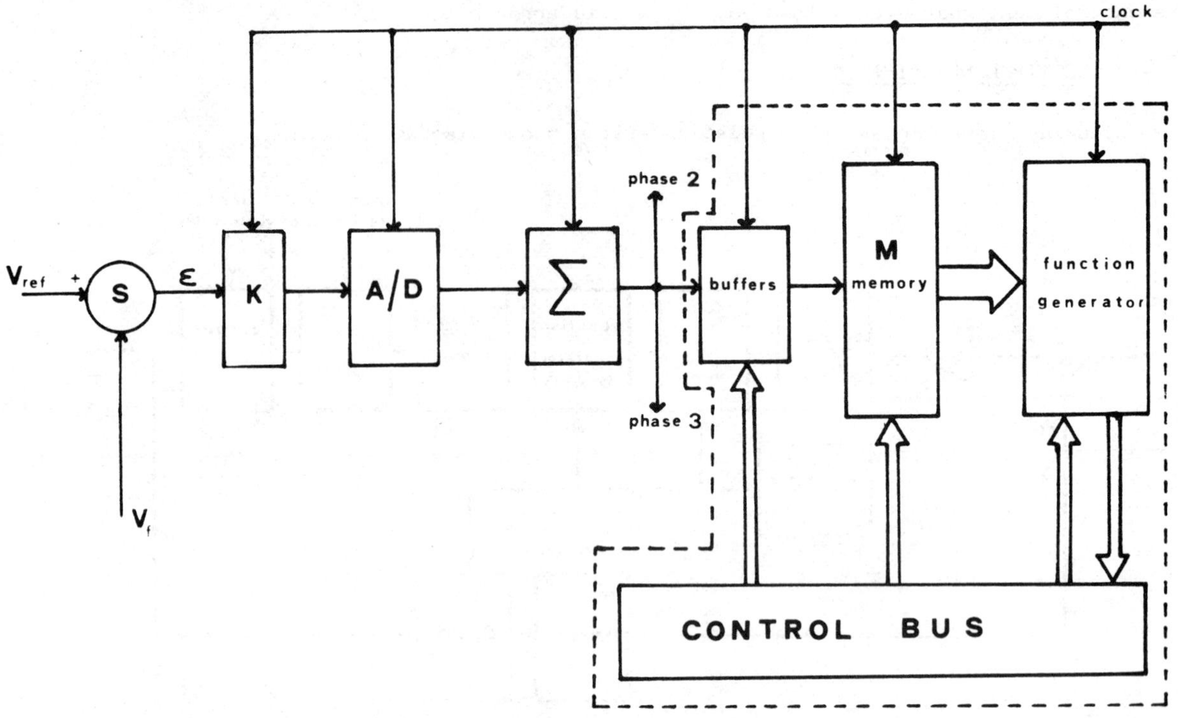

Fig. 6. Regulator Block Diagram

The feedback voltage compared with a reference generates the error signal (ε) which, properly amplified, is converted by an A/D into a digital signal. The subsequent values of ε are stored and summed-up by a full adder whose output represents directly the discrete value of ρ .
The range of ρ value has been divided into 128 steps: each step identifies a proper memory address. The division of ρ range has not been made linearly; this allowed us to obtain a relative truncation error which is constant and equal to 5‰.
A memory stack, of proper capacity, provides the right correspondence between the 128 steps and the partialization ratios. The values of such angles are converted by a decoder into binary code and transferred then, with the proper sequence,to the function generator; it processes this information and generates the output waveform logic levels (Fig. 2).
A bus controls the several functions of the whole system and more precisely enables the memory stack to feed the data into the function generator according to a prearranged sequence.
In Fig. 6 the dotted line regards the circuitry of only one phase; of course the circuit is repeated three times to obtain a complete three-phase system; proper synchronizing signals guarantee the right phase-difference among the three phases (120°).
The circuit providing the feedback voltage is sketched in Fig. 7.
The feedback signal is stored by three Sample & Hold (S & H) circuits which hold the peak value of the output voltage for the proper time .
Stored information feed an average circuit whose output goes through the summing point.
So acting new data are presented to the control circuits every 60 electrical degrees.
These new data generate the new ε value which updates the digital adder and, if necessary, sets a new memory address.

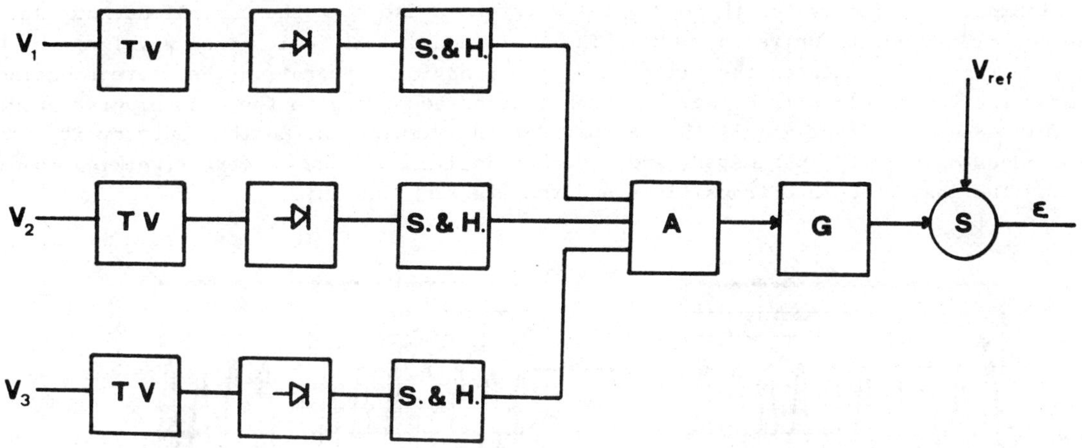

Fig. 7. Feedback Voltage Detector

Having used a clock frequency sufficiently high we have obtained a great precision in angle values; this fact allowed us to have small contents of THD according to the designing theories. The exact phase relations among the different signals and more precisely among the output waveforms, have brought us to take certain decisions for the hardware choice.
To provide such a function we have designed a priority network using only combinatory logic circuits, in order to obtain a high reliability in generating the output waveforms.
The circuitry that has been chosen is the logic family C - MOS and particularly the family 74C; only two blocks use different components; the memories and the displays.
The peculiar characteristics of this technology, such as the high voltage noise immunity, the wide supply voltage range, and the low power consumption makes the equipment safe and reliable. In fact all these properties allow the designers to override the serious problems in supply voltage encountered when using other logic families such as the TTL or the more updated Hi NIL.

SIMULATION

We have verified the stability and the features of the system through a partial simulation: the power block (battery + inverter + output filter + load) has been simulated through a minicomputer; on the contrary we have used an experimental prototype of the control to check the whole system behavior. The two blocks interact according to the diagram shown in Fig. 8, thus simulating the closed loop.

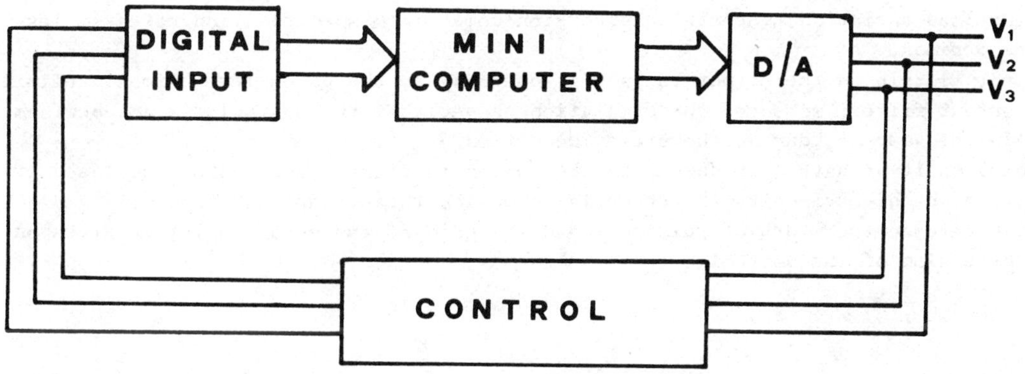

Fig. 8. Simulation Block Diagram

The minicomputer is fed by the three digital waveform present at the control output, designed, in the actual system, to drive the SCR's. The minicomputer multiplies this waveforms by the battery voltage and simulates the filter and load behavior, integrating the state equations. Of course the system clock frequency has been lowered according to the minicomputer processing time. This does not affect at all the validity of the simulation; on the contrary it allows a finer adjustment of the parameters and a deeper analysis of the different working conditions, including the load - no-load transitions and the starting instants.

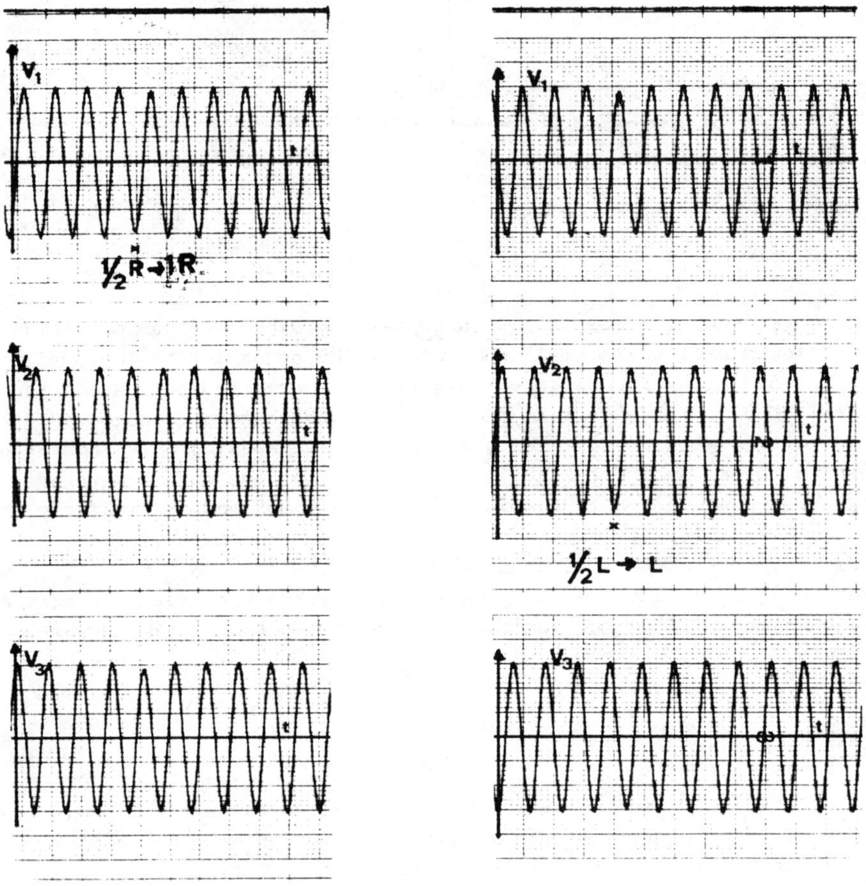

Fig. 9. Output line voltage changes due to load variations

Figure 9 shows the output voltage changes at the load varying conditions; it can be noted that after any load variation, the static precision comes into specification rates in less than two or three periods.

The ripple visible on the output voltage is due only to the uncomplete harmonic elimination by the output filter; moreover the simulation showed that the control did not perform any jitter in the address, during the steady condition.

To accomplish it we have introduced, in the analog to digital conversion, an insensitivity zone within which, obviously, the error falls during the steady condition.

The ratio between the width of this zone and the gain of the error amplifier gives out the static precision of the control.

CONCLUSIONS

The most peculiar characteristic of the system just described consists in the fact that the control has been carried out using the digital technology in almost every part of it.
Several are the advantages thus obtained: the accuracy achieved in the waveform generation, and thus the real harmonic cancellation; the complete simmetry of the three waveforms; the absolute repetitivity of the control performance.
Finally it is evident that an equipment of this type allows to increase by far the reliability of an U.P.S. system, compared with more traditional solutions.

REFERENCES

(1) H.S. Patel, R.G. Hoft, Generalized techniques of harmonic elimination and voltage control in thyristor inverter - Part II - Voltage control techniques, IEEE Trans. on Ind. Appl. 1A-10, 5 (1974).

HARMONIC ANALYZER FOR SUPPLY FREQUENCIES

B. Wehrli

Camille Bauer Measuring Instruments Ltd., CH - 5610 Wohlen/Switzerland

INTRODUCTION

In recent years, static converter technology has found a wide application field. The main reasons for this are the high reliability, freedom from maintenance and particularly high economy. However, there is one fundamental disadvantage: the current taken by a static converter is generally non-sinusoidal to the highest degree, which requires more reactive power in the first place and also causes harmonic voltages through the network impedance. These in turn may affect other users detrimentally, especially the operation of electrical machines, capacitors, meters, centralized telecontrol systems, computers and other electronic equipment. On account of this, standards and guidelines have been laid down, specifying the maximum harmonics that may be generated.

STANDARDS

The unclear situation with regard to reactions on the supply system has prompted CENELEC (European Committee for Electrotechnical Standardization) to draw up a standard which has been introduced with identical wording in all the eleven European member-countries (Ref. 1). However, it applies only to electrical domestic appliances, and therefore takes the form of an approval test. Among other things it states the maximum harmonics which an appliance may generate with a standardized supply impedance. It also includes exact regulations for the measuring instrument used. These deal with the frequency range to be measured, the maximum error admissible, the measuring bandwidth, time response and maximum overswing. Since this CENELEC standard is at present the most important and widerspread ruling on supply harmonics, it has the character of a guideline for major industrial installations and for regulations in supply contracts between power utilities and major consumers. Presently the IEC (International Electrotechnical Commission) is also engaged in establishing standards with wording similar to the CENELEC standard.

In September 1976 in Britain there appeared an Engineering Recommendation (Ref. 2) as a supplement to CENELEC standard EN 50.006, dealing with all cases of major installations in low and medium voltage networks having non-linear load charakteristics. Three cases are distinguished, depending on the gate power taken, the pulse number of the static converter, the harmonic level already existing at the point of common coupling, and the short-circuit power. For each of these groups the maximum harmonic currents taken and/or the harmonic voltages generated at the point of common coupling are laid down.

This development shows clearly that the measurement of supply harmonics is extremely important, whether for analyzing supply networks, for continuous monitoring at critical network points, or in production control when manufacturing equipment and systems embodying static converters.

EQUIPMENT SPECIFICATION

There are a number of different analysis procedures for measuring network harmonics. Depending on the requirement, one or the other method will be better suited for the specific task. In particular the following questions must be answered first:

- Is the information on both amplitude and phase required, or on the amplitude only?
- Does the entire spectrum have to be measurable simultaneously, or is a sequential measurement of the desired harmonics sufficient?
- Does the instrument have to be portable or not?
- Is the instrument intended more for laboratory duty or more for field work?
- How much is it allowed to cost?

In the present case the aim was to achieve a real operational measuring instrument with the following capabilities:

- Amplitude measurement without phase information.
- Suitable for prolonged duty without supervision too, especially with supply frequency fluctuations.
- Accommodation of the demands of power engineering with regard to input level and isolation of the input.
- Operation as simple, practical and foolproof as possible.
- Easy to transport.

It was also to satisfy all the requirements of the European standard EN 50.006.

The measuring system adopted works on the heterodyne principle. With suitable conception, this method of analysis meets all the above demands in elegant manner, guaranteeing high dependability and accuracy for reasonable outlay. The finished instrument is shown in Fig. 1.

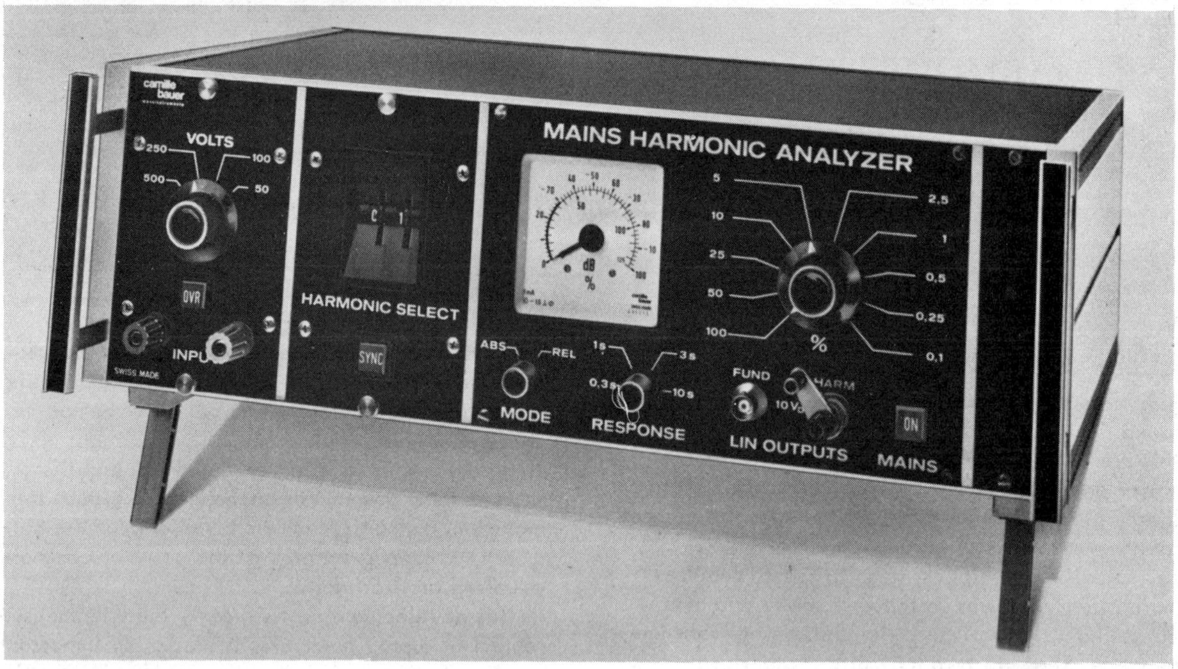

Fig. 1: Front view of the unit

MODE OF OPERATION

As already mentioned, the harmonics are analyzed on the heterodyne principle. This offers the advantage of constant absolute measuring bandwidth throughout the entire frequency range. Fig. 2 shows the block diagram for the instrument.

The measuring signal, whether current or voltage, passes first into a matching circuit, where it is brought to a level suitable for further processing. From here it passes on the one hand into a 50 Hz bandpass filter, which eliminates all harmonics. After subsequent rectification an impressed voltage is available, proportional to the fundamental oscillation of the input signal. On the other hand, from the matching circuit the measuring signal passes into a hole filter, which attenuates the fundamental oscillation so strongly that it has about the same amplitude as the harmonics. The low-pass filter following eliminates any very high frequency components still present, lying outside the frequency range to be measured. In the mixer A, the input spectrum is transposed into a higher frequency range by means of a suitable oscillator signal, as shown in Fig. 3. The mixer is able to form the sum and the difference of two input frequencies. In this case the entire input spectrum is mixed with a fixed oscillator frequency, giving rise to an upper and lower sideband. If the oscillator frequency is altered, the two sidebands move with it. This makes it possible to employ a fixed intermediate frequency filter with constant bandwidth, and to control the selection of the input frequencies to be measured with a suitable oscillator frequency. Filtering the intermediate frequency is performed by a crystal filter with suitable bandwidth. The signal is then amplified in steps (measuring range selection) and rectified. The result is an impressed voltage proportional to the amplitude of the harmonic to be measured. For absolute measurement of the harmonics, this signal is employed directly for the indication. If a relative measurement is required, as a percentage of the fundamental oscillation, the signal passes as numerator into a division stage. The denominator signal is provided by the dc voltage proportional to the fundamental oscillation mentioned above. This ensures that with this mode of operation a genuine percentage measurement of the harmonics is possible related to the fundamental oscillation, even with fluctuating input levels.

The measuring frequency for the analyzer is selected by adjusting a suitable oscillator frequency. This is processed in a Phase-Locked Loop so that operating is restricted to a minimum, and the measurement is independent of the exact frequency of the fundamental oscillation.

The following frequency must be supplied to the mixer A:

$$f_{OSZ} = f_{IF} - f_{HARM} = f_{IF} - N \cdot f_{FUND} \qquad (1)$$

f_{OSZ}: oscillator frequency for mixer A

f_{IF}: centre frequency of intermediate frequency filter

f_{HARM}: frequency of the harmonic to be measured

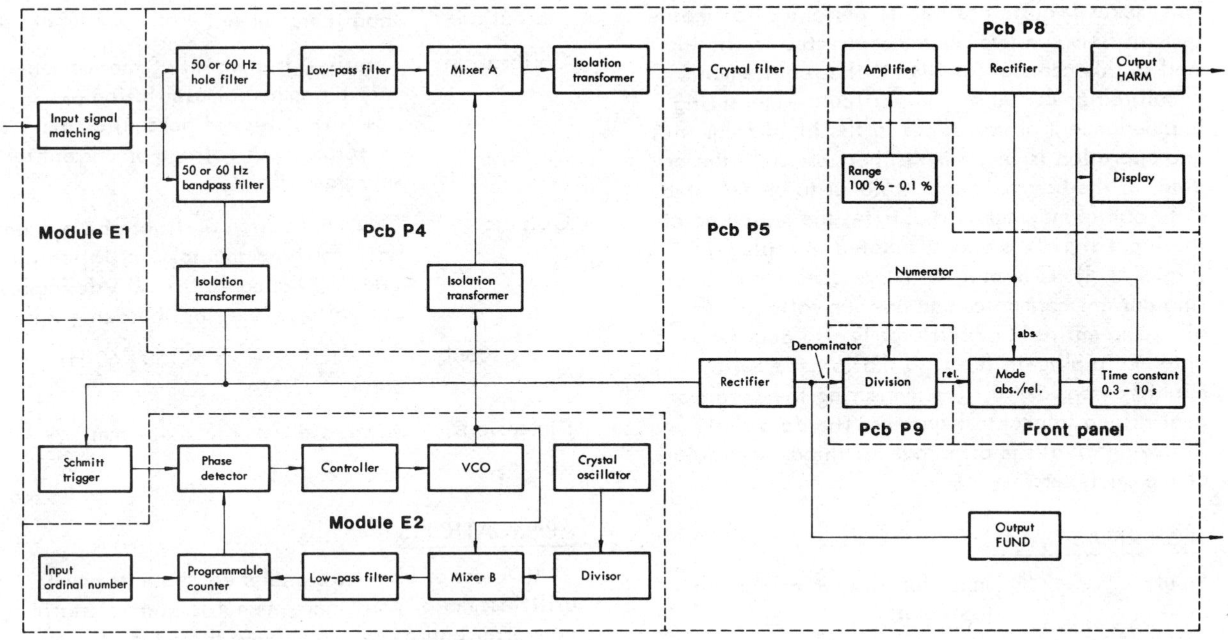

Fig. 2: Block diagramm of the instrument

f_{FUND}: frequency of the fundamental oscillation of the measuring signal

N: ordinal of the harmonic to be measured

This is generated in a voltage-controlled oscillator (VCO). Besides mixer A it is also led into a mixer B, where the two signals result:

$$f_{MB1} = f_{IF} + f_{OSZ} \tag{2}$$

$$f_{MB2} = f_{IF} - f_{OSZ} \tag{3}$$

f_{MB}: output frequency of mixer B

The signal f_{IF} is simulated in a crystal oscillator followed by a divider, and thus has the same stability as the intermediate frequency filter itself. The low-pass filter following mixer B eliminates the sum signal f_{MB1}. By combining equations (1) and (3) the difference signal may also be written:

$$f_{MB2} = f_{IF} - f_{OSZ} = N \cdot f_{FUND} \tag{5}$$

This signal is led to a programmable divider, for which the division factor can be adjusted on the front panel. If the ordinal N is put in as division factor, the frequency f_{FUND} appears as the output.

A phase detector now compares this signal with the fundamental oscillation obtained from the measuring signal. If there is a phase difference or even a frequency difference, the phase detector controls the integral controller in the appropriate sense, and the frequency of the VCO changes until the phase condition is satisfied. With that the control loop is closed, and the oscillator signal always has the right frequency, regardless of the frequency of the fundamental oscillation. The harmonic is selected simply by putting the ordinal number into the programmable divider.

From the block diagram in Fig. 2 it will be clear that the plug-in E 1 and the printed circuit board P 4 are isolated from the rest of the electronics.

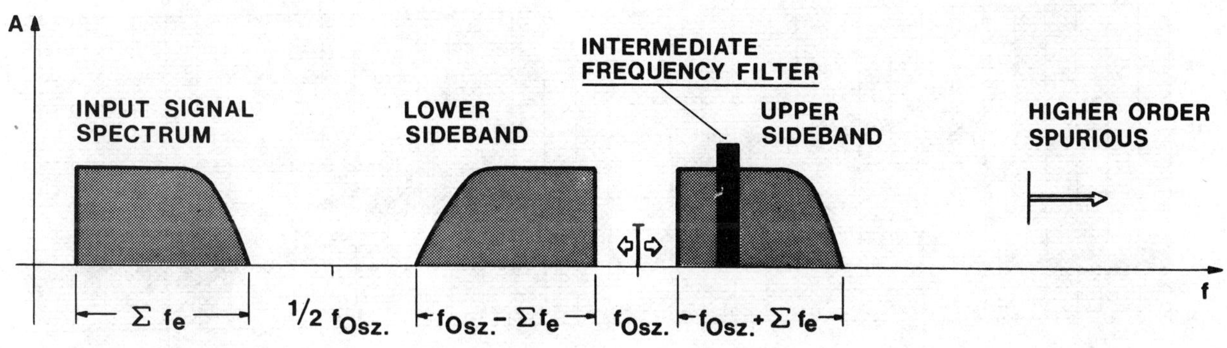

Fig. 3: Spectrum of mixer output

This makes it quite possible to perform also measurements between two phases connected to the input. And so this concept fulfills all the requirements stipulated at the outset. In particular, measuring is independent of variations in the input frequency and operation is very simple, because only the ordinal of the harmonic sought needs to be selected.

To obtain maximum flexibility, the matching of the input signals is accommodated in a plug-in (Fig. 2, E 1). One plug-in is provided for measuring current harmonics and one for voltage harmonics. The entire PLL oscillator is also accommodated on a plug-in (E 2), so that other applications might be considered, like measuring the levels of centralized telecontrol systems. The data sheet following gives the principal technical particulars of the analyzers.

TECHNICAL DATA

Input:	Input for voltage or current (optional)
Voltage module:	500/250/100/50/25/10 V_{rms}
	Input resistance 200 kΩ
Current module:	25/10/5/2,5/1 A_{rms}
	Inbuilt low-induction shunt approx. 16 mΩ for all ranges
Isolation:	2 kV_{rms}
Frequency range:	Fundamental oscillation (ordinal number 1), 50 Hz, optionally 60 Hz, harmonics of ordinal number 2 - 49 directly selectable by thumbwheel switch
Measuring range:	Selectable between 100 % and 0,1 % (FSD of instrument)
Attenuation against neighbouring harmonics:	≥ 60 dB
Accuracy:	= ± 5 % referred to momentarily measured value (absolute value)
Indication:	Inbuilt instrument
Calibration:	Relative, in percent of the momentary fundamental oscillation or absolute value, in percent of the selected input voltage or current by selector switch
Output:	One connection on front and one on back for fundamental oscillation and selected harmonic 0 - 10 Vdc impressed voltage, external burden ≥ 2 kΩ
Power supply:	110/220 V, ± 15 %, 50/60 Hz, approx. 25 VA
Dimensions:	Approx. 437 x 133 x 361 mm
Weight:	Approx. 10 kg

APPLICATIONS

All users of harmonic analyzers, such as power utilities, major consumers, manufacturers of static converter equipment etc., have their specific measuring problems. Four typical applications are given below as examples.

1. Direct measurement of the instantaneous value of the harmonics in the network, whether tracing faulty equipment, at the point of common coupling between generating station and major consumer, testing new systems in the laboratory, at final inspection in production or acceptance tests of entire installations. The very straightforward and foolproof operation is a big advantage here, especially the direct selection of the harmonic ordinal instead of the frequency.
2. Continuous measuring for monitoring and fault annunciation, on static converter systems for example. In some cases it is no straightforward matter to establish whether all components in a system are functioning properly. Often the malfunction reveals itself only by increased harmonics on the secondary side.
3. Use of the harmonic analyzer in conjunction with a recorder for plotting the contents over days and weeks. The instrument needs no attention or maintenance, because it is independent

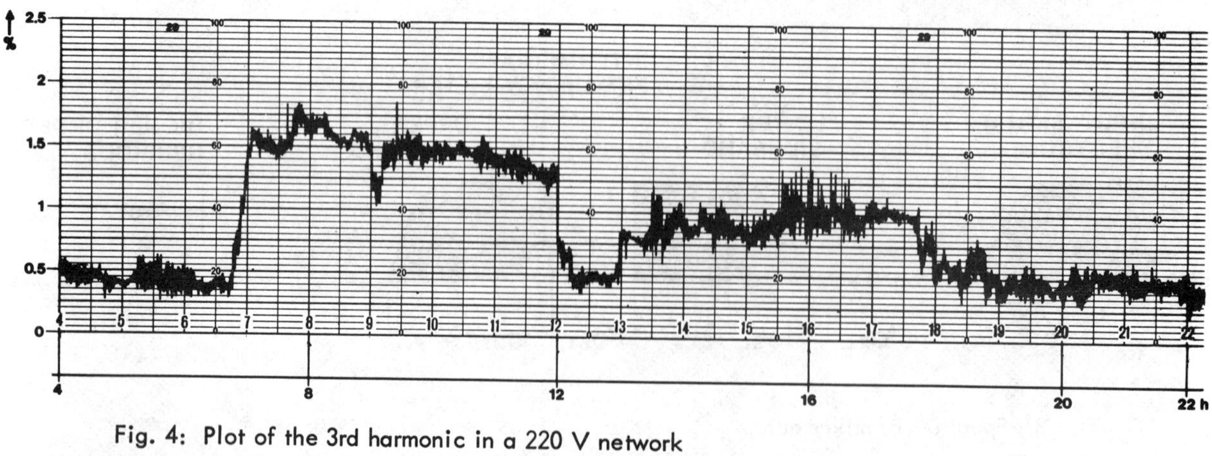

Fig. 4: Plot of the 3rd harmonic in a 220 V network

of the momentary supply frequency. Fig. 4 shows a plot of the 3rd harmonic in a 220 V network as an example. Very clear is the rise in the harmonic content in the morning, and the drop over midday.

4. Use of the analyzer with a high-speed recorder. This will often pinpoint individual faulty equipment, because the switch-on time can be compared with the plot. Fig. 5 shows a recording of a brief disturbance, in this case a photocopier which set up a considerable amount of harmonic voltages every time a copy went through. The copying machine was about 30 metres from the measuring location, and the analyzer was not even connected to the same mains feeders.

CONCLUSION

The demand that something should be done about the problem of network harmonics, including the laying down of maximum values, originates certainly from the power utilities, because they naturally want to be able to offer all their consumers a supply satisfying certain minimum quality standards. Generally it is they too who have to cover the increased reactive power demand and the higher transmission losses (Ref. 3). Power electronics are now in a position to offer circuitries with low network reaction, whether by increasing the pulse number, by employng choppers or by similar circuitry artifices. In any case, existing and additionally generated harmonics will have to be taken into account in the future, both in the development of equipment and in the control and supervision of supply networks. Here the measurement and recording of supply harmonics assumes crucial importance.

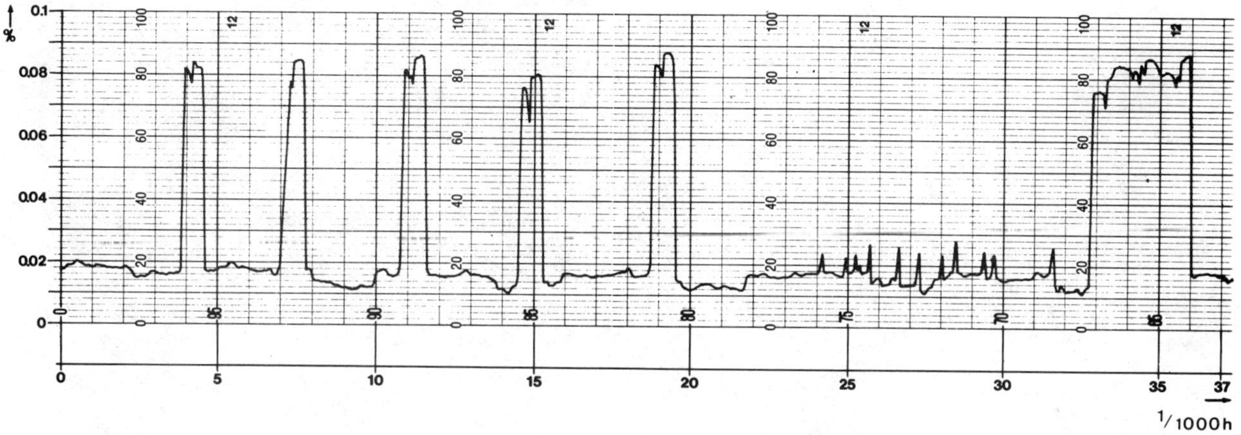

Fig. 5: Recording of the 21st harmonic produced by a photocopy machine

References:

(1) The Limitation of disturbances in electricity supply networks caused by domestic and similar appliances equipped with electronic devices. European Standard EN 50.006 CENELEC

(2) Limits for harmonics in the United Kingdom Electricity Supply System
Engeneering Recommendation G. 5/3
System Design and Development Committee

(3) Köntje, C Die wirtschaftliche Auswirkung phasenanschnittgesteuerter Verbraucher in Niederspannungsnetzen Dissertation (75) RWTH Aachen, Deutschland

POWER ELECTRONIC CIRCUIT ANALYSIS TECHNIQUES

Richard G. Hoft and Jordan B. Casteel

Electrical Engineering, University of Missouri-Columbia, Mo., USA

ABSTRACT

A number of methods of analyzing power electronic circuits are discussed and illustrated. Each of these methods has unique advantages and disadvantages. No one method appears to have all of the features necessary for it to become the preferable analysis approach. Thus, the search must continue for the ultimate power electronic circuit analysis technique.

INTRODUCTION

A block diagram of a general power converter is shown in Fig. 1. The purpose of the circuit is to efficiently convert an input voltage or set of voltages, $\vec{v}_{IN}$, having a particular waveform to another voltage or set of voltages, $\vec{v}_{OUT}$, having a different waveform. Usually $\vec{v}_{IN}$ is a dc, single-phase ac or polyphase ac voltage vector. The desired output is a dc voltage of a different magnitude or a single-phase or polyphase ac of a different voltage magnitude and frequency than the input. The conversion is accomplished by a set of semiconductor switches, usually thyristors or transistors, and filters connected between the input and the output. The switches are opened and closed in a particular manner controlled primarily by the input, $\vec{d}$.

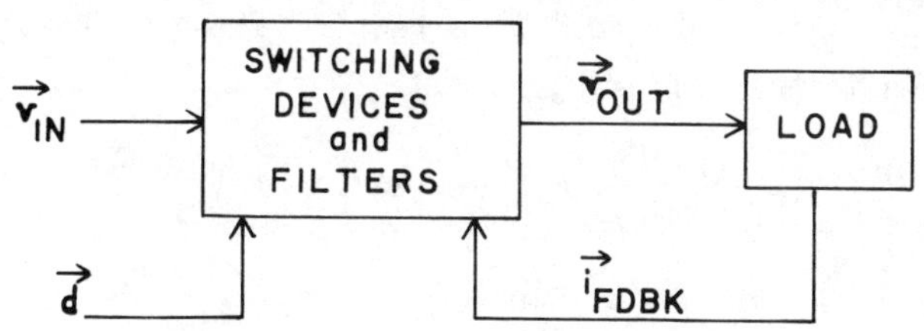

Fig. 1 General block diagram of power electronic converter

Unfortunately, the nature of the switches and the load may be such that the conduction intervals of the switches, and hence $\vec{v}_{OUT}$, also depend upon the load current. This complication is indicated in Fig. 1 by the signal $\vec{i}_{FDBK}$. There are four basic types of power electronic converters-namely, dc-to-dc converters, ac-to-dc rectifiers, dc-to-ac inverters, and ac-to-ac converters.

The analysis of power electronic switching circuits is difficult due to the complexity and nonlinearity of the equations which describe these systems. Perhaps the greatest source of difficulty in the circuit analysis is the discontinuous binary control signal, $\vec{d}$. This problem is compounded by the effect of the load current on switch conduction intervals and hence load voltage in many circuits. Frequently the load itself is nonlinear, as in the case of motor drives. Finally, the presence of many switches in the circuit results in many modes of operation during a single period of operation.

The ultimate goal of power electronic circuit analysis techniques is to determine exactly all significant variables in the circuit as a function of time. This is seldom feasible with presently known techniques. Instead, approximate solutions are obtained from simplified linear or nonlinear representations of the system. These solutions provide limited information such as waveforms or stability of the system under some operating conditions.

The purpose of this paper is to review presently used power electronic circuit analysis techniques. This is done by discussing the general methodology of the various techniques. In addition, more detailed derivations and application illustrations of the techniques are given for particular circuits.

ANALYSIS TECHNIQUES

Laplace Transform-Unit Step Function

Many power semiconductor circuits can be analyzed by a straightforward application of Laplace transforms, although the algebra usually becomes quite involved. Takeuchi [1] extensively developed this approach. It is most often applied to phase-controlled circuits, where the load voltage $\vec{v}_{OUT}$ consists of segments of sine waves. The key to the technique is using the right expression for $\vec{v}_{OUT}$. Rather than using a Fourier series, $\vec{v}_{OUT}$ is represented as an infinite series in which each term is the product of a sine wave and a switching function consisting of the difference of unit step functions, such that each term represents one period, or quasi-period, during transient states, of the voltage.

Once $\vec{v}_{OUT}(t)$ is determined, its Laplace transform is obtained by using the well known transform formulas and the time-domain shifting theorem. Then the equation for the load current $\vec{i}_{OUT}(t)$ as a function of $\vec{v}_{OUT}(t)$ is written and transformed to the s-domain. Using substitution the load current is found as a function of s and possibly other unknowns such as the thyristor extinction angle in line commutated circuits. Applying the inverse Laplace transformation yields an explicit function for $\vec{i}_{OUT}(t)$ as an infinite series. The main advantage of this expression is its compactness while retaining exactness even during the starting transient. Thus the approach is a more direct procedure for obtaining the current than a repeated solution of the circuit differential equation over each period.

Application to full wave rectifier. As an example of this technique, consider a single-phase full wave uncontrolled rectifier supplying a series LR load from a sinusoidal source, $E_m \sin(\omega t)$. The load voltage is thus the full wave rectification of this sinusoid, and may be expressed as discussed above in the following form.

$$v_{OUT}(t) = \sum_{n=0}^{\infty} [u(t-\frac{n\pi}{\omega})-u(t-\frac{(n+1)\pi}{\omega})]E_m \sin(\omega t-n\pi) \tag{1}$$

where u(t) is the usual unit step function. Since

$\sin(\omega t-n\pi) = -\sin(\omega t-n\pi-\pi)$, equation (1) becomes

$$v_{OUT}(t) = \sum_{n=0}^{\infty} E_m\left\{u(t-\frac{n\pi}{\omega})\sin[\omega(t-\frac{n\pi}{\omega})]+u(t-\frac{(n+1)\pi}{\omega})\sin[\omega(t-\frac{(n+1)\pi}{\omega})]\right\} \tag{2}$$

Using the following formulas

$$L[u(t)\ \sin(at)] = \frac{a}{s^2+a^2} \tag{3}$$

and
$$L[u(t-a)f(t-a)] = e^{-as}L[f(t)] \tag{4}$$

the transform of (2) is written

$$V_{OUT}(s) = \sum_{n=0}^{\infty} \frac{\omega E_m}{s^2+\omega^2} [e^{-\frac{n\pi}{\omega}s} + e^{-\frac{(n+1)\pi}{\omega}s}] \tag{5}$$

Assuming zero initial dc load currents in the circuit, its equation is

$$V_{OUT}(s) = (R + sL)\ I_{OUT}(s)$$

or
$$I_{OUT}(s) = \frac{1}{L}\frac{E_o(s)}{S+R/L} = \frac{E_m}{L}\sum_{n=0}^{\infty} \frac{\omega}{(s+R/L)(s^2+\omega^2)} \left[e^{-\frac{n\pi}{\omega}s} + e^{-\frac{(n+1)\pi}{\omega}s}\right] \tag{6}$$

Using (4) and the following formula

$$L^{-1}\left[\frac{\omega}{(s+R/L)(s^2+\omega^2)}\right] = \frac{\omega L^2}{Z^2} e^{-\frac{R}{L}t} + \frac{L}{Z}\sin(\omega t-\theta) \tag{7}$$

where $Z^2 = R^2 + \omega^2 L^2$ and $\theta = \tan^{-1}(\omega L/R)$, the inverse transform of (6) is written

$$i_{OUT}(t) = \frac{E_m}{L}\sum_{n=0}^{\infty}\left\{u(t-\frac{n\pi}{\omega})\left[\frac{\omega L^2}{Z^2}e^{-\frac{R}{L}(t-\frac{n\pi}{\omega})} + \frac{L}{Z}\sin(\omega t-n\pi-\theta)\right]\right.$$
$$\left. + u(t-\frac{(n+1)\pi}{\omega})\left[\frac{\omega L^2}{Z^2}e^{-\frac{R}{L}(t-\frac{(n+1)\pi}{\omega})} - \frac{L}{Z}\sin(\omega t-n\pi-\theta)\right]\right\} \tag{8}$$

Equation (8) shows how complicated the final result can get even for simple circuits. However, it is the complete time response of the circuit starting from rest.

Application to 3Ø induction motor. In the preceeding example, the analysis was simple because the load voltage was essentially known and the load volt-ampere relationship was simple. In another common problem, the inverter-fed 3Ø induction motor, the load voltages are often known but the load VA relationship is much more complicated. The following development from [2] of a 3Ø induction motor transfer function is included to illustrate how s-domain techniques may be applied to this type of problem.

Based on the assumptions of balanced three-wire, steady-state operation and completely known stator phase voltages, the analysis begins by writing the Laplace transform, V(s), of one of the phase voltages, as was done before. These voltages are balanced and periodic, so the Laplace transforms of the other two phase voltages can be expressed in terms of V(s) using the shifting theorem, (4). Taking the α-β transformation, both the α and β stator voltages can be expressed in terms of the single quantity V(s). Similarly, α and β stator and rotor currents can be expressed in terms of $I_s(s)$ and $I_r(s)$, which are the transforms of the unknown stator and rotor currents in one phase. Substituting the above expressions into the four α-β KVL equations of the motor leads to two equations in unknowns $I_s(s)$ and $I_r(s)$, assuming a short-circuited rotor. From these equations an equivalent circuit is drawn and a transfer function relating $I_s(s)$ to V(s) is derived. If the applied voltage is simple enough, an analytic solution may be carried out, but in most cases a numerical solution using fast Fourier transforms or numerical Laplace transform inversion is more practical.

To illustrate the approach in more detail, suppose the stator voltages are

$$\begin{aligned} v_1(t) &= v(t)\,u(t) \\ v_2(t) &= v(t-\tfrac{2\pi}{3\omega})u(t-\tfrac{2\pi}{3\omega}) \\ v_3(t) &= v(t-\tfrac{4\pi}{3\omega})u(t-\tfrac{4\pi}{3\omega}) \end{aligned} \tag{9}$$

Although (9) does not accurately represent the voltages near t=0, it results in simple expressions for the transforms.

$$\begin{aligned} V_1(s) &= V(s) \\ V_2(s) &= V(s)\,e^{-as} \\ V_3(s) &= V(s)\,e^{-2as} \end{aligned} \tag{10}$$

where $a=2\pi/(3\omega)$. Taking the α-β transformation of (10) results in

$$V_{s\alpha}(s) = \sqrt{\frac{2}{3}}\, V(s)(1 - \frac{1}{2}e^{-as} - \frac{1}{2}e^{-2as}) = \sqrt{\frac{1}{6}}\, V(s)(2 + e^{-as})(1 - e^{-as})$$

$$V_{s\beta}(s) = \sqrt{\frac{2}{3}}\, V(s)(-\frac{\sqrt{3}}{2}e^{-as} + \frac{\sqrt{3}}{2}e^{-2as}) = -\sqrt{\frac{1}{2}}\, V(s)\, e^{-as}\,(1-e^{-as}) \tag{11}$$

In a similar manner, expressions are written for the unknown currents:

$$I_{s\alpha}(s) = \sqrt{\frac{1}{6}}\, I_s(s)(2+e^{-as})(1-e^{-as})$$

$$I_{s\beta}(s) = -\sqrt{\frac{1}{2}}\, I_s(s)\, e^{-as}\,(1-e^{-as})$$

$$I_{r\alpha}(s) = \sqrt{\frac{1}{6}}\, I_r(s)(2+e^{-as})(1-e^{-as}) \tag{12}$$

$$I_{r\beta}(s) = -\sqrt{\frac{1}{2}}\, I_r(s)\, e^{-as}\,(1-e^{-as})$$

Combining (11) and (12) with the motor equations

$$\begin{bmatrix} R_s + sL_s & 0 & sM & 0 \\ 0 & R_s + sL_s & 0 & sM \\ sM & -\omega_r M & R_r + sL_r & -\omega_r L_r \\ \omega_r M & sM & \omega_r L_r & R_r + sL_r \end{bmatrix} \begin{bmatrix} I_{s\alpha}(s) \\ I_{s\beta}(s) \\ I_{r\alpha}(s) \\ I_{r\beta}(s) \end{bmatrix} = \begin{bmatrix} V_{s\alpha}(s) \\ V_{s\beta}(s) \\ 0 \\ 0 \end{bmatrix} \tag{13}$$

yields

$$\begin{bmatrix} R_s + sL_s & sM \\ sM & R_r \dfrac{2+e^{-as}}{2+e^{-as}(1+\sqrt{3}\omega_r/s)} + sL_r \end{bmatrix} \begin{bmatrix} I_s(s) \\ I_r(s) \end{bmatrix} = \begin{bmatrix} V(s) \\ 0 \end{bmatrix} \tag{14}$$

Eliminating $I_r(s)$ results in a transfer function in the following form

$$\frac{I_s(s)}{V(s)} \overset{\Delta}{=} Y(s) = \frac{L_r}{L_sL_r - M^2}\, H(s)$$

$$\text{where} \quad H(s) = \frac{2(s + \frac{1}{\tau}) + e^{-as}(s + \frac{1}{\tau} + \sqrt{3}\omega_r)}{2[s^2 + s(\frac{1}{\tau_1} + \frac{1}{\tau_2}) + \frac{1}{\tau\tau_1}] + e^{-as}[s^2 + s(\sqrt{3}\omega_r + \frac{1}{\tau_1} + \frac{1}{\tau_2}) + \frac{1}{\tau_1}(\frac{1}{\tau} + \sqrt{3}\omega_r)]} \tag{15}$$

For a wye motor, the transfer function can be expressed in terms of line quantities as follows

$$V_L(s) \overset{\Delta}{=} V_1(s) - V_2(s) = V(s)(1 - e^{-as}) \tag{16}$$

$$I_L(s) \overset{\Delta}{=} I_1(s) = \sqrt{\frac{2}{3}}\, I_{s\alpha}(s) = I_s(s)(2 + e^{-as})(1 - e^{-as})/3 \tag{17}$$

$$Y_\lambda(s) \overset{\Delta}{=} \frac{I_L(s)}{V_L(s)} = Y(s)\, \frac{2 + e^{-as}}{3} \tag{18}$$

Equation (18) is the transfer function in its final form.

Fourier Series Techniques

If the load of the switching circuit can be considered linear, and the voltages being applied to the load by the converter are known, then superposition holds and Fourier series may be used to determine the load currents. The analysis begins by finding the Fourier series representation of the load voltages and the equivalent circuit of the load at each significant frequency appearing in the series. For each harmonic, the current is found by solving the circuit which results from applying the harmonic voltage to the respective harmonic equivalent load circuit. These currents are summed to get the approximate load current waveform. If the load is truly linear, the load voltage is known accurately, and enough terms are retained in the series, the current waveform will be accurate.

The technique may be applied to circuits in which there remains a degree of uncertainty in the voltage waveform. For example, if the thyristor extinction angle β is unknown, β may be left as a variable in the Fourier series of the voltage. After applying the above technique, β is found by solving for the zero crossing the the resulting current waveform.

Application to 3Ø induction motor. To illustrate the application of Fourier series to a switching circuit, consider a 3Ø, three-wire induction motor fed from a six-step bridge inverter with 180° device conduction [3]. The motor voltages in this case can be considered to be square waves with only odd harmonics, of which only non-triplen harmonics need be considered. The well known fundamental frequency per phase equivalent circuit of the motor is shown in Fig. 2. Using this circuit to develop the harmonic equivalent circuit leads to the circuit of Fig. 3. Here R_2/s is neglected since the slip s is high at harmonic frequencies, the magnetizing reactance ωL_m is neglected since it is shunted by the much smaller secondary leakage reactance $n\omega L_2$, and the stator resistance R_1 is neglected in comparison with the leakage reactances at harmonic frequencies.

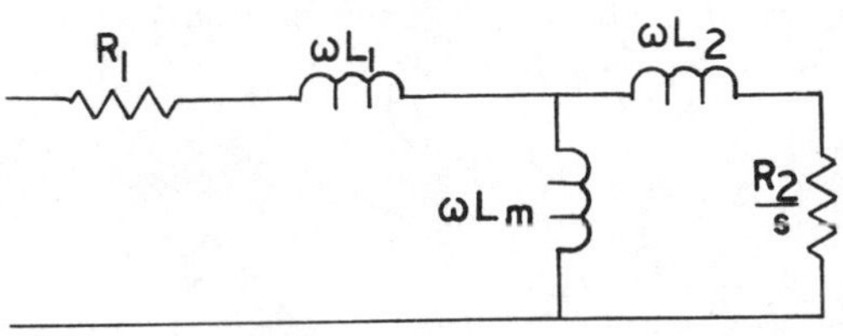

Fig. 2 Approximate fundamental frequency induction motor equivalent circuit.

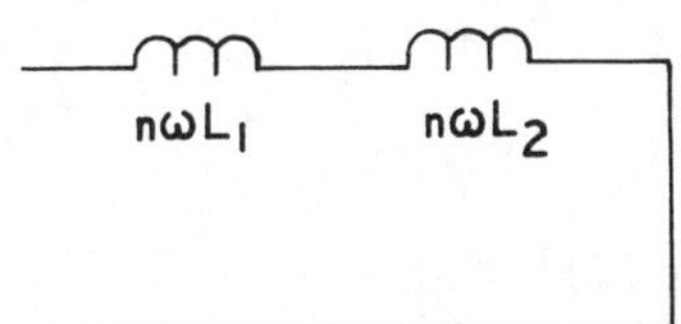

Fig. 3 Approximate nth harmonic frequency induction motor equivalent circuit, n = 5, 7, 11, 13, . . .

Assuming that the applied stator voltage is an odd function and that c_n is the signed magnitude of the nth harmonic voltage calculated in the Fourier series, and defining I_n and $\cos Ø_n$ to be the nth harmonic current amplitude and power factor,

$$v_s \triangleq \sum_{n=1,5,7,11,...}^{\infty} c_n \sin n\omega t \tag{19}$$

$$I_1 = \frac{c_1}{|Z_o|} \tag{20}$$

$$\cos Ø_1 = \frac{\mathrm{Re}(Z_o)}{|Z_o|} \tag{21}$$

where
$$Z_o = R_1 + j\omega L_1 + \frac{(R_2/s + j\omega L_2)\, j\omega Lm}{(R_2/s) + j\omega(L_2 + Lm)}$$

$$= R_1 + j\omega L_2 + \frac{(R_2/s)\omega^2 L_m^2 + j\omega L_m[(R_2/s)^2 + \omega^2 L_2(L_2 + L_m)]}{(R_2/s)^2 + \omega^2(L_2 + L_m)^2} \tag{22}$$

$$I_n = \frac{c_n}{n\omega(L_1 + L_2)}, \ \cos Ø_n = 0 \ ; \ n = 5, 7, 11, 13, . . . \tag{23}$$

If a four-wire induction motor were being used, so that triplen harmonics could flow, a reasonable assumption for the load impedance at such frequencies is $Z = R_1 + n\omega L_1$, i.e., the primary resistance and leakage reactance.

In [3], it is shown that the assumption of a square wave for the applied voltage in this analysis is adequate for predicting motor performance, but leads to discrepancies when used to predict current waveforms since it neglects device turn-off and dead-time effects.

Analysis by Neglecting Harmonics

In some switching circuits, a greatly simplified analysis of sufficient accuracy may be obtained by neglecting all harmonics produced by the converter, retaining only one term in the Fourier series expansion of the output waveform. For example, a rectifier might be described solely in terms of its average dc output, or a square-wave inverter output might be considered sinusoidal. This simplification is useful when the dominant performance characteristics are all that are required and harmonics do not greatly affect the load.

Neglecting harmonics in a 3Ø induction motor drive. As an example of this technique, a 3Ø rectifier-filter-inverter induction motor drive system is approximately represented in two different ways by neglecting harmonics. In one case, all harmonics due to rectifier action are neglected and only the average dc output of the rectifiers is used. In the second case, all harmonics due to the inverter are also neglected and only its fundamental output voltage is used. The d-q transformation is applied to simplify system representation. This example is taken from [4], in which analog computer representations are given and used to show that neglecting the rectifier harmonics does not seriously affect accuracy for most cases provided the rectifiers operate in a continuous conduction mode.

The system studied is shown in Fig. 4. The voltage equations for the filter during continuous operation, $I_R > 0$, are

$$V_R = V_I + (R_F + pL_F)I_R \qquad V_I = (I_R - I_I)/(pC_F) \tag{24}$$

where p indicates time differentiation and 1/p indicates integration. Since I_R cannot become negative, when I_R is not positive then $I_R = 0$ and $V_R = V_I$.

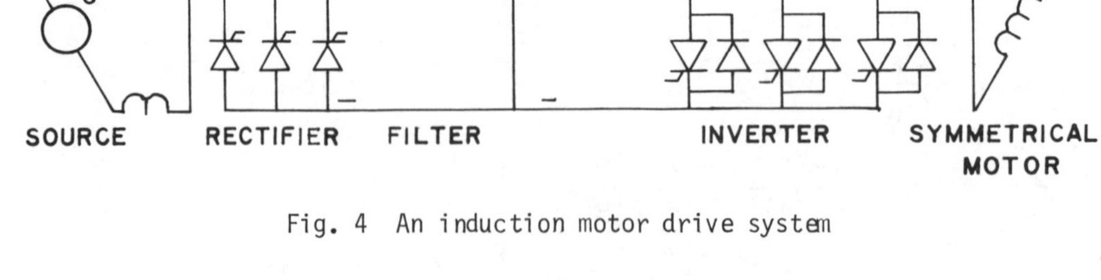

Fig. 4 An induction motor drive system

An approximate expression for the average output voltage of a six-pulse rectifier in a d-q representation, which neglects all harmonics as stated above, is derived in [5] and is

$$V_R = \frac{3\sqrt{3}}{\Pi} V_s \cos \alpha - \frac{3}{\Pi} X_{co} I_R \tag{25}$$

where V_S is the magnitude of the line to neutral source voltage, α the delay angle, and X_{CO} the commutating reactance, assumed to be small compared to the filter inductor reactance.

Assuming the inverter produces stepped waveforms with negligible commutation intervals, the waveforms applied to the stator phases of the motor may be approximated by Fourier series. For 120° conduction of the inverter devices, for example, v_{as} may be expressed as

$$v_{as} = \frac{2V_I}{\Pi} (\cos \omega_e t + \frac{1}{5} \cos 5 \omega_e t - \frac{1}{7} \cos 7 \omega_e t - \ldots) \tag{26}$$

Since V_I is not constant, each of the harmonics of the stator voltages form a balanced 3Ø set amplitude modulated by V_I. Transforming the phase voltages to a synchronously rotating reference frame of angular velocity ω_e leads to applied stator d-q voltages of the form

$$\begin{aligned} v_{qs}^e &= (2 V_I/\Pi)(1 + \frac{2}{35} \cos 6 \omega_e t - \frac{2}{143} \cos 12 \omega_e t + \ldots) \triangleq \frac{2V_I}{\Pi} g_{qs}^e \\ v_{ds}^e &= (2 V_I/\Pi)(\frac{12}{35} \sin 6 \omega_e t - \frac{24}{143} \sin 12 \omega_e t + \ldots) \triangleq \frac{2V_I}{\Pi} g_{ds}^e \end{aligned} \tag{27}$$

Here the superscript e represents the synchronous frame and new variables are defined to represent the series. Assuming there is no power loss in the inverter,

$$V_I I_I = \frac{3}{2} (v_{qs}^e i_{qs}^e + v_{ds}^e i_{ds}^e) \tag{28}$$

where the constant corrects for the power variant d-q transformation used. Using (27) in (28) gives

$$I_I = \frac{3}{\Pi} (i_{qs}^e{}_{qs}s^e + e_{ds}^e g_{ds}^e) \tag{29}$$

Equations (24)-(28) completely describe the drive system neglecting rectifier harmonics and may be combined into an equivalent circuit suitable for analog simulation or other analysis.

An even simpler representation of the system can be obtained by additionally neglecting the inverter produced harmonics appearing in (27). Thus in (27) and (29), let $g_{qs}^e = 1$ and $g_{ds}^e = 0$. This simplification results in neglecting the motor magnetizing currents [4]. Nevertheless, many of the dominant performance characteristics are preserved even though all harmonics are neglected.

State Variable Analysis

The generality of the state variable method of system representation makes it applicable to a wide class of power converter circuits, and much work has been done in this area. The basic principle of the state variable approach is that given the differential equations describing a system, and all the inputs $\vec{u}$ to the system, there is a minimum set of n-numbers, called the state vector $\vec{x}$, which completely determines the behavior of the system at all times. For example, if the effects of gravity and friction on a projectile are known and its position and velocity (the state variables) are known at some instant during its flight, its path can be completely determined.

The state representation of a system expresses the system equations in vector form as n first-order differential equations in the following form.

$$\frac{d\vec{x}}{dt} \triangleq \dot{\vec{x}} = A\vec{x} + B\vec{u} \tag{30}$$

where A and B are matrices. If the desired output $\vec{y}$ of the system is different than the state vector, a second equation is used

$$\vec{y} = C\vec{x} + D\vec{u} \tag{31}$$

where C and D are matrices.

Although the choice of variables for the state vector is not unique, in electrical systems it is common to choose inductor currents and capacitor voltages as the state variables. The advantage of the state variable approach lies in the fact that it is systematic and thus easy to adapt for computer simulation.

In switching circuits the state variable approach is complicated by the different modes caused by the switching of the devices. Thus a single circuit is actually many different systems depending on which devices conduct. The frequently unknown length of conduction intervals adds another complication. The usual approach is to choose a single state vector for all conduction modes and determine (30) and its solution for each mode. Continuity conditions at the switching instants are then used to piece together the solution. If the conduction intervals are unknown, iteration is normally used to solve the resulting transcendental equations. Another approach [6] reduces solution time by using the same A and B matrices in different modes and redefining the state and input vectors.

Two examples of state variable application are now given to illustrate the approach.

Application to 1Ø induction motor. This example, from [7], illustrates the straightforward application of state variable analysis to a single steady-state half cycle of operation of a single-phase induction motor whose stator voltage is controlled by a pair of back-to-back thyristors. It will be assumed the auxiliary winding is open, resulting in a third-order system of the form $\dot{\vec{x}} = A\vec{x} + B\vec{u}$, although in [7] a fifth order capacitor-run motor is also considered. The chief difficulty in the analysis is the unknown thyristor conduction period, which depends on the slip, firing angle, motor parameters, and stator voltage. To overcome this problem, the conduction period and slip are considered as independent variables and the triggering angle is found through iteration.

The first step in the analysis is to write the KVL equations for the motor windings--stator and rotor. Then the winding currents are chosen as the state variables and state equations are determined for both the conducting and blocking modes of the thyristor during a half cycle. Time-domain solutions for the state vector at the end of the half cycle are then found in terms of the initial state vector. These vectors are additive inverses due to steady-state continuity conditions and half-wave symmetry, so that an expression for the initial state $\vec{x}(0)$ is found in terms of slip, conduction period, and firing angle. For a given slip and conduction period, $\vec{x}(0)$ and the firing angle are found by assuming a firing angle, computing $\vec{x}(0)$, and iterating until an $\vec{x}(0)$ is found in which the main winding current is zero at the beginning of the blocking period.

In more detail, the KVL equations are given by [8] as

$$\begin{bmatrix} v_s \\ 0 \\ 0 \end{bmatrix} = \begin{bmatrix} R_s & 0 & 0 \\ 0 & R_r & -L_rS \\ MS & L_rS & R_r \end{bmatrix} \begin{bmatrix} i_s \\ i_{r\alpha} \\ i_{r\beta} \end{bmatrix} + \frac{d}{dt} \begin{bmatrix} L_s & M & 0 \\ M & L_r & 0 \\ 0 & 0 & L_r \end{bmatrix} \begin{bmatrix} i_s \\ i_{r\alpha} \\ i_{r\beta} \end{bmatrix} = R\vec{i} + \frac{d}{dt} L\vec{i} \tag{32}$$

where S = speed, L_s, L_r and M are inductances, and subscripts s and r refer to stator main and rotor winding quantities. In the conducting mode, letting $\vec{x} = \vec{i}$, the state equation becomes

$$\dot{\vec{x}} = -L^{-1}R\vec{x} + L^{-1}\begin{bmatrix}1\\0\\0\end{bmatrix}u = A_1\vec{X} + \vec{b}u = \begin{bmatrix}\frac{L_rR_s}{D} & \frac{-MR_r}{D} & \frac{ML_rS}{D}\\ \frac{-MR_s}{D} & \frac{L_sR_r}{D} & \frac{L_sL_rS}{D}\\ \frac{-MS}{L_r} & -S & \frac{R_r}{L_r}\end{bmatrix}\vec{x} + \begin{bmatrix}\frac{-L_r}{D}\\ \frac{M}{D}\\ 0\end{bmatrix}u \quad (33)$$

where u = applied voltage and $D = M^2 - L_sL_r$.

In the blocking mode, with $i_s = 0$, (32) reduces to

$$\begin{bmatrix}0\\0\end{bmatrix} = \begin{bmatrix}R_r & -L_rS\\ L_rS & R_r\end{bmatrix}\begin{bmatrix}i_{r\alpha}\\ i_{r\beta}\end{bmatrix} + \frac{d}{dt}\begin{bmatrix}L_r & 0\\ 0 & L_r\end{bmatrix}\begin{bmatrix}i_{r\alpha}\\ i_{r\beta}\end{bmatrix} \quad (34)$$

Using the same state vector as before, the state equation becomes

$$\dot{\vec{x}} = A_2\vec{x} = \begin{bmatrix}0 & 0 & 0\\ 0 & -R_r/L_r & S\\ 0 & -S & -R_r/L_r\end{bmatrix}\begin{bmatrix}i_s\\ i_{r\alpha}\\ i_{r\beta}\end{bmatrix} \quad (35)$$

Assuming t = 0 at the beginning of the blocking mode and $t = T_{OFF}$ at the end of this mode, we have from the from the solution of (35) that

$$\vec{x}(T_{OFF}) = \exp(A_2T_{OFF})\,\vec{x}(0) \quad (36)$$

During the conducting mode, solving (33) gives

$$\vec{x}(t) = \exp[A_1(t - T_{OFF})]\,\vec{x}(T_{OFF}) + \int_{T_{OFF}}^{t} \exp[A_1(t - m)]\vec{b}\,\mu(m)dm \quad (37)$$

At the end of this mode, when $t = T_{OFF} + T_{ON} = T$, we have in steady-state,

$$\vec{x}(T) = -\vec{x}(0) = \exp(A_1T_{ON})\,\vec{x}(T_{OFF}) + \int_{T_{OFF}}^{T} \exp[A_1(T-m)]\vec{b}u(m)\,dm \quad (38)$$

Combining (36) and (38) and using $u(t) = V\sin[\omega(t-T_{OFF}) + \alpha]$, where α is the firing angle, we have finally that

$$\vec{x}(0) = -C^{-1}\exp(A_1T)\int_{T_{OFF}}^{T} \exp(-A_1m)\vec{b}\,V\sin[\omega(m-T_{OFF}) + \alpha]dm \quad (39)$$

where $C = I + \exp(A_1T_{ON})\exp(A_2T_{OFF})$ and I = identity matrix.

The only unknowns on the right side of (39) are the speed S, conduction time T_{ON}, and firing angle α. Since it is known that $i_s(0) = 0$, for a given S and T_{ON} an α can be assumed and corrected through iteration.

Application to 3∅ induction motor. This example, from [9], which is similar to but more complicated than the preceeding example, uses the state approach in the analysis of a 120° conduction interval 3∅ inverter supplying an induction motor in steady-state. As shown in

Fig. 5, the system studied consists of an ideal dc source, an LC filter, an ideal inverter and an induction machine. Due to the half-wave and three-phase symmetry, the steady-state analysis only needs to extend over one-sixth of a cycle, as will be shown. However, current is discontinuous with 120° gating so that two modes are present during the interval studied.

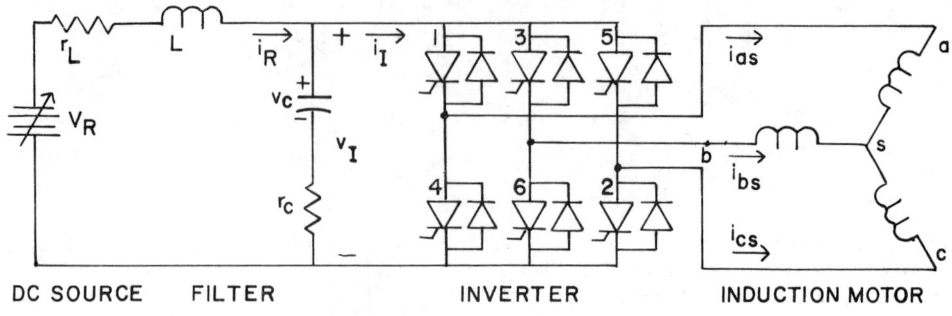

Fig. 5 Another induction motor drive system

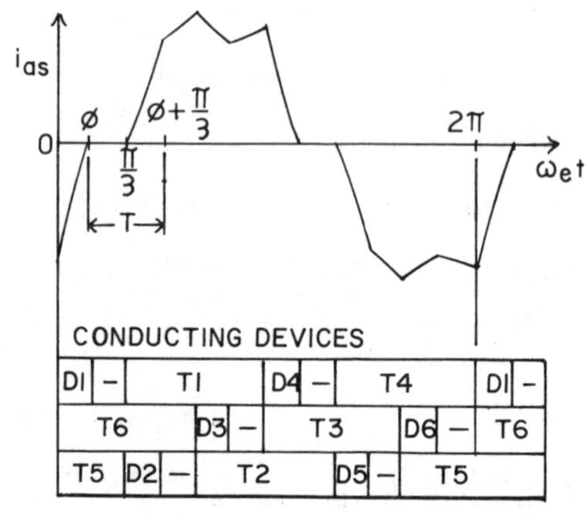

Fig. 6 Typical steady-state line current waveform and inverter operation

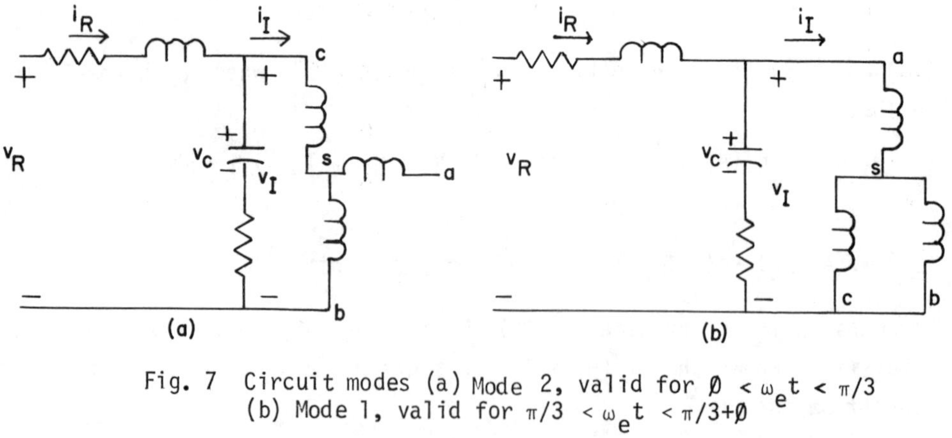

Fig. 7 Circuit modes (a) Mode 2, valid for $\emptyset < \omega_e t < \pi/3$ (b) Mode 1, valid for $\pi/3 < \omega_e t < \pi/3+\emptyset$

Some of the variables are defined in Fig. 5 and Fig. 6, and Fig. 7 shows equivalent circuits of the system during the two modes-mode 1, the beginning of positive conduction in phase a, and mode 2, the zero current interval for phase a. The d-q equations of the motor are

$$\begin{bmatrix} v_{qs}^s \\ v_{ds}^s \\ 0 \\ 0 \end{bmatrix} = \begin{bmatrix} r_s + L_s p & 0 & Mp & 0 \\ 0 & r_s + L_s p & 0 & Mp \\ Mp & -\omega_r M & r_r' + L_r' p & -\omega_r L_r' \\ \omega_r M & Mp & \omega_r L_r' & r_r' + L_r' p \end{bmatrix} \begin{bmatrix} i_{qs}^s \\ i_{ds}^s \\ i_{qr}'^s \\ i_{dr}'^s \end{bmatrix} \tag{40}$$

As indicated by the superscript s, (40) is written for d-q axes fixed in the stator. With the q-axis coincident with the phase a-axis and based on the absence of zero-sequence quantities, relations between phase and d-q variables are

$$v_{qs}^s = v_{as}, \quad v_{ds}^s = \frac{1}{\sqrt{3}}(v_{cs} - v_{bs})$$
$$i_{qs}^s = i_{as}, \quad i_{ds}^s = \frac{1}{\sqrt{3}}(i_{cs} - i_{bs}) \tag{41}$$

Equations for the filter are

$$i_I + pC\, v_c - i_R = 0$$
$$V_R = (r_L + pL)i_R + v_c + r_c(i_R - i_I) \tag{42}$$

During mode 1, when all phases carry current, as in Fig. 7(b), it follows that

$$v_{as} - v_{bs} = v_I, \; v_{bs} = v_{cs}, \; i_I = i_{as}$$
$$v_{qs}^s = \frac{2}{3}v_I = \frac{2}{3}[v_c + r_c(i_R - i_I)], \; v_{ds}^s = 0, \; i_I = i_{qs}^s \tag{43}$$

Combining (40), (42), and (43) into matrix form gives

$$\begin{bmatrix} 0 \\ 0 \\ 0 \\ 0 \\ 0 \\ V_R \end{bmatrix} = \begin{bmatrix} r_s + \frac{2}{3}r_c + L_s p & 0 & Mp & 0 & -\frac{2}{3} & -\frac{2}{3}r_c \\ 0 & r_s + L_s p & 0 & Mp & 0 & 0 \\ Mp & -\omega_r M & r_r' + L_r' p & -\omega_r L_r' & 0 & 0 \\ \omega_r M & Mp & \omega_r L_r' & r_r' + L_r' p & 0 & 0 \\ 1 & 0 & 0 & 0 & pC & -1 \\ -r_c & 0 & 0 & 0 & 1 & r_L + r_c + pL \end{bmatrix} \begin{bmatrix} i_{qs}^s \\ i_{ds}^s \\ i_{qr'}^s \\ i_{dr'}^s \\ v_c \\ i_R \end{bmatrix} \tag{44}$$

In state form, this becomes

$$\dot{\vec{x}} = \vec{A_1}\,\vec{x} + \vec{B_1 \mu} \tag{45}$$

where

$$\vec{x} = \begin{bmatrix} i_{qs}{}^{s} \\ i_{ds}{}^{s} \\ i_{qr'}{}^{s} \\ i_{dr'}{}^{s} \\ v_c \\ i_R \end{bmatrix}, \quad \vec{\mu} = \begin{bmatrix} 0 \\ 0 \\ 0 \\ 0 \\ 0 \\ V_R \end{bmatrix}, \quad B_1 = \begin{bmatrix} L_s & 0 & M & 0 & 0 & 0 \\ 0 & L_s & 0 & M & 0 & 0 \\ M & 0 & L_{r'} & 0 & 0 & 0 \\ 0 & M & 0 & L_{r'} & 0 & 0 \\ 0 & 0 & 0 & 0 & C & 0 \\ 0 & 0 & 0 & 0 & 0 & L \end{bmatrix}^{-1}$$

and

$$A_1 = -B_1 \begin{bmatrix} r_s + \frac{2}{3} r_c & 0 & 0 & 0 & -\frac{2}{3} & -\frac{2}{3} r_c \\ 0 & r_s & 0 & 0 & 0 & 0 \\ 0 & -\omega_r M & r_{r'} & -\omega_r L_{r'} & 0 & 0 \\ \omega_r M & 0 & \omega_r L_{r'} & r_{r'} & 0 & 0 \\ 1 & 0 & 0 & 0 & 0 & -1 \\ -r_c & 0 & 0 & 0 & 1 & r_L + r_c \end{bmatrix}$$

During mode 2, when $i_{as} = 0$, it follows that

$$v_{cs} - v_{bs} = v_I \text{ and } i_I = i_{cs} = -i_{bs} \tag{46}$$

so that from (41),

$$v_{ds}{}^{s} = \frac{1}{\sqrt{3}} v_I, \quad v_{qs}{}^{s} = Mp\, i_{qr'}{}^{s}$$
$$i_{qs}{}^{s} = 0, \text{ and } i_I = -\frac{1}{2} i_{qs}{}^{s} + \frac{\sqrt{3}}{2} i_{ds}{}^{s} \tag{47}$$

Combining (40), (42), and (47) into matrix forms yields

$$\begin{bmatrix} 0 \\ 0 \\ 0 \\ 0 \\ 0 \\ V_R \end{bmatrix} = \begin{bmatrix} r_s + L_s p & 0 & 0 & 0 & 0 & 0 \\ -\frac{1}{2\sqrt{3}} r_c & r_s + \frac{1}{2} r_c + L_s p & 0 & Mp & -\frac{1}{\sqrt{3}} & -\frac{1}{\sqrt{3}} r_c \\ Mp & -\omega_r M & r_r' + L_r' p & -\omega_r L_r' & 0 & 0 \\ \omega_r M & Mp & \omega_r L_r' & r_r' + L_r' p & 0 & 0 \\ -\frac{1}{2} & \frac{\sqrt{3}}{2} & 0 & 0 & pC & -1 \\ \frac{1}{2} r_c & -\frac{\sqrt{3}}{2} r_c & 0 & 0 & 1 & r_L + r_c + pL \end{bmatrix} \begin{bmatrix} i_{qs}{}^{s} \\ i_{ds}{}^{s} \\ i_{qr}{}'^{s} \\ i_{dr}{}'^{s} \\ v_c \\ i_R \end{bmatrix} \tag{48}$$

In state form this becomes

$$\dot{\vec{x}} = A_2\vec{x} + B_2\vec{\mu} \tag{49}$$

where

$$B_2 = \begin{bmatrix} L_s & 0 & 0 & 0 & 0 & 0 \\ 0 & L_s & 0 & M & 0 & 0 \\ M & 0 & L_{r'} & 0 & 0 & 0 \\ 0 & M & 0 & L_{r'} & 0 & 0 \\ 0 & 0 & 0 & 0 & C & 0 \\ 0 & 0 & 0 & 0 & 0 & L \end{bmatrix}^{-1} \text{and } A_2 = -B_2 \begin{bmatrix} r_s & 0 & 0 & 0 & 0 & 0 \\ -\frac{1}{2\sqrt{3}}r_c & r_s + \frac{1}{2}r_c & 0 & 0 & -\frac{1}{\sqrt{3}} & -\frac{1}{\sqrt{3}}r_c \\ 0 & -\omega_r M & r_{r'} & -\omega_r L_{r'} & 0 & 0 \\ \omega_r M & 0 & \omega_r L_{r'} & r_{r'} & 0 & 0 \\ -\frac{1}{2} & \frac{\sqrt{3}}{2} & 0 & 0 & 0 & -1 \\ \frac{1}{2}r_c & -\frac{\sqrt{3}}{2}r_c & 0 & 0 & 1 & r_L + r_c \end{bmatrix}$$

The solution of (49) during $\emptyset/\omega_e < t < \pi/(3\omega_e)$ is

$$\vec{x}(t) = e^{A_2(t-\frac{\emptyset}{\omega_e})}\,\vec{x}\left(\frac{\emptyset}{\omega_e}\right) + \lambda_2\left(t,\frac{\emptyset}{\omega_e}\right)B_2\vec{\mu} \tag{50}$$

where

$$\lambda_2\left(t,\frac{\emptyset}{\omega_e}\right) = \int_{\frac{\emptyset}{\omega_e}}^{t} e^{A_2(t-\tau)}d\tau = A_2^{-1}\left(e^{A_2(t-\frac{\emptyset}{\omega_e})} - I\right)$$

The solution of (45) during $\pi/(3\omega_e) < t < \emptyset/\omega_e + \pi/(3\omega_e)$ is

$$\vec{x}(t) = e^{A_1(t-\frac{\pi}{3\omega_e})}\,\vec{x}\left(\frac{\pi}{3\omega_e}\right) + \lambda_1\left(t,\frac{\pi}{3\omega_e}\right)B_1\vec{\mu} \tag{51}$$

where

$$\lambda_1\left(t,\frac{\pi}{3\omega_e}\right) = A_1^{-1}\left(e^{A_1(t-\frac{\pi}{3\omega_e})} - I\right)$$

Substituting (50) with $t = \pi/(3\omega_e)$ into (51) with $t = \emptyset/\omega_e + \pi/(3\omega_e)$ gives a relationship between $\vec{x}(\frac{\emptyset}{\omega_e})$ and $\vec{x}\,(\frac{3\emptyset+\pi}{3\omega_e})$.

$$\vec{x}\left(\frac{\emptyset}{\omega_e} + \frac{\pi}{3\omega_e}\right) = e^{A_1\frac{\emptyset}{\omega_e}}\,e^{A_2(\frac{\pi}{3\omega_e}-\frac{\emptyset}{\omega_e})}\,\vec{x}\left(\frac{\emptyset}{\omega_e}\right)$$

$$+\left[e^{A_1\frac{\emptyset}{\omega_e}}\lambda_2\left(\frac{\pi}{3\omega_e},\frac{\emptyset}{\omega_e}\right)B_2 + \lambda_1\left(\frac{\emptyset}{\omega_e}+\frac{\pi}{3\omega_e},\frac{\pi}{3\omega_e}\right)B_1\right]\vec{\mu} \tag{52}$$

It will now be shown that another relationship between the state vectors at these two instants can be derived based on symmetry relations in steady-state. Due to three-phase and half-wave symmetry, with $\vec{i}(\theta) = [i_{as}(\theta)\ i_{bs}(\theta)\ i_{cs}(\theta)]^T$, it follows that

$$\vec{i}\left(\omega_e t + \frac{\pi}{3}\right) = \begin{bmatrix} 0 & 1 & 0 \\ 0 & 0 & 1 \\ 1 & 0 & 0 \end{bmatrix} \vec{i}\left(\omega_e t + \frac{\pi}{3} + \frac{2\pi}{3}\right) = \begin{bmatrix} 0 & -1 & 0 \\ 0 & 0 & -1 \\ -1 & 0 & 0 \end{bmatrix} \vec{i}(\omega_e t) \tag{53}$$

Due to the absence of zero-sequence current, (53) reduces to

$$\begin{bmatrix} i_b(\omega_e t + \frac{\pi}{3}) \\ i_c(\omega_e t + \frac{\pi}{3}) \end{bmatrix} = \begin{bmatrix} 0 & -1 \\ 1 & 1 \end{bmatrix} \begin{bmatrix} i_b(\omega_e t) \\ i_c(\omega_e t) \end{bmatrix} \tag{54}$$

and also the d-q transformation with the q-axis coincident with the phase a-axis gives

$$\begin{bmatrix} i_q^s \\ i_d^s \end{bmatrix} = \begin{bmatrix} -1 & -1 \\ -\frac{1}{\sqrt{3}} & \frac{1}{\sqrt{3}} \end{bmatrix} \begin{bmatrix} i_{bs} \\ i_{cs} \end{bmatrix} \tag{55}$$

$$\text{Letting } T = \begin{bmatrix} -1 & 1 \\ -\frac{1}{\sqrt{3}} & \frac{1}{\sqrt{3}} \end{bmatrix} \text{ then } T^{-1} = -\frac{\sqrt{3}}{2} \begin{bmatrix} \frac{1}{\sqrt{3}} & 1 \\ \frac{1}{\sqrt{3}} & -1 \end{bmatrix} \tag{56}$$

Combining (54) and (55) using (56) gives

$$\begin{bmatrix} i_q^s(\omega_e t + \frac{\pi}{3}) \\ i_d^s(\omega_e t + \frac{\pi}{3}) \end{bmatrix} = T \begin{bmatrix} 0 & -1 \\ 1 & 1 \end{bmatrix} T^{-1} \begin{bmatrix} i_q^s(\omega_e t) \\ i_d^s(\omega_e t) \end{bmatrix} = \begin{bmatrix} \frac{1}{2} & \frac{\sqrt{3}}{2} \\ -\frac{\sqrt{3}}{2} & \frac{1}{2} \end{bmatrix} \begin{bmatrix} i_q^s(\omega_e t) \\ i_d^s(\omega_e t) \end{bmatrix} \tag{57}$$

This relation holds for both stator and rotor quantities. Also since v_C and i_R are sixth harmonics of the motor quantities,

$$v_c(\omega_e t + \frac{\pi}{3}) = v_c(\omega_e t) \text{ and } i_r(\omega_e t + \frac{\pi}{3}) = i_r(\omega_e t)$$

From these relations it follows finally that

$$\vec{x}\left(\frac{\emptyset}{\omega_e} + \frac{\pi}{3\omega_e}\right) = \begin{bmatrix} \frac{1}{2} & \frac{\sqrt{3}}{2} & 0 & 0 & 0 & 0 \\ -\frac{\sqrt{3}}{2} & \frac{1}{2} & 0 & 0 & 0 & 0 \\ 0 & 0 & \frac{1}{2} & \frac{\sqrt{3}}{2} & 0 & 0 \\ 0 & 0 & -\frac{\sqrt{3}}{2} & \frac{1}{2} & 0 & 0 \\ 0 & 0 & 0 & 0 & 1 & 0 \\ 0 & 0 & 0 & 0 & 0 & 1 \end{bmatrix} \vec{x}\left(\frac{\emptyset}{\omega_e}\right) \stackrel{\Delta}{=} D\vec{x}\left(\frac{\emptyset}{\omega_e}\right) \tag{58}$$

The unknown initial condition vector $\vec{x}(\emptyset/\omega_e)$ is now found by combining (52) and (58).

$$\vec{x}\left(\frac{\emptyset}{\omega_e}\right) = \left[D - e^{A_1 \frac{\emptyset}{\omega_e}} e^{A_2 \left(\frac{\pi}{3\omega_e} - \frac{\emptyset}{\omega_e}\right)}\right]^{-1} [C] \tag{59}$$

$$[C] = \left[e^{A_1 \frac{\emptyset}{\omega_e}} \lambda_2 \left(\frac{\pi}{3\omega_e}, \frac{\emptyset}{\omega_e}\right) B_2 + \lambda_1 \left(\frac{\emptyset}{\omega_e} + \frac{\pi}{3\omega_e}, \frac{\pi}{3\omega_e}\right) B_1\right] \vec{\mu}$$

The only unknown on the right side of (59) at a given steady-state rotor speed is $\emptyset$, the extinction angle defined in Fig. 6, which is known to lie in the range $0 < \emptyset < \pi/3$ for 120° inverter gating. The fact that $i_{qs}^s(\emptyset/\omega_e)$ must be zero on the left side of (59) provides a criterion for iterating to the proper value of $\emptyset$. Once $\vec{x}(\emptyset/\omega_e)$ is known, (50), (51), and (57) may be used to construct the complete waveforms during any period.

The state variable analysis of an inverter with 180° conduction is much simpler since only one mode exists during a 60° interval, and is considered in [9].

Analysis by Averaging Techniques

The method of averaging is useful for analyzing the low frequency, small-signal response of switching circuits. It is most applicable to circuits in which the power switching devices are operated with a switching period that is short compared to the response time of the output, and the effect low-frequency perturbations have on the output is desired. Thus it has been effectively applied to well-filtered high-frequency dc-to-dc power converters [10], [11], although it is not limited to these.

Referring to Fig. 1, the basic approach is to average the inputs, $\vec{v}_{IN}(t)$ and $\vec{d}(t)$, over a period of time that is short compared to the response time of the output, $\vec{v}_{OUT}(t)$. It is assumed that the load does not affect device conduction so that $\vec{i}_{FDBK}$ may be neglected. In the dc-to-dc converter this condition is met by assuming continuous current in the filter inductor and load. The system resulting from the averaging process may or may not be linear, and based on this the analysis proceeds by using standard applicable techniques such as Laplace transforms, state variables, small displacement theory, etc.

Two different approaches have been used to apply the averaging technique to dc-to-dc converters. In [10], the linear models from which transfer functions can be written are derived from a series of equivalent circuit manipulations. Since $\vec{v}_{IN}$, $\vec{d}$, and $\vec{v}_{OUT}$ reduce to scalars in this type of converter, the transfer functions developed are $V_{OUT}(s)/V_{IN}(s)$, the line to load transfer function, and $\hat{V}_{OUT}(s)/\hat{D}(s)$, the perturbed duty ratio to load transfer function. Although valid only for the low frequency response of the power converter, they are quite useful in studying average trends. For example, excellent correlation is obtained between theoretical and experimental response to step changes in duty ratio and in prediction of loop stability. The transfer functions also reveal interesting theoretical characteristics such as the dependence of the effective component values of the converter on the duty ratio.

The second approach [11] uses state variable techniques to develop the same two transfer functions as above. In addition, a general circuit model can be derived from the average state equations which describes the input-output and control properties of any dc-to-dc converter. This model is useful for studying the performace of the converter filter network throughout the range of duty cycle, optimizing the design, determining stability, and comparing various types of converters.

The averaged state equation is developed by first writing a pair of state equations for the two modes during a period T in which the duty ratio, defined as switch closed time over open time, is d.

$$\text{Switch closed:}\quad \dot{\vec{x}} = A_1\vec{x} + \vec{b}_1 v_{IN},\quad v_{OUT} = \vec{c}_1^{\,T}\vec{x} \tag{60}$$

$$\text{Switch opened:}\quad \dot{\vec{x}} = A_2\vec{x} + \vec{b}_2 v_{IN},\quad v_{OUT} = \vec{c}_2^{\,T}\vec{x} \tag{61}$$

Here the state equations are simplified from the form of (30) and (31) since the input and output are scalars. The averaged state equation is then written by taking a weighted average of the system matrices and the input and output vectors, based on the duty ratio d.

$$\text{Averaged model:}\quad \dot{\vec{x}} = A\vec{x} + \vec{b}\, v_{IN},\quad v_{OUT} = \vec{c}^{\,T}\vec{x} \tag{62}$$

$$\text{where}\quad A = dA_1 + d'A_2,\ \vec{b} = d\vec{b}_1 + d'\vec{b}_2,\ \vec{C} = d\vec{C}_1 + d'\vec{C}_2,\ \text{and } d' = 1-d \tag{63}$$

It may be shown that the approximation made by averaging is equivalent to the approximation $e^{A_1 t}e^{A_2 t} \approx I + (A_1 + A_2)t$.

Applying the usual techniques now, steady-state and line to load transfer functions are developed by assuming d is a constant, D, from period to period, and that $\vec{x}$ and v_{OUT} are perturbed by a small signal $\hat{v}_{IN}$ on the line. Making the substitutions d = D, $v_{IN} = V_{IN} + \hat{v}_{IN}$, $\vec{x} = \vec{X} + \vec{\hat{x}}$ and $v_{OUT} = V_{OUT} + \hat{v}_{OUT}$ in (62) and solving yields

$$\text{Steady state: } V_{OUT} = -\vec{C}^T A^{-1} \vec{b}\, \hat{V}_{IN} \tag{64}$$

$$\text{Line to output: } \hat{v}_{OUT}(s) = \vec{C}^T (sI - A)^{-1} \vec{b}\, \hat{v}_{IN}(s) \tag{65}$$

Next it is assumed that the duty ratio is also perturbed, so that $d = D + \hat{d}$. Adding this perturbation to (62) and neglecting second order perturbed terms yields the final dynamic small-signal model.

$$\dot{\hat{\vec{x}}} = A\vec{\hat{x}} + \vec{b}\,\hat{v}_{IN} + [(A_1 - A_2)\vec{X} + (\vec{b}_1 - \vec{b}_2) V_{IN}]\,\hat{d} \tag{66}$$

$$\hat{v}_{OUT} = \vec{C}^T \vec{\hat{x}} + (\vec{C}_1^{\,T} - \vec{C}_2^{\,T})\,\vec{X}\,\hat{d} \tag{67}$$

This linearized model can be used to simplify the analysis of a complete switching regulator circuit. The principal draw-back of the model is that it is only valid for small-signal, low frequency response. The averaging technique will now be illustrated in more detail.

<u>Averaging in a boost regulator by circuit manipulations</u>. To illustrate the equivalent circuit manipulations approach, the transfer functions for the so called boost regulator circuit of Fig. 8(a) are determined by averaging. First the switch and the diode of the original circuit are eliminated by splitting the circuit into two parts and introducing controlled sources. This is shown in Fig. 8(b), in which the inputs and outputs have also been averaged over a short interval by replacing v_{IN}, d, v_{OUT}, and i by their averages, $< v_{IN} >$, $<d>$, etc.

The line transfer function, $V_{OUT}(s)/V_{IN}(s)$, is developed by first assuming the averaged input $<d>(t)$ is a constant D, the duty ratio. Defining D' = 1 - D and replacing $<1-d>$ in Fig. 8(b) by D' yields the circuit of Fig. 8(c). The controlled sources enclosed by dotted lines in Fig. 8(c) have the effect of an ideal transformer of turns ratio D'. Normalizing this transformer to a unity turns ratio yields the circuit of Fig. 8(d), which immediately reduces to Fig. 8(e), from which the transfer function is written.

$$\frac{V_{OUT}(s)}{V_{IN}(s)} = \frac{D'}{R_D'^2 + R_1}\;\frac{1 + CR_c s}{1 + \frac{1}{Q}\frac{s}{\omega_0} + (\frac{s}{\omega_0})^2}, \text{ where}$$

$$Q = \frac{1}{\omega_0}\left[CR_c + \frac{CRR_1 + L}{D'^2 R + R_1}\right]^{-1} \text{ and } \omega_0 = \frac{1}{\sqrt{LC}}\sqrt{\frac{RD'^2 + R_1}{R + R_c}} \tag{68}$$

Equation (68) relates line and load for a constant duty ratio, and is also valid for a transient analysis following a step change in the duty ratio of the control input.

The perturbed control transfer function, $\hat{V}_{OUT}(s)/\hat{D}(s)$, is developed by assuming $< v_{IN} > = V_{IN}$, a constant, and that $<d>$, $<v_{OUT}>$, and $<i>$ are constants plus a perturbation, i.e., $<d> = D + \hat{d}(t)$, etc. Substituting into the averaged model of Fig. 8(b) results in the circuit of Fig. 9(a). The steady-state dc circuit, shown in Fig. 9(b), is evaluated and subtracted from Fig. 9(a). In addition the second order terms, $\hat{d}\hat{v}_{OUT}$ and $\hat{d}\hat{i}$, in the controlled sources are neglected, assuming $\hat{v}_{OUT} << V_{OUT}$, $\hat{i} << I$, and $\hat{d}$ is small. The result of these steps is shown in Fig. 9(c). Next each dependent source can be replaced by two sources, and as before, the circuit values are modified to yield a unity turns ratio transformer, as shown in Fig. 9(d). Reducing Fig. 9(d) to that of Fig. 9(e) and then combining the sources using Thevenin's theorem, the final linearized equivalent circuit for small amplitude control perturbations in a boost converter is shown in Fig. 9(f). The transfer function is

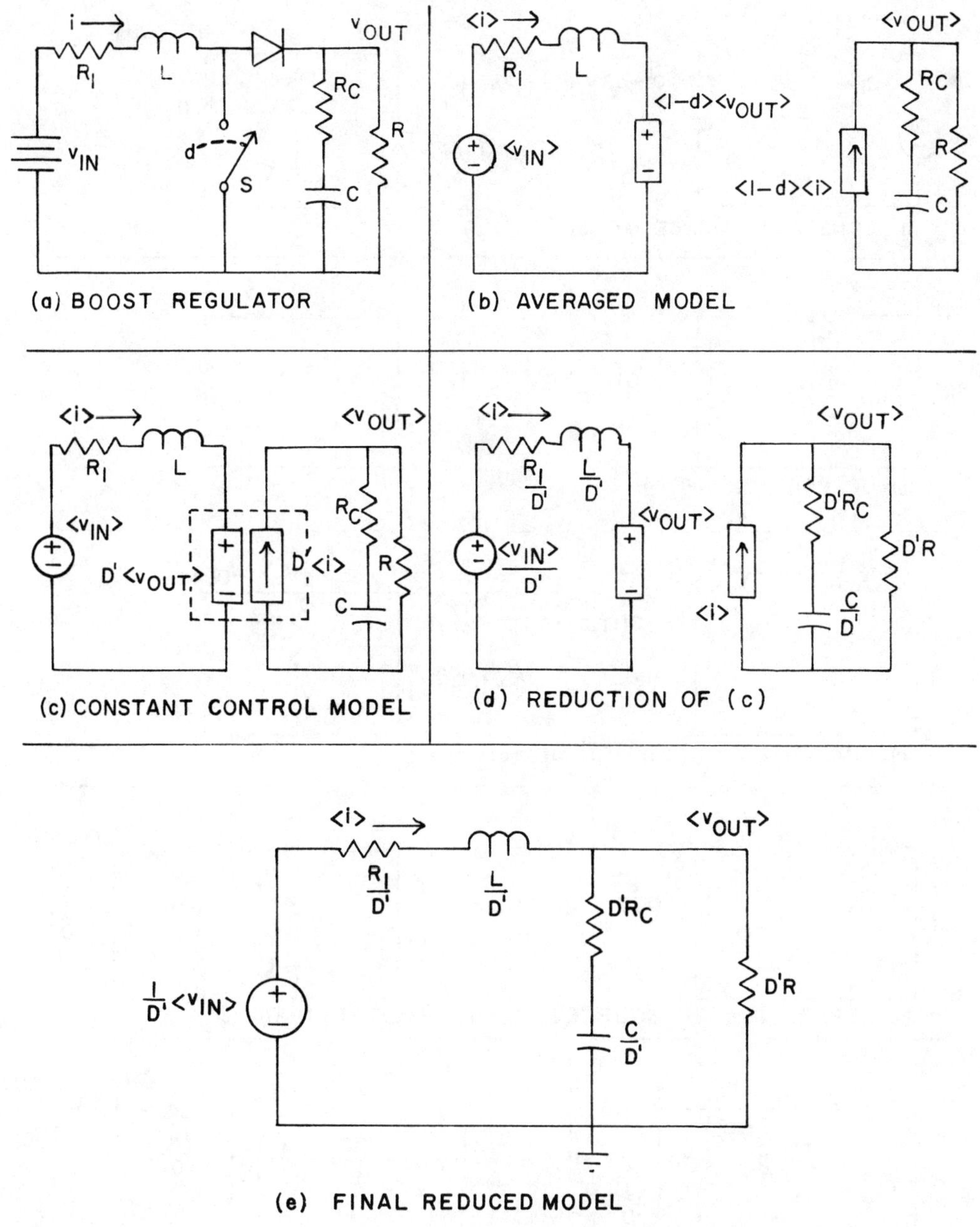

Fig. 8 Reduction of averaged boost regulator model for source variations and constant control

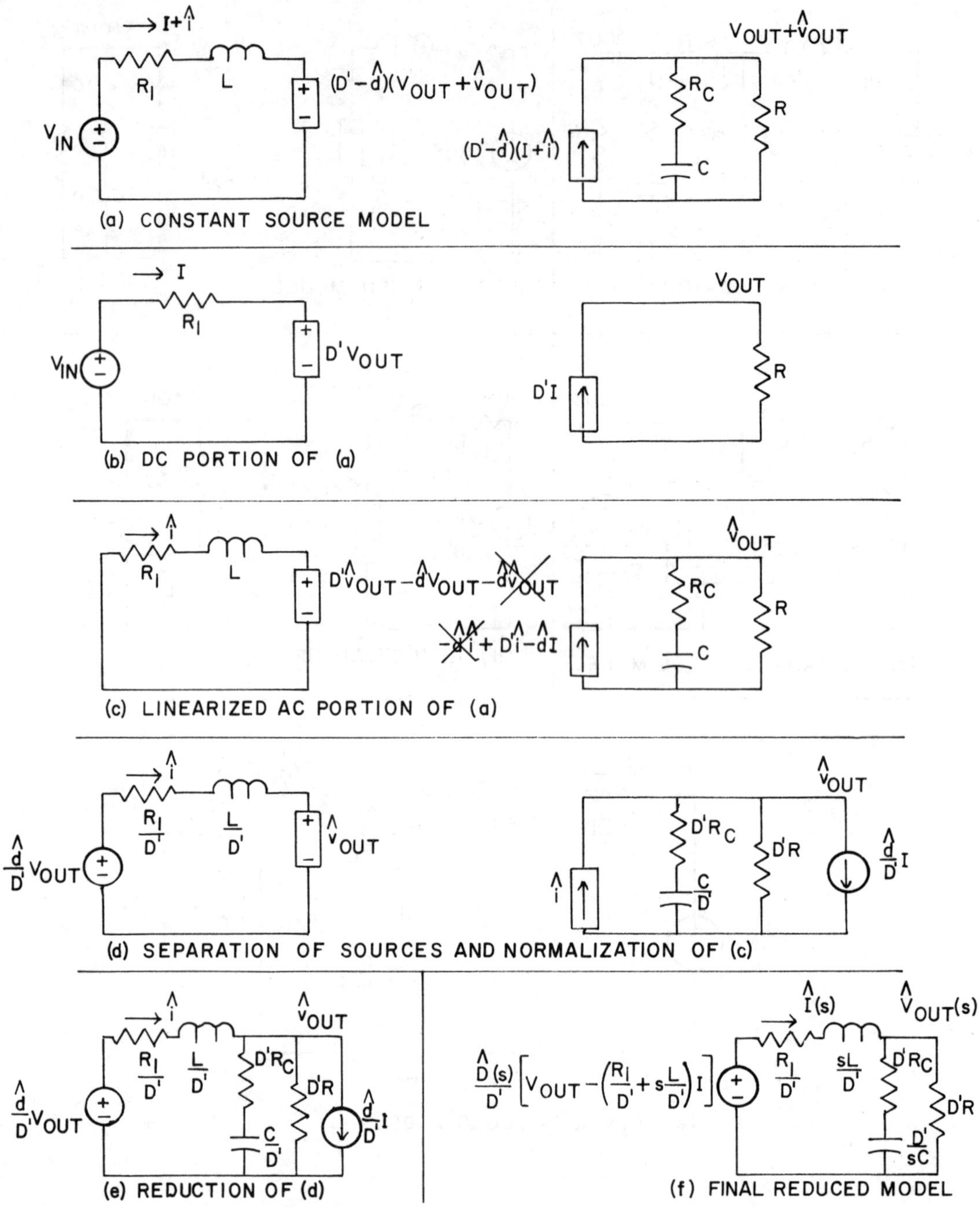

Fig. 9 Reduction of averaged boost regulator model for small control variations and a constant source

$$\frac{\hat{V}_{OUT}(s)}{\hat{D}(s)} = \frac{D'^2R - R_1}{D'^2R + R_1} \frac{V_{IN}}{D'} \left(1 - \frac{sL}{D'^2R - R_1}\right) \frac{V_{OUT}(s)}{V_{IN}(s)} \tag{69}$$

where the expression $V_{OUT}(s)/V_{IN}(s)$ is given by (68).

State variable averaging in a boost regulator. To illustrate the use of averaging by means of state variables, consider the boost converter of Fig. 10. The inductor current and capacitor voltage will be chosen at state variables, and the resistor voltage will be chosen as the output y, so that $[x_1 \ x_2] = [i \ v_{OUT}]$ and $y = v_{OUT}$.

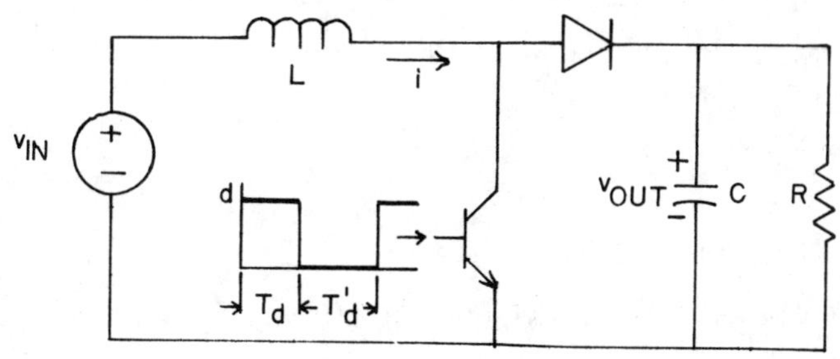

Fig. 10 A simple boost regulator circuit

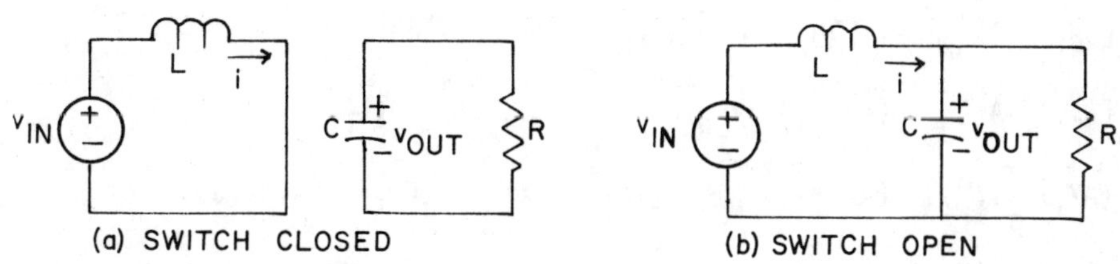

Fig. 11 The two modes of the boost regulator of Fig. 10

When the switch is closed, Fig. 11(a) applies and the circuit equations and state equations are

$$v_{IN} = L\frac{di}{dt} = L\dot{x}_1 \text{ and } \frac{v_{OUT}}{R} = -C\frac{d\,v_{OUT}}{dt} = -C\dot{x}_2 \tag{70}$$

$$\dot{\vec{x}} = \begin{bmatrix} 0 & 0 \\ 0 & \frac{-1}{RC} \end{bmatrix} \vec{x} + \begin{bmatrix} \frac{1}{L} \\ 0 \end{bmatrix} \vec{v}_{IN} \overset{\Delta}{=} A_1\vec{x} + \vec{b}_1 v_{IN} \tag{71}$$

$$y = [0\ 1]\,\vec{x} \overset{\Delta}{=} \vec{c}_1^{\,T}\,\vec{x} \tag{72}$$

When the switch is open, Fig. 11(b) applies and the circuit and state equations are

$$v_{IN} - v_{OUT} = L\frac{di}{dt} \text{ and } i - C\frac{dv_{OUT}}{dt} = \frac{v_{OUT}}{R} \tag{73}$$

$$\dot{\vec{x}} = \begin{bmatrix} 0 & \frac{-1}{L} \\ \frac{1}{C} & \frac{-1}{RC} \end{bmatrix} \vec{x} + \begin{bmatrix} \frac{1}{L} \\ 0 \end{bmatrix} v_{IN} \overset{\Delta}{=} A_2\vec{x} + \vec{b}_2\, v_{IN} \tag{74}$$

$$y = [0\ 1]\ \vec{x} \triangleq \vec{c}_2^{\,T}\ \vec{x} \tag{75}$$

Letting D = the steady-state duty ratio, T_d/T_d', in Fig. 10, and D' = 1 - D, the averaged state equation is written by using the following weighted averages.

$$\begin{aligned} A &= DA_1 + D'A_2 \\ \vec{b} &= D\vec{b}_1 + D'\vec{b}_2 \\ \vec{c}^{\,T} &= D\vec{c}_1^{\,T} + D'\vec{c}_2^{\,T} \end{aligned} \tag{76}$$

$$\dot{\vec{x}} = A\vec{x} + \vec{b}v_{IN} = \begin{bmatrix} 0 & -\frac{D'}{L} \\ \frac{D'}{C} & -\frac{1}{RC} \end{bmatrix} \vec{x} + \begin{bmatrix} \frac{1}{L} \\ 0 \end{bmatrix} v_{IN} \tag{77}$$

$$y = \vec{c}^{\,T}\ \vec{x} = [0\ 1]\ \vec{x} \tag{78}$$

Adding perturbations to the systems gives

$$(\dot{\vec{X}} + \dot{\hat{\vec{x}}}) = [(D + \hat{d})A_1 + (D' - \hat{d})A_2]\ (\vec{X}+\hat{\vec{x}}) + [(D + \hat{d})\vec{b}_1 + (D' - \hat{d})\vec{b}_2]\ (V_{IN} + \hat{v}_{IN}) \tag{79}$$

$$(Y + \hat{y}) = [(D + \hat{d})\vec{c}_1^{\,T} + (D' - \hat{d})\vec{c}_2^{\,T}]\ (\vec{X} + \hat{\vec{x}}) \tag{80}$$

or

$$\begin{aligned} \dot{\hat{\vec{x}}} &= [DA_1 + D'A_2]\vec{X} + [D\vec{b}_1 + D'\vec{b}_2]V_{IN} + [DA_1 + D'A_2]\hat{\vec{x}} + [D\vec{b}_1 + D'\vec{b}_2]\hat{v}_{IN} \\ &\quad + [(A_1 - A_2)\vec{X} + (\vec{b}_1 - \vec{b}_2)V_{IN}]\hat{d} + [(A_1 - A_2)\hat{\vec{x}} + (\vec{b}_1 - \vec{b}_2)\ \hat{v}_{IN}]\hat{d} \\ &\approx (A\vec{X} + \vec{b}V_{IN}) + (A\hat{\vec{x}} + \vec{b}\hat{V}_{IN}) + [(A_1 - A_2)\ \vec{X} + (\vec{b}_1 - \vec{b}_2)V_{IN}]\hat{d} \end{aligned} \tag{81}$$

The approximation in (81) linearizes the equation by assuming the perturbations are small and hence the second order perturbations are neglected. The equation now consists of essentially three parts: the dc term, the line variation term, and the duty ratio variation term. The output relation (80) is also simplified using this same assumption.

$$\begin{aligned} Y + \hat{y} &= \vec{c}^{\,T}\vec{X} + \vec{c}^{\,T}\hat{\vec{x}} + (\vec{c}_1^{\,T} - \vec{c}_2^{\,T})\ \vec{X}\hat{d} + (\vec{c}_1^{\,T} - \vec{c}_2^{\,T})\ \hat{\vec{x}}\hat{d} \\ &\approx \vec{c}^{\,T}\vec{X} + \vec{c}^{\,T}\hat{\vec{x}} + (\vec{c}_1^{\,T} - \vec{c}_2^{\,T})\ \vec{X}\hat{d} \end{aligned} \tag{82}$$

Equations (81) and (82) represent the dc and small-signal dynamic linearized system. Separating these two parts gives

$$\text{dc model:}\quad \vec{X} = -A^{-1}\vec{b}\ V_{IN},\ Y = \vec{c}^{\,T}\vec{X} = -\vec{c}^{\,T}A^{-1}\vec{b}\ V_{IN} \tag{83}$$

$$\text{dynamic model:}\quad \dot{\hat{\vec{x}}} = A\hat{x} + \vec{b}\ \hat{v}_{IN} + [(A_1 - A_2)\vec{X} + (\vec{b}_1 - \vec{b}_2)\ V_{IN}]\hat{d} \tag{84}$$

$$\hat{y} = \vec{c}^{\,T}\hat{\vec{x}} + (\vec{c}_1^{\,T} - \vec{c}_2^{\,T})\ \vec{X}\hat{d} \tag{85}$$

For the specific boost converter example being considered, (83) - (85) become

$$\vec{X} = \begin{bmatrix} \frac{L}{RD'^2} & -\frac{C}{D'} \\ \frac{L}{D'} & 0 \end{bmatrix} \begin{bmatrix} \frac{1}{L} \\ 0 \end{bmatrix} V_{IN} = \begin{bmatrix} \frac{1}{RD'^2} \\ \frac{1}{D'} \end{bmatrix} V_{IN} \tag{86}$$

$$Y = [0 \quad 1]\,\bar{X} = \frac{1}{D'}\,V_{IN} \tag{87}$$

$$\dot{\hat{x}} = \begin{bmatrix} 0 & -\frac{D'}{L} \\ \frac{D'}{C} & -\frac{1}{RC} \end{bmatrix} \hat{x} + \begin{bmatrix} \frac{1}{L} \\ 0 \end{bmatrix} \hat{v}_{IN} + \left\{ \begin{bmatrix} 0 & \frac{1}{L} \\ -\frac{1}{C} & 0 \end{bmatrix} \begin{bmatrix} \frac{1}{RD'^2} \\ \frac{1}{D} \end{bmatrix} V_{IN} + \begin{bmatrix} 0 \\ 0 \end{bmatrix} V_{IN} \right\} \hat{d}$$

$$= \begin{bmatrix} 0 & -\frac{D'}{L} \\ \frac{D'}{C} & -\frac{1}{RC} \end{bmatrix} \hat{x} + \begin{bmatrix} \frac{1}{L} \\ 0 \end{bmatrix} \hat{v}_{IN} + \begin{bmatrix} \frac{1}{LD'} \\ \frac{-1}{RCD'^2} \end{bmatrix} V_{IN}\,\hat{d} \tag{88}$$

$$\hat{y} = [0 \quad 1]\hat{x} \tag{89}$$

Equation (87) should be recognized as a well known result for boost regulators. From (88) and (89) the dynamic duty ratio to output and line voltage to output transfer functions may be found. The final result is indicated by following equation.

$$\hat{v}_{OUT}(s) = \frac{\frac{D'}{LC}}{s^2 + \frac{1}{RC}s + \frac{D'^2}{LC}}\,\hat{v}_{IN}(s) + \frac{\frac{-1}{RCD'^2}s - \frac{1}{(RCD')^2} + \frac{1}{LC}\,V_{IN}}{s^2 + \frac{1}{RC}s + \frac{D'^2}{LC}}\,\hat{d}(s) \tag{90}$$

Multiple Reference Frame Analysis

If the d-q transformation is applied to a set of balanced 3ϕ sinusoidal voltages with the d-q axis rotating at the electrical angular velocity ω_e, the d-q voltages will be constants. This suggests that for a balanced set of periodic but nonsinusoidal voltages, each of the harmonics will appear as constants in some properly chosen d-q reference frame. Thus if the load these voltages are supplying is linear, each harmonic current may be evaluated using dc circuit theory. The complete current waveform is then determined by transforming each of these currents back to a common reference frame and summing them.

This technique, referred to as multiple reference frames analysis, is developed in [12] and applied to the analysis of symmetrical induction motors. Since superposition is used, the circuit must be linear and the method only applies to the steady-state or constant speed transient analysis of the motor. However, the applied voltages do not have to be balanced for the method to work, so that unbalanced operation may be studied by this type of analysis provided the waveforms are periodic.

The equations necessary to apply this technique are derived in the following section.

Application to 3ϕ motor. The reference frames to be used in the analysis and the magnitudes of the dc voltages appearing in these frames will now be determined. It is assumed that the load is a 3ϕ, three-wire system. The applied d-q voltages in a stationary reference frame, denoted by $v_{ds}{}^s$ and $v_{qs}{}^s$, may be obtained from the phase voltages and since the phase voltages are assumed to be periodic, the d-q voltages may be expressed by Fourier series expansions.

$$\begin{aligned} v_{qs}{}^s &= v_{as} = \sum_{k=1}^{\infty} (V_{kq\alpha} \cos k\,\omega_e t + V_{kq\gamma} \sin k\omega_e t) \\ v_{ds}{}^s &= \frac{1}{\sqrt{3}}(-v_{bs} + v_{cs}) = \sum_{k=1}^{\infty} (V_{kd\alpha} \cos k\omega_e t + V_{kd\gamma} \sin k\omega_e t) \end{aligned} \tag{91}$$

Here k is the order of the harmonic, ω_e the angular velocity of the fundamental, and α and γ are used to distinguish the coefficients of the cosine and sine terms. The dc component, if any is present, should also be included in (91), but it is assumed to be zero for this analysis.

By applying the transformation

$$v_{qs}^{e} = v_{qs}^{s} \cos \omega_e t - v_{ds}^{s} \sin \omega_e t$$
$$v_{ds}^{e} = v_{qs}^{s} \sin \omega_e t + v_{ds}^{s} \cos \omega_e t \qquad (92)$$

to the Fourier expansions of (91), the applied voltages may be expressed in the synchronously rotating reference frame in an interesting form. It is shown in [12] that regardless of the form of the phase voltages, for each value of k in the series expansion of the d-q voltages in the synchronous frame, there is a balanced pair of d-q voltages rotating counterclockwise at a velocity $k\omega_e$ with respect to the stator and there is a balanced pair rotating clockwise at a velocity $-k\omega_e$ with respect to the stator. This property is very similar to the concept of symmetrical components, which allows unbalanced 3Ø waveforms to be expressed as the sum of balanced positive and negative sequence sets (when no zero-sequence terms are present, as is being assumed for this example).

Now if a balanced set of voltages appears in any reference frame there is another reference frame in which this balanced set will appear as constants rather than sinusoids. For example, for k=1, the voltages rotating at velocity $k\omega_e$ with respect to the stator, i.e., the fundamental frequency voltages, appear as constants in the above expansion in the synchronously rotating frame. Since superposition is applicable to the motor model at constant speed, the steady-state currents due to these voltages are also constants which may be found using dc circuit theory. Extending this idea, all the currents due to the series of balanced voltages may be found by transformations to two different reference frames, rotating at $\pm k\omega_e$, for each value of k in the truncated Fourier series, and application of dc circuit theory. The resulting currents can then be transformed back to the synchronously rotating reference frame and combined to give the constant-speed performance of the motor. This application of multiple reference frames permits a rigorous constant-speed electrical transient analysis or a steady-state performance analysis with stator phase voltages of any periodic waveform.

By performing transformations to each of the rotating frames, the dc load voltages which appear in each frame may be found as

$$v_{qs}^{+ke} = 0.5\,(V_{kq\alpha} - V_{kd\gamma})\,,\quad v_{ds}^{+ke} = 0.5\,(V_{kq\gamma} + V_{kd\alpha})$$
$$v_{qs}^{-ke} = 0.5\,(V_{kq\alpha} + V_{kd\gamma})\,,\quad v_{ds}^{-ke} = 0.5\,(V_{kq\gamma} - V_{kd\alpha}) \qquad (93)$$

where the quantities on the right are from (91). The superscripts +ke and -ke identify the speed and direction of the reference frame rotation with respect to the stator. For example, +1e denotes the synchronously rotating reference frame.

In the case of a symmetrical 3Ø induction motor, the constant speed electrical transient currents in each frame are found by substituting into the machine equations in an arbitrary frame with $\omega = \pm k\omega_e$. If only steady-state currents are desired, this results in a simple algebraic equation

$$\begin{bmatrix} v_{qs} \\ v_{ds} \\ 0 \\ 0 \end{bmatrix}^{\pm ke} = \begin{bmatrix} r_s & \pm k\omega_e L_s & 0 & \pm k\omega_e M \\ \mp k\omega_e L_s & r_s & \pm k\omega_e M & 0 \\ 0 & (\pm k\omega_e - \omega_r)M & r_r' & (\pm k\omega_e - \omega_r)L_r' \\ -(\pm k\omega_e - \omega_r)M & 0 & -(\pm k\omega_e - \omega_r)L_r' & r_r' \end{bmatrix} \begin{bmatrix} i_{qs} \\ i_{ds} \\ i_{qr}' \\ i_{dr}' \end{bmatrix}^{\pm ke} \qquad (94)$$

The solution of (94) is given in general in [12] along with the equations for transforming all the resulting currents back to the synchronous frame, the stationary frame, and finally back to actual phase currents. Also, a series expression for the torque is given in terms of the dc currents resulting from the solution of (94).

Complex Variable Analysis

In [13] it is shown that complex variables may be applied to simplify the analysis of an inverter-fed 3∅ induction motor because of the symmetry present in the system. Complex variables reduce the steady-state voltage source inverter system problem to a second order linear system, and the current source inverter problem to a first order linear system, and closed-form solutions are amenable to either hand or simple computer calculations. The representation of the motor, discussed in more detail below, is derived by writing the machine equations in the α-β axes and combining pairs of α-β variables into single complex variables. This combination is suggested by the quadrature relationship between α and β variables. A complex variable representation of inverters using the same ideas is discussed in depth in the section of this paper on switching function analysis.

Representation of a 3∅ induction motor. In the following derivation of the complex variable induction motor representation, it is assumed that the 3∅ motor has no neutral connection. Based on the reference axes shown in Fig. 12, the α-β equations of the motor are

$$\begin{bmatrix} v_{s\alpha} \\ v_{s\beta} \\ v_{r\alpha} \\ v_{r\beta} \end{bmatrix} = \begin{bmatrix} R_s + L_s p & 0 & Mp\cos\theta_r & -Mp\sin\theta_r \\ 0 & R_s + L_s p & Mp\sin\theta_r & Mp\cos\theta_r \\ Mp\cos\theta_r & Mp\sin\theta_r & R_r + L_r p & 0 \\ -Mp\sin\theta_r & Mp\cos\theta_r & 0 & R_r + L_r p \end{bmatrix} \begin{bmatrix} i_{s\alpha} \\ i_{s\beta} \\ i_{r\alpha} \\ i_{r\beta} \end{bmatrix} \tag{95}$$

The rotor quantities are expressed in terms of α-β axes fixed on the rotor and stator quantities in terms of stator-fixed axes. In this representation the torque equation becomes

$$T = 0.5\ MP\ [(i_{s\beta} i_{r\alpha} - i_{s\alpha} i_{r\beta})\cos\theta_r - (i_{s\alpha} i_{r\alpha} + i_{s\beta} i_{r\beta})\sin\theta_r] \tag{96}$$

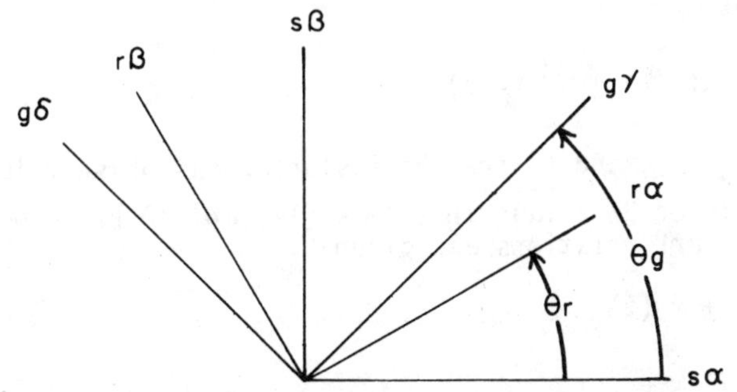

Fig. 12 Reference axes for 3∅ motor complex variable analysis

Using f to represent either v or i, complex variables are defined as

$$f_{sc} = f_{s\alpha} + jf_{s\beta} \text{ and } f_{rc} = f_{r\alpha} + jf_{r\beta} \tag{97}$$

Applying these definitions to the machine equations (95) and (96) gives the following complex machine equations.

$$\begin{bmatrix} v_{sc} \\ v_{rc} \end{bmatrix} = \begin{bmatrix} R_s + L_s p & Mp\,e^{j\theta_r} \\ Mp\,e^{-j\theta_r} & R_r + L_r p \end{bmatrix} \begin{bmatrix} i_{sc} \\ i_{rc} \end{bmatrix} \tag{98}$$

$$T = 0.5\ MP\ \text{Im}\ [i_{sc}(e^{j\theta_r}\, i_{rc})^*] \tag{99}$$

The relations between v_{sc} and the phase voltages are

$$v_{sc} = \sqrt{2/3}\,(v_a + av_b + a^2 v_c)\ , \quad a = e^{j2\pi/3}$$

$$v_a = \sqrt{2/3}\ \mathrm{Re}\,(v_{sc})\ , \quad v_b = \sqrt{2/3}\ \mathrm{Re}\,(a^2 v_{sc})\ , \quad v_c = \sqrt{2/3}\ \mathrm{Re}\,(av_{sc}) \tag{100}$$

Equations (98) and (99) may be rewritten in terms of a single pair of arbitrary axes $g\gamma$- $g\delta$ as shown in Fig. 12 by using the relations

$$f_{sg} = e^{-j\theta_g} f_{sc}$$

$$f_{rg} = e^{-j(\theta_g - \theta_r)} f_{rc} \tag{101}$$

where f denotes either v or i and the displacement θ_g of the new axes is defined in the figure. Letting $\omega_g = p\theta_g$, (98) and (99) become

$$\begin{bmatrix} v_{sg} \\ v_{rg} \end{bmatrix} = \begin{bmatrix} R_s + L_s\,(p + j\omega_g) & M(p + j\omega_g) \\ M[p + j(\omega_g - \omega_r)] & R_r + L_r[p + j(\omega_g - \omega_r)] \end{bmatrix} \begin{bmatrix} i_{sg} \\ i_{rg} \end{bmatrix} \triangleq \begin{bmatrix} Z_{sg} & Z_{mg} \\ Z_{dg} & Z_{rg} \end{bmatrix} \begin{bmatrix} i_{sg} \\ i_{rg} \end{bmatrix} \tag{102}$$

$$T = 0.5\ MP\ \mathrm{Im}\,[i_{sg}\, i_{rg}^*] \tag{103}$$

The complex impedances defined on the right side of (102) depend on the choice of ω_g for the desired reference frame—stationary, synchronous, rotor-fixed, or others. Equations (102) and (103) are the final complex variable representation of the motor and may be used in the study of steady-state performance.

Park Vector Analysis

The Park vector, defined as

$$f(t) = \frac{2}{3}\,[f_a(t) + af_b(t) + a^2\, f_c(t)] \tag{104}$$

where $a = \exp(j2\pi/3)$ and f_a, f_b, and f_c are the instantaneous phase values of a 3Ø system, has been applied to the analysis of 3Ø bridge inverters [14] and 3Ø line-commutated bridge converters [15]. The inverse Park relations are given by

$$f_a(t) = \mathrm{Re}\,[f(t)] + f_o(t)$$

$$f_b(t) = \mathrm{Re}\,[a^2 f(t)] + f_o(t) \tag{105}$$

$$f_c(t) = \mathrm{Re}\,[af(t)] + f_o(t)$$

where $f_o(t)$, the zero sequence quantity, is

$$f_o(t) = \frac{1}{3}\,[f_a(t) + f_b(t) + f_c(t)] \tag{106}$$

In the absence of zero sequence quantities, the Park vector transformation reduces the 3Ø system to a single phase complex variable system. It should be noted that the relations in (100) from the section on complex variable analysis, which were derived from α-β variables, are the same as the Park vector relations (104) and (105) when $f_o(t)$ is absent; hence, these two analysis methods are related.

One of the advantages of Park vectors is that the half-wave and three-phase symmetry which appear in many 3Ø converter systems appear as a hexagonal symmetry in the Park vector representation. Thus a complete cycle is easy to determine once one-sixth of a cycle has been investigated.

In [14], the technique is developed for the steady-state analysis of 3Ø six-step and PWM bridge inverters. The Park vector is determined over a 60° period for the inverter output voltages. This Park voltage is then applied to an equivalent single phase representation of the load and the Park current vector is found. Initial conditions are found based on the assumption of steady-state operation and hexagonal symmetry. Finally, actual output phase and inverter switch currents can be found from the Park current vector.

Application to 3Ø inverter with RL load. The three-phase six-step inverter and its conduction pattern are shown in Fig. 13. During the first conduction interval, the inverter output voltages are $v_a = E$ and $v_b = v_c = -E$. Hence from (104) the Park voltage vector during this interval is

$$v_{P1} = \frac{2}{3}(1 - a - a^2) E = \frac{4}{3} E \qquad (107)$$

During the second interval, the vector is

$$v_{P2} = \frac{2}{3}(1 + a - a^2) E = \frac{4}{3} e^{j\frac{\pi}{3}} E \qquad (108)$$

During successive intervals the vector magnitude remains unchanged but rotates 60° in the complex plane. Thus, six symmetrically spaced vectors on the complex plane completely represent the inverter output voltage, as shown in Fig. 14.

Applying any one of these voltage vectors to a series RL circuit results in the following equation and solution.

$$v_P = i_P R + L \frac{d}{dt} i_P \qquad (109)$$

$$i_P(t) = \frac{v_P}{R} + C e^{-\frac{t}{\tau}} \qquad (110)$$

where $\tau = L/R$ and C is a complex constant. Assuming $i_P(0) = I$, in steady-state $i_P(\pi/3\omega) = I \cdot \exp(j\pi/3)$ due to symmetry in the 3Ø circuit. Substituting these boundary conditions into (110) and eliminating C gives

$$I = \frac{v_P}{R} \frac{1 - e^{-\frac{\pi}{3\omega\tau}}}{e^{j\frac{\Pi}{3}} - e^{-\frac{\Pi}{3\omega\tau}}} \qquad (111)$$

and

$$i_P(t) = \frac{v_P}{R}(1 - e^{\frac{t}{\tau}}) + I e^{-\frac{t}{\tau}} \qquad (112)$$

It can be shown that (112) is a straight line in the complex plane with t as a parameter. The typical path of the current vector, $i_P(t)$, over 360° period is shown in Fig. 15 for both six-step and PWM inverters.

The phase currents may be found from (105), noting that the zero-sequence term is zero. The currents may also be found graphically projecting the instantaneous Park current vector onto the R, S, and T axes shown in Fig. 15.

Switching Function Analysis

Because of the repetitive manner in which the devices in a power converter are switched between conducting and blocking states, it seems reasonable that some sort of periodic function of rectangular pulses may simplify the converter representation. Such functions are called switching functions and may switch between a finite number of states or may be multiplied by sinusoids to produce segments of sine waves. Referring to Fig. 1, the basic approach is to

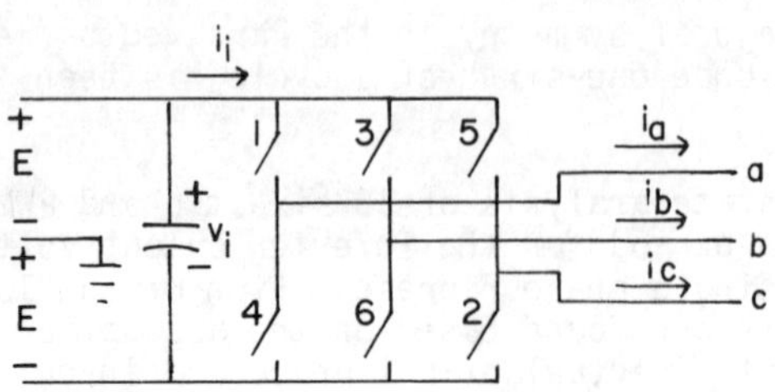

INTERVAL	1	2	3	4	5	6
SWITCHES CONDUCTING	6	1	2	3	4	5
	1	2	3	4	5	6
	2	3	4	5	6	1

Fig. 13 3∅ inverter and conduction pattern

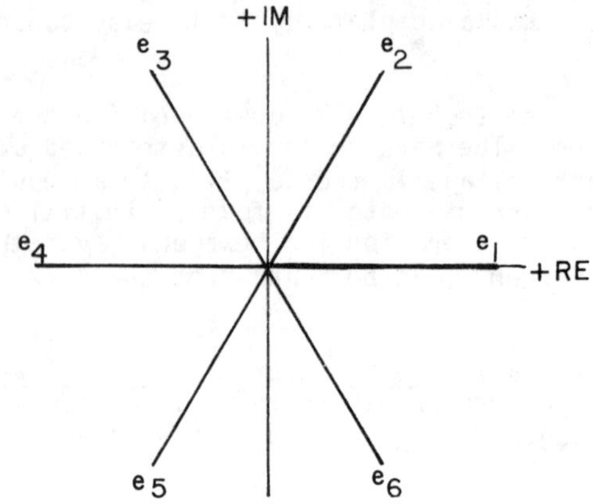

Fig. 14 Park vectors of inverter output voltage

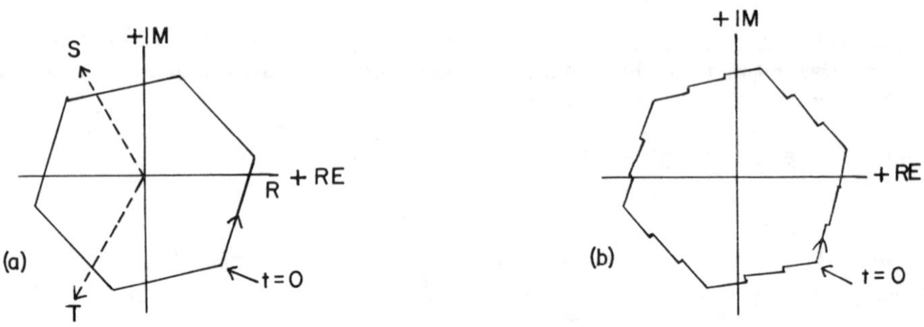

Fig. 15 Locus of load current Park vectors for series RL load supplied by (a) six-step inverter, and (b) two-notch PWM inverter

find a useful set of functions to represent the control input $\vec{d}$ so that when combined with the input $\vec{v}_{IN}$, a convenient formulation for $\vec{v}_{OUT}$ results. As with most analysis techniques, switching functions are not as easy to apply when the load current determines conduction through the signal $\vec{i}_{FDBK}$.

Switching functions have not been utilized to the fullest extent, and it is believed much useful work remains to be accomplished in this area. At present there are no general procedures for finding useful switching functions, and a whole new class of functions may be awaiting discovery. An excellent illustration of switching function application is found in [16], where the 36 equations describing the voltage and current transfer relationships of a six-step inverter are reduced to two or three equations which take advantage of the symmetry and redundancy present in the 36 equations. This paper is discussed in more detail below.

Application to a 3∅ inverter. In [16] both the voltage source and current source inverters with wye and delta loads are studied. Here only the development for the six-step VSI with a wye-connected load is repeated. The basic circuit and its six conduction intervals, or modes, are shown in Fig. 13. It is assumed that the load is balanced, commutation time is negligible and the inverter is lossless. The equations which characterize the inverter during modes 1, 2, and 3 are shown in Fig. 16. The equations during modes 4, 5, and 6 are found by multiplying the right side of the equations in Fig. 16 by minus one. The input-output voltage relationships are completely determined by these equations, but a degree of freedom exists in the current relations. Using switching functions and complex variables, a single equation will now be developed to represent all these current equations.

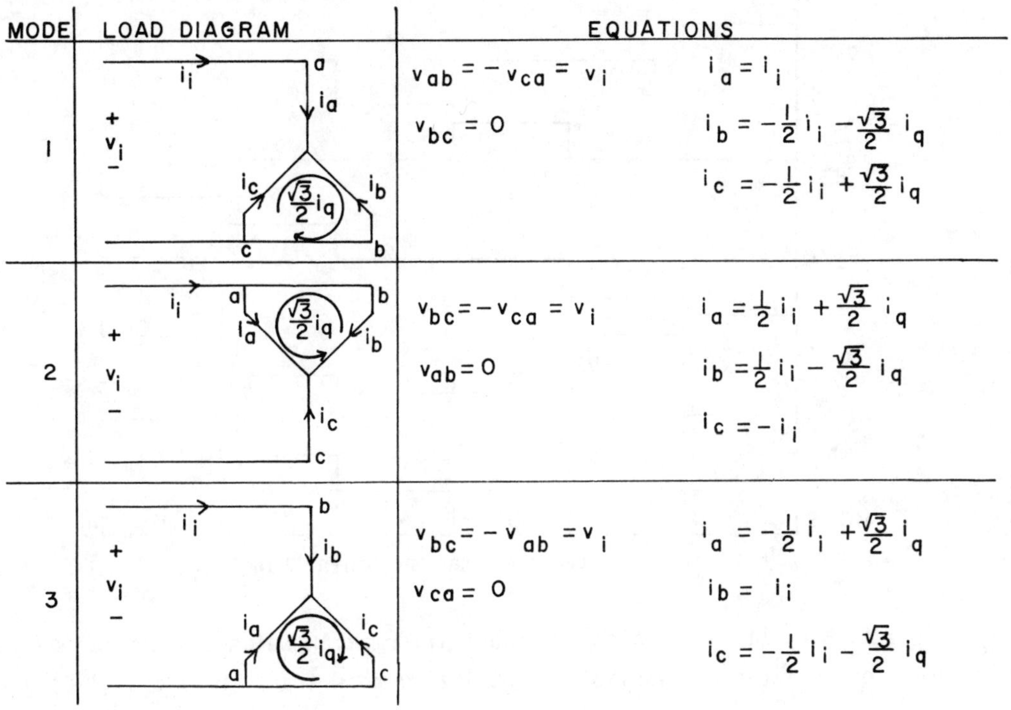

Fig. 16 VSI mode circuit diagrams and equations

The 3Ø switching functions g_a, g_b, and g_c are defined by Fig. 17. One phase of each of two other sets of functions is also shown in the figure. These additional sets of functions are defined in terms of the switching functions as follows

$$f_X = g_X - \frac{1}{2}(g_Y + g_Z) \tag{113}$$

$$g_{XY} = g_X - g_Y = \frac{2}{3}(f_X - f_Y) = \frac{2}{3}(f_Z + 2f_X) \tag{114}$$

Here X denotes a, b, or c and Y and Z then denote the cyclic permutations of a, b, and c. From Fig. 16 and Fig. 17, the line to line and line to neutral voltages may be written as

$$[v_{ab} \;\; v_{bc} \;\; v_{ca}]^T = [g_{ab} \;\; g_{bc} \;\; g_{ca}]^T \, v_i \tag{115}$$

$$[v_a \;\; v_b \;\; v_c]^T = \frac{2}{3}[f_a \;\; f_b \;\; f_c]^T \, v_i \tag{116}$$

The currents in Fig. 16 may also be described in terms of the switching functions. For example,

$$i_a = i_i g_a - (\frac{1}{2} i_i + \frac{\sqrt{3}}{2} i_q)\, g_c + (-\frac{1}{2} i_i + \frac{\sqrt{3}}{2} i_q)\, g_b \tag{117}$$

Writing similar relations for i_b and i_c and applying (113) and (114) gives

$$\begin{bmatrix} i_a \\ i_b \\ i_c \end{bmatrix} = \begin{bmatrix} f_a & \frac{\sqrt{3}}{2} \cdot g_{bc} \\ f_b & \frac{\sqrt{3}}{2} \cdot g_{ca} \\ f_c & \frac{\sqrt{3}}{2} \cdot g_{ab} \end{bmatrix} \begin{bmatrix} i_i \\ i_q \end{bmatrix} \tag{118}$$

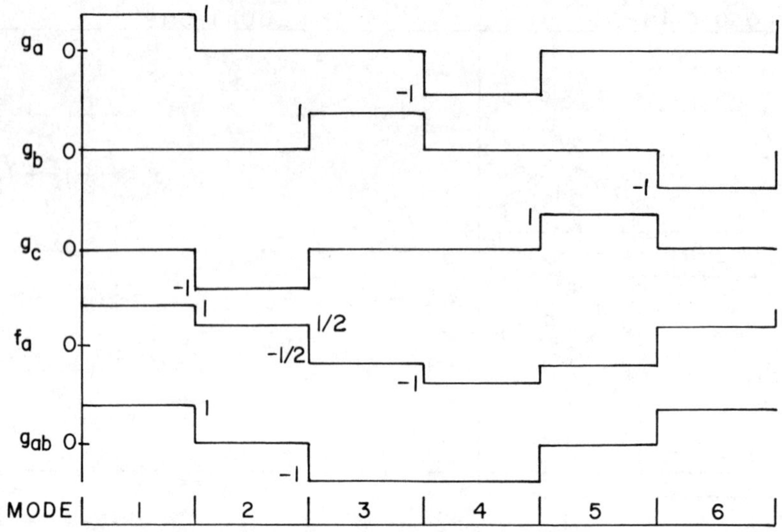

Fig. 17 Six-step inverter switching functions

Since $i_a + i_b + i_c = 0$, (118) contains redundant information which is eliminated by making the α-β transformation to two-phase variables. Utilizing this transformation, (116) and (118) become

$$[v_\alpha \;\; v_\beta]^T = \sqrt{\tfrac{2}{3}}\,[f_a \;\; \tfrac{\sqrt{3}}{2} g_{bc}]^T \, v_i \tag{119}$$

$$\begin{bmatrix} i_\alpha \\ i_\beta \end{bmatrix} = \sqrt{\tfrac{3}{2}} \begin{bmatrix} f_a & \tfrac{\sqrt{3}}{2} g_{bc} \\ \tfrac{\sqrt{3}}{2} g_{bc} & -f_a \end{bmatrix} \begin{bmatrix} i_i \\ i_q \end{bmatrix} \tag{120}$$

From (120) it follows that

$$\begin{bmatrix} i_i \\ i_q \end{bmatrix} = \sqrt{\tfrac{2}{3}} \begin{bmatrix} f_a & \tfrac{\sqrt{3}}{2} g_{bc} \\ \tfrac{\sqrt{3}}{2} g_{bc} & -f_a \end{bmatrix} \begin{bmatrix} i_\alpha \\ i_\beta \end{bmatrix} \tag{121}$$

since $f_a^{\,2} + \tfrac{3}{4} g_{bc}^2 = -1$.

Equation (121) is a fairly compact representation of the inverter transfer characteristic. However, by defining complex variables to represent pairs of two-phase variables, an even simpler representation results. The definitions are

$$\begin{aligned} v_s &= v_\alpha + j v_\beta \\ i_s &= i_\alpha + j i_\beta \\ i_{iq} &= i_i - j i_q \end{aligned} \tag{122}$$

Equations (119)-(121) then become

$$v_s = \sqrt{\tfrac{2}{3}}\, h \, v_i \tag{123}$$

$$i_s = \sqrt{\tfrac{3}{2}}\, h \, i_{iq} \tag{124}$$

$$i_{iq} = \sqrt{\frac{2}{3}}\, h^*\, i_s \qquad (125)$$

where $h = f_a + j\,\frac{\sqrt{3}}{2}\,g_{bc} = e^{\frac{j(m-1)\pi}{3}}$ m = mode number

and h* is the complex conjugate of h. The complex switching function h may be viewed as a vector which rotates 60° each time the inverter switches modes. It is interesting to note that thus v_S in (123) is equivalent to the Park vector of the inverter voltage as discussed earlier and shown in Fig. 14.

Equations (123) and (124) are equivalent to all the equations listed partially in Fig. 16, and hence are a compact description of the inverter. This inverter representation may be used with the motor representation given in the previous section of this paper on complex variable analysis to analyze a complete induction motor drive system.

Sampled-Data Modelling

Since the gate has no control over the thyristor once it has turned on, thyristor systems are basically discrete control systems. Hence discrete methods such as sampled-data modelling have been applied to various types of switching converters. Two examples of applications to controlled rectifiers are given below. In the first, from [17], the thyristor is modelled as a sample and hold device. This model is then linearized to a model which clearly shows the time lag introduced into the system by the thyristor. In the second example, from [18], the thyristor is modelled as a sampler of both the load current and the firing angle, and a linearized discrete model is developed.

Sample and hold model of a controlled rectifier. In [17], sampled-data control theory is used to develop a linear model for single-phase, phase-angle modulated thyristor systems by first representing the thyristor as a sample and zero-order hold device. The linear model then results from assumptions of a) a system bandwidth small relative to the sampling frequency, and b) limited excursion of the thyristor triggering angle. In this simplified model, the effect of the thyristor is to introduce into the control circuit a transport lag time of half the thyristor conduction period.

A typical circuit for which the model will be derived is shown in Fig. 18. In this circuit, the trigger angle α of the thyristors is varied to maintain v_0 constant. The waveform of v_i' shown in Fig. 19(a) suggests representing the rectifiers and diode by a sample and partial

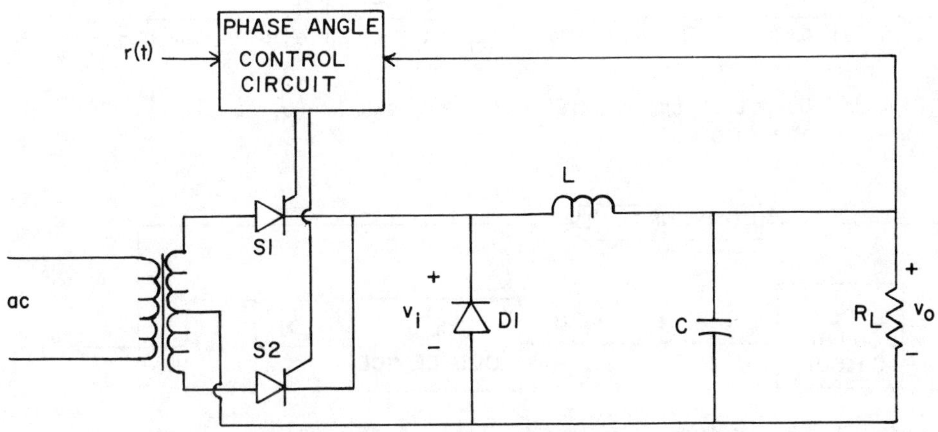

Fig. 18 A typical AC-to-DC thyristor controlled rectifier

hold model with waveforms as shown in Fig. 19(b). Although the hold or conduction period T_c is not constant, an average value will be used based on assumption b) above. The sampling period T is half the ac input period of $2\pi/\omega_L$.

Fig. 18 is represented by the block diagram of Fig. 20. A simplified representation of the open loop transfer function $G_T(s) = v_o(s)/R(s)$ will now be determined. The input to the hold device of Fig. 20 is a train of equivalent impulses, described by

$$\begin{aligned} V_c^*(s) &= L\left\{v_c(t)\sum_{n=0}^{N}\delta(t-nT)\right\} = L\left\{\sum_{n=0}^{N} v_c(nT)\,\delta(t-nT)\right\} \\ &= \sum_{n=0}^{N} v_c(nT)\, L\{\delta(t-nT)\} = \sum_{n=0}^{N} v_c\,(nT)\, e^{-nTs} \end{aligned} \tag{126}$$

where $v_c(t)$ is calculated such that the areas enclosed in Fig. 19(a) and Fig. 19(b) are the same. The output of the hold device is given by

$$\begin{aligned} v_i(s) &= L\left\{\sum_{n=0}^{N} v_c(nT)\,[\mu(t-nT) - \mu\,(t-nT-T_c)]\right\} \\ &= \sum_{n=0}^{N} v_c(nT)\left(\frac{e^{-nT_s}}{s} - \frac{e^{-nT_s - T_c s}}{s}\right) = \frac{1-e^{-T_c s}}{s}\sum_{n=0}^{N} v_c\,(nT)e^{-nT_s} \end{aligned} \tag{127}$$

The transfer function of the hold device, from (126) and (127), is

$$G_h(s) = \frac{v_i(s)}{V_c^*(s)} = \frac{1-e^{-T_c s}}{s} \tag{128}$$

Equation (128) is the transfer function of the rectifier model, which will now be simplified using the assumption of small bandwidth. Replacing s by $j\omega$ in (128)

$$G_h(j\omega) = \frac{1-e^{-T_c j\omega}}{j\omega} = \frac{2\,e^{-\frac{1}{2}j\omega T_c}\left(e^{\frac{1}{2}j\omega T_c} - e^{-\frac{1}{2}j\omega T_c}\right)}{2\omega j} = \frac{T_c \sin\frac{\omega T_c}{2}}{\frac{\omega T_c}{2}}\, e^{-j\frac{\omega T_c}{2}} \tag{129}$$

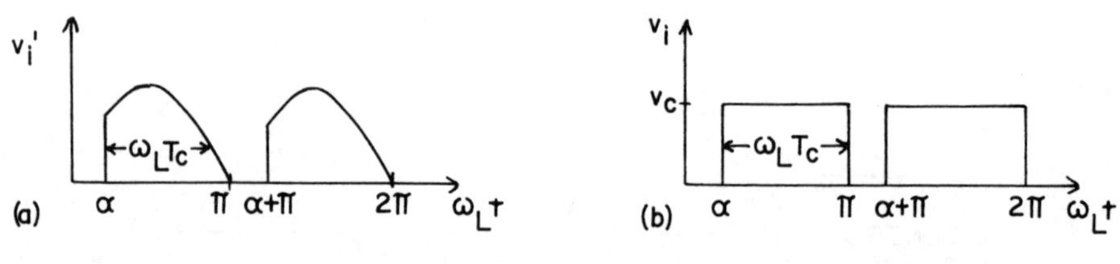

Fig. 19 Output of the rectifiers: (a) actual, (b) sample and hold model

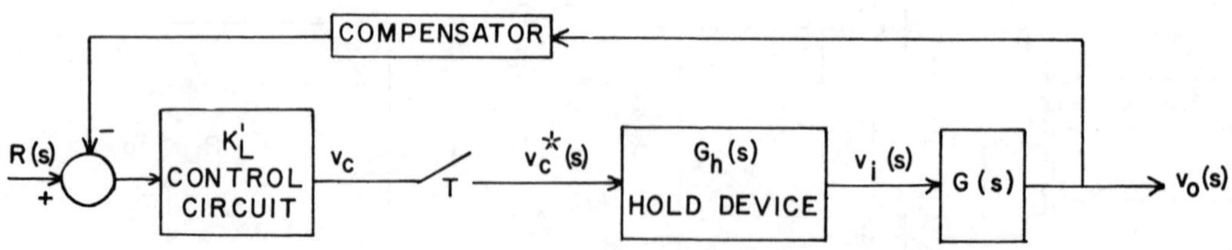

Fig. 20 Block diagram of the circuit of Fig. 18 using sampled-data model

In (129) the magnitude and phase components of $G_h(j\omega)$ are shown, and the phase $\phi_h(j\omega) = \frac{1}{2}\omega T_c$ makes it apparent that thyristor operation results in a delay time equal to half the conduction period. Phase lag is introduced in direct proportion to signal frequency, which suggests that a thyristor introduces phase shift into a system in a manner equivalent to transport lag.

If system bandwidth is small, $\sin(\frac{1}{2}\omega T_c)/(\frac{1}{2}\omega T_c) \approx 1$ and (124) simplifies to

$$G_h(j\omega) = T_c\, e^{-j\frac{\omega T_c}{2}} \tag{130}$$

which is the simplified thryristor model.

To further illustrate some of the methods of sampled-data theory, the open-loop system transfer function $G_T(s)$ is now developed. The transfer function is found by sampling the output v_o synchronously with the input. Thus

$$v_o^*(s) = L\{\delta_T(t)\, v_o(t)\} \tag{131}$$

where $\delta_T(t)$ is a train of impulses which may be represented as a complex Fourier series

$$\delta_T(t) = \sum_{n=0}^{\infty} \delta(t - nT) = \frac{1}{T}\sum_{n=-\infty}^{\infty} e^{jn\omega_L t} \tag{132}$$

where $\omega_L = 2\pi/T$ as before. Combining these equations gives

$$v_o^*(s) = \frac{1}{T}\sum_{n=-\infty}^{\infty} L\{v_o(t)\, e^{jn\omega_L t}\} = \frac{1}{T}\sum_{n=-\infty}^{\infty} V_o(s + jn\omega_L) \tag{133}$$

The transfer function may now be written as

$$G_T^*(s) = \frac{k_L'}{T}\sum_{n=-\infty}^{\infty} G_h(s + jn\omega_L)G(s + jn\omega_L) \tag{134}$$

Substituting (130) into (134) with $s = j\omega$ and $K_L = T_c K_L'/T$ = static loop gain, and assuming low pass filtering is present to reduce the system bandwidth, (134) becomes

$$G_T(s) \approx K_L\, G(j\omega)\, e^{-j\frac{\omega T_c}{2}} \tag{135}$$

Thus the sampled-data representation of Fig. 20 is reduced to a continuous representations containing a transport lag equal to half the conduction time.

Linearized discrete model of a rectifier. In a controlled rectifier in which the commutation overlap, μ, is not negligible, the unfiltered dc voltage, v_d, depends on the firing angle, α, and the dc current, i_d. In [18], a model for such a rectifier is developed which determines the relationship between small variations in α and i_d and the output v_d. It is shown that the rectifier behaves approximately as a sampler—i.e., the rectifier samples i_d and α and produces a Δv_d which depends on the variation of the samples from steady-state.

To illustrate the effect of $\Delta\alpha$ and Δi_d on Δv_d, consider a small variation in the firing angle with the output current kept constant. As shown in Fig. 21, the increment in output voltage will consist of two pulses in each firing period. For an infinitesimal change of α, the two pulses can be replaced by impulses having the same strength as the pulses they replace. This permits the SCR to be considered as an ideal sampler which samples the incremental firing angle and produces two impulses, one at the sampling instant and another delayed by the overlap interval, μ.

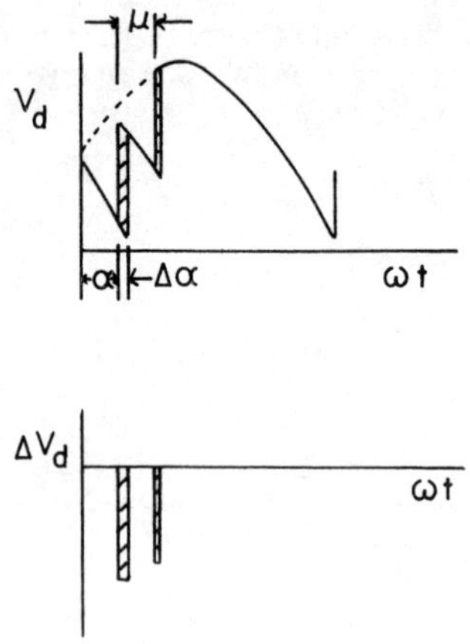

Fig. 21 Effect of α on v_d for a phase-controlled rectifier

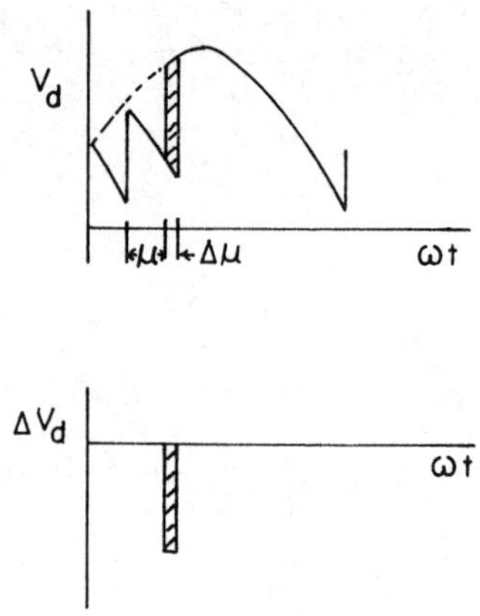

Fig. 22 Effect of i_d on v_d

If a small variation Δi_d of the output current is now considered, with the firing angle kept constant, the increment in output voltage will be one single pulse delayed by the interval μ with respect to the sampling instant as shown in Fig. 22. If the current variation during the commutation period is assumed to be negligible, the action of the rectifier can now be represented by a sampler which samples the current at the firing instants and produces impulses delayed by μ/ω seconds with respect to the firing instants. The area of each impulse is proportional to the deviation of current from the steady-state value.

These facts lead to a linearized discrete model of the rectifier in the following form.

$$\Delta v_d = G_1(s)\ \Delta I_d{}^* + G_2(s)\ \Delta\alpha^* \tag{136}$$

where the asterisks denote sampled variables. The transfer functions in (136) are derived in [18] and are

$$G_1(s) = -L_c e^{-sT_\mu} \tag{137}$$

$$G_2(s) = -\frac{1}{2} V_{do}\ T_s\ \sin\alpha^\circ\ [1 + e^{-sT_\mu}] \tag{138}$$

where L_c = commutating inductance per phase

T_μ = commutation time

V_{do} = dc component of rectifier output under no load and α = 0

T_s = firing or sampling period

α^o = steady-state firing angle

I_d = variation in dc component of rectifier output current.

A typical block diagram of a circuit using the linearized discrete model of (136) is shown in Fig. 23.

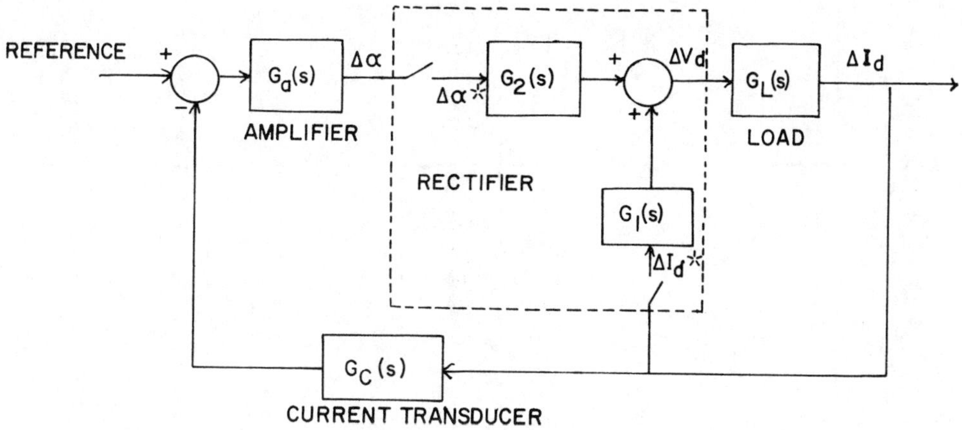

Fig. 23 Block diagram of linearized discrete model of a rectifier under current control

Analysis Using Boolean Algebra

Since the thyristor is generally modelled as a binary device, either conducting or blocking, Boolean algebra may sometimes be used to simplify waveform analysis or representation. In the following example from [19], Boolean algebra is used in the analysis of the harmonic content of cycloconverter and PWM waveforms. In the section of this paper on computer simulation, a second example is given in which a few simple rules and a Boolean equation are used to determine a conduction matrix which represents the instantaneous states of all the SCRs in a cycloconverter.

Application to a cycloconverter. The straightforward application of Fourier series to determine the harmonic content of a PWM or cycloconverter output waveform is generally complicated due to the presence of several pulses of different widths during a single period. In [19] a method is presented which may reduce the amount of effort required in the analysis by breaking the waveform up into a set of waveforms whose Fourier series representations are easy to determine. These waveforms are usually just pulses of constant width and period. The Fourier series representations are then truncated and recombined using Boolean algebra expressions to determine the overall harmonic content. The techniques assumes knowledge of the applied voltage source and the conduction intervals of the different switches in the converter system.

In [19] a step-up cycloconverter is considered. As another simple example, consider the output of a PWM, group-triggered 3Ø to 1Ø cycloconverter operating at an integral step-down ratio. A typical output waveform might appear as in Fig. 24(a). The harmonic content of this waveform is desired. Fig. 24(b) shows four simpler waveforms which, when properly combined, yield the output signal of Fig. 24(a). The Boolean relationship is

$$v_o = v_1\ v_2\ (\mu_1 + \mu_2) \tag{138}$$

The Fourier series for v_1, v_2, μ_1, and μ_2 are easy to determine. Once these series are truncated to an acceptable degree of accuracy, the harmonic content of v_o can be determined by substitution into (138). The amount of effort involved in multiplying out the expression for this example is likely greater than if the Fourier series of the original waveform had been determined directly, since the product of three series is called for in (138). However, for many waveforms, such as pulse-width modulation of a square wave, only two series have to be multiplied, and this analysis technique might result in a reduction of labor over the direct method.

Analog and Digital Computer Simulation

Although little reference to the use of computers has been made thus far in this paper, it is obvious that the digital computer is necessary to handle the calculations in most of the analysis techniques discussed. Analog, digital, and hybrid computer simulation techniques have

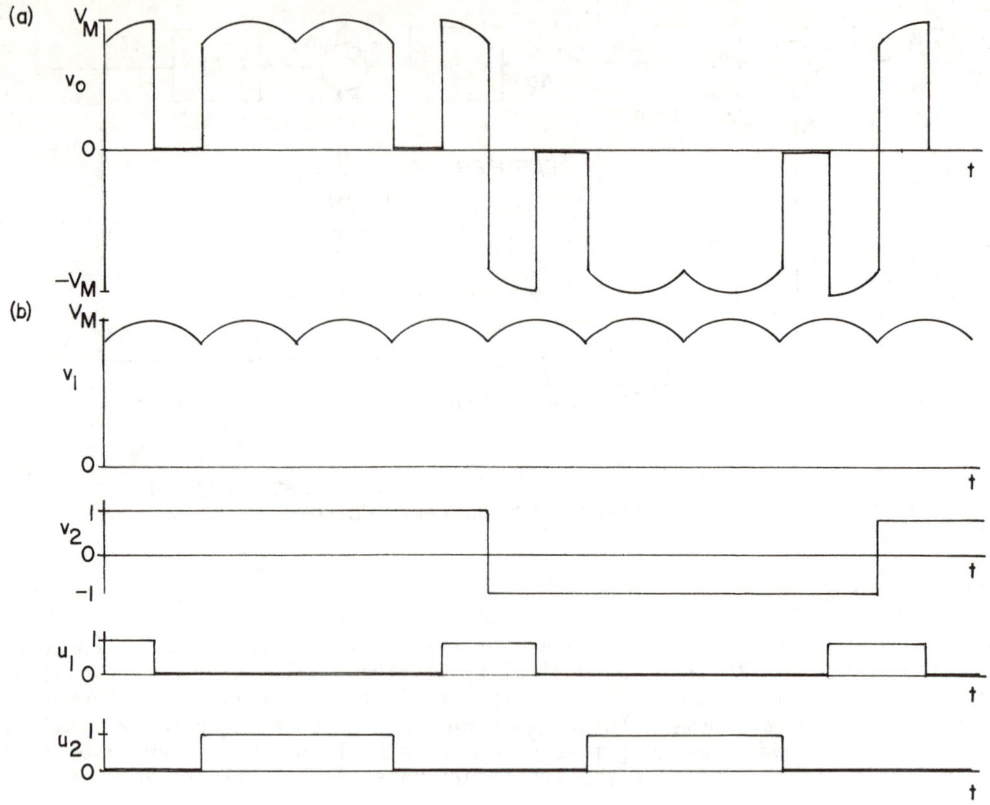

Fig. 24 A complex waveform (a) and its decomposition into simpler waveforms (b)

also been widely used as a complete and detailed analysis technique for switching converters. For complex circuits, such as those in which the conduction times of the devices are unknown, simulation may be the only feasible analysis technique.

References [20] and [21] discuss some of the general advantages and applications of analog computers in the analysis of switching circuits. Analog simulation can reduce the trial and error involved in circuit selection and design, is faster and less expensive than extensive digital computer runs, and is readily programmable to accomodate extremely complicated configurations. The usual outputs from an analog computer are the actual circuit waveforms, although the computer may be coupled with a digital computer to study harmonics, losses, etc. The switching devices are modelled as impedances whose magnitudes are controlled by the polarity of the applied voltages and by gating signals.

In digital simulation the problem is generally to determine a means of solution which maintains accuracy, saves computation time, and provides a maximum of information. A number of general circuit simulation programs such as ECAP and CSMP are available, but are not very efficient since they are designed to apply to a very large class of circuits with a minimum of effort by the programmer. Although no general simulation programs have appeared which have been designed especially for switching converter simulation, one of the most popular approachers is the state variable representation with numerical integration using Runge-Kutta methods.

One of the major problems encountered in converter analysis is that many switching devices may be present in the circuit, resulting in numerous different configurations during a single period of operation. Hence in writing a computer program, it may be difficult to keep track of which circuit equations apply at any instant. In [22], this problem is dealt with through a state of conduction matrix, which determines what devices are conducting at any instant. Each element in the matrix represents a particular switching device, and is binary valued to indicate whether the device is conducting or not at a given time. A general set of rules are used to determine when a device has changed states so that the matrix is updated when necessary. This approach is discussed in more detail for a cycloconverter in the following section.

Application of a conduction matrix to a cycloconverter. Suppose thyristors are driving an ac motor in a cycloconverter circuit. A particular thyristor is conducting if 1) a trigger signal is applied AND 2) its anode to cathode voltage is positive, OR 3) it was conducting during the previous step in the solution and no thyristors are triggered which will take its current. Thus if D is the conduction matrix as introduced above, the elements of D are described by the following Boolean matrix equation.

$$D = (D_1 \text{ AND } D_2) \text{ OR } D_3 \tag{139}$$

where D_1, D_2, and D_3 are each defined solely by the corresponding three rules above. D_1, which describes when the triggers are applied, depends solely on the desired voltage waveform and frequency at the motor terminals. D_2 depends on the instantaneous magnitude of the source voltages. D_3 depends on D during the previous interval and the solutions for the currents during the previous interval. If a judicious choice is made as to which elements of D represent each thyristor of the circuit, the elements of D_1, D_2, and D_3 may have a type of symmetry which makes them easier to determine.

In [22], a 36 thyristor cycloconverter is considered. As a second example, the 18 thyristor cycloconverter of Fig. 25 is considered. Note that there are six groups of thyristors with three devices in each group. The D matrix is thus chosen to be a 6 X 3 matrix, each element representing one thyristor. For example, if the third element of the fourth row is one, this implies $THY_{4,3}$ in Fig. 25 is conducting, and a zero would mean nonconduction. Suppose the group trigger mode is used to control the cycloconverter, i.e., all thyristors in a group are always triggered simultaneously. For example, during part of the positive half cycle of phase "a" of the motor, all three thyristors in group a^+ are triggered. The only one which conducts, however, is the one whose anode is connected to the most positive of the line to neutral source voltages. Fig. 26 illustrates typical motor voltage and current waveforms. During region 1, group a^+ is triggered. At $\theta = \alpha$, the trigger angle, triggering is removed. Region 2 then lasts until $\theta = \beta$, the extinction angle, when i_a decays to zero. Induced voltages then appear until the end of region 3, when the negative half cycle begins.

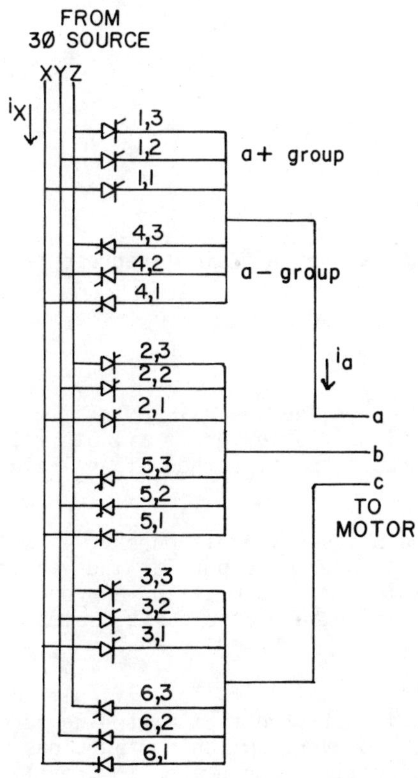

Fig. 25 An 18 thyristor cycloconverter

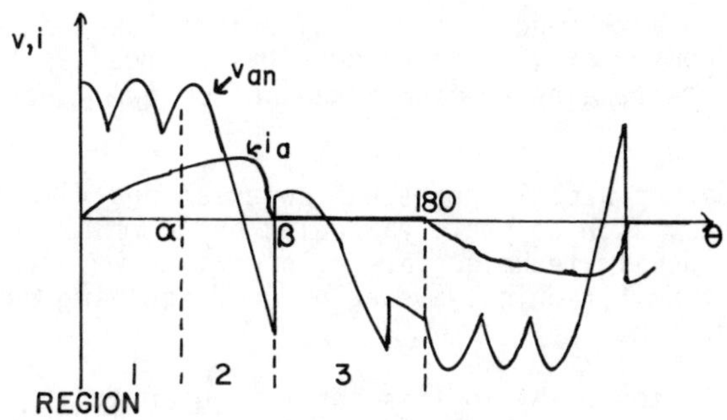

Fig. 26 Typical waveforms for the circuit of Fig. 25

From this information, matrices D_1, D_2, and D_3 can be defined. The first row of D_1, which indicates when triggers are applied to group a^{+}, is all ones when $0 < \theta < \alpha$ in Fig. 26. Row 4 is all ones when $180 < \theta < 180 + \alpha$, and the other rows are similar but shifted $\pm 120°$. D_2 indicates when anode to cathode voltages are positive. It is apparent that the first elements of column one are one when $v_X = \max(v_X, v_Y, v_Z)$, where v_X, v_Y, and v_Z are the line to neutral source voltages. The last three elements of column one are one when $v_X = \min(v_X, v_Y, v_Z)$, etc. Matrix D_3 is required to properly simulate region 2 in Fig. 26. At $\theta = \alpha$, prior to resetting row 1 of D_1, the information about which thyristors were conducting is transferred by equating D_3 to D. Thus (139) will ensure that whatever devices were conducting at the end of region 1, these same devices will continue to conduct for some interval. When the current goes to zero, D_3 is reset to zero.

The conduction matrix D can be used to define 3 X 3 connection matrices C_i and C_v which relate motor currents $\vec{i}_m$ and line to neutral voltages $\vec{v}_m$ to source currents $\vec{i}_s$ and line to neutral voltages v_s. Defining δ_{ij} to be the (i,j)th element of D, the relations are

$$\vec{v}_m = [C_v]\,\vec{v}_s \tag{140}$$

$$C_v = \begin{bmatrix} \delta_{11} + \delta_{41} & \delta_{12} + \delta_{42} & \delta_{13} + \delta_{43} \\ \delta_{21} + \delta_{51} & \delta_{22} + \delta_{52} & \delta_{23} + \delta_{53} \\ \delta_{31} + \delta_{61} & \delta_{32} + \delta_{62} & \delta_{33} + \delta_{63} \end{bmatrix}$$

$$\vec{i}_m = [C_i]\,\vec{i}_s \tag{141}$$

$$C_i = \begin{bmatrix} \delta_{11} - \delta_{41} & \delta_{12} - \delta_{42} & \delta_{13} - \delta_{43} \\ \delta_{21} - \delta_{51} & \delta_{22} - \delta_{52} & \delta_{23} - \delta_{53} \\ \delta_{31} - \delta_{61} & \delta_{32} - \delta_{62} & \delta_{33} - \delta_{63} \end{bmatrix}$$

These equations completely model the cycloconverter and are an efficient representation for use in a computer simulation.

CLOSING REMARKS

Although a tremendous amount of effort has been expended on the analysis of static switching power converters, no technique has emerged as being clearly superior to all others. Thus the challenge remains to find a new and uniquely advantageous general method of analyzing these circuits.

A number of useful approaches do exist and have been discussed in this paper. Time domain approaches, such as state variables and complex variables, are quite popular today although frequency domain techniqes are also well known and valuable. Many simplified approaches, such as neglecting harmonics, averaging, and applying superposition, are also quite useful to the designer.

As indicated early in this paper, the analysis of switching circuits is difficult due to the complexity and nonlinearity of the equations. It was illustrated that switching functions greatly reduced the complexity of an inverter system. However, switching functions have not been fully exploited and an entire new class of such functions may emerge to greatly simplify converter analysis.

Transformations of variables may also be used to greatly simplify nonlinear systems. A simple example of this is the system

$$[\dot{Y}_1 \quad \dot{Y}_2]^T = [3Y_1 + 3Y_1^{\frac{2}{3}} Y_2^{\frac{1}{5}} \quad 5Y_2^{\frac{4}{5}} Y_1^{\frac{1}{3}}]^T \tag{142}$$

By using the transformation

$$[Y_1 \quad Y_2]^T = [X_1^{\,3} \quad X_2^{\,5}]^T \tag{143}$$

in (142), a linear system,

$$[\dot{X}_1 \quad \dot{X}_2]^T = [X_1 + X_2 \quad X_1]^T \tag{144}$$

results. This linear system is referred to as topologically equivalent to the nonlinear system and leads to a simple solution of a system which would be difficult to solve if approached directly.

Another example of a useful transformation of variables is the representation of rotating machinery using two-axis reference frames. This transformation linearizes the induction motor equation in steady-state.

We believe that further research on power electronic circuit analysis techniques is essential. There is a need to fully exploit the advantages of each of the analytical methods discussed in this paper, and the search must continue for the ultimate power electronic analysis techniques.

REFERENCES

1. Takeuchi, T. J. (1968) Theory of SCR Circuit and Application to Motor Control, Tokyo Electrical Engineering College Press, Tokyo.

2. R. E. Bedford and V. D. Nene, Complex Frequency-Domain Analysis of Inverter-Fed Induction machines, IEEE Trans. on Ind. App. IA-8, No. 3, 269-77 (1972).

3. Linos Jacovides, Analysis of Induction Motor Drives with a Nonsinusoidal Supply Voltage Using Fourier Analysis, IEEE Trans. on Ind. App., IA-9, No. 6, 741-7 (1973).

4. P. C. Krause and T. A. Lipo, Analysis and Simplified Representations of a Rectifier-Inverter Induction Motor Drive, IEEE Trans. on Power App. and Sys., PAS-88, No. 5, 588-596 (1969).

5. H. A. Peterson and P. C. Krause, A D-Q Axis Representation of a Parallel AC and DC Power System, IEEE Trans. on Power App. and Sys., PAS-85, No. 3, 210-25 (1966).

6. P. H. Nayak and R. G. Hoft, Computer-Aided Steady-State Analysis of Thyristor DC Drives on Weak Power Systems, IEEE Conf. Rec. 1976 11th Annu. Meet. Ind. App. Soc., 835-47.

7. M. Ramamoorty and B. Ilango, Application of State-Space Techniques to the Steady-State Analysis of Thyristor-Controlled Single-Phase Induction Motors, Int. J. Control, 16, No. 2, 353-368 (1972.

8. D. W. Novotny and A. F. Fath, The Analysis of Induction Machines Controlled by Series Connected Semiconductor Switches, IEEE Trans. on Power App. and Sys., PAS-87, No. 2, 597-605 (1968).

9. T. A. Lipo and F. G. Turnbull, Analysis and Comparison of Two Types of Square-Wave Inverter Drives, IEEE Trans. on Ind. App., IA-11, No. 2, 137-47 (1975).

10. G. W. Wester and R. D. Middlebrook, Low-Frequency Characterization of Switched DC-DC Converters, IEEE Trans. on Aero. and Elec. Sys., AES-9, No. 3, 376-85 (1973).

11. R. D. Middlebrook and S. Cuk, A General Unified Approach to Modelling Switching-Converter Power Stages, Conf. Rec. 1976 IEEE Power Electronics Specialists Conference, 18-34.

12. Paul Krause, Method of Multiple Reference Frames Applied to the Analysis of Symmetrical Induction Machinery, IEEE Trans. on Power App. and Sys., PAS-87, No. 1, 218-27 (1968).

13. D. W. Novotny, Steady State Performance of Inverter Fed Induction Machines by Means of Time Domain Complex Variables, IEEE Trans. on Power App. and Sys., PAS-95, No. 3, 927-35 (1976).

14. K. R. Jardan, S. B. Dewan, G. R. Slemon, General Analysis of Three-Phase Inverters, IEEE Trans. on Ind. and Gen. App., IGA-5, No. 6, 672-9 (1969).

15. Gabor P. Kalman, Capacitor-Assisted Line-Commutated Converter, Conf. Rec. 1977 Meet. IEEE/IAS International Semiconductor Power Converter Conference, 442-52.

16. D. W. Novotny, Switching Function Representation of Polyphase Inverters, Conf. Rec. 1975 10th Annu. Meet. Ind. App. Soc., 823-31.

17. E. A. Parrish and E. S. McVey, A Theorectical Model for Single-Phase Silicon-Controlled Rectifier Systems, IEEE Trans. on Auto. Cont., AC-12, No. 5, 577-9 (1967).

18. J. P. Sucena-Paiva, R. Hernandez and L. L. Freris, Stability Study of Controlled Rectifiers Using a New Discrete Model, Proc. IEE, 119, No. 9, 1285-93 (1972).

19. S. B. Dewan and M. D. Kankam, A Method for Harmonic Analysis of Cycloconverters, IEEE Trans. on Ind. and Gen. App., IGA-6, No. 5, 455-62 (1970).

20. C. B. Mayer and T. A. Lipo, The Use of Simulation in the Design of an Inverter Drive, IEEE Conf. Rec. 1972 7th Annu. Meet. Ind. App. Soc., 745-52.

21. P. C. Krause and L. T. Woloszyk, Comparison of Computer and Test Results of a Static AC Drive System, IEEE Trans. on Ind. and Gen. App., IGA-4, No. 6, 583-8 (1968).

22. L. J. Jacovides, Analysis of a Cycloconverter-Induction Motor Drive System Allowing for Stator Current Discontinuities, IEEE Trans. on Ind. App., IA-9, No. 2, 206-215 (1973).

AUTHOR INDEX